Immature Insects

Immature Insects

Edited by
Frederick W. Stehr
Department of Entomology
Michigan State University

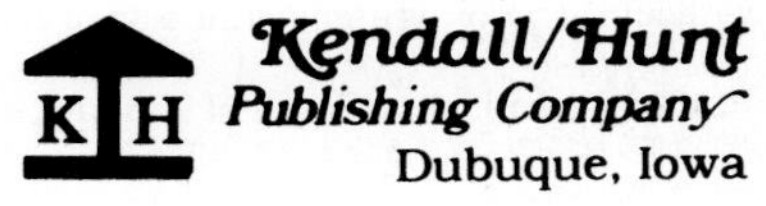

Credits

Figures in the Key to Orders identified with an Ⓟ are reproduced with permission from A. Peterson, *Larvae of Insects,* Vol. 1 (1948) and Vol. 2 (1951); those identified with "WCB" are from B. McDaniel, *How to Know the Mites and Ticks.* Copyright © 1979 by Wm. C. Brown Company Publishers, Dubuque, Iowa. All rights reserved. Reprinted by permission; those identified with a "CDC" are reproduced from *Pictorial Keys to Arthropods, Reptiles, Birds and Mammals of Public Health Significance* (1964), Communicable Disease Center, U.S. Public Health Service, Atlanta, Georgia; and those identified with "JL" are reproduced with permission from J. L. Luth.

Ephemeroptera Figures 9.1–9.45. Reprinted from *Mayflies of North and Central America* by G. Edmunds, S. Jensen, and L. Berner. University of Minnesota Press, Minneapolis. Copyright © 1976 by the University of Minnesota. Used with permission.

Ephemeroptera Figures 9.46 and 9.50. Reprinted from B. D. Burks, *The Mayflies of Illinois,* Illinois Natural History Survey Bulletin 26, Article 1. Used with permission.

Ephemeroptera Figure 9.57. Reprinted from Bulletin of Florida State Museum Biological Sciences 17:173. Used with permission.

Odonata Figures 10.1–5a, 10.18–28, 10.30, 10.32, 10.34a–37. From Ohio Journal of Science XLIV 94:151–166.

Odonata Figures 10.6, 10.16, 10.29, 10.33–34, 10.38. From R. Merritt, and K. Cummins, *An Introduction to the Aquatic Insects of North America.* Copyright © 1978 by Kendall/Hunt Publishing Company.

Anoplura Figures 23.1–2. From Systematic Entomology 3:254, 255. Blackwell Scientific Publications Limited. Used with permission.

Hymenoptera Figures 27.77–78, 27.82, 27.86, 27.88, 27.91–92, 27.94–98, 27.100. Trans. Soc. Br. Ent. 7:123–190. Redrawn with permission of the Royal Entomological Society of London.

Hymenoptera Figures 27.104–105, 27.107–108, 27.111–112, 27.122. Trans R. Ent. Soc. Lond. 103:27–84. Redrawn with permission of the Royal Entomological Society of London.

Hymenoptera Figures 27.79–81, 27.83–84, 27.87, 27.89, 27.99, 27.101–102, 27.106, 27.109–110, 27.113–121, 27.123–129a, 27.131, 27.132. Reproduced with permission of the Entomological Society of Canada.

Hymenoptera Figures 27.85, 27.90. Reproduced with permission of the Smithsonian Institution Press.

COVER: *Parasa indetermina* (Boisduval), the stinging rose caterpillar (Lepidoptera: Limacodidae). Most commonly collected on eastern redbud, *Cercis canadensis,* in August; also found on oak and sycamore. Photo by James E. Appleby, Section of Economic Entomology, Illinois State Natural History Survey, Champaign, Illinois.

Library of Congress Catalog Card Number: 85–81922

ISBN 0–8403–3702–7

Printed in the United States of America
10 9 8 7 6 5 4 3 2 1

Dedication

To the memory of Alvah Peterson (1888–1972), who first introduced immature insects to me, and whose pioneering books, Larvae of Insects, *have been of immeasurable value in stimulating and facilitating their study.*

Contents

Preface

The identification of immature insects and the location of information about them is a regular and continuing problem for many. The literature available to accomplish this is widely scattered, limited to certain groups (e.g., aquatics), outdated, difficult to use, or nonexistent. This book has been conceived to address all of these needs.

Specifically, it has been designed to serve as: (1) a textbook for courses on immature insects; (2) an introduction to and partial source of the literature on immature insects; (3) a means to identify the larvae of all orders to family; (4) a means to identify a number of common, economically important, or unusual species (largely North American) using the abundant illustrations, descriptions, and/or selected keys; and (5) a source of information on the biology and ecology of the families and selected important species. The emphasis is on larvae, with minimal coverage of eggs and pupae, since knowledge of the latter is *relatively* meager and the need to identify them is not great compared with larvae or adults.

All orders are covered, not just the Holometabola. Some nonholometabolous immatures can be identified using existing keys for adults, but others, especially groups where wing characters are used, cannot be identified satisfactorily. In addition, in existing keys to families for adults, the user is never informed whether the key will work for immatures.

The key to orders covers immature, brachypterous, and wingless adult insects, as well as other terrestrial and freshwater invertebrates that are likely to be collected with or possibly confused with immature insects. Keys to pupae are not included, but general descriptions are provided for many groups at the order or family level.

Volume 1 covers techniques, the key to orders, and the Protura, Collembola, Diplura, Microcoryphia, Thysanura, Ephemeroptera, Odonata, Blattodea, Isoptera, Mantodea, Grylloblattodea, Phasmatodea, Orthoptera, Dermaptera, Embiidina, Zoraptera, Plecoptera, Psocoptera, Mallophaga, Anoplura, Mecoptera, Trichoptera, Lepidoptera, and Hymenoptera. The techniques covered include collecting, rearing, proper killing, fixation and preservation, permanent storage, study methods, and shipping procedures.

Volume 2 will cover the Hemiptera, Homoptera, Thysanoptera, Megaloptera, Raphidiodea, Neuroptera, Coleoptera, Strepsiptera, Siphonaptera, and Diptera.

Quick recognition characters (diagnoses) for distinguishing between closely related or superficially similar groups are supplied for all taxa.

The descriptions provided for each family, along with the diagnoses and illustrations, are an important tool for confirming identifications made with the keys. They are particularly important for immatures, since many users will not have access to an adequate reference collection as they frequently do for adults. In addition, virtually all entomology textbooks emphasize adults and offer little help in recognition of immatures below the order level, although some contain general descriptive statements, and some information on biology and common species is often presented.

The area emphasized is America north of Mexico, although some of the coverage (e.g., Coleoptera, in volume 2) is worldwide. Although the keys may work reasonably well for many specimens from other parts of the world (especially Eurasia), there are taxa that do not occur or are minimally represented in America north of Mexico and that will not key satisfactorily.

In the keys, direct character comparisons between couplet halves are used, but supplementary information is added at times. Illustrations are used abundantly in the keys, and in the longer ones figures are inserted where referred to and duplicated if used again in the key.

Family coverage begins with a presentation of relationships and recognition characters (the diagnosis). Following this is a section on biology and ecology, a description of the family, and comments on a variety of topics such as economic importance, size, distribution, and state of knowledge. An author-year index to selected literature for the family follows or is incorporated in the comments. Illustrations provided for each family range from a single figure for small families to multiple figures for the larger ones.

The literature for each order is extensive but not complete; however, it is believed that sufficient publications have been included to enable the user to find the additional information necessary for further study.

Each volume contains a complete index and an independent glossary, which assumes a basic knowledge of insects and their structure.

Acknowledgments

At one time I had a co-editor, Lauren D. Anderson, professor emeritus of the Department of Entomology, University of California, Riverside, but after retiring in 1976, he wisely relinquished his duties in 1980. Nevertheless, we share the responsibility for coming up with the idea for this book longer ago than either of us cares to admit, and I thank him for his enthusiastic support and help when the prospects for completion and publication were somewhat dim.

Administrative support for a project such as this is absolutely essential, and I thank James E. Bath, Chairman, Department of Entomology, Michigan State University for his continued support and encouragement in every way. In addition, much of my work has been supported by the Michigan Agricultural Experiment Station and the College of Natural Science, Michigan State University.

The contributing authors have helped each other in many ways that are not acknowledged individually, and all have been patient and helpful to me.

On behalf of the contributing authors acknowledgement is made to all of our colleagues, students, and friends whose help has made a more complete and accurate publication possible. We deeply appreciate it.

Illustrations are perhaps the most important part of a book like this. They come from many sources. Contributing authors have been responsible for providing their own, either via originals or other arrangements. Some are reproduced with permission from other publications and are acknowledged where used. Some of the artists and photographers are acknowledged in the chapters or family write-ups, or with the individual figures. The line drawings for Davis's Lepidoptera families were drawn by Birute Akerbergs Hansen. A considerable number of Alvah Peterson's original drawings have been used (some with slight modifications) by agreement with his widow, Helen Peterson; they are marked with Ⓟ and/or acknowledged in the legends. Those figures in the Key to Orders by Joseph L. Luth are marked JL, those from the Centers for Disease Control, U.S. Public Health Service, with CDC, and those used with permission from William C. Brown Co., are marked WCB. Others are from the contributing authors. I would particularly like to thank Lana Tackett LT for her special touch and care in illustrating many of the Lepidoptera, for drawing or redrawing many of the figures for the keys to families of Lepidoptera and Hymenoptera, and for drawing, inking or re-inking assorted other drawings for some of the contributing authors.

Special thanks are due to D. M. Weisman, Systematic Entomology Laboratory, USDA and to J. A. Powell, Department of Entomology, University of California, Berkeley, who provided many of the specimens and substantial help in constructing the Lepidoptera key.

The seemingly endless task of reviewing the entire set of galley proofs has been accomplished by Lois Connington. Her careful attention has made the final product substantially more polished than if only the contributing authors and editor had checked them.

The following have made special efforts to collect or lend specimens, test keys, answer questions, provide information, and review manuscripts, or have been helpful to the contributing authors in other ways:

S. Allyson, Biosystematics Research Institute, Ottawa
R. C. Beckwith, Forest Service, USDA
G. Berkey, Ohio Agricultural Research and Development Center
N. L. Braasch, Southeast Missouri State University
T. J. Cohn, San Diego State University
C. H. Collison, Pennsylvania State University
E. F. Cook, University of Minnesota
T. E. Eichlin, California Department of Food and Agriculture
M. E. Farris, Illinois Natural History Survey
D. C. Ferguson, Systematic Entomology Laboratory, USDA
J. G. Franclemont, Cornell University
L. Gall, Yale University
J. R. Gorham, United States Food and Drug Administration
L. R. Grimes, Meredith College, North Carolina
A. B. Gurney, Systematic Entomology Laboratory, USDA
K. S. Hagen, University of California, Berkeley
R. W. and E. R. Hodges, Systematic Entomology Laboratory, USDA
T. H. Hubbell, University of Michigan
D. Jokinen, formerly New Mexico State University
R. D. and C. Kendall, San Antonio, Texas
D. K. McE. Keven, McGill University
P. D. Kinser, formerly North Carolina State University
W. E. LaBerge, Illinois Natural History Survey
J. D. Lafontaine, Biosystematics Research Institute, Ottawa
W. H. Lange, University of California, Davis
R. Levy, Lee County Mosquito Control District, Ft. Meyers, Florida
J. K. Liebherr, Cornell University
W. C. McGuffin, Biosystematics Research Institute, Ottawa
F. Mead, Florida Department of Agriculture
A. S. Menke, Systematic Entomology Laboratory, USDA
D. R. Miller, Systematic Entomology Laboratory, USDA
G. L. Motyka, American Museum of Natural History
W. L. Murphy, Beltsville Agricultural Research Center, USDA
G. Okumura, California Department of Food and Agriculture

D. Otte, Academy of Natural Sciences of Philadelphia
F. Pacheco Mendivil, CIANO, Ciudad Obregón, Sonora, Mexico
S. Passoa, University of Illinois
R. S. Peigler, Greenville, North Carolina
J. E. Rawlins, Carnegie Museum of Natural History
R. W. Rings, Ohio Agricultural Research and Development Center
L. M. Roth formerly, U.S. Army Natick Laboratories, Natick, Massachusetts
J. Scott, Lakewood, Colorado
D. J. Shetlar, formerly Pennsylvania State University
T. J. Spilman, Systematic Entomology Laboratory, USDA
D. L. Stephan, North Carolina State University
J. G. Sternburg, University of Illinois
J. G. Thomas, Texas A & M University
P. M. Tuskes, Houston, Texas
J. D. Unzicker, Illinois Natural History Survey
D. J. Voegtlin, Illinois Natural History Survey
D. M. Weisman, Systematic Entomology Laboratory, USDA
J. J. Whitesell, Valdosta State College, Georgia

And for those whom I may have overlooked, we appreciate your help just as much.

List of Contributors

R. K. ALLEN, 22021 Jonesport Lane, Huntington Beach, California

J. E. APPLEBY, Section of Economic Entomology, Illinois Natural History Survey, Champaign, Illinois

R. W. BAUMANN, Department of Biology, Brigham Young University, Provo, Utah

E. C. BERNARD, Department of Entomology and Plant Pathology, University of Tennessee, Knoxville, Tennessee

A. BRINDLE, Department of Entomology, Manchester Museum, University of Manchester, Manchester, England

J. W. BROWN, 791 My Way, San Diego, California

R. L. BROWN, Department of Entomology, Mississippi State University, Mississippi State, Mississippi

M. A. BRUSVEN, Department of Plant, Soil and Entomological Sciences, University of Idaho, Moscow, Idaho

G. W. BYERS, Department of Entomology, University of Kansas, Lawrence, Kansas

D. R. DAVIS, Department of Entomology, Smithsonian Institution, Washington, D.C.

J. P. DONAHUE, Natural History Museum of Los Angeles County, 900 Exposition Blvd., Los Angeles, California

J. C. DOWNEY, Graduate College, Northern Iowa University, Cedar Falls, Iowa

G. F. EDMUNDS, JR., Department of Biology, University of Utah, Salt Lake City, Utah

W. R. ENNS, Professor Emeritus, Department of Entomology, University of Missouri, Columbia, Missouri

H. E. EVANS, Department of Zoology and Entomology, Colorado State University, Fort Collins, Colorado

T. FINLAYSON, Professor Emerita, Department of Biological Sciences, Simon Fraser University, Burnaby, British Columbia, Canada

F. W. FISK, Professor Emeritus, Department of Entomology and Zoology, The Ohio State University, Columbus, Ohio

D. C. FRACK, 13506 Alanwood Rd., LaPuente, California

G. L. GODFREY, Section Faunistic Surveys and Insect Identification, Illinois Natural History Survey, Champaign, Illinois

D. H. HABECK, Department of Entomology and Nematology, University of Florida, Gainesville, Florida

D. J. HARVEY, Department of Zoology, University of Texas, Austin, Texas

J. B. HEPPNER, Florida State Collection of Arthropods, Bureau of Entomology, DPI, FDACS, P.O. Box 1269, Gainesville, Florida

S. B. HILL, Department of Entomology, Macdonald College of McGill University, St. Anne de Bellevue, Quebec, Canada

M. JEFFORDS, Section of Economic Entomology, Illinois Natural History Survey, Champaign, Illinois

K. C. KIM, The Frost Entomological Museum, Department of Entomology, Pennsylvania State University, University Park, Pennsylvania

N. MCFARLAND, P.O. Box 1404, Sierra Vista, Arizona

R. J. MCGINLEY, Department of Entomology, Smithsonian Institution, Washington, D.C.

P. J. MARTINAT, Division of Forestry, West Virginia University, Morgantown, West Virginia

W. W. MIDDLEKAUFF, Professor Emeritus, Department of Entomology, University of California, Berkeley, California

J. Y. MILLER, Allyn Museum of Entomology, Florida State Museum, 3701 Bay Shore Road, Sarasota, Florida

E. L. MOCKFORD, Department of Biological Science, Illinois State University, Normal, Illinois

H. H. NEUNZIG, Department of Entomology, North Carolina State University, Raleigh, North Carolina

D. A. NICKLE, Systematic Entomology Laboratory USDA, U.S. National Museum, Washington, D.C.

R. D. PRICE, Department of Entomology, University of Minnesota, St. Paul, Minnesota

G. T. RIEGEL, Department of Zoology, Eastern Illinois University, Charleston, Illinois

E. S. ROSS, Department of Entomology, California Academy of Sciences, Golden Gate Park, San Francisco, California

D. R. SMITH, Systematic Entomology Laboratory USDA, U.S. National Museum, Washington, D.C.

R. J. SNIDER, Department of Zoology, Michigan State University, East Lansing, Michigan

F. W. STEHR, Department of Entomology, Michigan State University, East Lansing, Michigan

M. E. TOLIVER, Department of Biology, Eureka College, Eureka, Illinois

S. L. TUXEN (deceased), Universitetes Zoological Museum, Universitetsparket, Copenhagen, Denmark

D. L. WAGNER, Department of Entomology, University of California, Berkeley, California

T. J. WALKER, Department of Entomology and Nematology, University of Florida, Gainesville, Florida

F. M. WEESNER, Department of Zoology and Entomology, Colorado State University, Fort Collins, Colorado

M. J. WESTFALL, Jr., Department of Zoology, University of Florida, Gainesville, Florida

G. B. WIGGINS, Royal Ontario Museum and University of Toronto, Toronto, Ontario, Canada

P. WYGODZINSKY, American Museum of Natural History, Central Park W. and 79th St., New York, New York

Introduction

1

Frederick W. Stehr
Michigan State University

Encounters with immature insects are relatively uncommon compared to those with adults, yet larvae undoubtedly cause more *direct* injury to crops and other materials valued by humans than do adults. This is true because most (sometimes all) feeding is done in the larval stage, whereas adults are primarily involved in dispersal and reproduction, although many do cause feeding or oviposition injury, and some transmit plant and animal diseases. Most larvae are not harmful; indeed, many appear to be "neutral" in not being obviously harmful or beneficial. Others are beneficial predators or parasitoids, or they are essential components in nutrient cycling and food webs. Accordingly, a knowledge of immatures and an ability to identify them is important to anyone concerned with natural, agricultural, or urban ecosystems and related subjects.

There are no books that provide relatively comprehensive coverage of immature insects for all orders, although some books cover the aquatic groups. Three recent ones are Merritt and Cummins (1984), Brigham et al. (1982) (eastern North America), and McCafferty (1981). See table 1A in Merritt and Cummins (1984) for a list of twenty-two books with some general coverage of aquatic groups, or in this book see the selected bibliographies for orders containing aquatic species for publications with more depth of coverage.

Practically nothing "comprehensive" has been published on terrestrial groups since 1954. Peterson's two volumes (1948, 1951) on the Holometabola have been available, as has the generalized *How to Know the Immature Insects* by Chu (1949). Some keys to larvae for selected orders are provided in Brues et al. (1954), but there is no family information. *The Manual of Nearctic Diptera,* Volume 1 (McAlpine et al. 1981), provides a key to families and substantial information and excellent illustrations on larvae of the lower Diptera; Volume 2 on the higher Diptera will soon be available. Other older publications, such as Böving and Craighead (1931) on Coleoptera larvae, are out of print and outdated, but still of use. This book is intended to provide relatively comprehensive coverage of all orders down to the family level and to varying levels below the family for selected taxa.

LARVAE AND NYMPHS

The term "immature insects" encompasses all life stages except adults, no matter how many specialized names are applied to the various developmental forms in the different orders. There is little difficulty in defining an egg or an adult, but problems arise when attempts are made to name and define the instars or stages that may occur between egg and adult. Some insects are larviparous, never depositing eggs, some multiply from a single egg by polyembryony, and some are sexually mature as immatures (paedogenesis or neoteny). Nevertheless, all of them undergo a series of molts as they grow.

Arthropods and many other animals (frogs and toads for example) go through a metamorphosis of some kind in their development; the changes may be rather minor or they may result in great differences in form. Some authors restrict the term "metamorphosis" to the major changes that take place at the last molt to the adult stage (e.g., Chapman 1982); others take a broader view and regard all changes in form throughout the life history as metamorphosis. That view is followed here.

In the Insecta (broadly speaking) metamorphosis ranges from ametabolous, in which the number of molts is indefinite and molting may continue throughout life (Apterygota except Protura), to holometabolous, in which the instars egg, larva, pupa, and adult occur. Between these extremes there is a series of different types of metamorphosis and terms that the user of this book has undoubtedly encountered. However, since there is considerable nonagreement on some terms, a brief review and definition of the major terms used for immatures is presented to clarify their use here.

The tendency in North America, and to a certain extent in Great Britain and elsewhere, since the publication of Comstock's (1918) paper, has been to restrict the term "larva" to juveniles of the holometabolous orders and to apply the term "nymph" to the juveniles of non-holometabolous orders. However, the term "larva" has very broad usage in invertebrate zoology, being applied to an assortment of forms (often dispersal in function) in virtually all invertebrate phyla.

In the Arthropoda other than insects, "larvae" is most often used for first instars, as in the mites and ticks, and for first instar "hexapod larvae" of millipedes. For many other arthropod groups the term "juvenile" is used.

In the insects, "larvae" is used in greatly different ways, including such diverse forms as immatures of Protura and of Hymenoptera. Among the insects, the termites present an interesting problem: some authors use "nymph" for all juvenile termites, whereas others use "larva" for those lacking wingpads, and "nymph" for those having wingpads. This is complicated by supplementary reproductives that may be wingless or bear wingpads, even though the two forms are functionally equivalent. Should they be "larvae" or "nymphs"?

The term "nymph" also has broad usage, being applied to second-stage mites and ticks in addition to most juvenile non-Holometabola, and in Continental Europe (especially France) a "nymphe" is a pupa. However, in English-speaking countries "pupa" is universally used for the stage between the larva and adult of Holometabola.

There has been a tendency to use "larva" for all immatures that are not eggs, pupae, or adults, as was done by Chapman (1982), partially by Richards and Davies (1977), and partially in *The Insects of Australia* (Waterhouse, et al. 1970). In North America "larva" has not yet prevailed, although there is movement in that direction. For example, many who work on the Ephemeroptera, Odonata, and Plecoptera use "larva" instead of "naiad" or "nymph". The contributors to this book are not in agreement regarding the use of "larva" and "nymph", but for consistency, larva has been adopted for all orders, sometimes with a parenthetical "nymph" added for clarification.

When "larva" is used in the comprehensive sense, the subcategories "exopterygote larva" and "endopterygote larva" are useful for pterygote juveniles. "Juvenile", a useful term equivalent to "larva" in the broad sense, can be used as a general term for nonadults of all orders.

Below are the commonly encountered terms used for different types of metamorphosis. "Hemimetabolous" as used here includes "paurometabolous".

Simple metamorphosis: a broad term covering all types of metamorphosis except holometabolous.

Anamorphosis: development with fewer body segments at hatching than when mature; found in the Protura, in which three abdominal segments are added as the individual develops to an adult.

Ametabolous: development with the major change being an increase in size until mature; the number of molts is indefinite, and molting may continue throughout life. Found in Apterygota (excluding the Protura).

Hemimetabolous: (gradual, incomplete, direct, paurometabolous) originally defined by the Europeans as including everything except ametabolous, anametabolous, and holometabolous, but restricted by Comstock (1918) to the aquatic orders Ephemeroptera, Odonata, and Plecoptera, in which he termed the juveniles "naiads". It is used here as originally broadly defined.

Paurometabolous: (gradual, incomplete, direct) Comstock's (1918) term for everything that did not fall under ametabolous, hemimetabolous as restricted by Comstock, or holometabolous. The juveniles were termed "nymphs".

Holometabolous: (complete, indirect) development through egg, larva, pupa, and adult.

Hypermetamorphosis (also known as heteromorphosis): a special kind of complete metamorphosis found in some parasitic insects in which the form of different instars varies greatly, with the first instar being active and seeking the host. Subsequent instars, in (or on) the host, become more grublike and rather immobile. Hypermetamorphosis is found in the Strepsiptera, Mantispidae (Neuroptera), Meloidae, Rhipiphoridae, and a few Staphylinidae (Coleoptera), some Hymenoptera (some Chalcidoidea), some Diptera (Acroceridae, Bombyliidae, Nemestrinidae), and even Lepidoptera (Epipyropidae).

The term "triungulin" has been used for the first instars of many hypermetamorphic species, but it should be restricted to first instar Meloidae (Coleoptera) since the meloid pretarsus bears a claw and a pair of flanking clawlike setae, thus giving a "three-clawed" (triungulin) appearance (fig. 1.9). Hypermetamorphic, campodeiform first instars of other groups do not have this three-clawed appearance and should be termed "triungulinids" or "triungulinlike" or "planidia" or other suitable terms.

There are some general terms used for holometabolous larvae that have broad application as follows:

Campodeiform: (figs. 1.1, 1.2, 1.9): body elongate, somewhat flattened, thoracic legs well developed, head directed forward (prognathous), no abdominal prolegs, antennae and cerci usually conspicuous as in *Campodea* (Diplura); common in the Coleoptera, Megaloptera, Neuroptera, and Raphidiodea.

Elateriform: (fig. 1.2): somewhat similar to a campodeiform larva, but body more elongate and subcylindrical, more heavily sclerotized, setae shorter, and cerci often replaced by urogomphi. Common in Elateridae and some other Coleoptera.

Scarabaeiform: (fig. 1.3): body C-shaped, thoracic legs well developed, head directed downward (hypognathous), abdominal prolegs absent, cerci absent, usually lies on its side when removed from its habitat. Prevalent in certain groups of Coleoptera, especially Scarabaeoidea. White grubs (Scarabaeidae) are the best example.

Eruciform: (figs. 1.4, 1.5, 1.6): Caterpillarlike, body cylindrical, thoracic legs well developed, head usually hypognathous, abdominal prolegs present. Common in the Lepidoptera, Mecoptera, and Hymenoptera (Symphyta).

Vermiform: (fig. 1.8): wormlike; an ill-defined term, but generally an elongate, legless larva with or without a conspicuous head.

Maggot: (fig. 1.7): restricted to larvae of higher Diptera; shape peglike, tapered anteriorly, legless, head greatly reduced, but with conspicuous mouthhook(s).

Grub: (figs. 1.10, 1.11, 1.12): another imprecise term, often applied to comma-shaped larvae with or without legs or having greatly reduced legs. Commonly applied to curculionoid and scarabaeoid Coleoptera and to many larvae of the higher Hymenoptera (Apocrita) with reduced heads and appendages.

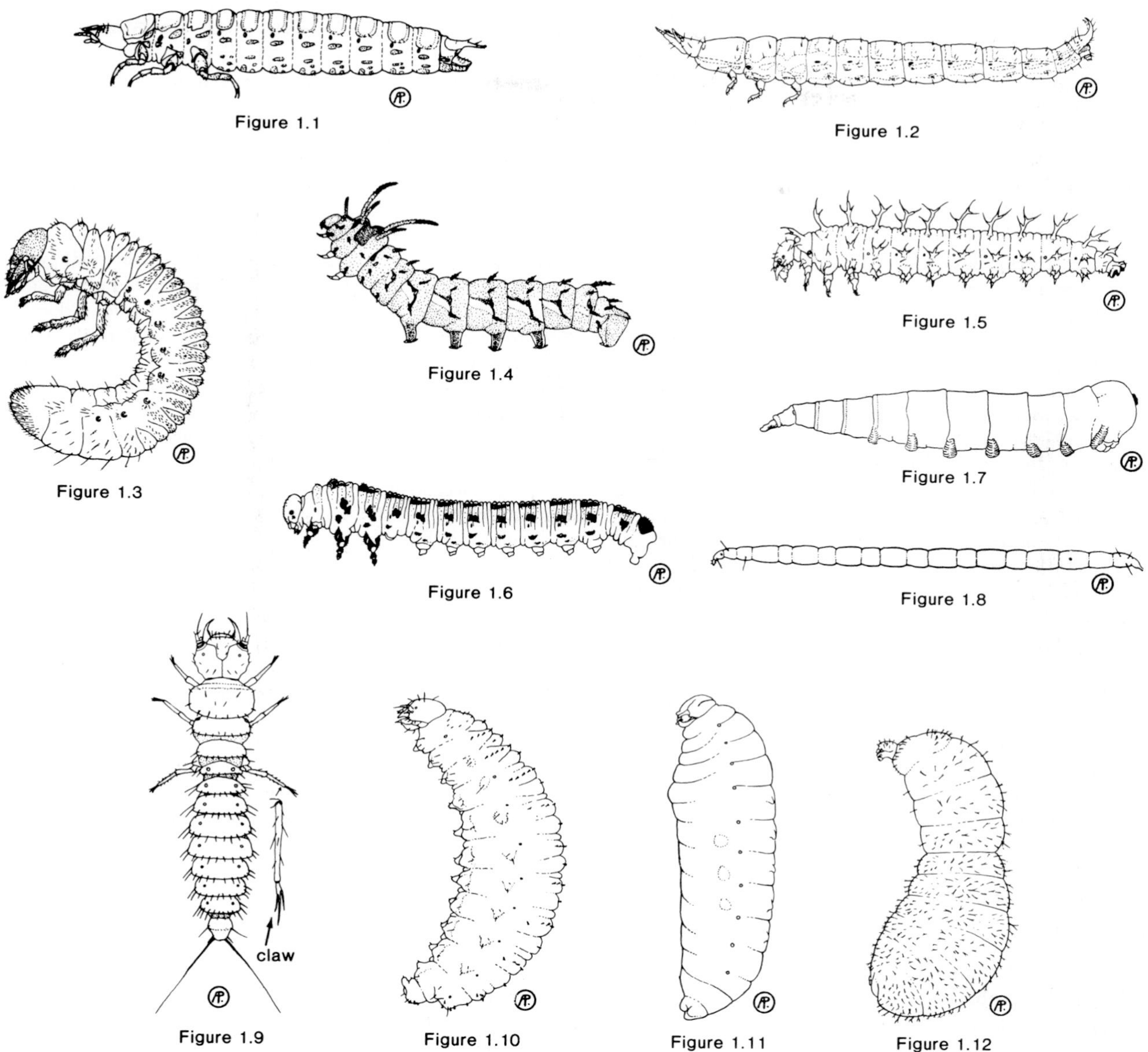

Figures 1.1–1.12. Types of Larvae (from Peterson 1948 and 1951).

Figure 1.1. Campodeiform larva: *Calosoma* (Coleoptera: Carabidae). See also figures **1.2** and **1.9.**

Figure 1.2. Elateriform larva (wireworm): *Ludius* (Coleoptera: Elateridae).

Figure 1.3. Scarabaeiform larva: *Phyllophaga* (Coleoptera: Scarabaeidae).

Figures 1.4–1.6. Eruciform larvae: **(1.4)** *Citheronia* (Lepidoptera: Saturniidae); **(1.5)** *Bittacus* (Mecoptera: Bittacidae); **(1.6)** *Neodiprion* (Hymenoptera: Diprionidae).

Figure 1.7. Maggot: *Haematobia* (Diptera: Muscidae).

Figure 1.8. Vermiform larva: *Scenopinus* (Diptera: Scenopinidae).

Figure 1.9. Triungulin: *Lytta*, first instar (Coleoptera: Meloidae).

Figures 1.10–1.12. Grubs: **(1.10)** *Hypera* (Coleoptera: Curculionidae); **(1.11)** *Vespa* (Hymenoptera: Vespidae); **(1.12)** Hymenoptera: Formicidae.

PUPAE

A "pupa" is understood to be that stage of holometabolous insects between the mature larva and the adult wherein major structural reorganization takes place. Pupation usually occurs in a protected situation in a cell or cocoon, but in some groups, such as many butterflies, the pupa (chrysalis) (fig. 1.20) is suspended openly and is usually cryptically colored.

There are two basic kinds of pupae, *exarate* and *obtect*. An exarate pupa (figs. 1.13–1.18) has free appendages and may be decticous or adecticous. Pupae may have articulated mandibles (*decticous*) or nonarticulated mandibles (*adecticous*). Decticous pupae, which are capable of chewing their way out of cells or cocoons, occur in the Mecoptera, Megaloptera, Neuroptera, Trichoptera, and primitive Lepidoptera. An obtect pupa (figs. 1.19–1.23) has the appendages adhering to the body wall and is always adecticous. Most Lepidoptera, most lower Diptera, some chrysomelid and staphylinid beetles, and many chalcidoid Hymenoptera have obtect pupae; nearly all other pupae are exarate. In the special case of the higher Diptera, the exarate pupa (also called a *coarctate* pupa) is enclosed in a delicate membrane within the hardened and barrel-shaped last (third) larval skin, which is termed the *puparium* (figs. 1.24–1.26).

A *prepupa* can be defined as the last-instar larva that has completed feeding. It may wander in search of a pupation site, but generally becomes nonmobile before pupation. It is easily observed in Lepidoptera such as butterflies, wherein the mature larva becomes shortened and the prolegs and crochets become progressively retracted before pupation. Some workers restrict the term "prepupa" or "propupa" to thrips, male scale insects, some gracillariids (Lepidoptera), and other groups where there is a separate, completely nonfeeding instar preceding the pupa.

Most pupae are inactive, their body movements often limited to the abdominal segments. However, pupae in some groups are capable of locomotion, and some have functional mandibles that enable them to cut their way out of the pupal cell, cocoon, or chamber. These active pupae are frequently referred to as pharate adults since the adult is still enclosed in the pupal cuticle. "Pharate" refers to any life stage that remains within and often visible beneath the cuticle of the preceding stage; the pharate life stage may be active or inactive and terminates at ecdysis to the next instar or emergence of the adult instar. Active pupae are common in the Megaloptera and Raphidiodea and in the Trichoptera as they prepare to emerge as adults. A few Diptera pupae (Chironomidae and Culicidae, fig. 1.22) are active.

Keys to pupae are not provided here, but keys to the order level are available in Peterson (1948, pp. 12, 14, 24) and Chu (1949, couplets 17 and 45). In addition, keys to the pupae of the major families of Lepidoptera are available in Mosher (1916) or as modified by Chu (1949). Keys to the pupae of the families of the lower Diptera (suborders Nematocera, Tabanomorpha, and Asilomorpha) are also available in Chu (1949). Selected keys to genera, some figures, and pupal descriptions are available for families of Diptera in McAlpine et al. (1981, vol. 1, and 198?, vol. 2), but there is no key to families.

SIMPLE EYES: OCELLI AND STEMMATA

The number, size, and arrangement of simple eyes are important diagnostic characters for many larvae. Until relatively recently the simple eyes of both larvae and adults have been generally termed "ocelli" (ocellus), although it has been recognized for some time that there are two different groups that are innervated from different parts of the brain (Snodgrass 1935 and earlier authors). See Paulus (1981) for an extensive survey and discussion of insect (and arthropod) eyes.

One group of simple eyes, found in adult insects and larvae (nymphs) of non-Holometabola, is termed "dorsal ocelli" or simply "ocelli". They are innervated dorsally from the protocerebrum between the optic lobes. There are basically four, but one pair is fused to form the median ocellus; thus, there are typically three ocelli located near the midline of the head, but the number varies from zero to three (eight in Collembola).

The second group, termed "lateral ocelli" or "*stemmata*" (stemma), is found in the larvae of Holometabola; they are innervated laterally from the optic lobes, and typically there is a group on each side of the head. There are at least five different types, all believed to have been derived from the typical insect ommatidium, but all five types are termed stemmata (Paulus 1981). The number of stemmata is variable, ranging from zero to seven (but see Mecoptera below), and the number and arrangement can be diagnostic for some taxa.

Although broad usage of "ocelli" has not appeared to cause confusion, there has been a tendency to adopt "stemmata" for the lateral simple eyes of holometabolous larvae. In line with Paulus (1981) and some current usage, *stemmata* is adopted here for the lateral simple eyes of holometabolous larvae. This can result in some changes in related nomenclature, such as the change in the Lepidoptera from ocellar setae (01, 02, 03) to stemmatal setae (S1, S2, S3), but consistency mandates such changes. The term *ocelli* is retained for the simple eyes of nonholometabolous larvae and for adults of all orders.

Larvae of Mecoptera may have faceted eyes with up to thirty *ommatidia* (Panorpidae), or as few as three (Boreidae and Panorpodidae), or none (Panorpodidae) that are nearly identical to adult insect ommatidia (Paulus 1981). For convenience, they are included under the term "stemmata" in the Key to Orders and elsewhere.

ABBREVIATIONS

The terms prothorax, mesothorax, and metathorax, and various combinations and numbers of the abdominal segments are repeatedly used, especially in the descriptive sections and keys. The following abbreviations for these terms have been used whenever it did not appear that confusion would result: prothorax (**T1**), mesothorax (**T2**), metathorax (**T3**), abdominal segments one through ten (**A1–10**), etc. A mature or full-grown larva is abbreviated (**f.g.l.**). Other abbreviations are defined in the sections where used.

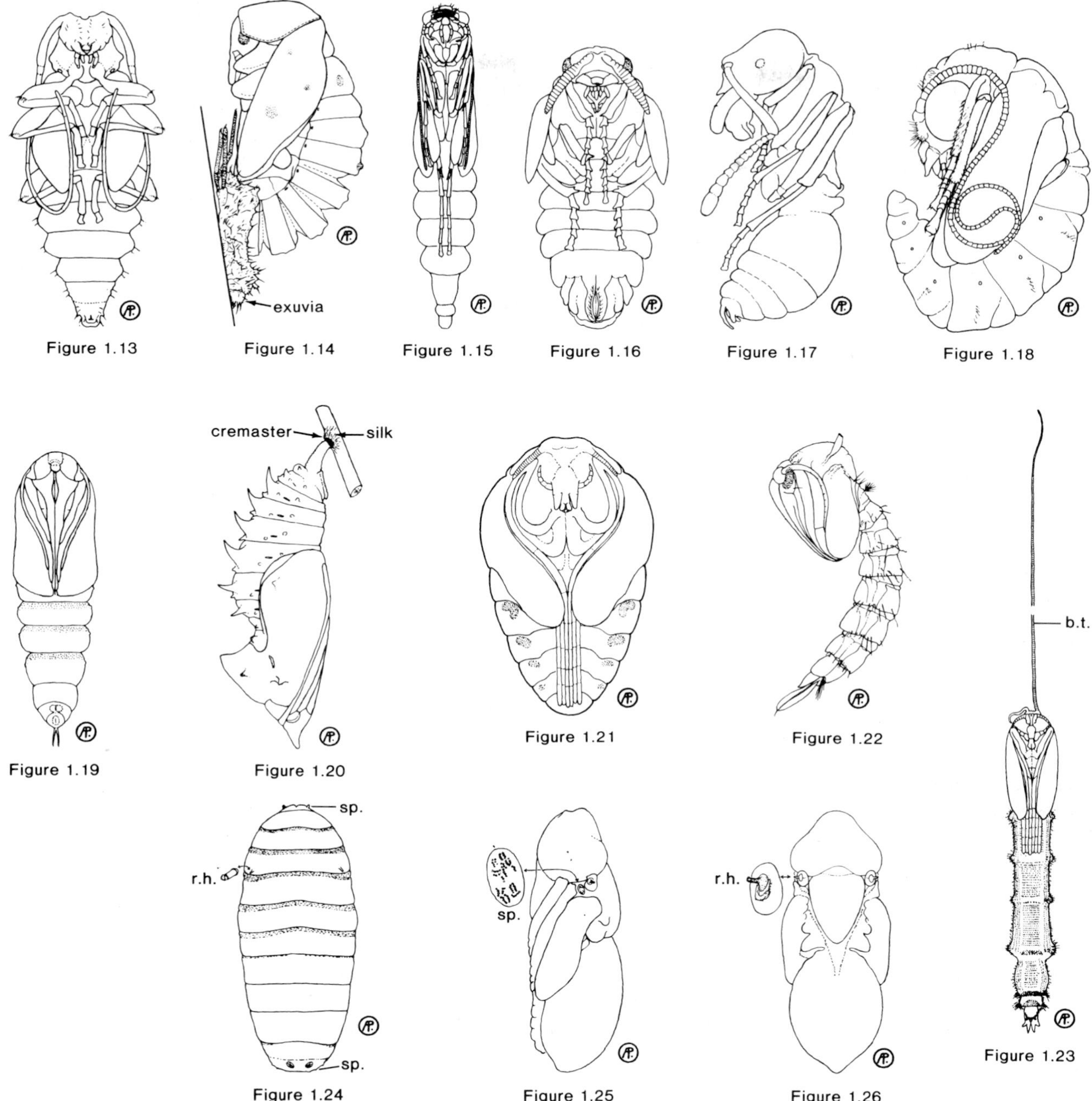

Figures 1.13–1.26. Types of pupae; figs. **1.13–1.18,** exarate; figs. **1.19–1.23,** obtect (from Peterson 1948 and 1951).

Figure 1.13. Exarate pupa, ventral (Coleoptera: Cerambycidae).

Figure 1.14. Exarate pupa, lateral (Coleoptera: Coccinellidae).

Figure 1.15. Exarate pupa, ventral (Trichoptera).

Figure 1.16. Exarate pupa, ventral (Hymenoptera: Diprionidae).

Figure 1.17. Exarate pupa, lateral (Hymenoptera: Formicidae).

Figure 1.18. Exarate pupa, lateral (Neuroptera: Chrysopidae).

Figure 1.19. Obtect pupa, ventral (Lepidoptera: Noctuidae).

Figure 1.20. Obtect chrysalis, hanging, lateral (Lepidoptera: Nymphalidae).

Figure 1.21. Obtect pupa, ventral (Diptera: Blephariceridae).

Figure 1.22. Obtect pupa, lateral (Diptera: Culicidae).

Figure 1.23. Obtect pupa with breathing tube (b.t.), ventral (Diptera: Ptychopteridae).

Figure 1.24–1.26. Puparium and pupa (Diptera:Calliphoridae). **(1.24)** puparium, dorsal; **(1.25)** lateral, and **(1.26)** dorsal, newly formed pupa extracted from puparium, with prothoracic and caudal spiracular scars (sp.) and respiratory horns (r.h.)

CLASSIFICATION

The recognition here of 34 orders (and four classes) does not duplicate any currently published list, but is more a synthesis of the views of the contributors. It is probably closest to that of Borror, Delong, and Triplehorn (1981), with Protura, Collembola, and Diplura recognized as classes (and orders), with their "Orthoptera" recognized as five orders (Blattodea, Mantodea, Grylloblattodea, Phasmatodea, and Orthoptera) and their "Neuroptera" as three orders (Megaloptera, Raphidiodea, and Neuroptera). For each order a synoptic listing of the classification used is presented near the key to families.

General discussion of philosophical and evolutionary questions relating to immatures is not presented; however, evolutionary considerations are frequently presented under "Relationships and Diagnosis" for orders and families.

SELECTED BIBLIOGRAPHY

Borror, D. J., D. M. DeLong , and C. A. Triplehorn. 1981. An introduction to the study of insects. 5th ed. Philadelphia, Penn.: Saunders College Publ., 827 pp.

Böving, A. G., and F. C. Craighead. 1931. Larvae of Coleoptera. Brooklyn Ent. Soc. 351 pp.

Brigham, A. R., W. U. Brigham, and A. Gnilka (eds.) 1982. The aquatic insects and oligochaetes of North and South Carolina. Midwest Aquatic Enterprises, Mahomet, Ill. 837 pp.

Brues, C. T., A. L. Melander, and F. M. Carpenter. 1954. Classification of insects. Bull. Mus. Comp. Zool. Cambridge, Mass.: Harvard University. 917 pp.

Chapman, R. F. 1982. The insects: structure and function. Cambridge, Mass.: Harvard Univ. Press. 919 pp.

Chen, S. H. 1946. Evolution of the insect larva. Trans. Roy. Ent. Soc. London 97:381–404.

Chu, H. F. 1949. How to know the immature insects. Dubuque, Iowa: Wm. C. Brown Co. 234 pp.

Comstock, J. H. 1918. Nymphs, naiads and larvae. Ann. Ent. Soc. Amer. 2:222–24.

Hinton, H. E. 1963. The origin and function of the pupal stage. Proc. Roy. Ent. Soc. London, A, 38:77–85.

Hinton, H. E. 1971. Some neglected phases in metamorphosis. Proc. Roy. Ent. Soc. London, C, 35:55–64.

McAlpine, J. R. et al. 1981. Manual of Nearctic Diptera, vol. 1. Res. Branch, Agr. Canada, Monograph No. 27. 674 pp.

McCafferty, W. P. 1981. Aquatic entomology. Boston, Mass., Science Books. Int. 448 pp.

Merritt, R. W., and K. W. Cummins (eds.). 1984. An introduction to the aquatic insects of North America. 2d ed. Dubuque, Iowa: Kendall/Hunt Publ. Co., 710 pp.

Mosher, Edna. 1916. A classification of the Lepidoptera based on characters of the pupa. Bull. Ill. St. Lab. Nat. Hist. Vol. 12:17–159.

Paulus, H. F. 1981. Eye structure and the monophyly of the Arthropoda, chap. 6, 299–383. *In* Gupta, A. P. 1981. Arthropod phylogeny. New York: Van Nostrand Reinhold Co. 726 pp.

Peterson, A. 1948. Larvae of insects. Part 1. Lepidoptera and Hymenoptera, 315 pp. Printed for the author by Edwards Bros., Ann Arbor, Mich.

Peterson, A. 1951. Larvae of insects. Part 2. Coleoptera, Diptera, Neuroptera, Siphonaptera, Mecoptera, Trichoptera, 416 pp. Printed for the author by Edwards Bros., Ann Arbor, Mich.

Richards, O. W., and R. G. Davies. 1977. Imm's general textbook of entomology. 10th ed. London: Chapman & Hall. New York: John Wiley & Sons. Vol. 1, Structure, physiology and development, pp. 1–418; vol. 2, Classification and biology, pp. 419–1354.

Snodgrass, R. E. 1935. Principles of insect morphology. New York: McGraw-Hill. 667 pp.

Snodgrass, R. E. 1954. Insect metamorphosis. Smithsonian Misc. Coll. 122, 124 pp.

Waterhouse, D. R., et al. 1970. Insects of Australia. Carlton, Victoria: Melbourne Univ. Press. 1029 pp. Supplement, 1974, 146 pp.

Wigglesworth, V. B. 1964. The hormonal regulation of growth and reproduction in insects. Adv. Ins. Physiol. 2:247–336.

Techniques for Collecting, Rearing, Preserving, and Studying Immature Insects

2

Frederick W. Stehr
Michigan State University

COLLECTING

The different places where immature insects are found and can be collected are best described by saying that they occur everywhere except in the air (normally) and in salt water (but see Cheng (1976) for a book on all insects associated in any way with the marine environment).

Many caterpillars, sawfly larvae, some beetle larvae, numerous nonholometabolous larvae (nymphs), and assorted others feed externally on plants and are easily collected by sweeping, beating, or searching for damaged or webbed parts of plants or for the larvae themselves. Although galls are formed by many species, one must be careful in assuming that the immature(s) found within galls caused formation of the gall, since parasitoids and hyperparasitoids are frequently present. Felt's (1918) well-illustrated book on American insect galls is still the best available aid for the identification of gall-formers, although Johnson and Lyon (1976) cover many of those found on trees and shrubs. Leaf miners are common on many plants. The mines are species-characteristic for some, but representatives of all four major orders are commonly encountered, with all but the Diptera being superficially quite similar in appearance. Be sure of the order before trying to key leaf miners to family.

Many other immatures are borers in all parts of living, dying, dead, decaying, and decayed plant parts. Borers are commonly indicated by dead twigs, buds, and cones, damaged fruits, exuding sap, and accumulations of frass. There is a whole succession of immatures that lives beneath the bark and in the wood of trees as they decay. These are best collected by peeling, prying, splitting, and chopping the wood apart. In addition, numerous species feed on or bore in all parts of the root system.

A great variety of immatures inhabiting ground litter and soil can be collected with anything from a hand trowel to a plow, but other more specialized and labor-saving methods such as Berlese or Tullgren funnels and pitfall traps are very effective for many groups. A pitfall trap with a flashlight bulb rigged up over it is very effective for collecting Mecoptera larvae (especially Bittacidae), which are otherwise almost impossible to spot in the litter because of their branched scoli and adhering debris.

Special microcosms containing an assortment of immatures include dung, fungi, and dead animals, all of which support an array of species that changes relatively rapidly over time. Many larvae immediately burrow into or retreat to burrows in the soil when dung or carcasses are turned over, so the soil should be dug up and examined. Nests and dens of birds and mammals are sources of assorted scavengers and are of course the major source of flea larvae.

The freshwater aquatic environment has its own multitude of species, with several orders and other taxa restricted to water in the larval stages. Other not truly aquatic forms bore in or feed on aquatic vegetation or live on the surface or shores. The oceans are nearly devoid of insect life, but beach debris at the driftline contains a great assortment of immatures. Special nets, screens, or grabs are needed for collecting some aquatics, but many can be obtained by simple examination of rocks, vegetation, mud, and other debris by hand or in a white pan.

Bee and wasp larvae must be obtained from their nests, which vary from being solitary and undefended to being colonial and well defended, making collection of some species difficult, if not hazardous.

Parasitoid larvae are relatively easy to collect by dissecting or holding the hosts, but because identification even to family may be difficult, rearing to adults is encouraged. First instars of parasitoids, either triungulins of hypermetamorphic species or those of many other species, may be very different from other instars; they frequently have mandibles and tails that are lost in later instars. These may not key out, but see the illustrations following the Key to Orders.

Larvae of predators can be found everywhere their prey is found, so one can expect to find them when collecting other immatures. Other predators that ambush or trap their prey may be found in special locations such as at the bottom of conical pits in sand or loose soil for antlion larvae or in vertical burrows in bare areas for tiger beetle larvae (indicated by round holes about the size of a pencil (spiders and bees make similar holes; ant holes have soil particles around them)). Predators are also commonly found near aggregations of prey such as aphid colonies; chrysopid, coccinellid, and cecidomyiid larvae are common, and unexpected ones like the larvae of the "Harvester", *Feniseca tarquinius,* a lycaenid predator of wooly aphids, are occasionally found.

REARING

Any student of immatures should have as an important objective the rearing to adults of as many unknown larvae as possible, since classifications have been largely based on adults, and the immature stages of many species are unknown or at least not associated with *reared* adults. Such reared material is essential for the advancement of our knowledge of immature insects. Natural enemies may also be reared; they should always be saved and properly labelled to associate them with the immatures they emerged from.

Although rearing may seem simple, quite often it isn't, since the proper food, temperature, humidity, and pupation sites must be provided in a relatively confined situation in the lab or in the field. Hence, the more mature the larvae are, the more successfully they are reared, but the early instars are obviously missed if only mature larvae are reared. Absolute conspecificity of the field-collected preserved larvae with reared adults is also frequently not possible (even though it may be probable) since similar larvae of different species may be collected together. Diapause can be a frustrating problem, too, in those species that overwinter as larvae, prepupae, and pupae. Diapausing stages usually must be held for three months or longer at temperatures cooler than 10°C to be reasonably sure of breaking diapause and obtaining adults. Numerous specialized techniques for rearing different groups are widely scattered through the literature, especially as "scientific notes" or in columns such as the "Culture Corner" of the *Lepidopterists' News*. Many rearing and collecting techniques that are broadly useful for many groups are covered by Peterson (1953), Martin (1977), Borror, Delong and Triplehorn (1981), and Merritt and Cummins (1984).

A better alternative to rearing late instars (except for the length of feeding necessary) is to obtain eggs from females and to rear larvae to maturity while preserving samples of all instars. This positively associates all instars with the female, which can usually be identified to species. And of course, if males are needed for identification, some will usually be obtained if larvae are reared through. When individuals are selected for preservation, a sample representative of the variability in size, structure, color, and any other diagnostic features should be taken. Don't, for example, select the largest ones, assuming them to be more mature, since female immatures are larger than males in many species.

Ideally, some specimens of all stages are preserved and some are reared to adults, but problems arise when only one specimen is collected. Obviously it can't be preserved, so it should be photographed if possible, and detailed notes and sketches should be made. In addition the cast skin should be preserved (in 70% ETOH) for later examination since all external features will be present and the skin can be softened for study or slide mounting. Some insects eat the cast skin immediately after molting, so unless the specimen is pupating the skin may not be available. Others molt in protected situations where the skin is hard to find.

Rearing and Transporting Living Material

Specialized containers have to be used for some groups, but polyethylene bags can be used for most groups in many different ways. Some immatures with chewing mouthparts will chew their way out so the heavier the material, the better. Polyethylene is ideal in many respects because it is light, compact, cheap, waterproof, and disposable, yet gases such as O_2 readily pass through it. Leaf-feeding insects can be maintained on cut foliage for several days in polyethylene bags, and rotten wood, fungi, and such materials can be kept moist. Some aquatic insects can be temporarily held in polyethylene bags filled with water (fish are often shipped this way), but bags of water in the field are not ideal. Holding larvae in bottles of water is unsatisfactory since the oxygen is rapidly used up or lost and cannot be replenished. Many aquatic forms can be kept alive in good condition for several days by holding them in a fairly tight container or polyethylene bag containing wet sphagnum moss or other suitable moist material, because the integument and respiratory surfaces remain moist amid an abundant oxygen supply. Avoid leaving closed polyethylene bags in the sun since inside temperatures may reach lethal limits for the organisms inside. For terrestrial insects, condensation in bags may be a problem if material is held for several days, but bags can be changed or reversed and paper towels can be added to absorb excess moisture.

Temperature regulation of the organisms during the collecting process is obviously desirable, but refrigeration is rarely available in the field. Ice chests are convenient and practical if ice is available, since specimens can be placed in just about any container (preferably dry but with obvious exceptions) and kept inactive. Plastic jugs or bottles partially filled with water and then frozen will eliminate the melt water problem, and they can be refrozen. Sealed, commercially available refreezable plastic and metal containers are equally useful. Ice chests are ideal, since no food is required, predators need not be isolated, and one need not be concerned with overheated bags and cars.

It is essential to maintain proper moisture and humidity, and too much water is as harmful as too little. Many larvae that live in soil, litter, logs, fungi, stems, under rocks, and in similar places are living under saturated or nearly saturated atmospheric conditions. In the field, if conditions become unfavorable they can move toward more optimum conditions. Such movement is impossible or greatly restricted in the lab since moisture gradients are not present in most cages, bags, or other containers, but a technique using plaster of Paris blocks has worked quite well for many groups. The following technique described by Snider et al. (1969) for soil arthropods should work for organisms that require relative humidity approaching 100%.

1. Mix dry: 50% by volume powdered activated charcoal or boneblack, 50% by volume plaster of Paris.
2. Add distilled water to make a slurry that can be poured into containers or molds.
3. Let air-dry completely.
4. Saturate with distilled water when needed for rearing. Add water as necessary during rearing. Once moistened, do not let the plaster dry out or it will separate from the sides of the container.
5. Containers with snap caps normally do not need any ventilating punctures. Containers with gaskets and screw caps need a very small hole in the lid to avoid CO_2 buildup.
6. The plaster can be cleaned by scraping the surface.

Steel (1970) describes the use of connected paired cells (one for food, one for cover) in plaster of Paris blocks for rearing staphylinid larvae from egg to adult and describes the methods used for preparation and examination of the larvae. H. H.

Neunzig (personal communication) recommends the use of cardboard food containers partially filled with moist sand, which keeps the plant food fresh, provides proper humidity, and provides pupation sites in the soil when needed.

Proper food is essential; the obvious choice is the same food larvae were feeding on when collected (if it can be determined). Leaf feeders are perhaps easiest to rear since cut foliage is usually satisfactory and readily available. Vegetation (especially the tougher, more leathery kinds) that may not be locally available can be misted lightly and held for one to several weeks in plastic bags under refrigeration. Solutions are also available that help maintain cut foliage in better condition (for example, Wellington 1965, p. 12), and satisfactory artificial diets are available for some species, with new ones continually being developed. One never knows if a diet is satisfactory until it is tried, so valuable material should be fed a "natural" diet if at all possible, although natural diets have their limitations, too (Scribner, 1977). For phytophagous species (especially Lepidoptera), artificial diets that contain some natural materials such as beans and wheat germ have been quite successful for a diversity of species (Hinks and Byers 1976; Neil 1984). Vanderzant (1974) has reviewed the subject of artificial diets for insects, and Singh (1977) provides a bibliography of artificial diets for insects and mites. King and Leppla (1984) have edited a comprehensive handbook on insect rearing that covers everything from rearing a few to full commercial production.

A simple technique that appears to work quite well for many woodborers and may work for a variety of other larvae that eat relatively solid materials is to feed them Post "Grape Nuts" cereal as it comes from the box. One to 2 cm of cereal in a container the larvae can't chew through works well. Control of humidity does not appear to be necessary unless the larvae actually feed on fungi associated with the food material, in which case humidity control and addition of some of the natural food material may be necessary.

The number of specimens per rearing container will be highly variable depending on the species, but predators and other aggressive species (such as various grubs) may have to be isolated to avoid loss, damage, or both.

KILLING, FIXATION, AND PRESERVATION

It should be noted that many of the materials used in fixatives, killing solutions, and preservatives are toxic, flammable, potentially explosive, or carcinogenic; **good judgment must be exercised in the use, storage, and disposal of all of them.**

Most immature insects are not satisfactorily preserved if pinned, so they must be killed and preserved in other ways (usually in fluids). The ideal technique should easily and permanently preserve the insect in a nearly natural condition with regard to color, shape, and flexibility of internal and external structures. No single technique currently meets these requirements, so one must select the most acceptable one(s) for particular needs and conditions. It is also evident that very little *research* has been done on the development of better methods of preservation. Kelsheimer's (1928) paper seems to have been a start, and Houyez (1978) has described a promising method for the injection and pinning of larvae of Lepidoptera (including green ones) with viscous materials. Currently, in North America, Dr. C. Romero-Sierra and the Biological Preservation Group, Department of Anatomy, Queen's University, Kingston, Ontario, are engaged in the development of methods of preservation of biological materials (see Romero-Sierra and Webb (1983) for a brief history and discussion of the potential for the science of preservation (diatirology)).

There are advantages in bringing live specimens to the laboratory for rearing, killing, and fixation, since just plain hot water is one of the best all-purpose killing agents. However this is not always practical, so cold solutions for use in the field have been developed. If one examines the ingredients in cold killing and fixing solutions recommended for immature insects and other groups, it is immediately obvious that most of them contain distilled water, acetic acid, and an alcohol of some kind (usually ethyl or isopropyl), plus an assortment of other ingredients. The acetic acid serves two functions: it stops enzymatic action and thus avoids darkening, and it helps retain flexibility. Alcohol is the preservative; usually the concentration is 70–80% in order to retain flexibility and avoid the shriveling that frequently occurs at higher concentrations. Acetic acid and alcohol alone make up acid alcohol (one part glacial acetic acid; nine parts 70% ETOH), which is a good killing agent by itself, provided the specimens are distended by killing in hot water or by injection orally or anally.

Distention is obtained in noninjected cold solutions by the addition of a penetrating agent such as kerosene, which in turn may require the addition of emulsifiers to make it miscible with the alcohol (95% is normally used to minimize the water). **NOTE:** For large or sclerotized larvae such as many macrolepidoptera, Corydalidae (Megaloptera) etc., anal or oral injection may be necessary to obtain properly distended larvae.

Table 2.1 summarizes by orders some commonly used killing and preservation solutions. Color is not retained satisfactorily by any of these materials.

Additional Killing, Preservation, and Examination Suggestions

1. Select larvae for preservation that are not newly molted or ready to molt, since the old skin and head capsule of premolting larvae may separate from the new one beneath, and newly molted larvae do not distend as well, making it more difficult to see setae and other structures.
2. One to five percent glycerin is often added to the alcohol to avoid the possibility of total dehydration, especially if vials are stoppered with corks or other materials that may not seal tightly. Cleared specimens of all orders can be stored in glycerin, as can uncleared specimens (but see text).
3. Write the killing technique and preservation fluid on the label; this lets one know what fluid to add or examine specimens in, but most important, it lets future workers determine if the methods used were suitable for long-term storage.

Table 2.1. Some commonly used killing and storage solutions.

Order	Killing Solution	Permanent Storage
Protura* Collembola*	95% ETOH	95% ETOH (colors fade in weaker solutions)
Diplura* Microcoryphia* Thysanura*	75–80% ETOH	75–80% ETOH Store in individual vials if cerci are fragile
Ephemeroptera Odonata Plecoptera	75–80% ETOH	75–80% ETOH Can be preserved directly in 95% ETOH
Grylloblattodea, Blattodea, Mantodea, Phasmatodea, Orthoptera, Dermaptera, Isoptera, Zoraptera, Psocoptera, Mallophaga, Anoplura,* Hemiptera, Embiidina,* Strepsiptera, Siphonaptera,* Homoptera (Auchenorrhyncha)	75–80% ETOH	75–80% ETOH
Homoptera (Sternorrhyncha)		
Many Coccoidea* Armored, pit, and soft scales	75–80% ETOH Attached to host plant, preserved dry in envelopes	75–80% ETOH
Aphidoidea	95% ETOH Preserve galls dry or in alcohol	95% ETOH
Aleyrodoidea	95% ETOH Keep puparia dry on host plant	95% ETOH
Psyllidae	95% ETOH Preserve galls dry	95% ETOH
Thysanoptera*	10% ETOH Adults—A.G.A.[a]	60% ETOH
Megaloptera	KAAD, acid alcohol[b] Inject large specimens orally	75–80% ETOH
Neuroptera Raphidiodea Mecoptera	KAAD; hot water	75–80% ETOH
Trichoptera	Kahle's; hot water; KAAD Remove some if in cases	75–80% ETOH
Coleoptera	KAAD; hot water; Acid alcohol[b] Avoid solutions containing formalin (Kahle's)	75–80% ETOH Acid alcohol[b]
Lepidoptera	KAAD; hot water Inject large ones orally; Acid alcohol[b] if injected. Remove some if in cases.	75–80% ETOH
Hymenoptera		
Symphyta	KAAD; hot water	75–80% ETOH
Apocrita*	Kahle's; hot water; KAAD	75–80% ETOH
Galls, leaf mines, and other damage	Dry preservation is superior to fluids in order to avoid discoloration or bleaching. Photographs are recommended.	

[a]A.G.A.: 60% ETOH, ten parts; glycerine, one part; glacial acetic acid, one part.
[b]Acid alcohol: nine parts 70% ETOH, one part glacial acetic acid.
*See text for these orders for additional details.

4. Specimens killed in KAAD can be preserved in 95% ETOH, but should be put in 75–80% ETOH if structures such as mouthparts will need to be moved.
5. Specimens killed in hot water should NOT be placed in 95% ETOH—they usually collapse. Put them in 75–80% ETOH or run them up gradually.
6. If ethyl alcohol is not obtainable (most private collections) 70% isopropyl alcohol for both killing agents and preserving is generally acceptable.
7. Large larvae of all orders are best preserved by oral or anal injection with a hypodermic.
8. Examine specimens in the same concentration of alcohol they are stored in—if they float, add 95% ETOH to the dish until they sink (or wait a while).
9. Submerge specimens completely for examination to avoid distortion and surface reflections.
10. Do not completely fill individual storage vials that cannot be totally submerged—three-fourths to four-fifths is best; many structures can be quickly examined without removing the specimen from the vial (a full vial creates distortions). Microvials or other small vials placed within storage vials or jars should be totally filled since they can be completely submerged for examination.

Inflation

Inflation can be used for medium to large larvae, and the technique has been extensively used in the past, primarily for Lepidoptera. Briefly described, first roll the internal contents out through a posterior cut, snugly clamp the larva on a tapered glass tube of suitable diameter, blow up the larva with a hand-operated rubber bulb provided with a rubber pressure reservoir, and maintain pressure until the larva is dried out over a small "oven" heated with an alcohol burner or hot plate (anything to provide heat, but keep direct flames away). Then glue inflated specimens to twisted wires, balsa wood, or other similar material, attach to insect pins, and handle like other pinned insects. Colors (except green, which is not retained well by any method) preserve well, but specimens are brittle, often "bloated," the posterior has been damaged, internal structures are gone, and further dissections of mouthparts and other structures are difficult. Therefore, as a method to prepare specimens for display, inflation has some merit (freeze drying is better—see below), but for the preservation of research specimens it is minimally acceptable (most useful for rapid examination of external structures). See Peterson (1948) or Martin (1977) for more details.

Injection

Houyez's (1978) injection method involves dehydration and elimination of gut contents and fats by refrigeration at 4°C for 3–21 days (depending on the species) before preservation; anal injection to maximum distention with 1–2% Titriplex III to prevent darkening (Merk Laboratories, always mixed and injected with plastic utensils); evisceration, followed by injection with one of two materials: (1) 50 ml each of bees wax and soft high-grade silicone vacuum grease mixed with three teaspoons of Aerosil (Edwards), an industrial emulsifier and stabilizer, or (2) aquarium sealer. Both can be tinted to desired colors. See the article (in French) for details, but freeze drying (see below) is the simplest method to preserve greens if the equipment is available.

Freeze Drying

Freeze drying takes advantage of the well-known phenomenon of freezer burn in the home freezer. Specimens are frozen in the desired position and the water is removed by sublimation under a partial vacuum. A good description of equipment and a discussion of principles is available in Dominick (1972) and Roe and Clifford (1976), and other papers on the subject are cited. Freeze drying has obvious advantages over inflation since colors are preserved as well or better (including greens to some degree), the internal contents remain, the posterior is not damaged, and the specimen is not distorted. Specimens are excellent for display purposes but are less satisfactory for detailed research because of the difficulty in making further dissections. The critical point process described on p. 16 for the preparation of scanning electron microscopy specimens can also be used, but color retention is poor.

Chemical Drying

Another method that produces specimens similar to the freeze-drying process (but with more loss of color) and that only requires access to a hood is what can be termed the "chemical drying method." It is very similar to the process used to prepare specimens for slide mounting in Canada balsam except for the final step, and the method is currently being used for the preparation and study of Lepidoptera larvae by P. T. Dang of Ottawa (personal communication). It is useful to take color photographs before proceeding as follows:

1. Use living specimens, or clean specimens thoroughly if removed from alcohol by cleaning in a detergent solution in an ultrasonic cleaner (place specimens in vials to reduce the vibration and possible loss of setae, especially if they have been in 95% ETOH).
2. Kill in KAAD of a suitable strength until distended as desired (usually 5–30 minutes).
3. Transfer to 95% ETOH for 24–48 hours, depending on size.
4. Transfer to absolute ETOH for 24–48 hours, depending on size.
5. Transfer to clean absolute ETOH for 24 hours since *all* water must be removed.
6. Transfer to xylene for up to 24 hours, depending on size.
7. Remove, place on paper towels, and dry in a hood or outdoors if hood is not available.
8. Pin as you would an adult insect, or glue the right side of the specimen to a pin with white glue (soluble in water if specimen must be removed).

If dissections must be made, use one of the methods for restoring dehydrated specimens (p. 15) to soften them. The advantages of preserving immatures by inflating, freeze drying, critical point or chemical drying, or injection-distention with pastes are several, including:

1. They can be stored with associated, reared, adults;
2. Each specimen has its own label;
3. External morphology is easier to study;
4. Many specimens can be examined quickly;

5. The possibility of total loss of preservative is avoided, but dermestids must be protected against; and
6. Some colors are preserved.

The major disadvantage is the need to soften specimens if dissections or manipulations of structures such as mouthparts must be made.

Cold Killing Solutions

KAAD was developed by Alvah Peterson in the 1940s for field use. It was originally composed of kerosene (K), ethyl alcohol (A), glacial acetic acid (A), and dioxane (D) (not dioxin, the carcinogen). Various modifications have been suggested by others, primarily regarding the replacement of dioxane with ionic detergents (D) or other alcohols since dioxane's function is to make the kerosene miscible with the alcohol. Dioxane presents some risk of explosion if stored in pure form longer than 12 months (*National Safety News,* March 1976).

Ethyl alcohol (95%) is used in KAAD, but other alcohols such as isopropyl have been used. Kerosene is the penetrating agent and distends the specimens. The amount of kerosene can vary considerably, depending on the specimens to be preserved. Up to three parts are used for large Lepidoptera, but as little as 0.1 part for many maggots is enough, with greater concentrations often bursting them. One part is used in standard solutions prepared for most uses. The lowest grade (light yellow) kerosene ("coal oil") is miscible with alcohol, but it is often difficult to get; more highly refined grades of kerosene and fuel oil may not be miscible.

The formula with some modifications for KAAD is:

Component	Parts
Kerosene (or coal oil)	1 (0.1 for Diptera)
Alcohol, 95% ethyl	10
Acetic acid, glacial	2
If the kerosene is not miscible with the alcohol, use one of the following ionic detergent emulsifiers (dioxane is *NOT* recommended):	
"Triton X-100," "Tween" or "Lubrol (X,W)"	0.2 or more
or	
Iso-butyl alcohol or sec-butyl alcohol	1–5

These "parts" are approximate. More or less may be needed, depending on the kerosene. Add and shake until the solution is uniform and clear.

KAAD cannot be made with 70% ETOH, because of the miscibility problem between the kerosene and the 30% water in the 70% ETOH. Isopropyl alcohol is more miscible with kerosene than ethyl alcohol and can be substituted if an emulsifier is not available. It does not penetrate as well or as fast, so specimens may have to be left in it longer; by the same token, it does not burst or explode larvae as readily as ethyl alcohol.

Specimens should be put in KAAD while alive since dead specimens usually do not distend satisfactorily (they should be injected). They should be left in KAAD until fully distended; this may vary from a few minutes to one-half hour or longer, depending on the size and condition of the specimens. Specimens should be removed within 24 hours, or sooner if clearing or separation of the cuticle occurs. However, it must be emphasized that no killing agent works satisfactorily for everything, and there is almost unlimited room for improvement. Separation of the cuticle or unsatisfactory distention can often be avoided by killing the larva in the solution and then immediately injecting it with the same or a weaker solution.

Peterson (1948) recommended transfer to 95% ethyl alcohol for permanent storage since specimens did not become brittle or shrivel after being killed in KAAD. It is true they do not become brittle, but the tissues are hardened enough so that movement and/or dissection of mouthparts and other structures is difficult. Therefore, specimens are best transferred to 70%–80% ETOH, despite the fact that partial collapse of some specimens may occur. If collapse is evident or external bubbles appear, immediate transfer to 95% ETOH is recommended. Do not pack specimens into vials; the body fluids dilute the alcohol and poor preservation results. The alcohol should be changed at least once before permanent storage, especially for large specimens.

Direct injection orally or anally with a hypodermic needle of a killing and preserving agent is without doubt the most foolproof method for moderate to large specimens (longer than 1 cm). Godfrey (1972) gives the following method used by J. G. Franclemont at Cornell (kerosene is not needed since injection provides distention).

1. Kill in a mixture of nine parts 70% ETOH and one part glacial acetic acid (= acid alcohol).
2. Inject anally with same solution until properly distended.
3. Return to original solution for 24 hrs.
4. Store in 70% ETOH.

(Wear rubber gloves if you inject, no matter what the solution.)

Ethyl alcohol (95%), although sometimes used "as is" for a few groups such as Plecoptera, Ephemeroptera, and Odonata, is unsatisfactory for most immatures (especially soft-bodied ones) because of dehydration and the resultant shrivelling and hardening of tissues, which makes examination and dissection difficult. If nothing else is available, a 70% mixture is far better.

Isopropyl alcohol (rubbing alcohol) can be used, but it may not be as satisfactory as ethyl (in addition, the odor is objectionable).

Formalin causes shrivelling and hardening similar to 95% ethyl alcohol, and in addition to being obnoxious to use, it is listed as a carcinogen. Moreover, tissues fixed in solutions containing formaldehyde may not dissolve well in caustic solutions such as KOH. It is used in some fixatives such as Kahle's and Pampel's (see below).

Kahle's (also called Dietrich's) is a good fixative for internal tissues of both immatures and adults. It is used for killing immatures such as wasp and bee larvae by some, and Trichoptera and Diptera by others, but it may not work well for other groups. The use of formalin is objectionable and questionable for the reasons given above, and proper precautions should be observed, but since specimens are normally transferred to 70–80% ethyl alcohol for permanent storage, exposure is reduced. Pampel's fluid is nearly identical to Kahle's (four parts of acetic acid instead of one). It can be used in the same way as Kahle's, but gives the best results when used as an intermediate fixative after specimens are killed in hot water and prior to permanent storage in 70–80% ethyl alcohol. This procedure is reported to be especially good for small larvae.

	Kahle's	Pampel's
Ethyl alcohol, 95%	15	15
Distilled water	30	30
Formalin (= 40% formaldehyde)	6	6
Acetic acid, glacial	1	4

Barber's has been used for killing and reclaiming heavily sclerotized or dried-out larvae (such as wireworms) and adults, primarily because it is a good relaxer. However, it contains benezene, a known carcinogen, so it is *not recommended.* See "Restoring Dehydrated Specimens", p. 15 for alternatives.

Special solutions are used for some taxa and numerous solutions for various purposes have been proposed. Many of them can be found in commonly available publications such as Martin (1977), Pennack (1978), and Borror, DeLong and Triplehorn (1981). Any special techniques for various orders are given in that section or in table 2.1. For valuable larvae or very large ones it may be desirable to puncture the integument for rapid penetration of the solution or to inject them until fully distended, using a hypodermic needle. Use good rubber gloves if your hands contact the solutions since most of the chemicals are irritating to some degree, and some individuals are hypersensitive.

Hot Killing Agents

Hot water: Variations of this technique are some of the most universally satisfactory methods for killing the widest diversity of specimens (without using injection, which is difficult for small specimens). Hot water also stops enzyme action, and nearly everything can be satisfactorily killed (and also cleaned of debris). Many different variations have been proposed, but two basic ones are apparent.

1. Drop the specimen in boiling (or near boiling) water and let it cool (preferred). Specimens will normally not be harmed by gentle boiling, but some may burst, especially larger, soft-bodied ones. They usually straighten out, but it may be necessary to shake the specimen or place it in the desired position until the tissues have been coagulated.
2. Drop the specimen in cold water and heat to a boil (or near boiling). Small amounts of ethyl alcohol can be added to enhance relaxation and straightening out. Specimens are normally transferred to 70% ethyl alcohol, but placing them in warm Kahle's or Pampel's promotes better fixation of internal tissues; they can also be transferred temporarily to more dilute alcohol if collapse occurs (rare).

Eggs and Pupae

Hinton (1979) provided an introduction via literature citations to all sorts of techniques involving eggs but did not cover preservation methods. In fact, in comparison to larvae, relatively little has been done on the preservation of eggs and pupae, and the advent of scanning electron microscopy (ideal for eggs) is probably going to result in the preservation of many "gold-plated" eggs, which of course are colorless. Good color photography would appear to be a useful part of SEM work.

Eggs can be freeze dried as can pupae. Those eggs with a rigid chorion can also be air dried if they are killed before hatching (freezing or heat of 125–135°F is good, but heat may collapse many of them (Peterson 1960)). The shells of hatched eggs may be nearly as good if the larvae don't eat them or if they can be prevented from eating them as McFarland (1972) has pointed out. Soft-shelled eggs can be preserved in solutions similar to those used for larvae; Peterson (1960) suggests using a standard KAAD mixture made with isopropyl alcohol and diluted four or five times to prevent the inner tissues from separating from the chorion. Kahle's is also suggested.

Mosher (1916) does not mention how the Lepidoptera pupae she studied were preserved; however, pupae are frequently preserved in the same ways as larvae, but without the distending agent and with the use of isopropyl alcohol. Laurent LeSage (personal communication) reports good results for beetle pupae by placing them in hot water for 10–30 seconds, transferring them directly to a large volume of pure isopropyl alcohol for 24 hours, and then storing them in pure isopropyl.

LABELING

Data labels and determination labels must always be separate since data labels are permanent records, whereas determination labels can be changed. Small labels like those used on most pinned insects are undesirable for use in vials since much of the time they end up facing each other or facing the specimen or tilting so they are impossible or difficult to read. Labels approximately the length of the inside of a two-dram vial (45 mm) and wide enough to curl one-third to one-half of the way around the inside of a vial (20 mm) are ideal, since they stay in place against the vial wall, are easily read,

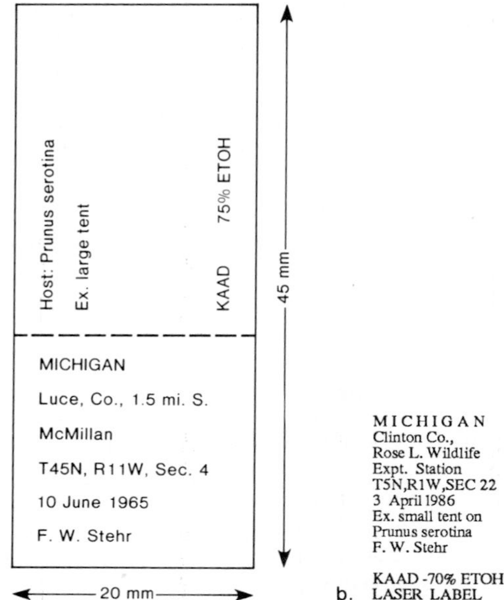

Figure 2.1. a. A label for vials that provides easy reading of locality data with the vial upright. b. A computer generated label printed with a desktop laser printer at 50% reduction. Long term durability is unknown, but seems OK.

and provide a uniform, light background to view specimens against. Determination labels of the same size can be curled back to back so the data label is facing outward and the determination label inward. The determination label should be the inner one since the inner label is more easily added or replaced. The same size of label can also be used in larger vials.

Labels should be printed or written on a high-quality rag paper in permanent ink that does not dissolve in the preservative fluid. Let the ink dry thoroughly before placing the label in the vial. Labels should read from the bottom to the top of the vial for ease in scanning a rack of vials.

A useful variation on data labels used by some institutions is to print the collection data horizontally across the bottom of the label so the label can be read when the vial is sitting upright in a rack, and to write habitat or ecological data and preservation data on the upper part of the label as in figure 2.1a. The new desktop laser printers make it possible to print immediately all data (fig. 2.1b). Township (T), Range (R) and Section (S) coordinates are "permanent" and locate the site within one square mile. They are not usually printed on ordinary highway maps because the scale is too small, but the coordinates can be obtained from county maps, topographic maps, and plat books. Roads and cities are subject to changes in their names, locations, and distances, so use both systems if possible.

PERMANENT STORAGE

When diluting alcohol and other solutions by adding water, always use distilled water if available (or rainwater, even though it may be acidic), because tap water may be quite hard or may contain other substances that react adversely. To dilute to a lower percent proceed as follows: to dilute 95% ETOH to 70% ETOH, take 70 units of 95% ETOH and add distilled water until 95 units of liquid is obtained (in other words, take 70 units of 95% and add 25 units of distilled water); the resulting solution is 70% ETOH.

Ethyl alcohol (70–95%) is the most widely used permanent storage fluid because it is relatively cheap, easy to get (except by private individuals), clear, nontoxic (externally), and nonodorous. The major problem is rapid evaporation in poorly sealed containers. Screw caps with cardboard liners, polyethylene snap caps, and corks are all relatively unsatisfactory because some of them will leak, requiring frequent checking and addition of alcohol, and risking the possible loss of some specimens due to complete drying. Glycerin (1–5%) can be added to the alcohol to keep the specimen moist and soft if all the alcohol evaporates. Ordinary black rubber stoppers are unsatisfactory because of excessive swelling and the high sulphur content, which badly discolors the alcohol (and specimens) with time. Neoprene rubber stoppers have been satisfactorily used for long-term storage, but they are not immune to swelling, since they may be affected to some degree over a period of time by many of the agents used in killing solutions, even if the quantities are not great. Even alcohol alone (plus water) may cause some swelling since stoppers used with 75% ethyl alcohol often show swelling whereas those used with pure 95% ethyl alcohol (with no remnants of killing agents present) swell only minimally or not at all. Stoppers should fit snugly so that a minimum of the stopper is exposed to the alcohol.

Screw caps with conical polyethylene liners (Polyseal® is one brand) are the best method to prevent loss of alcohol from individual vials, since these caps totally prevent evaporation when screwed down tightly, and the polyethylene is not affected by the various chemicals in killing solutions. Most immature insects are small enough to fit in two-dram vials, which most collections use as a standard size, with larger sizes used as needed.

Before the availability of neoprene stoppers and polyethylene-lined screw caps, many collections used shell vials filled with alcohol and plugged with cotton, with ten or more vials per larger screw cap bottle, so only the large bottles had to be replenished with alcohol. If this method is used, the vials should be inverted (with the cotton at the bottom) so the cotton does not act as a wick when the alcohol in the large bottle drops below the top of the vials. This method is inconvenient if specimens must be examined often.

P. T. Dang (1983) uses pure glycerin as a preservation fluid with satisfactory results. Color is retained quite well, but specimens cannot be killed in hot water and placed directly in glycerin because they shrivel and discolor rather quickly unless the following steps recommended by Dang are taken:

1. Kill the larva in near-boiling water (95–98°C) one to three minutes, depending on size.
2. Transfer to cold water (5–10°C) to stop the "cooking."
3. Leave the larva in the water and make one or two small holes (00 or 000 insect pin) at each intersegmental area so glycerin can penetrate more easily and shrivelling or collapse is prevented when the larva is transferred to pure glycerin for storage.

Specimens I have tried did not shrivel when transferred to glycerin from 70% ETOH, but did become somewhat darker than those retained in 70% ETOH. Specimens killed in KAAD, preserved in 95% ETOH, and transferred to glycerin did not shrivel or darken.

The advantages of glycerin are its lack of evaporation in storage, lack of fumes during examination, and slow leaching of colors. Specimens in glycerin need no special treatment for shipping because its relatively high viscosity limits damage—specimens don't "rattle around." Nematologists have been preserving nematodes for 50 years or more in glycerin with good success, so there is reason to believe it may work for insects. The major disadvantage is the viscosity and messiness in working with it, making the use of a "finger bowl" of water or alcohol for cleaning a necessity. A simple way to gradually transfer specimens to pure glycerin is to place them in a solution of glycerin (5%) and 70% ETOH (95%) and then permit the alcohol and water to evaporate, resulting in a gradual change to 100% glycerin.

Various kinds of vial racks have been devised, but the two most important considerations are visibility of specimens and labels without removing the vials and a way to keep vials from tipping over when the rack is not full. Plastic racks have been available commercially in two- and four-dram sizes, and although they are not cheap, they probably are the best long-term solution to these problems.

Restoring Dehydrated Specimens

Several techniques for restoring such specimens have been used, but if polyethylene-lined screw caps, neoprene stoppers, or glycerin have been used they should not be needed. Five possible techniques are:

1. A 0.5% solution of trisodium phosphate (Van Cleave and Ross, 1947); may take up to two days; can be transferred to 80% ethyl alcohol.
2. Weak KOH (3–5%). Carefully warm specimen until it is nearly normal in size and shape, or leave it overnight if only the chitinized exoskeletal parts are wanted (similar to a genitalic preparation). Stretching may be necessary.

   Rinse in distilled water for methods 1 and 2. Neutralize any remaining KOH by use of acetic acid (5–10%) for a few minutes. The specimen can often be transferred directly to 70% ethyl alcohol, but if shrivelling occurs, run it through 25% and 50% alcohol.
3. A 50–80% solution of warm lactic acid; leave in for two days, then place back in 70% ETOH.
4. Propylene glycol (Marhue 1983 and Thompson et al. 1966); soak in solution of 50% propylene glycol and 50% distilled H_2O for 24 hours, or longer if necessary; transfer to 70% ETOH for storage.
5. Surfactant; Banks and Williams (1972) soaked dried coccids, thrips, and other insects requiring the gentle softening of hardened or shrivelled tissues in Decon 90®[1] to soften them and to remove wax and other materials. A 1:1 solution of Decon 90 and 5% KOH was used when large amounts of resins were present.
   1. Soak overnight;
   2. Heat container in boiling water *bath* if specimen is not reinflated;
   3. Soak in cold distilled water until reinflated (usually about two hours).

1. Decon 90: Decon Laboratories, Ltd., Ellen St. Portslode, Brighton, Sussex, England, BNY 1E0. An alkaline detergent such as Multi-Terge®, available from American Scientific Products, may work equally well.

Cast skins may be the only material available in some situations, but the thorax and abdomen are usually crumpled together so structures cannot be seen. The above techniques may work, but Hinton (1956) suggests the following procedure for making mounts of cast skins (of Lepidoptera). Place the skin for two or three days in 10% KOH in an open watch glass. He says the solution absorbs CO_2 from the atmosphere, and some carbonates are formed. When the cuticle is soft, acetic acid (he doesn't say what strength) is added. The carbon dioxide evolved from the carbonates enclosed by the cuticle is supposed to blow the cuticle out to its original shape, and the thorax and abdomen can then be examined or mounted on a slide.

POSITIONING SPECIMENS IN FLUIDS FOR EXAMINATION

Specimens must be held firmly in diverse positions for study, and placing them in a dish of alcohol is not very satisfactory. Better methods are use of boric acid ointment in alcohol, or use of a small culture dish or stender dish partially filled with clean, washed sand and enough alcohol to cover the specimen being examined. The thinner the layer of alcohol covering the specimen, the better, since distortion is greater because of convection currents set up in deeper layers of alcohol by the heat of the light. The newer fiber optic lights avoid the heat problem. Specimens can be held in any desired position by pushing them far enough into the sand so they stay in place. Very delicate specimens also can be positioned, but a hole should be dug and sand filled around the specimen to avoid any possible damage. The best sand is white with rounded grains for easier movement of the specimen, but other kinds of clean sand will work, and dark sand is sometimes better for light specimens. Glycerin with small glass beads also works well, especially for persons who are irritated by alcohol fumes.

In cases where setae or other minute structures are difficult to see, the specimen should be cleared with 10% KOH and examined with light transmitted from below. A stain may show up minute structures even better. Stains such as 1% mercurochrome (in water) or carbol fucsin (buy prepared, or dissolve 4 gm basic fucsin and 8 ml phenol in 20 ml 95% ETOH, and add 100 ml H_2O) have worked well.

A method for converting small Lepidoptera larvae into "pelts" is as follows:

1. Slit larva on right side above the prolegs (larvae are normally examined from the left side). If too small to slit, perforate in several places with needle or pin.
2. Place in 10% KOH overnight.

3. Remove specimen to 70% ETOH for clearing. A small hypodermic can be used for internal flushing.
4. Transfer pelt to proper solvent for stain to be used. Chlorozol Black is used as an example below.
5. Add enough Chlorozol Black crystals to 70% ETOH (95% causes precipitation) to form deep blue solution.
6. Stain pelts for 5 to 30 minutes (or longer) until they are evenly stained. Do not overstain, since the pelts will not destain in alcohol.
7. Transfer to 70% ETOH for rinsing. Pelts can be stored in microvials with other larvae since they do not destain in 70% ETOH. [Courtesy of D. L. Wagner]

SLIDE MOUNTS

Small specimens of any order must often be slide-mounted for study (either temporarily or permanently), in order to use transmitted light with a compound or phase contrast microscope. Methods and materials are quite diverse, but there are three basic groups, based on degree of permanence.

1. *Temporary,* using water, alcohol, or preferably glycerin, which does not evaporate and is soluble in both water and alcohol.
2. *Questionably permanent,* but longer-lasting nonresinous media, such as Hoyer's and CMCP-9, which should be ringed for long-term storage.
3. "*Permanent*", using Canada balsam or synthetic media such as Permount®. Preparation of Hoyer's and CMCP-9 is given below, and slide-mounting techniques are given in the introductory parts for some orders. See especially Collembola, Thysanoptera, Anoplura, Coccoidea, and Siphonaptera.

Hoyer's Medium

Distilled H_2O	50 g
Powdered gum arabic (clear crystals)	30 g
Chloral hydrate	200 g
Glycerine	20 g

1. Dissolve gum Arabic in water (24–36 hrs.);
2. Filter at room temperature;
3. Add chloral hydrate and dissolve;
4. Add glycerine;
5. Filter through clean fiberglass gauze;
6. Store in glass-stoppered bottle.

NOTE: Hoyer's is hygroscopic. Under humid conditions cover slips will loosen and slip unless the mount is carefully ringed. Hoyer's is a mild clearing agent.

Preparation of CMCP-9

1. Mix 7.5 g PVA with 112 ml of distilled water.
2. Dissolve 44 g of phenol crystals (Reagent) in 44 ml of lactic acid (white USP) in hot water bath.
3. Add part "2" to "1" and heat in water bath.

NOTE: Do not use liquid phenol—it is highly toxic. Here PVA is Dupont Elvanol®. Formula courtesy of TURTOX/CAMBOSCO, MacMillan Science Co., Chicago, Illinois.

Specimens can be mounted directly in CMCP-9 from water, alcohols, glycerin, and solutions containing formalin. It can be thinned with water. CMCP-9 is also a clearing agent, clearing Collembola scales within 24 hours, a big advantage if species must be determined.

SCANNING ELECTRON MICROSCOPY

Scanning electron microscopy (SEM) is a very useful research and photographic tool for immature insects, but it has practical limitations for "routine" identifications and student use in systematics classes because of the time, expense, and technical knowledge involved in specimen preparation and SEM operation.

SEM is most useful for observing minute structures that were previously difficult or impossible to resolve. It is also very helpful in the photographic illustration of these very small structures and of larger structures that are visible with light microscopy but are difficult to draw because of the great depth of field or their complexity.

Most research workers and students can obtain access to an SEM and trained technicians to operate it. However, proper preparation of specimens for examination by SEM is essential, and many technicians may not have had experience with immature insects. Hence, the following techniques used by Don Davis, Department of Entomology, Smithsonian Institution, are provided (the specimens used for his excellent photographs in this book were prepared this way).

Try to select recently collected specimens (within the last five years) to increase the chance of clean material. Prior cleaning of larvae in an ultrosonic cleaner is usually not advisable because of the possible breakage of setae (but see "chemical drying method", p. 11.) Although simple air-dried samples of some structures are satisfactory (head capsules are frequently okay), larvae should be dried with the *critical point process* to prevent shrinkage. The advantage of this over freeze drying is that the critical point apparatus is much less expensive and usually provides faster results. It takes about 30 minutes to dry as many as four or five specimens once they have been placed in the critical point chamber. This method is particularly suitable for specimens preserved in alcohol.

First, run specimens through a step series (70–90) to absolute ETOH and then up through 50% ethyl alcohol/50% amyl acetate to 100% amyl acetate. Normally, this is done over a 12-hour period. Then quickly remove the specimens from amyl acetate and place them in small porous plastic containers, which are quickly loaded into a mesh basket and lowered into the critical point chamber.

Essentially, this process replaces all liquids in the specimens with liquid CO_2, which is then quickly brought to the critical point (about 1060 psi) by submerging the chamber in warm water. As the CO_2 gas is bled off above the critical temperature, the specimens remain unchanged and dry. Coat specimens with a thin layer of carbon by evaporation and sputter-coat them with a gold palladium alloy.

A method for the killing and preparation of immatures for SEM examination that have water- and alcohol-impermeable cuticles (for example, the sugar beet root maggot, *Tetanops myopaeformis* (Roder) (Diptera:Otitidae), and others with more permeable cuticles), is given by Bjerke, Freeman, and Anderson (1979), using acidified 2,2-dimethoxypropane (which they recommend should be used with a fume hood and other safety precautions since its toxicology is unknown).

SEM specimen storage: For convenience, glue SEM specimens to 12 mm glass coverslips, then lightly attach the coverslips with a carbon glue to 12 mm metal SEM stubs. After use, these are popped off the stubs with a razor blade and stored in cardboard paleontological slide mounts. These are well-slides of varying depths, with a plastic cover that slides over the well. Most specimens can be stored in regular slide collection trays. Retain larger specimens on stubs and store them in the usual bulky fashion in standard stub boxes.

ILLUSTRATING

Good illustrations are essential for good and usable systematics publications. Nearly all of the illustrations in this book are SEM photos, black and white photos, color slides printed as black and whites, or line drawings, preparation of all of which requires some special skills or equipment. There are almost limitless books on photography; there are also several books on scientific illustration that may be useful. See Zweifel (1961), Papp (1968) and Hodges (in press).

SHIPPING

The primary consideration in shipping specimens is to make sure they cannot be bounced around by air bubbles in the preservation fluid, which frequently results in broken or denuded specimens. Such damage can be prevented by enclosing them in smaller vials stoppered with cotton, by holding them in place with cotton wads, by removing all air bubbles, or by using a viscous fluid such as glycerin. Release air bubbles in rubber-stoppered vials by temporarily inserting a fine insect pin alongside the stopper as it is inserted. Tape or block stoppers in place since temperature or pressure changes can pop them out. Screwcap vials sealed with conical polyethylene liners should not contain bubbles if the vial is overflowing as the cap is screwed on. For shipment, vials must be wrapped in paper or foam rubber or otherwise padded or secured so they don't bump against each other. They should, of course, be sent in mailing tubes or packed in sturdy boxes with a minimum of two inches of padding. A technique used by Gary Ulrich, University of California, Berkeley, to keep a vial in the center of the loose polystyrene pieces commonly used for shipping is to wrap tape around the vial with the sticky side out; the polystyrene pieces adhere to the tape, assuring cushioning and preventing the vial from shifting to the edge of the carton.

ACKNOWLEDGMENT

The translation of the Houyez (1978) paper by Suzanne Allyson, Biosystematics Research Institute, Ottawa, is greatly appreciated.

SELECTED REFERENCES

Banks, H. J., and D. J. Williams. 1972. Use of the surfactant, Decon 90, in the preparation of coccids and other insects for microscopy. J. Australian Entomol. Soc. 11:347–48.

Bjerke, J. M., T. P. Freeman, and A. W. Anderson. 1979. A new method of preparing insects for scanning electron microscopy. Stain Technology 54:29–31.

Borror, D. J., D. M. Delong, and C. H. Triplehorn, 1981. An introduction to the study of insects. Fifth ed. New York: Saunders College Publishing. 827 pp.

Cheng, Lanna (ed.). 1976. Marine insects. New York: North Holland Publ. Co. (Am. Elsevier Publ. Co.) 581 pp.

Dang, P. T. 1983. Glycerin as a preservative for immature insects, p. 69. *In* Faber, D. J. (ed.) Proc. of 1981 Workshop on Care and Maintenance of Natural History Collections. Syllogeus, No. 44, 196 pp. Nat. Mus. of Canada, Ottawa.

Dominick, R. B. 1972. Practical freeze-drying and vacuum dehydration of caterpillars. J. Lepid. Soc. 26(2):69–79.

Felt, E. P. 1918. Key to American insect galls. N.Y. State Mus. Bull. 200, 310 pp.

Godfrey, G. L. 1972. A review and reclassification of larvae of the subfamily Hadeninae (Lepidoptera:Noctuidae) of America north of Mexico. USDA Tech. Bull. No. 1450. 265 pp.

Hinks, C. F., and J. R. Byers. 1976. Biosystematics of the genus *Euxoa* (Lepidoptera:Noctuidae). V. Rearing procedures, and life cycles of 36 species. Can. Entomol. 108:1345–57.

Hinton, H. E. 1956. The larvae of the species of Tineidae of economic importance. Bull. Ent. Research 47(2):251–346.

Hinton, H. E. 1979. Biology of insect eggs. Vol. 1, Chap. 12, Techniques. pp. 312–16.

Hodges, E. R. S. (ed.) The guild handbook of scientific illustration. New York: Van Nostrand Reinhold. In press.

Houyez, P. 1978. La preparation des chenilles et des larves. Annales Soc. Roy. Zool. Belg. 107:91–100. (in French).

Johnson, W. T., and H. H. Lyon, 1976. Insects that feed on trees and shrubs. Ithaca, New York: Cornell Univ. Press. 464 pp.

Kelsheimer, E. G. 1928. The preservation of immature insects. Ann. Ent. Soc. Amer. 21:436–44.

King, E. G., and N. C. Leppla. 1984. Advances and challenges in insect rearing. ARS, USDA, U.S. Govt. Printing Office, Washington, D.C. 306 pp.

McFarland, N. 1964. Notes on collecting, rearing, and preserving larvae of macrolepidoptera. J. Lepid. Soc. 18:201–10.

McFarland, N. 1965. Additional notes on rearing and preserving larvae of macrolepidoptera. J. Lep. Soc. 19:233–36.

McFarland, N. 1972. Notes on describing, measuring, preserving, and photographing the eggs of Lepidoptera. J. Res. Lepid. 10:203–14.

McFarland, N. 1973. Some observations on the eggs of moths and certain aspects of first instar larval behavior. J. Res. Lep. 12:199–208.

Marhue, L. 1983. Techniques to restore dried-up invertebrate specimens, pp. 175–177. *In* Faber, D. J. (ed.). Proc. of 1981 Workshop on care and maintenance of natural history collections. Nat. Mus. of Canada, Ottawa, Syllogeus, No. 44, 196 pp.

Martin, J. E. H. (ed.). 1977. The insects and arachnids of Canada. Pt. 1. Collecting, preparing, and preserving insects, mites, and spiders. Can. Dep. of Agr., Publ. 1643, 182 pp.

Merritt, R. W., and K. W. Cummins. 1984. An introduction to the aquatic insects of North America. Dubuque, Ia.: Kendall/Hunt Publ. Co. 710 pp.

Mosher, E. 1916. A classification of the Lepidoptera based on characters of the pupa. Bull. Ill. State Lab. Nat. Hist. 12:17–159. (27 pls.)

Neil, Kenneth, 1984. Lepidoptera reared on a simple wheat germ diet. J. Lep. Soc. 37:311–13.

Papp, C. S. 1968. Scientific illustration. Theory and practice. Dubuque, Ia.: Wm. C. Brown Co. 318 pp.

Pennack, R. W. 1978. Freshwater invertebrates of the United States. 2d. ed. New York: John Wiley and Sons. 803 pp.

Peterson, A. 1953. A Manual of Entomological Techniques, 367 pp. Printed for the author by Edwards Bros., Ann Arbor, Mich.

Peterson, A. 1948. Larvae of insects. Part 1. Lepidoptera and Hymenoptera, 315 pp. Printed for the author by Edwards Bros., Ann Arbor, Mich.

Peterson, A. 1960. Photographing eggs of insects. Florida Ent. 43:1–7.

Roe, R. M., and C. W. Clifford. 1976. Freeze-drying of spiders and immature insects using commercial equipment. Ann. Ent. Soc. Amer. 69:497–99.

Romero-Sierra, C., and J. C. Webb. 1983. The potentials for diatirology, p. 21–28. *In:* Faber, D. J. (ed.). Proc. of 1981 Workshop on care and maintenance of natural history collections. Nat. Mus. of Canada, Ottawa, Syllogeus, No. 44, 196 pp.

Scribner, J. M. 1977. Limiting effects of low leaf-water content on the nitrogen utilization, energy budget and larval growth of *Hyalophora cecropia* (Lepidoptera:Saturniidae). Oecologia 28:269–287.

Singh, P. 1977. Artificial diets for insects, mites, and spiders. New York: Plenum Publ. Corp. 606 pp.

Snider, R., J. Shaddy, and J. W. Butcher. 1969. Culture techniques for rearing soil arthropods. Mich. Ent. 1:357–62.

Steel, W. O. 1970. The larvae of the genera of the Omaliinae (Coleoptera:Staphylinidae) with particular reference to the British fauna. Trans. Roy. Ent. Soc. London 122(1):1–47.

Thompson, R. J., M. H. Thompson, and S. Drummond. 1966. A method for restoring dried crustacean specimens to taxonomically usable condition. Crustaceana 10:169.

Van Cleave, H. J., and J. A. Ross. 1947. A method for reclaiming dried zoological specimens. Science 105:318.

Vanderzant, E. S. 1974. Development, significance, and application of artificial diets for insects. Ann. Rev. Ent. 19:139–60.

Wellington, W. G. 1965. Some maternal influences on progeny quality in the western tent caterpillar, *Malacosoma pluviale* (Dyar). Can. Ent. 97(1):1–14.

Zweifel, F. W. 1961. A handbook of biological illustration. Chicago, Ill.: Univ. Chicago Press. 131 pp.

Key to Orders of Immature Insects and Selected Arthropods[1,2]

3 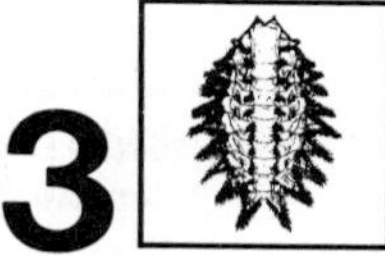

Stuart B. Hill,[3]
Macdonald College, McGill University

Frederick W. Stehr
Michigan State University

Wilbur R. Enns
University of Missouri

KEY TO ORDERS

You may start at the following couplets if the following is true:

0–3 pairs of true, segmented legs ... 1
4 or more pairs of true, segmented legs ... 76
Primitively wingless "insects" ... 6
Insects ectoparasitic on mammals, birds, or honeybees ... 11
Holometabolous insect larvae (and a few Homoptera) with:
  Thoracic legs on 2 or more segments ... 17
  Thoracic legs absent or 1 small pair ... 36
Hemimetabolous larvae (nymphs) and brachypterous or wingless adult insects[4] ... 54
Most Arthropoda other than insects ... 76
Arachnida ... 77

1. **Thoracic legs:** 0–3 pairs ... 2

1′. **Thoracic legs:** 4 or more pairs ... (most Arthropoda other than Insecta) 76

2(1). **Thoracic legs:** present on 2 or more segments, usually clearly segmented and with claws, but may be reduced to conical segmented or unsegmented clawless protuberances, or to minute segmented legs with claws ... 3

2′. **Thoracic legs:** absent, or 1 small pair (may have fleshy swellings in their place) or a single, sometimes branched prothoracic proleg ... 36

3(2). **Antennae:** branched ... (fig. 3.1) (most first instar) Class **PAUROPODA**

3′. **Antennae:** unbranched or absent ... 4

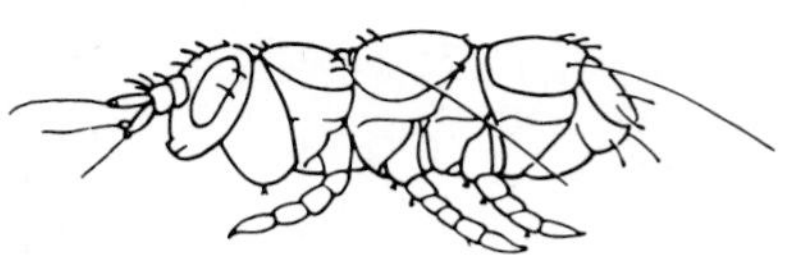

Figure 3.1

1. Including insect larvae, wingless adult insects, major groups of terrestrial and freshwater Arthropoda exclusive of plankton, and selected other invertebrates of America north of Mexico. Early instars of many will key out, but the earlier the instar, the greater the chance that it will not key correctly. Illustrations of selected first instar parasitoids are given at the end of the key (pp. 45, 46).

2. Usually larvae are easily distinguished from pupae, but some pupae, especially Neuroptera, Megaloptera, Trichoptera, Raphidiodea, and the most primitive Lepidoptera, have movable mouthparts, and some pupae can move about to a certain extent. Pupae of Raphidiodea, the most mobile, are capable of walking.

3. Expanded, modified, and reorganized by F. W. Stehr, with assistance by S. B. Hill and W. R. Enns, from a key to terrestrial, nonparasitic Canadian arthropods compiled by S. B. Hill.

4. Separation of brachypterous and wingless adults from immatures is possible in some taxa by nongenitalic characters, but examination for fully formed genitalia is necessary in others. See the individual order for details.

4(3′). **Head:** usually evident
Body segmentation: nearly always clearly evident; may not be evident if scalelike
Abdomen: distinguishable from rest of body; may not be if scalelike
Legs: present or absent
Antennae: present or absent
Size: minute to large 5

4′. **Head:** not evident, reduced to a capitulum (figs. 3.2, 3.3)
Body segmentation: usually not evident
Abdomen: broadly fused to rest of compact and ovoid body
Legs: 3 pairs
Antennae: absent
Size: minute .... (figs. 3.2, 3.3) (first instar larval ticks and mites) Class **ARACHNIDA,** Subclass **ACARINA**

5(4). **Segmented legs:** on 1st, 2nd, and 3rd segments behind the head (or only 2 pairs), 1st segment not as below 6

5′. **Segmented legs:** on 2nd, 3rd, and 4th segments behind the head, 1st body segment (collum) legless, wide dorsally and narrow ventrally .... (fig. 3.4) (first instar millipedes) Class **DIPLOPODA**

6(5). **Antennae:** absent
Forelegs: not used for walking but held forward like antennae
Abdomen: pairs of ventral appendages (styli) on first 3 segments; terminal appendages absent
Size: under 2 mm .... (fig. 3.5) immature and adult **PROTURA** (p. 47)

6′. **Antennae:** 1 pair, though may be greatly reduced or hidden beneath
Forelegs: used for walking (or rarely, for grasping); greatly reduced in some
Abdomen: ventral appendages usually absent; if present, then not on all of first 3 segments or not confined to them; terminal appendages present or absent
Size: often over 2 mm 7

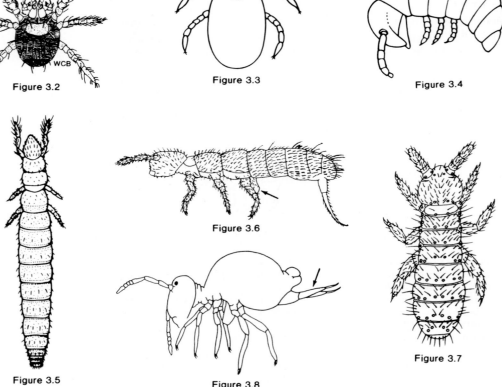

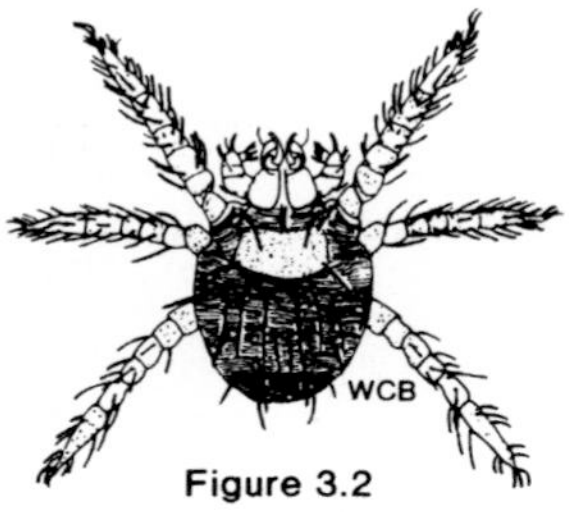

Figure 3.2

Figure 3.3

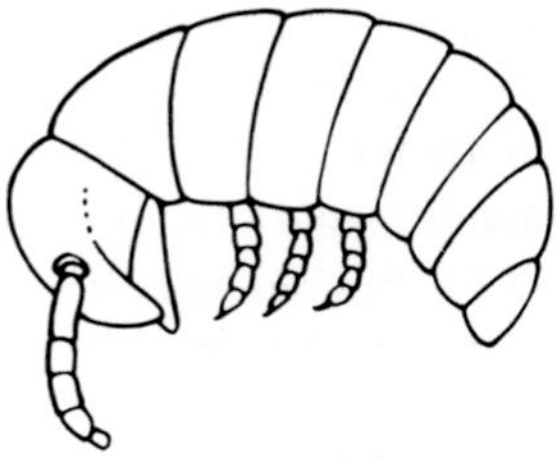

Figure 3.4

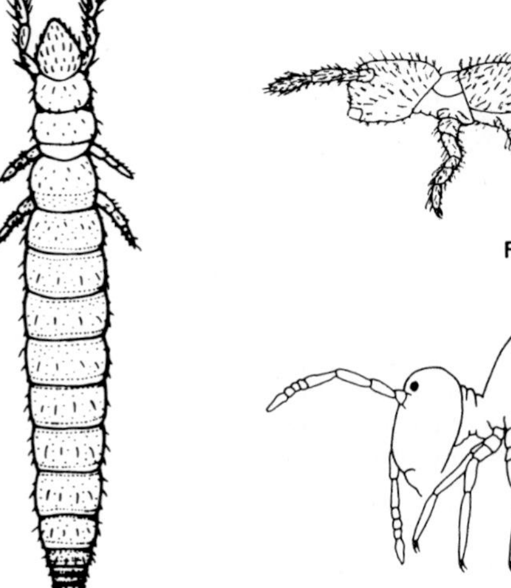

Figure 3.5

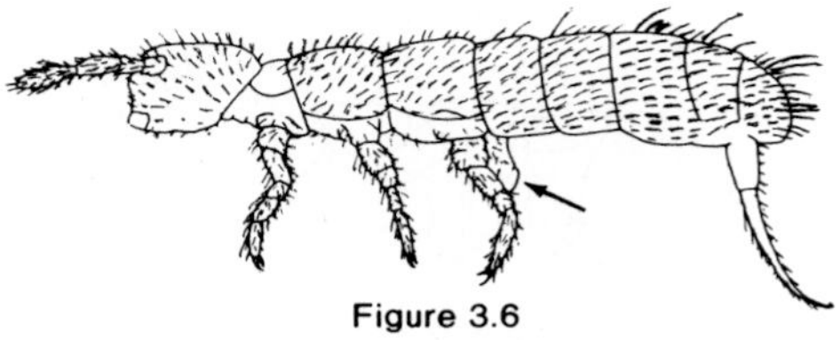

Figure 3.6

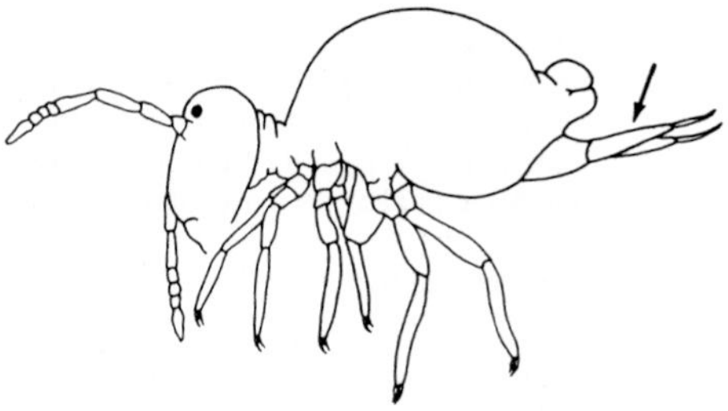

Figure 3.7

Figure 3.8

7(6′). **Abdomen:** with single, often forked, ventral tube (collophore) (fig. 3.6) on segment 1, and usually with forked jumping organ (furcula) on segment 4 or 5 (fig. 3.8); 6 segments (often indistinct due to fusion)
Size: small to minute (most under 6 mm) .... (figs. 3.6, 3.7, 3.8) immature and adult **COLLEMBOLA** (p. 55)

7′. **Abdomen:** collophore and furcula absent; more than 6 segments
Size: often over 6 mm 8

8(7′). **Abdomen:** 2 or more of segments 1–9 with pairs of ventral styli (fig. 3.9); 2 or 3 terminal appendages present (filaments (fig. 3.9) or forceps (fig. 3.10)) 9

8′. **Abdomen:** ventral styli absent (may have prolegs or appendagelike gills) or, more rarely, with 1 posterior pair of styluslike appendages (wingless Mantodea, Blattodea, Grylloblattodea, Phasmatodea, and Isoptera); terminal appendages present or absent 11

9(8). **Abdomen:** 2 terminal appendages (cerci) that may be forcepslike (Japygidae, fig. 3.10) or filamentous (Campodeidae, fig. 3.11); pairs of ventral styli on segments 1–7 or 2–7
Tarsi: 1 segment
Compound eyes: absent
Body: neither scaly nor patterned, usually white .... immature and adult **DIPLURA** (p. 65)

9′. **Abdomen:** 3 terminal filamentous appendages (2 cerci and a median caudal filament); pairs of ventral styli on segments 2–9, or 7, 8 and 9, or 8 and 9
Tarsi: 2 to 4 segments
Compound eyes: present (rarely absent—in Nicoletiidae)
Body: scaly and often patterned 10

10(9′). **Body:** somewhat cylindrical
Thorax: arched
Compound eyes: large, usually touching
Abdominal styli: on segments 2–9 .... (fig. 3.12) immature and adult **MICROCORYPHIA** (p. 68)

10′. **Body:** somewhat dorsoventrally flattened
Thorax: not arched
Compound eyes: small and widely separated or absent
Abdominal styli: variable in number and location .... (fig. 3.13) immature and adult **THYSANURA** (p. 71)

11(8′). **Habitat:** ectoparasites of birds, mammals, or honeybees, usually found on host
Body: usually strongly flattened dorsoventrally or laterally, more or less leathery 12

11′. **Habitat:** not ectoparasitic on birds, mammals, or honey bees (phoretic first instars (triungulins) of Meloidae, Rhipiphoridae, and Mantispidae may be tightly wrapped around honeybee hairs)
Body: usually not strongly flattened (may be if on plants) and not leathery 16

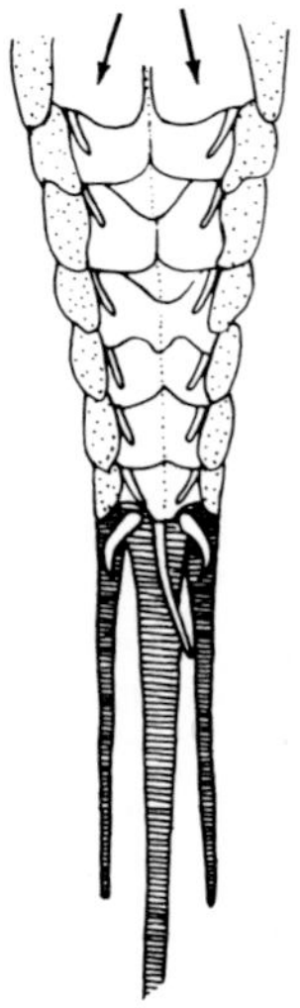

Figure 3.9

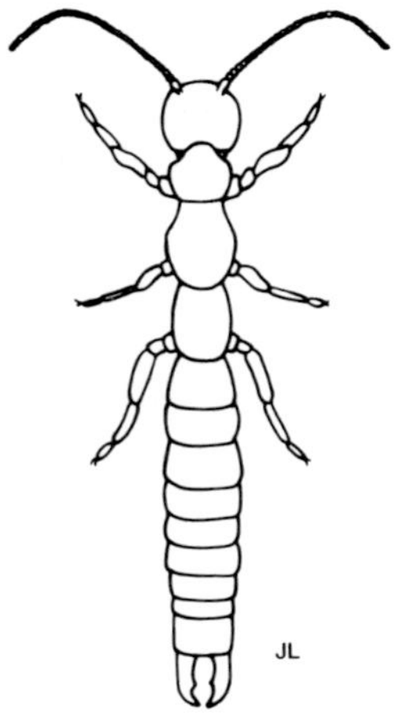

Figure 3.10

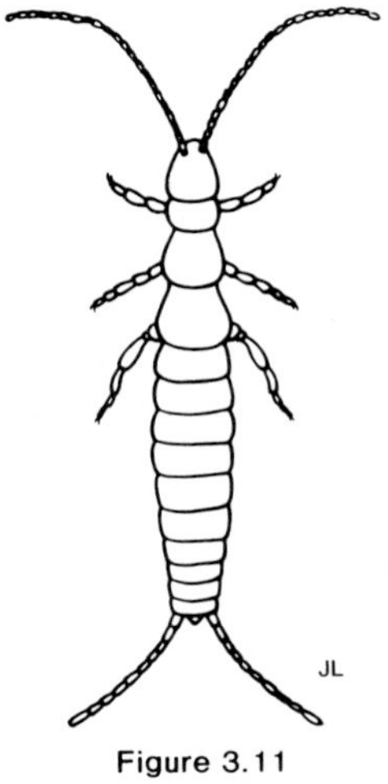

Figure 3.11

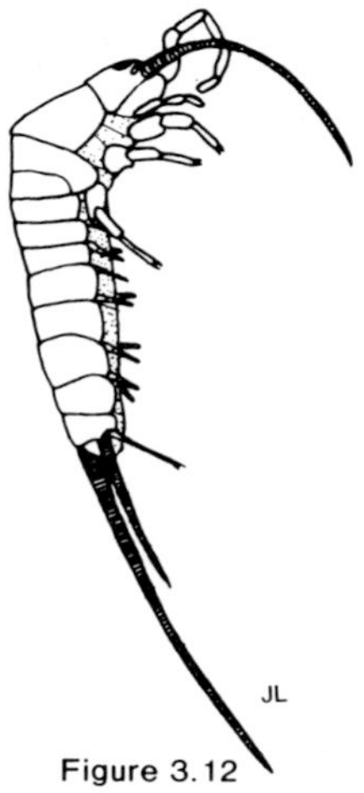

Figure 3.12

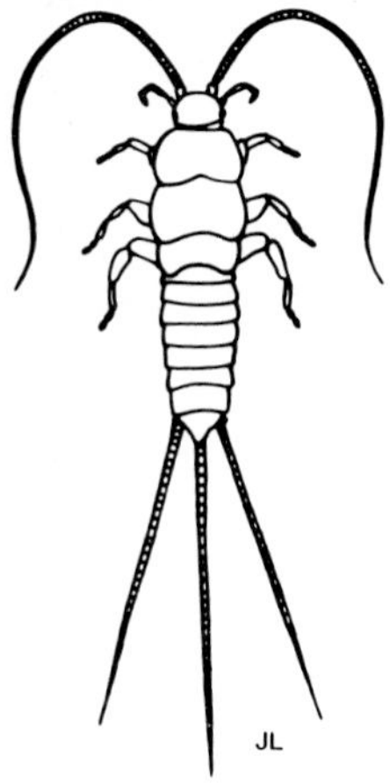

Figure 3.13

12(11). **Tarsi:** 5 segments
Antennae: short and usually concealed in grooves 13

12′. **Tarsi:** 1–4 segments
Antennae: variable 14

13(12). **Body:** flattened laterally
Legs: ventrally extended
Behavior: usually jumping .... (fig. 3.14) adult **SIPHONAPTERA**

13′. **Body:** flattened dorsoventrally
Legs: laterally extended
Behavior: not jumping .... (fig. 3.15) (few) adult **DIPTERA**
(bee lice (Braulidae), louse flies (Hippoboscidae), and bat flies (Nycteribiidae and Streblidae)) (larvae of all except Braulidae retained in uterus until mature)

14(12′). **Tarsi:** 1 segment
Antennae: shorter than head .... (lice) 15

14′. **Tarsi:** 3 segments
Antennae: longer than head ....
.... (figs. 3.16, 3.16a) (adult and immature bed, bat, and bird bugs) **HEMIPTERA** (Vol. 2)

15(14). **Head:** as wide as or wider than prothorax
Mouthparts: opposable, mandibulate (chewing)
Habitat: parasitic on birds (2 claws) or mammals (1 small or large claw) ....
.... (fig. 3.17) immature and adult **MALLOPHAGA** (p. 215)

15′. **Head:** narrower than prothorax
Mouthparts: piercing-sucking
Habitat: parasitic on mammals (1 large claw) .... (fig. 3.18) immature and adult **ANOPLURA** (p. 224)

16(11′). **Compound eyes:** absent
Wing pads: absent
Tarsi: 1 segment (most) or absent
Stemmata (simple eyes): often present (in a few there are as many as 30 pairs grouped into "pseudocompound" eyes) .... (holometabolous larvae and a few others) 17

16′. **Compound eyes:** usually present, rarely reduced or absent
Wing pads: often present (especially in later instars)
Tarsi: usually 2–5 segments (1 segment in mayflies and absent in immature thrips)
Stemmata: absent, athough may have dorsal and/or median ocelli ....
.... hemimetabolous larvae (nymphs) and wingless adult **INSECTA** 54

17(16). **Mouthparts:** chewing (rarely reduced to knobs on coarctate Meloidae) or mandibulo-suctorial (mandibles and maxillae long, usually sickle-shaped, and closely united to form sucking jaws arising from front of head, fig. 3.19); if mouthparts are styletlike, then the larvae are aquatic and associated with freshwater sponges (fig. 3.20); labial and/or maxillary palpi present
Habitat: variable 18

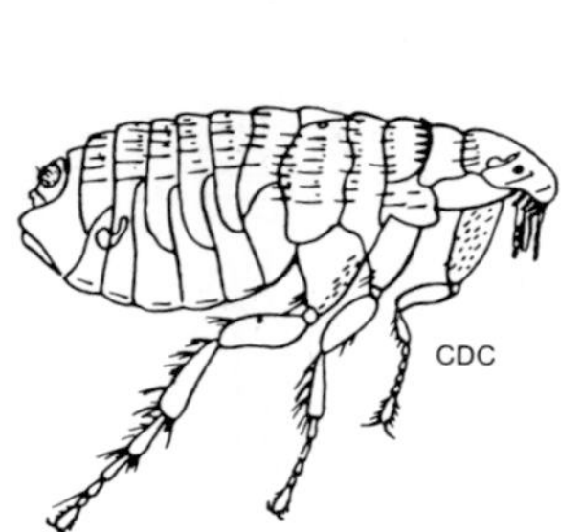

Figure 3.14

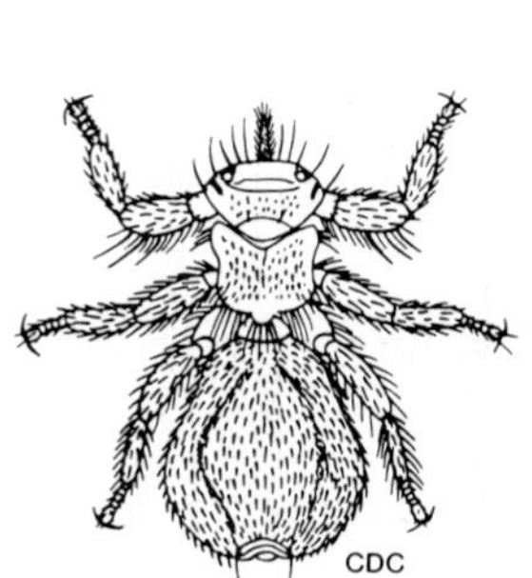

Figure 3.15

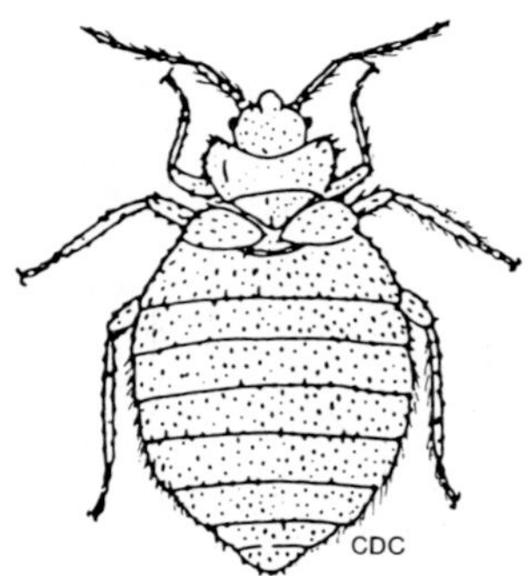

Figure 3.16

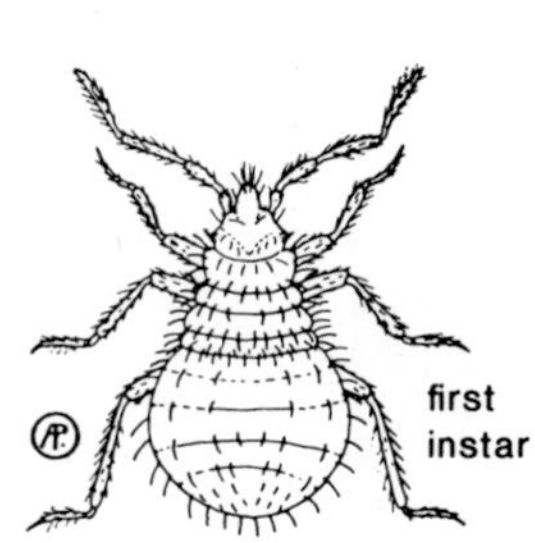

Figure 3.16a

17′. **Mouthparts:** piercing-sucking, with stylets arising from rear of head (fig. 3.21); labial and maxillary palpi absent
Habitat: on plants .... (legged immature Coccoidea) **HOMOPTERA** (Vol. 2)

18(17). **Abdomen:** with 2 or more pairs of soft, unsegmented, fleshy or peglike prolegs that may or may not bear crochets (hooks) (if fleshy prolegs appear to be absent, but crochets are present, prolegs are said to be present) 19

18′. **Abdomen:** with no fleshy prolegs or with 1 terminal pair that may bear 1–3 claws or setae or spines; proleg(s) rarely present on apparent 8th or last segment (a few Coleoptera); a few aquatic Coleoptera (Hydrophilidae, fig. 3.22) have ventral prolegs with masses of crochets on several segments; a few other Coleoptera found in moist, decaying wood (Oedemeridae) have nublike prolegs bearing multiple, irregularly arranged small spines (fig. 3.23) on abdominal segments 2, 3, 2–4, or 2–5 23

19(18). **Body:** sluglike, covered with jellylike tubercles (fig. 3.24)
Abdomen: segments 2 and 7 bearing 3–5 small crochets per "proleg" (look very closely) .... (few) larval (Dalceridae) **LEPIDOPTERA** (p. 288)

19′. **Body:** variable, without jellylike tubercles
Abdomen: segments 2 and 7 without crochets 20

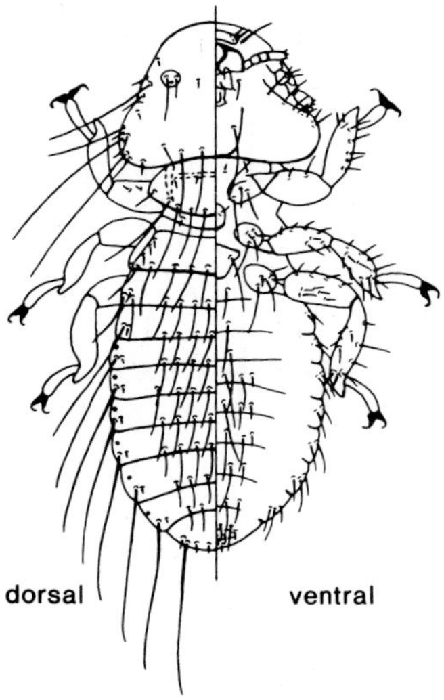

dorsal ventral

Figure 3.17

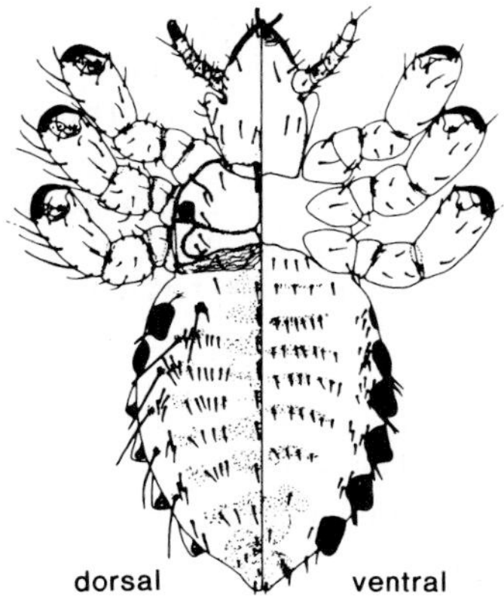

dorsal ventral

Figure 3.18

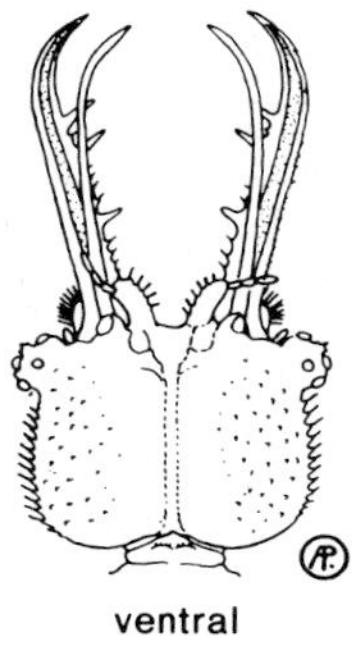

ventral

Figure 3.19

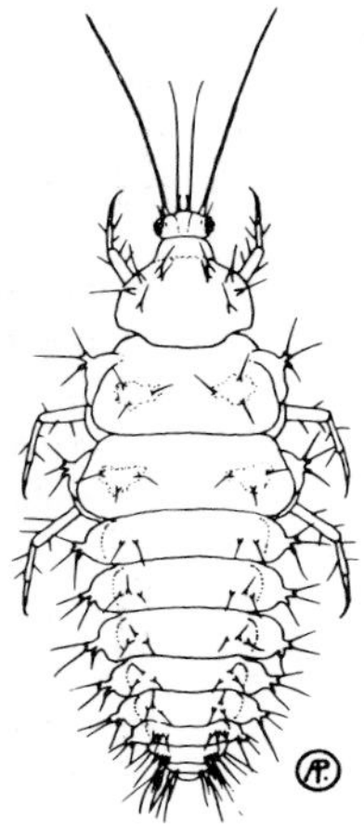

Figure 3.20

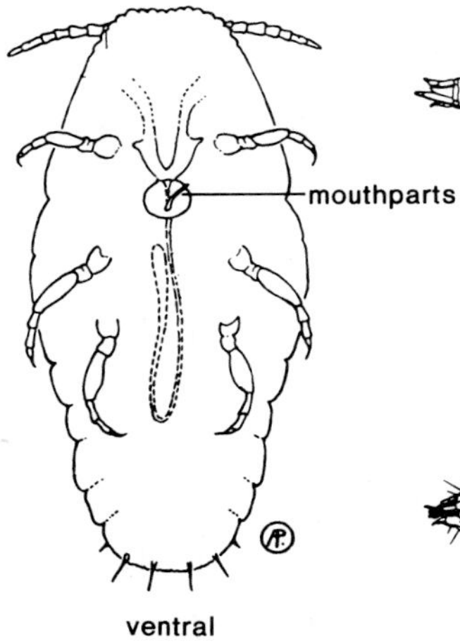

mouthparts

ventral

Figure 3.21

Figure 3.22

Figure 3.24

Figure 3.23

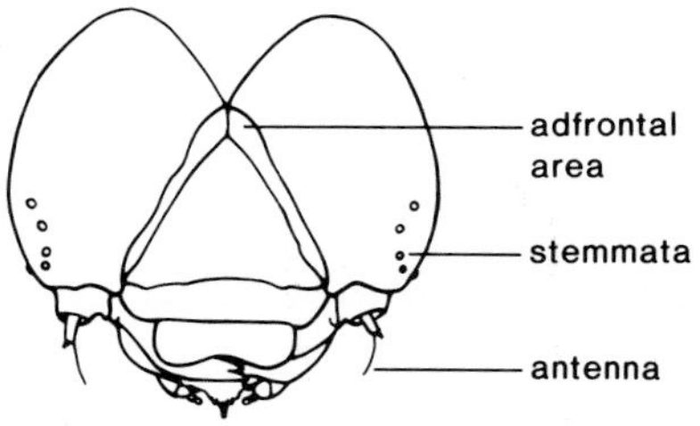

adfrontal area
stemmata
antenna

Figure 3.25

20(19′). **Prolegs:** usually 2–5 pairs on abdominal segments 3–6 and 10; if prolegs are also present on segments 2 and 7, then 2 and 7 without crochets (if crochets arise directly from the venter, prolegs are said to be present)
Crochets: present in rows or circles, usually hooked, sometimes more spinelike, especially if miners in plant tissues
Head: adfrontal areas usually clearly evident (fig. 3.25) .... (most) larval **LEPIDOPTERA** (p. 288)

20′. **Prolegs:** 6 or more pairs
Crochets: absent, though prolegs may bear 1 or 2 claws
Head: adfrontal areas absent 21

21(20′). **Body:** with 4 double or single rows of large, elongate or oval setae (fig. 3.26)
Size: up to 5 mm long
Head: retractile; a single medial seta midway between antennae
Prolegs: 8 pairs on first 8 abdominal segments, very short, with single "claw" .... (few) larval (Micropterigidae) **LEPIDOPTERA** (p. 288)

21′. **Body:** without rows of large, elongate, or oval setae
Size: usually over 5 mm long
Head: rarely retractile; medial seta absent
Prolegs: usually 7 or 8 pairs, although 4 to 10 pairs possible; short or long, with 1 or 2 claws that may be minute and spinelike or rarely absent 22

22(21′). **Stemmata (ommatidia):** 7 or more pairs; many have up to 30 pairs aggregated into a pair of "pseudocompound" eyes
Prolegs: nearly always present on abdominal segment 1; usually 8 pairs although 0–8 pairs possible
Abdomen: segments not secondarily annulated (ringed), often spiny
Terminal segment: often expanded into suctorial disc .... (fig. 3.27) (most) larval **MECOPTERA** (p. 246)

22′. **Stemmata:** 1 pair
Prolegs: absent on abdominal segment 1 (present but greatly reduced in Xyelidae); usually 7 or 8 pairs although 6–10 pairs possible
Abdomen: segments usually secondarily annulated (ringed), rarely spiny
Terminal segment: suctorial disc absent .... (some) larval **HYMENOPTERA** (p. 597) (fig. 3.28) (most sawflies: suborder Symphyta)

23(18′). **Head:** hypognathous (distance along middorsal line much greater than along midventral line (fig. 3.29)); usually cylindrical or depressed 24

23′. **Head:** prognathous (distance along middorsal and midventral lines nearly equal (fig. 3.30)); usually more rounded, boxlike 28

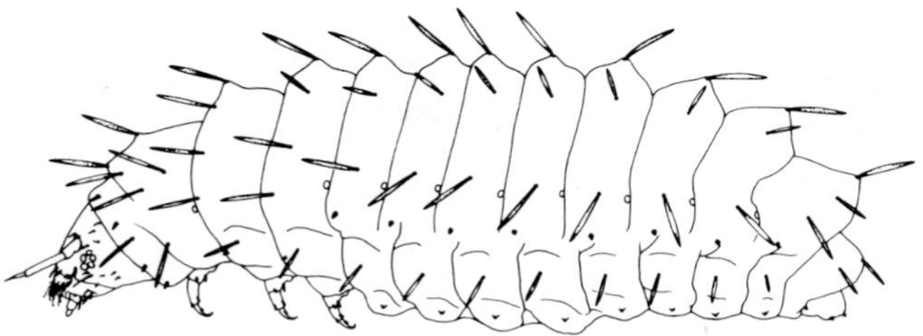
Figure 3.26

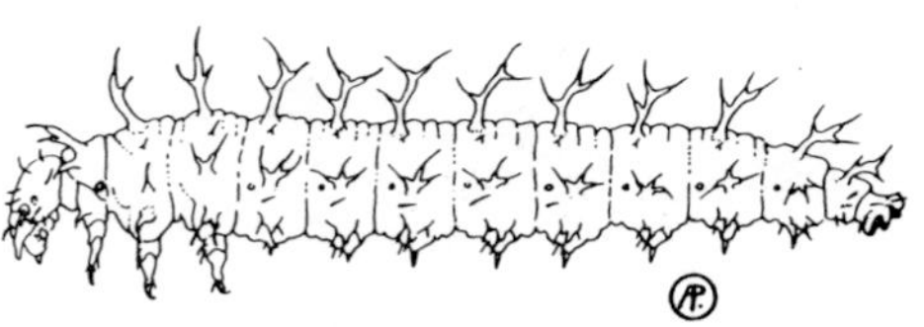
Figure 3.27

Figure 3.28

Figure 3.29

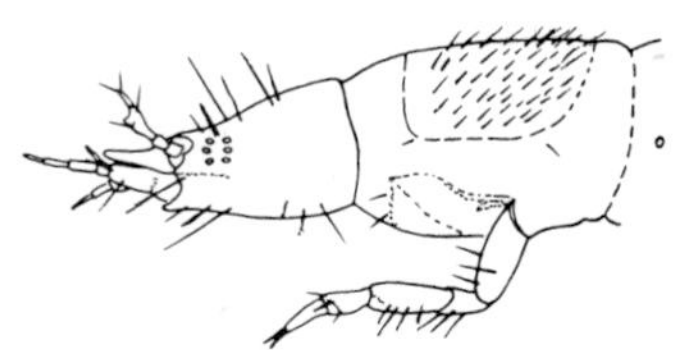
Figure 3.30

24(23). **Body form:** usually sluglike
Lateral stemmata: usually 6, arranged in semicircle (fig. 3.25)
Adfrontal areas: present (similar to fig. 3.25)
Antennae: arising from triangular or U-shaped membranous area adjacent to base of mandibles ..... (few) larval **LEPIDOPTERA** (p. 288)

24′. **Body form:** variable, but not sluglike
Lateral stemmata: 0–6 pairs (rarely 7), not arranged in semicircle
Adfrontal areas: absent
Antennae: arising from cranium or if at base of mandible, from a small, more circular area (often separated from mandible by sclerotized area or ridge) 25

25(24′). **Thoracic legs:** fore legs small, close together, projecting ventrally; mid and hind legs larger, farther apart, tending to project laterally (fig. 3.31); all legs of 3 segments; claws absent, although legs usually bearing fleshy terminal projection
Stemmata: with 3 pairs or none
Gills: absent ..... (few) larval (Boreidae and Panorpodidae) **MECOPTERA** (p. 246)

25′. **Thoracic legs:** 3 pairs similar in structure, position, and usually size; if fore legs smaller, then mid and hind legs not projecting laterally; legs usually of more or fewer than 3 segments; 1, 2 or no claws
Stemmata: up to 6 pairs
Gills: present or absent 26

26(25′). **Terminal segment:** usually with 1 pair of anal prolegs, each bearing 1–3 hooked claws
Tarsi: 1 claw (in some a spur creates an impression of 2 claws)
Body: frequently with gills, often enclosed in a case
Habitat: almost always aquatic, a few in continuously damp terrestrial areas ..... (figs. 3.32, 3.33) (caddisfly larvae) **TRICHOPTERA** (p. 253)

26′. **Terminal segment:** without prolegs bearing hooked claws (hookless subanal appendages may be present (fig. 3.36))
Tarsi: 1, 2 or no claws
Body: rarely with gills, hardly ever enclosed in a case, although may bear fecal material or other debris on back, or dorsum may be expanded
Habitat: terrestrial or aquatic 27

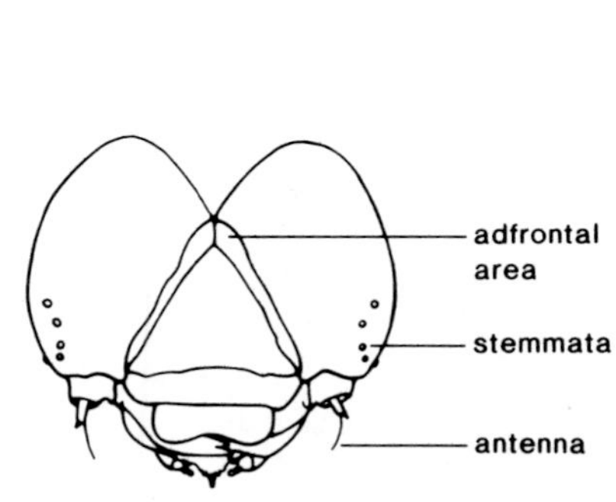

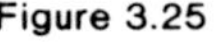
Figure 3.25

Figure 3.31

Figure 3.32

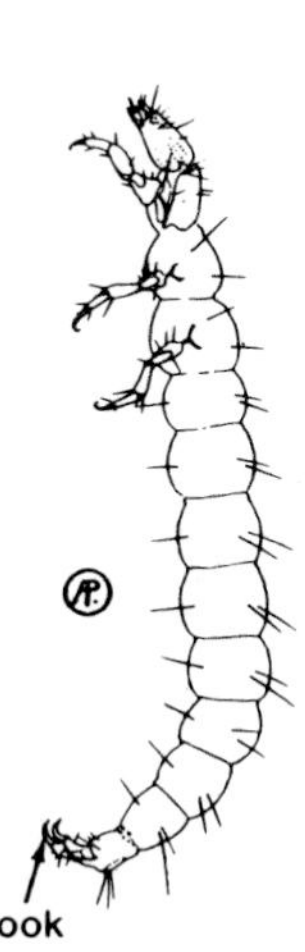

Figure 3.33

27(26′). **Stemmata:** 1 pair
Labial palpi: 3 segments (most), 2 on a few, especially some leafminers
Spinneret: (salivary opening)—often distinct (medial or paired)
Thoracic legs: not distinctly elbowed
Thoracic spiracles: on prothorax and mesothorax, or in a fold between pro- and mesothorax and in a fold between meso- and metathorax; rarely on prothorax alone
Body: elongate, not heavily sclerotized
Habitat: never truly aquatic .... (some) larval **HYMENOPTERA**[5] (p. 597) (figs. 3.34, 3.35, 3.36) (mostly suborder Symphyta)

27′. **Stemmata:** 2–6 pairs (most), absent or 1 pair (few)
Labial palpi: 1 or 2 segments, rarely more than 2 or vestigial
Spinneret: absent
Thoracic legs: usually elbowed at 1 or more joints, sometimes short and not elbowed
Thoracic spiracles: usually on mesothorax, sometimes on prothorax (notably Scarabaeoidea and Bostrichoidea) or gills present, or, if aquatic, spiracles lacking or vestigial
Body: may be curved like C or U, occasionally very heavily sclerotized
Habitat: terrestrial or aquatic .... (some) larval **COLEOPTERA**[5] (Vol. 2)

28(23′). **Body:** with 4 double or single rows of large, elongate, or oval setae (fig. 3.26)
Size: up to 5 mm long
Head: retractile; a single medial seta between antennae
Ventral prolegs: present on first 8 abdominal segments but very short, with single claw on each .... (few) larval (Micropterigidae) **LEPIDOPTERA** (p. 288)

28′. **Body:** without such setae
Size: often longer than 5 mm
Head: rarely retractile, usually conspicuous, but may be concealed beneath expanded dorsum
Ventral prolegs: absent (fewer than 8 pairs on a few Coleoptera and some leafminers) 29

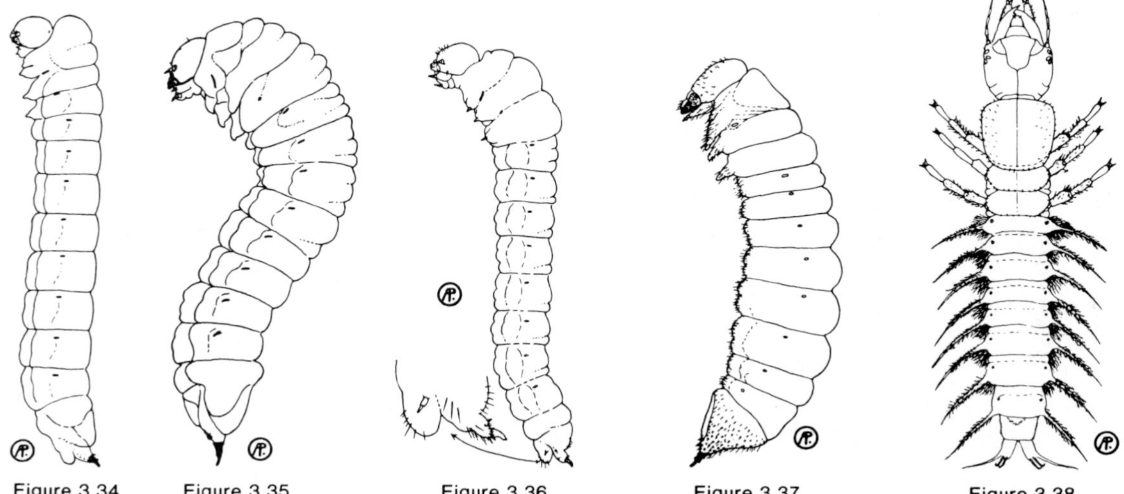

Figure 3.34 Figure 3.35 Figure 3.36 Figure 3.37 Figure 3.38

5. Some woodboring and stemboring Symphyta (especially Anaxyelidae, Siricidae (fig. 3.34), Xiphydriidae (fig. 3.35), and Cephidae (fig. 3.36)) and Coleoptera (some Mordellidae, fig. 3.37) are superficially very similar in appearance. Anaxyelids, cephids, siricids, and xiphydriids have oval to elongate spiracles located in or near the pro-mesothoracic and meso-metathoracic folds (meso-metathoracic spiracle is vestigial in Xiphydriidae and sometimes not easily seen on anaxyelids and siricids) and a median pore at tip of labium. Mordellids have a small, more circular spiracle in the middle or near the lower anterior margin of the mesothorax and no median labial pore. Cephids are most easily recognized by a pair of subanal appendages on the tenth segment (fig. 3.36).

29(28′). **Thoracic legs:** 1 claw (stout spines or claw with spur or teeth may create impression of 2 or 3 claws); rarely absent 34

29′. **Thoracic legs:** 2 claws 30

30(29′). **Mouthparts:** labrum and clypeus present and visible; if end of abdomen bears hooks, then with a *pair* of prolegs each bearing 2 hooks 31

30′. **Mouthparts:** labrum and/or clypeus absent, or fused so only one is apparent, or concealed beneath margin of head; if end of abdomen bears hooks, then without a *pair* of prolegs each bearing 2 hooks 33

31(30). **Abdomen:** 7 or 8 pairs of long, pointed, lateral processes
Habitat: aquatic (or near water if larvae have crawled out of water to pupate) .... (figs. 3.38, 3.39) larval **MEGALOPTERA** (Vol. 2)

31′. **Abdomen:** lateral processes absent
Habitat: terrestrial 32

32(31′). **Labial palpi:** 3 segments
Stemmata: 4–7 (1 or 2 may be difficult to see)
Abdomen: with 10 well-defined and rather sharply constricted segments, terminal segment with soft pads ventrally .... (fig. 3.40) larval **RAPHIDIODEA** (Vol. 2)

32′. **Labial palpi:** no more than 2 segments
Stemmata: 0–6 present
Abdomen: not as above .... (Cupedidae and Micromalthidae) larval **COLEOPTERA** (Vol. 2)

33(30′). **Maxillary palpi:** absent
Mandibles and maxillae: *united* to form piercing-sucking jaws
Habitat: terrestrial .... (figs. 3.41, 3.42) larval **NEUROPTERA** (Vol. 2)

33′. **Maxillary palpi:** present (fig. 3.43)
Mandibles and maxillae: distinctly *separate* mandible may have blood duct or groove along mesal edge and be used for piercing-sucking
Habitat: terrestrial or aquatic .... (some) larval **COLEOPTERA** (Vol. 2)

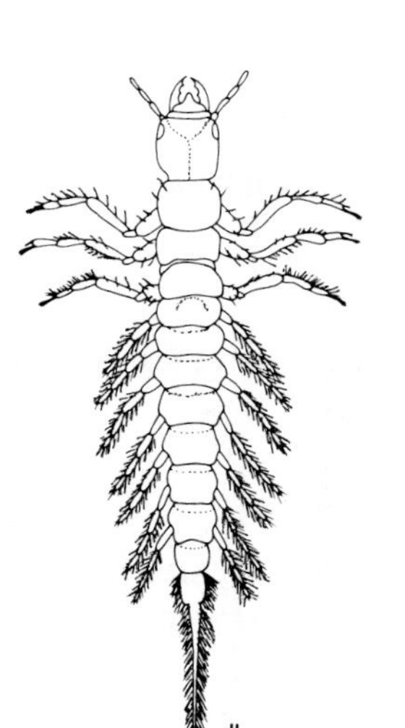

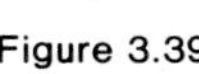

Figure 3.39

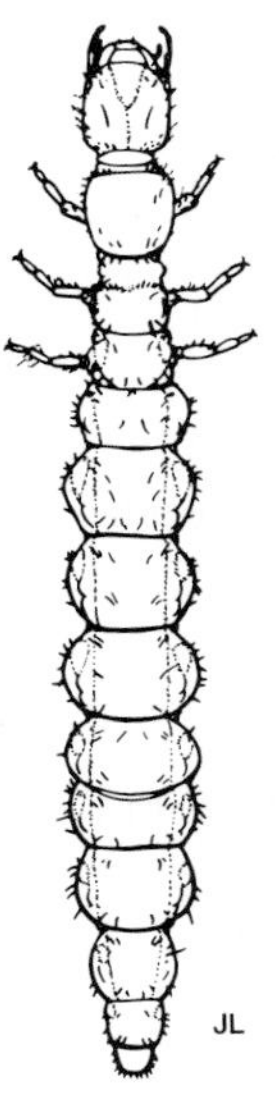

Figure 3.40

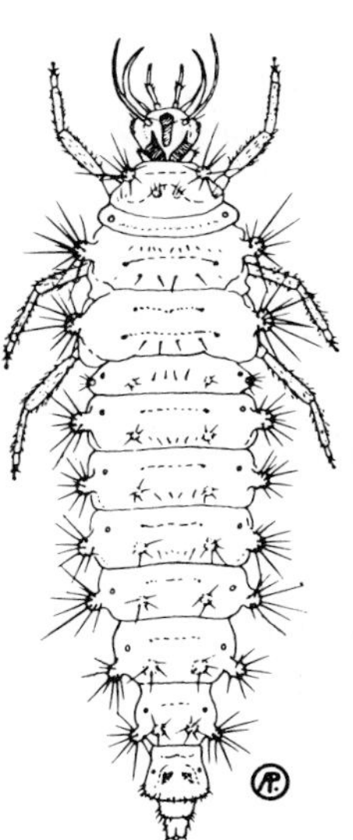

Figure 3.41

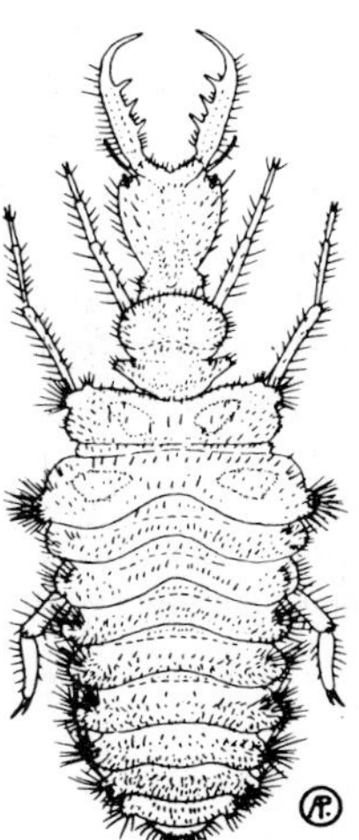

Figure 3.42

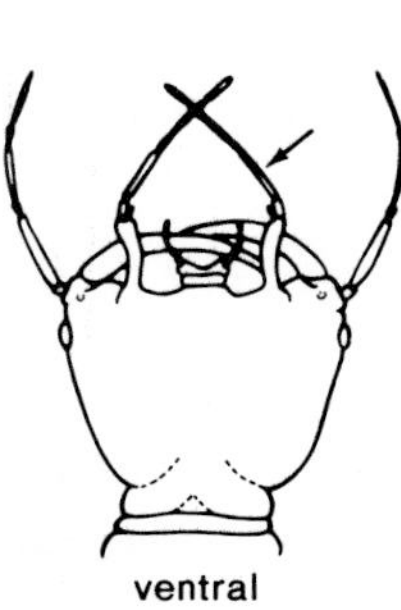

Figure 3.43

34(29). **Mandibles and maxillae:** modified into long needlelike structures
Tarsal claws: 1
Gills: folded beneath abdomen
Antennae: 5–16 segments
Habitat: in or on freshwater sponges .... (fig. 3.20) (Sisyridae) larval **NEUROPTERA** (Vol. 2)

34′. **Mandibles and maxillae:** not greatly modified, mandibles opposable, chewing
Tarsal claws: 1 or 2 (rarely absent)
Gills: if present, not folded beneath abdomen
Antennae: variable
Habitat: not associated with freshwater sponges (see third choice, 34″) 35

34″. **Mandibles and maxillae:** mandibles absent; maxillae represented by palpi at most
Tarsal claws: absent
Gills: absent
Antennae: bristlelike at most
Habitat: terrestrial, usually found clinging to insects .... first instar (triungulinids) **STREPSIPTERA** (Vol. 2)

35(34′). **Terminal segment:** 1 pair of anal prolegs, each bearing 1–3 hooked claws
Thoracic legs: 5 segments, trochanter often subdivided
Spiracles: absent
Gills: usually present
Body: often living in a case
Antennae: usually 1 segment, peglike, short, or as inconspicuous patches of tiny papillae .... (see figures at couplet 26) larval **TRICHOPTERA** (p. 253)

35′. **Terminal segment:** without prolegs bearing hooked claws (a median proleg, or a pair with spines, or 2 hooks above a flap are rarely present)
Thoracic legs: usually 4 segments, rarely 5
Spiracles: present, or, if aquatic, vestigial or absent
Gills: present or absent
Body: rarely living enclosed in a case; a few live as leaf miners (if a leaf miner, see 35 A, B, C below)
Antennae: usually 2 or more segments .... (most) larval **COLEOPTERA** (Vol. 2)
(and adult female larviform glowworms: Phengodidae)[6]

Leaf miners:

35A. **Labium:** *protruding* spinneret present
Thoracic spiracle: only on T1, not adjacent to T2
Adfrontal areas: may be evident
Stemmata (simple eyes): 0–6 pairs .... (some) larval **LEPIDOPTERA** (p. 288)

35B. **Labium:** spinneret absent
Thoracic spiracle: on T2
Adfrontal areas: absent
Stemmata (simple eyes): 0–6 pairs .... (some Chrysomelidae) larval **COLEOPTERA** (Vol. 2)

35C. **Labium:** *protruding* spinneret absent (may have median pore)
Thoracic spiracles: on T1, rather large and oval and often very near T2; T3 spiracle very small or vestigial and difficult to see
Adfrontal areas: absent
Stemmata (simple eyes): always 1 pair .... (some Tenthredinidae) larval **HYMENOPTERA** (p. 597)

36(2′). **Habitat:** aquatic .... (many) larval **DIPTERA** (Vol. 2)

36′. **Habitat:** terrestrial 37

37(36′). **Body:** covered with waxy material, frequently scalelike in form
Mouthparts: sucking, arising from rear of head region
Habitat: on plants, generally immobile .... (Coccoidea in part, Aleyrodoidea) **HOMOPTERA** (Vol. 2)

37′. **Not fitting the above combination** 38

6. Adult female larviform phengodids can be distinguished from larvae by the presence of a gonopore on the venter of A9 (it may be concealed by an arched fold or groove).

38(37′). **Head capsule:** distinct, partially or completely sclerotized, usually pigmented; may be embedded in prothorax 40

38′. **Head capsule:** not distinct, not sclerotized and not pigmented; at most only the mandibles and supporting structures sclerotized and pigmented; may be embedded in prothorax 39

39(38′). **Mandibles:** opposable, but substantially reduced, ranging from pointed mandibles to fleshy unpigmented structures
Body: often C-shaped, often widest in middle and tapered at both ends; usually lacking transverse rows of spines or microspines
Spiracles: 1–3 on thorax and 5–8 on abdomen, usually inconspicuous; caudal respiratory tube or prothoracic spiracular horn or rake absent
Prolegs: not present and crochets never present on body
Habitat: in cells or nests built by adults, as parasites in plant tissues (usually gall-formers); gall-formers always without a breastbone .... (figs. 3.44–3.47) (many, mostly Apocrita) larval **HYMENOPTERA** (p. 597)

39′. **Mandibles:** usually parallel, reduced to 1 or 2 mouth hooks, embedded within head to varying degrees; may be opposable if deeply embedded (may be difficult to locate); if specimens are small (less than 5 mm), the head is minute, and mandibles are difficult to discern, then usually a spatula (breastbone) present on thorax (fig. 3.50)
Body: highly variable, but rarely C-shaped; often peg-shaped; often bearing transverse rows of spines or microspines
Spiracles: usually on prothorax and at or near caudal end, or apparently absent; caudal respiratory tube and/or prothoracic spiracular horn or rake may be present
Prolegs: may be present and may bear crochets
Habitat: highly variable, but not in cells or nests built by adults; if gall-formers, then often with breastbone on thorax (fig. 3.50) .... (figs. 3.48–3.55)(many) larval **DIPTERA** (Vol. 2)

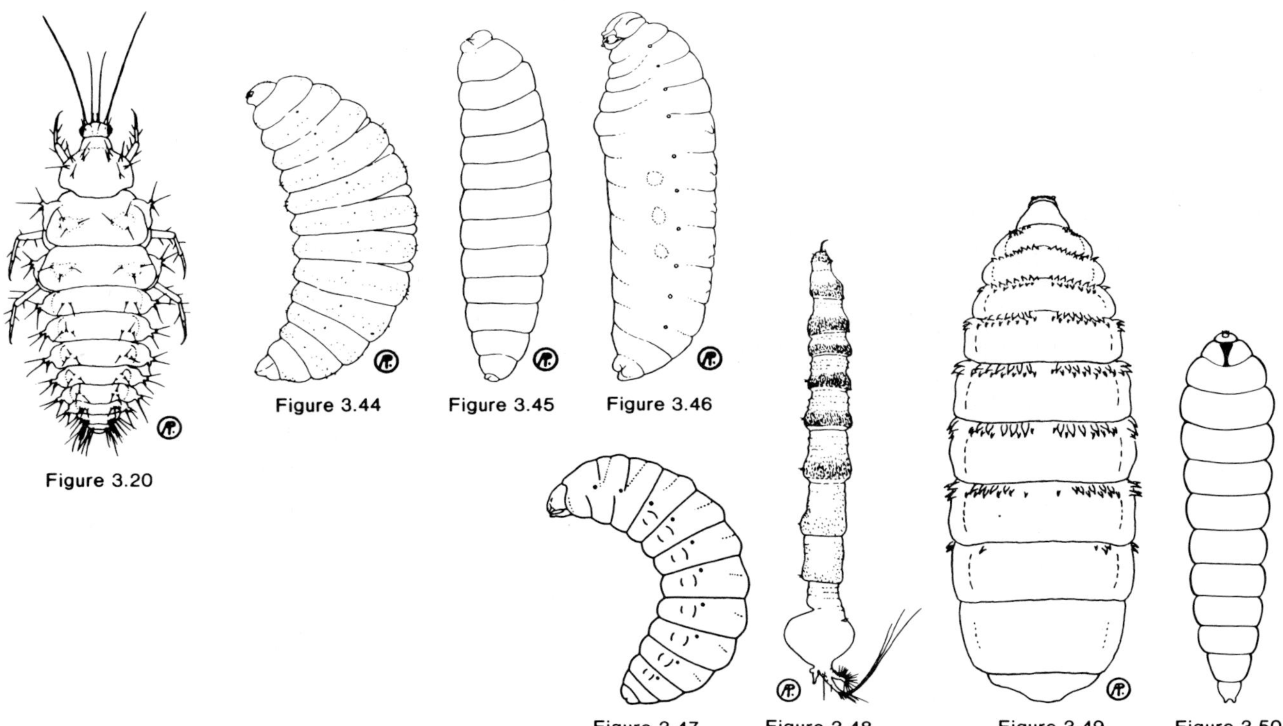

Figure 3.20

Figure 3.44

Figure 3.45

Figure 3.46

Figure 3.47

Figure 3.48

Figure 3.49

Figure 3.50

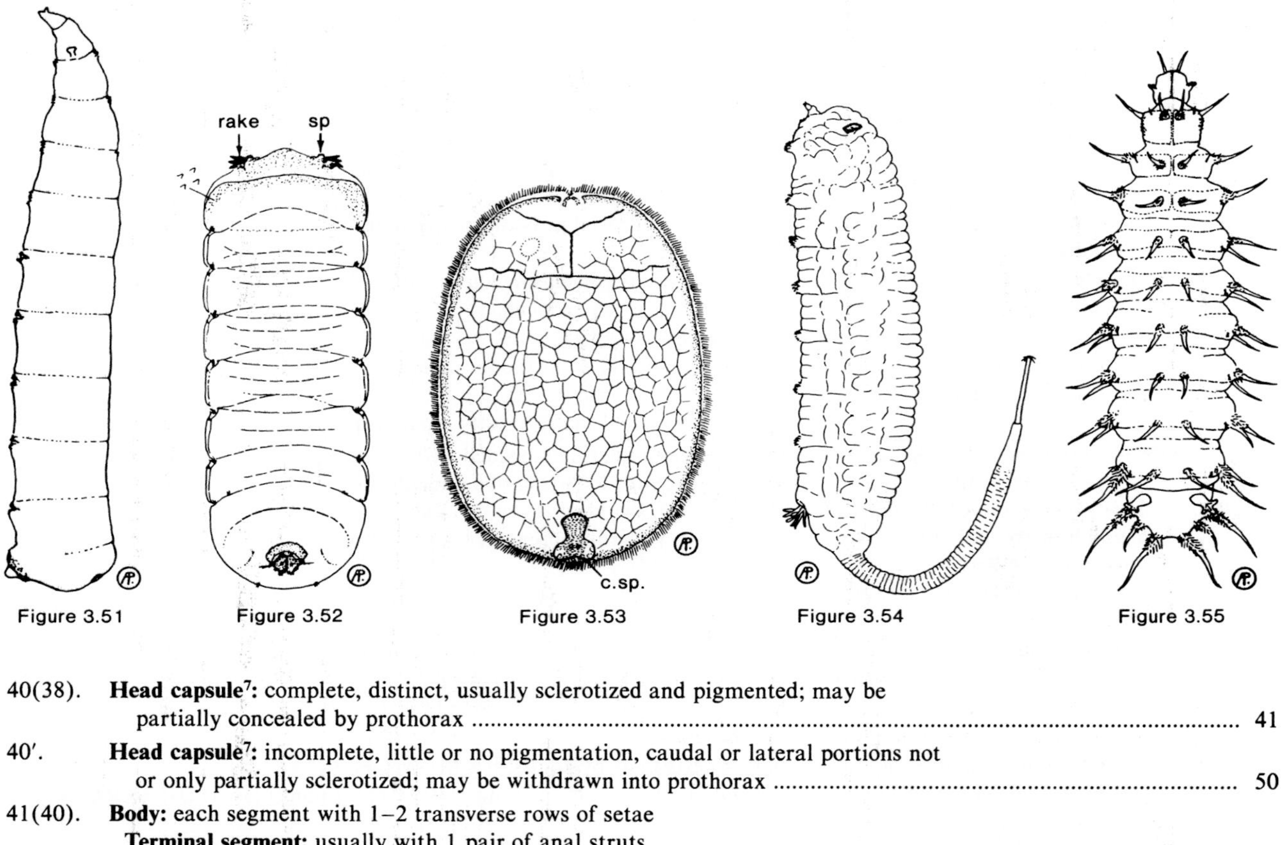

Figure 3.51 Figure 3.52 Figure 3.53 Figure 3.54 Figure 3.55

40(38). **Head capsule**[7]: complete, distinct, usually sclerotized and pigmented; may be partially concealed by prothorax 41

40′. **Head capsule**[7]: incomplete, little or no pigmentation, caudal or lateral portions not or only partially sclerotized; may be withdrawn into prothorax 50

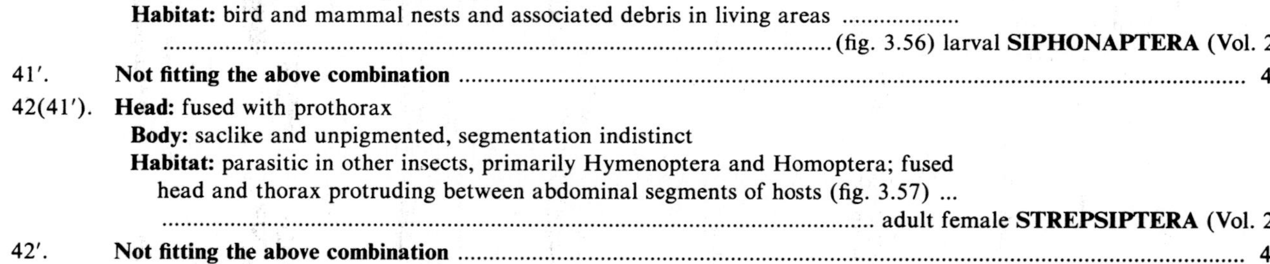

41(40). **Body:** each segment with 1–2 transverse rows of setae
Terminal segment: usually with 1 pair of anal struts
Habitat: bird and mammal nests and associated debris in living areas ....
.... (fig. 3.56) larval **SIPHONAPTERA** (Vol. 2)

41′. **Not fitting the above combination** 42

42(41′). **Head:** fused with prothorax
Body: saclike and unpigmented, segmentation indistinct
Habitat: parasitic in other insects, primarily Hymenoptera and Homoptera; fused head and thorax protruding between abdominal segments of hosts (fig. 3.57) ...
.... adult female **STREPSIPTERA** (Vol. 2)

42′. **Not fitting the above combination** 43

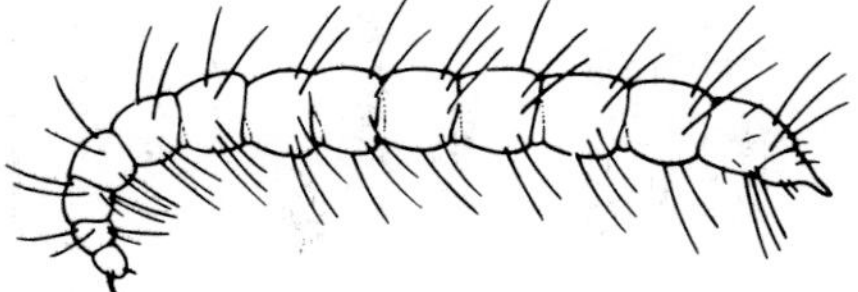

Figure 3.56

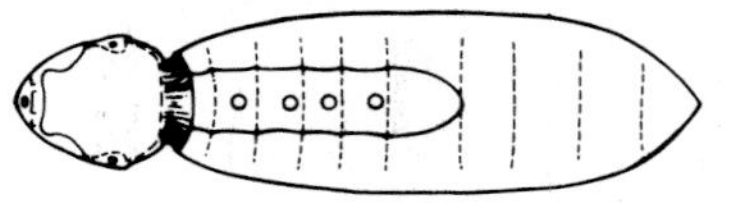

Figure 3.57

7. If head is questionably distinct, sclerotized or pigmented, try other half of couplet if specimen doesn't key well.

43(42′). **Head:** hypognathous (distance along middorsal line much greater than along midventral line, as in fig. 3.29) 46

43′. **Head:** prognathous (distance along middorsal and midventral lines nearly equal as in fig. 3.30) 44

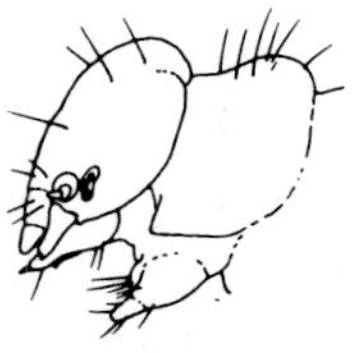

Figure 3.29

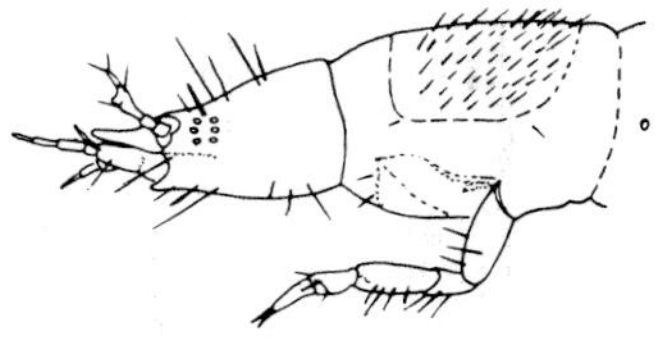

Figure 3.30

44(43′). **Prolegs:** often present, but if present, only on abdominal segments 3–6 and 10 (may be reduced to crochets arising directly from body wall or absent)
Labium: central, terminal spinneret present
Adfrontal areas: usually evident
Stemmata (simple eyes): 1–6 pairs or absent; often 1 pair—if several pairs, often arranged in semicircle
Epicranial notch: often very deep, especially if miners in plant tissues
Antennae: arising in triangular or U-shaped membranous area at base of mandibles .... (some) larval **LEPIDOPTERA** (p. 288)

44′. **Prolegs:** usually absent on abdominal segments 3–6 and 10—if present, without crochets and/or with some prolegs on other segments
Labium: spinneret absent
Adfrontal areas: never present
Stemmata: absent (most) or with 1–6 pairs—never arranged in semicircle
Epicranial notch: relatively shallow or not evident
Antennae: if evident, arising from head capsule or from a small, more circular area near mandibles 44a

44a(44′). **Median endocarina:** present and extending forward between epicranial sutures (fig. 3.57a) ....
(a few leafmining or leafgallforming Curculionidae and Chrysomelidae) .... **COLEOPTERA** (Vol. 2)

44a′. **Median endocarina:** absent 45

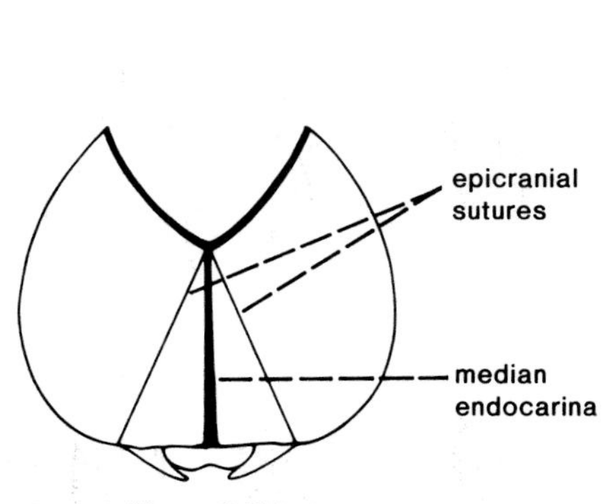

Figure 3.57a

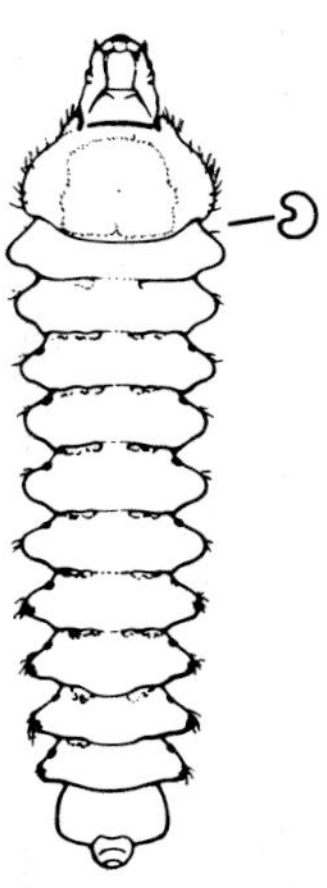

Figure 3.58

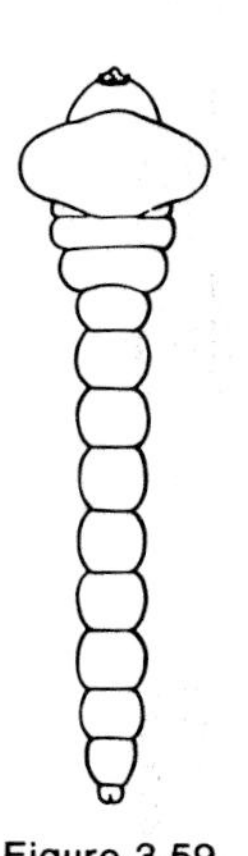

Figure 3.59

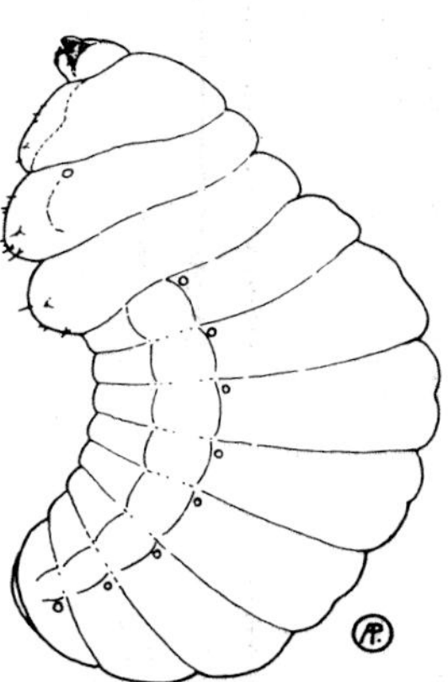

Figure 3.60

45(44a′). NOTE: COMPARE ALL 5 CHARACTERS: If none of the three options fits well and head is lightly pigmented, try couplet 48 (especially if from a plant gall).

Mouthparts: mandibles clearly opposable[8]

Spiracles: usually 1 pair on mesothorax and 8 pairs of equal size along abdomen, the last one no larger than the others; some spiracles always visible

Body: shape variable, but often stout and depressed or curved like a C; sometimes elongate; lacking knobs or fleshy protuberances on segments, but may have distinct constrictions between segments

Epicranial head sutures: may form a Y

Prothorax: sometimes 2 or more times as long as, and usually wider than, mesothorax or metathorax; never bearing a proleg .... (figs. 3.58, 3.59, 3.60) (some) larval **COLEOPTERA** (Vol. 2) (mostly wood and seed boring beetles; a few leaf miners)

45′. **Mouthparts:** mandibles opposable (but often difficult to see) or vertical mouth hooks

Spiracles: usually 1 pair on prothorax, and usually 1 terminal pair on abdomen, or if along abdomen, then terminal pair often largest; sometimes vestigial or apparently absent

Body: usually elongate; may bear knobs or fleshy protuberances on some body segments

Epicranial head sutures: rarely forming a Y (except some Bibionidae—they have 10 pairs of conspicuous spiracles), or rarely appearing somewhat Y-shaped as in fig. 3.64a

Prothorax: about same width and length as mesothorax or metathorax, or narrower; may bear a proleg .... (see third choice, 45″) (figs. 3.62–3.67) (many) larvae of "lower" **DIPTERA** (Vol. 2)

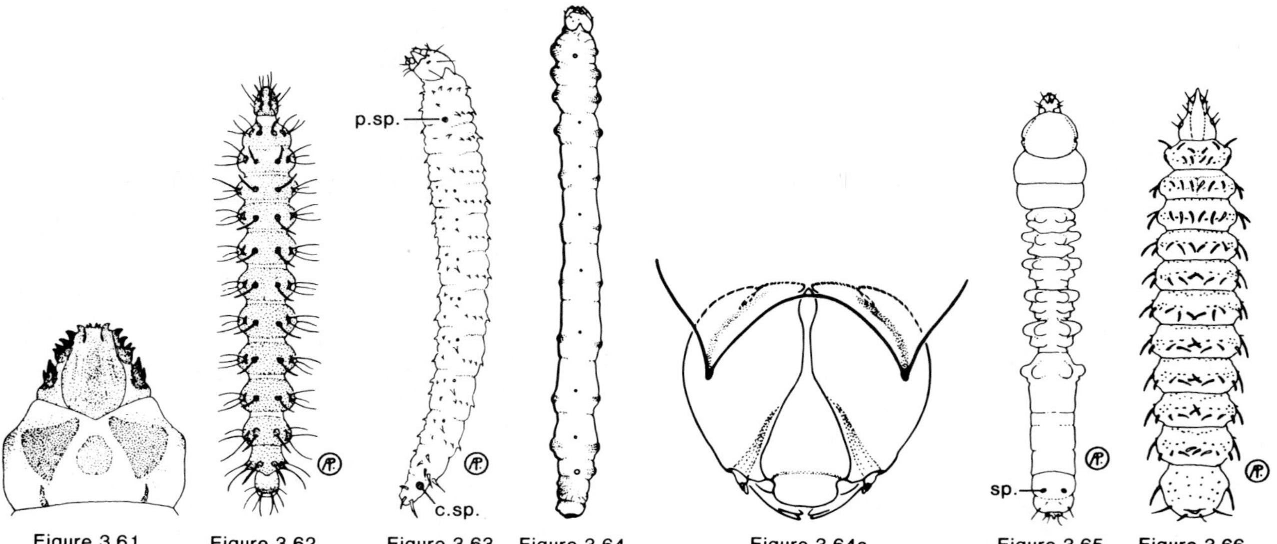

Figure 3.61 Figure 3.62 Figure 3.63 Figure 3.64 Figure 3.64a Figure 3.65 Figure 3.66

8. If head is strongly depressed and broad, the anterior margin may be serrated, the antennae and maxillae obsolete, and the mandibles serrate with teeth projecting laterally (fig. 3.61, Eucnemidae).

45″. **Mouthparts:** opposable, but may be highly modified for sap feeding as leaf miners
Spiracles: 1 pair on prothorax (never on mesothorax) and 8 pairs on abdominal segments 1–8, sometimes quite small or difficult to see on Al–8; those on prothorax and 8th abdominal segment usually somewhat larger than others (**Check:** if spiracle is on mesothorax, see Coleoptera)
Body: elongate; flattened dorsoventrally, usually with distinct constrictions between segments; lacking knobs or fleshy protuberances on segments
Epicranial head sutures: not forming a Y
Prothorax: about same width and length as meso- and metathorax; never bearing a proleg (fig. 3.68) .... (some) leafmining **LEPIDOPTERA** (p. 288)

46(43). **Thoracic spiracles:** 1 pair only, on mesothorax or prothorax
Spiracular openings: often biforous or annular biforous (figs. 3.69, 3.70, 3.71)
Body segments: often with strong transverse furrows (annuli)
Mouthparts: usually well developed; maxillary palpi of 2 or more segments
Body: often somewhat curved like C or U (fig. 3.72) .... (some) larval **COLEOPTERA**[9] (Vol. 2)
(most snout and bark beetles and a few others)

46′. **Thoracic spiracles:** 0–3 pairs, if 1 pair, then on prothorax, or if apparently on mesothorax, with blunt spur (postcornu) at tip of abdomen, or posterior pair of abdominal spiracles larger than others
Spiracular openings: single, or with 3 or more openings, very rarely with some biforous openings
Body segments: usually without annuli, but may bear protuberances
Mouthparts: sometimes reduced; maxillary palpi indistinct or of 1 or 2 segments
Body: curved or not 47

47(46′). **Body:** heavily sclerotized, brown, with segments fused to form rigid capsule (fig. 3.73) .... (few) larval (coarctate larvae of Meloidae) **COLEOPTERA** (Vol. 2)

47′. **Body:** not heavily sclerotized or brown, and without segments fused to form a rigid capsule 48

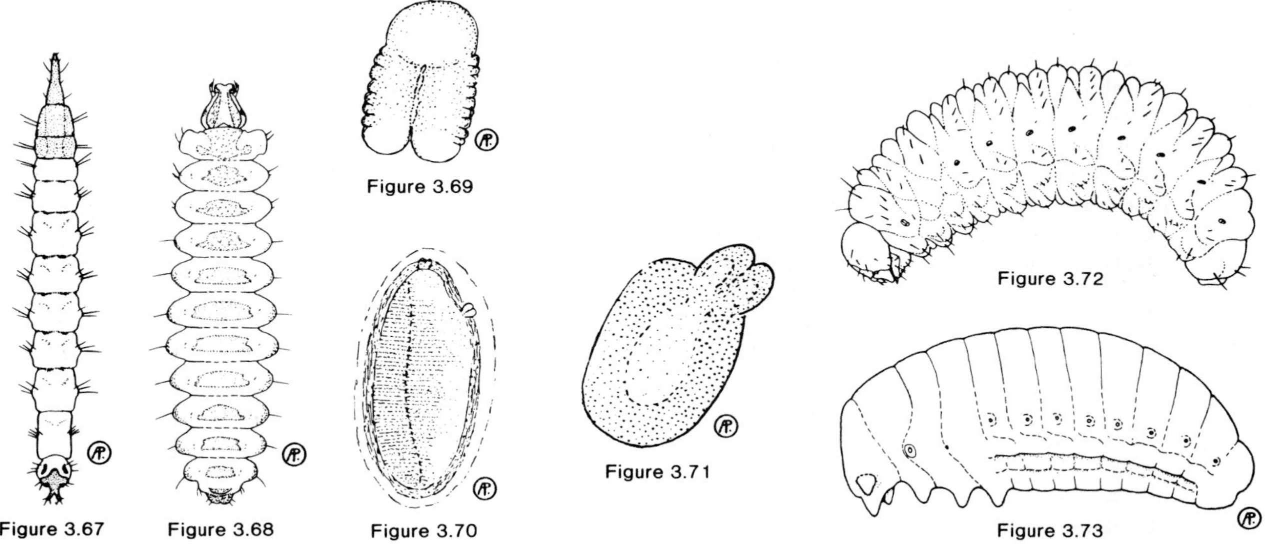

Figure 3.67 Figure 3.68 Figure 3.69 Figure 3.70 Figure 3.71 Figure 3.72 Figure 3.73

9. If larvae are about 2 mm long, have 7 stemmata, and are from the Pacific Northwest, see Boreidae (*Caurinus*), Mecoptera.

48(47′). **Labium:** bearing protruding, median, pointed spinneret
Antennae: arising from triangular or U-shaped membranous area at base of mandibles
Ventral prolegs: if evident, with crochets
Adfrontal areas: present, but usually inconspicuous .... (few) larval **LEPIDOPTERA** (p. 288)

48′. **Labium:** protruding, median, pointed spinneret absent; may have nonprotruding median opening, or protruding slit, or protruding paired spinnerets
Antennae: if evident, arising from cranium or from small, circular area near base of mandible
Ventral prolegs: if present, without crochets
Adfrontal areas: absent 49

49(48′). **Mandibles:** opposable, pointed, weakly to moderately (rarely strongly) sclerotized (some wasp larvae)
Spiracles: usually 2 pairs on thorax (rarely 1) and 8 pairs of equal size on abdomen, the posterior pair not larger than others
Habitat: in cells or nests built by adults, as parasitoids, or in plants .... (some) larvae of Apocrita **HYMENOPTERA** (p. 597)

49′. **Mandibles:** usually opposable, sometimes vertical mouthhooks; may be modified as "brushes"
Spiracles: None or 1 pair on thorax; usually fewer than 8 pairs on abdomen (rarely apparently as many as 9 pairs), with posterior pair often larger than others; sometimes vestigial or apparently absent
Habitat: diverse, often in damp or wet areas, as internal parasites or in plants or decaying materials .... (some) larval **DIPTERA** (Vol. 2)

50(40′). **Mouthparts:** mandibles opposable; maxillae, labrum and labium clearly visible and normally developed, sclerotized, pigmented, and segmented; not deeply embedded in prothorax
Antennae: usually distinct and usually with more than 1 segment, arising near the mandibles .... (few) larval **COLEOPTERA** (Vol. 2)

50′. **Mouthparts:** (a) mandibles reduced to weakly sclerotized and weakly pigmented, or unsclerotized and unpigmented opposable structures, OR
(b) mandibles reduced to 1 or 2 parallel mouthhooks, sometimes small, peglike, and roughened when single, OR
(c) mandibles opposable but strongly sclerotized (at least basally) and deeply embedded in prothorax; labrum often absent, labium and maxillae variable, but may be absent or greatly reduced
Antennae: usually not evident or absent, rarely 1- or 2-segmented and often arising on the face 51

51(50′). **Prosternum:** usually with sclerotized, anchor-shaped or notched spatula (fig. 3.50)
Color: often yellow to orange or red when alive
Body: spindlelike or peglike, and apparently with 13 segments behind the tiny head (thorax appears to have 4 segments)
Size: almost always under 4 mm long .... (few) larval **DIPTERA** (Vol. 2)
(gall-gnats, gall-midges: Cecidomyiidae)

51′. **Prosternum:** spatula absent
Color: usually pale cream when alive
Body: sometimes somewhat C- or U-shaped, usually with fewer than 13 segments behind head (if 13, thorax has 3 segments)
Size: often over 4 mm long 52

52(51′). **Mandibles:** entirely or partially embedded or withdrawn into prothorax, and reduced to 1 or 2 parallel mouthhooks, or reduced to sclerotized, opposable mandibles
Maxillae: variable, but often absent, or sometimes mouthhooks flanked by well developed maxillae that may appear mandiblelike (fig. 3.74) .... (some) larval **DIPTERA** (Vol. 2)

52′. **Mandibles:** opposable and reduced, but not embedded or withdrawn into prothorax
Maxillae: usually evident but reduced, or mouthparts reduced to a mouth opening with surrounding indistinct sensory structures 53

53(52′). **Maxillae:** reduced to fleshy lobes or conical protuberances, and not bearing conspicuous setae or other structures
Maxillary palpi: reduced to discs or conical protuberances
Antennae (or antennal areas): if present, on the face, may resemble "eyespots" (fig. 3.75)
Spinneret: (salivary opening)—may be distinct (medial or paired)
Abdomen: usually not annulated or fewer than 3 annuli per segment .... (most larval Apocrita) **HYMENOPTERA** (p. 597)
(ants (fig. 3.76), bees (fig. 3.47), wasps (fig. 3.46), parasitoids and gallmakers (fig. 3.44))

53′. **Maxillae:** may be small, but usually[10] distinctly segmented, and bearing setae or other structures
Maxillary palpi: usually[10] 2–5 segmented
Antennae: if present, near the mandibles
Spinneret: absent
Abdomen: often 3 or more annuli per segment .... (few) larval **COLEOPTERA** (Vol. 2)

54(16′). **Labium:** extensible (folded back on itself at midlength and held beneath fore legs), often covering face like a mask (figs. 3.77, 3.78)
Habitat: aquatic (a few live in continuously damp areas) .... larval damselflies and dragonflies **ODONATA** (p. 95)

54′. **Labium:** not extensible or folded back on itself, and not covering face like a mask
Habitat: aquatic or terrestrial 55

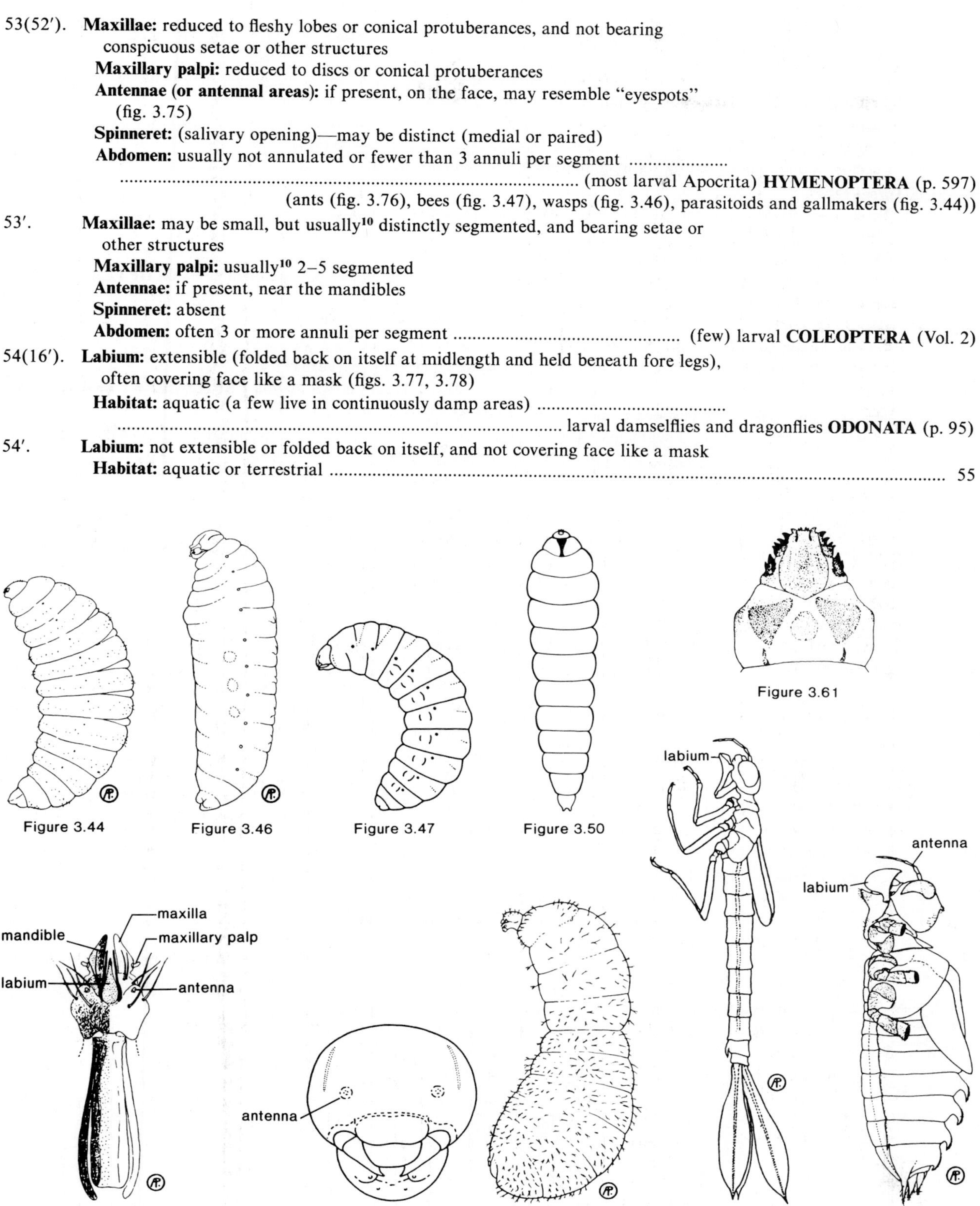

Figure 3.44 Figure 3.46 Figure 3.47 Figure 3.50 Figure 3.61 Figure 3.74 Figure 3.75 Figure 3.76 Figure 3.77 Figure 3.78

10. If head is strongly depressed and broad, the anterior margin may be serrated, and maxillae and palpi not evident or minimally evident (fig. 3.61)

55(54′). **Gills:** present (absent on some stonefly larvae similar in appearance to fig. 3.79)
Habitat: aquatic
Cerci: long and many-segmented 56

55′. **Gills:** absent; may have rows of gill-like plastron hairs if mouthparts are piercing-sucking
Habitat: terrestrial or aquatic
Cerci: variable, but absent or not long and many-segmented if aquatic 57

56(55). **Tarsal claws:** 2
Tarsi: more than 1 segment
Terminal filaments: 2
Gills: beneath thorax and fingerlike; can also be on head (inconspicuous at base of mandibles), abdomen and legs (fig. 3.79) .... larval stoneflies **PLECOPTERA** (p. 186)

56′. **Tarsal claws:** 1
Tarsi: 1 segment
Terminal filaments: most with 3 (some 2)
Gills: on abdomen, usually laterally or dorsally (filaments, plates or featherlike) (figs. 3.80, 3.81) .... larval mayflies **EPHEMEROPTERA** (p. 75)

57(55′). **Body:** strongly constricted between abdomen and thorax
Antennae: usually 12 or 13 segments, often elbowed .... wingless adult **HYMENOPTERA**

57′. **Body:** not strongly constricted between abdomen and thorax
Antennae: usually more or fewer than 12 or 13 segments, never elbowed 58

58(57′). **Thorax:** largely consisting of mesothorax to which prothorax and metathorax are fused, thus making them indistinct
Halteres: (clublike structures on caudal portion of thorax) may be present .... wingless adult **DIPTERA**

58′. **Thorax:** prothorax and/or metathorax distinct
Halteres: absent 59

59(58′). **Body:** thickly covered with flattened scales and hairs
Mouthparts: as coiled proboscis beneath head, with reduced maxillary palpi, or inconspicuous
Antennae: relatively long, of many segments .... wingless or brachypterous adult female **LEPIDOPTERA** (some female Geometridae, Lymantriidae, Psychidae, Pyralidae, Tortricidae)

59′. **Body:** not thickly covered with scales, and if hairy, only moderately so
Mouthparts: never as a coiled proboscis, may be inconspicuous
Antennae: variable 60

60(59′). **Mouthparts:** as triangular or styletlike beak for piercing and sucking, or rarely absent 61

60′. **Mouthparts:** always present as jaws for chewing; if beaklike, then with distinct, separate chewing mouthparts at apex of beak 63

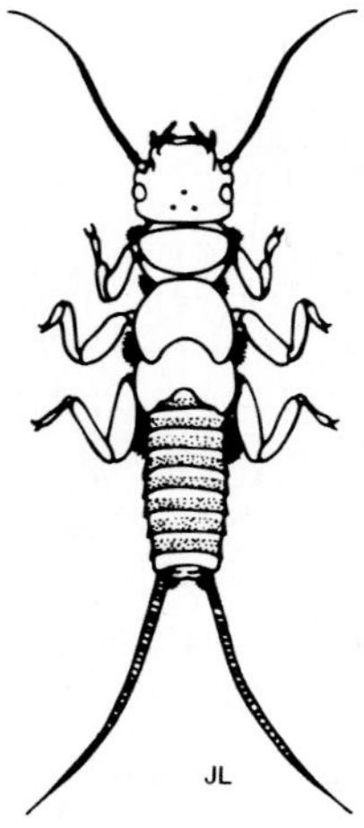

Figure 3.79

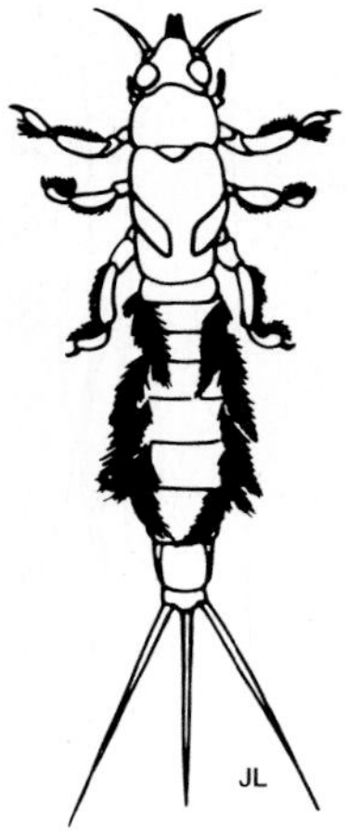

Figure 3.80

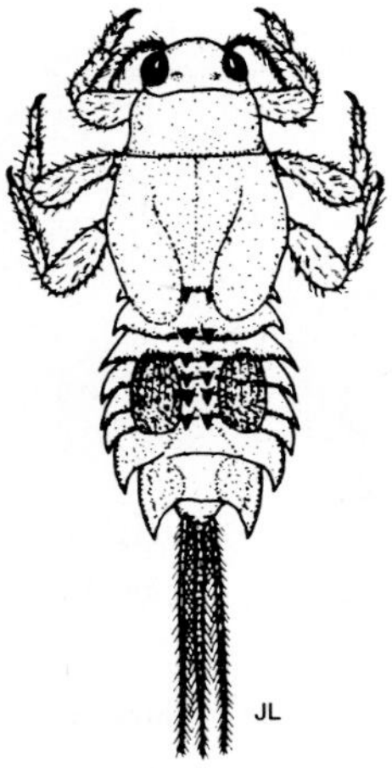

Figure 3.81

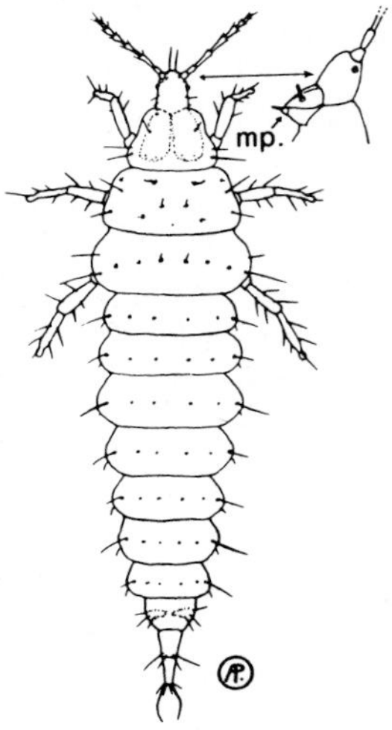

Figure 3.82

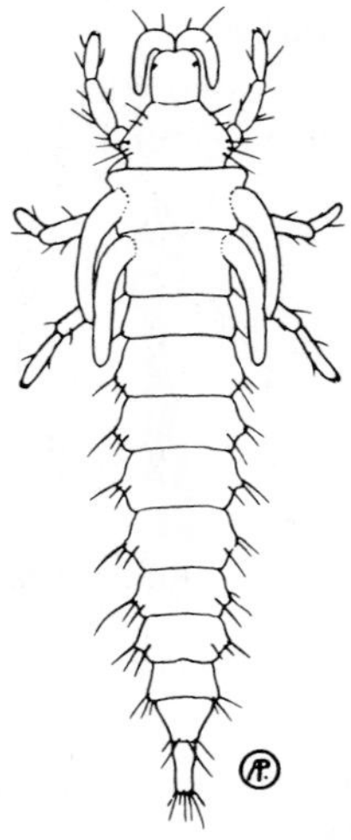

Figure 3.83

61(60). **Legs:** with or without claws and ending in protrusible bladder
Beak: unjointed, fleshy or horny and cone-shaped, palpi of 1 to 4 segments .... (figs. 3.82, 3.83) immature and wingless adult **THYSANOPTERA** (Vol. 2)

61′. **Legs:** with 2 claws (rarely 1), terminal protrusible bladder absent
Beak: triangular and without palpi, or of 4 needlelike components (mandibles and maxillae) that are usually enclosed in a troughlike, jointed labium lacking palpi, the whole structure being beaklike or styletlike 62

62(61′). **Beak:** appears to arise from front of head (more toward the rear on some aquatic forms)
Pronotum: usually prominent
Tarsi: usually 3 segments
Ocelli: present or absent (adults); always absent in immatures
Habitat: aquatic or terrestrial .... (figs. 3.84–3.87) immature and wingless adult **HEMIPTERA** (Vol. 2)

62′. **Beak:** appears to arise from posterior part of head
Pronotum: usually not especially prominent (except Membracidae)
Tarsi: 1–3 segments
Ocelli: present or absent (immatures and adults)
Habitat: terrestrial (not truly aquatic if associated with water) .... (figs. 3.88–3.91) immature and wingless adult **HOMOPTERA** (Vol. 2)

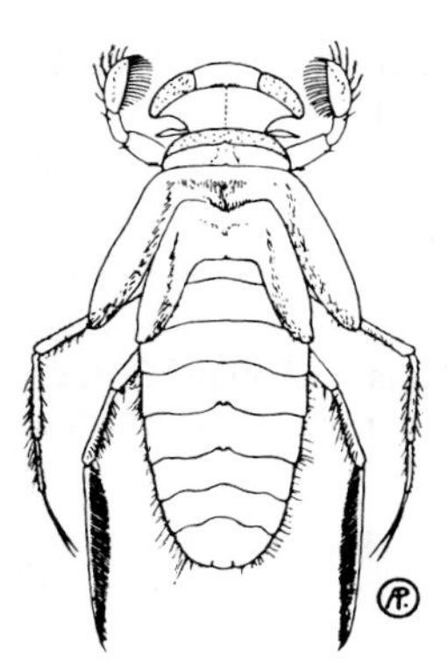
Figure 3.84

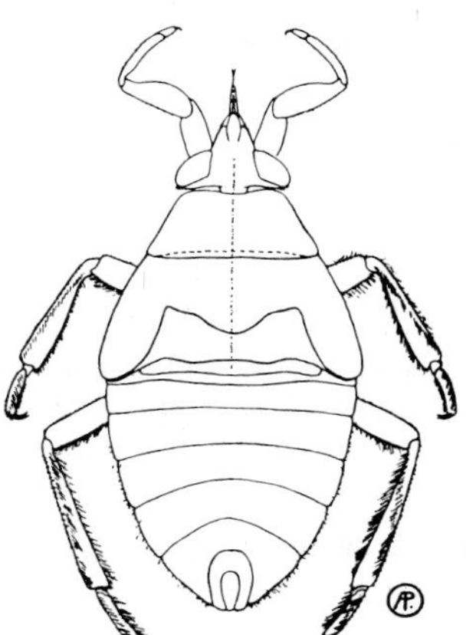
Figure 3.85

Figure 3.86

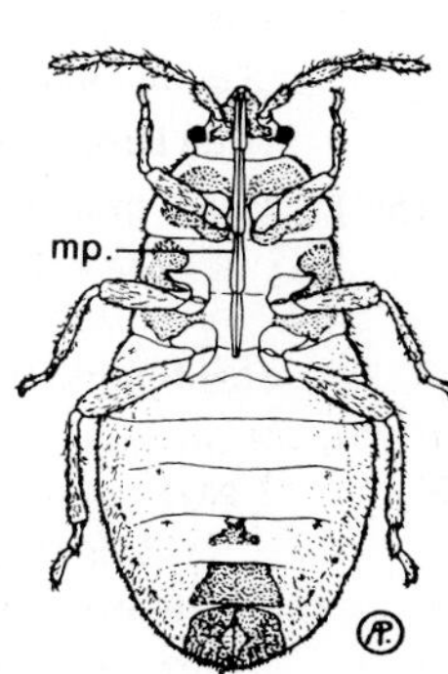

Figure 3.87

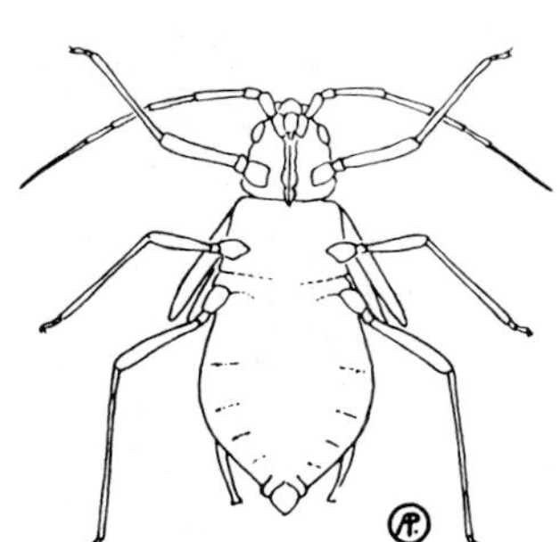
Figure 3.88

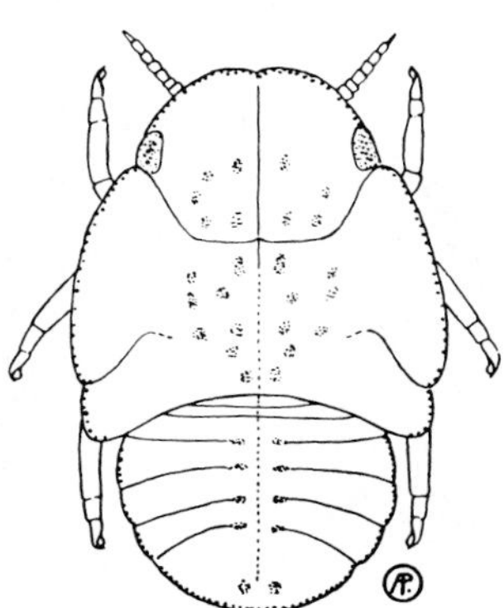
Figure 3.89

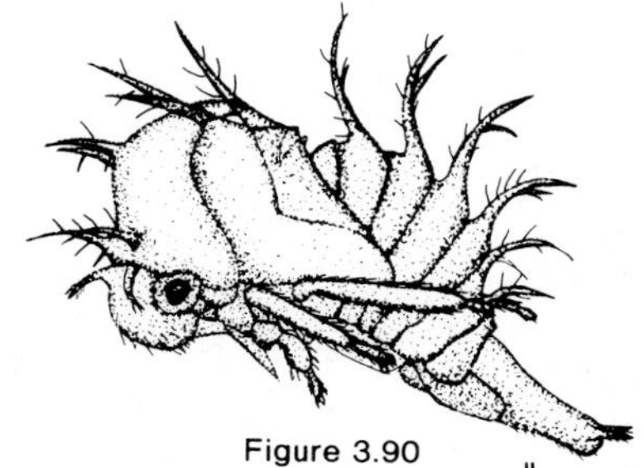

Figure 3.90

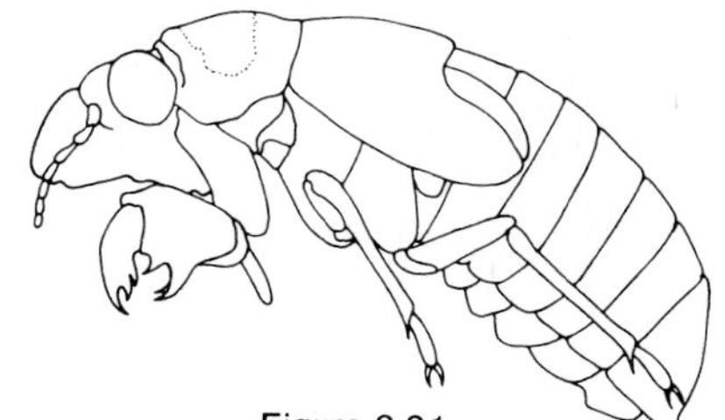

Figure 3.91

63(60′). **Head:** usually with conspicuous, outwardly-swelling postclypeus (area in front of or below the antennae)
Pronotum: usually inconspicuous, smaller than mesonotum or metanotum
Tarsi: 2 (immatures), 2 or 3 (adults) segments and with pair of claws
Cerci: absent .... (fig. 3.92) immature and wingless adult **PSOCOPTERA** (p. 196)

63′. **Head:** without conspicuous outward swelling in front of or below the antennae
Pronotum: conspicuous, usually approximately equal to, or sometimes larger than mesonotum
Tarsi: 1–5 segments, with or without claws
Cerci: often present 64

64(63′). **Cerci:** 1 segment but prominent, often with an apical spinelike process (fig. 3.93)
Tarsi: 2 segments
Antennae: 9 beadlike segments
Habitat: under bark, in sawdust piles .... (figs. 3.94, 3.95) immature and wingless adult **ZORAPTERA** (p. 184)

64′. **Cerci:** absent or, if present, often of 2 or more segments and without an apical spinelike process
Tarsi: usually 3 to 5 segments, at least the hind tarsi rarely of 2 segments
Antennae: variable
Habitat: variable 65

65(64′). **Front tarsi:** basal segment large and bulbous (fig. 3.96) .... immature and adult **EMBIIDINA (EMBIOPTERA)** (p. 179)

65′. **Front tarsi:** basal segment not large and bulbous 66

66(65′). **Hind legs:** femur enlarged for jumping .... (figs. 3.97, 3.98) most immature and wingless adult **ORTHOPTERA** (p. 147)

66′. **Hind legs:** femur not enlarged 67

67(66′). **Head:** prolonged downward into trunklike beak with terminal chewing mouthparts (fig. 3.99) .... wingless adult **MECOPTERA** (snow scorpionflies, Boreidae; and 1 species of hanging-fly, Bittacidae)

67′. **Head:** not prolonged into trunklike beak 68

68(67′). **Prothorax:** greatly lengthened, much longer than mesothorax
Fore legs: enlarged and modified for grasping prey .... (fig. 3.100) immature and wingless adult **MANTODEA** (p. 140)

68′. **Prothorax:** not greatly lengthened
Fore legs: not enlarged or modified for grasping prey 69

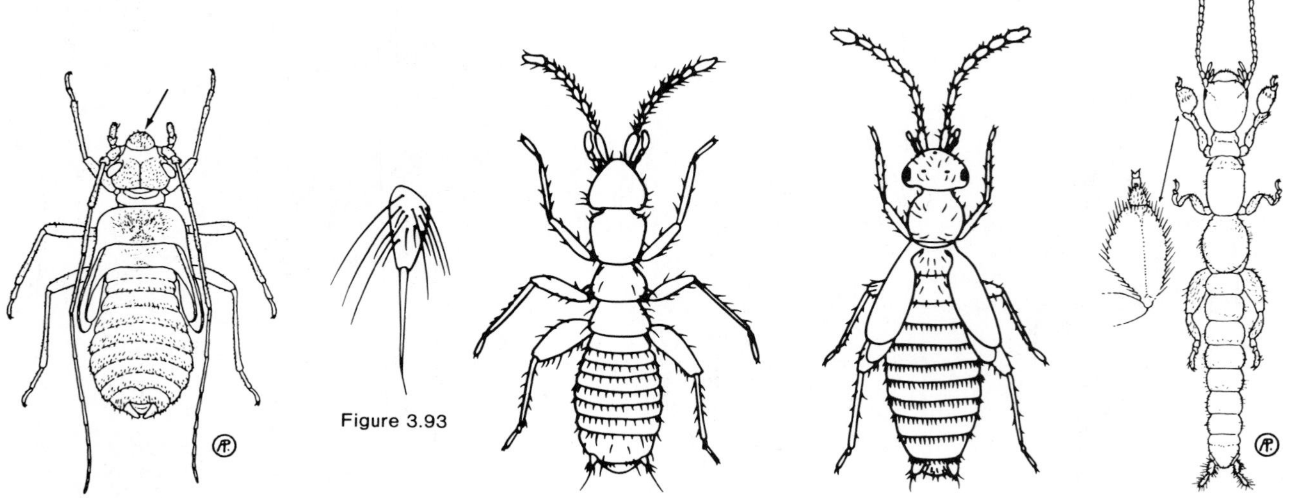

Figure 3.92 Figure 3.93 Figure 3.94 Figure 3.95 Figure 3.96

69(68′). **Abdomen:** without pair of terminal appendages, although may have single median process
Antennae: usually 11 or fewer segments, may be enlarged toward tip .... (some) adult **COLEOPTERA** (wingless female Dermestidae without elytra (*Thylodrias*) and a few Staphylinidae and Pselaphidae with very short elytra)

69′. **Abdomen:** with pair of terminal appendages (cerci) but without median process other than ovipositor
Antennae: usually more than 11 segments (if 11, or more rarely 10 or 9, then antennae not usually clublike or gradually enlarged toward tip) 70

70(69′). **Cerci:** 4 or more segments 71

70′. **Cerci:** 1–3 segments 73

71(70). **Head:** hypognathous or slanting posteriorly
Pronotum: shieldlike, with rounded outline, usually covering all or part of head
Body: dorsoventrally flattened and oval .... (fig. 3.101) immature and wingless adult **BLATTODEA** (p. 120)

71′. **Head:** prognathous
Pronotum: not shieldlike, and not covering head
Body: cylindrical and more elongate 72

72(71′). **Tarsi:** 5 segments
Antennae: filiform
Pronotum: roughly quadrate
Abdomen: females with prominent, sclerotized, sword-shaped ovipositor
Cerci: usually at least half length of abdomen and of 8 or 9 segments
Range: Northwestern U.S., Alberta, and British Columbia .... (fig. 3.102) immature and adult **GRYLLOBLATTODEA** (p. 143)

72′. **Tarsi:** 4 segments (an indistinct 5th may be present)
Antennae: moniliform to submoniliform (beadlike)
Pronotum: rarely quadrate
Abdomen: females lacking sclerotized ovipositor
Cerci: short, 4 to 8 segments
Range: Pacific coastal states, including British Columbia, Baja California, and Desert Southwest .... (fig. 3.103) (some) immature and wingless adult **ISOPTERA** (p. 132)

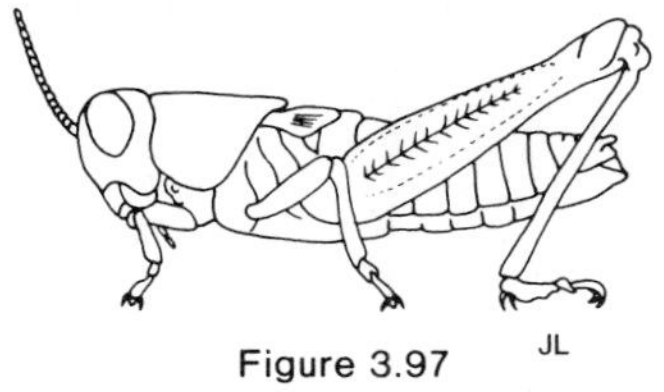

Figure 3.97

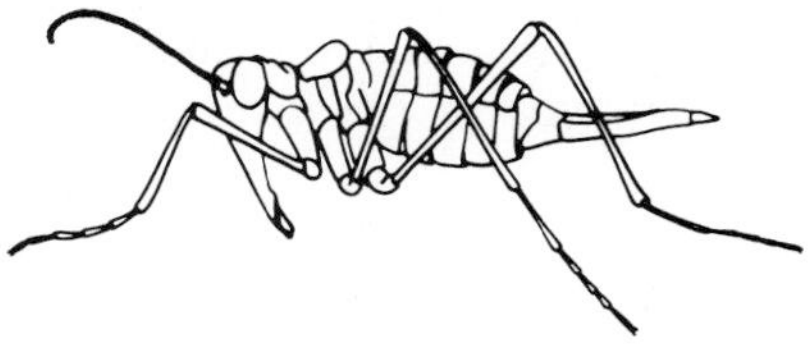

Figure 3.99

Figure 3.101

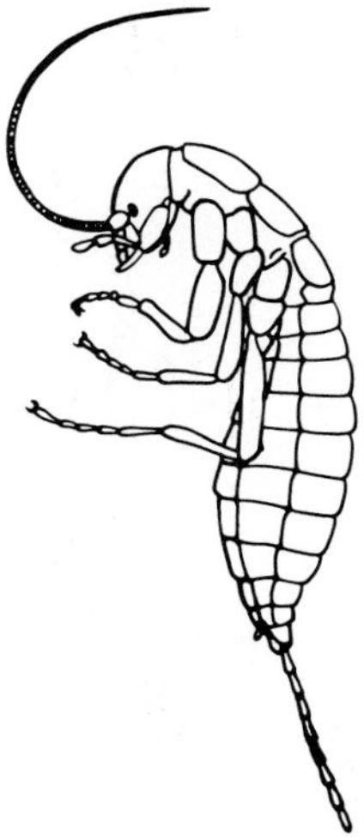

Figure 3.102

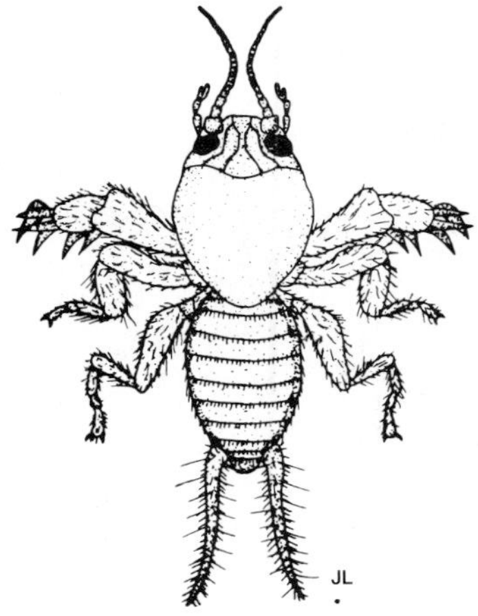

Figure 3.98

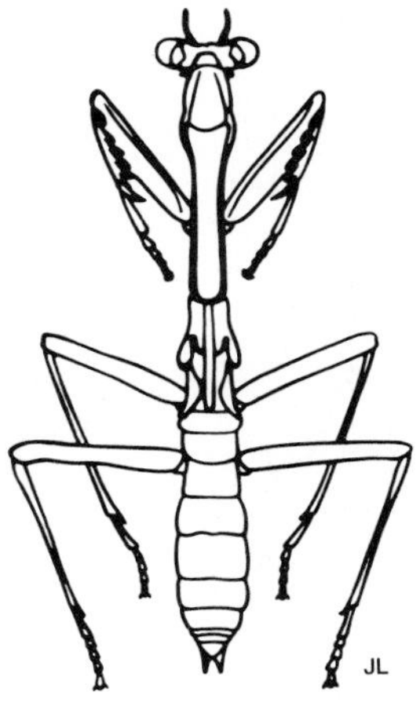

Figure 3.100

73(70′). **Antennae:** 11, or more rarely 10 or 9 segments; usually clublike or gradually enlarged toward tip
Abdomen: 6 or 7 (rarely 8) segments visible ventrally; tarsi usually 5-segmented .. .. (few) adult **COLEOPTERA**
(some rove beetles (not really wingless but have very short elytra) Staphylinidae)

73′. **Antennae:** usually more than 11 segments; threadlike or beadlike
Abdomen: 9–11 segments visible ventrally (if 6 or 7, then tarsi with 3 segments) 74

74(73′). **Tarsi:** 5 distinct segments
Body: sticklike .... (fig. 3.104) immature and wingless adult **PHASMATODEA** (p. 145)

74′. **Tarsi:** 2–4 segments (an indistinct 5th may be present)
Body: not sticklike 75

75(74′). **Terminal appendages:** very small, 1 to 3 segments and divergent
Tarsi: 4 segments (an indistinct 5th may be present)
Body: usually pale or white .... (fig. 3.105) (most) immature and wingless adult **ISOPTERA** (p. 132)

75′. **Terminal appendages:** large, 1 segment and somewhat convergent, becoming forcepslike in late instars
Tarsi: 3 segments (rarely 2)
Body: usually dark brown or black .... (fig. 3.106) immature and wingless adult **DERMAPTERA** (p. 171)

76(1′). **Antennae:** branched .... (fig. 3.107) (most first instar pauropods) Class **PAUROPODA**

76′. **Antennae:** not branched, or absent 77

77(76′). **Antennae:** absent, although other appendages may be used as antennae
Segmented legs or leglike appendages: 4 pairs (also with chelicerae and pedipalps, the latter sometimes appearing leglike) .... Class **ARACHNIDA** 78

77′. **Antennae:** present
Segmented legs: with 5 or more pairs 88

78(77). **Opisthosoma ("abdomen"):** segmentation absent (most) or only visible dorsally
Ventral spinning organs: present, or, if lacking, body usually minute 79

78′. **Opisthosoma:** segmentation distinct (ventrally and/or dorsally)
Ventral spinning organs: absent 80

79(78). **Opisthosoma ("abdomen"):** joined to prosoma (rest of body) by slender petiole
Spinning organs: present beneath the anus (fig. 3.108)
Chelicerae: single claw with venom duct opening .... (immature and adult spiders) **ARACHNIDA,** Order **ARANEAE**

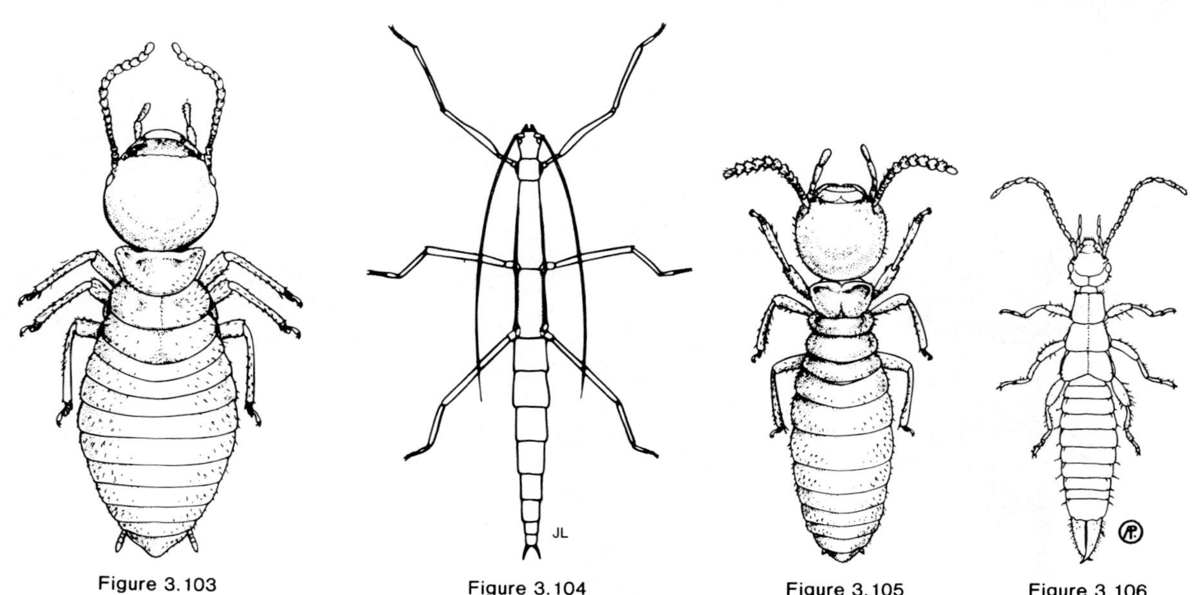

Figure 3.103 Figure 3.104 Figure 3.105 Figure 3.106

79′. **Opisthosoma:** broadly fused to prosoma, body often saclike
Spinning organs: absent
Chelicerae: usually with pincers .... (figs. 3.109–3.112) (nymphal & adult ticks, mites) **ARACHNIDA,** Order **ACARINA**

80(78′). **Opisthosoma ("abdomen"):** ending in tail-like or whiplike appendage (may be short) 81

80′. **Opisthosoma:** tail-like or whiplike appendage absent 85

81(80). **Opisthosoma ("abdomen"):** with long, thick, tail-like prolongation terminating in prominent, bulbous sting; with pair of comblike organs on second ventral segment .... (fig. 3.113) immature and adult scorpions **ARACHNIDA,** Order **SCORPIONIDA**

81′. **Opisthosoma:** without terminal sting, though may have prolongation; ventral comblike organs absent 82

82(81′). **Prosoma ("cephalothorax"):** movable flap concealing mouthparts and front of face (southern Texas) .... (fig. 3.114) immature and adult ricinuleids **ARACHNIDA,** Order **RICINULEIDA**

82′. **Prosoma:** movable flap absent 83

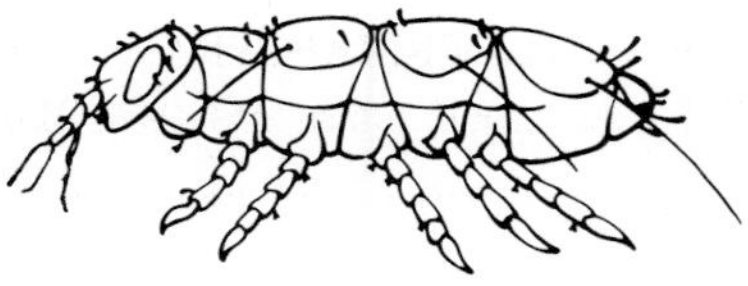

Figure 3.107

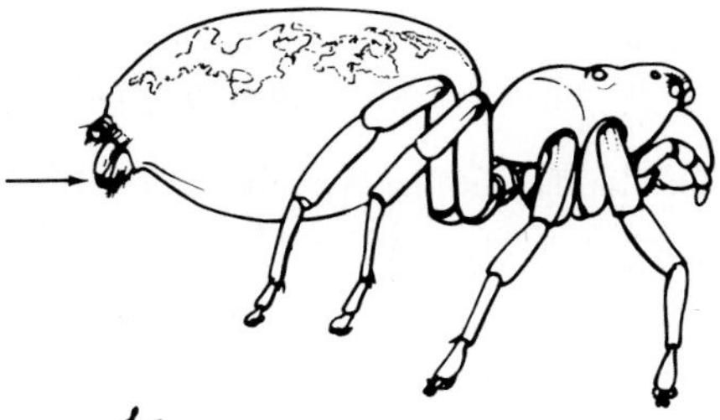

Figure 3.108

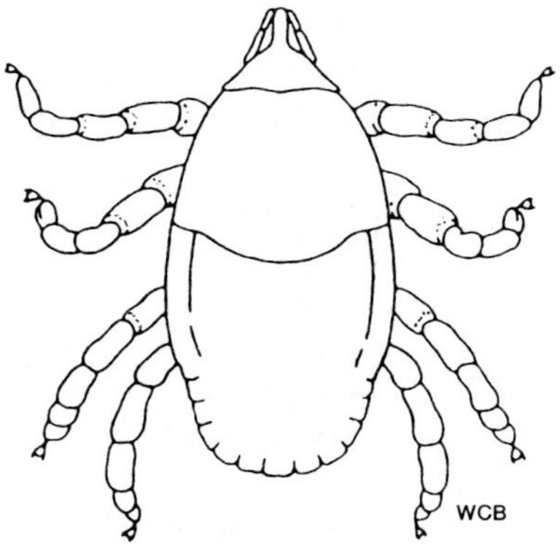

WCB

Figure 3.109

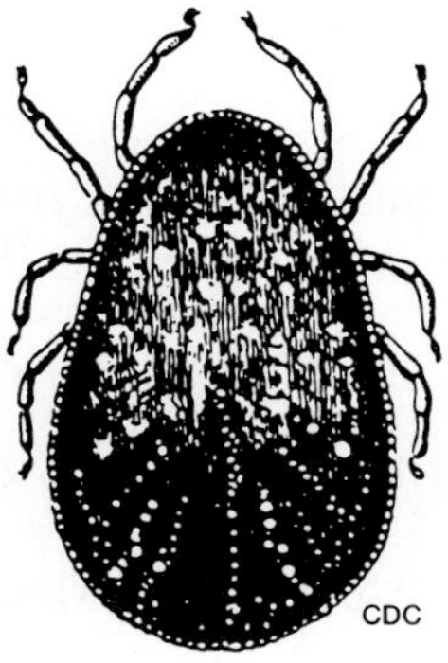

CDC

Figure 3.110

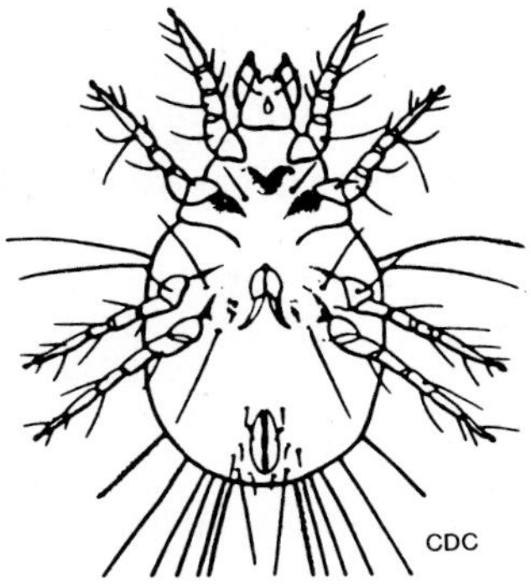

CDC

Figure 3.111

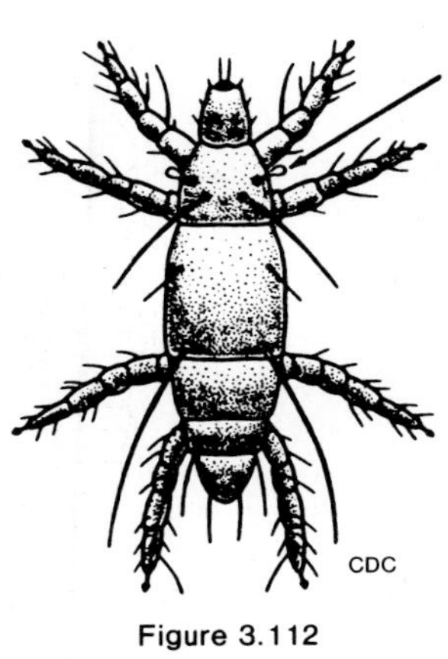

CDC

Figure 3.112

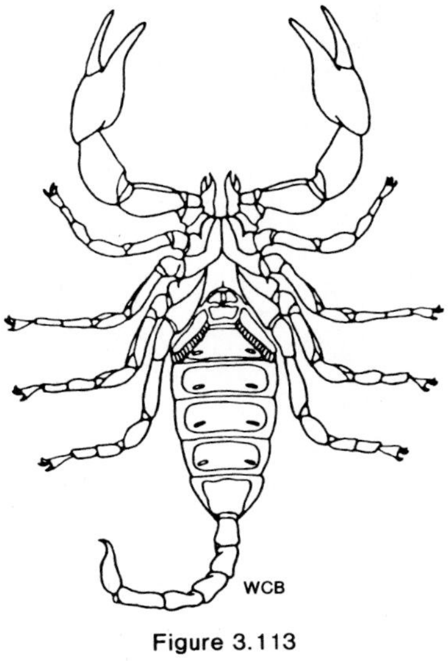

WCB

Figure 3.113

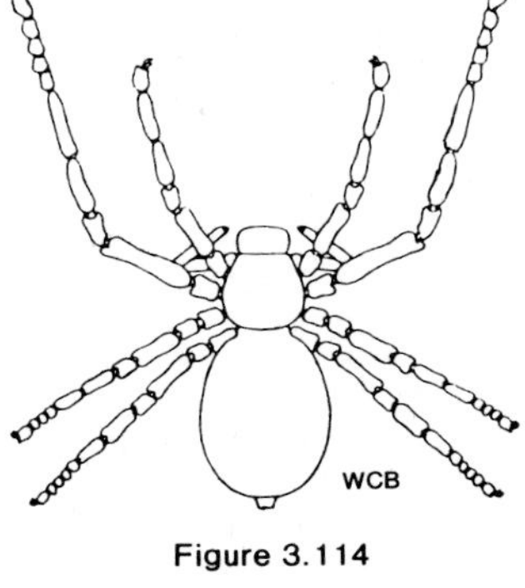

WCB

Figure 3.114

83(82′). **Pedipalps:** not stout and raptorial, leglike
Adult size: under 5 mm
U.S. Distribution: Texas to California .... (fig. 3.115) immature and adult micro whipscorpions **ARACHNIDA,** Order **PALPIGRADI**

83′. **Pedipalps:** stout and raptorial, not leglike
Adult size: most over 5 mm
U.S. Distribution: Florida to California 84

84(83′). **Opisthosoma ("abdomen"):** with short tail of 1–4 segments
Eyes: absent
Adult size: under 10 mm
U.S. Distribution: Texas to California .... (fig. 3.116) immature and adult short-tailed whipscorpions **ARACHNIDA,** Order **SCHIZOMIDA**

84′. **Opisthosoma:** with long tail of many segments
Eyes: present (2 median and 3 on each lateral margin)
Adult size: over 50 mm
U.S. Distribution: Florida to Arizona .... (fig. 3.117) immature and adult whipscorpions **ARACHNIDA,** Order **UROPYGIDA**

85(80′). **Chelicerae:** massive, projecting forward
Prosoma ("cephalothorax"): posterior 2 or 3 segments distinguishable (not fused with the rest)
U.S. and Canadian Distribution: Florida, and the West, north into British Columbia .... (fig. 3.118) the windscorpions **ARACHNIDA,** Order **SOLIFUGAE**

85′. **Chelicerae:** not greatly enlarged
Prosoma: all segments fused
U.S. and Canadian Distribution: widespread 86

86(85′). **Opisthosoma ("abdomen"):** narrowly joined to prosoma ("cephalothorax")
Legs I: long, thin, and antennalike
U.S. Distribution: Florida to Texas .... (fig. 3.119) immature and adult tailless whipscorpions **ARACHNIDA,** Order **AMBLYPYGIDA**

86′. **Opisthosoma:** broadly joined to prosoma
Legs I: quite similar to other legs
Distribution: widespread 87

87(86′). **Pedipalps:** pincerlike, long, and heavy
Eyes: absent or 1 or 2 pairs on anteriolateral margin (hard to see)
Body: dorsoventrally flattened; longer than wide .... (fig. 3.120) immature and adult pseudoscorpions **ARACHNIDA,** Order **PSEUDOSCORPIONIDA**

87′. **Pedipalps:** leglike, with claws, but shorter than legs
Eyes: 1 pair, usually medially located on ocularium (easy to see)
Body: round to oval .... (fig. 3.121) immature and adult harvestmen = daddy-long-legs **ARACHNIDA,** Order **PHALANGIDA**

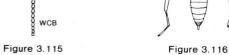

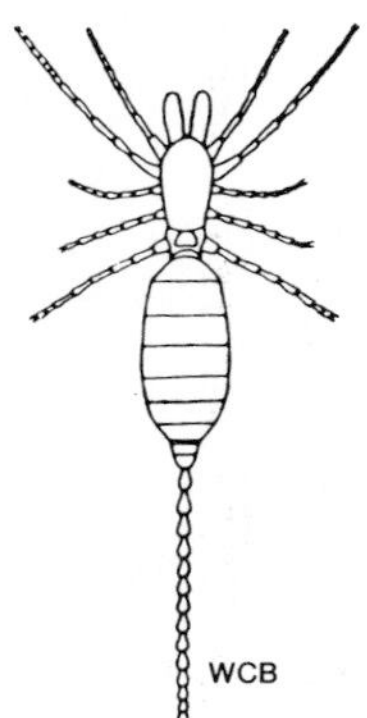

Figure 3.115

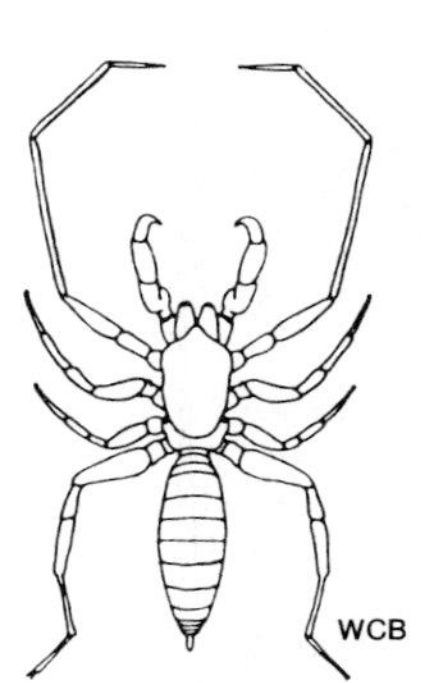

Figure 3.116

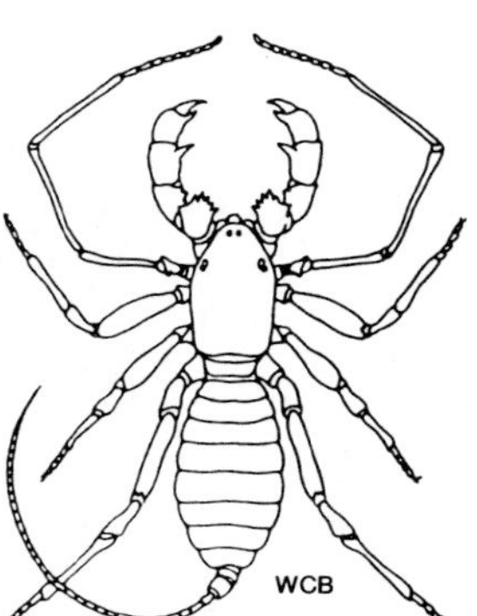

Figure 3.117

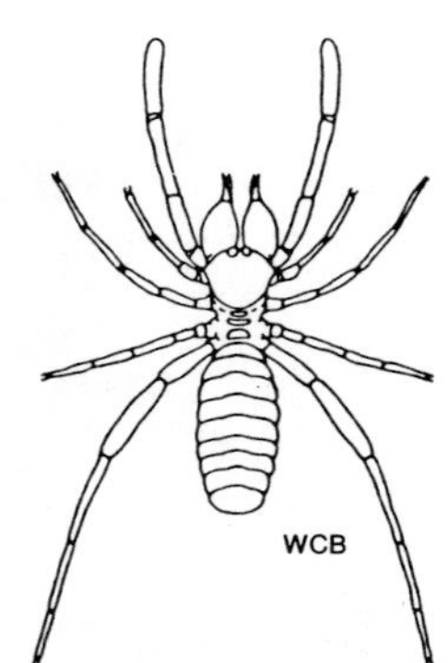

Figure 3.118

88(77′). **Appendages (excluding those of head):** 6 or 7 pairs of walking legs followed by 5 pairs of small flat plates (pleopods) and 1 pair of tail-like, biramous legs (uropods) ..... mostly isopods (fig. 3.122), amphipods (fig. 3.123), and decapods (fig. 3.124) Class **CRUSTACEA**

88′. **Appendages:** all, or all except first pair, as walking legs; distributed more or less uniformly along body 89

89(88′). **Legs I:** modified as large but short poison fangs (maxillipeds)
Legs: with 1 claw; 21 or more pairs of legs (hatch with full number) or 15 pairs (hatch with 7 pairs) .... (figs. 3.125, 3.126) immature and adult centipedes Class **CHILOPODA**

89′. **Legs I:** not as maxillipeds
Legs: with 1 or 2 claws; 4 or more pairs of legs (often over 30) 90

90(89′). **Body segments:** after segment 3, appear to have 2 pairs of legs per segment
Legs: with 1 claw; adult usually has 13 or more pairs of legs (hatch with 3 or rarely 4 pairs)
Color: rarely white (although some are pale)
Size: commonly over 10 mm .... (fig. 3.127) immature and adult millipedes Class **DIPLOPODA**

90′. **Body segments:** not fused in pairs, having 1 pair of legs per segment (3 or 4 segments lack legs)
Legs: with 2 claws; adult has 11 or 12 pairs of legs (usually hatch with 6 pairs)
Color: usually white
Size: rarely over 10 mm long .... (fig. 3.128) immature and adult symphylids Class **SYMPHYLA**

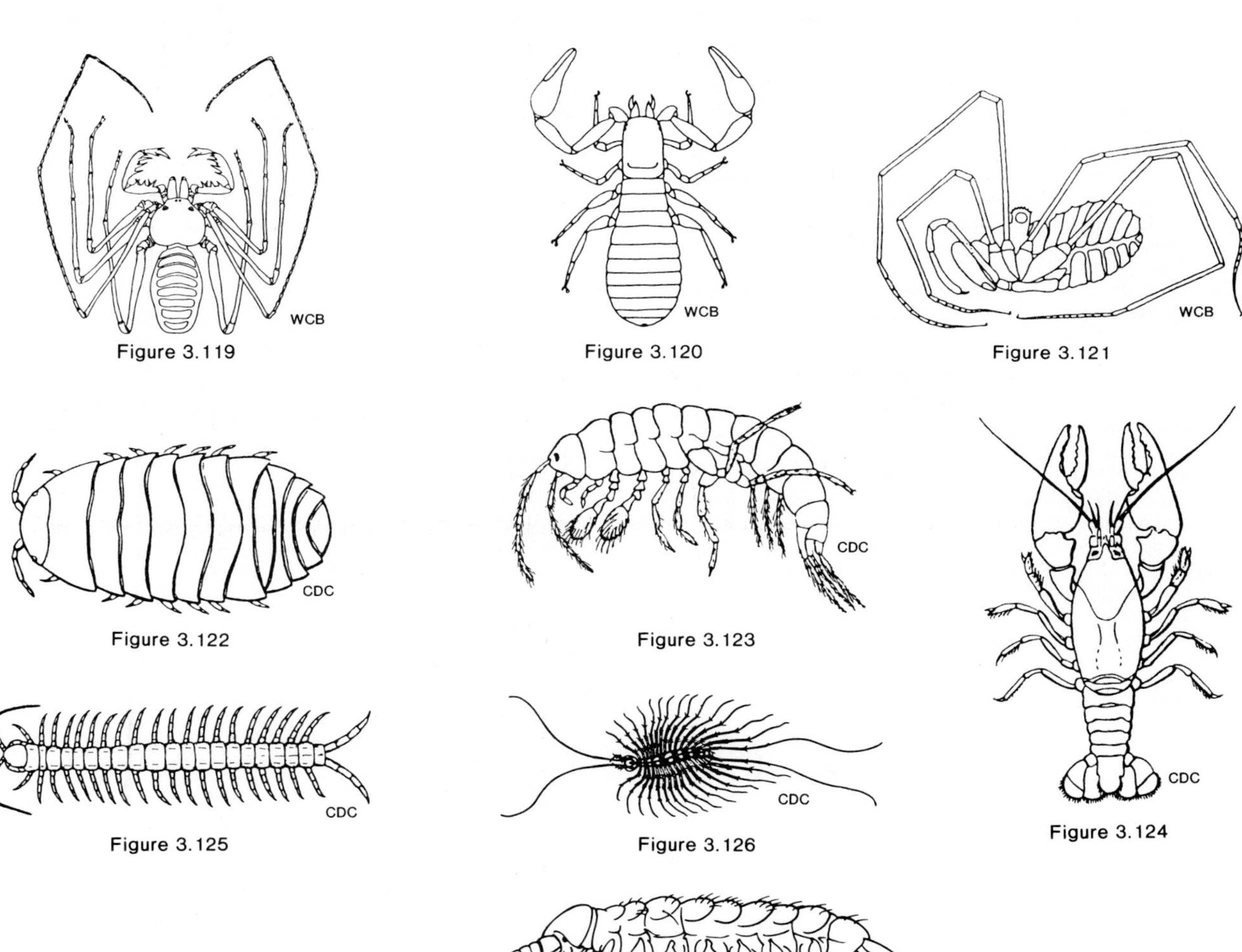

Figure 3.119

Figure 3.120

Figure 3.121

Figure 3.122

Figure 3.123

Figure 3.124

Figure 3.125

Figure 3.126

Figure 3.127

Figure 3.128

FIRST INSTAR PARASITOID LARVAE

The following figures give the user some idea of the types and diversity of forms found in first instar parasitoid larvae. They may be temporarily free living, attach externally to the host or other carrier, or be found internally. After the first molt, most become quite nondescript and grublike. See Clausen (1940) for the most comprehensive coverage of immature parasitic insects that is available. See Hagen (1964) for a more concise coverage of the developmental stages of parasitic insects.

SELECTED BIBLIOGRAPHY FOR PARASITOID LARVAE

Allen, H. W. 1958. Orchard studies on the effect of organic insecticides on parasitism of the oriental fruit moth. J. Econ. Ent. 51:82–87.

Balduf, W. V. 1928. Observations of the buffalo treehopper *Ceresia bubalus* Fabr. (Membracidae, Homoptera), and the bionomics of an egg parasite, *Polynema striaticorne* Girault (Mymaridae, Hymenoptera). Ann. Ent. Soc. Amer. 21:419–35.

Bradley, W. G., and E. D. Burgess. 1934. The biology of *Cremastus flavoorbitalis* (Cameron), an ichneumonid parasite of the European corn borer. USDA Tech. Bull. 441. 15 pp.

Bohart, R. M. 1941. A revision of the Strepsiptera with special reference to the species of North America. Univ. Calif. Publ. Ent. 7:91–160.

Clancy, D. W. 1946. The insect parasites of the Chrysopidae (Neuroptera). Univ. Calif. Publ. Ent. 7:403–496.

Clausen, C. P. 1923. The biology of *Schizaspidia tenuicornis* Ashm., a Eucharid parasite of *Camponotus*. Ann. Ent. Soc. Amer. 16:195–219.

Clausen, C. P. 1928. *Hyperalonia oenomaus* Rond., a parasite of *Tiphia* larvae. (Dipt.: Bombyliidae.) Ann. Ent. Soc. Amer. 21:461–72, 642–59.

Clausen, C. P. 1931. Biological notes on the Trigonalidae. Proc. Ent. Soc. Wash. 33:72–81.

Clausen, C. P. 1940. Entomophagous insects. New York: McGraw-Hill Book Co., Inc. 688 pp. (Reprinted by Hafner Publ. Co., New York, 1972).

Clausen, C. P., J. L. King, and C. Teranishi. 1927. The parasites of *Popillia japonica* in Japan and Chosen (Korea) and their introduction into the United States. USDA Agric. Bull. 1429. 55 pp.

Coppel, H. C. 1958. Studies on dipterous parasites of the spruce budworm, *Choristoneura fumiferana* (Clem.) (Lepidoptera: Tortricidae). VI. *Phorocera incrassata* Smith (Diptera: Tachinidae). Canadian J. Zool. 36:453–62.

Dowden, P. B. 1941. Parasites of the birch leaf-mining sawfly (*Phyllotoma nemorata*). USDA Tech. Bull. 757. 55 pp.

Ganin, M. 1869. Beiträge zur Erkenntniss der Entwicklungsgeschichte bei den Insecten. Zeitschr. f. wiss. Zool. 19:381–451.

Hagen, K. S. 1964. Developmental stages of parasites. pp. 168–246. *In:* Debach, P. (ed.). Biological control of insect pests and weeds. New York: Reinhold Publ. Corp.

Hidaka, T. 1958. Biological investigation on *Telenomus gifuensis* Ashmead (Hym.:Scelionidae), an egg-parasite of *Scotinophara lurida* Burmeister (Hem.:Pentatomidae) in Japan. Acta Hymenopterologica 1:75–93.

Kemner, N. A. 1926. Zur Kenntnis der Staphyliniden-larven. II. Die Lebensweise und die parasitische Entwicklung der echten Aleochariden. Ent. Tidskr. 47:133–70.

King, J. L. 1916. Observations on the life history of *Pterodontia flavipes* Gray. Ann. Ent. Soc. Amer. 9:309–321.

Linsley, E. G., J. W. MacSwain, and R. F. Smith. 1952. The life history and development of *Rhipiphorous smithi* with notes on their phylogenetic significance (Coleoptera: Rhipiphoridae). Univ. Calif. Publ. Ent. 9:291–314.

MacSwain, J. W. 1956. A classification of the first instar larvae of the Meloidae (Coleoptera). Univ. Calif. Publ. Ent. 12:1–182.

Morris, K. R. S., E. Cameron, and W. F. Jepson. 1937. The insect parasites of the spruce sawfly, *Diprion polytomum* in Europe. Bull. Ent. Res. 28:341–93.

Prescott, H. W. 1955. *Neorhynchocephalus sackenii* and *Trichopsidea clausa,* nemestrinid parasites of grasshoppers. Ann. Ent. Soc. Amer. 48:392–402.

Principi, M. M. 1947. Contributi allo studio dei "Neurotteri" Italiani V. Ricerche su *Chrysopa formosa* Brauer e su alcuni suoi parassiti. Boll. Instit. Ent. Univ. Bologna 16:134–75.

Schell, S. C. 1943. The biology of *Hadronotus ajax* Girault (Hymenoptera: Scelionidae) a parasite in the eggs of squash-bug (*Anasa tristis* DeGeer). Ann. Ent. Soc. Amer. 36:625–35.

Schlinger, E. I., and J. C. Hall. 1961. The biology, behavior, and morphology of *Trioxys* (*Trioxys*) *utilis,* an internal parasite of the spotted alfalfa aphid, *Therioaphis maculata* (Hymenoptera: Braconidae, Aphidiinae). Ann. Ent. Soc. Amer. 54:34–45.

Simmonds, F. J. 1952. Parasites of the frit-fly, *Oscinella frit* (L.), in eastern North America. Bull. Ent. Res. 43:503–42.

Thorpe, W. H. 1934. The biology and development of *Cryptochaetum grandicorne* (Diptera), an internal parasite of *Guerinia serratulae* (Coccidae). Quart. J. Micros. Sci. 77:273–304.

Townsend, C. H. T. 1908. A record of results from rearings and dissections of Tachinidae. U.S. Bur. Ent. Tech. Ser. 12, 6:95–118.

Tripp, H. A. 1961. The biology of a hyperparasite, *Euceros frigidus* Cress. (Ichneumonidae) and description of the planidial stage. Can. Ent. 93:40–58.

Wright, D. W., Q. A. Geering, D. G. Ashby. 1947. The insect parasites of the carrot fly, *Psila rosae* Fabr. Bull. Ent. Res. 37:507–29.

TRIUNGULIN AND TRIUNGULINID TYPES

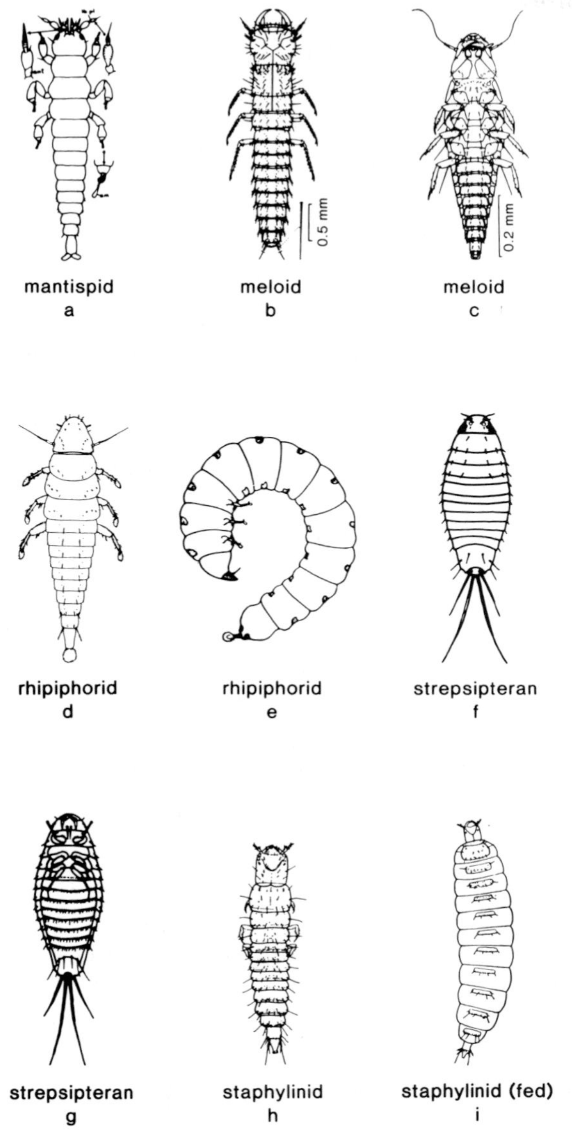

mantispid a — meloid b — meloid c — rhipiphorid d — rhipiphorid e — strepsipteran f — strepsipteran g — staphylinid h — staphylinid (fed) i

0.5 mm — 0.2 mm

Figures 3.129a-i

PLANIDIUM TYPES

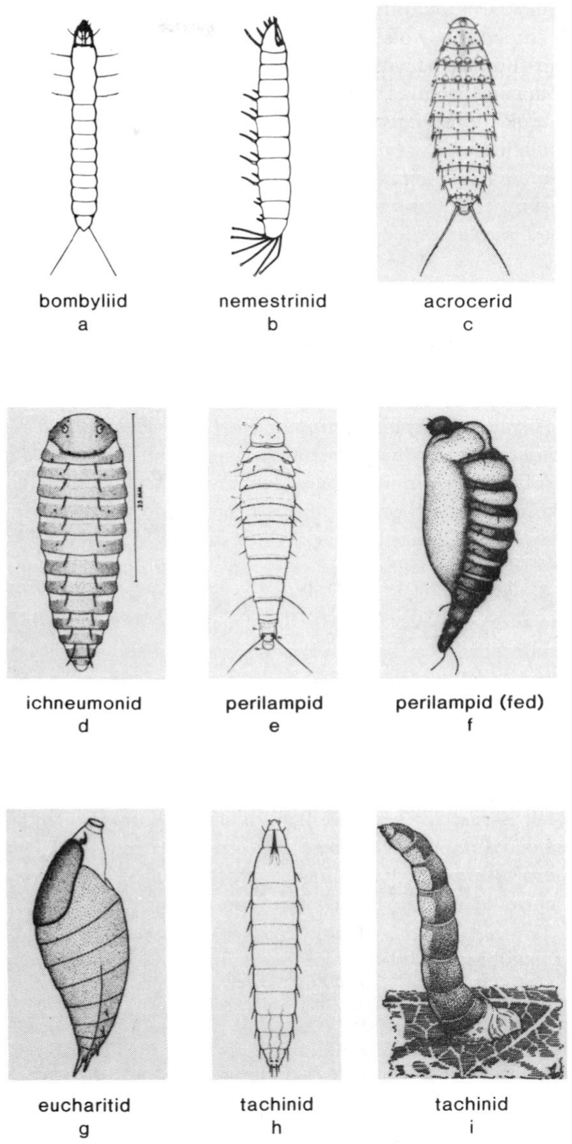

bombyliid a — nemestrinid b — acrocerid c — ichneumonid d — perilampid e — perilampid (fed) f — eucharitid g — tachinid h — tachinid i

Figures 3.130a-i

Figures 3.129a-i. Triungulin and triungulinid type first instar parasitoid larvae (**b-g** courtesy of Hagen 1964): **a.** Neuroptera. *Mantispa interrupta* Say (from Peterson 1951); **b.** Coleoptera. *Epicauta pardalis* (LeConte) (from MacSwain 1956); **c.** Coleoptera. *Zonitis bilineatus* Say, ventral view (from MacSwain 1956); **d.** Coleoptera. *Rhipiphorus smithi* Linsley and MacSwain (from Linsley, MacSwain, and Smith 1952); **e.** Coleoptera. *R. smithi* first instar endoparasitic phase (from Linsley, MacSwain, and Smith 1952); **f.** Strepsiptera. *Eoxenos laboulbenei* Peyer, dorsal view (from Bohart 1941); **g.** Strepsiptera. *E. laboulbenei,* ventral view (from Bohart 1941); **h.** Coleoptera. *Aleochara curtula* Goeze, newly hatched first instar larva (from Kemner 1926); **i.** Coleoptera. *A. curtula,* fully fed first instar larva (from Kemner 1926).

Figures 3.130a-i. Planidium-type first instar parasitoid larvae (courtesy of Hagen 1964): **a.** Diptera. *Hyperalonia oenomaus* Rondani (from Clausen 1928); **b.** Diptera. *Trichopsidea clausa* (Osten Sacken), lateral view (from Prescott 1955); **c.** Diptera. *Pterodontia flavipes* Gray, dorsal view (from King 1916); **d.** Hymenoptera. *Euceros frigidus* Cresson, dorsal view (from Tripp 1961); **e.** Hymentopera. *Perilampus laevifrons* Dal., dorsal view (from Principi 1947); **f.** Hymentopera. *Perilampus chrysopae* Crawford, engorged planidium, lateral view (from Clancy 1946); **g.** Hymenoptera. *Schizaspidia tenuicornis* Ashmead, lateral view (from Clausen 1923); **h.** Diptera. *Prosena siberita* (F.) (from Clausen, King, and Teranishi 1927); **i.** Diptera. *Tachina magnicornis* Zetterstedt, showing the egg chorion serving as an attachment "cup" to leaf (from Townsend 1908).

FIRST INSTAR PARASITOID LARVAE

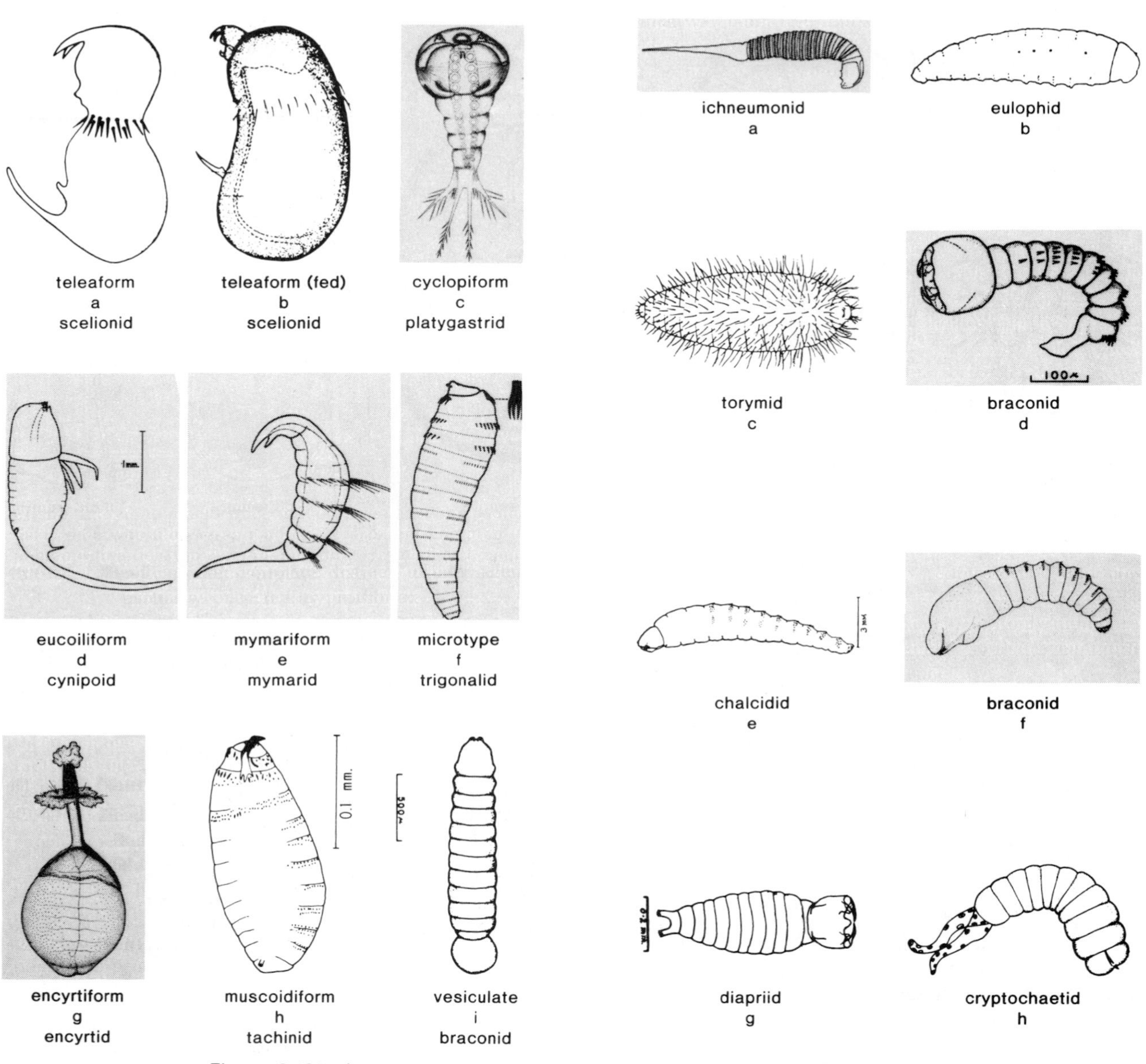

Figures 3.131a–i

Figures 3.132a–h

Figure 3.131a-i. Assorted types of first instar parasitoid larvae (courtesy of Hagen 1964): **a.** Hymenoptera. *Telonomus gifuensis* Ashmead (from Hidaka 1958); **b.** Hymenoptera. Engorged *Hadronotus ajax* Girault (from Schell 1943); **c.** Hymenoptera. Platygastrid (from Ganin 1869); **d.** Hymenoptera, Eucoilidae. *Hexacola sp.* (from Simmonds 1952); **e.** Hymenoptera. *Polynema striaticorne* Girault (from Balduf 1928); **f.** Hymenoptera. *Orthogonalys debilis* Teranishi (from Clausen 1931a); **g.** Hymenoptera. *Isodromus niger* Ashmead (from Clancy 1946); **h.** Diptera. *Phorocera incrassata* Smith (from Coppel 1958); **i.** Hymenoptera. Second stage of *Apanteles medicaginis* Muesebeck (from Allen 1958).

Figures 3.132a-h. Assorted first instar parasitoid larvae (courtesy of Hagen 1964): **a.** Hymenoptera. *Trathala flavoorbitalis* (Cameron), 96 hours after oviposition (from Bradley and Burgess 1934); **b.** Hymenoptera. *Sympiesis* sp. (from Dowden 1941); **c.** Hymenoptera. *Monodontomerus dentipes* (Dalman) (from Morris et al. 1937); **d.** Hymenoptera. *Apanteles medicaginis* Muesebeck, first instar (from Allen 1958); **e.** Hymenoptera. *Spilochalcis side* (Walker) (from Arthur 1958); **f.** Hymenoptera. *Trioxys utilis* Muesebeck, first instar (from Schlinger and Hall 1961); **g.** Hymenoptera. *Eutrias tritoma* (Thomson), first instar (from Wright et al. 1947); **h.** Diptera. *Cryptochaetum grandicorne* Rondani, late first instar (from Thorpe 1934).

Class and Order Protura

4

E. C. Bernard
University of Tennessee

S. L. Tuxen[1]
Zoologisk Museum, Copenhagen, Denmark

PROTURANS

Proturans comprise a primitive, wingless group of hexapods characterized by lack of antennae, fore legs modified for sensory purposes, and anamorphosis of the abdomen. They were first recognized and described at the beginning of the century (Silvestri 1907), and their morphology and anatomy were thoroughly and accurately described by Berlese in his monumental monograph of the order (1909), much of which was excerpted by Janetschek (1970). A number of species were described in the 50 years following Berlese's work, but not until 1964, when Tuxen published his comprehensive monograph on the groups, were the important diagnostic criteria firmly established. Although detailed treatises of European Protura (Nosek 1973) and Japanese Protura (Imadaté 1974) have been published, the only work covering a major portion of the North American fauna is that of Ewing (1940), whose species were redescribed by Bonet and Tuxen (1960).

Phylogenetically, Protura seem to bear closest relationships to Collembola (Tuxen 1970), but it is evident that these two have evolved independently for a very long time. Although no fossil Protura are known, they appear to date from the Devonian period (Tuxen 1978).

BIOLOGY AND ECOLOGY

Protura primarily inhabit organic litter layers, rotten logs, moss and soil in wooded areas, but they may be found also in meadows, agricultural soils, or turf. Lussenhop (1973) recovered specimens from grassy expressway margins in an urban area. Most Protura live near the soil surface, but some small forms may be found at considerable depths (Strenzke 1942, Tuxen 1949, 1977b, Price 1973, Yin et al. 1981). For instance, Price (1973) found specimens at a 15–25 cm depth.

The seasonal abundance of proturans apparently varies with the species. In general, densities of Eosentomidae drop to low levels in summer (Pearse 1946, Tuxen 1949, Price 1973, Walker and Rust 1975), but Acerentomidae increase during summer and decrease in winter and early spring (Tuxen 1949; Walker and Rust 1975). On studies of Japanese Protura, Imadaté (1974) observed that *Eosentomon sakura* peaked about April and October with severe declines in summer and winter, whereas *E. asahi* and *Berberentulus tosanus* fluctuated irregularly. The factors influencing population fluctuations and appearance of immature stages are nearly totally unknown.

1. Deceased, 15 June 1983.

The food of Protura is little known, a situation that more than any other has prevented successful laboratory culture of these animals. Sturm (1959) observed two species feeding upon the mantle and free hyphae of ectotrophic mycorrhizae of oak and hornbeam roots. No other observations of feeding have been published. Until food relationships and other environmental considerations can be worked out, workers will be hampered in their attempts to understand proturan development and life cycles.

Larval development among Protura is anamorphic, i.e., the larvae hatch with a few abdominal segments and the number increases with successive stages. Berlese (1909) recognized this type of development, but assumed that there were four stages with 9, 10, 11, and 12 abdominal segments. Tuxen (1949) was the first to establish the sequence prelarva–larva I–larva II–maturus junior–adult as the usual means of postembryonic development. François (1960) later substantiated the details of this work. Tuxen (1949) reported the actual development as follows:

1. A *prelarva* with 9 abdominal segments; mouthparts and other sclerotized structures weakly developed and different in appearance from following stages. It is largely an immobile stage.
2. A *larva I* with 9 abdominal segments, the shape of the ninth different from that of the prelarva.
3. A *larva II* with 10 abdominal segments. The new segment (segment X in the adult) arises between the eighth segment and the terminal one (which thus must be regarded as a telson).
4. A *maturus junior* with 12 abdominal segments. Copeland (1964) found that, at least in the genus *Eosentomon,* the two new segments arise on either side of the future adult segment X (thus becoming segments IX and XI in future stages).
5. A *preimago* with 12 abdominal segments and only a weakly developed squama genitalis. Known only for males of Acerentomidae.
6. The *adult* with 12 abdominal segments and fully developed squama genitalis.

Tuxen (1949) proved the above sequence by observing molting specimens of *Nosekiella danica* (Acerentomidae). Although these observations were performed only on this species, morphological and taxonomic studies by many subsequent authors have amply confirmed them. There is no stage with 11 abdominal segments, although most authoritative texts continue to follow the older sequence (e.g., Matsuda 1976).

Imadaté (1966) studied the development of the chaetotaxy and distinguished between primary setae (those of larva I), secondary setae (added in larva II), tertiary setae (added in the maturus junior), and complementary setae (added in the adult stage).

The egg stage, only recently described (Bernard 1976b, 1979), is poorly known but appears to resemble Collembola or tardigrade eggs in its gross morphology.

DESCRIPTION

Entognathous hexapods lacking eyes, antennae, and cerci; length less than 2 mm, bodies delicate, rarely strongly sclerotized. Head prognathous; mandible styliform; maxilla with elongated galea and lacinia, the lacinia produced into two pointed or hooked processes; labium apically bilobed, each piece triangular. Eyes absent; head with a dorsolateral pair of platelike organs, the pseudoculi, possibly homologous with the postantennal organs of Collembola. Prothorax reduced dorsally, enlarged ventrally, bearing fore legs specialized apparently to function as antennae. Spiracles on meso- and metanotum present or absent. First three abdominal segments bearing paired styluslike appendages ventrally, which may be one- or two-segmented. Male and female with internal, sclerotized genitalia (the squama genitalis), opening between the eleventh segment and telson.

COMMENTS

There are about 30 species of Eosentomidae known in North America, 4 of Protentomidae, and about 14 of Acerentomidae. Worldwide, there are nearly 400 described species.

Systematically, the Protura are divided into two superfamilies on the basis of development of the female squama genitalis. Eosentomoidea contains the large family Eosentomidae, in which the squama genitalis has a complex development of sclerotized areas and plates (Tuxen 1964). In contrast, the superfamily Acerentomoidea contains species in which the female squama genitalis does not possess large sclerotized areas but instead has large terminal acrostyli that are either pointed or rounded. This superfamily contains three families: Sinentomidae, Protentomidae, Acerentomidae. Species of the oriental family Sinentomidae have characters in common with both Eosentomidae and Protentomidae. One of us (SLT) feels strongly that sinentomids should be included in the Protentomidae (Tuxen 1977a).

TECHNIQUES

Protura are collected usually with Berlese-type funnels. Soil forms are more easily recovered by the centrifugation-sugar flotation technique (Jenkins 1964), but fore legs frequently are lost with this method, making identification difficult. Preservation is in 95% ethanol. Various mounting media have been used, including Swan's, Hoyer's and balsam. Tuxen (1964) and Nosek (1973) present information on preparing permanent mounts of Protura.

CLASSIFICATION

Class and order **PROTURA** (Myrientomata)
- Superfamily Eosentomoidea
  - Family Eosentomidae
- Superfamily Acerentomoidea
  - Family Acerentomidae
  - Family Protentomidae
  - Family Sinentomidae

KEY TO WORLD SUPERFAMILIES AND FAMILIES OF PROTURAN JUVENILES

1. Mouthparts weakly developed, fore tarsus with few or no sensillae (prelarvae) 2
1′. Mouthparts well developed, fore tarsus with most of the adult sensillae 3
2(1). Fore tarsus with a claw .... Acerentomoidea
2′. Fore tarsus without a claw but with an apical, triangular spur .... Eosentomoidea
3(1′). Meso- and metathoracic spiracles present (figs. 4.7 and 4.9) 4
3′. Spiracles absent (Acerentomoidea) 5
4(3). Appendages of abdominal segments II-III 2-segmented (as in fig. 4.40); body without rows of spines (Eosentomoidea) .... ***Eosentomidae*** (p. 49)
4′. Appendages of abdominal segments II-III 1-segmented (as in fig. 4.41); body adorned with rows of prominent spines .... ***(Sinentomidae)***[2] (p. 50)
5(3′). Appendages of abdominal segments I-II 2-segmented (fig. 4.41), sometimes all 3 sets 2-segmented .... ***Protentomidae*** (p. 50)
5′. Only the appendages of segment I 2-segmented .... ***Acerentomidae*** (p. 49)

2. The status of this family is uncertain. One of us (SLT) believes that *Sinentomon* (Sinentomidae) is really a protentomid and should be included in that family (Tuxen 1977a).

EOSENTOMIDAE

The Eosentomids

Figures 4.1–4.13

The species of this family are mostly very similar. One well-known species is *Eosentomon transitorium* Berlese (figs. 4.1–4.13) (Tuxen 1949).

The *prelarva* (figs. 4.1–4.6) possesses three pairs of abdominal appendages, all with terminal vesicles (similar to the adult). Mouthparts are weak and only partially developed, without labial palpi; the fore tarsus lacks sensillae and ends in a short spine, lacking a claw; tracheal system and spiracles are absent; the abdominal chaetotaxy is simple, with only one row of setae dorsally and ventrally and very few setae overall. This stage is easily recognizable by the rounded head and truncate terminal abdominal segment.

The *larva I* (figs. 4.7–4.8) assumes most of the adult characteristics: well-developed abdominal appendages and mouthparts, fore tarsi fully equipped with sensillae. The chaetotaxy, however, is not fully developed, lacking an anterior row of setae on abdominal terga I-III and possessing only two anterior and ten posterior setae on terga IV-VII. Sternal setae also are fewer in number, with two located anteriorly and four or six posteriorly.

The *larva II* (figs. 4.9–4.10) differs from the other stages only in the chaetotaxy. The abdominal terga have two, four, or six setae in the anterior row; the sterna have four setae.

The *maturus junior* (fig. 4.11) has all of the adult setae except on the sternite of segment XI, where it has four instead of eight setae (figs. 4.12–4.13). The squama genitalis is absent.

A *preimago* is unknown among the Eosentomidae. The prelarvae of several European and Asian species are known, and in the United States the prelarvae of *Eosentomon australicum* Womersley and *E. wheeleri* Silvestri have been observed (Bernard 1976b and unpublished observations). The development from larva I to adult has been followed in a number of European and Oriental species as well as in *E. rostratum* Ewing (Durey and Copeland 1977) and *E. pomari* (Bernard 1976a) from the United States.

Selected Bibliography

Bernard 1976b, 1979.
Bonet and Tuxen 1960.
Copeland 1964.
Imadaté 1974.
Nosek 1973.
Pearse 1946.
Tuxen 1949, 1964.
Yin et al. 1981.

ACERENTOMIDAE

The Acerentomids

Figures 4.14–4.31

Tuxen (1949) carefully studied the development in *Nosekiella danica* (Condé) (figs. 4.14–4.31).

The *prelarva* (figs. 4.14–4.22) has only the first pair of abdominal appendages (fig. 4.15), developed as in the three pairs of Eosentomidae. The second and third pairs are represented only by two setae. The fore tarsus has a claw and two claviform sensillae, xl and x2 (fig. 4.16), but these sensillae have not been homologized with those of later stages. The acerentomid prelarva differs from later stages in several ways: in the weakly developed mouthparts, though a labial palpus is present; the small number of fore tarsal sensillae; the absence of a transversely striate band and comb on abdominal segment VIII; and simple development of chaetotaxy. The abdominal terga lack anterior rows of setae, whereas the sterna lack setae on segments I-III and possess only two setae on segments IV-VII. As in eosentomid prelarvae, the head is anteriorly rounded and the terminal segment truncate.

The *larva I* (figs. 4.23–4.24) differs from the adult only in the chaetotaxy, with no anterior row of setae on the abdominal terga, no anterior setae on segments I-III of the sterna and only one on segments IV-VII.

The *larva II* (figs. 4.25–4.26) has anterior rows of setae on all abdominal sterna but lacks them on the terga except for segment VIII.

The *maturus junior* (figs. 4.27–4.28) possesses all the adult setae except for two fewer anterior setae on terga II-VII, and only two instead of six setae on sternum XI. The squama genitalis is absent.

The *preimago* male (fig. 4.29) lacks seta p1′ on terga I-VI and possesses a squama genitalis less developed than that of the adult male (compare fig. 4.30).

In addition to *Nosekiella danica,* the full development from prelarva to adult has been followed in *Acerentomon propinquum* Condé (François 1960) and *Nipponentomon uenoi paucisetosum* Imadaté (Imadaté 1974). Development of these latter species is quite similar to that of *N. danica.* The development from larva I is known for numerous European, Japanese, and Southeast Asian species (Imadaté 1965, 1966, 1974; Nosek 1973) and in North America for *Proacerella reducta* Bernard (Bernard 1976a).

Selected Bibliography

François 1960.
Imadaté 1974.
Nosek 1973.
Tuxen 1949, 1964.

PROTENTOMIDAE

The Protentomids

Figures 4.32-4.35, 4.40, 4.41

Only *Protentomon michiganense* Bernard (Bernard 1976a) has been followed from prelarva through adult. The prelarva of *Proturentomon discretum* Condé was mentioned by Condé (1961) and the fore tarsus described and drawn.

The *prelarva* cannot be separated morphologically from that of Acerentomidae except possibly by the pseudoculus. In *P. michiganense* there is a posterior prolongation that is absent in acerentomid prelarvae (figs. 4.18–4.19). This character probably does not hold true for protentomids whose adults do not possess the prolongation. In *P. discretum* the fore tarsi possess a claw and two sensillae (figs. 4.32–4.33), whereas in *P. michiganense* the sensillae are absent. Abdominal terga I-VIII of *P. michiganense* possess eight setae in a single row; sterna I-III lack setae and sterna IV-VIII have two each.

The *larva I* is known only for *P. michiganense.* Of the abdominal terga and sterna, only sternum I has an anterior row.

The *larva II* in *P. michiganense* resembles the larva I but the setal rows on the sterna and terga increase by one, two, or four setae. In *Condeellum ishiianum* Imadaté the tergal anterior setal rows are lacking (Imadaté 1965) (figs. 4.34–4.35).

Chaetotaxy of the *maturus junior* resembles the adult, except that there are four less setae on sternum XI, as in most acerentomids.

Selected Bibliography

Bernard 1976a.
Condé 1961.
Imadaté 1974.
Nosek 1973.
Tuxen 1964.

SINENTOMIDAE

The Sinentomids

Figures 4.36-4.39

Two species are known: *Sinentomon erythranum* Yin (Yin 1965) from China and *S. yoroi* Imadaté (Imadaté 1977) from Japan. All stages except the prelarva have been described for both species. In all forms the body is ornamented with numerous small spines.

The *larva I* has only one row of dorsal setae on abdominal segments I-VII, but the accessory setae are shifted posteriorly to give the appearance of two rows (Tuxen 1977a).

The *larva II* possesses anterior rows of setae on terga II-VII to give the appearance of three rows. However, the total number of setae is still less than in the adult.

The *maturus junior* has two setae on sternum XI (figs. 4.36–4.37), and the adult has six setae, exactly as in the acerentomids (figs. 4.38–4.39).

Selected Bibliography

Imadaté 1977.
Tuxen 1977a.
Yin 1965.

BIBLIOGRAPHY

Berlese, A. 1909. Monografia dei Myrientomata. Redia 6:1–182.

Bernard, E. C. 1976a. A new genus, six new species, and records of Protura from Michigan. Great Lakes Ent. 8:157–81.

Bernard, E. C. 1976b. Observations on the eggs of *Eosentomon australicum* (Protura:Eosentomidae). Trans. Amer. Micros. Soc. 95:129–30.

Bernard, E. C. 1979. Egg types and hatching of *Eosentomon* Berlese (Protura: Eosentomidae). Trans. Amer. Micros. Soc. 98:123–126.

Bonet, F., and S. L. Tuxen. 1960. Reexamination of species of Protura described by H. E. Ewing. Proc. U.S. Nat. Mus. 112:265–305.

Condé, B. 1961. Nouvelles recoltes de Protoures au Maroc. Bull. Mus. Hist. Nat. (2)33:495–99.

Copeland, T. P. 1964. New species of Protura from Tennessee. J. Tenn. Acad. Sci. 39:17–29.

Durey, R. A., and T. P. Copeland. 1977. Postembryonic development and synonymy for *Eosentomon rostratum* Ewing (Insecta-Protura). Proc. Biol. Soc. Wash. 90:778–87.

Ewing, H. E. 1940. The Protura of North America. Ann. Ent. Soc. Amer. 33:495–551.

François, J. 1960. Development postembryonnaire d'un protoure du genre *Acerentomon* Silv. Trav. Lab. Zool. Fac. Sci. Dijon 33:1–11.

Imadaté, G. 1965. Proturans-fauna of Southeast Asia. Nature and Life in SE Asia 4:195–302.

Imadaté, G. 1966. Taxonomic treatment of Japanese Protura. IV. The proturan chaetotaxy and its meaning to phylogeny. Bull. Nat. Sci. Mus., Tokyo 9:277–315.

Imadaté, G. 1974. Fauna Japonica. Protura (Insecta). Tokyo. 351 pp.

Imadaté, G. 1977. Occurrence of *Sinentomon* (Protura) in Japan. Bull. Nat. Sci. Mus., Tokyo, Ser. A (Zoology) 3:37–48.

Janetschek, H. 1970. Ordnung Protura. *In* Kukenthal et al. Handbuch der zoologie, Zweite Auflage. Berlin, IV(2)2–3:1–72.

Jenkins, W. R. 1964. A rapid centrifugal-flotation technique for separating nematodes from soil. Plant Dis. Rep. 48:692.

Lussenhop, J. 1973. The soil arthropod community of a Chicago expressway margin. Ecology 54:1124–37.

Matsuda, R. 1976. Morphology and evolution of the insect abdomen. New York: Pergamon Press. 534 pp.

Nosek, J. 1973. The European Protura. Mus. D'Hist. Naturelle, Geneva: 345 pp.

Pearse, A. S. 1946. Observations on the microfauna of the Duke Forest. Ecol. Monog. 16:127–50.

Price, D. W. 1973. Abundance and vertical distribution of microarthropods in the surface layers of a California pine forest soil. Hilgardia 42:121–48.

Silvestri, F. 1907. Descrizione di un novo genere di insetti apterigoti, rappresentante di un novo ordine. Boll. Lab. Zool. Portici 1:296–311.

Strenzke, K. 1942. Norddeutsche Proturen. Zool. Jahrb., Syst. 75:73–102.

Sturm, H. 1959. Die Nahrung der Proturen. Die Naturwiss. 46:90–91.

Tuxen, S. L. 1949. Über den Lebenszyklus und die postembryonale Entwicklung zweier dänischer Proturengattungen. Kgl. da. vid. Selsk. Biol. Skr. 6, 3, 49 p.

Tuxen, S. L. 1964. The Protura. Paris: Hermann. 360 pp.

Tuxen, S. L. 1970. The systematic position of entognathous apterygotes. Ann. Esc. Nac. Cienc. Biol. Mex. 17:65–79.

Tuxen, S. L. 1977a. The systematical position of *Sinentomon* (Insecta, Protura). Bull. Nat. Sci. Mus., Tokyo, Ser. A. (Zoology) 3:25–26.

Tuxen, S. L. 1977b. Ecology and zoogeography of the Brazilian Protura (Insecta). Stud. Neotrop. Fauna 12:225–47.

Tuxen, S. L. 1978. Protura (Insecta) and Brazil during 400 million years of continental drift. Stud. Neotrop. Fauna 13:23–50.

Walker, G. L., and R. W. Rust. 1975. Seasonal distribution of Protura in three Delaware forests. Ent. News 86:187–98.

Yin Wenying. 1965. [Studies on Chinese Protura II. A new family of the suborder Eosentomoidea.] Acta Ent. Sinica 14:186–95. [In Chinese]

Yin Wenying, Ren Bingfu, Jin Gentao, and Guo Peifu. 1981. [The ecological investigation of proturan fauna in the soil of bamboo forest, East Sheshan, Shanghai.] Acta Ecol. Sinica 1:126–135. [In Chinese]

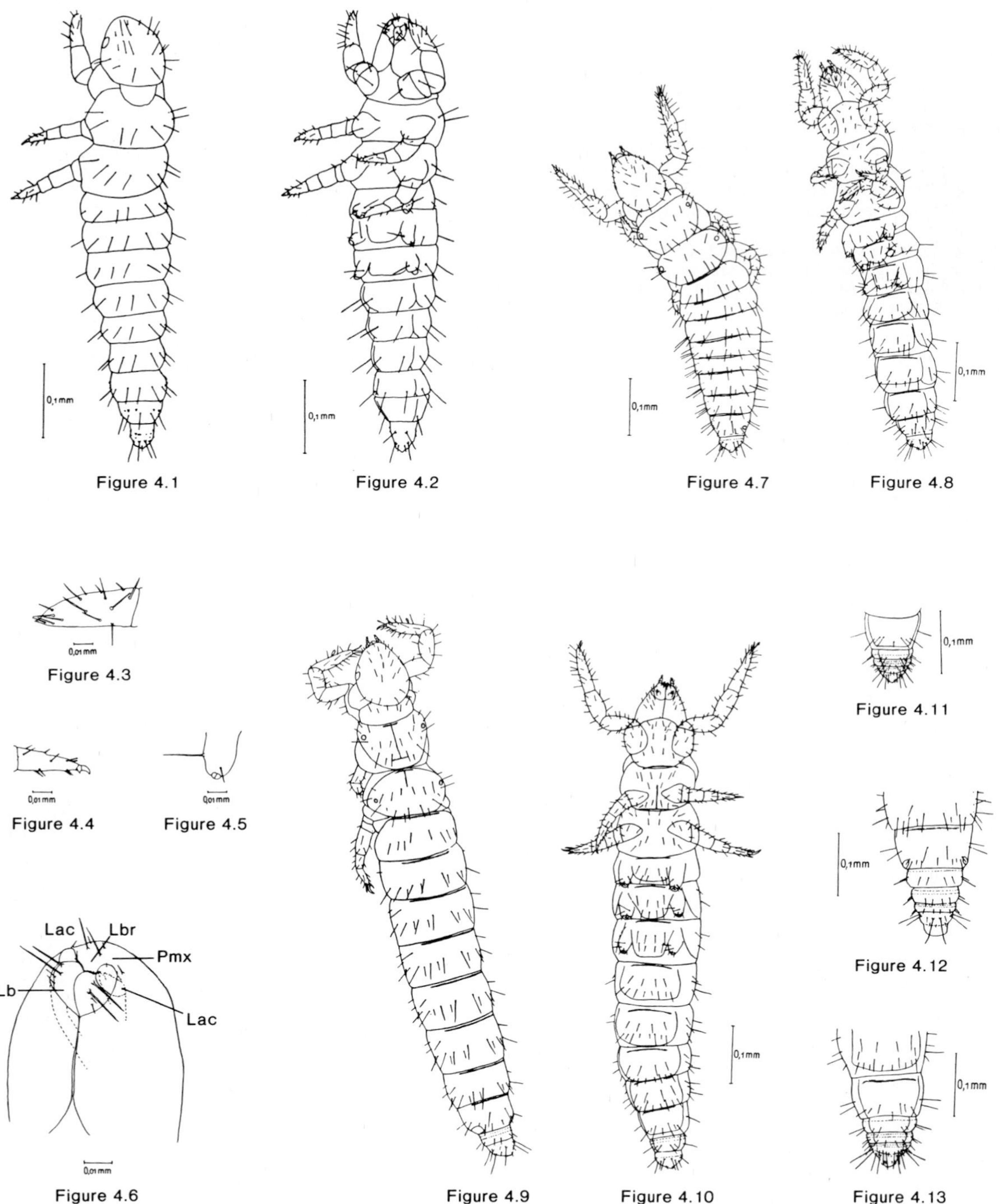

Figure 4.1 Figure 4.2 Figure 4.7 Figure 4.8

Figure 4.3 Figure 4.4 Figure 4.5 Figure 4.6 Figure 4.9 Figure 4.10 Figure 4.11 Figure 4.12 Figure 4.13

Figures 4.1–4.13. *Eosentomon transitorium* Berlese (Eosentomidae).

Figures 4.1–4.6. Prelarva: **(4.1)** dorsal; **(4.2)** ventral; **(4.3)** fore tarsus; **(4.4)** hind tarsus; **(4.5)** appendage of third abdominal segment; **(4.6)** mouthparts. *Lac:* maxilla, probably lacinia; *Lb:* labium; *Lbr:* labrum; Pmx: maxillary palpus.

Figures 4.7–4.8. Larva I: **(4.7)** dorsal; **(4.8)** ventral.

Figures 4.9–4.10. Larva II: **(4.9)** dorsal; **(4.10)** ventral.

Figure 4.11. Maturus junior, ventral view of abdominal segments VIII-XII.

Figures 4.12–4.13. Adult abdominal segments VIII-XII: **(4.12)** dorsal; **(4.13)** ventral. (Reprinted with permission from Tuxen 1949).

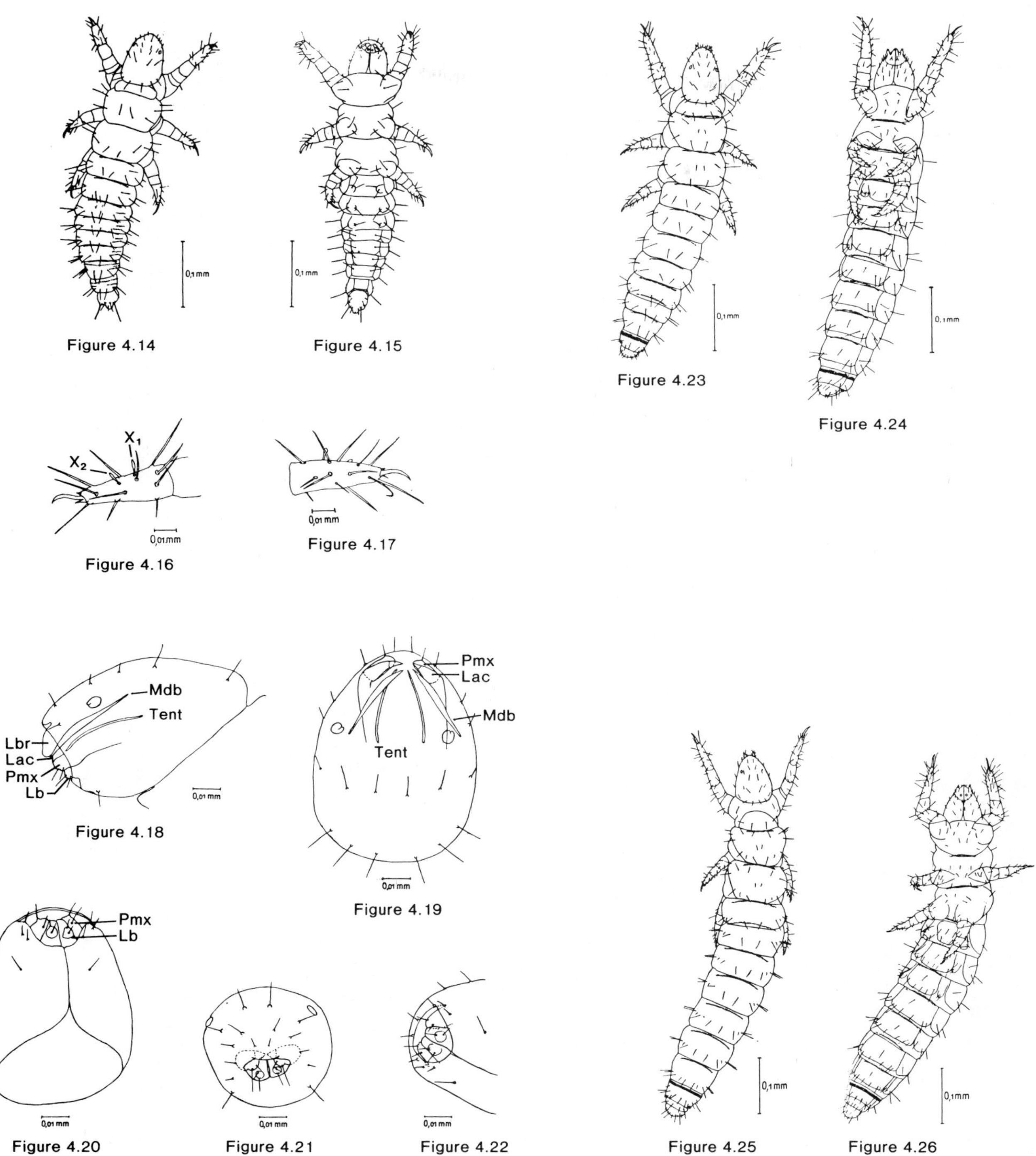

Figures 4.14–4.26. *Nosekiella danica* (Condé) Acerentomidae.

Figures 4.14–4.22. Prelarva: **(4.14)** dorsal; **(4.15)** ventral; **(4.16)** fore tarsus in exterior view (X_1, X_2-sensillae); **(4.17)** fore tarsus in interior view; **(4.18)** head in lateral view; **(4.19)** dorsal; **(4.20)** ventral; **(4.21)** anterior; **(4.22)** mouthparts, ventral. *Lac:* maxilla, probably lacinia; *Lb:* labium; *Lbr:* labrum; *Mdb:* mandible; *Pmx:* maxillary palpus; *Tent:* fulcrum.

Figures 4.23–4.24. Larva I: **(4.23)** dorsal; **(4.24)** ventral.

Figures 4.25–4.26. Larva II: **(4.25)** dorsal; **(4.26)** ventral. (Reprinted with permission from Tuxen 1949).

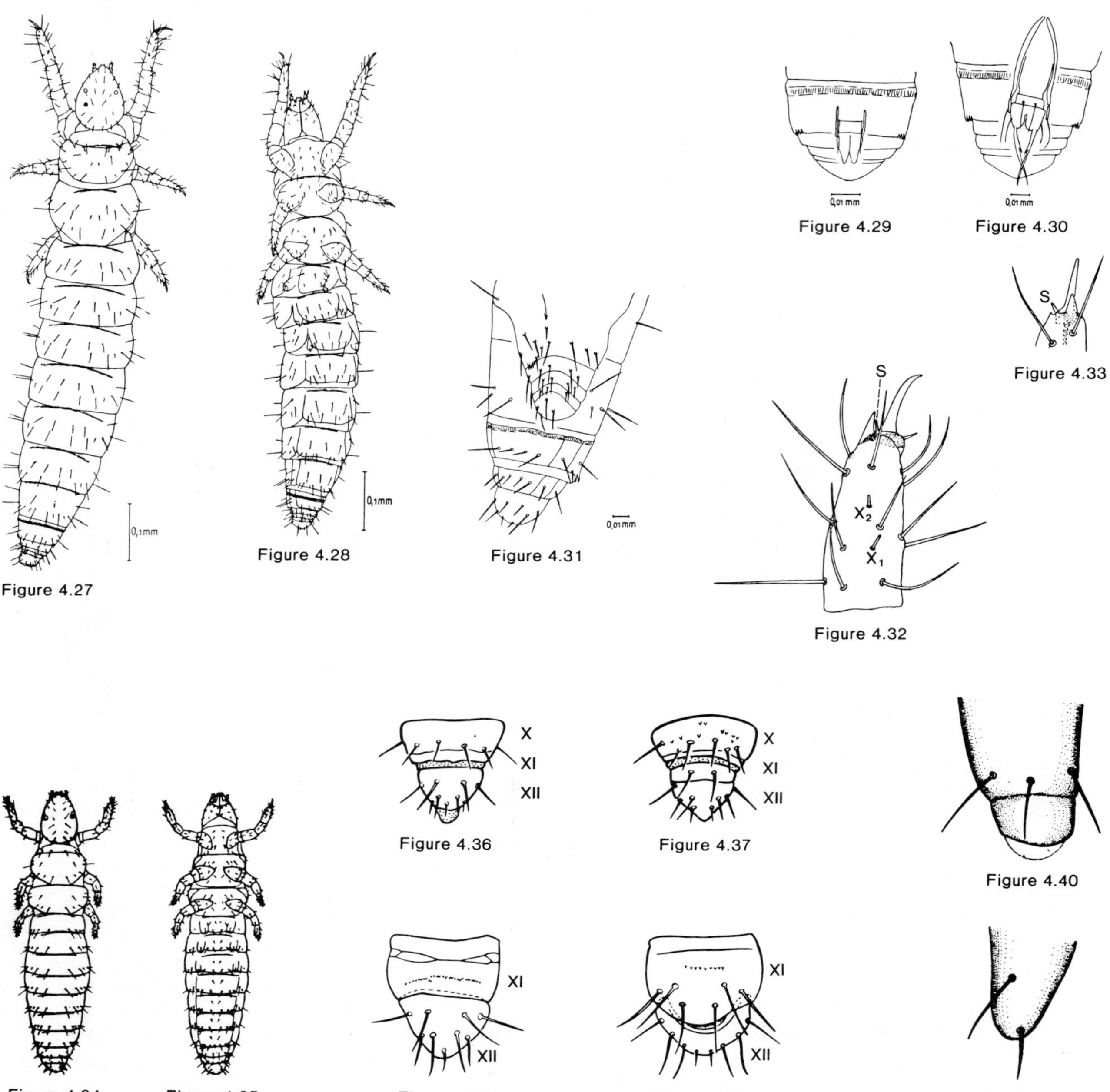

Figure 4.27 Figure 4.28 Figure 4.29 Figure 4.30 Figure 4.31 Figure 4.32 Figure 4.33 Figure 4.34 Figure 4.35 Figure 4.36 Figure 4.37 Figure 4.38 Figure 4.39 Figure 4.40 Figure 4.41

Figures 4.27–4.31. *Nosekiella danica* (condé) (Acerentomidae).

Figures 4.27–4.28. Maturus junior: **(4.27)** dorsal; **(4.28)** ventral.

Figure 4.29. Preimago male.

Figure 4.30. Adult male.

Figure 4.31. Larva II in molt to maturus junior. (Reprinted with permission from Tuxen 1949).

Figures 4.32–4.33. *Proturentomon discretum* Condé (Protentomidae), prelarva: **(4.32)** foretarsus in interior view, S, X_1, X_2: sensillae; **(4.33)** foretarsus in exterior view, S, sensilla. (Reprinted with permission from Condé 1961).

Figures 4.34–4.35. *Condeellum ishiianum* Imadaté (Protentomidae), larva II: **(4.34)** dorsal; **(4.35)** ventral. (Reprinted with permission from Imadate 1965).

Figures 4.36–4.39. *Sinentomon erythranum* Yin (Sinentomidae). **(4.36–4.37)** Abdominal segments X-XII of maturus junior: **(4.36)** dorsal; **(4.37)** ventral. **(4.38–4.39)** Abdominal segments XI-XII of adult male: **(4.38)** dorsal; **(4.39)** ventral.

Figures 4.40–4.41. Abdominal appendages of Protentomidae: **(4.40)** AI, 2-segmented; **(4.41)** AIII, 1-segmented. (Reprinted with permission from Tuxen 1949).

Class and Order Collembola

5

Richard J. Snider
Michigan State University

SPRINGTAILS

The Collembola, or "springtails," are of ancient lineage, evidenced by their occurrence among the fossils of the early or middle Devonian cherts (Hirst and Maulik 1926). Even then they were highly specialized and recognizable to subfamily level. Their distant origins, relatively small body size, and great numerical abundance have produced a group that is cosmopolitan today. In addition, humans have been responsible for transporting a large number of species from continent to continent during exploration and commerce.

Metamorphosis in the Collembola is gradual (ametabolous). Larvae, or juveniles, are distinguished by a relatively large head in proportion to the body. Color patterns often do not develop until the fifth or sixth instar. Likewise, chaetotaxy of body segments changes with each instar (molt) until the adult instar is reached. However, morphological characteristics used for determination of adults will, with few exceptions, also serve for identification of immatures to family and generic levels.

Of the numerous studies on the life history of Collembola, most report instar duration and egg production, but studies illustrating the immature stages of Collembola are scant (Davis and Harris 1936; Hale 1965; Snider 1977). Only recently has the study of phylogenetic relationships reawakened an interest in immature stages (Barra 1975; Szeptycki 1967, 1969, 1972, 1979).

BIOLOGY AND ECOLOGY

Most species of Collembola live in soil or soil-related habitats, and a few are found in semiaquatic situations. It is, therefore, not surprising that moisture is a primary factor in their habitat selection and distribution. Collembola commonly dwell in organic soils, leaf litter, mosses, under fallen logs, beneath stones, under loose bark, in dung, nests, burrows, and caves. Many are able to survive epigeally on vegetation where humid conditions prevail, some scaled and heavily setaceous forms can withstand the hostile climate of deserts, and some are associated with ice fields, freshwater ponds, and marine littoral waters.

Eggs are laid singly or in clumps, in crevices or pockets of the substrate. Collembola generally pass through five or six instars before they are sexually mature. Instars one through four are usually considered to be the immature stages, and progressive morphological changes take place during that period: the trunk increases in size relative to the head; dorsal setae are added with each molt; antennal segments lengthen; the genital openings may develop operculi and setal patterns; the furcula enlarges (in *Tomocerus* sp., the mucronal shape may become more complex); and color patterns develop (figs. 5.1a–e). Molting continues at fairly regular intervals throughout life, even after maximum size has been reached.

Collembola mouthparts conform to dietary habits, and some species have mandibles with molar surfaces for grinding, whereas others possess mandibles that lack a molar surface or have become styliform. In *Brachystomella* sp., a genus that is largely tropical with five Nearctic species, mandibles are lacking altogether. These varied adaptations permit collembolans to occupy a multitude of niches within the soil ecosystem, and many species are capable of feeding on fresh plant tissues as well as those just beginning to decay. Some ingest their own fecal pellets and those of other arthropods. Gut content analysis has revealed that fungal hyphae, spores, bacteria, and nematodes are commonly taken as food. Species dwelling on the surface may feed on algae, mosses, pollen, and plant fluids. A few species are carnivorous and even cannibalistic.

DESCRIPTION AND DIAGNOSIS

Collembola may be recognized by the following characteristics (figs. 5.2 and 5.3): body comprised of head, three-segmented thorax, and six-segmented abdomen; ocelli usually eight to a side, sometimes reduced or absent; one pair of four-segmented antennae, of which one and two may appear to be subsegemented and three and four subsegmented or annulated; postantennal organ (PAO, fig. 5.12) between base of antenna and eyepatch present or absent; mouthparts either prognathous or hypognathous, mandible with or without molar plate; legs comprised of precoxal, coxal, trochanteral, femoral, tibial and pretarsal segments, ending in a claw (unguis) with or without an opposable unguiculus. The first segment of the abdomen bears the collophore, containing a bifurcate filament, sometimes greatly reduced (for osmoregulation); third abdominal segment with a retinaculum consisting of a corpus and two rami; fourth abdominal segment with a furcula consisting of basal manubrium, dentes, and distal mucrones; fifth abdominal segment with the genital opening and sixth bearing the anus. Modification of the trunk is usually by reduction and fusion of various segments.

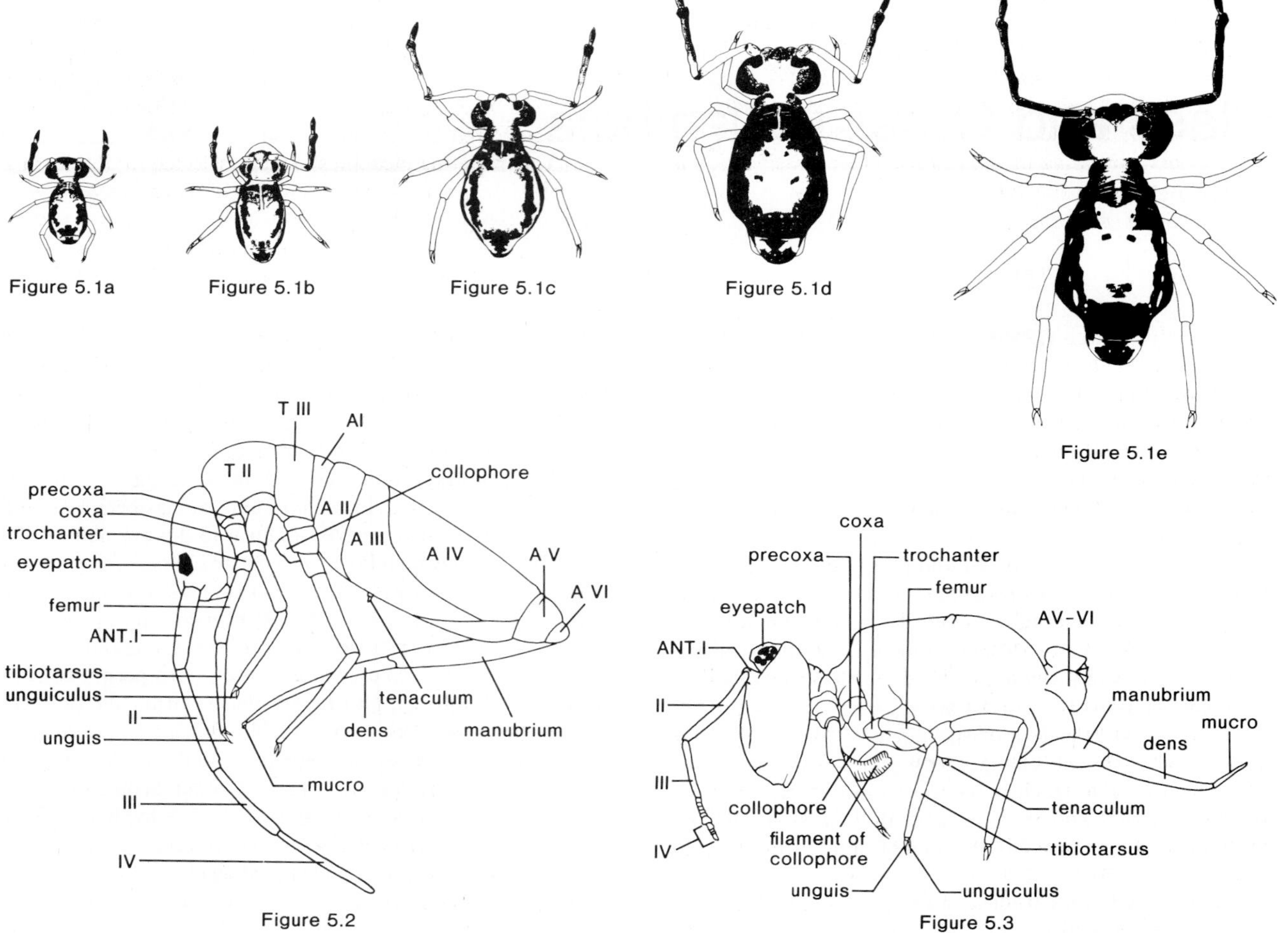

Figure 5.1a-c. Sminthuridae, *Dicyrtoma (Dicyrtoma) hageni* (Folsom). a. 1st instar; b. 2nd instar; c. 3rd instar; d. 4th instar; e. adult.

Figure 5.2. Body form of an arthropleonid collembolan.

Figure 5.3. Body form of a symphypleonid collembolan.

Body with setae and sometimes scales; anal horns sometimes present. Many species have pigment and color patterns in various configurations. Molting is continuous throughout life and there is no metamorphosis. Most species are small (1–2 mm); a few may reach larger sizes (8 mm).

COMMENTS

In general, the economic importance of Collembola has been reduced from the approximately 70 species once considered to be pests (Folsom 1933, Scott 1953). Today, *Sminthurus viridis* (L.), the "Lucerne flea," which feeds on leguminous crops and pastures, is one of a few prominent pest species with the potential of becoming a problem in the western U.S. and southern Canada. Currently only one record of its occurrence in North America (Canada) has been published Hammer (1953); however collophore setal pattern and gut content analysis of these specimens indicates that they are species *Sminthurus nigromaculatus* Tullberg, another European import that feeds on detritus. Collembola populations can also spell trouble for the mushroom industry. In normal practice, when a grower "cooks out" the culture medium, Collembola are destroyed by heat. If the procedure is ignored or shortened, outbreaks of soil species can occur, mushroom mycelia are soon devoured, and crop production severely reduced.

Identification of Collembola beyond the family level is greatly facilitated by publications that either have or soon will become standard references. The large index compiled by Salmon (1964) contains keys to the world genera, and Uchida (1971, 1972a,b) updated these keys and added new genera and illustrations. Stach (1947, 1949a,b, 1951, 1954, 1956, 1957, 1960, 1963) produced a series of monographs useful on a cosmopolitan basis. Gisin (1960) in his "Collembolenfauna Europas" prescribed the systematics currently in vogue.

In the United States the literature was limited to regional studies (Guthrie 1903, Mills 1934, Maynard 1951, Bellinger 1954, Snider 1967, Lippert and Butler 1976), until Christiansen and Bellinger (1980–81) produced a monograph on the Collembola of North America north of the Rio Grande, which for the first time coherently described our North American fauna in its entirety. Their classification is followed here and given below.

TECHNIQUES

The most common method used to collect Collembola is to place soil, leaf litter, or bark in Berlese funnels. Extraction may take several days to a week. More rapid extraction is achieved by the Tullgren funnel, in which heat from above is supplied by a light bulb (Kevan 1955). Collection should be made in alcohol (95%) and glycerine (1%), but when living specimens are needed, substitute water or a culture container.

Aspirators, which provide another rapid collection technique, are particularly convenient for keeping collections separate while sampling under bark of dead trees. Grasses and low vegetation may be sampled with a shallow enamel pan held at a 45° angle and swept forward. From these, specimens can be aspirated with ease.

Pit traps of various sizes are useful. The simplest consist of cottage cheese containers buried to just above their lip, with a small quantity of glycerine or ethylene glycol in the bottom. Traps should be emptied within 24 hours and the glycerine or glycol rinsed from the specimens before storage, otherwise specimens become very soft and unmanageable. A commonly used storage medium is 95% ethanol or isopropanol with 1% glycerine. Collembola have been stored for over 20 years in this mixture with good success.

Phase-contrast microscopy is usually necessary for identification of specimens to species. Several slides should be prepared. Remove the head and mount dorsal side up with antennae spread out. Mount the trunk dorsal side up with legs and furcula spread out; frequently the furcula must be mounted separately. While holding the trunk with watchmaker's forceps, insert a fine needle just under the base of the manubrium so the whole structure can be peeled away and transferred to a separate slide.

The mounting medium should be selected for its clearing properties. In the past, fine mounts have been achieved with xylol-to-balsam technique, often with KOH as a preliminary clearing agent, but this requires time and some skill. More recently, Hoyer's medium as well as various lactophenol mixtures are giving favorable results. One medium that satisfies broad requirements is CMCP–9® (see techniques), basically a polyvinyl alcohol medium with very positive clearing qualities. In practice, transfer the specimen from 95% alcohol to a drop of CMCP–9 on a slide and place a coverslip over it; then apply pressure with forceps to flatten the specimen and allow the mount to dry. Within 48 hours the specimen will generally be clear enough to identify. It is desirable to ring the coverslip with "glyptol" (General Electric insulating enamel) or asphaltum for long-term storage.

CLASSIFICATION

Class and Order **COLLEMBOLA**
- Suborder Arthropleona
  - Superfamily Poduroidea
    - Poduridae
    - Hypogastruridae
    - Onychiuridae
  - Superfamily Entomobryoidea
    - Isotomidae
    - Entomobryidae
- Suborder Symphypleona
  - Neelidae
  - Sminthuridae

KEY TO THE FAMILIES OF IMMATURE COLLEMBOLA

1.	Body elongate. Thoracic and abdominal segments distinct, with at most the last 2–3 segments fused (fig. 5.4)	Suborder ARTHROPLEONA	2
1′.	Body subglobate. Thoracic and first 4 abdominal segments fused (fig. 5.5)	Suborder SYMPHYPLEONA	7
2(1).	Trunk segments distinct; first thoracic segment with dorsal setae; scales absent; pseudocelli (fig. 5.6) present or absent	Poduroidea	3
2′.	Trunk segments dissimilar; first thoracic segment reduced or hidden beneath second segment (fig. 5.7), if visible, then without dorsal setae; scales present or absent; pseudocelli absent	Entomobryoidea	5
3(2).	Head hypognathous, first thoracic segment similar in size to second and third; dentes (fig. 5.8) 3 times as long as manubrium and elbowed in dorsal view	***Poduridae*** (p. 59)	
3′.	Head prognathous, first thoracic segment reduced compared to second and third; dentes 2.5 times (or less) as long as manubrium		4

4(3′). Pseudocelli present; eyes absent; sense organ of Ant. III complex (fig. 5.9), consisting of sense rods, clubs, cuticular papillae, and folds; pigment usually lacking .... ***Onychiuridae*** (p. 59)

4′. Pseudocelli absent; eyes present or absent; sense organ of Ant. III simple (fig. 5.10), consisting of sense rods or clubs; pigment usually present .... ***Hypogastruridae*** (p. 59)

5(2′). Postantennal organ (fig. 5.11) absent; Abd. IV usually longer than Abd. III .... ***Entomobryidae*** (p. 61)

5′. Postantennal organ (PAO) present (fig. 5.12) or absent: Abd. IV subequal to or shorter than Abd. III 6

6(5′). Scales present; dentes sometimes with multidentate spines (fig. 5.13) .... ***Entomobryidae*** (p. 61)

6′. Scales absent; dentes without such spines .... ***Isotomidae*** (p. 60)

7(1′). Antennae shorter than head (fig. 5.14), eyes absent .... ***Neelidae*** (p. 63)

7′. Antennae longer than head, eyes present or absent .... ***Sminthuridae*** (p. 63)

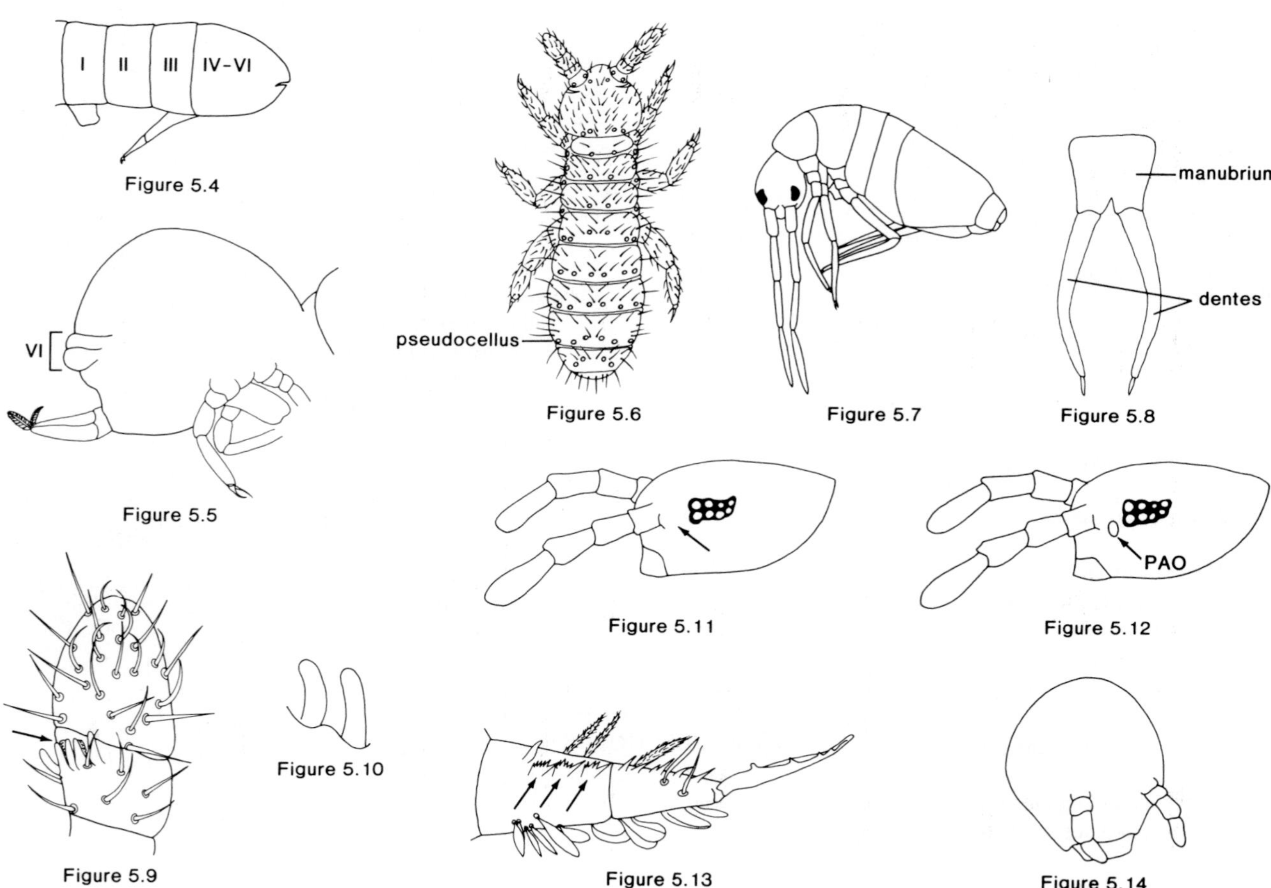

Figure 5.4. Fusion of posterior abdominal segments in subclass Arthropleona.

Figure 5.5. Fusion of abdominal segments in subclass Symphypleona.

Figure 5.6. Onychiuridae, *Onychiurus folsomi* (Schaeffer), diagram showing location of pseudocelli.

Figure 5.7. Entomobryidae, *Entomobrya (Entomobryoides) purpurascens* (Packard), diagram showing mesothorax obscuring prothorax.

Figure 5.8. Poduridae, *Podura aquatica* (Linnaeus), dorsal view of furcula.

Figure 5.9. Onychuridae, *Onychiurus similis* Folsom, 3rd and 4th antennal segments, arrow indicates sense organ.

Figure 5.10. Typical simple sense organ of antennal segment 3.

Figure 5.11. Diagram of head without a postantennal organ.

Figure 5.12. Diagram of head with a postantennal organ (PAO).

Figure 5.13. Entomobryidae, *Harlowmillsia oculata* (Mills), furcula showing multidentate spines.

Figure 5.14. Neelidae, head, antennae shorter than diameter of head.

Suborder ARTHROPLEONA

PODURIDAE

The Podurids

Figures 5.8, 5.15

This family is represented worldwide by one genus and species, *Podura aquatica* L. It is characterized by: hypognathous head; eyes 8 + 8 on dark patches; postantennal organ absent; antennae shorter than head; mandible with molar plate; short, stout trunk with all thoracic segments distinct; long furcula, extending beyond the ventral tube, dentes three times the length of the manubrium and strongly bowed in dorsal view. Because the head is hypognathous, some authors believe that the family is closely allied to the Symphypleona (Uchida 1971).

Podura aquatica is one of the most abundant and cosmopolitan species. It frequents standing waters of ponds and streams and is often found in large numbers on vegetation washed up on the edges.

Selected Bibliography

Christiansen and Bellinger 1980–81.
Maynard 1951.
Mills 1934.

HYPOGASTRURIDAE

The Hypogastrurids

Figures 5.16–5.21

A large family of diverse forms, the hypogastrurids are characterized by: prognathous head; eyes 8 + 8, reduced or absent; postantennal organ present or absent, when present represented by an oval with a few (usually four) peripheral tubercles (fig. 5.16); antennae in some species longer than head, third antennal segment sense organ simple (fig. 5.17); mandibles with or without molar plate (figs. 5.18–5.19), furcula not extending beyond posterior of abdomen (fig. 5.20); many species dorsoventrally depressed; and some species highly pigmented.

This family contains the bluish-black "snow flea," *Hypogastrura nivicola* (Fitch), found in swamps and woodlands during winter and late spring (fig. 5.21). Many species prefer habitats associated with the soil surface, i.e., leaf litter, mosses, under stones and logs, as well as loose bark, bracket fungi, pond margins, and tree holes.

Selected Bibliography

Christiansen and Bellinger 1980–81.
Gisin 1960.
Maynard 1951.
Mills 1934.
Salmon 1964.
Stach 1949a, 1949b, 1951.
Uchida 1971, 1972a.

ONYCHIURIDAE

The Onychiurids

Figures 5.6, 5.9, 5.22–5.25

Uchida (1971) considers Onychiuridae to be more highly evolved than Hypogastruridae, although both families arose from Poduroidea ancestors. Onychiurids are easily separated from hypogastrurids by the presence of pseudocelli (fig. 5.22), which are dorsal paramedial and lateral depressions of the exoskeleton appearing to have ridges or "combs" radiating from the outer edges to the median. These structures are considered to be sensory organs by some investigators (Hale 1969) and, in my own investigations, proved to be sites of autohaemorrhage when they were attacked by predators, an observation also confirmed by K. A. Christiansen (*in litt.*).

Eyes absent; postantennal organ usually present, with many tubercles in a long, parallel-sided groove (Fig. 5.23); third antennal segment with sensory organ consisting of two to three clubs, a pair of sense rods, and usually outer papillae with guard setae (fig. 5.24); mandible with well-developed molar plate (fig. 5.25), an exception being the genus *Tetrodontophora;* furcula absent; clavate tenent hairs lacking; pigment generally absent.

With few exceptions onychiurids are edaphic, often found in the soils of forests and old fields. They comprise a major part of the collembolan fauna of agricultural soils. A few species live under rotten logs, bark, and in the fruiting bodies of fungi. Occasionally they may become pests in mushroom houses and commercial worm cultures.

Selected Bibliography

Christiansen and Bellinger 1980–81.
Gisin 1960.
Hale 1965, 1969.
Mills 1934.
Salmon 1964.
Snider 1977.
Stach 1954.
Uchida 1972a.

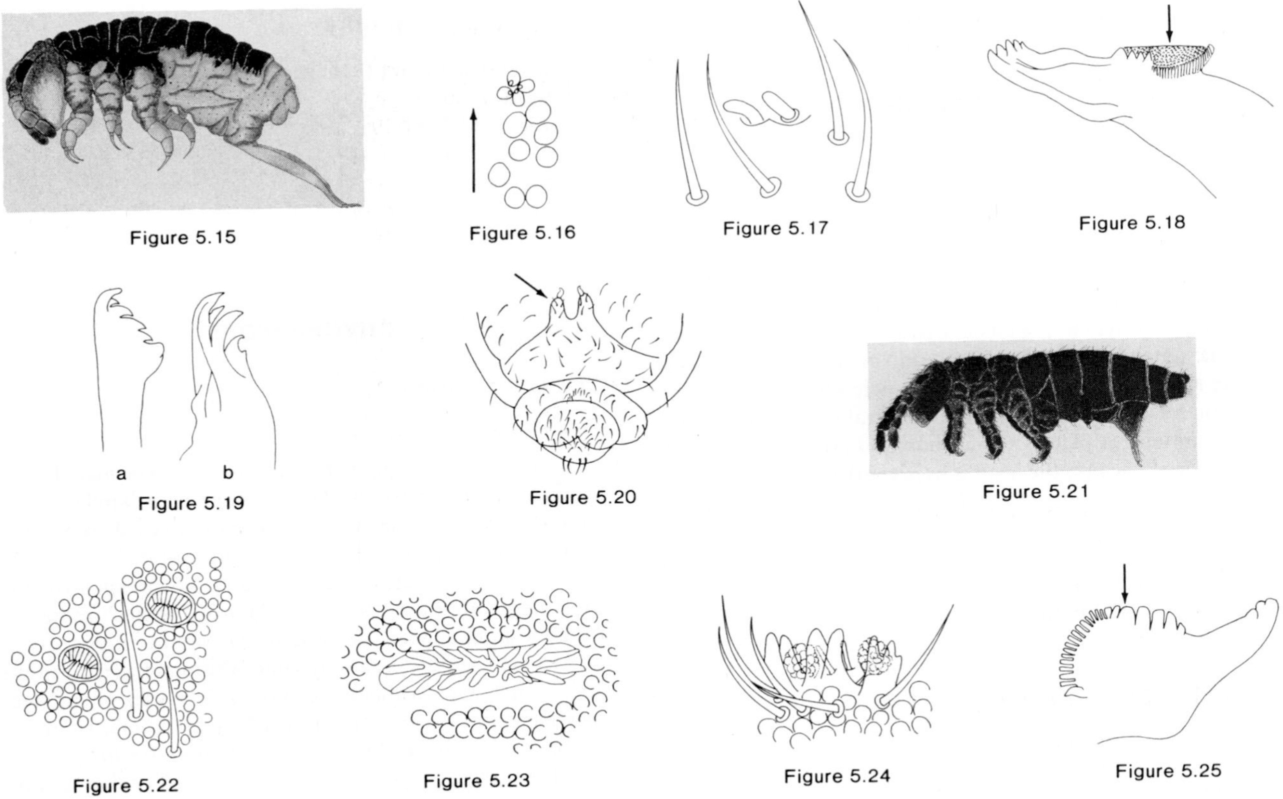

Figure 5.15. Poduridae, *Podura aquatica* (Linnaeus), habitus.

Figure 5.16. Hypogastruridae, *Hypogastrura* sp., ocellar pattern and postantennal organ.

Figure 5.17. Hypogastruridae, *Hypogastrura* sp., simple sense organ of antennal segment 3.

Figure 5.18. Hypogasruridae, *Hypogastrura nivicola* (Fitch), mandible showing molar plate.

Figure 5.19. Hypogastruridae, *Morulina crassa* Christiansen & Bellinger, (a) mandible, (b) maxilla.

Figure 5.20. Hypogastruridae, *Pseudachorutes* sp., furcula.

Figure 5.21. Hypogastruridae, *Hypogastrura nivicola* (Fitch).

Figure 5.22. Onychiuridae *Onychiurus folsomi* (Schaeffer), pseudocelli.

Figure 5.23. Onychiuridae, *Onychiurus reus* Christiansen and Bellinger, postantennal organ.

Figure 5.24. Onychiuridae, *O. reus*, detail of sense organ of antennal segment 3.

Figure 5.25. Onychiuridae, *Onychiurus pseudarmatus* Folsom, mandible showing molar plate.

ISOTOMIDAE

The Isotomids

Figures 5.26–5.34

According to Folsom (1937), the isotomids are postulated as originating from ancestors resembling hypogastrurids. Uchida (1971) places them between Entomobryidae and Hypogastruridae. The family is so diverse (figs. 5.26–5.29) that some of its representatives resemble hypogastrurids, even to the extreme reduction of the furcula; in these more specialized forms, the pronotum must be carefully examined for the absence of setae.

The large and diverse isotomid family has the following characteristics: eyes 8 + 8, reduced, or absent; antennae shorter, subequal, or longer than head, sense organ of third segment consisting of two simple papillae (fig. 5.30); postantennal organ simple, usually ovate (fig. 5.31); pronotum membranous, without setae, mesonotum not projecting over the head (fig. 5.32); furcula sometimes absent; third and fourth abdominal segments subequal (fig. 5.33); occasionally the last two or three abdominal segments fused (fig. 5.34); setae smooth, serrate, or fringed; scales absent.

The Isotomidae live within soil and litter and on vegetation. A number of species that dwell on the margins of ponds, streams, and seashores are considered semiaquatic.

Selected Bibliography

Christiansen and Bellinger 1980–81.
Folsom 1937.
Gisin 1960.
Guthrie 1903.
Maynard 1951.
Mills 1934.
Salmon 1964.
Stach 1949a.
Uchida 1972a.

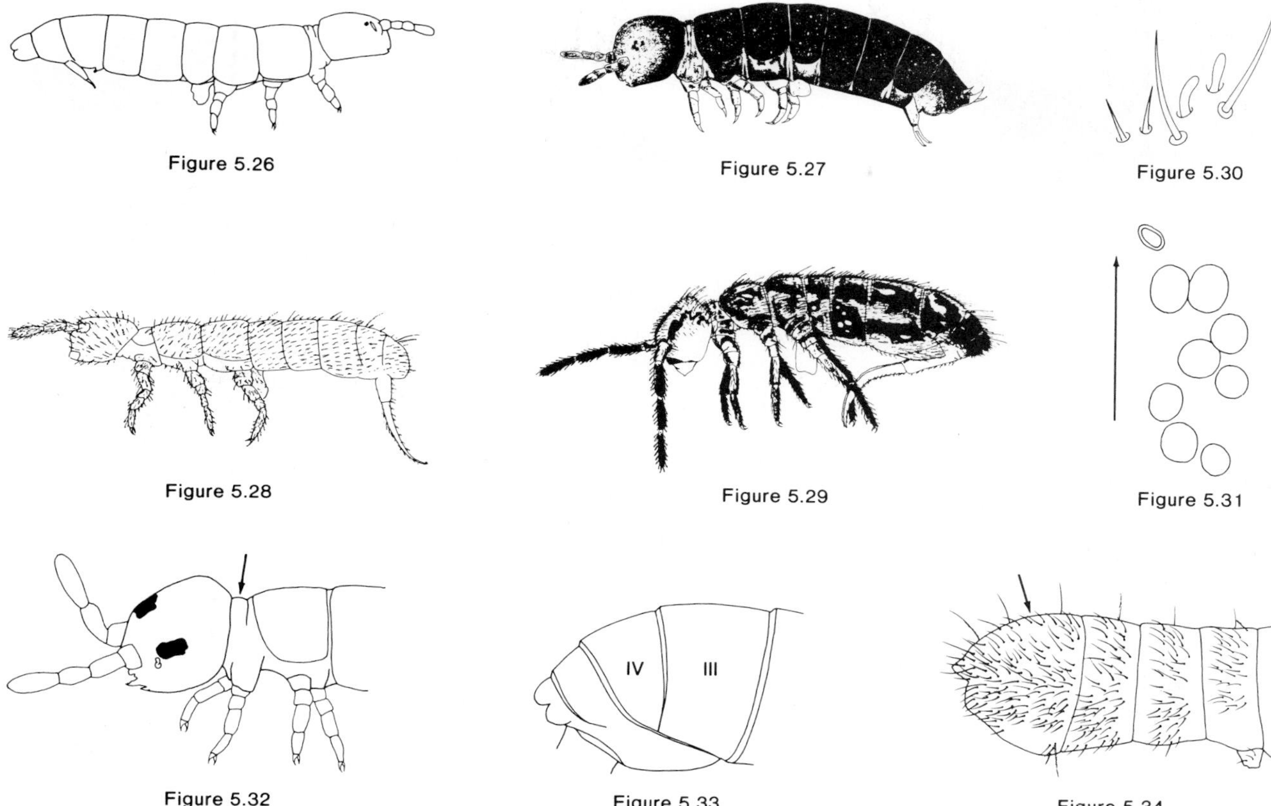

Figure 5.26. Isotomidae, *Folsomides americanus* Denis.

Figure 5.27. Isotomidae, *Tetracanthella bellingeri* Deharveng.

Figure 5.28. Isotomidae, *Isotomiella minor* (Schaeffer).

Figure 5.29. Isotomidae, *Isotomurus tricolor* (Packard).

Figure 5.30. Simple sense organ of antennal segment 3.

Figure 5.31. Postantennal organ and associated ocelli.

Figure 5.32. Isotomidae, *Cryptopygus thermopilus* (Axelson), mesothorax not projecting overhead.

Figure 5.33. Posterior abdominal segments of an isotomid.

Figure 5.34. Isotomidae, *Folsomia* sp., arrow indicates fusion of segments IV, V, and VI.

ENTOMOBRYIDAE

The Entomobryids

Figures 5.7, 5.13, 5.35–5.42

The large family Entomobryidae, the most highly specialized one in the suborder Arthropleona, contains a wide variety of genera that are phylogenetically dichotomous (figs. 5.35–5.38). In some genera color patterns are key characters but extremely variable, and a series of specimens should be examined when making determinations. Chaetotaxy is a powerful tool for identification at the species level. Unfortunately, a large number of genera still need investigation. Anomalies are encountered frequently in this family. Thus the genus *Orchesella* Templeton has developed subsegmentation of the first two antennal segments, giving an appearance of five or six segments, whereas the genus *Tomocerus* Nicolet has a very reduced fourth antennal segment that frequently breaks off, giving a semblance of three segments. In addition, the distal two segments of *Tomocerus* are finely subannulated and very flexible (fig. 5.39).

The entomobryids are characterized by: eyes 8 + 8, reduced or lacking; antennae longer than head, fourth segment usually longer and sometimes subannulated, first and second segments in some species apparently divided (fig. 5.40); postantennal organ seldom present; pronotum membranous, without setae and obscured by mesonotum (fig. 5.41); abdominal segment IV dorsally longer than III (fig. 5.42); furcula long, always reaching the ventral tube; smooth and fringed setae present; scales present or absent.

Entomobryids are found in a variety of habitats. Species with dense setation and/or scales can be found living under xeric conditions in leaf litter, under bark, or on vegetation. Some are inhabitants of ant and termite nests, and others live in semiaquatic habitats.

Selected Bibliography

Christiansen and Bellinger, 1980–81.
Gisin 1960.
Maynard 1951.

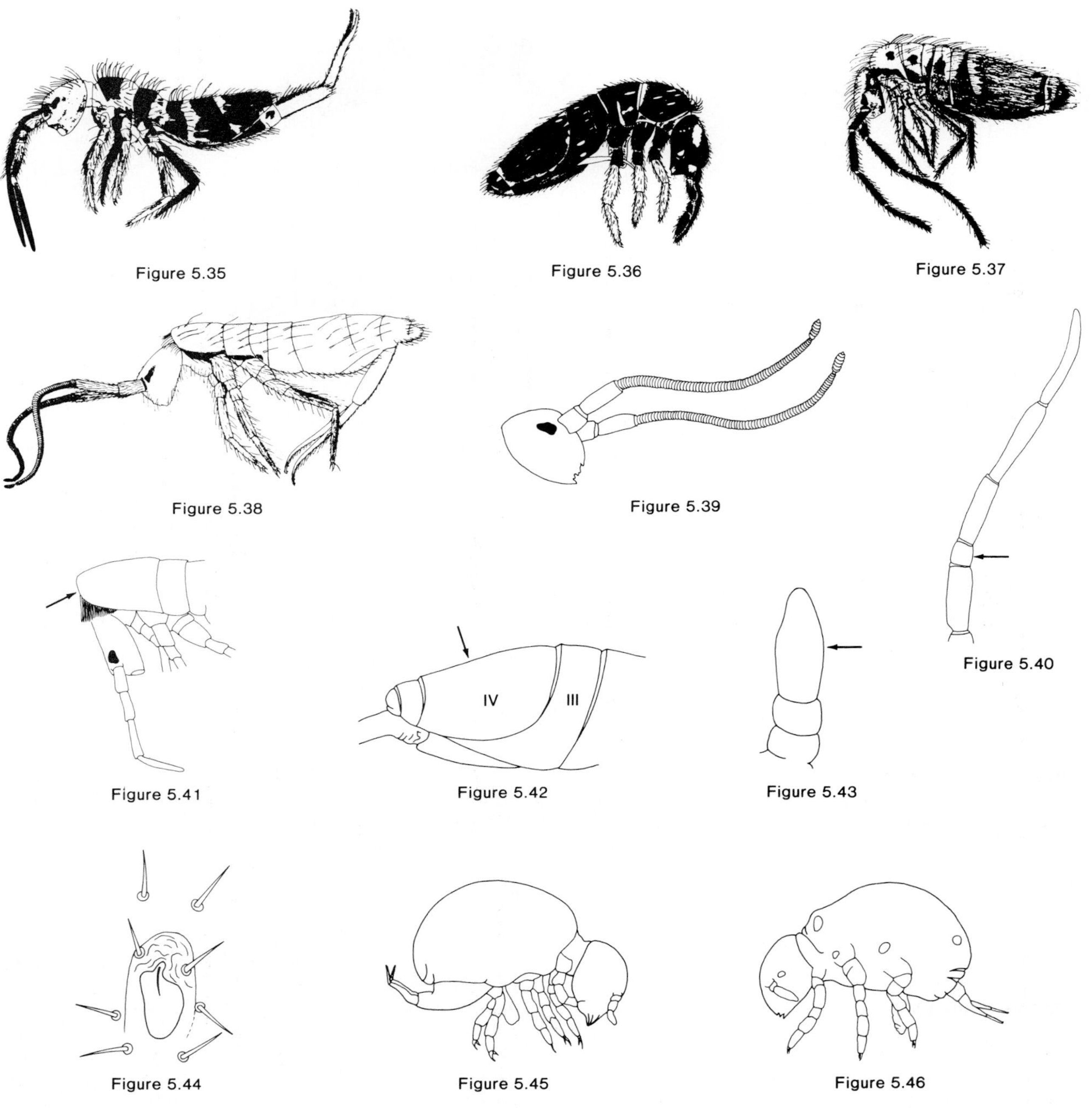

Figure 5.35. Entomobryidae, *Orchesella hexfasciata* Harvey.

Figure 5.36. Entomobryidae, *Lepidocyrtus violaceous* Fourcroy.

Figure 5.37. Entomobryidae, *Entomobrya (Entomobryoides) mineola* Folsom.

Figure 5.38. Entomobryidae, *Tomocerus flavescens* Tullberg.

Figure 5.39. Entomobryidae, *Tomocerus* sp., head.

Figure 5.40. Entomobryidae, *Orchesella* sp., antenna showing subsegmentation of segment 1.

Figure 5.41. Entomobryidae, *Lepidocyrtus paradoxus* Uzel, mesonotum obscuring pronotum.

Figure 5.42. Entomobryidae, Posterior abdominal segments of an entomobryid.

Figure 5.43. Neelidae, *Megalothorax incertus* Borner.

Figure 5.44. Neelidae, *M. incertus,* upper anterior sensory field.

Figure 5.45. Neelidae, *Neelides minutus* (Folsom).

Figure 5.46. Neelidae, *M. incertus.*

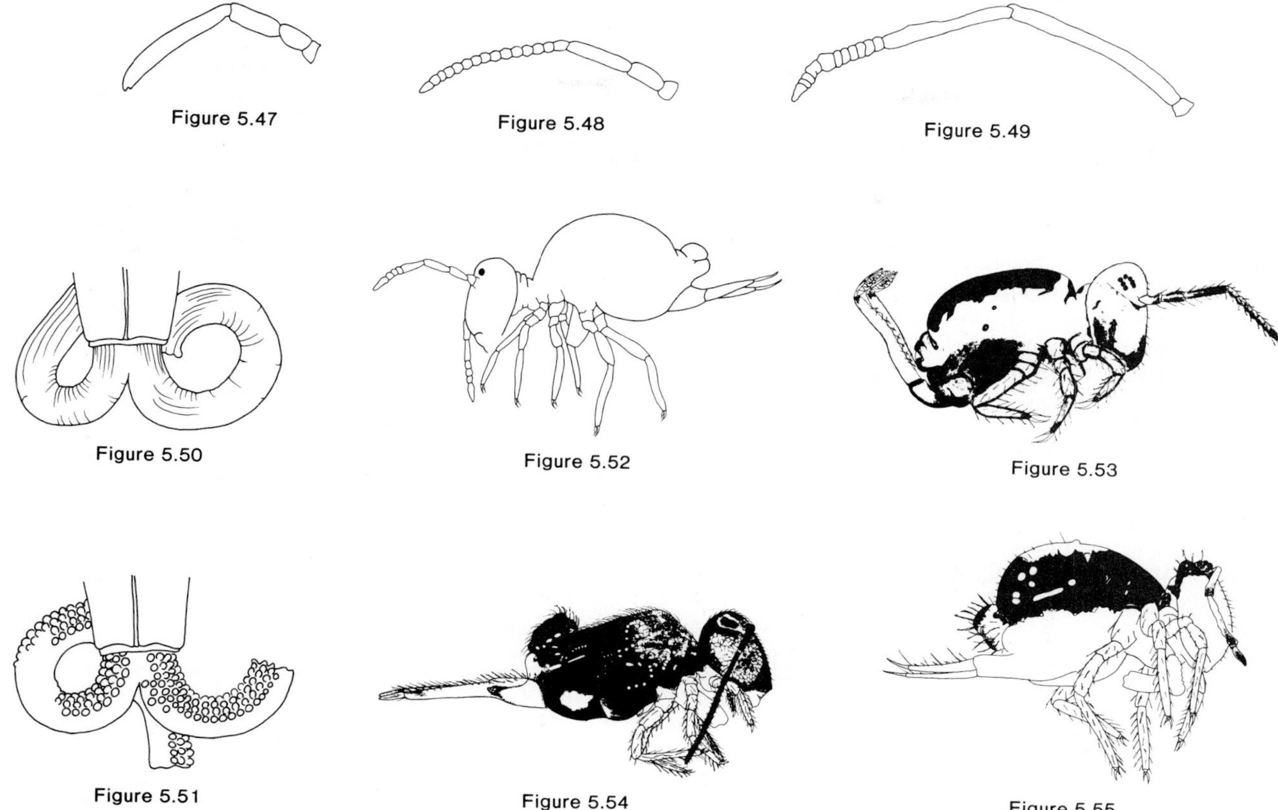

Figure 5.47. Sminthuridae, *Sminthurinus* sp., antenna.

Figure 5.48. Sminthuridae, *Sminthurus* sp., antenna.

Figure 5.49. Sminthuridae, *Dicyrtoma (Ptenothrix)* sp., antenna.

Figure 5.50. Collophore with smooth filaments.

Figure 5.51. Collophore with warty filaments.

Figure 5.52. Sminthuridae, *Arrhopalites* sp.

Figure 5.53. Sminthuridae, *Sminthurides malmgreni* (Tullberg).

Figure 5.54. Sminthuridae, *Bourletiella (Pseudobourletiella) spinata* (MacGillivray).

Figure 5.55. Sminthuridae, *Dicyrtoma (Dicyrtoma) hageni* (Folsom).

Mills 1934.
Salmon 1964.
Stach 1960, 1963.
Szeptycki 1967, 1969, 1972, 1979.
Uchida 1972b.

Suborder SYMPHYPLEONA

NEELIDAE

The Neelids

Figures 5.43–5.46

The family is described by the following characteristics: eyes absent; postantennal organ absent; antennae shorter than head, segments three and four may be fused (fig. 5.43); thorax longer than abdomen, metathorax indistinctly jointed to abdomen; abdominal segments fused, segments V and VI weakly sutured and partially concealed from above; midgut divided into four chambers; sensory fields present in the form of depressions resembling pseudocelli (fig. 5.44); setae sparse.

There are only three genera, *Neelides* Caroli, *Neelus* Folsom, and *Megalothorax* Willem. All species are very small, less than 1 mm long (fig. 5.45–5.46). Most frequently they are encountered in humus, rotten logs, bogs, and highly organic soils. Because of their small size they are generally overlooked in field collections.

Selected Bibliography

Christiansen and Bellinger 1980–81.
Gisin 1960.
Stach 1957.
Uchida 1972b.

SMINTHURIDAE

The Sminthurids

Figures 5.1a–e, 5.47–5.55

This large family is typically defined by: eyes 8 + 8, reduced or absent; postantennal organ absent; antennae longer than head, segments three and four sometimes subsegmented (figs. 5.47–5.49); with the exception of a few genera, thoracic segmentation obscure; abdominal segments I to IV fused, segments V and VI usually delimited by sutures or lines; exertile filaments of the collaphore smooth or warty (figs. 5.50–5.51); setae smooth, serrate, or fringed.

Sminthurids are perhaps the most interesting family in this suborder. Their globate shape (resembling small spiders, figs. 5.52–5.55), bright color patterns, and complex behavior have attracted the attention of many investigators. They are found in humus, under bark, in fungal fruiting bodies, on vegetation, and in semiaquatic situations. Sexual dimorphism is frequent and has occasionally led to species descriptions with the female as one species and the male as another. In some genera, the eggs are covered with fecal material to protect the developing embryo. In my cultures, *Dicyrtoma* (*Pterothrix*) *atra* (L.) produced juveniles that lined up single-file behind the female, following her wherever she went.

Selected Bibliography

Christiansen and Bellinger 1980–81.
Gisin 1960.
Guthrie 1903.
Maynard 1951.
Mills 1934.
Salmon 1964.
Stach 1956, 1957.
Uchida 1972b.

BIBLIOGRAPHY

Barra, J. A. 1975. Le developpement postembryonnaire de *Pseudosinella decipiens*. 1. Etudes morphologique et chaetotaxique (Collembola). Ann. speleol., Paris 30:173–86.

Bellinger, P. F. 1954. Studies of soil fauna with special reference to the Collembola. Conn. Agric. Exp. Sta. Bull. 583:1–67.

Boudreaux, H. B. 1979. Arthropod phylogeny with special reference to insects. New York: John Wiley & Sons. 320 pp.

Christiansen, K. A., and P. F. Bellinger. 1980–81. The Collembola of North America north of the Rio Grande. Grinnell, Ia.: Grinnell College. 1322 pp. (4 parts).

Davis, H., and H. M. Harris. 1936. The biology of *Pseudosinella violenta* (Folsom), with some effects of temperature and humidity on its life stages (Collembola:Entomobryidae). Iowa State Coll. J. Sci. 10(4):421–30.

Folsom, J. W. 1933. The economic importance of Collembola. J. Econ. Ent. 26:934–39.

Folsom, J. W. 1937. Nearctic Collembola or springtails of the family Isotomidae. Bull. U.S. Nat. Mus., No. 168. 144 pp.

Gisin, H. 1960. Collembolenfauna Europas. Mus. Hist. Nat., Geneve, 312 pp.

Guthrie, J. E. 1903. The Collembola of Minnesota. Rep. Geol. Nat. Hist. Surv. Minn., Zool. Ser. 4:1–110.

Hale, W. G. 1965. Post-embryonic development in some species of Collembola. Pedobiol. 5:228–43.

Hale, W. G. 1969. Preliminary stereoscan studies of the genus *Onychiurus* Gervais (Collembola:Onychiuridae), pp. 169–86. *In* J. G. Sheals, (ed.) The soil ecosystem. London: The Systematics Association.

Hammer, M. 1953. Investigations on the microfauna of northern Canada. Pt. II. Collembola Acta Arctica. Fasc. 6; 1–107. Copenhagen.

Hirst, S., and K. Maulik. 1926. On some arthropod remains from the Rhynie chert (old red sandstone). Geol. Mag. 63:69–71.

Kevan, D. K. McE. 1955. Soil zoology. London: Butterworths. 512 pp.

Lippert, G., and L. Butler. 1976. Taxonomic study of Collembola of West Virginia. West Virginia Univ. Agric. Exp. Sta. Bull. 643T, 27 pp.

Maynard, E. A. 1951. The Collembola of New York State. Comstock Publ. Co., New York. 339 pp.

Mills, H. B. 1934. A monograph of the Collembola of Iowa. Iowa State Coll., Monogr. no. 3, Div. Ind. Sci., 143 pp.

Salmon, J. T. 1964. An index to the Collembola. Bull. Roy. Soc. New Zealand, 1:1–144; 2:1–644.

Scott, D. B. 1953. The economic biology of Collembola. J. Econ. Ent., 46:1048–51.

Snider, R. J. 1967. An annotated list of the Collembola (springtails) of Michigan. Mich. Ent. 1(6):179–234.

Snider, R. J. 1977. Development of instar chaetotaxy of *Onychiurus* (*Onychiurus*) *folsomi*. Trans. Amer. Micros. Soc. 96(3):355–62.

Stach, J. 1947. The apterygotan fauna of Poland in relation to the world fauna of this group of insects. Family: Isotomidae. Pol. Akad. Sci. & Lett. Krakow. 488 pp.

Stach, J. 1949a. Ibid. Families: Neogastruridae and Brachystomellidae. Acta Mon. Mus. Hist. Nat. Poland. 341 pp.

Stach, J. 1949b. Ibid. Families: Anuridae and Pseudachorutidae. Ibid. 122 pp.

Stach, J. 1951. Ibid. Family: Bilobidae. Ibid. 97 pp.

Stach, J. 1954. Ibid. Family: Onychiuridae. Ibid. 219 pp.

Stach, J. 1956. Ibid. Family: Sminthuridae. Ibid. 287 pp.

Stach, J. 1957. Ibid. Families: Neelidae and Dicyrtomidae. Ibid. 113 pp.

Stach, J. 1960. Ibid. Tribe: Orchesellini. Ibid. 151 pp.

Stach, J. 1963. Ibid. Tribe: Entomobryini. Ibid. 126 pp.

Szeptycki, A. 1967. Morpho-systematic studies on Collembola. Part 1. Material to a revision of the genus *Lepidocyrtus* Bourlet, 1839 (Entomobryidae s.l.). Acta Zool. Cracov., Krakow 12:369–78.

Szeptycki, A. 1969. Morpho-systematic studies on Collembola. Part 2. Postembryonic development of the chaetotaxy in *Entomobryoides myrmecophila* (Reuter, 1886) (Entomobryidae). Acta Zool. Cracov., Krakow 14:136–72.

Szeptycki, A. 1972. Morpho-systematic studies on Collembola. Part 3. Body chaetotaxy in the first five instars of several genera of the Entomobryomorpha. Acta Zool. Cracov. Krakow 17:341–72.

Szeptycki, A. 1979. Morpho-systematic studies on Collembola. Part 4. Chaetotaxy of the Entomobryidae and its phylogenetical significance. Polska Akad. Nauk, Krakow. 218 pp.

Uchida, H. 1971. Tentative key to the Japanese genera of Collembola, in relation to the world genera of this order (1). Sci. Rep. Hirosaki Univ. 18(2):64–76.

Uchida, H. 1972a. Tentative key to the Japanese genera of Collembola, in relation to the world genera of this order (2). Sci. Rep. Hirosaki 19(1):19–42.

Uchida, H. 1972b. Tentative key to the Japanese genera of Collembola, in relation to the world genera of this order (3). Sci. Rep. Hirosaki 19(2):79–114.

Waldorf, E. 1971. Selective egg cannibalism in *Sinella curviseta* (Collembola: Entomobryidae). Ecology, 54(4):673–75.

Class and Order Diplura

6 

Pedro Wygodzinsky
The American Museum of Natural History

DIPLURANS

Diplurans are basically cryptic, found in the soil, under rocks, in leaflitter, under bark of dead or dying trees, and in rotten wood; some are cavernicolous. Some Diplura feed on organic detritus and soil microorganisms; others are predators. None are economically significant.

DESCRIPTION

Diplurans are primitively wingless, with mouthparts endognathous, mandibles monocondylous, eyes and ocelli absent, and with two paired caudal appendages; the abdomen consists of normally developed segments; abdominal styli and exsertile vesicles are present in most cases. The immatures are distinguished from adults by their smaller size, fewer setae, and less completely developed cerci. First instars (where known) are quiescent, not feeding, with rudimentary, nonfunctional mouthparts, and sparse body vestiture.

Body from subcylindrical to slightly flattened, invariably slender and elongate, generally without, rarely with, scales; color from whitish to yellow, in some cases (japygids) posterior body segments conspicuously darkened. Head prognathous. Mandibles and maxillae enclosed in pouches formed by fusion of cranial folds with sides of labium; only tips of mouthparts exposed anteriorly. Eyes and ocelli absent. Antennal flagellum moniliform, with intrinsic musculature, with or without trichobothria. Mandibles monocondylous. Maxillary palpi small or obsolescent; labial palpi reduced or absent. Tarsi not segmented. Pretarsus with two or three claws. Abdomen from subcylindrical to slightly flattened, with ten normally developed segments. Styli and in some cases exsertile vesicles present. Urosternites consisting each of a single plate. Cerci present, of different structure in the various families. Terminal filament not developed.

COMMENTS

Although the higher taxonomic categories within the Diplura are not agreed upon, there seem to be three monophyletic groups that are treated here as families. None have unequivocally been shown to be economically significant.

TECHNIQUES

Diplurans can be collected manually by careful search in their habitat, and mechanically using soil traps, flotation techniques, or Berlese-type funnels. They can be cultured in glass or plastic dishes in which the bottoms are covered with a plaster of Paris mixture (see chapt. 2) and fed, according to their respective requirements, with decaying organic matter, fungi and similar substances, or with small arthropods.

For study, specimens should be preserved in 70% ethyl alcohol, to be changed at least once. Specimens with fragile cerci (e.g., campodeids) must be preserved in individual small vials to assure correct association of specimens with their taxonomically important caudal appendages.

KEY TO THE FAMILIES OF DIPLURA, BASED UPON IMMATURES AND ADULTS

1. Cerci not segmented, forcepslike (fig. 6.1); mandibles without prostheca (Fig. 6.7) .... ***Japygidae*** (p. 67)

1′. Cerci segmented (figs. 6.2, 6.4); mandibles with prostheca (fig. 6.8) 2

2(1′). Cerci rigid, open apically (figs. 6.2, 6.3); first segment of antennal flagellum without trichobothria; lacinia with 1 or several pectinate appendages (similar to fig. 6.6); urosternite I with styli in addition to coxal appendages (fig. 6.11) .... ***Projapygidae*** (p. 67)

2′. Cerci flexible, moniliform, closed apically (figs. 6.4, 6.5); first segment of antennal flagellum with trichobothria (fig. 6.9); lacinia without pectinate appendages; urosternite I without styli, only with coxal appendages (fig. 6.10) .... ***Campodeidae*** (p. 67)

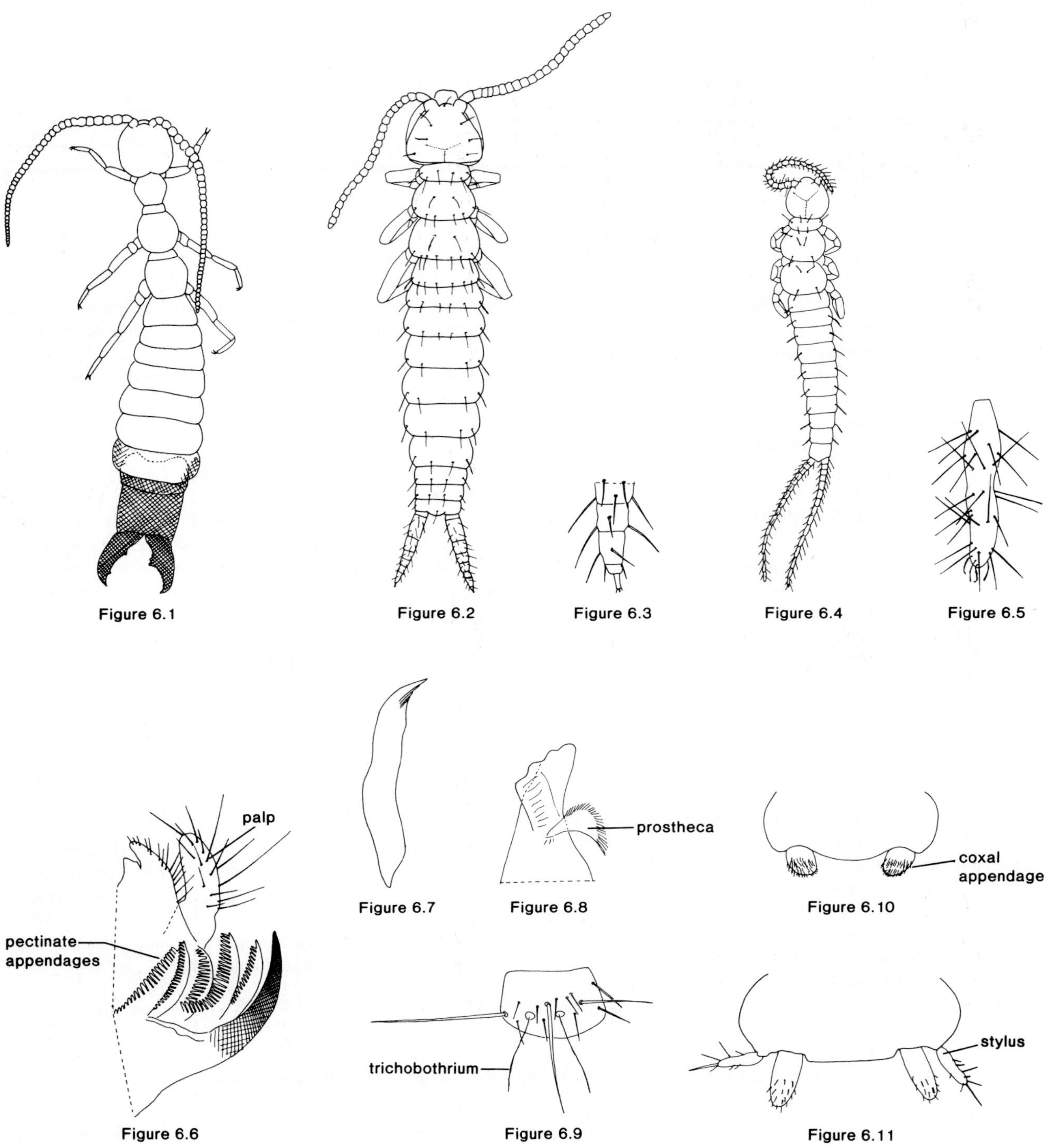

Figure 6.1. A japygid, dorsal view, schematic.

Figure 6.2. *Symphylurinus* sp. (Projapygidae), dorsal view, schematic.

Figure 6.3. *Symphylurinus* sp., apex of cercus.

Figure 6.4. A campodeid, dorsal view, schematic.

Figure 6.5. A campodeid, apical article of cercus.

Figure 6.6. A japygid, apex of maxilla.

Figure 6.7. A japygid, mandible.

Figure 6.8. A campodeid, apex of mandible with prostheca.

Figure 6.9. Campodeid, first segment of flagellum.

Figure 6.10. Campodeid, first urosternite, schematic.

Figure 6.11. *Symphylurinus* sp. (Projapygidae), first urosternite, schematic.

CAMPODEIDAE

The Campodeids

Figures 6.4, 6.5, 6.8–6.10

Relationships and Diagnosis: Campodeids differ from other Diplura by their multiarticulate, flexible cerci and by the absence of the maxillary palpus and pectinate appendages to the lacinia. Campodeids can be distinguished from "campodeiform" immature stages of other insects immediately by the combination of entotrophic mouthparts, absence of eyes and ocelli, and the multiarticulate moniliform cerci.

Biology and Ecology: Campodeids are found in the ground, among leaf litter, under the bark of trees, in rotten wood, and in similar situations. Many species are cavernicolous, and these generally have greatly elongated appendages. They feed on organic detritus and fungi.

Description: As described for the order. Scales absent or rarely present; body whitish or yellowish. Antennae with trichobothria from first to fourth flagellar segments. Lacinia without pectinate appendages; maxillary palpus absent; mandible with prostheca. First urosternite with coxal appendages but without styli. Cerci multiarticulate, moniliform, flexible, apically closed. First instar quiescent, with very short and sparse setae, cerci and antennae very short, feebly segmented; second instar from quiescent to mobile, more setose than first, but less than following instars; third and following instars closely resembling adults in shape and chaetotaxy.

Comments: This family has numerous genera and species in North America and Mexico; several species are cavernicolous. The early instars of the New World species have not been studied.

PROJAPYGIDAE

The Projapygids

Figures 6.2, 6.3, 6.11

Relationships and Diagnosis: Projapygids differ from all other Diplura by their short multiarticulate but rigid cerci.

Biology and Ecology: Found in the soil, among leaf litter and especially under large stones embedded in the ground. Food, where known, consists of microarthropods, mainly mites.

Description: As described for the order. Scales absent, body white or yellowish, antennal trichobothria present, beginning on second or third segment of flagellum. Lacinia with pectinate appendages. Maxillary palpus present. Mandible with prostheca. First urosternite with coxal appendages and with styli. Cerci multiarticulate but rigid, open apically, with silk gland. Immatures not described.

Comments: There is only one species in the United States, and four in Mexico, belonging to two genera.

JAPYGIDAE

The Japygids

Figures 6.1, 6.6, 6.7

Relationships and Diagnosis: Japygids differ from all Diplura by their cerci, which are transformed into forceps, and by having the terminal abdominal segments heavily sclerotized and conspicuously pigmented. Superficially similar to Dermaptera, they can of course be easily distinguished from the latter by their entotrophous mouthparts and the absence of eyes and ocelli.

Biology and Ecology: Japygids are soil inhabitants, only rarely found in caves. Some species produce tunnels in the soil. They feed on other arthropods which they grasp with their forceps.

Description: As for the order. Scales absent. Body from white to yellowish; posterior segments and forceps heavily pigmented. Antennae with or without trichobothria. Mandibles without prostheca. Lacinia with pectinate appendages. First urosternite without articulate coxal appendages, but with styli. Cerci not segmented, short, forcepslike. First instar white, quiescent, with very short and sparse setae. Cerci and antennae very short, the former of simple structure as compared to later instars, white, without denticles. Second instar from quiescent to mobile, more setose than first instar but less so than later instars; third and following instars closely resembling adults in shape, pigmentation and chaetotaxy. Pagès (1967) has given an excellent summary of our knowledge of immature japygids.

Comments: The Japygidae possess several genera and species in North America, but the early instars are known only from a few. The family has been divided into several family group taxa.

BIBLIOGRAPHY

Bareth, C., and B. Condé. 1965. La prélarve de *Campodea* (*C.*) *remyi*. Rev. Ecol. Biol. Sol. 2:397–401.

Condé, B. 1955. Matériaux pour une monographie des Diploures Campodéides. Mém. Mus. nat. Hist. natur., Paris (A), zool. 12:1–203.

Gyger, H. 1960. Untersuchungen zur postembryonalen Entwicklung von *Dipljapyx humberti* (Grassi). Verh. Naturf. Ges. Basel, 71: 29–95.

Orelli M. von. 1956. Untersuchungen zur postembryonalen Entwicklung von Campodea. Verh. Naturf. Ges. Basel, 67:501–74.

Pagès J. 1951. Contribution à la connaissance des Diploures. Bull. sci. Bourgogne, 13: suppl. mécan. no. 9, 97 pp.

Pagès, J. 1967. Données sur la biologie de *Dipljapyx humberti* (Grassi). Thesis, Limoges, 99 pp.

Silvestri, F. 1908. Ueber die Projapygiden und einige Japyx-Arten. Zool. Anz. 28:638–43.

Silvestri, F. 1928. On postembryonal development of Japygidae (Thysanura). Trans. 4th Int. Congress Ent., Ithaca, 2:905–908.

Smith, L. M. 1961. Japygidae of North America, 8. Postembryonal development of Parajapyginae and Evalljapyginae (Insecta, Diplura). Ann. Ent. Soc. Amer. 54:437–44.

Order Microcoryphia

7

Pedro Wygodzinsky
The American Museum of Natural History

JUMPING BRISTLETAILS

Most microcoryphians are found on cliffs, on rock outcroppings, or among boulders and smaller stones; species of some genera are collected only at or near the seashore (*Petrobius* Leach, *Neomachilis* Silvestri). A few can be found remote from rocks, on the ground or on trees. They occur in a wide range of habitats, from very moist coastal forests to semiarid areas. They feed on a wide variety of decaying organic matter, preferably vegetal, on lichens and mainly on terrestrial algae (Pleurococcales). Microcoryphians are generally crepuscular and nocturnal except the very early instars; the members of *Machilinus* Silvestri are diurnal.

DIAGNOSIS

Microcoryphians are primitively wingless with conspicuously arched thorax, ectognathous mouthparts, monocondylous mandibles, large contiguous eyes and three caudal appendages. This combination of characters serves to distinguish microcoryphians from all other insects. Immatures are distinguished from adults by smaller size and absent or not fully developed genitalia. First and second instars are invariably without scales but with regularly arranged short erect setae and with very short cerci (fig. 7.4). Lacinia of first instar with prominent apical spinelike tooth (egg-burster).

DESCRIPTION

Primarily wingless, ametabolous insects. Body subcylindrical, tapering posteriorly, thorax conspicuously arched. Body heavily scaled, but with setae only in first and second instars. Head hypognathous. Mouthparts exposed, rasping. Compound eyes large, contiguous dorsally. One median and 1 + 1 submedian or sublateral ocelli present. Antennal flagellum without intrinsic musculature. Mandibles monocondylous, incisor portion remote from molar area. Maxillary palp large, forwardly and upwardly arched, seven-segmented. Legs in some cases with coxal styli. Tarsi two or three segmented. Pretarsus with two simple claws. Urosternites 2–9 with one pair of styli each. Eversible vesicles present, variable in number, or absent. Abdominal coxites well developed, on pregenital segments either contiguous, with sternite strongly reduced, or separated by large triangular sternite. Three rigid caudal appendages, two cerci and one terminal filament.

COMMENTS

Microcoryphians are of no known economic importance.

TECHNIQUES

During the daytime, microcoryphians can be collected under stones and boulders, in rock crevices, on the walls of cave entrances, under bark of trees, or in the ground and among leaf litter. At dusk and during the night, they come to the surface and can be collected on the ground, on tree trunks and on rocks with the aid of a lantern. Most microcoryphians can be kept alive for a limited time in a container provided with bark covered with terrestrial algae, changed at regular intervals. Temperature and moisture should be kept within the limits of conditions found where the specimens were taken.

For taxonomic or morphological studies, preservation in 70% ethyl alcohol is adequate; it should be changed at least once. Kahle's is adequate for studies of gross internal anatomy.

KEY TO THE FAMILIES OF MICROCORYPHIA[1]

1. Urosternites with sternite very small (fig. 7.6); entire antennae, legs and abdominal styli without scales .... *Meinertellidae* (p. 69)

1′. Urosternites with sternite well developed (fig. 7.5); at least scapus of antennae as well as legs and abdominal styli with scales .... *Machilidae* (p. 69)

1. Based on characters of immature specimens (except first and second instars).

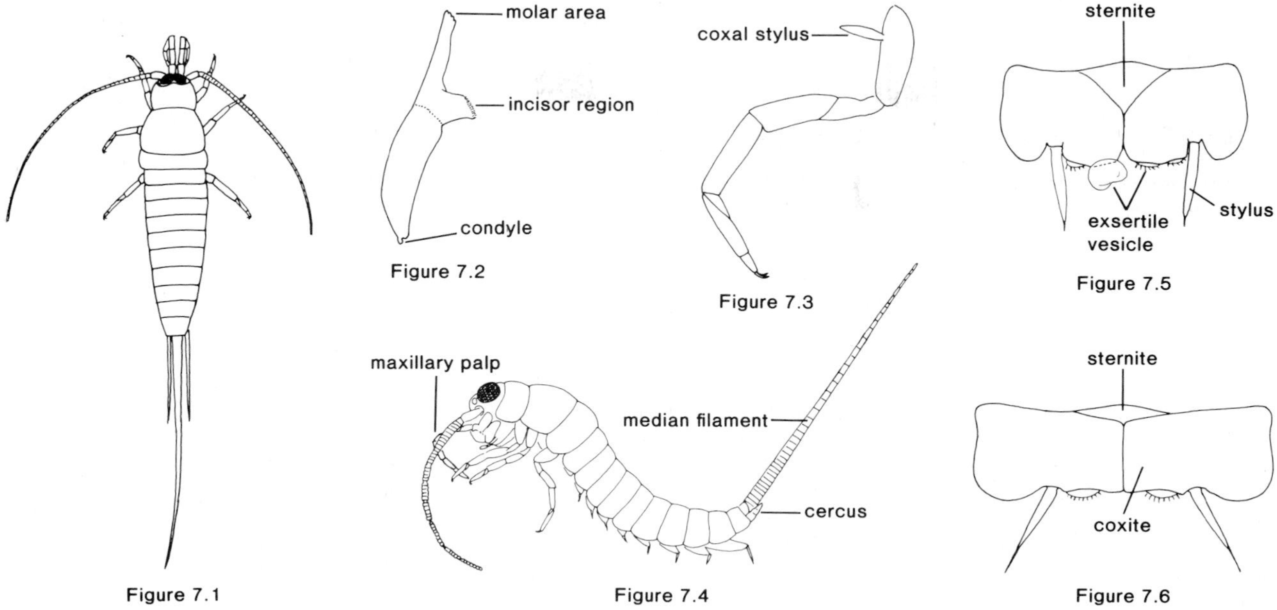

Figure 7.1. *Mesomachilis* sp. (Machilidae), dorsal, schematic.

Figure 7.2. *Mesomachilis* sp., mandible.

Figure 7.3. *Mesomachilis* sp., female, hind leg.

Figure 7.4. *Trigoniophthalmus alternatus* Silvestri, (Machilidae), second instar, side view, setae not shown.

Figure 7.5. *Mesomachilis* sp. (Machilidae), fifth urosternite.

Figure 7.6. *Machiloides* sp. (Meinertellidae), fifth urosternite.

MACHILIDAE

The Machilids, Bristletails

Figures 7.1–7.5

Relationships and Diagnosis: The machilids are very similar to the meinertellids, but they can easily be recognized by the well developed sternal plate of the urosternites, the presence of parameres in the juvenile and adult male, and the presence of scales at least on the scapus of the antennae and on the abdominal styli.

Biology and Ecology: As described for the order. All genera are basically crepuscular or nocturnal, except the very early instars (first to third or fourth).

Description: As described for the order. Scales present at least on scapus of antennae, on the legs, and on the styli. Coxites of pregenital segments with 1 + 1 or 2 + 2 exsertile vesicles. Males with parameres.

Comments: Six or seven valid genera have been described from the United States and Mexico, but several additional genera occur in the area. *Pedetontus* Silvestri is the most widely distributed genus, with several described and many undescribed species.

Delany (1959), Heymons (1906), and Verhoeff (1910, 1911) have provided some information on immatures of European genera; no work has been done on American forms.

MEINERTELLIDAE

The Meinertellids, Bristletails

Figure 7.6

Relationships and Diagnosis: The meinertellids are very similar to the machilids and are easiest to recognize by the extreme reduction of the sternite of the abdominal segments, the absence of parameres in the males, and the absence of scales on the entire antennae, the legs and the abdominal styli.

Biology and Ecology: As described for the order. One of the genera included here, *Machilinus* Silvestri, is diurnal, an exception for members of this order; all other genera are basically nocturnal, except the very early instars.

Description: As for the order. Scales absent on entire antennae, legs, and styli. Urosternites with sternite very small. Exsertile vesicles present or absent, but never more than 1 + 1 on each segment. Males without parameres.

Comments: Four genera occur in the United States and Mexico. *Machiloides* Silvestri is the most common eastern genus, *Machilinus* the main western genus.

No published work exists on the immatures of meinertellids.

BIBLIOGRAPHY

Delany, M. J. 1959. The life histories and ecology of two species of *Petrobius* Leach, *P. brevistylis* and *P. maritimus*. Trans. Roy. Soc. Edinburgh, 58:501–33, 37 figs.

Heymons, R. 1906. Über die ersten Jugendformen von *Machilis alternata* Silv. Ein Beitrag zur Beurteilung der Entwicklungsgeschichte bei den Insekten. Sitz. ber. Gesell. naturforschenden Freunde Berlin 1906 (1):253–59.

Verhoeff, K. W. 1910. Über Felsenspringer, Machiloidea. 3. Aufsatz: Die Entwicklungsstufen. Zool. Anz. 36:385–99.

Verhoeff, K. W. 1911. Über Felsenspringer, Machiloidea. 5. Aufsatz: Die schuppenlosen Entwicklungsstufen und die Orthomorphose. Zool. Anz. 38:254–63.

Order Thysanura 8

Pedro Wygodzinsky
The American Museum of Natural History

BRISTLETAILS

Thysanurans occupy a variety of habitats; most are edaphic, a few are domestic, others are myrmecophilous or termitophilous, and others still are troglobionts. They occur in many habitats, from moist leaf litter to the sands of the desert. Free-living and cave forms feed on organic detritus; domestic species, although thriving on starchy substances, also require protein. The feeding patterns of most myrmecophilous and termitophilous species are not well known, but some are cleptobiotic. All free-living and domestic species are nocturnal, even the early instars.

DIAGNOSIS

Primitively wingless insects, with thorax not arched, mouthparts ectognathous, mandibles dicondylous, eyes not contiguous or absent, three caudal appendages; abdomen in most cases with styli on pregenital and genital segments. This combination of characters serves to distinguish immature (and mature) thysanurans from all other insects. Immatures distinguished from adults by smaller size and absent or not fully developed genitalia. First instar not feeding. Head of first instar with dorsal egg-tooth (egg-burster).

DESCRIPTION

Primarily wingless, ametabolous insects; body from subcylindrical to flattened, elongate, or limuloid. Thorax not arched. Body with or without scales; the first two instars invariably without scales, only with setae. Head prognathous or hypognathous, invariably hypognathous in first instar. Mouthparts exposed, chewing. Eyes present or absent; when present, lateral, widely separated. Ocelli present or absent. Antennal flagellum without intrinsic musculature. Mandibles dicondylous, incisor portion close to molar area. Maxillary palpi small, not forwardly arched, five-segmented. Legs never with coxal styli. Pretarsus with one median and two lateral claws, simple or variously modified. Styli present on genital and in most cases pregenital segments, their number varied, very rarely absent. Eversible vesicles absent or present in variable number. Urosternites with coxites well developed, posterolateral or posterior, or absent. Three somewhat flexible caudal appendages, two cerci and one terminal filament.

COMMENTS

Several thysanurans are of economic importance as pests of household and stored products and of materials containing starch, such as paper and books.

TECHNIQUES

Thysanurans can be obtained by careful search in their respective habitat; an aspirator is useful. Domestic forms can be caught with oat-traps. Berlese-type funnels will produce thysanurans, especially immatures. Silverfish, firebrats, and related forms can be kept in the laboratory if provided with the correct degree of moisture and temperature and with adequate food (oatmeal, yeast, etc). Cultures can be easily established. Soil-inhabiting thysanurans of the family Nicoletiidae can be cultured in a glass or plastic dish, the bottom of which is covered with a plaster of Paris mixture (see chap. 2) and fed with small pieces of lettuce that are daily replaced. Preservation for study is the same as indicated for Microcoryphia.

KEY TO THE FAMILIES OF THYSANURA BASED ON CHARACTERS OF IMMATURE SPECIMENS

1. Compound eyes absent (fig. 8.15) .... ***Nicoletiidae*** (p. 72)
1′. Compound eyes present (figs. 8.14, 8.16) 2
2(1′). Ocelli present (fig. 8.14); scales absent .... ***Lepidotrichidae*** (p. 72)
2′. Ocelli absent (fig. 8.16); scales present .... ***Lepismatidae*** (p. 72)

LEPIDOTRICHIDAE

The Lepidotrichids

Figures 8.1, 8.7, 8.9, 8.12, 8.14

Relationships and Diagnosis: Lepidotrichids are superficially similar to certain lepismatids and nicoletiids but can be recognized easily by the combination of the lack of scales and the heavy overall hypodermal pigment, the presence of eyes and three ocelli, the five-segmented tarsi, and, on the abdomen, the large transverse sternites with posteriorly situated coxites.

Biology and Ecology: Found under rotten bark or in decaying wood of Douglas fir. Feeds on vegetable detritus, terrestrial algae, and similar substances.

Description: As for the order. Shape elongate, parallel-sided. Scales absent; hypodermal pigment covering entire body. Head hypognathous in all instars. Eyes lateral, well developed. One median and two lateral ocelli. Lacinia with several strongly sclerotized apical teeth and three pectinate and several simple appendages. Tarsi five-segmented. Pretarsus with one median and two lateral claws, all simple. Proventriculus absent. Caudal appendages longer than body. Sternites large, transverse; coxites posteriorly situated. Styli on abdominal segments 2–9, 1 + 1 functional exsertile vesicles on segments 2–7.

Comments: This relict family possesses a single recent genus and species, *Tricholepidion gertschi* Wygodzinsky, only known from the coastal ranges of northern California.

There is no information on immature lepidotrichids.

NICOLETIIDAE

The Nicoletiids

Figures 8.2, 8.3, 8.5, 8.10, 8.15, 8.18

Relationships and Diagnosis: Nicoletiids can be distinguished from the lepismatids and lepidotrichids by their invariably pale yellowish or whitish color, the absence of eyes and of ocelli, the presence of one pectinate appendage to the lacinia, and the lateral position of the coxites when present.

Biology and Ecology: Most nicoletiids live in the soil, under rocks, among leaf litter, and in similar situations (*Nicoletia* spp.). Others, which live in holes, in some cases have enormously elongate antennae and caudal appendages (*Texoreddellia* Wygodzinsky and others). Still others are myrmecophilous, with very short appendages (*Grassiella* spp.). Most are detritus or plant feeders; myrmecophilous species may have specialized feeding habits.

Description: As for the order. Shape from elongate parallel-sided to limuloid. Scales absent or present, invariably pale, translucent. Head hypognathous in first, prognathous in remaining instars. Eyes and ocelli absent. Lacinia with two heavily sclerotized, pointed apical teeth and with one pectinate and several simple, membranous appendages. Tarsi four-segmented. Pretarsus with one median and two lateral claws, simple or variously modified. Caudal appendages from much shorter to much longer than body. Proventriculus absent. Sternites fused with coxites, forming a transversal plate, or, when not fused to coxites, the latter posterolateral, with sternite subtriangular or trapezoidal. Number of abdominal styli variable; exsertible vesicles present.

Comments: This family has been divided into two subfamilies, the Nicoletiinae (elongate, slender) and the Atelurinae (limuloid shape), but on a world basis these subfamilies cannot yet be defined.

One species (*Nicoletia meinerti* Silvestri) is found occasionally in greenhouses, but damage has not been shown to occur; unidentified nicoletiids have been reported to damage roots of sugar cane.

The North American genera are *Nicoletia, Texoreddellia,* and *Grassiella;* additional genera occur in Mexico. There are no published studies on immature nicoletiids.

LEPISMATIDAE

The Lepismatids, Silverfish, Firebrats, and Allies

Figures 8.4, 8.6, 8.8, 8.11, 8.13, 8.16, 8.17

Relationships and Diagnosis: Lepismatids can be distinguished from lepidotrichids and nicoletiids by the absence of ocelli combined with the presence of eyes, the presence of scales frequently forming simple or intricate patterns, and the absence of pectinate appendages to the lacinia.

Biology and Ecology: Most lepismatids are free living but some are myrmecophilous. There are no cavernicolous species, but several species are adapted to life in houses and other buildings where they feed on stored products, food remainders, paper, and some fabrics. The free-living species feed on a variety of organic matter in and on the surface of the soil.

Description: As for the order. Shape from elongate, parallel-sided to limuloid. Scales invariably present except in the two or three early instars. Head hypognathous in first instar, prognathous in all others. Eyes well developed, lateral. Ocelli absent. Lacinia with two strongly sclerotized apical teeth and various simple membranous projections, none pectinate. Tarsi three- or four-segmented. Pretarsus with one median and two lateral claws, all simple. Proventriculus present. Caudal appendages from much shorter to longer than body. Sternites fused to coxites, forming a single plate (except genital segments). Number of abdominal styli varied. Exsertile vesicles absent. First two or three instars without scales and with appendages proportionately shorter than in adults.

Comments: There are several economically important lepismatid species that attack stored products; in our area, the following are actually or potentially involved: *Acrotelsa collaris* (F.); *Ctenolepisma lineata pilifera* (Lucas); *Ctenolepisma longicaudata* Escherich; *Lepisma saccharina* L., the silverfish; *Thermobia campbelli* (Barnhart), and *Thermobia domestica* (Packard), the firebrat. A key to the adults of these species (and all other Lepismatidae occurring in the area) has been given by Wygodzinsky (1972) and is applicable to immatures with caution.

There are several papers (Adams 1933, Cornwall 1915, Heymons 1897, Sahrhage 1952, and others) that provide information on the postembryonal development of *Lepisma saccharina* and *Thermobia domestica.*

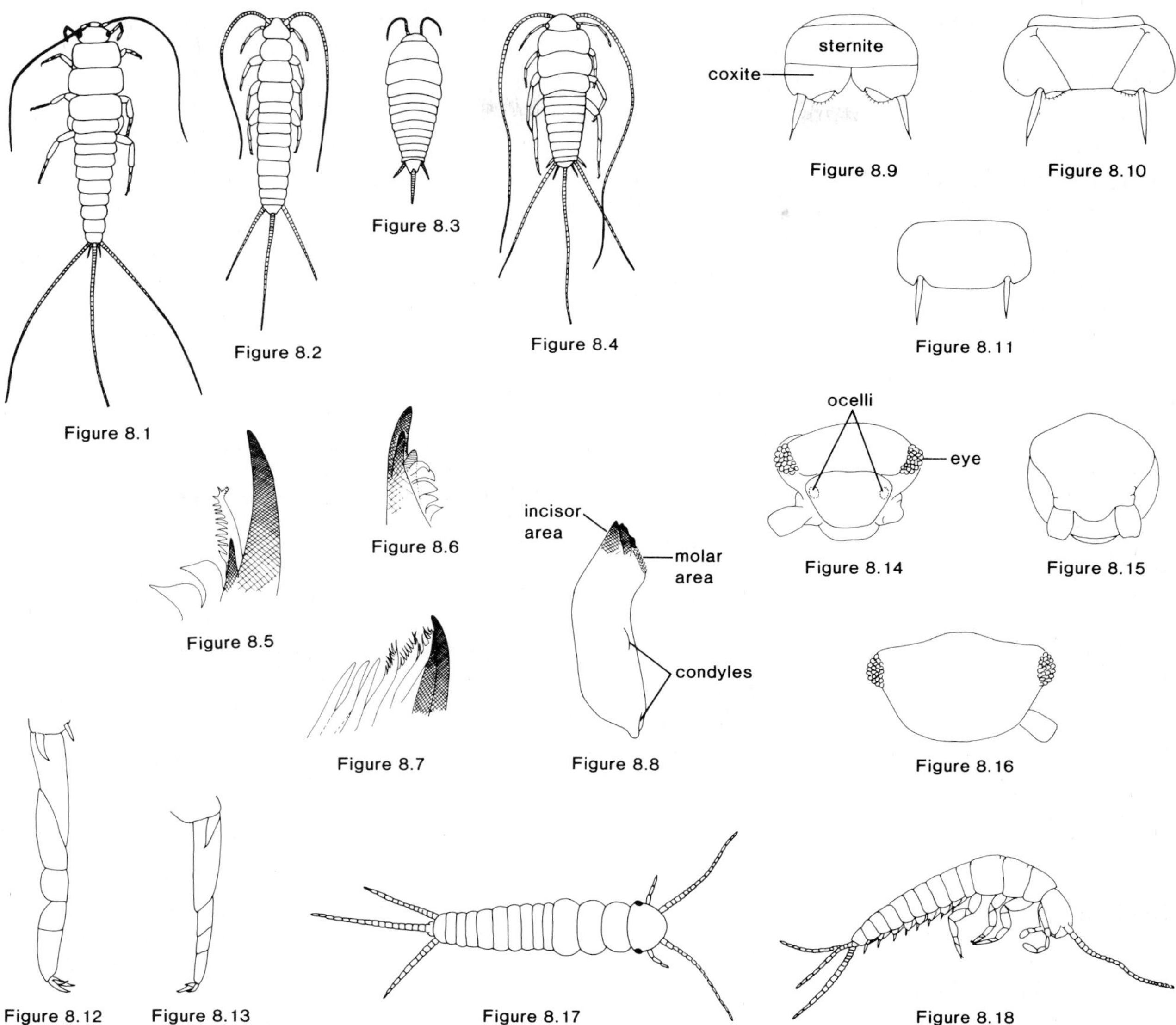

Figure 8.1 Figure 8.2 Figure 8.3 Figure 8.4 Figure 8.5 Figure 8.6 Figure 8.7 Figure 8.8 Figure 8.9 Figure 8.10 Figure 8.11 Figure 8.12 Figure 8.13 Figure 8.14 Figure 8.15 Figure 8.16 Figure 8.17 Figure 8.18

Figure 8.1. *Tricholepidion gertschi* Wygodzinsky (Lepidotrichidae), dorsal view, schematic.

Figure 8.2. *Nicoletia* sp. (Nicoletiidae), dorsal view, schematic.

Figure 8.3. *Grasiella* sp. (Nicoletiidae), dorsal view, schematic.

Figure 8.4. *Thermobia* sp. (Lepismatidae), dorsal view, schematic.

Figure 8.5. *Texoreddellia texensis* (Ulrich) (Nicoletiidae), apex of lacinia.

Figure 8.6. *Acrotelsa collaris* (Fabricius) (Lepismatidae), apex of lacinia.

Figure 8.7. *Tricholepidion gertschi* Wygodzinsky (Lepidotrichidae), apex of lacinia.

Figure 8.8. *Acrotelsa collaris* (Fabricius) (Lepismatidae), mandible.

Figure 8.9. *Tricholepidion gertschi* Wygodzinsky (Lepidotrichidae), fifth urosternite.

Figure 8.10. *Texoreddellia texensis* (Ulrich) (Nicoletiidae), fifth urosternite.

Figure 8.11. *Allacrotelsa spinulata* (Packard) (Lepismatidae), fifth urosternite.

Figure 8.12. *Tricholepidion gertschi* Wygodzinsky (Lepidotrichidae), tarsus.

Figure 8.13. *Acrotelsa collaris* (Fabricius) (Lepismatidae), tarsus.

Figure 8.14. *Tricholepidion gertschi* Wygodzinsky (Lepidotrichidae), head (median ocellus not visible), schematic.

Figure 8.15. *Nicoletia meinerti* Silvestri (Nicoletiidae), head, schematic.

Figure 8.16. *Allacrotelsa spinulata* (Packard) (Lepismatidae), head, schematic.

Figure 8.17. *Lepisma saccharina* L. (Lepismatidae), second instar, schematic.

Figure 8.18. *Nicoletia meinerti* Silvestri (Nicoletiidae), second instar, schematic.

BIBLIOGRAPHY

Adams, J. A. 1933. The early instars of the firebrat, *Thermobia domestica*. Proc. Iowa Acad. Sci. 40:217–19.

Cornwall, J. W. 1915. *Lepisma saccharina* (?): its life history and anatomy and its gregarine parasites. Indian J. Med. Res. 3:116–41.

Heymons, R. 1897. Entwicklungs geschichtliche Untersuchungen von *Lepisma saccharina* L. Ztschr. wiss. Zool. 62:583–631.

Sahrhage, D. 1952. Ökologische Untersuchungen an *Thermobia domestica* (Packard) und *Lepisma saccharina* L. Ztschr. wiss. Zool. 157:77–168.

Wygodzinsky, P. 1972. A review of the silverfish (Lepismatidae, Thysanura) of the United States and the Caribbean area. Amer. Mus. Novitates, 2481:1–26.

Order Ephemeroptera

9

George F. Edmunds, Jr.
University of Utah

Richard K. Allen
Huntington Beach, California

MAYFLIES

Ephemeroptera larvae are Exopterygota that have the claws single (lacking in Behningiidae and on the forelegs of *Homoeoneuria*), the mesothorax larger than the prothorax or metathorax, and abdominal gills on segments 4, 5, and 6 (primitively on 1–7, but may be absent at either or both ends of the series).

All mayflies are aquatic as larvae. They occupy an extreme range of aquatic habitats and feed primarily as scrapers and gatherers of vegetation and detritus; a few are omnivores. Those few that are carnivores are found most frequently in large rivers.

Although mayflies are rarely pests in the traditional sense, the adults of certain species become a nuisance when they are attracted to lights in large numbers, and fragments of their dried bodies may become allergens. At times *Hexagenia* adults interfere with Mississippi River boat traffic when attracted to the boat lights in great swarms. But mayfly larvae are generally regarded as being of positive economic importance. They play a significant role in the transfer of nutrients to fish and by emergence into the terrestrial segment of the environment, and thus are of considerable interest and importance to fly fishermen and aquatic biologists.

DESCRIPTION

Head (fig. 9.1)

Shape and deflection of the head variable. Primitively the head is grasshopperlike, short and hypognathous, with compound eyes anterolateral. A median ocellus and a pair of lateral ocelli usually located between the eyes. A pair of filiform antennae usually arises ventral to the eyes. Antennae may be short (less than width of head) to more than twice as long as width of head. In larvae with hypognathous heads the mouthparts are oriented vertically and directed ventrally. In some derived forms the head is flattened, eyes and antennae dorsal, and the head prognathous; in these the mouthparts are usually concealed beneath or behind the head capsule (although portions may be exposed). In larvae with prognathous heads the mouthparts are oriented horizontally to the substrate and directed anteriorly. In such cases the face is still regarded as anterior and the underside as posterior. The larval mouthparts are adapted for various feeding habits. The labrum varies from small and narrow to wider than the head (fig. 9.2). The mandibles differ in shape: left mandible with molar surface oriented somewhat parallel to lateral margin of mandible; molar surface of right mandible oriented at a right angle to lateral margin. Mandibular incisors and molar surfaces heavily sclerotized. In the Ephemeroidea (except Behningiidae) and some *Paraleptophlebia* (Leptophlebiidae) the lateral surfaces of the mandibles are enlarged into tusks. Maxillary palpi two- or three-segmented (rarely absent). In some genera the maxillae may bear gills. The body of the labium consists of a small submentum (postmentum) and a larger mentum (prementum). Arising from the mentum are two pairs of lobes, the glossae (mesal pair) and paraglossae (lateral pair). Also arising from the mentum is a pair of palpi. Each palpus may consist of two or three segments.

Thorax (fig. 9.1)

Pronotum relatively small, the largest thoracic segment being the mesonotum, bearing the fore wing pads. In mature larvae the wing pads extend from the pronotum posteriorly to abdominal segment two or beyond. In mature larvae of the suborder Schistonota, more than half of the wing pad is free; in the derived Pannota the wing pads are more or less fused, and less than half the wing pad is free. The metanotum, somewhat concealed dorsally by the mesonotum, especially in the Pannota, is much smaller than the mesonotum, and, except in those species whose adults lack hind wings, bears the hind wing pads. There is great variety in the structure of the legs; most have spines, tubercles, or setae. The legs of various genera are modified for such special functions as burrowing, filtering food, grooming, and gill protection. Fingerlike gills are found on the coxae of some larvae.

Abdomen (fig. 9.1)

All mayflies have ten-segmented abdomens, although some segments may be concealed beneath the mesonotum and segment 1 may be fused partially with the metanotum. The abdominal terga may have spines, tubercles, or both, and the posterolateral corners may be enlarged. Always count forward from segment 10 to determine the segment number.

Gills are the most variable structures in mayfly larvae. Gill position on the abdomen is also variable. In North American Oligoneuriinae, the gills on segment 1 are ventral, and those on succeeding segments dorsal; in *Dolania* and *Anepeorus* the gills are ventral. In most mayflies they are lateral

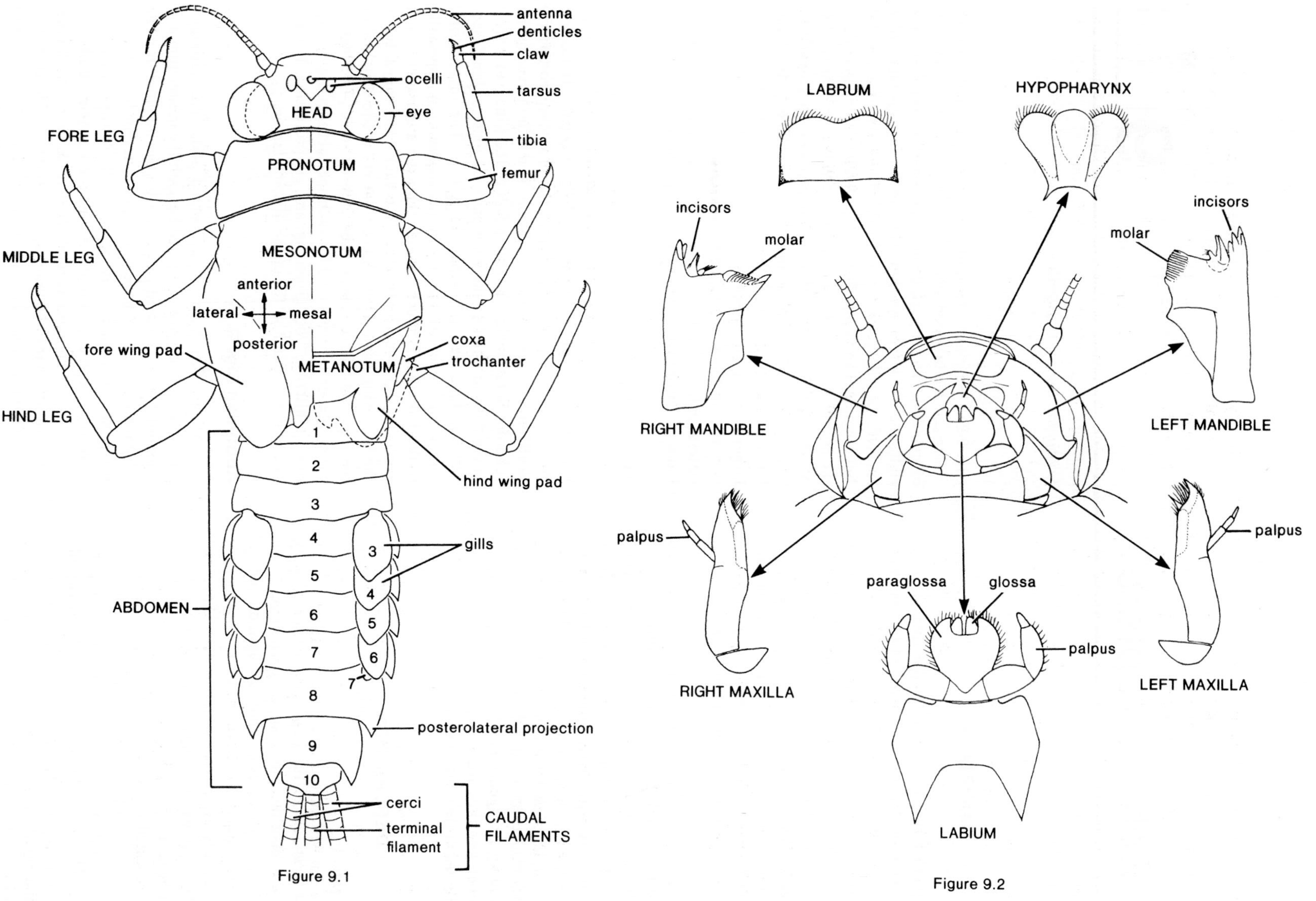

Figure 9.1. Dorsal aspect of mayfly larva. (Figures 9.1–9.45 reprinted with permission from Edmunds, Jensen, and Berner, 1976, *Mayflies of North and Central America.* Minneapolis: Univ. of Minnesota Press.)

Figure 9.2. Ventral aspect of mayfly mouthparts.

or dorsal. Gills may occur on abdominal segments 1 through 7 or be absent from one or more segments in various combinations. They may be absent from either or both ends of the series. In *Ephemerella* they are always absent from segment 2 and sometimes from segment 3, but vestigial gills may occur on segment 1. The structure of the gills is highly variable. Those of the middle abdominal segments may each consist of two platelike lamellae that may be of various shapes, two lamellae with the margins fringed or finely dissected, a dorsal platelike lamella with the ventral lamella dissected into a number of small lobes, a single platelike lamella, or other configurations (the illustrations show much of the variability.) Gills on abdominal segment 1, when present, may be rudimentary, or similar in shape to those of succeeding pairs of gills, but are usually smaller or different in shape in comparison to succeeding pairs. Gills on abdominal segment 7 are usually similar in shape to the preceding gills but smaller.

Most species have three caudal filaments (a terminal filament and two cerci). In some species only the cerci are well developed, the terminal filament being represented by a short rudiment or absent. The terminal filament, when present, may vary in length and thickness relative to the cerci. The length of the caudal filaments varies from shorter than the body to two or three times the length of the body.

COMMENTS

The larvae of at least one species of all North American genera of Ephemeroptera are now apparently known. Of the sixty genera reported to occur in America north of Mexico, 12 have only a single species (unnamed in the case of the group tentatively referred to *Homothraulus,* but actually closer to, or possibly congeneric with, *Farrodes* Peters of the West Indies). Good keys for larvae exist for many genera; references are given for these keys in the family accounts that follow, and a key to the genera by Edmunds is found *in* Merritt and Cummins (1984).

Substantial collections of reared but undescribed mayflies are deposited at several institutions or are in the hands of private individuals. Nevertheless, rearing adults to get the correct association with larvae remains important in practically all parts of North America. Revisional studies of the taxonomy of many genera with the publication of larval keys is essential before larval species identifications can be accurately accomplished for most North American mayflies.

The taxonomic arrangement followed here is a modification of that proposed by Edmunds and Traver (1954) and used by Edmunds et al. (1963). A few important changes were made by McCafferty and Edmunds (1976, 1979). The subfamily Pentageniinae was shifted from the Ephemeridae to the Palingeniidae, thus Palingeniidae is now represented in North America (McCafferty and Edmunds 1976). The subfamily Isonychiinae was transferred from the Siphlonuridae to the Oligoneuriidae (McCafferty and Edmunds 1979); because of this change we have included a couplet separating the larvae of the subfamilies Oligoneuriinae and Isonychiinae. McCafferty and Edmunds also divided the Ephemeroptera into two suborders; because ordinal and superfamily groups tend to range over a gradient of characters, they are difficult for keying purposes and are not used in the keys.

Landa and Soldán (1985) have proposed major changes in the higher classification of mayflies. The Oligoneuriidae and Heptageniidae have been removed from Baetoidea and placed in a superfamily Heptagenioidea, and the Baetiscidae are placed in Caenoidea. North American genera of Tricorythidae are now in the family Leptohyphidae. The Metretopodidae is placed as a subfamily of Siphlonuridae, and the subfamily Pseudironinae (*see* Edmunds et al. 1976) has been moved from Heptageniidae to Siphlonuridae.

CLASSIFICATION

Order **EPHEMEROPTERA**
- Suborder Schistonota
  - Superfamily Baetoidea
    - Siphlonuridae, primitive minnow mayflies
    - Ametropodidae, sand minnow mayflies
    - Baetidae, small minnow mayflies
    - Metretopodidae, cleftfooted minnow mayflies
    - Oligoneuriidae, brushlegged mayflies
    - Heptageniidae, flatheaded mayflies
  - Superfamily Leptophlebioidea
    - Leptophlebiidae, pronggilled mayflies
  - Superfamily Ephemeroidea
    - Behningiidae, tuskless burrowing mayflies
    - Potamanthidae, hacklegill mayflies
    - Polymitarcyidae, pale burrowing mayflies
    - Ephemeridae, common burrowing mayflies
    - Palingeniidae, spinnyheaded burrowing mayflies
- Suborder Pannota
  - Superfamily Ephemerelloidea
    - Ephemerellidae, spiny crawler mayflies
    - Tricorythidae, little stout crawlers
  - Superfamily Caenoidea
    - Neoephemeridae, large squaregills
    - Caenidae, small squaregills
  - Superfamily Prosopistomatoidea
    - Baetiscidae, armored mayflies

KEY TO THE FAMILIES OF EPHEMEROPTERA LARVAE

1. Thoracic notum enlarged to form a shield extended to abdominal segment 6, gills enclosed beneath shield (fig. 9.3) .... ***Baetiscidae*** (p. 93)

1′. Thoracic notum not enlarged as above; abdominal gills exposed 2

2(1′). Gills on abdominal segments 2–7 forked, with margins fringed (fig. 9.4); mandibles with large tusks projected forward and visible from above head (figs. 9.5–9.7), or head and pronotum with pads of spines on each side (fig. 9.8) 3

2′. Gills on abdominal segments 2–5, 2–6, or 2–7 variable; if gills forked, margins not fringed; tusks rarely present (figs. 9.19–9.25), head and prothorax without pads of spines 7

3(2). Head and prothorax with dorsal pads of long spines on each side (fig. 9.8); without mandibular tusks; gills ventral; North Carolina to Florida and Louisiana .... ***Behningiidae*** (p. 88)

3′. Head and prothorax without pads of spines; mandibular tusks conspicuous (fig. 9.3); gills lateral or dorsal 4

4(3′). Fore tibiae and tarsi cylindrical, unmodified (fig. 9.5); abdominal gills held laterally; Texas, Florida, north to Manitoba .... ***Potamanthidae*** (p. 88)

4′. Fore tibiae and tarsi more or less flattened, adapted for burrowing (fig. 9.9); abdominal gills held dorsally 5

5(4′). Ventral apex of hind tibiae projected into distinct acute point (fig. 9.11); mandibular tusks curved upward apically as viewed laterally (fig. 9.6) 6

5′. Ventral apex of hind tibiae rounded (fig. 9.10); mandibular tusks curved downward apically as viewed laterally (fig. 9.7); widespread .... ***Polymitarcyidae*** (p. 89)

6(5). Mandibular tusks with a distinct dorsolateral keel that is more or less toothed, with a line of spurs along the toothed edge (fig. 9.9); processes on front of head as in figure 9.9; central North America, Texas, and Florida north to Manitoba .... ***Palingeniidae*** (p. 90)

6′. Mandibular tusks more or less circular in cross section, without a distinct toothed keel (fig. 9.6); processes on front of head variable; widespread .... ***Ephemeridae*** (p. 89)

7(2′). Fore legs with double row of long setae on inner surface (figs. 9.12, 9.13); gills present at base of maxillae (fig. 9.12) .... ***Oligoneuriidae*** (p. 86) 8

7′. Fore legs with setae other than above; gills absent from bases of maxillae 9

8(7). Gills on abdominal segment 1 ventral (fig. 9.12); gills on abdominal segments 4–6 slender and lanceolate or about one-half as long as abdominal terga; widespread .... **Oligoneuriinae**

8′. Gills on abdominal segment 1 dorsal; gills on abdominal segments 4–6 as long or longer than abdominal terga; widespread .... **Isonychiinae**

9(7′). Gills on abdominal segment 2 (gill 2) operculate or semioperculate, covering succeeding pairs (fig. 9.14) 10

9′. Gills on abdominal segment 2 neither operculate nor semioperculate, either similar to those on succeeding segments or absent 12

10(9). Gills on abdominal segment 2 triangular, semitriangular, or oval, not meeting medially (fig. 9.14); gill lamellae on segments 3–6 simple or bilobed, without margins fringed; widespread .... ***Tricorythidae*** (p. 91)

10′. Gills on abdominal segment 2 quadrate (fig. 9.15a) meeting or almost meeting medially, gill lamellae on segments 3–6 with margins fringed (fig. 9.15b) 11

11(10′). Operculate gills fused medially; mesonotum with distinct rounded lobe on anterolateral corners (fig. 9.16); developing hind wing pads present; eastern Canada west to Michigan and south to Florida .... ***Neoephemeridae*** (p. 92)

11′. Operculate gills not fused medially; mesonotum without anterolateral lobes (fig. 9.17); developing hind wing pads absent; widespread .... ***Caenidae*** (p. 92)

12(9′). Gills on segments 3–7 or 4–7 consisting of anterior (dorsal) oval lamella and posterior (ventral) lamella with numerous lobes (fig. 9.18); gills absent on abdominal segment 2, rudimentary or absent on segment 1, and present or absent on segment 3; paired tubercles often present on abdominal terga (fig. 9.19); widespread .... ***Ephemerellidae*** (p. 90)

12′. Gills variable, not as above; gills present on abdominal segments 1–5, 1–7, or 2–7; paired tubercles rarely present on abdominal terga 13

13(12′). Larvae distinctly flattened; head prognathous, eyes and antennae dorsal (figs. 9.20, 9.21). [Note: Some Leptophlebiidae appear to be intermediate in flattening, but either half of this couplet leads to Leptophlebiidae.] 14

13′. Larvae not flattened, or only slightly flattened, being more cylindrical; head hypognathous, eyes or antennae, or both, lateral, anterolateral, or on front of head (figs. 9.22–9.25) 15

14(13). Mandibles concealed beneath flattened head capsule (Fig. 9.20); labial palpi two-segmented; labrum rarely visible; abdominal gills of single lamellae, usually with ventral fibrilliform tuft at base, rarely pointed or with narrow lanceolate ventral branch; gills sometimes ventral .... ***Heptageniidae*** (p. 86)

14′. Mandibles visible and forming part of upper surface of head (fig. 9.21); labial palpi three-segmented; labrum conspicuously visible; abdominal gills either forked (fig. 9.28), formed of lamellae with margins fringed (fig. 9.30) or terminated in filaments or points (figs. 9.31, 9.32), sometimes appearing as a tuft (fig. 9.29); gills never ventral .... (in part) ***Leptophlebiidae*** (p. 87)

15(13′). Claws of fore legs differing in structure from those on middle and hind legs (figs. 9.26, 9.27); claws of middle and hind legs long and slender, about as long as tibiae (figs. 9.26, 9.27) 16

15′. Claws of all legs similar in structure, usually sharply pointed, rarely spatulate; claws variable in length, if those of middle and hind legs long and slender, then usually shorter than tibiae (longer than tibiae in three rare genera) 17

16(15). Claws on fore legs simple, with long slender denticles; spinous pad present on fore coxae (fig. 9.26); western North America .... ***Ametropodidae*** (p. 84)

16′. Claws on fore legs bifid; without spinous pad on fore coxae (fig. 9.27); Alaska and Canada, midwestern and eastern United States .... ***Metretopodidae*** (p. 85)

17(15′). Abdominal gills on segments 2–7 either forked (fig. 9.28), in tufts (fig. 9.29), with all margins fringed (fig. 9.30), or with double lamellae terminated in filaments or points (figs. 9.31, 9.32) .... (in part) ***Leptophlebiidae*** (p. 87)

17′. Abdominal gills not as above; gills usually single and more or less ovate (figs. 9.33, 9.34) cordate (heartshaped) or subcordate, lamellae may be double or triple (figs. 9.35, 9.36); inner margin of gills usually entire, rarely finely dissected (fig. 9.37) 18

18(17′). Posterolateral projections present and prominent on abdominal segments 8–9 .... (in part) ***Siphlonuridae*** (p. 84)

18′. Posterolateral projections absent or small on abdominal segments 8–9 19

19(18′). Antennae short, about one and one-half times the width of head (figs. 9.24, 9.25); maxillae with broad apical crown of pectinate spines (fig. 9.25), gills obovate, with sclerotized band along outer margin and with similar band near or on inner margin (figs. 9.42, 9.43) .... (in part) ***Siphlonuridae*** (p. 84)

19′. Antennae usually over twice width of head (fig. 9.23); maxillae narrow at apex, without pectinate spines; gills other than above, but margins may be sclerotized .... ***Baetidae*** (p. 85)

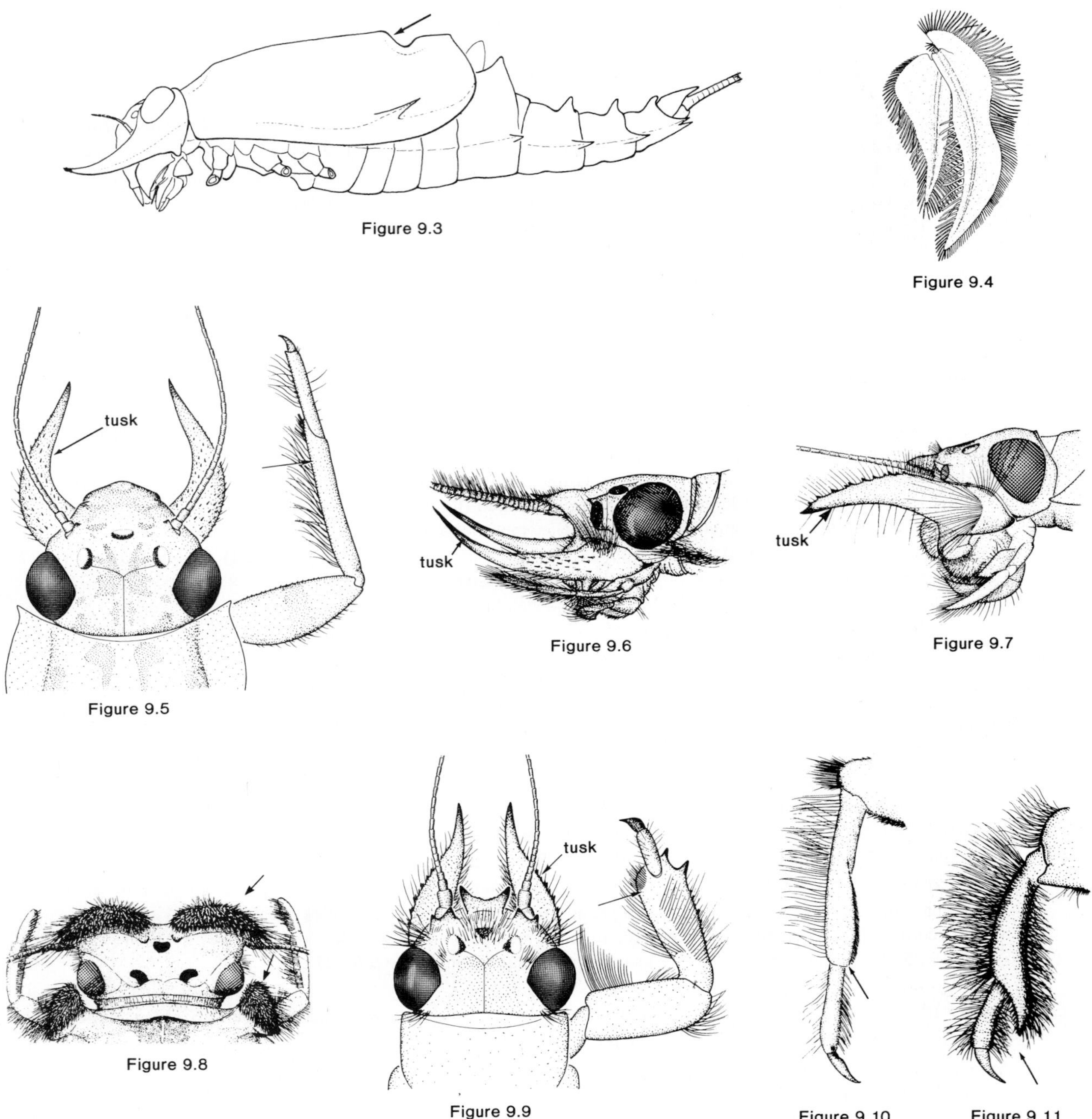

Figure 9.3. Lateral view of *Baetisca* (Baetiscidae) (from Edmunds, Jensen, and Berner 1976).

Figure 9.4. Gill 4, *Ephoron album* (Polymitarcyidae).

Figure 9.5. Head and fore leg, *Potamanthus* (Potamanthidae).

Figure 9.6. Head, *Ephemera* (Ephemeridae).

Figure 9.7. Head, *Ephoron* (Polymitarcyidae).

Figure 9.8. Head and prothorax, *Dolania* (Behningiidae).

Figure 9.9. Head and fore leg of *Pentagenia* (Palingeniidae).

Figure 9.10. Hind tibia and tarsus, *Ephoron* (Polymitarcyidae).

Figure 9.11. Hind tibia and tarsus, *Hexagenia* (Ephemeridae).

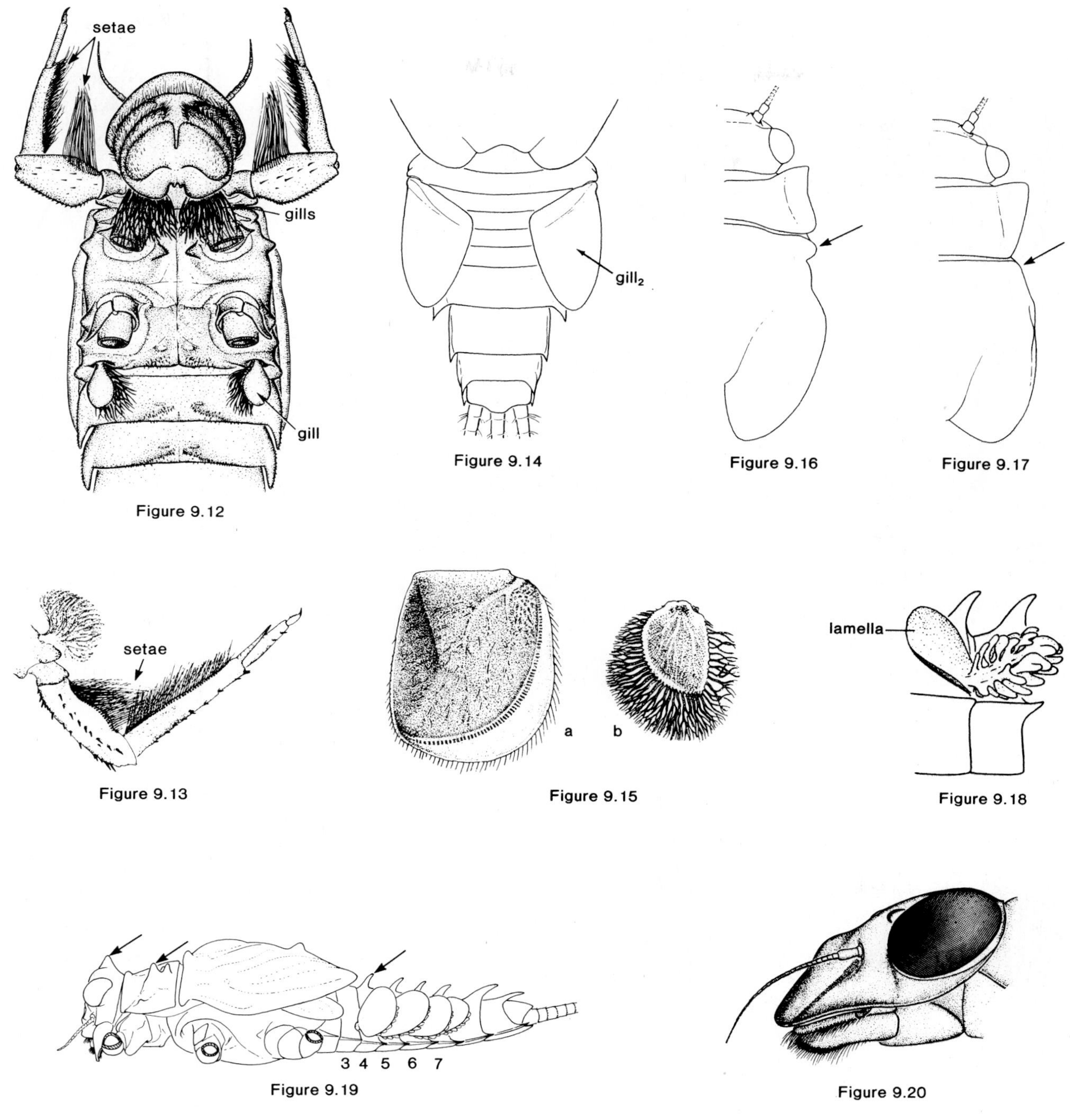

Figure 9.12. Ventral view, anterior half, *Lachlania* (Oligoneuriidae).

Figure 9.13. Fore leg *Isonychia* (Oligoneuriidae).

Figure 9.14. Abdomen with operculate gills on segment 2, *Tricorythodes* (Tricorythidae).

Figure 9.15a,b. Gills, segment 2 (9.15a) and 4 (9.15b), *Caenis* (Caenidae).

Figure 9.16. Head and thorax, *Neoephemera* (Neoephemeridae).

Figure 9.17. Head and thorax, *Caenis* (Caenidae).

Figure 9.18. Gill with lamella raised, *Drunella* (Ephemerellidae).

Figure 9.19. Lateral view, body, *Drunella* (Ephemerellidae).

Figure 9.20. Lateral view, head, *Heptagenia* (Heptageniidae).

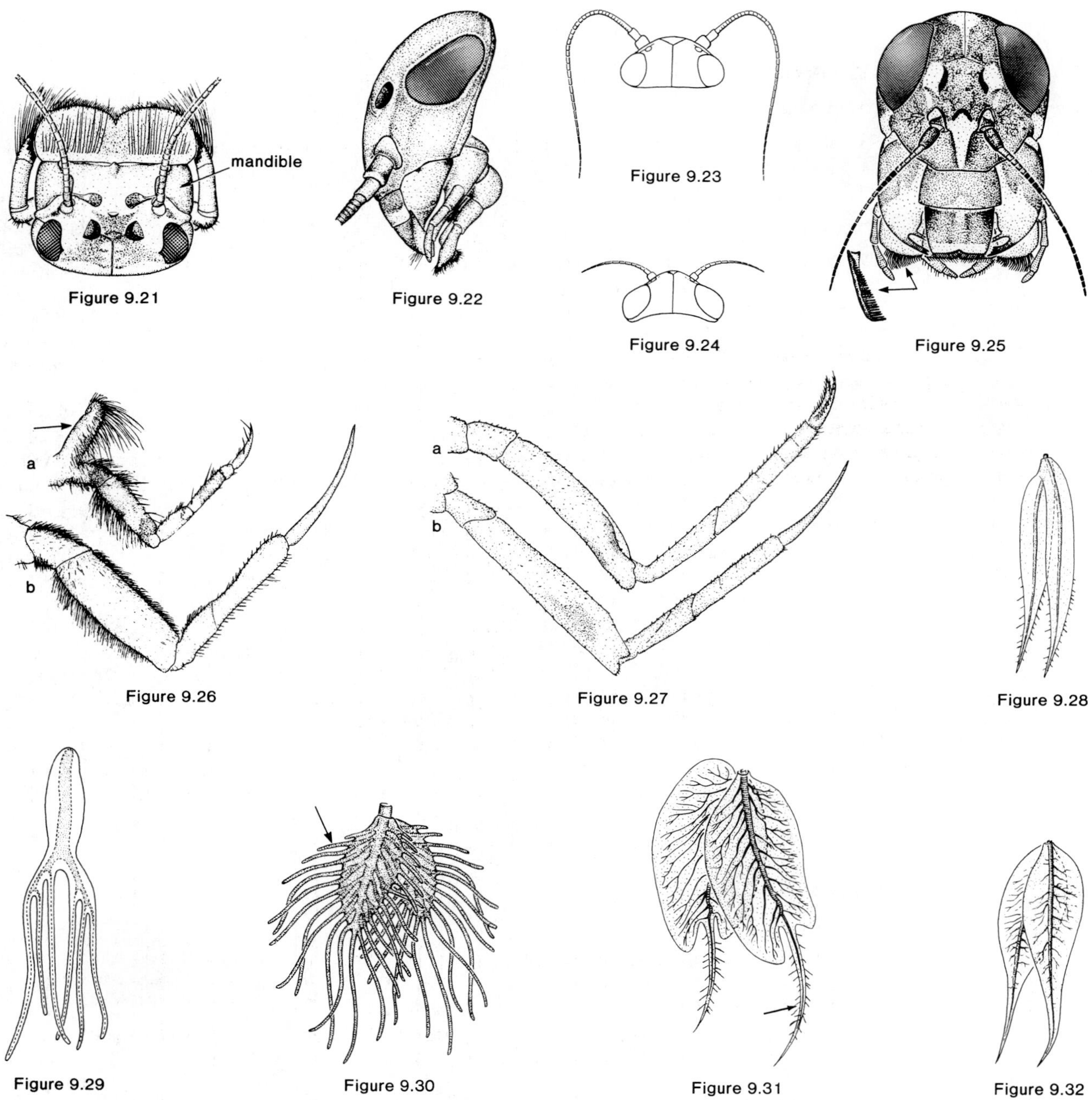

Figure 9.21. Dorsal view, head, *Traverella* (Leptophlebiidae).

Figure 9.22. Lateral view, head, *Baetis* (Baetidae).

Figure 9.23. Dorsal view, head, *Baetis* (Baetidae).

Figure 9.24. Dorsal view, head, *Siphlonurus* (Siphlonuridae).

Figure 9.25. Head, anterior view, *Ameletus,* showing pectinate spines of maxillae (Siphlonuridae).

Figure 9.26. Fore leg (a), hind leg (b) of *Ametropus* (Ametropodidae).

Figure 9.27. Fore leg (a), hind leg (b), *Siphloplecton* (Metretopodidae).

Figure 9.28. Gill 4, *Paraleptophlebia* (Leptophebiidae).

Figure 9.29. Gill 4, *Habrophlebia* (Leptophlebiidae).

Figure 9.30. Gill 4, *Traverella* (Leptophlebiidae).

Figure 9.31. Gill 4, *Leptophlebia* (Leptophlebiidae).

Figure 9.32. Gill 4, *Thraulodes* (Leptophlebiidae).

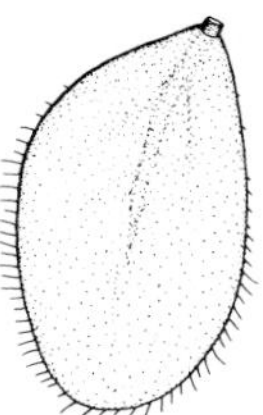
Figure 9.33

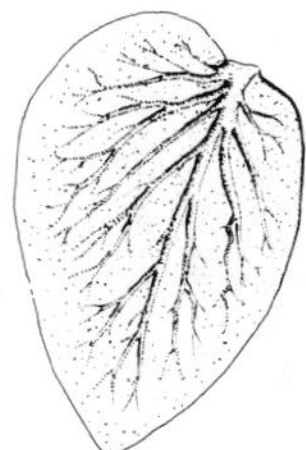
Figure 9.34

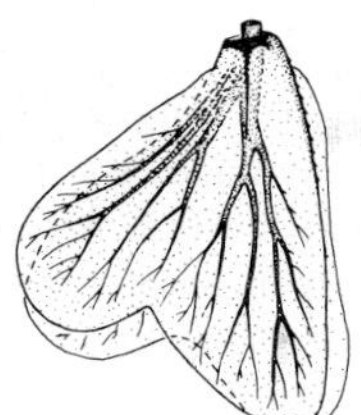
Figure 9.35

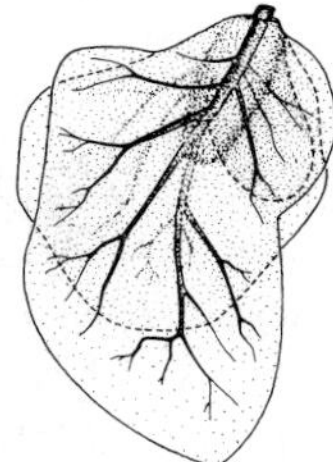
Figure 9.36

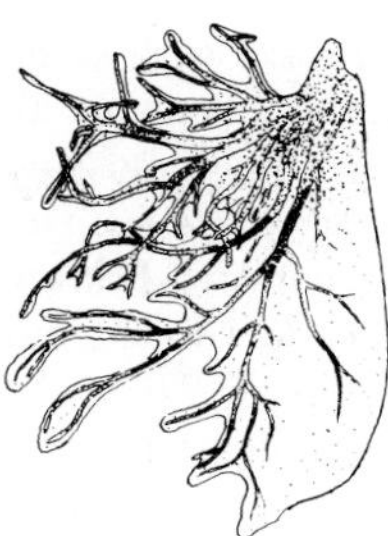
Figure 9.37

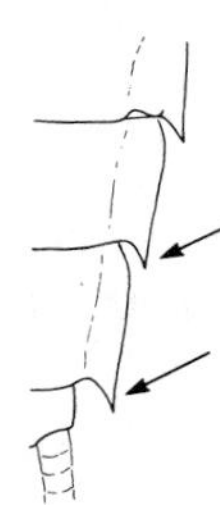
Figure 9.38

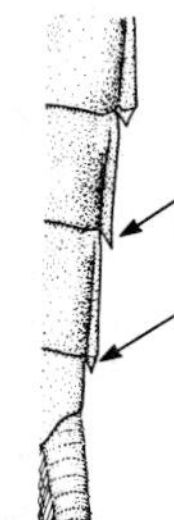
Figure 9.39

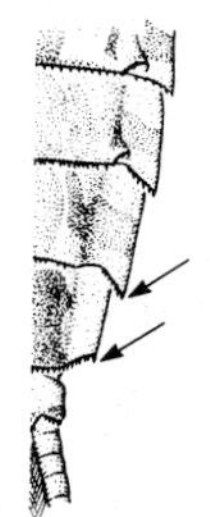
Figure 9.40

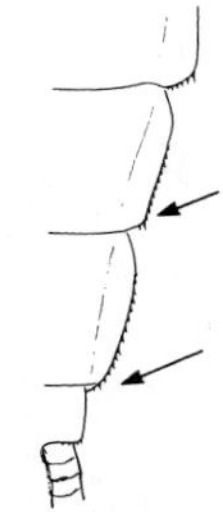
Figure 9.41

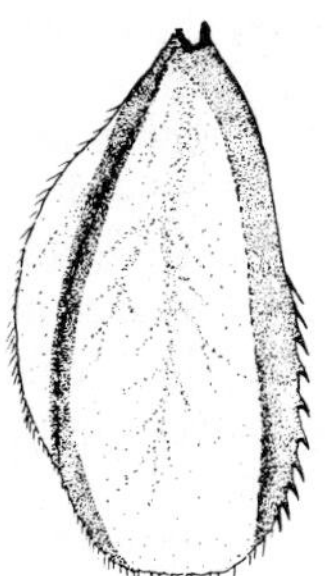
Figure 9.42

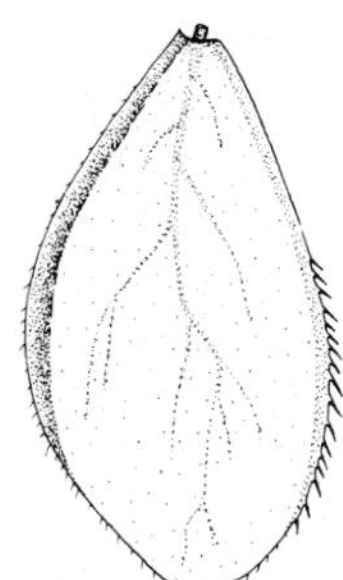
Figure 9.43

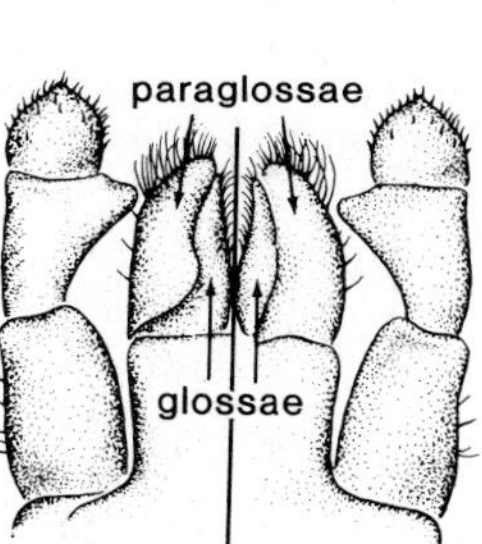

Figure 9.44

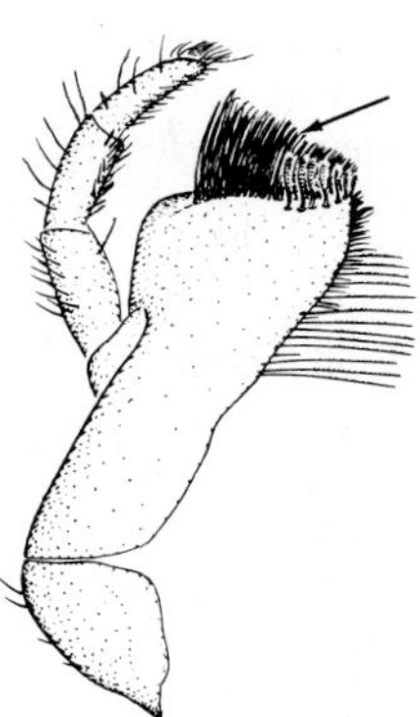
Figure 9.45

Figure 9.33. Gill 4, *Baetis* (Baetidae).

Figure 9.34. Gill 4,*Siphlonisca* (Siphlonuridae).

Figure 9.35. Gill 2, *Siphlonurus* (Siphlonuridae).

Figure 9.36. Gill 4, *Callibaetis* (Baetidae).

Figure 9.37. Gill 4, *Acanthametropus* (Siphlonuridae).

Figure 9.38. Abdominal segments 8-10, *Siphlonurus* (Siphlonuridae).

Figure 9.39. Abdominal segments 8-10, *Ameletus* (Siphlonuridae).

Figure 9.40. Abdominal segments 7-10, *Callibaetis* (Baetidae).

Figure 9.41. Abdominal segments 7-10, *Cloeon* (Baetidae).

Figure 9.42. Gill 4, *Ameletus* (Siphlonuridae).

Figure 9.43. Gill 4, *Ameletus* (Siphlonuridae).

Figure 9.44. Labium, *Baetis* (Baetidae).

Figure 9.45. Maxilla, *Leptophlebia* (Leptophlebiidae).

SUBORDER SCHISTONOTA

The wing pads of mature larvae are fused mesally to the mesonotum on their mesal edges for only one-eighth to less than one-half their length. Most larvae are quite active. Some are minnowlike strong swimmers, and others are active crawlers. The flattened Leptophlebiidae like *Traverella* and the Heptageniidae are also active. Slow-moving crawlers like *Lachlania* are rare in the group. Many larvae are burrowers, either as very young forms or throughout their larval development. There are three superfamilies in the suborder, Baetoidea (=Heptagenioidea), Leptophlebioidea, and Ephemeroidea.

SIPHLONURIDAE (BAETOIDEA)

The Siphlonurids, Primitive Minnow Mayflies

Figures 9.24, 9.25, 9.34, 9.35, 9.37–39, 9.42, 9.43, 9.46

Relationships and Diagnosis: The Siphlonuridae are the most primitive extant family of mayflies. All Baetoidea, except possibly the Ametropodidae, can be traced to a proto-siphlonuroid origin. The larvae are streamlined and minnowlike with the caudal filaments forming a paddle. They are distinguished from other minnowlike mayflies by the simple fore claws (not bifid or with 4–5 long spines), short antennae (figs. 9.24, 9.25) (less than twice as long as width of head), and with well-developed posterolateral projections on segments 8–9 (fig. 9.38) (except in *Ameletus* (fig. 9.39) whose maxillae (fig. 9.25) and gills (figs. 9.42, 9.43) are distinctive). The gills are usually single, more or less oval plates (fig. 9.34), but may be double (fig. 9.35) on segments 1–2 or 1–7 (and rarely of three lamellae) with the inner margin dissected in one rare genus (fig. 9.37).

Biology and Ecology: The larvae inhabit a wide variety of aquatic habitats from large silted rivers to tiny rivulets, ponds, and swamps. Most are general feeders, gathering plant, animal, and detrital food. At least one genus appears to feed primarily by scraping rocks and other surfaces. The rare Acanthametropodinae are carnivores of chironomid larvae. The larvae crawl from the water before the subimago emerges.

Several types of eggs are known (described by Smith 1935, Koss 1968, and Koss and Edmunds 1974).

Description: Mature larvae (fig. 9.46) small to large, 6–20 mm. Body cylindrical to moderately flattened, streamlined, minnowlike.

Head: Hypognathous, grasshopperlike. Maxillary palpi usually three-segmented, absent in *Acanthametropus*. Mandible usually with molar area, absent in Acanthametropodinae. Labial palpi three-segmented.

Thorax: Claws similar on all legs, sometimes longer and more slender on T2 and T3 legs.

Abdomen: Terga without dorsal tubercles, except *Acanthametropus* with single median hooklike tubercle on each tergum. Gills usually oval, entire, on segments 1–7; usually with one lamella, but all gills, or anterior two pairs, may have two or three lamellae. Segments 8–9 with prominent posterolateral projections, except in *Ameletus*. Three caudal filaments with long setae, thus forming a paddle; terminal filament with lateral long setae on both sides, cerci with row on inner side only.

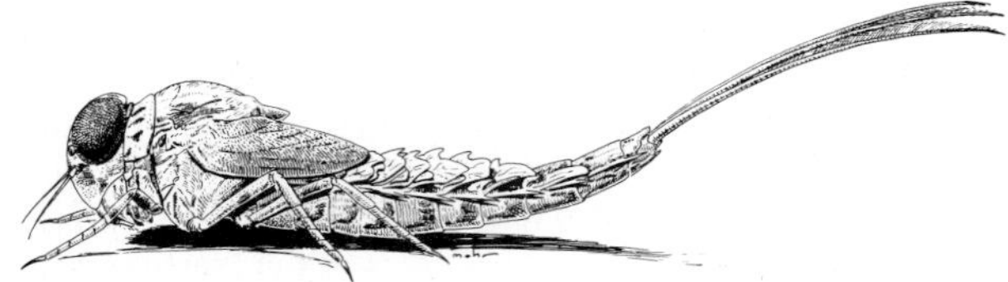

Figure 9.46. Siphlonuridae. *Siphlonurus*, from Burks (1953).

Comments: The family is of moderate size with seven genera and 56 species in North America. Edmunds et al. (1976) key the genera and list the species, and Edmunds and Koss (1972) reviewed *Acanthametropus* and *Analetris*. Demoulin (1974) placed these last-named genera in separate subfamilies (Acanthametropodinae and Analetridinae) but this change is not accepted by Edmunds et al. (1976). The family is widespread in cooler waters of North America; it is known as far south as Georgia, Missouri, Arizona, and California.

AMETROPODIDAE (BAETOIDEA)

The Ametropodids, Sand Minnow Mayflies

Figure 9.26

Relationships and Diagnosis: The Ametropodidae are Baetoidea that diverged early and are isolated within the superfamily. The larvae are distinctive, because the claws of the T1 legs are short, slender, and gently curved, and bear 4–5 long spines; spinous pads are present at the base of the coxae, and the T2 and T3 legs have long tarsal claws, subequal in length to the tarsi and twice as long as the tibiae (fig. 9.26).

Biology and Ecology: Larvae bury to the gills in firm, slightly silty sand, usually in large rivers, but rarely in small to medium sandy-bottomed rivers. The percentage of suitable bottom in a river is small and larvae are difficult to locate, even in rivers where they are known to occur. The larvae swim rapidly with the T2 and T3 legs trailing to the side of the body. The T1 legs are used primarily for grooming the head. The larvae feed principally on algae and other microbiota in the sand.

Eggs were described by Koss and Edmunds (1974); they have a single polar cap with many long, compactly coiled threads.

Description: Mature larvae medium to large, 11–24 mm. Body flattened.

Head: Hypognathous, grasshopperlike. Ridge between antennae. Eyes distinctly anterolateral. Mouthparts distinctive, maxillary and labial palpi three-segmented.

Thorax: T1 legs short, claws long and gently curved, with 4–5 long spines (fig. 9.26). T2 and T3 legs long; tibiae about half length of tarsi. Claws long and slender, subequal to tarsi.

Abdomen: Gills on segments 1–7, ovate, margined with long setae. Three caudal filaments, paddlelike with long setae, terminal filament with lateral rows, cerci with single mesal row.

Comments: This is a small family with one genus and three species in North America. Larvae of the North American species are keyed and figured by Allen and Edmunds (1976). *Ametropus* is found from British Columbia and Saskatchewan south to central California, New Mexico, and Colorado. See habitus figure 396 in Edmunds et al. (1976).

BAETIDAE (BAETOIDEA)

The Baetids, Small Minnow Mayflies

Figures 9.22, 9.23, 9.33, 9.36, 9.40, 9.41, 9.44, 9.47

Relationships and Diagnosis: The nearest relative of the Baetidae is the monobasic Siphlaenigmatidae of New Zealand, but baetids are also closely related to the siphlonurine Siphlonuridae. Larvae are small, streamlined, and usually minnowlike. Antennae are usually two or more times as long as width of head, and the claws are similar on all legs. The long narrow labial glossae and paraglossae (fig. 9.44) are diagnostic. Abdominal segments 8–9 usually without posterolateral projections (fig. 9.41); small to moderate projections present on some species (fig. 9.40).

Biology and Ecology: Most Baetidae are inhabitants of rivers and streams; some live in fast to moderate currents (most *Baetis, Pseudocloeon, Apobaetis, Baetodes, Dactylobaetis,* and *Heterocloeon*) whereas others are found frequently in slack current areas (*Centroptilum, Paracloeodes,* and *Cloeon*). Others (*Callibaetis,* some *Cloeon,* and *Centroptilum*) are found in lakes and ponds. All North American species are detritivores or herbivores, feeding largely as gatherers of detritus, scraping plant material from the surfaces of stones and higher aquatic plants, or feeding on filamentous algae. The larvae are, with few exceptions, able to swim rapidly, most making short dashes and then coming to rest; this is clearly a predator avoidance mechanism. *Baetodes* larvae are sprawlers on rocks, move slowly, and are feeble swimmers.

The eggs are described by Smith (1935), Koss (1968), and Koss and Edmunds (1974). They have tagenoform micropyles located midway between the equator and one pole. Some *Baetis* crawl underwater to lay their eggs and deposit them in distinct rows.

Description: Mature larvae (fig. 9.47) small to large, 3–12 mm. Body cylindrical to flattened, usually streamlined and minnowlike, moderately flattened in *Pseudocloeon* and *Baetodes.*

Head: Hypognathous, grasshopperlike (fig. 9.22). Antennae long, usually 2–3 times as long as width of head, labial glossae and paraglossae long and narrow (fig. 9.44).

Thorax: Legs usually long, slender; claws with or without denticles. Leg bases with fingerlike gills in *Heterocloeon,* some *Dactylobaetis.*

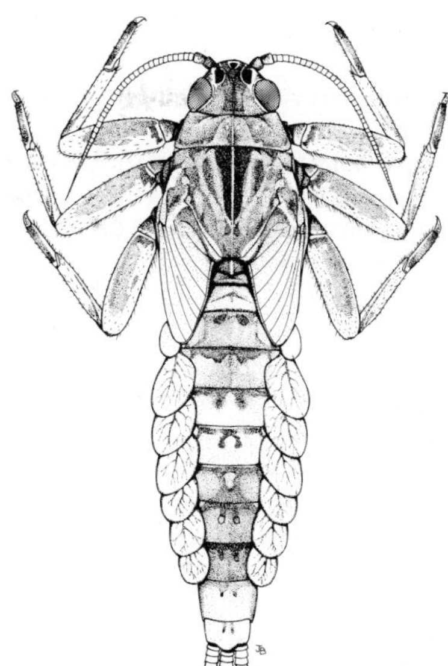

Figure 9.47. Baetidae. *Baetis.*

Abdomen: Gills usually on segments 1–7 or 2–7, 1–5 in *Baetodes;* gills variable (figs. 9.33, 9.36) usually dorsolateral, ventrolateral in *Baetodes.* Segments 8–9 without posterolateral projections or with small to moderate projections. Caudal filaments 2 or 3, glabrous to setaceous, nearly glabrous in *Baetodes* to paddlelike in *Callibaetis* and *Cloeon,* with long setae on both sides of terminal filament and inner side of cerci.

Comments: This is a relatively large family with 146 species and ten genera in America north of Mexico. The genera have been keyed and species listed by Edmunds et al. (1976). Species identifications of larvae are difficult for most genera. Müller-Liebenau (1974) keyed the larvae of *Heterocloeon* (as *Rheobaetis*), Traver and Edmunds (1968) keyed the known *Dactylobaetis* larvae, Cohen and Allen (1978) keyed the larvae of *Baetodes,* and Morihara and McCafferty (1979) keyed and revised the larvae of *Baetis.* Generic identifications may be difficult for some species from southwestern United States. The family is extremely widespread and is known from higher altitudes and latitudes than any other mayfly family.

METRETOPODIDAE (BAETOIDEA)

The Metretopodids, Cleftfooted Minnow Mayflies

Figure 9.27

Relationships and Diagnosis: The metretopodids presumably are derived from protosiphlonurine Siphlonuridae. Larvae are streamlined and minnowlike. They are unique in that the tarsal claws of the T1 legs are bifid (fig. 9.27).

Biology and Ecology: Larvae are found principally in slow-flowing water among marginal vegetation, but others occur in moderately strong current. *Siphloplecton* larvae have been collected by deep dredging in Lake Superior. The larvae are active, strong swimmers and difficult to collect by standard methods. Clifford (1976) has shown that results using emergence traps suggest that populations are larger than collections indicate.

The eggs are described by Koss (1968) and Koss and Edmunds (1974), and they differ in the two genera.

Description: Mature larvae medium, 9–16 mm. Body subcylindrical to moderately flattened.

Head: Hypognathous, grasshopperlike. Maxillary and labial palpi three-segmented.

Thorax: Tarsi longer than tibiae. Claws of T1 legs bifid, claws of T2 and T3 legs slender, longer than tibiae (fig. 9.27).

Abdomen: Gills single on segments 1–7; gills on 1–2 or 1–3 with recurved ventral flaps in some *Siphloplecton.* Three caudal filaments paddlelike, terminal filament with long setae on both sides, cerci with long setae on inner side.

Comments: This is a small family in North America with only six species in two genera. Keys for distinguishing the larvae of the genera are inadequate except for Berner (1978). The larvae of *Metretopus* and *Siphloplecton* have weak morphological differences, but the adults differ strikingly in wing venation. The family is known from Alaska, across Canada, and south to Michigan, Illinois, and Florida. See habitus figure 395 in Edmunds et al. (1976).

OLIGONEURIIDAE (BAETOIDEA)

The Oligoneuriids, Brushlegged Mayflies

Figures 9.12, 9.13, 9.48

Relationships and Diagnosis: The Oligoneuriidae are derived from siphlonuridlike ancestors. The double row of long setae on the fore legs of the larvae is diagnostic (figs. 9.12, 9.13). Two body forms are found, minnowlike (*Isonychia,* and the pale burrowing *Homoeoneuria*) or flattened and slow moving (*Lachlania*).

The Isonychiinae have traditionally been placed in the Siphlonuridae but it has long been recognized that proto-*Isonychia* gave rise to the Oligoneuriidae. Riek (1973) placed *Isonychia* within the Oligoneuriidae and we accept that classification. The decision to place interstitial groups with their derived relatives rather than with their ancestral-type relatives was explained by McCafferty and Edmunds (1976). The placement is further justified by observations on the behavior of *Elassoneuria* (*Madeconeuria*) *insulicola* in Madagascar. The larvae of this species are extremely *Isonychia*-like, being strong, active minnowlike swimmers. The North American Oligoneuriinae represent endpoints of two extreme specialized oligoneuriid lineages.

Biology and Ecology: Larvae are largely filter feeders. The three North American genera are strikingly different in larval ecology. *Isonychia* (Isonychiinae) usually occur on tangles of vegetation and debris, especially on branches and clusters of fallen leaves, but also on leaves and in gravel where there is strong-flowing water. *Lachlania* (Oligoneuriinae) larvae cling to twigs and other items, occasionally under rocks, in current where they filter feed. Submerged twigs in rapid current are often literally covered with several layers of larvae.

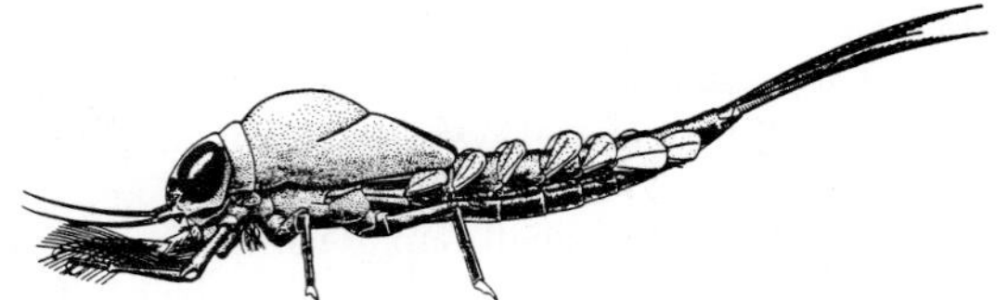

Figure 9.48. Oligoneuridae. *Isonychia,* from Needham (1920).

The robust larvae move slowly, and disturbed individuals frequently curl the tip of the abdomen upward. The pale larvae of *Homoeoneuria* (Oligoneuriinae) burrow beneath the sand surface where they filter food from the water in the interstices of the sand.

The eggs, described by Koss and Edmunds (1974), are similar to those of *Siphlonurus* (Siphlonuridae); the chorion is smooth, polar caps are absent, and a dense orderly layer of coiled fibers is over the egg. *Isonychia* eggs are spherical.

Description: Mature larvae medium, 8–17 mm. Body minnowlike in *Isonychia;* moderately flattened in *Lachlania;* cylindrical, elongate, and slender in *Homoeoneuria.*

Head: Hypognathous, grasshopperlike in *Isonychia,* moderately flattened in *Lachlania,* subspherical in *Homoeoneuria.* Maxillary and labial palpi two-segmented, gill tuft at maxillae bases, and at base of fore legs in *Isonychia.*

Thorax: Femora and tibiae of T1 legs with double rows of long setae on inner margins (figs. 9.12, 9.13).

Abdomen: Abdominal segments with posterolaterally projecting flanges in *Isonychia* and *Lachlania,* lateral margins appearing serrated in *Lachlania;* abdomen cylindrical in *Homoeoneuria.* Gills dorsal on segment 1 in *Isonychia;* ventral on segment 1 in Oligoneuriinae (fig. 9.13), dorsal in Isonychiinae, with fibrilliform tuft; gills on segments 2–7 slender lamellae in *Homoeoneuria,* small lamellae and fibrilliform tuft in *Lachlania* and *Isonychia.* Caudal filaments two in *Lachlania,* three in *Isonychia* and *Homoeoneuria.*

Comments: This is a moderate-sized family in North America with three genera and 30 species north of Mexico. The Oligoneuriinae were treated by Edmunds et al. (1958), larvae of the North America species of *Lachlania* were keyed by Koss and Edmunds (1970), and the larvae of *Homoeoneuria* were revised and keyed by Pescador and Peters (1980). The Oligoneuriinae represent North American extensions of neotropical genera as far north and east as Saskatchewan, Indiana, and South Carolina. The Isonychiinae are holarctic and extend south to Honduras. *Isonychia* larvae were keyed by Kondratieff and Voshell (1984).

HEPTAGENIIDAE (BAETOIDEA)

The Heptageniids, Flatheaded Mayflies

Figures 9.20, 9.49

Relationships and Diagnosis: The Heptageniidae are most closely related to the isonychiine Oligoneuriidae. Larvae are depressed, most of them strongly so, with flattened femora that are appressed to the substrate. The flat head capsule hides all mouthparts (fig. 9.20), except the labrum may show in a few. In *Arthroplea,* the only genus where the head

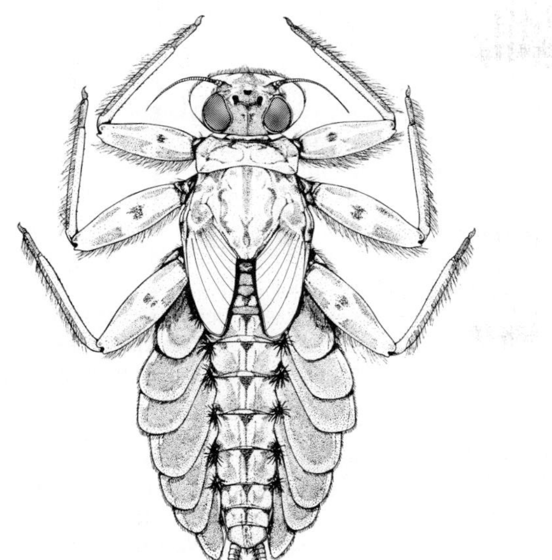

Figure 9.49. Heptageniidae. *Epeorus.*

is not strongly depressed, the very long maxillary palpi armed with long setae are visible at the sides or behind the head capsule.

Biology and Ecology: The Heptageniinae are flattened stream-inhabiting mayflies. They occur on rocks, wood, and vegetation, or in and on packs of leaves. Most larvae are capable of rapid movement; they scurry forward, backward, or laterally with equal facility. Many are adapted to cool water, but a number of *Stenacron, Stenonema, Heptagenia,* and some *Rhithrogena* are found in warmer waters. They are generally poor swimmers. Most feed by collecting detritus or scraping attached periphyton.

Arthroplea (Arthropleinae) inhabits ponds or slow current, and feeds by sweeping up algae, detritus, and small animals. The continuous motion of the palpi produces a vortex reminiscent of feeding rotifers or mosquito larvae. *Pseudiron* (Pseudironinae) are found in large rivers, are sand dwellers, strong swimmers, and carnivorous. *Anepeorus* (Anepeorinae) are found in rubble or gravel, and the larvae are rapidly moving active carnivores. *Spinadis* (Spinadinae) are known only from large rivers, and the larvae are carnivorous.

The rather diverse eggs are described by Smith (1935), Koss (1968), and Koss and Edmunds (1974).

Description: Mature larvae small to large, 5–20 mm. Body strongly to moderately depressed.

Head: Prognathous, strongly to moderately depressed, usually broader than long. Mandibles not visible in dorsal view; eyes and antennae dorsal. Maxillary and labial palpi two-segmented (except maxillary palpi three-segmented in *Pseudiron*).

Thorax: More or less depressed. Legs flat, modified to press flat against substrate; femora broad and flat.

Abdomen: Gills present on segments 1–7, usually with lamella and fibrilliform tuft. Tuft reduced or absent in *Cinygmula,* absent in *Arthroplea.* In *Pseudiron,* an auxiliary fingerlike appendage arises from each slender lamella. Gills usually inserted posterolaterally; gills ventral in *Anepeorus,* 1–2 ventral in *Spinadis,* 1 and 7 extend ventrally in *Rhithrogena* and some *Epeorus.* Three caudal filaments, except two in *Epeorus, Ironodes,* and *Spinadis;* caudal filaments with only short setae.

Comments: This family is moderately large in North America with 15 genera and 132 species. Edmunds et al. (1976) keyed the genera and listed species, and existing keys to larvae are those of Traver (1935) and Burks (1953). The larvae of *Stenacron* (as *Stenonema interpunctatum*-group) are keyed by Lewis (1974). Bednarik and McCafferty (1979) keyed the larvae and revised *Stenonema,* and Edmunds and Allen (1966) keyed the Rocky Mountain species of *Epeorus.* The family is widespread and abundant in North America, and four genera, *Epeorus, Heptagenia, Rhithrogena,* and *Stenonema,* occur as far south as southern Mexico and Central America.

LEPTOPHLEBIIDAE (LEPTOPHLEBIOIDEA)

The Leptophlebiids, Pronggilled Mayflies

Figures 9.21, 9.28–9.32, 9.45, 9.50

Relationships and Diagnosis: The Leptophlebiidae are a distinctive family that originated early and appear to have given rise to the entire Pannota and the Ephemeroidea. Despite the essential unity of this family and characters recognizable to persons familiar with mayflies, it is fairly difficult to characterize the family (even in North America) on characters other than mouthparts. The maxillae, with their broad apices with a dense brush of setae (fig. 9.45), are diagnostic of the family. Both gills and head shape vary from one genus to another. The gills of the middle segments of most genera are either forked (fig. 9.28) or of two lamellae, each of which ends in single (fig. 9.32) or multiple (fig. 9.31) points; in *Habrophlebia* the gills appear to be a tuft of fingerlike projections (fig. 9.29). The head varies from grasshopperlike to flattened; if the head is strongly flattened the mandibles form part of the upper surface (Fig. 9.21), and the labium is usually conspicuous at the front of the head.

Biology and Ecology: Larvae occupy a wide range of habitats from fast-running streams to large silt-laden rivers, slow-flowing stream margins, and lakes and ponds. They are found on a variety of substrates, from rocks and leafpacks to tangles of debris and roots of exposed plants. Most larvae feed on detritus and algae, but a few larger species are omnivorous. Most genera are gatherers, but some scrape food from rocks, and *Traverella* filters food from passing water. The larvae are poor swimmers. They usually crawl on surfaces or through the substrate material, perhaps penetrating several feet into the sides and bottom of the stream.

The eggs are described by Smith (1935), Koss (1968), and Koss and Edmunds (1974). They lack polar caps and have a funnelform micropyle; the surface sculpture and attachment structures are extremely diverse.

Description: Mature larvae small to medium, 4–15 mm. Body subcylindrical to strongly depressed.

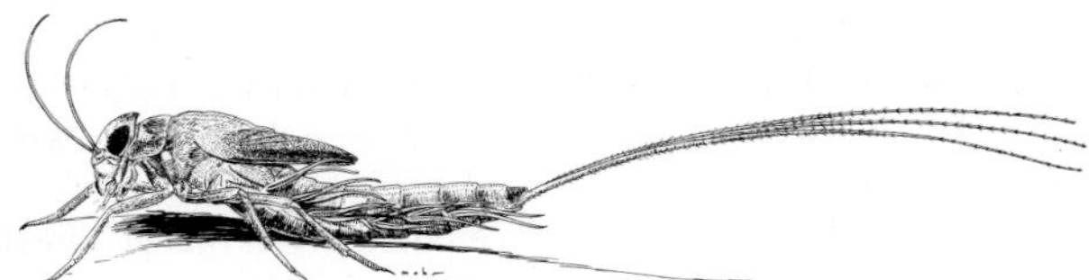

Figure 9.50. Leptophlebiidae. *Paraleptophlebia,* from Burks (1953).

Head: Hypognathous to prognathous, grasshopperlike to depressed. Mandibles visible in dorsal view if head depressed. Maxillae broad at apex with dense brush of setae (fig. 9.45). Maxillary and labial palpi three-segmented.

Thorax: Legs usually slender, femora often enlarged and flattened. Claws similar on all legs.

Abdomen: Gills forked, depth of fork varies from one-third to full length, often forming two separate lamellae (Fig. 9.28). Gills of segment one often reduced to single long, slender lamella; gills 2–7 variable, with a tuft of fingerlike lobes in *Habrophlebia* (fig. 9.29), broad lamellae in *Choroterpes, Leptophlebia* (fig. 9.31), *Thraulodes* (fig. 9.32), and *Traverella* (fig. 9.30). Gills acutely pointed in most *Thraulodes,* three lobes in *Leptophlebia* and *Choroterpes,* and each lamella fringed in *Traverella.* Three caudal filaments, whorls of setae at apex of each segment.

Comments: This is a large family with eight genera and 70 species in the region. It is diverse in America north of Mexico, as both southern and northern hemisphere lineages occur here. The genera are keyed and the species listed by Edmunds et al. (1976) except for the genus *Farrodes* (or a close ally) reported from Texas (but not keyed) as *Homothraulus.* Keys to the species are generally not available except those in standard works (Traver 1935, Berner 1950, Burks 1953). Allen (1973) has keyed the larvae of *Traverella,* and Allen and Brusca (1978) key the larvae of *Thraulodes.* The family is distributed in most of North America.

BEHNINGIIDAE (EPHEMEROIDEA)

The Behningiids, Tuskless Burrowing Mayflies

Figure 9.8

Relationships and Diagnosis: The Behningiidae appear to be derived as the earliest branch of the Ephemeroidea. The larvae are the most distinctive of all mayflies. A unique feature among mayflies are the paired crowns of spines on the anterior margin of the head and at the anterolateral corners of the pronotum (fig. 9.8). The fringed gills are ventral, gill 1 is single and longest, gills 2–7 bilamellate.

Biology and Ecology: Larvae live in large rivers or their tributaries in a substrate of deep loose sand. They are good swimmers and can burrow rapidly into the sand. They are carnivorous, feeding on chironomid and ceratopogonid larvae.

The eggs, described by Koss and Edmunds (1974), are ovoid, without polar caps, and very large (about 0.75 mm long).

Description: Mature larvae medium, about 13 mm. Body robust and subcylindrical.

Head: Antennae ventral. Prominent patch of spines on anterolateral margin (fig. 9.8). Mandibles small, without tusks. Maxillary and labial palpi large and prominent.

Thorax: Pronotum produced anterolaterally, covered with patch of spines and setae (fig. 9.8). Legs without tarsal claws. T1 legs palplike. T2 legs modified, tibiae and tarsi with spinous pads. Coxae of T3 legs enlarged, all segments bearing numerous spines and setae.

Abdomen: Segments with dense lateral setae. Gills ventral. Gills of segment 1 single, longer than others, gills 2–7 bilamellate, with fringed margins. Three caudal filaments, terminal filament with row of setae on each side, cerci with inner row only.

Comments: There is a single species in North America, *Dolania americana* Edmunds and Traver. Suitable habitat seems to be rare.

The species is known from North Carolina to Florida and Louisiana. A good population is found in the Blackwater River in northwestern Florida. See habitus figure 424 in Edmunds et al. (1976).

POTAMANTHIDAE (EPHEMEROIDEA)

The Potamanthids, Hacklegill Mayflies

Figure 9.5, 9.51

Relationships and Diagnosis: The Potamanthidae is a primitive family of Ephemeroidea; ancestral potamanthids probably gave rise to all other Ephemeroidea except Behningiidae. The combination of fringed gills, tusks, and the fore legs *not* modified for burrowing (fig. 9.5) is diagnostic.

Biology and Ecology: The Potamanthidae are the only North American Ephemeroidea that do not burrow as well-developed larvae. Young larvae burrow in silty substrates, but as they grow and mature they move to gravel substrates and to the surfaces of large rocks.

Eggs were described by Ide (1935), Smith (1935), and Koss (1968). They have a tuberculate chorion and two polar caps.

Description: Mature larvae medium to large, 8–15 mm. Body subcylindrical.

Head: Antennae with small scattered setae. Mandibles with distinct tusks, tusks with scattered spines and setae (fig. 9.5).

Thorax: Pronotum with distinct lateral flanges. T1 legs longer than T2 or T3 legs. Femora broad and flattened.

Abdomen: Gills on segments 1–7, gill 1 vestigial and single; gills on 2–7 forked and fringed. Three caudal filaments, a row of long setae on both sides of each filament.

Comments: There is a single North American genus, *Potamanthus,* with eight described species. McCafferty (1975) has reviewed the genus and provided a key to the larvae. They are widely distributed in eastern North America to Kansas and Nebraska.

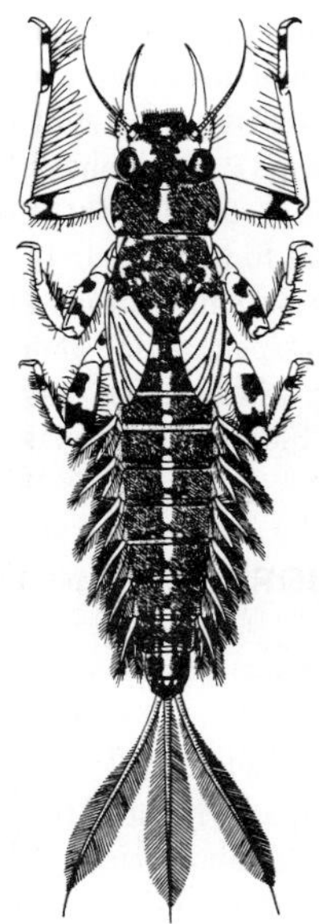

Figure 9.51. Potamanthidae. *Potamanthus,* from Needham (1920).

POLYMITARCYIDAE (EPHEMEROIDEA)

The Polymitarcyids, Pale Burrowing Mayflies

Figures 9.4, 9.7, 9.10

Relationships and Diagnosis: This family is most closely related to the tropical Euthyplociidae. Larvae are distinguished by the mandibular tusks that turn downward apically as viewed laterally (fig. 9.7); the gills have fringed margins (fig. 9.4).

Biology and Ecology: Larvae are burrowers in the banks or bottoms of rivers and canals. The burrows are U-shaped, principally in firm clay, in a variety of situations, beneath rocks, in river bottoms, on river banks, for instance. They appear to be filter feeders.

Eggs variable, those of *Ephoron* with two polar caps, those of *Tortopus* and *Campsurus* shaped somewhat like a deflated ball with one side pushed in (Ide 1935, Smith 1935, Koss 1968, Koss and Edmunds 1974).

Description: Mature larvae medium to large, 9–30 mm. The body is generally subcylindrical.

Head: Ovoid to elongate. Antennae with scattered short setae on flagellum; scape and pedicel with or without longer setae. Maxillary and labial palpi two-segmented; labial palpi held beneath and at right angles to glossae and paraglossae. Mandibular tusks present, curved downward apically (fig. 9.4); upper surface of tusks with numerous tubercles in *Ephoron* (Polymitarcyinae), or mesal surfaces with one or more tubercles in Campsurinae.

Thorax: Pronotum with projections at anterolateral corners. T1 legs fossorial, with numerous long setae; tibiae and tarsi fused in Campsurinae, separate in Polymitarcyinae. T2 and T3 legs with marginal long setae. Apex of tibia of T3 legs rounded at apex (fig. 9.10).

Abdomen: Gills dorsal, vestigial on segment 1 (single in Polymitarcyinae, double in Campsurinae); bilamellate on segments 2–7, margins fringed. Three caudal filaments, with rows of long setal tufts on both sides.

Comments: This is a small family with three genera and six species in North America. Genera and species are keyed as larvae by McCafferty (1975). Edmunds et al. (1976) key the genera and list the species. The family is fairly widespread in North America and frequently abundant; this abundance is most obvious from the mass flights of the pale adults, but the larvae are not collected frequently except by mayfly specialists. See habitus figures 431 and 432 in Edmunds et al. (1976).

EPHEMERIDAE (EPHEMEROIDEA)

The Ephemerids, Common Burrowing Mayflies

Figures 9.6, 9.11, 9.52

Relationships and Diagnosis: The Ephemeridae are closely related to the Potamanthidae and Palingeniidae. Ephemerids probably were derived from potamanthidlike ancestors and protoephemerids gave rise to the palingeniids. The larvae of the Ephemeridae have upturned mandibular tusks that are more or less circular in cross section (fig. 9.6) and lack a distinct spuriferous keel; the gills have fringed margins.

Biology and Ecology: Larvae are burrowers in soft substrates; *Hexagenia* forms U-shaped burrows in silt substrates, and *Ephemera* burrows freely in sand or silty sand. They are found in a wide variety of streams and lakes. *Hexagenia* may be extremely abundant; in clean lakes the numbers may reach 100/meter^2; in polluted habitats they have been reported as high as 9,000/meter^2. The larvae are only fair swimmers. They feed primarily on detritus and algae, although *Ephemera simulans* Walker is carnivorous; *Hexagenia* is largely a filter feeder.

Eggs of Ephemeridae are rather diverse; they have been described by Smith (1935), Hunt (1953), Koss (1968), and Koss and Edmunds (1974); the latter paper includes a key to eggs of genera.

Description: Mature larvae medium to large, 10–32 mm. The body is subcylindrical.

Head: With a prominent frontal process between the antennae. Mandibles with large, cylindrical, upcurved tusks, the tusks without a spuriferous keel. Galea-lacinia of maxillae relatively narrow, more than twice as long as wide; maxillary palpi three-segmented.

Thorax: T1 legs fossorial. Apex of tibia of T3 legs expanded ventroapically into acute process (fig 9.11).

Abdomen: Segments 2–7 expanded laterally. Gills dorsal; gills on segment 1 vestigial, single or forked; gills on segments 2–7 bilamellate with fringed margins. Three caudal filaments.

Comments: This is a small family in North America with three genera and 13 species. The conspicuously large and abundant species are well known. The family, reviewed by McCafferty (1975), is widespread and common.

Figure 9.52. Ephemeridae. *Ephemera,* from Needham (1920).

PALINGENIIDAE (EPHEMEROIDEA)

The Palingeniids, Spinyheaded Burrowing Mayflies

Figure 9.9

Relationships and Diagnosis: The Palingeniidae, a family derived from the Ephemeridae, are primarily Old World but with the most primitive genus, *Pentagenia,* in North America. McCafferty (1972) first recognized the correct phylogenetic position of *Pentagenia* and placed it in a separate family, Pentageniidae; Edmunds et al. (1976) placed the taxon as a subfamily of the Ephemeridae, and McCafferty and Edmunds (1976) placed the subfamily in the Palingeniidae. The larvae have unique upturned mandibular tusks with a distinct dorsolateral keel that is more or less toothed, with a line of spurs along the toothed edge (fig. 9.9). In the more derived genera (i.e., *Palingenia* of Eurasia), the toothed keel is greatly expanded.

Eggs of *Pentagenia* have been described by Smith (1935) and Koss and Edmunds (1974). They are unique in containing only one tagenoform micropyle.

Biology and Ecology: Larvae of *Pentagenia,* the only North American palingeniid, burrow in U-shaped tubes in hard clay banks of large rivers. They are most frequently collected in drift nets at night.

Description: Mature larvae large, 18–24 mm. Body subcylindrical.

Head: A bifid frontal process between the antennal bases (fig. 9.9). Mandibular tusks with a distinct dorsolateral keel that is variably toothed, a row of short stout spurs along this toothed edge. Galea-lacinia of maxillae slightly longer than wide. Maxillary palpi three-segmented, the segments broad and rounded at apex. Labial palpi two-segmented, broad and flat. Flagella of antennae without setae.

Thorax: T1 legs adapted for burrowing; femora and tibia flattened, with dense patches of setae ventrally. Apex of tibia of T3 legs expanded ventroapically into an acute process.

Abdomen: Segments 2–7 expanded laterally. Gills dorsal; on segments 1–7, gill 1 divided to base; gills with fringed margins. Three caudal filaments.

Comments: This is a small family in North America with a single genus and two species, one generally uncommon, but locally abundant, the other known only from the types. *Pentagenia* ranges from Florida and Texas north to Manitoba.

SUBORDER PANNOTA

The wing pads of mature larvae are fused mesally to the mesonotum for more than one-half their length; this trend culminates in *Baetisca* where the wing pads are fused into the mesothoracic carapace. Most of the larvae are slow crawlers, often camouflaged by detritus clinging to their setae. Many larvae bend the abdomen and caudal filaments up over the dorsum when disturbed. There are three superfamilies, Ephemerelloidea, Caenoidea, and Prosopistomatoidea.

EPHEMERELLIDAE (EPHEMERELLOIDEA)

The Ephemerellids, Spiny Crawler Mayflies

Figures 9.18, 9.19, 9.53

Relationships and Diagnosis: The Ephemerellidae, with the Tricorythidae, form the superfamily Ephemerelloidea of the Pannota. The Ephemerellinae, the only subfamily occurring in North America, is composed of eight genera. Larvae are distinguishable from all other mayflies by the distinctive dorsal gills on segments 3–7 or 4–7; the ventral lamella has numerous lobes (fig. 9.18).

Biology and Ecology: Larvae are found in a wide variety of running water and on wave-washed shores of lakes. Various species are found over a wide range of stream velocity and habitat. Some are found sheltered in moss, others in tangles of roots or debris or on or under rocks.

Larvae are poor swimmers and generally move slowly as they crawl around on the substrate. They are largely detritivores or omnivores, but some of the larger species are carnivores.

Eggs, described by Smith (1935), Koss (1968), and Koss and Edmunds (1974), are ovoid in most, with a single polar cap. Eggs of the subgenus *Eurylophella* are rectangular and without a polar cap.

Description: Mature larvae small to medium, 4–15 mm. Body variable in different genera. Subcylindrical in *Ephemerella, Serratella, Caudatella,* and most *Drunella,* depressed in *Attenella, Dannella,* and *Eurylophella,* and strongly depressed in *Timpanoga.* They often have paired tubercles of various sizes on head, thorax, and/or abdomen (fig. 9.19).

Head: Hypognathous, oval to strongly depressed. Maxillary palpi with zero to three segments, labial palpi three-segmented.

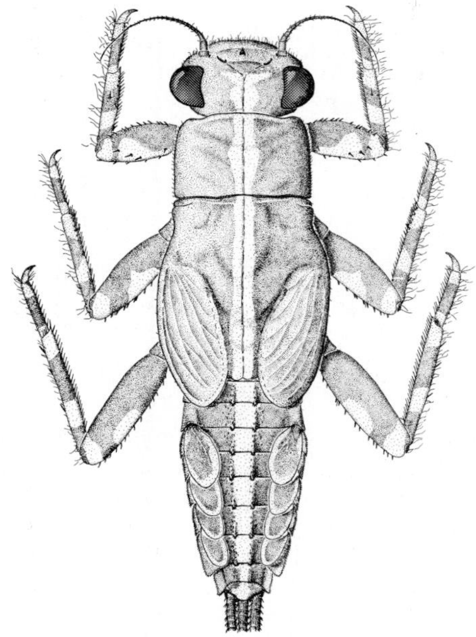

Figure 9.53. Ephemerellidae. *Serratella.*

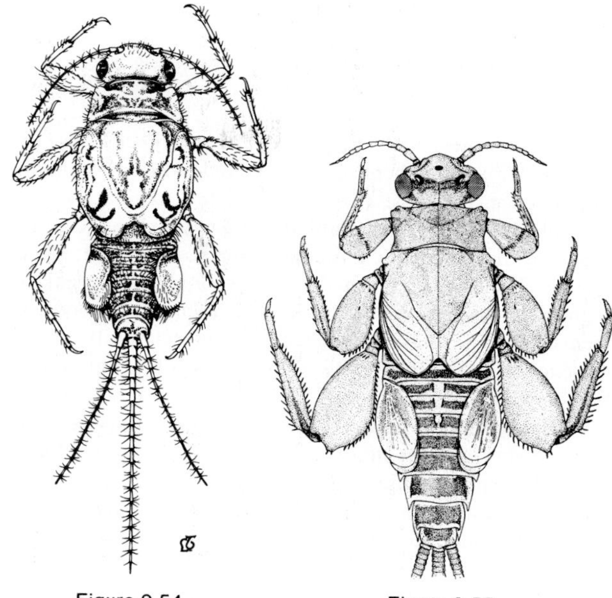

Figure 9.54 Figure 9.55

Figure 9.54. Tricorythidae. *Tricorythodes,* Gunnison Co., Colo. Drawing by L. Tackett, Courtesy F. W. Stehr.

Figure 9.55. Tricorythidae. *Leptohyphes.*

Thorax: Legs variable, long and slender to short and robust; anterior margin of femora often with large tubercles in genus *Drunella.*

Abdomen: Segments expanded laterally with posterolateral projections. Gills dorsal, vestigial or absent on segment 1. Gills on 3–7 in *Caudatella, Ephemerella* s.s., *Serratella,* and *Drunella* (fig. 9.19); on 4–7 in *Eurylophella, Dannella, Attenella,* and *Timpanoga.* Gills imbricated, or with gill 3 or 4 semioperculate or operculate. Three caudal filaments, usually subequal, cerci short in *Caudatella.*

Comments: This is a relatively large family in North America with 79 species. The genera, which have been regarded as subgenera in the past, are now considered to have full generic rank by Allen (1980) and many European workers. Allen and Edmunds (1959–65) and Allen (1968) keyed the larvae of the North American species. The genera are widely distributed in North America, with the greatest diversity of species in the northern United States and southern Canada. Species are largely absent from the plains region, and a few occur as far south as northern Mexico in western North America and central Florida in the East.

TRICORYTHIDAE (EPHEMERELLOIDEA)

The Tricorythids, Little Stout Crawlers

Figures 9.14, 9.54, 9.55

Relationships and Diagnosis: The Tricorythidae are most closely related to the Teloganodinae of the Ephemerellidae. Leptohyphinae, the only subfamily in North America, is represented by two genera. Gills are present on segments 2 to 5 or 6, and those on segment 2 are triangular (fig. 9.14) or suboval operculate.

Biology and Ecology: Larvae occur on a variety of substrates in streams of all sizes. They occur most frequently in warm streams, but *Tricorythodes* is sometimes abundant in trout streams. Larvae, being poor swimmers, move almost entirely by crawling in and among rocks and vegetation. They appear to feed on detritus and biota that grows on aquatic plants and on submerged objects.

Eggs have been described by Smith (1935), Koss (1968), and Koss and Edmunds (1974). They have a single polar cap and a chorionic sculpturing of overlapping layers of plates.

Description: Mature larvae small, 5–9 mm. Body subcylindrical, sometimes rather flattened.

Head: Hypognathous, usually ovoid. Eyes small in male and female, rarely larger in males.

Thorax: Legs slender to stout; femora with spines and long setae. Claws with or without denticles.

Abdomen: Segments 3–6 shorter than others, 8–9 usually longest. Gills absent on segment 1; subovate to triangular and operculate on segment 2. Gills of segments 3–6 double, ventral lamella lobed. Caudal filaments with whorls of setae at apex of each segment; caudal filament bases thick in males.

Comments: In North America this family includes two genera and 22 species. Allen keyed the larvae of some of the western North American *Tricorythodes* (1977) and the North American species of *Leptohyphes* (1978). *Tricorythodes* is widespread and abundant in North America, and *Leptohyphes* occurs north to Utah, Texas, and Maryland; both genera also occur in South America and appear to have dispersed northward.

NEOEPHEMERIDAE (CAENOIDEA)

The Neoephemerids, Large Squaregills

Figure 9.16

Relationships and Diagnosis: The neoephemerids, the most primitive members of the Caenoidea-Prosopistomatoidea lineage, are most closely related to Caenidae. Larval Neoephemeridae are very similar to *Caenis* larvae except for size and size-related characters. North American Neoephemeridae are distinguished by quadrate operculate gills on segment 2, fused at the midline and covering the gills on segments 3–6. They have a rounded lobe anterolaterally on the mesonotum (fig. 9.16), hind wing pads, and a fibrilliform portion on gills of segments 3–6.

Biology and Ecology: Larvae are inhabitants of moderate or slow current in a variety of rivers. They are found primarily among tangles of roots, branches, or debris, although some live in moss, and *Neophemera purpurea* Traver is reported under large flat rocks. They crawl slowly and are poor swimmers.

The eggs of *Neoephemera* are covered with a dense filamentous adhesive layer; they are described by Smith (1935) and Koss and Edmunds (1974).

Description: Mature larvae medium to large, 8–17 mm. Body moderately flattened.

Head: Subspherical; eyes large in male, moderate in female. Maxillary and labial palpi three-segmented.

Thorax: Legs long, the T3 legs longest. Claws without denticles.

Abdomen: Posterolateral projections well developed on segments 6–9. Gills on segment 1 vestigial; on segment 2 operculate, quadrate, and fused at midline; on segments 3–6 with fringed margins and fibrilliform tuft. Three caudal filaments, whorls of setae at apex of each segment.

Comments: This is a small family with one genus, *Neoephemera,* and four species in North America. The species are described and keyed by Berner (1956). One species occurs from Michigan to Quebec; the other three in the southeastern United States. See habitus figure 420 in Edmunds et al. (1976).

CAENIDAE (CAENOIDEA)

The Caenids, Small Squaregills

Figures 9.15, 9.17, 9.56

Relationships and Diagnosis: The caenids were derived from protoneoephemerids. They are distinguished by quadrate gills on segment 2 (fig 9.15a), not fused (often overlapping) at midline, and covering the gills on segments 3–6 (fig. 9.15a). Very similar to Neoephemeridae, but caenids have no hind wing pads, and no fibrilliform tuft on the gills of segments 3 to 6 (fig. 9.15b). The anterolateral corner of the mesonotum usually with no indication of a lobe (fig. 9.17), a very gentle lobe in one or more species.

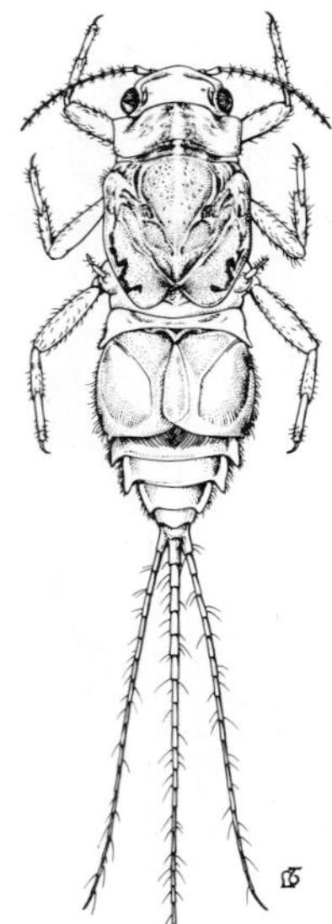

Figure 9.56. Caenidae. *Caenis,* Lake Winnipeg Beach, Manitoba. Drawing by L. Tackett, courtesy F. W. Stehr.

Biology and Ecology: The genus *Brachycercus* is found in rivers and on wave-washed shores of lakes. In most of the world *Caenis* larvae live in rivers, but in North America and Europe *Caenis* is predominantly a pond- or lake-dwelling genus. Some North American species are found in rivers, and some of the predominantly pond-dwelling species are found at the edges of rivers. The larvae are omnivorous; many feed on the epibiota of submerged plants. They are poor swimmers, usually crawling slowly over vegetation.

Eggs have one or two polar caps; they are described by Smith (1935), Koss (1968), and Koss and Edmunds (1974).

Description: Mature larvae small to medium, 3–8 mm. The body is slightly to moderately flattened.

Head: Ovoid, in *Brachycercus* there are three ocellar tubercles. Maxillary and labial palpi three-segmented in *Caenis,* two-segmented in *Brachycercus.*

Thorax: Legs moderately robust and claws strongly curved in *Caenis.* In *Brachycercus* T2 and T3 legs long and slender, T1 legs short; claws slender. Without T3 wing pads, lobe on the anterolateral corner of mesonotum absent or small.

Abdomen: More or less depressed; segments 2–5 short, and 7–9 elongate. Gills on segment 1 vestigial, on segment 2 quadrate, not fused, touching or overlapping at midline; lamellae of gills on segments 3–6 with fringed margins. Three caudal filaments, with a whorl of setae at apex of each segment; caudal filament bases thick in males.

Comments: The caenids, a small family of two genera and 18 species, are widespread and common. There are no usable keys to the larvae, although the family is reviewed by Edmunds et al. (1976).

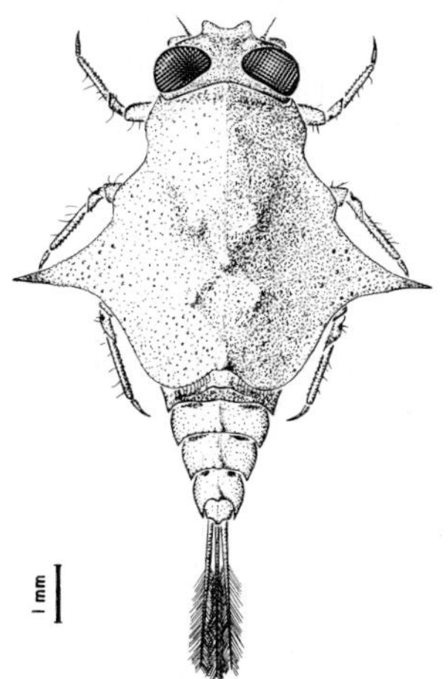

Figure 9.57. Baetiscidae. *Baetisca*, from Pescador and Peters (1974).

BAETISCIDAE (PROSOPISTOMATOIDEA)

The Baetiscids, Armored Mayflies

Figures 9.3, 9.57

Relationships and Diagnosis: The monotypic Baetiscidae are not closely related to any other family; their nearest relatives are another monotypic family, the Old World Prosopistomatidae. The large mesonotal shield that extends to tergum 6 is diagnostic in the American fauna (fig. 9.3).

Biology and Ecology: *Baetisca* larvae are found most frequently in small to moderate streams, but they also occur in large rivers and on wave-washed shores of lakes. They are found in a variety of microhabitats; most are found in rocks or gravel mixed with sand, but a few live in exposed sand or silt and sand. They may be partially buried in the substrate. They are fair swimmers; they pull the legs under the body and rapidly undulate the caudal filaments.

Description: Mature larvae small to medium, 4–14 mm. Body robust; the mesonotum fused with the pronotum and forming a carapace that reaches to abdominal segment 6.

Head: Tubercles between antennae and on genae.

Thorax: Mesonotum fused to pronotum and expanded to cover the gills. Legs short and stout.

Abdomen: Segments 1–5 shortened. Gills on segments 2–6, margins of most fringed. Caudal filaments short, forming a paddle, long setae on both sides of terminal filament and cerci.

Comments: This is a small endemic family of one genus that is most abundant and diverse in the southeast, but is widespread in the eastern and central U.S., as far west as Wyoming, Alberta, and Yukon Territory. One species is found in the Columbia River in Washington. Keys to larvae are given by Traver (1935), Burks (1953), Berner (1955), and Pescador and Berner (1981).

BIBLIOGRAPHY

Allen, R. K. 1968. New species and records of *Ephemerella* (*Ephemerella*) in western North America (Ephemeroptera: Ephemerellidae). J. Kansas Ent. Soc. 41:557–67.

Allen, R. K. 1973. Generic revisions of mayfly nymphs. I. *Traverella* in North and Central America (Leptophlebiidae). Ann. Ent. Soc. Amer. 66:1287–95.

Allen, R. K. 1977. A new species of *Tricorythodes* with notes (Ephemeroptera: Tricorythidae). J. Kansas Ent. Soc. 50:431–35.

Allen, R. K. 1978. The nymphs of North and Central American *Leptohyphes* (Ephemeroptera: Tricorythidae). Ann. Ent. Soc. Amer. 71:537–58.

Allen, R. K. 1980. Geographic distribution and reclassification of the subfamily Ephemerellinae (Ephemeroptera: Ephemerellidae), 71–91. *In* Flannagan, J. F., and K. E. Marshall, Advances in Ephemeroptera biology. New York: Plenum Publ. Corp. 552 pp.

Allen, R. K., and R. C. Brusca. 1978. Revisions of mayfly nymphs. II. *Thraulodes* in North and Central America (Leptophlebiidae). Can. Ent. 110:413–33.

Allen, R. K., and G. F. Edmunds, Jr. 1959. A revision of the genus *Ephemerella*. I. The subgenus *Timpanoga* (Ephemeroptera: Ephemerellidae). Can. Ent. 91:51–58.

Allen, R. K., and G. F. Edmunds, Jr. 1961a. A revision of the genus *Ephemerella*. II. The subgenus *Caudatella* (Ephemeroptera: Ephemerellidae). Ann. Ent. Soc. Amer. 54:603–12.

Allen, R. K., and G. F. Edmunds, Jr. 1961b. A revision of the genus *Ephemerella*. III. The subgenus *Attenuatella* (Ephemeroptera: Ephemerellidae). J. Kansas Ent. Soc. 34:161–73.

Allen, R. K., and G. F. Edmunds, Jr. 1962. A revision of the genus *Ephemerella*. IV. The subgenus *Dannella* (Ephemeroptera: Ephemerellidae). J. Kansas Ent. Soc. 35:333–38.

Allen, R. K., and G. F. Edmunds, Jr. 1963a. A revision of the genus *Ephemerella*. VI. The subgenus *Serratella* in North America (Ephemeroptera: Ephemerellidae). Ann. Ent. Soc. Amer. 56:583–600.

Allen, R. K., and G. F. Edmunds, Jr. 1963b. A revision of the genus *Ephemerella*. VII. The subgenus *Eurylophella* (Ephemeroptera: Ephemerellidae). Can. Ent. 95:597–623.

Allen, R. K., and G. F. Edmunds, Jr. 1965. A revision of the genus *Ephemerella*. VIII. The subgenus *Ephemerella* in North America (Ephemeroptera: Ephemerellidae). Misc. Publ. Ent. Soc. Amer. 4:243–82.

Allen, R. K., and G. F. Edmunds, Jr. 1976. A revision of the genus *Ametropus* in North America (Ephemeroptera: Ametropodidae). J. Kansas Ent. Soc. 94:625–35.

Bednarik, A. F. and W. P. McCafferty. 1979. Biosystematic revision of the genus *Stenonema* (Ephemeroptera: Heptageniidae). Can. Bull. Fish. Aquat. Sci. 201. 73 pp.

Berner, L. 1950. Mayflies of Florida. Univ. Florida Studies. Biol. Sci. Ser. 4:267 pp.

Berner, L. 1955. The southeastern species of *Baetisca* (Ephemeroptera: Baetiscidae). Quart. J. Fla. Acad. Sci. 18:1–19.

Berner, L. 1956. The genus *Neoephemera* in North America (Ephemeroptera: Neoephemeridae). Ann. Ent. Soc. Amer. 49:33–42.

Berner, L. 1978. A review of the mayfly family Metretopodidae. Trans. Amer. Ent. Soc. 104:91–137.

Burks, B. D. 1953. The mayflies or Ephemeroptera of Illinois. Bull. Ill. St. Nat. Hist. Surv. 26:1–216.

Clifford, H. F. 1976. Observations on the life cycle of *Siphloplecton basale* (Walker) (Ephemeroptera: Metretopodidae). Pan-Pac. Ent. 52:265–71.

Cohen, S. D., and R. K. Allen. 1978. Revision of mayfly nymphs. III. *Baetodes* in North and Central America (Baetidae). J. Kansas Ent. Soc. 51:253–69.

Demoulin, G. 1974. Remarques critiques sur les Acanthametropodinae et sur certaines formes affines (Éphémèropteres: Siphlonuridae). Bull. Inst. Roy. Sci. Nat. Belg. 59 (2):1–5.

Edmunds, G. F., Jr., and R. K. Allen. 1966. Rocky Mountain species of *Epeorus* (*Iron*) (Ephemeroptera: Heptageniidae). J. Kansas Ent. Soc. 37:275–88.

Edmunds, G. F., Jr., R. K. Allen and W. L. Peters. 1963. An annotated key to the nymphs of the families and subfamilies of mayflies (Ephemeroptera). Salt Lake City, Univ. Utah Press. 49 pp.

Edmunds, G. F., Jr., L. Berner, and J. R. Traver. 1958. North American mayflies of the family Oligoneuriidae. Ann. Ent. Soc. Amer. 51:375–82.

Edmunds, G. F., Jr., S. L. Jensen, and L. Berner. 1976. The mayflies of North and Central America. Minneapolis: Univ. Minn. Press, 330 pp.

Edmunds, G. F., Jr., and R. W. Koss. 1972. A review of the Acanthometropodinae with a description of a new genus (Ephemeroptera: Siphlonuridae). Pan-Pac. Ent. 48:136–44.

Edmunds, G. F., Jr., and J. R. Traver. 1954. An outline of a reclassification of the Ephemeroptera. Ent. Soc. Wash. 56:236–40.

Hunt, B. P. 1953. The life history and economic importance of a burrowing mayfly (*Hexagenia limbata*) in southern Michigan lakes. Bull. Mich. Inst. Fish. Res. 4:1–151.

Ide, F. P. 1935. Life history notes on *Ephoron, Potamanthus, Leptophlebia* and *Blasturus* with descriptions (Ephemeroptera). Can. Ent. 67:113–25.

Kondratieff, B. C. and J. R. Voshell, Jr. 1984. The North Central American species of *Isonychia* (Ephemeroptera: Oligoneuriidae). Trans. Amer. Ent. Soc. 110:129–244.

Koss, R. W. 1968. Morphology and taxonomic use of Ephemeroptera eggs. Ann. Ent. Soc. Amer. 61: 696–721.

Koss, R. W., and G. F. Edmunds, Jr. 1970. A new species of *Lachlania* from New Mexico with notes on the genus (Ephemeroptera: Oligoneuriidae). Proc. Ent. Soc. Wash. 72:55–65.

Koss, R. W., and G. F. Edmunds, Jr. 1974. Ephemeroptera eggs and their contribution to phylogenetic studies of the order. Zool. J. Linn. Soc. 55:267–349.

Lameere, A. 1917. Etude sur l'evolution des ephemeres. Bull. Soc. Zool. France 42:41–59, 61–81.

Landa, V. and T. Soldán. 1985. Phylogeny and higher classification of the order Ephemeroptera: a discussion from the comparative anatomical point of view. Studie ČSAV (Praha) 4–85. 121 pp.

Lewis, P. A. 1974. Taxonomy and ecology of *Stenonema* mayflies (Heptageniidae: Ephemeroptera) Environ. Monit. Ser. Report No. EPA–670/4–74–006. 81 pp.

McCafferty, W. P. 1972. Pentageniidae, a new family of Ephemeroidea (Ephemeroptera). J. Ga. Ent. Soc. 7:51–56.

McCafferty, W. P. 1975. The burrowing mayflies (Ephemeroptera: Ephemeroidea) of the United States. Trans. Amer. Ent. Soc. 101:447–504.

McCafferty, W. P., and G. F. Edmunds, Jr. 1976. Redefinition of the family Palingeniidae and its implication for the higher classification of Ephemeroptera. Ann. Ent. Soc. Amer. 69:486–90.

McCafferty, W. P., and G. F. Edmunds, Jr. 1979. The higher classification of the Ephemeroptera and its evolutionary basis. Ann. Ent. Soc. Amer. 72:5–12.

Merritt, R. W. and K. W. Cummins. 1984. An introduction to the aquatic insects of North America. Dubuque, Iowa, Kendall/Hunt Publ. Co. 722 pp.

Morihara, D. K. and W. P. McCafferty. 1979. The *Baetis* larvae of North America (Ephemeroptera: Baetidae). Trans. Amer. Ent. Soc. 105:139–221.

Müller-Liebenau, G. 1974. *Rheobaetis:* A new genus from Georgia (Ephemeroptera: Baetidae). Ann. Ent. Soc. Amer. 67:555–67.

Needham, J. G. 1920. Burrowing mayflies of our larger lakes and streams. Bull. U.S. Bur. Fish. 36:269–92. (12 pls.)

Needham, J. G., J. R. Traver and Y. C. Hsu. 1935. The biology of mayflies. Ithaca, N.Y.: Comstock. 759 pp.

Pescador, M. L. and L. Berner. 1981. The mayfly family Baetiscidae (Ephemeroptera). Part II. Biosystematics of the genus *Baetisca*. Trans. Amer. Ent. Soc. 107:163–228.

Pescador, M. L. and W. L. Peters. 1974. The life history of *Baetisca rogersi* Berner (Ephemeroptera: Baetiscidae). Bull. Fla. St. Mus. Biol. Sci. 17:151–209.

Pescador, M. L. and W. L. Peters. 1980. A revision of the *Homoeoneuria* (Ephemeroptera: Oligoneuriidae). Trans. Amer. Ent. Soc. 106:357–393.

Riek, E. F. 1973. The classification of the Ephemeroptera, pp. 160–78. *In* Proc. 1st Int. Conf. Ephemeroptera, 1970. Leiden. E. J. Brill.

Smith, O. R. 1935. The eggs and egg-laying habits of North American mayflies, p. 67–89. *In* Needham, J. G., J. R. Traver, and Y. C. Hsu. The biology of mayflies. Ithaca, N.Y.: Comstock. 759 pp.

Traver, J. R. 1935. Systematic, Part II, 267–739. *In* Needham, J. G., J. R. Traver, and Y. C. Hsu. The biology of mayflies. Ithaca, N.Y.: Comstock. 759 pp.

Traver, J. R., and G. F. Edmunds, Jr. 1968. A revision of the Baetidae with spatulate-clawed nymphs (Ephemeroptera). Pacific Insects 10:629–77.

Order Odonata

10 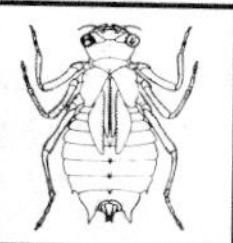

Minter J. Westfall, Jr.
University of Florida

DRAGONFLIES AND DAMSELFLIES

The order Odonata contains at least 510 species in our range of North America, including Alaska, Hawaii, the Mexican border states and the Greater Antilles. A great number of these species form an important link in food chains for fish and other aquatic vertebrates. They in turn destroy countless numbers of pest species, such as the larvae of midges and mosquitoes. A few large species have been known to pose a problem in fish hatcheries, where they may eat the fingerlings. On occasion, anglers have used some common species for fish bait.

The immatures, which undergo hemimetabolous (incomplete) metamorphosis, are referred to as larvae, nymphs, or naiads by different entomologists (the term larvae will be used here). The larvae are known for most North American species, but some have not yet been described and illustrated in detail. I am currently (1984) preparing descriptions for a number of the latter. The species of some genera are quite similar, and more study is needed to develop dependable keys. Very few species have been reared from egg to adult, and some key characters now used may not apply except in the last few instars. The classification for the Anisoptera is that used by Walker (1958), and Walker and Corbet (1975). The classification for the Zygoptera is that to be used in my *Manual of the Damselflies of North America* (in preparation).

Recent trends have been to raise to family rank groups of genera and species formerly considered as subfamilies, i.e., Macromiidae and Corduliidae. For the Macromiidae, Gloyd (1959) gave ample reasons for recognizing this family. For the Corduliidae, no single character or group of characters can now be given to separate all larvae of this family from all larvae of Libellulidae. For this reason, some prefer to recognize only one family, Libellulidae, with two subfamilies Corduliinae and Libellulinae. Because the adults appear more distinct, two families are recognized here, but in the larval key it has been necessary to use several characters with exceptions to separate them.

Odonata are aquatic, with wings developing externally in the larvae. There is a distinct head, thorax, and abdomen. The legs are free. The most unique characteristic is the peculiar labium or lower lip. It is folded upon itself at midlength and turned backward beneath the front legs (figs. 10.4, 10.21), which alone separates the larvae of Odonata from closely related aquatic orders.

Biology and Ecology

The larvae are all carnivorous, even cannibalistic, and most are found in permanent lakes, ponds, or streams, although some with a short life cycle have become adapted to temporary ponds. A few are terrestrial or semiterrestrial, living in bog moss or under damp leaves in seepage areas. At least one species of *Megalagrion* (Zygoptera) in Hawaii lives in damp, dead vegetation on the ground, far removed from bodies of water. Several species have become adapted for life in the water that collects in bromeliads and similar plants, and there are some species that are able to withstand long periods of desiccation.

The larval stage lasts from a few weeks to about five years, depending on the species, and the larvae pass through 10 to 15 instars. The wing cases appear after the third or fourth molt, becoming very swollen in the final instar, indicating imminent emergence. Some univoltine species have a synchronized emergence, after which it is difficult to collect the larvae until a new generation develops. The larvae of some Odonata serve as hosts for various water mites (*Hydracarina*) and other parasites. Informative accounts of the biology and ecology of Odonata are found in Needham and Heywood (1929), Walker (1953, 1958), Walker and Corbet (1975), Needham and Westfall (1955), and Corbet (1963).

DESCRIPTION

Head:

The unique labium or lower lip may be flat, as in the Gomphidae and Aeshnidae (fig. 10.18), or spoon-shaped, as in the Corduliidae and Libellulidae, covering much of the face like a mask (figs. 10.2, 10.21). The larger central portion is the prementum (formerly mentum); it usually bears important premental setae used in classification. The front margin of the prementum may have a central projection known as the ligula, which may be cleft medially (fig. 10.33). Hinged to the front of the prementum are a pair of strong, palpal lobes (formerly lateral lobes) armed with hooks, spines, teeth and usually raptorial setae, varying in different families. Each palpal lobe terminates in a strong movable hook, which may bear raptorial setae of importance in classification. Antennae with 4–7 segments, the shape and number of taxonomic importance (figs. 10.23, 10.24, 10.25, 10.26). Compound eyes may be high and capping the frontolateral margins of the head or lower and covering more of the sides. The Macromiidae have a distinct frontal horn projecting upward (fig. 10.35), and the Cordulegastridae have a rounded frontal shelf with taxonomic significance.

Thorax

The three segments are fused into a boxlike structure that bears legs and wing cases. The legs in some families have strong burrowing hooks, in which case the legs are short. In others, such as the Macromiidae and Calopterygidae, the legs are very long; these larvae may be called sprawlers. Fore and middle tarsi two- or three- segmented. In the Macromiidae, metasternum with a broad mesal tubercle near posterior margin. Wing cases usually parallel, but strongly divergent in some Gomphidae.

Abdomen

Composed of ten distinct segments, may be broad and flattened in some Anisoptera (fig. 10.19), or narrower than the head in the Zygoptera and other Anisoptera. The relative lengths of segments may be useful in classification. Some species have a series of mid-dorsal hooks and/or lateral spines on the segments (fig. 10.20). The Zygoptera have three external tracheal gills: a pair of lateral gills and a median gill (caudal lamellae of some authors) (figs. 10.3, 10.4). These vary in shape in different families, but are usually laterally flattened and used for swimming as well as respiration. Gill shape, markings, and number of marginal setae are used in identification. The abdomen of the Anisoptera ends in several sharp valves (anal appendages), a single dorsal one (epiproct), a pair of ventral ones (paraprocts), and a pair of lateral ones (cerci); the cerci are slower in developing to their full length (fig. 10.31).

TECHNIQUES

It is often desirable to rear larvae for positive identification. If an emerging adult is found, it should be placed with the exuvia in some container such as a paper bag, and allowed to mature for at least 24 hours, or even until it dies. Then it can be preserved with the exuvia in 75% ethyl alcohol. Larvae can be brought alive from the field in polyethylene bags or containers of water or damp material, making certain to separate individuals as much as possible to avoid cannibalism. Gomphines may require sand in which to burrow, with air bubbled into the water. Most Corduliidae, Libellulidae, and the Zygoptera do well in any container. Baby food jars with screen in the top and a strip of screen protruding above the water have served well. Unless the larva is very close to emergence, it will need to be fed. I use very successfully the small annelids, *Enchytreus,* reared in a soil culture, kept cool in a modified refrigerator, and fed with bread and powdered milk. Most Zygoptera need to be isolated to prevent one larva from nipping off the gills of another. Just before emergence, the gills can be removed to alcohol, as the gills of exuviae are often crumpled and specific characters are difficult to find. With some Anisoptera, color patterns are well preserved in the exuviae, but in others they are not. If one has a series of larvae of apparently the same species, several should be preserved in 75% alcohol, and the others reared.

Collecting can be done with any small-mesh dip net. A net such as the Needham scraper is valuable, especially for burrowing species. Often dead leaves and other vegetation from the bottom can be placed on a light-colored surface or sheet. When larvae move, they can be readily seen and collected.

GENERAL REFERENCES ON ANISOPTERA AND ZYGOPTERA

Bick 1957.
Byers 1930.
Cannings and Stuart 1977.
Corbet 1963, 1980.
Davies 1981
Garman 1927
Gloyd and Wright 1959.
Howe 1920.
Huggins and Brigham 1982.
Kennedy 1915, 1917.
Needham and Heywood 1929.
Needham and Needham 1962.
Paulson and Jenner 1971.
Smith and Pritchard 1956.
Westfall 1984.
Westfall and Roback 1967.
Wright 1943, 1946a.
Wright and Peterson 1944.
Wright and Shoup 1945.

GENERAL REFERENCES ON ZYGOPTERA ONLY

Garman 1917.
Johnson and Westfall 1970.
Needham 1903.
Walker 1953.

GENERAL REFERENCES ON ANISOPTERA ONLY

Dunkle 1980.
Louton 1982.
Musser 1962.
Needham 1901.
Needham and Hart 1901.
Needham and Westfall 1955.
Stanford 1977.

CLASSIFICATION

Following is a synopsis of the Odonata occurring in North America, including Alaska, Hawaii, the Mexican border states, and the Greater Antilles.

Order **ODONATA**
- Suborder Zygoptera, Damselflies
  - Superfamily Calopterygoidea
    - Calopterygidae, broad-winged damselflies
  - Superfamily Lestoidea
    - Lestidae, spread-winged damselflies
    - Synlestidae, the synlestids
    - Hypolestidae, the hypolestids
  - Superfamily Coenagrionoidea
    - Platystictidae, the platystictids
    - Protoneuridae, the protoneurids
    - Coenagrionidae, narrow-winged damselflies
- Suborder Anisoptera, Dragonflies
  - Superfamily Aeshnoidea
    - Aeshnidae, the darners
    - Petaluridae, the graybacks
    - Gomphidae, the clubtails
  - Superfamily Cordulesgastroidea
    - Cordulesgastridae, the biddies
  - Superfamily Libelluloidea
    - Macromiidae, the belted skimmers and river skimmers
    - Corduliidae, the green-eyed skimmers
    - Libellulidae, the common skimmers

KEY TO SUBORDERS OF ODONATA LARVAE

1. Larvae relatively stout, head usually narrower than thorax and abdomen, the latter never cylindrical but widening from base to middle or beyond; without external caudal tracheal gills, the anus surrounded by 3 short, stiff pointed valves (anal appendages) forming the anal pyramid at tip of abdomen (figs. 10.1, 10.2). .... **ANISOPTERA**

1′. Larvae slender, head wider than thorax and abdomen, the latter usually cylindrical, not widening behind the base; 3 long caudal tracheal gills at tip of abdomen in place of the anal pyramid (figs. 10.3, 10.4) .... **ZYGOPTERA**

KEY TO FAMILIES OF ZYGOPTERA LARVAE

1. First antennal segment greatly elongate, as long as the combined length of the remaining segments (fig. 10.5); lateral gills considered in cross section, triangular .... ***Calopterygidae*** (p. 99)

1′. First antennal segment not so elongate, distinctly less than the combined length of the remaining segments (figs. 10.3, 10.4); lateral gills flat or balloonlike 2

2(1′). Prementum distinctly petiolate and spoon-shaped, the narrow proximal part as long as or longer than the expanded distal part (fig. 10.6); movable hook of each palpal lobe with 2 or 3 setae (fig. 10.6); usually 5–8 premental setae present each side of median line (when only 4 or 5 prementals present, only 3 palpal setae present (fig. 10.6)) .... ***Lestidae*** (p. 105)

2′. Prementum not distinctly petiolate, more or less triangular or subquadrate in shape (fig. 10.7); movable hook of each palpal lobe without setae (fig. 10.7); premental setae 0–3 each side of median line (some Coenagrionidae with 4 or 5 in which case, 5–6 palpal setae are present) (fig. 10.7) 3

3(2′). Premental and palpal setae absent (fig. 10.8); margin of prementum distinctly cleft medially, although the cleft may be closed (fig. 10.8) 4

3′. Premental and palpal setae present (fig. 10.7) (*Argia* without prementals and sometimes without palpal setae); margin of prementum never cleft (fig. 10.7) 6

4(3). Gills flat with apices broadly rounded (fig. 10.9); pedicel of gills very distinct and separated from gill by a "breaking joint" (fig. 10.9); medial cleft of prementum extending proximally beyond base of palpal lobes (fig. 10.8) .... ***Synlestidae*** (p. 105)

4′. Gills balloonlike with apices long and sharply pointed (fig. 10.10); pedicel of gills inconspicuous (fig. 10.10); medial cleft of prementum not extending proximally beyond base of palpal lobes (fig. 10.11) 5

5(4′). Eyes forming large portion of lateral margin of head (fig. 10.12); prementum narrowed proximally (fig. 10.11); distal margin of each palpal lobe with three teeth (fig. 10.11); body with few, if any, long setae (fig. 10.12) .... ***Hypolestidae*** (p. 106)

5′. Eyes small and not forming a portion of lateral margin of head (fig. 10.13); prementum expanded proximally, with lateral margins convex (fig. 10.14); distal margin of each palpal lobe with one tooth (fig. 10.14); antennae, head and legs with numerous long setae (fig. 10.13) .... ***Platystictidae*** (p. 106)

6(3′). One premental seta each side of median line (fig. 10.15); palpal setae 3–5 (fig. 10.15); proximal portion of gills thickened and darkened, apical portion thinner and more lightly pigmented, thus nodus very distinctly delineated across entire width of gill (fig. 10.16) .... ***Protoneuridae*** (p. 107)

6′. Premental setae usually 3–5 each side of median line or absent (if only one present, gills not as above and usually 6 palpal setae present) (fig. 10.7); palpal setae 0–6 (fig. 10.7); proximal portion of gills usually not differing markedly from distal portion, thus nodus not delineated across entire width of gill (fig. 10.17) (*Nehalennia* and *Argiallagma* with gills in exception to this have 3–4 premental setae each side of median line) .... ***Coenagrionidae*** (p. 107)

KEY TO FAMILIES OF ANISOPTERA LARVAE

1. Prementum and palpal lobes flat or nearly so, dorsal surface without long, stout premental setae and usually without palpal setae (fig. 10.18) 2

1′. Prementum and palpal lobes together spoon-shaped or mask-shaped, covering the face to the base of the antennae; dorsal surface usually with long premental and always with palpal setae (figs. 10.21, 10.22) 4

2(1). Antennae 4-segmented (figs. 10.23, 10.24); mesotarsi 2-segmented (fig. 10.19); ligula (front margin of prementum) without a median cleft (fig. 10.18) .... ***Gomphidae*** (p. 109)

2′. Antennae 6- or 7-segmented (figs. 10.25, 10.26); mesotarsi 3-segmented (fig. 10.1); ligula with a median cleft (figs. 10.27, 10.28) 3

3(2′). Prementum with sides subparallel in distal three-fifths, then abruptly narrowed near basal hinge (fig. 10.29); each palpal lobe with a stout dorsolateral spur at base of movable hook (fig. 10.29); epiproct not bifid at tip (fig. 10.30); antennal segments short, thick, and hairy (fig. 10.26); a pair of latero-dorsal tufts of long black bristles on abdominal segments 2 or 3 to 9 (fig. 10.30) .... ***Petaluridae*** (p. 109)

3′. Prementum widest in distal half, much narrower in proximal half (fig. 10.28); no dorsolateral spur at base of movable hook (fig. 10.27); epiproct usually distinctly bifid at tip (fig. 10.31); antennal segments slender and bristlelike (fig. 10.25); no laterodorsal hair tufts on abdomen (fig. 10.32) .... ***Aeshnidae*** (p. 108)

4(1′). Distal edge of each palpal lobe deeply cut, forming large irregular teeth, without associated groups of setae (fig. 10.33); ligula represented by a toothlike process that is cleft (fig. 10.33) .... ***Cordulegastridae*** (p. 110)

4′. Distal edge of each palpal lobe smooth or evenly crenate, each crenation bearing 1 or more setae (fig. 10.34a); ligula without a median cleft, or ligula not recognizable (fig. 10.34) 5

5(4′). Head with a prominent, suberect thick frontal horn between the bases of the antennae, its width distinctly less than its length (figs. 10.21, 10.35); metasternum with a broad median tubercle (best seen in lateral view) near posterior margin; legs very long, the apex of each hind femur reaching to or beyond the hind margin of abdominal segment 8; abdomen strongly depressed, almost circular in outline when viewed from above .... ***Macromiidae*** (p. 111)

5′. Head without a prominent, suberect, thick frontal horn (fig. 10.36) (*Neurocordulia molesta* (Walsh) has a *flattened* triangular frontal shelf, its width at base greater than its length (figs. 10.37, 10.38)); metasternum without a tubercle near posterior margin; legs shorter, the apex of each hind femur usually not reaching to the hind margin of abdominal segment 8; abdomen less depressed, more cylindrical in outline 6

6(5′). Crenations on distal margins of palpal lobes of labium separated by deep notches, crenations usually one-fourth to one-half as long as they are broad (fig. 10.34a); cerci generally more than one-half as long as paraprocts; lateral spines of abdominal segment 9 usually longer than its middorsal length; middorsal hooks on abdomen often cultriform (sicklelike) .... ***Corduliidae**** (p. 112)

6′. Crenations on distal margins of palpal lobes of labium generally separated by shallow notches, crenations usually one-tenth to one-sixth as long as they are broad (fig. 10.34b); cerci generally not more than one-half as long as paraprocts; lateral spines of abdominal segment 9 usually shorter than its middorsal length, but if longer, then middorsal hooks on abdomen are not cultriform but more spinelike, stubby, or absent .... ***Libellulidae**** (p. 112)

*No single character will reliably separate all larvae of Corduliidae from those of Libellulidae. For this reason some specialists have recognized only one family, Libellulidae, and two subfamilies, Libellulinae, and Corduliinae. Specimens difficult to identify in couplet 6 may need to be compared with available figures of both families.

Suborder ZYGOPTERA

CALOPTERYGIDAE (CALOPTERYGOIDEA)

The Calopterygids, Broad-Winged Damselflies

Figures 10.5, 10.5a

Relationships and Diagnosis: This is the only one of six families of Calopterygoidea that occurs within our range. It is represented by two subfamilies, Calopteryginae and Hetaerininae, each with one genus, *Calopteryx* and *Hetaerina*, respectively.

This family stands apart from all other families considered here by having the first antennal segment as long as the six remaining segments together (fig. 10.5). The lateral caudal gills are thick and ridged longitudinally, appearing triangular in cross section. The deep median cleft of the prementum is diamond-shaped (fig. 10.5a). There are no raptorial setae on the labium. Some previous authors have treated this family as the Agrionidae or Agriidae.

Biology and Ecology: These long-legged, stiff insects cling to roots, stems, and trash at the edges of the current in slow-flowing streams. They move infrequently from one place to another. The elongate eggs are deposited singly, below the surface of the water, in various kinds of plant tissue, living or dead.

Description: Mature larvae quite large, 30–40 mm, including the gills. Body cylindrical, light brown to black and quite stiff. Color markings variable, even among larvae of the same species.

Head: Flat and pentagonal, antennae sometimes longer than head. Labial mask not very long, but prementum considerably produced and deeply cleft, each division with a small dorsal seta. Diamond-shaped median cleft may extend halfway to base and well beyond the articulation of the palpal lobe (fig. 10.5a). These palpal lobes end in three curved and acute processes, the middle one longest and representing the end-hook. Premental and palpal setae absent. Caudolateral margins of head forming blunt to sharp tubercle.

Thorax: Approximately twice as long as broad, often with a dark line extending along side. Lateral margins of pronotum usually noticeably elevated, scalloped, the margins produced at two points on each side to form tubercles. Legs long, slender, and stiff, without heavy setae sometimes marked with dark and light rings. Wing cases usually extending posteriorly about to abdominal segment 4 in mature larvae.

Abdomen: Subcylindrical and usually without a distinct lateral keel, but some species of *Hetaerina* may have small lateral spines on segments 8 and 9. Caudal gills long, lateral gills stiff, triquetral, longer and narrower than middle gill, which is flat and thin-walled. Gills usually with long slender setae around margin, intermixed in some species with sharp spines.

Comments: This is a small family in our range, consisting of only two genera and nine species, widespread in the United States and Canada. Several species are quite variable, and subspecies have been described. Larvae of all species are known through rearing, but a few are still awaiting adequate descriptions.

Selected Bibliography

Johnson 1973, 1974.
Martin 1939.
Needham 1911.
Provonsha and McCafferty 1973.
Wright 1946e.

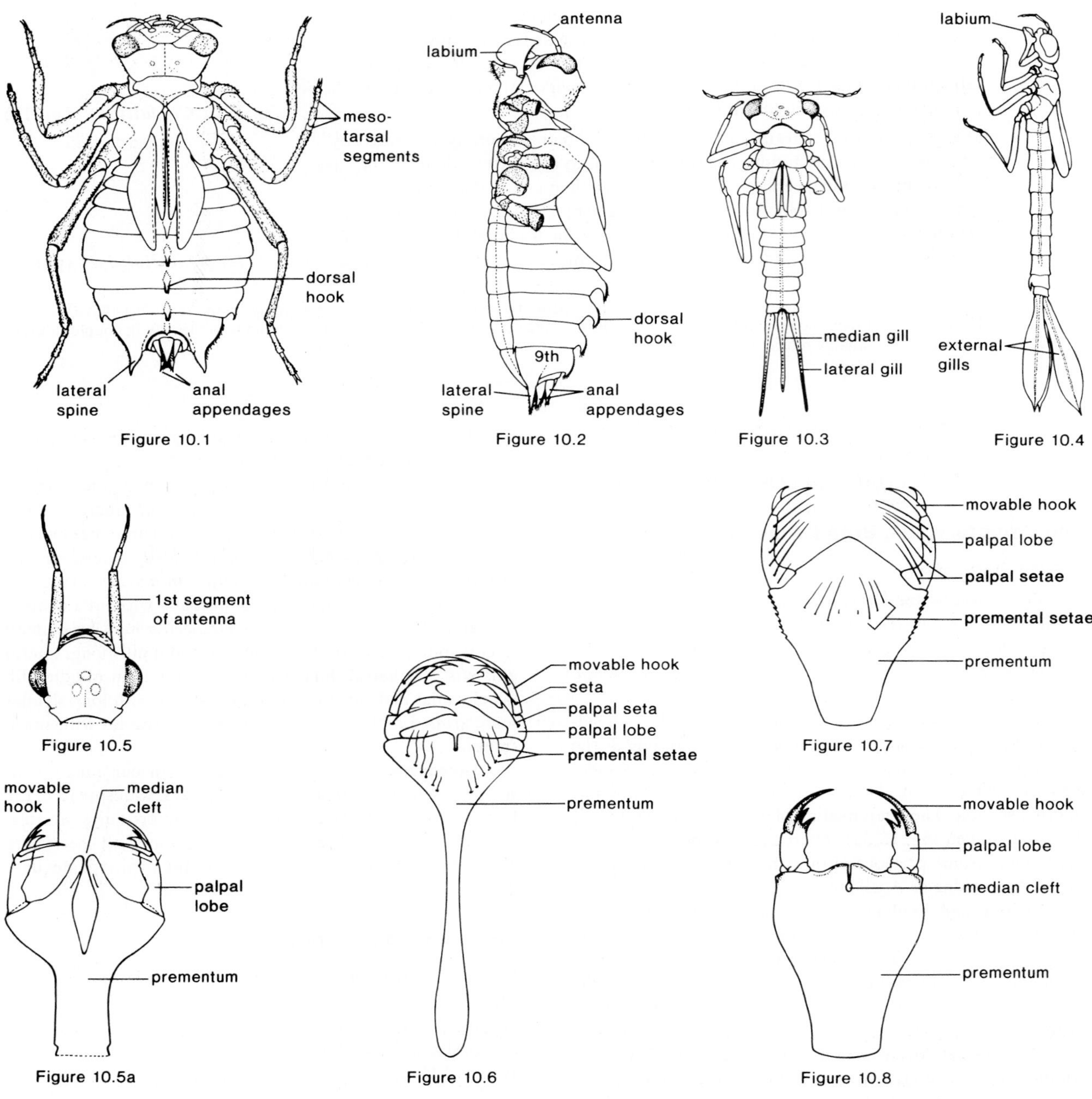

Figure 10.1. Dorsal view, larva of *Epicordulia princeps* (Hagen) (Corduliidae) (Anisoptera). (Redrawn from Wright and Peterson 1944.)

Figure 10.2. Lateral view, larva of *Epicordulia princeps* (Hagen) (Corduliidae) (Anisoptera). (Redrawn from Wright and Peterson 1944.)

Figure 10.3. Dorsal view of larva of Ceonagrionidae (Zygoptera). (Redrawn from Wright and Peterson 1944.)

Figure 10.4. Lateral view, larva of Coenagrionidae (Zygoptera). (Redrawn from Wright and Peterson 1944.)

Figure 10.5. Dorsal view, head of larva of *Calopteryx* (Calopterygidae). (Redrawn from Wright and Peterson 1944.)

Figure 10.5a. Dorsal view, prementum of labium of larva, *Calopteryx* (Calopterygidae). (Redrawn from Wright and Peterson, 1944.)

Figure 10.6. Dorsal view, prementum of larva of *Lestes* (Lestidae). (From Merritt and Cummins 1978.)

Figure 10.7. Dorsal view, prementum of labium of larva, *Argiallagma minutum* (Selys) (Coenagrionidae).

Figure 10.8. Dorsal view, prementum of labium of larva, *Phylolestes ethelae* Christiansen (Synlestidae).

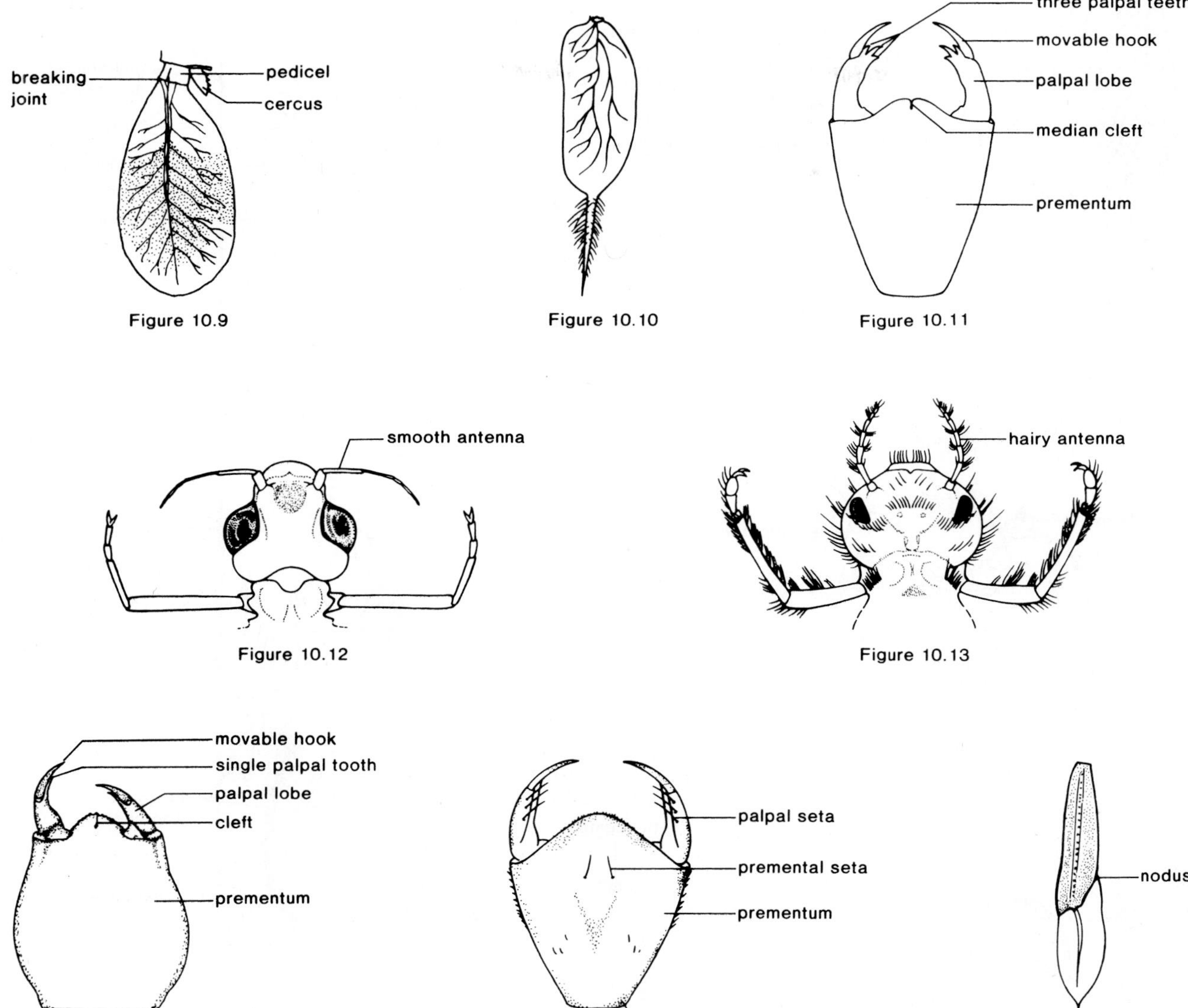

Figure 10.9. Lateral view, right caudal gill of larva, *Phylolestes ethelae* Christiansen (Synlestidae).

Figure 10.10. Lateral view, right caudal gill of larva, *Hypolestes clara* (Calvert) (Hypolestidae).

Figure 10.11. Dorsal view, prementum of labium of larva, *Hypolestes clara* (Calvert) (Hypolestidae).

Figure 10.12. Dorsal view, head and prothorax of larva,*Hypolestes clara* (Calvert) (Hypolestidae).

Figure 10.13. Dorsal view, head and prothorax of larva, *Palaemnema* (Platystictidae).

Figure 10.14. Dorsal view, prementum of labium of larva, *Palaemnema* (Platystictidae).

Figure 10.15. Dorsal view, prementum of labium of larva, *Neoneura maria* (Scudder) (Protoneuridae).

Figure 10.16. Lateral view, right caudal gill of larva, *Protoneura viridis* Westfall (Protoneuridae). (From Merritt and Cummins 1978.)

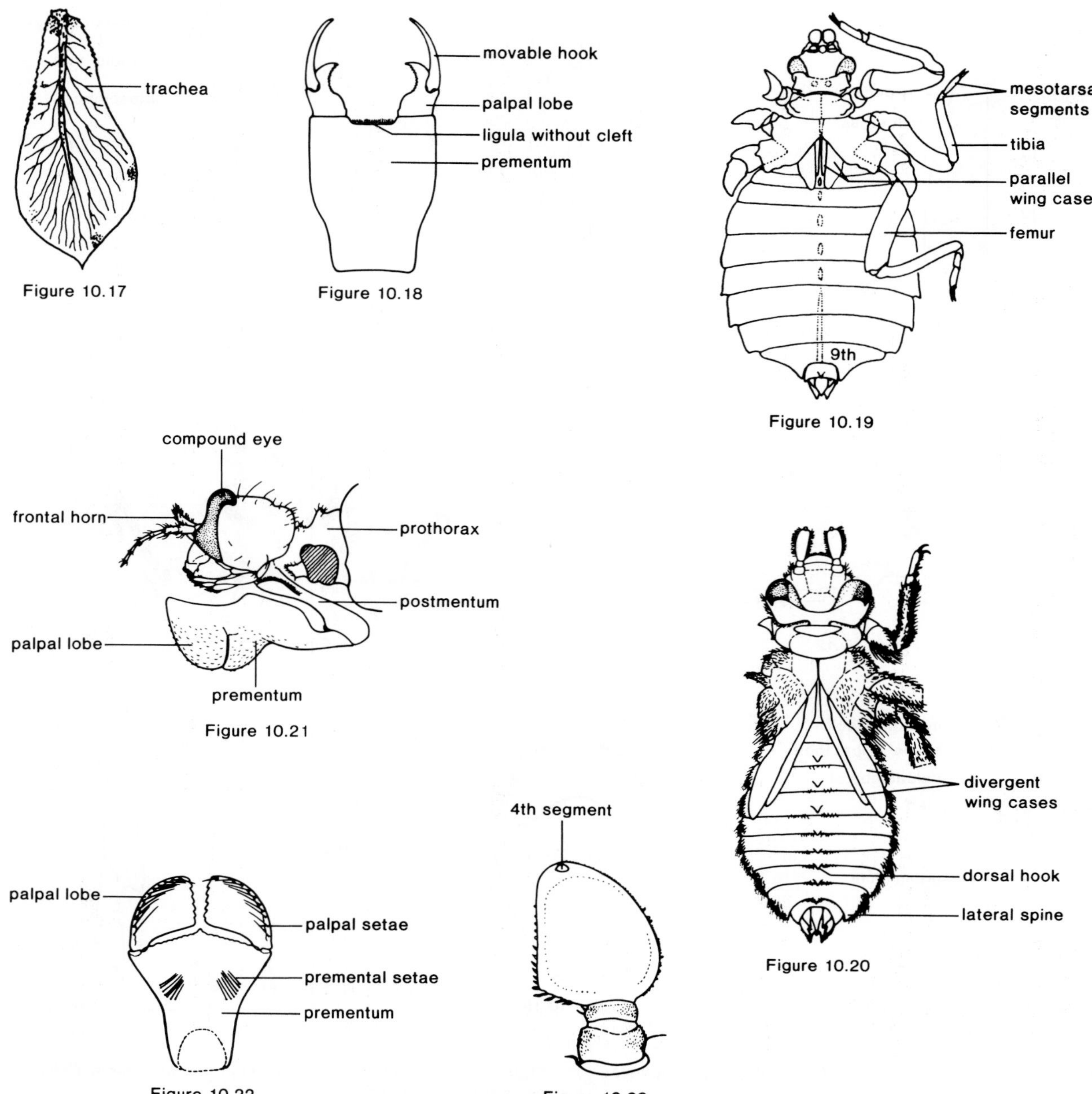

Figure 10.17. Lateral view, right caudal gill of larva, *Telebasis byersi* Westfall (Coenagrionidae).

Figure 10.18. Dorsal view, prementum of labium of larva, *Dromogomphus spinosus* Selys (Gomphidae). (Redrawn from Wright and Peterson 1944.)

Figure 10.19. Dorsal view, larva of *Hagenius brevistylus* Selys (Gomphidae). (Redrawn from Wright and Peterson 1944.)

Figure 10.20. Dorsal view, larva of *Ophiogomphus* (Gomphidae). (Redrawn from Wright and Peterson 1944.)

Figure 10.21. Lateral view of head and prothorax of larva, showing labium composed of postmentum, prementum and palpal lobes, *Macromia* (Macromiidae). (Redrawn from Wright and Peterson 1944.)

Figure 10.22. Dorsal view, prementum of labium of larva, *Plathemis lydia* (Drury) (Libellulidae). (Redrawn from Wright and Peterson 1944.)

Figure 10.23. Antenna of *Stylogomphus albistylus* (Hagen) (Gomphidae). (Redrawn from Wright and Peterson 1944.)

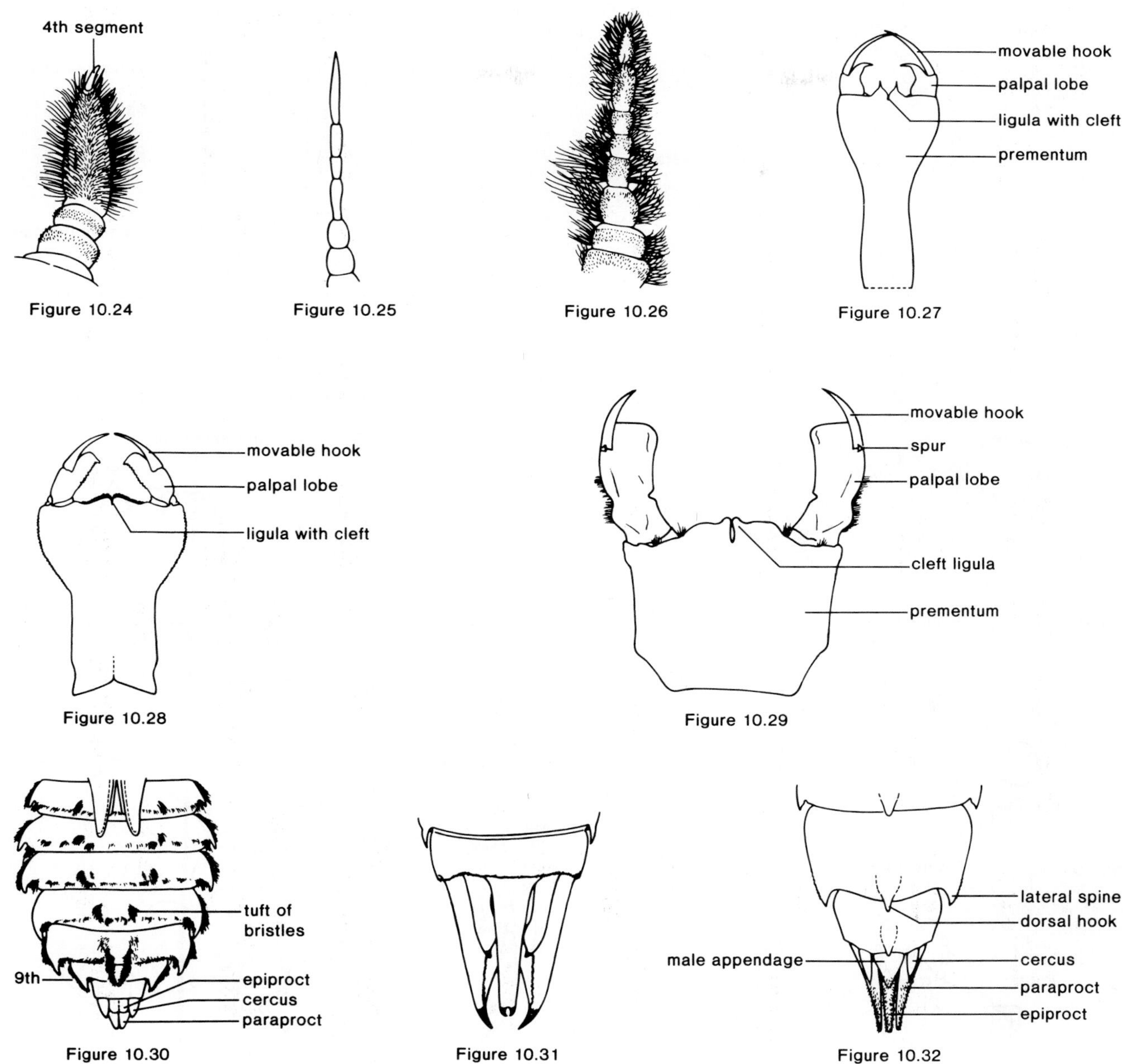

Figure 10.24. Antenna of larva of *Progomphus obscurus* (Rambur) (Gomphidae). (Redrawn from Wright and Peterson 1944.)

Figure 10.25. Antenna of larva of *Aeshna* (Aeshnidae). (Redrawn from Wright and Peterson 1944.)

Figure 10.26. Antenna of larva of *Tachopteryx thoreyi* (Hagen) (Petaluridae). (Redrawn from Wright and Peterson 1944.)

Figure 10.27. Dorsal view, prementum of labium of larva, *Coryphaeschna ingens* (Rambur) (Aeshnidae). (Redrawn from Wright and Peterson 1944.)

Figure 10.28. Dorsal view, prementum of labium of larva, *Aeshna* (Aeshnidae). (Redrawn from Wright and Peterson 1944.)

Figure 10.29. Dorsal view, prementum of labium of larva, *Tachopteryx thoreyi* (Hagen) (Petaluridae). (From Merritt and Cummins 1978.)

Figure 10.30. Dorsal view, abdomen of larva, *Tachopteryx thoreyi* (Hagen) (Petaluridae). (Redrawn from Wright and Peterson 1944.)

Figure 10.31. Dorsal view, tip of abdomen of larva, *Oplonaeschna armata* (Hagen) (Aeshnidae).

Figure 10.32. Dorsal view, tip of abdomen of larva, *Nasiaeschna pentacantha* (Rambur) (Aeshnidae). (Redrawn from Wright and Peterson 1944.)

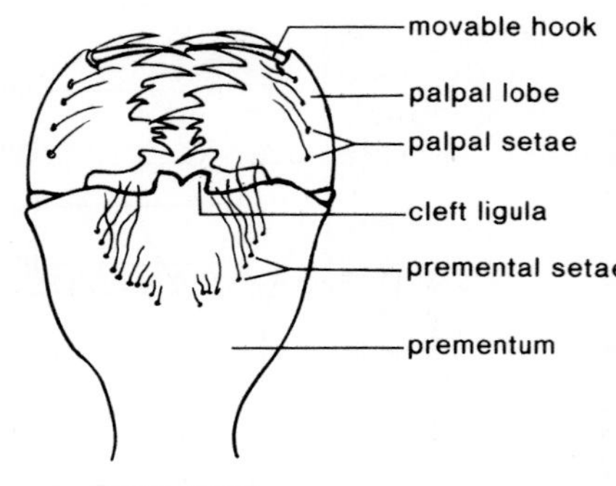

Figure 10.33

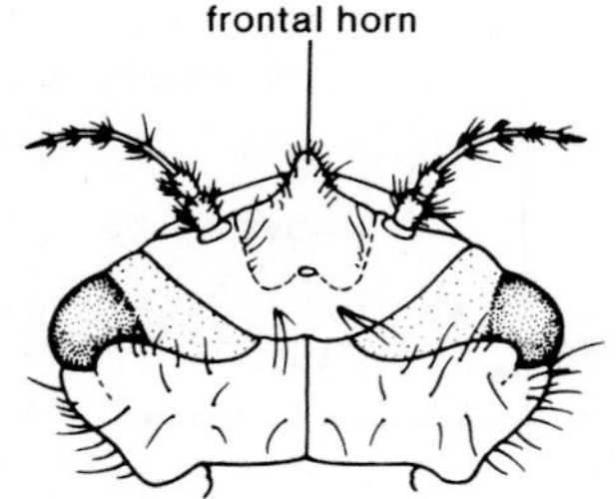

Figure 10.35

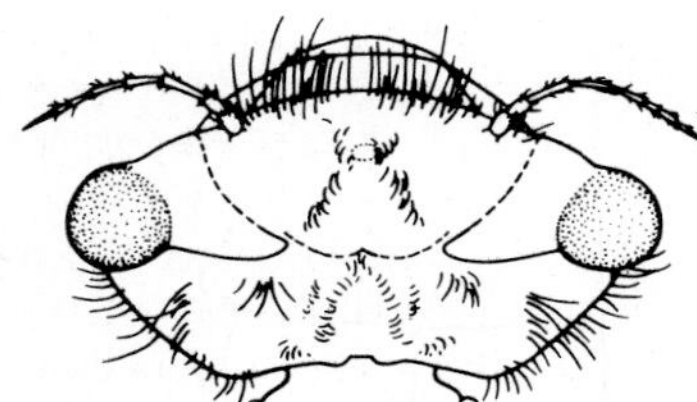

Figure 10.36

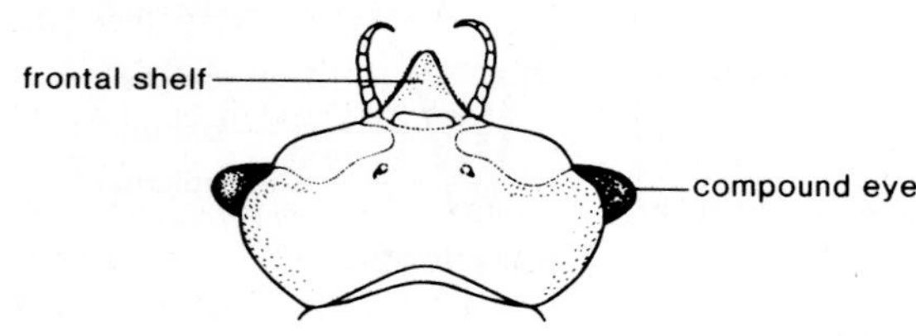

Figure 10.37

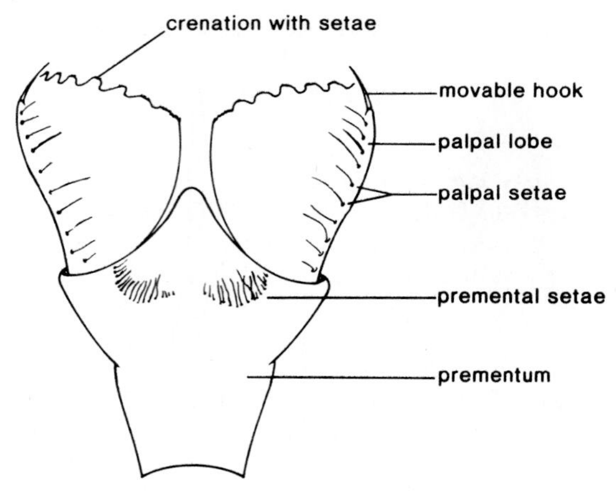

Figure 10.34

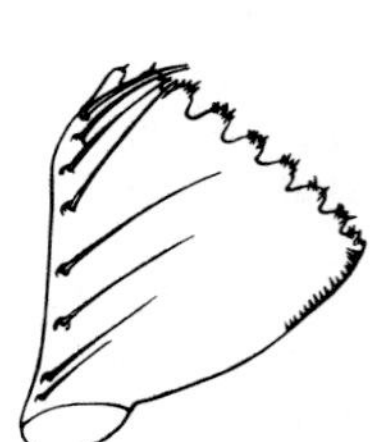

Figure 10.34a

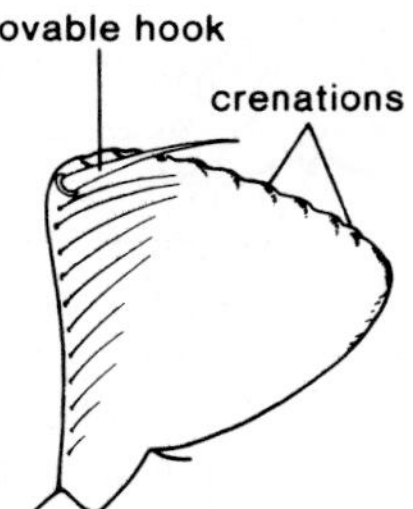

Figure 10.34b

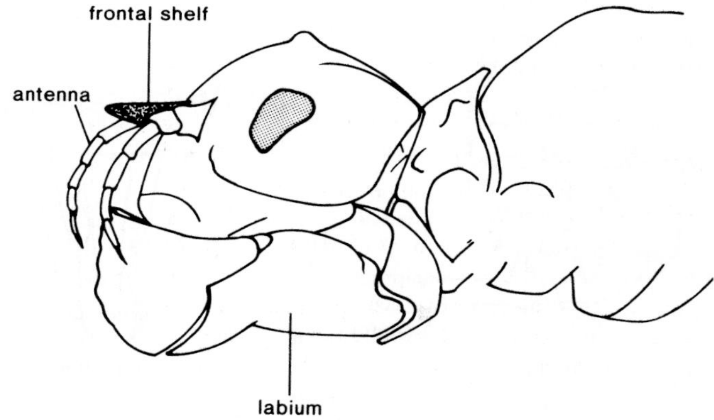

Figure 10.38

Figure 10.33. Dorsal view, prementum of labium of larva, *Cordulegaster* (Cordulegastridae). (From Merritt and Cummins 1978.)

Figure 10.34. Dorsal view, prementum of labium of larva, *Paltothemis lineatipes* Karsch (Libellulidae). (From Merritt and Cummins 1978.)

Figure 10.34a. Dorsal view, left palpal lobe of labium, *Somatochlora elongata* (Scudder) (Corduliidae). (Redrawn from Wright and Peterson 1944.)

Figure 10.34b. Dorsal view, left palpal lobe of labium of larva, *Tramea* (Libellulidae). (Redrawn from Wright and Peterson 1944.)

Figure 10.35. Dorsal view, head of larva, *Macromia* (Macromiidae). (Redrawn from Wright and Peterson 1944.)

Figure 10.36. Dorsal view, head of larva, *Libellula* (Libellulidae). (Redrawn from Wright and Peterson 1944.)

Figure 10.37. Dorsal view of head of larva, *Neurocordulia molesta* (Walsh) (Corduliidae). (Redrawn from Wright and Peterson 1944.)

Figure 10.38. Lateral view of head of larva, *Neurocordulia molesta* (Walsh) (Corduliidae). (From Merritt and Cummins 1978.)

LESTIDAE (LESTOIDEA)

The Lestids, Spread-Winged Damselflies

Figure 10.6

Relationships and Diagnosis: This is one of three families of the Lestoidea represented in our range. The others are the Synlestidae and Hypolestidae. The latter is considered by some authors as a subfamily of Pseudolestidae of Fraser (1957).

The larvae are unique in having the prementum distinctly petiolate and spoon-shaped, with the narrow proximal part as long as or longer than the expanded distal part (fig. 10.6). The movable hook of each palpal lobe has two to four long setae. There are usually five to eight (rarely four) premental setae. The gills are nearly parallel-sided, and the numerous tracheae are almost perpendicular to the long axis (fig. 10.6).

Biology and Ecology: These long, slender, climbing larvae are found in still, marshy, or bog-margined waters, usually in ponds or the edges of small sheltered lakes. They may also be found in the dammed-up portions of streams and in slow, weedy streams. When captured, they may flip about like small minnows. The elongate eggs are usually deposited well above the water line in standing aquatic plants. The larvae are usually common to abundant where found.

Description: Larvae long and slender, with long legs and noticeably stalked prementum. Length of mature specimens 22–38 mm, including gills. Color usually light green to brown, sometimes quite dark, often with little marking.

Head: Very wide, two or more times as wide as long, eyes prominent. Prementum noticeably spoon-shaped and stalked, the slender proximal part comprising up to three-fourths of total length and extending to apices of metacoxae. Prementum with short, closed median cleft. Palpal lobes trifid, with three to five strong raptorial setae, all but one on the movable hook. Four to eight premental setae on each side of midline (fig. 10.6).

Thorax: Usually slender, but may be as broad as long, with T2 and T3 considerably wider than T1. Legs long and slender. T3 wing cases extending to abdominal segment 3 or 4 in mature larvae, and parallel to each other.

Abdomen: Subcylindrical with lateral spines, usually on segments 6–9, some species with spines on more proximal segments also. Large caudal gills lacking a nodus or stiff marginal setae; tracheae numerous, closely aligned, nearly perpendicular to the axis, and branched near the margins.

Comments: As limited by Davies (1981), Lestidae contains 12 genera. Of these, only two, *Archilestes* and *Lestes,* are found within our range. Two species of *Archilestes* occur in the United States, but several other species are found in the New World tropics. *Lestes* is a large genus (about 105 species) with worldwide distribution. Within our range, only 18 species are present, and the larvae are known for all but one.

Selected Bibliography

Klots 1932.
Lutz 1968.
Lutz and Pittman 1968.
Needham, 1904, 1941b.
Walker 1914a, 1914b.
Westfall and Tennessen, 1973.

SYNLESTIDAE (LESTOIDEA)

The Synlestids

Figures 10.8, 10.9

Relationships and Diagnosis: This is one of three families of the Lestoidea occurring in our range and is considered to be very ancient. The others are the Lestidae and Hypolestidae. The latter is considered by some authors as a subfamily of Pseudolestidae. Within our range, the Synlestidae is represented by only one species, *Phylolestes ethelae* Christiansen of Hispaniola in the West Indies. The nearest relatives are in Africa.

P. ethelae is unique in our fauna by having the pedicel of each gill very distinct and separated from the blade of the gill by a "breaking joint" (fig. 10.9). Nevertheless, a larva is rarely seen with a missing gill. The medial cleft of the prementum extends beyond the base of the palpal lobes (fig. 10.8). The gills are flat with broadly rounded apices (fig. 10.9). Members of this family, along with those of Hypolestidae and Platystictidae, lack both premental and palpal setae. Mature larvae of our species are 30–34 mm, including the gills. The general color is light tan with darker brown markings. The legs are relatively long, and the short broad gills are not caducous.

Biology and Ecology: Larvae of *Phylolestes ethelae* were found in 1960 in a cold, clear stream on a small plateau at La Visite, near Seguin in the LaSelle Mountains of southern Haiti at an elevation of about 7000 ft. They were observed sitting in plain sight on large rocks in the very cold spring-fed stream. Abdominal segment 1, seen between the bases of the wing cases, was conspicuously white. Very few other odonates were found at this stream. There were many larvae of *Aeshna psilus* Calvert, thousands of mayflies, and some hydropsychid Trichoptera, a few water beetles, and small Hemiptera. When brought down to sea level, where it was much warmer, the larvae began vigorously swinging their abdomens from side to side. Adults have also been taken in high mountains of the Dominican Republic.

Comments: The nine genera of this family have worldwide distribution. The single species found in our range is a relic species with its nearest relatives in the genus *Chlorolestes* of Africa. It is confined to very cold mountain-top streams in Hispaniola, and the lack of many such habitats nearby apparently prevents it from spreading very far. Some authors have treated this family as Chlorolestidae.

Selected Bibliography

Christiansen 1947 (adults only).
Westfall 1976.

HYPOLESTIDAE (LESTOIDEA)

The Hypolestids

Figures 10.10, 10.11, 10.12

Relationships and Diagnosis: This is one of three families of the Lestoidea found in our range. The others are the Lestidae and the Synlestidae. As treated here, the discussion concerning Hypolestidae is based on the genus *Hypolestes,* which contains two or three species found in Cuba, the Dominican Republic, Haiti, and Jamaica.

The Hypolestidae, along with the Platystictidae and Synlestidae, always lack premental and palpal setae. An occasional species of *Argia* of the Coenagrionidae may also agree in this character. The Hypolestidae and Platystictidae have saccoid or balloonlike gills with the apices long and sharply pointed (fig. 10.10). Larvae of Hypolestidae may be distinguished from Platystictidae by the eyes forming a large portion of the lateral margin of the head, whereas the eyes of the Platystictidae are small and do not form such a part of the margin (figs. 10.12, 10.13). Also, the palpal lobes of Platystictidae are less than half as long as the prementum; in Hypolestidae, more than half. The distal margin of each palpal lobe has three hooks in addition to the movable hook; in Platystictidae there is only one. The mature larva is rather stout and somewhat depressed, about 20 mm in length, including the gills. The color is light tan with some darker markings on the abdomen.

Biology and Ecology: The genus *Hypolestes,* endemic to the West Indies, is found in small, swift streams. It may be sifted from among coarse gravel or small stones in the stream bed, or it may be found clinging to larger rocks when lifted from the stream. Individuals may emerge on large boulders in the stream bed a few inches above the water in the Dominican Republic. Needham (1941b) found the larvae associated with those of *Macrothemis celeno* (Selys) and *Scapanea frontalis* (Burmeister) of the Libellulidae. Although the larva has been described, very little is known of the behavior and ecology of the genus.

Comments: *Hypolestes* is put in Fraser's family Pseudolestidae by Davies (1981). The latter is admitted by Fraser (1957) to be an annectant family between the Coenagrionoidea and Calopterygoidea and served as sort of a catch-all for genera of doubtful relationship. Davies further uses a subfamily Hypolestinae to contain *Hypolestes* and three other extant genera not within our range. The genus *Hypolestes* has been collected in Cuba, the Dominican Republic, Haiti, and Jamaica.

Selected Bibliography

Needham 1941b.
Westfall 1964a.

PLATYSTICTIDAE (COENAGRIONOIDEA)

The Platystictids

Figures 10.13, 10.14

Relationships and Diagnosis: This is one of three families of the Coenagrionoidea occurring within our range. The others are Coenagrionidae and Protoneuridae. Two subfamilies, Palaemnematinae and Platystictinae, contain four genera. Palaemnematinae is restricted to the New World and is represented in our range by the single genus *Palaemnema.*

Larvae of this genus have been described as somewhat "termitelike" in general appearance, especially the head. The eyes are small, not forming a portion of the lateral margin of the head (fig. 10.13). The larva differs from the Calopterygidae by having a short first antennal segment, much shorter than the remaining segments together. Unlike the Lestidae, the prementum is as wide at the base as at the apex, and the gills are saccoid or balloonlike. There are no premental or palpal setae, which is also true only of Hypolestidae and Synlestidae of our range. Hypolestidae and Platystictidae are the only families in our range with saccoid gills. They are easily separated by the palpal lobes, which are more than half as long as the prementum in Hypolestidae, and which have three end hooks in addition to the movable hook. In Platystictidae the palpal lobes are small, less than half as long as the prementum and with only one distal hook in addition to the movable hook (fig. 10.14).

Biology and Ecology: The adults of *Palaemnema* are usually encountered in dense woods, but nothing has been published about the larvae or their habitat.

Description: So far as known, the larvae of *Palaemnema* of our range are unique in general appearance, having characters, especially of the head, suggesting a termite. The species known are not patterned with color but are rather uniformly light brown.

Head: Rounded, almost as long as wide. Eyes small, beady, not forming a part of the lateral margin of the head as in other families (fig. 10.13). Antennae six-segmented and hairy. Labium broad at base, almost as wide as at apical end. Sides of prementum strongly convex. Front border of labium strongly convex and with a closed median cleft. Each palpal lobe bearing a single pointed end hook and a small movable hook. No palpal or premental setae (fig. 10.14).

Thorax: Legs moderate in length, hind legs not reaching end of adbominal segment 10 and covered with numerous long setae. Wing cases parallel to each other.

Abdomen: Bearing three long, pointed, saccoid caudal gills, the median gill longest. Each gill with long, thin, hairlike setae near apex, but no coarse setae along the margins.

Comments: In our range, this family has only one genus, *Palaemnema;* a single species has been taken in Nuevo Leon, Mexico. There are some 30 species known in Central and South America, and many more species awaiting description. Much work needs to be done on this group, as larvae are almost unknown for any species. The other three genera contain over 100 species but all are East Palaearctic or Southeast Asian.

Selected Bibliography

Calvert 1931, 1934 (adults only).

PROTONEURIDAE (COENAGRIONOIDEA)

The Protoneurids

Figures 10.15, 10.16

Relationships and Diagnosis: The Protoneuridae is one of three families in the Coenagrionoidea occuring within our range. The others are Coenagrionidae and Platystictidae.

Larvae of Protoneuridae may be distinguished from those of Synlestidae, Hypolestidae, and Platystictidae by the presence of premental and palpal setae (fig. 10.15). These larvae differ from Calopterygidae larvae by the length of the first antennal segment, which is shorter than the combined length of the remaining segments. They also differ from the Lestidae by the shape of the prementum, which is not distinctly petiolate, and by the absence of setae on the movable hook of each palpal lobe. Larvae of Protoneuridae usually possess only one premental seta each side of the midline, whereas Coenagrionidae larvae usually possess three to five. The prementum of Platystictidae and Hypolestidae larvae (and certain larvae of Synlestidae) differs by the cleft in the anterior margin, a characteristic absent in the Protoneuridae (fig. 10.15).

Biology and Ecology: Larvae of these genera are primarily stream dwellers. They are often found in small, swift streams clinging to the underside of rocks. However, *Neoneura aaroni* Calvert has also been collected in a large river in Texas. Farther south, in Ecuador, adults of *Protoneura* were collected as mating pairs around the edge of a lake. *Microneura* larvae have been found in the same stream from which *Hypolestes* has been collected. Exuviae of *Microneura* were taken from rocks two inches above the water line.

Description: Larvae rather slender with thin lamellate gills. Known species have rather strongly banded legs and gills, which are conspicuously nodate, the proximal part of the gill being thicker and darker than the apical portion (fig. 10.16). Full-grown larvae are 12–17 mm in length, including the gills.

Head: Rather broad and flattened, widest across the very large eyes, which occupy more than half the lateral margin of the head and may protrude from the side. In *Microneura,* there are about 13 alternating light and dark dorsoventral bands on the eyes. Caudolateral angles of head rounded. Prementum of all known species only a little longer than wide, with a single premental seta each side of midline, and 3–5 palpal setae (fig. 10.15). Front margin of prementum entire. Distal margin of each palpal lobe with long, curved mesal end hook, and a lateral, shorter hook adjacent to the movable hook, which is very long in *Microneura.*

Thorax: Dorsum usually conspicuously patterned in light and darker brown. Legs strongly banded, the metathoracic legs usually reaching end of abdominal segment 10. Wing cases in a full-grown larva usually extending at least to end of abdominal segment 5 and in some species to half length of segment 6. In *Protoneura,* as far as known, wing pads almost transparent, so that abdominal segments can be seen through them, and pterostigma conspicuous.

Abdomen: Tapering from base to apex and usually noticeably patterned in dark and light brown. Gills unusual in that basal part of each gill is thicker and darker than apical portion, with a conspicuous node (fig. 10.16). Gills of known *Protoneura* longer than abdomen, but in *Neoneura* shorter.

Comments: As treated by Davies (1981), the Protoneuridae is a widespread family with 21 genera. Of these, only three, *Microneura, Neoneura,* and *Protoneura,* occur within our range. *Microneura* is limited to Cuba, and in the United States *Protoneura cara* Calvert and *Neoneura aaroni* Calvert are known only from Texas. *Protoneura* contains five species in our range, *Neoneura* has four, and *Microneura* a single species.

Selected Bibliography

Needham 1939.
Westfall 1964a, 1964b.

COENAGRIONIDAE (COENAGRIONOIDEA)

The Coenagrionids, Narrow-Winged Damselflies

Figures 10.3, 10.4, 10.7, 10.17

Relationships and Diagnosis: This is one of three families in the Coenagrionoidea occurring within our range. Numerically, this is the largest of the six families in the superfamily. Only two others occur in our range, the Platystictidae and Protoneuridae.

The larval gills differ from those of the Protoneuridae in that the proximal portion usually does not differ markedly from the distal portion, and therefore, the nodus is usually not delineated across the entire width of the gill (fig. 10.17). The family differs from the Platystictidae, as well as the Hypolestidae and Synlestidae, in having the margin of the prementum entire, without a median cleft, and in usually possessing premental and palpal setae (fig. 10.7). The first antennal segment is not as long as the combined length of the remaining segments, a character that distinguishes the family from the Calopterygidae. The Coenagrionidae lack the petiolate, spoon-shaped prementum of the Lestidae. The family also has gone under the name Coenagriidae.

Biology and Ecology: Larvae of this large family differ in habitat, but most are found in quiet waters of ponds or lakes. Most species of *Argia* occur in running water of rivers or smaller streams, sometimes clinging to rocks on the bottom. Eggs are usually inserted into plant tissues below the water line, the female at times descending to a depth of a foot or more for oviposition. Larvae of at least one species of *Megalagrion* in Hawaii are quite terrestrial, living far from standing water. Others, such as *Diceratobasis macrogaster* (Selys) in Jamaica, live in water in the bases of bromeliads.

Description: Larvae of the generalized slender form (figs. 10.3, 10.4). Some, such as species of *Argia*, rather short and stocky, with short legs. Color usually light green to very dark brown or almost black.

Head: Eyes wide apart as typical for the Zygoptera and may be characteristically marked in some genera and species. Prementum more or less triangular or subquadrate, its front margin entire, never with a median cleft. Almost always 3–5 premental setae each side of the midline, as well as 1–7 (10 in *Diceratobasis*) raptorial setae on each palpal lobe. Palpal lobes parallel-sided and terminating in not more than two hooklike processes. Movable hook arising nearer the midline than in the Lestidae and never bearing raptorial setae as it does in that family.

Thorax: Quite variable, legs short to medium in length, usually not extending to end of abdomen. Wing cases, when fully grown, extending about to middle of abdomen.

Abdomen: Only in a few genera are there any lateral setae on the abdomen, and never a middorsal ridge with spines on any segment. The caudal gills have the most diagnostic characters of the abdomen, with the basal part of the gill almost never thicker and darker than distal part. Tracheae forming an acute angle with the main trachea in middle of gill (fig. 10.17).

Comments: Davies (1981) lists this family as worldwide in distribution, with six subfamilies containing about 95 genera (although not all authors recognize some of these to be generically distinct). Four subfamilies are represented in our range by 20 genera, with approximately 146 species recorded in the literature, plus several species currently being described. The majority of species are placed in three genera: *Argia* (32); *Enallagma* (39); and *Megalagrion* of Hawaii (34). The larvae of most species have been reared but some have not been described.

Selected Bibliography

Baker 1981.
Corbet 1983.
Crowley 1979.
Garcia-Diaz 1938.
Geijskes 1943.
Klots 1932.
Needham 1904, 1941b, 1945a.
Pearse 1932.
Pritchard and Pelchat 1977.
Provonsha and McCafferty 1973.
Tennessen and Knopf 1975.
Walker 1914b.
Westfall 1957, 1964a, 1976.

Suborder ANISOPTERA

AESHNIDAE (AESHNOIDEA)

The Aeshnids, Darners

Figures 10.25, 10.27, 10.28, 10.31, 10.32

Relationships and Diagnosis: In our range the well-defined superfamily Aeshnoidea contains three families, Petaluridae, Gomphidae, and Aeshnidae, which includes two subfamilies. These families agree in having the labium flat, without premental setae, palpal lobes narrow with lateral and mesal edges subparallel, movable hook large, palpal setae almost always absent; ligula with or without a median cleft.

Larvae of Aeshnidae differ from those of Gomphidae in having slender, bristlelike six- or seven-jointed antennae, rather than thick, four-jointed antennae. The relatively long three-jointed tarsi differ from the short tarsi of Gomphidae in which the tarsi of the first two pairs of legs are two-jointed and are usually equipped with a burrowing hook. The slender antennae of the Aeshnidae differ from the thick and hairy antennae of the Petaluridae. The Petaluridae also have a sharp spur at the base of the movable hook of the palpal lobe, a structure not present in the Aeshnidae or any other family. The costal margins of the wing cases are parallel, as in the Petaluridae, whereas in Gomphidae some genera have them strongly divergent.

Biology and Ecology: Larvae are primarily climbers on stems, stumps, and other submerged surfaces. Habitats vary from green aquatic vegetation to dead stems along the water's edges. They are found in most bodies of permanent fresh water, occurring on all continents, but missing in Oceania and Madagascar (Malagasy Republic). The eggs are laid in various kinds of living or dead plant stems.

Description: Larvae have long, graceful bodies, and are generally larger than the Gomphidae, ranging from *Gomphaeschna* 30 mm long, to *Anax*, 40–62 mm in length. They are colored and patterned to match their environment.

Head: More or less flattened, eyes lateral. Hind angles of head rounded or angulate. Antennal segments long and slender. Eyes typically more prominent than those of Gomphidae and nearer front of the head. Labium long, flat, never masklike.

Thorax: Somewhat depressed, widening a little posteriorly. Underneath the prothorax is a pair of blunt supracoxal processes. Sides of synthorax sloping steeply down to leg bases. Legs long and thin, sometimes crossbanded with brown. Wing cases parallel, not divergent.

Abdomen: Dorsal hooks (except for low humps in *Nasiaeschna*) lacking. Lateral spines on segments 5–9 varying by genera and species. Epiproct usually distinctly bifid at tip (fig. 10.31).

Comments: These dragonflies are found almost worldwide. The larvae are extremely aggressive and voracious when prey is nearby, often attacking smaller individuals of their own species. They are long and graceful and include some of the largest species, e.g., *Anax walsinghami* McLachlan, and have been commonly called "darning needles" as adults. Some species have been destructive of small fingerlings in fish hatcheries and fish may have to be protected by screening tanks. Needham and Westfall (1955) used a "c" in the spelling of the family and genus (Aeschnidae and *Aeschna*) to agree with other generic names like *Basiaeschna, Gomphaeschna,* etc., but Walker has shown that an error was probably made when *Aeschna* was first described and the name should be *Aeshna,* and the family Aeshnidae, although the other generic names must retain their original spellings with the "c".

Selected Bibliography

Byers 1927b.
Cabot 1881.
Dunkle 1977b.
Garcia-Diaz 1938.
Kennedy 1936.
Klots 1932.
Needham 1904, 1945a.
Pritchard 1964.
Walker 1912, 1913, 1941, 1958.

PETALURIDAE (AESHNOIDEA)

The Petalurids, Graybacks

Figures 10.26, 10.29, 10.30

Relationships and Diagnosis: The Petaluridae is considered to be an ancient and primitive family in form and pattern, and is very local in distribution. In our range the well-defined superfamily Aeshnoidea contains the Petaluridae, Aeshnidae, and Gomphidae. All are alike in having the labium flat, without premental setae, palpal lobes narrow with lateral and mesal edges subparallel, movable hook large, palpal setae almost always absent; ligula with or without a median cleft. The subcylindrical body is suggestive of *Cordulegaster.*

Larvae of Petaluridae differ from those of Aeshnidae by having six- or seven-jointed thick antennae (fig. 10.26), and an epiproct that is smooth at the tip (fig. 10.30) and without sharp points, as it is in Aeshnidae. The antennae of Aeshnidae are quite thin in comparison, and those of Gomphidae have only four joints. There is a small, sharp spur at the base of the movable hook of the palpal lobes (fig. 10.29), not present in any other family. The costal margins of the wing cases are always parallel in the Petaluridae and Aeshnidae, but strongly divergent in some Gomphidae. Although the petalurids might resemble *Cordulegaster* in body shape, the flat labium, without premental setae, and the relatively narrow palpal lobes, also without setae, show a greater resemblance to the Aeshnidae and the Gomphidae. Only two genera, *Tachopteryx* and *Tanypteryx,* occur in our range.

Biology and Ecology: Both *Tanypteryx* and *Tachopteryx* are rather uncommon and are local in distribution. *Tanypteryx* is found in the mountains on the Pacific side of North America, and *Tachopteryx* occurs on the Atlantic. They appear to have been supplanted in their favorite habitats by more specialized dragonflies. Larvae of *Tanypteryx* have been collected from mountainous areas where rocks have been covered by various mosses kept moist by constant shallow streams of water. They construct burrows in this plant material and muck, as do members of other genera of Petaluridae in other areas of the world. *Tachopteryx* has not been found in such burrows. They have been taken in very small permanent pools, but in Florida are usually found in permanent, spring-fed hillside seeps in broad-leaved deciduous forests. They are usually near the uphill edges of the seep, and under, or between, wet leaves, but not in the stream into which the seeps empty.

Description: Primarily sluggish, rough-hewn larvae, dark and hairy. Antennae 6- or 7-jointed and eyes wide apart. Length when full-grown about 30–38 mm.

Head: Eyes small and wide apart, capping front angles of the head. Antennae thick and hairy, 6-jointed in *Tanypteryx* and 7-jointed in *Tachopteryx,* each segment about as wide as long. Prementum of labium fully as broad as long, and with median cleft on anterior margin. Palpal lobe somewhat curved and flat, without palpal setae, and truncate at distal end. Distal and mesal margins minutely and irregularly crenate. No end hook, and movable hook short compared with that of Aeshnidae and Gomphidae. A small spur at base of movable hook (fig. 10.29).

Thorax: Tarsi three-jointed, tibiae ending in several strong spurs. The stout carinated legs are characteristic of this group. Costal margins of wing cases parallel.

Abdomen: Spindle-shaped, broader than head and widest at middle segments. Lateral margins of segments strongly angulated, especially in *Tachopteryx,* and bearing very conspicuous tubercles with tufts of hair, or bristles, on abdominal segments each side of the midline, from segments 2 or 5 to 9 (fig. 10.30). No dorsal hooks on midline.

Comments: This family has relict members in Australasia, Chile, New Zealand, and the East Palaearctic region, as well as in North America, where they are local in distribution and usually rare in collections. Our two genera are placed in two subfamilies. *Tachopteryx,* with one species, is found only in North America, but *Tanypteryx* is represented by one species in North America and a second in Japan.

Selected Bibliography

Dunkle 1981.
Svihla 1958, 1959, 1975a, 1975b.
Walker 1958.
Williamson 1901.

GOMPHIDAE (AESHNOIDEA)

The Gomphids, Clubtails

Figures 10.18, 10.19, 10.20, 10.23, 10.24

Relationships and Diagnosis: The well-defined superfamily Aeshnoidea in our range contains three of the four recognized families—Gomphidae, Aeshnidae, and Petaluridae. All agree in having the labium flat, without premental setae, palpal lobes narrow with lateral and mesal edges subparallel, movable hook large, palpal setae almost always absent, ligula with or without a median cleft (fig. 10.18). The family Gomphidae is quite distinct from the others in both adult and larval stages.

Larvae of Gomphidae are distinguished from those of other families of Aeshnoidea by the flat head which bears thick, four-jointed antennae (figs. 10.23, 10.24), not six- or seven-jointed as in the others (figs. 10.25, 10.26). The front and middle tarsi are two-jointed (figs. 10.19, 10.20), not three-jointed (fig. 10.1) as in all other families of Anisoptera. Each tarsus is usually equipped with a burrowing hook in the Gomphidae, but not in the other families. The ligula is also different by being entire, lacking a median cleft (fig. 10.18). The costal margins of the wing cases are parallel or strongly divergent in Gomphidae (figs. 10.19, 10.20), always parallel in the other two families.

Biology and Ecology: The larvae are predominantly burrowers, with the exception of *Hagenius* (fig. 10.19), and are primarily stream dwellers. *Progomphus* digs like a mole in loose beds of drifting sand, leaving a trail that can be followed easily. The thick, short legs of most genera prevent the larvae from climbing up slender stems for transformation, therefore emergence usually takes place on the sand of the shore or on broad, rough surfaces of logs and stones, or on lily pads. Emergence of the entire brood may occur during a relatively short period of time.

Description: Mature larvae vary considerably in size from *Lanthus,* which is 20–23 mm long, to *Aphylla,* which is 42–68 mm. The body is predominantly flat, aiding the larvae in burrowing shallowly in bottom mud and sediment. Coloring is usually nondescript.

Head: Flat and wedge-shaped, with hind angles generally rounded. Antennae thick and 4-jointed (fig. 10.24), the fourth joint usually minute or vestigial, and the third joint sometimes greatly expanded laterally (fig. 10.23). Prementum and palpal lobes flat, or nearly so, dorsal surface lacking stout premental or palpal setae (fig. 10.18). Ligula entire, without a median cleft. Eyes not prominent and often placed near rear of head.

Thorax: In most genera legs short and thick. Front and middle tarsi 2-jointed (figs. 10.19, 10.20), their tips more or less flattened, and hooked outward for burrowing. Most with considerable hairiness about leg bases. *Hagenius* (fig. 10.19) with long legs, sprawling on bottom, and not having burrowing hooks. Wing cases parallel or divergent, depending on the genus (figs. 10.19, 10.20).

Abdomen: Shape varying from being nearly as broad as long in *Hagenius* (fig. 10.19), to being scarcely wider than the head in *Stylurus.* The relative length of segments is important in generic and specific determinations. Many genera with dorsal hooks on the midline (fig. 10.20); the nature of lateral spines important in identification.

Comments: These dragonflies are worldwide in distribution, and five subfamilies are recognized, of which three occur in North America. There are over 100 species within our range. Many have limited distributions and are not common in collections as adults or larvae. With the exception of *Hagenius* (a sprawler), all of our gomphids are burrowers. *Progomphus* digs a molelike trench in loose beds of drifting sand. *Aphylla* is distinguished by the elongation of segment 10, which is cylindrical and half as long as the entire abdomen, adapting these larvae for dwelling in deep, loose muck. The upturned end lifts the respiratory opening through silt to clear water. There are only a few species for which the larvae have not been described.

Selected Bibliography

Cabot 1872.
Calvert 1921.
Carle 1980, 1982.
Dunkle 1976.
Hagen 1885.
Kennedy 1921.
Kennedy and White 1979.
Knopf 1977.
Knopf and Tennessen 1980.
Needham 1904, 1940, 1941a, 1945b, 1948.
Walker 1928, 1932, 1933, 1958.
Westfall 1950, 1974.
Westfall and Tennessen 1979.
Westfall and Trogdon 1962.
Wright 1946c.

CORDULEGASTRIDAE (CORDULEGASTROIDEA)

The Cordulegastrids, Biddies

Figure 10.33

Relationships and Diagnosis: This is the only family of the Cordulegastroidea; it has been divided into two subfamilies, only one of which occurs within our range. Usually only one genus, *Cordulegaster,* is recognized in North America, although some authors have used other names, such as *Zoraena, Thecaphora,* or *Taeniogaster,* for certain species, whereas others have considered these as subgenera. In general form, these larvae superficially resemble some Petaluridae and some Libellulidae.

Compared with the other superfamilies of the Anisoptera, larvae of the Cordulegastroidea have a spoon-shaped labium covering the face up to the eyes and with premental setae as in the Libelluloidea, not flat and without premental setae as in the Aeshnoidea. The ligula is a small, cleft, tooth-like process; in the Aeshnoidea it is entire or cleft, though in the Libelluloidea a recognizable ligula is absent. The palpal lobes of the labium are narrow with the lateral and mesal edges subparallel in the Aeshnoidea, but broad and triangular in the Cordulegastroidea and the Libelluloidea. The distal margins of the palpal lobes of the Cordulegastroidea are so deeply cut as to form huge irregular teeth without setae on their margins (fig. 10.33), whereas in the Libelluloidea this edge is evenly crenate, with one seta or a group of them on each crenation, or if smooth, the edge has setae evenly spaced, singly or in groups (fig. 10.34). The movable hook (fig. 10.33) of the palpal lobe is smaller than in the Aeshnoidea but larger than in the Libelluloidea. The costal margins of the wing cases are slightly divergent, parallel in the Libelluloidea and parallel or strongly divergent in the Aeshnoidea. Although the general form of the Cordulegastroidea may resemble the Petaluridae, the shape of the labium is not flat as in that family and the antennae are slender (as in fig.

10.25), not thick and hairy as in the Petaluridae (fig. 10.26). The huge irregular teeth of the palpal lobes are probably the most distinctive feature identifying the Cordulegastroidea.

Biology and Ecology: Larvae are usually found buried in the sand or silt of small woodland streams, with only the tips of the eyes and the tip of the abdomen exposed. They do not burrow, but descend by raking the sand from beneath them by using sweeping, lateral movements of the legs. When deep enough, the sand is kicked up over the back. They may remain motionless for weeks until a prey organism approaches. Then the hidden labium springs into action and clutches the victim.

Description: Larvae are stout, conspicuously hairy, cylindrical, and tapering beyond the middle of abdomen to pointed apex. Longitudinal axis upcurved at both ends, with the tips of eyes and abdominal appendages the high points. Total length of mature larvae 31–45 mm.

Head: Eyes capping the angular anterolateral prominences of head. Hind angles of head rounded and posterior margin not noticeably concave. A large labium extending posteriorly between bases of middle legs, its spoon-shaped anterior part covering face to bases of antennae. A convex frontal shelf bears sensory hairs on its margin; its shape and the nature of hairs on its surface varying with different species.

Prementum triangularly widened beyond middle, and produced into a median tooth that is bifid on the median line (fig. 10.33). Palpal lobes broad, triangular, concave, and bearing a row of short raptorial setae near external margin, and a stouter, but not longer, movable hook at end of this row. On the opposed margins, palpal lobes deeply and irregularly cut to form huge, irregular, interlocking teeth. Antennae slender, with seven joints.

Thorax: Prothorax with dorsal, flattened extension on each side termed the epaulet. It is fringed on the margin with stiff hairs. Shape of epaulets and length of hairs can be used in identifying some species. Tarsi three-jointed. Wing cases a little divergent and reaching to about fourth abdominal segment in mature larvae.

Abdomen: Subcylindrical, arcuately upcurved toward tip. No dorsal hooks, but apices of segments bearing rows of very stiff hairs said to aid in holding a layer of sand, dirt, and other material around the body. Lateral spines absent in western U.S. species, small in eastern ones.

Comments: Although the family has a wide distribution in the Palaearctic region and Southeast Asia, only the genus *Cordulegaster* occurs in our range, with eight species. One of these, *C. sayi* Selys of Florida, is on the list of endangered species. The larvae are voracious feeders and are said to capture young brook trout. All our species have been reared.

Selected Bibliography

Cabot 1872.
Carle 1983.
Hagen 1885.
Walker 1958.

MACROMIIDAE (LIBELLULOIDEA)

The Macromiids, Belted Skimmers, River Skimmers

Figures 10.21, 10.35

Relationships and Diagnosis: The Macromiidae, Corduliidae, and Libellulidae in our range belong to the Libelluloidea, a well-defined superfamily. See the family Libellulidae for a discussion of relationships (p. 112). Some Libelluloidea may superficially resemble some species of the superfamily Cordulegastroidea. See p. 110 for a comparison.

In our range we have only two genera, *Didymops* and *Macromia*. The larvae have very long legs, with the hind femur extending as far posteriorly as abdominal segment 8. The abdomen is three-fourths to four-fifths as broad as long, and there is always an upcurved triangular projection between the antennae termed a frontal horn (figs. 10.21, 10.35). The only larva in our range with a similar structure is *Neurocordulia molesta* Walsh of the Corduliidae. Its triangular frontal shelf is not upturned but projects straight forward, its base is broader than its length, and the legs are much shorter (fig. 10.37). The Cordulegastridae also have a spoon-shaped labium, but the palpal lobes are more deeply cut on the opposing margins to form huge, irregular teeth (fig. 10.33). Also, the legs are much shorter in the Cordulegastridae and the abdomen subcylindrical, with the wing cases slightly divergent.

Biology and Ecology: These large larvae with long, spiderlike legs, lie sprawled on the bottom of small woodland streams or quiet stretches of rivers where sticks and leaves have accumulated. In Florida, one species, *Didymops floridensis* Davis, inhabits the deeper waters of sand-bottomed lakes and comes into shallow water when it is close to emergence. Larvae of both genera are excellent climbers, and the exuviae may be found as high as three meters on tree trunks or in bushes, and several meters from water.

Description: Larvae flattened with broadly oval or subcircular abdomen and very long legs. Color from light brown to almost black. Mature larvae 24–42 mm in length.

Head: Outline pentagonal. Slender antennae with seven segments, and pyramidal frontal horn between the antennae almost erect (fig. 10.21). Eyes capping anterodorsal angles of head, being very prominent (fig. 10.35). Prementum of labium very broadly triangular; palpal lobes also very wide, with distal margins deeply cut into two series of crenations, each usually bearing a large group of setae. Movable hook small as in other Libelluloidea. Posterolateral angles of head bearing an erect tubercle (fig. 10.35).

Thorax: Pronotum bearing a pair of small, erect tubercles laterally, adjacent to head tubercles. Depressed pterothorax at level of middle and hind legs decidedly wider than the head. Very long legs slender and spreading, with very long and simple tarsal claws. Wing cases parallel over the abdomen.

Abdomen: Very broadly oval or subcircular, somewhat more than two-thirds as wide as long. Most segments bearing rather high cultriform dorsal hooks on midline; lateral spines present only on segments 8 and 9. Anal pyramid short and broad.

Comments: The genus *Didymops* is represented by only two species and both occur in North America. *Macromia* is a very large genus of more than 100 species, almost worldwide in distribution. In North America there are about eight valid species. A third genus, *Epophthalmia,* contains about 11 species found in East and West Palearctic regions and in Southeast Asia.

Selected Bibliography

Cabot 1890.
Tillyard 1917.
Walker 1937.
Walker and Corbet 1975.
Williams 1978.

CORDULIIDAE (LIBELLULOIDEA)

The Corduliids, Green-Eyed Skimmers

Figures 10.1, 10.2, 10.34a, 10.37, 10.38

Relationships and Diagnosis: In our range, the Libelluloidea contains three families, the Macromiidae, Corduliidae, and Libellulidae. See the family Libellulidae for a discussion of relationships below.

In the Corduliidae, the cerci are generally more than half as long as the paraprocts, usually not more than half in the Libellulidae. Most Corduliidae larvae have the lateral spines of abdominal segment 9 longer than its middorsal length, whereas they are usually shorter in the Libellulidae. The middorsal hooks of the abdomen are often cultriform (sicklelike) in Corduliidae, but usually not in Libellulidae, and more spinelike, stubby, or absent. In the species of Corduliidae having middorsal hooks, they are always present on segment 9. In the Libellulidae, a number of species may possess middorsal hooks only on segments anterior to 9. The crenations of the palpal lobes are generally longer in the Corduliidae than in the Libellulidae. The Macromiidae stand apart from the other two families in a number of characters, such as the very long legs, the abdomen three-fourths to four-fifths as broad as long, and the almost erect frontal horn on the head.

Biology and Ecology: This family is predominantly northern. The larvae are inhabitants of small woodland (and often spring-fed) streams, ponds, and lakes, even living in sphagnum bogs in the North. They are chiefly sprawlers on the bottom with some climbing over bottom trash and waterweeds. Species of *Neurocordulia* may cling tightly to submerged sticks, logs, and stones, so that ordinary methods of collecting produce very few individuals. One must lift the sticks or stones from the water and remove the larvae by hand. One species, *Neurocordulia xanthosoma* (Williamson), feigns death when disturbed. In northwest Florida, where adults of six species of *Somatochlora* have been collected, some in large numbers, much intensive searching has produced almost no larvae of this genus. The larvae of *Tetragoneuria* (*Epitheca* of some authors) are more easily collected as they climb around in vegetation. These are sometimes attractively patterned in green and brown; the sprawlers (*Cordulia,* etc.) are often dull-colored and quite hairy.

Description: Medium-sized, from about 17–29 mm long. Some rather stockily built, such as *Neurocordulia* and *Helocordulia,* for example; others with a more tapered abdomen, rather elongate, and quite hairy, a characteristic associated with the sprawlers. These may easily be mistaken for larvae of Libellulidae.

Head: Usually about twice as broad as long. Eyes capping frontolateral angles of head, but not so highly elevated as in *Libellula* or *Orthemis* of the Libellulidae. Palpal lobes of spoon-shaped labium with crenations, usually one-fourth to one-half as long as broad, or longer, having the form of somewhat rounded scoops. Usually 5–7 (sometimes 4–10) palpal setae, and 7–15 (4–5 in *Epicordulia*) premental setae.

Thorax: Legs much shorter than in Macromiidae, the hind femora usually extending as far back as some part of abdominal segment 6. Wing cases with costal margins parallel.

Abdomen: Dorsal hooks often cultriform (sicklelike) and always present on segment 9 unless entirely absent, as in some *Somatochlora.* Lateral spines of segment 9 usually longer than its middorsal length. Some *Neurocordulia* with lateral spines of segment 8 quite divergent. Cerci generally more than half as long as epiprocts or paraprocts. In the sprawlers, such as the *Somatochlora,* the abdomen may be very hairy, accumulating silt and debris for camouflage.

Comments: This large family is worldwide. There are six recognized subfamilies and about 50 genera, with over 200 species. In our range, we have only eight genera and about 50 species. The larvae are known, with the exception of *Williamsonia fletcheri* Williamson and a few species of *Somatochlora.*

Selected Bibliography

Benke 1978.
Byers 1937.
Cabot 1890.
Dunkle 1977a.
Kennedy 1924.
Kormondy 1959.
Needham 1945a, 1945b.
Pritchard 1964, 1965.
Tennessen 1975, 1977.
Walker 1913, 1925.
Walker and Corbet 1975.
Westfall 1951.
White and Raff 1970.
Williams 1976, 1979.
Williams and Dunkle 1976.

LIBELLULIDAE (LIBELLULOIDEA)

The Common Skimmers

Figures 10.22, 10.34, 10.34b, 10.36

Relationships and Diagnosis: In our range the superfamily Libelluloidea contains the Macromiidae, Corduliidae, and Libellulidae. This last family contains more genera and species than the other two combined. All members of the superfamily possess a spoon-shaped or mask-shaped labium covering the face up to the eyes (as in figs. 10.21 and 10.38) with palpal setae and usually premental setae; recognizable ligula absent; palpal lobes broad and triangular, with the distal margins smooth or evenly crenate, the crenations each bearing one or more setae or, if crenations absent, then the margins bearing evenly spaced setae, singly or in groups; movable hook small; costal margins of wing cases parallel. Needham and Westfall (1955) treated these three groups as subfamilies of the Libellulidae. Gloyd (1959) gave full family rank to the Macromiidae, showing that they stood well apart from the other two groups. A single character or group of characters has not been found that will always separate all larvae of the Corduliidae from all larvae of the Libellulidae, but Walker (1958) recognized these two as separate families and subsequent authors have done likewise, although some think the subfamily rank is better. Some Libellulidae, as well as Corduliidae, superficially resemble species of Cordulesgastridae, which also have a spoon-shaped labium.

Larvae of the Libellulidae generally have the cerci not more than half as long as the paraprocts, but in the Cordullidae they are usually more than half as long. Most Libellulidae have the lateral spines of abdominal segment 9 shorter than its middorsal length, but in Corduliidae they are usually longer than the middorsal length of 9. The middorsal hooks of the abdomen are usually not cultriform (sicklelike) as they normally are in the Corduliidae, but are spinelike, stubby, or absent. In the species having middorsal hooks they are always present on abdominal segment 9 in Corduliidae, but some species of Libellulidae have them only on abdominal segments anterior to 9. The crenations of the palpal lobes are usually shorter in the Libellulidae than in the Corduliidae. The *Cordulegaster* larvae that superficially resemble some Libellulidae may be recognized at once by the palpal lobes, which have the distal margins cut so deeply as to form huge, irregular teeth (fig. 10.33). The prementum also has a definite ligula (fig. 10.33), and the costal margins of the wing cases are slightly divergent in the Cordulegastridae.

Biology and Ecology: The numerous genera are found in a great variety of habitats, usually inhabiting warmer, shallower, more eutrophic waters than do larvae of Corduliidae, and tending to be more active and to grow faster. Most occur in ponds and lakes, or slow-flowing streams, and a few, such as *Brechmorhoga* and *Paltothemis,* in streams with a moderate flow. Most are found in fresh water, but a few, such as *Idiataphe* and *Macrodiplax,* seem to prefer brackish water. *Erythrodiplax berenice* (Drury) is collected from strongly saline waters, whereas *Brachymesia gravida* (Calvert) is found in both fresh and brackish water of lakes and marshes. Libellulidae are characteristically sprawlers or crawlers over bottom silt and debris, or climbers in vegetation. They never burrow deeply, though they may be covered with silt. *Plathemis lydia* (Drury) placed in aquaria in the laboratory disappeared completely under white sand on the bottom. *Miathyria marcella* (Selys) in Florida is found associated with the roots of water hyacinths.

Description: This is a large and diverse family. The larvae of most species are stocky, short-bodied, and not much depressed. They vary in total length from the small *Nannothemis,* as small as 9 mm, to the large *Pantala* and *Tramea,* which may be 28 mm. The bottom sprawlers tend to be very hairy *(Libellula)* and covered with silt, with no distinctive color pattern. Those that climb around in beds of water weeds may have distinctive color patterns, including conspicuously banded legs.

Head: Generally blunt anteriorly, but may project in front of the eyes due to prominence of the labium. Eyes in some genera *(Libellula, Orthemis, Plathemis,* and *Ladona)* at frontolateral part of head and very prominent, so that a line drawn through their tips will not touch another part of the head. In others, eyes more broadly rounded and lateral in position. Labium spoon-shaped, narrow at its base and widened near palpal articulations. Its concave dorsal surface bearing 3–18 premental setae (formerly known as mental setae) on each side of midline (*Ladona* may have none). Beyond the articulations of the palpal lobes, prementum narrowing to an angular apex, but no recognizable ligula present; anterior margin smooth or markedly crenate. Palpal lobes broadly triangular, concave within, each with a much-reduced movable hook, followed along lateral margin with a series of 6–18 palpal setae (sometimes known as lateral or raptorial setae). Palpal lobes regularly scalloped on distal margins, each crenation bearing a series of spinelike setae on the curved surface.

Thorax: Legs may be relatively short, as in *Brechmorhoga,* or long, as in *Idiataphe* and *Lepthemis,* where femora may extend to abdominal segment 8 or beyond. Legs may be quite smooth, but in most sprawlers very hairy. No special burrowing hooks as in Gomphidae, and legs often encircled with several dark bands.

Abdomen: Not much depressed, may be quite blunt behind or tapered to the apex. Important in identification is the presence or absence of middorsal hooks, which when present vary in shape from genus to genus or species to species. The number of segments on which they are found is also important. Most of our species have lateral spines on abdominal segments 8 and 9, although in some they are very small or absent. Terminal abdominal appendages short, or quite long as in *Brachymesia* and *Idiataphe.*

Comments: This is a very large worldwide family. Davies (1981) lists 11 subfamilies, 147 genera, and 1028 species. In our range 31 genera and about 116 species are currently recognized. There is some disagreement as to whether certain groups should be given generic or subgeneric rank, and specific or subspecific rank.

Selected Bibliography

Benke 1978.
Bick 1951, 1953, 1955.
Byers 1927a, 1927b, 1936.
Cabot 1890.
Corbet 1983.
Garcia-Diaz 1938.
Klots 1932.

Leonard 1934.
Needham 1904, 1937, 1943, 1945a.
Needham and Fisher 1936.
Pearse 1932.
Pritchard 1964, 1965.
Walker 1913, 1917.
Walker and Corbet 1975.
Weith and Needham 1901.
Westfall 1953.
Wright 1946b, 1946d.

BIBLIOGRAPHY

Baker, R. L. 1981. Behavioural interactions and use of feeding areas by nymphs of *Coenagrion resolutum* (Coenagrionidae: Odonata). Oecologia (Berl) 49:353–58.

Baker, R. L., and H. F. Clifford. 1982. Life cycle of an *Enallagma boreale* Selys population from the boreal forest of Alberta, Canada (Zygoptera: Coenagrionidae). Odonatologica 11(4):317–22.

Benke, A. C. 1978. Interactions among coexisting predators—a field experiment with dragonfly larvae. J. Anim. Ecol. 47:335–50.

Bick, G. H. 1951. The nymph of *Libellula semifasciata* Burmeister (Odonata, Libellulidae). Proc. Ent. Soc. Wash. 53(5):247–50.

Bick, G. H. 1953. The nymph of *Miathyria marcella* (Odonata, Libellulidae). Proc. Ent. Soc. Wash. 55(1):30–36.

Bick, G. H. 1955. The nymph of *Macrodiplax balteata* (Hagen). Proc. Ent. Soc. Wash. 57:191–96.

Bick, G. H. 1957. The Odonata of Louisiana. Tulane Stud. Zool. 5:71–135.

Borror, D. J., D. M. DeLong, and C. A. Triplehorn. 1981. Odonata. Chap. 11, pp. 197–212. An introduction to the study of insects. 5th ed. New York: Saunders College Publishing. 827 pp.

Byers, C. F. 1927a. The nymph of *Libellula incesta* and a key for the separation of the known nymphs of the genus *Libellula* (Odonata). Ent. News 38:113–15.

Byers, C. F. 1927b. Notes on some American dragonfly nymphs (Odonata, Anisoptera). J. of N.Y. Ent. Soc. 35:65–74.

Byers, C. F. 1930. A contribution to the knowledge of Florida Odonata. Gainesville, Fla.: University of Florida Publ., vol. 1, no. 1. 310 pp.

Byers, C. F. 1936. The immature form of *Brachymesia gravida* with notes on the taxonomy of the group (Odonata: Libellulidae). Ent. News 47:35–37, 60–64.

Byers, C. F. 1937. A review of the dragonflies of the genera *Neurocordulia* and *Platycordulia.* Univ. Mich. Misc. Publ. Mus. Zool. 36:1–36.

Cabot, L., 1872. The immature state of the Odonata. Part 1. Subfamily Gomphina. Illus. Cat. Mus. Comp. Zool. 5:1–17.

Cabot, L. 1881. The immature state of the Odonata. Part II. subfamily Aeschnina. Mem. Mus. Comp. Zool. 8:35–39.

Cabot, L. 1890. The immature state of the Odonata. Part III. subfamily Cordulina. Illus. Cat. Mus. Comp. Zool. 17:37–41.

Calvert, P. P. 1921. *Gomphus dilatatus, vastus,* and a new species, *lineatifrons* (Odonata). Trans. Am. Ent. Soc. 47:221–32.

Calvert, P. P. 1931. The generic characters and the species of *Palaemnema* (Odonata: Agrionidae). Trans. Amer. Ent. Soc. 57:1–111 (adults only).

Calvert, P. P. 1934. Two Mexican species of *Palaemnema* (Odonata: Agrionidae). Trans. Amer. Ent. Soc. 59:377–81 (adults only).

Cannings, R. A. 1982. Notes on the biology of *Aeshna sitchensis* Hagen (Anisoptera: Aeshnidae). Odonatologica 11(3):219–23.

Cannings, R. A., S. G. Cannings and R. J. Cannings. 1980. The distribution of the genus *Lestes* in a saline lake series in central British Columbia, Canada (Zygoptera: Lestidae). Odonatologica 9(1):19–28.

Cannings, R. A., and K. M. Stuart. 1977. The dragonflies of British Columbia. British Columbia Prov. Mus., Victoria. Handbook No. 35. 256 pp.

Carle, F. L. 1980. A new *Lanthus* (Odonata: Gomphidae) from eastern North America with adult and nymphal keys to American octogomphines. Ann. Ent. Soc. Amer. 73:172–79.

Carle, F. L. 1982. *Ophiogomphus incurvatus:* a new name for *Ophiogomphus carolinus* Hagen (Odonata: Gomphidae). Ann. Ent. Soc. Amer. 75(3):335–39.

Carle, F. L. 1983. A new *Zoraena* (Odonata: Cordulegastridae) from eastern North America, with a key to the adult Cordulegastridae of North America. Ann. Ent. Soc. Amer. 76(1):61–68.

Christiansen, K. A. 1947. A new genus and species of damselfly from Southern Haiti (Odonata). Psyche 54(4):256–62.

Cook, C. and J. J. Daigle. 1985. *Ophiogomphus westfalli* Spec. Nov. from the Ozark region of Arkansas and Missouri, with a key to the *Ophiogomphus* species of eastern North America (Anisoptera: Gomphidae). Odonatol. 14(2):89–99.

Corbet, P. S. 1963. A biology of dragonflies. Chicago: Quadrangle Books, Inc. 247 pp.

Corbet, P. S. 1980. Biology of Odonata. Ann. Rev. Ent. 25:189–217.

Corbet, P. S. 1983. Odonata in phytotelmata. *In* Frank, J. H., and L. P. Lounibos (eds.). Phytotelmata: terrestrial plants as hosts of aquatic insect communities. Medford, N.J.: Plexus Publ., Inc. 293 pp.

Crowley, P. H. 1979. Behavior of zygopteran nymphs in a simulated weed bed. Odonatologica 8:91–101.

Davies, D. A. L. 1981. A synopsis of the extant genera of the Odonata. Soc. Int. Odon. Rapid Comm. No. 3. 59 pp.

Dunkle, S. W. 1976. Larva of the dragonfly *Ophiogomphus arizonicus* (Odonata: Gomphidae). Fla. Ent. 59(3):317–20.

Dunkle, S. W. 1977a. The larva of *Somatochlora filosa* (Odonata: Corduliidae). Fla. Ent. 60(3):187–92.

Dunkle, S. W. 1977b. Larvae of the genus *Gomphaeschna* (Odonata: Aeshnidae). Fla. Ent. 60(3):223–25.

Dunkle, S. W. 1980. Second larval instars of Florida Anisoptera (Odonata). Phd. Dissertation, University of Florida. 123 pp.

Dunkle, S. W. 1981. The ecology and behavior of *Tachopteryx thoreyi* (Hagen) (Anisoptera: Petaluridae). Odonatologica 10:189–99.

Dunkle, S. W. 1984. Novel features of reproduction in the dragonfly genus *Progomphus* (Anisoptera:Gomphidae). Odonatol. 13(3):477–80.

Dunkle, S. W. 1985. Larval growth in *Nasiaeschna pentacantha* (Rambur) (Anisoptera:Aeshnidae). Odonatol. 14(1):29–35.

Fraser, F. C. 1957. A reclassification of the order Odonata. Royal Zoological Soc. of New South Wales. pp. 1–133. (Mostly adults)

Garcia-Diaz, J. 1938. An ecological survey of freshwater insects of Puerto Rico. J. Agr. Univ. Puerto Rico 22(1):43–97. (7 pls.)

Garrison, R. W. 1981. Description of the larva of *Ischnura gemina* with a key and new characters for the separation of sympatric *Ischnura* larvae. Ann. Ent. Soc. Am. 74:525–530.

Garrison, R. W. 1984. Revision of the genus *Enallagma* of the United States west of the Rocky Mountains and identification of certain larvae by discriminant analysis (Odonata:Coenagrionidae). Univ. Calif. Publ. in Entomology, v. 105, Berkeley, Calif.: Univ. Calif. Press. 129 pp.

Garrison, R. W. and J. E. Hafernik, Jr. 1981. Population structure of the rare damselfly, *Ischnura gemina* (Kennedy) (Odonata: Coenagrionidae). Oecologia (Berl) (1981) 48:377–384.

Garman, P. 1917. The Zygoptera, or damsel-flies of Illinois. Bull. Ill. State Lab. Nat. Hist. 587 pp. (16 pls.)

Garman, P. 1927. Guide to the insects of Connecticut. V. The Odonata or dragonflies of Connecticut. State Geol. & Nat. Hist. Surv., Hartford. 331 pp. (22 pls.)

Geijskes, D. C. 1943. Notes on Odonata of Surinam IV. Nine new or little known zygopterous nymphs from the inland waters. Ann. Ent. Soc. Amer. 36(2):165–84. (7 pls.)

Gloyd, L. K. 1959. Elevation of the *Macromia* group to family status (Odonata). Ent. News 70(8): 197–205. (4 figs.)

Gloyd, L. K., and M. Wright. 1959. Odonata, pp. 917–26. *In* Ward, H. B., and G. C. Whipple (W. T. Edmondson, ed.). Fresh water biology, 2d ed. New York: John Wiley & Sons, Inc. 1248 pp.

Hagen, H. A. 1885. Monograph of the earlier stages of the Odonata (subfamilies Gomphina and Cordulesgastrina). Trans. Amer. Ent. Soc. 12:249–91.

Halverson, T. G. 1984. Autecology of two *Aeshna* species (Odonata) in western Virginia. Can. Ent. 116(4):567–78.

Howe, R. H. 1920. Manual of the Odonata of New England. Part VI. Larvae or Nymphs, pp. 95–149. *In* Thoreau Mus. of Nat. Hist. Memoir II. Concord, Massachusetts. 149 pp.

Huggins, D. G., and W. U. Brigham. 1982. Odonata, pp. 4.1–4.100. *In* Brigham, A. R., W. U. Brigham, and A. Gnilka (eds.). Aquatic insects and oligochaetes of North and South Carolina. Mahomet, Ill.: Mid-west Aquatic Enterprises. 837 pp.

Huggins, D. G., and M. B. DuBois. 1982. Factors affecting microdistribution of two species of burrowing dragonfly larvae with notes on their biology (Anisoptera: Gomphidae). Odonatologica 11(1):1–14.

Johnson, C. 1973. Distributional patterns and their interpretation in *Hetaerina* (Odonata: Calopterygidae). Fla. Ent. 56(1):24–42. (Adults only)

Johnson, C. 1974. Taxonomic keys and distributional patterns of *Calopteryx* damselflies. Fla. Ent. 57(3):231–248. (Adults only)

Johnson, C., and M. J. Westfall, Jr. 1970. Diagnostic keys notes on the damselflies (Zygoptera) of Florida. Bull. Fla. State Mus., Biol. Sci., 15(2):45–89.

Kennedy, C. H. 1915. Notes on life history and ecology of the dragonflies of Washington and Oregon. Proc. U.S. Nat. Mus., vol. 49, pp. 259–345.

Kennedy, C. H. 1917. Notes on the life history and ecology of the dragonflies (Odonata) of central California and Nevada. Proc. U.S. Nat. Mus. vol. 52, pp. 482–635.

Kennedy, C. H. 1921. Some interesting dragonfly naiads from Texas. Proc. U.S. Nat. Mus., 59(2390):595–98. (1 pl.)

Kennedy, C. H. 1924. Notes and descriptions of naiads belonging to the dragonfly genus *Helocordulia.* Proc. U.S. Nat. Mus. 64(12):1–4. (1pl.)

Kennedy, C. H. 1936. The habits and early stages of the dragonfly *Gomphaeschna furcillata* (Say). Proc. Indiana Acad. Sci. 45:315–22.

Kennedy, J. H. and H. B. White, III. 1979. Description of the nymph of *Ophiogomphus howei* (Odonata: Gomphidae). Proc. Ent. Soc. Wash. 81:64–69. (2 figs.)

Klots, E. B. 1932. Scientific survey of Puerto Rico and the Virgin Islands, vol. 14, pt. 1. New York Acad. Sci., 107 pp. (7 pls.)

Knopf, K. W. 1977. Protein variation in *Gomphus* (Odonata: Gomphidae). Unpubl. Ph.D. dissertation, University of Florida, Gainesville, 112 pp.

Knopf, K. W., and K. J. Tennessen. 1980. A new species of *Progomphus* Selys, 1854, from North America (Anisoptera: Gomphidae). Odonatologica 9:247–52.

Kormondy, E. J. 1959. The systematics of *Tetragoneuria,* based on ecological, life history, and morphological evidence (Odonata: Corduliidae). Univ. Michigan Mus. Zool., Misc. Publ. No. 107, 79 pp. (4 pls.)

Leonard, J. W. 1934. The naiad of *Celithemis monomelaena* Williamson (Odonata: Libellulidae). Occ. Pap. Univ. Mich. Mus. Zool. No. 297. 5 pp.

Louton, J. A. 1982. Lotic dragonfly (Anisoptera: Odonata) nymphs of the southeastern United States: identification, distribution, and historical biogeography. Ph. D. Dissertation, University of Tennessee, Knoxville, Tennessee. 357 pp. (236 figs., 11 tables)

Louton, J. A. 1982. A new species of *Ophiogomphus* (Insecta: Odonata: Gomphidae) from the western highland rim in Tennessee. Proc. Biol. Soc. Wash. 95(1):198–202.

Lutz, P. E. 1968. Life history studies on *Lestes eurinus* Say (Odonata). Ecology 49:576–79.

Lutz, P. E., and A. R. Pittman. 1968. Oviposition and early developmental stages of *Lestes eurinus* (Odonata: Lestidae). Amer. Midl. Nat. 80:43–51.

Martin, R. D. C. 1939. Life histories of *Agrion aequabile* and *Agrion maculatum* (Agriidae: Odonata). Ann. Ent. Soc. Amer. 32:601–19.

Merritt, R. W., and K. W. Cummins. 1984. An introduction to the aquatic insects of North America, 2d ed. Dubuque, Ia.: Kendall/Hunt Publ. Co. 710 pp.

Musser, R. J. 1962. Dragonfly nymphs of Utah (Odonata: Anisoptera). Univ. of Utah Biol. Ser. 12(6):1–66. (5 pls.)

Needham, J. G. 1901. Aquatic insects of the Adirondacks (Odonata). Bull. New York State Mus. 47:429–540.

Needham, J. G. 1903. Life histories of Odonata, suborder Zygoptera damselflies, pp. 218–279. *In* Aquatic insects in New York State. New York State Mus. Bull. 68. (15 figs., 9 pls.)

Needham, J. G. 1904. New dragonfly nymphs in the United States National Museum. Proc. U.S. Nat. Mus. 137, pp. 685–720. (7 pls.)

Needham, J. G. 1911. Descriptions of dragonfly nymphs of the subfamily Calopteryginae (Odonata). Ent. News 22:151 ("unknown nymph from Jamaica" is *Hypolestes*).

Needham, J. G. 1937. The nymph of *Pseudoleon superbus* Hagen. Pamona Coll. J. Ent. Zool. 29:107–109. Claremont, Calif.

Needham, J. G. 1939. Nymph of the protoneurine genus *Neoneura* (Odonata). Ent. News 50(9):241–45.

Needham, J. G. 1940. Studies on Neotropical gomphine dragonflies (Odonata). Trans. Amer. Ent. Soc. 65:363–94. (3 pls.)

Needham, J. G. 1941a. Life history studies on *Progomphus* and its nearest allies (Odonata: Aeschnidae). Trans. Amer. Ent. Soc. 67:221–45.

Needham, J. G. 1941b. Life history notes on some West Indian coenagrionine dragonflies (Odonata). J. Dept. Agr. Puerto Rico 25(3):1–18.

Needham, J. G. 1943. Life history notes on *Micrathyria* (Odonata). Ann. Ent. Soc. Amer. 36(2):185–89.

Needham, J. G. 1945a. Notes on some dragonflies of southwest peninsular Florida. Bull. Brooklyn Ent. Soc. 40(4):104–10.

Needham, J. G. 1945b. Tracking dragonfly nymphs. Ent. News 56(6):141–43.

Needham, J. G. 1948. Studies on the North American species of the genus *Gomphus* (Odonata). Trans. Amer. Ent. Soc. 73:307–39.

Needham, J. G., and E. Fisher. 1936. The nymphs of North American libelluline dragonflies (Odonata). Trans. Amer. Ent. Soc. 62:107–16.

Needham, J. G., and C. A. Hart. 1901. The dragonflies (Odonata) of Illinois. Pt. 1. Petaluridae, Aeshnidae and Gomphidae. Bull. Ill. St. Lab. of Nat. Hist., vol. 6, art. 1, 94 pp.

Needham, J. G., and H. B. Heywood. 1929. A handbook of the dragonflies of North America. Springfield, Ill.: C. C. Thomas. 378 pp.

Needham, J. G., and P. R. Needham. 1962. A guide to the study of fresh water biology, 5th ed. San Francisco: Holden-Day. 108 pp.

Needham, J. G., and M. J. Westfall, Jr. 1955. A manual of the dragonflies of North America. (Anisoptera) including the Greater Antilles and the provinces of the Mexican border. Berkeley, Calif.: Univ. Calif. Press. 615 pp.

Paulson, D. R., and C. E. Jenner. 1971. Population structure in overwintering larval Odonata in North Carolina in relation to adult flight season. Ecology 52:96–107.

Pearlstone, P. S. M. 1973. The food of damselfly larvae in Marion Lake, British Columbia. Syesis 6:33–39.

Pearse, A. S. 1932. Animals in brackish water ponds and pools at Dry Tortugas. Pap. Tortugas Lab. Carnegie Inst. Washington 28(435):125–42. (3 pls.)

Pilon, J.-G., and M. J. Masseau. 1983. External morphology of the larval stages of *Enallagma hageni* (Walsh) (Zygoptera: Coenagrionidae). Odonatologica 12(2):125–40.

Pritchard, G. 1964. The prey of dragonfly larvae (Odonata: Anisoptera) in ponds in northern Alberta. Can. J. Zool. 42:785–800.

Pritchard, G. 1965. Prey capture by dragonfly larvae (Odonata: Anisoptera). Can. J. Zool. 43:271–89.

Pritchard, G. 1980. The life cycle of *Argia vivida* Hagen in the northern part of its range (Zygoptera: Coenagrionidae). Odonatologica 9(1):101–106.

Pritchard, G., and B. Pelchat. 1977. Larval growth and development of *Argia vivida* (Odonata: Coenagrionidae) in warm sulphur pools at Banff, Alberta. Can. Ent. 109:1563–70.

Provonsha, A. V., and W. P. McCafferty. 1973. Previously unknown nymphs of western Odonata (Zygoptera: Calopterygidae) Proc. Ent. Soc. Wash. 75:449–54.

Smith, R. F., and A. E. Pritchard. 1956. Odonata, pp. 106–153. *In* Usinger, R. L. (ed.) Aquatic insects of California. Los Angeles: Univ. Calif. Press. 508 pp.

Snodgrass, R. E. 1954. The dragonfly larva. Smithsonian Misc. Coll. (Publ. 4175) 123(2):1–38.

Stanford, D. F. 1977. The immature dragonflies (Odonata: Anisoptera) of North Mississippi. Masters Thesis, University of Mississippi. 98 pp.

Svihla, A. 1958. The nymph of *Tanypteryx hageni* Selys (Odonata). Ent. News. 69(10):261–66.

Svihla, A. 1959. The life history of *Tanypteryx hageni* Selys (Odonata). Trans. Amer. Ent. Soc. 85:219–32. (1 pl., 4 figs.)

Svihla, A. 1975a. Another locality of the larvae of *Tanypteryx hageni* Selys in Washington. Tombo 18(1–4):43–44.

Svihla, A. 1975b. Adverse factors affecting the distribution of *Tanypteryx hageni* Selys. Tombo 18(1–4):44–45.

Tennessen, K. J. 1975. Description of the nymph of *Somatochlora provocans* Calvert (Odonata: Corduliidae). Fla. Ent. 58(2):105–10. (7 figs., 1 table)

Tennessen, K. J. 1977. Rediscovery of *Epitheca costalis* (Odonata: Corduliidae). Ann. Ent. Soc. Amer. 70(2):267–77 (11 figs.)

Tennessen, K. J. 1979. Distance traveled by transforming nymphs of *Tetragoneuria* at Marion County Lake, Alabama, United States (Anisoptera: Corduliidae). Notul. Odonatol. 1(4):63–65.

Tennessen, K. J., and K. W. Knopf. 1975. Description of the nymph of *Enallagma minusculum* (Odonata: Coenagrionidae). Fla. Ent. 58(3):199–201. (1 fig.)

Tillyard, R. J. 1917. The biology of dragonflies (Odonata or Paraneuroptera). Cambridge: Cambridge Univ. Press. 396 pp.

Walker, E. M. 1912. The North American dragonflies of the genus *Aeshna.* Univ. of Toronto Studies, Biol. Ser. (11):1–213. (28 pls.)

Walker, E. M. 1913. New nymphs of Canadian Odonata. Can. Ent. 45:161–70. (2 pls.)

Walker, E. M. 1914a. The known nymphs of the Canadian species of *Lestes* (Odonata). Can. Ent. 46:189–200.

Walker, E. H. 1914b. New and little known nymphs of Canadian Odonata. Can. Ent. 46:349–57; 369–77. (1 pl.)

Walker, E. H. 1917. The known nymphs of the North American species of *Sympetrum* (Odonata). Can. Ent. 49:409–18.

Walker, E. H. 1925. The North American dragonflies of the genus *Somatochlora.* Univ. of Toronto Studies, Biol. Ser. (26):1–202. (17 text figs., 35 pls.)

Walker, E. H. 1928. The nymphs of the *Stylurus* group of the genus *Gomphus* with notes on the distribution of this group in Canada (Odonata). Can. Ent. 60:79–88.

Walker, E. H. 1932. The nymph of *Gomphus quadricolor* Walsh. Can. Ent. 64:270–73.

Walker, E. H. 1933. The nymphs of the Canadian species of *Ophiogomphus* (Odonata, Gomphidae). Can. Ent. 65:217–29.

Walker, E. H. 1937. A new *Macromia* from British Columbia (Odonata: Corduliidae). Can. Ent. 69:5–13.

Walker, E. H. 1941. The nymph of *Aeschna verticalis* Hagen. Can. Ent. 73:229–31.

Walker, E. H. 1953. The Odonata of Canada and Alaska. Vol. 1. Toronto: Univ. of Toronto Press. 292 pp.

Walker, E. H. 1958. The Odonata of Canada and Alaska. Vol. 2. Toronto: Univ. of Toronto Press. 318 pp. (64 pls.)

Walker, E. H., and P. S. Corbet. 1975. The Odonata of Canada and Alaska. Vol. 3. Toronto: Univ. of Toronto Press. 307 pp. (45 pls.)

Weith, R. J., and J. G. Needham. 1901. The life history of *Nannothemis bella* Uhler. Can. Ent. 33:252–55.

Westfall, M. J., Jr. 1950. Nymphs of three species of *Gomphus* (Odonata). Fla. Ent. 33(1):33–39. (7 figs.)

Westfall, M. J., Jr. 1951. Notes on *Tetragoneuria sepia* Gloyd, with descriptions of the female and nymph. Fla. Ent. 34(1):9–14. (1 pl.)

Westfall, M. J., Jr. 1953. The nymph of *Miathyria marcella* Selys (Odonata). Fla. Ent. 36(1):21–25. (1 pl.)

Westfall, M. J., Jr. 1957. A new species of *Telebasis* from Florida (Odonata: Zygoptera). Fla. Ent. 40(1):19–27. (2 pls.)

Westfall, M. J., Jr. 1964a. Notes on the Odonata of Cuba. Qtly. J. Fla. Acad. Sci. 27(1):67–85. (3 figs.)

Westfall, M. J., Jr. 1964b. A new damselfly from the West Indies (Odonata:Protoneuridae). Qtly. J. Fla. Acad. Sci. 27(2):111–119.

Westfall, M. J., Jr. 1974. A critical study of *Gomphus modestus* Needham, 1942, with notes on related species (Anisoptera: Gomphidae). Odonatologica 3(1):63–73. (3 figs.)

Westfall, M. J., Jr. 1976. Taxonomic relationships of *Diceratobasis macrogaster* (Selys) and *Phylolestes ethelae* Christiansen of the West Indies as revealed by their larvae (Zygoptera: Coenagrionidae, Synlestidae). Odonatologica 5(1):65–76. (3 figs.)

Westfall, M. J., Jr. 1984. Odonata, pp. 126–176. *In* An introduction to aquatic insects of North America, 2d ed. R. W. Merritt and K. W. Cummins (eds.). Dubuque, Ia.: Kendall/Hunt Publ. Co., 710 pp.

Westfall, M. J., Jr., and S. S. Roback. 1967. New records of Odonata nymphs from the United States and Canada, with water quality data. Trans. Amer. Ent. Soc. 93:101–24.

Westfall, M. J., Jr., and K. J. Tennessen. 1973. Description of the nymph of *Lestes inaequalis* (Odonata: Lestidae). Fla. Ent. 56(4):291–93.

Westfall, M. J., Jr., and K. Tennessen. 1979. Clarification within the genus *Dromogomphus* Selys (Odonata: Gomphidae). Fla. Ent. 62(3):266–73.

Westfall, M. J., Jr., and R. P. Trogdon. 1962. The true *Gomphus consanguis* Selys (Odonata: Gomphidae). Fla. Ent. 45(1):29–41.

White, H. B. III and R. A. Raff. 1970. The nymph of *Williamsonia lintneri* (Hagen) (Odonata: Corduliidae). Psyche 77(2):252–57.

White, T. R., K. J. Tennessen, R. C. Fox, and P. H. Carlson. 1980. The aquatic insects of South Carolina. Pt. I: Anisoptera (Odonata). South Carolina Agric. Exp. Sta. Bull. 632. 153 pp.

White, T. R., K. J. Tennessen, R. C. Fox, and P. H. Carlson. 1983. The aquatic insects of South Carolina. Part II: Zygoptera (Odonata). South Carolina Agric. Exp. Sta. Bull. 648. 72 pp.

Williams, C. E. 1976. *Neurocordulia (Platycordulia) xanthosoma* (Williamson) in Texas (Odonata: Libellulidae: Corduliinae). Great Lakes Ent. 9(1):63–73.

Williams, C. E. 1978. Notes on the behavior of the late instar nymphs of four *Macromia* species under natural and laboratory conditions (Anisoptera: Macromiidae). Notul. Odonatol. 1:27–28.

Williams, C. E. 1979. Observations on the behavior of the nymph of *Neurocordulia xanthosoma* (Williamson) under laboratory conditions (Anisoptera: Corduliidae). Notul. Odonatol. 1:44–46.

Williams, C. E., and S. W. Dunkle. 1976. The larva of *Neurocordulia xanthosoma* (Odonata: Corduliidae). Fla. Ent. 59(4):429–33.

Williamson, E. B. 1900. The dragonflies of Indiana. Indiana Dept. Geol. Nat. Res., 24th Ann. Rept., pp. 229–333, 1003–11.

Williamson, E. B. 1901. On the manner of oviposition and on the nymph of *Tachopteryx thoreyi* (order Odonata). Ent. News 12:1–3.

Wright, M. 1943. The effect of certain ecological factors on dragonfly nymphs. J. Tennessee Acad. Sc. 18(2):172–96.

Wright, M. 1946a. The economic importance of dragonflies (Odonata). J. Tennessee Acad. Sci. 21(1):60–71.

Wright, M. 1946b. A description of the nymph of *Sympetrum ambiguum* (Rambur), with habitat notes. J. Tennessee Acad. Sci. 21:135–38.

Wright, M. 1946c. Taxonomic notes on the nymphs of the dragonfly genus *Dromogomphus* Selys. J. Tennessee Acad. Sci. 21:183–86.

Wright, M. 1946d. Notes on nymphs of the genus *Tarnetrum*. J. Tennessee Acad. Sci. 21(2):198–200.

Wright, M. 1946e. A description of the nymph of *Agrion dimidiatum* (Burmeister). J. Tennessee Acad. Sci. 21:336–38.

Wright, M., and A. Peterson. 1944. A key to the genera of anisopterous dragonfly nymphs of the United States and Canada (Odonata, suborder Anisoptera). Ohio J. Sci. 44(4):151–66.

Wright, M., and C. S. Shoup. 1945. Dragonfly nymphs from the Obey River drainage and adjacent streams in Tennessee. J. Tennessee Acad. Sci. 20(3):266–78.

An Introduction to the Orthopteroid Insects

David A. Nickle
Systematic Entomology Laboratory, ARS, USDA

Several diverse but related groups of insects, associated historically at one time or another with the order Orthoptera, are known today as the orthopteroid insects. They include the cockroaches, mantids, termites, walkingsticks, earwigs, crickets, katydids, grasshoppers, and several interesting but lesser known groups. Several shared features prompted early entomologists to combine them: they are generalized insects with chewing mouthparts and cerci; all have two pairs of wings or are derived from insects with two pairs of wings, in which the anterior pair usually is thickened and leathery, whereas the posterior pair is membranous with an expanded anal lobe; and all undergo metamorphosis in which larvae (nymphs) resemble the adults and wings develop externally as wing pads. However, in spite of these unifying characters, each group possesses a combination of features uniquely characterizing it as a separate entity.

Although the earwigs and the termites were each removed as a separate order by early in this century, the other groups remained together until quite recently. Brues et al. (1954) and Chopard (1949) were among the first to divide Orthoptera into orders generally recognized today. Others to follow the expanded classification of the orthopteroids include Mackerras in *The Insects of Australia* (1970) and, with some deviations, Kevan (1976, 1982).

Following a more conservative approach of orthopterists such as Blatchley (1920), Morse (1920), and Rehn and Hebard (both of whom dominated the world orthopteran systematics from 1903 to 1965), American entomologists have been slow to accept the growing trend to redefine the orthopteroid orders. As recently as 1981, most of the orthopteroid groups appeared as families of Orthoptera in widely used textbooks such as Borror and White (1971) and Borror, DeLong, and Triplehorn (1981).

Within the last 20 years there has been a renewed interest in the higher systematics of the orthopteroids. Studies by Blackith and Blackith (1968) and Kamp (1973), and a review by Kevan (1976) convinced many American entomologists to adopt an expanded classification of the orthopteroids. In this volume we recognize seven orders of orthopteroid insects: the Dermaptera (earwigs), Phasmatodea (walkingsticks), Grylloblattodea or Notoptera (rock crawlers or icebugs), Isoptera (termites), Blattodea (cockroaches), Mantodea (praying mantids), and Orthoptera (the saltatorial orthopteroids, including the crickets, katydids, and grasshoppers). An eighth order, the Zoraptera, possesses several orthopteroid characteristics but also has been associated with the Psocoptera. Until its relationships have been established by a more thorough investigation, it seems best not to include it in the following discussion.

In addition to the renewed interest in the higher classification of the orthopteroids, there has been a trend in insect pest management for seasonal monitoring of insect populations as a biorational approach to pest population management. This trend has prompted more research on all stages of development of insect pests. The orthopteroids contain many pest species, including termites, cockroaches, and rangeland grasshoppers and katydids. Because little has been written about the immature stages of most of these species, it seems appropriate that attention is given to both the higher classification and the identification of the immatures.

It is not possible here to discuss all the features that define each order. In a numerical taxonomic analysis, Blackith and Blackith (1968) used 92 characters to evaluate affinities of the orthopteroids. Kamp (1973) used 84 and 80 characters, respectively, in two separate evaluations. It is easy to relate the cockroaches with their specialized relatives, the mantids and termites, although problems arise in assigning the categories to which they belong. Less obvious are the affinities of the walkingsticks, earwigs, icebugs, and the saltatorial Orthoptera. The male cerci of the Orthoptera, Phasmatodea, and Dermaptera all are unsegmented, suggesting a closer affinity of these groups than with the others in which cerci contain two or more segments. Unlike all the other orthopteroids, only in the Orthoptera and most Phasmatodea are the male internal genitalia bilaterally symmetrical (they are asymmetrical, however, in the walkingstick family Timematidae). Kamp (1973) showed that Dermaptera, Grylloblattodea, and Phasmatodea shared many features suggesting a relationship among them. Orthoptera contains several diverse families, which are united primarily on the basis of the enlarged hind legs and characters involving the genitalia of both sexes.

In this book, the cockroaches, mantids, and termites are regarded as separate orders. Placing all these as suborders of the order Dictyoptera, or placing the mantids and cockroaches as suborders of the order Dictyoptera and the termites as a separate order Isoptera, are two alternatives that are less acceptable for several reasons. The first alternative would upset the stability and wide acceptance of an old established order, the Isoptera. The phylogeny of genera of this order is better understood than that of most animal groups of comparable size. The biology of termites has been studied extensively, and because it contains many pest species, research on the control of pest termites is enormous. Termites were removed as a separate order because of their universal, highly specialized adaptation of sociality. The mantids are also a highly specialized group. All are predaceous insects with well developed raptorial fore legs. The mantids in fact evolved from the cockroaches before the termites. Although both groups share features with the stem group, the cockroaches, they are distinctively specialized, and it seems more logical to rank them at the same hierarchical level of "order."

Vickery et al. (1974) and Kevan (1976) recently divided the Orthoptera into two orders, the Grylloptera (crickets, katydids, and their relatives) and the Orthoptera *sensu stricto* (grasshoppers and their relatives). They reasoned that the relationship based on saltatorial hind legs has been too heavily weighted in comparison with other characters, such as position of the tympanum, shape of the ovipositor, and number of tarsal segments. Although there is merit in this line of reasoning, it has not been widely accepted, possibly because it disproportionately increases the number of higher categories without substantially contributing to the understanding of the lower levels. Orthoptera in this book includes all saltatorial forms.

Selected Bibliography

Complete citations begin on p. 167.

Blackith and Blackith 1968.
Blatchley 1920.
Borror and White 1971.
Borror et al. 1981.
Brues et al. 1954.
Chopard 1949.
Kamp 1973.
Kevan 1976, 1982.
Mackerras 1970.
Morse 1920.
Vickery et al. 1974.

Order Blattodea

11

Frank W. Fisk[1]
The Ohio State University

See the "Introduction to the Orthopteroid Insects," p. 118.

COCKROACHES

Cockroaches are small to very large insects with chewing mouthparts, simple metamorphosis, body flattened and the head more or less hidden from above by the pronotum. Tegmina and wings are generally present in adults but may be reduced or lacking, especially in females; when present they develop as external wing pads in older larvae (nymphs). Legs are fitted for running, the tarsi five-segmented. Cerci usually with eight or more segments. Princis (1962–71) lists about 460 genera and 3680 described species worldwide in 28 families, of which less than one percent are household pests. However, McKittrick (1964) recognized five families, all represented by United States species. McKittrick's classification is supported by studies of several biological systems (Roth 1970), and her general conclusions have been accepted in the classification used here. Some information is provided here on oöthecae (egg cases), but see Roth (1968a, 1968b, 1971) for an excellent discussion of the evolution of ovoviviparity and viviparity in cockroaches, and for illustrations of the oöthecae of many species.

One of the characteristics of the Blattodea is the formation of an oötheca or egg case that may contain only a few or many eggs, depending on the species. Paired accessory or colleterial glands secrete the materials that form the oötheca. The completed oötheca of oviparous species almost invariably contains two rows of eggs surrounded by a hard protective coat.

During the evolution of ovoviviparity and viviparity in the Blattodea, the oötheca has changed from a hard rigid structure, completely enclosing the eggs (Blattidae), that was deposited relatively shortly after its formation, to a soft, flexible transparent membrane, often incompletely enclosing the eggs, that was retracted and incubated in a uterus or brood sac (Blaberidae).

The oöthecae of more than 200 Blattodea have been illustrated (Roth 1968a, 1971); those of oviparous species are often distinctive and some may be identified to genus and even to species. Sweetman (1965), Cornwell (1968), Bennett (1977), and Gurney and Fisk (1986) have devised keys to the oöthecae of a few household species.

Many oviparous females protect their oöthecae by concealing them in crevices or by preparing a hole in the substrate and covering the deposited oötheca with substrate. McKittrick (1964) has summarized this aspect of oviposition behavior in several species.

Oöthecae of oviparous species that are collected in the field may be parasitized by wasps that belong to at least six families of Hymenoptera (Roth and Willis 1960). Empty oöthecae that have one or more holes in the wall are those from which parasitoids have emerged. All the Evaniidae (ensign wasps) are presumed to be parasitic on cockroach eggs and their presence in dwellings indicates the presence of cockroaches.

LARVAE

Since cockroaches are exopterygotes, the larvae (nymphs) show many similarities to the adults and can usually be recognized and identified when found in association with them. There are also obvious differences. External genitalia develop slowly, showing the greatest change with the final molt. Hence, species diagnostic characters of the genitalia are only useful for adults. The same is true of diagnostic characters of the tegmina and hind wings (venation, size, coloration and folding), which are not expressed in the wing pads of even the last instar larvae. The morphology of the remaining structures, especially the arrangement of spines and setae on the legs, is more stable and useful. The proventriculus (between crop and midgut) is constant, irrespective of age and sex. Dissection and microscopic examination of proventriculi may offer excellent characters for determination to family and sometimes to genus (McKittrick 1964; Miller and Fisk 1971).

Reliance upon the stability of coloration and cuticle texture may be either misleading or most helpful, depending on the species. Many rather uniformly colored species (e.g., *Blatta*) change little from early instars to adults. In others having color patterns, these show up at an early stage and continue to develop through to the adult (e.g., *Supella, Blattella*) so that most larval stages can be determined. Patterns on the face, pronotum, and legs are helpful with this group.

Some genera (e.g., *Eurycotis*) show distinctive color patterns in the young that disappear as the adult stage is reached, but often just the opposite is the case as in *Neostylopyga*. In a relatively few genera (e.g., *Panchlora, Phortioeca*) there are dramatic changes in color and pattern at the final molt so that adults appear very unlike any of the earlier instars.

1. The assistance of L. M. Roth in providing information on oöthecae, oviposition, and photos of larvae is gratefully acknowledged.

In collecting immatures, one should remember that they are usually nocturnal and hide by day in loose aggregations, often in the company of adults. So, if a single roach is encountered, there will often be others of the same species nearby. If adults are collected, their identity will help in determining the larvae. Alternatively, it may be possible to rear the larvae and identify the adults.

In the keys and descriptions, *size, wherever mentioned, refers to the length of the adult.* For example, cockroaches less than 8 mm long are designated as small, those 8–20 mm medium, and those over 20 mm are large. Obviously the larvae are smaller than their respective adults, their sizes varying with the instar.

CLASSIFICATION

Order **BLATTODEA**
- Superfamily Blattoidea
  - Cryptocercidae
  - Blattidae
- Superfamily Blaberoidea
  - Polyphagidae
  - Blattellidae
  - Blaberidae

FAMILY KEY TO THE LARVAE OF BLATTODEA[2]

1. Compound eyes very small, ovoid, or absent in very early instars; antennae relatively short, about one-fourth body length; tibiae spiny, but mid and hind femora lacking spines; short cerci usually concealed by supra-anal plate except in youngest ones. Proventriculi with narrow bladelike teeth (fig. 11.6) and the pads (pulvilli) bearing obvious tubercles. No trace of wing pads or wings in either larvae or adults (fig. 11.7) .... ***Cryptocercidae*** (p. 124)

1′. Compound eyes present, larger, kidney-shaped; antennae short or long; spines on tibiae and mid and hind femora may be numerous or nearly lacking; cerci short and ovoid or long and tapering but never completely concealed by supra-anal plate. Proventriculi various, but never with obvious tubercles on the pads. Older larvae usually with wing pads 2

2(1′). Antennae short, less than one-half body length; legs relatively short; mid and hind femora with spines lacking except at distal end; cerci stout and ovoid, often partly concealed; proventriculi with teeth reduced to tubercles or entirely lacking 3

2′. Antennae longer, more than one-half body length; legs relatively long and slender; mid and hind femora with numerous strong spines along ventral margins; cerci long and tapering, carried at an angle to the body; proventriculi with teeth variously shaped, but always obvious 4

3(2). Head with thickened postclypeus (fig. 11.1), thorax conspicuously hairy (setaceous), especially along outer margins of the terga; proventriculi with dental tubercles minute or lacking; small- to medium-sized species .... ***Polyphagidae*** (p. 128)

3′. Head without thickened postclypeus; thorax not conspicuously hairy; proventriculi with dental tubercles obvious, minute, or lacking; medium to large species .... ***Blaberidae*** (p. 129)

4(2′). (The spination of femora on very young larvae is not well developed and may not agree with the statements in this couplet.) Ventroanterior margins of front femora with an unbroken row of spines ending apically in 2 or 3 larger spines; the row of spines equal in size or decreasing very slightly toward the distal end (fig. 11.2). (In *Eurycotis,* the row of spines may be interrupted by occasional tiny spines before the 2 apical spines are reached.) Proventriculi with large, heavily sclerotized, sculptured teeth, one of them dumbbell-shaped (fig. 11.5); medium to large species .... ***Blattidae*** (p. 127)

4′. Ventroanterior margins of front femora with (a) a row of spines that definitely, but gradually, decrease in size before the 2 or 3 large apical spines (fig. 11.4), or (b) the spines decrease abruptly in length and diameter before the apical spines are reached (fig. 11.3). Proventriculi with slender, bladelike teeth of various shapes (fig. 11.6); small to medium species .... ***Blattellidae*** (p. 128)

2. Intended primarily for the United States species.

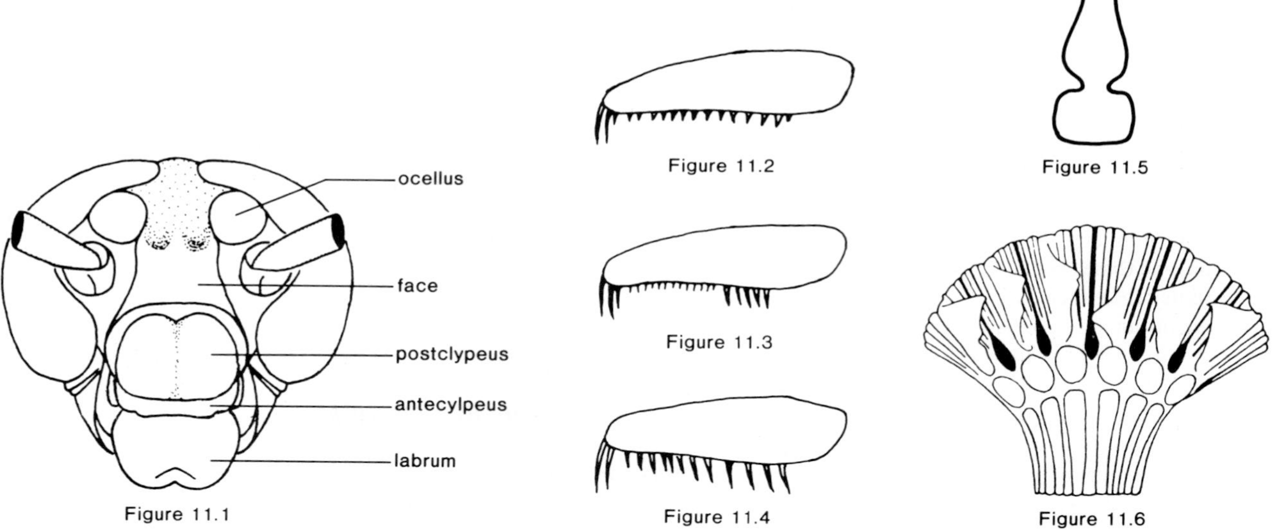

Figure 11.1. Polyphagidae. Front of head of *Arenivaga* sp., mole. (Drawing by A. D. Cushman, courtesy of J. R. Gorham.)

Figure 11.2–11.4. Ventro-anterior margins of front femora. **(11.2)** Blattidae. *Periplaneta* sp.; **(11.3)** Blattellidae. *Parcoblatta* sp.; **(11.4)** Blattellidae. *Supella longipalpa* (F.).

Figure 11.5. Proventricular tooth from blattid proventriculus. (Redrawn by Fisk from McKittrick 1964.)

Figure 11.6. Generalized blattellid proventriculus. (Drawing by Fisk.)

KEY TO THE LARVAE OF PEST COCKROACHES IN THE UNITED STATES[3]

In several species where color patterns of early instar larvae differ markedly from late instars and are keyed out separately, note that changes in color pattern are seldom completed in a single molt so instars with "intergrade" patterns will be encountered.

1. Mid and hind femora with numerous strong spines along ventral margins (in youngest ones these may be reduced to setae but are still obvious) 6

1′. Mid and hind femora without strong spines except at distal end (do not confuse with scattered fine setae that may be present) 2

2(1′). Ventroanterior margins of front femora with several short stout spines; body shape ovoid and flattened (fig. 11.20). Outdoors, southern Florida .... *Blaberus craniifer* Burmeister (p. 129)

2′. Ventroanterior margins of front femora with a few setae or very fine spines 3

3(2′). Dorsal surface of thorax and abdomen rough, covered with short, microscopic spines, especially along posterior margins of segments; color olive grey; larger larvae have obvious yellowish spots laterally on each abdominal tergum (fig. 11.21). Primarily outdoors, Florida, Hawaii, and New York City .... *Leucophaea maderae* (Fabricius) (p. 130)

3′. Dorsal surface of thorax and abdomen smooth (except distal abdominal terga in *Pycnoscelus*); color reddish brown 4

4(3′). Dorsal surface of abdomen roughened posterior to third segment, each tergum with several rows of low but strong tubercles; remainder of abdomen and thorax smooth; mid tibiae very stout and spiny, about 3× longer than wide (fig. 11.22). Mostly outdoors, Hawaii, Florida, Louisiana, Texas .... *Pycnoscelus surinamensis* (L.) (p. 130)

4′. Entire abdomen and thorax smooth; posterior abdominal terga each with single row of obvious tubercles (*Panchlora*) or indications of several rows of weak tubercles (older *Nauphoeta*); mid tibae spiny but less stout, at least 4× longer than wide 5

3. Dr. Ashley B. Gurney's assistance is gratefully acknowledged.

5(4′). Terga of abdomen with single well developed row of tubercles along posterior margin. Brown larvae may be associated with bright green adults. Outdoors, Gulf Coast states .... *Panchlora nivea* (L.) (p. 130)

5′. Terga of abdomen without tubercles except in older larvae—several rows of weak tubercles may be indicated, but none on posterior margins. Brown larvae may be associated with mottled grey adults. Outdoors, Florida, Hawaii .... *Nauphoeta cinerea* (Olivier) (p. 130)

6(1). Dorsal surface of pronotum with obvious pattern of contrasting dark and light colors 7

6′. Dorsal surface of pronotum uniformly colored or with different shades of color which blend without forming a contrasting pattern 14

7(6). Pronotal pattern extending to meso- and metanotum and consisting of 2 longitudinal dark stripes separated from lateral margins and from each other by pale background color (fig. 11.16) 8

7′. Pronotal pattern not as above 9

8(7). Dark longitudinal stripes (fig. 11.16) less widely separated than in *vaga* below; if face is dark, the color more generally distributed. See couplet 11. Indoors, U.S. .... *Blattella germanica* (L.) (p. 129)

8′. Dark longitudinal stripes well separated; if face is dark, the colored portion limited to a wide median stripe (fig. 11.17). Outdoors, southwestern states .... *Blattella vaga* Hebard (p. 129)

9(7′). Pronotal pattern complex and distinctive, including a dark transverse band along posterior margin (also posterior margins of meso- and metanotum), and 2 large irregularly shaped ringlike splotches that contrast with light areas outside and within the rings, the splotches linked to the posterior margin by a median dark stripe; tarsi with very large arolia (fig. 11.11). Outdoors, Texas .... *Neostylopyga rhombifolia* (Stoll) (p. 127)

9′. Pronotal pattern not as above 10

10(9′). Pronotum solidly dark in youngest larvae but developing pale areas that become more extensive and contrasting with each molt as follows: (a) lateroposterior spots, which extend (with successive molts) to a solid transverse band separating the disc from the equally dark posterior marginal band, and (b) a pale anterior marginal band, which extends around the lateral margins of the pronotum until it joins the pale posterior band by the time the adult stage is reached (fig. 11.13). See couplets 18 and 21. Outdoors and indoors, South .... *Periplaneta australasiae* (Fabricius) (p. 127)

10′. Pronotal pattern not as above 11

11(10′). Dorsal surface of body brown except for pale lateral margins of thorax and a broad median stripe on meso- and metanotum that just reaches posterior margin of pronotum (fig. 11.16) (first instar larvae not over 4 mm long). See couplet 8. Indoors, U.S. .... *Blattella germanica* (L.) (p. 129)

11′. Dorsal surface of body not colored as above 12

12(11′). Pronotum dark except for transparent lateral margins, but most of meso- and metanotum (and part of abdomen) contrastingly light (figs. 11.18 and 11.19). Indoors, U.S. .... *Supella longipalpa* (Fabricius) (p. 129)

12′. Body solidly dark except for lateral margins of pronotum (outdoor species) 13

13(12′). Ventroanterior margins of front femora with several stout spines followed distally by a row of very small, slender setae before the 3 large apical spines; tarsi with arolia relatively small. Mostly outdoors, eastern U.S. .... *Parcoblatta pennsylvanica* (DeGeer) (p. 129)

13′. Ventroanterior margins of front femora with continuous row of stout spines before the 2 large apical spines; tarsi with arolia large (in older nymphs, fig. 11.9), the pale yellowish lateral margins of pronotum reduced in successive instars to a pair of lateroposterior pale spots, which then disappear altogether in the subsequent molt. See couplet 15. Outdoors, Gulf Coast states .... *Eurycotis floridana* (Walker) (p. 127)

14(6′). Dorsal surface of body uniformly black or very dark brown or mahogany colored 15

14′. Dorsal surface of body not uniformly dark; predominant color reddish brown, often with pale transverse markings laterally. (Small early instars with white-banded antennae and pale transverse bands on mesonotum and second abdominal tergum key out here) 17

15(14). Seventh abdominal tergum with lateroposterior angles acute, spinelike; tarsi with large arolia. (Younger larvae (fig. 11.8) show cream-colored spots on lateral margins of pronotum.) See couplet 13. Outdoors, Gulf Coast states .... *Eurycotis floridana* (Walker) (p. 127)

15′. Seventh abdominal tergum with lateroposterior angles less acute, not spinelike; tarsi with arolia present or absent 16

16(15′). Tarsi with arolia absent (barely visible in oldest larvae); cerci relatively stout; all instars uniformly colored. Mostly indoors, U.S. .... *Blatta orientalis* L. (p. 127)

16′. Arolia obvious; cerci slender; older larvae (>8 mm) reddish brown to mahogany colored; see couplet 18. More often outdoors, southeastern states .... *Periplaneta fuliginosa* (Serville) (p. 127)

17(14′). Very young larvae (<8 mm long) 18

17′. Older larvae (>8 mm long) 20

18(17). Antennae unbanded; body uniformly colored (except for pale lateral margins on thoracic terga of *Neostylopyga*) 19

18′. Antennae banded with white near base; transverse pale bands on mesonotum and second abdominal tergum .... early instars of *Periplaneta australasiae* (Fabricius), *P. brunnea* Burmeister, or *P. fuliginosa* (Serville) (p. 127)

19(18). Arolia large; pale lateral margins on thoracic terga (fig. 11.10). See couplet 9 .... early instar *Neostylopyga rhombifolia* (Stoll) (p. 127)

19′. Arolia absent; thoracic and abdominal terga uniformly colored. See couplet 20 .... *Periplaneta americana* (L.) (p. 127)

20(17′). Pronotum suffused with light tan markings especially near posterior margin; arolia very small, barely visible; cerci relatively slender (fig. 11.12). See couplet 19. Indoors, U.S.; also outdoors in South .... *Periplaneta americana* (L.) (p. 127)

20′. Pronotum may or may not have lighter markings; tarsi with obvious medium-sized arolia; cerci slender or stout 21

21(20′). Pronotum not showing lighter spots or bands but whole pronotum may be pale compared with *P. americana;* obvious lateral pale spots on sixth abdominal tergum; cerci stout (fig. 11.14). See couplet 18. Indoors and outdoors, South .... *Periplaneta brunnea* Burmeister (p. 127)

21′. Pronotum developing well-marked lateroposterior pale spots (and, later, other markings as described in couplet 10); contrasting lateral pale markings develop on abdominal terga (but not on sixth tergum as noted for *P. brunnea*); cerci slender (fig. 11.13). See couplets 10 and 18. Indoors and outdoors, South. .... *Periplaneta australasiae* (Fabricius) (p. 127)

CRYPTOCERCIDAE

The Cryptocercids, Wood-Feeding Cockroaches

Figure 11.7

Relationships and Diagnosis: This small family is believed to be the most primitive and has been shown to have many characters in common with termites (Cleveland et al. 1934; McKittrick 1964). It is most closely linked to the large family Blattidae. Cryptocercid larvae are solid ivory color when very young or freshly molted, becoming reddish-brown to dark brown when older. Diagnostic characters are the small ovoid compound eyes (except in very young larvae), no ocelli, short cerci usually concealed by the supra-anal plate (except in very young larvae), and no trace of wing pads at any age. The proventriculi, which resemble those of the termite genus *Zootermopsis,* have narrow bladelike teeth with the pads (pulvilli) below them bearing obvious tubercles.

Biology and Ecology: Wood-feeding roaches are always found in family groups (adults, larvae and oöthecae) or small colonies tunneling in moist fallen logs or rotting stumps. As shown by Cleveland et al. (1934), they are capable of digesting pure cellulose with the help of symbiotic protozoa (order Hypermastigina) present in their hind intestines. Adults are apterous, appearing like larger, darker larvae. There is no evidence of the polymorphism, caste system, or complex nest construction characteristic of the Isoptera. Growth and metamorphosis are slow. Cleveland et al. (1934) suggested a single molt per year, but more recent observations indicate several instars per year.

Description: Body smooth, slender, but not flat. Adults medium-sized. In addition to the characters noted in the diagnosis, they have short legs with spiny tibiae but the mid and hind femora are without spines.

Figure 11.7

Figure 11.8

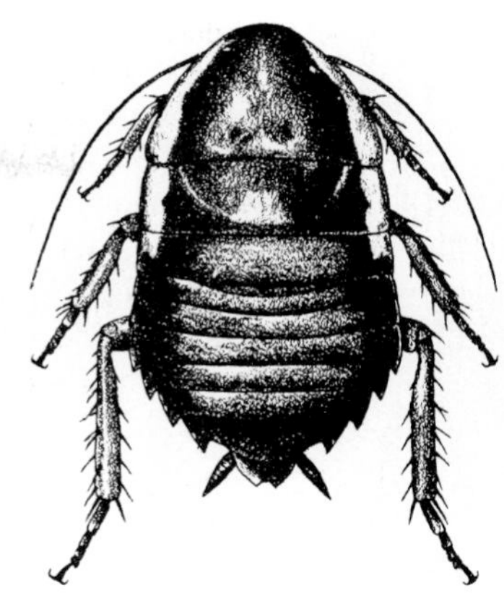
Figure 11.9

Figure 11.10

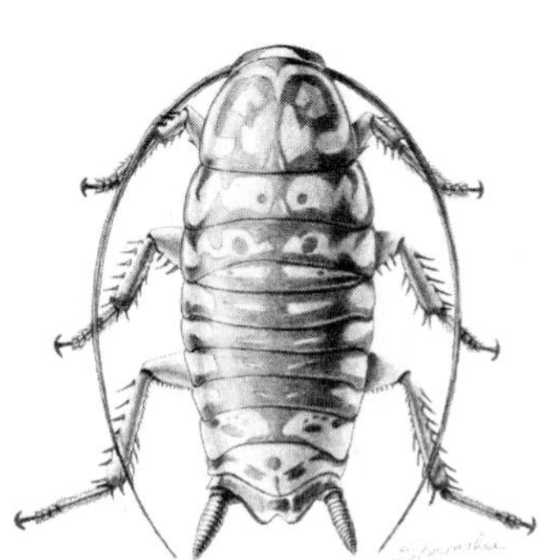
Figure 11.11

Figure 11.12

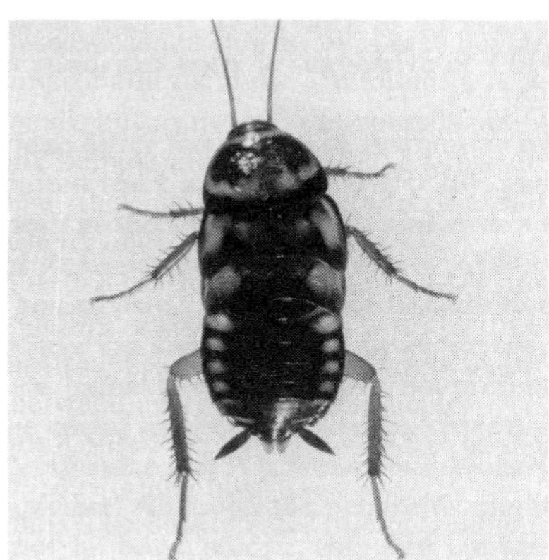
Figure 11.13

Figure 11.14

Figure 11.7. Cryptoceridae. *Cryptocercus punctulatus* Scudder, late instar. (Roth photo.)

Figure 11.8. Blattidae. *Eurycotis floridana* (Walker), Florida stinkroach, 1st instar. (Roth photo.)

Figure 11.9. Blattidae. *Eurycotis floridana* (Walker), late instar. (Drawing by A. D. Cushman, courtesy of J. R. Gorham.)

Figure 11.10. Blattidae. *Neostylopyga rhombifolia* (Stoll), harlequin cockroach, 3rd instar. (Roth photo.)

Figure 11.11. Blattidae. *Neostylopyga rhombifolia* (Stoll), harlequin cockroach, late instar. (Drawing by L. Shoemaker.)

Figure 11.12. Blattidae. *Periplaneta americana* (L.), American cockroach, late instar. (Roth photo.)

Figure 11.13. Blattidae. *Periplaneta australasiae* (F.), Australian cockroach, late instar. (Roth photo.)

Figure 11.14. Blattidae. *Periplaneta brunnea* Burmeister, brown cockroach, late instar. (Roth photo.)

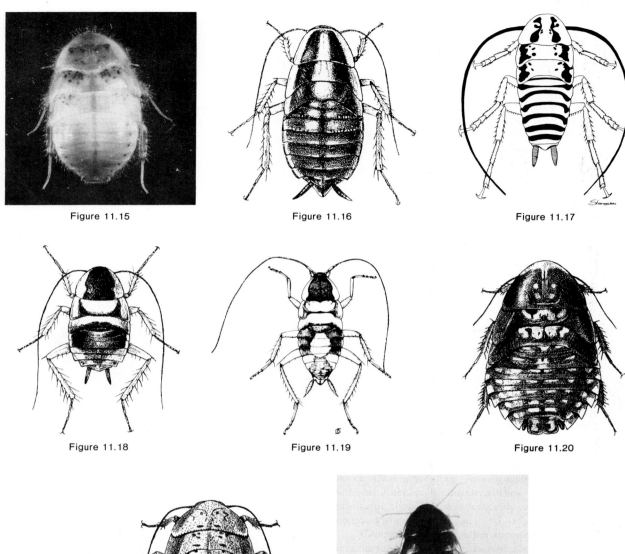

Figure 11.15

Figure 11.16

Figure 11.17

Figure 11.18

Figure 11.19

Figure 11.20

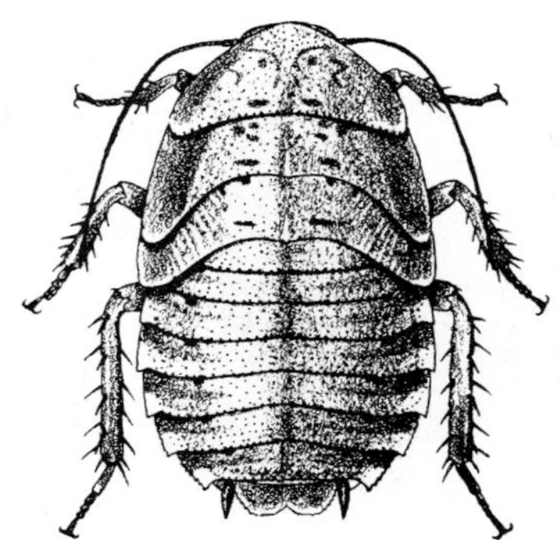

Figure 11.21

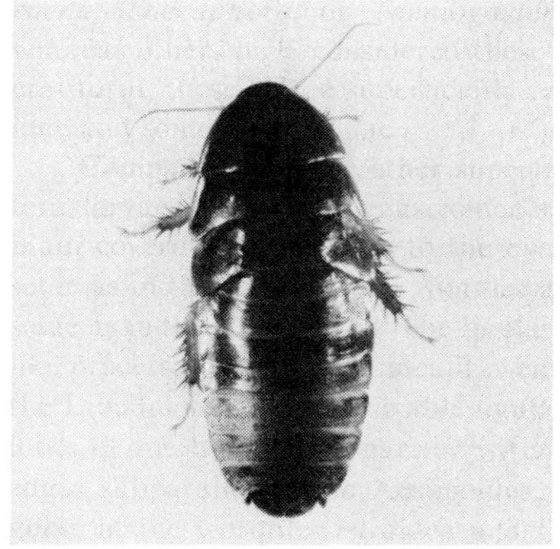

Figure 11.22

Figure 11.15. Polyphagidae. *Arenivaga investigata* Friauf and Edney, late instar male. (Roth photo.)

Figure 11.16. Blattellidae. *Blattella germanica* (L.), German cockroach, late instar. (Drawing by A. D. Cushman, courtesy of J. R. Gorham.)

Figure 11.17. Blattellidae. *Blattella vaga* Hebard, late instar. (Drawing by L. Shoemaker.)

Figure 11.18. Blattellidae. *Supella longipalpa* (F.), brownbanded cockroach, early instar. (Drawing by A. D. Cushman, courtesy of J. R. Gorham.)

Figure 11.19. Blattellidae. *Supella longipalpa* (F.), brownbanded cockroach, late instar. (Drawing by L. Tackett, courtesy of F. W. Stehr.)

Figure 11.20. Blaberidae. *Blaberus craniifer* Burmeister, late instar. (Drawing by A. D. Cushman, courtesy of J. R. Gorham.)

Figure 11.21. Blaberidae. *Leucophaea maderae* (F.), Madeira cockroach, late instar. (Drawing by A. D. Cushman, courtesy of J. R. Gorham.)

Figure 11.22. Blaberidae. *Pycnoscelus surinamensis* (L.), Surinam cockroach, late instar. (Roth photo.)

Comments: The family is represented by a single genus and species, *Cryptocercus punctulatus* Scudder, in the U.S. and two other *Cryptocercus* species in China. The American species is found in hilly or mountainous Appalachian forests and the Pacific Coastal foothills from Washington to northern California, frequently in association with termites. Larvae can be identified by the hidden cerci, in addition to habitat and locality information. They are not economically important since they usually prefer damp, rotting, chestnut, hemlock, oak, pine, spruce or arborvitae.

Selected Bibliography

Cleveland et al. 1934.

BLATTIDAE

The Blattids

Figures 11.2, 11.5, 11.8–11.14

Relationships and Diagnosis: This family of medium- to large-sized roaches includes some familiar pests. The closest affinities of the blattids are to the very primitive wood-feeding roaches (Cryptocercidae) (McKittrick and Mackerras 1965), but immature members are more likely to be confused with large larvae of the Blattellidae, which also have heavily spined mid and hind femora. They can be separated from blattellids by the spination of the ventroanterior margin of the front femora. In blattids there is an unbroken row of numerous similar-sized spines ending apically in 2 or 3 larger spines (fig. 11.2), whereas in blattellids the row of spines is less crowded and either decreases gradually in size or decreases abruptly to a row of fine piliform setae before the 2 or 3 large apical spines are reached (figs. 11.3 and 11.4). Adult female blattids are readily separated from other families because the subgenital plate is keel-shaped and divided at the midline. However, this character is not obvious in the younger larvae. Blattid males and larvae are distinguished by their simple, slender, symmetrical styles and relatively long, tapering cerci. Members of other families have more robust, asymmetrical, and differently shaped styles, and (except for Blattidae) their cerci are not long and tapering. The character is not useful for all larvae because the younger larvae of both sexes may have short, symmetrical styles, as illustrated by Lawson and Lawson (1965). Only blattids have a proventriculus with broad, sculptured, heavily sclerotized teeth. In U.S. blattids, 1 proventricular tooth is always constricted midway between the anterior and posterior ends to form a dumbbell shape (fig. 11.5) (Miller and Fisk 1971). Tarsal arolia are lacking or minute in *Blatta,* small in *Periplaneta americana,* medium-sized in *Neostylopyga* and most *Periplaneta,* and large in *Eurycotis.*

Biology and Ecology: The life cycles and habits of several pest species are fully discussed by Mallis (1982), Cornwell (1968, 1976), and Ebeling (1975); Roth and Willis (1960) discuss the biology of both outdoor and household species. All blattids are oviparous, with the oöthecae being buried or hidden in a variety of ways. Cockroaches in the genus *Periplaneta* mix particles of their environment and excreta with a secreted glue to form a claylike substance (superficially like the tubes of termites) with which they cover their oöthecae and portions of their daytime resting places. Females of *Eurycotis* bury their oöthecae in sand using a gluelike secretion to cement the sand particles around each oötheca. *Eurycotis* adults, but not the larvae, are well known for their ill-smelling defensive secretion, 2-hexenal (Roth et al. 1956).

Description: The general references noted in the previous paragraph may be consulted for descriptions of the larvae. It may be noted that the first stage larvae of *Periplaneta* (except *P. americana*) have conspicuous white bands at the bases and tips of the antennae and on certain terga. White-banded antennae are observed on first stage larvae of certain species of cockroaches in other families as well. Generally, they are lost soon after the first molt. Wing pads develop in older blattid larvae; the adults may be fully winged in both sexes (*Periplaneta*), short-winged in the female and some males (*Blatta*), or short-winged in both sexes (*Eurycotis* and *Neostylopyga*).

Comments: There are about 44 genera and 560 species worldwide. The U.S. species are primarily ubiquitous house pests that have spread over warm temperate and tropical areas of the world from origins in Africa or Asia. They include the following (with probable origins listed in parentheses):

Eurycotis floridana (Walker), the Florida stinkroach or palmettobug (U.S. native) (figs. 11.8, 11.9): most of the more than 30 species of *Eurycotis* occur in Cuba and South America; *E. lixa* is rare in Florida.

Blatta orientalis Linnaeus, the Oriential cockroach (Asia, a moisture-loving species).

Neostylopyga rhombifolia (Stoll), the harlequin cockroach (Asia) (figs. 11.10, 11.11).

Periplaneta americana (Linnaeus), the American cockroach (Africa) (fig. 11.12).

Periplaneta australasiae (Fabricius), the Australian cockroach (Africa) (fig. 11.13).

Periplaneta brunnea Burmeister, the brown cockroach (Africa) (fig. 11.14).

Periplaneta fuliginosa (Serville), the smokybrown cockroach (Asia).

To this list one may add two potential pests: *Blatta lateralis* (Walker) from Asia and reported in California (Gurney 1978); and *Periplaneta japonica* Karny from Asia, which is in laboratory cultures at several U.S. localities.

Selected Bibliography

Gurney 1978.
McKittrick and Mackerras 1965.
Roth et al. 1956.

POLYPHAGIDAE

The Polyphagids

Figures 11.1, 11.15

Relationships and Diagnosis: This family is considered by McKittrick (1964) to be a primitive branch of the Blaberoidea. Larvae and apterous females of the U.S. genera (*Arenivaga, Compsodes, Eremoblatta, Holocompsa*) can be recognized by the postclypeus, a bulging area above the clypeus (fig. 11.1) and by the presence of numerous scattered setae on the thoracic and abdominal terga. The short, stout legs with very spiny tibiae are fitted for digging. The proventriculi are weakly sclerotized, with the teeth reduced to minute tubercles or lacking entirely, and the remaining proventricular structures hardly discernable. Only the male larvae develop wing pads as they mature. Unfortunately, the above characters are not all shared by *Holocompsa* and members of the family found outside the U.S. (e.g., *Latindia, Homoegamia,* etc.), and therefore they cannot be used to categorize the family worldwide.

Biology and Ecology: The U.S. polyphagids are all desert species. Like most cockroaches, they are nocturnal. By day they hide under rocks and other debris, in the tunnels of small mammals, or deep in loose sand. Their unusual ability to retain body water or even absorb water from the atmosphere has been studied by Edney (1966). Females are oviparous, but only oöthecae of *Arenivaga* are known in the U.S. The oöthecae they produce are clasped by the tip of the abdomen and carried about with them for a day or more. Adult males are fully winged, with very large ocelli and compound eyes, and are attracted to lights. Edney et al. (1974) have studied the ecology of *Arenivaga investigata,* and Hawke and Farley (1973) reported on an unidentified species of the genus.

Species of *Arenivaga* are difficult to culture because, for unknown reasons, the eggs usually fail to hatch when deposited in the laboratory. However, Cochran (1979) has succeeded in culturing *Arenivaga tonkawa* and suggested that his simple method might be applicable to other species of the genus.

Description: Body setaceous above, almost circular in shape in female *Arenivaga,* more ovoid in *Compsodes* and *Eremoblatta.* Head with compound eyes and ocelli, antennae relatively short, front with enlarged clypeal shield. Legs short, with tibiae spiny and remaining parts (including exposed parts of coxae) armed with backward-projecting setae or tiny spines. Tarsi with simple symmetrical claws but no pulvilli or arolia. *Compsodes* species are small, species of *Arenivaga* and *Eremoblatta* are medium-sized.

Comments: This is a small family with about 175 species distributed worldwide. In the U.S., the largest genus is *Arenivaga* (subgenus *Arenivaga*) with about nine described species; *Compsodes* and *Eremoblatta* have two species each. Except for *Arenivaga floridensis* Caudell, all the species are limited to the arid Southwest, from Texas to California. Identification of the immatures depends upon collecting them with identifiable adults, preferably males. None of the polyphagids is of economic importance.

Selected Bibliography

Cochran 1979.
Edney 1966.
Friauf and Edney 1969.
Hebard 1920.

BLATTELLIDAE

The Blattellids

Figures 11.3, 11.4, 11.6, 11.16–11.19

Relationships and Diagnosis: This very large family of mostly medium-sized oviparous cockroaches is placed by McKittrick (1964) between the Polyphagidae and the livebearing Blaberidae. Worldwide it includes very small cockroaches (*Anaplecta, Attaphila*) and very large ones (*Nyctibora, Megaloblatta*). The U.S. species discussed here are characterized by having the mid and hind femora with numerous strong spines along the ventral margins and the front femora with the ventroanterior margin bearing a row of spines that decreases in size gradually or abruptly before the 2 or 3 large apical spines are reached (figs. 11.3, 11.4). The proventriculi bear obvious bladelike teeth (fig. 11.6), except in *Ectobius* where the teeth are very fine and other proventricular structures poorly defined. The proventricular teeth show considerable variation in shape among genera. Generally, blattellids are slender-bodied, long-legged, fast-moving cockroaches that carry the slender cerci at about a 45° angle to the body. These characters are shared by the genus *Periplaneta* (especially in the males) in the Blattidae, but the older larvae of *Periplaneta* species can be separated from blattellids by their large size and different spination of the front femora.

Biology and Ecology: Most species are oviparous; females carry the oötheca just a few days while it is being formed and then drop it, but *Blattella germanica* females retain the fully formed, partially expelled oötheca until about 24 hours before the young hatch, supplying moisture and protection to the incubating eggs. Some other blattellids (*Ectobius*) drop the oötheca promptly, but unless the developing eggs have access to moisture sufficient to cause them to swell, they will not hatch (Roth and Willis 1957). Life cycles are comparatively short in the household pest species, 60 days for German and 100 days for brown-banded females under optimum conditions, and a given female can produce several oöthecae (with a single mating) during her lifetime.

Species of *Blattella* carry their eggs externally for the full gestation period and may be links between oviparous Blattellidae and ovoviviparous and viviparous Blaberidae (Roth 1976b). The oöthecae of all species of *Blattella* are very similar (Roth 1971). They are lightly sclerotized and thinner at their anterior ends, so that water is able to pass through the oötheca from the female to the eggs.

Species of the genus *Parcoblatta* live outdoors, often in or near wooded areas, beneath loose bark or in decomposing logs where the oöthecae are deposited. Females are almost all brachypterous, cannot fly, and rarely become established in houses, although they may be introduced with firewood.

Description: For descriptive details see the general references (Cornwell 1976, Ebeling 1975, Mallis 1982) in addition to Roth (1985) (*Blattella vaga*), Gurney (1953, 1968) (*Ectobius*), and Roth and Willis (1957) (*Ectobius*).

Comments: This is the largest family, numbering about 1850 species in five subfamilies worldwide (Princis 1962–71). In the United States there are about 28 species in 13 genera, representing three subfamilies. The genera *Parcoblatta, Ectobius, Cariblatta, Chorisoneura, Euthlastoblatta, Ischnoptera, Latiblattella, Plectoptera, Pseudomops,* and *Symploce* are usually strictly outdoor, seldom-encountered, noneconomic cockroaches.

Attaphila fungicola has been considered to be a polyphagid (Princis 1960 and others), but McKittrick (*in* Roth 1968b) placed it in the Blattellidae. It lacks the bulging postclypeus of the Polyphagidae. The ovaries of *A. fungicola* are more like those of a blattellid (Roth 1968b). Its oötheca, which lacks an anterior flange (characteristic of polyphagids) and has a greatly reduced keel, is also more like a blattellid than polyphagid. It lives in the fungus gardens of the ant *Atta texana* (Buckley) in Texas and Louisiana (Wheeler 1900), and oöthecae can be found in their fungus gardens (J. C. Moser, personal communication). Princis (1963) erected the family Attaphilidae for *Attaphila,* but further study may require its inclusion as a subfamily of the Blattellidae.

Other blattellids, most of them mentioned in the key to larvae of pest species, are noted individually as follows, with probable origin given in parentheses.

Supella longipalpa (Fabricius) (figs. 11.18, 11.19), the brownbanded cockroach (Africa), is difficult to control. Both brownbanded and German cockroaches may occur in the upper floors of houses or of high rise apartments or office buildings. The contrasting bands from which it gets its common name are obvious distinguishing characters in both larvae and adults.

Blattella germanica (Linnaeus) (fig. 11.16), the German cockroach (Africa), is the most important house-infesting pest insect, not only in the U.S. but probably worldwide. In addition to its short life cycle and great fecundity, this species has become quite resistant to most insecticides, making the elimination of infestations difficult, even for pest control operators.

Blattella vaga Hebard, the field cockroach (Asia) (fig. 11.17; see also Roth (1985) fig. 52A), is an outdoor species that occasionally enters houses during the dry season where it may be confused with the German cockroach but can be separated by the characters given in the key. In southern Arizona it occasionally damages newly emerged seedlings.

Blattella lituricollis (Walker), the false German cockroach (Asia), is an outdoor species reported from Hawaii (Zimmerman 1948) that seldom enters houses but may be attracted to lights at night. The larva is described and illustrated by Roth (1985). Adult males are required to distinguish it from the German cockroach with certainty.

Parcoblatta pennsylvanica (DeGeer), the Pennsylvania wood cockroach, occasionally enters buildings. Males are fully winged and often fly to lights and enter buildings in rural areas. Oöthecae are parasitized by several species of wasps (Roth and Willis 1960).

Ectobius pallidus (Olivier), the spotted Mediterranean cockroach (native to areas bordering the Mediterranean Sea), is primarily an outdoor species that has become established in Michigan and Massachusetts, but it has the potential for increasing its distribution (Gurney 1953, 1968). Although nondomiciliary it occasionally invades houses. The eggs are deposited under leaf litter and overwinter as an oötheca. Eggs appear to pass through a dormant period before absorbing water and completing development (Roth and Willis 1957).

The subfamily Nyctiborinae consists of eight genera, principally found in the American tropics. None is established in the United States but *Nyctibora noctivaga* Rehn and *Nyctibora laevigata* (Beauvois) have occasionally been taken in shipments of bananas (Hebard 1917).

Selected bibliography

Anonymous 1976.
Eye 1973.
Gurney 1953.
Roth 1985.
Roth and Willis 1957.
Zimmerman 1948.

BLABERIDAE

The Blaberids

Figures 11.20–11.22

Relationships and Diagnosis: This large and varied group, considered a superfamily by some authors, includes on a worldwide basis, large, medium and (a very few) small cockroaches. According to McKittrick (1964): "the Blaberidae are the most recently evolved family among cockroaches. They constitute a large, highly complex assemblage that has undergone extensive adaptive radiation. . . ." The family is characterized by basic similarities of genitalia and proventriculi and the fact that all species are live-bearing and do not deposit oöthecae, but hold them in a brood sac and incubate the eggs internally. Forerunners of this adaptation have been noted in certain polyphagids (*Homoeogamia, Therea,* and *Arenivaga*) where the oöthecae are carried externally by the female from one to several days, depending on the species, and in *Blattella germanica* (Blattellidae) which carry the eggs externally until they are nearly ready to hatch.

Diagnostic characters that apply to both larvae and adults include relatively short antennae (usually less than half the body length), body shape rather ovoid, legs relatively short, mid and hind femora lacking spines except at distal end, cerci stout and partly concealed by supra-anal plate, proventriculi with teeth reduced to tubercles or lacking altogether (*Panchlora*).

Biology and Ecology: Blaberids found in the U.S. are subtropical or tropical outdoor species that are localized near ports of entry where they have been introduced, or, in the case of *Blaberus craniifer* and *Panchlora nivea,* are native species at the northern limit of their range. They may be found in greenhouses, heated food warehouses, grocery stores, and in the South in outdoor vegetable trash rather than inside houses or apartments.

Although most blaberids are ovoviviparous, *Diploptera punctata* (Eschscholtz), the Pacific beetle cockroach, is the only viviparous cockroach known (Roth and Willis 1955). The beetle cockroach does not occur in continental U.S. but is found in Hawaii, Australia, and the Orient. Another reproductive specialization occurs in *Pycnoscelus surinamensis* where parthenogenesis occurs. Life cycles vary from about 150 days in *Nauphoeta* to over a year in the very large *Blaberus* species.

Description: For descriptive details see the general references (Cornwell 1976, Ebeling 1975, Mallis 1982) in addition to Gurney (1953) (*Leucophaea*) and Roth and Willis (1958) (*Panchlora*).

Comments: The blaberids comprise about 1,020 species and about 155 genera worldwide, but in the United States we have seven species as listed below. Their probable origin is listed in parentheses and their U.S. distribution is noted in the key to the larvae of pest species.

Blaberus craniifer Burmeister, sometimes called the "death's head cockroach" because of the skull-like design on its pronotum (fig. 11.20), is a Neotropical species established but uncommon in the Florida Keys. Because of its large size, long adult life, and ease of rearing, it is also found in several invertebrate physiology laboratories around the country.

Leucophaea maderae (Fabricius), the Madeira cockroach (Africa) (fig. 11.21), is a large species found in tropical and subtropical localities worldwide. Adults produce an objectionable odor from the second abdominal spiracles, and both sexes can stridulate by rubbing the pronotum against the tegmina (Hartman and Roth 1966). It often infests food stores and is abundant in countries south of the United States bordering the Caribbean Sea. It has also become established in heated buildings in New York City (Gurney 1953).

Nauphoeta cinerea (Olivier), the cinereous cockroach (Africa), is sometimes called the "lobster cockroach" from the design on the adult's pronotum. Males of *Nauphoeta* are also able to stridulate (Hartman and Roth 1966). It has become established in food storage and grain milling plants in Florida (Gurney 1953).

Panchlora nivea (Linnaeus), the Cuban cockroach, is a Neotropical species that has become established in the Gulf Coast states. The bright green adults are easily recognized and fly readily, temperature permitting.

Phoetalia pallida (Brunner), a circumtropical species, is limited in the U.S. to the Florida Keys, where it is very scarce. Dorsal surface of the body shining brown; pronotum with narrow anterior and lateral yellow margins.

Pycnoscelus surinamensis (Linnaeus), the Surinam cockroach (fig. 11.22), although described by Linnaeus from the Dutch colony of Surinam in South America, is a parthenogenetic species probably of Southeast Asian origin. Other members of the genus are bisexual. Larvae of *Pycnoscelus* may be confused with *Panchlora* and *Nauphoeta* but can be distinguished by the posterior abdominal terga being dull, not shiny, and other characters as noted in the key. It is established outdoors in several Gulf States and has been found in greenhouses in the northern U.S. and northern Europe (Cornwell 1968).

Selected Bibliography

Gurney 1953.
Hartman and Roth 1966.
Roth and Willis 1955.

BIBLIOGRAPHY

Anonymous. 1976. Cockroaches: Their identification, biology and habits. Appendix III, National Pest Control Association, Vienna, Vir. ESPC 031101. 37 pp.

Bennett, G. W. 1977. The domestic cockroach and human bacterial disease. Pest Control 45:22–24, 44.

Cleveland, L. R., S. R. Hall, E. P. Sanders, and J. Collier. 1934. The woodfeeding roach *Cryptocercus*, its protozoa, and the symbiosis between protozoa and roach. Mem. Amer. Acad. Arts & Sci. 17:185–342.

Cochran, D. G. 1979. A method for rearing the sand cockroach, *Arenivaga tonkawa* (Dictyoptera:Polyphagidae). Proc. Ent. Soc. Wash. 81 (4):580–82.

Cornwell, P. B. 1968. The cockroach. Vol. 1. A laboratory insect and an industrial pest. London: Hutchinson. 391 pp.

Cornwell, P. B. 1976. The cockroach. Vol. 2. Insecticides and cockroach control. London: Associated Business Programmes. 557 pp.

Ebeling, Walter. 1975 (slightly revised 1978). Pests on or near food, pp. 217–44. *In* Urban entomology. Los Angeles: Univ. Calif. Div. Agric. Sci.

Edney, E. B. 1966. Absorption of water vapor from unsaturated air by *Arenivaga* sp. (Polyphagidae, Dictyoptera). Comp. Biochem. Physiol. 19:387–408.

Edney, E. B., S. Haynes, and D. Gibo. 1974. Distribution and activity of the desert cockroach, *Arenivaga investigata* (Polyphagidae), in relation to microclimate. Ecology 55:420–27.

Eye, J. G., Jr. 1973. The comparative morphology of the male genitalia of the cockroach genus *Parcoblatta* (Dictyoptera, Blattellidae, Blattellinae). Masters thesis, Ohio State University, Columbus. 76 pp. (15 pls.)

Flock, R. A. 1941. The field roach, *Blattella vaga*. J. Econ. Ent. 34:121.

Friauf, J. J., and E. B. Edney. 1969. A new species of *Arenivaga* from desert sand dunes in southern California (Dictyoptera:Polyphagidae). Proc. Ent. Soc. Wash. 71:1–7.

Gordh, G. 1973. Biological investigations on *Comperia merceti* (Compere), an encyrtid parasite of the cockroach *Supella longipalpa* (Fabricius). J. Ent. (A)47:115–23.

Gurney, A. B. 1953. Distribution, general bionomics and recognition characters of two cockroaches recently established in the United States. Proc. U.S. Nat. Mus. 103:39–56.

Gurney, A. B. 1968. The spotted Mediterranean cockroach, *Ectobius pallidus* (Olivier) (Dictyoptera, Blattaria, Blattellidae), in the United States. U.S. Dept. Agr. Coop. Econ. Ins. Rep. 18:684–86.

Gurney, A. B. 1978. A cockroach, *Blatta lateralis* (Walker). Coop. Plant Pest Rept. 3 (25):295.

Gurney, A. B., and F. W. Fisk. 1986. Cockroaches. Chap. 2. *In* Gorham, J. Richard (ed.). Insects and mites in foods. Washington, D.C. USDA.

Hartman, H. B., and L. M. Roth. 1966. Stridulation by the cockroach, *Nauphoeta cinerea*, during courtship behaviour. J. Insect Physiol. 13:579–86.

Hawke, S. D., and R. D. Farley. 1973. Ecology and behavior of the desert burrowing cockroach *Arenivaga* sp. (Dictyoptera, Polyphagidae). Oecologia 11:263–79.

Hebard, M. 1917. The Blattidae of North America north of the Mexican boundary. Mem. Amer. Ent. Soc. 2:1–284.

Hebard, M. 1920. Revisionary studies in the genus *Arenivaga* (Orthoptera, Blattidae, Polyphaginae). Trans. Amer. Ent. Soc. 56:197–217.

Lawson, F. A. 1951. Structural features of the oöthecae of certain species of cockroaches (Orthoptera, Blattidae). Ann. Ent. Soc. Amer. 44:269–85.

Lawson, F. A. 1967. Ecological and collecting notes on eight species of *Parcoblatta* (Orthoptera:Blattidae) and certain other cockroaches. J. Kansas Ent. Soc. 40:267–69.

Lawson, F. A. 1968. Structural features of cockroach egg capsules. VI. The oötheca of *Ischnoptera deropeltiformis* (Orthoptera:Blattidae). J. Kansas Ent. Soc. 41:419–24.

Lawson, F. A. and E. Q. Lawson. 1965. Sexing first instar cockroaches. J. Kansas Ent. Soc. 38:408–10.

Mackerras, M. J. 1970. Blattodea (cockroaches), pp. 262–74. *In* the Insects of Australia. Melbourne: Melbourne Univ. Press.

Mallis, Arnold. 1982. Cockroaches. *In* Handbook of pest control, 6th ed. Cleveland, Ohio: Franzak & Foster. 1101 pp.

McKittrick, F. A. 1964. Evolutionary studies of cockroaches. Cornell Univ. Agric. Exp. Sta. Mem. 389. 197 pp.

McKittrick, F. A. 1965. A contribution to the understanding of cockroach-termite affinities. Ann. Ent. Soc. Amer. 58:18–22.

McKittrick, F. A., and M. J. Mackerras. 1965. Phyletic relationships within the Blattidae. Ann. Ent. Soc. Amer. 58:224–30.

Miller, H. K., and F. W. Fisk. 1971. Taxonomic implications of the comparative morphology of cockroach proventriculi. Ann. Ent. Soc. Amer. 64:671–87.

Piper, G. L., G. W. Frankie, and J. Loeher. 1978. Incidence of cockroach egg parasites in urban environments in Texas and Louisiana. Env. Ent. 7:289–93.

Princis, K. 1960. Zur Systematik der Blattarien. Eos. 36:427–49.

Princis, K. 1962–71. Blattaria. *In* M. Beir (Ed.) Orthopterorum Catalogus. Pars 3,4,6,7,8,11,13,14. Gravenhage: W. Junk. 1224 pp.

Rehn, J. A. G. 1945. Man's uninvited fellow traveler—the cockroach. Sci. Monthly 61:265–76.

Ross, M. H., and D. G. Cochran. 1960. A simple method for sexing German cockroaches. Ann. Ent. Soc. Amer. 53:550–51.

Roth, L. M. 1967. Water changes in cockroach oöthecae in relation to the evolution of ovoviviparity and viviparity. Ann. Ent. Soc. Amer. 60:928–46.

Roth, L. M. 1968a. Oöthecae of the Blattaria. Ann. Ent. Soc. Amer. 61:83–111.

Roth, L. M. 1968b. Ovarioles of the Blattaria. Ann. Ent. Soc. Amer. 61:132–40.

Roth, L. M. 1970. Evolution and taxonomic significance of reproduction in Blattaria. Ann. Rev. Ent. 15:75–96.

Roth, L. M. 1971. Additions to the oöthecae, uricose glands, ovarioles, and tergal glands of Blattaria. Ann. Ent. Soc. Amer. 64:127–41.

Roth, L. M. 1985. A taxonomic revision of the genus *Blattella* Caudell (Dictyoptera, Blattaria:Blattellidae) Ent. Scandinavica. Suppl. 22. 221 pp.

Roth, L. M., W. D. Niegisch, and W. H. Stahl. 1956. Occurrence of 2-hexenal in the cockroach, *Eurycotis floridana*. Science 123:670–71.

Roth, L. M., and E. R. Willis. 1954. The reproduction of cockroaches. Smithsonian Misc. Coll. 122:1–49.

Roth, L. M., and E. R. Willis. 1955. Intra-uterine nutrition of the "beetle-roach" *Diploptera dytiscoides* (Serv.) during embryogenesis, with notes on its biology in the laboratory (Blattaria:Diplopteridae). Psyche 62:55–68.

Roth, L. M., and E. R. Willis. 1957. Observations of the biology of *Ectobius pallidus* (Olivier) (Blattaria, Blattidae). Trans. Amer. Ent. Soc. 83:31–37.

Roth, L. M., and E. R. Willis. 1958. The biology of *Panchlora nivea* with observations on the eggs of other Blattaria. Trans. Amer. Ent. Soc. 83:195–207.

Roth, L. M., and E. R. Willis, 1960. The biotic associations of cockroaches. Smithsonian Misc. Coll. 141:1–439.

Sweetman, H. L. 1965. Recognition of structural pests and their damage. Dubuque, Ia.: W. C. Brown. 371 pp.

Wheeler, W. M. 1900. A new myrmecophile from the mushroom gardens of the Texas leaf-cutting ant. Amer. Nat. 34:851–62.

Zimmerman, E. C. 1948. Insects of Hawaii. Vol. 2. Honolulu: University of Hawaii Press. 475 pp.

Order Isoptera

12 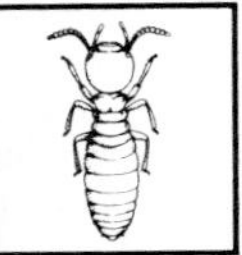

Frances M. Weesner[1]
Colorado State University

See the "Introduction to the Orthopteroid Insects," on p. 118.

Termites, White Ants

A considerable number of isopteran species are of great economic importance because they feed upon any material consisting of cellulose, including structural wood. This is a relatively small order with fewer than 2000 described species (Snyder 1949), with some variation in the assignment of family status to certain groups. In general there are two major views: that advocated by Emerson (see Krishna 1970) and that by Grassé (1949). The major difference concerns the primitive damp-wood-dwelling Termopsinae, the primitive soil-dwelling Porotermitinae and Stolotermitinae, and the harvester termites, Hodotermitinae. The latter are soil-dwelling forms appearing to have a definitive, pigmented, worker caste that harvests grasses above ground and then stores then in underground galleries. In view of the fossil record, Emerson considers all four of these to be members of the Hodotermitidae. Grassé, however, assigns them to two separate families, the Hodotermitidae and the Termopsidae (including the wood-dwelling Termopsinae and the primitive soil-dwelling forms). The Emerson system is followed here, although many investigators follow the Grassé view.

Of the six (or seven) generally recognized families, the Mastotermitidae is represented by a single living species in the Australian region. The Serritermitidae of Brazil is monotypic. The Nearctic fauna includes the only three living species of the damp-wood genus *Zootermopsis* (Termopsinae). Eight genera of dry-wood termites (Kalotermitidae) are encountered in the southern and coastal areas of the Nearctic, including three monospecific genera in the Desert Southwest (*Marginitermes, Paraneotermes,* and *Pterotermes*) and one introduced species in the southeastern and coastal areas (*Cryptotermes brevis* (Walker)). The most widespread and economically important genus in the family Kalotermitidae is *Incisitermes.* Four genera of subterranean termites of the family Rhinotermitidae are encountered in the Nearctic, including the introduced *Coptotermes formosanus* Shiraki in the gulf coastal region. The only wide-ranging genus in the Nearctic is *Reticulitermes,* which probably is responsible for the greatest economic damage of any Nearctic termite. Two subfamilies of the Termitidae are represented in the fauna of the contiguous U.S.; these are limited to the Desert Southwest and, like the Rhinotermitidae, are subterranean in habit.

The termites of Hawaii include five kalotermitids: two species of *Incisitermes,* two of *Neotermes,* and the introduced *Cryptotermes brevis.* The greatest economic loss may be attributed to the introduced rhinotermitid, *Coptotermes formosanus. Coptotermes vastator* Light also appears to have been introduced into Hawaii. There are apparently no termites endemic to or established in Alaska, although the detection of transported groups, particularly of *Zootermopsis* and any of the kalotermitids, is to be expected.

1. Frances W. Lechleitner

DIAGNOSIS

Soft-bodied, hemimetabolous insects, generally unpigmented in immature forms. Head dorsoventrally flattened; mouthparts prognathous; mandibles with two or three marginal teeth on the left mandible and two on the right. Each mandible has a basal molar plate (figs. 12.1–12.4). Antennae moniliform with 10 to 33 segments, arising near the base of the mandibles. Most immature individuals and workers lack compound eyes or ocelli, although they are present in well-developed brachypterous larvae (nymphs). Ocelli, if present, consisting of single pair.

The thorax consists of three distinct segments with the tergites subequal in most forms; it is broadly joined to the abdomen, never "wasp-waisted" as in the Formicidae (Hymenoptera) as might be concluded from the term "white ant." Legs similar in size and shape, lacking the various specializations encountered in many other orders (e.g., the enlarged first tarsal segment of the fore legs in the Embiidina or the greatly enlarged hind legs of the Orthoptera, etc.). The tarsi are four- (or five-) segmented, with the basal three (or four) segments short and together approximating the length of the terminal segment, a situation not present in many orders. An arolium may be present in many Isoptera but is generally absent in immatures. In a similar manner, tibial spines or spurs may be present in soldiers or alates but are generally inconspicuous in immature forms. Wing pads, when present, are about equal in size on the meso- and metathorax. The smallest wing pads tend to project laterally. As they become longer during subsequent stages they overlap one another and the abdomen, with the posterior pair just projecting beyond the upper pair. In the penultimate stage they may be fully half the length of the abdomen, and as with the adult wings, they lie over one another and flat on the abdomen (fig. 12.8).

BIOLOGY AND ECOLOGY

Termites are all colonial insects, living in family groups that may consist of only a few hundred individuals in some kalotermitids to many thousands, or even millions, in others. Colonies consist of individuals in many stages of development, all of which are active. The makeup of these individuals is complicated by the presence of several distinct morphological and functional types: the "workers," soldiers and reproductives.

At any time of the year, the majority of the individuals encountered in a termite colony usually consists of unspecialized forms, the so-called "workers," which lack wing pads or any other marked structural specialization. Workers may be either functional or definitive. The functional workers remain capable of further molts and growth or differentiation into other types. Such individuals are termed apterous larvae, since they lack any indication of wing pad development. The definitive workers, in contrast, have lost the capacity for further molting or differentiation. Definitive workers are typical of the Termitidae and are frequently pigmented.

Soldiers have structural specialization and behavioral adaptations focused toward defense of the colony. In most soldiers the head is greatly enlarged, usually elongated and heavily sclerotized, and the mandibles are enlarged and variously modified. In other soldiers the head is phragmotic or "stopperlike," and used to physically plug openings in the workings. In still others, soldiers are specialized to eject sticky and/or repulsive fluids from the frontal gland through a more or less elongated nasus that projects from the front of the head capsule. Space does not permit a consideration of the various modifications of structure and function, but many accounts are available (see Krishna and Weesner 1969 and 1970). The terminal soldier is always preceded by a presoldier stage, which possesses the general features of the mature individual. These presoldiers have been termed white-soldiers or soldier-nymphs. Young stages preceding the soldier-nymphs are termed apterous larvae (nymphs) and may be functional workers.

The reproductives represent a third type of individual in the colonies. The development of winged adults (alates) is seasonal, and a number of different stadia are involved with increasing development of the wing pads. Such immatures are termed brachypterous larvae (nymphs). Most frequently the alates are in the seventh or eighth stadium. Various types of neotenic reproductives may be produced but they are not pertinent to our consideration. Such reproductives may be apterous or brachypterous, depending upon the type of immatures from which they develop.

Termite eggs are elongate and slightly bean-shaped with a smooth surface. They may be as much as 1 mm in length although they are usually smaller. They are deposited singly (expect in the Mastotermitidae) but are stacked together in an apparently random fashion by the workers. Desert termites move eggs daily from subsurface workings to under-rock chambers and back again. If galleries containing eggs are exposed, the workers frantically carry them to inner chambers. No pupal stages are present in these hemimetabolous insects.

The colony workings are generally inconspicuous in Nearctic forms, being confined to wood or soil, depending upon the species. Termites are most frequently encountered when infested wood is broken into or subterranean workings are exposed beneath surface rocks, wood, or debris, or when foraging groups are encountered in or under herbivore dung. Certain species build extensive covered passageways to reach wood above ground; others, such as *Gnathamitermes,* construct extensive earthen sheets over materials they are feeding upon such as fence posts, shrubs, or grass. *Cryptotermes* may first be detected by accumulations of fecal pellets (hexagonal in cross section) that are expelled from the workings and collect upon undisturbed surfaces below.

The presence of even large colonies of termites is frequently not suspected until a flight of alates is observed. Such flights are staged once or twice a year, under conditions typical for particular species. Flights of individual species tend to be synchronized from many colonies at the same time. Crepuscular and nocturnal swarmers may be encountered around lights.

DESCRIPTION

Soft-bodied insects. Most immature stages lack pigmentation although some may have a yellowish cast. Among the Nearctic species, only the workers and nasuti of *Tenuirostitermes* (Termitidae) are dark brown. The abdomen may have a mottled appearance of varying color because gut contents are visible through the thin exoskeleton. Generally small, ranging in size from 1 mm, or even less, in recently hatched first instar young to as much as 15 mm in penultimate instars in some species.

Head

Dorsoventrally flattened; mouthparts prognathous with chewing mandibles. Dentition generally consisting of rather sharply defined teeth along each mandible, with two or three marginal teeth on the left and two on the right. The second on the right mandible may not project conspicuously from the mandibular face, and the right mandible may also have a small subsidiary tooth at the anterior base of the first marginal (figs. 12.2 and 12.4).

The terminology used for a designation of the dentition follows that put forth by Emerson (1933). The terminal tooth is referred to as the apical, and the marginal teeth are numbered in sequence (1 through 3) from the apical toward the mandibular base. Posterior to the marginal teeth on each mandible is a wedge-shaped molar plate. On the left mandible, as viewed from above, the wide, roughened surface of this plate may be obscure. It is important not to mistake the anterior termination of this plate for a marginal tooth. The roughened molar plate of the right mandible is usually the most conspicuous. It should be noted that the terminology used here does not indicate possible homologies between teeth in different families. Thus Ahmad (1950) considers the apparent second marginal in the Termitidae to be homologous to the third in the Hodotermitidae and Rhinotermitidae, and Krishna (1970) considers the first marginal of the left mandible in the Kalotermitidae to be homologous to the first plus second of the other two families.

It should be noted that the medial, sclerotized articulations of the mandibles, which frequently appear as two brownish spots, one to either side of the postclypeus, are often mistaken for eyes.

Maxilla with a hooded galea, toothed lacinia and five-jointed palp. Labium with four-lobed ligula and three-jointed palps. Antennae moniliform, with 10–33 segments. Clypeus subdivided into anterior and posterior portions. In the Rhinotermitidae and Termitidae, the postclypeus is somewhat swollen and weakly or strongly subdivided medially into right and left portions. Eyes and ocelli generally lacking in apterous larvae although well-developed eyes may be present in large individuals of this type in *Zootermopsis*. Compound eyes are generally present in older brachypterous larvae (nymphs) and may be pigmented in premolt, penultimate individuals. Ocelli absent, even in mature hodotermitids. A fontanelle is present in the rhinotermitids and termitids but is usually not detectable in immatures. It is, however, present and even conspicuous in soldiers of these two families.

Thorax

Thoracic tergites free. In many the pronotum is slightly larger than the meso- and metanotum. In a few the mesonotum is largest. In general the tergites are flat although the pronotum is saddle-shaped in the termitids. Legs with four (or five) tarsal segments of which the basal three (or four) are short and together approximate the length of the terminal segment. Tarsi terminating in two small claws. Arolium generally lacking in immature forms although present in some soldiers and reproductives in the hodotermitids and kalotermitids. Tibial spines and spurs usually lacking or inconspicuous in immatures.

Abdomen

Abdomen ten-segmented and somewhat flattened dorsoventrally although much less so than the head capsule. It is very flat-appearing in desiccated individuals. The only abdominal appendages are small styles situated on the ninth sternite (these are generally lacking in the Termitidae), and the cerci, on or to the sides of the tenth sternite. Cerci are usually two-segmented but have more than two segments (usually four) in the Hodotermitidae. In *Mastotermes darwinensis* Froggatt (Mastotermitidae) of the Australian Region, they are eight-segmented.

COMMENTS

Identification of termites is usually based upon an examination of alates and/or soldiers, and many keys are available to assist in the identification of various Nearctic and Hawaiian forms on this basis (Banks and Snyder 1920, Kofoid et al. 1934, Miller 1949, Weesner 1965). There are occasions when neither of these types is present in a collection. It is still possible to identify such material to the level of family (and even to genus) on the basis of the structure of immature forms, particularly the mandibular dentition. This aspect of termite morphology and its implications in terms of phylogeny was extensively investigated by Ahmad (1950). The dentition of the soldier mandibles is highly modified (Hare 1936) and is not included here.

TECHNIQUES

Since the Isoptera are soft-bodied insects, they are best preserved in 70% ethanol or some other suitable fluid, rather than being pinned. Usually a group, rather than single individuals, is encountered, so most collections include a number adequate to permit examination of the mandibles. Mandibular structure can be best observed in larger individuals by removal from the head. In small forms, the head capsule, or even the entire insect, may be placed ventral side down on a slide. If the specimen is flooded with alcohol or other fluid and a coverglass applied, pressure exerted on the cover glass will force the mandibles out to the sides of the labrum and clypeus so that mandibular dentition can be readily observed.

CLASSIFICATION

Order **ISOPTERA**

Hodotermitidae (may be split into the Hodotermitidae and Termopsidae), dampwood termites
Kalotermitidae, drywood termites, furniture termites
Mastotermitidae (living forms limited to the Australian Region)
Rhinotermitidae, subterranean termites
Termitidae, higher termites
Termopsidae, see Hodotermitidae

KEY TO FAMILIES OF IMMATURE ISOPTERA

Based upon the characters of the immatures (including workers) of the Nearctic, Hawaiian, and Mexican species.

1. Cerci of more than 2 segments. Tarsi with 4 short (imperfectly separated) basal segments and a longer terminal segment .... ***Hodotermitidae*** (p. 135)

1′. Cerci of 2 segments. Tarsi with 3 short basal segments and a longer terminal segment 2

2(1′). Pronotum flat although the lateral margins may be depressed; usually as wide as, or wider than, the head. Left mandible with 2 conspicuous marginal teeth, the second of which is situated at, or anterior to, the midline of the mandibular face (fig. 12.1). Postclypeus not subdivided medially .... ***Kalotermitidae*** (p. 136)

2′. Pronotum flat or saddle-shaped; usually narrower than head. Left mandible with 2 or 3 marginal teeth; if there are but 2, the second is often reduced and is usually situated posterior to the midline of the mandibular face. Postclypeus subdivided medially by a groove, although this might not be conspicuous 3

3(2′). Left mandible with 3 marginal teeth (fig. 12.2). Right mandible with 2 marginal teeth and a small subsidiary tooth at the anterior base of the first .... ***Rhinotermitidae*** (p. 136)

3′. Left mandible with 2 marginal teeth (fig. 12.3), the second of which is often reduced and situated posterior to the midline of the mandibular face. The right mandible with 2 marginal teeth and lacking a subsidiary tooth at the anterior base of the first .... ***Termitidae*** (p. 137)

HODOTERMITIDAE

The Hodotermitids, Dampwood Termites

Figures 12.4 and 12.5

Relationships and Diagnosis: The Nearctic dampwood termites are members of the subfamily Termopsinae, which includes three genera, all of which are large, wood-dwelling forms. The genus *Zootermopsis,* which is limited to the Nearctic, occurs in southwestern (coastal) Canada, the western and southwestern United States, and into Baja California. All are large termites, with well-developed immatures that are larger (10–15 mm or more) than any other Nearctic forms with the possible exception of the kalotermitid genus *Neotermes.* The latter are also dampwood-dwelling species, but their distribution is limited in the Nearctic to coastal Florida. The immatures of *Zootermopsis* tend to have a distinct yellowish color, rather than the whiteness or mottled appearance of immatures of other families. Their most distinctive and readily observed characteristic is the relatively long, three- to five- (usually four-) segmented cerci, which are two-segmented in other families encountered in the Nearctic and Hawaii.

Biology and Ecology: The dampwood termites are all wood-dwelling forms whose colony workings are restricted to the wood in which it is established or to wood in immediate contact with the colony. They do not penetrate the soil to reach new wood. The two coastal species inhabit relatively damp standing or felled dead trees and, when conditions are favorable, structural wood. Mature colonies are large, often comprising many thousands of individuals, and are usually encountered in rotten wood. These coastal termites, however, actually colonize relatively sound wood, usually dead logs with the bark intact (Weesner 1965). The third species occurs in the Desert Southwest, usually in dead wood of standing trees along waterways or ponds (Nutting 1965).

Description: Well-developed immatures more than 10 mm long; yellowish. In some well-developed individuals, pale compound eyes may be observed, which may be pigmented in premolt penultimate larvae. Lacking ocelli or fontanelle. Postclypeus not subdivided medially. Antennae with 20 or more segments (fewer in young stages). Left mandible with three marginal teeth; right mandible with two marginal teeth and a small subsidiary tooth at the anterior base of the first marginal (fig. 12.4). Pronotum broader in front than behind. Mesonotum broader behind than in front and strongly overlapping the metanotum. Metanotum also broader posteriorly and somewhat overlapping the first abdominal segment (fig. 12.5). Tarsi imperfectly five-segmented; arolium absent in immature forms. In older individuals heavy spines may be present at the distal margin of the tibia or even on the shaft. Abdomen rather broad. Relatively large styles present on ninth sternite. Cerci relatively conspicuous, plainly visible with the unaided eye and usually four-segmented (never two-segmented).

Comments: Although the coastal species are generally associated with damp wood, often with rotten wood, they can survive for extended periods of time in relatively dry, sound wood. This is attested to by the fact that they are frequently encountered in finished lumber shipped throughout the U.S. and to foreign ports (Gay 1969). They are most frequently encountered in 10.16 × 10.16 cm, or larger, posts. Frequently the individuals intercepted in such shipments are somewhat desiccated, which enhances the dorsoventral flattening of the abdomen and gives them a rather roachlike appearance. They are not known to have become established in any of the areas into which they have been introduced outside of their normal range.

Selected Bibliography

Emerson 1933.
Nutting 1965.

KALOTERMITIDAE

The Kalotermitids, Drywood Termites, Furniture Termites

Figures 12.1 and 12.6

Relationships and Diagnosis: The Kalotermitidae includes a diversity of forms in both the Nearctic and Hawaii. No subfamilies have been designated for this group (Krishna 1961). As in the Hodotermitidae, the postclypeus is not divided medially. The tarsi are four-jointed and the cerci two-jointed as in other Nearctic and Hawaiian forms, except in Hodotermitidae. The most reliable diagnostic character for immature forms is mandibular structure. The left mandible has only two marginal teeth, rather than three as in the Hodotermitidae and Rhinotermitidae. The second marginal is as conspicuous as the first (fig. 12.1), not reduced as in the Termitidae. The right mandible also has only two marginal teeth, which is true of termites generally. There is no subsidiary tooth at the base on the first marginal tooth of the right mandible, as in the Hodotermitidae and Rhinotermitidae. The body tends to have a more blockish appearance, with head, thorax, and abdomen being more similar in width than is true of other termites (fig. 12.6).

Biology and Ecology: Despite the common designation of the Kalotermitidae as dry-wood termites, they occupy many different ecological situations. Some are limited to very dry, even protected, wood, including furniture (*Cryptotermes*); many inhabit relatively dry wood in dead trees, or dead portions of living trees, or structural wood (*Calcaritermes, Incisitermes, Kalotermes, Marginitermes, Pterotermes*); others inhabit relatively damp wood or dead wood in living trees (*Neotermes*). One species (*Paraneotermes simplicornis* Banks of the Desert Southwest) penetrates the soil to reach wood in contact with or buried in the ground (Light 1937). With the exception of this species, all the kalotermitids are strictly wood dwellers and do not penetrate the soil.

Description: Well-developed immatures may be 6–8 mm long (larger in *Neotermes*). They lack a fontanelle, even in mature forms. Ocelli present in adults but not usually detectable in immatures. Compound eyes may be observed in brachypterous larvae (nymphs) (fig. 12.6) and are present in some soldiers but generally absent in others. Postclypeus not subdivided medially. Antennae relatively short, usually with fewer than 20 segments. Both left and right mandibles with two marginal teeth (fig. 12.1). Pronotum variable, usually as broad as, or even broader than, the head; usually much wider than long. Mesonotum and metanotum usually similar in size and not wider behind than in front, and not strongly overlapping one another nor the first abdominal segment. Head, thorax, and abdomen tending to be about the same width, resulting in a blockish appearance. Tarsi four-segmented. An arolium may be present, but generally absent in immatures. Tibial spines and spurs also generally lacking or inconspicuous in immatures. Styles present; cerci two-segmented and inconspicuous.

Comments: Some members of this group, particularly species of *Incisitermes* (formerly *Kalotermes*) are of major economic importance throughout the southern and coastal regions of the Nearctic. Many are readily transported in household goods and have become established far beyond their usual range. The "furniture" termites (*Cryptotermes*) are the most prone to transport and are a serious pest in Hawaii and in the Gulf Coast states. Although their colonies are relatively small, usually consisting of only a few hundred individuals, they can impart considerable damage. In areas of the Southwest, particularly in Arizona, *Marginitermes* replaces *Incisitermes* as the major dry-wood pest.

Selected Bibliography

Krishna 1961.
Light 1937.

RHINOTERMITIDAE

The Rhinotermitids, Subterranean Termites

Figures 12.2, 12.7, 12.8

Relationships and Diagnosis: The Rhinotermitidae is represented in the Nearctic by three native genera and one introduced genus. This latter (*Coptotermes*) is abundant in Hawaii and has become established in limited areas of the Gulf Coast (Texas and Louisiana) in addition to being detected elsewhere. This genus represents the monogeneric Coptermitinae. Species of the Holarctic genus *Reticulitermes* are widespread through the contiguous United States and are encountered in southern Canada. *Heterotermes,* which with *Reticulitermes,* constitutes the subfamily Heterotermitinae, is present in the Desert Southwest and ranges southward into Mexico. *Prorhinotermes,* which is encountered in a limited area of coastal Florida (Dade County), is considered to be the most primitive genus of the large subfamily Rhinotermitinae.

In all instances the postclypeus is divided medially, a division that may not be as conspicuous as in the Termitidae. (The postclypeus is simple in the Hodotermitidae and Kalotermitidae.) The dentition of the mandibles represents that of the Hodotermitidae (*Zootermopsis*) with three marginal teeth on the left mandible and two on the right, compared with two on each in the Kalotermitidae and Termitidae. As in *Zootermopsis,* there is a small subsidiary tooth at the anterior margin of the first marginal of the right mandible, although it may be quite inconspicuous. Rhinotermitids can be readily distinguished from hodotermitids on the basis of the two-segmented (Rhinotermitidae) rather than more than two-segmented cerci. Immature stages are much smaller than comparable stages of hodotermitids or most kalotermitids. The largest apterous larvae are usually 5–7 mm in total length. Brachypterous larvae (nymphs) of the penultimate stage may be fully 10 mm.

Biology and Ecology: Colonies of Rhinotermitidae, which are subterranean in habit, are usually established and centered in wood buried in the ground, although these termites are capable of colonizing damp wood that is isolated from the ground. Wood and other cellulose materials above ground may be attacked via extensions of the workings through wood or by building passageways over the exterior surface. *Coptotermes* frequently colonizes wood above ground and may build extensive structures in wall voids. *C. formosanus* has long been known to colonize ship timbers and is consequently widely distributed to many seaports and established in some (Gay 1969).

Description: Postclypeus subdivided medially, although this may be inconspicuous. Antennae 10–17 segments, depending upon stage and species. Mandibles with three distinct marginal teeth on the left, the most posterior of which projects backward (fig. 12.2). Right mandible with two marginal teeth and a minute subsidiary tooth at the anterior base of the first marginal. Tergites of the thorax subequal. Pronotum somewhat larger in most, but the mesonotum larger in *Prorhinotermes*. Pronotum flat in most but saddle-shaped in *Prorhinotermes*. A pair of inconspicuous styles present on the ninth sternite; the cerci, associated with the tenth abdominal sternite, two-segmented and inconspicuous. Styles absent in functional females.

Comments: Species of *Reticulitermes* (in the continental U.S.) and *Coptotermes* (in Hawaii) are responsible for millions of dollars in damage annually to structures (and their contents) in their areas of occurrence. The activities of *Reticulitermes* are most pronounced in the eastern and Gulf coastal regions of the continent (primarily *R. flavipes* (Kollar)) and the western coastal states (primarily *R. hesperus* Banks). *R. tibialis* Banks is quite common throughout mid North America and overlaps the ranges of the other species to the east and west.

TERMITIDAE

The Termitids, Higher Termites, Desert Termites

Figure 12.3

Relationships and Diagnosis: The Nearctic Termitidae are limited to the Desert Southwest (hence the common name of desert termites) and range southward into Mexico. This is by far the largest family of Isoptera, encompassing about 75% of the described species. Species occurring in the Nearctic include members of two of the four subfamilies. The Amitermitinae are represented by *Amitermes, Gnathamitermes,* and *Anoplotermes;* the Nasutitermitinae by *Tenuirostritermes*. Both *Gnathamitermes* and *Tenuirostritermes* are restricted to this region. No termitids are encountered in Hawaii.

Apterous larvae of these species are quite small (5–7 mm in well-developed forms), unlike many of the numerous tropical members of this family in which they are often larger. Brachypterous larvae (nymphs) are conspicuously larger than apterous larvae, workers, or soldiers. In *Tenuirostritermes* this is particularly true. The soldiers in this genus are of the nasute type and are much smaller than the workers or brachypterous larvae (nymphs). They are dark brown and possess rather long legs. It is unusual to encounter a collection of these termites without nasuti, since they constitute a fair proportion of the colony.

The postclypeus in the termitids is conspicuously divided into right and left portions by a medial groove. In general this is much more conspicuous here than in the rhinotermitids and there is no such division in the hodotermitids or kalotermitids. As in the Kalotermitidae, each mandible possesses two marginal teeth. The second marginal of the right mandible does not project conspicuously from the mandibular face. The second marginal of the left mandible is generally greatly reduced and is situated posterior to the midpoint of the mandibular face (fig. 12.3), unlike the more anterior position of this tooth in the Kalotermitidae. There is no subsidiary tooth at the anterior base of the first marginal on the right mandible as occurs in the hodotermitids and rhinotermitids. As in the other termites, except the Mastotermitidae and Hodotermitidae, the tarsi are four-segmented.

The thorax is narrower than the head or abdomen and the prothorax is saddle-shaped. In all other families, the prothorax is flattened, although the lateral margins may be depressed. In the rhinotermitid, *Prorhinotermes,* the prothorax of the workers is saddle-shaped, however the mandibular dentition is clearly rhinotermitid in nature.

Styles are generally lacking in this group although their points of origin may be detectable on imagos. Styles are present in other termites but are inconspicuous. The cerci are two-segmented as in the kalotermitids and rhinotermitids, never of more than two, as in the hodotermitids.

Biology and Ecology: All of the Nearctic termitids are subterranean in habit with the colony workings situated under surface rocks. None of the species encountered in the Nearctic build conspicuous mounds or arboreal nests usually associated with tropical members of this family. Species of *Gnathamitermes* build earthen sheets over surface wood and grass upon which they feed. *Amitermes* is more prone to actively penetrate surface or subsurface wood. The workers of *Tenuirostritermes* cut short lengths of grass and small shrubs and carry these back to the workings where small stacks of such material may be encountered. Foraging apparently occurs at night or on heavily overcast days when the workers, escorted by nasuti, may occasionally be encountered foraging above ground. All of these termites, as is the case with most subterranean forms, may be encountered attacking herbivore dung.

Description: Apterous larvae and workers, as well as nasute soldiers, small. Workers and nasuti of *Tenuirostritermes* dark brown. The brachypterous larvae (nymphs) may be much larger than the other forms. Compound eyes and even ocelli are detectable in brachypterous larvae (nymphs) only. Although a fontanelle is present in mature forms, it is generally not detectable in immatures. Mandibles with two marginal teeth on each. Second marginal of the left mandible greatly reduced and situated posterior to the middle of the mandibular face. Prothorax saddle-shaped; thorax much narrower than head or abdomen. Tarsi four-segmented. Styles usually lacking; cerci very inconspicuous, two-segmented.

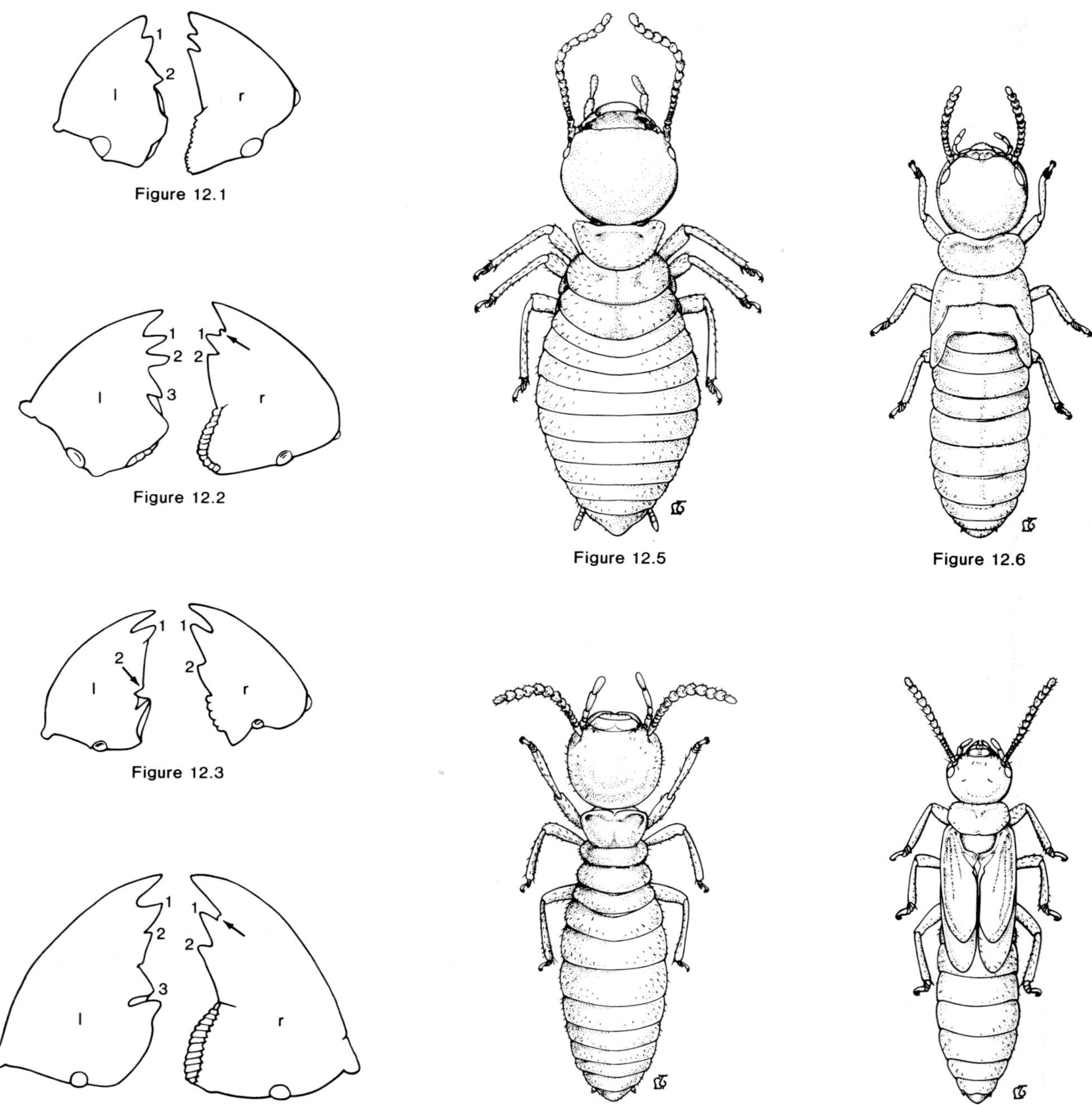

Figure 12.1

Figure 12.2

Figure 12.3

Figure 12.4

Figure 12.5

Figure 12.6

Figure 12.7

Figure 12.8

Figures 12.1.–12.4. Mandibular structure typical of various families or subfamilies of Isoptera: **(12.1)** *Incisitermes* (Kalotermitidae); **(12.2)** *Reticulitermes* (Rhinotermitidae) (In *Heterotermes* and *Coptotermes* the first marginal tooth of the left mandible is shorter than the apical or second marginal tooth); **(12.3)** *Gnathamitermes* (Termitidae); **(12.4)** *Zootermopsis* (Termopsinae, Hodotermitidae). All are in dorsal view.

Figures 12.5–12.8. General structure of immature individuals of different types and families: **(12.5)** Well-developed apterous larva of *Zootermopsis angusticollis* (Hagen) (Hodotermitidae). The spines on the shaft of the tibia are only present in very large individuals; **(12.6)** brachypterous larva (nymph) of *Cryptotermes brevis* (Kalotermitidae); **(12.7)** Functional worker of *Reticulitermes tibialis* Banks (Rhinotermitidae); **(12.8)** Penultimate brachypterous larva (nymph) of the same species. (Figs. 12.5–12.8, drawings by L. Tackett, courtesy of F. W. Stehr.)

Comments: The termitids of the Desert Southwest are rather inconspicuous but very abundant. In many areas it is difficult to locate a surface rock that does not cover the workings of one or more colonies. Considering their abundance, very little economic damage is attributed to termitids in the Nearctic, most of the damage being due to the activities of the kalotermitids and rhinotermitids.

BIBLIOGRAPHY

Ahmad, M. 1950. The phylogeny of termite genera based on imago-worker mandibles. Bull. Amer. Mus. Nat. Hist. 95:41–86.

Banks, N., and T. E. Snyder. 1920. A revision of the Nearctic termites with notes on biology and geographic distribution. Bull. U.S. Nat. Mus. 108. 228 pp.

Emerson, A. E. 1933. A revision of the genera of fossil and recent Termopsinae (Isoptera). Univ. California Publ. Ent. 6:165–96.

Gay, F. J. 1969. Species introduced by man, pp. 459–94. *In* Krishna, K., and F. M. Weesner (eds.). Biology of termites, vol. 1. New York and London: Academic Press. 598 pp.

Grassé, P. P. 1949. Ordre des Isoptères ou termites, pp. 408–544. *In* Grassé, P. P. (ed.). Traité de Zoologie, vol. 9. Masson et Cie, Paris. 1117 pp.

Hare, L. 1936. Termite phylogeny as evidenced by soldier mandible development. Ann. Ent. Soc. Amer. 37:459–86.

Kofoid, C. A. et al. (eds.). 1934. Termites and termite control. Second edition. Berkeley, Calif.: Univ. of California Press. 795 pp.

Krishna, K. 1961. A generic revision and phylogenetic study of the family Kalotermitidae (Isoptera). Bull. Amer. Mus. Nat. Hist. 122:307–408.

Krishna, K. 1970. Taxonomy, phylogeny and distribution of termites, pp. 127–152. *In* Krishna, K., and F. M. Weesner (eds.). Biology of termites, vol. 2. New York and London: Academic Press. 643 pp.

Krishna, K., and F. M. Weesner (eds.). 1969. Biology of termites, vol. 1. New York and London: Academic Press. 598 pp.

Krishna, K., and F. M. Weesner (eds.). 1970. Biology of termites, vol. 2. New York and London: Academic Press. 643 pp.

Light, S. F. 1937. Contributions to the biology and taxonomy of *Kalotermes (Paraneotermes) simplicornis* Banks (Isoptera). Univ. California Publ. Ent. 6:423–64.

Miller, E. M. 1949. A handbook on Florida termites. Coral Gables, Fla.: Univ. Miami Press. 30 pp.

Nutting, W. L. 1965. Observations on the nesting site and biology of the Arizona damp-wood termite *Zootermopsis laticeps* (Banks) (Hodotermitidae). Psyche 72:113–25.

Snyder, T. E. 1949. Catalog of the termites (Isoptera) of the world. Smithsonian Misc. Collections, No. 112. 496 pp.

Snyder, T. E. 1956. Annotated subject-heading bibliography of termites, 1350 B.C. to A.D. 1954. Smithsonian Misc. Collections, No. 130. 305 pp.

Snyder, T. E. 1961. Supplement to the annotated subject-heading bibliography of termites, 1955 to 1960. Smithsonian Misc. Collections, No. 143. 137 pp.

Weesner, F. M. 1960. Evolution and biology of the termites. Ann. Rev. Ent. 5:153–170.

Weesner, F. M. 1965. The termites of the United States, a handbook. Elizabeth, N.J.: Natl. Pest Control Assoc. 71 pp.

Order Mantodea

13 

David A. Nickle
Systematic Entomology Laboratory, ARS, USDA

See the "Introduction to the Orthopteroid Insects," p. 118.

MANTIDAE

Praying Mantids

Figures 13.1–13.7

The more than 1800 species of praying mantids are exclusively predaceous. The distribution is worldwide, but most species are tropical. In the United States mantids have perhaps a unique distinction among insects in commanding an almost universal respect for their well-being. Folk legend is such that nearly everyone who is familiar with praying mantids thinks (erroneously) that they are protected by law and that killing them is wrong. In fact, the only such law is in Florida, enacted in 1979, that declares "the praying mantid" the state insect and offers it the same level of protection afforded other fauna and flora so designated by the state. However, the folk legend provides essentially the same degree of protection for these insects nationwide.

RELATIONSHIPS AND DIAGNOSIS

In the United States and Canada there are 20 species of mantids. Traditionally, mantids have been considered a nonsaltatorial family of the Orthoptera, but current interpretations of the higher classification of the Orthoptera and its relatives have resulted in treating mantids as a suborder, Mantodea, of the Dictyoptera (Chopard 1949), an order it shares with the cockroaches, or as a separate order, the Mantodea (Mackerras 1970). The mantids are closely related to the cockroaches and termites. These groups share several characters: hemimetabolous metamorphosis, hypognathous chewing mouthparts, segmented cerci, asymmetrical male genitalia that are concealed from view, the absence of an external ovipositor, and hindwings with large anal lobes. Mantids and cockroaches are related to each other and separated from the termites by the fact that they lay eggs in egg cases or oöthecae, and their tarsi are five-segmented, not four-segmented (rarely five) or fewer as in the termites. From these groups the mantids differ in the shape and function of their fore legs, which are clearly raptorial and modified for grasping.

BIOLOGY AND ECOLOGY

Mantids are a predominantly tropical group. The only two species that occur in Canada, *Mantis religiosa* L. and *Tenodera aridifolia sinensis* Saussure, are introduced, so there are no endemic species north of the United States. In the United States and Canada praying mantids overwinter exclusively in the egg stage. During courtship and mating, the male approaches the female with caution and mounts her. In some cases the female in copulation may turn and devour the male. Females of some species use a pheromone to attract potential mates (Robinson 1979).

Eggs are laid in groups of about 12 to 400 in an egg case (oötheca), which begins as a frothy liquid but hardens into a tough fibrous protective structure after the eggs have been deposited in it. The shape and placement of the oötheca is characteristic for various species (Breland and Dobson 1947). Some species, such as *Tenodera aridifolia sinensis* and all species of *Stagmomantis,* place their oöthecae on goldenrod, blackberry, or other weeds. The shape of the oötheca is often species specific. In contrast to species that place their oöthecae on weeds, the grizzled mantis, *Gonatista grisea* (F.) deposits the oötheca in crevices on the bark of trees.

In mid-spring the larvae emerge through holes they chew in the wall of the oötheca. The average number of instars is about seven (range four to nine). All species in the United States are univoltine.

Most mantids are cryptic in some way. Most are green and leaflike. Others are sticklike (e.g., species of ***Brunneria***) or resemble the bark of trees (e.g., *Gonatista grisea*). Several tropical species are flower mimics (Varley 1939).

Although praying mantids are exclusively predaceous, their feeding habits are generalized. Few, if any species limit their diet to only one kind of host; possible exceptions are some of the flower-mimicking species of the tropics, where hosts could be limited to those species of insects that have co-evolved to pollinate the flower that the mantid mimics. As generalized predators, mantids are not considered very useful biological control agents, although they may be welcome inhabitants of home gardens.

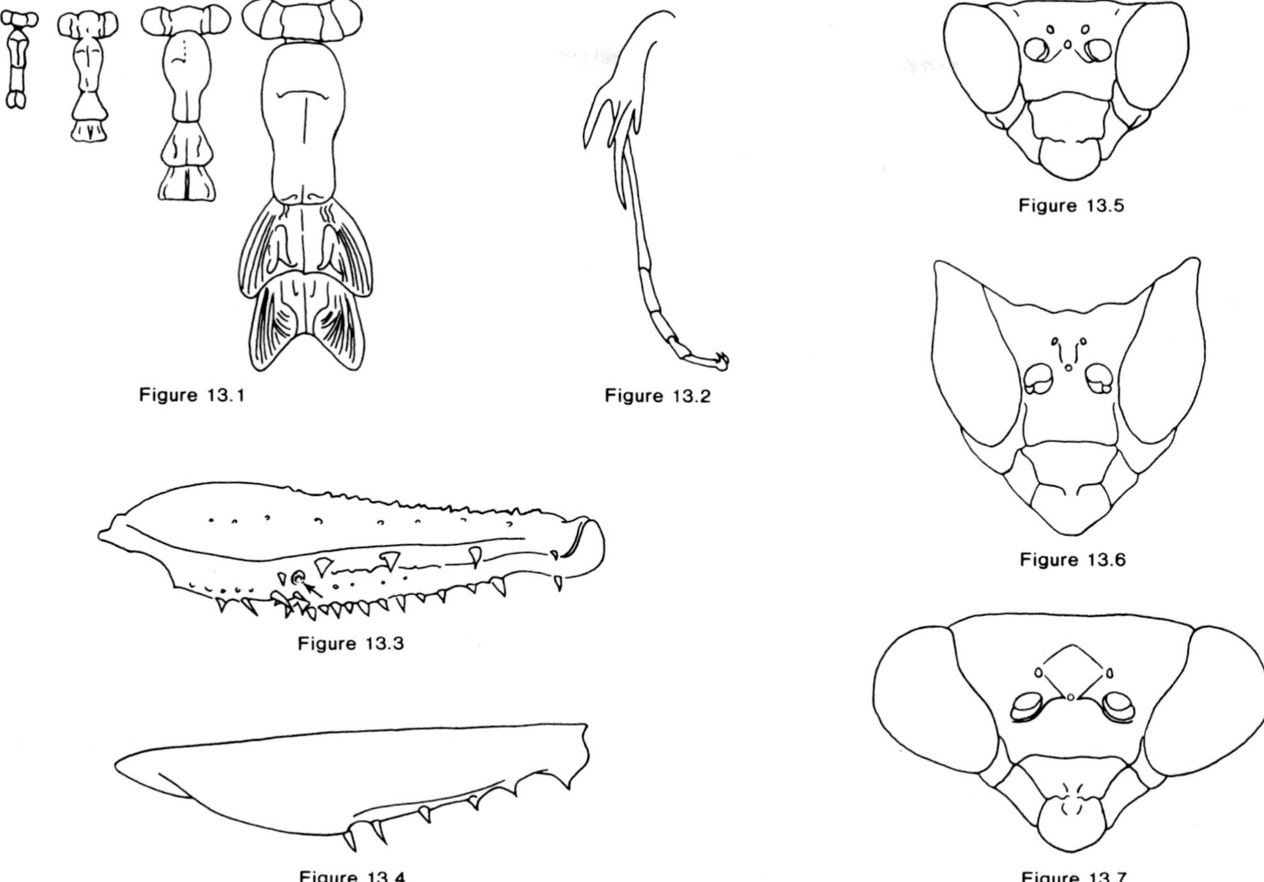

Figure 13.1. Development of wing pads in larvae of *Gonatista grisea* (F.), dorsal view.

Figure 13.2. Left fore tibia of larva (instar 3) of *Thesprotia graminis* (Scudder), median view.

Figure 13.3. Right fore femur of larva of *Gonatista grisea,* lateral view. Arrow indicates position of fovea, a pit between the first and second spine.

Figure 13.4. External ventral margin of right fore femur of larva (instar 4) of *Iris oratoria* (L.).

Figure 13.5–13.7. Faces of larvae, frontal view: **(13.5)** *Litaneutra minor* (Scudder); **(13.6)** *Yersiniops solitarium* (Scudder); **(13.7)** *Mantis religiosa* L.

DESCRIPTION

Small to very large insects (2–100 mm), hemimetabolous, with hypognathous chewing mouthparts. Mid to late instar larvae similar to adults except for development of wings, genitalia, and antennae. External wing pads visible from fourth instar, increasing in size in subsequent moults (fig. 13.1).

Head

Triangular in frontal view (figs. 13.5–13.7), laterally broad, freely movable to pivot around junction with thorax. Eyes large, ovoid, elliptical, or pointed, widely spaced; ocelli three, arranged on a triangular elevation just above antennal insertions.

Thorax

Pronotum several times longer than wide for most species (only slightly longer than wide in one U.S. species, *Mantoida maya* Saussure and Zehntner), with broadest region in anterior half just above coxal attachment of fore legs.

Legs

Mid and hind legs very similar (figs. 13.2–13.4), adapted for walking. Fore legs raptorial; coxa greatly enlarged; femur usually with several stout spines on ventral margins; tibia also bearing several spines and a long, sharp terminal spine; tarsi long, five-segmented.

Wings

Usually alate as adults; some species with tegmina and wings of female reduced; some species apterous. Fore wings or tegmina thickened. Hind wings membranous, clear or colored, with or without dark markings. Size of anal lobe of hind wing variable but enlarged.

Abdomen

Male: Elongate, dorsoventrally compressed. Ten terga; tenth tergum narrowing apically, triangular. Cercus elongate, segmented with 10–20 articles. Subgenital plate (sternum 9) elongate, spatulate, apically usually bearing two articulating styles; in most species, subgenital plate partly or completely concealing internal genitalia, which are a complex of asymmetrical sclerites (see Beier 1970).

Female: Elongate, dorsoventrally compressed, usually broader and more robust than that of conspecific male. Cercus elongate, segmented, similar to male. Subgenital plate (sternum 7) elongate, ensheathing short blunt ovipositor. Ovipositor comprised of three pairs of valves.

COMMENTS

The praying mantids belong to the single family Mantidae (with 13 subfamilies), which in the United States is represented by 7 subfamilies totaling 20 species. Three species have been introduced into the United States from Europe or Asia: *Mantis religiosa, Tenodera aridifolia sinensis,* and *Iris oratoria* (L.). The subfamilies are separated on the basis of pronotal shape, ornamentation of the head (figs. 13.5–13.7), and spination of the raptorial front legs (figs. 13.3, 13.4). The only world treatment of mantids is the work of Giglio-Tos (1927), whose divisions of the group remain basically uncontested. Additional information about mantid relationships has been gleaned from chromosome studies (Hughes-Schrader 1950). In the keys of Giglio-Tos and Brues, Melander, and Carpenter (1954), several subfamilies are divided in several couplets. Useful keys to adult mantids that may be used to identify larvae to species include Blatchley (1920), Gurney (1951), Hebard (1943), and Helfer (1963). Most keys to adults will fail to assist in the determination of some larvae, however, because important couplets often use only wing shape and coloration.

Selected Bibliography

Complete citations begin on p. 167.

Beier 1934, 1935, 1937, 1970.
Blatchley 1920.
Brues et al. 1954.
Breland and Dobson 1947.
Chopard 1949.
Giglio-Tos 1927.
Gurney 1951.
Helfer 1963.
Hughes-Schrader 1950.
Mackerras 1970.
Rehn 1911.
Robinson 1979.
Varley 1939.

Order Grylloblattodea (Notoptera)

14

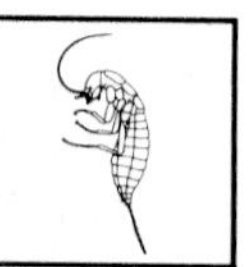

David A. Nickle
Systematic Entomology Laboratory, ARS, USDA

See the "Introduction to the Orthopteroid Insects," p. 118.

GRYLLOBLATTIDAE

Rock Crawlers, Icebugs

Figures 14.1–14.6

The Grylloblattodea are an unusual group of orthopteroid insects in that the individuals of this order display many characters considered primitive for the orthopteroid orders; they are probably relict insects, and are geographically restricted to remote glacial regions in northwestern United States, Canada, and northeastern Asia.

RELATIONSHIPS AND DIAGNOSIS

Until recently this group was placed as a family of Orthoptera (Grylloblattidae), and many of the standard entomology textbooks still treat rock crawlers as such, but they are now considered to represent a separate order (Kamp 1973). They are of great interest in that they possess a mosaic of features found in other orthopteroid orders, as well as unique features considered primitive or ancestral for the orthopteroid lineage. They share several morphological characters with the Dermaptera; also, the legs, tarsi, and cerci of both sexes are similar to the Blattodea and Mantodea. On the other hand, the ovipositor has six valves and is similar in form to the Tettigonioidea (Orthoptera). Using a numerical taxonomic analysis of the orthopteroids, Kamp (1973) demonstrated that the Grylloblattodea should be treated as an order and that its closest affinities are to the Dermaptera. Moreover, it shares with that order a more distant relationship with the Phasmatodea. The higher classification of the orthopteroid orders has undergone major revisions recently (see Kevan 1976), and there is now general agreement among orthopterists that Kamp's recognition of the rock crawlers as an order is correct. Interestingly enough, although the grylloblattids were first given ordinal status by Crampton (1915), who named it Notoptera, it was not well received, although it was used at least once in a review by Chopard (1949). However, the term Grylloblattodea, proposed by Brues and Melander (1932) and used by Essig (1942), has gained general acceptance.

The Grylloblattodea consists of one small family containing three genera: *Galloisiana* (with ten species from Japan, Korea and USSR), *Grylloblattina* (with a single species from Siberia), and *Grylloblatta* (with ten species from western Canada and northwestern United States).

The rock crawlers are all small apterous campodeiform insects, about 20–30 mm in length. Eyes are small, absent in some species; ocelli are absent. The antennae are long, filiform, with 30–40 articles. The prothorax is quadrate, longer than wide, longer than the meso- and metathorax. Tarsi are five-segmented; those of adult males bear a pair of membranous lobes between each segment. The abdomen is elongated, and the cerci are long, filamentous, with eight to nine segments. The ovipositor is elongated, sword-shaped, consisting of three paired valves (Scudder 1970).

BIOLOGY AND ECOLOGY

Although little is known about the biology, their geographic distribution and tolerance to cold temperatures have attracted the interest of numerous biologists (see Kamp 1979, see also H. Ando, ed. 1982). Their common name, "icebugs", is derived from the remarkable phenomenon of most species being active at temperatures that cause most insects to be totally inactive. The optimum temperature for U.S. species appears to be about 2°C (35°F).

Eggs are laid in moss and decaying logs, between rocks, or in moist soil. They are about 3 mm long, oval, and black. The number of larval instars is between eight and nine. They are very slow to mature and may require two or three years to reach adulthood.

Grylloblattids are either predators or scavengers, but stomach contents of species from the United States nearly always contain fragments of insects and spiders (Ford 1926). Activity is confined to periods of moderate cold, between 30–50°F, and most captured specimens have died on prolonged exposure to direct sunlight. Those that have been observed in nature have been found crawling on rocks or in damp decaying logs.

DESCRIPTION

Grylloblattids are secondarily apterous in all stages of development, and larvae resemble adults except for size and degree of development of the genitalia.

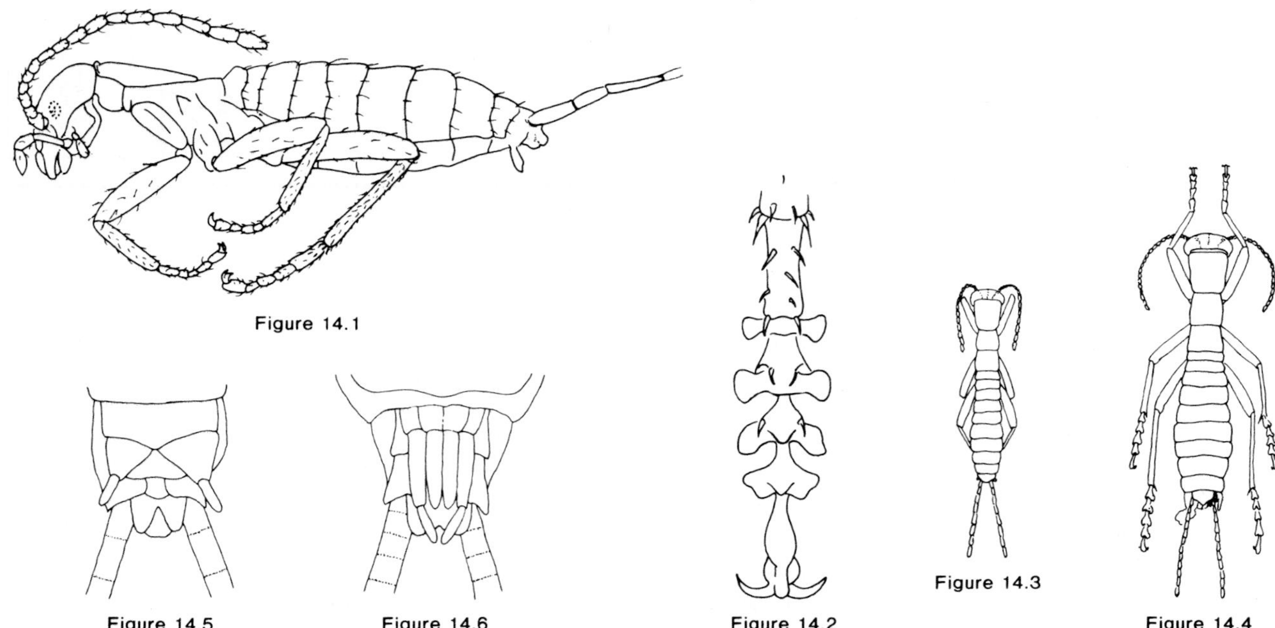

Figure 14.1. *Grylloblatta campodeiformis* Walker, habitus of larva (instar 3), left lateral view.

Figure 14.2. Hind tarsus of *G. campodeiformis,* male larva (instar 3), ventral view.

Figures 14.3–14.4. Habitus, *G. campodeiformis,* dorsal view (modified from Walker 1919): **(14.3)** larva (instar 3); **(14.4)** larva (instar 5.)

Figures 14.5–14.6. Apex of abdomen of *G. campodeiformis,* ventral view; **(14.5)** male larva (instar 3); **(14.6)** female larva (instar 3). (Modified from Walker 1919).

Head

Similar in form to Dermaptera. Prognathous. Compound eyes oval, reduced to about 60 ommatidia. Ocelli absent. Antennae filiform, nearly half as long as body, with 20–40 articles (figs. 14.1, 14.3, 14.4).

Thorax

Pronotum small, rectangulate, not extending over meso- and metanotum. Thoracic sterna separate, not fused.

Legs

Similar in form to Blattodea. Tarsi five-segmented. Pulvilli and arolia absent. In males, a pair of membranous lobes is present between each tarsal segment (fig. 14.2).

Abdomen

Male external genitalia asymmetrical; ninth sternum bearing articulating styles (fig. 14.5). Cerci long, filiform, with eight or nine segments. Ovipositor elongate, composed of three paired valves (fig. 14.6).

COMMENTS

Walker (1914, 1919) described the first rock crawler and immediately recognized its primitive features among the Orthoptera. Gurney (1937, 1953, 1961) reviewed most of what is known on the systematics of the Grylloblattodea. Kamp (1973) provided an excellent analysis of the order, using numerical taxonomic techniques. Additional information on phylogenetic relationships of this order was provided by Walker (1937, 1956) and Scudder (1970).

Selected Bibliography

Complete citations begin on p. 167.

Ando 1982.
Brues and Melander 1932.
Chopard 1949.
Crampton 1915.
Essig 1942.
Ford 1926.
Gurney 1937, 1953, 1961.
Kamp 1973, 1979.
Kevan 1976.
Scudder 1970.
Walker 1914, 1919, 1937, 1956.

Order Phasmatodea

15

David A. Nickle
Systematic Entomology Laboratory, ARS, USDA

See the "Introduction to the Orthopteroid Insects," p. 118.

PHASMATIDAE

Walkingsticks

Figures 15.1–15.4

The Phasmatodea (walkingsticks) are an unusual group of insects that traditionally has been included as a family of Orthoptera (Brunner and Redtenbacher 1906–1908, Caudell 1903). Several names have been used for the order, including Phasmida (Leach 1815, Beier 1968), Phasmatoptera (as a suborder of Orthoptera, Rehn and Grant 1961), and Cheleutoptera (Crampton 1939, Kevan 1976), but there seems to be growing acceptance for the name Phasmatodea as a distinct order (Jacobsen and Bianchi 1902, Gunther 1953, Bradley and Galil 1977). Of the more than 2500 species that have been described, most are tropical, and only about 30 occur in the United States. Most species resemble twigs (fig. 15.1) or, in some cases, leaves. All species are vegetarians, but only a few are considered pests. Some species are among the largest of insects: *Hermarchus pythonius* (Westwood) attains a length of more than 260 mm; most species, however, average no more than 70 mm in length.

The order consists of six families, of which four occur in the United States and Canada. Although many walkingsticks have fully developed wings and are capable of flight, in the United States all but one species (*Aplopus mayeri* Caudell) are apterous.

RELATIONSHIPS AND DIAGNOSIS

The Phasmatodea are members of that orthopteroid line in which the external male genitalia are concealed and asymmetrical. This line includes the Grylloblattodea, Mantodea, and Blattodea. However, the cerci of walkingsticks are unsegmented, unlike the grylloblattids, mantids, and cockroaches, which have segmented cerci.

Phasmatodea mouthparts are hypognathous or prognathous. All legs are similar, and tarsi are five-segmented (except in Timematidae, with three-segmented tarsi; see figs. 15.2 and 15.3). Both male and female genitalia are usually concealed. Unlike the Orthoptera, wing pads in larvae, when present, are not reversed. These insects lack both stridulatory organs and auditory tympana.

BIOLOGY AND ECOLOGY

Walkingsticks are exclusively plant feeders. Aspects of their behavior have attracted biologists for many years. For example, some species are parthenogenetic (Bergerard 1962), several species possess toxin-producing prothoracic glands, and they are one of the few groups of insects that can regenerate lost legs.

Walkingsticks have evolved several strategies in defense against predators. Most species display a habit of rocking back and forth while perched on a branch which they often continue for several minutes at a time. Presumably this rocking motion enhances their mimicry to leaves or twigs swaying in the breeze. Other strategies include passive mimicry or crypsis, in which the insect resembles its background environment; catalepsy or feigning death, in which the insect becomes immobile for a prolonged period of time; warning displays, in which coloration patterns are abruptly presented to a would-be predator; and secretion of toxic or noxious substances.

Walkingsticks generally freely drop their eggs as they walk or remain motionless on a branch. However, some species deposit their eggs in the soil or in other substrates. The oviposition behavior of *Anisomorpha buprestoides* is unusual. The female digs a hole with her fore and mid legs, and while still in the hole she recurves her abdomen over the body to her head and deposits several eggs in the hole. She then backs out and covers up the hole (Hetrick 1949). Species of *Timema* and *Pseudosermyle* glue their eggs in clusters or singly to the surface of leaves.

Eggs in general are oval or round with hard shells. They have useful characters for species identification, although only about five percent of the species have been described (Clark 1976, Coleman 1942). Most species of walkingsticks in North America overwinter in the egg stage. In many species two periods of diapause occur within egg development, and many species require two years for eggs to hatch. Usually there are five molts to adulthood, although females of many species molt six times (Coleman 1942, Savage 1957, Stringer 1970). During mating, the male mounts the female, coupling occurs, and the pair may remain attached for extended periods of time.

A list of natural enemies of phasmatids has been compiled by Bedford (1978). In the United States only one species, *Diapheromera femorata,* is an occasional economically important pest. It attacks hardwood forests, and several papers describe its biology and damage potential in northern forests (Severin and Severin 1911, Eaton 1952). Environmental factors affecting outbreaks of pest walkingsticks in Australia have been reviewed by Campbell (1974).

DESCRIPTION

Larvae resemble adults, differing mainly in size and development of genitalia. All but one species in the United States are apterous.

Head

Prognathous or hypognathous. Antennae long, stout. Eyes small; ocelli often absent.

Thorax

Pronotum short, seldom exceeding twice the length of the head. Mesonotum and metanotum usually very long in comparison to pronotum. Wings absent in U.S. species (although short tegmina are present in *Aplopus*). Legs long, slender, nearly equal in size, usually exceeding four times the length of head. Apices of mid and hind tibiae with or without an areola (fig. 15.4). Tarsi usually five-segmented, terminating in two claws. Arolium present.

Abdomen

Basal segment closely adherent to metathorax, the two often fused. Ten abdominal sclerites. Male and female genitalia usually concealed. Cerci of both sexes unsegmented.

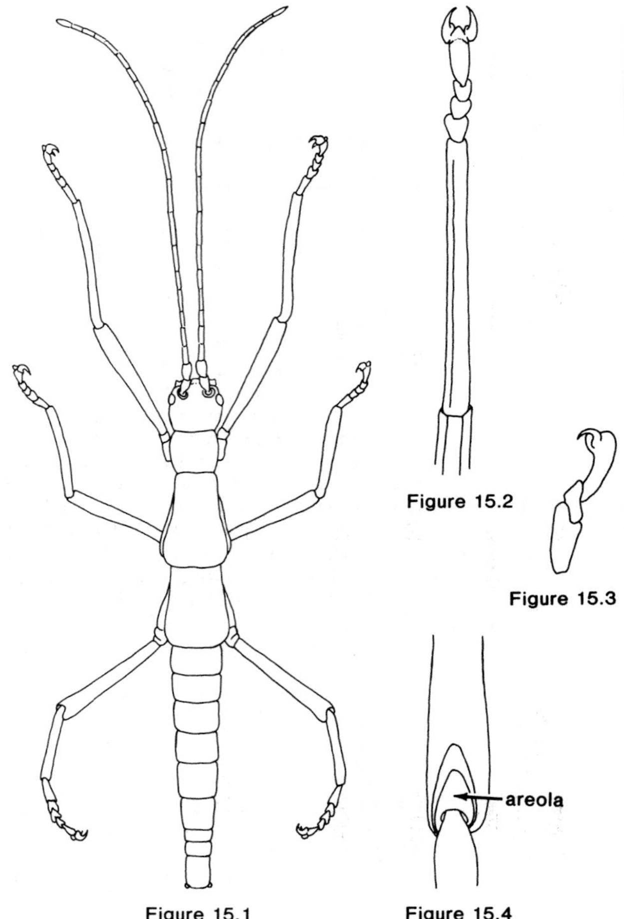

Figure 15.1. Habitus of a larva (instar 4) of *Anisomorpha buprestoides* (Stoll) (Pseudophasmatidae).

Figures 15.2–15.3. Tarsal segments of larvae (instar 3) of walkingsticks: **(15.2)** *Anisomorpha buprestoides* (Pseudophasmatidae); **(15.3)** *Timema* sp. (Timematidae).

Figure 15.4. Apex of mid tibia of larva of *Anisomorpha buprestoides*, ventral view (Pseudophasmatidae).

CLASSIFICATION

Order **PHASMATODEA**
- Heteronemiidae
- Phasmatidae
- Timematidae
- Bacillidae
- Pseudophasmatidae
- Phylliidae

COMMENTS

Several keys for adult phasmatids are available for identifying larvae at least to genus. The most useful of these are Helfer (1963) and Hebard (1943).

Selected Bibliography

Complete citations begin on p. 167.

Bedford 1978.
Beier 1968.
Bergerard 1962.
Bradley and Galil 1977.
Brunner von Wattenwyl and Redtenbacher 1906–1908.
Campbell 1974.
Caudell 1903.
Clark 1976.
Coleman 1942.
Crampton 1939.
Eaton 1952.
Gunther 1953.
Hebard 1943.
Helfer 1963.
Hetrick 1949.
Jacobson and Bianchi 1902.
Kevan 1976.
Leach 1815.
Rehn and Grant 1961.
Savage 1957.
Severin and Severin 1911.
Stringer 1970.

Order Orthoptera

16

David A. Nickle
Systematic Entomology Laboratory, ARS, USDA

Thomas J. Walker
University of Florida

M. A. Brusven
University of Idaho

See the "Introduction to the Orthopteroid Insects," p. 118.

GRASSHOPPERS, CRICKETS, KATYDIDS AND OTHERS

Order **ORTHOPTERA**
- Suborder Ensifera
  - Superfamily Grylloidea
    - Gryllidae, crickets
    - Gryllotalpidae, mole crickets
  - Superfamily Gryllacridoidea
    - Gryllacrididae, leaf-rolling crickets
    - Rhaphidiphoridae, cave and camel crickets
    - Stenopelmatidae, Jerusalem or stone crickets
  - Superfamily Tettigonioidea
    - Haglidae, hump-winged katydids
    - Tettigoniidae, katydids, long-horned grasshoppers
- Suborder Caelifera
  - Superfamily Acridoidea
    - Tetrigidae, pygmy or grouse locusts
    - Tridactylidae, pygmy mole crickets
    - Eumastacidae, monkey grasshoppers
    - Tanaoceridae, desert long-horned grasshoppers
    - Acrididae, short-horned grasshoppers

KEY TO THE FAMILIES OF IMMATURE ORTHOPTERA

1. Fore legs fossorial, enlarged, modified for digging 2

1′. Fore legs not fossorial, not enlarged, similar in form to mid legs 3

2(1). Fore and mid tarsi 2-segmented; hind tarsi 1-segmented or absent; 3 ocelli; tympanum on side of abdomen; small insects, less than 10 mm long .... ***Tridactylidae*** (p. 158)

2′. Fore and mid tarsi 3-segmented; 2 ocelli; tympanum on fore tibia of late instars and adults; larger insects, greater than 18 mm long when wing pads are present .... ***Gryllotalpidae*** (p. 148)

3(1′). Pronotum elongated, extending to or exceeding tip of abdomen; fore and mid tarsi 2-segmented, hind tarsi 3-segmented; arolia absent .... ***Tetrigidae*** (p. 157)

3′. Pronotum saddle-shaped, but not greatly elongated; tarsi 3- or 4-segmented; arolia present 4

4(3′). Antenna filamentous, usually as long as or longer than body; tympanum, if present, located at base of fore tibia (fig. 4); dorsal and ventral valves of ovipositor articulated along entire length, forming a single ensiform or stilettiform structure (figs. 1, 2) 5

4′. Antenna usually short, seldom exceeding half of body length; tympanum, if present, located on side of first abdominal segment; dorsal and ventral valves of ovipositor consisting of 2 pairs of separate pronglike valves with a basal hinge articulation (fig. 3) 7

5(4). Tarsi 3-segmented .... ***Gryllidae*** (p. 148)

5′. Tarsi 4-segmented 6

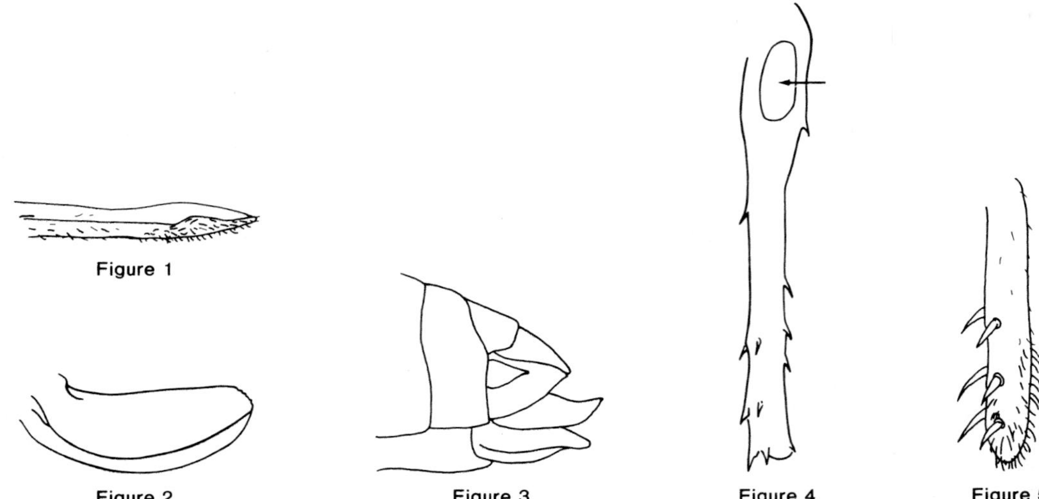

6(5′). Tympana present, at least on mid instars; pronotum usually with lateral carinae; fore and mid tibiae lacking ventral articulated spines. (fig. 4) .... (including ***Haglidae***) ***Tettigoniidae*** (p. 153)

6′. Tympana absent; pronotum without lateral carinae; fore and mid tibiae usually with ventral articulated spines (fig. 5) .... ***Gryllacridoidea*** (p. 151)

7(4′). Antenna shorter than fore femur; tympana absent; small wingless species .... ***Eumastacidae*** (p. 160)

7′. Antenna longer than fore femur; tympana present or absent 8

8(7′). Antenna very long, in males, longer than body, in females, about half as long; tympana absent; males with a stridulatory ridge on third abdominal segment; frontal ridge forming a single carina beneath median ocellus .... ***Tanaoceridae*** (p. 160)

8′. Antenna less than half as long as body; tympana on sides of first abdominal segment; males lacking a stridulatory ridge on third abdominal segment; frontal ridge broad or with 2 carinae beneath median ocellus .... ***Acrididae*** (p. 162)

Suborder ENSIFERA

Superfamily GRYLLOIDEA

Thomas J. Walker, *University of Florida*

GRYLLIDAE, CRICKETS

(Including GRYLLOTALPIDAE, Mole Crickets)

Figures 16.1–16.11

Relationships and Diagnosis: Some authors assign all crickets to the family Gryllidae, but the more usual treatment is to place mole crickets in a separate family, the Gryllotalpidae. The closest relatives of crickets are the Tettigoniidae and Gryllacrididae. With these families they share hind legs that are adapted to jumping (enlarged hind femora) and long, threadlike antennae. They differ in having all tarsi three-segmented and in bearing club-shaped sensory setae on the basal inner surface of the cerci (fig. 16.11); these apparently act as gravity receptors (Bischof 1975), and have been named tricholiths (Hartman et al. 1979).

Biology and Ecology: Crickets are most diverse in warm, moist regions, but some species occur in deserts and others in southern Canada. Juveniles live in the same habitats as adults; they are found in soil, on the ground, in and under litter, on herbaceous plants, and on leaves, twigs, and trunks of trees and shrubs. Four North American species live in the nests of ants, and a larger number construct their own burrows. A few are found almost exclusively on one species or genus of plants, ranging from sphagnum moss to pine trees. Crickets are generally omnivorous but few details are known as to the mixes of foods consumed under natural circumstances by even the most common species, either as juveniles or adults.

Description: Juveniles resemble adults in color and proportions; length of body less than 27 mm (see figs. 16.1–16.7).

Head: Hypognathous or prognathous. Antennae longer than head and thorax combined, except in mole crickets. Ocelli present or absent.

Thorax: Last two juvenile instars with wing pads. All tarsi three-segmented. Tibial tympana generally not developed (except in mole crickets).

Abdomen: Cerci tapering and slender, bearing clublike setae at base (fig. 16.11).

Comments: Crickets and all their major subgroups are found on all continents. Approximately 3000 species are known, 130 occurring in America north of Mexico. Many of our species are difficult to identify as adults, and practically no attention has been given to identification of juveniles. Fortunately, juvenile crickets can generally be identified to subfamily by the same features that are useful in adults (see key below).

The species of greatest economic importance are the house cricket (*Acheta domesticus*), southern mole cricket (*Scapteriscus acletus*), and tawny mole cricket (*S. vicinus*) (Walker 1984). House crickets, reared now by the millions for fish bait and animal food, were introduced from the Old World and are feral in North America only in the immediate vicinity of buildings or areas of intense human disturbance. The juveniles resemble those of *Gryllus* spp. (fig. 16.5), but they are mottled brown rather than black and have a conspicuous dark bar across the vertex. Southern mole crickets and tawny mole crickets damage pastures, turf, and vegetable crops in the Southeast by feeding and tunneling, and annual losses amount to tens of millions of dollars. Juveniles of *Scapteriscus* differ from those of other mole crickets in having two tibial dactyls (figs. 16.8, 16.9), rather than four (fig. 16.10). The tibial dactyls of *S. vicinus* nearly touch at their bases (fig. 16.9), whereas those of *S. acletus* are separated at their bases by a gap nearly equalling the width of one (fig. 16.8). The northern mole cricket, *Neocurtilla hexadactyla*, (fig. 16.10), is a common species throughout eastern United States. It occurs along the margins of ponds and streams and is seldom of economic consequence.

A variety of other ground-dwelling crickets are of occasional economic importance. Foremost are field crickets (*Gryllus* spp., fig. 16.5), which sometimes damage crops such as cotton, strawberries, alfalfa, and wheat. They also cut the twine on bales of hay, and the adults sometimes fly to lights in such numbers that householders feel besieged and restaurant owners lose business. On the positive side, they may be important predators of the apple maggot (Monteith 1971). In the Southeast, the short-tailed cricket (*Anurogryllus arboreus*) damages pine plantings (Weaver and Sommers 1969) and leaves unsightly mounds of dirt on golf greens and lawns. Juveniles (fig. 16.6) resemble those of field crickets except that they are brown rather than black and have a more rounded frons and shorter hind tibiae. Ground crickets (Nemobiinae; see key below) are often abundant and occasionally of economic importance. The striped ground cricket, *Allonemobius fasciatus*, has been implicated as a destroyer of white clover seedlings (Nielsson and Bass 1967), but it also feeds on the pupae of horn flies and apple maggots (Bourne and Nielsson 1967, Monteith 1971). Juveniles are 10 mm or less and have orange heads and dark longitudinal stripes laterally and mediodorsally (fig. 16.7). Juveniles of field crickets and ground crickets sometimes climb to feed but are characteristically restricted to the litter and close-to-ground vegetation.

Tree crickets (Oecanthinae; see key below) are the crickets most frequently collected from vegetation and are numerous on herbs as well as trees and shrubs. They often occur on plants of value to human use and are beneficial in that they eat phytophagous insects such as aphids and scales; they are harmful in that they feed on flowers, fruits, and foliage, kill or weaken stems by ovipositing in them, and provide access for plant pathogens (Parrot and Fulton 1914). Juveniles (fig. 16.3) are slender, usually pale green to white, and have the mouthparts directed forward. They can often be identified to species by the dark marks on the first two antennal segments that are used in identifying adults (Walker 1962, 1963). Larger bush crickets (Eneopterinae; see key below) are found principally on foliage and concealed under bark. The restless bush cricket, *Hapithus agitator*, (fig. 16.4) occurs in eastern U.S. south of the fortieth parallel and is a minor pest of citrus (Bullock 1973).

KEY TO THE MAJOR SUBFAMILIES OF IMMATURE CRICKETS[1]

(Principal genera listed in parentheses)

1. Fore legs conspicuously modified for digging .... mole crickets (figs. 16.1, 16.8–16.10) (*Scapteriscus, Neocurtilla*), **Gryllotalpinae** *or* ***Gryllotalpidae***

1′. Fore legs not conspicuously modified for digging 2

2(1′). Body covered with minute scales; clypeus bulging .... scaly crickets (fig. 16.2)(*Cycloptilum, Hoplosphyrum*) **Mogoplistinae**

2′. Body not covered with scales; clypeus not bulging 3

3(2′). Upper margins of hind tibiae with rows of spines with teeth between (figs. 16.3, 16.4) (rarely with neither spines nor teeth); usually on plants 4

3′. Upper margins of hind tibiae with rows of prominent spines with no teeth between (fig. 16.5–16.7); usually on the ground 5

1. Three U.S. subfamilies are omitted: Trigonidiinae, Myrmecophilinae, and Pentacentrinae. Juveniles of these are tiny and/or seldom collected.

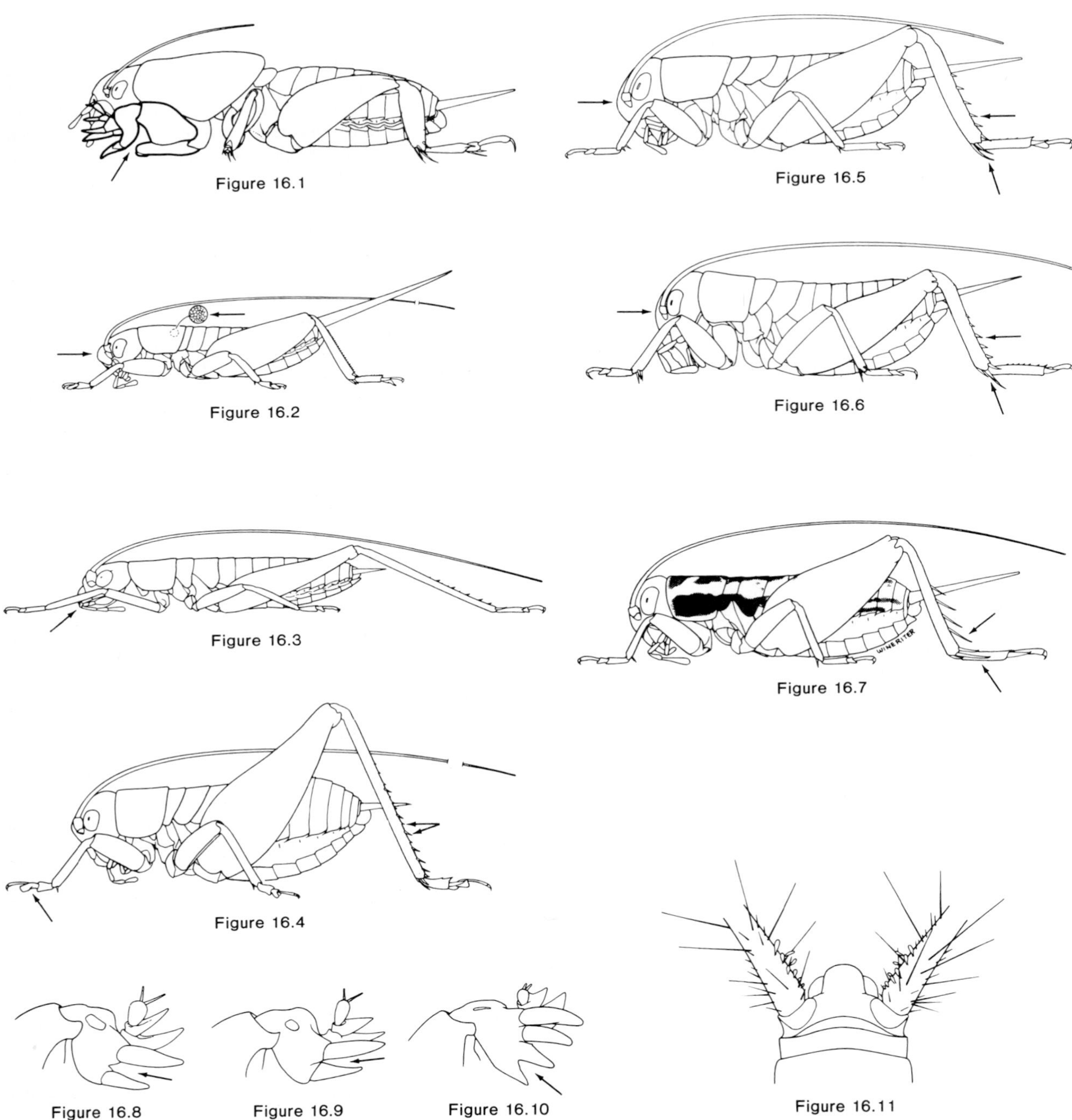

Figures 16.1–16.7. Middle stage juveniles of representative crickets: **(16.1)** Mole cricket, Gryllotalpinae, *Scapteriscus acletus* Rehn and Hebard (Gainesville, Fla.); **(16.2)** Scaly cricket, Mogoplistinae, *Cycloptilum* sp. (Dry Tortugas, Fla.); **(16.3)** Tree cricket, Oecanthinae, *Oecanthus celerinictus* Wallker (Tallahassee, Fla.); **(16.4)** Larger bush cricket, Eneopterinae, *Hapithus agitator* Uhler (Key Largo, Fla.); **(16.5)** Field cricket, Gryllinae, *Gryllus rubens* Scudder (Gainesville, Fla.); **(16.6)** Short-tailed cricket, Gryllinae, *Anurogryllus arboreus* Walker (Gainesville, Fla.); **(16.7)** Ground cricket, Nemobiinae, *Allonemobius fasciatus* De Geer (Sanford, Fla.).

Figures 16.8–16.10. Fore legs of mole cricket juveniles: **(16.8)** *Scapteriscus acletus* Rehn and Hebard; **(16.9)** *S. vicinus* Scudder; **(16.10)** *Neocurtilla hexadactyla* (Perty).

Figure 16.11. Cercal bases of juvenile cricket (*Allonemobius* sp.) showing club-shaped sensory setae (dorsal view).

4(3). Slender, prognathous, usually pale green; second tarsal segment cylindrical .... tree crickets (fig. 16.3) (*Oecanthus, Neoxabea*) **Oecanthinae**

4′. Robust, hypognathous, never pale green; second tarsal segment depressed and heart-shaped .... larger bush crickets (fig. 16.4) (*Hapithus, Orocharis*) **Eneopterinae**

5(3′). Spines along upper margins of hind tibiae stout; apical hind tibial spines less than half as long as first tarsal segment .... field crickets (figs. 16.5, 16.6) (*Gryllus, Acheta, Anurogryllus, Miogryllus*) **Gryllinae**

5′. Spines along upper margins of hind tibiae slender, apical hind tibial spines more than half as long as first tarsal segment .... ground crickets, (fig. 16.7) (*Allonemobius, Eunemobius, Neonemobius, Pictonemobius*) **Nemobiinae**

Selected Bibliography

Bischof 1975.
Blatchley 1920.
Bourne and Nielsson 1967.
Bullock 1973.
Folsom and Woke 1939.
Fulton 1915.
Hartman et al. 1979.
Monteith 1971.
Nielsson and Bass 1967.
Parrot and Fulton 1914.
Walker 1962, 1963, 1984.
Weaver and Sommers 1969.

Superfamily GRYLLACRIDOIDEA

David A. Nickle, *Systematic Entomology Laboratory, ARS, USDA*

GRYLLACRIDIDAE, Leaf-Rolling Crickets

RHAPHIDOPHORIDAE, Cave and Camel Crickets

STENOPELMATIDAE, Jerusalem or Stone Crickets

Figures 16.12–16.18

The superfamily Gryllacridoidea is a group of orthopterans that possesses some characters of the crickets and the katydids. They are usually small to medium (length 5–40 mm) and usually brown or gray. All species in the United States are wingless, although they have relatives in tropical regions of the world that are fully alate. The group is worldwide, with about 1000 described species (Karny 1937). It contains six families, three of which are represented in the United States and Canada: the Gryllacrididae (leaf-rolling crickets), Rhaphidophoridae (cave and camel crickets), and Stenopelmatidae (Jerusalem or stone crickets).

Relationships and Diagnosis: The Gryllacridoidea are closely related to the Grylloidea and Tettigonioidea. They are defined here as jumping orthopterans with long, filamentous antennae and (with rare exceptions) four-segmented tarsi, but which lack auditory organs and sound-producing apparatus. Females have ovipositors with six valves. All species in the United States and Canada are apterous and superficially resemble crickets in their general morphology and behavior but are probably more closely related to the katydids (Karny 1937).

The three families of Gryllacridoidea in the United States and Canada are disproportionately represented. The Rhaphidophoridae are extensively represented with 17 genera and about 100 species, most of which, however, belong to the single genus *Ceuthophilus* (Hubbell 1936); the next largest genus, *Pristoceuthophilus,* is confined to the Pacific states and western desert regions. In contrast, the Gryllacrididae are known only from the single species *Camptonotus carolinensis* (Gerstaecker), which occurs in the eastern United States. The Stenopelmatidae are represented by two genera, in which all species are found in dry environments from the Great Plains to the Pacific coast, with the greatest number in California.

Biology and Ecology: The biology of most of the gryllacridoids is unknown, but several species have highly specialized habits and life histories. As with crickets and katydids, juveniles and adults of gryllacridoids are similar in form and occupy the same habitats. Nearly all species are nocturnal or occupy dark niches such as leaf litter, tree holes, caves, and animal burrows. Members of the commonest genus, *Ceuthophilus,* hide during the day under logs, rocks, and leaf litter and wander about at night. All five species of *Hadenoecus* and two of the four species of *Euhadenoecus* are exclusively cavernicolous (Hubbell and Norton 1978); *Typhloceuthophilus floridanus* Hubbell occurs only in the burrows of pocket-gophers in northern Florida (Helfer 1963). To some degree most species require moist conditions to survive. Desert species of several genera (including some *Ceuthophilus, Ammobaenetes, Daihinibaenetes, Utabaenetes, Macrobaenetes,* and *Daihiniodes*) exist in parts of western United States, but little is known of them other than that they are usually nocturnal and are effective burrowers (Tinkham 1962a,b, 1968, 1970a,b).

Most species probably have a one-year life cycle, but life history studies indicate that some species require several years to mature. Overwintering is usually in the egg and mid to late instars. At least seven moults are required to reach the adult stage. The stenopelmatid, *Ammopelmatus kelsoensis* Tinkham, may require four years to mature, though other stenopelmatids usually develop within two or three years (Tinkham and Rentz 1969).

The leaf-rolling behavior of the gryllacridid, *Camptonotus carolinensis,* has attracted the attention of many naturalists in eastern United States. This insect cuts incisions along the edges of a leaf with its mandibles, then pulls the opposing edges of the leaf together and unites them with silk threads produced by its labial glands, all within five minutes. The rolled leaf is used as a protective chamber during the day when they are not active (Blatchley 1920).

Most gryllacridoids are scavengers, although several species have been observed to feed on other insects. Most are not economically important, but a few are considered pests. One such species, introduced from Europe or Asia, is the greenhouse stone cricket, *Tachycines asynamorus* Adelung, which invades homes in the late autumn and frequently is destructive in greenhouses in various localities in the United States. Some species, notably *Pristoceuthophilus pacificus* Thomas, feed on commercially grown mushrooms and are a major pest on that crop.

Description: Larvae resemble adults, differing in size and incomplete development of genitalia.

Head: Head globose. Eyes round, elliptical, or in some species absent. Antennae long, filamentous, Palpi long and slender to short and stout.

Thorax: Pronotum rounded; lateral carinae absent in all species. Mesonotum and metanotum exposed, not concealed beneath pronotum.

Wings: Absent in species in the United States and Canada.

Legs: Variable, ranging from long and slender in most Rhaphidophoridae to short and robust in other Rhaphidophoridae and in Stenopelmatidae. Auditory tympana absent. All tarsi four-segmented (with rare exceptions in Rhaphidophoridae), either compressed (Stenopelmatidae and Rhaphidophoridae) (Fig. 16.18) or more or less depressed (Gryllacrididae) (Fig. 16.17). Hind tibia simple, with small fixed teeth only, or heavily armed with long spurs (with or without smaller teeth between them) which may be crowded at apex to form a "sand basket". Spurs on apex of hind tibia modified into broad calcars used for digging (Stenopelmatidae), or spinelike or rarely bladelike (Rhaphidophoridae), or greatly reduced (Gryllacrididae).

Abdomen. Male. Tergal surface smooth or tuberculate, in some species with one or more well developed protuberances on middle terga. Cercus simple, one-segmented, often long, either tapering and flexible with long setae, or with a rigid base and flexible tip (Rhaphidophoridae), or short and bearing only a few short setae (Stenopelmatidae). *Female.* Cercus simple, similar to males. Ovipositor with six valves; long and bladelike, in adults usually serrated or toothed along apical margins (Rhaphidophoridae and Gryllacrididae), or very small, reduced to three pairs of lobes equal in length to cerci (Stenopelmatidae).

Comments: The following key for larvae of U.S. Gryllacridoidea is modified from keys for adults. Additional papers helpful in the identification of larvae at least to genus are Caudell (1916), Blatchley (1920), and Helfer (1963).

KEY TO IMMATURE GRYLLACRIDOIDEA OF THE UNITED STATES AND CANADA

1. Antennae continguous at base or nearly so (fig. 16.15) .... (*Ammobaenetes, Ceuthophilus* (fig. 16.12), *Daihinia, Daihinibaenetes, Daihiniella, Daihiniodes, Euhadenoecus, Gammarotettix, Hadenoecus, Macrobaenetes, Phrixocnemis, Pristoceuthophilus, Rhachocnemis, Tachycines, Tropidischia, Typhloceuthophilus, Styracoseceles, Udeopsylla, Utabaenetes*) .... ***Rhaphidophoridae***

1′. Antennae separated at base by a distance equal to or greater than first antennal segment length (fig. 16.16) 2

2(1′). Tarsi somewhat flattened horizontally, second and third segment with lateral lobes and more or less depressed (fig. 16.17); tibiae narrow, cylindrical; tibial spurs small, inconspicuous; eastern United States .... (*Camptonotus,* fig. 16.13) ***Gryllacrididae***

2′. Tarsi more or less flattened vertically, without lateral lobes (fig. 16.18); tibiae broad, robust; tibial spurs well developed, tooth-like, modified for digging; western United States .... (*Ammopelmatus, Cnemotettix, Stenopelmatus,* fig. 16.14) ***Stenopelmatidae***

Selected Bibliography

Blatchley 1920.
Borror et al. 1981.
Caudell 1916.
Helfer 1963.
Hubbell 1936.
Hubbell and Norton 1978.
Karny 1937.
Rentz and Weissman 1973.
Tinkham 1962a,b, 1968, 1970a,b.
Tinkham and Rentz 1969.

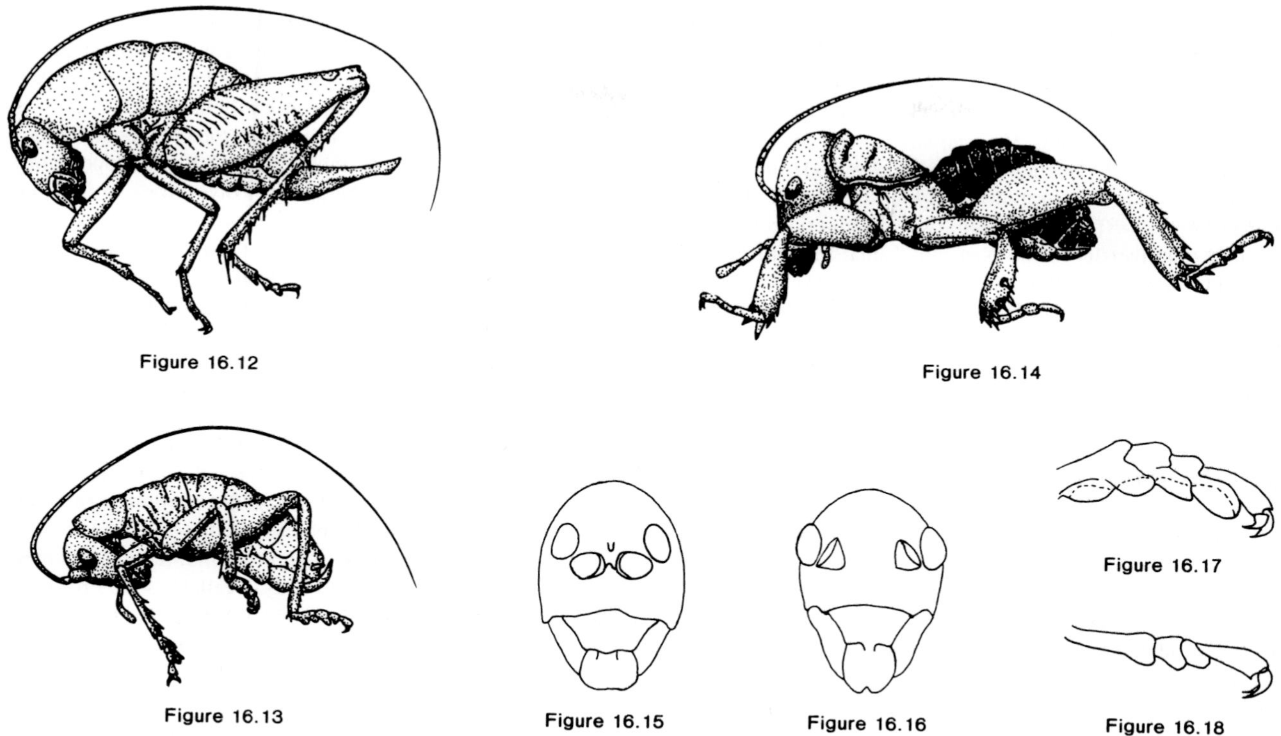

Figure 16.12. *Ceuthophilus nigricans* Scudder (Rhaphidophoridae), larva (instar 4), left lateral view.

Figure 16.13. *Camptonotus carolinensis* (Gerstaeker) (Gryllacrididae) larva (instar 3), left lateral view.

Figure 16.14. *Stenopelmatus* sp. (Stenopelmatidae), larva (instar 3), left lateral view.

Figures 16.15–16.16. Gryllacridoid faces, frontal view: **(16.15)** *Ceuthophilus nigricans;* **(16.16)** *Camptonotus carolinensis.*

Figures 16.17–16.18. Gryllacridoid tarsi, left lateral view: **(16.17)** *Camptonotus carolinensis;* **(16.18)** *Stenopelmatus* sp.

Superfamily TETTIGONIOIDEA

David A. Nickle, *Systematic Entomology Laboratory, ARS, USDA*

HAGLIDAE, Hump-Winged Katydids

TETTIGONIIDAE, Katydids, Long-Horned Grasshoppers

Figures 16.19–16.33

Relationships and Diagnosis: Although there is some controversy concerning the systematic limits of the Tettigonioidea, the superfamily is defined here as those saltatorial (jumping) Orthoptera having four-segmented tarsi, usually long, filamentous antennae, auditory organs or tympana on the fore tibiae (fig. 16.32), sound-producing or stridulatory apparatus on the tegmen of males, and ovipositors with six valves. Their closest relatives are the Grylloidea (crickets and mole crickets) and the Gryllacridoidea (cave and camel crickets and their relatives). Although all three groups share the characters of enlarged hind legs used for jumping and long filamentous antennae, the Grylloidea differ in having three-segmented tarsi and ovipositors with four valves; and the Gryllacridoidea, although they have four-segmented tarsi and ovipositors with six valves, universally lack auditory organs on the fore tibiae and stridulatory apparatus on the tegmina of males. Several authors have disputed the placement of certain species within these groups. For an interesting but unresolved discussion of this controversy, one may refer to several reviews of the subject (Ander 1939, Sharov 1968, Ragge 1977, Rentz 1979).

Two families of katydids are recognized in the United States and Canada: the Haglidae and the Tettigoniidae. Members of the Haglidae are the source of most of the above controversy. The Haglidae are represented in northwestern United States and Canada by the genus *Cyphoderris* with three species. These are called the "hump-winged katydids" and are more similar in shape and most characters to the gryllacridids, the leaf-rolling crickets, except that males and females have auditory organs on their fore tibiae and males produce sound.

The family Tettigoniidae, which includes the katydids or long-horned grasshoppers, is represented in the United States and Canada by more than 250 species (*s*), which are divided into nine subfamilies with a total of 49 easily recognizable genera (*g*). The subfamilies include the Phaneropterinae (*g*=14, *s*=66), known as the bush katydids; Pseudophyllinae (*g*=3, *s*=6), the true katydids; Copiphorinae (*g*=4, *s*=21) the cone-headed katydids; Conocephalinae (*g*=3, *s*=39), the meadow katydids; Decticinae (*g*=22, *s*=124), the shield-backed katydids; Listrocelidinae (*g*=1, *s*=2); and Tettigoniinae (*g*=1, *s*=1). In addition, two introductions have increased the diversity of the Nearctic fauna: the matriarchal katydid, *Sago pedo* (Pallas) (Saginae), and the drumming katydid, *Meconema thalassinum* (DeGeer) (Meconematinae).

Biology and Ecology: Within the United States and Canada, diversity increases from north to south. In the Atlantic Provinces of Canada, for example, there are 15 species of tettigoniids, compared with 86 species in eastern United States. As with crickets, juveniles and adults both occupy the same habitats; most species are found on foliage of herbaceous plants, trees, or shrubs, or in grasses. Some species of decticine katydids (e.g., most species of *Atlanticus*) are found on the ground in leaf litter in forests or on the edge of forests. Although many species of katydids are found on a variety of species of plants, some are host-specific. For example, three species of the bush katydid genus *Inscudderia* are quite host specific: *I. strigens* (Hebard) is found on St. John's wort (*Hypericum* sp.), and *I. taxodii* Caudell and *I. walkeri* (Hebard) occur only in the canopy of baldcypress (*Taxodium distichum* (L.)). The creosote bush katydid, *Insara covilleae* Rehn and Hebard, an attractive desert katydid with pearly white markings on its tegmina and thorax, is restricted to creosote bush (*Larrea divaricata* Cav.). A related species, *I. juniperi* (Scudder), is found only on a juniper (*Juniperus monosperma* (Engelm.), whereas *I. elegans* (Scudder) is found on mesquite (*Prosopis juliflora* (Swartz)).

Katydids are generally considered to be primarily plant feeders; the Phaneropterinae, for example, are exclusively phytophagous. Most other groups are predominantly dendrophagous (Pseudophyllinae, Tettigoniinae), omnivorous (Decticinae, Conocephalinae), or seminivorous (Copiphorinae); the Listrocelidinae are generally carnivorous. For more information on the feeding habits of the katydids, one may refer to several articles (Gangwere 1961, 1965, 1967).

In the United States and Canada, most species are univoltine, but in the southern regions of the United States there is a trend toward bivoltinism. In general, katydids overwinter in the egg stage, become adults by mid to late summer, and lay eggs that diapause through the winter. Species are rather specific in their oviposition behavior. Some species oviposit in the soil, some in the mesophyll tissue of leaves or grasses, some in the crevices of bark, and some on the surface or edge of leaves or stems of woody plants. Tettigoniid eggs are generally elongate and oval. The eggs of the bush katydids (Phaneropterinae), in addition, are laterally flattened.

The juveniles or larvae of the tettigonioids resemble their adults in general features, lacking only fully developed wings and external genitalia. The greatest divergence in form between larvae and adults in all species occurs at the first instar. This is particularly true of the phaneropterine or bush katydids. In this group the head shape of early instars of several species differs dramatically from adults in both morphology and color pattern. By the third to the final (usually fifth) instar, the juveniles resemble the adults, and identification at least to genus is usually possible using published keys for adults (Blatchley 1920).

Description: Juveniles similar to adults (figs. 16.19–16.21) especially in later instars; first and second instars, especially among phaneropterine katydids, may differ radically from adults in head shape and coloration.

Head: Hypognathous. Antennae filamentous, in most cases as long as or longer than body. Ocelli present. Frons and fastigium (figs. 16.22, 16.23) of vertex variable in shape, broad and flat to pointed and lamellate, and in relation to each other either convergent and attingent or nonconvergent.

Thorax: Pronotum saddle-shaped, similar in form to adults. Last two instars with wing pads. All tarsi four-segmented. Tympana not exposed or apparent until fourth or fifth instar.

Abdomen: External genitalia undeveloped in early instars, becoming more similar to adults with each successive moult. Cerci modified in males as claspers, usually with one or more sharp teeth at or near apex; cerci short, cylindrical, and unspecialized in females. Ovipositor bladelike (fig. 16.33) or swordlike for most species, apically serrate or smooth.

Comments: Although many katydids are described as phytophagous, few are known to be serious pests in the United States and Canada. Of those considered pests, two decticines are major agricultural pests: the coulee cricket, *Paranabrus scabricollis* (Thomas) (Melander and Yothers 1917), and the Mormon cricket, *Anabrus simplex* Haldemann (Cowan 1929, Cowan and Wakeland 1955, Wakeland 1959). In addition, phaneropterine species of the genera *Scudderia* and *Microcentrum* occasionally cause damage to ornamentals and fruit trees.

Although it is not difficult to identify katydid larvae to genus, the current state of katydid taxonomy almost precludes identification of larvae (especially early instars) to species. Rearing katydids and recording their developmental changes are necessary to obtain adequate keys at the species level for the immatures.

KEY TO THE MAJOR SUBFAMILIES OF IMMATURE KATYDIDS[2]

(Principal genera listed in parentheses)

1. Head globose, face not usually slanted or frontally flattened; prosternum unarmed; ovipositor usually short, upturned, laterally flattened, with serrated or toothed margins (figs. 16.21, 16.28–31) .... (*Amblycorypha, Arethaea, Dichopetala, Insara, Inscudderia, Scudderia, Microcentrum*) **Phaneropterinae**

1′. Head and face slanted, conical, or frontally flattened; if globose, then prosternum usually armed with 2 spines or nodes; ovipositor long, straight or bladelike, nearly always smooth and lacking serrated or toothed margins 2

2(1′). Margins of antennal sockets inflated, well developed (fig. 16.24); prothoracic spiracle small, inconspicuous, located below ventral margin of pronotum (fig. 16.24) .... (*Paracyrtophyllus, Lea, Pterophylla*) **Pseudophyllinae**

2′. Margins of antennal sockets not inflated, weakly developed (fig. 16.25); prothoracic spiracle large, conspicuous, partly concealed beneath pronotum or behind posterior margin of lateral lobe of pronotum (fig. 16.25) 3

3(2′). Fastigium very narrow, less than half as broad as width of first antennal segment, laterally compressed into a narrow fold above frons; fore tibia armed with 6–7 long movable spines; spines on fore tibia longer than those on hind tibia .... (*Neobarrettia*) **Listrocelidinae**

3′. Fastigium at least as broad as first antennal segment; spines on fore tibia as small as or smaller than those on hind tibia 4

4(3′). Head usually rounded; fastigium broad, in most species at least as broad as width of antennal scape (fig. 16.22); plantula of hind basitarsus very well developed, at least half as long as basitarsus (fig. 16.26); pronotum usually well developed; pronotal disc elongated posteriorly in most species; humeral sinus of posterior margin of lateral lobe of pronotum usually not emarginate .... (*Aglaothorax, Anabrus, Atlanticus, Capnobotes, Idiostatus, Metrioptera, Steiroxys*) **Decticinae**

4′. Head slightly to acutely angulate; fastigium either broad, obtuse or conical, elongated; if broad, obtuse, then not as broad as width of antennal scape; plantula of hind basitarsus weakly or at most moderately developed, but less than half as long as basitarsus (fig. 16.27); pronotum saddle-shaped, but disc not elongated posteriorly, and humeral sinus of posterior margin of pronotal lobe almost always emarginate 5

5(4′). Fore and mid femora unarmed; fastigium broad, obtuse, with rounded or truncate untoothed ventral margin (fig. 16.22); dorsum of body often with reddish stripes extending from pronotum to tip of abdomen (fig. 16.20) .... (*Conocephalus, Orchelimum*) **Conocephalinae**

5′. Fore and mid femora armed with spines; fastigium conical, narrow, in some cases extremely elongated, with a tooth or tubercle extending from ventral margin in many species (figs. 16.19, 16.23); dorsum of body without reddish stripes (fig. 16.23) .... (*Belocephalus, Neoconocephalus*) **Copiphorinae**

Selected Bibliography

Ander 1939.
Blatchley 1920.
Cantrall 1943.
Cohn 1965.
Cowan 1929.
Cowan and Wakeland 1955.
Gangwere 1961, 1965, 1967.
Melander and Yothers 1917.
Morris and Gwynne 1978.
Morris and Walker 1976.
Ragge 1977.
Rehn and Hebard 1914a,b,c,d, 1915a,b, 1916.
Rentz 1973, 1979.
Rentz and Birchim 1968.
Sharov 1968.
Vickery et al. 1974.
Wakeland 1959.

2. The subfamilies Saginae, Meconematinae, and Tettigoniinae and the family Haglidae are omitted from the key; larvae as well as adults are uncommonly encountered and their representation in the United States and Canada is geographically restricted.

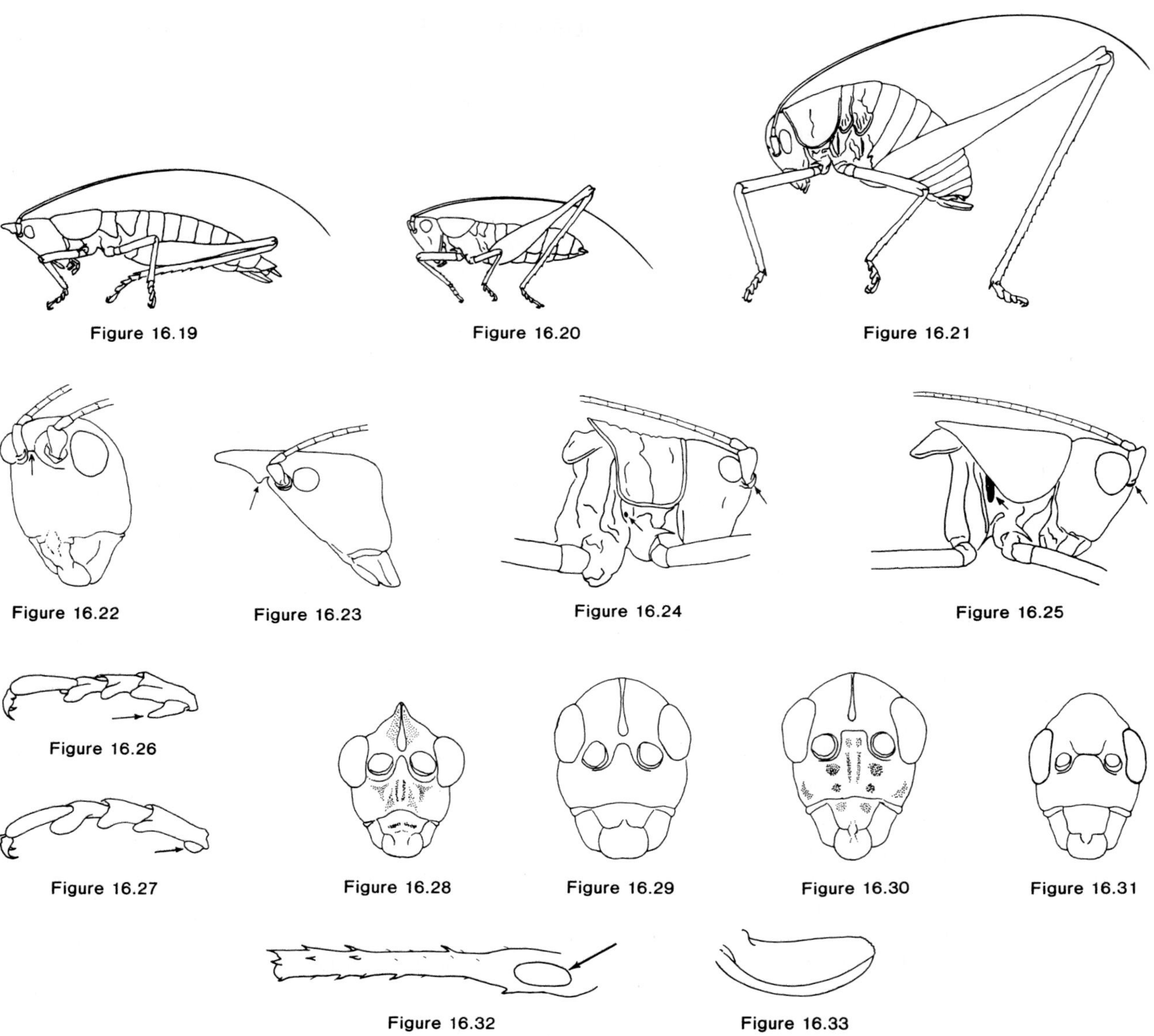

Figures 16.19–16.21. Middle stadia larvae of representative katydids: **(16.19)** Broad-tipped conehead, *Neoconocephalus triops* (L.), Copiphorinae (Gainesville, Fla.); **(16.20)** A meadow katydid, *Conocephalus* sp., Conocephalinae (Gainesville, Fla.); **(16.21)** Florida oval-winged katydid, *Amblycorypha floridana* Rehn and Hebard, Phaneropterinae (Gainesville, Fla.).

Figures 16.22–16.23. Margins of fastigia: **(16.22)** Meadow katydid, *Orchelimum* sp. Arrow stresses truncate ventral margin of fastigium; **(16.23)** Conehead katydid, *Neoconocephalus* sp. Arrow stresses tubercle on ventral margin of fastigium.

Figures 16.24–16.25. Prothoracic spiracles and margins of antennal sockets: **(16.24)** Northern true katydid, *Pterophylla camellifolia* (F.), Pseudophyllinae (Annandale, Va.). Arrows stress inconspicuous prothoracic spiracle and well-developed margin of antennal socket; **(16.25)** Meadow katydid, *Conocephalus* sp. Arrows stress large prothoracic spiracle and weakly developed margin of antennal socket.

Figures 16.26–16.27. Metathoracic tarsi. Arrows stress degree of development of plantula: **(16.26)** *Idiostatus californicus* Pictet, Decticinae (Sissou, Calif.); **(16.27)** *Conocephalus* sp. (Annandale, Va.).

Figures 16.28–16.31. Faces of representative first instar larvae of phaneropterine katydids: **(16.28)** *Scudderia furcata* Brunner (Annandale, Va.); **(16.29)** *Microcentrum rhombifolium* (Saussure) (Washington, D.C.); **(16.30)** *Stilpnochlora couloniana* (Saussure) (Paradise Key, Fla.); **(16.31)** *Amblycorypha floridana* Rehn and Hebard (Gainesville, Fla.).

Figure 16.32. Fore tibia of larva (instar 4) of *Scudderia furcata*, showing tympanum.

Figure 16.33. Ovipositor of larva (instar 4) of *Scudderia furcata.*

Suborder CAELIFERA

Superfamily ACRIDOIDEA

TETRIGIDAE

David A. Nickle, *Systematic Entomology Laboratory, ARS, USDA*

Pygmy or Grouse Locusts

Figures 16.34–16.42

Relationships and Diagnosis: The Tetrigidae (grouse or pygmy locusts) are among the smallest orthopterans, ranging from less than 7 mm to slightly more than 30 mm. There are about 200 genera worldwide with nearly 2000 species or subspecies. The most distinctive feature is the pronotum, which usually extends beyond the tip of the abdomen. Tegmina are always reduced or absent and are nonfunctional, and the hind wings are usually well developed and efficient flight organs. A series of revisions of North American tetrigids provides us with most of the information available on this group (Grant 1955, 1956, Rehn and Grant 1955, 1958, 1961).

Several characters when used in combination distinguish the Tetrigidae from other acridoid families. These include two-segmented fore and mid tarsi and a pronotum that extends posteriorly over the abdomen. The pronotum is short in early instars but nevertheless extends over the first abdominal terga, becoming longer in subsequent instars and surpassing the tip of the abdomen in the final instars of most North American species. In addition, the prosternum is developed into a sternomentum, which forms a collar around the lower face. Although acridoid affinities are reflected in the short antennae and external male and female genitalia of adults, the tetrigids probably evolved as a separate branch of Acridoidea early in the Miocene epoch. Their closest relatives, members of the Tridactylidae, also have two-segmented fore and mid tarsi, and both families lack auditory tympana, but the tridactylids lack the well developed pronotum, and male genitalia of most species are much more complex than those of tetrigids.

Biology and Ecology: Tetrigids occur worldwide and are more diverse in form and greater in number of species in tropical regions. In some tropical forms (e.g., *Zaphyllonotum rhombeum* (Selton)), the median carina of the pronotum may be cristate or foliacious, developed into a large crescent and similar in appearance to the bark of trees on which they rest. Several species are dimorphic with respect to pronotal length. These populations display constant proportions of long-pronotum vs. short-pronotum forms. Many species exhibit polyornatism, that is, several distinct color patterns are present within a population.

Almost all tetrigids in temperate regions of the world diapause not only in the egg stage, but as late instar larvae and adults. Adults usually live more than one year. Larvae and adults bury themselves under the debris of dead leaves, twigs, mosses, or bark and break diapause early in the spring. They live on the ground and seem to require a moist environment, since they are usually found on the forest understory, near water, or in boggy or marshy places.

Tetrigids feed primarily on decaying vegetable matter, molds, or algae mixed with soil. They are also known to feed on lichens, sprouting grasses and sedges, or other tender vegetative material. Hancock (1898) noted the occurrence of mud eating among North American tetrigids, in which mud mixed with algae and other debris formed the major diet of species of *Paratettix*.

In Canada and most areas of the United States, the first eggs are laid in the ground in mid-May. The offspring of this brood continue to develop until late fall and diapause through the winter as last instars or more commonly as adults. Later in the summer, any clutch of eggs to develop invariably produces individuals that pass through the winter as late instars.

Description: Usually five instars. Juveniles resemble adults except for degree of development of pronotum, wings, and genitalia, and the number of antennal articles.

Head: Hypognathous. Antenna in first instar usually having 10 articles, increasing in number in subsequent instars, with final instar having 12–15 articles in the Tetriginae (fig. 16.37) and 19–22 in the Batrachideinae (fig. 16.38). Ocelli three, median ocellus located in lower portion of grooved frontal costa; lateral ocelli situated on either side of frontal costa, either behind or above and behind median ocellus (figs. 16.41, 16.42).

Thorax: Median carina on dorsum of pronotum well developed, usually more so than on respective adults for most North American species, (figs. 16.39, 16.40). Pronotum compact in first instar but elongated posteriorly in all stages over part or entire abdomen. Prosternum produced anteriorly and dorsally into a sternomentum, forming a shield or collar around lower face. Wing pads absent until third or fourth instar, frequently concealed beneath pronotum. In final instar, tegmen moves from dorsum of mesonotum to a lateral position, exposed behind tegminal sinus of pronotum.

Legs: Fore and mid tarsi two-segmented; hind tarsi three-segmented. Arolia absent between tarsal claws. Posterior margins of hind tibiae often broadened, lamellate as a modification for surface swimming.

Abdomen: Similar in later instars to adults. Auditory tympanum absent on proximal abdominal tergum. Cerci of males and females unsegmented, unmodified. Ovipositor consisting of three pairs of valves, dorsal and ventral pairs serrated on dorsal and ventral margins, respectively.

Comments: The most recent revision of North American Tetrigidae of is that of Rehn and Grant (1961). They list 17 species of Tetriginae and 6 species of Batrachideinae in the United States and Canada. Other references on the systematics of this group include Bolivar (1887), Bruner (1910), and Hancock (1902). Research on the genetic inheritance of polymorphism in the Tetrigidae of North America is extensive (Nabours 1919, Nabours and Stebbins, 1950).

Although tetrigids are not economically important in agriculture, they are biologically interesting organisms; some species, for example, are facultatively parthenogenetic (Nabours 1925).

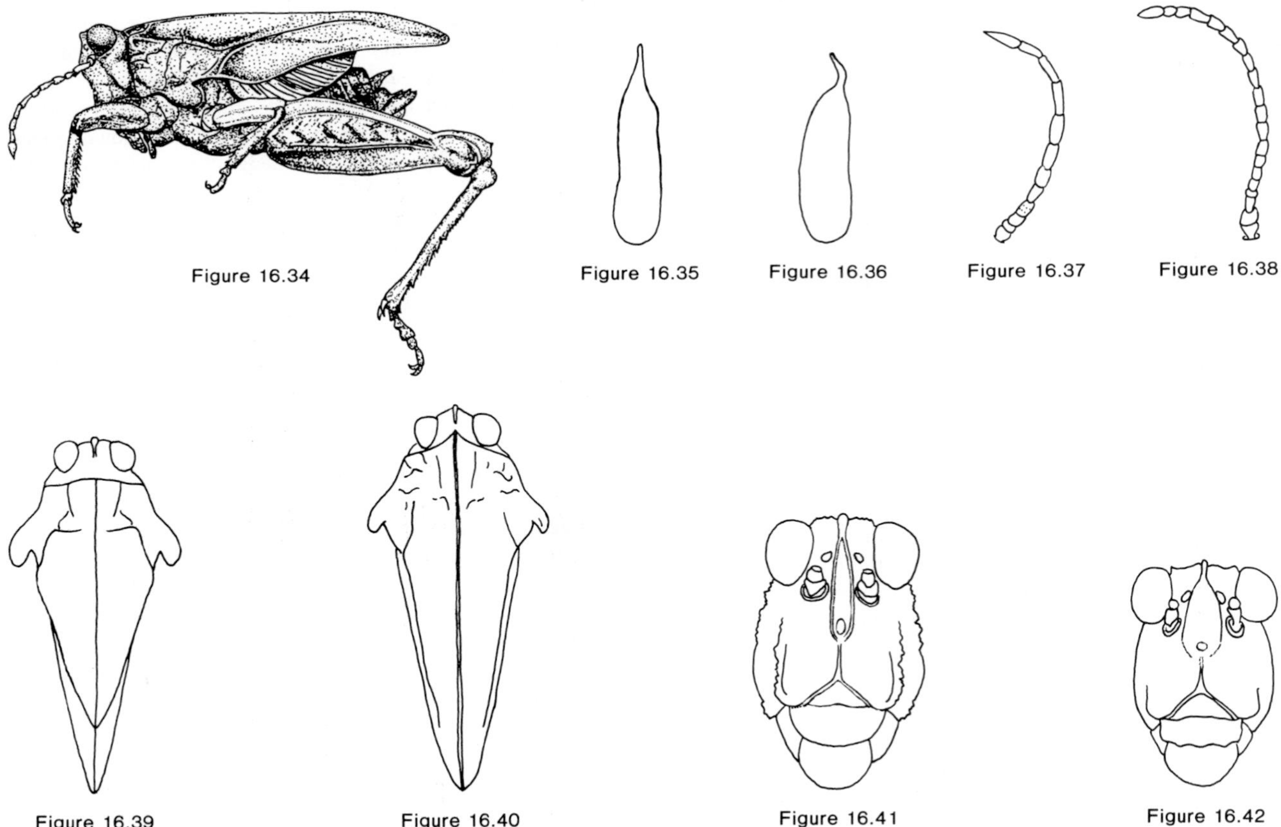

Figure 16.34. *Tetrix arenosa* Burmeister, penultimate instar, lateral view (Tetrigidae).

Figures 16.35–16.36. Tetrigid eggs, lateral view: **(16.35)** *Tetrix* sp. (Tetriginae); **(16.36)** *Tettigidea* sp. (Batrachideinae).

Figures 16.37–16.38. Antennae of larvae (instar 3) of tetrigids: **(16.37)** *Tetrix* sp.; **(16.38)** *Tettigidea* sp.

Figures 16.39–16.40. Pronotum of larvae (instar 4) of tetrigids, dorsal view: **(16.39)** *Tetrix* sp.; **(16.40)** *Tettigidea* sp.

Figures 16.41–16.42. Faces of tetrigid larvae (instar 3): **(16.41)** *Paratettix* sp.; **(16.42)** *Neotettix* sp.

Selected Bibliography

Bolivar 1887.
Bruner 1910.
Grant 1955, 1956.
Hancock 1898, 1902.
Nabours 1919, 1925.
Nabours and Stebbins 1950.
Rehn and Grant 1955, 1958, 1961.

Superfamily ACRIDOIDEA

TRIDACTYLIDAE

David A. Nickle, *Systematic Entomology Laboratory, ARS, USDA*

Pygmy Mole Crickets

Figures 16.43–16.46

Relationships and Diagnosis: The placement of the Tridactylidae within the systematic hierarchy of the Orthoptera was uncertain for a long time. Much of the early literature placed them as a subfamily of the Gryllidae or the Gryllotalpidae. This placement originated because they superficially resemble the gryllotalpids, both being burrowing insects with front legs adapted for digging. Chopard (1943, 1949),

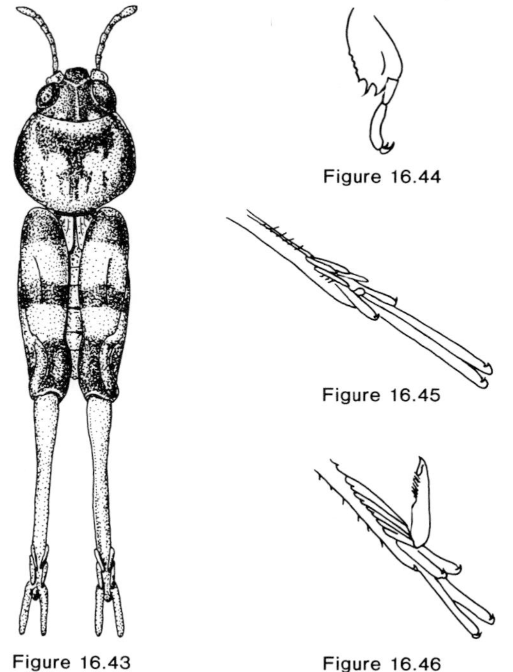

Figure 16.43

Figure 16.44

Figure 16.45

Figure 16.46

Figure 16.43. *Ellipes minutus* (Scudder) (Tridactylidae), larva (instar 3), dorsal view.

Figure 16.44. Fore tibia and tarsus of *Neotridactylus apicalis* (Say) (Tridactylidae) larva (instar 4), left lateral view.

Figures 16.45–16.46. Tridactylid hind tarsi, right lateral view: **(16.45)** *Ellipes minutus;* **(16.46)** *Neotridactylus apicalis.*

following the earlier work of Ander (1939), established the superfamily Tridactyloidea of the Acridoidea, with one family, the Tridactylidae. The short antennae and morphology of the external genitalia, especially the female genitalia, place the Tridactylidae with the acridoids, and their closest relatives within the Acridoidea are the tetrigids. Both groups have two-segmented fore and midtarsi, and both lack auditory tympana. The tridactylids, however, have a short, rounded pronotum, and fore legs are modified for digging. Today, the superfamily Tridactyloidea comprises three families, Cylindrachetidae (known only from Australia, Chile, and Argentina), Rhipipterygidae (from Central and South America), and Tridactylidae (worldwide). Only the Tridactylidae are represented in the United States and Canada, and all further remarks pertain only to this family.

Members of the family are small, measuring 3–6 mm. The antennae are short, widely separated at the base, with 11 distinct articles. There are three very small ocelli. The pronotum is short, rounded, or convex laterally. Tegmina are reduced to small lobes, whereas the hind wings are well developed with scalloped margins. The fore legs are modified for digging; the femur is thickened, and the tibia is expanded into several lobes. Fore and mid tarsi are two-segmented. As with all Acridoidea, hind legs are modified for jumping.

Biology and Ecology: Pygmy mole cricket biology is poorly known, but they live dual lives above and below the ground. They apparently feed on vegetation, algae, and plant debris during the day but often stay within the safety of shallow burrows when not feeding. They prefer sandy areas in close proximity to ponds, lakes, or rivers. They are capable of swimming, and in several species the hind tibiae are flared posteriorly, serving as paddles when the insect kicks across the surface.

Urquhart (1937) observed several burrows of *Tridactylus* (= *Neotridactylus*) *apicalis* Say. In January the burrows were deep, 1.5–2.0 ft. below the surface, and contained clusters of hibernating larvae, which were still active when disturbed. In August, burrows were shallow, less than three inches deep, with clusters of 10–27 eggs, apparently guarded by the female. In some cases the young were found crowded with the female. This small species, only 4.5 mm long, produces very large eggs, about 1.5 mm long, often in batches of as many as 27. It is not known how many molts occur, but typically young and adults are similar in form and habits.

Description: Larvae and adults are similar in morphology and behavior. Larvae differ from adults in degree of development of the genitalia and wings.

Head: Hypognathous. Shape of head broad, dorsoventrally compressed. Eyes well developed, oval. Three ocelli present, arranged nearly on a line connecting ventral margins of compound eyes. Antennae short, with 10–12 articles.

Thorax: Pronotum rotund, oval as seen from above. Prosternal spine present. Tegmen in adult males short, elytriform, with a stridulatory organ near apex. Hind wing of adults well developed, with scalloped margin.

Legs: Fore legs modified for digging (fig. 16.44); coxa, femur, and tibia much enlarged; tibia variable in shape but always expanded with two to four dactyl-like expansions. Fore and mid tarsi two-segmented; hind tarsi one-segmented, without claws. Mid leg unmodified. Hind leg saltatorial; posterior margin of hind tibia with one to four pairs of long, flat plates near apex.

Abdomen: Slender, with terminal tergum enlarged, deeply cleft. Cercus slender, bristly two-segmented. A second pair of one-segmented appendages situated ventrad of cerci.

Comments: Nearly 200 species have been described worldwide. Two genera are represented in the United States and Canada. *Ellipes* (figs. 16.43, 16.45) with three species and *Neotridactylus* (fig. 16.46) with one (see Gunther (1980) for a species list). Most of our knowledge is from the work of Gunther (1969, 1975, 1977) and several earlier workers (see Blatchley 1920, pp. 654–58).

Selected Bibliography

Ander 1939.
Blatchley 1920.
Chopard 1943, 1949.
Gunther 1969, 1975, 1977, 1980.
Urquhart 1937.

Superfamily ACRIDOIDEA

David A. Nickle, *Systematic Entomology Laboratory, ARS, USDA*

EUMASTACIDAE, Monkey Grasshoppers

TANAOCERIDAE, Desert Long-Horned Grasshoppers

Figures 16.47–16.54

The distinctive monkey grasshoppers are worldwide in distribution, and are highly specialized in habitat preference and behavior, but in the United States, they are represented by members of a single endemic subfamily, the Morsinae (Rehn 1948b). The six species in this subfamily are included in the genera *Morsea, Psychomastax,* and *Eumorsea.*

The tanaocerids are a distinctive family of nocturnal grasshoppers that occur only in the southwestern deserts of the United States (Grant and Rentz 1967, Rehn 1948a). Two genera are represented, *Tanaocerus* with two species (only one of which occurs in the United States, the other restricted to Baja California) and *Mohavacris* with a single species.

Relationships and Diagnosis: *Eumastacidae.* Eumastacids share the following characteristics: they are small to medium, not exceeding 35 mm in length; all North American species are apterous (although species in other parts of the world include alate, brachypterous, and apterous forms). Although the family has characters that distinguish it from the Tetrigidae and Tanaoceridae on the one hand and from the Acrididae on the other, there is no single set of characters that uniquely characterizes the Eumastacidae. They have three-segmented tarsi and arolia between their tarsal claws (as in Acrididae) but lack auditory tympana on the abdomen (present in most Acrididae). They agree with the Tetrigidae in having a definite supraclypeal triangle (absent in Acrididae) and in having the abdominal spiracles located in the laterodorsal membrane (in the Acrididae they are located directly in the terga). They are the only acridoids in which fingerlike diverticula or caecae are present in the hind gut of the female. Finally, the Eumastacidae possess a simpler phallic complex than do the Tanaoceridae and Acrididae.

Tanaoceridae. The three species of tanaocerids are medium sized grasshoppers, lacking wings and auditory tympana, with filiform antennae that in the males are longer than the body, the first two articles being very thick. A stridulatory organ is present, consisting of a row of curved ringlets or teeth on the third abdominal tergum and a scraper or short ridge near the lower internal basal margin of the hind femur. The tarsi are three-segmented. The internal genitalia of males are more specialized than in the Eumastacidae but less so than in the Acrididae. The presence of an external apical spine on the hind tibia is similar to the Romaleinae, but internal male and female genitalia of these two groups indicate that they are distinct (Dirsh 1955, Tinkham 1955).

Biology and Ecology: *Eumastacidae.* Little is known of the biology of the eumastacids. In the United States, members are found in a number of diverse habitats, from the hot dry regions in the Mohave Desert of California through various chaparral areas in the Coast Range of California, to their apparently most common habitat of timbered sections of the Sierra Nevada, San Jacinto, and San Bernardino Mountains, the Kaibab Plateau, and certain mountainous areas of Nevada, Arizona, and Utah (Hebard 1934). Apparently all North American species overwinter in the egg stage. Individuals usually become adults in June or July, and there is only one generation per year. Several species occur at high altitudes; for example, species of *Psychomastax* have been found as high as 11,500 feet on Charleston Peak, Nevada.

Most species are thamnophilous and assume characteristic positions with their hind femora extended at right angles to the body. They are usually found on the exposed upper surface of leaves of bushes or occasionally on the ground, and several species appear to be host specific. Rentz (*in* Strohecker et al. 1968:19) mentioned that eumastacids, like their relatives the tanaocerids, are nocturnal. *Morsea californica tamalpaisensis* Rehn and Hebard occurs generally on chamise (*Adenostoma fasciculatum*) and manzanita (*Arctostaphylos granulosa*), whereas the preferred host of *Morsea californica piute* Rehn and Grant is bitterbrush (*Purshia tridentata*).

Tanaoceridae. The tanaocerids are endemic to a restricted area in the Upper and Lower Sonoran Life Zones of Utah, Nevada, Arizona, California, and Baja California. All are nocturnal desert species, found on sagebrush, *Artemisia tridentata,* and chamise. Males are active at night, moving about on branches, whereas females usually remain motionless on branches with their heads pointed toward the ground.

Apparently the overwintering stage is the late instar larva and possibly the adult, at least for *Mohavacris timberlakei,* since adults are active in early spring and have been collected in March and April.

Description: *Eumastacidae.* Larvae similar to adults. In North American species, wing pads absent in larvae and adults; the only external differences between larvae and adults are the general size and degree of development of the external genitalia; also, adults are more sclerotized than larvae.

Head: Enlarged; occiput projecting dorsad beyond anterior margin of pronotum (fig. 16.47). Eyes large, elongated, nearly half as long as face for most species (fig. 16.48). Frontal costa narrow, sulcate, with well developed carinate margins. Three ocelli. Antennae with 12 or 13 articles in adults, fewer in larvae; shorter than combined length of head and pronotum.

Thorax: Pronotum short, with shallow lateral lobes; lateral carinae obsolete or weakly expressed. Meso- and metasternum united into a single sclerite. All species apterous within the United States.

Legs: All legs long and slender. Tarsi three-segmented. Tarsal arolia well developed. Hind femora without dorsoproximal overhanging lobe found in most acridids. Hind tibiae bearing two pairs of distal spurs.

Abdomen: Abdomen elongate, narrow in males, more robust in females. Auditory tympanum absent on proximal abdominal tergum. Spiracles situated in laterodorsal membrane between terga. External genitalia similar to those in acridids.

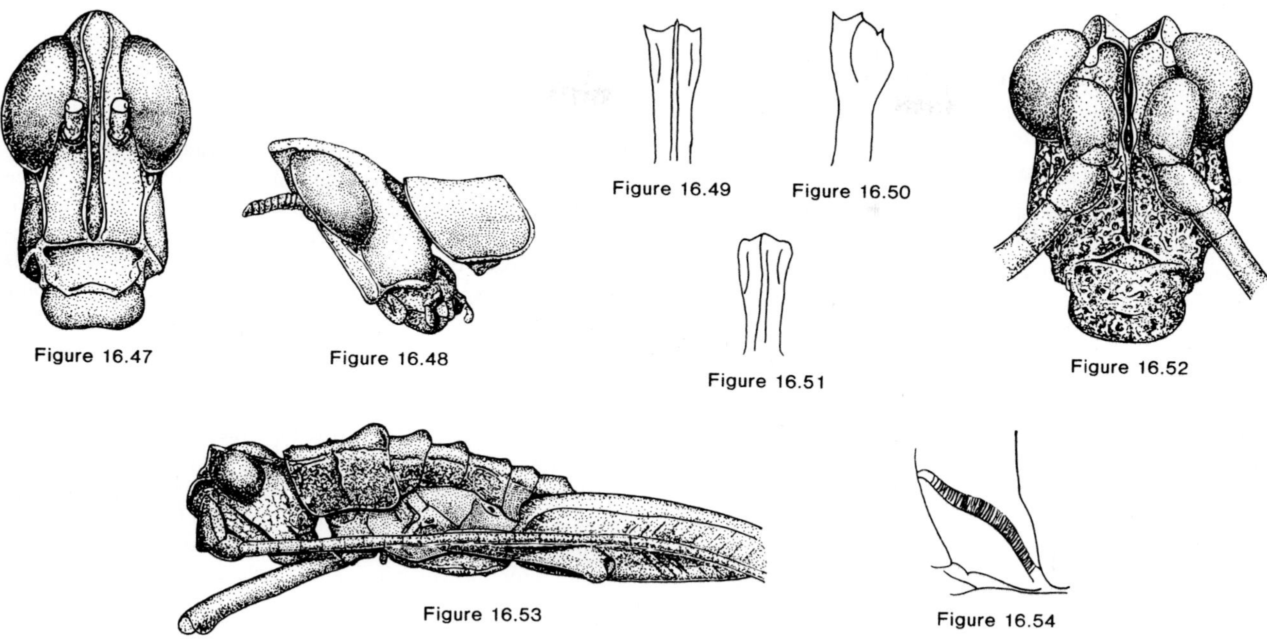

Figures 16.47–16.51 / Figure 16.47 / Figure 16.48 / Figure 16.49 / Figure 16.50 / Figure 16.51 / Figure 16.52 / Figure 16.53 / Figure 16.54

Figures 16.47–16.51. Morphological features of larvae of eumastacid grasshoppers (Eumastacidae).

Figure 16.47. *Psychomastax psylla psylla* Rehn and Hebard, face of late instar larva, frontal view.

Figure 16.48. Same specimen, head and pronotum, left lateral view.

Figures 16.49–16.50. Apex of hind femur of larva of *Morsea* sp.: **(16.49)** dorsal view; **(16.50)** left lateral view.

Figure 16.51. Apex of hind femur of larva of *Eumorsea pinaleo* Rehn and Grant, dorsal view.

Figures 16.52–16.53. Morphological features of larva of the long-horned grasshopper *Mohavacris timberlakei* (Tanaoceridae): **(16.52)** face of late instar larva, frontal view; **(16.53)** same specimen, left lateral view.

Figure 16.54. Stridulatory field on third abdominal tergum of male fourth instar larva of *Tanaocerus koebelei*, left lateral view.

Tanaoceridae. Larvae similar to adults. All species apterous, the only differences between larvae and adults are general size differences and degree of development of genitalia.

Head: Broad, spherical. Frontal costa with two lateral carinae narrowing ventrad of median ocellus to a single carina (fig. 16.52). Eyes oval or nearly spherical, situated high on face, in length less than 0.25 times facial length. Antennae long, filiform; thick at base, tapering apically.

Thorax: Pronotum slightly longer than longitudinal axis of head; median carina weakly expressed; lateral carinae absent. Posterior margin of pronotum with several shallow crenulations.

Legs: Hind femora without dorsoproximal overhang. Tarsi three-segmented. Arolia present. Stridulatory ridge of male located near lower internal basal margin of hind femur.

Abdomen: Stridulatory ridge on third abdominal tergum of male (fig. 16.54), absent on female. Male subgenital plate composed of two or three sclerotic plates separated by a membrane. Other structures on males and females similar to Acrididae.

Comments: In general, identification keys for adults of North American eumastacids serve as well for the larvae (at least for late instars). See the following keys: Rehn and Grant 1961: 127–54; Strohecker et al. 1968: 16, 20.

Selected Bibliography

Dirsh 1955.
Grant and Rentz 1967.
Hebard 1934.
Rehn 1948a,b.
Rehn and Grant 1961.
Rehn and Hebard 1918.
Strohecker et al. 1968.
Tinkham 1955.

Superfamily ACRIDOIDEA

M. A. Brusven,[3,4] ***University of Idaho***

ACRIDIDAE, Short-Horned Grasshoppers

Figures 16.55–16.85

Relationships and Diagnosis: Larval Acrididae are distinguished from other closely related members in the suborder Caelifera by a combination of characters. Acridid larvae have three-segmented tarsi on all legs, differing from Tridactylidae and Tetrigidae larvae which have front and middle tarsi two-segmented. The antennae of larval Acrididae and Eumastacidae are typically shorter than the head and pronotum combined, whereas in Tanaoceridae the antennae are longer than the head and pronotum. All instars of North American Eumastacidae lack wing pad development; most species of Acrididae have wing pads.

Biology and Ecology: Most grasshoppers have a single generation per year and overwinter in the egg stage. In warmer climates, some species have more than one generation per year and lack a distinct diapause; in colder, more rigorous climates, complete development may require more than one year. Middle and late instars overwinter in a few species, particularly members of Gomphocerinae and Oedipodinae.

Oviposition is usually done in the soil, although plant material may be used by some species. Soil particles adhere to a material that the female secretes around the eggs to form an encasement, which together form an egg pod, holding from 3 to 50 eggs per pod. A female may oviposit several times during the course of her life. A brief substage called a "vermiform larva" hatches from the egg and is terminated by the first molt. The general body form of the vermiform stage is more embryonic than larval and quite different from that of the subsequent instars. Anatomical changes among instars are gradual, with the greatest change occurring during the last molt when the wings and genitalia are fully formed in the adult.

Ecologically, grasshoppers are a diverse, successful group of insects. Many species are habitat-specific, thus restricted in distribution; others have a wide range of adaptation and are found in many habitats. Food habits range from monophagous to polyphagous. The Gomphocerinae and Acridinae are principally grass feeders; the other subfamilies are largely general feeders, often preferring broad-leaved plants.

Description and Development: Five instars are recognized for most North American grasshoppers. Fewer or greater number of instars have been reported by Dirsh (1968) with females occasionally having one or two more instars than males. For purposes of this text five instars are recognized (figs. 16.59–16.63). The shape of the pronotum and size and position of the wing pads are the best criteria for instar differentiation for species having moderate to well developed wings as adults. Several species, especially in the subfamilies Melanoplinae and Romaleinae, are brachypterous or apterous as adults, which poses special difficulty for instar differentiation. Here, the number of antennal segments, body size, and general shape of the pronotum are useful characters. Because of considerable species variation within and among subfamilies, detailed rearing studies must be conducted to establish species-specific instar differences. In the typical five-instar sequence of species having moderate to well developed wings as adults, the wing pads develop as extensions or lobes of the meso- and metanotum. The wing pads are essentially undetectable in the first instar; they become moderately lobed, directed downward, and weakly tracheated in the second instar; and prominently lobed, heavily tracheated, and directed posteriorly in the third instar. The fourth instar has the wing pads inverted and extending nearly to the posterior margin of the first abdominal segment. The fifth instar has wing pads extending to the third abdominal segment in long-winged forms. If an extra instar exists, it typically occurs between the normal third and fourth instars. The wing pads differ from the normal third instar in that they are more prominently lobed and tracheated and have a pronounced posteroventral extension.

Comments: Larval grasshopper taxonomy has fallen far behind that of the adults, and several works are important for the serious student. Uvarov's 1966 study treats the biology and taxonomy of larval and adult grasshoppers; the postembryonic ontogeny of Acridomorpha is explicitly detailed by Dirsh (1968). No general keys to larval grasshoppers for North America are in existence; however, several studies are of importance at the regional level: Criddle (1926), immature grasshoppers of Manitoba; Hanford (1946), *Melanoplus* nymphs of Manitoba; Brusven (1967), immature Acridinae of Kansas; Scoggan and Brusven (1972), immature Gomphocerinae and Oedipodinae of Idaho; and Brusven (1972), immature Melanoplinae (Catantopinae) and Cyrtacanthacridinae of Idaho.

Taxonomy of Subfamilies: The subfamily classification of Acrididae has undergone several revisions (Rehn and Grant 1960, Dirsh 1961, Uvarov 1966, and Jago 1969). The classification used here is that of the United States National Museum of Natural History. The following subfamilies are recognized from North America: Acridinae, Cyrtacanthacridinae, Gomphocerinae, Melanoplinae, Oedipodinae, and Romaleinae.

Identification of larval grasshoppers to the subfamily level poses a challenge because many of the characters used for identifying adults are either lacking or poorly defined (e.g., genitalic features, tegmina and pronotal sulci and carinae). As a general rule, the later the instar, the easier it is to identify because the aforementioned characters become progressively more developed and the overall appearance becomes more like the adult.

Great detail in terminology of characters for identifying larvae at the subfamily level is not included here; rather, I have relied heavily upon illustrations for character interpretation in the keys. General body shape, profile of the head, shape and coloration of the eye, type of antenna, shape of the pronotum, and spines on the hind tibia are especially useful characters.

3. Published with approval of the director of the Agricultural Experiment Station, University of Idaho, as Research Paper No. 7862.

4. I wish to acknowledge Mr. Marvin Hanks, Department of Entomology, University of Idaho, who prepared the illustrations of Acrididae.

KEY TO SUBFAMILIES OF IMMATURE ACRIDIDAE

1. Hind tibia with both inner and outer immovable apical spines (fig. 16.64), (external apical spine absent in *Spaniacris deserticola* (Bruner), a desert grasshopper from California); form mostly robust (figs. 16.78,16.79) .... (lubber grasshoppers) **Romaleinae** (p. 164)

1′. Hind tibia without both inner and outer immovable apical spines (fig. 16.65); form typically more slender except Oedipodinae (figs. 16.80–16.83) 2

2(1′). Prosternum with a distinct median spine, at least in instars 2–5 (fig. 16.58); pronotum with median carina absent or weak (distinct in *Schistocerca*) 3

2′. Prosternum without a distinct median spine (fig. 16.57); pronotum with median carina absent to prominent 4

3(2). Pronotum with a distinct median carina; compound eye elongately ovate, anterior half usually with vertical pale banding, at least in instars 3–5 (fig. 16.69); body color mostly pale green, occasionally light brown, most species without conspicuous contrasting dark brown or black markings (fig. 16.80). *Schistocerca* the only North American genus .... (bird grasshoppers) **Cyrtacanthacridinae** (p. 163)

3′. Pronotum with median carina indistinct or weak; compound eye broadly oblong, anterior half without pale vertical banding, may be variously speckled, mottled, concolorous or with transverse median band (figs. 16.70, 16.71); body color variable, mostly cream to light brown, occasionally green; ground color often interrupted with conspicuous brown or black markings (figs. 16.82, 16.83). Many North American species .... (spur-throated grasshoppers) **Melanoplinae** (p. 163)

4(2′). Head moderately to strongly slanted when viewed in profile (figs. 16.72,16.73); form mostly slender (fig. 16.81); pronotum with median carina absent or shallow, except *Acrolophitus;* antennae mostly filiform (threadlike), occasionally ensiform (sword-shaped) or clavate (clubbed) (figs. 16.66–16.68). Includes subfamily Acridinae—represented by the North American genus *Metaleptea* (formerly *Truxalis*) .... (slant-faced grasshoppers) **Gomphocerinae** and **Acridinae** (p. 164)

4′. Head mostly vertical, never strongly slanted when viewed in profile (figs. 16.75, 16.76); form mostly robust, occasionally slender (figs. 16.84, 16.85); pronotum with median carina absent to prominent; antennae filiform (fig. 16.66) .... (band-winged grasshoppers) **Oedipodinae** (p. 164)

Subfamily Melanoplinae

Figures 16.74, 16.82, 16.83

The subfamily Melanoplinae (spur-throated grasshoppers), represented by over 300 species in North America, is the largest subfamily of Acrididae. Larvae lack the anatomical diversity of the Oedipodinae and Romaleinae, although body coloration and markings are often extreme and contrasting. A conspicuous prosternal spine readily distinguishes this subfamily from other subfamilies except Cyrtacanthacridinae, which shares this character. Brachyptery is common, aptery is rare; thus, adults of some species appear as larvae to the beginning student.

Larvae are prevalent during spring, summer, and early fall. Most species have a single generation per year. In warmer climates, some species are nondiapausing and have several generations per year; others require more than one year to complete their life cycle. Species occur in a wide variety of habitats ranging from desert to arctic tundra, and from semibarren mountain tops to valley floors.

The Melanoplinae are represented by some of the most destructive species in North America, causing damage to cropland and rangeland. The lesser migratory grasshopper, *Melanoplus sanguinipes* (Fabricius) (Fig. 16.83) is one of the most widespread and destructive species in North America. Ravenous appetites and the migratory behavior of this species have necessitated extensive control programs over the years. The red-legged grasshopper, *Melanoplus femurrubrum* (DeGeer) (fig. 16.82) is also widespread, common, and sometimes destructive.

Subfamily Cyrtacanthacridinae

Figures 16.69, 16.80

The subfamily Cyrtacanthacridinae (bird grasshoppers) is represented by the single genus *Schistocerca* (fig. 16.80) in North America. Six species are recognized. The presence of a prosternal spine readily distinguishes the group from other subfamilies except Melanoplinae, which shares this character. Earlier classifications placed all spur-throated grasshoppers in the subfamily Cyrtacanthacridinae. The common name "bird grasshopper" is applied to the group because of the strong powers of flight of the adult. The distinctive body appearance and vertical banding of the compound eye, at least in middle to late instars of most species, makes the larvae readily identified.

Larvae of most species are present during spring and summer. *Schistocerca americana* (Drury) has continuous generations in southeastern United States and is seasonal farther north. The greatest species diversity occurs in the southeast and southcentral United States and Mexico. Most species are associated with old fields, roadsides, and rank vegetation, but they are occasionally found in xeric habitats. Most species are general feeders, preferring forbs, and although they seldom attain densities large enough to be of economic importance, *S. americana* is occasionally destructive.

Subfamilies Gomphocerinae and Acridinae

Figures 16.72, 16.73, 16.81

The Gomphocerinae and Acridinae (slant-faced grasshoppers) are treated together because of the lack of diagnostic characters to satisfactorily separate larvae of these two subfamilies. Earlier classifications placed all slant-faced grasshoppers in the single subfamily Acridinae. North American Acridinae is represented by the genus *Metaleptea* (formerly *Truxalis* from the northeastern United States). Gomphocerinae includes numerous genera and is represented by some 90 species. Larvae of most species can be readily recognized by the moderately to strongly slanting head.

The slant-faced grasshoppers are widely distributed across North America, exhibiting their greatest diversity and abundance in grasslands. Unlike the other subfamilies, slant-faced grasshoppers feed almost exclusively on grasses and sedges and often cause serious damage to rangeland. *Aulocara elliotti* (Thomas) is especially destructive in the Midwest where it feeds extensively on wheatgrasses.

Most species overwinter in the egg stage; however, *Psoloessa delicatula* (Scudder) overwinters as a late instar. Larvae are prevalent during spring and summer, adults in the fall.

Subfamily Oedipodinae

Figures 16.75, 16.76, 16.84, 16.85

The subfamily Oedipodinae (band-winged grasshoppers) is extremely diverse. It has approximately 200 species recognized for North America, which occur in virtually all regions of North America but exhibit greatest diversity in desert and grassland-shrub areas. Because of extreme anatomical variations among species, many of their taxonomic characteristics are present in other subfamilies of grasshoppers and cause difficulty in identification to the beginning student. Most members of this subfamily can be recognized by their stout, robust appearance and prominent median and lateral pronotal carinae. The pronotum and head are often heavily ridged and pitted.

Band-winged grasshoppers occasionally become abundant and attain economic importance. *Camnula pellucida* (Scudder) (fig. 16.85) is subject to periodic outbreaks and is especially injurious to the High Plains rangeland. Most band-winged grasshoppers are general feeders, feeding on grasses, shrubs, and forbs. They usually overwinter in the egg stage; larvae are prevalent in spring and early summer.

Subfamily Romaleinae

Figures 16.64, 16.77, 16.78, 16.79

The Romaleinae (lubber grasshoppers) are a small but diverse subfamily composed of 7 genera and 9 species in North America. The greatest diversity of species occurs in desert and semidesert regions of southcentral and southwestern United States and Mexico. ***Brachystola magna*** (Girard) occurs throughout the Great Plains from North Dakota to Mexico; *Romalea guttata* (Houttuyn) is principally eastern. Species of this subfamily seldom become abundant enough to cause economic damage; however, *R. guttata* is occasionally destructive to truck crops in the Southeast.

The large and often bizzare shape of larvae and adults make many romaleine grasshoppers a prize to collectors. The presence of both inner and outer immovable spines on the apex of the hind tibia of all species except *Spaniacris* permits recognition of larvae of this subfamily. Larvae of most species can be collected during the spring and early summer months.

Comments: The family Acrididae contains over 600 species in North America. Most occur in terrestrial habitats, reaching their greatest diversity and abundance in grassland, desert, and semidesert regions. Heavily forested areas, alpine areas, and tundra have sparse grasshopper faunas, although grasshoppers may be found over wide elevational gradients.

Although grasshoppers have often been universally condemned as being destructive, most of the serious damage is caused by relatively few species that are capable of rapid population increase under favorable conditions, possess exceptional abilities for dispersal, and feed ravenously on native and planted vegetation. Many North American species are rare, occurring only in very specific habitats and are seldom if ever of economic importance.

Several references on adult grasshoppers give accounts of the biology, ecology, and keys to identification and are useful study aids for larvae in a specific region (Morse 1920, 1921; Hebard 1925, 1928, 1929, 1931, 1936; Ball et al. 1942; Helfer 1963; Froeschner 1954; Brooks 1958; Strohecker et al. 1968; Dakin and Hays 1970; Otte 1981, 1984).

Selected Bibliography

Ball et al. 1942.
Brooks 1958.
Brusven 1967, 1972.
Criddle 1926.
Dakin and Hays 1970.
Dirsh 1961, 1968.
Froeschner 1954.
Hanford 1946.
Hebard 1925, 1928, 1929, 1931, 1936.
Helfer 1963.
Jago 1969.
Morse 1920, 1921.
Otte 1981, 1984.
Rehn and Grant 1960.
Scoggan and Brusven 1972.
Strohecker et al. 1968.
Uvarov 1966, 1977.

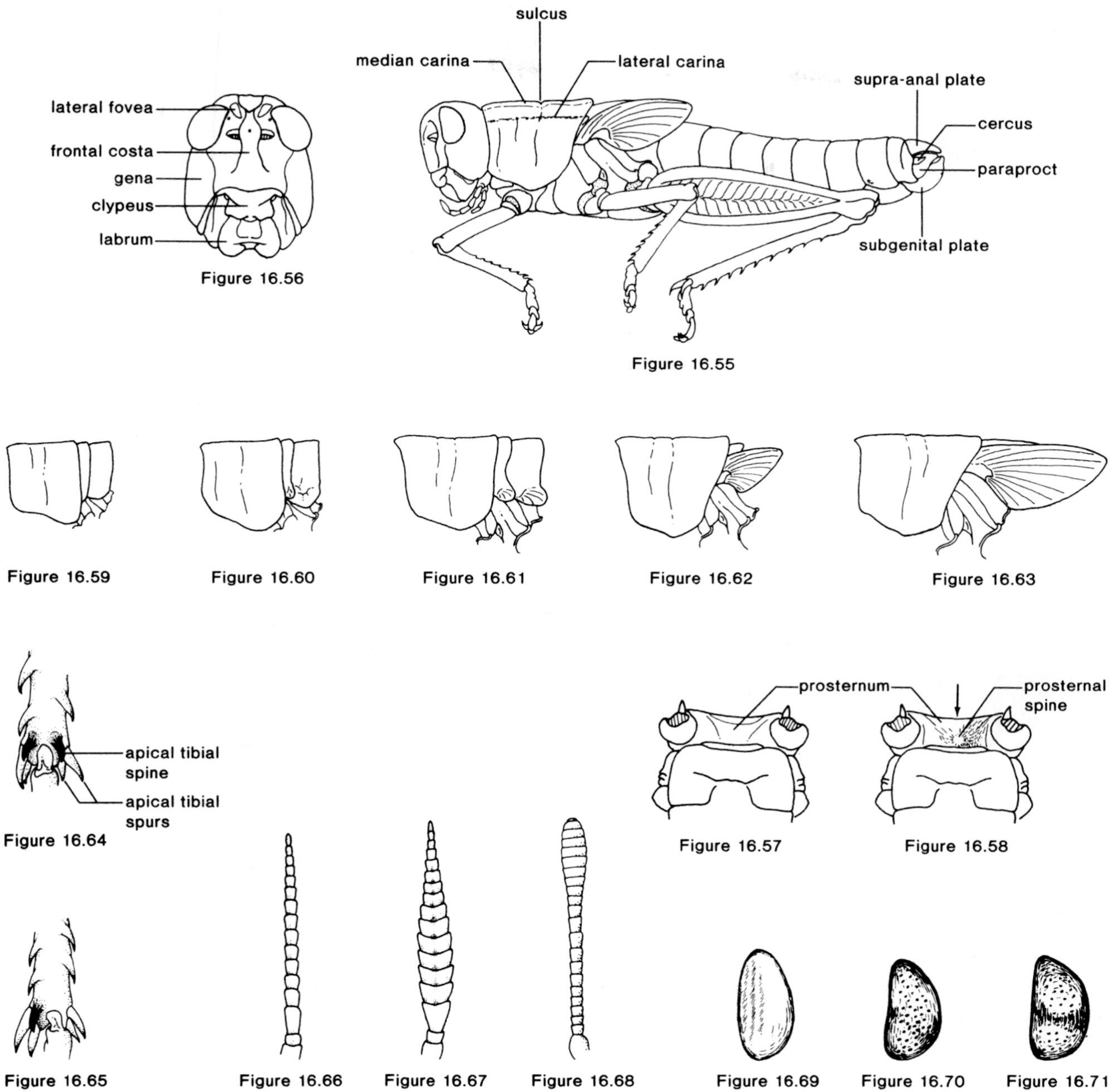

Figure 16.55. Generalized lateral view of fifth instar.

Figure 16.56. Frontal view of head.

Figure 16.57. Prosternum without spine.

Figure 16.58. Prosternum with spine.

Figure 16.59. Wing pads, first instar.

Figure 16.60. Wing pads, second instar.

Figure 16.61. Wing pads, third instar.

Figure 16.62. Wing pads, fourth instar.

Figure 16.63. Wing pads, fifth instar.

Figure 16.64. Hind tibia with 2 apical spines (Romaleinae).

Figure 16.65. Hind tibia with 1 apical spine.

Figure 16.66. Filiform antenna.

Figure 16.67. Ensiform antenna.

Figure 16.68. Clavate antenna.

Figure 16.69. Compound eye of *Schistocerca* (Cyrtacanthacridinae), with vertical bands.

Figure 16.70. Compound eye, unbanded.

Figure 16.71. Compound eye, with transverse band.

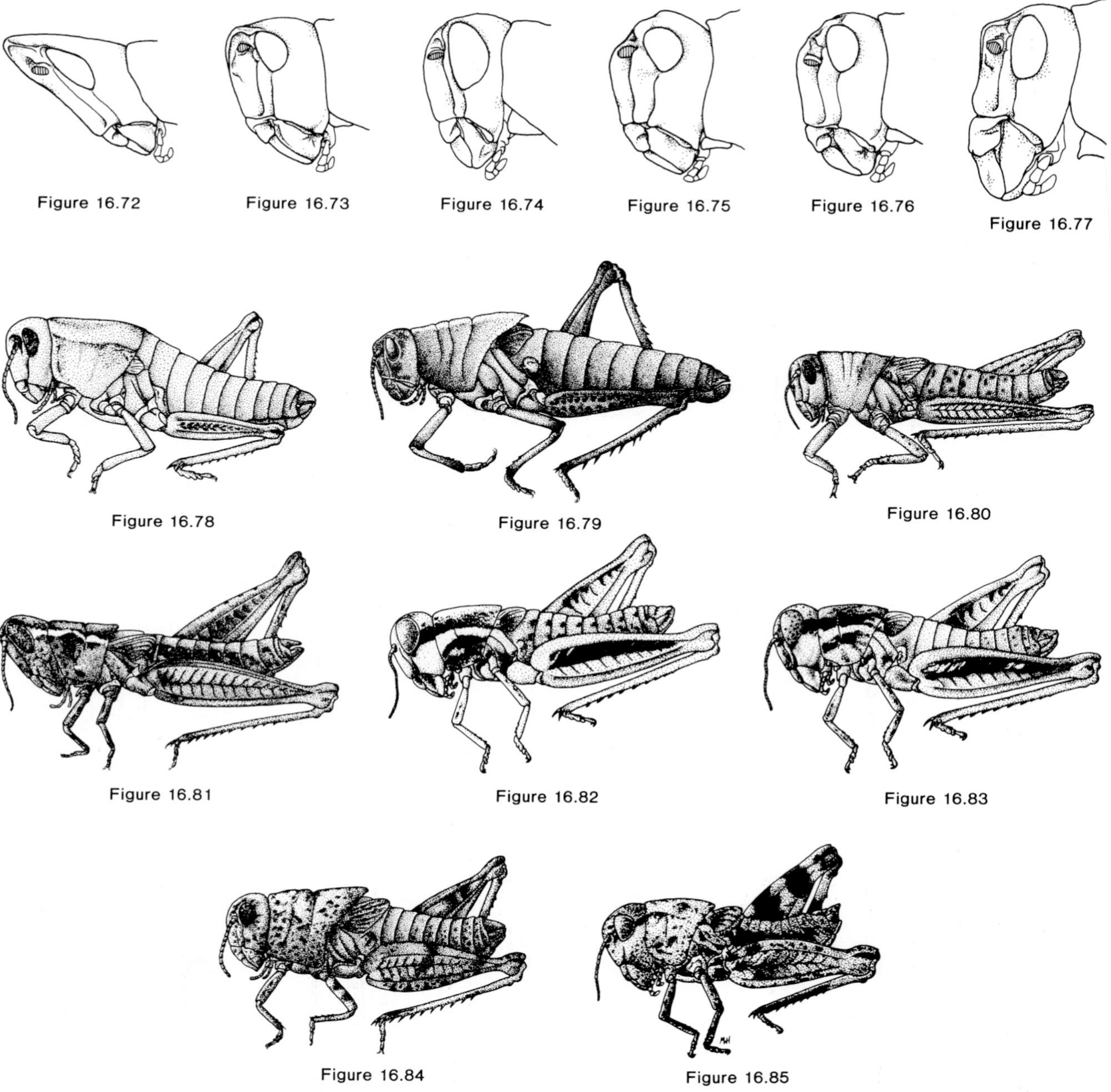

Figures 16.72–16.77. Heads of Acrididae.

Figure 16.72. *Chloealtis brachyptera* (Scudder) (Gomphocerinae).

Figure 16.73. *Brunneria brunnea* (Thomas) (Gomphocerinae).

Figure 16.74. *Melanoplus femurrubrum* (De Geer) (Melanoplinae).

Figure 16.75. *Tachyrhacis kiowa* (Thomas) (Oedipodinae).

Figure 16.76. *Dissosteira carolina* (Linnaeus) (Oedipodinae).

Figure 16.77. *Taeniopoda eques* (Burmeister) (Romaleinae).

Figure 16.78. Fourth instar, *Brachystola magna* (Girard) (Romaleinae).

Figure 16.79. Fourth instar, *Taeniopoda eques* (Burmeister) (Romaleinae).

Figure 16.80. Fourth instar, *Schistocerca lineata* Scudder (Cyrtacanthacridinae).

Figure 16.81. Fourth instar, *Ageneotettix deorum* (Scudder) (Gomphocerinae).

Figure 16.82. Fourth instar,*Melanoplus femurrubrum* (De Geer) (Melanoplinae).

Figure 16.83. Fourth instar, *Melanoplus sanguinipes* (Fabricius) (Melanoplinae).

Figure 16.84. Fourth instar, *Cratypedes neglectus* (Thomas) (Oedipodinae).

Figure 16.85. Fourth instar, *Camnula pellucida* (Scudder) (Oedipodinae).

BIBLIOGRAPHY MANTODEA, GRYLLOBLATTODEA, PHASMATODEA AND ORTHOPTERA

Ander, K. 1939. Vergleichend-anatomische und phylogenetische Studien uber die Ensifera (Saltatoria). Opusc. Entomol. Suppl. 2:1–306.

Ando, H. (ed.). 1982. Biology of the Notoptera. Nagano, Japan: Kashiyo-Insatsu Co., Ltd. 194 pp.

Ball, E. D., E. R. Tinkham, R. Flock, and C. T. Vorheis. 1942. The grasshoppers and other Orthoptera of Arizona. Ariz. Agr. Exp. Sta. Tech. Bull. 93:257–373.

Bedford, G. O. 1978. Biology and ecology of the Phasmatodea. Ann. Rev. Ent. 23:125–50.

Beier, M. 1934. Mantodea. Family Mantidae; Subfamily Hymenopodinae. *In* Genera Insectorum, No. 196. 37 pp.

Beier, M. 1935. Mantodea. Family Mantidae; Subfamily Thespinae. *In* Genera Insectorum, No. 200. 31 pp.

Beier, M. 1937. Mantodea. Family Mantidae; Subfamily Mantinae. *In* Genera Insectorum, No. 203. 146 pp.

Beier, M. 1968. Phasmida (Stab-order Gespenstheuschrecken). *In* Kukenthal's Handbuch der Zoologie. pp. 1–56.

Beier, M. 1970. Chapter 7. Dictyoptera (Blattoidea et Mantoidea). *In* Tuxen, S. L., (ed.). Taxonomist's glossary of genitalia in insects. Copenhagen: Munksgaard. 359 pp.

Bergerard, J. 1962. Parthenogenesis in the Phasmidae. Endeavour 21: 137–43.

Bischof, Hans-Joachim. 1975. Die keulenförmigen Sensillen auf den Cerci der Grille *Gryllus bimaculatus* als Schwerezeptoren. J. Comp. Physiol. 98:277–88.

Blackith, R. E. and R. M. Blackith. 1968. A numerical taxonomy of orthopteroid insects. Austr. J. Zool. 16:111–31.

Blatchley, W. S. 1920. The Orthoptera of Northeastern America. Indianapolis: Nature Publishing Co., 784 pp.

Bolivar, I. 1887. Essai sur les Acridiens de la Tribu des Tettigidae. Ann. Soc. Ent. Belg. 31:175–313.

Borror, D. J. and R. E. White. 1971. A field guide to the insects of America north of Mexico. Boston, Mass.: Houghton Mifflin. 404 pp.

Borror, D. J., D. M. DeLong, and C. A. Triplehorn. 1981. An introduction to the study of insects. 5th ed. New York: Holt, Rinehart & Winston. 827 pp.

Bourne, J. R. and R. J. Nielsson. 1967. *Nemobius fasciatus*—a predator on horn fly pupae. J. Econ. Ent. 60:272–4.

Bradley, J. C. and B. S. Galil. 1977. The taxonomic arrangement of the Phasmatodea with keys to the subfamilies and tribes. Proc. Ent. Soc. Wash. 79:176–208.

Breland, O. P. and J. W. Dobson. 1947. Specificity of mantid oöthecae. Ann. Ent. Soc. Amer. 40:557–75.

Brooks, A. R. 1958. Acridoidea of southern Alberta, Saskatchewan, and Manitoba (Orthoptera). Can. Ent. Suppl. 9. 92 pp.

Brues, C. T. and A. L. Melander. 1932. Classification of insects. A key to the known families of insects and other terrestrial arthropods. Bull. Mus. Comp. Zool. 73:1–672.

Brues, C. T., A. L. Melander, and F. M. Carpenter. 1954. Classification of the insects. Bull. Mus. Comp. Zool. 108:1–917.

Bruner, L. 1910. VII. South American Tetrigidae. Ann. Carnegie Mus. 7:89–143.

Brunner von Wattenwyl, K., and J. Redtenbacher. 1906–1908. Die insekten Familie der Phasmiden. Leipzig. 590 pp.

Bullock, R. C. 1973. Toxicity of selected insecticides to the restless bush cricket, a minor pest of citrus in Florida. J. Econ. Ent. 66:559–60.

Brusven, M. A. 1967. Differentiation, ecology, and distribution of immature slant-faced grasshoppers (Acridinae) in Kansas. Kansas Agr. Exp. Sta. Tech. Bull. 149. 59 pp.

Brusven, M. A. 1972. Differentiation and ecology of common Catantopinae and Crytacanthacridinae nymphs (Orthoptera: Acrididae) of Idaho and adjacent areas. Melanderia 9:1–31.

Campbell, K. G. 1974. Factors affecting the distribution and abundance of the three species of phasmatids which occur in plague numbers in the forests of southeastern Australia. J. Ent. Soc. Austr. 8:3–6.

Cantrall, I. J. 1943. The ecology of the Orthoptera and Dermaptera of the George Reserve, Michigan. Misc. Publ. Mus. Zool., Univ. Mich. No. 54. 184 pp.

Caudel, A. N. 1903. The Phasmidae, or walking sticks of the U.S. Proc. U.S. Nat. Mus. 26:863–85.

Caudell, A. N. 1916. The genera of the tettigoniid insects of the subfamily Rhaphidophorinae found in America north of Mexico. Proc. U.S. Nat. Mus. 49:655–90.

Chopard, L. 1943. Orthopteroides de l'Afrique du Nord. Paris. 450 pp.

Chopard, L. 1949. Ordre des Dictyoptères, 9:355–407; ordre des Notoptères, 9:587–93; ordre des Orthoptères, 9:617–722. *In* Grassé, P. P., (ed.). Traité de Zoologie Anatomie, Systematique, Biologie. Insectes.

Clark, J. T. 1976. The eggs of stick insects: a review with descriptions of the eggs of eleven species. Syst. Ent. 1:95–105.

Cohn, T. J. 1965. The arid-land katydids of the North American genus *Neobarrettia* (Orthoptera: Tettigoniidae): their systematics and a reconstruction of their history. Misc. Publ. Mus. Zool., Univ. Mich. No. 126. 179 pp.

Coleman, E. 1942. Notes on the great brown stick insect. Part I. Development of eggs and young. Part 2. Development of eggs and nymphs. Vict. Nat. 59:46–8; 62–5.

Cowan, F. T. 1929. Life history, habits, and control of the Mormon cricket. USDA, Tech. Bull. No. 161. 28 pp.

Cowan, F. T. and C. Wakeland. 1955. Mormon crickets—how to control them. USDA, Farmers' Bull. No. 2081. 13 pp.

Crampton, G. C. 1915. The thoracic sclerites and the systematic position of *Grylloblatta campodeiformis* Walker, a remarkable annectant "Orthopteroid" insect. Ent. News 26:337–50.

Crampton, G. C. 1939. The interrelationships and lines of descent of living insects. Psyche 45:165–82.

Criddle, N. 1926. Studies of the immature stages of Manitoba Orthoptera. Trans. Royal Soc. Can. Sect. V, Ser. 3, 20:505–525.

Dakin, M. E. and K. L. Hays. 1970. A synopsis of Orthoptera of Alabama. Auburn Univ. Exp. Sta. Bull. 404. 118 pp.

Dirsh, V. 1955. Tanaoceridae and Xyronotidae: two new families of Acridoidea (Orthoptera). Ann. Mag. Nat. Hist. 12:285–8.

Dirsh, V. 1961. A preliminary revision of the families and subfamilies of Acridoidea (Orthoptera, Insecta). Bull. Br. Mus. (Nat. Hist.) Ent. 10:351–419.

Dirsh, V. 1968. The post-embryonic ontogeny of Acridomorpha. Eos 43:413–514.

Eaton, C. B. 1952. Walkingstick damage in hardwood stands and its control with DDT aerial sprays in 1950–51. Progr. Rept., USDA, Agr. Res. Serv., Milwaukee. 17 pp.

Essig. E. O. 1942. College Entomology New York: Macmillan. 900 pp.

Ford, N. 1926. On the behavior of *Grylloblatta.* Can. Ent. 58:66–70.

Folsom, J. W. and P. A. Woke. 1939. The field cricket in relation to the cotton plant in Louisiana. USDA, Tech. Bull. 642:1–28. (Photographs of all instars of *Gryllus* sp.)

Froeschner, R. C. 1954. The grasshoppers and other Orthoptera of Iowa. Iowa State College J. Sci. 29:163–354.

Fulton, B. B. 1915. The tree crickets of New York: life history and bionomics. New York Agr. Exp. Sta. Tech. Bull. 42:1–47. (Drawings of all instars of *Oecanthus fultoni.*)

Gangwere, S. K. 1961. A monograph on food selection in Orthoptera. Trans. Amer. Ent. Soc. 87:67–230.

Gangwere, S. K. 1965. The structural adaptations of mouthparts in Orthoptera and allies. Eos 40:67–85.

Gangwere, S. K. 1967. The phylogenetic development of food selection in certain orthopteroids. Eos 42:383–392.

Giglio-Tos, E. 1927. Orthoptera, Mantidae. *In* Das Tierreich. Berlin: Walter de Gruyter & Co. 707 pp.

Grant, H. J., Jr. 1955. Systematics of the Neotropical batrachideine genera *Plectronotus, Halmatettix* and *Cranotettix* gen. nov. (Orthoptera; Acridoidea; Tetrigidae). Proc. Acad. Nat. Sci. Philadelphia 107:57–74.

Grant, H. J., Jr. 1956. The taxonomy of *Batrachidea, Puiggaria, Lophoscirtus, Eutettigidea,* and *Rehnidium* n. gen. (Orthoptera; Acridoidea; Tetrigidae). Trans. Amer. Ent. Soc. 82:67–108.

Grant, H. J., Jr. and D. C. Rentz. 1967. A biosystematic review of the family Tanaoceridae including a comparative study of the proventriculus (Orthoptera: Tanaoceridae). The Pan-Pacific Ent. 43:65–74.

Gunther, K. K. 1953. Über die taxonomische Gliederrung und die geographische Verbreitung der Insektenordnung der Phasmatodea. Beitr. Ent. 3:541–63.

Gunther, K. K. 1969. Revision der Familie Rhipipterygidae Chopard, 1949. Mitt. Zool. Mus. Berlin. 45:259–424.

Gunther, K. K. 1975. Das Genus *Neotridactylus* Gunther, 1972. Mitt. Zool. Mus. Berlin. 51:305–65.

Gunther, K. K. 1977. Revision der Gattung *Ellipes* Scudder, 1902. Dtsch. Entomol. Z., N.F. 24:47–122.

Gunther, K. K. 1980, Katalog der Caelifera-Unterordnung Tridactyloidea. Dtsch. Entomol. Z., N.F. 27:149–78.

Gurney, A. B. 1937. Synopsis of the Grylloblattidae with the description of a new species from Oregon. *G. barberi*. Pan-Pacif. Ent. 13:159–71.

Gurney, A. B. 1951. Praying mantids of the United States: native and introduced. Smithsonian Report for 1950:339–62.

Gurney, A. B. 1953. Recent advances in the taxonomy and distribution of the Grylloblattidae. J. Wash. Acad. Sci. 43:325–32.

Gurney, A. B. 1961. Further advances in the taxonomy and distribution of the Grylloblattidae. Proc. Biol Soc. Wash. 74:67–76.

Hancock, J. 1898. The food habits of the Tettigidae. Ent. Record 10:6–7.

Hancock, J. 1902. The Tettigidae [*sic*] of North America. Chicago. 188 pp.

Hanford, R. H. 1946. The identification of nymphs of the genus *Melanoplus* of Manitoba and adjacent areas. Sci. Agr. 26:147–80.

Hartman, H. B., W. W. Walthall, L. P. Bennett, and R. R. Stewart. 1979. Giant interneurons mediating equilibrium reception in an insect. Science 205:503–5.

Hebard, M. 1925. The Orthoptera of South Dakota. Proc. Acad. Nat. Sci. Phila. 77:33–155.

Hebard, M. 1928. The Orthoptera of Montana. Proc. Acad. Nat. Sci. Phila. 80:211–306.

Hebard, M. 1929. The Orthoptera of Colorado. Proc. Acad. Nat. Sci. Phila. 81:303–425.

Hebard, M. 1931. The Orthoptera of Kansas. Proc. Acad. Nat. Sci. Phila. 83:119–227.

Hebard, M. 1934. Studies in Orthoptera which occur in North America north of the Mexican boundary. 1. *Psychomastax,* a genus of grasshoppers of the western United States. Trans. Amer. Ent. Soc. 59:363–70.

Hebard, M. 1936. Orthoptera of North Dakota. N.D. Agr. Coll. Exp. Sta. Tech. Bull. 284. 69 pp.

Hebard, M. 1943. The Dermaptera and orthopterous families Blattidae, Mantidae, and Phasmidae of Texas. Trans. Amer. Ent. Soc. 68:239–311.

Helfer, J. R. 1963. How to know the grasshoppers, cockroaches, and their allies. Dubuque, Ia.: Wm. C. Brown Co. Publishers. 353 pp.

Hetrick, L. A. 1949. The oviposition of the two-striped walkingstick *Anisomorpha buprestoides* (Stoll). Fla. Ent. 32:74–77.

Hubbell, T. H. 1936. A monographic revision of the genus *Ceuthophilus* (Orthoptera: Gryllacrididae, Rhaphidophorinae). Univ. Fla. Publ. Biol. Sci. Series 2:1–551.

Hubbell, T. H. and R. M. Norton. 1978. The systematics and biology of the cave-crickets of the North American tribe Hadenoecini (Orthoptera Saltatoria: Ensifera: Rhaphidophoridae: Dolichopodinae). Misc. Publ. Mus. Zool., Univ. Mich. No. 156. 124 pp.

Hughes-Schrader, S. 1950. The chromosomes of mantids (Orthoptera: Manteidae [*sic*]) in relation to taxonomy. Chromosoma 4:1–55.

Jago, N. D. 1969. A revision of the systematics and taxonomy of certain North American gomphocerine grasshoppers (Gomphocerinae, Acrididae, Orthoptera). Proc. Acad. Nat. Sci. Phila. 121:229–335.

Jacobson, G. G. R. and V. L. Bianchi. 1902. Orthoptera and Pseudoneuroptera of the Russian Empire. St. Petersburg. 432 pp.

Karny, H. 1937. Orthoptera, family Gryllacrididae, subfamiliae omnes. *In* Genera Insectorum 206:1–317.

Kamp, J. W. 1973. Numerical classification of the orthopteroids, with special reference to the Grylloblattodea. Can. Ent. 105:1235–49.

Kamp, J. W. 1979. Taxonomy, distribution, and zoogeographic evolution of *Grylloblatta* in Canada (Insecta: Notoptera). Can. Ent. 111:27–38.

Kevan, D. K. McE. 1976. The higher classification of the orthopteroid insects: a general view. *In* Kevan, D. K. McE., (ed.). XVth Int. Congr. Ent. 1976. Mem. Lyman Ent. Mus. Res. Lab. 4:1–31.

Kevan, D. K. McE. 1982. Orthoptera, pp. 352–83. *In* Synopsis and classification of living organisms. New York: McGraw-Hill Book Company, Inc.

Leach, W. E. 1815. Zoological Miscellany. London. 3v. Orthoptera. *In* vol. 2. pp. 57–172.

Mackerras, M. J. 1970. Evolution and classification of the insects, pp. 152–67. *In* The insects of Australia. A textbook for students and research workers. Division of Entomology. CSIRO, Canberra, Australia. Melbourne: Melbourne University Press, 1029 pp.

Morse, A. P. 1920. Manual of the Orthoptera of New England, including the locusts, grasshoppers, crickets, and their allies. Proc. Boston Soc. Nat. Hist. 35:197–556.

Morse, A. P. 1921. Orthoptera of Maine. Maine Agr. Exp. Sta. Bull. 296. 36 pp.

Melander, A. L. and M. A. Yothers. 1917. The coulee cricket. Wash. Agr. Exp. Sta. Bull. No. 137. 57 pp.

Monteith, L. G. 1971. Crickets as predators of the apple maggot, *Rhagoletis pomonella* (Diptera: Tephritidae). Can. Ent. 103:52–8.

Morris, G. K. and D. T. Gwynne. 1978. Geographic distribution and biological observations of *Cyphoderris* (Orthoptera; Haglidae) with the description of a new species. Psyche 85:147–67.

Morris, G. K. and T. J. Walker. 1976. Calling songs of *Orchelimum* meadow katydids (Tettigoniidae). I. Mechanism, terminology, and geographic distribution. Can. Ent. 108:785–800.

Nabours, R. K. 1919. Parthenogenesis and crossing-over in the grouse locust *Apotettix*. Amer. Nat. 53:131–43.

Nabours, R. K. 1925. Studies in the inheritance and evolution in Orthoptera V. The grouse locust, *Apotettix eurycephalus* Hancock. Tech. Bull. Kans. Agric. Coll. No. 17. 231 pp.

Nabours, R. K. and F. M. Stebbins. 1950. Cytogenetics of the grouse locust *Apotettix eurycephalus* Hancock. Kansas Agricult. Exp. Sta. Tech. Bull. 67. 116 pp.

Nielsson, R. J. and M. H. Bass. 1967. Seasonal occurrence and number of instars of *Nemobius fasciatus,* a pest on white clover. J. Econ. Ent. 60:699–701.

Otte, D. 1981. The North American grasshoppers. I. Acrididae: Gomphocerinae and Acridinae. Cambridge, Mass.: Harvard Univ. Press. 275 pp.

Otte, D. 1984. The North American grasshoppers. II. Acrididae: Oedipodinae. Cambridge, Mass.: Harvard Univ. Press. 366 pp.

Parrot, P. J. and B. B. Fulton. 1914. Tree crickets injurious to orchard and garden fruits. New York Agr. Exp. Sta. Bull. 338:417–61.

Ragge, D. R. 1977. Classification of Tettigonioidea. *In* The higher classification of the orthopteroid insects. D. K. McE. Kevan (ed.) XVth Int. Congr. Ent. 1976. Mem. Lyman Entomol. Mus. Res. Lab. 4:44–46.

Rehn, J. A. G.1911. Orthoptera, family Mantidae; subfamily Vatinae. *In* Genera Insectorum. No. 119. 28 pp.

Rehn, J. A. G. 1948a. The locust genus *Tanaocerus* as found in the United States, and the description of a related new genus (Orthoptera: Acridoidea). Proc. Acad. Nat. Sci. Phila. 100:1–22.

Rehn, J. A. G. 1948b. The acridoid family Eumastacidae (Orthoptera). A review of our knowledge of its components, features, and systematics, with a suggested new classification of its major groups. Proc. Acad. Nat. Sci. Phila. 100:77–139.

Rehn, J. A. G. and H. J. Grant, Jr. 1955. The North American tetrigid genus *Nomotettix* (Orthoptera; Acridoidea; Tetrigidae). Proc. Acad. Nat. Sci. Phila. 107:1–34.

Rehn, J. A. G. and H. J. Grant, Jr. 1958. The Batrachideinae (Orthoptera; Acridoidea; Tetrigidae) of North America. Trans. Amer. Ent. Soc. 84:13–103.

Rehn, J. A. G. and H. J. Grant, Jr. 1960. A new concept involving the subfamily Acridinae (Orthoptera: Acridoidea). Trans. Amer. Ent. Soc. 86:173–85.

Rehn, J. A. G. and H. J. Grant, Jr. 1961. A monograph of the Orthoptera of North America (north of Mexico). Volume 1. Acridoidea in part, covering the Tetrigidae, Eumastacidae, Tanaoceridae, and Romaleinae of the Acrididae. Monogr. Acad. Nat. Sci. Phila., no. 12. 257 pp. 8 pl.

Rehn, J. A. G. and M. Hebard. 1914a. A study of the species of the genus *Dichopetala* (Orthoptera: Tettigoniidae). Proc. Acad. Nat. Sci. Phila. 66:64–160.

Rehn, J. A. G. and M. Hebard. 1914b. A revision of the orthopterous group Insarae (Tettigoniidae, Phaneropterinae). Trans. Amer. Ent. Soc. 40:37–184.

Rehn, J. A. G. and M. Hebard. 1914c. Studies in American Tettigoniidae (Orthoptera). I. A synopsis of the species of the genus *Scudderia*. II. A synopsis of the species of the genus *Amblycorypha* found in America north of Mexico. Trans. Amer. Ent. Soc. 40:271–344.

Rehn, J. A. G. and M. Hebard. 1914d. Studies in American Tettigoniidae (Orthoptera). III. A synopsis of the species of the genus *Neoconocephalus* found in North America north of Mexico. Trans. Amer. Ent. Soc. 40:365–413.

Rehn, J. A. G. and M. Hebard. 1915. Studies in American Tettigoniidae (Orthoptera). IV. A synopsis of the species of the genus *Orchelimum*. Trans. Amer. Ent. Soc. 41:11–83.

Rehn, J. A. G. and M. Hebard. 1918. A study of the North American Eumastacidae. Trans. Amer. Ent. Soc. 44:223–50.

Rentz, D. C. 1973. The shield-backed katydids of the genus *Idiostatus*. Mem. Amer. Ent. Soc. 42:33–100.

Rentz, D. C. 1979. Comments on the classification of the orthopteran family Tettigoniidae, with a key to subfamilies and description of two new subfamilies. Austr. J. Zool. 27:991–1013.

Rentz, D. C. and J. D. Birchim. 1968. Revisionary studies in the Nearctic Decticinae. Mem. Pac. Coast Ent. Soc. 3. 173 pp.

Rentz, D. C. and D. B. Weissman. 1973. The origins and affinities of the Orthoptera of the Channel Islands and adjacent mainland California. Part I. The genus *Cnemotettix*. Proc. Acad. Nat. Sci. Phila. 125:89–120.

Robinson, M. 1979. By dawn's early light: matutinal mating and sex attractants in a Neotropical mantid. Science 205:825–27.

Savage, A. A. 1957. The identification of the nymphal and imaginal stages of the stick insect *Carausius morosus* Brunner. Proc. Leeds Phil. Lit. Soc. Sci. Sect. 7:29–33.

Scoggan, A. C. and M. A. Brusven. 1972. Differentiation and ecology of common immature Gomphocerinae and Oedipodinae (Orthoptera: Acrididae) of Idaho and adjacent areas. Melanderia 8:1–76.

Scudder, G. G. E. 1970. Chapter 11. Grylloblattaria, pp. 55–58. *In* Tuxen, S. L., (ed.), Taxonomist's glossary of genitalia in insects. Copenhagen: Munksgaard. 359 pp.

Severin, H. H. P. and H. C. Severin. 1911. The life history of the walking-stick *Diapheromera femorata* Say. J. Econ. Ent. 4:307–20.

Sharov. A. G. 1968. Phylogeny of the Orthopteroidea. Acad. Sci. USSR, Trans. Inst. Paleontol. 118. 251 pp.

Stringer, I. A. N. 1970. The nymphal and imaginal stages of the bisexual stick insect *Clitarchus hookeri*. New Zealand Ent. 4:85–95.

Strohecker, H. F., W. W. Middlekauff, and D. C. Rentz. 1968. The grasshoppers of California (Orthoptera: Acridoidea). Bull. Cal. Insect Surv. vol. 10. 177 pp.

Tinkham, E. R. 1955. Description of the female *Mohavacris timberlakei* in a new family and subfamily of the Acridoidea. Bull. S. Calif. Acad. Sci. 54:156–60.

Tinkham, E. R. 1962a. Studies in Nearctic desert sand dune Orthoptera. Part V. A new genus and two new species of giant sand treader camel crickets with keys and notes. Great Basin Nat. 22:12–29.

Tinkham, E. R. 1962b. Studies in Nearctic desert sand dune Orthoptera. Part VI. A new genus and three new species of large sand-treader camel crickets from the Colorado Desert with keys and notes. Bull. S. Calif. Acad. Sci. 61:89–111.

Tinkham, E. R. 1968. Studies in Nearctic desert sand dune Orthoptera. Part XI. A new arenicolous species of *Stenopelmatus* from Coachella Valley with key and biological notes. Great Basin Nat. 28:124–31.

Tinkham, E. R. 1970a. Studies in Nearctic desert sand dune Orthoptera. Part XII. A remarkable new genus and species of Stenopelmatinae crickets from the Viscaino Desert, Baja California, Mexico, with key. Great Basin Nat. 30:173–79.

Tinkham, E. R. 1970b. Studies in Nearctic sand dune Orthoptera. Part XIII. A remarkable new genus and species of giant black sand treader camel cricket from the San Rafael Desert with key and notes. Great Basin Nat. 30:242–49.

Tinkham, E. R. and D. C. Rentz. 1969. Notes on the bionomics and distribution of the genus *Stenopelmatus* in central California with the description of a new species (Orthoptera: Gryllacridinae). Pan-Pac. Ent. 45:4–14.

Uhler, P. R. 1884. Natural history of the arthropods, bugs. Stand. Nat. Hist., 2:263.

Urquhart, F. A. 1937. Some notes on the sand cricket (*Tridactylus apicalis* Say). Can. Field-Nat. 51:28–29.

Uvarov, B. P. 1966. Grasshoppers and locusts. Anti Locust Res. Cntr. Publ., Cambridge Press. 481 pp.

Uvarov, B. P. 1977. Grasshoppers and locusts. II. Centre Overseas Pest Res. Cambridge Press. 613 pp.

Varley, G. C. 1939. Frightening attitudes and floral simulations in praying mantids. Proc. Roy. Ent. Soc. London 14:91–96.

Vickery, V. R., D. E. Johnstone, and D. K. McE. Kevan. 1974. The orthopteroid insects of Quebec and the Atlantic Provinces of Canada. Lyman Ent. Mus. and Res. Lab. Mem. No. 1 (Spec. Publ. No. 7). 204 pp.

Wakeland, C. 1959. Mormon crickets in North America. USDA, ARS Tech. Bull. No. 1202. 77 pp.

Walker, E. M. 1914. A new species of Orthoptera, forming a new genus and family. Can. Ent. 46:93–99.

Walker, E. M. 1919. On the male and immature state of *Grylloblatta compodeiformis* Walker. Can. Ent. 51:131–39.

Walker, E. M. 1937. *Grylloblatta,* a living fossil. Trans. Roy. Soc. Can., Sec. V, Ser. III, 26:1–10.

Walker, E. M. 1956. Grylloblattaria, pp. 47–49. *In* Tuxen, S. L., (ed.). Taxonomist's glossary of genitalia in insects. Copenhagen: Munksgaard. 359 pp.

Walker, T. J. 1962. The taxonomy and calling songs of United States tree crickets (Orthoptera: Gryllidae: Oecanthinae). I. The genus *Neoxabea* and the *niveus* and *varicornis* groups of the genus *Oecanthus.* Ann. Ent. Soc. Amer. 55:303–22.

Walker, T. J. 1963. The taxonomy and calling songs of United States tree crickets (Orthoptera: Gryllidae: Oecanthinae). II. The *nigricornis* group of the genus *Oecanthus.* Ann. Ent. Soc. Amer. 56:772–89.

Walker, T. J. 1984. Mole crickets in Florida. Fla. Agr. Expt. Sta. Bull. 846. 54 pp.

Weaver, J. R. and R. A. Sommers. 1969. Life history and habits of the short-tailed cricket, *Anurogryllus muticus,* in central Louisiana. Ann. Ent. Soc. Amer. 62:337–42.

Order Dermaptera

17

Alan Brindle[1]
University of Manchester, England

EARWIGS

The Dermaptera, or earwigs, form a small but well-defined order of about 1500 world species. They are mainly tropical or subtropical and few species occur in temperate countries. The order has consisted of three very unequal suborders—the Arixeniina, the Hemimerina, and the Forficulina, the last containing the majority of the species and representing the Dermaptera as generally understood.

The Arixeniina contains a single genus, *Arixenia* Jordan, that is associated with bats in Malaysia, Indonesia, and Sarawak, and whose biology is described by Cloudsley-Thompson (1957), among others. The Hemimerina also contains a single genus, *Hemimerus* Walker, that is associated with the African giant rat, *Cricetomys gambianus;* recent papers on the biology of this genus include Giles (1961), Davies (1966), and Ashford (1970). Popham (1961) gave the Hemimerina ordinal status, although Giles (1974) has published a case for its reinstatement within the Dermaptera.

The Forficulina, which are regarded here as representing the Dermaptera, are very similar in external structure, and although they are thus easily recognizable as such, the taxonomy of the order is necessarily based on less prominent external characters. The most certain and most generally adopted method of classification is by the structure of the male genitalia, with the males usually having more distinctive external characters. Females are often difficult to name unless associated with males, and many are impossible to determine; it follows that immature specimens are usually impossible to name with any certainty unless associated with adults, although there are exceptions. In the small fauna of the United States, including Hawaii, and Canada, however, it is possible to name the families of larvae, and sometimes to name them to genus and species.

Few species are of any economic importance, and the life histories of few species have been adequately described, but the life history does show the unusual feature of maternal care for the eggs and young.

DIAGNOSIS

Dermaptera are easily recognized as adults by the paired forceps at the end of the abdomen; each branch of the forceps consists of a single structure. Though the dipluran family Japygidae has forcepslike cerci, that group is separated from the Dermaptera by the one-segmented tarsi (three-segmented in Dermaptera) and the lack of compound eyes; compound eyes are absent from only four known Dermaptera species. A few other insects have paired caudal forcepslike structures, but these are different and there is rarely any confusion in making the ordinal placement of earwig specimens. The forceps of adult male earwigs may be highly specialized in shape, whereas those of females and larvae are usually of a more uniform shape, the branches being simple. In one subfamily of the Pygidicranidae (Karschiellinae—African) the forceps of the larvae may be partially segmented, and in the subfamily Diplatyinae, Pygidicranidae (or family Diplatyidae), which is circumtropical, the forceps of the larvae are represented by long, multisegmented cerci.

BIOLOGY AND ECOLOGY

The white eggs, large for the size of the insect, are laid in a batch in a dark moist situation, such as beneath the bark of trees, under stones, in the basal leaves of plants, or in burrows in the soil. The number of eggs varies in different species from 15–80. Earwigs are strongly thigmotactic, and the choice of a nest situation depends largely on the presence of tactile contact between the earwig and the enclosing soil or other substance. Humidity and darkness also seem to be necessary. The excavation of the nest changes the behavior of the female, and she will try to attack any moving object that comes near; and the male, if present, often seems to be ejected or sometimes, apparently, killed, although there is evidence that the male sometimes may be associated with the female in the nest. Laying the first egg brings into operation two further responses, licking the eggs and rolling them; they are kept in a batch and will be collected into a heap if they become scattered. Licking and rolling the eggs remove fungal spores and other extraneous material, and they soon become moldy if removed from the female. If one artificially cleans and rolls the eggs with a fine brush, however, a portion of the eggs may be hatched without the presence of the female (Buxton and Madge 1974).

1. I am grateful to Dr. Ashley B. Gurney, USDA, for photocopies of publications and for information on the common names of the American Dermaptera, as well as for other details; and to Professor D. K. McE. Kevan, McGill University, Quebec, for information on the Canadian species and papers concerning these species.

Larvae usually pass through four instars, but five have been recorded in one Australian species (Giles 1953). The period from hatching to last ecdysis varies from 40–165 days in recorded species. In temperate countries, the life cycle may follow an annual pattern, although there may be more than one brood produced by a single female. In tropical climates there is likely to be continuous breeding with a shorter life cycle unless such a feature as aestivation plays a part during hot dry periods.

Adult earwigs range from 2.5 mm to nearly 80 mm in total length, and most seem to be nocturnal in habit, hiding by day in dark sheltered situations, preferably in narrow crevices or beneath bark or stones. A few are known to be diurnal. They are probably mainly omnivorous, feeding on various parts of plants, acting as scavengers, or feeding on other insects. Recent studies tend to show that a carnivorous habit may be dominant in some species or genera, and may be more widespread than previously thought. A few species are known to fly readily, mainly species of the family Labiidae, but most earwigs, although they may be fully winged, only fly occasionally and for short distances. Earwigs are attracted to light and can climb rough surfaces to a considerable height, even the trunks of palm trees.

Earwigs occur from sea level to 4000 or 5000 meters, but montane species are in a minority. Adults and larvae occur in humus, vegetation, under bark or stones, or in various kinds of debris. They are most common in equatorial rain forests, the favoured habitats being moist and warm, with only a few species recorded from dry or semidesert areas. There is a good deal of uncertainty about many of the habits and life histories, and much more research is necessary before we can have any real conception about their way of life. The lack of research is partly due to their small economic importance, but it has been found that certain species may be useful in controlling some insect pests (Terry 1905, Williams 1931). It has also been found that the results recorded by one author may differ from those recorded by another; authors have reported that few males of the common European earwig, *Forficula auricularia,* survive the winter in temperate countries, whereas others state that males can be quite common in spring. There are similar discrepancies as to the part played by the male in the nest, the kind of food usually taken, and other features.

The most generally useful summaries of the known habits, ecology, biology, and structure of earwigs, together with a summary of the taxonomy of the order, are those in Beier (1959) and Guenther and Herter (1974), which also have extensive bibliographies. Chopard (1949) gives a much shorter account, and short summaries occur in most entomological textbooks. Zimmerman (1948) gives an account of the earwigs of Hawaii, and Gurney (1950, 1959) gives a largely up-to-date list of the earwigs of the United States except for a few name changes, and the addition of one species by Nutting and Gurney (1961). Langston and Powell (1975) give an account of the earwigs of California.

DESCRIPTION

Larvae of Dermaptera resemble adults in general structure except for the absence of elytra and wings and the presence of simpler forceps. Adults often have no elytra or wings, and the forceps of females are usually more or less simple. Recognizing a specimen as being immature, therefore, is sometimes difficult, and the most practical method is by the less strongly sclerotized cuticle of the larvae, the cuticle distorting when dried. This also happens in teneral adults, where the antennal segments and the legs may become flattened. Antennal segments of larvae are less numerous than in adults and they are perhaps shorter and more cylindrical. This is particularly prominent in younger larvae of the Pygidicranidae, Labiduridae, and Carcinophoridae, and is less obvious in the Labiidae, Chelisochidae, and Forficulidae. In *Euborellia annulipes* (Lucas) (Carcinophoridae) for example, the antennal segments, especially the basal ones, are more or less cylindrical and are separated from each other by white membraneous cuticle (fig. 17.4) which contrasts with the uniformly dark basal segments of the adult (fig. 17.5). This is particularly useful in the Carcinophoridae, because so many of the adults lack elytra and wings.

Head

Prognathous, mouthparts of a biting type similar in most respects to those of the Orthoptera (Popham 1959). Compound eyes of variable size almost always present, but ocelli absent. Epicranial sutures distinct or not. First antennal segment (scape, figs. 17.6,17.7) is longer than other segments but shorter than in adults; second segment (pedicel, figs. 17.6, 17.7) always short and often transverse; remaining segments, forming the flagellum, shorter and more cylindrical than in adults and vary according to the particular instar, earlier instars having fewer segments than later ones. The third antennal segment, or meriston (Henson 1947, Chapman 1982), which is the first of the flagellum, is the segment that repeatedly divides during larval life to form additional distal segments, so this segment may vary in length, depending on whether an additional segment has recently been formed (figs. 17.6, 17.7). Although the number of the antennal segments is a guide to the particular instar, it has been shown that growth of the head capsule may be a more reliable guide (Giles 1952).

Thorax

Pronotum a single sclerotized plate, usually more or less rectangular and prominent; mesonotum shorter and transverse; metanotum is shorter still and with posterior margin concave. In the last instar the mesonotum may be inflated, indicating development of the elytra, and if lateral longitudinal ridges occur on the elytra of the adult, these ridges are indicated on the sides of the mesonotum of the larva. Metanotum, in those species fully winged as adults, strongly produced backward in later instars, and in the last instar forming two large platelike structures on which rudiments of the venation are visible.

Each leg is comprised of a relatively large coxa, a trochanter, a femur which is usually broader medially and sometimes compressed, a shorter and more slender tibia, and a 3-segmented tarsus bearing two curved claws, between which may be an arolium.

Abdomen

Abdomen usually more or less parallel-sided and depressed, less often strongly widened, each segment consisting of a dorsal tergite that curves ventrally and overlaps the ventral sternite. The number of visible segments varies and some are often hidden, but nine are usually visible and ten complete segments may be distinct. The adult male has ten visible tergites, adult females only eight, females being characterized, in most species, by retraction of segments 8 and 9. In many species there is a tubercle toward the lateral margin of the third and fourth tergites; these are variously referred to as scent glands, repugnatorial glands, or simply as lateral tubercles. In all cases where they exist, the tubercles on the third tergite are smaller than those on the fourth. They are absent, or generally less conspicuous, in the more primitive families and may be large and prominent in the Forficulidae. At the end of the abdomen may be a visible pygidium between the bases of the branches of the forceps (figs. 17.10, 17.11, 17.12). This represents the eleventh segment and strictly consists of the epiprocts and paraprocts (Richard and Davies 1977), the former being divisible in their most complete condition by three sclerites—the pygidium, the metapygidium, and the telson. In practice it is usual to term the entire structure the pygidium.

Forceps

Each branch of the forceps, except in the Karschiellinae and Diplatyinae, consisting of a single sclerotized structure, typically broader at the base and narrowed distally; articulated with the last segment to allow mainly lateral movement, although restricted vertical movement possible. In most cases the base is trigonal, i.e., with a triangular cross section, and with a dorsal ridge at the base, but the forceps of some larvae are cylindrical throughout. Although the term "forceps" is generally applied to these processes (Richards and Davies 1977), the term "cerci" is also used (Chapman 1982).

Most earwigs are more or less glabrous, few being strongly pubescent, and even in these the hairs are short; the distal part of the tibiae and ventral part of the tarsi usually have short dense hairs and the latter may have short peglike processes in addition. Stronger hairs may occur laterally or on the posterior margins of the abdominal tergites and forceps, and in some earwigs the whole body may have very short, thick, and often truncate setae. The term "seta" is generally used to indicate stronger or thicker hairlike structures and not necessarily setae in the strict sense of the term. Research on the location of sensory setae in earwigs is most desirable.

COMMENTS

The classification adopted is basically that of Hincks (1955, 1959), and the more recent, but broadly similar classifications of Brindle (1973) and Steinmann (1975). This divides the order into 4 superfamilies and 7 families, although the Diplatyinae are sometimes removed from the Pygidicranidae and given family status, thus making 8 families.

Popham (1965) has proposed a different classification, the main difference being that the positions of the Carcinophoridae and the Labiidae are interchanged, but this really needs further study. Sakai's (1982) classification changes the family Carcinophoridae to Anisolabididae (Anisolabioidea). All the above families are represented in the United States, Canada, or Hawaii, with the exception of the Apachyidae. The Pygidicranidae are represented only by 1 adventive species in the United States, and the Chelisochidae are found under natural conditions only in Hawaii, being adventive in the United States but not in Canada.

TECHNIQUES

Collecting and killing methods generally used for Coleoptera are suitable for Dermaptera. Adults may be dried, pinned or mounted on cards, or preserved in alcohol or other preservative liquid. Larvae tend to distort if dried directly after capture, so are better preserved in liquid unless treated with a degreasing and hardening agent before pinning.

CLASSIFICATION

Order **DERMAPTERA**
- Suborder Arixeniina
  - Arixeniidae
- Suborder Hemimerina
  - Hemimeridae
- Suborder Forficulina
  - Superfamily Pygidicranoidea
    - Pygidicranidae
  - Superfamily Carcinophoroidea
    - Carcinophoridae
    - Labiduridae
  - Superfamily Apachyoidea
    - Apachyidae
  - Superfamily Forficuloidea
    - Labiidae
    - Chelisochidae
    - Forficulidae

KEY TO FAMILIES OF LARVAL DERMAPTERA

1. Second tarsal segment simple (fig. 17.1) 2

1′. Second tarsal segment not simple, the ventral part prolonged beneath the third distal segment (figs. 17.2, 17.3) 5

2(1). Antennal segments more numerous, never fewer than 15 and most instars with 20 or more; forceps as long as the abdomen in earlier instars, and at least three-quarters as long in later instars (fig. 17.8) 3

2′. Antennal segments less numerous, usually fewer than 15 and rarely more than 20 in all instars; forceps not more than half as long as abdomen in all instars or only slightly more (figs. 17.9, 17.10) 4

3(2). Head strongly transverse, usually with short thick setae; antennae with basal flagellar segments strongly transverse (fig. 17.6); cuticle with at least scattered short dark setae .... (**Pyragrinae**) ***Pygidicranidae*** (p. 174)

3′. Head less transverse and without short thick setae; antennae with basal flagellar segments less transverse (fig. 17.7); cuticle without short dark setae .... ***Labiduridae*** (p. 175)

4(2′). Forceps with branches relatively short and broad, inner margins smooth or almost so, branches close together and pygidium not visible (fig. 17.9); fourth instar larvae almost always without wing sheaths .... (mainly **Carcinophorinae***) ***Carcinophoridae*** (p. 174)

4′. Forceps with branches longer and more slender, inner margins often dentated or at least crenulated, and branches well separated from each other; pygidium usually prominent (fig. 17.10); fourth instar larvae almost always with wing sheaths .... ***Labiidae*** (p. 175)

5(1′). Second tarsal segment prolonged beneath third segment as narrow flattened lobe, the lobe usually closely pressed against lower surface of third segment (fig. 17.2); basal tarsal segment shorter than third (distal) in hind legs or only as long .... ***Chelisochidae*** (p. 175)

5′. Second tarsal segment prolonged beneath third segment as wide flattened lobe, the lobe well separated from lower surface of third segment (fig. 17.3); basal tarsal segment longer than third (distal) in hind legs .... ***Forficulidae*** (p. 177)

*Species of the subfamily Brachylabidinae occur in the Caribbean and Mexico; they are easily distinguished from other Dermaptera larvae by their very thick antennal segments, small head, fusiform abdomen, and long legs. Their forceps are slender, usually cylindrical, and pointed distally.

Pygidicranidae (Pygidicranoidea)

Figure 17.6

Diagnosis: Of several distinctive subfamilies, only the Pyragrinae is recorded from the continental United States (not from Hawaii or Canada). Larvae are similar to those of the Labidurinae (Labiduridae) in general appearance; but usually separable by the darker and more uniform coloration, by the short basal antennal segments (fig. 17.6), and by the presence of numerous and rather conspicuous short dark setae, most prominent laterally and along the posterior margins of the abdominal tergites. The biology is unknown.

Comments: The species of this subfamily, together with the closely similar Pygidicraninae, have distributions in South and Central America, including the Caribbean; the Pyragrinae are entirely American in distribution. The only species recorded from the United States is *Pyragropsis buscki* (Caudell), which has occurred as an adventive (Gurney 1959). The subfamily Diplatyinae (regarded as a separate family by some) also occurs in Central America, but the larvae are easily distinguished by the forceps being long, multisegmented cerci that may be longer than the body, which is very slender.

Carcinophoridae (Carcinophoroidea)

Figures 17.4, 17.5, 17.9

Diagnosis: Larvae (fig. 17.9) are usually recognizable by their shining and glabrous cuticle, depressed body, relatively short and broad branches of the forceps that tend to be close together, and by the rather short legs. Small specimens are similar to larvae of the Labiidae, and both have simple second tarsal segments, but the absence of wing sheaths in the fourth instar of most species will clearly separate the larger larvae from those of the Labiidae.

Biology and Ecology: Larvae of some species seem to be mainly carnivorous and wide-ranging, feeding on various soft-bodied insects. *Euborellia annulipes* (Lucas), the ring-legged earwig (U.S. including Hawaii, and Canada) has been recorded from beneath stones, debris, or litter of various kinds, in compost heaps or dung in which it may be numerous, or in fruits, among vegetation, or in houses. *Anisolabis maritima* (Bonelli in Gene), the maritime earwig (U.S. including Hawaii, and Canada) is typically a coastal species, hiding beneath boards, seaweed, or other litter on beaches and in mangrove swamps; it has been recorded as entering water and swimming freely; Guppy (1950) gives an account of its life history.

Comments: *Euborellia annulipes* has been noted as an important control on certain leafhopper pests (Terry 1905, Williams 1931, Hincks, 1947) and the less common *E. stali* (Dohrn) (United States) may be similarly useful in certain areas. Both of them have some distal antennal segments white that will separate them from the larvae of *A. maritima,* in which all the antennal segments are dark. Hincks (1947) summarizes the distinctions between the larval instars of *Euborellia annulipes,* based on the number of antennal segments.

Selected Bibliography

Giles 1952, 1953.
Guppy 1950.
Hincks 1947.
Terry 1905.

Labiduridae (Carcinophoroidea)

Figures 17.7, 17.8

Only one species of this family, *Labidura riparia* (Pallas), the shore or striped earwig, is found in the United States including Hawaii, but not in Canada.

Diagnosis: This large earwig is distinctive; the larva (fig. 17.8) can be recognized by the rather long and simple forceps, the long antennae and large eyes, the absence of setae, and by the coloration, which is usually yellowish with a median pattern of brown; some variation in color does occur.

Biology and Ecology: This species occurs both in dry sandy areas where the color tends to be paler and where it may be the dominant earwig, and in more humid habitats, where the color is darker. It is especially common on sandy beaches near the coast or sandy margins of rivers and lakes, or on mud flats or salt pans. Nest sites may be in deep tunnels in the ground. The food appears to be largely insects or other small animals.

Comments: This species is likely to have some importance, where common, in pest control; specimens examined in connection with surveys of crop pests in Pakistan have shown that it seems to be involved in controlling some lepidopterous pests of cotton crops (not published), and Terry (1905) records it as feeding on psyllids in Hawaiian sugarcane fields. In Australia a closely similar species, *L. truncata* (Kirby) has been recorded attacking codling moth larvae on apple trees (Giles 1970).

Labiidae (Forficuloidea)

Figures 17.1, 17.10

This is one of the largest families and contains mainly smaller species; a size range of 5–10 mm is usual, although some quite large species also occur.

Diagnosis: Usually small, characterized in the Forficuloidea by the simple second tarsal segment (fig. 17.1). The branches of the forceps are slender and the inner margins often serrate. Sometimes the body is pubescent and with longer hairs. Fourth instar larvae have wing sheaths in almost all cases (fig. 17.10).

Biology and Ecology: These earwigs seem to be more active than many others and may fly readily and at some height, having been taken in light traps in the tree canopy in Panama in large numbers. Recorded habitats include under bark, stones, in dead vegetation, rotten wood, dung, or compost heaps, in the tops of palm trees, in decaying stems, and in houses. *Marava arachidis* (Yersin) (U.S. including Hawaii) is apparently ovoviviparous, the larvae hatching soon after the eggs are laid, and the female helping to remove egg shells from the larvae (Beier 1959).

Comments: Hincks (1947) comments on the instars and habitats of *Labia curvicauda* (Motschulsky) (U.S. including Hawaii) and *Marava arachidis,* both of which are thought to have some value in controlling insect pests. The small earwig, *Labia minor* (L.) (U.S. and Canada) is frequent around and in compost heaps in company with staphylinid beetles and is thought to be predaceous. Terry (1905) records *Labia dubronyi* Hebard as feeding on psyllids in Hawaiian sugarcane fields.

Selected Bibliography

Hincks 1947.
Nutting and Gurney 1961.
Terry 1905.

Chelisochidae (Forficuloidea)

Figures 17.2, 17.11

Two species of this family occur in Hawaii, *Hamaxas nigrorufus* (Burr) and the black earwig, *Chelisoches morio* (Fabricius), but only the latter occurs in the continental United States, and that only as an occasional introduction.

Diagnosis: Second tarsal segment prolonged ventrally beneath the third distal segment as a narrow lobe (fig. 17.2); this characteristic clearly separates these larvae from all other earwig families.

Biology and Ecology: *C. morio* is diurnal and active on vegetation generally; it flies readily and feeds on leafhoppers, coccids, and larvae of beetles and moths among other insects. The eggs are laid in a heap in leaf sheaths at the bases of such plants as sugarcane, and it is common in the cane fields of Hawaii. Adults have also been recorded from the inflorescences of *Pandanus,* decaying banana stalks, under coconut shells or other debris on beaches. Nothing is known regarding the biology of *H. nigrorufus.*

Comments: *C. morio* seems to be well established as a control on leafhoppers in Hawaii (Terry 1905; Williams 1931). Terry (1905) and Hincks (1947) give details to separate the various instars. Larvae of *C. morio* can be separated from *H. nigrorufus* by the coloration, the former being shining black, often with contrasting whitish marks on the pronotum and elsewhere, whereas the antennae have some distal segments whitish; *nigrorufus* is more or less uniform dark brown or brown, and the antennae are unicolorous.

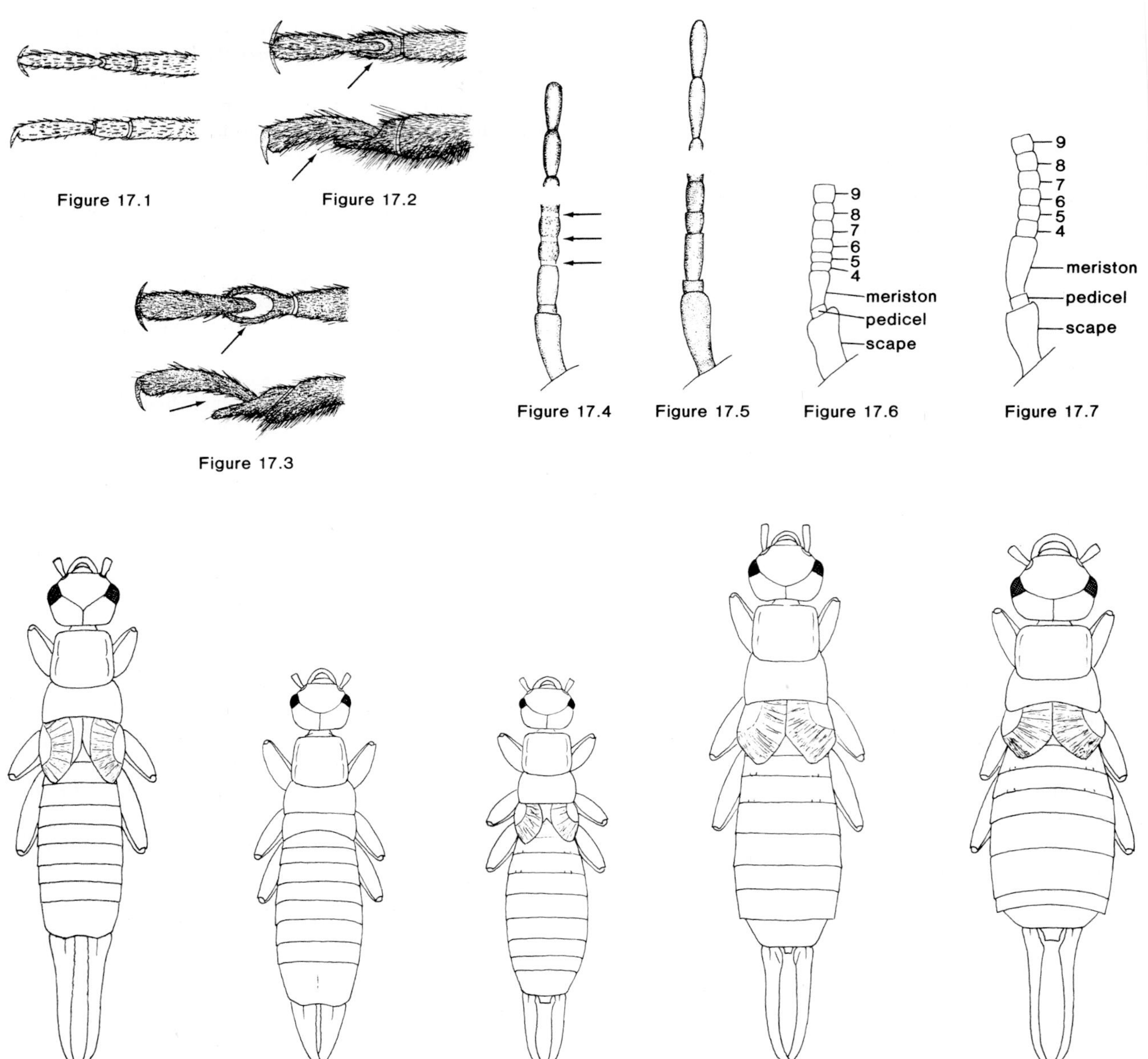

Figures 17.1–17.3. Tarsi, dorsal and lateral views.

Figure 17.1. *Labia minor* (L.), tarsus (Labiidae).

Figure 17.2. *Chelisoches morio* (Fabricius), tarsus (Chelisochidae).

Figure 17.3. *Forficula auricularia* L., tarsus (Forficulidae).

Figure 17.4. *Euborellia annulipes* (Lucas), antenna of larva (Carcinophoridae).

Figure 17.5. *Euborellia annulipes*, antenna of adult (Carcinophoridae).

Figure 17.6. *Pyragra*, basal antennal segments (Pygidicranidae).

Figure 17.7. *Labidura*, basal antennal segments (Labiduridae).

Figure 17.8. *Labidura riparia* Pallas, fourth instar larva (Labiduridae).

Figure 17.9. *Euborellia annulipes*, fourth instar larva (Carcinophoridae).

Figure 17.10. *Labia curvicauda* (Motschulsky), fourth instar larva (Labiidae).

Figure 17.11. *Chelisoches morio*, fourth instar larva (Chelisochidae).

Figure 17.12. *Forficula auricularia*, fourth instar larva (Forficulidae).

Forficulidae (Forficuloidea)

Figures 17.3, 17.12

Only one species of this family occurs in Hawaii, and few occur in the continental United States and Canada. The native species belong to the genus *Doru,* which is mainly southern, although *Doru aculeatum* (Scudder) occurs rarely as far north as the southern part of Ontario. The European earwig, *Forficula auricularia* L., however, is now well established in the continental United States and Canada and has become established in Hawaii (Dr. F. G. Howarth, *in litt.*).

Diagnosis: Distinguished by the broad flattened second tarsal segment from all other earwig families (fig. 17.3).

Biology and Ecology: The European earwig usually lays her eggs under a stone, under bark, or in burrows in the soil in autumn and winter in North Temperate countries; the young become adult towards midsummer, and more than one brood may be produced per year. Its diet is diverse, including various kinds of caterpillars, larvae, aphids, and other insects; but the earwig also acts as a scavenger, feeding on flowers and fruit. This latter habit seems to be mainly a nuisance, although Tillyard (1925) regarded this species as a major pest in New Zealand, and Fulton (1924) mentions reports of various crops receiving serious damage in the U.S.S.R. and in Europe. Crumb, Eide, and Bonn (1941) give an excellent account of *F. auricularia* in the United States, and Canadian papers on its biology include Lamb (1975, 1976) and Lamb and Wellington (1974, 1975). *Doru* adults and larvae are often associated with bromeliads and also with flowering heads of grasses and cacti, but nothing is certain about their food.

Comments: Larvae of the European earwig can be distinguished from those of *Doru* by the almost smooth inner margin of the forceps; in *Doru* the inner margins are crenulated with a regular row of distinct toothlike projections. Other species are intercepted from time to time, largely from imported plants. Some belong to *Metresura* and other genera; these can usually be separated as larvae from *Forficula* or *Doru* by their mainly black color and much longer and more slender forceps.

Selected Bibliography

Buxton and Madge 1974.
Crumb, Eide, and Bonn 1941.
Fulton 1924.
Henson 1947.
Lamb 1975, 1976.
Lamb and Wellington 1974, 1975.
Popham 1959.
Tillyard 1925.

BIBLIOGRAPHY

Ashford, R. W. 1970. Observations on the biology of *Hemimerus talpoides* (Insecta: Dermaptera). J. Zool. Lond. 162:413–18.

Beier, M. 1959. *In* Bronn. Klassen und Ordnungen des Tierreichs (Dermaptera) 5:455–585.

Brindle, A. 1973. The Dermaptera of Africa. Ann. Mus. Roy. Afr. Centr., Tervuren, in–8° Zool. 205:1–335.

Buxton, J. H., and D. S. Madge. 1974. Artificial incubation of eggs of the common earwig, *Forficula auricularia* L. Ent. Mon. Mag. 110:55–57.

Chapman, R. F. 1982. The insects—structure and function. 3rd ed. Cambridge, Mass.: Harvard Univ. Press. 919 pp.

Chopard, L. 1949. *In* Grassé. Ordre des Dermaptères. Traité de Zoologie. 9:745–70.

Cloudsley-Thompson, J. L. 1957. On the habitat and growth stages of *Arixenia esau* Jordan and *A. jacobsoni* Burr (Dermaptera: Arixenoidea), with descriptions of the hitherto unknown adults of the former. Proc. Roy Ent. Soc. Lond. (A) 32:1–12.

Crumb, S. E., P. M. Eide, and A. E. Bonn. 1941. The European earwig. Tech. Bull. U.S. Dept. Agric. 766:1–76.

Davies, R. G. 1966. The postembryonic development of *Hemimerus vicinus* Rehn and Rehn (Dermaptera: Hemimeridae). Proc. Roy. Ent. Soc. Lond. (A) 41:67–77.

Fulton, B. B. 1924. The European earwig. Bull. Ore. Agric. Exp. Sta. 207:1–29.

Giles, E. G. 1952. The growth of the head capsule and antennae of *Anisolabis littorea* (White) (Dermaptera: Labiduridae). Proc. Roy. Ent. Soc. Lond. (A) 27:91–98.

Giles, E. G. 1953. The biology of *Anisolabis littorea* (White) (Dermaptera, Labiduridae). Trans. Roy. Soc. New Zealand 80:383–98.

Giles, E. G. 1961. Further studies on the growth stages of *Arixenia esau* Jordan and *Arixenia jacobsoni* Burr (Dermaptera, Arixeniidae), with a note on the first instar antennae of *Hemimerus talpoides* Walker (Dermaptera, Hemimeridae). Proc. Roy. Ent. Soc. Lond. (A) 36:21–26.

Giles, E. G. 1970. Dermaptera, pp. 306–13. *In* The insects of Australia. Melbourne: Melbourne University Press. 1029 pp.

Giles, E. G. 1974. The relationship between the Hemimerina and the other Dermaptera: a case for reinstating the Hemimerina with the Dermaptera, based on a numerical procedure. Trans. Roy. Ent. Soc. Lond. 126:189–206.

Guppy, R. 1950. Biology of *Anisolabis maritima* (Gene), the seaside earwig, on Vancouver Island (Dermaptera: Labiduridae). Proc. Ent. Soc. Brit. Columbia 46:14–18.

Guenther, K., and K. Herter. 1974. *In* Beier, M. Handbuch der Zoologie (Dermaptera) 4 (part 2):1–158. Berlin.

Gurney, A. B. 1950. An African earwig new to the United States with a corrected list of the Nearctic Dermaptera. Proc. Ent. Soc. Wash. 52:200–203.

Gurney, A. B. 1959. New records of Orthoptera and Dermaptera from the United States. Florida Ent. 42:75–80.

Henson, H. 1947. The growth and form of the head and antennae in the earwig (*Forficula auricularia* L.). Proc. Leeds Phil. Soc. 5:21–32.

Hincks, W. D. 1947. Preliminary notes on Mauritian earwigs (Dermaptera). Ann. Mag. Nat. Hist. (11) 14:517–40.

Hincks, W. D. 1955. 1959. A systematic monograph of the Dermaptera of the world, Pt. 1, pp. 1–132; Pt. 2, pp. 1–218. British Mus. (Natural History).

Lamb, R. J. 1975. Effects of dispersion, travel, and environmental heterogeneity on populations of the earwig, *Forficula auricularia* L. Can. J. Zool. 53:1855–67.

Lamb, R. J. 1976. Parental behaviour in the Dermaptera, with special reference to *Forficula auricularia* (Dermaptera: Forficulidae). Can. Ent. 108:609–19.

Lamb, R. J. and W. G. Wellington. 1974. Techniques for studying the behavior and ecology of the European earwig, *Forficula auricularia* (Dermaptera: Forficulidae). Can. Ent. 106:881–88.

Lamb, R. J., and W. G. Wellington. 1975. Life history and population characteristics of the European earwig, *Forficula auricularia* (Dermaptera: Forficulidae) at Vancouver, British Columbia. Can. Ent. 107:819–24.

Langston, R. L., and J. A. Powell. 1975. The earwigs of California. Bull. Calif. Insect Surv. 20:1–25.

Nutting, W. L., and A. B. Gurney. 1961. A new earwig in the genus *Vostox* (Dermaptera: Labiidae) from the southwestern United States and Mexico. Psyche 68:46–52.

Popham, E. G. 1959. The anatomy in relation to feeding habits of *Forficula auricularia* L., and other Dermaptera. Proc. Zool. Soc. Lond. 133:251–300.

Popham, E. G. 1961. On the systematic position of *Hemimerus* Walker—a case for ordinal status. Proc. Roy. Ent. Soc. Lond. (B) 30:19–25.

Popham, E. G. 1965. The functional morphology of the reproductive organs of the common earwig (*Forficula auricularia*) and other Dermaptera, with reference to the natural classification of the order. J. Zool. Lond. 146:1–43.

Richards, O. W., and R. G. Davies. 1977. Imm's general textbook of entomology. 10th ed. London: Chapman and Hall. 2 vols. pp. 1–1354.

Sakai, S. 1982. A new proposed classification of the Dermaptera with special reference to the Check List of the Dermaptera of the World. Bull. Daito Bunka Univ. 20:1–108.

Steinmann, H. 1975. Suprageneric classification of Dermaptera. Acta Zool. Acad. Hung. 21:195–220.

Terry, T. W. 1905. Leafhoppers and their natural enemies, V. Forficulidae. Hawaiian Sugar Planter's Assoc. Exp. Sta. Div. Ent. Bull. 1 (5):163–74.

Tillyard, J. 1925. Insects in relation to the New Zealand food supply. Mid-Pacific Mag. 29:671–72.

Williams, F. X. 1931. Handbook of the insects and other invertebrates of Hawaiian sugarcane fields. Hawaiian Sugar Planters' Assoc. Expt. Sta. Honolulu. 400 pp.

Zimmerman, E. C. 1948. Insects of Hawaii. Vol. 2:197–212.

Order Embiidina (Embioptera) 18

Edward S. Ross
California Academy of Sciences

WEB-SPINNERS

Embiidina, commonly called embiids, or web-spinners, are small, elongate, orthopteroid insects (fig. 18.1). They are essentially tropical, and thus the few species occurring in temperate regions are limited to localities lacking prolonged cold periods. Poor dispersal ability and vulnerability to predators may account for the spotty occurrence of the order in suitable environments.

DIAGNOSIS

Embiids can be recognized at a glance by the large, bulbous, basal tarsomere of each fore leg (fig. 18.1). This unique character holds for all immature stages, as well as adults of both sexes. Except for short dispersal movements of adults, all embiid activities are confined to labyrinths of silk galleries spun with the fore tarsi, and all order-defining characters are adaptations to the physical features of this self-created microenvironment.

BIOLOGY AND ECOLOGY

The silk galleries mainly serve as coverways to reach food that usually consists of outer bark surfaces, dead leaf litter, lichens, and moss. Such protection from predators, desiccation, and excessive moisture is increased when the galleries extend into bark or soil crevices. When a predator is encountered (usually at the periphery of a gallery), an embiid uses the runway as a predetermined escape route, and the most effective movement is rapidly backward.

Embiids are seldom noticed, even by entomologists, except for males of certain species attracted to lights on warm nights. Males of species that disperse diurnally, or lack wings, are rarely encountered. Females are universally apterous, neotenic, and, except for larger size and darker color, they are very similar to their juveniles. The adult male's life is short, but adult females may live at least 6 months.

The spread of a species is limited to the short distances females can walk outside of galleries without being killed by predators. A species' range may be extended, however, as substrate objects are moved about by wind, rafting, and human commerce. Human introduction accounts for the presence of five vigorous, Old World species of the family Oligotomidae in North America and Hawaii.

In regions with summer rainfall, embiid colonies usually are found on and in bark and weathered fence post surfaces. In seasonally arid regions, soil cracks are used as retreats and the galleries are best located by turning stones or by searching leaf litter. A species may range over a variety of microclimates and thus may occur in both subterranean and aboveground habitats.

Embiid eggs (fig. 18.2) are elongate-oval, slightly curved, with a slanted, disclike, rimmed operculum. The eggs are generally laid in a single-layered cluster within the galleries and usually imbedded in a matrix of finely masticated leaf or bark fragments. Some species, however, deposit loose clumps of uncovered eggs in the galleries.

The parent female remains near her eggs (fig. 18.2) and early instar brood, and reacts defensively against approaching predators and other disturbances. After a period of clumping, the young radiate out in their own galleries, which are spun just large enough to accommodate their body size. Most species exhibit very close mutual contact and share a common system of galleries. Such species may be termed subsocial, communal insects, and a colony usually consists of the brood of one female. In certain Old World genera, however, individuals must develop in separate galleries because of mutual antagonism.

Embiids have only one generation per year except in the families Teratembiidae and Oligotomidae, which seem to have continuous, overlapping development of broods. The number of instars is difficult to determine because of the gregarious, gallery-confined life. Development is basically similar to that of other orthopteroid insects.

First instars are delicate miniatures of adult females but have disproportionately large heads, and their antennae have only nine segments. In the second instar, by division of the third segment into four segments, the antennae become twelve-segmented. Female juveniles simply increase in size with each ecdysis, and adults are recognized by their relatively darker pigmentation and the opening of the genital aperture, which often has associated sclerotic plates and small flaps. Apterism through neoteny obviously facilitates quick movement in the galleries. The trend also occurs in males of many species on various evolutionary lines.

Male juveniles destined to develop into apterous adults (*Anisembia texana* may have apterous and alate males in the same colony) develop in the manner of females. However, they are recognized by their smaller size and more slender, flattened bodies. Apterous adult males are neotenic only in lacking wings; their head, genitalic structures and pigmentation are fully developed.

Juveniles destined to become alate males will, when about half grown, develop small wing pads on the caudal angles of the meso- and metathoracic scuta (fig. 18.3a). These pads gradually enlarge and elongate (fig. 18.3b). Finally, in the penultimate instar, the adult wing venation is evidenced by rows of macrotrichiae (fig. 18.3c). Just prior to the final moult, the pads greatly thicken. At this time the cerci of the adult can be seen retracting within the cuticle of the juvenile cerci.

Soon after each moult an embiid eats its cast skin. In the case of adult males this may be the only ingestion and thus a lack of adult nutrition may in part explain their short life.

Ideally, identifications should be based on characters of adult males but all too often the collector kills and preserves only juvenile specimens. Adult males usually are present in the colonies for a limited period each year; therefore, it is desirable to rear field-encountered embiids to maturity.

COMMENTS

Because living plant tissue is seldom eaten, embiids have little or no economic importance. However, there is occasional public concern when galleries are conspicuous on the trunks of shade trees and around foundations.

Culturing

A culture container can be a straight-sided glass or plastic vial, ideally about one inch in diameter and four to six inches long, stoppered with a firm cotton plug. I prefer a closure consisting of a slip-on polyethylene cap with a large circular vent punched out of its center. The embiids are confined by a disc of fine-mesh brass or aluminum wire screen cut to fit like a gasket within the cap.

When a colony is encountered, fill the vial half full of habitat material before introducing the delicate embiids. Such media should be firmly pushed into the vial to eliminate the lethal consequences of rattling during transport of the culture. Suitable habitat material consists of outer bark flakes or crumpled dead leaves, which serve as food as well as support for the new galleries that will be immediately spun.

While collecting, it may be necessary to pry open bark crevices with a tool such as a screw driver. Soil-inhabiting species must often be excavated if the season is dry. They are introduced into the vial by trapping them in sections of field galleries or by teasing them in with forceps. Avoid picking them up with forceps, but, if necessary, grip them in the abdominal region.

Captive embiids will quickly spin new galleries and continue their development. If a culture is well stocked and free of disease, it is possible to rear an indefinite number of generations. Of course, as a culture grows it must be transferred into larger containers; some of my cultures, at times started with a single gravid female, eventually had to be kept in wide-mouthed gallon jars.

Routine maintenance consists of pipetting in water to keep the media slightly moist. A bottom zone will become an uninhabitable damp wick and most embiid activity will be confined to upper levels. As the medium becomes thoroughly webbed and crowded, a sheet of fresh lettuce can periodically be added to the surface as it is consumed. This replenishable, fresh, diet supplement sustains larger populations than otherwise would be possible in a small container.

When enlarging a culture, it is not necessary to use original habitat food. Regardless of their regional origin or identity, embiids will eat any dead bark or dead leaves, later supplemented by fresh lettuce. I have adopted dead, fragmented live-oak leaves as a standard medium for all species.

Adult males and females may be collected as they venture into upper galleries of the culture. They may be caught by blocking the galleries behind them with a hooked teasing needle. Transfer the entangled specimens into an empty dry watchglass and allow them to disengage themselves from the silk before killing and preservation in 70% alcohol. For advanced research, cleared slide preparations of representative adult males are required. These should be made by a specialist if the species is likely to be undescribed.

Recognition of Taxa.

Because of the universal physical uniformity of the galleries, there has been no adaptive radiation of body form and general facies, even between totally unrelated families. However, the great age and taxonomic differentiation of the order is evident in adult males, especially in the complex diversification of their terminalia. Juveniles and adult females throughout the order are difficult to identify without associated males, but a person familiar with the few species in any local fauna may learn to make accurate sight identifications of at least some species on the basis of size, color, and general appearance.

The following key clearly distinguishes *Haploembia* juveniles. However, it is unavoidably vague for recognition of juveniles of other genera. Couplets based on indefinite characters, such as size and coloration, must be supplemented by consideration of geographic range and reference to the discussion of families.

KEY TO IMMATURE EMBIIDINA OF THE UNITED STATES

1. All juveniles and adults with 2 papillae ("bladders") on ventral surface of basitarsus of hind leg (fig. 18.5). Parthenogenetic except in a limited region just south of San Francisco, Calif. .... (*Haploembia solieri*) ***Oligotomidae*** (p. 183)

1′. All individuals with 1 hind basitarsal papilla (fig.18.4). Never parthenogenetic 2

2(1′). General color of juveniles pink to rust-red, more reddish with age; occasionally with whitish coxae and pale zones between thoracic segments. Limited to regions west of (or near) Mississippi River .... ***Anisembiidae*** (p. 181) 3

2′. General color of juveniles uniformly pale tan to brownish. Ranging across southern and southwestern U.S.A. 4

3(2). Occurring east of Continental Divide. Juveniles robust and moderately large (may exceed 7 mm in length). Maturing juveniles with pale mid and hind coxae and a light band between each thoracic segment. Head of adults golden .... *Anisembia texana* (p. 181)

3′. Occurring west of Continental Divide (except in Sacramento Mts., New Mexico). Small (less than 7 mm in length), slender; late instar juveniles distinctly reddish, without pale coxae and thoracic bands. Head of adult females red, that of adult males black .... *Dactylocerca rubra* complex (p. 181)

4(2′). Late instar juveniles moderately large, at least 10 mm in length. Occurring at low elevations across south and southwestern U.S., and in Hawaii. Adult males with wing vein MA (R_{4+5}) unbranched .... (*Oligotoma*) ***Oligotomidae*** (p. 183)

4′. Late instar juveniles small, never more than 6 mm in length. Confined to lowlands of south and southeastern U.S. Adult males with wing vein MA branched .... ***Teratembiidae*** (p. 181)

CLASSIFICATION

Order **EMBIIDINA**
- Anisembiidae (endemic)
- Teratembiidae (=Oligembiidae) (endemic)
- Oligotomidae (introduced)

Anisembiidae

Diagnosis: Living juveniles of this family can be distinguished by their pink to rust-red color, which intensifies with age, and by means of other details mentioned below. Living juveniles of other families are light tan to brownish. The family is also largely confined to regions west of the Mississippi, where it extends southward into South America.

Comments: Only two species, or species complexes, are known from the United States. *Anisembia texana* (Melander) is recorded from western Mississippi, Louisiana, Arkansas, southern Oklahoma, Texas, and into northeastern Mexico. In drier regions it is found under stones, but it also colonizes fence posts and bark of living and dead trees, especially in damper environments. Adults are recognized by their relatively large size (females 10 mm, males 7 mm), dark brown body with pale intersegmental bands on thorax, and golden head. Adult females also have pale mid and hind coxae, a unique character also found in juveniles of both sexes. Adult males have nondentate mandibles and wing vein MA is unbranched. They are often apterous, especially in arid environments.

The family is also represented by species of the *Dactylocerca rubra* complex which occurs from southern California into Arizona, Nevada, Utah and New Mexico, and extends southward into Mexico. Colonies of these rarer species occur under stones in undisturbed environments, especially juniper-grasslands. They are quite small (adult females are 7 mm, males 6 mm), juveniles are unicolorous pink, and females reddish; the reddish adult males have jet-black heads and terminalia and are always alate. Males fly diurnally, whereas those of all other alate U.S. embiids (except *Oligembia melanura*) are nocturnal. A few adult males have been collected by sweeping vegetation, but they are most readily collected in their colonies.

Teratembiidae

Diagnosis: Living juveniles of this family are pale tan in color and similar to early instar juveniles of *Oligotoma* that, in disturbed areas, may occur on the same trees. Unlike *Oligotoma,* which spin larger and more extensive galleries, teratembiids spin small galleries that are usually partially obscured in bark crevices and under flakes. Small embiid galleries found in tree bark in areas with natural vegetation are likely to have been spun by species of this family. *Diradius vandykei,* however, commonly inhabits shade tree bark in southeastern lowlands.

Comments: The five species occurring in the United States cannot be distinguished from one another as juveniles. However, some can be identified according to locality. *Oligembia hubbardi* (Hagen) is apparently confined to Florida and Bimini; but its range overlaps *Diradius vandykei* (Ross) and, in such regions, juveniles of the two species cannot be distinguished; *vandykei,* however, ranges much farther north, along the Gulf coast to the Mississippi delta and along the lowlands of Georgia, South Carolina, and North Carolina. *Diradius caribbeanus* (Ross) appears to be limited to the Florida Keys and Cuba. *Diradius lobatus* (Ross) is known only from the lower Rio Grande of Texas but is widespread in Mexico. *Oligembia melanura* Ross, essentially a Mexican species, ranges as far north as Austin, Texas, and east to the lower Mississippi region of Louisiana where it occurs on baldcypress bark.

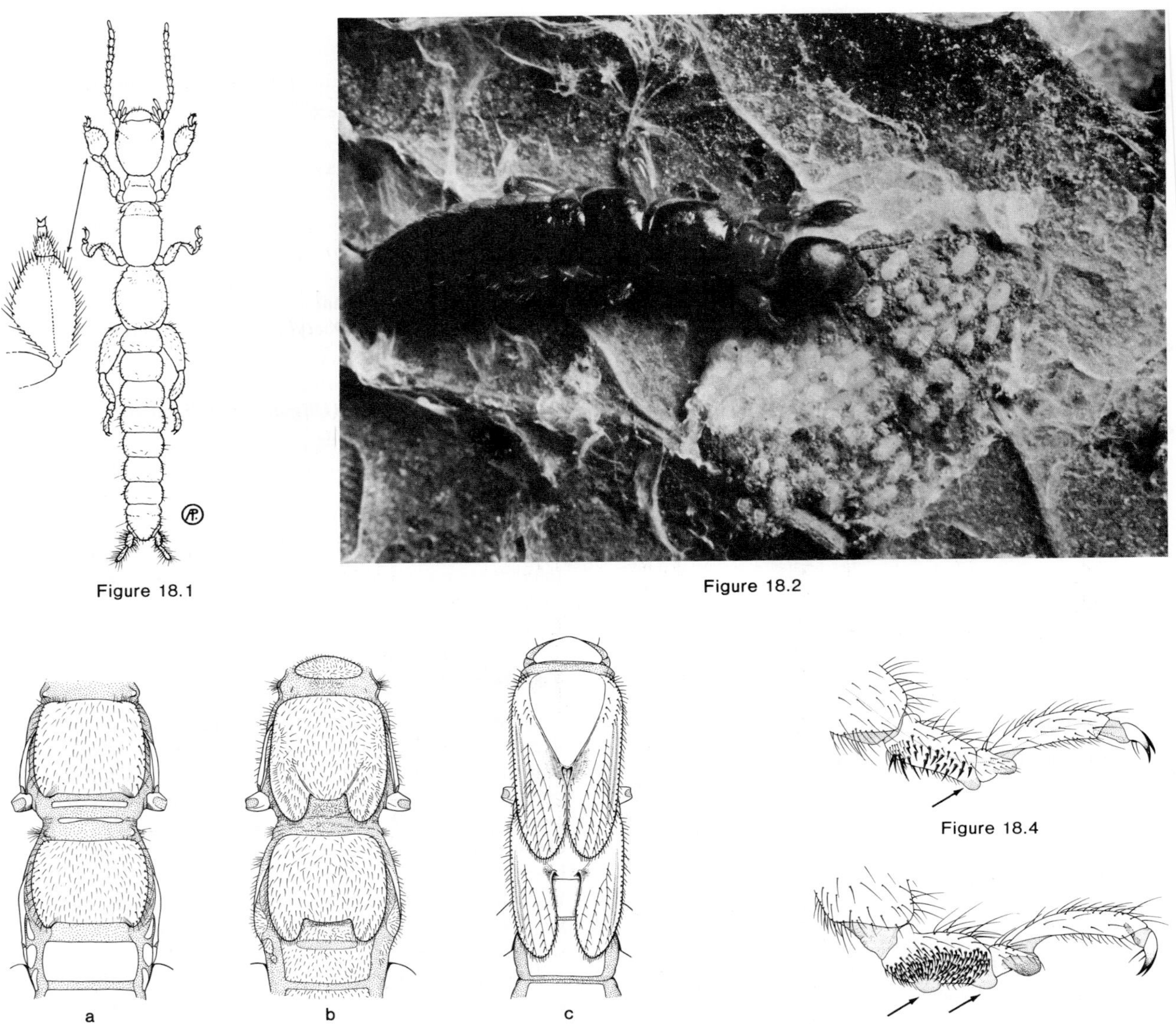

Figure 18.1

Figure 18.2

Figure 18.3

Figure 18.4

Figure 18.5

Figure 18.1. *Haploembia solieri* (Rambur) (Oligotomidae). Juvenile. (From Peterson 1948.)

Figure 18.2. Adult female embiid guarding eggs. Females are always wingless.

Figure 18.3a-c. *Oligotoma saundersii* (Westwood) (Oligotomidae). Wing pad development of the last 3 juvenile stages of male.

Figure 18.4. Single basitarsal papilla ("bladder") on hind tarsus of *Oligotoma nigra* (Hagen) (Oligotomidae).

Figure 18.5. Two basitarsal papillae on hind tarsus of *Haploembia solieri* (Rambur) (Oligotomidae).

Oligotomidae

Figures 18.1-18.5

Diagnosis: Living juveniles of this Old World family tend to be light tan to brownish in color and are difficult to distinguish from those of teratembiids in their early instars. However, late stage juveniles of oligotomids become much larger than those of Teratembiidae, which never exceed 6 mm. The much larger diameter and extent of the silk galleries of oligotomids is the best clue to family identification in the field. The galleries of teratembiids are small and largely obscured by bark flakes and crevices.

Comments: Four introduced species are known from the United States. *Haploembia solieri* (Rambur), which never has alate males and is usually parthenogenetic, can be recognized in all stages by the presence of a second, or medial, papilla on the ventral surface of the hind basitarsus (fig. 18.5). The bisexual form has recently been found near San Francisco, apparently recently introduced.

All stages of the other continental U.S. oligotomids, *Oligotoma saundersii* (Westwood) and *Oligotoma nigra* (Hagen), have only one papilla (fig. 18.4), and adult males are present and always alate. Juveniles of these two species are indistinguishable but can, except in an overlap area around San Antonio, Texas, be identified by distribution; *saundersii* being confined to southeastern United States, and *nigra* to the Southwest, including southern California.

Haploembia solieri almost always colonizes soil crevices. *Oligotoma saundersii* usually is found in or on bark surfaces of shade trees, while *Oligotoma nigra* is generally found in or near the ground, usually under stones, but also in the bases of cultivated palms.

In Hawaii, *Aposthonia oceania* (Ross), an introduction by ancient Polynesians throughout Oceania, occurs in trail banks away from settled areas, and *Oligotoma saundersii* is common in disturbed lowland environments. Juveniles and adult females of *A. oceania* have pale mid and hind coxae and a distinct pale band between the meso- and metathorax, whereas these areas are dark in *saundersii*. Adult males have distinct terminalia, *saundersii* being characterized by a sickle-shaped hook beneath the hypandrium process. This hook is very short in *A. oceania*.

BIBLIOGRAPHY

Hepner, L. 1948. Note on Embioptera in Kansas. J. Kansas Ent. Soc. 21:35. [*Oligotoma saundersii* record]

Peterson, A. 1948. Larvae of Insects. Part I. Lepidoptera, Hymenoptera. 315 pp. Printed by Edwards Bros., Ann Arbor, Mich., for the author.

Ross, E. S. 1940. A revision of the Embioptera of North America. Ann. Ent. Soc. Amer. 33:629–76.

Ross, E. S. 1944. A revision of the Embioptera, or web-spinners of the New World. Proc. U.S. Nat. Mus. 94:401–504.

Ross, E. S. 1952. The identity of *Teratembia geniculata* Krauss, and a new status for the family Teratembiidae (Embioptera). Wasmann J. Biol. 10:225–34.

Ross, E. S. 1957. The Embioptera of California. Bull. Calif. Insect Surv. 6:51–57.

Ross, E. S. 1970. Biosystematics of the Embioptera. Ann. Rev. Ent. 15:157–72.

Ross, E. S. 1984. A synopsis of the Embiidina of the United States. Proc. Ent. Soc. Wash. 86(1):82–93.

Sanderson, M. W. 1941. The order Embioptera new to Arkansas. J. Kansas Ent. Soc. 14:60.

Order Zoraptera

19

Garland T. Riegel
Eastern Illinois University

ZORAPTERANS

This small orthopteroid group has one family, Zorotypidae, containing a single genus, *Zorotypus*. The only common name suggested for Zoraptera is *Bodenläuse* (Weidner 1970). It does not seem particularly appropriate.

DIAGNOSIS AND DESCRIPTION

The ordinal (and familial) diagnosis given by Gurney (1974) is as follows: "mandibulate mouthparts, maxillary palpi five-segmented, labial palpi three-segmented; nine-segmented moniliform antennae; legs adapted for running, tarsi two-segmented; cerci present, one-segmented; wings present or absent, shed by indefinite basal fracture, venation simple, hind wing smaller than front wing. Size small, body length ranging from about 1.5 mm to 2.5 mm. There are two types of individuals, both of which have sexually productive males and females: (1) wingless, pale general color, eyes absent or rudimentary (this type most numerous); (2) winged, dark pigmented, fully developed compound eyes and ocelli (these individuals less numerous)." Some apterous males of some species have a cephalic fontanelle. P. H. Darst and C. M. Cooper (*in litt.*) first noted the fontanelle on males of *Zorotypus hubbardi* Caudell. It has not been observed on *Z. snyderi* Caudell.

Zoraptera are gregarious. The best way to distinguish the juveniles is by association with the easily identified adult males. Adult male *hubbardi* have a dark band on the ventral side of the abdomen near the caudal end. Adult male *snyderi* have the easily seen "horseshoe-shaped" plate illustrated by Gurney (1938).

The juveniles of *snyderi* tend to be somewhat larger, darker, and more setaceous than those of *hubbardi,* but for the purposes of identification, and in the usual absence of comparison material, these facts are of little value. The use of femoral spines to distinguish juveniles of the two species is not too reliable because of considerable variation as Gurney (1938) notes.

Zorotypus hubbardi appears to have four juvenile instars (Riegel and Eytalis 1974). The same may be true for other species. The first instar (fig. 19.2) is approximately 1 mm long and has 8-segmented antennae. Instars 2 and 3 also have eight-segmented antennae, and this count sometimes occurs in the fourth stage. Usually however, fully grown juveniles of either the apterous form (fig. 19.3) or the alate form (fig. 19.4) will have nine-segmented antennae as do the adults.

Eggs of zorapterans are relatively large. All known eggs have a reticulated chorion (fig. 19.1).

BIOLOGY AND ECOLOGY

In the south these insects are normally found in rotting logs and stumps and under loose bark of trees. Farther north they have been found in logs and stumps but are more easily taken in decaying sawdust piles at sawmills. At the proper stage of decay these heaps are quite warm, allowing activity all winter. Zorapterans primarily eat fungus spores and hyphae but also are known to scavenge their own dead, mites, and probably other small organisms.

Ecological requirements are similar to those of subterranean termites and some ants. They need warmth, moisture, and a supply of fungus-infested rotting wood. Besides termites and ants, Zoraptera are often found associated with Collembola, enicocephalids, immatures of wax filament-bearing fulgoroids, mites, pseudoscorpions, small fungus-feeding beetles, and occasionally *Campodea.*

COMMENTS

Zoraptera are most easily collected with an aspirator, and then transferred to alcohol. If collecting in a sawdust pile, inspect the underside of half-buried slabs of wood and pieces of bark. Specimens can be studied in alcohol, on microscope slides, or prepared for SEM study.

Of the approximately 25 species known, three are found in the United States, one of these in Hawaii. One species has been described from Mexico, and several from the West Indies, Central and South America. The rest are found in various parts of the eastern hemisphere, including Tibet (Hwang 1974, 1976).

The Hawaiian species (*Zorotypus swezeyi* Caudell) is known from eight adult specimens. Presumably others taken on Oahu or Kauai would be of this same species. Apparently the juveniles are unknown.

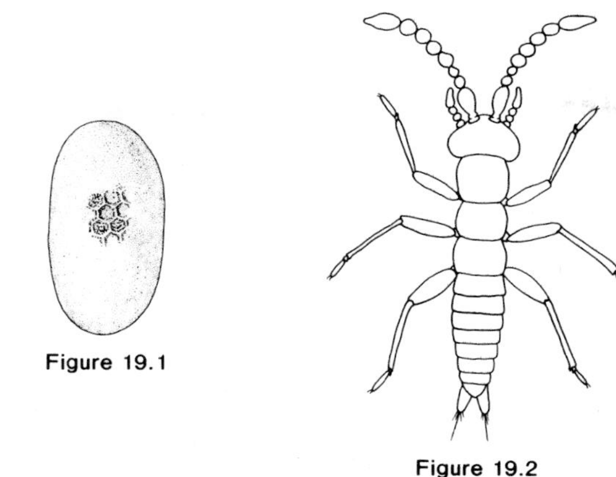

Figure 19.1

Figure 19.2

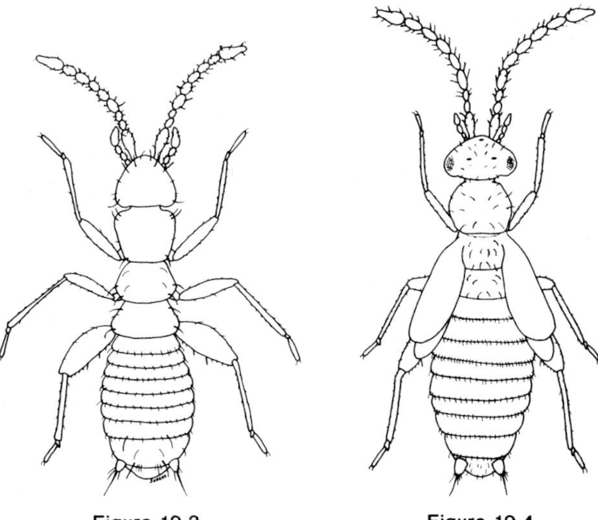

Figure 19.3

Figure 19.4

Figures 19.1–19.4. *Zorotypus hubbardi* Caudell: **(19.1)** egg; **(19.2)** first instar; **(19.3)** last instar apterous juvenile; **(19.4)** last instar alate juvenile.

In the continental United States the two species are *Z. hubbardi* Caudell and *Z. snyderi* Caudell. In the U.S. *snyderi* has been taken only in Florida but also is known from Jamaica. *Zorotypus hubbardi* is found over most of the deciduous forest region in the southeastern portion of the country. This is the area bounded by Florida north to southern Pennsylvania, west to southern Iowa, and south through eastern Kansas, Oklahoma and Texas.

Zoraptera are of no direct economic importance, although there is a record of *hubbardi* from fungus-damaged timbers of a house in Georgia (R. T. Franklin, *in litt.*).

The recent papers cited below, plus the bibliographies of the older works listed, will cover the world literature on Zoraptera.

BIBLIOGRAPHY

Cooper, C. M. 1976. New Alabama records for *Zorotypus hubbardi* Caudell (Zoraptera). Ent. News 87 (1 & 2):57.

Darst, P. H. et al. 1974. New Mississippi records for *Zorotypus hubbardi* (Zoraptera). Ent. News 85 (7 & 8):224.

Gurney, A. B. 1938. A synopsis of the order Zoraptera, with notes on the biology of *Zorotypus hubbardi* Caudell. Proc. Ent. Soc. Wash. 40 (3):57–87.

Gurney, A. B. 1974. Class Insecta, Order Zoraptera. *In* Coaton, W. G. H. (ed.). Status of the taxonomy of the Hexapoda of Southern Africa. RSA Dept. Agr. Tech. Serv., Ent. Mem. 38:32–34.

Hwang, F. 1974. *Zorotypus chinensis,* a new species from China. Acta Ent. Sinica 17 (4):423–27.

Hwang, F. 1976. A new species of Zoraptera. Acta Ent. Sinica 19 (2):225–27.

Mignot, E. C., and J. O. Sillings. 1969. Winter occurrence of *Zorotypus hubbardi* (Zoraptera) in Tippecanoe County, Indiana. Mich. Ent. 2(3 & 4):72–73.

Mizell, R. F., and T. E. Nebeker. 1976. New Mississippi records of *Zorotypus hubbardi* Caudell (Zoraptera). Ent. News 87 (1 & 2): 58.

New, T. R. 1978. Notes on Neotropical Zoraptera, with descriptions of two new species. Syst. Ent. 3:361–70.

Riegel, G. T. 1963. The distribution of *Zorotypus hubbardi* (Zoraptera). Ann. Ent. Soc. Amer. 56(6):744–47.

Riegel, G. T. 1969. More Zoraptera records. Proc. North Central Branch, Ent. Soc. Amer. 23 (2):125–26.

Riegel, G. T. 1978. Feature photograph. Ann. Ent. Soc. Amer. 71 (6):iii.

Riegel, G. T., and S. J. Eytalis. 1974. Life history studies on Zoraptera. Proc. North Central Br., Ent. Soc. Amer. 29:106–107.

Shetlar, D. J. 1967 (1969). New distribution records of Zoraptera (Insecta) in Oklahoma. Proc. Okla. Acad. Sci. 48:104–105.

Shetlar, D. J. 1978(1979). Biological observations on *Zorotypus hubbardi* Caudell (Zoraptera). Ent. News 89 (9 & 10):217–23.

Slifer, E. H., and S. S. Sekhon. 1978. Structures on the antennal flagellum of *Zorotypus hubbardi* (Insecta, Zoraptera). Notulae Naturae (Phila.) No. 453:1–8.

Weidner, H. 1970. 15. Ordnung Zoraptera (Bodenläuse). *In* Helmeke, J.-G. et al. (eds.). Handb. Zool., Berlin 4 (2) 2/15:1–12.

Weidner, H. 1976. Eine neue *Zorotypus*-art von den Galapagosinseln, *Zorotypus lelupi* sp. n. (Zoraptera). Mission Zoologique Belge aux Iles Galapagos et en Ecuador (N. et J. Leleup 1964–1965) 3:161–76.

Order Plecoptera

20

Richard W. Baumann
Brigham Young University

STONEFLIES

Immature stoneflies are usually found in clear, cold, unpolluted rivers, streams, and lakes with high levels of dissolved oxygen, although a few species are able to survive in habitats that warm up or dry up during part of the year. The special names "salmonfly" and "troutfly" for adults and "hellgrammite" for larvae (often called "nymphs" or "naiads") have been used by anglers for certain species in different parts of North America, although "hellgrammite" is commonly used for the larvae of Corydalidae (Megaloptera).

DIAGNOSIS

The combination of two long cerci and two tarsal claws distinguishes stonefly larvae from those of other aquatic orders.

BIOLOGY AND ECOLOGY

Larvae are herbivores, detritivores, or carnivores, feeding on aquatic plants, algae or detritus, or on other insects and small animals. The larvae of herbivorous families are rounded or robust (figs. 20.18–20.23), whereas those of carnivorous families are dorsoventrally flattened (figs. 20.15–20.17). Larvae go through 12–24 instars, depending on species, sex, size, and habitat conditions. Most North American species are univoltine, but the larger species commonly take two or three years to complete development.

DESCRIPTION

Stonefly larvae have a distinct head, thorax, and abdomen composed of ten segments (fig. 20.1). Long, filiform antennae and cerci are completely developed in all families. Well-developed legs composed of a coxa, trochanter, femur, tibia and three tarsal segments are present (fig. 20.2). Gills arise individually or are tufted or branched, or absent in some species; they arise on various parts of the body, including the neck, thorax, leg bases, abdomen, and anal area. Wing pads are not evident in earlier instars, appearing in the middle instars, and becoming conspicuous in late instars.

COMMENTS

Since most species were described from adults, study of the larvae has progressed more slowly. Beginning with the Claassen monograph in 1931, several studies have concentrated on the larvae of a certain taxon or geographic region (Frison 1935, Ricker 1952, Hitchcock 1974, Stark and Gaufin 1976a and b, Baumann et al. 1977, Harper and Hynes 1971a–d, Szczytko and Stewart 1979, Fullington and Stewart 1980). However, much more work needs to be done because the larvae of many genera are still undescribed. The classification used here is that of Illies (1966) and Zwick (1973). Harper and Stewart (1984) provide a key to North American genera.

TECHNIQUES

Larvae are most easily collected by dislodging substrates and capturing the drifting organisms in a net or on a screen. Specimens are best preserved in 75% ethyl alcohol.

CLASSIFICATION

Order **PLECOPTERA**
- Suborder Arctoperlaria, Northern Hemisphere families
  - Group Euholognatha, true herbivores
    - Nemouridae, brown stoneflies
    - Taeniopterygidae, spring stoneflies
    - Capniidae, winter stoneflies
    - Leuctridae, rolled-winged stoneflies
  - Group Systellognatha, carnivores, omnivores and herbivores
    - Peltoperlidae, roachlike stoneflies
    - Pteronarcyidae, giant stoneflies
    - Perlodidae, patterned stoneflies
    - Perlidae, common stoneflies
    - Chloroperlidae, yellow or green stoneflies

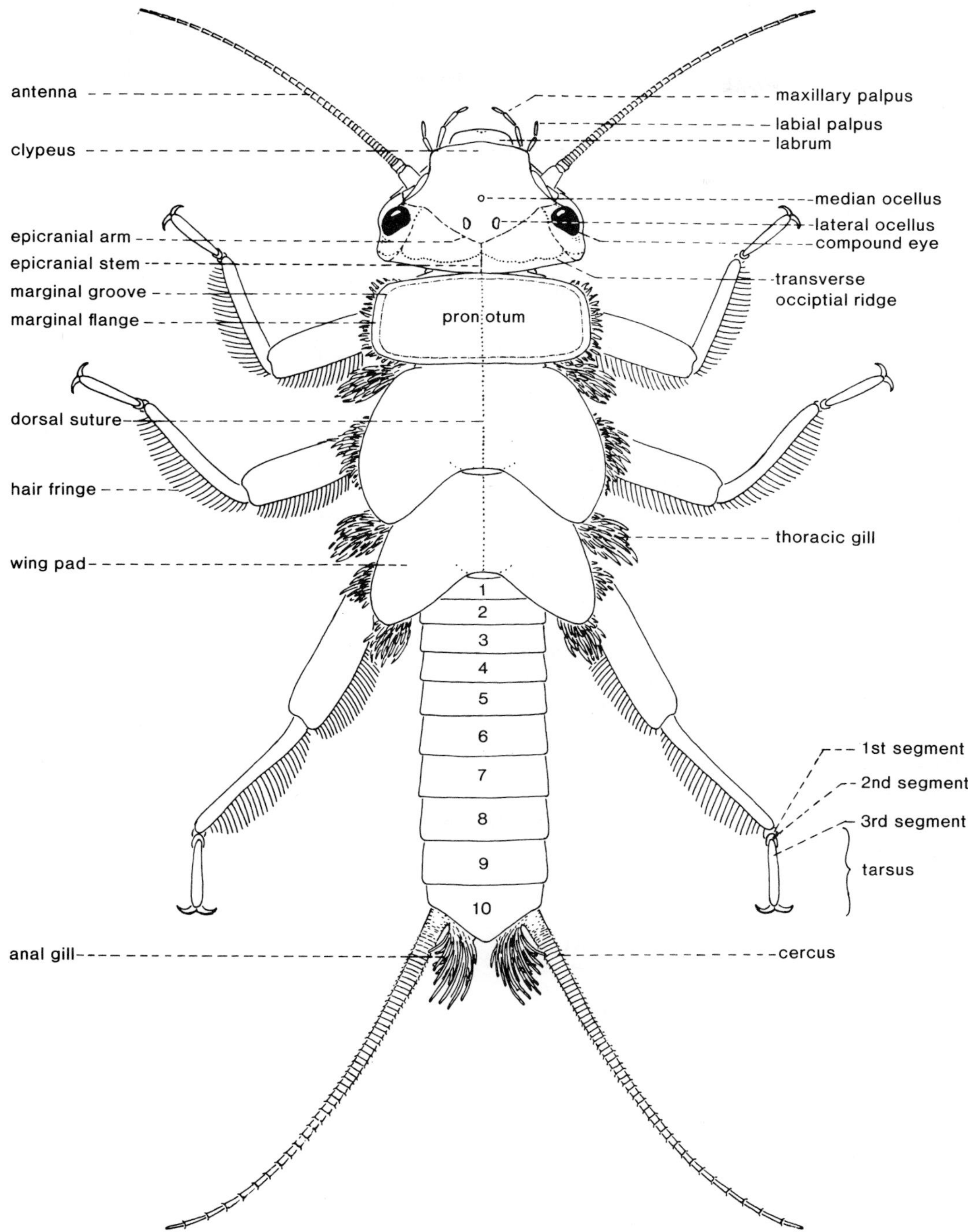

Figure 20.1. Plecoptera larval diagram, dorsal. (From Baumann et al. 1977.)

KEY TO FAMILIES OF MATURE PLECOPTERA LARVAE

1. Branched gills present on abdominal sterna 1, 2 and sometimes 3 (fig. 20.9) .... *Pteronarcyidae* (p. 191)

1′. Branched gills absent from anterior abdominal sterna 2

2(1′). Head strongly depressed (fig. 20.18); thoracic sterna overlapping (fig. 20.12) .... *Peltoperlidae* (p. 192)

2′. Head not depressed; thoracic sterna not overlapping 3

3(2′). Paraglossae and glossae subequal in length (fig. 20.6); dorsal surface concolorous, yellow or brown (figs. 20.18–20.23) 4

3′. Paraglossae longer than glossae (figs. 20.7, 20.8); dorsal surface usually patterned with contrasting light and dark areas (figs. 20.15–20.17) 7

4(3). Extended hind legs exceeding apex of abdomen; wing pads divergent from body axis (figs. 20.19, 20.20) 5

4′. Extended hind legs not exceeding apex of abdomen; wing pads nearly parallel to body axis (figs. 20.22, 20.23) 6

5(4). Second tarsal segment shorter than first (fig. 20.3); posterior margin of ninth sternum not produced (fig. 20.10) .... *Nemouridae* (p. 190)

5′. Second tarsal segment as long as first (fig. 20.2); posterior margin of ninth sternum produced, forming rounded plate (fig. 20.11) .... *Taeniopterygidae* (p. 190)

6(4′). Plane of dorsal surface of front wing pads level (fig. 20.22); abdominal segments 8 and 9 divided by membranous fold laterally (fig. 20.5) .... *Capniidae* (p. 190)

6′. Plane of dorsal surface of front wing pads tilted downward laterally (fig. 20.23); abdominal segments 8 and 9 not divided by membranous fold laterally (fig. 20.4) .... *Leuctridae* (p. 191)

7(3′). Extended hind legs not exceeding apex on abdomen; cerci not more than three-fourths as long as abdomen; wing pads rounded laterally (fig. 20.17) .... *Chloroperlidae* (p. 194)

7′. Extended hind legs exceeding apex of abdomen; cerci more than three-fourths as long as abdomen; wing pads nearly straight laterally (figs. 20.15, 20.16) 8

8(7′). Branched gills present on venter of thorax (fig. 20.14); anal gills often present (fig. 20.15); paraglossae rounded at apex (fig. 20.8) .... *Perlidae* (p. 192)

8′. Branched gills absent from thorax, anal gills absent; paraglossae pointed at apex (fig. 20.7) .... *Perlodidae* (p. 192)

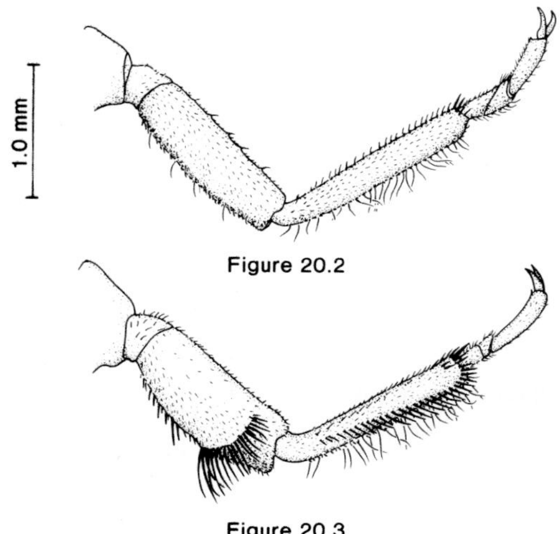

Figure 20.2–20.3. Right fore leg: **(20.2)** *Taeniopteryx* (Taeniopterygidae); **(20.3)** *Zapada* (Nemouridae).

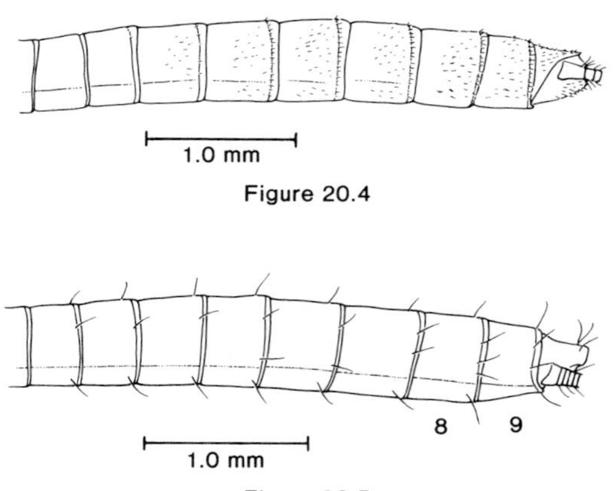

Figures 20.4–20.5. Abdomen, lateral: **(20.4)** *Leuctra* (Leuctridae); **(20.5)** *Capnia* (Capniidae).

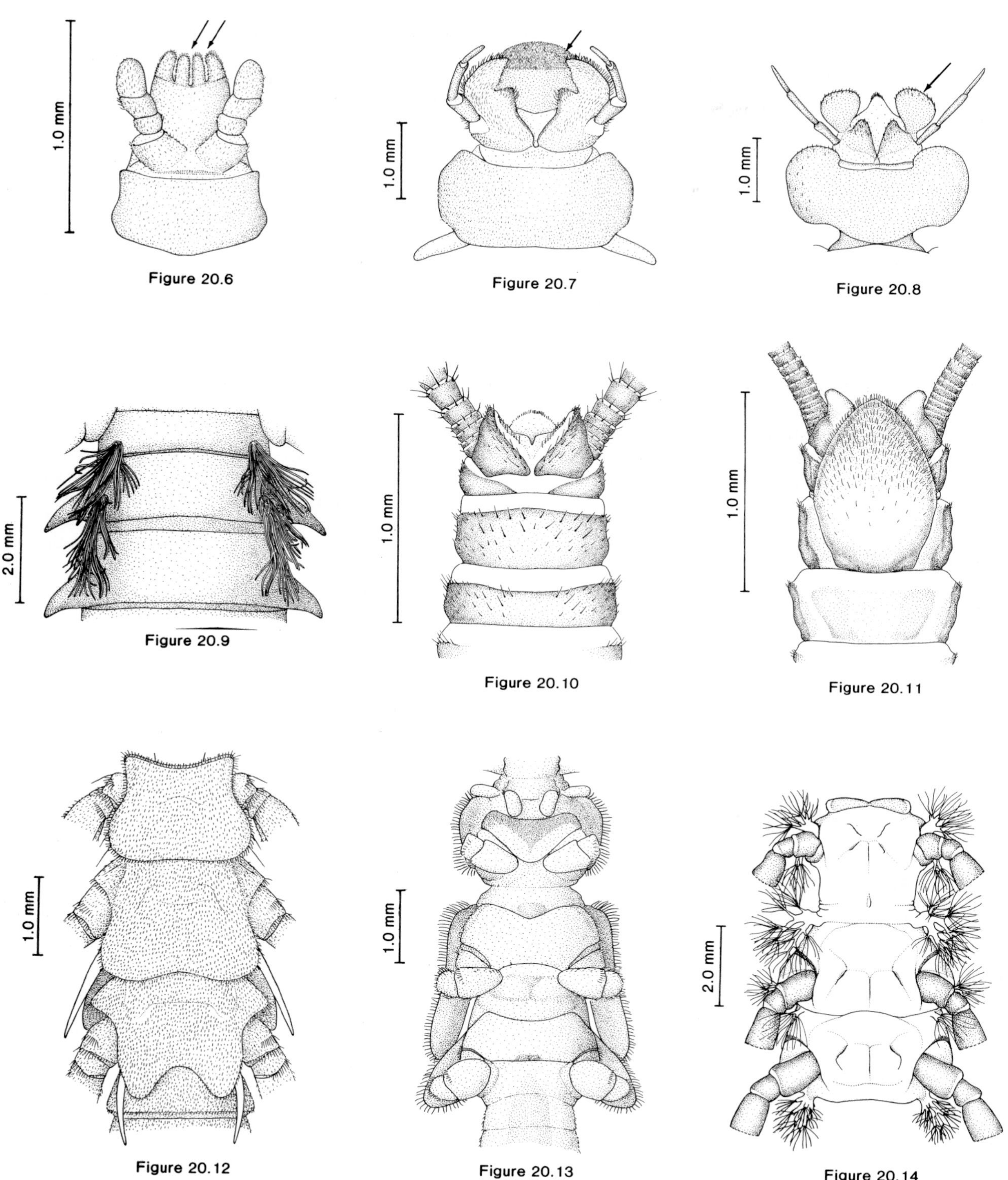

Figures 20.6–20.8. Labium: **(20.6)** *Zapada* (Nemouridae); **(20.7)** *Isogenoides* (Perlodidae); **(20.8)** *Acroneuria* (Perlidae).

Figure 20.9. Anterior abdominal sterna, *Pteronarcys* (Pteronarcyidae).

Figures 20.10–20.11. Posterior abdominal sterna: **(20.10)** *Zapada* (Nemouridae); **(20.11)** *Taenionema* (Taeniopterygidae).

Figures 20.12–20.14. Thoracic sterna: **(20.12)** *Peltoperla* (Peltoperlidae); **(20.13)** *Zapada* (Nemouridae); **(20.14)** *Acroneuria* (Perlidae).

Nemouridae
Brown Stoneflies

Figures 20.3, 20.6, 20.10, 20.13, 20.19

Relationships and Diagnosis: Nemouridae superficially resemble the Taeniopterygidae (fig. 20.20) in their stout, robust general shape. They are also characterized by well-developed spines on the legs, thorax, and abdomen.

They can be separated from Taeniopterygidae by the difference in size ratio of the tarsal segments (fig. 20.3) and by the absence of a ventral plate on sternum nine in mature larvae (fig. 20.10). They can be distinguished from Capniidae and Leuctridae because their wing pads are divergent and not parallel to the body axis.

Biology and Ecology: Nemouridae feed on vegetative material and detritus and occur in a wide variety of habitats from lakes to rivers and springs. Most species emerge in spring or summer but some emerge in fall or winter. Some, like *Zapada cinctipes* (Banks), are known to emerge during all twelve months in some systems (Hales and Gaufin 1971).

Description: All larvae brown and small to medium (5–20 mm). Head bearing three ocelli and the genera *Amphinemura, Malenka, Visoka,* and *Zapada* possessing gills in neck region (fig. 20.13). Prothorax often fringed with stout spines, as are the posterior thoracic segments. Mature larvae with wing pads that diverge greatly from the body axis (fig. 20.19). Legs of species in some genera with distinctive patterns of spines and/or hairs (fig. 20.3). The cerci also bear spines.

Comments: Nemouridae once included the other three families of Euholognatha: Taeniopterygidae, Capniidae, and Leuctridae. They were separated by Illies (1966) who also raised the subgenera to genera. The family was revised at the generic level by Baumann (1975).

Species of Nemouridae are widely distributed in North America. They occur from coast to coast and from the Arctic as far south as Mexico City.

Selected Bibliography

Baumann 1975.
Hales and Gaufin 1971.
Harper and Hynes 1971d.
Lemkuhl 1971.
Ricker 1952.

Taeniopterygidae
Spring Stoneflies

Figures 20.2, 20.11, 20.20

Relationships and Diagnosis: Larvae of this family are always brown. They are similar to the other Euholognatha, especially the Nemouridae (fig. 20.19).

They can be differentiated by segment 2 of each tarsus, which is as large or larger than segment 1 (fig. 20.2). In the other related families, the second tarsal segment is smaller than the first (fig. 20.3). Mature larvae possess a large rounded plate on sternum 9 (fig. 20.11).

Biology and Ecology: Taeniopterygidae are usually found in flowing water of relatively high quality where they feed on detritus or vegetative materials. However, the genera *Taeniopteryx* and *Oemopteryx* contain species that occur in large rivers of marginal quality. In fact, one species, *Taeniopteryx lonicera* Ricker and Ross, occurs in bayous along the Gulf Plain. Larvae of some species move down into the hyporheic zone during part of their life cycle to escape poor environmental conditions. For this reason, larvae are often not collected except near the time of adult emergence.

Description: Larvae vary in length from 10–20 mm. Antennae and cerci of most species quite long when compared to those of many families. Morphology of head and mouthparts typical. However, thorax relatively naked of spines when compared to the Nemouridae. Wing pads divergent from body axis (fig. 20.20). One genus, *Taeniopteryx,* with a single telescopic gill arising from base of each leg. Abdomen bearing short spines along posterior margin of the terga. Some species have larvae with well developed swimming hairs near apex of cerci (fig. 20.20). Several species in the genus *Taeniopteryx* exhibit a distinctive, light middorsal line whose size and area of occurrence are useful for taxonomic purposes.

Comments: Larvae preserved in alcohol often curl up in a very characteristic fashion that is accentuated by the long antennae and cerci when they overlap.

Taeniopterygidae occur throughout North America, although there are some distinctive eastern and western genera that do not occur on the other half of the continent (Ricker and Ross 1968, 1975). Adults have been implicated in agricultural damage to peach blossoms in the Pacific Northwest (Newcomer 1918).

Selected Bibliography

Fullington and Stewart 1980.
Harper and Hynes 1971c.
Newcomer 1918.
Ricker and Ross 1968, 1975.

Capniidae
Winter Stoneflies

Figures 20.5, 20.22

Relationships and Diagnosis: Capniidae larvae are very similar to Leuctridae larvae (fig. 20.23). Both are small, elongate and somewhat delicate. Their wing pads lie parallel to the long axis of the body and the abdomen is longer than the extended hind legs (fig. 20.22).

Capniidae differ from Leuctridae in the place of attachment of the anterior wingpads, which attach dorsally in Capniidae (fig. 20.22) and dorsolaterally in Leuctridae (fig. 20.23). In addition, the eighth and ninth sterna and terga are divided by membranous lateral folds in Capniidae (fig. 20.5) but not in Leuctridae (fig. 20.4).

Biology and Ecology: Winter stoneflies are detritivores that spend most of their life cycle in the hyporheic zone or possibly developing very quickly after a summer diapause.

They can occur in intermittent streams that do not have surface water for more than six months during the year. They emerge from late fall to early spring, with many species active on the snow. Since the adults are black they are easily spotted against the white background. Several species occur in lakes and other lentic habitats (Lemkuhl 1971).

Description: Larvae usually small (5–15 mm) but members of the genus *Isocapnia* reach lengths of 25 mm or more. Most species yellow or brown as larvae, without a distinctive color pattern. Head well developed, but thorax rather small compared to most other stoneflies. Wing pads well developed in most species, however, almost absent in species where microptery or aptery occur. Thorax, legs, and abdominal region covered by short hairs or spines. Antennae and cerci rather long and about equal length (fig. 20.22).

Comments: This family is the largest in North America, with species that occur from the Arctic to Mexico and in every state and province. One species undergoes its complete life cycle underwater in Lake Tahoe (Jewett 1963, 1965).

Selected Bibliography

Baumann and Gaufin 1970.
Hanson 1946.
Harper and Hynes 1971b.
Jewett 1963, 1965.
Nebeker and Gaufin 1965.
Ross and Ricker 1971.

Leuctridae
Rolled-winged Stoneflies

Figures 20.4, 20.23

Relationships and Diagnosis: This is the sister family of the Capniidae (fig. 20.22). They are difficult to separate in the larval stage even when mature. Lack of a lateral membranous fold on abdominal segments 8 and 9 in the Leuctridae, which the Capniidae possess, has been used to separate these families in the past. This character state works if the specimens are fully mature and well preserved; otherwise it is almost impossible to delineate. However, the fact that the Leuctridae have their fore wing pads attached dorsolaterally (fig. 20.23), instead of dorsally, as in the Capniidae (fig. 20.22), is an excellent means of separation. This is based on the reality that in the Leuctridae the adult wings are indeed attached more laterally, thus giving the appearance that the wings are rolled instead of flat.

Biology and Ecology: Leuctridae are detritivores that occur in small creeks, springs, and seeps. They are not commonly collected as larvae because they are not usually common in the benthic community except around the time of emergence. Their long, thin shape and parallel wing pads make them well suited for burrowing in the hyporheic zone.

Description: Larvae usually white or yellow. Head distinctive and bearing long antennae. Wing pads attached more laterally than in other families. Abdomen long, thin, and nearly twice as long as head and thorax combined in the genus *Paraleuctra*. Cerci long and well developed. Larvae of all genera small and delicate except *Megaleuctra*, where they are rather stout like the Nemouridae.

Comments: Larvae of several genera have not yet been described, primarily because of the very specialized habitats and behavior. The family is widespread, but most genera are regionally restricted.

Selected Bibliography

Harper and Hynes 1971a.

Pteronarcyidae
Giant Stoneflies

Figures 20.9 and 20.21

Relationships and Diagnosis: This primitive family is most closely related to the Peltoperlidae in North America. They were both placed in the Euholognatha (Filipalpia) until the order was completely revised (Zwick 1973). Since both families are functionally herbivores, their mouthparts (glossae and paraglossae) are modified accordingly. This led to them being incorrectly grouped with the true herbivores.

They can be distinguished from other families by their anterior abdominal gills (fig. 20.9), large size, and lateral projections on the thorax and abdomen of most species (fig. 20.21).

Biology and Ecology: Pteronarcyids are detritivores that are usually found in creeks and rivers, with *Pteronarcys californica* Newport being restricted to rivers in western North America. The *Pteronarcella* species and *Pteronarcys princeps* Banks are, however, usually found in small streams in mountainous areas.

The eggs of Pteronarcyidae are rounded, highly sculptured, and possess a large cap. They are heavily sclerotized and distinctively shaped in each species (Knight et al. 1965a).

Description: Pteronarcyidae are the largest stoneflies (15–70 mm). They are dark brown and adults have numerous crossveins in their wings. Head somewhat depressed but not as much as in Peltoperlidae. All thoracic segments large and well developed, often exhibiting lateral projections. Abdominal segments similar in size and shape except number 10, which shields the developing genitalia. Some species such as *Pteronarcys comstocki* Smith have large, laterally produced flanges on most abdominal segments.

Comments: Pteronarcyidae are distributed throughout North America, with two species in eastern Asia. Larvae of all North American species are known. Two genera, *Pteronarcella* and *Pteronarcys*, are presently recognized. Larvae of large *Pteronarcys* species have a multiyear life cycle (Branham and Hathaway 1975). This family includes the salmonflies and troutflies, upon which, when they are emerging in early summer, fish feed voraciously.

Selected Bibliography

Branham and Hathaway 1975.
Nelson and Hanson 1971.
Ricker 1952.

Peltoperlidae
Roachlike Stoneflies

Figures 20.12, 20.18

Relationships and Diagnosis: The Peltoperlidae are quite distinctive and cannot be confused except possibly with small pteronarcyid larvae. They can be distinguished from other North American families by their strongly depressed head (fig. 20.18), overlapping thoracic sterna (fig. 20.12) and roachlike appearance (fig. 20.18).

Biology and Ecology: Members of this family are detritivores that are usually restricted to springs or streams that are heavily influenced by springwater sources. Peltoperlid eggs are different from those of the other North American taxa. They are very flat and limpet-shaped with ornate dorsal sculpturing (Stark and Stewart 1981).

Description: Head strongly depressed and bent downward under prothorax. All thoracic segments large and bearing pairs of flat single gills ventrally (fig. 20.12). Thoracic sterna developed into large, flat plates that overlap each other. Legs short, stout, and covered by fine spines. Mature larvae varying from 5–15 mm.

Comments: This is a small family with six genera recognized from North America. Representatives occur in both eastern and western North America but are absent from the central portion (Stark and Stewart 1981).

Selected Bibliography

Ricker 1952.
Stark and Stewart 1981.

Perlodidae
Patterned Stoneflies

Figures 20.7, 20.16

Relationships and Diagnosis: The Perlodidae have very distinctly marked larvae similar to those in the Perlidae (fig. 20.15). They usually have a light pattern that is imposed on a dark background. The most comprehensive study on the North American species is by Ricker (1952).

They can be separated from the Perlidae by the absence of branched gills on the thorax and abdomen and by the fact that the paraglossae are pointed (fig. 20.7) instead of rounded (fig. 20.8).

Biology and Ecology: Larvae are generally predators that occur in all types of cold, flowing water, and also cold lakes. Some genera such as *Hydroperla* and *Isogenoides* contain species that live in large, rather warm rivers. Their eggs are heavily sclerotized and distinctly shaped (Knight et al. 1965a and b). The taxonomy of the females has been greatly improved since egg characters have been used.

Description: Larvae from 10–50 mm and exhibiting contrasting dark and light patterns on their dorsal surface that range from bands and stripes to almost checkerboard patterns (fig. 20.16). Head, thorax, and abdomen dorsoventrally flattened and inner margin of wing pads diverging laterally (fig. 20.16). All leg pairs relatively long, and extended hind legs reaching beyond apex of abdomen. Gills single, when present, and can occur on submentum, cervix, thorax, and anterior abdominal sterna.

Comments: Although they are large and distinctively patterned, the larvae of some North American species have not been associated with the adults. They are all carnivorous and quite active. Perlodidae, which occur from coast to coast and from Alaska to northern Mexico in the Nearctic region, contains many small genera (Stewart and Stark 1984).

Selected Bibliography

Ricker 1952.
Stewart and Stark 1984.
Szczytko and Stewart 1979.

Perlidae
Common Stoneflies

Figures 20.8, 20.14, 20.15

Relationships and Diagnosis: The Perlidae generally resemble the Perlodidae (fig. 20.16) since they are strongly dorsoventrally flattened and exhibit distinctive dark and light color patterns dorsally (fig. 20.15). They can be distinguished by their dorsal color pattern, branched thoracic and sometimes anal gills, and rounded paraglossae.

Biology and Ecology: Perlidae larvae are predators that occur in running water, ranging from small streams to large rivers. Some species, such as *Hesperoperla pacifica* (Banks) and *Acroneuria abnormis* (Newman), are extremely vagile and are widely distributed in North America in a broad range of lotic habitats.

Description: Mature larvae vary from 20–50 mm. Dorsal portion often covered with small hairs or spines in addition to being contrastingly marked. Head large and mouthparts well developed for predation. Highly branched gills occur on thorax (fig. 20.14) and sometimes between the cerci. A fringe of long hairs found along hind margin of legs of many species (fig. 20.15).

Comments: Perlids are relatively large, and since the adults often fly to lights, they are the stoneflies that are most often collected by general collectors. The larvae of all the large North American species are known, but those of several smaller species in genera such as *Neoperla* and *Perlesta,* still need to be associated. They are often used by anglers for bait or imitated with artificial flies (Schwiebert 1973). The North American species are well known because of recent studies by Stark and Gaufin (1976a and b). They are widespread, with more genera in the East.

Selected Bibliography

Stark and Baumann 1978.
Stark and Gaufin 1976a, 1976b.
Stark and Szczytko 1981.

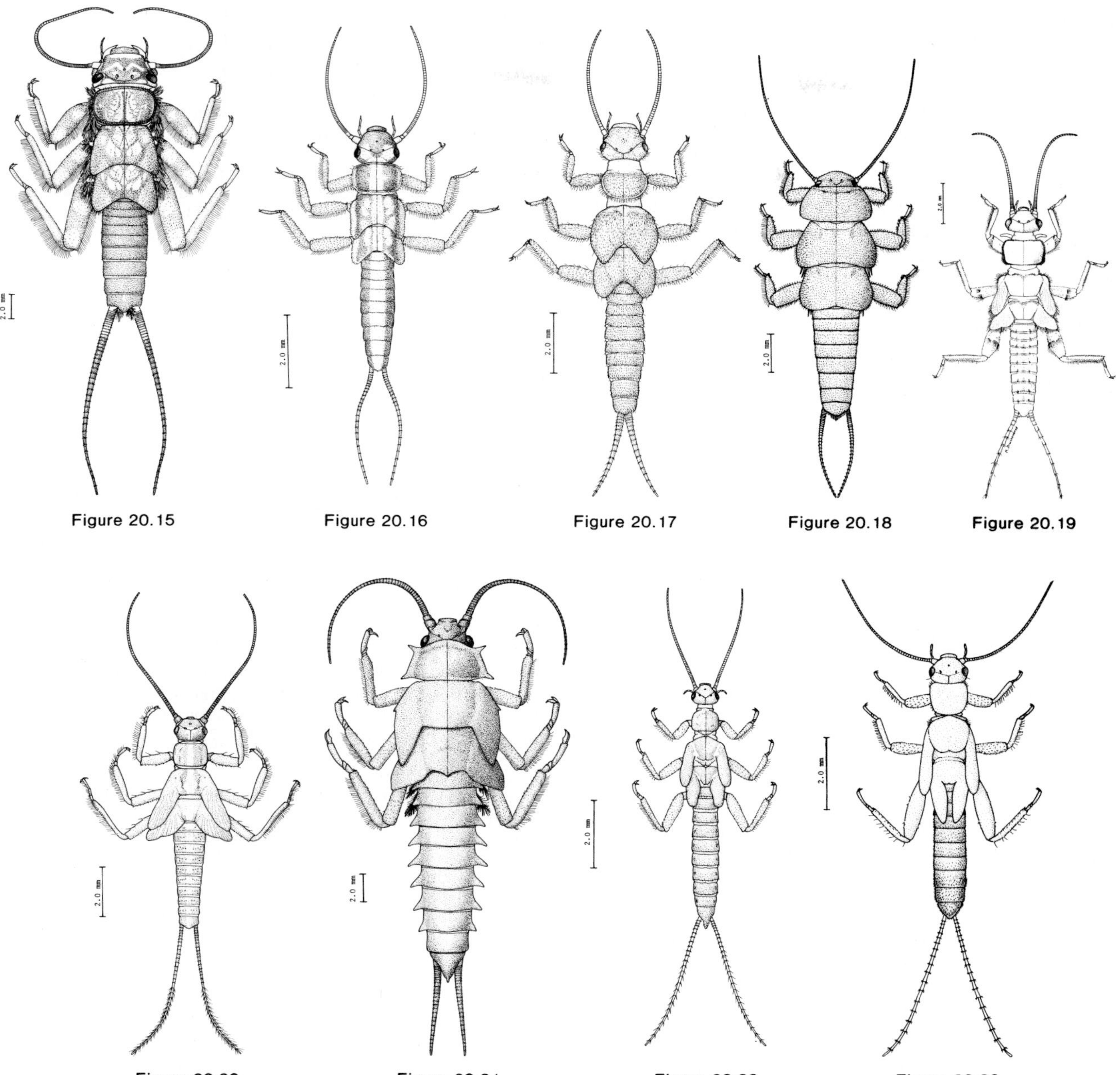

Figures 20.15–20.23. Habitus larvae: **(20.15)** *Claassenia sabulosa* (Perlidae); **(20.16)** *Isoperla fulva* (Perlodidae); **(20.17)** *Sweltsa* (Chloroperlidae); **(20.18)** *Peltoperla* (Peltoperlidae); **(20.19)** *Zapada* (Nemouridae) (courtesy of Smithsonian Institution); **(20.20)** *Taeniopteryx* (Taeniopterygidae) **(20.21)** *Pteronarcys* (Pteronarcyidae); **(20.22)** *Mesocapnia* (Capniidae); **(20.23)** *Leuctra* (Leuctridae).

Chloroperlidae
Yellow or Green Stoneflies

Figure 20.17

Relationships and Diagnosis: The Chloroperlidae are most similar to the Perlidae (fig. 20.15) and Perlodidae (fig. 20.16) in general appearance. They are, however, usually mostly brown, and if they do exhibit a dorsal pattern it is much more subtle than in the other two families. They can be distinguished by the fact that they lack gills, possess wing pads that are rounded laterally, and have very short legs and cerci (fig. 20.17).

Biology and Ecology: This family is quite restricted ecologically, occurring only in clean, cool, running waters. Some species, such as *Paraperla frontalis* (Banks), also live in the wave-washed shores of northern lakes. The larvae are carnivorous when mature but omnivorous when young (Richardson and Gaufin 1971).

Description: Most species quite small, 10–20 mm in length, but some members of the subfamily Paraperlinae reach 30–40 mm. Body parts all well developed and well delineated from each other. Larvae with short legs and cerci, enabling them to move through the substrate in a snakelike manner.

Comments: Chloroperlidae larvae are not often collected because they are small and inconspicuous. Adults, however, are often noticed because they are usually bright pastel yellow or green and are widespread.

Selected Bibliography

Ricker 1952.
Surdick 1985.

BIBLIOGRAPHY

Baumann, R. W. 1973. Studies on Utah stoneflies (Plecoptera). Great Basin Natur. 33:91–108.

Baumann, R. W. 1975. Revision of the stonefly family Nemouridae (Plecoptera): A study of the world fauna at the generic level. Smithsonian Contr. Zool. 211:1–74.

Baumann, R. W., and A. R. Gaufin. 1970. The *Capnia projecta* complex of western North America (Plecoptera: Capniidae). Trans. Amer. Ent. Soc. 96:435–68.

Baumann, R. W., A. R. Gaufin, and R. F. Surdick. 1977. The stoneflies (Plecoptera) of the Rocky Mountains. Mem. Amer. Ent. Soc. 31:1–208.

Branham, J. M., and R. R. Hathaway. 1975. Sexual differences in the growth of *Pteronarcys californica* Newport and *Pteronarcella badia* (Hagen) (Plecoptera). Can. J. Zool. 53:501–506.

Claassen, P. W. 1931. Plecoptera nymphs of America (north of Mexico). Thomas Say Found. Ent. Soc. Amer. 3. 199 pp.

Dosdall, L., and D. M. Lemkuhl. 1979. Stoneflies (Plecoptera) of Saskatchewan. Quaest. Ent. 15:1–116.

Frison, T. H. 1935. The stoneflies, or Plecoptera of Illinois. Bull. Ill. Natur. Hist. Sur. 20:281–471.

Frison, T. H. 1942. Studies of North American Plecoptera, with special reference to the fauna of Illinois. Bull. Ill. Natur. Hist. Sur. 22:235–355.

Fullington, K. E., and K. W. Stewart. 1980. Nymphs of the stonefly genus *Taeniopteryx* (Plecoptera: Taeniopterygidae) of North America. J. Kansas Ent. Soc. 53:237–59.

Gaufin, A. R., A. V. Nebeker, and J. Sessions. 1966. The stoneflies (Plecoptera) of Utah. Univ. Utah Biol. Ser. 14:1–93.

Gaufin, A. R., W. E. Ricker, M. Miner, P. Milam, and R. A. Hays. 1972. The stoneflies (Plecoptera) of Montana. Trans. Amer. Ent. Soc. 98:1–161.

Hales, D. C., and A. R. Gaufin. 1971. Observations on the emergence of two species of stoneflies. Ent. News 82:107–109.

Hanson, J. F. 1946. Comparative morphology and taxonomy of the Capniidae (Plecoptera). Amer. Midl. Natur. 35:193–249.

Harper, P. P., and H. B. N. Hynes. 1971a. The Leuctridae of eastern Canada (Insecta; Plecoptera). Can. J. Zool. 49:915–20.

Harper, P. P., and H. B. N. Hynes. 1971b. The Capniidae of eastern Canada (Insecta; Plecoptera). Can. J. Zool. 49:921–40.

Harper, P. P., and H. B. N. Hynes. 1971c. The nymphs of the Taeniopterygidae of eastern Canada (Insecta; Plecoptera). Can. J. Zool. 49:941–47.

Harper, P. P., and H. B. N. Hynes. 1971d. The nymphs of the Nemouridae of eastern Canada (Insecta; Plecoptera). Can. J. Zool. 49:1129–42.

Harper, P. P., and K. W. Stewart. 1984. Plecoptera. Chap. 13. *In* Merritt, R. W. and K. W. Cummins (eds.). An introduction to the aquatic insects of North America, 2nd ed. Dubuque, Ia.: Kendall/Hunt Publ. Co. 722 pp.

Hitchcock, S. W. 1974. Guide to the insects of Connecticut. VII. The Plecoptera or stoneflies of Connecticut. Bull. Conn. Geol. Natur. Hist. Surv. 107:1–262.

Hynes, H. B. N. 1976. Biology of Plecoptera. Ann. Rev. Ent. 21:135–53.

Illies, J. 1966. Katalog der rezenten Plecoptera. Das Tierreich. Berlin: Walter de Gruyter and Co. 82. 632 pp.

Jewett, S. G., Jr. 1959. The stoneflies (Plecoptera) of the Pacific Northwest. Oregon State Monogr. Stud. Ent. 3:1–95.

Jewett, S. G., Jr. 1960. The stoneflies (Plecoptera) of California. Bull. California Insect Sur. 6:125–77.

Jewett, S. G., Jr. 1963. A stonefly aquatic in the adult stage. Science 139:484–85.

Jewett, S. G., Jr. 1965. Four new stoneflies from California and Oregon. Pan-Pac. Ent. 41:5–9.

Knight, A. W., A. V. Nebeker, and A. R. Gaufin. 1965a. Description of the eggs of common Plecoptera of western United States. Ent. News 76:105–11.

Knight, A. W., A. V. Nebeker, and A. R. Gaufin. 1965b. Further descriptions of the eggs of Plecoptera of western United States. Ent. News 76:233–39.

Lemkuhl, D. M. 1971. Stoneflies (Plecoptera:Nemouridae) from temporary lentic habitats in Oregon. Amer. Midl. Natur. 85:514–15.

Nebeker, A. V., and A. R. Gaufin. 1965. The *Capnia columbiana* complex of North America (Capniidae:Plecoptera). Trans. Amer. Ent. Soc. 91:467–87.

Needham, J. G., and P. W. Claassen. 1925. A monograph of the Plecoptera or stoneflies of America north of Mexico. Thomas Say Found. Ent. Soc. Amer. 2. 397 pp.

Nelson, C. H., and J. F. Hanson. 1971. Contribution to the anatomy and phylogeny of the family Pteronarcidae (Plecoptera). Trans. Amer. Ent. Soc. 97:123–200.

Newcomer, E. J. 1918. Some stoneflies injurious to vegetation. J. Agr. Res. 13:37–42.

Richardson, J. W., and A. R. Gaufin. 1971. Food habits of some western stonefly nymphs. Trans. Amer. Ent. Soc. 97:91–121.

Ricker, W. E. 1952. Systematic studies in Plecoptera. Indiana Univ. Publ. Sci. Ser. 18:1–200.

Ricker, W. E. 1965. New records and descriptions of Plecoptera (Class Insecta). J. Fish. Res. Bd. Can. 22:475–501.

Ricker, W. E., and H. H. Ross. 1968. North American species of *Taeniopteryx* (Plecoptera, Insecta). J. Fish. Res. Bd. Can. 25:1423–39.

Ricker, W. E., and H. H. Ross. 1975. Synopsis of the Brachypterinae (Insecta: Plecoptera: Taeniopterygidae). Can. J. Zool. 53:132–53.

Ross, H. H., and W. E. Ricker. 1971. The classification, evolution, and dispersal of the winter stonefly genus *Allocapnia*. Illinois Biol. Monog. 45:1–166.

Schwiebert, E. 1973. Nymphs. A complete guide to naturals and imitations. Winchester Press, New York. 339 pp.

Stark, B. P., and R. W. Baumann. 1978. New species of Nearctic *Neoperla* (Plecoptera: Perlidae), with notes on the genus. Great Basin Natur. 38:97–114.

Stark, B. P., and A. R. Gaufin. 1976a. The Nearctic species of *Acroneuria* (Plecoptera: Perlidae). J. Kansas Ent. Soc. 49:221–53.

Stark, B. P., and A. R. Gaufin. 1976b. The Nearctic genera of Perlidae (Plecoptera). Misc. Publ. Ent. Soc. Amer. 10:1–77.

Stark, B. P., and A. R. Gaufin. 1979. The stoneflies (Plecoptera) of Florida. Trans. Amer. Ent. Soc. 104:391–433.

Stark, B. P., and K. W. Stewart. 1981. The Nearctic genera of Peltoperlidae (Plecoptera). J. Kansas Ent. Soc. 54:285–311.

Stark, B. P., and S. W. Szczytko. 1981. Contributions to the systematics of *Paragnetina* (Plecoptera: Perlidae). J. Kansas Ent. Soc. 54:625–48.

Stewart, K. W., R. W. Baumann, and B. P. Stark. 1974. The distribution and past dispersal of southwestern United States Plecoptera. Trans. Amer. Ent. Soc. 99:507–46.

Stewart, K. W., and B. P. Stark. 1984. Nymphs of North American Perlodinae genera (Plecoptera: Perlodidae) Great Basin Natur. 44:373–415.

Surdick, R. F. 1985. Nearctic genera of Chloroperlinae (Plecoptera: Chloroperlidae). Ill. Biol. Monog. 54:1–146.

Surdick, R. F., and K. C. Kim. 1976. Stoneflies (Plecoptera) of Pennsylvania, a synopsis. Bull. Penn. Agr. Exp. Station 808:1–73.

Szczytko, S. W., and K. W. Stewart. 1977. The stoneflies (Plecoptera) of Texas. Trans. Amer. Ent. Soc. 103:327–78.

Szczytko, S. W., and K. W. Stewart. 1979. The genus *Isoperla* (Plecoptera) of western North America; holomorphology and systematics, and a new stonefly genus *Cascadoperla*. Mem. Amer. Ent. Soc. 32:1–120.

Zwick, P. 1973. Insecta: Plecoptera, phylogenetisches System und Katalog. Das Tierreich. Berlin: Walter de Gruyter and Co. 94. 465 pp.

Order Psocoptera

21

Edward L. Mockford
Illinois State University

PSOCIDS, BOOKLICE

Psocids are small, rather delicate insects of limited economic importance. Some 50 species are known to occur in human habitations or in stored grain. The smaller species, often called booklice, may become very abundant under ideal conditions of plentiful food and high relative humidity. Booklice are contaminants of human food and possible causative agents of asthmatic reactions (Spieksma and Smits 1975). A few species occurring on rangeland grasses have been incriminated as possible intermediate hosts of a tapeworm of sheep (Allen 1973).

Psocoptera exhibits its greatest diversity in the tropics. Several families that are very poorly represented in temperate North America are highly diverse in tropical areas.

DIAGNOSIS

Antennae elongate, filiform; head rounded or somewhat depressed; mandibles chewing, laciniae elongate, slender (fig. 21.45); tarsi two-segmented (two- or three-segmented in adults), terminating in pair of claws, wing pads usually present in instars beyond first; abdomen weakly sclerotized; cerci absent.

BIOLOGY AND ECOLOGY

Approximately 500 species occur in North America, including Mexico. The great majority dwell out-of-doors on trunks and branches of trees and shrubs, on rock outcrops, cave walls, dead persistent leaves on plants, dead leaves in ground litter, and on the larger grasses with persistent dead stems. In these habitats they feed on algae, fungi, lichens, and various forms of organic debris. Some species become associated with bird and mammal nests (references in Mockford 1967b, 1971a) and a few species have been found to be phoretic on birds (Mockford 1967b).

DESCRIPTION

Last instars small, 1–5 mm. Body, viewed laterally, depressed (fig. 21.1) to normal (fig. 21.2); viewed dorsally, slender (head including eyes wider than greatest width of abdomen) to robust (head including eyes narrower than greatest width of abdomen). [Note: these terms will be used without added definition in the family descriptions]. Head prognathous (fig. 21.1) to hypognathous (fig. 21.2), well sclerotized. Compound eyes prominent and many faceted to greatly reduced with few facets; postclypeus bulging, generally set off by sutures; mandibles stout, with distinct molar lobe; galea short and broad; lacinia elongate, rodlike (fig. 21.45), generally bearing a few tines on distal end; hypopharynx with basal sitophore sclerite connected by two filaments to the pair of oval lingual sclerites (fig. 21.4). Prothorax small. Meso- and metathorax generally separate, each considerably larger (broader and higher) than prothorax. Wing pads held horizontally (above thorax and abdomen, fig. 21.1) to vertically (at sides of thorax and abdomen figs. 21.2, 21.3). Abdomen weakly sclerotized; in dorsal view, usually fusiform (fig. 21.3), or broad at base, parallel-sided, tapering or rounded near apex (Family Liposcelidae only); the first segment reduced; terga 8–10 or 9–10 partially or completely fused, the terminal segment (11) composed of epiproct and paraprocts.

Figure 21.1

Figure 21.2

Figure 21.3

Figure 21.1. *Embidopsocus needhami* (Enderlein) (Liposcelidae), last instar larva of macropterous female, lateral view (legs removed).

Figure 21.2. *Neurostigma* sp. (Neurostigmatidae). Lateral view (legs removed).

Figure 21.3. *Neurostigma* sp. (Neurostigmatidae). Dorsal view (legs removed).

COMMENTS

Perhaps 20 percent of the North American species of psocids are identifiable as immatures by association with adults in field collections. Only some 10 species have been associated by rearing. In the present work, only late instars are covered. In general, early instars and eggs cannot yet be determined for North American species. Immature stages of some North American psocids have been described by Dunham (1972), Eertmoed (1966), Mockford (1957, 1974b, 1977, 1979, 1984b), and Sommerman (1943a,b,c, 1944).

It even remains difficult to obtain reliable determinations of adults in some taxa of North American psocids. Chapman's paper (1930) is still an important reference, including slightly over 25 percent of the species known at present from the United States. For particular taxa in North America, papers by Eertmoed (1973), Mockford (1951, 1953, 1955, 1959, 1963, 1965, 1966, 1967a, 1969, 1971b, 1974a, 1984a), Mockford and Wong (1969), and Sommerman (1946, 1948, 1956, 1957) are important.

The classification followed is essentially that of Badonnel (1951), which is followed by most specialists at present, although several families have been discovered or recognized as families since Badonnel's review.

TECHNIQUES

Psocids are best reared in cotton-stoppered glass tubes kept at high relative humidity but below saturation (about 80%) by use of appropriate saturated salt solutions (Winston and Bates 1960). In general, the natural substrate is the best diet, and spare substrate for maintenance of a culture must be kept refrigerated. Specimens for rearing are best collected with an aspirator, the body of which is made from soft clear plastic tubing. If glass or hard plastic is used, many are killed or injured. Specimens are best transported in the glass rearing tubes together with pieces of natural substrate. Care must be taken that none of the substrate placed in the tubes is damp, so in wet weather, dry bark and twigs must be found. In the field, tubes with specimens may be kept in plastic bags with wet paper towelling, but in the laboratory, or during an extended stay in the field, avoid letting the tubes contact wet surfaces. This results in condensation inside the tube that may result in drowning and will certainly encourage growth of mold.

Psocids are best killed and preserved in 75–80 percent ethanol. Superficial observations may be made in this solution; detailed observations must be done on slide-mounted whole specimens or parts. For observation of cuticular structures, macerate in lactophenol for 24–36 hours. If the cuticle is unpigmented, stain lightly by adding a small amount of acid fuchsin crystals to the lactophenol, heat the specimen in this solution in a water bath to the boiling point of the water, then allow it to cool slowly. The specimen should be passed through distilled water and mounted whole or in parts in Hoyer's medium.

Selected Bibliography

Badonnel 1951.
Chapman 1930.
Mockford 1967b, 1971a,c.
New 1977.
Pearman 1928.
Smithers 1972.
Sommerman 1942.

CLASSIFICATION

Order **PSOCOPTERA**
- Suborder Trogiomorpha
  - Group Atropetae
    - Lepidopsocidae
    - Trogiidae (= Atropidae)
    - Psoquillidae
  - Group Psocatropetae
    - Psyllipsocidae (= Psocatropidae)
    - Prionoglaridae
- Suborder Troctomorpha
  - Group Nanopsocetae
    - Pachytroctidae
    - Sphaeropsocidae
    - Liposcelidae
  - Group Amphientometae
    - *Musapsocidae
    - *Troctopsocidae (= Plaumanniidae)
    - *Manicapsocidae
    - *Compsocidae
    - Amphientomidae
- Suborder Psocomorpha (= Eupsocida)
  - Group Epipsocetae
    - Epipsocidae
    - *Dolabellopsocidae
    - *Neurostigmatidae
    - Ptiloneuridae
    - *Spurostigmatidae
  - Group Caecilietae
    - Superfamily Asiopsocoidea
      - Asiopsocidae
    - Superfamily Caecilioidea
      - Caeciliidae
      - Amphipsocidae (including Polypsocidae)
  - Group Homilopsocidea
    - Elipsocidae
    - Mesopsocidae
    - Philotarsidae
    - Lachesillidae
    - Peripsocidae
    - Pseudocaeciliidae
    - Trichopsocidae
    - Ectopsocidae
    - Archipsocidae
  - Group Psocetae
    - Hemipsocidae
    - Myopsocidae
    - Psocidae

*Not recorded north of Mexico (1985).

KEY TO FAMILIES OF NORTH AMERICAN PSOCOPTERA LARVAE[1]

1. Antennae with more than 20 segments, hypopharyngeal filaments separate throughout their length (fig. 21.4) .... Suborder **TROGIOMORPHA** 3

1′. Antennae with 17 or fewer segments; hypopharyngeal filaments fused proximally (fig. 21.5) 2

2(1′). Antennae 13-segmented. Sitophore sclerite lacking anterior arms as continuous regions (fig. 21.6) or with only very short ones arising laterally from an anterior shelf (fig. 21.7) .... Suborder **PSOCOMORPHA** 12

2′. Antennae 11-, 12-, 13-, 15-, or 17-segmented. Sitophore sclerite with elongate anterior arms arising close together (fig. 21.5) ...... Suborder **TROCTOMORPHA** 6

3(1). Legs and antennae long and slender; femora not over 2× as wide as greatest width of tibiae. Shortest flagellomere at least 3× longer than broad. Posterior first tarsomere about twice length of second .... Group PSOCATROPETAE .... ***Psyllipsocidae*** (p. 204) and ***Prionoglaridae*** (p. 204)

3′. Legs and antennae relatively shorter; femora about 3× as wide as greatest width of tibiae. Shortest flagellomere not over 2× longer than broad. Posterior first tarsomere about 1.5× length of second 4

4(3′). Body and legs densely hairy. Face generally patterned (fig. 21.8). Labial palpi broadly joined to labium, not prominent. Slender wing pads present, pointed at their apices, variable in length .... ***Lepidopsocidae*** (p. 203)

4′. Body and legs, if hairy, not densely so. Face patterned or not. Labial palpi variable. Wing pads broad basally, pointed or not; if pointed, not or barely reaching abdomen 5

5(4′). Distal segment of maxillary palpus hatchet-head-shaped (fig. 21.9). Wing pads rounded distally .... ***Psoquillidae*** (p. 203)

5′. Distal segment of maxillary palpus rounded or truncated (fig. 21.10). Wing pads pointed distally .... ***Trogiidae*** (p. 203)

6(2′). Posterior femora broader than others. Body dorsoventrally at least slightly depressed (fig. 21.1). Head prognathous .... ***Liposcelidae*** (p. 205)

6′. Posterior femora not broader than others. Body not dorsoventrally depressed. Head not distinctly prognathous 7

7(6′). Distance between eye and postclypeus greater than greatest diameter of eye (fig. 21.11). Antennae longer than body .... ***Pachytroctidae*** (p. 204)

7′. Distance between eye and postclypeus less than greatest diameter of eye (fig. 21.12). Antennae variable .... Group AMPHIENTOMETAE 8

8(7′). Dorsal surfaces of body and wing pads beset with stout upright setae that are truncated or knobbed distally; terminal flagellomere ending in an elongate, slender, aristalike process (fig. 21.13) .... ***Musapsocidae*** (p. 205)

8′. Dorsal surfaces of body and wing pads lacking stout upright setae; terminal flagellomere pointed or not, but never ending in an elongate, aristalike process 9

9(8′). Antenna essentially bare or beset with very short and/or not more than a total of 3 long setae 10

9′. Antennal flagellum regularly beset with long setae, 2 or more of these per flagellomere, the longest ones at least half the length of their flagellomere 11

10(9). Row of denticles present on anterior carina of fore femur (fig. 21.14). Lateral ocellar anlagen separated (as in fig. 21.30) by not over 2× width of median ocellar anlage .... ***Compsocidae*** (p. 205)

10′. Row of denticles absent on anterior carina of fore femur, or, if present, lateral ocellar anlagen separated by about 6× width of (minute) median ocellar anlage .... ***Troctopsocidae*** (p. 205)

1. Excluding Sphaeropsocidae for which immatures were not available.

11(9′). Row of denticles present on anterior carina of fore femur .... ***Amphientomidae*** (p. 206)

11′. Row of denticles absent from anterior carina of fore femur .... ***Manicapsocidae*** (p. 205)

12(2). Labrum internally with 2 anteroposteriorly directed, strongly sclerotized ridges clearly showing through cuticle (fig. 21.15) .... Group EPIPSOCETAE 28

12′. Labrum internally with only a small parenthesislike mark bearing a sclerotized tubercle on either side bordering anterior margin (fig. 21.16). In a few forms a pair of pigment bands running anteroposteriorly through labrum 13

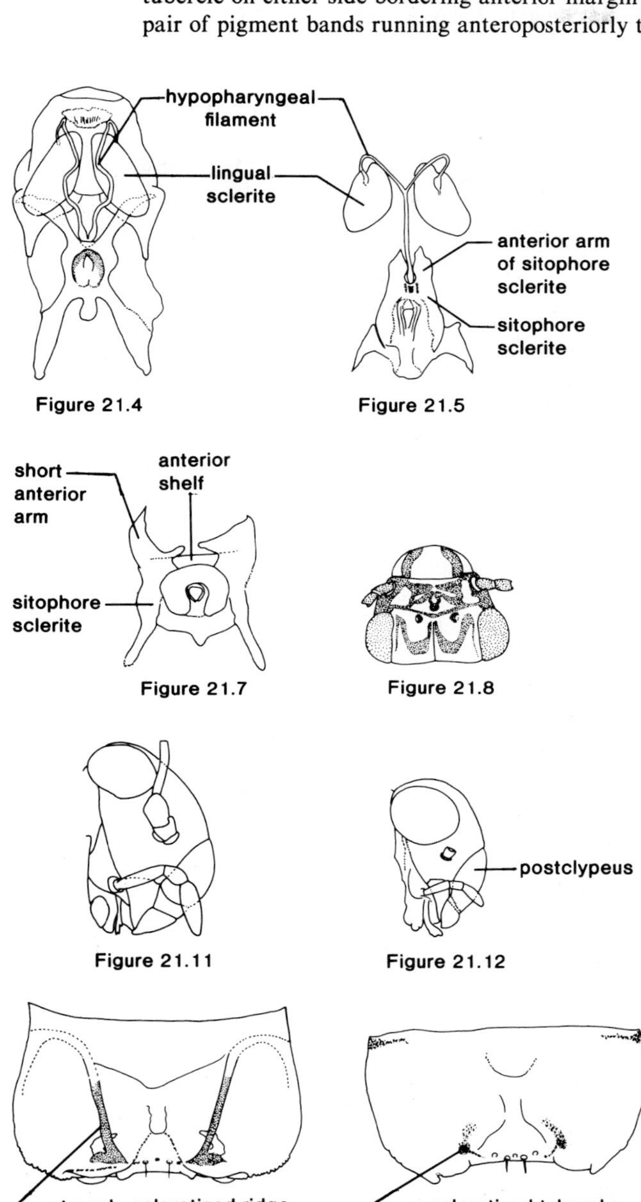

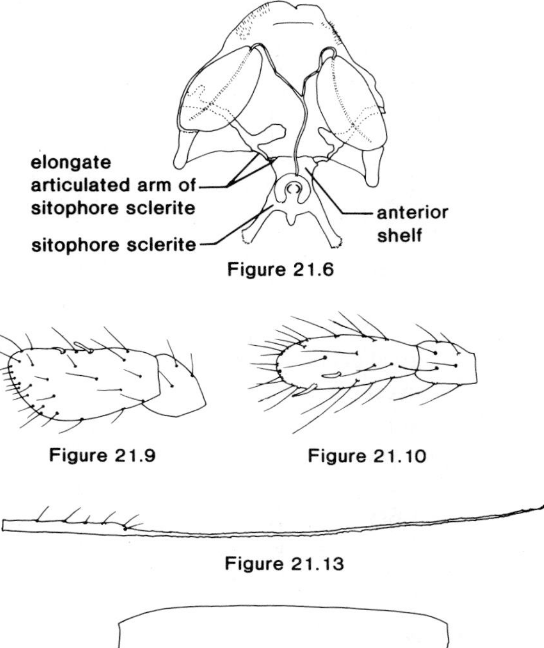

Figure 21.4. *Dorypteryx* sp. (Psyllipsocidae). Hypopharynx.

Figure 21.5. *Liposcelis bostrychophilus* Badonnel (Liposcelidae). Sitophore and lingual sclerites and hypopharyngeal filaments.

Figure 21.6. *Aaroniella maculosa* (Aaron) (Philotarsidae). Hypopharynx.

Figure 21.7. *Caecilius totonacus* Mockford (Caeciliidae). Sitophore sclerite.

Figure 21.8. *Echmepteryx hageni* (Packard) (Lepidopsocidae). Head, dorsal view.

Figure 21.9. *Psoquilla marginepunctata* (Hagen) (Psoquillidae). Distal two segments of maxillary palpus.

Figure 21.10. *Lepinotus reticulatus* Enderlein (Trogiidae). Distal two segments of maxillary palpus.

Figure 21.11. *Tapinella maculata* Mockford (Pachytroctidae). Head, lateral view.

Figure 21.12. *Seopsis* sp. (Amphientomidae). Head, lateral view.

Figure 21.13. *Musapsocus* sp. (Musapsocidae). Terminal flagellomere.

Figure 21.14. *Compsocus elegans* Banks (Compsocidae). Anterior femur showing row of denticles.

Figure 21.15. *Epipsocus* sp. (Epipsocidae). Labrum.

Figure 21.16. *Psocus leidyi* Aaron (Psocidae). Labrum.

13(12′). Sitophore sclerite with short, broad anterior arms continuous with anterior shelf (fig. 21.7). Pretarsal claw without a denticle .... Group CAECILIETAE 14

13′. Sitophore sclerite with elongate arms, slender at their bases, articulated with anterior shelf (fig. 21.6). Pretarsal claw with or without a denticle 17

14(13). Labrum not conforming closely to contour of mandibles (fig. 21.18), its heavy distal sclerotizations not in the form of a pair of parentheses, 1 on each side of median sensillar field (figs. 21.17 and 21.20). Mandibles elongate, hollowed out posteriorly to accommodate bulging galeae (fig. 21.18) 15

14′. Labrum held closely against mandibles (as in fig. 21.19), its heavy distal sclerotizations in the form of a pair of parentheses, 1 on each side of median sensillar field (fig. 21.16). Mandibles shorter, not hollowed out posteriorly, the galeae relatively flat (as in fig. 21.19) .... Genus *Notiopsocus* (currently placed in the ***Asiopsocidae***) (p. 207)

15(14). Ventral abdominal vesicles absent. A distinct pair of pigmented bands running length of labrum (fig. 21.20) .... ***Asiopsocidae*** (p. 207)

15′. Ventral abdominal vesicles present (1–3) (fig. 21.21). Pigment bands of labrum usually absent, obscure if present 16

16(15′). Body robust (i.e., greatest width of abdomen more than that of head including eyes). Width of thorax including fore wing bases greater than width of head including eyes .... ***Amphipsocidae*** (p. 209)

16′. Body slender (i.e., greatest width of abdomen less than width of head including eyes). Width of thorax including wing bases about equal to width of head including eyes .... ***Caeciliidae*** (p. 207)

17(13′). Preapical denticle present on pretarsal claw (fig. 21.22) 18

17′. Preapical denticle absent on pretarsal claw 25

18(17). Distal margin of paraproct with hyaline cone variously developed, flanked by 2 heavy setae, 1 above cone, other below (figs. 21.23–21.25) 19

18′. Distal margin of paraproct with hyaline cone flanked by a single heavy seta located ventral to it (figs. 21.26–21.28) 23

19(18). Ocellar anlagen elongate, the laterals a pair of curved pigment lines resembling parentheses facing outward; the median somewhat anterior to the laterals (figs. 21.29, 21.30) 20

19′. Ocellar anlagen compact, the laterals short, the median between them anteriorly; in some forms anlagen obscured by surrounding dark brown pigment spot (fig. 21.31) 21

20(19). Compound eyes decidedly prominent laterally on wide head (fig. 21.29). A broad sclerotized line between bases of paraglossae on ventral surface of labium (fig. 21.32) .... ***Mesopsocidae*** (p. 209)

20′. Compound eyes not as prominent laterally (fig. 21.30). Head not decidedly wide. A narrow sclerotized line between bases of paraglossae on ventral surface of labium (fig. 21.33) .... ***Elipsocidae*** (p. 209)

21(19′). Lacinial tip approximately bifid, i.e., lateral and median tines about equal in size (fig. 21.34) .... ***Peripsocidae*** (p. 210)

21′. Lateral tine of lacinial tip much larger than median (fig. 21.35) 22

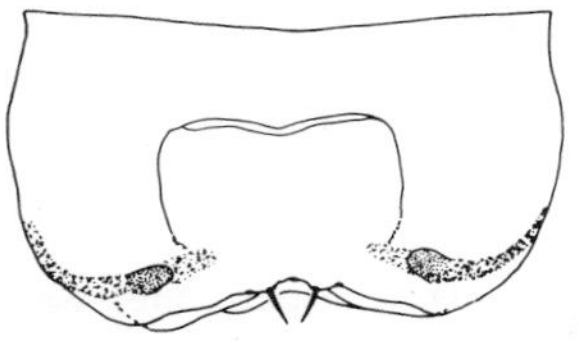

Figure 21.17

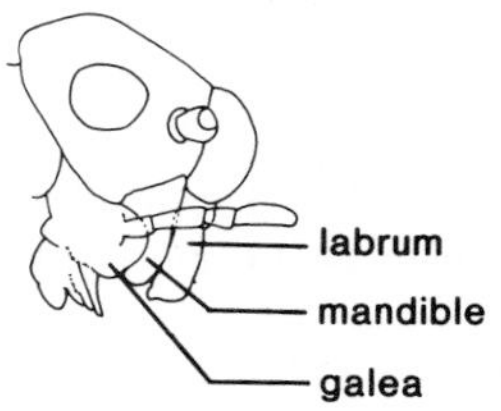

Figure 21.18

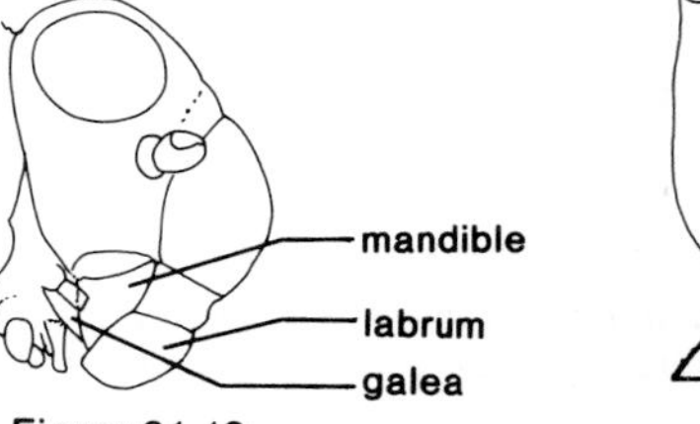

Figure 21.19

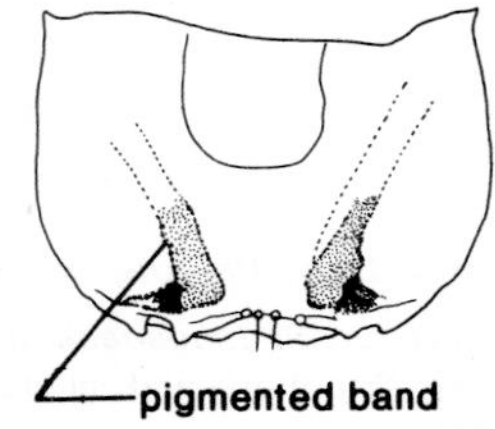

Figure 21.20

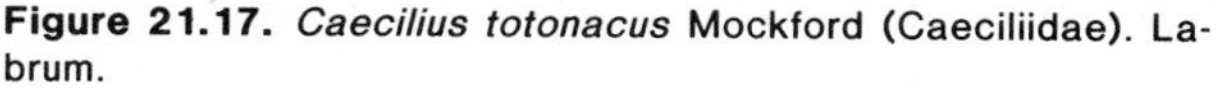

Figure 21.17. *Caecilius totonacus* Mockford (Caeciliidae). Labrum.

Figure 21.18. *Polypsocus corruptus* (Hagen) (Amphipsocidae). Head, lateral view.

Figure 21.19. *Psocus leidyi* Aaron (Psocidae). Head, lateral view.

Figure 21.20. *Asiopsocus sonorensis* Mockford and Garcia Aldrete (Asiopsocidae). Labrum.

22(21′). Pulvillus bent near its base (fig. 21.36). Labium with a distinct sclerotized line on outer surface along midline between bases of paraglossae (as in fig. 21.33) .... ***Myopsocidae*** (p. 212)

22′. Pulvillus straight or slightly curved (fig. 21.22). Labium with or without a sclerotized line on outer surface along midline between bases of paraglossae .... ***Psocidae*** (p. 212)

23(18′). Pulvillus broad, bent near its base (fig. 21.37) .... ***Hemipsocidae*** (p. 212)

23′. Pulvillus narrow, straight or curved 24

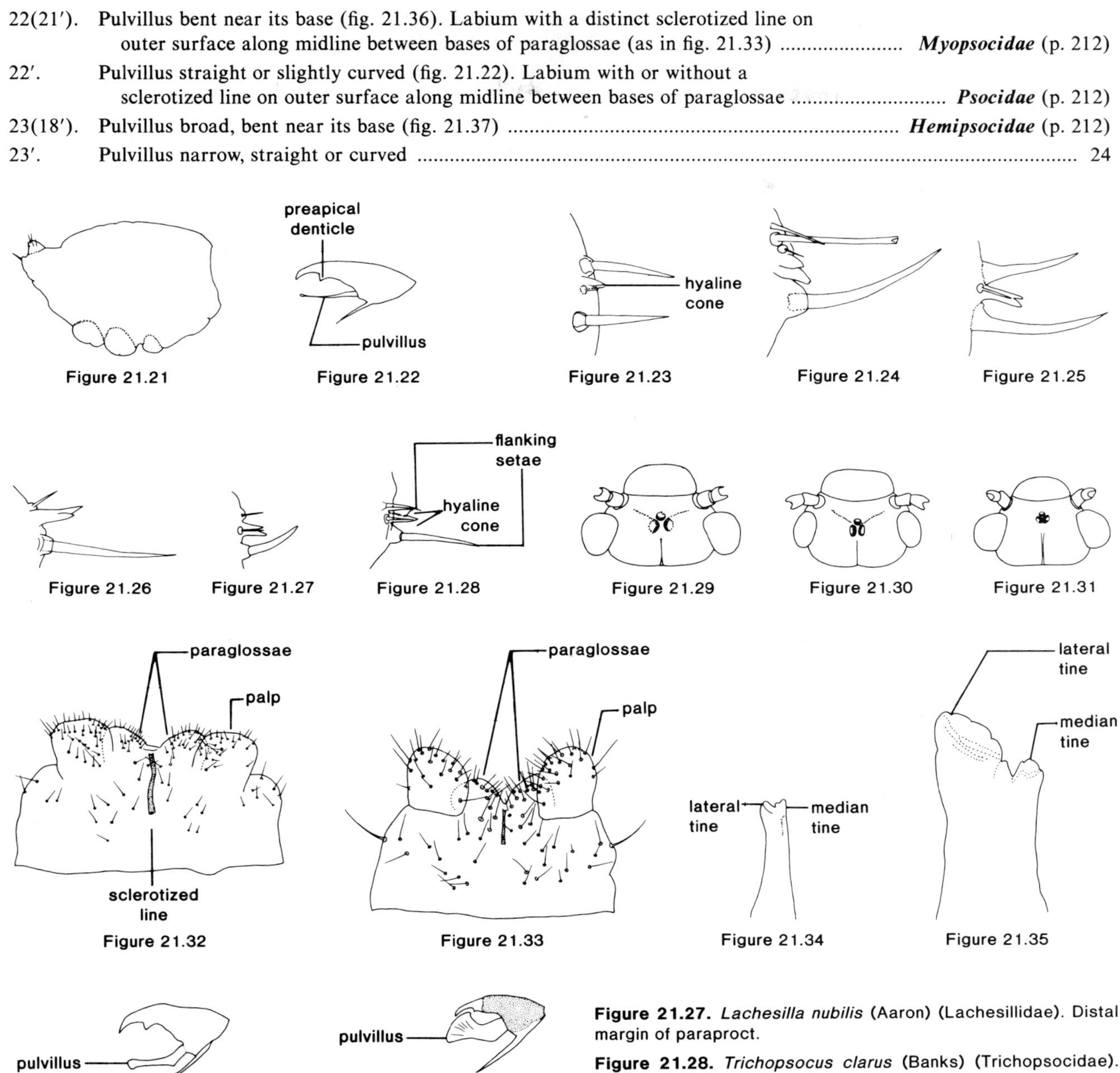

Figure 21.21. *Polypsocus corruptus* (Hagen) (Amphipsocidae). Abdomen, lateral view.

Figure 21.22. *Psocus leidyi* Aaron (Psocidae). Pretarsal claw.

Figure 21.23. *Mesopsocus unipunctatus* (Müller) (Mesopsocidae). Distal margin of paraproct.

Figure 21.24. *Aaroniella* sp. (Philotarsidae). Distal margin of paraproct.

Figure 21.25. *Peripsocus madidus* (Hagen) (Peripsocidae). Distal margin of paraproct.

Figure 21.26. *Hemipsocus africanus* Enderlein (Hemipsocidae). Distal margin of paraproct.

Figure 21.27. *Lachesilla nubilis* (Aaron) (Lachesillidae). Distal margin of paraproct.

Figure 21.28. *Trichopsocus clarus* (Banks) (Trichopsocidae). Distal margin of paraproct.

Figure 21.29. *Mesopsocus unipunctatus* (Müller) (Mesopsocidae). Head, dorsal view.

Figure 21.30. *Elipsocus guentheri* Mockford (Elipsocidae). Head, dorsal view.

Figure 21.31. *Psocus leidyi* Aaron (Psocidae). Head, dorsal view.

Figure 21.32. *Mesopsocus unipunctatus* (Müller) (Mesopsocidae). Labium, ventral view.

Figure 21.33. *Elipsocus guentheri* Mockford (Elipsocidae). Labium, ventral view.

Figure 21.34. *Peripsocus madidus* (Hagen) (Peripsocidae). Lacinial tip.

Figure 21.35. *Psocus leidyi* Aaron (Psocidae). Lacinial tip.

Figure 21.36. *Myopsocus* sp. (Myopsocidae). Pretarsal claw.

Figure 21.37. *Hemipsocus africanus* Enderlein (Hemipsocidae). Pretarsal claw.

24(23′). Labium with a distinct, broad, sclerotized line on outer surface along midline between bases of paraglossae (fig. 21.38). Lacinial tip with lateral tine broader than median (fig. 21.39). .... ***Philotarsidae*** (p. 209)

24′. Labium with only a very thin line or none on outer surface along midline between bases of paraglossae. Lacinial tip with lateral and median tines of approximately equal size (fig. 21.40) .... ***Lachesillidae*** (p. 210)

25(17′). Ventral abdominal vesicles present (as in fig. 21.21). Distal flagellomere acuminately tipped (fig. 21.41) 26

25′. Ventral abdominal vesicles absent. Distal flagellomere not acuminately tipped 27

26(25). Hyaline cone of paraproct minute, flanked ventrally by a relatively long, slender seta, the latter subtended by another of similar proportions (fig. 21.42) .... ***Pseudocaeciliidae*** (p. 210)

26′. Hyaline cone of paraproct minute, flanked ventrally by a relatively short, heavy seta, the latter well isolated from other setae below it .... ***Trichopsocidae*** (p. 211)

27(25′). Paraproctal hyaline cone absent. Usually colonial forms in webs .... ***Archipsocidae*** (p. 211)

27′. Paraproctal hyaline cone present, flanked dorsally by a single, short, slender seta (fig. 21.43). Insects dwelling singly or in groups of 2–3 in small webs or none .... ***Ectopsocidae*** (p. 211)

28(12). First hind tarsomere slightly over 3× length of second. Setae sparse on dorsal surfaces .... ***Epipsocidae*** (p. 206)

28′. First hind tarsomere less than 3× length of second. Setae conspicuous on dorsal surfaces 29

29(28′). Stout, upright setae in tufts on dorsal surfaces of head and thorax. Abdomen dorsally and laterally with small tubercles, each bearing a single seta longer than surrounding ones (figs. 21.2, 21.3) .... ***Neurostigmatidae*** (p. 206)

29′. No tufts of setae on head or thorax. Abdomen lacking tubercles 30

30(29′). All (3) ocellar anlagen visible. First flagellomere about 1.5× length of pedicel .... ***Spurostigmatidae*** (p. 207)

30′. Only the lateral ocellar anlagen or none visible. First flagellomere 3–4× length of pedicel 31

31(30′). Lateral ocellar anlagen visible. First flagellomere about 3× length of pedicel. Setae of upper body surface upright, curving slightly backward, many serrate .... ***Ptiloneuridae*** (p. 206)

31′. No ocellar anlagen visible. First flagellomere about 4× length of pedicel. Setae of upper body surface remaining close to body, smooth .... ***Dolabellopsocidae*** (p. 206)

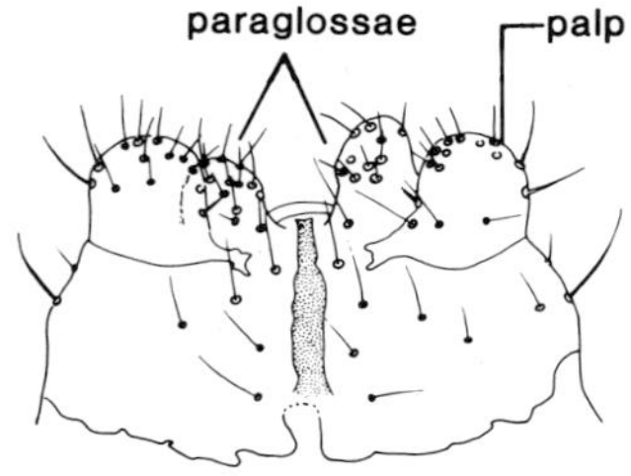

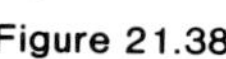
Figure 21.38

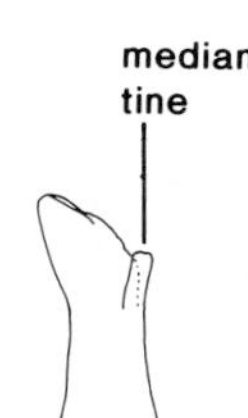

Figure 21.39

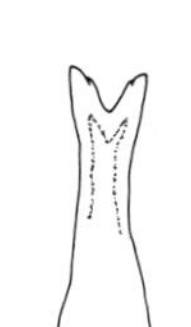
Figure 21.40

Figure 21.41

Figure 21.38. *Aaroniella maculosa* (Aaron) (Philotarsidae). Labium, ventral view.

Figure 21.39. *Aaroniella maculosa* (Aaron) (Philotarsidae). Lacinial tip.

Figure 21.40. *Lachesilla nubilis* (Aaron). (Lachesillidae) Lacinial tip.

Figure 21.41. *Pseudocaecilius citricola* (Ashmead) (Pseudocaeciliidae). Terminal flagellomere.

Figure 21.42. *Pseudocaecilius citricola* (Ashmead) (Pseudocaeciliidae). Distal margin of paraproct.

Figure 21.43. *Ectopsocus meridionalis* Ribaga (Ectopsocidae). Distal margin of paraproct.

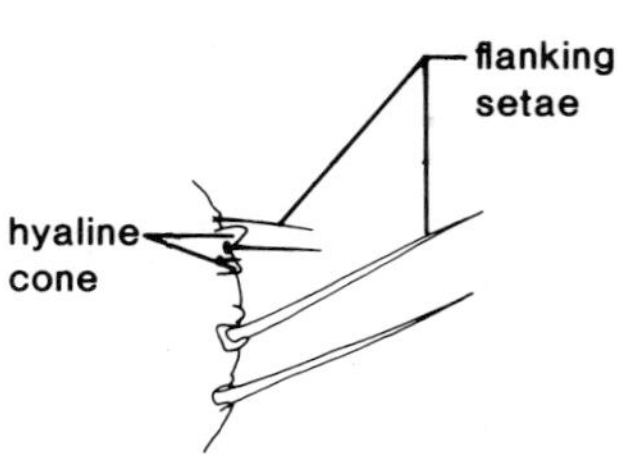

Figure 21.42

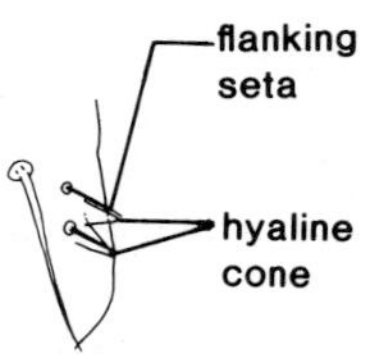

Figure 21.43

SUBORDER TROGIOMORPHA, GROUP ATROPETAE

Lepidopsocidae

Figures 21.8, 21.50

Relationships and Diagnosis: This family, in which adults generally have scaly wings and body, is probably most closely related to Trogiidae. Larvae of both families are similar in appearance, sharing the character of pointed wing pads.

Lepidopsocids differ consistently from trogiids by larger size (of adults and mature larvae). Most species, even as immatures, have characteristic facial markings (fig. 21.8) that are generally absent in Trogiidae.

Biology and Ecology: Found on trunks and branches of trees and shrubs, rock outcrops, dead, persistent palm leaves, bromeliads (probably dead leaves), sugarcane and other large grasses, and dead leaves of ground litter. They feed on algae, lichens, leaf fungi, and a variety of organic debris. A reared bark-inhabiting species underwent six instars and deposited its eggs in crevices in bark. Several species are parthenogenetic (Mockford 1971c).

Description: Mature larvae 2–3 mm, relatively slender, slightly depressed with head semiprognathous. Compound eyes well developed, with numerous facets. Antennae not quite as long as body. Wing pads of long-winged forms pointed apically, held semihorizontally, extending at angle of about 10° from body to about end of distal three/fifths of abdomen. Hairs of various lengths abundant on all body surfaces. Color from creamy yellow to medium brown; pale forms generally with darker markings on face, legs, and sides of body.

Comments: The family is primarily tropical and subtropical. Fourteen species occur in North America but only three reach northern United States and Canada. *Echmepteryx hageni* (Packard) occurs on tree trunks and branches throughout eastern United States and southeastern Canada. Two closely related species occur in Florida (Mockford 1974a). Species of *Echmepteryx, Proentomum,* and *Thylacella* occur on palms, bromeliads, and large grasses in coastal plain areas of Mexico. Species of *Soa, Lepolepis, Nepticulomima,* and *Cyptophania* occur in ground litter in Mexico and Florida; *Lepolepis occidentalis* Mockford occurs in ground litter throughout eastern United States and southeastern Canada. They are of no known economic importance.

Selected Bibliography

Mockford 1974a.

Trogiidae

Figures 21.10, 21.44

Relationships and Diagnosis: This family is probably most closely related to Lepidopsocidae. Larvae of both families are similar in appearance, generally sharing the character of pointed wing pads.

Adults are neotenic with winglets not reaching beyond the second abdominal segment and ocelli completely lacking. Mature larvae differ from those of Lepidopsocidae by smaller size, never being over 1.5 mm, and by shorter wing pads, these never reaching beyond the extreme base of the abdomen. Larvae, except of the genus *Cerobasis* (fig. 21.44), do not show facial markings.

Biology and Ecology: Most species occur in semiarid areas on junipers, pines, small palms, yuccas, and ground litter. A few species occur in more humid areas on tree trunks and rock outcrops. Several species are domestic, occurring on dusty furniture tops, in old papers, and occasionally in stored grain. Females of some species make sounds by tapping the abdomen on a surface (Pearman 1928). Most are presumably feeders on small fungi and organic debris.

Description: Mature larvae 1.0–1.5 mm, relatively slender, slightly depressed with head semiprognathous to hypognathous. Compound eyes well developed with numerous facets. Antennae about two-thirds length of body. Wing pads extending barely to base of abdomen or represented by minute swellings. Some hairs present on body surface. Color uniform creamy yellow or variegated with darker markings in reticulate pattern over much of body surface.

Comments: Nine species are known in North America. Domestic species of the genera *Trogium, Lepinotus,* and *Cerobasis* are widely distributed, probably by transport in human commerce. Species of *Cerobasis* occur on juniper, pines, and yuccas in semiarid areas of southwestern United States and Mexico. *Lepinotus reticulatus* Enderlein is widely distributed in ground litter and occasionally infests stored grain. It has been incriminated as a possible intermediate host of the fringed tapeworm of sheep (Allen 1973).

Selected Bibliography

Mockford 1971a.

Psoquillidae

Figures 21.9, 21.52

Relationships and Diagnosis: This family is rather closely related to Lepidopsocidae and Trogiidae. Larvae in general resemble those of both families.

They differ from both lepidopsocids and trogiids in that wing pads of mature larvae are rounded rather than pointed. They also differ from trogiids in the somewhat larger size and relatively longer wing pads. They differ from lepidopsocids in relatively less body hair.

Biology and Ecology: Most species occur in dead persistent leaves and in dead leaves of ground litter. A few species occur under bark of trees, and a few are known from human habitations, occasionally infesting stored mealy foods. Most species probably feed on small fungi and organic debris.

Description: Mature larvae 1.5–2 mm, relatively slender, slightly depressed with head semiprognathous. Compound eyes well developed with numerous facets. Antennae about as long as body or slightly shorter. Wing pads of long-winged forms held horizontally, extending at an angle of about 15° from body to end of basal two-thirds of abdomen. Color uniform creamy yellow, sometimes with darker lateral markings.

Comments: The family is primarily tropical and subtropical. Nine species occur in North America, none reaching beyond the Mississippi Embayment in southern Illinois and the Atlantic coastal plain in Maryland. Species of *Rhyopsocus* inhabit persistent dead leaves of palms, yuccas, bromeliads, and broad-leaved trees in general. Two species sometimes invade houses in southeastern United States (Sommerman 1956). Several brachypterous species occur in ground litter in Mexico and southwestern United States. *Psoquilla marginepunctata* Hagen occurs under bark of trees and logs in southern Mexico, and it is probably established in houses in southern Florida. These insects feed on small fungi and organic debris. *Rhyopsocus squamosus* Mockford was implicated as a possible intermediate host of the fringed tapeworm of sheep (Allen 1973). Otherwise, except as occasional household pests in southeastern United States, they are of no economic importance.

Selected Bibliography

Broadhead 1961.
Sommerman 1956.

GROUP PSOCATROPETAE

Psyllipsocidae

Figures 21.4, 21.45, 21.49

Relationships and Diagnosis: This family is closely related to Prionoglaridae, based on adult characters of venation and genitalia. Larvae of the two families share the proportion characters used in the key (couplet 3).

Mature larvae are distinguishable from those of the North American species of Prionoglaridae by size, psyllipsocids being not over 2 mm, the prionoglarids being about 4 mm in body length.

Biology and Ecology: Found in caves, cellars, on rock outcrops, on persistent dead leaves of yuccas in semiarid areas, and occasionally on tree trunks in tropical forests. They probably feed on small fungi and organic debris.

Description: Mature larvae about 2 mm. Body relatively slender, normal, head hypognathous. Compound eyes variable. Antennae about length of body. Wing pads of long-winged forms held horizontally at angle of about 20° from body, attaining end of about basal third of abdomen, but most individuals brachypterous with much shorter wing pads. Proportions of leg parts as mentioned in key (couplet 3). Color white to creamy yellow.

Comments: The family is primarily tropical and subtropical. Of the 11 species known in North America, only 3 occur north of the southern tier of states in the United States. *Psyllipsocus ramburi* Selys Longchamps is frequently encountered in caves and cellars throughout eastern and central United States and in Mexico. *Psocatropos microps* Enderlein occurs on walls in houses in Florida. *Psyllipsocus oculatus* Gurney occurs commonly on persistent dead leaves of yuccas in semiarid areas of northern Mexico, southern Texas, and southern New Mexico. Except as occasional household pests, they are of no economic importance.

Selected Bibliography

Gurney 1943.

Prionoglaridae

Relationships and Diagnosis: This family is closely related only to Psyllipsocidae, as discussed under that family.

Mature larvae distinguishable from psyllipsocids by size, being about twice as large (4 mm).

Biology and Ecology: The North American species occur in caves, sheltered rock outcrops, and skirts of desert fan palms in southwestern United States.

Description: Mature larvae about 4 mm. Body robust, normal, head hypognathous. Compound eyes small, with few facets, set decidedly low on sides of head. Antennae extremely slender, somewhat longer than body. Wing pads held horizontally, sloping slightly downward toward their tips, at angle out from body of about 30°, attaining about end of basal third of abdomen. Proportions of leg parts as in Psyllipsocidae, but legs longer. Color white.

Comments: Species of *Speleketor* are known from Arizona, Nevada, and southern California; specimens were collected in caves, in a sheltered canyon, and on dead skirts of desert fan palms. Gurney (1943) and Mockford (1984a) discussed the relationships of this genus.

Selected Bibliography

Gurney 1943.
Mockford 1984a.

SUBORDER TROCTOMORPHA
GROUP NANOPSOCETAE

Pachytroctidae

Figure 21.11

Relationships and Diagnosis: Pachytroctids are most closely related to Sphaeropsocidae and Liposcelidae. They share with both families small size and male winglessness, with Sphaeropsocidae general body shape, and with Liposcelidae the adult character of wings folding flat over the back at rest.

They differ from sphaeropsocids by better development of compound eyes (in adults, fore wings not elytriform), and from liposcelids in the body being much less depressed with head hypognathous.

Biology and Ecology: Found on dead persistent leaves of palms, yuccas, and large grasses, in dead leaves of ground litter, and occasionally on walls in human dwellings. They probably feed on small fungi and organic debris.

Description: Mature larvae 1–1.5 mm. Body slender, slightly depressed, head semiprognathous. Compound eyes well developed with numerous facets. Antennae about length of body to slightly longer. Wing pads of long-winged forms held horizontally, extending straight backward, to a little less than end of basal half of abdomen. Color white or creamy yellow to medium brown, some forms with darker lateral marks.

Comments: The family is primarily tropical and subtropical. Twelve species occur in North America but only one occurs north of the coastal plain of the Gulf of Mexico. Species of *Tapinella* are common on dead persistent leaves of palms and large grasses in Florida, Mexico, and southern Texas. *Nanopsocus oceanicus* Pearman occurs commonly on walls in buildings in Florida and on dead persistent leaves there and in Mexico. It was taken in a sawdust pile in southern Illinois. Except as occasional household pests in Florida, they are of no economic importance.

Sphaeropsocidae

No immatures are available for study. From adult characters one can state with a high degree of probability that mature larvae are somewhat smaller than those of pachytroctids, with compound eyes rather poorly developed and having few facets. The body is probably shaped as in Pachytroctidae, hence much less depressed than in liposcelids. They are rare in North America, where only two species have been found. Both were taken in ground litter. They are of no known economic importance.

Liposcelidae

Figures 21.1, 21.5, 21.54

Relationships and Diagnosis: Relationships are as indicated for Pachytroctidae (above). The decidedly depressed body, prognathous head (fig. 21.1), laterally sprawling legs with relatively broad posterior femora mark this family.

Biology and Ecology: Found under bark of trees and shrubs, in dead persistent leaves of yuccas, palms, bromeliads, and grasses, in leaves of ground litter, in nests of birds and mammals, and in human dwellings and stored grain. Species of subfamily Embidopsocinae show wing dimorphism, with wingless males and long-winged and wingless females. Species of subfamily Liposcelinae are wingless. They pass through three to four larval instars.

Description: Mature larvae 0.6–1.5 mm, body relatively robust, depressed, (fig. 21.1), head prognathous. Compound eyes of wingless forms greatly reduced, with not over eight facets; those of winged forms much larger with many more facets. Antennae slightly less than half length of body. Wing pads of winged forms held horizontally (fig. 21.1), extending straight backward to about third abdominal segment. Abdomen broad basally, parallel sided, except tapering or rounded at distal end. Color white to pale brown, a few species with some darker markings.

Comments: This family has not yet been well studied in North America, where 26 described species are known, and perhaps as many more remain undescribed or unrecorded. Subfamily Embidopsocinae is largely tropical, with only two species recorded north of the southern tier of states in the United States. Subfamily Liposcelinae, represented by the large genus *Liposcelis*, is common and diverse northward into Canada. These are the true booklice. Several of its species commonly occur in human dwellings and infest stored grain. Certain household species have been implicated as possible causative agents of allergic reactions to house dust in humans (Spieksma and Smits 1975). These insects are possible intermediate hosts of the fringed tapeworm of sheep (Allen 1973).

Selected Bibliography

Broadhead 1950.
Mockford 1963, 1971a.
Sommerman 1957.
Spieksma and Smits 1975.

GROUP AMPHIENTOMETAE

Musapsocidae (Tropical)

Figure 21.13

Found on persistent dead leaves of musaceous plants, especially banana (*Musa sativa*) and *Heliconia* spp. in tropical forests and adjacent plantations. They are strictly tropical, reaching north to about central Veracruz, Mexico. Only two species are known in North America, and although they commonly occur on banana trees, they probably feed only on small fungi on the dead leaves and are of no known economic importance.

Selected Bibliography

Mockford 1967a.

Troctopsocidae (Tropical)

Inhabiting dead persistent leaves of ferns, club mosses, and flowering plants in tropical forests. One species occurs in dead leaves of ground litter. They are strictly tropical and subtropical. Only three species occur in North America, and none reach the United States. They are of no known economic importance.

Selected Bibliography

Mockford 1967a.

Manicapsocidae (Tropical)

Found on trunks of large trees in tropical forests. They run rapidly when disturbed. Most species appear to be dimorphic, with long-winged males and brachypterous females. They are strictly tropical and temperate South American. The single North American species *Nothoentomum tuxtlarum* (Mockford) inhabits forests of southern Mexico. They are of no known economic importance.

Selected Bibliography

Mockford 1967a.

Compsocidae (Tropical)

Figure 21.14

Found on trunks of large trees in tropical forests. They run rapidly when disturbed. They are strictly tropical. The two species known from North America both inhabit forests of southern Mexico. They are of no known economic importance.

Selected Bibliography

Mockford 1967a.

Amphientomidae

Figure 21.12

Relationships and Diagnosis: This family appears to be most closely related to Compsocidae, with which it shares color banding of the compound eyes, genitalic features of adults of both sexes, and presence of a row of denticles along the anterior carina of the anterior femur.

It differs from the previous four tropical families by certain venational features of adults (subcosta not looping posteriorly to join R, IA and IIA not joining together before reaching wing margin, and by the presence in adults of scales over body and wing surfaces. The larvae differ from those of Musapsocidae and Manicapsocidae by the presence of a row of denticles on the anterior carina of the fore femur. They differ from those of Troctopsocidae and Compsocidae by having the antennae beset with elongate setae, some approaching the length of their flagellomere (antennae essentially bare in Compsocidae, with very few long setae if any in Troctopsocidae).

Biology and Ecology: Found on trunks of large trees and rock outcrops. They run rapidly when disturbed.

Description: Mature larvae 2–3 mm, body relatively robust, normal, head hypognathous. Compound eyes well developed with numerous facets. Antennae half to about two-thirds the length of body. Wing pads held vertically, extending posterolaterally at angle of about 10° from body to slightly beyond end of basal half of abdomen; somewhat tapering to blunt point distally. Color creamy white with brown markings on face, sides, and legs.

Comments: This family is primarily tropical. Of the approximately six species in North America, only one—apparently an introduction from Asia—occurs north of the southern tier of states of the United States. The others inhabit rock outcrops and trunks of forest trees throughout Mexico and southern Texas. They are of no known economic importance.

SUBORDER PSOCOMORPHA, GROUP EPIPSOCETAE

Epipsocidae

Figure 21.15

Relationships and Diagnosis: Among the five families designated by Eertmoed (1973) in group Epipsocetae, relationships remain poorly understood. On Eertmoed's phenogram, this family is placed closest to Neurostigmatidae.

The larvae differ from those of other Epipsocetae by possessing only sparse, inconspicuous setae on all dorsal body surfaces and in having the first posterior tarsomere slightly over 3× the length of the second.

Biology and Ecology: Found on tree trunks, rock outcrops, and in ground litter in tropical and montane forests of Mexico and Central America and in the eastern deciduous forest of United States and southeastern Canada.

Description: Mature larvae 2–3 mm, body relatively robust, normal, head hypognathous. Compound eyes well developed with numerous facets. Antennae longer than body. Three ocellar anlagen visible. Wing pads held vertically, close to sides of body, extending to about distal end of basal half of abdomen. Color creamy white with brown or reddish-brown markings on body and appendages.

Comments: Numerous species occur in the American tropics, most of them still undescribed. Northward, the family falls off rapidly in diversity, and only two species occur in the United States. Both of these are widely distributed, reaching southeastern Canada. They are of no known economic importance.

Selected Bibliography

Eertmoed 1973.

Dolabellopsocidae (Tropical)

Found in dead persistent leaves and ground litter in tropical forests and adjacent plantations. There are 15 known species, most of them from the American tropics. Only a single species reaches north to southern Mexico. They occur in persistent dead leaves and are of no known economic importance.

Selected Bibliography

Eertmoed 1973.

Neurostigmatidae (Tropical)

Figures 21.2, 21.3

Found on trunks and branches of trees in tropical forests. They are strictly American Tropical, with 1 species reaching southern Mexico. They are of no known economic importance.

Selected Bibliography

Eertmoed 1973.

Ptiloneuridae

Relationships and Diagnosis: On Eertmoed's phenogram (1973) this family stands closest to Spurostigmatidae.

Larvae differ from those of other Epipsocetae by having only the lateral ocellar anlagen conspicuous, the median being minute. They differ from those of all families but Spurostigmatidae in possessing regularly spaced, upright, curved, serrate setae dorsally on head, thorax, and abdomen.

Biology and Ecology: Found on tree trunks and rock outcrops in tropical and montane forests.

Description: Mature larvae 3–4 mm, body robust, normal, head hypognathous. Compound eyes well developed with numerous facets. Antennae slightly shorter than body. Lateral ocellar anlagen conspicuous, median minute. Wing pads held vertically, projecting posterolaterally at angle of about 20° from body, reaching distal end of third abdominal segment. Color (on single species at hand) creamy white extensively marked with dark brown.

Comments: This is a small family primarily of the American tropics. A few species reach northern Mexico and southern Arizona in montane localities. The family is of no known economic importance.

Selected Bibliography

Eertmoed 1973.

Spurostigmatidae (Tropical)

Found on tree trunks and occasionally rock outcrops in tropical forest edges, woodlands, and isolated trees. Individuals spin a small dwelling web. Most species of this small family are South American and Antillean, and a single species is known from Mexico. The family is of no known economic importance.

Selected Bibliography

Eertmoed 1973.

GROUP CAECILIETAE

SUPERFAMILY ASIOPSOCOIDEA

Asiopsocidae

Figure 21.20

Relationships and Diagnosis: This group appears to be most closely, but rather distantly, related to Caeciliidae and Amphipsocidae. They share with the latter two families absence of a preapical denticle of pretarsal claw and the adult character of near or complete glabrousness of the third ovipositor valvula.

Both adults and larvae are readily distinguishable from Caeciliidae and Amphipsocidae by their lack of ventral abdominal vesicles (fig. 21.21).

Biology and Ecology: These insects inhabit dead branches of trees. The two genera currently included, *Asiopsocus* and *Notiopsocus,* both show dimorphism, *Asiopsocus* with long-winged males and wingless females, *Notiopsocus* with wingless males and long-winged females.

Description: Mature larvae 1.5–2.5 mm, body relatively robust, normal, head hypognathous with postclypeus decidedly bulging anteriorly. Compound eyes well developed, with numerous facets. Antennae slightly more than half length of body (*Notiopsocus*) to about length of body (*Asiopsocus*). Wing pads held vertically, extending posterolaterally at about 15° angle from body to about end of basal half of abdomen (*Asiopsocus* ♂) or slightly less (*Notiopsocus* ♀). Color creamy white with contents of digestive tract showing through darkly in abdomen.

Comments: The three (?) North American species of this family show a seemingly relictual distribution. The single species of *Asiopsocus, A. sonorensis* Mockford and Garcia Aldrete, is apparently restricted to northern Sonora and adjacent parts of Arizona. The species of *Notiopsocus* (one or two) are restricted to southeastern Mexico and south-central Florida. These insects are of no known economic importance.

Selected Bibliography

Mockford 1977.

SUPERFAMILY CAECILIOIDEA

Caeciliidae

Figures 21.7, 21.17

Relationships and Diagnosis: This family is closely related and very similar to Amphipsocidae. The two families share an "open-mouthed" appearance due to a flat, broad labrum and bulging galeae (fig. 21.18). In both families two or three ventral abdominal vesicles are present (fig. 21.21).

Caeciliid larvae differ from those of amphipsocids (at least, North American) in having the head as broad as the thorax at attachment of fore wing bases, including the wing bases. The thorax is decidedly wider in amphipsocid larvae. Larvae have the flagellum beset with rather fine hairs directed more distally than outward and each one little, if any, longer than the width of the flagellomere from which it arises. Amphipsocid larvae have flagellar hairs somewhat coarser, directed more outward than distally, and most are longer than the width of the flagellomere from which each arises.

Biology and Ecology: Caeciliids primarily inhabit living leaves of trees and shrubs, including foliage of conifers. They may also be found on dead persistent leaves and dead leaves of ground litter. They feed on small fungi. Eggs of broadleaf-inhabiting species of the temperate area overwinter in ground litter, where the first generation larvae remain.

Description: Mature larvae 2–3 mm, body relatively slender, normal, head hypognathous with vertex and frons directed upward, clypeus decidedly bulging forward. Compound eyes well developed with numerous facets. Antennae about length of body. Wing pads held vertically, extending posterolaterally at angle of about 15°, the anteriors and posteriors held tightly together, with tips of anteriors extending slightly beyond posteriors to about distal end of basal half of abdomen. Color creamy yellow, some species with darker markings; contents of digestive tract showing through darkly in abdomen.

Comments: The family is abundant and diverse in all parts of North America but shows a tendency to fall off in diversity northward. They are of no known economic importance.

Selected Bibliography

Dunham 1972.
Mockford 1965, 1966, 1969.
Sommerman 1943a.

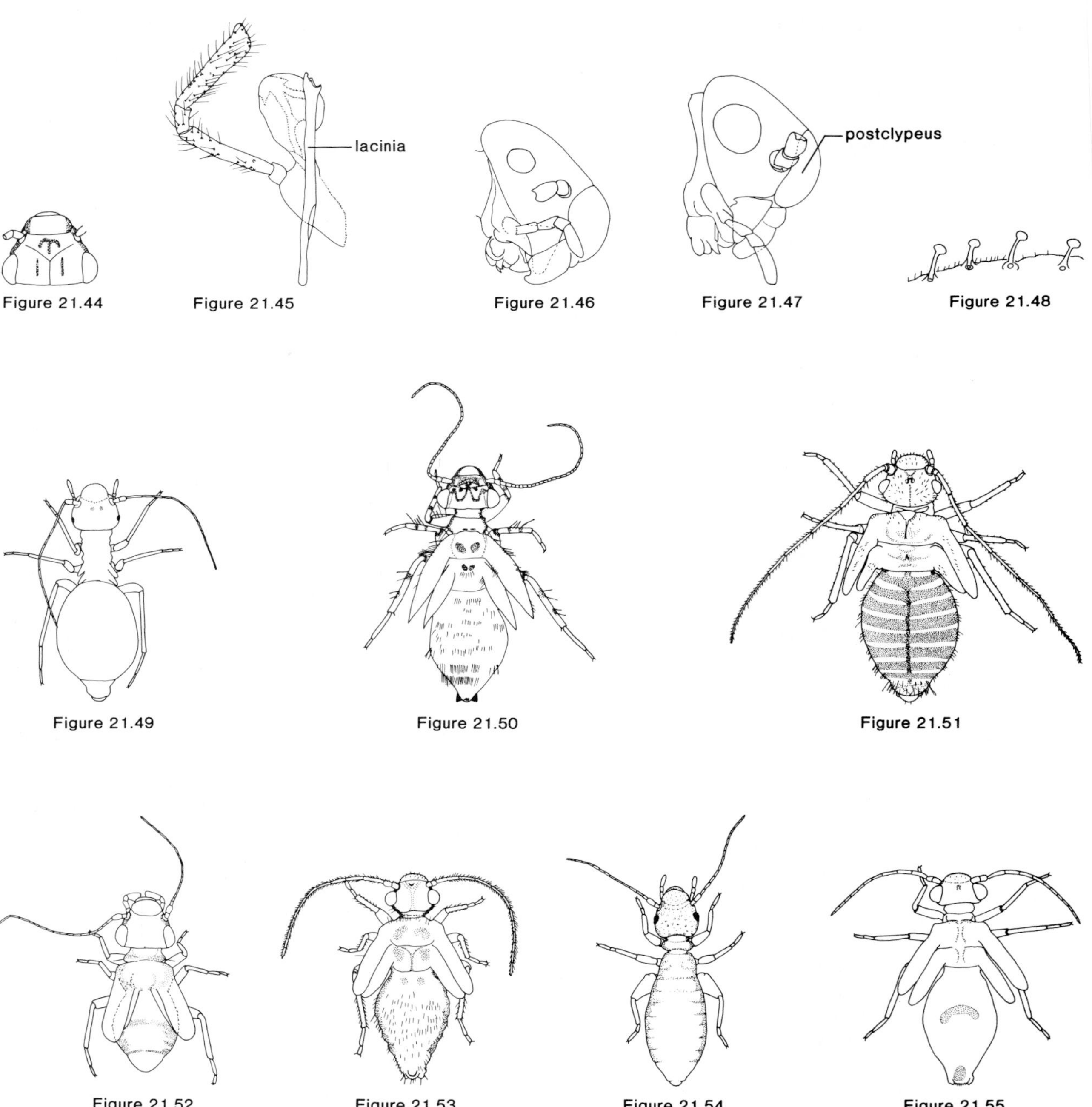

Figure 21.44. *Cerobasis guestfalica* (Kolbe) (Trogiidae). Head, dorsal view.

Figure 21.45. Dorypteryx sp. (Psyllipsocidae). Maxilla showing typical structure for Psocoptera.

Figure 21.46. *Archipsocus floridanus* Mockford (Archipsocidae). Head, lateral view.

Figure 21.47. *Ectopsocus meridionalis* Ribaga (Ectopsocidae). Head, lateral view.

Figure 21.48. *Psocus leidyi* Aaron (Psocidae). Gland hairs on surface of wing pad.

Figure 21.49. ***Psyllipsocus ramburi* Selys Longchamps (Psyllipsocidae)**

Figure 21.50. ***Echmepteryx hageni* (Packard) (Lepidopsocidae)**

Figure 21.51. ***Cerastipsocus venosus* Burmeister (Psocidae)**

Figure 21.52. ***Psoquilla marginepunctata* Hagen (Psoquillidae)**

Figure 21.53. ***Polysocus corruptus* Hagen (Amphipsocidae)**

Figure 21.54. ***Liposcelis bostrychophilus* Badonnel (Liposcelidae)**

Figure 21.55. ***Lachesilla nubilis* (Aaron) (Lachesillidae)**

Amphipsocidae (including Polypsocidae)

Figures 21.18, 21.21, 21.53

Relationships and Diagnosis: As discussed under Caeciliidae (above).

Biology and Ecology: Found on living leaves of trees and shrubs, including foliage of conifers. They may also be found on dead persistent leaves and dead leaves of ground litter.

Description: Mature larvae 3–3.5 mm, body robust, normal, head hypognathous with vertex and frons directed upward, clypeus decidedly bulging forward. Compound eyes well developed with numerous facets. Antennae slightly over half length of body to about length of body. Wing pads held vertically, extending posterolaterally at angle of about 25° from body to about distal end of third abdominal segment. Color creamy white to dull gray with darker markings as in adults of the species.

Comments: The family is primarily Old World and tropical. The genera *Polypsocus* and *Dasydemella* are strictly American and have attained a fair diversity. Five of the seven North American species are in these two genera. Only two species reach the United States, both extending to southern Canada. These insects are of no known economic importance.

Selected Bibliography

Mockford 1978.

GROUP HOMILOPSOCIDEA

Elipsocidae

Figures 21.30, 21.33

Relationships and Diagnosis: This family appears to be most closely related to Mesopsocidae, with which adults in general share several characters of wing venation, male genitalia, and ovipositor valvulae. Larvae share with those of Mesopsocidae approximate orientation of wingpads (see description below), presence of a dark sclerotized semicircle on labrum bordering anterior margin, and presence of a hyaline cone and two large flanking setae on distal margin of the paraproct.

Larvae are distinguishable from those of Mesopsocidae by their considerably less prominent (i.e., less laterally protruding) compound eyes.

Biology and Ecology: Found on trunks and branches of trees and rock outcrops. Some forms spin webs in which they dwell singly or in small groups. Some genera show dimorphism, with long-winged males and micropterous females or the reverse.

Description: Mature larvae 1–3 mm, body relatively slender, normal, head hypognathous. Compound eyes usually well developed with numerous facets, occasionally much reduced. Antennae half to about entire length of body. Wing pads of long-winged forms held semivertically, extending posterolaterally at angle of about 30° from body, reaching nearly to fourth abdominal segment with tips of posteriors extending beyond tips of anteriors. Color creamy yellow to orange-brown, often with darker markings.

Comments: The family was divided into four subfamilies by Smithers (1964); each is rather distinctive in appearance, habits, and distribution. A few species occur in Florida, several on the West Coast, with one extending into British Columbia; others are found in the Rockies and central United States. These insects are of no known economic importance.

Selected Bibliography

Mockford 1955.
Smithers 1964.

Mesopsocidae

Figures 21.23, 21.29, 21.32

Relationships and Diagnosis: See discussion of relationships under Elipsocidae, above.

Biology and Ecology: Found on branches of broad-leaved and coniferous trees, feeding on lichens and algae. Some species are dimorphic with long-winged males and wingless females.

Description: Mature larvae 4 mm, body relatively robust, normal, head hypognathous. Compound eyes well developed with numerous facets. Antennae slightly shorter than body. Wing pads of long-winged forms held semivertically to vertically, extending posterolaterally at an angle of 20–25° from body, reaching nearly to fourth abdominal segment with tips of posteriors extending slightly beyond tips of anteriors. Color dull yellowish- to grayish-white marked with various shades of brown.

Comments: In North America this family is represented by three species of *Mesopsocus* with northern (Holarctic) distributions. Apparently none occur south of southern Illinois and central Colorado. They are of no known economic importance.

Philotarsidae

Figures 21.6, 21.24, 21.38, 21.39

Relationships and Diagnosis: This family is probably closest to Mesopsocidae, with which it shares certain adult features of external genitalia, and, in adults and larvae, the presence of a longitudinal dark, sclerotized band along midline of ventral surface of labium in groove between bases of paraglossae (fig. 21.38).

Larvae are readily distinguishable from those of Mesopsocidae by their less prominent compound eyes, possession of two flanking setae, the dorsal one small and the ventral one large, both pointed (*Philotarsus*, as in fig. 21.28), or the dorsal one a gland hair (*Aaroniella*, fig. 21.24) (two pointed flanking setae in Mesopsocidae), and presence of pointed setae of at least moderate length on all dorsal body surfaces, those on flagellum longer than width of flagellomere from which each arises (only minute truncate setae present in Mesopsocidae; those of flagellum never as long as half width of flagellomere from which each arises).

Biology and Ecology: Found on trunks and branches of trees and rock outcrops. Some forms spin webs in which they dwell singly or in small groups.

Description: Mature larvae 2.5–3 mm, body relatively robust, slightly depressed with head semiprognathous (*Aaroniella*) to normal with head hypognathous (*Philotarsus*). Compound eyes well developed with numerous facets. Antennae about two-thirds body length. Wing pads held vertically, extending posterolaterally at angle of about 20° from body, reaching not quite to middle of abdomen. Color grayish- or greenish-white with medium brown markings (*Aaroniella*) or pale tan with medium brown markings (*Philotarsus*).

Comments: Of approximately nine species that occur in North America, five show tropical or subtropical distributions and the others temperate. Species of *Aaroniella* occur in southern and gulf-coastal Mexico, throughout Florida, and eastern United States. They spin small dwelling webs on tree trunks, branches, and stone outcrops. *Philotarsus kwakiutl* Mockford is common on coniferous trees along the Pacific coast from central California north to British Columbia. An undetermined species of *Philotarsus* is common on conifers and oaks in the Mexican highlands. These insects are of no known economic importance.

Selected Bibliography

Mockford 1951, 1979.

Lachesillidae

Figures 21.27, 21.40, 21.55

Relationships and Diagnosis: This family appears to be most closely, though rather distantly, related to Peripsocidae, with which larvae and adults share a bifid lacinial tip, the two tines being subequal in length.

Larvae are readily distinguishable from those of Peripsocidae by their less robust body form and presence of a small hyaline cone of the paraproct flanked ventrally by a single stout seta (fig. 21.27) (the somewhat larger hyaline cone of Peripsocidae has a stout seta above and below, fig. 21.25).

Biology and Ecology: Found on dead persistent leaves of all kinds and occasionally in dead leaves of ground litter. They feed primarily on small fungi. Eggs are laid singly and bare on the substrate.

Description: Mature larvae 2–2.5 mm, body relatively slender, normal, head hypognathous, the postclypeus occupying much of anterior surface of head but not very bulging. Compound eyes well developed with numerous facets. Antennae about two-thirds the length of body. Wing pads held semivertically, extending posterolaterally at angle of about 30° from body with a slight space between anteriors and posteriors; tips of posteriors extending slightly beyond anteriors and reaching end of about basal third of abdomen. Color creamy yellow with darker marks present in some species but restricted to abdominal annulations and slight pigmentation along sutures.

Comments: This appears to be the largest family of psocids in North America, with close to 200 species, most of them currently undescribed (Garcia Aldrete 1974). Although abundant and diverse in all parts of North America, the group tends to fall off in diversity northward. *Lachesilla pedicularia* (L.) occasionally invades houses and stored food materials. Aside from that, the family is of no known economic importance.

Selected Bibliography

Garcia Aldrete 1974.
Sommerman 1943b, 1944, 1946.

Peripsocidae

Figures 21.25, 21.34

Relationships and Diagnosis: See this section under Lachesillidae, above.

Biology and Ecology: Found on trunks and branches of trees and shrubs and on rock outcrops, feeding on algae and lichens. Eggs are laid singly or in small groups and are overplastered.

Description: Mature larvae 2–2.5 mm, body relatively robust, normal, head hypognathous, the postclypeus occupying much of anterior surface of face and somewhat bulging. Compound eyes well developed with numerous facets. Antennae slightly over half length of body. Wing pads held vertically, extending posterolaterally at about 30° angle from body, the tips of posteriors extending slightly beyond those of anteriors, not quite to end of basal third of abdomen. Color creamy white with medium to dark brown marks on dorsal surfaces of head, thorax, and abdomen.

Comments: The family is represented in North America by two genera (*Kaestneriella* and *Peripsocus*) and about 28 species. There is a decline of diversity northward. *Kaestneriella* is mostly confined to Mexico, with only one species reaching the United States. *Peripsocus* occurs throughout North America, but only five species reach Canada. This family is of no known economic importance.

Selected Bibliography

Eertmoed 1966.
Mockford 1971b.
Mockford and Wong 1969.

Pseudocaeciliidae

Figures 21.41, 21.42

Relationships and Diagnosis: This family appears to be most closely, though rather distantly, related to Trichopsocidae, with which they share in both larva and adult the presence of ventral abdominal vesicles (only one in Trichopsocidae) and an acuminate tip of the distal flagellomere.

Larvae of Pseudocaeciliidae differ from those of Trichopsocidae by possessing (usually) two, rather than one, ventral abdominal vesicles and by virtual or complete lack of the hyaline cone of the paraproct with flanking setae relatively slender (fig. 21.42) (in Trichopsocidae a well-developed cone present with a heavy ventral and weaker dorsal flanking seta, fig. 21.28).

Biology and Ecology: Found on living and persistent dead leaves of trees and shrubs, probably feeding on small fungi.

Description: Mature larvae 2.5–3 mm, body relatively slender, normal, head hypognathous, the postclypeus occupying much of anterior surface of face and somewhat bulging. Compound eyes well developed with numerous facets. Abundant and conspicuous hairs present on all dorsal body surfaces and on appendages. Antennae about three-fourths length of body. Wing pads held vertically, extending posterolaterally at 15–20° from body, the tips of posteriors not exceeding those of anteriors and reaching not quite to end of basal third of abdomen. Color creamy yellow with dark brown lateral marks; ground color probably darker in some species.

Comments: This primarily Old World family is represented in North America by only four species, all of them tropical or subtropical. *Pseudocaecilius citricola* (Ashmead) is common on leaves of citrus in Florida, Mexico, and southern Texas. These insects are of no known economic importance.

Trichopsocidae

Figure 21.28

Relationships and Diagnosis: See this section under Pseudocaeciliidae, above.

Biology and Ecology: These insects inhabit living and persistent dead leaves of trees and shrubs.

Description: Mature larvae 2–2.5 mm, body relatively slender, normal, head hypognathous, the postclypeus relatively smaller and less bulging than in Pseudocaeciliidae. Compound eyes well developed with numerous facets. Abundant and conspicuous hairs present on all dorsal body surfaces and on antennae. Antennae about three-fourths length of body. Wing pads held vertically, extending from body at about 30° angle, tips of anteriors slightly exceeding those of posteriors and reaching not quite to end of basal half of abdomen. Color creamy yellow with dark brown lateral marks.

Comments: This family contains the single genus *Trichopsocus,* represented in North America by only two species. *Trichopsocus clarus* (Banks) occurs along the Pacific Coast of the United States. Another, as yet undetermined, species has been taken in northeastern Mexico. The family is of no known economic importance.

Ectopsocidae

Figures 21.43, 21.47

Relationships and Diagnosis: This family is probably most closely related to the preceding two and to Archipsocidae. All of these families share the absence of the subapical denticle of the pretarsal claw and abundant and conspicuous hairs on dorsal body surfaces and antennae.

Ectopsocid larvae differ from pseudocaeciliids and trichopsocids by lacking ventral abdominal vesicles. They differ from archipsocids by possessing a distinct hyaline cone of the paraproct (fig. 21.43), totally absent in archipsocids.

Biology and Ecology: Ectopsocids occur primarily on persistent dead leaves of trees and shrubs. Some species with short-winged morphs occur in dead leaves of ground litter. A few species have invaded human dwellings and stored foods.

Description: Mature larvae 1.5–2.5 mm, body relatively robust, slightly depressed, head hypognathous, vertex and frons directed upward, postclypeus decidedly bulging forward. Compound eyes well developed with numerous facets. Abundant and conspicuous hairs present anteriorly and dorsally on head and dorsally on thorax; those dorsally on abdomen shorter and less conspicuous. Antennae about two-thirds length of body. Wing pads held vertically, extending from body at about 20° angle, tips of posteriors slightly exceeding those of anteriors, and reaching not quite to end of basal half of abdomen. Color pale yellow-brown with medium brown abdominal annulations; contents of digestive tract showing through abdominal wall darkly.

Comments: This family contains 17 species in North America, most of them confined to the tropical and subtropical portions. *Ectopsocopsis cryptomeriae* (Enderlein) is common in dead persistent leaves throughout eastern United States. It occasionally invades households and stored foods. *Ectopsocus vachoni* Badonnel occurs in ground litter across southern United States and northern Mexico. It was incriminated as a possible vector of the fringed tapeworm of sheep (Allen 1973).

Selected Bibliography

Mockford 1959.
Sommerman 1943c.

Archipsocidae

Figure 21.46

Relationships and Diagnosis: This family appears to be rather closely related to the three preceding ones, sharing with these the absence of the subapical denticle of the pretarsal claw and abundant and conspicuous hairs on dorsal body surfaces and antennae.

Archipsocid larvae differ from pseudocaeciliids and trichopsocids by lack of ventral abdominal vesicles. They differ from ectopsocids by total lack of the paraproctal hyaline cone and by the postclypeus being restricted to the ventral half of the head, well below the compound eye (fig. 21.46), extending more dorsally in Ectopsocidae (fig. 21.47).

Biology and Ecology: Archipsocids are for the most part colonial forms which spin extensive webs on trunks and branches of trees, sometimes on leaves of palms and other trees. They feed on lichens and small fungi. Most species are dimorphic with long- and short-winged females. Males are rarely long winged.

Description: Mature larvae 1.5–2 mm, body relatively slender, normal, head hypognathous, postclypeus decidedly bulging anteroventrally. Compound eyes small, with relatively few facets to rather well developed with numerous facets. Abundant and conspicuous hairs present on all dorsal body surfaces and on antennae. Antennae less than half length of body. Wing pads of long-winged forms pointed at tips, held semihorizontally to semivertically, extending from body at about 10–20° angle, tips of posteriors slightly exceeding those of anteriors in some forms, not in others, and reaching about end of basal half of abdomen or slightly beyond. Color pale to medium reddish brown on body; wings, legs, and antennae creamy white.

Comments: This family is strictly tropical and subtropical. They occur throughout Florida, coastal Georgia, the Gulf coast, and throughout the tropical lowlands of Mexico. The number of North American species, not known at present, probably does not exceed 20. *Archipsocus nomas* Gurney and *A. floridanus* Mockford spin extensive webs on shade trees in central Florida, their dense webbing sometimes covering the entire trunk. Although the sight of their webs on city trees occasionally is objectionable, they are of negligible economic importance.

Selected Bibliography

Gurney 1939.
Mockford 1953, 1957

GROUP PSOCETAE

Hemipsocidae

Figures 21.26, 21.37

Relationships and Diagnosis: This family appears to be most closely, though rather distantly, related to Psocidae and Myopsocidae, with which it shares the adult characters of narrow mesoprecoxal bridge, wide-based mesotrochantin, flagellar sculpture, shape of female paraproct, and presence of male paraproctal prong. It shares with Myopsocidae the adult and larval character of pulvillus bent near its base.

Larvae of this family differ from those of both Psocidae and Myopsocidae by possession of conspicuous hairs on dorsal surfaces of body (only short hairs in latter two families), possession of broad pulvilli (fig. 21.37) (narrow in the latter two), and possession of a conspicuous paraproctal hyaline cone flanked (ventrally) by a single heavy hair (the cone present or absent in the other two families, but two heavy flanking setae always present).

Biology and Ecology: Primarily inhabitants of persistent dead leaves, including dead and dying palm leaves. They run rapidly when disturbed.

Description: Mature larvae 2–2.5 mm, body relatively slender, normal, head hypognathous, vertex and frons directed dorsally, postclypeus somewhat bulging anteriorly. Compound eyes well developed, with numerous facets. Conspicuous upright hairs present anteriorly and dorsally on head and dorsally on thorax and abdomen. Antennae about 1⅓× as long as body. Wing pads held vertically, extending from body at about 20° angle, tips of anteriors exceeding those of posteriors and reaching slightly less than basal half of abdomen. Color creamy white to slightly darker.

Comments: This family is tropical and subtropical in distribution. Three species are known in North America. *Hemipsocus chloroticus* (Hagen) has been introduced into Florida and occurs northward on the Atlantic Coast to Charleston, South Carolina. It also is known from southeastern Mexico. *H. pretiosus* Banks occurs in understory plants of pine lands in southern Florida. *H. africanus* Enderlein is common on native and introduced oil palms in southeastern Mexico. The latter species has been taken in stored foods in Africa. Aside from that, the family is of no known economic importance.

Myopsocidae

Figure 21.36

Relationships and Diagnosis: This family is most closely related to Psocidae, with which it shares adult characters of flagellar sculpture, Cu_{1a} joined to M in fore wing, shape of female paraproct, presence of a prong on male paraproct, and presence of a specialized row of ctenidia-based setae running the length of the posterior tibia. The characters shared with Hemipsocidae are discussed under that family.

Larvae of Myopsocidae differ consistently from those of Psocidae in having the pulvillus bent at an angle near its base (fig. 21.36) (pulvillus in Psocidae straight or slightly curved). They share this character with Hemipsocidae but differ markedly from larvae of that family in having the body much more robust and in having antennae shorter than the body.

Biology and Ecology: Found on tree trunks and shaded rock outcrops, usually in loose groups. They probably feed on algae.

Description: Mature larvae 2.5–3.5 mm, body robust, normal, head hypognathous, postclypeus somewhat bulging anteriorly. Compound eyes well developed with numerous facets. Short, pointed hairs present on postclypeus; other body hairs short and knobbed (gland hairs, as in fig. 21.48). Antennae about three-fourths length of body. Wing pads held vertically, extending from body at about 25° angle; tips of anteriors exceeding those of posteriors and reaching about end of basal third of abdomen. Color creamy white with medium to dark brown markings on body and wing pads.

Comments: This family is primarily tropical and subtropical. Of approximately 17 species that occur in North America, 12 occur in the United States, but only 5 north of Florida. The family is of no known economic importance.

Psocidae

Figures 21.16, 21.19, 21.22, 21.31, 21.35, 21.48, 21.51

Relationships and Diagnosis: See this section under Myopsocidae, above.

Biology and Ecology: Found on trunks and branches of trees and shrubs and on shaded rock outcrops. They feed on lichens and algae. Eggs are laid singly or in clusters and are overplastered with a cementlike material. In some forms a few strands of silk are spun over the eggs. The large species of *Cerastipsocus* and closely related genera are gregarious as larvae and young adults, occurring in herds of up to several hundred.

Description: Mature larvae 2–5 mm, body robust, normal, head hypognathous, postclypeus slightly bulging anteriorly. Compound eyes well developed with numerous facets. Short, pointed hairs abundant on postclypeus; other body hairs—located on dorsal surfaces of head, thorax, and abdomen, and sparsely on ventral surface of abdomen—short and knobbed (gland hairs, subfamilies Psocinae and Amphigerontiinae, fig. 21.48), or short to moderate and pointed or blunt-tipped (Subfamily Cerastipsocinae). Antennae from about ⅔ to about 1 ⅓× length of body. Wing pads held vertically, extending from body at about 15–25° angle; tips of anteriors exceeding those of posteriors, or the reverse, reaching about to fourth abdominal segment. Color variable; creamy white with dark brown marks in some forms, dull yellow-brown with reddish brown abdominal annulations in others.

Comments: This family is abundant and diverse in all parts of North America. The precise number of North American species remains unknown but is probably about 80. Of this number, fully half occur north of the subtropics, and many reach Canada. They are of no known economic importance.

Selected Bibliography

Mockford 1974b, 1984b.
Sommerman 1948.

BIBLIOGRAPHY

Allen, R. W. 1973. The biology of *Thysanosoma actinioides* (Cestoda: Anoplocephalidae), a parasite of domestic and wild ruminants. N. Mex. St. Univ. Ag. Exp. Sta. Bull. 604:1–68.

Badonnel, A. 1951. Ordre des Psocoptères, V. 10, Fasc. 2:1301–40. *In* Grassé, P. P. (ed). Traité de Zoologie. Paris: Masson et Cie. 17 vols.

Broadhead, E. 1950. A revision of the genus *Liposcelis* Motschulsky with notes on the position of this genus in the order Corrodentia and on the variability of ten *Liposcelis* species. Trans. Roy. Ent. Soc. Lond. 101:335–88.

Broadhead, E. 1961. Biology of *Psoquilla marginepunctata* (Hagen) (Corrodentia, Trogiidae). Trans. Soc. Brit. Ent. 14:223–36.

Chapman, P. J. 1930. Corrodentia of the United States of America: I. Suborder Isotecnomera. J. N.Y. Ent. Soc. 38:219–90, 319–403.

Dunham, R. S. 1972. A life history study of *Caecilius aurantiacus* (Hagen) (Psocoptera: Caeciliidae). Great Lakes Ent. 5:17–27.

Eertmoed, G. 1966. The life history of *Peripsocus quadrifasciatus* (Psocoptera: Peripsocidae). J. Kansas Ent. Soc. 39:54–65.

Eertmoed, G. 1973. The phenetic relationships of the Epipsocetae (Psocoptera): the higher taxa and the species of two new families. Trans. Amer. Ent. Soc. 99:373–414.

Garcia Aldrete, A. N. 1974. A classification above species level of the genus *Lachesilla* Westwood (Psocoptera: Lachesillidae). Folia Ent. Mex. 27:1–88.

Gurney, A. B. 1939. Nomenclatorial notes on Corrodentia with descriptions of two new species of *Archipsocus*. J. Wash. Acad. Sci. 29:501–15.

Gurney, A. B. 1943. A synopsis of psocids of the tribe Psyllipsocini, including the description of an unusual new genus from Arizona (Corrodentia: Empheriidae: Empheriinae). Ann. Ent. Soc. Amer. 36:195–220.

Mockford, E. L. 1951. On two North American philotarsids (Psocoptera). Psyche 58:102–107.

Mockford, E. L. 1953. Three new species of *Archipsocus* from Florida (Psocoptera: Archipsocidae). Fla. Ent. 36:113–24.

Mockford, E. L. 1955. Studies on the Reuterelline psocids. Proc. Ent. Soc. Wash. 57:97–108.

Mockford, E. L. 1957. Life history studies on some Florida insects of the genus *Archipsocus* (Psocoptera). Bull. Fla. St. Mus. 1:253–74.

Mockford, E. L. 1959. The *Ectopsocus briggsi* complex in the Americas. Proc. Ent. Soc. Wash. 61:260–66.

Mockford, E. L. 1963. The species of Embidopsocinae of the United States (Psocoptera: Liposcelidae). Ann. Ent. Soc. Amer. 56:25–37.

Mockford, E. L. 1965. The genus *Caecilius* (Psocoptera: Caeciliidae). Part I. Species groups and the North American species of the *flavidus* group. Trans. Amer. Ent. Soc. 91:121–66.

Mockford, E. L. 1966. The genus *Caecilius* (Psocoptera: Caeciliidae). Part II. Revision of the species groups, and the North American species of the *fasciatus, confluens,* and *africanus* groups. Trans. Amer. Ent. Soc. 92:133–72.

Mockford, E. L. 1967a. The electrentomoid psocids (Psocoptera). Psyche 74:118–65.

Mockford, E. L. 1967b. Some Psocoptera from plumage of birds. Proc. Ent. Soc. Wash. 69:307–09.

Mockford, E. L. 1969. The genus *Caecilius* (Psocoptera: Caeciliidae. Part III. The North American species of the *alcinus, caligonus,* and *subflavus* groups. Trans. Amer. Ent. Soc. 95:77–151.

Mockford, E. L. 1971a. Psocoptera from sleeping nests of the dusky-footed wood rat in southern California (Psocoptera: Atropidae, Psoquillidae, Liposcelidae). Pan-Pac Ent. 47:127–40.

Mockford, E. L. 1971b. *Peripsocus* species of the *alboguttatus* group (Psocoptera: Peripsocidae). J. N.Y. Ent. Soc. 79:89–115.

Mockford, E. L. 1971c. Parthenogenesis in psocids (Insecta: Psocoptera). Amer. Zool. 11:327–39.

Mockford, E. L. 1974a. The *Echmepteryx hageni* complex (Psocoptera: Lepidopsocidae) in Florida. Fla. Ent. 57:255–67.

Mockford, E. L. 1974b. *Trichadenotecnum circularoides* (Psocoptera: Psocidae) in southeastern United States, with notes on its reproduction and immature stages. Fla. Ent. 57:369–70.

Mockford, E. L. 1977. *Asiopsocus sonorensis* (Psocoptera: Asiopsocidae): a new record, augmented description, and notes on reproductive biology. Southw. Nat. 22:21–29.

Mockford, E. L. 1978. A generic classification of the family Amphipsocidae (Psocoptera: Caecilietae). Trans. Amer. Ent. Soc. 104:139–90.

Mockford, E. L. 1979. Diagnoses, distribution, and comparative life history notes on *Aaroniella maculosa* (Aaron) and *A. eertmoedi* n. sp. (Psocoptera: Philotarsidae). Great Lakes Ent. 12:35–44.

Mockford, E. L. 1984a. Two new species of *Speleketor* from southern California, with comments on the taxonomic position of the genus (Psocoptera: Prionoglaridae). S.W. Nat. 29:169–179.

Mockford, E. L. 1984b. A systematic study of the genus *Camelopsocus,* with descriptions of three new species (Psocoptera: Psocidae). Pan-Pac. Ent. 60:193–212.

Mockford, E. L. and Wong Siu Kai. 1969. The genus *Kaestneriella* (Psocoptera: Peripsocidae). J. N.Y. Ent. Soc. 77:221–49.

New, T. R. 1977. Notes on the identification of nymphs of the British Psocoptera. Ent. Gazette 28:61–71.

Pearman, J. V. 1928. On sound production in the Psocoptera and on a presumed stridulatory organ. Ent. Mon. Mag. 64:179–86.

Smithers, C. N. 1964. Notes on the relationships of the genera of Elipsocidae (Psocoptera). Trans. Roy. Ent. Soc. Lond. 116:211–24.

Smithers, C. N. 1972. The classification and phylogeny of the Psocoptera. Australian Mus. Mem. 14:1–349.

Sommerman, K. M. 1942. Rearing techniques for Corrodentia. Ent. News 53:259–61.

Sommerman, K. M. 1943a. Description and bionomics of *Caecilius manteri* n. sp. (Corrodentia). Proc. Ent. Soc. Wash. 45:29–39.

Sommerman, K. M. 1943b. Bionomics of *Lachesilla nubilis* (Aaron) (Corrodentia, Caeciliidae). Can. Ent. 75:99–105.
Sommerman, K. M. 1943c. Bionomics of *Ectopsocus pumilis* (Banks) (Corrodentia, Caeciliidae). Psyche 50:53–64.
Sommerman, K. M. 1944. Bionomics of *Amapsocus amabilis* (Walsh) (Corrodentia, Psocidae). Ann. Ent. Soc. Amer. 37:359–64.
Sommerman, K. M. 1946. A revision of the genus *Lachesilla* north of Mexico. Ann. Ent. Soc. Amer. 39:627–61.
Sommerman, K. M. 1948. Two new Nearctic psocids of the genus *Trichadenotecnum* with a nomenclatural note on a third species. Proc. Ent. Soc. Wash. 50:165–73.
Sommerman, K. M. 1956. Two new species of *Rhyopsocus* from the U.S.A., with notes on the bionomics of one household species. J. Wash. Acad. Sci. 46:145–49.
Sommerman, K. M. 1957. Three new species of *Liposcelis* (= *Troctes*) (Psocoptera) from Texas. Proc. Ent. Soc. Wash. 59:125–29.
Spieksma, F. Th. M., and C. Smits. 1975. Some ecological and biological aspects of the booklouse *Liposcelis bostrychophilus* Badonnel 1931 (Psocoptera). Netherl. J. Zool. 25:219–30.
Winston, P. W., and D. H. Bates. 1960. Saturated solutions for the control of humidity in biological research. Ecology 41:232–37.

Order Mallophaga

22

Roger D. Price
University of Minnesota

CHEWING LICE

Mallophaga are permanent ectoparasites, primarily on birds, but with a small percentage of species on mammals.

The economic importance of the vast majority of the chewing lice is not known; however, a few species found on poultry and livestock have been reported to cause irritation, loss of weight or production, and to otherwise contribute to an unhealthy condition of the host.

DIAGNOSIS

Small, somewhat flattened individuals, with mandibulate mouthparts. Body divided into distinct head, thorax, and abdomen; head wider than pronotum; three pairs of segmented thoracic legs. Antennae short, of only three to five segments. Tarsi one-segmented, with zero to two claws. Abdomen of eight or nine apparent segments.

Immatures are much like adults except for smaller size, reduced chaetotaxy and sclerotization, and absence of genitalic features. A progressive increase in size, number of setae, and degree of sclerotization occurs from early to late instars, as shown in figs. 22.11–22.13 and 22.26–22.28.

Most work to date on the Mallophaga has been taxonomic, this being almost exclusively based on adult forms. As a result, immatures have for the most part been disregarded or discarded. Information is limited to occasional descriptions of an immature or part thereof. Nothing is known of the taxonomic characters of immatures and the possibility of species identification in the absence of adults. Workers should be encouraged to collect early instars whenever possible so that collections can gradually be built up. Ultimately, in this manner, a comprehensive treatise may be possible.

BIOLOGY AND ECOLOGY

Biological information is available for a relatively small number of species. From what is known, lice deposit their eggs on the host, attaching them directly to the feather or hair, or in some cases depositing them within the shafts of primary feathers. Development is hemimetabolous. There are three immature instars, with all stages permanently associated with the host; lice can live only a short time away from the host. Their food consists of bits of feathers or hair, or other particulate matter associated with the skin; some species of Menoponidae and Ricinidae ingest blood as a regular part of their diet.

Although it has often been stated that the chewing lice are highly host specific and that each host taxon has its own unique louse taxa, such a generalization is open to challenge. Work based on morphological evidence suggests that some lice are indeed quite specific, others not so much so, and at least a few may be found on a wide spectrum of hosts. The practice of some workers in describing new louse taxa solely on the fact that they are taken from hosts not previously yielding lice is to be condemned. Lice must be recognized as distinct on the basis of characteristics of the lice and not solely because of their host association.

DESCRIPTION

Head distinct from thorax, as wide or wider than prothorax, with pair of mandibles situated ventrally. Antennae three- to five-segmented, clubbed or filiform. Eyes reduced or absent.

Prothorax usually distinct from mesothorax; meso- and metathorax often closely associated to form pterothorax. Three pairs of distinctly segmented legs; tarsi one-segmented, either with zero, one, or two terminal claws.

Abdomen distinctly segmented, usually with at most only weak sclerotization. With eight or nine apparent segments, these being 2–8 or 1–8, with terminal portion beyond 8 representing fusion of segments 9 and beyond. With six, less often zero to five pairs of spiracles, these associated when present with segments 3–8.

COMMENTS

The classification followed here for the higher taxa is essentially that of Hopkins and Clay (1952), in which the Mallophaga are divided into three suborders, two of which occur in North America: the Amblycera with seven families, six of which are represented in the North American fauna; and the Ischnocera, with two of three families in North America.

TECHNIQUES

The preparation of chewing lice for microscopic study is essentially the same as for other small soft-bodied insects. Although staining may facilitate character observation, a phase contrast microscope is essential for detailed study.

Rearing methods for some of the few successful attempts to date may be found in Stenram (1956) for *Columbicola columbae* Linnaeus; Nelson (1971) for *Colpocephalum turbinatum* Denny; Hopkins (1970) for *Bovicola ovis* (Schrank); and Hopkins and Chamberlain (1969, 1972) for *B. crassipes* (Rudow), *B. limbata* (Gervais), and *B. bovis* (Linnaeus). For additional techniques, see the order Anoplura.

KEY TO THE FAMILIES OF IMMATURE NORTH AMERICAN MALLOPHAGA

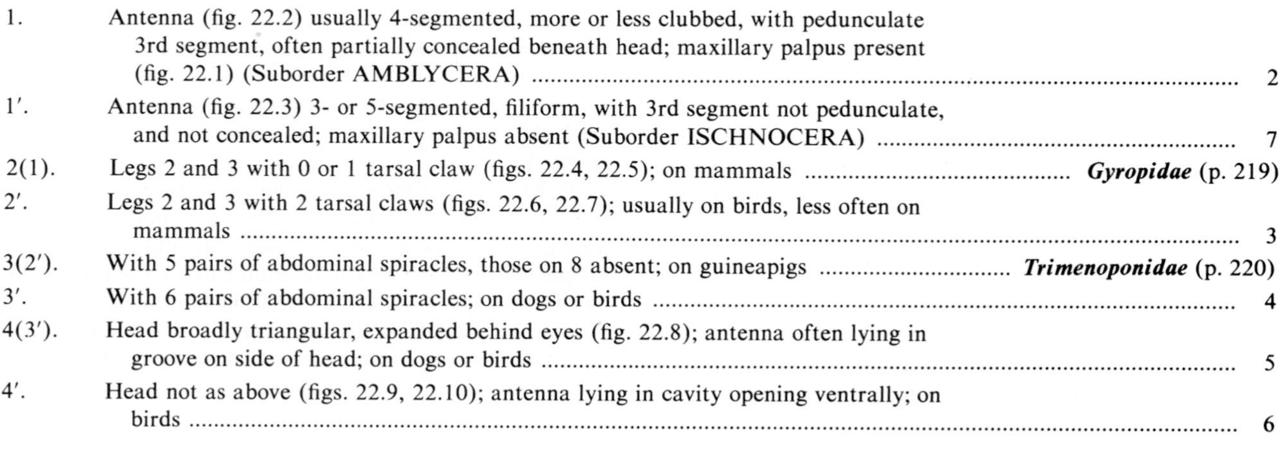

1. Antenna (fig. 22.2) usually 4-segmented, more or less clubbed, with pedunculate 3rd segment, often partially concealed beneath head; maxillary palpus present (fig. 22.1) (Suborder AMBLYCERA) 2

1′. Antenna (fig. 22.3) 3- or 5-segmented, filiform, with 3rd segment not pedunculate, and not concealed; maxillary palpus absent (Suborder ISCHNOCERA) 7

2(1). Legs 2 and 3 with 0 or 1 tarsal claw (figs. 22.4, 22.5); on mammals .... ***Gyropidae*** (p. 219)

2′. Legs 2 and 3 with 2 tarsal claws (figs. 22.6, 22.7); usually on birds, less often on mammals 3

3(2′). With 5 pairs of abdominal spiracles, those on 8 absent; on guineapigs .... ***Trimenoponidae*** (p. 220)

3′. With 6 pairs of abdominal spiracles; on dogs or birds 4

4(3′). Head broadly triangular, expanded behind eyes (fig. 22.8); antenna often lying in groove on side of head; on dogs or birds 5

4′. Head not as above (figs. 22.9, 22.10); antenna lying in cavity opening ventrally; on birds 6

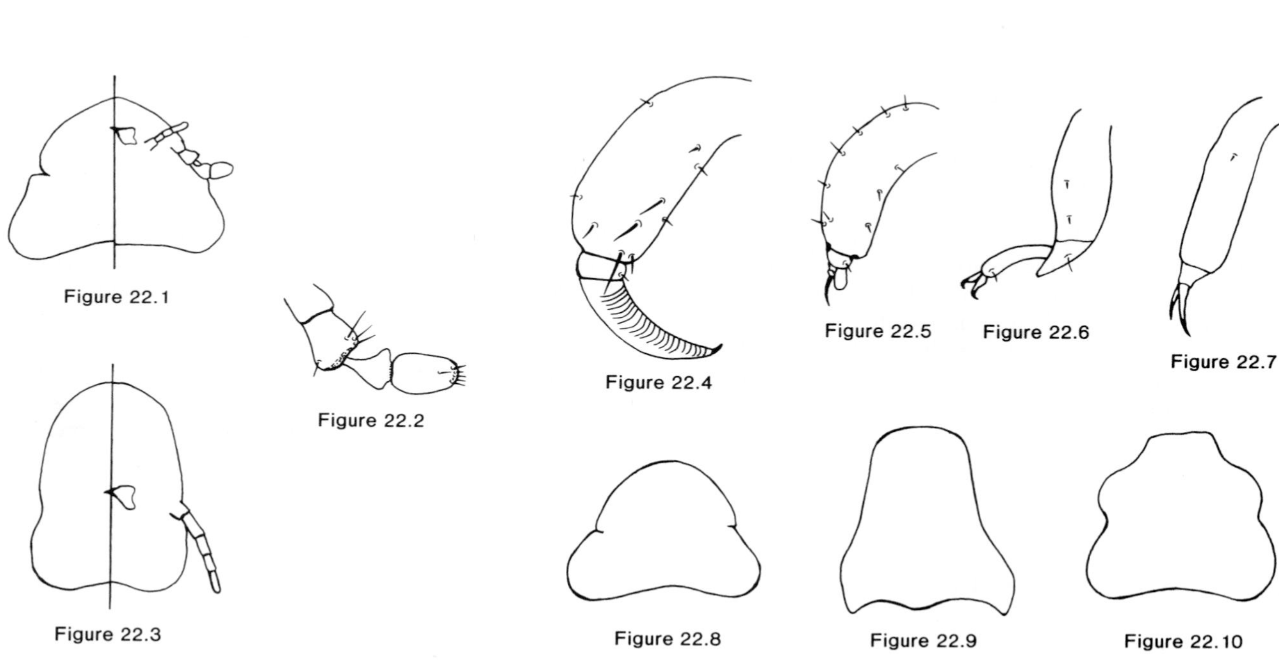

Figures 22.1–22.3. Head: **(22.1)** Head outline of *Trimenopon hispidum* (Burmeister) (Trimenoponidae); **(22.2)** Antenna of *Menopon gallinae* (L.) (Menoponidae); **(22.3)** Head outline of *Lipeurus caponis* (L.) (Philopteridae).

Figures 22.4–22.7. Distal leg 3 of: **(22.4)** *Gyropus setifer* (Ewing) (Gyropidae); **(22.5)** *Gliricola echimydis* (Gyropidae); **(22.6)** *Menopon gallinae* (L.) (Menoponidae); **(22.7)** *Lipeurus caponis* (L.) (Philopteridae).

Figures 22.8–22.10. Head outline of: **(22.8)** *Menopon gallinae* (L.) (Menoponidae), shaft louse; **(22.9)** *Ricinus fringillae* De Geer (Ricinidae); **(22.10)** *Laemobothrion maximum* (Laemobothriidae).

5(4). Mesonotum with pair of stout setae each borne on protuberance (fig. 22.18); abdominal tergum 1 fused with metanotum, with 1st abdominal spiracle on apparent 2nd segment; on dogs .... ***Boopiidae*** (p. 219)

5′. Mesonotum without such stout setae or protuberances; abdominal tergum 1 not fused with metanotum, with 1st abdominal spiracle on apparent 3rd segment; on birds .... ***Menoponidae*** (p. 217)

6(4′). Side of head with conspicuous swelling in front of eye at base of antenna (fig. 22.10); ventral temple (fig. 22.15) with sculpturing of outer rows of peglike projections; venter of 3rd femur with microtrichial patch (fig. 22.16) .... ***Laemobothriidae*** (p. 218)

6′. Side of head without such swelling (fig. 22.9); without peglike sculpturing on temple or microtrichial patch on 3rd femur .... ***Ricinidae*** (p. 219)

7(1′). With 2 tarsal claws; antenna with 5 segments; on birds .... ***Philopteridae*** (p. 220)

7′. With only 1 tarsal claw; antenna usually with 3 segments; on mammals .... ***Trichodectidae*** (p. 220)

CLASSIFICATION

Order **MALLOPHAGA**
- Suborder Amblycera
  - Menoponidae, avian body lice
  - Laemobothriidae, on birds
  - Ricinidae, on birds
  - Boopiidae, on mammals
  - Gyropidae, rodent chewing lice
  - Trimenoponidae, on mammals
- Suborder Ischnocera
  - Philopteridae, feather chewing lice
  - Trichodectidae, mammal chewing lice

HOST LISTS

Since so much Mallophaga identification is based on host lists rather than on formal keying of the lice themselves, a set of four publications by Emerson (1972a, b, c, d) is indispensable for anyone attempting to affix a specific name to a North American chewing louse. The first two works list the species of lice by family and genus for each suborder, giving known hosts for each species. The last two works present a mammal and bird host list, giving the lice known from each host.

Selected Bibliography

General

Hopkins and Clay 1952.
Keler 1960 (adults only).

Rearing

Stenram 1956.
Nelson 1971.
Hopkins 1970.
Hopkins and Chamberlain 1969, 1972.

Suborder AMBLYCERA

Relationships and Diagnosis

An assemblage of six North American families, three of which are restricted to birds and three to mammals. Specimens are readily separated from those of the suborder Ischnocera by having short, four-segmented, clubbed antennae (fig. 22.2) and maxillary palps (fig. 22.1). A discussion of the relationships among the families of Amblycera may be found in Clay (1970).

MENOPONIDAE

The Menoponids, Avian Body Lice

Figures 22.2, 22.6, 22.8, 22.11–22.14

Relationships and Diagnosis: This family of Amblycera contains by far the largest number of species of the suborder, with over 250 North American species distributed in about 35 genera, all restricted to a wide spectrum of bird hosts.

Separable from the other families of amblyceran lice by the broadly triangular head and presence of nine apparent abdominal segments bearing six pairs of spiracles on 3–8.

Biology and Ecology: Members of this family are believed to ingest blood fairly often in the course of their feeding and, by so doing, have the potential to transmit pathogens among their avian hosts.

Description: In size, adults most commonly from 1–3 mm long, but occasionally up to 7–8 mm; generally somewhat ovoid slender specimens, often brownish or gray. Head roughly triangular, with temple variably wider than anterior head; each tarsus with two claws; abdomen with nine apparent segments, with six pairs of spiracles laterally on 3–8.

Comments: Clay (1969) has applied numbers to most of the head and pronotal setae and has otherwise discussed in detail the morphology of adult menoponids. At least some of this will be pertinent to immature descriptions following study to establish appropriate homologies.

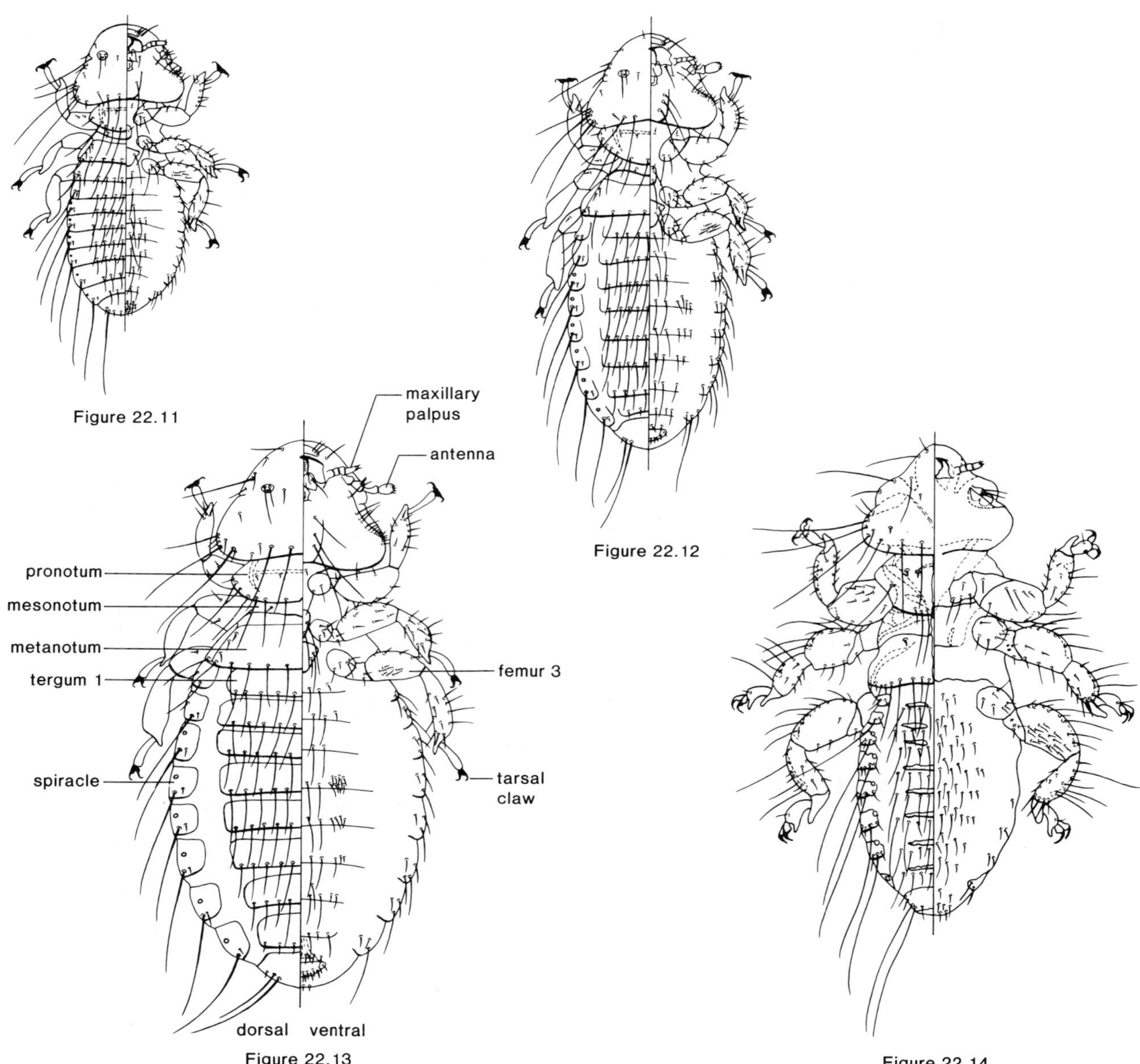

Figures 22.11–22.13. *Menopon gallinae* (L.) (Menoponidae): **(22.11)** first instar; **(22.12)** second instar; **(22.13)** third instar.

Figure 22.14. *Trinoton querquedulae* (L.), large duck louse (Menoponidae).

The most common economically important lice are those associated with poultry: *Menacanthus stramineus* (Nitzsch), the chicken body louse, and *Menopon gallinae* (L.), the shaft louse (figs. 22.11–22.13). Several of the largest members of the family (*Trinoton querquedulae* (L.), the large duck louse (fig. 22.14), and *T. anserinum* (J. C. Fabricius), the goose body louse) are associated with ducks and geese and are often encountered by hunters handling freshly killed game.

Selected Bibliography

Clay 1969.
Emerson 1956, 1962 (adults only).

LAEMOBOTHRIIDAE

The Laemobothriids

Figures 22.10, 22.15, 22.16

Relationships and Diagnosis: The largest of all Mallophaga (adults to 12 mm long), but with only eight North American species, four of these found on various falconiform hosts and the others on gallinules, limpkins, coots, and grebes.

The large size and unique head shape (fig. 22.15), with the swelling at the side in front of the eye at the antennal base, separate laemobothriids from the other amblyceran bird lice.

Description: Adults up to 12 mm long, with moderately heavy ovoid bodies, usually tan to dark brown. Head angular (fig. 22.15), with conspicuous swelling at side in front of eye near antennal base; antennae mostly concealed in ventral concavity; ventral temple with sculpturing of rows of peglike projections; prothorax as for ricinids; each tarsus with two claws; venter of femur 3 (fig. 22.16) with patch of microtrichia; abdomen with nine apparent segments; six pairs of spiracles laterally on 3–8.

Comments: Essentially nothing has been done on the immatures of this family or on the adults of the nonfalconiform hosts. Nelson and Price (1965) have discussed the adult taxonomy of the four species found on the Falconiformes.

Selected Bibliography

Nelson and Price 1965 (adults only).

RICINIDAE

The Ricinids

Figures 22.9, 22.17

Relationships and Diagnosis: Only about 30 species in three genera that are restricted in distribution to the songbirds and hummingbirds. Readily distinguished from both the Menoponidae and Laemobothriidae by the head shaped much as in figure 22.17 and by lacking a conspicuous swelling at side in front of the eye near antennal base, and from the Menoponidae by the clubbed antennae lying in a cavity opening ventrally.

Biology and Ecology: Although little is known other than the generalizations given earlier, an excellent summary of available information is given by Nelson (1972). He compiles details of life cycle, sex ratio, intensity of infestation, rate of incidence, seasonal abundance, oviposition sites, dispersion on the host, and blood feeding. In this last regard, it should be noted that ricinids apparently regularly feed on blood and serum and, by so doing, have the potential for transmitting avian infectious agents.

Description: Adults from 2–6 mm long, rather elongate specimens with parallel sides, and pale colored. Head uniquely shaped (fig. 22.17), lacking conspicuous swelling at side in front of eye near antennal base; antennae concealed in concavity opening ventrally; mandibles reduced and inconspicuous; prothorax rectangular; each tarsus with two claws; abdomen with only eight apparent segments, segment 1 absent; six pairs of spiracles laterally on 3–8.

Comments: No members of this family are considered of economic importance, since they all occur on nondomestic hosts.

Selected Bibliography

Nelson 1972.
Rheinwald 1968 (adults only).

BOOPIIDAE

The Boopiids

Figure 22.18

Relationships and Diagnosis: With a number of features in common with, and possibly closely related to, the Menoponidae. Found only on mammals, with only one North American species.

Separated from other amblycerans by the presence of a seta on a prominent protuberance on each side of mesonotum (fig. 22.18) in combination with the head shape, presence of two tarsal claws, and only eight apparent abdominal segments.

Description: Adults about 2.5 mm long, generally ovoid and brownish. Head roughly triangular; with ventral postpalpal spinous process; mesonotum with seta-bearing protuberance on each side; quadrate metathorax; each tarsus with two claws; abdomen with only eight apparent segments bearing six pairs of spiracles laterally on 3–8.

Comments: Only one species, *Heterodoxus spiniger* (Enderlein) (fig. 22.18), is known in North America; it has been found on the domestic dog, coyote, and wolf.

Selected Bibliography

Emerson and Price 1975, 1981 (adults only).
Hopkins 1949 (adults only).
Keler 1971 (adults only).
Werneck 1948 (adults only).

GYROPIDAE

The Rodent Chewing Lice

Figures 22.4, 22.5, 22.19

Relationships and Diagnosis: Clay (1970) finds this family having the fewest characters in common with the Menoponidae of all Amblycera; the diversity within the Gyropidae even suggests it may be polyphyletic. There are four North American species, each placed in a separate genus.

The gyropids are the only Amblycera having only zero to one tarsal claws on legs 2 and 3, and this claw often is extremely modified and large (fig. 22.4) or quite small (fig. 22.5).

Description: Small, slender lice, only 1.0–1.5 mm long, and whitish. Head rather broad, angular; legs often disproportionately long, with no or single claw on 2 and 3, either very large (fig. 22.4) or inconspicuously small (fig. 22.5); abdomen with eight to nine apparent segments, with five to six pairs of spiracles.

Comments: Found only on mammals, the most likely species of this family to be found in North America are associated with guineapigs in laboratories—*Gliricola porcelli* (Schrank), the slender guineapig louse, and *Gyropus ovalis* Burmeister, the oval guineapig louse. In addition to these, a louse from the nutria and one from the collared peccary are the only other ones likely to be found.

Selected Bibliography

Clay (1970).
Emerson and Price 1975, 1981 (adults only).
Hopkins 1949 (adults only).
Werneck 1948 (adults only).

TRIMENOPONIDAE

The Trimenoponids

Figures 22.1, 22.20

Relationships and Diagnosis: Close to the Menoponidae and Boopiidae in some respects, but with distinctive features. Found on mammals, with only one possible North American species.

Separated from other amblyceran lice by having a broadly triangular head, abdominal tergite 1 reduced, and with nine apparent abdominal segments, but with only five pairs of spiracles, those on 8 absent.

Description: Head roughly triangular; ovoid prothorax; each tarsus with two small claws; abdomen with nine apparent segments, with first reduced and much shorter than others; with only five pairs of spiracles, those on 8 being absent.

Comments: Only one species, *Trimemopon hispidum* (Burmeister) (fig. 22.20), may be found in North America; it occurs on guineapigs, but has not been reported to date north of Panama.

Selected Bibliography

Emerson and Price, 1975, 1981 (adults only).
Hopkins 1949 (adults only).
Werneck 1948 (adults only).

Suborder ISCHNOCERA

Relationships and Diagnosis

There are only two North American families in this suborder, with one of these restricted to birds and the other to mammals. Specimens are easily separated from those in the Amblycera by having short three- or five-segmented filiform antennae (fig. 22.3) and lacking maxillary palpi.

PHILOPTERIDAE

The Feather Chewing Lice

Figures 22.3, 22.7, 22.21–22.25

Relationships and Diagnosis: This family has the largest number of species of any in the Mallophaga, with over 450 North American species distributed in about 60 genera, all found on birds. It and the Trichodectidae represent the only North American families of Ischnocera and thus are presumably closely related.

The Philopteridae are distinguished by having an antenna of five segments and two tarsal claws on each leg.

Description: A wide variety of sizes and shapes, with adults usually from 2–4 mm long, and broad (fig. 22.22) to slender (fig. 22.21) individuals. Head from narrowly rounded to broadly tapered anteriorly; antenna five-segmented; tarsi with two claws; abdomen with eight apparent segments; with six pairs of spiracles laterally on 3–8.

Comments: Found only on birds. As would be expected, a family of this size includes a number of parasites of poultry and other domestic birds. Commonly occurring on chickens are *Cuclotogaster heterographus* (Nitzsch), the chicken head louse (fig. 22.23); *Goniodes dissimilis* Denny, the brown chicken louse; *G. gigas* (Taschenberg), the large chicken louse; *Goniocotes gallinae* (De Geer), the fluff louse (fig. 22.24); and *Lipeurus caponis* (L.), the wing louse (fig. 22.21). Several species from turkeys are *Chelopistes meleagridis* (L.), the large turkey louse (fig. 22.25); and *Oxylipeurus polytrapezius* (Burmeister), the slender turkey louse.

Selected Bibliography

Emerson 1956, 1962 (adults only).

TRICHODECTIDAE

The Mammal Chewing Lice

Figures 22.26–22.30

Relationships and Diagnosis: The only family of Ischnocera containing mammal-infesting lice; closely related to the Philopteridae. There are somewhat more than 50 North American species occurring in ten genera.

Recognized by having an exposed antenna, usually of three segments, and only one tarsal claw on each leg.

Description: A variety of sizes and shapes, much as for the Philopteridae. Antenna usually three-segmented; tarsi each with one claw; abdomen with nine apparent segments; with variable number of spiracles, from zero to six pairs.

Comments: Restricted to mammals. The genus *Bovicola* contains a number of important livestock pests, including *B. bovis* (L.), the cattle biting louse (figs. 22.26–22.28); *B. caprae* (Gurlt), the goat biting louse; *B. equi* (Denny), the horse biting louse; *B. limbata* (Gervais), the Angora goat biting louse; and *B. ovis* (Schrank), the sheep biting louse. Lice found on domestic pets around the home include *Felicola subrostratus* (Burmeister), the cat louse (fig. 22.30); and *Trichodectes canis* (De Geer), the dog biting louse (fig. 22.29).

Selected Bibliography

Emerson and Price 1975, 1981 (adults only).
Hopkins 1949 (adults only).
Werneck 1948, 1950 (adults only).

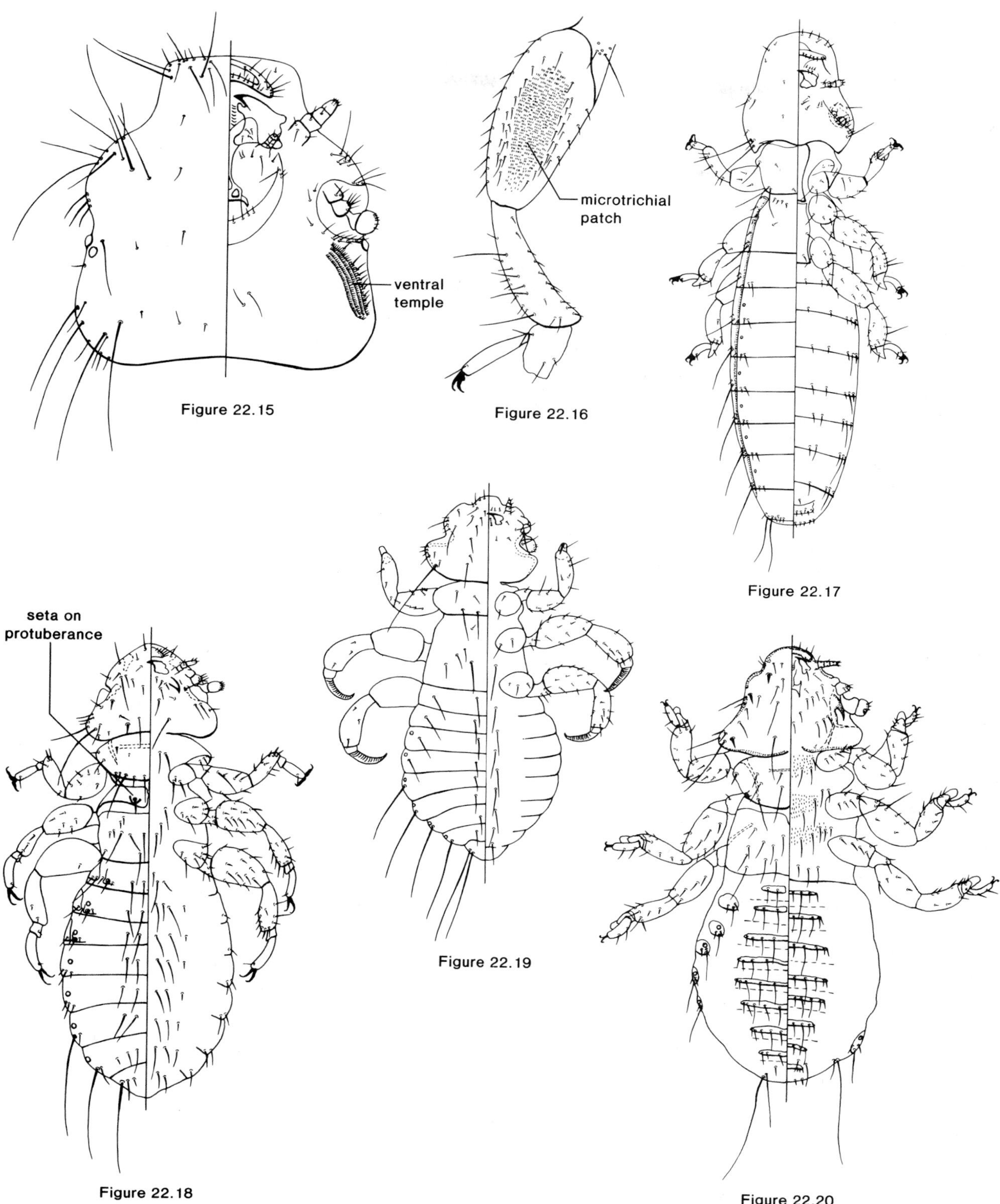

Figure 22.15

Figure 22.16

Figure 22.17

Figure 22.18

Figure 22.19

Figure 22.20

Figures 22.15–22.16. *Laemobothrion maximum* (Scopoli) (Laemobothriidae): **(22.15)** head; **(22.16)** ventral leg 3.

Figure 22.17. *Ricinus fringillae* De Geer (Ricinidae).

Figure 22.18. *Heterodoxus spiniger* (Enderlein) (Boopiidae).

Figure 22.19. *Gyropus setifer* (Ewing) (Gyropidae).

Figure 22.20. *Trimenopon hispidum* (Burmeister) (Trimenoponidae).

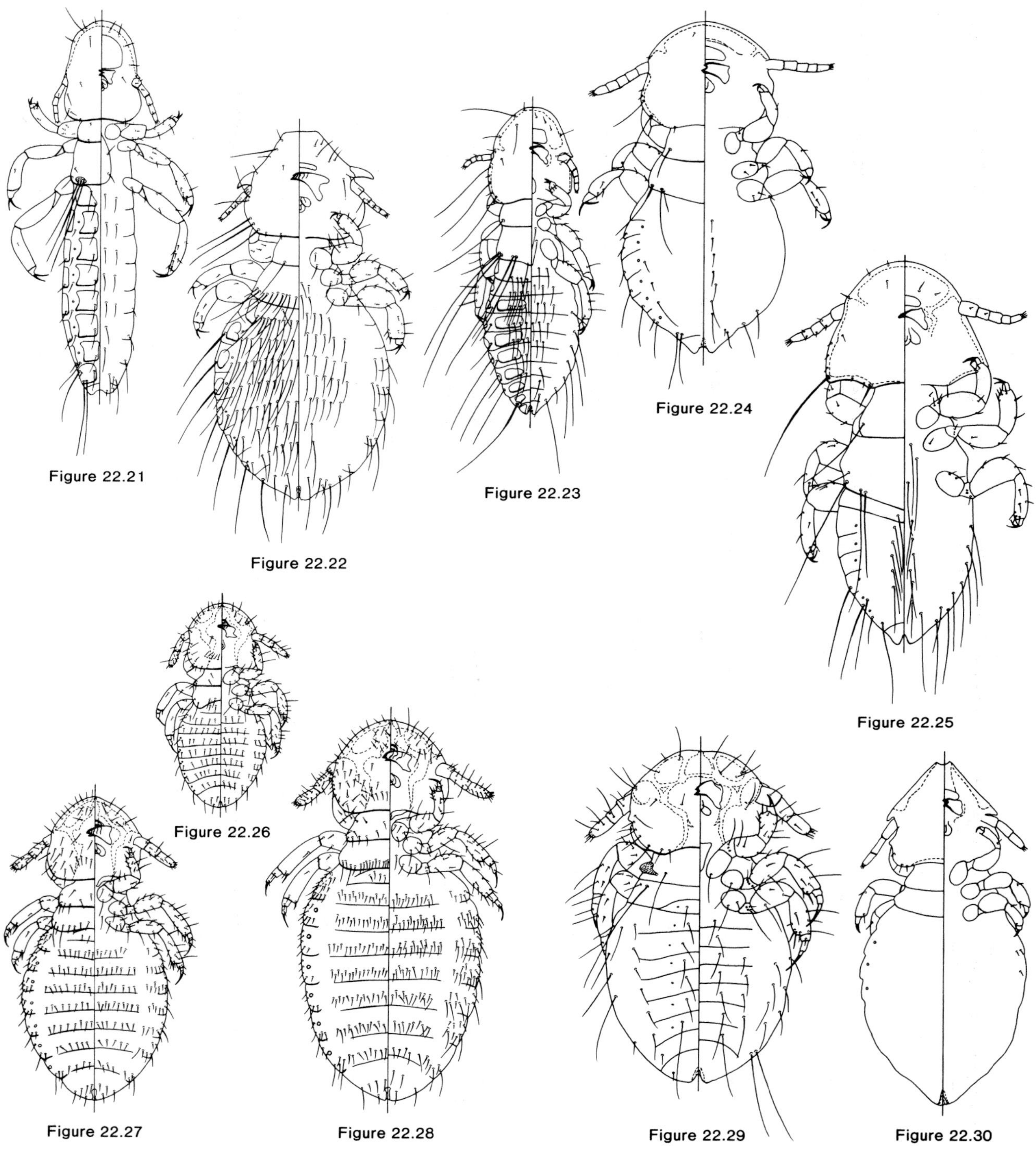

Figures 22.21–22.22. (22.21) *Lipeurus caponis* (L.), wing louse (Philopteridae); **(22.22)** *Philopterus ocellatus* (Scopoli) (Philopteridae).

Figures 22.23–22.25. (22.23) *Cuclotogaster heterographus* (Nitzsch), chicken head louse (Philopteridae); **(22.24)** *Goniocotes gallinae* (De Geer), fluff louse (Philopteridae); **(22.25)** *Chelopistes meleagridis* (L.), large turkey louse (Philopteridae).

Figures 22.26–22.28. *Bovicola bovis* (L.), cattle biting louse (Trichodectidae): **(22.26)** first instar; **(22.27)** second instar; **(22.28)** third instar.

Figures 22.29–22.30. (22.29) *Trichodectes canis* (De Geer), dog biting louse (Trichodectidae); **(22.30)** *Felicola subrostratus* (Burmeister), cat louse (Trichodectidae).

BIBLIOGRAPHY

Clay, T. 1969. A key to the genera of the Menoponidae (Amblycera: Mallophaga: Insecta). Bull. Brit. Mus. (Nat. Hist.) Ent. 24:1–26.

Clay, T. 1970. The Amblycera (Phthiraptera: Insecta). Bull. Brit. Mus. (Nat. Hist.) Ent. 25:73–98.

Emerson, K. C. 1956. Mallophaga (chewing lice) occurring on the domestic chicken. J. Kansas Ent. Soc. 29:63–79 (adults only).

Emerson, K. C. 1962. Mallophaga (chewing lice) occurring on the turkey. J. Kansas Ent. Soc. 35:196–201 (adults only).

Emerson, K.C. 1972a. Checklist of the Mallophaga of North America (north of Mexico). Part I. Suborder Ischnocera. Deseret Test Center, Dugway, Utah. 200 pp.

Emerson, K. C. 1972b. Ibid. Part II. Suborder Amblycera. 118 pp.

Emerson, K. C. 1972c. Ibid. Part III. Mammal host list. 28 pp.

Emerson, K. C. 1972d. Ibid. Part IV. Bird host list. 216 pp.

Emerson, K. C., and R. D. Price. 1975. Mallophaga of Venezuelan mammals. Brigham Young Univ. Sci. Bull. Biol. Series 20(3):1–77 (adults only).

Emerson, K. C., and R. D. Price. 1981. A host-parasite list of the Mallophaga on mammals. Misc. Publ. Ent. Soc. Amer. 12(1):1–72 (adults only).

Hopkins, D. E. 1970. *In vitro* colonization of the sheep biting louse, *Bovicola ovis*. Ann. Ent. Soc. Amer. 63:1196–97.

Hopkins, D. E., and W. F. Chamberlain. 1969. *In vitro* colonization of the goat biting lice, *Bovicola crassipes* and *B. limbata*. Ann. Ent. Soc. Amer. 62:826–28.

Hopkins, D. E., and W. F. Chamberlain. 1972. *In vitro* colonization of the cattle biting louse, *Bovicola bovis*. Ibid. 65:771–72.

Hopkins, G. H. E. 1949. The host-associations of the lice of mammals. Proc. Zool. Soc. London 119:387–604 (adults only).

Hopkins, G. H. E., and T. Clay. 1952. A check list of the genera and species of Mallophaga. London. British Museum. 362 pp.

Keler, S. von. 1960. Bibliographie der Mallophagen. Mitteilungen aus dem Zoologischen Museum in Berlin 36:145–403 (adults only).

Keler, S. von. 1971. A revision of the Australasian Boopiidae (Insecta: Phthiraptera), with notes on the Trimenoponidae. Australian J. Zool., Suppl. Series, Suppl. No. 6:1–126 (adults only).

Nelson, B. C. 1971. Successful rearing of *Colpocephalum turbinatum* (Phthiraptera). Nature New Biology 232:255.

Nelson, B. C. 1972. A revision of the New World species of *Ricinus* (Mallophaga) occurring on Passeriformes (Aves). Univ. Calif. Publ. in Entomol. 68. 180 pp.

Nelson, R. C., and R. D. Price. 1965. The *Laemobothrion* (Mallophaga: Laemobothriidae) of the Falconiformes. J. Med. Ent. 2:249–57.

Rheinwald, G. 1968. Die Mallophagengattung *Ricinus* De Geer, 1778. Revision der ausseramerikanischen Arten. Mitteilungen aus dem Zoologischen Staatsinstitut und Zoologischen Museum in Hamburg 65:181–326 (adults only).

Stenram, H. 1956. The ecology of *Columbicola columbae* L. (Mallophaga). Opusc. Ent. 21:170–90.

Werneck, F. L. 1948. Os malofagos de mamiferos. Parte I: Amblycera e Ischnocera (Philopteridae e parte de Trichodectidae). Revista Brasileira de Biologia, Rio de Janeiro. 243 pp. (adults only).

Werneck, F. L. 1950. Os malofagos de mamiferos. Parte II: Ischnocera (continuacao de Trichodectidae) e Rhyncophthirina. Instituto Oswaldo Cruz, Rio de Janeiro. 207 pp. (adults only).

Order Anoplura[1]

23

Ke Chung Kim
Pennsylvania State University

SUCKING LICE

The sucking lice are obligatory, wingless, hemimetabolous permanent ectoparasites of eutherian mammals. They have adapted successfully to the microenvironment of the host body surface, the fur environment, and coevolved with the mammalian hosts. Accordingly, the sucking lice are relatively host specific; often a species of Anoplura is restricted to a single host species. They spend their entire life on the individual host, and they have developed many unique adaptive traits for an ectoparasitic mode of life through a long association with mammals. Their body is dorsoventrally flattened and the head is equipped with a protrusible proboscis and sucking mouthparts. They are exclusively blood suckers. Naturally, sucking lice are closely associated with transmission of certain mammalian diseases. *Pediculus humanus* L. is a well-known vector of epidemic relapsing fever caused by *Borrelia recurrentis* and the sole agent in the transmission among people of typhus fever caused by *Rickettsia prowazeki*.

The immature stages of the sucking lice are poorly known. Ferris (1951) illustrated several, but until 1959 there was no serious effort to study them. For the first time, Cook and Beer (1959) made an attempt to systematically study *Hoplopleura* larvae in North America. Following Cook and Beer's pioneering work, the larvae of numerous taxa have been studied by Kim (1965, 1966a, b, c, 1968, 1971, 1975), yet our present knowledge is scanty. A systematic treatment is for the first time presented here. The study of larvae is important in biological and ecological research, useful in recognizing sibling species, and provides additional information on anopluran phylogeny.

The present state of the knowledge of the Anoplura larvae does not warrant a serious attempt to generalize structural patterns of different stages in depth and to describe them for every higher taxon. Furthermore, the third stage larvae are considerably different from the first stages, which makes a general description rather difficult and superfluous at best.

DIAGNOSIS

Body dorsoventrally flattened with a prognathous head and segmented abdomen. Head conical with peculiar piercing-sucking mouthparts in a small snout-like proboscis. Thorax relatively small and completely fused, usually with a pair of mesothoracic spiracles. Legs strongly developed with delicate modifications of tibia and tarsus, and the tarsus one-segmented with a strong claw.

Larvae are generally similar to adults in major morphological characters, but they are usually not heavily sclerotized, and they go through some definite changes in three larval stages. These changes are principally associated with the growth of body dimensions, body proportion, development of the thoracic sternites and paratergites, a progressive change of chaetotaxy, and occasionally antennal segmentation. The three stages are easily recognized when a series of specimens is available. In any long series of specimens at least a few are at the point of moulting. At this time, the succeeding instar is usually visible within the skin of the preceding stage.

BIOLOGY AND ECOLOGY

Information on biology of the sucking lice is limited to a handful of species, such as *Pediculus humanus* L. (Buxton 1946), *Haematopinus suis* (L.) (Florence 1921), and *H. eurysternus* Denny (Matthysse 1944, 1946). Murray and his colleagues worked on population biology of the sucking lice on seals (Murray 1958; Murray and Nicholls 1965; Murray, Smith and Soucek 1965), on mice and sheep (Murray 1960, 1961, 1963a, b). Kim (1971, 1972, 1975) published on population ecology of *Antarctophthirus callorhini* (Osborn) and *Proechinophthirus fluctus* (Ferris) on northern fur seals. Factors influencing the distribution of the sucking lice on the host animal were studied by Murray (1960, 1961, 1963a, b) and by Jensen and Roberts (1966).

DESCRIPTION

Immatures are naturally smaller than adults, with a relatively small abdomen. All are quite similar to adults in head shape, eyes, ocular points, thoracic dorsum, legs, and basic chaetotaxy.

Body dorsoventrally flattened with prognathous head and oval abdomen.

Head generally conical with definite proboscis in which are located peculiar piercing-sucking mouthparts. No tentorium. Antennae short, filiform, usually five-segmented, occasionally reduced to three or fourth segments. Fourth and

1. Authorized for publication as paper no. 5713 in the journal series of the Pennsylvania Agricultural Experiment Station.

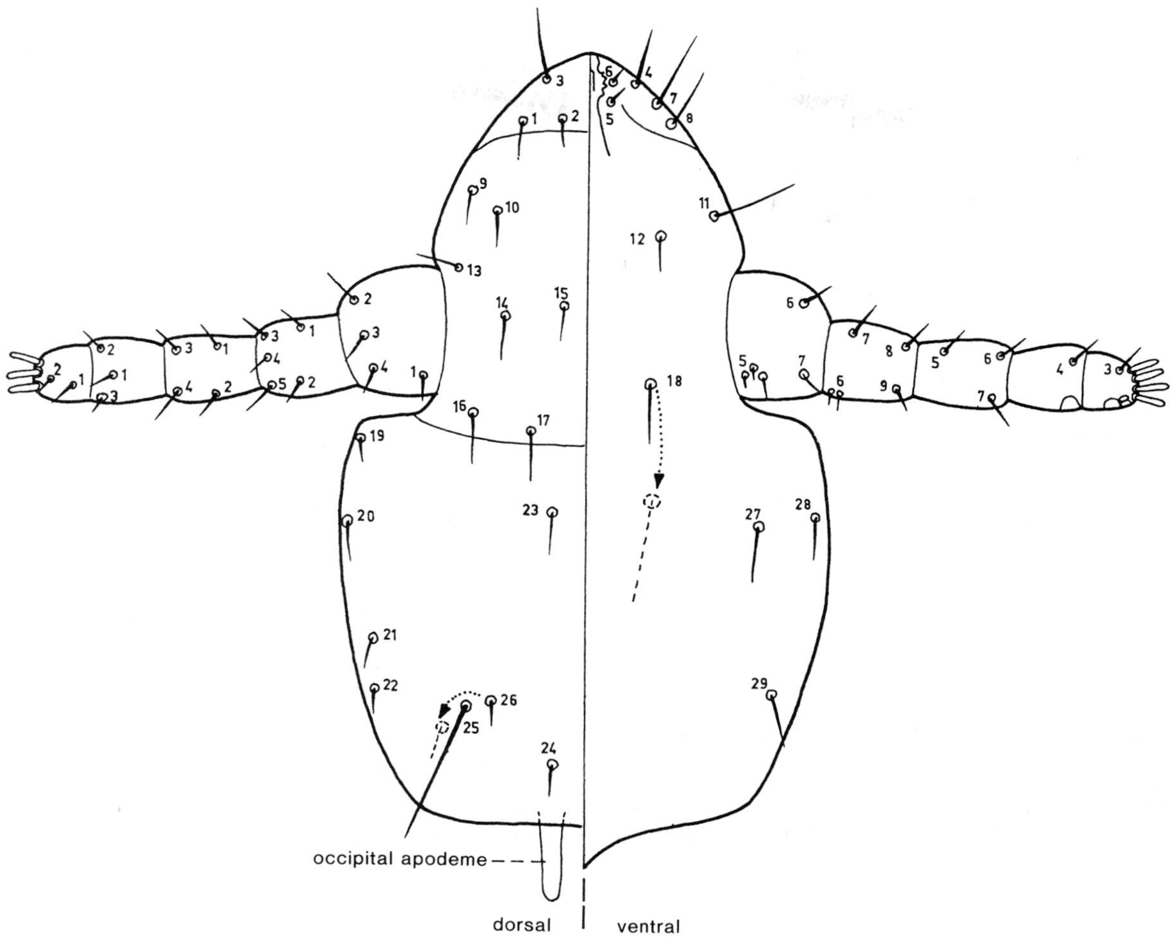

Figure 23.1. Standardized chaetotaxy of Anoplura (generalized), setae on head, thorax, and abdomen are continuously numbered dorsally and ventrally. **HEAD:** Setae on each antennal segment are separately numbered without specific positional designation; 1,2: dorsal anterior head setae (DAnHS); 3,4: apical head setae (ApHS); 5,6: oral setae (OrS); 7,8: anterior marginal head setae (AnMHS); 9,10: dorsal preantennal lateral head setae (DPaLHS); 11,12: ventral preantennal head setae (VPaHS); 13: dorsal preantennal head setae (DPaHS); 14: supraantennal head setae (SpAtHS); 15: supraantennal central head setae (SpAtCHS); 16,17: sutural head setae (SuHS); 18: ventral principal head seta (VPHS); 19, 20, 21, 22: dorsal marginal head setae (DMHS); 23: dorsal anterior central head setae (DAnCHS); 24: dorsal posterior central head setae (DPoCHS); 25: dorsal principal head setae (DPHS); 26: dorsal accessory head setae (DAcHS); 27: ventral lateral head setae (VLHS); 28: ventral anterior marginal head setae (VAnMHS); 29: ventral posterior marginal head setae (VPoMHS) (after Kim and Ludwig 1978).

fifth segments each bearing a tuft organ and often additional pore and peg organs. No ocelli present. Compound eyes usually reduced or absent.

Thorax narrower than abdomen, with a pair of mesothoracic spiracles. Thoracic dorsum mostly composed of subcoxal or pleural components. Venter generally membranous but occasionally with a median sternal plate in second and third stages. Legs strongly developed, with a modification of the tibia and tarsus for effective holdfast. Tarsus one-segmented, with a strong claw.

Abdomen of ten segments, of which only nine are easily identified, with terga and sterna usually not sclerotized. Paratergites often developed in the second and third stages. Abdomen normally bearing six pairs of spiracles on lateral part of segments 3–8. In the first instar abdominal spiracles often indistinct. Sexual determination can often be made in the third and rarely in the second stage.

Morphology and Chaetotaxy

The head is of a generalized type usually found in adults (fig. 23.1) but shows several unique characters in some taxa. Larvae of *Hoplopleura* and *Pterophthirus* have small unpigmented tubercles on the ventral side of the head, antennae, and even on the coxae (figs. 23.18, 23.38, 23.40, 23.41), whereas those of *Proechinophthirus* have numerous heavily sclerotized spiniform setae on the head (figs. 23.10, 23.21–23.23). The number of antennal segments is usually five as in the adults except for *Latagophthirus rauschi* Kim and Emerson, which has three segmented antennae in all stages. In some taxa, however, larvae have fewer segments than the adults. Larvae of *Antarctophthirus callorhini* (Osborn) have four-segmented antennae, whereas the adults have five.

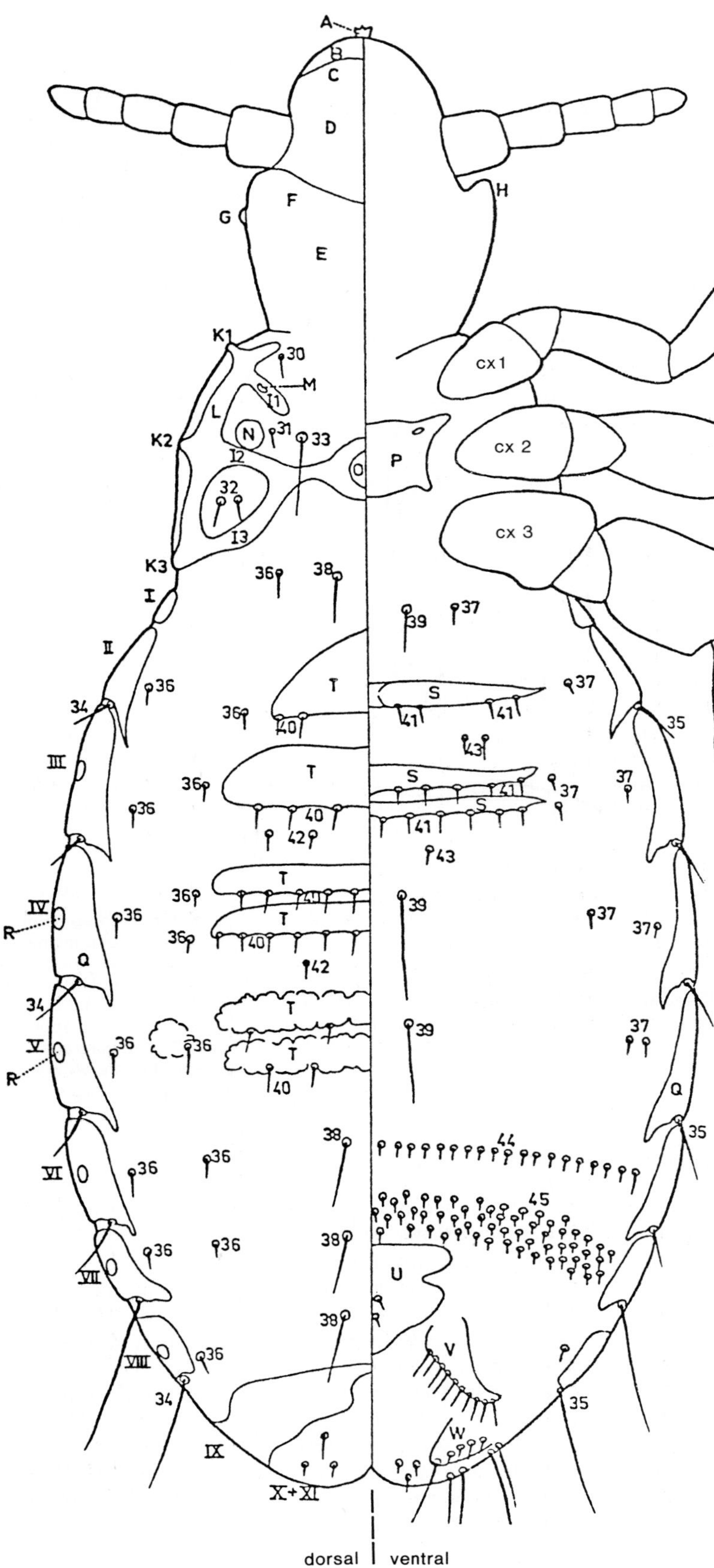

Figure 23.2. Anatomy and standardized chaetotaxy of Anoplura (generalized); setae on head, thorax, and abdomen are continuously numbered dorsally and ventrally. **ANATOMY:** A: haustellum; B: labrum; C: clypeus; D: forehead; E: hindhead; F: clypeofrontal suture; G: eye; H: ocular point; I1: prothoracic pleural apophysis; I2: mesothoracic pleural apophysis; I3: metathoracic pleural apophysis; K1: prothoracic coxal process; K2: mesothoracic coxal process; K3: metathoracic coxal process; L: longitudinal pleural bar; M: pleural apophysial pit; N: mesothoracic spiracle; O: notal pit; Q: paratergites; R: spiracles; T: tergites; S: sternites; U: subgenital plate: V: gonopod VIII; W: gonopod IX; Cx 1,2,3 = Coxae 1,2,3; I − X + XI − Number of Abdominal segments. **CHAETOTAXY:** 30: dorsal prothoracic setae (DPtS); 31: dorsal mesothoracic setae (DMsS); 32: dorsal metathoracic setae (DMtS); 33: dorsal principal thoracic setae (DPTS); 34: dorsal marginal abdominal seta (DMAS) = dorsal paratergal setae (DPrS); 35: ventral marginal abdominal seta (VMAS) = ventral paratergal setae (VPrS); 36: dorsal lateral abdominal setae (DLAS); 37: ventral lateral abdominal setae (VLAS); 38: dorsal central abdominal setae (DCAS); 39: ventral central abdominal setae (VCAS); 40: tergal abdominal setae (TeAS); 41: sternal abdominal setae (StAS); 42: intertergal abdominal setae (InTeAS); 43: intersternal abdominal setae (InStAS); 44: transverse setae row; 45: transverse setal field (after Kim and Ludwig 1978).

Thoracic sternal plate usually undeveloped in the larvae and always lacking in the first stage. In some taxa, the sternal plate is already developed in the second stage; for example, *Enderleinellus* (figs. 23.30, 23.31).

The abdomen is proportionally small and usually composed of ten segments of which the last segment may not be distinct (fig. 23.2). Each segment can be identified by the position of spiracles and setal rows. In *Hoplopleura onycomydis* Cook and Beer and *H. oenomydis* Ferris, the abdominal segmentation is evident in the first stage (figs. 23.39a, b). The spiracles are usually evident in the second and third stages but are occasionally seen in the first stage, as in *Hoplopleura arboricola* Kellogg and Ferris and *H. sciuricola* Ferris. The spiracles are often associated with paratergites, which usually become sclerotized in the second stage, yet are distinctly visible in all stages of some species as in *Haematopinus quadripertusus* Fahrenholz and *H. eurysternus* Denny (figs. 23.34–23.37).

The primary pattern of chaetotaxy on the head, antennae (fig. 23.1), thorax, and legs is already developed in the first stage and does not change throughout postembryonic development. New setae are added to the primary chaetotaxy or each seta becomes longer and larger as it goes through successive stages to the adult. No setae of the primary chaetotaxy are lost during metamorphosis, although some setae in the larval instars are replaced by other types or become greatly reduced in subsequent instars. The most striking difference between larvae and adults is in the abdominal chaetotaxy. In more generalized taxa, such as *Haematopinus* and *Solenopotes,* the adult chaetotaxy is already mostly developed in the second stage (figs. 23.35, 23.49). In specialized taxa, exemplified by *Hoplopleura* and *Polyplax,* larvae lack the extensive chaetotaxy of the adult (figs. 23.38–23.43, 23.57–23.59). In many species larvae I and II simply have one or two rows of central abdominal setae (CAS) for each side (dorsal and ventral) and one to three marginal abdominal setae (MAS) on abdominal segments 7 and 8, and sometimes on segment 9 (fig. 23.2).

EXPLANATION OF CHAETOTAXAL TERMS

Position	*Structure*
Ac—accessory	A—abdominal
An—anterior	At—antennal
Ap—apical	G—genital
C—central	H—head
D—dorsal	Ms—mesothoracic
I—inner	Mt—metathoracic
If—infra	Oc—occipital
Im—intermedian	Or—oral
In—inter	Pa—preantennal
M—marginal	Pr—paratergal
Md—median	Pt—prothoracic
L—lateral	S—setae
O—outer	St—sternal
P—principal	Su—sutural
Po—posterior	T—thoracic
Sp—supra	Te—tergal
V—ventral	

Examples:
DAnHS = dorsal anterior head setae
VLAS = ventral lateral abdominal setae
SpAtHS = supraantennal head setae

COMMENTS

Sucking lice are widely distributed throughout the world (Hopkins 1949; Ludwig 1968) wherever host animals exist. The anopluran fauna is especially rich in the Ethiopean region. Ludwig (1968) recorded 34.4 percent of the total, or 135 species, from the Ethiopean region, 18.3 percent from the Palaearctic, 13.3 percent from the Oriental, 10.2 percent from the Nearctic, and 11.0 percent from the Neotropical. Currently, 76 species of sucking lice (9 families, 19 genera) are known from North America (Kim et al. 1986).

The diversity of the Anoplura is not fully known as yet. At present there are 486 species known from approximately 840 species (241 genera) of mammals (Kim and Ludwig 1978). They are parasitic on diverse groups of mammals but are apparently absent in several major mammalian taxa; Monotremata, Marsupialia, Edentata, Pholidota, Chiroptera, Cetacea, Proboscida, Sirenia, and most terrestrial Carnivora (Kim 1985).

The suprageneric classification of Anoplura by Kim and Ludwig (1978), which is adopted here, contains 15 families and 42 genera.

COLLECTION AND PRESERVATION TECHNIQUES

Sucking lice are small and often difficult to find on the host animals. Most immature lice are very small and usually escape visual examination of the host skin, unless special care is taken. When sucking lice are found on freshly killed animals, all specimens should be preserved in 75–80 percent ethyl alcohol. They can be collected from museum skins by combing; these specimens should be relaxed in Barber's before preservation in alcohol (but the benzene in Barber's is a known carcinogen, so use proper precautions or another relaxing technique).

To collect all stages, the host animal may be trapped in the field and skinned. Each skin or hide is placed in a separate plastic bag in order to prevent contamination and kept frozen for future study. To extract lice, the entire skin or a small piece of a large hide is placed in a beaker with 1 percent trypsin (certified 1:250) buffered to pH 8.3 ± with sodium diphosphate and kept at ambient temperature for 10 to 24 hours. After the initial digestion period, an equal amount of 10 percent KOH solution is added to the beaker containing the digested skin. This mixture is then boiled for several minutes or until hairs and tissues are completely dissolved. The cleared specimens are then strained out of the resulting solution through an 80-mesh screen. This process recovers the entire louse population including eggs (Cook and Beer 1959; Kim 1972). The concentration of KOH and the length of the digestion period may vary with the size and thickness of skin and the amount of blubber or fat material.

To study lice, specimens must be slide mounted. Clear specimens by placing them in a small dish containing 10% KOH for about 10 hours. To speed up this process warm the beaker on a hot plate for a few minutes (if specimens are collected by the digestion process, they need not be cleared this way). After clearing, wash specimens thoroughly in distilled water and dehydrate them by going through a 30–50–75–95% to absolute ethyl alcohol series. Leave specimens in each solution 5 minutes or more before transferring to creosote, where they can remain until mounted. Center the specimen on a microscope slide, ventral side up, one specimen per slide. Place a small drop of Canada balsam on the specimen, add the cover glass, and press gently to spread the legs and antennae properly. Slides should be completely dry before permanent labels are attached, with collection label on the right and identification label on the left.

CLASSIFICATION OF NORTH AMERICAN FAMILIES

Order **ANOPLURA**
- Echinophthiriidae, the echinophthiriids
- Haematopinidae, wrinkled sucking lice
- Enderleinellidae, the enderleinellids
- Hoplopleuridae, the hoplopleurids
- Linognathidae, smooth sucking lice
- Pecaroecidae, the pecaroecids
- Pediculidae, human lice
- Polyplacidae, the polyplacids
- Pthiridae, crab lice

HOST LIST OF NORTH AMERICAN ANOPLURA

Order **Insectivora,** insectivores
- Family Talpidae, moles .... *Haematopinoides*

Order **Lagomorpha,** lagomorphs
- Family Leporidae, hares and rabbits .... *Haemodipsus*

Order **Rodentia,** rodents
- Family Sciuridae, squirrels .... *Enderleinellus*, *Microphthirus*, *Linognathoides*, *Neohaematopinus*, *Hoplopleura*
- Family Heteromyidae, heteromyids .... *Fahrenholzia*
- Family Cricetidae, New World rats and mice .... *Hoplopleura*, *Neohaematopinus*, *Polyplax*
- Family Muridae, Old World rats and mice .... *Hoplopleura*, *Polyplax*

Order **Carnivora,** carnivores
- Family Canidae, canids .... *Linognathus*
- Family Mustelidae, mustelids .... *Latagophthirus*

Order **Pinnipedia,** pinnipeds
- Family Otariidae, eared seals .... *Antarctophthirus*, *Proechinophthirus*
- Family Odobenidae, walrus .... *Antarctophthirus*
- Family Phocidae, haired seals .... *Antarctophthirus*, *Echinophthirus*

Order **Artiodactyla,** even-toed ungulates
- Family Tayassuidae, peccaries .... *Pecaroecus*
- Family Cervidae, cervids .... *Solenopotes*
- Family Bovidae, bovids .... *Linognathus*, *Haematopinus*
- Family Suidae, pigs .... *Haematopinus*

Order **Perissodactyla,** odd-toed ungulates
- Family Equidae, horses .... *Haematopinus*

Order **Primates,** primates
- Family Hominidae, humans .... *Pediculus*, *Pthirus*

KEY TO THE FAMILIES OF NORTH AMERICAN IMMATURE ANOPLURA

1. Head with distinct eyes (fig. 23.3) or subacute ocular points (fig. 23.4) on lateral margins posterior to antennae 2

1′. Head without external eyes or prominent ocular points (fig. 23.5) 5

2(1). Head with prominent ocular points but without eyes (fig. 23.4); on bovids, pigs and horses .... ***Haematopinidae*** (p. 233)

2′. Head with eyes having a distinct lens but without ocular points (fig. 23.3); on other hosts 3

3(2′). Head long and slender, much longer than thorax (fig. 23.6); large louse with narrowly elliptical abdomen; on peccaries .... ***Pecaroecidae*** (p. 238)

3′. Head about as long as thorax (figs. 23.7, 23.8); small lice with oval or elliptical abdomen; on humans 4

4(3′). Compact, with body less than 2× as long as wide (fig. 23.7); thorax very wide; fore legs slender, mid and hind legs very stout, each with stout claw; on humans .... ***Pthiridae*** (p. 241)

4′. Slender, with body more than 2× as long as wide (fig. 23.8); abdomen long, wider than thorax; all legs slender, each with an acuminate claw; on humans .... ***Pediculidae*** (p. 239)

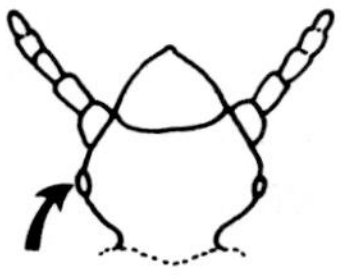

Figure 23.3

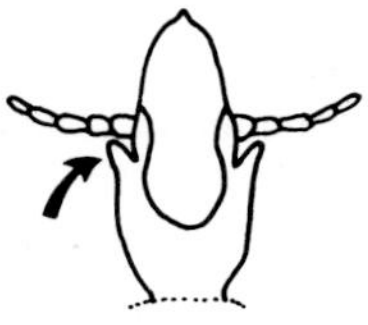

Figure 23.4

Figure 23.5

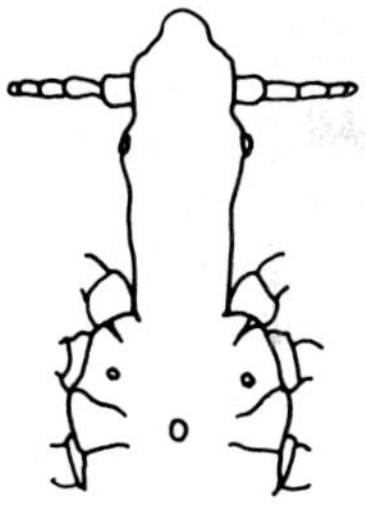

Figure 23.6

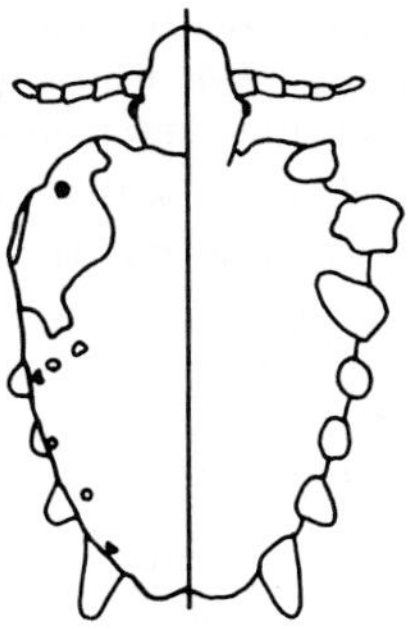

Figure 23.7

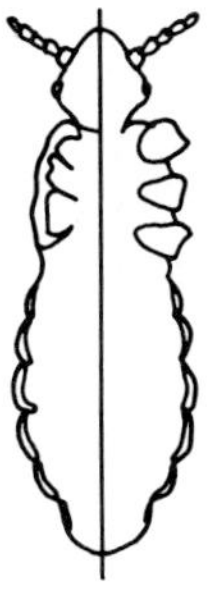

Figure 23.8

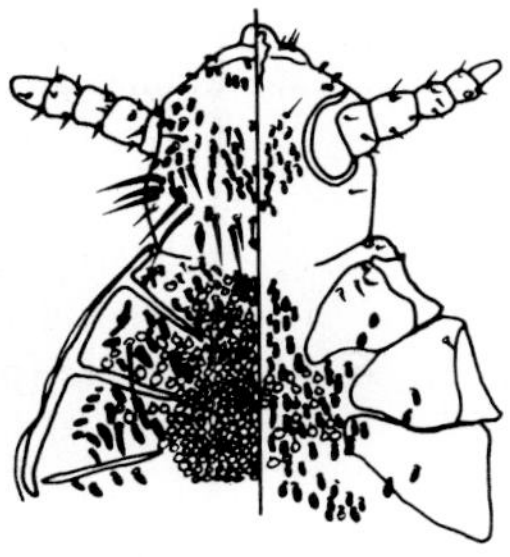

Figure 23.9

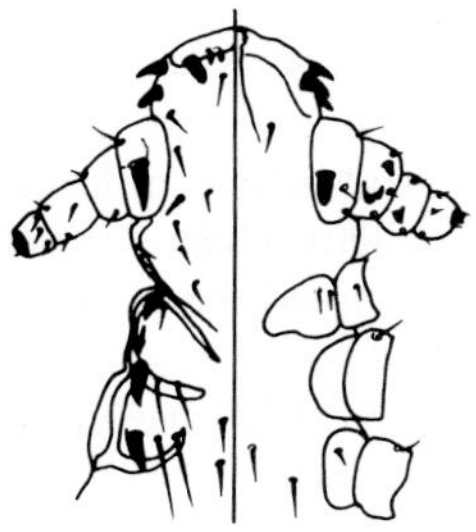

Figure 23.10

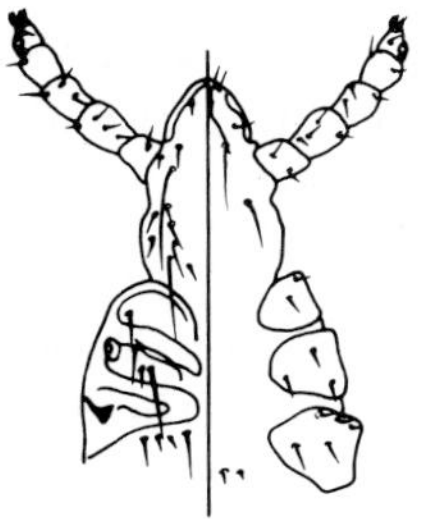

Figure 23.11

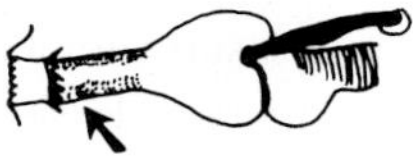

Figure 23.11a

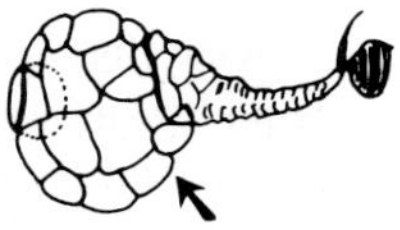

Figure 23.11b

fore legs

mid legs

hind legs

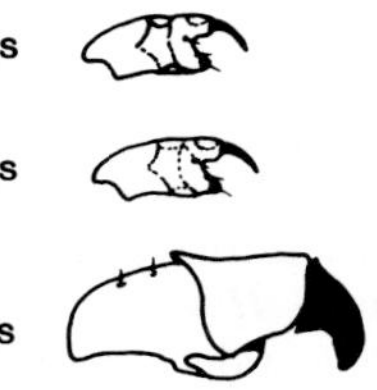

Figure 23.12

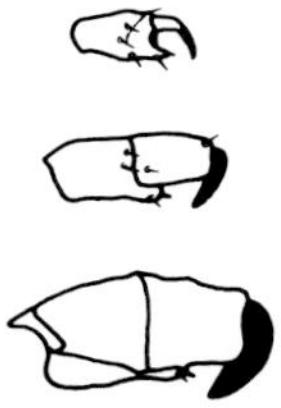

Figure 23.13

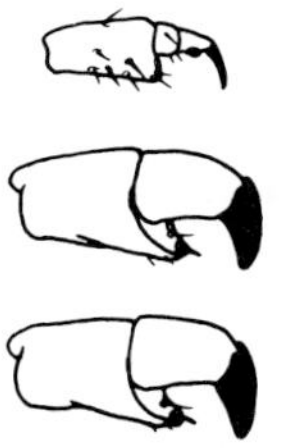

Figure 23.14

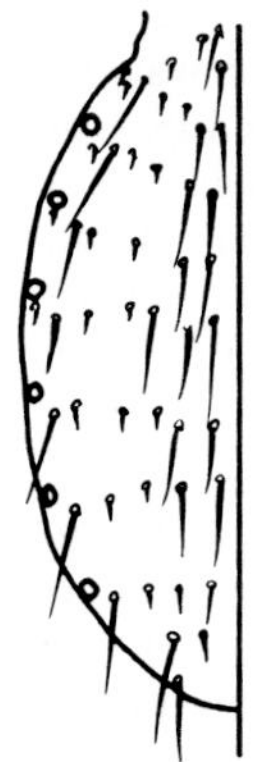

Figure 23.15

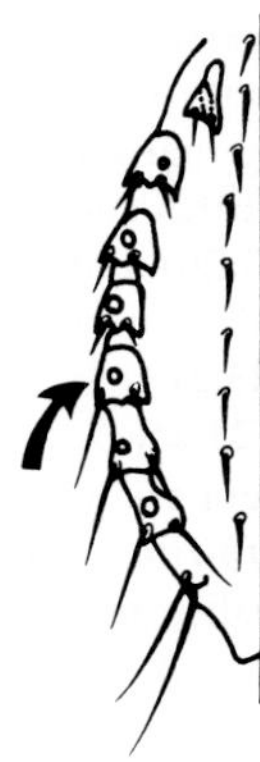

Figure 23.16

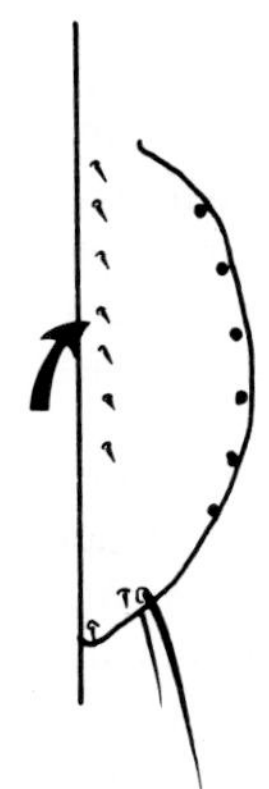

Figure 23.17

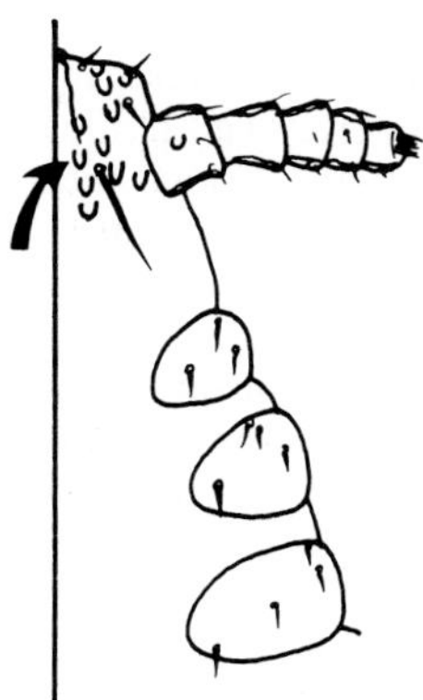

Figure 23.18

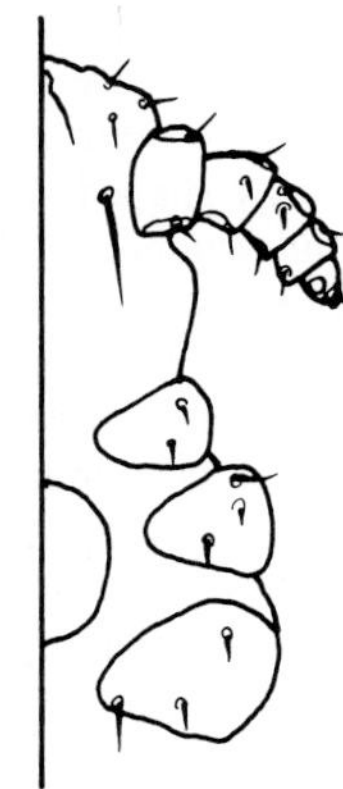

Figure 23.19

5(1′). Head and thorax thickly covered with various setae (fig. 23.9) or with strong spiniform setae (fig. 23.10); spiracular atrium tubular (fig. 23.11a); on seals, sea lions, walrus, and river otter .... ***Echinophthiriidae*** (p. 230)

5′. Head and thorax with only a few setae (fig. 23.11); spiracular atrium bulbous (fig. 23.11b); on terrestrial mammals 6

6(5′). Fore legs subequal to mid legs in size and shape, both more slender and smaller than hind legs, each with acuminate claw (fig. 23.12); on squirrels .... ***Enderleinellidae*** (p. 233)

6′. Fore legs smallest of the 3 pairs (fig. 23.13, 23.14); mid legs usually subequal to hind legs in size and shape or at least somewhat larger than fore legs, each with larger and stouter claws 7

7(6′). Abdomen without any evidence of paratergites in all stages (fig. 23.15); first instar with 4 rows[2] of VCAS (ventral central abdominal setae) (figs. 23.44, 23.48); on even-toed ungulates and canids .... ***Linognathidae*** (p. 237)

7′. Abdomen usually with paratergites in second or third larval stage (fig. 23.16); first instar with no, or at most, 2 rows[2] of VCAS (fig. 23.17); on small mammals 8

8(7′). Hind legs largest of three pairs, stout, each with a stout, blunt claw (fig. 23.13); small unpigmented tubercles densely distributed on ventral surface of head, first antennal segment and coxae (fig. 23.18); parasitic on rodents (**Hoplopleurinae**), or if absent, abdomen with 4 rows[2] of CAS (central abdominal setae) in third stage (fig. 23.43), parasitic on moles (**Haematopinoidinae**) .... ***Hoplopleuridae*** (p. 233)

8′. Hind legs usually subequal to mid legs in shape and size (fig. 23.14); head, basal antennal segment and coxae usually without such tubercles (fig. 23.19); all instars usually with more than 4 rows of CAS; on rodents .... ***Polyplacidae*** (p. 241)

2. The figures show only half the rows since the left half of each figure is a dorsal and the right half is a ventral view.

ECHINOPHTHIRIIDAE

The Echinophthiriids

Figures 23.20–23.27

Relationships and Diagnosis: The echinophthiriids are rather unique lice, exclusively parasitic upon aquatic carnivores, namely Pinnipedia and aquatic Mustelidae. They are somewhat related to the Haematopinidae. All the echinophthiriids are easily recognized by the presence of unique spiracles with a tubular atrium (Key fig. 23.11a) and variously shaped setae on the head and thorax; tuberculiform and setaceous setae of various sizes (*Antarctophthirus, Echinophthirus,* and *Latagophthirus,*) or strong spiniform setae (*Proechinophthirus*).

Biology and Ecology: All instars inhabit the fur and skin of the host and feed on blood. All larvae tend to aggregate on particular areas. *Antarctophthirus* seems to prefer the areas of the body surface devoid of dense fur, such as the base of the flipper, eyelids, and the area surrounding the genital opening, but *Proechinophthirus* inhabits the area of dense fur (Kim 1972, 1975).

Eggs are usually oblong with a number of small knoblike tubercules (5–15) on the operculum (figs. 23.20, 23.24). The first instar does not have scales on the abdomen and ventral surface of the thorax.

Description: Body size variable. The third-stage larvae 1.5–2.7 mm in length. Body usually covered with setae of various shapes and sizes. Primary setae distinct in the first stage. *Head* usually about as wide as long or much longer than wide (*Proechinophthirus*); anterior margin rounded; postantennal angles usually developed; antennae usually four-segmented, five-segmented (many *Antarctophthirus*) or rarely three-segmented (*Latagophthirus*). *Thorax* with phragmata well developed; mesothoracic spiracles small, with specialized closing apparatus; no sternal plate. *Abdomen* oval, with six small spiracles, each with specialized closing apparatus; no tergal, sternal, or paratergal sclerites developed; with numerous spiniform setae, scales and regular setae in *Antarctophthirus* (figs. 23.25, 23.26) and *Latagophthirus;* or with numerous spiniform and regular setae in *Echinophthirus* and *Proechinophthirus* (figs. 23.21–23.23) and in the first instars of *Antarctophthirus* (fig. 23.25).

Comments: Five genera are presently recognized; *Proechinophthirus, Echinophthirius, Lepidophthirus,* and *Antarctophthirus* on Pinnipedia, and *Latagophthirus* on river otter (Mustelidae, Carnivora). All genera except *Lepidophthirus* occur in North America.

Selected Bibliography

Ferris 1951.
Kim 1971, 1972, 1975.
Kim and Emerson 1974.

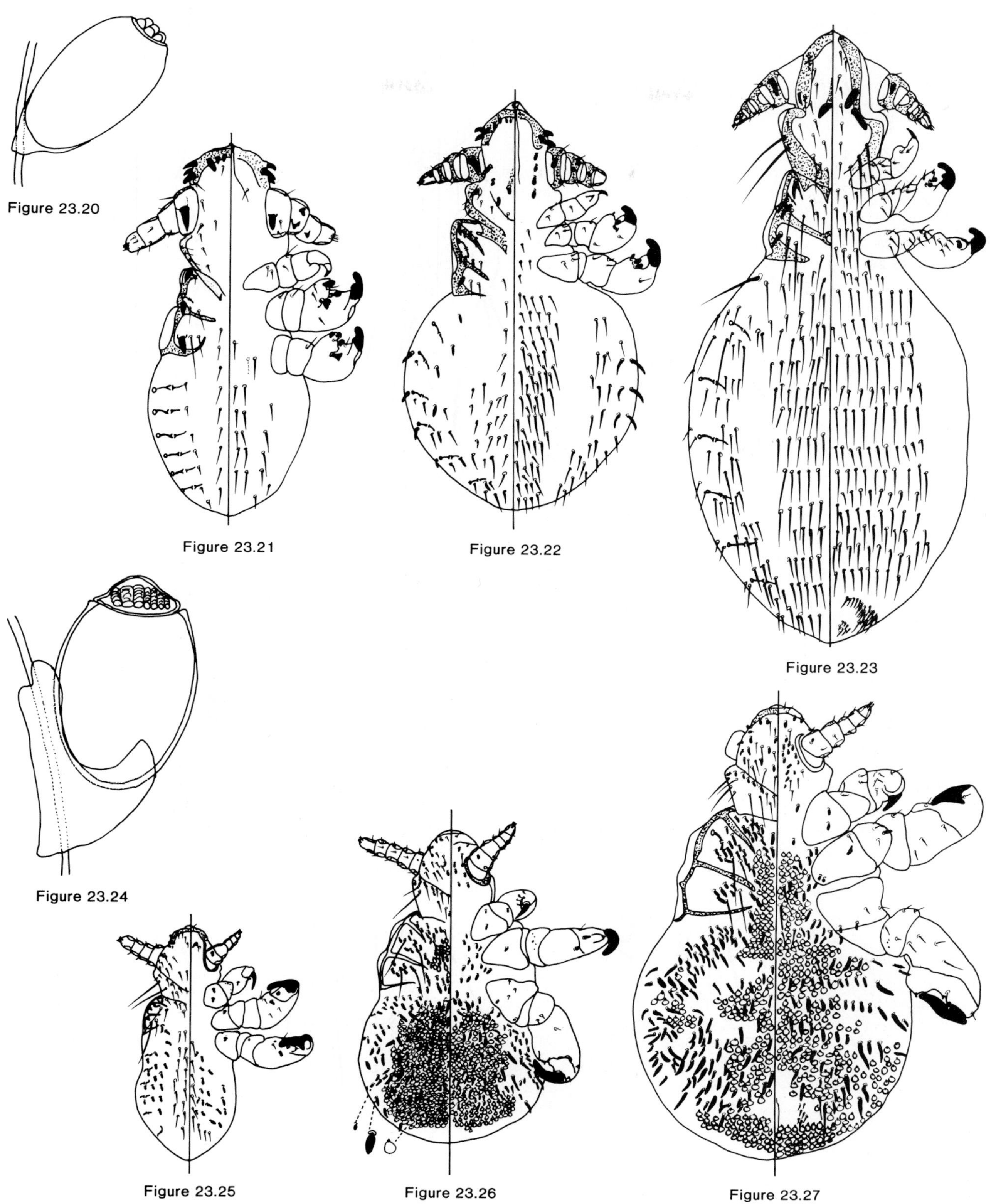

Figures 23.20–23.23. Echinophthiriidae. *Proechinophthirus fluctus* (Ferris); **(23.20)** egg; **(23.21)** larva I; **(23.22)** larva II; **(23.23)** larva III.

Figures 23.24–23.27. Echinophthiriidae. *Antarctophthirus microchir* (Trouessart and Neumann); **(23.24)** egg; **(23.25)** larva I; **(23.26)** larva II; **(23.27)** larva III.

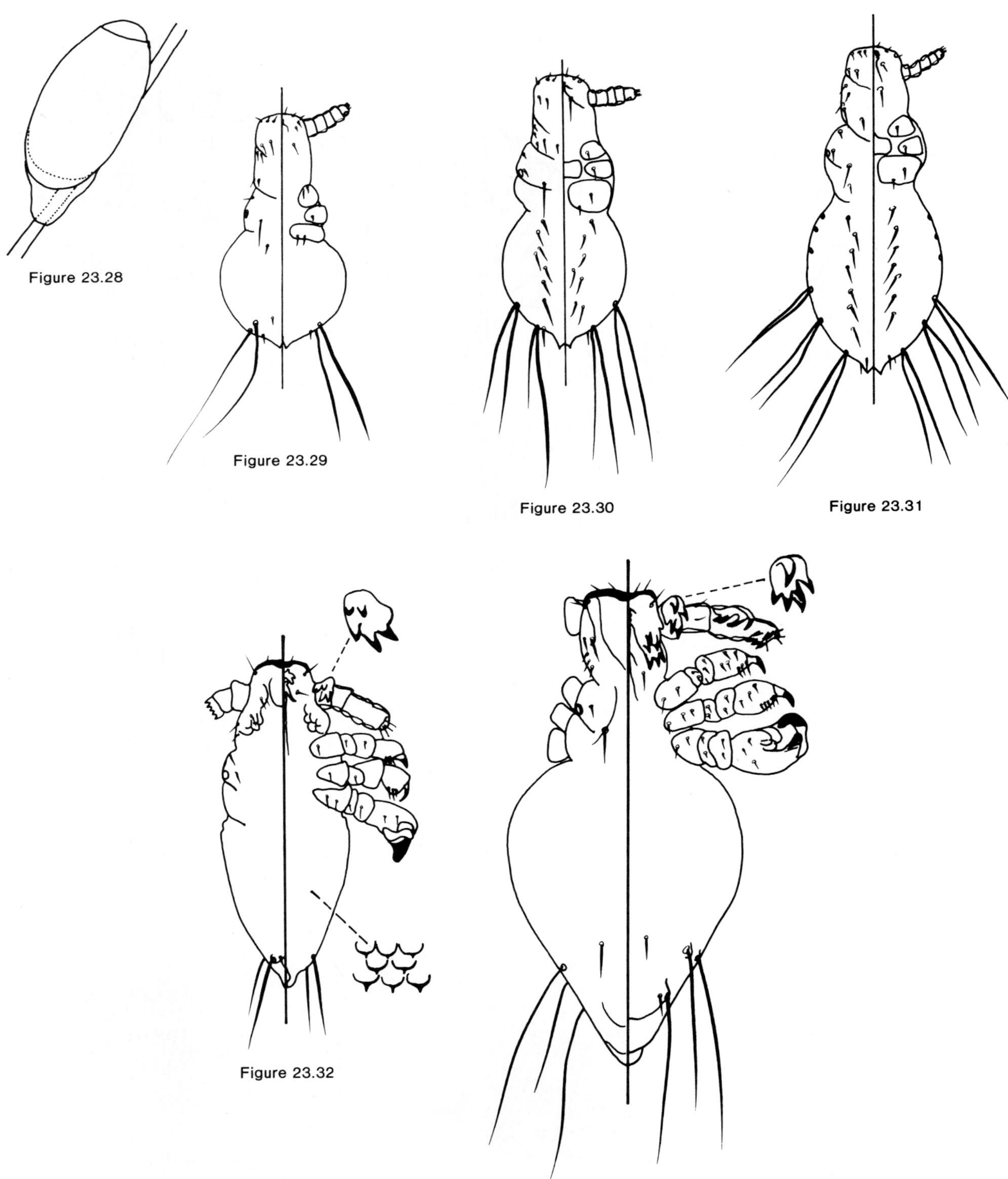

Figure 23.28

Figure 23.29

Figure 23.30

Figure 23.31

Figure 23.32

Figure 23.33

Figures 23.28–23.31. Enderleinellidae. *Enderleinellus;* **(23.28)** egg of *E. osborni* Kellogg and Ferris; **(23.29–23.31)** *E. longiceps* Kellogg and Ferris, legs removed; **(23.29)** larva I; **(23.30)** larva II; **(23.31)** larva III.

Figures 23.32–23.33. Enderleinellidae. *Microphthirus uncinatus* (Ferris); **(23.32)** larva I; **(23.33)** larva III.

ENDERLEINELLIDAE

The Enderleinellids

Figures 23.28-23.33

Relationships and Diagnosis: Very small lice, related to the hoplopleurids, but easily distinguished from other lice by having the fore legs subequal to the mid legs in size and shape, both small and slender. DPHS are very small but VPHS are distinct and large in all larval stages.

Biology and Ecology: The enderleinellids are exclusively parasitic on squirrels (Sciuridae).

Description: Third stage larvae about 0.5 mm in length. *Head* with anterior margin rounded or truncated; antennae five-segmented; clypeofrontal suture distinct; DPHS very small but VPHS distinct and long. *Thorax* with comparatively large DPTS; thoracic sternal plate generally absent. *Abdomen* with two central rows of DCAS and VCAS and three visible pairs of spiracles in larva III.

Differences between species and between the early stages are striking in abdominal morphology (figs. 23.29–23.31). The first stage has only two pairs of MAS. The second stage has four pairs of MAS and five to seven or more pairs of DCAS and VCAS, and usually three or more paratergites on each side. The third stage has six pairs of MAS and numerous LAS in addition to CAS. On the anterior half of the abdomen there are four or more paratergal plates on each side. In *E. longiceps* Kellogg and Ferris there is no evidence of paratergites but the thoracic sternal plate is developed in the second and third stages (figs. 23.30, 23.31).

Comments: Five genera are presently recognized: *Enderleinellus* (43 species), *Werneckia* (3 species), *Microphthirus* (1 species), *Phthirunculus* (1 species), and *Atopophthirus* (1 species) (Kim and Ludwig 1978). *Enderleinellus* (figs. 23.28–23.31) and *Microphthirus* (figs. 23.32, 23.33) occur in North America.

Selected Bibliography

Ferris 1951.
Kim 1966a, 1966b, 1977.
Kuhn and Ludwig 1965.

HAEMATOPINIDAE

Haematopinids, Wrinkled Sucking Lice

Figures 23.34-23.37

Relationships and Diagnosis: The haematopinids are among the most devastating ectoparasites of domesticated animals. They are easily distinguished from other lice by their large size, the presence of prominent ocular points posterior to the antennae, all legs subequal in size and shape, and abdominal cuticula leathery and minutely wrinkled.

Biology and Ecology: The life cycle takes three to five weeks from eggs to eggs. The incubation period for eggs takes 10–17 days and each stage requires three to seven days before moulting (Matthysse 1946). However, the total period for the life cycle varies by season and host species.

Each species prefers specific parts of the host body by season. During August and September *Haematopinus eurysternus* Denny is found in the ears near the tips but the main area of infestation is the top of the neck in the winter in New York state (Matthysse 1946). *H. suis* L. frequents the folds of skin on the neck and the jowl, the inside and the base of the ears, the inside of the legs, flanks, and, in smaller numbers, the back (Florence 1921). This distribution pattern may be altered by seasonal temperature changes.

H. suis feeds readily on humans and other hosts and thus is well adapted for experimental work. Hosts of *Haematopinus* are Suidae, some Bovidae, Cervidae (Artiodactyla), and Equidae (Perissodactyla).

Description: Third instars 2–2.8 mm long. *Head* with distinct ocular points; antennae five-segmented; primary chaetotaxy distinct. *Thorax* much wider than head, heavily pigmented, with distinct notal pit and large mesothoracic spiracles; no evidence of sternal plate. All legs subequal in size and shape, each leg with strong acuminate claw. *Abdomen* membranous, leathery and wrinkled, with distinct paratergites and spiracles on segments 3–8; with numerous sclerotic plates; with 9 DCAS, 2–6 DInAS, 3–7 VMdAS on segments 1–7, 1–3 VLAS on each side (fig. 23.37).

Larva II (fig. 23.35) is very similar to larva III in general appearance, but smaller. Larva I (fig. 23.34) is naturally smaller, with a reduction in chaetotaxy and fusion of sclerotic plates, and with no mesothoracic seta.

Comments: This monotypic family, with 22 known species, is widely distributed throughout the world. *Haematopinus eurysternus* Denny, *H. quadripetusus* Fahrenholz, and *H. tuberculatus* (Burmeister) are important cattle lice (Meleney and Kim 1974). *H. suis* and *H. apri* Goureau are lice of swine; *H. asini* (L.) is an important parasite of horses.

Selected Bibliography

Bruce 1947.
Chaudhuri and Kumar 1961.
Craufurd-Benson 1941.
Florence 1921.
Matthysse 1946.
Meleney and Kim 1974.
Roberts 1953.
Stimie and van der Merwe 1968.

HOPLOPLEURIDAE

The Hoplopleurids

Figures 23.38-23.43

Relationships and Diagnosis: The hoplopleurids are specialized lice somewhat similar to Linognathidae and Polyplacidae. Larvae of Hoplopleuridae can be distinguished from the linognathids and the polyplacids by having the hind legs stout and usually larger than the mid and fore legs, and each leg with a stout and blunt claw. Larvae of the subfamily Hoplopleurinae have numerous tubercles on the ventral side of the head and antennae, whereas the Haematopinoidinae larvae lack ventral tubercles on the head and usually have distinct paratergites.

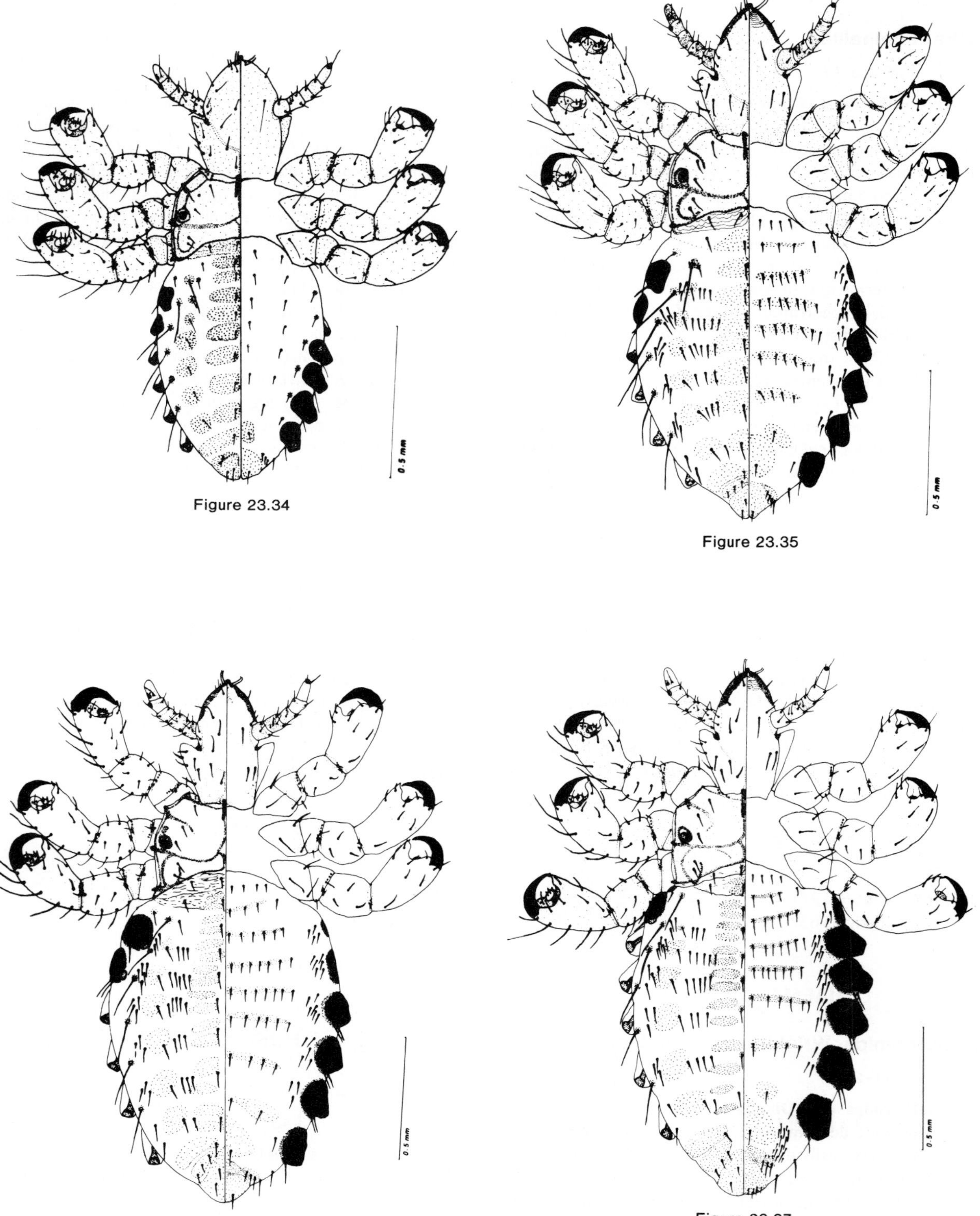

Figure 23.34

Figure 23.35

Figure 23.36

Figure 23.37

Figures 23.34–23.37. Haematopinidae. *Haematopinus eurysternus* Denny; **(23.34)** larva I; **(23.35)** larva II male; **(23.36)** larva III male; **(23.37)** larva III female.

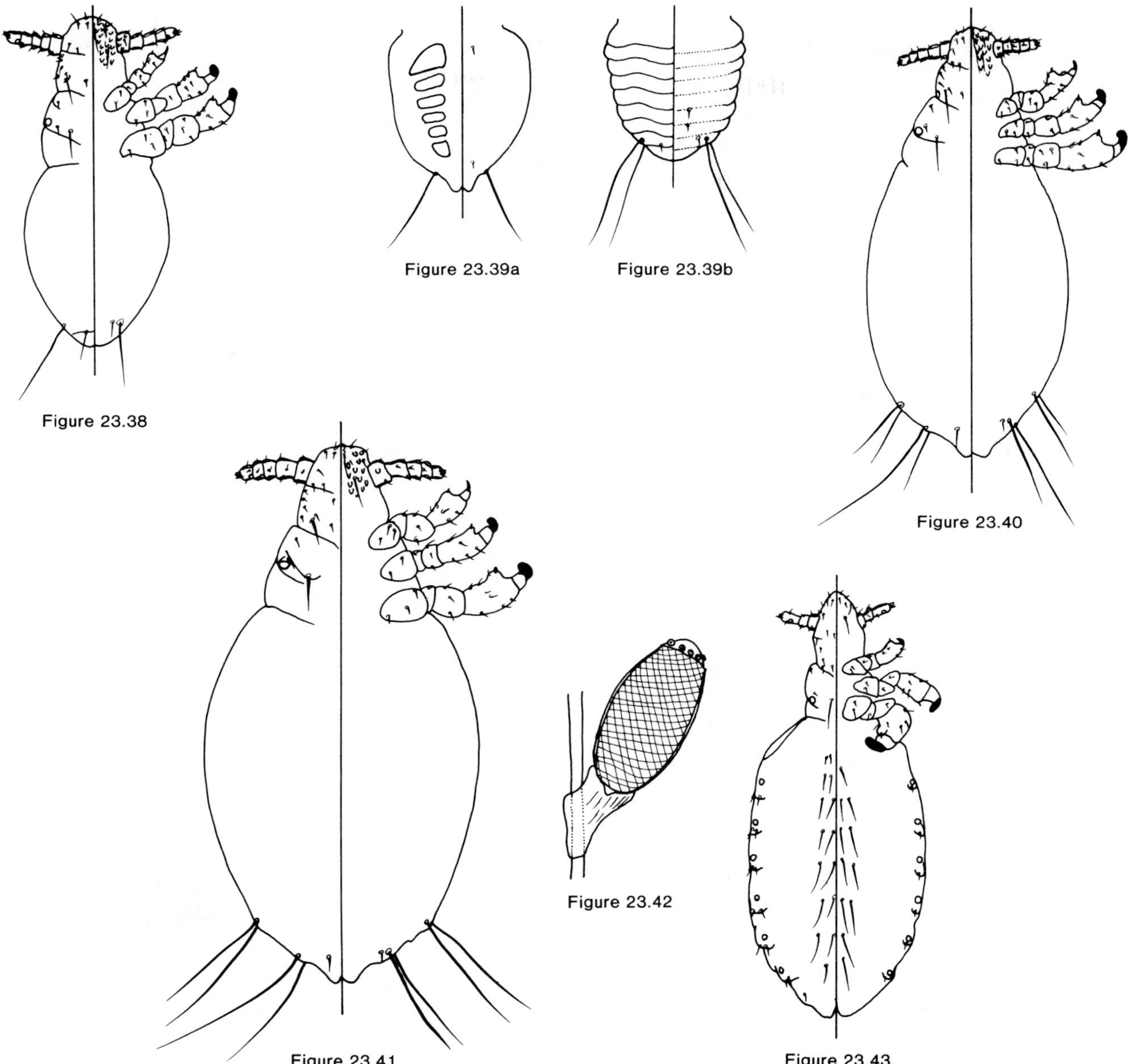

Figures 23.38–23.43. Hoplopleuridae. **(23.38)** *Hoplopleura acanthopus* (Burmeister), larva I; **(23.39a)** abdomen of *H. pacifica* Ewing, larva I; **(23.39b)** abdomen of *H. onychomydis* Cook & Beer, larva I; **(23.40)** *H. acanthopus* (Burmeister), larva II; **(23.41)** *H. acanthopus* (Burmeister), larva III; **(23.42)** egg of *H. pacifica* Ewing; **(23.43)** *Haematopinoides squamosus* Osborn, larva III.

Biology and Ecology: Hoplopleurinae are parasitic on rodents and pikas (Ochotonidae, Lagomorpha), and Haematopinoidinae are parasites of moles and shrews (Talpidae and Soricidae; Insectivora) and myomorph rodents (Gliridae and Zapodidae).

Description: Small lice; larva I 0.30–0.50 mm long, larva II 0.35–0.70 mm, and larva III 0.60–0.90 mm. *Head* with anterior margin irregularly rounded or truncated and without external eyes or ocular points; antennae usually five-segmented or rarely four-segmented (*Ancistroplax* and *Haematopinoides,* both on Insectivora); numerous tubercles usually present on the ventral head (*Hoplopleura* and *Pterophthirus*). *Thorax* gradually larger than head; thoracic sternal plate usually lacking except *Schizophthirus;* mesothoracic spiracles small; no notal pit. Fore legs always small, each with an acuminate claw; mid legs usually larger than fore legs, although similar in shape; hind legs usually largest, each with a stout claw and highly developed tibial thumbs. *Abdomen* usually with one or more pairs of MAS and two central rows of DCAS and VCAS.

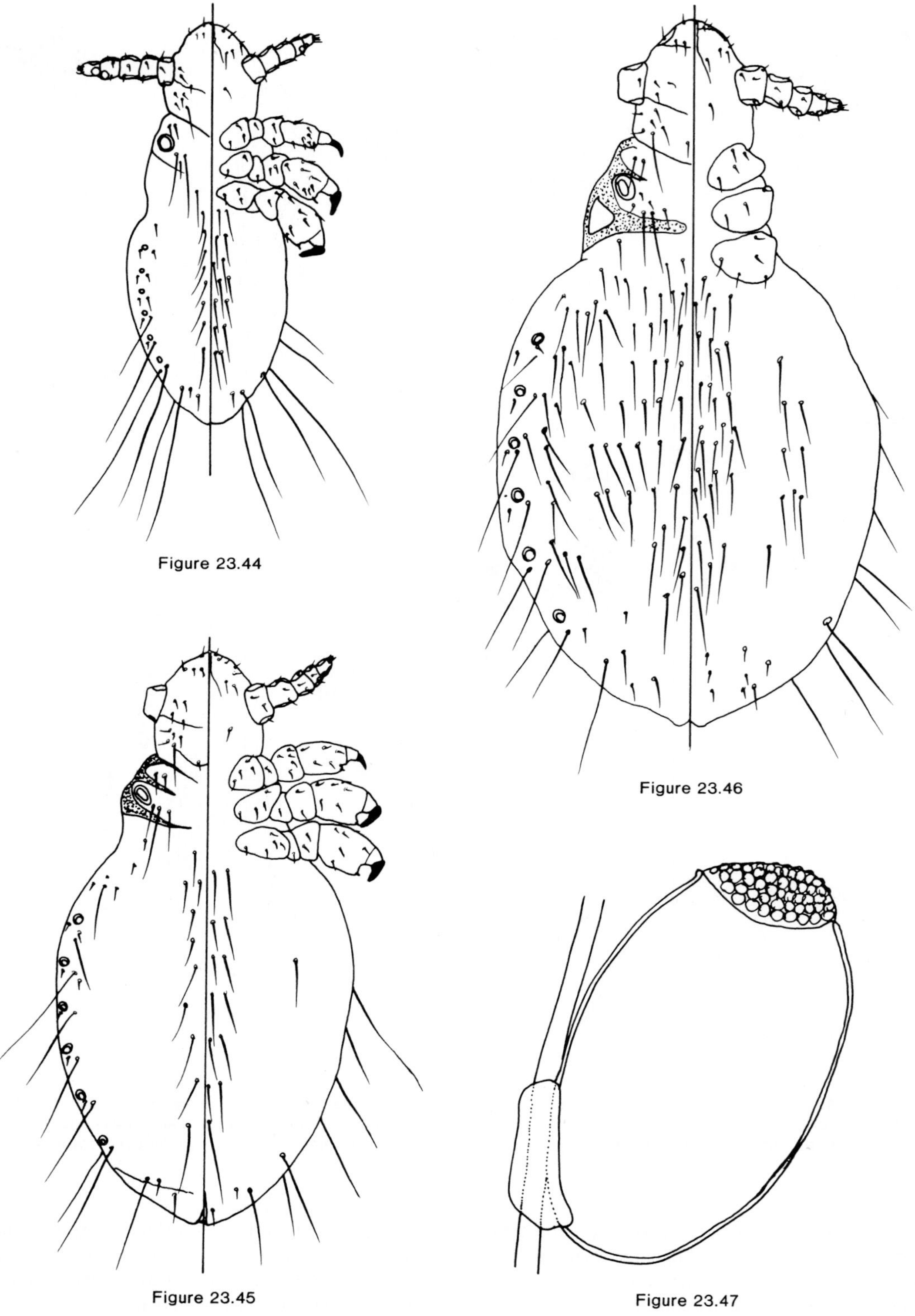

Figures 23.44–23.47. Linognathidae. *Linognathus;* **(23.44–46)** *L. setosus* (von Olfers): **(23.44)** larva I; **(23.45)** larva II; **(23.46)** larva III, legs removed; **(23.47)** egg of *L. pedalis* (Osborn).

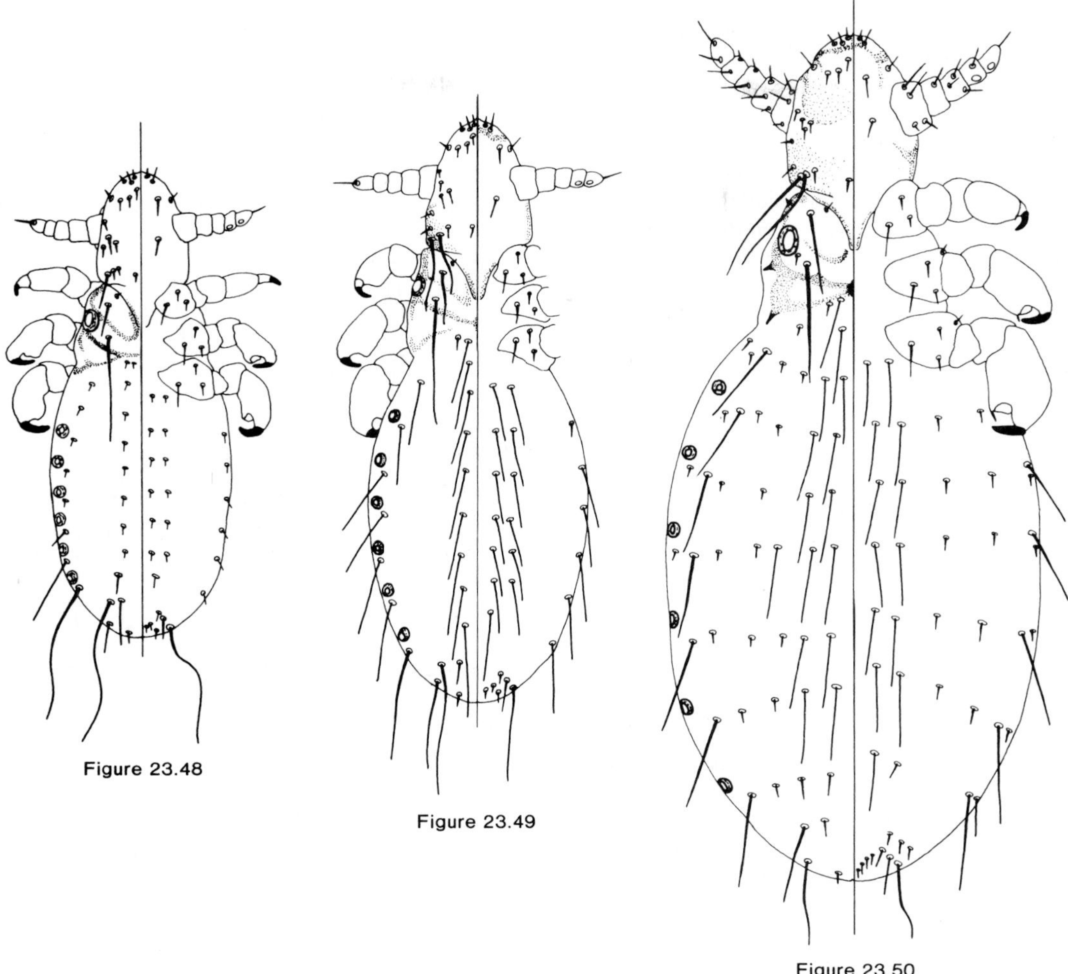

Figures 23.48–23.50. Linognathidae. *Solenopotes tarandi* (Mjöberg); **(23.48)** larva I; **(23.49)** larva II; **(23.50)** larva III.

Comments: Two subfamilies and five genera are recognized: Hoplopleurinae—*Hoplopleura* (117 species) (figs. 23.38–23.42) and *Pterophthirus* (5 species), and Haematopinoidinae—*Haematopinoides* (1 species) (fig. 23.43), *Ancistroplax* (2 species), and *Schizophthirus* (7 species). Of these, *Hoplopleura* and *Haematopinoides* occur in North America.

Selected Bibliography

Cook and Beer 1959.
Ferris 1951.
Johnson 1972.
Kim 1965.
Pratt and Karp 1953.

LINOGNATHIDAE

Linognathids, Smooth Sucking Lice

Figures 23.44–23.50

Relationships and Diagnosis: The linognathids are a rather homogeneous group, closely related to the Ratemiidae (Old World zebra and ass lice) and the Polyplacidae. Linognathids can be distinguished by having four rows of VCAS and DLAS in larva I, the mid legs and hind legs subequal and larger than the fore legs, and no paratergites in larvae II and III.

Biology and Ecology: The life cycle usually takes 21–30 days from eggs to eggs. Each species infests particular parts of the host animal. *Linognathus vituli* (L.) is abundant on the dewlap and shoulders, although it is also found on the sides of the neck, the rump, sides of the body, topline, udder, perineum, and belly. *Solenopotes capillatus* Enderlein is usually found on the neck and head of infested animals (Matthysee 1946). Seasonal fluctuations in the *Solenopotes capillatus* population are caused by air temperature changes in the louse habitat and hair shedding (Jensen and Roberts 1966). As the temperature moves above or below the optimum (31–33°C), less favorable environmental conditions are available for reproduction and population maintenance.

Linognathus is primarily parasitic on Bovidae and Giraffidae (Artiodactyla) and has expanded its distribution to Canidae (Carnivora). *Solenopotes* is parasitic on Bovidae and Cervidae (Artiodactyla), and *Prolinognathus* is found exclusively on Procaviidae (Hyracoidea).

Description: Medium lice; larva I 0.80–1.25 mm, larva II 1.10–1.80 mm, larva III 1.70–2.10 mm. The abdominal chaetotaxy develops gradually in size and number from the first to the third stage. *Head* usually cone-shaped, without external eyes or ocular points; antennae usually five-segmented or rarely four-segmented (*Prolinognathus*); DPoMHS and DPHS distinct. *Thorax* without sternal plate; DPtS and DPTS distinct; mesothoracic spiracles usually large (or small in *Prolinognathus*). Legs relatively short; fore legs smallest, mid legs much larger, and hind legs largest; fore coxae separated widely from each other. *Abdomen* elliptical, without any indication of paratergites, sternites or tergites; spiracles usually visible; abdominal chaetotaxy with mostly minute DMdAS and VMdAS between larger LAS and CAS.

Comments: Three genera are recognized: *Linognathus* (51 species), *Solenopotes* (10 species), and *Prolinognathus* (8 species). Some species of *Linognathus* (figs. 23.44–23.47) and *Solenopotes* (figs. 23.48–23.50) are found in North America. *Linognathus vituli* and *Solenopotes capillatus* are important cattle lice.

Selected Bibliography

Jensen and Roberts 1966.
Kim and Weisser 1974.
Matthysse 1946.
Weisser and Kim 1973.

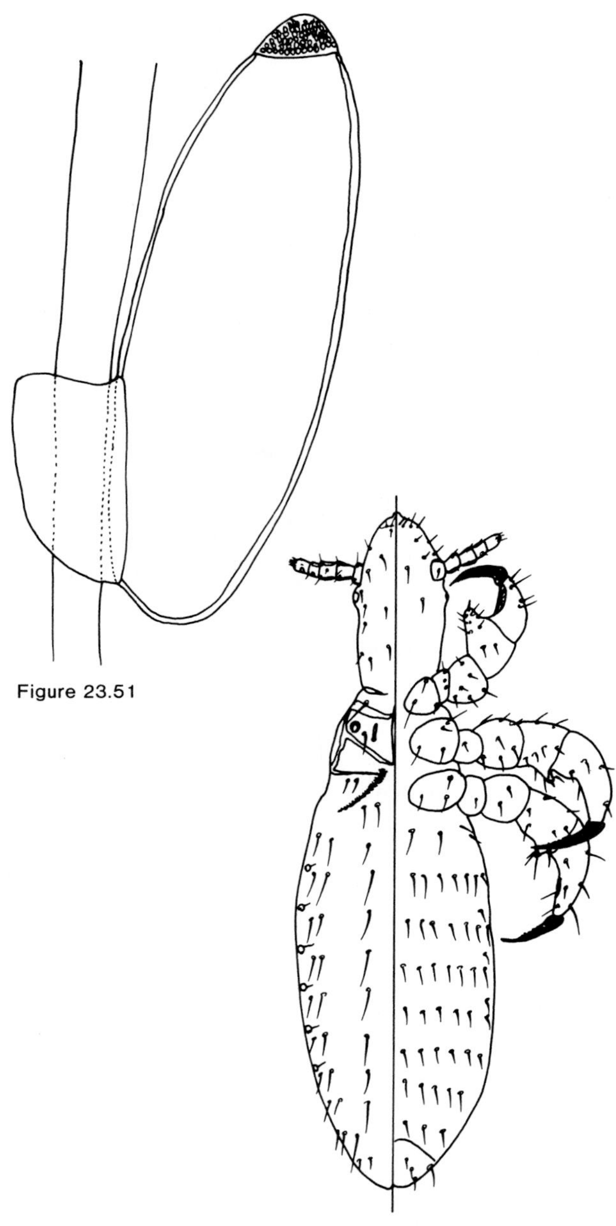

Figures 23.51–23.52. Pecaroecidae. *Pecaroecus javalli* Babcock & Ewing; **(23.51)** egg; **(23.52)** larva II.

PECAROECIDAE

The Pecaroecids

Figures 23.51, 23.52

Relationships and Diagnosis: Pecaroecids are superficially similar to *Haematopinus* and are distinguished from other lice by the long slender head and distinct eyes.

Biology and Ecology: Pecaroecids are parasites of the peccary (Tayassuidae).

Description: Large lice with long, slender body. *Head* long and slender, with clearly evident eyes represented by a lens; antennae five-segmented. *Thorax* relatively short and heavily sclerotized, with distinct notal pit; no sternal plate; mesothoracic spiracles distinct. All legs subequal in size and shape but fore legs with enlarged tibial thumb. *Abdomen* long and narrowly elliptical; derm finely wrinkled; segmental setae short and arranged in transverse rows.

Comments: The single known species, *Pecaroecus javalii* Babcock and Ewing (1938), is distributed in the southwestern United States.

Selected Bibliography

Babcock and Ewing 1938.
Ferris 1951.

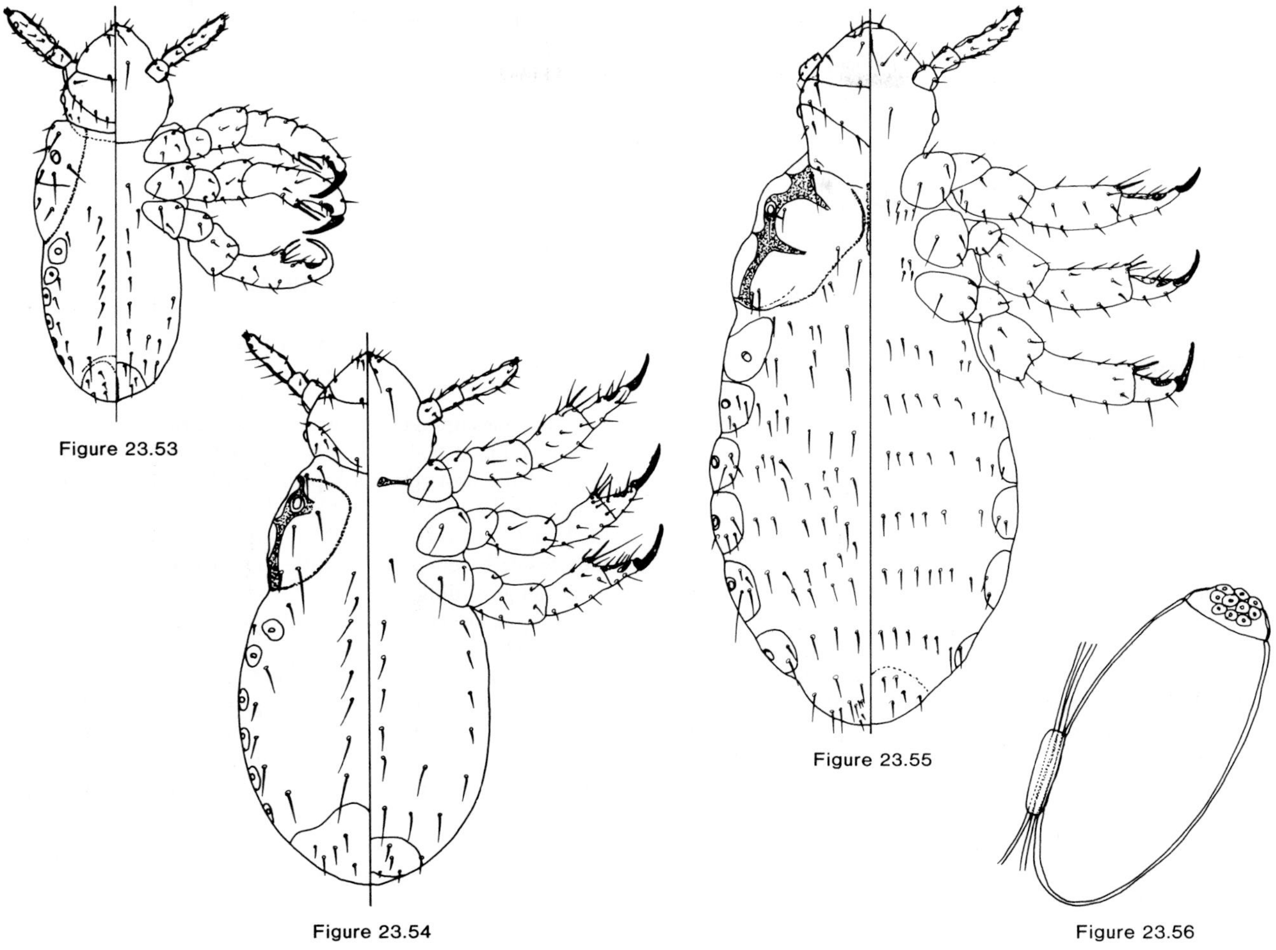

Figures 23.53–23.56. Pediculidae. *Pediculus humanus humanus* L. **(23.53)** larva I; **(23.54)** larva II; **(23.55)** larva III; **(23.56)** egg.

PEDICULIDAE

Pediculids, Human Lice

Figures 23.53–23.56

Relationships and Diagnosis: The pediculids, although highly specialized as are the Pecaroecidae and the Haematopinidae, retain many primitive characters. Larvae of the pediculids can be distinguished from other lice by the presence of external eyes, distinct notal pit and all legs subequal in size and shape.

Biology and Ecology: *Pediculus* is parasitic upon New World monkeys (Cebidae), gibbons and great apes (Pongidae), and humans (Hominidae). The human louse, *Pediculus humanus* L., has two subspecies, *P. h. humanus* L. and *P. h. capitis* De Geer. The life cycle takes 18–20 days from eggs to eggs. The eggs (fig. 23.56) take 8–9 days at 30° C. to hatch, and larval stages last 12–21 days. *Pediculus h. capitis* (head louse) is able to feed at any time, but *P. h. humanus* (body louse) can only feed undisturbed when the host is at rest (Buxton 1946). The eggs of head lice are cemented to hairs whereas body louse eggs are glued to fibers of the clothing.

Description: Medium lice, often translucent; larva I 0.90–1.30 mm long (fig. 23.53); larva II 1.35–1.58 mm long (fig. 23.54); larva III 2.00–2.70 mm long (fig. 23.55). *Head* relatively short, abruptly constricted posteriorly into a short neck, with eyes externally represented by pair of distinct lenses and pigmentation on the lateral lobes; antennae five-segmented (terminal segments often fused). *Thorax* with well-developed phragmata and notal pit; no sternal plate; mesothoracic spiracles distinct. All legs subequal in shape and size and each with a long acuminate claw; tibial thumbs developed. *Abdomen* membranous, with lateral margins more or less lobed and six pairs of spiracles; segmental setae distinct, arranged in transverse fields.

Comments: Many species had formerly been recognized for *Pediculus*. Currently only two species are accepted as distinct and others as subspecies or infraspecific variants: *Pediculus humanus* on humans and New World monkeys (Cebidae), and *P. schaeffi* Fahrenholz on gibbons and great apes (Pongidae).

Selected Bibliography

Buxton 1946.
Ferris 1951.
Kim and Emerson 1968a.

POLYPLACIDAE

The Polyplacids

Figures 23.57-23.73

Relationships and Diagnosis: The polyplacids are a rather heterogeneous group and somewhat related to the Linognathidae and the Hoplopleuridae. They can be distinguished from other lice by having no or two rows of VCAS and no DLAS in larva I, and mid legs usually subequal to the hind legs in size and shape, and paratergites in larvae II and III.

Larvae are superficially similar among *Polyplax* (figs. 23.57–23.60), *Neohaematopinus* (figs. 23.64–23.67), *Proenderleinellus, Linognathoides* (figs. 23.61–23.63), *Fahrenholzia* (figs. 23.68–23.70), and *Haemodipsus* (figs. 23.71–23.73), but larvae are quite different among *Lemurphthirus, Lemurpediculus, Phthirpediculus, Eulinognathus, Ctenophthirus, Scipio, Sathrax, Johnsonpthirus,* and *Docophthirus.* However, the larvae are very similar to their adults in general morphology and primary chaetotaxy.

Biology and Ecology: Polyplacids are parasitic on Rodentia, Lagomorpha, Insectivora, and Prosimian Primates. *Polyplax* (76 species) and *Neohaematopinus* (30 species) are the two largest genera, primarily parasitic on rodents and occasionally infesting Insectivora. *Sathrax* (1 species) and *Docophthirus* (1 species) are found on Tupaiidae (Primates), and three genera are parasitic on Prosimian Primates: *Lemurphthirus* (2 species) on Lorisidae, *Lemurpediculus* (2 species) on Lemuridae, and *Phthirpediculus* (2 species) on Indridae. *Haemodipsus* (6 species) are parasites of rabbits (Leporidae, Lagomorpha).

Description: Small lice; body size variable. *Head* with antennae five-segmented; head about as long as wide; VPHS long; DPHS, MHS, DAnCHS, DPoCHS distinct; some species with sclerotized ventral tubercles (e.g., *Fahrenholzia microcephala*). *Thorax* wider than head; DPtS and DPTS distinct; sternal plate usually lacking in the first stage but frequently present in the second and third stages. Fore legs always small and slender, each with an acuminate claw; mid legs subequal to hind legs in size and shape, or hind legs larger than mid legs. *Abdomen* oval or elliptical, with six pairs of small spiracles; paratergites often present in larvae II and III; MAS and CAS distinct; usually larva I with two rows of CAS and two pairs of MAS, larva III with two or four rows of CAS and four or more pairs of MAS, and larva III with four or more rows of CAS and usually six pairs of MAS.

Comments: *Polyplax, Neohaematopinus, Linognathoides, Fahrenholzia,* and *Haemodipsus* are commonly found in North America.

Selected Bibliography

Ewing 1927.
Ferris 1951.
Johnson 1969.
Kim and Adler 1982.
Kim and Emerson 1968b, 1973.
Pratt and Karp 1953.

PTHIRIDAE

Pthirids, Crab Lice, Pubic Lice

Figures 23.74-23.77

Relationships and Diagnosis: The pthirids are unique lice with a compact body, wide thorax, and short abdomen. Because of the host relationships, *Pediculus* and *Pthirus* have been considered closely related and have been grouped into the family Pediculidae by many workers (Ferris 1951), but they basically represent two different lineages with numerous morphological differences (Kim and Ludwig 1978).

Larvae are very similar to the adults except for size and setal density.

Biology and Ecology: The entire life cycle of *Pthirus pubis* L. takes 13–17 days from eggs to eggs at skin temperature. Eggs hatch 7–8 days after oviposition (Buxton 1946). *Pthirus pubis* infests the pubic regions particularly, but also the armpits and more rarely the mustache, beard, eyelashes, and eyebrows.

Description: Medium lice with compact body; larva I 0.63–0.85 mm long, larva II 0.9–1.2 mm, and larva III 1.10–1.50 mm. *Head* short, much narrower than thorax, with distinct eyes; antennae five-segmented. *Thorax* short and wide, without notal pit or sternal plate; mesothoracic spiracles distinct. Fore legs slender, with pointed, acuminate claws; mid legs subequal to hindlegs in size and shape, very large, with stout claws. *Abdomen* relatively small, as broad basally as the posterior part of the thorax, with six pairs of large spiracles, the first three being crowded together and the first two displaced toward the dorsal meson.

Comments: Two species of *Pthirus* are so far known: *Pthirus pubis* (crab louse) on humans and *P. gorillae* Ewing on the gorilla. *Pthirus pubis* is distributed worldwide. Crab lice are usually transmitted from one person to another during sexual activity. They may also be spread on loose hairs transferred by infested persons to such items as towels and bedding.

Selected Bibliography

Buxton 1946.
Ewing 1927.
Kim and Emerson 1968a.
Piotrowski 1961.

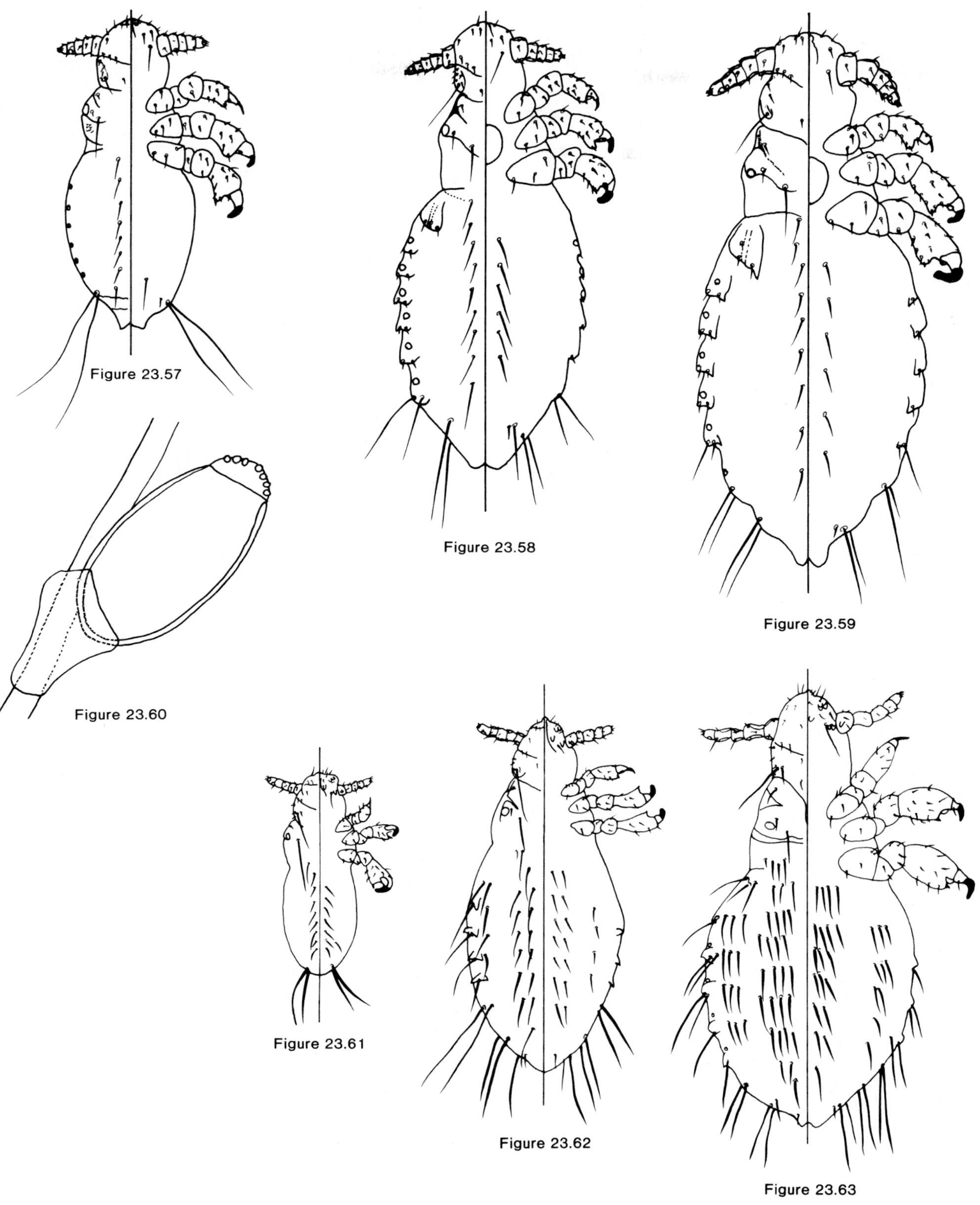

Figures 23.57–23.60. Polyplacidae. *Polyplax spinulosa* (Burmeister); **(23.57)** larva I; **(23.58)** larva II; **(23.59)** larva III; **(23.60)** egg.

Figures 23.61–23.63. Polyplacidae. *Linognathoides marmotae* (Ferris); **(23.61)** larva I; **(23.62)** larva II; **(23.63)** larva III.

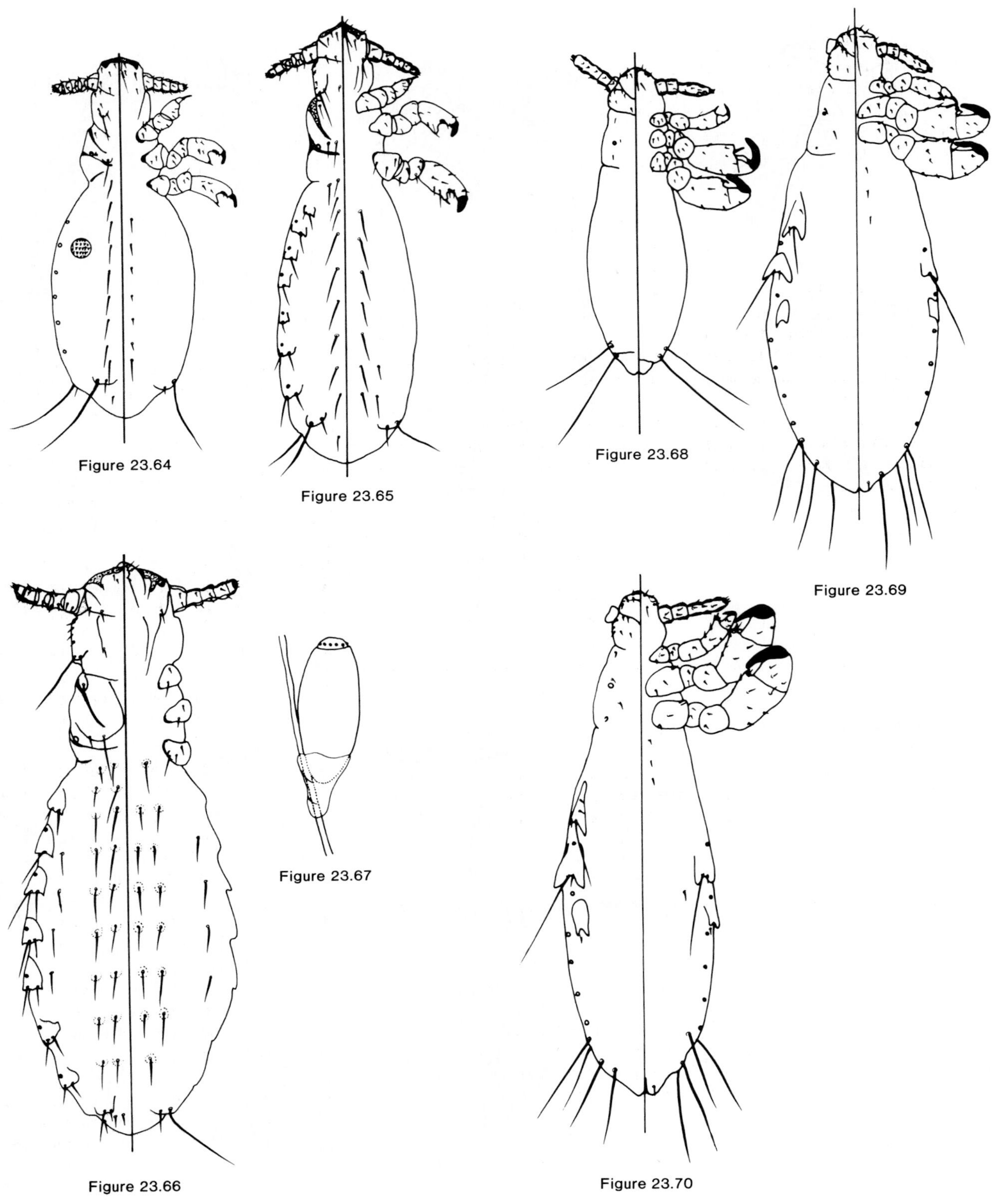

Figures 23.64–23.67. Polyplacidae. *Neohaematopinus;* **(23.64–66)** *N. sciuropteri* (Osborn); **(23.64)** larva I; **(23.65)** larva II; **(23.66)** larva III, legs removed; **(23.67)** *N. sciuri* Jancke, egg.

Figures 23.68–23.70. Polyplacidae. *Fahrenholzia fairchildi* Johnson; **(23.68)** larva I; **(23.69)** larva II; **(23.70)** larva III.

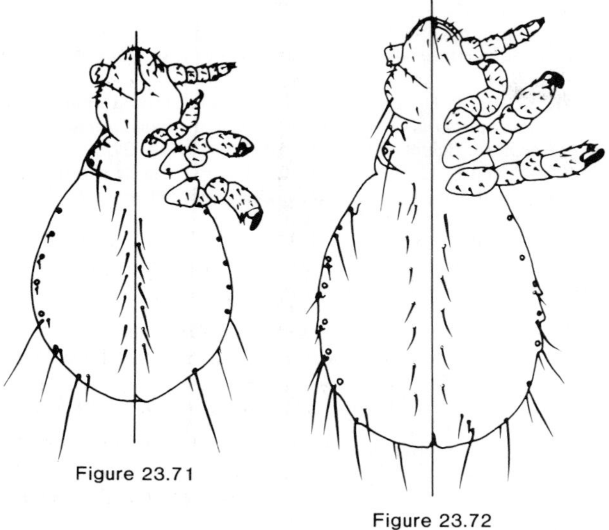

Figure 23.71

Figure 23.72

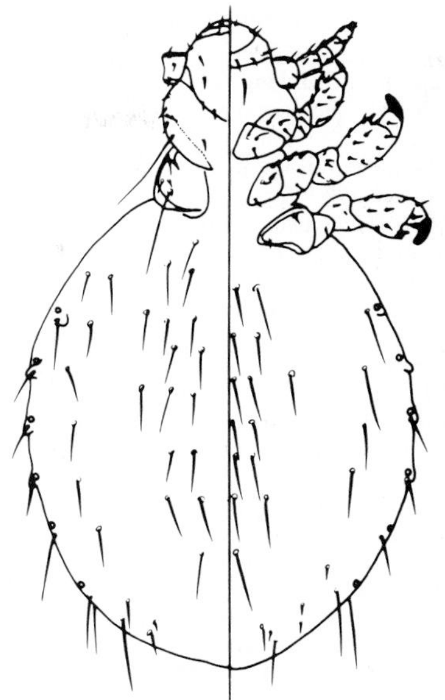

Figure 23.73

Figures 23.71–23.73. Polyplacidae. *Haemodipsus setoni* Ewing; **(23.71)** larva I; **(23.72)** larva II; **(23.73)** larva III.

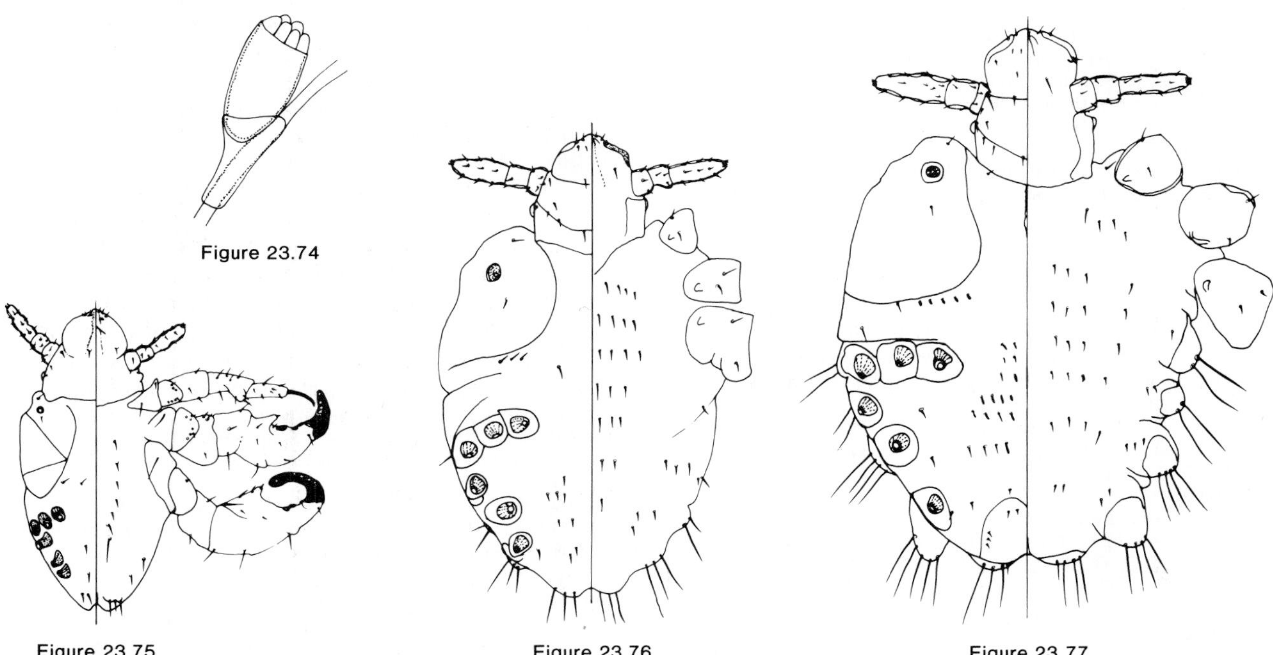

Figure 23.74

Figure 23.75

Figure 23.76

Figure 23.77

Figures 23.74–23.77. Pthiridae. *Pthirus pubis* (L.); **(23.74)** egg; **(23.75)** larva I; **(23.76)** larva II, legs removed; **(23.77)** larva III, legs removed.

BIBLIOGRAPHY

Babcock, O. G., and H. E. Ewing. 1938. A new genus and species of Anoplura from the peccary. Proc. Ent. Soc. Wash. 40(7):197–201.

Bruce, W. G. 1947. The tail louse, a new pest of cattle in Florida. J. Econ. Ent. 40:590–99.

Buxton, P. A. 1946. The louse. An account of the lice which infest man. Their medical importance and control. London: Edward Arnold and Co. 164 pp.

Chaudhuri, R. P., and P. Kumar. 1961. The life history and habits of the buffalo louse, *Haematopinus tuberculatus* (Burmeister) Lucas. Indian J. Vet. Sci. 31:275–87.

Cook, E. F., and J. R. Beer. 1959. The immature stages of the genus *Hoplopleura* (Anoplura: Hoplopleuridae) in North America, with descriptions of two new species. J. Parasit. 45:405–16.

Craufurd-Benson, H. J. 1941. The cattle lice of Great Britain. I. Biology with special reference to *Haematopinus eurysternus*. II. Lice populations. Parasitology 33:331–42, 342–58.

Ewing, H. E. 1927. Descriptions of three new species of sucking lice, together with a key to some related species of the genus *Polyplax*. Proc. Ent. Soc. Wash. 29:118–21.

Ferris, G. F. 1951. The sucking lice. Mem. Pacific Coast Ent. Soc. 1:1–320.

Florence, L. 1921. The hog louse, *Haematopinus suis* Linné: its biology, anatomy, and histology. Mem. Cornell Univ. Agr. Expt. Sta. 51:642–744.

Hopkins, G. H. E. 1949. The host-associations of the lice of mammals. Proc. Zool. Soc. Lond. 119:387–604.

Jensen, R. E., and J. E. Roberts. 1966. A model relating microhabitat temperatures to seasonal changes in the little blue louse (*Solenopotes capillatus*) population. Tech. Bull. N.S. 55, Ga. Agr. Expt. Sta. Univ. Ga. Coll. Agr., 22 pp.

Johnson, P. T. 1969. *Hamophthirius galeopitheci* Mjöberg rediscovered; with the description of a new family of sucking lice (Anoplura: Hamophthiriidae). Proc. Ent. Soc. Wash. 71(3):420–28.

Johnson, P. T. 1972. Sucking lice of Venezuelan rodents, with remarks on related species (Anoplura). Brigham Young Univ. Science Bull., Biol. Ser. 17(5):1–62.

Kim, K. C. 1965. A review of the *Hoplopleura hesperomydis* complex. J. Parasit. 51:871–87.

Kim, K. C. 1966a. The nymphal stages of three North American species of the genus *Enderleinellus* Fahrenholz (Anoplura, Hoplopleuridae). J. Med. Ent. 2(4):327–30.

Kim, K. C. 1966b. The species of *Enderleinellus* (Anoplura, Hoplopleuridae) parasitic on the Sciurini and Tamiasciurini. J. Parasit. 52(5):988–1024.

Kim, K. C. 1966c. A new species of *Hoplopleura* from Thailand, with notes and descriptions of nymphal stages of *Hoplopleura captiosa* Johnson (Anoplura). Parasitology 56:603–12.

Kim, K. C. 1968. Two new species of the sucking lice (Hoplopleuridae, Anoplura) from *Rattus* (Muridae, Rodentia) in Thailand. 58(3):701–707.

Kim, K. C. 1971. The sucking lice (Anoplura, Echinophthiriidae) of the northern fur seal; descriptions and morphological adaptation. Ann. Ent. Soc. Amer. 64(1):280–92.

Kim, K. C. 1972. Louse populations of the northern fur seal (*Callorhinus ursinus*). Amer. J. Vet. Res. 33(10):2027–36.

Kim, K. C. 1975. Ecology and morphological adaptation of the sucking lice (Anoplura: Echinophthiriidae) on the northern fur seal. Rapp. P.-v. Reun. Cons. Int. Explor. Mer. 169:504–15.

Kim, K. C. 1977. *Atopopthirus emersoni,* new genus and new species (Anoplura: Hoplopleuridae) from *Petaurista elegans* (Sciuridae, Rodentia), with a key to the genera of Enderleinellinae. J. Med. Ent. 14(4):417–20.

Kim, K. C. 1985. Chap. 5. Evolution and host associations of Anoplura. pp. 197–231 *in* Kim, K. C. (ed.). Coevolution of parasitic arthropods and mammals. New York: John Wiley and Sons.

Kim, K. C., and P. H. Adler. 1982. Taxonomic relationships of *Neohaematopinus* to *Johnsonpthirus* and *Linognathoides* (Anoplura: Polyplacidae). J. Med. Ent. 19:615–27.

Kim, K. C., and K. C. Emerson. 1968a. Descriptions of two species of Pediculidae (Anoplura) from great apes (Primates, Pongidae). J. Parasit. 54(4):690–95.

Kim, K. C., and K. C. Emerson. 1968b. New records and nymphal stages of Anoplura from Central and East Africa, with description of a new *Hoplopleura* species. Rev. Zool. Bot. Afr. 78(1–2):1–45.

Kim, K. C., and K. C. Emerson. 1973. Anoplura from Mozambique with descriptions of a new species and nymphal stages. Rev. Zool. Bot. Afr. 87(3):425–55.

Kim, K. C., and K. C. Emerson. 1974. *Latagophthirus rauschi,* new genus and new species (Anoplura: Echinophthiriidae) from the river otter (Carnivora: Mustelidae). J. Med. Ent. 11(4):442–46.

Kim, K. C., and H. W. Ludwig. 1978. The family classification of the Anoplura. Syst. Ent. 3:249–84.

Kim, K. C., and C. F. Weisser. 1974. Taxonomy of *Solenopotes* Enderlein, 1904, with redescription of *Linognathus panamensis* Ewing (Linognathidae: Anoplura). Parasitology 69:107–35.

Kim, K. C., H. D. Pratt, and C. J. Stojanovich. 1986. The sucking lice of North America: an illustrated manual for identification. The Penn. St. Univ. Press, University Park, PA.

Kuhn, H. J., and H. W. Ludwig. 1965. *Phthirunculus sumatranus* n. gen., n. sp., eine Lause des Flughornchens *Petaurista petaurista.* Sencken. Biol. 46:245–50.

Ludwig, H. W. 1968. Zahl, Vorkommen und Verbeitung der Anoplura. Z. f. Parasitenk. 31:254–65.

Matthysse, J. G. 1944. Biology of the cattle biting louse and notes on cattle sucking lice. J. Econ. Ent. 37:436–42.

Matthysse, J. G. 1946. Cattle lice: their biology and control. Bull. Cornell Univ. Agr. Expt. Sta. 832:1–67.

Meleney, W. P., and K. C. Kim. 1974. A comparative study of cattle-infesting *Haematopinus,* with redescription of *H. quadripertusus* Fahrenholz, 1916 (Anoplura: Haematopinidae). J. Parasit. 60(3):507–22.

Murray, M. D. 1958. Ecology of the louse *Lepidophthirus macrorhini* Enderlein 1904 on the elephant seal, *Mirounga leonina* L. Nature 182:404–405.

Murray, M. D. 1960. The ecology of lice on sheep. I. The influence of skin temperature on populations of *Linognathus pedalis* (Osborn). Aust. J. Zool. 8:349–56.

Murray, M. D. 1961. The ecology of the louse *Polyplax serrata* (Burm.) on the mouse, *Mus musculus* L. Aust. J. Zool. 9:1–13.

Murray, M. D. 1963a. The ecology of lice on sheep. III. Differences between the biology of *Linognathus pedalis* (Osborn) and *L. ovillus* (Neumann). Aust. J. Zool. 11:153–56.

Murray, M. D. 1963b. The ecology of lice on sheep. IV. The establishment and maintenance of populations of *Linognathus ovillus* (Newmann). Aust. J. Zool. 11:157–72.

Murray, M. D., and D. G. Nicholls. 1965. Studies on the ectoparasites of seals and penguins. I. The ecology of the louse *Lepidophthirus macrorhini* Enderlein on the southern elephant seal, *Mirounga leonina* (L.). Aust. J. Zool. 13:437–54.

Murray, M. D., M. S. R. Smith, and Z. Soucke. 1965. Studies on the ectoparasites of seals and penguins. II. The ecology of the louse *Antarctophthirus ogmorhini* Enderlein on the Weddell seal, *Leptonychotes weddelli* Lesson. Aust. J. Zool. 13:761–71.

Piotrowski, F. 1961. The nymphs of crab-louse *Phthirus pubis* L. (Anoplura). Polsk. Pismo Entomolog. 31(22):321–34.

Pratt, H. D., and H. Karp. 1953. Notes on the rat lice *Polyplax spinulosa* (Burmeister) and *Hoplopleura oenomydis* Ferris. J. Parasit. 39(5):495–504.

Roberts, F. H. S. 1952. Insects affecting livestock, with special reference to important species occurring in Australia. Sydney: Angus and Robertson 267 pp.

Stimie, M., and S. van der Merwe. 1968. A revision of the genus *Haematopinus* Leach (Phthiraptera: Anoplura). Zool. Anz. 180:183–220.

Weisser, C. F., and K. C. Kim. 1973. Rediscovery of *Solenopotes tarandi* (Mjoberg 1915) (Linognathidae: Anoplura), with ectoparasites of the Barren Ground caribou. Parasitology 66:123–32.

Order Mecoptera

24

George W. Byers
University of Kansas

SCORPIONFLIES AND RELATIVES

Mecoptera are generally associated with forested parts of the temperate and tropical regions. A few are grassland species; some are subarctic. Larval Mecoptera mostly occur in or on the soil or in mosses or leafy liverworts, usually in humid environments; some bittacids and boreids survive in semiarid habitats. No Mecoptera, either immature or adult, are of real economic importance; however, since adult Bittacidae are predaceous, and larvae of both Bittacidae and Panorpidae are scavengers, they may be regarded as beneficial.

Mecoptera have been evolving for some 280 million years, and many of the evolved taxa are now extinct. As a result, we see very distinct, taxonomically isolated family groups and, accordingly, it is difficult to give a useful description of the larvae of the entire order.

The classification adopted here (Byers 1965) recognizes the family Panorpodidae as separate from the Panorpidae, largely on the basis of larval morphology. Larvae and pupae of Meropeidae are still unknown; this is probably the greatest gap remaining in our knowledge of North American Mecoptera.

DIAGNOSIS

Head well developed and sclerotized, without adfrontal areas; mouthparts hypognathous, mandibles and maxillae relatively large (26–38% of total head length, lateral aspect). Eyes apparently "compound" in Panorpidae, or of 3 to 30 closely set stemmata, or absent. Antennal pedicel as long as or longer than scape, antennal base above anterior mandibular articulation. Thoracic legs subconical, thick and fleshy at base; eight pairs of prolegs without crochets, or none. If prolegs present, eyes with either 7 stemmata in a ring or 24–30 apparent ommatidia in a tight cluster; if prolegs absent, eyes reduced to 3–7 stemmata or absent, and meso- and metathoracic legs extending laterad or rudimentary; prothoracic legs extending ventrad.

Selected Bibliography

Byers 1954.
Byers and Thornhill 1983.
Carpenter 1931.
Peterson 1951.

KEY TO FAMILIES AND GENERA OF NORTH AMERICAN MECOPTERA LARVAE[1]

1. Larva eruciform; small prolegs on first 8 abdominal segments; prominent setiferous projections on at least abdominal terga 8 and 9 2

1′. Larva modified scarabaeiform, with meso- and metathoracic legs directed laterad, or curculioniform; no prolegs or conspicuous projections on any abdominal segments 5

2(1). Prominent dorsal and lateral, branched, fleshy projections (scoli) on thoracic and abdominal segments (Bittacidae) 3

2′. Thoracic and first 7 abdominal segments without fleshy, branched projections but bearing setiferous pinacula .... (*Panorpa*) ***Panorpidae*** (p. 247)

3(2). Seta on middle branch of each dorsal projection of abdomen terminal (fig. 24.4) .... (*Bittacus*) ***Bittacidae*** (p. 248)

3′. Seta on middle branch of each dorsal projection of abdomen subterminal (fig. 24.3) 4

4(3′). Posterior branch of dorsal projection longer than middle branch on A1–7 (fig. 24.5); projections wider than high (measured between apices of anterior and posterior branches, setae excluded) (fig. 24.5) .... (*Apterobittacus*) ***Bittacidae*** (p. 248)

4′. Posterior branch of dorsal projection shorter than middle branch on A1–7; projections higher than wide (fig. 24.3) .... (*Hylobittacus*) ***Bittacidae*** (p. 248)

1. Larvae of Meropeidae are unknown.

5(1′). Eyeless; no stemmata present .... *(Brachypanorpa)* ***Panorpodidae*** (p. 248)
5′. Cluster of 3–7 stemmata surrounded by black pigmentation near base of each mandible (Boreidae) 6
6(5′). Body curculioniform; legs rudimentary; integument in thick folds .... *(Caurinus)* ***Boreidae*** (p. 249)
6′. Body scarabaeiform; meso- and metathoracic legs directed laterad; integument not in thick folds 7
7(6′). Transverse row of setae on each of abdominal terga 1–9 .... *(Boreus)* ***Boreidae*** (p. 249)
7′. Transverse band of tiny denticles on abdominal terga 1–5; transverse row of setae on abdominal terga 1 and 6–9 .... *(Hesperoboreus)* ***Boreidae*** (p. 249)

CLASSIFICATION

Order **MECOPTERA**
- Suborder Protomecoptera
  - Meropeidae, meropeids
- Suborder Eumecoptera
  - Panorpidae, scorpionflies
  - Bittacidae, hangingflies
  - Panorpodidae, panorpodids
  - Boreidae, snow scorpionflies

PANORPIDAE

Scorpionflies, Panorpids

Figures 24.1, 24.2, 24.13

Relationships and Diagnosis: In North America, only the Panorpidae and Bittacidae have eruciform larvae; this may not indicate close relationship. The Panorpidae probably more nearly resemble the closest common ancestor. Although larval panorpids and bittacids at first appear utterly different, there is a general correspondence of setal patterns, with most setae in the Panorpidae arising from sclerotized pinacula that have positions equivalent to those of the fleshy projections on Bittacidae.

Larvae of *Panorpa* (figs. 24.1, 24.2) are eruciform and sparsely hairy, somewhat resembling certain Lepidoptera caterpillars and sawfly larvae. Unlike these, larval panorpids have compound eyes (or large numbers of closely grouped stemmata) and long, thickened setae borne on subconical, fleshy projections that are paired on abdominal terga 8 and 9, and single and median on tergum 10. They do not have conspicuous, branched projections on the thoracic dorsum or on abdominal terga 1–7, as do the Bittacidae (figs. 24.6, 24.7).

Biology and Ecology: Larvae of Panorpidae occur near the surface in humic forest soil, where they are somewhat gregarious and therefore unevenly distributed. They wander in search of food during the first three stadia and into the early fourth. At such times, they can be collected by Berlese funnel extraction, in pitfall traps, or at suitable baits. Their diet consists mainly of dead insects but may include other dead animal matter. Eggs of most species are subspherical, the chorion covered by a pattern of polygonal depressions. They attain the fourth instar within a month of hatching, and shortly enter a lengthy prepupal diapause.The pupa (fig. 24.13) has the shape of the adult, except that the wings are compressed and the rostrum is much shorter until just before emergence. There are one or two generations per year, depending on the species, possibly also on climatic factors. Larva, pupa, egg and general life history of *Panorpa nuptialis* Gerstaecker have been described by Byers (1963) and Gassner (1963).

Description: Mature larva (figs. 24.1, 24.2); about 12 mm (e.g., *Panorpa nebulosa* Westwood) to 25 mm (*P. nuptialis*) long, depending on species. Color in life sordid grayish yellow with diffuse pinkish gray areas on terga; head glossy dark brown, eyes black; preserved larvae paler, with more distinct tergal markings of dull pink or pale purple; pinacula light yellowish brown, pronotal shield glossy brown. Uneven transverse row of setae on paired dorsal sclerites of T2, T3, and A1–7; other setae on smaller dorsal and lateral pinacula. Paired, long, thick, annulated setae on fleshy, subconical projections on abdominal terga 8 and 9; one such annulated seta on posterodorsal midline of tergum 10. Head hypognathous, mandibles large, clypeus and labrum distinct, maxillae prominent; antennae with short scape, elongate pedicel widest subapically, and slender, short flagellum. Eyes comprising 24–30 facets (or closely packed, nearly hexagonal stemmata, giving appearance of compound eyes). Thoracic legs three-segmented with lightly sclerotized terminal claw; conical, fleshy prolegs on A1–8. Large thoracic spiracle enclosed in posterolateral corner of pronotal shield; smaller spiracles each enclosed by small pinaculum near anterior edge of A1–8. Terminal part of A10 membranous, eversible to form a holdfast organ, sometimes used in locomotion. Larvae of several North American species have been differentiated by use of chaetotaxy (Boese 1973), using the system of Yie (1951).

Comments: The largest family of Mecoptera, the Panorpidae, is widely distributed in temperate Eurasia and North America, entering the tropics in the mountains of Indo-China, Indonesia, and Mexico. *Panorpa,* the only genus found in the Nearctic region, where it is represented by 45 species, is also found in Europe and Asia.

Selected Bibliography

Boese 1973.
Brauer 1863.
Byers 1963.
Felt 1896.
Gassner 1963.
Mampe and Neunzig 1965.
Miyake 1912.
Potter 1938.
Shiperovitsh 1925.
Steiner 1930.
Yie 1951.

BITTACIDAE

Hangingflies, Bittacids

Figures 24.3-24.7

Relationships and Diagnosis: The caterpillarlike larvae of Bittacidae resemble those of Panorpidae, and bittacids are probably more closely related to Panorpidae than to Boreidae or Panorpodidae.

Larval bittacids are eruciform (figs. 24.6, 24.7), with conspicuous, branched, fleshy scoli (setiferous projections) (figs. 24.3–24.5) on the dorsum and pleura of all thoracic and abdominal segments. They differ from larvae of Panorpidae (fig. 24.2) by these scoli, and by the eye, which has only seven stemmata arranged in a ring. They bear a superficial resemblance to some Lepidoptera larvae, notably Nymphalidae, because of the scoli, but they differ in having conical prolegs without crochets, eyes with seven virtually contiguous stemmata, and other details.

Biology and Ecology: Larvae of Bittacidae are usually found on the surface of the soil but beneath leaf litter or other organic debris. Apparently they crawl about chiefly at night, seeking food, and therefore they may be collected the way panorpids are. Richard J. Sauer (personal communication to F. W. Stehr) has noted that *Bittacus* larvae were abundantly collected in North Dakota in pitfall traps equipped with a flashlight bulb connected by wires to a single "C" size battery clipped to a sheet metal cover. Larvae of *Hylobittacus* have been observed foraging on the surface of moss during daylight (Byers 1954, as *Bittacus apicalis* Hagen). It is common for bittacids to have the more anterior scoli thickly coated with soil. Soil is ingested and excreted with the larva's feces onto the dorsum of the anterior segments. In life, the larva usually has a thin covering of soil particles over most of its skin. The diet consists of dead animal matter. Setty (1940) reared larvae of several species of *Bittacus* on dead blow flies and chopped earthworms, but found that they would also feed on finely ground beef.

Eggs of *Bittacus* and *Apterobittacus* are cuboidal or polyhedral, with slightly impressed surfaces when deposited, later enlarging and becoming subspherical; those of *Hylobittacus* are at first nearly spherical. There are four larval instars, and Setty (1940) has described an additional molt between the fourth larval and the prepupal stadia. The exarate pupa generally resembles the adult, except for the wings being tightly compressed. There is one generation per year.

Description: Larva with conspicuous, branched scoli on dorsum and pleura of all thoracic and abdominal segments (figs. 24.6, 24.7). Length of mature larva 10–15 mm. Body dull grayish yellow, head brown to dark yellowish brown, strongly sclerotized; pronotal shield glossy brown; small median tergal sclerites brownish. Most projections branched, bearing setae at or near apices of branches; some setae tapering, others clavate or bladelike, their form and distribution varying with species. Scolus on tergum 10 of abdomen single, median. Dorsal scoli paired on T2, T3, and A1–9. Upper pleural scoli on these segments at level of abdominal spiracles, those on abdomen bearing three setae each; lower pleural scoli above bases of thoracic legs and abdominal prolegs. Head hypognathous, with conspicuous mandibles and maxillae; antennae comprising short, subcylindrical scape, elongate pedicel widest slightly before apex, and short, slender flagellum. Eyes of seven stemmata in a broadly elliptical or circular ring. Thoracic legs subconical, indistinctly three-segmented, with terminal clawlike portion; abdominal prolegs conical, curved, positioned near ventral midline. Thoracic spiracles adjacent to posterolateral edges of pronotal shield or enclosed by shield; abdominal spiracles smaller, near anterior edge of A1–8. Terminal portion of abdomen membranous, eversible, forming a holdfast structure.

Comments: Most widespread of the Mecoptera, the Bittacidae are found in all temperate and tropical regions. There are 14 described genera and others still unnamed. Only seven species of *Bittacus,* one of *Apterobittacus,* one of *Hylobittacus,* and one of *Orobittacus* are known in America north of Mexico.

Selected Bibliography

Applegarth 1939.
Setty 1931, 1939, 1940, 1941.

PANORPODIDAE

The Panorpodids

Figures 24.11, 24.12

Relationships and Diagnosis: Until recently, the Panorpodidae were included in the Panorpidae because of various structural similarities in the adults. Profound differences in adult and larval structures, however, suggested that the genera involved merited recognition as a separate family. Larval morphology indicates a level of relationship to the Boreidae that was not suspected before the discovery of the larvae of *Brachypanorpa.*

Larvae of *Brachypanorpa* (fig. 24.11) are modified scarabaeiform, the body strongly curved, abdominal prolegs absent, and the T2 and T3 legs extended somewhat laterad, the T1 legs ventrad, as in the Boreidae. Eyes are totally lacking (reduced in Boreidae), and sternal and pleural setae are longer than their counterparts in Boreidae.

Biology and Ecology: Larvae and pupae of three species of *Brachypanorpa* have so far been found. These were in forest soil at varying depths to several centimeters and were usually closely associated with roots of grasses and other herbaceous plants. Eggs are whitish and subspherical. There are probably four larval stages. The larval diet is not known but is assumed to be plant matter. In the laboratory, young larvae fed on soft, decayed wood but did not survive; they did not accept various plant rootlets, dead insects, or other foods offered. The pupa (fig. 24.12) has the form of the adult except for the compressed, ensheathed wings. The life cycle is apparently completed in a year.

Description: Mature larva (fig. 24.11) about 11.5 mm long; abdomen strongly curved, whitish to cream-colored; head, pronotal shield, and cervical sclerites pale yellowish-brown. Transverse bands of minute denticles on anterior 1/3 of each abdominal tergum 2–10, less dense band of similar denticles across posterior 1/3 of terga 1–7; setae in short transverse rows or small clusters on all terga, sterna, and pleura. Thoracic spiracle at posterolateral corner of pronotal shield; smaller spiracles on A1–8. Mouthparts hypognathous; clypeus and labrum distinct; maxillae large, lightly sclerotized; mandibles large, strongly sclerotized; antennae three-segmented. Eyes absent. T1 legs extend ventrad, T2 and T3 legs directed mainly laterad. Legs thick, subconical, indistinctly segmented except terminal segment abruptly more slender than more basal portion and cylindrical. No abdominal prolegs.

Comments: Only four species of *Brachypanorpa* in North America and five of *Panorpodes* in eastern Asia comprise this small family. Two species of *Brachypanorpa* occur in the southern Appalachian Mountains and the other two are found in the western states.

Selected Bibliography

Byers 1976.

BOREIDAE

Boreids, Snow Scorpionflies

Figures 24.8–24.10

Relationships and Diagnosis: On the basis of the adults, the Boreidae are a very distinct family that appears to have no near relatives. The larvae, however, indicate some common ancestry with the Panorpodidae, although the extent of this relationship has not yet been investigated in detail.

Larvae of most boreids (figs. 24.8, 24.9) can be differentiated from those of other Mecoptera by their somewhat scarabaeiform shape, laterally extended T2 and T3 legs, absence of abdominal prolegs, and small eyes. The only similar larvae are those of *Brachypanorpa* (Panorpodidae, fig. 24.11), which are eyeless, much larger when mature, and occur in a different habitat (humic forest soil). Larvae of the western genus *Caurinus* (one species) are curculioniform, with rudimentary legs and wrinkled or thickly folded integument.

Biology and Ecology: Larval boreids are usually found in small, oblong, earthen cells among the rhizoids of terrestrial mosses, particularly in compact, cushion-forming mosses such as *Dicranella* and *Tortula*. Their diet, as far as known, consists of fragments of the rhizoids and lower leaves. Cooper (1974) found larvae of *Hesperoboreus* in dry mosses such as *Grimmia*, growing on small amounts of trapped soil on boulders. Larvae of *Caurinus* occur in epiphytic mosses and leafy liverworts (Russell 1979, 1982). The number of larval instars is inferred to be four. Temperate North American species probably have a two-year life cycle, but subarctic species may require longer to reach maturity. The pupa of *Boreus brumalis* Fitch was described by Williams (1916).

Description: Mature larvae attain a length of 4.8 mm in such species as *Boreus brumalis* (eastern U.S.) and *B. coloradensis* Byers (figs. 24.8, 24.9) (Rocky Mountain region), and 3.7 mm in *Hesperoboreus notoperates* Cooper. They are normally strongly curved and were described as scarabaeiform by Peterson (1951). The curculioniform, slightly curved larva of *Caurinus* is about 2.5 mm in length.

Larva whitish to cream-colored, with pale brownish yellow head; smooth-skinned with a transverse row of eight pale setae on each abdominal tergum in *Boreus; Hesperoboreus* with a transverse band of minute denticles on terga 1–5 and a row of setae on terga 1 and 6–9 (fig. 24.10), as well as less conspicuous pleural and sternal setae, groups of setae on thoracic dorsum, and setae on the head. Mouthparts hypognathous with distinct clypeus and labrum and conspicuously developed maxillae; antennae with subconical basal segment, a slender, cylindrical apical segment set in a nearly circular socket; eyes comprising three (*Boreus, Hesperoboreus*) to seven (*Caurinus*) stemmata in an arc, surrounded by intense black pigmentation. T1 legs projecting ventrad; T2 and T3 legs extending more nearly laterad; legs rudimentary in *Caurinus*. Basal portion of leg thick, fleshy, rounded; second segment subconical; terminal segment abruptly more slender, cylindrical. T1 notum not sclerotized. Thoracic spiracle minute; abdominal spiracles barely evident in *Hesperoboreus,* not visible at ordinary magnifications in *Boreus*. First eight abdominal segments distinct, last three more or less merged to form bluntly rounded cauda without projections.

Comments: This is a small, circumboreal family comprising ten species of *Boreus* in North America and eleven species in Europe and Asia, two species of *Hesperoboreus* in western North America (Penny 1977), and one species of *Caurinus* in western Oregon and Washington (Russell 1979).

Selected Bibliography

Cooper 1974.
Penny 1977.
Potter 1938.
Russell 1979, 1982.
Williams 1916.

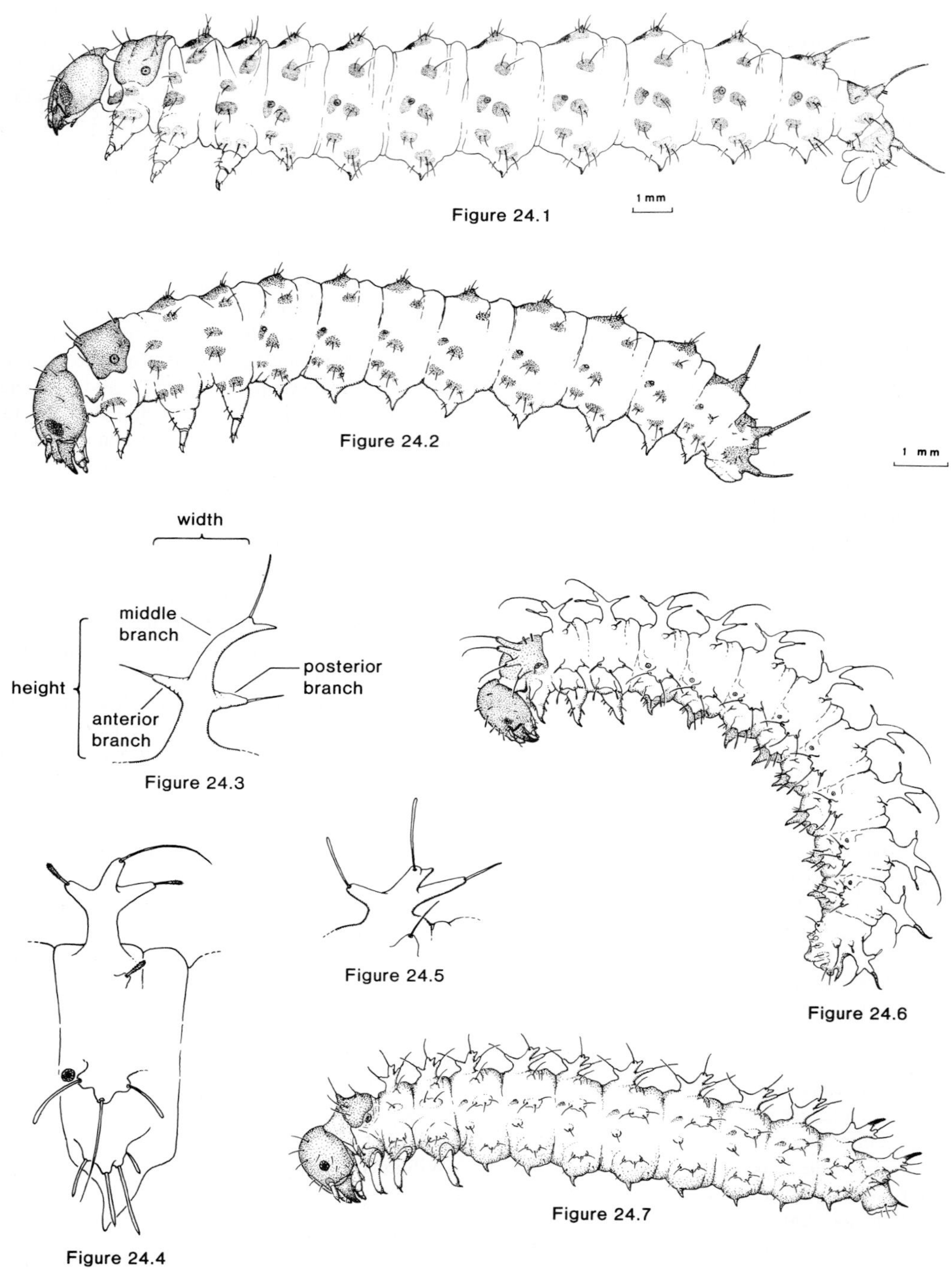

Figure 24.1. Panorpidae. *Panorpa nuptialis* Gerstaecker. f.g.l. ca. 24mm.

Figure 24.2. Panorpidae. *Panorpa helena* Byers. f.g.l. ca. 14mm.

Figure 24.3. Bittacidae. *Hylobittacus apicalis* (Hagen). Left lateral view of A6 dorsal scolus, showing measurements.

Figure 24.4. Bittacidae. *Bittacus stigmaterus* Say. Left lateral view of A5.

Figure 24.5. Bittacidae. *Apterobittacus apterus* (MacLachlan). Left lateral view of A2 scolus.

Figure 24.6. Bittacidae. *Bittacus stigmaterus* Say. f.g.l. ca. 12mm.

Figure 24.7. Bittacidae. *Apterobittacus apterus* (MacLachlan).

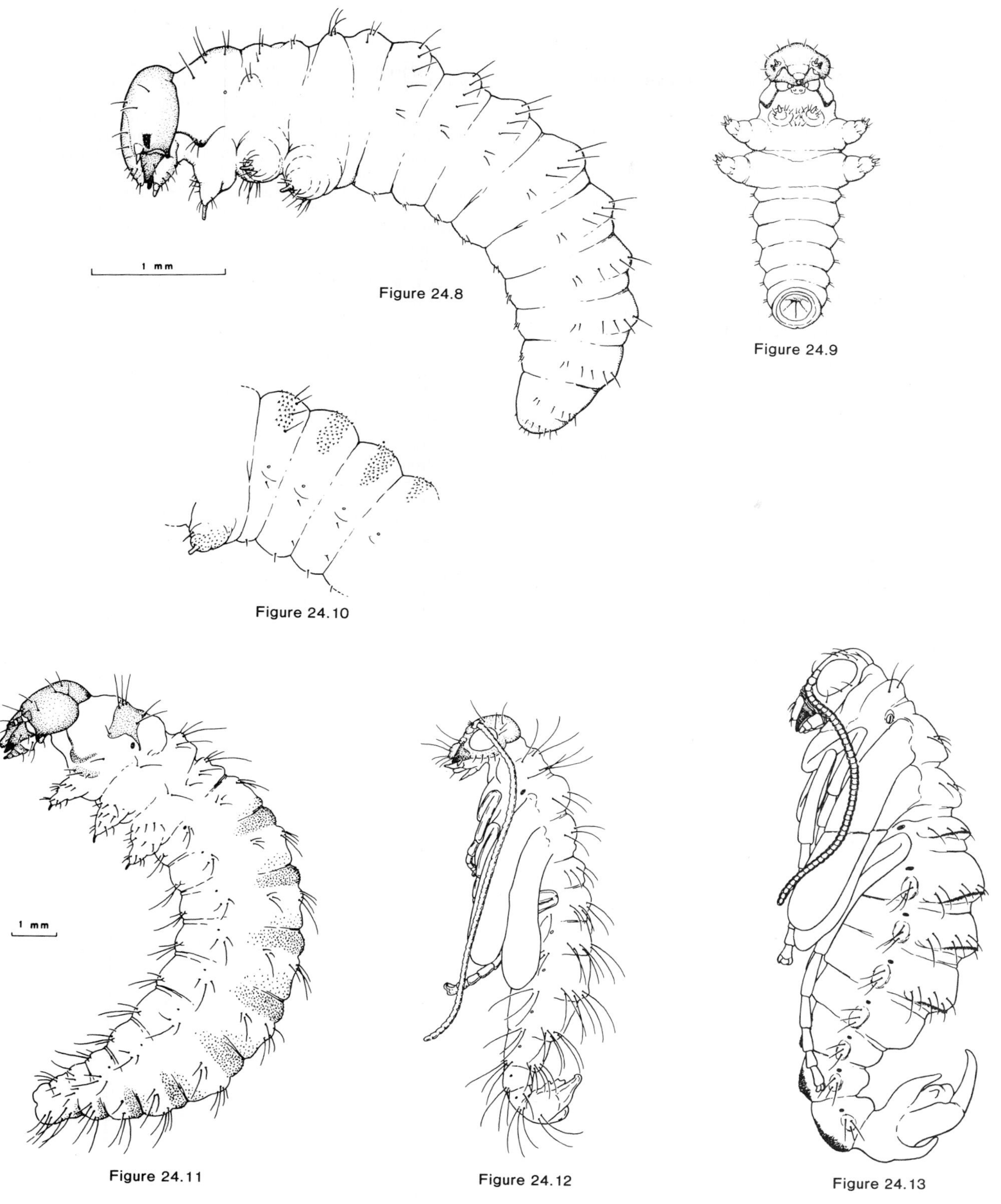

Figure 24.8

Figure 24.9

Figure 24.10

Figure 24.11

Figure 24.12

Figure 24.13

Figures 24.8–24.9. Boreidae. *Boreus coloradensis* Byers. Lateral and ventral respectively. f.g.l. ca. 4.8mm.

Figure 24.10. Boreidae. *Hesperoboreus notoperates* Cooper. metathorax and A1-4. f.g.l. ca. 3.7mm.

Figures 24.11–24.12. Panorpodidae. *Brachypanorpa* sp. Larva and pupa. f.g.l. ca. 11mm.

Figure 24.13. Panorpidae. *Panorpa nuptialis* Gerstaecker. Pupa. (From Byers 1963.)

BIBLIOGRAPHY

Applegarth, A. G. 1939. The larva of *Apterobittacus apterus* MacLachlan (Mecoptera: Panorpidae). Microentomology 4:109–20.

Boese, A. E. 1973. Descriptions of larvae and key to fourth instars of North American *Panorpa* (Mecoptera: Panorpidae). Univ. Kansas Sci. Bull. 50:163–86.

Brauer, F. 1863. Beiträge zur Kenntnis der Panorpiden-Larven. Verh. zool.-bot. Ges. Wien 13:307–24.

Byers, G. W. 1954. Notes on North American Mecoptera. Ann. Ent. Soc. Amer. 47:484–510.

Byers, G. W. 1963. The life history of *Panorpa nuptialis* (Mecoptera: Panorpidae). Ann. Ent. Soc. Amer. 56:142–49.

Byers, G. W. 1965. Families and genera of Mecoptera. Proc. XII Int. Congr. Ent., London, 1964:123.

Byers, G. W. 1976. A new Appalachian *Brachypanorpa* (Mecoptera: Panorpodidae). J. Kans. Ent. Soc. 49:433–40.

Byers, G. W., and R. Thornhill. 1983. Biology of the Mecoptera. Ann. Rev. Ent. 28:203–28.

Carpenter, F. M. 1931. The biology of the Mecoptera. Psyche 38:41–55.

Cooper, K. W. 1974. Sexual biology, chromosomes, development, life histories and parasites of *Boreus*, especially of *B. notoperates*, a southern California *Boreus*. II. (Mecoptera: Boreidae). Psyche 81:84–120.

Felt, E. P. 1896. The scorpion-flies. New York State Ent. Rept. No. 10:463–80.

Gassner, G. 1963. Notes on the biology and immature stages of *Panorpa nuptialis* Gerstaecker (Mecoptera: Panorpidae). Texas J. Sci. 15:142–54.

Mampe, C. D., and H. H. Neunzig. 1965. Larval descriptions of two species of *Panorpa* (Mecoptera: Panorpidae) with notes on their biology. Ann. Ent. Soc. Amer. 58:843–49.

Miyake, T. 1912. The life-history of *Panorpa klugi* MacLachlan. Jour. Coll. Agr., Imper. Univ. Tokyo 4:117–39.

Penny, N. D. 1977. A systematic study of the family Boreidae (Mecoptera). Univ. Kansas Sci. Bull. 51:141–217.

Peterson, A. 1951. Larvae of insects. Part II. Coleoptera, Diptera, Neuroptera, Siphonaptera, Mecoptera, Trichoptera. Printed for the author by Edwards Bros., Ann Arbor, Mich. 421 pp.

Potter, E. 1938. The internal anatomy of the larvae of *Panorpa* and *Boreus* (Mecoptera). Proc. Roy. Ent. Soc. London (A) 13:117–30.

Russell, L. K. 1979. A new genus and a new species of Boreidae from Oregon (Mecoptera). Proc. Ent. Soc. Wash. 81:22–31.

Russell, L. K. 1982. The life history of *Caurinus dectes* Russell, with a description of the immature stages (Mecoptera: Boreidae). Ent. Scand. 13:225–35.

Setty, L. R. 1931. Biology of *Bittacus stigmaterus*. Ann. Ent. Soc. Amer. 24:467–84.

Setty, L. R. 1939. The life history of *Bittacus strigosus* with a description of the larva. J. Kansas Ent. Soc. 12:126–28.

Setty, L. R. 1940. Biology and morphology of some North American Bittacidae (Order Mecoptera). Amer. Midland Nat. 23:257–353.

Setty, L. R. 1941. Description of the larva of *Bittacus apicalis* and a key to bittacid larvae (Mecoptera). J. Kansas Ent. Soc. 14:64–65.

Shiperovitsh, V. J. 1925. Biologie und Lebenszyklus von *Panorpa communis* L. Rev. Russe d'Ent. 19:27–40.

Steiner, P. 1930. Studien an *Panorpa communis* L. Zeitschr. Morphol. und Ökol. Tiere 17:1–67.

Williams, F. X. 1916. The pupa of *Boreus brumalis* Fitch. Psyche 23:36–39.

Yie, S.-T. 1951. The biology of Formosan Panorpidae and morphology of eleven species their immature stages. Mem. Coll. Agr., Nat. Taiwan Univ. 2:1–111.

Order Trichoptera

25

Glenn B. Wiggins[1]
Royal Ontario Museum and University of Toronto

CADDISFLIES

The Trichoptera or caddisflies are closely related to the Lepidoptera, but larvae of all species now extant are aquatic, with a few secondarily terrestrial larvae clearly derived from aquatic ancestors. Unlike Lepidoptera, larvae of most species do not feed on living vascular plants and are not generally regarded as destructive insects. Larval Trichoptera consume decaying organic materials of plant and animal origin, assimilating mainly the associated fungi and bacteria, and thereby contributing to the reduction of large organic pieces such as leaves into small faecal particles that are themselves colonized by fungi and consumed by other aquatic larvae including Trichoptera. They also consume algae of all types; they scrape diatoms from rock substrates and the Hydroptilidae use filamentous algae by puncturing individual cells and sucking out their contents. Some Trichoptera are predaceous on other insects and small invertebrates. Thus, trichopteran larvae are involved in the transfer of energy and nutrients from the degradation process through to higher trophic levels.

Because Trichoptera are one of the most diverse and abundant groups of organisms in all types of freshwater habitats, their contribution to these fundamental ecological roles is extremely important. Lacking the food specificity to particular plants that has characterized evolution of the Lepidoptera, trichopteran diversity derives from differences in the behavior of larvae through which simliar food resources are obtained in different ways. It seems clear that this evolutionary thrust in diversifying larval behavior has been greatly enhanced by the ability of larvae to produce silk (Wiggins and Mackay 1978), which stems from the common ancestor of both Trichoptera and Lepidoptera.

The behavior of caddisfly larvae, manifest in the remarkable range of retreats, capture nets, and portable cases that they build, is an integral part of both their ecology and taxonomy. Larvae of Hydropsychoidea use silk, for example, to construct the fine-meshed, fixed filter nets by which they strain food particles from the current and fashion protective tunnels on rocks and tubes in soft sediments. Larvae of Limnephiloidea and some Rhyacophiloidea use silk to build portable tubular cases of mineral and plant pieces, which serve to protect the larva from predators and strong currents, and to enhance its respiratory efficiency by channeling the ventilatory current of water set up by the undulating abdominal movements of the larva within the case. Respiration based on this self-generated water circulation system seems to have released larvae from dependence on the natural currents of highly oxygenated stream water and to have been largely responsible for exploitation of the resources of standing waters by Trichoptera (see Wiggins 1977).

Trichopteran larvae are found in most types of freshwater habitats. All North American families are represented in cool flowing waters, the least demanding respiratory environment, and probably the primitive aquatic habitat for the order (Ross 1956). Waters of reduced current and higher temperature have fewer genera, and standing waters of lakes and marshes have fewer still. Problems of periodic drought in transient waters have been overcome by only a few species, largely through diapause (Novák and Sehnal 1963) and specialized oviposition (Wiggins 1973a, Wiggins et al. 1980).

Larvae usually pass through five instars during development, sometimes up to seven in certain genera. Final instars seal themselves within a pupal enclosure where metamorphosis is usually completed within two to three weeks. The pharate adult within the pupal integument is equipped with sharp mandibles, with which it cuts the sieve membrane and swims to the surface. Eclosion takes place on some emergent object or at the surface in open water. Most species in temperate latitudes are univoltine, although some complete more than one generation and others emerge as two discrete cohorts in one year. Larvae in a number of genera require two years to complete development. Diapause may be imposed at any stage to suspend normal development until the appropriate environmental cue is received, usually photoperiod and/or temperature.

More than 1250 species of Trichoptera are now known in North America north of the Rio Grande, representing approximately 145 genera in 22 families. Diagnoses of larvae are available for most genera (Wiggins 1977) but identification of larvae to species is possible for no more than one-third of the Nearctic fauna. Given the importance in ecological work of distinguishing larvae of different species, there is a continuing priority for further work on taxonomic discrimination of larvae. Systematic work of this kind is most useful when done for an entire genus or for geographic components of large genera.

Taxonomic references are widely scattered in the literature. In a comprehensive synthesis of generic taxonomy and biology of larvae by Wiggins (1977), references are given for taxonomic papers to species level for larvae and also for adults.

1. Acknowledgment for illustrations is made to Anker Odum and Zile Zichmanis.

For the adult stage, the basic reference to species, particularly for eastern North America, remains the classic work by Ross (1944); a recent summary of genera of the Canadian fauna was made by Schmid (1980). Wiggins (1984) provided keys to all stages of the North American families, along with references to specialized works dealing with particular parts of the fauna; Morse and Holzenthal (1984) provided generic keys to the larvae. Betten (1934) and Balduf (1939) have reviewed the early literature on morphology and biology of larval Trichoptera.

DIAGNOSIS

The Trichoptera are neopterous, endopterygote insects, with larvae and pupae mainly found in freshwater, but a few secondarily terrestrial or marine, and adults entirely terrestrial. Larvae are generally similar to those of the Lepidoptera but with abdominal prolegs only on the apical segment and terminating in a hooklike anal claw rather than crochets. Larvae range in length from 2 to 40 mm.

Head with eyes of seven or fewer stemmata grouped closely together; mandibles usually with discrete toothlike points but with an entire scraping edge in groups feeding mainly on diatoms; antennae short in most families of the Limnephiloidea and Rhyacophiloidea, longer in the Leptoceridae and Hydroptilidae, very small in the Hydropsychoidea.

Thorax with three well-developed segments, each with a pair of segmented, sclerotized walking legs; pronotum and sometimes mesonotum covered with a pair of large sclerotized plates, mesonotal sclerites subdivided, reduced or lacking in some families; metanotum covered with paired sclerotized plates in some families, more often with small and usually paired sclerites or entirely membranous.

Abdomen almost entirely membranous, with small sclerites sometimes present on segments 1 and 9; filamentous tracheal gills present in some genera; terminal anal prolegs variously shaped.

DESCRIPTION

Head

The head capsule (fig. 25.7) comprises a mid-dorsal frontoclypeal apotome separated from the two lateral parietal sclerites by frontoclypeal sutures that converge posteriorly to meet the median coronal suture; the sutures are the dorsal ecdysial lines along which the head sclerites separate at ecdysis. Ventrally the parietals meet along the ventral ecdysial line, but they may be partially or entirely separated by the median ventral apotome (fig. 25.8), which may have separate anterior and posterior components (fig. 25.9). The exterior surface of the head often bears a dorsolateral carina and may be roughened by spines, pebbling, or other sculpturing. Internal muscle attachments (fig. 25.7) show externally as scars that give a characteristic pattern in some larvae. Primary setae of the head are as illustrated (fig. 25.7; see Williams and Wiggins (1981) for a comprehensive system of larval chaetotaxy), but secondary setae occur in some genera. The eyes consist of seven or fewer stemmata grouped together. Antennae in the Rhyacophiloidea (Glossosomatidae and Hydroptilidae) and Limnephiloidea are rodlike and usually short, although much longer in many Leptoceridae and Hydroptilidae. In the Hydropsychoidea and Rhyacophilidae antennae are very small and not easily distinguished. Mandibles have a cutting edge of pointed teeth, usually surrounding a central concavity (fig. 25.10); but in some groups where larvae feed principally by scraping diatoms and fine organic particles from rocks, the teeth are replaced by an entire, uniform edge (fig. 25.11). Each maxilla bears a fingerlike palp, and in the Hydropsychoidea and Rhyacophilidae a well-developed mesal lobe (fig. 25.12, lacinia); this mesal lobe is shortened and barely recognizable in most Limnephiloidea (fig. 25.13). The mesal lobe in Trichoptera is considered a fusion product of the galea and lacinia (Matsuda 1965), but is most often derived mainly from the lacinia. Silk is emitted from an opening at the tip of the labium (fig. 25.13).

Thorax

The thorax (figs. 25.2, 25.7) always has a sclerotized pronotum, with two sclerites closely appressed along the mid-dorsal ecdysial line. In several families of the Limnephiloidea, a membranous prosternal horn (fig. 25.2) is present. The trochantin (fig. 25.2), a derivative of the prothoracic pleuron, is characteristically shaped in some families. The mesonotum may be largely sclerotized with two primary sclerites appressed along the midline (fig. 25.7), or these sclerites may be subdivided (fig. 25.38), very small (fig. 25.24), or entirely absent (fig. 25.32). Sclerotization of the metanotum is equally diverse. On both mesonotum and metanotum setae are arranged in three primary setal areas, *sa 1, sa 2 and sa 3* (figs. 25.2, 25.7); these setal areas range from a separate sclerite bearing many setae to a single seta. Each thoracic segment bears a well-developed pair of legs, the relative lengths of the legs characteristic of particular families and genera. The legs bear setae and spines, diversely modified: spurs are very stout setae, usually at the distal end of the tibia, and often paired; the tarsal claw bears a basal seta. At the base of the meso- and metathoracic legs is the pleuron (fig. 25.2), a sclerite consisting of an anterior episternum, and a posterior epimeron, separated by a darkened infolding of the sclerite, the pleural suture.

Abdomen

The abdomen (fig. 25.1) is almost entirely membranous except in early instars of the Hydroptilidae, where abdominal sclerites may be present; a dorsal sclerite occurs on segment 9 in many families. Segment 1 in the Limnephiloidea may bear small, irregular sclerites and usually has a dorsomedian and two lateral expansile protuberances.

The anal prolegs exhibit significant structural diversity, with homologies designated here for component parts largely as proposed by Ross (1964). The condition in the Hydropsychoidea (fig. 25.5) is considered to represent the primitive type for the Trichoptera; in the net-spinning families and also in the free-living Rhyacophilidae, the anal prolegs are elongate, separate, and mobile. In the derivative condition seen in the Limnephiloidea, the prolegs are considered to have become short and thick, their bases swollen and contiguous. The Glossosomatidae of the Rhyacophiloidea (fig. 25.4) are considered to represent an intermediate condition in which partial shortening of the proleg and reduction in the size of the anal claw are apparent. The anal prolegs are considered a derivative of segment 10 (Matsuda 1976).

A line of fine filaments termed the lateral fringe (fig. 25.1) usually extends along each side of most segments in the Limnephiloidea. In many Limnephiloidea there is also a row of tiny, forked, sclerotized lateral tubercles (fig. 25.3) on each side of certain abdominal segments; in the Limnephilidae, Lepidostomatidae, and Brachycentridae lateral tubercles occur on most abdominal segments, but in most other genera tubercles are confined to segment 8 and are absent from the Phryganeidae. Segments frequently bear tracheal gills, sometimes as single filaments (fig. 25.1), sometimes branched (fig. 25.3), but often entirely lacking; gill filaments occasionally occur on thoracic segments. Arrangement of the gills is in three longitudinal series—dorsal, lateral, and ventral—with an anterior and posterior position on each segment. Gills may not occur at all positions and their arrangement is of taxonomic value, although the gill complement increases in successive instars and is often variable in the final instar.

Osmotic regulation in several families of Trichoptera is mediated through anal papillae (fig. 25.6), fingerlike lobes best seen in living specimens when they are everted by pressure of the haemolymph. Often termed blood gills in the literature, anal papillae occur in at least some genera of all families of the Hydropsychoidea and Rhyacophiloidea. Tracheoles in the anal papillae of glossosomatid larvae (Nüske and Wichard 1972) suggest that in this family at least, the structures serve as both respiratory and ion transporting epithelia. Other structures involved in ion absorption for osmoregulation are the chloride epithelia of the Limnephilidae, Goeridae, Hydroptilidae, and Molannidae (Wichard 1976). In larvae of the Limnephilidae, usually on the venter of abdominal segments 2–7, chloride epithelia are seen as ovoid areas of modified cuticle bordered by a thin sclerotized line (fig. 25.1); these have been termed oval sclerotized rings in the taxonomic literature. In some genera, especially of the tribe Limnephilini, similar areas are present dorsally and laterally on most of these same segments. These rings enclose areas of the hypodermis specialized for ionic transport (Wichard and Komnick 1973); in living larvae, chloride ions are absorbed from water passing through the case, accumulated in the cuticle, and transmitted to the body through the chloride epithelium in osmoregulatory compensation for ions lost through renal excretion.

TECHNIQUES

The data points for larval taxonomy are specimens of proven identity, obtained by rearing or by collecting in the field. For Trichoptera in general, adult males are the best material for unequivocal identification of a species. Laboratory rearing of part of a conspecific series of larvae to the adult stage can be done with rather simple equipment and techniques; some examples are available in papers by Anderson (1974), Bjarnov and Thorup (1970), Hiley (1969), Mackay (1981), Philipson (1953), and Wiggins (1959, 1977). Eggs collected from a female of known identity can be reared to final instar larvae (see Resh 1972). Field collecting to associate larvae with adults depends on the fact that the larval exuviae can be retrieved from pupal cases in which the pharate adult has all of the genitalic features necessary for identification to species. Larvae collected earlier from the same site as the pupae can be confirmed with reasonable assurance as conspecific with the adult by comparison with the exuvial sclerites. Known as the metamorphotype method, this technique is useful in most families except the Leptoceridae and Molannidae where the larval exuviae are expelled from the case after ecdysis. Pupae collected in the field but not developed sufficiently to be taxonomically useful will continue development to eclosion if maintained on moist leaves, moss, or paper towel in a jar kept in an ice chest or refrigerator.

For preservation, the best study material of Trichoptera is obtained from larvae killed in Kahle's fluid (see section on techniques for formula) and left for fixation for two or three weeks. Since the formalin in Kahle's fluid has an objectionable odour, specimens are transferred to 80 percent alcohol for long-term storage. Initial fixation in ethyl or isopropyl alcohol does not provide adequate preservation of internal tissues unless the fluid is changed several times during the first few days. Kahle's fluid always yields superior larval specimens for all groups of Trichoptera; it has the added advantage that even in jars filled with larvae the fluid need not be changed during the fixation period. Adequate fixation of material is of prime importance in the taxonomic study of larval Trichoptera.

Selected General References

Balduf 1939.
Betten 1934.
Bjarnov and Thorup 1970.
Coffman et al. 1971.
Crichton 1978.
Fischer 1960–73.
Hiley 1969.
Lepneva 1964, 1966.
Malicky, 1973, 1974.
Matsuda 1965.
Merrill 1969.
Moretti, 1981.
Morse 1984.
Resh 1972.
Ross 1944, 1956, 1964, 1967.
Schmid 1980.
Wichard 1976.
Wichard and Komnick 1973.
Wiggins 1959, 1973a, 1977, 1981, 1982, 1984.
Wiggins and Mackay 1978.
Wiggins et al. 1980.
Williams and Wiggins 1981.

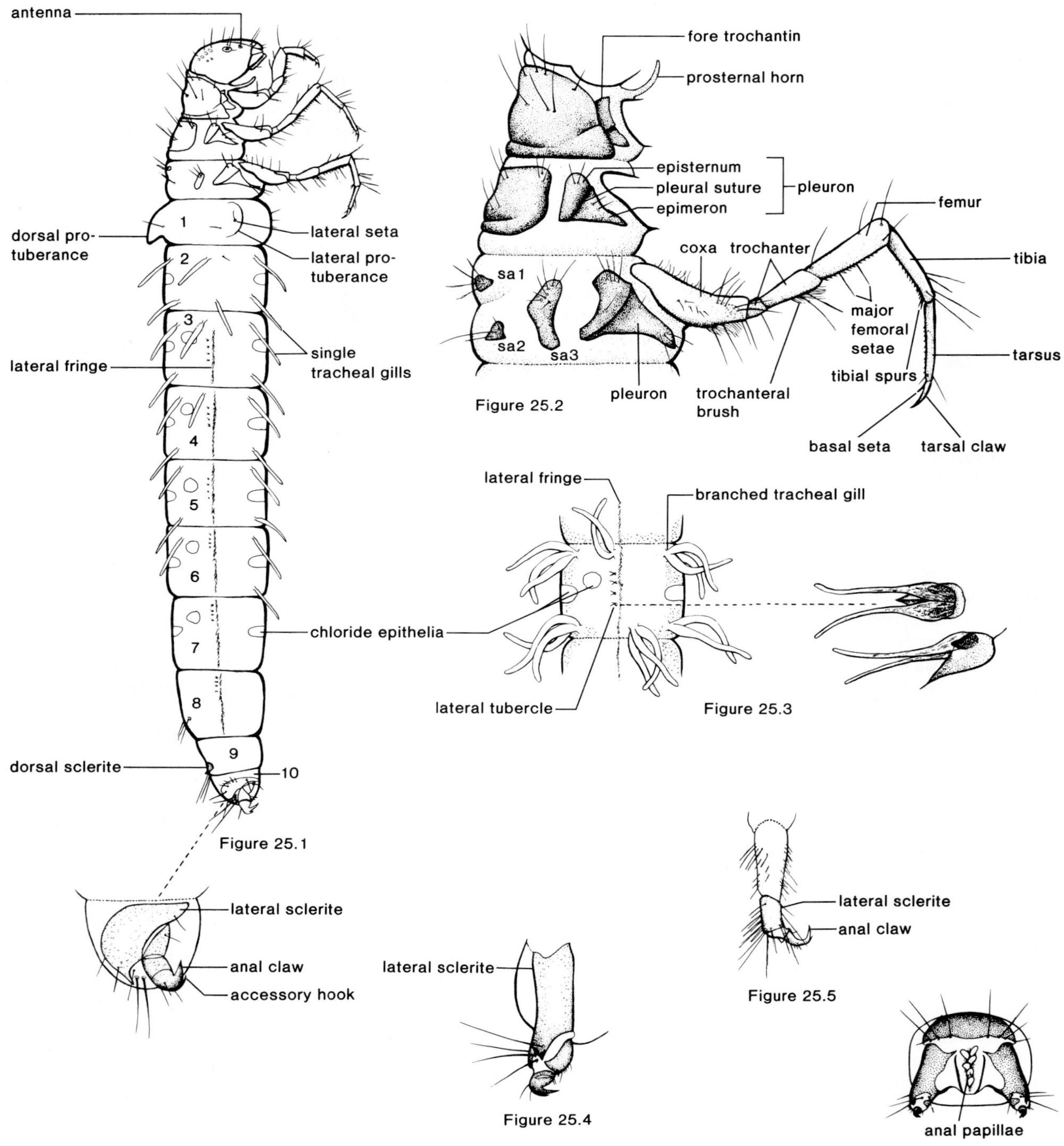

Figure 25.1. Lateral view of larva (Limnephilidae), abdominal segments numbered, detail of anal proleg.

Figure 25.2. Lateral view of larval thorax and hind leg (Limnephilidae).

Figure 25.3. Lateral view of abdominal segment of larva (Limnephilidae) bearing branched gills, detail of lateral tubercle from lateral and profile aspects.

Figure 25.4. Lateral view of anal proleg of larva (Glossosomatidae).

Figure 25.5. Lateral view of anal proleg of larva (Philopotamidae).

Figure 25.6. Caudal view of segment 9 and anal prolegs of larva (Glossosomatidae).

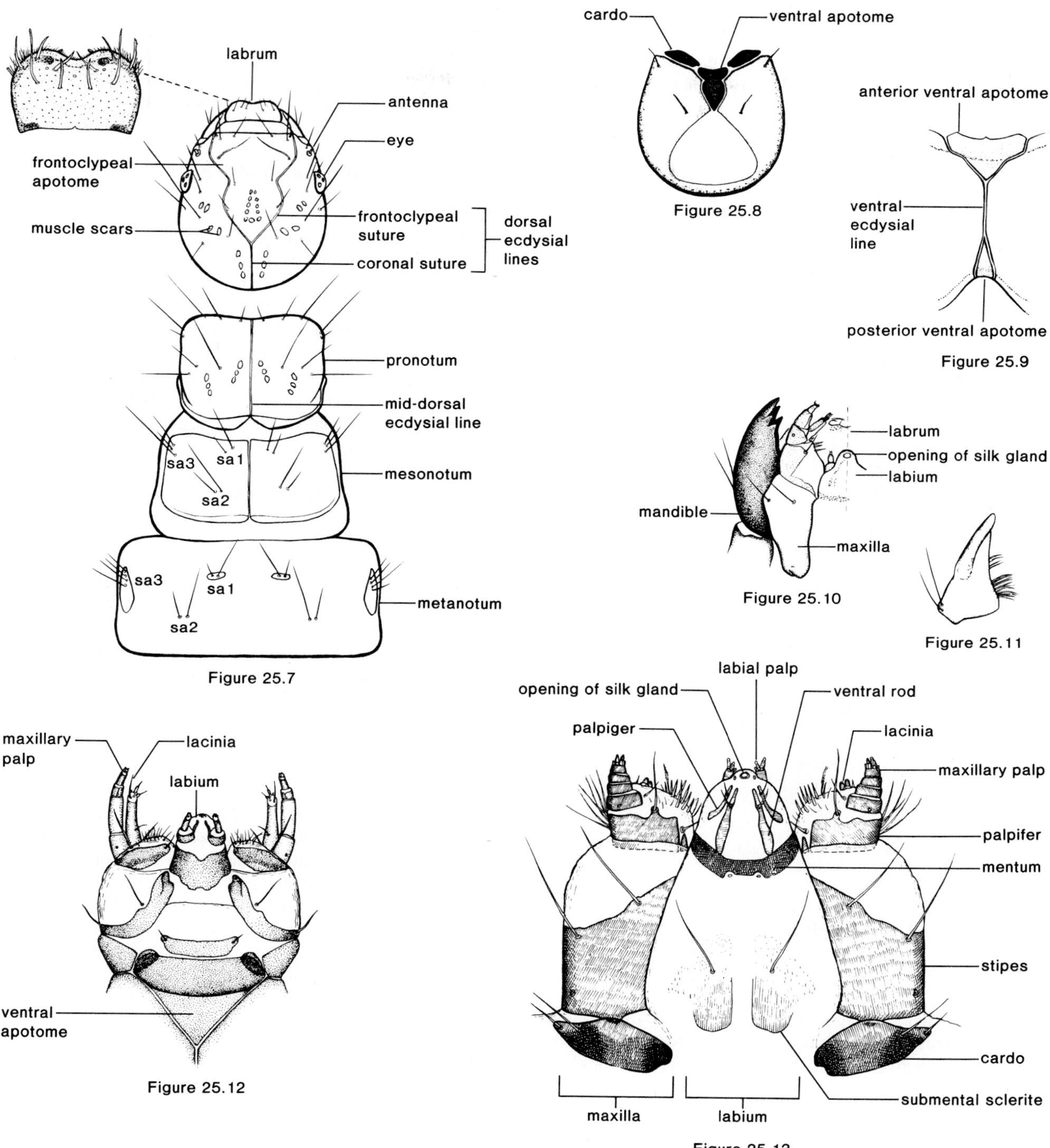

Figure 25.7

Figure 25.8

Figure 25.9

Figure 25.10

Figure 25.11

Figure 25.12

Figure 25.13

Figure 25.7. Dorsal view of larval head and thorax (Limnephilidae), detail of labrum.

Figure 25.8. Ventral view of larval head (Limnephilidae).

Figure 25.9. Ventral view of median line of larval head (Hydropsychidae: Diplectroninae).

Figure 25.10. Ventral view of larval mouthparts (Leptoceridae).

Figure 25.11. Ventral view of larval mandible (Limnephilidae: Neophylacinae).

Figure 25.12. Ventral view of larval mouthparts (Rhyacophilidae).

Figure 25.13. Ventral view of larval mouthparts (Limnephilidae: Apataniinae).

CLASSIFICATION OF NEARCTIC TRICHOPTERA

As proposed by Ross (1967) and adapted by Wiggins (1977, 1982), the classification followed here is based on three superfamilies. Traditional subdivision into two suborders, Annulipalpia (Rhyacophiloidea, Hydropsychoidea), and Integripalpia (Limnephiloidea) is advocated by Schmid (1980), but differing interpretations have been proposed on the relationships of the Rhyacophiloidea (cf. Ross 1967, Schmid 1980).

Order **TRICHOPTERA**
- Superfamily Rhyacophiloidea
  - Rhyacophilidae
  - Hydrobiosidae
  - Glossosomatidae
  - Hydroptilidae
- Superfamily Hydropsychoidea
  - Philopotamidae
  - Hydropsychidae
  - Polycentropodidae
  - Ecnomidae*
  - Psychomyiidae
  - Xiphocentronidae
- Superfamily Limnephiloidea
  - Limnephilid Branch
    - Phryganeidae
    - Brachycentridae
    - Lepidostomatidae
    - Limnephilidae
    - Uenoidae
    - Goeridae
  - Leptocerid Branch
    - Beraeidae
    - Sericostomatidae
    - Helicopsychidae
    - Odontoceridae
    - Molannidae
    - Leptoceridae
    - Calamoceratidae

*See footnote 2, p. 259.

Superfamily Rhyacophiloidea

Free-Living, Saddle-Case, and Purse-Case Caddisflies

Larvae are mobile for at least part of the life cycle, using silk to construct portable cases or fixed retreats; campodeiform or eruciform. Head with antennae not apparent or close to anterior margin of the head capsule. Thorax with only pronotum sclerotized, or with all segments bearing sclerites in the Hydroptilidae; legs usually of equal length, but middle and hind legs longer in some genera of the Hydroptilidae. Abdomen lacking dorsal and lateral protuberances on segment 1, a lateral fringe, and usually tracheal gills; dorsal sclerite present on segment 9.

Superfamily Hydropsychoidea

Net-Spinning Caddisflies

Larvae are mostly stationary, using silk to construct fixed shelters, tubes, or filter nets; campodeiform and prognathous. Head with antennae very small, maxillary palp and lacinia usually well developed and fingerlike. Thorax usually with only the pronotum sclerotized, but all segments sclerotized in the Hydropsychidae; legs generally of equal length. Abdomen without dorsal and lateral protuberances on segment 1, and without dorsal sclerite on 9; tracheal gills lacking except in the Hydropsychidae; anal prolegs elongate and highly mobile, anal claw large.

Superfamily Limnephiloidea

Tube-Case Caddisflies

Larvae mobile, using silk to construct portable cases of tubular, tapering form from plant and mineral materials; mainly eruciform and hypognathous. Head with peglike antennae, maxillary palp and lacinia frequently short and thick, lacinia sometimes hardly apparent. Thorax with pronotum and mesonotum usually sclerotized (except for Phryganeidae where the mesonotum is entirely or largely membranous); prosternal horn present in some families, absent in others; metanotum usually mainly membranous, with up to three pairs of small sclerites; fore legs shorter than others, middle or hind legs the longest. Abdomen with segment 1 usually bearing dorsal and lateral protuberances (lacking entirely in Brachycentridae); filamentous tracheal gills frequently present, single or branched, variously arranged on segments; midlateral fringe of small filaments usually present along each side; lateral row of small, bifurcate tubercles frequently present; segment 9 usually with a dorsal sclerite; anal prolegs short, fused for the most part with the abdomen, the terminal anal claw usually bearing small dorsal accessory hook(s).

KEY TO FAMILIES OF TRICHOPTERA LARVAE[2]

1. Anal claw comb-shaped (fig. 25.14); larvae (fig. 25.92) constructing portable cases of sand grains or small rock fragments, coiled to resemble a snail shell (fig. 25.93). Widespread in rivers, streams and wave-washed shorelines of lakes .... ***Helicopsychidae*** (p. 281)

1′. Anal claw hook-shaped (figs. 25.15, 25.18); larval case straight or nearly so, not resembling a snail shell, or larvae not constructing portable cases 2

2(1′). Dorsum of each thoracic segment covered by sclerites, usually closely appressed along the middorsal line (fig. 25.16), sometimes subdivided with thin transverse sutures (fig. 25.17), or some sclerites undivided 3

2′. Metanotum and sometimes mesonotum entirely membranous (fig. 25.32), or largely so and bearing several pairs of smaller sclerites (figs. 25.19, 25.20) 4

3(2). Abdomen with ventrolateral rows of branched gills, and with prominent brush of long hairs at base of anal claw (fig. 25.18); posterior margin of meso- and metanotal plates lobate (fig. 25.17); larvae (fig. 25.66) constructing fixed retreats (e.g., fig. 25.67). Widespread in rivers and streams, occasionally along rocky shores of lakes .... ***Hydropsychidae*** (p. 272)

3′. Abdomen lacking ventrolateral gills, and with only 2 or 3 hairs at base of anal claw (fig. 25.21); posterior margin of meso- and metanotal plates usually straight (fig. 25.16); minute larvae usually less than 6 mm long (fig. 25.64), constructing portable case of sand (e.g. fig. 25.65) or algae, or fixed cases of silk. Widespread in rivers, streams, and lakes .... ***Hydroptilidae***[2] (p. 269)

4(2′). Antennae very long and prominent, at least 6 times as long as wide (fig. 25.22); and/or sclerites on mesonotum lightly pigmented except for a pair of dark curved lines on posterior half (fig. 25.23); larvae (fig. 25.94) constructing portable cases of various materials (e.g., figs. 25.95, 25.96). Widespread in lakes and rivers .... ***Leptoceridae*** (p. 281)

4′. Antennae of normal length, no more than 3 times as long as wide (fig. 25.24); or not apparent; mesonotum never with a pair of dark curved lines as above 5

5(4′). Mesonotum largely or entirely membranous (fig. 25.24), or with small sclerites covering not more than half of notum (fig. 25.19); pronotum never with an anterolateral lobe (fig. 25.24) 6

5′. Mesonotum largely covered by sclerotized plates, variously subdivided and usually pigmented (fig. 25.20) although sometimes lightly; pronotum sometimes with a transverse carina terminating in prominent anterolateral lobes (fig. 25.26) 13

6(5). Abdominal segment 9 with a dorsal sclerite (fig. 25.27) 7

6′. Abdominal segment 9 with dorsum entirely membranous (fig. 25.28) 10

7(6). Metanotal *sa* 3 usually consisting of a cluster of setae arising from a small rounded sclerite (fig. 25.24); prosternal horn present (fig. 25.25); larvae (fig. 25.78) constructing tubular portable cases, mainly of plant materials (e.g., fig. 25.79). Widespread in lakes, ponds, and slow streams .... ***Phryganeidae*** (p. 276)

7′. Metanotal *sa* 3 consisting of a single seta not arising from a sclerite (fig. 25.19); prosternal horn absent; larvae either constructing a tortoiselike case of stones (fig. 25.63) or free-living 8

2. Larvae of the family Ecnomidae were found in Texas in 1977 (Waltz & McCafferty 1983), the first American record of this family north of Mexico. Although the Ecnomidae are related to the Polycentropodidae, larvae will key to Hydroptilidae here because each of the thoracic segments has a sclerotized notum. Ecnomidae can be readily distinguished from Hydroptilidae by the larger size, abdomen not swollen, depressed rather than compressed body form, elongate anal prolegs, and head 3–4 times longer than the prothorax. As in all hydropsychoid families, ecnomid larvae construct fixed shelters on substrates, and do not have portable cases at any time during development. The larva from Texas has been identified as *Austrotinodes* sp.; an illustration and description of *Austrotinodes* larvae were given by Flint (1973).

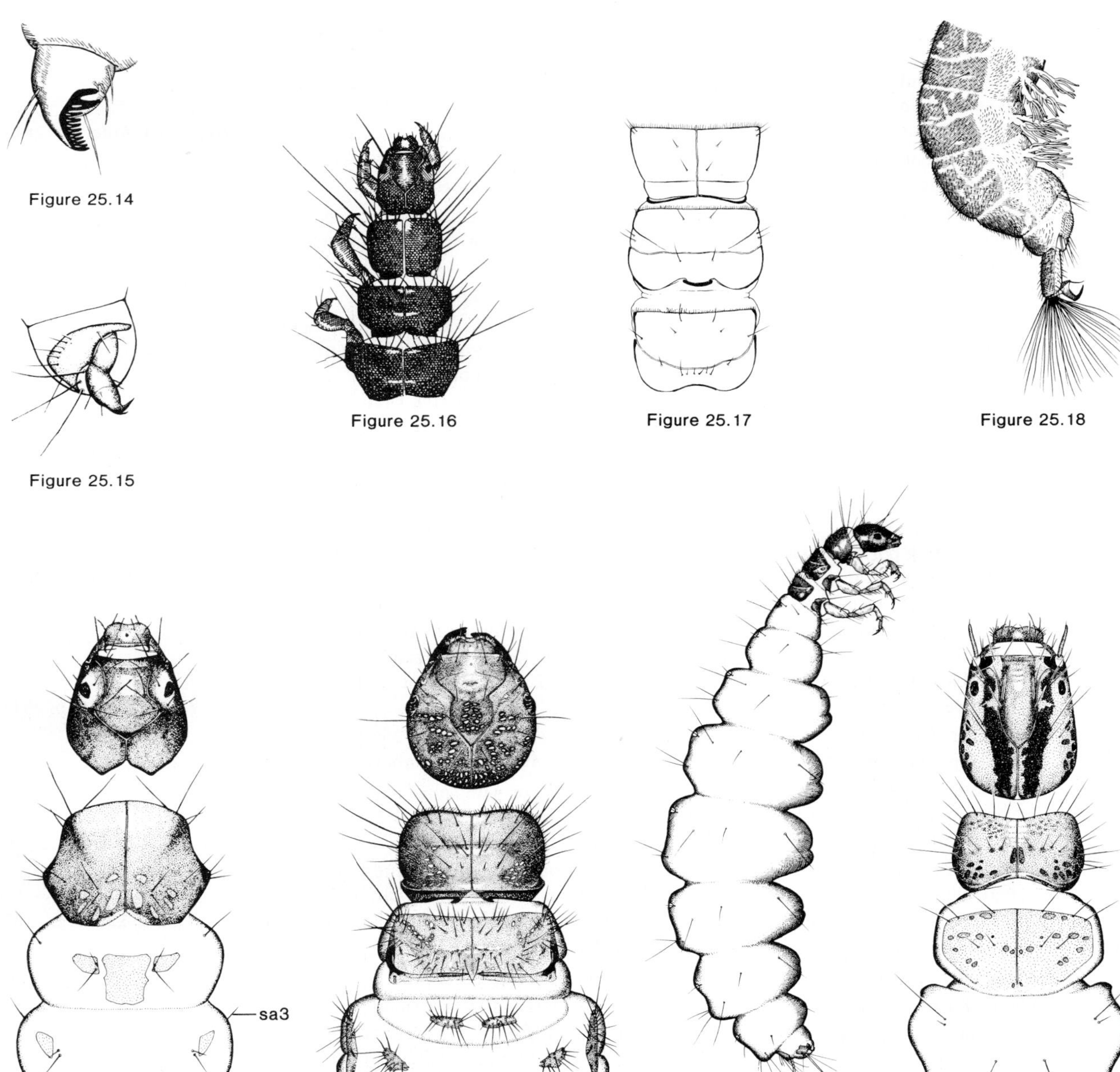

Figure 25.14

Figure 25.15

Figure 25.16

Figure 25.17

Figure 25.18

Figure 25.19

Figure 25.20

Figure 25.21

Figure 25.22

Figure 25.14. Lateral view of anal proleg of *Helicopsyche* (Helicopsychidae) larva.

Figure 25.15. Lateral view of anal proleg of larva (Limnephilidae).

Figure 25.16. Dorsal view of head and thorax of *Ochrotrichia* (Hydroptilidae) larva.

Figure 25.17. Dorsal view of thorax of *Homoplectra* (Hydropsychidae) larva.

Figure 25.18. Lateral view of posterior abdominal segments of *Hydropsyche* (Hydropsychidae) larva.

Figure 25.19. Dorsal view of head and thorax of *Protoptila* (Glossosomatidae) larva.

Figure 25.20. Dorsal view of head and thorax of *Homophylax* (Limnephilidae) larva.

Figure 25.21. Lateral view of *Hydroptila* (Hydroptilidae) larva.

Figure 25.22. Dorsal view of head and thorax of *Triaenodes* (Leptoceridae) larva.

8(7′). Basal half of anal proleg broadly joined with segment 9, anal claw with at least one dorsal accessory hook (fig. 25.29); larvae (fig. 25.62) constructing tortoiselike portable cases of small stones (fig. 25.63). Widespread in rivers and streams .... ***Glossosomatidae*** (p. 269)

8′. Most of anal proleg free from segment 9, anal claw without dorsal accessory hooks (fig. 25.27); larvae free-living without cases or fixed retreats until pupation 9

9(8′). Tibia, tarsus, and claw of fore leg articulating against ventral lobe of femur to form a chelate leg (figs. 25.30, 25.61). Southwestern, running waters .... ***Hydrobiosidae*** (p. 268)

9′. Fore leg normal, not chelate as above (figs. 25.31, 25.60). Widespread in running waters .... ***Rhyacophilidae*** (p. 268)

10(6′). Labrum membranous and T-shaped (fig. 25.32), often withdrawn from view in preserved specimens; larvae (fig. 25.69) constructing fixed sack-shaped nets of silk (fig. 25.70). Widespread in rivers and streams .... ***Philopotamidae*** (p. 271)

10′. Labrum sclerotized, rounded, and articulated in normal way (fig. 25.33) 11

11(10′). Mesopleuron extended anteriorly as a lobate process, tibiae and tarsi fused together on all legs (fig. 25.34); larvae (fig. 25.68) constructing fixed tubes of sand in small streams. Southern Texas .... ***Xiphocentronidae*** (p. 274)

11′. Mesopleuron not extended anteriorly, tibiae and tarsi separate on all legs (fig. 25.35) 12

12(11′). Trochantin of prothoracic leg with apex acute, fused completely with episternum without separating suture (fig. 25.36); larvae (figs. 25.73, 25.76) constructing exposed funnel-shaped capture nets (fig. 25.74), flattened retreats (fig.25.75), or tubes buried in loose sediments (fig. 25.77). Widespread in most types of aquatic habitats .... ***Polycentropodidae*** (p. 274)

12′. Trochantin of prothoracic leg broad and hatchet-shaped, separated from episternum by dark suture line (fig. 25.35); larvae (fig. 25.71) construct tubular retreats on rocks and logs (e.g., fig. 25.72). Widespread in running waters .... ***Psychomyiidae*** (p. 272)

13(5′). Abdominal segment 1 lacking both dorsal and lateral protuberances (fig. 25.37); each metanotal *sa* 1 usually lacking entirely (fig. 25.38), or if represented only by a single seta without a sclerite, mesonotal sclerites subdivided similar to fig. 25.38; larvae (fig. 25.80) construct portable cases of various materials and arrangements (e.g., fig. 25.81). Widespread in running waters .... ***Brachycentridae*** (p. 276)

13′. Abdominal segment 1 always with a lateral protuberance on each side, although not always prominent, and with (fig. 25.39) or without (fig. 25.56) a median dorsal protuberance; metanotal *sa* 1 always present, usually represented by a sclerite bearing several setae (fig.25.40) but with at least a single seta; larvae construct portable cases of various materials and arrangements 14

14(13′). Tarsal claw of hind leg modified to form a short setose stub (fig. 25.41), or a slender filament (fig. 25.42); larvae (fig. 25.103) constructing cases of sand grains with lateral flanges (fig. 25.104). Transcontinental through Canada to Alaska, and eastern, in lakes and large rivers, less commonly in spring streams .... ***Molannidae*** (p. 282)

14′. Tarsal claws of hind legs no different in structure from those of other legs (fig. 25.43) 15

15(14′). Labrum with transverse row of approximately 16 long setae across central part (fig. 25.44); larvae (fig. 25.97) use a hollowed twig as a case, or construct cases of leaves (fig. 25.98) and bark variously arranged. Eastern and western streams .... ***Calamoceratidae*** (p. 282)

15′. Labrum with no more than 6 long setae across central part (fig. 25.45) 16

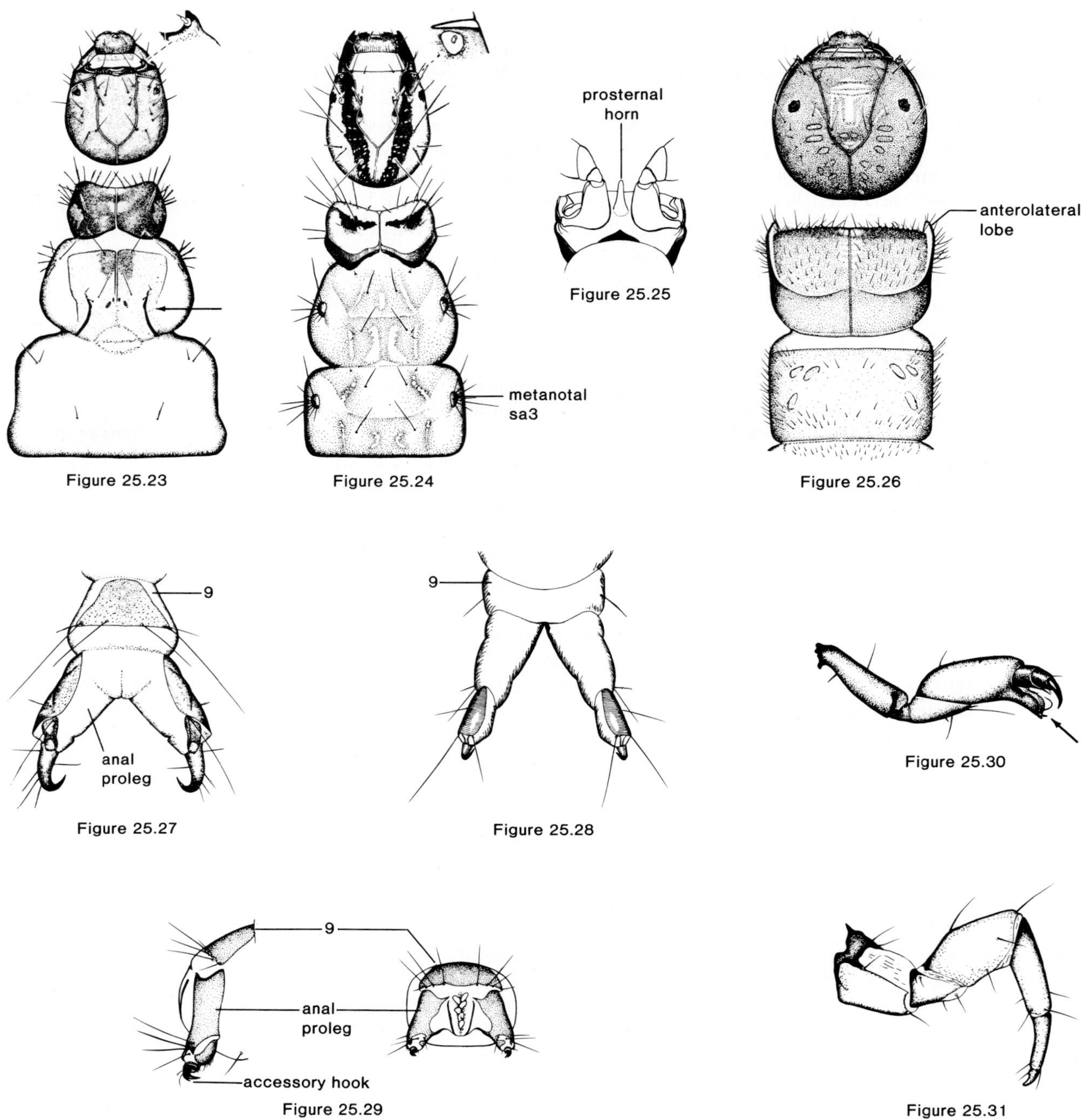

Figure 25.23. Dorsal view of head and thorax of *Ceraclea* (Leptoceridae) larva, detail of antenna.

Figure 25.24. Dorsal view of head and thorax of *Ptilostomis.* (Phryganeidae) larva, detail of antenna.

Figure 25.25. Ventral view of prothorax of *Ptilostomis* (Phryganeidae).

Figure 25.26. Dorsal view of head and thorax of *Beraea* (Beraeidae).

Figure 25.27. Dorsal view of segment 9 and anal prolegs of *Rhyacophila* (Rhyacophilidae) larva.

Figure 25.28. Dorsal view of segment 9 and anal prolegs of *Chimarra* (Philopotamidae) larva.

Figure 25.29. Lateral and caudal view of segment 9 and anal prolegs of *Glossosoma* (Glossosomatidae) larva.

Figure 25.30. Lateral view of fore leg of *Atopsyche* (Hydrobiosidae) larva.

Figure 25.31. Lateral view of fore leg of *Rhyacophila* (Rhyacophilidae) larva.

16(15′). Anal proleg with lateral sclerite much reduced in size and produced posteriorly as a lobe from which a stout apical seta arises (fig. 25.46); base of anal claw with ventromesal membranous surface bearing a prominent brush of 25–30 fine setae (fig. 25.47); larvae (fig. 25.101) constructing cases of sand grains (fig. 25.102). Eastern, exceedingly local, in wet muck of spring seepage areas .... ***Beraeidae*** (p. 280)

16′. Anal proleg with lateral sclerite not produced posteriorly as a lobe around base of apical seta (fig. 25.48); base of anal claw with ventromesal surface lacking prominent brush of fine setae (fig. 25.49), although setae may be present dorsally (fig. 25.51) 17

17(16′). Antenna situated at or very close to the anterior margin of the head capsule (fig. 25.50); prosternal horn lacking (fig. 25.53); larval cases mainly of rock fragments 18

17′. Antenna removed from the anterior margin of the head capsule and approaching the eye (fig. 25.52); prosternal horn present although sometimes short (fig. 25.52); larval cases of rock fragments or of plant materials 19

18(17). Anal proleg with dorsal cluster of approximately 30 or more setae posteromesad of lateral sclerite (fig. 25.51); fore trochantin relatively large, the apex hook-shaped (fig. 25.53); larvae (fig. 25.90) constructing cases mainly of sand (fig. 25.91). Widespread in running waters and along lake shorelines .... ***Sericostomatidae*** (p. 280)

18′. Anal proleg with no more than 3 to 5 dorsal setae posteromesad of lateral sclerite, sometimes short spines (fig. 25.55); fore trochantin small, the apex not hook-shaped (fig. 25.54); larvae (fig. 25.99) constructing cases mainly of small rock fragments (fig. 25.100). Widespread in running waters .... ***Odontoceridae*** (p. 283)

19(17′). Antenna situated close to the anterior margin of the eye, median dorsal protuberance of abdominal segment 1 lacking (fig. 25.56); larvae (fig. 25.84) constructing cases of various materials and arrangements, frequently four-sided (e.g., fig. 25.85). Widespread, mainly in lotic habitats .... ***Lepidostomatidae*** (p. 277)

19′. Antenna situated approximately halfway between the anterior margin of the head capsule and the eye, median dorsal protuberance of abdominal segment 1 almost always present (fig.25.52) 20

20(19′). Mesopleuron modified, usually extended anteriorly as a prominent process (fig. 25.57); larvae (fig. 25.88) constructing cases of mineral materials (e.g., fig. 25.89). Widespread in running waters or spring seepage .... ***Goeridae*** (p. 278)

20′. Mesopleuron unmodified (fig. 25.59) 21

21(20′). Larvae very slender, pronotum longer than wide, metanotal *sa* 1 a single seta (fig. 25.58); larvae (fig. 25.86) constructing very slender cases of fine mineral materials (fig. 25.87), sometimes of silk alone. Western montane streams .... ***Uenoidae*** (p. 278)

21′. Larvae not as slender as above, pronotum usually wider than long, metanotal *sa* 1 with more than one seta (fig. 25.59); larvae (fig. 25.82) constructing cases of many types using plant materials and rock fragments, usually of rough, irregular texture (e.g., fig. 25.83). Widespread in all types of aquatic habitats .... ***Limnephilidae*** (p. 277)

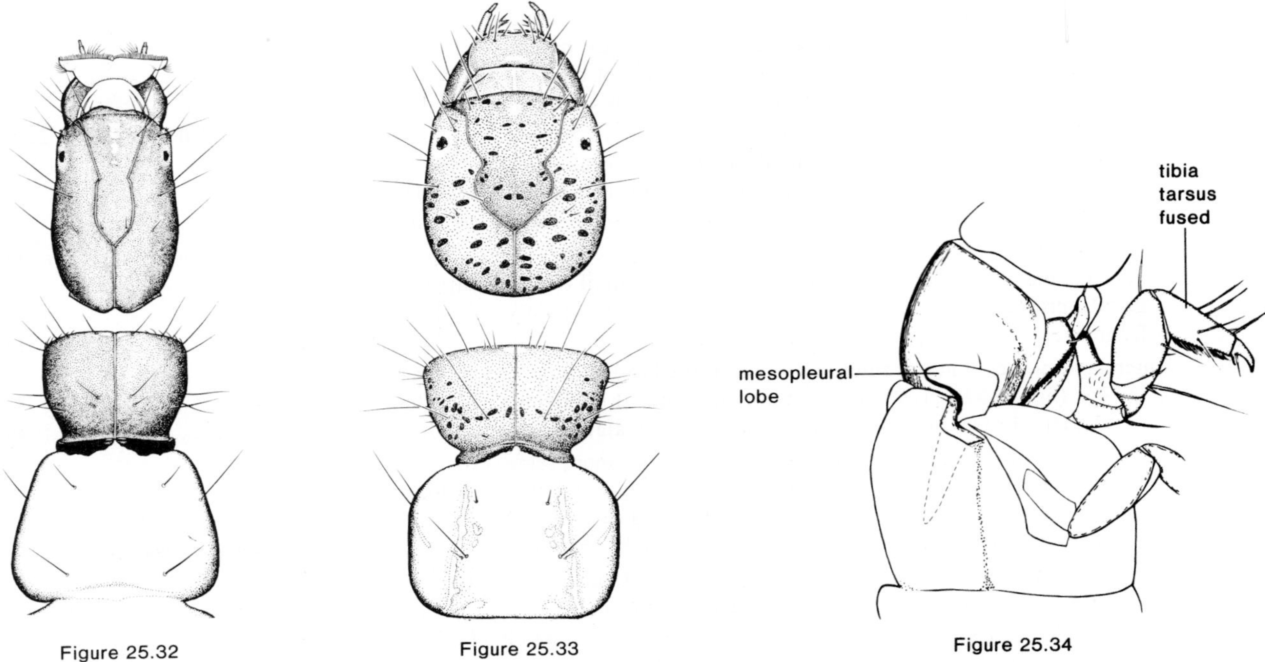

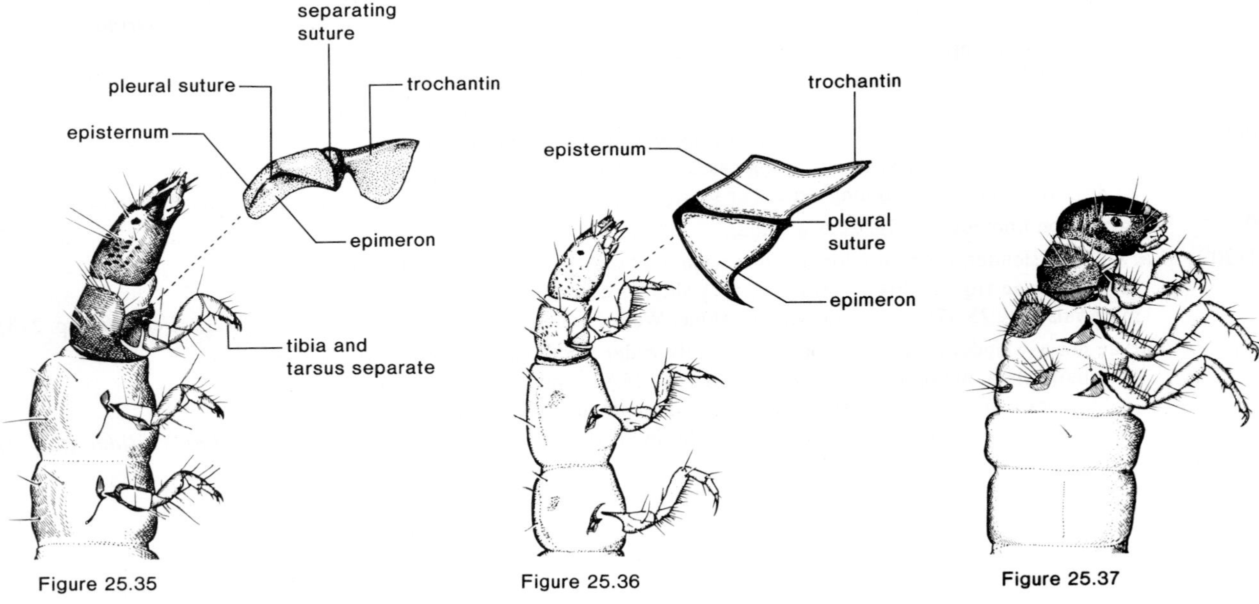

Figure 25.32. Dorsal view of head and thorax of *Dolophilodes* (Philopotamidae) larva.

Figure 25.33. Dorsal view of head and thorax of *Polycentropus* (Polycentropodidae) larva.

Figure 25.34. Lateral view of prothorax and mesothorax of *Xiphocentron* (Xiphocentronidae) larva.

Figure 25.35. Lateral view of head and thorax of *Tinodes* (Psychomyiidae) larva, detail of trochantin.

Figure 25.36. Lateral view of head and thorax of *Neureclipsis* (Polycentropodidae) larva, detail of trochantin.

Figure 25.37. Lateral view of *Micrasema* (Brachycentridae) larva.

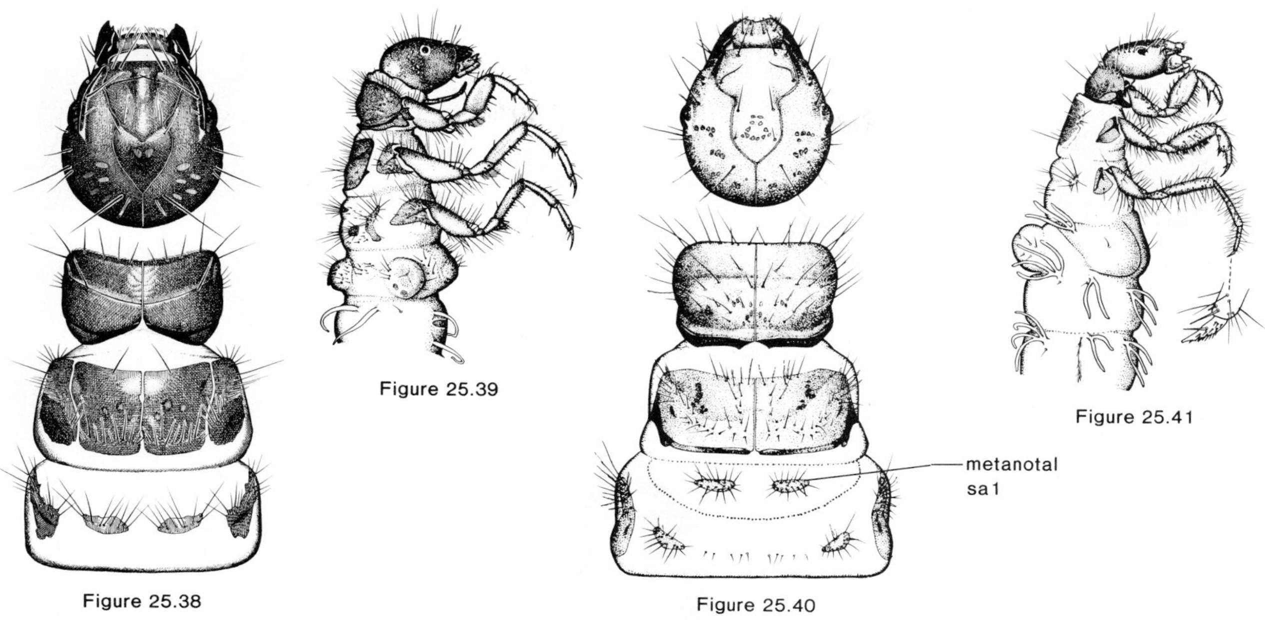

Figure 25.38 Figure 25.39 Figure 25.40 Figure 25.41

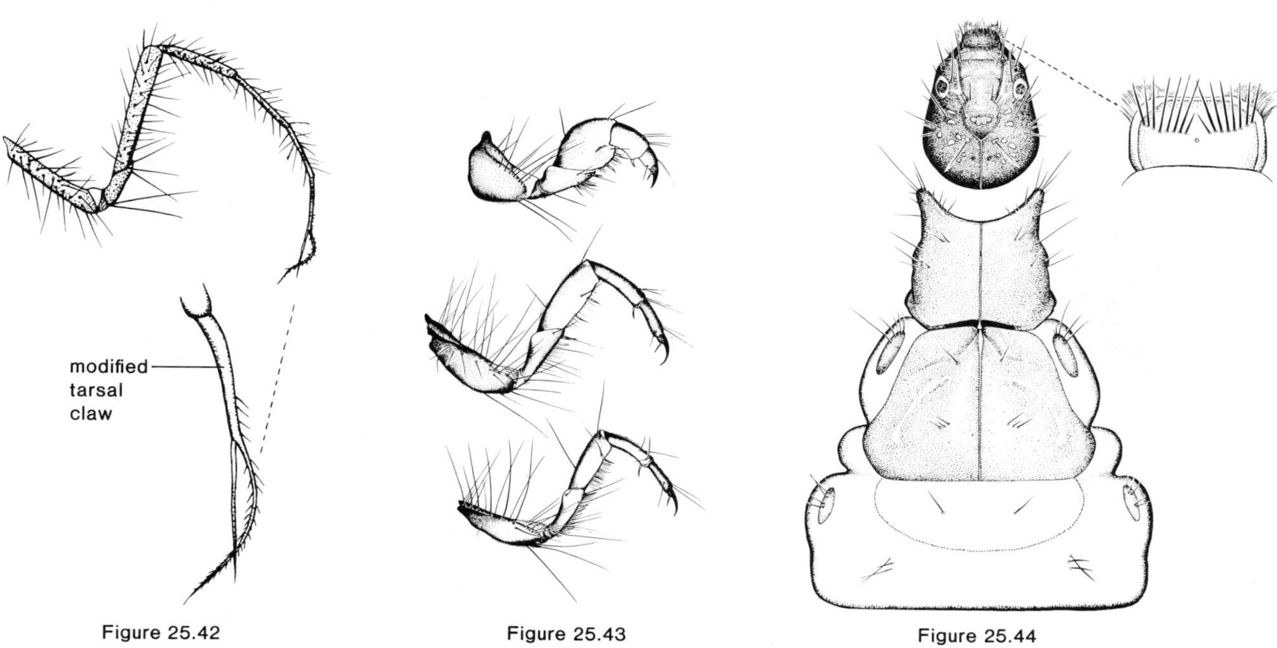

Figure 25.42 Figure 25.43 Figure 25.44

Figure 25.38. Dorsal view of head and thorax of *Brachycentrus* (Brachycentridae) larva.

Figure 25.39. Lateral view of *Desmona* (Limnephilidae) larva.

Figure 25.40. Dorsal view of head and thorax of *Pseudostenophylax* (Limnephilidae) larva.

Figure 25.41. Lateral view of *Molanna* (Molannidae) larva, detail of hind tarsal claw.

Figure 25.42. Lateral view of hind leg of *Molannodes* (Molannidae) larva, detail of tarsal claw.

Figure 25.43. Lateral view of legs of *Eobrachycentrus* (Brachycentridae) larva.

Figure 25.44. Dorsal view of head and thorax of *Anisocentropus* (Calamoceratidae) larva, detail of labrum.

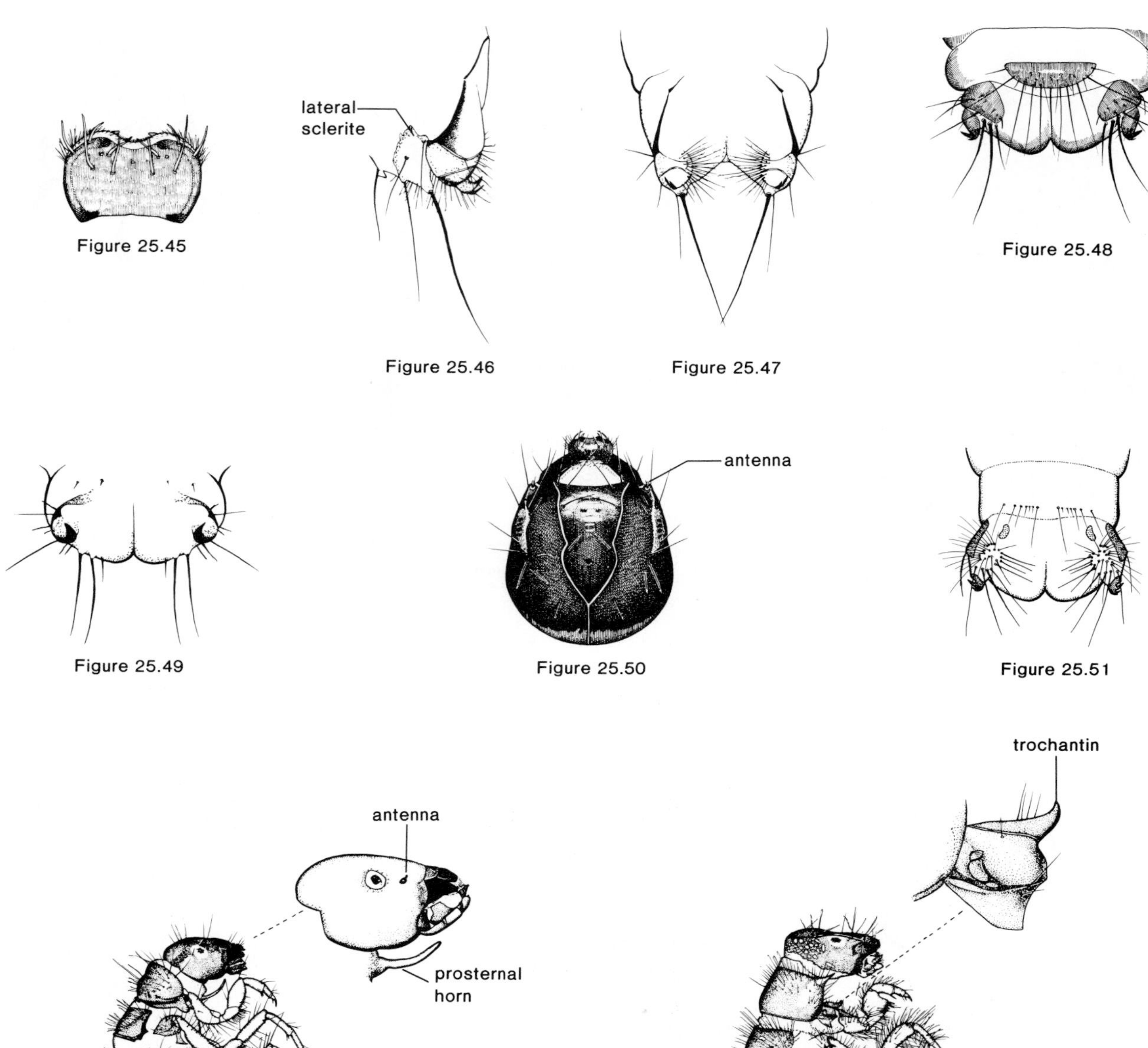

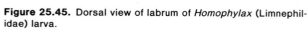

Figure 25.45. Dorsal view of labrum of *Homophylax* (Limnephilidae) larva.

Figure 25.46. Lateral view of anal proleg of *Beraea* (Beraeidae) larva.

Figure 25.47. Ventral view of anal prolegs of *Beraea* (Beraeidae) larva.

Figure 25.48. Dorsal view of segment 9 and anal prolegs of *Homophylax* (Limnephilidae) larva.

Figure 25.49. Ventral view of anal prolegs of *Homophylax* (Limnephilidae) larva.

Figure 25.50. Dorsal view of head of *Agarodes* (Sericostomatidae) larva.

Figure 25.51. Dorsal view of segment 9 and anal prolegs of *Agarodes* (Sericostomatidae) larva.

Figure 25.52. Lateral view of head and thorax of *Pseudostenophylax* (Limnephilidae) larva, detail of head.

Figure 25.53. Lateral view of *Fattigia* (Sericostomatidae) larva, detail of trochantin.

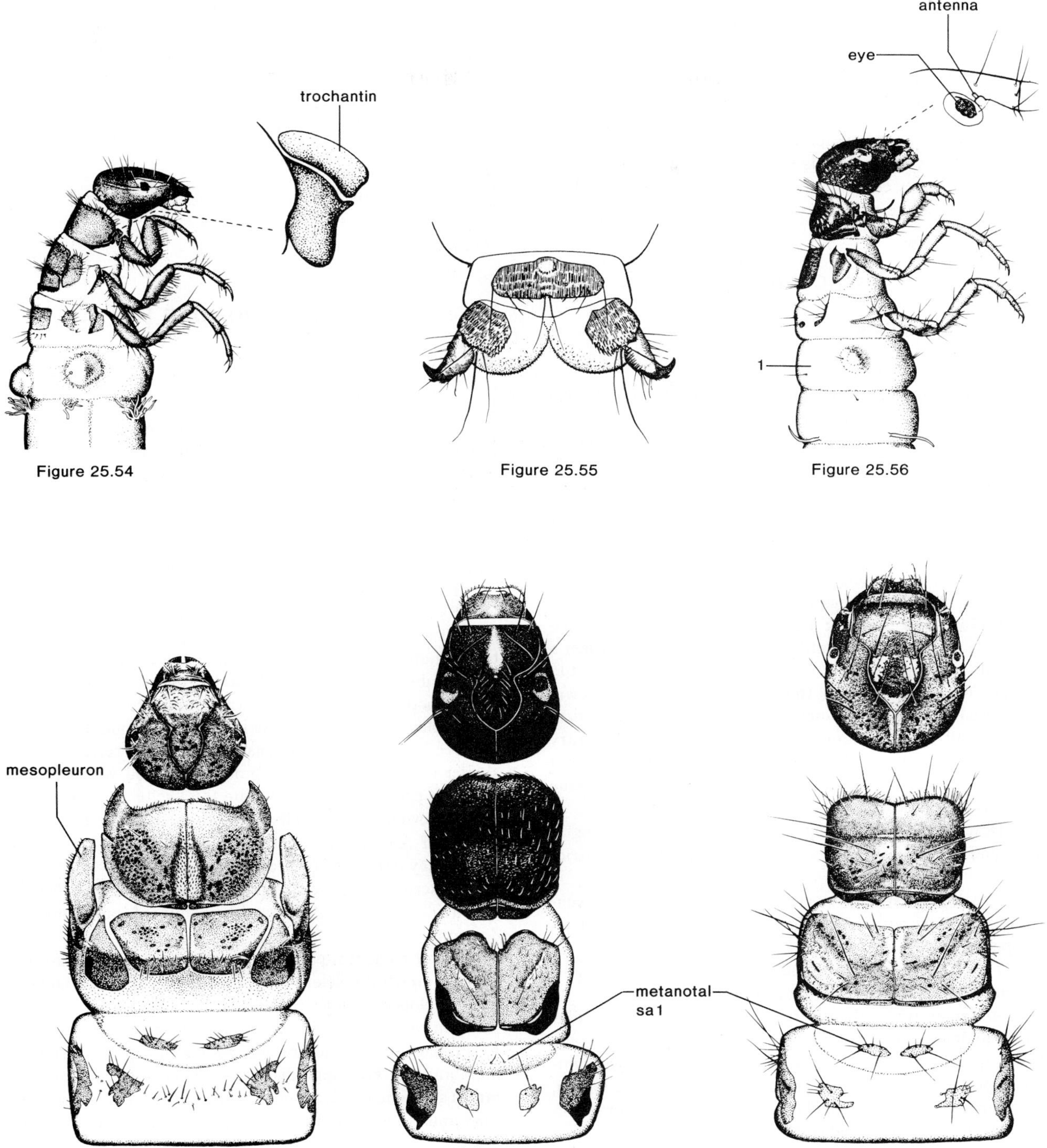

Figure 25.54. Lateral view of *Marilia* (Odontoceridae) larva, detail of trochantin.

Figure 25.55. Dorsal view of segment 9 and anal prolegs of *Pseudogoera* (Odontoceridae) larva.

Figure 25.56. Lateral view of *Lepidostoma* (Lepidostomatidae) larva, detail of antenna.

Figure 25.57. Dorsal view of head and thorax of *Goera* (Goeridae) larva.

Figure 25.58. Dorsal view of head and thorax of *Neothremma* (Uenoidae) larva.

Figure 25.59. Dorsal view of head and thorax of *Philarctus* (Limnephilidae) larva.

SUPERFAMILY RHYACOPHILOIDEA

Rhyacophiloid caddisflies, believed to be the most primitive of the Trichoptera, comprise three distinct groups. Rhyacophilidae and Hydrobiosidae, entirely free living as larvae, are confined to running waters. Larvae of the Glossosomatidae construct portable saddle cases and also are confined to running waters. Hydroptilidae live in all types of permanent water, where they construct portable cases that are purselike and flasklike and even some that are fixed in position.

RHYACOPHILIDAE

The Rhyacophilids

Figures 25.12, 25.27, 25.31, 25.60

Relationships and Diagnosis: Rhyacophilids are believed to be the most primitive living family of Trichoptera and to have their closest relationships with the Glossosomatidae. Their traditionally close relationship with the Hydrobiosidae has been questioned by Schmid (1980). Rhyacophilid larvae (fig. 25.60) are similar in general appearance to those of the Hydrobiosidae, Philopotamidae, and Polycentropodidae, all of which have a membranous mesonotum but can be distinguished by the sclerotized labrum, sclerite on segment 9, and (fig. 25.31) fore legs similar in structure to the others.

Biology and Ecology: Larvae are free living and construct no shelter or case until just prior to pupation when a crude dome-shaped shelter of rock fragments is fastened to a rock. Within this shelter the larva spins a tough, brown, ovoid cocoon of silk where metamorphosis takes place. Larvae occur in all types of running waters, including temporary streams. Most rhyacophilids are free-living predators, but some species are herbivorous, feeding on vascular plant tissue both living and dead, and on algae (Thut 1969).

Description: Larvae (fig. 25.60) up to 32 mm in length, agile and free living with prominent intersegmental constrictions.

Head prognathous, parietals largely in contact ventrally; maxillary palpi and laciniae both elongate and well developed.

Thorax with only the pronotum sclerotized; all legs similar in length and structure, short and unmodified.

Abdomen with gills diverse, ranging from simple filaments to dense clusters of filaments, or lacking; lateral fringe lacking; segment 9 with a dorsal sclerite; anal prolegs prominent, projecting freely from the body for their entire length, each terminating in a large claw; lateral sclerite in some species developed as a curved spike.

Comments: Rhyacophilidae in North America comprise two genera, *Rhyacophila* with more than 100 species and *Himalopsyche* with only one; both genera are widely distributed in the Northern Hemisphere, and *Rhyacophila* with more than 500 species is the largest genus in the Trichoptera. *Rhyacophila* larvae are especially characteristic of running waters in montane areas, where several species frequently occur at the same site; species are more numerous in the West.

A key to genera was given by Wiggins (1977); diagnoses to larvae of the eastern North American species by Flint (1962), Weaver and Sykora (1979), and to western species by Smith (1968b, 1984a,b) and Peck and Smith (1978).

Selected Bibliography

Flint 1962.
Peck and Smith 1978.
Smith 1968b, 1984a,b.
Thut 1969.
Weaver and Sykora 1979.

HYDROBIOSIDAE

The Hydrobiosids

Figures 25.30, 25.61

Relationships and Diagnosis: Hydrobiosidae have traditionally been held to be closely related to the Rhyacophilidae, and they were treated as a subfamily of that group until Schmid's (1980) suggestion that they showed some relationships with the Hydropsychoidea. In any case, hydrobiosids resemble in general appearance larvae of the Rhyacophilidae, Philopotamidae, and Polycentropodidae.

Hydrobiosid larvae are distinguished primarily by a chelate fore leg in which the shortened tibia and tarsus close against a concave extension of the femur (fig. 25.30). Only the pronotum is sclerotized, the mesonotum and metanotum are membranous (fig. 25.61); the labrum is fully sclerotized and unmodified, and the prosternum is sclerotized.

Biology and Ecology: As in the Rhyacophilidae, larvae are free living and do not construct a case or retreat until a shelter is formed of rock fragments fastened with silk to a rock just prior to pupation; within this shelter the larva spins a brown ovoid cocoon of silk in which metamorphosis occurs. In North America larvae are confined to running waters, where they appear to be predators.

Description: Larvae (fig. 25.61) up to 22 mm in length, free living and agile, slender with prominent intersegmental divisions.

Head elongate, parietals contiguous ventrally; labrum rounded and fully sclerotized; maxillary palpi and laciniae well developed as in the Rhyacophilidae.

Thorax with only the pronotum sclerotized, a large prosternal sclerite also present; mesonotum and metanotum membranous. Fore legs distinctive in having the tibia and tarsus shortened and closing against a distal concave lobe of the femur to form a chelate leg; other legs short and unmodified.

Abdomen lacking gills and lateral fringe; segment 9 with dorsal sclerite; anal prolegs well developed, projecting freely from the body for their entire length, each terminating in a large claw.

Comments: Hydrobiosidae in North America comprise only the single genus *Atopsyche,* a large Neotropical genus of which three species have been recorded north of the Rio Grande, but only in the extreme southwestern part of the United States.

Selected Bibliography

Schmid 1980.

GLOSSOSOMATIDAE

The Glossosomatids

Figures 25.4, 25.6, 25.19, 25.29, 25.62, 25.63

Relationships and Diagnosis: Glossosomatids are most closely related to the Rhyacophilidae and Hydroptilidae; they were included within the Rhyacophilidae until family status was proposed by Ross in 1956. Their larvae do not resemble those of any other family.

Larvae are distinctive in being hypognathous and rather thick-bodied, with only the pronotum entirely sclerotized (fig. 25.62), but with very small sclerites on the mesonotum and metanotum in some genera (fig. 25.19); abdominal gills are lacking, segment 9 bears a dorsal sclerite (fig. 25.29), and only the apical half of each anal proleg is free from the abdomen.

Biology and Ecology: Larvae of all instars construct distinctive tortoiselike cases, ovoid flat-bottomed domes of tiny rock fragments (fig.25.63); a broad strap across the ventral surface leaves an opening at each end through which head and legs can be extended. Larvae graze diatoms and fine organic particles on the upper exposed surfaces of rocks in running waters and occasionally along the wave-washed margins of large lakes. Crawling over the rock, they feed concealed beneath the domed top of the case. Spaces between the rock fragments allow the water current to bathe the larva continuously. Larvae readily abandon their cases under stress, building new ones. For pupation the ventral strap is cut away, and the dome fastened with silk to a rock; within the dome the larva spins a brown ovoid cocoon of silk, as in the Rhyacophilidae.

Description: Larvae (fig. 25.62) up to 10 mm in length, thick bodied and slow moving.

Head hypognathous and uniformly dark brown; parietals contiguous ventrally; mandibles with an entire scraping edge; labrum with a membranous outer fringe.

Thorax with a pair of dark brown sclerotized plates covering the pronotum; mesonotum and metanotum with small sclerites in some genera (entirely membranous in others); prosternum largely sclerotized; legs short, all similar in length and structure.

Abdomen with gill filaments and lateral fringe lacking; segment 1 unmodified; segment 9 with a dorsal sclerite; anal prolegs mostly sclerotized, basal half fused with abdomen, terminating in a small anal claw with one or more dorsal accessory hooks; anal papillae present.

Comments: Approximately 80 species in six genera are known in America north of Mexico. Glossosomatids occur in most lotic habitats throughout the continent, including temporary streams, with different genera characteristic of particular conditions of current and temperature. A key to Nearctic genera was given by Wiggins (1977), but little taxonomic work has been done at the species level.

Selected Bibliography

Anderson 1974.
Nüske and Wichard 1972.

HYDROPTILIDAE

The Hydroptilids, Micro-Caddisflies

Figures 25.16, 25.21, 25.64, 25.65

Relationships and Diagnosis: Hydroptilids are held to be most closely related to the Glossosomatidae, but their larvae are very distinctive and share with the Hydropsychidae sclerotization of each thoracic notum. Since hydroptilids are small and lack segmentally arranged abdominal gills (fig. 25.64), confusion is possible only with very early instars of Hydropsychidae before their gills develop, when the straight posterior margins of thoracic sclerites in Hydroptilidae (fig. 25.16) are readily distinguished from the lobate margins of Hydropsychidae (fig. 25.17).

Biology and Ecology: Hydroptilids exhibit larval heteromorphosis, unique in the Trichoptera; they pass through the first four instars as tiny, free-living larvae, deferring case construction until the beginning of the fifth instar. The early instars resemble small coleopteran larvae with prominent sclerites on a slender abdomen. At the beginning of the fifth instar the abdomen is membranous and small, but it soon becomes distended with food reserves (fig. 25.64). Larval cases constructed at the beginning of the fifth instar are purse-shaped and portable in most genera—commonly in the form of a silken bivalve to which sand grains or algae are attached (e.g., fig. 25.65), but resembling a flattened silken flask or a stout barrel in some genera. In the Leucotrichiini, free-living larvae take up a sedentary existence in the fifth instar by fixing to rocks their silken cases resembling the egg cocoon of a leech; larvae extend the slender head and thorax through a small hole at each end to graze diatoms and organic particles from the surrounding area. Hydroptilids feed on algae of all kinds, mainly filamentous forms which they pierce and suck out the cell contents. Biology of Nearctic species of Hydroptilidae has been little studied, with most information still derived from Nielsen's work (1948) on European species. One of the few examples of food specificity implicit in Trichoptera is the monotypic genus *Dibusa* of eastern North America, in which larvae have been found only on the freshwater red alga *Lemanea* (Wiggins 1977). Habitats for hydroptilids range from cold springs and running waters of all kinds to lakes.

Description: Mature larvae (e.g., fig. 25.64) unusually diverse in structural detail among genera; small, 2 mm in length in some genera, up to 6 mm in others; strongly compressed in most genera, depressed or cylindrical in a few.

Head tending to be prognathous, extended and snoutlike in a few genera; antennae relatively long, as in Leptoceridae, in most genera.

Thorax with the notum of each segment entirely covered by sclerotized plates (fig. 25.16), subdivided along the median line in most genera, but the mesonotal and metanotal plates not subdivided in the Leucotrichiini; posterior margins of segments straight or nearly so (fig. 25.16). Legs short, all

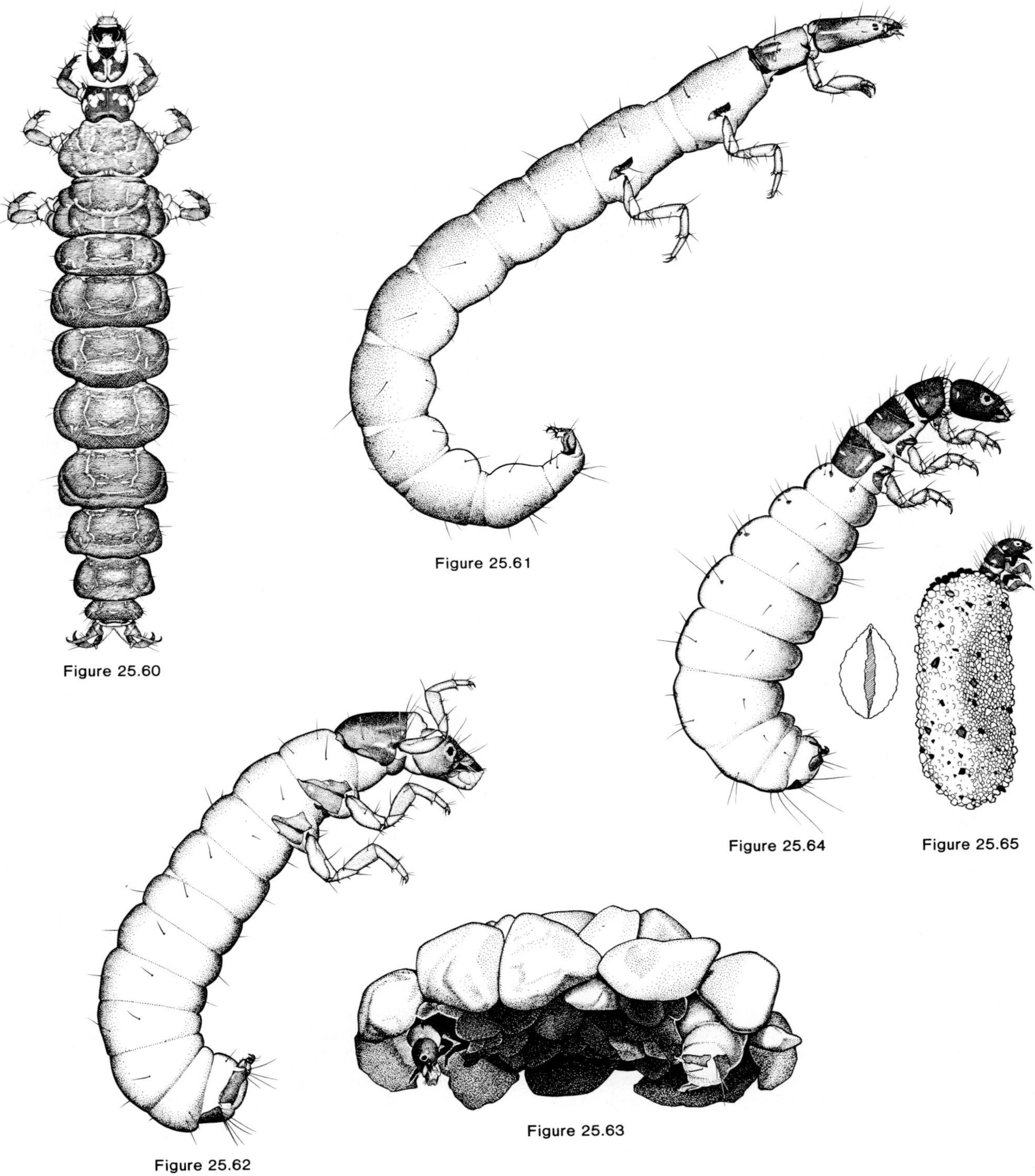

Figure 25.60. Dorsal view of *Rhyacophila* (Rhyacophilidae) larva.

Figure 25.61. Lateral view of *Atopsyche* (Hydrobiosidae) larva.

Figure 25.62. Lateral view of *Glossosoma* (Glossosomatidae) larva.

Figure 25.63. Ventrolateral view of *Glossosoma* (Glossosomatidae) case with larva.

Figure 25.64. Lateral view of *Ochrotrichia* (Hydroptilidae) larva.

Figure 25.65. Lateral view of *Ochrotrichia* (Hydroptilidae) case with larva, detail of end view.

similar in size and structure, with tibiae and tarsi short and stout in some genera, but with the mesothoracic and metathoracic legs much longer than the prothoracic legs, and tibiae and tarsi long and slender in other genera. Tarsal claws short and stout in some genera, long and slender in others.

Abdomen with intersegmental divisions very prominent in some genera, much less so in others, segments extended into prominent dorsal and ventral lobes in *Ithytrichia*. Chloride epithelia present dorsally in some genera; segment 1 not modified; segmentally arranged gills lacking but slender filaments on segment 9 in some genera; lateral fringe lacking in most genera. Anal prolegs short and lacking accessory hooks in most genera, longer in genera with portable tubular cases.

Comments: Hydroptilidae are an important component of the Trichoptera, the Nearctic fauna now comprising 14 genera with nearly 200 species. Many more are undoubtedly still to be discovered. Little work has been done on larval taxonomy to species but a key to Nearctic genera is available (Wiggins 1977). Larvae of world genera were reviewed by Marshall (1979).

Selected Bibliography

Marshall 1979.
Nielsen 1948.

SUPERFAMILY HYDROPSYCHOIDEA

Hydropsychoid caddisfly larvae are stationary during their development, constructing fixed filter nets and shelters, and most depending on food carried by the current. They do not have well-developed walking legs, all thoracic legs being of approximately the same length, and they lack the prominent protuberances on the first abdominal segment found in Limnephiloids. Long anal prolegs are an asset in maintaining position in strong currents. For respiration, almost all of them are entirely dependent on the currents of lotic waters, and apart from the Hydropsychidae, none have tracheal gills; a few species of the Polycentropodidae have penetrated standing waters.

PHILOPOTAMIDAE

The Philopotamids

Figures 25.5, 25.28, 25.32, 25.69, 25.70

Relationships and Diagnosis: The Philopotamidae are closely related to the Stenopsychidae, a family of Asia and the Southern Hemisphere that is absent from North America. The two are regarded as the most primitive living families of the net-spinning Hydropsychoidea. Philopotamids are similar in general appearance to polycentropodids, hydrobiosids, and some rhyacophilids.

Philopotamid larvae are distinguished by a broad, membranous, somewhat T-shaped labrum (fig. 25.32), modified from the normal sclerotized rounded labrum found in larvae of all other families of Trichoptera. Because the membranous labrum is so flexible, it is often withdrawn entirely beneath the anterior margin of the head capsule in preserved specimens, frequently leading to misinterpretation of the anterior margin of the frontoclypeus as a sclerotized labrum. Philopotamid larvae (fig. 25.32) usually have a uniformly brownish orange head and pronotum, without spots, and a prominent black band along the posterior border of the pronotum; membranous segments of the thorax and abdomen are yellowish in life, and usually white in preserved specimens.

Biology and Ecology: Philopotamid larvae live in running waters where they spin fixed, sacklike silken nets (fig. 25.70) approximately 25 to 60 mm long and 2.5 to 5 mm in diameter on the under sides of rocks (Wallace and Malas 1976); the mesh openings are the smallest known in Trichoptera, as small as $0.4 \times 3.7\ \mu m$ for final instars. Constructed in relatively reduced currents under rocks, the nets filter very fine detrital particles and diatoms; the larva living inside the net uses its broad membranous labrum to sweep accumulating particles from the meshes (Philipson 1953). The nets collapse into an amorphous silt-laden mass when the rock is removed from water.

Description: Larvae (fig. 25.69) up to 17 mm in length, slender and agile.

Head elongate (fig. 25.32), uniform brownish orange; anterior margin of frontoclypeus asymmetrical in some genera; parietal sclerites entirely in contact along the midventral line. Labrum membranous, the anterior margin straight and wide, tapering behind to its articulation, the entire labrum often retracted beneath the anterior margin of the head capsule.

Thorax with only the pronotum sclerotized, brownish orange, posterior border bounded with a prominent black line; mesonotum and metanotum entirely membranous. Legs unmodified, short, all similar in size and structure.

Abdomen lacking gills and lateral fringe; anal papillae present; yellowish in living specimens, white in preservation; segment 9 lacking dorsal sclerite. Anal prolegs prominent, basal segment long, anal claw small (fig. 25.28).

Comments: Three genera of Philopotamidae are known in North America: *Chimarra, Dolophilodes,* and *Wormaldia,* with a total of approximately 50 species. All three genera are widespread and common, with species diversity somewhat greater in the West but decreasing generally at higher latitudes. Wiggins (1977) provided a key to genera, and some species diagnoses for larvae are found in Ross (1944).

Selected Bibliography

Philipson 1953.
Wallace and Malas 1976.

HYDROPSYCHIDAE

The Hydropsychids

Figures 25.9, 25.17, 25.18, 25.66, 25.67

Relationships and Diagnosis: The hydropsychids are not closely related to any other family of the Hydropsychoidea. The European tradition of treating the subfamily Arctopsychinae as a distinct family, followed by Schmid (1980), has not been accepted by most North American workers (see Wiggins 1981).

All hydropsychid larvae can be readily distinguished by the unique combination of thickly branched abdominal and thoracic gills and sclerotized plates covering the nota of all three thoracic segments (fig. 25.66).

Biology and Ecology: Larvae are confined to running waters, with a few species finding sufficient movement from the wave action along the shore of very large lakes. Current is essential for the operation of the silken filter nets (e.g., fig. 25.67) constructed by hydropsychid larvae, carrying the particulate organic matter and small insects that they eat. The larva rests in an adjacent tubular retreat of plant and rock fragments, coming out periodically to remove food and debris accumulated by the net. Mesh sizes in the silken filter nets are different (Wallace 1975a, b). The largest meshes are made by the Arctospychinae, large predaceous larvae living in fast, cold headwater streams. The smallest meshes are made by the Macronematinae, living in slower currents of large rivers where small organic particles comprise the larger part of the suspended materials. Larvae of the Hydropsychinae and Diplectroninae build filter nets of a wide range of intermediate mesh sizes, with different species occurring under particular combinations of river size, current, and temperature. By rubbing a series of ridges on the ventrolateral surface of the head against a tubercle on the femora of the fore legs, *Hydropsyche* larvae produce sound (Johnstone 1964). Since larvae in all hydropsychid genera have head ridges of slightly different configuration (Wiggins 1977), all of them probably produce sound (and perhaps different sounds). The sound is evidently part of the defensive behavior of larvae in protecting their retreat against congeneric intruders (Jansson and Vuoristo 1979), but whether this protective role extends to predators generally is not known.

Description: Larvae (fig. 25.66) up to 30 mm in length, agile.

Head quadrate or rounded in dorsal aspect, sometimes carinate, secondary setae abundant; parietals entirely separated by the rectangular ventral apotome (Arctopsychinae), or ventral apotome subdivided into anterior and posterior triangular sections of approximately equal length between which the parietals are contiguous (Diplectroninae) (fig. 25.9), or the posterior ventral apotome minute (Hydropsychinae), and minute or lacking (Macronematinae). Labrum with dense lateral brushes of setae. Ventral surface of head with two elongate patches of closely spaced transverse ridges, differing in subfamilies (see Wiggins 1977), which function in sound production when rubbed against the fore femur.

Thorax with nota of all three segments entirely covered by sclerotized plates; plates of the pronotum subdivided along the median line, with those of the mesonotum and metanotum subdivided transversely (Arctopsychinae, Diplectroninae) or undivided (Hydropsychinae, Macronematinae). Transverse prosternal plate usually present; ventral gills present on last two thoracic segments.

Abdomen with many secondary setae that are frequently modified as scale hairs, thickly branched ventral gills on most segments, paired ventral sclerites bearing stout setae on segments 8 and 9, anal prolegs with a prominent brush of setae at apex of lateral sclerite.

Comments: The Hydropsychidae are one of the dominant groups of aquatic insects in running waters. Riffle areas of rivers support extremely high populations of the larvae, particularly below the effluent of lakes from which planktonic organisms are carried downstream to their filter nets. Approximately 150 Nearctic species are known, representing 11 genera. Diagnoses for larvae of the genera were given by Wiggins (1977). Larvae of species in the Arctopsychinae were treated by Flint (1961), Givens and Smith (1980), and Smith (1968a); in the genus *Hydropsyche* by Schuster and Etnier (1978) and Schefter and Wiggins (1986) and in *Aphropsyche* by Weaver et al. (1979).

Selected Bibliography

Flint 1961.
Givens and Smith 1980.
Jansson and Vuoristo 1979.
Johnstone 1964.
Schefter and Wiggins, 1986.
Schuster and Etnier 1978.
Smith 1968a.
Wallace 1975a, 1975b.
Weaver et al. 1979.

PSYCHOMYIIDAE

The Psychomyiids

Figures 25.35, 25.71, 25.72

Relationships and Diagnosis: The Psychomyiidae are closely related to the Xiphocentronidae and also bear a general resemblance to some Polycentropodidae. They are distinguished from other larvae in which only the pronotum is sclerotized by the broad, hatchet-shaped trochantin of the prothorax (fig. 25.35).

Biology and Ecology: Psychomyiid larvae live in cool running waters for the most part, although some *Tinodes* live in isolated stream pools in western North America and along lake margins in Europe. Larvae live on rocks and logs in individual meandering silken tubes a centimeter or more long that are covered with sand and fine detritus (e.g., fig. 25.72). They do not filter food particles from the current as do most Hydropsychoidea; rather they feed on diatoms, on fine organic material from the substrate around the openings to their tube, and sometimes on insect larvae (Coffman et al. 1971).

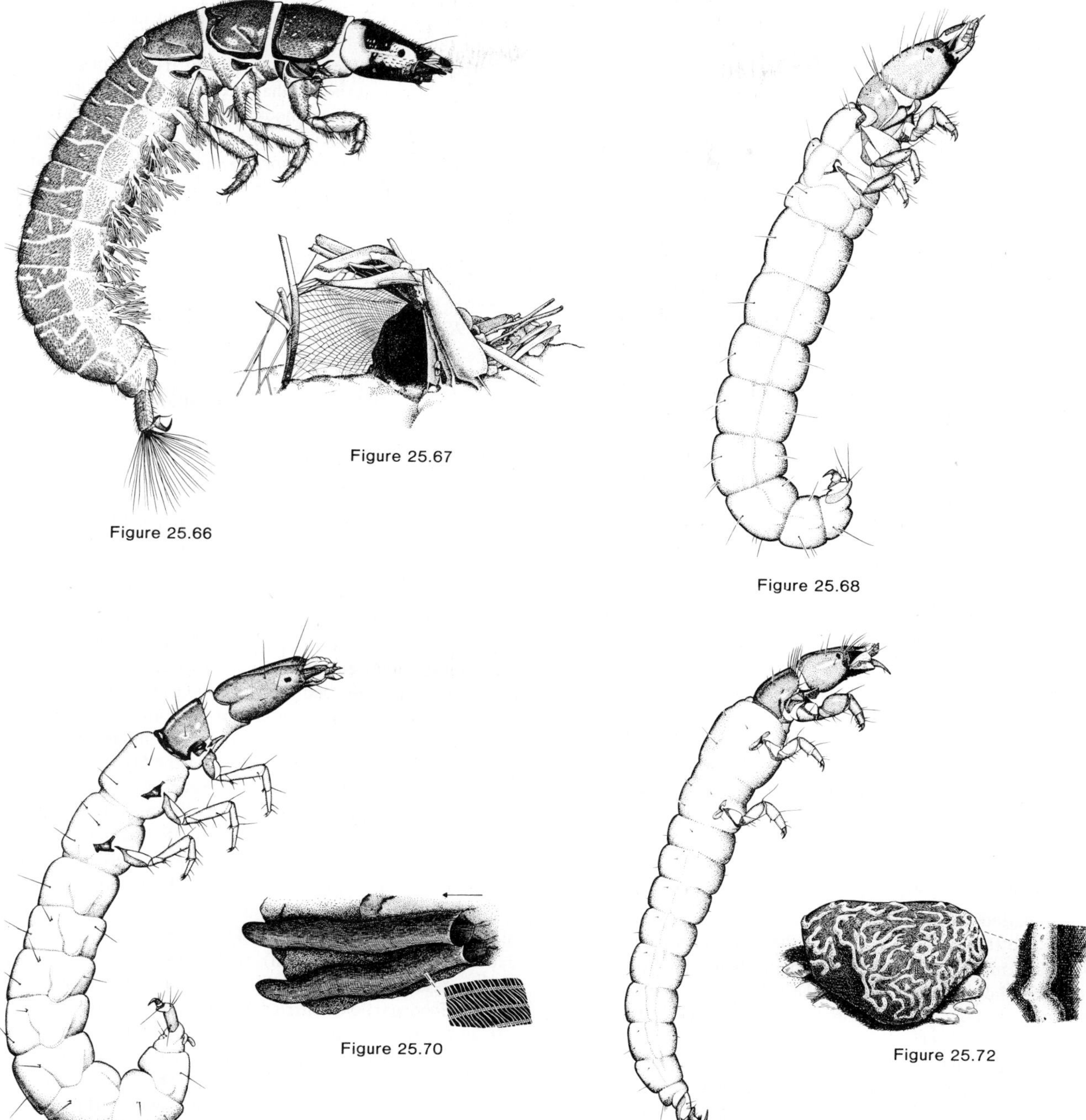

Figure 25.66

Figure 25.67

Figure 25.68

Figure 25.69

Figure 25.70

Figure 25.71

Figure 25.72

Figure 25.66. Lateral view of *Hydropsyche* (Hydropsychidae) larva.

Figure 25.67. Filter net and retreat of *Hydropsyche* (Hydropsychidae) larva.

Figure 25.68. Lateral view of *Xiphocentron* (Xiphocentronidae) larva.

Figure 25.69. Lateral view of *Dolophilodes* (Philopotamidae) larva.

Figure 25.70. Two filter nets of *Dolophilodes* (Philopotamidae) larvae, detail of mesh, arrow indicates direction of current.

Figure 25.71. Lateral view of *Psychomyia* (Psychomyiidae) larva.

Figure 25.72. Rock with sand-covered tubes of *Psychomyia* (Psychomyiidae) larvae, detail of tube.

Description: Larvae (fig. 25.71) up to 15 mm in length.

Head rather short, parietals entirely in contact along the midventral line; labium slender, pointed and extending well beyond the head. Labial palpi lacking; mandibles toothed; submental sclerites much enlarged in *Psychomyia.*

Thorax with only the pronotum sclerotized; fore trochantin (fig. 25.35) broad and hatchet-shaped, clearly separated from the episternum by a dark suture; mesonotum membranous, swollen and larger than other segments; metanotum membranous. Legs short, those of the prothorax sometimes larger than the others.

Abdomen lacking gills and lateral fringe; anal papillae usually present; basal segment of anal proleg shorter than distal segment. Anal claw moderately large and lacking dorsal accessory hooks, but in *Psychomyia* with teeth on the concave base.

Comments: Two subfamilies are known in North America: Psychomyiinae are widely distributed over the continent, with three genera—*Lype, Psychomyia,* and *Tinodes*—containing approximately 15 species, and Paduniellinae with one species of *Paduniella* known only from Arkansas and for which no larva is yet known. Keys to genera were given by Wiggins (1977) and Flint (1964a), and some information on species diagnosis is also available in these references. In many of the taxonomic references in North America, the Polycentropodidae and Xiphocentronidae have been treated as subfamilies of the Psychomyiidae.

Selected Bibliography

Flint 1964a.

XIPHOCENTRONIDAE

The Xiphocentronids

Figures 25.34, 25.68

Relationships and Diagnosis: Xiphocentronids are closely related to the Psychomyiidae and have often been included as a subfamily of that group. Although their larvae resemble psychomyiids, they are readily distinguished by the fused tibiae and tarsi of all legs and by the lobate process of the mesopleuron (fig. 25.34).

Biology and Ecology: Larvae live on rocks in cool streams, where they build tubes of fine sand held together with silk, frequently extending the tubes several centimeters above the water surface on wet substrates; tubes up to 5 cm in length and 2.5 mm in diameter have been recorded (Edwards 1961; Flint 1964b).

Description: Larvae (fig. 25.68) up to 8 mm in length, generally similar to psychomyiids.

Head short, parietals entirely in contact along the midventral line; labium slender and pointed, extending well beyond the head, labial palpi lacking; mandibles toothed.

Thorax (fig. 25.34) with only the pronotum sclerotized; fore trochantin a small lobe separated from the episternum by a suture, but not expanded as in the Psychomyiidae. Mesonotum membranous, a unique upturned lobe on each side, arising from an invagination in the mesopleuron; metanotum membranous. All legs short, with tibia and tarsus fused into a single segment.

Abdomen lacking gills and lateral fringe; anal prolegs short, basal segment shorter than the distal segment, anal claw lacking teeth or accessory hooks; anal papillae present.

Comments: The Xiphocentronidae in the New World comprise species of *Xiphocentron* in Mexico, Central America, and the Caribbean Islands. North of the Rio Grande one species has been recorded from Texas (Edwards 1961).

Selected Bibliography

Edwards 1961.
Flint 1964b.

POLYCENTROPODIDAE

The Polycentropodids

Figures 25.33, 25.36, 25.73-25.77

Relationships and Diagnosis: The Polycentropodidae do not share close relationships with any other family, but their larvae are somewhat similar in general appearance to philopotamids, hydrobiosids, and some rhyacophilids. They lack a dorsal sclerite on segment 9, have a rounded sclerotized labrum (fig. 25.33), and unmodified legs (fig. 25.36).

Biology and Ecology: Polycentropodids are biologically diverse and have invaded the slower warmer waters of lakes and large rivers to a greater extent than any other net-spinning Trichoptera. Some *Polycentropus* live in temporary pools (Wiggins 1973a), the only members of the Hydropsychoidea known to do so. In genera such as *Neureclipsis,* larvae construct large trumpet-shaped nets of silk in slow currents to filter suspended organic materials (fig. 25.74). In others such as *Nyctiophylax* (fig. 25.75) and some *Polycentropus,* larvae live in a flattened tubular retreat of silk, darting out like a spider to capture any small creature that strikes the threshold of silken strands. The genus *Phylocentropus* is the most unusual (see comments section), with larvae burrowing in soft sediments and constructing silken tubes covered with sand (fig. 25.77); food particles carried by the current enter through one end of the tube projecting above the bottom and are filtered out by a mass of silken threads in a bulbous section, the water passing out again through the other end of the tube (Wallace et al. 1976). The larva remains within the tube, removing organic particles from the filter, probably maintaining the circulatory current through abdominal undulations.

Description: Larvae (figs. 25.73, 25.76) up to 20 mm in length.

Head generally short and rounded, parietals usually entirely in contact along the midventral line; mandibles elongate (short and broad in *Phylocentropus,* with thick mesal brushes of setae). Labrum short, labial palpi small (labium very long and slender, labial palpi lacking in *Phylocentropus*); maxillary palpi and laciniae well developed (laciniae flattened and setate in *Phylocentropus*).

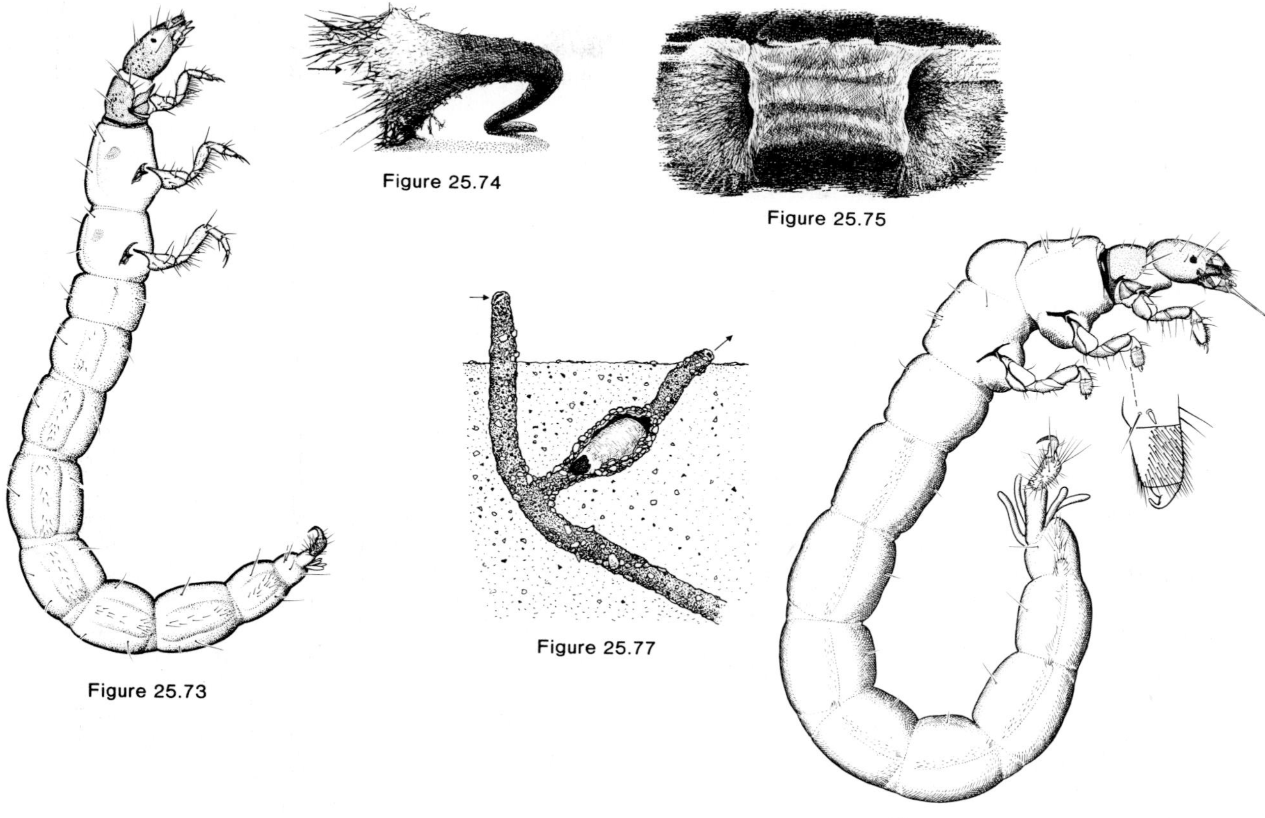

Figure 25.73. Lateral view of *Neureclipsis* (Polycentropodidae) larva.

Figure 25.74. Filter net of *Neureclipsis* (Polycentropodidae) larva, arrow indicates direction of current.

Figure 25.75. Retreat of *Nyctiophylax* (Polycentropodidae) larva on rock.

Figure 25.76. Lateral view of *Phylocentropus* (Polycentropodidae: Dipseudopsinae) larva.

Figure 25.77. Dwelling tube of *Phylocentropus* (Polycentropodidae: Dipseudopsinae) larva, ends projecting above sediment, filter exposed, arrows indicating direction of current.

Thorax with only the pronotum sclerotized, mesonotum and metanotum entirely membranous. All legs short and similar in structure (tibia and tarsus short and flattened in *Phylocentropus*).

Abdomen lacking gills but usually with a lateral fringe; anal papillae present in at least some genera; segment 9 without a dorsal sclerite. Anal prolegs prominent, extending freely from the abdomen, anal claw in some genera with a row of comblike points on the concave margin and an accessory hook.

Comments: The seven Nearctic genera of the Polycentropodidae include approximately 70 species north of the Rio Grande. All are assigned to the subfamily Polycentropodinae except for *Phylocentropus,* which was placed in the Dipseudopsinae (Ross and Gibbs 1973), a group otherwise confined to the Old World tropics. *Phylocentropus* has also been assigned to the Hyalopsychidae (Schmid 1980), but evidence in support of either position is not yet complete. In any case, recognition of *Phylocentropus* at the familial level seems justified. Keys to genera were given by Wiggins (1977); the diagnosis for *Cernotina,* which was not then known, was provided by Hudson et al. (1981).

Selected Bibliography

Flint 1964a.
Hudson et al. 1981.
Ross and Gibbs 1973.
Wallace et al. 1976.

SUPERFAMILY LIMNEPHILOIDEA

Limnephiloid caddisfly larvae walk, or in a few genera swim, actively with their cases, and hence are able to move from one food resource to another. The middle and hind legs, which are longer than the fore legs, serve as walking legs. Short, stout anal prolegs with claws bearing accessory hooks hold firmly to the silken lining of the case. Portable cases confer protection from predators as well as respiratory independence from the stream current. The expansile protuberances on abdominal segment 1 are believed to space the larva evenly from all sides of the case, allowing uniform and unimpeded respiratory exchange. Filamentous gills in most larvae extend the respiratory surfaces into the current passing through the case. All families of the Limnephiloidea are represented in cool, running waters and several (Limnephilidae, Leptoceridae, Phryganeidae, Molannidae) are well represented in standing waters of lakes, ponds and marshes.

The Limnephiloidea comprise two groups of families—the limnephilid branch in which most larvae have a prosternal horn (Phryganeidae, Brachycentridae, Lepidostomatidae, Limnephilidae, Goeridae, and Uenoidae); and the leptocerid branch in which the prosternal horn is lacking (Beraeidae, Sericostomatidae, Helicopsychidae, Leptoceridae, Calamoceratidae, Molannidae, and Odontoceridae). Function of the prosternal horn is not known.

PHRYGANEIDAE

The Phryganeids

Figures 25.24, 25.25, 25.78, 25.79

Relationships and Diagnosis: Members of the limnephilid branch, the Phryganeidae are distinguished by the absence of mesonotal and metanotal *sa* 1 and *sa* 2 sclerites, although small mesonotal sclerites are present in one or both of these positions in a few genera, and especially by the small rounded *sa* 3 sclerites on the mesonotum and metanotum (fig. 25.24).

Biology and Ecology: Phryganeids live in lakes, marshes, temporary pools, and slow-flowing parts of rivers and streams. Larval cases are usually constructed of uniform pieces of leaves, bark, or stems fastened together in a continuous band (fig. 25.79) or in discrete cylindrical sections joined end to end; occasionally the pieces are arranged irregularly, and in the western genus *Yphria,* rock fragments and irregular pieces of bark are used. Larval cases are large, up to 60 mm in length. Phryganeids are unusual among the Limnephiloidea in abandoning their cases when picked up; but correspondingly they show a greater ability to re-enter abandoned cases than do larvae in other families (Merrill 1969). In the spiral cases of *Phryganea,* dextral and sinistral spirals occur within the same species. Larvae feed on plant detritus but may also be strongly predaceous.

Description: Larvae (fig. 25.78) up to 43 mm long, suberuciform, active, slender and somewhat depressed, with prominent intersegmental constrictions. Head and pronotum usually with conspicuous dark bands on a yellowish brown ground colour.

Head prognathous, elongate, parietals entirely separated by an elliptical ventral apotome, except in *Yphria* where they are separated only anteriorly by a small triangular ventral apotome; antennae at anterior margin of head capsule (fig. 25.24); maxillae with lacinia slender and fingerlike.

Thorax (fig. 25.24) with prosternal horn well developed; mesonotum and metanotum usually lacking *sa* 1 and *sa* 2 sclerites, although small mesonotal *sa* 1 sclerites are present in *Oligostomis* and mesonotal *sa* 1 and *sa* 2 sclerites in *Yphria; sa* 3 sclerites on mesonotum and metanotum small and rounded; hind and mid legs approximately same length.

Abdomen with protuberances of segment 1 prominent; gills single and long; lateral fringe well developed; lateral tubercles absent; dorsal sclerite on segment 9.

Comments: Although not a large family, the Phryganeidae are common in northern latitudes and in western mountain lakes. Ten genera with 27 species are known in North America; a number are northern and transcontinental and several are Holarctic in distribution. The subfamily Yphriinae was established for the aberrant western species *Yphria californica* Banks (Wiggins 1962); all others are assigned to the Phryganeinae. Larval taxonomy was summarized (Wiggins 1977), and diagnoses for a number of species given by Wiggins (1960).

Selected Bibliography

Wiggins 1960, 1962.

BRACHYCENTRIDAE

The Brachycentrids

Figures 25.37, 25.38, 25.43, 25.80, 25.81

Relationships and Diagnosis: Although members of the limnephilid branch, the Brachycentridae are not particularly closely related to any other family. Larvae are distinguished from all others in the Limnephiloidea by the absence of both dorsal and lateral protuberances on the first abdominal segment (fig. 25.37). Most of them are further characterized by a transverse pronotal carina and by secondary subdivisions in the mesonotal sclerites (fig. 25.38).

Biology and Ecology: Brachycentridae are confined to running waters, ranging from cold springs to slow moving, marshy rivers. Larvae of *Brachycentrus* occur in large rivers where they filter food materials from the current by extending their unusually long legs; they also graze algae from surrounding substrates. Larvae of some other genera (*Micrasema, Eobrachycentrus, Adicrophleps*) live in submerged moss, feeding on these plants and on algae and fine detrital particles. Larval cases, constructed of moss or other plant materials, sand, or occasionally mainly of silk alone, are four-sided in some genera (fig. 25.81), cylindrical in others; cases range up to 18 mm in length.

Description: Larvae (fig. 25.80) up to 13 mm in length, sclerotized areas generally uniform dark brown to black, occasionally reddish or with contrasting light and dark markings.

Head (fig. 25.38) hypognathous, usually round in dorsal aspect, somewhat flattened dorsally and frequently with an anterolateral carina; parietals entirely separated ventrally by a quadrate ventral apotome; antennae close to the anterior margin of the head capsule.

Thorax (fig. 25.38) with pronotum usually bearing a transverse carina or sulcus, short prosternal horn present in some genera; mesonotum sclerotized, the two primary sclerites usually subdivided longitudinally; metanotum with small setate sclerites at *sa* 2 and *sa* 3, but with only a single seta or none at all at *sa* 1. Middle legs usually the longest, both middle and hind legs much elongated in the filter-feeding genus *Brachycentrus.*

Abdomen (fig. 25.37) with expansile protuberances of segment 1 entirely absent; gills single or lacking; lateral fringe much reduced or absent; lateral tubercles on several segments.

Comments: The Brachycentridae comprise five genera with approximately 30 species in North America; representatives occur in most parts of the continent. Larvae are known for all of the Nearctic genera (Flint 1984, Wiggins 1977), two of which are monotypic, and one is represented by only a single Nearctic species. Larvae of most Nearctic species in the widespread and common genus *Brachycentrus* can be identified in Flint (1984). Diagnoses for *Micrasema* larvae have been worked out (Chapin 1978).

Selected Bibliography

Chapin 1978.
Flint 1984.

LEPIDOSTOMATIDAE

The Lepidostomatids

Figures 25.56, 25.84, 25.85

Relationships and Diagnosis: Lepidostomatidae belong to the limnephilid branch; they are discrete and readily separated from other families by the location of the antenna close to the eye and by the absence of a median dorsal protuberance on abdominal segment 1 (fig. 25.56).

Biology and Ecology: Lepidostomatids typically occur in cool running water, but some are found in temporary streams and in the littoral zone of lakes. Larvae frequent accumulations of plant debris, and the few species that have been studied are detritivorous (e.g., Anderson and Grafius 1975). Larval cases are constructed of plant materials or sand grains fastened together in various forms; a common type is a four-sided case of bark fragments (fig. 25.85), but early instars in these species construct a cylindrical case of sand. Cases range up to 15 mm in length.

Description: Larvae (fig. 25.84) up to 13 mm in length, sclerotized areas most often dark with large, light muscle scars on the head posterolaterally; some larvae reddish brown.

Head (fig. 25.56) hypognathous, usually rounded in dorsal aspect, sometimes inflated dorsally or flattened and with a dorsal carina; parietal sclerites partially separated anteriorly by a wedge-shaped ventral apotome; antennae close to the eye.

Thorax (fig. 25.56) with prosternal horn prominent; mesonotum covered by two large, quadrate, sclerotized plates without secondary subdivision; metanotum mainly membranous with three pairs of small sclerites bearing setae. Middle and hind legs of approximately same length.

Abdomen with segment 1 lacking median dorsal protuberance (fig. 25.56); gills single or absent; lateral fringe sparse or lacking; lateral tubercles on most segments; segment 9 with dorsal sclerite; anal papillae usually present.

Comments: The Lepidostomatidae are a moderate-sized family with some 70 species known in North America. These are currently assigned to two genera, *Lepidostoma* and *Theliopsyche*, but generic placement has fluctuated considerably in the past. The majority of Nearctic species are unknown in the larval stage, but the considerable structural and behavioral diversity known to exist among their larvae is an important source of data in resolving the taxonomy (see Wiggins 1977).

Selected Bibliography

Anderson and Grafius 1975.

LIMNEPHILIDAE

The Limnephilids

Figures 25.1–25.3, 25.7, 25.8, 25.11, 25.13, 25.15, 25.20, 25.39, 25.40, 25.45, 25.48, 25.49, 25.52, 25.59, 25.82, 25.83

Relationships and Diagnosis: The family belongs to the limnephilid branch, with genera assigned to the Goeridae as its nearest relatives. Limnephilid larvae have antennae located midway between the eye and the anterior margin of the head capsule (fig. 25.52), unmodified mesopleura, and metanotal *sa* 1 usually represented by a small sclerite bearing setae (fig. 25.59) but, in a few genera, by several setae without a sclerite.

Biology and Ecology: Limnephilid larvae occur in a wider range of habitats than any other family, with genera characteristic of lakes, marshes, rivers, springs including at least one restricted to the wet organic muck of seepage areas, and temporary pools. Larvae of a few species are secondarily terrestrial (Anderson 1967, Williams and Williams 1975). Although many limnephilid genera are confined to lotic habitats, the family is the dominant group in standing waters of lakes, ponds, and marshes, particularly at higher latitudes and elevations. They are also the most abundant caddisflies in temporary pools, a demanding habitat made accessible to them by delayed oviposition through diapause (Novák and Sehnal 1963) and by unusual features of oviposition and eggs (Wiggins 1973a, Wiggins et al. 1980). Larval cases constructed of both plant and mineral materials (e.g., fig. 25.83), are highly diverse in form and range in size from 7 to 76 mm. Most larvae in the Dicosmoecinae, Pseudostenophylacinae, and Limnephilinae have toothed mandibles (as in fig. 25.10) and are detritivores or omnivores; larvae in the Apataniinae and Neophylacinae (fig. 25.11) have mandibles with entire edges and feed mainly by scraping diatoms and fine organic particles from exposed rock surfaces.

Description: Larvae (fig. 25.82) up to 35 mm in length, eruciform, sclerotized areas diverse in color but often with contrasting markings.

Head rounded in dorsal aspect, a dorsolateral carina in some species; parietals entirely separated by the ventral apotome (fig. 25.8). Secondary setae present in some genera; antennae located midway between the eye and the anterior margin of the head capsule (fig. 25.52); mandibles with separate teeth in most genera, entire edges in a few (fig. 25.11).

Thorax with prosternal horn well developed (fig. 25.52); mesonotum usually with a single pair of large sclerites; metanotum usually with primary setal areas represented by small paired sclerites (fig. 25.59), but each *sa* 1 by a few setae in Neophylacinae, and fused as a transverse band of setae in Apataniinae; middle and hind legs longer than fore legs.

Abdomen with gills branched, single or lacking; lateral fringe usually well developed; lateral tubercles on most segments; dorsal sclerite on segment 9; ventral chloride epithelia (fig. 25.1) usually present on most segments, often on dorsal and lateral areas as well.

Comments: The Limnephilidae are by far the largest family of Trichoptera in the Northern Hemisphere. The Nearctic fauna comprises more than 300 species in some 50 genera. Species occur throughout most of North America, but the family is particularly dominant at higher latitudes and elevations. Five subfamilies are represented in North America (see above). Larvae of most Nearctic genera can be identified (Wiggins 1977); diagnoses for larvae of some species were given by Flint (1956, 1960), Wiggins and Anderson (1968), Wiggins and Richardson (1982), Parker and Wiggins (1985), but the majority of larvae are still unknown.

Selected Bibliography

Anderson 1967.
Flint 1956, 1960.
Novák and Sehnal 1963.
Parker and Wiggins 1985.
Wiggins 1973b, 1976.
Wiggins and Anderson 1968.
Wiggins and Richardson 1982.
Williams and Williams 1975.

GOERIDAE

The Goerids

Figures 25.57, 25.88, 25.89

Relationships and Diagnosis: Status of this group as an independent family or as a subfamily of the Limnephilidae remains unresolved because some genera have characters shared by the traditional diagnoses for the two families (Wiggins 1973b, 1981). Review of larvae of the world genera, necessary for resolution of the problem, is under way. For the present review, the traditional diagnosis of larvae can be adopted through which most goerids can be recognized by a modified mesopleuron, usually as a long process directed anteriorly (fig. 25.57).

Biology and Ecology: Larvae live in running waters, from small cold spring streams to rivers; but a few occur only in the wet organic muck of spring seepage areas (Wiggins 1973b, 1976). Larval cases are always constructed of rock pieces, a simple curved and tapered cylinder in most genera, but with lateral ballast stones in *Goera* (fig. 25.89) and *Goeracea.* Larvae in most genera have mandibles with entire edges, feeding primarily by scraping diatoms and fine organic particles from rocks. Those living in wet organic muck have toothed mandibles and appear to feed mainly on detritus of higher plants. The mesopleural processes and enlarged pronotum of typical goerids fit together to form a remarkably close-fitting operculum that seals the anterior opening of the case. The head in these larvae folds beneath the pronotum.

Description: Larvae (fig. 25.88) up to 11 mm long, eruciform, sclerotized areas light to dark brown, head and pronotum without distinctive color markings.

Head (fig. 25.57) short, anteriorly narrowed or rounded in dorsal aspect, frequently with an anterolateral carina; parietals separated by a wedge-shaped ventral apotome; antennae midway between the eye and anterior margin of head capsule, frequently in a concavity. Secondary setae often present dorsally; mandibles usually with edges entire, toothed in *Lepania* and *Goereilla.*

Thorax (fig. 25.57) with pronotum usually enlarged and thickened laterally, often with a pronounced dorsal hump and extended anterolaterally, prosternal horn well developed; mesonotum usually with a pronounced transverse ridge, each primary sclerite subdivided into two or three smaller sclerites; mesopleuron modified as a spinose lobe or most often produced anteriorly as a prominent process; metanotum usually with all three setal areas represented by small paired sclerites, *sa* 1 a single seta in *Goeracea;* middle legs longer than hind legs.

Abdomen with gills branched, single or lacking; lateral fringe variable; lateral tubercles present on several segments; dorsal sclerite on segment 9; anal prolegs usually lacking accessory hook.

Comments: As defined here, the Goeridae are a rather small family of five genera and 12 species in North America: *Lepania, Goereilla,* and *Goeracea* in the West; *Goerita* in the East; and *Goera* in both East and West. Larval taxonomy at the generic level was summarized by Wiggins (1977), with additional species diagnoses given by Flint (1960) and Wiggins (1973b).

Selected Bibliography

Wiggins 1977.

UENOIDAE

The Uenoids

Figures 25.58, 25.86, 25.87

Relationships and Diagnosis: Because the Nearctic genera *Farula* and *Neothremma* show the traditional diagnostic characters of the Limnephilidae, they have been identified as members of that family in most keys. Evidence from the larval stages reveals clearly, however, that the true affinity of these groups is with the Asian family Uenoidae (Wiggins et al. 1985). Uenoid larvae have a very characteristic slender form (fig. 25.86), the pronotum longer than wide

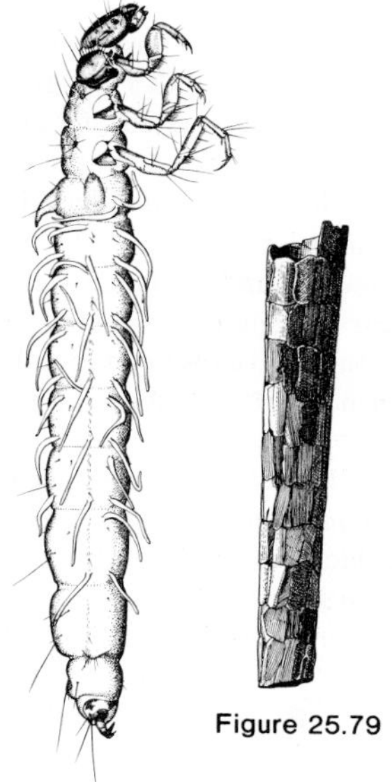

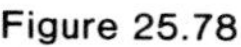

Figure 25.78 Figure 25.79

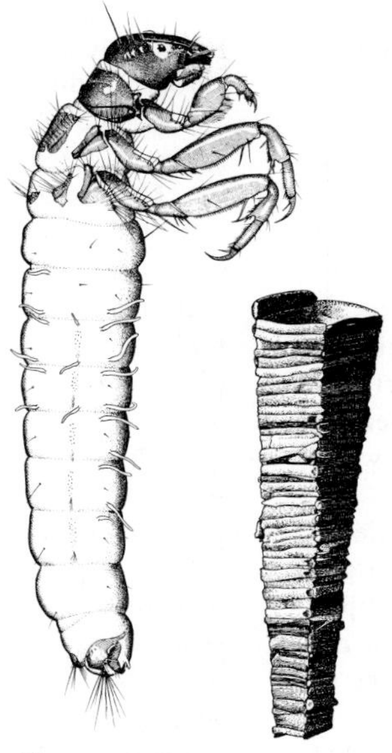

Figure 25.80 Figure 25.81

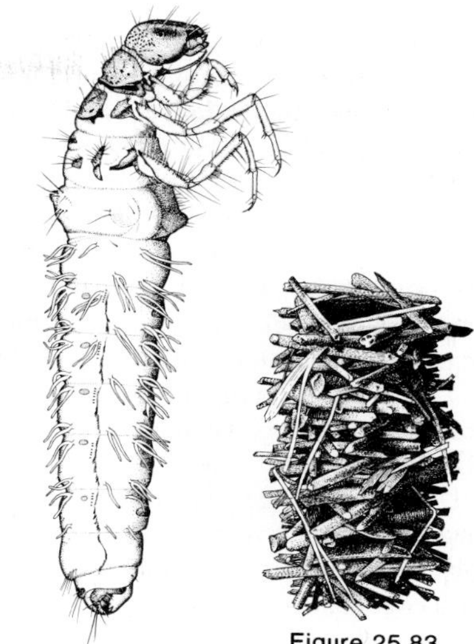

Figure 25.82 Figure 25.83

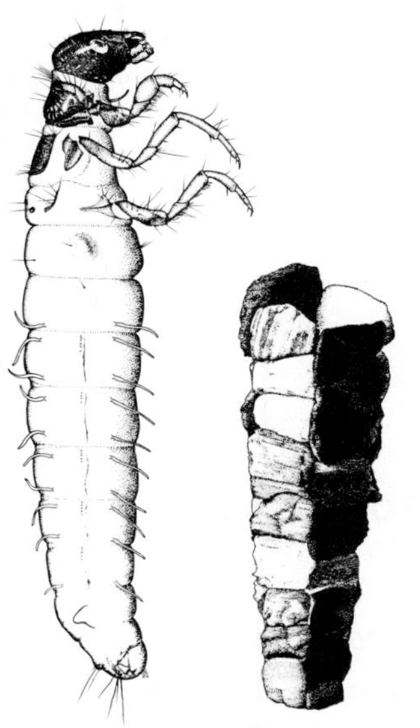

Figure 25.84 Figure 25.85

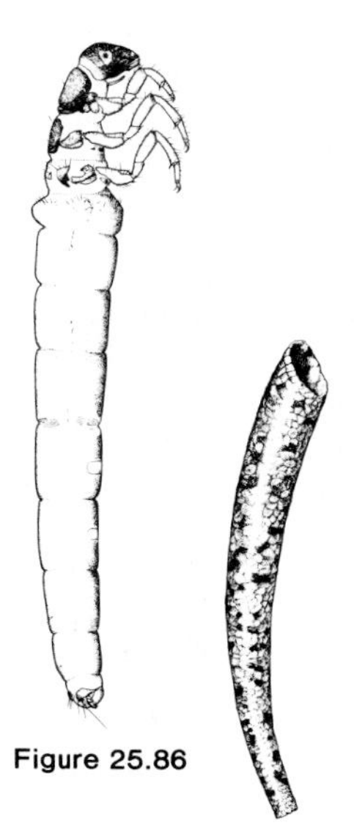

Figure 25.86 Figure 25.87

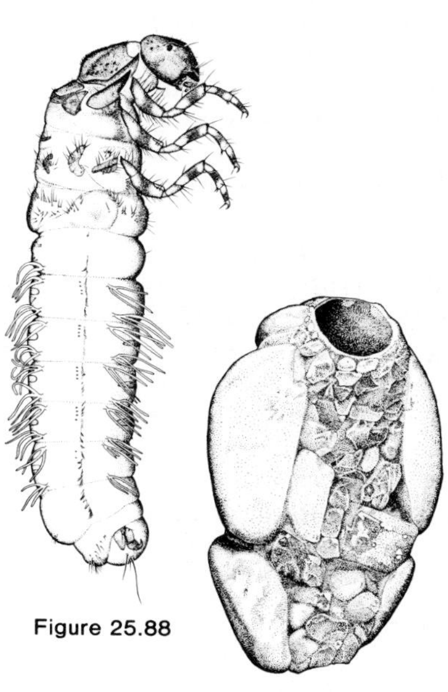

Figure 25.88 Figure 25.89

Figure 25.78. Lateral view of *Agrypnia* (Phryganeidae) larva.

Figure 25.79. Larval case of *Agrypnia* (Phryganeidae).

Figure 25.80. Lateral view of *Brachycentrus* (Brachycentridae) larva.

Figure 25.81. Larval case of *Brachycentrus* (Brachycentridae).

Figure 25.82. Lateral view of *Platycentropus* (Limnephilidae) larva.

Figure 25.83. Larval case of *Platycentropus* (Limnephilidae).

Figure 25.84. Lateral view of *Lepidostoma* (Lepidostomatidae) larva.

Figure 25.85. Larval case of *Lepidostoma* (Lepidostomatidae).

Figure 25.86. Lateral view of *Neothremma* (Uenoidae) larva.

Figure 25.87. Larval case of *Neothremma* (Uenoidae).

Figure 25.88. Lateral view of *Goera* (Goeridae) larva.

Figure 25.89. Larval case of *Goera* (Goeridae).

and metanotal *sa* 1 a single seta (fig. 25.58); the slender larval cases (fig. 25.87) are also unlike any in the Limnephilidae. Uenoidae belong to the limnephilid branch.

Biology and Ecology: Uenoid larvae are confined to small, cold and turbulent montane streams. Their cases (fig. 25.87), so slender that they superficially resemble coniferous needles, are constructed of fine sand covered with silk, and sometimes largely or entirely of dark brown silk alone. Larvae are usually abundant where they occur, and pupate in aggregations on rocks. They have scraping mandibles with the edge entire, and they feed on algae and fine organic particles.

Description: Larvae (fig. 25.86) up to 9 mm in length, eruciform and very slender, sclerotized parts dark brown to black, head and pronotum without markings.

Head (fig. 25.58) hypognathous, elongate parietals largely in contact ventrally except anteriorly where the tip of the triangular ventral apotome lies between them; mandibles with edges entire; antennae located midway between the eye and the edge of the head capsule; many secondary setae on the frontoclypeal apotome.

Thorax (fig. 25.58) with pronotum longer than wide, rather uniformly covered with short setae; prosternal horn well developed; mesonotal sclerites each longer than wide, anterior margins curved convexly, forming a median notch; metanotal *sa* 1 a single seta with no sclerite, *sa* 3 sclerites quadrate and three to four times larger than *sa* 2 sclerites; small sternal sclerites present on all three segments; legs not greatly different in length.

Abdomen with many segments longer than wide; gills absent; lateral fringe reduced; ventral chloride epithelia present on some segments; dorsal sclerite on segment 9.

Comments: The Uenoidae are a small family represented in Japan and the Himalayas, and to which the North American genera *Neothremma* and *Farula* were recently transferred (Wiggins et al. 1985); a third Nearctic genus, *Sericostriata,* was also discovered. Approximately 14 species are known, all from western montane parts of the continent. Most larvae in *Neothremma* and *Farula* have yet to be identified specifically.

Selected Bibliography

Wiggins et al. 1985.

BERAEIDAE

The Beraeids

Figures 25.26, 25.46, 25.47, 25.101, 25.102

Relationships and Diagnosis: Beraeids belong to the leptocerid branch and are generally held to be closely related to the Sericostomatidae. Larvae are unlike all others in having a transverse pronotal carina terminating in a rounded lobe at each anterolateral corner (fig. 25.26), and an anal proleg with reduced lateral sclerite and ventral brush of setae (figs. 25.46, 25.47).

Biology and Ecology: Larvae are known for two species in North America, and both live in the wet organic muck of spring seepage areas (Wiggins 1954, Hamilton 1985). What little evidence is available suggests that larvae are detritivorous, feeding on coarse and fine plant materials (Wiggins 1977). The larval case (fig. 25.102) is constructed of sand grains, tapered, slightly curved and up to 6 mm in length.

Description: Larvae (fig. 25.101) up to 6 mm in length, sclerotized parts brownish orange in color.

Head (fig. 25.26) hypognathous, round in dorsal view, with a prominent anterolateral carina; antennae located at the end of the carina at the edge of the head capsule; parietals entirely in contact ventrally with no visible midventral ecdysial line.

Thorax (fig. 25.26) with a characteristic transverse pronotal carina extended anterolaterally into a rounded lobe at each corner; prosternal horn lacking; mesonotum only lightly sclerotized and pigmented, with no middorsal ecdysial line; metanotum membranous with an anterior patch of short setae.

Abdomen with a lateral line of sparse, feathery setae; single gills on segments 2 and 3; lateral tubercles present on segment 8; segment 9 without a dorsal sclerite. Anal prolegs (figs. 25.46, 25.47) distinctive in having a brush of 25–30 setae arising mesally from the bulbous base, and with the lateral sclerite strongly reduced in area but extended posteriorly as a lobe giving rise to a stout black seta.

Comments: Only the single genus *Beraea* occurs in North America, where three species are described from the eastern part of the continent. Colonies are rare and exceedingly local. Larvae are most effectively located by washing samples of muck in a strainer until all fine suspended materials are removed, and then sorting the remaining components in a clean basin.

Selected Bibliography

Hamilton 1985.
Wiggins 1954.

SERICOSTOMATIDAE

The Sericostomatids

Figures 25.50, 25.51, 25.53, 25.90, 25.91

Relationships and Diagnosis: The Sericostomatidae belong to the leptocerid branch and, as larvae, bear superficial resemblance to some odontocerids. Larvae are clearly distinguished from all other families by the dorsal cluster of setae on the anal proleg (fig. 25.51) and by the hook-shaped fore trochantin (fig. 25.53).

Biology and Ecology: Sericostomatid larvae live in spring streams, larger rivers, and along the shorelines of large lakes. Larval cases (fig. 25.91) are of sand, the external surface uniform, and range up to 30 mm in length. Six larval instars have been found in some European species (Elliott 1969), compared to five in most families. Larvae show a tendency to burrow beneath bottom materials; in the eastern genera, larvae are seldom seen unless sand and gravel deposits are sifted. Vascular plant detritus is the principal food.

Description: Larvae (fig. 25.91) up to 19 mm long, eruciform, sclerotized parts uniform dark or reddish brown, head lacking pigmentation ventrally, occasionally with prominent light markings dorsally.

Head (fig. 25.50) hypognathous, broad and rounded in dorsal aspect; anterolateral carina usually present; parietals almost entirely separated by a wedge-shaped ventral apotome; antennae at the anterior margin of the head capsule.

Thorax (fig. 25.53) with pronotum frequently extended into an anterolateral point; fore trochantin rather large and hook-shaped; mesonotum entirely sclerotized, even posteriorly from the two primary sclerites; *sa* 3 sclerite subdivided from the *sa* 1—*sa* 2 sclerite in *Gumaga.* Metanotum with primary setal areas usually represented by indistinct paired sclerites, but in *Gumaga* metanotal *sa* 2 a single seta. Hind legs longer than middle legs.

Abdomen with lateral protuberances of segment 1 bordered by a circular sclerite basally; gills single or branched; lateral tubercles present on segment 8; dorsal sclerite lacking on 9. Anal proleg (fig. 25.51) with lateral sclerite somewhat reduced dorsally, the membrane bearing a dorsal cluster of approximately 30 small setae, accessory hook of anal proleg frequently bearing another small accessory hook.

Comments: The Sericostomatidae are a small family, and in North America 12 species are known in three genera: *Agarodes* in the East, *Fattigia* in the Southeast, and *Gumaga* in the West (Ross and Wallace 1974). Larval taxonomy was summarized by Wiggins (1977), and diagnoses for *Agarodes* larvae given by McEwen (1980).

Selected Bibliography

Elliott 1969.
McEwen 1980.
Ross and Wallace 1974.

HELICOPSYCHIDAE

The Helicopsychids

Figures 25.14, 25.92, 25.93

Relationships and Diagnosis: The Helicopsychidae are a distinctive family, not closely related to any other in the Trichoptera, although they are members of the leptocerid branch. Apart from their behavior in constructing a very distinctive, helical case (fig. 25.93), larvae of this family are readily recognized by their comblike anal claw (fig. 25.14).

Biology and Ecology: Helicopsychids live in moving water, usually in streams, but also on rocks subject to wave action in the littoral zone of lakes. Larval cases of sand grains fastened together with silk (fig. 25.93) are essentially tubular, but coiled into a typical gastropod form; in some exotic species, the coil is open rather than closed, and in others small rock fragments are fastened around the periphery. Cases of Nearctic species range up to 7 mm in diameter. Larvae graze mainly on diatoms and fine organic particles from the upper exposed surfaces of rocks while concealed beneath the anterior cowled lip of the case.

Description: Larvae (fig. 25.92) lie on one side within the case, with the abdomen coiled somewhat dorsoventrally; sclerotized areas dark mottled in color, head light ventrally.

Head hypognathous, dorsolaterally carinate, dorsum flattened; parietals separated by triangular ventral apotome; antennae midway between eye and margin of head capsule; mandibles with edges entire.

Thorax with pronotum bearing stout, pointed setae along anterior margin; prosternal horn lacking; trochantin long and slender; mesonotum entirely sclerotized; metanotum largely sclerotized, somewhat membranous posteromesally. Hind legs longer than middle legs.

Abdomen with lateral protuberances of segment 1 bearing many small spines; gills short and branched, confined to anterior segments; lateral fringe lacking; lateral tubercles present on segment 8; segment 9 lacking dorsal sclerite; anal claw comblike (fig. 25.14).

Comments: Only the single genus *Helicopsyche* occurs in North America, with six species recognized north of the Rio Grande; one of these, *H. borealis* (Hagen) is widespread and common over much of the continent (Wiggins 1977).

Selected Bibliography

Hamilton and Holzenthal 1984.

LEPTOCERIDAE

The Leptocerids

Figures 25.10, 25.22, 25.23, 25.94–25.96

Relationships and Diagnosis: The family belongs to the leptocerid branch but has no close relatives. Most Leptoceridae larvae have long antennae (fig. 25.22), which clearly distinguish them from other Nearctic Limnephiloidea; but some species of *Ceraclea* with short antennae (fig. 25.23) can be distinguished by the pair of characteristic dark curved lines on the mesonotal sclerites. Leptocerid larvae are further distinguished by unpigmented secondary ecdysial lines on the head and in some genera on the thorax, too (figs. 25.22, 25.23).

Biology and Ecology: The Leptoceridae are a dominant family in lakes and in larger, warmer rivers. Larvae in most genera seem to be omnivorous, and predation is common; *Oecetis* larvae with specialized bladelike mandibles are evidently mainly predaceous; and some *Ceraclea* larvae feed on freshwater sponges; some *Triaenodes* feed on green plants. Larvae of *Leptocerus* and *Triaenodes* swim through the water carrying their cases and live mainly in beds of aquatic plants well above the bottom substrate. Casemaking behavior is diverse: in several genera, rock and plant pieces are used to make cylindrical cases (e.g., fig. 25.95), which often have a small twig extending beyond the anterior end; cases of *Ceraclea* are of sand grains with an overhanging cowl (fig. 25.96) as in *Molanna,* although the sponge-feeding species of *Ceraclea* construct tapered, cylindrical cases entirely of silk; cases of the free-swimming *Leptocerus* are also of silk, but in *Triaenodes* plant pieces are fashioned in a slender spiral. Larval cases range in length from 8 to 33 mm.

Leptocerids share with molannids the unusual behavior of ejecting the larval exuviae from the pupal case.

Description: Larvae (fig. 25.94) up to 15 mm in length, sclerites of head and pronotum frequently marked with contrasting dark and light patches or bands.

Head hypognathous, elongate, unpigmented secondary ecdysial lines extending longitudinally (fig. 25.23). Parietal sclerites separated entirely by a quadrate ventral apotome in some genera, or incompletely by a wedge-shaped apotome. Antennae long in most genera (fig. 25.22), at least six times longer than wide, but short in sponge-feeding *Ceraclea* larvae (fig. 25.23).

Thorax (figs. 25.22, 25.23) with pronotum showing unpigmented secondary ecdysial lines in some genera; mesonotum largely covered by pair of large but lightly sclerotized plates; metanotum largely membranous, small sclerites bearing setae sometimes present. Hind legs much longer than middle legs, often with many setae and especially so in genera with swimming larvae (*Leptocerus* and *Triaenodes*); all segments except coxa lengthened; femur, tibia, and sometimes tarsus subdivided.

Abdomen with lateral protuberances of segment 1 located ventrolaterally, bearing a spinose sclerite in some genera. Gills usually single, sometimes branched or absent; lateral fringe sparse but usually present; lateral tubercles on segment 8.

Comments: The Leptoceridae are a large family, with approximately 100 Nearctic species in seven genera. They are widespread and often abundant. A key to larvae of the Nearctic genera was given by Wiggins (1977), and keys to some species by Haddock (1977, *Nectopsyche*), Resh (1976, *Ceraclea*), Ross (1944), and Yamamoto and Wiggins (1964, *Mystacides*). Approximately half of the Nearctic species are still unidentified as larvae.

Selected Bibliography

Haddock 1977.
Resh 1976.
Yamamoto and Wiggins 1964.

CALAMOCERATIDAE

The Calamoceratids

Figures 25.44, 25.97, 25.98

Relationships and Diagnosis: A member of the leptocerid branch, this family is not closely related to any other. The larvae are readily distinguished by a transverse row of approximately 16 long setae across the central part of the labrum and by the separation of mesonotal *sa* 3 from the main sclerite (fig. 25.44).

Biology and Ecology: Larvae are mainly detritivorous (Anderson et al. 1978, Patterson and Vannote 1979), occurring in pools and slower sections of streams where plant detritus accumulates. *Heteroplectron* larvae are unusual in that they use a piece of wood as a case, excavating a central cavity. Larvae of *Anisocentropus* (fig. 25.98) and *Phylloicus* fasten a few pieces of leaves or bark together to form a central chamber. Cases range up to at least 40 mm in length.

Description: Larvae (fig. 25.97) up to 25 mm in length, sclerotized parts dark or light brown, head without distinctive markings.

Head (fig. 25.44) hypognathous, labrum with anterolateral patches of fine setae, distinctive in having approximately 16 stout setae in a central transverse row; antennae situated approximately midway between eye and anterior margin of head capsule; parietal sclerites separated anteriorly by a triangular ventral apotome.

Thorax (fig. 25.44) with sclerotized plates of pronotum extended anterolaterally, as prominent attenuate lobes in *Anisocentropus* and *Phylloicus:* fore trochantin hook-shaped. Mesonotum with small *sa* 3 sclerites separate from the larger sclerotized plate incorporating *sa* 1 and *sa* 2. Metanotum mainly membranous, a single seta at *sa* 1, one to several setae at *sa* 2, small sclerite bearing several setae at *sa* 3. Hind legs much longer than middle legs in *Anisocentropus,* and the tibia with a secondary subdivision; the two legs approximately equal in length in the other two genera.

Abdomen with lateral protuberances of segment 1 arising more ventrally than in other families; gills single or branched; lateral fringe prominent; lateral tubercles restricted to segment 8.

Comments: The Calamoceratidae are a small and mainly subtropical family. Three genera with five species extend into North America: *Anisocentropus* in the East, *Phylloicus* in the Southwest, and *Heteroplectron* with one eastern and one western species. Larval taxonomy was summarized by Wiggins (1977).

Selected Bibliography

Anderson et al. 1978.
Patterson and Vannote 1979.

MOLANNIDAE

The Molannids

Figures 25.41, 25.42, 25.103, 25.104

Relationships and Diagnosis: Members of the leptocerid branch, molannids are readily distinguished by their specialized hind tarsal claws: in *Molanna* the hind claw is a short, stout, setose lobe with basal seta short (fig. 25.41); in *Molannodes,* the claw is a long, slender filament with a long basal seta (fig. 25.42).

Biology and Ecology: Most molannids occur in lakes or the slower sections of rivers, but *Molanna blenda* is characteristic of pools in small spring runs. The flattened tubular larval case, up to 27 mm in length, is constructed mainly of rock fragments and has a characteristic wide lateral flange extended as a cowl overhanging the anterior opening (fig. 25.104). Concealed by the cowl, the larva can extend its head and legs and move about to feed; thus the insect is difficult to detect until the case moves. Some species in the leptocerid genus *Ceraclea* construct somewhat similar cases (fig. 25.96).

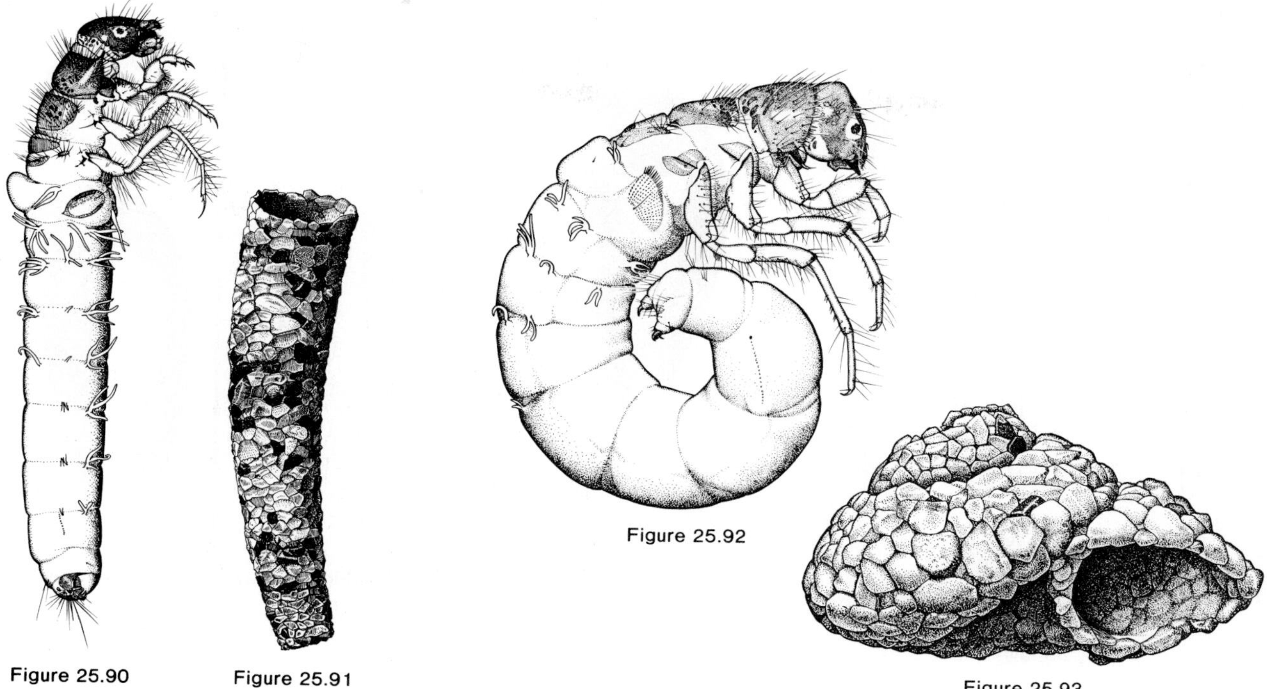

Figure 25.90. Lateral view of *Agarodes* (Sericostomatidae) larva.

Figure 25.91. Larval case of *Agarodes* (Sericostomatidae).

Figure 25.92. Lateral view of *Helicopsyche* (Helicopsychidae) larva.

Figure 25.93. Larval case of *Helicopsyche* (Helicopsychidae).

Larvae burrow into soft substrates for pupation. Molannids share with leptocerids the unusual behavior of expelling the larval exuviae through the posterior opening of the pupal case. The larvae feed on diatoms, filamentous algae, vascular plant detritus, and small invertebrates.

Description: Larvae (fig. 25.103) up to 19 mm in length, head and pronotum usually with dark markings on a lighter background.

Head (fig. 25.41) hypognathous, elongate; parietal sclerites entirely separated by a trapeziform ventral apotome. Tentorial pits of the frontoclypeal sutures clearly visible; antennae close to the anterior margin of the head capsule.

Thorax with a weakly sclerotized mesonotal plate, subdivided by a thin suture either medially or transversely, in the latter case with the anterior sclerite further subdivided by a median suture. Metanotum mostly membranous but with setae in the primary setal areas. Hind legs much longer than the others, the tarsal claw in final instars modified as a short stout lobe (*Molanna*) or a long slender filament (*Molannodes*); tibia of fore leg with an elongate process forming base of distal spur.

Abdomen with protuberances of segment 1 prominent; gills branched or single; lateral fringe well developed; lateral tubercles on segment 8; segment 9 with dorsal sclerite.

Comments: The Molannidae are a small family of two genera in North America: *Molanna* with five species chiefly eastern and northern, and *Molannodes* with one species of sporadic occurrence from Alaska to Ontario. Larval taxonomy is now fairly well established (Wiggins 1977, Sherberger and Wallace 1971).

Selected Bibliography

Sherberger and Wallace 1971.

ODONTOCERIDAE

The Odontocerids

Figures 25.54, 25.55, 25.99, 25.100

Relationships and Diagnosis: Within the leptocerid branch, some larvae of the Odontoceridae are similar to the Sericostomatidae, but they can be distinguished by the lack of a dorsal cluster of setae on the anal proleg (fig. 25.55) and by the absence of an apical hook on the fore trochantin (fig. 25.54). In other features, odontocerid larvae are rather diverse structurally.

Biology and Ecology: Larvae live in running waters, where they burrow in silt or gravel deposits. Odontocerid cases (fig. 25.100) are of sand or rock fragments, and in some genera the silken lining characteristic of most limnephiloid families is replaced by silken brace-bands between adjacent rock pieces. Smaller pieces are also fitted into the interstices from the inside, resulting in a case with unusual resistance to crushing in *Psilotreta*. Larvae of *Psilotreta* pupate in large aggregations on rocks. Food of odontocerids is algae, vascular plant detritus, and small invertebrates.

Description: Larvae (fig. 25.99) up to 20 mm in length, with sclerotized parts usually uniform dark brown in color, the head with contrasting markings in some species.

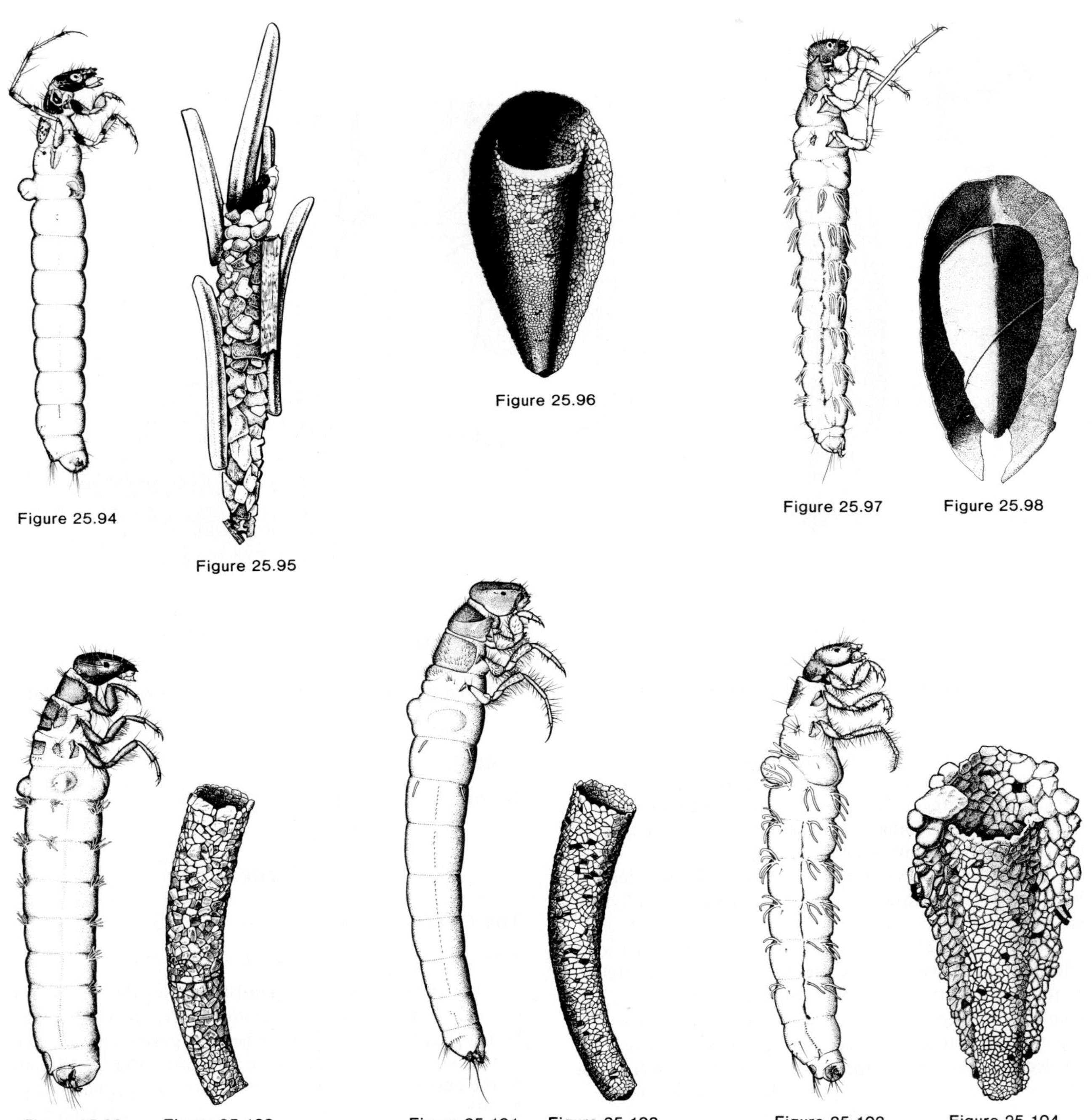

Figure 25.94. Lateral view of *Mystacides* (Leptoceridae) larva.

Figure 25.95. Larval case of *Mystacides* (Leptoceridae).

Figure 25.96. Larval case of *Ceraclea* (Leptoceridae).

Figure 25.97. Lateral view of *Anisocentropus* (Calamoceratidae) larva.

Figure 25.98. Larval case of *Anisocentropus* (Calamoceratidae).

Figure 25.99. Lateral view of *Marilia* (Odontoceridae) larva.

Figure 25.100. Larval case of *Marilia* (Odontoceridae).

Figure 25.101. Lateral view of *Beraea* (Beraeidae) larva.

Figure 25.102. Larval case of *Beraea* (Beraeidae).

Figure 25.103. Lateral view of *Molanna* (Molannidae) larva.

Figure 25.104. Larval case of *Molanna* (Molannidae).

Head hypognathous, elongate in most genera, dorsal aspect rounded in others, an anterolateral ridge variously developed; parietals entirely separated by a vase-shaped ventral apotome in some genera, but in others only partially separated by a triangular apotome, which is minute in *Pseudogoera.*

Thorax (fig. 25.54) with pronotum extended as anterolateral points in some genera, normally rounded in others; large prosternal sclerites; fore trochantin not hooked apically; mesonotum comprising two heavily sclerotized plates in most genera, each plate subdivided in *Marilia* into three small sclerites; mesosternal sclerites present in some genera. Metanotum with small sclerites on primary setal areas variously shaped, discrete in some genera, *sa* 1 and *sa* 2 sclerites fused into single transverse sclerites in others.

Abdomen with gills in tufts; lateral fringe often prominent; lateral tubercles on segment 8; dorsal sclerite on 9. Anal claw not as in other Limnephiloidea, apex less acute, not as strongly hooked, accessory hook usually replaced by several straight spines.

Comments: The Odontoceridae are a small family of 12 species in six genera in North America, three of which are monotypic. Subfamily Pseudogoerinae comprises the unusual southeastern monotypic genus *Pseudogoera* (Wallace and Ross 1971). Larval taxonomy for the Nearctic fauna was summarized by Wiggins (1977), and for larvae of the genus *Psilotreta* by Parker and Wiggins (in press).

Selected Bibliography

Parker and Wiggins in press.
Wallace and Ross 1971.

BIBLIOGRAPHY

Anderson, N. H. 1967. Life cycle of a terrestrial caddisfly, *Philocasca demita* (Trichoptera: Limnephilidae), in North America. Ann. Ent. Soc. Amer. 60:320–23.

Anderson, N. H. 1974. The eggs and oviposition behavior of *Agapetus fuscipes* Curtis (Trich., Glossosomatidae). Ent. Mon. Mag. 109:129–31.

Anderson, N. H., and E. Grafius. 1975. Utilization and processing of allochthonous material by stream Trichoptera. Verh. Int. Verein. Limnol. 19:3083–88.

Anderson, N. H., J. R. Sedell, L. M. Roberts, and J. F. Triska. 1978. The role of aquatic invertebrates in processing of wood debris in coniferous forest streams. Amer. Midl. Nat. 100:64–82.

Balduf, W. V. 1939. The bionomics of entomophagous insects, Pt. II. St. Louis: John S. Swift. 384 pp.

Betten, C. 1934. The caddis flies or Trichoptera of New York State. Bull. N.Y. St. Mus. 292. 576 pp.

Bjarnov, N., and J. Thorup. 1970. A simple method for rearing running-water insects, with some preliminary results. Arch. Hydrobiol. 67:201–209.

Chapin, J. W. 1978. Systematics of Nearctic *Micrasema* (Trichoptera: Brachycentridae). Ph.D. diss., Clemson University, Clemson, S. C.

Coffman, W. P., K. W. Cummins, and J. C. Wuycheck. 1971. Energy flow in a woodland stream ecosystem: I. Tissue support trophic structure of the autumnal community. Arch. Hydrobiol. 68:232–76.

Crichton, Ian M. (ed.) 1978. Proceedings of the Second International Symposium on Trichoptera, University of Reading, England. The Hague: W. Junk. 359 pp.

Fischer, F. C. J. 1960–73. Trichopterorum catalogus. Amsterdam: Nederlandse Entomologische Vereeniging (vol. 1, 1960; vol. 2, 1961; vol. 3, 1962; vol. 4, 1963; vol. 5, 1964; vol. 6, 1965; vol. 7, 1966; vol. 8, 1967; vol. 9, 1968; vol. 10, 1969; vol. 11, 1970; vol. 12, 1971; vol. 13, 1972; vol. 14, 1972; vol. 15, 1973.)

Edwards, S. W. 1961. The immature stages of *Xiphocentron mexico* (Trichoptera). Tex. J. Sci. 13:51–56.

Elliott, J. M. 1969. Life history and biology of *Sericostoma personatum* Spence (Trichoptera). Oikos 20:110–18.

Flint, O. S. 1956. The life history and biology of the genus *Frenesia* (Trichoptera: Limnephilidae). Bull. Brooklyn Ent. Soc. 51:93–108.

Flint, O. S. 1960. Taxonomy and biology of Nearctic limnephilid larvae (Trichoptera), with special reference to species in eastern United States. Entomologica Am., n.s. 40:1–117.

Flint, O. S. 1961. The immature stages of the Arctopsychinae occurring in eastern North America (Trichoptera: Hydropsychidae). Ann. Ent. Soc. Am. 54:5–11.

Flint, O. S. 1962. Larvae of the caddis fly genus *Rhyacophila* in eastern North America (Trichoptera: Rhyacophilidae). Proc. U.S. Nat. Mus. 113:465–93.

Flint, O. S. 1964a. Notes on some Nearctic Psychomyiidae with special reference to their larvae (Trichoptera). Proc. U.S. Nat. Mus. 115:467–81.

Flint. O. S. 1964b. The caddisflies (Trichoptera) of Puerto Rico. Univ. Puerto Rico, Agric. Exp. Sta., Tech. Pap. 40.

Flint, O. S. 1973. Studies of Neotropical caddisflies, XVI: the genus *Austrotinodes* (Trichoptera: Psychomyiidae). Proc. Biol. Soc. Wash. 86:127–42.

Flint, O. S. 1984. The genus *Brachycentrus* in North America, with a proposed phylogeny of the genera of Brachycentridae (Trichoptera). Smithsonian Contr. Zool. No. 398, 58 pp.

Givens, D. R., and S. D. Smith. 1980. A synopsis of the western Arctopsychinae (Trichoptera: Hydropsychidae). Melanderia 35:1–24.

Haddock, J. D. 1977. The biosystematics of the caddis fly genus *Nectopsyche* in North America with emphasis on the aquatic stages. Am. Midl. Nat. 98:382–421.

Hamilton, S. W. 1985. The larva and pupa of *Beraea gorteba* Ross (Trichoptera: Beraeidae). Proc. Ent. Soc. Wash. 87:783–89.

Hamilton, S. W., and R. W. Holzenthal. 1984. The caddisfly genus *Helicopsyche* in America north of Mexico (Trichoptera: Helicopsychidae). [Abstract]. p. 167 *In* Morse, J. C. (ed.) 1984.

Hiley, P. D. 1969. A method of rearing Trichoptera larvae for taxonomic purposes. Ent. Mon. Mag. 105:278–79.

Hudson, P. L., J. C. Morse, and J. R. Voshell. 1981. Larva and pupa of *Cernotina spicata.* Ann. Ent. Soc. Am. 74(5):516–19. (Polycentropodidae).

Jansson, A., and T. Vuoristo. 1979. Significance of stridulation in larval Hydropsychidae (Trichoptera). Behaviour 71:167–86.

Johnstone, G. W. 1964. Stridulation by larval Hydropsychidae (Trichoptera). Proc. Roy. Ent. Soc. Lond. (A) 39:146–50.

Lepneva, S. G. 1964. Fauna SSSR, Rucheiniki, vol. 2, no. 1. Lichinki i kukolki podotryada kol'chatoshchupikovykh. Zoologicheskii Institut Akademii Nauk SSSR, n.s. 88. 560 pp. (In Russian. Translated into English as: Fauna of the U.S.S.R.; Trichoptera, vol. 2, no. 1. Larvae and Pupae of Annulipalpia. Published by the Israel Program for Scientific Translations, 1970.)

Lepneva, S. G. 1966. Fauna SSSR, Rucheiniki, vol. 2, no. 2. Lichinki i kukolki podotryada tsel'noshchupikovykh. Zoologicheskii Institut Akademii Nauk SSSR, n.s. 95. 560 pp. (In Russian. Translated into English as Fauna of the U.S.S.R.; Trichoptera, vol. 2, no. 2. Larvae and Pupae of Integripalpia. Published by the Israel Program for Scientific Translations, 1971.)

Mackay, R. J. 1981. A miniature laboratory stream powered by air bubbles. Hydrobiologia, 83:383–85.

Malicky, H. 1973. Trichoptera (Köcherfliegen). Handb. Zool. 4(2). 114 pp.

Malicky, H. 1974 (ed.) Proceedings of the First International Symposium on Trichoptera. Lunz am See, Austria. The Hague: W. Junk. 213 pp.

Marshall, J. E. 1979. A review of the genera of the Hydroptilidae (Trichoptera). Bull. Brit. Mus. (Nat. Hist.) 39:135–239.

Matsuda, R. 1965. Morphology and evolution of the insect head. Mem. Amer. Ent. Inst. 4, 334 pp.

Matsuda, R. 1976. Morphology and evolution of the insect abdomen, with special reference to developmental patterns and their bearing on systematics. Pergamon Press, New York. 532 pp.

McEwen, E. 1980. Larvae of the genus *Agarodes* (Trichoptera: Sericostomatidae). M.Sc. thesis. Clemson Univ., Clemson, S.C.

Merrill, D. 1969. The distribution of case recognition in ten families of caddis larvae (Trichoptera). Anim. Behav. 17:486–93.

Moretti, G. (ed.). 1981. Proceedings of the Third International Symposium on Trichoptera, University of Perugia, Italy. Series Entomologica, vol. 20, The Hague: W. Junk. 472 pp.

Morse, J. C. (ed.). 1984. Proceedings of the Fourth International Symposium on Trichoptera. Clemson University, S.C. The Hague: W. Junk. 486 pp.

Morse, J. C. and R. W. Holzenthal. 1984. Trichoptera genera, Chap. 17. *In* Merritt, R. W. and K. W. Cummins (eds.). An introduction to the aquatic insects of North America. 2nd ed. Kendall/Hunt Publishing Company, Dubuque, Iowa. 722 pp.

Nielsen, A. 1948. Postembryonic development and biology of the Hydroptilidae. Kgl. Danske Vidensk. Selsk. Biol. Skr. 5. 200 pp.

Novák, K., and F. Sehnal. 1963. The development cycle of some species of the genus *Limnephilus* (Trichoptera). Cas. Cs. Spol. Ent. 60:68–80.

Nüske, H., and W. Wichard. 1972. Die Analpapillen der Köcherfliegenlarven. II. Feinstruktur des ionen-transportierenden und respiratorischen Epithels bei Glossosomatiden. Cytobiologie 6:243–249.

Parker, C. R., and G. B. Wiggins. 1985. The Nearctic caddisfly genus *Hesperophylax* Banks (Trichoptera: Limnephilidae). Can. J. Zool. 63:2443–72.

Parker, C. R., and G. B. Wiggins. In press. Revision of the caddisfly genus *Psilotreta* (Trichoptera: Odontoceridae). Life Sci. Contr., Roy. Ont. Mus.

Patterson, J. W., and R. L. Vannote. 1979. Life history and population dynamics of *Heteroplectron americanum*. Environ. Ent. 8:665–669.

Peck, D. L., and S. D. Smith. 1978. A revision of the *Rhyacophila coloradensis* complex (Trichoptera: Rhyacophilidae). Melanderia 27:1–24.

Philipson, G. N. 1953. The larva and pupa of *Wormaldia subnigra* McLachlan (Trichoptera: Philopotamidae). Proc. Roy. Ent. Soc. Lond. (A) 28:57–62.

Resh, V. H. 1972. A technique for rearing caddisflies (Trichoptera). Canad. Ent. 104:1959–61.

Resh, V. H. 1976. The biology and immature stages of the caddisfly genus *Ceraclea* in eastern North America (Trichoptera: Leptoceridae). Ann. Ent. Soc. Amer. 69:1039–61.

Ross, H. H. 1944. The caddis flies, or Trichoptera, of Illinois. Bull. Ill. Nat. Hist. Surv. 23. 326 pp.

Ross, H. H. 1956. Evolution and classification of the mountain caddisflies. Urbana, Ill.: Univ. Illinois Press. 213 pp.

Ross, H. H. 1964. Evolution of caddisworm cases and nets. Amer. Zoologist 4:209–20.

Ross, H. H. 1967. The evolution and past dispersal of the Trichoptera. Ann. Rev. Ent. 12:169–206.

Ross, H. H., and D. G. Gibbs. 1973. The subfamily relationships of the Dipseudopsinae (Trichoptera, Polycentropodidae). J. Georgia Ent. Soc. 8:312–16.

Ross, H. H., and J. B. Wallace, 1974. The North American genera of the family Sericostomatidae (Trichoptera). J. Georgia Ent. Soc. 9:2–48.

Schefter, P. W., and G. B. Wiggins. 1986. Larval systematics of the Nearctic species in the *Hydropsyche morosa* group (Trichoptera: Hydropsychidae). Life Sci. Misc. Publ., Roy. Ont. Mus.

Schmid, F. 1980. Genera des Trichoptères du Canada et des États adjacents. Les insectes et arachnides du Canada, Partie 7. Agric. Canada Publ. 1692. 296 pp.

Schuster, G. A., and D. A. Etnier. 1978. A manual for the identification of the larvae of the caddisfly genera *Hydropsyche* Pictet and *Symphitopsyche* Ulmer in eastern and central North America (Trichoptera: Hydropsychidae). U.S. Environ. Prot. Agency 600/4–78–060, Cincinnati. 129 pp.

Sherberger, F. F., and J. B. Wallace. 1971. Larvae of the southeastern species of *Molanna*. J. Kans. Ent. Soc. 44:217–24.

Smith, S. D. 1968a. The Arctopsychinae of Idaho (Trichoptera: Hydropsychidae). Pan-Pac. Ent. 44:102–12.

Smith, S. D. 1968b. The *Rhyacophila* of the Salmon River drainage of Idaho with special reference to larvae. Ann. Ent. Soc. Amer. 61:655–74.

Smith, S. D. 1984a. Larvae of Nearctic *Rhyacophila,* part I: *acropedes* group. Aquatic Insects. 6:37–40.

Smith, S. D. 1984b. Larvae of Nearctic *Rhyacophila* (Trichoptera: Rhyacophilidae) II: *Rhyacophila lieftincki* group. J. Kansas Ent. Soc. 57:540–42.

Thut, R. N. 1969. Feeding habits of larvae of seven *Rhyacophila* (Trichoptera: Rhyacophilidae) species with notes on other life-history features. Ann. Ent. Soc. Amer. 62(4):894–98.

Wallace, J. B. 1975a. The larval retreat and food of *Arctopsyche;* with phylogenetic notes on feeding adaptations in Hydropsychidae larvae (Trichoptera). Ann. Ent. Soc. Amer. 68:167–73.

Wallace, J. B. 1975b. Food partitioning in net-spinning Trichoptera larvae: *Hydropsyche venularis, Cheumatopsyche etrona,* and *Macronema zebratum* (Hydropsychidae). Ann. Ent. Soc. Amer. 68:463–72.

Wallace, J. B., and D. Malas. 1976. The fine structure of capture nets of larval Philopotamidae (Trichoptera), with special emphasis on *Dolophilodes distinctus.* Can. J. Zool. 54:1788–1802.

Wallace, J. B., and H. H. Ross. 1971. Pseudogoerinae: a new subfamily of Odontoceridae (Trichoptera). Ann. Ent. Soc. Amer. 64:890–94.

Wallace, J. B., W. R. Woodall, and A. A. Staats. 1976. The larval dwelling-tube, capture net and food of *Phylocentropus placidus* (Trichoptera: Polycentropodidae). Ann. Ent. Soc. Amer. 69:149–54.

Waltz, R. D., and W. P. McCafferty. 1983. *Austrotinodes* Schmid (Trichoptera: Psychomyiidae), a first U.S. record from Texas. Proc. Ent. Soc. Wash. 85:181–82.

Weaver, J. S., B. G. Swegman, and J. L. Sykora. 1979. The description of immature forms of *Aphropsyche monticola* Flint (Trichoptera: Hydropsychidae). Aquatic Insects 1:143–48.

Weaver, J. S., and J. L. Sykora. 1979. The *Rhyacophila* of Pennsylvania, with larval descriptions of *R. banksi* and *R. carpenteri* (Trichoptera: Rhyacophilidae). Ann. Carnegie Mus. 48:403–23.

Wichard, W. 1976. Morphologische Komponenten bei der Osmoregulation von Trichopterenlarven, pp. 171–177. *In* Malicky, H. (ed.). Proc. First Int. Symp. Trichoptera. Lunz am See (Austria), 1974. The Hague: W. Junk.

Wichard, W. and H. Komnick. 1973. Fine structure and function of the abdominal chloride epithelia in caddisfly larvae. Z. Zellforsch. 136:579–90.

Wiggins, G. B. 1954. The caddisfly genus *Beraea* in North America (Trichoptera). Contr. Roy. Ont. Mus. Zool. Palaeont. 39. 13 pp.

Wiggins, G. B. 1959. A method of rearing caddisflies (Trichoptera). Can. Ent. 91:402–05.

Wiggins, G. B. 1960. A preliminary systematic study of the North American larvae of the caddisfly family Phryganeidae (Trichoptera). Can. J. Zool. 38:1153–70.

Wiggins, G. B. 1962. A new subfamily of phryganeid caddisflies from western North America (Trichoptera: Phryganeidae). Can. J. Zool. 40:879–91.

Wiggins, G. B. 1973a. A contribution to the biology of caddisflies (Trichoptera) in temporary pools. Life Sci. Contr., Roy. Ont. Mus. 88. 28 pp.

Wiggins, G. B. 1973b. New systematic data for the North American caddisfly genera *Lepania, Goeracea* and *Goerita* (Trichoptera: Limnephilidae). Life Sci. Contr., Roy. Ont. Mus. 91. 33 pp.

Wiggins, G. B. 1976. Contributions to the systematics of the caddis-fly family Limnephilidae (Trichoptera). III: The genus *Goereilla,* pp. 7–19. *In* Malicky, H. (ed.). Proc. First Int. Symp. Trichoptera. Lunz am See (Austria), 1974. The Hague: W. Junk.

Wiggins, G. B. 1977. Larvae of the North American caddisfly genera (Trichoptera). Toronto: Univ. Toronto Press. 401 pp.

Wiggins, G. B. 1981. Considerations on the relevance of immature stages to the systematics of Trichoptera, pp. 397–409. *In* Moretti, G. P. (ed.). Proc. Third Int. Symp. Trichoptera. Series Entomologica 20. The Hague: W. Junk.

Wiggins, G. B. 1982. Trichoptera, pp. 599–612. *In* Parker, S. P. (ed.). Synopsis and classification of living organisms. New York: McGraw-Hill.

Wiggins, G. B. 1984. Trichoptera. Chap. 16. *In* Merritt, R. W. and K. W. Cummins (eds.). An introduction to the aquatic insects of North America, 2d ed. Dubuque, Ia.: Kendall/Hunt Publishing Company. 722 pp.

Wiggins, G. B., and N. H. Anderson. 1968. Contributions to the systematics of the caddisfly genera *Pseudostenophylax* and *Philocasca* with special reference to the immature stages (Trichoptera: Limnephilidae). Can. J. Zool. 46:61–75.

Wiggins, G. B., and R. J. Mackay. 1978. Some relationships between systematics and trophic ecology in Nearctic aquatic insects, with special reference to Trichoptera. Ecology 59:1211–20.

Wiggins, G. B., R. J. Mackay, and I. M. Smith. 1980. Evolutionary and ecological strategies of animals in annual temporary pools. Arch. Hydrobiol./Suppl. 58:97–206.

Wiggins, G. B., J. S. Weaver and J. D. Unzicker. 1985. Revision of the caddisfly family Uenoidae (Trichoptera). Can. Ent. 117:763–800.

Wiggins, G. B., and J. S. Richardson. 1982. Revision and synopsis of the caddisfly genus *Dicosmoecus* (Trichoptera: Limnephilidae; Dicosmoecinae). Aquatic Insects 4:181–217.

Williams, N. E., and G. B. Wiggins. 1981. A proposed setal nomenclature and homology for larval Trichoptera, pp. 421–29. *In* Moretti, G. P. (ed.). Proc. Third Int. Symp. Trichoptera. Series Entomologica 20. The Hague: W. Junk.

Williams, D. D., and N. E. Williams. 1975. A contribution to the biology of *Ironoquia punctatissima* (Trichoptera: Limnephilidae). Can. Ent. 107:829–32.

Yamamoto, T., and G. B. Wiggins. 1964. A comparative study of the North American species in the caddisfly genus *Mystacides* (Trichoptera: Leptoceridae). Can. J. Zool. 42:1105–26.

Order Lepidoptera[1]

26

Frederick W. Stehr, Coordinator
Michigan State University

BUTTERFLIES, SKIPPERS, AND MOTHS

The order Lepidoptera contains well over 100,000 species and more than 100 families worldwide. In America north of Mexico there are over 11,000 species and about 80 families, according to the "Check List of the Lepidoptera of America North of Mexico" (Hodges et al. 1983).

Considering their relatively large size and abundance, Lepidoptera larvae (generally known as caterpillars) are not well known, ranging from families like the better-known Saturniidae, Sphingidae, Nymphalidae, and Papilionidae that contain many large and conspicuous species whose larvae are reliably associated with the adults, to many families of the microlepidoptera in which considerably fewer larvae are associated with adults.

Proper preservation is critical because the numbers and relationships of setae and crochets are difficult or impossible to see on shrivelled or darkened larvae; the use of KAAD or hot water for killing, and other solutions and procedures as outlined in the "techniques" section that distend and straighten larvae, is essential. The colors of living, externally feeding caterpillars are frequently striking and often useful in identification, but, unfortunately, current preservation methods (except for freeze-drying) destroy or obscure most colors.

DIAGNOSIS

The external structures of Lepidoptera larvae are diverse, yet throughout the order there is a remarkable adherence to a fundamental plan that enables one to easily place most larvae to order. As a general model, larvae of Lepidoptera possess a distinct head, chewing mouthparts, one pair of short antennae, six stemmata (simple eyes), adfrontal areas, a protruding labial spinneret, three pairs of thoracic legs, ten abdominal segments with crochet-bearing prolegs on 3, 4, 5, 6, and 10, and spiracles on the prothorax and abdominal segments 1–8.

1. The comments and suggestions of John G. Franclemont on the entire Lepidoptera section exclusive of the family figures and those of George L. Godfrey on the introductory part are gratefully acknowledged.

BIOLOGY AND ECOLOGY

Caterpillars are perhaps the most commonly encountered immature insects since they are primarily phytophagous and many are exposed feeders. Virtually all plants are attacked (including poison ivy), and many species are important pests. In addition to chewing, boring, mining, or forming galls on living plants, some are scavengers on dead plant materials, and some consume stored products. A few groups (especially Tineidae) feed on animal materials such as feathers, wool, or fur. Epipyropidae are ectoparasitoids on Homoptera. A few are predators, with the largest number in the Lycaenidae, but with the most unusual ones being the Hawaiian geometrids that strike backward, seizing with their enlarged thoracic legs any prey that touches their rear end (Montgomery 1982).

Exposed feeders are heavily parasitized and preyed upon; as a result many unusual adaptations and habits have evolved. These range from simple cryptic coloration that matches the background, to such things as the very realistic twig mimicry and behavior found in some geometrids (complete with "bark irregularities," "scars," "stipules," etc.). Other caterpillars may be warningly colored and/or bear diverse lobes, spines, horns, knobs, and urticating hairs or spines. Many of the hairy or spiny exposed-feeding macrolepidoptera are physically or chemically irritating or cause allergic reactions in susceptible individuals, whereas most of the microlepidoptera feed in protected or concealed situations and do not possess these defenses. The ill-smelling secretions of the eversible, dorsal prothoracic glands (osmeteria) of papilionids and parnassians are well known, as are the irritating properties of saddleback caterpillars (Limacodidae) and the apparent distastefulness of many monarch larvae to birds. Certain notodontids spray acid from a ventral prothoracic gland toward predators and parasitoids, and the dorsal glands of lymantriids are believed to be protective. Some caterpillars rear up and shake their heads vigorously from side to side when disturbed; others drop to the ground or spin down on silken threads. In general, the structural and behavioral adaptations of caterpillars are superficially known, and even in the relatively conspicuous, abundant, and important groups, much remains to be discovered.

Caterpillars spin silk through a conspicuous labial spinneret. Species in many different taxa use silk in diverse ways—from webbing together leaves or other materials (Tortricidae,

Pyralidae, and others) to constructing large "tents" (*Malacosoma* (Lasiocampidae)) or webs (fall webworm (Arctiidae)) or silken tube shelters (Crambinae (Pyralidae), Acrolophinae (Tineidae)); some even make complex bags or shelters that they carry with them (Coleophoridae, Psychidae, Mimallonidae). Many caterpillars spin cocoons in which they pupate, ranging from the tight and commercially valuable silkworm cocoons, to those of the gypsy moth, which are at best a few strands of silk. Others pupate in the ground or litter (most Sphingidae, Notodontidae, Citheroniinae (Saturniidae), many Noctuidae and others). Butterfly pupae (chrysalids) frequently are naked and suspended from a silken pad by the hooked cremaster. In some species the chrysalis is held upright by a band of silk around the middle.

Most Lepidoptera overwinter in immature stages, with a few butterflies (some nymphalids and the monarch, which migrates south) and some moths (especially noctuids) going through as adults. Most overwinter as various larval instars, prepupae, or pupae, but some as eggs.

The eggs of Lepidoptera are diverse in structure, abundance, and the ways they are laid. Some, such as certain Australian hepialids, are numerous and scattered on the ground by the females in flight (Common 1970); others are laid singly on carefully selected parts of hosts (many papilionids). However, most species lay one or more egg masses that range from loose aggregations to tight masses laid in regular formation that may or may not be protected in some way on hosts. A few lay egg masses in the soil. Considerably less is known about the eggs of Lepidoptera than about the larvae, and although the eggs of selected groups are relatively well known, no comprehensive study of eggs has been made in North America. Hinton's (1979) 1500-page monograph on insect eggs is barely an introduction to Lepidoptera eggs, with only eight pages. Doring's (1955) work on the eggs of central European Lepidoptera and that for Belgium by Sarlet (1964 and other papers since 1949) are much more comprehensive. Salkeld (1983, 1984) has recently illustrated the eggs of 131 and 124 Canadian Geometridae and Noctuidae respectively.

Pupae of Lepidoptera were studied by Mosher (1916a), but nothing comprehensive has been done since then. All except the more primitive groups have obtect pupae and are without functional mandibles. The more primitive groups may have exarate pupae, but only in the suborders Zeugloptera and Dacnonypha are functional (decticous) mandibles found that assist in emergence from the cocoon. Many larvae spin cocoons of various kinds, often incorporating leaves or other materials into them; others pupate in earthen cells or protected locations where little or no cocoon is spun. Emergence from cocoons is commonly through prespun one-way escape hatches, through caplike flaps, by secretion of a fluid that softens or partially dissolves part of the cocoon, or by cutting or forcing through the wall of the cocoon with sharp structures on the pupal head that are moved by the pharate adult inside the pupal skin. Generally, if the pupa moves, it only partially protrudes from the cocoon, being held in place by forward-projecting spines near the rear that anchor it in the cocoon, enabling the adult to pull out of the pupal skin more easily.

DESCRIPTION

Head

The caterpillar head (figs. 26.1, 26.2, 26.3) is a sclerotized capsule with a large occipital foramen strengthened dorsally by a ⅄ -shaped internal ridge. Hinton's (1947) interpretation of the head sutures is followed. The anterior arms of the ⅄ -shaped ridge extend anteriorly and are represented externally by the lateral adfrontal sutures. The stem, or epicranial suture (=median adfrontal suture) extends dorsally to the epicranial notch (=vertical notch or vertical triangle). In all but the last instar the head capsule is shed whole at ecdysis, but at pupation the last instar head capsule splits along the ecdysial lines of weakness lateral to the adfrontal sutures. The areas between the ecdysial lines and the adfrontal sutures are the adfrontal areas. Between the lateral adfrontal sutures is a triangular combined frons and clypeus, or frontoclypeus. The frontoclypeal suture is sometimes distinct, in which case the frons (=front) and clypeus can be distinguished. Anterior to the frontoclypeus is the anteclypeus, which articulates with the labrum.

The mandibles are opposable, toothed, and conspicuous in the vast majority of the Lepidoptera, but along with the other mouthparts, they may be greatly modified in the sap-feeding instars of leaf miners. The maxilla is well developed, usually consisting of a cardo, stipes, and a three-segmented palpus that terminates in a medial lobe provided with several sensory cones and an antennalike apical lobe. The labium usually consists of three distinguishable parts, the basal submentum (=postmentum), the mentum, and the prementum, which bears the median spinneret distally and the minute, usually two-segmented labial palpi laterally (sometimes absent). The spinneret is absent in Micropterigidae and some internal feeders. The prementum joins with the hypopharynx at the front of the mouth (fig. 26.4).

Antennae are usually quite short (longer in Micropterigidae, Limacodidae, and others with retractile heads, and reduced in some internal feeders), three-segmented, and usually arise from a U-shaped or △-shaped membranous area adjacent to the base of the mandibles.

The relative length of the epicranial suture is related to the orientation of the mouthparts. Typical of many macrolepidoptera, the mouthparts in the hypognathous head are directed downward, perpendicular to the long axis of the body, and the epicranial suture is long relative to the length of the frontoclypeus. In the semiprognathous head, typical of most internally feeding microlepidoptera, the mouthparts are directed more forward, at a shallow angle to the long axis of the body, and the frontoclypeus is longer than the epicranial suture. In the prognathous head, typical of leaf miners, the mouthparts are directed forward, parallel to the long axis of the body, and the apex of the frontoclypeus may reach the epicranial notch, resulting in loss of the epicranial suture. In some leaf miners the change is such that the frontoclypeus assumes a square or rectangular shape. The mouthparts, especially the mandibles and hypopharynx (fig. 26.4), provide many structures that are primarily useful below the family level.

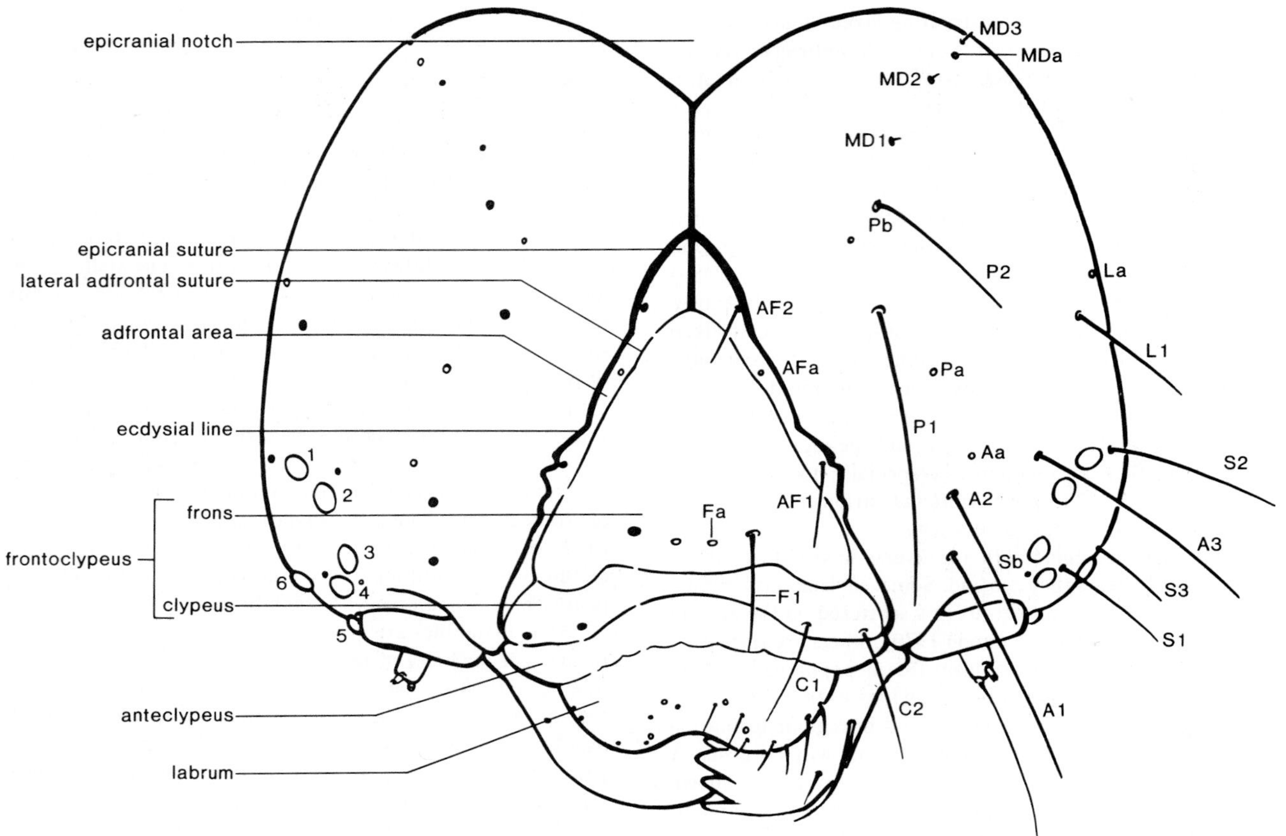

Figure 26.1. Head capsule, frontal view. See pages 299-301 and table 26.2.

Caterpillars typically have six stemmata (formerly termed ocelli by many workers and ommatidia by others) on each side of the head, arranged in a semicircle adjacent to the antennal sockets and numbered 1–6 beginning dorsally (figs. 26.1, 26.2, 26.3). Two common variations in the arrangement of the stemmata are as follows: stemmata 1 and 2 are sometimes close together and separated from the others (many Lyonetiidae, Gracillariidae, Elachistidae, and Tineidae); stemma 5 is sometimes separated from the others, as in many macrolepidoptera, and may be difficult to see when it is very low. Usually, all six stemmata are nearly equal in size, but in some butterflies such as the Satyrinae (Nymphalidae), the third stemma is about twice the size of the others. In other butterfly larvae stemmata 2, 3, 4, and 5 may be elevated, and in the moth family Apatelodidae stemma 3 is sometimes conspicuously more convex and distinct than the others.

Reductions in number of stemmata are common in leaf-mining families, and Nepticulidae, Eriocraniidae, and Opostegidae have only one on each side. Reductions are also found in other families, including the Heliozelidae, Elachistidae, Oinophilidae, and Gracillariidae. Many Prodoxidae, Lyonetiidae, and Tineidae have only five stemmata, and a few Tineidae have none.

Thorax

The prothorax (T1) is fundamentally different from the meso- (T2) and metathorax (T3), the most striking differences being the presence of a spiracle and the T1 shield. Setal differences also exist and are described under the section on chaetotaxy.

Thoracic Legs

Caterpillars typically have three pairs of thoracic legs, each leg consisting of a coxa, trochanter, femur, tibia, tarsus, and claw. Some workers (e.g. Snodgrass, 1954) believe they are homologous with the adult legs, but Birket-Smith (1984) terms them prolegs, believes they are segmentally homologous with the abdominal prolegs, and have nothing to do with adult legs.

In families with specialized habits (leaf miners), thoracic legs are rudimentary or lost. In the Micropterigidae the coxa, trochanter, and femur are fused and the leg appears three-segmented. In some other families, the trochanter may be fused with the femur so the leg appears four-segmented.

Abdomen

The abdomen consists of ten segments, which can be divided into groups of roughly similar segments (A1 and A2), (A3–A6), (A7 and A8), (A9) and (A10). A1–8 bear spiracles. The T1 and A8 spiracles are often larger than the others

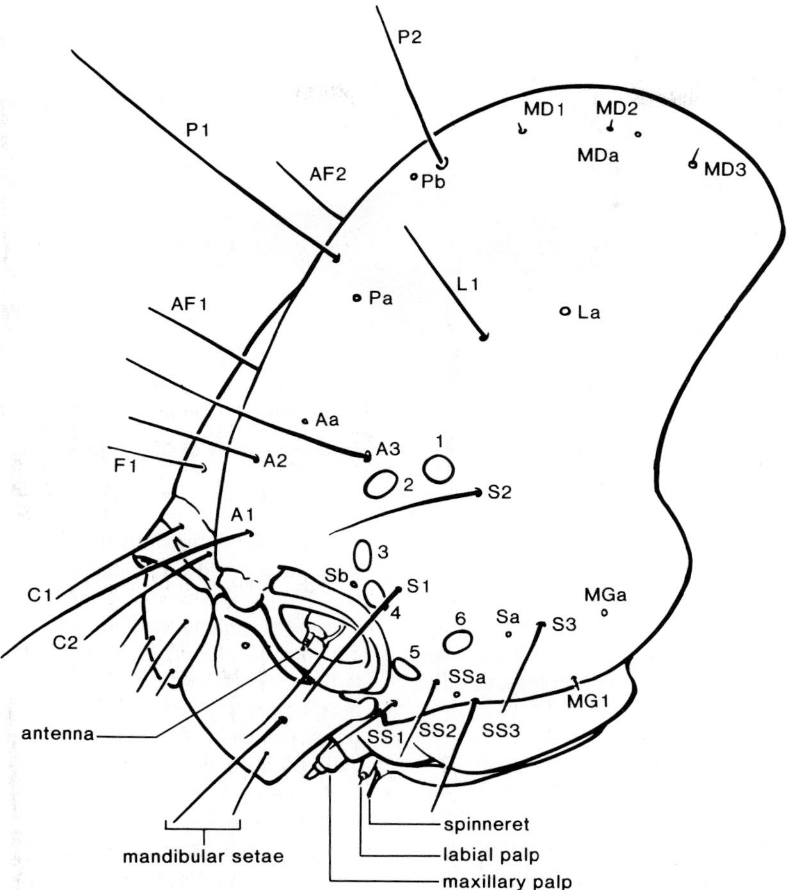

Figure 26.2. Head capsule, lateral view. See pages 299–301 and table 26.2.

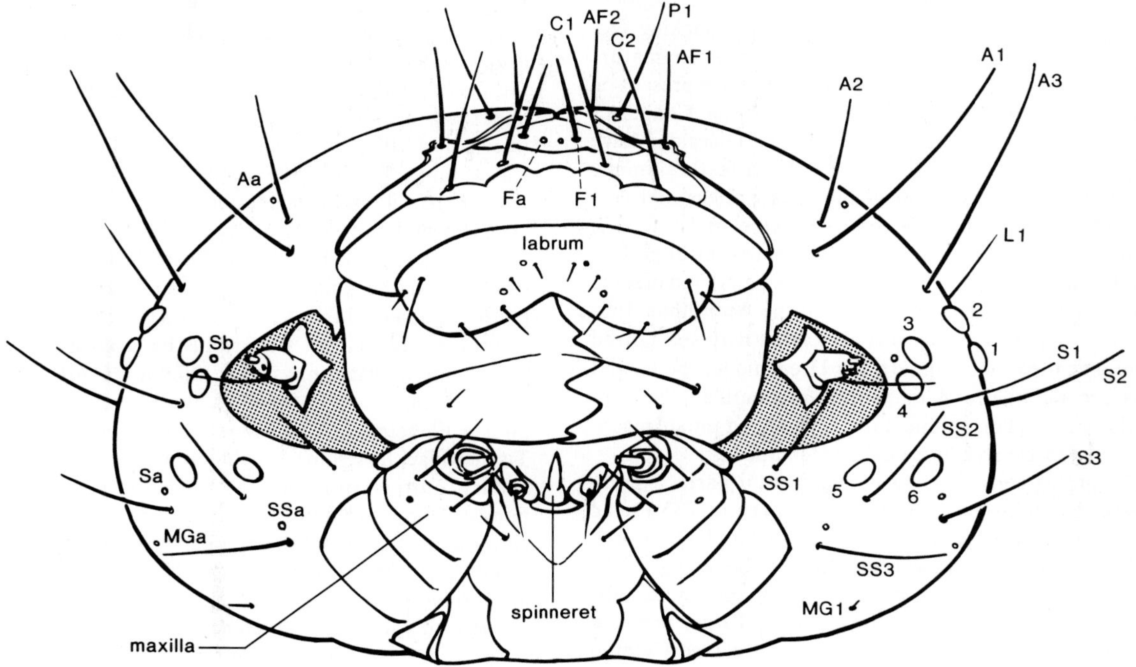

Figure 26.3. Head capsule, ventral mouth area. See pages 299–301 and table 26.2.

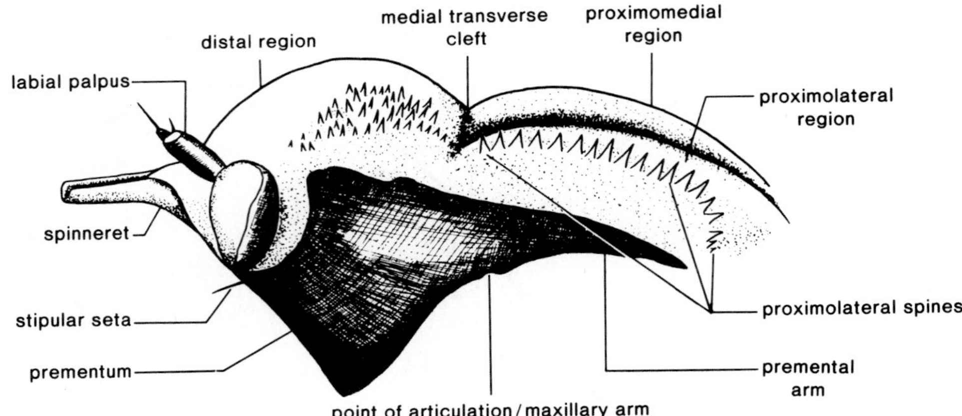

Figure 26.4. Hypopharyngeal complex. Courtesy of G. L. Godfrey (1972).

and similar in size. A1 and A2 usually do not bear prolegs; A1 does not bear crochets, and only in the later instars of the family Dalceridae does A2 bear any crochets; A3–6 bear the ventral prolegs, usually with crochets, A9 never bears prolegs, and only in the later instars of the family Dalceridae does A7 bear crochets; A10 bears the caudal prolegs, usually with crochets.

Prolegs

Prolegs are paired ventral muscled outgrowths of the body wall. Only the most primitive family, the Micropterigidae, have unmuscled prolegs. Most Lepidoptera larvae have prolegs with crochets on A3–6 (ventral prolegs) and A10 (caudal prolegs), but there are important exceptions. Prolegs show a wide range of development. In many leaf miners and internally feeding groups, they are absent or rudimentary, and the crochets may also be present or absent. In the Nepticulidae prolegs are lost, but "ambulatory warts" lacking crochets have evolved on A2–7 (Hinton 1955a). In mature gracillariid larvae, prolegs are lost on A6 but are present on earlier instars. The Megalopygidae have lobes on A2 and A7 resembling prolegs but lack crochets. The Dalceridae have small prolegs with crochets on A2 and A7, a feature unique among the Lepidoptera. Limacodids lack prolegs and crochets, and some have median ventral suckers on A1–7. The Nolidae lack prolegs on A3. The loss of some ventral prolegs in most Geometridae (usually the first three pairs) and in some Noctuidae (the first one or two pairs in the Acontiinae, Plusiinae, and some Catocalinae) is associated with specialized locomotion called "looping." In the Drepanidae, the caudal prolegs are nublike, and in many Notodontidae they are somewhat reduced or rudimentary, or modified into elongate alarm or defensive structures called stemapoda.

The proleg consists of two fundamental parts, a proximal base, distinguished by bearing the subventral (SV) setae (and sometimes secondary setae), and a distal planta, *which never bears setae* and from which the crochets arise (fig. 26.5). Contraction of retractor muscles inserted on the planta pulls its center inward, disengaging the crochets. Turgor pressure pushes the planta out once the muscles are relaxed (Hinton 1955a).

Two fundamentally different kinds of prolegs are found, depending on the relationship between the proleg base and the planta. In the first kind (found in groups such as the Tineoidea, Yponomeutoidea, Cossoidea, Tortricoidea, Gelechioidea, Copromorphoidea, and Pyraloidea) the proleg base is little more than a ring encircling the planta. The planta may be greatly reduced so the crochets appear to arise directly from the venter (fig. 26.6), or it may be elongate and cylindrical (fig. 26.7), but the *planta never bears setae.* In this kind of proleg, the crochets are usually in transverse bands, a complete or incomplete circle, or a mesopenellipse. In the second kind (found in groups such as the Papilionoidea, Geometroidea, Bombycoidea, Sphingoidea, and Noctuoidea) the proleg base is elongate and forms the greater part of the proleg (fig. 26.5). It bears the SV group (fig. 26.8) and/or numerous secondary setae (fig. 26.5) and often has a partially sclerotized area laterally. The planta is reduced to a lobe at the distal end of the proleg base, with the crochets usually in a mesoseries. These two kinds of prolegs parallel the feeding habits. The first (especially those with crochets appearing to arise directly from the venter) is typical of many internally feeding caterpillars that bore, live in nests, folded leaves, mines, etc., and the second is typical of many external feeders that climb on vegetation.

Crochets are small hooks arranged in rows or circles around the edge of the planta. Their development probably began from the enlargement of cuticular granules which gradually assumed the shape of hooks (Gerasimov 1952). The primitive arrangement would thus be large spines arranged in circular multiple rows (figs. 26.9m, 26.9d). This multiserial arrangement is found in the Gracillariidae, Acrolophinae (Tineidae), Yponomeutidae, Hepialidae, Adelidae, Tischeriidae, and some Castniidae where they may be secondary. Further specialization has led to a reduction in the number of rows to a uniserial (single row) circle and differentiation in size. If all the crochets in a circle are the same length, they are uniordinal (fig. 26.9a; two alternating lengths are termed biordinal (fig. 26.9b); three alternating lengths are termed triordinal (fig. 26.9c).

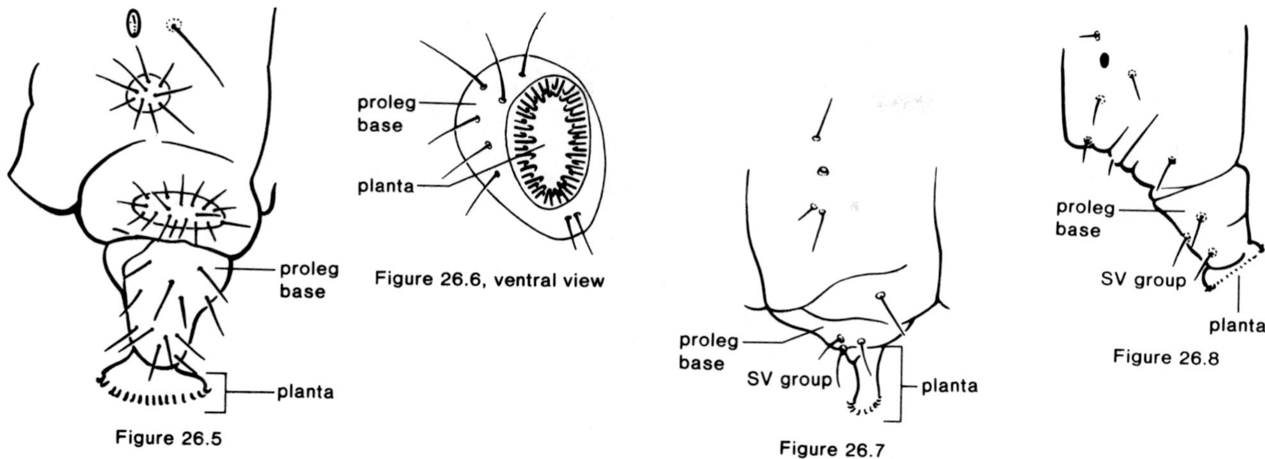

Figures 26.5–26.8. Abdominal prolegs, showing variation in proleg base and planta.

The circles of crochets are often broken into rows, semicircles, bands, etc. An incomplete circle has a small gap in an otherwise continuous arc. If the crochets are interrupted mesally and are continuous laterally, the arrangement is a lateropenellipse (fig. 26.9h). If the crochets are interrupted laterally and are continuous mesally, it is a mesopenellipse (fig. 26.9g). If a circle has a gap both laterally and mesally, the crochets are in two transverse bands (fig. 26.9e). If a single transverse row of crochets is present, the arrangement is a single transverse band. Most leaf feeders have a mesoseries (fig. 26.9i,j), which is a single longitudinal row of crochets on the mesal side of the planta, parallel to the meson. When the crochets are almost alike in structure and size throughout, they are termed a homoideous mesoseries (fig. 26.9i). If the crochets in the center of a mesoseries are *abruptly* longer than those at either end (most Arctiidae, Ctenuchidae, Pericopidae), it is a heteroideous mesoseries (fig. 26.9j). If two longitudinal parallel rows occur (fig. 26.9k), the outer is a lateroseries and the inner is a mesoseries. In addition to being broken by gaps, rows of crochets may be interrupted by structures such as a plantar "sucker" in some geometrids (Forbes 1948) and a spatulate fleshy lobe in most lycaenids and some riodinids (fig. 26.9o).

ADDITIONAL EXTERNAL STRUCTURES

In addition to those discussed above, the more significant external structures that are of taxonomic importance include glands, fleshy lobes, tonofibrillary platelets, setal bases, and setae.

Glands and Lobes

Eversible glands are commonly found in families such as the Lyonetiidae, Yponomeutidae, Papilionidae, Lycaenidae, Lymantriidae, Notodontidae, Noctuidae, and others. In many groups there is a midventral cervical gland anterior to the prothoracic legs. It secretes noxious defensive fluids in some notodontids (Herrick and Detweiler 1919, Eisner et al. 1972, and others), and its function in other caterpillars is probably similar. Papilionidae have an eversible middorsal Y-shaped gland (osmeterium) on T1 that is also defensive (Eisner et al. 1970), and the Lymantriidae have middorsal glands on A6 (may be absent) and A7 that are most likely defensive. The Danainae (Nymphalidae) and some Papilionidae, Noctuidae, Geometridae, and others have fleshy lobes or filaments on various segments.

Tonofibrillary Platelets

Snodgrass (1935) defined tonofibrillae as "cuticular fibrils connecting the muscle fibers with the inner surface of the cuticula." These tonofibrillae frequently penetrate to the outer part of the cuticula where they may form a small, conspicuous, external, pigmented, frequently depressed area on the body termed a "tonofibrillary platelet" by Neunzig (1979). These tonofibrillary platelets are useful taxonomic characters below the family level in some taxa and have been most used in the Pyralidae. The terminology of Mutuura (1980) as simplified by Stehr and Neunzig (1981) is used when needed.

Setae

Most setae are simple (i.e., hairlike), but in some taxa, setae that are plumose, knobbed, spinelike, or spatulate are found, and some are poisonous or urticating. A seta arises from a small sclerotized ringlike papilla on the integument and internally connects with at least one hypodermal cell. A sclerotized area around the base of one or more setae is a pinaculum (figs. 26.20–26.25, dotted circles). If the "pinaculum" is distinctly elevated, cone-shaped, and bears a single seta, it is a chalaza (figs. 26.10, 26.11). If the "chalaza" is a larger structure bearing setae or branching spines, it is a scolus (figs. 26.12, 26.13). Setae may also be grouped on nonsclerotized fleshy lobes, or on rather flat, disclike verricules with parallel setae (fig. 26.14) or on more convex verrucae with divergent setae (figs. 26.15, 26.16, 26.17). Some structures, such as the middorsal horn or tubercle on A8 of Sphingidae, are thought to be modified setae (Gerasimov 1952). Other

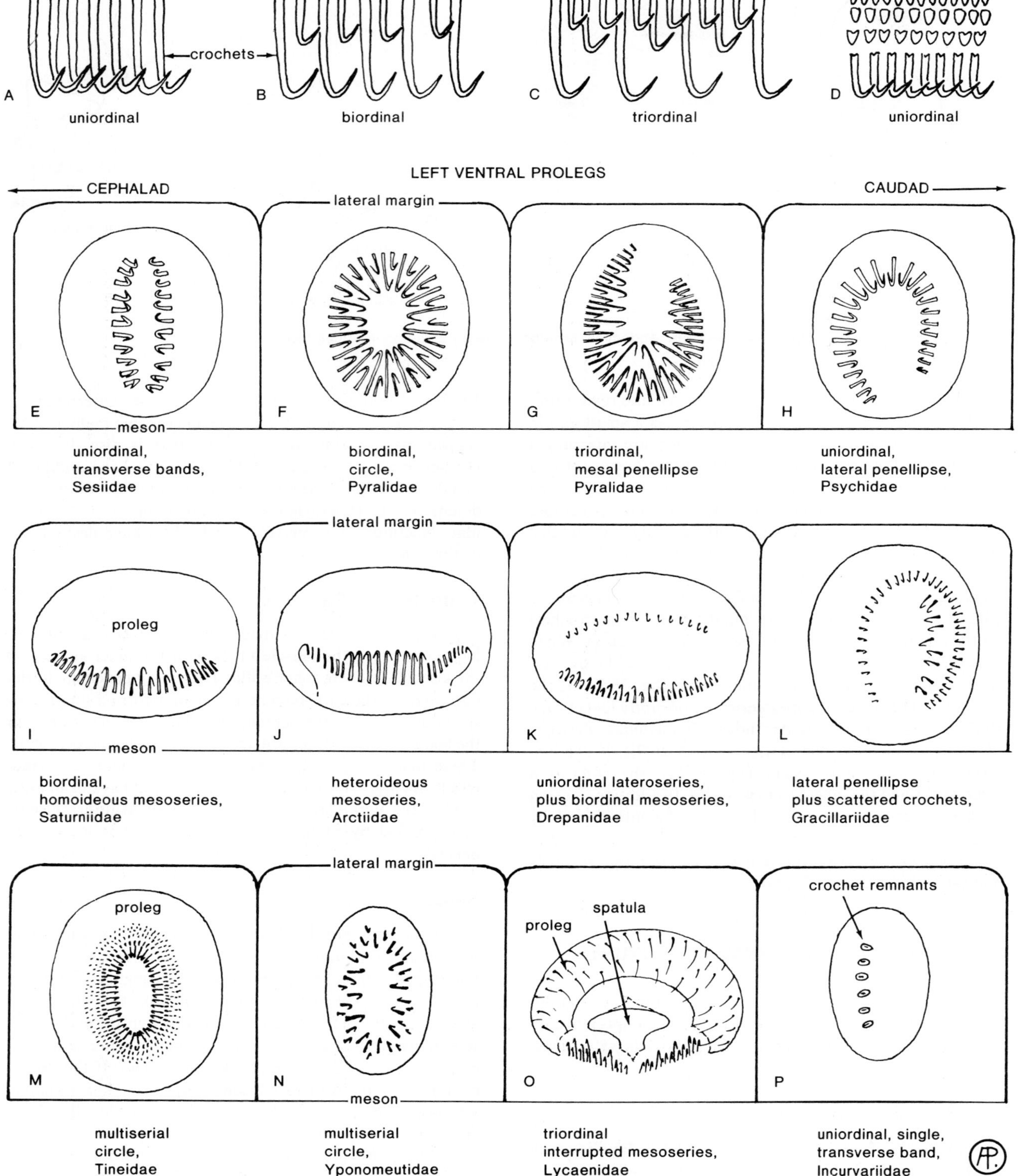

Figures 26.9a-p. Crochet arrangements and terminology (from Peterson 1948).

structures resembling horns, tubercles, and antlers are found on some Nymphalidae, Notodontidae, and Bombycidae. The anal comb or fork (fig. 26.18) found on many Oecophoridae, Gelechiidae, Tortricidae, Thyatiridae, and Hesperiidae, and used for flipping frass (fecal pellets), probably originated from modification of cuticular granules (Gerasimov 1952).

Such features as setae, spines, bristles, and horns vary greatly throughout the order, but the location and arrangement of the primary setal groups are critical to family and lower taxa recognition, and the terminology must be known. The currently used system of Hinton (1946) is not entirely satisfactory from the standpoint of homologies, but as an information system of names for taxonomic purposes, it serves quite well.

Fracker (1915) distinguished three kinds of setae on caterpillars: primary, subprimary, and secondary. Primary setae are sensory in function and are present on all instars of all but a few groups. They are relatively constant in number and position. Because primary setae occur throughout the Lepidoptera, their presence is a primitive condition (Hinton 1946). As has long been recognized, there are two fundamentally different kinds of primary setae. The microscopic or proprioceptor setae (prefixed "M") are very small and are located near the anterior margins of body segments, making sensory contact with the immediately anterior segment as it folds over or rides against them. The microscopic setae on the head are located posteriorly and make contact with the prothorax. The remaining and more visible long tactile setae function as receptors of external stimuli. Subprimary setae are like primaries but appear at the second instar. Primaries and subprimaries have been named and are reasonably well homologized throughout the order. Variations in position, length, structure, and number provide useful taxonomic characters. Primary and subprimary setae are both termed "primaries" in later instars because of difficulty in distinguishing between them. Verrucae and verricules that occur in primary positions are equivalent to primary setae and are named accordingly (see "secondary setae" and "pinacula and verrucae").

Secondary setae vary in number, size, and position within as well as between taxa, and they usually are not present until after the first instar (some Lasiocampidae and others have secondary setae on the first instar). They usually occur in one of two patterns: as numerous setae scattered over the cuticle, or as dense tufts (verrucae) or hair pencils (verricules). Verrucae and verricules usually occur in the same positions as the primary setae; hence most verrucae and verricules are homologous with primary setae and are named the same way as primary setae are. The setae comprising verrucae or verricules can be regarded as secondary since there is more than one seta and the number and position is usually not constant between specimens of the same species.

The appearance of secondary setae probably happened as a protective and/or sensory adaptation to an external feeding habit, since they are usually absent on internal feeders. Small secondaries scattered over the cuticle as in the Hesperioidea and Bombycoidea probably originated from modifications of cuticular granules (Gerasimov 1952).

Secondary setae are highly variable in number, size, and position. Many Tortricidae, Noctuidae, Lyonetiidae, and Pterophoridae have spinules (unarticulated minute cuticular spines that may represent intermediate conditions in the development of secondary setae (fig. 26.19)). Primaries are obscured by secondaries in most butterfly larvae, but in the Megathymidae (giant skippers) primaries stand out above the surrounding secondaries. Some Notodontidae, Noctuidae, Ethmiinae (Oecophoridae), Scythrididae, and Pterophoridae have extra setae on the primary pinacula. Many Cossidae, Noctuoidea, and Geometroidea have extra setae in "fixed" positions remote from the primaries. These setae resemble primaries in size and have occasionally been given names of various sorts, but their occurrence is inconsistent at the family level. See the discussion on segment A10 in the section on chaetotaxy for a suggested system to name them as the need arises.

SETAL MAPS

A setal map (figs. 26.20–26.25) is a diagram showing the relative positions and sizes of setae and other structures on the left side (by convention) of the thorax and abdomen. Each segment except A10 (and sometimes A9) is drawn as a rectangle proportional to the shape of the segment and covering the area between the middorsal and midventral lines. A10 is drawn more similar to the A10 shape since it is so different from the other segments. All segments are sometimes drawn, but more commonly T1, T2 (T3 is similar to T2), A1 (A2 is similar to A1), A3 or A6 (A3–6 are similar) and A7–10 are drawn (such deletions are also frequent for conventional drawings). Setal maps are used because (1) they are easier and faster to draw, (2) a standard outline can be used, and (3) setal relationships can be shown just as well or better on a flat surface (the relationships of the more dorsal and more ventral setae are difficult to show in a conventional drawing). Setal maps are most useful in comparing larvae that are closely related or closely similar in shape. However, it is difficult to get an overall image of the whole larva from a setal map. Hence, in this book conventional drawings have been widely used to give the user a better "feel" for the species or family except where detailed differences must be illustrated to distinguish closely similar characters or taxa as in the keys. Body setae are labelled on figures 26.50c and 26.52a, b, and c).

Setal maps are also used for head structures, using either a face-on or side view (or both) to show relationships of sutures, pores, setae, stemmata, and patterns.

CHAETOTAXY

Numerous systems for naming the primary setae have been proposed. The most well known are those of Dyar (1894a, 1896b), Fracker (1915), Heinrich (1916), Forbes (1910a, 1923), Gerasimov (1935), Hinton (1946), and Mutuura (1956), although Müller (1886) apparently was the first to

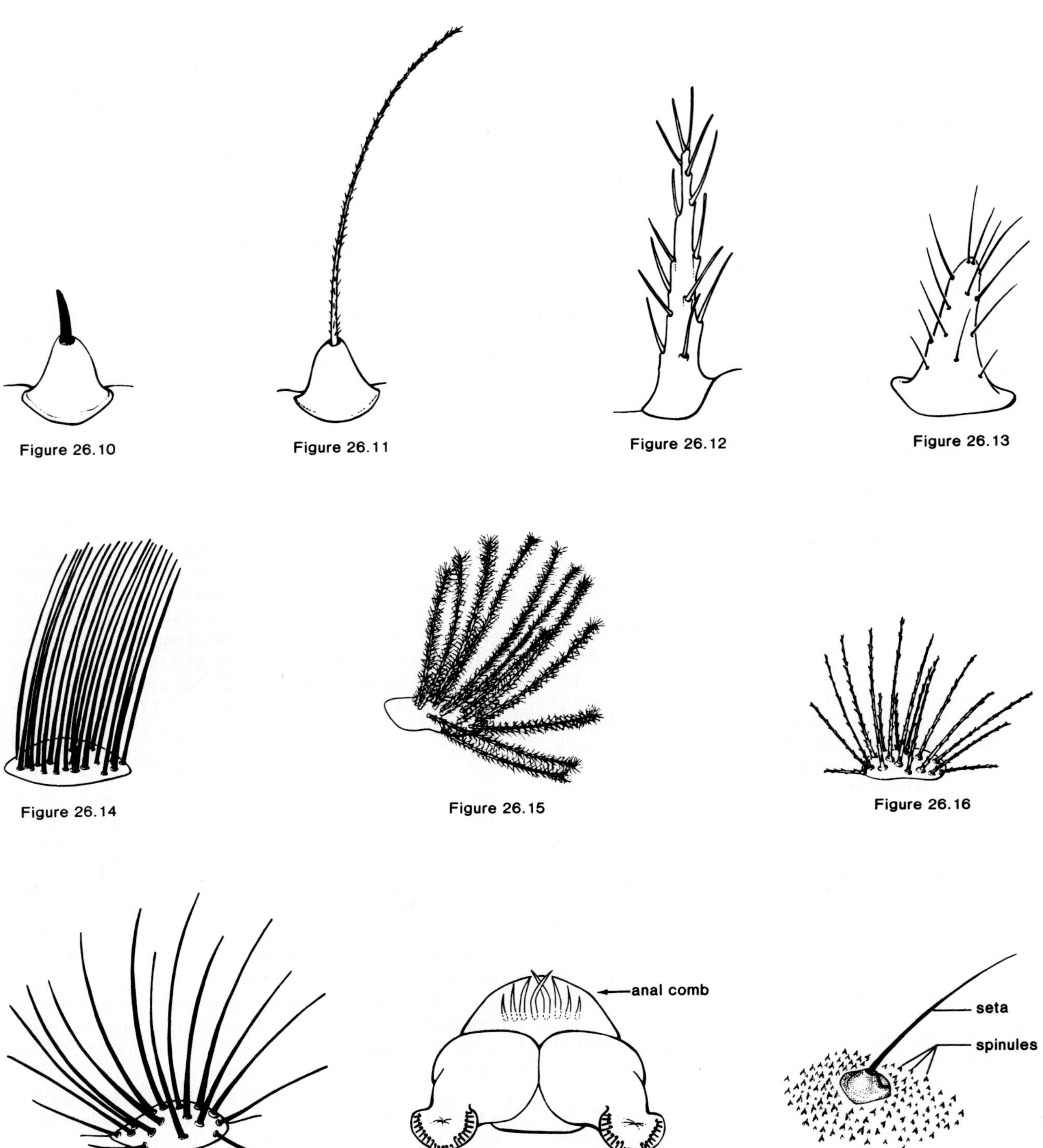

Figures 26.10–26.19. External structures. Figures 26.10, 26.11. chalazae; Figures 26.12, 26.13. scoli; Figure 26.14. verricule; Figures 26.15–26.17. verrucae; Figure 26.18. anal comb or fork; Figure 26.19. cuticular spinules.

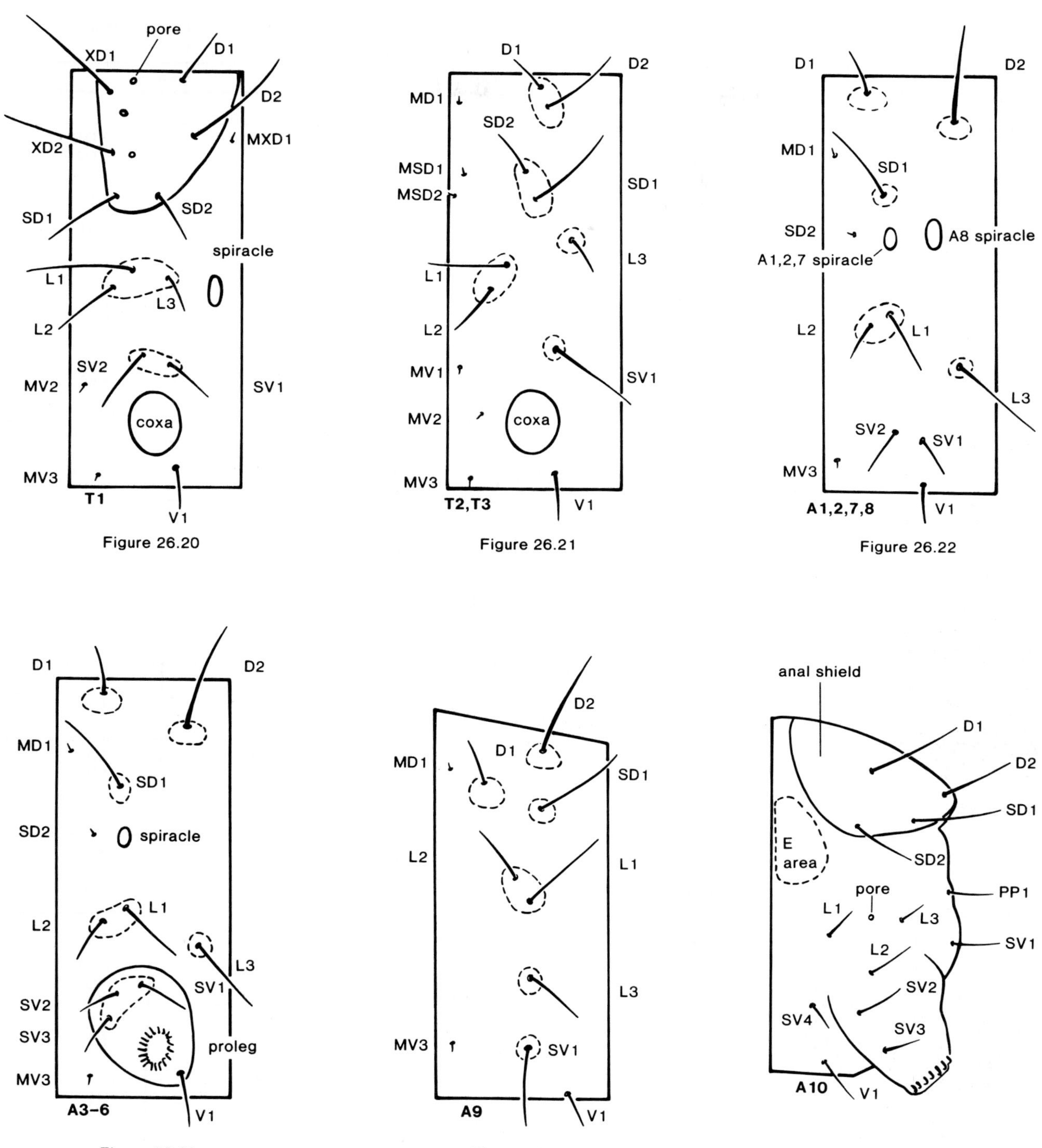

Figures 26.20–26.25. Setal maps: **26.20.** T1; **26.21.** T2, T3; **26.22.** A1, A2, A7, A8; **26.23.** A3-6; **26.24.** A9; **26.25.** A10. See pages 299-304 and tables 26.1 and 26.3-26.5.

Table 26.1. The Names of the Setae of the Thorax and Abdomen According to Fracker (1915), Gerasimov (1935), and Hinton (from Hinton (1946))

	Fracker		Gerasimov		Hinton	
	Jugatae	Frenatae	Hepialidae	Others		
Prothorax	Gamma	Alpha	IX	X	Tactile	XD1
"	Epsilon	Gamma	IIIa	IX	"	XD2
"	Alpha	Beta	X	I	"	D1
"	Beta	Delta	I	II	"	D2
"	Rho	Epsilon	II	IIIa	"	SD1
"	Delta	Rho	III	III	"	SD2
"	Eta	Kappa	V	IV	"	L1
"	Kappa	Eta	IV	V	"	L2
"	Theta	Theta	VI	VI	"	L3
"	Pi	Pi	VIIa	VIIa	"	SV1
"	Nu	Nu	VIIb	VIIb	"	SV2
"	Sigma	Sigma	VIII	VIII	"	V1
"	—	—	—	Xa	Proprioceptor	MXD1
"	Tau	—	VIIc	VIIc	"	MV2
"	Phi	—	VIId	VIId	"	MV3
Meso- and metathorax	Alpha	Alpha	I	I	Tactile	D1
"	Beta	Beta	II	II	"	D2
"	Rho	Rho	III	III	"	SD1
"	Delta	Epsilon	IIIa	IIIa	"	SD2
"	Kappa	Kappa	V	IV	"	L1
"	Epsilon	Eta	IV	V	"	L2
"	Theta	Theta	VI	VI	"	L3
"	Pi	Pi	VIIa	VIIa	"	SV1
"	—	Nu	—	—	"	SV2
"	Sigma	Sigma	VIII	VIII	"	V1
"	Gamma	—	Xa	Xa	Proprioceptor	MD1
"	—	—	—	Xb	"	MD2
"	Gamma	—	IXa	IXa	"	MSD1
"	Gamma	—	IXb	IXb	"	MSD2
"	Nu	—	VIIb	VIIb	"	MV1
"	Tau	—	VIIc	VIIc	"	MV2
"	Omega	Omega	VIId	VIId	"	MV3
Abdomen, 1–9	Alpha	Alpha	I	I	Tactile	D1
"	Beta	Beta	II	II	"	D2
"	Rho	Rho	III	III	"	SD1
"	Epsilon	Epsilon	IIIa	IIIa	"	SD2
"	Theta	Kappa	IV	IV	"	L1
"	Kappa	Eta	V	V	"	L2
"	Eta	Mu	VI	VI	"	L3
"	Pi	Pi	VIIa	VIIa	"	SV1
"	Nu	Nu	VIIb	VIIb	"	SV2
"	Tau	Tau	VIIc	VIIc	"	SV3
"	Sigma	Sigma	VIII	VIII	"	V1
"	—	—	X	Xa	Proprioceptor	MD1
"	—	—	—	Xb	"	MD2
"	Omega	—	VIId	VIId	"	MV3

suggest in his work on Nymphalidae that there was a common pattern of setae. Hinton (1946) gives other references and Fracker (1915) gives an account of the early work. The Dyar-Heinrich-Forbes system was the basis for Gerasimov's system and was used by Crumb (1956) for Noctuidae. Godfrey (1972) used Hinton's system for Noctuidae and gives a table of Dyar-Heinrich-Forbes equivalents for Noctuidae. Mutuura's system is most similar to Hinton's, and he provides tables of equivalents in Issiki et al. (1965) for Dyar, Forbes, Fracker, Gerasimov, and Hinton. Hinton's system has been adopted by most workers and is used here. Table 26.1 (Hinton 1946) compares Fracker's (1915–Greek letters) Gerasimov's (1935–Roman numerals (adapted from Dyar-Forbes)) and Hinton's (Roman letters and Arabic numerals) systems.

Head Chaetotaxy and Olfactory Pores (punctures)[2]

Head setae and pores are little used at the family level, but they can be important diagnostic structures for lower taxa. Dyar (1896b) was the first to name the head setae, being followed by others including Forbes (1910a), Dampf (1910), and Fracker (1915) before Heinrich (1916) came up with the basic system used today, with some modifications by Gerasimov (1935) and Hinton (1946), and one adopted here to replace the confusing term "vertical" (see below). The important feature of Heinrich's system is that most of the names chosen incorporated location information (for example AF1 = adfrontal seta 1), in contrast to the totally arbitrary systems that others had proposed. Unfortunately, a similar location system for body setae was not proposed until Hinton's (1946) system, which has now been generally adopted. See Hinton (1946) for a more complete discussion of the history and development of Lepidoptera head (and body) chaetotaxy.

Consistent use of the terms "anterior" (=in front of), "posterior" (=behind, rear), "dorsal" (=above), and "ventral) (=below) requires consistent orientation of the objects being described. There is no problem with the thorax and abdomen of caterpillars, but the orientation of heads varies from prognathous to hypognathous. For purposes of describing the location of the head setae and pores in table 26.2, the use of these terms as applied to a hypognathous head is used since the hypognathous or semihypognathous orientation is by far the most common. Hence, heads with prognathous or intermediate orientations should be viewed as if they were hypognathous. In other words, the frons or frontoclypeus is anterior, the epicranum (vertex) is dorsal, and the mouthparts are ventral.

The use of the term "vertical triangle" for the notch formed by the dorsal epicranial halves and the dorsoanterior margin of the prothorax is potentially confusing. Although "vertical" is the adjectival form of vertex and is correct, in everyday English the word "vertical" means oriented "up and down" and is not related to the vertex. Because of this potential confusion in meanings, the term "epicranial notch" is adopted here in place of "vertical triangle."

The use of "V" for the three proprioceptor setae on the epicranium, as proposed by Gerasimov (1935) and followed by Hinton (1946), is also confusing for the same reason, and also because Hinton used "V" for the ventral tactile setae on the body. For these reasons, and because these three setae and the "G" setae on the head are proprioceptors, they should be named in accordance with Hinton's system for the rest of the body. Hence, V1, V2, and V3 are renamed microdorsal setae (MD1, MD2, and MD3), and G1 and G2 are renamed microgenal setae (MG1 and MG2) (table 26.2, figs. 26.1, 26.2, 26.3).

The basic number of primary setae on the *head* of Lepidoptera larvae (excluding the mouthparts) (table 26.2) is 17 long tactile setae (A1, A2, A3, S1, S2, S3, SS1, SS2, SS3, L1, P1, P2, F1, AF1, AF2, C1, and C2), and four (or five) minute proprioceptor setae (MD1, MD2, MD3, (=V)) and MG1 (MG2 usually absent) located on those parts of the head that contact the prothorax or can be withdrawn into the prothorax.

McIndoo (1919) reported from 12–26 olfactory pores on the head capsule of 30 species in 21 different families, but 9–13 rather conspicuous pores are commonly found per side of the head capsule (table 26.2 and figs. 26.1, 26.2, 26.3). Hinton (1946) believed pores Sb(=Ob), SSa(=SOa), SSb(=SOb), and SSc(=SOc) on the head were of such little value in taxonomy that he chose not to name them. Gerasimov (1935) did name them, and they are included in table 26.2. In addition to the head capsule, pores are present on the antennae and mouthparts.

Body Chaetotaxy, Setal Bases, and Pores

Because of their small size, the microscopic proprioceptor setae have been little used and are not described here, but they are plotted on the setal maps (figs. 26.20–26.24) and listed in table 26.3. The primary tactile setae are divided into six groups based on their location (figs. 26.20–26.24). The following statements apply to one side only (the left side is used by convention). Segment A10 (fig. 26.25) is discussed separately below since it is so different.

XD Group

The two XD setae are only found on the anterior margin of the T1 shield, are nearly equal in length, are longer than D1, and are as long as D2. XD2 is below XD1.

Dorsal (D) Group

Two D setae are usually present near the dorsal midline on all segments. D1 is shorter than D2. On T1, both are on the shield, and D1 is usually above D2. On T2 and T3, D1 is above D2, and they are usually close together. On A1–8, D1 is anterior to D2. On A9, D2 is above or posterior to D1.

2. The term "pore" is more correct (as opposed to "puncture"), because a puncture is a depression that may not penetrate the cuticle, whereas a pore is a small sensory opening going through the cuticle (McIndoo 1919; Gerasimov 1952).

Table 26.2. Head Setae and Pores: Locations and Relative Lengths (modified from Hinton (1946); see figs. 26.1–26.3)

Setal Group	Location	Relative Lengths of Setae
TACTILE SETAE		
Anterior (A) Setae	Area in front of the stemmata	
A1	In front of stemma 2 or 3 and the most ventral A seta	Long as A3
A2	In front of stemma 2 and directly above seta A1	Shorter than A1 & A3
A3	Highly variable, but usually above or behind and closest to seta A2 and stemma 1; sometimes closer to L1	Long as A1
Pore Aa	Usually above and closest to seta A2	
Stemmatal (S) Setae	Area within and behind the stemmatal semicircle, including stemmata 1,2,3,4,6	
S1	Usually close to stemmata 3 and 4 and within the stemmatal semicircle	Short as A2
S2	Usually near the opening of the stemmatal semicircle	Longer than S1 & S3
S3	Usually behind stemma 6	Short as S1 or shorter
Pore Sa	Close to stemma 6 (often between 6 and seta S3)	
Pore Sb (often absent)	Close to stemma 4 (often between 3 and 4)	
Substemmatal (SS) Setae	Area below the stemmata, but including stemma 5	
SS1	Most ventral SS seta (closest to mandible)	
SS2	Variable, but between SS1 and SS3	Usually similar lengths
SS3	Most posterior SS seta (toward the genal area)	
Pore SSa	Within the substemmatal area	
Pore SSb (often absent)	Variable	
Pore SSc (often absent)	Variable	
Frontal (F) Setae	Area of the frons	
F1	Only seta on the frons	Same as C setae
Pore Fa	Only pore on the frons, mesad of F1	
Adfrontal (AF) Setae	Area between adfrontal suture and ecdysial line	
AF1	Ventral seta	
AF2	Dorsal seta	Shorter than C setae
Pore AFa	Usually between setae AF1 and AF2	
Clypeal (C) Setae	Area of the clypeus	
C1	Medial seta	Nearly equal
C2	Lateral seta	
Lateral (L) Setae	Area above the stemmata and on the *side* of the head	
L1	Above and somewhat behind stemma 1; often near A3	Shorter than A3
Pore La	Near and usually above seta L1	

Setal Group	Location	Relative Lengths of Setae
Posteriodorsal (P) Setae	Area on the upper *face* of the head between the lateral area and the upper adfrontal area	
P1	Ventral and longer seta, near adfrontal suture	Largest head seta
P2	Dorsal and shorter seta	Much shorter than P1
Pore Pa	Ventral pore and closest to P1 or L1	
Pore Pb	Dorsal pore and usually closest to P2	
PROPRIOCEPTOR SETAE		
Dorsal (MD) Setae	Area of the dorsal epicranium (vertex) (vertical setae of Hinton)	
MD1	Ventral seta	
MD2	Middle seta	All short
MD3	Dorsal seta	
Pore MDa	Between MD2 and MD3	
Genal (MG) Setae	Area at the lower, rear portion of the head	
MG1	Farthest from the stemmata	
MG2 (often absent)	Closest to the stemmata	Both short
Pore MGa	Closer to MG1	

Subdorsal (SD) Group

Two SD setae are present on all segments except A9 where SD2 is absent, although SD2 is usually minute and sometimes absent on A1–8. SD1 is longer than SD2. On T1 they are usually on the T1 shield below D1, D2, XD1, and XD2. SD1 is usually below XD2 and anterior to SD2. Sometimes (in many Noctuoidea) SD1 and SD2 are very close together and are located below the T1 shield. On T2 and T3, SD1 is below SD2, and D1, D2, SD2, and SD1 often form a nearly straight vertical line with D1 and D2 frequently on one pinaculum and SD1 and SD2 on another. On A1–8, SD2 is usually minute (sometimes absent) and anterior or anterodorsal to the spiracle; SD1 is usually above the spiracle. On A9, SD2 is absent and SD1 is below D1.

Lateral (L) Group

Three L setae are usually present on all segments. On T1 they are anterior to the spiracle, usually on a single pinaculum. L1 is the longest, and L2 the next longest. L1 is between L2 and L3. L2 is usually anterior to L1, and L3 is posterior and closest to the spiracle. In nearly all Copromorphoidea, Noctuoidea, and Pyraloidea, L3 is missing on T1. On T2 and T3, L1 is primary, L2 and L3 subprimary; L1 is in the middle, L3 is caudodorsad of L1, and L2 is anteroventrad of L1; usually L1 and L2 share a pinaculum, and L3 is on its own. On A1–8, L1 and L2 are close together and below the spiracle in most Gelechioidea, Tortricoidea, and Pyraloidea; but they are far apart, with L1 often behind the spiracle, in most Tineoidea, Yponomeutoidea, Geometroidea, and Noctuoidea; L3 is caudoventrad and farther from the spiracle than L1 and L2 and is often the longest seta. On A9 the L group is often unisetose or bisetose; L1 is always present. If all three are present on A9, they are usually arranged vertically with L1 in the middle and L3 below.

Subventral (SV) Group

The SV group is usually bisetose on T1 and unisetose on T2 and T3 (bisetose in Thyrididae, Pterophoridae, some Tineidae). They are above the coxa and are often on large pinacula. On A1–8, the SV group may be unstable in number, even on opposite sides of the same segment. In most microlepidoptera and cutworm-type macrolepidoptera, three is the maximum number of SV setae on A1–8. In many Oecophoridae, Pterophoridae, Scythrididae, and others, secondary setae are present in the SV group, and in most externally feeding macrolepidoptera, the ventral prolegs bear many secondary setae, whereas other abdominal segments may have a normal SV group, or an SV verucca, or the SV group may be obscured by numerous secondaries.

Ventral (V) Group

A single V seta is usually present on all segments but on rare occasions a secondary seta(e) may be present (some Scythrididae, for example). It is closer to the ventral meson than the SV group. On T1–3 it is on the underside of the coxa; on A3–6 it is on the inside of the proleg.

Table 26.3. Basic Occurrence of Primary Setae on Thorax and Abdomen of Lepidoptera Larvae (modified from Hinton, 1946)

Setae	Segment: T1	T2,3	A1,2	A3–6	A7	A8	A9	A10	
Tactile:									
XD1	+	0	0	0	0	0	0	0	
XD2	+	0	0	0	0	0	0	0	
D1	+	+	+	+	+	+	+	+	anal shield
D2	+	+	+	+	+	+	+	+	anal shield
SD1	+	+	+	+	+	+	+	+	anal shield
SD2	+	+	+	+	+	+	0	+	anal shield
L1	+	+	+	+	+	+	+	+	anal proleg
L2	+	+	+	+	+	+	+*	+	anal proleg
L3	+/0	+	+	+	+	+	+*	+	anal proleg
SV1	+	+	+	+	+	+	+	+	anal proleg
SV2	+	+/0	+/0	+/0	+/0	+/0	+/0	+	anal proleg
SV3	0	0	+/0	+/0	+/0	0	0	+	anal proleg
SV4	0	0	0	0	0	0	0	+	anal proleg
V1	+	+	+	+	+	+	+	+	
PP1	0	0	0	0	0	0	0	+	
Proprioceptor:									
MXD1	0	+/0	0	0	0	0	0	0	
MD1	0	+	+	+	+	+	+	+	
MD2	0	+/0**	+	+	+	+	+	+	
MSD1	0	+	0	0	0	0	0	0	
MSD2	0	+	0	0	0	0	0	0	
MV1	0	+	0	0	0	0	0	0	
MV2	+	+	0	0	0	0	0	0	
MV3	+	+	+	+	+	+	+	+	

+ = present
0 = absent
+/0 = present or absent
*frequently missing
**always absent on T2

The 9th Segment (A9)

The setae of A9 (table 26.4, fig. 26.4) are the most reduced in number, commonly reduced to D1 and D2, SD1, L1 (sometimes L2 and L3), SV1 (sometimes SV2) and V1. The relationships of D1, D2, and SD1 are frequently used in keys, with D1 usually located below D2, and sometimes located on the same pinaculum with SD1.

The Anal Segment (A10)

The chaetotaxy of A10 (fig. 26.25) is the most confusing of all segments, since the segment has a basically different shape. It is minimally used at the family level, although it has proved to be useful at lower levels. In fact, it was believed by Hinton (1946) and Mutuura (1956) to be so seldom useful that they did not attempt to name the A10 setae.

Table 26.4. Setae of Abdominal Segment 9

Seta	Presence	Location and Size
D2	Always present	Usually dorsal to D1, longer seta
D1	Always present	Usually ventral to D2, shorter seta
SD1	Always present	Below the D group
SD2	Never present	
L2	Sometimes absent	Above L1, short
L1	Always present	In the middle if there are 3 L's, long
L3	Frequently absent	Below L1, shorter than L1
SV1	Always present	Above V1, longer than SV2
SV2	Rarely present	
V1	Always present	Close to midventral line

Table 26.5 summarizes the systems of nomenclature for A10 as used by various authors since Ripley (1923) first proposed the use of Greek letters derived from Fracker (1915) for Noctuidae. Gerasimov (1952), the only one to consider all families in designing a system, used Roman numerals, as did Hasenfuss (1960) for Pyralidae, whereas Singh (1953), Dugdale (1961), and McGuffin (1967, 1972, 1977) used modifications of Hinton's (1946) system for Geometridae (which are unusual in some ways). Allyson (1976) followed Hasenfuss (1960) for Pyralidae, who followed Gerasimov (1952), but she substituted Hinton's (1946) terminology. Gerasimov's system with Hinton's terminology (table 26.5, Allyson) is consistent in that it uses the terminology widely adopted for the other segments and attempts to homologize setal groups for the whole segment, instead of regarding the anal shield and anal proleg as rather discrete units with different terminologies as Singh, Dudgale, and McGuffin have done for geometrids.

When the A10 segment is viewed as an anal shield and an anal proleg area (fig. 26.25) there are fundamentally four setae on the anal shield (D1, D2, SD1, SD2) and nine on the anal proleg area (L1, L2, L3, SV1, SV2, SV3, SV4, V1 and PP, the paraproct seta). The paraproct seta is abbreviated PP as Dugdale (1961) did, rather than "sppr" as Hasenfuss (1960) or "Sppr" as Allyson (1976) have done, since PP is shorter and distinguishes it from the posterior (P) setae on the head.

In addition, extra setae may be found on both the anal shield and/or the anal proleg or other segments. Gerasimov (1952), who used Roman numerals, tried naming a couple of them IX and X, and Singh (1953) called them "extra" setae. They are all secondary setae and do not occur frequently enough to warrant specific names. Singh's suggestion to call them "extra" setae when necessary and identify them by reference to nearby setal groups seems reasonable and is adopted here, using the next higher number in the setal *group* and using the prefix "EX" (rather than "X" to distinguish from XD1 and XD2 on the prothorax, which are not "extra" at all). Hence, an extra seta near D1 or D2 on the anal shield would be "EXD3" (not EXD1 or EXD2); a group of three setae near the D group (D1 and D2) would be EXD3, EXD4, and EXD5. Obviously, illustrations should be used to accurately locate any EX setae since their occurrence is sporadic and more variable than the regular setae.

Pinacula and Verrucae

Pinacula are usually named according to the primary setae occurring on them. Seta D1 is on pinaculum D1. If D1 and D2 setae are on a single pinaculum, that pinaculum is D. Verrucae are also labeled in this manner, since they are modified pinacula. If two separate dorsal verrucae are present, they are D1 and D2, but if only one dorsal verucca (fused) is present it is verucca D. If two verrucae of different primary groups are fused (for example, the fused D and SD verrucae on T2 and T3 in the Arctiidae), the verruca is D + SD.

Pores

In addition to the head, pores are present on the prothorax, all three pairs of thoracic legs, anal shield, and anal prolegs. Primary pores are present in first instars and the positions of some may vary considerably between taxa. Secondary pores appear in later instars, especially in the more advanced groups. There are no subprimary pores. Pores are seldom used in family classification, but *should not be confused with seta ring bases, which they may superficially resemble when setae are broken off.* The pore most likely to be mistaken for a seta base is the one usually found on the upper, lateral part of the A10 proleg (fig. 26.25). Sclerotization on the inside of a pore is usually heavier than on the inside of a setal ring. Pores are named for areas and/or nearby structures and are distinguished by adding a lower case letter (La is the lateral pore).

Table 26.5. Setae of Abdominal Segment 10 as Named by Various Authors (see fig. 26.25)

Noctuidae	Geometridae			All Families	Pyralidae		All Familes	Locations	Type of Seta
Ripley 1923	Singh 1953	Dugdale 1961	McGuffin 1977	Gerasimov 1952	Hasenfuss 1960	Allyson 1976	This Book		
Anal shield									
alpha	D1	D1	D1	I	I	D1	D1	Anterior from D2 and on disc of the shield	Primary
beta	L1	D2	D2	II	II	D2	D2	Posterior seta on edge of shield	Primary
kappa	L2	L1	L1	III	III	SD1	SD1	Middle seta along edge of shield	Primary
rho	L3	SD1	SD1	IIIa	IIIa	SD2	SD2	Anterior seta along edge of shield	Primary
Anal proleg area									
eta	L1	L3	LG3	IV	IV	L1	L1	Most anterior L seta (*dorsal to proleg*)	Primary
tau	L2	L2	LG2	V	V	L2	L2	Most ventral L seta (*dorsal to proleg*)	Primary
mu	L3	L1	LG1	VI	VI	L3	L3	Most posterior L seta (L2 sometimes more posterior) (*dorsal to proleg*)	Subprimary
pi	CD2	CD2	CD2	VIIa	VIIa	SV1	SV1	First seta below PP1 (second seta below anus)	Primary
epsilon	CP1	CP1	CP1	VIIb	VIIb	SV2	SV2	*Vertical pair on anterior half of the proleg; SV2 is dorsal, SV3 is ventral*	Primary
omega	CP2	CP2	CP2	VIIc	VIIc	SV3	SV3	*Vertical pair on anterior half of the proleg; SV2 is dorsal, SV3 is ventral*	Subprimary
—	not named	not named	not named	VIId	VIId	SV4	SV4	Most anterior seta above V1 (*anterior to SV2, SV3*)	Subprimary
sigma	not named	not named	not named	VIII	VIII	V1	V1	Most ventral, anterior seta (*anterior to SV2, SV3*)	Primary
nu	CD1	PP	CD1	s. ppr.	sppr	Sppr	PP1	First seta below anus	Primary
—	EXTRA		LG4	IX[1]			EXL4		Probably secondary
—				X[2]			EXL5		Probably secondary
				——[3]					

IX[1] = extra (above L group)
X[2] = intersegmental
——[3] = subanal (below SV1)

INTERNAL STRUCTURES

The internal structures of caterpillars have been essentially ignored as far as *identification* is concerned for obvious reasons. They should be used more in building classifications, as should characters of all immature stages. The alimentary canal is usually relatively straight and simple, with six Malpighian tubules. The paired labial (modified salivary) silk glands are usually large and conspicuous and may extend backward from the spinneret for a considerable distance. The nervous, respiratory, and circulatory systems are not unusual.

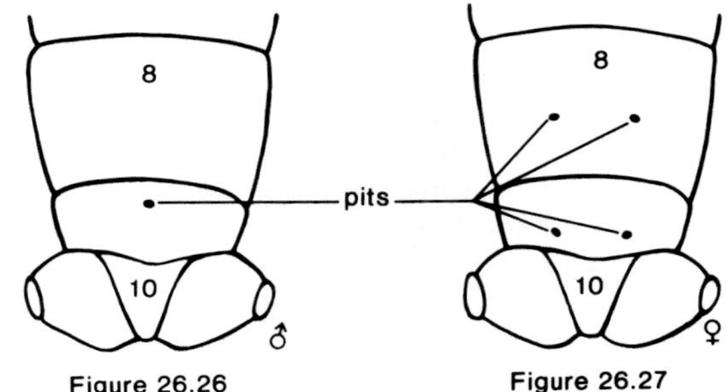

Figure 26.26. Sexing larva, male.

Figure 26.27. Sexing larva, female.

The ovaries or testes are located in the fifth abdominal segment and are frequently visible in light-colored larvae (especially males).

Caterpillars can often be easily sexed by the gonads visible through the integument or by the external pits marking the internal location of the genitalic histoblasts on segments A8 and A9 (Gerasimov 1952; Stehr and Cook 1968, and others). Males have a single median pit on A9 that may be difficult to see (fig. 26.26), whereas females have a more visible pair of pits on A8 and a pair on A9 (fig. 26.27).

CLASSIFICATION[3]

Order **LEPIDOPTERA** (11313)
- Suborder Zeugloptera (2)
  - Micropterigoidea (2)
    - Micropterigidae (2)
- Suborder Dacnonypha (15)
  - Eriocranioidea (15)
    - Eriocraniidae (12)
    - Acanthopteroctetidae (3)
- Suborder Exoporia (20)
  - Hepialoidea (20)
    - Hepialidae (20)
- Suborder Monotrysia (224)
  - Nepticuloidea (89)
    - Nepticulidae (82)
    - Opostegidae (7)
  - Tischerioidea (48)
    - Tischeriidae (48)
  - Incurvarioidea (87)
    - Heliozelidae (31)
    - Adelidae (18)
    - Incurvariidae (22)
    - Prodoxidae (16)
- Suborder Ditrysia (11080)
  - Tineoidea (598)
    - Tineidae (174)
    - Psychidae (26)
    - Ochsenheimeriidae (1)
    - Lyonetiidae (122)
    - Gracillariidae (275)
  - Gelechioidea (1460)
    - Oecophoridae (225)
    - Elachistidae (57)
    - Blastobasidae (121)
    - Coleophoridae (169)
    - Momphidae (37)
    - Agonoxenidae (6)
    - Cosmopterigidae (180)
    - Scythrididae (35)
    - Gelechiidae (630)
  - Copromorphoidea (66)
    - Copromorphidae (5)
    - Alucitidae (1)
    - Carposinidae (11)
    - Epermeniidae (11)
    - Glyphipterigidae (38)
  - Yponomeutoidea (166)
    - Plutellidae (54)
    - Yponomeutidae (32)
    - Argyresthiidae (52)
    - Douglasiidae (5)
    - Acrolepiidae (3)
    - Heliodinidae (20)
  - Sesioidea (142)
    - Sesiidae (113)
    - Choreutidae (29)
  - Cossoidea (45)
    - Cossidae (45)
  - Castnioidea (0)
    - Castniidae (0)
  - Tortricoidea (1164)
    - Tortricidae (1164)
  - Hesperioidea (290)
    - Hesperiidae (262)
    - Megathymidae (28)
  - Papilionoidea (470)
    - Papilionidae (33)
    - Pieridae (63)
    - Lycaenidae (136)
    - Riodinidae (25)
    - Libytheidae (3)
    - Nymphalidae (210)
  - Zygaenoidea (87)
    - Zygaenidae (22)
    - Megalopygidae (11)
    - Limacodidae (52)
    - Epipyropidae (1)
    - Dalceridae (1)
  - Pyraloidea (1387)
    - Pyralidae (1374)
    - Hyblaeidae (1)
    - Thyrididae (12)
  - Pterophoroidea (146)
    - Pterophoridae (146)
  - Drepanoidea (21)
    - Thyatiridae (16)
    - Drepanidae (5)
  - Geometroidea (1414)
    - Geometridae (1404)
    - Epiplemidae (8)
    - Sematuridae (1)[4]
    - Uraniidae (1)[4]
  - Mimallonoidea (4)
    - Mimallonidae (4)
  - Bombycoidea (109)
    - Apatelodidae (5)
    - Bombycidae (1)
    - Lasiocampidae (35)
    - Saturniidae (68)
  - Sphingoidea (124)
    - Sphingidae (124)
  - Noctuoidea (3359)
    - Notodontidae (135)
    - Dioptidae (2)
    - Doidae (1)
    - Pericopidae (10)
    - Arctiidae (225)
    - Ctenuchidae (29)
    - Lymantriidae (32)
    - Nolidae (16)
    - Noctuidae (2909)

3. Slightly modified from Hodges et al. (1983), Check List of the Lepidoptera of America North of Mexico. The number following the taxon shows the species recognized north of Mexico.

4. Not covered, only adults are recorded north of Mexico.

LEPIDOPTERA
Key to Families of Larvae

Frederick W. Stehr, ***Michigan State University***
Peter J. Martinat, ***West Virginia University***

This key is artificial; it is designed for identification—not to show relationships, although it will do so to some extent.

In a number of places it is possible to make a "wrong" choice and arrive at the correct family via a different route through the key or by being directed back to the correct couplet.

An attempt has also been made to minimize conditional statements (hypothetical example: A3 with a group of 4D setae, but if with 1 or 2, then on a verruca, or if with 3, then with the middle seta longer). The reduction of conditional statements leads to additional couplets and to families keying out at additional different places, but makes the couplet pairs easier to use. Because of this, the total length is greater than "usual," but the actual path through the key may be no longer than usual and perhaps will be easier or more accurate. In addition, it may not be possible to determine family characters by checking through the key; consult the diagnoses and descriptions for such information.

The key will work best with mature larvae. Key earlier instars with caution, since some differ considerably from mature larvae. Prepupae should key out nearly as well as mature larvae *if the crochets can be accurately observed* (the prolegs and crochets are usually retracted and may need to be dissected).

You may start at couplet 39 if there are crochets on A3, 4, 5, 6, *and* 10, and three pairs of thoracic legs with claws.

ORIENT LARVA WITH HEAD TO THE LEFT

1. Antennae nearly as long as width of head, and inserted above the stemmata (fig. 1); setae modified as thickened or elongate "scales"; ventral prolegs present on A1–8 (fig. 2); mature larvae less than 5 mm, sluglike (found in mosses, liverworts, and lichens, or in litter and duff) .... ***Micropterigidae*** (p. 341)

1′. Antennae much shorter than width of head, and inserted between the stemmata (if present) and base of mandibles (fig. 3); setae rarely modified as above; complement of ventral prolegs not as above 2

2(1′). Externally affixed to Fulgoroidea adults until mature, pupate on foliage; primary setae apparently absent; secondary setae minute; body stout, humped, with head retracted into prothorax; stemmata clumped into an "eyespot"; thoracic legs present but rudimentary; prolegs absent, but uniordinal crochets arranged in a complete circle present on A3–6, and A10 (fig. 4) .... ***Epipyropidae*** (p. 456)

2′. Never externally affixed to Fulgoroidea; primary setae present or absent; not exactly fitting combination of other characters 3

3(2′). Crochets or transverse rows of multiserial microspines present on the venter of at least 1 segment of A1–9 (look very carefully if caterpillar is sluglike) 4

3′. Crochets absent on A1–9 (*irregularly spaced* spines (fig. 5) (may be present on ventral prolegs if mature larvae are large subtropical or tropical borers in stems of banana and other plants) .... **(see third choice)** 5

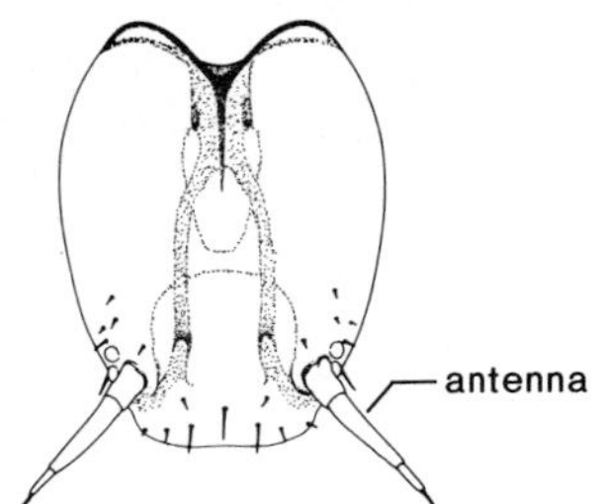

Figure 1

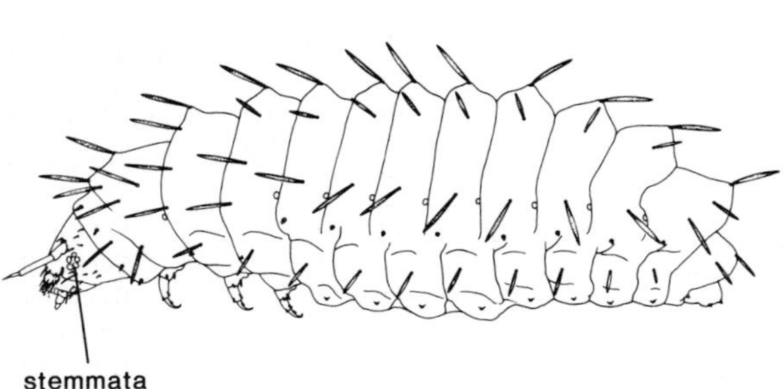

Figure 2

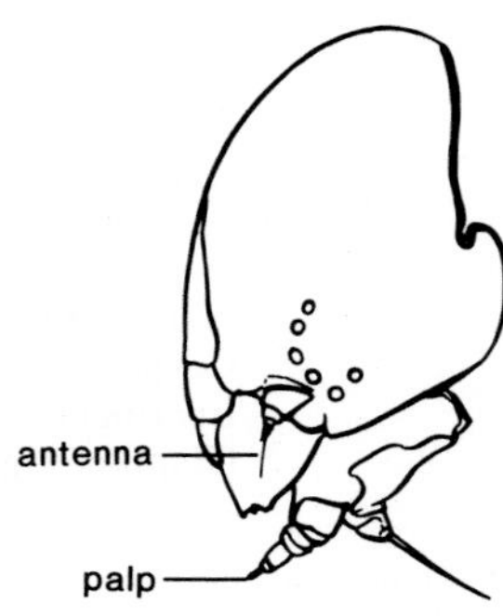

Figure 3

3″. Crochets *AND* spines present on the venter of at least 1 segment of A1–9 (fig. 6); mature larvae large (borers in stems of banana and other plants; tropical and subtropical) .... (part) ***Castniidae*** (p. 417)

4(3). Crochets present on A10 (at least 1 remnant of a crochet present) (check carefully if proleg is not fully distended) 34

4′. Crochets absent on A10 26

5(3′). At least 1 pair of segmented, thoracic legs *with claws* present (may be very small, look carefully) 14

5′. Segmented, thoracic legs with claws absent (knobs present at most) 6

6(5′). Frontoclypeus with posterior margin "squared" off (fig. 7), curved (figs. 8, 9), or open (figs. 10, 11) (frontoclypeal shape basically rectangular or circular) 7

6′. Frontoclypeus with posterior margin more acute (figs. 12, 13, 14); never open (frontoclypeal shape basically triangular or pentagonal) 11

7(6). Frontoclypeus less than 2× as high as widest part (figs. 8, 9); body cylindrical; prolegs, if well developed, present on A2–7; T2 and T3 usually with pairs of ventral fleshy lobes (fig. 15) (miners in leaves, bark, or fruits, or forming galls in twigs or petioles) .... ***Nepticulidae*** (p. 350)

7′. Frontoclypeus at least 2× as high as widest part (figs. 7, 10, 11); complement of prolegs not as above; ventral fleshy lobes on thoracic segments present or absent 8

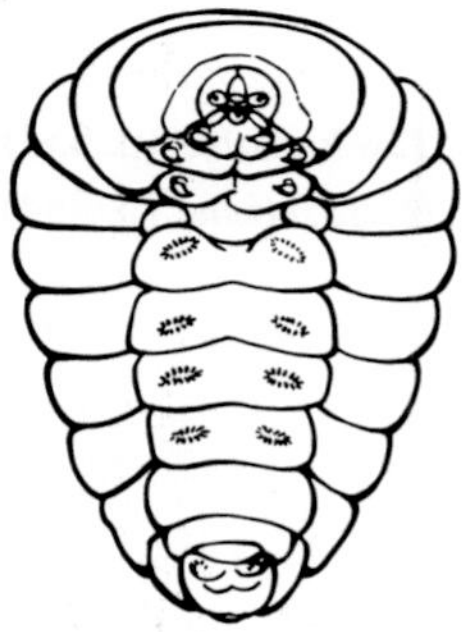

Figure 4

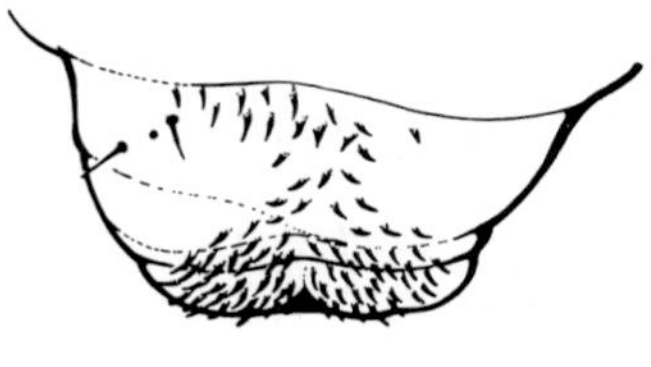

Figure 5

Figure 6

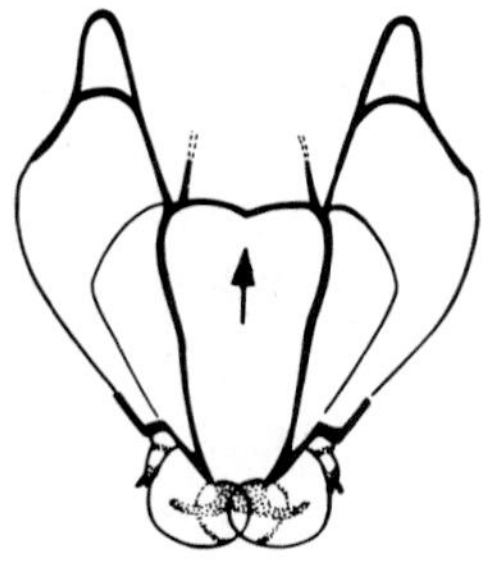

Figure 7

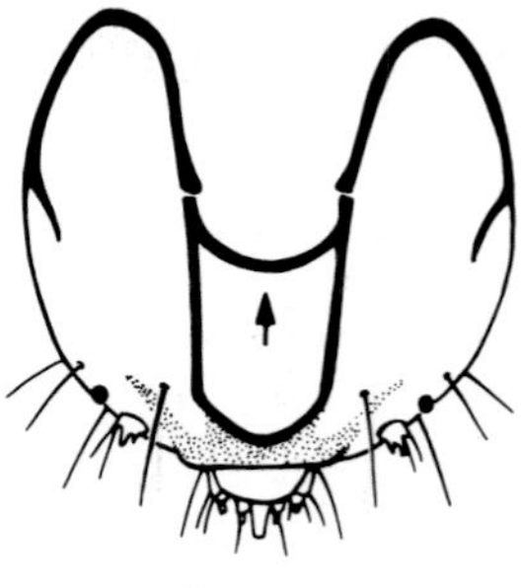

Figure 8

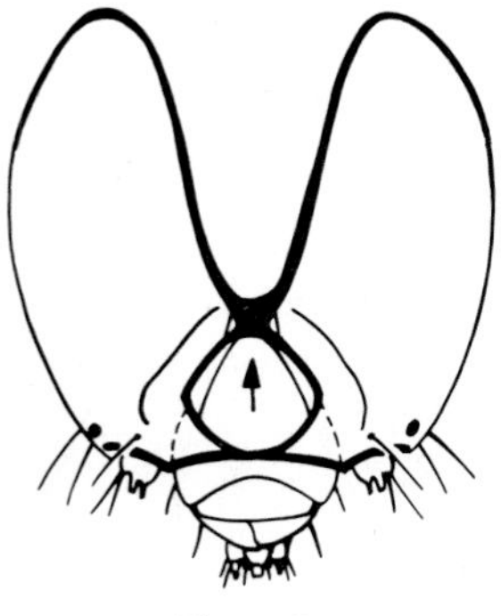

Figure 9

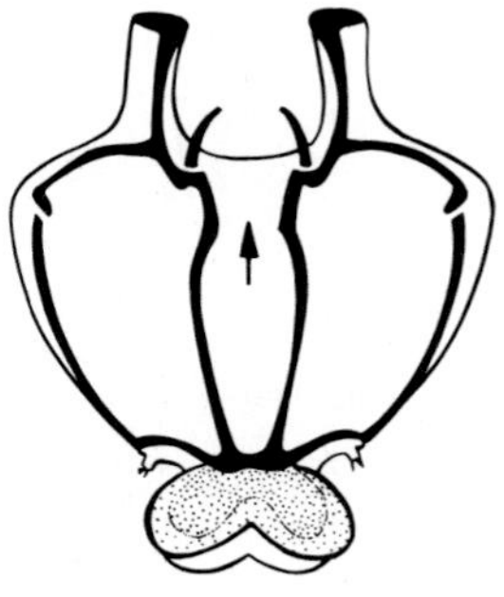

Figure 10

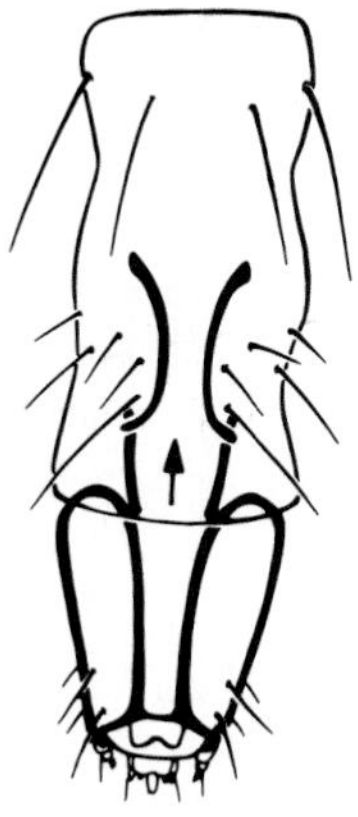

Figure 11

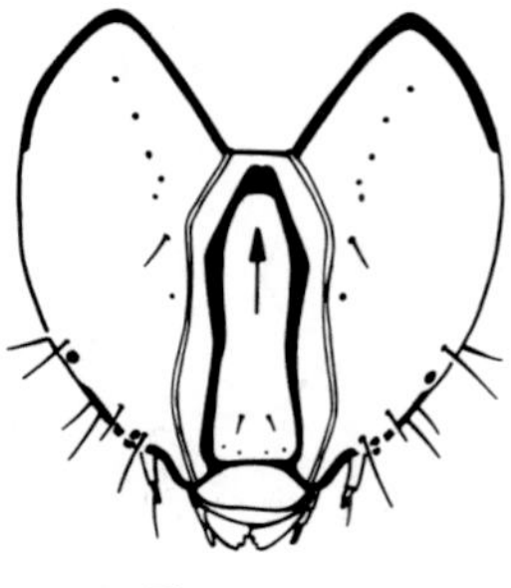

Figure 12

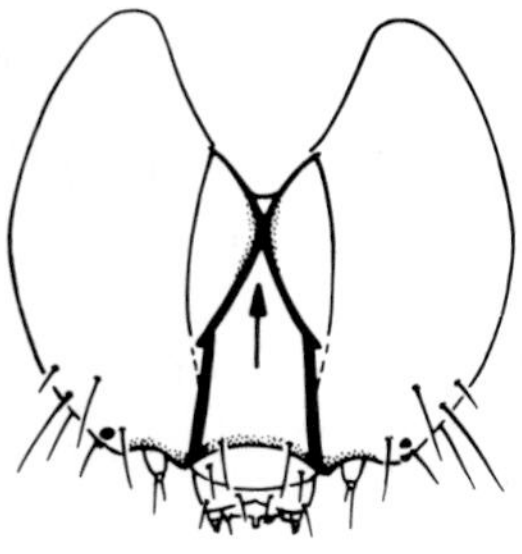

Figure 13

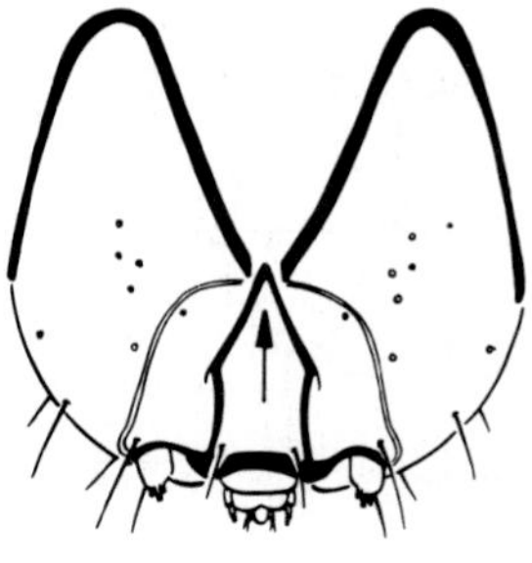

Figure 14

Figure 15

8(7′). Frontoclypeus closed behind (fig. 7) 9

8′. Frontoclypeus open behind (figs. 10, 11) 10

9(8). Four to six stemmata (simple eyes) per side .... (part) ***Tischeriidae*** (p. 354)

9′. Stemmata absent or fewer than 4 per side .... (part) ***Gracillariidae*** (p. 372)

10(8′). Body long and slender, about 10× longer than wide; T2 and T3 each with a pair of ventral fleshy lobes which are absent on T1; setae long and conspicuous and arranged in vertical rows on A1–8; labrum as wide as frontoclypeus; mouthparts not extremely modified (fig. 11) (stem borers) .... ***Opostegidae*** (p. 351)

10′. Body not long and slender, less than 10× longer than wide, usually dorsoventrally flattened with prominent lateral lobes; all 3 thoracic segments without ventral fleshy lobes, or, if present, on all 3 thoracic segments; setae minute or obscure; labrum much wider than frontoclypeus; mouthparts extremely modified for sap feeding (fig. 10) (leaf miners) .... (early instars) ***Gracillariidae*** (p. 372)

11(6′). Body swollen, cyphosomatic (figs. 16, 17); 3 stemmata (simple eyes) on each side; prolegs absent (borers in stalks or fruit of *Agave* or *Yucca*) .... (part) ***Prodoxidae*** (p. 361)

11′. Body not swollen, more cylindrical or dorsoventrally flattened; number of stemmata variable (often 2 or 4 per side); prolegs present or absent 12

12(11′). Head capsule with the somewhat irregular ecdysial lines (sometimes faint) terminating outside *lateral* margins of antennal sockets (fig. 18); head with a single stemma on each side which may be faint; leaf miners .... ***Eriocraniidae*** (p. 344)

12′. Head capsule with ecdysial lines terminating between antennal sockets and mouthparts (fig. 19); number of stemmata variable 13

13(12′). Crochets present on A10 .... (part) ***Tischeriidae*** (p. 354)

13′. Crochets absent on A10 .... (part) ***Heliozelidae*** (p. 354)

14(5). Secondary setae numerous; prominent scoli (figs. 20, 21) or hairy protuberances often present; suckerlike structures present on venter of abdominal segments .... (part) ***Limacodidae*** (p. 456)

14′. Secondary setae and scoli absent (plates or hairless lobes may be present); venter of abdominal segments variable 15

15(14′). Distinct (but sometimes short) fleshy prolegs present on A3–6 16

15′. Distinct fleshy prolegs absent on A3–6 (small bumps or suckers may be present) 21

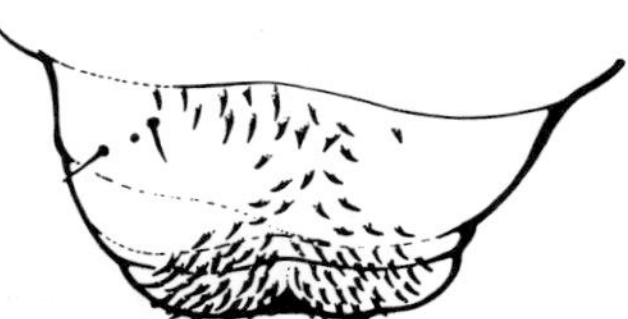

Figure 5

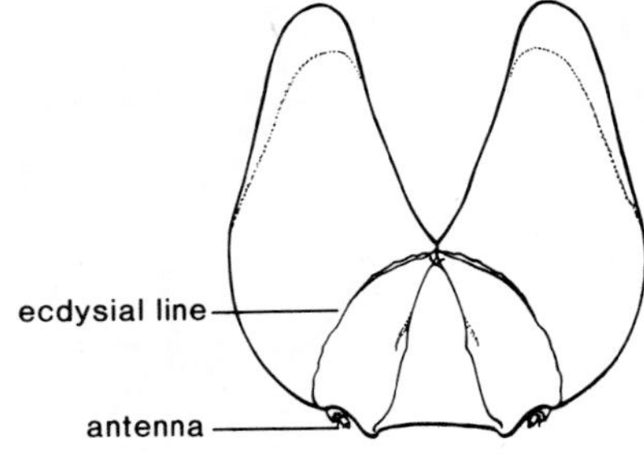

Figure 18

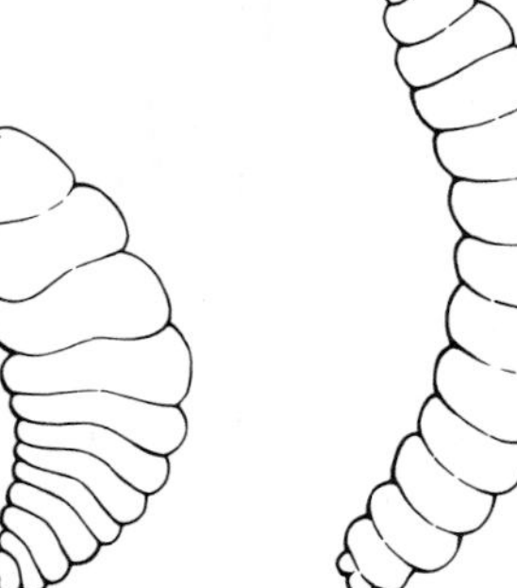

Figure 16 Figure 17

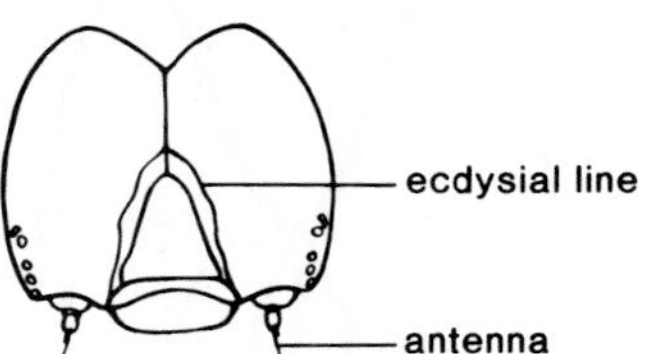

Figure 19

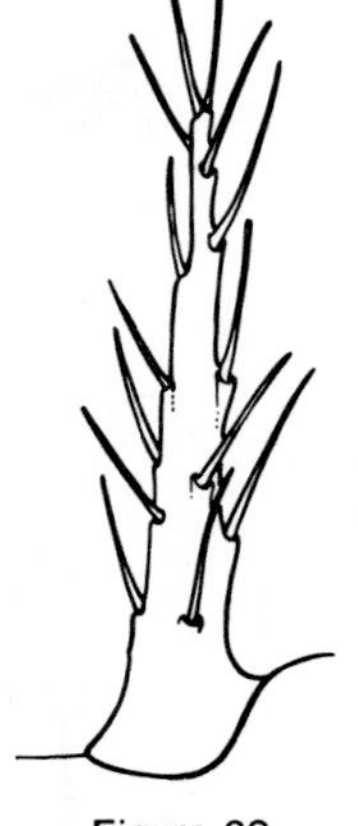

Figure 20

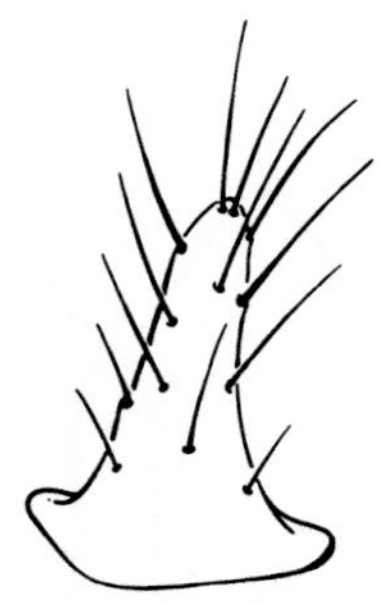

Figure 21

⟵ Figs. 7, 10, 11

16(15). Many spines (fig. 5) (not true hooked crochets) present on ventral part of prolegs; dorsal ambulatory ridges with many spines present on some species; mature larvae large (borers in stems of banana and other plants; tropical and subtropical) .... (part) ***Castniidae*** (p. 417)

16′. Spines absent on ventral part of proleg (setae may be present); dorsal ambulatory ridges lacking spines 17

17(16′). Abdominal spiracles surrounded by a sclerotized spot which includes the SD2 seta base (seta may be absent) (fig. 22) .... (part) ***Ochsenheimeriidae*** (p. 370)

17′. Abdominal spiracles not surrounded by a sclerotized spot that includes the SD2 seta base (if present) 18

18(17′). Spiracles oval, mature larvae greater than 2 cm long; not leaf miners . . . (Hadeninae, *Lasionycta* in part) .... (part) ***Noctuidae*** (p. 549)

18′. Spiracles circular, mature larvae less than 1 cm long; may be leaf miners 19

19(18′). A8 spiracles raised and caudo-projecting .... (part) ***Glyphipterigidae*** (p. 403)

19′. A8 spiracles not raised and caudo-projecting 20

20(19′). A1–8 with L1 and L2 well separated (fig. 23) or L2 difficult to locate; crochets difficult to see at 50× or less, but actually present; (may be leaf miners; some *Bucculatrix, Bedellia* and *Paraleucoptera*) .... (part) ***Lyonetiidae*** (p. 370)

20′. A1–8 with L1 and L2 close together; crochets totally absent (leaf miners) .... ***Douglasiidae*** (p. 408)

21(15′). Body appearing "armored" with platelike areas; head withdrawn beneath prothorax .... (part) ***Limacodidae*** (p. 456)

21′. Body not appearing "armored" with platelike areas; head may or may not be withdrawn beneath prothorax 22

22(21′). *Mesothoracic* shield distinctly sclerotized (case bearers) .... (part) ***Coleophoridae*** (p. 387)

22′. *Mesothoracic* shield not distinctly sclerotized 23

23(22′). Body covered with jellylike, conical tubercles; head withdrawn beneath prothorax .... (part) ***Dalceridae*** (p. 460)

23′. Body not covered with jellylike, conical tubercles; head not withdrawn beneath prothorax 24

24(23′). L1 caudad to spiracle on A3–6 (fig. 24); head seta S2 closer to stemma 6 than 1 (fig. 25) (if leaf miners, go to couplet 25) 25

24′. L1 ventrad to spiracle on A3–6 (fig. 26); head seta S2 closer to stemma 1 than 6 (not leaf miners) .... (a few) ***Gelechiidae*** (p. 394)

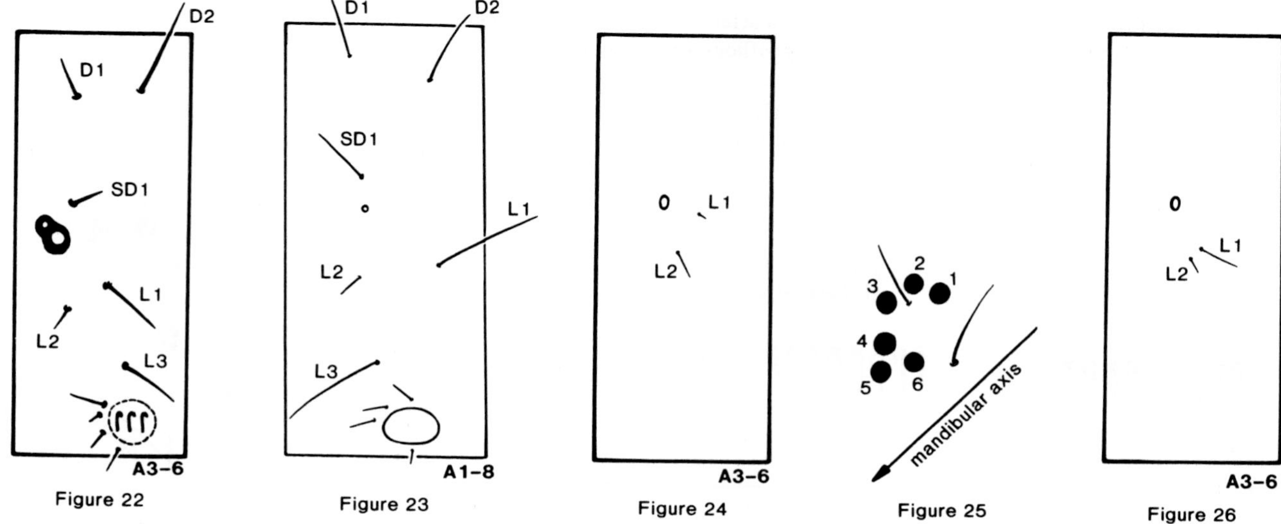

Figure 22 Figure 23 Figure 24 Figure 25 Figure 26

⟵ Fig. 5

25(24). Larvae 15–27 mm, usually pink or red when preserved (borers in *Yucca* pods) .... (part) ***Prodoxidae*** (p. 361)

25′. Larvae 5–6 mm, white (leaf miners) .... ***Acanthopteroctetidae*** (p. 347)

26(4′). Crochets absent on A6 27

26′. Crochets present on A6 28

27(26). Head flattened; stemmata in a nearly straight line along lateral margin of head .... (part) ***Gracillariidae*** (p. 372)

27′. Head not flattened; stemmata in a more normal C-shaped arrangement .... (part) ***Sesiidae*** (p. 411)

28(26′). Crochets on A2–7, those on A2 and A7 smaller and fewer (look very closely); body covered with jellylike, conical tubercles .... (part) ***Dalceridae*** (p. 460)

28′. Crochets absent on A2 and A7; body not covered with jellylike conical tubercles 29

29(28′). All prolegs rudimentary or absent; crochets on A3–6 in transverse rows or reduced in number; (mature larvae internal feeders in leaves, needles, stems, flowers, and roots, or external feeders bearing portable cases) 30

29′. Only caudal prolegs rudimentary or absent, ventral prolegs well developed; crochets on ventral prolegs in a mesoseries (fig. 27) or meso- and lateroseries (fig. 28); (mature larvae external feeders on deciduous trees and shrubs, not bearing cases) 33

30(29). Thoracic legs absent; crochets in a longitudinal small group .... (part) ***Gelechiidae*** (p. 394)

30′. Thoracic legs present; crochets in transverse rows or bands 31

31(30′). Ventral crochets in 2 transverse rows; anal plate with 4 pairs of setae; T1 with SD2 almost dorsad to SD1 and nearly as close to XD2 as to SD1 (fig. 84) .... (few) ***Sesiidae*** (p. 411)

31′. Ventral crochets in 2 transverse rows; anal plate with 4 pairs of setae; T1 with SD2 almost caudad to SD1 and never as close to XD2 as to SD1 (fig. 85) .... (**see 3rd choice**) (few) ***Cossidae*** (p. 416)

31″. Ventral crochets in multiserial rows of spinelike crochets; anal plate with 3 pairs of setae 32

32(31″). Procoxae clearly fused into a single plate; 6 stemmata per side .... ***Adelidae*** (p. 358)

32′. Procoxae not clearly fused into a single plate; no more than 5 stemmata per side .... (part) (*Heliozela aesella*) ***Heliozelidae*** (p. 354)

33(29′). Ventral crochets in uniordinal or biordinal mesoseries plus a uniordinal lateroseries (fig. 28); anal *shield* usually modified into spinose caudoprojecting elongate process .... ***Drepanidae*** (p. 501)

33′. Ventral crochets in uniordinal mesoseries only (fig. 27), no lateroseries present; anal shield never modified into caudoprojecting process (although rudimentary *caudal prolegs* may be caudoprojecting and a pair of caudoprojecting processes may be *below* the anus) .... (part) ***Notodontidae*** (p. 524)

34(4). Thoracic segments without segmented sclerotized legs bearing a claw (ventral knobs may be present); head prognathous and extremely dorsoventrally flattened (fig. 29) 35

34′. Thoracic segments with segmented sclerotized legs bearing a claw on at least 1 thoracic segment; head prognathous or not 36

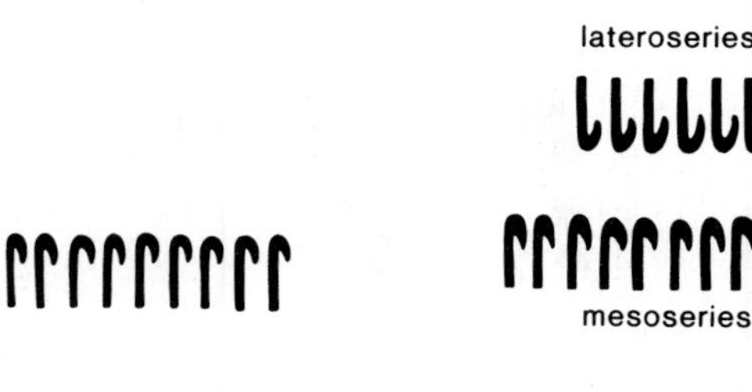

Figure 27 Figure 28

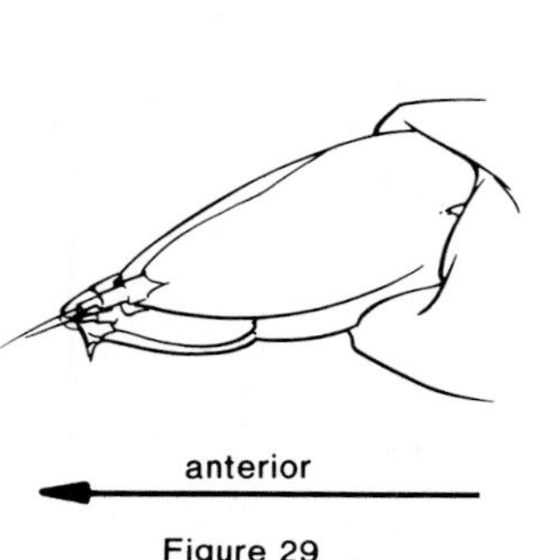

Figure 29

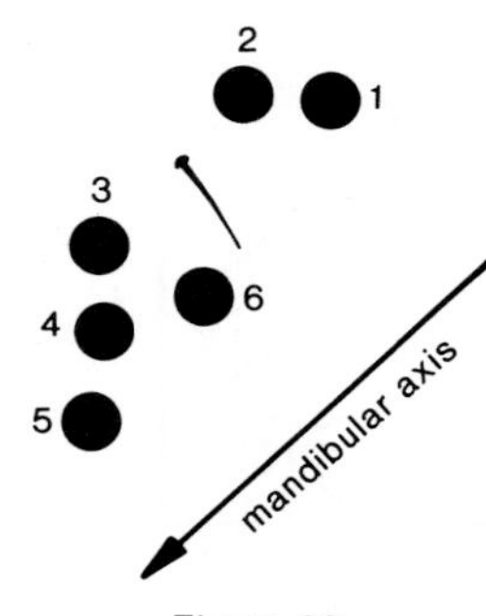

Figure 30

Figs. 84, 85 ⟶

35(34). Crochets absent on A6 .... (part) ***Gracillariidae*** (p. 372)

35′. Crochets present on A6 .... (part) ***Tischeriidae*** (p. 354)

36(34′). Crochets absent on A6 37

36′. At least a few crochets on A6 39

37(36). Six stemmata (simple eyes) per side 38

37′. Fewer than 6 stemmata per side .... (late instar) ***Gracillariidae*** (p. 372)

38(37). Stemmata 1, 2, 3, 4, 5 relatively evenly spaced (casebearers) .... (a few) ***Coleophoridae*** (p. 387)

38′. Stemmata not evenly spaced (1 and 2 close together and separated from 3 by at least 3× the distance between 1 and 2 (fig. 30) (external feeders on foliage, may construct folded or rolled leaf shelter) .... (late instars) ***Gracillariidae*** (p. 372)

39(36′). **Be sure to locate the spiracles: setal bases sometimes resemble spiracles when the setae are broken off. Count all setae on the prolegs.**

Secondary setae absent on A6 (7 or fewer total setae below a line drawn through the dorsal margins of spiracles A5, A6, and A7, and usually 3 or fewer setae in SV position). [Do not count MV3 or SD2 if visible (fig. 31) or spinules (they lack ring bases, fig. 32); be sure to count L1, which rarely may be caudad and very near dorsal margin of the spiracle] 44

39′. Secondary setae (which may be short) present on A6 (8 or more total setae below a line drawn through the dorsal margins of spiracles A5, A6, and A7, and usually 4 or more setae in SV position). [Do not count MV3 or SD2 if visible; be sure to count L1 which rarely may be caudad and very near dorsal margin of the spiracle] 40

40(39′). T2 and A8 (and sometimes other segments) bearing a single pair of subdorsal to dorsal filaments (fig. 33); remainder of head and body essentially bare and vertically striped with combinations of black, yellow, and white; crochets triordinal; all setae very short above the prolegs; no osmeterium on T1 (commonly found on milkweed, oleander, and fig) .... (**Danainae**) ***Nymphalidae*** (p. 448)

40′. Lacking the above combination of characters 41

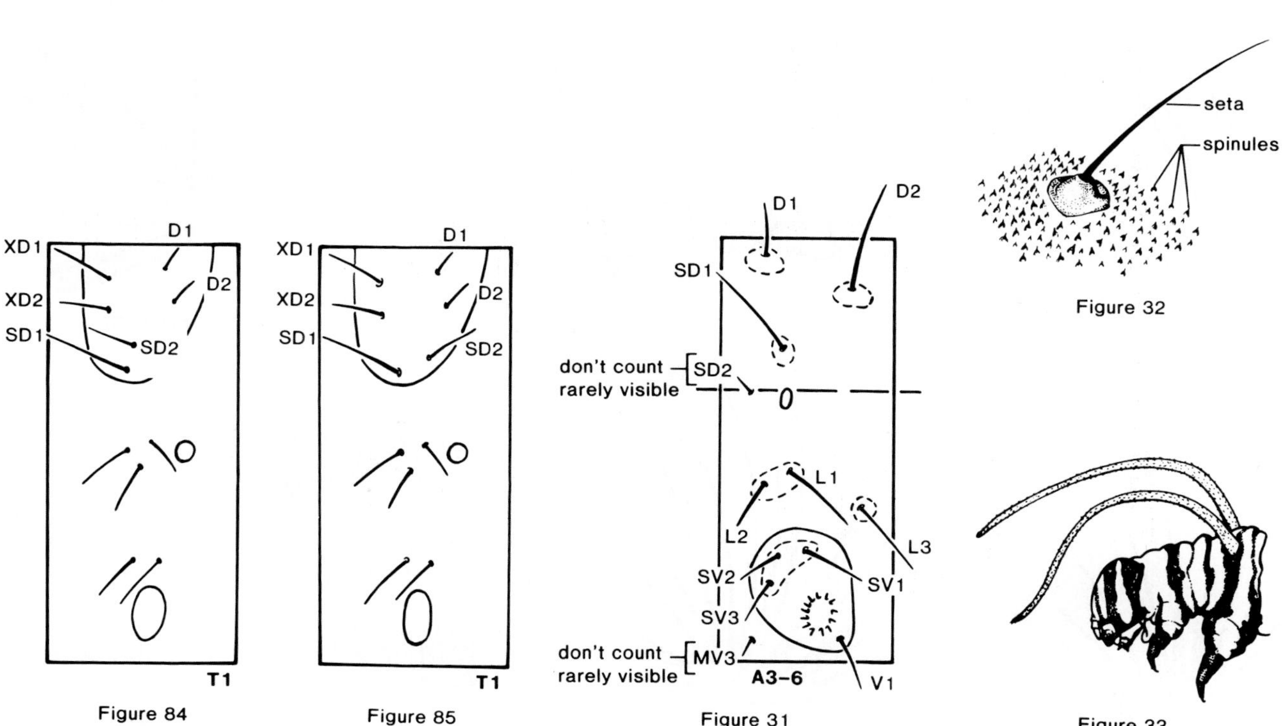

Figure 84

Figure 85

Figure 31

Figure 32

Figure 33

41(40′). T1 through A8 (***examine* all 11 segments!**) with area between spiracles and dorsal midline bearing *conspicuous **seta-bearing** protuberance(s)* or tufts of various kinds on at least 1 segment (the setae may be quite small (fig. 34) or spinelike, but they are ***articulated***) (scoli (figs. 20, 21, 34), chalazae (figs. 35, 36), verrucae (figs. 37, 38, 39), verricules (fig. 40), knobs, fleshy lobes, etc. are frequently present; if the verrucae or verricules are flat and resemble pinacula, then with 7 or more *prominent* setae per pinaculum 130

41′. T1 through A8 with area between spiracles and dorsal midline *not* bearing *conspicuous seta-bearing protuberances* or tufts on at least 1 segment (horns or other protuberances that do *not* bear *articulated* setae may be present); secondary setae not grouped onto verrucae or verricules, although setae may be dense; if flat pinaculalike structures are present, then with 6 or fewer *prominent* setae per pinaculum (secondary setae may be totally absent above the spiracles) 42

42(41′). Osmeterium present on dorsum of T1 (fig. 33a) (it may be invaginated: look for elongate transverse middorsal slit on anterior part) .... ***Papilionidae,*** couplet 174 (p. 438)

42′. Osmeterium not present on dorsum of T1 43

43(42′). None or only a few secondary setae above dorsal margin of spiracle on A3–6, with *most of those present located on primary pinacula* (D1, D2, SD1), and often nearly as long as the primary setae) 196

43′. Many secondary setae above dorsal margin of spiracle on A3–6, with *most of them not located on primary pinacula* (*if present*), and usually shorter than the primary setae (if the primary setae can be distinguished) (secondary setae may be minute) 161

Figure 20

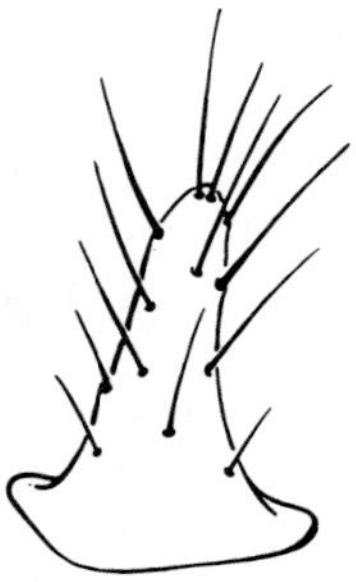

Figure 21

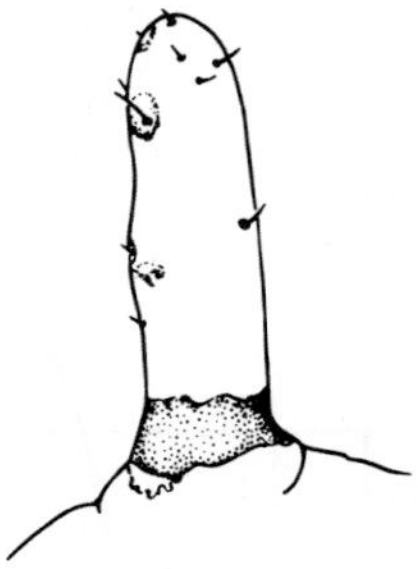

Figure 34

Figure 35

Figure 36

Figure 37

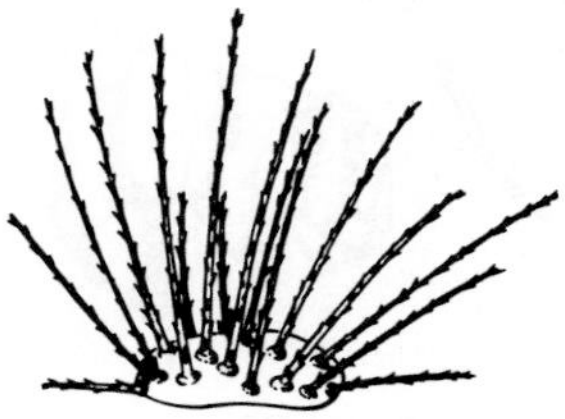

Figure 38

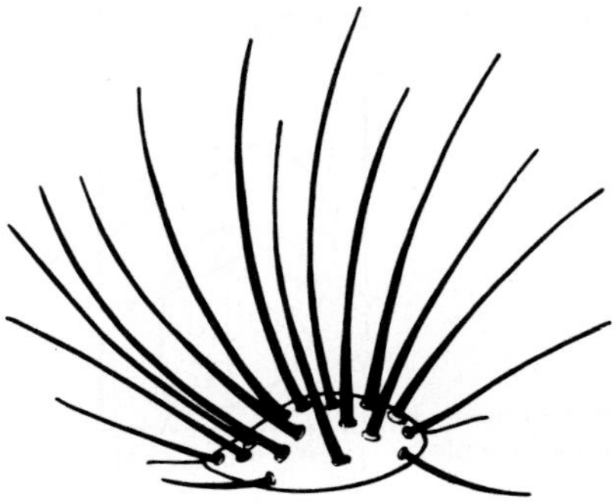

Figure 39

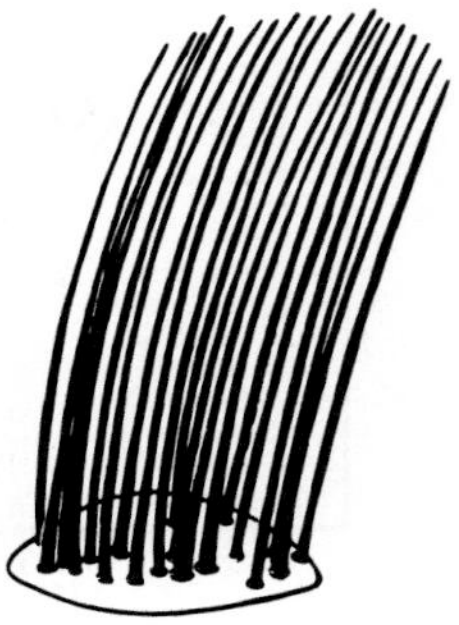

Figure 40

Figure 33a

44(39). Crochets on ventral prolegs arranged in a circle (fig. 41), mesopenellipse (fig. 42), lateropenellipse (fig. 43), transverse bands (fig. 44), or few in number, *never in a mesoseries;* if arranged as a mesopenellipse the gap free of crochets is usually distinctly less than one third the circumference of the projected circle, and the entire series of crochets on each caudal proleg is more nearly *perpendicular* than parallel to the meson 45

44′. Crochets on ventral prolegs *arranged in a mesoseries* (fig. 27) or partial circle, never in a lateropenellipse; if arranged as a partial circle or ellipse, then the gap free of crochets is equal to or greater than one-third the circumference of the projected circle (fig. 45), and the entire series of crochets on each caudal proleg is *usually* (not always) more nearly *parallel* than perpendicular to the meson 123

45(44). A3–6 crochets bi- or triordinal (at least in part) (figs. 46, 47); crochet *bases* arising in a single row throughout (if more than 1 row, go to couplet 46, but check the *bases,* which may be more lightly pigmented than the tips) 96

45′. A3–6 crochets uniordinal *throughout OR* arranged quite irregularly if of more than 1 length; crochet bases arising in 1 or more rows; crochets in transverse bands may be longer in one band than the other 46

46(45′). T1 with L (prespiracular) group trisetose (fig. 48) (if L group is combined with prothoracic shield there are 3 L's) (check both sides if fewer than 3 and check very carefully from several angles; if setae are very difficult to see on small casebearers, regard them as having 3 L's) 58

46′. T1 with L (prespiracular) group unisetose or bisetose (check both sides if fewer than 3 and check very carefully from several angles) 47

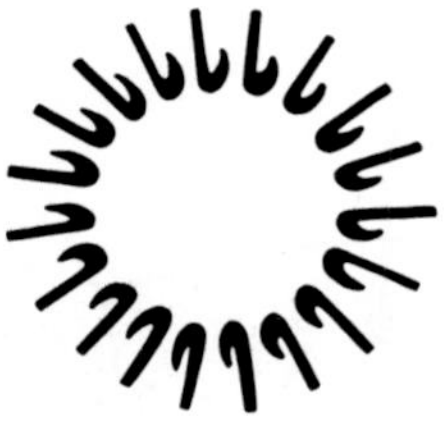

Figure 41

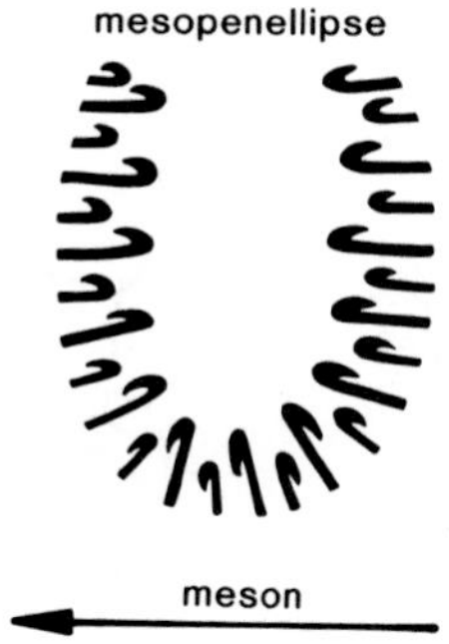

mesopenellipse

meson

Figure 42

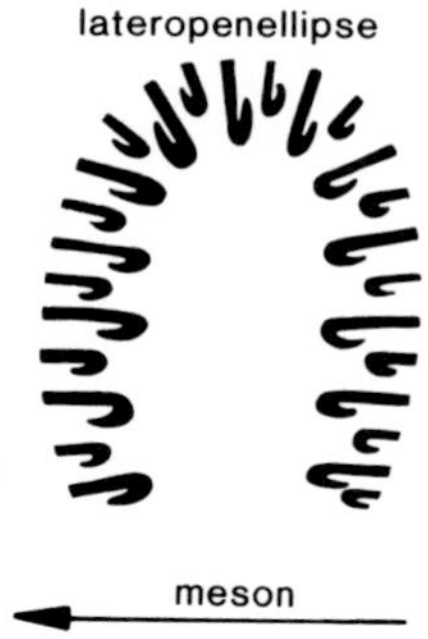

lateropenellipse

meson

Figure 43

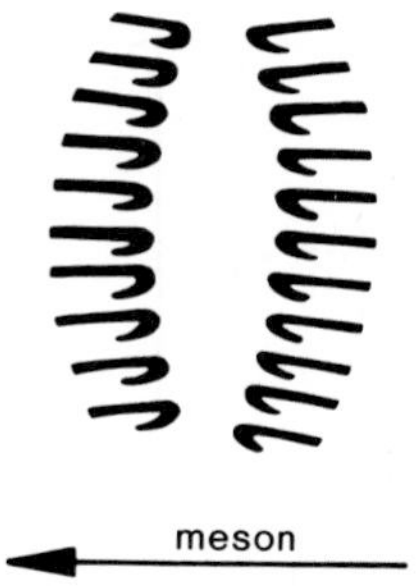

meson

Figure 44

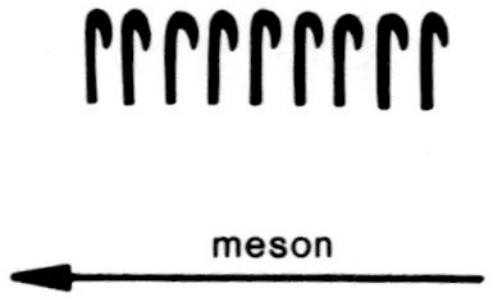

meson

Figure 27

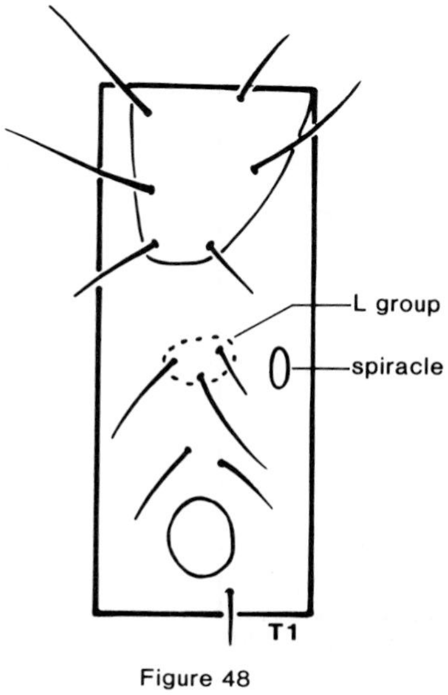

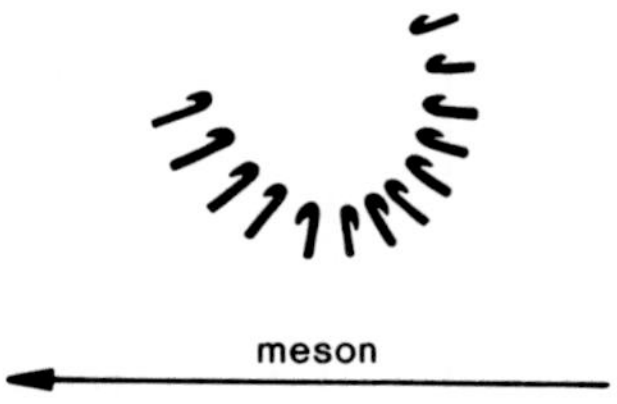

meson

Figure 45

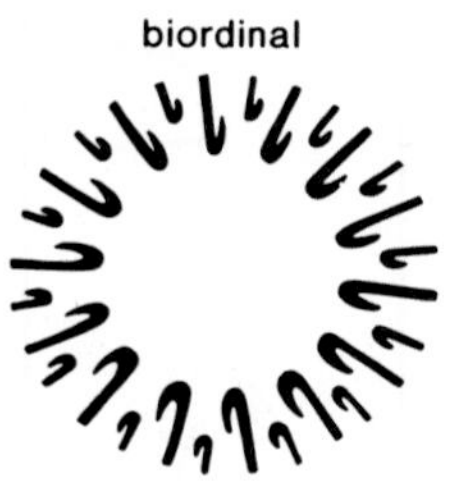

biordinal

Figure 46

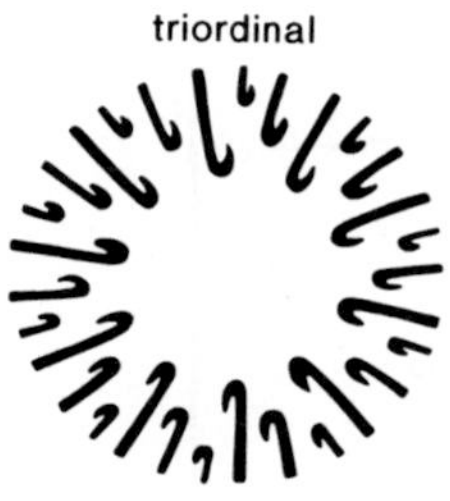

triordinal

Figure 47

L group

spiracle

T1

Figure 48

47(46′). **If neither couplet fits, go to couplet 90**

T2 and T3 with SV group bisetose (fig. 49), and with L3 absent; head with seta S1 usually between stemmata 2 and 3 or more cephalad than 2 and 3 (fig. 50); (leaf rollers and stem borers) .... ***Thyrididae*** (p. 495)

47′. T2 and T3 with SV group unisetose, and with L3 usually present; head with seta S1 usually behind stemmata 2 and 3 (fig. 51) 48

48(47′). **The relationships of these setae are usually most easily observed by standing the larva on its head and observing them from the rear with the setae silhouetted.**

A9 with *D1* (*which may be very small and short*) closer to SD1 than to D2 (D1 and SD1 often on the same pinaculum) (figs. 52, 53) 49

48′. A9 with D1 (*which may be very small and short*) more equidistant between SD1 and D2 or closer to D2 (D1 and SD1 never on the same pinaculum) (figs. 54, 55, 56) 54

49(48). A8 with SD1 cephalad or cephalodorsad of spiracle (fig. 57); T1 with SD2 almost directly dorsad of SD1 (fig. 58) 50

49′. A8 with SD1 more nearly dorsad of spiracle (fig. 59); T1 with SD2 more caudad than dorsad of SD1 (fig. 60) 51

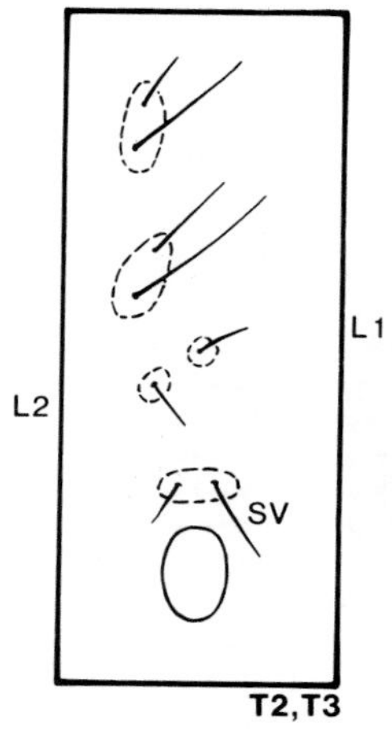

Figure 49

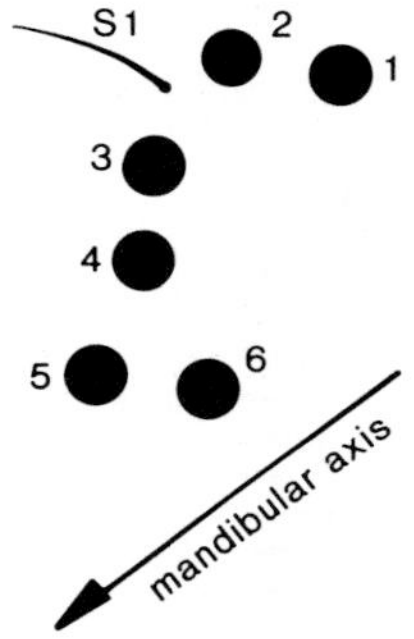

Figure 50

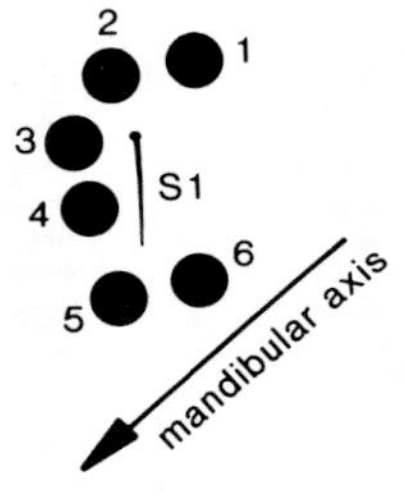

Figure 51

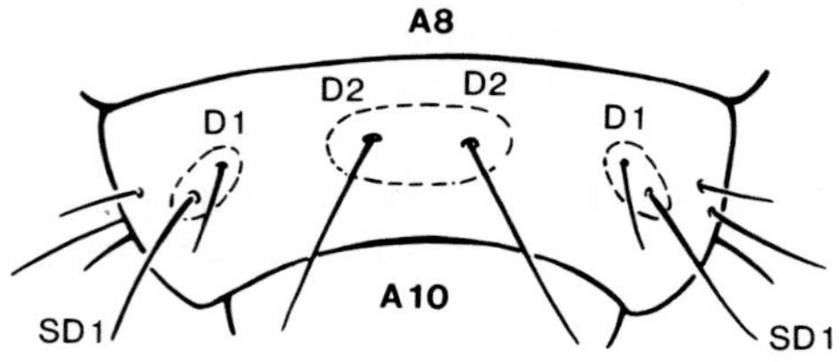

Figure 52

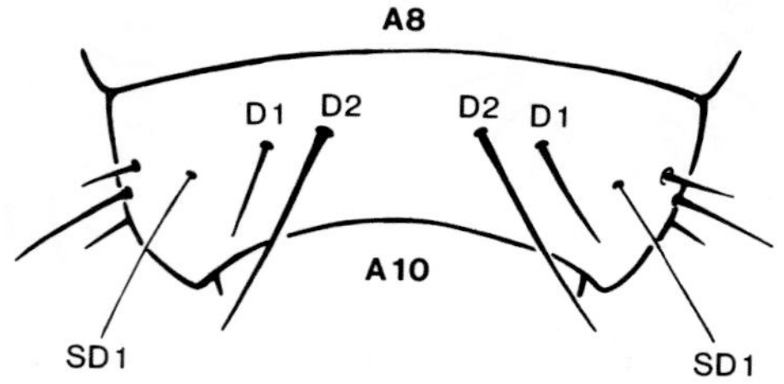

Figure 56

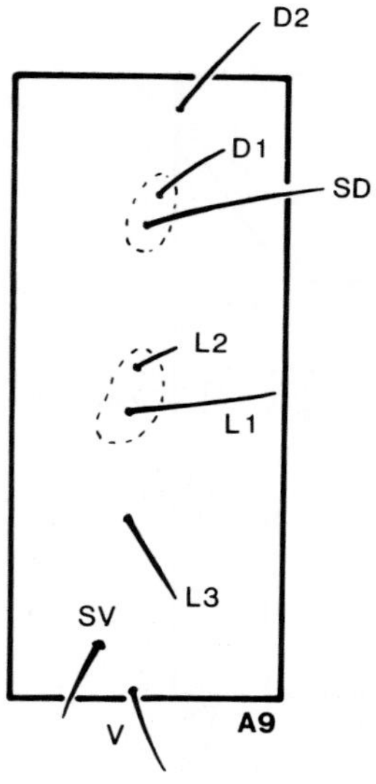

Figure 53

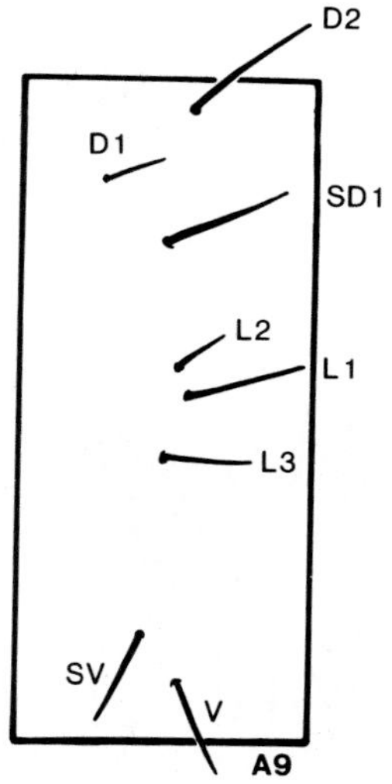

Figure 54

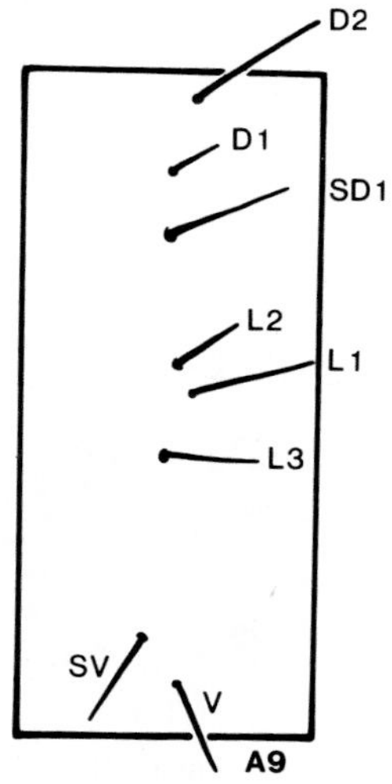

Figure 55

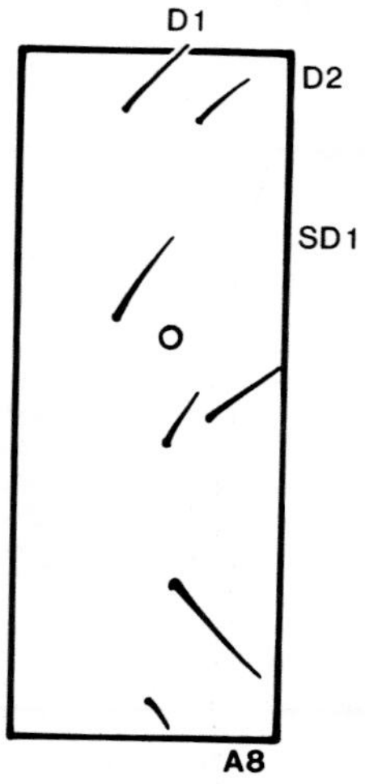

Figure 57

50(49). Spiracles circular; SV3 on A3–6 anterolaterad of crochets (fig. 61) .... ***Epermeniidae*** (p. 402)

50′. Spiracles more oval; SV3 on A3–6 directly in front of crochets (fig. 62) .... (*Eorema*, **Crambinae**) ***Pyralidae*** (p. 462)

51(49′). Spiracles on A8 distinctly closer to dorsomeson than on preceding abdominal segments (fig. 63), and caudoprojecting .... (part) ***Glyphipterigidae*** (p. 403)

51′. Spiracles on A8 on about same level as spiracles on preceding abdominal segments and never caudoprojecting 52

52(51′). A10 prolegs with 2 rows of crochets .... (part) ***Copromorphidae*** (p. 399)

52′. A10 prolegs with a single row of crochets 53

53(52′). Five or six stemmata per side—not grouped as below; (some Chrysauginae, feeders in pods of trumpet-creeper; some Crambinae, feeders in roots and stems of grasses; and Odontiinae, in seed capsules of malvaceous plants) .... (part) ***Pyralidae*** (p. 462)

53′. Five stemmata per side, grouped into 2 pairs, and with the fifth one nearly hidden beneath the antenna (from *Polyporus* (shelf) fungi) .... (part) ***Tineidae*** (p. 362)

54(48′). A1–7 with L1 caudad of and at same level as spiracle, and as close or closer to spiracle than it is to L2 (fig. 64); fungus feeders .... (part) ***Tineidae*** (p. 362)

54′. A1–7 with L1 lower than spiracle and closer to L2 than L1 is to spiracle, or L2 absent (fig. 65) 55

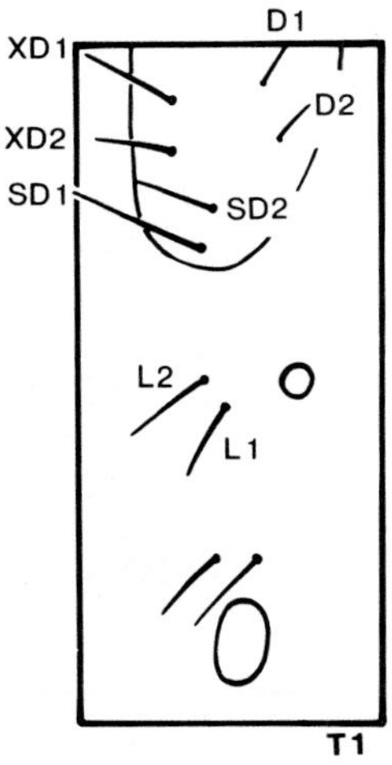

Figure 58

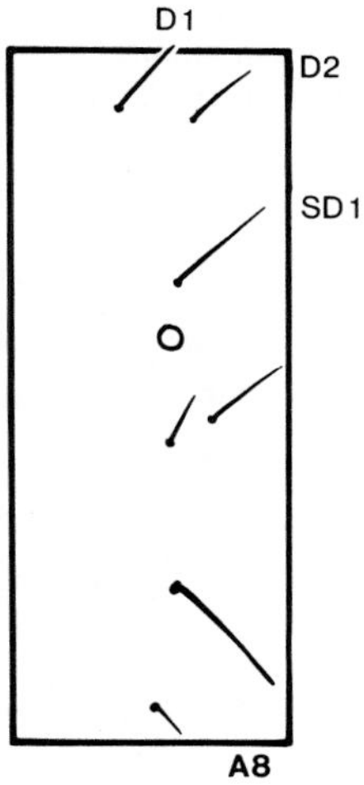

Figure 59

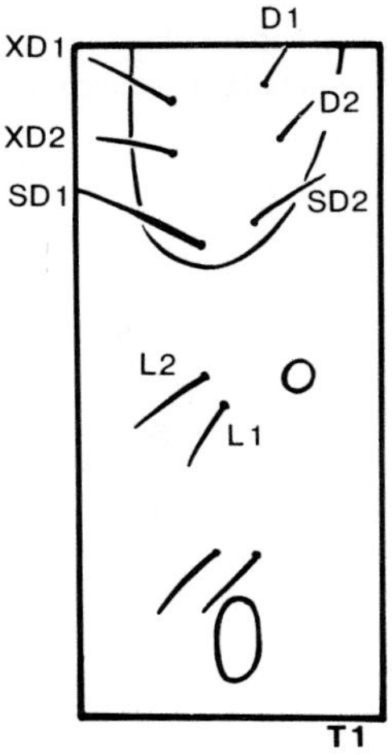

Figure 60

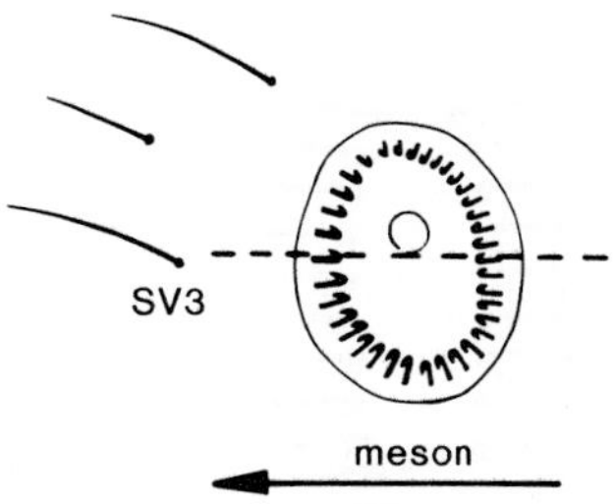

Figure 62

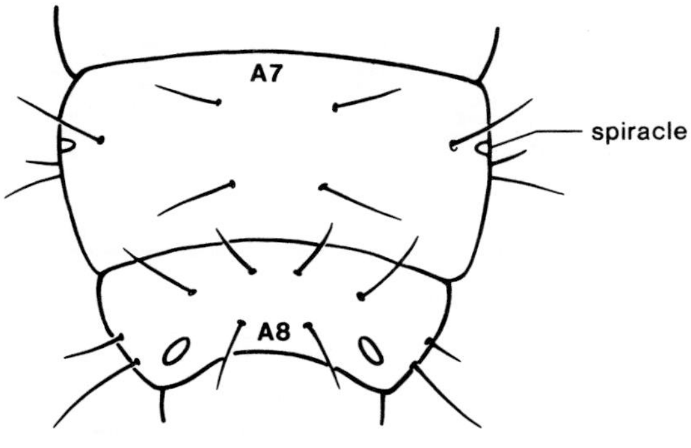

Figure 63

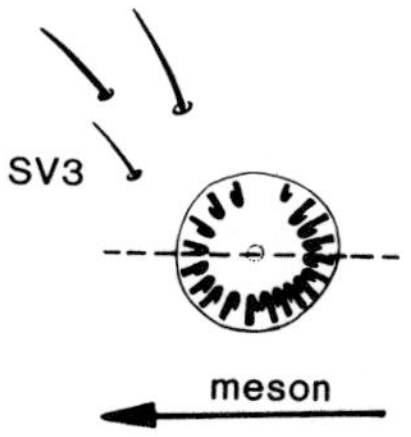

Figure 61

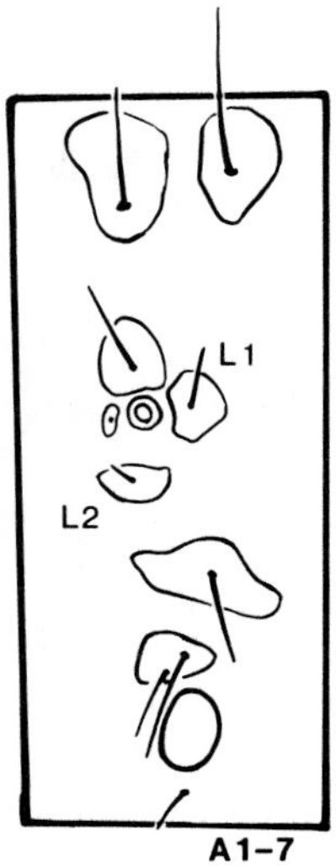

Figure 64

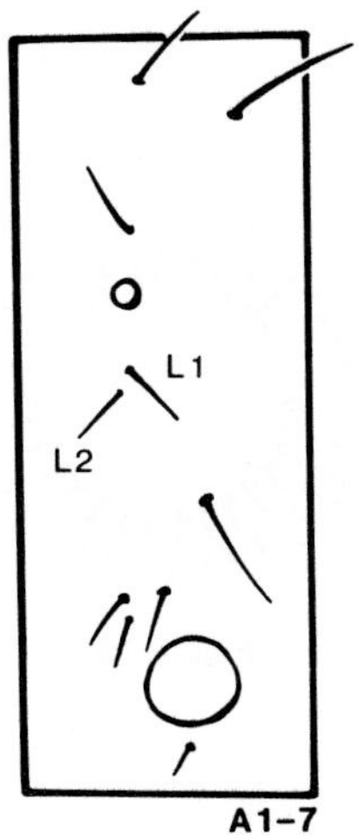

Figure 65

55(54′). A3–6 with L1 and L2 about the width of a spiracle apart (fig. 66), and with a line drawn through them more nearly vertical than horizontal 56

55′. A3–6 with L1 and L2 distinctly farther apart than the width of a spiracle, and with a line drawn through them more nearly horizontal than vertical, or L2 missing 57

56(55). A3–6 with L2 (short seta) dorsad from L1 (long seta) 56a

56′. A3–6 with L2 (short seta) ventrad from L1 (long seta) .... ***Alucitidae*** (p. 401)

56a(56). A8 spiracle in line with the spiracles anterior to it, and laterally oriented; spiracles usually oval .... (a few) ***Pyralidae*** (p. 462)

56a′. A8 spiracle distinctly more dorsal than the spiracles anterior to it, and more posteriorly oriented; spiracles circular .... (part) ***Carposinidae*** (p. 401)

57(55′). Stemmata arranged in relatively normal arc; if stemmata arranged similar to arrangement below, then L2 missing on A3–6 .... ***Momphidae*** (p. 389)

57′. Stemmata arranged differently—1 and 2 separated from 3 and 4, with 5 beneath the antenna, and 6 separated caudally (fig. 67); L2 always present on A3–6 (from shelf fungi) .... (part) ***Tineidae*** (p. 362)

58(46). Ventral crochets uniserial (fig. 44), no small hookless crochets on planta (sometimes with a *few* extra crochets *scattered* posteriorly (fig. 68) that are no longer than the surrounding ones) 63

58′. Ventral crochets biserial or multiserial (figs. 69, 70), *OR* with a single short row of well-developed unscattered crochets plus one or several rows of smaller crochets or small hookless crochets that encircle or are in front of the well developed crochets (fig. 71) 59

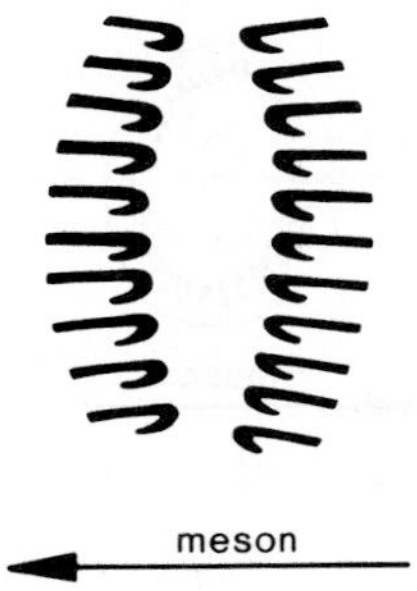

Figure 44

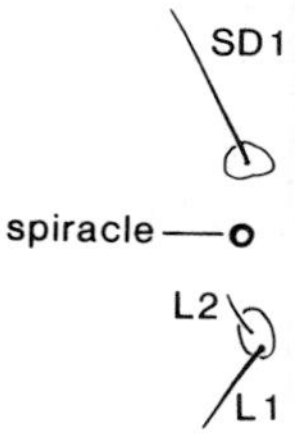

Figure 66

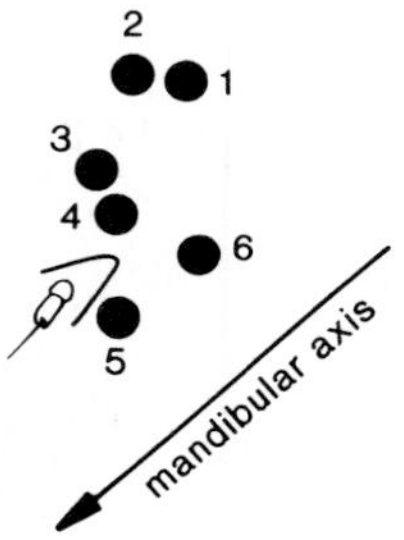

Figure 67

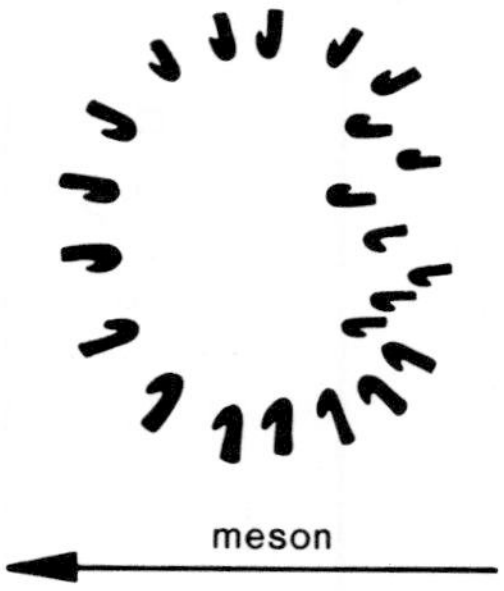

Figure 68

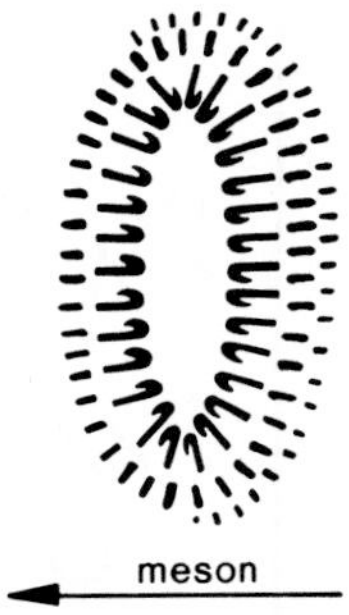

Figure 69

Figure 70

Figure 71

Figure 76

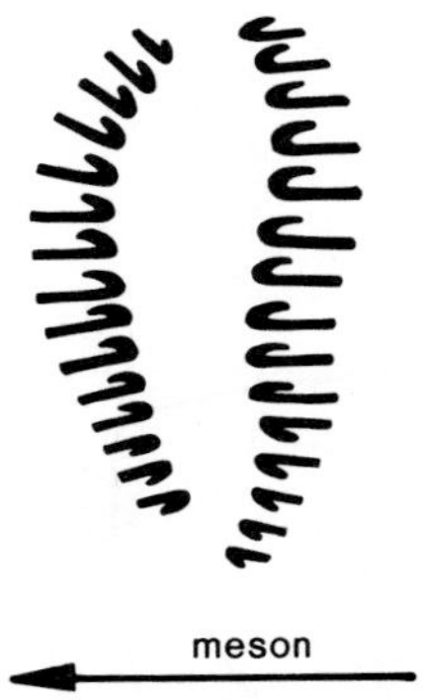

Figure 77

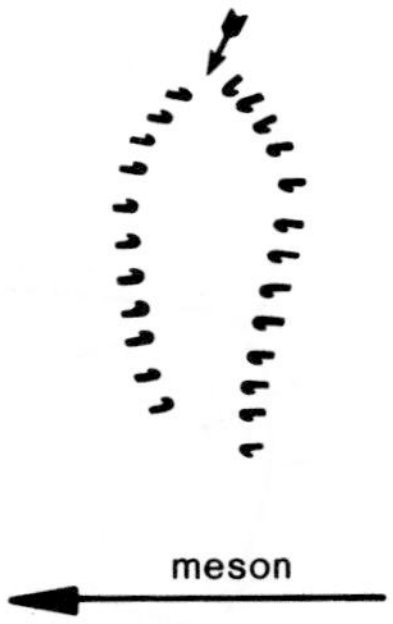

Figure 78

Figs. 72-75 ⟶

59(58′). Prothoracic spiracle included on (surrounded by) the *distinct* L pinaculum (which may be continuous with prothoracic shield (fig. 72)) (often on clover and in roots of grasses, may construct silken galleries and webs) .... (**Acrolophinae,** a few **Tineinae**) ***Tineidae*** (p. 362)

59′. Prothoracic spiracle excluded from the L pinaculum or all pinacula indistinct 60

60(59′). Four SV setae on prolegs; 3 pairs of setae on anal plate; L1 and L2 essentially posterior to spiracle on A3–6 (fig. 73) .... (part) ***Hepialidae*** (p. 347)

60′. Three SV setae on prolegs; 4 pairs of setae on anal plate (look closely); L1 and L2 not essentially posterior to spiracle on A3–6 (at least 1 of them (L2) ventral or somewhat anterior to spiracle) 61

61(60′). A3 with L1 *distinctly* closer to spiracle than L2 is to spiracle (fig. 74) .... (part) ***Yponomeutidae*** (p. 406)

61′. A3 with L1 and L2 equidistant from spiracle (fig. 75) (if L1 is *slightly* closer, then L1 is closer to L2 than it is to spiracle) 62

62(61′). A9 with 8 or 9 setae per side .... (part) ***Plutellidae*** (p. 404)

62′. A9 with 6 setae per side .... ***Acrolepiidae*** (p. 409)

63(58). Crochets on ventral prolegs in transverse band(s) (figs. 76, 77) or reduced in number (less than 5 on a proleg), or if arranged as a semicircle, then crochets make up less than one-half the circumference of the projected circle 64

63′. Crochets on ventral prolegs in a complete or incomplete circle or a penellipse, and not reduced or little reduced in number (5 or more per proleg); if an incomplete circle, or less than 5 crochets are present, then crochets make up more than one-half the circumference of the projected circle (if proleg is somewhat collapsed, "crochets in a circle" may appear to be 2 transverse bands: also, if "transverse bands" could be regarded as an incomplete circle or penellipse by the addition of *one* crochet at one end (fig. 78), proceed to couplet 75 75

64(63). T1 with legs absent or distinctly smaller than legs on T2 and T3; head prognathous, dorsoventrally flattened, elongate; (leaf miners on grasses and sedges) .... (part) ***Elachistidae*** (p. 383)

64′. T1 with legs never absent or smaller than legs on T2 and T3; head semiprognathous or hypognathous, not extremely dorsoventrally flattened 65

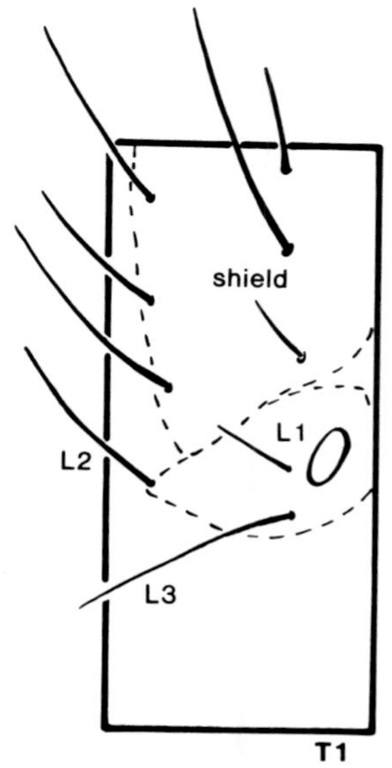

Figure 72

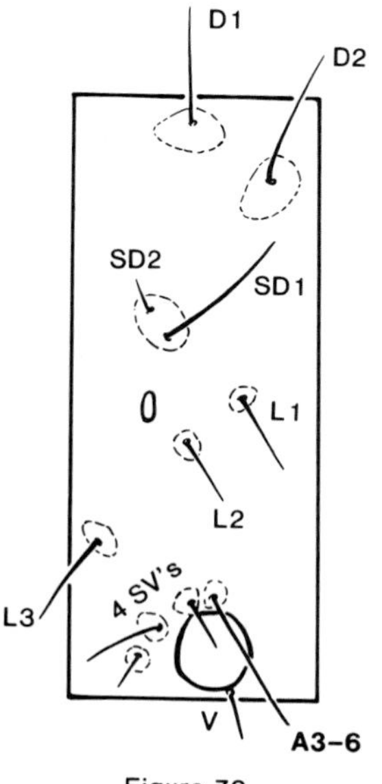

Figure 73

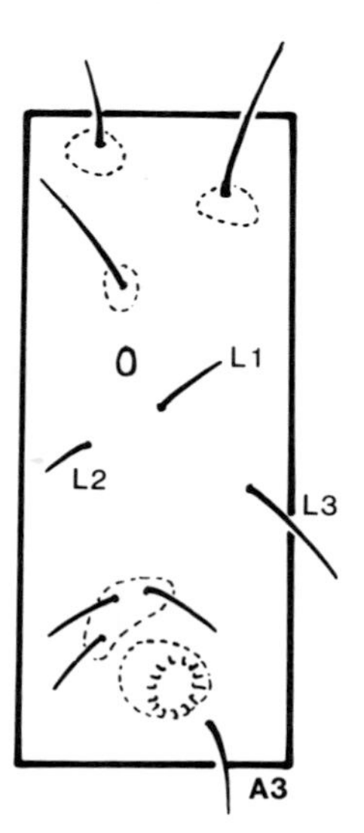

Figure 74

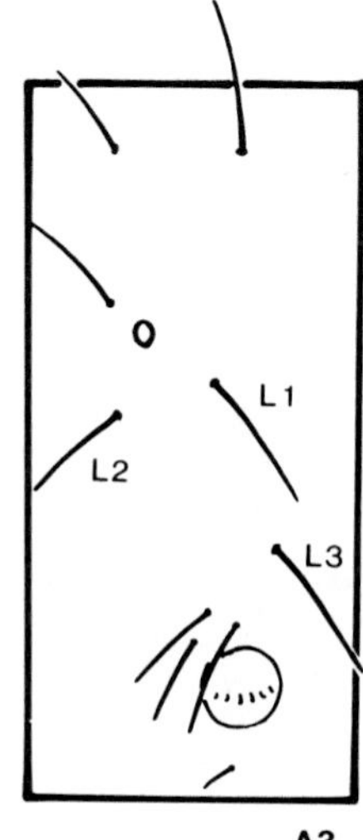

Figure 75

←—— Figs. 76–78

65(64′). Ventral prolegs with both crochets and spines, the spines continuous between the prolegs (fig. 79) (borers in stems of banana and other plants, tropical and subtropical) .... (part) ***Castniidae*** (p. 417)

65′. Ventral prolegs with crochets only 66

66(65′). T2 and T3 with L1 equidistant between L2 and L3 or L1 closer to L3 (fig. 80); stemmata 3, 4, and 5 not equally spaced, distance between 4 and 5 usually (not always) 2× the diameter of 4 (fig. 81); crochets in 2 transverse bands or a complete ellipse (medium to large wood or stem borers) 67

66′. T2 and T3 with L1 closer to L2 than to L3 (fig. 82) or only 2 L setae present; stemmata 3, 4, and 5 more equally spaced, distance between 4 and 5 less than 2× the diameter of 4 (fig. 83); crochet arrangement variable (if setae are so faint or short they are nearly impossible to see, go to couplet 68) (small to medium larvae) 68

67(66). T1 with SD2 almost dorsal to SD1 and nearly as close to XD2 as to SD1 (fig. 84) .... (part) ***Sesiidae*** (p. 411)

67′. T1 with SD2 almost posterior to SD1 and never as close to XD2 as to SD1 (fig. 85) .... (part) ***Cossidae*** (p. 416)

68(66′). Abdominal spiracles surrounded by a sclerotized spot which includes the SD2 seta base (fig. 22); from 1–14 crochets arranged in a lateroseries (borers in stems of wheat and other grasses) .... (part) ***Ochsenheimeriidae*** (p. 370)

68′. Abdominal spiracles not surrounded by a sclerotized spot that includes the SD2 seta base; usually less than 5 crochets arranged in transverse bands on each proleg 69

69(68′). Two setae caudad of spiracle on A3–6 (fig. 86); each proleg area with up to 12 crochets in a single transverse row .... (*Paraclemensia, Vespina*) ***Incurvariidae*** (p. 358)

69′. No more than 1 seta caudad of spiracle on A3–6; crochets variable 70

70(69′). A3 with L1 and L2 closer to each other than either is to spiracle; L1 ventrad or caudoventrad of the spiracle (fig. 87) (if spiracles and/or setae are so faint or short they are nearly impossible to see, go to couplet 71) 71

70′. A3 with L1 and L2 as far from each other or farther from each other than either is from spiracle (fig. 88); L1 sometimes nearly caudad of the spiracle .... (part) ***Lyonetiidae*** (p. 370)

71(70). A10 with crochets divided by a gap into 2 groups (sometimes each group may contain only a single crochet) (fig. 89) .... (part) ***Gelechiidae*** (p. 394)

71′. A10 with crochets never divided by a gap into 2 groups 72

72(71′). Ventral prolegs distinct and peglike .... (part) ***Heliodinidae*** (p. 410)

72′. Ventral prolegs rudimentary 73

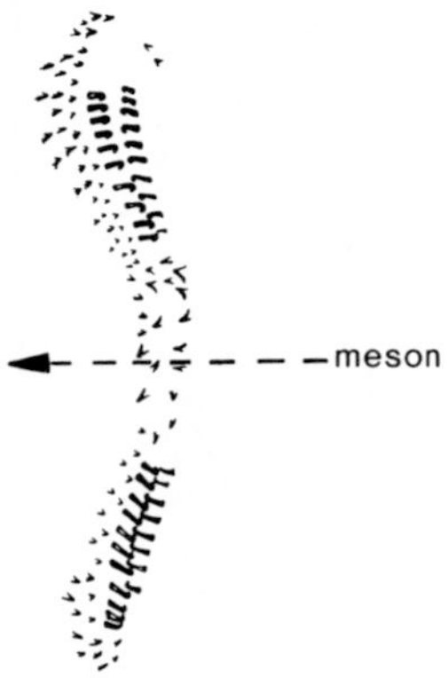

Figure 79

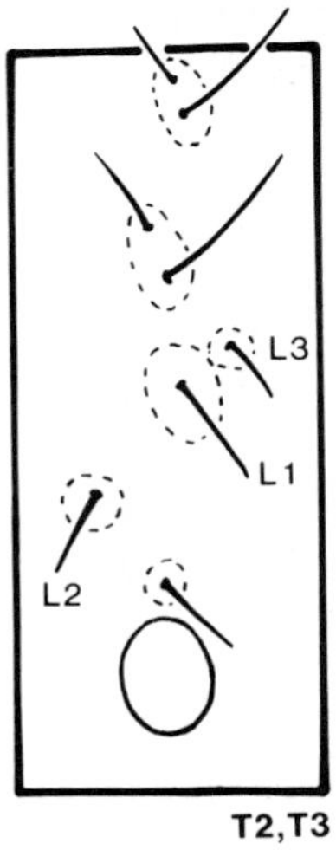

Figure 80

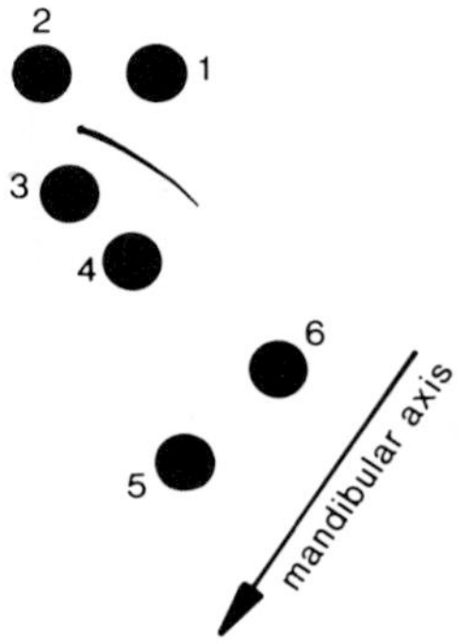

Figure 81

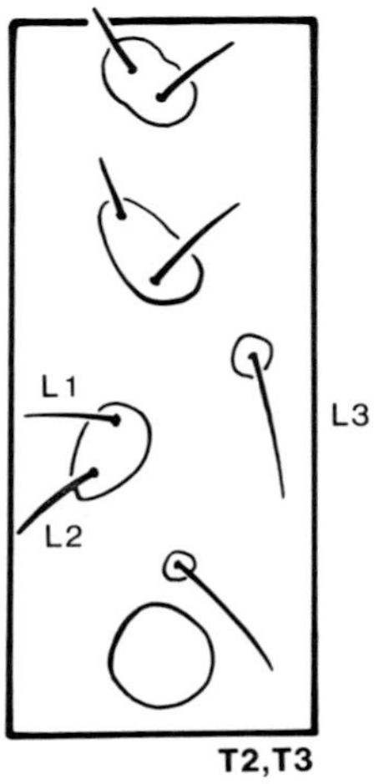

Figure 82

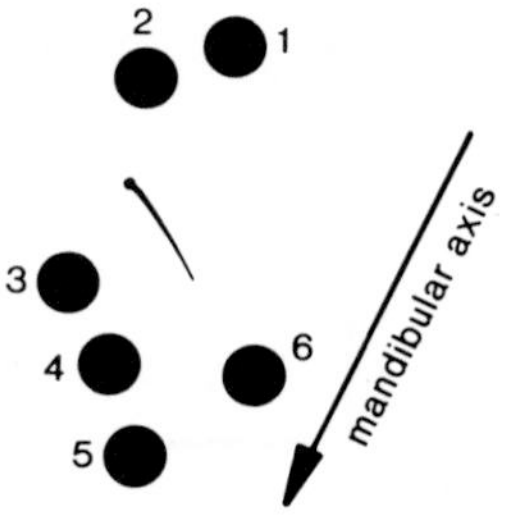

Figure 83

Fig. 22 ⟶

73(72′). A3–6 prolegs with 2 rows of crochets on each 74
73′. A3–6 prolegs with 1 row of crochets on each .... (part) ***Gelechiidae*** (p. 394)
74(73). Frontoclypeus extending to epicranial notch, which is very deep (fig. 90) (case-bearers) .... (part) ***Coleophoridae*** (p. 387)
74′. Frontoclypeus not reaching epicranial notch (the adfrontal areas may), which is more shallow (fig. 91) (not case-bearers) .... (part) ***Argyresthiidae*** (p. 407)
75(63′). T2 and T3 with SV group (above the leg) unisetose 78
75′. T2 and T3 with SV group bisetose 76
76(75′). Prothoracic shield including (surrounding) the spiracle and L (prespiracular) group (fig. 92); all 6 stemmata present and easily visible, with stemmata 1–5 evenly spaced; thoracic legs large and heavily sclerotized (polyphagous feeders constructing portable bags) .... ***Psychidae*** (p. 366)
76′. Prothoracic shield excluding (not surrounding) the spiracle and the L group (which is often on its own well-sclerotized pinaculum); usually with at least 3, sometimes all 6 stemmata missing; if more than 3 are present, then 1 and 2 are conspicuously separated from 3, 4, and 5 (fig. 83); thoracic legs not conspicuously large 77

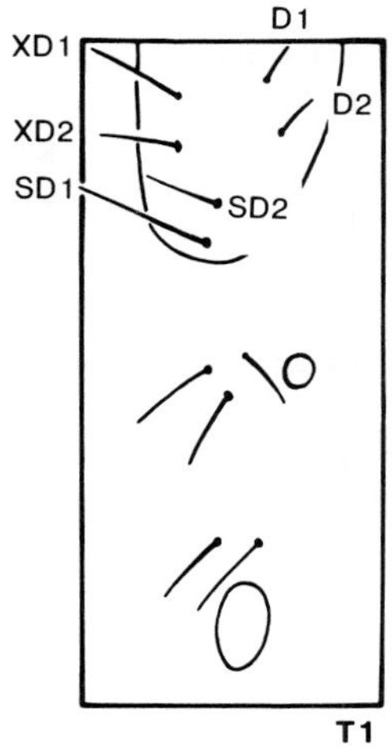

Figure 84

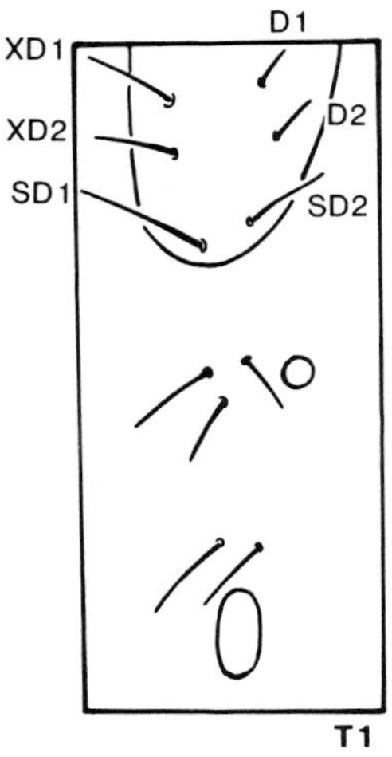

Figure 85

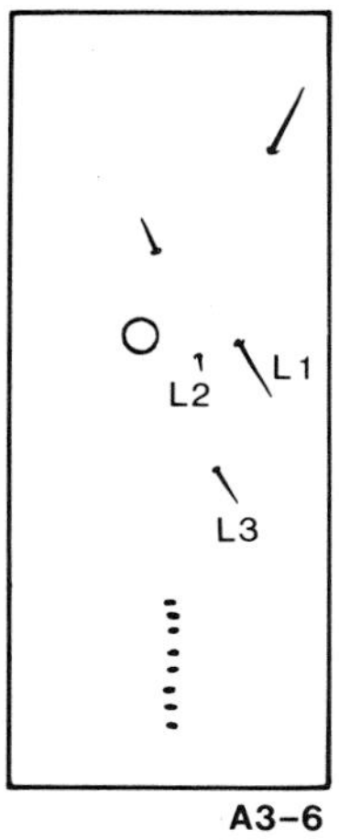

Figure 86

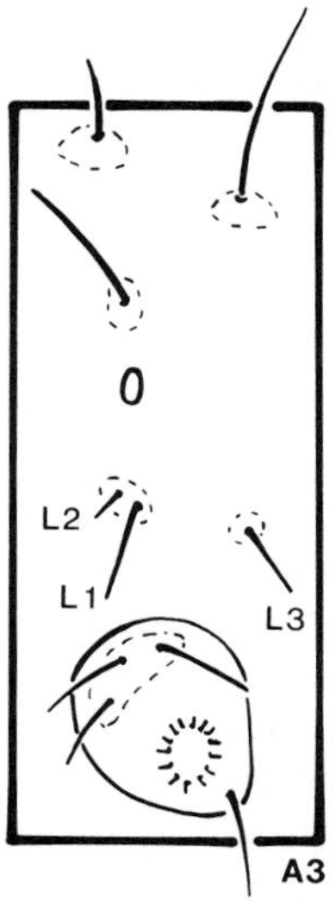

Figure 87

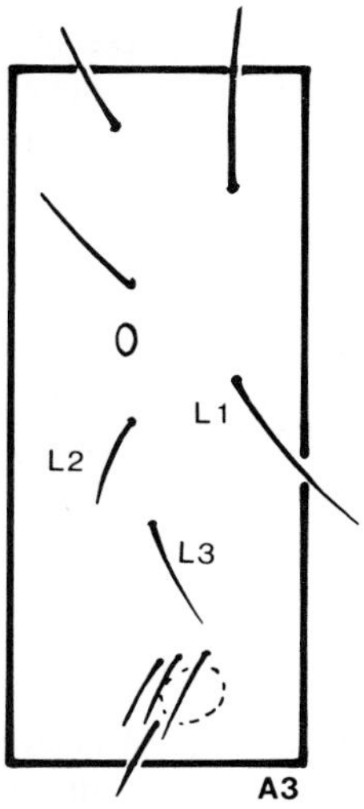

Figure 88

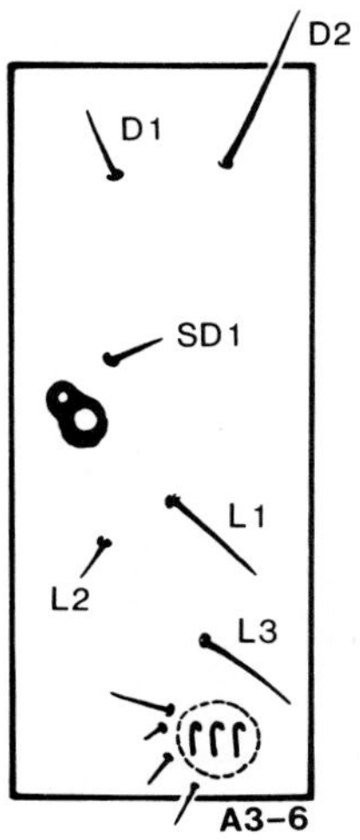

Figure 22

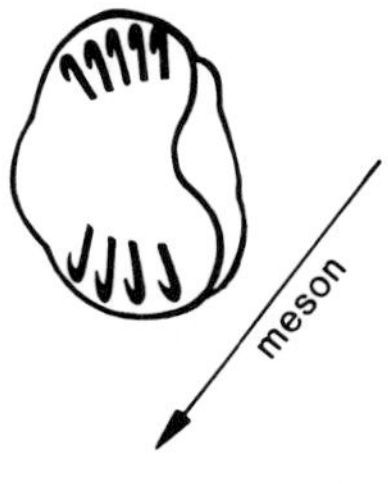

Figure 89

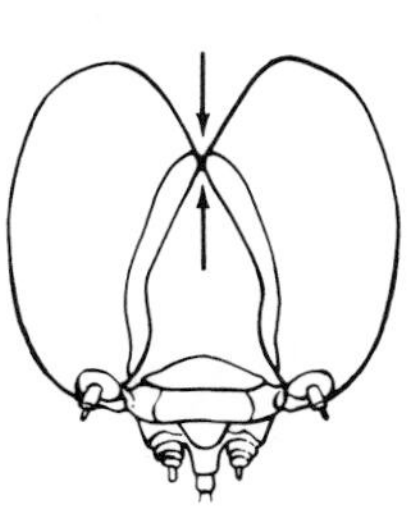

Figure 90

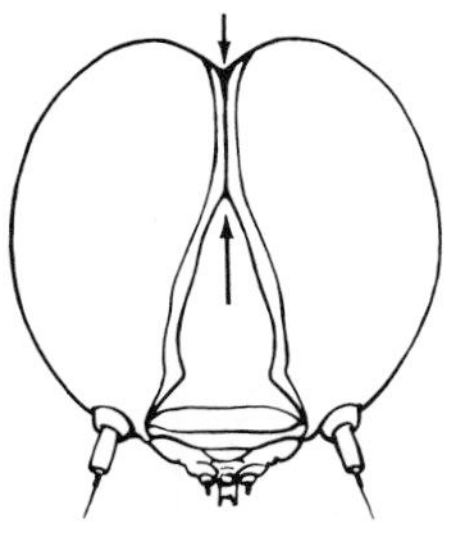

Figure 91

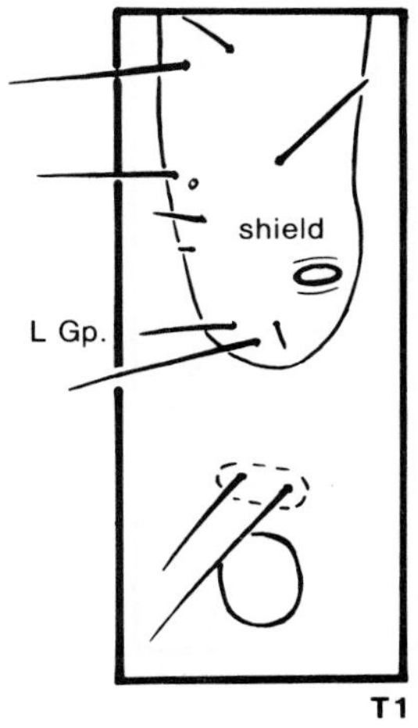

Figure 92

77(76′). Spinules abundant on body (resembling secondary setae); (feeding or boring in living plants, sometimes forming webs) .... (part) ***Pterophoridae*** (p. 497)

77′. Spinules absent on body; (feeding on dried animal and vegetable matter, sometimes constructing cases) .... (part) ***Tineidae*** (p. 362)

78(75). A3 with L1 or L2 closer to spiracle than L1 and L2 are to each other (fig. 93), or distances equal (fig. 94), or L2 absent 79

78′. A3 with L1 and L2 closer to each other (sometimes only slightly so) than either is to spiracle (fig. 87) 85

79(78). With 4 or fewer stemmata on each side of head 80

79′. With at least 5 stemmata on each side of head (check beneath the antenna for one which may be present) 81

80(79). A3 with SD2 almost as large as SD1 and with L1 much closer to spiracle than L2 is (fig. 95); body elongate and covered with minute spinules arranged in vertical rows; (scavengers; may construct silk galleries in fungus) .... (1 species in U.S., *Oinophila v-flavum*) (**Oinophilinae**) ***Tineidae*** (p. 362)

80′. A3 with SD2 much smaller (and sometimes hard to see) than SD1 and with L1 and L2 more nearly equidistant from spiracle (fig. 96); body not fitting above description (mostly feeders in fungus or decaying wood, occasionally found in dried food products or nests of birds, mammals, insects, etc.) .... (part) ***Tineidae*** (p. 362)

81(79′). Distance between stemmata 2 and 3 more than 2× the greatest diameter of stemma 2 (fig. 97) .... (part) ***Tineidae*** (p. 362)

81′. Distance between stemmata 2 and 3 less than 2× the greatest diameter of stemma 2 82

82(81′). A3 with L1 and L2 present and conspicuous (fig. 98) 83

82′. A3 with L1 or L2 absent or very inconspicuous 84

83(82). A3–6 with SD2 present above and in front of spiracle (fig. 98) .... (a few) ***Tineidae*** (p. 362)

83′. A3–6 with SD2 not evident .... (part) ***Plutellidae*** (p. 404)

84(82′). Thoracic legs with the claw elongate and narrow (fig. 99); prolegs on A3 smaller and farther apart than those on A4–6 (leaf miners) .... (part) ***Lyonetiidae*** (p. 370)

84′. Thoracic legs with the claw relatively short (fig. 100); prolegs on A3 similar to those on A4–6 .... (part) ***Gelechiidae*** (p. 394)

85(78′). T3 legs with distance between coxal bases *less than* 1.5× the maximum diameter (any direction) of a (projected) coxal base (fig. 101) 92

85′. T3 legs with distance between coxal bases *greater than* 1.5× the maximum diameter (any direction) of a (projected) coxal base 86

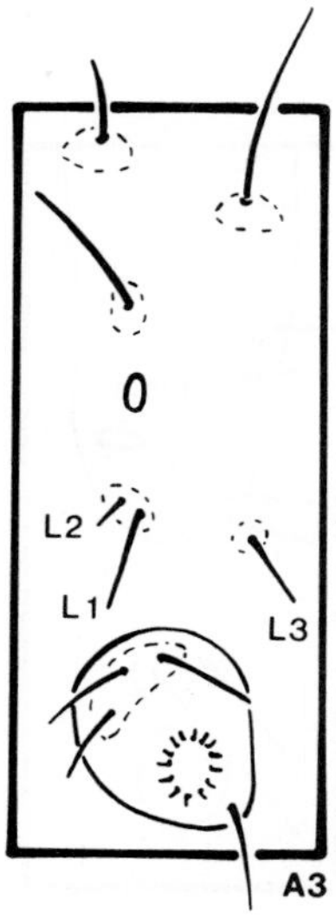

Figure 87

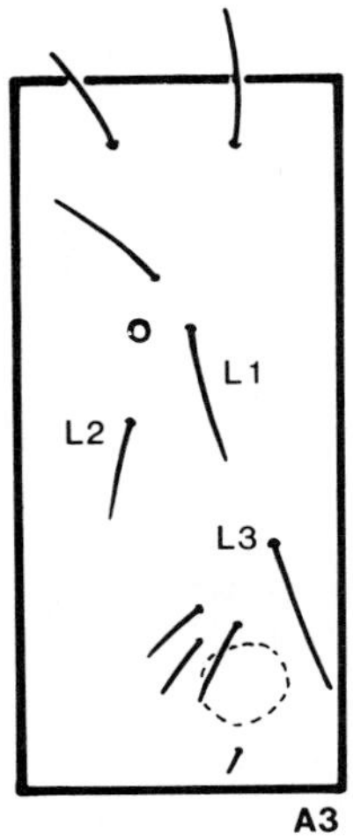

Figure 93

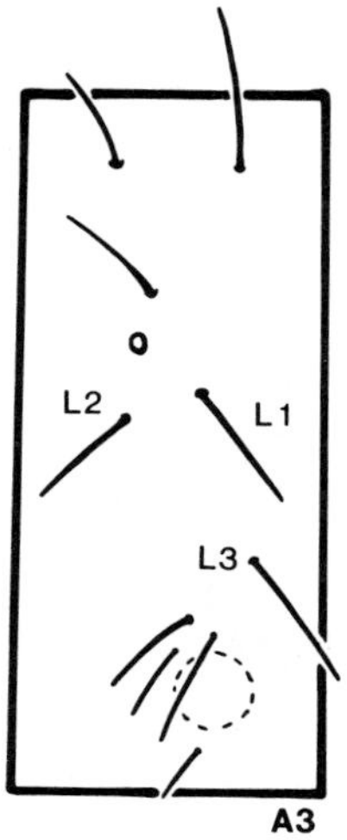

Figure 94

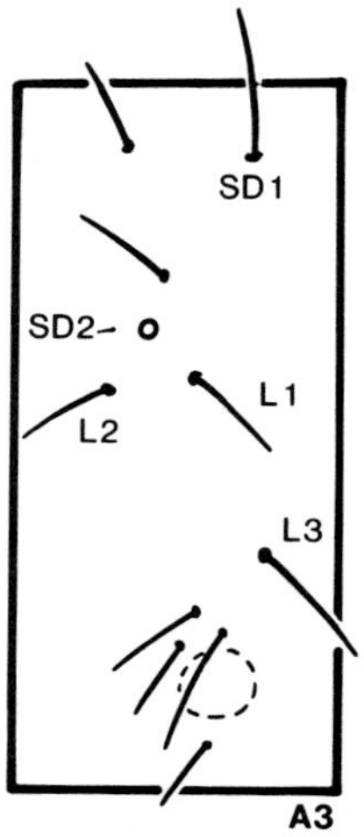

Figure 96

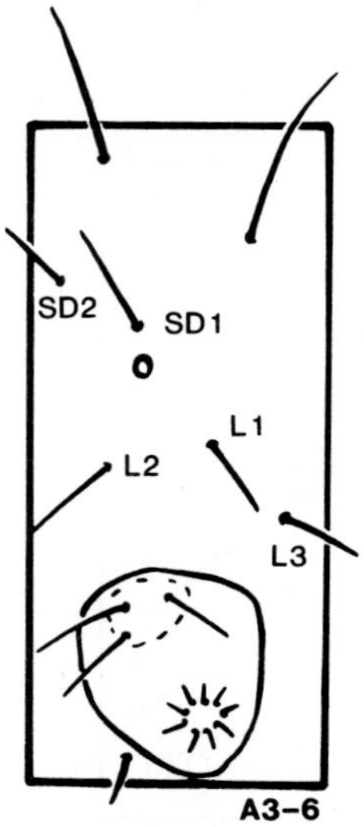

Figure 98

Figs. 95, 97 ⟶

86(85′). Stemmata "overlapping" and difficult to count, but head usually appearing to have 2 or 3 "eyespots" on each side (fig. 102) (leaf miners) .... (part) ***Lyonetiidae*** (p. 370)

86′. Stemmata usually easy to count, and with at least 5 on each side; if stemmata are faint and difficult to count, then 2 or 3 "eyespots" are *not present* on each side of head 87

87(86′). A3 with D1 and D2 adjacent, distance between the two D1 setae at least 4× distance between D1 and D2 (fig. 103) (leaf miners) (*Lampronia*) .... (part) ***Prodoxidae*** (p. 361)

87′. A3 with D1 and D2 farther apart, distance between the two D1 setae less than 4× distance between D1 and D2 (fig. 104) (if D1 is absent go to couplet 112) 88

88(87′). Venter of T1 with darkened, sclerotized spots .... (part) ***Elachistidae*** (p. 383)

88′. Venter of T1 without darkened sclerotized spots 89

89(88′). A3–6 with L2 (short seta) more anterior than dorsal of L1 (fig. 105) 91

89′. A3–6 with L2 absent or more dorsal than anterior of L1 90

90(89′). Crochets on A3–6 in transverse bands or lateropenellipse .... (part) ***Cossidae*** (p. 416)

90′. Crochets on A3–6 not in transverse bands or lateropenellipse 109

91(89). A3–6 with distance between L1 and L2 greater than diameter of an A3–6 planta .... (part) ***Heliodinidae*** (p. 410)

91′. A3–6 with distance between L1 and L2 less than diameter of an A3–6 planta ..... (a few) ***Gelechiidae*** (p. 394)

92(85). Prolegs elongate, peglike, at least 3× longer than greatest planta diameter (fig. 106) 93

92′. Prolegs not conspicuously elongate or peglike, less than 3× longer than greatest planta diameter 101

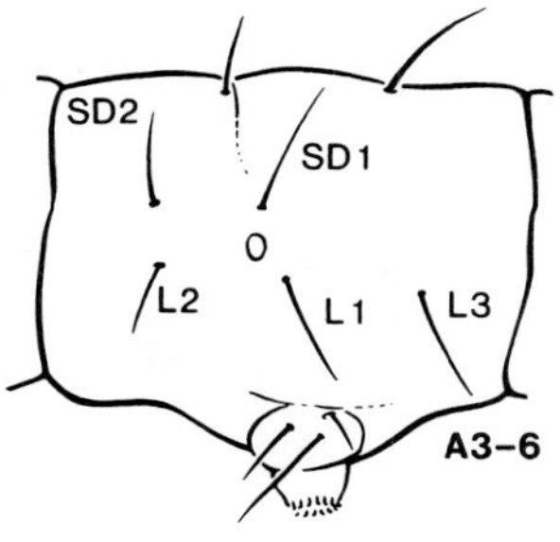

Figure 95

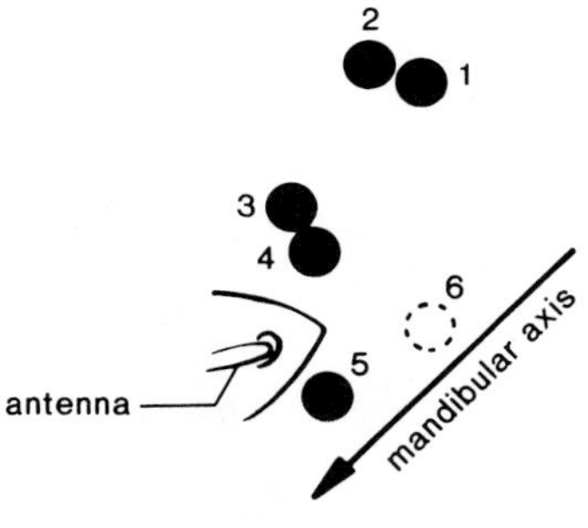

Figure 97

Figure 99

Figure 100

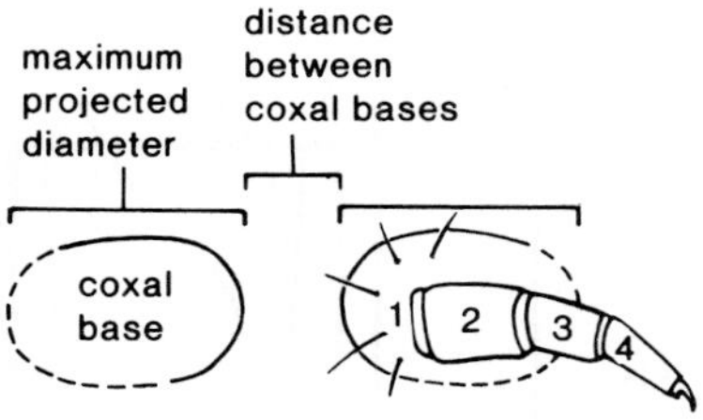

Figure 101

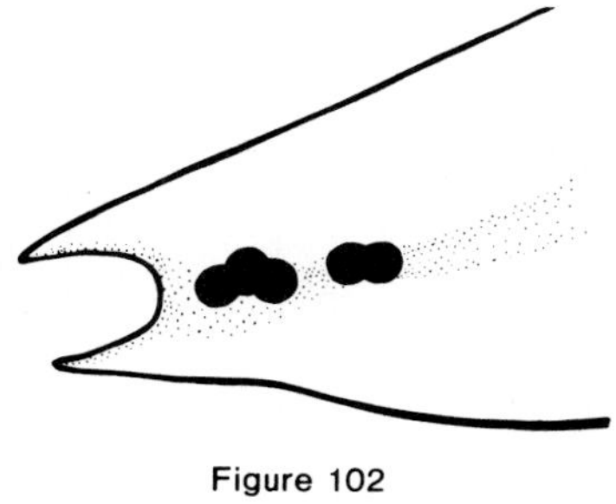

Figure 102

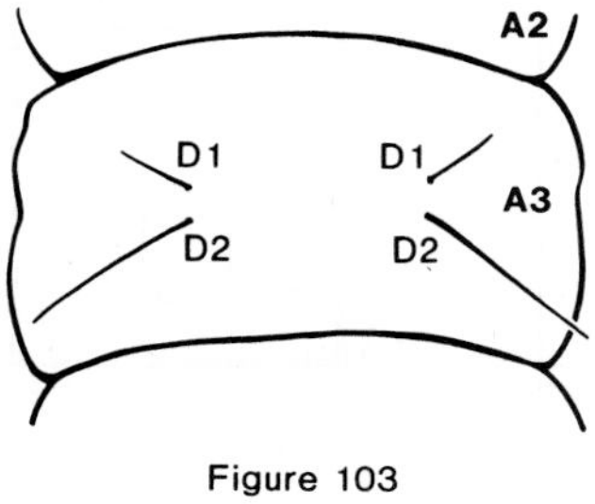

Figure 103

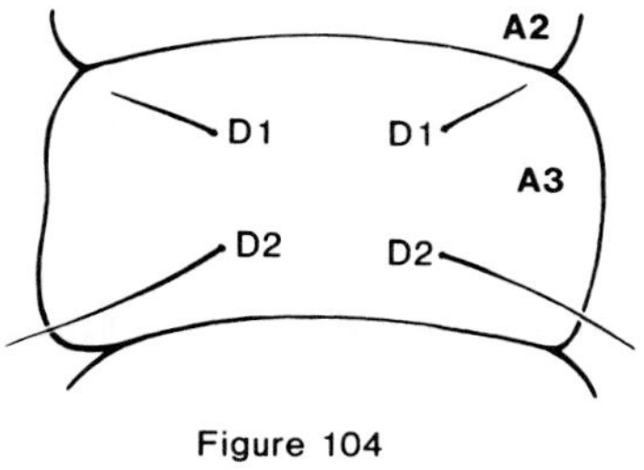

Figure 104

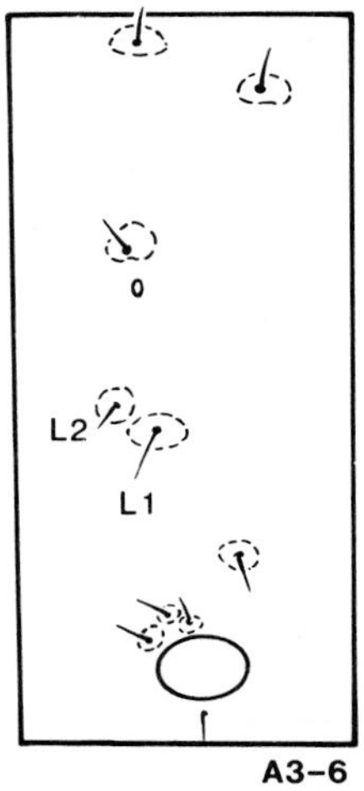

Figure 105

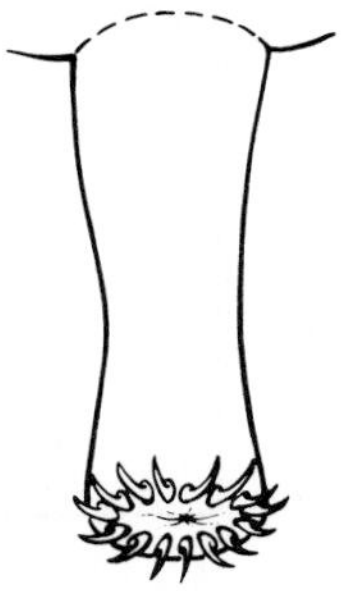

Figure 106

93(92). Tarsi conspicuously slender and elongate, slightly more than 1× to 1.5× the length of the tibia (fig. 107); (skeletonizers on rolled, folded, or webbed leaves) .... ***Choreutidae*** (p. 414)

93′. Tarsi not conspicuously slender and elongate, about equal to tibia in length or less (fig. 108) 94

94(93′). Head setae A1, A2, and A3 (above the stemmata) forming an angle less than 90° (fig. 109) 95

94′. Head setae A1, A2 and A3 (above the stemmata) forming an angle of 90° or more (fig. 110) 101

95(94). A8 spiracle at least 2× diameter of A7 spiracle .... (part) ***Yponomeutidae*** (p. 406)

95′. A8 spiracle nearly same diameter as A7 spiracle .... (part) ***Heliodinidae*** (p. 410)

96(45). T1 with L (prespiracular) group trisetose (if L group is combined with prothoracic shield there are 3 L's) 99

96′. T1 with L (prespiracular) group unisetose or bisetose (check both sides if fewer than 3 setae) (if neither couplet fits, go to couplet 41; if setae are very hard to see and head is flattened, go to couplet 99) 97

97(96′). SD1 about 1/5 as long on T3 as on T2 (fig. 111) (if SD1 setae are broken off, the SD1 base is distinctly smaller on T3) (only 1 species, *Hyblaea puera,* in U.S., on tropical plants) .... ***Hyblaeidae*** (p. 494)

97′. SD1 seta equal in size on T2 and T3 or slightly smaller on T3 (if SD1 setae are broken the SD1 seta bases are about equal in size, or the SD1 base is slightly smaller on T3) 98

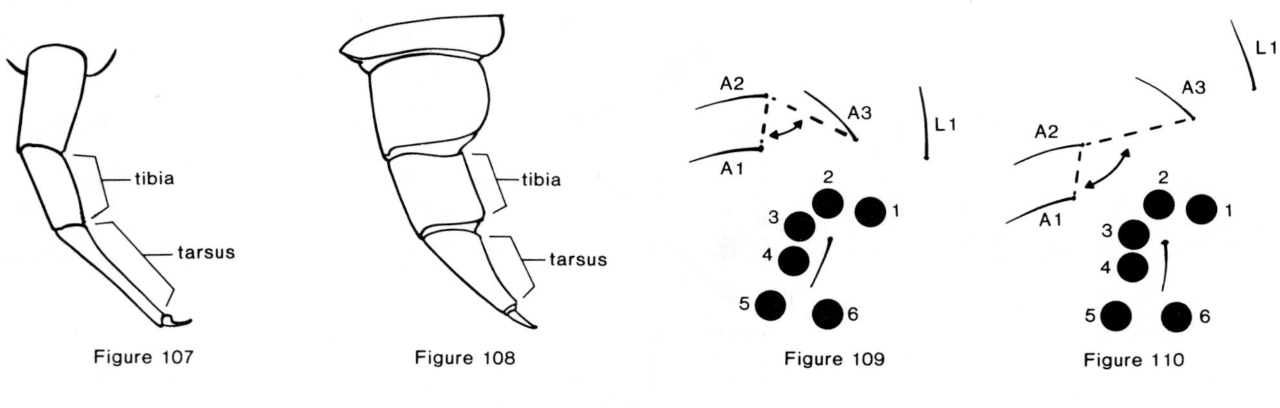

Figure 107 Figure 108 Figure 109 Figure 110

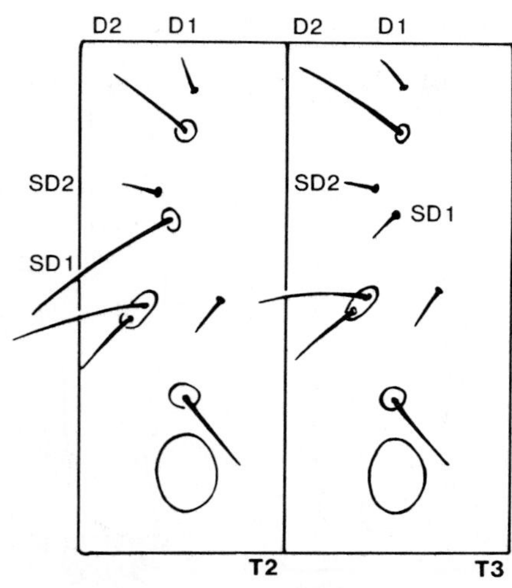

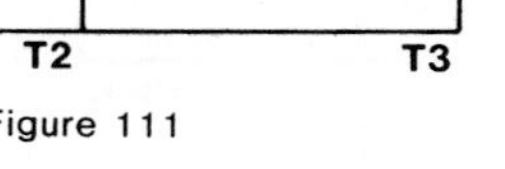

Figure 111

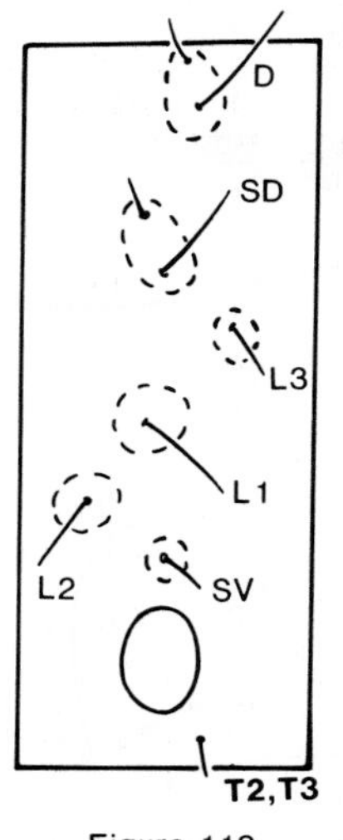

Figure 112

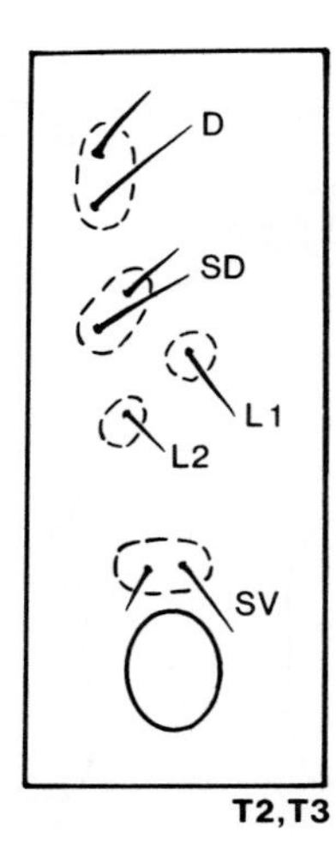

Figure 113

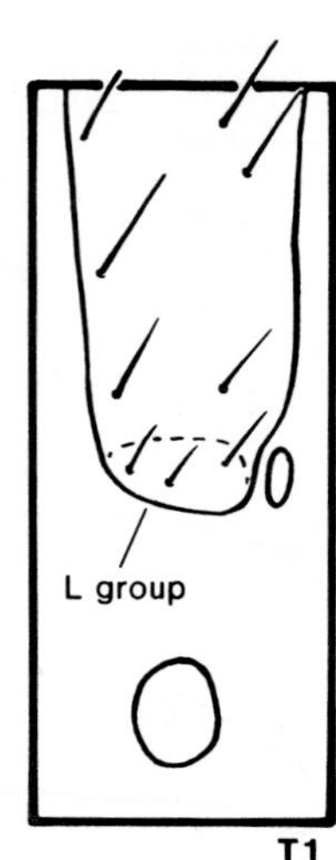

Figure 114

98(97′). T2 and T3 with 3 L setae (fig. 112) (some aquatic or semiaquatic pyralids (Nymphulinae and Schoenobiinae) have L3 missing or very small) .... (most) ***Pyralidae*** (p. 462)

98′. T2 and T3 with 2 L setae (fig. 113) (not aquatic) .... (part) ***Thyrididae*** (p. 495)

99(96). A3–6 with 4 SV setae and 2 conspicuous SD setae (fig. 73); 2 SV setae on T1 .... (part) ***Hepialidae*** (p. 347)

99′. A3–6 with 3 SV setae and 1 conspicuous SD seta (a tiny second SD seta rarely present); 2 SV setae on T1 .... **(see third choice)** 100

99″. A3–6 with SD seta "impossible" to see and only 2 SV setae; 1 SV seta on T1 ..... (a few) ***Elachistidae*** (p. 383)

100(99′). Prothoracic shield surrounding the L (prespiracular) group (fig. 114) .... (part) ***Plutellidae*** (p. 404)

100′. Prothoracic shield not surrounding the L (prespiracular) group, which is often on its own well-sclerotized pinaculum 101

101(92′, 94′, 100′). Abdomen variable, but segments shorter or little longer than high, and A3–6 with L1 and L2 *more ventral* than anterior to spiracle (at the most, slightly anterior to spiracle) (fig. 115) 102

101′. Abdominal segments much longer than high (usually at least twice as long), and A3–6 with L1 and L2 far *more anterior* than ventral to spiracle (fig. 116) .... (*Symmoca*, **Symmocinae**) (part) ***Oecophoridae*** (p. 379)

102(101). T2 and T3 with L1 closer to L2 than L3 (fig. 117) (setae L2 and L3 may be minute and not easily seen) 104

102′. T2 and T3 with L1 equidistant between L2 and L3 or closer to L3 (fig. 118), or L3 absent (setae may be minute on small larvae) 103

103(102′). T1 shield with SD2 approximately caudad to SD1 (fig. 119); prothoracic shield often elevated, bearing a roughened caudal area with cornicula; A8 with spiracle often distinctly near caudal margin of segment (mature larvae may exceed 50 mm; borers in wood, roots, stems) .... (part) ***Cossidae*** (p. 416)

103′. T1 shield with SD2 approximately dorsad to SD1 (fig. 120) prothoracic shield never elevated or bearing cornicula; A8 with spiracle usually located near center of segment, or only slightly caudad of center (mature larvae under 10 mm; feeders in fruits, berries, buds). .... (part) ***Argyresthiidae*** (p. 407)

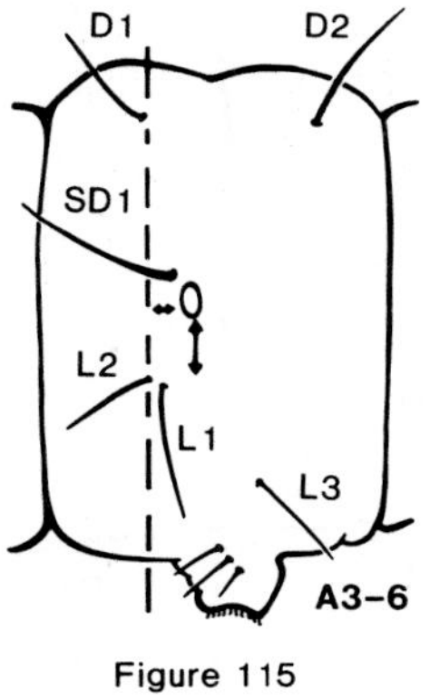

Figure 115

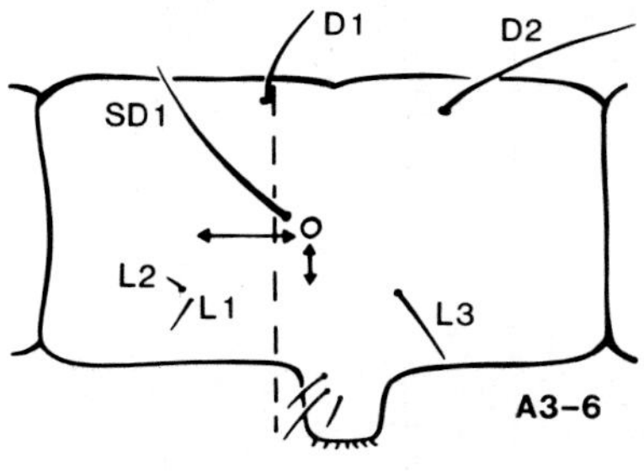

Figure 116

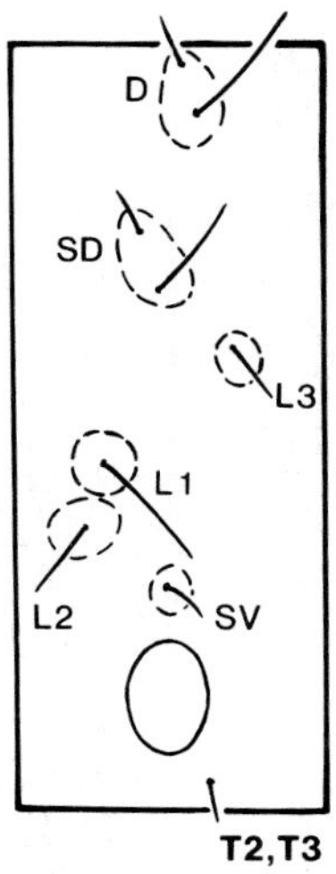

Figure 117

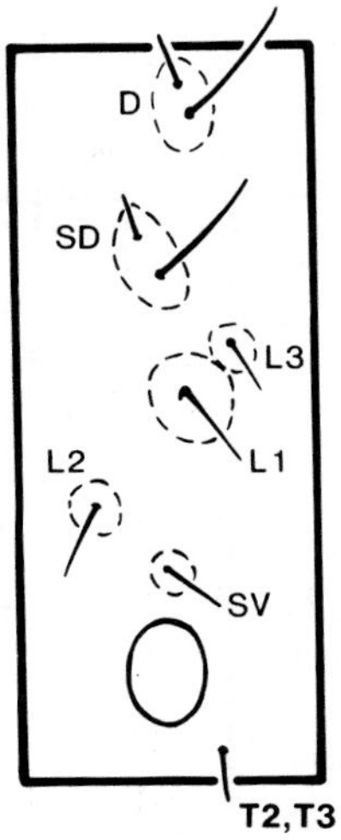

Figure 118

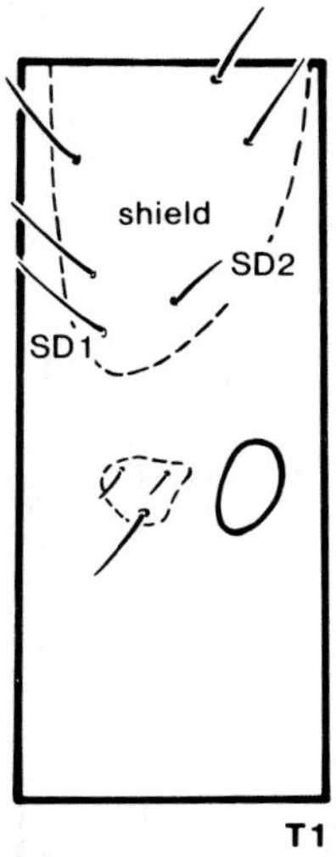

Figure 119

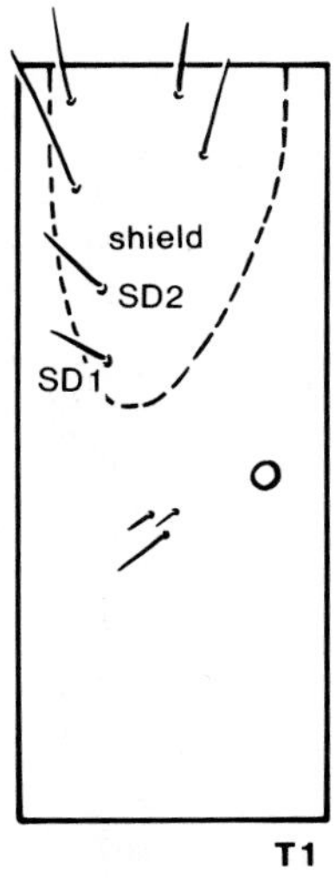

Figure 120

Fig. 73 ⟶

104(102). A10 prolegs with crochets divided by gap into 2 groups (forming a broken row, which may contain as few as a single crochet at each end) *OR* with the central crochets much shorter than those at either end (fig. 121) 105

104′. A10 prolegs with crochets in continuous row, and with central crochets same length or longer than those at either end 107

105(104). Frontoclypeus extending halfway or less to epicranial notch (fig. 122) 106

105′. Frontoclypeus extending more than halfway to epicranial notch .... (part) ***Gelechiidae*** (p. 394)

106(105). Anal comb present (fig. 121) .... (part) ***Gelechiidae*** (p. 394)

106′. Anal comb absent .... (*Setiostoma*, **Stenomatinae**) (part) ***Oecophoridae*** (p. 379)

107(104′). A9 with 3 setae grouped together (usually on raised bases, fig. 123) at D position .... (part) ***Heliodinidae*** (p. 410)

107′. A9 with 2 or fewer setae grouped at D position and bases not raised 108

108(107′). **Note: Larvae may not have all five of characters, a, b, c, d, e, but usually have 4 of 5 (rarely only 3).**

a. SD1 on A8 usually anterior or somewhat anterodorsal or anteroventral from spiracle (fig. 124);
b. A9 with D2 setae usually on a common middorsal pinaculum and closer to each other than each is to its own D1 (fig. 52);
c. A9 with D2's usually closer together than D1's (the anterior setae) on A8;
d. A9 with D1 usually closer to SD1 than to D2 (fig. 52), and D1 and SD1 frequently on same pinaculum (fig. 52);
e. A9 with SD1 never hairlike (i.e. gradually tapering from base toward tip the same way as other setae do) .... (including **Cochylinae** and **Olethreutinae**) (most) ***Tortricidae*** (p. 419)

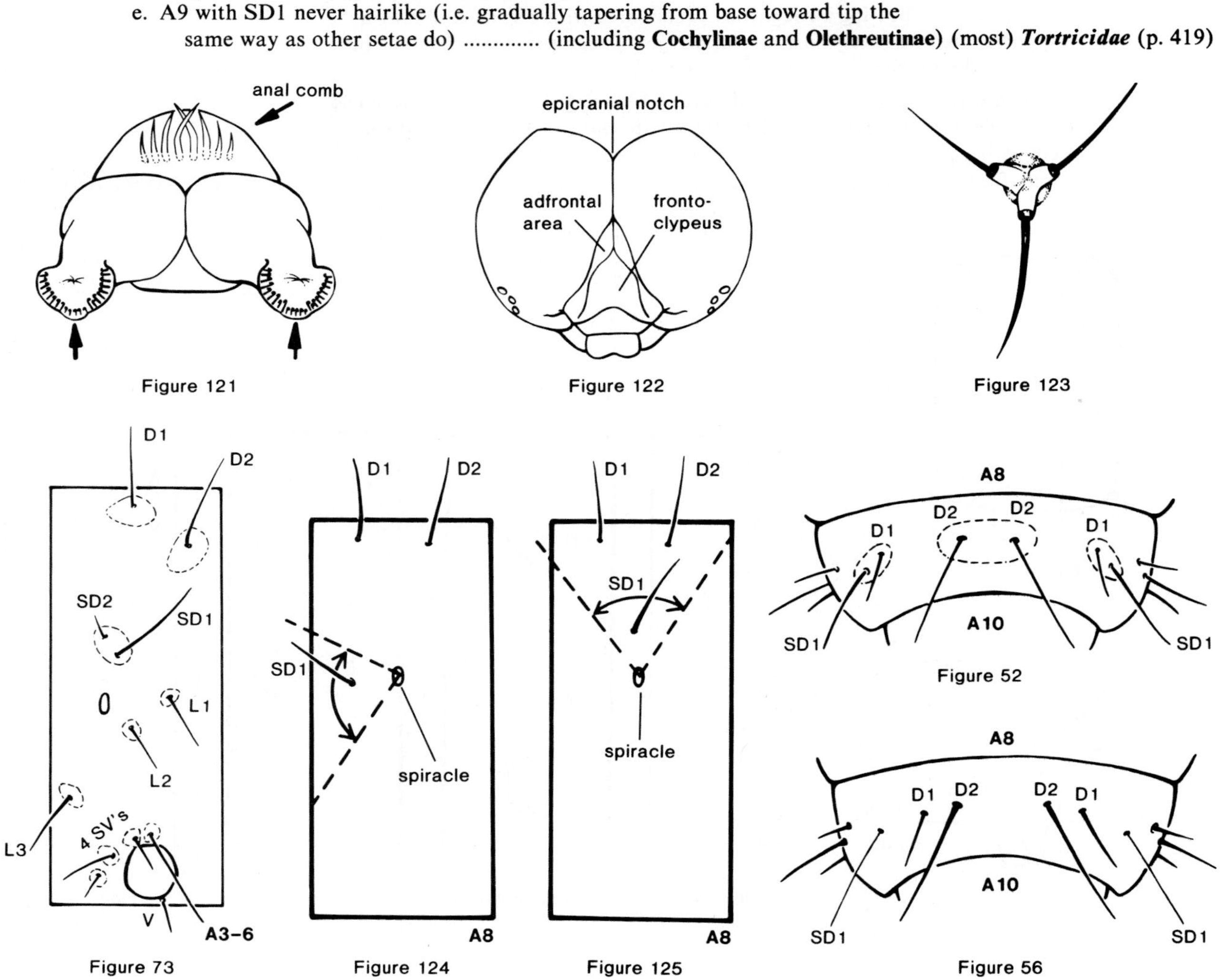

Figure 121

Figure 122

Figure 123

Figure 52

Figure 73

Figure 124

Figure 125

Figure 56

108′. a′. SD1 on A8 usually dorsal or dorsoanterior or dorsoposterior from spiracle (fig. 125);
b′. A9 with D2 setae rarely on a common middorsal pinaculum, and usually farther from each other than each is from its own D1 (fig. 56);
c′. A9 with D2's usually as far apart or farther apart than D1's (the anterior setae) on A8; if the prolegs are quite long, go to the next couplet;
d′. A9 with D1 usually closer to D2 than to SD1 or equidistant between them (fig. 56), and D1 and SD1 rarely on same pinaculum;
e′. A9 with SD1 frequently hairlike (i.e. essentially uniform in diameter from base for much of its length and *finer than other setae*) (fig. 56) 109

109(90′, 108′). Submentum with a distinct pit (fig. 126); seta SD1 on A1–7 often with a sclerotized pinaculum ring around its base (fig. 127) 110

109′. Submentum without a pit; seta SD1 on A1–7 usually without a sclerotized pinaculum ring around its base 112

110(109). A1 with 3 or more setae in SV group 111

110′. A1 with 2 setae in SV group .... some *Batrachedra, Homaledra* (part) ***Oecophoridae*** (p. 379)

111(110). Secondary setae absent; stemmata 3 and 4 nearly touching, and usually separated from 5 by space about as wide as stemma 4 (fig. 128) .... (part) ***Blastobasidae*** (p. 385)

111′. Secondary setae present (but you overlooked them at couplet 39); stemmata 3, 4, *and* 5 equally close together, the space between 4 and 5 much less than width of stemma 4 .... (a few) ***Scythrididae*** (p. 393)

112(109′). Abdominal segments with D1 or D2 apparently absent and with the remaining visible D seta small 113

112′. Abdominal segments with D1 and D2 clearly present (make sure they aren't broken off) 114

113(112). Stemmata grouped into 2 or 3 spots; crochets on A3–6 arranged in ovals, with distance between the 2 ovals on A4 less than or equal to maximum diameter of an oval .... (part) ***Lyonetiidae*** (p. 370)

113′. Stemmata grouped "normally", not in 2 or 3 spots; crochets on A3–6 arranged in circles, with distance between the 2 circles on A4 greater than maximum diameter of a circle .... (some *Cosmopterix*) ***Cosmopterigidae*** (p. 391)

114(112′). Frontoclypeus extending halfway or less to epicranial notch (fig. 122) .... (part) ***Oecophoridae*** (p. 379)

114′. Frontoclypeus extending more than halfway to epicranial notch 115

115(114′). T1 with L group (prespiracular) included on (surrounded by) prothoracic shield .... (part) **Plutellidae** (p. 404)

115′. T1 with L group separated from prothoracic shield 116

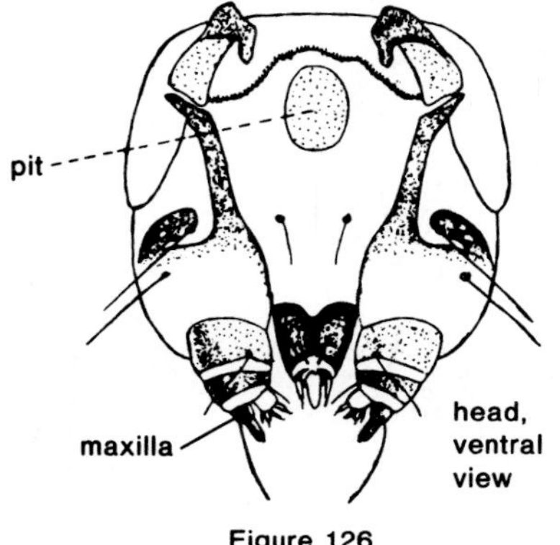

Figure 126

Figure 127

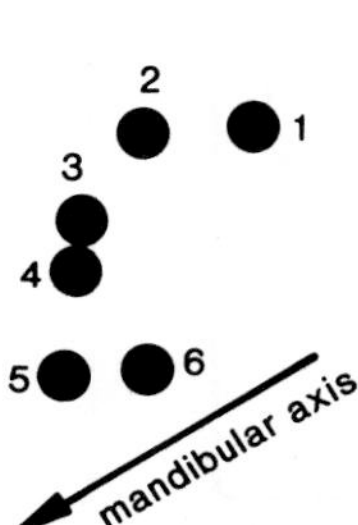

Figure 128

←—— Fig. 56

116(115′). T1 shield with SD2 approximately caudad to SD1 (fig. 119) 117

116′. T1 shield with SD2 approximately dorsad to SD1 (fig. 120) .... (most) ***Argyresthiidae*** (p. 407)

117(116). Head seta L1 farther from A3 than A3 is from A2 (fig. 129) (do not mistake pores for setal bases) 118

117′. Head seta L1 closer to A3 than A3 is to A2, or distances equal 119

118(117). A9 with D1 setae (the shorter, smaller ones) farther apart than D2 setae (fig. 130) .... (many) ***Gelechiidae*** (p. 394)

118′. A9 with D1 setae (the shorter, smaller ones) closer together than D2 setae (fig. 131) .... (a few) ***Heliodinidae*** (p. 410)

119(117′). A9 with SD1 hairlike (i.e. essentially uniform in diameter for much of its length and finer than adjacent setae (fig. 132). (If both SD1 setae are broken off and most D and L setae are still present assume SD1 is hairlike) 120

119′. A9 with SD1 setalike (i.e. gradually tapering from base toward tip the same way as adjacent setae) .... (most) ***Cosmopterigidae*** (p. 391)

120(119). A9 with D1 setae (the shorter, smaller ones) closer together than D2 setae (fig. 131) 121

120′. A9 with D1 setae (the shorter, smaller ones) farther apart than D2 setae 122

121(120). SD1 on A8 anterodorsad to spiracle (fig. 133); crochets in an incomplete circle, or if complete, then crochets biordinal .... (a few **Ethmiinae**) ***Oecophoridae*** (p. 379)

121′. SD1 on A8 dorsad and slightly caudad to spiracle; crochets in a complete circle and uniordinal .... (part) ***Heliodinidae*** (p. 410)

122(120′). A1 with 3 SV setae (check both sides) (fig. 134) .... (part) ***Oecophoridae*** (p. 379)

122′. A1 with 2 or fewer SV setae (check both sides) .... ***Assorted Gelechioidea*** (Gelechiidae, Oecophoridae, possibly Cosmopterigidae, and probably some others)

123(44′). T1 with L (prespiracular) group trisetose (fig. 135) 126

123′. T1 with L group unisetose or bisetose 124

124(123′). A5 with prolegs absent or conspicuously smaller than A6 prolegs .... (part) ***Geometridae*** (p. 502)

124′. A5 prolegs as large as or almost as large as A6 prolegs 125

125(124′). A3–6 crochets heteroideous (abruptly shorter at both ends) .... (*Utetheisa*) ***Arctiidae*** (p. 538)

125′. A3–6 crochets homoideous (essentially the same length) .... (**see 3rd choice**) (part) ***Noctuidae*** (p. 549)

125″. A3–6 crochets biordinal (two different lengths) .... (part) ***Thyatiridae*** (p. 500)

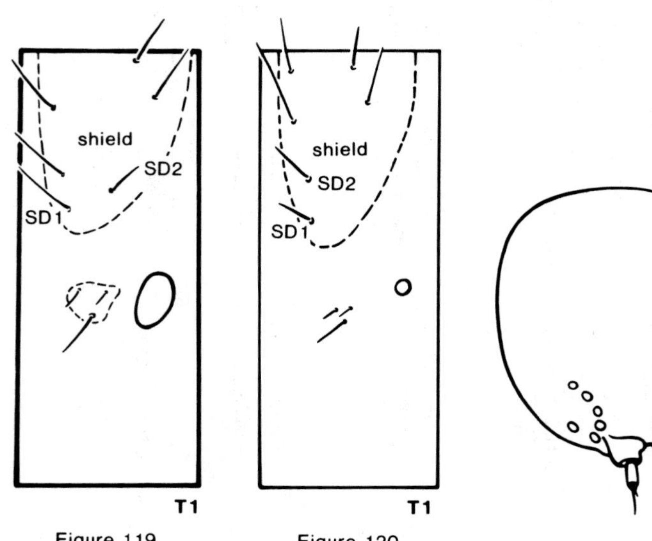

Figure 119 Figure 120

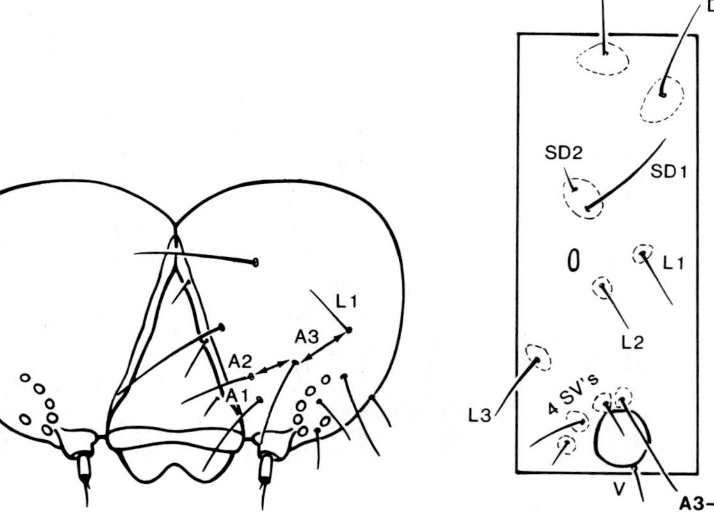

Figure 129 Figure 73

126(123). A3–6 bearing 2 SD setae and 4 SV setae (fig. 73) .... (part) ***Hepialidae*** (p. 347)

126′. A3–6 bearing 1 SD seta and 3 SV setae 127

127(126′). A3 with L1 or L2 closer to spiracle than to each other or distances equal (fig. 75); crochets uniordinal .... (a few) ***Plutellidae*** (p. 404)

127′. A3 with L1 and L2 closer to each other than either is to spiracle; crochets uniordinal, biordinal, or triordinal 128

128(127′). T2 and T3 with SV group (above the leg) unisetose 129

128′. T2 and T3 with SV group at least bisetose .... (part) ***Pterophoridae*** (p. 497)

129(128). A9 with D1 and D2 on same pinaculum (fig. 136); A10 with secondary setae present (10 or more below the anal shield) .... (a few **Ethmiinae**) ***Oecophoridae*** (p. 379)

129′. A9 with D1 and D2 on separate pinacula; A10 with no secondary setae (9 or fewer below the anal shield) .... (part) ***Yponomeutidae*** (p. 406)

130(41). A2 and A7 with pairs of ventral lobes or "pads" resembling prolegs but lacking crochets; head retracted into prothorax; setae long, in tufts (foliage feeders on shrubs and trees) .... ***Megalopygidae*** (p. 454)

130′. A2 and A7 without conspicuous ventral lobes or "pads"; head retracted or not; secondary setae variable 131

131(130′). Prolegs and crochets absent on A3 132

131′. Prolegs and crochets present on A3 133

132(131). Prolegs present on A4, 5, 6, and 10 .... ***Nolidae*** (p. 548)

132′. Prolegs present on A6 and A10 .... **(see 3rd choice)** (a few) ***Geometridae*** (p. 502)

132″. Prolegs present on A5, 6 and 10 .... (a few) ***Noctuidae*** (p. 549)

133(131′). Ventral crochets biordinal or triordinal (i.e. *alternating* 2 or 3 different lengths) (if a mesoseries, check *very* carefully on the inside) 147

133′. Ventral crochets uniordinal (i.e. all the same length) or heteroideous (longer in the middle than at the ends) 134

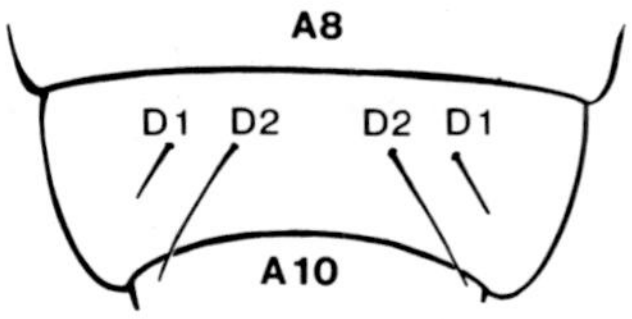

Figure 130

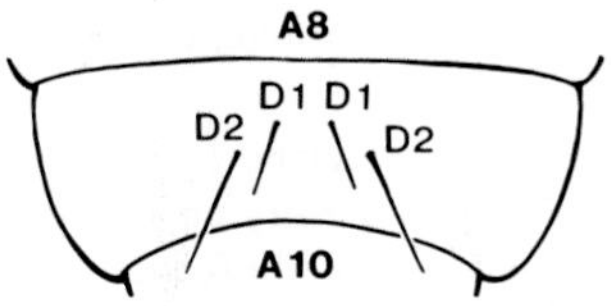

Figure 131

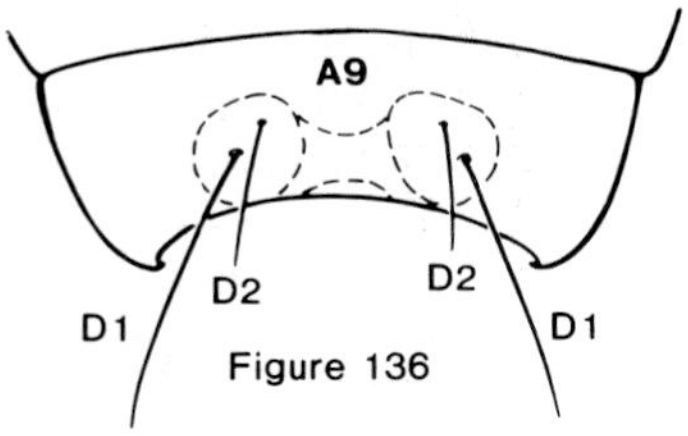

Figure 136

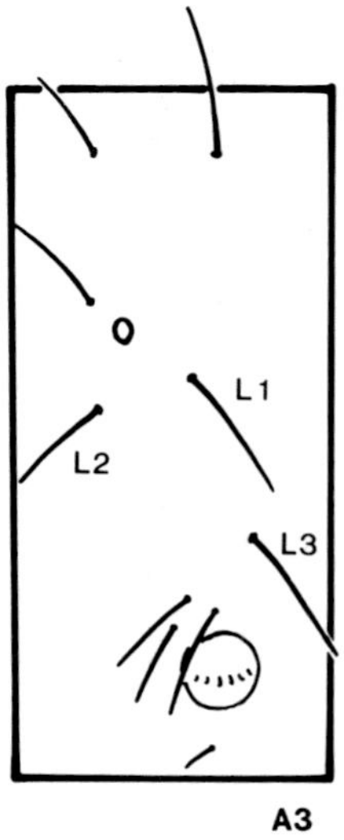

Figure 75

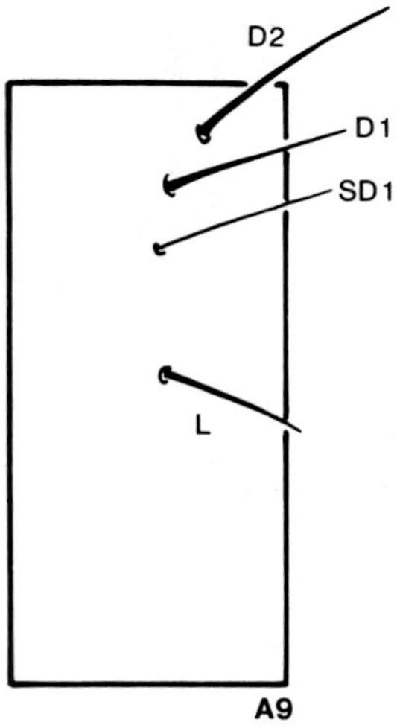

Figure 132

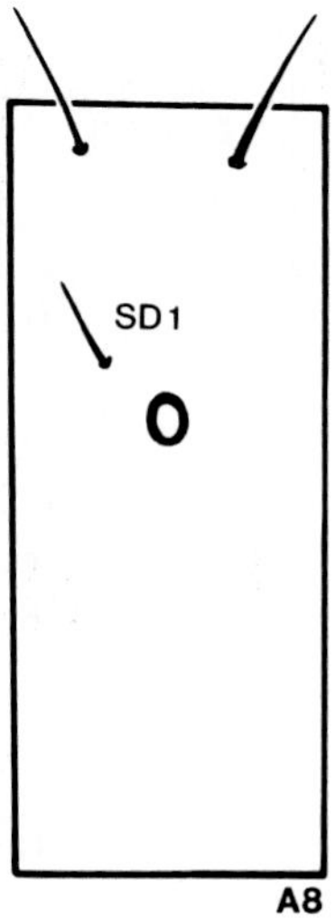

Figure 133

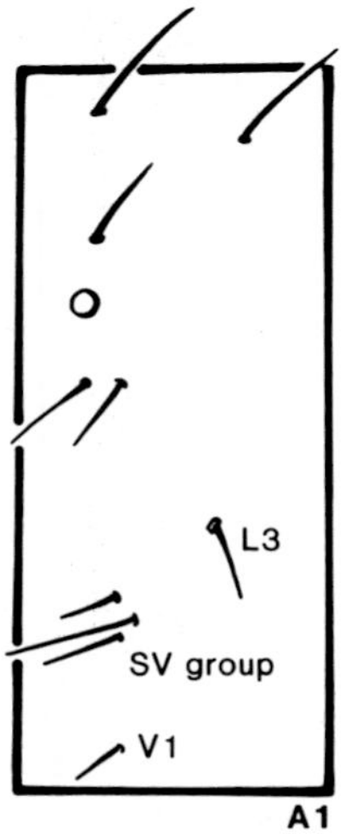

Figure 134

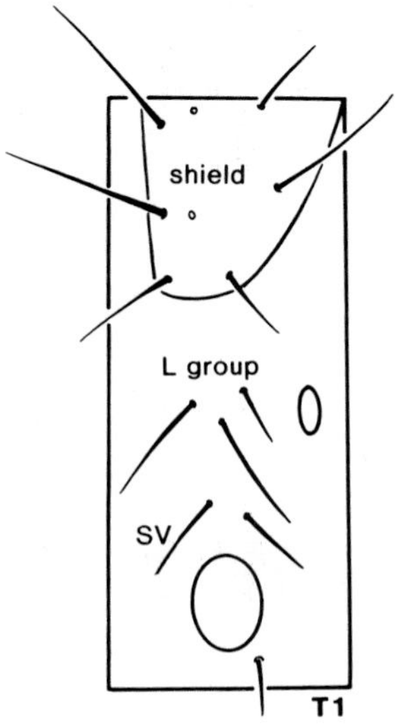

Figure 135

← Fig. 73

134(133′). Ventral crochets in a heteroideous mesoseries (fig. 137) (examine the ends of the row carefully for *abruptly* shorter crochets) 135

134′. Ventral crochets in a homoideous mesoseries or mesopenellipse 139

135(134). T2 and T3 with at least 4 verrucae between coxa and middorsal line (i.e. D and SD verrucae separated, fig. 138) (foliage feeders on wide variety of plants, favoring herbaceous plants) .... (part) ***Arctiidae*** (p. 538)

135′. T2 and T3 with only 3 verrucae between coxa and middorsal line (i.e. D and SD verrucae fused, fig. 139) 136

136(135′). Verruca L1 on A7 the same distance or only slightly farther from spiracle as verruca L1 on A1–6 (fig. 140) .... (many) ***Ctenuchidae*** (p. 542)

136′. Verruca L1 on A7 distinctly farther from spiracle than verruca L1 on A1–6 (fig. 141) 137

137(136′). A6 with several to many setae per verruca above the spiracle 138

137′. A6 with a single seta per verruca above the spiracle .... (*Utetheisa*) ***Arctiidae*** (p. 538)

138(137). Head with only 6 prominent, long setae (and no short ones) on face above the stemmatal arc (setae A1, A2, A3, L1, P1, P2, fig. 142); A3–6 prolegs usually with only a single prominent, isolated seta (V1) on the inside .... ***Pericopidae*** (p. 536)

138′. Head with more than 6 (usually many more) relatively shorter setae on face above the stemmatal arc; A3–6 prolegs usually with many, nearly uniform setae on the inside .... (*Ctenucha, Cisseps*) (part) ***Ctenuchidae*** (p. 542)

139(134′). Spiracles on A8 elliptical and with long axis nearly vertical 142

139′. Spiracles on A8 more nearly circular than elliptical; if elliptical (a few pterophorids) then long axis more nearly horizontal 140

140(139′). A3–6 with only 1 seta per "D" verruca .... (part) ***Arctiidae*** (p. 538)

140′. A3–6 with more than 1 seta per "D" verruca 141

141(140′). A2 and A7 with distinct gland adjoining the spiracle (compare with A3–6 spiracles); crochets in mesoseries on fan-shaped planta; (skeletonizers on deciduous plants including Virginia creeper and grape) .... ***Zygaenidae*** (p. 453)

141′. A2 and A7 without distinct gland adjoining the spiracle; crochets usually in mesopenellipse on cylindrical, elongate planta .... (part) ***Pterophoridae*** (p. 497)

142(139). A7 (and sometimes A6) with middorsal eversible gland (fig. 143); secondary setae usually in tufts of different lengths and density on verrucae; often with middorsal parallel pencils of setae on A1–4 (usually feeders on deciduous trees, many are forest defoliators) .... ***Lymantriidae*** (p. 544)

142′. A7 without middorsal eversible gland; secondary setae variable; if on verrucae, then tufts usually of similar length; middorsal pencils rarely present on A1–4 143

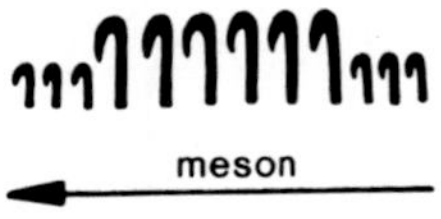

meson

Figure 137

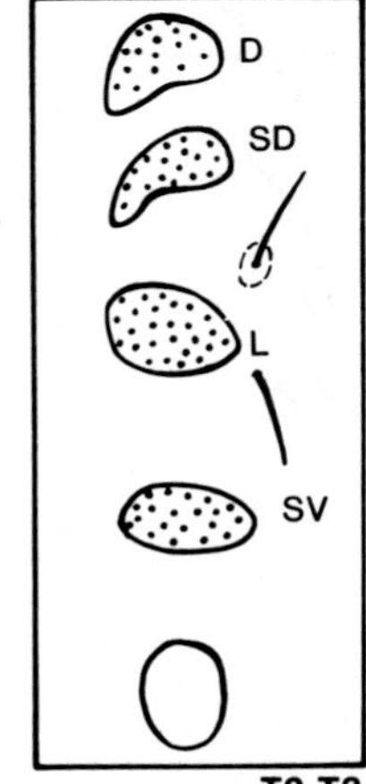

Figure 138

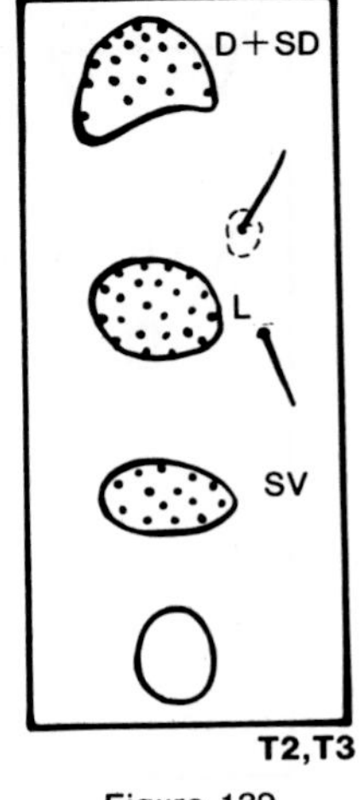

Figure 139

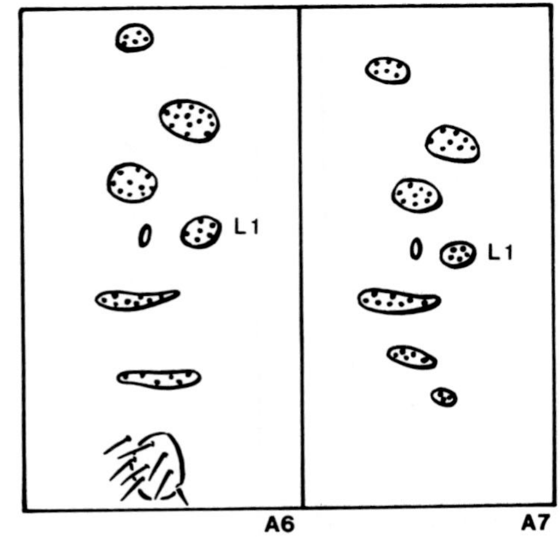

Figure 140

143(142′). Labral notch shallow, extending less than half the distance to base of labrum (fig. 144); spinules *never* conspicuous on cuticle (make sure they are *not* present) .... (part) ***Arctiidae*** (p. 538)

143′. Labral notch usually deeper (not always), usually extending at least one-half the distance to base of labrum (fig. 145); spinules often conspicuous on cuticle (fig. 146) 144

144(143′). Labral notch V-shaped and usually continued as groove to base of labrum (fig. 147); anal prolegs usually somewhat caudo-projecting and somewhat smaller than ventral prolegs, or greatly modified into elongate stemapoda .... (part) ***Notodontidae*** (p. 524)

144′. Labral notch more nearly U-shaped and not continued as groove to base of labrum (fig. 145); anal prolegs never caudo-projecting and rarely somewhat smaller than ventral prolegs 145

145(144′). A8 with a *small* verruca (smaller than the one above it) *directly behind* spiracle (fig. 148) .... (**Pantheinae,** many **Acronictinae**) ***Noctuidae*** (p. 549)

145′. A8 without a *small* verruca *directly behind* spiracle (a large one about same size as that above may be present) 146

146(145′). Verruca or seta L1 same distance from spiracle on A7 as on A6 (fig. 140); verruca L1 on A6, 7, and 8 similar in size .... (a few) ***Arctiidae*** (p. 538)

146′. Verruca or seta L1 farther from the spiracle on A7 than on A6 (fig. 141); verruca L1 on A7 sometimes reduced to 1 or a few setae or absent .... (a few) ***Noctuidae*** (p. 549)

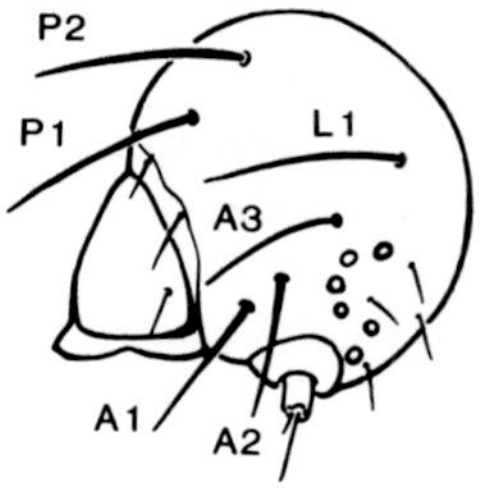

Figure 142

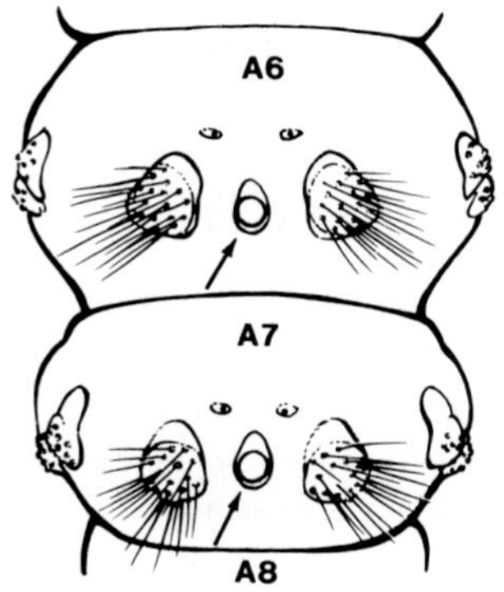

Figure 143

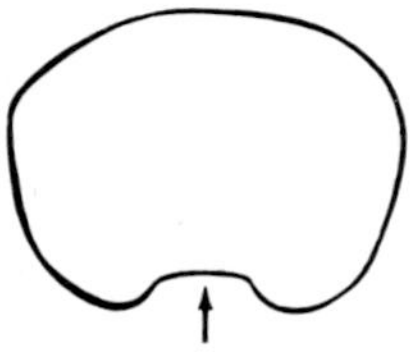

Figure 144

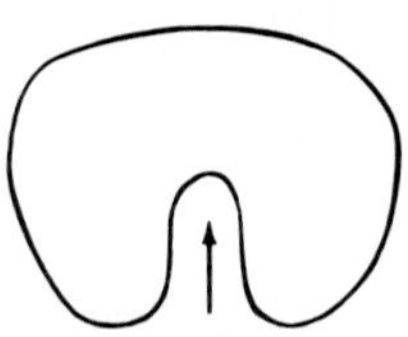

Figure 145

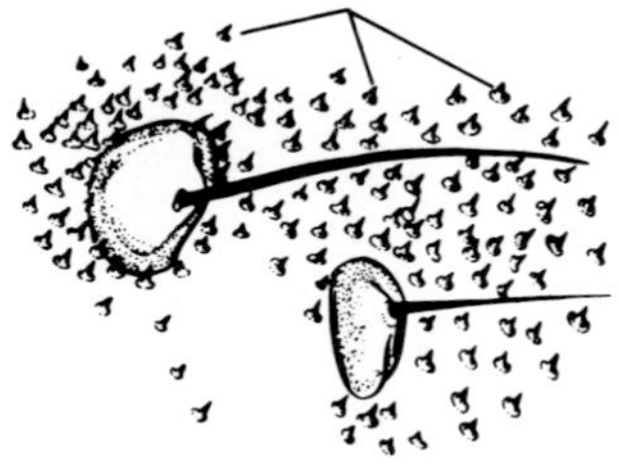

Figure 146

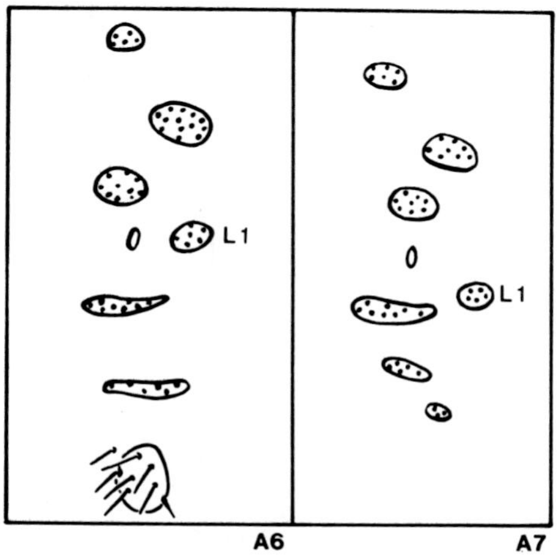

Figure 141

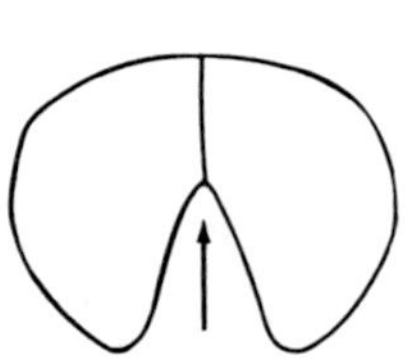

Figure 147

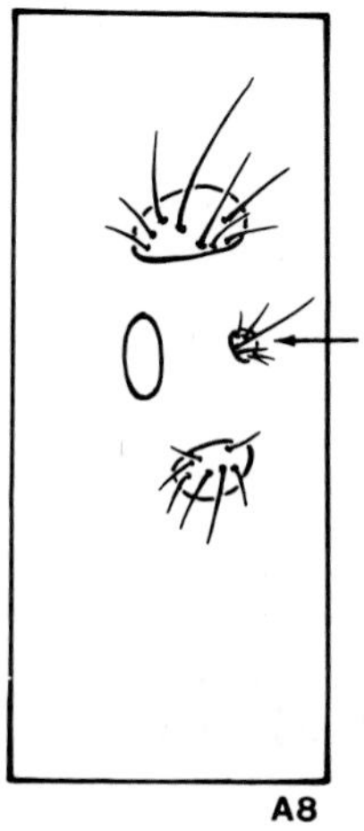

Figure 148

147(133). T1 with middorsal protruding fleshy osmeterium (osmeterium may be invaginated into prothorax; look for elongated transverse middorsal slit on anterior part) .... (a few) ***Papilionidae*** (p. 438)

147′. T1 without osmeterium or transverse middorsal slit 148

148(147′). A8 with at least 1 *middorsal* horn, scolus, chalaza, tubercle, verruca or tuft (do not include larvae with closely spaced addorsal tufts) 149

148′. A8 without *middorsal* horn, scolus, chalaza, tubercle, verruca or tuft (may be present on other segments) 154

149(148). A7 with middorsal *scolus;* crochets usually triordinal, sometimes biordinal (feeders on a wide variety of herbaceous and woody plants) .... (most) ***Nymphalidae*** (p. 448)

149′. A7 without middorsal *scolus* (verrucae or tufts may be present); crochets biordinal, rarely partially "triordinal" 150

150(149′). A8 with middorsal scolus, horn, button or rounded pinaculum; most setae similar in length and little or no longer than body diameter; stemma 3 usually not distinctly more convex and larger than stemmata 2 and 4 151

150′. A8 without middorsal scolus, horn, button or rounded pinaculum—a middorsal tuft usually present; some setae may be much longer than others and much longer than body diameter; stemma 3 often distinctly more convex and somewhat larger than stemmata 2 and 4 .... (part) ***Apatelodidae*** (p. 509)

151(150). A8 with middorsal horn or button (fig. 149) or rounded pinaculum; scoli usually absent on other abdominal segments 152

151′. A8 with middorsal scolus; scoli usually present on other abdominal segments (see couplet 153 if other scoli are lacking) .... (part) ***Saturniidae*** (p. 513)

152(151). Abdominal segments divided into 6–8 annulets .... (part) ***Sphingidae*** (p. 521)

152′. Abdominal segments divided into 3 annulets at most 153

153(152′). Thorax (especially T2 and T3) conspicuously humped (fig. 150) .... (domestic silkworm) ***Bombycidae*** (p. 510)

153′. Thorax not conspicuously humped .... (a few) ***Saturniidae*** (p. 513)

154(148′). Scoli or chalazae (figs. 20, 21, 35, 36) present; verrucae (fig. 38), verricules (fig. 40), and pinacula usually absent 156

154′. Scoli and chalazae absent; either verrucae, verricules, pinacula or setae that arise from no definite structures present 155

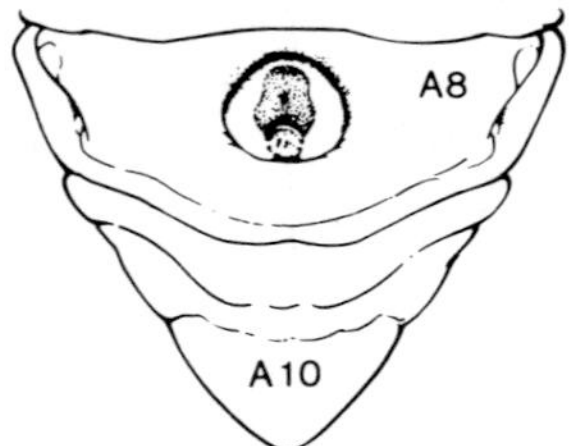

Figure 149

Figure 150

Figure 20

Figure 21

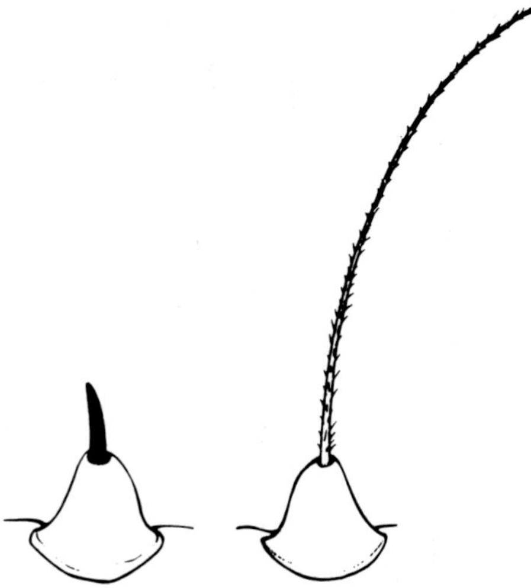

Figure 35

Figure 36

Figs. 38, 40 ⟶

155(154′). Ventral crochet rows with a distinct median fleshy lobe which may interrupt or reduce the crochets in size (fig. 151) 160

155′. Ventral crochet rows without such a fleshy lobe 185

156(154). A9 with middorsal scolus .... (part) ***Saturniidae*** (p. 513)

156′. A9 without middorsal scolus 157

157(156′). A3–6 with many more than 10 setae below spiracle (including those on scoli or pinacula) 158

157′. A3–6 with fewer than 10 setae below spiracle .... (*Lactura*) ***Yponomeutidae*** (p. 406)

158(157). Body setae very irregular in length, some often much longer than others; fleshy setiferous lobes bearing many setae usually present above legs; head without scoli .... (part) ***Lasiocampidae*** (p. 511)

158′. Body setae rather uniform in length; fleshy, setiferous lobes bearing many setae usually absent above legs; head sometimes with scoli 159

159(158′). Body bearing conspicuous scoli with several to many stout setae or spines (fig. 20); body segments without annuli .... (**Heliconiinae** and a few others) ***Nymphalidae*** (p. 448)

159′. Body not bearing conspicuous scoli, but bearing rather abundant setae that may be borne on distinct chalazae in some species (fig. 35); chalazae, when evident, almost invariably bearing only a single seta; body segments frequently with annuli 161

160(155). A1 spiracle distinctly lower than A2 spiracle; mandible bearing more than 2 setae (look near the ventral articulation); head capsule usually quite hairy and usually wider than half greatest body width .... (part) ***Riodinidae*** (p. 446)

160′. A1 spiracle essentially in line with A2 spiracle; mandible bearing no more than 2 setae; head capsule usually rather bare and usually less than half greatest body width .... (part) ***Lycaenidae*** (p. 443)

161(43′, 159′). *Planta* of *ventral* prolegs (planta never bears setae), cylindrical, and usually much longer than wide (figs. 152, 153, 154); mature larvae small, rarely over 20 mm 162

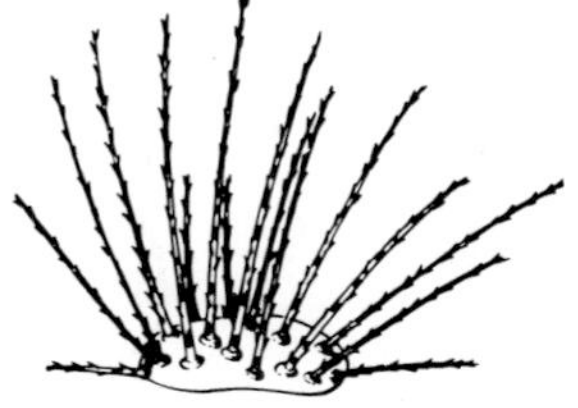

Figure 38

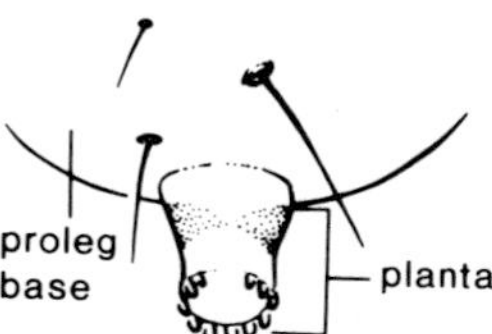

Figure 154

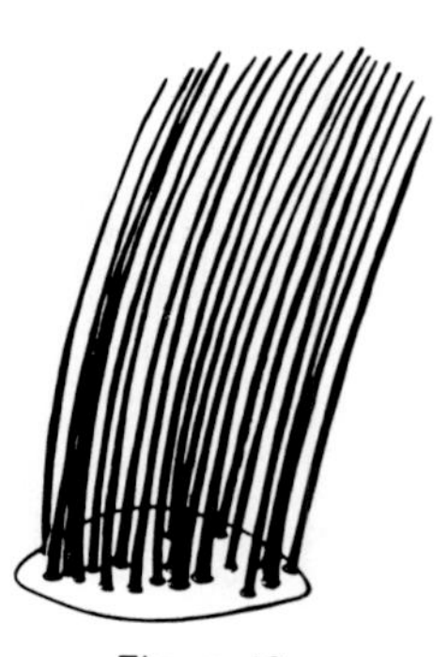

Figure 40

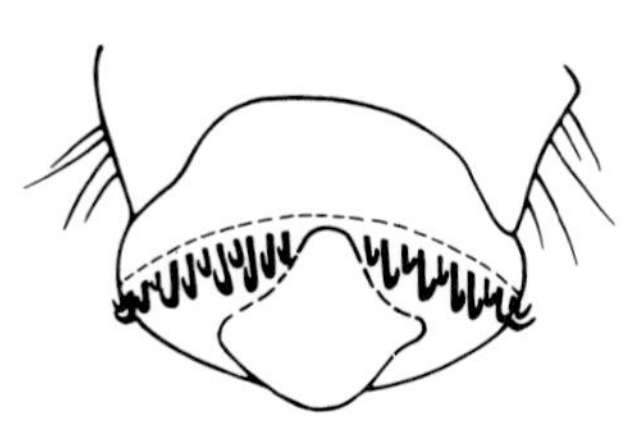

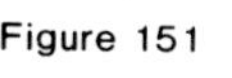

Figure 151

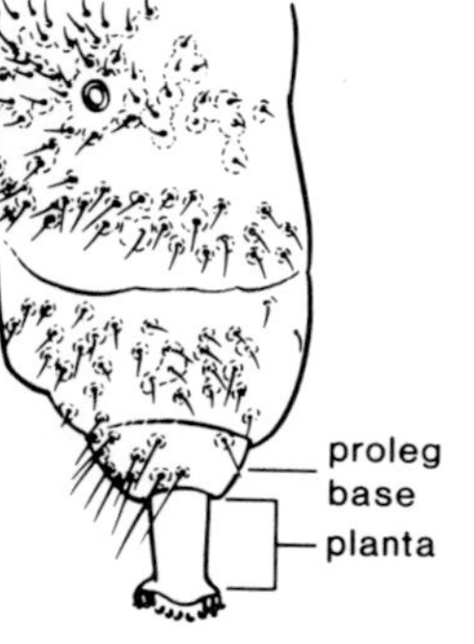

Figure 152

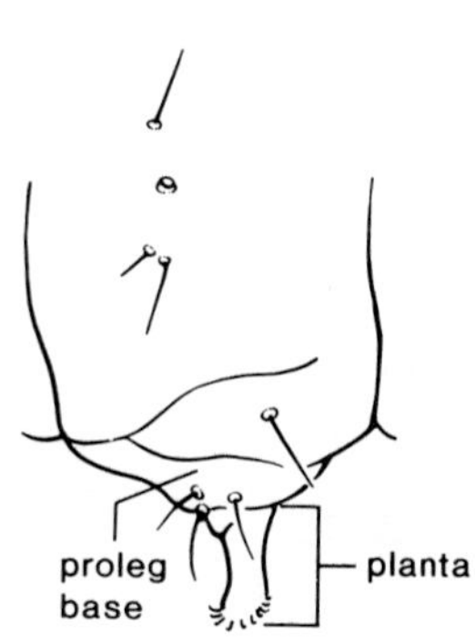

Figure 153

←— Figs. 20, 35

161′. *Planta* of *ventral* prolegs never long and cylindrical, either reduced or rudimentary (fig. 155, 156) or short and fan-shaped, or oval at distal end of the proleg base (figs. 157, 158); mature larvae variable, but frequently over 20 mm 165

162(161). Ventral crochet rows with a distinct median fleshy lobe that may interrupt or reduce the crochets in size (fig. 151) .... (part) ***Riodinidae*** (p. 446)

162′. Ventral crochet rows without such a fleshy lobe 163

163(162′). Ventral crochets in a *meso*penellipse or mesoseries 164

163′. Ventral crochets in a *closed* ellipse (fig. 159) .... (part) ***Scythrididae*** (p. 393)

164(163). Thoracic segments with SV setae obscured by numerous secondary setae of same length (if SV setae are conspicuous they are unisetose on T2 and T3 (fig. 160)); (feeders in fruits, stems, berries, seed pods, and with no webbing) .... (part) (**Blastodacninae**) ***Agonoxenidae*** (p. 390)

164′. Thoracic segments with SV setae longer (and sometimes paler) than surrounding secondary setae, and bisetose on T2 and T3; (stem borers or foliage feeders within a webbed mass) .... (part) ***Pterophoridae*** (p. 497)

165(161′). Only 2 pairs of prolegs (A6 and A10) .... (a few) ***Geometridae*** (p. 502)

165′. More than 2 pairs of prolegs 166

166(165′). Ventral crochets in complete circle or ellipse, or with 1 or 2 small gaps in an otherwise continuous arc (gaps always less than one-third the circumference of projected circle (fig. 161)) 167

166′. Ventral crochets in mesoseries or mesoseries plus lateroseries; if mesoseries resembles a mesopenellipse, then gap free of crochets is at least one-third the circumference of projected circle (disregard a few widely spaced isolated crochets on lateral side of planta) 168

167(166). Head nearly always distinctly larger than prothorax in profile (figs. 162, 163); anal comb present (fig. 164) (mature larvae usually less than 40 mm; external feeders, often on grasses, usually constructing shelters) .... ***Hesperiidae*** (p. 434)

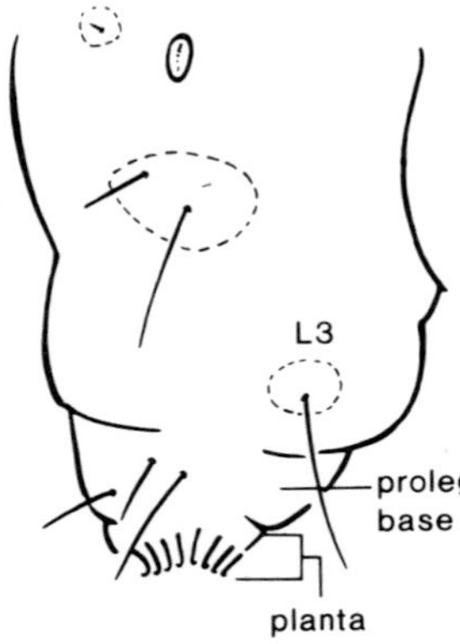

Figure 155

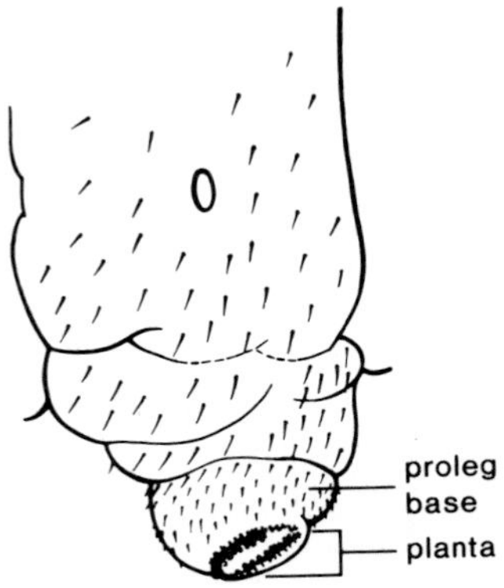

Figure 156

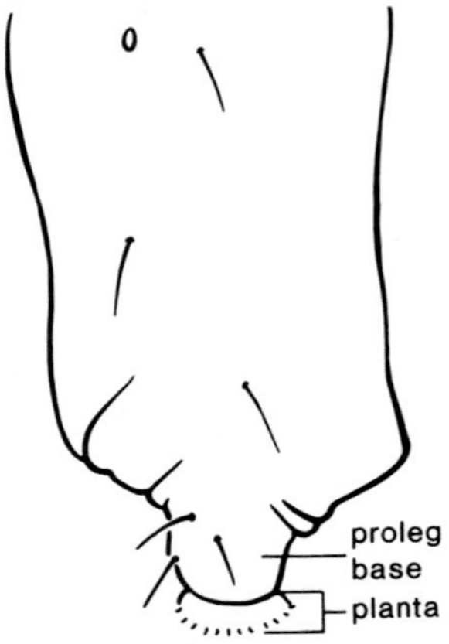

Figure 157

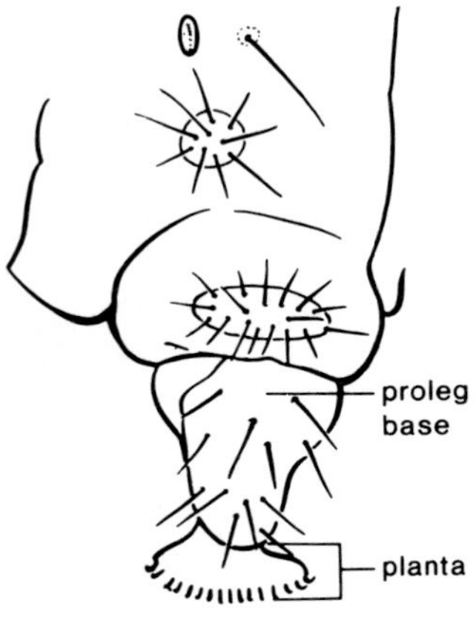

Figure 158

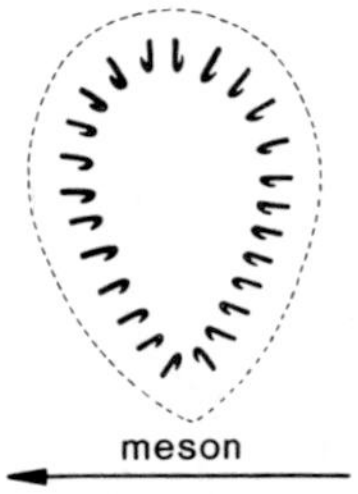

Figure 159

Figure 160

Figure 162

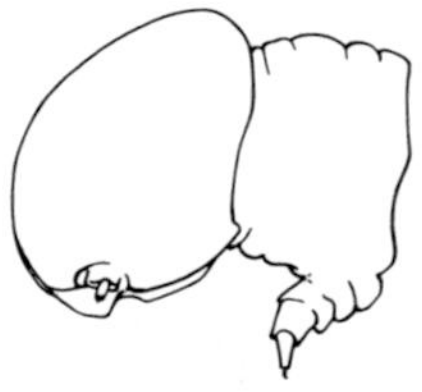

Figure 163

Figure 164

⟵ Fig. 151

Fig. 161 ⟶

167′. Head equal to or smaller than prothorax in profile; anal comb absent or rudimentary (mature larvae large, 50 mm or more; bore in *Yucca* roots and leaves and caudex of *Agave* and *Manfreda* (all Agavaceae); southwestern U.S.) .... ***Megathymidae*** (p. 436)

168(166′). Stemmata 1, 2, and 3 grouped together, with stemma 3 about 1.5–2× larger than the others, and with 4, 5, and 6 usually spaced farther apart (fig. 165) (if stemmata 1 and 6 are not clearly visible, look for dark neutral roots of 1 and 6) 169

168′. Stemmata not arranged as above, and stemma 3 not 1.5–2× larger than others 170

169(168). Labral notch very deep; body segments clearly distinct dorsally and without annulae; anal plate not forked (feeders on leaves of trees and shrubs) .... (a few) ***Saturniidae*** (p. 513)

169′. Labral notch shallow; body segments relatively indistinct dorsally and with annulae; anal plate usually forked (feeders on grasses) .... (**Satyrinae**) ***Nymphalidae*** (p. 448)

170(168′). Ventral crochets uniordinal (all the same length) 172

170′. Ventral crochets biordinal or triordinal (i.e. regularly alternating 2 or 3 different lengths) (check *very* carefully on the inside) 171

171(170′). A8 with median dorsal horn or button that does not bear setae (fig. 149) .... (part) ***Sphingidae*** (p. 521)

171′. A8 without median dorsal horn or button (a small hump bearing secondary setae may be present) 173

172(170). Labral notch V-shaped, and usually continued as a groove to base of labrum (fig. 147); anal prolegs usually smaller than ventral prolegs, often somewhat caudo-projecting, and sometimes without crochets .... (part) (*Closterna, Datana,* possibly others) ***Notodontidae*** (p. 524)

172′. Labral notch U-shaped with parallel sides, and never continued as a groove to base of labrum (fig. 145); anal prolegs not conspicuously smaller (rarely somewhat thinner) than ventral prolegs, not at all caudo-projecting, and always with crochets .... (many **Acronictinae** and a few others) ***Noctuidae*** (p. 549)

173(171′). Thorax with fleshy filaments or an osmeterium above the spiracular line (osmeterium may be inverted: look for elongate transverse middorsal slit on anterior part of prothorax) 174

173′. Thorax without filaments or an osmeterium above the spiracular line (no middorsal slit on anterior part of prothorax) 177

174(173). T2 with at least 1 pair of fleshy filaments; osmeterium present or absent on T1; lateroseries of crochets never present 175

174′. T2 without fleshy filaments; osmeterium always present on T1; lateroseries of crochets sometimes present (fig. 166) 176

175(174). Osmeterium absent; vertical black, yellow, and white stripes on all segments and head; bearing *dorsal* or *subdorsal* filaments (common on milkweed, oleander, and fig) .... (**Danainae**) ***Nymphalidae*** (p. 448)

175′. Osmeterium present; black with dorsal rows of white spots; bearing *lateral* filaments on thoracic segments and smaller *lateral* and/or *subventral* filaments on abdominal segments (on pipe-vine) .... (*Battus*) ***Papilionidae*** (p. 438)

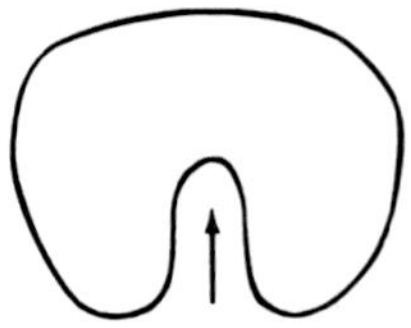

Figure 145

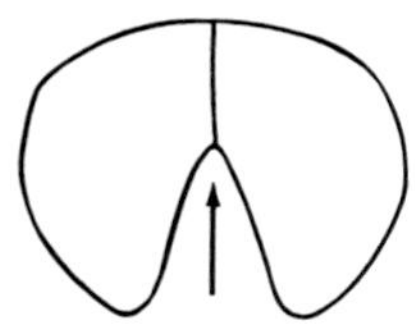

Figure 147

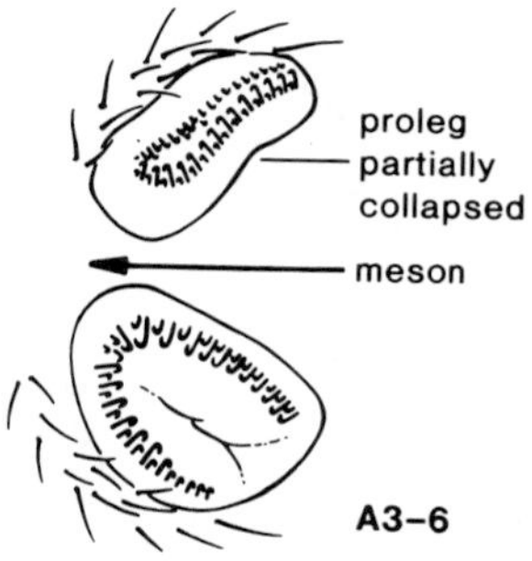

proleg partially collapsed
meson
A3–6
Figure 161

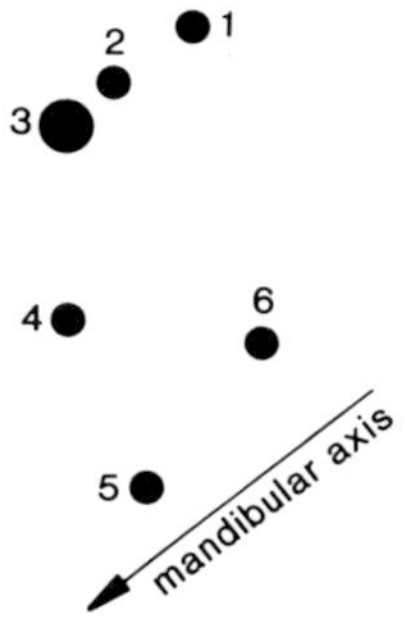

mandibular axis
Figure 165

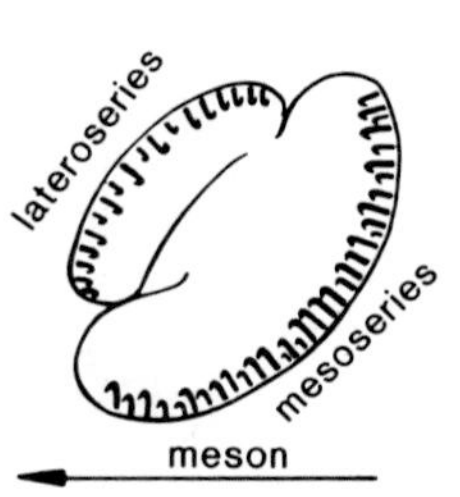

lateroseries
mesoseries
meson
Figure 166

Fig. 149 ⟶

176(174′). Thoracic and abdominal segments bearing pinacula in the positions of primary setae, with each pinaculum bearing 4–6 short setae; larva black with dorsal and subdorsal rows of white spots (on *Dicentra* and *Sedum* (stonecrop); mountainous regions of western North America) .... (*Parnassius*) ***Papilionidae*** (p. 438)

176′. All segments lacking pinacula; coloration variable, may be aposematic or cryptic (wide variety of hosts, wide distribution) .... (part) ***Papilionidae*** (p. 438)

177(173′). Stemmata 2, 3, 4, and 5 all distinctly different from and *elevated above* stemmata 1 and 6 (fig. 167) 178

177′. Three or fewer of stemmata 2, 3, 4, and 5 distinctly different from and elevated above stemmata 1 and 6 185

178(177). Head with pair of protuberances nearly as long as or longer than frontoclypeus .... (part) (*Asterocampa*, **Apaturinae**) ***Nymphalidae*** (p. 448)

178′. Head without pair of long protuberances (small ones may be present) 179

179(178′). A8 with short median horn .... (domestic silkworm) ***Bombycidae*** (p. 510)

179′. A8 without median horn 180

180(179, 201′). T1 with middorsal osmeterium (may be inverted into prothorax; look for elongate transverse middorsal slit on anterior part) .... (part) ***Papilionidae*** (p. 438)

180′. T1 without osmeterium or transverse middorsal slit 181

181(180′). A3–6 crochets in weak circle (made up of semicircular mesoseries plus 4–6 widely spaced crochets making up rest of circle); A3–6 segments with 4 distinct and equal annulets each .... (part) ***Libytheidae*** (p. 446)

181′. A3–6 crochets in mesoseries; A3–6 annulets, if present, variably distinct or more than 4 per segment 182

182(181′). A2 spiracle distinctly more dorsal than A3 spiracle; head distinctly larger than prothorax in profile .... (part) (*Anaea*, **Apaturinae**) ***Nymphalidae*** (p. 448)

182′. A2 spiracle essentially in line with A3 spiracle; head same size or smaller than prothorax in profile 183

183(182′). Body segments with distinct annulets (usually 6) (fig. 168); stemmata 2, 3, 4, and 5 usually distinctly different from 1 and 6 .... ***Pieridae*** (p. 441)

183′. Body segments without annulets; stemmata 2, 3, 4, 5 similar to 1 and 6 184

184(183′). A1 spiracle distinctly lower than A2 spiracle; head capsule usually quite hairy and usually wider than half the greatest body width .... (a few) ***Riodinidae*** (p. 446)

184′. A1 spiracle essentially in line with A2 spiracle; head capsule usually rather bare and usually narrower than half the greatest body width .... (part) ***Lycaenidae*** (p. 443)

185(155′, 177′). Ventral crochets reduced in size or interrupted near center by conspicuous fleshy pad (fig. 151); head smaller than prothorax and retractable (may be greatly protruded in preserved specimens) (external feeders on foliage, myrmecophilous, or predaceous on Homoptera) .... (part) ***Lycaenidae*** (p. 443)

185′. Ventral crochets not interrupted by fleshy pad; head not retractable 186

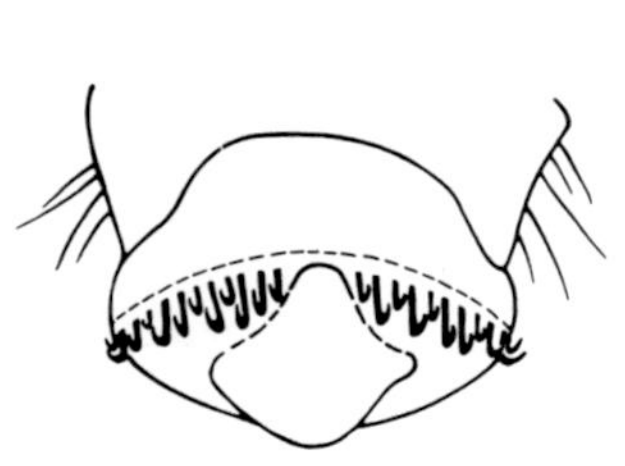

Figure 151

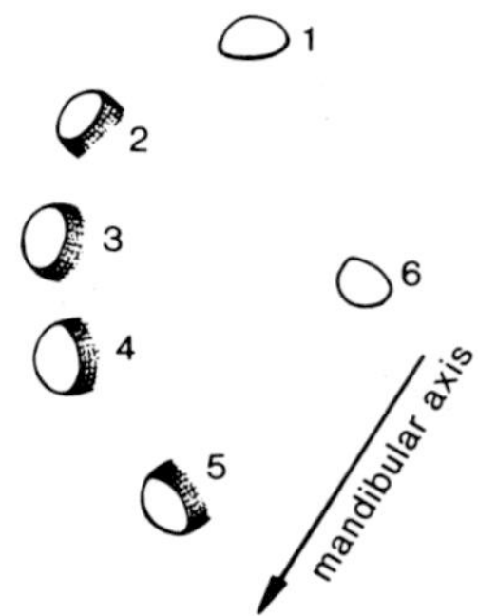

Figure 167

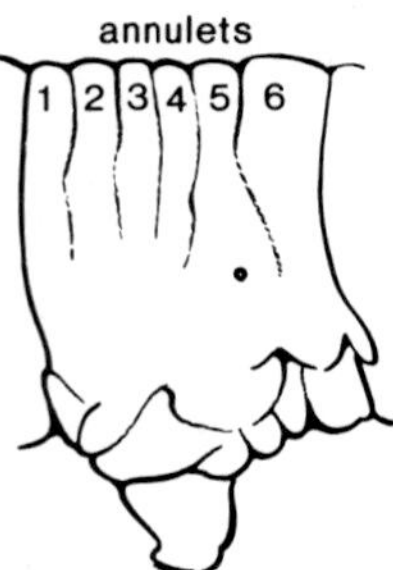

Figure 168

Fig. 149 ⟶

186(185′). Stemma 3 distinctly more convex and more distinct than the others (examine carefully, fig. 169); conspicuous dorsal tufts of setae usually present (fig. 170); labrum with moderately deep V-shaped notch .... (part) ***Apatelodidae*** (p. 509)

186′. Stemma 3 not more convex and distinct than the others; conspicuous dorsal tufts usually absent; labrum usually (not always) with shallower, more U-shaped notch 187

187(186′). T1 with L (prespiracular) group bisetose and with at least one of the L setae more than twice the length of the spiracle; body covered with dense spinules (fig. 171) 188

187′. T1 with L (prespiracular) group (if distinguishable) with more than 2 setae, or if only 2 are present, they are less than twice the length of the spiracle; body without dense spinules (numerous short secondary setae may be present) 189

188(187). T2 and T3 with 1 SV seta .... (a few) ***Noctuidae*** (p. 549)

188′. T2 and T3 with 2 or more SV setae .... **(see 3rd choice)** (*Doa*) ***Doidae*** (p. 534)

188″. T2 and T3 with SV setae not clearly evident .... (part) ***Libytheidae*** (p. 446)

189(187′). Verrucae or groups of setae present on body, the body rather setiferous 192

189′. Verrucae or groups of setae absent on body, the body rather "bare" 190

190(189′). A3–6 crochets in mesoseries 191

190′. A3–6 crochets in pseudocircle comprised of curved mesoseries, plus a few scattered crochets laterally .... **(see 3rd choice)** (part) ***Libytheidae*** (p. 446)

190″. A3–6 crochets in mesopenellipse .... (a few **Ethmiinae**) (part) ***Oecophoridae*** (p. 379)

191(190). A8 with no trace of middorsal button; body setae rarely quite short; abdomen terminating normally; larvae usually less than 25 mm when mature .... (part) ***Lycaenidae*** (p. 443)

191′. A8 with *very inconspicuous* middorsal button (fig. 149); body setae quite short, abdomen terminating in 3 angular lobes; larvae usually greater than 25 mm when mature .... (part) ***Sphingidae*** (p. 521)

192(189). A3–6 with verrucae above the spiracles pin-cushionlike, very conspicuous, and with more than 20 setae each 193

192′. A3–6 with verrucae above the spiracles not pin-cushionlike, less conspicuous, and with fewer than 20 setae each 194

193(192). A1 spiracle distinctly lower than A2 spiracle; mandible with more than 2 setae (look near ventral articulation); head capsule usually quite hairy, and usually wider than half the greatest body width .... (part) ***Riodinidae*** (p. 446)

193′. A1 spiracle essentially in line with A2 spiracle; mandible with no more than 2 setae; head capsule usually rather bare, and usually less than half the greatest body width .... (part) ***Lycaenidae*** (p. 443)

194(192′). Those secondary setae above prolegs distributed over entire body, many of them not grouped 195

194′. Those secondary setae above prolegs not distributed over entire body, all or most of them grouped 196

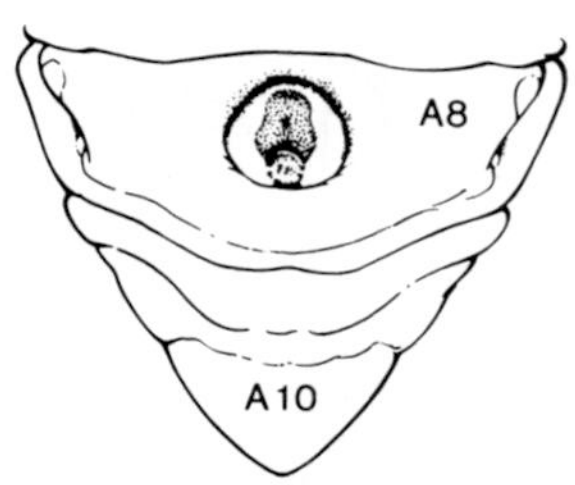

Figure 149

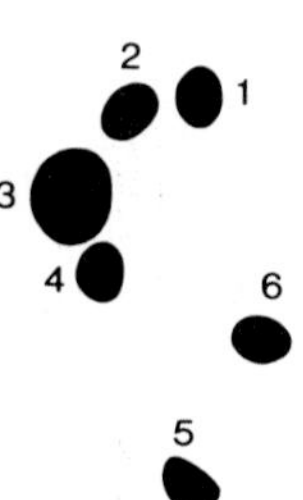

Figure 169

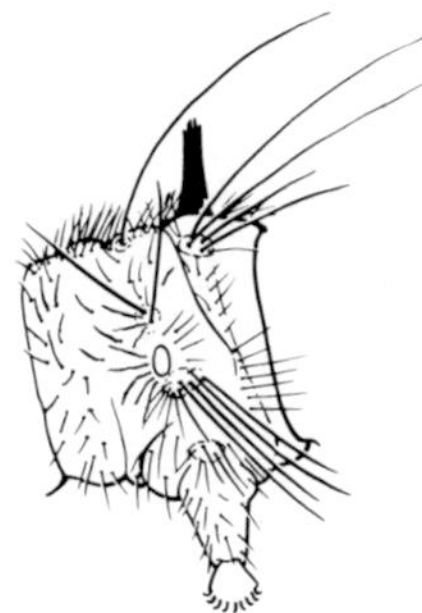
Figure 170

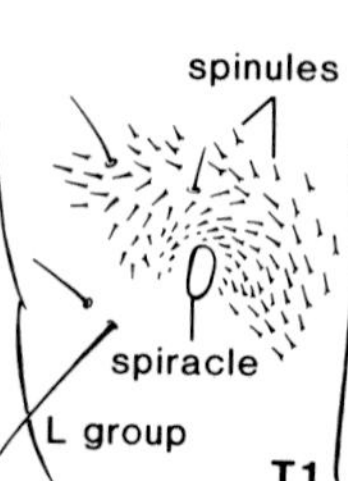

Figure 171

195(194). All secondary setae relatively stout and nearly *equal* in length and thickness; setae relatively sparse and easy to count in stemmatal/antennal area; fleshy lobes not present above legs .... (part) ***Saturniidae*** (p. 513)

195′. Secondary setae of different lengths and thicknesses; setae relatively abundant and difficult to count in stemmatal/antennal area; fleshy lobes frequently present above legs .... (part) ***Lasiocampidae*** (p. 511)

196(43, 194′). A6 with proleg base (part bearing the SV group and any secondary setae) usually extensively developed, forming greater part of well-formed prolegs (figs. 158, 172). Planta usually laterally flattened and fan shaped (fig. 158), rarely somewhat peglike and cylindrical, but never greatly reduced so crochets appear to arise directly from venter of body; crochets usually in mesoseries (mostly external feeding macrolepidoptera) 197

196′. A6 with proleg base (part bearing the SV group and any secondary setae) not forming more than one-half of proleg (figs. 155, 173); planta peglike and cylindrical (long or short) (figs. 106, 154) *OR* greatly reduced so crochets appear to arise directly from venter of body (fig. 174); crochets in complete circle, penellipse or transverse bands (mostly internal feeding microlepidoptera) 212

197(196). Prolegs only on A6 and 10, if prolegs are present on A5–6, A4–6 or A3–6 they are progressively smaller toward A3 and have fewer crochets than A6 .... (most) ***Geometridae*** (p. 502)

197′. Prolegs on A3–6 and 10, and with A3–6 prolegs nearly equal in size and with similar complements of crochets 198

198(197′). A8 with middorsal horn or button (fig. 149) 199

198′. A8 without middorsal horn or button, although other humps of various kinds may be present 200

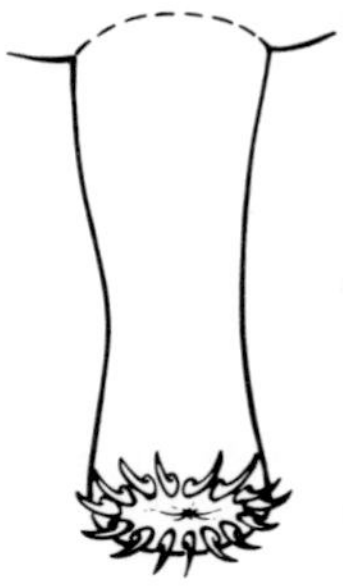

Figure 106

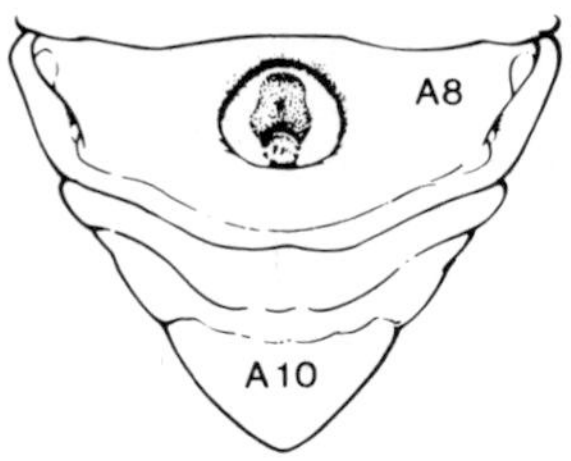

A8
A10

Figure 149

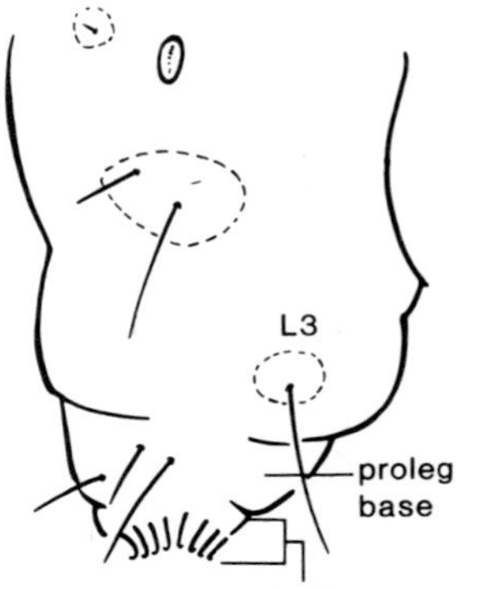

L3
proleg base
planta

Figure 155

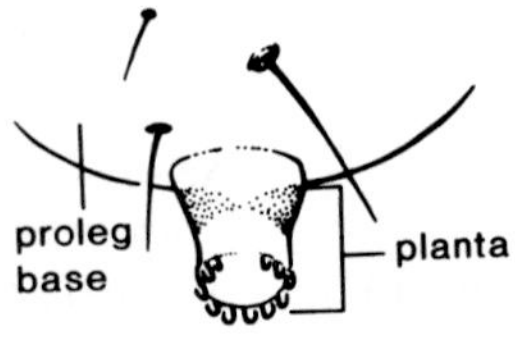

proleg base
planta

Figure 154

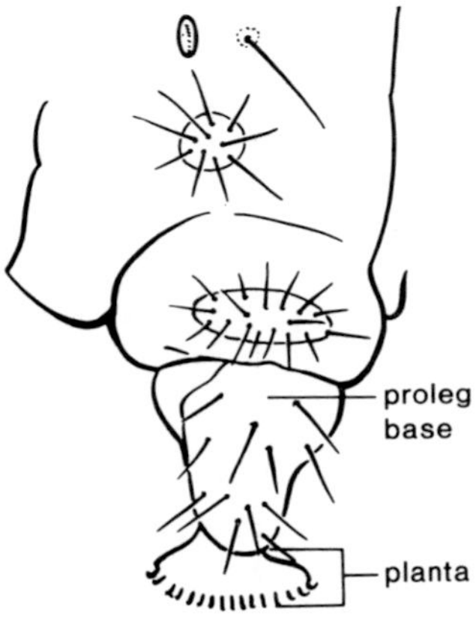

proleg base
planta

Figure 158

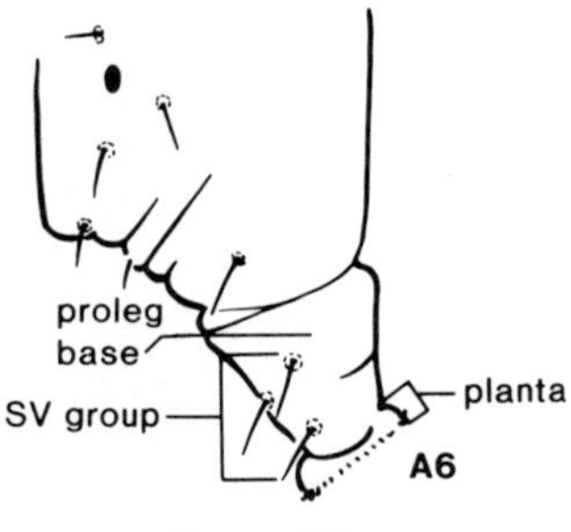

proleg base
SV group
planta
A6

Figure 172

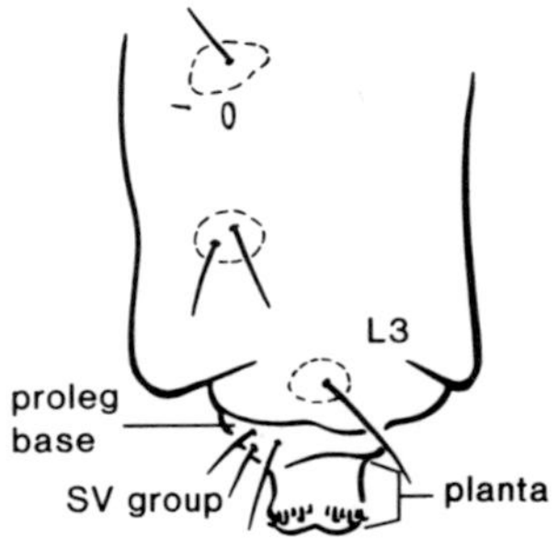

L3
proleg base
SV group
planta

Figure 173

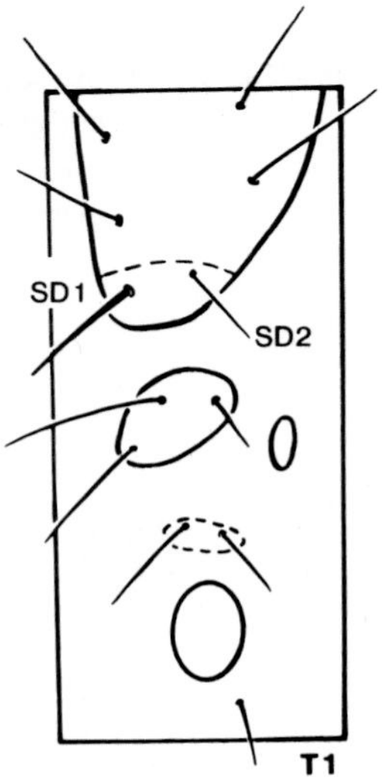

SD1
SD2
T1

Figure 178

Fig. 174 ⟶

199(198). Abdomen with distinct annuli; A10 terminating in 3 angular lobes (fig. 182); caudal prolegs similar in size to ventral prolegs .... (part) ***Sphingidae*** (p. 521)

199′. Abdomen without annuli or only very weak ones; A10 not terminating in 3 angular lobes; caudal prolegs smaller than ventral prolegs .... (a few) ***Notodontidae*** (p. 524)

200(198′). At least *some* primary setae visible and distinguishable from secondary setae on body; pinacula often distinct 202

200′. *No* primary setae visible and distinguishable from secondary setae on body; pinacula absent 201

201(200′). A3–6 crochets reduced in size or interrupted near center by conspicuous fleshy pad (fig. 151); head smaller than prothorax and retractable (may be greatly protruded in preserved specimens) .... (part) ***Lycaenidae*** (p. 443)

201′. A3–6 crochets not reduced in size or interrupted near center by conspicuous fleshy pad; head not retractable 180

202(200). A7 with L1 seta or L1 pinaculum (may be secondary setae on the pinaculum) farther from spiracle than on A1–6, or more caudo-ventrad on A7 than on A1–6 (figs. 175, 176); T1 with SD1 usually thinner and/or shorter than SD2 (fig. 177) 203

202′. A7 with L1 seta or L1 pinaculum (may be secondary setae on the pinaculum) nearly the same distance from spiracle and in same position on A7 as on A1–6; T1 with SD1 never thinner or shorter than SD2, usually larger than SD2 (fig. 178). (If primary setae or pinacula are difficult to locate, go to couplet 207) 208

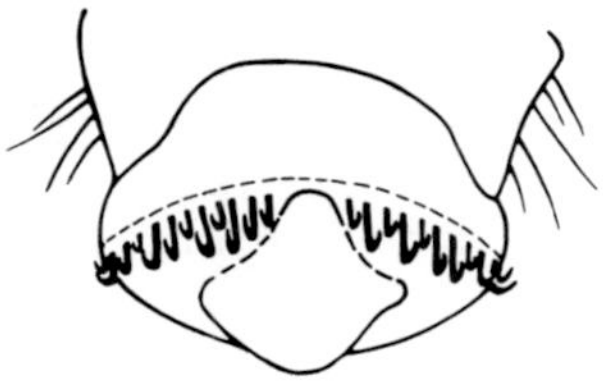

Figure 151

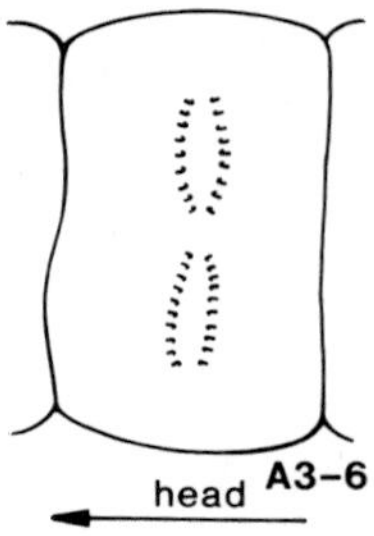

head A3–6

Figure 174

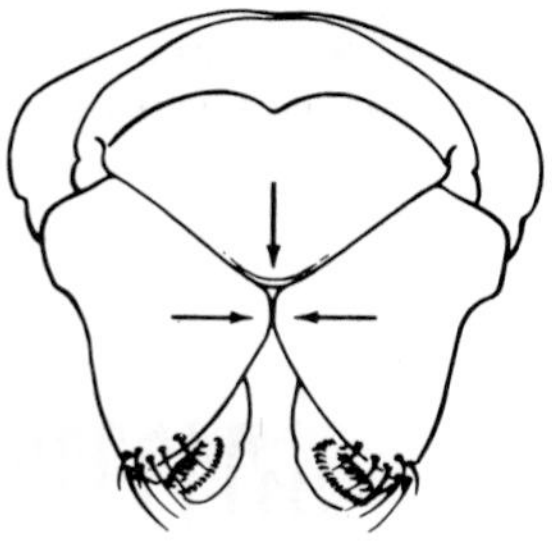

Figure 182

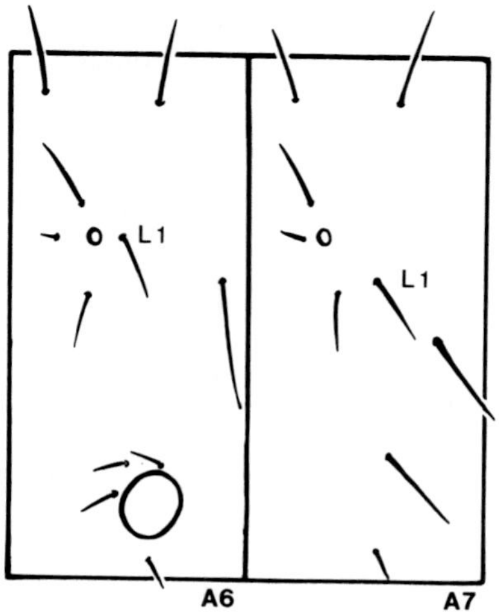

L1 L1 A6 A7

Figure 175

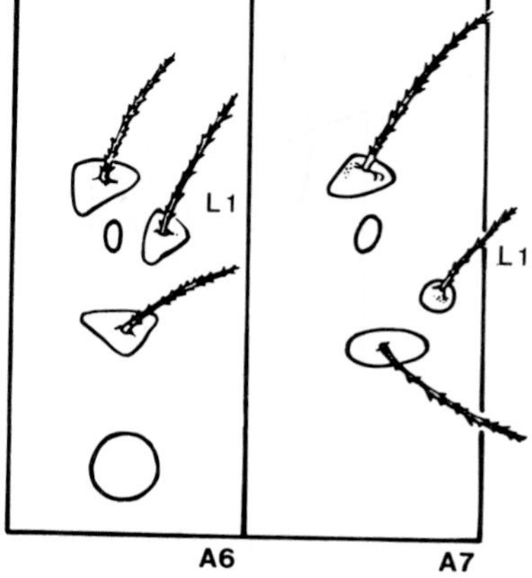

L1 L1 A6 A7

Figure 176

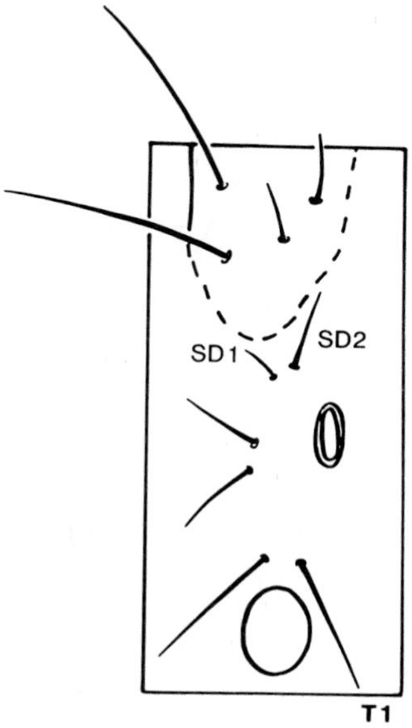

SD1 SD2 T1

Figure 177

← Fig. 178

203(202). T1 with SD1 or SD2, or both, excluded from (not surrounded by) prothoracic shield (which may be faint) (fig. 177), or shield absent (SD setae may be on their own pinaculum which may also have a few extra setae) 204

203′. T1 with SD1 and SD2 included on (surrounded by) prothoracic shield (fig. 179) .... (part) ***Arctiidae*** (p. 538)

204(203). Ventral crochets in heteroideous mesoseries (fig. 137) .... (*Utetheisa*) ***Arctiidae*** (p. 538)

204′. Ventral crochets in homoideous mesoseries 205

205(204′). A10 with 1–4 setae in area anterior to anal shield or on "anterior corner of anal shield" (anal shield normally bears 8 setae, 4 per side), and located approximately as high as SD1 on A9 (fig. 180); labral notch usually acute, V-shaped, and continued as groove to base of labrum (fig. 147); caudal prolegs often caudo-projecting or modified into caudo-projecting processes 207

205′. A10 with no setae in above areas; labral notch never V-shaped or continued as groove to base of labrum; caudal prolegs not caudo-projecting 206

206(205′). Mandible with 3 large, *adjoined, inner* teeth (*Cargida pyrrha,* southeast Arizona, possibly southwest New Mexico and west Texas) .... (part) ***Notodontidae*** (p. 524)

206′. Mandible without 3 large, adjoined inner teeth; widespread .... (some) ***Noctuidae*** (p. 549)

207(205). T1 with shield absent or not darkly sclerotized, and with posterior pair of setae on each side shorter than anterior pair; caudal prolegs often modified into caudo-projecting processes or smaller than ventral prolegs (totally reliable characters separating dioptids and notodontids are lacking, check descriptions); (widespread distribution and many common species; some are forest defoliators) .... (part) ***Notodontidae*** (p. 524)

207′. T1 with darkly sclerotized shield bearing 4 conspicuous setae of approximately equal size per side (8 total setae); caudal prolegs never modified as above, but somewhat smaller than ventral prolegs (distribution in the U.S. confined to Oregon and California (and possibly the southwestern U.S.)) .... ***Dioptidae*** (p. 533)
1 common species, the California oakworm (*Phryganidia californica*)

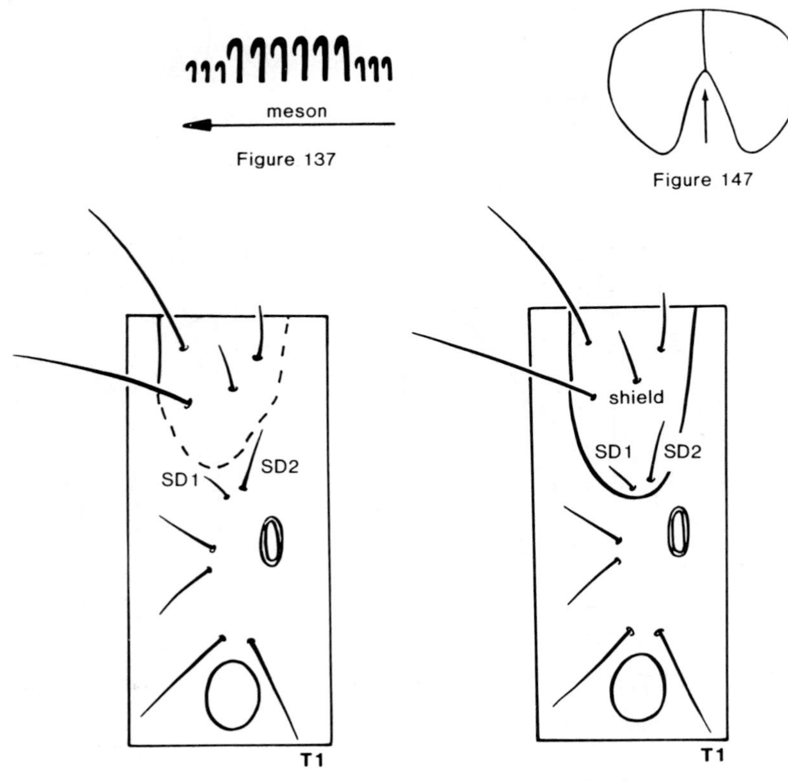

Figure 137

Figure 147

Figure 177

Figure 179

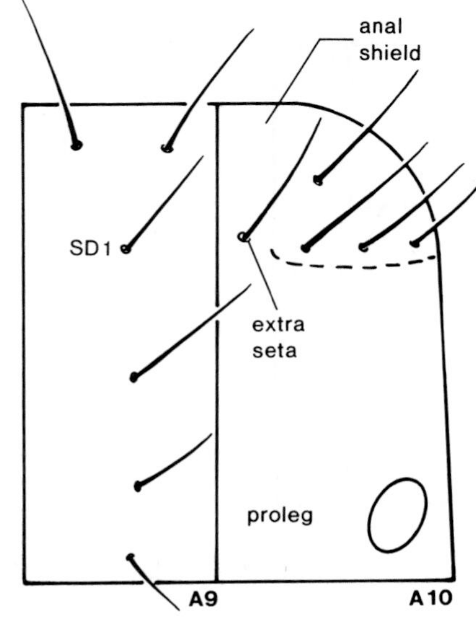

Figure 180

208(202′). A3–6 usually with 2 to 3 prominent setae in area caudad of spiracle, and with a group of 4–12 prominent setae in the SV1 area just above and distinct from the multisetiferous proleg base .... (*Doa*) ***Doidae*** (p. 534)

208′. A3–6 with extra setae not located in positions as above or not as numerous or not distinguishable 209

209(208′). A10 crochets in complete circle or oval (look very carefully if proleg is not fully distended) .... (part) ***Mimallonidae*** (p. 508)

209′. A10 crochets not in complete circle or oval 210

210(209′). Setae L1 and L2 (if distinguishable) similarly spaced on A1–3 and A4–8; ventral crochets uniformly biordinal (if crochets are uniordinal go to couplet 202) 211

210′. Setae L1 and L2 close together on A1–3, but about twice as far apart on A4–8; ventral crochets uniordinal in center, biordinal at both ends (fig. 181) .... ***Epiplemidae*** (p. 507)

211(210). A3–6 prolegs with many setae; abdomen terminating in 3 sharply pointed lobes (fig. 182); head triangular, and with some knobby seta bases that somewhat resemble stemmata .... (*Lapara*) ***Sphingidae*** (p. 521)

211′. A3–6 prolegs with 4 setae; abdomen terminating smoothly; head rounded, and without knobby seta bases .... (part) ***Thyatiridae*** (p. 500)

212(196′). A3–6 crochets multiserial (fig. 69), biserial (fig. 183) or partially biserial (fig. 184) and in complete circle 213

212′. A3–6 crochets in single row (uniserial) in complete or incomplete circle 215

213(212). Anal shield with 3 pairs of setae .... (part) ***Hepialidae*** (p. 347)

213′. Anal shield with 4 pairs of setae or secondary setae present 214

214(213′). T1 with 2 L setae .... (part) ***Copromorphidae*** (p. 399)

214′. T1 with 3 L setae .... (part) ***Agonoxenidae*** (p. 390)

215(212′). Stemmata 4 and 5 farther apart than stemmata 3 and 4 (fig. 81), distance between 4 and 5 *usually much greater* than diameter of stemma 4 216

215′. Stemmata 4 and 5 not farther apart than stemmata 3 and 4 (fig. 83), distance between 4 and 5 *usually equal to or less* than diameter of stemma 4 219

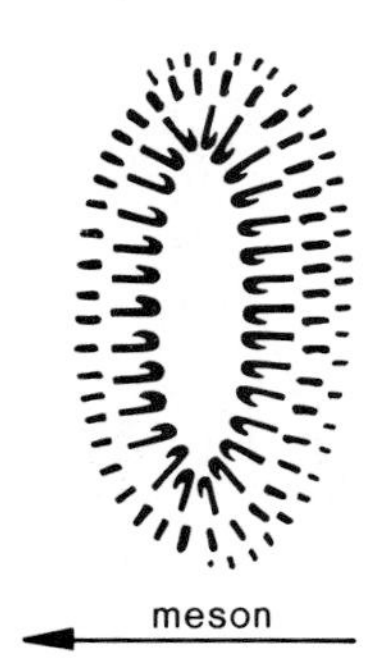

Figure 69

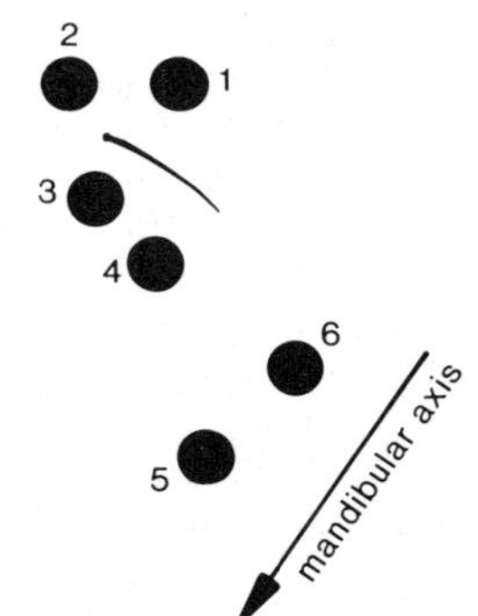

Figure 81

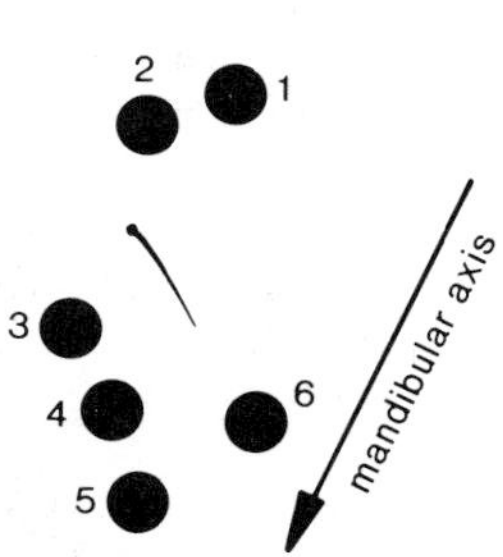

Figure 83

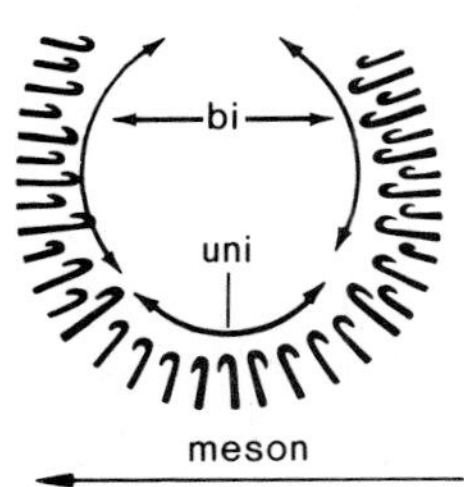

Figure 181

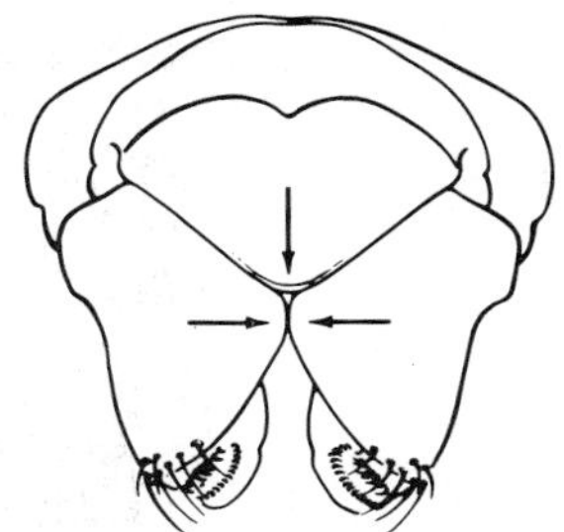
Figure 182

Figure 183

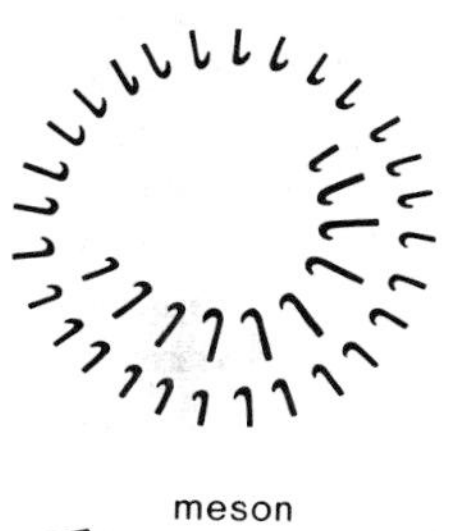

Figure 184

216(215). A3–6 crochets in mesoseries .... (part) ***Thyatiridae*** (p. 500)
216′. A3–6 crochets not in mesoseries 217
217(216′). T1 with L (prespiracular) group trisetose (check both sides) (fig. 185) .... (part) ***Cossidae*** (p. 416)
217′. T1 with L (prespiracular) group bistose (check both sides) 218
218(217′). A10 crochets in complete circle or oval (if proleg not fully distended, crochets may appear to be 2 transverse bands); no setae caudad of spiracle on A3–6; spiracles elliptical .... (part) ***Mimallonidae*** (p. 508)
218′. A10 crochets not in complete circle or oval; at least 1 seta in area caudad of spiracle on A3–6; spiracles essentially round .... (a few **Scardiinae**) ***Tineidae*** (p. 362)
219(215′). Ventral crochets uniordinal 220
219′. Ventral crochets biordinal (sometimes weakly so or triordinal) 224
220(219). T1 with L (prespiracular) group bisetose (check both sides) 221
220′. T1 with L (prespiracular) group trisetose (check both sides) 222
221(220). A3–6 crochets in circle .... (part) ***Carposinidae*** (p. 401)
221′. A3–6 crochets in mesopenellipse .... (*Lotisma trigonana*) ***Copromorphidae*** (p. 399)
222(220′). T2 and T3 with SV group bisetose or trisetose; prolegs often longer than wide 223
222′. T2 and T3 with SV group unisetose; prolegs usually shorter than wide .... (part) ***Cosmopterigidae*** (p. 391)
223(222). Ventral crochets in complete circle or *closed* ellipse .... (part) ***Scythrididae*** (p. 393)
223′. Ventral crochets in mesopenellipse .... (part) ***Pterophoridae*** (p. 497)
224(219′). T2 dorsum with transverse band or patches of spinulae (fig. 186). .... (part) ***Agonoxenidae*** (p. 390)
224′. T2 dorsum without such a band or patches 225
225(224′). SD1 on A1–7 with base usually set in conspicuous pinaculum ring (fig. 187); SD1 (just above spiracle) on A8 usually finer than SD1 on A1–7 .... (part ***Scythrididae*** (p. 393)
225′. SD1 on A1–7 with base not set in pinaculum ring, but sometimes on well sclerotized pinaculum; SD1 on A8 not finer than SD1 on A1–7 .... (part) ***Oecophoridae*** (p. 379)

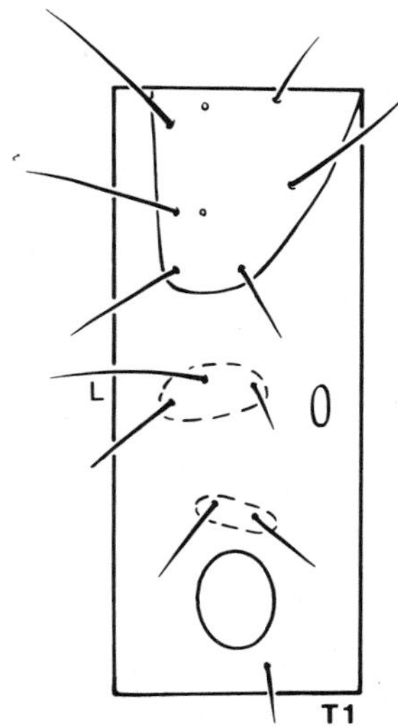

Figure 185

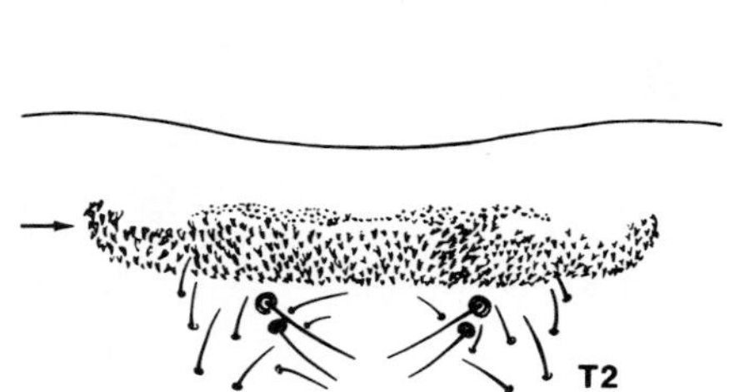

Figure 186

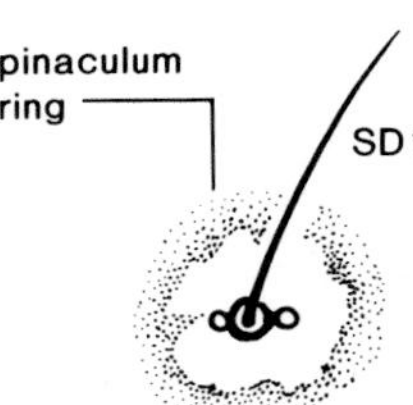

Figure 187

SUBORDER ZEUGLOPTERA

MICROPTERIGIDAE (MICROPTERIGOIDEA)

Donald R. Davis, *Smithsonian Institution*

Micropterigids

Figures 26.28a-g, 26.29a-l

Relationships and Diagnosis: Relatively isolated systematically and now recognized to be the only member of the suborder Zeugloptera, the Micropterigidae is the most archaic extant family of Lepidoptera. The primitive nature of the mandibulate adults was early recognized by Packard (1895) who proposed a separate suborder, Protolepidoptera. Chapman (1917) further distinguished the group by erecting a new order, Zeugloptera. Largely influenced by the very unusual morphology of the larva, Hinton (1946) reaffirmed Chapman's decision, adding that the Zeugloptera were more primitive than either the Trichoptera or Lepidoptera. Kristensen (1971, 1984), however, enumerated the close affinities of the adult Micropterigidae to the Lepidoptera and once again recognized them as a suborder of Lepidoptera. These varying opinions concerning the proper placement of the Micropterigidae reflect a basic problem of congruency. The adult and pupal stages, although quite primitive, are decidedly lepidopterous. The larval stage, however, demonstrates several plesiomorphic features not present in any member of the Lepidoptera Glossata, as well as some apomorphies not known to occur in any other family of insects. Although differing in most other respects, the larvae of some Micropterigidae superficially resemble those of Panorpidae (Mecoptera), particularly in the possession of eight pairs of nonmuscular prolegs on the first eight abdominal segments.

Functional mandibles in the adult accompanied by an absence of a haustellum also occurs in at least two other families of Lepidoptera, the Australian Agathiphagidae, and the South American Heterobathmiidae. The latter, at least on the basis of adult characters, appears to be the family most closely allied to Micropterigidae. The highly divergent larvae of both the Agathiphagidae and Heterobathmiidae have required separate subordinal placements for each (i.e., the suborders Aglossata and Heterobathmiina respectively, Kristensen and Nielsen 1983). The larvae of Agathiphagidae are borers in the seeds of *Agathis* and those of *Heterobathmia* are leaf miners in deciduous *Nothofagus* (Kristensen and Nielsen 1983).

The single endemic North American genus, *Epimartyria,* demonstrates the principal diagnostic features of the family and suborder: antennae three-segmented, relatively simple and elongate; a single medial seta (M1); internal adfrontal ridge absent; absence or vestigial condition of the adfrontal sutures and ecdysial lines; absence of tubular spinneret; tentorium well developed with tentorial bridge short and broad; hypostomal bridge complete; body hexagonal in cross section; generally reduced chaetotaxy; thoracic legs four-segmented; and eight pairs of ventral prolegs on first eight abdominal segments.

Biology and Ecology: Eggs of *Micropterix* are laid singly or in small clumps of 2–45 (Lorenz 1961; Tuskes and Smith 1984) on the food substrate.

Larvae have been reported feeding upon lower forms of plant life such as liverworts (Tillyard 1923; Tuskes and Smith 1984), leaf detritus, and possibly fungal hyphae (Heath 1976b). Lorenz (1961) reared larvae of *Micropterix calthella* (L.) on decayed plant detritus as well as upon fresh leaves of *Veronica agrestis* L. (Scrophulariaceae). Luff (1964) found larvae of *Micropterix* in tussocks of cocksfoot grass (*Dactylis glomerata* L.) in Berkshire, England, and suggested that they fed on dead leaves. Carter and Dugdale (1982) reported that successful rearing of two species of British *Micropterix* was dependent upon a supply of fresh, photosynthetic angiosperm tissue, particularly chickweed (*Stellaria media* (L.). Although external feeders, the larvae are secretive, preferring to burrow through thick mats of mosses and lichens or leaf litter. Heath (1976b) reports larvae of *Micropterix* at depths down to 10 cm. in loose soil. Lorenz (1961) noted that the larvae he studied required from 132 to 141 days for development, but he was uncertain as to the number of instars. Three larval instars have been suggested by Heath (1976b), with the last instar overwintering. Three instars have also been reported for *Epimartyria pardella* (Tuskes and Smith 1984).

Larvae of the eastern North American *Epimartyria auricrinella* Wlsm. require two seasons to mature, as evidenced by larval collections in the field. Tuskes and Smith (1984) observed a two-year life cycle for the western *E. pardella* (Wlsm.).

The pupa is decticous with articulated mandibles and all appendages free. Pupation occurs in early spring within a tough silken cocoon in the ground or host substrate. Adults emerge in the spring.

Description: Mature larva small, not exceeding 5 mm. Body color varies from whitish with greenish brown markings (Tillyard 1923) to entirely dark brown. Body typically hexagonal in cross section with ribbed, bulbous, or hairlike to claviform setae. Cuticle rugose; dorsal surface devoid of spinules but heavily ridged; ventral surface and intersegmental areas densely covered by small papillae. Interpapillar area sometimes densely supplied with numerous, ultrascopic projections, with those of *Epimartyria* probably functioning in part as a plastron (Davis in prep.). Spiracles peripneustic, conical, with filamentous walls in *Epimartyria.*

Head: Somewhat prognathous, capable of being retracted into prothorax. Antenna elongate, slender, three-segmented, arising well posterior of clypeal margin and above stemmata; all segments without sensory setae except for peglike to elongate terminal spine. Five stemmata, arranged in a closed circle. Adfrontal sutures vestigial, not extending to vertex. Adfrontal ridges similarly undeveloped. Ecdysial lines externally indistinct. Cranial setae reduced. A single medial seta (M) arising midway between antennae. Stemmatal setae absent.

Thorax: T1 with as many as 11 pairs of setae; in *Epimartyria,* anterior vertical row of four setae of uncertain homology and greatly reduced. Lateral series apparently unisetose; L1 far anterior of spiracle. T2 and T3 with six pairs

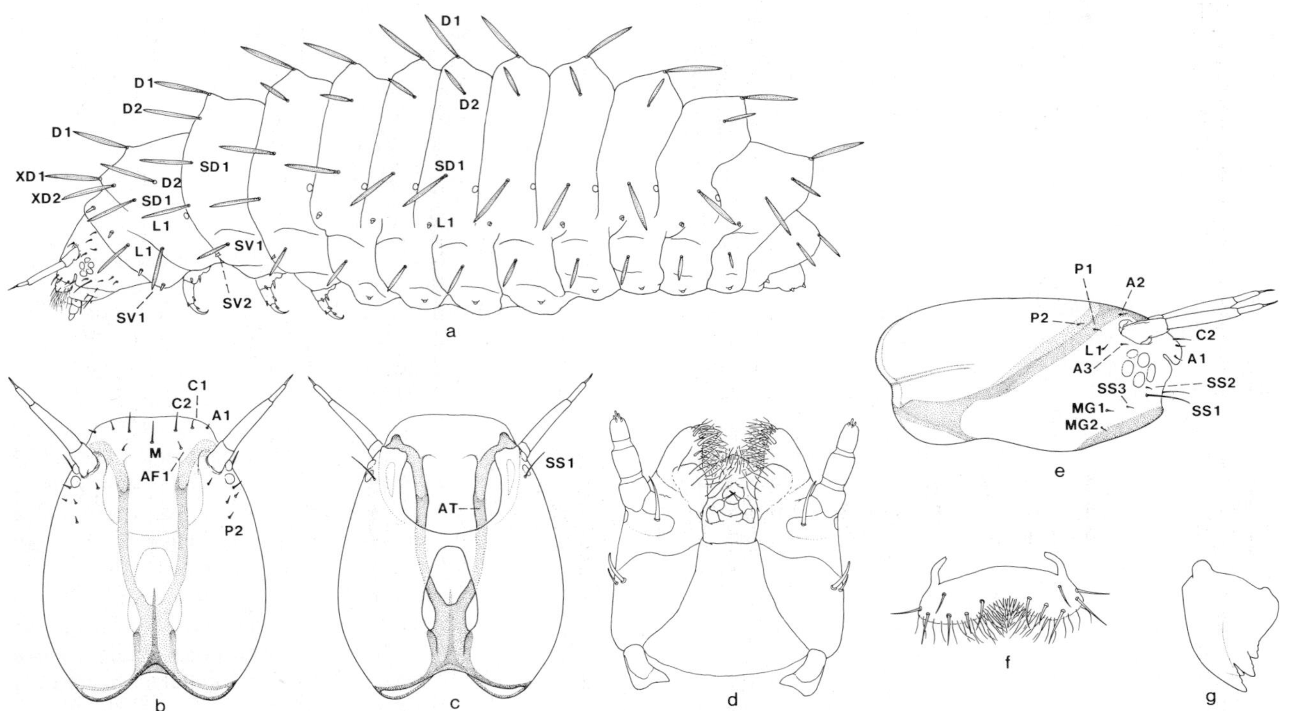

Figure 26.28a-g. Micropterigidae, *Epimartyria auricrinella* Walsingham. **a.** lateral; **b.** head, dorsal (M = medial seta); **c.** head, ventral (AT = anterior arms of tentorium); **d.** maxilla, labium, ventral; **e.** head, lateral; **f.** labrum, dorsal; **g.** mandible.

of primary setae; SV2 greatly reduced in *Epimartyria.* Legs four-segmented, including large pretarsal segment; coxa reduced, femur and trochanter fused, as well as tibia and tarsus.

Abdomen: Usually six pairs of primary setae on A1–8; A9 with three pairs and A10 with two pairs; L1 and V1 greatly reduced, peglike in *Epimartyria.* One pair of ventral prolegs on A1–8; prolegs without muscles (Hinton 1958), well developed in *Micropterix* but reduced in all other genera; venter of A9–10 with a trilobed suction organ in *Micropterix* (Gerasimov 1952).

Comments: The Micropterigidae currently consists of nine extant genera and approximately 100 species (Gibbs 1983). The group is poorly represented in the Western Hemisphere, with the greatest concentration of species in the Palearctic and Australian Regions. The family is an ancient one, with fossil records from the lower Cretaceous (Whalley 1978).

Although Gerasimov (1952) and Hinton (1958) believed the chaetotaxy of Micropterigidae to be entirely different from the higher Lepidoptera, I have attempted to homologize these setae, according to the body areas from which they arise. Some uncertainties persist, particularly with regard to the homology of the unique medial head seta, present in all genera studied. Similarly, the homologies of the anterior vertical row of prothoracic setae arising on the prothoracic membrane in *Epimartyria* are unknown. Possibly they may represent proprioceptor setae, although the pronotal seta (MXD1) has never been reported anterior to the pronotum. It is further possible that a few of these anterior setae may represent other primary setae (e.g., L2 and SV2), or extra setae, like the medial seta above, that are unique for the family.

Hinton (1958) further stated that the tracheal system in the larva of Micropterigidae was holopneustic, with functional spiracles on both T1 and T2 and first seven abdominal segments. The spiracles are normally situated deep in the intersegmental folds, frequently being hidden or difficult to examine in larvae not fully expanded. Thus, it is not obvious on which segment they actually arise. It appears more likely that micropterigid larvae are peripneustic, as is typical for the Lepidoptera, with the spiracles of the abdominal segments having migrated slightly forward to the intersegmental fold.

Selected Bibliography

Carter and Dugdale 1982.
Chauvin and Chauvin 1980.
Gerasimov 1952.
Gibbs 1983.
Heath 1976b.
Hinton 1958.
Kristensen 1971, 1984.
Kristensen and Nielsen 1979, 1983.
Mutuura 1956.
Lorenz 1961.
Luff 1964.
Martinova 1950.
Tillyard 1923.
Tuskes and Smith 1984.
Whalley 1978.
Yasuda 1962.

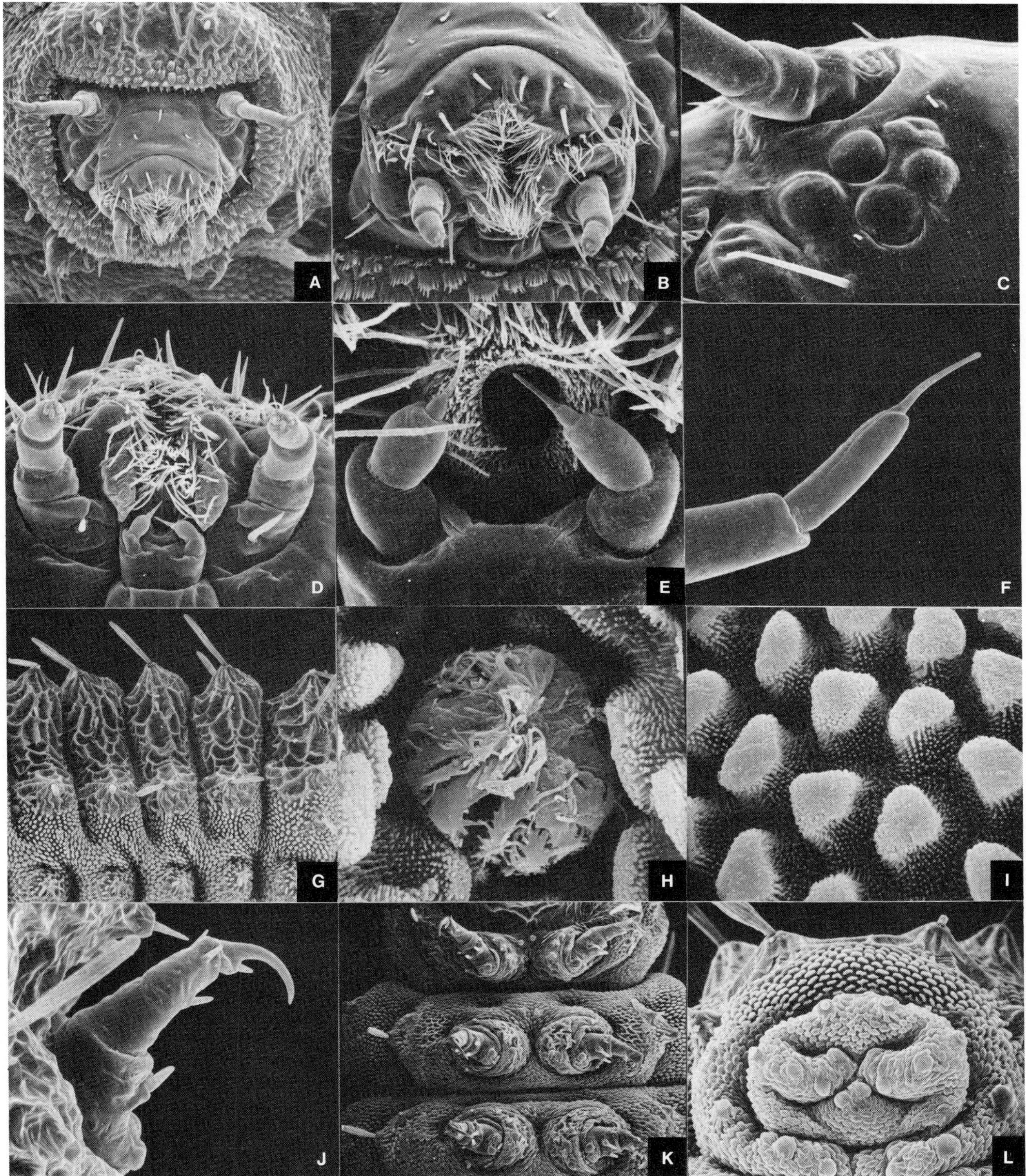

Figure 26.29a-l. Micropterigidae, *Epimartyria auricrinella* Walsingham. a. head ×153; b. detail of mouthparts, labrum and maxilla, ×319; c. stemmata, ×354; d. ventral view of mouthparts, maxilla and labium, ×463; e. opening to labial gland, labial palpi, ×1180; f. apex of antenna, ×826; g. lateral view of abdomen, A1-5, ×71; h. spiracle, A4, ×1652; i. plastron surface in subventral region of A4, ×1180; j. thoracic leg, T3, ×384; k. ventral view of thorax, ×57; l. ventral view of A10, ×118.

SUBORDER DACNONYPHA

ERIOCRANIIDAE (ERIOCRANIOIDEA)

Donald R. Davis, *Smithsonian Institution*

The Eriocraniids

Figures 26.30a-e, 26.31a-c, l

Relationships and Diagnosis: The Eriocraniidae, Acanthopteroctetidae, and possibly the endemic Australian family Lophocoronidae comprise the Eriocranioidea, currently the only superfamily in the suborder Dacnonypha. The Acanthopteroctetidae demonstrate major apormorphies that will probably necessitate a new superfamily and possibly subordinal placement. The immature stages of the Lophocoronidae are unknown; consequently, our knowledge of the preadult stages for the suborder is based upon what is known for the Acanthopteroctetidae and Eriocraniidae. The known larvae are leaf miners, either completely apodal or with thoracic legs. The pupae of Acanthopteroctetidae and Eriocraniidae are exarate with large movable mandibles.

The larvae of Eriocraniidae possess several features that easily distinguish them from most, if not all, Lepidoptera. The most diagnostic character of the head is the very distinct ecdysial lines. These extend anteriorly around the antennae and terminate on the posterior rim of the antennal socket. The adfrontal sclerite is sufficiently broad to include seta Al, if present. A single pair of rudimentary stemmata is present immediately beneath the antenna. The body is cylindrical and without legs, prolegs, or crochets. Paired ambulatory calli are present on the meso- and metathorax.

Biology and Ecology: All species are univoltine leaf miners with adults emerging in late winter or early spring. Eggs are oval and smooth with a thin chorion and are inserted singly into young leaf tissue by a piercing ovipositor. The larval mine at first is serpentine, becoming a large, full-depth blotch with loose, thin, stringy frass. Four larval instars are believed to exist. The preferred hosts include *Alnus, Betula, Carpinus,* and *Corylus* (Betulaceae) as well as *Castanea* and *Quercus* (Fagaceae). Larval development is rapid and 7–10 days after hatching the larva abandons the mine, burrows into the soil 2–10 inches and overwinters inside a silken cocoon. Pupation probably does not occur until the following spring.

The pupa is decticous and exarate with extremely large, articulated mandibles. The apices of the mandibles are broadly truncate and partially surrounded by four sharp spines. The pupal cuticle is thin and flexible with all appendages free and probably capable of some movement. Another feature of the eriocraniid pupa is the elongate, spinose projection of the frontal ridge.

Description: Mature larva small, usually not exceeding 11 mm. Body mostly whitish in color, cylindrical, and somewhat fusiform in outline; apodal. Cuticle appearing smooth, but densely covered by microscopic spines.

Head: Prognathous, partially retracted into thorax, usually pale brown to stramineous. Vertex deeply divided to frontoclypeus. Adfrontal sclerite broad with ecdysial line terminating at anterior margin of head lateral to antennae; ecdysial lines very well defined, broad, irregular, sometimes enclosing P1. A single rudimentary stemma present near anterior margin of head immediately ventrad to antenna.

Thorax: Usually whitish except for large, brownish pronotal and prosternal plates. D1 and D2 on all segments; D1 smaller and more anterior. SD1 caudad and more ventral to SD2. L group trisetose on all segments. SV group either bisetose or trisetose on T1, bisetose on T2 and T3. Ventral ambulatory calli on T2 and T3; dorsal calli also sometimes evident on T2 and T3.

Abdomen: Whitish. D1 and D2 on all segments; D1 reduced and situated more cephalad. SD group bisetose on segments 1–6, unisetose on 7–9, rarely bisetose on 9. L group bisetose or trisetose; if L1 present then situated high on pleuron above and posterior to spiracle and SD1. SV group either unisetose or bisetose on 1–7 and 9; unisetose on 8. Segment 9 usually with seven primary setae, rarely with ten. Tenth segment usually with anal shield poorly defined; only two pairs of dorsal setae and a pair of small, elongate spots or brace rods situated lateroposteriorly between D1 and D2.

Comments: The Eriocraniidae is a small, archaic family of homoneurous moths with somewhat specialized larvae. It is reasonable to assume that the present day eriocraniid larva descended from forms possessing thoracic legs and possibly abdominal prolegs with crochets, and that the recent apodal condition was an evolutionary adaptation associated with its leaf-mining habit. However, it is uncertain at what point in the hierarchy of Lepidoptera that crochets originated, or if this feature evolved more than once. Because of the enormous inadequacies of the fossil record, these queries will probably never be answered. All evidence suggests that crochets did not appear until after the separation of the higher Lepidoptera from the Zeugloptera stem. In the homoneurous families they are first encountered in the Hepialidae where they are multiserial and uniordinal. In the heteroneurous groups they first appear in the Incurvarioidea and are of a primitive form, resembling enlarged, slightly modified cuticular spines arranged in transverse rows.

The family is restricted to the Holarctic region where less than 30 species have been described; 16 of those are known from the United States and Canada. The most complete review of the group is by Davis (1978a). The British species were treated by Heath (1976a).

Selected Bibliography

Davis 1978a.
Davis and Faeth 1986.
Gerasimov 1952.
Heath 1976a.

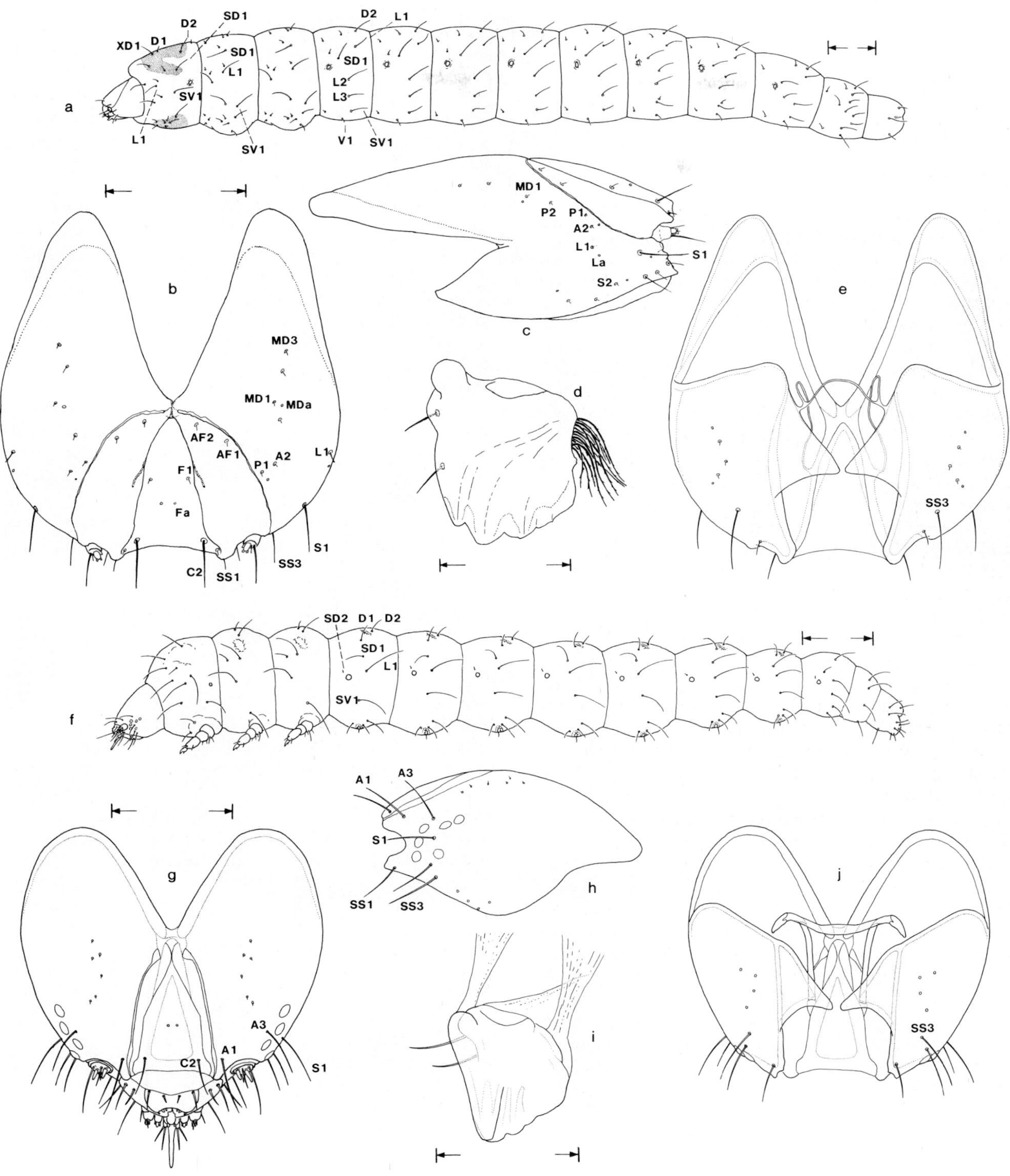

Figure 26.30a-j. Eriocraniidae, a–e; Acanthopteroctetidae, f–j. **a-e.** *Dyseriocrania griseocapitella* (Walsingham). a. lateral view, (scale = 0.5 mm); b. head, dorsal view (scale = 0.5 mm); c. head, lateral; d. mandible (scale = 0.1 mm); e. head, ventral; **f-j.** *Acanthopteroctetes unifascia* Davis. f. lateral (scale = 0.5 mm); g. head, dorsal (scale = 0.5 mm); h. head, lateral; i. mandible (scale = 0.1 mm); j. head, ventral.

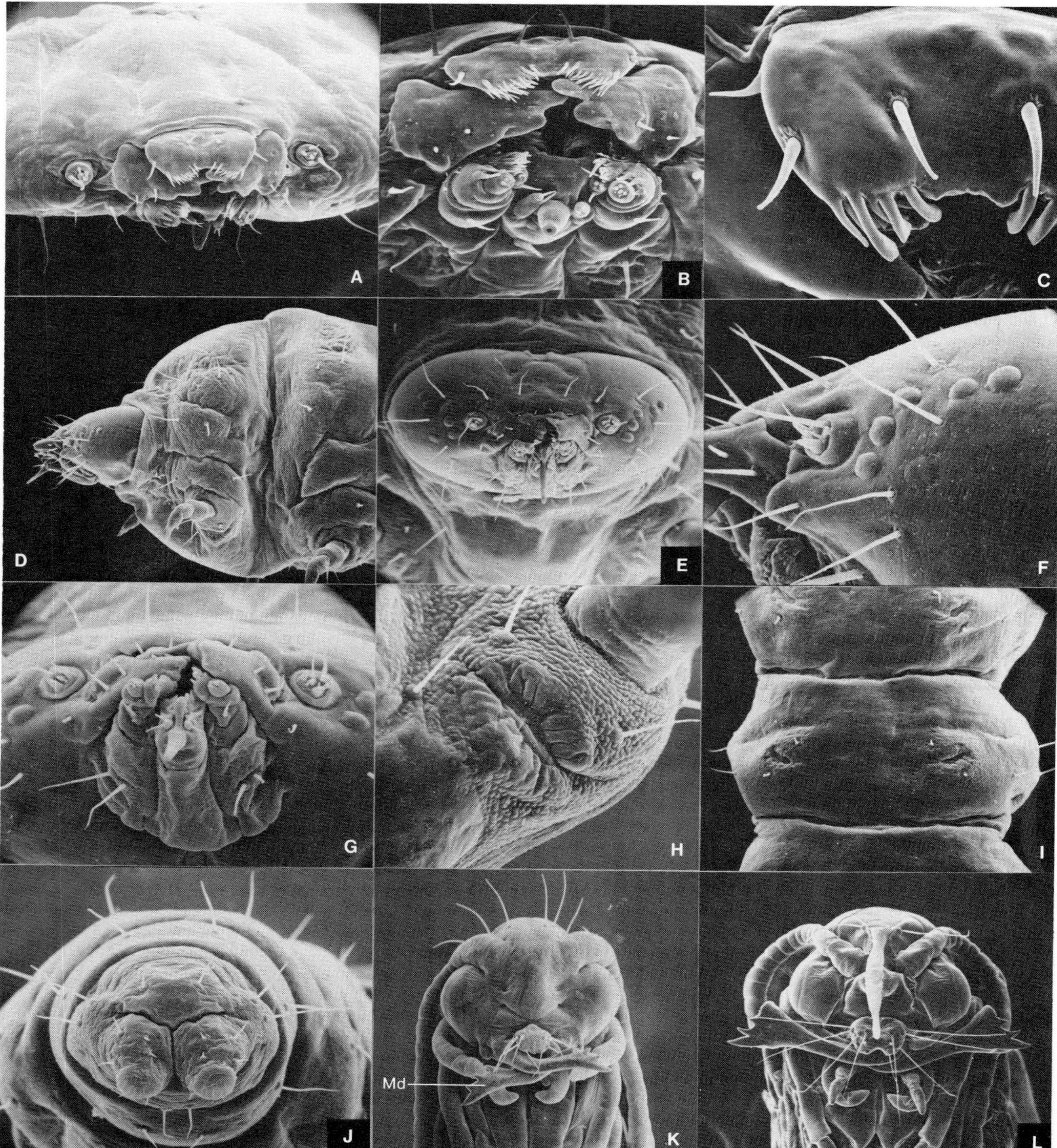

Figure 26.31a-l. Eriocraniidae, a–c, l; Acanthopteroctetidae, d–k. **a,b.** *Dyseriocrania griseocapitella* (Walsingham): a. head, ×128; b. head, detail of mouthparts, ×209; **c.** *Eriocraniella ?longifurcula* Davis, labrum, ×638; **d-h.** *Acanthopteroctetes unifascia* Davis. d. head and thorax, lateral, ×49; e. head, ×93; f. stemmata, ×261; g. head, detail of mouthparts, ×174; h. ambulatory callus, coxa T3, ×232; i. dorsal ambulatory calli, A4, ×58; j. anal prolegs, A10, ×79; k. head of pupa, ×50 (Md = mandibles); **l.** *Eriocrania ?semipurpurella* (Stephens), pupa, ×38.

ACANTHOPTEROCTETIDAE (ERIOCRANIOIDEA)

Donald R. Davis, *Smithsonian Institution*
Donald C. Frack, *La Puente, California*

The Acanthopteroctetids

Figures 26.30f-j, 26.31d-k

Relationships and Diagnosis: The Acanthopteroctetidae is a small, rarely encountered family, presently including a single genus and four species from the western United States (Davis 1978a). Although currently associated with the Eriocraniidae in the suborder Dacnonypha, recent research suggests a new superfamily and possibly subordinal placement for the family.

Recent discovery of the immature stages of *Acanthopteroctetes unifascia* Davis by D. C. Frack has made it possible to present this information for the first time. Further details will be presented in another paper. The larva of this species is a blotch leaf miner with slender thoracic legs and no crochets; a single pair of anal prolegs is present. The coxal segments of T1–3 are unusual in possessing suctionlike ambulatory calli; similar paired, sunken areas are also present on the dorsum of T1–3 and both the dorsal and ventral surfaces of A1–7. The lateral setae are bisetose on T1. Of the very few larvae bearing only thoracic legs, this family alone has seta L1 directly caudad to the spiracle on A1–8 and has leaf-mining habits until maturity.

Biology and Ecology: *Acanthopteroctetes* is a full-depth blotch miner on the leaves of various species of *Ceanothus* (Rhamnaceae) in southern California. The eggs are inserted singly into the upper epidermis of the leaf by the piercing ovipositor. Larval feeding begins during summer and continues into fall with the half-grown larva overwintering. The relatively large frass pellets are packed into the older portions of the mine. Feeding continues the following spring until maturity, at which time the larva exits through the lower epidermis and constructs a cocoon in debris beneath the host plant.

The pupa (fig. 26.31k) is decticous and exarate, with hypertrophied mandibles similar to those of Eriocraniidae. The mandibles of *A. unifascia* differ from the latter in being more slender and acute. The species is univoltine.

Description: Mature larva small, 5–6 mm. Body white except for brownish tergal and sternal plates on T1. Head slightly depressed, body subcylindrical. Chaetotaxy reduced, with only the lateral setae well developed. Spiracles round.

Head: Prognathous, partially retracted into prothorax. Stemmata 6, arranged in a sharply angulate, almost rectangular line with the sixth stemmatal lens reduced. Ecdysial lines close to adfrontal sutures, terminating anteriorly well mesad of antennae and converging caudally near prominent epicranial notch. Adfrontal sclerite well defined, narrow, not bearing A1. Frontoclypeus triangular, longer than wide, converging slightly before epicranial notch and near ecdysial lines.

Thorax: Prothorax with brownish tergal and sternal plates. D group unisetose on T1, bisetose on T2–3. L group bisetose on T1, unisetose on T2–3. Legs present, relatively short and slender, widely separated, and with indistinct coxae. Paired suctionlike discs (probably ambulatory calli) present on dorsum of T1–3 and on coxal membrane of T1–3.

Abdomen: Without pigmented areas or distinct pinnacula. D group bisetose on A1–7, unisetose on A8–9. SD group bisetose on A1–8 with SD2 minute and immediately dorsad to spiracle. L group bisetose on A1–8, unisetose on A9, with L1 directly caudad to spiracle. SV group trisetose on A1–7. Prolegs absent except for anal pair. Crochets completely absent. Paired, sunken, suctionlike, elliptical discs present on both dorsum and venter of A1–7.

Comments: As may be observed from a comparison of the family descriptions, the larvae of Acanthopteroctetidae and Eriocraniidae bear almost no close similarity to one another, in contrast to their very similar pupae. The most unusual feature of the larva of *Acanthopteroctetes unifascia* is the development of paired, probably locomotory organs on both the dorsal and ventral surfaces of the first ten body segments in addition to the presence of five-segmented thoracic legs. The large number of calli in this family all appear to be expansible and may be used in a manner similar to those described for some Heliozelidae; that is, to assist the larva in maintaining its position between the two walls of its mine. Another peculiar locomotory specialization in *A. unifascia* is the absence of all crochets and the absence of all prolegs except the anal pair. The ventral ambulatory calli on the first seven abdominal sterna are not believed to be homologous to prolegs. The discussion under the Eriocraniidae regarding the first appearance of crochets in the Lepidoptera and their supposed "loss" in leaf-mining groups is equally pertinent to the Acanthopteroctetidae.

Selected Bibliography

Davis 1978a.

SUBORDER EXOPORIA

HEPIALIDAE (HEPIALOIDEA)

David Wagner, *University of California, Berkeley*

Swift or Ghost Moths

Figures 26.31t-26.31z

Relationships and Diagnosis: Hepialids belong to the archaic suborder Exoporia, members of which possess an oviduct and copulatory orifice that is not internally interconnected by a ductus seminalis; rather, sperm must pass externally from the bursa copulatrix to the common oviduct to effect fertilization. Exoporian families are primarily restricted to the southern continents; hepialids are the only common Holarctic members of the suborder. Relationships among the families have been discussed by Kristensen (1978).

Hepialid larvae are among the most distinct lepidopteran larvae, possessing many unique characteristics. On the head capsule (fig. 26.31y) AFa is absent, Ga is absent, Fa is laterad of F1, and two microgenal setae are present. Prothoracic shield (fig. 26.31z) enlarged with L setae included or with L3 nearly included. Abdominal segments 1–8 with SD2 conspicuous and L2 anteroventrad of spiracle. MV3 enlarged and adjacent to SV setae on A2–7. Abdominal prolegs with crochets (second to final instars) in multiserial, uniordinal ellipses.

Biology and Ecology: Eggs are broadcast aerially, often indiscriminately, as females fly over suitable habitat. Females are extremely fecund and one Australian species may lay upwards of 18,000 ova (Common 1970). The eggs of North American species are shiny black, without ornamentation, and nearly spherical.

The feeding biology of early instars is not known for North American species. Later instars tunnel in woody tissues above and below ground or feed on roots externally. All Holarctic species are believed to be polyphagous. *Sthenopis* Packard larvae (fig. 26.31u-x) have been recovered from the lower trunks and roots of alder, birch, and willow. *Hepialus* F. larvae (fig. 26.31t) are known to feed on a wide range of hosts including ferns, legumes, rosaceous plants, composites, and grasses.

Prior to pupation, the larval tunnel is lined with silk; within the tunnel an elongate cocoon is spun in which the pupa is afforded a great deal of mobility. The pupal shell is extruded from one end of the tunnel prior to emergence. Our smallest species, *Hepialus hectoides* Boisduval is univoltine, whereas larger species, e.g., *Sthenopis,* may take several years to mature (Swaine 1909).

Adults of most species are crepuscular. Courtship activity may be exceptionally brief, often lasting only 20–30 minutes following sunset. Many of our montane and boreal species are diurnal and display more extensive periods of flight activity.

Description: Mature larvae 28–75 mm. Body elongate and cylindrical; ground color whitish, often with conspicuous pinacula. Microtrichia often conspicuous on first instar and on abdominal segments of later stages of some West Coast species. Spiracles round in first instar; elliptical, often concave, and with a distinct peritreme in second to final instars.

Head: (fig. 26.31y) Nearly prognathous. Head capsule deep reddish brown in late instars. Ga absent, AFa absent, Fa laterad of F1. Two microgenal setae; MG1, MG2, and SS3 nearly aligned; MG2 bisecting the distance between MG1 and SS3. A1, A2, and A3 in a line; A3 caudad of P1. Stemmata forming two vertical "rows".

Thorax: (fig. 26.31z) D1 on anterior margin of cervical shield and in line with XD1 and XD2. L setae included on the shield or nearly included. SD1 and SD2 (D2 of Hinton 1946) diagonally positioned and thin in comparison to D2 (SD2 of Hinton). SD1 and SD2 on darkened pinacula that are enlarged in some species to include D2 in a heavily pigmented sensory pit (fig. 26.31x). T1 with two SV setae; MV3 elongated.

T2 and T3 with SD2 directly above SD1 and on the same pinaculum that is sometimes confluent with L3 and D2 pinacula. MSD1, MSD2, and MD1 conspicuous, on a single pinaculum anteriorad of SD2; MD1 often elongate. Small pits located ventrocaudad of D2 and immediately caudad of L1. One SV seta.

Abdomen: SD1 and SD2 on a single pinaculum above spiracle; SD2 conspicuous, at least one-third length of SD1. L2 caudad of L1 and spiracle; L1 ventrocaudad of spiracle; L3 below and just anteriorad of spiracle on A1–8. MV3 lengthened and grouped with three SV setae on A2–7. A9 with nine setae; A10 with six setae on anal plate and eight above each proleg.

Crochets uniserial, uniordinal in a circle in first instars; multiserial uniordinal ellipses with crochet size increasing toward the center in second to final instars; up to six rows in *Sthenopis.*

Comments: Hepialids comprise a worldwide family with over 500 species in 80 genera. Diversity is greatest in the Indo-Australian Region where adults of some species are large and brilliantly colored. The North American fauna has 20 recognized species in the genera *Hepialus* and *Sthenopis.*

Although hepialids may be locally common, adult or larval collections are rare, primarily because of the brief flight activity of adults and the concealed habits of larvae. Hepialids are found in northern, montane, or coastal regions from Newfoundland to Alaska, south along the Appalachians to North Carolina, south to the mountains of Arizona in the Rocky Mountains, and along the Pacific coast into southern California.

Because of their polyphagous feeding habits, hepialids occasionally damage cultivated plants. For example, larvae of *Hepialus californicus* Boisduval (=H. *sequoiolus* Behrens) have been found tunneling in apple, blackberry, and peonies (Essig 1958).

Selected Bibliography

Aikenhead and Baker 1964.
Common 1970.
Essig 1958.
Hinton 1946.
Kristensen 1978.
Swaine 1909.

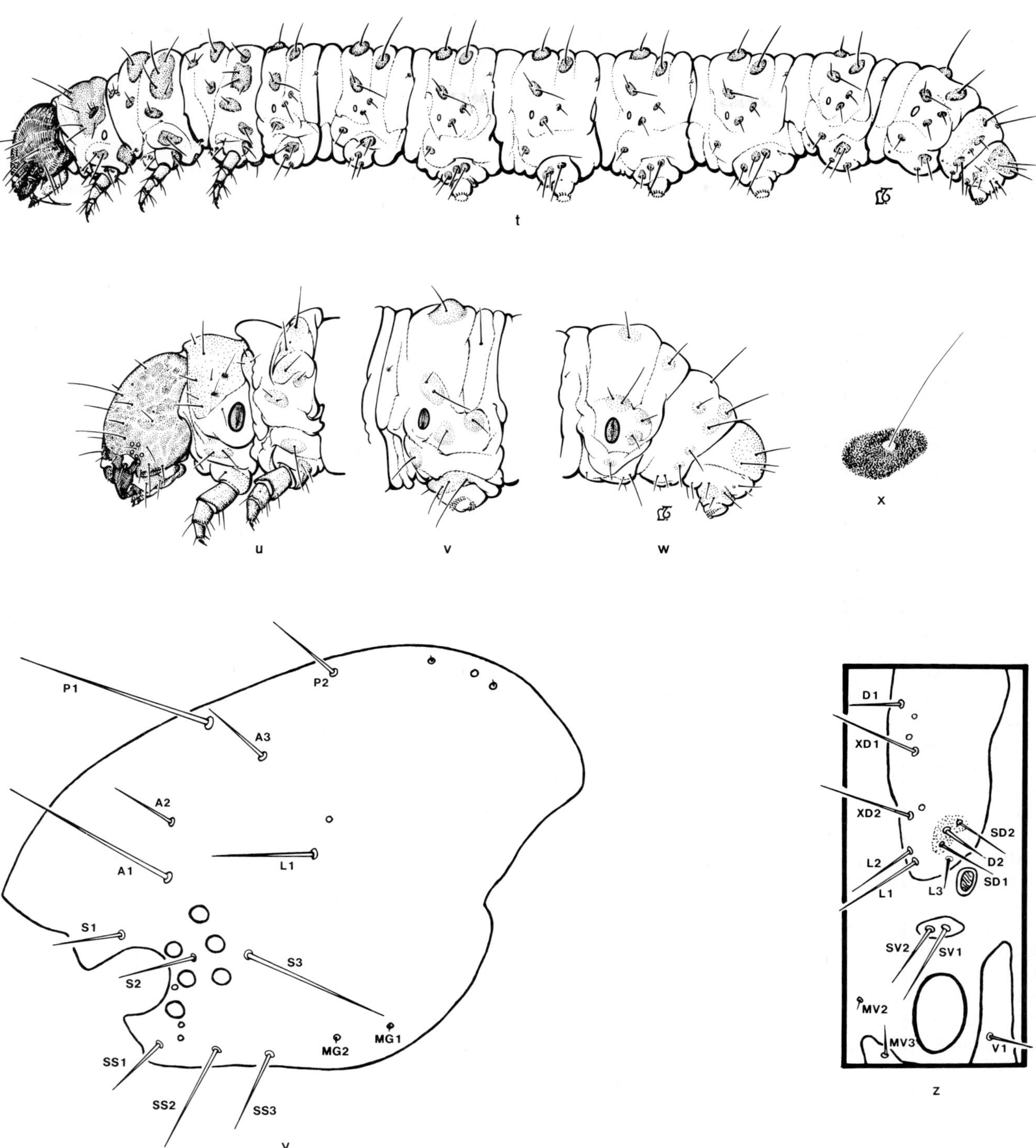

Figure 26.31t. Hepialidae. *Hepialis* sp. f.g.1. 50± mm. Bodega Head, Sonoma Co., Calif. July 1967, ex. *Lupinus arboreus.* Univ. Calif., Berkeley.

Figure 26.31u-x. Hepialidae. *Sthenopis quadriguttatus* (Grote)? f.g.1. 60-70± mm. n. head, T1, T2; o. A3; p. A8-A10; q. SD1 on T1 in heavily pigmented sensory pit. Syracuse, N.Y., 10 Jan 1932, in white birch butt. U.S. Natl. Mus.

Figure 26.31y. Hepialidae. *Hepialis californicus* Boisduval. Head setae left side. Pt. Reyes Nat. Seashore, Marin Co., Calif., boring in trunks of *Lupinus arboreus.* Univ. Calif. Berkeley.

Figure 26.31z. Hepialidae. *Hepialis hectoides* Boisduval. T1 setae, penultimate instar. San Bruno Mtns., San Mateo Co., Calif., feeding externally on *Baccharis pilularis* roots. Univ. Calif. Berkeley.

SUBORDER MONOTRYSIA

NEPTICULIDAE (NEPTICULOIDEA)

Donald R. Davis, *Smithsonian Institution*

The Nepticulids

Figures 26.32a-g, 26.33a-e

Relationships and Diagnosis: The Nepticulidae and Opostegidae, which constitute the Nepticuloidea, have larvae that are all highly specialized miners. Although the larvae and adults of the two families differ considerably from one another, several similarities indicate that they are closely related.

The adults are monotrysian, with a soft ovipositor, reduced wing venation, and greatly enlarged antennal scapes (i.e., eye-caps). Heinrich (1918) noted certain similarities among the larvae, particularly in the anterior development of the hypostoma, with that of the Opostegidae the most displaced and that of Nepticulidae the least.

In the Nepticulidae, the hypostoma has moved far forward under the head, just short of the anterior edge of the clypeus. Other features distinguishing the larvae are the minute size, single stemma, extremely deep epicranial notch, and complete absence of legs and prolegs, although paired ambulatory protuberances are sometimes present on the venter of T2–3 and A2–7.

Biology and Ecology: The larvae of Nepticulidae are miners throughout their life, usually in leaves, more rarely in the thin bark of woody twigs, or in fruit. Some species of *Ectoedemia* produce twig galls on Fagaceae. More than 20 plant families serve as hosts, particularly the Betulaceae, Fagaceae, Rosaceae, and Salicaceae.

The flat oval eggs are laid singly and attached to the substratum by a glistening viscid fluid. Upon eclosion the larva bores directly into the host, usually initiating a slender, serpentine mine. The mine may continue in a linear fashion, with frass often being deposited in a dark median line, or it may abruptly enlarge into a blotch. It is interesting that the larvae of *Stigmella* tend to feed in a normal position (i.e., their dorsal surface toward the upper surface of the leaf) in contrast to the larvae of many *Ectoedemia,* which feed in an inverted position (with their dorsal surface toward the venter of the leaf) (Gustafsson 1981). Development of the larva is sometimes very rapid, with one European species reportedly requiring only 36 hours (Gerasimov 1952). Most species probably complete their larval period in about two weeks. The number of generations varies, with up to three being reported.

Pupation almost always occurs outside the mine, usually in detritus, bark crevices or upon the soil, and in an oval, flattened cocoon of densely woven brown to yellowish silk. The pupa usually protrudes through a slitlike aperture from the cocoon immediately prior to adult emergence.

Description: Mature larva extremely small, usually less than 5 mm, sometimes up to 8 mm. Body slightly depressed, essentially apodal, and densely covered with microtrichia or granules. The cuticle may be partially transparent, thus exposing the internal organs, or unicolorous and ranging in color from bright green to bright yellow. The spiracles are round.

Head: Strongly flattened, usually longer than broad, prognathous, usually retracted up to two-thirds its length into prothorax. Epicranial notch deep, extending approximately half the length of head; posterior lobes (vertex) of head well developed, extending deep into prothorax. Posterior cranial setae greatly reduced or lost. Frontoclypeus triangular to rectangular, usually longer than wide, extending to epicranial notch; adfrontal ridge joining to epistomal ridge to form a conspicuous rectangular to triangular internal ridge. Ecdysial lines indistinct. A single stemma located posterior to the antenna which is very reduced in length, with only one segment apparent.

Thorax: T1 typically with an elongate pair of dorsal sclerites and a broad to narrow sternal plate; no other markings normally evident. D1 and D2 present on all segments. L group trisetose on T1, bisetose on T2–3. Legs absent; paired, ventral, ambulatory calli present on T2–3; calli often prominent in *Stigmella.*

Abdomen: Chaetotaxy greatly reduced. D group unisetose on A1–9, with D1 absent on all segments. SD group bisetose on A1–8. L group unisetose on A1–8. SV group unisetose on A1–9. V1 present on A1–8 and rarely A9. A10 with a pair of slender, lateral sclerites, or brace rods; one to three pairs of setae dorsad to brace rods. Prolegs and crochets absent. Paired, ventral ambulatory calli usually present on A1–7, sometimes absent on A1.

Comments: The Nepticulidae, in general, are the smallest Lepidoptera known, with the wing expanse of some adults no more than 3 mm. As is true for all leaf miners, the best means of collecting them is by locating the mines and carefully rearing the larvae. Unfortunately, rearing Nepticulidae can be difficult, although Emmet (1976) provides helpful advice.

The family is cosmopolitan and relatively small with just over 400 described species, including approximately 100 for North America. However, due to inadequate collecting in most areas of the world, far more undiscovered species probably remain than are presently described.

Only a few species of Nepticulidae are of serious economic concern in North America. The larva of the hard maple budminer, *Obrussa ochrefasciella* (Cham.), bores into the petioles and buds of sugar maple and *Acer nigrum,* frequently destroying the terminal bud. *Obrussa sericopeza* (Zell.) mines the leaf petioles and fruit (samara) of Norway maple, sometimes causing premature leaf fall. Cotton is occasionally damaged by the cotton leaf miner, *Stigmella gossypii* (Forbes and Leonard), and the leaves of pecan are often infested by the pecan serpentine leaf miner, *S. juglandifoliella* (Clem.).

Selected Bibliography

Braun 1917.
Emmet 1976.
Gerasimov 1952.
Gustafsson 1981.
Heinrich 1918.
Hering 1951.
Jayewickreme 1940.
MacKay 1972.
Newton and Wilkinson 1982.
Scoble 1983.
Trägardh 1913.
Wilkinson and Newton 1981.
Wilkinson and Scoble 1979.

OPOSTEGIDAE (NEPTICULOIDEA)

Donald R. Davis, *Smithsonian Institution*

The Opostegids

Figures 26.32h-k, 26.33f-l

Relationships and Diagnosis: The Opostegidae present a curious blend of very primitive and highly specialized features. Although it possesses several unusual or unique characters in both larval and adult stages, this family demonstrates definite affinities to the Nepticulidae. One synapomorphy that has long been recognized as linking the adults of these two families is the greatly enlarged scape, or "eyecap." The larvae of the two families are similar in possessing a single pair of stemmata and in having the hypostoma displaced far forward. The anterior development of the latter is greatest in Opostegidae where it comes to underlie the epistoma, and together the two structures form a kind of sheath enclosing the mouthparts (fig. 26.33g). The pupa of Opostegidae resembles that of Nepticulidae closely in general form, and both are characterized by the broadly exposed coxae.

The larvae of Opostegidae are highly specialized miners, easily distinguished from all other Lepidoptera. In addition to the unusually developed hypostoma, other diagnostic features include an extremely long, slender, and cylindrical body that is apodal except for paired ventral ambulatory calli on T2 and T3. The head is very depressed and wedge-shaped with a pair of articulated cranial apophyses (fig. 26.32h) projecting deeply into the prothorax. The single mandibular seta is unique in being spinose (fig. 26.32j).

Biology and Ecology: Because few species have been reared or studied in any detail, relatively little is known about the biology of Opostegidae. Larvae have been reported mining the leaves of *Pelea* (Rutaceae) in Hawaii (Zimmerman 1978), the stems of *Ribes* (Saxifragaceae) in the United States (Forbes 1923), the flower stalks of *Caltha palustris* L. (Ranunculaceae) in Europe (Stainton 1872), and the cambium layer in *Nothofagus* bark (Fagaceae) in Chile (Carey et al. 1978).

The few records to date indicate the group to be univoltine, with the pupa overwintering outside the mine in a flat, oval, and relatively tough cocoon of brownish silk located sometimes in leaf litter.

Description: Mature larvae small to medium, 10–20 mm. Body whitish (in alcohol), extremely elongate and slender, cylindrical, and apodal. Spiracles circular.

Head: Prognathous, extremely depressed, usually whitish with reddish brown longitudinal lines corresponding to sclerotized internal ridges. Epicranial notch only slightly developed. Adfrontal sutures (and ridges) and frontoclypeus diverging posteriorly to vertex. Ecdysial line obsolete and adfrontal sclerite not defined. A well-developed lateral ridge present, with dorsal branch continuing around lateral margin of cranium to termination of adfrontal ridges on vertex, and ventral branch turning anteriorly to hypostoma. A pair of prominent cranial apophyses extending internally from posterior end of adfrontal ridge into prothorax. Only one pair of stemmata, situated far forward.

Thorax: Pronotum with a pair of elongate, sclerotized bars longitudinally extended near dorsal midline. D group bisetose on T1–3. L group trisetose on T1, bisetose on T2–3. One coxal (SV) seta on T1, two on T2 and T3. Legs absent; a pair of ventral ambulatory calli on T2–3.

Abdomen: Cuticle smooth, naked except for primary setae and minute spinules. Pinnaculae absent to poorly defined. Chaetotaxy reduced; D1 and L3 absent on A1–10; SD2 and L2 absent on A9; SV2 absent on A1–10; SV1 either present or absent on A9. Dorsum of anal segment either with two pairs of setae or none. Prolegs and crochets absent.

Comments: The Opostegidae is a small cosmopolitan family of approximately 85 species, of which 7 occur in the continental United States and Canada. Although only one genus is currently recognized, research now in progress will result in new genera being described, including one for North America.

Selected Bibliography

Carey et al. 1978.
Eyer 1963.
Forbes 1923.
Gerasimov 1952.
Heinrich 1918.
Stainton 1872.
Zimmerman 1978.

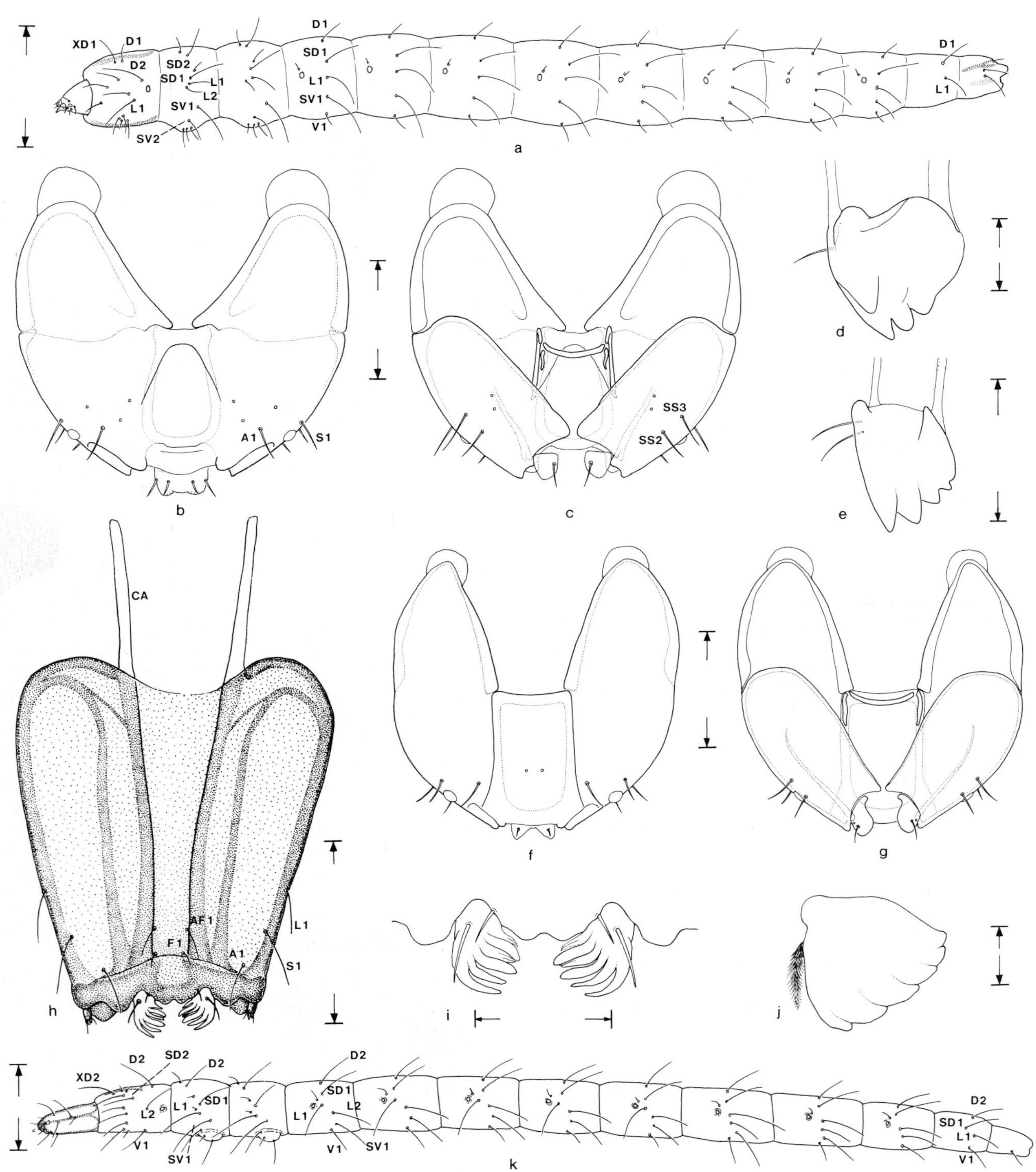

Figure 26.32a-k. Nepticulidae, a–g; Opostegidae, h–k. **a-d.** *Ectoedemia phleophaga* Busck. a. lateral (scale = 1 mm); b. head, dorsal (scale = 0.2 mm); c. head, ventral; d. mandible (scale = 0.05 mm); **e–g.** *Stigmella ?slingerlandella* (Kearfott). e. mandible (scale = 0.05 mm); f. head, dorsal (scale = 0.2 mm); g. head, ventral; **h-k.** *Opostega scioterma* Meyrick. h. head, dorsal (CA = cranial apophysis, scale = 0.2 mm); i. labrum, dorsal (scale = 0.1 mm); j. mandible (scale = 0.05 mm); k. lateral view (scale = 1 mm).

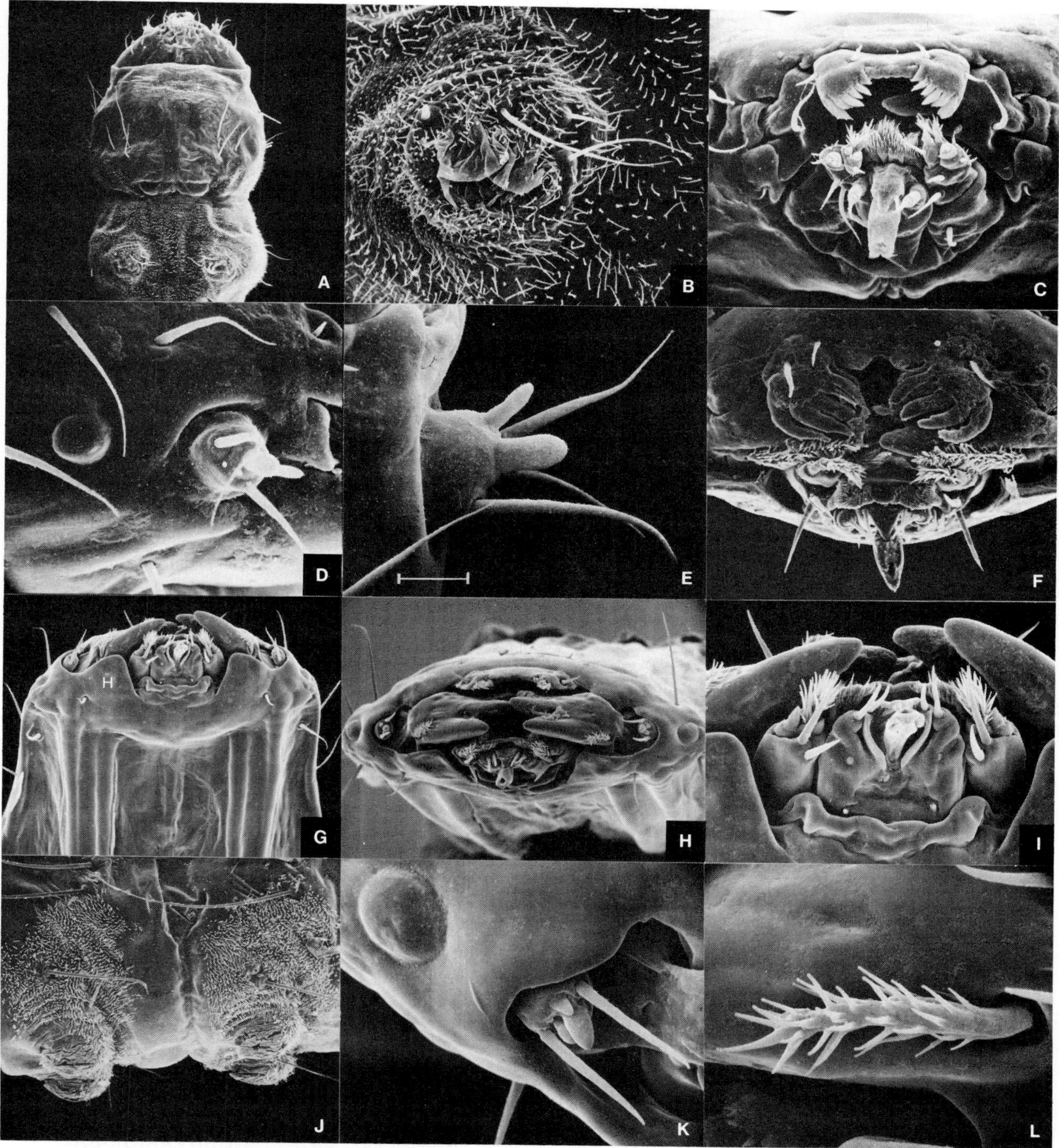

Figure 26.33a-l. Nepticulidae, a–e; Opostegidae, f–l. **a-e.** *Ectoedemia phleophaga* Busck. a. head and thorax, ventral, ×44; b. ambulatory callus, T2 ventral, ×227; c. mouthparts, ×248; d. stemma and antenna, ×372; e. antenna (scale = 20 mm); f. *Opostega scioterma* Meyrick, mouthparts, ×295; **g-l.** *Notiopostega atrata* Davis. g. head, ventral, ×89 (H = hypostoma); h. mouthparts, ×112; i. labium, ventral, ×245; j. ventral ambulatory calli, T2-3, ×53; k. stemma and antenna, ×434; l. mandibular seta, ×767.

TISCHERIIDAE (TISCHERIOIDEA)

Donald R. Davis, *Smithsonian Institution*

The Tischeriids, Trumpet Leaf Miners

Figures 26.34a-f, 26.35a-f

Relationships and Diagnosis: Because the adults exhibit certain problematic morphological features that presently are difficult to interpret, the relationships of the Tischeriidae are uncertain. The females are believed to be monotrysian (Braun 1972; Dugdale 1974) and perhaps distantly related to the Nepticulidae. However, because of their very distinct male and female genitalia (Braun 1972), absence of antennal eye-caps, and a unique haustellum bearing scutiform plates, a separate superfamily placement for the Tischeriidae appears justified. The adults are further recognized by other features of the head, including the roughly scaled vertex with forward-projecting scales; frons smooth and tapering; maxillary palpi minute, one-segmented; and labial palpi short, slender and drooping.

Heinrich (1918) associated the Tischeriidae with the Nepticuloidea largely on the basis of the forward development of the hypostoma. Heinrich's interpretation of this feature is questionable, however, as the tischeriid hypostoma appears to have developed no farther forward than that of several other families (e.g., Heliozelidae). Larval Tischeriidae are best characterized by a strongly depressed head, deep epicranial notch, and four to six stemmata typically arranged in a horizontal line along the lateral edge of the head. Thoracic legs are absent, but crochets are usually present and arranged in multiserial bands or incomplete ellipses. Setae D1 and D2 typically arise extremely close together (fig. 26.34a) on A3–6, but are well separated elsewhere.

Biology and Ecology: The larvae are leaf miners throughout the larval period and pupate within the mine, usually in a flimsy cocoon. Prior to adult emergence, the pupa partially protrudes from the mine or nidus. The egg is deposited singly, often on the upper surface of the leaf. Upon eclosion the larva bores directly into the leaf, eventually forming an upper surface or full-depth blotch mine lined with silk. In those species that feed in a single direction, the mine becomes trumpet-shaped. Frass is normally expelled from the mine through a single hole. The larva actively spins throughout its existence, perhaps the only leaf-mining group to do so (Trägardh 1913). Some species construct a firm, circular, silk-lined nidus within the mine into which they often retreat when alarmed or not feeding. Although as many as four generations may occur, the different broods are usually not well defined (Braun 1972). Hosts include the families Asteraceae, Fagaceae, Rhamnaceae, and Rosaceae. A few Tischeriids are known to feed on annual herbs (certain Asteraceae), one of the few leaf-mining Lepidoptera to do so.

Description: Mature larvae very small, not exceeding 6 mm. Body depressed, relatively broad, segments sometimes strongly rounded laterally. Color pale green to yellow. Spiracles round.

Head: Strongly depressed, prognathous, and partly retactible. Epicranial notch deep. Frontoclypeus narrow, elongate, usually equalling epicranial notch in length, converging with ecdysial lines at apex of epicranial notch. Stemmata four to six, often arranged in a nearly straight line along lateral margin of head in two groups of three stemmata each; posterior group sometimes reduced.

Thorax: Pronotum and prosternum well defined, brown to nearly black. Legs greatly reduced to absent, either with two vestigial segments (Emmet 1976) or with small unsegmented pads. Paired, dorsal ambulatory calli also present on T1–3 (fig. 26.34b). L group on T1 trisetose, separate from pronotum. Coxal plates poorly defined or absent.

Abdomen: Anal plate usually distinct, darkly pigmented, with three pairs of dorsal setae. Cuticle smooth, minutely granular; pinnacula indistinct. D1 and D2 usually contiguous on A3–6 (fig. 26.34a), well-separated on other segments. Prolegs poorly developed but with simple spinelike crochets present on A3–6 and A10; crochets usually arranged in multiserial and most frequently two biserial (fig. 26.35f) bands or partial ellipses on A3–6; A10 with two to three rows of crochets.

Comments: The Tischeriidae constitute a small, primarily Holarctic family of approximately 70 species, all included in one genus, *Tischeria*. The greatest concentration of species occurs in the United States (47) with only a few described from the Neotropical, Ethiopian, and Indo-Malayan regions, and none from the Australian region or Oceania. Only one species normally causes much damage in the United States; the appleleaf trumpet miner, *Tischeria malifoliella* Clemens (figs. 26.34a-f, 26.35a-f).

Larval Tischeriidae present an interesting exception to a statement made by Hinton (1955a:508) that prolegs are lost before true legs in the Lepidoptera and that no larva is known with prolegs but without true legs. This statement, although generally true, is somewhat puzzling since Hinton later in the same paper notes the absence of legs and presence of crochet-bearing prolegs in the Tischeriidae.

Selected Bibliography

Braun 1972.
Dugdale 1974.
Emmet 1976.
Gerasimov 1952.
Grandi 1932–33.
Heinrich 1918.
Hering 1951.
Hinton 1955a.
Jayewickreme 1940.
Trägardh 1913.

HELIOZELIDAE (INCURVARIOIDEA)

Donald R. Davis, *Smithsonian Institution*

The Heliozelids, Shield Bearers

Figures 26.34g-k, 26.35g-l

Relationships and Diagnosis: The Heliozelidae, along with the Adelidae, Crinopterygidae, Incurvariidae, Cecidosidae, and Prodoxidae, comprise the monotrysian superfamily Incurvarioidea (Nielsen and Davis 1985). The Crinopterygidae (one species) contains a case-bearing larva and is confined to south-central Europe. The Cecidosidae are an ancient and remarkable gall-making family restricted to the Southern Hemisphere, with at least one genus present in South Africa and four genera in southern South America. Their larvae are apodal with numerous secondary setae. The Heliozelidae, Adelidae, Incurvariidae, and Prodoxidae are more cosmopolitan and all are represented in the United States. The larvae are either internal borers or leaf miners, or construct portable, lenticular cases and feed externally.

The Heliozelidae are closely related to the Adelidae, both in morphology and biology. The larvae are typically apodal leaf miners until the final instar. Some species of *Heliozela,* however, possess either rudimentary or well-developed thoracic legs and multiserial crochets very similar to those of Adelidae. The rudimentary prolegs of all three North American genera are unusual in being fused into one transversely elongate callus situated midventrally (figs. 26.34h, 26.35j) with only those of *H. aesella* bearing crochets. In addition to those four calli, both the dorsum and venter of T2–3 and A8 of *Coptodisca* bear paired calli, as well as T2–3 and the middorsum of A8 of *Antispila.* Also, all known heliozelid larvae possess fewer than six stemmata and have either vestigial or separated coxal plates.

Biology and Ecology: The eggs are inserted singly into plant tissue by means of a slender, piercing ovipositor. North American host records include 12 plant families with some preference shown for the families Betulaceae, Cornaceae, Ericaceae, Rosaceae, and Vitaceae. The larvae usually commence as serpentine miners, but soon enlarge the mine into a small, full-depth blotch with the frass being retained inside the mine. All instars are miners except the last, which constructs a flat, oval case by cutting sections from the upper and lower surfaces of the mine and attaching them together with silk. Most larvae then release a silk strand to lower themselves to the ground. Although usually apodal, some species (e.g., *Coptodisca splendoriferella* Clem.) are capable of considerable movement not only within the mine or case but also of dragging the case some distance before pupating. Snodgrass (1922) reports the presence of paired, inflatable "adhesive discs" on the tergites and sternites of T2, T3, and A8. Similar but unpaired transversely elongate areas are also present on the sternites of A3–6. The ventral calli are used for dragging the larval case, and both dorsal and ventral calli are used for movement within the mine. Snodgrass further suggests that the larva can maintain a fixed position inside the mine by using the full complement of calli as adjustable hydraulic wedges. A similar complement of ambulatory calli exists for *C. arbutiella* Bsk., but, in *Antispila nysaefoliella* Clem., A8 bears only a single, middorsal callus.

After the case is constructed, no further feeding is believed to occur. The case is lined with silk to form a firm cocoon. The larva overwinters inside the case and most likely pupates a few weeks prior to adult emergence. Most species are probably univoltine; some, however (e.g., *C. splendoriferella*) are known to be at least bivoltine.

Some *Heliozela* mine the twigs, petioles, or leaf ribs of their host. *Heliozela aesella* Cham. from the southern United States initiates petiole and leaf galls on *Vitus* in which the larva feeds (Needham et al. 1928, McGiffin and Neunzig 1985). Later, after the gall has been largely consumed, the larva constructs a portable case of the remains.

Description: Mature larva very small, 4–6 mm. Body slightly to moderately depressed, relatively broad; cuticle smooth, minutely granular, without accessory setae or processes, usually white to pale yellow or green and with a brownish to black pronotum, sometimes with a series of median spots dorsally; anal plate occasionally dark; lateral setae usually long and complete; dorsal setae reduced. Spiracles round.

Head: Mostly prognathous, slightly to strongly depressed. Frontoclypeus elongate, extending to or nearly to epicranial notch. Ecdysial lines converging at epicranial notch. Stemmata two to five pairs. Mandibles of normal chewing type, usually with four large, acute cusps. Labrum relatively large, with median notch extending about one-third the length of labrum in *Antispila* and *Heliozela;* labrum more complex in *Coptodisca,* with lateral lobes deeply subdivided into three bladelike lobes. Hypostoma either separate or contiguous.

Thorax: Pronotum well defined, dark, not fused to prespiracular plate. L group trisetose on T1. Legs usually absent, but well developed and five-segmented in *Heliozela aesella.* Coxal plates, if present, separate. At least some species of *Antispila* and *Coptodisca* with paired ventral and dorsal ambulatory calli on T2–3.

Abdomen: Pinnacula usually indistinct; anal plate typically well defined, with three pairs of setae. Ventral prolegs either absent or nearly so; crochets usually absent, present on A3–6 in *Heliozela aesella* and consisting of multiserial rows of simple, enlarged spines resembling those of Adelidae except contiguous (fig. 26.35i). Crochets always absent on A10. At least some *Antispila* and *Coptodisca* with fused ventral ambulatory calli on A3–6; paired dorsal and ventral calli also present on A8 in *Coptodisca,* but only a single middorsal callus on A8 in *Antispila.*

Comments: The Heliozelidae is a small family of largely cosmopolitan distribution consisting of nine genera and approximately 100 species (Gerasimov 1952). Three genera and 31 species have been reported from the United States and Canada. Only two species are usually encountered in such numbers to be of economic concern. The tupelo leaf miner, *Antispila nysaefoliella* Clem. attacks tupelo (sour gum), *Nyssa sylvatica* Marsh. The resplendent shield bearer, *Coptodisca splendoriferella* Clem., at times can be a major pest of apple, with as many as 30 mines reported in a single leaf (Needham et al. 1928).

All previous diagnoses have stated that the larvae of Heliozelidae are apodal or nearly so (Forbes 1923; Gerasimov 1952; Emmet 1976; etc.). The North American *Heliozela aesella* Cham. is therefore of some interest in possessing well-developed thoracic legs and adelidlike, multiserial crochets. These features, among others, demonstrate the family's close affinities to the Adelidae. The fused condition of the ventral prolegs is a significant apomorphy for the family.

Selected Bibliography

Emmet 1976.
Forbes 1923.
Gerasimov 1952.
McGiffin and Neunzig 1985.
Needham et al. 1928.
Peterson 1948.
Snodgrass 1922.

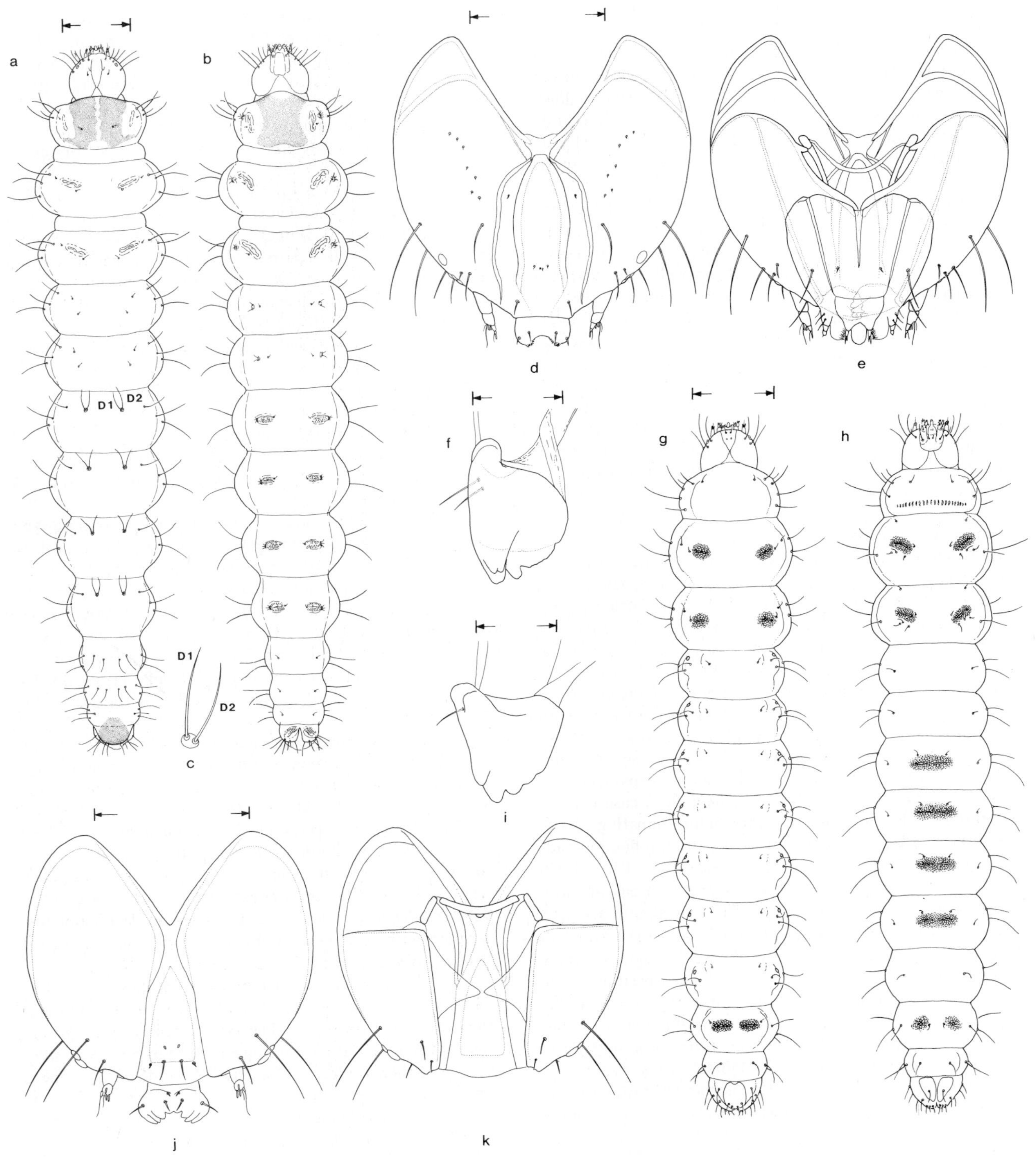

Figure 26.34a-k. Tischeriidae, a-f; Heliozelidae, g-k. **a-f.** *Tischeria malifoliella* Clemens. a. dorsal (scale = 0.5 mm); b. ventral; c. abdominal setae D1-2 enlarged; d. head, dorsal (scale = 0.2 mm); e. head, ventral; f. mandible (scale = 0.05 mm); **g-k.** *Coptodisca arbutiella* Busck. g. dorsal (scale = 0.05 mm); h. ventral; i. mandible (scale = 0.05 mm); j. head, dorsal (scale = 0.2 mm); k. head, ventral.

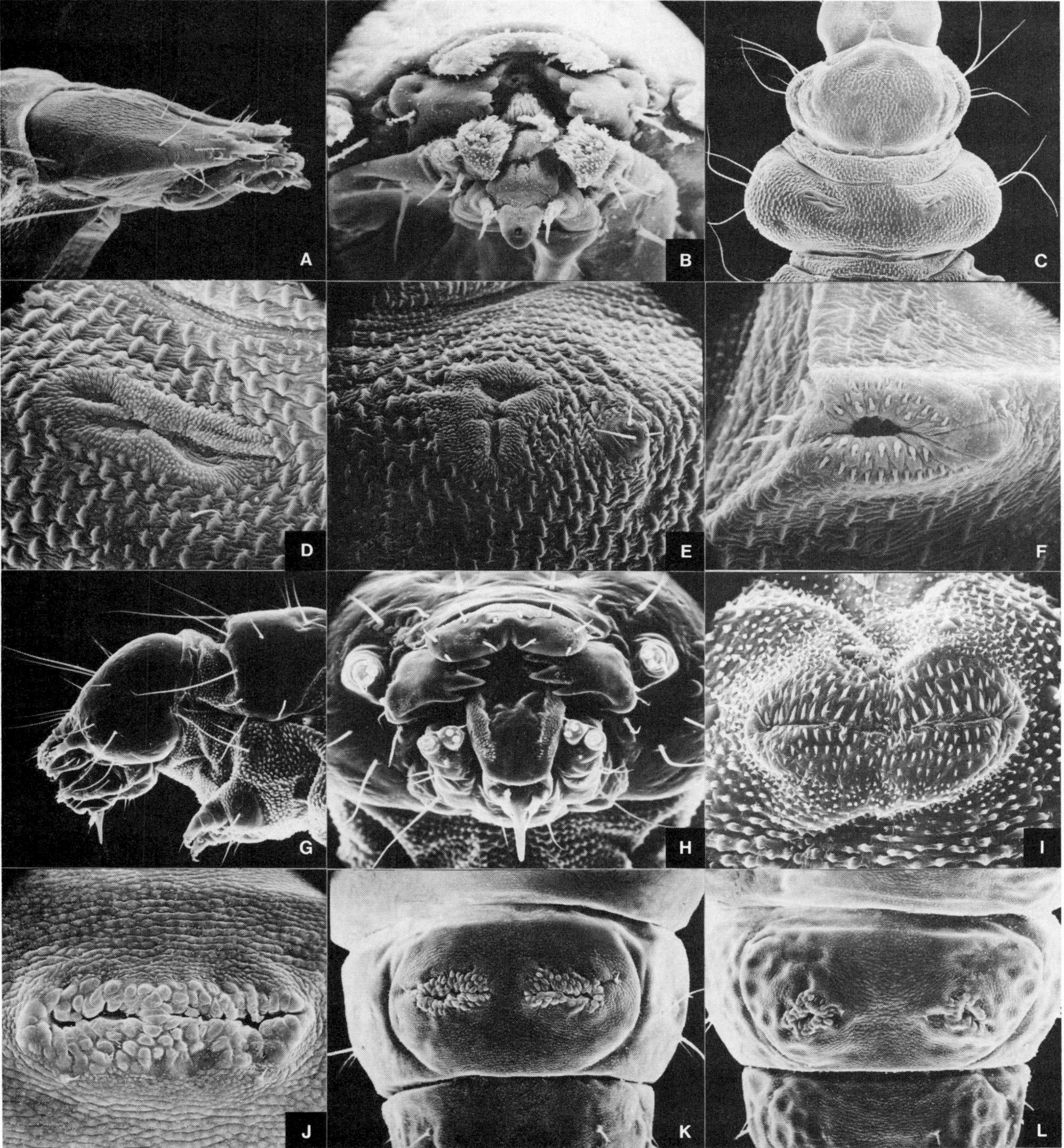

Figure 26.35a-l. Tischeriidae, a–f; Heliozelidae, g–l. **a-f.** *Tischeria malifoliella* Clemens. a. head, lateral, ×103; b. mouthparts, ×371; c. thorax, dorsal of T1-2, ×43; d. dorsal ambulatory callus, T2, ×228; e. ventral ambulatory callus, T2, ×214; f. crochets, A5, ×285; **g-i.** *Heliozela aesella* Chambers. g. head and T1, lateral, ×77; h. mouthparts, ×171; i. crochets, A3, ×285; **j-l.** *Coptodisca arbutiella* Busck. j. ventral ambulatory callus, A5, ×271; k. dorsal ambulatory callus, A8, ×131; l. ventral ambulatory callus, A8, ×131.

ADELIDAE (INCURVARIOIDEA)

Donald R. Davis, *Smithsonian Institution*

The Adelids, Fairy Moths

Figures 26.36f, 26.37g

Relationships and Diagnosis: The Adelidae are a relatively homogenous family closely related to the Heliozelidae and Incurvariidae. Larvae are similar to those of Incurvariidae in bearing cases through most of the later instars. They differ primarily in possessing multiserial crochets. Other features include the presence of six stemmata per side, adfrontal sclerite extending to epicranial notch, thoracic legs present, coxae fused, and crochets always absent on A10.

Biology and Ecology: The eggs are inserted singly into the host by the piercing ovipositor. The first instar usually begins as either a leaf miner or borer in the ovaries of various plants. After leaving the primary host on which oviposition occurred, the young (probably second instar) larva then constructs a portable, lenticular case of various plant material or soil debris. For the remainder of its larval existence, it feeds on low, living vegetation or in leaf litter, often of its primary host, on the ground. Because most larval feeding is believed to occur on fallen leaves, specific host data for adelids are meager or questionable. Oviposition records, sometimes associated by first instar feeding, have been noted on Brassicaceae, Dipsacaceae, Lamiaceae, Salicaceae, and Scrophulariaceae. The last instar larva (rarely pupa) usually overwinters, but some species may feed through the winter (Heath and Pelham-Clinton 1976) whenever conditions permit and continue feeding into the next season before pupating.

Pupation always occurs inside the larval cocoon on the ground in leaf litter. The pupal exuvia is partially extended from the larval case prior to adult emergence. All species are believed to be univoltine.

Description: Mature larva small, length 7–12 mm. Body slightly depressed, white to green in color with darkly pigmented, sclerotized plates and head.

Head: Mostly prognathous. Frontoclypeus relatively large, approximately two-thirds the length of head. Adfrontal sclerite elongate, narrow, terminating at deeply incised epicranial notch. Labrum broad, width exceeding length of mandible. Six stemmata per side, arranged in an uneven, sometimes compressed circle. Seta P1 adjacent to ecdysial suture. Seta AF1 arising rather high on adfrontal sclerite near apex of frontoclypeus; AF2 absent.

Thorax: Dorsal plates dark, well defined on T1–3. Prespiracular sclerite fused to pronotum; spiracle usually free from prespiracular sclerite. Coxal plates fused, at least on T1. L group trisetose on T1; SV group bisetose on T1, unisetose on T2–3. Legs well developed; tarsi without enlarged, squamiform seta.

Abdomen: Prolegs greatly reduced, multiserial crochets present on A3–6 and absent on A10; crochets arranged in several more or less definite transverse rows that gradually decrease in size from center so that peripheral crochets hardly distinct from normal cuticular spines. SD2 absent. SV group bisetose or trisetose. A10 with dark dorsal and ventral plates; dorsal plate with three setal pairs.

Comments: The Adelidae, Incurvariidae, and Prodoxidae have been variously treated as separate families or subfamilies encompassing different assemblages of genera. Whenever these three groups have been considered as subfamilies, the name Incurvariidae Spuler (1910) has most frequently been used for the familial name, as has Incurvarioidea for the superfamily. In recent years, however, a few authors (Common 1975, Razowski and Wojtusiak 1978a, b), sometimes citing Adelidae Heinemann (1870) as the oldest name, and priority as the rule, have reverted to using Adelidae and Adeloidea respectively. Actually, Nemophoridae Leach (1815) is the oldest available name, which emphasizes again, for the sake of stability and familiarity, the preference for the continued use of Incurvariidae and Incurvarioidea.

The Adelidae is a widely distributed, nearly cosmopolitan group. The principal genera of the northern hemisphere are *Adela, Cauchas, Nematopogon,* and *Nemophora.* The family is represented in the southern hemisphere by the genus *Ceromitia,* with nearly 100 species in Africa and South America.

Selected Bibliography

Common 1975.
Gerasimov 1952.
Heath and Pelham-Clinton 1976.
Heinemann 1870.
Kuroko 1961.
Leach 1815.
Nielsen 1980, 1985.
Razowski and Wojtusiak 1978a, 1978b.
Spuler 1910.

INCURVARIIDAE (INCURVARIOIDEA)

Donald R. Davis, *Smithsonian Institution*

The Incurvariids

Figures 26.36c-e, 26.37a-f

Relationships and Diagnosis: The Incurvariidae are most closely allied to the Prodoxidae and Adelidae. The larvae are very similar to Adelidae in general appearance and biology. The first instar is a leaf miner, which usually soon changes to a case-bearing form, feeding on or beneath its host. Thoracic legs are present in all genera except *Phylloporia,* which also lacks crochets. Prolegs are vestigial with usually a single transverse band of 4–7 crochets on A3–6. Crochets are usually absent on A10 but present in *Paraclemensia* and *Vespina.* Other diagnostic features include the extension of the adfrontal sclerite to the epicranial notch, the presence of 6 stemmata, and the separated condition of the coxae.

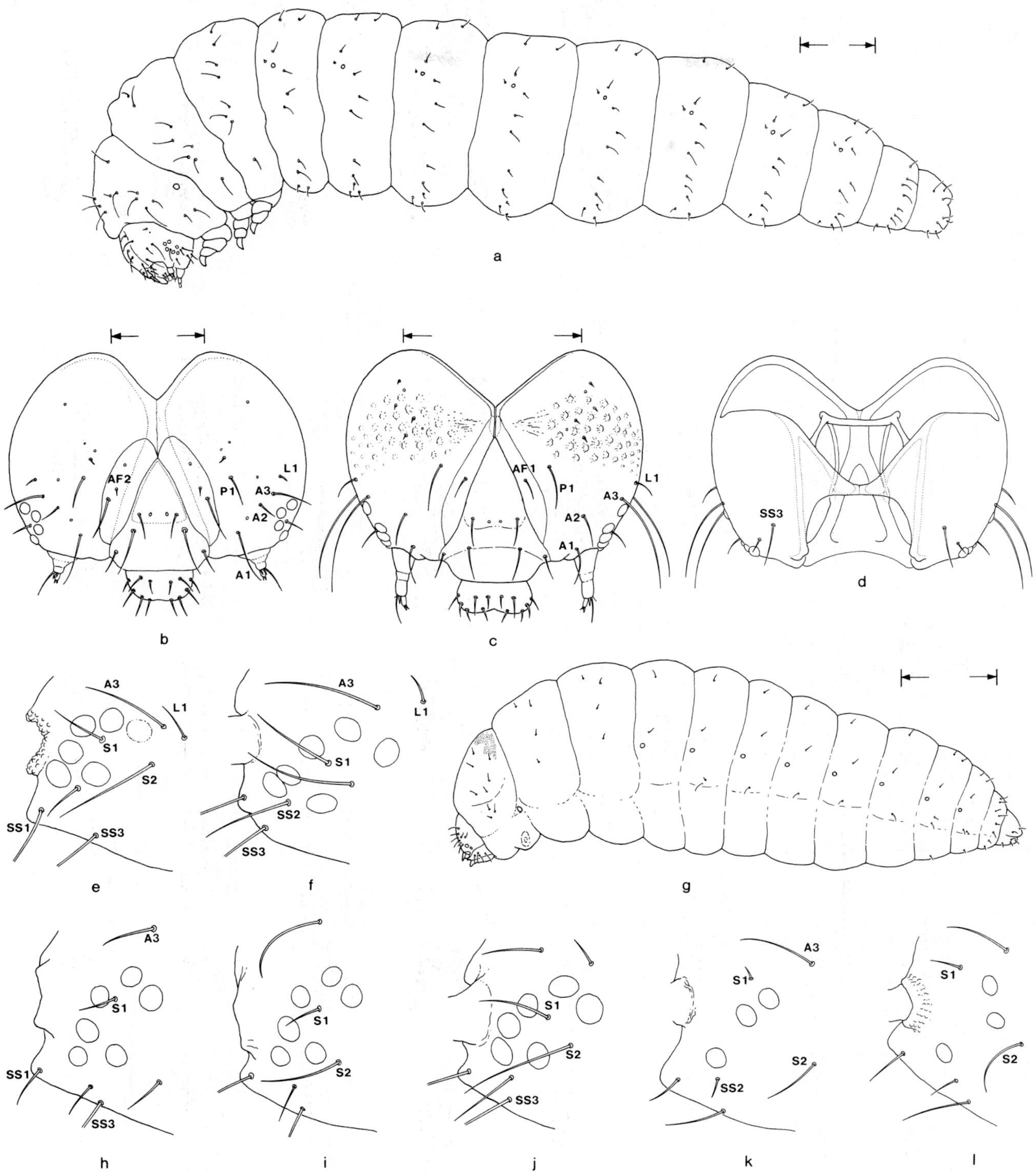

Figure 26.36a-j. Prodoxidae, a-b, g-l; Incurvariidae, c-e; Adelidae, f. **a-b.** *Tegeticula yuccasella* (Riley). a. lateral (scale = 1.0 mm); b. head, dorsal (scale = 0.5 mm); **c-e.** *Paraclemensia acerifoliella* (Fitch). c. head, dorsal (scale = 0.5 mm); d. head, ventral; e. stemmatal area; **f.** *Adela ?septentrionella* (Walsingham), stemmatal area; **g.** *Prodoxus quinquepunctellus* (Chambers), lateral (scale = 1.0 mm); **h.** *Tegeticula yuccasella* (Riley), stemmatal area; **i.** *Parategeticula pollenifera* Davis, stemmatal area; **j.** *Mesepiola specca* Davis, stemmatal area; k. *Prodoxus quinquepunctellus* (Chambers), stemmatal area; l. *Agavenema pallida* Davis, stemmatal area.

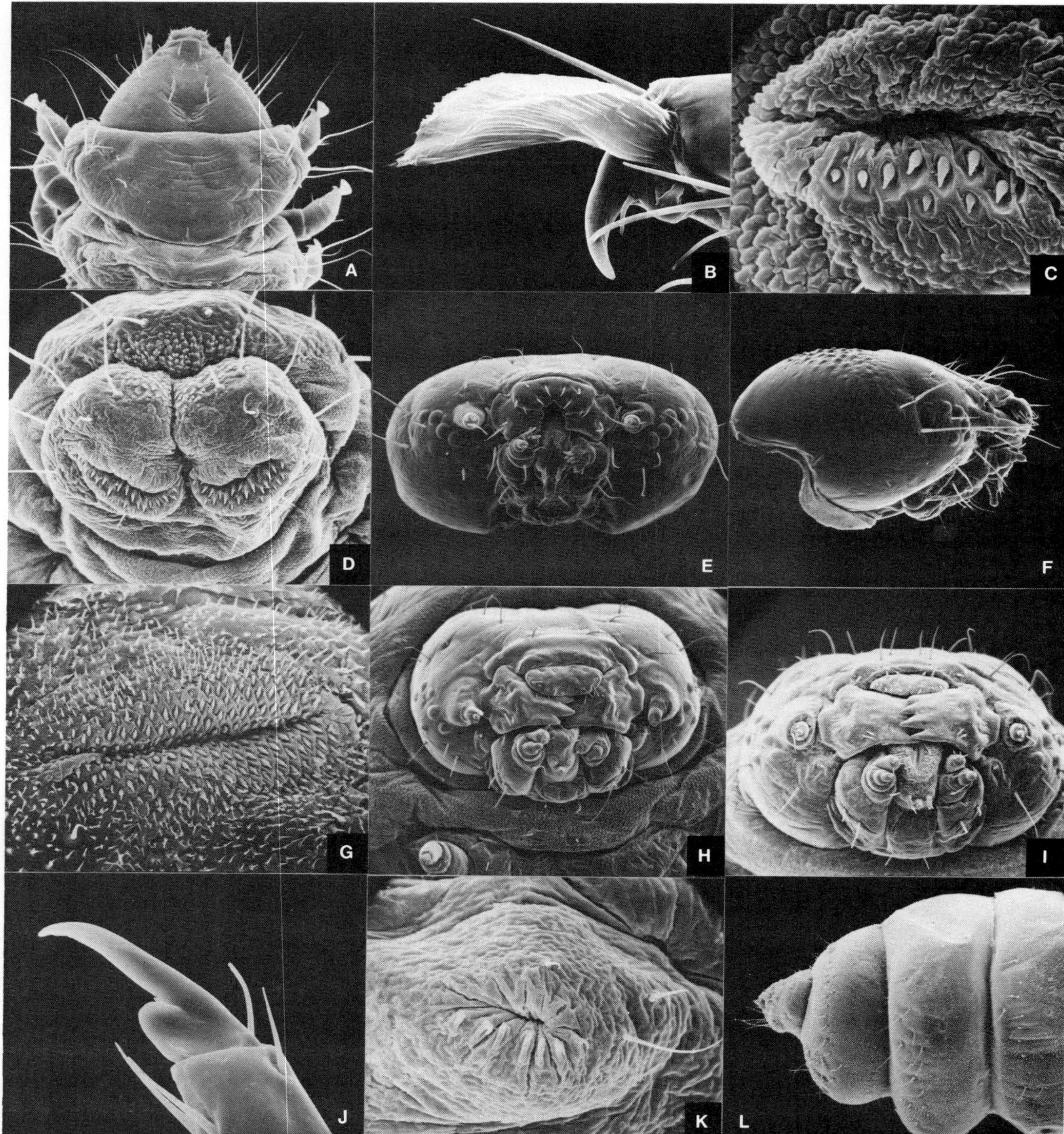

Figure 26.37a-l. Incurvariidae, a–f; Adelidae, g; Prodoxidae, h–k; Cecidosidae, l. **a-d.** *Vespina quercivora* (Davis). a. head and T1, dorsal, ×40; b. tarsus of T1, ×485; c. crochets, A4, ×570; d. anal prolegs, A10, ×114; **e-f.** *Paraclemensia acerifoliella* (Fitch). e. head, frontal, ×86; f. head, lateral, ×74; **g.** *Adela ?septentrionella* Walsingham, crochets, A5, ×114; **h.** *Tegeticula yuccasella* (Riley), head, frontal, ×29; **i.** *Parategeticula pollenifera* Davis, head, frontal, ×32; **j-k.** *Mesepiola specca* Davis. j. tarsus, T1, ×499; k. crochets, A5, ×114; **l.** *Eucecidoses minutanus* Brèthes, head and thorax, lateral, ×23.

Biology and Ecology: Eggs are frequently inserted into the underside of the leaves of the host where the first instar develops an irregular blotch mine. At the completion of the mining stage, most larvae cut through the upper and lower epidermal layers of the mine and construct an oval, lenticular case of the two sections by sewing the edges together. Dragging its case, the larva then feeds externally on living leaves or descends to the ground to feed on dead leaves and detritus. As the larva develops, it enlarges the case by cutting oval sections from leaves and adding these to the top and bottom of the smaller, older case. A few genera (e.g., *Perthida, Phylloporia,* and *Protaephagus*) continue to mine throughout their larval development and do not abandon the mine to construct a case until ready to pupate. Pupation occurs within the larval case, either attached to the host or on the ground in leaf litter. The pupal exuvia is partially extruded from the larval case prior to adult emergence. All species are believed to be univoltine.

Description: Mature larva small, 5–9 mm long. Body slightly depressed, variable in color (white, pale yellow, or gray) with darkly pigmented plates and head.

Head: Mostly prognathous. Frontoclypeus relatively large, approximately two-thirds the length of head. Adfrontal sclerite elongate, narrow, terminating at deeply incised epicranial notch. Labrum broad, width exceeding length of mandible. Six stemmata in an ellipse. Seta P1 separated from ecdysial suture approximately the width of adfrontal sclerite. Seta AF1 arising from middle of sclerite, AF2 absent.

Thorax: Dorsal plates brown to black, much better defined on T1 than on T2–3. Sternal plates reduced, coxae separate. Legs well developed; tarsi sometimes with an enlarged, squamiform seta adjacent to claw (fig. 26.37b). Prespiracular plate bearing the spiracle usually separate from pronotum. L group trisetose; SV group bisetose on T1, unisetose on T2–3.

Abdomen: Prolegs greatly reduced; crochets usually present on A3–6 (rarely absent) and arranged in a uniserial, transverse band; A10 usually without crochets but present in bi- to triserial rows in *Paraclemensia* and *Vespina.* SD2 either minute or absent. SV group trisetose on A3–6, bisetose on remaining segments. A10 often with a brownish dorsal plate bearing three setal pairs.

Comments: The Incurvariidae is a small, cosmopolitan family perhaps best represented in the Australian region. It is poorly represented in Africa (Scoble 1980) and South America (Nielsen and Davis 1981). Four genera, each represented by a single species, occur in North America. Few species have been reported of economic importance, but in the northeastern United States the maple leaf cutter, *Paraclemensia acerifoliella* (Fitch) (figs. 26.36c-e, 26.37e-f) is a serious pest of sugar maple (Ross 1958).

The frequent absence of anal prolegs and crochets throughout the Incurvarioidea, although most unusual, has probably received more emphasis than warranted. Hinton (1955a) was forced to some length to explain such a phenomenon, even suggesting that this group evolved prolegs independently from those of all other Lepidoptera. The fact that some genera of Incurvariidae do possess anal crochets, thereby indicating the plesiomorphic condition for the family, was apparently unknown by Hinton.

Selected Bibliography

Davis 1972.
Gerasimov 1952.
Heath and Pelham-Clinton 1976.
Hinton 1955a.
MacKay 1972.
Nielsen and Davis 1981.
Nielsen 1982b.
Ross 1958.
Scoble 1980.

PRODOXIDAE (INCURVARIOIDEA)

Donald R. Davis, *Smithsonian Institution*

The Prodoxids, Yucca Moths

Figures 26.36a-b, g-l; 26.37h-l

Relationships and Diagnosis: The Prodoxidae is believed to be one of the most advanced families in the Incurvarioidea. Together with the Cecidosidae, it differs from all other Incurvarioidea in being strictly endophagous.

The larvae are best characterized by their completely endophagous habit, boring into stems, flower stalks, and fruits. Thoracic legs may be present or absent. Crochets, if present, are arranged in a uniserial (rarely biserial) transverse band on A3–6; anal crochets always absent. Those larvae with legs typically possess six pairs of stemmata, those without legs only three pairs.

Biology and Ecology: The Prodoxidae may be divided into two groups (i.e., Lamproniinae and Prodoxinae) with regard to their general biology (Frack in prep.). In the Lamproniinae, the eggs are usually inserted singly into the young flower buds of the host (predominantly Rosaceae and Saxifragaceae). Young larvae feed on the developing seeds or on the fruit receptacle. By the third instar, the larvae abandon this site and overwinter in a firm silken hibernaculum, usually at the base of the host plant or in the soil. Early the following spring, the half-grown larvae emerge from this shelter, ascend the host and bore into new buds, often attacking more than one bud (Hill 1952). The European *Lampronia fuscatella* (Tengst.) exhibits a somewhat aberrant life history in that it produces and feeds within galls on young birch twigs (Heath and Pelham-Clinton 1976). Pupation usually occurs within the feeding galleries in developing shoots but may also occur in a cocoon outside the host. Adult emergence is in late spring, and there is one generation per year.

The larvae of Prodoxinae are mainly seed or stem borers in the monocot family Agavaceae, particularly the genus *Yucca*. The latter host harbors the well-known yucca moths, the biology of which comprises one of the most fascinating relationships in Lepidoptera. The interesting life histories and symbiotic relationships of this group have been investigated by Riley (1892b), Davis (1967), Powell and Mackie (1966), Powell (1984), and Frack (in prep.).

Adult females of *Tegeticula* and *Parategeticula* are known to pollinate their host immediately before or after oviposition. This insures the development of seeds upon which their larvae will feed. The larvae of *Prodoxus* bore in the flower stalks or sterile fruit tissue of *Yucca*, and those of *Agavenema* in the flower stalks of *Agave*. The larvae of *Mesepiola* bore in the seeds of *Nolina* and *Dasylirion* (Frack in prep.). Pupation may occur either outside the host in a subterranean earthen cocoon (*Mesepiola, Tegeticula,* and *Parategeticula*) or within the larval galleries within the host (*Agavenema, Prodoxus*). Last instar larvae of some Prodoxinae are capable of prolonged diapause, a xerophytic adaptation that is closely associated with the irregular flowering period of the host (Riley 1892b). Powell (1984) has succeeded in rearing adults from larvae that underwent diapause for as long as ten years.

The prodoxid pupa bears abdominal spines that are best developed in those species emerging from subterranean cocoons and least developed in the stalk-boring genera that pupate within the host plant. A prominent frontal beak is developed in those species pupating inside the host plant (*Agavenema, Prodoxus*). This is used by the pharate adult to push through the outer walls of the plant immediately prior to adult emergence (Davis 1967).

Description: Mature larvae small to medium, 6–22 mm long. Body cylindrical or fusiform, broadest anteriorly; color white, green, or red, with or without pigmented plates. Body setae greatly reduced in size.

Head: Mostly prognathous, usually withdrawn partially into prothorax. Frontoclypeus variable, one-half to two-thirds the length of head. Adfrontal sclerite variable, either slender and extending to epicranial notch or shortened and ending well before notch. Labrum usually reduced in size, width often less than length of mandibles. Stemmata either six (in podal forms) or three per side (in apodal forms). Seta P1 variable in position. Setae AF1 and AF2 arising close together at or below middle of adfrontal sclerite.

Thorax: Pigmented, sclerotized plates absent. Thoracic legs present or absent; when absent, replaced by moderately developed ambulatory calli. L group variable in number, usually bisetose or trisetose on T1 and uni-, bi-, or trisetose on T2–3. SV group bisetose or trisetose.

Abdomen: Prolegs greatly reduced or absent; crochets usually absent, with a single (rarely biserial) transverse row on A3–6 in *Lampronia* and *Mesepiola;* crochets always absent on A10. SD2 present, minute. L group variable, usually bi- or trisetose. SV group variable, usually trisetose on A3–6, bisetose on A1, 2, 7–8. Anal shield indistinct, with two or three pairs of setae.

Comments: The Prodoxidae are almost exclusively restricted to the Northern Hemisphere and are best represented in North America where ten genera and nearly 40 species occur. A new genus and species has recently been discovered in southern South America, the first record for the Southern Hemisphere (Nielsen and Davis 1985). Only one species in our fauna, *Lampronia rubiella* (probably introduced from Europe), has been reported as a major pest, the larvae sometimes being very destructive to raspberry.

Selected Bibliography

Davis 1967, 1978c.
Gerasimov 1952.
Heath and Pelham-Clinton 1976.
Hill 1952.
MacKay 1972.
Nielsen 1982.
Nielsen and Davis 1985.
Powell 1984.
Powell and Mackie 1966.
Riley 1892b.

SUBORDER DITRYSIA

TINEIDAE (TINEOIDEA)

Donald R. Davis, *Smithsonian Institution*

The Tineids, Carpet Moths, Clothes Moths, Grain Moths

Figures 26.38a-q, 26.39a-l

Relationships and Diagnosis: Since the time of Linnaeus, the Tineoidea has served as a disposal area for numerous, little-known and often unrelated taxa of microlepidoptera. Currently ten families are placed within Tineoidea (Heppner in prep.). The most archaic member of the superfamily is the Tineidae, a large cosmopolitan family in great need of revision. No comprehensive key has been compiled to distinguish the more than 15 subfamilies that have been proposed; consequently, numerous generic placements within the family remain uncertain.

In relation to the size and diversity of the family, the immature stages of only a few Tineidae have been described. For some subfamilies no information is available. Major features apparently shared by most, if not all, tineid larvae are: head with ecdysial lines terminating at epicranial notch; frontoclypeus usually large, extending almost to epicranial notch in most genera but sometimes only one-half to two-thirds the distance; usually with zero, one, or six stemmata per side, rarely two or five; three pairs of epipharyngial setae present; and hypostoma either narrowly separated or contiguous. The thorax has the prespiracular (L) sclerite typically separate from pronotum; L group usually trisetose (bisetose in Scardiinae); and legs present. The abdomen possesses four pairs of ventral prolegs and one pair of anal prolegs; crochets always present and arranged in a uniserial circle or ellipse, rarely in a mesal penellipse.

Biology and Ecology: Tineid eggs are typically oval and slightly flattened, and are usually deposited singly into crevices in the larval substrate by means of an elongate, telescoping ovipositor possessed by most females.

The feeding habits of tineid larvae vary considerably, with certain preferences dominating within some of the major subfamilies. The Nemapogoninae and Scardiinae typically feed on fungi or in decayed wood infested by fungi. A few Nemapogoninae are major pests of stored grains, and some infest wine corks. The larvae of a few *Acrolophus* (Acrolophinae) construct long silken, subterranean tubes and feed on the roots of grasses or other vegetation. This habit has generally been attributed to the entire group, although recent discoveries indicate a more varied diet. For example, at least one West Indian *Acrolophus* is known to be coprophagous, a habit prevalent in another acrolophine genus, *Amydria*. The larvae of Tineinae tend to be either coprophagous or keratophagous. The latter habit is particularly true for the clothes moths, which are among the few insects capable of digesting keratin (Hinton 1956). The Setomorphinae tend to feed on dried plant material and are often reported as pests of stored fruit, seeds, tobacco, and grain. The Hieroxestinae feed mostly on fungi or dead plant debris, but a few species of *Opogona* are serious pests of sugarcane in the field or stored tubers (Davis 1978b). Some groups of Tineidae (e.g., Acrolophinae and Scardiinae) tend to feed within silken tunnels constructed through, or attached to, the host substrate. A few species are myrmecophiles and feed on detritus inside ant nests. Many species, particularly in the Tineinae, construct flattened portable cases, in which pupation eventually occurs. Immediately prior to adult emergence, the pupa is extended about halfway free of its cocoon or pupal case.

Description: Mature larva small to large, ranging from 6–50 mm. Body cylindrical, moderately slender, and naked except for primary setae. Cuticle appearing relatively smooth but normally with minute spinules. Body usually whitish, except for brownish to nearly black head and thoracic plates. Spiracles round, with those on T1 slightly larger and more oval.

Head: Hypognathous. Frontoclypeus usually longer than wide and extending nearly four-fifths to epicranial notch; less elongate in some species and extending no more than one-half. Ecdysial lines extending beyond apex of frontoclypeus to epicranial notch. Stemmata variable, usually 1 or 6 per side, sometimes 0, 2, or 5.

Thorax: Pronotum usually distinct. Prespiracular plate usually separate from pronotum; partially fused in Acrolophinae. Procoxal plates usually separate or contiguous, rarely fused. D1 and D2 on all segments, D2 ventral to D1. XD1, XD2 and SD1 in nearly a straight vertical line. L group usually trisetose on T1, bisetose in Scardiinae. SV group bisetose on T1, usually unisetose on T2 and T3 but bisetose in Tineinae. Legs always well developed.

Abdomen: D1 and D2 on all segments, with D2 more caudad. SD1 on all segments. SD2 usually small, on all segments except A9. L group typically trisetose. SV group trisetose on A1–6, usually bisetose on A7–9, but normally unisetose on A9 in Tineinae. Anal plate with four pairs of setae. Prolegs present on A3–6 and A10. Ventral crochets almost always in a uniserial ellipse or circle, but in a mesal penellipse in *Oinophila*. Cuticular spines around planta variable, either present or absent, and if present, then sometimes slightly enlarged (Acrolophinae). Anal crochets in a curved uniserial row.

Comments: The family is moderately large, including nearly 3000 named species, mostly concentrated in the tropics. Although only 173 species have recently been listed for North America north of Mexico (Davis 1983), more than 70 additional species are known and awaiting description.

Although information on the immature stages of Tineidae is sparse, the most useful reference to date for the larvae is that of Hinton (1956) which treats 32 species, including most of those of major economic importance. Unfortunately, as MacKay (1963) points out, Hinton diverged somewhat in that paper in interpreting the nomenclature of the pronotal setae as compared to his original 1946 work. Because this divergence seems unjustified, later authors, including myself, have not followed this revised nomenclature. Hinton (1955b, 1956) has provided a key to four principal subfamilies of Tineidae. Zimmerman (1978) revised Hinton's key in order to accommodate three additional subfamilies present in Hawaii.

Several species of Tineidae are of serious economic concern. According to the *1952 Yearbook of Agriculture* (p. 146), clothes moths and similar pests are responsible for 200–500 million dollars worth of damage each year to products of animal origin. Much of the damage in this country is caused by *Tineola bisselliella* (Hum.) (figs. 26.38j, p, q), with various members of the *Tinea pellionella* group (figs. 26.38i, 26.39n, o) (Robinson 1979) also involved. The poultry house moth, *Niditinea fuscella* (L.) is commonly found in hen houses and bird nests where the larvae feed on feathers, hair, and possibly bird guano. Other species, particularly *Nemapogon granella* (L.) (figs. 26.38e, g, h, k-l, 26.38a-c) and *Setomorpha rutella* Zell. (figs. 26.39d-f), are important pests of stored plant material including grains.

The crochet arrangement of the ventral prolegs of *Acrolophus* has previously been described as a multiserial circle or ellipse. Actually, the crochets of *Acrolophus* are in a uniserial arrangement, as they are in all Tineidae. Scattered rows of slightly enlarged and moderately curved cuticular spines encompass the planta of *Acrolophus,* and should not be interpreted as crochets. Similar spines occur in *Amydria* and *Setomorpha* and are at least partially developed in several other genera.

Selected Bibliography

Davis 1978b, 1983.
Fracker 1915.
Gozmany and Vari 1973.
Hinton 1955b, 1956.
Lawrence and Powell 1969.
MacKay 1963.
Peterson 1948.
Robinson 1979.
Zagulajev 1960, 1964, 1973, 1975.
Zimmerman 1978.

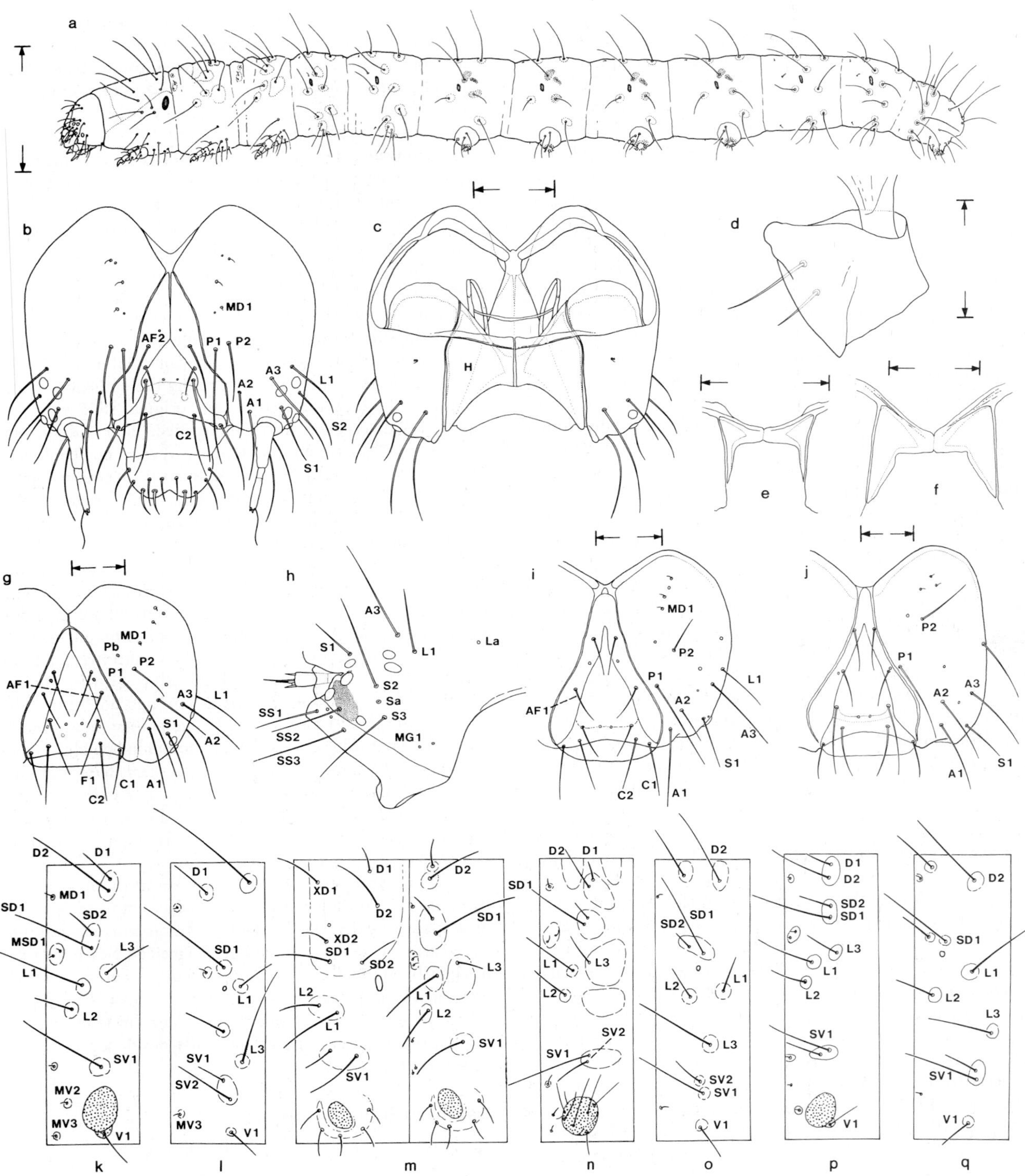

Figure 26.38a-q. Tineidae. **a-d.** *Acrolophus arcanella* (Clemens). a. lateral (scale = 5 mm); b. head, dorsal; c. head, ventral (H = hypostoma, scale = 1 mm); d. mandible (scale = 1.0 mm); **e.** *Nemapogon granella* (L.), hypostoma (scale = 0.5 mm); **f.** *Amydria arizonella* Dietz, hypostoma (scale = 0.5 mm); **g-h.** *Nemapogon granella* (L.). g. head, dorsal (scale = 0.2 mm); h. Stemmatal region; **i.** *Tinea pellionella* (L.), head, dorsal (scale = 0.2 mm); **j.** *Tineola bisselliella* (Hummel), head, dorsal (scale = 0.2 mm). **k-l.** *Nemapogon granella* (L.). k. mesothorax; l. A7; m. *Fernaldia anatomella* Grote, T1 and T2; **n-o.** *Tinea pellionella* (L.). n. mesothorax; o. A7; **p-q.** *Tineola bisselliella* (Hummel): p. T2; q. A7. (Figs. k-l, n-q redrawn from Hinton 1956.)

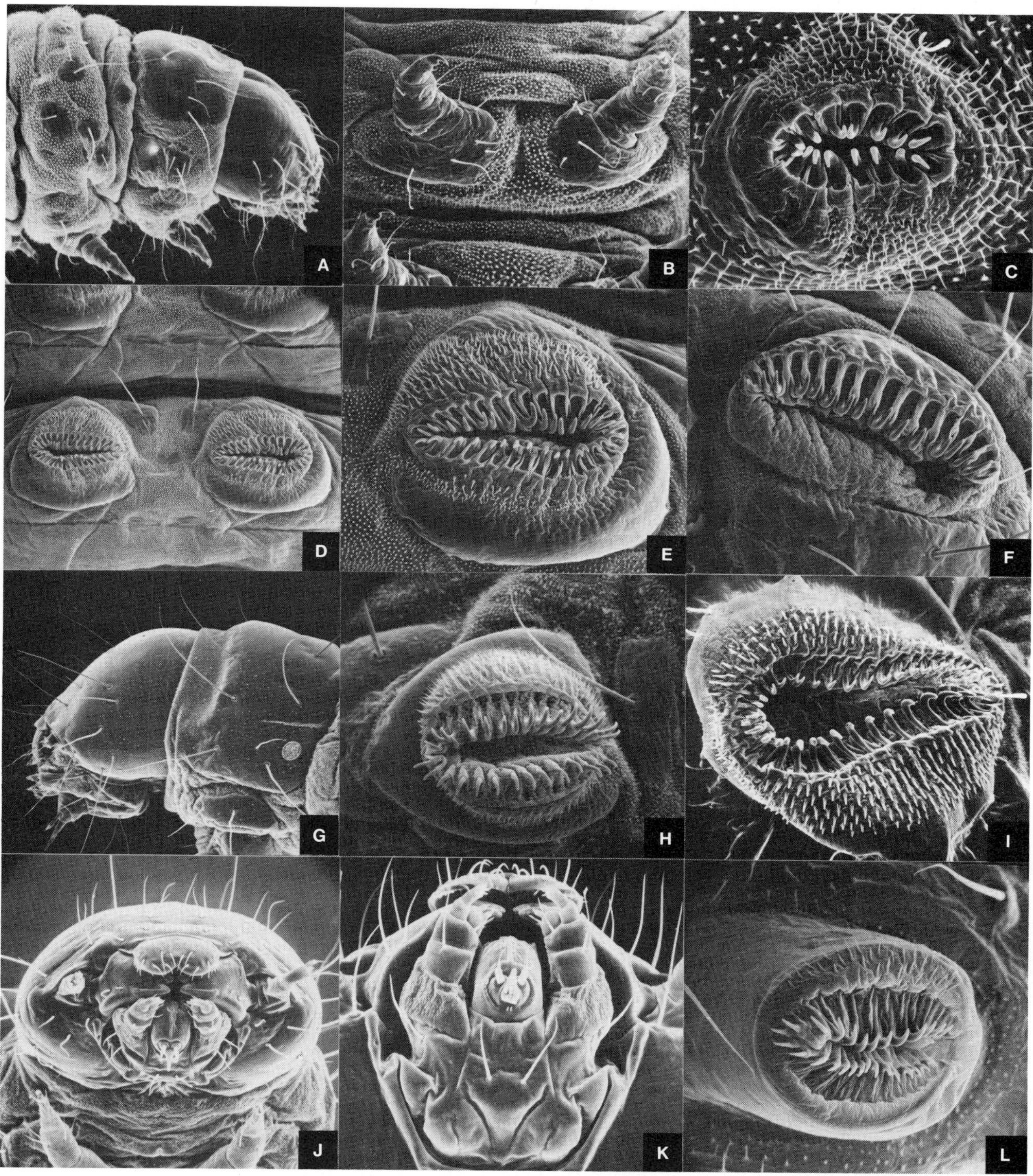

Figure 26.39a-l. Tineidae. **a-c.** *Nemapogon granella* (L.). a. head and thorax, lateral, ×41; b. T1 legs, ×84; c. crochets, A4; **d-f.** *Setomorpha rutella* Zeller. d. prolegs, A4, ×49; e. crochets, A4, ×110; f. crochets, A10, ×212; **g-h.** *Amydria arizonella* Dietz. g. head and T1, lateral ×32; h. crochets, A5, ×116; i. *Acrolophus arcanella* (Clemens). crochets, A4, ×81; **j-k.** *Tinea translucens* Meyrick. j. head, frontal, ×73; k. labium, ventral, ×131; **l.** *Opogona omoscopa* (Meyrick), crochets, A4,×203.

PSYCHIDAE (TINEOIDEA)

Donald R. Davis, *Smithsonian Institution*

The Bagworm Moths

Figures 26.40a-f, 26.41a-g, 26.42a-q

Relationships and Diagnosis: The Psychidae represent a specialized branch of the family Tineidae. Although the most primitive genera possess fully winged females, the great majority have either brachypterous or apterous females; the latter have usually undergone further specialization and are larviform.

The larvae construct sturdy silken cases usually adorned with fragments of the plant host or substrate. The anterior end of the case is used for feeding, with excrement being ejected from the posterior end. The larvae possess variously pigmented heads, well-developed thoracic legs, and a full complement of prolegs with uniordinal crochets arranged in a lateral penellipse. Psychid larvae are further distinguished from most other larvae including Tineidae by the four pairs of epipharyngeal setae on the venter of the labrum, by having the prespiracular pinaculum containing all three lateral setae and the spiracle fused to the T1 shield, and by the fusion or near fusion of the thoracic coxal plates.

Biology and Ecology: The eggs, numbering anywhere from 200 to over 13,000, are deposited inside the larval case, which also functions as a pupal cocoon. Dense, hairlike setae from the body of the female are shed and intermixed among the eggs. Soon after hatching, the first instar larvae begin to construct small cases, continually adding to these throughout their life. Because of the rather conservative chaetotaxy of the larvae, the sometimes bizarre larval cases frequently are the best means for identifying the immature stages (figs. 26.42a-q). Nearly all species are external feeders; the most primitive genera feed on lichens and the more advanced groups feed on the leaves of woody trees and shrubs. Pupation occurs inside the case, and the last instar larva attaches the case to some support and then inverts itself with the head pointing down. In the more advanced genera (i.e., the larviform species), the female does not leave the case except to die after mating and egg deposition. The females of a few species are parthenogenetic. All species are believed to be univoltine with an unusually long larval period and a brief, nonfeeding adult period.

Description: Mature larvae small to large, 8 to over 50 mm. Body cylindrical with well developed thoracic legs, four pairs of abdominal prolegs, and one pair of anal prolegs. Cuticle relatively smooth, without secondary setae, bumps, or processes, but minutely granular and sometimes wrinkled. Spiracles usually oval but sometimes round in more primitive forms; if prothoracic spiracle oval, then typically positioned horizontal to body axis.

Head: Mostly hypognathous; slightly retractible into prothorax; variously pigmented with brown or black. Frontoclypeus slightly longer than wide, extending approximately one-half the distance to epicranial notch. Ecdysial lines usually close to adfrontal sutures and meeting at or just above frontoclypeal apex. Six stemmata per side, 1–5 evenly spaced, either in a semicircle or angular line.

Thorax: Prothoracic shield well developed, fused to prespiracular pinaculum, thus bearing lateral setae as well as spiracle; variously pigmented. D1 and D2, SD1 and SD2 on all segments. XD1, XD2, and SD1 usually in a straight vertical line. L group trisetose on T1–3. SV bisetose on all segments.

Abdomen: Cuticle variously pigmented from white to nearly black. D1 and D2 on all segments, with D1 more dorsad and caudad except on A9 and A10 where D1 is either directly above or more cephalad. SD1 on all segments. SD2 greatly reduced on all segments except A9 and A10. L group trisetose on all segments. SV usually bisetose on A1 and A2, trisetose on A3–6, unisetose on A7–9. Anal shield with three to four pairs of setae. Crochets of abdominal prolegs in a uniordinal lateral penellipse.

Comments: The family is relatively small, with 26 species known from the United States and Canada and approximately 600 species for the world. Of these, about 500 species occur in the Old World. An excellent biological review of the family is provided by Kozhanshikov (1956); the New World species are treated by Davis (1964, 1975b). The systematic importance of larval morphology has been reviewed for the Palearctic psychids by Gerasimov (1937), a reference that also includes a larval key to the Palearctic genera.

Several species of Oiketicinae are major pests of both deciduous and evergreen trees and shrubs around the world. In the eastern United States, the evergreen bagworm, *Thyridopteryx ephemeraeformis* (Haworth) (figs. 26.41a-e, 26.42q) is a serious defoliator of conifers, particularly ornamentals such as aborvitae. *Oiketicus abbotii* Grote (fig. 26.42l) is occasionally a pest on maples, oaks, and sycamore. *Apterona helix* (Siebold) (fig. 26.42c) is a relatively recent introduction from Europe that has spread as a pest on numerous forage and vegetable crops in the eastern and particularly western United States. The female is parthenogenetic with no males known.

Selected Bibliography

Davis 1964, 1975b.
Gerasimov 1937.
Jones 1927.
Jones and Parks 1928.
Kozhantshikov 1956.

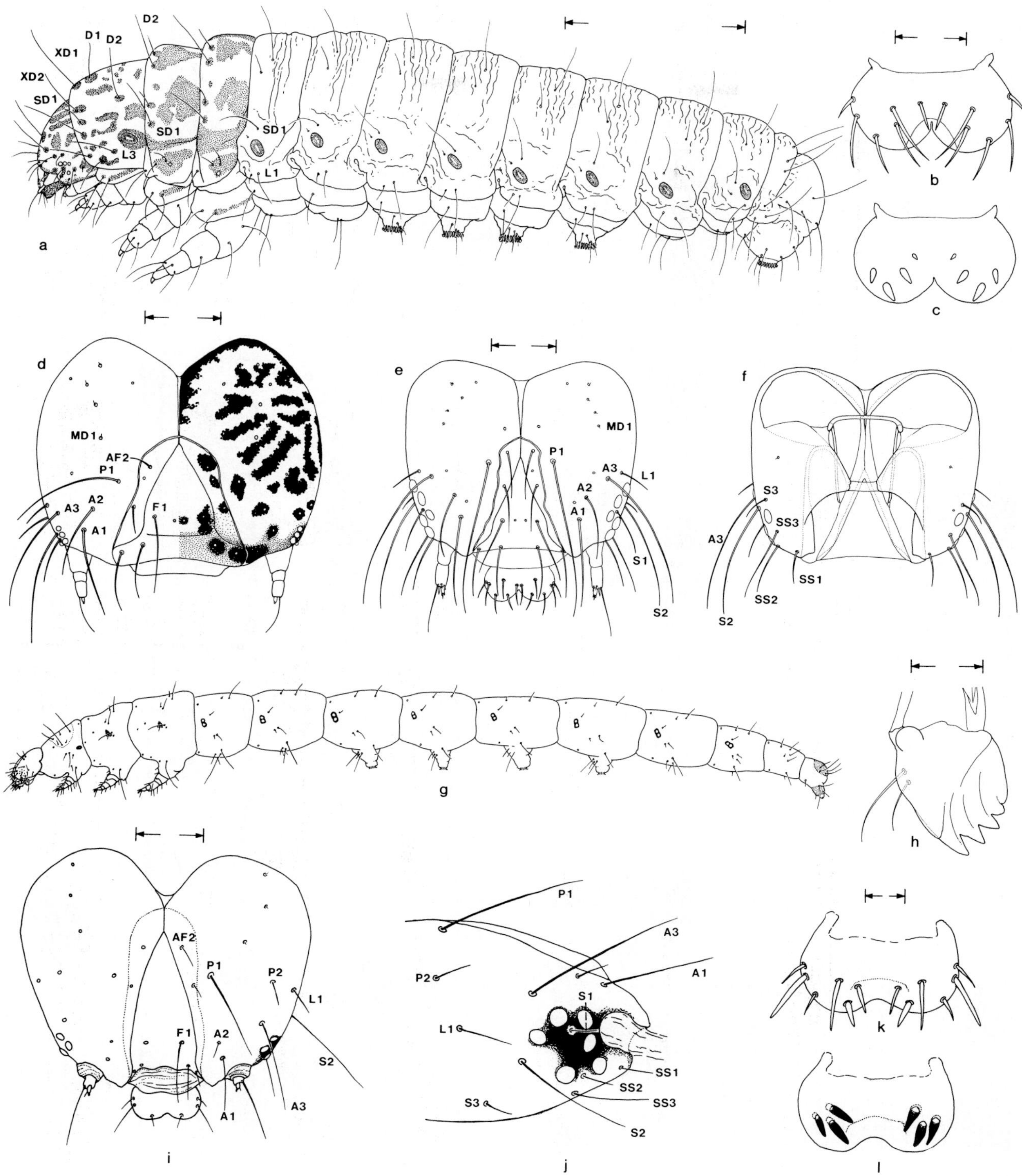

Figure 26.40a-l. Psychidae, a-f; Ochsenheimeriidae, g-l. **a-d.** *Thyridopteryx ephemeraeformis* (Haworth). a. lateral (scale = 5 mm); b. labrum, dorsal (scale = 0.05 mm); c. labrum, ventral; d. head, dorsal (scale = 1 mm); **e-f.** *Solenobia walshella* Clemens. e. head, dorsal (scale = 0.2 mm); f. head, ventral; **g-l.** *Ochsenheimeria* sp. g. lateral; h. mandible (scale = 0.1 mm); i. head, dorsal (scale = 0.2 mm); j. head, lateral; k. labrum, dorsal (scale = 0.1 mm); l. labrum, ventral.

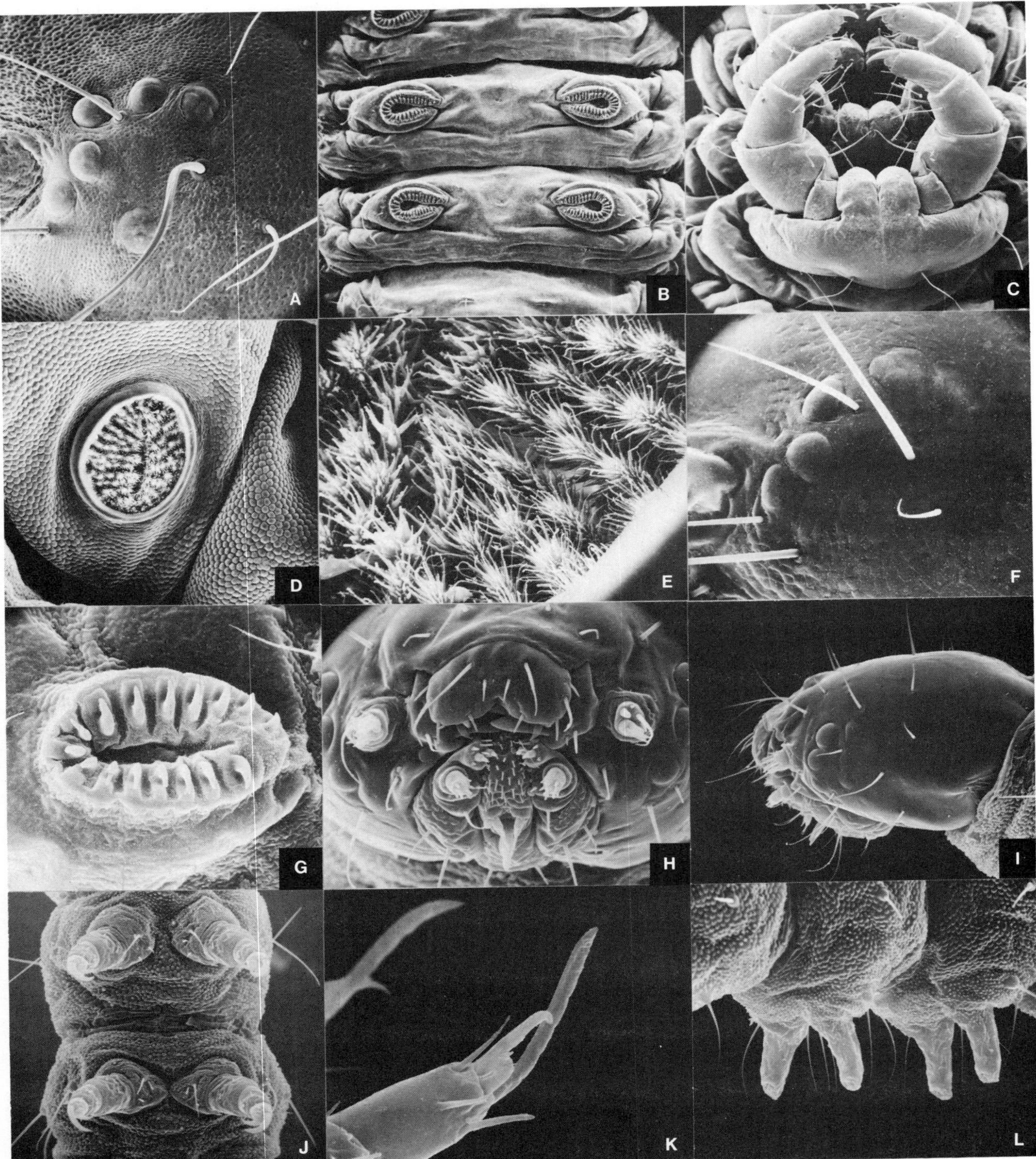

Figure 26.41a-l. Psychidae, a–g; Ochsenheimeriidae, h–l. **a.-e.** *Thyridopteryx ephemeraeformis* (Haworth). a. stemmatal area, ×93; b. prolegs, A5–6, ×22; c. T3 legs, ×21; d. spiracle, A1, ×102; e. spiracular filter, A1, ×580; **f-g.** *Solenobia walshella* Clemens. f. stemmatal area, ×209; g. crochets, A3, ×365; **h-l.** *Ochsenheimeria vacculella* F. R., first instar. h, head, frontal, ×493; i. head, lateral, ×319; j. legs, T2–3, ×276; k. tarsus, T3, ×1218; l. prolegs, A4–5, ×368.

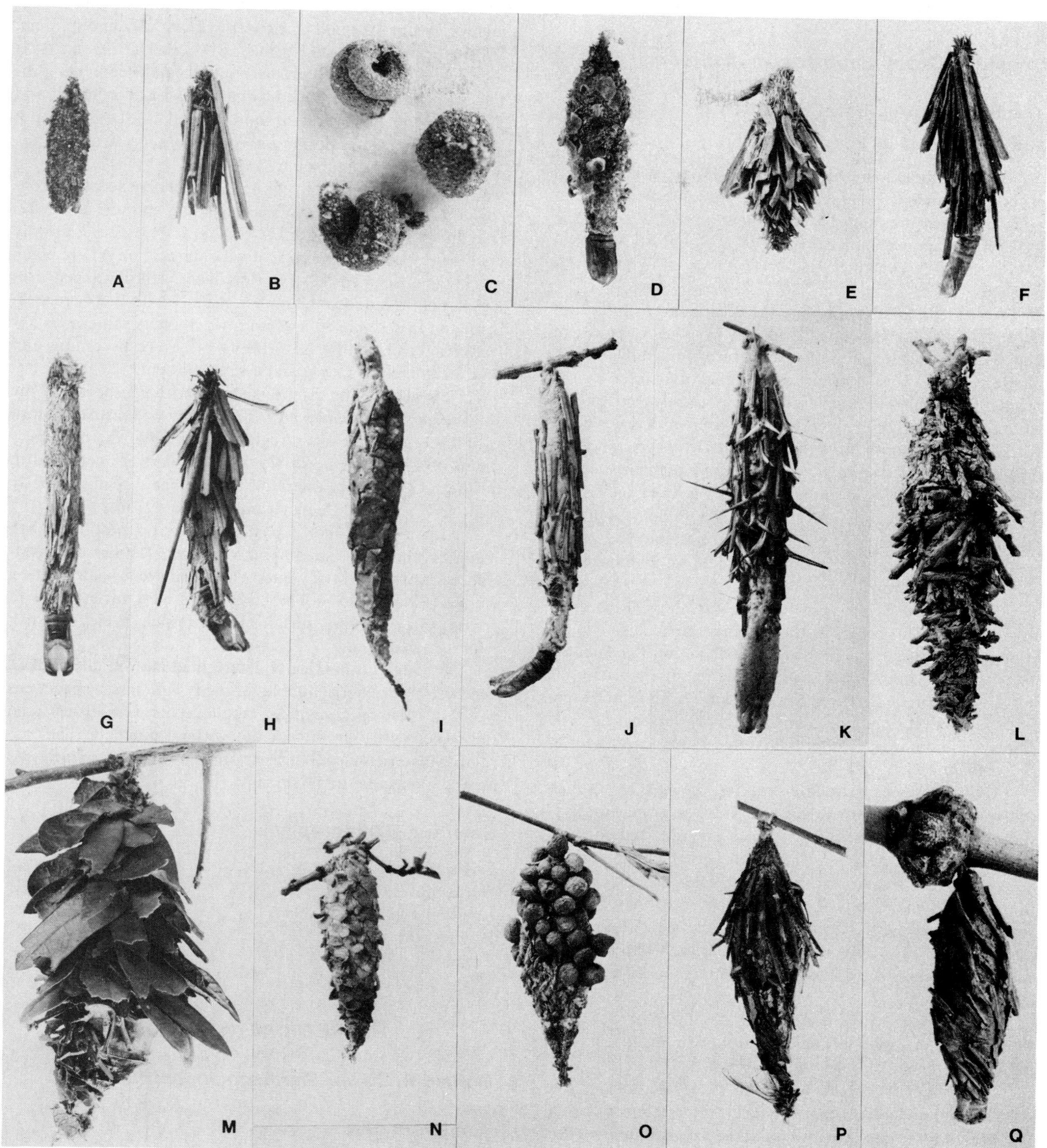

Figure 26.42a-q. Psychidae, larval cases. **a.** *Solenobia walshella* Clemens, length 8 mm; **b.** *Psyche casta* (Pallas), length 14 mm; **c.** *Apterona helix* (Siebold), greatest diameter 3 mm; **d.** *Cryptothelea gloverii* (Packard), length 15 mm; **e.** *Cryptothelea nigrita* (Barnes and McDunnough), length 17 mm; **f.** *Astala confederata* (Grote and Robinson), length 17 mm; **g.** *Astala edwardsi* (Heyl.), length 26 mm; **h.** *Basicladus tracyi* (Jones), length 27 mm; **i.** *Oiketicus toumeyi* Jones, length 90 mm; **j.** *Oiketicus toumeyi* Jones, length 70 mm; **k.** *Oiketicus townsendi* Townsend, length 85 mm; **l.** *Oiketicus abbotii* Grote, length 80 mm; **m.** *Oiketicus abbotii* Grote, length 80 mm; **n.** *Thyridopteryx meadii* Hy. Edwards, length 30 mm; **o-q.** *Thyridopteryx ephemeraeformis* (Haworth). o. length 52 mm; p. length 55 mm, with ♂ in mating position; q. length 48 mm, twig damage caused by persistent attachment of larval case.

OCHSENHEIMERIIDAE (TINEOIDEA)

Donald R. Davis, *Smithsonian Institution*

The Stem Moths, Stem Borers

Figures 26.40g-l, 26.41h-l

Relationships and Diagnosis: The family relationships of the Ochsenheimeriidae remain somewhat questionable, and a few adult features suggest yponomeutoid affinities. On the basis of most morphological features, particularly wing venation, the Ochsenheimeriidae appear most akin to the Tineoidea.

Very little is known about the morphology of the immature stages. As explained by Davis (1975a), our present knowledge is largely limited to one mature, unidentified larva. Consequently, as other species are studied, diagnoses for the family could change considerably. As presently understood, the major features of the larva are the extremely slender body, with a mostly smooth, unpigmented cuticle except for a lateral series of black spiracular spots on all segments except A9 and A10 (may be absent in first instar larvae); on T2 and T3 the spots are near L1 and L2. A full complement of legs and prolegs is present, the crochets reduced in number or absent; if present, then they number from 2–14, arranged in a single transverse row.

Biology and Ecology: All known larvae of Ochsenheimeriidae are stem borers in the grasslike monocots Cyperaceae and Juncaceae, and in the grasses, Poaceae. The Poaceae are of greatest importance, with a wide variety of grasses as hosts—the genera *Bromus, Secale,* and *Triticum* are especially favored. It should be pointed out, however, that the food plants are unknown for approximately half of the species of *Ochsenheimeria.*

Oviposition occurs in late summer; the eggs are laid on overwintering grasses or in such habitats as granaries, haystacks, or refuse piles. Some species are reported to overwinter as first instar larvae, others as eggs that contain formed larvae. Larvae become active in the spring, usually mining the leaves of their host. Within a week or so they become stem borers, capable of passing from one stem to another. Normally there is only one generation per year with pupation occurring in early summer and adult emergence in middle to late summer.

Description: Maximum length 18–27 mm. Body extremely slender, essentially naked and whitish (in alcohol) with indistinct pinnacula.

Head: Entirely dark brown to pale stramineous and largely unpigmented except for black interstemmatal area. Six stemmata, arranged in a relatively close but irregular circle. Frontoclypeus relatively narrow, length approximately twice the width, apex extending 0.8 the distance to epicranial notch. Ecdysial lines close to adfrontal sutures and joining above apex of frontoclypeus near vertex. Labrum moderately bilobed; three pairs of epicranial setae present. Mandible with four well-developed cusps.

Thorax: Pronotum either dark brown or largely unpigmented except for small black area around XD2 and spiracle. D1 smaller than D2. XD1, XD2, SD1, and L1 arranged almost in a vertical row. L group trisetose on all segments. Prothoracic spiracle surrounded by narrow black border. T1 and T2 with small black area around L1 and L2. Legs well developed, tarsal claws simple and relatively short. Coxal plates indistinct, not contiguous.

Abdomen: Uniformly white (in alcohol) except for conspicuous black, oblong spots surrounding spiracle and SD2. D1 smaller and anterior to D2. SD2 greatly reduced, usually enclosed by black pigment spot with spiracle on A1–8; black markings absent on A9–10. Relatively well-developed prolegs; crochets reduced in number, usually with 2–14 crochets arranged in a single transverse row; crochets sometimes absent. Anal shield either dark brown or barely discernible and pale stramineous; four pairs of setae present.

Comments: The family is small, with 23 species in the single genus, *Ochsenheimeria.* The family was restricted until recently to the Palearctic region, but in 1964 the first specimens of *Ochsenheimeria vacculella* F.R. were collected in northern Ohio. Since then, this species has spread gradually into New York and Pennsylvania (Davis 1975a).

Although still not reported as a crop pest in North America, the cereal stem borer, *O. vacculella,* has been considered a serious pest of winter wheat and rye in southwestern Russia (Pavlov 1961). The larva feeds on a variety of wild grasses and apparently sustains itself on such species in North America at present.

Although authentically determined mature larvae of *O. vacculella* were not available for study, first instar specimens were. In these specimens, spiracular spots were absent, and the prolegs were prominent but without crochets. The thoracic legs were unusual in possessing an elongate, curved seta arising alongside the tarsal claw (fig. 26.41k).

Selected Bibliography

Davis 1975a.
Gerasimov 1952.
Karsholt and Nielsen 1984.
Pavlov 1961.
Zagulajev 1971.

LYONETIIDAE (TINEOIDEA)

Donald R. Davis, *Smithsonian Institution*

The Lyonetiids

Figures 26.43a-o, 26.44a-i

Relationships and Diagnosis: The Lyonetiidae are believed to be more related to the Gracillariidae than to any other family of Lepidoptera. The relationship appears to be a tenuous one, with few, if any, unquestionable synapomorphies uniting the two groups. Superficially their adults are similar in being slender-bodied with narrow fore wings and lanceolate hind wings. Although the larvae demonstrate major morphological differences, the habits of both groups agree in being predominantly leaf miners, at least in their early instars. Lyonetiid pupae in some instances differ considerably from those of Gracillariidae (and *Bucculatrix*), which possess a few movable abdominal segments, free appendages, and dorsal abdominal spines. The lyonetiid pupa (except *Bucculatrix*) lacks abdominal movement and dorsal spines, and all appendages are firmly fused to the body.

Placement of the genus *Bucculatrix* has varied for many years between a subfamily of Lyonetiidae and a separate monotypic family. Most authors, largely influenced by similarities in the adult and to a lesser extent in the larva, regard the genus as no more than a subfamily. If pupal characters are emphasized, then *Bucculatrix*, based upon present knowledge, should be treated as a distinct family, although one problem with this emphasis is that the pupae of relatively few "lyonetiid" genera have been studied. Kuroko (1964), for example, in his review of the Lyonetiidae makes almost no reference to pupal structure. Similarly, if such specializations as the reduced tentorium in larval *Bucculatrix* or the reduced prementum and associated expansion of the hypopharynx in all Lyonetiidae except *Bucculatrix* are found to occur consistently, then this would indicate ample family distinction between the two groups. However, *Bedellia* shows some intermediacy regarding certain features, and it seems likely that genera whose larvae are at present unknown could demonstrate even closer relationships to *Bucculatrix* (e.g., Scoble and Scholtz 1984). Primarily because of these inadequacies and the uncertainties they present, *Bucculatrix* has been retained as a subfamily of Lyonetiidae in the current treatment.

The larvae of Lyonetiidae are typically cylindrical with relatively depressed heads. Little or no hypermetamorphism is evident except that the early leaf-mining instars of some genera are apodal. Mature larvae always bear thoracic legs (reduced in some *Lyonetia*) and a full complement of five pairs of prolegs with crochets usually arranged in a uniordinal, uniserial circle, sometimes biserial in caudal half of circle, or in one to two transverse bands; prolegs usually short, but long and slender in *Bedellia* and *Bucculatrix*. Stemmata usually six per side, the sixth sometimes reduced or lost. Frontoclypeus terminating anywhere from epicranial notch to two-thirds the distance. Prothorax with three prespiracular (L group) setae.

Biology and Ecology: The larvae of most Lyonetiidae are leaf miners, at least in the early instars. Some species are stem miners or stem borers and a few are gall feeders. The eggs, typically flat and oval and variously sculptured, are deposited singly on the host plant, usually cemented to some crevice or adjacent to a leaf rib. The emerging larva bores directly into the host, normally initiating extremely narrow, serpentine mines, depositing the frass in a thin median line. All instars are tissue feeders, and there is no early sap-feeding stage. The early instar larvae of many, if not all, Lyonetiidae are apodal. By the third instar the larvae of all species possess a full complement of legs and prolegs. From this stage on, the larva either continues mining, usually enlarging the mine to form large blotches and, in some cases (e.g., in *Bedellia*) leaving the original mine to initiate new mines, or it may abandon the mining habit altogether to feed externally. The latter is true for most species of *Bucculatrix*, which leave the mine after the second instar and spin a flat "molting cocoon" within which they molt (Braun, 1963). A second molting cocoon similar to the first but slightly larger is spun at the end of the fourth instar. Larvae of *Bucculatrix* that feed externally generally skeletonize the undersides of leaves, occasionally creating holes over much of the leaf. A few species continue as leaf miners or stem borers until mature or feed throughout their larval life within galls formed most often on Asteraceae. At least one gall feeder, *Bucculatrix needhami* Braun, is interesting in that a nonfeeding instar reportedly occurs between the last feeding instar and the pupal stage (Needham et al. 1928).

Kuroko (1964) summarizes the host records of the Leucopterinae and Lyonetiinae for the world. Over 20 plant families are listed including one monocot (Liliaceae), with the Betulaceae, Fabaceae, and Rosaceae particularly favored. Braun (1963) reports over 25 host families for *Bucculatrix*, the Asteraceae serving as hosts for nearly two-thirds of the species in the United States.

Pupation occurs outside the mine or gall, often in a spindle-shaped cocoon of white silk. This is either suspended in a silken hammocklike support or attached to some part of the host. In a few genera (e.g., *Bedellia* and some *Lyonetia*) the pupa is naked. The cocoons of *Bucculatrix* are characteristically ribbed (fig. 26.44e) and intricately constructed (Jäckh 1955, Snodgrass 1922) with an encircling stockade of erect silken fibers. The pupa is obtect and complete or incomplete. The more primitive, gracillariidlike pupa of *Bucculatrix* is incomplete, having dorsal abdominal spines, most appendages free, and segments A3–7 moveable in the male and A3–6 in the female. On emergence, the pupa projects more than half its length through the anterior end of the cocoon. In sharp contrast, the pupae of all other known Lyonetiidae are much more advanced and complete. In those species that construct cocoons, the pupa may not protrude prior to adult emergence, although this needs to be confirmed.

From one to three generations per year have been recorded for various species. Braun (1963) reports that most *Bucculatrix* have but one generation each year, with two or three occurring in some of the oak or Asteraceae feeding species.

Description: Mature larva small, usually 5–10 mm. Body somewhat flattened in early instars, cylindrical when mature; color variable, ranging from white to greenish to purplish red, occasionally with pale intersegmental bands and dark, longitudinal stripes dorsally. Pronotum usually brown; pinnacula small, sometimes distinct and bulbous. Spiracles round.

Head: Prognathous to semihypognathous, slightly to strongly depressed, more round in *Bucculatrix*. Epicranial notch deep. Frontoclypeus variable, terminating anywhere from epicranial notch to two-thirds the distance. Ecdysial lines often indistinct, converging at epicranial notch or below. Stemmata typically six, with sixth sometimes reduced or absent; fifth stemma rarely reduced; arrangement of stemmata variable depending upon species. Labrum usually deeply bilobed, often with accessory spines from venter of anterior margin; three pairs of epipharyngeal setae present. Mandibles variable, sometimes flattened, usually with six or more cusps; two rows of prominent cusps present in *Bedellia somnulentella* Zell. (fig. 26.43g). Most genera (except *Bucculatrix*) have premental lobe receded so that spinneret and labial palpi are situated far posterior of anterior margin of head (fig. 26.43m); hypopharynx conversely enlarged and thrust far forward; mouthparts of *Bucculatrix* normal, not highly modified. Tentorial bridge greatly reduced in *Bucculatrix*, normal in all other genera examined.

Thorax: Pronotum usually well defined, dark brown. L group trisetose on T1, bi- or trisetose on T2 and T3. Legs present in mature larvae, rarely reduced (two-segmented in some species of *Lyonetia*), frequently absent in first two instars; a pair of large, truncate setae situated on either side of tarsal claw in most *Bucculatrix* (fig. 26.44d).

Abdomen: D group bisetose on A1–9, D1 either above or below level of D2. SD group often bisetose on A1–8 but with SD2 very small, indistinct, and perhaps lacking on some segments. L series usually trisetose on A1–8, L2 absent on A1–2 in some *Leucoptera;* L2 absent on A1–8 in *Bedellia somnulentella;* L1 and L2 on A1–8 widely separated in *Bucculatrix*, much closer together in Leucopterinae and Lyonetiinae. SV group either uni- or bisetose on A1–2, usually trisetose on A3–6. Prolegs usually absent in first two instars, present on A3–6 and 10 in mature larvae; prolegs usually short but elongate and slender in *Bucculatrix* and on A4–6 and 10 in *Bedellia somnulentella*. Crochets usually in a uniordinal, uniserial circle, sometimes biserial in caudal half of circle or in one to two transverse bands; anal prolegs often in a single band, which may be reduced to a single hook in *Bucculatrix thurberiella* Bsk.

Comments: The Lyonetiidae comprises a relatively small cosmopolitan family consisting of less than 350 named species, almost equally divided in number between the Old and New World. Some genera, however, are disproportionately distributed. Most species of *Bucculatrix*, for example, are endemic to the continental United States. The family is in serious need of revision, because the systematic relationships of most genera are poorly understood.

Several species of *Bucculatrix* are of economic importance, perhaps the most important being the widespread cotton leaf perforator, *B. thurberiella* Bsk. (figs. 26.43a-f, 26.44a-f), which attacks several varieties of cotton throughout the southwestern United States, Mexico, and Hawaii. *Bedellia somnulentella* (Zell.) (figs. 26.43g-j) is a cosmopolitan pest of morning glories (*Ipomaea* sp.) and possibly of sweet potato (*Ipomaea batatas*). The latter is also the primary host for the sweet potato leaf miner, *Bedellia orchilella* Wlsm., presently known only from Hawaii (Zimmerman 1978). The South American coffee leaf miner, *Perileucoptera coffeella* (Guer.-Men.) now occurs almost everywhere in the western hemisphere where coffee is cultivated. At least one species of Lyonetiidae has been used as a weed control agent. *Leucoptera spartifoliella* (Hbn.) was successfully introduced into the western United States (Frick 1964, Parker 1964) to combat the spread of scotch broom (*Cytisus scoparius* Link.).

Selected Bibliography

Braun 1963.
Forbes 1923.
Fracker 1915.
Frick 1964.
Grandi 1932–33.
Jäckh 1955.
Jayewickreme 1940.
Kuroko 1964.
MacKay 1972.
Mosher 1916a.
Needham, Frost and Tothill 1928.
Parker 1964.
Scoble and Scholtz 1984.
Snodgrass 1922.
Zimmerman 1978.

GRACILLARIIDAE (TINEOIDEA)

Donald R. Davis, *Smithsonian Institution*

The Gracillariids, Blotch Leaf Miners

Figures 26.44j-l, 26.45, 26.46

Relationships and Diagnosis: The Gracillariidae and Lyonetiidae have questionably been placed in the Tineoidea by most authors or treated as members of a separate superfamily, Gracillarioidea (Mosher 1916, Zimmerman 1978). Until the relationships of these two families have been more thoroughly analyzed on a global basis, not only in comparison with the other Tineoidea but between themselves as well, it seems preferable to follow a more conservative approach. One problem is that the morphological definitions of adult Gracillariidae and Lyonetiidae remain rather vague, so that certain intermediate or aberrant genera, particularly *Phyllocnistis* and *Bucculatrix*, have been difficult to place. One solution has been to erect monotypic families for each (Spuler 1910, Fracker 1915). In this treatment *Phyllocnistis*, largely on the basis of its hypermetamorphic, sap feeding larva, has been considered a subfamily of Gracillariidae, and *Bucculatrix*, primarily on adult similarities, a subfamily of Lyonetiidae.

Although the adults of these two families are difficult to define on a world basis, their immature stages demonstrate major differences. The larvae of Gracillariidae are hypermetamorphic with two distinct forms and habits: (1) an early sap-feeding stage possessing a flattened, apodal body and highly specialized mouthparts that lack a spinneret, and (2) a later tissue-feeding and/or spinning stage typified by a more generalized cylindrical body usually possessing legs and functional spinneret. An additional nonfeeding, quiescent stage is also present in the genera *Chrysaster, Dendrorycter*, and *Marmara* (Kumata 1978). The larvae of Gracillariidae are further characterized by two prespiracular (L group) setae on T1, and prolegs, if present, occurring on A3–5, and 10 and almost always absent on A6 (present in *Artifodina*).

Biology and Ecology: The Gracillariidae constitute the principal family of plant-mining Lepidoptera. The early instars of most species are leaf miners with the others mining the cambium layer of herbaceous or woody stems or fruit. The later instars of the most primitive genera in the Gracillariinae feed externally but are concealed in folded or rolled leaves.

The eggs are laid singly and cemented to the surface of the host plant. Upon emergence the larva bores through the lower wall of the egg directly into the plant epidermis. The early instar larvae, typically sap feeders, sheer through a few layers of subepidermal parenchyma cells and suck their liquid contents without ingesting any solid tissue. Because vertical space is severely restricted, the entire body of the apodal larva has become highly modified and flattened. The head capsule is a triangular wedge consisting essentially of a pair of saw-toothed, scissorlike blades enclosed by two sheaths: a dorsal, usually enlarged labrum and a ventral plate consisting primarily of a variously modified hypopharynx, maxilla, and labium. A spinneret is absent. The Gracillariidae, including the genus *Phyllocnistis,* is the only family of Lepidoptera possessing a sap-feeding larval stage.

The number of larval instars varies from 4 to 11, depending upon the genus and sometimes species. Commencing no earlier than the third instar (i.e., second larval molt) but sometimes as late as the final instar (fourth to ninth instar), a major change in morphology and feeding behavior occurs. The larva usually reverts to the typically lepidopterous form, which is cylindrical with a round head, chewing mouthparts, and a functional spinneret. The transition to this tissue-feeding stage occurs at different instars according to genus and with varying degrees of completeness. In the more primitive genera (e.g., *Aristaea, Caloptilia, Macrostola,* and *Parornix*) there occur only two sap feeding instars, the third instar being cylindrical and possessing legs and normal chewing mouthparts. The fourth and fifth instars of *Caloptilia* and *Macarostola* are also tissue feeding like the third, but these larvae feed outside the mine, usually in a rolled-over portion of the leaf. In *Phyllonorycter,* the first three to four instars are sap feeding, the fourth and fifth (or fifth to sixth) tissue feeding. The larvae of *Cameraria* are sap feeders through the third instar, as in *Phyllonorycter,* and the larval form continues to be somewhat flat through the sap feeding fourth and fifth stages (Needham et al. 1928). The sixth and/or seventh instars are more cylindrical, nonfeeding and possess vestigial mouthparts, but with a well-developed spinneret. Thus, the principal function of the final instars is to spin a silken pupal chamber within the mine. Some species of *Marmara,* a genus of mostly stem and bark miners, may undergo as many as six to nine sap feeding instars, as well as one quiescent instar, before changing into the final nonfeeding, spinning stage (Fitzgerald and Simeone 1971). *Phyllocnistis* is somewhat similar in having only sap feeding instars (only three in this genus) before the final (fourth) nonfeeding, spinning instar. The latter, however, is the most specialized larva in the family and in the entire order, in having lost all legs, stemmata, and mouthparts except for the functional spinneret (fig. 26.45l).

The shape and location of the larval mine, as well as the method of frass deposition and the identity of the host, are all important specific characteristics. Although they probably first initiate a slender linear mine, the great majority of gracillariid larvae eventually construct blotch leaf mines (stigmatonomes). Most species of *Marmara* form rather broad serpentine stem or bark mines (ophionomes). The leaf mine of *Phyllocnistis* is usually a slender, very elongate, upper-surface mine containing a dark median frass line, the mine terminating in a small blotch, usually near the edge of the leaf in which the fourth instar pupates. *Phyllonorycter* larvae most frequently devour the parenchyma tissue in the lower half of the leaf, thereby creating a blotch mine visible only from the lower leaf surface. Their frass is typically dry and loose and at times may be incorporated into the cocoon. *Cameraria* construct upper surface blotch mines in which the dark, viscid frass dries to form tar-like spots over the ventral surface of the mine. The spinning larvae of both *Cameraria* and *Phyllonorycter* usually line the epidermis of the blotch mine with silk, causing it to constrict, thus producing a slightly deeper tentlike mine.

Several genera of Gracillariinae pupate in a cocoon outside the mine, often in a rolled-over leaf. This habit varies according to genus, the Lithocolletinae and Phyllocnistinae typically pupating inside the mine, either with or without forming a silken cocoon. Depending upon the species, pupation in *Marmara* occurs either in a widened area inside the mine, or outside the mine in some protected crevice partially covered by silk and usually ornamented by frothy globules. The larvae of *Marmara* and *Dendrorycter* are unusual in undergoing a quiescent pseudopupal stage prior to the final spinning instar (DeGryse 1916, Kumata 1978). The formation of the cylindrical pseudopupal stage occurs inside the body of the last mining instar without eclosing. The final instar, or true spinning larva, is formed within the body of the pseudopupa so that when growth is completed, the spinning larva sheds the exuviae of both the mining and the pseudopupal stages simultaneously.

A majority of the Nearctic Gracillariidae probably undergo some kind of diapause, most frequently in the last larval, over-wintering instar. Several species, particularly those feeding on evergreen leaves or as bark miners, apparently experience no true diapause but feed whenever temperatures or other conditions permit. Although the number varies, sometimes for a given species, most Nearctic Gracillariidae appear to pass through one to three generations per year.

Description: Mature larva small, usually not exceeding 10 mm. Body varying from white to green and usually immaculate. *Cameraria* larvae typically with a dorsal row of dark spots. Cuticle smooth except for primary setae. Larvae hypermetamorphic, with two distinct body forms; early instars extremely flat, apodal, with highly specialized head capsule, mouthparts (figs. 26.45a, b), and reduced chaetotaxy; later, or at least, last instar cylindrical, usually with legs, generalized head and spinneret, and with body setae more developed. Spiracles round.

Head: Early instars (from 1–2 or 1–8) extremely flat and prognathous. Stemmata highly variable, four to six in a linear series, indiscernible or reduced to a single pigmented spot. Antennae typically without elongate macrosetae. Labrum enlarged, bi- or trilobed and with entire or serrated margins. Mandible flattened, scissorlike, either with a few distinct cusps and/or minutely serrated. Maxilla with galea and palpus greatly reduced and usually absent but with enlarged stipes. Hypopharynx usually broadly laminate, densely pubescent, and largely forming the prominent lower lip of the buccal cavity (Trägardh 1913). Labium reduced with labial palpus and spinneret absent or vestigial.

Later tissue and/or spinning instars (commencing with third to tenth instar) cylindrical with head more round. Frontoclypeus elongate in both larval forms and usually extending to epicranial notch. Usually with four to six well-developed stemmata. Antennae with long macrosetae. Labrum reduced, usually bilobed. Mandible of normal chewing type, with four to six distinct cusps. Maxilla with normal galea and palpus. Labium with palpus and spinneret. Last (spinning) instar of *Phyllocnistis* highly specialized, cylindrical, but with all mouthparts lost except for spinneret (fig. 26.45l); stemmata absent; antennae greatly reduced and one-segmented.

Thorax: Most setae generally greatly reduced or absent in early sap-feeding instars and more developed in later tissue-feeding stages, although greatly reduced in last instar of *Phyllocnistis*. Mature larva usually with D group bisetose on T1–3 but setae sometimes greatly reduced in size on T1. SD group bisetose on T1–2 but sometimes absent on T1. L group bisetose on T1, uni- or trisetose on T2–3. SV group bisetose on T1, uni- or bisetose on T2–3. Legs generally present in later instars (absent or reduced in *Phyllocnistis* and *Cameraria,* respectively) and vestigial to absent in earlier sap-feeding instars.

Abdomen: Mature larva with D group usually bisetose on A1–8. SD group bisetose on A1–8 with SD1 usually minute and occasionally absent on some segments. L group unisetose on all segments. SV either uni- or bisetose on A1–2, trisetose on A3–5, uni- or bisetose on A6–7, and unisetose on A8–9. Anal plate with three pairs of setae. Prolegs present on A3–5 and A10 only, usually absent in early sap-feeding instars and in last instar (fourth) of *Phyllocnistis,* and reduced in last instar of *Cameraria*. Crochets uniordinal, typically arranged in a lateral penellipse, irregular bands, or circles, and usually biserial with inner series incomplete. Crochets absent in early instars and in all stages of *Phyllocnistis.*

Comments: The Gracillariidae constitute the largest family of plant-mining Lepidoptera, with over 1600 species known and countless others remaining to be discovered; 271 species have been reported from America north of Mexico (Davis 1983). The principal success of this group derives from its ability to exploit a wide range of plant hosts. Within the continental United States alone, over 40 plant families are represented as hosts, with the Asteraceae, Betulaceae, Fagaceae, Fabaceae, Rosaceae, and Salicaceae being slightly favored.

Larval Gracillariidae frequently are serious pests of orchards and nurseries. *Phyllonorycter blancardella* (F.), the spotted tentiform leaf miner (figs. 26.45a-c, 26.46a-c) and *P. crataegella* (Clem.) (figs. 26.45d, 26.46j-m) produce tentiform blotch mines on the underside of apple leaves in northeastern North America. At times their mines can become so numerous that premature fruit drop occurs. Similar injury caused by *P. elmaella* Dog. and Mut. has been reported from northwestern North America. Apple trees may also be heavily infested by the apple bark miner, *Marmara elotella* (Bsk.) and the apple leaf miner *M. pomonella* (Bsk.). The azalea leaf miner *Caloptilia azaleella* (Brants) at first mines and later skeletonizes the under surfaces of both deciduous and particularly evergreen azaleas. North American oak trees are frequently attacked by several species of *Phyllonorycter* and especially *Cameraria.* In the eastern United States the solitary oak leaf miner, *C. hamadryadella* (Clem.) (figs. 26.45e-l, 26.45n-q) and the gregarious oak leaf miner *C. cincinnatiella* (Cham.) can be particularly numerous on white oaks. *Cameraria agrifoliella* (Braun) is a major pest of California live oak (Opler and Davis 1981).

Selected Bibliography

Braun 1908.
Condrashoff 1962.
Davis 1983.
DeGryse 1916.
Fitzgerald and Simeone 1971.
Forbes 1923.
Fracker 1915.
Grandi 1932–33.
Heinrich and DeGryse 1915.
Hering 1951.
Jayewickreme 1940.
Kumata 1978.
MacKay 1972.
Mosher 1916.
Needham et al. 1928.
Opler and Davis 1981.
Peterson 1948.
Pottinger and LeRoux 1971.
Spuler 1910.
Still and Wong 1973.
Trägardh 1913.
Vari 1961.
Zimmerman 1978.

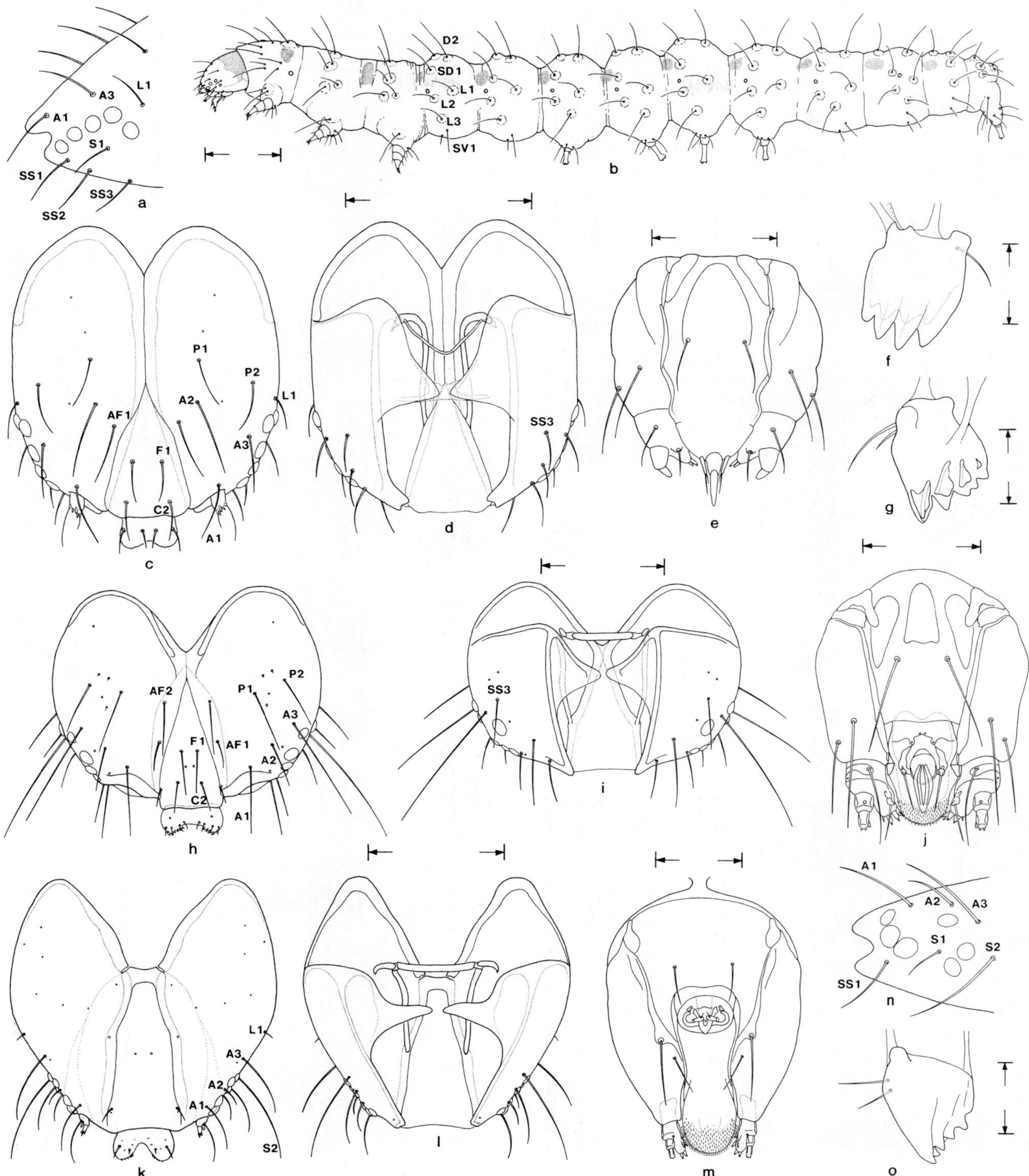

Figure 26.43a-o. Lyonetiidae. **a-f.** *Bucculatrix thurberiella* Busck. a. stemmatal area; b. lateral (scale = 0.5 mm); c. head, dorsal; d. head, ventral (scale = 0.2 mm); e. labium and maxilla, ventral (scale = 0.1 mm); f. mandible (scale = 0.05 mm); **g-j.** *Bedellia somnulentella* (Zeller), redrawn in part from Grandi 1931: g. mandible (scale = 0.05 mm); h. head, dorsal; i. head, ventral (scale = 0.2 mm); j. labium and maxillae, ventral (scale = 0.1 mm); **k-o.** *Leucoptera laburnella* (Stn.), redrawn in part from Grandi 1931. k. head, dorsal; l. head, ventral (scale = 0.2 mm); m. labium and maxillae, ventral (scale = 0.1 mm); n. stemmatal area; o. mandible (scale = 0.05 mm).

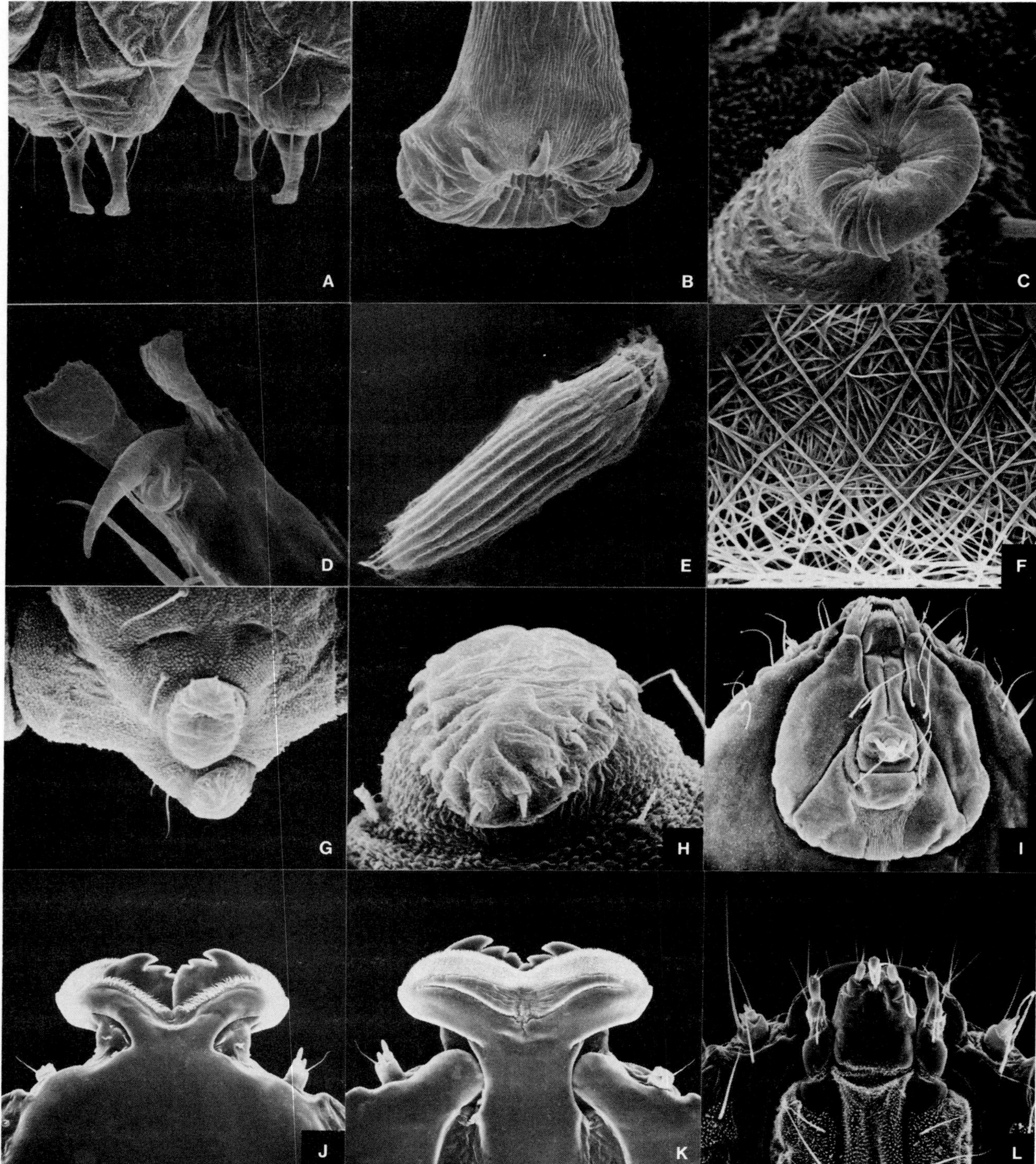

Figure 26.44a-l. Lyonetiidae, a–i; Gracillariidae, j–l. **a-f.** *Bucculatrix thurberiella* Busck. a. prolegs, A4–5, ×93; b. crochets, A5, ×870; c. crochets, A5, ×696; d. tarsus, T2, ×870; e. cocoon, ×12; f. detail of cocoon between longitudinal ribs, ×290; **g-i.** *Lyonetia speculella* Clemens. g. prolegs, A6, ×139; h. crochets, A6, ×464; i. *Leucoptera laburnella* (Stn.), head, ventral, ×168; **j-l.** *Marmara* species. j. head of sap feeding larva, dorsal, ×157; k. ventral view of fig. j, ×157; l. head of spinning larva, ventral, ×187.

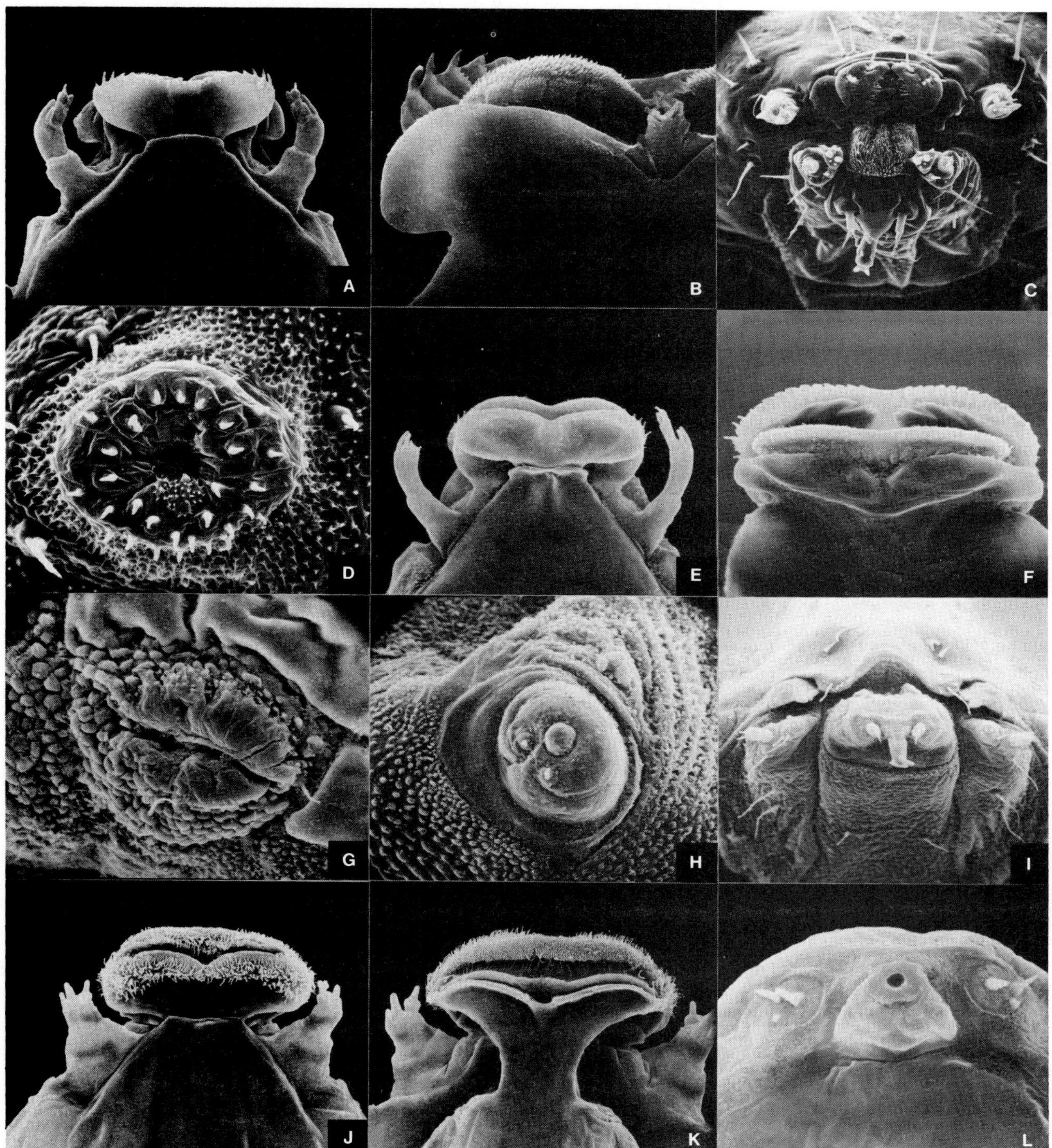

Figure 26.45a-l. Gracillaridae. **a-c.** *Phyllonorycter blancardella* (Fab.). a. head of sap-feeding larva, dorsal, ×341; b. ventral view of fig. a, ×893; c. head of tissue-feeding larva, anterior, ×232; **d.** *Phyllonorycter crataegella* (Clemens), crochets of A5, ×377; **e-i.** *Cameraria hamadryadella* (Clemens). e. head of sap feeding larva, dorsal, ×203; f. anteroventral view of fig. e, ×319; g. ventral ambulatory callus (proleg) of sap feeding larva, A4, ×377; h. vestigial leg of spinning larva, T3, ×377; i. head of spinning larva, anterior, ×194; **j-l.** *Phyllocnistis insignis* F. & B. j. head of sap-feeding larva, dorsal, ×232; k. ventral view of fig. j, ×261; l. head of spinning larva, fourth instar, ventral, ×696.

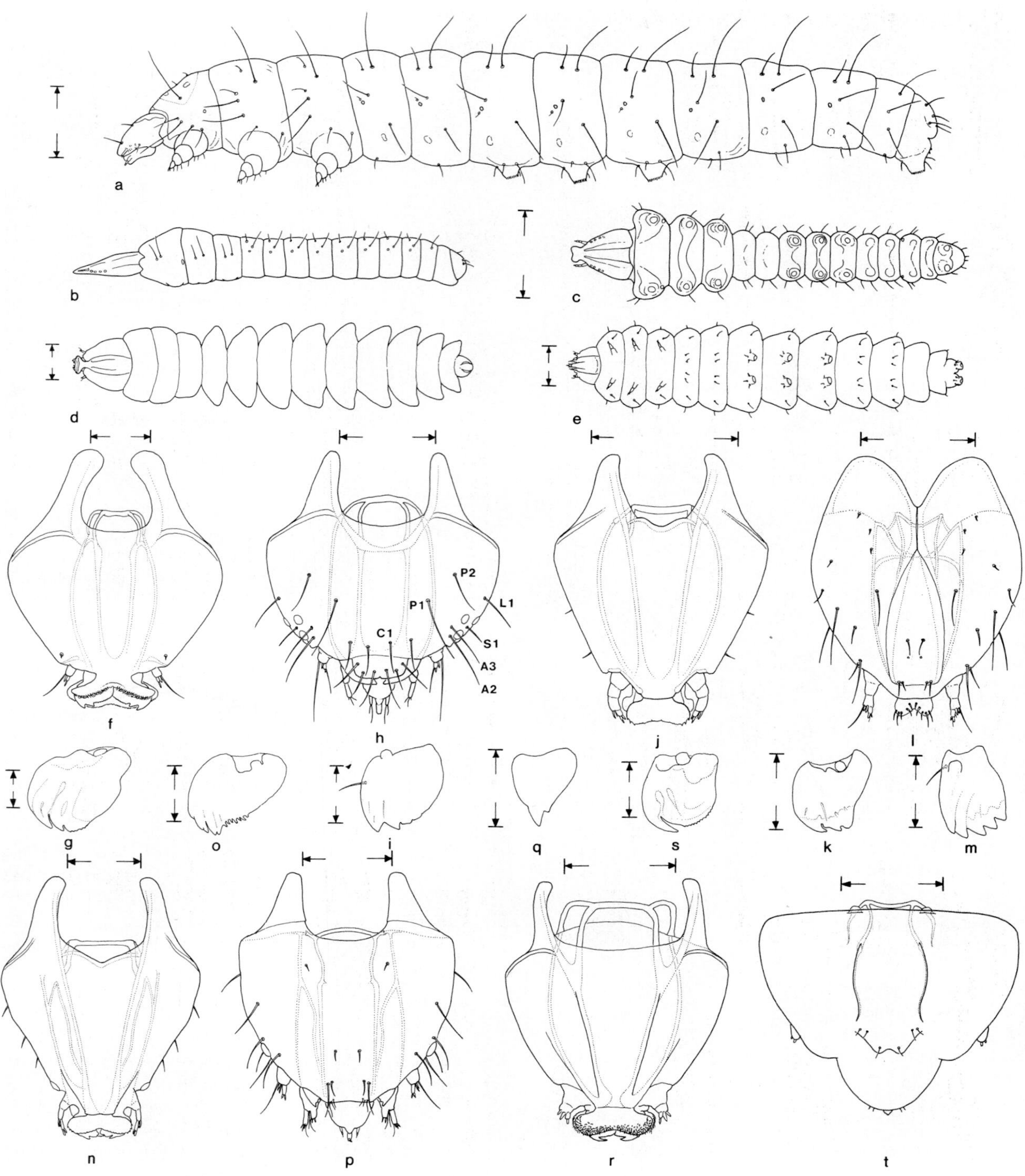

Figure 26.46a-t. Gracillariidae. **a-c.** *Phyllonorycter blancardella* (Fab.), redrawn from Pottinger and LeRoux 1971. a. fifth instar, lateral (scale = 0.5 mm); b. third instar, lateral (scale = 0.5 mm); c. third instar, ventral; **d-i.** *Marmara* species. d. sap feeding larva, ventral (scale = 0.5 mm); e. spinning larva, ventral (scale = 0.5 mm); f. head of sap feeding larva, dorsal (scale of all heads shown = 0.2 mm); g. mandible of sap feeding larva (scale of all mandibles shown = 0.05 mm); h. head of spinning larva, dorsal; i. mandible of spinning larva; **j-m.** *Phyllonorycter crataegella* (Clemens). j. head of third instar, dorsal; k. mandible of third instar; l. head of fifth instar, dorsal; m. mandible of fifth instar; **n-q.** *Cameraria hamadryadella* (Clemens). n. head of sap feeding larva, dorsal; o. mandible of sap feeding larva; p. head of spinning larva, dorsal; q. mandible of spinning larva; **r-t.** *Phyllocnistis insignis* F. & B. r. head of sap feeding larva, dorsal; s. mandible of sap feeding larva; t. head of spinning larva, dorsal.

SUPERFAMILY GELECHIOIDEA[4]

Frederick W. Stehr, *Michigan State University*

As workers have realized for a long time, the delimitation of families within the gelechioids is difficult at best, and next to impossible at worst, even when a relatively restricted geographical region is considered. Even though Fracker (1915) originally proposed the Gelechioidea based on larval characters, a comparison of the diverse groupings of genera into families by Forbes (1923), McDunnough (1939), Zimmerman (1978), and Hodges et al. (1983), as well as the family groupings given in tables 1 and 2 of Hodges (1978) helps explain why construction of a satisfactory family key for the gelechioid larvae is still not possible. And it is likely to remain so in the forseeable future, even when non-adult parameters can be more extensively used in arriving at a more satisfactory classification. The current classification also shows a considerable amount of discordance between adults and larvae, leading one to speculate that the Gelechioidea may be one group, which, because of its size, diversity, variation, and rapid evolution may defy comfortable "pigeonholing."

Most gelechioid larvae are small (5–15 mm), and many are active wrigglers when disturbed. They occupy nearly every kind of ecological niche, but even partial information on the biologies of species north of Mexico is known for half at best.

OECOPHORIDAE (GELECHIOIDEA)

Frederick W. Stehr, *Michigan State University*

The Oecophorids

Figures 26.47-26.59

Relationships and Diagnosis: In the Check List of the Lepidoptera of America North of Mexico (Hodges et al. 1983), the Ethmiidae and Stenomatidae are included in the Oecophoridae as subfamilies, thereby making a more diverse family of the Oecophoridae, and also making general statements more difficult, if not less accurate. Oecophorids are most likely to be confused with Gelechiidae and Cosmopterigidae, and there is no known way to positively identify these families, as they are relatively large and diverse (especially the Gelechiidae). In general, on A9 oecophorids have D1 more forward than D2, so D2, D1, and SD1 do not line up as well as they do in gelechiids and cosmopterigids. In addition, SD1 on A9 is usually hairlike for oecophorids and gelechiids, but is usually not hairlike in cosmopterigids. Also, Ethmiinae (Oecophoridae) have secondary setae in the SV group on the abdomen (especially on the prolegs or A9 (where they may be few and minute). Secondary setae are usually absent in the gelechiids, cosmopterigids, and most other oecophorids (present in *Apachea*). The genus *Symmoca* (Symmocinae) (figs. 26.59a-d) is also included here because they lack a submental pit (excluding them from Blastobasidae), the frontoclypeal area is similar to many oecophorids, and D1 is almost directly in front of D2 on A9, an oecophorid tendency. The genera *Batrachedra* (fig. 26.56) and *Homaledra* are placed here for the reasons given in the relationships and diagnosis for Coleophoridae, although they don't fit well here either. More study of all the immatures is definitely needed.

Biology and Ecology: The Depressariinae feed on living leaves, flowers, or seeds, and make shelters by rolling or tying them together, and one is even a leaf miner (Clarke 1941). Many of the flower feeders occur on Apiaceae. In contrast, the Oecophorinae feed almost exclusively on dead plant materials and/or associated fungi, and rarely on living plants (Hodges 1974). Ethmiinae are largely restricted to Boraginaceae and Hydrophyllaceae (Powell 1973) where they feed on flowers and buds, or in slight webs on leaves. Stenomatinae occur primarily in Central and South American wet forests on trees and shrubs; in North America most species feed on *Quercus,* a few on shrubs, and one on grasses (Duckworth 1964). Many Stenomatinae live in shelters formed by tying leaves together.

Description: Most mature larvae medium, 10–25 mm. A normal complement of primary setae present except for secondaries in most Ethmiinae. Pinacula variable, but most distinct in Stenomatinae. Spiracles circular to oval, those on T1 and A8 somewhat to 2× larger.

Head: Hypognathous to semiprognathous, heavily pigmented and smooth to rugose. Frontoclypeus longer than wide and extending one-half to three-fourths to epicranial notch except in Stenomatinae, where it extends less than halfway to epicranial notch. Ecdysial lines faint to distinct, running from one-half (some Stenomatinae) to all the way to epicranial notch. Usually six stemmata (apparently four or five in *Hofmannophila* and two in *Endrosis*), stemmata 1–4 (and sometimes 5) in a normal arc, 5 sometimes beneath the antennal socket, 6 caudad.

Thorax: T1 shield variably pigmented and with standard complement of setae. L group trisetose and linearly to triangularly arranged on a variably distinct pinaculum. SV group with two setae on T1 and 1 on T2 and T3. T2 and T3 with D1 and D2 close together and SD1 and SD2 close together, and with L1 closer to L2 than L3.

Abdomen: D1 and D2 quite wide of the midline, with D2 usually farther away. SD1 on A1–7 dorsad to somewhat anterodorsad of spiracle. L1 and L2 below the spiracle and close together. L3 located more or less above caudal margin of proleg. SV group bi- or trisetose on A1 and A7, trisetose on A2–6 and bisetose on A8 except for most Ethmiinae, which usually have secondary setae in the SV groups (*Ethmia charybdis* Powell and a few others have only a couple of minute secondaries on A9, and *Pyramidobela* has secondaries on A10). SD on A8 not hairlike and usually dorsad to dorsoanterad from spiracle, but directly anterior on *Psilocorsis* spp. and nearly so on *Setiostoma* (Stenomatinae). A9 usually with D1 more anterior than D2, and D1's farther apart than D2's except for most Ethmiinae where the D1's are closer together than the D2's (an exception is *Ethmia semiombra* Dyar). SD1 on A9 usually hairlike (*Hofmannophila* an exception). L1 and L2 together on A9 with L3 below, except Stenomatinae where all three are usually grouped together on the same pinaculum. SV with one seta on A9.

4. The comments and suggestions of Ron Hodges and Don Weisman, USDA Systematic Entomology Laboratory, and Jerry Powell, University of California, Berkeley, on the gelechioid families are greatly appreciated.

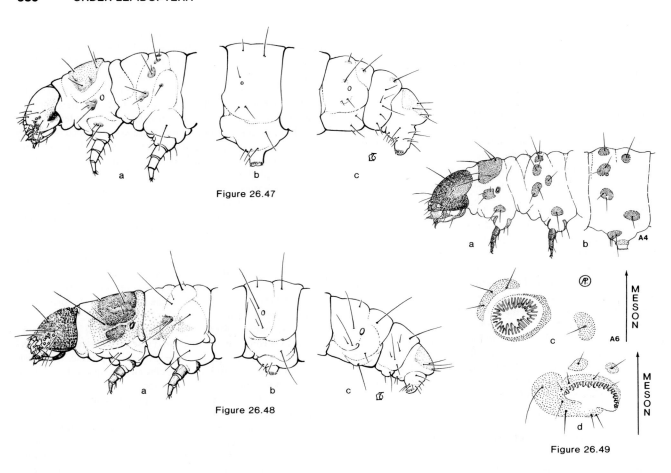

a b c

Figure 26.47

a b A4

c A6 MESON

d MESON

Figure 26.49

a b c

Figure 26.48

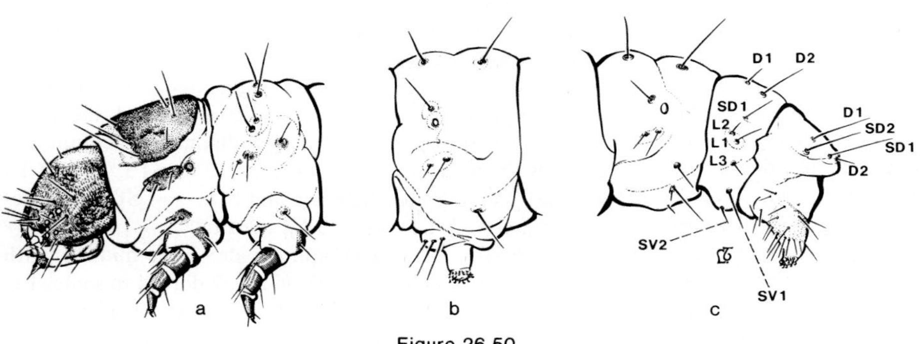

Figure 26.50

Figure 26.47a-c. Oecophoridae, Depressarinae. *Agonopterix dimorphella* Clarke. a. head, T1, T2; b. A3; c. A8-A10. f.g.l. 15± mm. Seward, Neb., 16 May 1960. Food plant: *Amorpha fruticosa.* U.S. Natl. Mus.

Figure 28.48a-c. Oecophoridae, Depressarinae. *Psilocorsis cryptolechiella* (Chambers). a. head, T1, T2; b. A3; c. A8-A10. f.g.l. 12± mm. Kirsten, N.C., 12 July 1961. A leaftier on willow oak. U.S. Natl. Mus.

Figure 26.49a-d. Oecophoridae, Depressarinae. *Depressaria pastinacella* (Duponchel), parsnip webworm. a. head, T1, T2; b. A4; c. A6 crochets; d. A10 crochets. f.g.l. 15± mm. Brown to black spotted, yellowish-green with grayish blue dorsum. *Hosts:* parsnip, celery, wild parsnip, wild carrot, and related weeds. Webs together and feeds on leaves and flowers. Pupates within stems (figs. from Peterson 1948.)

Figure 26.50a-c. Oecophoridae, Ethmiinae. *Pyramidobela angelarum* Keifer. a. head, T1, T2; b. A3; c. A8-A10. f.g.l. 10± mm. Fresno, Calif., 15 May 1945. *Host: Buddleia,* they roll and skeletonize the leaves. Found from the San Francisco Bay area south. Univ. Calif., Berkeley (Powell 1973). U.S. Natl. Mus.

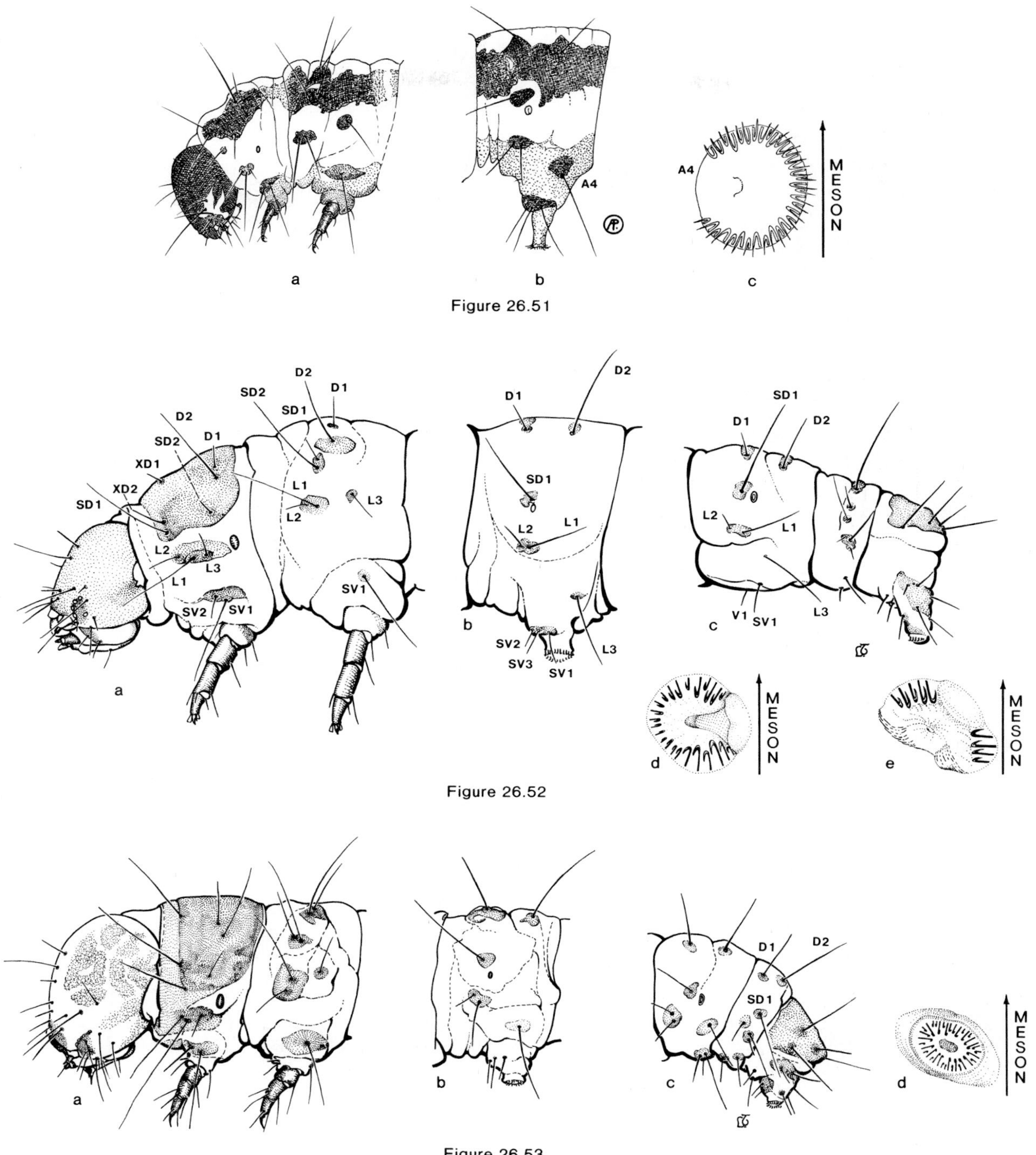

Figure 26.51a-c. Oecophoridae, Ethmiinae. *Ethmia monticola fuscipedella* (Walsingham). a. head, T1, T2; b. A4; c. A4 crochets. f.g.l. 20± mm. With 2 wide supraspiracular near black bands and black pinacula. *Food plant: Lithospermum gmelini* and *L. canescens.* Eastern North America (figs. from Peterson 1948). Powell 1973).

Figure 26.52a-e. Oecophoridae, Stenomatinae. *Setiostoma xanthobasis* Zeller. a. head, T1, T2; b. A3; c. A8-A10; d. A3 crochets; e. A10 crochets. f.g.l. 12± mm. Lady Lake, Lake Co., Fla., 3 June 1973, D. H. Habeck. *Food plant: Quercus laurifolia.* Univ. Florida.

Figure 26.53a-d. Oecophoridae, Stenomatinae. *Gonioterma mistrella* (Busck). a. head, T1, T2; b. A3; c. A8-A10; d. A3 crochets. Tall Timbers Res. Sta., Leon Co., Fla., 28 April 1970, D. H. Habeck. Ex. ova. Univ. Florida.

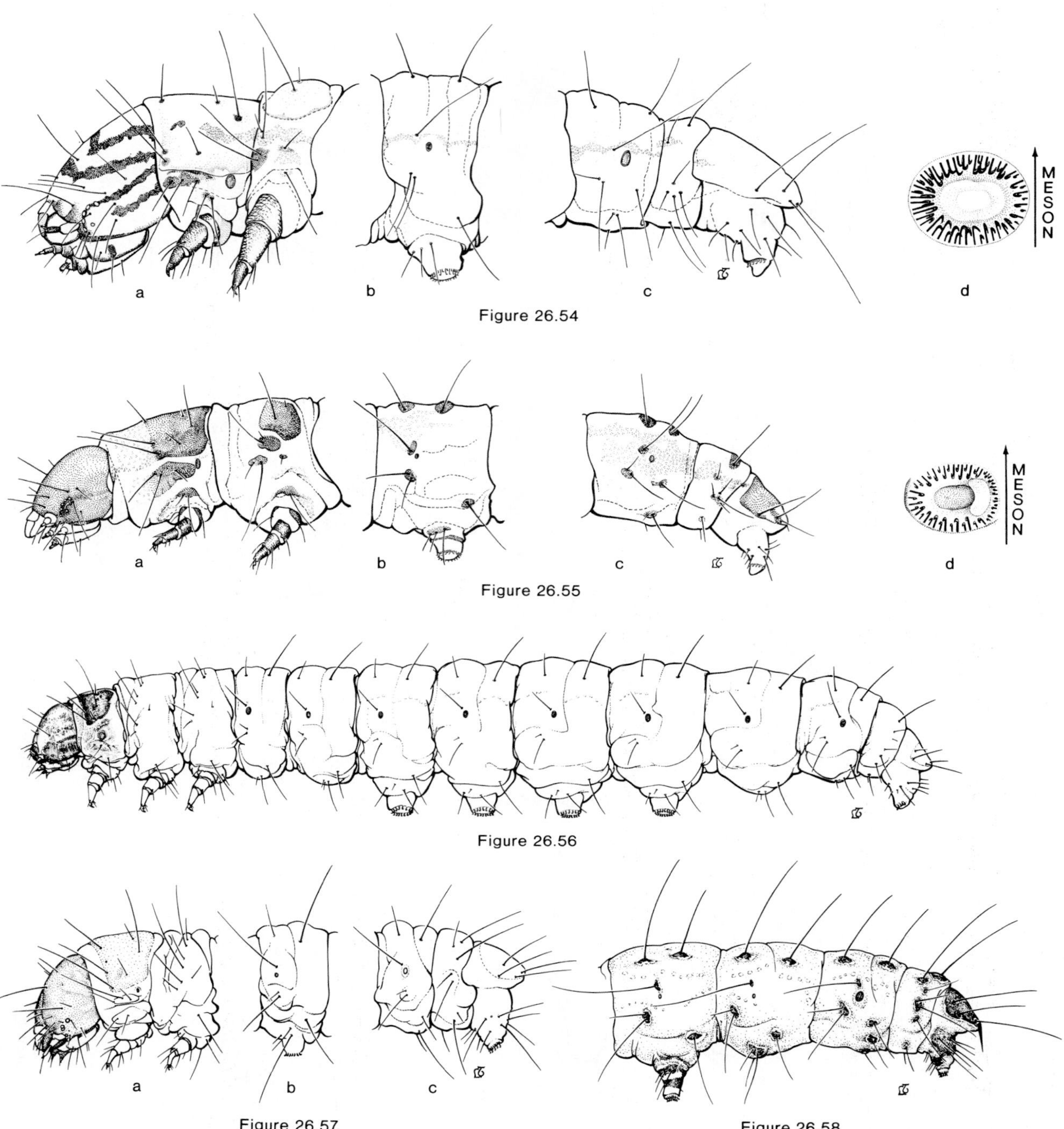

Figure 26.54a-d. Oecophoridae, Stenomatinae. *Anteotricha schlaegeri* (Zeller). a. head, T1, T2; b. A3; c. A8-A10; d. A3 crochets. f.g.l. 12± mm. Gainesville, Alachua Co., Fla., 30 October 1971, D. H. Habeck. *Food plant:* reared on oak. Univ. Florida.

Figure 26.55a-d. Oecophoridae, Stenomatinae. *Anteotricha humilis* (Zeller). a. head, T1, T2; b. A3; c. A8-A10; d. A3 crochets. f.g.l. 10± mm. Gainesville, Alachua Co., Fla., 23 September 1971, D. H. Habeck. *Food plant: Quercus nigra.* Univ. Florida.

Figure 26.56. Oecophoridae, Batrachedrinae. *Batrachedra linaria* Clarke. f.g.l. 15± mm. Baja Calif., 17 July 1952, on *Agave* leaf. U.S. Natl. Mus.

Figure 26.57a-c. Oecophoridae, Oecophorinae. *Hofmannophila pseudospretella* (Stainton), brown house moth. a. head, T1, T2; b. A3; c. A8-A10. f.g.l. 15-20± mm. New Zealand, 18 May 1937, USDA Cat. No. 37-15351, in packing for gladiolas. Host materials: see text. U.S. Natl. Mus.

Figure 26.58. Oecophoridae, Symmocinae. *Glyphidocera* sp.; A6-A10. f.g.l. 12± mm. Hamilton, Monroe Co., Miss., 26 August 1981, D. Tatum. Food plant: "Gold Coast" juniper. Note downward-projecting, tusklike setae on A10. U.S. Natl. Mus.

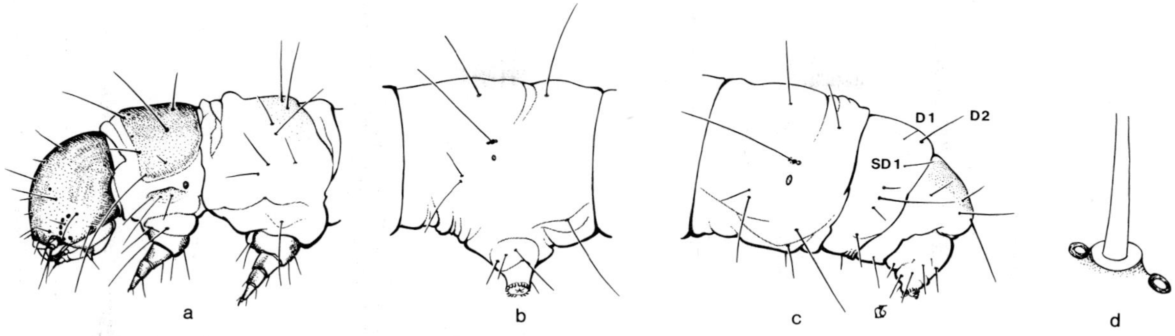

Figure 26.59a-d. Oecophoridae, Symmocinae. *Symmoca signatella* (H.-S.); a. head, T1, T2; b. A3; c.A8-A10; d. SD1 seta base on A3. f.g.l. 15± mm. Caruthers, Fresno Co., Calif. 9 September 1969, D. Rush, CDA #69I-11-11. Food plant: *Acer.* Univ. Calif., Berkeley.

A3–6 prolegs usually short, with crochets in a biordinal (rarely triordinal) circle or ellipse; in some Stenomatinae (*Setiostoma*) the prolegs are longer and the crochets are in a uni- to biordinal *latero*penellipse; in some *Anteotricha* the mesal part of the crochet ring is uniordinal. In most Ethmiinae the prolegs are fairly long and peglike, with the crochets in a biordinal *meso*penellipse.

A10 with standard complement of setae except in *Pyramidobela,* which has secondaries. Prolegs normal, with biordinal crochets varying from a transverse arc to almost a complete circle (*Depressaria*).

Comments: The Oecophoridae are most extensively developed in Australia (Common 1970), occurring in diverse climatic conditions, and they are abundant throughout North America (Clarke 1941, Hodges 1974). The Stenomatinae are primarily Neotropical and are most abundant in wet forest conditions (Duckworth 1964), whereas the Ethmiinae are, in contrast, most abundant in seasonally dry areas, especially microphyllous thorn forest (Powell 1973). In North America Stenomatinae are concentrated in the Southeast and Ethmiinae in the Southwest, with very few species of either group in Canada.

Most oecophorids are of little or no economic importance, but a few Depressariinae and Oecophorinae and one introduced Chimabachinae can be serious pests. The most important plant feeder is the parsnip webworm, *Depressaria pastinacella* (Duponchel) (figs. 26.49a-d), which damages flowers and developing seeds. *Cheimophila salicella* (Hübner) (Chimabachinae) is a pest on highbush blueberries in British Columbia (Raine 1966). *Psilicorsis* species (figs. 26.48a-c) attack beech, oak, and hickory, and pecan in the South where *P. reflexella* Clemens can be damaging. In the Oecophorinae, the brown house moth, *Hofmannophila pseudospretella* (Stainton) (figs. 26.57a-c) which is most common in the West, attacks a great variety of products, including stored cereals, dried fruits, bulbs, furs, wool products, wine corks, dried meats, nuts, seeds, etc. The white-shouldered house moth, *Endrosis sarcitrella* (L.) is a minor pest in houses on various organic products worldwide (Woodroffe 1951a, 1951b; Hodges 1974). *Ethmia nigroapicella* (Saalmuller) was a serious defoliator of *Cordia subcordata* in Hawaii in the past (Zimmerman 1978), but currently is quite rare. *Homaledra sabalella* (Chambers) skeletonizes palm fronds in Florida (Johnson and Lyon 1976).

Selected Bibliography

Powell (1971) gives details on life histories of *Ethmia,* and his 1973 monograph on New World Ethmiinae summarizes much of the information on larvae. Clark (1941) and Hodges (1974) offer information on larvae of Oecophorinae and Depressarinae.

Clarke 1941, 1952.
Duckworth 1964, 1973.
Embree 1958.
Heinrich 1921a,b.
Hodges 1974, 1978.
Johnson and Lyon 1976.
Lawrence and Powell 1969.
Powell 1971, 1973.
Pritchard and Powell 1959.
Raine 1966.
Woodroffe 1951a, 1951b.
Zimmerman 1978.

ELACHISTIDAE (GELECHIOIDEA)

David Wagner, *University of California, Berkeley*

The Elachistids

Figures 26.60–26.63

Relationships and Diagnosis: Elachistids are members of the Gelechioidea, in which the relationships among the families are poorly understood. Braun (1948) suggests that the Elachistidae have close affinity with the Oecophoridae and Scythrididae. Traugott-Olsen and Nielsen (1977) note that the Elachistidae, together with the Oecophoridae, Coleophoridae, and Agonoxenidae, retain several plesiomorphic characters and suggest that these taxa were among the first to have diverged from the ancestral gelechioid stock.

Characterization of larvae is difficult because of the diverse body forms within the family. These range from the generalized oecophoridlike larvae of *Coelopoeta* (figs. 26.63a-d) to the highly specialized leaf miners on grasses, e.g., *Dicranoctetes* (fig. 26.61). Nevertheless, most larvae may be recognized by the following combination of characters: small

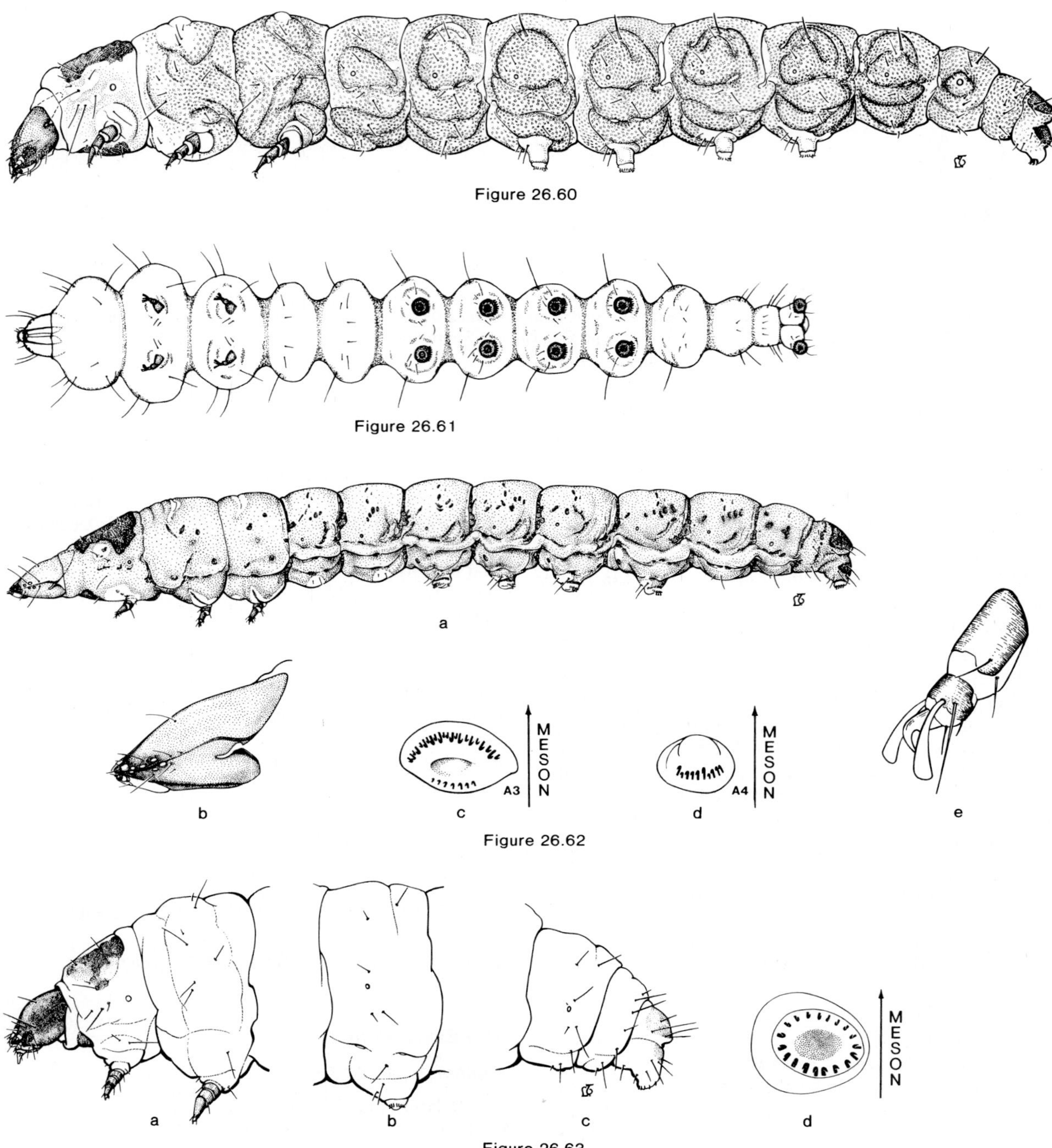

Figure 26.60. Elachistidae, Elachistinae. *Onceroptila cygnodiella* (Busck) f.g.l. 6± mm. Near Wilsonville, Clackamas Co., Ore., 16 June 1983. A blotch miner on *Symphoricarpos rivularis.* Univ. Calif., Berkeley.

Figure 26.61. Elachistidae, Elachistinae. *Dicranoctetes saccharella* (Busck). Venral view. f.g.l. 7± mm. Jamaica, July 1959, USDA 59-20500. A leaf miner in sugarcane. Note absence of T1 legs.

Figure 26.62a-e. Elachistidae, Elachistinae. *Elachista* sp. a. entire larva; b. head; c, d. A3 and A4 crochets, note 2 transverse bands on A3; e. tarsus with thoracic claw showing flangelike setae. f.g.l. 6± mm. Pygmy Forest, Van Damme St. Park, Mendocino Co., Calif., 13 April 1981. Food plant: *Hierochloe occidentalis.* Univ. Calif., Berkeley.

Figure 26.63a-d. Elachistidae, Coelopoetinae. *Coelopoeta glutinosi* Walsingham. a. head, T1, T2; b. T3; c. A8-A10; d. A3 crochets. f.g.l 8± mm. Horse Bridge, 1.5 mi. S.W. Arroyo Seco G. Sta., 1300′, 3-7 May 1975. *Food plant: Eriodictyon californicum.* Univ. Calif., Berkeley.

size, i.e., less than 9 mm; body dorsoventrally flattened and fusiform, widest at the pro- or mesothorax, and with deep intersegmental constrictions (except *Coelopoeta*); head prognathous, flattened, and recessed into T1; thoracic legs well developed, tarsal claw with a flangelike pretarsal seta (fig. 26.62e) on either side in grass- and sedge-feeding species; SV2 absent on T1 in Elachistinae (virtually all Lepidoptera larvae have two SV setae on T1). The grass-feeding species are not likely to be confused with any other leaf miners with the possible exception of *Cosmopterix*. In contrast to elachistids, *Cosmopterix* larvae are cylindrical, widest in the abdominal segments, lack a strongly sclerotized sternal plate, and possess two SV setae on T1.

Biology and Ecology: Elachistids are predominantly leaf miners, although a few are known to bore into stems. The more primitive genera are leaf miners in dicots; *Coelopoeta* is a western Nearctic genus restricted to the Hydrophyllaceae, *Onceroptila* (fig. 26.60) and related taxa feed on members of the Caprifoliaceae and Calycanthaceae, and *Stephensia* has been reared from *Cunila* in the Lamiaceae. The majority of elachistids are grass-, sedge- and rush-feeders, and include the North American genera *Elachista* (figs. 26.62a-e), *Cosmiotes, Biselachista* and *Dicranoctetes*.

Larvae form a full-depth blotch mine by consuming all the host tissues between the epidermal leaf surfaces. Typically, frass is retained within the mine. Larvae may abandon mined leaves and initiate new mines, especially in those species that overwinter as larvae within the mine.

Pupation usually occurs outside the mine, although *Coelopoeta glutinosi* Wlsm. frequently pupates within the mine. The larvae of *Onceroptila* and related taxa tunnel beneath bark or into unmined leaf tissue prior to pupation. Many of the grass- and sedge-feeding species pupate in a fashion similar to butterflies; the cremaster is well developed and a girdle of silk is fastened around the thorax. Other species pupate under a loose cocoon. The pupa is often ridged and beset with tubercles and spines.

Description: Length less than 9 mm. Body dorsoventrally flattened with deep intersegmental constrictions (except in *Coelopoeta*). Integument color variable, ranging from whitish to greenish and gray-brown, often darkening after leaving the mine. Primary setae only, inconspicuous; pinacula undifferentiated.

Head: Prognathous, heavily sclerotized, retracted into prothorax. Grass- and sedge-feeding species with strongly compressed head capsules, often bearing a genal ridge. Frontoclypeus extending to epicranial notch. Five or six stemmata. Labrum with strong bristles.

Thorax: Body widest at T1 or T2; T1 shield divided; T1 spiracle similar in size to abdominal spiracles; thoracic legs well developed (T1 legs absent in *Dicranoctetes*), grass- and sedge-feeding species with a flattened, clavate seta on either side of the claw (fig. 26.62e). SV2 on T1 absent in the Elachistinae; SD2 absent or minute on T1 and T2.

Abdomen: SD1 above spiracle; SD2 absent or minute on A1–8. Often only two L setae on A1–8. Prolegs on A3–6 and A10; crochets on A3–6 uniordinal, rarely biordinal, arranged in a circle or transverse bands.

Comments: Elachistids are well represented in the Holarctic region. The North American fauna consists of 57 recognized species in eight genera, although a large portion of the fauna is undescribed. Very little is known about the more primitive dicotyledonous-feeding species. Traugott-Olsen and Nielsen (1977) removed *Coelopoeta* from the Elachistidae because of numerous specializations of the genitalia, larva, and biology, and regarded *Coelopoeta* as a member of a monobasic taxon related to the Elachistidae and Oecophoridae.

When elachistid larvae are prodded on their dorsum, they bend their heads backward over the abdomen until the head touches A8 or A9, and then rapidly snap out of this position. In *Coelopoeta* this behavior is weakly developed. In other elachistids the snapping behavior may be especially animated, with larvae repeating the response up to a dozen times upon a single contact.

Elachistids are not known to inflict economic injury to any grain crops in North America, although *Dicranoctetes saccharella* (Busck) has been reported as a pest on sugarcane in Cuba (Busck 1934).

Selected Bibliography

Braun 1948.
Busck 1934.
Hering 1951.
Traugott-Olsen and Nielsen 1977.

BLASTOBASIDAE (GELECHIOIDEA)

Frederick W. Stehr, *Michigan State University*

The Blastobasids

Figure 26.64

Relationships and Diagnosis: The relationships of blastobasids are uncertain; in structure and habits the larvae are similar to scavenging oecophorids, but they can be distinguished from most of them by their submental pit (family key fig. 126). Known exceptions are the oecophorid genera *Endrosis* and *Martyringa*, a few Gelechiinae (Gelechiidae), and some non-American (Asian/Oceanic) Autostichinae (Oecophoridae) (Hodges 1978). The submental pit is also present in some species of other groups such as the Scythrididae and the exotic Xyloryctidae. At least some species of *Homaledra* and of *Batrachedra* (Oecophoridae?) have a pit, but they can be distinguished by having only two SV setae on A1 (three in blastobasids). Blastobasids and many scythridids have a variably distinct pinaculum ring around seta SD1 on A1–7 (as do some pyralids); some scythridids also have a submental pit, but scythridids always have secondary setae that blastobasids lack. The extra L seta(e) anterior to the L3 group on the abdomen of scythridids also distinguishes them from blastobasids. Scythridid larvae also have stemmata 3, 4 and 5 tight together, whereas blastobasids have a space the diameter of stemma 4 between 4 and 5.

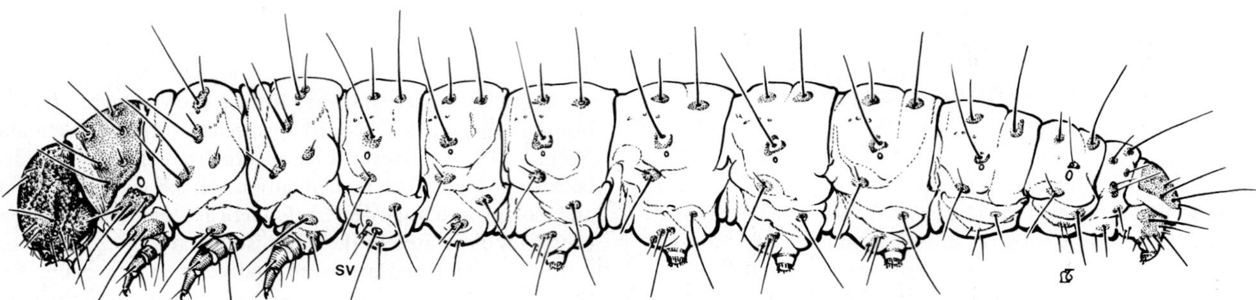

Figure 26.64. Blastobasidae. *Eubolepia gargantuella* Heinrich. f.g.l. 12± mm. Sabino Canyon, Santa Catalina Mountains, Ariz. 1-2 February, 1921. In cynipid galls on *Quercus oblongifolia.* U.S. Natl. Mus.

Biology and Ecology: Powell (1976b) reported on two species of giant blastobasids (*Holcocera*) that fed in flowers of *Agave* and in the open green and old fruit of *Yucca* spp. He summarized the known habits of blastobasids as feeding in a wide variety of situations, mostly associated with detritus in the nests of external-feeding caterpillars, in flower or seed heads, in the galleries of insect borers, or even on the ground or in subterranean situations (Pigritinae). He concluded that larvae of most species are opportunists, basically established as scavengers, but feeding in undamaged plant tissue when convenient. They even act as predators when susceptible prey such as scale insects, beetle larvae, and moth pupae are encountered. It is apparent that some species may live in a great diversity of habitats and eat a wide variety of foods, although others are relatively specific plant feeders (*Yucca* and acorns).

Description: Mature larvae small to medium, ranging up to 25 mm, but many about 10–15 mm. A normal complement of setae present, pinacula sometimes quite distinct. Spiracles small and circular to oval, those on T1 and A8 somewhat larger.

Head: Hypognathous to semiprognathous, rather uniformly pigmented. Frontoclypeus longer than wide and extending one-half to two-thirds to epicranial notch. Ecdysial lines distinct, closely parallel to adfrontal sutures, and extending to epicranial notch. Six stemmata in species examined; 1, 2 and 3 separated from each other by up to a stemma's width, 3 and 4 tight together, with 5 separated by approximately stemma 4's width, and 6 caudad. Submental pit present.

Thorax: T1 shield heavily pigmented, with standard complement of setae. L group trisetose and horizontally arranged on a distinct, elongate pinaculum anterioventrad of the spiracle. SV group on T1 with two setae and with one seta on T2 and T3. T2 and T3 with D1 and D2 on same pinaculum and SD1 and SD2 on same pinaculum; L1 closer to L2 than L3 and usually on same pinaculum.

Abdomen: D1 and D2 both quite wide of the midline and about the same distance from it. SD1 on A1–7 above spiracle and in a variably distinct pinaculum ring. L1 and L2 beneath and slightly ahead of spiracle, with L2 above L1. L3 located more or less above caudal margin of proleg. SV group trisetose on A1–6, bisetose on A7 and A8. SD1 on A8 hairlike and located on a variably shaped, elongated pinaculum. A9 with D1 more anterior than D2 and approximately equidistant between D2 and SD1. SD1 hairlike. L setae nearly vertically aligned, with upper two on same pinaculum. SV with one seta on A9.

A3–6 prolegs with a short, peglike planta bearing an oval of uni- or biordinal crochets.

A10 with a standard complement of setae. Prolegs normal, with uni- or biordinal crochets transverse to the meson.

Comments: There are about 120 species in 20 genera described, and another hundred undescribed (Powell 1976b). They are under revision by Adamski at Mississippi State. In the East, the larva of the acorn moth, *Valentinia glandulella* (Riley), is commonly found in acorns infested by the acorn weevil. *Holcocera lepidophaga* Clarke is reported to damage the flowers and young cones of slash and longleaf pines in Florida (Clarke 1960; U.S.D.A. 1985).

Selected Bibliography

Clarke 1960.
Dietz 1910.
Heinrich, 1921a.
MacKay 1972.
Powell 1976b.
Powell and Mackie 1966.
U.S.D.A. 1985.

COLEOPHORIDAE (GELECHIOIDEA)

Frederick W. Stehr, ***Michigan State University***

The Casebearers

Figures 26.65–26.68

Relationships and Diagnosis: In North America in the past the family Coleophoridae has contained 145 species in the single genus *Coleophora,* whose larvae were readily recognized as being leaf miners as first instars, and in the later instars being case-bearing, "external" leafminers or skeletonizers, rarely chewing through the entire leaf. In the genus *Coleophora,* cases are often recognizable to species, the larvae have reduced and short setae, and the ventral crochets (sometimes absent) are in two transverse bands. Other casebearers, like small Psychidae, have two SV setae on T2 and T3 (coleophorids have 1) and the A3–6 crochets are not in two transverse bands. Coleophorids also have on T2 and T3 a rather conspicuously sclerotized lateral patch (which is not a pinaculum since the L setae are almost invariably adjacent to, but not on it).

The genera *Batrachedra* and *Homaledra,* placed in the Coleophoridae in the subfamily Batrachedrinae by Hodges (1978), are placed in the Oecophoridae here because most are apparently scavengers and a few may be predators instead of leaf feeders as *Coleophora* species are. Hodges reports that a few *Batrachedra* make cases, but case making of various kinds is widely scattered among the Lepidoptera. In addition, the setae of *Batrachedra* and *Homaledra* are "normal" (no casemakers have been examined), and their crochets are in uni- or biordinal circles. *Batrachedra* and *Homaledra* are tentatively keyed as Oecophoridae because SD1 on A9 is hairlike, as are most species of oecophorids (SD is not hairlike in most Cosmopterigidae, another family where they have been placed). They are not placed in the Momphidae because the L group on T1 is trisetose (bisetose in Momphidae).

Biology and Ecology: The first instars of known species of *Coleophora* are leaf miners. As second instars they construct a portable case from bits of leaves, frass, and silk. The form of the case may be characteristic for the species. Second and later instars of some species can be called "external" leaf miners, because the case and posterior of the larva remain at the leaf surface while the larva extends itself inside the leaf to feed. Others are skeletonizers or feed more conventionally. In many (or most) species winter is passed as partially grown larvae inside the cases attached to the host. In such examples most damage occurs in the spring when the later instars complete feeding before pupating in the case. Eggs are laid on the leaves.

Description: Mature larvae small, 5–10 mm. Casebearers after the first instar. Head and T1 shield usually well sclerotized, T2 and T3 variable; thoracic legs well developed, but abdomen usually pale, with reduced numbers and size of setae except for the anal plate and lateral A10 proleg, which are usually well sclerotized. Cuticle frequently granulated, especially dorsally.

Head: Semiprognathous. Frontoclypeus longer than wide, extending to epicranial notch, which extends forward considerably between the cranial lobes (usually concealed by the thorax). Ecdysial lines very distinct and meeting the epicranial notch. Six stemmata in a normal arc with stemma 6 caudad. Head setae distinct.

Thorax: T1 normal, with three L setae that may be difficult to see on some species. T1 shield and sometimes T2 and T3 sclerotized to various degrees. L area of T1 and sometimes of T2 and T3 variably sclerotized. T2 and T3 with L2 closer to L1 than to L3. T2 and T3 with a rather conspicuously sclerotized lateral patch that is not a pinaculum, since the L setae are almost invariably adjacent to but not on it.

Abdomen: Some or most setae may be absent, short or difficult to see. D1 shorter and closer to midline than D2. SD1 caudodorsad of spiracle. A3–6 with L1 and L2 together, ventral and slightly posterior of vertical line through spiracle; L3 more ventral, minute, and in line with a vertical line through spiracle. Two SV setae (one minute) separated and above the third SV, which is adjacent to the crochets and minute. V seta in front of crochets and minute. Prolegs on A3–6 reduced to lumps bearing transverse bands of small uniordinal crochets, varying in number and sometimes absent. Crochets sometimes absent on A6 even though present on A3–5. The transverse bands of crochets vary from adjacent to the midline to rather widely separated. SD1 on A8 anterodorsad. A9 setae in a nearly vertical row. A10 with anal plate usually sclerotized. A10 proleg usually short, sclerotized laterally, and usually bearing a single row of well-developed crochets (sometimes reduced and rarely absent), even when A3–6 crochets are reduced. Spiracles circular, A8 spiracle simiiar to T1 spiracle, and somewhat to 2× larger than A1–7 spiracles.

Comments: The genus *Coleophora* contains about 100 species in America north of Mexico. Little has been done with the larvae other than descriptions of the cases of numerous species. A number of species are fairly common and several can be pests, including the pistol casebearer, *C. malivorella* Riley (figs. 26.67a-e) on apple and other fruit trees, the larch casebearer, *C. laricella* Hübner (figs. 26.66a-e), the elm casebearer, *C. ulmifoliella* McDunnough, the birch or cigar casebearer, *C. serratella* (L.)., the pecan cigar casebearer, *C. laticornella* Clemens, on pecan, walnut and hickory, and the cherry casebearer, *C. pruniella* Clemens (figs. 26.68a-e) on cherry, apple, plum and other hosts. *Coleophora parthenica* Meyrick from Pakistan and Egypt, and *C. klimesckiella* Toll from Pakistan have been introduced as potential biocontrol agents against Russian thistle, *Salsola iberica* (J. Coulson, personal communication).

Selected Bibliography

Bryant and Raske 1975.
Gould 1931.
Heinrich (*in* Forbes 1923).
Powell 1980.
Raske 1976.
U.S.D.A. 1985.

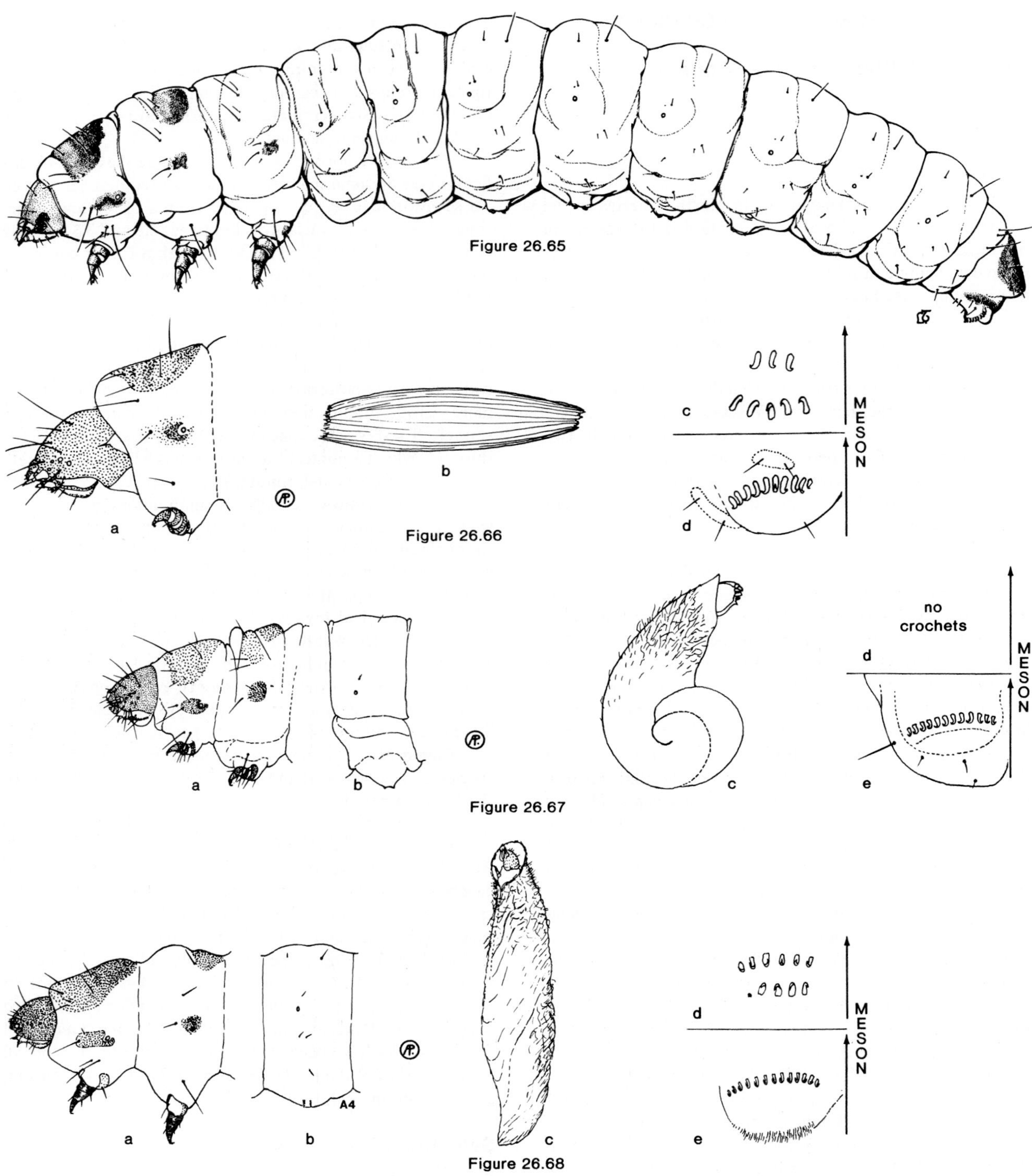

Figure 26.65. Coleophoridae. *Coleophora acordella* Walsingham. f.g.l. 5± mm. Seattle, Wash., 5 July 1943. *Food plant:* willow leaves. U.S. Natl. Mus.

Figure 26.66a-d. Coleophoridae. *Coleophora laricella* (Hübner), larch casebearer. f.g.l. 5± mm. a. head, T1; b. case; c. A6 crochets; d. A10 crochets. Hibernating larvae often found in cases in axils of buds. In spring it increases the size of the case, feeding on needles of European or American larch (figs. from Peterson 1948).

Figure 26.67a-e. Coleophoridae. *Coleophora malivorella* Riley, pistol casebearer. f.g.l. 5± mm. a. head, T1, T2; b. A4; c. case; d. A6 proleg area; e. A10 crochets. Constructs coiled case resembling a pistol. Crochets absent on A6. Feeds on leaves, buds, and fruit of apple, pear, quince, cherry, plum, hawthorne, and related plants (figs. from Peterson 1948).

Figure 26.68a-e. Coleophoridae. *Coleophora pruniella* Clemens, cherry casebearer. f.g.l. 5± mm. a. head, T1, T2; b. A4; c. case; d. A6 crochets; e. A10 crochets. Feeds primarily on foliage of apple, cherry, plum and other plants; may attack buds, stems, and young fruit (figs. from Peterson 1948).

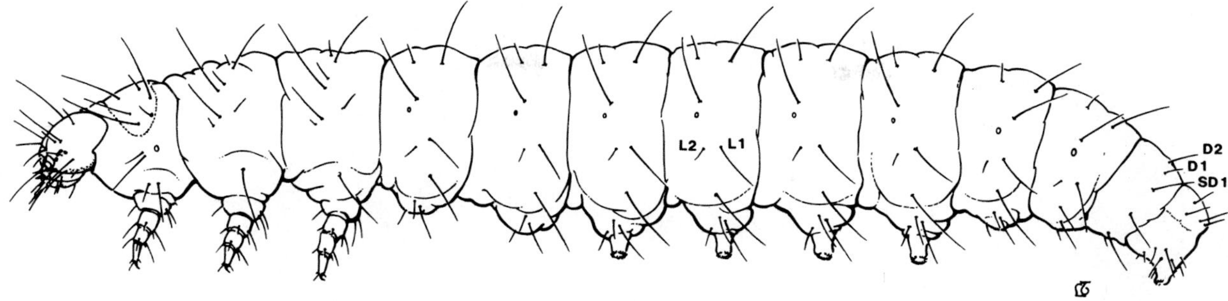

Figure 26.69. Momphidae. *Mompha* sp. f.g.l. 8± mm. Troutdale, Ore., 22 July 1944, ex. *Oenothera muricata* flower buds. U.S. Natl. Mus.

MOMPHIDAE (GELECHIOIDEA) (COPROMORPHOIDEA?)

Frederick W. Stehr, ***Michigan State University***

The Momphids

Figure 26.69

Relationships and Diagnosis: Hodges (1978) has placed the momphids in the Gelechioidea, although stating that they are somewhat puzzling, because they feed on living tissue, whereas the adults are somewhat similar to the Blastobasidae, most of which are scavengers.

Only the genus *Mompha* (Hodges 1978) and possibly *Chrysoclista* and *Synallagma* remain in the family, although Hodges et al. (1983) has placed *Chrysoclista* in the Agonoxenidae under the name *Glyphipteryx* (not *Glyphipterix* of Glyphipterigidae in the checklist). The larvae do not fit in Agonoxenidae comfortably, because they lack secondary setae and the *L group on T1 is bisetose.* The four species of *Mompha* examined here (identified as *stellella, eloisella?* and two as *Mompha* sp.) have only two L setae on T1, thereby distinguishing and tending to exclude them from the Gelechioidea, which have three, and possibly placing them in the Copromorphoidea (which have two and are plant feeders). The larvae key out near the copromorphoid families Carposinidae and Alucitidae and can be distinguished from both by having L1 and L2 on A1–8 distinctly farther apart than the width of the spiracle (L2 missing on *Chrysoclista*) and by having a line drawn through them more nearly horizontal than vertical (carposinids and alucitids have L1 and L2 about the width of a spiracle apart and have a line drawn through them more nearly vertical than horizontal). Larvae of more *Mompha* species must be examined to be certain that all species have only two L setae on T1.

Biology and Ecology: Although very little is known, *Mompha* have been collected from the stems and flower buds of *Oenothera,* and *Chrysoclista linneella* (Clerck), a European import, has been collected from European linden bark in the Northeast. Worldwide, a high percentage use Onagraceae, the only microlepidoptera so specialized (Powell 1980).

Description: Mature larvae small, not exceeding 15 mm. Body cylindrical, primary setae only and pinacula obscure.

Head: Semiprognathous to prognathous; frontoclypeus higher than wide, and extending about three-fourths to epicranial notch; ecdysial sutures inconspicuous. Six stemmata in a normal arc with stemma 6 caudad, but with larger gaps in *Chrysoclista.* Head setae conspicuous.

Thorax: T1 with standard complement of setae, but with only two L setae; SD2 caudad to SD1 and both on the weakly to moderately sclerotized shield. T2 and T3 normal, with 1 SV, and L2 closer to L1 than L3. T1 spiracle circular and somewhat larger than abdominal spiracles.

Abdomen: D1, D2, and SD1 normal. Except for *Chrysoclista,* which lacks L2, L1 and L2 are relatively close together, with L2 more or less anterior to L1; L3 caudoventrad of the spiracle; SV group trisetose on A3–6 proleg base; plantae short and peglike, and relatively far apart, with uniordinal crochets in a circle; or plantae shorter and nearly fused medially, with uniordinal crochets in a mesopenellipse. A9 with D1 short and about midway between D2 and SD1 (which is *not* hairlike); one or two L setae (*Chrysoclista* with two close together, one minute), one SV seta. A10 crochets in uniordinal transverse series. SD1 on A8 dorsad of the spiracle. A1–8 spiracles circular, with A8 spiracle little larger than A1–7 spiracles.

Comments: The examined specimens of this family were not reared from eggs of positively identified females, but misidentification is unlikely. However, considerable rearing and additional study of more species is needed to clarify the status and position of the "Momphidae."

Selected Bibliography

Hodges 1962, 1978.
Hodges et al. 1983.
Powell 1980.

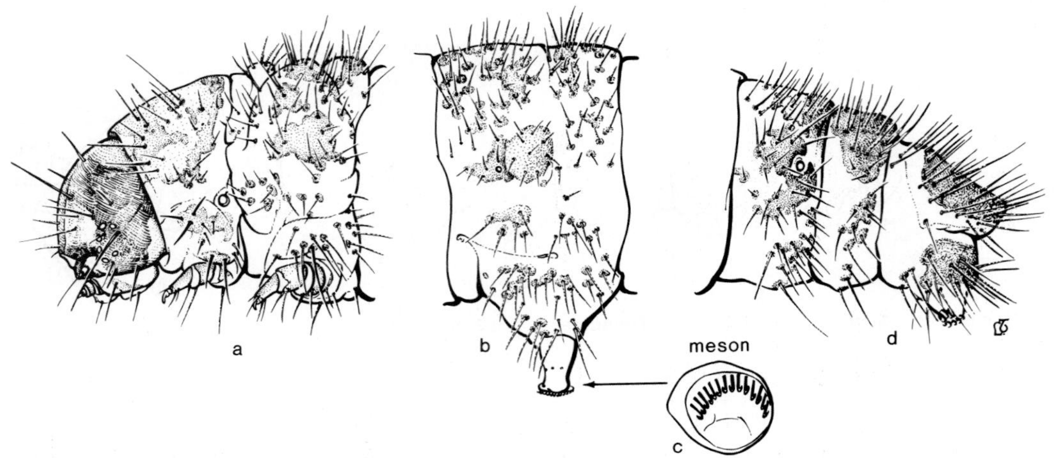

Figure 26.70a-d. Agonozenidae. *Blastodacna curvilineella* (Chambers). f.g.l. 7± mm. a. head, T1, T2; b. A3; c. crochets, A3; d. A8-A10. Ancaster, Ontario, 10 October 1962, ex. fruit of *Crataegus* sp. U.S. Natl. Mus.

AGONOXENIDAE (GELECHIOIDEA)

Frederick W. Stehr, *Michigan State University*

The Agonoxenids

Figure 26.70

Relationships and Diagnosis: The agonoxenids are represented in continental America north of Mexico only by the subfamily Blastodacninae, containing three genera and six species. They are easily distinguished from other gelechioids by the abundant short secondary setae distributed over the entire body. They are superficially similar to those pterophorid larvae that have abundant secondary setae, but they differ in having a single SV seta on T1 and T2 (if it can be distinguished from the secondary setae) (pterophorids usually have two clearly distinct SV's), and by not having any of the secondary setae grouped into verrucae, etc., as some pterophorids do.

The other subfamily, Agonoxeninae, contains four species in the genus *Agonoxena,* which are distributed from Hawaii to Australia, primarily on coconut palm (Bradley 1966). Bradley indicates that only one species, *A. argaula* Meyrick, bears secondary setae, but specimens of *A. pyrogramma* Meyrick in the USNM clearly bear secondary setae on the prolegs and A10, so secondary setae may be usual in the subfamily. The two species of Agonoxeninae examined also differ from Blastodacninae by having the L group of T1 included on the shield (separate from the shield in Blastodacninae).

Biology and Ecology: Known species of Blastodacninae are borers or gall-formers in the leaves, stems, fruits, berries, and seeds of *Malus, Crataegus, Aronia,* and *Croton* (Forbes 1923). Agonoxeninae are primarily leaf feeders on coconut palm (Bradley 1966).

Description: The following description applies to Blastodacninae; Agonoxeninae differ in having fewer secondary setae, in having the crochets in a biserial circle with the crochets alternately displaced (family key, fig. 183), and in having a nearly prognathous head. Mature larvae small, 10–15 mm, cylindrical; sometimes with small, pigmented pinaculalike areas of cuticle around many setae that may merge to form large areas. Primary setae present but frequently obscured by abundant, relatively short, secondary setae.

Head: Hypognathous to semiprognathous, frequently with dark areas dorsolaterally; frontoclypeus longer than wide and extending one-third to one-half to epicranial notch. Ecdysial sutures meeting shortly beyond apex of frontoclypeus. Six stemmata in a normal arc, with the fifth somewhat lower.

Thorax: T1 with relatively fewer secondary setae than the rest of the body; when visible, L group with three or four setae and SV group with two setae; T1 shield conspicuous, but primary setae difficult to discern. T2 and T3 with many secondary setae, SV with one seta when discernable; some other primary setae sometimes evident.

Abdomen: Many secondary setae; primaries usually not evident. Prolegs with short to moderately long cylindrical planta, crochets of A3–6 and A10 uniordinal and in a mesopenellipse to mesoseries. Spiracles small and circular, those on T1 and A8 slightly larger.

Comments: This is a small family, with six species in three genera in the continental U.S. One species, *Agonoxena argaula,* the coconut flat or leaf moth, is reported to be a pest on coconut palms in Fiji (Bradley 1966) and is also a pest in Hawaii (Zimmerman 1978).

Selected Bibliography

Bottimer 1926.
Bradley 1966.
Forbes 1923.
Hodges 1978.
Zimmerman 1978.

COSMOPTERIGIDAE (GELECHIOIDEA)

Frederick W. Stehr, *Michigan State University*

The Cosmopterigids

Figures 26.71–26.75

Relationships and Diagnosis: Cosmopterigids are apparently most closely related to the Gelechiidae; the larvae are difficult to characterize or distinguish from other gelechioids. The small size of some species makes the observation of setae and other characters more difficult than for many other gelechioids, and there are no simple recognition characters. Most species examined have SD1 on A9 *not* hairlike, in contrast to most gelechiids and oecophorids, where it is hairlike. However, one of the most commonly collected species, *Limnaecia phragmitella* Stainton (fig. 26.71), from cattail heads, has a short SD1 seta on A9 that varies toward being hairlike on some specimens. USNM specimens of *Triclonella pergandeella* Busck definitely have SD1 on A9 hairlike—perhaps it should be returned to the Oecophoridae, where Forbes (1923) and McDunough (1939) placed it. At least, SD1 on A9 being nonhairlike should be more fully examined as a delimiting or distinguishing character for cosmopterigids. The grass-mining *Cosmopterix* (fig. 26.72) may be confused with grass-mining elachistids, but *Cosmopterix* are cylindrical, widest in the abdominal segments, lack a strongly sclerotized sternal plate, and have two SV setae on T1. Grass-mining elachistids are widest in the thorax, have strongly sclerotized sternal plates, and have a single SV seta on T1.

Biology and Ecology: Hodges (1978) has recognized three subfamilies, one of which (Antequerinae) contains only three species, two in the genus *Antequera* which are parasitic on *Kermes* scales on oaks. The larvae of Cosmopteriginae are usually leaf miners in monocots (all *Cosmopterix* are miners), but some are seed feeders, stem miners or scavengers. The habits of the Chrysopeleinae are quite diverse, including species reported to be needle and stem miners, gall makers, leaf tiers, leaf and flower feeders, and fruit borers.

Description: Most mature larvae small, 5–10 mm. A normal, although sometimes reduced or minute complement of primary setae present, except for a few species with secondaries on the prolegs. *Anoncia sphacelina* (Keifer) has secondaries, but specimens I have examined of three of the other five species that Hodges (1978) reported as having secondaries do not have them (*Limnaecia phragmitella, Pyroderces rileyi* (Walsingham), and *Triclonella pergandeella* Busck).

Head: Semiprognathous to somewhat hypognathous; basically heart-shaped and usually not heavily sclerotized. Frontoclypeus 2× as long as wide and extending two-thirds to four-fifths to the epicranial notch. Ecdysial lines faint to distinct and meeting "at" the epicranial notch. Six stemmata in a normal arc, with the sixth caudad.

Thorax: T1 shield variably pigmented and with standard complement of setae. L group trisetose on a variably distinct pinaculum. SV group with two setae on T1 and one on T2 and T3. T2 and T3 with D1 and D2 close together, SD1 and SD2 closer together, and with L1 closer to L2 than L3.

Abdomen: D1 and D2 both quite wide of the midline, with D2 somewhat farther away. SD1 on A1–7 dorsad to slightly anteriodorsad of spiracle. L1 and L2 below spiracle and close together. L3 more or less above posterior margin to middle of proleg. SV group usually with two setae on A1 and A7, three on A2 and one on A9, but sometimes with three on A1, two on A2, and one on A7 (or more than three if secondaries are present, especially on A3–6). SD1 on A8 not hairlike, usually dorsad to anterodorsad of the spiracle, but anterior on a few such as the genus *Pyroderces*. A9 with D2, D1 and SD in a nearly straight line except for *Triclonella pergandeella,* where D1 is more anterior (tending to resemble oecophorids). SD1 on A9 generally *not* hairlike. L1 and L2 close together on A9, L3 below or absent. A3–6 prolegs short to flush with body and with uniordinal or biordinal crochets in a circle or oval (rarely a mesopenellipse). A10 with standard complement of setae; prolegs normal, with uni- or biordinal crochets in a transverse arc.

Comments: Some 26 genera and 180 species are recognized north of Mexico, although the family is most abundant in Oceania, with over 500 species in Hawaii and more than 350 in the genus *Hyposmocoma* alone (Zimmerman 1978). Most are of little or no economic importance. *Periploca nigra* Hodges was reported by Koehler and Tauber (1964) to cause dieback of junipers in California. *Walshia miscecolorella* (Chambers), the sweetclover root borer, attacks some legumes (Manglitz et al. 1971). *Ithome concolorella* (Chambers), native to the southern half of the U.S., is an introduced pest in Hawaii on *Acacia farnesiana* and on a South American species of mesquite, *Prosopis chilensis,* where it reduces flowers and consequently nectar production (Namba 1956). The pink scavenger moth, *Pyroderces rileyi* (Walsingham) (fig. 26.75), is a common scavenger in old and damaged bolls, and in a variety of other situations in cotton areas from Virginia to Arizona. It should not be confused with the pink bollworm, *Pectinophora gossypiella* (Saunders), a primary gelechiid pest of cotton. It is easily separated by having the A3–6 crochets in a circle (the pink bollworm has a mesal penellipse).

Selected Bibliography

Brandhorst 1962.
Classen 1921.
Heinrich 1921a,b.
Hodges 1962, 1964, 1978.
Koehler and Tauber 1964.
Lindquist and Bowser 1966.
Manglitz et al. 1971.
Namba 1956.
Opler 1974.
Powell 1963.
Zimmerman 1978.

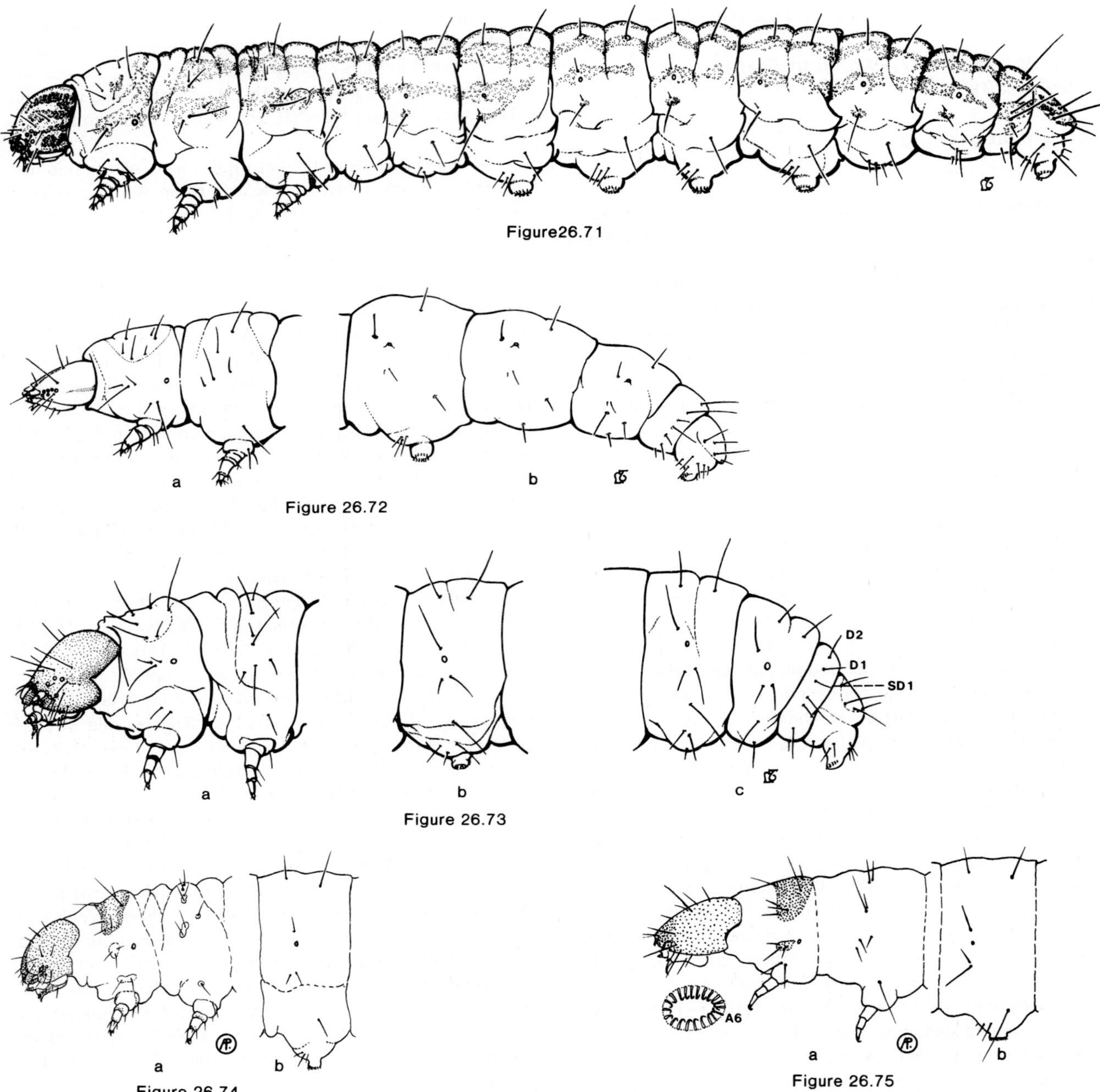

Figure 26.71. Cosmopterigidae. *Lymnaecia phragmitella* Stainton. f.g.l. 8± mm. Berrien Co., Mich., T55, R19W, Sec. 29, 20 May 1966, J. P. Donahue, ex. head of cattail, *Typha latifolia.* Mich. State Univ.

Figure 26.72a,b. Cosmopterigidae. *Cosmopterix clandestinella* Busck. f.g.l. 5± mm. a. head, T1, T2; b. A6-A10. Falls Church, Va., 25 September 1914. Leaf miner in grass (*Panicum*). U.S. Natl. Mus.

Figure 26.73a-c. Cosmopterigidae. *Sorhagenia nimbosa* (Braun). f.g.l. 6± mm. a. head, T1, T2; b. A3; c. A7-A10. Inverness Ridge, 2 mi. S. E. Inverness, Marin Co., Calif., 700-1100′, 15 May 1970, on *Rhamnus californica.* Univ. Calif., Berkeley.

Figure 26.74a,b. Cosmopterigidae. *Walshia amorphella* Clemens. f.g.l. 12± mm. a. head, T1, T2; b. A4. Ex. stem galls on *Amorpha fruticosa* (figs. from Peterson 1948).

Figure 26.75. Cosmopterigidae. *Pyroderces rileyi* (Walsingham), pink scavenger caterpillar. f.g.l. 8± mm. Pink or reddish body. A scavenger in cotton bolls, corn husks, dried and decayed fruit, and other injured or damaged plants (fig. from Peterson 1948).

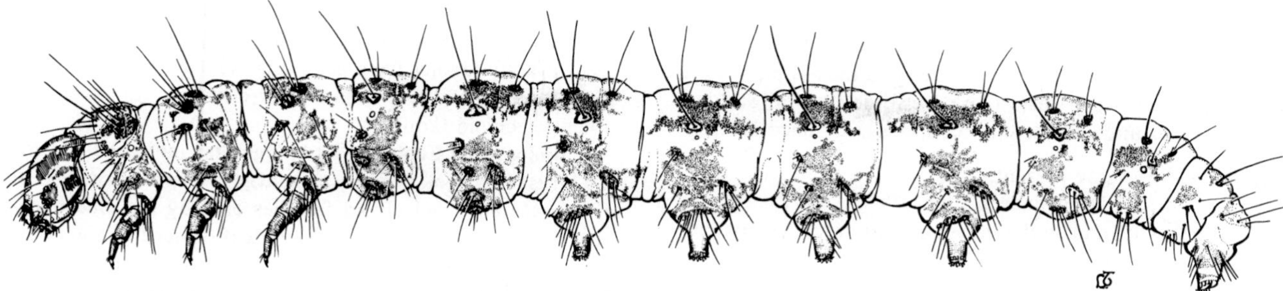

Figure 26.76. Scythrididae. *Scythris eboracensis* (Zeller). f.g.l. 12± mm. Falls Church, Va., July 1951. Reared from *Cirsium discolor,* J. F. G. Clarke. U.S. Natl. Mus.

SCYTHRIDIDAE (GELECHIOIDEA)

Frederick W. Stehr, ***Michigan State University***

The Scythridids

Figure 26.76

Relationships and Diagnosis: Powell (1976a) summarized the variable placement by previous workers of the genus *Scythris* and related genera in yponomeutoid or gelechioid families. Mosher (1916a) suggested placement as a distinct family in the Gelechioidea based on pupal characters. Keifer (1937) suggested placement there based on larval characters, and this placement has been widely followed since Common (1970) placed it there. The scythridids may be the most highly evolved gelechioid family.

All scythridid larvae examined possess secondary setae on the prolegs or body, although they may be difficult to see on some. This separates those scythridids having a submental pit and a pinaculum ring on A1–7 from those Blastobasidae that are similar, because blastobasids never have secondary setae. In addition, the scythridid larvae examined have little or no space between stemmata 4 and 5, whereas blastobasids that have been examined have a space about the diameter of stemma 4 between 4 and 5.

Biology and Ecology: According to Powell (1976a) scythridids usually live in slight webs, feeding externally on buds or leaves. Forbes (1923) reports *S. graminvorella* Braun to be a leaf miner in grasses, and *S. magnatella* Busck folds a leaf of *Epilobium* together to form a shelter. Several eastern species feed on Asteraceae, but the western ones have been reported from diverse families, including Solanaceae, Apiaceae, Cactaceae, Hydrophyllaceae, Onagraceae, Polygonaceae, Lamiaceae, Fabaceae, and Asteraceae. Powell (1976a) has described an unusual California dune-dwelling, flightless scythridid, *Areniscythris brachypteris* Powell, whose larvae are quite elongate, have short setae, and live in sand-covered silken tubes attached to buried green parts of diverse plant species at the edge of active, moving sand dunes.

Description: Mature larvae small, ranging from 5–15 mm. Body usually rather elongate, but occasionally rather stubby. Secondary setae usually conspicuous, especially on proleg base, and usually present in most other setal groups. Body sometimes striped or with pigmented patches. Spiracles small and circular, those on T1 and A8 about twice as large as the others.

Head: Hypognathous to semiprognathous, frequently mottled. Frontoclypeus longer than wide and extending one-half to four-fifths to epicranial notch. Ecdysial lines bent or straight, and usually meeting near epicranial notch. Six stemmata, 1, 2, and 3 usually separated from each other by width of a stemma, 3, 4, 5 close together, 6 caudad of 4 and 5. A submental pit present or absent.

Thorax: T1 shield variably distinct, with standard complement of primary setae that may be difficult to distinguish from secondaries. L group trisetose, usually with secondaries present. SV group on T1 usually with three or more setae. T2 and T3 with D1 and D2 usually on same pinaculum, SD1 and SD2 on the same pinaculum, and L1 and L2 on the same pinaculum. Secondary setae usually present. L3 isolated and almost never with a secondary seta. SV seta group on T2 and T3 usually with 3 or more setae.

Abdomen: D1 and D2 on all segments, usually accompanied by secondary setae. SD setae located above the spiracle and in a pinaculum ring on A1–7, with SD1 long and SD2 minute or absent (with only the seta base adjacent to SD1). L1 below L2 and on the same pinaculum anteroventrad of the spiracle, with secondary setae never present; an extra L seta nearly directly above the proleg and sometimes with a second seta. Caudad to this group is the L3 group of several setae. Several to many setae on proleg base, but always more than three. Setal group V with one or two setae. SD1 on A8 usually very fine and located on a small, irregular pinacular area. A9 with D2 more mesad than D1, SD1 weak and often very caudad, L setae usually nearly vertically aligned, SV group with several setae, V with one or two setae.

A3–6 proleg usually with a peglike, cylindrical planta bearing triordinal (but sometimes uni- or biordinal) crochets in a circle that may be weaker laterally.

A10 with a standard complement of setae, sometimes with some secondary setae. A10 prolegs normal to somewhat caudo-projecting; crochets uni-, bi- or triordinal and transverse to the meson in a partial circle.

Comments: The family contains 35 described species in the genus *Scythris,* 1 in *Areniscythris,* and upwards of 100 undescribed species (Powell, 1976a); most of the larvae are unknown. Jean-Francois Landry, Biosystematics Research

Institute, Ottawa, is revising them. With the exception of *Areniscythris brachypteris* and possibly other dune-adapted species in which the setae are reduced and the SD pinaculum ring is absent or very weak, there are several overall similarities that are quite consistent in the species examined, namely:

1. Secondary setae present.
2. A1–7 with SD pinaculum ring.
3. L1 below L2 and on the same pinaculum, with secondary setae never present in this group.
4. Stemmata 3, 4, 5 close together.
5. The extra L group of setae located anterior to the L3 group on the abdomen.
6. A3–6 crochets in a circle that is somewhat weaker laterally.

For a gelechioid family this is rather consistent, but only eight species have been examined. Scythridids are not reported to cause economic damage.

Selected Bibliography

Forbes 1923.
Powell 1976a, 1980.

GELECHIIDAE (GELECHIOIDEA)

Frederick W. Stehr, *Michigan State University*

The Gelechiids

Figures 26.77-26.89

Relationships and Diagnosis: This is a large and diverse family, and as indicated in the key, some gelechiids are not separable from cosmopterigids and oecophorids (and possibly some others) in the current state of our knowledge. Perhaps they never will be, but according to Don Weisman of the USDA Systematic Entomology Laboratory, there are certain characters and combinations that help to define gelechiid larvae, even though species in other families may have some of the same characters.

One of these is the presence of two SV setae on A1—gelechiids almost never have one or three SV's on A1, but species in other families may also have two. Another trait common in the gelechiids is the hairlike condition of SD1 on A9 (fig. 26.83c), a character also common in blastobasids and oecophorids, among others, but rare in cosmopterigids. Another is the presence of an anal fork—gelechiids are the only Gelechioidea that have an anal fork (fig. 26.89c, d) but relatively few species possess one. Among other families that may resemble gelechiids, tortricids commonly have an anal fork, but SD1 on A9 is tapered and never hairlike. In addition, SD1 on A8 is usually more anterior to the spiracle on tortricids and more dorsal in gelechiids (see couplet 108 in the family key). Blastobasids always have a submental pit (rare in gelechiids). In oecophorids the D1 seta on A9 is usually forward of the other setae so the A9 setae tend not to line up in a vertical row (in gelechiids D1 is usually more in line with the other setae, fig. 26.78). On the head, head seta L1 is usually farther from head seta A3 than A3 is from A2 (family key, fig. 129); it is usually closer or equidistant in other gelechioid families).

Biology and Ecology: As Hodges has indicated (1966), the larval habits of a family this large are varied, rivalled in diversity among gelechioid families only by the Oecophoridae. Species include ones that are leaf or needle miners for part or all of their development, leaf rollers or leaf tiers, external leaf feeders, stem borers, gall makers, feeders in flower and seed heads, twig, bark or seed feeders, communal feeders on vegetation beneath loose webs, and a few scavengers. The biology of perhaps half the species is unknown, but that of some of the economically important ones such as the pink bollworm is relatively well known.

Description: Mature larvae small to medium, most ranging from small leaf miners to 10–15 mm. A normal complement of primary setae present but sometimes reduced in size or difficult to see on leaf miners and seed or gall inhabitors. Secondary setae usually absent, but present on A3–6 prolegs and/or A10 in a few species such as *Dichomeris marginella* (Fabricius) (fig. 26.89) and the peach twig borer, *Anarsia lineatella* Zeller (fig. 26.86). Integument usually smooth, but sometimes granulated. Spiracles circular, usually small and distinct, T1 and A8 spiracles usually slightly larger, but up to 3× larger when A1–7 spiracles are minute.

Head: Semihypognathous to prognathous (leaf miners), usually heavily pigmented and smooth. Frontoclypeus longer than wide, usually extending one-half to four-fifths to epicranial notch, but all the way in some prognathous species. Ecdysial lines usually distinct, close to adfrontal sutures anteriorly, angling outward posteriorly, usually meeting at epicranial notch. Six stemmata arranged in a normal arc, with 6 caudad. Head seta L1 usually farther from head seta A3 than A3 is from head seta A2.

Thorax: T1 shield variably pigmented, but usually distinctly so; a standard complement of setae present. L group on T1 trisetose, the middle seta usually distinctly lower. SV group with two setae on T1, and one on T2 and T3. T2 and T3 with setae normally arranged, with L1 closer to L2 than L3.

Abdomen: D1 and D2 quite wide of the midline, with D2's usually farther away. SD1 on A1–7 dorsad to somewhat anterodorsad of spiracle. L1 and L2 below spiracle and close together. L3 more or less above caudal margin of proleg. SV group almost invariably bisetose on A1, trisetose on A2–6, bisetose on A7, and usually unisetose on A8 (rarely bisetose). SD1 on A8 usually dorsad or anterodorsad to spiracle, rarely anterior as on *Dichomeris marginella.* A9 with D1 below and usually lined up with D2 and SD1, which may be hairlike; L1 and L2 close together on the same pinaculum, with L3 close to distant or absent; SV rarely absent.

A3–6 prolegs short to 2× as long as width of planta at crochets; crochets in a uni-, partially bi- or biordinal circle or penellipse, rarely reduced in number, or to mere nubs (*Sitotroga cerealella* (Olivier)), or almost to two transverse rows (*Exoteleia* sp.), or absent (*Metzneria* sp. and *Isophrictus* sp.).

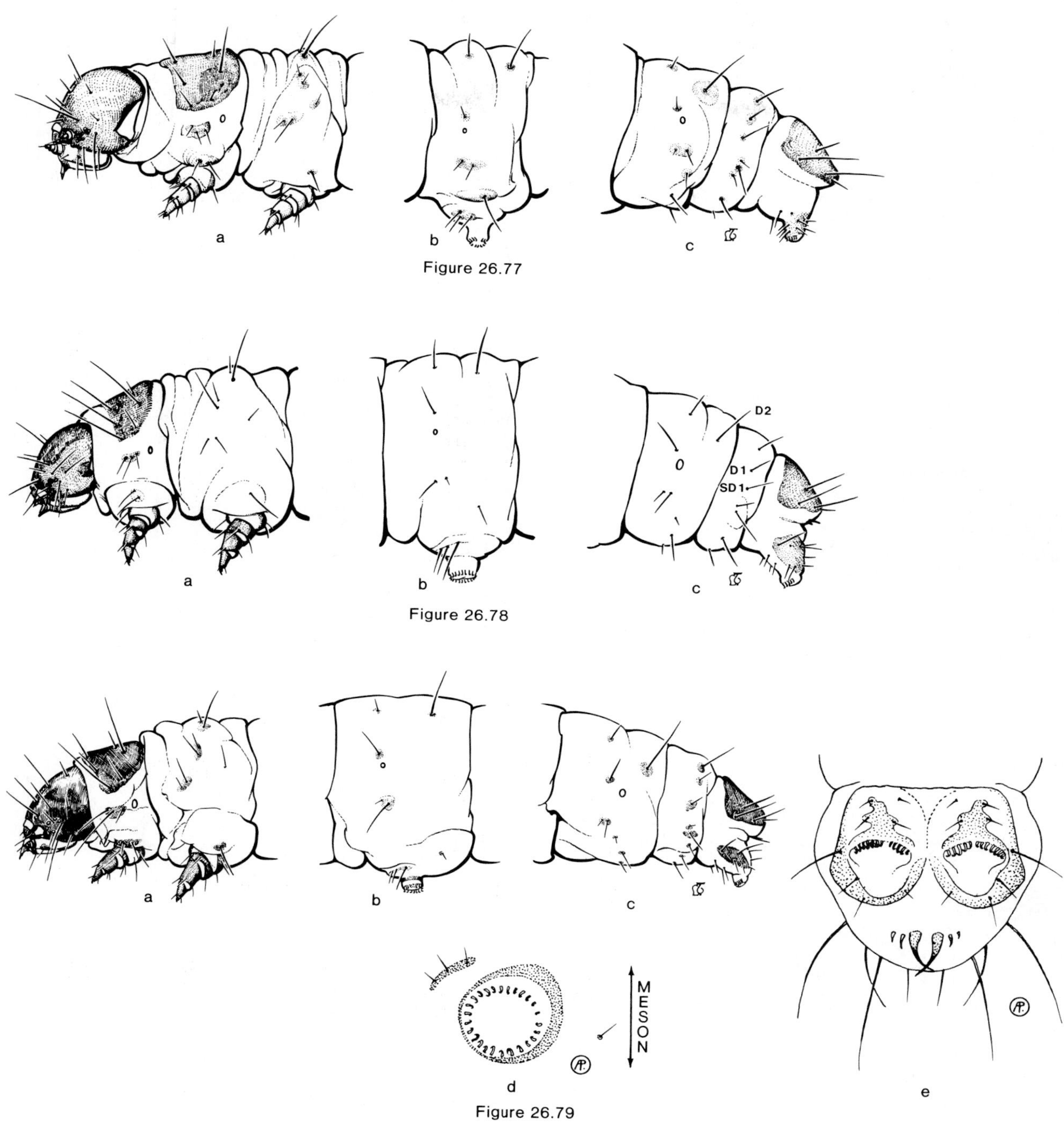

Figure 26.77a-c. Gelechiidae, Anomologinae. *Monochroa fragariae* (Busck), strawberry crown miner. f.g.l. 10± mm. a. head, T1, T2; b. A3; c. A8-A10. Woodburn, Ore., 29 October 1926. Ex. strawberry crowns. U.S. Natl. Mus.

Figure 26.78a-c. Gelechiidae, Gelechiinae. *Recurvaria nanella* (Denis and Schiffermüller), lesser budmoth. f.g.l. 10± mm. a. head, T1, T2; b. A3; c. A8-A10. Dalles Co., Ore., 8 May 1981, cherry foliage. U.S. Natl. Mus.

Figure 26.79a-e. Gelechiidae, Gelechiinae. *Coleotechnites milleri* (Busck), lodgepole needleminer. f.g.l. 7± mm. Yosemite Nat. Park, Calif., 25 September 1914. U.S. Natl. Mus. a. head, T1, T2; b. A3; c. A8-A10; d. planta A6 proleg; e. venter of A10. Larva yellow to orange with darker red line along dorso-meson. Anal comb with 2 prominent spines (figs. d and e from Peterson 1948).

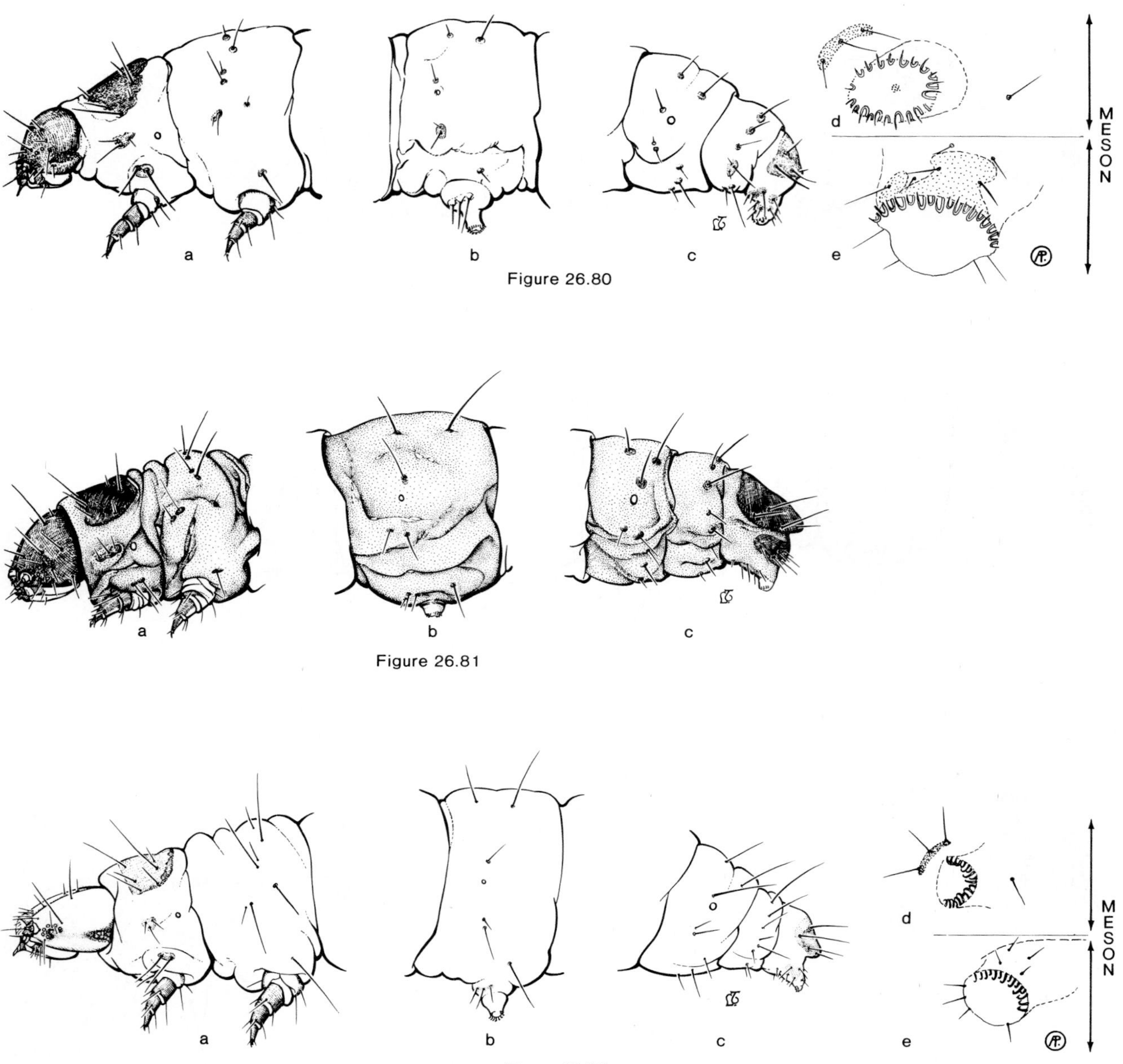

Figure 26.80

Figure 26.81

Figure 26.82

Figure 26.80a-e. Gelechiidae, Gelechiinae. *Phthorimaea operculella* (Zeller), potato tuberworm. f.g.l. 12± mm. N.Y., in potatoes from Chile, 26 July 1923. U.S. Natl. Mus. a. head, T1, T2; b. A3; c. A8-A10; d. A6 proleg; e. A10 proleg. Slightly fusiform, creamy white, greenish or pinkish-white with fuscous head, near-black cervical shield and yellowish anal plate. No anal comb. In the field burrows in leaves, petioles or stems of potato, tobacco, tomato, eggplant and solanaceous weeds. May also riddle tubers in the field or storage with slender, silk-lined burrows filled with excrement (figs. d, e from Peterson 1948).

Figure 26.81a-c. Gelechiidae, Gelechiinae. *Exoteleia pinifoliella* (Chambers), pine needleminer. f.g.l. 4± mm. a. head, T1, T2; b. A3; c. A8-A10; Bent Cr. Expt. For., N.C., May-June 1953, pitch pine. U.S. Natl. Mus.

Figure 26.82a-e. Gelechiidae, Gelechiinae. *Keiferia lycopersicella* (Walsingham), tomato pinworm. f.g.l. 8± mm. a. head, T1, T2; b. A3; c. A8-A10; d. A6 planta; e. A10 proleg. Kirkwood, Mo., 9 September 1941, tomato leaves in greenhouse. Yellowish, gray or green, usually with purple spots. Head yellowish with dark band from stemmata to caudal margin. Yellowish cervical shield with dark caudal margin. Setae light, arising from inconspicuous pinacula. Produces serpentine mines or blotches in tomato leaves in greenhouses or outdoors. Leaves may be folded and held together (figs. d, e from Peterson 1948).

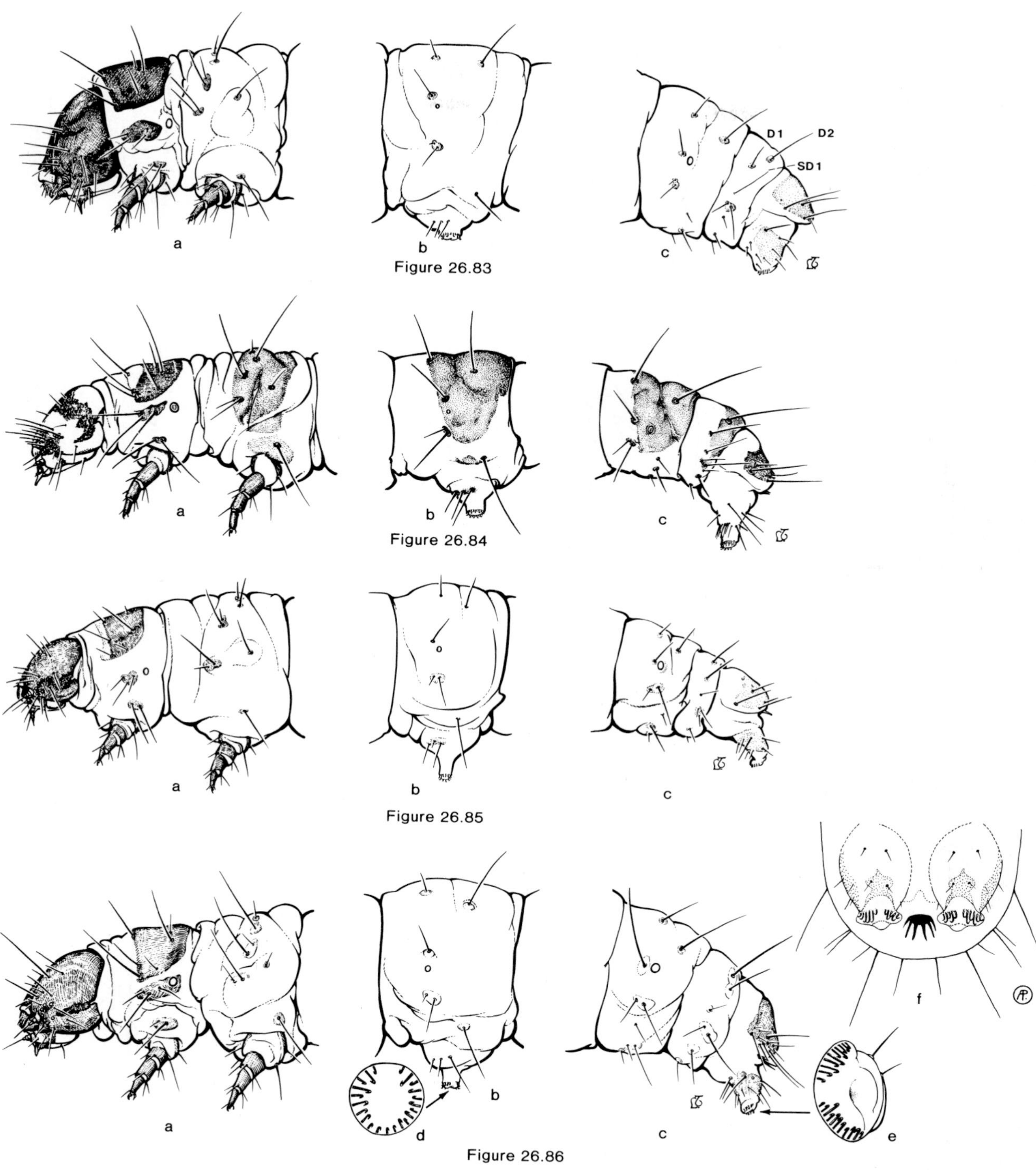

Figure 26.83a-c. Gelechiidae, Gelechiinae. *Aroga websteri* Clarke, sagebrush defoliator. a. head, T1, T2; b. A3; c. A8-A10. Jerome, Id., 18 June 1953, sagebrush. L.W. Orr. f.g.l. 12± mm. U.S. Natl. Mus.

Figure 26.84a-c. Gelechiidae, Gelechiinae. *Fascista cercerisella* (Chambers), redbud leafroller. a. head, T1, T2; b. A3; c. A8-A10. f.g.l. 15± mm. Near Marlboro, Md., 30 June 1928, on *Cercis canadensis,* C. Heinrich. U.S. Natl. Mus.

Figure 26.85a-c. Gelechiidae, Gelechiinae. *Stegasta bosqueella* (Chambers), rednecked peanutworm. f.g.l. 8± mm. a. head, T1, T2; b. A3; c. A8-A10. Angleton, Tex., 2 September 1944, in stems, buds, webbed leaves of peanut. U.S. Natl. Mus.

Figure 26.86a-f. Gelchiidae, Chelariinae. *Anarsia lineatella* Zeller, peach twig borer. a. head, T1, T2; b. A3; c. A8-A10; d. A3 crochets; e. A10 crochets; f. venter of A10. f.g.l. 10± mm. Stockton, Calif., September 1918, on peach. U.S. Natl. Mus. Light to reddish brown with near-black head, cervical shield and anal plate. Anal comb 6± prongs. Crochets on anal proleg broken into 2 groups. Infests tender twigs or fruit of peach, plum, apricot, almond (fig. f from Peterson 1948).

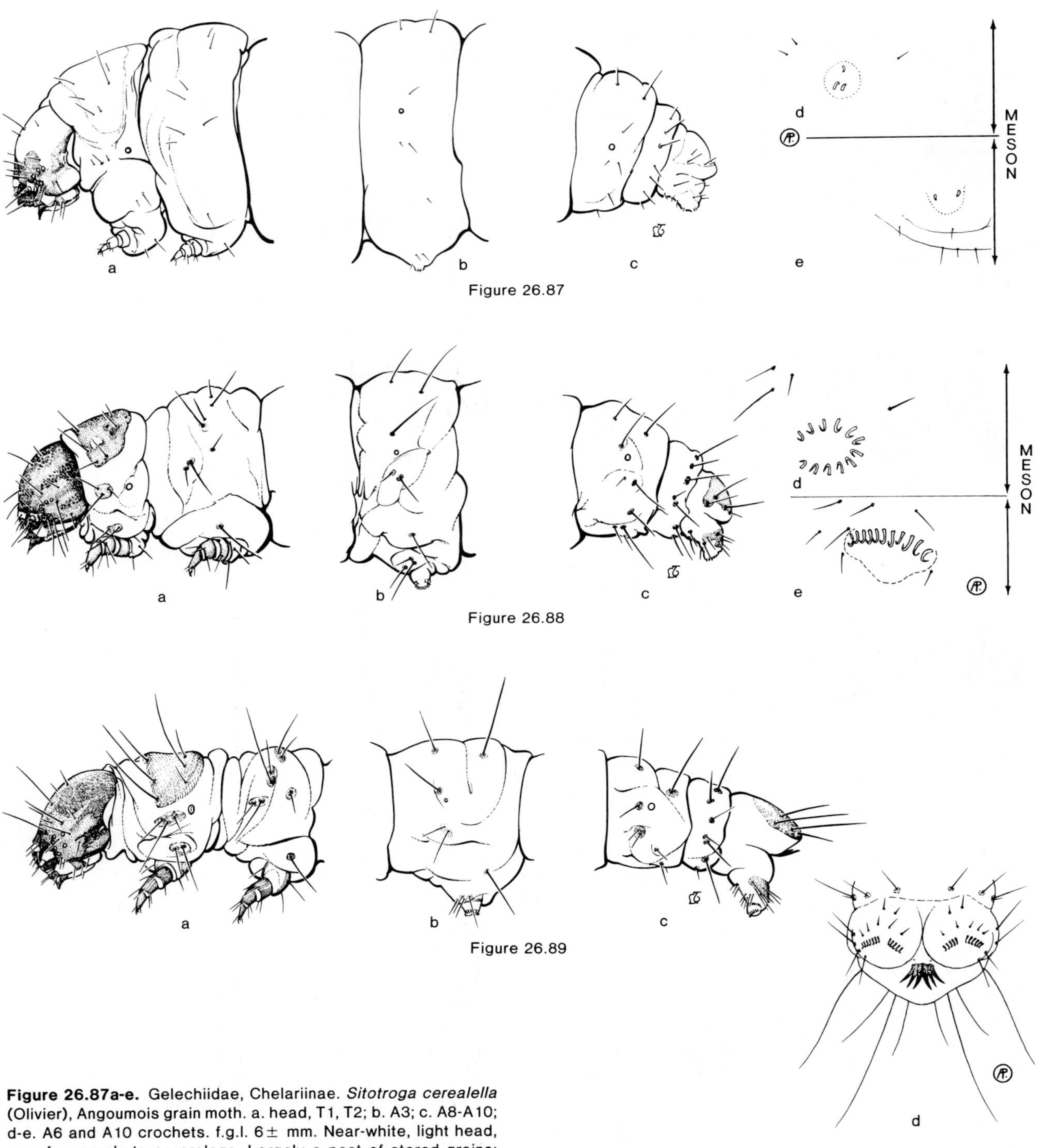

Figure 26.87a-e. Gelechiidae, Chelariinae. *Sitotroga cerealella* (Olivier), Angoumois grain moth. a. head, T1, T2; b. A3; c. A8-A10; d-e. A6 and A10 crochets. f.g.l. 6± mm. Near-white, light head, very few crochets on prolegs. Largely a pest of stored grains; occasionally a pest of corn and grains in the field in more southern areas. U.S. Natl. Mus. (figs. d, e from Peterson 1948).

Figure 26.88a-e. Gelechiidae, Chelariinae. *Pectinophora gossypiella* (Saunders), pink bollworm. a. head, T1, T2; b. A3; c. A8-A10; d-e. A6 and A10 crochets. f.g.l. 11± mm. Tlahualila, Durango, Mexico, 20 September 1918, cotton bolls. U.S. Natl. Mus. Pink with yellowish brown head, cervical shield and anal plate. No anal fork. Bores into squares, bolls, or seeds of cotton, also attacks other malvaceous plants. Busck 1917, Heinrich 1921 (figs. d-e from Peterson 1948).

Figure 26.89a-d. Glechiidae, Dichomeridinae. *Dichomeris marginella* (Fabricius), juniper webworm. a. head, T1, T2; b. A3; c. A8-A10; d. venter A10. f.g.l. 15± mm. Biltmore, N.C., 27 April 1927, F. C. Craighead. U.S. Natl. Mus. Light brown with median reddish-brown line paralleled by 2 wider dark brown stripes. Most setae near-white, arising from pigmented pinacula. Anal comb 6± prongs. Feeds on and webs together terminal shoots of *Juniperus* varieties (fig. d from Peterson 1948).

A10 usually with standard complement of setae (secondaries present in a few species, such as the peach twig borer, *Anarsia lineatella*). Prolegs normal, with transverse row of uni- or biordinal crochets that may be divided by a gap, but rarely reduced or absent. Anal fork sometimes present.

Comments: In America north of Mexico, the Gelechiidae is the largest family of the Gelechioidea, with some 630 species in 91 genera in six subfamilies. Most are of no economic importance, but 30–40 species are occasional pests, and a few are continually troublesome.

Probably the most important gelechiid to agriculture is the pink bollworm, *Pectinophora gossypiella* (Saunders) (fig. 26.88), a primary pest of cotton, that is distributed nearly worldwide. Another important pest is the potato tuberworm, *Phthorimaea operculella* (Zeller) (fig. 26.80), a borer in stems and leaves of solanaceous plants (potato, tomato, eggplant, tobacco, and weeds), and also a borer in potato tubers in the field or in storage. The lodgepole needle miner, *Coleotechnites milleri* (Busck) (fig. 26.79), has a long history of serious damage to mature lodgepole pine and other hosts on the west slope of the Sierra Nevada in California, and other species of *Coleotechnites* can damage other conifers. *Dichomeris marginella,* the juniper webworm (fig. 26.89), is an introduced European pest on *Juniperus* spp., including red cedar; it mines needles as young larvae and feeds externally while communally webbing foliage together in later instars. Other troublesome *Dichomeris* species are *D. acuminatus* (Staudinger), the alfalfa leaftier, and *D. ligulella* Hübner, the palmerworm, a pest on apple and oak. Several species in the genus *Exoteleia,* including the pine needleminer, *E. pinifoliella* (Chambers) (fig. 26.81), and *E. nepheos* Freeman, the pine candle moth, attack *Pinus* spp., especially under plantation or ornamental conditions, causing damage to both needles and buds, resulting in stunted or bushy growth.

Furniss and Carolin (1977) and U.S.D.A. (1985) summarize the life cycles of important forest species. *Aroga websteri* Clarke, (fig. 26.83), the sagebrush defoliator, is a sporadic and sometimes lethal pest on western ranges. *Anarsia lineatella,* the peach twig borer (fig. 26.86), damages twigs and fruit in backyard, neglected and unsprayed orchards. *Keiferia lycopersicella* (Walsingham), the tomato pinworm (fig. 26.81), is a leaf miner or drills small holes in buds, fruits, or stems in the field or greenhouse. Another European import, the Angoumois grain moth, *Sitotroga cerealella* (fig. 26.87), is a pest of stored grain throughout North America. The larvae feed within whole kernels and are not a problem in milled grain.

Selected Bibliography

Bailey and Kok 1982.
Burdick and Powell 1960.
Busck 1917.
Capps 1946, 1958.
Carroll et al. 1979.
Finnegan 1965.
Furniss and Carolin 1977.
Hain and Wallner 1973.
Heinrich 1921a,b.
Hodges 1966, 1978 (Gelechioidea discussion).
Hodges et al. 1983.
Khare and Mills 1968.
McLeod 1969.
Powell and Mackie 1966.
Stark 1954.
U.S.D.A. 1985.
Valley and Wheeler 1976.
Wilson 1974.

COPROMORPHIDAE (COPROMORPHOIDEA)[5]

J. B. Heppner, ***Florida State Collection of Arthropods***

The Copromorphids, Tropical Fruit Worms

Figures 26.90, 26.91

Relationships and Diagnosis: The Copromorphidae, Alucitidae, Carposinidae, Epermeniidae, and Glyphipterigidae make up the Copromorphoidea, the superfamily distinguished by the larvae that are borers or feed externally on flowers and fruits (sometimes leaves), as far as is known. The L group prespiracular setae on the prothorax are usually bisetose. The L group appears to be secondarily bisetose since one compromorphid (*Isonomeutis*) has a trisetose L group and one New Zealand glyphipterigid appears to have a vestigial third seta (Dugdale, personal communication). The Epermeniidae and Glyphipterigidae have been recently added to the superfamily from their previous placement in Yponomeutoidea (Heppner 1977).

The North American copromorphids are *Lotisma trigonana* (Walsingham) (fig. 26.90) and *Ellabella* species (fig. 26.91). Larvae resemble those of Carposinidae. The larvae of *Lotisma* and *Ellabella* may be distinguished by the absence of D1 on A9 (figs. 26.90, 26.91), a character that varies in other genera. The submentum of *Lotisma* also has paired flaplike protrusions that are unforked protrusions in Carposinidae and less developed in Alucitidae although forked; submental setae occur on small tubercles.

Biology and Ecology: *Lotisma* larvae have been reared as external feeders or tunnelers of flowers and fruits of *Arbutus menziesii, Vaccinium ovatum, Gaultheria shallon,* and *Arctostaphylos* spp. (Ericaceae), but little other biological information is known. The larvae of one *Ellabella* species are leaf tiers of *Mahonia pinnata* (Berberidaceae). Pupae do not have spines on the abdominal dorsum and are not protruded at adult eclosion. *Isonomeutis amauropa* Meyrick, from New Zealand, bores in bark and sooty mold of bark (Dugdale, personal communication).

Description: Mature larvae small, approximately 10 mm (some tropical larvae may approach 45 mm). Primary setae only.

5. The comments and additions of John DeBenedictis are greatly appreciated.

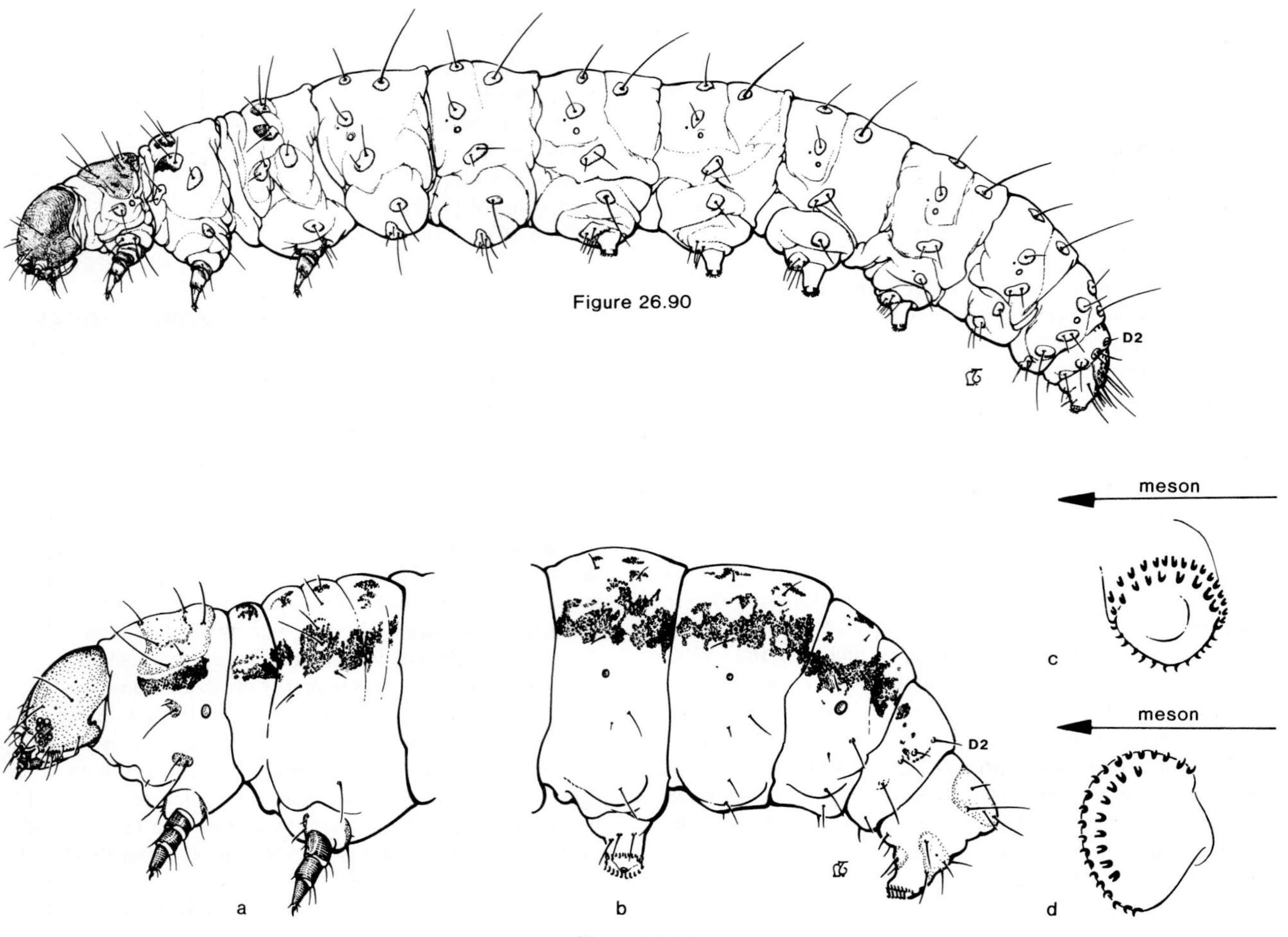

Figure 26.90. Copromorphidae. *Lotisma trigonana* (Walsingham). f.g.l. 10± mm. Red Bay, Prince of Wales Is., Alaska, 25 July 1946, huckleberries, R. L. Furniss. U.S. Natl. Mus. (FWS).

Figure 26.91a-d. Copromorphidae. *Ellabella bayensis* Heppner. a. head, T1, T2; b. A6-A10; c. A6 crochets; d. A10 crochets. f.g.l. 12± mm. San Bruno Mountain, San Mateo Co., Calif., 5-14 May 1983. J. A. DeBenedictis. Univ. Calif., Berkeley (FWS).

Head: Hypognathous. Frontoclypeus higher than wide, extending two-thirds to three-fourths to epicranial notch. Six stemmata; in *Lotisma* 1–5 in evenly spaced semicircle and 6 somewhat separated from 1–5; in *Ellabella* 5 displaced toward the mouth. Head setae A1, A2 and A3 as obtuse triangle with A2 most distant from stemmata. Submentum of *Lotisma* with pair of posterior flaplike protrusions (lacking in *Ellabella*); submental setae borne on tubercles anteriorly.

Thorax: T1 shield evident. Pinacula usually inconspicuous. T1 L group usually bisetose. Spiracle somewhat enlarged on T1.

Abdomen: L2 anteroventrad of L1 on A1–8; close together. SD1 anterodorsad to spiracle on A1–8. D1 absent on A9 (minute and on SD1 pinaculum in *Isonomeutis*). Spiracle on A8 somewhat enlarged; sometimes on tubercle.

Prolegs slender, somewhat long, or shorter; present on A3–6 and A10. Crochets in uniordinal mesal penellipse or circle (uniordinal or partly biordinal).

Comments: This small and relatively unstudied family is almost exclusively tropical, centered in Indo-Australia and Oceania. Only *Lotisma trigonana* and four species of *Ellabella* are known for North America, ranging from British Columbia to northern Mexico. Another species of *Lotisma* is known from Costa Rica but no *Ellabella* are known south of the United States. *Lotisma* and *Ellabella* have until recently been placed in Glyphipterigidae (*sensu* Meyrick), but have proved to belong to the copromorphids (Heppner 1978, 1984). The systematics of the family has been restricted to isolated species descriptions and no comprehensive revision has been attempted. No species have been reported to be of economic importance in North America, but tip damage to figs has been reported in India (Fletcher 1933).

Selected Bibliography

Busck 1909.
DeBenedictis 1984.
Fletcher 1933.
Heppner 1977, 1978, 1984, 1986.

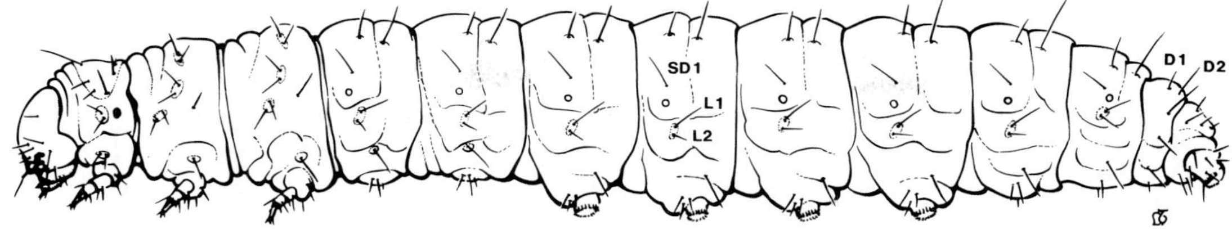

Figure 26.92. Alucitidae. *Alucita* sp. Ocampo, Chihuahua, Mexico, 7 May 1961, in *Lonicera* flowers. f.g.l. 8± mm. El Paso 58127. U.S. Natl. Mus. (FWS).

ALUCITIDAE (COPROMORPHOIDEA)

J. B. Heppner, ***Florida State Collection of Arthropods***

The Alucitids, Many-Plumed Moths

Figure 26.92

Relationships and Diagnosis: Alucitid moths usually have both fore and hind wings split into six or seven lobes, and thus are superficially similar to Pterophoridae which usually have two to four lobes, but they are phylogenetically related to Copromorphidae. Larvae conform to the characteristics of the superfamily Copromorphoidea.

Alucitid larvae are similar to both Copromorphidae and Carposinidae in having submental protrusions like the former, only less developed, and with seta L2 anteroventrad of L1 on the abdominal segments (fig. 26.92) rather than basically dorsad of L1 as in Carposinidae. The larvae are distinguished from copromorphids by possessing seta D1 on segment A9 (absent in copromorphids).

Biology and Ecology: Little is known of the biology of alucitids. European species are known to bore in flowers and buds, sometimes causing galls. *Alucita hexadactyla* Linnaeus bores in *Lonicera* (Ford 1954); an exotic species feeds on coffee (Viette 1958).

Description: Mature larvae small, not exceeding 10 mm. Integument rugose; primary setae only.

Head: Hypognathous. Frontoclypeus higher than wide, extending three-fourths to epicranial notch. Six stemmata; 1–5 in evenly spaced semicircle, with 6 closely dorsad of 5. Head setae A1, A2 and A3 in nearly right triangle with A3 almost between A2 and stemmata. Submentum with paired posterior flaplike protrusions, less forked than in Copromorphidae.

Thorax: T1 shield evident. Pinacula not conspicuous. T1 L group bisetose. T1 spiracle enlarged.

Abdomen: L2 anteroventrad of L1 on A1–8 and close together. SD1 dorsad of spiracle on A1–8. D1 present on A9. Spiracle on A8 enlarged. Prolegs short, present on A3–6 and 10. Crochets few but large, in uniordinal circle.

Comments: The family is small with about four genera and 130 described species, distributed in all faunal regions. Most described species are Indo-Australian. There is one described North American species, *Alucita hexadactyla* L. that is also found in Europe, and one in Hawaii; there appear to be at least another two undescribed species in North America. One species has been reported to be economically damaging to coffee in Madagascar (Viette 1958). The name *Alucita* has been erroneously applied to Pterophoridae in the past.

Selected Bibliography

Ford 1954.
Viette 1958.
Zimmerman 1958, 1978.

CARPOSINIDAE (COPROMORPHOIDEA)

J. B. Heppner, ***Florida State Collection of Arthropods***

The Carposinids, Fruit Worms

Figure 26.93

Relationships and Diagnosis: The carposinids are very similar to known copromorphid larvae and to alucitids. The North American species are largely unknown except for the economic species that have had some biological investigation.

The larvae may be distinguished by prolegs shorter than in copromorphids, but longer than in alucitids; in possessing D1 on segment A9; and in having flap-like protrusions* without a posterior fork on the submentum (fig. 26.93b).

Biology and Ecology: The few known larvae are borers in fruits, stems, stem galls, or bark; some are leaf miners. Hosts are known in Campanulaceae, Epacridaceae, Ericaceae, Fagaceae, Myrtaceae, and Rosaceae, with major economic species on apple and peach. The stem galls apparently caused by some carposinid larvae appear to actually be the result of secondary invasions of host cells by an ascomycete

*These "flap-like protrusions" vary considerably in the taxa examined, ranging from an apparently medial submental structure on USDA specimens identified as the copromorphid *Lotisma trigonana,* to conspicuous paired flaplike protrusions arising from the stipes on USDA specimens intercepted at Detroit in 1929 on *Crataegus* and identified by Heinrich as the carposinid *Carposina* sp. (*persicae?*) (fig. 26.93b). The protrusions on both specimens of this carposinid have internal tissues that appear to be muscular, although they could also be sensory or glandular. Their function may be related to the spinneret since on both specimens the spinneret projects caudad and the tip appears capable of nesting between the flap-like protrusions. SEM work and behavioral observations on living larvae (if possible) seem necessary for final determination of structure and function. [FWS]

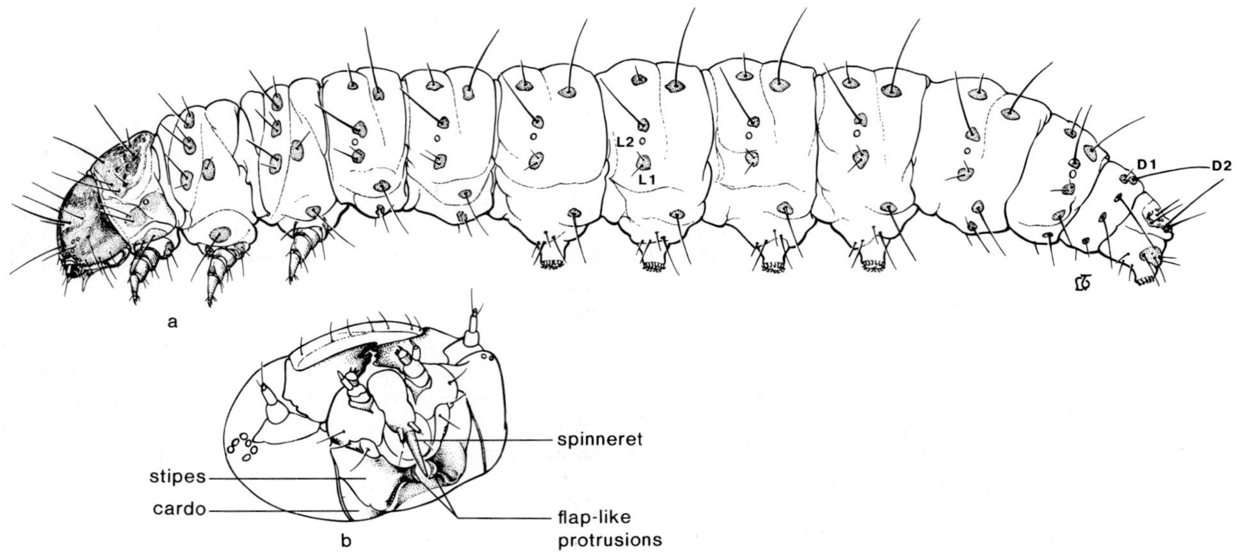

Figure 26.93a–b. Carposinidae. *Bondia comonana* (Kearfott). f.g.l. 11± mm. a. larva; b. mouthparts, ventral, showing flap-like protrusions. Walla Walla, Wash., 15 July 1944, ex. *Prunus virginiana.* U.S. Natl. Mus. (FWS).

fungus, *Plowrightia morbosa*, known as the black knot disease of North American *Prunus*, which follows the damage produced by the larva. This information has been established for *Bondia comonana* (Kearfott) (Davis 1969). A similar biology is known for a Japanese species. Pupae are without dorsal abdominal spination.

Description: Mature larvae small, not exceeding 12 mm (some tropical species may approach 30 mm). Primary setae only.

Head: Hypognathous. Frontoclypeus somewhat higher than wide, extending somewhat more than one-half to epicranial notch. Six stemmata; 1–5 in semicircle with gap between 2 and 3, with 6 forming acute angle with 3–5. Head setae A1, A2 and A3 forming right angle with A2 most distant from stemmata. Submentum with pair of posterior flap-like protuberances (not forked as in Copromorphidae or Alucitidae).

Thorax: T1 shield evident. Pinacula evident, not conspicuous. T1 L group bisetose. Spiracle on T1 enlarged.

Abdomen: L2 anterodorsad of L1 on A1–8 and close together. SD1 basically dorsad to spiracle on A1–9. D1 present on A9. Spiracle on A8 enlarged.

Prolegs slender, shorter than in Copromorphidae; present on A3–6 and A10. Crochets as uniordinal circles.

Comments: The Carposinidae comprise a small family of approximately 24 genera and 250 species worldwide, with most species known from pantropical regions. North of Mexico there are 11 described species with at least two currently undescribed. Thirty-seven described species are known from Hawaii. The immature stages are poorly known except for the economic species: *Carposina niponensis* Walsingham on peach and apple in Japan and North America (Riley 1889, Hukusima 1953, Miyashita et al. 1955); *Carposina fernaldana* Busck on currant (Davis 1969); and *Bondia comonana* (Kearfott) (fig. 26.93) on *Prunus* limbs (Keifer 1943). Swatschek (1958) and Yano (1959) published information on the larvae of Palearctic species; Diakonoff (1954) provided a key to genera for adults.

Selected Bibliography

Davis 1969.
Diakonoff 1954.
Hukusima 1953.
Keifer 1943.
Miyashita et al. 1955.
Riley 1889.
Swatschek 1958.
Yano 1959.

EPERMENIIDAE (COPROMORPHOIDEA)

J. B. Heppner, *Florida State Collection of Arthropods*

The Epermeniids

Figure 26.94

Relationships and Diagnosis: The Epermeniidae are most closely related to Glyphipterigidae, less so to Carposinidae. Previous classifications considered them to be related to families in Yponomeutoidea, but this has been shown to be erroneous, especially when characters of the immature stages are considered (Heppner 1977).

Epermeniid larvae are distinguished by SD1 being nearly directly above the spiracle on segments A1–8, together with medium-length prolegs. The spiracles are somewhat enlarged and on small conelike protuberances, especially on T1 and A8, but less so than in Glyphipterigidae.

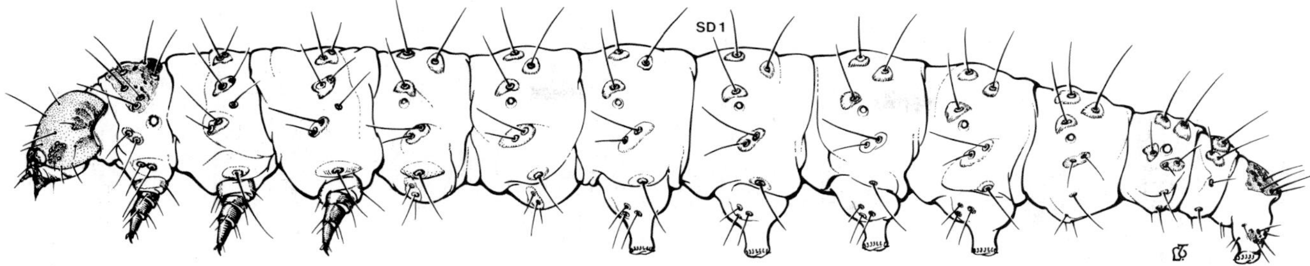

Figure 26.94. Epermeniidae. *Epermenia* sp. f.g.l. 9± mm. 3 mi NNE Soquell, Santa Cruz Co., Calif., El 130–400′, 222 April 1972, J. A. Powell. Univ. Calif., Berkeley (FWS).

Biology and Ecology: Only a few species have host associations and very little has been published on their biologies. Known species are borers of buds, fruits and seeds, and some are leaf miners with late instars going to external leaf feeding. Hosts have been recorded from Araliaceae, Ammiaceae, Loranthaceae, Olacaceae, Santalaceae, and Apiaceae (= Umbelliferae). *Epermenia cicutaella* Kearfott is the only North American species with some published description of the life history (Kearfott 1903). *Epermenia albapunctella* Busck is known from *Ligusticum* (Apiaceae)(MacKay 1972).

Description: Mature larvae small, not exceeding 10 mm. Primary setae only.

Head: Hypognathous. Frontoclypeus as wide as high, reaching somewhat beyond one-half to epicranial notch. Six stemmata in semicircle with gaps between 2 and 5, and 5 and 6, with 6 forming acute angle to 3–5. Head setae A1, A2 and A3 form a right triangle with A2 reduced and most distant from stemmata. Submentum with slight posterior unforked protuberances.

Thorax: T1 shield and pinacula evident. T1 L group bisetose. Spiracles enlarged and in part elevated on small protrusions.

Abdomen: L2 anteroventrad to L1 on A1–8. SD1 nearly dorsad to spiracle on A1–8. D1 present on A9. Spiracles on A8 enlarged more than others and somewhat elevated on protrusions.

Prolegs medium length, present on A3–6 and 10. Crochets in uniordinal circles.

Comments: This family is distributed worldwide, comprising about 10 genera and 80 described species. The world fauna for the adults has been revised by Gaedike in several papers, including one for the New World (Gaedike 1977). North of Mexico there are 11 described species. The reported trisetose T1 L group (Forbes 1923) may be erroneous (MacKay 1972).

Selected Bibliography

Gaedike 1966, 1977, 1978.
Heppner 1977.
Kearfott 1903.
MacKay 1972.
Murtfeldt 1900.

GLYPHIPTERIGIDAE (COPROMORPHOIDEA)

J. B. Heppner, ***Florida State Collection of Arthropods***

The Glyphipterigids, Sedge Moths

Figure 26.95

Relationships and Diagnosis: Glyphipterigids are most closely related to Epermeniidae and not to yponomeutoid families as previously thought (Heppner 1977). The superfamilies, however, appear to be closely related via Plutellidae and Glyphipterigidae.

Larvae are distinguished by the lack of distinct prolegs (usually without many or any crochets), and by the unusual protruded spiracles, especially on T1 and A8. Larval characteristics noted for this family in Peterson (1948) correspond only to Choreutidae.

Biology and Ecology: Most larvae of Glyphipterigidae are seed and stem borers, borers in leaf axils, rarely leaf miners. The majority of known hosts are in Juncaceae and Cyperaceae but several other plant families have been recorded as hosts, including Poaceae, Araceae, Piperaceae, Urticaceae, and Crassulaceae in various parts of the world.

Eggs are deposited singly on the host plant. Larvae of known species pupate within the larval chambers. Pupae are devoid of dorsal abdominal spination and are not protruded at adult eclosion. Pupae with elevated spiracles; without dorsal spines. Adults are diurnally active and most species often push their wings up with the hind legs (perhaps only males as a mating signal).

Description: Mature larvae small, usually not exceeding 8 mm (some tropical species should have larvae approaching 35 mm). Primary setae only.

Head: Hypognathous. Frontoclypeus higher than wide, reaching two-thirds to epicranial notch. Six stemmata in semicircle with gaps between 2 and 3, and 3 and 4. Head setae A1, A2, and A3 as an obtuse or right triangle with A3 closest to stemmata. Submentum without posterior protuberances.

Thorax: T1 shield and pinacula evident. T1 L group bisetose. Spiracles enlarged and on conelike protrusions, especially on T1.

Abdomen: L2 anterior to L1 on A1–8. SD1 anterodorsad to spiracle on A1–8. D1 present on A9 (absent in some

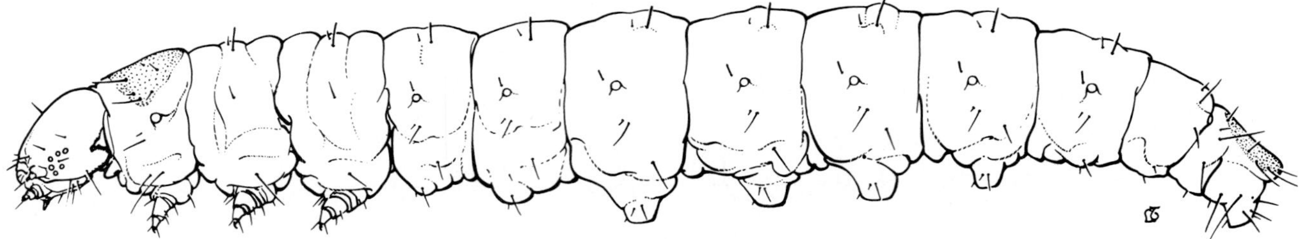

Figure 26.95. Glyphipterigidae. *Diploschizia habecki* Heppner. f.g.l. 5± mm. Highlands Hammock St. Park, Fla. 4 May 1974. J. Heppner. Note crochets absent. Fla. St. Coll. Arthropods, Gainesville. (FWS).

species). Spiracles on conelike protuberances, with A8 spiracle on an especially large cone. A10 usually with large sclerotized tergal plate, often concave, with stout posterior setae.

Prolegs vestigial on A3–6 and A10. Crochets absent or as sparse uniordinal lateral penellipse.

Comments: The family is relatively small, but with 360 described species it is the largest family of the superfamily. Thirty-eight species in 5 genera are known north of Mexico. Species can be locally common, usually in the vicinity of the host plant.

The biology of very few species is known. The only economic species are the cocksfoot moth of Europe, *Glyphipterix simpliciella* (Stephens), where the larvae bore in seeds of *Dactylis*, and the New Zealand cocksfoot stemborer, *Glyphipterix achyloessa* (Meyrick), a stem borer of *Dactylis* (Penman 1978). In North America the commonest species is *Diploschizia impigritella* (Clemens), which is a borer in the stem and leaf axils of *Cyperus*. *Diploschizia habecki* Heppner (fig. 26.95) can be locally common on *Rhynchospora* in Florida and Georgia (Heppner 1980). Two revisions are recent: Diakonoff (in preparation) for the Palearctic fauna and Heppner (1985) for the Nearctic fauna.

Selected Bibliography

Chopra 1925.
Empson 1956.
Heppner 1977, 1980, 1985.
Kodama 1961.
Penman 1978.

PLUTELLIDAE (YPONOMEUTOIDEA)

J. B. Heppner, ***Florida State Collection of Arthropods***

The Plutellids, Diamondback Moths

Figures 26.96, 26.97

Relationships and Diagnosis: The Plutellidae, Yponomeutidae, Argyresthiidae, Douglasiidae, Acrolepiidae, and Heliodinidae make up the Yponomeutoidea. The extra-limital Amphitheridae apparently are tineoid, not yponomeutoid. Plutellidae are closely related to Yponomeutidae and some authors include the group as a subfamily of the latter, but larval and adult characters warrant separation at the family level. Until a world generic revision has been completed, some genera of Yponomeutoidea will remain questionably placed to family; larval characters for the Plutellidae and other families of the superfamily are therefore based on the major genera of each family.

Larvae of plutellids are similar to Yponomeutidae but have uniserial crochets, noticeably elongate prolegs, and L1 and L2 separate on A9 (but not in the Palearctic genus *Prays*).

Biology and Ecology: Plutellid larvae are usually leaf feeders, often tying host plant leaves together with loose silk webbings and skeletonizing the leaves. Some species are borers in buds and petioles (Oku 1964). The main economic species are pests on various Brassicaceae; other recorded hosts are found in Caprifoliaceae, Celastraceae, Fagaceae, Juglandaceae, Oleaceae, Pinaceae, Rutaceae, and Ulmaceae.

Pupae usually do not have dorsal abdominal spination except among a few genera (e.g., *Homadaula*), and often have protruding spiracles. Pupation is usually on the host leaf, often in a delicate lacelike cocoon. Some larvae drop from the host and pupate in crevices. Adults are predominately nocturnal; some are crepuscular or even diurnal.

Description: Mature larvae small, usually less than 12 mm but some species should have larvae to 25 mm. Integument varing from yellow to green or whitish, sometimes marked with brown or orange striping. Primary setae only.

Head: Hypognathous. Frontoclypeus higher than wide, extending about two-thirds to epicranial notch. Six stemmata in a semicircle but with stemma 5 sometimes more ventrad. Head setae A1, A2 and A3 as obtuse triangle with A2 most distant from stemmata.

Thorax: T1 shield evident, usually variously colored, often brown. Pinacula evident and colored. T1 L group trisetose (bisetose in first instar larvae), rarely bisetose (*Rhabdocosma* and *Orthotaelia* from Palearctic). L1 and L2 approximate, distant from L3 on T2 and T3; rarely all subequal (*Rhabdocosma*).

Abdomen: L2 antero-ventrad of L1 on A1–8, usually approximate and on a single pinaculum (except for *Plutella xylostella,* fig. 26.97), distant from L3. SD1 antero-dorsad to spiracle on A1–8. D1 setae closer together than D2 setae on A9, but nearly subequal in many species. L1 and L2 separate on A9. Anal shield evident on A10.

Prolegs long and slender; present on A3–6 and A10. Crochets variable, mostly uniordinal but often somewhat biordinal, as uniserial circles or slightly incomplete circles, sometimes biserial in part of circle.

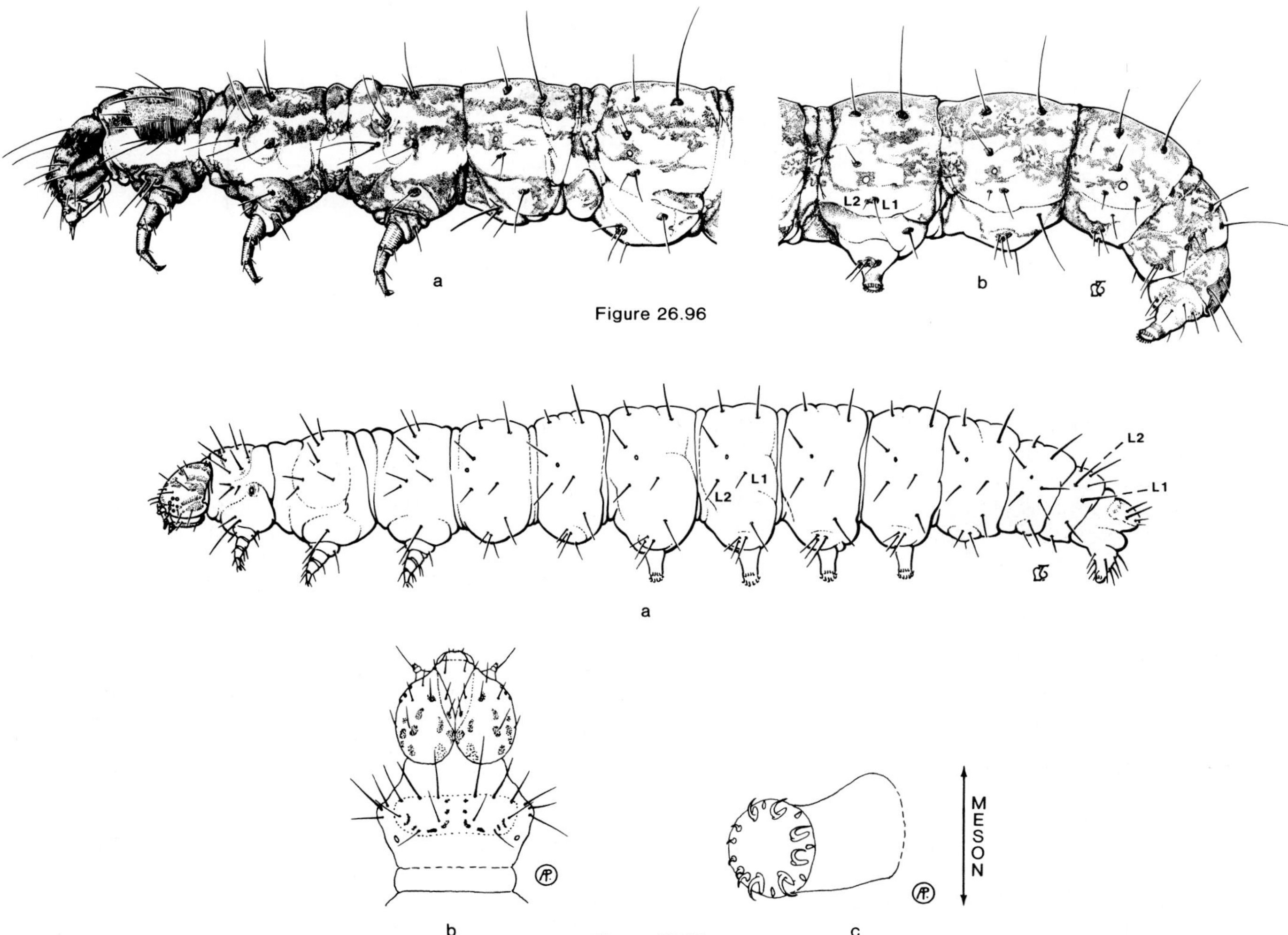

Figure 26.96a,b. Plutellidae. *Homadaula anisocentra* Meyrick, mimosa webworm. f.g.l. 15± mm. a. head, T1, T2, A1, A2; b. A6-A10. Atlanta, Ga., Ainsley Park area, on mimosa, 22 August 1947. U.S. Natl. Mus. Introduced from China. Can be a pest on mimosa (*Albizia*) and honeylocust (*Gleditsia*); larvae web foliage together (FWS).

Figure 26.97a-c. Plutellidae. *Plutella xylostella* (L.), diamondback moth. f.g.l. 9± mm. a. larva; b. dorsum of head, T1; c.A3 crochets. Pale green to cream, with fairly conspicuous erect, dark setae. Head yellowish mottled with dark brown spots. Feeds under leaves, riddling foliage. When disturbed, drops and suspends itself by silken thread. *Food plants:* cabbage, cauliflower, brussels sprouts, horseradish, kale, mustard, radish, rape, turnip, and watercress. Can be a pest in the greenhouse on a variety of plants (figs. b, c from Peterson 1948) (FWS).

Comments: The plutellids comprise approximately 390 species from all faunal regions, with nine genera and 54 species north of Mexico. The major economic species include the diamondback moth, *Plutella xylostella* (Linnaeus) (fig. 26.97, formerly *Plutella maculipennis* (*Curtis*); the horseradish webworm, *Plutella armoraciae* Busck; the European honeysuckle leafroller, *Ypsolopha dentella* (Fabricius) (formerly *Harpipteryx xylostella* of authors); and the mimosa webworm, *Homadaula anisocentra* Meyrick (fig. 26.96). The latter species was formerly in Glyphipterigidae but is neither in that family nor a choreutid and appears to belong in Plutellidae, although together with related genera of the plutellid subfamily Galacticinae, it is rather unusual. In Europe and Asia *Prays citri* Millière, the citrus flower moth, is a pest on citrus, and *Prays oleae* (Bern.), the olive kernel borer, is a pest of olives.

The Plutellidae are relatively closely related to Glyphipterigidae (Copromorphoidea) and Acrolepiidae (Yponomeutoidea), at least in pupal characters, and to the Yponomeutidae, so the separation of Copromorphoidea and Yponomeutoidea, which appears valid and useful, involves a narrow gap among various taxonomic characters.

Selected Bibliography

Clarke 1943.
Harcourt 1956, 1957.
Heppner and Dekle 1975.
Marsh 1913a.
Martouret et al. 1966.
Moriuti 1963, 1977.
Oku 1964.
Werner 1958.

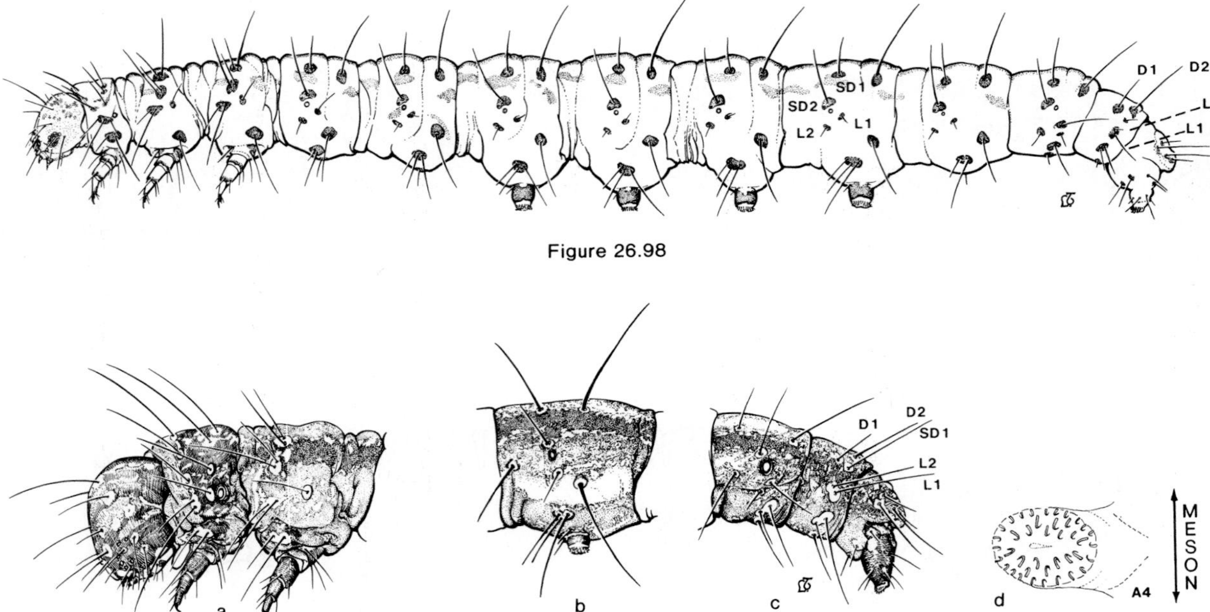

Figure 26.98. Yponomeutidae. *Yponomeuta multipunctella* (Clemens). f.g.l. 20± mm. Richmond, Va., 15 May 1962, on *Euonymus americanus.* U.S. Natl. Mus. (FWS).

Figure 26.99a-d. Yponomeutidae. *Atteva punctella* (Cramer). ailanthus webworm. f.g.l. 25–30± mm. a. head, T1, T2; b. A3; c. A8-A10. Slender, reddish to olive brown, with long, light setae arising from near-white pinacula. Fairly conspicuous, dark stripe bordered by white specks above the spiracles and lighter area below. Head, prothorax deeply pigmented. Ventral prolegs with multiserial or scattered crochets (fig. d). *Food plant:* tree of heaven, *Ailanthus.* Gregarious within thin web that encloses foliage. Most abundant in the South (fig. d from Peterson 1948) (FWS).

YPONOMEUTIDAE (YPONOMEUTOIDEA)

J. B. Heppner, *Florida State Collection of Arthropods*

The Yponomeutids, Ermine Moths

Figures 26.98, 26.99

Relationships and Diagnosis: The yponomeutids, which are closely related to Plutellidae and Argyresthiidae, often have been one subfamily of a broad concept of the family, encompassing groups now considered as distinct families.

Larvae are similar to Plutellidae but usually have multiserial crochets on shorter prolegs than are found among the plutellids. Both families, however, have anomalous genera that do not conform to some larval characters common to each family. The yponomeutids are distinctive, though, in all having L1 and L2 approximate on segment A9, the converse of what is found among plutellid larvae.

Biology and Ecology: Yponomeutid larvae are mostly leaf feeders, tying leaves together in loose webs as do the related Plutellidae. Only the anomalous genera (e.g., *Ocnerostoma*) are borers, and all known species of these genera have host plants among the Coniferales where the larvae mine the needles of Pinaceae and Cupressaceae. Host records of the external leaf feeders include the plant families Aceraceae, Betulaceae, Celastraceae, Corylaceae, Crassulaceae, Empetraceae, Ericaceae, Fagaceae, Lauraceae, Oleaceae, Rhamnaceae, Rosaceae, Salicaceae, Santalaceae, Saxifragaceae, and Simaroubaceae.

Eggs are deposited singly or in clusters. Larvae are sometimes communal when external feeders. Pupation is on the host plant in the larval chambers for borers, or in a delicately lacelike cocoon, often on a slender stalk, among the external feeders. The pupa is without dorsal abdominal spination and is not protruded upon adult eclosion. Adults are crepuscular and nocturnal.

Description: Mature larvae small, averaging 15 mm, with some species ranging to 30 mm. Integument of varying coloration, from yellow to green or whitish, often with markings or stripes of brown. Head capsule sometimes marked. Primary setae only.

Head: Hypognathous. Frontoclypeus higher than wide, extending two-thirds to epicranial notch. Six stemmata in oblique rectangular arrangement with 1 and 2 dorsad and 5 and 6 ventrad of seta S1; sometimes a gap between 2 and 3. Head setae A1, A2 and A3 as obtuse triangle with A2 most distant from stemmata.

Thorax: T1 shield and pinacula conspicuous, variously colored, often brown. T1 L group trisetose, rarely bisetose (*Ocnerostoma*). L1 and L2 approximate on T2 and T3, distant from L3; rarely with L2 and L3 absent (*Ocnerostoma*).

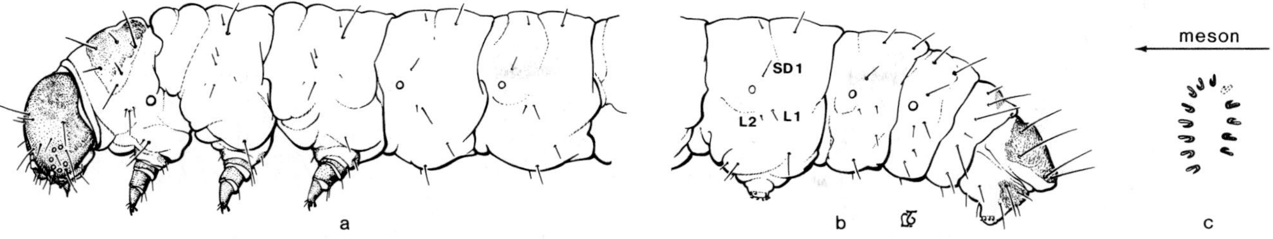

Figure 26.100

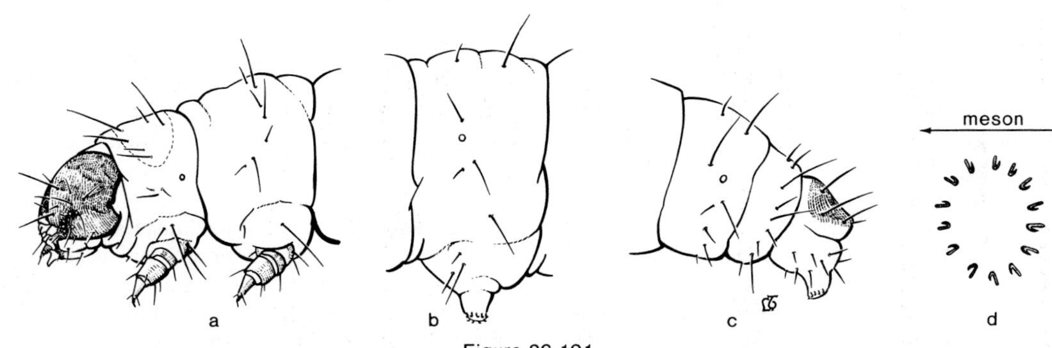

Figure 26.101

Figure 26.100a-c. Argyresthiidae. *Argyresthia thuiella* (Packard), arborvitae leaf miner. f.g.l. 4± mm. a. head, T1-A2; b. A6-A10; c. A6 crochets, open laterally, sometimes weak mesally. Falls Church, Va., 15 June 1952, ex. arborvitae. U.S. Natl. Mus. Mines the foliage of *Thuja* (white cedar, arborvitae) in eastern U.S. and Canada (FWS).

Figure 26.101a-d. Argyresthiidae. *Argyresthia conjugella* Zeller, apple fruit moth, f.g.l. 7± mm. a. head, T1; b. A3; c. A8-A10; d. A3 crochets, in complete circle. From Norway, ex. mountain ash N.Y. no. 172084, 31 August 1961. U.S. Natl. Mus. (FWS).

Abdomen: L2 anteroventrad of L1 on A1–8, distant from each other and on separate pinacula, and distant from L3. SD1 anterodorsad to spiracle on A1–8. D2's closer together than D1's on A9. L1 and L2 approximate on A9, on one pinaculum (L2 absent in *Ocnerostoma*). Anal shield evident on A10.

Prolegs normal (shorter than thoracic legs); present on A3–6 and A10. Crochets uniordinal in triserial to multiserial circles; rarely uniserial (*Ocnerostoma*) or biserial (*Swammerdamia*).

Comments: The Yponomeutidae encompass about 540 species worldwide, mainly in North and South Temperate regions but with a sizable tropical component. There are ten genera and 32 species known for America north of Mexico. The family does not have many economic species. In North America, the pine needle miners have some importance: the pine needle sheath miner *Zelleria haimbachi* Busck, and other *Zelleria* species; *Ocnerostoma piniariella* Zeller, on several *Pinus* species in western North America and Europe; and *Ocnerostoma strobivorum* Freeman, on eastern white pine. Common yponomeutids in North America are *Atteva punctella* (Cramer) (fig. 26.98, formerly *Atteva aurea* Fitch), on *Ailanthus* trees and other Simaroubaceae; and *Urodus parvula* (H. Edwards) in the southeastern states on *Persea* (Lauraceae); both genera are of tropical origin and are not typical yponomeutids. The true ermine moths are of the genus *Yponomeuta* (fig. 26.99). *Yponomeuta padellus* (L.) has been reported as a pest on apple in Europe.

Selected Bibliography

Duckworth 1965.
Dyar 1913.
Freeman 1960.
Friese 1960.
Frost 1972.
Herrebout et al. 1976.
Menken 1981.
Povel and Beckers 1982.
Powell et al. 1973.
Werner 1958.

ARGYRESTHIIDAE (YPONOMEUTOIDEA)

J. B. Heppner, *Florida State Collection of Arthropods*

The Argyresthiids, Head-Standing Moths

Figures 26.100, 26.101

Relationships and Diagnosis: Argyresthiids are a homogeneous family related to Yponomeutidae. Whereas yponomeutid larvae are predominately leaf feeders, argyresthiid larvae are only known to be miners of various plant parts, mainly on conifers.

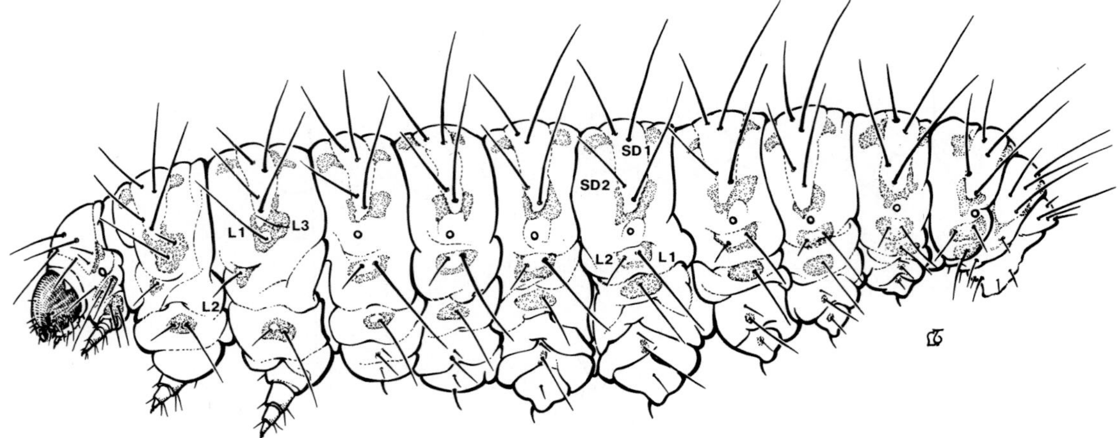

Figure 26.102. Douglasiidae. *Tinagma balteolellum* (Fischer von Röslerstamm). f.g.l. 4± mm. No data (European). U.S. Natl. Mus. (FWS).

Larvae of argyresthiids are distinguished by their uniordinal crochets on normal prolegs (subequal to thoracic legs) and in having L1 and L2 approximate on A1–8. Plutellid larvae are very similar to argyresthiids but argyresthiid larvae have SD1 dorsad of the spiracle on A1–8 instead of anterodorsad and the prolegs are not markedly long and slender.

Biology and Ecology: Argyresthiid larvae are miners of buds, twigs, cones, fruits, and leaves of conifers and some deciduous plants. The pine needle miners are the main economic pests in North America when locally abundant, but most argyresthiids are not common. Host plant records include Betulaceae, Cupressaceae, Ericaceae, Fagaceae, Pinaceae, Rosaceae, and Salicaceae.

Pupae lack dorsal spination and are not protruded upon adult eclosion. The cocoon is a loose webbing among pine needles tied together or in the larval chamber in fruits, buds, or twigs. Adults appear to be crepuscular and nocturnal. They have a distinctive resting position whereby they appear to be standing on their heads with the body inclined upward at nearly a 60° angle.

Description: Mature larvae small, about 5–16 mm. Primary setae only.

Head: Hypognathous. Frontoclypeus higher than wide, extending two-thirds to three-fourths to epicranial notch. Six stemmata in an oblique rectangular arrangement with 1 and 2 dorsad and 5 and 6 ventrad of seta S1; usually a small gap between 2 and 3. Head setae A1, A2 and A3 as obtuse triangle with A3 closest to stemmata.

Thorax: T1 shield evident, yellow to brown. Pinacula evident but not conspicuous. T1 L group trisetose. L1 and L2 approximate on T2 and T3, distant from L3.

Abdomen: L2 anteroventrad of L1 on A1–8, approximate and on a single pinaculum, distant from L3. SD1 dorsad of spiracle on A1–8. D1's closer together than D2's on A9. L1 and L2 separate on A9. Anal shield usually evident.

Prolegs normal; present on A3–6 and A10. Crochets in uniordinal, uniserial circles, sometimes with incomplete circles; rarely in transverse bands.

Comments: The Argyresthiidae are a small family of about 150 species, mostly in only two genera (*Argyresthia* and *Paraargyresthia*), but known worldwide. There are 52 species, all in *Argyresthia*, described from America north of Mexico, but there are at least as many undescribed species known. Few species are abundant enough to be of economic concern but those that are of some economic interest have been reviewed for North America by Freeman (1960; 1972). The most important of these species is the arborvitae leaf miner, *Argyresthia thuiella* (Packard) (fig. 26.100), a pest on eastern white cedar (*Thuja*). In Europe *Argyresthia andereggiella* (Duponchel) and *Argyresthia conjugella* Zeller, the apple fruit moth (fig. 26.101), are sometimes pests on apples and other temperate fruits like plums and cherries; the latter species also occurs in North America.

Selected Bibliography

Busck 1907.
Freeman 1960, 1972.
Friese 1969.
MacKay 1972.
Moriuti 1965, 1969, 1977.

DOUGLASIIDAE (YPONOMEUTOIDEA)

J. B. Heppner, *Florida State Collection of Arthropods*

The Douglasiids

Figure 26.102

Relationships and Diagnosis: The Douglasiidae have been so little studied that their closest affinities are uncertain. They once were thought to be related to Glyphipterigidae but are currently placed near Argyresthiidae. They have also been allied to Elachistidae (Forbes 1923) but they are not gelechioid.

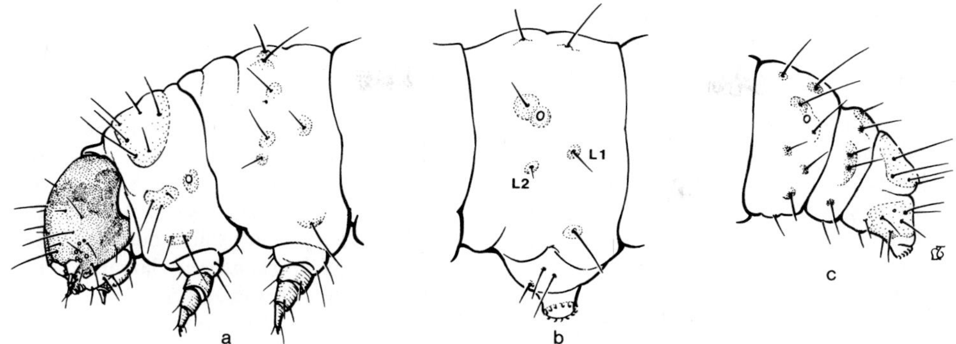

Figure 26.103a-c. Acrolepiidae. *Acrolepiopsis assectella* (Zeller), leek moth. f.g.l. 10± mm. From France, Philadelphia. No. 47900, 4 November 1960, from leek leaves. U.S. Natl. Mus. (FWS).

Douglasiids are modified as true borers, with reduced prolegs and an inflated grublike body. The larvae are unusual in having very long setae; some of the longest are flattened and silky. Crochets appear to be absent. Chaetotaxal distinctions include SD1 and SD2 long and dorsad of the abdominal spiracles, L1 and L2 approximate on abdominal segments, and L1 closer to L3 than to L2 on T2 and T3.

Biology and Ecology: Little is known of douglasiid biologies, and few species have been reared. Larvae have been reported as leaf miners, flower petiole miners, and stem borers. Host plants are known in Boraginaceae and Rosaceae for the most part, with two species of the Palearctic genus *Klimeschia* on Lamiaceae (= Labiatae).

Adults may be crepuscular or partially diurnal but are also collected at lights. Pupae were not available for study but are reported to remain in the pupal chamber at adult eclosion.

Description: Mature larvae small, about 5–8 mm. Integument white to yellow, sometimes with red-orange markings somewhat like stripes or elongate spots on large pinacula. Body grublike with intersegmental indentations and segmentally inflated. Primary setae only, but very long and some are flattened.

Head: Hypognathous. Frontoclypeus higher than wide, extending two-thirds to three-fourths to epicranial notch. Six stemmata in dorsally rounded oblique rectangle with 5 and 6 forming sharp angle at base. Head setae A1, A2, and A3 as obtuse triangle with A2 most distant from stemmata.

Thorax: T1 shield evident. Pinacula present, usually well developed and partly colored; sometimes less conspicuous. T1 L group trisetose. T2 and T3 with L1 and L3 approximate on merging pinacula, distant from L2.

Abdomen: L2 anterior to L1 on A1–8, approximate on merging pinacula, distant from L3. SD1 and SD2 dorsad of spiracle on A1–8. D1's closer together than D2's on A9. L1 and L2 approximate on A9, on separate pinacula. Anal shield evident on A10.

Prolegs reduced, present on A3–6 and A10. Crochets not visible, apparently absent.

Comments: The douglasiids are widely distributed over the world but contain only about 25 described species in two genera. Only five species of *Tinagma* (formerly *Douglasia*) occur in North America and none are commonly collected. *Tinagma obscurofasciella* Chambers has been reared from *Potentilla*. No species have been reported to be of economic significance, but the European *Tinagma perdicellum* Zeller is a leaf miner of strawberries and a few other Rosaceae.

The larval description is based on specimens of the European species, *Tinagma balteolellum* (Fischer von Röslerstamm) (fig. 26.102) and *Tinagma ocnerostomellum* (Stainton).

Selected Bibliography

Forbes 1923.
Gaedike 1974.

ACROLEPIIDAE (YPONOMEUTOIDEA)

J. B. Heppner, ***Florida State Collection of Arthropods***

The Acrolepiids

Figure 26.103

Relationships and Diagnosis: Acrolepiidae have sometimes been included in Plutellidae but the family is distinct and actually rather isolated in the Yponomeutoidea. It is possible that the family is closest to Heliodinidae.

The larvae are recognized by a combination of characters: L1 and L2 are approximate on abdominal segment 9; L2 anteroventrad of L1 on A1–8; SD1 anterodorsad to the spiracles on A1–8; and short prolegs with uniordinal crochets in circles, sometimes with a few extras inside the circle. They are easily distinguished from plutellids by having only six setae on A9 (plutellids have 8 or 9).

Biology and Ecology: Acrolepiid larvae, although mostly leaf miners, are also reported as borers of stems, flower buds, seeds, and bulbs. Host records include the plant families Asteraceae, Dioscoreaceae, Lamiaceae (= Labiatae), Liliaceae, and Solanaceae. One European species, *Acrolepia pygmeana* (Haworth), can tolerate the deadly alkaloids of poisonous belladonna (*Atropa*).

Pupae are without dorsal abdominal spination and are not protruded at adult eclosion. The cocoon is lacelike to some extent as in plutellids and yponomeutids. Adults probably are crepuscular but are collected at lights as well. Frediani (1954) recorded five or six generations annually in Italy for *Acrolepiopsis assectella* (Zeller) (fig. 26.103); also found in Hawaii.

Description: Mature larvae small, about 7–13 mm. Integument sometimes with dorsal markings. Primary setae only.

Head: Hypognathous. Frontoclypeus higher than wide, extending three-fourths to epicranial notch. Six stemmata in an oblique rectangle but dorsally rounded with a gap between 2 and 3. Head setae A1, A2 and A3 as obtuse triangle with A3 closest to stemmata.

Thorax: T1 shield and pinacula evident. T1 L group trisetose. L1 and L2 approximate on T2 and T3, on merging pinacula, and L1 nearly equidistant from L2 and L3.

Abdomen: L2 anteroventrad of L1 on A1–8, distant and on separate pinacula; both distant from L3. SD1 anterodorsad to spiracle on A1–8. D1's closer together than D2's on A9. L1 and L2 approximate on A9, on merging pinacula. Anal shield evident on A10. Spiracles small.

Prolegs short, present on A3–6 and A10. Crochets uniordinal in uniserial circles, sometimes with a few biserial crochets in part of circle.

Comments: Acrolepiids comprise only about 80 species worldwide, mainly from the Holarctic region. Only three species are described for North America. The main economic species, the leek moth, *Acrolepiopsis assectella* (Zeller) (fig. 26.103), is a pest on leeks and onions; it was introduced to Hawaii from Europe. An excellent paper on the biology and morphology of *A. assectella* by Frediani (1954) is in Italian. The North American *Acrolepiopsis incertella* (Chambers) is reported (Forbes 1923) on leaves of *Smilax* or bulbs of *Lilium,* although this report may represent two species. The European *Acrolepiopsis alliella* (Semenov and Kuznetsov) is a pest of onions. Two Japanese species feed on yams. The genus *Antispastis* from Argentina, for which Bourquin (1951) provided biological notes, also is an acrolepiid.

Selected Bibliography

Bourquin 1951.
Forbes 1923.
Frediani 1954.
Gaedike 1970.
Moriuti 1961.
Werner 1958.

HELIODINIDAE (YPONOMEUTOIDEA)

J. B. Heppner, *Florida State Collection of Arthropods*

The Heliodinids, Sun Moths

Figures 26.104–26.106

Relationships and Diagnosis: Heliodinids appear to be the most advanced yponomeutoid moths, perhaps most closely related to Acrolepiidae. The family is now restricted to true heliodinids. Several genera, formerly in Heliodinidae, are now placed in Oecophoridae (Stathmopodini), as well as Glyphipterigidae; one genus is in Cosmopterigidae. The moths appear superficially similar but the Stathmopodini have a scaled haustellum, whereas true Heliodinidae do not.

Included here in Heliodinidae is the genus *Schreckensteinia* (subfamily Schreckensteiniinae), which conforms to adult heliodinid characters but has tibial spines on the hind legs of the adult. Some specialists refer the genus to Epermeniidae due to the leg spines but this relationship needs further study. [*Schreckensteinia* larvae do not conform well to other heliodinid larvae and have spiracles like epermeniids, but the T1 L group is trisetose (FWS).]

Larvae of Heliodinidae are similar to acrolepiid larvae but some have very short setae. The spiracles are usually minute and difficult to see except on A8. SD1 is dorsad of the spiracles on A1–8, instead of anterodorsad as in Acrolepiidae.

Biology and Ecology: Larvae are leaf miners, leaf skeletonizers, and fruit raceme borers. Hosts have been recorded in Anacardiaceae, Chenopodiaceae, Nyctaginaceae, Poaceae (*Panicum*), Rosaceae, and Scrophulariaceae (*Orthocarpus*).

Pupae are without dorsal spination and are not protruded from the cocoon at adult eclosion. Pupae are often dorsally flattened and have lateral margin angular ridges. The cocoon is usually lacelike to some extent or is a looser webbing. Adults are diurnal but are also collected at lights.

Biologies that have been reported for genera now in the Oecophoridae (formerly Stathmopodidae, now a tribe of oecophorids) or that of *Euclemensia,* now in Cosmopterigidae, have been excluded here as well as in the larval description.

Description: Mature larvae small, to about 12 mm. Integument often green, mostly unmarked except for scattered brown pinacula. Head capsule yellow or amber. Primary setae only, often short.

Head: Hypognathous. Frontoclypeus higher than wide, extending three-fourths to epicranial notch. Six stemmata in oblique semicircle, with stemma 6 usually reduced. Head setae A1, A2 and A3 in obtuse triangle with A3 closest to the stemmata.

Thorax: T1 shield evident, amber. Pinacula present, brown. T1 with L group trisetose. T2 and T3 with L1 and L2 approximate on merging pinacula and relatively close to L3.

Abdomen: L2 anteroventrad and apart from L1 on A1–8, both distant from L3. SD1 dorsad of spiracle on A1–8. D1 setae closer together than D2 setae on A9. L1 and L2 approximate on A9, on separate pinacula. Spiracles very small, difficult to see. Anal shield evident on A10, but not well developed, amber.

Prolegs small, slender but relatively short, sometimes reduced; present on A3–6 and A10. Crochets in uniordinal circles.

Comments: The Heliodinidae are known from all faunal regions and comprise about 55 described species, of which 20 are from America north of Mexico. The numerous tropical species previously considered to be heliodinids are mostly stathmopodine oecophorids. No heliodinid has been reported to be of economic importance. The eastern *Heliodines nyctaginella* Gibson (fig. 26.105) is a leaf miner of *Oxybaphus* (Nyctaginaceae) and *Schreckensteinia erythriella* Clemens

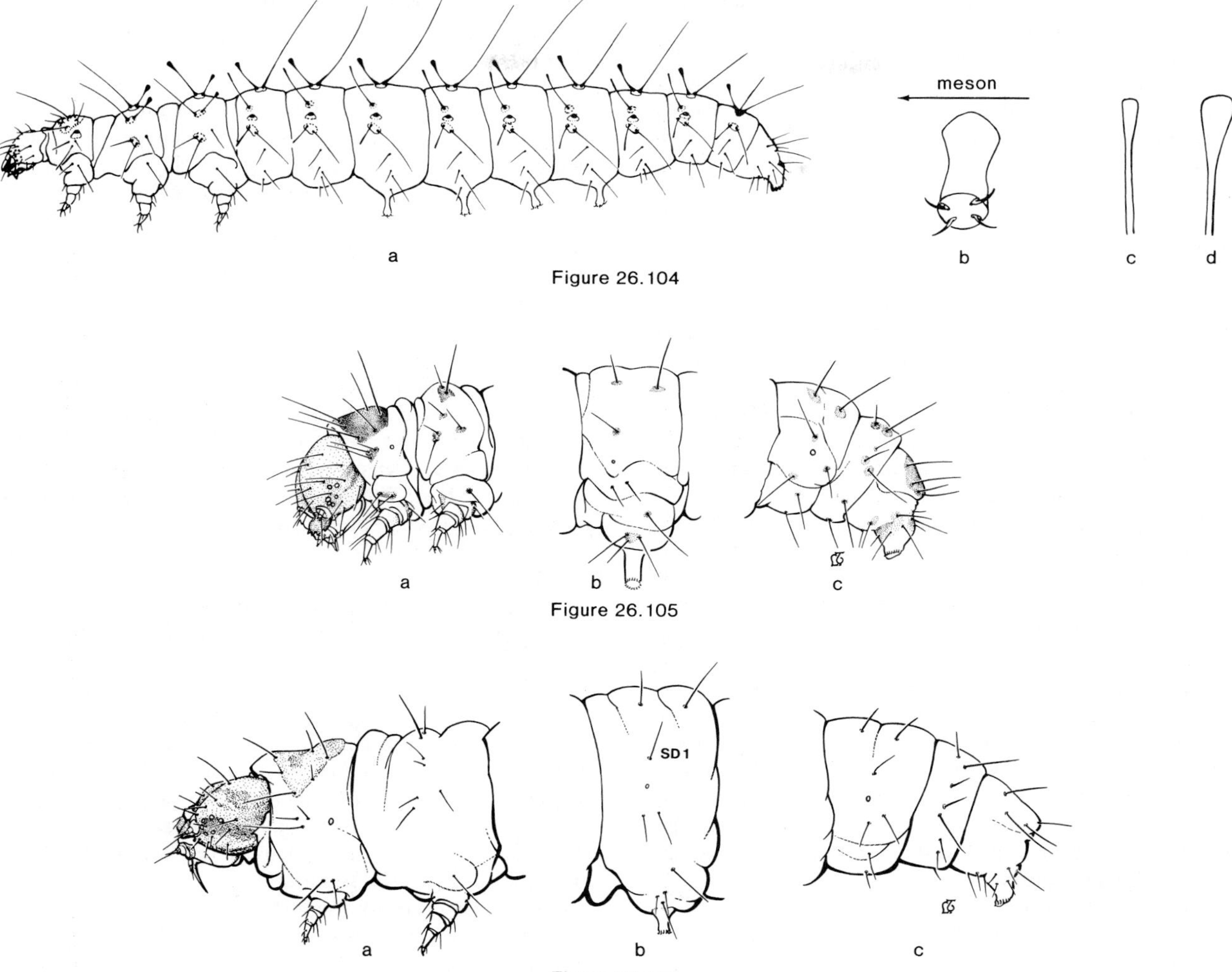

Figure 26.104a-d. Heliodinidae. *Schreckensteinia erythriella* Clemens. f.g.l. 10± mm. a. larva; b. A6 crochets; c-d. spatulate body setae. Falls Church, Va., 6 September 1915, racemes of *Rhus copallina,* C. Heinrich. U.S. Natl. Mus. (FWS).

Figure 26.105a-c. Heliodinidae. *Heliodines nyctaginella* Gibson. f.g.l. 10± mm. a. head, T1, T2; b. A3; c. A8-A10. Decatur, Ill., September 1920, A. W. Lindsey. U.S. Natl. Mus. (FWS).

Figure 26.106a-c. Heliodinidae. *Lithariapteryx abromiella* Chambers. f.g.l. 11± mm. a. head, T1, T2; b. A3; c. A8-A10. North Beach, Pt. Reyes Nat. Seashore, Marin Co., Calif., 11 May 1974, J. A. Powell. Univ. Calif., Berkeley (FWS).

(fig. 26.104) feeds in the fruit racemes of sumac. The western *Lithariapteryx mirabilinella* Comstock is a leaf miner on *Mirabilis* (Nyctaginaceae).

Larval characteristics noted for the family by Forbes (1923) and in Peterson (1948) are in part applicable to genera that are not heliodinids. *Euclemensia* is a cosmopterigid and *Idioglossa* and *Cyphacma* are in Oecophoridae (Stathmopodini).

Selected Bibliography

Comstock 1940.
Forbes 1923.

SESIIDAE (SESIOIDEA)

J. B. Heppner, *Florida State Collection of Arthropods*

The Sesiids, Clearwing Moths

Figures 26.107-26.112

Relationships and Diagnosis: The Brachodidae (formerly Atychiidae), Sesiidae, and Choreutidae make up the Sesioidea. Brachodidae are Palearctic and pantropical and are not represented in America north of Mexico. Sesiidae are most closely related to Brachodidae, the group presumed to resemble sesiid precursors prior to their evolution toward hymenopterous mimicry (Heppner 1979a, Heppner and Duckworth 1981).

Sesiid larvae are distinguished by the trisetose T1 L group (one of the three setae is often very small and inconspicuous), the short or reduced prolegs having crochets in transverse bands, and the oblique sclerotized ridges on the T1 shield (absent in Cossidae). Abdominal segments have L2 anterodorsad of L1, whereas in Brachodidae L2 is anterior to L1. D2 setae often as far apart as D1's on A6–8; in Brachodidae they are twice as far apart as D1's.

Biology and Ecology: Larvae are borers in tree trunks, bark, branches or roots, in woody shrubs and vines, and sometimes in roots and stems of herbaceous plants. A few species have evolved as inquiline borers in galls of woody and herbaceous plants, and two species are predaceous on scale insects (Duckworth 1969). Biological information is extensive for Holarctic sesiids, although most life histories have been studied for the economically important species. Life histories of most tropical species and many Holarctic noneconomic species are unknown. Primary summaries of biological information for North American sesiids can be found in Engelhardt (1946) and Duckworth and Eichlin (1977, 1978).

Most sesiids have a narrow host preference but the range of hosts for the whole family is quite large, with hosts recorded from at least 32 plant families.

Eggs are deposited singly on or near the host plant. Larvae construct galleries, forming an exit area prior to pupation that is covered with a thin layer of plant tissue through which the pupa pushes at adult eclosion. Pupation occurs in the larval gallery. Pupae have two rows of dorsal abdominal spines on segments 2–7. Root borers, especially among Tinthiinae, sometimes have galleries in the soil next to the host roots, similar to what is reported for Brachodidae from Europe. Most sesiids have an annual life cycle but some require two, possibly three years.

Sesiids are known from diverse habitats, but in North America extensive speciation has occurred in the arid regions of the Southwest and Great Basin. Sesiid adults are diurnally active, mainly in the morning hours and this, together with rapid adult flight and isolated colonies near hosts, has hindered research on the group because specimens are difficult to find. Development of sesiid synthetic pheromones now allows easier field location of colonies and greatly enhances the research possibilities.

Description: Mature larvae small to large, approximately 12 to perhaps 70 mm, averaging about 25 mm. Integument usually white with amber head capsule, usually with deep intersegmental indentations common to boring grublike larvae. Primary setae only.

Head: Somewhat prognathous. Frontoclypeus higher than wide, extending two-thirds to epicranial notch. Six stemmata arranged as a trapezoid with gaps between 2 and 3, and 4 and 5. Head setae A1, A2 and A3 as obtuse triangle with A3 closest to stemmata.

Thorax: T1 shield evident, usually with only the two oblique ridges that are convergent posteriorly colored brown. Pinacula inconspicuous. T1 L group trisetose.

Abdomen: L2 anterodorsad and approximate to L1 on A1–8. SD2 (if visible) subequal in distance to spiracle and SD1 on A1–6. SV setae more or less in a vertical line on A2. Prolegs short, reduced, present on A3–6 and A10. Crochets in uniordinal transverse bands (somewhat reduced in *Pennisetia*).

Comments: The family includes over 1,000 species and 123 genera worldwide, divided into three subfamilies: Tinthiinae, Paranthreninae, and Sesiinae (Heppner and Duckworth 1981). A tribal classification is being developed. All faunal regions are represented, with extensive speciation in pantropical regions and arid areas. Only the Holarctic fauna is relatively well known but new species are currently being discovered, especially since the advent of synthetic sex pheromones that have allowed fugitive species to be located. North of Mexico 113 valid species are currently recognized in 19 genera (Duckworth and Eichlin 1977).

Although most sesiids are not economically important on a world basis, many species are very destructive to economically important plants and all species do extensive damage to hosts, often compounded by the presence of many larvae in one host. The economic species in North America include the following: the raspberry crown borer, *Pennisetia marginata* (Harris); the oak clearwing moth, *Paranthrene asilipennis* (Boisduval); the western poplar clearwing, *Paranthrene robiniae* (H. Edwards) (fig. 26.108); the grape root borer, *Vitacea polistiformis* (Harris); the squash vine borer, *Melittia cucurbitae* Harris (fig. 26.107); the [European] hornet moth, *Sesia apiformis* (Clerck) (fig. 26.109); the American hornet moth, *Sesia tibialis* (Harris); the maple callus borer, *Synanthedon acerni* (Clemens); the strawberry crown moth, *Synanthedon bibionipennis* (Boisduval); the peach tree borer, *Synanthedon exitiosa* (Say) (fig. 26.111); the lesser peach tree borer, *Synanthedon pictipes* (Grote & Robinson) (fig. 26.112); the Douglas fir pitch moth *Synanthedon novaroensis* (H. Edwards); the apple bark borer, *Synanthedon pyri* (Harris); the pitch mass borer, *Synanthedon pini* (Kellicott); the sequoia pitch moth, *Synanthedon sequoiae* (H. Edwards); the rhododendron borer, *Synanthedon rhododendri* (Beutenmüller); the dogwood borer, *Synanthedon scitula* (Harris); the [European] currant borer, *Synanthedon tipuliformis* (Clerck); the banded ash clearwing, *Podosesia aureocincta* Purrington and Nielsen; the lilac borer, *Podosesia syringae* (Harris); and the persimmon borer, *Sannina uroceriformis* Walker.

The world fauna of Sesiidae is reviewed by Heppner and Duckworth (1981). Popescu-Gorj et al. (1958), Naumann (1971), and Fibiger and Kristensen (1974) provide reviews for European sesiids; those of Australia and New Zealand have been reviewed by Duckworth and Eichlin (1974).

Selected Bibliography

Duckworth 1969.
Duckworth and Eichlin 1974, 1977, 1978.
Engelhardt 1946.
Fibiger and Kristensen 1974.
Heppner 1979a.
Heppner and Duckworth 1981.
MacKay 1968.
Naumann 1971.
Popescu-Gorj et al. 1958.

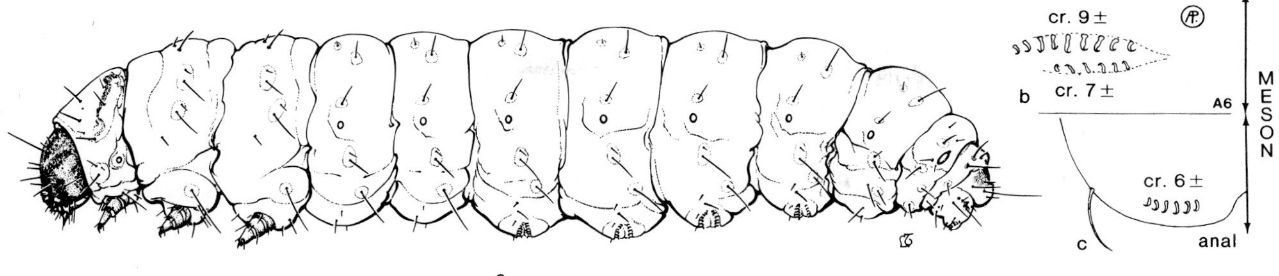

Figure 26.107

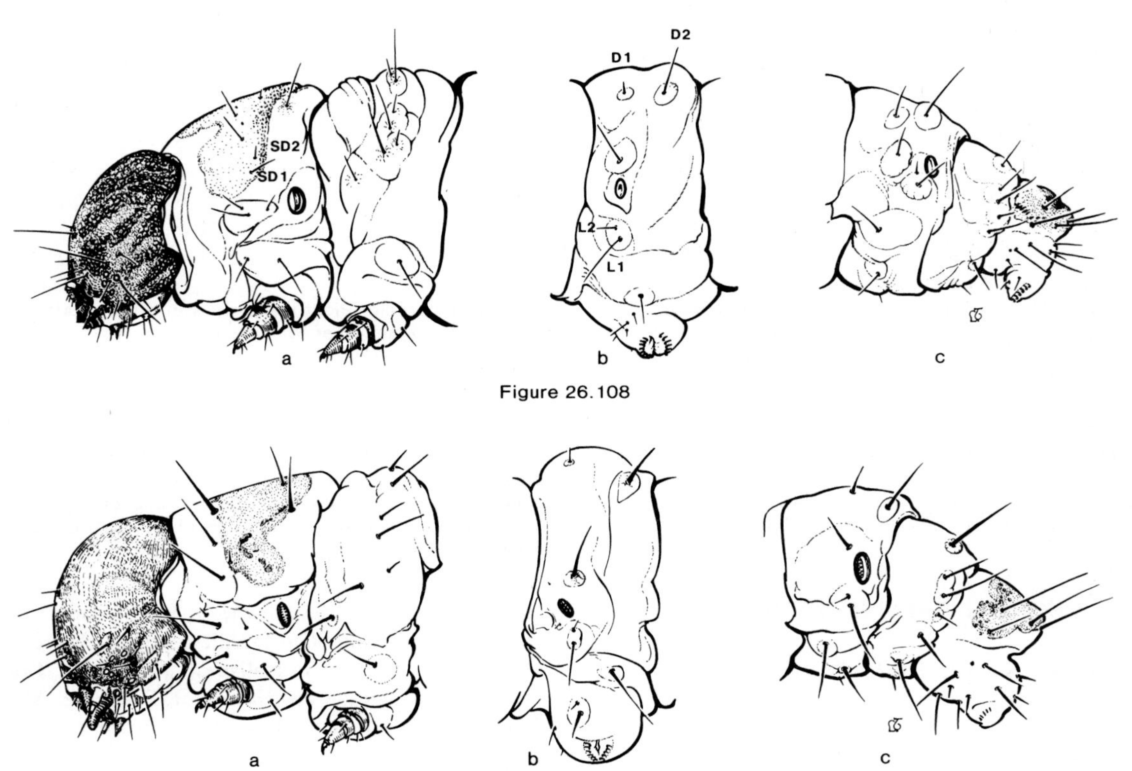

Figure 26.108

a b c

Figure 26.109

Figure 26.107a-c. Sesiidae. *Mellittia cucurbitae* (Harris), squash borer. a. larva; b. A6 crochets; c. A10 crochets. f.g.l. 35± mm. Near-white borer with brownish head, yellowish cervical shield, and 6-10 crochets per row on each A6 proleg. Infests stems (runners) of squash, pumpkins, muskmelons, cucumbers, gourds, and other cucurbits. U.S. Natl. Mus. (figs. b-c from Peterson 1948) (FWS).

Figure 26.108a-c. Sesiidae. *Parenthrene robiniae* (Henry Edwards), western poplar clearwing. a. head, T1, T2; b. A3; c. A8-A10. f.g.l. 35± mm. Arroyo Seco, S. Pasadena, Calif., 1 July 1928, cottonwood. Bores in trunks and branches of poplars, willows, and ornamental birches. U.S. Natl. Mus. (FWS).

Figure 26.109a-c. Sesiidae. *Sesia apiformis* (Clerck), hornet moth. f.g.l. 30± mm. a. head, T1, T2; b. A3; c. A8-A10. White Plains, N.Y., 21 April 1938, in roots of aspen. U.S. Natl. Mus. An introduced Palearctic species, boring in the trunks and exposed roots of poplars and willows (FWS).

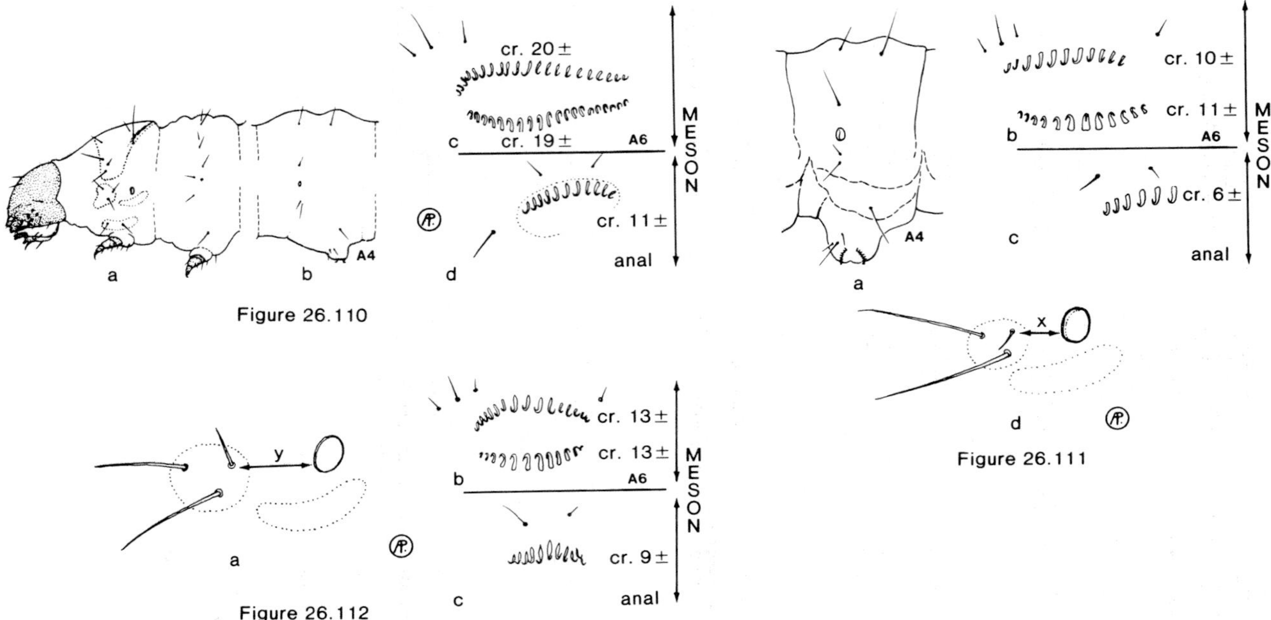

Figure 26.110

Figure 26.111

Figure 26.112

Figure 26.110a-d. Sesiidae. *Podosesia syringae* (Harris), lilac borer/ash borer. a. head, T1, T2; b. A4; c. A6 crochets; d. A10 crochets. f.g.l. 18± mm. 18-20 crochets per row on A6. Infests lilac and ash; ash may also be infested by *Podosesia aureocincta* Purrington and Nielsen, the banded ash clearwing (figs. from Peterson 1948) (FWS).

Figure 26.111a-d. Sesiidae. *Synanthedon exitiosa* (Say), peachtree borer. f.g.l. 30± mm. a. A4; b. A6 crochets; c. A10 crochets; d. T1 spiracle and L group, showing distance (x) between spiracle and nearest seta to be less than 2× smallest diameter of spiracle. Right mandible with 4 teeth. Attacks crown area of peach and other *Prunus* species (see fig. 26.112, lesser peachtree borer. (figs. from Peterson 1948) (FWS).

Figure 26.112. Sesiidae. *Synanthedon pictipes* (Grote and Robinson), lesser peachtree borer. f.g.l. 20± mm. (a) T1 spiracle and L group, showing distance (y) between spiracle and nearest seta to be greater than 3× smallest diameter of spiracle. Right mandible with 5 teeth. Attacks trunks and large branches of peach, plum, cherry, and other *Prunus* species, especially where injured by fruit harvest shakers (see fig. 26.111, peachtree borer. (figs. from Peterson 1948) (FWS).

CHOREUTIDAE (SESIOIDEA)

J. B. Heppner, *Florida State Collection of Arthropods*

The Choreutids, Metalmark Moths

Figures 26.113, 26.114

Relationships and Diagnosis: The Choreutidae are related to Sesiidae but have external feeding larvae. Adults are also diurnally active like sesiids, but have a scaled tongue. The family was previously incorporated in the Glyphipterigidae but the two families are unrelated.

Choreutid larvae are distinguished by unusually long, slender prolegs and long primary setae, especially in *Brenthia* (fig. 26.114) except Millieriinae with short prolegs. The larvae have a curious habit of very rapid movement, generally backward, when disturbed, as do many tortricid and pyralid larvae. Larval characters for Glyphipterigidae in Peterson (1948) correspond exclusively to Choreutidae.

Biology and Ecology: Virtually all known choreutid larvae are external leaf feeders. A few are known to feed inside hollow-stemmed plants such as thistles or in leaf buds of *Ficus*, and one European species feeds inside seeds, but these larvae are not modified as borers. Life histories of most species, especially the tropical ones, are unknown, but a few host records and life histories are known. Host records are known among the following plant families; Aristolochiaceae, Betulaceae, Boraginaceae, Bromeliaceae, Asteraceae, Ericaceae, Fagaceae, Lamiaceae (= Labiatae), Fabaceae, Moraceae, Arecaceae (= Palmae), Rosaceae, Salicaceae, Scrophulariaceae, Ulmaceae, Apiaceae (= Umbelliferae), and Urticaceae.

Eggs are deposited singly on host leaves or in protected places on the host plant. Larvae tie leaves together and form a loose larval gallery of silk; a Florida species makes a case. Pupation usually occurs in the host leaf-tie or in the larval gallery. The cocoon is often a complex three-layered white web with an inner fluted layer holding the pupa. Pupae typically have a single row of spines on the dorsum of A2–7 and they are protruded at adult eclosion, but the pupae of Millieriinae have double spine rows. Many temperate species are bivoltine but in moderate or tropical climates, species are known to be multivoltine. Although diurnally active, adults sometimes come to lights at night. They often hold the wings elevated from the body when perching and appear to hop from leaf to leaf.

Description: Mature larvae small, not exceeding about 20 mm, pale green, yellow-white, or brownish, with amber or yellow head capsule, sometimes marked. Primary setae only, especially long in *Brenthia*.

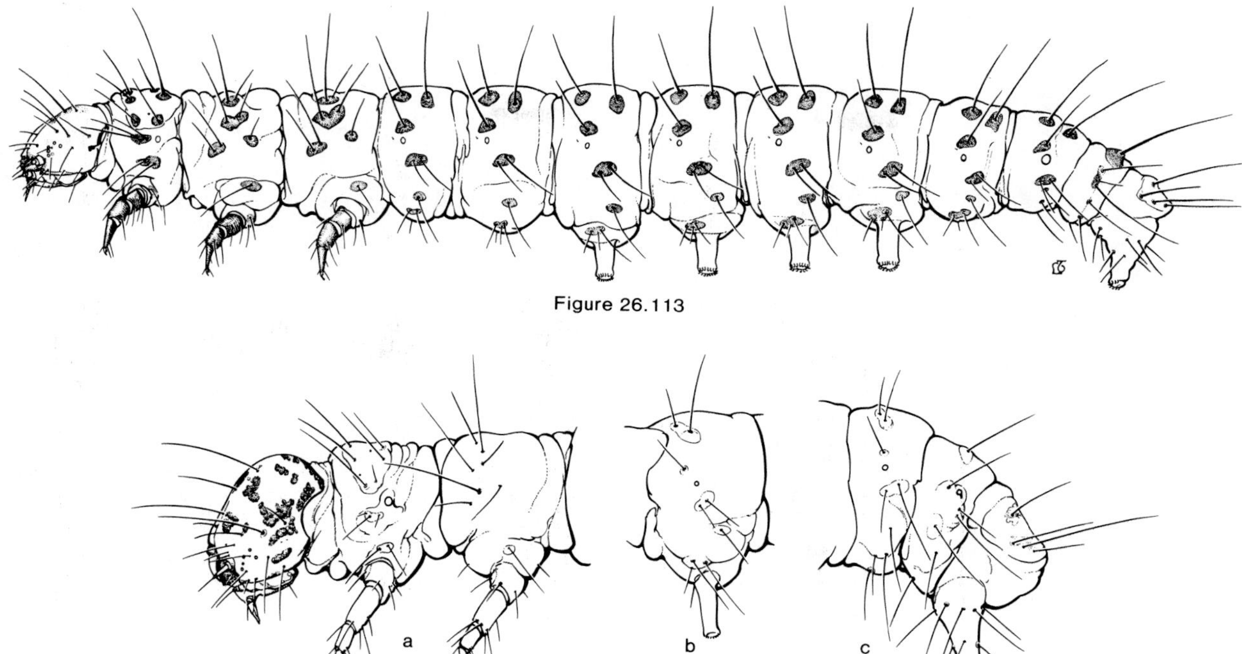

Figure 26.113

Figure 26.114

Figure 26.113. Choreutidae. *Choreutis pariana* (Clerck), apple-and-thorn skeletonizer. f.g.l. 10± mm. Romeo, Van Dyke, and 29 Mile Rd., McComb Co., Mich., 31 July 1980, apple, K. Strickler, R. Nielsen. Michigan State Univ. (FWS).

Figure 26.114. Choreutidae. *Brenthia pavonacella* Clemens. f.g.l. 10± mm. W. Falls Church, Va., 25 July 1914. U.S. Natl. Mus. (FWS).

Head: Hypognathous, sometimes very large (*Brenthia*) (fig. 26.114). Frontoclypeus higher than wide, extending from half to three-fourths to epicranial notch. Six stemmata in an evenly spaced rectangular semicircle but with stemma 6 often very reduced, giving the appearance of only five stemmata present. Head setae A1, A2, and A3 as very obtuse triangle and in a more posterior position than usual in Lepidoptera larvae, with A2 closest to stemma 2.

Thorax: T1 shield well developed, yellow-brown to nearly black. Pinacula distinctly marked, yellow-brown to brown-black, well developed. T1 L group trisetose, but bisetose in Millieriinae.

Abdomen: All segments with well-developed and distinctly marked pinacula. L2 anteroventrad and approximate to L1 on A1–8. SV setae on A2 in triangular arrangement. D2's more approximate than D1's on A8 (subequal in Millieriinae).

Prolegs very long and slender (short in Millieriinae), present on A3–6 and A10. Crochets in uniordinal (rarely biordinal) circle, sometimes in lateral penellipse (*Hemerophila*).

Comments: The choreutids are a relatively small family with 350 described species in 18 genera worldwide. Most work has been restricted to isolated species descriptions. Currently two revisions are being completed for the Holarctic fauna (Diakonoff, in prep.; Heppner in prep.). Some species in North America can be locally common but they are not generally encountered. They tend to congregate near their hosts or on flowers nearby. Only one species is of economic significance in North America, the introduced European apple-and-thorn skeletonizer, *Choreutis pariana* (Clerck) (fig. 26.113), especially more recently in British Columbia (Heppner 1979b). Other common species include *Caloreas multimarginata* (Braun) (formerly *Choreutis melanifera* Keifer), on sagebrush and artichoke (Lange 1941) in California, *Tebenna carduiella* (Kearfott), on thistles, and in Florida species of *Tortyra* and *Hemerophila* on *Ficus* leaves or in leaf buds.

Selected Bibliography

Cuscianna 1927.
Felt 1917.
Heppner 1979b.
Keifer 1937.
Lange 1941.
MacKay 1972.
Peiu and Patrascanu 1970.
Porter and Garman 1923.
Real and Balachowsky 1966.
Wille 1952.
Williams 1951.

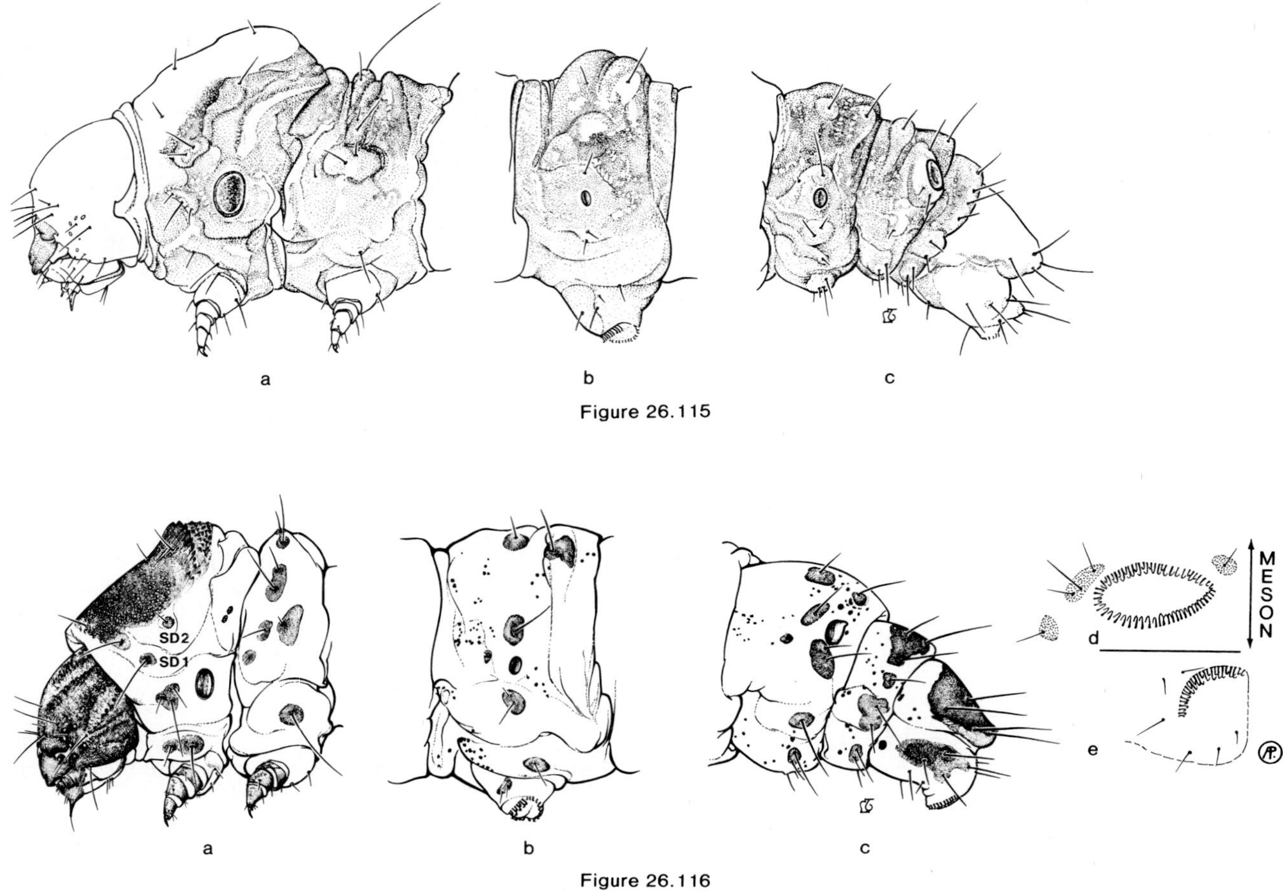

Figure 26.115a-c. Cossidae. *Cossula magnifica* (Strecker), pecan carpenterworm. a. head, T1, T2; b. A3; c. A8-A10. f.g.l. 40± mm. Florence, S.C., 15 November 1929, ex. pecan, O. L. Cartwright. U.S. Natl. Mus. Bores in pecan, oak, and hickory branches and trunks. Body pinkish. Apparently 1 generation per year.

Figure 26.116a-e. Cossidae. *Zeuzera pyrina* (L.), leopard moth. a. head, T1, T2; b. A3; c. A8-A10; d-e. A6 and A10 crochets. f.g.l. 50± mm. Brooklyn, N.Y. U.S. Natl. Mus. Introduced from Europe, bores in branches and trunks of over 125 species of deciduous trees, including ash, beech, cherry, and other fruit trees, elm, maple, oak, poplar, willow, walnut, and woody shrubs. Body cream to yellowish to pinkish. Two years per generation (figs. d and e from Peterson 1948).

COSSIDAE (COSSOIDEA)

Frederick W. Stehr, *Michigan State University*

The Cossids, Carpenter Moths and Leopard Moths

Figures 26.115–26.117a–e

Relationships and Diagnosis: Cossids are superficially similar to sesiid larvae, with which they share a wood- or stem-boring habit, but they are more closely related to the Psychidae and Tortricidae (Common 1970). They can be distinguished from sesiids by having SD2 nearly caudad of SD1 on T1 (sesiids have SD2 nearly dorsad of SD1 on T1) and by having uni-, bi-, or triordinal crochets in an ellipse, penellipse, or transverse bands (sesiids have uniordinal crochets in transverse bands, as do some cossids). Many cossids also have a rugose or humped caudal area on the T1 shield (fig. 26.116) (oblique, sclerotized ridges are usually present in sesiids).

Biology and Ecology: Cossids primarily consume woody tissues, usually boring in branches and trunks of a great diversity of hosts, but they occasionally feed beneath the bark, externally on roots, and sometimes in stems of less woody plants such as cacti. They attack a great diversity of deciduous hosts and some conifers. Cossids lay eggs in protected bark crevices, wounds, among lichens, etc., and larvae bore in upon hatching, usually boring deeper in later instars, and maturing in one to four years. Frass and debris are pushed to the outside. Pupation takes place in the tunnel, but before adult emergence the pupa protrudes from the exit hole.

Description: Mature larvae medium to large, 20–75 mm. Body basically cylindrical and pale, white, yellow, or pink, sometimes dusky dorsally, and often with T1 humped and/ or rugose. A standard complement of setae is present, but as

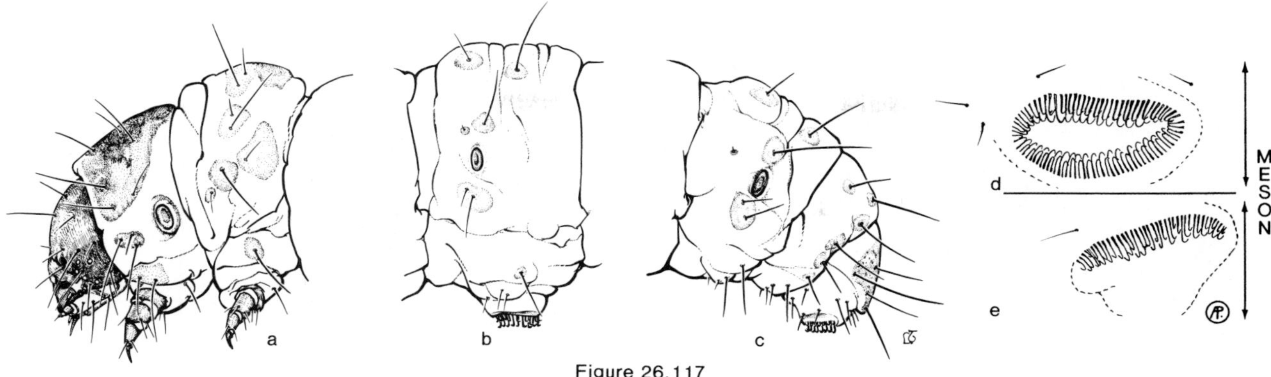

Figure 26.117

Figure 26.117a-e. Cossidae. *Prionoxystus robiniae* (Peck), carpenterworm. a. head, T1, T2; b. A3; c. A8-A10; d-e. A6 and A10 crochets. f.g.l. 65± mm. Twin Falls, Id., 31 December 1957, ex. willow. U.S. Natl. Mus. Bores in large branches and trunks of many hardwoods, especially ash, elm, black locust, maple, poplar, oak, willow, and large shrubs. Body cream to yellowish. One year in the South to 4 years in the North per generation (figs. d and e from Peterson 1948).

many as four extra setae may be present dorsocaudally from the spiracle and sometimes extras are present in the SV group and elsewhere.

Head: Semiprognathous to hypognathous. Frontoclypeus usually extending about halfway to epicranial notch but sometimes considerably less. Ecdysial lines bowed slightly inward near the center and meeting about halfway between apex of frons and epicranial notch. Six stemmata, evenly spaced in semicircle except for 5 which is "beneath" the antenna.

Thorax: T1 dorsum frequently humped and rugose or spined. SD2 nearly caudad to SD1 on T1. L group trisetose on T1, T2, and T3, L1 closer to L3 than L2 on T2 and T3, SV group bisetose on T1, unisetose on T2 and T3 (rarely bisetose).

Abdomen: D2's usually father apart than D1's on A1–7, often closer together on A8 and A9 and sometimes closer together on the more anterior segments. SD1 large and above spiracle on A1–7, SD2 small to minute and in front of spiracle on A1–7. SD1 and SD2 positions more variable on A8. L1 and L2 on same pinaculum below spiracle on A1–8. L3 closer to SV group on A1–8 than to L1 and L2 group. SV group usually bisetose on A1 and A8, trisetose on A2–7. All A9 setae in vertical line except D1, which is anterior. A10 shield bearing four setae per side. A10 crochets sometimes uniordinal.

Prolegs on A3–6 and A10 short or rudimentary. Crochets uni-, bi-, or triordinal, and arranged in an ellipse, penellipse, or transverse band(s), Spiracles large and round to oval, with T1 and A8 spiracles sometimes two or three times larger than A1–7 spiracles, and the A8 spiracle sometimes caudo-projecting and closer to midline than other spiracles.

Comments: About 45 species in 11 genera occur in Canada and the United States. Some cause economic injury, primarily by degrading timber quality through their borings in otherwise healthy trees. *Prionoxystus robiniae* (Peck), the carpenterworm (fig. 26.117) is probably the most common and destructive species, attacking oaks and many other hardwoods from coast to coast. The pecan carpenterworm, *Cossula magnifica* (Strecker) (fig. 26.115) attacks pecan, oak, and hickory in the South; the introduced leopard moth, *Zeuzera pyrina* (L.) (fig. 26.116) attacks hardwoods in the Northeast and Midwest. *Acossus populi* (Walker) attacks *Populus* spp., and *Givita lotta* Barnes and McDunnough attacks ponderosa pine in the Southwest.

Selected Bibliography

Barnes and McDunnough 1911.
Britton and Crombie 1911.
Furniss and Carolin 1977.
Johnson and Lyon 1976.
Solomon 1966, 1967, 1968.
U.S.D.A. 1985.

CASTNIIDAE (CASTNIOIDEA)

J. Miller, *Allyn Museum of Entomology, Florida State Museum*

The Castniids

Figure 26.117x-2

Relationships and Diagnosis: The Castniidae are closely related to the Sesiidae, Tortricidae, and especially Cossidae. The number of genera recognized in the Neotropics is variable, ranging from 29 (J. Miller, 1986) to 33 (Houlbert, 1918), with *Synemon* and *Tascinia* (=*Neocastnia*) being the sole counterparts in Australia and tropical Asia.

Late instar larvae bear only primary setae and may be distinguished from their close relatives by the poorly developed prolegs with reduced crochets (A3–6), and by the rudimentary prolegs on A10 with or without a few crochets.

Biology and Ecology: All larvae are borers and construct tunnels, sometimes loosely webbed with silk, but without the white powdery consistency shown by the Megathymidae (Hesperioidea), as suggested by Seitz (1913). Cocoons, fabricated from a combination of host plant debris and leaf litter

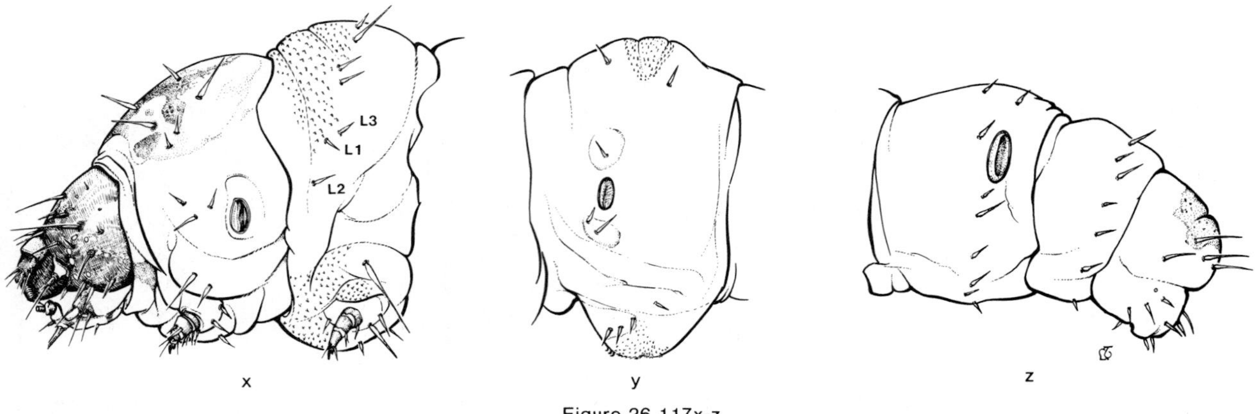

Figure 26.117x-z

Figure 26.117x-z. Castniidae. *Castnia humboldti* Boisduval. x. head, T1, T2; y. A3; z. A8-A10. f.g.l. 85± mm. Coto, Costa Rica, 26 December 1955, boring in banana. Cornell Univ. Note: Setae are wide at base, taper abruptly toward tip, and are less distinct than in this drawing (FWS).

and loosely webbed together with silk, are variable in hardness and length (13–95 mm). Neotropical species are catholic feeders primarily on palms, bromeliads, sugarcane and bananas; those of *Synemon* in Australia infest the roots of low-growing sedges. (Tindale 1928, Common and Edwards 1981).

Eggs are white to green upon deposition and change color to grey, yellow, green, or rose upon maturation. All known eggs are wheat-shaped with the exception of *Corybantes pylades* (Stoll), which is spherical (Moss 1945). Eggs are generally 0.5–0.7 mm. long with five raised longitudinal ribs bearing aeropyles. A sulcate micropyle is present on one end with a roughened area on the opposite pole.

Description: Late instar larvae (fig. 26.117) variable in size (15–90 mm). Body cylindrical and fairly stout with head two-thirds diameter of prothorax. Color variable in first instar larvae, ranging from white to rose to green; white to yellow in later instars. Phillipi (1863) and Moss (1945) indicated that first instar larvae might have additional setae; however, further studies indicate that setae, especially D1 and D2 on A6–A10, are elongate in the first and second instars, perhaps related to the caterpillar's free-living status. Cuticle generally smooth or granular and may have transverse patches of short spinules dorsally and ventrally. Spiracles elliptical, with those of T1 and A8 enlarged.

Head: Hypognathous. Frontoclypeus extending one-half to two-thirds the distance to the epicranial notch and almost as wide as long. Ecdysial sutures quite wide of the frontoclypeus and meeting near epicranial notch. Epicranium lightly sclerotized and usually concealed by T1. Mandibles prominent and heavily sclerotized. Six stemmata, 1–4 evenly spaced in semicircle with seta S1 between 2 and 3; 5 and 6 separated from 1–4; 6 caudad to 5.

Thorax: T1 shield poorly to moderately defined in early instar larvae, more prominent in later instars. D1 and D2 on all segments, with D2 more laterad of D1 on T2 and T3. L group trisetose on T1; T2 and T3 with L1 closer to L3 than L2. SV group bisetose on T1 and unisetose on T2 and T3.

Abdomen: Dorsal transverse patches of reduced spinules on A1–6 in later instar larvae. D1 and D2 on all segments, with D2 ventrocaudad of D1 except on A9, where D1 is ventrad to D2 and reduced. SD1 dorsoanteriad and closer to spiracle on A8 than on A7. L group trisetose on A1–8, with L1 and L2 ventrad of spiracle and L3 caudoventrad. SV group trisetose on A1–7, bisetose on A8, and unisetose on A9.

Prolegs variable, with early instar larvae possessing one or two uniserial bands. Mature larvae vary between species from no crochets to multiserial crochets on A3–6, which may be interrupted mesially by a fleshy lobular groove and laterally by a T-shaped groove. A10 prolegs rudimentary, without crochets or with only a few. Suranal plate poorly developed, bearing four setae per side (eight total).

Comments: The family is relatively small with almost 200 species described. Adult records taken in Mexico near the U.S. include two specimens of *Castnia escalantei* J. Miller from Primavera, Chihauahua, and specimens of *Castniomera atymnius futilis* (Walker) at Tampico, Tamaulipas. The states of Morelos, Oaxaca, and Veracruz provide most of the closest records, with a single outlier species, *Iricila hecate* (Herrich-Schaeffer), found in the Dominican Republic.

A few life histories of the Castniidae have been published (Korytkowski and Ruiz 1980a, 1980b; Esquivel 1980), but there is little identified material available for chaetotaxic study.

Most Castniidae are uncommon and economically unimportant with the exception of three genera, *Castniomera, Leucocastnia,* and *Eupalamides*. Marlatt (1905) reported on a severe infestation of *L. licus* (Fabricius) on sugarcane in Georgetown, British Guiana. Similarly, Salt (1929) published an economic study of *C. humboldti* (Boisduval) on bananas in the Santa Marta Region of Colombia. More recently, Korytkowski and Ruiz (1980a, 1980b) discussed the impact of *Eupalamides cyparissias* (Fabricius) [= *dedalus* (Cramer)] on various oil palm plantations. In general, young plants may be killed by the boring larva, but older plants may show little outward damage. On the other hand, the natives of Chile rear specimens of *Castnia eudesmia* Gray, the larvae of which are used as fish bait (Steinhauser, personal communication).

Selected Bibliography

Bourquin 1933.
Common and Edwards 1981.
Esquivel 1980.
Grünberg 1909.
Houlbert 1918.
Jörgensen 1930.
Korytkowski and Ruiz 1980a, 1980b.
Marlatt 1905.
Miller, J. 1980, 1986.
Moss 1945.
Philippi 1863.
Salt 1929.
Seitz 1913.
Tindale 1928.

TORTRICIDAE (TORTRICOIDEA)

Richard L. Brown, *Mississippi State University*

The Tortricids

Figures 26.118–26.136

Relationships and Diagnosis: The Tortricidae are composed of three subfamilies that include over 5000 species of small- to moderate-sized moths. Although ancestral relationships of Tortricidae are questionable, the family appears to be most closely related to Cossoidea and Sesioidea. In the United States and Canada, Tortricidae are represented by 370 species of Tortricinae, three species of Chlidanotinae, and 800 species of Olethreutinae. Cochylini and Olethreutinae have been given family status by some workers; the Cochylini have been treated as a tribe of Tortricinae by Kuznetsov and Stekolnikov (1973, 1977) and Razowski (1976).

Phylogenetic relationships of the tribes and subfamilies of Tortricidae remain unresolved, partly because various workers have placed emphasis on different types of characters and because plesiomorphic and apomorphic states of these characters have not been determined or clearly defined. MacKay (1959, 1962) concluded that less intraspecific variation of larval characters occurs in Tortricinae than in Olethreutinae, that the setal arrangement was more advanced in Olethreutinae, and that the Tortricinae was the evolutionarily older and less plastic of the two subfamilies. Within Oletreutinae, MacKay (1959) considered Olethreutini to be "the most evolved" group and the Grapholitini to be "the most primitive." The sequence of tribes given by MacKay (1962) for Tortricinae (Tortricini-Cnephasiini-Archipini-Sparganothini) was considered to reflect "increasing specialization and perhaps a true phylogenetic order". Note, too, that MacKay examined several species in genera of Cochylini, but included these genera in her key to species groups of Olethreutinae because their characteristics were so similar to those of some olethreutines.

Powell (1964) considered genital structures, as used by Heinrich (1923, 1926), to be the most rapidly evolving characteristics and to be of less value than biological characteristics in showing group relationships. Because Cossoidea and primitive groups of Lepidoptera are primarily internal feeders, Powell considered the externally feeding Tortricinae (excluding Cochylini) to be derived from an internal feeding olethreutoid ancestor. Powell's delineation of biological characteristics of Tortricinae support larval characteristics in suggesting that Tortricini and Cnephasiini are the most ancestral and Archipini and Sparganothini are the most recently evolved tribes.

Based on genitalic morphology with emphasis on musculature, Kuznetsov and Stekolnikov (1973, 1977) and Razowski (1976) concluded that Olethreutinae evolved from an external-feeding tortricine ancestor. These authors considered Olethreutini to be the most ancestral tribe and Grapholitini to be the most recently derived tribe within Olethreutinae, but they do not agree on relationships of the higher groups of Tortricinae.

The ancestor of Tortricidae is considered here to have the tortricine form of larva with external feeding habits and the tortricine form of genitalia. Internal feeding behaviors, e.g., stem and root boring, fruit feeding, and leaf mining, are considered to be secondarily derived and not homologous with internal feeding behaviors in Cossoidea, Gelechioidea, and other more primitive Lepidoptera. Internal feeding habits, as in some Chlidanotinae, Cochylini, and Olethreutinae, are thought to have evolved independently within the three subfamilies.

Most tortricid larvae can be recognized by the following set of characters: L1 and L2 adjacent on A1–8; SD1 on A8 usually anterior, anterodorsal or anteroventral to the spiracle (family key, fig. 124); A9 with D2 setae usually on a common middorsal pinaculum and closer to each other than each is to its own D1 (family key, fig. 52); A9 with D1 usually closer to SD1 than to D2 (family key, fig. 52), and D1 and SD1 frequently on the same pinaculum.

Tortricid larvae (except for known Hilarographini) have a trisetose T1 L group to distinguish them from pyralids. Crochets on the ventral prolegs are uniserial and in a complete circle, and uni-, bi-, or triordinal; crochets on the A10 prolegs are in a continuous row, whereas they are divided in some Gelechiidae.

Biology and Ecology: Parallel feeding habits have developed in Tortricidae; they are not always a tribal character but are often characteristic of species groups or genera and may vary with larval age or with generations. For example, second instar *Choristoneura fumiferana* (Clemens), mine needles whereas later instars are external feeders. Although dicotyledons are hosts to most Tortricidae, grasses and sedges are hosts to some species, e.g., *Clepsis* (Archipini), and *Bactra* (Olethreutini).

Tortricinae: Species are monophagous, oligophagous, or polyphagous; most feed externally on leaves that are rolled, crumpled, or tied together. With the exception of Tortricini, most species overwinter as eggs or larvae in hibernacula.

Tortricini. Eggs are usually laid singly and larvae are usually oligophagous. Host plants of most species are in Rosaceae, Salicaceae, Betulaceae, Ericaceae, and Fagaceae. A few species, including *Acleris gloverana* (Walsingham) and *A. variana* (Fernald), feed on Pinaceae. Larvae feed mainly on leaves, but some European species feed on fruits. Larvae of the African genus *Polemograptis* are known to feed on Coccidae living on the cocoa tree. Many Holarctic species of *Acleris*, the largest genus of the tribe, have two generations, with second-generation adults hibernating (Razowski 1966).

Cnephasiini. The biology is very poorly known for most species. Eggs are laid singly or in small groups. Although feeding habits are diverse, larvae tend to be polyphagous within a host family or among several families of trees and herbaceous plants, especially Rosaceae, Asteraceae (= Compositae), Betulaceae, and Fagaceae. Some species mine the leaf in early instars and later feed externally among webbed leaves. Many British species feed on flower heads, drawing together petals and eating immature seeds. Some feed on fibrous roots of grasses and herbaceous plants (Bradley et al. 1973). The form of larval shelter is species specific in some cases (Powell 1964, Razowski 1965).

Archipini. Eggs are deposited in masses of imbricated rows. Larval habits and host specificity are variable. Early instars of most Archipini are oligophagous; later instars are polyphagous, feeding on various secondary hosts (Chapman and Lienk 1971, Razowski 1977). *Choristoneura rosaceana* (Harris) and *Argyrotaenia citrana* (Fernald) have been recorded from over 50 and 100 hosts, respectively. In contrast, some species in the *C. fumiferana* complex have been separated by host specificity, e.g., *C. pinus* Freeman (jack pine budworm). Pinaceae are hosts to relatively more Archipini than to other tribes of Tortricinae. Archipini have developed the greatest specialization toward characteristic shelters (Powell 1964). Larvae of some *Archips*, e.g., *A. cerasivorana* (Fitch), form tents of dense silk that cover entire branches.

Sparganothini. Eggs are deposited in imbricated groups; many species overwinter as early instar larvae in hibernacula. Most known larvae are polyphagous on a wide range of hosts. Larvae roll leaves or spin them together, eat into shoots or flowers, or bore into buds, seeds, fruits, and stems. Larvae of some species are known to live as small groups in tubes of silk (MacKay 1962).

Cochylini. Eggs are deposited singly. Larvae of most species are oligophagous, generally feeding on flowers, seeds, stems and roots of Asteraceae (= Compositae) or Apiaceae (= Umbelliferae); other species feed on diverse hosts that are not used frequently by other Tortricinae or Olethreutinae. Larvae of many species feed initially on flowers or seeds before boring into the stem or root to complete development. Early generations that feed in flower heads may feed exclusively in stems during later generations. Overwintering occurs in the pupal state, and pupation may occur on the host plant or in ground litter. In some species, larvae leave their habitation in the flower head and bore into the stem or root to pupate.

Chlidanotinae: *Hilarographini.* Although larvae are unknown for the three Nearctic species, the larva of *Thaumatographa eremnotorna* Diakonoff and Arita, a Japanese species, differs from all other tortricid larvae in having a bisetose L group on the prothorax (Diakonoff and Arita 1981). *T. regalis* has been reared from cambium of *Pinus* in California and the other two Nearctic species have been collected in or near pine forests (Heppner 1982a).

Olethreutinae: Eggs are laid singly. Larvae of most species are oligophagous or monophagous. Many Olethreutinae are external feeders on leaves; certain species and genera considered to be derived are internal feeders in stems, roots, seeds, or fruits. Overwintering occurs in all stages, although larval and pupal diapause are most common.

Olethreutini. Polyphagy is more common in this tribe than in other Olethreutinae, although most species tend to be oligophagous leaf rollers and webbers. The various species use a wider diversity of host families than any other Nearctic group of Tortricidae. Rosaceae, Betulaceae, Salicaceae, and Ericaceae serve as frequent hosts; Pinaceae and Fabaceae (= Leguminosae) are hosts to a few species.

Eucosmini. Most species in primitive genera (e.g., *Ancylis, Epinotia, Gypsonoma*) are external leaf webbers and rollers on Rosaceae, Betulaceae, Salicaceae, Ericaceae, and Fagaceae. Recently derived genera are mostly internal feeders in stems and roots of Asteraceae (= Compositae), e.g., *Eucosma* and *Epiblema*, or in shoots or cones of Pinaceae, e.g. *Petrova, Rhyacionia*, and *Zeiraphera*. Stem galls are formed by many species of *Epiblema*.

Grapholitini. Monophagy is relatively more common in this tribe than in other Olethreutinae. Most species feed in fruits and seeds of Pinaceae, Asteraceae, or Fabaceae, e.g., *Cydia* and *Grapholita;* most *Dichrorampha* feed in roots and stems of Asteraceae; and many *Pammene* are gall inquilines.

Description: Mature larvae small, 8–25 mm, varying from slender (most external feeders) to rather stout (most internal feeders); body more or less uniform in color, near white, pink, green, purple, or brown; secondary setae absent.

Head: Hypognathous or semiprognathous; often darkly pigmented in early instars, lighter in final instar; adfrontals reaching or nearly reaching epicranial notch; epicranial suture usually short; six stemmata, usually unevenly distributed, with 6 always close to 4 and 5.

Thorax: T1 shield variably pigmented, often with distinct medial sulcus; T1 with trisetose L group (bisetose in Hilarographini); T2 and T3 with SV group unisetose or bisetose, L1 closer to L2 than L3.

Abdomen: L group setae adjacent to each other and obliquely or vertically placed; A3 with L1 and L2 closer to each other than either is to spiracle; SD1 on A8 usually anterior or somewhat anterodorsal or anteroventral from spiracle; A9 with D2's usually on same pinaculum and closer together than D2's on A8; D1 usually closer to SD1 than D2, frequently on same pinaculum as SD1. Spiracles broadly elliptical, with eighth pair usually somewhat larger and more dorsad. Prolegs on A3–6 and A10; uniordinal, biordinal, or triordinal crochets in complete circles on A3–6 prolegs; A10 prolegs with crochets in continuous row, and with central crochets same length as those at either end. Anal shield variably pigmented. Anal fork present or absent.

Comments: The Olethreutinae and Tortricinae cannot be separated by larval characters at present. Keys to tribes, as given by Swatschek (1958) and MacKay (1962), are of limited use. Keys for separating species of major economic importance on Pinaceae, legumes, apple, peach, and generically related fruits, provided below, are based largely on the treatments of selected tortricid larvae by MacKay (1959, 1962) and are, therefore, limited by the scope of these earlier works. Emphasis has been placed on species occurring in eastern North America, especially for those species feeding on Pinaceae. Chapman and Lienk (1971) have provided keys to larvae on apple, although emphasis is on color and pattern of nonpreserved specimens. Descriptions of larvae and figures of larval feeding damage have been provided for British species by Bradley et al. (1973, 1979).

Selected Bibliography

Bradley et al. 1973, 1979.
Chapman and Lienk 1971.
Diakonoff and Arita 1981.
Heinrich 1923, 1926.
Heppner 1982a.
Kuznetsov and Stekolnikov 1973, 1977.
MacKay 1959, 1962.
Powell 1964.
Razowski 1965, 1966, 1976, 1977.
Swatschek 1958.

KEY TO SOME COMMON TORTRICID LARVAE ON PINACEAE

1. Anal fork absent 2

1′. Anal fork present 13

2(1). SV group on A3–6 always with 4, often 5 setae; borer in bark and stems of pines and spruces, forming a nodule of pitch and frass ..... *Petrova* spp.

2′. SV group on A3–6 with 3 setae or fewer; larval habits variable, but not forming nodules 3

3(2′). SD group on A1–8 with extra seta posterodorsal, posterior, or posteroventral to spiracle (fig. 1); spinneret spatulate; borer in buds and new growth of pines (figs. 26.119, 26.120) ..... *Rhyacionia* spp. (p. 421)

3′. SD group on A1–8 without extra seta; spinneret not spatulate; habits variable 4

4(3′). A8 and A9 with SV group unisetose 5

4′. A8 and usually A9 with SV group bisetose or trisetose 7

5(4). A3–6 prolegs with 15–25 crochets, A10 prolegs with 3–6 crochets; usually in cones of spruce ..... *Cydia strobilella* (L.)

5′. A3–6 prolegs with 25–40 crochets, A10 prolegs with 7–12 crochets; usually in cones of pines 6

6(5′). SD1 on T1 equidistant from SD2 and XD2 or farther from SD2 than from XD2 (fig. 2) ..... (longleaf pine seedworm) *Cydia ingens* (Heinrich)

6′. SD1 on T1 much closer to SD2 than to XD2 ..... *Cydia piperana* Kearfott
Cydia injectiva (Heinrich)

7(4′). Spinneret with tip distinctly bifurcate (fig. 3); SD1 on A8 anterior to spiracle; borer in cones of spruce, balsam, Douglas fir ..... (Douglas fir cone moth) *Barbara colfaxiana* (Kearfott)

7′. Spinneret with tip not bifurcate; SD1 on A8 anterior, anterodorsal, or dorsal to spiracle (if anterior to spiracle, then not a borer in cones) 8

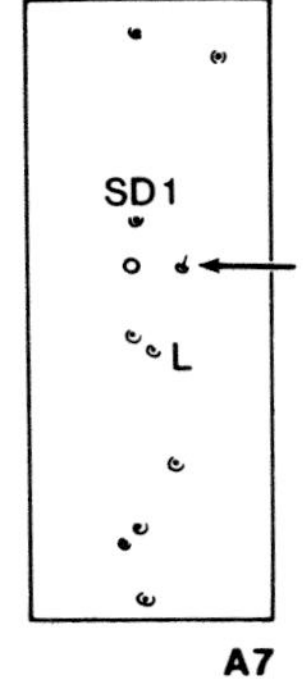

Figure 1

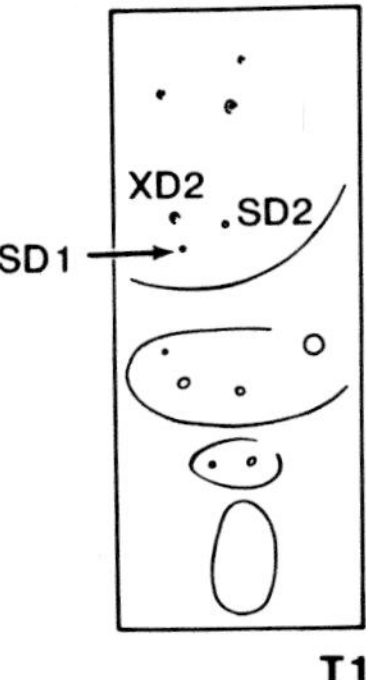

Figure 2

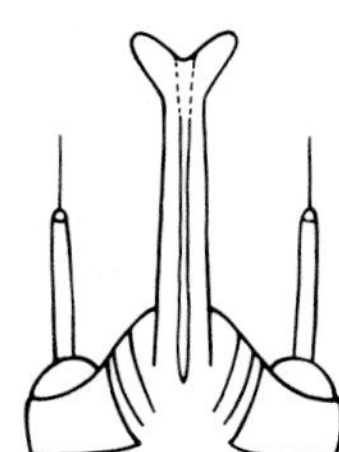
Figure 3

8(7′). Leaf miner or webber of spruce and balsam (sometimes feeding in opening buds); SD1 on A8 usually anterior to spiracle 9

8′. Borer in cones and shoots of spruce and pine; SD1 on A8 dorsal or anterodorsal to spiracle (fig. 4) 11

9(8). L group distinctly anterior to a vertical line through the spiracle on A2–8 (fig. 5); D1 often separate from SD1 on A9; V1s on A9 the same distance apart or slightly closer than those on A8 .... (spruce needleminer) *Endothenia albolineana* (Kearfott)

9′. L group not anterior to spiracle on A2–8; D1 and SD1 on same pinaculum; V1's on A9 farther apart than those on A8 (fig. 16) 10

10(9′). Head, thoracic shield, legs, and anal shield yellow; webbing opening buds and young needles .... (spruce bud moth) *Zeiraphera canadensis* Mutuura and Freeman

10′. Head, thoracic shield, legs, and anal shield dark brown; webbing and mining old needles .... *Epinotia nanana* (Treitschke)

11(8′) SV group on A3–6 bisetose (fig. 6); shoot borer .... (eastern pine shoot borer) *Eucosma gloriola*[5] Heinrich

11′. SV group on A3–6 trisetose; cone borer .... 12[6]

12(11′). SV group on A7 bisetose; anal shield pale; on spruce .... (white pine cone borer) *Eucosma tocullionana* Heinrich

12′. SV group on A7 trisetose; anal shield brown; on pine .... *Eucosma rescissoriana* Heinrich

13(1′). A9 with D1 and SD1 on same pinaculum or on approximate pinacula (fig. 7) 14

13′. A9 with D1 and SD1 on well separated pinacula (fig. 12) 15

14(13). SV group on A7 trisetose; in cones .... *Endopiza piceana* (Freeman)

14′. SV group on A7 bisetose; in buds .... *Zeiraphera* spp.

15(13′). SV group on A7 bisetose 16

15′. SV group on A7 trisetose 17

16(15). Occurring west of the continental divide .... (western blackheaded budworm) *Acleris gloverana* (Walsingham)

16′. Occurring east of the continental divide .... (eastern blackheaded budworm) *Acleris variana* (Fernald)

17(15′). Dorsal, and often subdorsal, pinacula on T2 elongated posteriorly (fig. 8) (difficult to observe in some *Argyrotaenia*, see couplet 21) 18

17′. Dorsal and subdorsal pinacula on T2 not elongated 21

18(17). Head seta P1 closer to P2 than to AF2 (fig. 9); V1's on A9 about 1½—2 times farther apart than those on A8 .... *Sparganothis tristriata* Kearfott

18′. Head seta P1 closer to AF2 than to P2 or equidistant from both (fig. 23); V1's on A9 only slightly farther apart than those on A8 19

19(18′). Head brownish yellow with distinct dark brown pattern; needle webber on spruce, not within a tube of webbed needles .... *Argyrotaenia occultana* Freeman

19′. Head without a distinct dark brown pattern; larva within tube of pine needles webbed together 20

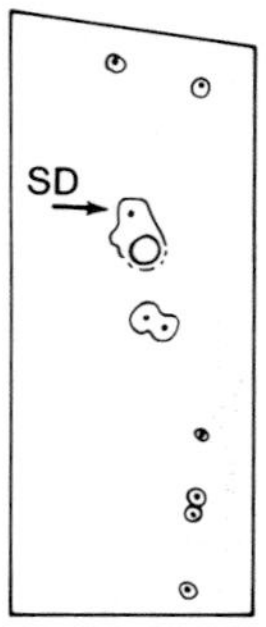

Figure 4

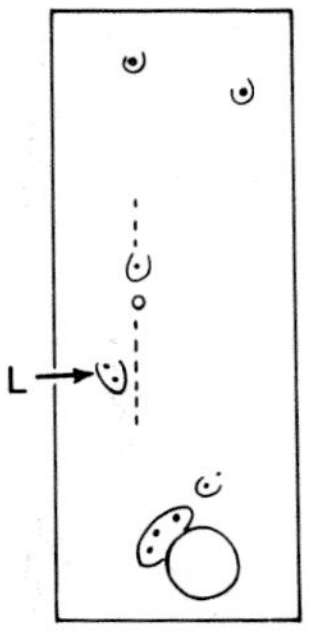

Figure 5

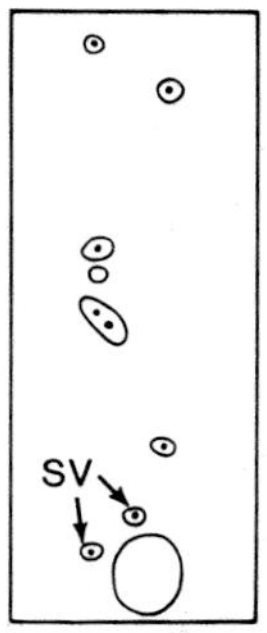

Figure 6

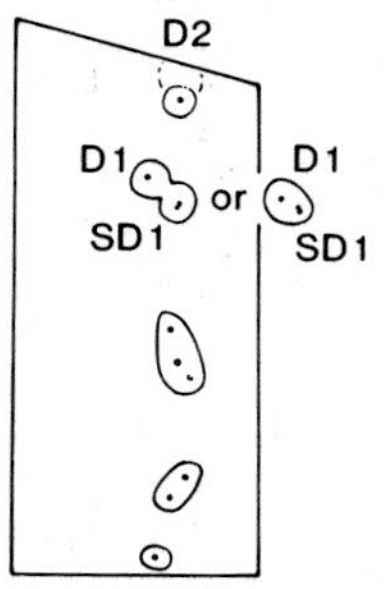

Figure 7

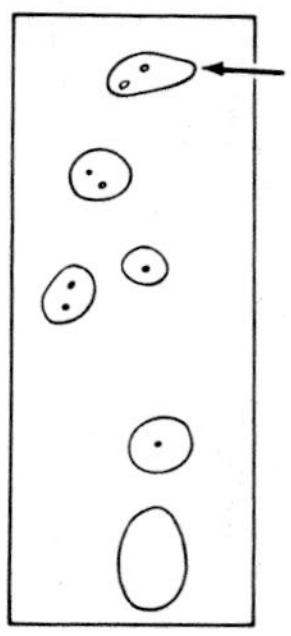

Figure 8

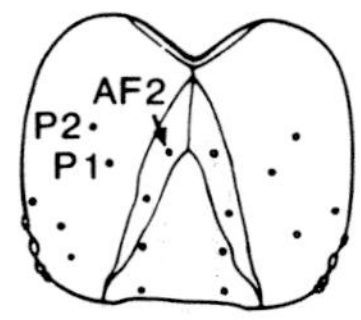

Figure 9

5. May also include *Eucosma sonomona* Kearfott.

6. May also include *Eucosma bobana* Kearfott, *E. monitorana* Heinrich, *E. cocana* Kearfott, *E. ponderosa* Powell and *E. monoensis* Powell (all on pine) and *E. siskiyouana* (Kearfott) (on spruce (*Abies*)).

Figs. 16, 23 ⟶

20(19′). Head brownish yellow, suffused with brown on dorsal area; thoracic legs light brown, without contrasting tarsi; on white pine .... (pine tube moth) *Argyrotaenia pinatubana* (Kearfott)

20′. Head pale brownish yellow without darker suffusion on dorsal area; thoracic legs light brown with dark brown tarsi; on jack pine, lodgepole pine, yellow pine .... *Argyrotaenia tabulana* Freeman

21(17′). Spiracles on A2–7 small, often the size of a setal base; head seta P1 closer to AF2 than to P2 or equidistant from both and at the apex of a right or acute angle (*Argyrotaenia* spp.) 19

21′. Spiracles on A2–7 larger than the size of a setal base; head seta P1 closer to P2 than to AF2 or equidistant from both and at the apex of a right or obtuse angle (fig. 9) 22

22(21′). On pines 23

22′. On spruces and firs 24

23(22). Occurring east of Rocky Mountains .... (jack pine budworm) *Choristoneura pinus* Freeman

23′. Occurring in Rocky Mountains and west .... *Choristoneura lambertiana* (Busck)

24(22′). Dorsal pinacula moderately large, D2 pinaculum on A1–3 with mesal margin on about same level as D1 seta base (fig. 10) .... *Choristoneura* spp.[7]

24′. Dorsal pinacula moderately small, D2 pinaculum on A1–3 with its mesal margin below that of lateral margin of D1 pinaculum (fig. 11) 25

25(24′). Spinules and thoracic legs of body color .... *Archips packardiana* (Fernald)

25′. Spinules and thoracic legs darker than body color 26

26(25′). Thoracic shield dark brown .... *Archips alberta* (McDunnough)

26′. Thoracic shield pale, bordered with dark brown .... *Archips striana* (Fernald)

7. *C. fumiferana* (Clemens), p. 431, *C. retiniana* (Walsingham), *C. occidentalis* Freeman, *C. biennis* Freeman, *C. orae* Freeman, *C. carnana* (Barnes and Busck), *C. spaldingiana* Obraztsov.

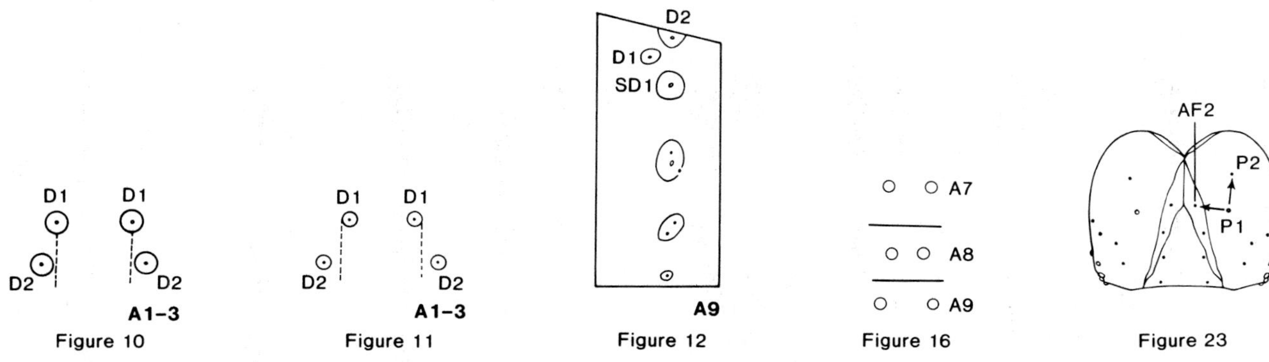

Figure 10 Figure 11 Figure 12 Figure 16 Figure 23

KEY TO SOME COMMON TORTRICID LARVAE ON SOYBEANS, ALFALFA, AND OTHER CULTIVATED LEGUMES

1. A9 with D1 and SD1 on same pinaculum (fig. 7) 2

1′. A9 with D1 and SD1 on separate pinacula (fig. 12) 7

2(1). L group on A9 bisetose 3

2′. L group on A9 trisetose 4

3(2). Stemma 3 prominent on raised area; L1 and L2 on A2–8 ventral to spiracle .... (omnivorous leaftier) *Cnephasia longana* (Haworth)

3′. Stemma 3 not more prominent than other stemmata; L1 and L2 on A2–8 anterior to a vertical line through spiracle (fig. 5) .... *Cnephasia interjectana* (Haworth)

4(2′). SV group on A9 unisetose and usually unisetose on A8 5

4′. SV group on A9 and A8 bisetose 6

5(4). SD pinaculum on A1–7 usually fused with sclerotized area surrounding spiracle (fig. 13); SV group on A2 bisetose; in pods and seeds of peas and beans .... (pea moth) *Cydia nigricana* (Fabricius)

5′. SD pinaculum on A1–7 not fused with sclerotized area surrounding spiracle; SV group on A2 trisetose; in flower heads and seeds of clover .... (clover head caterpillar) *Grapholita interstinctana* (Clemens)

6(4′). D1's on anal shield almost as long as SD2's (fig. 14); anal fork moderately developed (fig. 15) .... *Epinotia aporema* (Walsingham)

6′. D1's on anal shield much shorter than SD2's, scarcely longer than D2's; anal fork poorly developed .... *Epinotia perplexana* (Fernald)

7(1′). Dorsal, and often subdorsal, pinacula on T2 elongated posteriorly (fig. 8) 8

7′. Dorsal and subdorsal pinacula not elongated 14

8(7). Head seta P1 closer to AF2 than to P2 or equidistant from both; V1's on A9 rarely farther apart those on A8 and A7 9

8′ Head seta P1 closer to P2 than to AF2 (fig. 9); V1's on A9 always farther apart than those on A8 and A7 (1½–2 times) (fig. 16) 10

9(8). Thoracic shield with 2 brown areas anteromedially (fig. 17); anal shield with some brownish pigment .... *Choristoneura parallela* (Robinson)

9′. Thoracic shield without brown areas anteromedially; anal shield without brownish pigment (see MacKay 1962) .... (obliquebanded leafroller) *Choristoneura rosaceana* (Harris)
(fig. 26.129) *Choristoneura obsoletana* (Walker) (p. 431)

10(8′). Dorsal pinacula on A1–8 slightly elongated longitudinally and often cream-colored (as in fig. 8) 11

10′. Dorsal pinacula on A1–8 circular, never cream-colored 12

11(10). Head and thoracic shield brown, overlaid with darker brown pattern .... *Platynota idaeusalis* (Walker)

11′. Head and thoracic shield brownish yellow, without darker brown pattern, except laterally and posteriorly on thoracic shield .... (fig. 26.133) *Platynota stultana* Walsingham (p. 432)

12(10′). Head pale, except for dark pigmentation surrounding stemmata 3, 4, and sometimes 5; thoracic shield and legs pale and unicolorous .... *Sparganothis unifasciana* (Clemens)

12′. Head brownish-yellow with dark stemmatal area; thoracic shield edged laterally with brown; legs yellowish-brown or brown, or pale with brown-yellow tarsi 13

13(12′). Spinules and their bases dark on dorsum; thoracic legs pale with brownish-yellow tarsi; length of last instar larva less than 17 mm .... *Sparganothis sulfureana* (Clemens)

13′. Spinules and their bases pale; thoracic legs yellowish-brown or brownish-yellow, unicolorous; length of last instar larva over 18 mm .... *Platynota nigrocervina* Walsingham

14(7′). Stemmata 3, 4, and 5 strongly convex (fig. 18); anal shield weakly tapered posteriorly, with SD1's twice as long as anal segment ...... (three-lined leafroller) *Pandemis limitata* (Robinson)

14′. Stemmata not strongly convex; anal shield strongly tapered, with SD1's less than 1½ times as long as anal segment (fig. 19) 15

15(14′). Larva pale, unicolorous .... *Xenotemna pallorana* (Robinson)

15′. Larva with dark pigment on stemmatal area, thoracic, and anal shields .... *Aphelia alleniana* (Fernald)

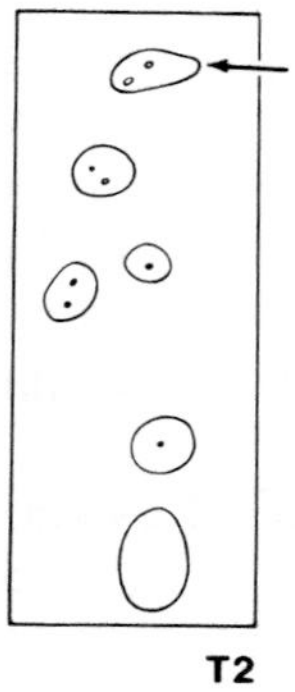

T2

Figure 8

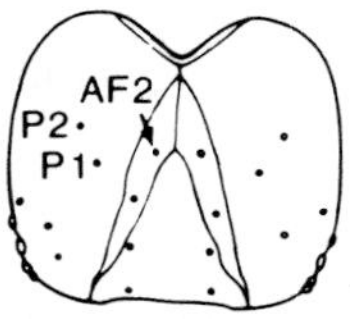

AF2 P2 P1

Figure 9

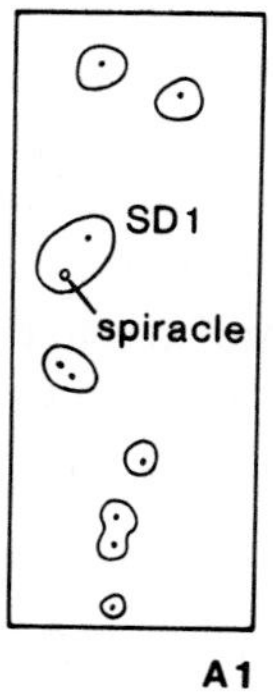

SD1 spiracle A1

Figure 13

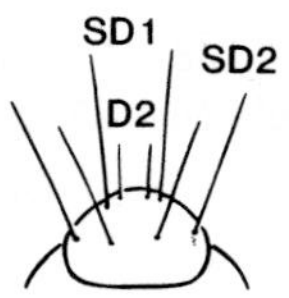

SD1 SD2 D2

Figure 14

Figure 15

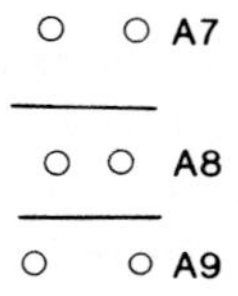

A7 A8 A9

Figure 16

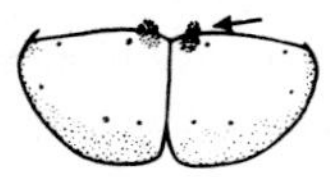

Figure 17

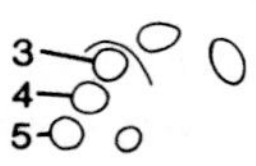

3 4 5

Figure 18

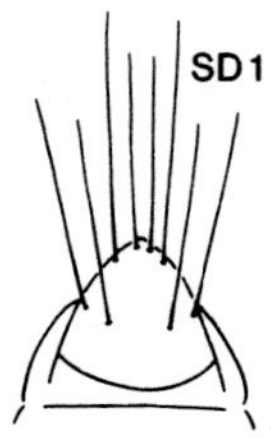

SD1

Figure 19

KEY TO SOME COMMON TORTRICID LARVAE ON APPLE, PEACH AND GENERICALLY RELATED FRUITS

1. A9 with D1 and SD1 on separate pinacula (fig. 12) 2

1′. A9 with D1 and SD1 on same pinaculum (fig. 7) 19

2(1). SV group on T2 bisetose; anal fork absent; larvae gregarious on host .... (uglynest caterpillar, fig. 26.132) *Archips cerasivorana* (Fitch) (p. 432)

2′. SV group on T2 unisetose; anal fork present; larvae not gregarious 3

3(2′). SV group on A7 bisetose 4

3′. SV group on A7 trisetose 6

4(3). V1's on A9 twice as far apart as those on A8 (fig. 20) .... *Acleris* species[8]

4′. V1's on A9 and A8 equal distance apart 5

5(4′). SD2 anterodorsal to SD1 on T2 (fig. 21); pinacula black .... (green budworm) *Hedya nubiferana* (Haworth)

5′. SD2 dorsal to SD1 on T2; pinacula greenish-amber .... (part) *Choristoneura fractivittana* (Clemens)

6(3′). Dorsal, and often subdorsal, pinacula on T2 elongated posteriorly (fig. 8) (difficult to observe in some *Argyrotaenia,* see couplet 14) 7

6′. Dorsal and subdorsal pinacula on T2 not elongated 13

7(6). Head seta P1 closer to AF2 than to P2 or equidistant from both; V1's on A9 rarely farther apart than those on A8 and A7 8

7′. Head seta P1 closer to P2 than to AF2 (fig. 9); V1's on V9 always farther apart than those on A8 and A7 (1½–2 times) (fig. 16) 11

8(7). Spiracle on abdomen of last instar larva always much larger than a setal base; last instar over 19 mm long .... (obliquebanded leafroller) *Choristoneura rosaceana* (Harris)

8′. Spiracle on abdomen of last instar larva about the size of a setal base or slightly larger; last instar less than 18 mm long 9

9(8′). Head yellowish or brownish yellow, last instar 13–18 mm long ....
.... (eastern North America, fig. 26.136, redbanded leafroller) *Argyrotaenia velutinana* (Walker) (p. 433)
.... (western North America, orange tortrix) *Argyrotaenia citrana* (Fernald)

9′. Head pale, unicolorous with body, last instar 12–23 mm long 10

10(9′). Spinules slender, darker than body color; pinacula of D1's on A1–8 about twice the diameter of a pinaculum apart; last instar over 17 mm .... (graybanded leafroller) *Argyrotaenia mariana* (Fernald)

10′. Spinules not apparent; pinacula of D1's on A1–8 about 4 times the diameter of a pinaculum apart (fig. 22); last instar less than 17 mm .... *Argyrotaenia quadrifasciana* (Fernald)

11(7′). T1 pinacula brown; head black, reddish-brown or brown with darker brown pattern .... *Platynota idaeusalis* (Walker)

11′. T1 pinacula pale; head brownish-yellow with brown restricted to stemmatal area 12

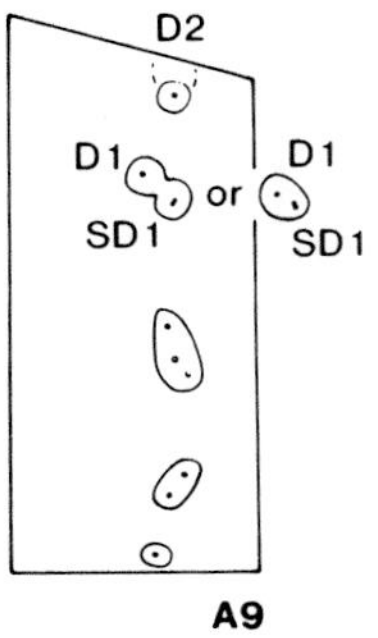

Figure 7

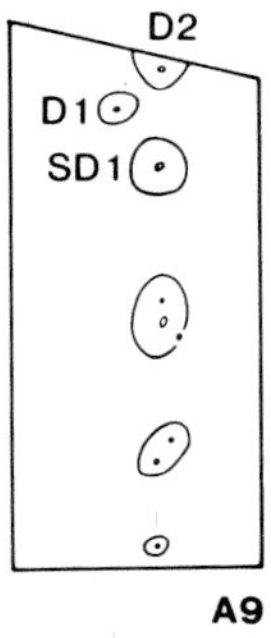

Figure 12

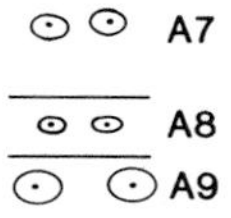

Figure 20

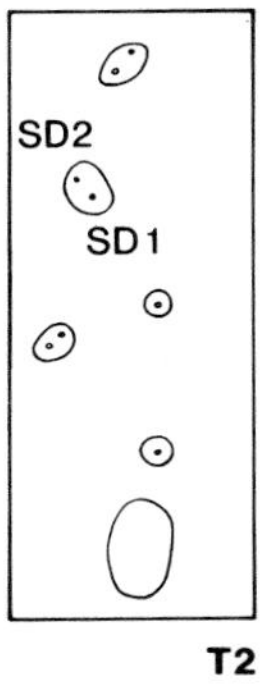

Figure 21

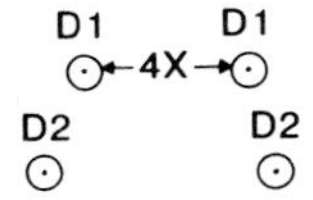

Figure 22

8. *Acleris minuta* (Robinson) is the major pest species in this genus; other apple-feeding species include *A. flavivittana* (Clemens) and *A. nivisellana* (Walsingham).

←— Figs. 8, 9, 16

12(11′) Anal shield brownish-yellow with large area of brown laterally .... *Sparganothis reticulatana* (Clemens)

12′. Anal shield yellowish, without lateral dark areas .... *Sparganothis sulfureana* (Clemens)

13(6′). SD2 dorsal to SD1 on T2; D2 pinaculum on A1 with mesal margin below lateral margin of D1 pinaculum (fig. 11) .... (part) *Choristoneura fractivittana* (Clemens)

13′. SD2 distinctly anterodorsal to SD1 on T2 (fig. 21); D2 pinaculum on A1 with mesal margin below or above lateral margin of D1 pinaculum 14

14(13′). Head seta P1 equidistant from P2 and AF2, or closer to AF2, and at apex of a right or acute angle formed with P2 and AF2 (fig. 23); spiracles small, often the size of a setal base on A2–7; thoracic shield usually without dark pigment ....(See couplet 9) *Argyrotaenia* spp.

14′. Larva without above combination of characters 15

15(14′). Length of coronal suture equal to or less than width of adfrontal area at apex of clypeus; D1 dorsal to D2 on T2 16

15′. Length of coronal suture nearly twice width of adfrontal area at apex of clypeus (fig. 24); D1 posterodorsal to D2 on T2 (fig. 25) 17

16(15′). Body greenish-yellow; thoracic and anal shields without dark markings; feeding on lower surface of leaf .... *Ancylis fuscociliana* (Clemens)

16′. Body olive-green; thoracic and anal shields with dark markings; feeding on upper surface of leaf .... *Ancylis nubeculana* (Clemens)

17(15′). Setal bases dark, usually ringed with brown pigment; thoracic legs brown, (not graduated from a dark anterior to a light posterior pair) .... *Archips rosana* (L.)

17′. Setal bases not dark and not ringed with brown; thoracic legs pale, concolorous with body, or graduated from a dark anterior pair to a light posterior pair 18

18(17′). Anal setae long, SD1's over twice as long as anal segment, D1's ¾ the length of SD1's .... (three-lined leafroller) *Pandemis limitata* (Robinson)
.... *P. pyrusana* Kearfott
.... *P. lamprosana* (Robinson)

18′. Anal setae shorter, SD1's less than twice as long as anal segment, D1's only slightly less than the length of SD1's ... (fig. 26.131, fruittree leafroller) *Archips argyrospila*[9] (Walker) (p. 432)

19(1′). Anal comb absent .... (fig. 26.126, codling moth) *Cydia pomonella* (L.) (p. 430)

19′. Anal comb present 20

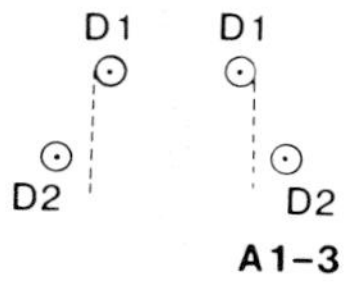

Figure 11

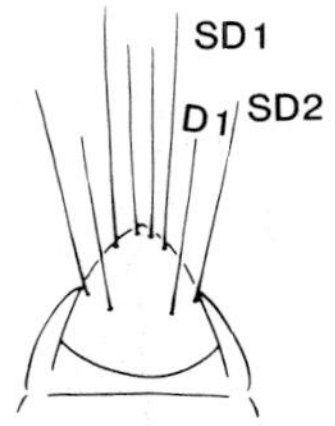

Figure 19

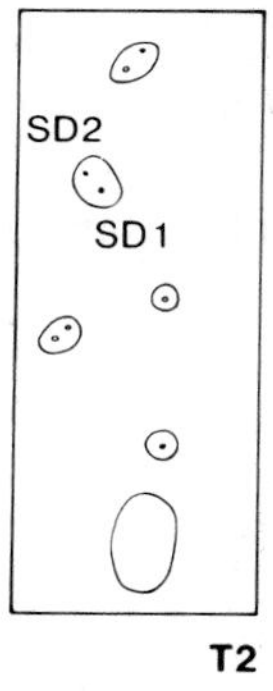

Figure 21

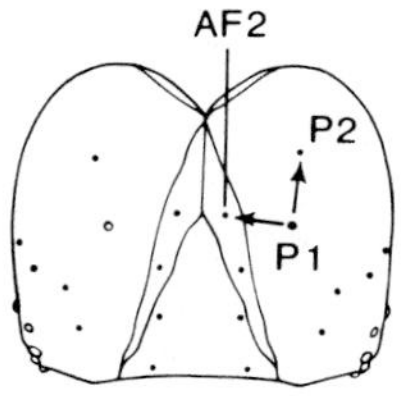

Figure 23

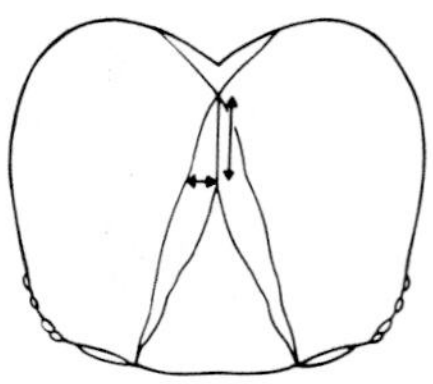

Figure 24

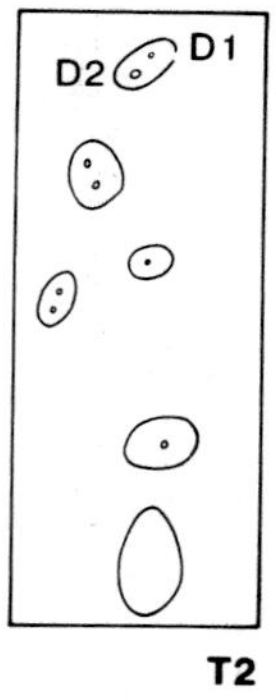

Figure 25

9. Possibly more than one species involved; see MacKay, 1962.

20(19′). SV group on A2 and L group on A9 bisetose .... *Hedya chionosema* (Zeller)

20′. SV group on A2 and L group on A9 trisetose 21

21(20′). SV group on A7 trisetose; D1 posterodorsal to D2 on T2 .... *Episimus tyrius* Heinrich

21′. SV group on A7 bisetose; if trisetose, D1 dorsal to D2 on T2 22

22(21′). SV group on A9 usually unisetose; SD2 on T2 usually anterodorsal to SD1; fruit, bud and twig borers 23

22′. SV group on A9 bisetose; SD2 on T2 usually dorsal to SD1, bud borers and leaf webbers 25

23(22). Length and width of head more than 9 and 9.5 mm respectively .... (fig. 26.124, Oriental fruit moth) *Grapholita molesta* (Busck) (p. 430)

23′. Length and width of head less than 9 and 9.5 mm respectively 24

24(23′). Pinacula on posterior segments suffused with reddish pigment; body retaining pinkish color in 70% alcohol .... (lesser appleworm) *Grapholita prunivora* (Walsh)

24′. Pinacula on posterior segments not suffused with reddish pigment; body not retaining pinkish color in 70% alcohol .... (cherry fruitworm) *Grapholita packardi* Zeller

25(22′). D1's on A8 farther apart than D2's; pinacula conspicuous, darker than body color .... (fig. 26.122, eyespotted budmoth) *Spilonota ocellana* (Denis and Schiffermüller) (p. 429)

25′. D1's on A8 closer together than D2's; pinacula inconspicuous, concolorous with body .... *Pseudexentera mali* Freeman

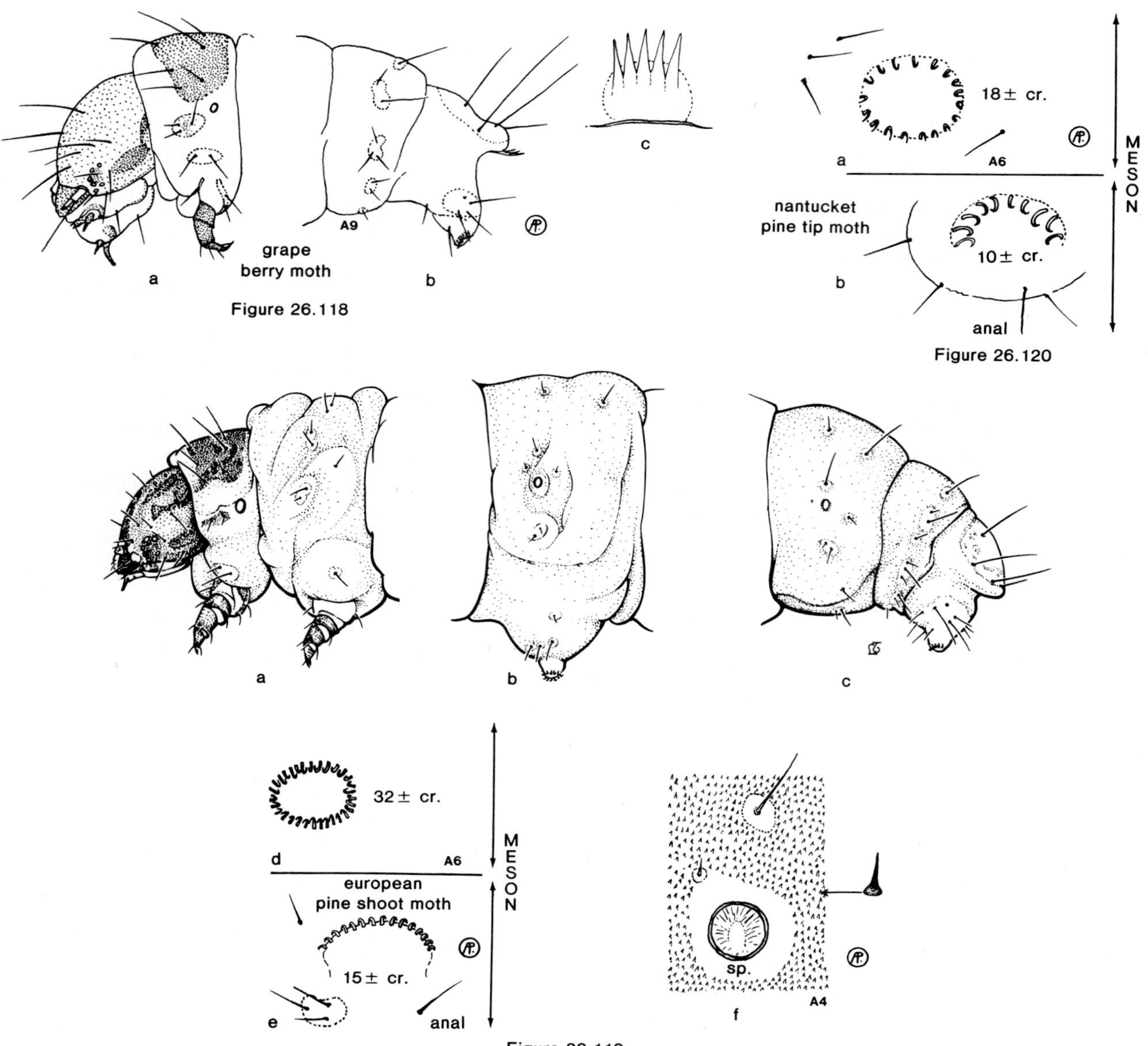

Figure 26.118a-c. Tortricidae. *Endopiza viteana* Clemens, grape berry moth. a. head, T1; b. A9-A10; c. anal comb. f.g.l. 10± mm. Greenish to gray to purple with light brown head and dark T1 shield. Anal comb, 5± prongs, inconspicuous. Infests cultivated and wild grapes, feeding externally on young fruit and bores within mature berries. On foliage it cuts out and folds over a small flap and spins a cocoon within (figs. a-c from Peterson 1948).

Figure 26.119a-f. Tortricidae. *Rhyacionia buoliana* (Denis and Schiffermüller), European pine shoot moth. a. head, T1, T2; b. A3; c. A8-A10; d. A6 crochets; e. A10 crochets; f. A4 spiracle and spinules. f.g.l. 14± mm. Dark brown with head, cervical shield, and thoracic legs near-black and yellowish brown speckled anal plate. Cuticle finely stippled with tiny microspines. Infests terminal shoots of many pines. Especially heavily attacked are Scotch, red, Austrian, and mugo; also attacks Japanese red, Japanese black, ponderosa, eastern white, longleaf, pitch, Virginia, jack and lodgepole (figs. d-f from Peterson 1948).

Figure 26.120a,b. Tortricidae. *Rhyacionia frustrana* (Comstock), nantucket pine tip moth. f.g.l. 9± mm. Yellowish, with light brown head and shield. Setae more difficult to see in comparison with European pine shoot moth (fig. 26.119). Spins web around terminal bud and mines the twig and bases of needles, especially young trees. Infests most eastern pines except longleaf and eastern white (figs. a, b from Peterson 1948).

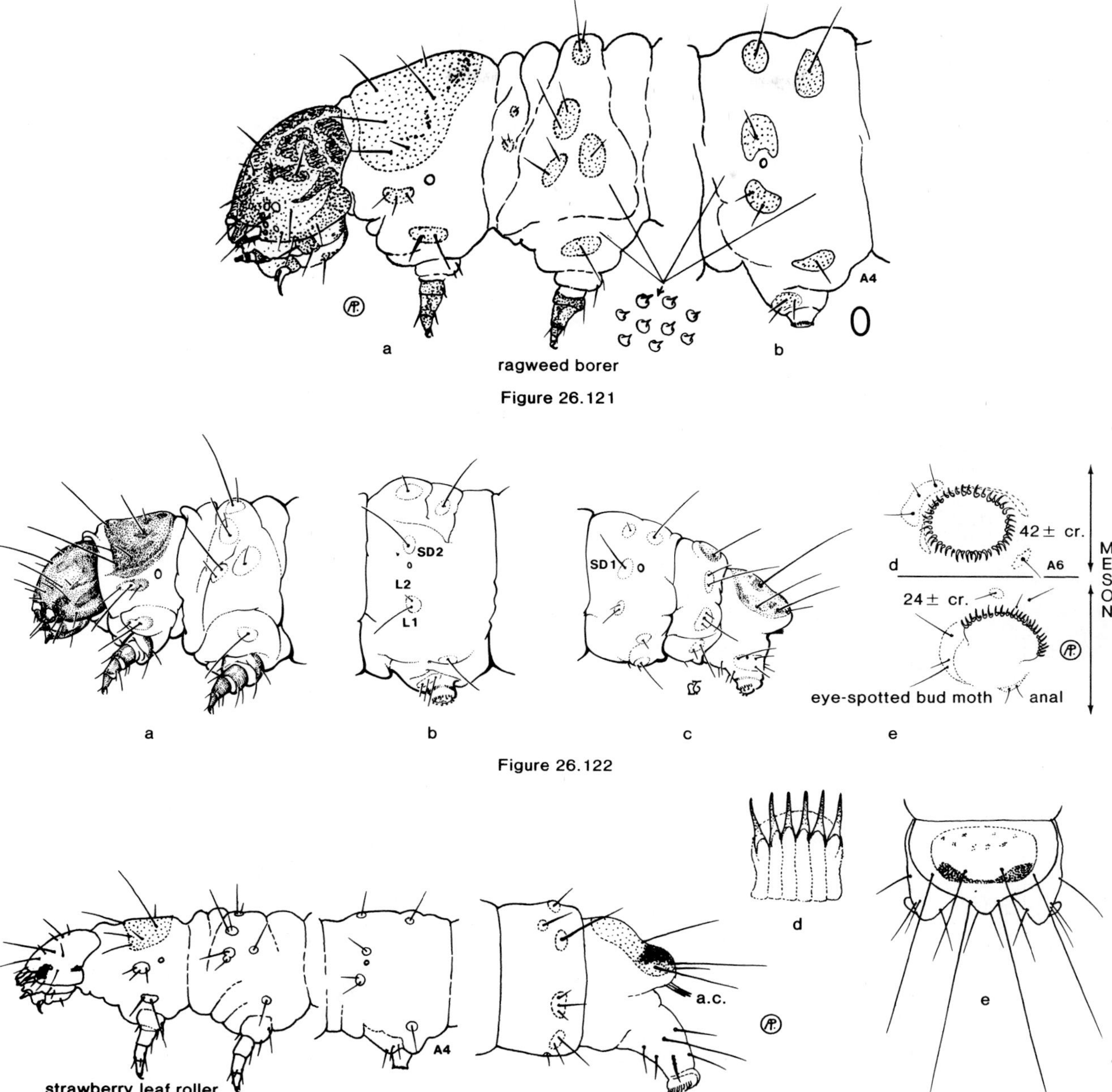

Figure 26.121

Figure 26.122

Figure 26.123

Figure 26.121a,b. Tortricidae. *Epiblema strenuana* (Walker), ragweed borer. a. head, T1, T2; b. A4. f.g.l. 20± mm. Dirty white to yellowish. Head, cervical shield, and anal plate deep brown. Sclerotized, brown pinacula conspicuous. Produces stem galls or enlargements on ragweed (figs. a, b from Peterson 1948).

Figure 26.122a-e. Tortricidae. *Spilonota ocellana* (Denis and Schiffermüller), eyespotted bud moth. a. head, T1, T2; b. A3; c. A8-A10; d. A6 crochets; e. A10 crochets. f.g.l. 9-14 mm. Head and thoracic shield varying from reddish brown to almost black, pinacula and thoracic legs dark brown, spinules of integument not apparent but with dark and large bases. Holarctic, feeding primarily on Rosaceae, especially stone and pome fruits (figs. d, e from Peterson 1948).

Figure 26.123a-e. Tortricidae. *Ancylis comptana* (Frölich), strawberry leafroller. f.g.l. 12± mm. a. head, T1, T2; b. A4; c. A9-A10; d. anal comb; e. anal plate. Greenish to bronze, with shining brown head and T1 shield. Caudal margin of anal plate deeply pigmented. Anal comb 6± prongs. Feeds on leaves of strawberry, blackberry, and raspberry. Folds leaf longitudinally, living and feeding within (figs. a-e from Peterson 1948).

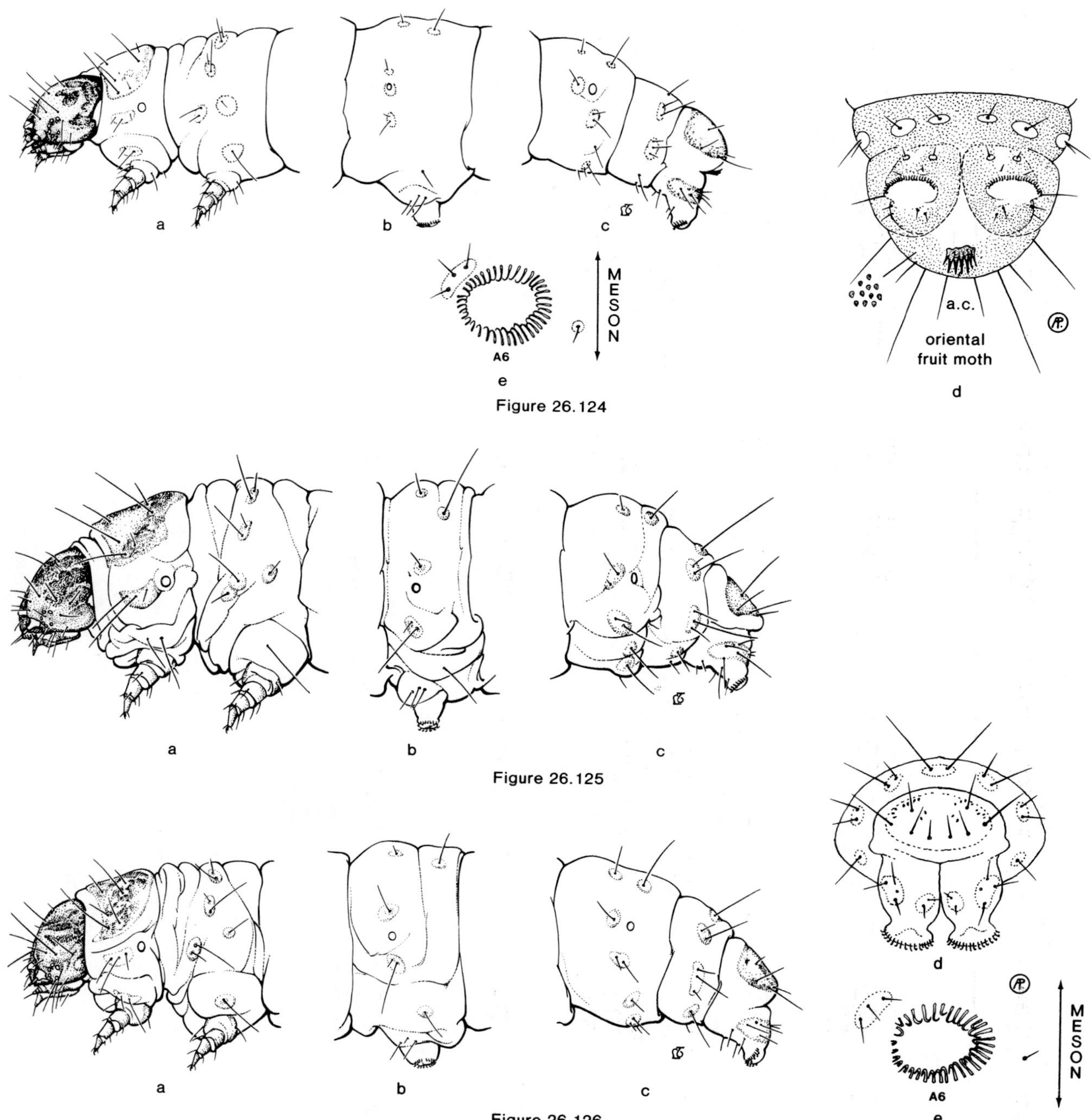

Figure 26.124

Figure 26.125

Figure 26.126

Figure 26.124a-e. Tortricidae. *Grapholita molesta* (Busck), Oriental fruit moth. f.g.l. 9–13 mm. a. head, T1, T2; b. A3; c. A8-A10; d. A9-A10 venter; e. A6 crochets. Pinkish to near-white with light brown head; thoracic and anal shields lightly sclerotized, pinacula large and pale; spinules of integument distinct, slender and darker than body color. Boring in shoot terminals or into fruits of peach, apricot, plum, cherry, apple, pear, and other fruit trees. Occurs in all peach growing districts of the United States, Canada, Eurasia, and South America (figs. d, e from Peterson 1948).

Figure 26.125a-c. Tortricidae. *Cydia latiferreana* (Walsingham), filbertworm. f.g.l. 12–15 mm. a. head, T1, T2; b. A3; c. A8-A10. Head yellowish brown to reddish, thoracic and anal shields yellowish or brownish, pinacula light brown or of pale body color, spinules of integument long and dark. In acorns, beech nuts, filbert nuts, chestnut burrs, and oak galls. Throughout the United States and Canada.

Figure 26.126a-e. Tortricidae. *Cydia pomonella* (L.), codling moth. f.g.l. 14–18 mm. a. head, T1, T2; b. A3; c. A8-A10; d. A9, A10, caudal view; e. A6 crochets. Head yellow-brown often overlaid with darker brown pattern, prothoracic and anal shields with dark speckling, spinules of integument distinct. Primarily on apple, occasionally on pear, apricot, and Persian walnut throughout most of the temperate regions of the world. Anal comb absent (figs. d, e from Peterson 1948).

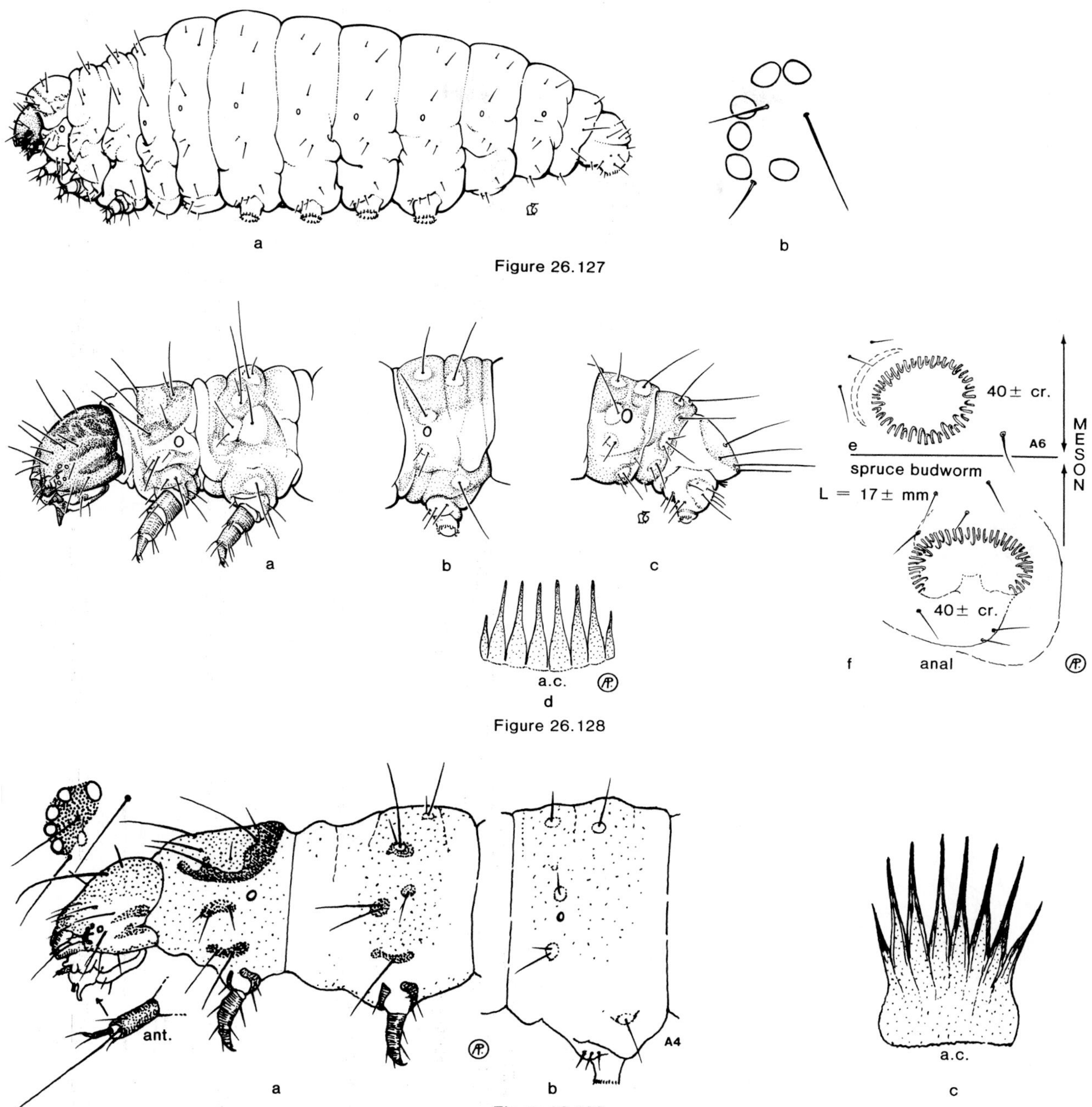

Figure 26.127

Figure 26.128

Figure 26.129

Figure 26.127a,b. Tortricidae. *Cydia deshaisiana* (Luc.), Mexican jumping bean larva. f.g.l. 8± mm. b. stemmata on left side. Near-white, stubby larva with brown head. Lives within seeds of *Sebastiana* spp. Larva feeds within the seed, and when nearly mature causes the seed to "jump" by quick internal movements. Sold commercially.

Figure 26.128a-f. Tortricidae. *Choristoneura fumiferana* (Clemens), spruce budworm. f.g.l. 20-22 mm. a. head, T1, T2; b. A3; c. A8-A10; d. anal comb; e. A6 crochets; f. A10 proleg. Head usually almost entirely brown, prothoracic shield usually brownish-yellow diffused with some brown pigment, occasionally entirely dark brown; anal shield brownish-yellow, spinules of integument dense, short, with dark bases on dorsum and venter, pale bases in spiracular and subspiracular areas. On balsam fir and, to a lesser degree, on spruce, larch, and pine. East of Rocky Mountains in Canada and northern United States (figs. d, e, f from Peterson 1948).

Figure 26.129a-c. Tortricidae. *Choristoneura obsoletana* (Walker). f.g.l. 22-25 mm. a. head, T1, T2; b. A4; c. anal comb. Olive-green with yellow to light brown head. T1 shield lightly pigmented except for narrow brown band at caudal and ventral margin. Thoracic legs brown, anal plate yellow. Anal comb conspicuous with 8± prongs. This strawberry pest folds the leaves or webs them together with the young fruit and feeds on them (figs. a-c from Peterson 1948).

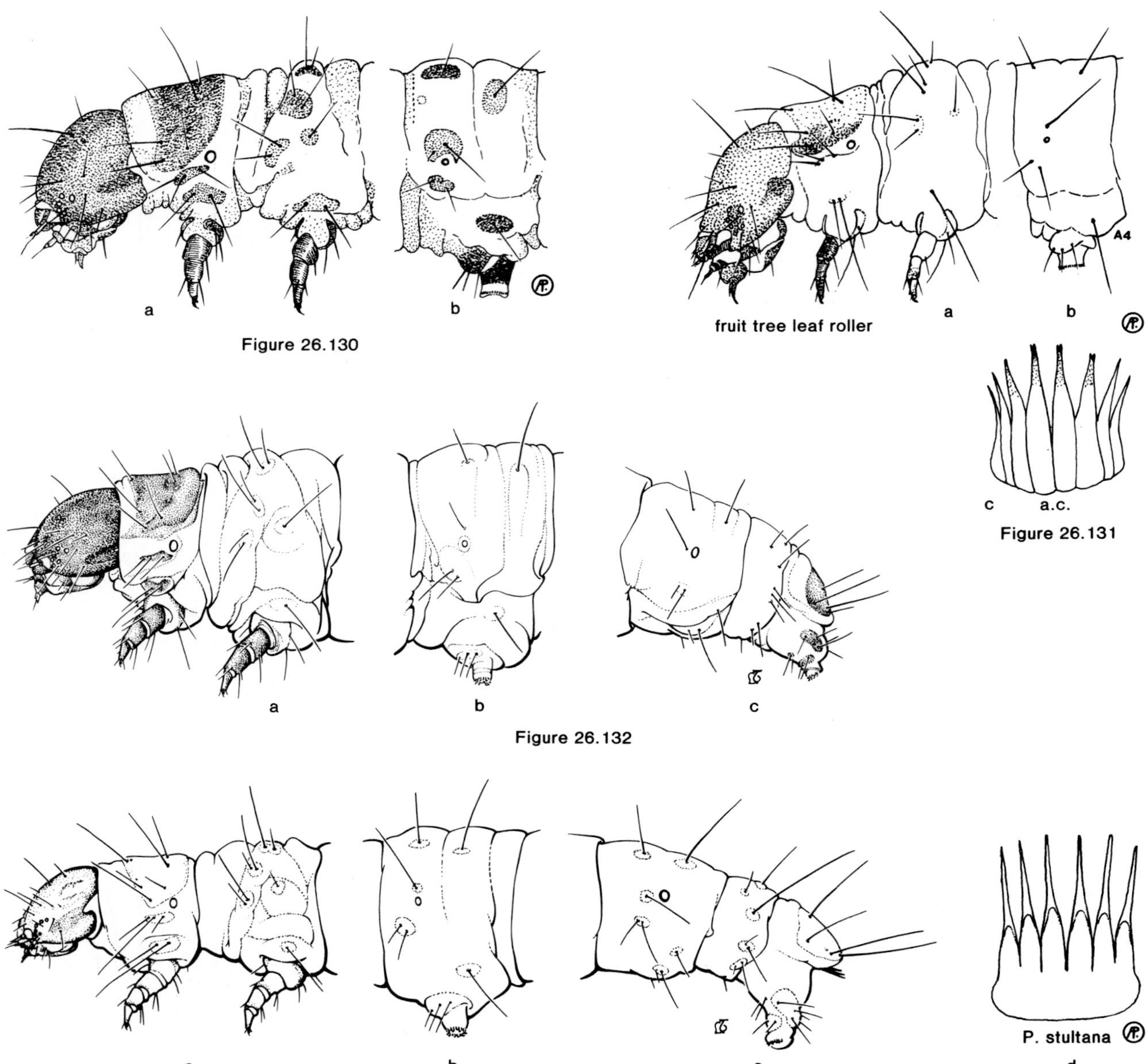

Figure 26.130a,b. Tortricidae. *Archips rileyana* (Grote). f.g.l. 23± mm. a. head, T1, T2; b. A4. Greenish-yellow with near-black head, deep brown to black T1 shield, deep brown to black anal plate, thoracic legs and prolegs; conspicuous brown to black pinacula. Microspines numerous, pigmented in folds of cuticle. Feeds in clusters on the foliage of buckeye, hickory, and walnut (figs. a, b from Peterson 1948).

Figure 26.131a-c. Tortricidae. *Archips argyrospila* (Walker), fruittree leafroller. f.g.l. 15-18 mm. a. head, T1, T2; b. A4; c. anal comb. Green to greenish-brown with light to dark brown head and light yellowish T1 shield. Most specimens without pigmented anal plate or pinacula. Feeds on buds, leaves, and small fruits that may be rolled and drawn together. Omnivorous, attacking apple, pear, cherry, plum, apricot, quince, currant, raspberry, and gooseberry, several woodland trees, roses, and other herbaceous plants. Entire United States and southern Canada (figs. a-c from Peterson 1948).

Figure 26.132a-c. Tortricidae. *Archips cerasivorana* (Fitch), ugly nest caterpillar. f.g.l. 19-26 mm. a. head, T1, T2; b. A3; c. A8-A10. Head dark brown or black, surface rough, thoracic and anal shields dark brown, pinacula inconspicuous, spinules of integument long, pale. Larvae feed gregariously within web around terminal growth of chokecherry and, during outbreaks, on apple, black cherry and other hardwoods. Across southern Canada and northern half of the United States.

Figure 26.133a-d. Tortricidae. *Platynota stultana* Walsingham. f.g.l. 12-18 mm. a. head, T1, T2; b. A3; c. A8-A10; d. anal comb. Early instars with black heads; last instar with brownish-yellow head and thoracic shield; prothoracic pinacula yellowish or brown, other pinacula of body color, spinules of integument small, with dark bases. A leaf-tier or roller and fruit-scarring larva, feeding on wide range of field and greenhouse plants. Southern United States, Arizona and California (fig. d from Peterson 1948).

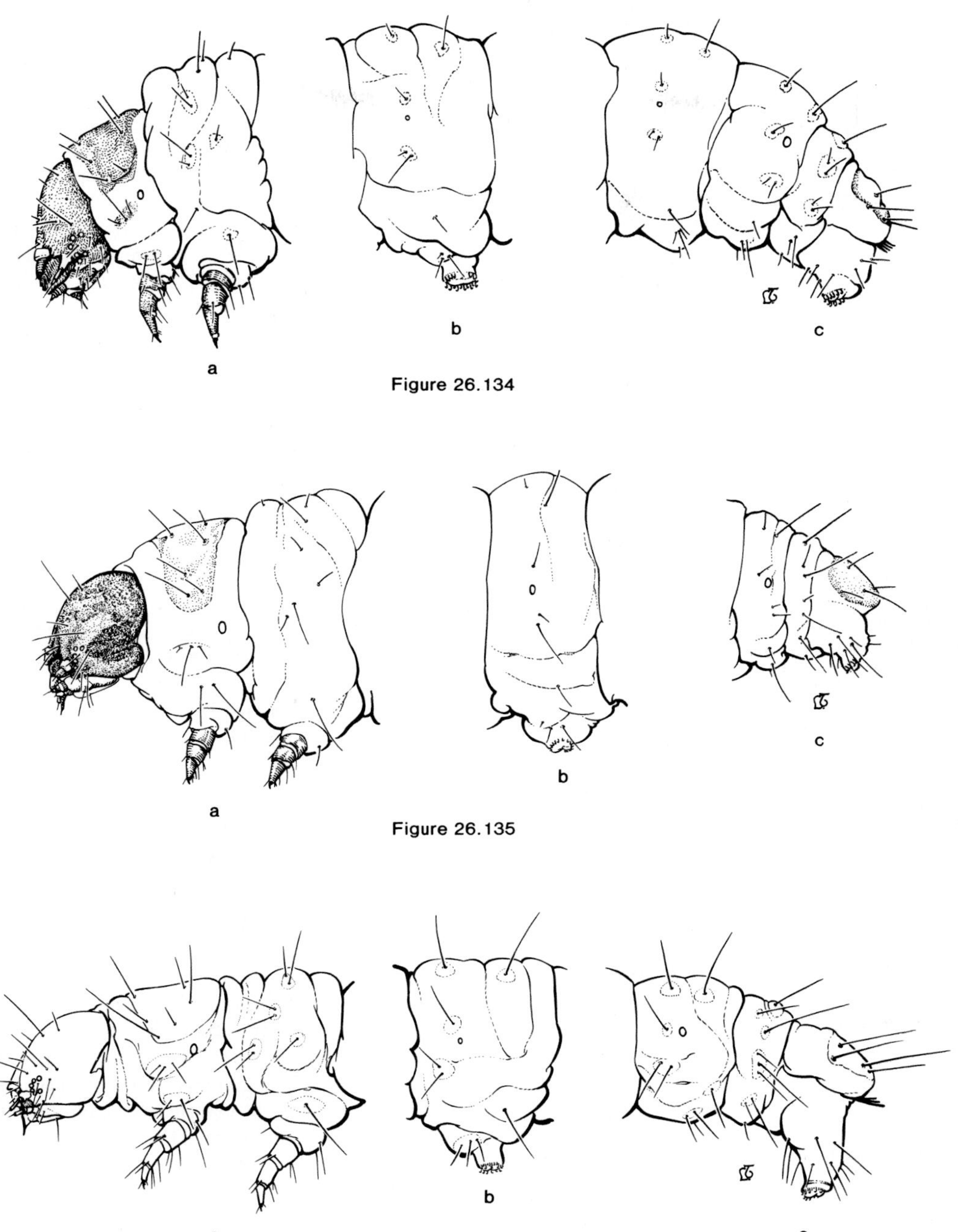

Figure 26.134a-c. Tortricidae (Cochylini). *Henricus macrocarpanus* (Walsingham). f.g.l. 13± mm. a. head, T1, T2; b. A3; c. A8-A10. Nearly white, light brown head, T1 shield and thoracic legs. Lobos Creek, San Francisco, Calif., ex. weekly samples, 1965, G. Frankie, cones of *Cupressus macrocarpa*. Univ. Calif., Berkeley.

Figure 26.135a-c. Tortricidae (Cochylini). *Aethes* (formerly *Phalonia*) sp. f.g.l. 12± mm. a. head, T1, T2; b. A3; c. A8-A10. Pale white, pale brown head. Paicenes, San Benito Co., Calif., 10 April 1967. Univ. Calif., Berkeley.

Figure 26.136a-c. Tortricidae. *Argyrotaenia velutinana* (Walker), redbanded leafroller. f.g.l. 13-18 mm. a. head, T1, T2; b. A3; c. A8-A10. Body pale greenish to yellowish, without markings except for dark pigment in stemmatal area; spinules of integument small, darker than body color. Feeds as a leaf-tier on fruit surfaces in contact with leaves on a wide range of plants, including apple and other fruits. Eastern United States and southern Canada.

HESPERIOIDEA AND PAPILIONOIDEA

Michael Toliver, *Eureka College*

Skippers and Butterflies

Introduction: We know more about butterflies than any other comparable group of insects, yet much remains to be learned about their immature stages. Many species, including common ones, have never been reared, and it is very difficult (often impossible) to identify a field-collected larva to species without rearing the specimen. Furthermore, until now, no one has carefully compared known immatures with the aim of formulating a classification of the groups, although Ehrlich (1958) has done this for adults. Kristensen (1976) has discussed the family level phylogeny.

Listed below are some references of particular importance for studying immatures of this group. These provide information on host plants, larval morphology, and larval habits, as well as access to the literature.

Two early works, unfortunately hard to obtain, provide a wealth of information about our species: Edwards (1868–1897) and Scudder (1889). Some more modern (and easier to obtain) works that give at least some information about immatures include Holland (1931), Howe (1975), Klots (1951), Ferris and Brown (1981), Pyle (1981), and Opler and Krizek (1984). Bibliographies of life histories include Davenport and Dethier (1937), Dethier (1946), Edwards (1889), and Tietz (1972). Use caution in consulting most of these works, especially the early ones and Tietz (1972), because many incorrect food plants have become enshrined in the literature, and species concepts have changed. A procedure for recording larval food plants is described by Shields et al. (1969 (1970)).

HESPERIIDAE (HESPERIOIDEA)

Michael Toliver, *Eureka College*

The Skippers

Figures 26.137–26.141

Relationships and Diagnosis: The diagnosis of Hesperioidea is discussed under the Megathymidae. Hesperiid larvae may be recognized by the characteristic constriction behind the head giving the appearance of a "neck." The head is larger than the prothorax. Other distinguishing characters include the presence of numerous secondary setae on the head and body (although the body may appear bare without magnification), bi- or triordinal crochets arranged in a circle, an anal comb, and the body usually tapering anteriorly and posteriorly.

Three distinctive subfamilies are currently recognized in North America. The Hesperiinae have larvae with relatively narrow heads that may be subconical at the vertex, and elongate, tapered bodies. The egg is spherical with a flattened base, with few or no reticulations, grooves or ribs (MacNeill 1975). North American hesperiines feed exclusively on monocotyledons, mostly grasses and sedges. Some authors (e.g. Miller and Brown 1981) place the genera *Carterocephalus* and *Piruna* in the subfamily Heteropterinae because the adults seem somewhat intermediate in morphology to the hesperiines and the next subfamily. What little is known of their early stages indicates closer affinities to the hesperiines, and they are included in that group here. The second subfamily, Pyrginae, has larvae with rounded, heart-shaped heads and robust bodies. Eggs of this group are strongly ribbed and/or reticulate (MacNeill 1975). Species in this group feed almost entirely on a variety of dicotyledons. The third subfamily, Pyrrhopyginae, is limited (in America north of Mexico) to one species found in Arizona, New Mexico, and west Texas. Larvae of *Pyrrhopyge araxes arizonae* Godman & Salvin are reddish with narrow bright yellow stripes, and are covered with scattered long white hairs (Comstock 1956a). They feed on oaks (at least *Quercus arizonica;* Burns 1964).

Biology and Ecology: Eggs are deposited singly on the larval food plant by almost all species. Most larvae form shelters by spinning silken tubes in clumps of grass, tying leaves of the host together, or folding a single leaf and tying it with silk. Many species feed only at night, remaining hidden in their shelters during the day (MacNeill 1975). They may "powder" their shelters with a white, waxy substance secreted from abdominal glands in the same manner as megathymid larvae (Dethier 1942a). A silken "girdle" supports pupae inside the former larval shelter (Emmel and Emmel 1973) or in a shelter spun specifically for pupation (MacNeill 1964). Pupal morphology is discussed by Mosher (1916a).

Description: Mature larva medium to large (20 to 50 mm +) (fig. 26.138). Body tapering from middle to either end; prominent "neck" results from the head being larger than T1. Cuticle variously wrinkled, with an abundance of secondary setae, particularly dorsally and ventrally. The location of primary setae is apparently indicated by rounded or irregularly shaped dark-rimmed plates. Body color some shade of green in many species, brown or white in others, often with longitudinal stripes of a darker or lighter color. Such coloration may help conceal the larvae when feeding.

Head: Hypognathous. Frontoclypeus as long or longer than wide, straight margins extending about one-half to the epicranial notch. Ecdysial lines meet about one-fourth to one-half the length of the frontoclypeus above its apex. Six stemmata, 1–4 evenly spaced in semicircle, 5 separated from 1–4 near antennal base; location of 6 variable, always caudad of 4 but may be located dorsad of 4; 6 may be farther than, equidistant from, or closer to 5 than 4 (see figs. 26.139, 26.140a). Primary setae obscured by an abundance of secondary setae, which may be as long as the longest body setae. The secondary setae are so numerous in some species as to make the head appear "fuzzy." Coloration variable, usually brown or black, often marked with vertical stripes or with dots lighter or darker than ground color. Cuticle pitted, rugose.

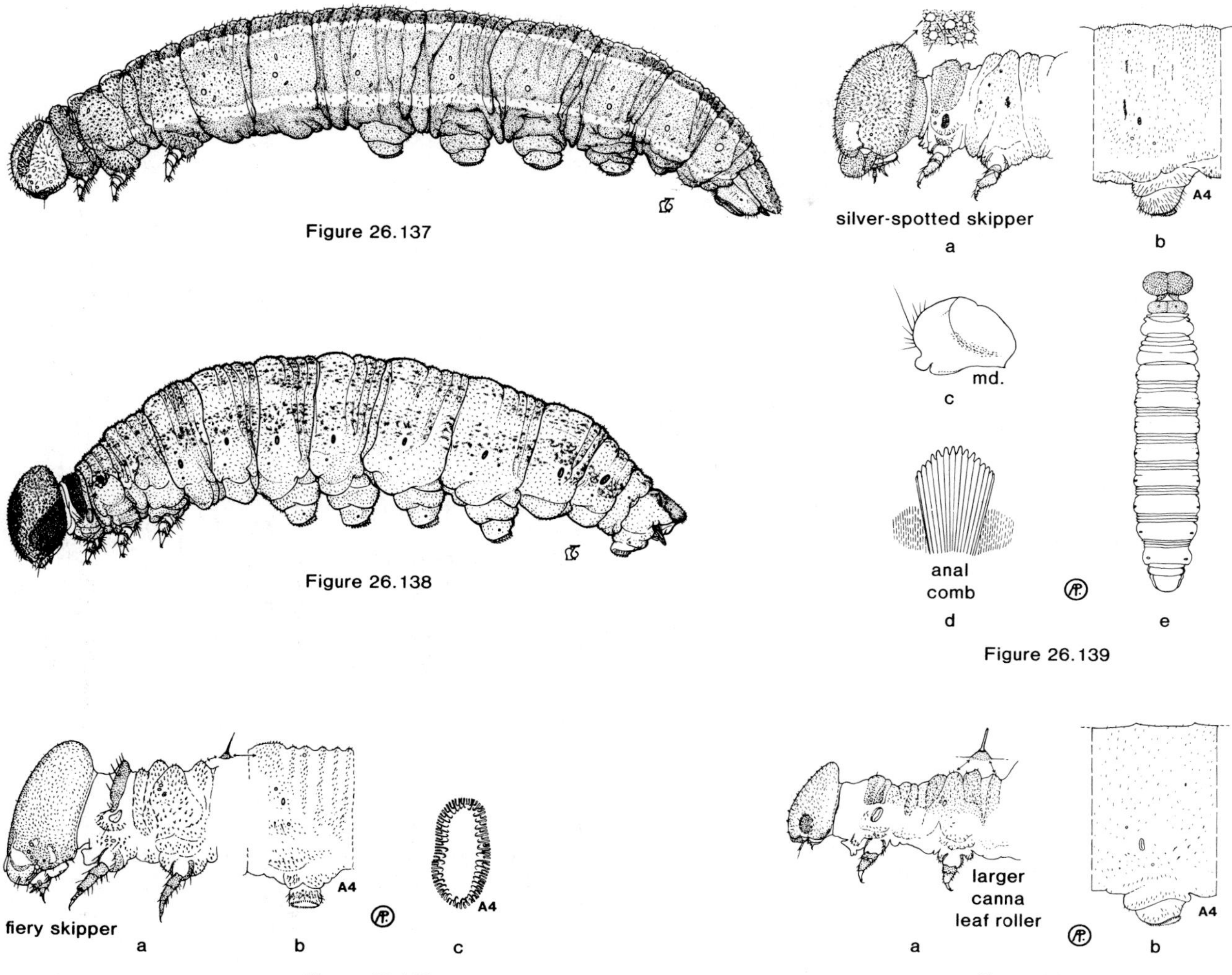

Figure 26.137

Figure 26.138

Figure 26.139

Figure 26.140

Figure 26.141

Figure 26.137. Hesperiidae. *Thymelicus lineola* (Ochsenheimer), European skipper, a "friendly" skipper, the adults frequently alighting on people. f.g.l. 21± mm. E. Lansing, Ingham Co., Mich., MSU campus, 13 June 1974. Michigan State Univ. Yellow green, with 2 addorsal and 1 subspiracular whitish strip on each side. Head with conspicuous pair of white vertical stripes bordered laterally by darker stripe. Head nearly same diameter as prothorax in contrast to most skipper larvae. On grasses, including small grains and corn; can be a pest in grass hay, especially timothy. Abundant in eastern Canada and northern United States west to Minnesota; also British Columbia; still expanding its distribution (FWS).

Figure 26.138. Hesperiidae. *Urbanus proteus* (L.), bean leafroller, long-tailed skipper. f.g.l. 30± mm. Miami, Fla., 18 January 1944, on bean foliage. U.S. Natl. Mus. Head brown and yellow. Green, with dark middorsal line, flanked by yellow lateral line and 2 pale green lines below. Areas between lines dotted with black and yellow on gray-green background. Reddish front and rear. Common in South on beans; also on *Wisteria,* other Fabaceae and crucifers. Northern United States (uncommon) to South America (FWS).

Figure 26.139a-e. Hesperiidae. *Epargyreus clarus* (Cramer), silverspotted skipper, f.g.l. 35–40 mm. a. head, T1, T2; b. A4; c. mandible; d. anal comb; e. dorsal outline. Green with conspicuous reddish-brown head possessing eyelike yellow or orange spots adjacent to stemmata. Lives in nest of foliage on legumes, especially *Wisteria* and black locust, but also on honey locust and many other legumes. Southern Canada (east to west) into Mexico (figs. from Peterson 1948) (FWS).

Figure 26.140a-c. Hesperiidae. *Hylephila phyleus* (Drury), fiery skipper. f.g.l. 20± mm. a. head, T1, T2; b. A4; c. A4 crochets. Greenish brown to tan with 3 dark stripes, dark brown head, T1 shield and thoracic legs. Numerous, dark, secondary setae arise from small pigment spots on all segments. A pest of bent grass, but feeds on many other grasses. Forms its shelter among the roots. Northern U.S. (uncommon) to South America, common in the South (figs. from Peterson 1948) (FWS).

Figure 26.141a,b. Hesperiidae. *Calpodes ethlius* (Stoll), larger canna leafroller. f.g.l. 60± mm. a. head, T1, T2; b. A4. Greenish with dark stemmatal area on yellowish head; dark spot on extreme lateral portions of T1 shield. Body gray-green with whitish line each side of dorsum. Feeds on and rolls leaves of *Canna* in the Southeast where it can be damaging. South to South America (figs. from Peterson 1948) (FWS).

Thorax: T1 much narrower right behind head, forming a "neck." T1 shield usually distinct, divided mid-dorsally; with numerous long setae that may be concentrated at anterior edge. Clear spots present in shield that may indicate former location of primary setae (fig. 26.139a). T1 spiracle elliptical, much larger (2–3×) than abdominal spiracles. Spiracle may be on plate. One or two rounded or irregular plates above leg, probably indicating former location of SV1 and SV2 (fig. 26.139a); other plates indicating former location of other primary setae located in vicinity of spiracle. T2 and T3 relatively simple in structure; some plates present, particularly subdorsally, which may indicate position of primary setae. Numerous long setae on thoracic legs. All thoracic segments divided into annulets.

Abdomen: Spiracles on A1–8 round or elliptical, much smaller than T1 spiracle. Spiracle on A8 larger than, and located dorsad of, spiracles on A1–7. Rounded or irregular dark-rimmed plates present on all segments, apparently indicating former location of primary setae. Secondary setae abundant, often more so on dorsal and ventral surfaces and on prolegs. Crochets on A3–6 bi- or triordinal, arranged in complete circle (fig. 26.140c), although there may be lateral gaps much less than one-third circumference of circle. Crochets on A10 bi- or triordinal, with a caudal gap in an otherwise complete circle. Dorsal surface of A10 often clothed with numerous long secondary setae. Anal comb present (figs. 26.138, 26.139d). All segments divided into annulets. Glands present on venter of A7 and A8, which may or may not be apparent.

Comments: The most diverse butterfly family in North America with over 230 species reported by MacNeill (1975), and over 260 species reported by Miller and Brown (1981). Many collectors ignore this interesting family because the adults are not brightly colored, and moth collectors ignore them because they are not moths, consequently comparatively little is known about the early stages. Nevertheless, MacNeill (1975) provides larval descriptions and/or food plant information for 147 species. Other important references for life history information include Edwards (1868–1897), Scudder (1889), Comstock (1927), Tietz (1972), and Stanford (1981). Emmel and Emmel (1973) provide an especially valuable regional work, illustrating the early stages of many species of hesperiids, quite a few of which occur outside the area covered by this book. Valuable generic revisions describing immatures are Burns (1964) and MacNeill (1964). General references include Lindsey, Bell and Williams (1931), and Evans (1951–1955).

Few skippers are economic pests. An introduced species that can cause extensive damage in timothy hay, especially that used for seed, is *Thymelicus lineola* (Ochsenheimer), the European skipper (fig. 26.137). This species is especially abundant in Canada. The bean leafroller, *Urbanus proteus* (L.) (fig. 26.138), occasionally damages beans and crucifers in the southeastern U.S. Some tropical species that barely enter our area (e.g., *Nyctelius nyctelius* (Latreille)) are pests of sugarcane and rice in areas south of the U.S., but these do little or no damage to crops in the U.S. *Atalopedes campestris* (Boisduval) has been reported as an economic pest of sod in Arkansas.

Selected Bibliography

Burns 1964.
Comstock 1927, 1956a.
Dethier 1942.
Edwards 1868–1897.
Emmel and Emmel 1973.
Evans 1951, 1955.
Lindsey, Bell and Williams 1931. (description, taxonomy).
MacNeill 1964, 1975.
Miller and Brown 1981.
Mosher 1916a.
Scudder 1889.
Tietz 1972.

MEGATHYMIDAE (HESPERIOIDEA)

Michael Toliver, *Eureka College*

The Giant Skippers

Figures 26.142, 26.143

Relationships and Diagnosis: The Megathymids are members of the Hesperioidea, a superfamily distinguished by the grublike appearance of the medium to large larvae, by crochets arranged in a circle, and by the presence of a large number of short secondary setae covering the entire body. In many species, the distal ends of these setae may be flattened.

Mature megathymids may be recognized by their large size (50 mm or more), head smaller than the prothorax, no anal comb, and larval host associations (borers in Agavaceae).

Biology and Ecology: All North American larvae are borers in roots of *Yucca,* leaves or caudex of *Agave* or both, and leaves or caudex of *Manfreda* (all Agavaceae) (Roever 1975). *Megathymus* larvae (*Yucca* feeders) make extensive silk-lined tunnels in the root systems, at some point in their development forming a characteristic "tent" over the tunnel exit. These tents are prominent silk tubes covered with bits of foodplant, dirt, and frass, projecting from the center or sides of the *Yucca* rosette or from the ground several inches away from the infested *Yucca. Megathymus* species are mostly single-brooded, emerging throughout the year depending on species and location, though they are primarily spring fliers. *Stallingsia maculosus* (Freeman) (a *Manfreda* feeder) also constructs tents over the burrow opening. This species is double-brooded, flying in April–May and September–October. *Agathymus* larvae (*Agave* feeders) initially burrow into the upper third of an *Agave* leaf, directing their tunnels toward the apex. They pass through the first and second instars in these apical tunnels, then migrate to the basal third of the leaf during the third instar, where they construct a burrow directed toward the leaf base in which they spend the rest of their larval life. Before pupating, *Agathymus* larvae build a characteristic silken "trap door" over the burrow exit. The location of the exit and the color of its trap door covering varies among *Agathymus* species (Roever 1964a). *Agathymus* are single-brooded, emerging August–November.

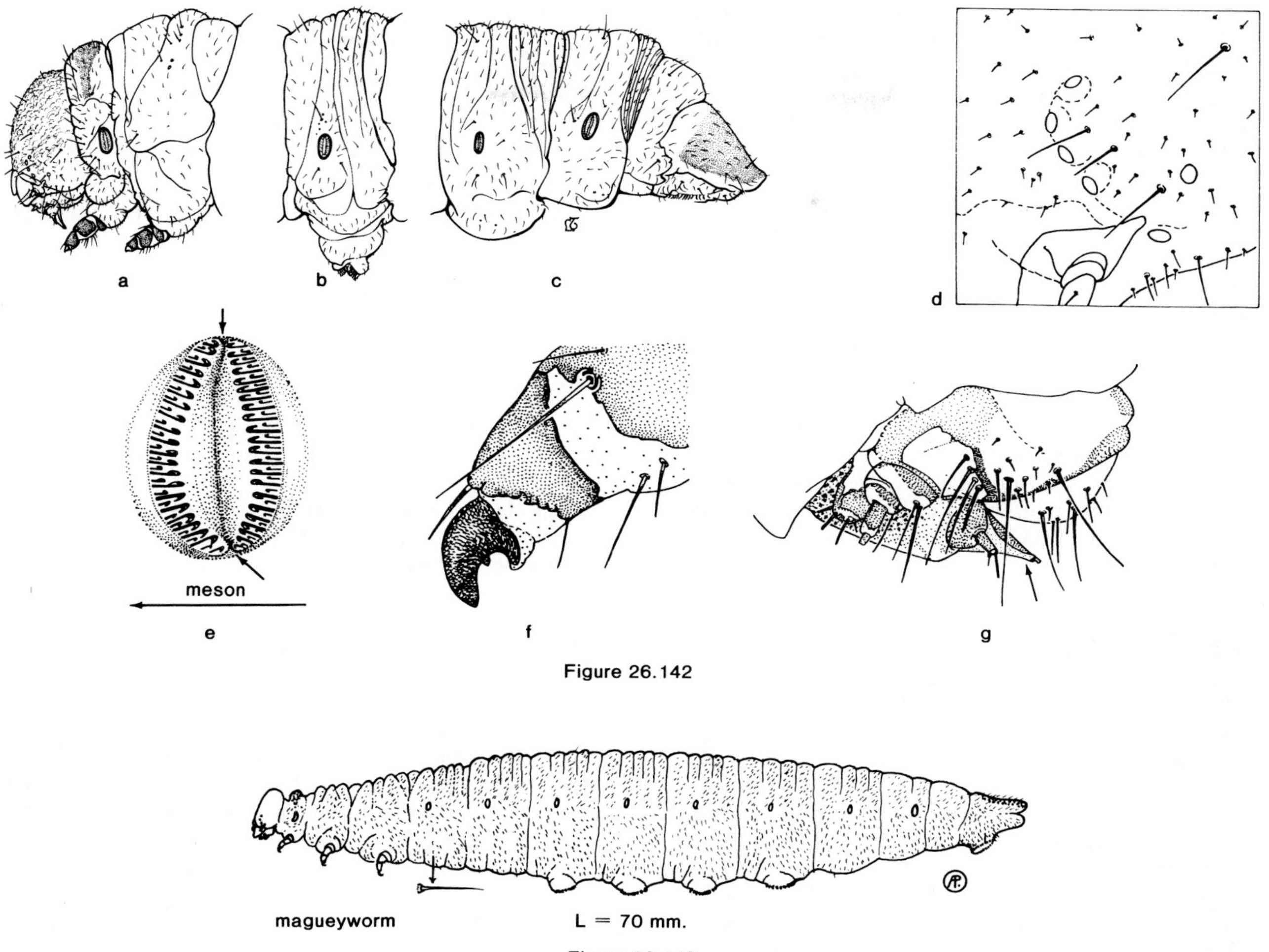

Figure 26.142

Figure 26.143

Figure 26.142a-g. Megathymidae. *Megathymus coloradensis kendalli* H. A. Freeman. f.g.l. 70± mm. a. head, T1, T2; b. A3; c. A6-A10; d. stemmatal area of head; e. A3 crochets, arrows indicate gaps; f. T1 claw; g. rearward projecting spinneret. Bexar Co., Tex., 1 February 1958, R. O. and C. A. Kendall. Ex. *Yucca constricta*.

Figure 26.143. Megathymidae. *Aegiale hesperiaris* (Walker), magueyworm. f.g.l. 70–75 mm. Fully distended larva. Feeds in the leaves and heart of the maguey or century plant (*Agave*) in Mexico where it is commonly placed in bottles of tequila (fig. from Peterson 1948) (FWS).

Most North American megathymid larvae "powder" their burrows, at least before pupation, with a hydrofugic substance, from glands on the ventral surface of A7 and A8, the specific composition and purpose of which is unknown (Roever 1964a). Megathymid larvae are host specific, so it is possible to determine the species by the host species and stage of larval development.

Megathymus eggs, round and somewhat flattened, are glued to *Yucca* leaves. *Agathymus* eggs, which are hemispherical, lack an adhesive and usually fall to the ground or lodge between *Agave* leaf bases after oviposition (Roever 1964a).

Pupae achieve a great deal of movement by rotating the abdomen and bracing the cremaster against the tunnel sides. They may move in response to temperature differences or disturbances (Roever 1964a). Pupation takes place in the larval tunnel. Pupal structure is discussed by Mosher (1916a).

Description: Mature larvae large, usually over 40 mm in length. Body cylindrical, grublike; cuticle variously wrinkled. Body whitish in most species, bluish in some. An abundance of secondary setae of the same length as primary setae (generally both are relatively short) makes the discernment of primary setae difficult.

Head: Hypognathous. Frontoclypeus as wide as long, extending about half way to epicranial notch. Ecdysial lines meeting approximately one-quarter the length of frontoclypeus above its apex. Six stemmata, 1–4 evenly spaced in semicircle, 5 separated from 1–4 near antennal base, 6 caudad of 4, about equidistant from 4 and 5 (fig. 26.142d). Head covered with numerous secondary setae that may be as long as primary setae. Head dark brown to black; cuticle rugose.

Thorax: T1 shield well defined. All segments with numerous generally distributed secondary setae that may be as long as (and therefore obscure) the primary setae. Spiracle on T1 elliptical.

Abdomen: Numerous secondary setae generally distributed on all segments. Spiracles on abdominal segments elliptical, similar in size to spiracle on T1. Prolegs on A3–6 and A10 may be indicated by crochets alone, especially on A10. Crochets biordinal, arranged in a circle (fig. 26.142e) that may be broken mesally and/or laterally, especially on A10. A10 may bear an ill-defined anal shield with numerous long stout setae. No anal comb.

Comments: A small family (15 North American species, Roever 1975) that is found primarily in the southwestern United States. Two species are found in the Southeast. Despite the fact that megathymids are best collected as immatures, and the host plants of all North American species are known, not a single species has had its life history thoroughly described. Wielgus and Wielgus (1974) and Wielgus and Stallings (1974) do provide abbreviated descriptions of all instars of *Megathymus yuccae "albasuffusa"* Wielgus and Wielgus and the two subspecies of *Megathymus streckeri* (Skinner) respectively. The most thorough description of megathymid immatures is that of *M. yuccae* (Boisduval and LeConte) by Riley over 100 years ago (1876). This lack of thorough descriptions may be attributed to the concealment of larvae in their burrows. Roever (1964a (1965)) has provided methods for instar determination of field-collected larvae and Petterson and Wielgus (1973 (1974)) and Wielgus (1974 (1975)) have provided methods of rearing larvae on artificial diets. These methods should allow more complete life history descriptions in the future. Other important references on the biology of this family include Roever (1964a, 1975), Ferris (1981a), and Freeman (1969). Comstock (1956b) includes a bibliography of life histories and notes on immatures.

Megathymids have no economic importance in the United States, but in Mexico the larvae of *Agave* borers (primarily the genus *Aegiale*) are considered a delicacy and are canned and sold as "gusanos de maguey". Larvae (fig. 26.143) are often included in bottles of tequila and mescal to certify the brew's authenticity and quality.

Selected Bibliography

Comstock 1956b.
Freeman 1969.
Mosher 1916a.
Petterson and Wielgus 1973 (1974).
Riley 1876.
Roever 1964a,b, 1975.

PAPILIONIDAE (PAPILIONOIDEA)

Michael Toliver, *Eureka College*

The Swallowtails and Parnassians

Figure 26.144–26.149

Relationships and Diagnosis: The swallowtails are one of six families of Lepidoptera comprising the Papilionoidea, or true butterflies (Ehrlich 1958). The larvae of this superfamily are diverse; no single characteristic serves to distinguish them from other Lepidoptera.

Papilionids, on the other hand, are unique among Lepidoptera in the possession of a dorsal prothoracic osmeterium, apparently used to repel enemies (Eisner et al. 1970). In addition, papilionids are large (usually 40 mm +) and most appear smooth. In fact, all our species possess numerous secondary setae, which may be so short as to be visible only under relatively high magnification (40×). Some species, especially *Aristolochia* feeders and Parnassiinae, possess filaments, verrucae or pinacula.

Two distinct subfamilies occur in our region. The Parnassinae (sometimes considered a distinct family) are found in the high mountains of the western U.S., in lowland rainforests of the Pacific Northwest, in western Canada, Alaska, and the Badlands of the Dakotas. The larvae possess conspicuous setae, often grouped onto pinacula or verrucae, which are velvety black with rows of subdorsal yellow spots (fig. 26.144). The Papilioninae are conspicuous, widely distributed insects that epitomize the word "butterfly". Most papilionine larvae appear smooth, with expanded thoracic segments imparting a "humped" appearance to these insects.

Biology and Ecology: The hemispherical eggs are usually deposited singly on larval food plants by females, but a few species (*Battus*) lay their eggs in small clusters. Parnassines may lay their eggs haphazardly in the vicinity of food plants.

Parnassine larvae feed on *Sedum, Dicentra,* and possibly *Corydalis* in our region, although they are mainly *Aristolochia* feeders elsewhere (Shepard and Shepard 1975, Tyler 1975). Papilionine larve feed on a greater variety of plants, but tend to feed on Apiaceae (= Umbelliferae), Rutaceae, Magnoliaceae, Lauraceae, Annonaceae, and *Aristolochia* (Tyler 1975). All are external feeders, although a few (e.g. *Papilio troilus* L.) tie the edges of a leaf together with silk to form a "leaf roll" in which they rest. In early instars, most species resemble bird droppings. In later instars, different species may be ornamented with large eyespots (e.g. *P. troilus,* fig. 26.149), conspicuously banded (e.g. *P. polyxenes* F., fig. 26.148), conspicuously colored (*Battus philenor* (L.), fig. 26.146), or cryptic (*P. glaucus* L., fig. 26.147). Presumably, such coloration helps to reduce predation by vertebrate predators.

Parnassine pupae are formed in loose cocoons on the ground (Shepard and Shepard 1975); papilionine pupae are usually suspended head up by a silk girdle and a silk pad to which the cremaster is attached. Papilionine pupal structure is discussed by Mosher (1916a).

Description: Mature larvae medium to large (30–50 + mm). Body frequently widest at Al, tapering abruptly to head and gradually toward A10. Head somewhat retractile, smaller than prothorax. Cuticle generally appearing smooth, but numerous secondary setae present that may be conspicuous or inconspicuous. Some species possess filaments, verrucae or pinacula. Color variable, usually some shade of brown, green, or black. Primary setae variable: absent, present but inconspicuous, or indicated by pinacula or verrucae. Osmeterium on dorsum of T1.

Head: Hypognathous. Frontoclypeus as wide as or wider than long, margins straight and extending from one-fourth to one-half to the epicranial notch. Ecdysial lines meeting some distance above frontoclypeal apex, usually at least a quarter the length of the frontoclypeus beyond its apex. Labrum

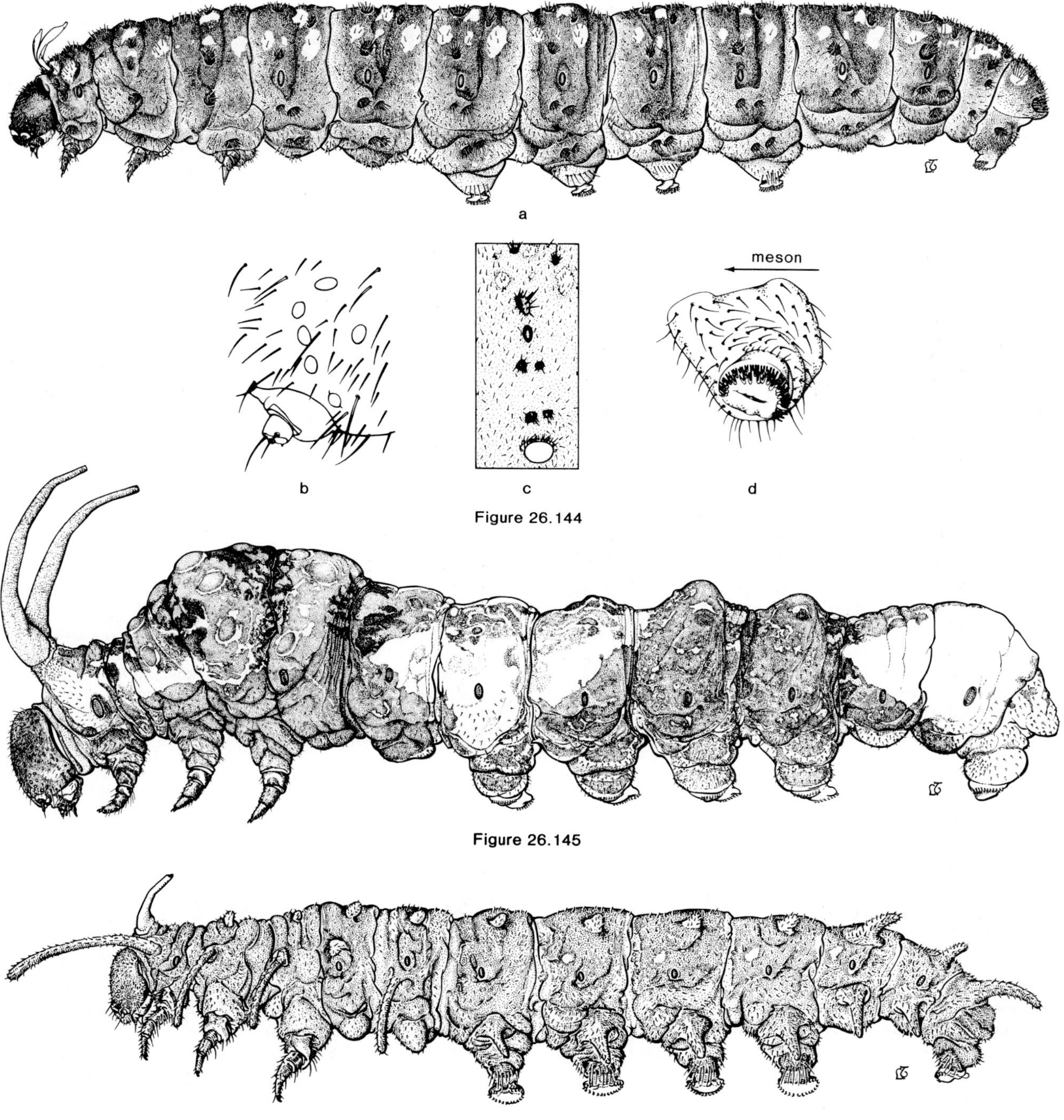

Figure 26.144

Figure 26.145

Figure 26.146

Figure 26.144a-d. Papilionidae, Parnassiinae. *Parnassius phoebus* (Fabricius). f.g.l. 35± mm. a. larva; b. stemmatal area; c. A3 setal map; d. A3 crochets. Stikine Plateau, near Tohltan, B.C. 23 June 1973. Jon Shepard and J. Gordon. Michigan State Univ. Arctic tundra and alpine clearings in the West; on stonecrops (*Sedum* spp.) (FWS).

Figure 26.145. Papilionidae. *Papilio (Heraclides) cresphontes* Cramer, orangedog or giant swallowtail. f.g.l. 55± mm. Vestaberg, Montcalm Co., Mich., 9 July 1959. Michigan State Univ. Resembling bird dropping, irregularly brown to black with conspicuous buff to white patches. Locally abundant on citrus in the South, northward to S. Canada; also on prickley-ash (*Zanthoxylum*), hoptree (*Ptelea*) and rue (*Ruta*) (FWS).

Figure 26.146. Papilionidae. *Battus philenor* (L.), pipevine swallowtail. f.g.l. 50± mm. John Bryan St. Pk., Greene Co., Ohio, 21 June 1965. Michigan State Univ. Black to reddish-black with black to red knobs and fleshy filaments, longest on T1. On pipevines (*Aristolochia* spp.) throughout most of the U.S. except Rocky Mountain region, and S.E. Canada (FWS).

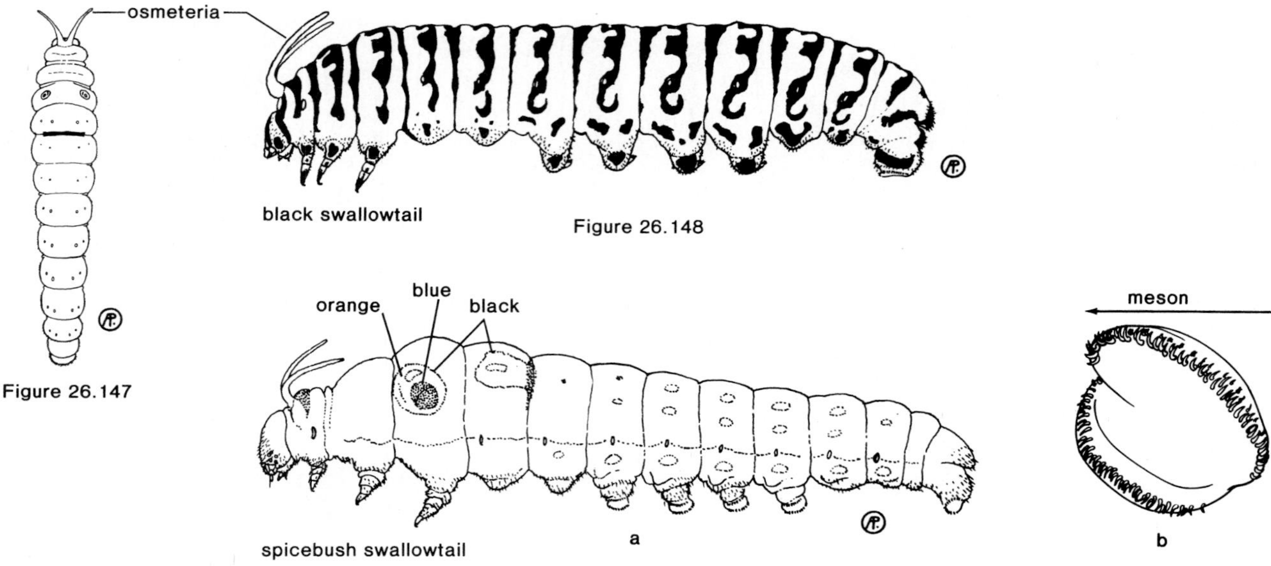

Figure 26.147. Papilionidae. *Papilio glaucus* L., tiger swallowtail. f.g.l. 50± mm. Greenish-yellow with orange and black "eyespots" on T3 and dark band between A1 and A2. On wide variety of trees and shrubs, including willow, birch, ash, cherry. Common east of the Rockies to Alaska (fig. from Peterson 1948) (FWS).

Figure 26.148. Papilionidae. *Papilio polyxenes asterias* Stoll, black swallowtail or parsleyworm. f.g.l. 45± mm. Green to yellowish-green to whitish-green with irregular black cross bands that partially surround yellowish to orange spots. Common on parsley, carrot, celery, Queen Anne's lace, and other relatives; also on some Rutaceae. From eastern Rockies eastward through southern Canada and U.S. (fig. from Peterson 1948) (FWS).

Figure 26.149. Papilionidae. *Papilio troilus* L., spicebush swallowtail. f.g.l. 45± mm. a. larva; b. A3 crochets. Green with pair of yellow-orange "eyespots" rimmed with black on T3 and A1, those on T3 with blue centers. Feeds on spicebush (*Lindera benzoin*), *Sassafras,* and *Persea.* Eastern U.S. deciduous forest, uncommon along streams through the Great Plains to Rocky Mountains (fig. 26.149a from Peterson 1948) (FWS).

notched. Six stemmata present. In *Parnassius,* 1 and 5 are slightly separated from 2–4, and 6 is widely separated and caudad from the rest. Five is near the antennal base (fig. 26.144b). In papilionines, 1–4 and 6 form a relatively compact group, often on a raised or darkened portion of cuticle, with 5 usually widely separated from the rest near the antennal base; 1–4 form a semicircle with 6 caudad of 3 and 4. Secondary setae numerous, primary setae obscured by secondary setae or absent. Cuticle smooth, slightly rugose or reticulate. Color variable; usually some shade of black, brown, or green; may be banded with black, yellow, or green.

Thorax: T1 wider than head. T1 shield present, usually indistinct, divided middorsally by a thin light line. Shield often covered with secondary setae. An osmeterium is located anterior to the shield in all our species.

In *Parnassius,* verrucae mark the location of primary setae. On T1 dorsal verrucae are posterior to the shield, whereas SD, L, and SV verrucae are arranged in a relatively straight vertical line ventrad to the shield. The spiracle is elliptical and somewhat smaller than the A1 spiracle. T2 and T3 have verrucae arranged in a straight vertical line. All segments are clothed with relatively long secondary setae with tuberculate bases. The body is cylindrical, not humped (fig. 26.144).

Papilionine swallowtails lack any indication of the primary setae on the thorax. In most species, T1 is slightly larger than the head, T2 is larger than T1, and T3 is still larger than T2, giving a humped appearance to the larvae. Spiracle on T1 elliptical or bean-shaped, approximately the size of spiracle on A1 or slightly larger. Our two species of *Battus* are quite distinct from other papilionine larvae. In *Battus,* T1 has a long fleshy filament clothed with setae arising just below the ventral margin of the T1 shield. T2 and T3 have long lateral filaments and a dorsal tubercle, all clothed with setae. The body is more or less cylindrical, not humped (fig. 26.146).

Abdomen: Spiracles on A1–8 elliptical, usually all the same size (A8 bigger in a few species). A1 usually largest body segment. Secondary setae numerous on all segments, especially on the ventral surface. Crochets triordinal mesoseries on A3–6 (fig. 26.144d), triordinal circle with caudal gap on A10. In many species, a weakly developed biordinal lateroseries is present on A3–6 (fig. 26.149b). In *Parnassius,* D, SD, L, SV, and V primary setae are apparently indicated by verrucae (fig. 26.144c). In all other species, primary setae are absent or weakly present (L1? on A3–7 of *P. cresphontes* Cramer). Dorsum of A10 with numerous relatively long secondary setae. *Battus* (fig. 26.146) possess numerous tubercles and filaments as follows: dorsal row of tubercles on all segments enlarging toward caudal end into filaments (A7–9); A1 tubercle just dorsad of spiracle; A2 long filament ventrad of spiracle covered with setae, and a ventral protuberance covered with setae; A3–8 with subventral filaments (near proleg base on A3–6). *Battus* also possesses an obvious anal plate.

Comments: A small but conspicuous and well-known family. Around 30 species are known from America north of Mexico, and we know at least something about the early stages of all of them. Although it would provide valuable information about the evolution and classification of the group, a careful comparison of the early stages has not been done. Tyler (1975) provides an excellent general reference for the North American species, including figures of adults and abbreviated descriptions of the immatures. The chapter on papilionids in Howe (1975) is also useful. Pyle (1981) figures immatures of eight species with excellent color photographs and describes the larvae of most North American species. Emmel and Emmel (1973) provide photographs and drawings of larvae of nine of our species. Fisher (1981) and Ferris (1981c) treat most western species. Munroe (1961) reviews the classification of the family.

Few swallowtails are of economic importance. *Papilio polyxenes asterius* Stoll (fig. 26.148) is mentioned as an occasional pest in garden umbellifers such as carrots and parsley, but it is mainly noticed because of its conspicuous appearance rather than the damage it does. *Papilio (Heraclides) cresphontes* Cramer (fig. 26.145) and its relatives are sometimes considered pests of *Citrus,* but it is unlikely that they do any real damage in the U.S.

Selected Bibliography

Ehrlich 1958
Emmel and Emmel 1973.
Fisher 1981.
Ferris 1981c.
Igarishi 1979.
McCorkle and Hammond 1985.
Mosher 1916a.
Munroe 1961.
Opler and Krizek 1984.
Pyle 1981.
Shepard and Shepard 1975.
Tyler 1975.

PIERIDAE (PAPILIONOIDEA)

Michael Toliver, *Eureka College*

The Whites and Sulphurs

Figures 26.150–26.153

Relationships and Diagnosis: The pierids are most closely related to the papilionids, but the larvae are abundantly distinct. Pierid larvae may be distinguished from other Lepidoptera by their abdominal segments (A2–6) being divided into five or more annulets (usually six), by an abundance of secondary setae often borne on chalazae, and by the absence of a prothoracic "neck" and osmeterium. They also lack filaments, scoli or protuberances of any type other than chalazae.

There are two subfamilies in America north of Mexico (Pierinae or whites and Coliadinae or sulphurs), with a third (Dismorphiinae) questionably recorded from the southern U.S. Attempts to place the genera *Euchloe* and *Anthocharis* (including *Falcapica*) in the separate subfamily Anthocharinae (Miller and Brown 1981) are not supported by morphology and foodplant choice or by biochemistry (Geiger "1980" (1981) and they are included in the Pierinae here.

Biology and Ecology: The spindle-shaped, ribbed eggs (unusual in butterflies) are deposited singly on the larval food plants by most species, but a very few (e.g. *Neophasia menapia* (Felder and Felder)) will lay several eggs in one clutch.

Pierine larvae feed mainly on Brassicaceae, including some cultivated crucifers, whereas Coliadinae larvae feed mainly on Fabaceae, including some cultivated legumes. A few species use other food plants, including pines, *Vaccinium, Salix,* and Asteraceae. All are external feeders, although some (e.g. *Phoebis sennae* (L.)) may form shelters from food plant leaves during the day (Pyle 1981). Most are some shade of green, and many are longitudinally striped with white, all of which may help to conceal them, although some *Eurema, Pontia, Phoebus* and others are conspicuously banded with dark colors.

Pupae are secured to the substrate by a cremastral silk pad (button) and a silk girdle. Pupal morphology is discussed by Mosher (1916a).

Description: Mature larvae medium, 20 mm–40 mm. Body cylindrical, slender with no protuberances except chalazae. Annulets (usually six on each abdominal segment) on all body segments, often emphasized by rows or bands of setae. Secondary setae abundant, often of different lengths, often on conspicuous chalazae. Primary setae obscured by secondary setae or absent. Body color usually some shade of green; often striped on banded.

Head: Hypognathous. Frontoclypeus longer or as long as wide, extending one-half to two-thirds to epicranial notch. Ecdysial lines usually meeting very near the frontoclypeal apex, occasionally meeting an eighth to a third the length of frontoclypeus above its apex. Six stemmata; 1–5 arranged in a semicircle with 2–5 elevated on dark papillae and thus appearing conspicuously different from 1 and 6 (fig. 26.150c); stemma 6 usually caudad of 3. Labrum notched, relatively large. Secondary setae abundant, usually long and often on chalazae. Primary setae lacking or obscured. Cuticle rugose. Color usually brown or green, occasionally banded.

Thorax: T1 shield usually indistinct. T1 spiracle larger than or equal to A1 spiracle; elliptical in coliadines, more rounded in pierines. Ventral T1 gland present, anterior to T1 legs. T1 annulets relatively indistinct, often no more than two. Abundant secondary setae, often on prominent chalazae, often of different lengths (fig. 26.152a). T2 and T3 with distinct annulets (usually five), often emphasized by rows or bands of secondary setae. Secondary setae particularly long and numerous in vicinity of thoracic legs.

Abdomen: A1 spiracle often ventrad of spiracles on A2–8, often slightly larger than spiracles on A2–7, about size of spiracle on A8. Spiracles elliptical (Coliadinae) to more rounded (Pierinae). Abdominal segments all divided into annulets, usually six, with the number decreasing caudally (A8

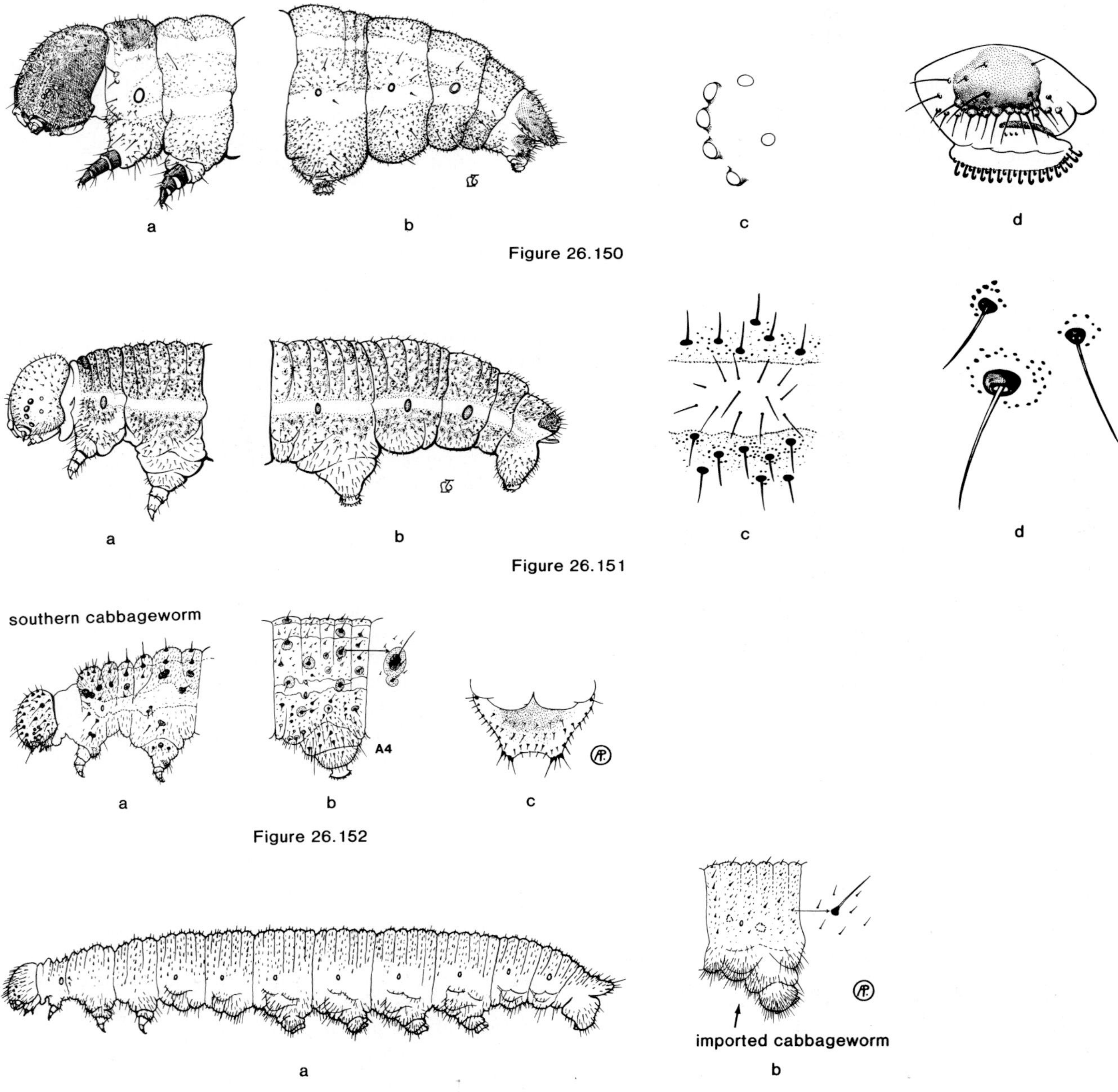

Figure 26.150a-d. Pieridae. *Neophasia menapia* (C. and R. Felder), pine butterfly. f.g.l. 25± mm. a. head, T1, T2; b. A6-A10; c. stemmata left side; d. left lateral view A6 crochets; note row of stout setae and sclerotized area above crochets. Boise Co., Id., 1953-54. U.S. Natl. Mus. Found throughout the mountainous West; especially destructive to ponderosa pine in Idaho, Washington, Oregon, and British Columbia. Also found on western white and lodgepole pine, Douglas-fir and larch in outbreak areas (FWS).

Figure 26.151a-d. Pieridae. *Colias eurytheme* Boisduval, alfalfa caterpillar, orange sulfur. f.g.l. 25± mm. a. head, T1, T2; b. A6-A10; c. enlargement of spiracular stripe area; d. enlargement of body setae. E. Lansing, Mich., July 1981, on alfalfa. Michigan State Univ. See text (FWS).

Figure 26.152a-c. Pieridae. *Pontia protodice* (Boisduval and Leconte), southern cabbageworm. f.g.l. 30± mm. a. head, T1, T2; b. A4; c. venter of A10, no anal comb. Purplish green with 2 greenish-yellow stripes per side, speckled with chalazae. On many cultivated and wild crucifers. Most abundant in South and West, ranging from southern Canada into Mexico except for Pacific Northwest (figs. from Peterson 1948) (FWS).

Figure 26.153a,b. Pieridae. *Pieris (Artogeia) rapae* (L.), imported cabbageworm. f.g.l. 25± mm. a. larva; b. A4. See text. Throughout North America (figs. from Peterson 1948) (FWS).

and A9 with fewer annulets than A2–7). A1 may have a reduced number of annulets as well. Secondary setae abundant, especially ventrally, often borne on prominent chalazae. Annulets marked with rows or bands of these setae. Long setae on dorsum of A10. Prolegs on A3–6 with mesoseries of bi- or (usually) triordinal crochets. Anal prolegs with bi- or triordinal caudally open circle of crochets. Anal proglegs may have a lateral pinaculum. Anal comb present or absent (figs. 26.151b, 26.152c). Anal shield obvious or indistinct. Shield may be produced into two projections (e.g. *Neophasia menapia, Pontia protodice* Boisduval and LeConte, see fig. 26.152c).

Comments: The family is relatively diverse (for butterflies) in North America, with about 60 species reported north of Mexico. Three species which are agricultural pests and often extremely abundant, not only in agricultural fields but in urban habitats as well, are *Pieris* (*Artogeia*) *rapae* (L.) (fig. 26.153), *Colias eurytheme* Bdv. (fig. 26.151), and *C. philodice* Godart, the cabbage white, alfalfa butterfly and clouded sulphur respectively.

P. rapae (fig. 26.153), which damages commercial and garden crucifers, such as cabbage (including broccoli, Brussels sprouts, etc.), turnips, radishes, and nasturtiums, was introduced from Europe around 1860. *P. rapae* larvae are light to medium green, with a dorsal yellow stripe and occasionally a lateral yellow stripe. The secondary setae give this caterpillar a "downy" appearance. Another pest from Europe, *Pieris* (*Pieris*) *brassicae* (L.), has recently been introduced into the Western Hemisphere in Chile (Gardiner 1974) and may spread from its introduction site. Gardiner provides information on the life cycle and potential threat of *P. brassicae. C. eurytheme* (fig. 26.151) and *C. philodice* are both pests of legumes, particularly alfalfa and clover. The larvae of these two species cannot easily be distinguished: both are yellowish green with white side stripes. The lateral stripes may be edged with black and contain a pink line. A darker middorsal line is also present. *C. eurytheme* feeds mainly on alfalfa, *C. philodice* feeds mainly on clover. They frequently hybridize in areas of overlap. A fourth species of economic importance is *Neophasia menapia* (fig. 26.150), the pine white. This species attacks pines, particularly ponderosa pine, in the western U.S. and Canada, and occasionally becomes numerous enough to kill them (Furniss and Carolin 1977). The caterpillars are green with two lateral white stripes.

General information about this family may be obtained from Howe (1975), Ferris (1981d) and Pyle (1981). Klots (1933) discusses the classification. Emmel and Emmel (1973) figure larvae of 15 of our species.

Selected Bibliography

Emmel and Emmel 1973.
Ferris 1981d.
Furniss and Carolin 1977.
Gardiner 1974.
Geiger 1980.
Howe 1975.
Klots 1933.
Miller and Brown 1981.
Opler and Krizek 1984.
Pyle 1981.

LYCAENIDAE (PAPILIONOIDEA)

John Downey, *University of Northern Iowa*

The Lycaenids, Blues, Coppers and Hairstreaks

Figures 26.154–26.157

Relationships and Diagnosis: Clench (1955, 1965) grouped several related families into the superfamily Lycaenoidea, including the Lycaenidae, Riodinidae, Liptenidae, and Liphyridae. Ehrlich (1958) however, recognized only one family, the Lycaenidae, with three subfamilies, the Styginae (monogeneric), the Lycaeninae, and the Riodininae. Authors have hoped that recent studies of the immature stages, including eggs, larvae and pupae, might shed some light on the higher categories, and in particular, answer the question whether the Riodinidae are confamilial with the Lycaenidae or represent a separate but closely related family. However, it has been my experience, from a modest sampling of the immature stages of both groups, that they likewise only confirm the close relationship of the two groups. In other words, although we know a little more about both groups now than we did two or three decades ago, the exact nature of their phylogenetic relationship has yet to be determined. For this reason, we are treating them as two families.

Riodinids are distinct from the lycaenids in that they have more than two mandibular setae, although the paucity of described riodinid larvae suggests continued observation of this character in new descriptions. Perhaps the best diagnostic character to separate lycaenids and riodinids is the arrangement of abdominal spiracles; the A1 and A2 spiracles are essentially in line in the Lycaenidae; in the Riodinidae the A1 spiracle is distinctly lower.

Biology and Ecology: Most larvae are plant feeders, and the early instars in particular eat their way into tender flower buds, flowers, fruits or pods, and become "internal" feeders. Later instars may insert their head and thorax into plant parts so that only the posterior, frass-depositing end is visible. Although they feed on a wide variety of plants, the Fabaceae are strong favorites, with a surprisingly large group feeding on Loranthaceae and even lichens! Many larvae are associated with ants, which is quite facultative in most cases, but has evolved to an obligate relationship in some tropical and semitropical species. Larvae have developed special exudate glands (Newcomer's organ) on the middorsal region of A7 (fig. 26.157a). Ants tend the larvae and have a repertoire of behavioral responses by which they "induce" the larva to exude a droplet of material from the gland, which they assiduously devour. Larvae also possess protrusible tentacles laterally on A8, which they are able to evert by fluid pressure from within the body. In most temperate species the tentacles bear apical setae that have a "whipping motion;" they may emit chemicals for attraction (or repellence). In most North American larvae the ants appear to be agitated by the tentacles and diverted from their attention to the honey gand, thus serving as a tempering device to help control ant behavior when the honey gland is nonproductive or overworked. Not all species possess one or both of the ant relational organs, and in those that do, the organs usually do not function in the first two instars. Apparently all larvae possess microscopic pore cupolas (= Malicky's glands) near the exudate gland, and these tiny structures are also involved in attracting the attention of ants.

The evolution of myrmecophily in both the lycaenids and riodinids is most interesting and exceedingly complex. In most cases mutualism is involved, with the larvae having a substantially lower rate of parasitoid infestation with the presence of attendant ants. However, some lycaenids, e.g., *Curetis*, have permanently everted tentacles, no honey gland, and are not associated with ants. In others, the intimate association with ants has lead to their becoming intranidal. In that situation some species have developed obligate carnivorous propensities and in later instars feed on ant larvae. Some species (*Liphyra*) have become completely carnivorous on ants, with eggs being laid directly on the ant nests. A few other species (mostly African) feed on scale insects, aphids, membracids and jassids. All instars of our *Feniseca tarquinius* (Fabr.), the "Harvester," feed on wooly aphids.

Description: Mature larvae small, 12–20 mm. Body dorsoventrally compressed, onisciform (i.e. sowbuglike), slightly attenuated at both ends. Integument thick, unsclerotized, often covered with minute, short, stout setae that impart a roughened appearance. In addition to imparting a texture (velvety, warty, crenulate) to the surface, the small setae, often with stellate bases, (fig. 26.157b), may impart color. The lateral body wall may be extended, flangelike, or the margins of segments (particularly T1) may be swollen and corrugated. Larvae move slowly with a gliding motion; legs and head are hidden.

Head: Small, generally 50 percent or less (often one-third) of greatest body width. With some exceptions (*Eumaeus*, some *Callophrys*, which may reach 54% head/body ratio), this characteristic helps separate lycaenids from riodinids. The latter group has a head capsule ratio that usually exceeds half the body width (a few exceptions in *Calephelis*). Cranial setae much reduced, particularly on the vertex and lateral areas where they may be absent; setae mainly clustered around mouthparts. Narrow, glabrous head capsule may be retracted into a recess in ventral T1, often hidden beneath T1 when feeding. A highly membranous cervix permits extension of the head capsule in a wide arc well beyond margins of T1 and for insertion of the head capsule into plant tissues where larvae may feed internally for at least an additional head capsule width around the original head-sized entrance hole.

Thorax: T1 shield usually unsclerotized and poorly defined. T1 usually 1.5–2× as long as other thoracic segments.

Abdomen: Crochets an interrupted mesoseries (a few exceptions and distinguished only with care in others, i.e., some *Satyrium*); crochets uniserial and triordinal to multiordinal. Often a uniordinal lateroseries (esp. *Lycaena*) on anterior prolegs. Riodinids may have an interrupted or noninterrupted mesoseries, the latter often occuring on the margin of a spatulate fleshy lobe; they also usually have a lateroseries. In the lycaenids, the fleshy lobe may arise within the crochets of the mesoseries or it may be laterad. Anal shield usually poorly defined and unsclerotized.

Three sets of myrmecophilous organs are usually present on abdominal segments of mature larvae (fig. 26.157a): Newcomer's organ, eversible tentacles and pore cupolas. Newcomer's organ is a dorsal transverse unpaired, slitlike structure on A7, from one-fifth to two-fifths the transverse diameter of the segment. The lips of the organ purse and open when a droplet of fluid is exuded, and minute setae help to confine the liquid to the gland site until it is collected by attendant ants. A pair of eversible tentacles are dorsolateral on A8, posterior to the last abdominal spiracle. These may be everted singly or together, and when fully extended they may be 2–3 mm high and contain a corona of setae radiating upward and outward from the tip. Pore cupolas are almost microscopic, round, pimplelike organs overlaying simple subcutaneous glands, which are concentrated on the lips and inner margin of Newcomer's organ. They may also occur near the bases of the eversible tentacles, or on the anal shield, or are located singly at other integumental sites.

Characters of taxonomic importance include head and body chaetotaxy (Clark and Dickson 1956b, 1971, urge using the "conservative" seventh abdominal segment in setal descriptions); prothoracic and anal shield and associated tentacles and setae; stemmatal arc and position of fifth stemma; type of setae; presence or absence of ant relational organs (honey gland, eversible tentacles, dew spots); occurrence of verrucae; labral notches and setae; crochet numbers (instars can be identified by the number of crochets). There may be both segmental and bilateral variance in crochet numbers.

As with most families, marked differences between first instar and mature larvae occur in all species. These include: head to body width; head and body chaetotaxy (particularly secondary setae); cross-sectional shape; presence of honey glands and tubercles; segmental subdivisions, proportional size of spiracles, etc.

Comments: From 35 to 40% of the world's butterflies (circa 5500 species) belong to this family, making it the largest family, exceeding the rather diverse Nymphalidae (excluding the satyrids) and Hesperiidae. It is worldwide, mainly tropical, and reaches its greatest diversity in the tropical rain forests of the African and Oriental realms.

It is estimated that for over three-fourths of the species the life cycles are not known, and most of the remaining fourth are only partially described. Entire lifetimes can be devoted to studying life histories of lycaenids, as Clark and Dickson (1971) have so admirably shown for South Africa.

Larvae should be killed in hot fluids, to prevent muscle contraction, curling and head withdrawal, and placed in KAAD for at least 24 hours, with storage in 80–95% alcohol. Leaching of some color components may occur, but distention of the cervix and head from its retracted position under T1 and expansion of the prolegs for crochet and setal studies are well worth the technique. Freeze-dry dispatching of mature larvae will distend the eversible tentacles for morphological and biochemical studies.

Very few lycaenids are economic pests because their population densities are low, and there is a tendency for most species to lay their eggs singly and broadly dispersed. *Strymon melinus* has been termed the "bean lycaenid" but also has been reported damaging cotton bolls and hibiscus buds. Other species feed on pineapples, pomegranates, and other tropical fruits, but their densities are too low to consider them major pests.

Selected Bibliography

Association with Ants

Hinton 1951.
Malicky 1970.
Farquharson 1922.
Pierce and Mead 1981.

Larval Exudate Glands

Clark and Dickson 1956a, 1956b.
Ehrhardt 1914.
Eltringhan 1921.
Guppy 1904.
Harrison 1908.
Kitching 1983.
Malicky 1970.
Newcomer 1912.

Chaetotaxy

Clark and Dickson 1971.
Downey and Allyn 1979.
Lawrence and Downey 1966.

Taxonomy

Clench 1955, 1965.
Ehrlich 1958.
Opler and Krizek 1984.

Figure 26.154

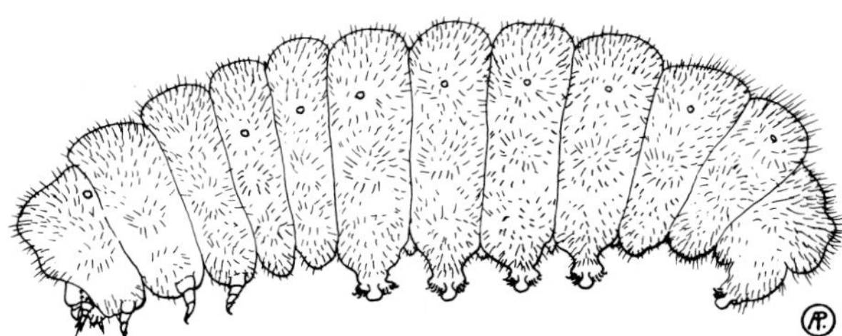

Figure 26.155

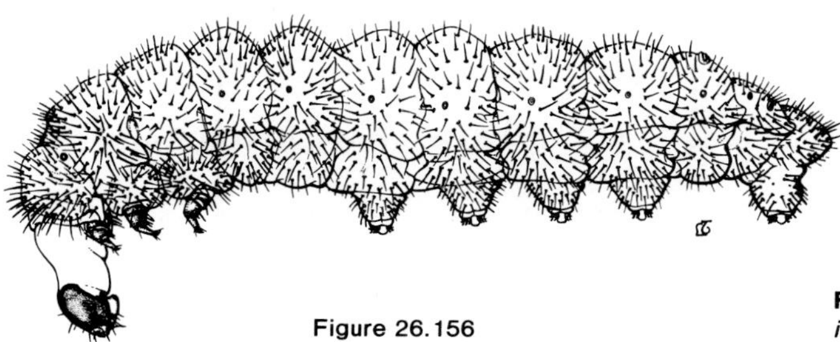

Figure 26.156

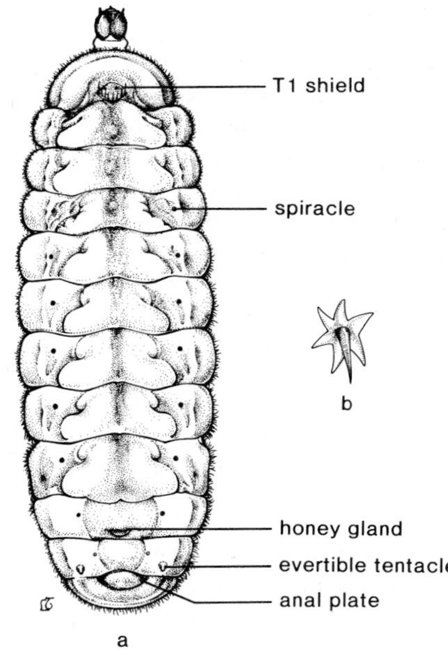

Figure 26.157

Figure 26.154. Lycaenidae, Liphyrinae. *Feniseca tarquinius* (Fabricius), the harvester. f.g.l. 13± mm. Green-brown, hides among its dead and/or living prey, woolly aphids (Eriosomatidae), covering itself with white aphid "wool" and other debris. Silver Spring, Md., 21 September 1931. Eastern N. America, mainly near wet areas. U.S. Natl. Mus. (FWS).

Figure 26.155. Lycaenidae, Eumaeinae. *Strymon melinus* Hübner, cotton square borer, or gray hairstreak. f.g.l. 15± mm. The highly extensible head region is retracted into a fold beneath the prothorax. Variable shades of green, sometimes with diagonal lateral stripes. Feeds on great diversity of trees, shrubs, and herbs, an occasional pest on cotton, beans, hops, and okra in the South. Throughout N. America except the extreme North (fig. from Peterson 1948) (FWS).

Figure 26.156. Lycaenidae, Polyommatinae. *Lycaenides melissa* (Edwards), the melissa blue. f.g.l. 13± mm. Delta, Utah, 6 August 1958, alfalfa. Head greatly extended from normal retracted position. Green with short, brownish setae. On lupine in the East, lupine, alfalfa, and other legumes throughout the West. In the East restricted to northern states where it is found in isolated pockets in New England, New York, and the Lake States area; northern Great Plains and much of the mountainous West, south into Mexico. U.S. Natl. Mus. (FWS).

Figure 26.157a,b. Lycaenidae, Polyommatinae. *Leptodes marina* (Reakirt), the marine blue. f.g.l. 12± mm. Tempe, Arizona, on alfalfa. a. dorsum, showing structures and glands that are difficult to observe on some preserved lycaenids; b. seta with stellate base. Color variable, green to brown. On leadwort (*Plumbago*) and assorted legumes, especially in disturbed areas. Resident from Texas to the Southwest, northward in warm weather. U.S. Natl. Mus. (FWS).

RIODINIDAE (PAPILIONOIDEA)

Donald J. Harvey, *University of Texas*

The Metalmarks

Figures 26.158–26.161

Relationships and Diagnosis: Ehrlich (1958) placed three subfamilies in the Lycaenidae: Styginae, Lycaeninae, and Riodininae. Many authors consider the Lycaeninae (*sensu* Ehrlich) to be of family status (e.g., Eliot 1973), with the riodinids a closely related but distinct family. The most recent classification of the Riodinidae (Stichel 1930–1931) is followed here with slight modification. Most New World taxa belong in the Riodininae (including *Helicopis*) or Euselasiinae. *Corrachia* and *Styx* are monotypic, tropical relicts whose immature stages are unknown. The majority of species, including all those in North America, are members of the Riodininae. Most genera are tropical, and only *Apodemia* and *Calephelis* have large ranges north of Mexico.

Larvae of Riodinidae are recognized by the presence of more than two mandibular setae (fig. 26.160b), in contrast to the Lycaenidae where only two are known to occur. In addition, in the subfamily Riodininae, the spiracle on A1 is ventrad and cephalad of its position on A2 (figs. 26.158, 26.159, 26.160a). Euselasiinae have only slight displacement of this spiracle, but they are unique in bearing spatulate-tipped setae mesally on the thoracic tibiae (fig. 26.161x). Lycaenids have the A1 and A2 spiracles essentially in line.

Biology and Ecology: All known larvae feed on Angiosperms, although some tropical taxa may eventually be found to feed on ant brood or Homoptera, as do some lycaenids (see Cottrell 1984). Host records include many plant families, with few generalizations yet possible. Most riodinids feed on foliage and, unlike lycaenids, they rarely feed on plant reproductive tissues.

Larvae of Riodininae are usually solitary, and some construct simple leaf shelters. Most Euselasiinae are gregarious and processional and pupate in groups.

Ant associations similar to those of lycaenids, although not known for genera north of Mexico, are widespread in tropical riodinines. Ants obtain fluid from paired glands on A8 (fig. 26.159) and may herd larvae either into chambers constructed at the base of the host plant during the day or into the ant nest. These relationships range from facultative attendance by several ant species to apparently species-specific interactions.

The few eggs known are morphologically diverse. Examples of most genera found in North America have been illustrated (Downey and Allyn 1980).

Description: Mature larvae small to medium, 12–35 mm. Body cylindrical to very flattened. Primary setae difficult to locate; secondary setae abundant, sometimes long and in dense clusters on verrucae, or evenly distributed, extremely short and inconspicuous dorsally (in ant-associated taxa) with only a fringe of long, lateral setae. Larvae conspicuous or quite cryptic on hosts.

Head: Hypognathous. Six stemmata, 1–4 usually evenly spaced in semicircle, 1 sometimes separated from 2–4, 5 usually separated from 1–4, 6 about equidistant from 1 and 4. Outline of head roughly circular in Riodininae, laterally compressed in Euselasiinae. Secondary setae present.

Thorax: T1 shield distinct, bearing abundant setae, sometimes projecting forward above head. Clusters of bladderlike setae present in some tropical genera (e.g., *Helicopis, Theope*). Paired, rodlike structures ("vibratory papillae" of Ross 1966) present on the anterior margin of T1 shield in many ant-associated taxa (fig. 26.159). Ventral T1 gland absent.

Abdomen: Prolegs present on A3–6 and A10, with uniserial, uni- to multiordinal mesoseries of crochets, interrupted by planta (continuous in *Apodemia*). Uni- to multiordinal lateroseries present (absent in *Apodemia*). Spiracle on A1 usually displaced ventrally (figs. 26.158, 26.159, 26.160a), often below lateral group of setae. Spiracles on A3–7 also displaced ventrally in some tropical groups (fig. 26.159).

Comments: The family reaches its greatest diversity (over 1000 species) in Neotropical rainforest habitats, with only 24 species recorded from North America. The Riodinidae is biologically the most poorly known family of butterflies. Hosts are known for most North American species, but those of tropical taxa (including the extent and nature of ant relationships) are largely unknown. The only economically important riodinids are some gregarious *Euselasia* species that occasionally defoliate cultivated Myrtaceae, *Eucalyptus* spp. and *Psidium* sp. (guava).

Selected Bibliography

Callaghan 1977.
Cottrell 1984.
Downey and Allyn 1980.
Ehrlich 1958.
Eliot 1973.
Guppy 1904.
Kendall 1976.
Opler and Krizek 1984.
Stichel 1930–1931.

LIBYTHEIDAE (PAPILIONOIDEA)

Michael Toliver, *Eureka College*

The Snout Butterflies

Figure 26.162

Relationships and Diagnosis: Snout butterflies are most closely related to the nymphalids, and although they have been considered a subfamily of that group (Kristensen 1976), the larvae are relatively distinctive, so I follow Ehrlich (1958) in considering them a family.

Libytheids are distinguished from other larvae by their abundance of relatively short secondary setae (which are not grouped on verrucae or pinacula, fig. 26.162a, b), crochets in a pseudocircle consisting of a biordinal mesoseries and a few uniordinal crochets in a lateroseries; T2 and T3 expanded, giving the larvae a humped appearance (see below); and most abdominal segments divided into four annulets. A10 is truncate.

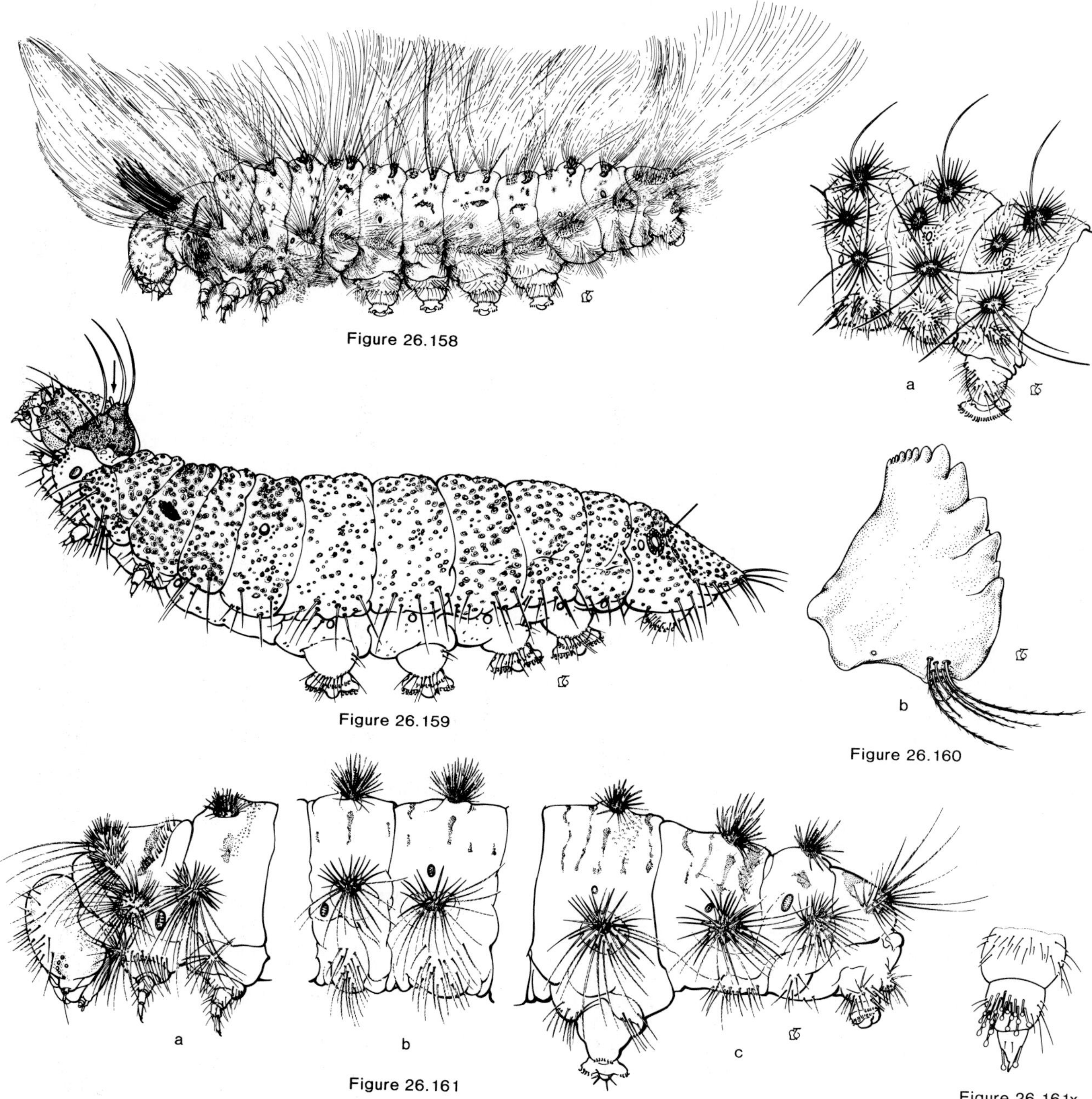

Figure 26.158

Figure 26.159

Figure 26.160

Figure 26.161

Figure 26.161x

Figure 26.158. Riodinidae, Riodininae. *Calephelis wrighti* (Holland). Wright's metalmark. f.g.l. 20± mm. Greenish-gray with dark spots and long, fine, white setae. On *Bebbia juncea* (Asteraceae) in S. California, S.W. Arizona and adjoining Mexico.

Figure 26.159. Riodinidae, Riodininae. *Calospila cilissa* (Hewitson). f.g.l. 10± mm. Lower abdomen extended to show A1 and A3-7 spiracles that are normally concealed beneath the edge. Note gland on A8 and "vibratory papilla" on T1 (arrows). Tan, with brown prothorax and white fringing setae. On *Stigmaphyllon lindenianum* (Malpighiaceae), Osa Peninsula, Costa Rica, 7 March 1982. Tended by ants (*Crematogaster brevispinosus*). Nicaragua to Panama.

Figure 26.160a,b. Riodinidae, Riodininae. *Apodemia mormo* (C. and R. Felder). a. A1-A3; b. mandible with 5 plumose setae. f.g.l. 22± mm. Gray-violet with dark spots and stiff, dark setae. On buckwheats (*Eriogonum* spp., including *fasciculatum, latifolium, elongatum,* and *wrighti*) (Polygonaceae). Most of United States west of Great Plains and northern Mexico.

Figure 26.161a-c. Riodinidae, Riodininae. *Melanis pixe* (Boisduval). f.g.l. 22± mm. a. head, T1, T2; b. A1, A2; c. A6-10. White and yellow, with dark markings; long white setae laterally, anteriorly (T1) and posteriorly (A9). On *Pithecellobium dulce* (Mimosaceae) in Southern Texas and Mexico, *Albizia* sp. (Mimosaceae) in Costa Rica. Extreme Southern Texas to Panama.

Figure 26.161x. Riodinidae. Euselasiinae. *Hades noctula* Westwood. f.g.l. 18± mm. Left T3 leg, mesal view, showing spatulate-tipped setae. Body light brown, with darker brown markings and brown setae. On *Spondias mombin* (Anacardiaceae) in Costa Rica. Mexico to northern South America. Drawing by D. J. Harvey.

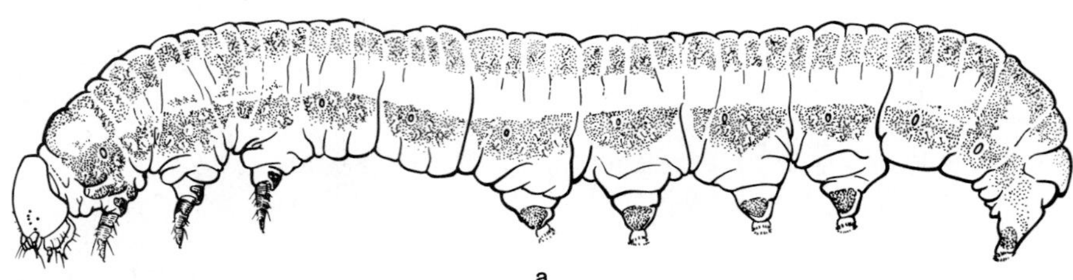

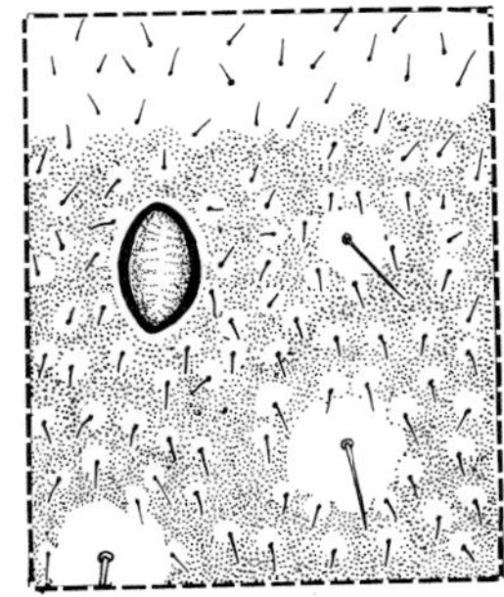

Figure 26.162a,b. Libytheidae. *Libytheana bachmanii* (Kirtland), snout butterfly. f.g.l. 25± mm. b. enlargement A3 spiracular area. Weslaco, Tex., September 1951. U.S. Natl. Mus. Host: *Celtis* spp. See text (FWS).

Biology and Ecology: Pale green eggs are somewhat barrel-shaped and ornamented with vertical ridges of varying length; they are deposited singly on the host plant. Larvae feed on terminal young leaves of *Celtis,* with additional reports of feeding on *Symphoricarpos.* When disturbed, they rear up in the manner of sphingid caterpillars, so that the expanded thoracic segments are maximally exposed. Complete development may take place in as little as 15 days (Edwards 1881a). Pupae are suspended from a silk pad by the cremaster (there is no silken girdle), and they superficially resemble nymphalid pupae.

Description: Mature larvae medium (30 + mm) (fig. 26.162a). Body more or less cylindrical, but with T2 and T3 swollen in our species. Cuticle granular, with an abundance of secondary setae of different lengths (fig. 26.162b), the longer ones located in lighter circular areas. Some setae may be borne on very small chalazae. Primary setae obscured or absent. Color usually dark green dorsally, pale green ventrally, with middorsal yellow stripe. Lateral white or yellow stripe just dorsad of spiracles also present.

Head: Hypognathous. Frontoclypeus as wide as long, extending one-half to epicranial notch. Ecdysial lines meeting approximately one-fourth length of frontoclypeus above its apex. Six stemmata, 1–5 in semicircle with 5 separated from 1–4; 2–4 on papillae and thus more conspicuous than 1 and 6; 5 also on a papilla, but its location near antennal base makes it inconspicuous. Six caudad of 3 and 4, and the most difficult to locate. Six about 2× as far from 5 as 5 is from 4. Labrum large, notched. Cuticle slightly rugose, with numerous secondary setae. Primary setae obscured or absent. Color green to brown.

Thorax: Prothoracic shield indistinct; dorsum with some relatively longer setae. T1 spiracle oval, slightly larger than A1 spiracle. Ventral prothoracic gland present. Prothorax divided into two obvious annulets. T2 and T3 expanded, T2 with four and T3 with five annulets. Secondary setae especially numerous around thoracic legs. Primary setae absent or obscured.

Abdomen: Spiracle on A2 slightly higher than other abdominal spiracles. All spiracles a rounded ellipse, with A8 larger than others. A1–8 divided into four annulets each. Secondary setae abundant, particularly on the venter, although they may be short and relatively inconspicuous elsewhere. Prolegs on A3–6 somewhat conical in shape; planta short and ovoid. Crochets in a biordinal mesoseries and a sparse uniordinal lateroseries, forming a pseudocircle. A10 proleg with biordinal penellipse or crochets broken caudolaterally. Dorsum of A9 and A10 curved abruptly ventrad, giving caudal end a truncate appearance.

Comments: Only one species, *Libytheana bachmanii* (Kirtland), is widespread in our area; reports of two others in the southern U.S. are inadequately substantiated or extremely rare. Adults often migrate in huge swarms, especially in the southwestern U.S. *L. bachmanii* is of no economic importance. Edwards (1881a) describes the life history.

Selected Bibliography

Edwards 1881a.
Ehrlich 1958.
Ferris and Brown 1981.
Klots 1951.
Kristensen 1976.
Opler and Krizek 1984.

NYMPHALIDAE (PAPILIONOIDEA)

Michael Toliver, *Eureka College*

The Brush-footed Butterflies

Figures 26.163–26.174

Relationships and Diagnosis: A single classification of the nymphaloid butterflies is by no means universally accepted. For example, Miller and Brown (1981) recognize (in America north of Mexico) the "families" Heliconiidae, Nymphalidae (with five subfamilies), Apaturidae (two subfamilies), Satyridae (two subfamilies) and Danaidae (two subfamilies); whereas Ehrlich (1958) recognizes only the Nymphalidae, with Danainae, Satyrinae, and Charaxinae (=Apaturinae) included as subfamilies (Heliconiidae are considered no more than a tribe). Ehrlich's (1958) classification is followed because he is the only worker to apply modern systematic techniques to an analysis of the higher classification of butterflies. Using many characters (including some from immatures), Ehrlich attempted to make the higher classification of butterflies comparable to higher classifications of other insects. A recent cladistic reanalysis of Ehrlich's data largely supports his conclusions (Kristensen 1976).

Such a classification includes a diverse array of larvae in a single family, so that no single character may be used to distinguish them from other Lepidoptera larvae. This situation is not unique to the nymphalids. Nymphalids' closest relatives are the libytheids, which may be distinguished from nymphalids by possessing a pseudocircular arrangement of the crochets. Nymphalids universally possess a uni-, bi- or triordinal mesoseries. This characteristic also serves to separate them from hesperiids, which many nymphalids resemble because they have a head much larger than the prothorax. Most species of nymphalids possess scoli, resembling saturniids, but nymphalids typically lack a middorsal scolus on A9 *and* have a middorsal scolus on A7 (fig. 26.167b), as well as having triordinal rather than biordinal crochets. Other characteristics that distinguish nymphalids include numerous secondary setae both dorsally and ventrally (these may be very short and inconspicuous, e.g., Danainae); stemmata conspicuously different from one another (either 2–5 on papillae and more prominent than 1 and 6, or 3 much larger than the others); heads in many species with prominent protuberances or scoli; no osmeterium; and a bifurcate anal plate in some (Apaturinae, Satyrinae, and some Nymphalinae).

Biology and Ecology: As might be expected in a group with larvae of diverse morphology, the biology and ecology are diverse as well. Typical nymphalines often lay their spherical, barrel-shaped or conical eggs in clusters, sometimes numbering several hundred eggs per cluster. When the larvae hatch, they are often gregarious, at least for the first few instars, and they may spin a silken web encompassing one to many leaves of the host. They may take shelter, rest, or feed in this web. Larval host plants of this group include a wide variety of trees, shrubs, and herbaceous plants. North American apaturines feed on *Celtis* or *Croton* and hatch from clusters or singly laid eggs. Some of them form larval shelters by tying the edges of a leaf together. The satyrines are grass feeders, hatching from eggs laid singly. The larvae are usually longitudinally striped, which may help to conceal them as they feed or rest in their grass clump homes. Danaines have spindle-shaped eggs, resembling those of pierids, which they lay individually on their hosts (Asclepidaceae). The brightly colored (aposematic) larvae sequester toxins from their host plants, rendering them distasteful to vertebrate predators. Depending on location and species, nymphalids may diapause as eggs, larvae, pupae or adults.

Pupae are hung from a variety of substrates from a silken pad in which the cremaster is hooked. There is no silken girdle. Pupal morphology is discussed by Mosher (1916a).

Description: Mature larvae medium to large, ranging from about 20 mm to over 60 mm. Body various; cylindrical or tapering to either end, often with head larger than T1, often with scoli, humps, protuberances or filaments. Cuticle smooth to granular, often with annulets on each segment. Secondary setae abundant, particularly on venter, usually conspicuous but occasionally inconspicuous (Danainae). Spinules sometimes present. Spiracles elliptical, T1 and A8 larger than others. Color variable, usually some shade of brown, black or green; may have transverse (vertical) bands of yellow, white, green or orange; may be longitudinally striped with white or yellow; may have orange or white spots. In some cases the larvae are quite conspicuous in their natural habitats; in others they are well camouflaged.

Head: Hypognathous. In **nymphalines** frontoclypeus as long or longer than wide, margins straight, extending from one-fourth to two-thirds to epicranial notch. Ecdysial lines meet from one-fourth to entire length of frontoclypeus above its apex. Six stemmata, with 2–5 on papillae and more prominent than 1 and 6. Spacing of stemmata variable; 1–5 may form relatively evenly spaced semicircle, 1 and 5 may be separated from 2–4, or 5 may be near antennal base and widely separated from 1–4. Six may be farther from 5 than 5 is from 4, or it may be closer. Stemmata may be on darkened patch of cuticle (figs. 26.163a, d) or relatively inconspicuous on a unicolorous background (fig. 26.169). Labrum relatively large, deeply notched. Numerous secondary setae on head, often on prominent chalazae (fig. 26.174c). Head frequently bilobed, and scoli, knobs or prominent chalazae may be present on apex of each lobe (fig. 26.163d). Cuticle smooth, granulate, punctate or rugose.

In **apaturines** (fig. 26.165), the frontoclypeus is longer than wide, reaching one-third to epicranial notch. Ecdysial lines meet about one-fourth length of frontoclypeus above its apex. Six stemmata, 1–4 in semicircle with 2–4 on papillae and more prominent than 1 and 6; 5 also on a papilla near antennal base; 6 caudad of 4, closer to 4 than 5. Labrum small, notched. Numerous secondary setae on head, often making it appear "fuzzy"; often with scoli present. Head bilobed, often with extremely long scoli (longer than frontoclypeus) at apex of each lobe (*Asterocampa,* fig. 26.165). In *Anaea,* no scoli present; instead apex of each lobe has obvious knobs (much shorter than frontoclypeus). Cuticle punctate or tuberculate. Head larger than T1.

In **satyrines** (fig. 26.174), frontoclypeus longer than wide, extending one-third to one-half to epicranial notch. Ecdysial lines meet relatively close to frontoclypeal apex, about one-eighth length of frontoclypeus above it. Six stemmata, with 1–3 grouped together and 3 much larger than others (fig. 26.174b); 4, 5 and 6 scattered; 5 near antennal base, and half again as far from 4 as 4 is from 3; 6 caudad of 4, about as far from 4 as from 5. Labrum relatively small, notched. Numerous evenly spaced secondary setae on head with tuberculate bases (fig. 26.174c). Head rounded or bilobed; horns may be present on apex of each lobe. Cuticle smooth to tuberculate. Color usually some shade of green and may be longitudinally striped.

Danaines (fig. 26.173) have the frontoclypeus wider than long, apex about one-half to epicranial notch. Ecdysial lines meet about one-sixth length of frontoclypeus above its apex. Six stemmata, 1–4 grouped in semicircle with 5 widely separated near antennal base; 2–5 on papillae, more prominent than 1 and 6; 6 caudad of 4, closer to 4 than 5. Labrum large, notched. Secondary and primary setae inconspicuous or absent. Head slightly bilobed, smaller than T1. Color light green, vertically banded with black.

Thorax: Ventral T1 gland present. Secondary setae generally longer and more conspicuous ventrally. In **nymphalines** T1 frequently narrows right behind head, forming a "neck" much as in hesperiids. Most nymphalines have the T1 shield indistinct or indicated by long setae on prominent chalazae.

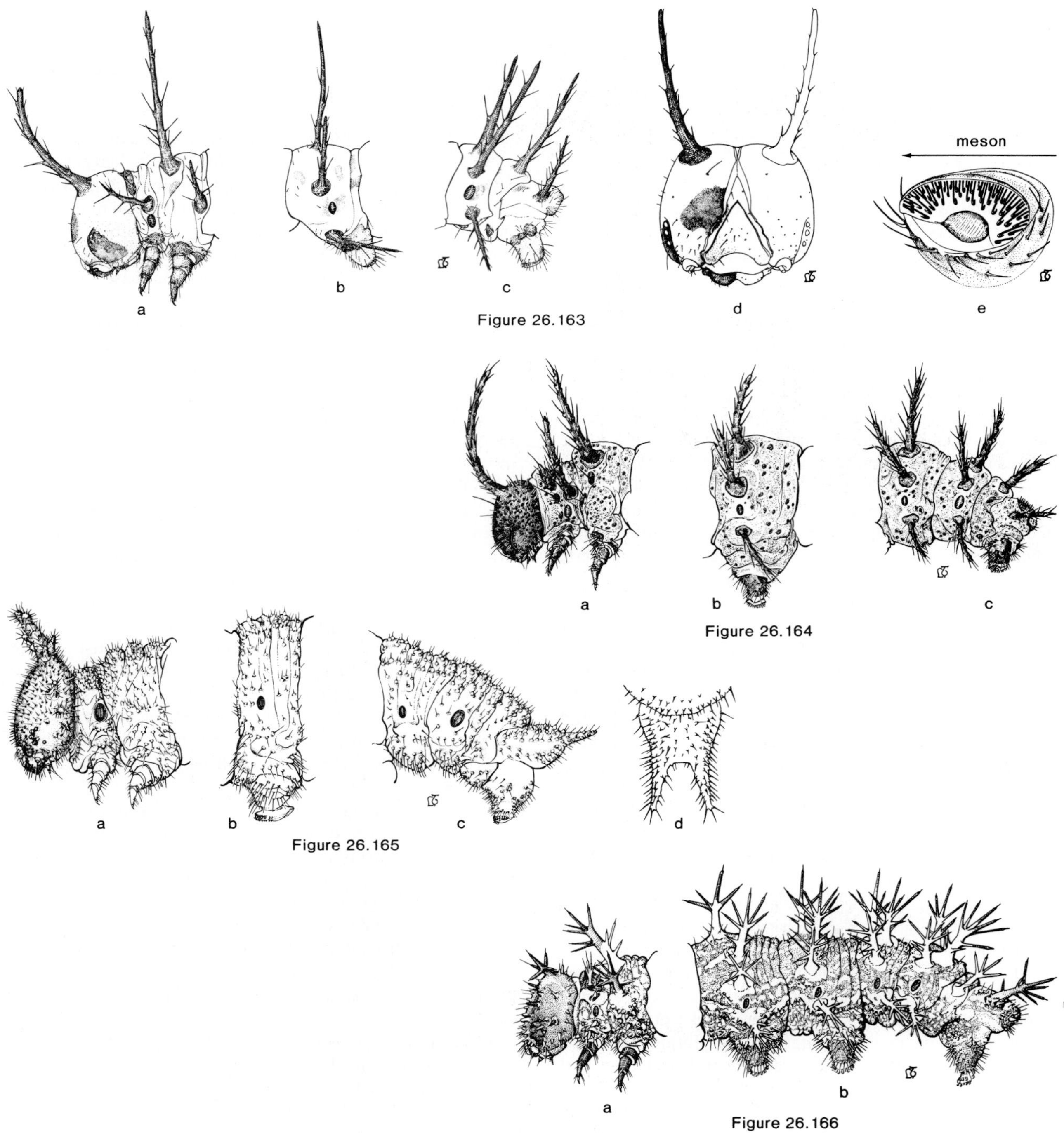

Figure 26.163a-e. Nymphalidae, Nymphalinae. *Heliconius charitonius* (L.), zebra butterfly, f.g.l. 40± mm. a. head, T1, T2; b. A3; c. A8-A10; d. head, front; e. A3 crochets. White, with slender, black spines. Florida City, Fla., 29 March 1936, on *Passiflora*. Cornell Univ. Larva 20 mm, instar uncertain. Texas to S. Carolina (FWS).

Figure 26.164a-c. Nymphalidae, Nymphalinae. *Agraulis vanillae* (L.) gulf fritillary. f.g.l. 35± mm. a. head, T1, T2; b. A3; c. A7-A10. Brownish, with rust stripes (not evident on preserved larvae or drawing) on black spines. St. Thomas, St. John, Virgin Islands, 20 October 1954, *Passiflora* vines. USNM. Southern U.S. (FWS).

Figure 26.165a-d. Nymphalidae, Apaturinae. *Asterocampa clyton* (Boisduval and Leconte), tawny emperor. f.g.l. 30± mm. a. head, T1, T2; b. A3; c. A7-A10; d. A10 dorsum. Brownsville, Tex., 24 January 1945, ex. *Celtis*. U.S. Natl. Mus. Bright green with yellow, bluish and tan stripes. Ranges to the Rockies and north to border states (fig. 26.165d from Peterson 1948) (FWS).

Figure 26.166a,b. Nymphalidae, Nymphalinae. *Polygonia comma* (Harris). hop merchant. f.g.l. 25± mm. a. head, T1, T2; b. A5-A10. Green to brown with numerous scoli. On hops (*Humulus*), nettles (*Urtica, Boehmeria*), and elm. East of the Rockies from southern Canada to Kansas and east to N. Carolina (FWS).

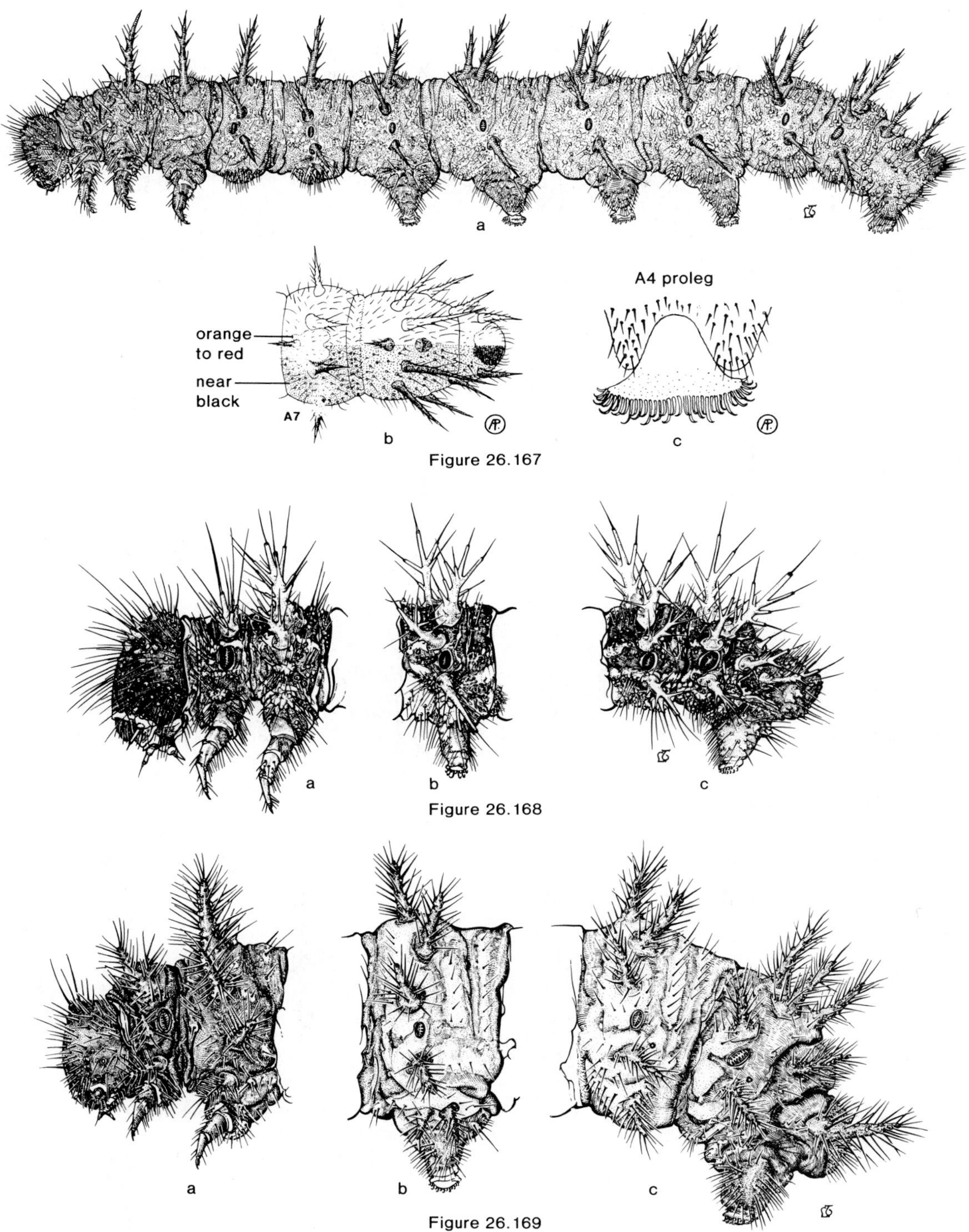

Figure 26.167

Figure 26.168

Figure 26.169

Figure 26.167a,b. Nymphalidae, Nymphalinae. *Nymphalis antiopa* (L.), mourning cloak. f.g.l. 50± mm. a. Larva: b. dorsum A7-A9; c. A4 proleg. Black with white flecks; red spots down back and many black scoli, prolegs reddish. On elm, willow, hackberry, cottonwood, rose, and others. Feeds gregariously throughout N. America where food plants occur (figs. 26.167b,c from Peterson 1948) (FWS).

Figure 26.168a-c. Nymphalidae, Nymphalinae. *Vanessa cardui* (L.), painted lady. f.g.l. 30± mm. a. head, T1, T2; b. A3; c. A7-A10. Green to purple with black mottled areas to pale yellow lateral stripe; yellowish scoli. Prefers thistle (*Cirsium*) but on great variety of other plants. Throughout N. America (FWS).

Figure 26.169a-c. Nymphalidae, Nymphalinae. *Euphydryas phaeton* (Drury), Baltimore. f.g.l. 25± mm. a. head, T1, T2; b. A3; c. A7-A10. Black and orange vertical stripes, many black scoli. Turtlehead (*Chelone glabra*) a preferred food; also on ash, honeysuckle, plaintain. Southern Canada east of Great Plains to northern parts of Gulf Coast states (FWS).

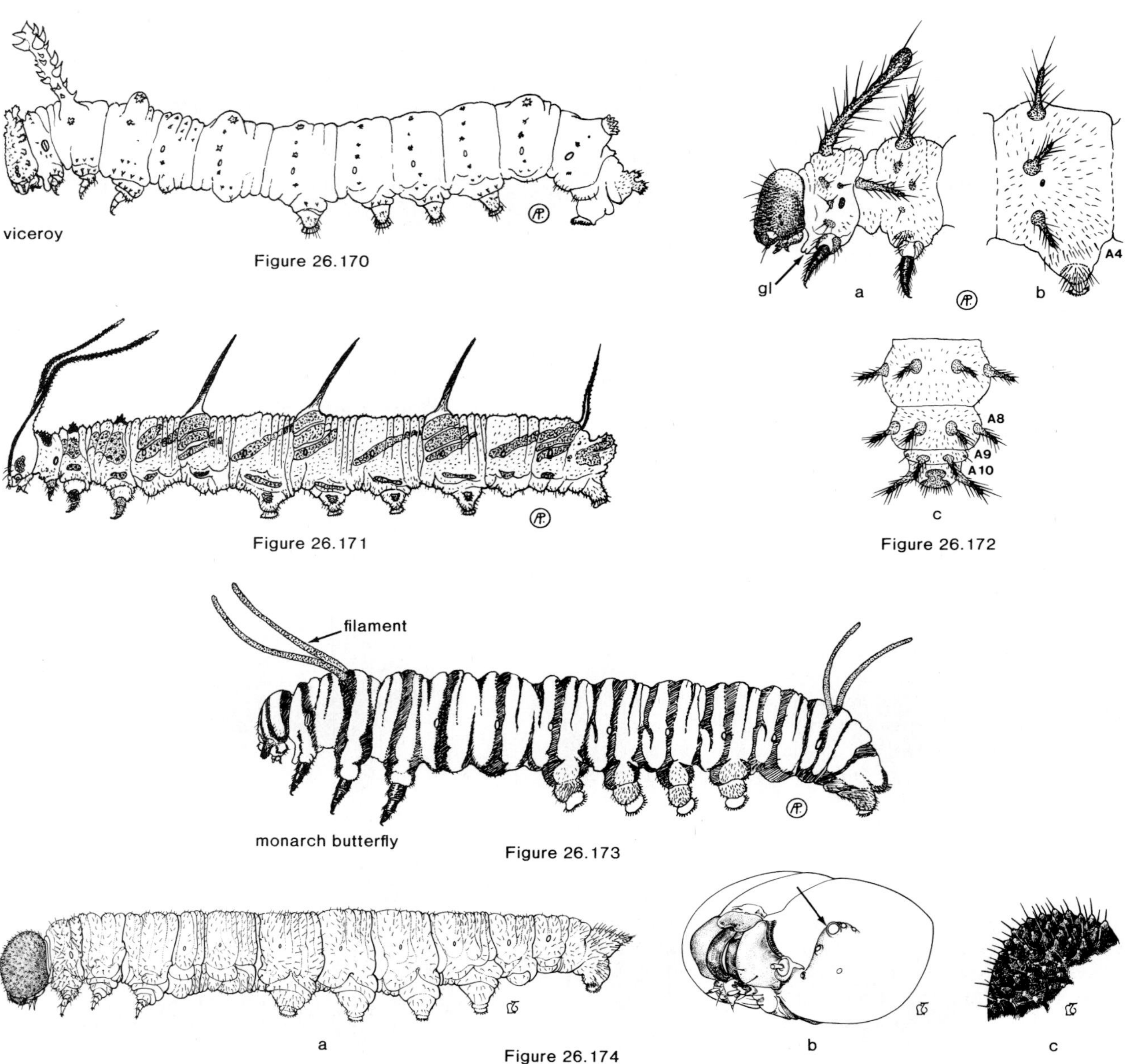

Figure 26.170. Nymphalidae, Nymphalinae. *Limenitis archippus* (Cramer), viceroy. f.g.l. 35± mm. Mottled brown to olive with cream patches laterally and a band across center of abdomen; pair of scoli on T2. On willows, poplars, and fruit trees. North America west to eastern parts of Pacific states (from Peterson 1948) (FWS).

Figure 26.171. Nymphalidae, Nymphalinae. *Marpesa petreus* (Cramer), daggerwing. f.g.l. 35± mm. Rust-colored with yellow to pink areas between the single black-tipped spine on A1, A4, A6 and A8; diagonal dark area below spine; pair of prominent scoli on head. On fig and cashew, Florida to Texas (fig. from Peterson 1948) (FWS).

Figure 26.172a-c. Nymphalidae, Nymphalinae. *Euptoieta claudia* (Cramer), variegated fritillary. f.g.l. 32± mm. a. head, T1, T2; b. A4; c. dorsum A7-A10. Orange-red with white lateral spotty lines; 6 rows of black scoli. Wide range of foodplants; including violets, passion flower, stonecrop (*Sedum*), flax (*Linum*), pansies, plaintain, and others. Southern two-thirds of U.S. west to California (from Peterson 1948) (FWS).

Figure 26.173. Nymphalidae, Danainae. *Danaus plexippus* (L.), monarch. f.g.l. 50± mm. Combination of whitish to pale greenish, yellow, and black cross stripes; pairs of black filaments on T2 and A8. Most commonly on milkweeds, occasionally on dogbane (*Apocynum*). Continent-wide (fig. from Peterson 1948) (FWS).

Figure 26.174a,b. Nymphalidae, Satyrinae. *Cercyonis pegala* (F.), the wood nymph. f.g.l. 30± mm. a. larva; b. stemmata showing enlarged third stemma (details omitted); c. portion of cranium. Yellow-green, fuzzy, with 4 lengthwise yellowish stripes. A10 dorsum forked and reddish. On assorted grasses (FWS).

Scoli or verrucae usually present; sometimes the most prominent scoli located addorsally on T1 (fig. 26.172a). Secondary setae prominent, often on chalazae; may be of different sizes, with long ones on chalazae and short ones scattered about between chalazae. T2 may have longest scoli present on body in addorsal position (fig. 26.170). Scoli, verrucae or pinacula usually present on T2 and T3.

In **apaturines** the T1 shield is indistinct. All segments may have annulets. Secondary setae on chalazae, often in prominent groups. Long secondary setae present on venter. Scoli absent.

In **satyrines** the T1 shield is indistinct and the segments have annulets, as in apaturines. The secondary setae are shorter and borne on smaller chalazae, and are evenly scattered. Spinules may be present. Scoli absent. Both satyrines (fig. 26.174) and apaturines (fig. 26.165) have T1 narrowed behind the head forming a conspicuous "neck," as in the hesperiids.

The T1 plate is also indistinct in **danaines.** Dorsal secondary setae are very difficult to see, so the cuticle appears bare without magnification. Setae are much less numerous in this subfamily. A pair of long fleshy filaments is located addorsally on T2; additional pairs may be present on other segments.

Abdomen: Prolegs present on A3–6 and A10; crochets in a uni-, bi- or (usually) triordinal mesoseries. Suranal plate on A10 truncate, sloping abruptly ventrad; or bifurcate and projecting caudally in two conical or scoliform processes. In **nymphalines,** A1–10 usually possess scoli, verrucae, and/or pinacula. When scoli are present, many species have up to 12 per segment on A3–6 (fig. 26.169b). Secondary setae abundant, frequently on chalazae. Primary setae absent, obscured or possibly indicated by setose tubercles. Abdominal and thoracic segments frequently annulate. Suranal plate usually truncate; may be bifurcate (fig. 26.165d).

Apaturines lack scoli, but abundant secondary setae are on chalazae. Abdominal segments divided into annulets (usually five). Suranal plate usually obviously bifurcate, with two long conical projections (fig. 26.165d), but *Anaea* has the plate only slightly bifurcate or simply truncate. Body usually widest in vicinity of A1 and tapering toward either end.

Satyrines also lack scoli. Secondary setae not on prominent chalazae, evenly distributed. Abdominal segments annulate. A10 usually with suranal plate bifurcate (*Erebia* is truncate).

Danaines lack scoli. Secondary setae very inconspicuous dorsally. One of our two common species possesses a pair of fleshy filaments on A2 (*Danaus gilippus* (Cramer), the queen), and both it and the monarch possess a pair of fleshy filaments on T2 and A8. Both species are conspicuously colored, with bright green and black transverse (vertical) bands on each segment (fig. 26.173).

Comments: The family is large, with over 200 species in America north of Mexico. Some of our most familiar butterflies are nymphalids, such as the mourning cloak (*Nymphalis antiopa* (L.), fig. 26.167), the viceroy (*Limenitis archippus* (Cramer), fig. 26.170), and the monarch (*Danaus plexippus* (L.) fig. 26.173). Despite this, the life histories of many of our species are totally unknown, and we possess inadequate knowledge of the majority of the rest. Useful information on immatures can be obtained from Holland (1931), Howe (1975), Miller (1981), Ferris (1981b), Ferris and Eff (1981) and Pyle (1981). Pyle provides color photographs of 18 species of our commonest nymphalid larvae and describes other larvae in the text. Holland (1931) reproduces Scudder's (1889) two plates of butterfly larvae (73 figures) and two plates of pupae (134 figures). Emmel and Emmel (1973) figure larvae of 31 of our species. Comstock (1927) also provides useful information, either as descriptions and figures or as references of life history descriptions. Edwards (1868–1897) and Scudder (1889) remain the best primary sources for life history information.

There are few nymphalids of economic importance. *Vanessa cardui* (L.) (fig. 26.168) can significantly damage soybeans in some midwestern states. The larva is figured in color by Pyle (1981). *Polygonia comma* (Harris) (fig. 26.166) and the related *P. interrogationis* (Fabricius) have been known to damage commercial hops. The buckeye, *Precis (Junonia) coenia* (Hübner) has been suggested as a possible biological control agent against certain weeds (*Seymeria*) in southern pine forests.

Selected Bibliography

Comstock 1927.
Edwards 1868–1897.
Ehrlich 1958.
Emmel and Emmel 1973.
Ferris 1981b.
Ferris and Eff 1981.
Holland 1931.
Howe 1975.
Kristensen 1976.
Miller 1981.
Miller and Brown 1981.
Mosher 1916a.
Opler and Krizek 1984.
Pyle 1981.
Scudder 1889.

ZYGAENIDAE (ZYGAENOIDEA)

Frederick W. Stehr, ***Michigan State University***

The Smoky Moths and Burnets

Figure 26.175

Relationships and Diagnosis: Zygaenids are most closely related to Limacodidae, Megalopygidae, and Epipyropidae, which all possess rather unusual body shapes and readily distinguishing characteristics. Zygaenids are easily recognized by the unique, mouthlike "gland" near the spiracle on A2 and A7 of our species, the conspicuous "gland" at the base of the T1 leg on many species, the tufts of nonplumose or minutely ringed setae arranged in transverse rows, crochets in a uniordinal mesoseries, head semiretracted, cuticle covered with short, dense spinules (difficult to see when the same color as the cuticle), and by the bright colors (often yellow and black).

Figure 26.175a,b. Zygaenidae. *Harrisina brillians* (Barnes and McDunnough), western grapeleaf skeletonizer. f.g.l. 17± mm. b. A2 post-spiracular gland, also one on A7. Also a gland dorsad of T1 leg. Sluglike, yellowish, with transverse rows of verrucae; verrucae above spiracles with black setae except on A3 and A5 where setae are light. Head nearly black and retractile. Younger larvae gregarious, feeding side by side and skeletonizing cultivated and wild grape and Virginia creeper. Larger larvae consume leaves and sometimes grapes. In California and the Southwest. The grapeleaf skeletonizer, *Harrisina americana* (Guerin) is very similar in appearance and habits, but smaller (f.g.l. 12± mm); and mature larvae have a transverse row of black verrucae bordered by white and yellow on each segment; younger larvae with light verrucae, but black on A2 and A8. Eastern North America.

Arctioidea and other caterpillars with verrucae do not have the A2 and A7 glands, the head is usually not retracted, and the verrucae and colors are different.

Biology and Ecology: Zygaenids are exposed feeders, with at least some species of the genus *Harrisina* lining up side-by-side and backing up together as they devour the edge of the leaf. They, and many other Zygaenoidea, are presumably aposematic. Vitaceous vines are preferred hosts, and both grape and Virginia creeper are common hosts. Eggs are laid in groups beneath the leaves and pupation is in tough, flattened cocoons.

Description: Mature larvae small to medium, 10 to 25 mm, superficially resembling some Arctiidae. Primary setae not evident, with secondary setae arranged in conspicuous verrucae, and with some very long setae present dorsally on T1, T2, and A9, and laterally on T2 thru A9.

Head: Semiconcealed, bearing primary setae only. Frontoclypeus about as wide as high; clypeus fairly distinctly delineated, and with retractile muscles attached near or at the apex of the frons. Frontoclypeus extending to long and narrow epicranial notch. Ecdysial sutures evident, but not sharply defined. Six stemmata in a normal arc, but with 6 closer to 1 than 5, or equidistant between them. Antennae quite long (as is usually the case in larvae with retracted or semiretracted heads).

Thorax: All thoracic legs with one or two spatulate setae near the claw. T1 with shield, L and SV verrucae. T2 and T3 with five verrucae.

Abdomen: Verrucae D, SD, L, and SV distinct on A1–9, but the SV group weaker on A7–9. A10 with tufts of setae on anal plate and proleg. Spiracles round, with a conical central protrusion that is quite conspicuous on some species. "Mouthlike glands" (fig. 26.175b) adjacent to the spiracle on A2 and A7. Prolegs on A3–6 and 10 bearing uniordinal crochets in a mesoseries, with the A10 mesoseries angled at 45° to the midline. Anal fork present in some genera.

Comments: The family is poorly represented in North America, with 22 species in six genera known. They are of little economic importance, but the grapeleaf skeletonizer, *Harrisina americana* (Guerin), and the western grapeleaf skeletonizer, *Harrisina brillians* Barnes and McDunnough (fig. 26.175), are occasionally pests on grape.

Selected Bibliography

Forbes 1923.
Johnson and Lyon 1976.
Langston and Smith 1953.
Smith and Langston 1953.

MEGALOPYGIDAE (ZYGAENOIDEA)

Frederick W. Stehr, *Michigan State University*

Puss Caterpillars (Flannel Moths)

Figures 26.176, 26.177

Relationships and Diagnosis: Megalopygids are related to Zygaenidae, Epipyropidae, Dalceridae and the Limacodidae, which they most resemble. They are easily recognized by their unusual complement of seven pairs of prolegs (fig. 26.177b), with A3–6 and A10 bearing crochets, and A2 and A7 bearing no crochets and only a pair of stout anterior SV setae (which are also present on A3–6). They have peglike glandular protuberances beside each spiracle on T1 and A2–8, bear urticating setae beneath long, dense, silky tufts that blend together to form a "tailed coat" (*Megalopyge*, fig. 26.177), or they bear more conventional verrucae (*Norape*) which have some long central setae (fig. 26.176).

Biology and Ecology: Megalopygids are leaf feeders on a variety of shrubs and trees, including redbud, beech, and hackberry (*Norape*), and elm, maple, sycamore, black locust, birch, cherry, pecan, oak, apple, blackberry, raspberry, and rose (*Megalopyge, Lagoa*). Young larvae may feed gregariously as zygaenids do. Pupation takes place in a very tough, tapered, brownish cocoon attached to and blended into a twig or branch, or in a barrel-shaped cocoon with a lid. Eggs are

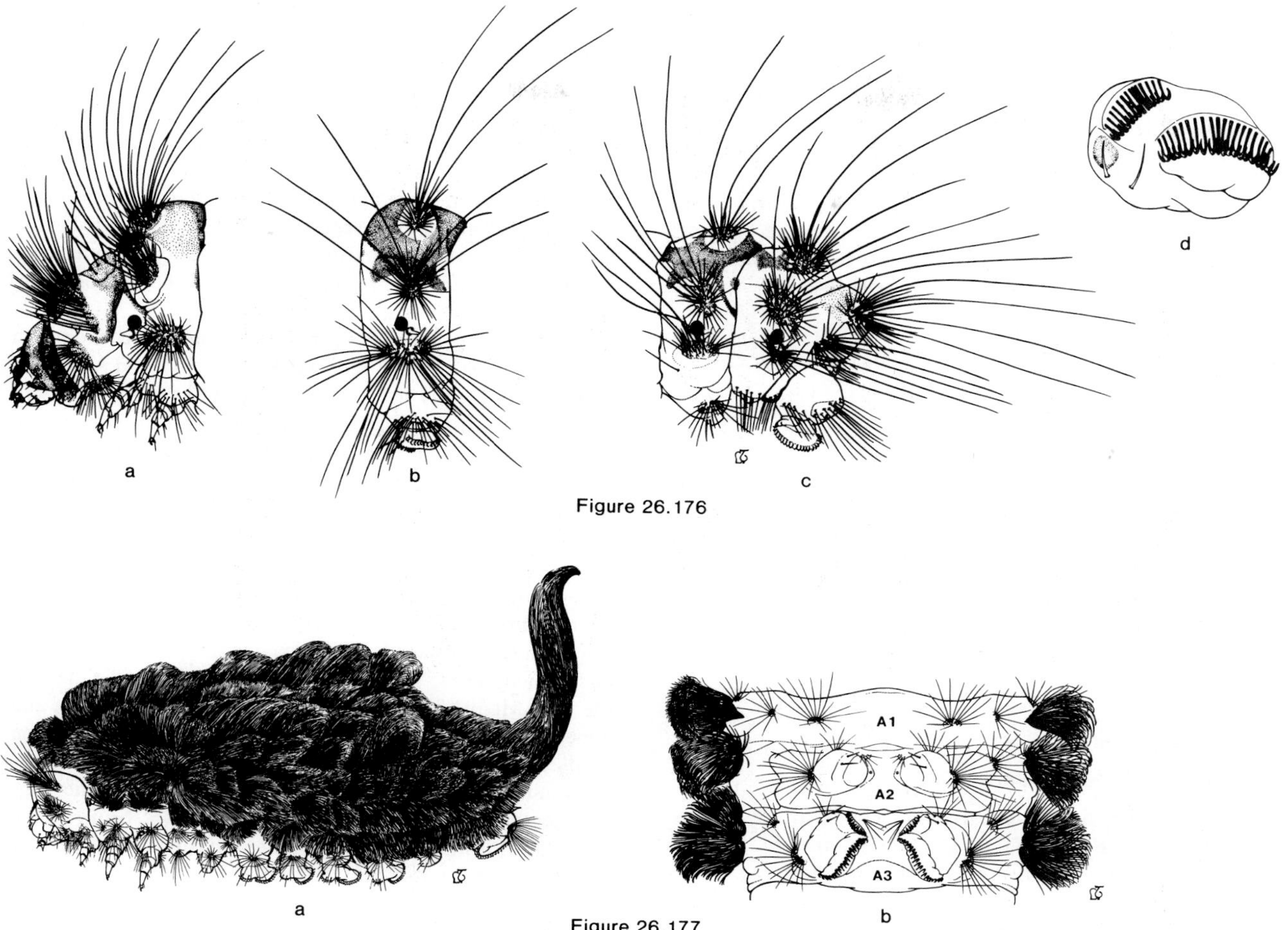

Figure 26.176a-d. Megalopygidae. *Norape ovina* (Sepp). a. head, T1, T2; b. A3; c. A7-A10; d. A3 crochets. f.g.l. 25± mm. Mt. Vernon, Va., 19 August 1921, on redbud. U.S. Natl. Mus. Preserved larvae with gray back and light verrucae. On redbud, mimosa, beech, hackberry. Ohio Valley southward.

Figure 26.177a,b. Megalopygidae. *Megalopyge opercularis* (J. E. Smith), puss caterpillar. f.g.l. 22± mm. LaCrosse, Va., 19 September 1961, on pecan. U.S. Natl. Mus. b. venter of A1, A2, A3 showing small A2 prolegs lacking crochets and A3 prolegs with crochets. Note anterior pair of stout proleg setae on A2 and A3 (also on *Norape,* fig. 26.176d). Densely covered with long yellow to brown to gray hairs with middorsal tufts and "tail". Ohio Valley southward on various shrubs and trees, including elm, hackberry, maple, oak, sycamore, and others.

laid on leaves and covered with hair. Larvae are commonly collected in late summer or early fall, with one generation in the North, and two generations possible in the South.

Description: Mature larvae medium, 15–30 mm, with two body types: the puss type (*Megalopyge*), and the arctiid type (*Norape*). Most setae are minutely plumose, possibly contributing to their urticating properties.

Head: Head capsule complete, but nearly always concealed, the anterior lobes of the prothorax capable of being folded together across the front of the head, totally obscuring it when the head is withdrawn. Frontoclypeus difficult to observe without dissection, but clearly divided into a frons and clypeus. Frons short, bluntly pointed dorsally and extending about one-third the distance to the epicranial notch. Ecdysial sutures difficult to see except on the lower sclerotized face. Six stemmata in a tight arc, except for 5, which is substantially lower.

Thorax: Thoracic legs normal. Prothorax extended forward around the head, the spiracle at the extreme posterior margin so it appears to be on T2. T1 with shield, L and SV verrucae; T2 with D1, D2, SD, L, and SV verrucae; T3 with D, SD, L, and SV verrucae.

Abdomen: A1–8 with D, SD, L, and SV verrucae. A9 and A10 unusual—A9 appearing to be the terminal segment, but with A10 tucked beneath it and anal plate area reduced to a small lobe. A9 bearing three verrucae on dorsal half and with ventral half reduced to a very narrow band. Proleg of A10 normal in size and bearing two verrucae. In effect, A9 and A10 have been "combined" into one segment, with A9 taking over the top half and A10 the bottom half. A3–6 and A10 prolegs with uniordinal crochets in a mesoseries, with the crochets on each A3–6 proleg nearly divided, or divided into two groups (fig. 26.176d). A2 and A7 bearing slightly smaller prolegs without crochets. A2–7 prolegs with a pair of stout setae near the anterior margin of the planta and with the area near the setae sometimes padlike, especially anteriorly. Spiracles round, with conical central protrusion. A peglike, glandular protuberance lies immediately adjacent to the spiracle (ventroanterior on T1 and ventroposterior on A1–8).

Comments: The family is poorly represented in North America, with 11 species in four genera known, primarily from the South and Southwest. They occasionally cause defoliation in limited areas and are sometimes a nuisance because of their urticating setae mixed with long hairs.

Selected Bibliography

Baerg 1924.
Bishop 1923.
Klots 1966.
Packard 1894.
U.S.D.A. 1985.

LIMACODIDAE (ZYGAENOIDEA)

Frederick W. Stehr, *Michigan State University*

Slug Caterpillars

Figures 26.178–26.183 and Cover

Relationships and Diagnosis: Limacodids are related to the Zygaenidae, Epipyropidae, Dalceridae, and the Megalopygidae, which they most closely resemble. They are readily recognized by their unusual A1–7 and A10 prolegs. Each pair of A1–7 prolegs is modified into a single, median, suckerlike, oval lobe that lacks crochets (fig. 26.182b); A10 is modified into a pair of smaller lobes that also lack crochets. Megalopygids have prolegs on A2–7 and 10, and have crochets on A3–6 and A10. The overall appearance of limacodids varies considerably, ranging from the hairy, multilobed hagmoth larvae to the spiny saddlebacks, to the smoother types that do not bear lobes or spines, but may be somewhat "armor-plated."

Biology and Ecology: Slug caterpillars are solitary leaf feeders on a great diversity of shrubs and trees. Some species such as the saddleback caterpillar, *Sibine stimulea* (Clemens), may be found on nonwoody plants (corn). Pupation and overwintering is in a tough, brown, oval cocoon provided with an emergence lid. Eggs are usually flat, and laid singly or in groups on leaves.

Description: Mature larvae medium, 10–30 mm, with three basic body types: those with hairy, detachable lobes (such as the hagmoth larva, *Phobetron* (figs. 26.178a, b)), those with spiny, nondetachable lobes (e.g. the saddleback, *Sibine stimulea* (fig. 26.180)), and the smoother larvae (e.g. *Prolimacodes* (fig. 26.181)).

Head: Head capsule complete, but nearly always concealed, the anterior sides of the prothorax capable of being folded forward so as to obscure the head when it is withdrawn. Frontoclypeus difficult to observe without dissection, but divided into frons and clypeus. Frons short, bluntly pointed dorsally, and extending about one-third the distance to the epicranial notch. Ecdysial sutures difficult to see except on the lower scelerotized face. Six stemmata in a tight arc, except for 5, which is lower.

Thorax: Thoracic legs normal but small. Prothorax extended forward around the head, with the spiracle at the extreme posterior margin and sometimes obscured by the other thoracic and abdominal segments if they are "fused" and hoodlike. T1 without armature, and T2 usually extended over T1. T2 and T3 armature usually similar to that of abdomen, but often larger.

Abdomen: Abdominal armature variable, depending on the species. Segment A8 reduced ventrally. A9 and A10 quite unusual; A9 greatly reduced ventrally, A10 small and tucked beneath A9, (essentially replacing A9 ventrally) with A10 proleg reduced to a small pair of lobes or suckers, and without crochets. A1–7 prolegs reduced to a single, oval, median sucker without crochets. A1–8 with round, conspicuous spiracles. A1 spiracle usually higher than A2–8 spiracles, sometimes visible dorsally in flattened larvae like *Isa* (= *Sisyrosea*) or hidden between the spines in others.

Comments: This family contains about 50 species in 21 genera in America north of Mexico, most of them being rather uncommon and of no economic importance in Canada and the U.S. Many of the larvae have a bizarre appearance, and some species possess urticating spines and hairs. The common saddleback caterpillar, *S. stimulea* (fig. 26.180), bears many poisonous spines that can cause severe irritation. The hagmoth caterpillar, *Phoebetron pithecium* (J. E. Smith) (fig. 26.178a, b) is more interesting than irritating, but does bear irritating hairs on its fleshy lobes.

Selected Bibliography

Dyar 1895–1899.
Harrison 1963.
U.S.D.A. 1985.

EPIPYROPIDAE (ZYGAENOIDEA)

Donald R. Davis, *Smithsonian Institution*

The Epipyropids

Figures 26.184a-i

Relationships and Diagnosis: Larvae of Epipyropidae are ectoparasitic on larvae and adults of certain Homoptera. Although family affinities are poorly understood, they appear most closely allied to the Australian Cyclotornidae, whose first instar larvae are also ectoparasitic on Homoptera. Larval Epipyropidae (fig. 26.184h) superficially resemble Limacodidae and Megalopygidae, particularly with their small head, stout body, reduced prolegs, and secondary setae. They differ from these groups in possessing piercing mandibles, crochets arranged in a complete circle on segments A3–6, and the absence of verrucae or elongate hairs. Also, the later instars of Epipyropidae are unique in secreting a thick, white covering of paraffin over most of their body.

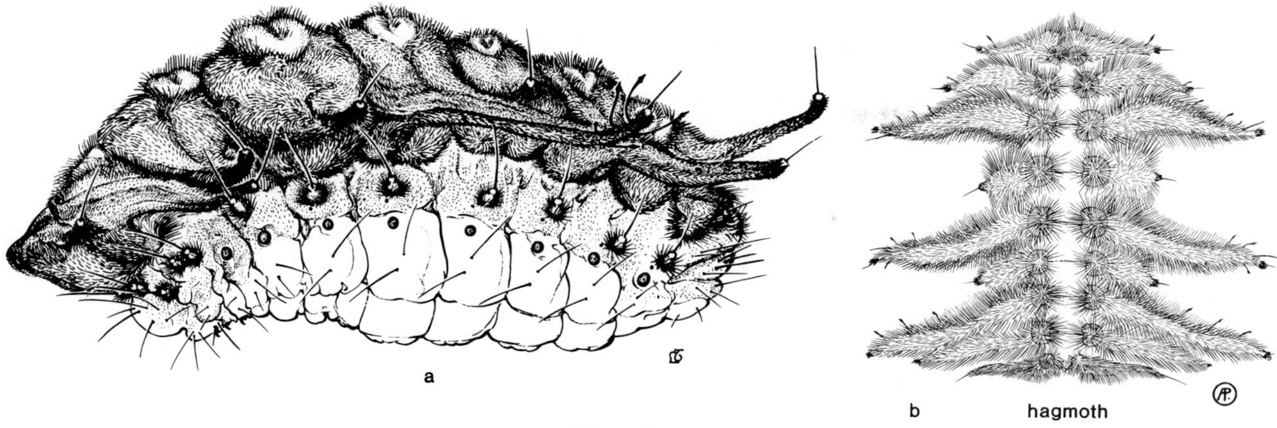

Figure 26.178

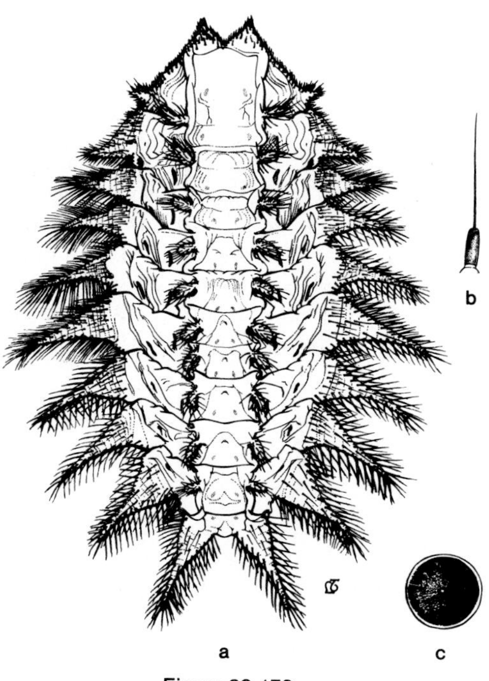

Figure 26.179

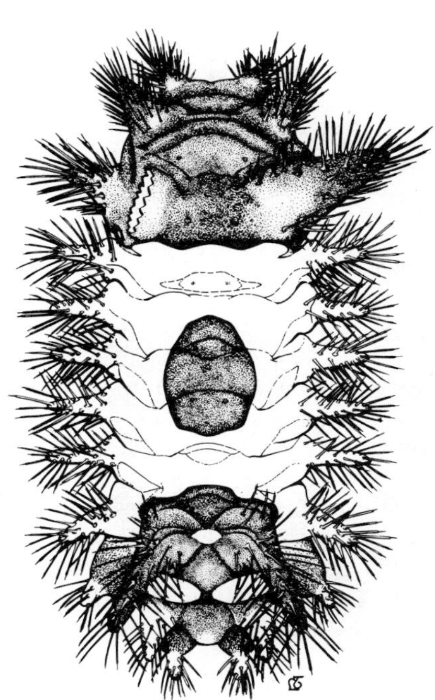

Figure 26.180

Figure 26.178a,b. Limacodidae. *Phobetron pithecium* (J. E. Smith), the hagmoth. a. lateral; b. dorsal. f.g.l. 16± mm. Brown, with 9 pairs of easily detached fleshy processes, the third, fifth and seventh ones much longer and all bearing short stinging hairs. On assorted shrubs and trees from southern Canada southward, east of Great Plains (fig. 26.178b from Peterson 1948).

Figure 26.179a-c. Limacodidae. *Isa textula* (Herrich-Schaeffer). f.g.l. 15± mm. a. dorsum; b. T1 seta; c. A1 conical spiracle. Green, on cherry, oak, Norway maple, and other trees and shrubs in U.S. mixed deciduous forest area.

Figure 26.180. Limacodidae. *Sibine stimulea* (Clemens), saddleback caterpillar. f.g.l. 25± mm. Brown anteriorly and posteriorly, with bright, purple-brown spot in center of green saddle between bases of prominent, reddish-brown scoli. On basswood, chestnut, cherry, oak, plum, and other trees; also on corn. Throughout eastern mixed deciduous forest area.

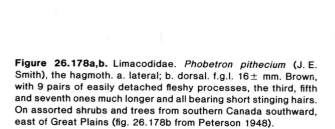

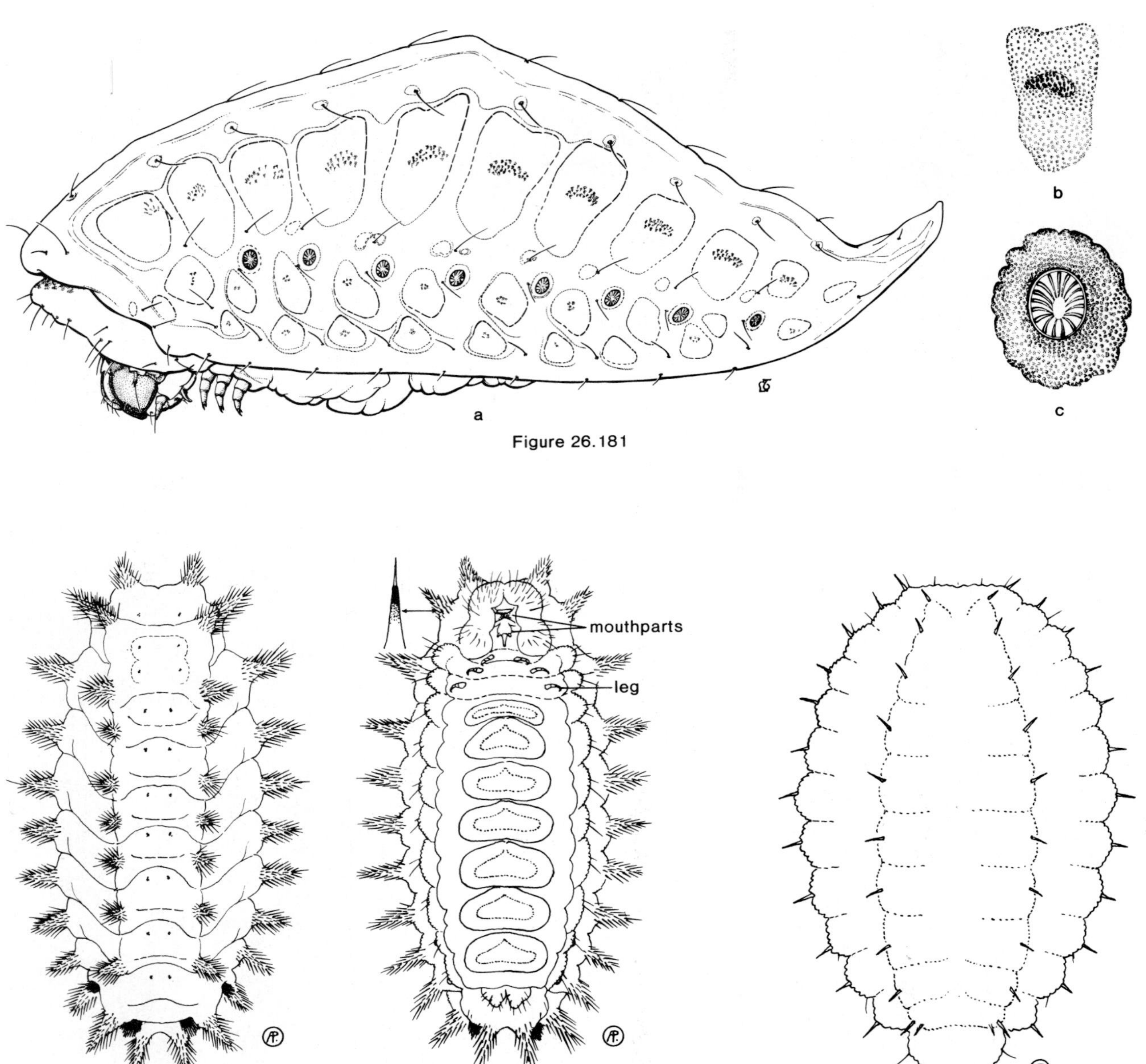

Figure 26.181a-c. Limocodidae. *Prolimacodes badia* (Hübner), skiff caterpillar. f.g.l. 13± mm. a. lateral; b. A4 lateral plaque; c. A4 spiracular area. Pale green, with dorsolateral lengthwise keel on each side; mature larvae with brown area on back. On assorted trees and shrubs, including oak, sycamore, beech, and black cherry. Southern Canada, eastern U.S., west to Arizona.

Figure 26.182a,b. Limacodidae. *Eudea delphinii* (Boisduval). a. dorsal; b. ventral. f.g.l. 18± mm. Pale yellow-green, with 4 patches of dark spines at bases of caudal scoli. On oak, pear, willow, wild cherry, and others. Southern Canada southward in eastern mixed deciduous forest area (figs. from Peterson 1948).

Figure 26.183. Limacodidae. *Tortricidia* sp., f.g.l. 8± mm. Green, with subdorsal yellow lines and red on back. On wide variety of trees and shrubs. Eastern U.S. and Canada (fig. from Peterson 1948).

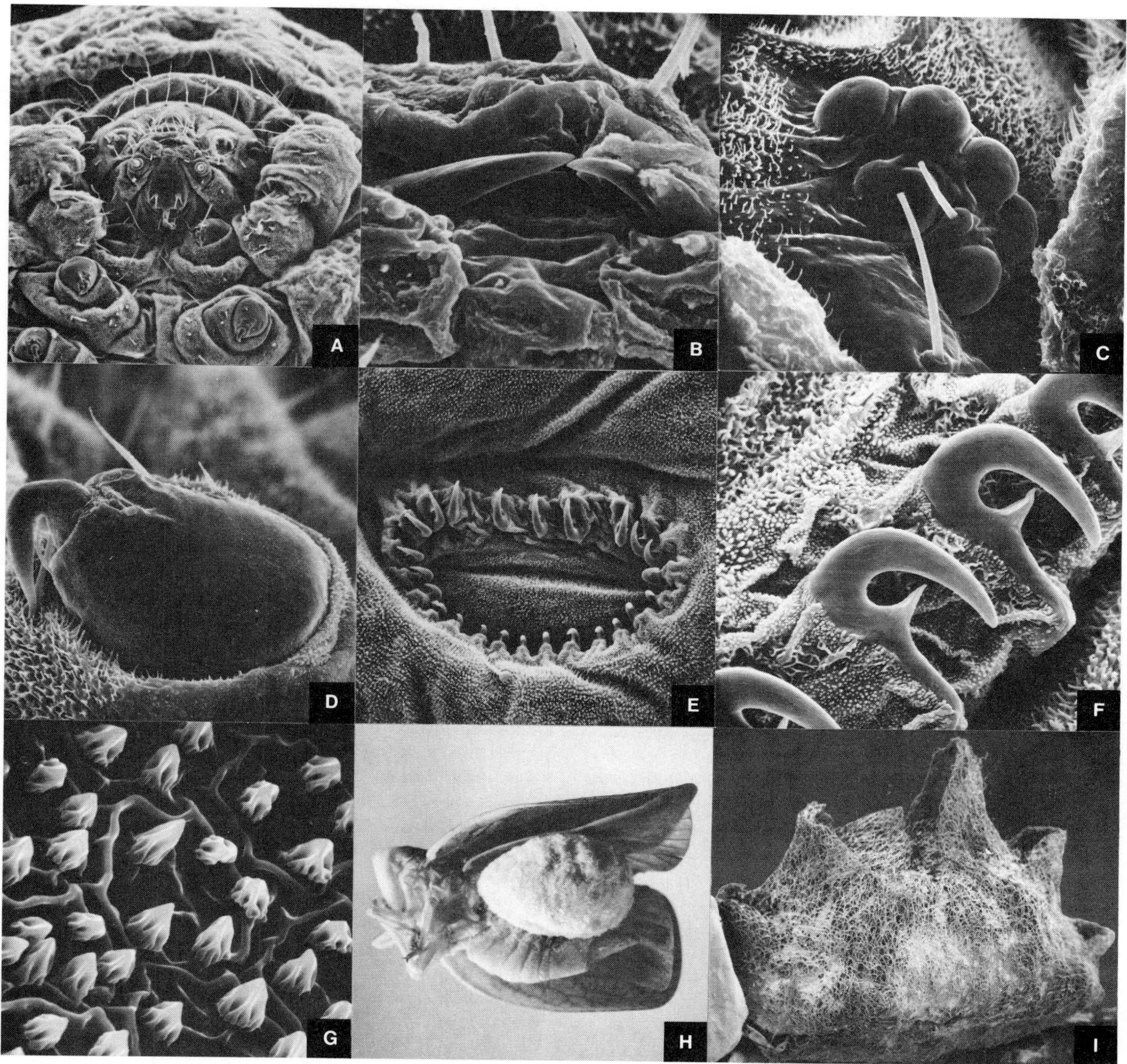

Figure 26.184a-i. Epipyropidae *Fulgoraecia exigua* (Hy. Edwards). a. head, frontal view, ×64; b. mandibles, ×673; c. stemmata, ×483; d. tarsus, T2, e. crochets, A5, ×176; f. crochets, showing retainer spines, ×799; g. integumental papillae and pores, A2, dorsal, ×4095; h. full-grown larva on abdomen of *Acanalonia acuta* (Say); i. cocoon, ×13. Also see Lepidoptera, key to families, fig. 4, venter of *Fulgoraecia exigua.*

Biology and Ecology: The eggs are usually deposited in clusters on the food plants of the homopteran hosts. As many as 3000 eggs are laid by a single female over a 24-hour period (Kirkpatrick 1947). The eggs are smooth, 0.25 mm to 0.50 mm long and half as wide, with a truncate micropylar end. Eclosion apparently varies considerably among species, with Krishnamurti (1933) reporting six to eight days following oviposition and Kirkpatrick (1947) observing a relatively steady rate of emergence from a single egg mass over a one-year span. The latter is probably an adaptation to increase the likelihood of at least some progeny locating suitable hosts.

Five larval instars have been reported (Krishnamurti 1933). Larval development is hypermetamorphic, with the triungulinid first instar actively seeking a host. The larva frequently assumes an erect searching posture, weaving side to side, while awaiting a host. They attack several families of Fulgoroidea as well as the Cicadellidae and Cicadidae. All later instars are passed on the body of a single host, where they feed on the haemolymph lapped up from small punctures caused by slender, styletlike mandibles (fig. 26.184b). By the second instar they are incapable of parasitizing a second host should they become dislodged. The body becomes considerably stouter and the legs and head greatly reduced. Attachment to the host is by means of powerful tarsal claws (fig. 26.184d) and specialized crochets (fig. 26.184e, f), which firmly grasp silk cinch lines laid down over the host. Each hook of the ventral crochets possesses a small retainer spine at its base, which serves to prevent slippage of the silk cinch. By the third instar the larva becomes even more sessile and has usually migrated to the dorsum of the host's abdomen. Also by this stage, a white waxy covering begins to cover the dorsal and lateral surfaces of the larva. The chemical composition of this substance is that of a very inert, long chain paraffin (Marshall et al. 1974), which is extruded as minute tubules through thousands of densely packed integumental pores (fig. 26.184g). The effects of the parasite on the host varies from having little or no effect in some species (Krishnamurti 1933) to causing death within 24 hours after larval departure in others (Kirkpatrick 1947).

The entire larval period requires from 4 to 6.5 weeks, after which the larva drops from its host to pupate in an intricately constructed, usually white silken cocoon attached to some plant support. Frequently the cocoon possesses dorsal flutes (fig. 26.184i) or rosettelike folds overlaid by the powdery paraffin cast off by the final instar. Immediately prior to adult emergence, the pupa is extruded partially out of the cocoon through a transverse slit at the anterior end. Adults are univoltine and short-lived, seldom surviving more than ten days.

Description: Hypermetamorphic; first instar an active triungulinid, campodeiform. Mature larvae nearly sessile, small to medium, 5–10 mm. Body immaculate, usually white, very stout, greatest diameter more than 0.6 the length; cuticle finely wrinkled, covered with relatively thick (0.2–0.3 mm) layer of flocculent paraffin filaments secreted from thousands of microscopic pores; pores (and paraffin) absent from head and venter of body. Numerous secondary setae finely scattered over body; all setae reduced. Spiracles round.

Head: Reduced in size and only partially sclerotized, deeply retracted into T1. Six stemmata, clustered closely together. Mandibles extremely slender, styletlike, minutely bidentate. Labrum reduced, poorly defined.

Thorax: Pronotal plate and pinacula absent. Legs greatly reduced in later instars, actually five-segmented but with only three segments evident; tarsal claw well developed, elongate, closing into a longitudinal grove on tarsus. Coxae indistinct.

Abdomen: Pinnacula and anal plate absent. Prolegs greatly reduced; crochets on A3–6 and 10; ventral crochets arranged in a uniordinal, uniserial circle; each crochet hook with a small basal spine. Crochets of A10 similar but in a nearly complete circle, interrupted posteriorly.

Comments: The Epipyropidae constitutes a small but widely dispersed and homogenous family consisting of nine genera and 30 species. Nine additional species will be proposed in a monograph of the family now in preparation (Davis, in prep.). The group is primarily pantropical with its greatest development in the Indo-Australian area. A single species, *Fulgoraecia exigua* (Hy. Edwards), occurs in North America in the area roughly south of a line from Connecticut to Ohio to Arizona.

Selected Bibliography

Clausen 1940.
Fracker 1915.
Kato 1940.
Kirkpatrick 1947.
Krishnamurti 1933.
Marshall 1970.
Marshall et al. 1974.
Perkins 1905.

DALCERIDAE (ZYGAENOIDEA)

Frederick W. Stehr, *Michigan State University*

Noel McFarland, *Sierra Vista, Arizona*

The Dalcerids or Jelly Slug Caterpillars

Figures 26.185a-c

Relationships and Diagnosis: Dalcerids are a New World family largely restricted to the Neotropics, with only a single species, *Dalcerides ingenitus* (Hy. Edwards) (fig. 26.185a), recorded north of Mexico (from southern Arizona). The following is based on this species, plus one penultimate and one antepenultimate larva from Costa Rica identified as *Acraga coa* Schaus (USNM specimens).

They appear to be most closely related to the Limacodidae and Megalopygidae. The larvae of limacodids and dalcerids are both sluglike in their movements, whereas megalopygids may move in a sluglike fashion (*Megalopyge* and *Lagoa*) or in a more conventional way (*Norape*). However, dalcerid larvae differ in substantially significant ways from both limacodids and megalopygids. One of the most interesting features is the presence of easily dislodged, probably defensive, jelly-like tubercles that completely cover all but the venter of the body and segments T1 and A10 (which are concealed beneath T2 and A9). Known limacodids and megalopygids do not have such jelly-like tubercles, and most megalopygids are densely hairy.

The most significant feature of dalcerids (at least for *D. ingenitus*) is the presence of crochets on abdominal segments 2, 3, 4, 5, 6 and 7 of the last two instars (Stehr and McFarland 1985). To our knowledge, crochets on A2 and A7 (fig. 26.185b, c) are not known for any other family of Lepidoptera. The crochets on A3–6 are slightly larger and more numerous (usually 7–10) than those on A2 and A7 (usually 3–5). Antepenultimate instars only have crochets on A3–6 and earlier instars have none. The prolegs on A2–7 are quite small and not readily discernable on preserved larvae. On living larvae they are clearly evident, and appear to be capable of functioning in a "normal" way, although the sluglike movement is clearly predominant on surfaces ranging from leaves to twigs and branches. Crochets are absent on A10, and the prolegs are essentially reduced to lobes bearing a few setae and a pair of peglike setae that have been observed to assist in flipping frass, even though they are ventral to the anus (in contrast to an anal comb which is dorsal).

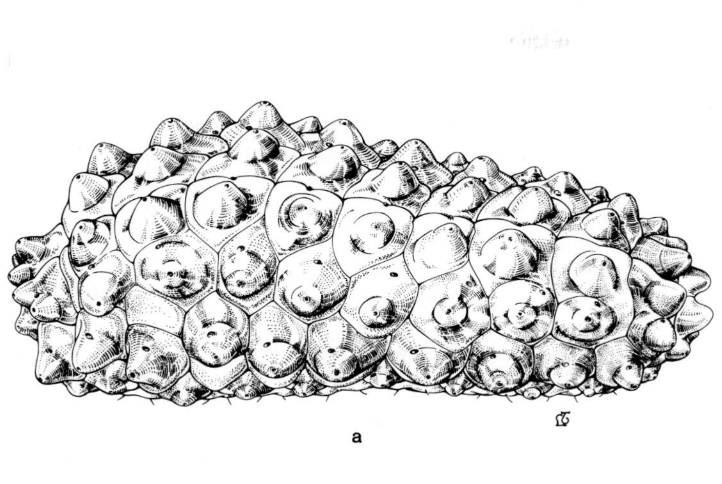

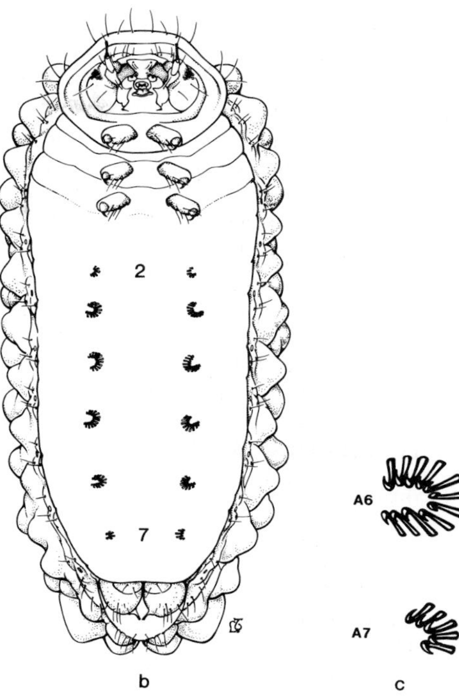

Figure 26.185a-c. Dalceridae. *Dalcerides ingenitus* (Henry Edwards). a. larva, dorsolateral, showing jellylike tubercles; b. venter, showing crochets on A2-A7; c. enlargement of A6 and A7 crochets. Color variable, from whitish (especially early instars) to orange-brown, each jellylike translucent mound enclosing a yellow inner tubercle that may be secretory. Each mound with apical depression. The 2 subdorsal rows of mounds have dorsolateral oval structures (openings?) of unknown function. Venter transparent, making observation of internal systems possible. Prepupae turning bright green, with middorsal area becoming transparent so beating heart is visible. On manzanita (*Arctostaphylos pungens*) and oaks in S.E. Arizona (Huachuca Mountains) and neighboring areas.

This unique arrangement of crochets and prolegs is not a totally unexpected phenomenon, when one considers that the megalopygids have prolegs on A2–7, but with the crochets absent on A2 and A7. It suggests an evolutionary progression from an ancestor with equal-sized prolegs with crochets on A2–7, to the dalcerids with smaller prolegs and reduced numbers of crochets on A2 and A7, to the megalopygids with smaller prolegs and no crochets on A2 and A7, to the limacodids with all prolegs reduced to suckers and crochets totally absent.

Biology and Ecology: *Dalcerides ingenitus* is abundant in the mountains of southern Arizona near the Mexican border, where its native Arizona hosts are a species of manzanita, *Arctostaphylos pungens,* and the new leaves of *Quercus* in early spring. It was also successfully reared on some other Ericaceae (such as blueberry), provided that the newly hatched larvae were started on the host.

The very soft eggs are laid singly or in small groups, close together but not in contact. Larvae hide beneath the leaves while feeding. They skeletonize (with a distinctive pattern of grooves) in the early instars, feeding on the edge of the leaf in later instars. They prefer the thick *older* leaves; hence the larvae are not easily located in the field, since they are less conspicuous among the old leaves that cover all but the tips of the branches. The cocoon is spun of loose, light yellow silk on or between contacting leaves of the host plant. There are at least two generations per year in southern Arizona (near Sierra Vista), with the first flying from mid-May to the end of June; the second begins to appear in late July, reaches a peak in mid-August, and tapers off in late September. They probably overwinter as larvae, based upon the field capture of a mature larva on *Q. emoryi* in early spring.

Description: Mature larvae 10–15 mm long, sluglike, and covered with colorless jellylike tubercles. Spiracles circular, subequal in size throughout, and on the abdomen located just below the lowest jellylike tubercles.

Head: Concealed, quite pale except for anterior margin. Frontoclypeus slightly longer than wide, extending somewhat less than halfway to epicranial notch, which is not visible without dissection. Ecdysial sutures only visible at anterior margin. Six stemmata, 1–4 in a tight arc, with 4 at the lateral apex of the antennal triangle, 5 below the apex, and 6 laterad to 4.

Thorax: T1 concealed beneath T2 and not bearing jellylike tubercles; T1 setae XD1, XD2, D1, D2, SD1, and SD2 near the anterior margin; L group trisetose; SV setae not evident on any thoracic segement. T2 and T3 bearing jellylike tubercles. Setae D1 and D2 present on T1, absent on T2. Some, but not all, of the other primary setae present on T2 and T3 along the margin below the jellylike tubercles.

Abdomen: All segments except A10 with conspicuous jellylike tubercles, and with three setae near each spiracle that are arranged similarly on A1 and A8 (with the lower seta anteroventrad of the spiracle), but with the lower seta posterio-ventrad of the spiracle on A2–A7. SV and V setae absent.

A9 with 3 setae in a nearly horizontal row beneath the jellylike tubercles. A10 lacking jellylike tubercles and prolegs; anal "plate" with four setae per side and numerous microspines beneath; A10 proleg reduced to small ventral lobe bearing two stout setae, and a larger more dorsal lobe bearing six stout setae, some microspines and a single large, stout, peglike seta that assists in flipping frass. Prolegs and crochets on A2–7 as described under "relationships and diagnosis".

Comments: One species has been reported as a minor pest on *Cacao* in Trinidad (Kirkpatrick 1953) but outside of the observations made by us in 1983, there appears to be very little known about dalcerid larvae, biology or ecology. Mr. Scott Miller, doctoral student at Harvard University, is revising them.

Selected Bibliography

Most of the following citations were furnished by Scott Miller.

Burmeister 1879.
Comstock 1959.
Costa Lima 1945.
Dyar 1910, 1925.
Gomes and Reiniger 1939.
Hopp 1928.
Jones and Moore 1882.
Kirkpatrick 1953.
Stehr and McFarland 1985.

PYRALIDAE (PYRALOIDEA)

H. H. Neunzig, *North Carolina State University*

The Pyralids, Webworms, Waxworms, Cereal Worms, Dried Fruit Worms, Casebearers, etc.

Figures 26.186–26.260

Relationships and Diagnosis: Larvae of the Pyraloidea (Pyralidae, Hyblaeidae, Thyrididae) have the body typically somewhat cylindrical and tapered anteriorly and posteriorly, with the usual complement of prolegs (prolegs usually well developed, at times considerably reduced, but crochets present). The crochets are arranged in a circle or a modified circle. Well-developed primary setae are usually present, frequently arising from distinct pinacula or occasionally on chalazae. The L (prespiracular) setae are bisetose on the prothorax.

The Pyralidae is the largest family within the superfamily Pyraloidea. Last-stage pyralid larvae have the crochets of the ventral prolegs usually triordinal or biordinal (sometimes uniordinal), and in the form of circles, ellipses, mesopenellipses, or ellipses open mesially and laterally (transverse bands of some authors).

Sixteen to seventeen subfamilies of North American Pyralidae are presently recognized. Hasenfuss (1960) has constructed keys for larvae of European pyralids that will also separate relatively well the major North American subfamilies (tribes of Hasenfuss). Allyson (1976, 1977a, 1977b) also discussed means of separating the larvae of some subfamilies.

Biology and Ecology: Almost all pyralid larvae are concealed feeders. Many feed within or upon a living host plant or occur near the host within a place of concealment. Some feed on quiescent or nonliving plant materials (frequently seeds or fruits, or similar materials, that have been stored), and a few are predaceous, occurring within nests of the host or under silk coverings constructed near the prey. Most pyralids are terrestrial, but aquatic forms exist, sometimes highly modified for such an existence.

The larval behavior and hosts of the various subfamilies, as presently known in North America, are briefly as follows: **Chrysauginae**—feed within seed capsules of Bignoniaceae or are leaf feeders on other groups of plants; **Crambinae**—bore in stems, roots and leaves of grasses (Poaceae), or form silk tubes in the soil and feed on grasses; **Epipaschiinae**—many are gregarious web makers on leaves of legumes (Fabaceae), oak (Fagaceae), pine (Pinaceae), and other hosts (some species feed within a mass of frass and silk); **Evergestinae**—web makers that feed mostly on leaves of crucifers (Brassicaceae) or capers (Capparaceae); **Galleriinae**—feed within nests of Hymenoptera (sometimes feeding on wax, etc.), predators of scale insects or are pests of stored fruit or other stored products; **Glaphyriinae**—webbers and borers of mostly crucifers (Brassicaceae), or feeders (within cases) on lichens, or apparently predators of insect larvae; **Nymphulinae**—borers in stems of ferns (Aspidiaceae), or eaters of aquatic or semiaquatic vascular plants in lentic environments, or alga eaters and webspinners on rocks in lotic situations; **Odontiinae**—leaf webbers of crucifers (Brassicaceae) and capers (Capparaceae), or leaf miners of composites (Asteraceae), or flower and seed feeders of Malvaceae; **Peoriinae**—primarily stem borers of grasses (Poaceae); **Phycitinae**—feed within stored products (meal, dried fruit, chocolate, tobacco, etc.), or tiers, webbers, rollers of leaves or borers of buds, shoots, stems, petioles, fruits, seeds, etc., of many plants, particularly legumes (Fabaceae), cacti (Cactaceae), pines (Pinaceae), Juglandaceae, Betulaceae, Fagaceae, Rosaceae, Ericaceae, and Asteraceae, or predaceous on scale insects; **Pyralinae**—mostly feed within stored products (meal, hay, etc.), or general scavengers; **Pyraustinae**—mostly folders or webbers of leaves, or stem or fruit borers, of mints (Lamiaceae), grasses (Poaceae), legumes (Fabaceae), cucurbits (Cucurbitaceae), and many other families; **Schoenobiinae**—borers in aquatic and semiaquatic plants (mostly grasses (Poaceae)); **Scopariinae**— borers in mosses, and clubmosses (Lycopodiaceae).

Pyralids have one to several generations each year. Some species in warmer parts of the United States such as Florida and Hawaii, and some infesting stored products, have a more or less continuous sequence of generations. Most North American species overwinter either as larvae or pupae. Larvae diapause usually when full grown, but a few species overwinter as small larvae; these small larvae usually construct hibernacula on the host.

Eggs are laid on, or near, the host plant, or on or near the stored product or refuse, or near the prey. Several kinds of eggs occur in the family Pyralidae. Peterson (1963a) has described and illustrated the major types and Arbogast et al.

(1980) have given detailed descriptions of eggs of some pyralids associated with stored products. Scalelike, distinctly flattened eggs are deposited by some Pyraustinae, Chrysauginae, and Crambinae. Usually these ova are placed in dense clusters with the individual eggs overlapping like shingles. Other eggs are ovoid or cylindrical; these are usually deposited singly or in small clusters by species of the subfamilies Pyraustinae, Galleriinae, Epipaschiinae, Crambinae, Phycitinae, and Schoenobiinae. The more or less cylindrical eggs of some of the cactus-feeding phycitines are glued together to form a remarkable, relatively long, egg "stick".

The chorion of the eggs is always colorless and transparent or translucent. Usually the surface of the eggs is relatively smooth; sometimes it may appear reticulate. In the genus *Crambus* the eggs possess distinct longitudinal ridges and furrows. All ova of pyralids are adhesive at the time they are deposited, except those of *Crambus*. The adhesive coating is usually transparent, and the egg masses are exposed. However, in the Schoenobiinae, after the eggs are deposited, they are covered with a layer of hair- or needlelike wax.

Most larvae, upon hatching, bore quickly into the host, or form shelters of silk, parts of the plant or plant products and sometimes frass. Some enter the soil near the host and construct silk tubes in close association with the host. Large larvae of some *Acrobasis* (Phycitinae) construct unique, relatively rigid, silk and frass tubes, and the leaf shelter, etc. of the late instars of many species, in general, is very characteristic. Pupation occurs within a silk-lined enclosure usually on the host, in litter, on the soil or in the soil. Some aquatic species pupate under water.

Pupae are typical lepidopteran, usually with lobes indicating the presence of pilifers. Maxillary palpi are present. The integument of most pupae is relatively smooth and simple.

Description: Last stage larvae small to medium, 8–40 mm. Head smooth to reticulate rugose. Body usually more or less cylindrical and slender to robust, sometimes bearing tracheal gills (Nymphulinae), or a bag-like structure or gibbosity between the thoracic legs (Schoenobiinae). Prolegs usually well developed on segments A3–6 and A10. Integument of body smooth to slightly granular, usually more or less unicolorous, frequently with venter slightly paler; some species have longitudinal stripes, a few are transversely banded (for example, some cactus-feeding Phycitinae), and some have contrasting dark pinacula. Tonofibrillary platelets are associated with the integument and constitute a distinctive pigmented feature on the head, prothoracic and anal shields of many pyralids; these structures are infrequently also very pronounced elsewhere on the body as in some *Dioryctria* and *Herculia*. The usual complement of primary (and subprimary) setae is present in most species (reduction in number of setae in a few species). Many Phycitinae, Pyralinae, Chrysauginae, Epipaschiinae, and Galleriinae have a pinaculum ring formed by a pinaculum with an unpigmented or unsclerotized center at the base of some setae. Spiracles usually elliptical, with those on T1 and A8 frequently larger.

Head: More or less evenly rounded, and semiprognathous in most species (prognathous and flattened in leaf miners (a few Odontiinae) and some Nymphulinae and Phycitinae). Six stemmata, occasionally fewer (Galleriinae). Mandibles usually simple, but sometimes with extra distal teeth or small to large retinacula or other modifications. Maxillae with sensilla trichodea usually simple, but sometimes forked.

Thorax: Prothorax with shield usually distinct. Characteristically six setae on each side of shield. Almost all species with the two L setae on a prespiracular plate of T1 (some Nymphulinae with one of the L setae very small and easily overlooked), but some (Galleriinae: *Macrotheca*) with L setae not on the prespiracular plate but on the lower margin of the prothoracic shield. D1 and D2 present and usually adjacent on T2 and also on T3 (D1 and D2 somewhat remote in Galleriinae). SD1 and SD2 present and adjacent on T2 and also on T3. Base of SD1 on T2 (most Phycitinae) or D2 on T3 (? most Chrysauginae) with a sclerotized, usually darkly pigmented, pinaculum ring. L group usually trisetose on T2 and T3 (apparently bisetose in some Nymphulinae and Schoenobiinae). SV group bisetose on T1, and unisetose or bisetose on T2 and T3. A few (some Nymphulinae) with simple or compound gills on T2 and T3, and the schoenobiines with anteromesial sacs or gibbosities associated with the thoracic legs.

Abdomen: D1 and D2 usually on all segments, with D1 more dorsad and cephalad, except on A9 where D1 is usually ventrad of D2 (D1 or D2 very small or missing on abdominal segments of some Nymphulinae and Schoenobiinae). SD1 present on all segments (small on some Nymphulinae) (Galleriinae sometimes with pinaculum ring around base of SD1 on A1). On A8, SD1 sometimes (most Phycitinae, Epipaschiinae, Chrysauginae, Galleriinae, Pyralinae) with a sclerotized, usually darkly pigmented pinaculum ring (for a few species of these subfamilies, ring not evident, but dark neural connection usually present at base of seta). SD2 on all segments except A9 and A10, but very small. On A1–8, L1 and L2 usually approximate and more or less ventrad of the spiracle (in some Nymphulinae and possibly some Schoenobiinae L1 (or L2) strongly reduced or absent). On A9, L2 usually missing in the Pyraustinae, Glaphyriinae, Odontiinae, Evergestinae, Crambinae, Nymphulinae, Schoenobiinae, Scopariinae, and usually present in the Epipaschiinae, Phycitinae, Chrysauginae, Galleriinae, and Pyralinae. On A1–8, L3 below L1 and L2. On A9, L3 missing in same subfamilies lacking L2; in other subfamilies where L3 is present on A9 it is in close proximity to L1 and L2. SV bisetose or trisetose on A1–6. A7–9 with SV unisetose to trisetose. V1 on all segments; relative distance between the V setae of each caudal segment differs in some subfamilies. Prolegs on segments 3–6 with crochets uni- to triordinal. Crochets most often arranged in the form of circles, ellipses, or mesopenellipses, infrequently as incomplete ellipses open laterally and mesially (transverse bands of some authors; a few Chrysauginae and Nymphulinae). A10 usually with a relatively indistinct anal shield bearing usually eight setae. Proleg on A10 with crochets usually bi- to triordinal. Typically, nine setae on and near the anal proleg. Most setae of A10 have been named for Pyralidae by Hasenfuss (1960) and Allyson (1976), but some questions remain as to homologies with the more anterior segments. Some nymphulines with simple or compound gills on some or all abdominal segments.

Comments: The Pyralidae is the third largest family in the Lepidoptera, with about 1400 known species in the United States and Canada. As is apparent by consulting the contributions of Munroe (1972–1976), the immature stages of many species are still unknown. Early significant studies on North American pyralid immatures were made by Dyar (1901, 1902), Fracker (1915), Forbes (1910b, 1923), Mosher (1916a), and Heinrich (1919, 1921a, 1921b). Larvae of Pyralidae that are stored-products pests have been given rather detailed treatment in important papers by Hinton (1942, 1943) and Aitken (1963). More recently, relatively detailed studies have been made of the immature stages of some of the North American Pyraustinae, and Epipaschiinae, by Allyson (1976, 1977a, 1981a,b, 1984), of some North American Phycitinae by Allyson (1980), Corrette and Neunzig (1979), Doerksen and Neunzig (1975), Grimes and Neunzig (1984), Neunzig (1972, 1979), Neunzig and Merkel (1967) and Neunzig et al. (1964), and of three Galleriinae (Macrothecini) by Allyson (1977b) and Liebherr (1977).

This family includes some of the most important pests of stored products, such as the meal moth (*Pyralis farinalis* (L.), fig. 26.228); the Indian meal moth (*Plodia interpunctella* (Hübner), fig. 26.257); the Mediterranean flour moth (*Anagasta kuehniella* (Zeller), fig. 26.260); and the almond moth (*Cadra cautella* (Walker), fig. 26.259). A key to aid in the identification of the larvae of these and other pyralids found in stored materials follows. Also, field crops such as corn and sugarcane are attacked by a group of pyralids including the lesser cornstalk borer (*Elasmopalpus lignosellus* (Zeller), fig. 26.246); the European corn borer (*Ostrinia nubilalis* (Hübner), fig. 26.206); and several species of *Diatraea.* A key to these and other species associated with these hosts follows. Numerous other crops, ornamental plants, etc. are attacked by one or more species of Pyralidae, making this family of major importance economically. *Cactoblastis cactorum* (Berg) (fig. 26.254) also belongs to this family; it is the species that has been used with such remarkable success in the biological control campaign against the prickly pear cactus in Australia.

KEY TO SOME LEPIDOPTERA LARVAE (LAST STAGE) INJURIOUS TO STORED FOOD PRODUCTS

(Pyralidae except where indicated; based, in part, on Hinton (1943) and Aitken (1963))

1. Prolegs weakly developed (fig. 1); each proleg with only a few crochets (fig. 3) .... (fig. 26.87) Angoumois grain moth (*Sitotroga cerealella* (Olivier) (Gelechiidae)) (p. 398)

1′. Prolegs well developed (fig. 2); each proleg with many crochets (fig. 4) 2

2(1′). T2 without broad, sclerotized pinaculum ring at base of seta SD1 (fig. 5) (pale pinaculum present, but not forming a ring); anal proleg with seta L3 closer to paraproct seta than to seta L1, or distance subequal (fig. 7) 3

2′. T2 with broad sclerotized pinaculum ring at base of seta SD1 (fig. 6) (ring pale yellow to black) (see figs. 19 and 20 for additional examples of pinaculum rings); anal proleg with seta L3 distinctly closer to seta L1 than to paraproct seta (fig. 8) 5

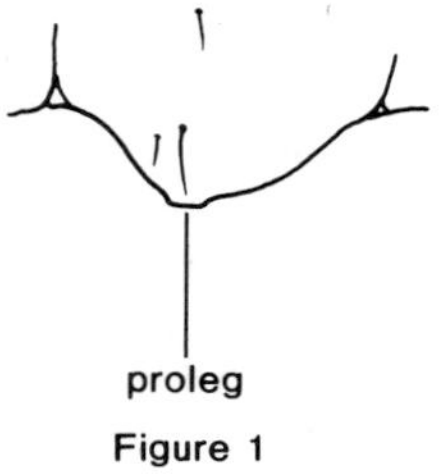

Figure 1

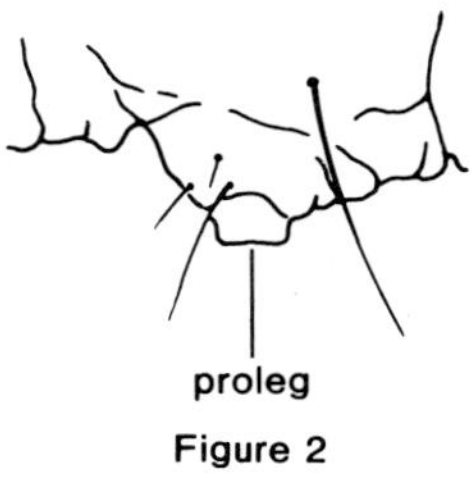

Figure 2

Figure 3

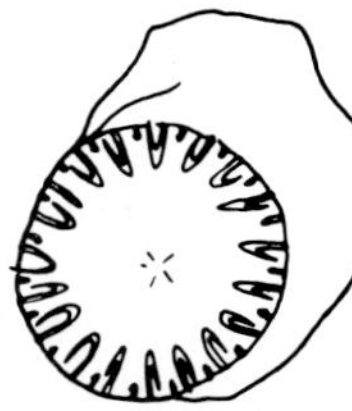

Figure 4

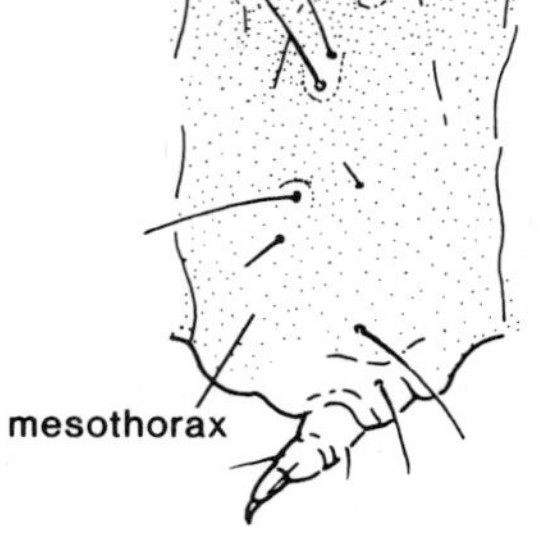

Figure 5

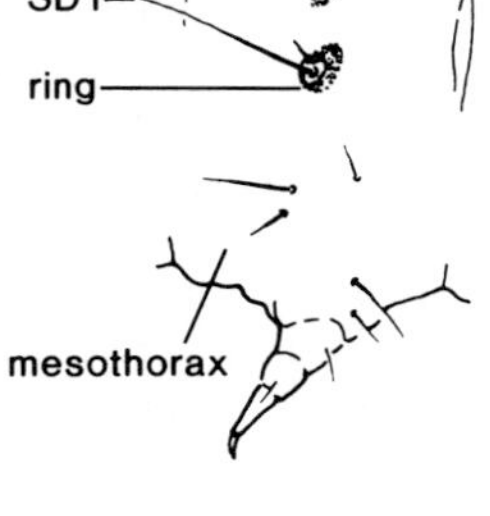

Figure 6

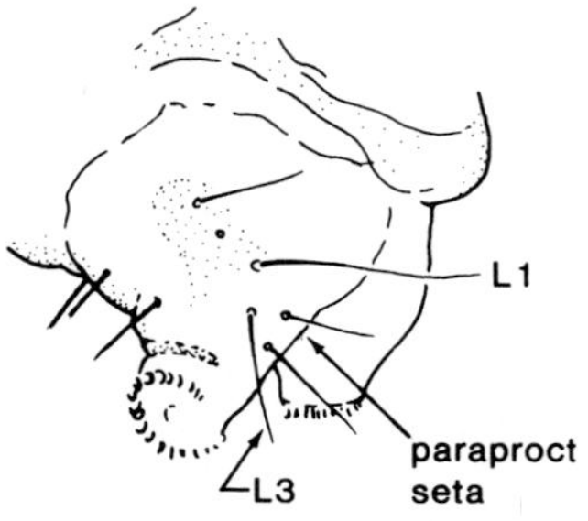

Figure 7

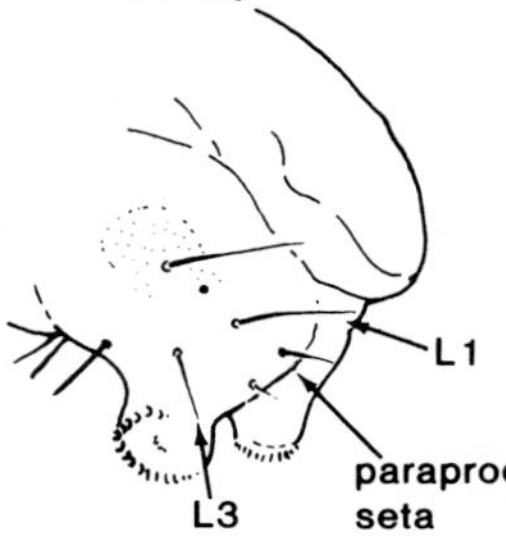

Figure 8

3(2). T1 with L (prespiracular) group trisetose .... (figs. 26.38, 26.39) European grain moth (*Nemapogon granella* (L.) (Tineidae)) (p. 363)

3′. T1 with L (prespiracular) group bisetose 4

4(3′). SV group of A9 unisetose (fig. 9); only 3–4 stemmata on each side of head (fig. 11); body relatively pale, except for some pigmentation on thorax and caudal abdominal segments .... (fig. 26.228) meal moth (*Pyralis farinalis* (L.)) (p. 483)

4′. SV group of A9 bisetose (fig. 10); 5–6 stemmata on each side of head (fig. 12); body dark .... (fig. 26.225) murky meal caterpillar (*Aglossa caprealis* (Hübner)) (p. 482)

5(2′). Setae of A1–9 not arising from pinacula (fig. 13); head with frontoclypeus extending two-thirds, or more, distance to epicranial notch (fig. 15) .... (fig. 26.257) Indian meal moth (*Plodia interpunctella* (Hübner)) (p. 492)

5′. Most dorsal setae of A1–9 with small, sometimes weakly developed, pinacula (fig. 14) (pinacula usually most apparent on posterior segments); head with frontoclypeus extending less than two-thirds, usually little more than half, the distance to epicranial notch (fig. 16) (except in some specimens of *A. kuehniella* that resemble *P. interpunctella* in this character) 6

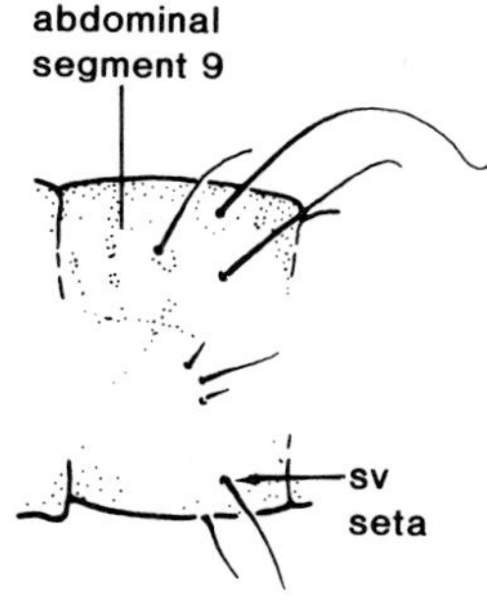

Figure 9

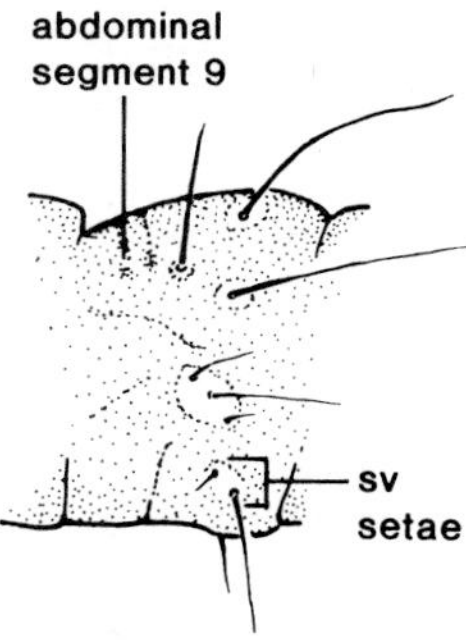

Figure 10

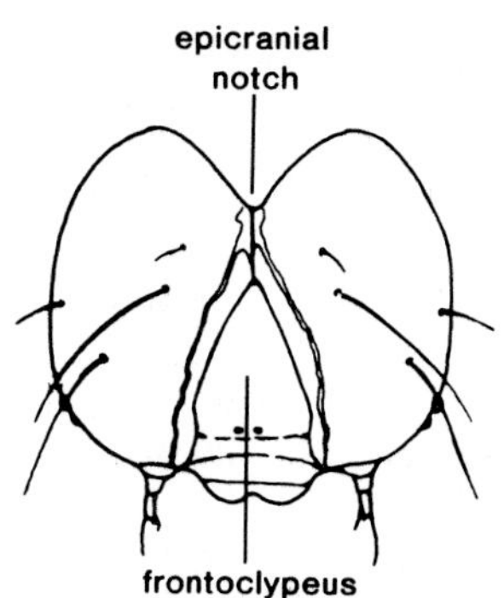

Figure 15

Figure 11

Figure 12

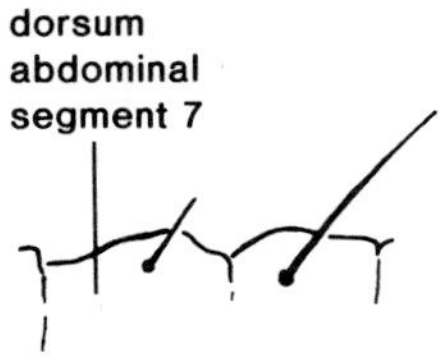

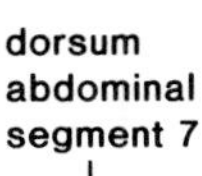

Figure 13

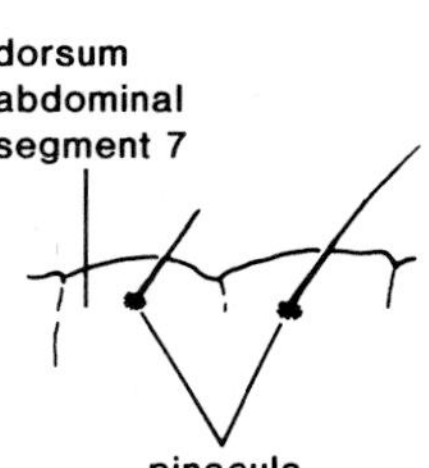

Figure 14

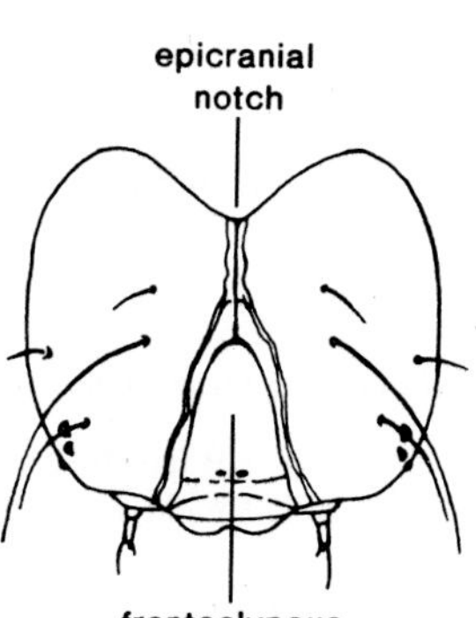

Figure 16

6(5′). A8 with seta SD2 separated from spiracle by distance equal to 2–3.5× diameter of spiracle (fig. 17) 7

6′. A8 with seta SD2 separated from spiracle by distance equal to, distinctly less, or slightly more than diameter of spiracle (fig. 18) 8

7(6). A8 with spiracle as large as, or slightly larger than area enclosed by SD1 pinaculum ring (fig. 19) .... (fig. 26.260) Mediterranean flour moth (*Anagasta kuehniella* (Zeller)) (p. 493)

7′. A8 with spiracle distinctly smaller, ⅔ or less as broad as area enclosed by SD1 pinaculum ring (fig. 20) .... (fig. 26.258) tobacco moth (*Ephestia elutella* (Hübner)) (p. 493)

8(6′). A8 with D2 only 2–2.5× as long as D1 (fig. 21) ... (fig. 26.259) almond moth (*Cadra cautella* (Walker)) (p. 493)

8′. A8 with D2 3–5.5 × longer than D1 (fig. 22) ...... (fig. 26.256) raisin moth (*Cadra figulliella* (Gregson)) (p. 492)

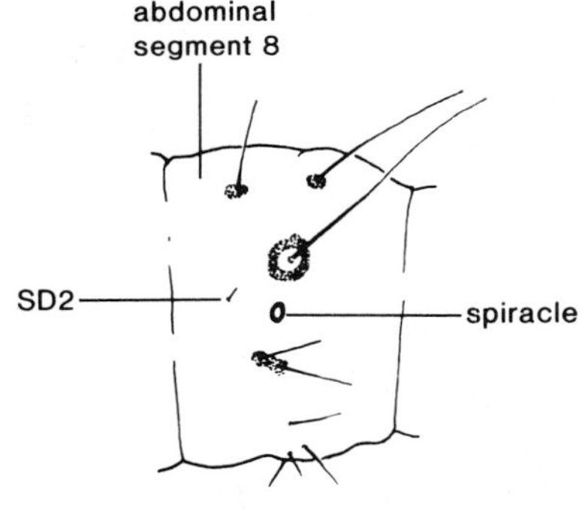

Figure 17

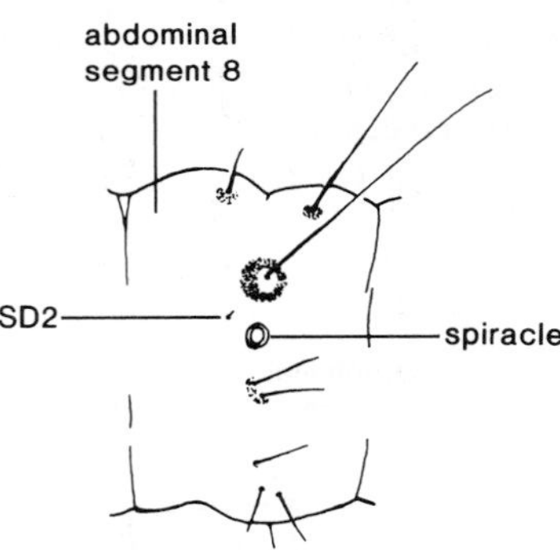

Figure 18

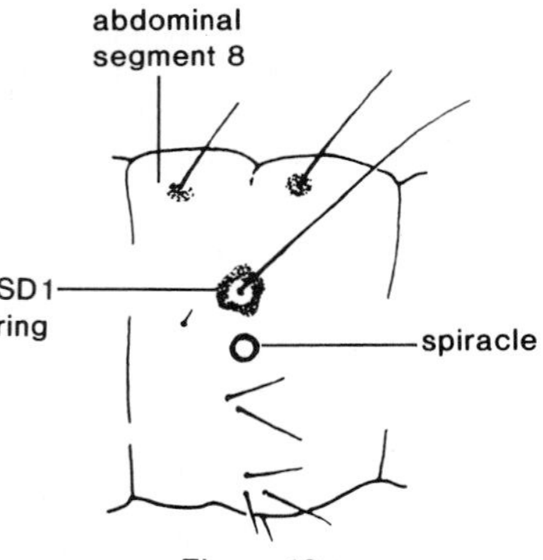

Figure 19

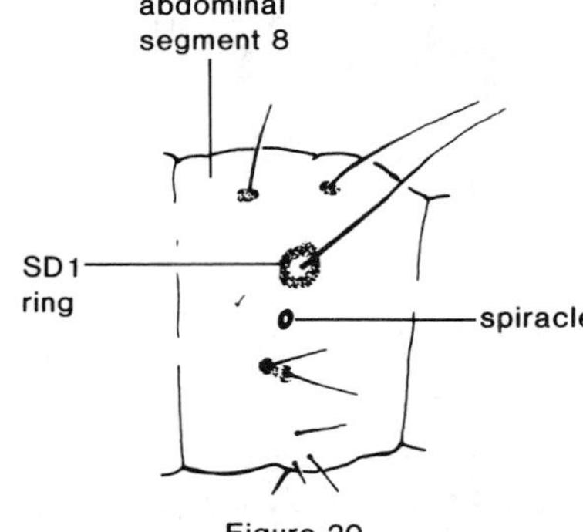
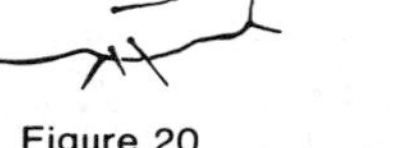

Figure 20

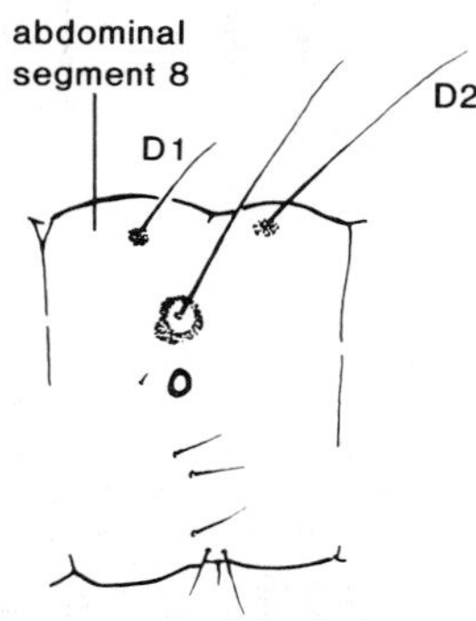

Figure 21

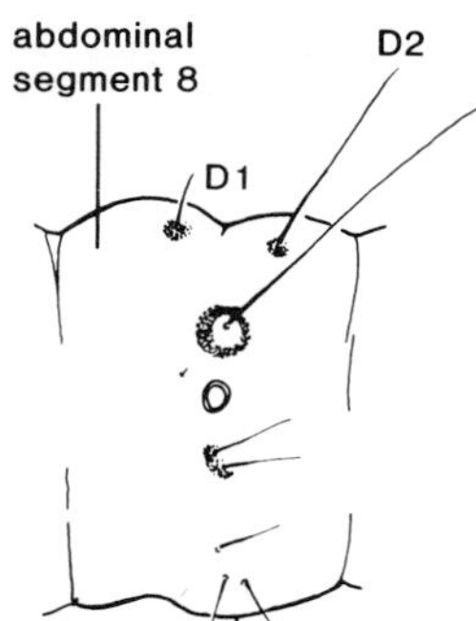

Figure 22

KEY TO SOME PYRALID LARVAE (LAST STAGE) INJURIOUS TO CORN, SUGARCANE, AND ASSOCIATED PLANTS

(Based in part on Peterson 1948)

Some diapausing larvae may be difficult to key out because of reduced pigmentation.

1. L group of A9 trisetose (fig. 23); head with small, pale, more or less circular areas at base of most setae (those associated with P setae about as large as stemmata (fig. 25) .... (fig. 26.246) lesser cornstalk borer (*Elasmopalpus lignosellus* (Zeller)) (p. 489)

1′. L group of A9 unisetose (fig. 24), sometimes bisetose; head without pale areas associated with setae 2

2(1′). V setae of A10 farther apart than V setae of A9 (fig. 26) 3

2′. V setae of A10 about one-half as far apart as V setae of A9 (fig. 27); V setae on A10 sometimes very small 5

3(2). Mandibles distinctly longer than wide (fig. 28) .... lotus borer (*Ostrinia penitalis* (Grote))

3′. Mandibles about as wide as long (fig. 29) 4

4(3′). Cranial setae A1 and A2 and pore Aa forming distinct angle, with Aa mesiad of line drawn through A1 and A2 (fig. 30) .... (fig. 26.205) smartweed borer (*Ostrinia obumbratalis* (Lederer)) (p. 475)

4′. Cranial setae A1 and A2 and pore Aa usually located in a straight line, or Aa somewhat laterad of A2 (fig. 31) .... (fig. 26.206) European corn borer (*Ostrinia nubilalis* (Hübner)) (p. 475)

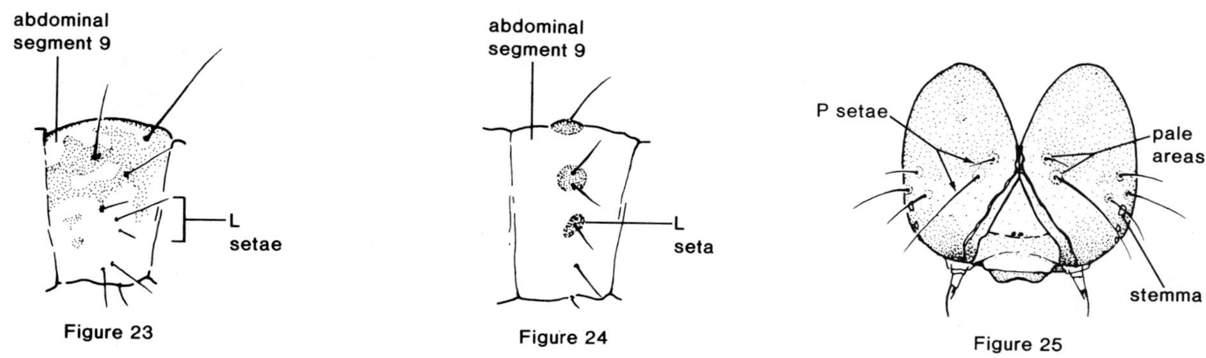

Figure 23 Figure 24 Figure 25

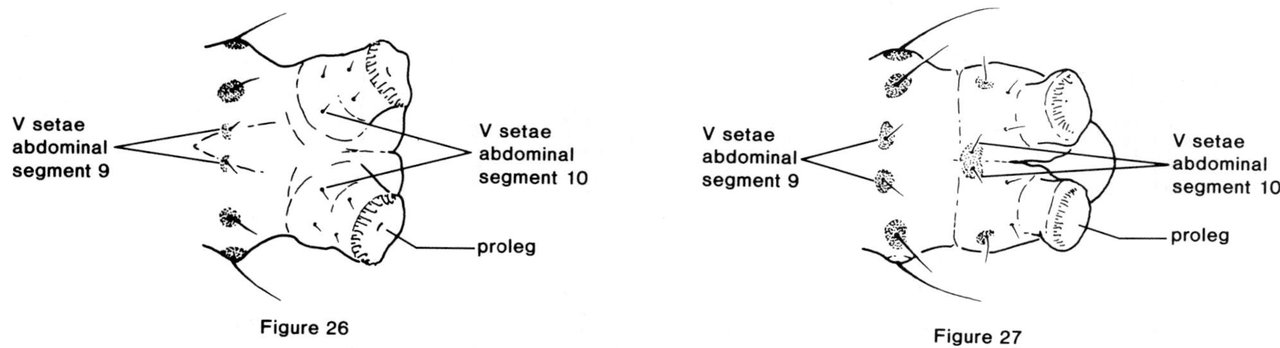

Figure 26 Figure 27

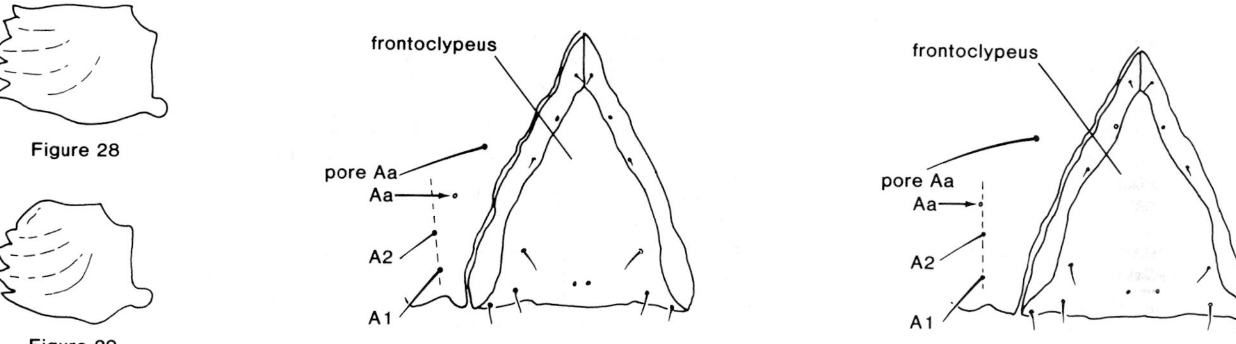

Figure 28 Figure 29 Figure 30 Figure 31

5(2′). Tonofibrillary platelet posterior to SD1 and spiracle of A3–6 distinct (fig. 32) .... (fig. 26.197) corn root webworm (*Crambus caliginosellus* (Clemens)) (p. 472)

5′. Tonofibrillary platelet posterior to SD1 and spiracle of A3–6 indistinct (fig. 33) 6

6(5′). Pinaculum of seta SD1 of A1–8 elongate, usually slightly angled and partially embracing spiracle (fig. 34); larvae relatively large, ca. 30–40 mm long; mostly southwestern United States .... (fig. 26.193) southwestern corn borer (*Diatraea grandiosella* (Dyar)) (p. 471)

6′. Pinaculum of seta SD1 of A1–8 more circular (figs. 35 and 36); larvae smaller, ca. 20–25 mm long; southeastern United States 7

7(6′). Mandibles with a distinct extra tooth associated with inner ridge of tooth 1 (fig. 37); dorsal pinacula surrounded by light brown or pink pigmentation (not always evident in preserved larvae); primarily a pest of sugarcane .... (fig. 26.194) sugarcane borer (*Diatraea saccharalis* (Fab.)) (p. 471)

7′. Mandibles simple, or with an indistinct tooth associated with ridge of tooth 1 (fig. 38); dorsal pinacula distinctly contrasting with surrounding integument, without associated brown or pink pigmentation; primarily a pest of corn .... (fig. 26.192) southern cornstalk borer (*Diatraea crambidoides* (Grote)) (p. 471)

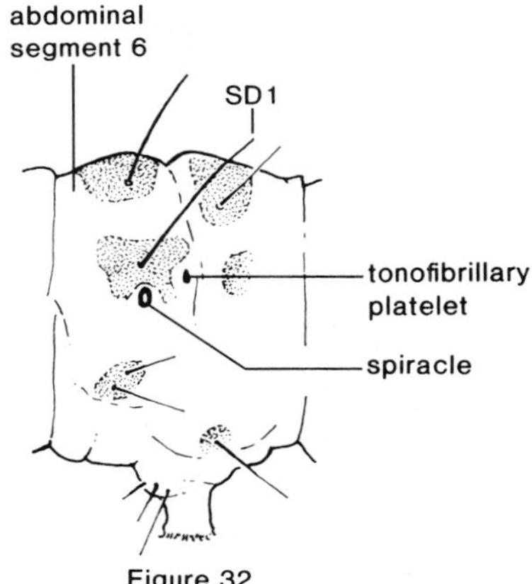

Figure 32

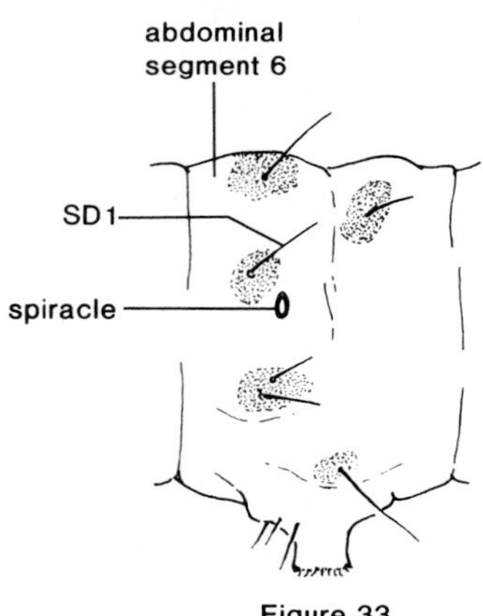

Figure 33

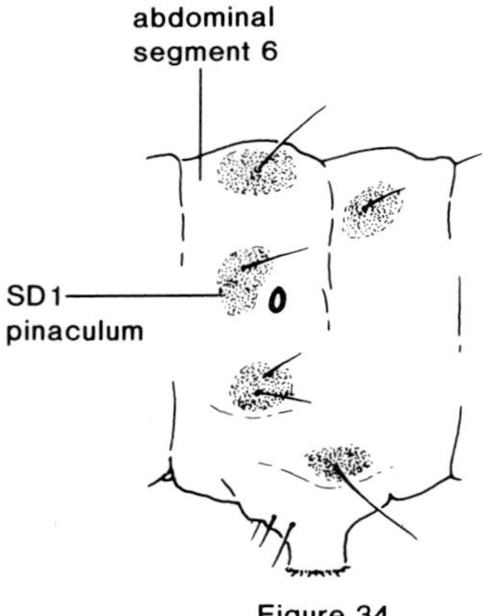

Figure 34

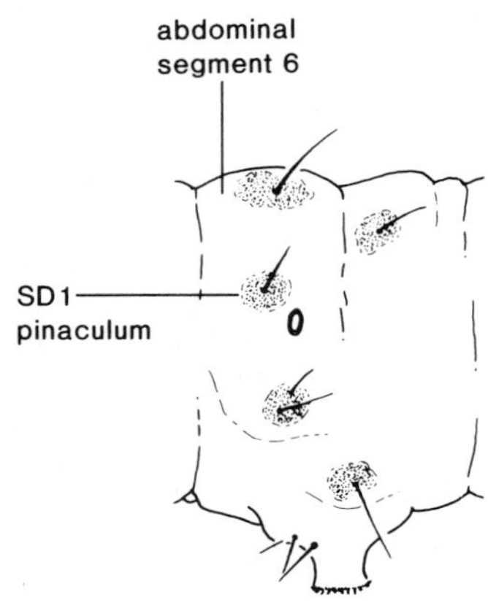

Figure 35

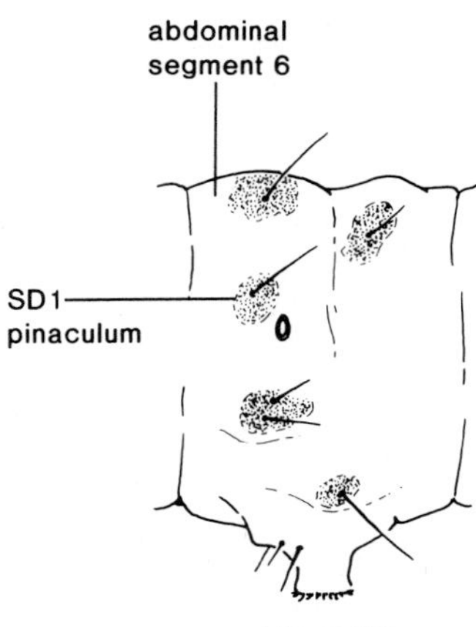

Figure 36

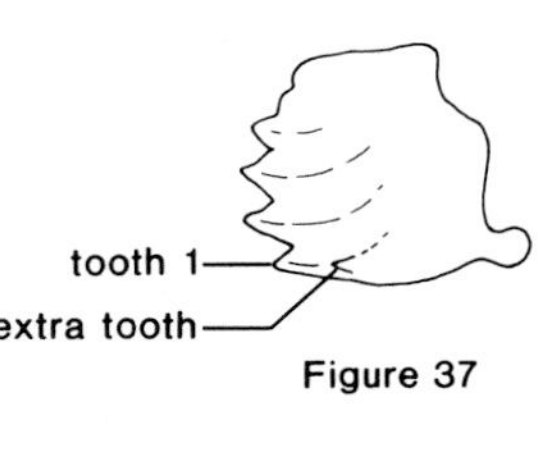

Figure 37

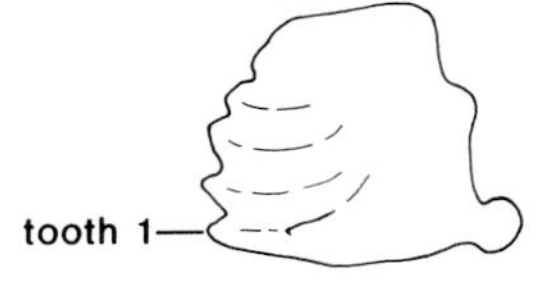

Figure 38

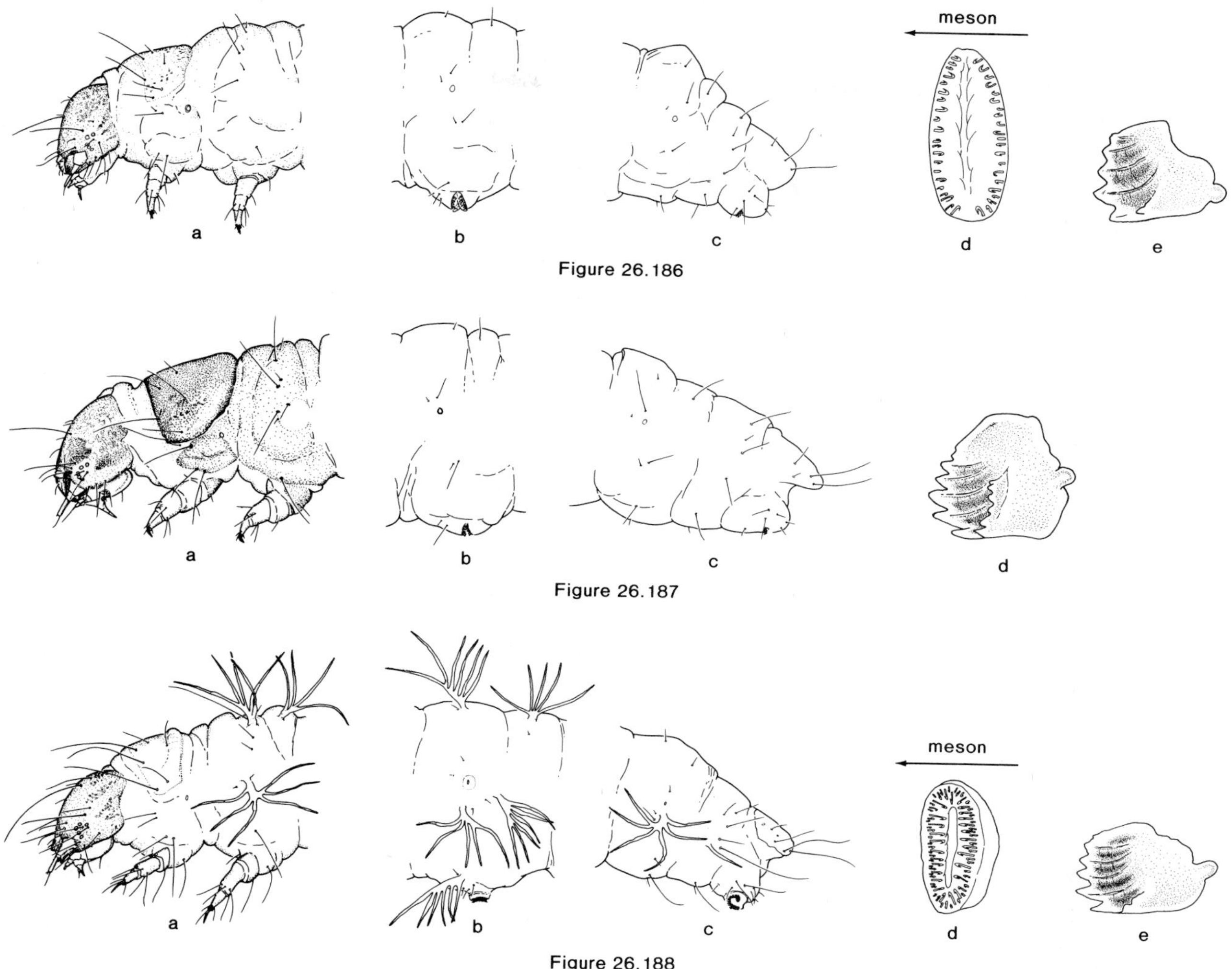

Figure 26.186a-e. Pyralidae, Nymphulinae. *Munroessa gyralis* (Hulst). f.g.l. 18± mm. a. head, T1, T2; b. A3; c. A8-10; d. A3 crochets; e. mesal view right mandible. Head pale yellowish white with slightly darker platelets. Mandibles with an additional small tooth (or 2 teeth) at base of tooth 1. T1 shield yellowish white, indistinct. Most of T2, T3 and abdomen yellowish white to gray. A3-6 prolegs rudimentary with biordinal crochets in narrow, incomplete ellipses open laterally and mesally (transverse bands). Kenilworth, District of Columbia, waterlily, May 3, 1955. *Hosts:* white waterlily (*Nymphaea* spp.), and yellow waterlily or spatterdock (*Nuphar luteum*). Feeds primarily on leaves, usually forming a retreat in stem.

Figure 26.187a-d. Pyralidae, Nymphulinae. *Synclita obliteralis* (Walker). Waterlily leafcutter. f.g.l. 13± mm. a. head, T1, T2; b. A3; c. A8-10; d. mesal view right mandible. Head pale brown to brown with indistinct to distinct darker platelets and bands or patches of darker brown maculation. Mandibles with enlarged ventral area and with a multitoothed slightly curved retinaculum that is continuous with the basic complement of distal teeth. T1 shield brown to dark brown with pale brown to pale yellow posterior, and sometimes lateral, margins. Most of remainder of T1 white to pale yellow with large brown lateral patches of hydrophil papillae ventral to the shield. T2 and T3 with brown papillae. Abdomen brownish white with brown papillae most evident anterodorsally. A3-6 prolegs rudimentary with biordinal crochets in narrow incomplete ellipses usually open laterally and mesally (transverse bands). Cumberland Co., N.C., Texas Pond. Ft. Bragg, on golden club (*Orontium aquaticum*), August 10, 1978. *Hosts:* duckweed (*Lemna* spp.), waterlilies, (*Nymphaea* and *Nuphar*), pondweed (*Potamogeton* spp.), golden club (*Orontium*), and other aquatic and semiaquatic plants. Lives in a flattened, oval-sided case made of part of host plant.

Figure 26.188a-e. Pyralidae, Nymphulinae. *Parapoynx allionealis* Walker. f.g.l. 14-18 mm. a. head, T1, T2; b. A3; c. A8-10; d. A3 proleg and crochets; e. mesal view right mandible. Head pale yellowish brown with indistinct slightly darker platelets. Mandibles with small additional teeth associated with tooth 1. T1 shield whitish yellow, distinctly elevated posteriorly. T2, T3 and abdomen mostly white to yellowish white. T2, T3 and A1-7 with several groups of white, compound gills arising from a common base. A8 usually with single, compound gill ventral to spiracle. More weakly developed, simple or compound gills also sometimes present on other segments. A3-6 crochets biordinal to triordinal ellipses. Maxton, N.C., pond adjacent to Lumber River, on *Utricularia* sp., April 5, 1966. *Hosts:* bladderwort (*Utricularia*), pondweed (*Potamogeton*), parrot-feather (*Myriophyllum*), and many other aquatic plants. Lives completely submerged on host within case made of fragments of plant.

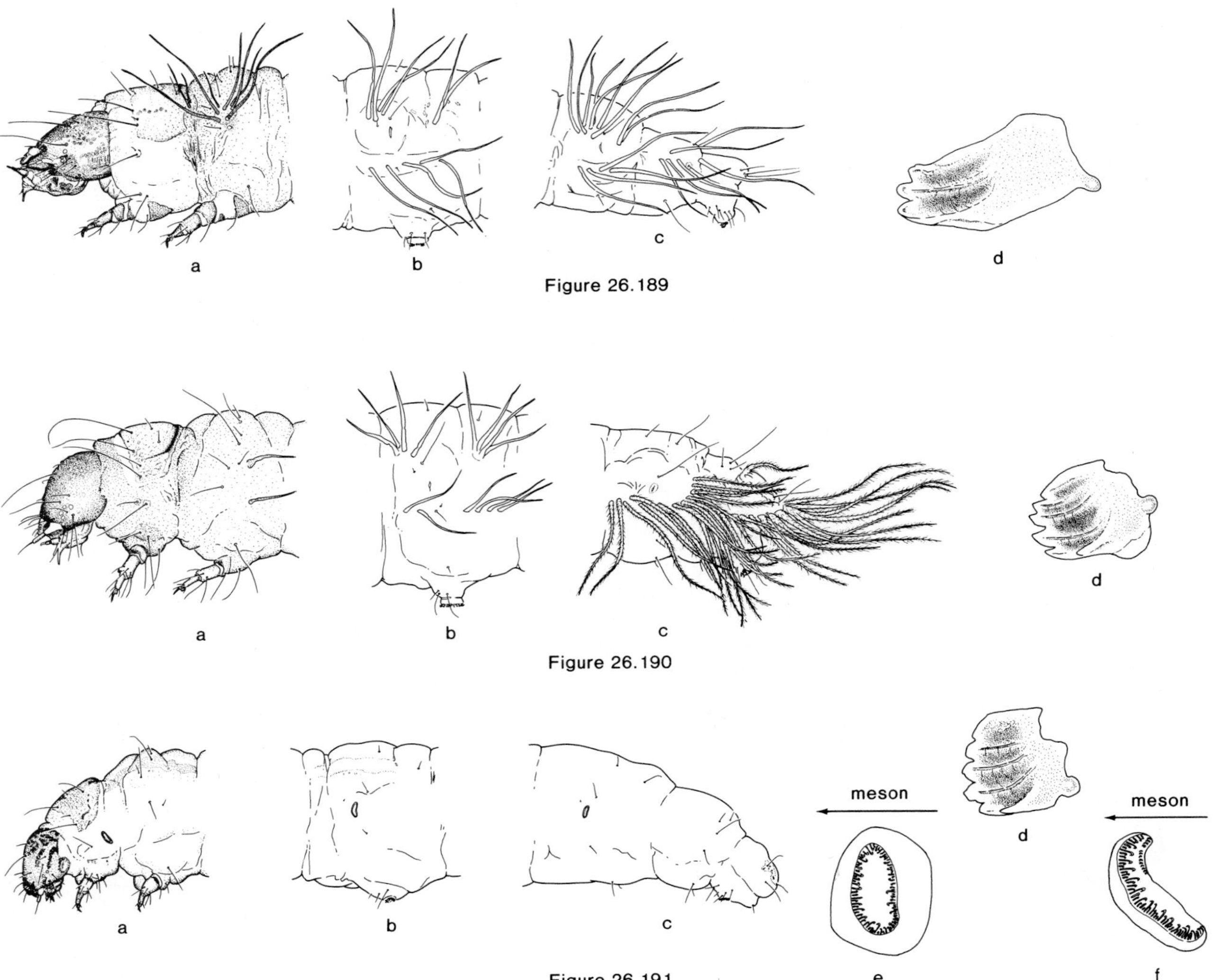

Figure 26.189a-d. Pyralidae, Nymphulinae. *Petrophila santafealis* (Heppner). f.g.l. 12± mm. a. head, T1, T2; b. A3; c. A8-10; d. mesal view right mandible. Head distinctly prognathous, finely granulate, brown, paler brown to white laterally and usually posteriorly. Mandibles distinctly elongate. Body somewhat flattened dorsoventrally. T1 shield pale brown with darker brown barlike patches dorsoposteriorly. T2, T3, and abdomen mostly pale brown. T2, T3 and A1-9 with several groups of brown, simple gills. A3-6 prolegs short with crochets in biordinal ellipses. Columbia Co., Fla., Santa Fe River, on rocks, May 8, 1974. *Host:* algae. Feeds under thin silk mat principally on algae growing on rocks in streams and rivers.

Figure 26.190a-d. Pyralidae, Nymphulinae. *Eoparargyractis plevie* (Dyar). f.g.l. 11± mm. a. head, T1, T2 (T2 sometimes with a pair of lower gills and an upper gill rather than a single lower gill and an upper gill); b. A3; c. A8-10; d. mesal view right mandible. Head rugulose, yellowish white suffused with brown. Mandibles with additional small tooth at base of tooth 1 and an enlarged ventral area. T1 shield rugulose, mostly pale yellowish white with brown posterior margin. T2, T3, and abdomen mostly white to brownish white (internal contents imparting a pale green color in living larva) and with several groups of grayish white, mostly unbranched or simple, gills. Gills of caudal segments strongly developed and plumose. A3-6 crochets in biordinal ellipses. Grafton Co., N.H., Mirror Lake, December 11, 1975. *Hosts:* lobelia (*Lobelia dortmanna*), quillwort (*Isoetes tuckermani* and *Isoetes muricata*). Occurs at base of rosette of host plant in protective case of silk and detritus; feeds on leaves of host.

Figure 26.191a-f. Pyralidae, Schoenobiinae. *Donacaula maximella* (Fernald). f.g.l. 35± mm. a. head, T1, T2 (note anteromesal sacs or gibbosities associated with thoracic legs, typical of larvae of this subfamily); b. A3; c. A8-10; d. mesal view right mandible; e. A3 proleg and crochets, f. A10 proleg and crochets. Head brownish white with darker brown platelets, and dark maculation near mouth parts. Mandibles with additional tooth at base of tooth 1. T1 shield pale brown with several groups of small brown platelets and maculae. T2, T3, and abdomen mostly yellowish white with dorsal pale brown to brown longitudinal stripes and a narrower ventromesal pale brown to brown stripe. T1, T2 and T3 legs with an associated anteromesal sac or gibbosity. A3-6 prolegs rudimentary with mostly bi- to triordinal crochets in ellipses. Anal proleg with a row of bi- to triordinal crochets plus an additional group of small uniordinal crochets. Thompsons, Tex., St. George Lake, in stems of southern wildrice. *Hosts:* southern wild rice (*Zizaniopsis miliacea*), bulrush (*Scirpus validus*) and other semiaquatic plants. Bores into stems of plants.

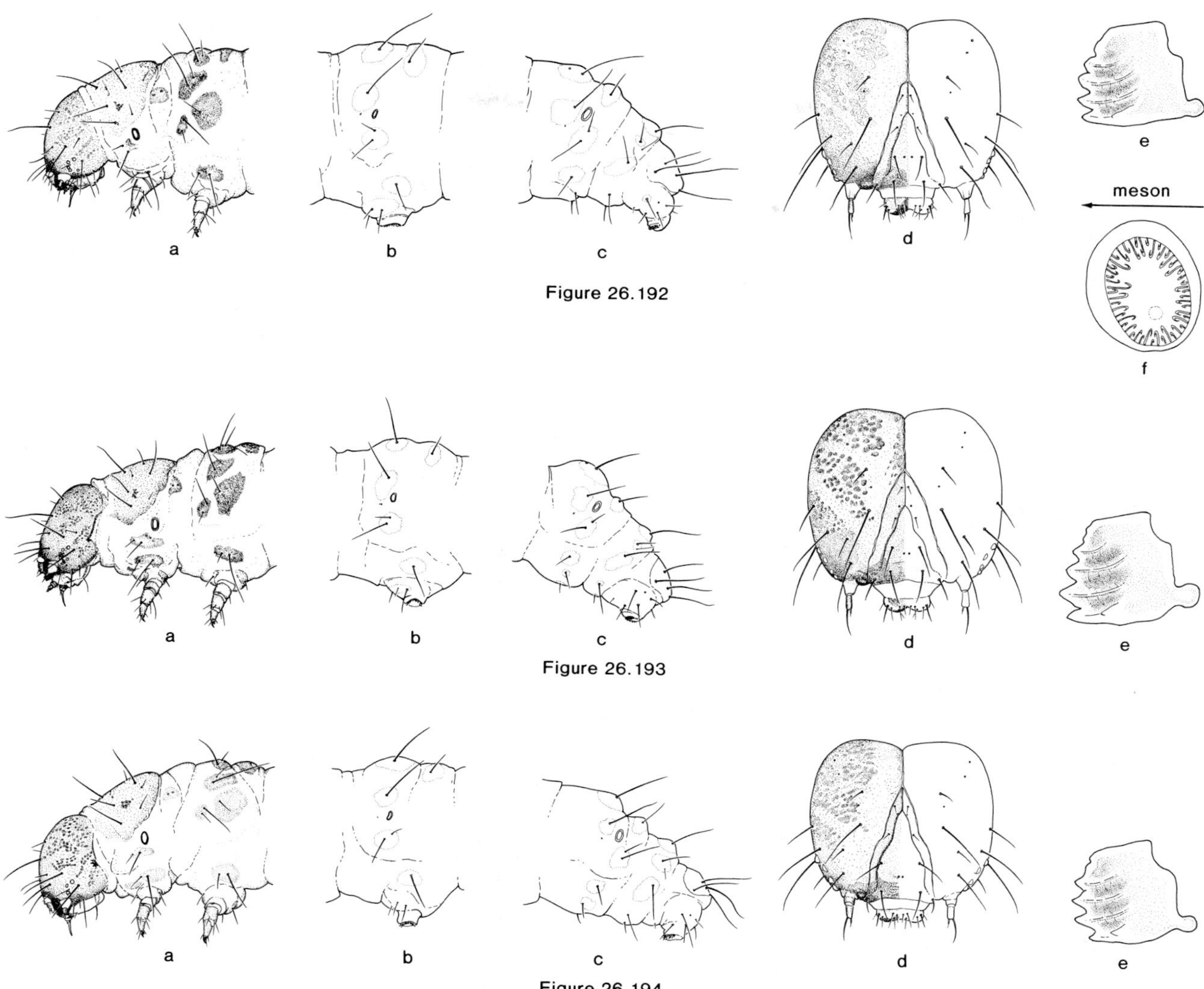

Figure 26.192a-f. Pyralidae, Crambinae. *Diatraea crambidoides* (Grote). Southern cornstalk borer. f.g.l. 25± mm. a. head, T1, T2; b. A3; c. A8-10; d. head; e. mesal view right mandible; f. A3 proleg and crochets. Head mostly whitish to brownish yellow with indistinct pale brown platelets, darker maculation near mandibles. Mandibles simple, or with indistinct tooth associated with inner ridge of tooth 1. Prothorax yellowish brown to brown. T2, T3, and abdomen mostly white to grayish or yellowish white with large brown to black pinacula (diapausing larvae yellowish white with pinacula and T1 shield pale, not contrasting with remainder of integument). Pinaculum of SD1 of A1-8 more or less circular. A3-6 crochets biordinal in ellipses or circles. Rocky Mount, N.C., in corn stalks, August 27, 1974. *Hosts:* primarily corn, also sorghum, *Paspalum, Panicum* and related plants. Sugarcane is also occasionally attacked. Feeds mostly within stems of hosts.

Figure 26.193a-e. Pyralidae, Crambinae. *Diatraea grandiosella* Dyar. Southwestern corn borer. f.g.l. 30-40 mm. a. head, T1, T2; b. A3; c. A8-10; d. head; e. mesal view right mandible. Head mostly brownish yellow with numerous brown to dark reddish brown relatively distinct, platelets and dark brown to black maculation near mandibles. Mandibles simple or with indistinct tooth associated with inner ridge of tooth 1. T1 shield smooth, yellowish brown to brown. T2, T3 and abdomen usually mostly grayish white with large brown to black pinacula (diapausing larvae with general color of integument grayish to yellowish white with pinacula indistinct (about same color as remainder of integument)). Pinaculum of SD1 of A1-8 elongate and usually slightly angled. A3-6 crochets biordinal in ellipses or circles. Caddo Parish, La., corn stalks. September 16, 1958. *Hosts:* primarily corn, also sorghum, sugarcane, rice, *Paspalum, Panicum* and related plants. Feeds at times on leaves, but usually in whorl; mostly tunnels in stems of hosts.

Figure 26.194a-e. Pyralidae, Crambinae. *Diatraea saccharalis* (Fabricius). Sugarcane borer. f.g.l. 22± mm. a. head, T1, T2; b. A3; c. A8-10; d. head; e. mesal view right mandible. Head mostly brownish yellow with brown platelets, dark brown to black near mandibles; frequently also with dark spot laterally between caudal margin and stemmata. Mandibles with a distinct, additional tooth associated with inner ridge of tooth 1. Prothoracic shield pale brown usually tinged with dark brown to black. T2, T3 and abdomen mostly white to brownish white with large brown pinacula (living larva with D pinacula usually surrounded by pale brown or pink pigmentation). (Diapausing larva yellowish white, with pinacula and T1 shield with little pigmentation). Pinaculum of SD1 of A1-8 more or less circular. A3-6 crochets biordinal in ellipses or circles. Baton Rouge, La., sugarcane, May 7, 1974. *Hosts:* primarily sugarcane; also, corn, sorghum, rice, *Paspalum, Panicum, Andropogon,* and related plants. Feeds mostly within stem of hosts.

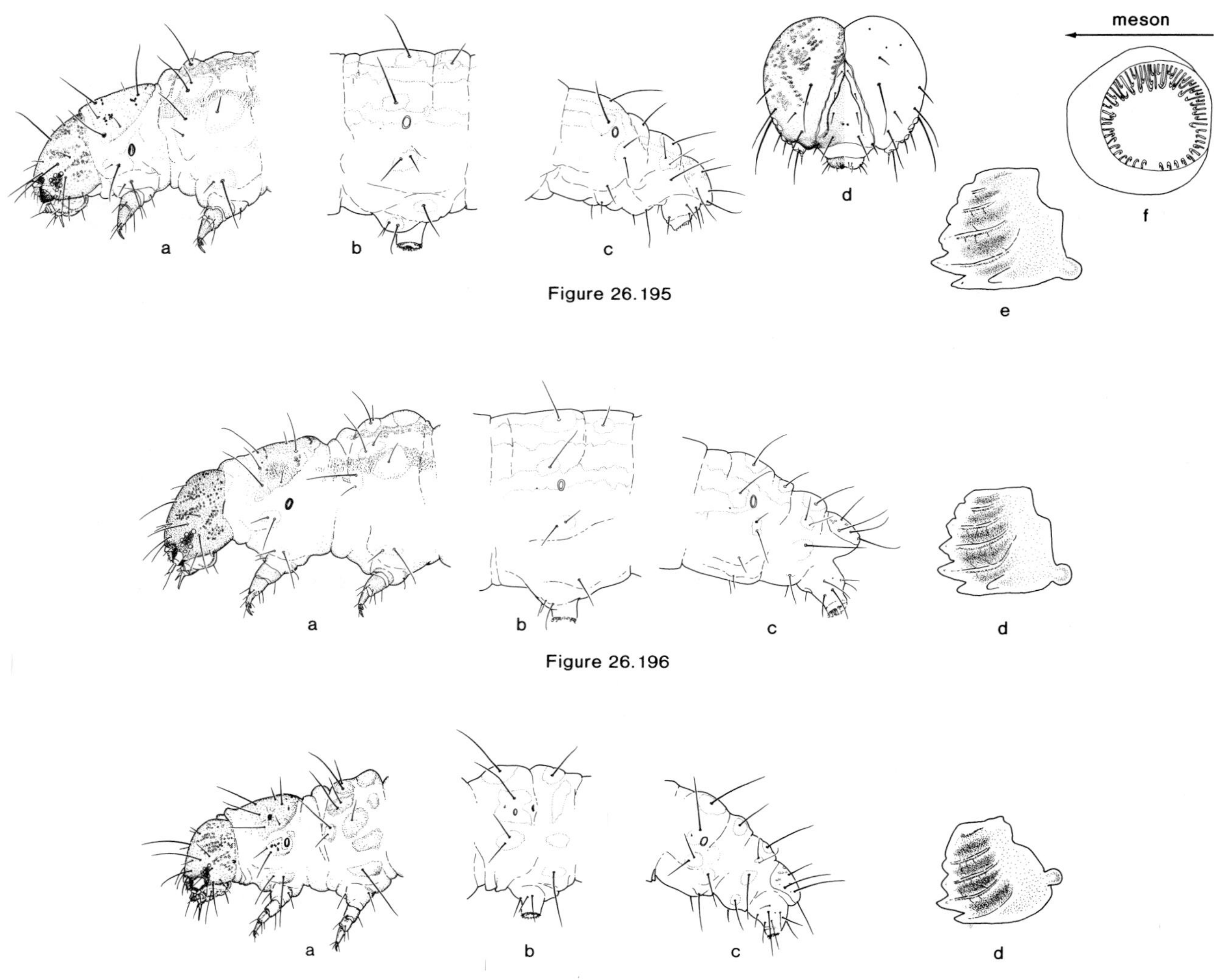

Figure 26.195

Figure 26.196

Figure 26.197

Figure 26.195a-f. Pyralidae, Crambinae. *Chilo suppressalis* (Walker). Asiatic rice borer. f.g.l. 25± mm. a. head, T1, T2; b. A3; c. A8-10; d. head; e. mesal view right mandible; f. A3 proleg and crochets. Head mostly brownish yellow with brown platelets. Distinct brown spot associated with stemmata. Mandibles with additional, distinct tooth at base of tooth 1. T1 shield whitish yellow with several groups of distinct, contrasting dark brown platelets. Most of T2, T3 and abdomen yellowish white with relatively distinct purplish brown longitudinal stripes, including a narrow middorsal stripe. Pinacula large, indistinct, brownish yellow. A3-6 crochets in circles or ellipses or partial circles or penellipses that are narrowly incomplete laterally; crochets small and more or less uniordinal laterally, becoming longer and distinctly multiordinal mesally. Seattle, Wash., in rice stems from Japan, March 15, 1929. *Hosts:* rice, other *Oryza* spp., northern wild rice (*Zizania aquatica*), *Phragmites* spp., common cattail and many other grasses. Feeds primarily within stem of host.

Figure 26.196a-d. Pyralidae, Crambinae. *Chilo plejadellus* Zincken. Rice stalk borer. f.g.l. 30± mm. a. head, T1, T2; b. A3; c. A8-10; d. mesal view right mandible. Head mostly yellowish brown with reddish brown platelets. Reddish brown spot associated with stemmata. Mandibles with additional, distinct tooth at base of tooth 1. T1 shield whitish yellow to yellowish brown suffused with brown and with several groups of brown platelets. Most of T2, T3 and abdomen white with purplish brown stripes; without a middorsal stripe (sometimes a few small dorsomesal patches of pigmentation). Pinacula large, indistinct, brownish yellow. A3-6 crochets in circles or ellipses, or partial circles or penellipses that are narrowly incomplete laterally; crochets small and more or less uniordinal laterally, becoming longer and distinctly multiordinal mesally. La., rice, October 1936. *Hosts:* rice, northern wild rice (*Zizania aquatica* L.), southern wild rice (*Zizaniopsis miliacea*), and *Spartina* spp. Small larvae feed on leaves; most damage and growth of larva is within stem of host.

Figure 26.197a-d. Pyralidae, Crambinae. *Crambus caliginosellus* Clemens. Corn root webworm. f.g.l. 20± mm. a. head, T1, T2; b. A3; c. A8-10; d. mesal view right mandible. Head pale yellow to yellowish brown with pale brown to brown platelets. Mandibles enlarged dorsally. T1 shield yellowish brown with only a few brown platelets. T2, T3 and abdomen mostly yellowish white with large, pale yellow to brown, pinacula. A3-6 crochets in biordinal to triordinal ellipses. Hickory Point, Tenn., roots of tobacco. Hosts: corn, tobacco, various grasses and other plants, especially asters, daisies (*Chrysanthemum* spp.), plantain (*Plantago* spp.), and wild carrot. Feeds on and within roots, and sometimes on stems and leaves. Forms loose tube of silk in soil at base of hosts.

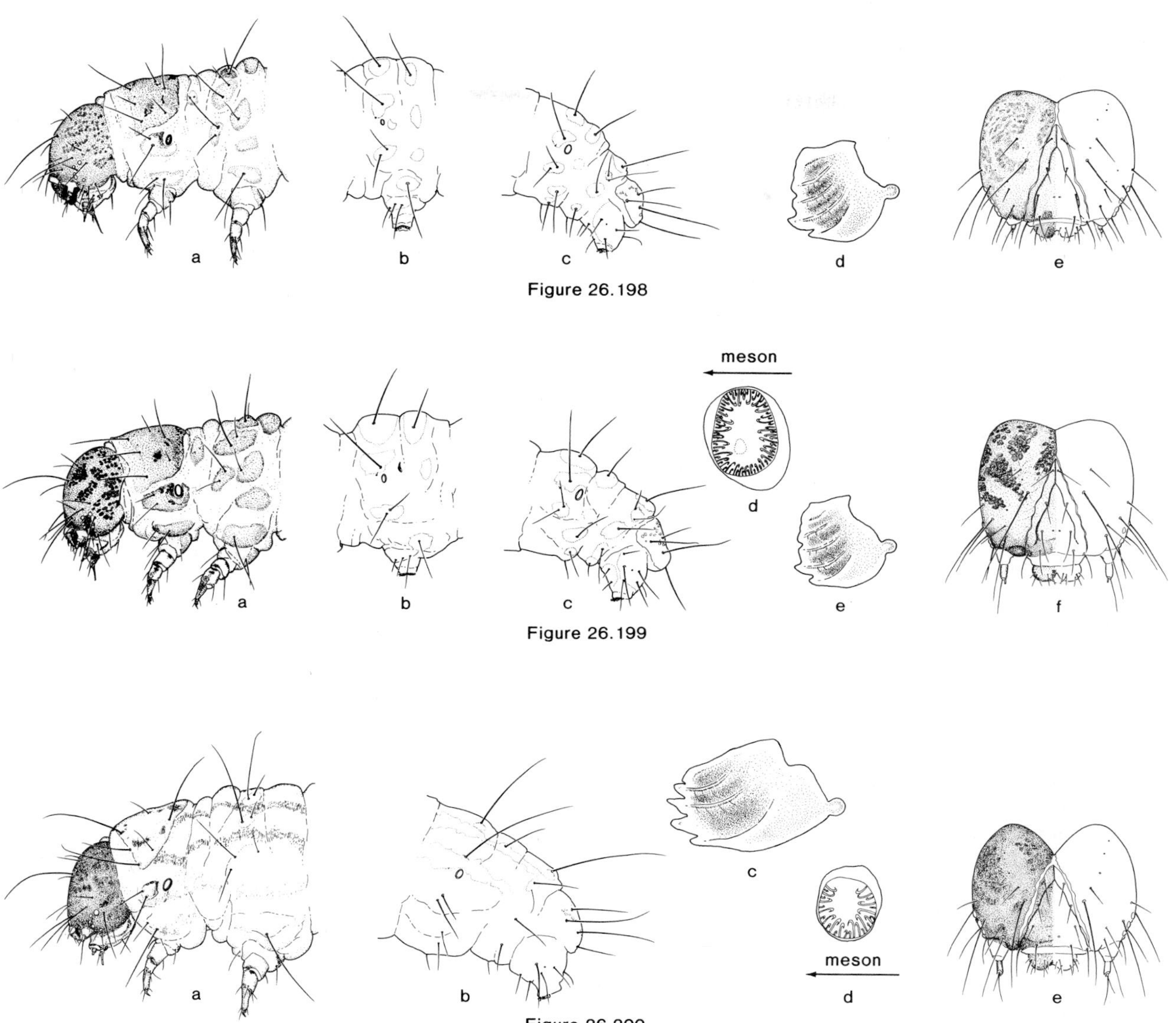

Figure 26.198

Figure 26.199

Figure 26.200

Figure 26.198a-e. Pyralidae, Crambinae. *Agriphila vulgivagella* (Clemens). Vagabond crambus. f.g.l. 17± mm. a. head, T1, T2; b. A3; c. A8-10; d. mesal view right mandible; e. head. Head pale reddish brown with reddish brown platelets. Mandibles enlarged dorsally. T1 shield yellowish brown with pale brown to brown platelets and sometimes additional brown suffusions. T2, T3 and abdomen mostly yellowish white with large pale brown to brown pinacula. A3-6 crochets mostly triordinal ellipses. Trentwood, Wash., creeping red fescue sod, May 4, 1958. *Hosts:* various grasses. Feeds in or at base of host, mostly on leaves.

Figure 26.199a-f. Pyralidae, Crambinae. *Parapediasia teterrella* (Zincken). Bluegrass webworm. f.g.l. 20± mm. a. head, T1, T2; b. A3; c. A8-10; d. A3 proleg and crochets; e. mesal view right mandible; f. head. Head yellowish brown sometimes with distinct reddish brown platelets. Mandibles enlarged dorsally. T1 shield reddish brown with dark brown platelets. T2, T3 and abdomen mostly whitish to brownish yellow with large brown to reddish brown pinacula. A3-6 crochets mostly triordinal ellipses. Gainesville, Fla., in grass, November 24, 1954. *Hosts:* Kentucky bluegrass and many other grasses. Feeds in or at base of host, mostly on leaves.

Figure 26.200a-e. Pyralidae, Glaphyriinae. *Hellula rogatalis* (Hulst). Cabbage webworm. f.g.l. 15± mm. a. head, T1, T2; b. A8-10; c. mesal view right mandible; d. A3 proleg and crochets; e. head. Head reddish brown with indistinct dark brown platelets (head appearing almost black in living larva). T1 shield mostly white usually with some pale brown suffusions and with brown platelets (shield of living larva purplish gray, marked with brown). T2, T3 and abdomen mostly yellowish white with purplish brown longitudinal stripes; stripes sometimes fade in preservative (living larva mostly grayish yellow with dark purplish brown longitudinal stripes). Midabdominal crochets biordinal and in shape of partial circles or penellipses that are incomplete laterally. Spring, Tex., cabbage, November 6, 1944. *Hosts:* cabbage, mustard, radish, other Brassicaceae, beet, purslane (*Portulaca* sp.), and other Amaranthaceae and Portulacaceae. Feeds mostly on leaves.

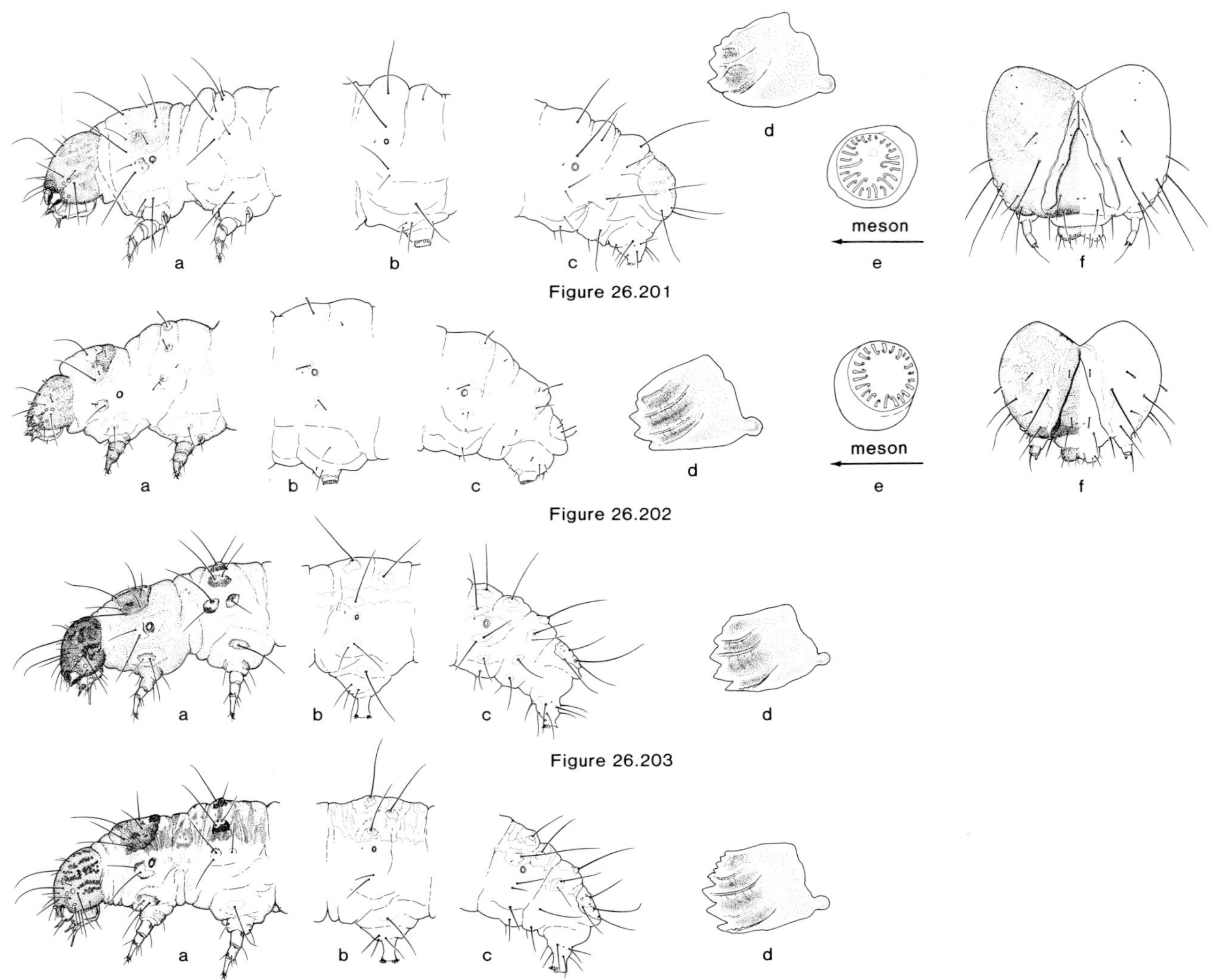

Figure 26.201a-f. Pyralidae, Glaphyriinae. *Dicymolomia julianalis* (Walker). f.g.l. 12± mm. a. head, T1, T2; b. A3; c. A8-10; d. mesal view right mandible; e. A3 proleg and crochets; f. head. Head yellowish brown with indistinct reddish brown platelets. T1 shield brownish yellow with pale brown platelets and maculation. T2, T3 and abdomen mostly whitish yellow to pink. A3-6 crochets mostly biordinal ellipses (usually weak laterally, sometimes forming mesal penellipses). Olmita, Tex., in gin trash, October 4, 1941. *Food:* not known. Possibly feeds on dead or live insects. Occurs in bagworm cases (*Thyridopteryx*) and seed capsules or seed clusters of cotton, cattail, thistle, milk-vetch (*Astragalus*), cactus, and other plants.

Figure 26.202a-f. Pyralidae, Glaphyriinae. *Calcoela pegasalis* (Walker). f.g.l. 12± mm. a. head, T1, T2; b. A3; c. A8-10; d. mesal view right mandible; e. A3 proleg and crochets; f. head. Head yellow to brownish yellow with pale yellowish white markings; frontoclypeus and adfrontals usually brownish yellow. T1 shield about same color as head. XD2 absent on shield. T2, T3 and abdomen mostly yellowish white. Many setae of head and body short and sometimes slightly enlarged distally. A3-6 crochets in uni- to biordinal ellipses. Durham, N.C., in nest of *Polistes annularis,* June 19, 1971. *Food:* probably *Polistes* larvae and associated materials.

Figure 26.203a-d. Pyralidae, Evergestinae. *Evergestis pallidata* (Hufnagel). Purplebacked cabbageworm. f.g.l. 20± mm. a. head, T1, T2; b. A3; c. A8-10; d. mesal view right mandible. Head brown with darker brown platelets. Mandibles with a series of small accessory teeth associated with tooth 1. T1 shield yellowish brown to brown with some darker maculation and yellowish white to white mostly along meson. T2 and T3 mostly brownish white. Abdomen brownish white with longitudinal purplish brown stripes that are most pronounced on some caudal segments. Most setae on chalazae (dorsal chalazae more elevated and brown to black; some chalazae partially pale). A3-6 crochets biordinal in partial circles or partial ellipses that are broadly incomplete laterally. Knappton, Wash., horseradish, October 8, 1943. *Hosts:* cabbage, horseradish and other Brassicaceae. Feeds primarily on foliage.

Figure 26.204a-d. Pyralidae, Evergestinae. *Evergestis rimosalis.* (Guenée). Cross-striped cabbageworm. f.g.l. 20± mm. a. head, T1, T2; b. A3; c. A8-10; d. mesal view right mandible. Head yellowish white with pale brown platelets (head with green undertones in living larva). Mandibles with series of small accessory teeth associated with tooth 1 and above tooth 4. T1 shield brown to black with large yellowish white to white patches. T2, T3 and abdomen mostly brownish white with broad purplish brown longitudinal stripes connected by purplish brown transverse stripes (living larva mostly greenish white to bluish gray with dark purplish brown longitudinal stripes, associated primarily with the SD setae, 3 or 4 dark purplish brown transverse stripes on dorsum of each segment interspersed with white, a bright yellow and white longitudinal band at the spiracles, and additional patches of white laterally and ventrally). Most setae arising from chalazae (dorsal chalazae well developed and brown to dark brown (black in living larva)) frequently with pale centers. A3-6 crochets bi- to triordinal in partial circles or partial ellipses that are broadly incomplete laterally. Taylorsville, N.C., on cabbage, October 21, 1949. *Hosts:* cabbage and other Brassicaceae. Usually eats holes in leaves, sometimes concealed within a folded leaf.

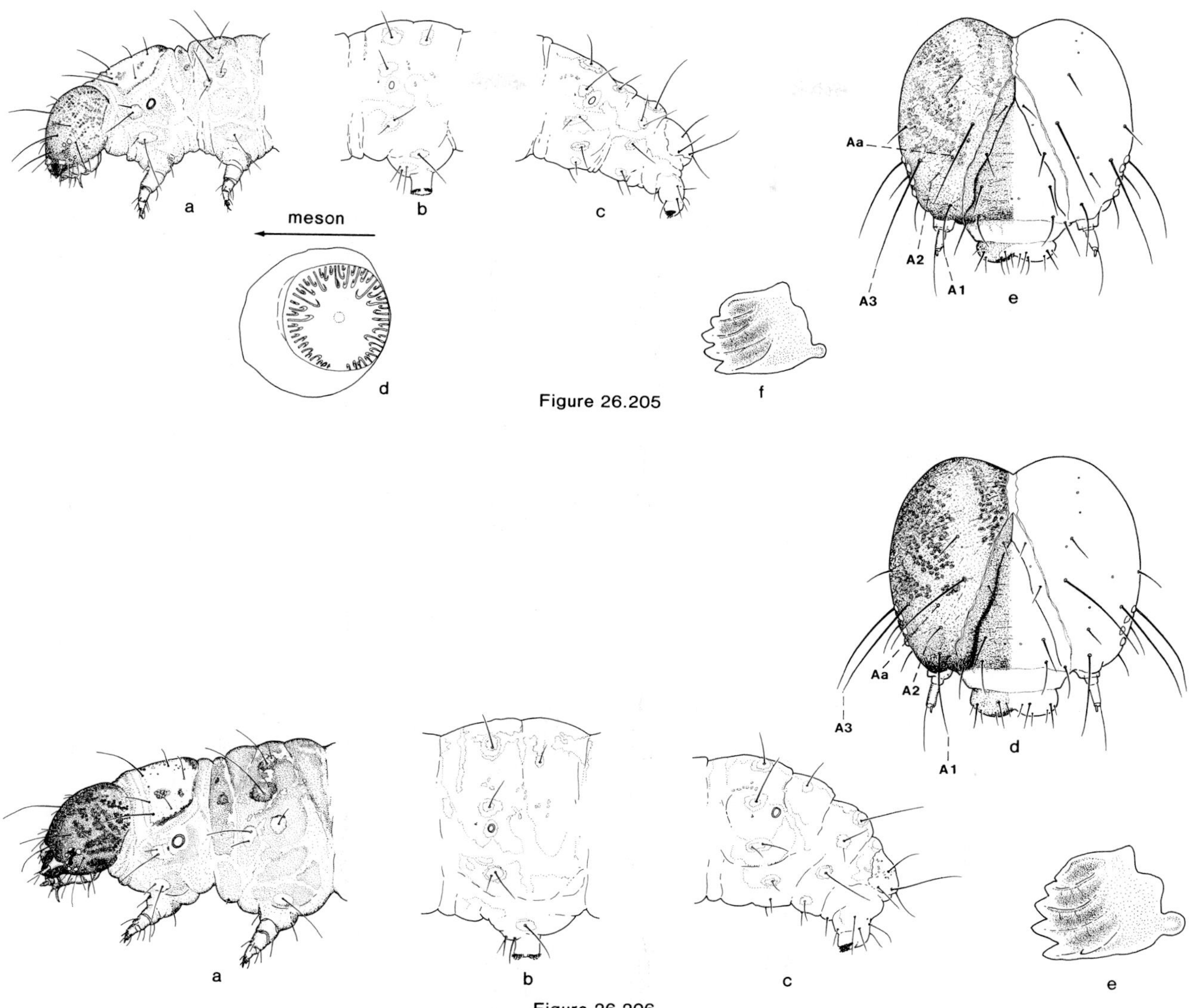

Figure 26.205

Figure 26.206

Figure 26.205a-f. Pyralidae, Pyraustinae.*Ostrinia obumbratalis* (Lederer). Smartweed borer. f.g.l. 22± mm. a. head, T1, T2; b. A3; c. A8-10; d. A3 proleg and crochets; e. head (note position of setae A1 and A2 and pore Aa); f. mesal view right mandible. Head yellowish to reddish brown with darker reddish brown platelets, reddish brown near mandibles. Setae A1 and A2 and pore Aa forming a distinct angle. T1 shield yellowish white with pale brown spots and usually pale brown along lateral margins. T2, T3 and abdomen mostly brownish white with finely granulate pale brown patches. Some dorsal patches forming vague longitudinal stripes. Pinacula pale brown dorsally, slightly paler ventrally, some with pale centers. A3-6 crochets bi- to triordinal, becoming smaller laterally and in partial circles or penellipses that are narrowly incomplete laterally. Urbana, Ill., in smartweed, September 1919. *Hosts:* prefers smartweed (*Polygonum* spp.), but also feeds on cocklebur, ragweed, Joe-pye weed, cattail, lamb's-quarters and many others. Bores principally in stems of hosts. Also sometimes found late in season in corn, cotton, etc.; these are not hosts, but so-called shelter plants selected for overwintering.

Figure 26.206a-e. Pyralidae, Pyraustinae. *Ostrinia nubilalis* (Hübner). European corn borer. f.g.l. 28± mm. a. head, T1, T2; b. A3; c. A8-10; d. head (note position of setae A1 and A2 and pore Aa); e. mesal view right mandible. Head yellowish to reddish brown with darker platelets, dark reddish brown to black near mandibles. Head setae A1 and A2 and pore Aa in line or almost a straight line. Mandibles sometimes with indistinct accessory tooth at base of tooth 1. T1 shield brownish yellow with reddish brown spots and pale brown along lateral and sometimes caudal margins. T2, T3 and abdomen mostly brownish white with a relatively distinct, finely granulate, dorsomesal, brown to gray longitudinal stripe and other broad, somewhat fragmented and diffuse, less distinct, finely granulate, brown to gray patches, vaguely forming, in part, into broad longitudinal stripes on the abdomen (living larva with granulate patches and stripes brown, gray or pink). Pinacula brown usually with pale centers dorsally, and pale brown with pale centers ventrally. A3-6 crochets mostly triordinal becoming smaller laterally and in partial circles or penellipses that are narrowly incomplete laterally. Emporia, Va., corn stalks, September 1972. *Hosts:* many herbaceous plants (occasionally will be secondarily associated with woody plants). In the U.S. and Canada feeds primarily on corn. Mostly within stems of herbaceous hosts; fruits and terminals of woody plants.

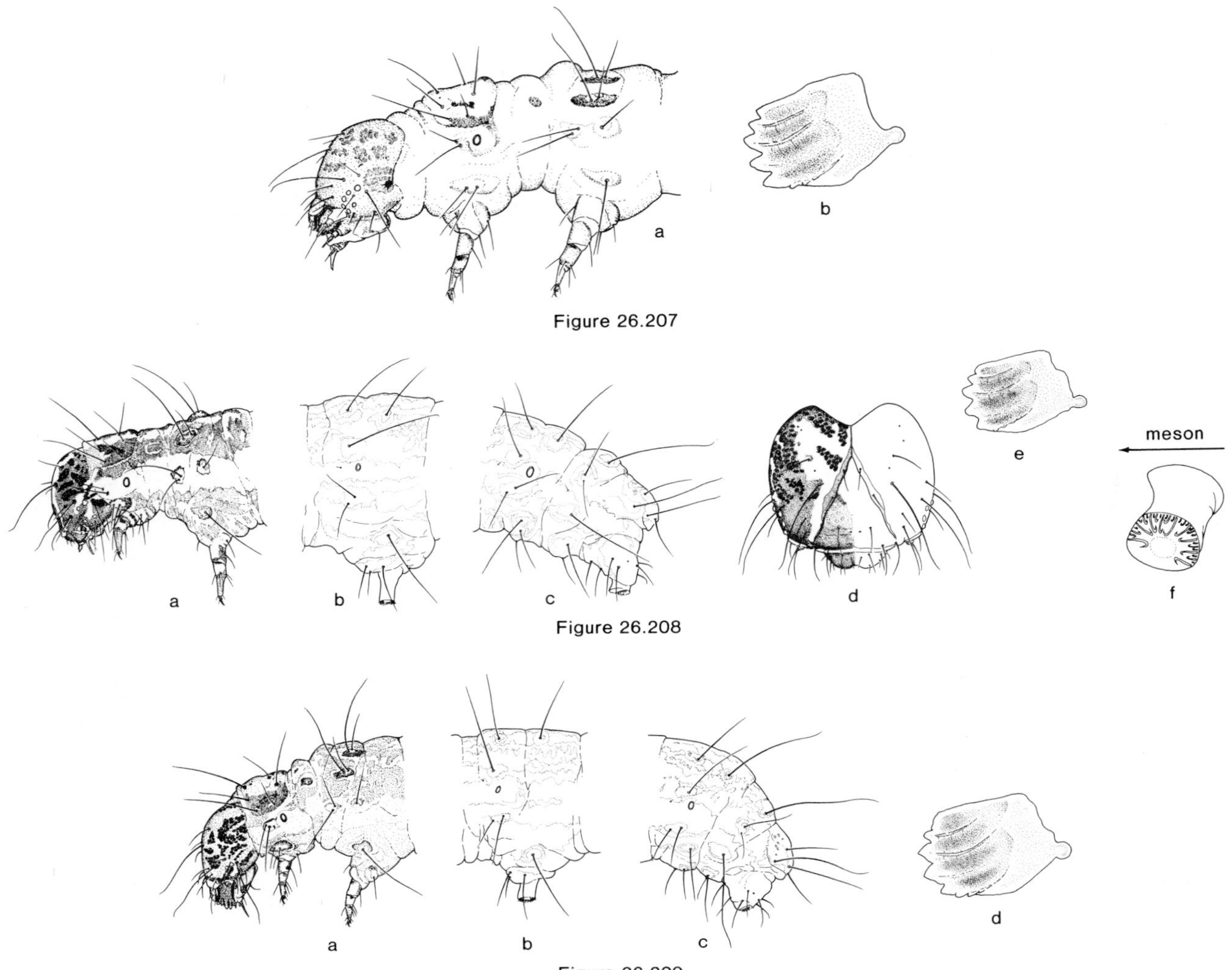

Figure 26.207

Figure 26.208

Figure 26.209

Figure 26.207a,b. Pyralidae, Pyraustinae. *Achyra rantalis* (Guenée). Garden webworm. f.g.l. 25± mm. a. head, T1, T2; b. mesal view right mandible. Head brownish white with pale brown platelets and a small fuscous posterolateral patch. Mandibles with series of small accessory teeth associated with tooth 1. T1 shield yellowish white with a brown band along lateral margin and other less obvious brown markings. Posterolateral angle of T1 shield extended under spiracle to fuse with prespiracular plate. T2, T3, and abdomen mostly brownish white (living larva green to very dark green). Pinacula dark brown to black; those above level of spiracles uniformly dark colored or with pale centers; those below spiracles with marked reduction in, or absence of, pigmentation. (Some pinacula, particularly those on anterior of body distinctly elongate). A3-6 crochets mostly triordinal and in partial circles, or partial ellipses that are broadly incomplete laterally. Newark, N.J., *Atriplex patula* L. August 8, 1944. *Hosts:* alfalfa, clover, beans, soybeans, cowpeas, sugarbeets, peas, strawberries, corn, cotton, and many uncultivated plants. Usually feeds under silk canopy on the leaves; sometimes found in silk-lined tunnel on or in the soil.

Figure 26.208a-f. Pyralidae, Pyraustinae. *Loxostege sticticalis* (L.). Beet webworm. f.g.l. 24± mm. a. head, T1, T2; b. A3; c. A8-10; d. head; e. mesal view right mandible; f. A3 proleg and crochets. Head reddish brown with distinct white lateral, dorsal and anterior streaks and brown to black platelets. Mandibles with series of small accessory teeth associated with tooth 1. T1 shield brown to black with mid-dorsal and sub-dorsal white streaks and brown to black platelets. T2, T3, and abdomen mostly white with several complete or fragmented brown to dark brown longitudinal stripes, including a middorsal stripe from T2 to A9 (living larva yellowish white or greenish white to very dark green with complete or fragmented, broad or narrow, brown to black, longitudinal stripes). Pinacula brown to black usually with white centers (some pinacula predominately white). A3-6 crochets triordinal in partial circles or partial ellipses, broadly open laterally. Stillwater, Okla., Russian thistle, May 21, 1935. *Hosts:* prefers sugar beet, but feeds on many other truck crops, forage crops and weeds. Consumes almost entire leaf, with exception of larger veins and petioles.

Figure 26.209a-d. Pyralidae, Pyraustinae. *Loxostege cereralis* (Zeller). Alfalfa webworm. f.g.l. 28± mm. a. head, T1, T2; b. A3; c. A8-10; d. mesal view right mandible. Head yellowish white with brown to dark brown platelets. Mandibles with series of small accessory teeth associated with tooth 1. T1 shield mostly yellowish white with a narrow dorsal streak, a very broad, lateral, brown to dark brown band and dark brown platelets and other maculae. T2, T3, and abdomen usually mostly white to brownish white (green in living larva) with fragmented broad or narrow pale brown to dark brown longitudinal stripes (dorsomeson without dark stripe from T2 to A9). Pinacula brown to dark brown usually with pale centers (some pinacula predominately pale). A3-6 crochets triordinal in partial circles or partial ellipses, broadly open laterally. Ft. Collins, Colorado, on alfalfa, June 17, 1920. *Hosts:* alfalfa, celery, beet, and many weeds. Feeds on leaves.

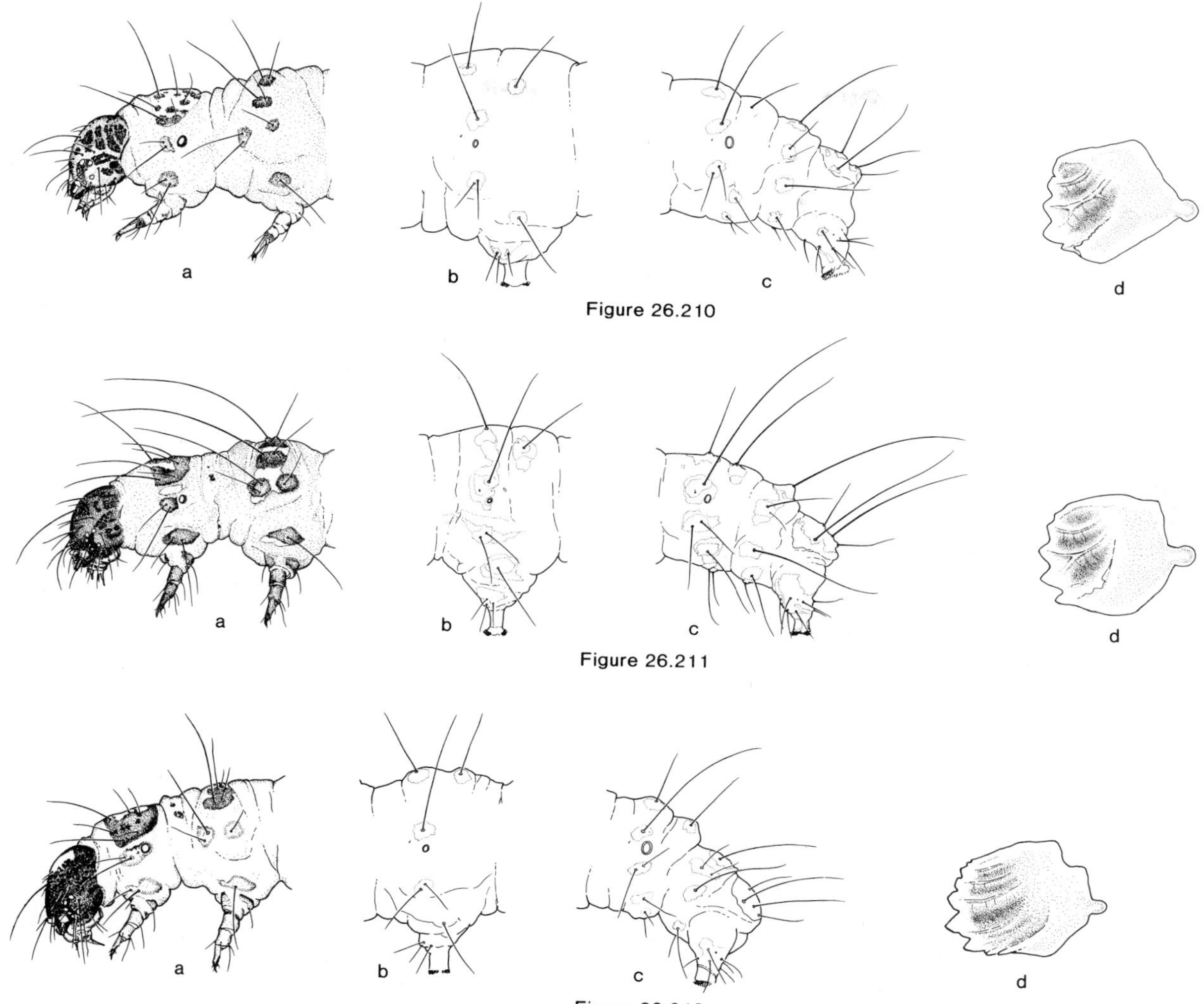

Figure 26.210a-d. Pyralidae, Pyraustinae. *Saucrobotys futilalis* (Lederer). f.g.l. 30± mm. a. head, T1, T2; b. A3; c. A8-10; d. mesal view right mandible. Head brownish white with dark brown to black platelets and additional dark markings, sometimes covering most of the head. Mandibles with ridge of small accessory teeth near tooth 1. T1 shield white with black platelets and markings. T2, T3, and abdomen mostly white to brownish or orangish white (living larva orange or greenish orange, with white on venter). Pinacula brown to black (those on dorsum with some associated white pigmentation). A3-6 crochets triordinal in partial circles or partial ellipses, broadly open laterally. Raleigh, N.C., on Indian hemp, July 21, 1964. *Hosts:* milkweeds, including Indian hemp (*Apocynum cannabinum* L.) and butterfly-weed (*Asclepias tuberosa* L.). Feeds on foliage and lives gregariously during most of larval stage in conspicuous nests made of silked-together leaves.

Figure 26.211a-d. Pyralidae, Pyraustinae. *Uresiphita reversalis* (Guenée). Genista caterpillar. f.g.l. 30± mm. a. head, T1, T2; b. A3; c. A8-10; d. mesal view right mandible. Head reddish brown to dark reddish brown with inconspicuous dark brown to black platelets, or head entirely black. Mandibles with irregular accessory ridge of small teeth starting near base of tooth 1 and extending dorsally. T1 shield mostly dark reddish brown to black with a yellowish white dorsomesal stripe and a usually truncated yellowish white patch between the XD setae and D2. T2, T3 and abdomen mostly yellowish to brownish white with patches or streaks of white that are mostly associated with pinacula or chalazae (white patches or streaks not always evident in preserved larvae). Pinacula and chalazae distinct, dark reddish brown to black. Setae pale, unusually long. A3-6 crochets triordinal in partial circles or partial ellipses, broadly open laterally. St. Augustine, Fla., *Lupinus* sp., April 1, 1972. *Hosts:* broom (*Genista* spp.), lupines, wild indigo (*Baptisia tinctoria*) and other herbaceous plants, and also some shrubs, especially Fabaceae. Feeds gregariously in silk enclosures, mostly on leaves, sometimes on other parts.

Figure 26.212a-d. Pyralidae, Pyraustinae. *Herpetogramma bipunctalis* (Fabricius). Southern beet webworm. f.g.l, 22-26 mm. a. head, T1, T2; b. A3; c. A8-10; d. mesal view right mandible. Head mostly pale brown to dark reddish brown (sometimes almost black) with distinct to indistinct darker platelets and maculae (darker larvae usually with distinct pale patches on vertex and frontoclypeus). Mandibles usually with 8-10 teeth, and enlarged ventrally. T1 shield dark reddish brown and broadly brownish white to white along dorsomeson. Spiracle broadly fused to prespiracular plate. T2, T3 and abdomen mostly brownish white (green in living larva). Most pinacula pale brown with slightly darker margins; a few pinacula, particularly on T2 darker (brown to reddish brown or black). A3-6 crochets mostly triordinal in partial circles or penellipses, narrowly incomplete laterally. Union Co., N.C., *Amaranthus* sp., September 6, 1955. *Hosts:* amaranth or pigweed, beet and many other plants. Feeds on webbed together foliage; sometimes on developing seeds.

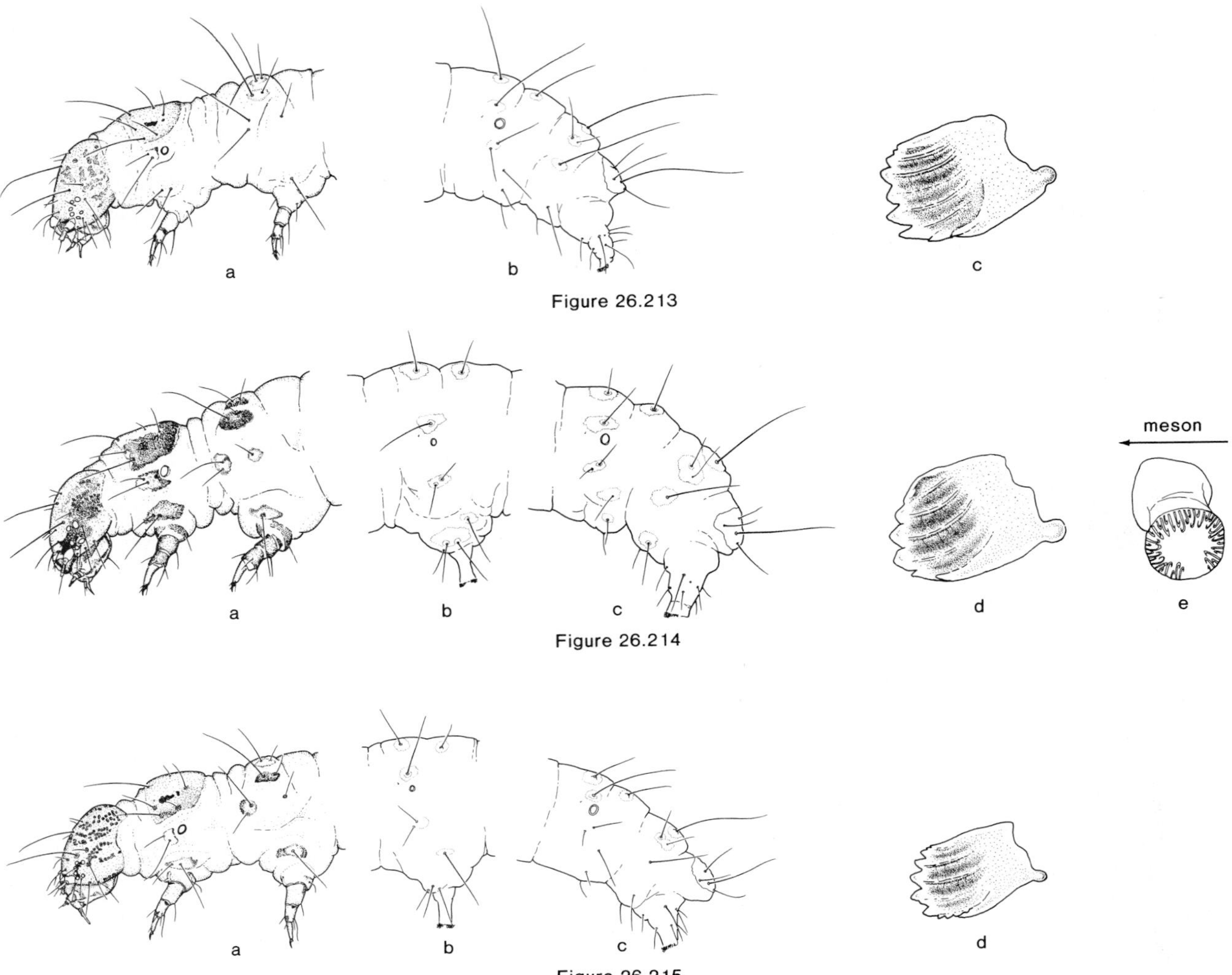

Figure 26.213a-c. Pyralidae, Pyraustinae. *Udea rubigalis* (Guenée). Celery leaftier or greenhouse leaftier. f.g.l. 18± mm. a. head, T1, T2; b. A8-10; c. mesal view right mandible. Head yellowish white with slightly darker indistinct platelets. Mandibles usually with 8-10 teeth. T1 shield whitish yellow, only slightly darker than surrounding integument with a distinct dark spot laterodorsally. T2, T3, and abdomen mostly yellowish white (living larva green; integument translucent so that portions of the trachea, particularly near the spiracles, can be seen; some larvae have a darker green stripe on dorsomeson and 2 whitish green subdorsal stripes). Pinacula about same color as surrounding integument, with slightly darker margins. A3-6 crochets triordinal mesally, becoming uniordinal laterally in circles or ellipses. Raleigh, N.C., March 1, 1970, on strawberry foliage in greenhouse. *Hosts:* chrysanthemum, rose, snapdragon, geranium, celery, cabbage, beets, lettuce, strawberry, and many other flowers and garden crops. Feeds primarily on the leaves, frequently on undersurface, or in webbed-together leaves.

Figure 26.214a-e. Pyralidae, Pyraustinae. *Desmia funeralis* (Hübner). Grape leaffolder. f.g.l. 22± mm. a. head, T1, T2; b. A3; c. A8-10; d. mesal view right mandible; e. A3 proleg and crochets. Head mostly yellowish white with large pale brown to brown areas laterally, usually darkest at stemmata, base of antennae and mandibles (head mostly pale brown with dark brown to black lateral patches in living larva). Mandibles usually with 5-6 teeth. T1 shield yellowish white dorsally with posterior and ventral areas broadly marked with brown (shield pale brown with black markings in living larva). T2, T3 and abdomen mostly white (integument somewhat transparent in living larva with body fluids and organs imparting a yellowish to whitish green color). Some anterior pinacula brown to dark brown (black in living larva), but most pinacula concolorous with integument or only slightly darker. A3-6 crochets bi- to triordinal, in circles or ellipses, or slightly incomplete circles or partial ellipses, smaller and fewer laterally. Hyattsville, Md., Virginia creeper, July 11, 1915. *Hosts:* grape, Virginia creeper and related plants. Feeds on leaves of host; forms a rolled leaf shelter.

Figure 26.215a-d. Pyralidae, Pyraustinae. *Spoladea recurvalis* (Fabricius). Hawaiian beet webworm. f.g.l. 20± mm. a. head, T1, T2; b. A3; c. A8-10; d. mesal view right mandible. Head yellowish or greenish white with brown platelets. Mandibles with small accessory teeth near tooth 1 and 5. T1 shield yellowish to greenish white suffused with brown laterally and with a few distinct dark brown to black platelets. T2, T3 and abdomen mostly yellowish to brownish white (living larva mostly greenish or reddish white). Most pinacula concolorous with remainder of integument; a few anterior pinacula, particularly SD on T2, dark and distinct. A3-6 crochets triordinal, in partial ellipses or incomplete circles open laterally. Norfolk, Va., feeding on spinach. October 12, 1926. *Hosts:* mostly beet, swiss chard, spinach, and several weeds such as *Amaranthus* and *Chaenopodium*. Feeds on leaves, sometimes resting under sparse silk covering.

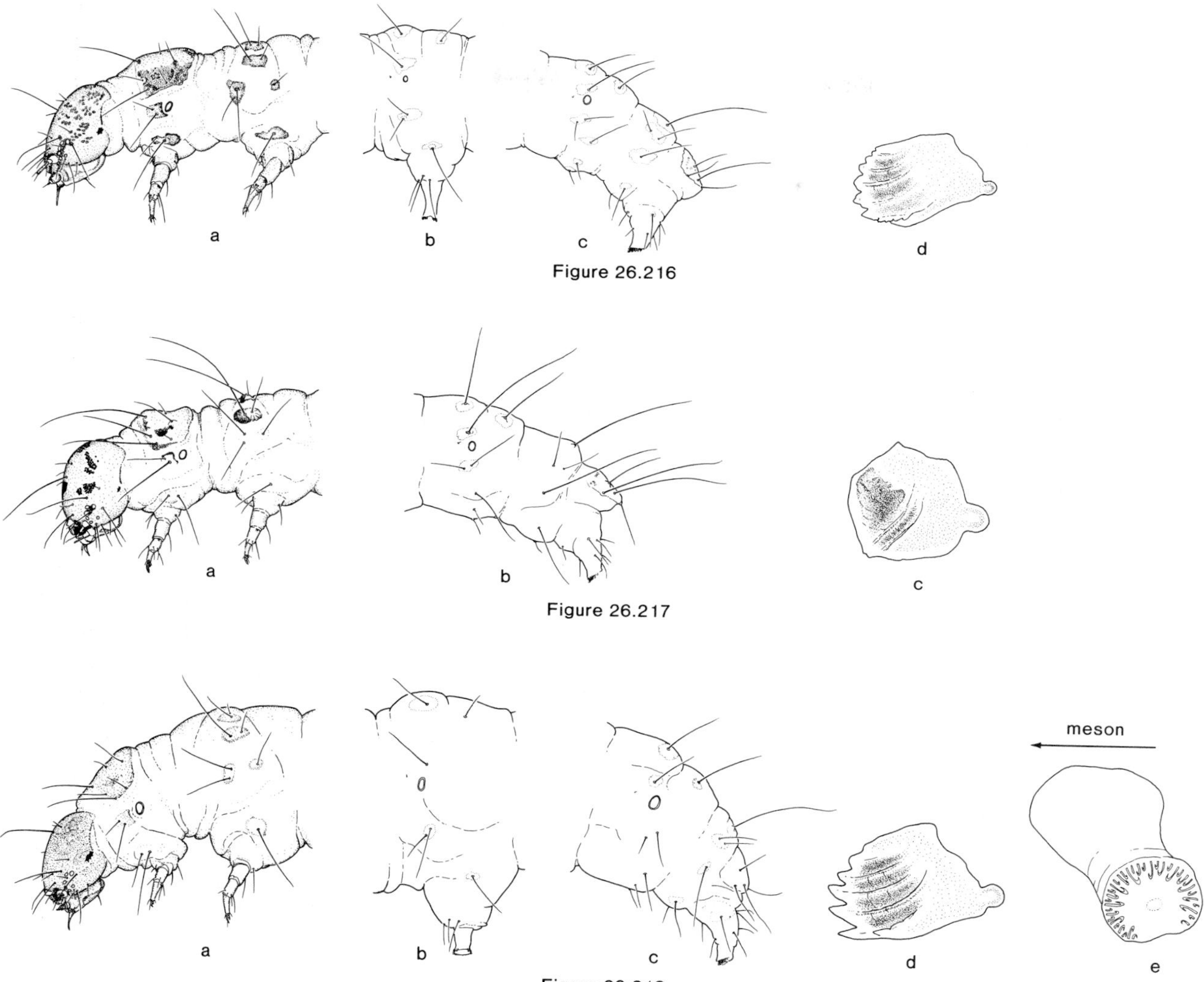

Figure 26.216a-d. Pyralidae, Pyraustinae. *Hymenia perspectalis* (Hübner). Spotted beet webworm. f.g.l. 18± mm. a. head, T1, T2; b.A3; c. A8-10; d. mesal view right mandible. Head yellowish white (greenish white in living larva) with brown to purplish brown platelets. Mandibles with small accessory teeth near tooth 1 and 5. T1 shield yellowish white, broadly margined laterally with brown to black. T2, T3 and abdomen mostly yellowish to brownish white (living larva mostly green). Pinacula relatively distinct, brown or partially brown to black; those associated with SD1 and SD2 on thorax and SD1 on abdomen darkest and most distinct. A3-6 crochets triordinal, in partial ellipses or incomplete circles open laterally. Goulds, Fla., spinach, March 20, 1945. *Hosts:* beet and swiss chard, spinach, *Alternanthera* sp. and other members of the Polygonaceae, Amaranthaceae and Chenopodiaceae. Feeds on leaves. Usually conceal themselves at base of plant during day.

Figure 26.217a-c. Pyralidae, Pyraustinae. *Omiodes blackburni* (Butler). Coconut leafroller. f.g.l. 32-35 mm. a. head, T1, T2; b. A8-10; c. mesal view right mandible. Head yellowish white to brownish yellow with several distinct brown to black groups of platelets; group of platelets just below the P1 setae usually darkest. Mandibles without distinct distal teeth. T1 shield white to yellowish white, brown to black anteriorly and laterally and with dark brown to black spots posterior to XD2. T2, T3 and abdomen mostly white to yellowish white (living larva mostly green with pale dorsal longitudinal stripes). Some setae, particularly those on dorsum, on chalazae. Most pinacula and chalazae yellowish white, sometimes margined with brown. Chalazae on dorsum of T2 and T3, in part, dark brown to black. A3-6 crochets mostly triordinal, in partial ellipses or incomplete circles, broadly open laterally. Larva illustrated penultimate stage. Kaaawa, Oahu, Hawaii, coconut palm leaves. April 28, 1959. *Hosts:* prefers coconut, occasionally on banana and palms other than coconut. Feeds mostly on young leaves.

Figure 26.218a-e. Pyralidae, Pyraustinae. *Diaphania nitidalis* (Stoll). Pickleworm. f.g.l. 25-30 mm. a. head,T1, T2 (note dark spot on lower half of caudal margin of head); b. A3; c. A8-10; d. mesal view right mandible; e. A3 proleg and crochets. Head yellowish white to brownish white with mostly indistinct pale platelets, and a distinct brown to black spot on the lower half of its caudal margin. Mandibles with an accessory ridge of teeth near base of tooth 1. T1 shield yellowish white without distinct maculation. T2, T3 and abdomen mostly yellowish white (mostly greenish white to greenish yellow in living larva). Pinacula indistinct, only slightly darker than surrounding integument. A3-6 crochets mostly triordinal, in partial ellipses or incomplete circles broadly open laterally. Brownsville, Tex., squash, April 7, 1938. *Hosts:* cantaloupe, cucumber, squash, and other cucurbits. Mostly feeds within fruit; sometimes feeds on blossoms, buds, stems, or petioles.

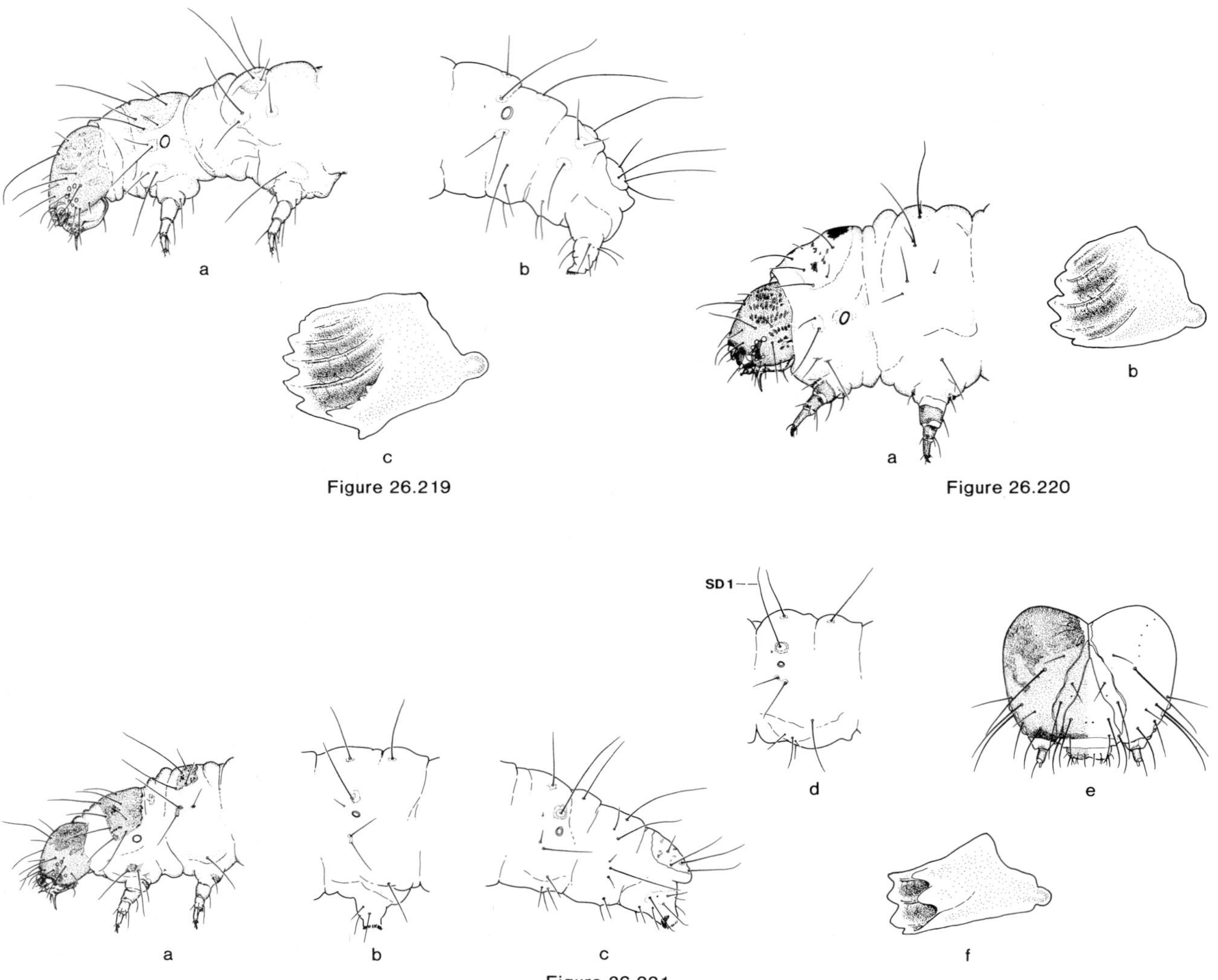

Figure 26.219

Figure 26.220

Figure 26.221

Figure 26.219a-c. Pyralidae, Pyraustinae. *Diaphania hyalinata* (L.). Melonworm. f.g.l. 25-30 mm. a. head, T1, T2 (note absence of dark spot on lower half of caudal margin of head; b. A8-10; c. mesal view right mandible. Head yellowish white with indistinct pale platelets; no distinct dark spot on lower half of caudal margin. Mandibles with an accessory ridge of teeth near base of tooth 1, and an additional dentate projection on the ventral margin. T1 shield yellowish white without distinct maculation. T2, T3 and abdomen mostly yellowish white (mostly greenish white to greenish yellow in living larva). Pinacula indistinct, only slightly darker than surrounding integument. A3-6 crochets mostly triordinal, in partial ellipses or incomplete circles, broadly open laterally. Savannah, Ga., okra, August 16, 1944. *Hosts:* cucurbits and occasionally other plants. Feeds on leaves and fruit of hosts.

Figure 26.220a,b. Pyralidae, Odontiinae. *Mimoschinia rufofascialis* (Stephens). Barberpole caterpillar. f.g.l. 15± mm. a. head, T1, T2; b. mesal view right mandible. Head mostly brownish yellow, with reddish brown to dark brown or black near mandibles, reddish brown at stemmata and at distal part of frontoclypeus, and with reddish brown relatively indistinct platelets. T1 shield white to yellowish white with many small brown platelets and maculae, and a distinct large dark brown to black fused group of spots near the posterior meson. T2, T3 and abdomen mostly yellowish white to brownish white (living larva with thoracic segments and anterior half of A1 mostly red; the remaining abdominal segments also partially encircled by a broad band of red (the red fades in preservative and the barberpole coloration is sometimes difficult to see). A3-6 crochets few in number, uniordinal and in mesal penellipses. Brownsville, Tex., *Pseudabutilon lozanii* pods, July 1951. *Hosts:* malvaceous plants such as Indian mallow (*Abutilon* spp.), *Pseudabutilon, Malvastrum, Wissadula* and *Sida.* Feeds mostly within seed pods of host.

Figure 26.221a-f. Pyralidae, Galleriinae. *Macrotheca unipuncta* Dyar. f.g.l. 14± mm. a. head, T1, T2; b. A3; c.A8-10; d. A1 (note pigmented ring at base of SD1); e. head; f. mesal view right mandible. Head brownish yellow to brown near mouthparts with indistinct pale brown platelets and maculation. Mandibles elongate with large, inner toothlike projections. T1 shield pale brown with brown platelets; shield extending laterally so as to include L1 and L2. T2 with a smaller, paler shield. T2, T3 and abdomen mostly white. Most pinacula about same color as surrounding integument, a few pale brown. SD1 of A1 and A8 with pale brown pinacula rings. A3-6 crochets mostly biordinal and in ellipses. Berrien Co., Mich., under bark of *Vitis,* June 24, 1977. *Food:* mealybugs and associated materials, under loose bark of grape canes.

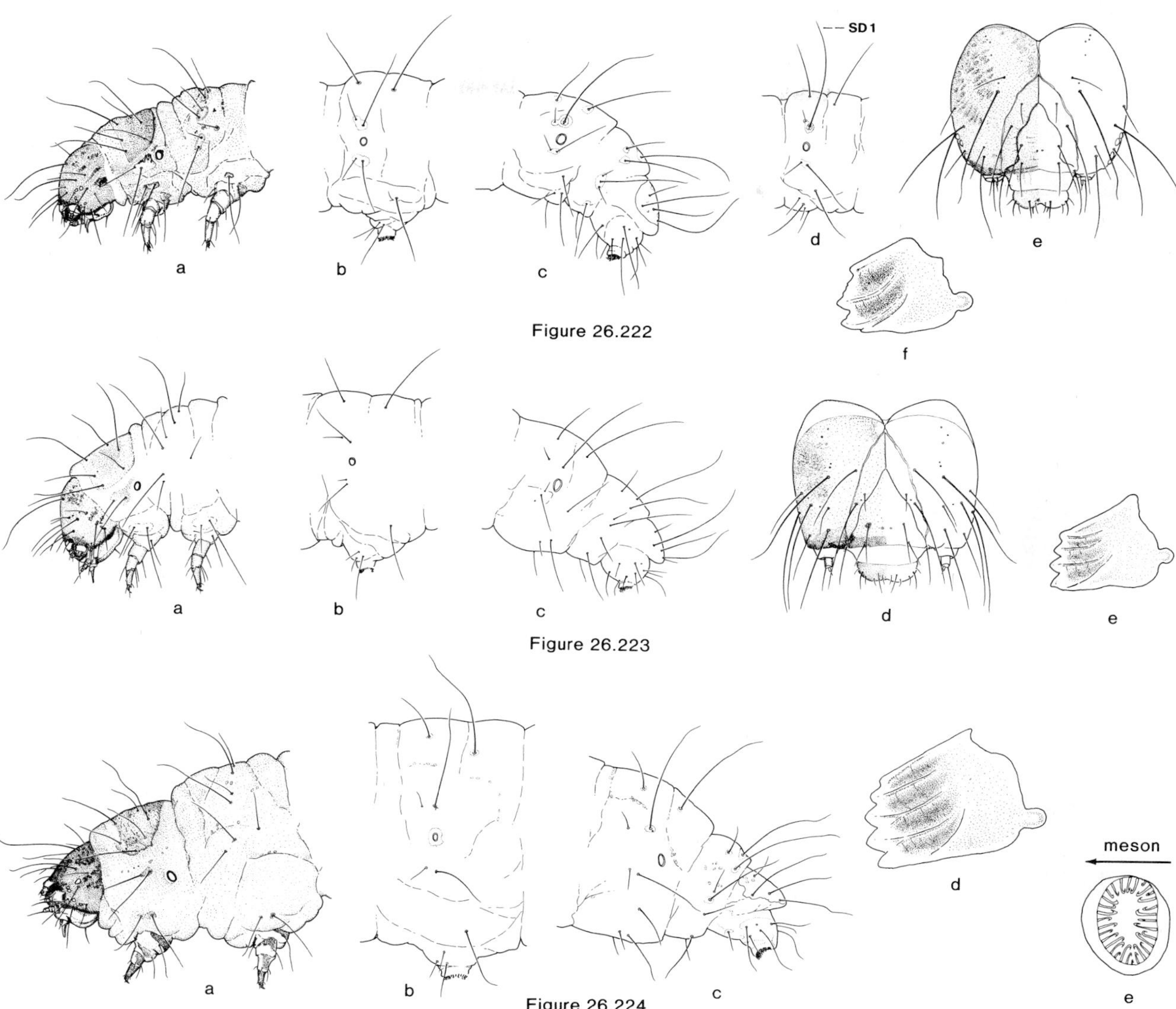

Figure 26.222

Figure 26.223

Figure 26.224

Figure 26.222a-f. Pyralidae, Galleriinae. *Paralipsa gularis* (Zeller). Stored nut moth. f.g.l. 14± mm. a. head, T1, T2; b. A3; c. A8-10; d. A1 (note pigmented ring at base of SD1); e. head; f. mesal view right mandible. Head yellowish brown to reddish brown with indistinct platelets, and a dark reddish brown spot latero-posteriorly (head mostly dark brown to black in living larva). T1 shield mostly reddish brown, slightly paler anteriorly (mostly dark brown to black in living larva). Most of T2, T3 and abdomen yellowish white with minute granules of integument pale brown (most of body sometimes appearing green in living larva). SV unisetose on T2 and T3. Pinacula pale brown to brown. SD1 of A1 and A8 with pale brown to brown pinacula rings with white centers. A3-6 crochets in bi-triordinal ellipses. Hawaii, in dry persimmon from Japan, June 10, 1930. *Food:* almonds, walnuts, soybeans, flax seed, dried fruit, and related materials. As common name implies, larvae are pests of stored foods.

Figure 26.223a-e. Pyralidae, Galleriinae. *Achroia grisella* (Fabricius). Lesser wax moth. f.g.l. 15± mm. a. head, T1, T2; b. A3; c. A8-10; d. head; e. mesal view right mandible. Head brownish yellow becoming reddish brown near mouthparts and with pale brown indistinct platelets. Stemmata absent, or a few rudimentary stemmata present. Mandibles protuberant immediately distad of condyle. T1 shield brownish yellow. T2, T3 and abdomen mostly brownish white or pink (mostly grayish or purplish white with a dark middorsal longitudinal line in living larva). No apparent pinacula. SD1 of A1 and A8 usually with neural connection at base of seta but pinacula rings very weak to not visible. A3-6 crochets mostly biordinal in ellipses. Clemson, S.C., honeybee combs, August 1935. *Food:* wax in beehives, prefering old rather than fresh wax. Also feeds on dead insects, and dried fruit.

Figure 26.224a-e. Pyralidae, Galleriinae. *Galleria mellonella* (L.). Greater wax moth. f.g.l. 28± mm. a. head, T1, T2; b. A3; c. A8-10; d. mesal view right mandible; e. A3 proleg and crochets. Head yellowish to reddish brown with slightly darker platelets and dark reddish brown near mouthparts. Some stemmata fused and some absent. T1 shield mostly yellowish to reddish brown, paler anteriorly. Most of T2, T3 and abdomen white with many minute pale brown to brown integument granules dorsally and laterally (most of body pink, yellowish white or gray in living larva). Pinacula relatively obvious on prothorax, weakly developed to absent elsewhere. SD1 of A1 and A8 with very weakly developed pinacula rings; neural connections associated with these setae more obvious than rings. A3-6 crochets mostly biordinal in ellipses. Memphis, Tenn., bee hive, October, 1960. *Food:* wax, and associated materials in bee hives. Feeds on honeycomb, prefering old combs, which it riddles with silk-lined burrows.

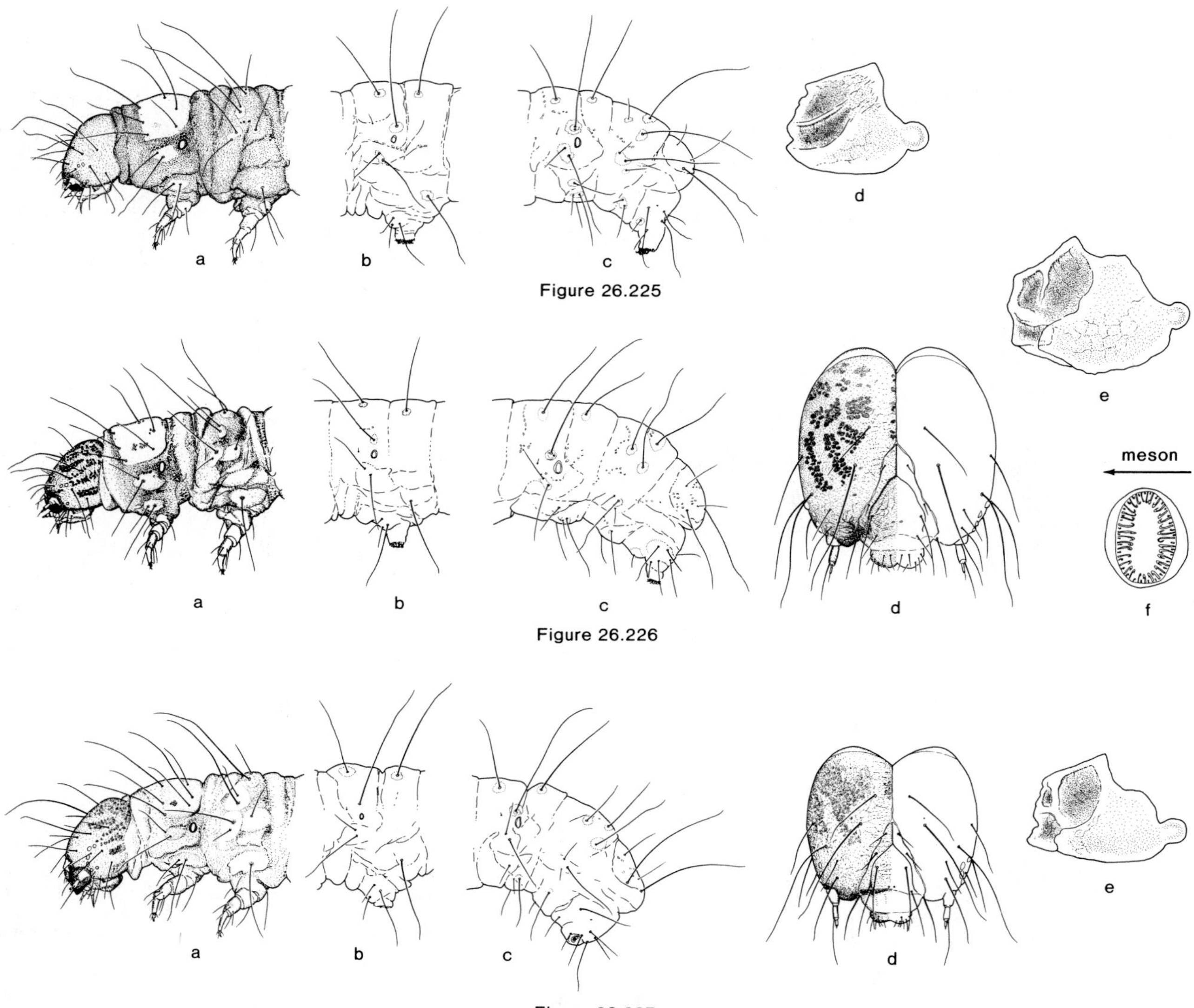

Figure 26.225a-d. Pyralidae, Pyralinae. *Aglossa caprealis* (Hübner). Murky meal caterpillar. f.g.l. 20± mm. a. head, T1, T2; b. A3; c. A8-10; d. mesal view right mandible. Head reddish brown (dark chestnut red in living larva). Mandibles with distal teeth not well developed, and with faint reticulated pattern on lower mesal surface. T1 shield yellowish brown. Most of T2, T3 and abdomen darkened with brown to dark brown integument granules; folding or wrinkling of integument distinct, most obvious at slender anterior body segments, and emphasized by darker pigmentation in the folds (living larva mostly bronzy black, slightly tinged with green). Pinacula yellowish brown. SD1 of A8 with yellowish brown pinaculum ring with pale center. A3-6 crochets biordinal in ellipses (very small crochets alternate with much longer ones, giving the appearance of uniordinal). Pleasant Garden, N.C., in feed room, October 28, 1962. *Food:* refuse in grain warehouse and similar structures. Inhabits silken, refuse-covered shelters in and under food. Also occurs in rodent nests.

Figure 26.226a-f. Pyralidae, Pyralinae. *Herculia sordidalis* Barnes and McDunnough. f.g.l. 20± mm. a head, T1, T2; b. A3; c. A8-10; d. head; e. mesal view right mandible; f. A3 proleg and crochets. Head rugose anteriorly becoming relatively smooth posteriorly, yellowish brown usually with distinct reddish brown platelets. Mandibles without well-developed teeth and with faint reticulated pattern on lower mesal surface. T1 shield yellowish brown with small brown platelets. Most of T2, T3 and abdomen yellowish brown with colorless to dark brown integument granules; darker granules mostly in folds or wrinkles of integument, particularly on thorax. Pinacula pale brown. SD1 of A8 with pale brown pinaculum ring with pale center. A3-6 crochets biordinal (long and very short) in ellipses. Lake Alfred, Fla., peanuts, September 12, 1929. *Food:* dried peanut plants and similar materials in the field and in storage.

Figure 26.227a-e. Pyralidae, Pyralinae. *Hypsopygia costalis* (Fabricius). Clover hayworm. f.g.l. 18± mm. a. head, T1, T2; b. A3; c. A8-10; d. head; e. mesal view right mandible. Head rugulose, mostly yellowish brown, sometimes with darker reddish brown platelets. Mandibles without well-developed teeth and with faint reticulated pattern on lower mesal surface. T1 shield mostly yellowish brown. Most of T2, T3 and abdomen yellowish white with pale brown to brown integument granules. Pinacula yellowish white to pale brown, indistinct. SD1 of A8 with yellowish brown indistinct pinaculum ring. A3-6 crochets biordinal (long and very short) in ellipses. Silver Creek, N.Y., refuse in barn, November 14, 1928. *Food:* stored clover and grass hay, and dead herbaceous materials in the field. Sometimes found in bumblebee nests.

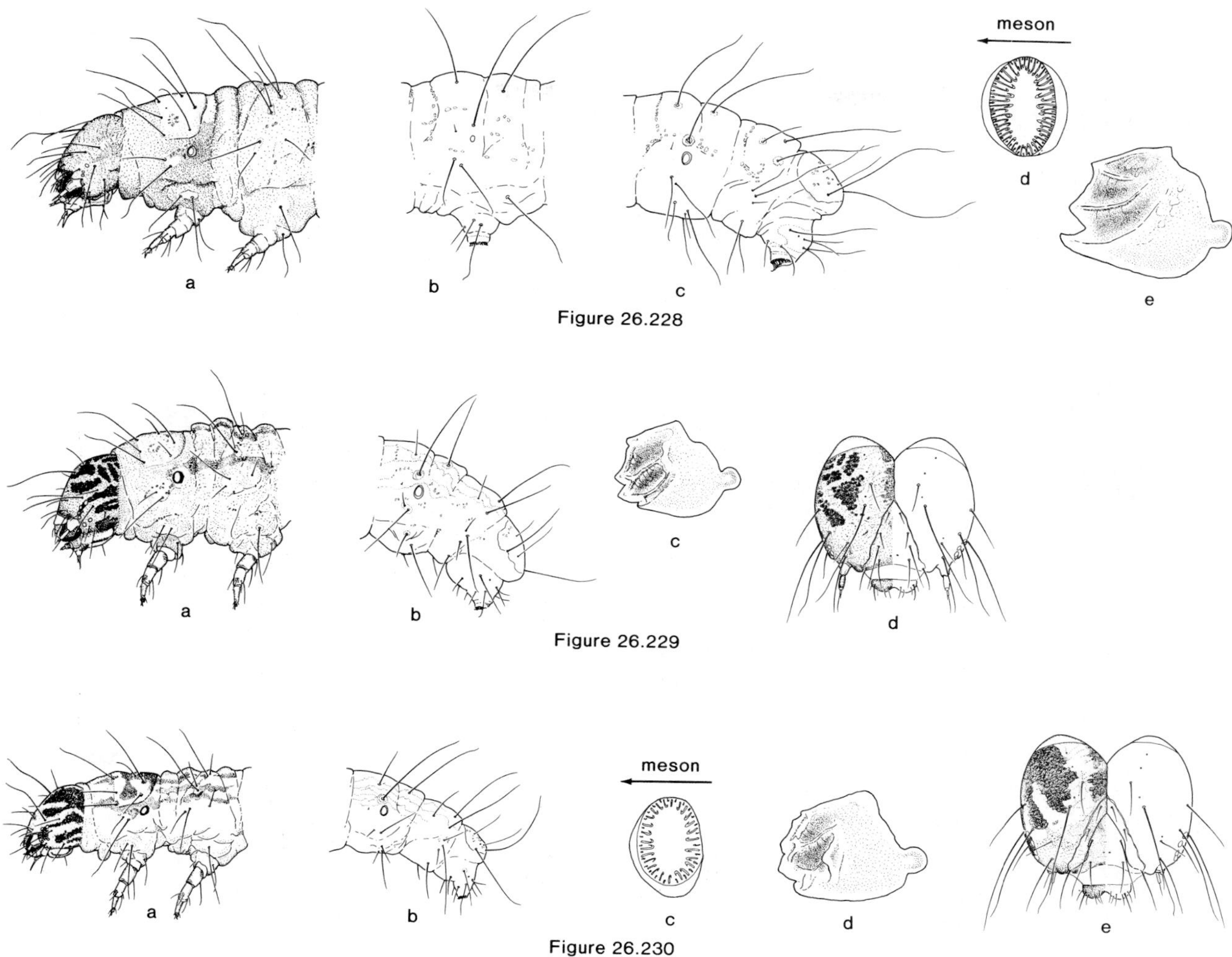

Figure 26.228a-e. Pyralidae, Pyralinae. *Pyralis farinalis* L. Meal moth. f.g.l. 18-22 mm. a. head, T1, T2; b. A3; c. A8-10; d. A3 proleg and crochets; e. mesal view right mandible. Head rugulose, yellowish to reddish brown. Only 3-4 stemmata on each side of head (usually 3 distinct and 1 weakly developed). Mandibles with faint reticulated pattern on mesal surface. T1 shield mostly yellowish brown. T2, T3 and abdomen mostly ochreous white with a dusky tinge on the thorax and caudal abdominal segments because of brown to dark brown integument granules anteriorly and posteriorly. Pinacula yellowish white, indistinct. SD1 of A8 with pale brown indistinct pinaculum ring. A3-6 crochets biordinal (long and short) in ellipses. Columbus, Ohio, chicken mash, April 3, 1958. *Food:* stored cereals and cereal refuse. Inhabits a relatively long silken tube attached to substrate. Found mostly in grain warehouses, mills, and barns. Prefers damp or moldy grain.

Figure 26.229a-d. Pyralidae, Epipaschiinae. *Tetralopha robustella* Zeller. Pine webworm. f.g.l. 18-20 mm. a. head, T1, T2; b. A8-10; c. mesal view right mandible; d. head. Head yellowish to brownish white with distinct reddish brown to black platelets. A narrow band of dark brown to black usually associated with stemmata. T1 shield without dark pigmentation, yellowish to brownish white (with green undertones in living larva) and with small pale brown platelets. Most of T2, T3, and abdomen yellowish white with pale brown to brown integument granules; darker granules most apparent in form of longitudinal stripes (living larva tan to greenish or pinkish tan, with darker brown to black stripes dorsally (minute pigmented granules of integument contributing to the brown or black color and producing distinct dark stripes particularly along the D setae on the abdomen)). Pinacula yellowish white to black. SD1 of A8 with brown to black pinaculum ring with pale center. A3-6 crochets biordinal in ellipses. Falls Church, Va., pine, October 8, 1915. *Food:* pine needles. Forms a conspicuous, more or less oval, frass and silk structure on pine shoot at base of fascicles. Several larvae usually in a single frass structure.

Figure 26.230a-e. Pyralidae, Epipaschiinae. *Tetralopha scortealis* (Lederer). Lespedeza webworm. f.g.l. 17± mm. a. head, T1, T2; b. A8-10; c. A3 proleg and crochets; d. mesal view right mandible; e. head. Head white with dark reddish brown to black platelets. Upper subdorsal groups of platelets suffused with dark pigmentation forming large dark patch. Adfrontals usually brown. Dark brown to black patch associated with stemmata. Mandibles with transverse retinaculum. T1 shield white with large pale brown to black patches. Most of T2, T3 and abdomen yellowish white with brown to pink stripes or patches and pale brown to brown integument granules; darker granules most apparent on thorax (living larva mostly yellowish to greenish white with brown, reddish brown or purplish brown stripes dorsally). Pinacula yellowish white to dark brown. SD1 of A8 with brown to black pinaculum ring with pale center. A3-6 crochets biordinal in ellipses. Spring Grove, Va., lespedeza, July 13, 1944. *Food: Lespedeza* spp. and possibly other Fabaceae. Lives in silk, frass, and mostly dead leaflet enclosures attached to stem of host.

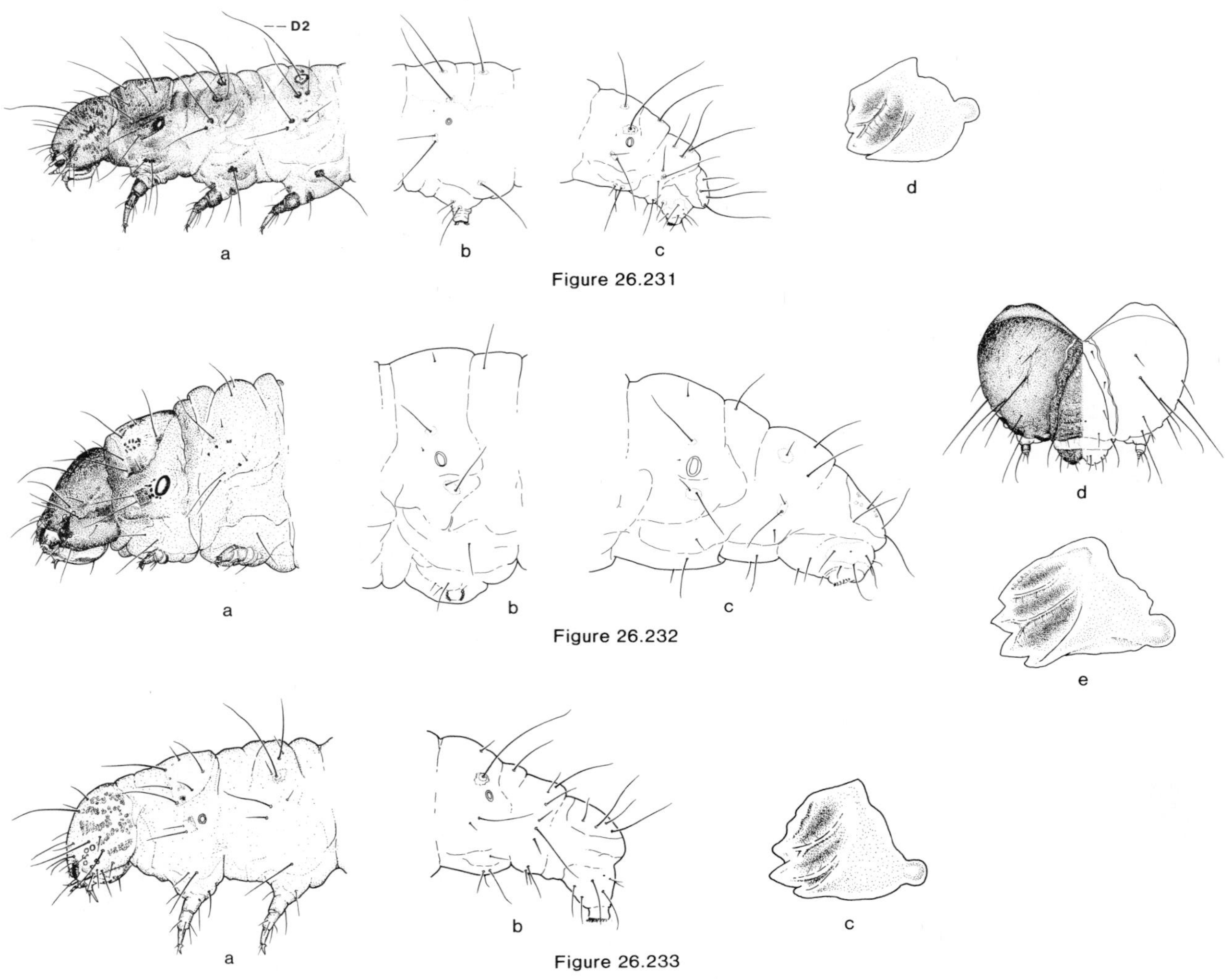

Figure 26.231a-d. Pyralidae, Chrysauginae. *Galasa nigrinodis* (Zeller). f.g.l. 18± mm. a. head, T1, T2, T3 (note pigmented ring at base of D2 on T3); b. A3; c. A8-10; d. mesal view right mandible. Head reddish brown with brown to dark brown platelets. Dark brown within arc of stemmata. T1 shield rugulose (with minute ridges), reddish brown to black. Prespiracular plate relatively large, elongate, concolorous with shield. Most of T2, T3, and abdomen darkened with brown to dark brown integument granules and purplish brown pigmentation; granules particularly dark and evident on T1 and T2, and usually forming on thorax a vague longitudinal stripe along dorsomeson, and sometimes indistinct longitudinal stripes or blotches on remainder of thorax (living larva with dorsum of T2 almost black and the rest of the thorax and abdomen purplish brown). Most pinacula pale brown, and narrowly dark brown to black near base of seta; those on thorax darker. D2 of T3 and SD1 of A8 with dark brown to black pinacula rings with pale centers. A3-6 crochets biordinal in ellipses. Roanoke Rapids, N.C., boxwood, April 19, 1974. *Hosts:* boxwood and possibly other woody plants. Forms concealed, loose, silk, frass and leaf enclosures on host. Apparently feeds only on dead leaves.

Figure 26.232a-e. Pyralidae, Chrysauginae. *Clydonopteron tecomae* Riley. f.g.l. 17± mm. a. head,T1, T2; b. A3; c. A8-10; d. head; e. mesal view right mandible. Head somewhat uneven (with the P setae arising from shallow depressions) and mostly reddish brown, usually with darker adfrontals and frontoclypeus. T1 shield yellowish brown suffused with varying amounts of reddish brown and with numerous, small, dark reddish brown platelets. Most of T2, T3 and abdomen yellowish white with hyaline to dark brown integument granules; darker granules concentrated on the thorax, particularly T1. Pinacula weakly developed laterally on thorax, more or less concolorous with integument; no pinacula or pinacula rings evident elsewhere. Prolegs rudimentary. A3-6 crochets few, uniordinal, in incomplete ellipses open laterally and mesally (transverse bands). Fayetteville, N.C., trumpet vine seed capsules, April 7, 1975. *Host:* apparently restricted to trumpet vine. Feeds on seeds within mature seed capsules.

Figure 26.233a-c. Pyralidae, Phycitinae. *Acrobasis vaccinii* Riley. Cranberry fruitworm. f.g.l. 12-16 mm. a. head, T1, T2; b. A8-10; c. mesal view right mandible. Head mostly yellowish white with pale brownish yellow platelets. T1 shield yellowish white, usually with a pale brown to brown spot posterior to SD1. T2, T3 and abdomen mostly yellowish white (living larva with dorsum of T2, T3 and abdomen yellowish green to reddish yellow; venter green to yellow green). Pinacula indistinct. SD1 of T2 and A8 with very faint brownish yellow pinacula rings. Spiracles on A8 slightly directed dorsocaudad. A3-6 crochets biordinal in broad ellipses. Magnolia, N.C., in blueberry fruit, May 28, 1968. *Hosts:* blueberry, cranberry and huckleberry. Feeds within developing fruit.

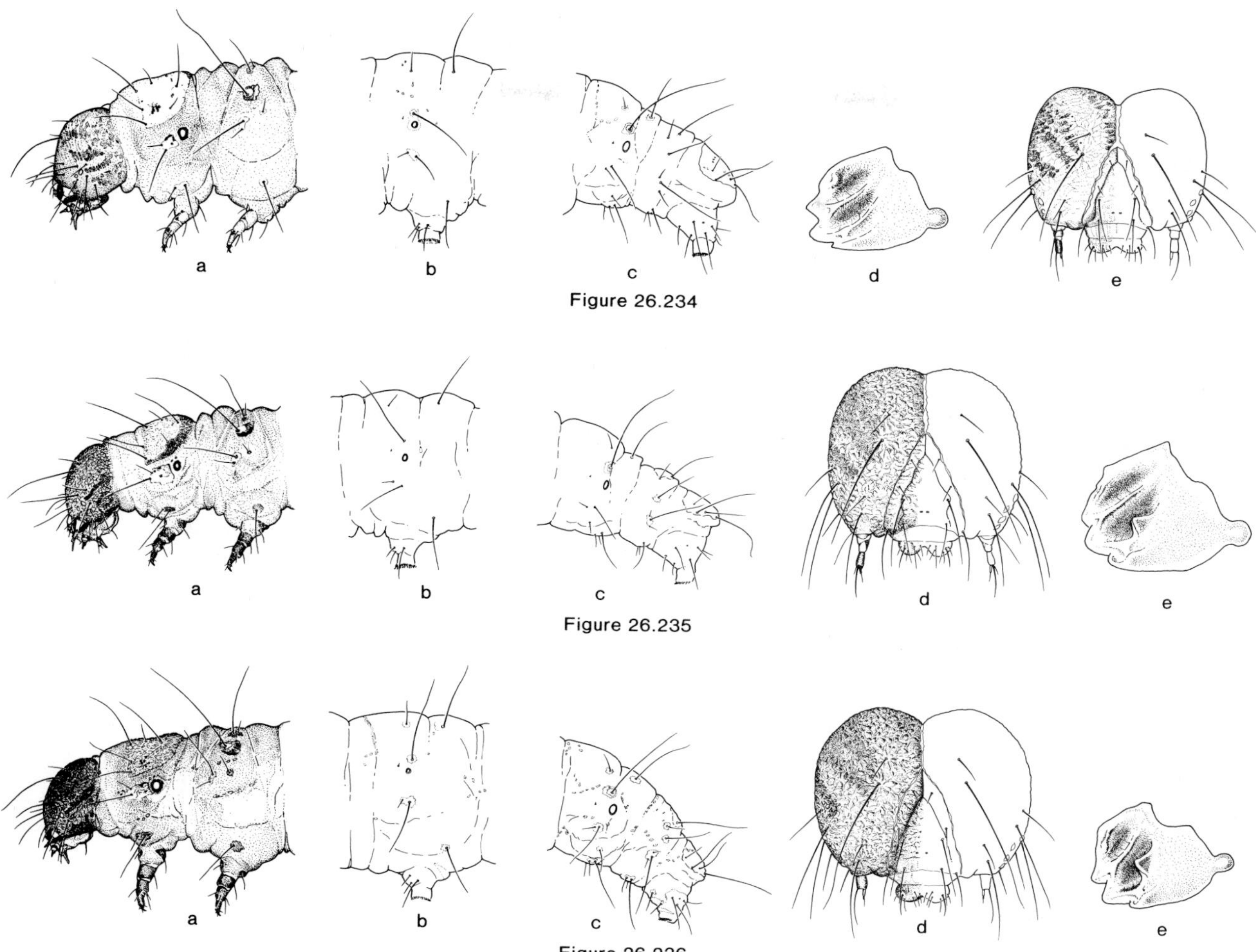

Figure 26.234

Figure 26.235

Figure 26.236

Figure 26.234a-e. Pyralidae, Phycitinae. *Acrobasis nuxvorella* Neunzig. Pecan nut casebearer. f.g.l. 10-17 mm. a. head, T1, T2; b. A3; c. A8-10; d. mesal view right mandible; e. head. Head yellowish brown with indistinct to distinct brown to reddish brown platelets. T1 shield mostly whitish yellow (usually distinctly paler than head), with pale brown to brown platelets and sometimes additional darker maculation. Most of T2, T3 and abdomen white with pale brown to brown integument granules (darker granules on thorax) (living larva with dorsum of body pale purplish brown to purplish brown with green undertones and with venter usually pale green). Pinacula brownish white. SD1 of T2 and A8 with pinacula rings; ring on T2 mostly brown with white center and usually pale yellow posteriorly. A3-6 crochets biordinal in broad ellipses. Bryan, Tex., in pecan nut, May 2, 1961. *Host:* pecan. Overwintering generation feeds early in spring in elongating shoots, and subsequent generations damage nuts.

Figure 26.235a-e. Pyralidae, Phycitinae. *Acrobasis indigenella* (Zeller). Leaf crumpler f.g.l. 14-18 mm. a. head, T1, T2; b. A3; c. A8-10; d. head; e. mesal view right mandible. Head distinctly reticulate rugose, mostly reddish or yellowish brown, usually with indistinct darker reddish brown platelets. Mandibles with large dentiform retinaculum. T1 shield distinctly and evenly curved from anterior to posterior margin, with minute reticulations, yellowish brown with broad area of brown to dark brown along lateral and posterior margins and usually small brown or dark brown platelets near meson. Most of T2, T3 and abdomen white to yellowish white with pale brown to brown integument granules; darker granules particularly evident on T1 (living larva mostly grayish green with varying amounts of purple overtones). Pinacula relatively indistinct, pale brown, usually with paler margins. SD1 of T2 and A8 with brown to dark brown pinacula rings; ring on T2 usually extending posterodorsally. A3-6 crochets biordinal in broad ellipses. Colchester, Conn. on apple, June 12, 1969. *Hosts:* apple, quince, cherry, plum, *Cotoneaster, Pyracantha,* hawthorn, and loquat (*Eribotrya japonica*). Feeds primarily on leaves and constructs sinuous 30-40 mm. long silk and frass tubes.

Figure 26.236a-e. Pyralidae, Phycitinae. *Acrobasis betulella* Hulst. Birch tubemaker. f.g.l. 16-23 mm. a. head, T1, T2; b. A3; c. A8-10; d. head; e. mesal view right mandible. Head reticulate rugose, reddish brown to dark reddish brown usually with indistinct darker brown platelets. Mandibles with large dentiform retinaculum. T1 shield yellowish brown to dark reddish brown with minute reticulations, and usually with transverse gibbosities near posterior margin. Most of T2, T3 and abdomen darkened with brown to dark brown integument granules; granules darkest and most evident on thorax (living larva with dorsum mostly dark purple or greenish purple). Pinacula mostly brown. SD1 of T2 and A8 with brown to dark brown pinacula rings with pale centers. A3-6 crochets biordinal in broad ellipses. Rutland, Mass. on birch, June 26, 1967. *Hosts:* birch, (*Betula populifolia, Betula papyrifera* and possibly other species of *Betula* in Canada and the northern United States). Large larvae occur within silk and frass tubes in a shelter composed of several host leaves. Leaves of, or near, the shelter consumed. An ovoid pupal chamber of silk and frass is made on the host.

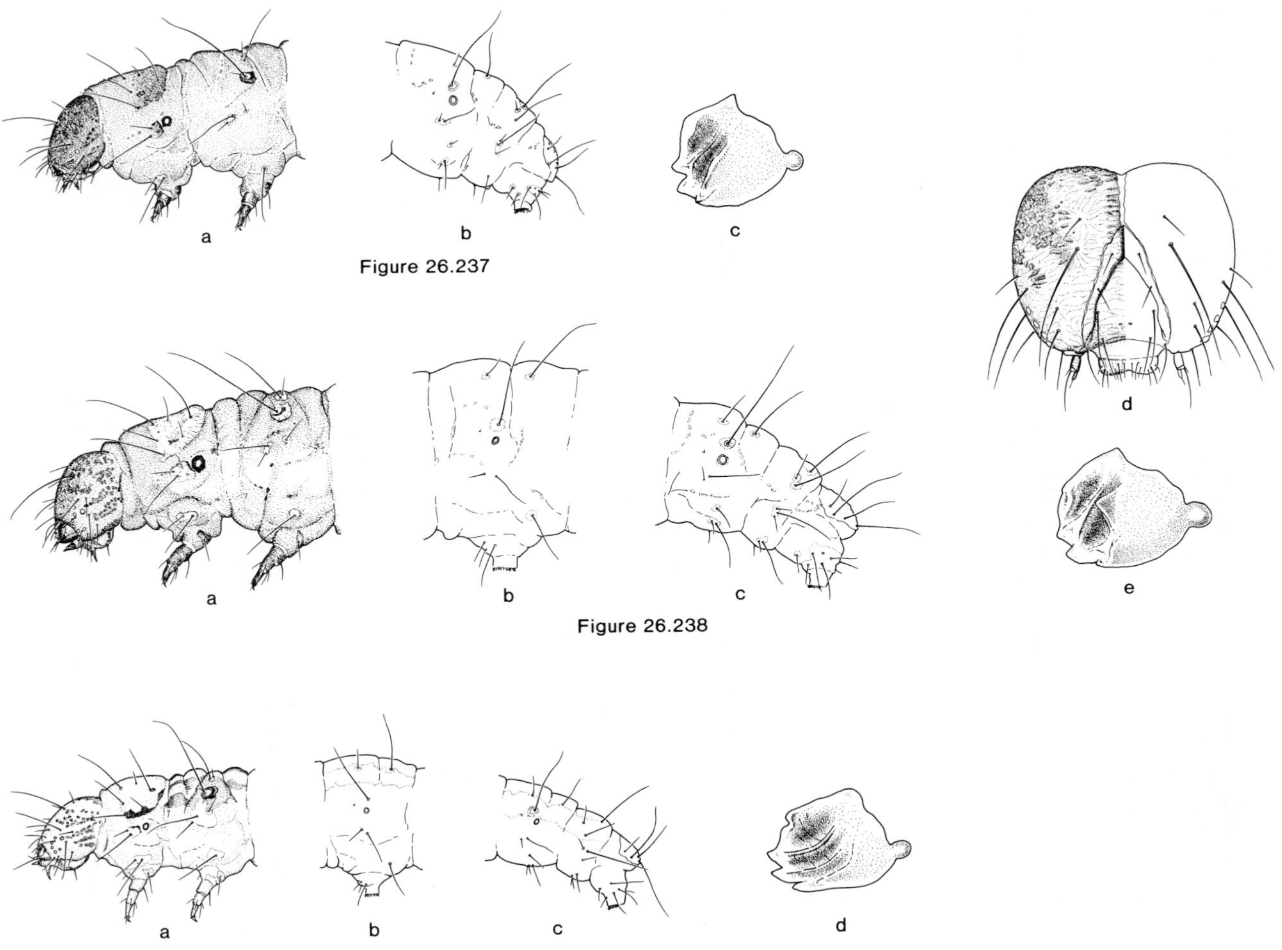

Figure 26.237a-c. Pyralidae, Phycitinae. *Acrobasis juglandis* (LeBaron). Pecan leaf casebearer. f.g.l. 12–18 mm. a. head, T1, T2; b. A8–10; c. mesal view right mandible. Head reddish brown to dark reddish brown (in living larva almost black), usually with indistinct darker platelets. T1 shield brownish yellow, sometimes brown to dark brown posteriorly and laterally, and with small brown platelets. Most of T2, T3 and abdomen darkened with brown to dark brown integument granules (living larva with dorsum of thorax blackish green and dorsum of abdomen olive green with very pale purple overtones). Pinacula mostly pale brown. SD1 of T2 and A8 with brown pinacula rings with pale centers. A3-6 crochets biordinal in broad ellipses. Albany, N.Y., black walnut, June 23, 1967. *Hosts:* black walnut, butternut, and pecan. Feeds primarily on foliage. Constructs elongate, more or less straight, rigid, silk and frass tube attached to the rachis and usually concealed by silked-together leaflets.

Figure 26.238a-e. Pyralidae, Phycitinae. *Acrobasis comptoniella* Hulst. Sweetfern leaf casebearer, f.g.l. 16–25 mm. a. head, T1, T2; b. A3; c. A8–10; d. head; e. mesal view right mandible. Head pale yellow to pale reddish brown with brown to reddish brown platelets. Mandibles with large dentiform retinaculum. T1 shield with surface wrinkled, usually brownish yellow with a few brown platelets and with transverse gibbosities near posterior margin. T2, T3 and abdomen darkened with pale brown to dark brown integument granules; granules dark brown over most of body, pale granules forming streaks dorsally and laterally (living larva mostly pale to dark purplish brown with meson and lateral aspects of thorax and anterior abdominal segments tan). Pinacula pale brownish yellow to brownish yellow. SD1 of T2 and A8 with pinacula rings that are usually mostly brownish yellow with a pale center. A3-6 crochets biordinal in broad ellipses. Rapid River, Mich., on sweetfern. July 11, 1950. *Hosts:* sweetfern (*Comptonia perigrina*) and sweet gale (*Myrica gale*). Feeds primarily on leaves and constructs silk and frass tube and relatively large (11–17 mm long) ovoid pupal chamber, usually including silked-together leaves, on the host.

Figure 26.239a-d. Pyralidae, Phycitinae. *Anabasis ochrodesma* (Zeller). f.g.l. 8–10 mm. a. head, T1, T2; b. A3; c. A8–10; d. mesal view right mandible. Head yellowish white with pale brown platelets. Mandibles simple, or with a small, extra, indistinct, ventromesal tooth (similar in appearance to distal teeth). T1 shield yellowish white with contrasting brown to dark brown spots at SD1, SD2, and usually at D2 and posterior to D2 on shield margin (spots at SD1 and SD2 frequently characteristically fused together along shield margin). T2, T3 and abdomen mostly yellowish white with pale brown stripes (stripes sometimes difficult to detect in preserved specimens) (living larva with dark green or grayish green stripes, pale green to yellowish green or yellow between stripes, and green to pale green on venter; some larvae with dark purplish stripes, reddish purple between stripes and pink and yellow venter). Pinacula pale brown. SD1 of T2 and A8 with pale brown to brown pinacula rings. A3-6 crochets biordinal in broad ellipses. Jensen Beach, Fla., on candlestick, May 18, 1955. *Hosts: Cassia* spp., including candlestick, emperor's candlestick, or king-of-the-forest (*Cassia alata*) and sicklepod (*Cassia obtusifolia*). Bores into host terminals and also feeds between silked-together leaflets.

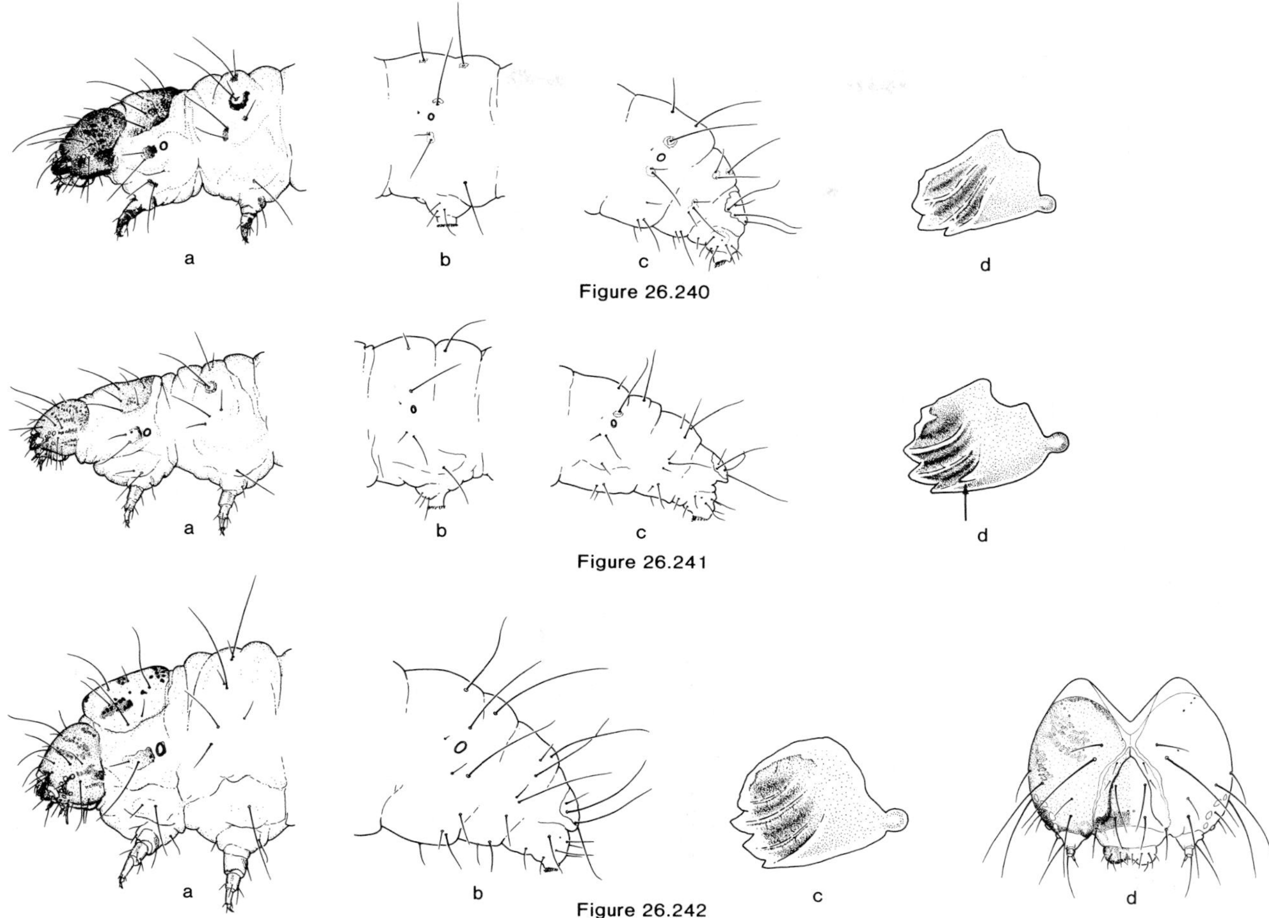

Figure 26.240

Figure 26.241

Figure 26.242

Figure 26.240a-d. Pyralidae, Phycitinae. *Amyelois transitella* (Walker). Naval orangeworm. f.g.l. 11-23 mm. a. head,T1, T2; b. A3; c. A8-10; d. mesal view right mandible. Head pale brown to dark brown with relatively indistinct pale brown, pale orange or black platelets; head frequently also with other indistinct pigmentation overlaying platelets giving it a mottled appearance, and with a distinct dark brown to black streak at posterolateral margin. T1 shield pale brown to brown usually with darker platelets and other dark brown maculation. T2, T3 and abdomen mostly yellowish white (living larva mostly white suffused with pink; pink particularly noticeable at intersegmental folds). Pinacula pale brown to brown. SD1 of T2 and A8 with pale brown to dark brown pinacula rings. SD1 of A1-A7 also usually partially encircled with pale brown to dark brown ring fragments (appearing as small, faint, brown patches at base of SD1 in many preserved specimens). A3-6 crochets biordinal in ellipses. Riverside, Calif., in walnuts, October 28, 1945. *Hosts:* fruits of walnut, orange, grapefruit, apple, peach, date, fig, almond, honey locust, and others. Feeds within mature fruit of host, preferring injured or diseased fruit.

Figure 26.241a-d. Pyralidae, Phycitinae. *Fundella pellucens* Zeller. Caribbean pod borer. f.g.l. 9-16 mm. a.head, T1, T2; b. A3; c. A8-10; d. mesal view right mandible (note small "extra" inner tooth at base of tooth 1). Head yellowish to greenish white with pale brown platelets (head pale brown with green undertones at adfrontals and indistinct brown markings in living larva). Mandibles with a small additional tooth (similar to distal teeth) at base of tooth 1. T1 shield mostly yellowish white (pale brown with green or purple undertones in living larva). T2, T3 and abdomen mostly yellowish white (living larva mostly green or green and purple). SD1 of T2 and A8 with pale yellowish brown relatively indistinct pinacula rings. A3-6 crochets biordinal in ellipses. Riviera Beach, Fla., in legumes of *Vigna,* May 12, 1973. *Hosts:* cowpea or black-eyed pea, crab's eyes *(Abrus),* pigeon pea *(Cajanus),* lima bean, swordbean (*Canavalia* spp.), and other Fabaceae. Large larva usually found within seed pods of host. Smaller larvae sometimes feed on flowers or flower buds.

Figure 26.242a-d. Pyralidae, Phycitinae. *Etiella zinckenella* (Treitschke). Limabean pod borer. f.g.l. 11-19 mm. a. head, T1, T2 (note horizontal, or approximately horizontal, arrangement of L setae on T1 (prespiracular setae), and lack of SD1 ring on T2); b. A8-10; c. mesal view right mandible; d. head. Head yellowish white with indistinct slightly darker platelets. (Head in living larva usually pale brown, partially transparent with some green showing through integument.) T1 shield yellowish to white with patches of distinct brown tonofibrillary platelets (TP T1C 1, TP T1C 2, TP T1C 3 and TP T1FA 1) (in living larva shield yellowish brown with green to purple undertones and distinct brown to black tonofibrillary platelets. T2, T3 and abdomen mostly yellowish white (living larva mostly pale green with indistinct dark greenish gray or pink stripes, or a more uniform (without discernible stripes) purple or reddish purple on dorsum and greenish white, bluish green or pink or purple on venter). L setae of T1 arranged in a horizontal or approximately horizontal configuration. No SD1 pinacula rings apparent. A3-6 crochets biordinal in broad ellipses. San Juan, Puerto Rico, pigeon pea, October 23, 1939. *Hosts:* lima bean and many other leguminous plants. With most hosts, larvae enter legume to feed on developing seed. With hosts having small legumes, larvae will silk several seed capsules together to form a shelter from which to feed.

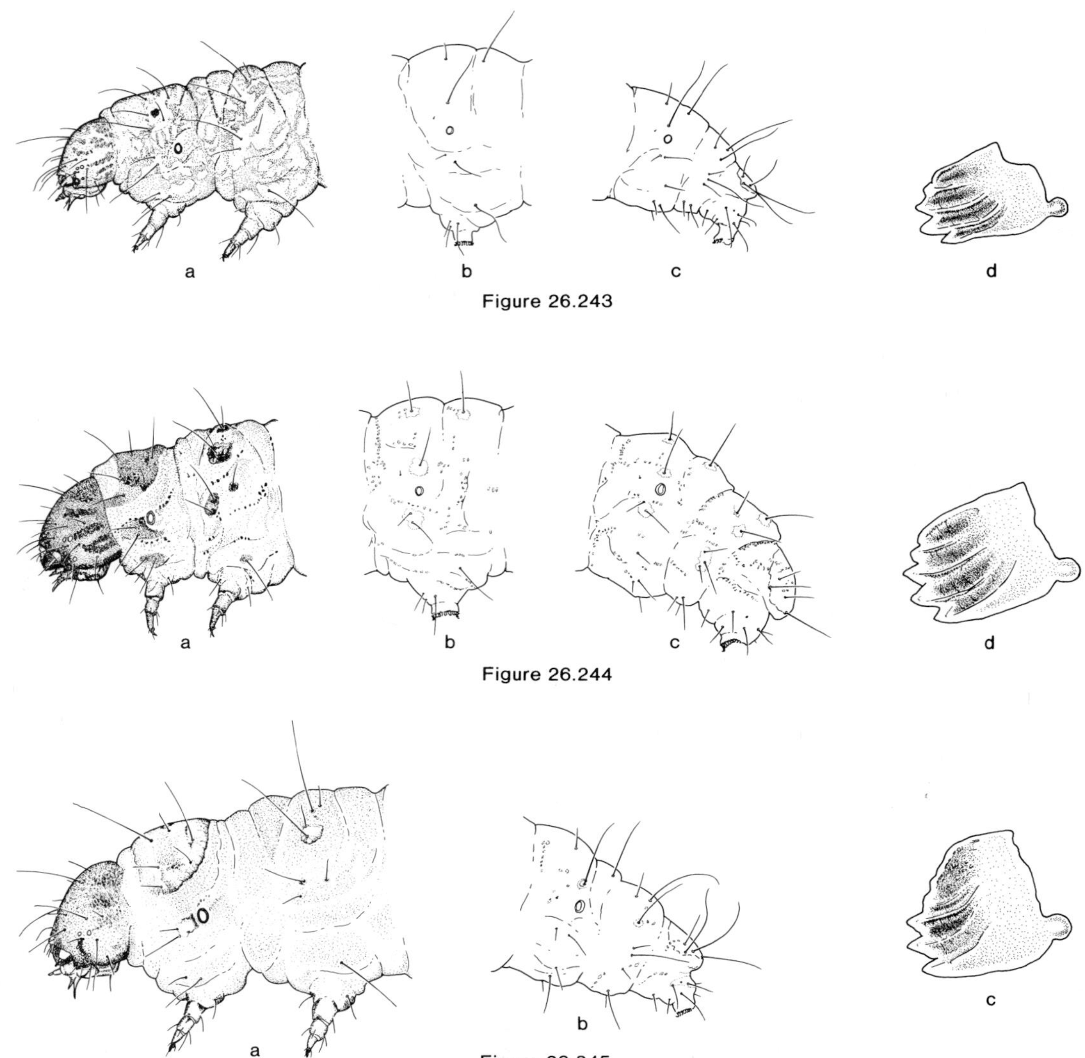

Figure 26.243

Figure 26.244

Figure 26.245

Figure 26.243a-d. Pyralidae, Phycitinae. *Pima albiplagiatella occidentalis* Heinrich. f.g.l. 13-21 mm. a. head, T1, T2; b. A3; c. A8-10; d. mesal view right mandible. Head brownish white with pale brown platelets (in living larva white with relatively indistinct green or greenish yellow platelets). T1 shield yellowish white with TP TsT and TP TTr brown to black and forming a distinct more or less circular spot (most of shield in living larva greenish white to white). T2, T3 and abdomen mostly yellowish white (living larvae mostly pale green, heavily overlaid with chalky white; white spots and patches sometimes forming vague, irregular, longitudinal stripes). Very weak SD1 pinacula rings at T2 and A8 (can't be detected on most specimens). A3-6 crochets biordinal in broad ellipses. Alamogordo, N. Mex., in legumes of locoweed, April 26, 1974. *Hosts:* locoweed (*Astragalus* spp.). Feeding occurs usually within seed pods of host. Most larvae after reaching maximum size do not pupate but form a silk hibernaculum; about 10-11 months later they pupate and emerge as an adult.

Figure 26.244a-d. Pyralidae, Phycitinae. *Dioryctria zimmermani* (Grote). Zimmerman pine moth. f.g.l. 18-25 mm. a. head, T1, T2; b. A3; c. A8-10; d. mesal view right mandible. Head yellowish brown, darker near mouthparts, reddish brown platelets. T1 shield reddish brown with brown platelets. Most of T2, T3 and abdomen white with hyaline to pale brown integument granules (living larva mostly pale gray, greenish or brownish white or pink). Pinacula pale brown to reddish brown. SD1 of T2 with a somewhat irregular, indistinct pinaculum ring; SD1 of A8 without pinaculum ring. Tonofibrillary platelets (muscle attachment markings) of body distinct, dark brown to black. A3-6 crochets biordinal (long ones plus very short, rudimentary ones) in broad ellipses or circles. Oregon, Ill., white pine whorl, July 10, 1950. *Hosts:* pine. Feed mostly within shoots, and cambium of trunks of host.

Figure 26.245a-c. Pyralidae, Phycitinae. *Monoptilota pergratialis* (Hulst). Limabean vine borer. f.g.l. 17-25 mm. a. head, T1, T2; b. A8-10; c. mesal view right mandible. Head pale brown to brown with indistinct reddish brown platelets. T1 shield yellowish brown frequently with brown to dark brown suffusions. T2, T3 and abdomen mostly white or pink with minute, relatively indistinct, pale brown to brown integument granules (living larva mostly blue, usually suffused with purple). SD1 of T2 and A8 with weakly developed, pale brown, pinacula rings. A3-6 crochets biordinal in broad ellipses. Auburn, Ala., in lima bean stem, September 14, 1963. *Hosts:* lima bean is principal host. Very occasionally occurs on other plants such as snap bean or black-eyed pea when these plants are growing adjacent to a heavily infested planting of lima bean. Relatively large, conspicuous galls form on plant where stem is infested.

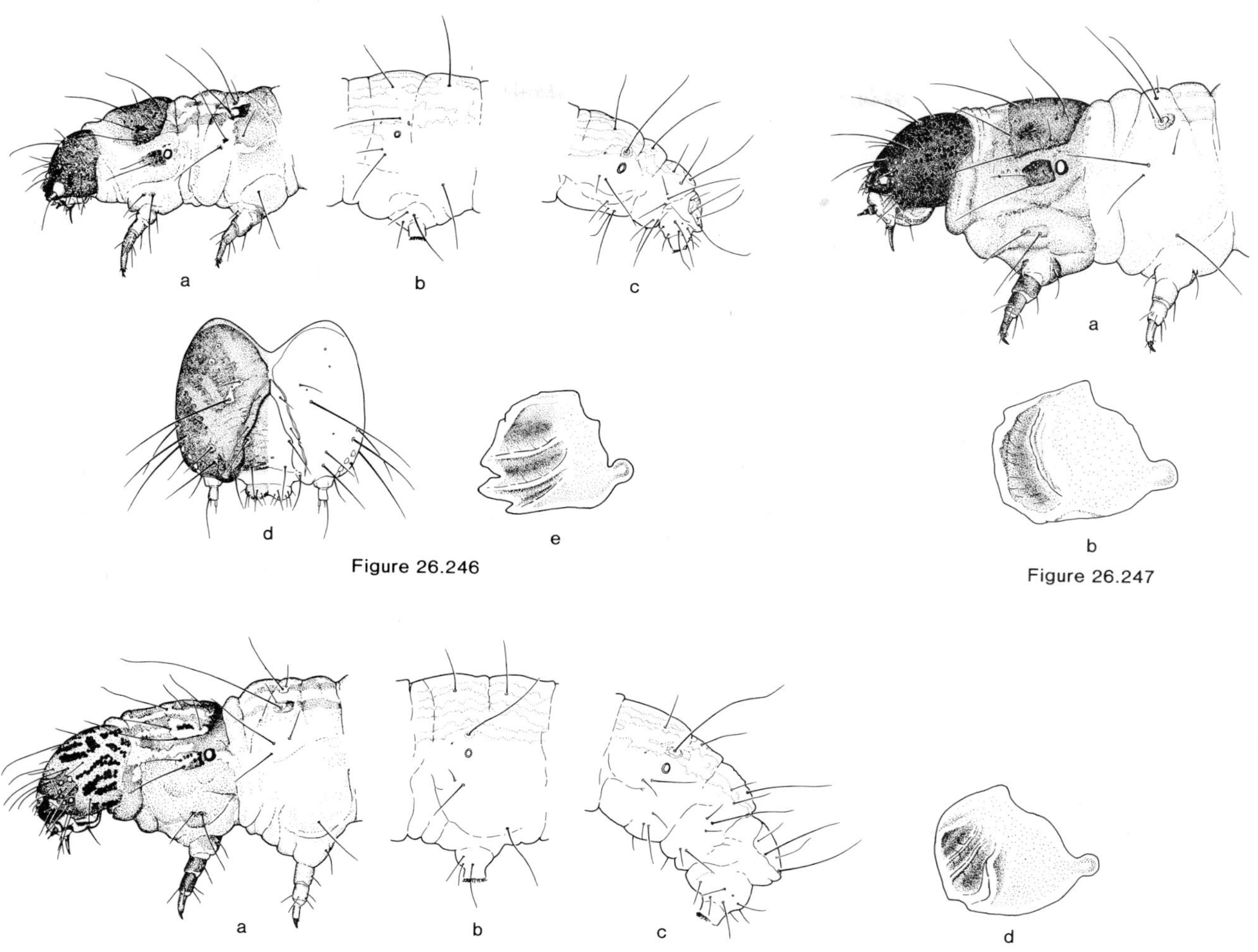

Figure 26.246

Figure 26.247

Figure 26.248

Figure 26.246a-e. Pyralidae, Phycitinae. *Elasmopalpus lignosellus* (Zeller). Lesser cornstalk borer. f.g.l. 10–17 mm. a. head, T1, T2; b. A3; c. A8–10; d. head (note pale areas at base of most setae and pores); e. mesal view right mandible. Head brown to very dark brown (almost black) with indistinct slightly darker platelets. Integument surrounding alveoli rings of many of the setae of head characteristically pale (some of the pale areas as large as, or larger than, the stemmata). T1 shield brown to very dark brown (or black) sometimes with darker platelets. Most of T2, T3 and abdomen white with pale brown streaks and patches (living larva mostly green and white suffused with varying amounts of purple; purple pigmentation usually forming longitudinal, partially fused, stripes and blotches). SD1 of T2 and A8 with dark brown to black pinacula rings (rings on T2 elongate). A3-6 crochets biordinal in ellipses. Charleston, S.C., snap beans, September 14, 1968. *Hosts:* feeds on a large variety of plants including many legumes, however, larvae are most closely associated with plants of the grass family (Poaceae) such as, corn. Larvae feed primarily in host stems at or below surface of soil. Usually tubes of soil, silk and frass are formed by larvae in soil and attached to host near feeding site.

Figure 26.247a,b. Pyralidae, Phycitinae. *Nephopterix subcaesiella* (Clemens). Locust leafroller. f.g.l. 16–24 mm. a. head, T1, T2; b. mesal view right mandible. Head reddish brown, with some darker maculation, to almost black. Mandibles with prominent transverse retinaculum, and with distal teeth and ridges indistinct. T1 shield reddish brown to almost black. Most of remainder of T1 darkened with pale brown, dark brown or gray integument granules, making T1 distinctly darker than rest of body. T2, T3 and abdomen mostly white with vague pale brown, longitudinal stripes (living larva mostly green with darker green longitudinal stripes). SD1 of T2 and A8 with pale brown, relatively inconspicuous, pinacula rings. A3-6 crochets biordinal in broad ellipses. Highlands, N.C., black locust, June 18, 1971. *Hosts:* black locust, other *Robinia* spp., and wisteria. Feeds on foliage, usually concealed between two silked-together leaflets.

Figure 26.248a-d. Pyralidae, Phycitinae. *Nephopterix celtidella* (Hulst). f.g.l. 17–26 mm. a. head, T1, T2; b. A3; c. A8–10; d. mesal view right mandible. Head brownish white to white with brown near stemmata and mouthparts, and with distinct dark brown to black platelets. Mandibles with a retinaculum extending obliquely from anteroventral margin, and with distal teeth and ridges indistinct. T1 shield white to yellowish white with dark brown to black platelets and other dark pigmentation, mostly along posterior margin and between XD2 and SD2. Most of remainder of T1 darkened with brown to dark purple integument granules, making T1 distinctly darker than rest of body. T2, T3 and abdomen mostly white, sometimes with indistinct pale brown stripes (living larva mostly pale green with gray longitudinal stripes). SD1 of T2 and A8 with pale brown to brown, relatively indistinct, pinacula rings (rings on A8 sometimes distinct). A3-6 crochets biordinal in broad ellipses. San Antonio, Tex., hackberry leaves, June 20, 1972. *Host:* hackberry. Feeds on foliage, frequently living as large larva in rolled leaf.

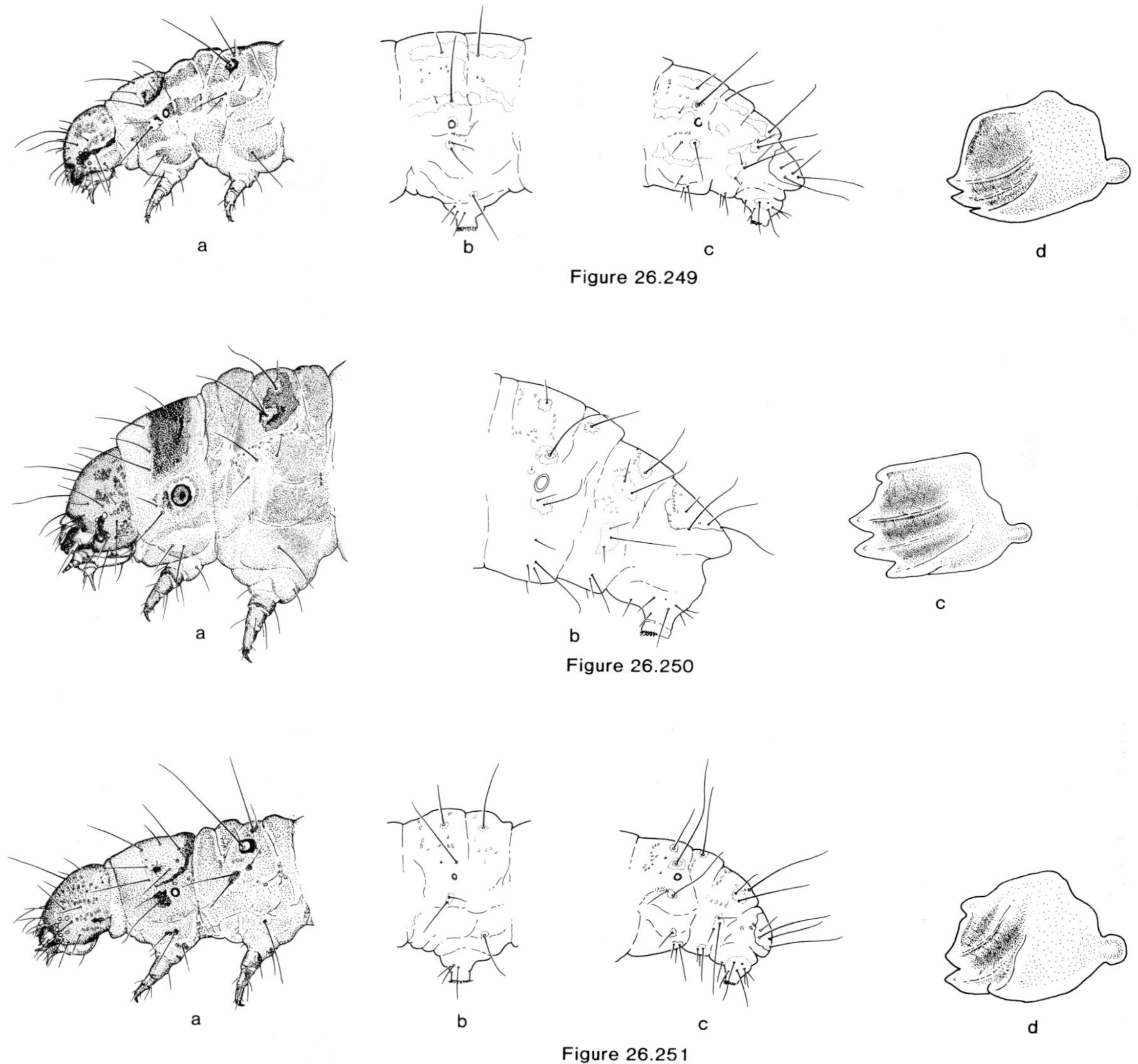

Figure 26.249

Figure 26.250

Figure 26.251

Figure 26.249a-d. Pyralidae, Phycitinae. *Homoeosoma electellum* (Hulst). Sunflower moth. f.g.l. 16–20 mm. a. head, T1, T2; b. A3; c. A8–10; d. mesal view right mandible. Head brownish yellow, darker near mouthparts and usually with dark streak between lateroposterior margin and stemmata; sometimes with indistinct reddish brown platelets. T1 shield mostly brownish yellow with dark brown lateroposterior margin and pale brown to dark brown subdorsal platelet group. Most of T2, T3 and abdomen white with brown to dark brown granules forming patches and longitudinal stripes (living larva mostly yellowish to bluish green with purple or reddish brown patches or longitudinal stripes). SD1 of T2 and A8 with brown to dark brown pinacula rings. A3–6 crochets biordinal in broad ellipses. Kingsbury, Tex., sunflower, June 7, 1972. *Hosts:* sunflower and many other Asteraceae. Feeds mostly on flowers and developing seeds of host.

Figure 26.250a-c. Pyralidae, Phycitinae. *Melitara prodenialis* Walker. Blue cactus caterpillar. f.g.l. 25–38 mm. a. head, T1, T2; b. A8–10; c. mesal view right mandible. Head yellowish brown to dark reddish brown or black near mouthparts and at stemmata and with reddish brown platelets. T1 shield mostly dark brown to black. T2, T3 and abdomen mostly whitish yellow with brown integument granules on dorsum (living larva with T2, T3 and abdomen mostly blue). Spiracles black. SD1 of T2 and A8 with dark brown, broad pinacula rings (ring somewhat diffuse and indistinct on T2). A3–6 crochets biordinal in broad ellipses or circles. Abbeville, Ala., cactus, April 23, 1969. *Hosts:* cactus (*Opuntia compressa, Opuntia drummondi*). Feeds gregariously within flattened stems (cladodes) of host.

Figure 26.251a-d. Pyralidae, Phycitinae. *Moodna ostrinella* (Clemens). f.g.l. 12± mm. a. head, T1, T2; b. A3; c A8–10; d. mesal view right mandible. Head yellowish to reddish brown, darker near mouthparts, with indistinct pale brown platelets, and usually a dark streak lateroposteriorly. T1 shield brownish yellow, usually dark brown posteriorly and with a subdorsal cluster of platelets between XD2-SD1 and SD2. Most of T2, T3, and abdomen darkened with brown to dark brown integument granules (living larva mostly dark purplish brown). SD1 of T2 and A8 with distinct dark brown pinacula rings with pale centers. A3–6 crochets biordinal in broad ellipses or circles. Clayton, N.C., peach mummies, winter 1973. *Hosts:* dried fruits, buds, seeds and similar materials. Found only outdoors. Common in dried clusters of sumac seed.

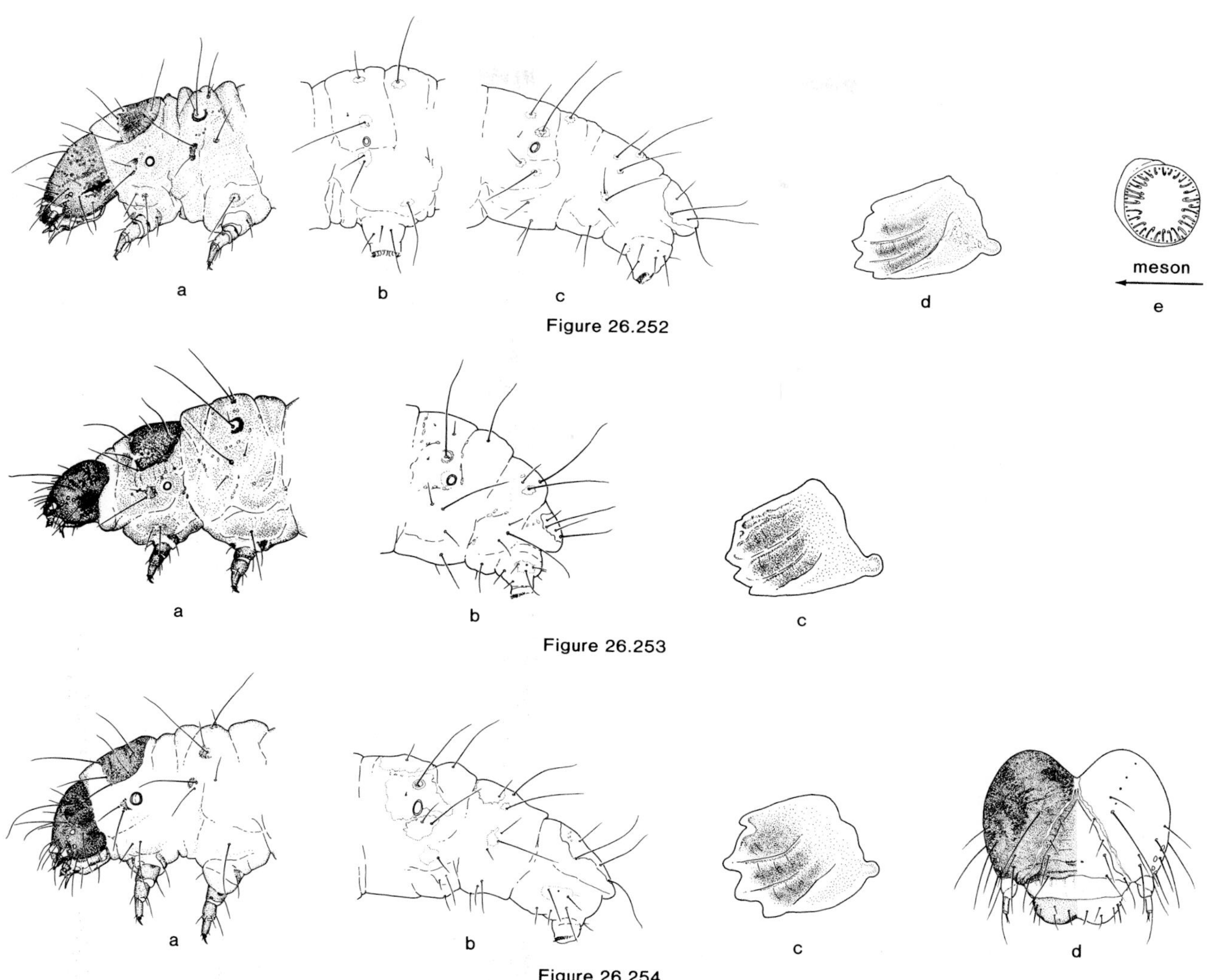

Figure 26.252a-e. Pyralidae, Phycitinae. *Euzophera semifuneralis* (Walker). American plum borer. f.g.l. 18± mm. a. head, T1, T2; b. A3; c. A8-10; d. mesal view right mandible; e. A3 proleg and crochets. Head yellowish brown to reddish brown, darker at adfrontals and frontoclypeus and near mouthparts, with indistinct slightly darker platelets, and a distinct dark brown to black streak lateroposteriorly. Mandibles somewhat elongate. T1 shield pale yellowish brown with dark brown to black platelets and pigmentation. T2, T3 and abdomen mostly white to yellowish white or pink with brown to dark brown integument granules (living larva with body mostly grayish white, pink or brownish green). SD1 of T2 and A8 with distinct brown pinacula rings. SD1 of A1-7 with brown neural connections, with or without distinct rings. A3-6 crochets biordinal to triordinal in broad ellipses or circles. Austin, Tex., girdle wound on pecan tree, July 6, 1939. *Hosts:* plum, walnut, pecan, many other trees, and sometimes stored somewhat woody plant material such as sweet potatoes. Usually associated with damaged or weakened trees or plant material. Larva bores into living tissue usually at a scar, wound or crevice, forming broad, shallow, irregular galleries just beneath surface.

Figure 26.253a-c. Pyralidae, Phycitinae. *Laetilia coccidivora* (Comstock). f.g.l. 12± mm. a. head, T1, T2; b. A8-10; c. mesal view right mandible. Head reddish brown to brown, darker at mouthparts, with dark brown, relatively indistinct platelets and dark brown to black patch at lateroposterior aspect. T1 shield mostly dark brown, sometimes paler toward meson and posterior to XD2. Most of T2, T3, and abdomen darkened with brown to dark brown integument granules. SD1 of T2 and A8 with dark brown pinacula rings. A3-6 crochets bi-triordinal in broad ellipses or circles. Melbourne, Fla., associated with scale insects, August 28, 1956. *Hosts:* scale insects. Usually forms a silk enclosure adjacent to the host insects to protect itself from ants tending the scales.

Figure 26.254a-d. Pyralidae, Phycitinae. *Cactoblastis cactorum* (Berg). Cactus moth. f.g.l. 35± mm. a. head, T1, T2; b. A8-10; c. mesal view right mandible; d. head. Head dark reddish brown to almost black, darker at adfrontals and frontoclypeus and near mouthparts, with darker, relatively indistinct platelets. T1 shield mostly dark reddish brown to almost black with relatively indistinct dark platelets. Mesothorax mostly yellowish white to brownish white; most of T3 and most of abdomen yellowish or brownish white with large brown spots encircling the dorsolateral part of each segment (living larva mostly bright orange or red with large dark spots). SD1 of T2 and A8 with brown to black pinacula rings (ring incomplete on T2 and obscured by large pigmented spot on A8). A3-6 crochets biordinal in broad ellipses or circles. Makapuu Pt., Oahu, Hawaii, in cactus, February 1970. *Host:* cactus (*Opuntia* spp.). Feeds within stems (cladodes) of host. This is the species known for its effectiveness as a biological control agent of cactus, particularly in Australia. Native to Argentina.

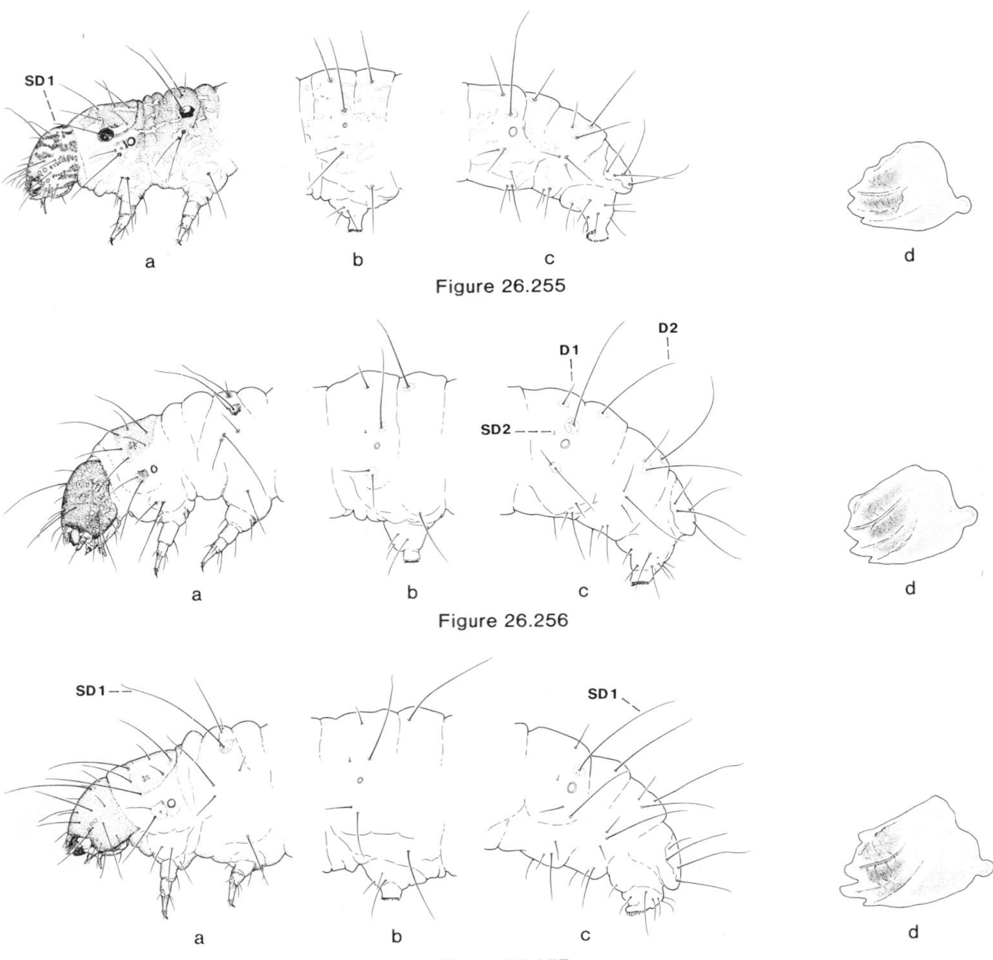

Figure 26.255

Figure 26.256

Figure 26.257

Figure 26.255a-d. Pyralidae, Phycitinae. *Psorosina hammondi* (Riley). Appleleaf skeletonizer. f.g.l. 12± mm. a. head, T1, T2 (note the distinctive, more or less circular, spot at SD1 on the T1 shield); b. A3; c. A8-10; d. mesal view right mandible. Head yellowish white to white with brownish yellow platelets. T1 shield mostly brownish yellow with paler streaks and spots and a very distinct, more or less circular dark brown to black spot at SD1. T2, T3 and abdomen mostly pale brown to pink with pale streaks, spots and tonofibrillary platelets (pale streaks, spots and platelets sometimes difficult to detect in preserved specimens) (living larva mostly greenish brown to reddish brown above with white streaks, spots and platelets; venter pale green to purple). Pinacula dark brown to black with pale centers. SD1 of T2 and A8 with distinct dark brown to black pinacula rings with pale centers. A3-6 crochets biordinal in broad ellipses or circles. Elizabethtown, N.C., *Crataegus,* July 3, 1972. *Hosts:* apple, hawthorn, pear, quince, and beach plum. Usually feeds under a very sparse silk covering on upper epidermis and mesophyll, leaving the leaf veins and veinlets.

Figure 26.256a-d. Pyralidae, Phycitinae. *Cadra figulilella* (Gregson). Raisin moth. f.g.l. 12± mm. a. head, T1, T2; b. A3; c. A8-10 (note distance between SD2 and spiracle, size of spiracle, and relative sizes of D1 and D2 on segment 8); d. mesal view right mandible. Head brown with indistinct paler platelets. Prothoracic shield brown, usually with slightly darker platelets. T2, T3 and abdomen mostly white (living larva, at times, somewhat pink). Pinacula pale brown to brown. SD1 of T2 and A8 with distinct brown pinacula rings. A8 with SD2 separated from spiracle by distance equal to, distinctly less, or slightly more than diameter of spiracle, and with D2 3-5.5 × longer than D1. A3-6 crochets biordinal in broad ellipses or circles. Fresno, Calif., raisins, May 1949. *Food:* raisins, carobs *(Ceratonia siliqua)* and similar fruit. According to Aitken (1963) feeds only on fallen fruit or freshly harvested crop.

Figure 26.257a-d. Pyralidae, Phycitinae. *Plodia interpunctella* (Hübner). Indian meal moth. f.g.l. 12± mm. a. head, T1, T2 (note that except for ring around SD1 on T2 and the slightly darker T1 shield and prespiracular plate, no pigmentation is evident around setal bases); b. A3 (note absence of pigmentation around base of setae); c. A8-10 (note that except for SD1 ring on A8, no pigmentation is evident around base of setae); d. mesal view right mandible. Head brownish yellow, darker at adfrontals, frontoclypeus and near mouthparts, and with indistinct platelets. Frontoclypeus extending two-thirds, or more, the distance to epicranial notch. T1 shield brownish yellow with pale brown to brown platelets. T2, T3 and abdomen usually mostly white (living larva sometimes mostly pink). No pinacula evident, except SD1 of T2 and A8 with relatively indistinct brownish yellow pinacula rings. A3-6 crochets biordinal in broad ellipses or circles. Beaufort, N.C., stored corn, December 15, 1971. *Food:* all kinds of grain, including oats, wheat, and corn, dried fruits, nuts, seeds, meal, dried roots and herbs and many other materials of vegetable origin, as well as some animal materials such as dried insects. In addition to feeding damage, larvae contaminate materials with silk and frass.

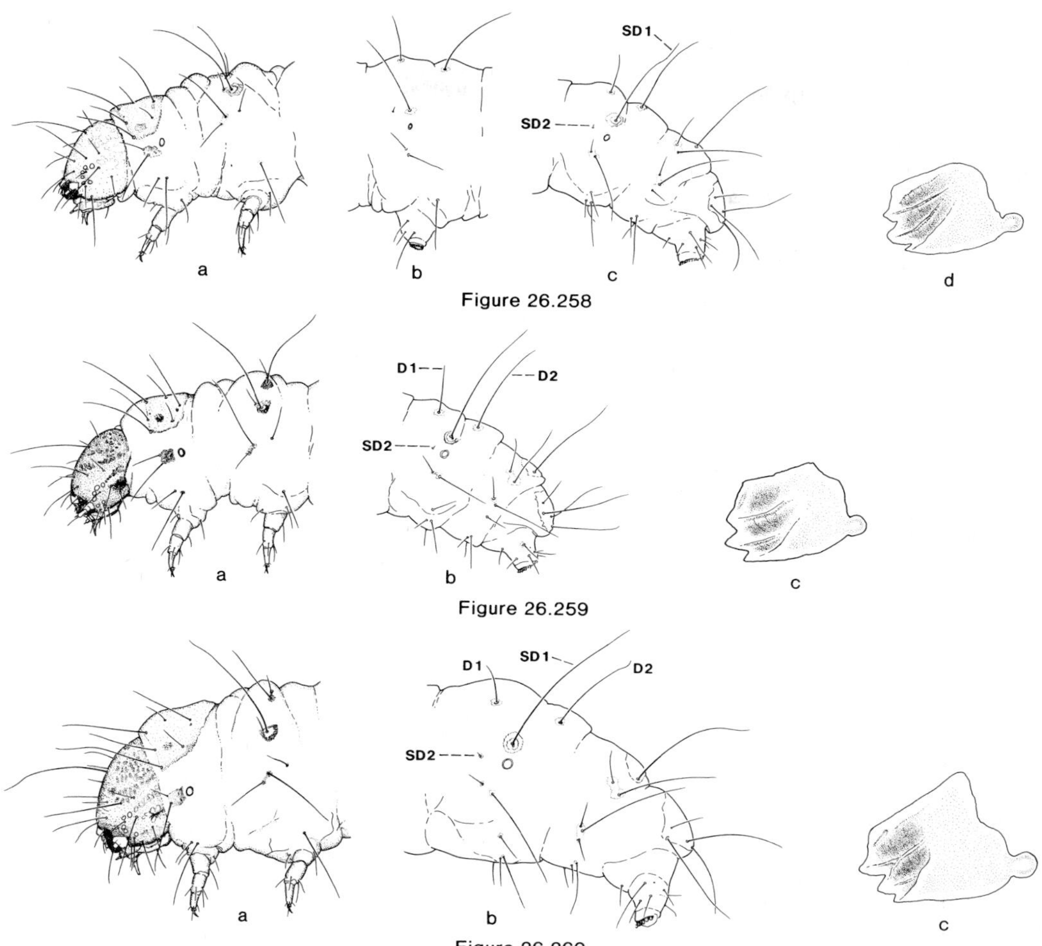

Figure 26.258

Figure 26.259

Figure 26.260

Figure 26.258a-d. Pyralidae, Phycitinae. *Ephestia elutella* (Hübner). Tobacco moth. f.g.l. 12± mm. a. head, T1, T2; b. A3; c. A8-10 (note distance between SD2 and spiracle, size of spiracle, and size of area enclosed by SD1 ring on A8); d. mesal view right mandible. Head pale brownish yellow, slightly darker at adfrontals, frontoclypeus and near mouthparts, with indistinct platelets. T1 shield pale brownish yellow with slightly darker brown platelets. Remainder of T1, T2, T3 and abdomen mostly white (living larvae sometimes pink). Pinacula pale brown to brown. SD1 of T2 and A8 with yellowish brown pinacula rings. A8 with SD2 separated from spiracle by distance equal to 2-3.5 × diameter of spiracle, and with spiracle ⅔ or less as broad as area enclosed by SD1 ring. A3-6 crochets biordinal in broad ellipses or circles. Richmond, VA., stored tobacco, June, 1936. *Food:* tobacco, cacao and chocolate, cereals and cereal products, dried fruit and vegetables, nuts, various seeds, cayenne pepper, other vegetable products and occasionally dried animal materials. In addition to consuming tobacco, etc., larvae contaminate materials with silk and frass.

Figure 26.259a-c. Pyralidae, Phycitinae. *Cadra cautella* (Walker). Almond moth. f.g.l. 12± mm. a. head, T1, T2; b. A8-10 (note distance between SD2 and spiracle, size of spiracle, and relative sizes of D1 and D2 on segment 8); c. mesal view right mandible. Head mostly brownish yellow to yellowish brown with some darker brown or reddish brown pigmentation; most noticeable dark patch is at the posterolateral aspect of the head (associated with TP Man Ab 1). T1 shield yellowish brown with brown to dark brown platelets and maculae. T2, T3 and abdomen mostly white to yellowish white (living larva sometimes pink). Pinacula brown to dark brown. SD1 of T2 and A8 with brown to dark brown pinacula rings. A8 with SD2 separated from spiracle by distance equal to, distinctly less, or slightly more than diameter of spiracle, and with D2 only 2-2.5 × as long as D1. A3-6 crochets biordinal, in broad ellipses or circles. Boiling Springs, N.C., stored cotton seed, September 14, 1971. *Food:* primarily dried fruits and nuts, but also feeds on all kinds of stored vegetable products. Most damage occurs through contamination of stored products with silk and frass.

Figure 26.260a-c. Pyralidae, Phycitinae. *Anagasta kuehniella* (Zeller). Mediterranean flour moth. f.g.l 16± mm. a. head, T1, T2; b. A8-10 (note distance between SD2 and spiracle, size of spiracle, and size of area enclosed by SD1 ring on segment 8); c. mesal view right mandible. Head mostly yellowish brown with dark streaks at posterolateral aspect and with relatively indistinct brown platelets. T1 shield yellowish brown, sometimes with relatively indistinct brown maculation. T2, T3 and abdomen mostly white to yellowish or brownish white (living larvae, at times, somewhat pink). Pinacula pale brown to brown. SD1 of T2 and A8 with brown to dark brown pinacula rings. A8 with SD2 separated from spiracle by distance equal to 2-3.5 × diameter of spiracle, and spiracle as large as, or slightly larger than area enclosed by SD1 ring. SD2 frequently with small pinaculum. A3-6 crochets biordinal in broad ellipses or circles. Hoboken, N.J., in oats from Germany. *Food:* primarily wheat flour, but also other flour, cereals, nuts, other vegetable products and occasionally dried animal materials. In addition to eating flour, etc., larvae contaminate materials with frass and silk.

Selected Bibliography

Allyson 1976, 1977a, 1977b, 1980, 1981a, 1981b, 1984.
Aitken 1963.
Arbogast et al. 1980.
Brigham and Herlong 1982.
Capps 1939, 1967.
Corrette and Neunzig 1979.
Doerksen and Neunzig 1975.
Dyar 1901, 1902.
Ellis 1925.
Farrier and Tauber 1953.
Fiance and Moeller 1977.
Forbes 1910b, 1911b, 1923.
Fracker 1915.
Grimes and Neunzig 1984.
Hasenfuss 1960.
Heinrich 1919, 1921a, 1921b, 1956.
Hinton 1942, 1943.
Holloway et al. 1928.
Kinser and Neunzig 1981.
Lange 1956a, 1956b, 1984.
Liebherr 1977.
Lloyd 1914.
MacKay 1943.
Matheny and Heinrichs 1972.
Mathur 1954.
Mathur and Singh 1956.
Mosher 1916a.
Munroe 1972–1976.
Mutuura 1980.
Neunzig 1972, 1979.
Neunzig et al. 1964.
Neunzig and Merkel 1967.
Peterson 1948, 1961, 1963a.
Ross 1959.
Weisman 1986.

Acknowledgments

Chien C. Chang, Shu-ling Tung, and Lana Tackett assisted with drawing and inking the figures.

HYBLAEIDAE (PYRALOIDEA)

H. H. Neunzig, *North Carolina State University*

The Hyblaeids, Leafrollers

Figure 26.261

Relationships and Diagnosis: Last-stage hyblaeid larvae (based on the single species occurring in the United States) have the L group bisetose on T1 and the crochets of A3–A6 mostly triordinal in a circle or ellipse. Hyblaeids appear most closely allied to the Pyralidae. Superficially, larvae (and adults) resemble many noctuids, but they are easily separated from noctuid larvae by the crochet arrangement.

Hyblaeid and pyralid larvae can be separated in that hyblaeids have SD1 on T3 reduced and much (about five times) smaller than SD1 on T2. In Pyralidae SD1 of T2 and T3 are usually about the same size; sometimes SD1 of T2 is enlarged in the pyralids, resulting in it being distinctly longer than SD1 of T3; however, under these circumstances SD1 of T3 remains relatively long and the difference in size between the two setae is considerably less than five-fold. Also, in hyblaeids D1 and D2 on T2 and T3 are remote with D1 about as close to the dorsomeson as to D2, whereas in pyralids D1 and D2 of T2 and T3 are usually closely associated, many times being on the same pinaculum or fused pinacula. In addition, hyblaeid spiracles appear to be generally larger than the spiracles on most pyralids.

Biology and Ecology: Only a single introduced species, *Hyblaea puera* (Cramer), occurs in the United States, being restricted to Florida. Larval host plants in Florida include sausage tree (*Kigelia pinnata*), calabash tree (*Crescentia cujete*), *Tabebuia avellanedae,* and possibly *Tecoma capensis.* All known hosts in the United States belong to the Bignoniaceae. The larvae form silk tunnels on the host and consume foliage. Pupation occurs in silken cocoons. Adults of *Hyblaea puera* have been collected in Florida every month of the year; apparently, therefore, larvae could be expected year round.

Description: Last stage larvae medium, 25–35 mm. Body more or less cylindrical and robust. Prolegs on A3–A6 and A10 well developed. Head relatively smooth. Integument of body minutely roughened and longitudinally striped. The usual complement of primary setae present (except as noted below). Spiracles large.

Head: Semiprognathous to hypognathous. Six stemmata. Mandibles with inner ridges of tooth 2 and 3 strongly developed and produced into two secondary teeth just basad of the distal teeth. Maxillae with sensilla trichodea simple.

Thorax: T1 with distinct shield, bearing six setae per side; T1 with two L setae on the prespiracular plate. D1 and D2 remote on T2 and T3; D1 on T2 and T3 about as close or closer to dorsomeson than to D2. SD1 and SD2 on fused pinacula on T2, and relatively close but on separate pinacula on T3. SD1 of T3 much smaller than SD1 of T2. L group trisetose on T2 and T3. SV group bisetose on T1 and unisetose on T2 and T3.

Abdomen: D1 and D2 on all segments with D1 more dorsad and cephalad except on A9 where D1 is ventrad of D2. SD1 long; SD2 very small, present on all segments except A9 and A10. On A1–8, L1 and L2 approximate and ventrad of spiracle. On A9, L2 and L3 missing and L1 weakly developed. SV bisetose on A1, trisetose on A2–6. A7 with SV bisetose. A8–9 with SV unisetose. V1 on all segments. Prolegs on segments 3–6 with crochets mostly triordinal and arranged in broad ellipses or circles. A10 with a relatively indistinct anal shield with eight setae. Proleg on A10 with crochets mostly triordinal. Nine setae on or near the anal proleg.

Comments: *Hyblaea puera* is known in the literature (Common 1970) as the teak leaf roller. It appears to be of some economic importance as a defoliator of *Tectona grandis* L. in Java and New Guinea. In North America injury is restricted to a few imported tropical ornamentals.

Selected Bibliography

Common 1970.

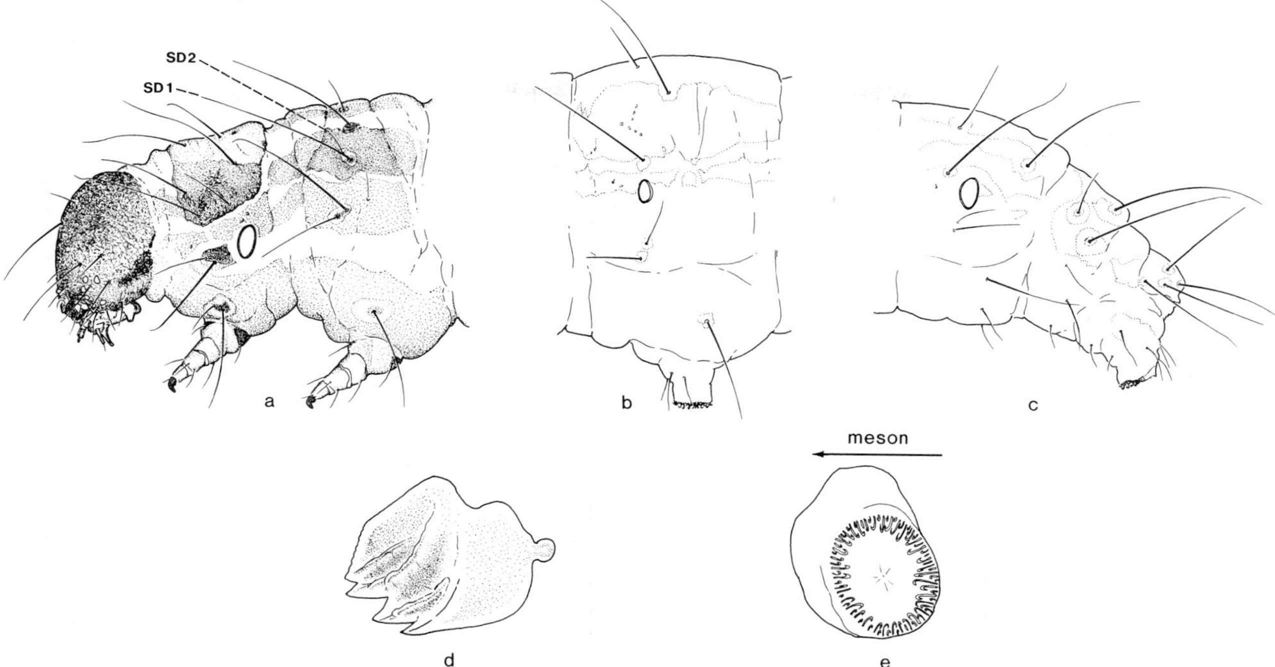

Figure 26.261a-e. Hyblaeidae. *Hyblaea puera* (Cramer). f.g.l. 35± mm. a. head, T1, T2; b. A3; c. A8-10; d. mesal view right mandible; e. A3 proleg and crochets. Head rugulose, usually reddish brown, with indistinct dark reddish brown platelets, or entirely black; head sometimes yellowish brown with darker pigmentation at adfrontals, frontoclypeus, near stemmata and mouthparts, and at posterolateral aspect. Mandibles with inner ridges of tooth 2 and 3 strongly developed and sometimes produced into 2 secondary teeth just basal to distal teeth. T1 shield relatively smooth, usually mostly brown with dark brown or black laterally and mostly brownish yellow to orange dorsally; sometimes entire shield dark except for a few pale spots. T2, T3 and abdomen mostly white with orange, brown or black granules forming broad pigmented stripes dorsally (some larvae without dorsomesal stripes); T2 and T3 sometimes with patches of brown to black granules ventrally. Pinacula hyaline, brown, or dark brown, sometimes very distinct because they are ringed with white. SD1 of T3 greatly reduced, about one-fifth length of SD1 of T2. Spiracles unusually large. A3-6 crochets mostly triordinal in broad ellipses or circles. Miami, Fla., on *Kigelia pinnata*, April 12, 1949. *Hosts:* sausage tree *(Kigelia pinnata)*, calabash tree *(Tabebuia avellanedae)* (Bignoniaceae), and other tropical plants. Feeds on foliage, forming silk enclosures on the host.

THYRIDIDAE (PYRALOIDEA)

H. H. Neunzig, *North Carolina State University*

The Thyridids, Gallworms

Figures 26.262-26.263

Relationships and Diagnosis: A family apparently closely related to the Pyralidae. Last-stage thyridid larvae have the L group bisetose on T1 and the crochets of the ventral prolegs uni- to biordinal in a circle or ellipse.

Thyridid larvae can be easily separated from nearly all other pyraloids in that only two L setae occur on each side of T2 and T3; the majority of pyraloids have three setae on these segments; a few (some nymphulines and schoenobiines (aquatic or semiaquatic species)) have the L setae of T2 and T3 bisetose.

Biology and Ecology: Larvae occur within stems or rolled leaves of hosts. In our area, known host plants include: grape, wild cotton (*Gossypium hirsutum*), waxmallow (*Malaviscus arboreus*), hibiscus (*Hibiscus* spp.), sea grape (*Coccoloba uvifera*), wild dilly (*Manilkara bahamensis*), sapodilla (*Manilkara zapoda*), and white indigo berry (*Randia aculeata*). Many form galls or live within galls on hosts. *Dysodia* has been reported as injurious to beans.

Description: Last-stage larvae 12–24 mm. Body more or less cylindrical, sometimes (*Hexeris*) with thoracic region slightly inflated. Prolegs well developed on A3–6 and A10. Head generally smooth, sometimes inconspicuously pitted (*Dysodia*). The usual complement of primary setae (except as noted previously and in the following description). Spiracles average in size, sometimes (*Hexeris*) those on A8 directed posteriorly.

Head: Semiprognathous to hypognathous. Six stemmata. Mandibles simple. Sensilla trichodea simple.

Thorax: T1 with distinct shield. Six setae on each side of the shield (SD2 usually more anteriorly situated than in other pyraloids; almost in line with XD1, XD2, and SD1). Two L group setae on T1. D1 and D2 present and more or less adjacent on T2 and T3. SD1 and SD2 present and more or less adjacent on T2 and also on T3. L group bisetose on T2 and T3. SV group bisetose on T1 and also on T2 and T3.

Abdomen: D1 and D2 on A1–8 with D1 more dorsad and cephalad. On A9, D1 sometimes ventrad of D2 and sometimes greatly reduced (*Hexeris*). SD1 usually strong, but

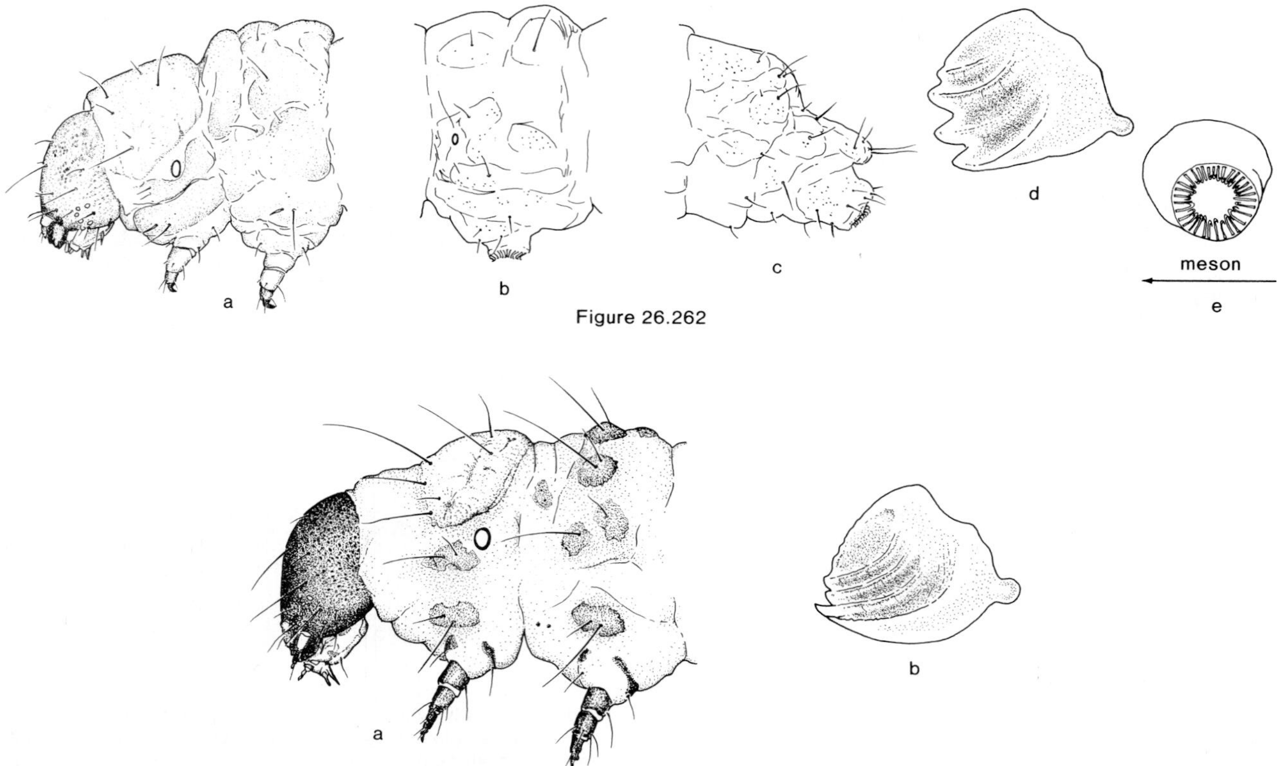

Figure 26.262a-e. Thyrididae. *Hexeris enhydris* Grote. Sea grape gallworm. f.g.l. 20± mm. a. head, T1, T2; b. A3; c. A8-10; d. mesal view right mandible; e. A3 proleg and crochets. Head mostly brownish white, reddish brown near mouthparts, with indistinct pale brown to gray platelets. Thoracic region slightly inflated. T1 shield brownish white, usually indistinctly flecked with small, mostly circular purplish brown hypodermal spots. T2, T3 and abdomen mostly white with brownish white pinacula (pinacula mostly large, elongate and ampullae-like) (living larvae mostly white with pink cast; pinacula shiny pale brown to brown); numerous small, usually circular, purplish-brown spots just under surface of integument. Setae short. T2 and T3 with only 2 L setae. Posterior part of A8 rolled inwardly so spiracles on A8 are in shallow pocket and directed posteriorly. Prolegs short. A3-6 crochets uniordinal to biordinal in broad ellipses or circles. St. Petersburg, Fla., sea grape, October 15, 1974. *Host:* sea grape *(Cocoloba uvifera)*. Feeds within stem of host, usually causing it to enlarge into a gall.

Figure 26.263a,b. Thyrididae. *Dysodia oculatana* Clemens. f.g.l. 17± mm. a. head, T1, T2; b. mesal view right mandible. Head inconspicuously pitted, dark reddish brown. T1 shield relatively smooth, brownish yellow (distinctly contrasting with head), usually narrowly margined with brown near spiracle. T2, T3 and abdomen mostly white to yellowish white with some granules of integument pale brown. Pinacula large, distinct, reddish brown to dark reddish brown; some pinacula elevated, some fused. T2 and T3 with only 2 L setae. A3-6 biordinal in broad ellipses or circles. Wake Co., N.C., September 15, 1950. *Hosts:* according to Forbes (1923) the larvae feed on Fabaceae. Occur within rolled leaves of host.

in *Hexeris,* on A9, greatly reduced with only the ring of the alveolus present. SD2 very small, present on all segments except A9 and A10. On A1–8, L1 and L2 ventrad of spiracle, approximate or somewhat separated, and more or less in a horizontal line. On A9 L is either unisetose or bisetose. SV bisetose on A1, trisetose on A2–6. A7 with SV bisetose. A8–9 with SV unisetose. V1 on all segments. Posterior part of A8, including spiracle, sometimes (*Hexeris*) directed posteriorly. Prolegs on segments 3–6 with crochets uni- to biordinal and arranged in circles or broad ellipses. A10 proleg with crochets uni- to biordinal. A10 shield bearing eight setae. Nine setae on or near anal proleg.

Comments: In North America, the family Thyrididae is poorly represented; only about 12 species, belonging to three subfamilies, are found in our fauna.

Selected Bibliography

Forbes 1923.
Fracker 1915.
Heinrich 1921a,b.

PTEROPHORIDAE (PTEROPHOROIDEA)

H. H. Neunzig, *North Carolina State University*

The Pterophorids, Plume Moths

Figures 26.264-26.270

Relationships and Diagnosis: This family has usually been included in the Pyraloidea. As early as 1926, Tillyard suggested that the group constituted a separate superfamily. Recent authors follow Tillyard's classification. Major differences in the immature stages of the Pterophoridae on the one hand, and the Pyralidae, Thyrididae, and Hyblaeidae on the other, support this separation.

Last-stage larvae of the Pterophoridae have the L setae on T1 trisetose (not bisetose, as stated in most of the literature) (or have an L verruca incorporating three L setae), usually small spiracles, and frequently long prolegs with the crochets of the ventral prolegs few in number, uniordinal, and arranged in a mesopenellipse. Pterophorids have the usual complement of primary setae, sometimes with a few localized secondary setae and/or scattered spinules (secondary setae of previous authors). Also many pterophorids having an L verruca on T1 in place of the L setae, have many additional verrucae elsewhere on the body and sometimes additional, localized secondary setae and/or scattered spinules.

Three subfamilies of North American pterophorids with approximately 150 species, have been established, and they seem to occur throughout our region. Forbes (1923) gave some brief information relative to the separation of the larvae of the more common genera. Yano (1963) also recognized three subfamilies in Japan and gave keys to separate both larvae and pupae.

Biology and Ecology: Larvae feed in or on stems, leaves, buds, flowers, or unripened seeds. Infested stems can be recognized by signs of wilt and by frass extruding from the stem. With some species, the leaves are mined when the larva is small. Large larvae usually feed concealed within rolled or tied-together leaves; some species feed in the open, depending on their cryptic coloration and/or cryptic setal and spinule covering for protection.

About 15 families of plants are hosts in North America, but the more common ones are the Asteraceae, Scrophulariaceae, Lamiaceae, and Geraniaceae.

Most species have several generations each year. In areas where over-wintering is necessary, the adult or larval stage undergoes diapause. The elongate-oval eggs are laid singly on the host. Larvae bore into the host or feed externally. Pupation usually occurs in the open on the host plant, although some species pupate in stems or silk cocoons. Most pupae are slender, with spinous projections, setae and spinules.

Description: Last-stage larvae rather small, usually under 15 mm. Body more or less cylindrical. Prolegs well developed on A3–6 and A10, frequently long and slender, particularly on A3–6. Head relatively smooth. Usual complement of primary setae, sometimes with a few localized secondary setae and/or numerous, scattered spinules; or in other species verrucae present in place of the usual primary setae, with or without additional secondary setae and/or spinules. A few species (such as *Oidaematophorus grandis* (Fish)) with rugosities (short, broad spinules) on the thorax and abdomen, and with hornlike structures on the anal shield. Spiracles usually small and circular.

Head: Usually hypognathous. Six stemmata. Mandibles simple, usually with five teeth. Maxillae with sensilla trichodea simple.

Thorax: T1 with or without distinct shield. Six setae or several verrucae on each side of shield (or on T1 in the position where shield is usually located). Three L group setae on the prespiracular plate of T1 (L3 sometimes very small) or an L verruca occurring in place of the L setae. D1 and D2 present and adjacent, or there are D1 and D2 verrucae, or a fused D1-D2 verruca, on T2 and T3. SD1 and SD2 present and adjacent, or there is an SD verruca, on T2 and T3. L group trisetose on T2 and T3, or 1–2 L verruca(e) present. SV bisetose on T1, T2, and T3, or an SV verruca is located where SV setae usually occur. Additional secondary setae and/or scattered spinules also frequently present on some or all thoracic segments. Primary and secondary setae and spinules sometimes modified (blunt, forked, inflated, etc.).

Abdomen: D1 and D2, or D1 and D2 verrucae, or a fused D1–D2 verruca on segments 1–8 (D1, or D1 verruca, more dorsad and cephalad than D2, or D2 verruca). On A9, D1 ventrad of D2, or D verruca present. SD1, or SD1 verruca, present on all segments. SD2 present and always small. On A1–8, L1 and L2 approximate, or an L verruca present and ventrad of spiracle. On A9, L3 is missing. SV bisetose, or an SV verruca on A1–A2. SV trisetose, or an SV verruca on A3–6. A7 with SV bisetose or an SV verruca. A8–9 with SV unisetose or an SV verruca. V1 on all segments. Prolegs on A3–6 with crochets uniordinal and in mesopenellipses. A10 with or without distinct anal shield. Proleg on A10 with crochets uniordinal. Additional secondary setae and/or scattered spinules also frequently present on some or all abdominal segments. Primary and secondary setae and spinules sometimes modified (blunt, forked, inflated, etc).

Comments: Apparently only a few species of pterophorids are of economic importance. The grape plume moth, *Geina periscelidactyla* (Fitch), is relatively common in vineyards in the United States and Canada, but the damage is usually negligible. In California, Lange (1950) reported that the artichoke plume moth, *Platyptilia carduidactyla* (Riley), at times causes severe damage to artichoke heads, and that several other species of *Platyptilia* and the snapdragon plume moth, *Stenoptilodes antirrhina* (Lange), infest snapdragon, geranium, calendula, and sage.

Selected Bibliography

Barnes and Lindsey 1921.
Common 1970.
Forbes 1923.

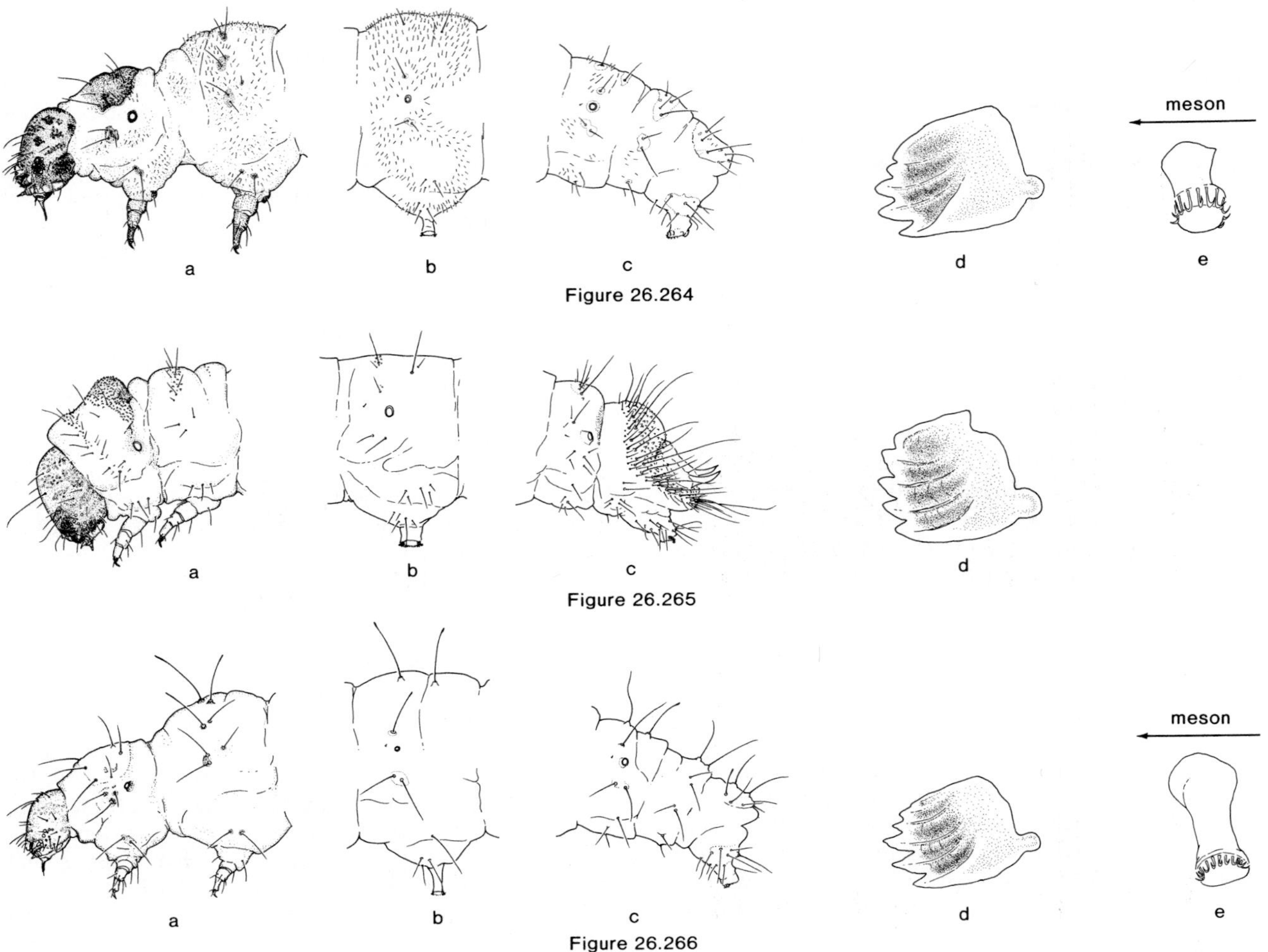

Figure 26.264a-e. Pterophoridae. *Platyptilia carduidactyla* (Riley). Artichoke plume moth. f.g.l. 13± mm. a. head, T1, T2; b. A3; c. A8-10; d. mesal view right mandible; e. A3 proleg and crochets. Head yellowish brown with dark brown to black platelets and dark brown to black spot at stemmata (platelets, particularly at the posterolateral margin of head, usually darkened by additional suffusions of brown or black). T1 shield dark brown to black; T2, T3 and abdomen mostly white with minute granules of integument hyaline to pale brown (most of body of living larva yellowish white to pink). Pinacula brown to black. Most thoracic and abdominal segments with numerous very short spinules. Spinules few on T1 and A9, absent on A10. One to several secondary setae associated with L3 on most abdominal segments, and sometimes on T3. A3-6 crochets few and in uniordinal ellipses that are broadly open laterally. Castroville, Calif., on artichoke, May 7, 1977. *Hosts:* primarily cultivated artichoke and artichoke thistle or cardoon *(Cynara cardunculus).* Feeds externally or internally on almost all parts above soil. Most important injury is on, and in, floral heads.

Figure 26.265a-d. Pterophoridae. *Oidaematophorus grandis* (Fish). f.g.l. 18± mm. a. head, T1, T2; b. A3; c. A8-10; d. mesal view right mandible. Head relatively large, yellowish white with brown platelets, dark reddish brown to black near mouthparts. Tiny brown rugosities also sometimes present on head. T1 shield yellowish white with numerous reddish brown rugosities. Posterior aspect of shield inflated. Approximately 20 setae on each side of shield. T1, T2 and abdomen mostly white (living larva yellowish white or pink with reddish brown or purplish brown markings). T2, T3 and A1-8 with small reddish brown dorsal rugosities. A9 with dorsum somewhat gibbous, more heavily sclerotized (usually reddish brown, particularly posteriorly) and with numerous setae and small rugosities. A10 with reddish brown anal shield bearing various sized rugosities and armed posteriorly with 2, relatively large, distinctive, stout, horn-like processes. In addition to secondary setae on T1 shield, and on A9 and A10, a few secondary setae occur in association with most primary setae elsewhere on the body. A3-6 crochets uniordinal and in mesal penellipses. Half Moon Bay, Calif., *Baccharis pilularis,* October 25, 1937. *Hosts: Baccharis* spp. Bores in stems.

Figure 26.266a-e. Pterophoridae. *Stenoptilodes antirrhina* (Lange). Snapdragon plume moth. f.g.l. 11± mm. a. head, T1, T2; b. A3; c. A8-10; d. mesal view right mandible; e. A3 proleg and crochets; Head relatively small, yellowish white with pale brown indistinct platelets (head mostly greenish white or purplish white in living larva). T1 shield yellowish white (greenish white in living larva), relatively indistinct. T2, T3 and abdomen mostly yellowish white (living larva greenish white or purplish white with white, interrupted, longitudinal stripes; fig. 26.266 is based on a preserved larva in which stripes are not apparent). Chalazae brown. Most larvae with only primary setae, some of which are slightly clubbed; some individuals with a few clubbed spinules associated with dorsal chalazae. A3-6 crochets uniordinal in mesal penellipses. San Leandro, Calif., snapdragon, September 22, 1941. *Hosts:* snapdragon and other Scrophulariaceae. Early stage larvae are leaf miners; late instars feed on flowers or bore into stems or seed capsules. Especially damaging in nurseries.

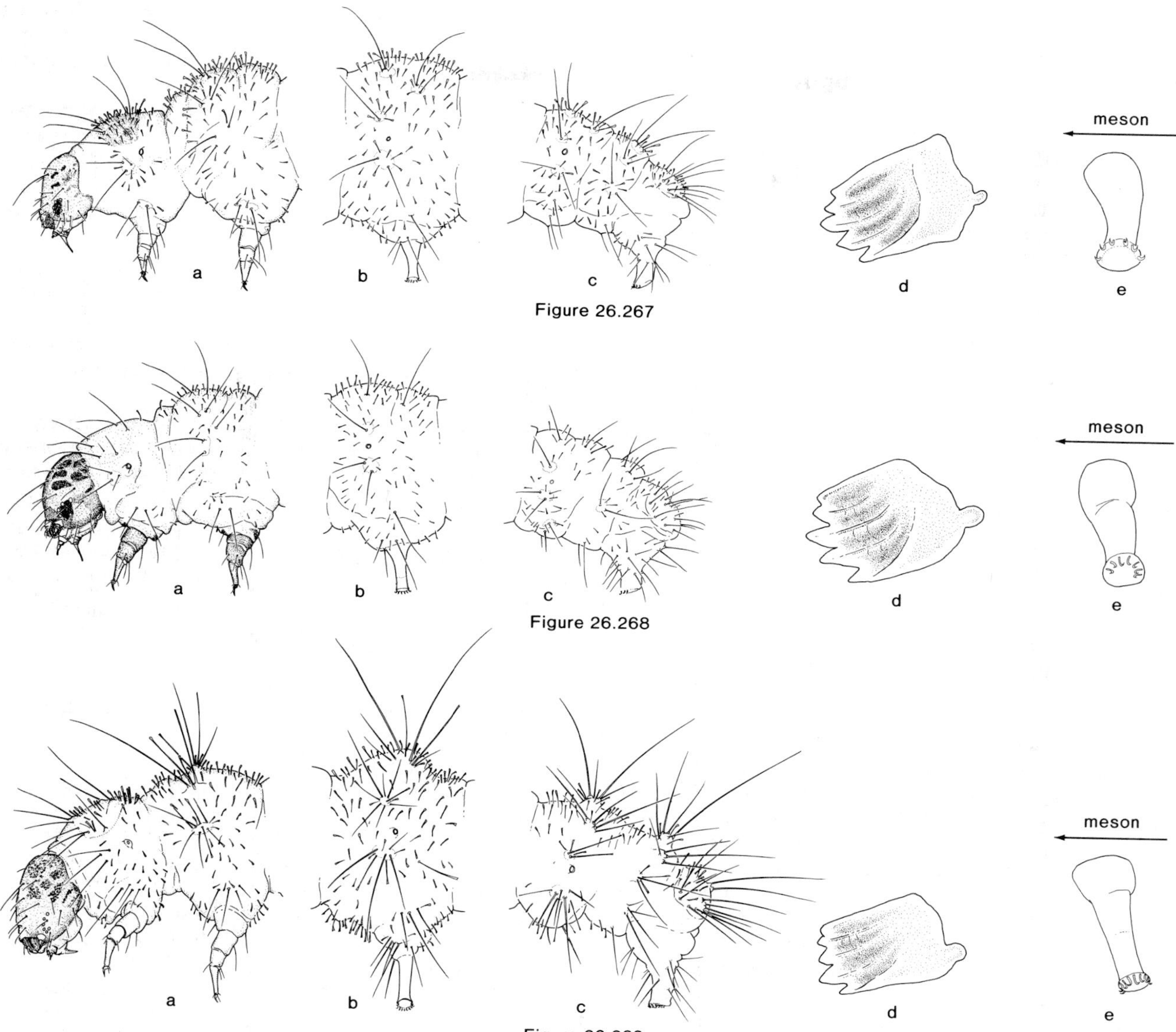

Figure 26.267

Figure 26.268

Figure 26.269

Figure 26.267a-e. Pterophoridae. *Stenoptilodes auriga* (Barnes and Lindsey). f.g.l. 12± mm. a. head, T1, T2; b. A3; c. A8-10; d. mesal view right mandible; e. A3 proleg and crochets. Head yellowish white with a brown spot at stemmata and pale brown to brown platelets (platelets at posterior of head usually pale). T1 shield not clearly delinated, yellowish white, with many, short distinctly clubbed spinules. T2, T3 and abdomen mostly yellowish white with numerous short, strongly clubbed spinules. Thoracic legs yellowish white. Chalazae pale brown to hyaline. A3-6 crochets uniordinal in mesal penellipses. Putnam Co., Ill., *Gerardia* sp., May 7, 1941. *Hosts:* species of Asteraceae.

Figure 26.268a-e. Pterophoridae. *Amblyptilia pica* prob. *crataea* (Fletcher). f.g.l. 10± mm. a. head, T1, T2; b. A3; c. A8-10; d. mesal view right mandible; e. A3 proleg and crochets. Head yellowish white with dark brown spot at stemmata and pale brown to dark brown platelets (platelets about as dark posteriorly as anteriorly, sometimes suffused with additional brown pigmentation). T1 shield not clearly delineated, yellowish white, without spinules. T2, T3 and abdomen mostly yellowish white, with many, short spinules, usually with slightly swollen apices (living larva mostly greenish yellow or pink). Thoracic legs brown or mostly brown. Chalazae pale brown to hyaline. California, on geranium. *Hosts:* snapdragon and geranium are the most common cultivated plants. Feeds on tender, terminal leaves which they web together, and bores into green seed pods, flower buds and flowers. Occurs in the field in California and in greenhouses elsewhere where geranium cuttings have been received from California.

Figure 26.269a-e. Pterophoridae. *Geina periscelidactyla* (Fitch). Grape plume moth. f.g.l. 12-16 mm. a. head, T1, T2; b. A3; c. A8-10; d. mesal view right mandible; e. A3 proleg and crochets. Head mostly yellowish white (greenish white in living larva) with pale brown to black platelets and usually with pale patches extending from near P1 along adfrontals to, and including, the area around the stemmata. T1 shield not clearly delineated, yellowish to brownish white with long primary and secondary setae, and short, clubbed spinules. T2, T3, and abdomen mostly yellowish to brownish white (yellowish green in living larva). Verrucae present, bearing long yellowish white primary and secondary setae and short clubbed spinules. Numerous, generally distributed spinules also present. A3-6 crochets uniordinal in mesal penellipses. Columbus, Ohio, grape, May 20, 1941. *Host:* grape. Feeds on young terminal foliage within webbed leaves.

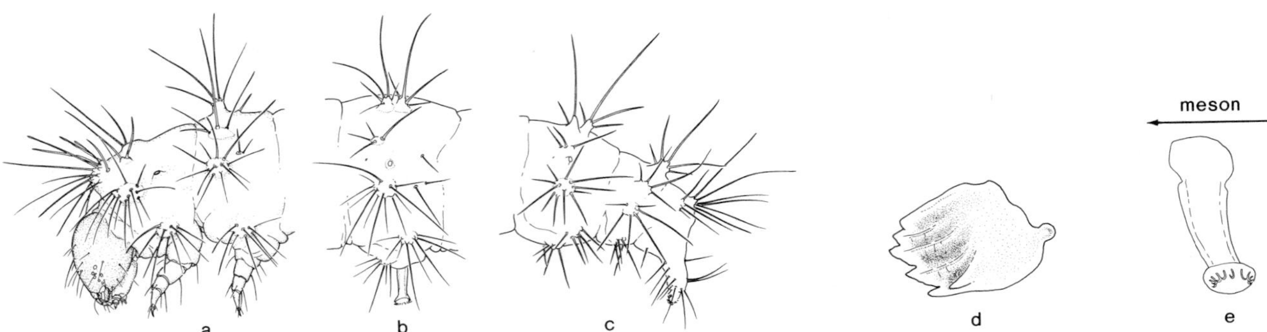

Figure 26.270a-e. Pterophoridae. *Oidaematophorus eupatorii* (Fernald). f.g.l. 10± mm. a. head, T1, T2; b. A3; c. A8–10; d. mesal view right mandible; e. A3 proleg and crochets. Head brownish to orangish white with indistinct brown platelets. Mandibles with extra tooth at base of tooth 1. T1 shield whitish yellow (usually with green undertones in living larva), somewhat elevated anteriorly, not clearly delineated ventrally and posteriorly. T2, T3 and abdomen mostly yellowish white (living larva mostly pale green to reddish green). Verrucae present, with setae very weakly plumose; setae on dorsum darker (smoky brown to black) and more robust than those more ventral. No spinules on verrucae or elsewhere. A3-6 crochets unidordinal in mesal penellipses. Baddeck Falls, Nova Scotia, *Eupatorium* sp., June 26, 1936. *Host:* Joe-pye weed *(Eupatorium purpureum)* and probably other species of *Eupatorium*. Larvae gregarious, feed mostly on foliage.

Heinrich 1921b.
Lange 1950.
McDunnough 1927, 1936.
Mosher 1916a.
Peterson 1948, 1961, 1967.
Pinhey 1975.
Tillyard 1926.
Valley et al. 1981.
Yano 1963.

Acknowledgments

Chien C. Chang assisted with drawing and inking the figures.

THYATIRIDAE (DREPANOIDEA)

Frederick W. Stehr, *Michigan State University*

The Thyatirids

Figures 26.271–26.272

Relationships and Diagnosis: Thyatirids are believed to be most closely related to the Drepanidae, which are easily recognized by their caudally projecting, nublike A10 prolegs that lack crochets, and the short to long median, suranal process. In contrast, the A10 prolegs of thyatirids are little, if any, reduced and they lack a median suranal process. Some thyatirids have 2 SV setae on T1, T2 and T3 as do the Drepanidae and the Epiplemidae (Geometroidea); others have two SV's on T1 and one on T2 and T3 as do the geometrids.

One of the more distinctive features of the thyatirids examined is the number and arrangement of the abdominal setae, with all setae "normal" in position (see fig. 26.272) except for L3, which is quite anterior to SV4, and for the two extra setae, one dorsocaudad and one ventrocaudad of the spiracle. There are also two extra setae dorsocaudally from the SV setae on T2 and T3, and SD1 and SD2 are not included on the T1 shield.

Biology and Ecology: The caterpillars reside in loosely folded leaves of bushes, shrubs, and trees; they have been reported from blackberry, thimbleberry, oak, and dogwood (Forbes 1923). There apparently are one or two generations per year.

Description: Mature larvae medium, 20–30 mm. Primary and a few secondary setae present. Pinacula indistinct or absent. Spiracles oval, T1 and A8 slightly larger to 2× larger than the other abdominal spiracles.

Head: Hypognathous. Wider than high. Frontoclypeus longer than wide and extending about halfway to epicranial notch. Ecdysial sutures meeting just beyond apex of frons. Six stemmata; 1–4 evenly spaced in an arc; 5 somewhat below 4; 6 caudad and nearly equidistant from 4 and 5.

Thorax: T1 shield distinct to indistinct. XD1 and XD2 quite low on shield; SD1 and SD2 not on shield and near the spiracle; 2 L and 2 SV setae; an extra seta located caudally about midway between the L and SV groups. T2 and T3 with D1, D2, SD1 and SD2 in a nearly vertical line; L2 equidistant between L1 and L3; SV group bisetose on T1 and uni- or bisetose on T2 and T3; two extra setae located caudally about midway between L3 and the SV setae on T2 and T3.

Abdomen: Setal pattern "normal" except as noted in the diagnosis. Prolegs normal; crochets on A3–6 and A10 in a biordinal mesoseries.

Comments: This small family contains six genera and 16 species in the U.S. and Canada. They are neither abundant nor damaging.

Selected Bibliography

Forbes 1923.

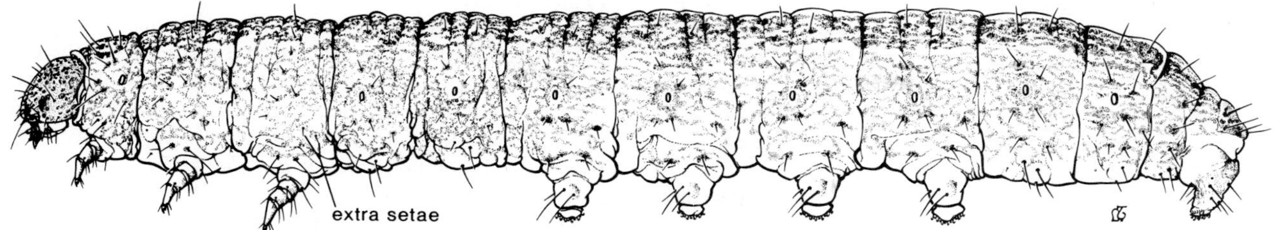

Figure 26.271. Thyatiridae. *Psuedothyatira cymatophoroides* (Guenée). f.g.l. 30± mm. Puyallup, Wash., 27 September 1928. Salmonberry, alder. U.S. Natl. Mus.

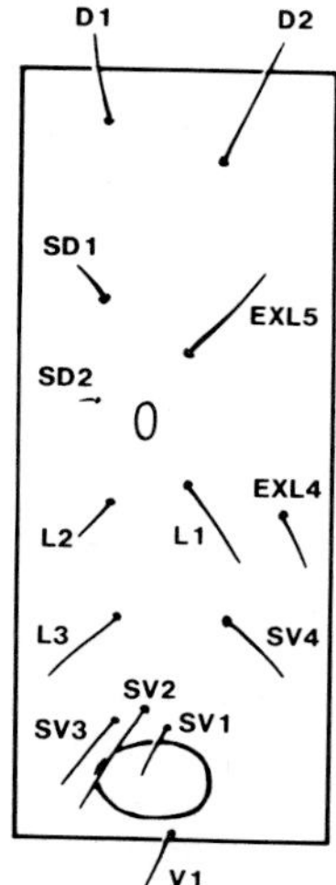

Figure 26.272. Thyatiridae. A3 setal map.

DREPANIDAE (DREPANOIDEA)

Frederick W. Stehr, *Michigan State University*

The Hook-tip Moths

Figures 26.273–26.274

Relationships and Diagnosis: Drepanids are believed to be most closely related to the Thyatiridae. They are easily recognized by their caudally projecting, nublike A10 "prolegs," which lack crochets (but do bear setae), and a short to long median knob or process on the suranal plate, both lacking in thyatirids and Geometroidea. Some notodontids have the A10 prolegs modified into long, paired processes (stemapoda), but they lack the median suranal process.

Most drepanids have some conspicuous secondary setae on the lower body and prolegs, and some have numerous spinules and/or a few tubercles on the upper body; *Eudeilinea* is nearly bare. They may have extra setae in fixed positions as do some Geometroidea (especially Geometridae). Drepanids also have a rather uncommon crochet arrangement, having a uni- or biordinal mesoseries and a shorter, uniordinal lateroseries on A3–6, usually with a small lobe between.

Biology and Ecology: They are leaf feeders on birch, alder, dogwood, and viburnum (Forbes 1923).

Description: Mature larvae small to medium, 10–30 mm. Vesiture variable, ranging from nearly bare to bearing all primary setae, plus some primarylike secondaries, spinules and some tubercles. Spiracles oval, with T1 and A8 slightly larger.

Head: Hypognathous. Variable in detail, but vertex usually bilobed, with epicranial suture meeting epicranial notch behind (over the top of) the vertex. Frontoclypeus extending one-third to halfway up the face. Head usually granulate, setae and adfrontal areas conspicuous or inconspicuous, the ecdysial sutures meeting about one-third distance between apex

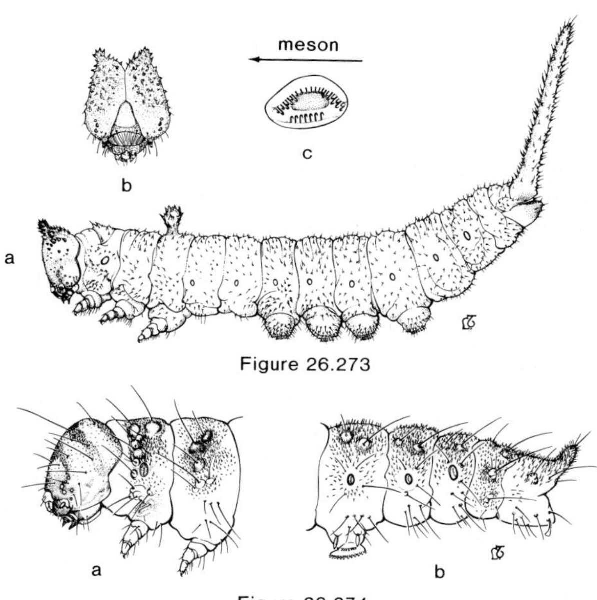

Figure 26.273a-c. Drepanidae. *Oreta rosea* (Walker). f.g.l. 25–30± mm. a. not mature larva; b. head; c. A3 proleg. U.S. Natl. Mus. Feeds on viburnum.

Figure 26.274a,b. Drepanidae. *Drepana arcuata* Walker. f.g.l. 25–30± mm. a. head, T1, T2; b. A6-A10. Falls Church, Va., 8 October 1915, on alder. U.S. Natl. Mus. Also on birch.

of frons and epicranial notch. Six stemmata, 1–4 evenly spaced in an arc; 5 considerably below 4; 6 weakest and caudad to 4; 3 larger than the others.

Thorax: With paired or median verruca(e) on one or more segments, or "bare." Setae variable, depending on the species. T1, T2, and T3 with two or more SV setae and two extra setae above them.

Abdomen: Variable, ranging from "bare" to having conspicuous setae, spinules and/or verrucae.

Prolegs normal on A3–6, but reduced on A10 to small, nublike caudo-projecting, seta-bearing structures that lack crochets. Crochets on A3–6 in a uni- or biordinal mesoseries, plus a short, uniordinal lateroseries. Suranal plate with median, short-to-long knob or process.

Comments: This small family contains three genera and five species in the U.S. and Canada, none of which occur in damaging numbers.

Selected Bibliography

Dyar 1895.
Forbes 1923.

GEOMETRIDAE (GEOMETROIDEA)

Frederick W. Stehr, *Michigan State University*

The Geometrids, Inchworms, Measuring Worms, Loopers

Figures 26.275–26.289

Relationships and Diagnosis: The Geometridae, Thyatiridae, and Epiplemidae make up the Geometroidea, a reasonably well-defined superfamily distinguished by most species having a fourth SV seta on A6 (which resembles an L seta by being on the body wall), by having L1 and L2 widely separated on A1–8 (widely separated only on A4–8 in Epiplemidae), and by having L3 more anterior than the usual posterior position on Lepidoptera (fig. 26.286c). Many geometrid larvae also have a secondary SV seta beneath the three primary SV setae on the A6 proleg, giving the appearance of four SV setae on the proleg and one above on the body (fig. 26.286c).

All but a few Geometridae can be recognized by their two pairs of prolegs (A6 and A10). In those genera with more than two pairs (*Archiearis, Leucobrephos* (Archiearinae)), *Alsophila* (Oenochrominae), and *Campaea* (Ennominae), the additional prolegs are always smaller than those on A6, and the crochets are reduced in number and/or size. In noctuids with prolegs only on A5, 6 and 10, the A5 proleg is the same size as A6.

Biology and Ecology: All larvae are "exposed" feeders (not borers or miners), although some live in folded leaves or cover themselves with debris of various kinds (Forbes 1948). They primarily feed on deciduous trees and shrubs, although some feed on ferns, coniferous foliage, and even coniferous seeds (McGuffin 1958b). All larvae except Archiearinae (which move in a modified looping manner like noctuid loopers) move in the characteristic looping manner by bringing the A6 and A10 prolegs forward to a place behind the thoracic legs before moving the anterior part of the body forward or sweeping it around in search of a new footing. The twig-mimicry and stipule-mimicry resting positions of many members of the Ennominae (*Biston, Euchlaena, Hesperumia,* and *Pero*), some Larentiinae (*Lygris, Plemyria,* and some *Eupithecia*), *Chlorochlamys* (Geometrinae) and doubtless many more genera where the body is held out at an angle from the twig also appear to be adaptations related to the proleg reduction.

Eggs variable, but usually oblong or flat with or without sculpture, laid singly or in groups on bark, twigs, or leaves of the host, or even dropped on the ground. No single character or combination of characters is known to separate geometrid eggs from other Lepidoptera eggs (McGuffin 1972). Salkeld (1983) presents scanning electron photographs of the eggs of 131 species from the Ottawa, Ontario vicinity.

Pupa typically lepidopteran, but lacking exposed maxillary palpi and mandibles (Mosher 1916a). Pupation usually in the soil or litter, rarely on the host, with little or no cocoon spun. More details (including keys) are given by Forbes (1948) and McGuffin (1967, 1972).

Description: Mature larvae small to medium, 15–60 mm. Body cylindrical, usually slender but occasionally stout and rarely flattened, sometimes bearing processes, humps, protuberances, or filaments; cuticle smooth or granular and often wrinkled in annulets on each segment. Body often longitudinally striped or cryptically or disruptively colored in a way that enhances concealment in the resting position. A standard complement of primary (including subprimary) setae is present. In addition, a fourth SV seta is present on the body wall of the abdominal segments, and secondary seta(e) (which may resemble the primaries in size and constancy of position) are present but restricted to the SV area on the prolegs, where a group of four setae composed of three SV setae and one secondary seta is often found. There is also an extra seta above the spiracle in a few genera (*Melanolophia, Eufidonia, Ectropis*) (McGuffin 1964) and below the spiracle in *Phaeoura* and *Gabriola* (McGuffin 1981). Spiracles elliptical, with those on T1 and A8 larger.

Head: Hypognathous. Frontoclypeus as wide as long or wider than long, margins straight and extending no more than two-thirds to the epicranial notch. Ecdysial lines usually very close to frontoclypeal margins and meeting just above the frontoclypeal apex. Six stemmata, 1–4 evenly spaced in semicircle, 5 separated from 1–4, 6 caudad of 4 and closer to 4 than 5.

Thorax: T1 shield usually indistinct. D1 and D2 on all segments. SD2 caudad and very close to SD1. XD1, XD2, and SD1 usually in a nearly straight vertical line. SD2 sometimes paler and with greater diameter than SD1. L group bisetose on T1, trisetose on T2 and T3. SV group bisetose on T1, unisetose on T2 and T3.

Abdomen: D1 and D2 on all segments, with D1 more dorsad and cephalad except on A9 where D1 is usually ventrad to D2. SD1 on all segments. SD2 on all segments except 9, but quite small. L1 and L2 far apart, with L1 nearly caudad

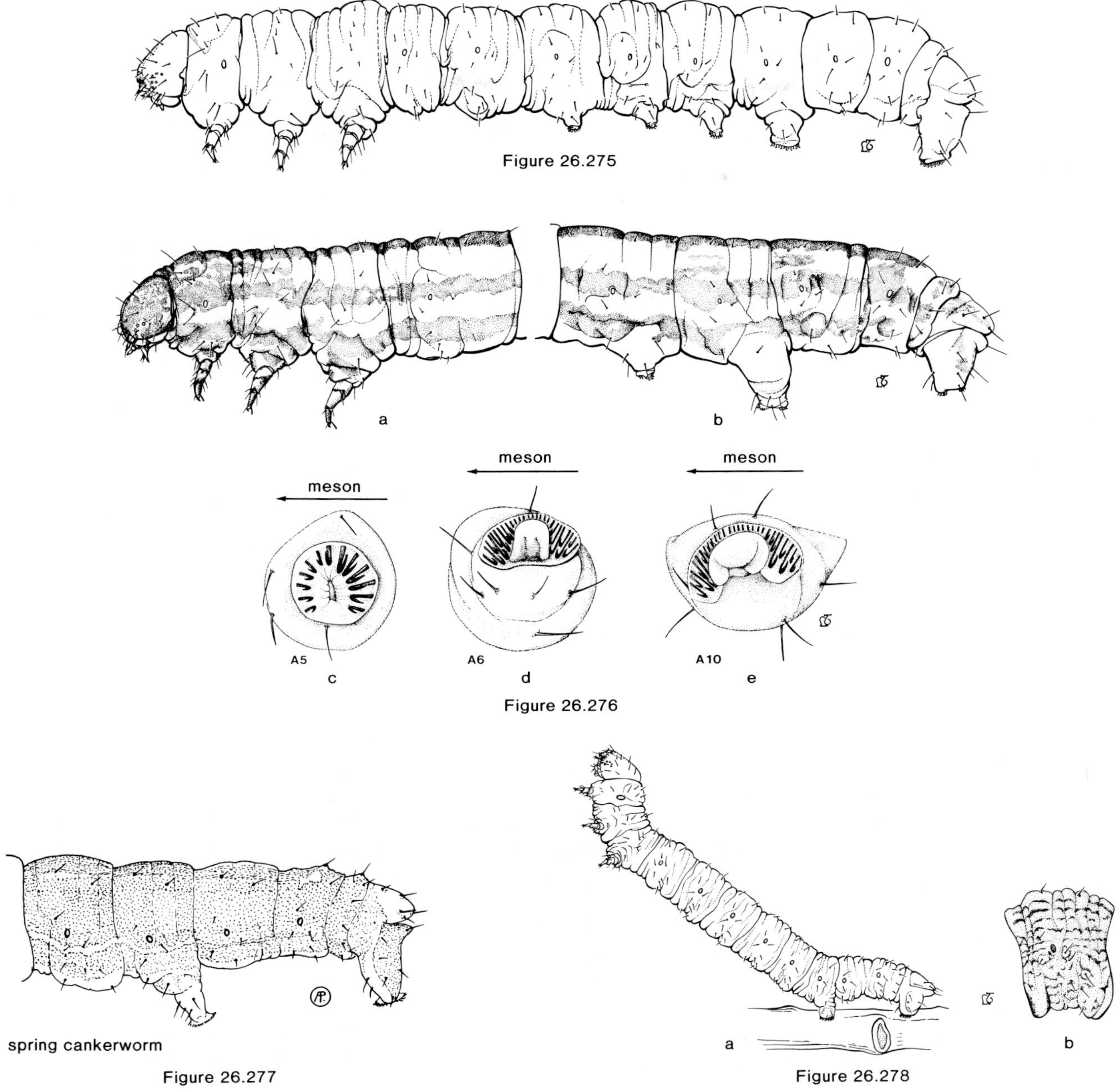

Figure 26.275. Geometridae, Archiearinae. *Archiearis infans* (Möschler). f.g.l. 25± mm. Green, with fine white stripes. One of our primitive species with 4 pairs of ventral prolegs, gradually reduced in size from A6 to A3. On white and gray birch and poplar. Northern U.S. and Canada west to Alaska.

Figure 26.276a-e. Geometridae, Oenochrominae. *Alsophila pometaria* (Harris), fall cankerworm. f.g.l. 25± mm. a. head, T1, T2, T3, A1; b. A5-A10; c-e. A5, A6 and A10 crochets, those on A6 and A10 divided into 2 groups. Color variable, green to brown; a middorsal dark stripe, with 3 narrow, whitish, longitudinal stripes above spiracles and yellow one below. Well-developed A6 prolegs and reduced pair on A5. Many shade and fruit trees, especially elm and apple. Adult emergence and oviposition in both late fall and early spring. N. Carolina to Maritimes, west through shelterbelts and river bottoms into Alberta, the Rocky Mountain states, Utah, and California. Can cause localized defoliation.

Figure 26.277. Geometridae, Ennominae. *Paleacrita vernata* (Peck), spring cankerworm, A5-A10. f.g.l. 25± mm. Color variable, light brown or gray to quite dark, with yellowish subspiracular stripe and greenish-yellow stripe on venter. One pair of ventral prolegs on A6, crochets on A6 and A10 not divided. Adult emergence in spring. Essentially the same hosts, distribution, and defoliation potential as fall cankerworm, but extends farther south and west into Texas, (from Peterson 1948).

Figure 26.278a,b. Geometridae, Ennominae. *Lycia ursaria* (Walker). f.g.l. 35± mm. a. larva drawn in the twig-mimic posture often assumed by species in this subfamily (they usually stretch out more so the skin is much smoother); b. detailed pattern on A3. Gray, interrupted with purplish or red-brown markings; T1 plate with yellow anterior margin and dark spots. On various trees and shrubs across northern U.S. and southern Canada.

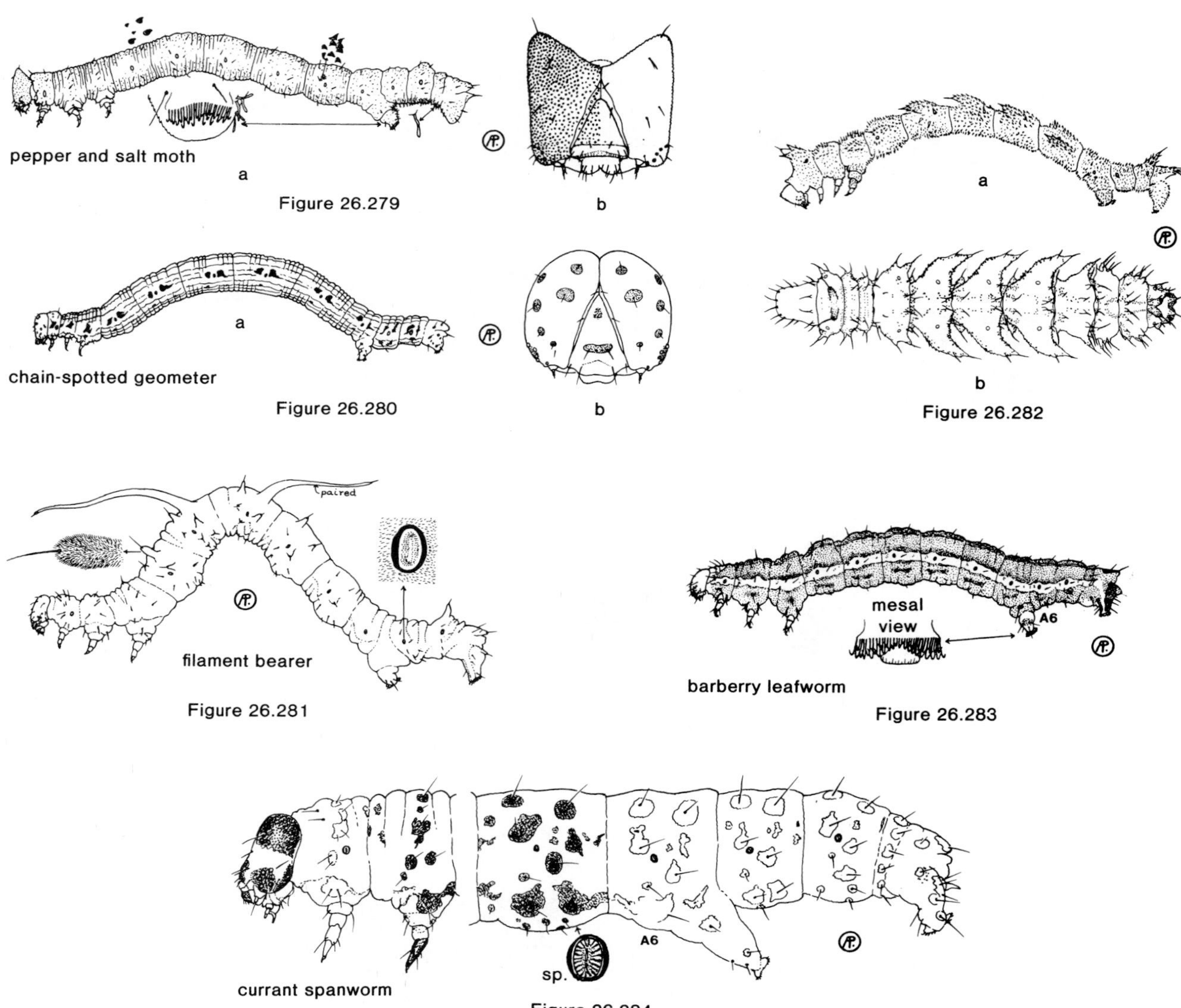

Figure 26.279a,b. Geometridae, Ennominae. *Biston betularia cognataria* (Guenée), pepper and salt moth. f.g.l. 40–60 mm. a. larva; b. head. Gray to green to brown, sometimes mottled, head brown to dark brown, with red-brown "horns". On willow, alder, birch, poplar, cherry, maple, larch, and numerous other trees and shrubs. Throughout northern North America, south to N. Carolina, Tennessee and into Mexico (figs. a, b from Peterson 1948).

Figure 26.280a,b. Geometridae, Ennominae. *Cingilia catenaria* (Drury), chainspotted geometer. f.g.l. 45± mm. a. larva; b. head. Straw yellow, spiracles located in white areas flanked front and rear with conspicuous black marks. On a great variety of deciduous trees and shrubs; also on some conifers when intermixed. Common in eastern Canada and northeastern U.S., a local defoliator (figs. a, b from Peterson 1948).

Figure 26.281. Geometridae, Ennominae. *Nematocampa limbata* (Haworth), filament bearer. f.g.l. 18± mm. Green-brown with red-brown head and conspicuous long pairs of light-tipped brown to black filaments on A2 and A3. On hemlock, fir, Douglas-fir, and many hardwoods coast to coast. Locally abundant (fig. from Peterson 1948).

Figure 26.282a,b. Geometridae, Geometrinae. *Nemoria rubrifrontaria* (Packard). f.g.l. 18± mm. a. lateral; b. dorsal, many spiny projections omitted. Yellowish to greenish-brown, with prominent reddish brown lateral extensions; body roughened with blunt, spinelike projections. Spiracles on A2, 3, 4, 5, 7 and 8 quite dorsal in position. On sweet-fern, *Comptonia perigrina,* whose young leaves it mimics well. Eastern North America where sweet-fern grows. (fig. 26.282a from Peterson 1948, fig. 26.282b from Dethier 1942).

Figure 26.283. Geometridae, Larentiinae. *Coryphista meadi* (Packard), barberry leafworm. f.g.l. 25± mm. Brown to purple-black, with orange head, broad, light spiracular stripe, and 4 narrow white dorsal lines. On barberry and *Mahonia.* Northern two-thirds of U.S. and southern Canada (fig. from Peterson 1948).

Figure 26.284. Geometridae, Oenochrominae. *Itame ribearia* (Fitch), currant spanworm. f.g.l. 28± mm. Longitudinally striped with white and yellow; pinacula and spots dark brown to near black. Head with near-black frons, stemmatal area and dorsolateral areas. On current, gooseberry, blueberry and huckleberry. Southern Canada and northern U.S. west to Colorado (fig. from Peterson 1948).

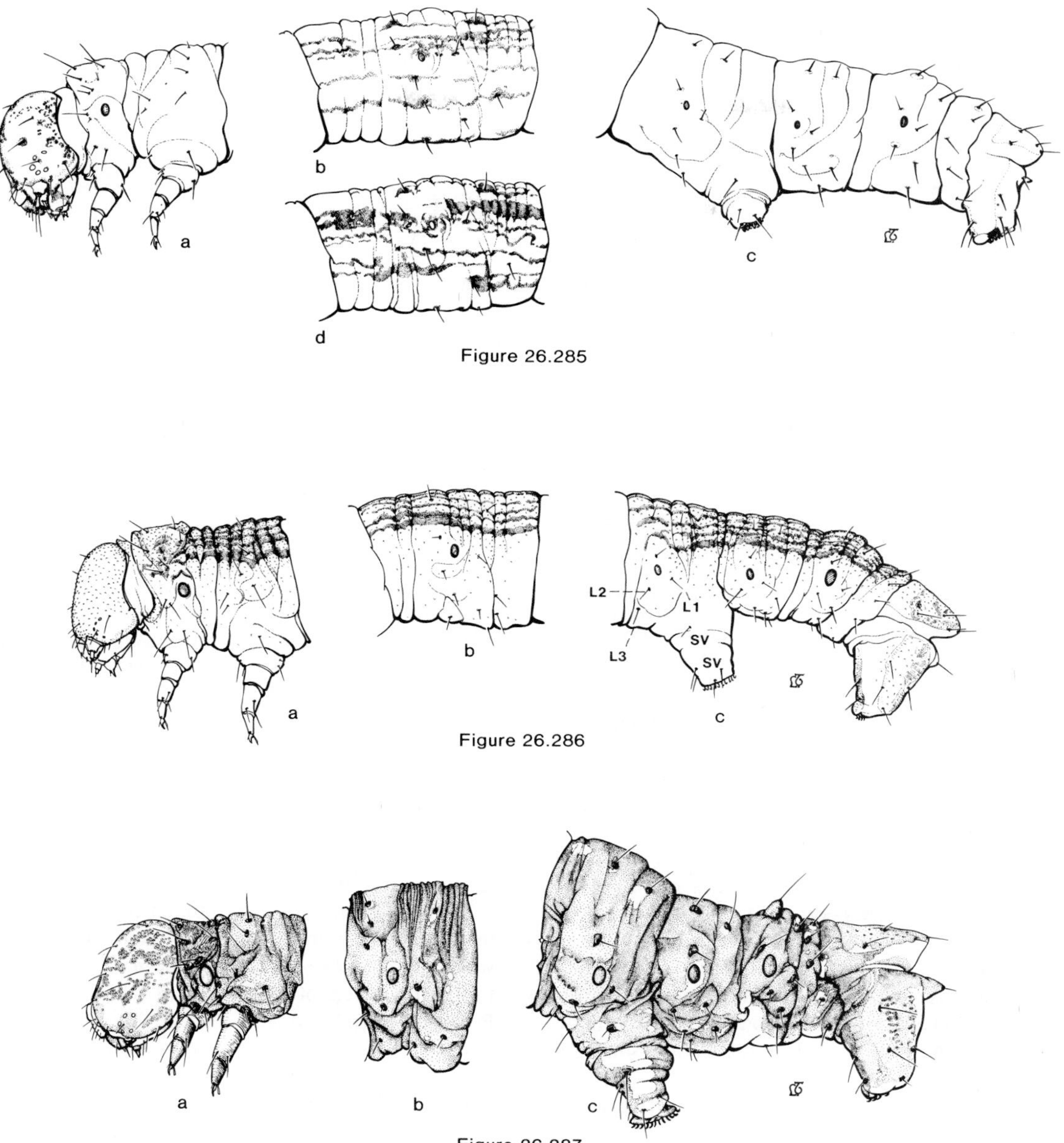

Figure 26.285

Figure 26.286

Figure 26.287

Figure 26.285a-d. Geometridae, Ennominae. *Lambdina fiscellaria lugubrosa* (Hulst), western hemlock looper. f.g.l. 28± mm. a. head,T1, T2; b. A3; c. A6-A10; d. A3 of *L. fiscellaria somniaria,* with more extensive pattern. Pale yellow brown to gray brown, with variably complex markings laterally and 4 dark spots per abdominal segment dorsally. *L. fiscellaria* occurs coast to coast where suitable hosts grow, especially northern U.S. and Canada to Alaska. On fir, Douglas-fir, spruce, and eastern and western hemlock and many others, including deciduous species during outbreaks. *L. fiscellaria lugubrosa* destructive to western hemlock and associated conifers in Pacific Northwest coastal forests. *L. fiscellaria somniaria* a local defoliator of Oregon white oak in Pacific Northwest.

Figure 26.286a-c. Geometridae, Ennominae. *Erannis tiliaria tiliaria* (Harris), linden looper. Two subspecies, *tiliaria* in the East, and *vancouverensis* Hulst in the West. f.g.l. 40± mm. a. head, T1, T2; b. A3; c. A6-A10. Yellow to red-brown, with rusty, rugulose head, antenna yellow, 10 wavy black lines on dorsum of abdomen. On variety of hardwoods, particularly basswood, maple, oak, birch, elm, hickory, willow, and others. Coast to coast in northern half of U.S. and southern Canada.

Figure 26.287a-c. Geometridae, Ennominae. *Ennomos subsignaria* (Hübner), elm spanworm. f.g.l. 50± mm. a. head, T1, T2; b. A3, c. A6-A10. Rather long, gray-green to gray to dull black, with a pair of pointed knobs on A8 and rusty head and tail. On hickory, oak, ash, elm, and many other hardwoods; has caused severe defoliation, especially in southern Appalachians. Eastern half of U.S. and southern fringe of Ontario.

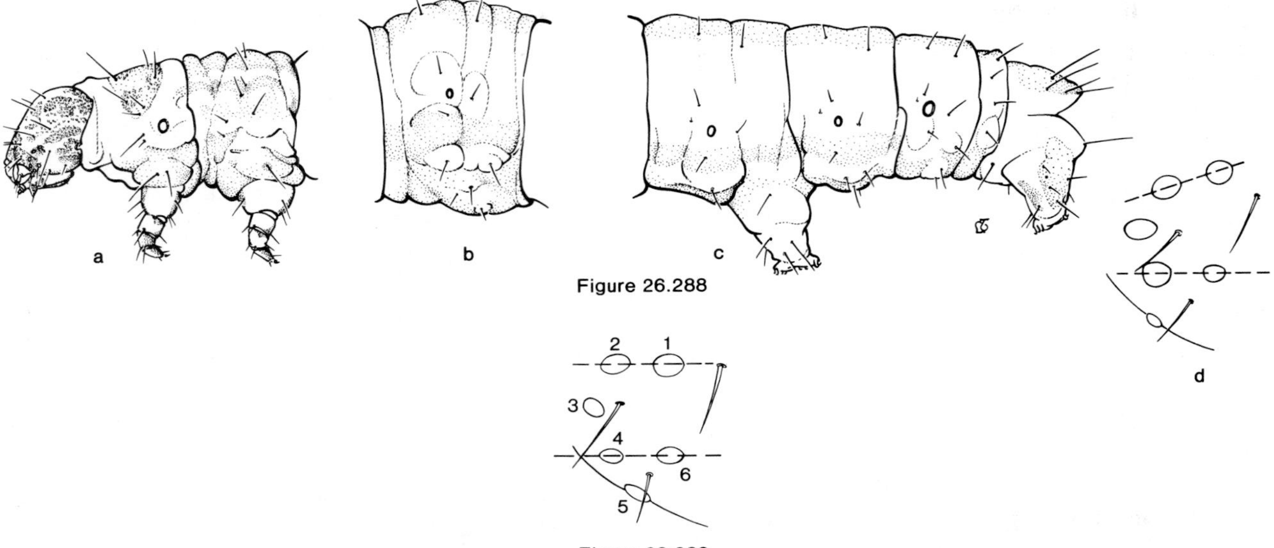

Figure 26.288a-d. Geometridae, Larentiinae. *Operophtera bruceata* (Hulst), Bruce spanworm. f.g.l. 18± mm. a. head, T1, T2; b. A3; c. A6-A10; d. stemmata. Stout, light green, with darker dorsum and 3 yellow lines per side; anal plate often dark. Stemmata as in figure 26.288d. Quite similar to winter moth, *O. brumata,* but see figure 26.289. On trembling aspen, willow, maple, birch, beech, alder, apple, cherry, and other trees and shrubs; an extensive defoliator in Canada and parts of U.S. Throughout Canada, the Lake States and the Northeast.

Figure 26.289. Geometridae, Larentiinae. *Operophtera brumata* (L.). Stemmata 1, 2, and 4, 6 in parallel rows in contrast to those of *O. bruceata* (fig. 26.288d), which are angled. *O. brumata* is introduced from Europe and found in the Maritime provinces, (figs. 26.288d and 26.289 redrawn from Eidt and Embree, 1968).

of the spiracle and L2 nearly below it. L3 below L1 and L2, near the proleg, and more anterior than usual for Lepidoptera larvae. SV group trisetose on A1–5 (bisetose in Larentiinae), but often with a fourth seta; A6 with at least four SV setae (SV1, 2, 3 on proleg, SV4 on body wall), and often with one secondary SV seta that is similar to SV1, 2 and 3 and grouped with them. There may also be many secondary setae on the A6 proleg. A7 and A8 with SV group uni-, bi- or trisetose. D2, D1, SD1, L1 and SV on A9 arranged in a nearly straight vertical line. V1 on A9 close to midventral line and small (or absent).

Prolegs in all but one subfamily (Archiearinae, two genera in North America (*Archiearis* and *Leucobrephos*)) are absent on A3 and A4 and absent or rudimentary on A5. A6 and A10 prolegs normal, with bi- or triordinal crochets in a mesoseries on a lobate planta and frequently interrupted or reduced near the middle by a fleshy lobe or "sucker" (Forbes 1948).

A10 with anal shield dorsal, often triangular, and bearing four setae per side. Anus bounded by an epiproct dorsally, paraprocts laterally, and a hypoproct ventrally that may be well developed. The setae of the anal plate and proleg have been named (Singh 1953; McGuffin 1958b; Dugdale 1961), but inconsistencies exist and the characters are generally more useful below the family level so they are not given here.

Comments: The family is large, with some 250 genera, and more than 1400 species are known from the U.S. and Canada. Our knowledge of the larvae is more comprehensive than for many other families, due to the monographic coverage of Canadian geometrids by McGuffin (1958b, 1967, 1972, 1977, 1981) and the more limited coverage by Forbes (1948) for northeastern U.S. Other important publications on geometrids that provide an entry into the literature for larvae and adults are given below.

Many geometrids are common but rarely economically important, with important exceptions. These include the winter moth, *Operophtera brumata* (L.) (fig. 26.289), a former defoliator of hardwoods in the Canadian Maritime Provinces introduced from Europe (Embree 1965); Bruce's spanworm, *Operophtera bruceata* (Hulst) (fig. 26.288), a northern defoliator of hardwoods (especially poplar and apple); the spring cankerworm, *Paleacrita vernata* (Peck) (fig. 26.277); and the fall cankerworm, *Alsophila pometaria* (Harris) (fig. 26.278). The latter two are both widely distributed in the U.S. and Canada, frequently causing severe defoliation of many hardwoods, especially in urban areas. Other defoliators include the elm spanworm, *Ennomos subsignaria* (Hübner) (fig. 26.287), the linden looper, *Erannis tiliaria* (Harris) (fig. 26.286), and *Phigalia titea* (Cramer), all three being defoliators of elm, oak, maple, and other hardwoods in both urban and forested parts of eastern North America. In addition, the hemlock looper, *Lambdina fiscellaria* (Guenée) (fig. 26.285), is a serious pest on conifers in the Pacific Northwest and in northern and eastern North America that also consumes hardwood foliage during outbreaks. Other occasionally abudant species in forest situations are covered in U.S.D.A. (1985) and Furniss and Carolin (1977).

Selected Bibliography

Dugdale 1961.
Dyar 1899–1905.
Eidt and Embree 1968.
Embree 1965.
Ferguson 1969, 1985a.
Forbes 1948.
Franklin 1948.
Furniss and Carolin 1977.
Johnson and Lyon 1976.
McGuffin, 1958b, 1964, 1967, 1972, 1977, 1981.
Mosher 1916a.
Salkeld 1983.
Singh 1953, 1956.
U.S.D.A. 1985.

EPIPLEMIDAE (GEOMETROIDEA)

Frederick W. Stehr, *Michigan State University*

The Epiplemids

Figure 26.290

Relationships and Diagnosis: The epiplemids are believed to be most closely related to the Uraniidae, a family containing many large, brightly colored, dayflying, papilionidlike moths that do not occur in the U.S. Epiplemids are unique in having L1 and L2 close together on the same pinaculum on A1–3, although having them widely separated and on different pinacula on A4–8. Epiplemids are superficially most similar to thyatirids, but they also differ in having L1 closer to L2 than to L3 on T2 and T3 (midway in thyatirids), by lacking the extra seta on T1, by lacking the two extra setae on T2 and T3 located caudodorsad from the SV group in the thyatirids, and by lacking the extra seta caudodorsad of the spiracle on the abdomen (see thyatirid diagnosis and fig. 26.272). They do have L3 (quite anterior) and SV4 (posterior) on the abdomen as do most other geometroids.

Biology and Ecology: Young larvae of some species live in a common web; larvae of some species hide at the base of the host when not feeding. Larvae are reported from *Viburnum prunifolium* and *Lonicera dioica* (Forbes 1923).

Description: Mature larvae of our species small (10–15 mm), but much larger species occur in the tropics. Primary and a few secondary setae present. Pinacula distinct and may be elevated or rounded. Spiracles broadly oval to round, those of T1 and A8 at least 2× those of A1–7.

Head: Hypognathous. Wider than high. Frontoclypeus longer than wide and extending about halfway to epicranial notch. Ecdysial sutures strongly bowed out about halfway to apex of frons, and meeting nearly halfway from apex of frons to epicranial notch. Six stemmata; 1–4 nearly evenly spaced in an arc; 5 distinctly lower; 6 caudad to 4 and nearly equidistant from 4 and 5.

Thorax: T1 with standard complement of setae, with 2 L setae and SD1 and SD2 on the shield. T2 and T3 standard except for 2 SV setae on each; L2 closer to L1 than L3.

Abdomen: D1, D2, SD1 normal. L1 and L2 close together on the same pinaculum on A1–3, but at least twice as far apart on separate pinacula on A4–8. L3 (anterior) and SV4 (posterior) midway between the L and SV groups, and separated about the same as the corresponding L1 and L2 setae on A1–8. Prolegs normal, but with four SV setae on each. Crochets on A3–6 and A10 in a semicircular mesoseries, uniordinal in the middle and biordinal at both ends in at least *Callizzia*, but reported as uniformly biordinal by Fracker (1915) and Peterson (1948).

Comments: This family is poorly represented in the U.S. and Canada, only seven genera and eight species being known, none of them common or damaging. The family is primarily tropical.

Selected Bibliography

Forbes 1923.
Fracker 1915.
Peterson 1948.

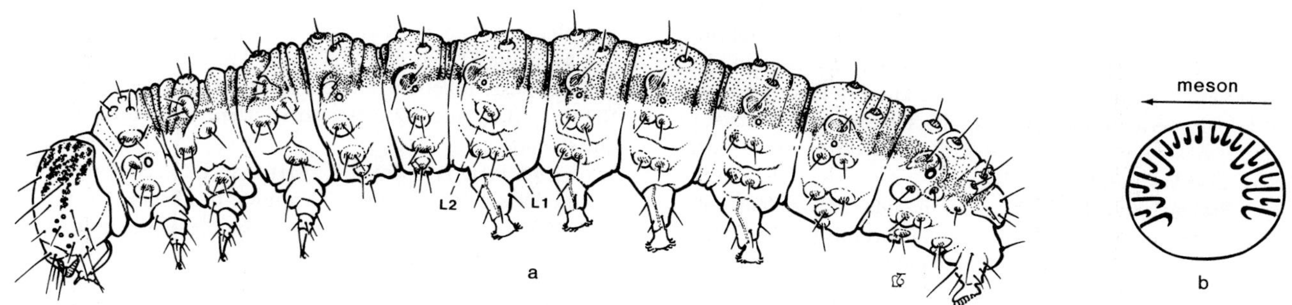

Figure 26.290a,b. Epiplemidae. *Callizzia amorata* Packard. f.g.l. 15± mm. a. larva; b. A4 crochets. Brown dorsally with black lateral band, lighter below the spiracular line; head pale with brown areas; pinacula humped. On honeysuckle, *Lonicera dioica*. Northern half of U.S. and southern Canada.

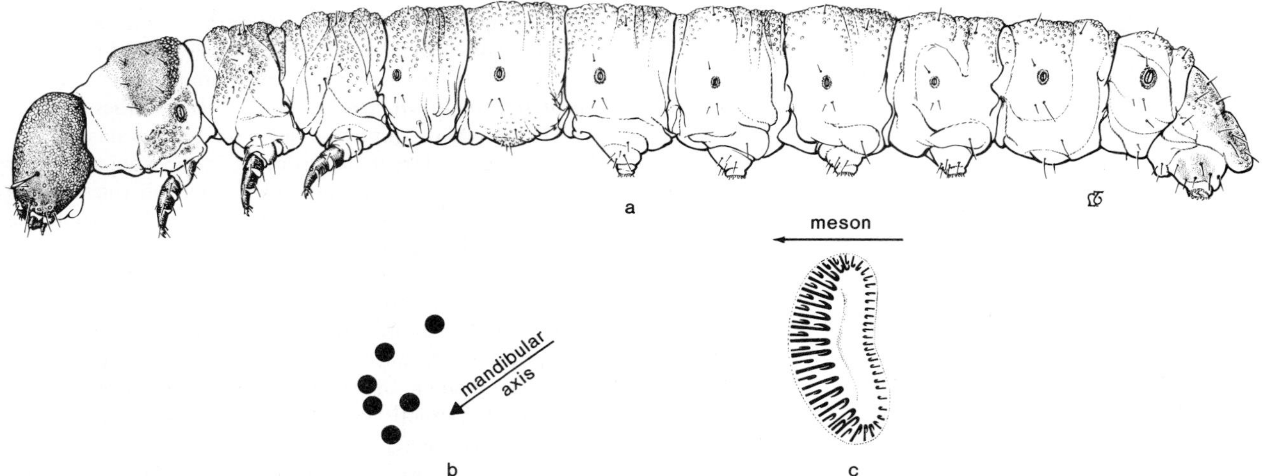

Figure 26.291a-c. Mimallonidae. *Cicinnus melsheimeri* (Harris). f.g.l. 45± mm. a. larva; b. stemmata; c. A10 crockets. In a strong case of leaves and silk open at both ends. Head heavily sclerotized and rugose, with a clubbed seta near the stemmata on each side. On scrubby oaks from the Northeast to Colorado, south into Florida and into Mexico.

MIMALLONIDAE (MIMALLONOIDEA)

Frederick W. Stehr, *Michigan State University*

The Sack-bearers

Figure 26.291

Relationships and Diagnosis: According to Franclemont (1973), the mimallonids are a small, wholly American family of uncertain relationship, primarily Neotropical, with only four species north of Mexico. Mimallonid larvae are superficially similar to larvae of a number of other families that are nearly devoid of secondary setae and have two L setae on T1. However, North American mimallonids I have seen (*Lacosoma chiridota* Grote and *Cicinnus melsheimeri* (Harris)) have a very heavily sclerotized rugose or punctate head, have only four to seven SV setae on the A3–6 prolegs, and—apparently unique to mimallonids—have the A10 crochets arranged in an *unbroken* transverse oval, with biordinal crochets anteriorly, and considerably smaller, biordinal or uniordinal crochets posteriorly (fig. 26.291a). In larvae that are not fully distended, the oval may appear to be two transverse rows, but the rows are connected at the ends. I know of no other larvae of Lepidoptera with the A10 crochets in an unbroken oval or circle. A single larva from Brazil and a dozen from Costa Rica also have unbroken ovals.

Biology and Ecology: Mimallonids are called sack-bearers because the later instars build portable cases of leaves that are open at both ends. Franclemont (1973) notes that some of the tropical species are reported to use their frass pellets as "bricks" in constructing their cases. Forbes (1923) describes the behavior as follows: "the caterpillar, when young, makes a lacelike nest between two leaves, later it forms a portable house of two pieces of leaf, which is lined with heavy silk, very roomy inside, and has a circular opening at each end. When not moving about, it anchors the case with silk, often in a slight nest formed of leaves drawn together, closing one opening of the case with its head, and the other with the circular thickened posterior end of its body." All of our species are reported to feed on oaks.

Description: Mature larvae medium, 15–25 mm, the head heavily sclerotized and projecting conspicuously downward and forward. A9 and A10 apparently combined to form a sloping anal plate. Spiracles ovoid, peritreme heavily sclerotized, T1 and A8 spiracles somewhat to 2× larger than others.

Head: Heavily sclerotized, rugose or punctate, projecting anteroventrad and longer than wide. Frontoclypeus almost twice as long as wide, extending nearly halfway to epicranial notch. Ecdysial sutures faint (*Cicinnus*) to distinct (*Lacosoma*), meeting slightly beyond apex of frons. Six stemmata characteristically spaced (fig. 26.291b), with stemma 1 about twice its diameter from 2, with 3 and 4 closest together, and with 2, 5, and 6 at least one diameter from nearest neighbor.

Thorax: T1 shield well sclerotized, with the standard complement of setae. L group on T1 bisetose. SV group with two setae on T1 and one on T2 and T3. T2 and T3 with D1 and D2 closer together than on T1 or the abdomen. SD1 and SD2 together, and L2 closer to L1 than L3, which may be quite small.

Abdomen: D2's farther apart than D1's. SD1 dorsad to anterodorsad of spiracle, SD2 minute or absent. L1 and L2 close together, horizontally in line, and beneath spiracle. L3 fairly low, above proleg or in line with rear margin. SV group bisetose on A1 and A7, trisetose on A2, 4–7 setae on A3–6 prolegs, bi- or unisetose on A8 and A9. SD1 on A8 nearly anterior to the top margin of the spiracle (*Lacosoma*) to anterodorsad (*Cincinnus*) to posterodorsad (Brazilian specimen). The dorsal, posterior portion of segment A8 (bearing D2) is sharply "separated" from the anterior part so as to appear to be A9. A9 is nearly fused with A10 to form what appears to be a large, sloping, sclerotized "anal plate" bearing seven setae (14 total). A9 with D2 behind D1 or slightly more laterad; SD1 on A9 in line with A8 spiracle; L1 and L2 together, L3 absent.

A3–6 prolegs short, crochets in biordinal, transverse oval. A10 prolegs short, A9–10 plate extending down and beyond them caudally; crochets in an unbroken transverse oval, biordinal in front and with smaller, uni- or biordinal crochets behind.

Comments: According to Franclemont (1973), four species in three genera occur north of Mexico; *Lacosoma chiridota* and *Cicinnus melsheimeri* (fig. 26.291) are widely distributed in eastern North America from New England to Texas, with *melsheimeri* extending to Colorado and Arizona. The larva of *Naniteta* is apparently unknown. They are of no economic importance.

Selected Bibliography

Dyar 1900a.
Forbes 1923.
Franclemont 1973.
Packard 1895.
U.S.D.A. 1985.

APATELODIDAE (BOMBYCOIDEA)

Frederick W. Stehr, *Michigan State University*

The Apatelodids

Figures 26.292–26.293

Relationships and Diagnosis: This is a small American family, probably most closely related to the Old World families Bombycidae, Anthelidae, and Eupterotidae, with two genera and five species north of Mexico (Franclemont 1973). Apatelodids are superficially similar to some acronictine noctuids, some notodontids, a few arctiids, some lymantriids, and many lasiocampids. They are readily distinguished from all but the lasiocampids by their biordinal crochets (uniordinal in acronictines, lymantriids, and notodontids; uniordinal or uniordinal, heteroideous in arctiids). The Lasiocampidae are bombycoids and have biordinal crochets; perhaps the best character to distinguish apatelodids from them is the lack of an "anal point" beneath the anus and between the A10 prolegs; all of our lasiocampids examined have an anal point (fig. 26.301c). Other characters are the tapered, middorsal tufts on mature larvae and the moderately deep, V-shaped labral notch. Our lasiocampids do not have such tufts and have a shallow U-shaped labral notch.

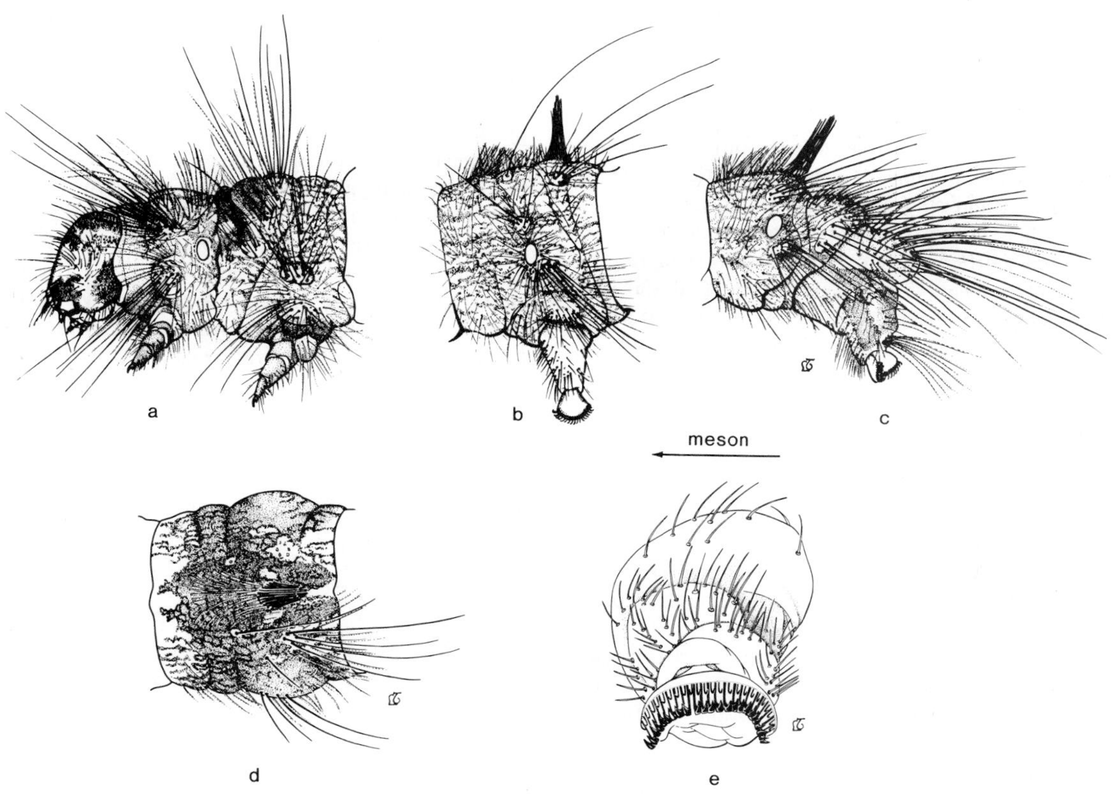

Figure 26.292a-e. Apatelodidae. *Olceclosterna ceraphica* (Dyar). f.g.l. 40± mm. a. head, T1, T2; b. A3; c. A8-A10; d. A3 dorsum; e. A3 proleg viewed from inside. Culberson Co., Tex., 1 October 1966, Cornell Univ. Gray with crinkled pattern; rests flattened against bark; head a patchwork of dark and light areas. Verrucae with long dark setae and shorter white setae. A1-A8 each with anterior, middorsal tuft of short white setae, followed by a longer, median tuft of black setae (shorter on A1). On desert willow, *Chilopsis linearis* var. *glutinosa* in southern-southwestern Texas.

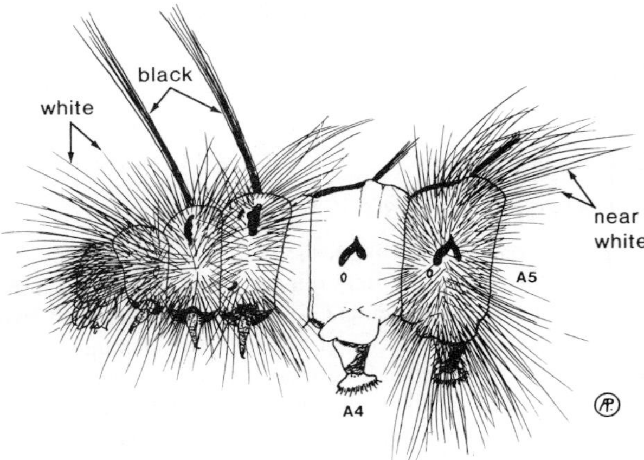

Figure 26.293. Apatelodidae. *Apatelodes torrefacta* (J. E. Smith). Head, T1, T2, T3, A4 and A5. f.g.l. 40± mm. Densely hairy, yellow to gray, with a long, black median hairpencil present on T2, T3 and A8, and a shorter median black tuft on A2-A7. Dark chevron-shaped spots above the spiracles. Head near-black. On a wide variety of trees and shrubs throughout the eastern deciduous forest into southern Ontario (fig. from Peterson 1948).

Biology and Ecology: Our species are exposed leaf feeders on a variety of trees and shrubs. They pupate in the ground, in contrast to many tropical species, which spin well-formed cocoons (Franclemont 1973). The caterpillars are apparently capable of flattening themselves against twigs or bark (Forbes 1923), as many lasiocampids and catocaline noctuids do.

Description: Mature larvae moderate in size, 25–40 mm, densely hairy and with conspicuous, tapered tufts and dorsal, dark hair pencils (*Apatelodes*). Spiracles elliptical, with a dark peritreme, T1 and A8 spiracles slightly larger.

Head: Rather square at the genae and flat in front, hypognathous, and mottled. Frontoclypeus nearly equilateral, extending about halfway to epicranial notch. Ecdysial sutures distinct, close to adfrontal sutures and meeting at apex of frons. Six stemmata in a normal arc, except for 5 which is about two stemmatal diameters away from 4 and 6. Stemma 3 sometimes conspicuously more convex and distinct than the others (especially in *Apatelodes*).

Thorax: T1 shield conspicuous, lateral area dominated by two lobes bearing numerous long, dark setae and shorter, light setae. T2 and T3 (and A8) bearing a long, medial pencil of dark setae in *Apatelodes,* and subdorsal verrucae with moderately long black setae in *Olceclosterna;* numerous light setae present in rows or tufts. A lobe above T2 and T3 prolegs, which is most conspicuous in *Olceclosterna.*

Abdomen: A1–8 with conspicuous middorsal tufts on posterior of each segment, the A1 tuft usually smaller, the A3–5 tufts the largest, and the A8 tuft very long in *Apatelodes. Olceclosterna* with primary setal groups D1, D2, SD and L conspicuous on A1–7, consisting of long black setae, with D1 single and the others multiple. A8, A9 and A10 with numerous, long, dark setae, especially in the SD position on A8 and A9.

A3–6 and A10 prolegs moderate in length, and capable of being spread out against the host. Proleg base with numerous secondary setae, and all crochets in a biordinal mesoseries.

Comments: The family contains two genera and five species in our area; only two species are widely distributed north of Mexico. *Apatelodes torrefacta* (J. E. Smith) feeds on a variety of trees and shrubs in the eastern deciduous forest; and *Olceclosterna angelica* (Grote) is recorded from ash and lilac in the northern two-thirds of that area (Franclemont 1973). They are of no economic importance.

Selected Bibliography

Forbes 1923.
Franclemont 1973.
Packard 1895.

BOMBYCIDAE (BOMBYCOIDEA)

Frederick W. Stehr, ***Michigan State University***

The Silkworms

Figure 26.294

The bombycids comprise a small family native to the southeastern Palearctic and the Oriental regions (Franclemont 1973). The only species in North America is the domestic silkworm, *Bombyx mori* L. It is not likely to be collected in the wild, since it has apparently become so dependent on humans through domestication that it is no longer able to survive in the wild, even in Asia, its native home. It resembles sphingids in having a small middorsal horn on A8, and some saturniids in various other ways, but its horn is rather characteristic (see fig. 26.294). In addition, it has an enlarged meso- and metathorax with distinctive patches of slightly longer setae; it also bears rather evenly distributed short secondary setae on the head and the rest of the body except for the ventrolateral area of A3–10, which bears somewhat longer setae. A3–6 and A10 prolegs normal, with many secondary setae, and crochets in a biordinal mesoseries. The preferred food is mulberry.

Selected Bibliography

Franclemont 1973.

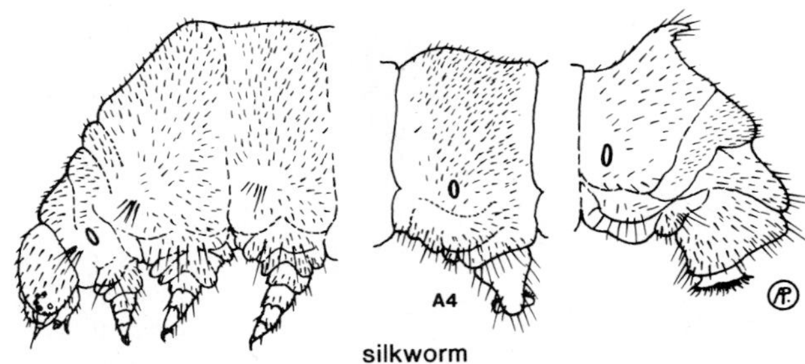

Figure 26.294. Bombycidae. *Bombyx mori* (L.), silkworm. f.g.l. 50± mm. Pale gray-green-white, thorax swollen and short horn on A8. See text (figs. from Peterson 1948).

LASIOCAMPIDAE (BOMBYCOIDEA)

Frederick W. Stehr, *Michigan State University*

Tent Caterpillars and Lappet Moths

Figures 26.295–26.301

Relationships and Diagnosis: This moderately large, worldwide family is primarily tropical and subtropical, but some important genera and species occur in temperate areas. Larvae are either flattened (lappets) or cylindrical; they may superficially resemble some hairy acronictine noctuids, some notodontids, some lymantriids, a few arctiids, and apatelodids. They are easily distinguished from all except the apatelodids by their biordinal crochets (uniordinal in noctuids, lymantriids, and notodontids, and uniordinal or heteroideous uniordinal in arctiids). Mature larvae of apatelodids have tapered *middorsal* tufts and a moderately deep, V-shaped labral notch; our lasiocampids do not have such tufts (they may have small *addorsal* ones), and have a shallower U-shaped labral notch. Another apparently reliable character that is easier to use for separating lasiocampids from apatelodids is the presence of a conspicuous caudally directed "anal point" located beneath the anus and between the A10 prolegs of lasiocampids (fig. 26.301c), which is absent on apatelodids. This anal point seems to be consistently present on all lasiocampid larvae examined, but its presence or absence in other bombycoid and noctuoid taxa has not been determined for taxa other than the Apatelodidae.

Biology and Ecology: Lasiocampid species feed on a wide variety of deciduous and coniferous shrubs and trees, and, especially as last instars, some species such as the forest tent caterpillar, *Malacosoma disstria* Hübner, will eat the leaves of nearly all trees and shrubs and even herbs when food is in short supply. Red maple is a notable exception that it won't touch. Oviposition is much more restricted. Eggs are laid singly, in small groups, or in larger masses of several hundred eggs. Winter can be spent in the egg, larval, or pupal stages, depending on the genus. The larvae may feed singly, in small groups, semisocially, or socially in large tents or webs. Pupation is in cocoons concealed in leaves, debris, bark crevices, old tents, and similar places. Most of our species have a single generation per year, but a few in the South have two or possibly three generations.

Despite the fact that lasiocampids are primarily tropical and subtropical, there are groups such as the tent caterpillars (*Malacosoma*) that are restricted to temperate regions. In North America, *Malacosoma* does not occur south of Sarasota, Florida, nor in lowland Mexico and south of Chiapas, Mexico, apparently because winter temperatures are not consistently cool enough to break the egg diapause.

Description: Mature larvae moderate to large, 20–80 mm, moderately to densely hairy along the sides, but varying dorsally from being nearly bare to having rather dense setae of varying lengths, some of which may be grouped on small verrucae or pinacula. Lappets variable, depending on the genus. Spiracles elliptical and somewhat larger on T1 and A8.

Head: Rounded, rather flat in front, hypognathous. Frons nearly equilateral, extending a fourth to halfway to epicranial notch. Ecdysial sutures highly variable, from "indistinguishable" (*Tolype, Artace, Gloveria*) to moderately distinct (*Phyllodesma, Malacosoma*), but short, and when visible, meeting only a little beyond apex of the frons.

Thorax: Variable, *Tolype* and *Artace* being the most flattened, with *Tolype* having larger lateroventral body lobes than *Artace*. *Phyllodesma, Malacosoma,* and *Gloveria* nearly cylindrical; *Phyllodesma* with two conspicuous T1 lobes per side and other body lobes moderate in size; *Malacosoma* and *Gloveria* with all body lobes minimal. *Tolype, Artace,* and *Phyllodesma* with rather long, abundant setae on the lappets, which enhance concealment as they rest flattened on bark or branches; *Tolype* and *Artace* with "fan" setae (fanlike tip); *Phyllodesma* without fan setae. *Malacosoma* and *Gloveria* without fan setae, with relatively shorter setae, and the larva not flattened when at rest.

Tolype, Artace and *Phyllodesma* have one or two transverse, color band(s) (which may be groovelike) with associated setae or scales that are concealed when the larva is not disturbed or alarmed. In the past these color bands have been described as intersegmental, but they are midsegmental on *Tolype* (dark band on T3, fig. 26.300) and *Phyllodesma* (orange-red bands on T2 and T3) and intersegmental only on *Artace* (orange band between T3 and A1). There are probably glands associated with them that produce defensive materials. Some Palearctic genera such as *Dendrolimus* have more conspicuous color bands that are more groovelike and lined with very conspicuous setae and scales.

Abdomen: Essentially as described for the thorax, but with smaller lobes and verrucae (if present). Left and right prolegs widely separated on *Tolype* and *Artace,* less so on *Phyllodesma,* closer together on *Gloveria,* and "normal" on *Malacosoma.*

Comments: Thirty-five species in 11 genera are known north of Mexico, the most commonly encountered genera being *Malacosoma, Tolype, Phyllodesma,* and *Gloveria.* Larvae of these genera and *Artace* were the only ones available for study. There are more lasiocampid species in the Southwest, but species of *Malacosoma* (the tent caterpillars) are widespread and abundant throughout the United States and Canada.

Except for *Malacosoma,* our lasiocampids are of no economic importance, although *Gloveria* and apparently *Quadrina* larvae are irritating when handled, and the powder of *Malacosoma* cocoons can cause allergic reactions. All species of *Malacosoma* can be abundant and destructive at times, but the forest tent caterpillar *M. disstria* Hübner (fig. 26.295) (which does not build a tent, but only spins a mat of silk to molt on) is a common defoliator of aspen in Canada and the northern United States, as well as maple, sweetgum, oak, and other hardwoods in other regions, especially the East and South. The eastern tent caterpillar, *M. americanum* (fig. 26.296) is common on *Prunus,* apple, and other rosaceous hosts and can be an important defoliator of commercial black cherry, as well as a nuisance on ornamentals. See Dethier (1980) for a delightful synopsis of the eastern tent caterpillar.

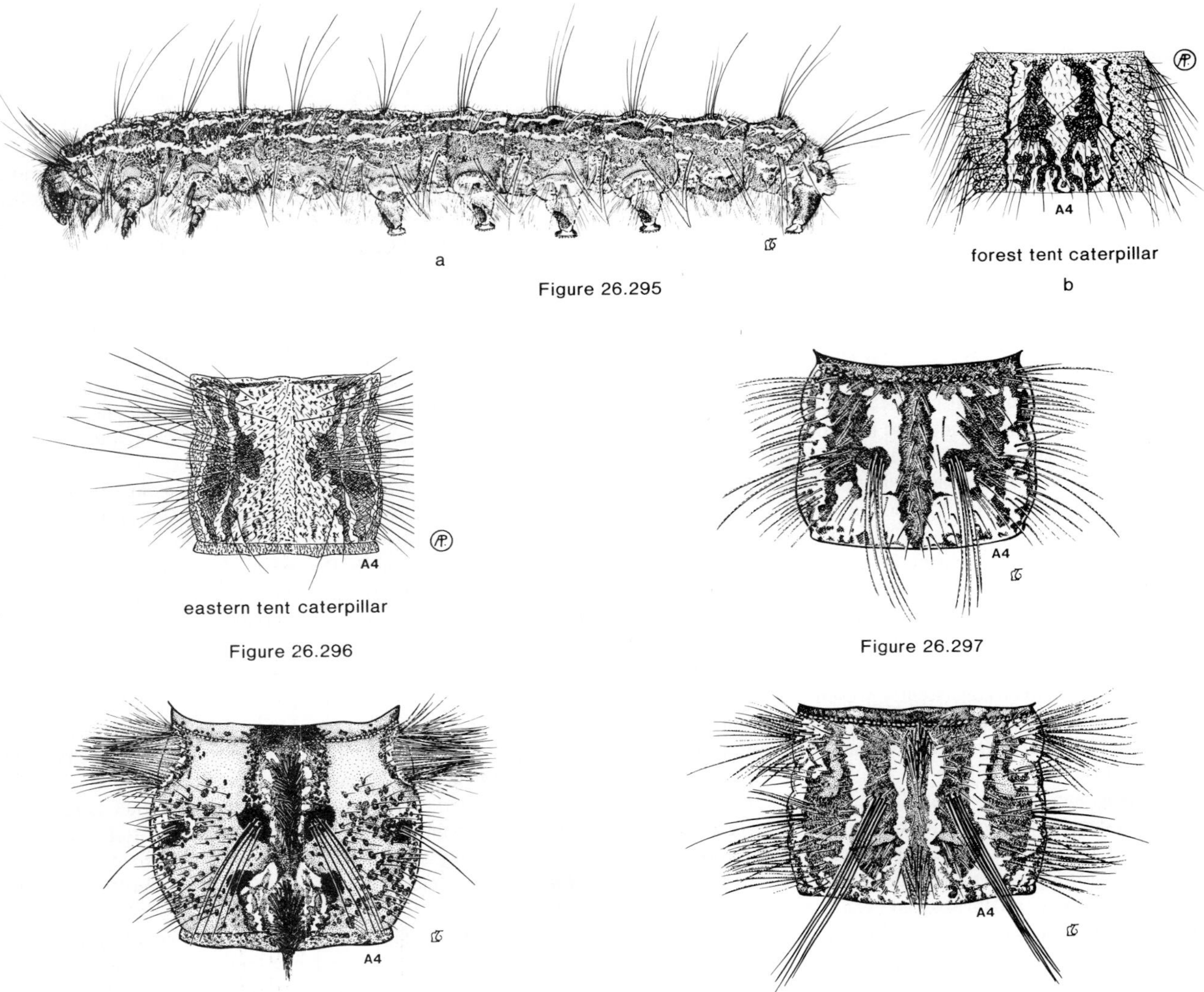

Figure 26.295

Figure 26.296

Figure 26.297

Figure 26.298

Figure 26.299

Figure 26.295a,b. Lasiocampidae. *Malacosoma disstria* Hübner, forest tent caterpillar. f.g.l. 50 ± mm. a. larva; b. A4 dorsum, showing keyhole-shaped middorsal spot that is diagnostic (sometimes only the anterior oval part is present). Head light blue, mottled with black. Body bluish to blue-black above spiracles, with an irregular reddish brown line and a similar, lower, yellowish line evident. Blue-gray to dark gray below spiracles. D1 setal group usually with 4 prominent setae (other species usually with 5). On most hardwoods, but particularly abundant and damaging on aspen in the North, tupelo gum, black gum and sweetgum in the South, oak in Texas and maple in New England. Will not eat red maple. Does not build a tent (fig. 26.295b from Peterson 1948).

Figure 26.296. Lasiocampidae. *Malacosoma americanum* (Fabricius), eastern tent caterpillar. f.g.l. 50± mm. A4 dorsum showing the diagnostic *continuous* intersegmental middorsal yellow-white stripe (no other American species has a *continous* stripe). Head coal black. Prefers black cherry, but also common on other *Prunus,* apple, hawthorn, and related plants. Found east of the central Great Plains, a nuisance on ornamentals, and damaging to commercial black cherry. Builds large tent (fig. from Peterson 1948).

Figure 26.297. Lasiocampidae. *Malacosoma californicum* (Packard), western tent caterpillar. f.g.l 50± mm. A4 dorsum showing a common black and orange pattern of subspecies *pluviale.* A highly variable species found west of central Great Plains and across Canada and northern fringe of U.S. On great diversity of trees and shrubs. Builds large tent; usually not damaging. *M. incurvum* (Henry Edwards) is common on cottonwoods and willows along rivers in southern Utah and Arizona.

Figure 26.298. Lasiocampidae. *Malacosoma constrictum* (Henry Edwards), Pacific tent caterpillar. f.g.l. 50± mm. A4 dorsum. Setae conspicuously tufted dorsally and laterally (more than other species), no conspicuous middorsal markings. Predominantly blue, black and white, with white lateral setae north of Los Angeles area (subspecies *constrictum*); predominantly orange and black with orange lateral setae south of Los Angeles area (subspecies *austrinum*). On deciduous and live oaks from southern Washington into Baja California west of the Cascades and Sierra Nevada. Builds only a small molting tent for crochet attachment; rarely damaging.

Figure 26.299. Lasiocampidae. *Malacosoma tigris* (Dyar), Sonoran tent caterpillar. f.g.l. 50± mm. A4 dorsum showing longitudinal orange and black (tiger) lines. Best recognized by the largely black dorsal and lateral areas of A8 and the half-moon-shaped subdorsal spots on T2 and T3. Like its close relative *constrictum,* it may be largely orange and black with orange lateral setae (Texas) or largely blue and black with white lateral setae (Utah). On deciduous or live oak from eastern Texas to Colorado, Arizona and south to Chiapas in Mexico. Builds only a small molting tent for crochet attachment; has defoliated scrub oaks in Texas.

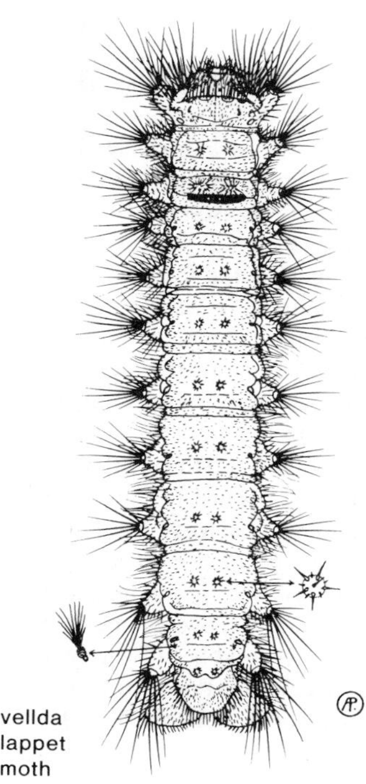

Figure 26.300

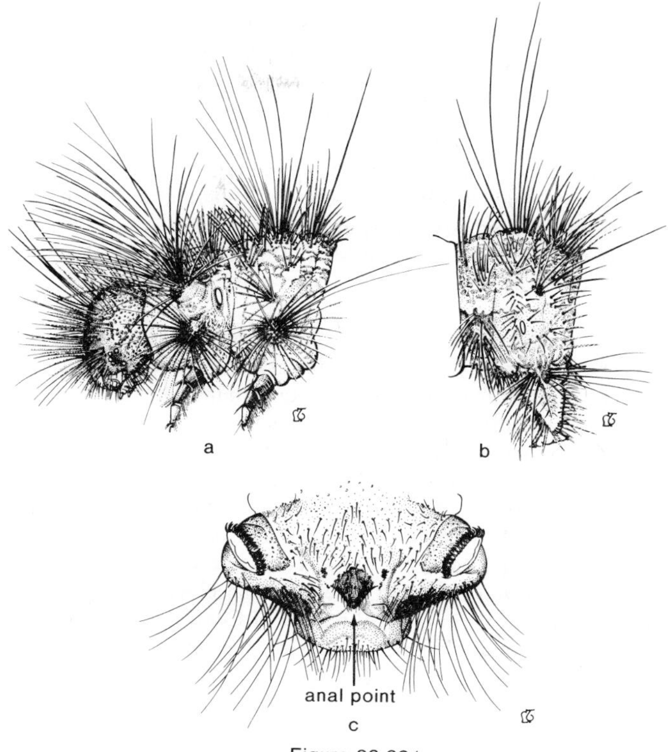

Figure 26.301

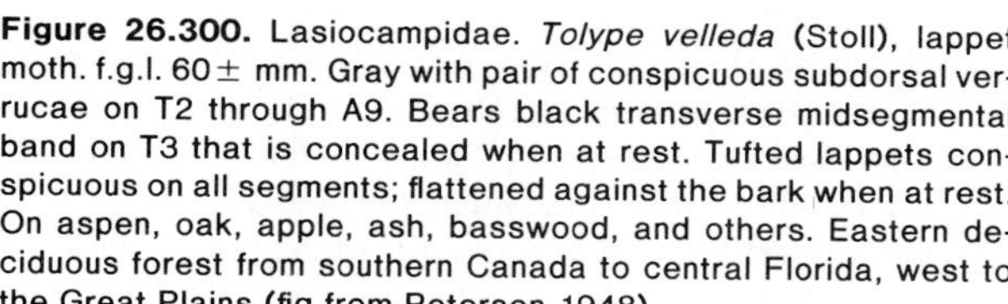
Figure 26.300. Lasiocampidae. *Tolype velleda* (Stoll), lappet moth. f.g.l. 60± mm. Gray with pair of conspicuous subdorsal verrucae on T2 through A9. Bears black transverse midsegmental band on T3 that is concealed when at rest. Tufted lappets conspicuous on all segments; flattened against the bark when at rest. On aspen, oak, apple, ash, basswood, and others. Eastern deciduous forest from southern Canada to central Florida, west to the Great Plains (fig from Peterson 1948).

Figure 26.301a,b. Lasiocampidae. *Gloveria* probably *gargamelle* (Strecker). f.g.l. 80± mm. a. head, T1, T2; b. A4; c. A10 venter, showing "anal point" between prolegs, a helpful diagnostic structure for North American lasiocampid larvae (see text and Apatelodidae). Mature larva mottled brownish-gray with relatively wide middorsal stripe bordered by irregular narrow darker stripe. Covered with dense silky hair, which is urticating to sensitive skin such as between the fingers. On *Quercus* sp., Guadelupe Mtns., Culberson Co., Tex., 6 Apr. 1961, Michigan State Univ. Central Arizona to southwest Texas on oaks.

Tent caterpillars, because of their destructiveness, have been the subject of many population studies and behavioral studies. In many respects they are ideal subjects, since those that build tents are easily located in the field in all life stages, and they are also easily reared in the laboratory. Details on all species of *Malacosoma* and color plates of the larvae can be found in Stehr and Cook (1968). Franclemont (1973) describes the known larvae of other genera and reproduces the *Malacosoma* color plates of Stehr and Cook (1968).

Selected Bibliography

Dethier 1980.
Franclemont 1973.
Furniss and Carolin 1977.
Ives 1973.
Johnson and Lyon 1976.
Stehr and Cook 1968.
U.S.D.A. 1985.

SATURNIIDAE (BOMBYCOIDEA)

George L. Godfrey,
Michael Jeffords,
James E. Appleby, ***Illinois Natural History Survey***

Royal and Giant Silkworms

Figures 26.302–26.325

Relationships and Diagnosis: Saturniids belong in the Bombycoidea (Ferguson 1971–1972), but because of superficial similarities, larval saturniids or "wild silkworms" are more often mistaken for larvae of Nymphalidae than they are for other bombycoids. Bodies of both groups generally have prominent chalazae or scoli (enlarged tubercles) with branching setae or spines, and possess varying amounts of secondary setae on the primary tubercles and/or elsewhere.

A reliable distinguishing feature that separates these groups is the head, which is smooth and rounded in the Saturniidae, but angulate or bearing dorsal tubercles or scoli in the Nymphalidae. The combined characters of the chalazae or scoli (at least rudimentary), smooth hypognathous head, and fully developed prolegs on A3–6 and A10 with strong biordinal crochets arranged in a homoideous mesoseries distinguish ultimate saturniid larvae from other Lepidoptera. North American saturniid larvae may attain lengths in excess of 100 mm (*Citheronia regalis* (Fabricius) (fig. 26.303)), but the majority of the species are medium-sized, measuring ca. 40–70 mm.

Two important aspects of saturniid chaetotaxy are that some have secondary setae as first instars, and that all exhibit fusion of some tubercles in many setal groups. Detailed information and discussions of the chaetotaxy and characteristics of saturniid larvae, including coverage of the first instars, phylogenetic implications, and other points may be found in Packard (1905, 1914), Forbes (1923), Munroe (1949), Michener (1952), Pease (1961), and Ferguson (1971–1972).

Biology and Ecology: Of the 1100 worldwide species of saturniids that represent 125 genera, 600 species occur in the New World, mostly in tropical regions of Central and South America. In America north of Mexico, the saturniid fauna consists of 65 species in 18 genera. Twenty-one species range northward into Canada, but none occur there exclusively (Ferguson 1971–1972).

Saturniidae are warm season, single-, double-, or even triple-brooded (*Sphingicampa bicolor* (Harris) (fig. 26.305)) medium- to large-sized moths. Certain species, such as *Coloradia pandora* Blake (fig. 26.311), have a two-year life cycle with the young larvae overwintering as second instars on the host plant (Carolin 1971). Others, like *Callosamia angulifera* (Walker) (fig. 26.322), are single-brooded in the North and double-brooded in the South (Eliot and Soule 1902).

Saturniid larvae are foliage feeders, principally on deciduous trees and shrubs. However, notable exceptions in host plant preference occur in all three subfamilies. The hemileucine, *Coloradia pandora,* feeds on many species of pine in the Rocky Mountain region, whereas *Hemileuca oliviae* Cockerell (fig. 26.312) uses many grass species found in the southwestern U.S. (Ferguson 1971–1972). In the Citheroniinae, *Citheronia sepulcralis* Grote and Robinson (fig. 26.304) has been reported from *Pinus strobus, P. rigida,* and *P. caribaea* (Packard 1905, Kimball 1965). *Hyalophora columbia* (S. I. Smith) (fig. 26.324), a member of the Saturniinae, feeds on larch. Approximately one-third of North American saturniid larvae are polyphagous (e.g., *Anisota*); the remainder are monophagous or relatively so (e.g., *Callosamia angulifera*). In general, saturniid larvae consume the entire leaf, leaving only the petiole. Feeding behavior ranges from gregarious through most of larval life (e.g., *Hemileuca oliviae, Anisota senatoria* (J. E. Smith) (fig. 26.309) to solitary (e.g., *Hyalophora cecropia* (Linnaeus) (fig. 26.323)).

Saturniid larvae are usually perceived as medium-sized to large-sized sluggish caterpillars that do not disperse any great distance from the host plant. Most species have well-developed crochets and grip the host plant tightly, which makes beating plants for collecting or sampling unproductive. The larvae of *Hemileuca oliviae,* the range caterpillar, do disperse widely during the penultimate or ultimate instar (Ainslie 1910).

Saturniid eggs are of two distinct types. Citheroniine eggs have nearly transparent shells, which give them a translucent appearance. No reddish-brown adhesive is present like that found on the opaque, whitish eggs of the other two North American subfamilies (Ferguson 1971–1972). Eggs are elongate, upright, and may be laid singly, in small groups, or in large clusters that encircle the food plant (*Coloradia, Hemileuca*). The latter pattern is associated with overwintering hemileucine eggs. Citheroniinae and Saturniinae overwinter as pupae, with the exception of the montane *Saturnia albofasciata* (Johnson), which overwinters as an egg (Hogue et al. 1965).

Pupation occurs in an earthen cell, or enclosed in some type of cocoon, either attached to the host plant or loose in the litter. Citheroniine larvae leave the host and enter the soil where they construct a cell and molt to the pupa. No cocoon is formed. A characteristic citheroniine pupa is covered with numerous spines and has a moveable but nontelescoping abdomen. The spines presumably aid the pupa in emergence from the soil prior to eclosion. Pupae of *Citheronia* differ from the norm by being smooth, not spinose (Mosher 1914). The Hemileucinae usually do not enter the soil, but pupate at or near the surface under litter or debris. The pupa is usually without spines. The cocoon may be present as only a few strands of silk (*Hemileuca*), or as a well-formed structure (*Automeris*) (Ferguson 1971–1972). Saturniine larvae usually pupate on the host plant in a well-formed, often double-walled cocoon that may remain attached to the host plant (e.g., *Callosamia promethea* (Drury)), or fall with the leaves (e.g., *Antheraea polyphemus* (Cramer)). Pupae of *Actias* are not formed on the plant, but among the debris on the ground. Cocoons are relatively distinctive for each species, with the cocoon of *Actias* considered to be the most primitive (Ferguson 1971–1972).

Defense tactics of saturniid larvae are not as well developed as in other families of Lepidoptera. The Hemileucinae (figs. 26.311–26.315) and a few Saturniinae (*Saturnia,* fig. 26.316) are covered with numerous, quill-like hairs that possess an extremely irritating toxin (Comstock and Dammers 1938). The hairs are often barbed and break off, but the defense seems to be primarily chemical. The Citheroniinae appear to be without defenses except for crypsis and disruptive coloration. The sphinx caterpillarlike behavior of *Antheraea* larvae may be a startle response against predators, but this is unconfirmed.

Description: Mature larva medium to large, 35–140 mm. Head hypognathous, exposed, sometimes slightly withdrawn under T1 shield, e.g., *Callosamia* (fig. 26.322) and *Hyalophora* (fig. 26.323); junction of adfrontal ecdysial lines distantly cephalad of epicranial notch; six stemmata per side. Body cylindrical and appearing stout. Integument conspicuously setose (fig. 26.302), granulate (fig. 26.306), or smooth (fig. 26.317). Chalazae or scoli and branching spines or setae

Figure 26.302

Figure 26.303

Figure 26.304

Figure 26.305

Figure 26.302. Saturniidae, Citheroniinae. *Eacles imperialis* (Drury), imperial moth, green form. f.g.l. 75-100 mm. Dark green or brown. Feeds on many broadleafed trees and some conifers, especially *Pinus*. Nominate subspecies occurs throughout most of the eastern U.S. and southern Ontario. Replaced northward by *Eacles imperialis pini* Michener, which feeds only on *Pinus;* other subspecies southward. Univoltine northward, bivoltine southward in North America. Eliot and Soule (1902), Packard (1905), Ferguson (1971-1972), and Janzen (1982). Photo by J.E. Appleby.

Figure 26.303. Saturniidae, Citheroniinae. *Citheronia regalis* (Fabricius), hickory horned devil. f.g.l. 125-140 mm. Basically olive and bluish-green or brownish with pale green, blue, or yellow lateral, oblique stripes on A2-A8. Scoli on T2 and T3 orange, black tipped, but majority of scoli black. Arboreal, on broadleafed species, especially butternut and walnut, pecan and other hickories, sweetgum *(Liquidambar),* and various sumachs. Eastern half of the U.S. from Missouri to Massachusetts, south to Texas and central Florida. Univoltine (possible occasional second larval generation in the South), matures in late summer. Eliot and Soule (1902), Packard (1905), and Ferguson (1971-1972). Photo by M. Jeffords.

Figure 26.304. Saturniidae, Citheroniinae. *Citheronia sepulcralis* Grote and Robinson, pine-devil moth. f.g.l. 90-100 mm. Brownish, black on dorsum of T2 and T3, shaded with black elsewhere. T1-T3 each with single pair of long, yellowish-brown, dorsal scoli; A8 and A9 each with single, middorsal scolus. Feeds on needles of various pines: *Pinus rigida, P. strobus,* and *P. caribaea.* Larval distributional records scarce; basically restricted to Atlantic coastal states (Maine to Florida), also reported from Alabama, Mississippi, Tennessee, and Kentucky. Univoltine northward (mature larva in July-August); bi-, possibly trivoltine southward. Packard (1905), Peterson (1948), Kimball (1965), Horn (1969), and Ferguson (1971-1972). Photo by G. R. Carner, courtesy of R. S. Peigler.

Figure 26.305. Saturniidae, Citheroniinae. *Sphingicampa bicolor* (Harris), honey locust moth. f.g.l. 40-60 mm. Green with slight bluish cast. Lateral area: pale blue line level with dorsal half and red line level with ventral half of spiracles, paralleled below by wide, white line. Subventer and venter of A1-A8 spattered with heterogeneous-sized silvery-white and black granules. Apparently restricted to foliage of Fabaceae: honey locust and Kentucky coffee-tree. Occurs between the Appalachian Mountains and eastern edge of Great Plains, especially Ohio and Kentucky, west to Kansas. Trivoltine. Packard (1905) and Ferguson (1971-1972). Photo by J. E. Appleby.

Figure 26.306

Figure 26.307

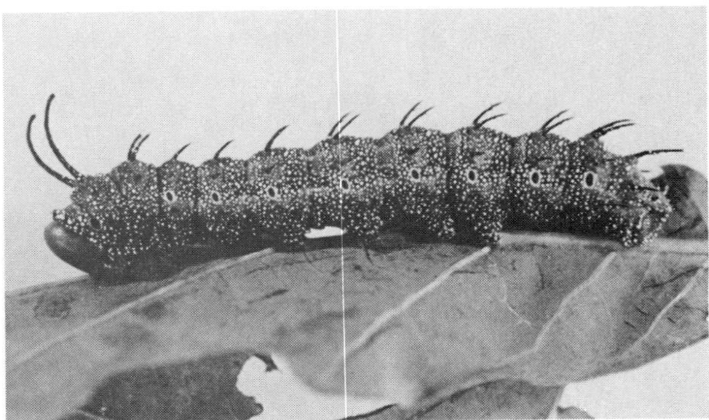

Figure 26.308

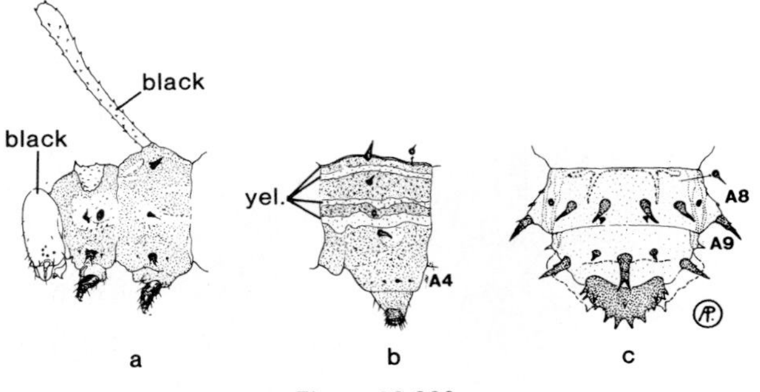

Figure 26.309

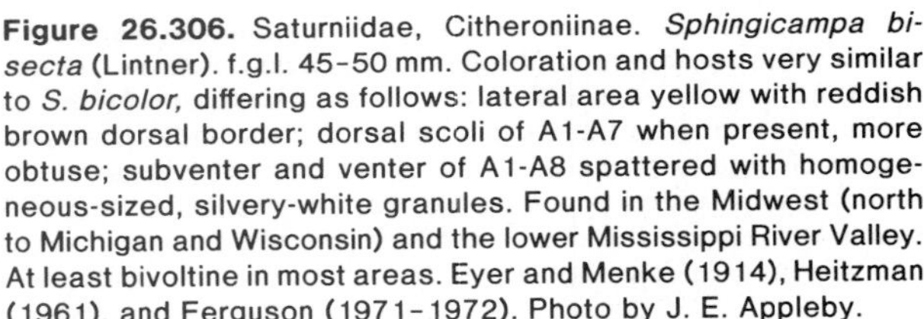

Figure 26.306. Saturniidae, Citheroniinae. *Sphingicampa bisecta* (Lintner). f.g.l. 45-50 mm. Coloration and hosts very similar to *S. bicolor,* differing as follows: lateral area yellow with reddish brown dorsal border; dorsal scoli of A1-A7 when present, more obtuse; subventer and venter of A1-A8 spattered with homogeneous-sized, silvery-white granules. Found in the Midwest (north to Michigan and Wisconsin) and the lower Mississippi River Valley. At least bivoltine in most areas. Eyer and Menke (1914), Heitzman (1961), and Ferguson (1971-1972). Photo by J. E. Appleby.

Figure 26.307. Saturniidae, Citheroniinae. *Dryocampa rubicunda* (Fabricius), greenstriped mapleworm. f.g.l. 40-50 mm. Body stripes: dark to blackish green and pale green or black and yellow; lateral area of A7-A9 red. T2 with single pair long black scoli. Lateral scoli on T2-A8, single, black, short, thornlike. Pair of similar scoli on dorsum of A8 and single, middorsal scolus on A9. Body otherwise covered with rounded granules. Principally feeds on maples (especially red, sugar, and silver) and less frequently on oak; an occasional defoliator. Eastern half of U.S. and adjacent Canada. Univoltine northward, bivoltine (possibly trivoltine) southward. Eliot and Soule (1902), Packard (1905), Howard and Chittenden (1909), and Ferguson (1971-1972). Photo by J. E. Appleby.

Figure 26.308. Saturniidae, Citheroniinae. *Anisota stigma* (Fabricius), spiny oakworm. f.g.l. 45-60 mm. Brownish orange, extensively covered with whitish granules; spiracles black, each enclosed by white ellipse. Scoli black. T2 dorsally with single pair scoli (length ca. 7 mm); T3-A8 armed with similar but shorter and sharper pairs of scoli dorsally, subdorsally, and laterally. Head and thoracic legs brown. See Riotte and Peigler [1980(1981)] for key to and diagnostic characters of North American *Anisota* larvae. Defoliater of various oaks [Eliot and Soule (1902) mentioned "low-growing oaks"], occasionally reported on hazel *(Corylus),* once reported from Chinese chestnut *(Castanea mollissima).* Massachusetts to central Florida, west to Minnesota, Kansas, and Arkansas; southern Ontario. Univoltine in the North; undefined southward. Eliot and Soule (1902), Packard (1905), Ferguson (1971-1972), and Riotte and Peigler [1980(1981)]. Photo by G. L. Godfrey.

Figure 26.309a-c. Saturniidae, Citheroniinae. *Anisota senatoria* (J. E. Smith), orangestriped oakworm. a. head, T1, T2; b. A4; c. A8-10, dorsal aspect. f.g.l. 35-45 mm. Dark brown to black with 4 thin, yellow or orange lines per side. T2 with single pair of thin dorsal scoli (length ca. 4 mm); soli on remainder of body short, thornlike. May be confused with *A. finlaysoni* Riotte except T2 dorsal scolli of the latter very short (ca. 1 mm), see Ferguson (1971-1972) and Riotte and Peigler [1980(1981)] for further discussion. Defoliator of various oaks, occasionally reaching pest status. Most of eastern U.S. but not north of Massachusetts, rarely in Florida, west to Minnesota and Texas; southern Ontario. Univoltine northward, undefined southward. Packard (1905), Peterson (1948), Hitchcock (1958, 1961), Ferguson (1971-1972), and Riotte and Peigler [1980(1981)]. (fig. from Peterson 1948).

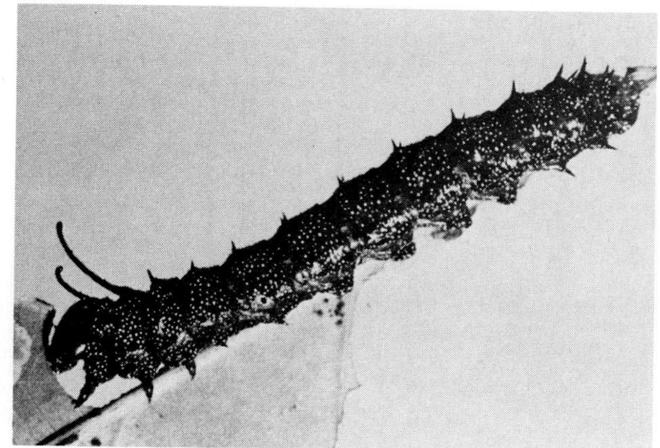

Figure 26.310

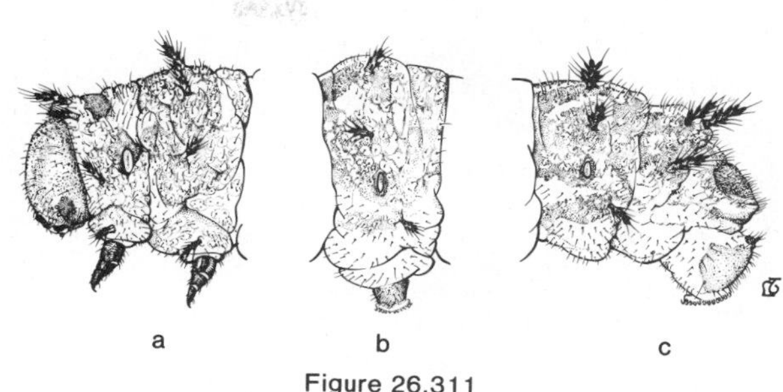

Figure 26.311

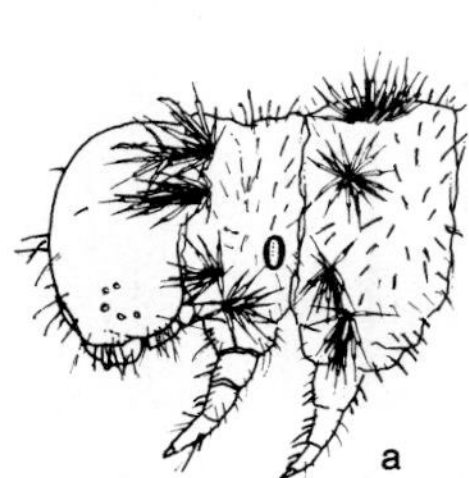

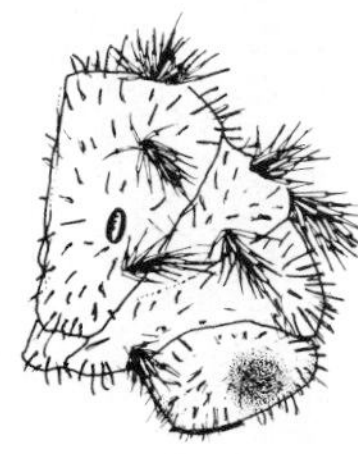

Figure 26.312

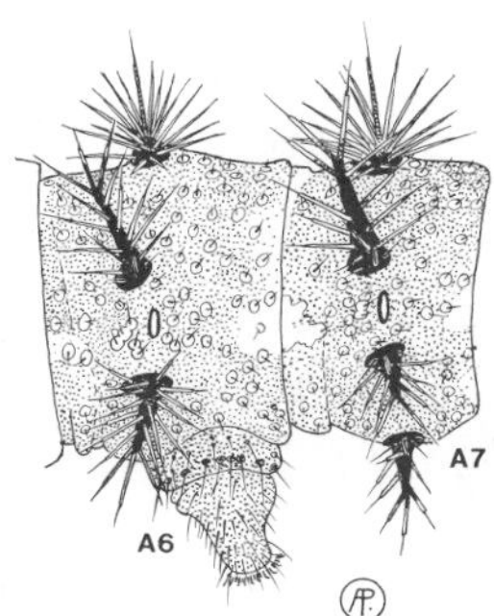

Figure 26.313

Figure 26.310. Saturniidae, Citheroniidae. *Anisota virginiensis* (Drury), pinkstriped oakworm. f.g.l. 40–55 mm. Pale olive green, speckled with white granules; subdorsal and lateral areas pink. T2 with single pair dorsal scoli (length ca. 6–8 mm). Spiracles black, within white ellipses. Head olive green. Various hosts have been reported, but primarily a defoliator of oaks. Mainly from Virginia to Arkansas, northward to Nova Scotia and Manitoba. Univoltine. Packard (1905), Ferguson (1971–1972), and Riotte and Peigler [1980(1981)]. Photo by J. E. Appleby.

Figure 26.311a-c. Saturniidae, Hemileucinae. *Coloradia pandora* Blake, pandora moth. a. head, T1, T2; b. A4; c. A8–10. f.g.l. 60–70 mm. Adequate descriptions of coloration and markings are lacking. Reportedly yellow-green (living) and pale brown (preserved) with pale middorsal line. Stemmatal area darkly pigmented on brownish head. Body scoli subequal in length; dorsal and lateral scoli bearing stubby branches. At times a serious defoliator of *Pinus ponderosa* and *P. jeffreyi.* The nominate subspecies occurs in the pine areas of Colorado, western S. Dakota, Wyoming, and northern Utah, whereas *C. p. lindseyi* Barnes and Benjamin is known from the mountain ranges of Oregon and California, especially the Cascade and Sierra Nevada. A 2-year life cycle, larvae present from about early August (overwinters as young larva) to late June or early July the next season, then pupates and overwinters before emerging the succeeding June or July. Patterson (1929), Wygant (1941), Carolin (1971), Ferguson (1971–1972), and Tuskes (1976), 1984 (1985). Drawings by L. Tackett, courtesy of F. W. Stehr.

Figure 26.312a-c. Saturniidae, Hemileucinae. *Hemileuca oliviae* Cockerell, range caterpillar. a. head, T1, T2; b. A6; c. A8–10. f.g.l. 50–60 mm. Mottled, yellowish brown to black with numerous whitish yellow, secondary setae. Surface of body slightly "warty" at 25X. Head reddish brown to black. Primary setae urticaceous. Gregarious, young larvae exhibit processional behavior. A major pest of rangeland, feeds on native grasses, especially gramma grasses, *Bouteloua* spp. Endemic, northeastern New Mexico and adjacent areas of Colorado, Texas, and western Oklahoma. Univoltine, larvae from late May to mid-September. Ainslie (1910), Ferguson (1971–1972), and Watts and Everett (1976). Drawings by J. L. Wissmann, I.N.H.S.

Figure 26.313. Saturniidae, Hemileucinae. *Hemileuca maia* (Drury), buck moth, A6–7. f.g.l. 45–60 mm. Yellow brown to purplish black, finely dotted with pale yellow spots marking bases of secondary setae, spots more numerous in lateral area. Reddish scoli bearing pale yellow and black, urticaceous spines. Pale specimens are similar to larvae of *H. nevadensis* Stretch, a western species, but lateral area more distinct and spines usually longer than in *nevadensis.* See Ferguson (1971–1972) for details. Young larvae gregarious. Mainly a defoliator of oaks, e.g., *Quercus ilicifolia* and *O. virginiana,* but willow a reported host in parts of the Midwest. Eastern U.S. from about New Hampshire to central Florida, westward to Illinois and Texas. Univoltine, feeds from late April to midsummer. Packard (1914), Holland (1903), Peterson (1948), and Ferguson (1971–1972). (fig. from Peterson 1948).

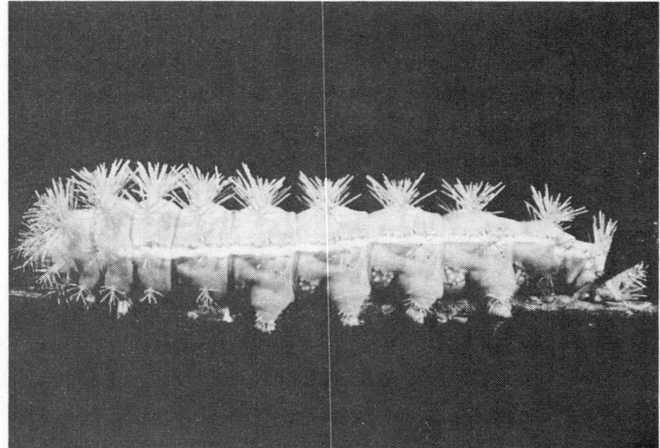

Figure 26.314

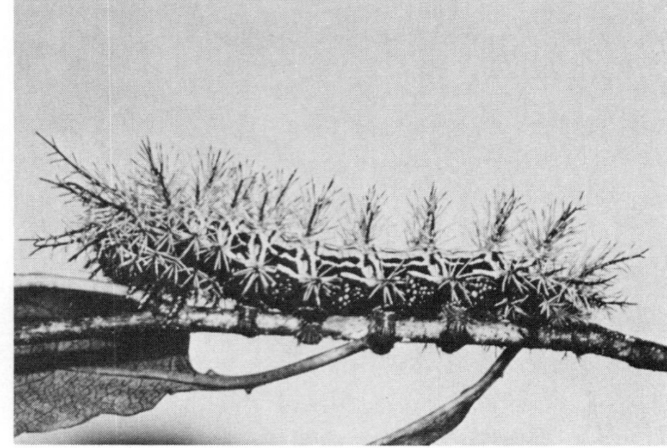

Figure 26.315

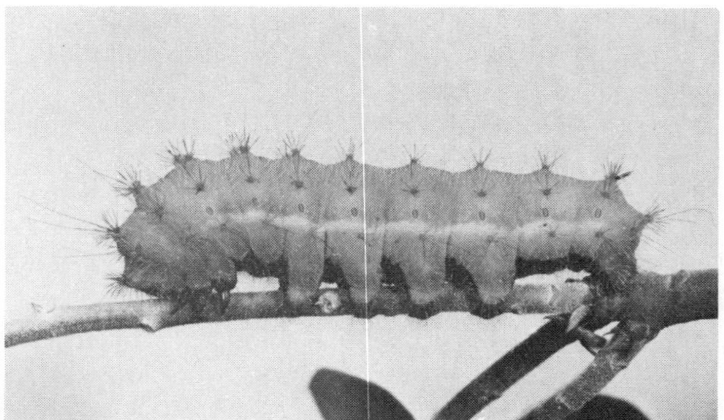

Figure 26.316

Figure 26.317

Figure 26.314. Saturniidae, Hemileucinae. *Automeris io* (Fabricius), io moth. f.g.l. 60± mm. Head and body yellowish green; lateral area white with red dorsal and ventral borders; spiracles white; thoracic legs and abdominal prolegs red. Spines urticaceous. Young larvae gregarious. Polyphagous, principally broadleaf shrubs and trees, some herbs, including large grasses, but not on conifers. Known throughout most of eastern North America from Canada to Costa Rica. Univoltine northward, bi- or trivoltine southward. Packard (1914), Ferguson (1971-1972), and Janzen (1982). Photo by M. Jeffords.

Figure 26.315. Saturniidae, Hemileucinae. *Automeris zephyria* Grote. f.g.l. 64-70 mm. Body (dorsal half) yellow with 4 black lines per side, middorsal line white; (ventral half) black with reddish speckles and 2 thin, interrupted white lines. Scoli yellow; urticating spines yellow, tips black. Head and prolegs reddish brown. Tends to aggregate at night on branches and trunk of host tree, *Quercus gambelii.* Reared on *Prunus serotina* in confinement (R. S. Peigler, pers. comm.). New Mexico and Guadalupe Mountains of Texas. Univoltine, larva in mid- to late summer. Snow (1885), Tietz (1972), and Ferguson (1971-1972). Photo by G. R. Carner, courtesy of R. S. Peigler.

Figure 26.316. Saturniidae, Saturniinae. *Saturnia mendocino* Behrens. f.g.l. 65± mm. Pale yellowish green, lateral area below spiracles pale yellow; D, SD, L scoli of most segments reddish orange; setae (some urticaceous) mainly white, terminal setae on scoli black. Head (covered with whitish secondary setae) and venter of body purplish brown. See Tuskes (1976) for contrasting characters of *S. mendocino, S. walterorum,* and *S. albofasciata* (Johnson). Feeds on *Arctostaphylos* and *Arbutus.* Occurs in chaparral, pine, or mixed woodland from extreme southern Oregon, southward to Santa Cruz and Mariposa counties, California. Ferguson (1971-1972) and Tuskes (1976). Photo by P. M. Tuskes.

Figure 26.317. Saturniidae, Saturniinae. *Antheraea polyphemus* (Cramer), polyphemus moth. f.g.l. 60-75 mm. Body translucent yellowish green; dorsal, subdorsal, and lateral scoli yellowish orange to red; lateral aspects of dorsal and subdorsal scoli on A1-A8 silvery white; spiracles orange; segmental series of yellow lines on A2-A7 extending from subdorsal to lateral scoli and touching posterior margin of spiracles. Head brownish. General feeder, especially on broad-leafed trees and shrubs; common at times on birch. Transcontinental, southward into Mexico. Univoltine northward, bivoltine southward. Eliot & Soule (1902), Packard (1914), and Ferguson (1971-1972). Photo by P. M. Tuskes.

Figure 26.318

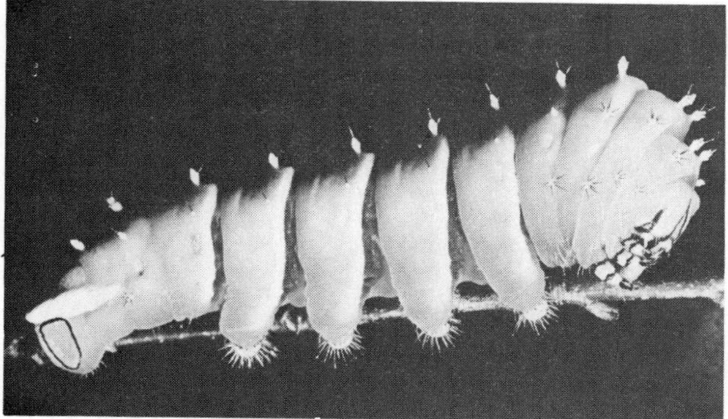

Figure 26.319

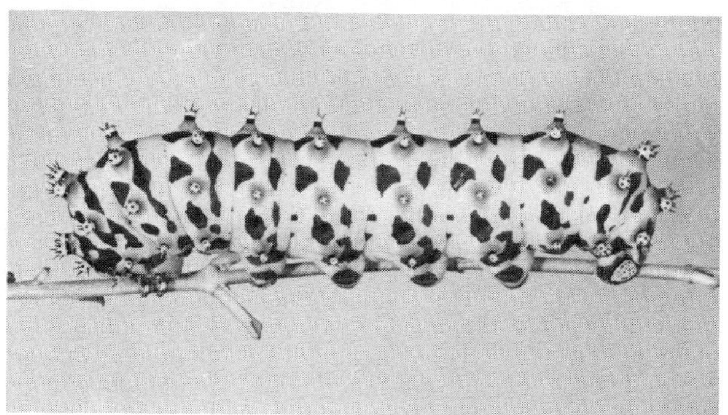

Figure 26.320

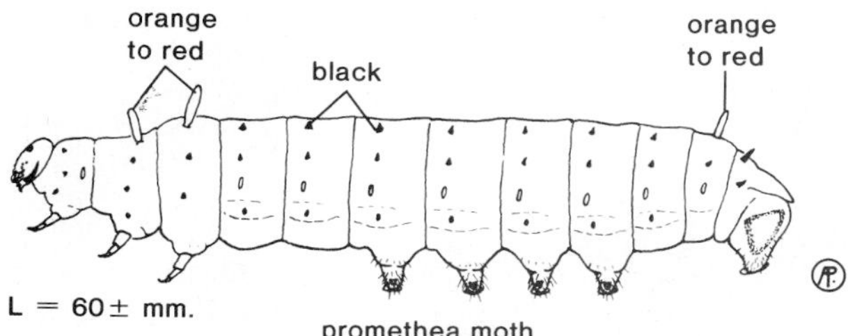

Figure 26.321

Figure 26.318. Saturniidae, Saturniinae. *Actias luna* (L.), luna moth f.g.l. 55-70 mm. Translucent green; T2 and T3 dorsal (D2) scoli, reddish, larger than other scoli which are white to red; narrow yellow area extending below orange spiracles; segmental series of yellow lines on posterior region of A1-A7 extending from lateral area to dorsum, not touching spiracles. Head brown to green. General feeder on broadleaf trees, reported from nearly 20 genera. Eastern North America from southern Canada to southern U.S. Univoltine northward, trivoltine southward. Eliot and Soule (1902), Packard (1914), and Ferguson (1971-1972). Photo by J. E. Appleby.

Figure 26.319. Saturniidae, Saturniinae. *Rothschildia forbesi* Benjamin. f.g.l. 60± mm. Translucent yellowish green; anterior portions of A2-A7 translucent bluish gray followed by narrow white, transverse band; spiracles and scoli yellow. Head greenish with black lateral stripes. Apparently a general feeder on broadleaf shrubs and trees; oviposition and cocoons have been observed on *Citrus,* ash, and willow, all implied hosts, as is cat's claw or esperanza *(Acacia).* Texas (Lower Rio Grande Valley) and Mexico (Monterey). Bivoltine. Collins and Weast (1961). Photo by J. G. Sternburg.

Figure 26.320. Saturniidae, Saturniinae. *Eupackardia calleta* (Westwood). f.g.l. 65-70 mm. Pale green with slight bluish cast, extent of black variable (black dorsal transvere bands may occur on anterior of abdominal segments, extending to level of spiracles); dorsal scoli stout, orange to red basally, black medially, blue distally; spiracles black. Young larvae solitary. Apparently a general feeder, reported from cat's claw *(Acacia),* ceniza *(Leucophyllum frutescens),* mesquite *(Prosopis),* and ocotillo *(Fouquieria splendens);* reared on other plants. Southern Texas, southern Arizona, and Mexico. Bivoltine in Texas, uncertain in Arizona. Packard (1914), Collins and Weast (1961), and Ferguson (1971-1972). Photo by P. M. Tuskes.

Figure 26.321. Saturniidae, Saturniinae. *Callosamia promethea* (Drury), promethea moth. f.g.l. 60-77 mm. Pale green with slight bluish cast, especially at bases of scoli. Five prominent dorsal scoli, pairs on T2 and T3 orange to red, single middorsal A8 scolus yellow (all with black basal ring and devoid of distal spines or setae); other scoli black, reduced to short tubercles. Head with 2 black facial spots at level of frontocylpeal apex. Young larvae gregarious. General feeder on broadleaf, woody hosts, reported from 12+ families. Eastern half of U.S. and contiguous parts of Canada. Univoltine northward, bivoltine southward. Eliot and Soule (1902), Packard (1914), and Ferguson (1971-1972). (fig. from Peterson 1948).

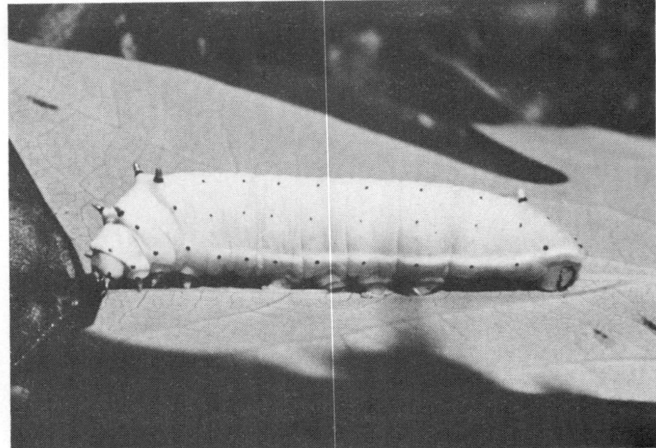

Figure 26.322

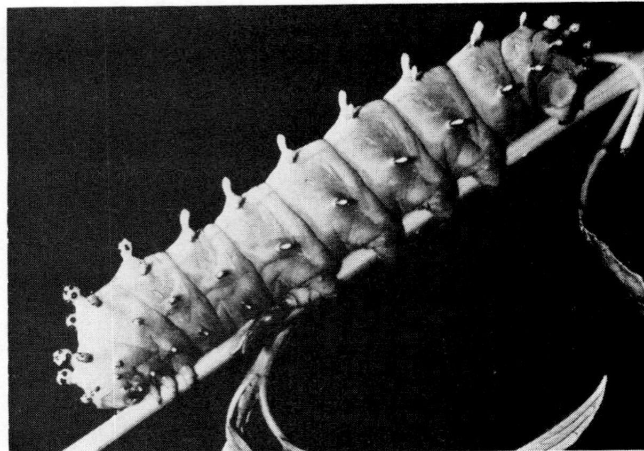

Figure 26.323

Figure 26.324

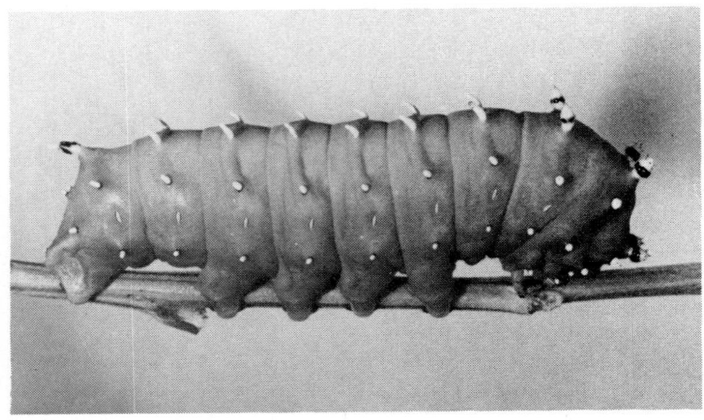

Figure 26.325

Figure 26.322. Saturniidae, Saturniinae. *Callosamia angulifera* (Walker). f.g.l. 70± mm. Similar to *C. promethea* but chalky or whitish green, all tubercles generally shorter, and head lacking 2 black facial spots. Principally feeds on tulip tree. East of the Mississippi River, south from Massachusetts, New York, Ontario, and southern Michigan, to northern Florida and northern Mississippi. Univoltine northward, bivoltine southward. Eliot and Soule (1902), Packard (1914), and Ferguson (1971-1972). Photo by J. G. Sternburg.

Figure 26.323. Saturniidae, Saturniinae. *Hyalophora cecropia* (Linnaeus), cecropia moth. f.g.l. 85-100 mm. Green with bluish tinge. Dorsal scoli on T2 and T3 yellow to red and on A1-A8 yellow; subdorsal and lateral scoli blue with pale blue or white tips. Differs from *H. columbia* (S. I. Smith) by presence of only 2 pairs of red dorsal scoli (T2 and T3) (*H. columbia* has 3 pairs (T2-A1) or by having all dorsal scoli yellow); if scoli yellow, *H. cecropia* separable from *H. euryalus* Boisduval by lacking black rings on dorsal scoli of T2, T3, and A1 and differs from *H. gloveri* (Strecker) by lacking black basal rings on lateral abdominal scoli. General feeder on broadleaf shrubs and trees, including at least 16 genera in 9 families. Known from most of southern Canada (Nova Scotia to Alberta) and most of the eastern U.S., ranging west to Washington, Utah, and Texas. Univoltine, but moths may have 2 emergence peaks. Eliot and Soule (1902), Packard (1914), Ferguson (1971-1972), and Sternburg et al. (1981). Photo by W. D. Zehr, I.N.H.S.

Figure 26.324. Saturniidae, Saturniinae. *Hyalophora columbia* (S. I. Smith). f.g.l. 75± mm. Similar to *H. cecropia* but slightly paler green, distinguishable from it and other *Hyalophora* by presence of 3 pairs of red, dorsal scoli (T2, T3, and A1). Principally restricted to larch. Occurs in boggy habitats, Nova Scotia to Manitoba and New Hampshire to northern Wisconsin and into southern Michigan. Univoltine. Caulfield (1878), Fernald (1878), Packard (1914), and Ferguson (1971-1972). Photo by J. G. Sternburg.

Figure 26.325. Saturniidae, Saturniinae. *Hyalophora euryalus* (Boisduval), ceanothus silkmoth. f.g.l. 85± mm. Pale green, dorsal scoli yellow, those on T2, T3, and A1 bearing black rings; subdorsal and lateral scoli whitish with black basal rings. The yellow scoli of T2-A1 distinguish this species from *H. columbia;* black rings on same scoli separate it from *H. gloveri* and *H. cecropia.* Common on *Ceanothus,* but also reported from various other genera of trees and shrubs from several families. Occurs in chaparral, oak, and pine forests from British Columbia to Baja California, occasionally eastward to western Montana and western Nevada. Number of generations unclear, thought to be univoltine. Edwards (1873), Packard (1914), Ferguson (1971-1972), and Tuskes (1976). Photo by P. M. Tuskes.

arising from structures variously prominent; dorsal spines urticating in Hemileucinae (figs. 26.311–26.315) and some Saturniinae. Body coloration cryptic (fig. 26.304), disruptive (fig. 26.305), or gaudy (fig. 26.320). Spiracles elliptical, conspicuous (fig. 26.302) to indistinct (fig. 26.322). A10 prolegs generally appearing larger than prolegs on A3-A6 because of presence and ventral extension of lateral plate.

Comments: Typically, most saturniid larvae seldom come into conflict with humans or their activities. However, exceptions are found in the Citheroniinae and Hemileucinae. The orange-striped oakworm, *Anisota senatoria* (J. E. Smith) (fig. 26.309), has been reported to cause serious defoliation of oaks during outbreak years (Packard 1905). Hitchcock (1958, 1961) considered its importance and impact second only to that of the gypsy moth, *Lymantria dispar* (L.) (Lymantriidae). In the Southwest, particularly New Mexico and Colorado, *Hemileuca oliviae* (fig. 26.312) undergoes massive, periodic outbreaks and can cause severe damage to pastureland (Ainslie 1910). During a 1970 outbreak, over 750,000 acres were treated with toxaphene for *H. oliviae* control (Ferguson 1971–1972). Another hemileucine, *Coloradia pandora* (fig. 26.311), which causes considerable defoliation in the pine forests of Colorado (Wygant 1941), reaches epidemic proportions every 20 to 30 years (Patterson 1929).

Although it appears that most of the larger saturniids have not adapted particularly well to humans and urbanization (Ferguson 1971–1972), at least one species, *Hyalophora cecropia* (fig. 26.323), may actually be on the increase. Sternburg et al. (1981) found that in new urban and suburban areas, *H. cecropia* populations were higher than in rural or wooded areas. Much of this increase is attributed to the large differential in predation pressure between these habitats, particularly from mice that prey on cocoons near the ground in nonurban areas.

Selected Bibliography

Ainslie 1910.
Carolin 1971.
Comstock and Dammers 1938.
Caulfield 1878.
Collins and Weast 1961.
Edwards 1873.
Eliot and Soule 1902.
Eyer and Menke 1914.
Forbes 1923. (general)
Ferguson, D. C., *in* Dominick et al. 1971–1972.
Fernald 1878.
Heitzman 1961.
Hitchcock 1958, 1961.
Hogue et al. 1965[1966].
Holland 1903. (general)
Horn 1969.
Howard and Chittenden 1909.
Janzen 1982.
Kimball 1965. (general)
Michener 1952.
Mosher 1914, 1916b.
Munroe 1949.
Packard 1905, 1914.
Patterson 1929.
Pease 1961.
Peterson 1948. (general)
Riotte and Peigler 1980(81).
Snow 1885.
Sternburg et al. 1981.
Tietz 1972.
Tuskes 1976, 1984 (1985).
Watts and Everett 1976.
Wygant 1941.

Acknowledgments

J. L. Wissmann graciously provided illustrative work. We are appreciative of the Dixon Springs Agriculture Center, Simpson, Illinois and The Morton Arboretum, Lisle, Illinois, for providing research facilities. Work was funded in part by grants from the Joyce Foundation and by Illinois Agricultural Experiment Station Project 12–361 Biosystematics of Insects.

SPHINGIDAE (SPHINGOIDEA)

Frederick W. Stehr, *Michigan State University*

The Hornworms, Hummingbird Moths, Hawkmoths, Sphinx Moths

Figures 26.326-26.333

Relationships and Diagnosis: Sphingids, some of our more highly advanced moths, are generally placed between the Bombycoidea and Noctuoidea in their own superfamily. The vast majority of both the adults and larvae are readily recognizable to family, the larvae by their relatively naked appearance (some have short or stubby setae), A10 terminating in three angular lobes, distinct body annuli, and a usually conspicuous (rarely small) middorsal A8 horn. Some mature larvae have a "button" instead of a horn (figs. 26.332, 26.333a; earlier instars nearly always have a horn), and the genus *Lapara* has neither a horn nor a button, but the head is rather high and triangular (figs. 26.329a,b). The domestic silkworm, a couple of notodontids and possibly a saturniid are the larvae most likely to be confused with sphingid larvae since they have an A8 horn or scolus, but sphingid larvae have 6–8 annulets per segment, whereas silkworms (fig. 26.294), notodontids (fig. 26.340), and saturniids (figs. 26.302–26.305) have none.

Biology and Ecology: Sphingid eggs are, with few exceptions, laid singly, but as with many Lepidoptera, eggs may be laid near one another and several to many larvae can often be collected in the same vicinity. The larvae of many species have striking patterns, even though many are rather hard to spot on their hosts and some are very cryptic. Hosts include a great diversity of shrubs, vines, and trees, and less commonly herbs. Pupation usually takes place in earthern cells, a loose cocoon at the soil surface, or rarely in a loose cocoon above the surface (Hodges 1971). Overwintering is in the pupal cell or cocoon.

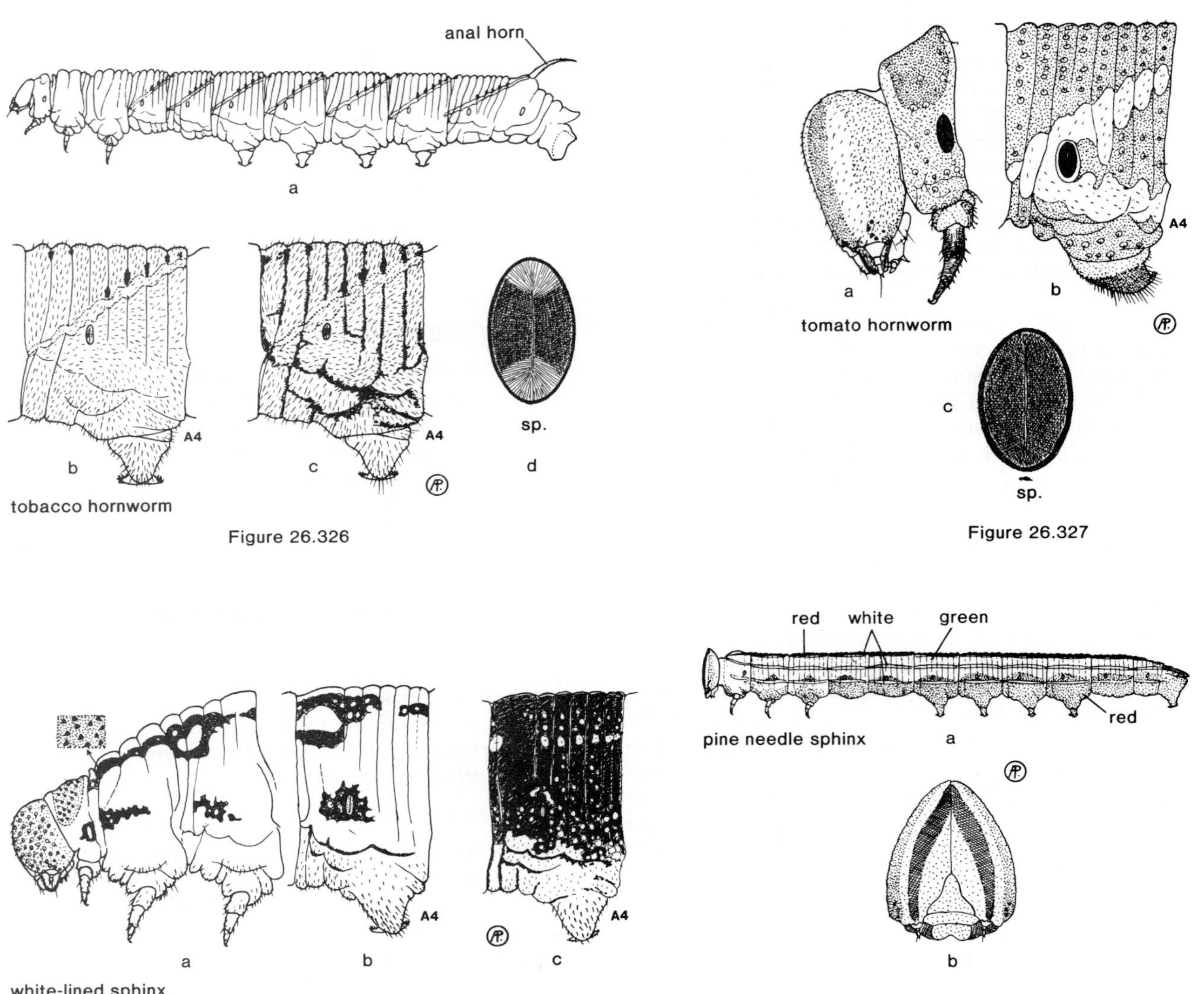

Figure 26.326

Figure 26.327

Figure 26.328

Figure 26.329

Figure 26.326a-d. Sphingidae, Sphinginae, *Manduca sexta* (L.), tobacco hornworm. a. larva; b. A4 with minimal black; c. A4 with greater black; d. A4 spiracle, light at top and bottom. f.g.l. 85± mm. Green, occasionally dark green-black, with a narrow diagonal whitish line above the spiracles on A1-A7 bordered dorsally by variable black dashes found in the body folds; horn reddish. A pest on tobacco and sometimes tomato; also on potato and other solanaceous plants. Eastern North America from southern Ontario to Florida, west into southern Arizona, and California into the San Joaquin Valley. (figs. from Peterson 1948).

Figure 26.327a-c. Sphingidae, Sphinginae, *Manduca quinquemaculata* (Haworth), tomato hornworm. a. head, T1; b. A4; c. A4 spiracle, totally black. f.g.l. 90± mm. Green, rarely tending toward blackish, with an irregular diagonal whitish line above the spiracles on A1-A7, connecting in front of spiracle with a similar horizontal line beneath the spiracle; horn green and black. On tomato, potato, and other solanaceous plants. Eastern North America into southern Ontario, more abundant in the north. (figs. from Peterson 1948).

Figure 26.328a-c. Sphingidae, Macroglossinae, *Hyles lineata* (Fabricius), whitelined sphinx. a. head, T1, T2; b. A4, lightly pigmented form; c. A4, deeply pigmented form. f.g.l. 80± mm. Yellow-green to nearly black, often with eyelike spots subdorsally. Head, T1 and anal shields, and lateral areas of prolegs yellowish with white pinacula. Horn yellow. On diverse plants, including purslane *(Portulaca)*, 4 o'clock *(Mirabilis)*, apple, and other herbs and trees. North America into southern Canada, more common toward the south. (figs. from Peterson 1948).

Figure 26.329a,b. Sphingidae, Sphinginae, *Lapara bombycoides* Walker, pine needle sphinx. a. larva; b. triangular head, face on. f.g.l. 60± mm. Green to yellow-green, longitudinally striped with variable amounts of red to red-purple, white and greens. Lacks a horn or even a remnant in all instars. On white, red, jack, pitch, and possibly other pines from Saskatchewan to Nova Scotia, northern U.S., and south along the Appalachians to Georgia. (figs. from Peterson 1948).

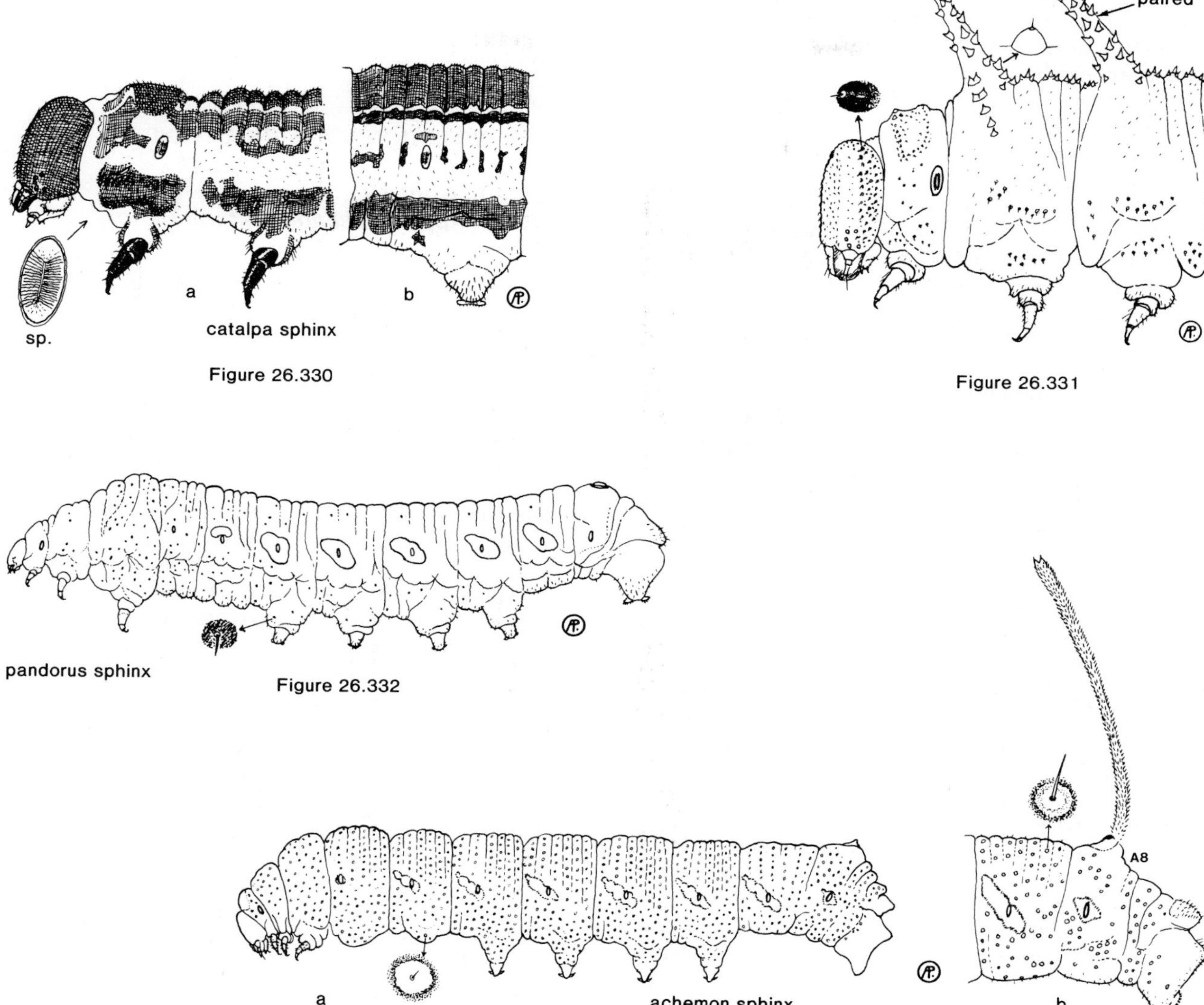

Figure 26.330

Figure 26.331

Figure 26.332

Figure 26.333

Figure 26.330a,b. Spingidae, Sphinginae, *Ceratomia catalpae* (Boisduval), catalpa sphinx. a. head, T1, T2; b. A4. f.g.l. 70± mm. Variable, but often black to striped black and yellow dorsally, with yellowish lateral areas bearing irregular vertical black lines and bordered dorsally by a narrow black line and ventrally by a wider black band above the proleg bases. Head and horn coal black. Common on *Catalpa* spp., in the eastern U.S., north to southern Michigan and New York; may cause compete defoliation in more southerly areas. (figs. from Peterson 1948).

Figure 26.331. Sphingidae, Sphiniginae, *Ceratomia amyntor* (Geyer), elm or four-horned sphinx. a. head and thorax. f.g.l. 75± mm. Green to rarely brownish, marked with 6 or 7 oblique white lines laterally, but distinguished by the pair of horn-like scoli on T2 and T3. On elm, basswood, and birch in eastern North America from southern Canada southward. (figs. from Peterson 1948).

Figure 26.332. Sphingidae, Macroglossinae, *Eumorpha pandorus* (Hubner), pandora sphinx. Last instar; lacking horn. f.g.l. 80± mm. Green to red-brownish, with oval, oblique cream patches above the spiracle on A2 and around the spiracles on A3-A7. Horn reduced to a hard shining "button" or "eyespot". On grape, Virginia creeper and *Ampelopsis* sp. Eastern U.S. into southern Canada, more common southward. (figs. from Peterson 1948).

Figure 26.333a,b. Sphingidae, Macroglossinae, *Eumorpha achemon* (Drury), achemon sphinx. a. last instar, lacking horn; b. A7-A10 of earlier instar, bearing long horn. Yellow to reddish-brown, covered with small black spots or rings around the secondary setae, and with elongate, oblique, somewhat beadlike spiracular spots. Horn reduced to "button" or "eyespot" on last instar. On grape, *Ampelopsis* sp., and possibly Virginia creeper from the northern U.S. south to Florida, and west into much of California. (figs. from Peterson 1948).

Description: Mature larvae moderate to very large, to nearly 100 mm. Primary setae not distinct. In first instars the head is generally rounded and the A8 horn quite long, in contrast to later instars of some species where the head may become more angular and/or knobby, and the horn may be shorter or reduced to a button. Spiracles elliptical, T1 and A8 spiracles about same size as A1–7 spiracles.

Head: Rounded to bumpy to angular. Frons clearly distinct from clypeus and nearly equilateral, extending ca. one-third to epicranial notch. Ecdysial sutures distinct and meeting just above apex of frons. Six relatively small stemmata in a normal arc, with stemma 5 below, and stemma 6 behind 4.

Thorax: T1 shield usually sclerotized and sometimes conspicuous. T2 and T3 usually like rest of the body, but occasionally with protuberances. Thoracic integument similar to abdomen.

Abdomen: Integument basically "naked," but some species with abundant, minute setae or stubby setae, or even small bumps bearing a single seta. Assorted stripes, spots, bars, diagonal lines, etc. usually present. A8 usually with conspicuous horn, sometimes short or reduced to a button; absent in *Lapara.* A10 terminating in three angular lobes, one dorsal and two lateral.

A3–6 and A10 prolegs moderate in length, bearing numerous short secondary setae that are longer than the body setae. Crochets in a biordinal mesoseries.

Comments: The family is moderate in size, with 124 species in 42 genera north of Mexico. Both Hodges (1971), for the entire area north of Mexico, and Forbes (1948), for the Northeast and Midwest, give considerable information on host plants and larval color and variation. Hodges (1971) provides a key to about half the genera based on Forbes (1911a) and Forbes (1948) provides a field key to 40 species in the Northeast. Hodges recognizes two subfamilies, the Sphinginae and the Macroglossinae. The Sphinginae, according to Hodges, in which the larvae are less diverse, have diagonal lines running from the anteroventral area of a segment to the posterodorsal area, and most have a well-developed, slightly curved, caudally directed, often roughened horn. In the Macroglossinae, a more variable group, "the larvae are variously modified with the third [T3] or fourth [A1] body segment being swollen and tapering rapidly to the head; the caudal horn is often replaced with a buttonlike structure, particularly in the last instar; often there are eyespots on the third or fourth body segments; and often the paired, oblique, lateral lines run from the posteroventral to anterodorsal margins of a segment."

Most sphingids are not of economic importance, but the tomato hornworm, *Manduca quinquemaculata* (Haworth) (figs. 26.327a,b), and especially the tobacco hornworm, *M. sexta* (L.) (figs. 26.326a-d), can be pests on solanaceous crops such as tomato, potato, and tobacco, particularly in the South. In addition, the catalpa sphinx, *Ceratomia catalpae* (Boisduval) (figs. 26.330a,b), is capable of totally defoliating catalpa trees. It is also commonly used for fish bait in the Southeast.

Selected Bibliography

Forbes 1911a, 1948.
Hodges 1971.
Madden and Chamberlin 1945.

NOTODONTIDAE (NOCTUOIDEA)

**George L. Godfrey,
James E. Appleby,** ***Illinois Natural History Survey***

The Notodontids and Prominents

Figures 26.334-26.354

Relationships and Diagnosis: Larvae of North America Notodontidae are medium-sized, generally measuring about 40 mm from the head to the tip of the anal shield (disregarding the A10 prolegs). They basically resemble most other Noctuoidea, especially many Noctuidae and Lymantriidae and the Dioptidae, because of the conspicuous, hypognathous head and subequally developed A3–6 prolegs, which bear uniordinal crochets, rarely biordinal (Common 1970), arranged in a homoideous mesoseries. However, beyond this basic concept, notodontid larvae possess many specializations, some quite unique, that are useful in distinguishing them from other noctuoids and, frequently, from each other.

Modified A10 prolegs and extra or secondary setae, at least on the A3–6 prolegs, have been used to characterize the notodontids (Fracker 1915, Forbes 1923, 1948). The A10 prolegs, frequently elevated when the larva is resting, usually are smaller than the A3–6 prolegs (figs. 26.335, 26.353) or modified as stemapods, which may be peglike structures (fig. 26.344) or very thin and elongate (fig. 26.341) (Packard 1890b, 1895). Crochets on the A10 prolegs are less numerous than on A3–6 or absent. The dioptid, *Phryganidia californica* Packard (California oakworm), has somewhat reduced A10 prolegs and, like the notodontids, generally elevates them during resting periods (Kellogg and Jack 1896).

The secondary setae of most notodontids are in the SV group on the lateral aspects of the A3–6 prolegs (fig. 26.348a), but in *Clostera* (fig. 26.334) and some species of *Datana* (figs. 26.335, 26.337), the body is extensively covered with uniformly distributed setae. Notodontids generally do not have paired, multisetose, dorsal verrucae on body segments T1-A9, as do most arctiids, lymantriids, and certain noctuids, e.g., *Panthea, Charadra,* and some *Acronicta.* Known exceptions are *Skewesia* and *Dunama. Skewesia, Clostera,* and *Dunama* have secondary head setae, whereas the other North American notodontid genera have only primary setae on the head.

Hinton (1946) noted that first instar notodontids have two MD setae (primary microscopic setae) on A1. All other noctuoids that he examined with the exception of *Acronicta* have one.

The full complement of prolegs on A3–6 separates the notodontids from the Nolidae, which have but three pairs, and from elements of the Noctuidae, which have two or three pairs. The homoideous crochets on the A3–6 prolegs are useful in distinguishing the notodontids from the Arctiidae and Pericopidae, which usually have heteroideous crochets. North American notodontids, like the noctuids, lack middorsal glands on the abdominal segments, unlike the Nearctic Lymantriidae, which have them on A7 and sometimes A6.

There is some difficulty in distinguishing the dioptid, *Phryganidia californica,* from some notodontid larvae. Fracker (1915) and Forbes (1948) used the shagreened integument (a dense covering of spinules visible at 50×) as a characteristic of the dioptids. One North American notodontid genus, *Clostera,* has this same characteristic, but it is readily separable from *Phryganidia* by its secondary head setae and conspicuous secondary body setae (fig. 26.334), which hide the spinules on the body. Also see the diagnosis for Dioptidae.

Biology and Ecology: Notodontid larvae are external foliage feeders. The principal hosts in Canada and the U.S. are dicotyledonous trees and woody shrubs. Exceptions are the various herbaceous and semiwoody legumes reported for *Dasylophia anguina* (J. E. Smith) (fig. 26.344) and *Euphorbia,* the only known host of *Theroa zethus* (Druce) (Franclemont, pers. comm.). McFarland (1979) collected and reared *Antimima cryptica* Turner and *A. corystes* Turner on semiwoody legumes in Australia. Notodontid-monocot associations are very uncommon. Larvae of *Dunama mexicana* Todd were collected in Mexico on *Chamaedorea* sp., a palm genus, but the Panamanian record of *Dunama angulinea* on banana is doubtful (Todd 1976). Some Asiatic species feed on bamboos, palms and grasses (see Gardner 1943, Issiki et al. 1969, and Holloway 1983). There are no records from conifers. A few notodontids are monophagous, about half are oligophagous, and the others are polyphagous.

Notodontid larvae typically exhibit two modes of feeding. First instars skeletonize the leaf blade (Riotte 1969), usually on the lower surface. Commencing with the second or third instar, the larvae eat most of the leaf except for the larger veins.

There is variation from solitary to social larval behavior in the Notodontidae, especially regarding feeding and molting. Contrasting extremes are *Nerice bidentata* (fig. 26.342), which feeds singly throughout its larval stages, and *Datana integerrima* Grote and Robinson (walnut caterpillar) (fig. 26.337), a gregarious species. Young larvae of *D. integerrima* feed in conspicuous clusters (Beutenmüller 1888, Packard 1890b), eating their way across a leaf in rank formation, but upon attaining the fifth instar, they begin separating into smaller groups (Farris and Appleby 1980 (1981)). Prior to molting, *D. integerrima* larvae move downward to larger branches or the trunk and aggregate. After molting in a compact mass, they ascend to resume feeding (Beutenmüller 1898). Synchrony is developed to the point in *Schizura concinna* (fig. 26.353) that whole broods may disappear during the same night into the leaf litter or soil and spin cocoons (Packard 1895). A few of the social species, e.g., *Clostera inclusa* (fig. 26.334) (Abbot and Smith 1797), form communal, silk-lined nests within folded or webbed leaves for resting places.

Notodontid larvae are well adapted to arboreal habitats (Packard 1890b) and have a tremendous capacity to grip leaves, but generally they are not capable of rapid or agile locomotion (Ferguson 1963). However, the mature larva of *Symmerista albifrons* (J. E. Smith) (fig. 26.343) will drop from its feeding or resting site to the ground if physically disturbed (pers. obs., G.L.G.). This rather sedentary behavior presents certain ecological risks, such as increased vulnerability to predators and parasitoids, which are avoided to some degree by disruptive and cryptic coloration combined with body form, plus, in some species, chemical and/or physical defenses.

The overall appearance of some species resembles whole leaves, as seen in *Nerice bidentata* (fig. 26.342), which has dorsal tubercles patterned after the double-serrate leaf margin of its host, *Ulmus.* Packard (1905) remarked that *Schizura ipomoeae* (Doubleday) (figs. 26.351a,b), *S. unicornis* (J. E. Smith) (fig. 26.352), and *S. leptinoides* (Grote), with their movable, dorsal abdominal humps, resemble partly eaten, twisted or tattered leaves. Some *Heterocampa* larvae have extremely variable and highly cryptic color patterns, e.g., *Heterocampa umbrata pulverea* Grote and Robinson (Klots 1967a).

Most North American notodontid genera have a T1 (cervical) gland that opens midventrally just behind the head capsule. Of the 27 genera examined, only *Clostera* lacks it. This structure may also occur in other Lepidoptera families. Several species of notodontids, both solitary and gregarious, are able to spray formic acid and certain accessory compounds (Eisner et al. 1972, Percy and MacDonald 1979) from this gland, but its presence does not imply a capacity to spray. Of the species that are able to emit an acidic spray, *Schizura concinna* (fig. 26.353) has been most thoroughly studied. See Detwiler (1922), Percy and MacDonald (1979), and Weatherston et al. (1979) for details.

The spray has a defensive function and can be ejected rapidly (Hintze 1969) and aimed at parasitoids and predators. A fully grown larva of *S. concinna* is able to spray nearly eight inches (15 cm) (Herrick and Detwiler 1919). The spray of *Lochmaeus manteo* is an effective deterrent against lycosid spiders (Eisner et al. 1972) and can severely irritate human skin (Kearby 1975).

Some *Datana* bob or jerk their heads and tails in unison (Comstock and Dammers 1938) in response to certain disturbances. *Datana* larvae may arch their heads and tails upward (Abbott 1927), and the last instar of *Cerura scitiscripta multiscripta* Riley extends its long stemapods, expands its prothorax, and thrashes its anterior and posterior ends in response to auditory stimuli (Klots 1969). However, the effect of these behaviors on avoiding parasitism or predation is questionable.

Notodontid eggs are an upright type (Forbes 1948, Common 1970); that is, the micropyle ordinarily is at the top of the egg (Franclemont pers. comm.). In side view, the eggs are hemispherical to nearly spherical. They are shiny or chalky (Peterson 1963b), indistinctly reticulated, and various shades of white, green, blue, or black. They usually are deposited singly on the host leaf or in single layers varying from a small, loose cluster to a large, compact cluster.

The pupae are similar to those of the Noctuidae (Forbes 1948). No single character can be used to separate them from the other noctuoids and some bombycoids, although two important characters are the absence of visible labial palpi and the presence of short maxillae that do not extend to the distal wing pad margin. The labial palpi, when visible, are extremely minute, and the maxillae are about 0.67 times the wing pad length (Mosher 1916a, 1917).

Pupation usually occurs within an earthen cell or in a cocoon at ground level, frequently in the leaf litter. Larvae of *Cerura* and *Furcula* pupate in shallow cavities that they chew into wood and conceal with tough, silken covers. The prepupa or pupa is the overwintering stage of North American notodontids. In a few species, the emergence of the moth from the pupa may be delayed until after the second winter (Millers and Wallner 1975, Surgeoner and Wallner 1978), whereas emergence after a single season is usually normal.

Description: Mature larva medium- to large-sized, total length 24–95 mm (head to tip of anal shield). Head rounded (fig. 26.338) or appearing flattened with slightly elongated epicranial vertices (fig. 26.354). Body cylindrical (fig. 26.338a, b), or with distinct horn (fig. 26.340), humps, and tubercles (figs. 26.342, 26.351a, 26.353, 26.354), or tapering caudad because of stemapods (figs. 26.341, 26.347); integument smooth, bearing indistinct pavement granules and sometimes glossy, or covered with spinules; extra setae on lateral aspects of prolegs A3–6 (fig. 26.348a) with single, primary setae dorsad (figs. 26.338–26.354) or secondary setae uniformly abundant dorsad (figs. 26.335, 26.337); rarely with multisetose, paired verrucae dorsad of prolegs. Body variously patterned: boldly spotted (figs. 26.336), boldly or indistinctly striped (fig. 26.338a), or cryptically marked (figs. 26.351a, 26.352); some species nearly solidly colored. Spiracles elliptical, usually larger on T1 and A8 than on A1–7.

Head: Hypognathous. Frontoclypeus roughly triangular to bell-shaped with apex at 0.28–0.42 of distance between base of frontoclypeus and epicranial notch. Junction of adfrontal ecdysial lines distantly cephalad of epicranial notch. Six stemmata, 1–4 in semicircle, 5 and 6 forming a line ventrocaudad of 1–4.

Thorax: T1 shield variously developed; distinctly pigmented, large, and glossy (e.g., *Datana ministra*); color and texture slightly differentiated, but shield small (e.g., *Lochmaeus manteo*); or shield not discernible (e.g., *Symmerista canicosta* Franclemont). Chaetotaxy basically noctuoid. SV group uni- or bisetose on T2 and T3; MD bisetose on T3 (Hinton 1946).

Abdomen: Chaetotaxy essentially noctuoid, but MD bisetose on A1 (Hinton 1946) and SV group usually bears more than three setae on prolegs A3–6.

Prolegs subequally developed on A3–6, bearing homoideous, uniordinal crochets in mesoseries. A10 prolegs generally reduced, shorter or peglike, with fewer crochets than on A3–6, or tapering and much elongated (fig. 26.341) with no crochets.

Dorsa of A1–8 unmodified or with at least one or several segments bearing single or double tubercles and/or humps; humps may be very large and movable (Packard 1895).

A10 with an extra seta near anterolateral corner of anal shield (fig. 180 in key to families) (absent in *Cargida* and undiscernible in some last instar *Datana* and *Clostera*, which have abundant secondary setae). Anal shield generally convex, posterior margin rounded in dorsal view; variously flattened and somewhat angulated in dorsal view in some species with long filamentous A10 prolegs; anal shield of *Pheosia* greatly sclerotized and rugose with hard, dorsal medial tubercle.

The preceding description pertains to the ultimate instar larva, which for some species differs remarkably from the first instar. Perhaps the best example of this among the North American notodontids is exhibited by *Heterocampa guttivitta* (Walker). In the first instar, this species is adorned with nine pairs of large antlerlike tubercles (fig. 26.348a), which are not present in the second and subsequent instars (Packard 1895) (figs. 26.348b,c). Interested readers may consult Packard (1890b, 1895, 1905) for additional information on the phenomenal changes that some notodontids undergo during larval development.

Comments: This is a relatively large family with approximately 2,000 described species (Forbes 1948). The Notodontidae occur worldwide in the temperate and tropical areas but are absent in New Zealand (Imms 1970) and are most diverse in the Neotropics (Franclemont, pers. comm.). Forty-one genera and 136 species are known from America north of Mexico (Franclemont 1983), and the greatest number are in the East.

A well-known transcontinental species is the redhumped caterpillar, *Schizura concinna* (fig. 26.353). This highly gregarious species is at times abundant on rosaceous and salicaceous plants. Another reasonably common species is the yellownecked caterpillar, *Datana ministra* (Drury) (fig. 26.335), which occasionally defoliates oak, crabapple, and apple. Other species of *Datana* have varying degrees of economic importance, including *D. major* (fig. 26.336), known to attack cultivated azalea, and *D. integerrima,* the walnut caterpillar (fig. 26.337), which can affect production of black walnuts and pecans. These last two species only occur in the eastern half of the continent, and *D. integerrima* ranges into Texas.

Notable defoliators of forested areas include the saddled prominent, *Heterocampa guttivitta* (figs. 26.348a-c), the variable oakleaf caterpillar, *Lochmaeus manteo* (fig. 26.349), and species of *Symmerista,* especially *S. albifrons* (fig. 26.343), the redhumped oakworm, and *S. leucitys* Franclemont, the orangehumped mapleworm.

Heterocampa guttivitta has attracted attention in southeastern Canada and the northeastern U.S. because of its damage to beech and sugar maple (Collins 1926, McDaniel 1933). Allen and Grimble (1970) observed a decrease in maple syrup production following heavy, successive defoliations of sugar maples by *H. guttivitta. Lochmaeus manteo, Symmerista albifrons,* and *S. canicosta* show considerable preference for oaks, whereas *S. leucitys* feeds primarily on sugar maple (Allen 1979). In addition to exerting physiological stress and promoting secondary pathological problems on the hosts, the defoliating populations of *Heterocampa guttivitta* and *Symmerista canicosta* may create problems in

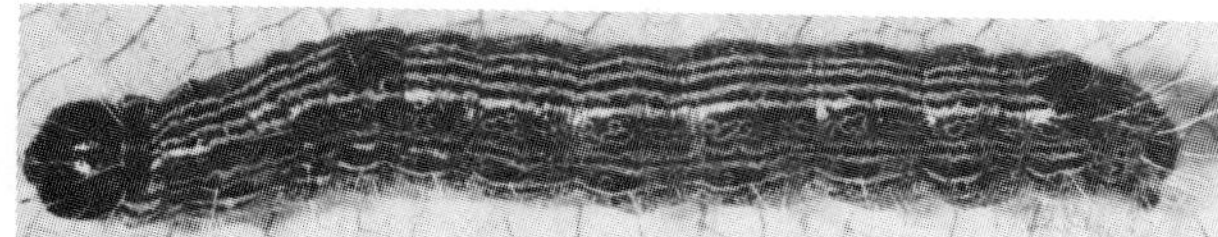

Figure 26.334

Figure 26.335

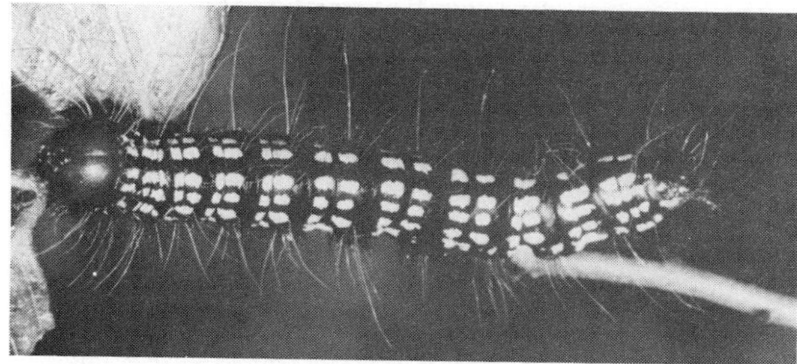

Figure 26.336

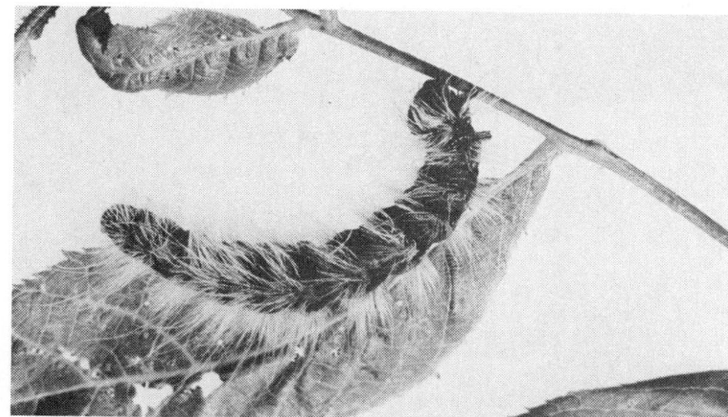

Figure 26.337

Figure 26.334. Notodontidae. *Clostera inclusa* (Hübner), poplar tentmaker. f.g.l. 28–44 mm. Grayish lavender; dorsal area with 4 distinct, yellow lines; D1 tubercles on A1 and A8 conspicuously enlarged, black, spinulate; primary setae long, white; A10 prolegs slightly reduced. Head black, secondary setae present. The secondary head setae in combination with the large D1 tubercles on A1 and A8 and the 4 prominent yellow dorsal lines distinguish this species. Larvae gregarious; each colony in silk-lined tent of single folded leaf or several webbed leaves; disperse prior to spinning cocoons among fallen leaves. Overwinters as pupa. On various species of poplar and willow. New Brunswick to southern Alberta and east of the Rocky Mountains in U.S., southward to Texas and Florida. Bivoltine (may be univoltine in Canada); larvae generally present May-July and August-October. Packard (1895, 1905), Forbes (1948), Prentice (1962), and McFarland (1967 (1968)). Photo by J. E. Appleby.

Figure 26.335. Notodontidae. *Datana ministra* (Drury), yellow-necked caterpillar. f.g.l. 38–55 mm. Recognizable by the distinctive "yellow" T1 shield, continuous yellow-white lines on black body, long, white, sparse secondary body setae, and black head. A10 prolegs stubby. First instars in dense colonies on lower leaf surface. Older instars may become solitary. Larvae crowd together when resting with heads and tails arched upward; together they move to lower branch or trunk to molt. Overwinters as subterranean pupa. A general feeder reported from more than 40 broadleaf shrubs and trees. Nova Scotia to interior British Columbia; generally east of the Rocky Mountains in the U.S. as far south as Florida and Texas but more common in Midwest and Northeast; also known from central and southern California. Univoltine; larvae from July to September northward and as late as November in Florida. Packard (1895), Forbes (1948), Prentice (1962), and Kendall and Kendall (1971). Photo by W. D. Zehr, I.N.H.S.

Figure 26.336. Notodontidae. *Datana major* Grote & Robinson. f.g.l. 45–60 mm. Black with 4 rows of distinct white or pale yellow spots between middorsal line and A3-6 prolegs. Head, T1 and anal shields, and prolegs bright red; primary body setae long, white; A10 prolegs stubby. The bright red head and T1 and anal shields plus the spotted body distinguish this species. Gregarious, but later instars feed singly more than other *Datana* species. Overwinters as subterranean pupa. Principally on the Ericaceae, especially *Lyonia ligustrina* and *L. mariana;* also on *Azalea* and *Vaccinium.* Infrequently reported from witch hazel, sumac, and apple. Southern Maine to Florida, west to Kansas and Arkansas. Most sources state that it is univoltine, feeding in August and September. Packard (1895), and Forbes (1948). Photo by J. W. and R. R. Graber, I.N.H.S.

Figure 26.337. Notodontidae. *Datana integerrima* Grote & Robinson, walnut caterpillar. f.g.l. 33–55± mm. Solid black or with traces of thin, longitudinal lines; primary and secondary setae white, long, shaggy; A10 prolegs stubby. Head dark brown to black. The shaggy appearance is characteristic. In conspicuous colonies with 5th instars tending to separate into small groups. Larvae crawl from feeding branch to larger branch or trunk to molt in masses 2 or 3 layers deep on communal, silken pad or within silken nest. Overwinters as pupa under leaf litter or below soil surface; does not spin a cocoon. Mainly occurs on walnuts and hickories. May defoliate isolated trees or small clumps. Also reported on azalea, beech, oak, ash, hawthorn, and apple, but Farris and Appleby (1980 (1981)) maintained that it survives only on the Juglandaceae. Maine to Florida, west to the eastern Great Plains in the U.S. and southern Ontario in Canada. Univoltine northward, larvae from mid-July to mid-September. As many as 2–3 broods southward. Grote and Robinson (1866), Packard (1890b, 1895), Hixson (1941), Forbes (1948), Prentice (1962), and Farris and Appleby (1980 (1981)). Photo by L. Farlow, I.N.H.S.

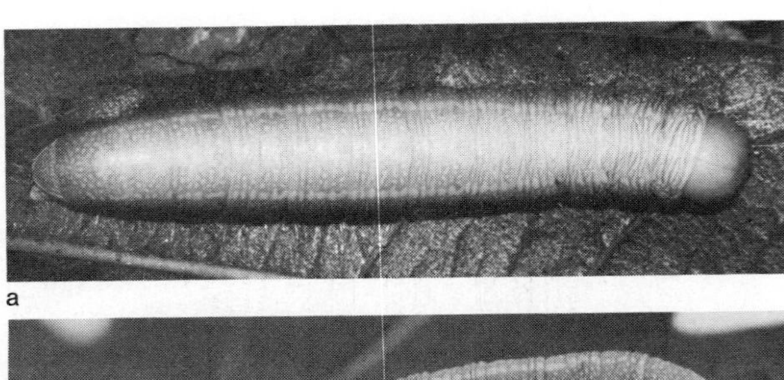
a

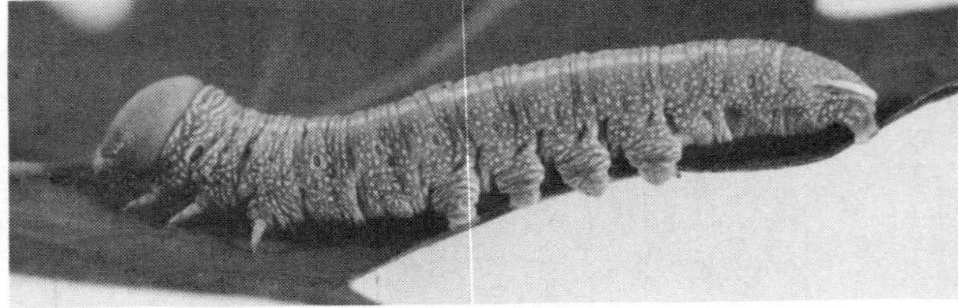
b

Figure 26.338

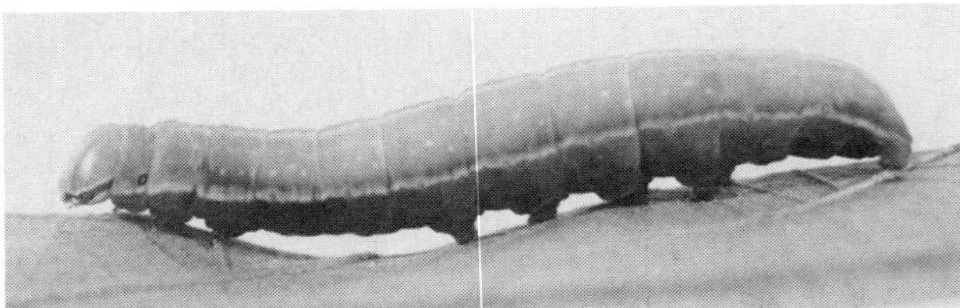
Figure 26.339

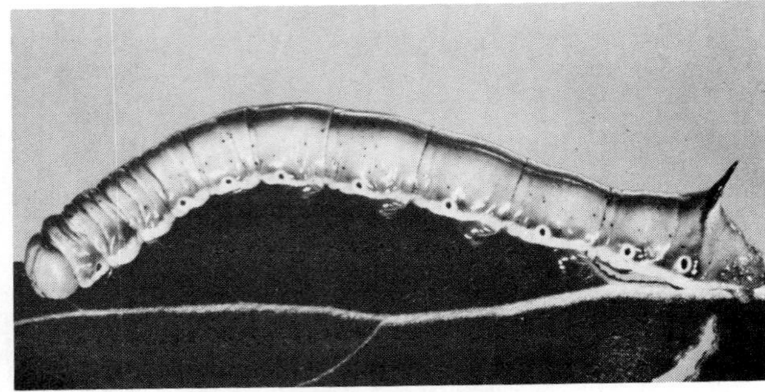
Figure 26.340

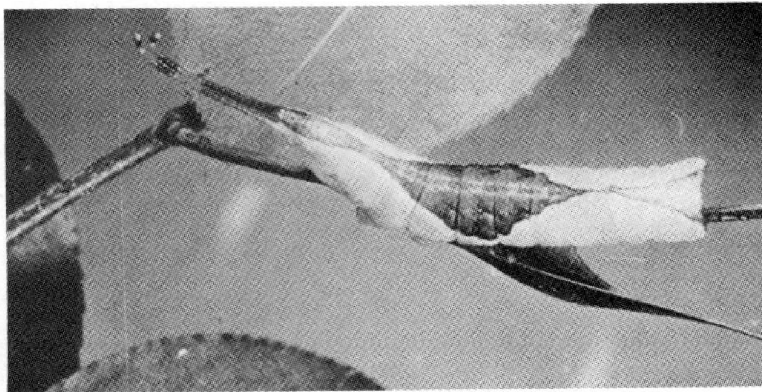
Figure 26.341

Figure 26.338a,b. Notodontidae. *Nadata gibbosa* (J. E. Smith). f.g.l. 33-38 mm. Pale, whitish green with thin, white subdorsal line and numerous small white dots; posterior margin of anal shield yellow, delineated cephalad by small ridge; A10 prolegs slightly reduced. Head whitish green. Distinguished by its relatively large, rounded head, and stout, "chalky," pale green body. A solitary defoliator. Overwinters as pupa in the soil or under leaf litter within soft cocoon of a few strands of silk. A general feeder, especially on various species of birch and oak. Other host records are maple, alder, hazelnut, beech, cherry, serviceberry, poplar, and willow. Newfoundland to British Columbia, south to Florida, Texas, and California. Bivoltine, larvae in June-July and August-September. Abbot and Smith (1797), Packard (1895), Beutenmüller (1898), Forbes (1948), and Prentice (1962). Fig. 26.338a photo by J. W. and R. R. Graber, I.N.H.S.; fig. 26.336b photo by G. L. Godfrey.

Figure 26.339. Notodontidae. *Peridea angulosa* (J. E. Smith). f.g.l. 38-40 mm. Three color phases: medium pastel green, pinkish, and intermediate (Riotte, 1969); middorsal (double) lines and subdorsal lines faint, whitish green; subdorsal area with few scattered yellow flecks; narrow, brownish to pinkish and yellow lateral line on T1-3, pink and yellow continuing to A10 on some specimens; spiracle on T1 with contrasting black peritreme; A10 prolegs slightly reduced. Head pale yellowish green to green with pinkish or brownish lateral stripe extending dorsocaudad. The lateral head stripe distinguishes *Peridea* from other North American notodontids with generalized body form. Riotte (1969) provided characters for separating *P. angulosa* from *P. basitriens* (Walker) and *P. ferruginea* (Packard). Overwinters as pupa in slight cocoon of silk and dirt on or beneath soil surface. Hosts: maple (J.E.A.), oaks, and pecan. From Maritime provinces to southern Ontario and Manitoba, south to Florida and Texas. Bivoltine in southern states, univoltine (larvae in August-October) in the North. Abbot and Smith (1797), Packard (1890a, 1893, 1895, 1905), Dyar (1893a), Forbes (1948), Prentice (1962), Riotte (1969), and Ferguson (1975). Photo by G. L. Godfrey.

Figure 26.340. Notodontidae. *Pheosia rimosa* Packard. f.g.l. 38-50 mm. Grayish green phase. Body glossy. Two color phases: grayish green with tinges of lavender or extensively lavender; lateral area of grayish green phase mostly yellow, but white on T2 and T3 and below A1-8 spiracles; abdominal spiracles black, surrounded by white; A8 with black, middorsal horn; anal shield hard, rugose, bearing middorsal tubercle; A10 prolegs about half the size of A3-6 prolegs. Head greenish or lavender. The middorsal horn on A8, the middorsal tubercle on the anal shield, and the glossy integument distinguish this species. Overwinters as pupa in silk-lined, earthen cell or in frail cocoon among fallen leaves. Reported only from the Salicaceae: *Populus balsamifera, P. tremuloides, P. trichocarpa,* and *Salix* sp. Newfoundland to British Columbia, south to N. Carolina, Illinois, the mountains of northern Arizona (Franclemont, pers. comm.), and Yosemite Valley, California. Reported feeding from June to October, apparently bivoltine. Packard (1890a, 1893, 1895), Dyar (1891b, 1892b), Forbes (1948), and Prentice (1962). Photo by J. E. Appleby.

Figure 26.341. Notodontidae. *Furcula borealis* (Guér.-Méneville). f.g.l. head to tip of anal shield, 26-45 mm; length of flagellar stemapods 10-20 mm. Yellow dorsally with brownish triangle over thoracic segments joining brownish saddle of abdominal segments that narrowly extends to A10 (fig. 26.341); stemapods extensible. Head brownish, partially concealed by enlarged and truncated T1 dorsum (fig. 26.341). Larvae of *Furcula* and *Cerura* similar in form. Forbes' (1948) key permits identification of the 6 eastern species of these genera. *Furcula borealis* differs from the others by its smooth T2 dorsum, continuous dorsal pattern from T1-A10, and specificity to *Prunus serotina.* Reports of other hosts are likely incorrect. Larva chews a depression into wood or bark and makes a firm silk and wood cocoon. Overwinters as a pupa. Southeastern Canada, south to Florida, west to Minnesota and Illinois. Bivoltine, feeds during May-June and July-September. Packard (1895), Dyar (1891a), Beutenmüller (1898), and Forbes (1948). Photo by W. D. Zehr, I.N.H.S.

Figure 26.342

Figure 26.343

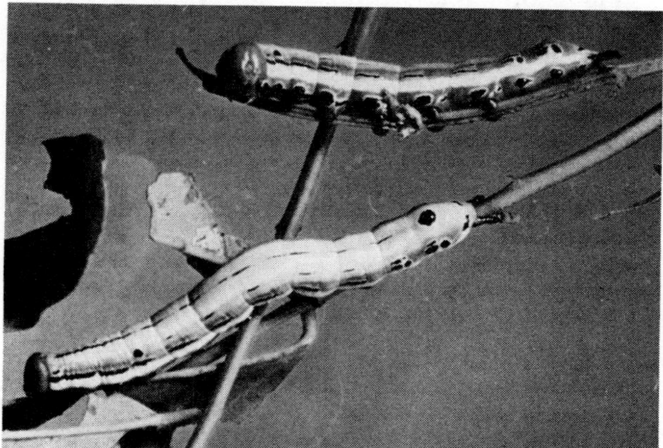

Figure 26.344

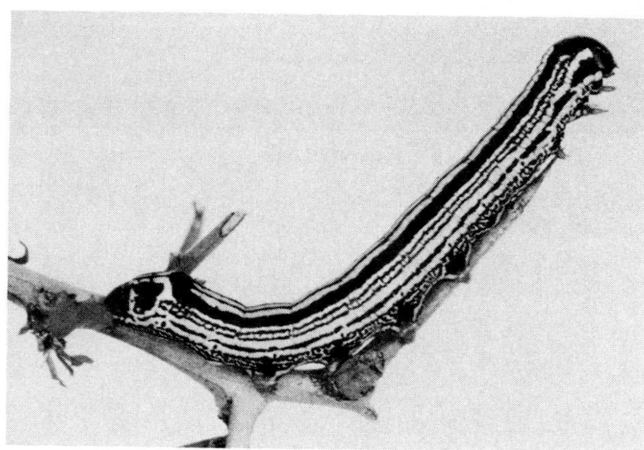

Figure 26.345

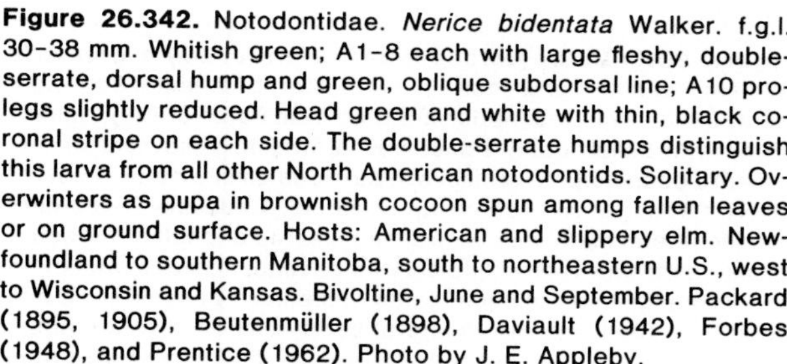

Figure 26.342. Notodontidae. *Nerice bidentata* Walker. f.g.l. 30-38 mm. Whitish green; A1-8 each with large fleshy, double-serrate, dorsal hump and green, oblique subdorsal line; A10 prolegs slightly reduced. Head green and white with thin, black coronal stripe on each side. The double-serrate humps distinguish this larva from all other North American notodontids. Solitary. Overwinters as pupa in brownish cocoon spun among fallen leaves or on ground surface. Hosts: American and slippery elm. Newfoundland to southern Manitoba, south to northeastern U.S., west to Wisconsin and Kansas. Bivoltine, June and September. Packard (1895, 1905), Beutenmüller (1898), Daviault (1942), Forbes (1948), and Prentice (1962). Photo by J. E. Appleby.

Figure 26.343. Notodontidae. *Symmerista albifrons* (J. E. Smith), redhumped oakworm. f.g.l. 35-50 mm. Body shiny; subdorsal line, lateral area, and dorsal hump on A8 orange to dark orange; remainder of body white or bluish white with thin, black, longitudinal lines; A10 prolegs about half the length of A3-6 prolegs. Head bright orange. Distinguished by 5 slightly wavy, bold, black dorsal lines on A1-7. Because of the similarities of this larva and that of *S. canicosta,* it is difficult to ascertain the taxon discussed by some earlier writers. Principally on bur and white oak; also reported from maple and white birch. May cause localized defoliation. Eastern half of North America, north to Nova Scotia. Bivoltine in the South, univoltine in the North. Young larvae gregarious, older instars tend to separate. Overwinters as pupa in whitish cocoon spun on the ground among fallen leaves. Abbot and Smith (1797), Packard (1890a, b, 1905), Franclemont (1946, 1948), and Prentice (1962). Photo by G. L. Godfrey.

Figure 26.344. Notodontidae. *Dasylophia anguina* (J. E. Smith). f.g.l. 45-55 mm. Ground color, bluish white or pinkish orange. White phase: thin black lines (sometimes broken) delineate middorsal line and dorsal and ventral margins of subdorsal area; subdorsal area yellow; lateral area white with yellow dorsal border. Single, medial, shiny, black rounded tubercle (button) on A8. Subventrally a series of shiny, black tubercles on T1-A9. A10 prolegs narrow, black stemapods, subequal in length to A3-6 prolegs, projecting caudad. Head brownish orange. The black, shiny button on the middorsum of A8 and the affinity of the larva for leguminous hosts are distinctive. Overwinters as pupa in thin cocoon of silk and leaves formed on the ground. On herbaceous dicots in addition to woody, broadleaf plants, but restricted to the Fabaceae: clover (unspecified), *Robinia* (black locust), *Lespedeza* (bush clover), *Baptisia* (false indigo), *Amorpha* (lead plant), and *Desmodium* (tick trefoil). Southern Maine and Quebec west to Montana; Colorado, Texas, and Florida. Bivoltine, spring and fall broods. Abbot and Smith (1797), Packard (1895), Forbes (1948), and Tietz (1972). Photo by W. D. Zehr, I.N.H.S.

Figure 26.345. Notodontidae. *Cargida pyrrha* (Druce). f.g.l. 30-38 mm. Velvety black, marked with narrow yellow and white stripes; all legs orange; prolegs with black patches at bases. Head black. This coloration is distinctive. Early instars gregarious, older instars tending to separate. Host: *Condalia lycioides* (Rhamnaceae). Mexico; southeastern and south-central Arizona. Univoltine, larvae in July-August. Comstock (1959a) and Godfrey (1984). Photo courtesy of J. G. Franclemont, Cornell University.

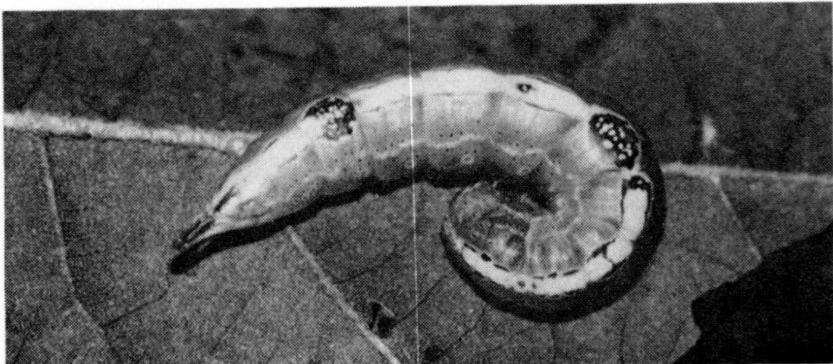

Figure 26.346

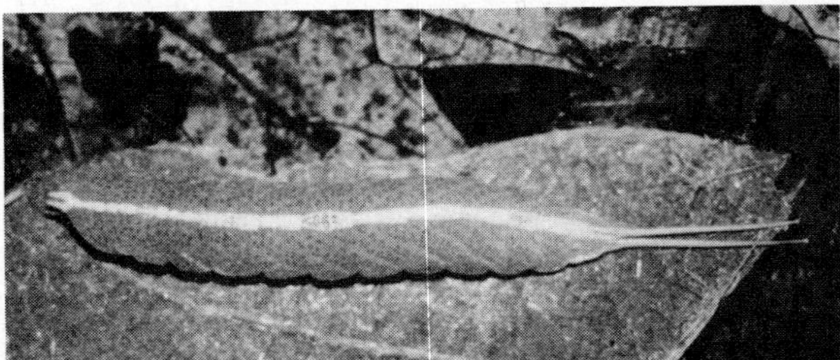

Figure 26.347

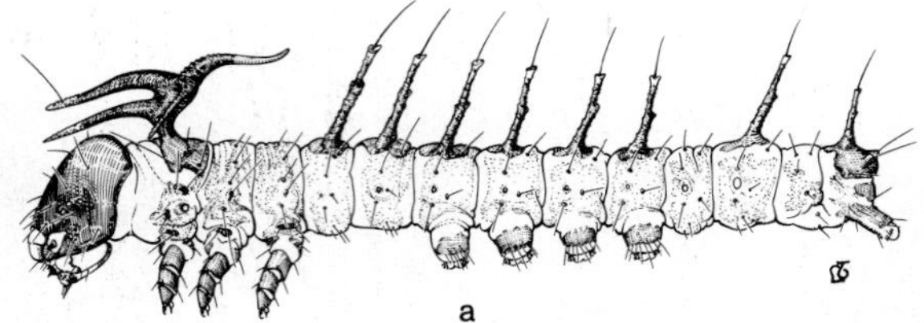

a

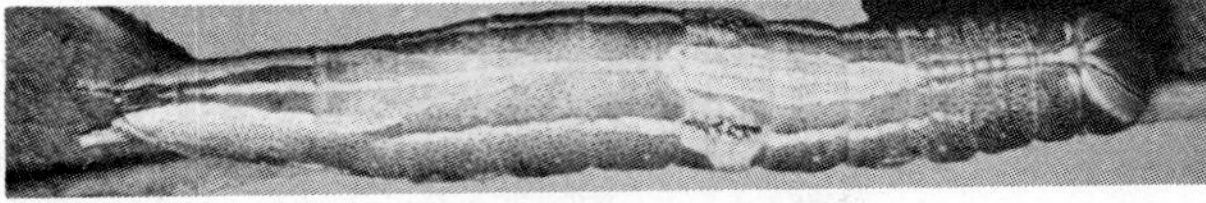

b

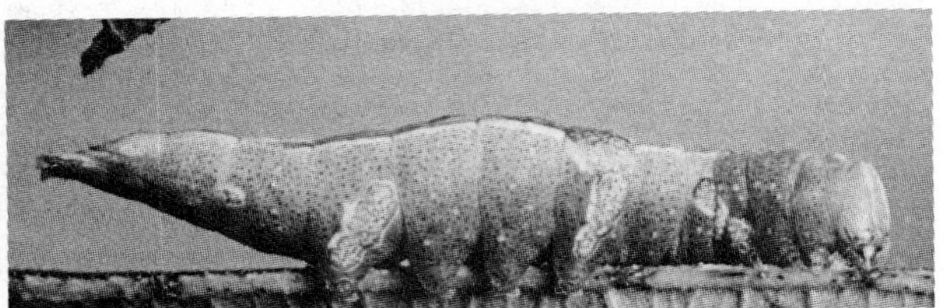

c

Figure 26.348

Figure 26.346. Notodontidae. *Misogada unicolor* (Packard). f.g.l. 43–50 mm. Translucent green; wide, white dorsal band interrupted on A3 and A7 by brownish red, white-spotted discs; lateral margins of white dorsal band yellow and brownish red; subdorsal area with 2 indistinct, yellow lines; lateral aspects of body bear numerous, small, reddish spots; spiracles orange, encircled by yellow; A10 prolegs narrow, reddish stemapods, projecting caudad, subequal in length to A3-6 prolegs. Head sordid green with wide, reddish, reticulated coronal stripes subtended laterally by white margins. This larva's distinctive dorsal band and restriction to sycamore makes it readily recognizable. First instars gregarious. Overwinters as larva within thin semitransparent cocoon spun at ground level, under or among leaves. Massachusetts to Virginia, west to Illinois, Kansas, south to Alabama, Arkansas, and Texas. Bivoltine, mature larvae in July-early August and late August. Popenoe et al. (1889), Dyar (1891c), Packard (1895), and Forbes (1948). Photo by J. W. and R. R. Graber, I.N.H.S.

Figure 26.347. Notodontidae. *Macrurocampa marthesia* (Cramer). f.g.l. 40–42 mm. Penultimate instar (last instar similar but with shorter stemapods). Green; whitish middorsal stripe extends from paired, middorsal tubercles on anterior edge of T1 shield to A10; middorsal stripe edged with yellow and reddish brown, reddish brown intensified on A3 and A7. Lateral aspects bear numerous dark specks and series of oblique, pale yellow lines. A10 prolegs short stemapods. Distinguished by short, paired, forward-projecting tubercles on anterior edge of T1 shield and oblique, lateral lines on body. Solitary defoliator capable of spraying acid from midventral, cervical gland. Overwinters as pupa in thin, transparent, cocoon spun amongst leaves on the ground. On beech and oak. Also reported on maple, sycamore, and poplar. Maine to Wisconsin, Illinois, and Missouri, south to Florida and central Mexico; Quebec and Ontario; Manitoba (subspecies *manitobensis* McDunnough). Number of broods uncertain, perhaps univoltine (August-October) in the North; indications of 2 broods southward. Packard (1895, 1905), Gibson (1902), Forbes (1948), and Prentice (1962). Photo by J. W. and R. R. Graber, I.N.H.S.

Figure 26.348a-c. Notodontidae. *Heterocampa guttivitta* (Walker), saddled prominent. a. first instar; b, c. last instar. f.g.l. instar 24–39 mm. Description adapted from Patch (1908): Usually green with bluish cast, sometimes brownish, yellowish, or lavender to deep purple; dorsum wth pale design in whitish yellow or heavy marking in reddish brown or purple; T3 usually bears dark oblique lateral mark with ventral end slanted cephalad; A6 bears similar mark with ventral end slanted caudad; similar marks often on other segments; oblique marks sometimes absent on T3 and A6. Head green with broad lateral band composed of 4 stripes: black, white, pink, and yellow. A10 prolegs distinctly smaller than A3-6 prolegs. The 4-colored lateral head band and 2 small, dorsal tubercles on anterior edge of T1 help diagnose this larva. Overwinters as pupa in thin cocoon beneath leaf mold under host. General feeder, sugar maple, and beech the principal hosts. A serious defoliator; effects of its feeding on sugar maples known to lower syrup production. Also reported from 20 other genera of trees and shrubs. Nova Scotia to southeastern Manitoba, south throughout eastern half of the U.S., including Florida, Texas, Minnesota, Nebraska, and Colorado. Most common in the northeastern U.S. Univoltine, larva from late June to mid-September. Packard (1895, 1905), Patch (1908), Forbes (1948), Prentice (1962), Allen and Grimble (1970), Grimble and Newell (1972a, b, 1973), Allen (1973), Grimble and Kasile (1974), and Loesch and Foran (1979). (fig. 26.348a drawing by L. M. Tackett, courtesy of F. W. Stehr; fig. 26.348b, c Photos by W. D. Zehr, I.N.H.S.

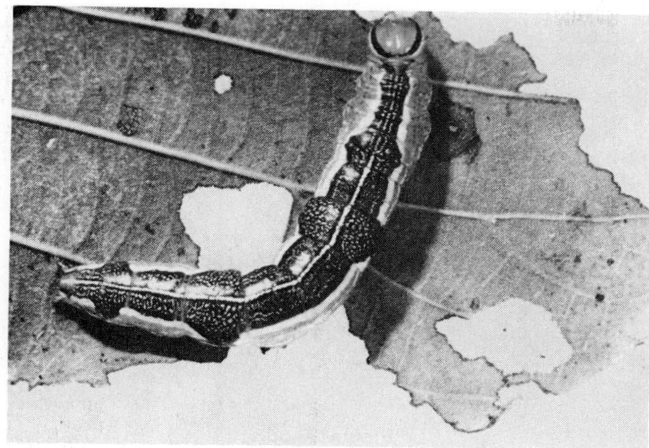

Figure 26.349

Figure 26.350

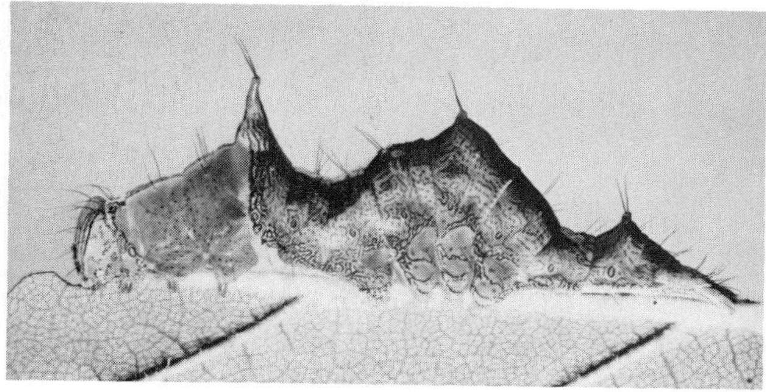

a

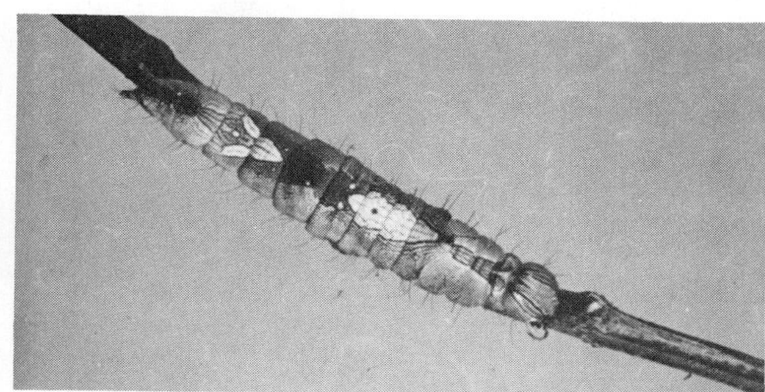

b

Figure 26.351

Figure 26.349. Notodontidae. *Lochmaeus manteo* Doubleday, variable oakleaf caterpillar. f.g.l. 30–40 mm. Ground color green; middorsal line white or yellow, narrow but distinct on A2-A8 of all color phases (lost in prepupal stage); dorsal area filled with varying amounts of reddish brown or purple; subdorsal lines white, wide, may be bent toward and touching spiracles on A3 and A6. Bases of D1 setae slightly tuberculate on A1 and A8. A10 prolegs about half the size of A3-6 prolegs. Head greenish with wide, black arcuate stripe (subtended by white). *L. manteo* and the following species are very similar. *L. manteo* has a distinct middorsal line in the various forms and the dorsal area of the green form is concolorous with the sides of the body or the sides are white (Dyar 1893a). The wide, black arcuate head stripe is characteristic. Young larvae gregarious, older instars tend to separate. Larvae able to eject a formic acid-acyclic ketone spray capable of causing skin blisters and lesions. Overwinters as prepupa within weak cocoon in the topsoil or leaf mold. Some prepupae remain in diapause for 2 years. General feeder on broadleaf trees, commonly reported on oak and basswood. Nova Scotia, to southern Ontario, South Dakota and Colorado, south to Florida and Texas. Univoltine in the North (July-early October), bivoltine southward. Dyar (1893c), Packard (1895), Beutenmüller (1898), Hooker (1908), Forbes (1948), Prentice (1962), Eisner et al. (1972), Kearby (1975), Surgeoner and Wallner (1975, 1978), Staines (1977), and Wilson and Surgeoner (1979). Photo by J. E. Appleby.

Figure 26.350. Notodontidae. *Lochmaeus bilineata* (Packard). f.g.l. 30–40 mm. Reddish brown form. Body and head color similar to those of *L. manteo.* The middorsal line of *L. bilineata* is not present on the brown (reddish brown to purple) form and the dorsal space (area) of the green form is white (Dyar 1893c); its black, arcuate head stripe is narrower than *L. manteo*'s. Solitary. Overwinters as prepupa within cocoon shallowly situated in the soil (Daviault 1942) or as pupa (Prentice 1962). Mainly on American and slippery elm, but also reported on white birch and beech, and seen on wild black cherry (G.L.G.). Nova Scotia to southeastern Manitoba, south to Colorado and Florida (subspecies *exsanguinis* Dyar). Usually univoltine, feeding from early July to mid-September, occasionally a second brood. French (1886), Dyar (1893c), Beutenmüller (1898), Daviault (1942), Forbes (1948), and Prentice (1962). Photo by W. D. Zehr, I.N.H.S.

Figure 26.351a,b. Notodontidae. *Schizura ipomoeae.* Doubleday, false unicorn caterpillar. f.g.l. 28–35 mm. Large fleshy, bifurcate, slightly protrusible hump on A1, bearing approximate D1 tubercles; smaller fleshy humps on A5 and A8, but bases of D1's disjunct. T2 and T3 green subdorsally and laterally, remainder of body grayish or reddish brown, profusely mottled; prominent white or flesh-colored, dorsal chevron on A6–7; a patch of same color sometimes present on A1–2. Primary setae long, stout, dark brown or black. Head gray or tan, each side with 2 black parallel lines. The distinct, parallel black lines on each side of the head and the prominent black primary setae separate *S. ipomoeae* from the following 2 species of *Schizura.* A solitary defoliator. Spins cocoon covered with dirt on soil surface. General feeder, reported on about 20 genera of broadleaf shrubs and trees. Newfoundland to British Columbia, south to Florida, Texas, and Arizona. Number of broods uncertain. Larvae from May-October. Packard (1890a & b, 1895, 1905), Townsend (1893), Beutenmüller (1898), Forbes (1948), and Prentice (1962). Fig. 26.351a, photo by J. E. Appleby; fig. 26.351b photo by W. D. Zehr, I.N.H.S.

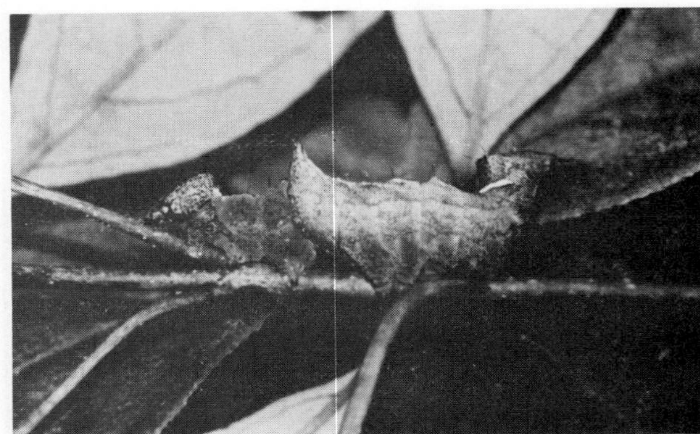
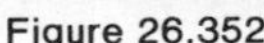

Figure 26.352

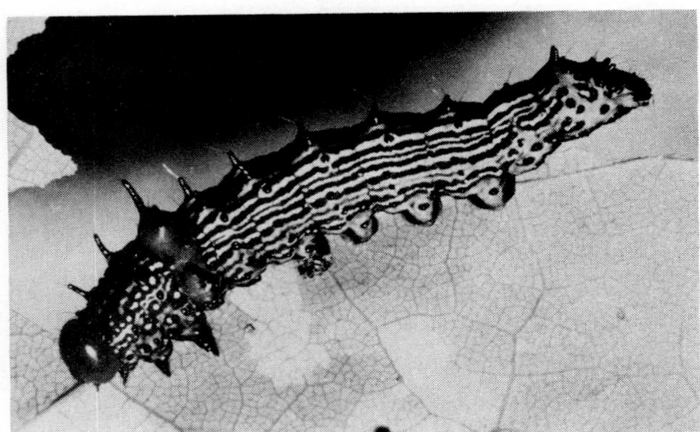

Figure 26.353

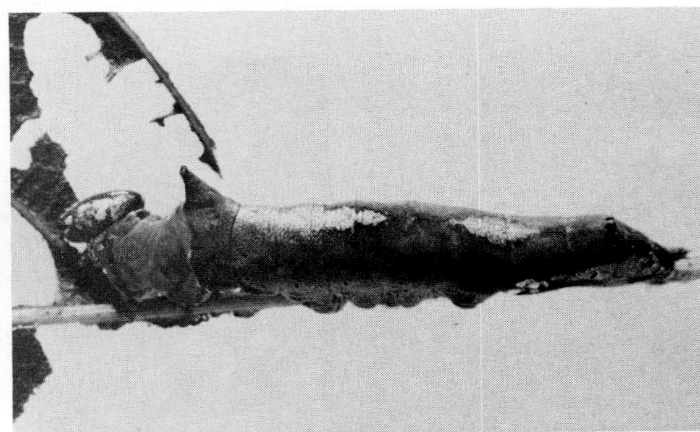

Figure 26.354

Figure 26.352. Notodontidae. *Schizura unicornis* (J. E. Smith), unicorn caterpillar. f.g.l. 25–35 mm. Fleshy humps similar to those of the preceding species. Grayish brown, heavily mottled; T2 and T3 green subdorsally and laterally; prominent white chevron on dorsum of A6–7. Primary setae short on A2–8. Head gray with blackish mottling; single, weakly defined black line on each half of head or none. The head pattern and short primary body setae readily distinguish it from *S. ipomoeae.* Solitary or in small, loose groups (3–4 larvae). Semitransparent cocoon spun on ground, covered with pieces of leaves. General feeder on about 20 genera of broadleaf shrubs and trees. Newfoundland to British Columbia, north to central Alberta, Manitoba, and Saskatchewan, south to northern California, throughout most of the eastern half of the U.S. Larvae reported from May-October. Two broods in the South, sometimes 2 in the North. Abbot and Smith (1797), Packard (1895, 1905), Beutenmüller (1898), Day (1912), Forbes (1948), and Prentice (1962). Photo by J. W. and R. R. Graber, I.N.H.S.

Figure 26.353. Notodontidae. *Schizura concinna* (J. E. Smith), redhumped caterpillar. f.g.l. 30–40 mm. Armed with elongate, black tubercles, tipped with primary setae; A1 red and swollen; A2–7 dark yellow dorsally and laterally, white subdorsally, striped with numerous, thin, black longitudinal lines. Head bright red. This species is unmistakable. Young larvae in conspicuous groups, older ones with tendency to separate. Nearly entire colony may synchronously abandon host to spin thin, parchmentlike cocoons in leaf litter or below soil surface. Overwinters as prepupa. General feeder, common at times on willow and trembling aspen, and the Rosaceae, especially apple. Reported from 26 additional genera of broadleaf shrubs and trees. Nova Scotia to southeastern British Columbia, north to central Manitoba and central Saskatchewan, south to Florida, Texas, and central California. Uni- or bivoltine; larvae spring and midsummer to September; spring brood apparently absent in far north. Abbot and Smith (1797), Packard (1895), Lugger (1898), Forbes (1948), Prentice (1962), and Ferguson (1975). Photo by J. E. Appleby.

Figure 26.354. Notodontidae. *Oligocentria lignicolor* (Walker). f.g.l. 35–38 mm. Pale form. Large fleshy hump on A1 bears contiguous D1 tubercles; remainder of body smooth except for very small dorsal hump on A8. T1–3 green but T1 with cream-colored to dark brown middorsal band and gray or brown lateral and ventral aspects. Green subdorsal patch on A3 and dorsal saddle on A4–10, saddle widest on A7; remainder of abdomen cream to purplish brown. A7 spiracle encircled by 4–5 white patches. Head cream with brown to dark brown bands on each half, with innermost margin of each band bent toward center. The circle of white patches around the A7 spiracle is quite distinctive. Solitary feeder. Overwinters as prepupa in parchmentlike cocoon spun between fallen leaves or under soil surface. Feeds on boxelder, white and yellow birch, beech, red and white oak, and chokecherry. Maine and Quebec to southeast Manitoba and S. Dakota, south to Florida and Texas. Apparently bivoltine, larvae from mid-June to October. Packard (1890b, 1895), Forbes (1948), and Prentice (1962). Photo by J. E. Appleby.

recreational areas (Grimble and Newell 1972b, Millers and Wallner 1975) by the mere presence of large numbers of crawling caterpillars and copious, dropping frass.

Selected Bibliography (General Papers)

Abbot and Smith 1797.
Beutenmüller 1898.
Common 1970.
Ferguson 1975.
Forbes 1923, 1948.
Fracker 1915.
Grote and Robinson 1866.
Hinton 1946.
Issiki et al. 1969.
Kellogg and Jack 1896.
Lugger 1898.
McDaniel 1933.
McFarland 1967(1968), 1979.
Mosher 1916a.
Peterson 1948, 1963b.
Prentice [coordinator] 1962.
Tietz 1972.

Selected Bibliography (Papers with Restricted Coverage)

Abbott 1927.
Allen 1973, 1979.
Allen and Grimble 1970.
Beutenmüller 1888.
Collins 1926.
Comstock 1959a.
Comstock and Dammers 1938.
Daviault 1942.
Day 1912.
Detwiler 1922.
Dyar, 1891a,b,c, 1892b, 1893a,c.
Eisner et al. 1972.
Farris and Appleby 1980 (1981).
Ferguson 1963.
Franclemont 1946, 1948, 1983.
French 1886.
Gardner 1943.
Gibson 1902.
Godfrey 1984.
Grimble and Kasile 1974.
Grimble and Newell 1972a,b, 1973.
Herrick and Detwiler 1919.
Hintze 1969.
Hixson 1941.
Holloway 1983.
Hooker 1908.
Imms 1970.
Kearby 1975.
Kendall and Kendall 1971.
Klots 1967a, 1969.
Loesch and Foran 1979.
Millers and Wallner 1975.
Mosher 1917.
Packard 1890a,b, 1893, 1895, 1905.
Patch 1908.
Percy and MacDonald 1979.
Popenoe et al. 1889.
Riotte 1969.
Staines 1977.
Surgeoner and Wallner, 1975, 1978.
Todd 1976.
Townsend 1893.
Weatherston et al. 1979.
Wilson and Surgeoner 1979.

Acknowledgments

The completion of this chapter was made possible by the enthusiasm and information generously provided by J. G. Franclemont and J. W. and R. R. Graber. We are grateful to the Dixon Springs Agricultural Center, Simpson, Illinois, and The Morton Arboretum, Lisle, Illinois, for the use of their research facilities. Work was funded in part by grants from the Joyce Foundation and by Illinois Agricultural Experiment Station Project 12–361 Biosystematics of Insects.

DIOPTIDAE (NOCTUOIDEA)

Frederick W. Stehr, ***Michigan State University***

The Dioptids

Figure 26.355

Relationships and Diagnosis: This small family is closely related to the Notodontidae and is sometimes included as a subfamily. Since the California oakworm, *Phryganidia californica* Packard, is the only known species in the U.S., this writeup is based on it. Totally reliable characters separating dioptid larvae from all notodontid larvae are not available, but the California oakworm does not have the A10 prolegs greatly reduced in size or modified into stemapoda as most notodontids do, although they are slightly reduced and may be elevated when the larva is at rest. Dioptids (at least the California oakworm) can be separated from similar "naked" notodontids that have anal prolegs bearing crochets by the presence of a darkly sclerotized T1 shield that bears four setae of nearly equal size per side (eight total setae); similar naked notodontids do not have a T1 shield, or if one is evident, it is not darkly sclerotized and the posterior pair of setae on each side is considerably shorter than the anterior pair; or if the shield is dark and the setae are nearly equal in length, the extra seta on A10 is absent (*Cargida pyrrha*). A better diagnostic character may be that the A2–7 spiracles of the California oakworm are nearly circular and the A7 spiracle is twice as large as the A2–6 spiracles (most notodontids have distinctly elliptical spiracles and the A7 spiracle is only slightly larger than the A2–6 spiracles (including *Cargida pyrrha*)).

Biology and Ecology: The California oakworm, the well-known defoliator of oaks in California, is more important toward the coast since it does not survive the winters of interior

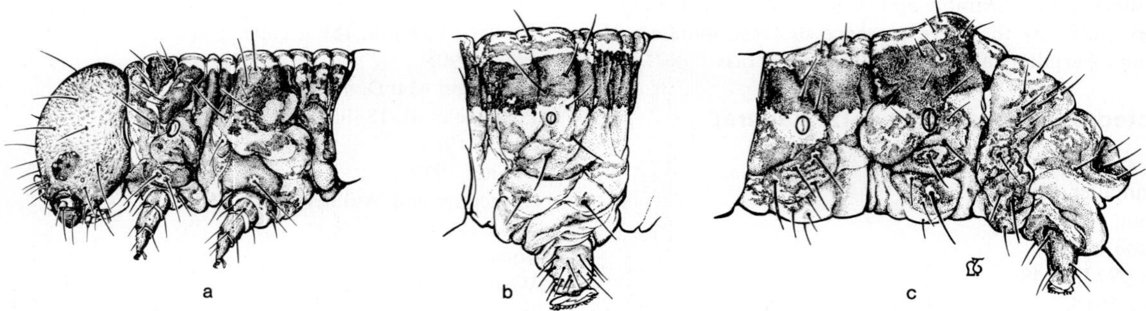

Figure 26.355a-c. Dioptidae. *Phryganidia californica* Packard, California oakworm. f.g.l. 25± mm. Olive green with black and yellow longitudinal stripes. Head dull reddish-orange. Reported from Oregon to San Diego, primarily on live oaks (see text).

California, even though it can cause damage when it spreads into that area in the summer. It is most important on the live oaks *Q. agrifolia* and *Q. dumosa,* but it attacks all oaks and occasionally wanders onto other shrubs and trees. There are two or three generations per year, and the summer populations are the most damaging.

Description: Mature larvae medium, 20–40 mm, essentially "naked", with most secondary setae confined to ventral prolegs. Body cylindrical, with a distinct dorsal hump on A8 and anal prolegs slightly smaller than "normal." A2–6 spiracles nearly circular; A7 spiracle about 2× larger than A2–6 spiracles, and nearly circular. T1 spiracle more elliptical than A2–6 spiracles and about the size of A7 spiracle. A8 spiracle elliptical and about 1.5× the size of T1 spiracle.

Head: Hypognathous, globose, larger than prothorax and nearly smooth. Frontoclypeus nearly equilateral, extending about one-fourth to epicranial notch, which is on the rear of the head. Ecdydial sutures distinct and meeting just above apex of the frons. Six stemmata in a normal arc except for 5, which is beneath lateral apex of the antennal socket. Stemma 3 somewhat larger than the others.

Thorax: T1 shield divided in the middle, but well sclerotized and bearing four conspicuous setae of approximately equal size on each half. SD, L and SV pinacula each with two setae. T2 and T3 with D2 nearly directly ventrad to D1; L1 slightly closer to L2 than L3; SV unisetose.

Abdomen: All segments with a standard complement of setae, plus a few extra ones that are not located on the prolegs and whose presence and locations are not very consistent. A10 with an extra seta just anterior to the lower edge of the anal plate (similar to many notodontids). A10 with a single prominent seta on the conical protrusion between each proleg and the anus (a similar structure is found on many notodontids). A3–6 prolegs with about 15 setae, with crochets in a uniordinal mesoseries; A10 prolegs slightly smaller.

Comments: The T1 shield and spiracular diagnostic characters given may be applicable only to the California oakworm; hence they should be checked on other "dioptids."

Selected Bibliography

Brown and Eads 1965.
Furniss and Carolin 1977.
Kellogg and Jack 1896.
Powell and Hogue 1979.
Wickman 1971.

DOIDAE (NOCTUOIDEA)

Julian P. Donahue, *Natural History Museum of Los Angeles County*
John W. Brown, *University of California, Berkeley*

The Doids or Euphorbia Moths

Figure 26.356

Relationships and Diagnosis: The Doidae, only recently elevated to family rank (Donahue, in prep.), is a small, primarily Neotropical group consisting of two or three genera and about seven to nine species. Mostly Mexican in distribution, two species occur in the southwestern U.S. Although doids share many individual character states with a number of families, the combination of character states will distinguish the Doidae (all earlier classifications were based entirely on the adult stage). That the doids do not readily "fit" in existing families is reflected by a history of diverse and uncertain classifications: they have previously been treated as Geometridae, Lymantriidae, Hypsidae, Arctiidae, Pericopidae, and Dioptidae. Most recently, Franclemont (1983) created for this group a new tribe, Doaini [sic], in the Arctiidae: Pericopinae.

Larvae of the Doidae appear to be unique in the following combination of character states: crochets biordinal, in a homoideous mesoseries; head small, thorax swollen (thus resembling a sawfly larva); secondary setae few, solitary above spiracle line on T2, T3, A7, and A8, coalescing to form small

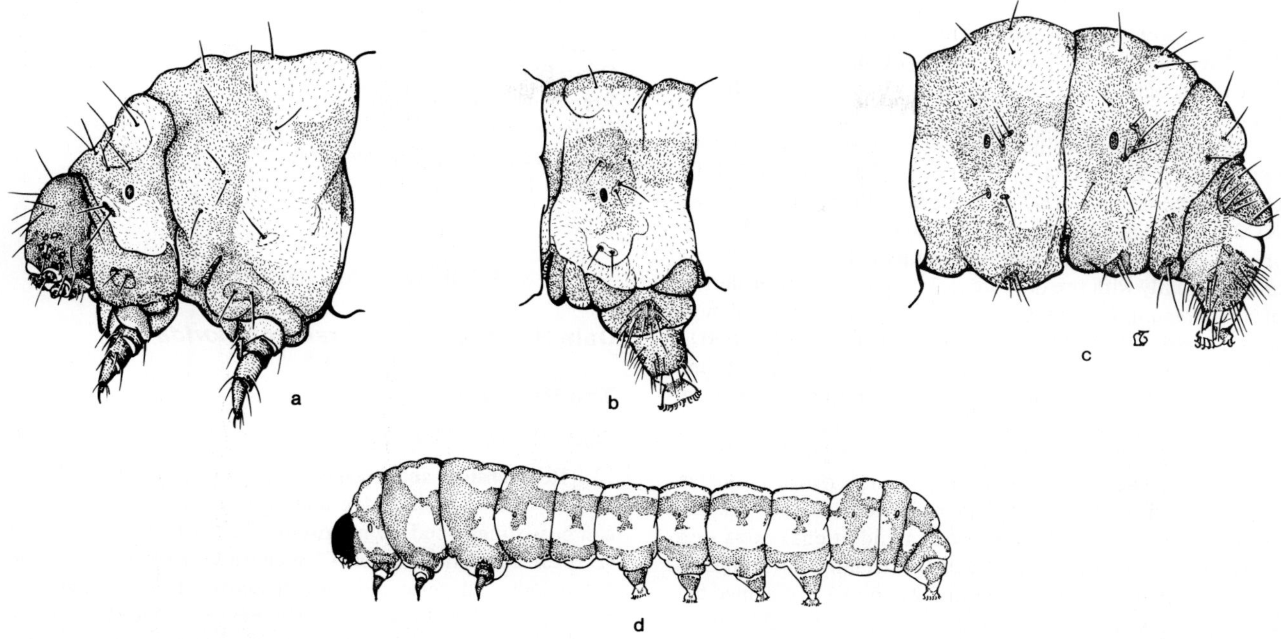

Figure 26.356a-d. Doidae. *Doa ampla* Dyar. f.g.l. 20± mm. a. head, T1, T2; b. A3; c. A7-10; d. entire larva (setae omitted). Head blackish. Color pattern bipolar, shaded areas black, with large white spots on T1, T2, A7-10, the intervening segments (T3-A6) with bold yellow stripes. See text.

verrucae in L and SV groups on abdomen; two L setae on T1; two SV setae on T2–3; spiracles small, those on T1 and A8 distinctly larger; L1 group retaining same relation to spiracle on A7 as on A3–6; mandible with prominent inner tooth. Perhaps sharing more larval characteristics with the Thyatiridae than any other family, doid larvae that have been examined (*D. ampla* and *D. dora*) are readily distinguished by their conspicuous integumentel spinules (absent in the thyatirids that have been examined).

Biology and Ecology: The early stages have only been described for two species, *Doa ampla* (Grote) (Dyar 1912) (figs. 26.356a-d) and *Doa raspa* (Druce) (Dyar 1911), the latter a species from Mexico and Central America; we have also examined larvae of the Mexican *Doa dora* Neumoegen and Dyar. The early stages of *Leuculodes,* the only other genus currently in Doidae, have not been described. The yellow eggs of *Doa ampla* are oblong and ovaloid, the entire surface covered with fine punctations. Eggs usually laid in a cluster of 15–35, in neat parallel rows of four to eight eggs per row. Early instar larvae moderately gregarious. Winter is probably spent as partially grown larvae, diapause being induced, at least in part, by the poor quality or insufficient quantity of the host plant. Pupation occurs in debris at the base of the host plant, in a single-layered, wiry, open-meshed brown cocoon. Adults are at least partly diurnal and are not readily attracted to ultraviolet light; their flight is weak and fluttering. All known larval hosts are in the Euphorbiaceae: *Doa ampla* has been reared on *Euphorbia robusta* in Colorado (Cockerell 1911), *E. incisa* and *E. lurida* in Arizona (R. Wielgus, label data; N. McFarland, pers. comm.), and *Stillingia texana* in Texas (R. Kendall, pers. comm.); *Doa dora* has been reared from *Euphorbia misera* in Baja California, Mexico (J. Brown, unpublished ms.).

Description (primarily based on *Doa ampla*): Mature larva medium, 18–22 mm; thorax swollen, larva appearing "humpbacked" (height of T2 in lateral view about 1.25× height of A2). Head small, smooth, brown, heavily mottled with black (*Doa ampla*) or orange with black "eye-patch" and labrum (*D. dora*). Crochets biordinal, in a homoideous mesoseries. Skin with numerous spinules, especially dense on pigmented areas; unpigmented skin actually colorless and transparent, revealing color of underlying tissue. Spiracles elliptical, peritreme well sclerotized, small yet conspicuous, those on T1 and A8 distinctly larger. All setae simple, pinacula small or absent. Secondary setae usually associated with L1, SV group (always on A3–6), and SV2 (always on A1 and A2), the pinacula then usually united to form small verrucae. Secondary setae may also occur elsewhere, their distribution and occurrence irregular and often bilaterally asymmetrical on a single larva. (Secondary setae in *Doa dora* usually restricted to prolegs.)

Head: Smooth, anteriorly quite flat or slightly concave, epicranial notch shallow. Frontoclypeus height = width, extending 0.55 distance to epicranial notch, the ventral margin broadly concave to more than half its width. Ecdysial sutures prominent, meeting about 0.35 distance between apex of frons and epicranial notch. Labrum with broad, rounded V-shaped notch. Six stemmata, 1 and 6 similar in size and larger than 2–5. Stemmata 1–4 about equally spaced; stemma 5 about 3× its diameter from both 4 and 6, stemma 6 equidistant from 5 and 4.

Thorax: T1 shield discontinuous across meson, reduced to a narrow transverse strip bearing only XD1 and XD2. D1's closer to meson than XD1's, D2's slightly farther apart than XD2's. SD1 dorsad of spiracle; SD2 small, between XD2 and spiracle. L group on T1 and SV group on T1–3 bisetose. T2 and T3 usually with two or more extra setae or verrucae dorsad of SV (*Doa ampla*), obscuring the homologies. Anterior-most dorsal setae (D1's?) 1.2× farther apart than the D2's. SD2 ventrad of D2, SD1 anteroventrad of SD2 (on T2, on a line between SD2 and T1 spiracle). L1 a group of two or more setae (unisetose in *D. dora*), remote or adjacent on a small verruca, in line with spiracles. L2 anteroventrad of L1 group, L3 posterad and/or dorsoposterad of L1 verruca. An extra seta posterad to L2 and slightly more ventrad than L2, directly dorsad of leg; *D. ampla* with an additional extra seta posterad of a line between L1 and L3 (usually closer to L1).

Abdomen: Distance between D2s up to 2× distance between D1's. A7 and A8 with extra seta dorsad of D2, giving appearance of 4 (total) equally spaced D2 setae in transverse line across dorso-meson. SD1 dorsad and slightly anterad of spiracle. SD2 minute, anterad of spiracle. L1 a group of 1–6 setae (one in *D. dora,* usually two in *D. ampla*), irregularly arranged posterad of spiracle on A1–8, pinacula of those on A7 and A8 tending to coalesce to form a rudimentary verruca in *D. ampla;* one (rarely two) L1 setae on A9. L2 and L3 about halfway between spiracle and SV1 group, about one spiracle height apart, L3 slightly lower than L2. SV1 group unisetose (rarely bisetose) on A1, A2, and A7–9, and 4–9 setose (usually bisetose in *D. dora*) on A3–6 prolegs. SV2 5–9 setose (2–3 setose in *D. dora*) on a small verruca on A1 and 2, unisetose (occasionally bi- or trisetose) on A7 and 8, absent on A9. SV3 unisetose on A1 and 2, absent on A7–9. SV2–3 group multisetose on A3–6 prolegs. A10 with 13–16 anal plate setae (26–32 total; fewer in *D. dora*) and many on the proleg. V1 apparently absent on A3–6, but undoubtedly grouped with other setae on inside of proleg.

Comments: *Doa ampla* is the most widely distributed species in the family, at least in North America, occurring from Colorado to the Mexican states of Durango and Nuevo Leon, and from Texas to Arizona. *Leuculodes lacteolaria* (Hulst), apparently scarce in the type locality of Arizona, is more common in western Mexico; we have examined specimens from Sinaloa, Nayarit, Morelos and Baja California Sur. The early stages of *Leuculodes* have not been described.

Although this small group appears morphologically distinct from any other American moth family, its affinities and higher classification are unclear. A separate family within the Noctuoidea seems to be a reasonable interim solution.

Selected Bibliography

Cockerell 1911.
Dyar 1911, 1912.
Franclemont 1983.
Holland 1903.
Neumoegen and Dyar 1893–94.
Schaus 1927.

PERICOPIDAE (NOCTUOIDEA)

Dale H. Habeck, *University of Florida*

The Pericopids

Figures 26.357-26.360

Relationships and Diagnosis: The Pericopidae are closely related to the Ctenuchidae and Arctiidae. Pericopid larvae can be distinguished from most arctiids by having only three verrucae above the coxa on T2, a characteristic shared by the ctenuchids. All pericopids except *Daritis* have a single V1 seta on the inside of the prolegs whereas ctenuchids have many secondary setae there. Pericopids also have dark and shiny (sometimes metallic-appearing) verrucae on certain thoracic and abdominal segments.

Biology and Ecology: Most larvae are brightly colored and they feed exposed on their food plants. Eggs spherical, with irregular or hexagonal reticulations, laid singly or in clusters. Pupae often brightly colored within a flimsy cocoon. The larvae presumably are unpalatable. Larvae of some species are gregarious. Plant families recorded as hosts are Fabaceae (= Leguminosae), Asclepidaceae, Apocyanaceae, Boraginaceae, and Asteraceae (= Compositae).

Description: Mature larvae 35–50 mm. Body cylindrical, prolegs on A3–6 and A10. Body may be striped longitudinally or transversely. Some thoracic and/or abdominal segments with verrucae contrastingly colored, usually dark brown, but shiny and often steely blue in live specimens. Verrucae of preserved specimens also exhibit this characteristic but less conspicuously. All setae barbed and on verrucae except for those on prolegs. Secondary setae on prolegs confined to lateral, anterior, and posterior aspects, with a single V1 seta on inner side of prolegs except in *Daritis,* which has four to five setae on inner side. Spiracles elliptical, those on T1 and A8 larger. Head without secondary setae.

Head: Hypognathous. Frontoclypeus about as wide as long, margins straight and extending one-half to two-thirds to epicranical notch. Ecdysial lines meeting not far above frontoclypeal apex. Six stemmata in semicircle, 1–4 nearly equidistant from each other, 5 separated from 1–4, 6 caudad of 4 and closer to 4 or equidistant from 4 and 5.

Thorax: T1 shield usually distinct. Two clumps of setae (XD1 and SD1?) on anterior margin. D1 and D2 generally

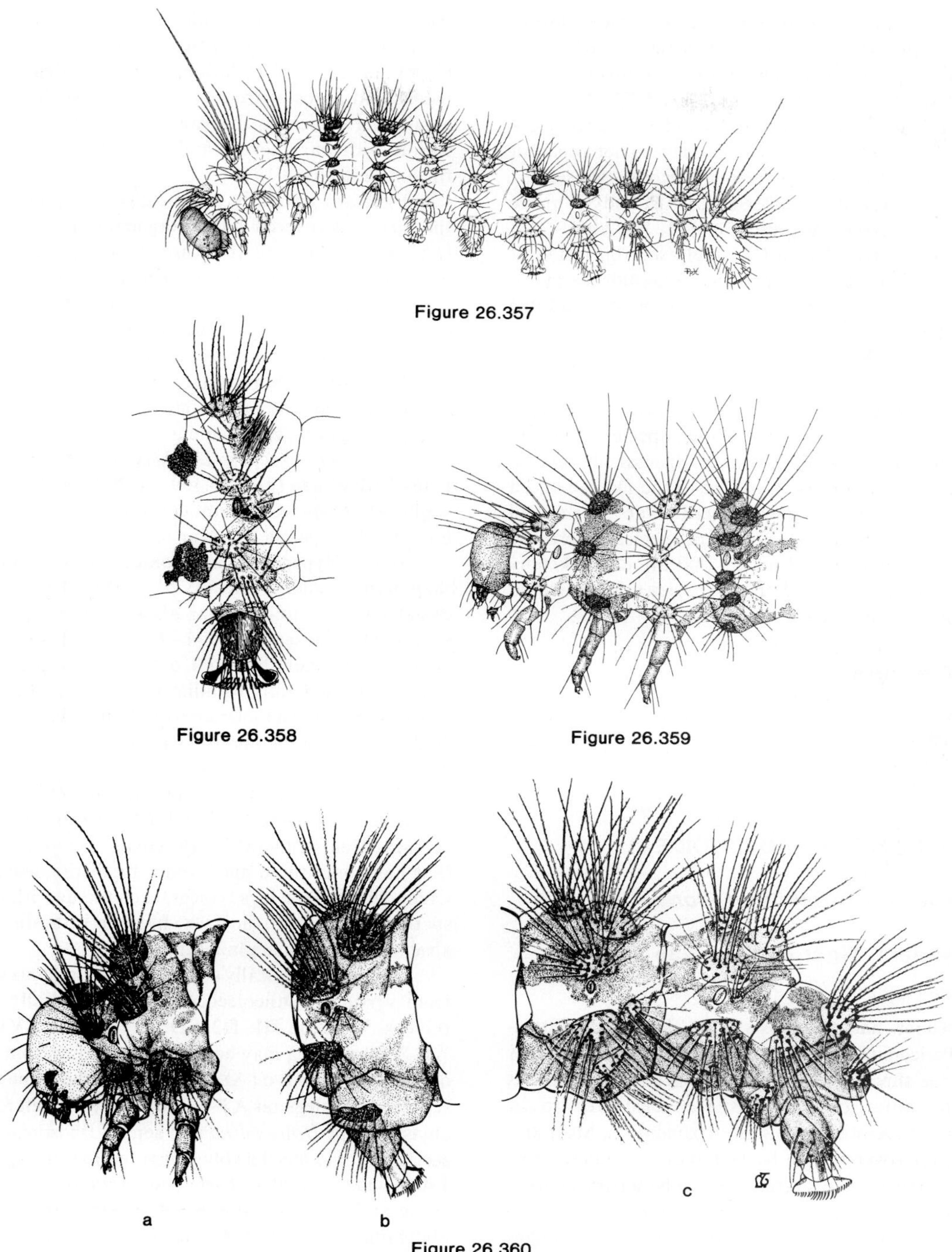

Figure 26.357

Figure 26.358

Figure 26.359

Figure 26.360

Figure 26.357. Pericopidae. *Composia fidelissima* Herrich-Schaeffer. f.g.l. 36± mm. Head and body bright red with metallic blue verrucae on T1-2, A6-7 and less so on A8. Other verrucae pale yellow. Feeds on *Echites*. Florida.

Figure 26.358. Pericopidae. *Daritis howardi* Hy. Edwards. Lateral view of A4. Head and legs dark brown. f.g.l. 50-55 mm. Body segments T1-2, A1-2, 5, 7, 9-10 maroon with dark brown shiny verrucae with a bluish cast. Verrucae on other segments yellow. Ventral portion of D2 verrucae on A1-8 with patch of dense short setae. Feeds on *Brickellia*. Texas.

Figure 26.359. Pericopidae. *Pericopis leucophaea* Walker. Head, T1-3, Al. f.g.l. 40± mm. Head and legs dark brown. Verrucae on T2, A1, A9 and D1, D2 and SD2 verrucae on A7 shiny dark brown with bluish cast. Other verrucae yellow to yellowish-brown. Feeds on *Bidens, Pluchea*. Mexico.

Figure 26.360. Pericopidae. *Gnophaela latipennis* Boisduval. a. head, T1-2; b. A3; c. A7-10. f.g.l. 40-45 mm. Verrucae on all segments dark brown with bluish cast. Feeds on *Cynoglossum*. Southwest U.S.

represented by only one or two setae, SD2 with two setae on a separate sclerite posteroventrad to T1 shield. L and SV verrucae in front of spiracle and above coxa, respectively. V represented by a single seta. Verrucae on T2 and T3: D and SD (combined), L, and SV. In *Composia* D and SD verrucae are adjacent but distinctly separated, which is often not apparent when viewed laterally. The SD verruca has only two setae.

Abdomen: Verrucae D1, D2, SD, L1, L2, and SV above proleg on A3–6. Verruca V present on A1, A2, and A7–9, but may be represented by only a single seta on inside of proleg. Crochets arranged in uniordinal heteroideous mesoseries. The longer central crochets range from about a third to a half of the total number of crochets. Verruca L1 more ventral on A7 and 2× farther from spiracle than on A6 or A8.

Comments: The family is small with less than ten species known from the United States. One species occurs in Florida and the others are southwestern. Species of *Doa* previously included here have been moved to the Doidae based on larval characters and other evidence. Larvae are not well known and there are no keys to genera or species. Literature on larvae consists of scattered papers that are often sketchy and without illustration. None of the species are considered to be of economic importance.

Selected Bibliography

Cockerell 1889.
Dyar 1896, 1899, 1914.
McFarland 1961.

ARCTIIDAE (NOCTUOIDEA)

Dale H. Habeck, *University of Florida*

The Woolybears, Tiger Moth Larvae

Figures 26.361-26.371

Relationships and Diagnosis: The Arctiidae, Ctenuchidae, and Pericopidae are generally distinguished from other members of the superfamily by the presence of heteroideous crochets. The latter two families are considered to be subfamilies of the Arctiidae by some taxonomists. Most arctiids are densely covered with barbed and/or plumose setae arising from verrucae. Most Arctiidae can be separated from the Ctenuchidae and Pericopidae by having four verrucae above the coxae on T2 and T3 whereas the latter families have only three. Some arctiids of the subfamily Lithosiinae have homoideous crochets and many are without verrucae, having single setae arising from pinacula. Some noctuids resemble arctiids, but arctiids never have integumental spicules, which similar noctuids usually have.

Biology and Ecology: Many arctiid larvae are inactive and hide during the day. All larvae feed exposed except for the fall webworm, *Hyphantria cunea* (Drury), and *Tyria jacobaeae* (Linn.), which live gregariously within a web until they seek pupation sites. Other species may be in loose colonies during the early stages, dispersing in later stages. Larvae feed on a wide variety of plants, from grasses to deciduous trees and shrubs. Many of the larger species crawl very rapidly. Larvae of the Lithosiinae feed on algae epiphytic on trunks and limbs of trees or on stones, or they may eat the algal layer of lichens. Larvae of a species of *Crambidia* associate with ants, move in and out of the nests, and pupate in the ant nest (Ayre 1958).

Eggs are generally dome-shaped or almost round, and the surface is covered with hexagonal reticulations. Eggs are fairly uniform in structure throughout the family.

Pupae stout, with a weak or no cremaster. Verrucae scars are surrounded by fine setae. Cocoons generally very flimsy, composed mainly of the larval setae incorporated with a little silk. Arctiids overwinter as pupae or partially grown larvae.

Description: Mature larvae small to large, 10–80 mm, cylindrical and generally stout, bearing verrucae and many setae. Standard complement of verrucae present (figs. 26.367–26.368) except in *Utetheisa* (fig. 26.363), *Tyria,* and many Lithosiinae (figs. 26.369–26.371), which usually have single setae. Spiracles elliptical, those on T1 and A8 larger, but circular in some Lithosiinae.

Head: Hypognathous. Frontoclypeus as wide or wider than high, extending halfway or less to epicranial notch. Ecdysial lines meeting not far above frontoclypeal apex. Six stemmata in a semicircle, 1–4 near each other, with 2–3 nearer each other than to 1 or 4; 5 separated from 1–4, 6 caudad and generally equidistant from 4 and 5. The Lithosiinae always with a molar area on the mandible (figs. 26.370b, 26.371b), which distinguishes them from the Arctiinae (fig. 26.361c, 26.364c).

Thorax: T1 shield present but often obscured by setae. Verrucae conspicuous. T1 and T2 with full complement of verrucae (four or five above the coxae) except in a few genera that have only three and cannot be distinguished from the Ctenuchidae. True legs typical, each tarsus with two or three spatulate setae near the apex. In some Lithosiinae, verrucae absent, but SV pinacula bisetose.

Abdomen: Typically covered with numerous setae arising from verrucae. Other secondary setae usually confined to prolegs. Verrucae D1, D2, SD, L1, L2, and SV1 present on A1–8. L1 verruca may be distinctly farther from the spiracle on A7 than on A6 or A8 or it may be the same on all three segments. Prolegs on A3–6 and A10 with heteroideous crochets (except *Holomelina* and some Lithosiinae) in a mesoseries. Planta considerably elongated in most species. Many Lithosiinae, as well as *Tyria* and *Utetheisa,* are without verrucae and bear only simple setae. SD pinacula bisetose on abdominal segments of Lithosiinae.

Comments: The family contains about 225 species in 67 genera north of Mexico. Although most larvae have been described, descriptions are often inadequate or without illustrations. Fracker (1915) and Forbes (1960) present keys to the genera of larvae and Mosher (1916a) and Forbes (1960) provide keys to the genera of pupae.

Although some arctiids are common, few are of economic importance. The saltmarsh caterpillar, *Estigmene acrea* (Drury) (fig. 26.361), feeds on a wide range of crops and weeds. The yellow woolybear, *Spilosoma virginica* (Fabricius), and the banded woolybear, *Pyrrharctia isabella* (J. E.

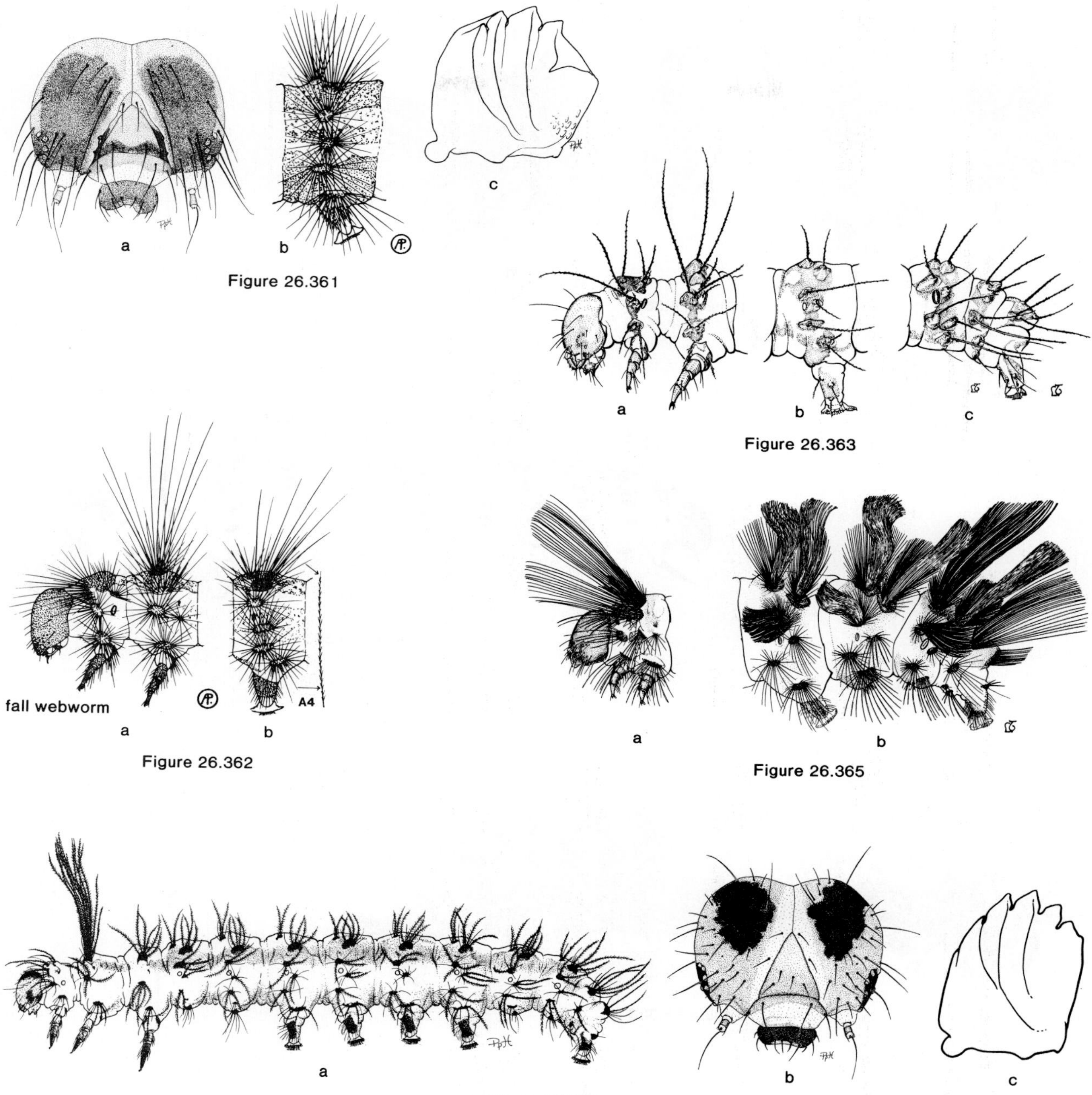

Figure 26.361a-c. Arctiidae, Arctiinae. *Estigmene acrea* (Drury), salt marsh caterpillar. f.g.l. 45–50 mm. a. head; b. lateral view of A4; c. mandible. Highly variable in color, pale yellow to dark brown, but readily identifiable by the yellow streak through the front of the head. Polyphagous. Widespread, common. (fig. 26.361b from Peterson 1948).

Figure 26.362a-b. Arctiidae, Arctiinae. *Hyphantria cunea* (Drury) fall webworm. f.g.l. 30–35 mm. a. head, T1–2; b. A4. The red- and black-headed forms exhibit differences in web formation and host preferences, and almost certainly represent distinct species. Black-headed form prefers willow, sweetgum, and other plants in wet areas; red-headed form prefers pecan, persimmon, and plants in drier areas although there is some overlap in hosts. Widespread, common. (figs. from Peterson 1948).

Figure 26.363a-c. Arctiidae, Arctiinae. *Utetheisa ornatrix* (Linnaeus). f.g.l. 30–35 mm. a. head, T1–2; b. A4; c. A8–10. Verrucae absent. Setae on chalazae. Feeds on *Crotalaria*. Southern Florida, common.

Figure 26.364a-c. Arctiidae, Lithosiinae. *Pagara simplex* Walker. f.g.l. 15–20 mm. a. larva; b. head; c. mandible. Hair pencil on D verrucae of T2. Head with 2 prominent dark spots. Feeds on plantain (in lab). Eastern U.S., widespread but uncommon.

Figure 26.365a,b. Arctiidae, Arctiinae. *Euchaetes egle* (Drury). f.g.l. 25± mm. a. head, T1–2; b. A6–10. Black, white and yellow. Pencils and tufts irregular in length. Feeds on milkweeds. Eastern U.S.

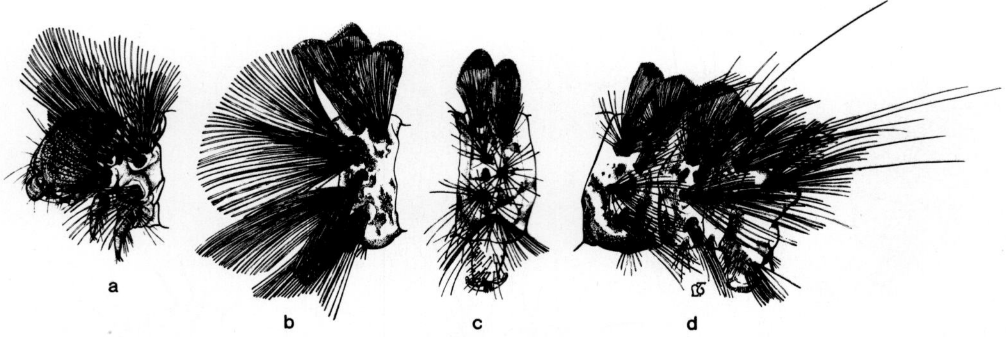

Figure 26.366

Figure 26.367

Figure 26.368

Figure 26.366a-d. Arctiidae, Arctiinae. *Lophocampa caryae* Harris. hickory tussock moth. f.g.l. 35-40 mm. a. head, T1-2; b. A1; c. A3; d. A7-10. Black and white. Black hair pencils on A1 and A7, white pencils on A8. Various trees. Widespread.

Figure 26.367. Arctiidae, Arctiinae. *Holomelina laeta* Guerin-Meneville. f.g.l. 15-20 mm. Setae removed to show verrucae. Mouse gray, host unknown, probably various low herbs, reared on dandelion, plantain and artificial diet. Eastern U.S.

Figure 26.368. Arctiidae, Arctiinae. *Ecpantheria scribonia* (Stoll). f.g.l. 75± mm. Setae removed to show verrucae. Black, bristly with red intersegmental areas. Setae uniform length. Feeds on various low herbs. Eastern U.S.

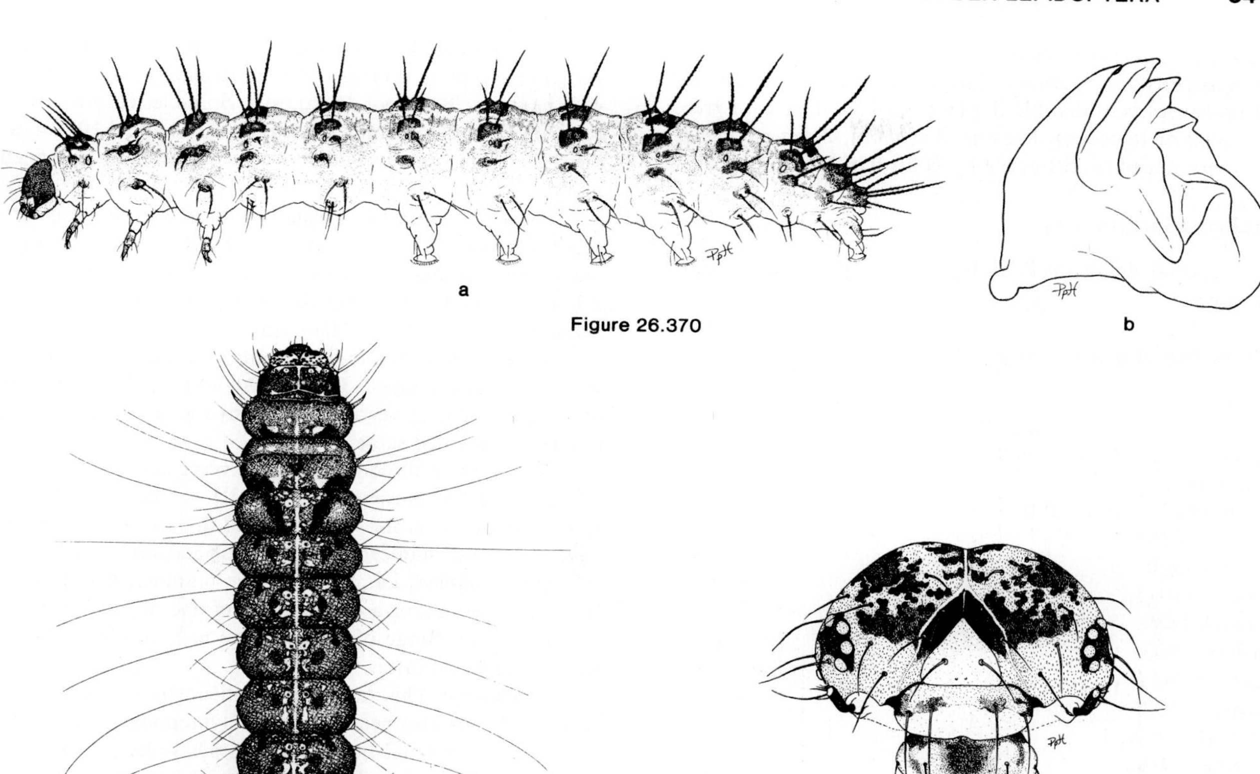

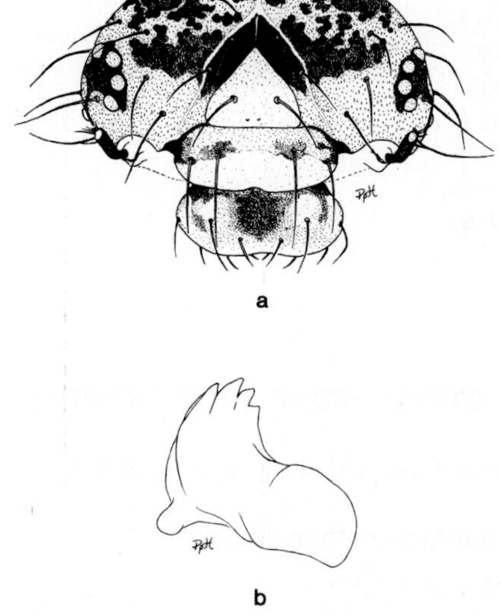

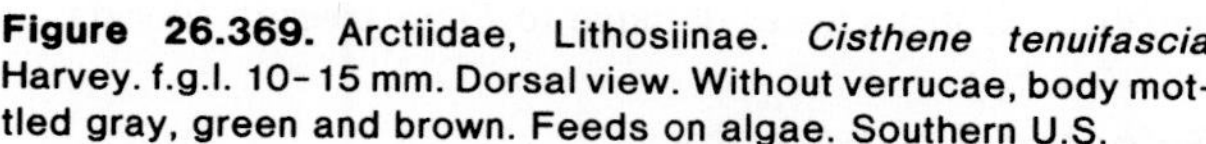

Figure 26.369. Arctiidae, Lithosiinae. *Cisthene tenuifascia* Harvey. f.g.l. 10–15 mm. Dorsal view. Without verrucae, body mottled gray, green and brown. Feeds on algae. Southern U.S.

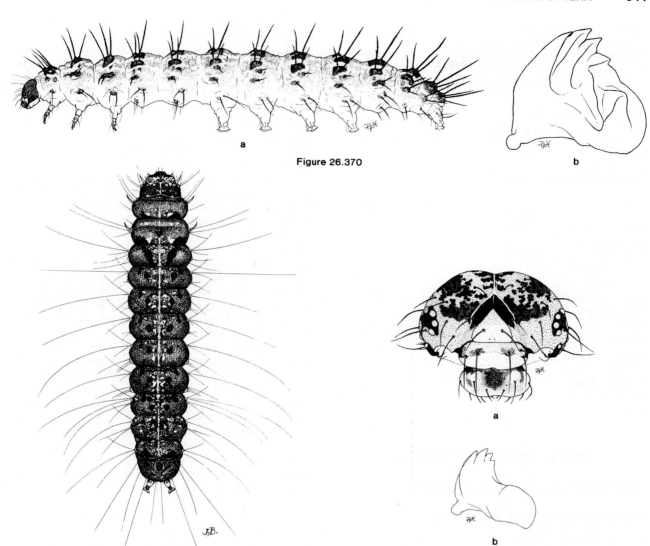

Figure 26.370a,b. Arctiidae, Lithosiinae. *Hypoprepia fucosa* Hübner. f.g.l. 25± mm. a. larva; b. mandible with enlarged mola. Without verrucae, mottled green and brown. Feeds on algae and lichens. Eastern U.S.

Figure 26.371a,b. Arctiidae, Lithosiinae. *Clemensia albata* Packard. f.g.l. 12–15 mm. a. head; b. mandible with enlarged mola. Without verrucae, mottled green, gray and brown. Feeds on lichens. Widespread.

Smith), are occasionally of some concern as are some species of *Apantesis* and *Grammia*. The fall webworm (fig. 26.362) often causes considerable damage to shade and fruit trees, particularly to pecans. The cinnabar moth, *Tyria jacobaeae*, was introduced from Europe for control of tansy ragwort.

Acknowledgments

Special thanks to P. Habeck and J. Bacheler for illustrations.

Selected Bibliography

Ayre 1958.
Barnes and McDunnough 1912.
Comstock and Henne 1967.
Drew 1962.
Dyar 1897a, 1900b, 1902.
Ferguson 1985b.
Forbes 1960.
Fracker 1915.
French 1889.
Gibson 1903.
LaFontaine et al. 1982.
Learned 1926.
McCabe 1981.
Mosher 1916a.
Oliver 1963.
Peterson 1948, 1963.

CTENUCHIDAE (NOCTUOIDEA)

Dale H. Habeck, *University of Florida*

The Ctenuchids, Wasp Moths

Figures 26.372–26.375

Relationships and Diagnosis: Ctenuchids are closely related to Arctiidae and Pericopidae and are considered by many taxonomists to be a subfamily of the former. They can be distinguished from most arctiids by having only three verrucae above the coxa on T2 and from pericopids by the absence of distinctly different-colored verrucae on several body segments. The inner sides of the prolegs have numerous secondary setae, whereas in all pericopids except *Daritis* there is only a single V1 seta. Most ctenuchids are more or less densely covered with setae arising from verrucae.

Biology and Ecology: Larvae feed exposed and usually singly, but oleander caterpillars, *Syntomeida epilais* (Walker), are gregarious. Most are host specific or feed on only a few closely related species. Eggs of ctenuchids are spherical and deposited singly or in clusters. The surface is marked with faint irregular or hexagonal reticulations. Pupae are stout, sometimes brightly colored and indistinguishable as a group from arctiid pupae. Cocoons are flimsy and composed mainly of larval setae and a small amount of silk.

Description: Mature larvae 15–55 mm. Setae limited to verrucae except on prolegs and head (usually). Many species with hair pencils and/or dorsal tufts. Spiracles elliptical with those on T1 and A8 larger. Prolegs on A3–6 and A10. Uniordinal, heteroideous crochets in a mesoseries on an extended planta.

Head: Hypognathous, usually with numerous secondary setae. Frontoclypeus extending about halfway to epicranial notch. Six stemmata, 2–4 usually closer to each other than 2 is to 1. Five distant from 4, and 6 nearer to 5 than to 4. Stemma 3 usually larger than others.

Thorax: Prothoracic shield generally distinct. L and SV verrucae well developed. T1 and T2 with only three verrucae above coxa. Dorsal verrucae may have long hair pencils or otherwise modified setae.

Abdomen: Full complement of verrucae: D1, D2, SD, L1, L2, and SV present on A4–6. L1 verruca usually distinctly farther from the spiracle on A7 than it is on A6 or A8. Secondary setae numerous on prolegs. Setae on L1 of A1 and A7 often modified, i.e., longer, more plumose, denser, contrastingly colored, or a combination of one or more of these characteristics. Setae of D1 verrucae may combine to give rise to middorsal tufts.

Comments: This family is small north of Mexico (29 species). It has also been known as Syntomidae, Euchromiidae, and Amatidae. No characteristics have been found that will separate all ctenuchids from arctiid larvae. The ctenuchid characteristic of having only three verrucae above the coxa on T2 is shared by some arctiids. Most of the larvae are known and described, but the only keys to ctenuchid larvae are by Fracker (1915) and Forbes (1906). The family is best represented in the Neotropical area and only two species (*Ctenucha virginica* (Esper) (fig. 26.372) and *Cisseps fulvicollis* (Hübner) (fig. 26.373)) occur in the more northern part of the United States.

The family is of little economic importance in the United States. The oleander caterpillar, *Syntomeida epilais* (Walker) (fig. 26.374), and the spotted oleander caterpillar, *Empyreuma affinis* Rothschild (fig. 26.375), are pests of oleander in Florida. The Edwards wasp moth caterpillar, *Lymire edwardsii* (Grote), causes damage to *Ficus* trees in Florida.

Selected Bibliography

Bratley 1932.
Forbes 1906.
Fracker 1915.
Genung 1959.
Habeck and Mead 1982.
Mosher 1916a.
Patch 1921.
Peterson 1948.
Remington 1963.

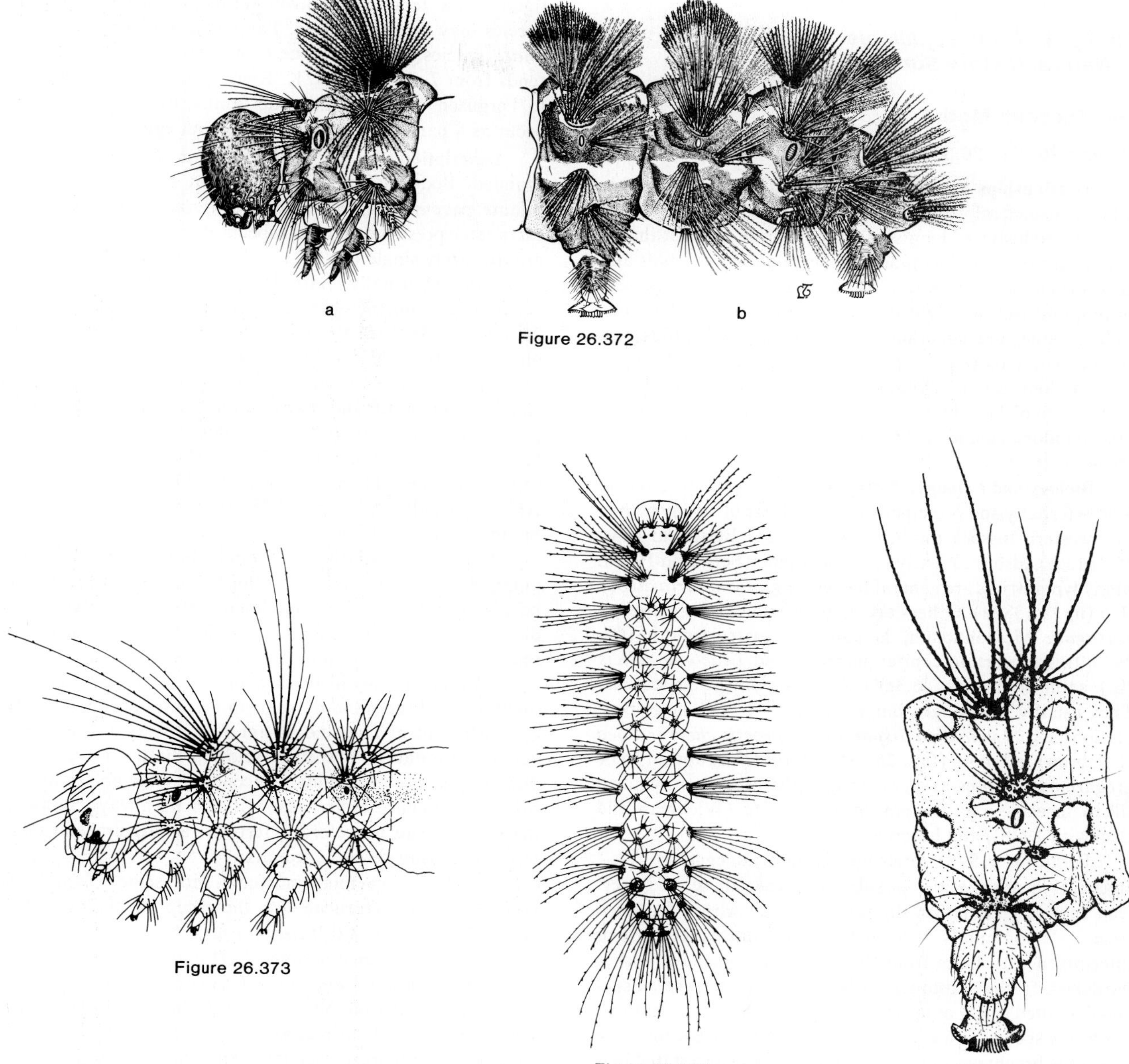

Figure 26.372

Figure 26.373

Figure 26.374

Figure 26.375

Figure 26.372a,b. Ctenuchidae. *Ctenucha virginica* (Esper). f.g.l. 30–35 mm. a. head, T1–2; b. A6–10. Head reddish brown and black. Body brown, lighter dorsally, a broad darker brown subdorsal stripe including SD verruca and spiracles, bordered dorsally and ventrally by narrow white stripe. D1 verrucae on abdomen close together forming short dorsal tufts of black setae. D2 verrucae on abdomen with creamy-white setae. Feeds on grasses. Eastern U.S. but not southern.

Figure 26.373. Ctenuchidae. *Cisseps fulvicollis* (Hübner). f.g.l. 25± mm. Head, T1–3, A1. Head and body yellow-brown, body with narrow middorsal dark stripe and wide darker lateral stripe encompassing SD verruca and spiracle. Setae relatively sparse, brown or white. Feeds on grasses. Eastern U.S.

Figure 26.374. Ctenuchidae. *Syntomeida epilais* (Walker), oleander caterpillar. f.g.l. 30–35 mm. Dorsal view. Head and body orange with dark hair pencils. Feeds on oleander. Florida.

Figure 26.375. Ctenuchidae. *Empyreuma affinis* Rothschild, spotted oleander caterpillar. f.g.l. 25–30 mm. Lateral view of A4. Similar to oleander caterpillar but with prominent white spots. Sparse black hair pencils on D verruca of T2 and D2 verruca of A8. Feeds on oleander. Florida.

LYMANTRIIDAE (NOCTUOIDEA)

George L. Godfrey, *Illinois Natural History Survey*

The Tussock Moths

Figures 26.376–26.385

Relationships and Diagnosis: North American lymantriid larvae generally are about 30–40 mm long but may reach 65 mm (exclusive of long setal tufts) during the last instar. Larvae are basically noctuoid with hypognathous heads and fully developed prolegs on A3–6 and A10. They possess paired, multisetose verrucae, and, because of abundant secondary setae, are superficially similar to the arctiids and "hairy" noctuids (e.g., *Acronicta, Charadra,* and *Panthea*) and notodontids (e.g., *Datana*). However, they can be distinguished from these by the presence of a single, fleshy, eversible, middorsal gland on A6 and usually on A7 (figs. 26.379a, 26.381b, 26.382).

Biology and Ecology: North American lymantriid larvae feed externally and principally on the foliage of woody plants. The western tussock moth, *Orgyia vetusta* Boisduval (figs. 26.381a,b), also frequently feeds on lupine (Ferguson 1978). Many lymantriids are general feeders, e.g., *Lymantria dispar* (L.) (fig. 26.383a). Others are more restricted: *Orgyia pseudotsugata* McDunnough, Douglas fir tussock moth (fig. 26.380), eats certain coniferous trees, and *Leucoma salicis* (L.), satin moth (figs. 26.384a,b), occurs on the Salicaceae. They usually feed singly, but some occur in such large populations, e.g., *Lymantria dispar* and *Euproctis chrysorrhoea* (L.), browntail moth (figs. 26.385a,b), that they appear to be gregarious. First instar larvae frequently disperse by wind on silken threads, e.g., *Orgyia pseudotsugata* (Wickman and Beckwith 1978), and become solitary.

Ferguson (1978) described lymantriid eggs as ". . . spherical, hemispherical, or subcylindrical, [those] of American species globular or slightly depressed with a flattened base, . . . commonly deposited in large masses, covered or intermixed with hairs from the female abdomen, or with a hardened, frothy substance, or both. Eggs of *Dasychira* species deposited singly or in small groups without covering; those of *Orgyia* species in a mass on the surface of the cocoon from which the brachypterous female emerged." He also stated that the surfaces of the eggs are unsculptured, but in most *Orgyia* they are sculptured, especially around the micropyle (Beckwith, pers. comm.).

Lymantriid pupae characteristically have long, coarse setae around the scars of the larval verrucae; labial palpi visible (usually concealed in *Lymantria*); epicranial suture absent; maxillae short, not exceeding two-fifths of the wing length; T2 legs never reaching caudal margin of wings; antennae pectinate; cremaster present, bearing hooks (Mosher 1916a); T1 femora not visible, a character shared with the Arctiidae (Ferguson 1978). Cocoons often contain some larval setae (Forbes 1948).

Lymantriids in southern Canada and the northern contiguous U.S. tend to be univoltine, but southward the same species may be multivoltine. The larval duration of *Gynaephora* species, at least in the Canadian Arctic Islands, extends from 7–11 years (J. K. Ryan, pers. comm. according to Ferguson 1978). Within the family, overwintering may occur as a partially grown larva or as an egg.

Description: Mature larvae medium, 20–65 mm. Head rounded. Body cylindrical; integument smooth or bearing minute pavement granules (e.g., *Euproctis chrysorrhoea*); extra setae present on lateral aspects of prolegs; primary setal groups rarely single (e.g., D1's of T2, T3 and A1–8 of *Leucoma salicis*), usually multisetose on distinct verrucae; setae occur as radiating tufts (figs. 26.376, 26.378), dense tufts (figs. 26.379b, 26.382) (dense tufts especially distinctive on dorsa of A1–4 in *Dasychira* and *Orgyia*), or hair pencils (especially on T1, A8, and A9) (figs. 26.377a–26.382). Setae either simple (may be stiff and sharp) or plumose. Body indistinctly patterned or marked with varying degrees of cream-colored to yellow spots and lines (figs. 26.382, 26.384); blue, orange, or red verrucae; and yellow to bright red middorsal glands on A6 and usually A7 (fig. 26.381b). Spiracles narrowly to broadly elliptical, larger on T1 and A8 than on A1–7.

Head: Hypognathous. Frontoclypeus roughly triangular, with apex at 0.24–0.45 of distance from frontoclypeal base to epicranial notch. Junction of adfrontal ecdysial lines distinctly cephalad of epicranial notch. Six stemmata, 1–4 in semicircle, 5 and 6 forming line ventrocaudad of 1–4.

Thorax: T1 shield texture similar to adjacent integument, quite discernible by color in certain genera, e.g., *Orgyia.* Midventral T1 gland absent. Chaetotaxy noctuoid, but actual delimitation of specific setae in mature larvae often difficult because of abundant secondary setae and tendency toward confluence of verrucae. L1 and L2 always confluent in mature larvae, henceforth referred to as prespiracular setae in the following species accounts.

Abdomen: Chaetotaxy noctuoid. Most setae (D's and L's) borne on distinct verrucae with the exception of single D1's on T2, T3, and A1–8 of *Leucoma salicis* and bi- or trisetose L1's of *Euproctis chrysorrhoea.* SD1 and L1 on A1–8 of *Lymantria dispar* and *Leucoma salicis* confluent. Prolegs subequally developed on A3–6 and A10, bearing homoideous, uniordinal crochets in mesoseries. A6 and usually A7 each with single middorsal, eversible gland, usually bright yellow to red.

Comments: Thirty-two species are known from Canada and the continental U.S. (Ferguson 1983). The diversity in the Neotropics is much greater, but the greatest number of species occurs in the Old World. Three species in the North American fauna are introductions from the Old World: *Lymantria dispar, Leucoma salicis,* and *Euproctis chrysorrhoea.*

Lymantriids are especially noted for their capacity to defoliate enormous areas of forested regions, for their negative aesthetic effects on wooded residential areas, and for the urticating properties of the setae and spines of certain species, especially the three aforementioned introduced species.

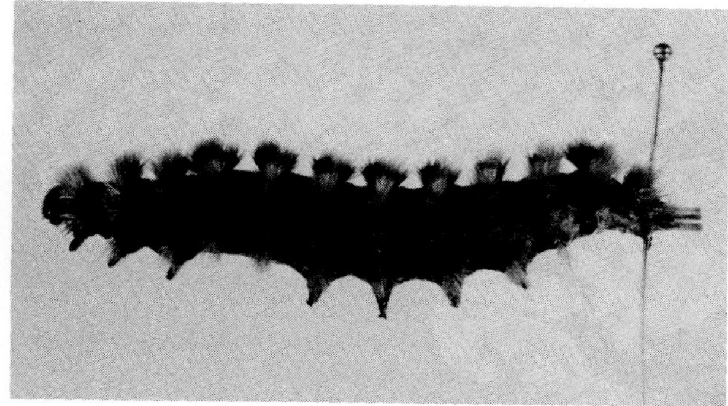

Figure 26.376

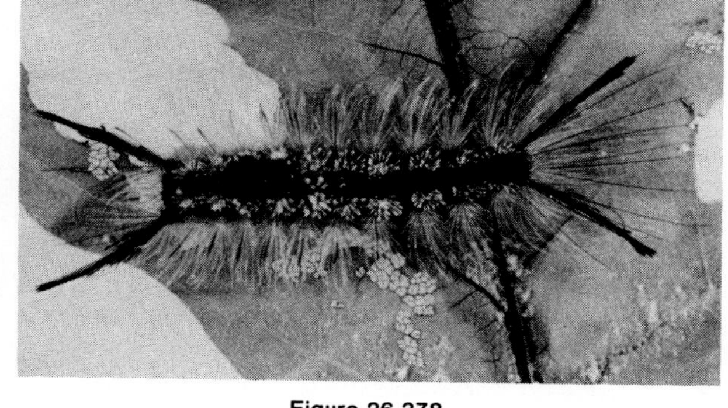

Figure 26.378

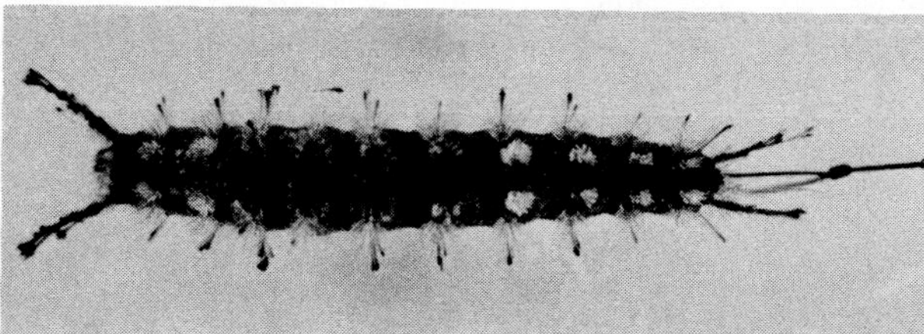

a

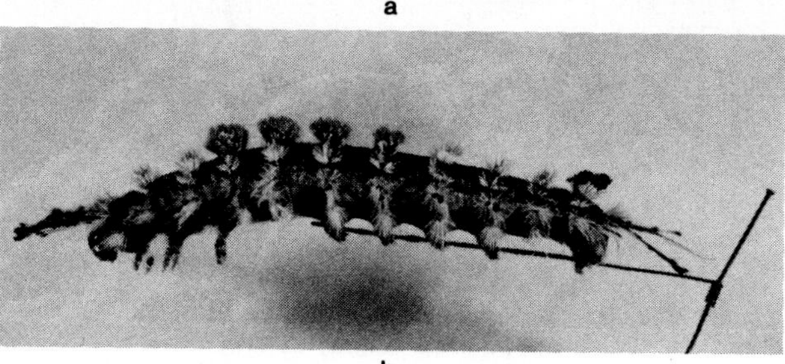

b

Figure 26.377

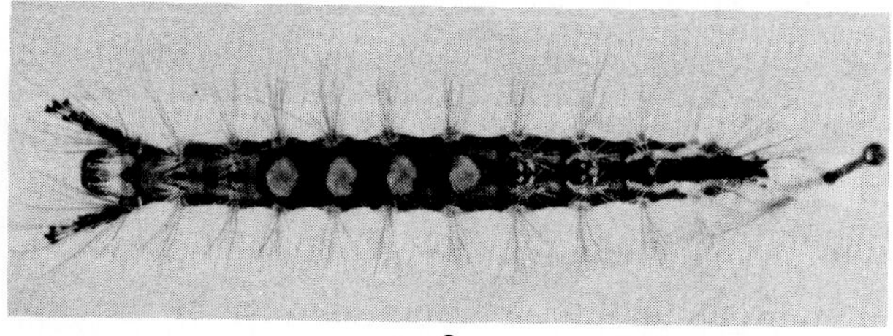

a

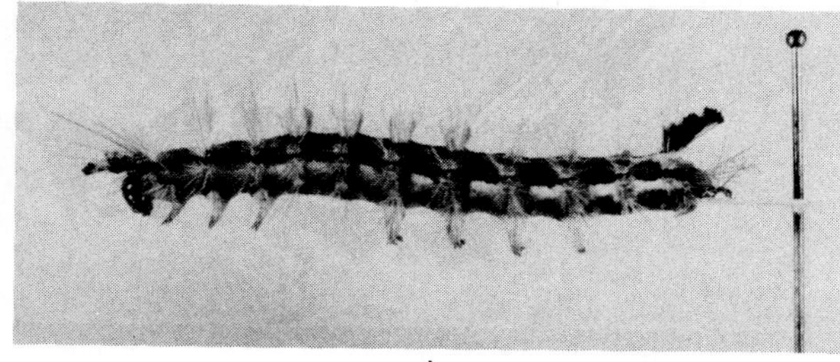

b

Figure 26.379

Figure 26.376. Lymantriidae. *Gynaephora rossii* (Curtis). f.g.l. 35–43 mm (blown specimen). Grayish orange to black; setae gray or yellow, dorsal thoracic setae equal mixture of gray and yellow; abdominal setae uniformly dark gray for majority of width, yellow laterally. Head brown to black, facial aspects (excluding frons and adfrontal areas) granulose, dull. Differs from other North American lymantriid genera by the 8 short, dense, dorsal abdominal setal tufts and absence of hair pencils. Larval duration may last 7–11 years. Occurs in dry and exposed conditions in nearly treeless arctic and alpine zones. Reported from Rosaceae (*Dryas* and *Saxifraga*) and Willow; 1 specimen (U.S. National Museum) labeled "spruce". Holarctic, in North America from the tundra of Labrador to Alaska, south to Colorado (higher elevations); above tree line in mountains of Quebec, New Hampshire, and Maine. Active soon after snow melt. Dyar (1897b), Oliver et al. (1964), and Ferguson (1978). Photo by G. L. Godfrey.

Figure 26.377a,b. Lymantriidae. *Dasychira vagans vagans* (Barnes and McDunnough). f.g.l. 43 mm (blown specimen). Grayish with black mottling. T1 and A9 each with pair of black hair pencils (long clavate, plumose setae). Five dense, dorsal setal tufts (fig. b): tufts on A1-4 brownish to black with scattering of white, plumose setae; tuft on A8 blackish medially, white laterally; all tufts longer than dorsal setae on other body segments. Middorsal glands on A6 and A7 coral red (living specimens). Head brownish red, shiny. The red middorsal glands distinguish *D. vagans* from *D. basiflava* (Packard) (**fig. 26.378**), which has white or yellow glands. See Ferguson (1978) for additional characters. Solitary defoliator on white birch, apple, balsam poplar, trembling aspen, and other deciduous trees. Ferguson (1978) considered coniferous hosts (Prentice 1962) as "highly improbable." The combined distribution of *D. vagans vagans* and *D. vagans grisea* (Barnes and McDunnough) extends across southern Canada and contiguous U.S. (*vagans* found to eastern edge of the Great Plains and *grisea* from there westward) and southward into montane areas. Larvae apparently occur from late June to mid-May, overwintering partially grown. Barnes and McDunnough (1913), Brittain and Payne (1919), Prentice (1962), and Ferguson (1978). Photos by G. L. Godfrey.

Figure 26.379a,b. Lymantriidae. *Orgyia antiqua* (L.). rusty tussock moth, dark phase. f.g.l. 30–40 mm. Mostly gray to black (darkest on dorsum of A1-8) with yellow to orange verrucae. Black hair pencils (clavate, plumose setae) on T1 and A2 verrucae (hair pencil on A2 absent in *O. antiqua badia* Hy. Edwards); large, medial, black hair pencil (clavate, plumose setae) on A8. Dense, white to yellow, dorsal setal tufts on A1-4. T1 shield yellowish orange with medial, dark brown or black triangular patch that distinguishes *antiqua* from other North American *Orgyia*. Middorsal glands on A6 and A7 red. Head shiny, mostly black. Solitary feeder on many deciduous shrubs and trees and most conifers except *Juniperus*. *Orgyia antiqua nova* Fitch occurs across Canada, northwest to central Alaska, south to Pennsylvania and Iowa. Replaced in southern British Columbia to northern California by *O. antiqua badia*. Overwinters as an egg; feeds from late spring to mid- or late summer. Prentice (1962), Tietz (1972), and Ferguson (1978). Photos by G. L. Godfrey (blown specimens).

Figure 26.380

Figure 26.382

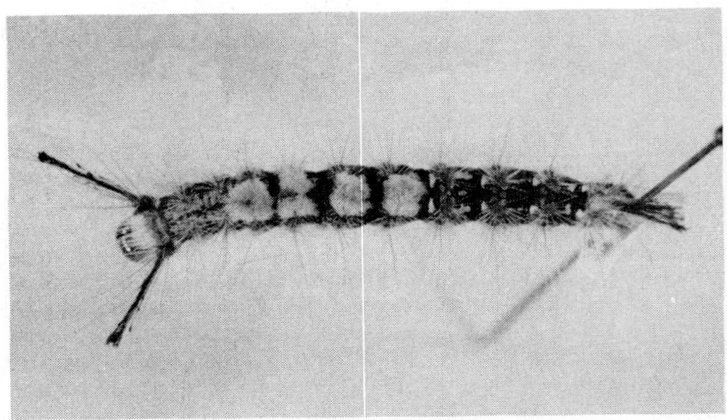

a

b

Figure 26.381

Figure 26.380. Lymantriidae. *Orgyia pseudotsugata* (McDunnough), douglas-fir tussock moth. f.g.l. 30± mm. Pale tan or dark brown (2 color phases), mottled with black, marked as follows: T1 shield red; verrucae red; black hair pencils of clavate, plumose setae on T1 prespiracular verrucae. Single black, dorsal hair pencil on A8; dense, whitish, dorsal setal tufts with tan tips on A1-4 and 1 small dense, tan tuft at anterior base of dorsal hair pencil on A8; tufts on A1-4 with long, dark setae concentrated on lateral edges; L2 abdominal verrucae, in addition to radiating clusters of white setae, bear 1-5 very long (length ca. 2× body width), dark brown or black setae. Middorsal glands on A6 and A7 red. Head shiny, orange-brown, infuscated with dark brown. The long, dark L1+L2 setae on the abdominal segments separate this species from the *O. vetusta* complex. Feeds singly. Early instars may aggregate on silken masses at tree tops prior to dispersing by wind. Major pest of Douglas fir, grand fir, and white fir. Also reported from other coniferous genera, occasionally from nonconiferous trees. A western species, abundant west of Continental Divide in areas where primary hosts occur from southern British Columbia to Arizona. Feeds for about 60 days during late May to September (rarely early October); univoltine. McDunnough (1921), Balch (1932), Eaton and Struble (1957), Sugden (1957), Prentice (1962), Mason and Baxter (1970), Beckwith (1978a, b), Ferguson (1978), and Wickman and Beckwith (1978). U.S.D.A., Forest Service photo, courtesy of R. C. Beckwith.

Figure 26.381a,b. Lymantriidae. *Orgyia "vetusta-cana"* complex, western tussock moth. a. blown specimen; b. living specimen. f.g.l. 30-40 mm (blown specimens). Blackish, marked as follows: T1 and anal shields red; D1 and D2 verrucae on A5-8 and SD1 verrucae red; dorsal area of T2 and T3 yellow to orange with black middorsal line; yellow to orange spots cephalad and caudad of D2 verrucae on A5-7; subdorsal area of some specimens with large yellow patches, some confluent with yellow dorsal spots; lateral area yellow to orange. Black hair pencils of clavate, plumose setae on T1 prespiracular verrucae; single black, dorsal hair pencil on A8. Dense, pale tan to gray setal tufts on A1-4; small dense tuft at anterior base of A8 hair pencil; tufts contain a few scattered, barbed, brown setae slightly longer than paler setae in tufts. Abdominal L1+L2 verrucae bear long, white, plumose setae in radiating clusters. Head shiny, orange-brown, dark brown infuscation on frontal aspects. The uniformly white setae on abdominal L1+L2 verrucae and bright yellow or orange dorsal, abdominal spots separate species of this complex from *O. pseudotsugata*. General feeders, principally on deciduous shrubs and trees. From Oregon to Montana, southward to northern Baja, California, Mexico and Utah. Edwards (1881b), Dyar (1892a, 1893b, 1896a), Tietz (1972), and Ferguson (1978). Fig. 26.381a photo by G. L. Godfrey; b, California Dept. Food and Agric. courtesy of T. D. Eichlin.

Figure 26.382. Lymantriidae. *Orgyia leucostigma* (J. E. Smith), whitemarked tussock moth. f.g.l. 30-34 mm. Mostly yellow or yellow and gray: T1 shield red; dorsal area of A1-4 black (interrupted by 4 dense, white, setal tufts); subdorsal line yellow, sometimes accented by grayish subdorsal area. Black hair pencils of clavate, plumose setae on T1 prespiracular verrucae; single black, dorsal hair pencil on A8. Whitish dorsal setal tufts on A1-4 frequently with single long, black seta at lateral boundaries. Middorsal abdominal glands on A6 and A7 red (blackish in some subspecies). The black dorsal area on A5-8 contrasting with the continuous, vivid yellow subdorsal lines distinguishes *O. leucostigma* from other *Orgyia* species. General, solitary feeder on trees and shrubs, including several coniferous species; occasionally feeds on herbs. *O. leucostigma leucostigma* occurs from S. Carolina to Florida and throughout the Gulf Coast to southern Texas. *O. leucostigma intermedia* Fitch ranges from Maine to Virginia, west to Alberta and Kansas. Overwinters in egg stage. Generally 2 broods. Ferguson (1978). Photo by G. L. Godfrey.

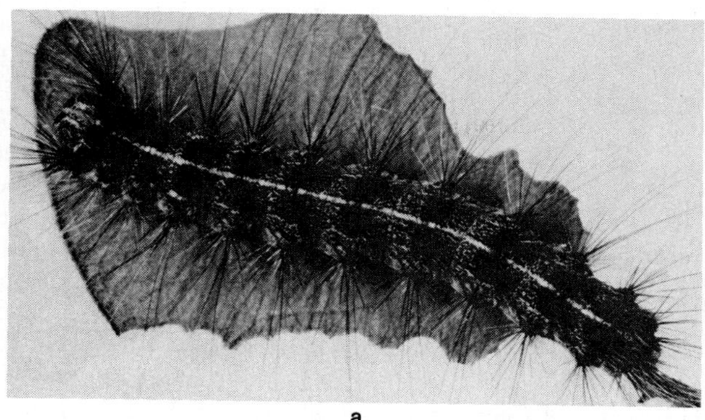

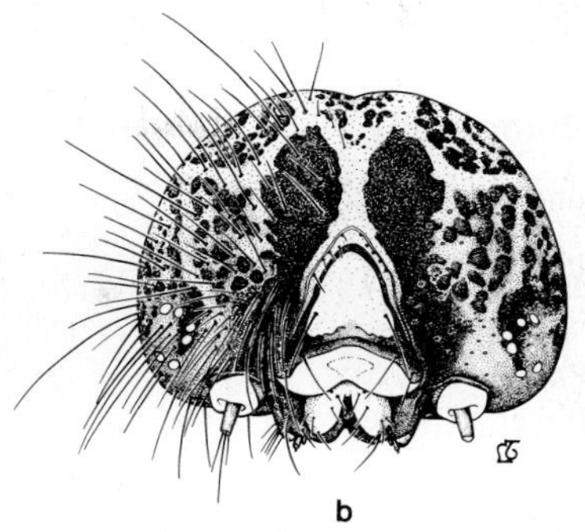

a b
Figure 26.383

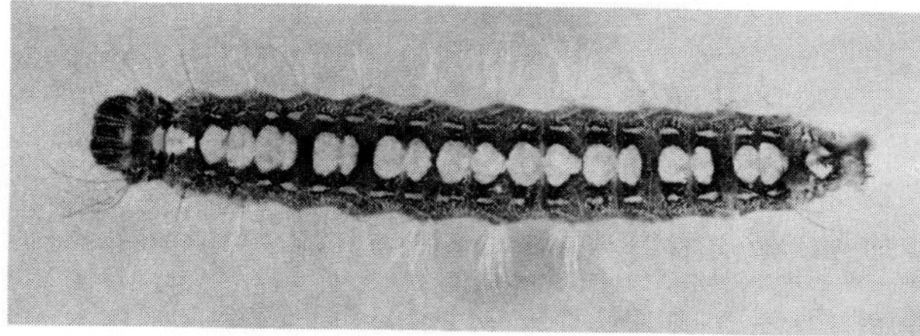

a

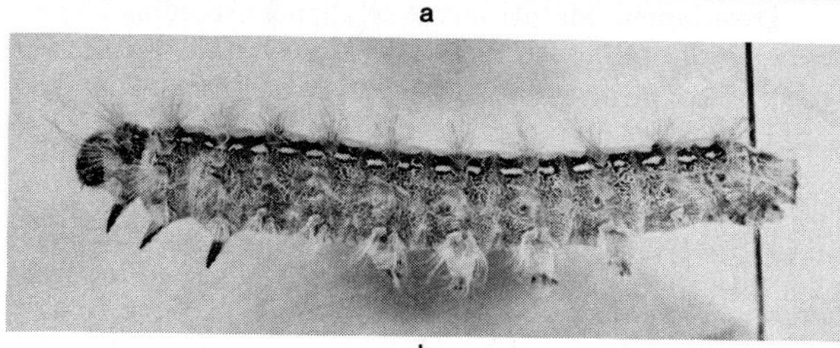

b
Figure 26.384

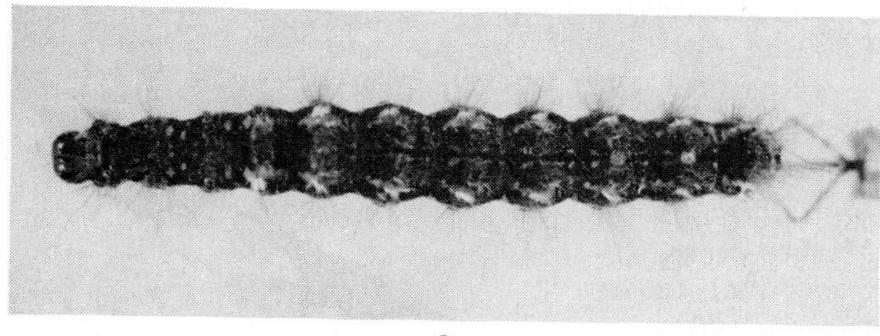

a

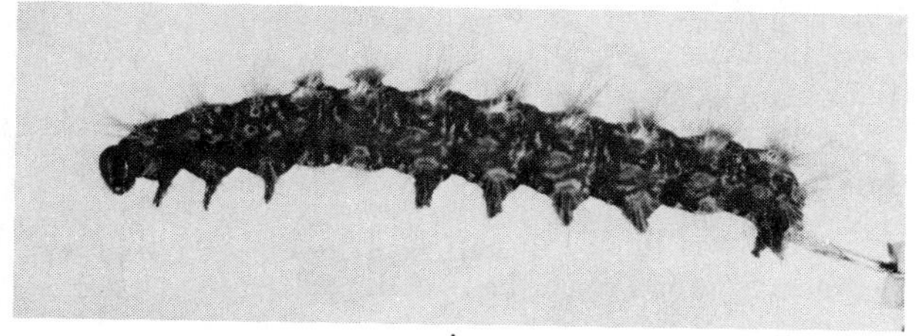

b
Figure 26.385

Figure 26.383a,b. Lymantriidae. *Lymantria dispar* (L.), gypsy moth. a. mature larva; b. head. f.g.l. 31–65 mm. Yellow brown, extensively mottled with shades of brown and black. Lacks hair pencils and dense, dorsal setal tufts. Dorsal verrucae on T1–3 and A1–2 blue, red on A3–8. Verrucae SD and L1 confluent on A1–8. Middorsal glands on A6 and A7 red. Dorsal setae stiff, sharp. Head yellow brown, marked with dark brown to black coronal stripes and numerous brown freckles (fig. 26.383b). Identifiable by the dark coronal stripes and the blue and red verrucae of the body. Solitary defoliator but commonly very numerous on a given tree. Hosts: feeds on many deciduous and coniferous trees, but prefers oaks, especially *Quercus alba* Linnaeus (see Abrahamson and Klass 1982 for details). Introduced from Europe; frequently abundant in the northeastern U.S. During the 1970s it spread to the Midwest and other areas, including the West Coast. Overwinters in egg stage; feeds for about 1 month in the spring or early summer. Campbell (1975), Ferguson (1978), and Abrahamson and Klass (1982). Photo by J. E. Appleby, I.N.H.S.; 26.383b drawing by L. Tackett, courtesy F. W. Stehr.

Figure 26.384a,b. Lymantriidae. *Leucoma salicis* (L.), satin moth (blown specimen). f.g.l. 35–50 mm. Conspicuously marked dorsally with series of large, cream-colored spots (each spot constricted laterally) set in black dorsal area; subdorsal line cream colored, interrupted by rust brown or orange D2 verrucae; remainder of body bluish gray. No dense tufts or hair pencils; body setae stiff. SD1 and L1 verrucae confluent on A1–8. Double medial dorsal glands (black) on A1 and A2, single yellow or brown middorsal glands on A6 and A7. Head black. The large cream-colored, dorsal spots distinguish *L. salicis* from all other North American lymantriids. Solitary feeder on poplar and willow, but may occur in large numbers. Introduced from temperate Eurasia. Massachusetts to all of Atlantic Canadian provinces, west to eastern Ontario and Utica, N.Y., in the East; Vancouver Island to southern Interior of British Columbia, south through western Washington and Oregon to Modoc County, California in the West. (Note: single female moth collected in Santa Rosa Range, Humboldt County, Nevada, August 1983). Feeds from mid- to late summer, overwinters under silken web in bark crevices, resumes and completes feeding in the spring. Burgess and Crossman (1927), Prentice (1962), and Ferguson (1978). Photos by G. L. Godfrey.

Figure 26.385a,b. Lymantriidae. *Euproctis chrysorrhoea* (L.), browntail moth (blown specimen). f.g.l. 28–35 mm. Dark brown to black with orange brown verrucae, interrupted subdorsal lines, and small irregular spots. D1+D2 and SD1 verrucae of A1–8 with mixture of radiating tufts of long, slightly plumose, tannish setae and compact tufts of shorter tannish and some white plumose setae (especially on SD1's adjacent to subdorsal lines). No dense tufts or hair pencils. Middorsal glands of A6 and A7 reddish. Head brownish with numerous dark brown to black freckles. The very short compact tufts of tannish and radiating plumose setae on the D1+D2 and SD verrucae differ from those of other North American lymantriids. Young larvae gregarious leaf skeletonizers in late summer, principally on rosaceous trees and shrubs but also reported from *Myrica* (bayberry), oak and willow. Overwinters as partially grown larva in communal nests of silk and leaves. Univoltine. Introduced European species, presently restricted to islands in Casco Bay, Maine and beach areas of Cape Cod, Massachusetts. Burgess and Baker (1938) and Ferguson (1978). Photos by G. L. Godfrey.

Larval accounts of representative species of all North American lymantriid genera are included in the illustrations.

Selected Bibliography

Abrahamson and Klass 1982.
Balch 1932.
Barnes and McDunnough 1913.
Beckwith 1978a,b.
Brittain and Payne 1919.
Burgess and Baker 1938.
Burgess and Crossman 1927.
Campbell 1975.
Dyar 1892a, 1893b, 1896a, 1897b.
Eaton and Struble 1957.
Edwards 1881b.
Ferguson 1978, 1983.
Forbes 1948.
Mason and Baxter 1970.
McDunnough 1921.
Mosher 1916a.
Oliver et al. 1964.
Prentice 1962.
Sugden 1957.
Tietz 1972.
Wickman and Beckwith 1978.

Acknowledgments

Work was funded in part by Illinois Agricultural Experiment Station Project 12–361 Biosystematics of Insects.

NOLIDAE (NOCTUOIDEA)

Frederick W. Stehr, *Michigan State University*

The Nolids

Figure 26.386

Relationships and Diagnosis: The nolids are superficially similar to some Arctiidae, Pericopidae, Ctenuchidae, and Noctuidae, but are believed to be most closely related to the Sarrothripinae (Noctuidae), with which they intergrade in the Old World tropics (Franclemont *in* Forbes 1960). (Franclemont *in* Forbes 1960) recognized two genera in the Northeast, *Nola* (=*Celama*) and *Meganola;* they are placed in the Nolinae (Noctuidae) by Franclemont in the checklist (Hodges et al. 1983). *Nigetia* (formerly included in Nolidae) is placed in the Hypenodinae (Noctuidae) in the checklist.

The larvae of our species can be distinguished from similar arctiids, pericopids, ctenuchids, noctuids, and other families by the absence of prolegs on A3. A few noctuids lack prolegs on A3 and have prolegs A4–A6 of equal size, but these species lack conspicuous secondary setae above the spiracles; some other noctuids may lack prolegs on A3, but if so, the prolegs of A4–A6 are reduced or modified. In addition, mature larvae are 20 mm or less in length; mature larvae of the other similar groups are usually larger.

Biology and Ecology: Known larvae usually live in a weak shelter formed from a folded leaf or webbed terminals. Cocoons are well-formed, boat-shaped, with an anterior ridge forming a valvelike slit. Hosts for species of *Meganola* include oak and apple; those of *Nola* include oak, witch hazel, *Clethra,* and grasses.

Description: Mature larvae small, not exceeding 20 mm, resembling Arctiidae. Primary setae obscured by tufts of secondary setae of irregular length on verrucae.

Head: Hypognathous, semihypognathous, bearing primary setae only. Frontoclypeus as wide as high, extending one-half to three-fourths to the epicranial notch. Six stemmata, 1–4 evenly spaced in a semicircle and 5 conspicuously separated from 1–4; 6 may be close to 4.

Thorax: T1 shield poorly to moderately defined. Verrucae conspicuous, sometimes extended as fleshy lobes. Verrucae on T1 are D (sometimes with seta D2 separate and distinct posteriorly), SD (small or reduced to a single seta), L and SV. Verrucae on T2 and T3 are D, SD, L, and SV. Ventral T1 gland present or absent.

Abdomen: Verrucae D, SD, and L well developed on A1–9; verruca SV less well developed. Spiracles oval, slightly larger to 2× larger (*Nola*) on T1 and A8.

Prolegs absent on A3, well developed on A4–A6 and A10, with secondary setae, and bearing uniordinal crochets in a mesoseries.

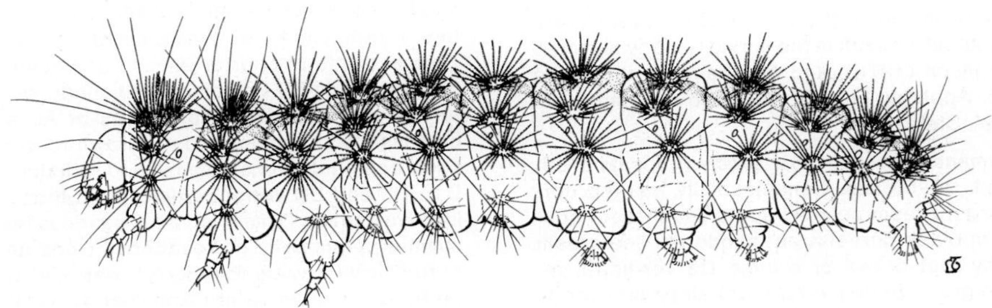

Figure 26.386. Nolidae. *Nola sorghiella* (Riley), the sorghum webworm. f.g.l. 12± mm. Greenish, with 4 red to brown longitudinal stripes dorsally. From the Great Plains eastward, more abundant in the South where multiple generations occur. See text.

Comments: The family is small, but widely distributed, with 16 species in two genera north of Mexico. They are occasionally abundant, but of little economic importance except for *Nola sorghiella* (Riley) (fig. 26.386), whose larvae live gregariously in webs on grass heads and are sometimes injurious to sorghum.

Selected Bibliography

Franclemont *in* Forbes 1960, p. 50.
Hinton 1943.
Reinhard 1938.

NOCTUIDAE (NOCTUOIDEA)

George L. Godfrey, *Illinois Natural History Survey*

Cutworms

Figures 26.387–26.440

Relationships and Diagnosis: The Noctuidae comprise a very large family, with the majority of the described larvae being medium-sized (25–40 mm), having hypognathus heads, lacking secondary setae, and possessing subequally developed prolegs on A3–6 and A10 that bear crochets in homoideous mesoseries. These include the frequently observed armyworms and cutworms (e.g., figs. 26.402, 26.414, 26.425). However, throughout the family there are many elements that show numerous exceptions to this generalized diagnosis: the presence of secondary setae (fig. 26.391), reduced numbers of prolegs (figs. 26.398, 26.432, 26.437a), vestigial prolegs, absence of crochets, and heteroideous crochets. There is no single character that can be used to separate *all* noctuid larvae from *all* other North American Noctuoidea (Nolidae, Notodontidae, Dioptidae, Lymantriidae, Arctiidae, Ctenuchidae, and Pericopidae) and even some Geometroidea.

North American noctuid larvae may have two, three, or four pairs of abdominal prolegs in addition to the caudal pair. Regardless of the number, the pair on A5 is subequal to A6 and can be used to separate the semilooping noctuids from the Geometridae, in which the prolegs on A5 are totally absent or distinctly smaller than the pair on A6. Nolid larvae have no prolegs on A3 as most noctuids do. However, the noctuids with no prolegs on A3 lack secondary setae, which nolids have. The caudal prolegs of noctuids generally are amply developed and bear crochets. In many notodontids the caudal prolegs are reduced or form stemapods that project caudally, sometimes resembling cerci, and lacking crochets. Many notodontids have a medial groove extending from the labral notch to the base of the labrum (Lepidoptera family key, fig. 147); this groove is absent in noctuids. Noctuids with secondary setae lack extra setae in the anterolateral area of A10, (Lepidoptera family key, fig. 180), the area where some notodontids and dioptids have one to four extra setae. The homoideous mesoseries crochet arrangement is quite reliable for distinguishing noctuids from the Arctiidae, Ctenuchidae, and Pericopidae, in which the crochets usually are in a heteroideous mesoseries. The few noctuids with heteroideous crochets possess only primary setae (e.g., *Paectes*). The absence of middorsal glands on A6 and A7 separates the North American noctuid larvae from the Lymantriidae.

Biology and Ecology: Feeding habits of noctuid larvae are quite diverse. Most of the known larvae are phytophagous; a few are carnivorous, including species that opportunistically feed on other caterpillars, intra- as well as interspecifically, and those that prey on scale insects (Coccidae) (see Forbes 1954, Common 1970). Sukhareva (1979) mentioned that some species of Eurasian *Orthosia* become predaceous on other caterpillars as a response to a moisture deficit. The plant feeders traditionally have been treated as cutworms, climbing cutworms, armyworms, green fruitworms, bollworms, earworms, borers, semiloopers, and loopers, the last adding confusion because the same term frequently is applied to the geometrids. The terms largely are based on behavioral characteristics relating to feeding or walking patterns. They primarily pertain to the last larval instar and need to be used cautiously because there may be behavioral variation between larval instars of the same species. For example, the black cutworm, *Agrotis ipsilon* (Hufnagel) (fig. 26.404), feeds on foliage in its early instars without necessarily cutting off stems or leaves (Ruesink, pers. comm.) and later becomes a true cutworm that cuts off plant material and pulls it into its burrow. The same species will occasionally climb more mature plants to get at suitable leaves and sometimes becomes carnivorous (Crumb 1929). Various correlations between the morphology of the larval noctuids and the factors of feeding and burrowing or climbing behaviors were discussed by Ripley (1923).

In addition to the phytophagous and carnivorous species there are also decomposers. Several species of *Idia* feed on decaying vegetation. *I. gopheri* (Smith) is more extreme in that it feeds on the excrement of the gopher tortoise, *Gopherous polyphemous* (Daudin) (Crumb 1956).

Relatively few noctuids construct nests. Those that do, form them out of folded leaves of their hosts (e.g., *Hexorthodes accurata* (Hy. Edwards) (see Comstock 1947, Godfrey 1972). Certain genera characteristically are borers in plant roots and stems, e.g., *Hydraecia* (fig. 26.419a) and *Papaipema* (fig. 26.422x). Leaf miners are uncommon in the Noctuidae. However, one example is the mining behavior of the first 3 instars of *Bellura gortynoides* Walker in the leaves of yellow water lily (Levine and Chandler 1976).

Eggs of noctuids basically are spherical, bearing parallel, vertical ridges that extend from the base and converge on the micropylar area. Parallel, but less conspicuous, horizontal ridges may be present (see Peterson 1961, Chu et al. 1963, Merzheyevskaya 1967, Angulo and Weigert 1975a, 1975b, Salkeld 1975, 1984, Weigert and Angulo 1977 for more details). Eggs are laid singly or in clusters that usually are one layer deep but may be two or three layers in some species (Peterson 1964). The eggs may be found completely exposed on leaves and stems, concealed in various plant parts, e.g., composite flowers (*Feltia jaculifera* (Guenée)) and grass sheaths (*Leucania pseudargyria* Guenée), or deposited loosely at bases of host plants. Pruess (1961) demonstrated that the eggs of *Euxoa auxiliaris* (Grote) are laid beneath the surface of sandy soil, and Parker et al. (1921) found eggs of *Agrotis orthogonia* Morrison as deep as a half inch in loose, mellow soil. Eggs deposited on leaf surfaces usually are naked but occasionally are densely covered with the anal scales from the abdomen of the ovipositing moth, e.g., *Lophoceramica artega* (Barnes) (see Godfrey 1972) and *Bellura gortynoides* Walker (Heitzman and Habeck 1981), or various secretions from the female (Peterson 1964).

Mosher (1916a) characterized noctuid pupae as having the labial palpi visible and maxillae extending to the caudal margin of the wing. Pupation may take place in the ground within earthen cells with little or no cocoon formation (typical of armyworms, climbing cutworms, and cutworms) or in cocoons of varying degrees of strength in leaf litter and rotting wood. Larvae of *Raphia abrupta* Grote were observed crawling into and pupating inside hanging, wooden, nesting traps set out to attract solitary-nesting Hymenoptera near Ware, Illinois; adults of *Bomolocha baltimoralis* (Guenée) and *Acronicta connecta* Grote also were reared from such traps (Braasch, pers. comm.). Perhaps these species seek out loose bark or the emergence holes of woodboring insects for natural pupation sites. Forbes (1954) noted that, unlike arctiid and lymantriid cocoons, larval setae are not incorporated into the cocoons of noctuids.

Description: Mature larvae small to medium, 12–70 mm. Head rounded. Body of the normal lepidopteran type, usually cylindrical and stout; A8 angulate dorsally in lateral view in some groups; transverse ridges, middorsal humps, and/or lateral fringes (noctuid key, fig. 91) may be present on abdominal segments; cuticle smooth or bearing granules (noctuid key figs. 18, 19, 20) or spinules (noctuid key fig. 16). Primary and subprimary setae present, secondary setae also present in a few subfamilies. Body concolorous, marked with chevrons or triangles, or longitudinally striped; transverse bands or lines common in the Agaristinae but rare in other subfamilies. Spiracles basically elliptical, sometimes with emarginate ends; those on T1 and A8 usually larger than those on A1–7.

Head: Hypognathous to semihypognathous. Frontoclypeus triangular with lateral margins slightly incurved to excurved and extending 0.50 to 0.88 of distance from base of frontoclypeus to epicranial notch. Adfrontal ecdysial lines close to frontoclypeal margins, merging cephalad with them; meeting each other above apex of frontoclypeus (noctuid key fig. 66), sometimes near or at epicranial notch (noctuid key fig. 35); six stemmata, 1–4 in semicircle with variable spacing from species to species, 5 and 6 removed from 1–4 semicircle, line extending through 5 and 6 not passing through semicircle.

Thorax: T1 shield usually well formed, may be concolorous with rest of body, in which case it may be indistinct. T1: setae XD1 and XD2 positioned near anterior margin and D1 and D2 close to posterior margin of T1 shield; in *Rivula* D1 and D2 closer to anterior than to posterior margin (Beck 1960). SD1 and SD2 approximate to each other, SD1 usually thinner than SD2. Only one SD seta present in *Nephelodes minians* Guenée (noctuid key fig. 36) (Crumb 1926a) and some Herminiinae (e.g., *Renia* (Godfrey 1980)). L1 and L2 arise close to each other, cephalad of spiracle; L2 usually thinner than L1, only one L seta present in some Herminiinae (Crumb 1934, Godfrey 1980). SV1 and SV2 present above leg. T2 and T3: D1, D2, and SD2 insertions approximate a straight line; SD1 usually hairlike, situated below and usually separated from SD2, both setae rarely on a common tubercle; L1, L2, and L3 present; an extra L2 seta occasionally present on T3. SV group usually unisetose. The Old World genus *Brithys* possesses 2 SV setae (Gardner 1946, 1947, 1948) as does *Xanthopastis* in the New World (Crumb 1956).

Abdomen: D1 and D2 present, D1 dorsocephalad of D2 on A1–8; on A9 D1 ventrocephalad of D2. SD1 basically dorsad and SD2 cephalad of spiracle; SD2 absent on A9. L group with three setae (four in *Acronicta*) on A3–6. L1 directly caudad of spiracle on A3–6, but usually ventrocaudad of spiracle on A7. SV group with two or three setae on A1, three on A2, three or more on A3–6, one on A7–9 except two in *Acronicta* and *Rivula* (Beck 1960) and in Agaristinae and Bagisarinae (Crumb 1956). A9 with only one seta in L group (two in Ufeiinae, see Crumb 1956). A10 bearing anal shield, sometimes concolorous with body.

Prolegs on A3–6 and A10, A4–6 and A10, or A5–6 and A10. A5 and A6 prolegs subequal; pairs on A3 and A4, when present, may decrease in size cephalad. Crochets uni- or less commonly biordinal, usually in homoideous mesoseries, rarely absent.

Comments: The family is large and widely distributed. North of Mexico nearly 3000 species have been described from adult specimens, but larvae are known for only about 800. Many of the documented larvae are so poorly understood that redescriptions, more host plant data, and direct, field behavioral observations are needed. Reliable keys with sufficient illustrations for localized faunae would greatly aid identification efforts.

Some of the more common armyworms include the armyworm, *Pseudaletia unipuncta* (Haworth) (fig. 26.414), a serious pest of grasses and grains, the wheathead armyworm, *Faronta diffusa* (Walker), which also is destructive to grains, and several species of *Spodoptera*, including *exigua* Hübner, *eridania* (Cramer), and *ornithogalli* (Guenée) (figs. 26.425a-c), which cause serious problems to many truck and field crops, especially in the southern United States. Perhaps the most serious cutworm is the black cutworm, *Agrotis ipsilon* (Hufnagel) (fig. 26.404), a cosmopolitan species especially destructive to seedling row crops. The corn earworm, *Heliothis zea* (Boddie) (fig. 26.394), feeds on the fruiting bodies of a wide variety of plants (e.g., corn, tomato, cotton, etc.) and may hold title within the North American Noctuidae to having had the most research dollars spent on it. Kogan et al. (1978) may be consulted for an exhaustive bibliography of the corn earworm and the tobacco budworm, *H. virescens* (Fabricius). Noctuids do not as commonly attract attention from pomologists and silviculturists as do other Lepidoptera families (e.g., Tortricidae, Lymantriidae, etc.). However, the green fruitworms that are climbing cutworms, typified by *Orthosia hibisci* (Guenée) (figs. 26.413a,b), occasionally cause appreciable damage to pome crops and definitely contribute to the defoliation of forested areas during late spring. Two loopers frequently encountered are the cabbage looper, *Trichoplusia ni* (Hübner), and the soybean looper, *Pseudoplusia includens* (Walker) (fig. 26.432). They are not restricted to specific plants as their common names imply, but will feed on a wide variety of herbs. Eichlin and Cunningham (1978) provide detailed information about these and other plusiines.

The following key is designed to acquaint the user, especially the student, with many of the economically important noctuid larvae plus larval representatives of the frequently encountered subfamilies occurring in the U.S. and Canada. It is beyond the scope of the key to treat every economic species, but it is hoped that the key plus the selected bibliography will aid in developing a knowledge of this diverse group. The key extensively employs adaptations from Crumb (1929, 1932, 1956), Walkden (1950), Beck (1960), Godfrey (1972), Levy and Habeck (1976), and Eichlin and Cunningham (1978). Most of the characters can be seen quite easily without any dissections except for certain ones on the mandible, hypopharyngeal complex, and head capsule. It usually is necessary to dissect these structures following the procedures given by Godfrey (1972) and below, if the associated characters are to be seen adequately.

1. Remove the left mandible by disjoining its anterior and posterior articulations from the head capsule with a minute insect pin or surgical scissors. Next cut between the outer base of the mandible and the antennal socket before pulling the mandible free with a fine jeweler's forceps. Use caution so that neither the outer cutting teeth nor the inner tooth (if present) are broken.
2. The hypopharyngeal complex, i.e., the hypopharynx, prementum, labial palpi, and spinneret (fig. 26.4), is dissected next. Cut the points of articulation between each premental arm and maxillary palpus with a slightly hooked minute insect pin, and then separate the prementum from the postmentum with the same instrument. Carefully but firmly grasp the hypopharyngeal complex along the length of the premental arms with a jeweler's forceps and pull out the entire structure. It may be necessary to clear the hypopharyngeal complex in a 10% solution of potassium hydroxide, stain it with mercurochrome, and view it in glycerine in order to see the more membranous structures.
3. Remove the head by cutting between it and the prothorax with small dissecting scissors.
4. Place the detached structures in a microvial and store with the larva.

KEY TO SOME REPRESENTATIVE LAST INSTAR NOCTUID LARVAE

1. One (fig. 1) or more (fig. 2) spatulate D2 setae on A8 2
1′. D2 setae on A1-A8 setose (fig. 3) or swollen (fig. 4), not spatulate 3
2(1). Single (one per side) spatulate D2 seta on A8 (fig. 1) .... (fig. 26.389) *Acronicta funeralis* (p. 563)
2′. Multiple spatulate D2 setae on A8 (fig. 2) .... (fig. 26.388) *Panthea furcilla* (p. 563)
3(1′). Proleg on A6 with more than 5 setae in SV group (fig. 5) 4
3′. Proleg on A6 with 3 setae in SV group (fig. 6), rarely 4 6

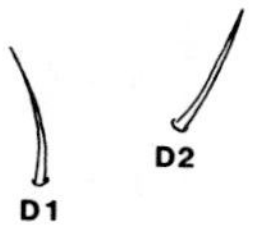

Figure 3

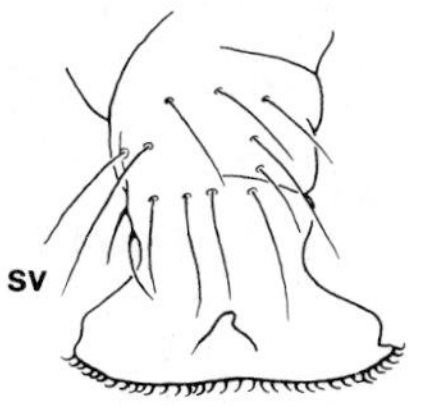

Figure 5

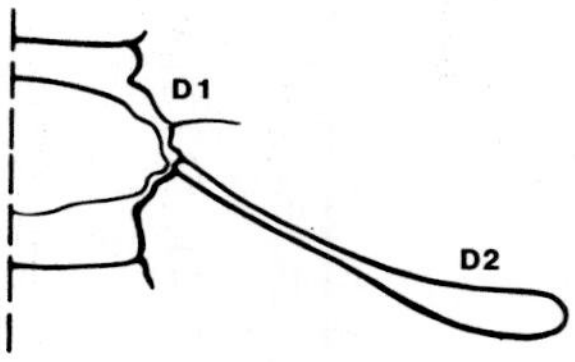

Figure 1

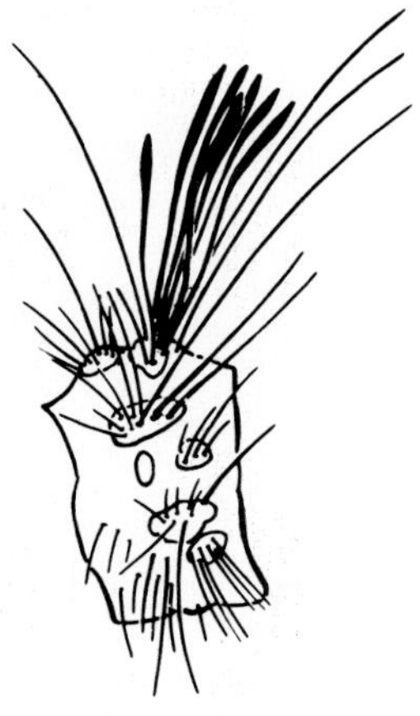

Figure 2

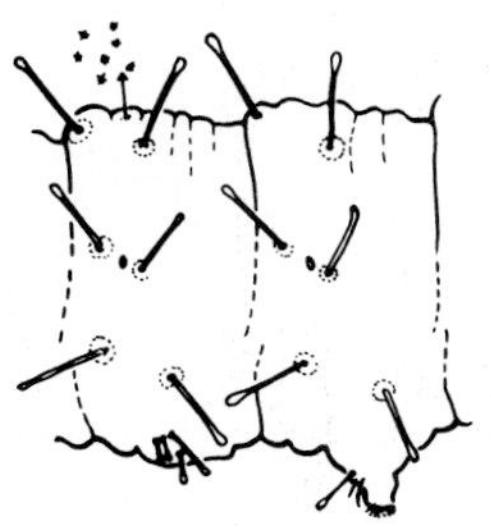

Figure 4

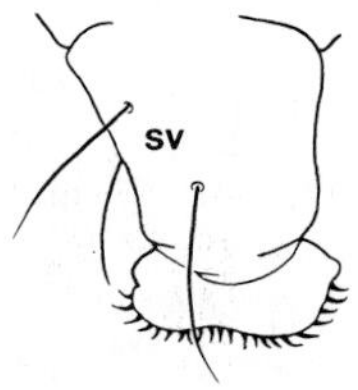

Figure 6

4(3). Secondary setae on dorsum of abdomen very conspicuous, black hair pencils present on A1, A3, and A8 (fig. 26.391) .... *Acronicta americana* (p. 564)

4′. Secondary setae on dorsum of abdomen absent or inconspicuous, no black hair pencils (fig. 26.392a) 5

5(4′). Head with contrasting pale coronal stripe including base of seta P2 (fig. 26.392b) .... (fig. 26.392a) *Simyra henrici* (p. 564)

5′. Head without pale stripe including base of seta P2, either unicolorous or base of seta P2 included in dark coronal stripe .... (fig. 26.390) *Acronicta oblinita* (p. 563)

6(3′). Two SV setae on A1 (fig. 7) 7

6′. Three SV setae on A1 (fig. 8) 63

7(6). Subanal setae and median posterior anal setae modified to form anal fork (fig. 9) .... *Acontia dacia* (p. 565)

7′. Subanal setae and median posterior anal setae unmodified 8

8(7′). Prolegs present on A3–6, bearing crochets 13

8′. Prolegs absent on A3 or A3 and A4 (fig. 26.432); if present on A3 and A4, then vestigial and lacking crochets (fig. 10) 9

9(8′). Vestigial prolegs present on A3 and A4 (fig. 10) 10

9′. Vestigial prolegs absent on A3 and A4 (fig. 11) 11

10(9). Ridges 2 and 3 of mandible terminating in processes before reaching cutting edge (fig. 12) .... (fig. 26.432) *Pseudoplusia includens* (p. 575)

10′. Ridges 2 and 3 of mandible not terminating in processes but continuing to cutting edge (fig. 13) .... *Trichoplusia ni* (p. 576)

11(9′). At least one ridge of mandible terminating in process before reaching cutting edge (fig. 12) 12

11′. Ridges of mandible not terminating in processes but continuing to cutting edge (fig. 13) .... *Autographa californica* (p. 576)

12(11). Pinacula SV1 and SV2 at least partially fused on A2 (fig. 14) .... (fig. 26.434) *Autographa precationis* (p. 576)

12′. Pinacula SV1 and SV2 separate on A2 (fig. 11) .... *Autoplusia egena* (p. 576)

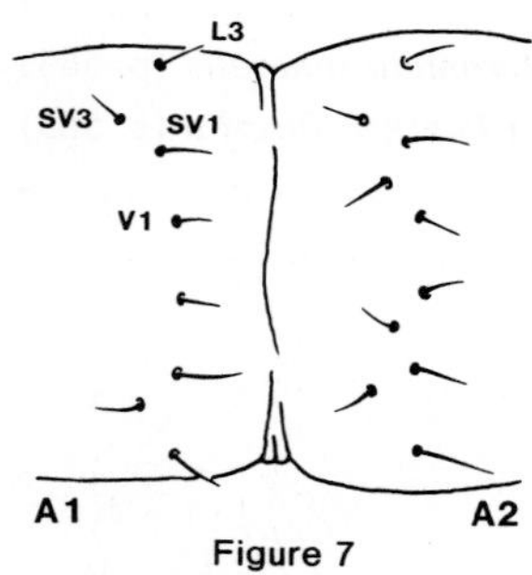

Figure 7

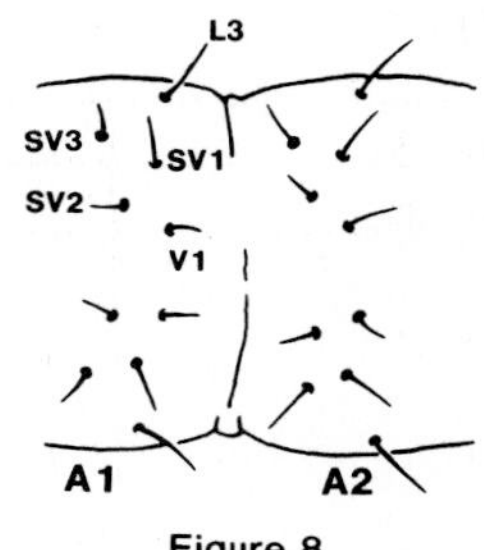

Figure 8

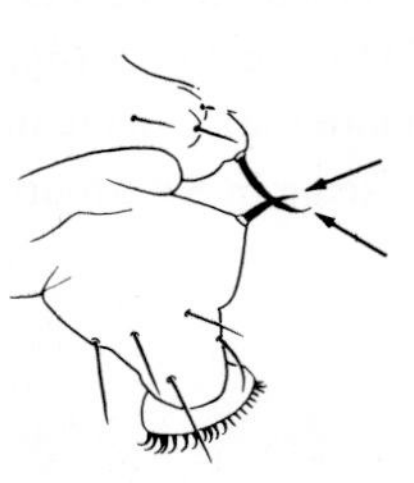
Figure 9

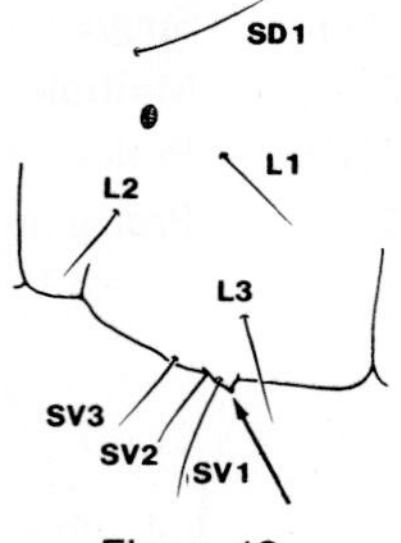

Figure 10

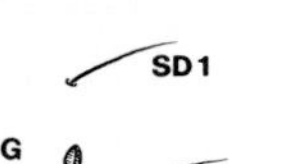

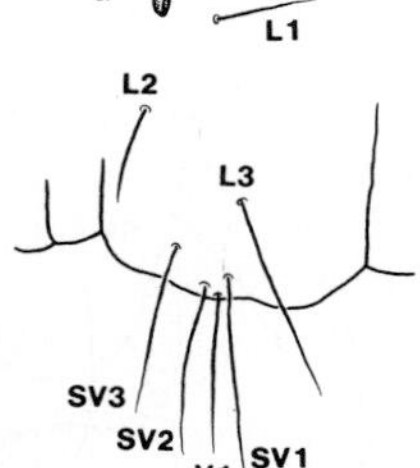

Figure 11

Figure 12

Figure 13

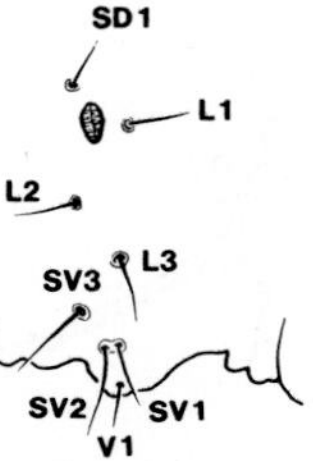

Figure 14

13(8). Body with large black tubercles, usually transversely striped (fig. 26.387); head with 5 or more contrasting black spots per side associated with setal bases (fig. 15) (Agaristinae) .... *Alypia octomaculata* (p. 563)

13′. Body not with above combination of characters; if body transversely striped then head immaculate, freckled, reticulate, or with 3 or fewer contrasting spots associated with setal bases 14

14(13′). Body covered with spinules (fig. 16); crochets of A3–6 prolegs weakly biordinal (fig. 17) (Heliothinae) 15

14′. Body smooth or covered with pavement, convex or conical granules (figs. 18, 19, 20); crochets of A3–6 prolegs uniordinal (fig. 21) 17

15(14). Subdorsal lines white, prominent, continuous (fig. 22); no tubercles associated with D1 and D2 on abdominal segments (fig. 23) .... (fig. 26.393) *Heliothis ononis* (p. 564)

15′. Subdorsal lines absent or inconspicuous (fig. 26.394); D1 and D2 borne on tubercles on abdominal segments (fig. 24) 16

16(15′). D1 tubercles on A1, A2, and A8 without spinules (fig. 25); mandible without or with but slight inner tooth (fig. 26) .... (fig. 26.394) *Heliothis zea* (p. 565)

16′. D1 tubercles on A1, A2, and A8 with spinules (fig. 27); mandible with distinct inner tooth (fig. 28) or scar of inner tooth (fig. 29) .... (fig. 26.395) *Heliothis virescens* (p. 565)
.... (fig. 26.396) *H. subflexus* (p. 565)

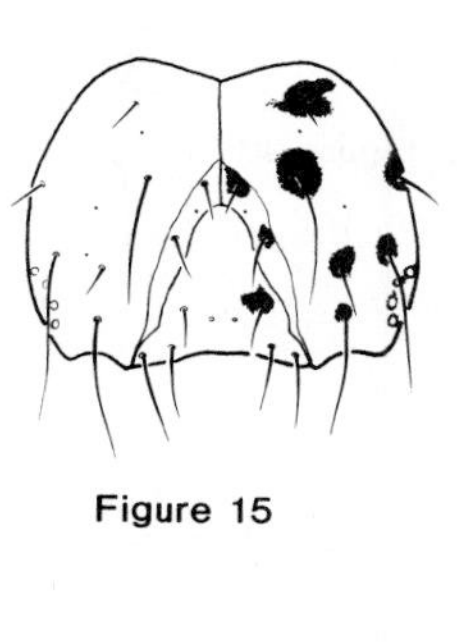
Figure 15

Figure 16

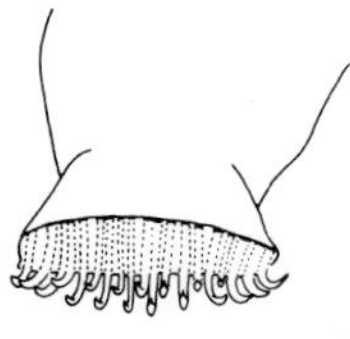
Figure 17

Figure 18

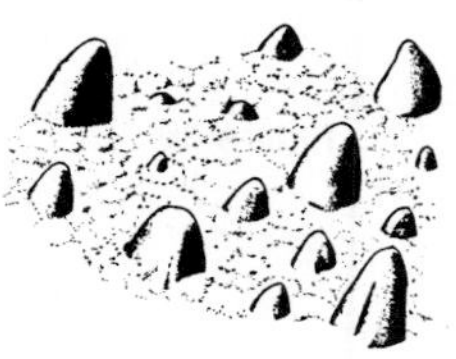
Figure 19

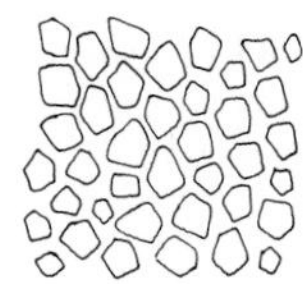
Figure 20

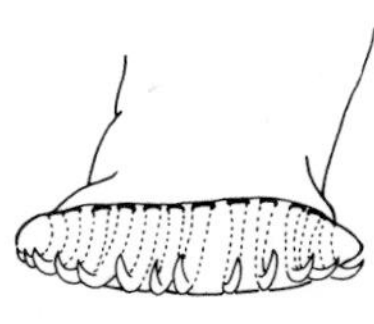
Figure 21

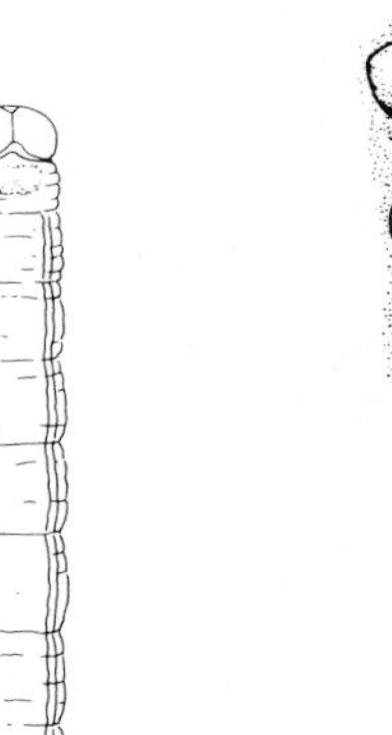
Figure 22

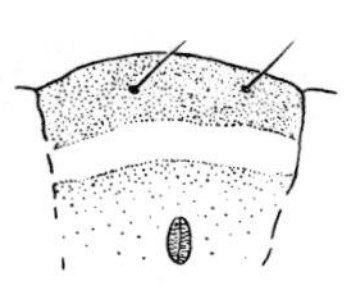
Figure 23

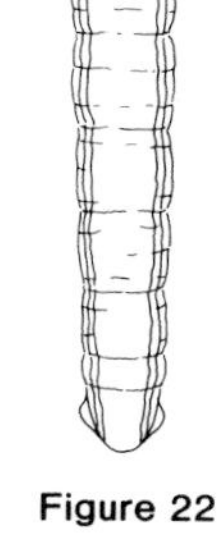
Figure 24

Figure 25

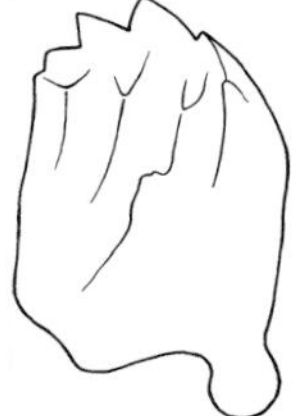
Figure 26

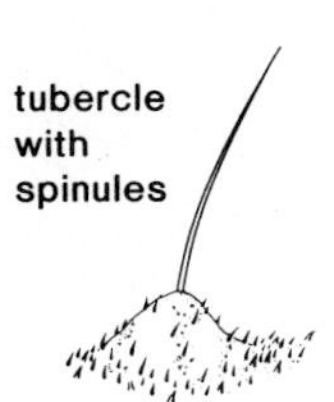

Figure 27

Figure 28

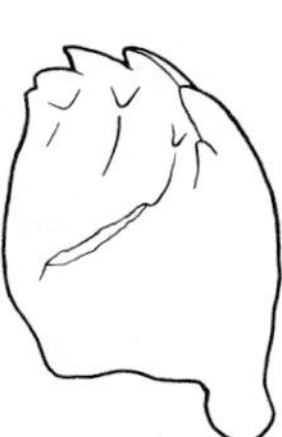
Figure 29

17(14′). SD1 on A9 setaceous, as strong or stronger than D1 (fig. 30); spiracular line extends caudad to form lateral border of anal plate (fig. 31) (Cuculliinae) .... (fig. 26.397) *Cucullia asteroides* (p. 565)

17′. SD1 on A9 hairlike, weaker than D1 (fig. 32); spiracular line when present extends downward to distal end of anal proleg (fig. 33) (Noctuinae; Hadeninae; Amphipyrinae; Lithacodiinae, in part) 18

18(17′). Adfrontal ecdysial lines not reaching epicranial notch, or distance from epicranial notch to apex of ecdysial lines (A) equal to or greater than distance from apex of ecdysial lines to apex of frontal sutures (B) (fig. 34) 19

18′. Adfrontal ecdysial lines reaching epicranial notch, or distance from epicranial notch to apex of ecdysial lines (A) less than distance from apex of ecdysial lines to apex of frontal sutures (B) (fig. 35) 55

19(18). T1 with 1 SD seta (fig. 36) .... (fig. 26.409) *Nephelodes minians* (p. 568)

19′. T1 with 2 SD setae, one may be indicated merely by minute papilla (fig. 37) 20

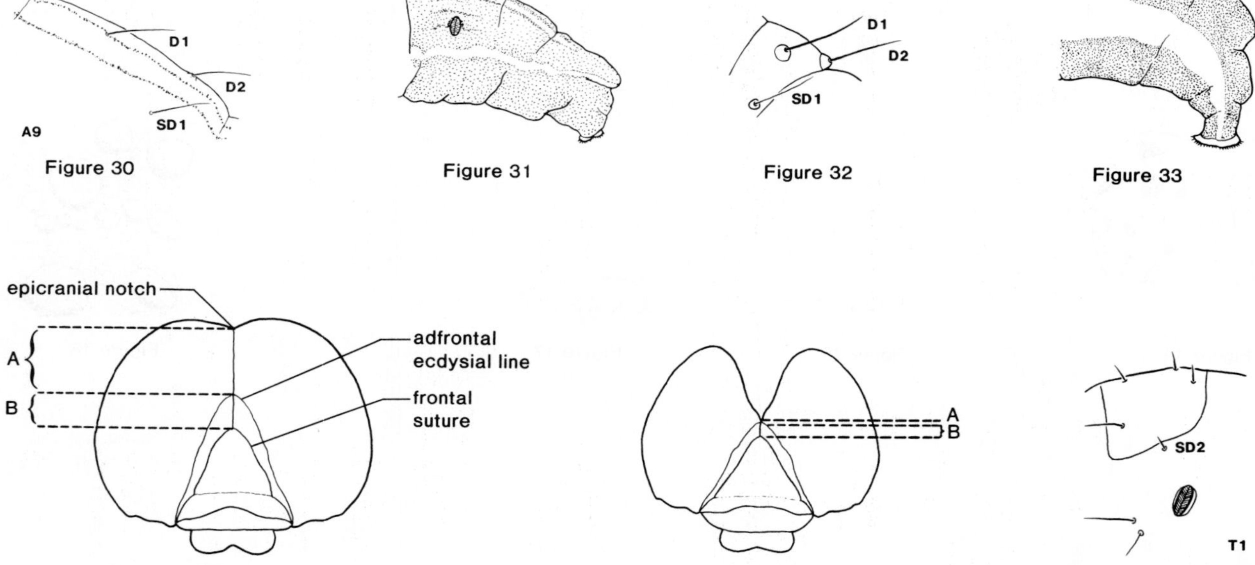

Figure 30 Figure 31 Figure 32 Figure 33

Figure 34 Figure 35 Figure 36

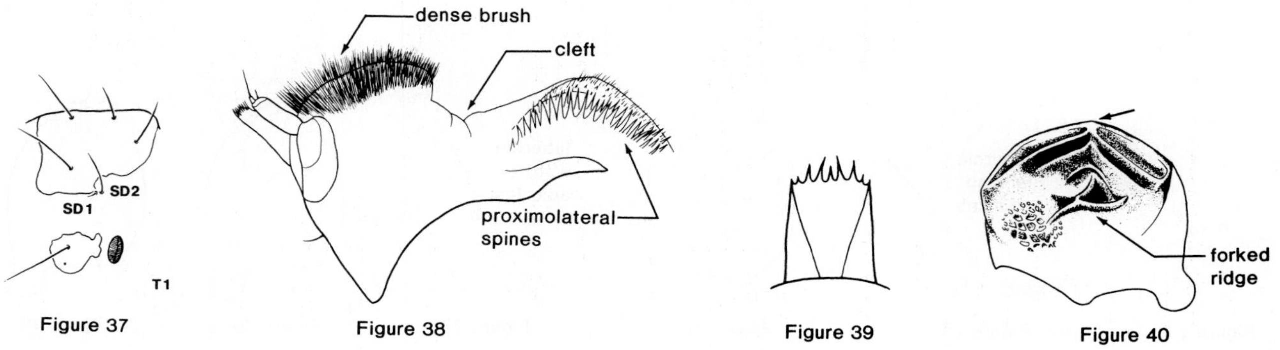

Figure 37 Figure 38 Figure 39 Figure 40

20(19′). Distal region of hypopharynx with thin spines forming dense brush and separated from proximal region by distinct, medial transverse cleft (fig. 38), *and* distal lip of spinneret fringed (fig. 39); mandible with cutting edge smooth or bearing two medial teeth (fig. 40); external feeders on grasses 21

20′. Not with the above combination of characters; distal region with thin spines *and* no medial, transverse cleft (fig. 41); *or* distal region with scattered stout spines (fig. 42) *and* distal lip of spinneret entire (fig. 43), lobed (fig. 44), spinose (fig. 45) or fringed; mandible with 4–12 teeth on cutting edge (fig. 46); inner surface with simple ridges (fig. 46) or bearing 1–2 teeth (figs. 47, 48); various host associations 22

21(20). Fa's of frons below line F1—F1 (fig. 49) .... (fig. 26.414) *Pseudaletia unipuncta* (p. 570)

21′. Fa's of frons on line F1—F1 (fig. 50) .... (fig. 26.415) *Leucania phragmatidicola* (p. 570)

22(20′). Spiracles on A8 positioned dorsocaudad (fig. 51) .... (fig. 26.431) *Bellura gortynoides* (p. 575)

22′. Spiracles on A8 positioned laterad (fig. 52) 23

23(22′). Mandible with 1 outer seta (fig. 53) 24

23′. Mandible with 2 outer setae (fig. 54) 26

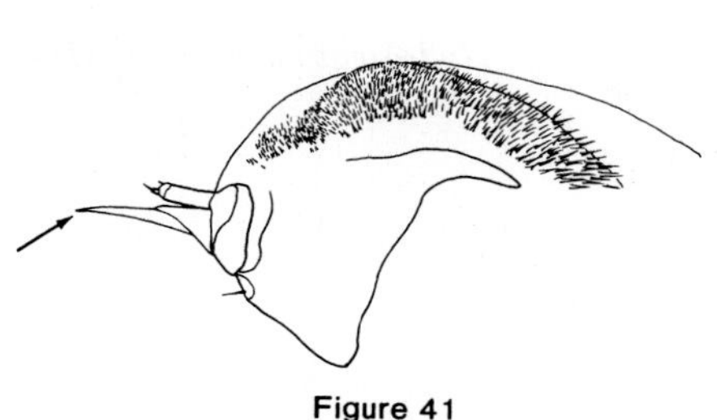

Figure 41

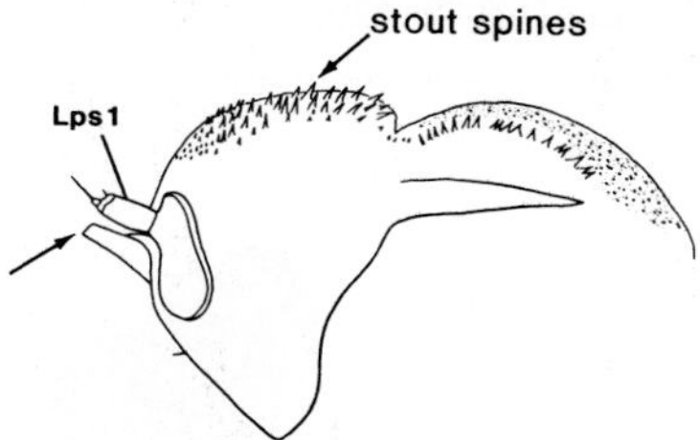

Figure 42

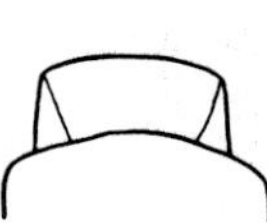

Figure 43

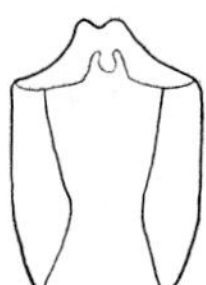

Figure 44

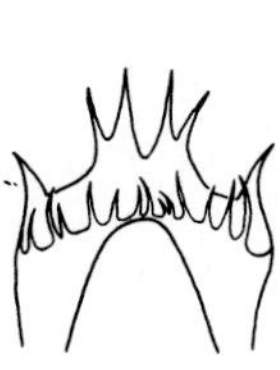

Figure 45

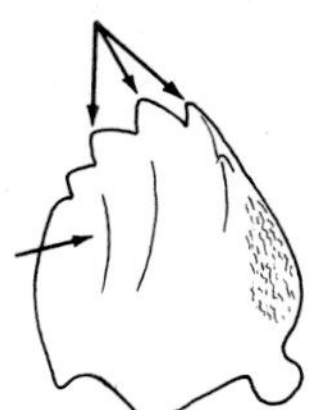

Figure 46

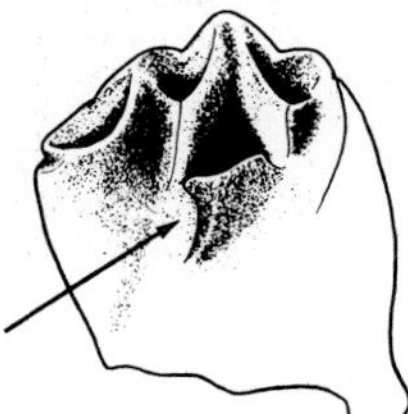

Figure 47

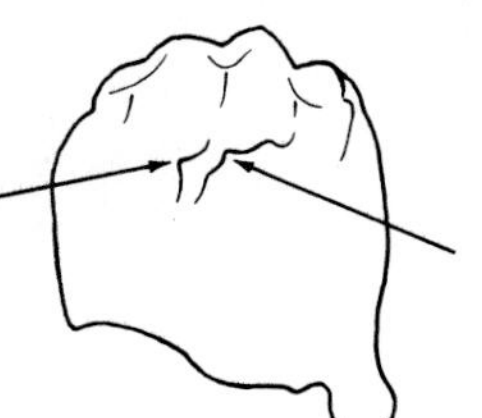

Figure 48

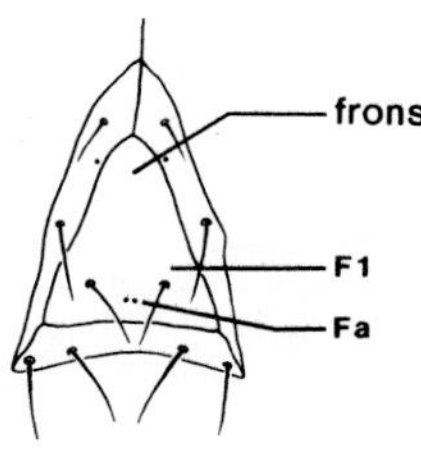

Figure 49

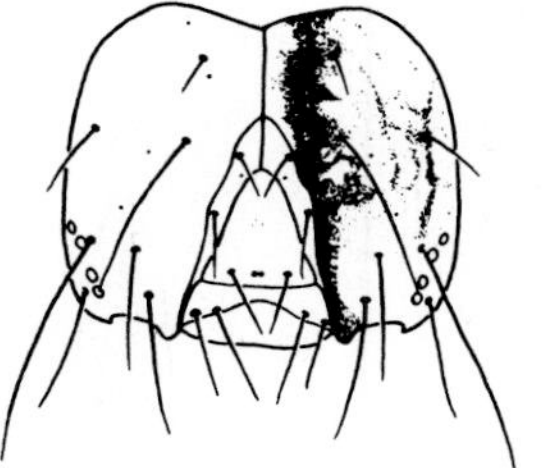

Figure 50

Figure 51

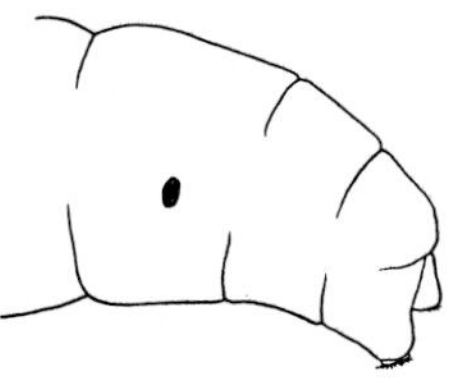

Figure 52

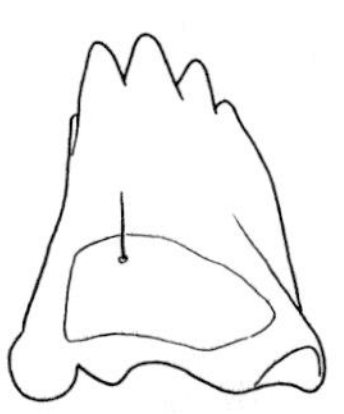

Figure 53

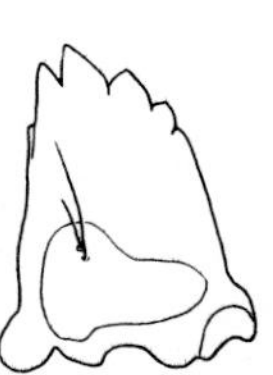

Figure 54

24(23). Pale subdorsal line present on anterior abdominal segments .... *Papaipema cataphracta* (p. 572)

24′. Pale subdorsal line absent on anterior abdominal segments (fig. 55) 25

25(24′). L1 setal base on A7 below level of spiracle; polyphagous .... (fig. 26.422) *Papaipema nebris* (p. 572)

25′. L1 setal base horizontally caudad of spiracle; columbine borer .... *Papaipema leucostigma* (p. 572)

26(23′). Posterior margin of anal shield distinctly lobed or tuberculate (fig. 56) 27

26′. Posterior margin of anal shield evenly convex, not lobed or tuberculate (fig. 57) 28

27(26). Anal shield black, distinctly rugose, 6 tubercules on posterior margin (fig. 56) .... *Achatodes zeae* (p. 575)

27′. Anal shield yellow, scarcely rugose, 5 lobes on posterior margin .... *Luperina stipata* (p. 571)

28(26′). Spiracles conspicuously narrowed, height 3.5 × width (fig. 58) .... *Macronoctua onusta* (p. 572)

28′. Spiracles not as above 29

29(28′). Mandible with more than 8 triangular-shaped outer teeth (fig. 59) .... *Callopistria floridensis* (p. 572)

29′. Mandible with 4–6 reduced or triangular-shaped outer teeth (fig. 60) 30

30(29′). Two outer mandibular setae closely spaced to each other (fig. 54); borers of stems and roots of herbs 31

30′. Two outer mandibular setae distantly spaced from each other (fig. 61); external feeders, may occasionally bore into fruiting structures 32

31(30). Dorsum of abdominal segments with violet patches separated by mid- and subdorsal lines (figs. 26.419a, b); distal lip of spinneret sharply pointed (fig. 41) ... *Hydraecia immanis* (p. 571)

31′. Dorsum of abdominal segments each with continuous transverse reddish-tinged band, mid- and subdorsal lines absent (fig. 26.420); distal lip of spinneret rounded (fig. 62) .... *Hydraecia micacea* (p. 571)

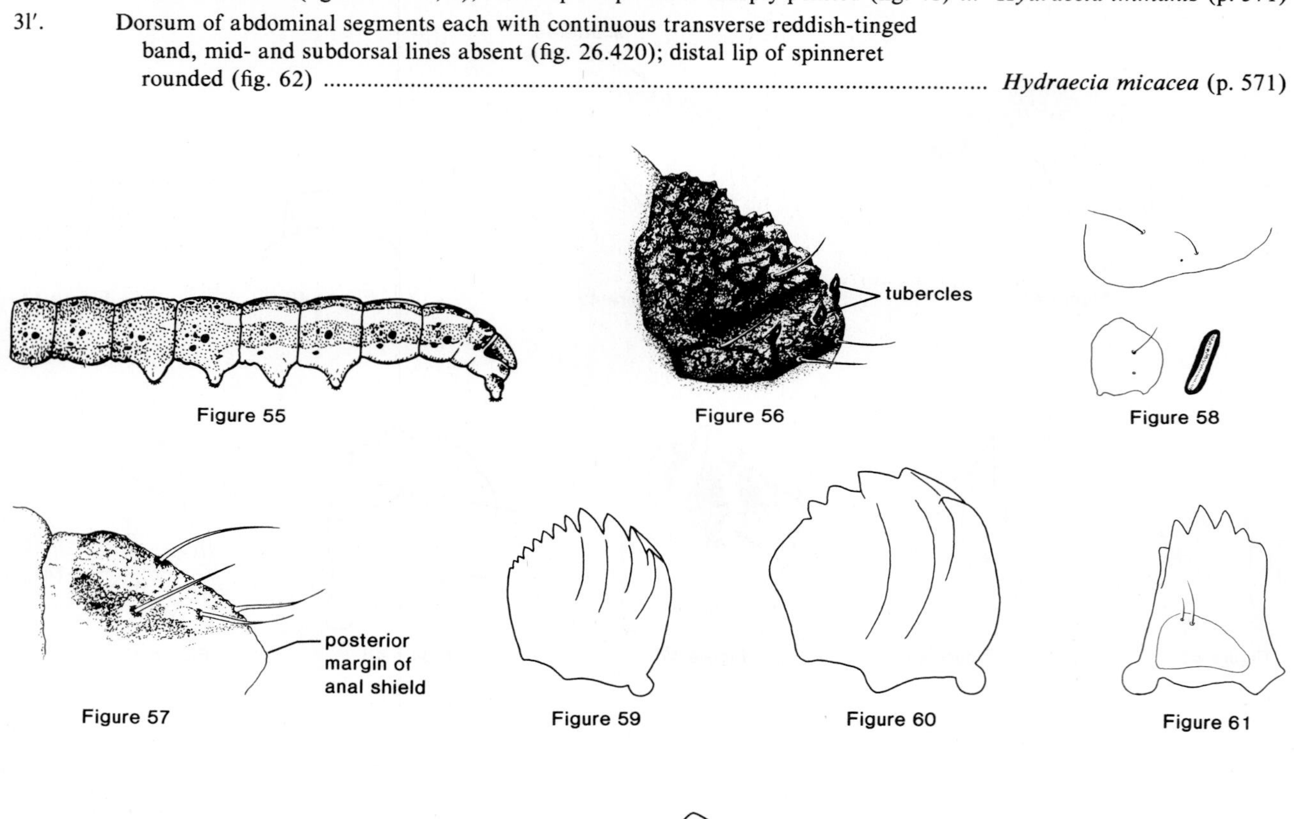

Figure 55

Figure 56

Figure 58

Figure 57

Figure 59

Figure 60

Figure 61

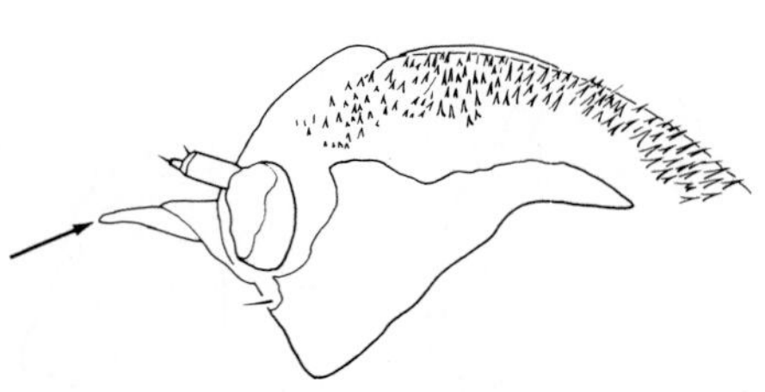

Figure 62

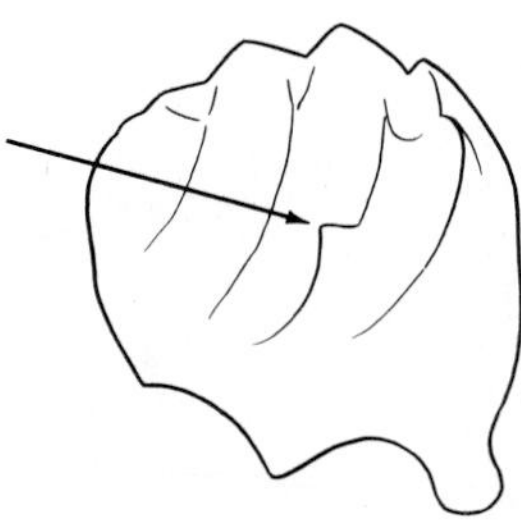

Figure 63

Figure 64

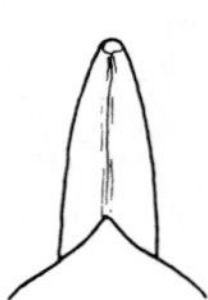

Figure 65

32(30′). Mandible lacking inner tooth, inner ridges not swollen or raised basad (fig. 60) 33

32′. Mandible with quadrate or triangular inner tooth associated with first inner ridge (figs. 47, 63) or first inner ridge swollen basad and transversely connected to second inner ridge (fig. 64) 49

33(32). Spinneret long and narrow, its length about 2–3× its width, distal lip entire (fig. 65) 34

33′. Spinneret short and broad, its length less than 2× its width, distal lip variable 38

34(33). Convex granules distinct (at 12× or more) between D1 and D2 tubercles on A1–8 35

34′. Body lacking granules 36

35(34). Abdominal dorsum with segmental series of diamond markings; dorsal setae appear stiff .... (fig. 26.416) *Lacinipolia renigera* (p. 570)

35′. Abdominal dorsum lacking such markings; dorsal setae inconspicuous, not stiff .... *Proxenus mindara* (p. 572)

36(34′). Head seta P1 shorter than width of frons (fig. 66) .... *Dargida procincta* (p. 569)

36′. Head seta P1 equal to or longer than width of frons (fig. 67) 37

37(36′). Spinneret only slightly surpassing tip of seta on second segment of labial palpus (Lp2) at best (fig. 68) .... *Dicestra trifolii* (p. 568)

37′. Spinneret distinctly surpassing tip of seta on second segment of labial palpus (Lp2) (fig. 69) .... *Lithacodia carneola* (p. 577)

38(33′). Proximolateral spines of hypopharynx absent or inconspicuous (fig. 70) 39

38′. Proximolateral spines of hypopharynx distinct (fig. 38) 46

39(38). Midventral muscle attachments between prolegs on A3–6 forming a "Y" (fig. 71) 40

39′. Midventral muscle attachments between prolegs on A3–6 in a line (fig. 72) 41

40(39). Pavement granules (fig. 20) visible on dorsum of abdomen at 25× or more .... (fig. 26.429) *Spodoptera frugiperda* (p. 574)

40′. Integument of abdomen smooth, no granules .... (fig. 26.430) *Spodoptera exigua* (p. 575)

41(39′). Lateral area distinct, interrupted on A1 by dark patch (fig. 26.427) .... *Spodoptera eridania* (p. 574)

41′. Lateral area, if distinct, not interrupted on A1 by dark patch (fig. 26.425a) 42

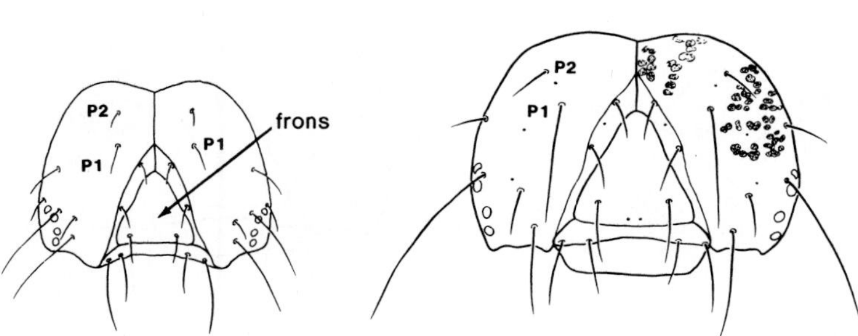

Figure 66

Figure 67

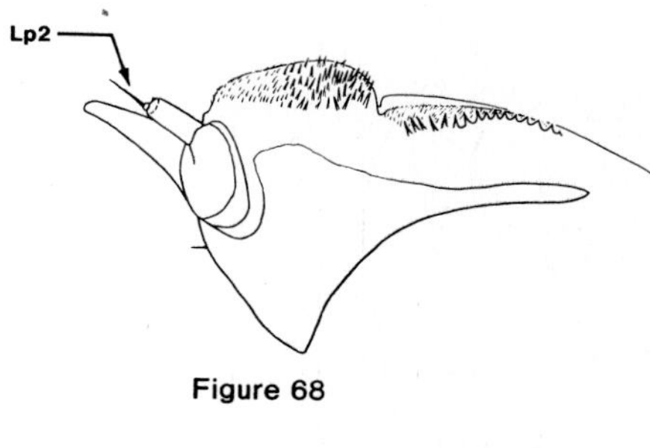

Figure 68

Figure 71

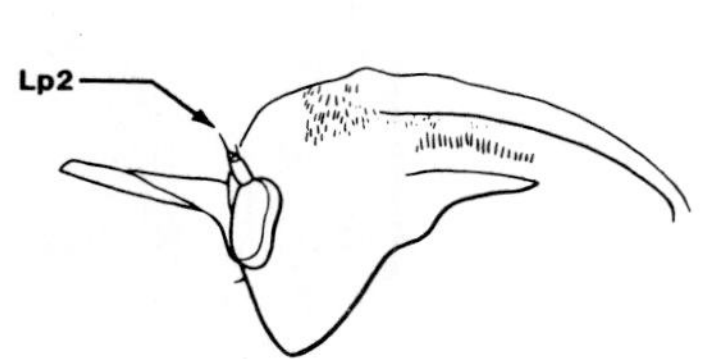

Figure 69

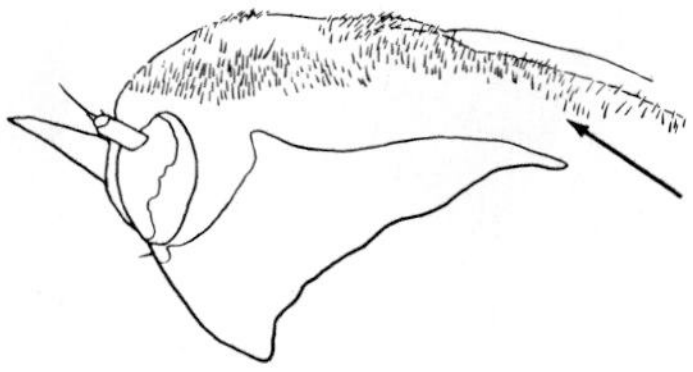

Figure 70

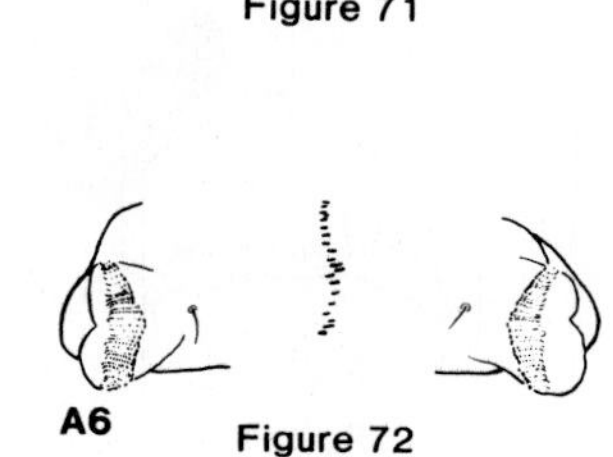

Figure 72

42(41′). Dark, dorsal patches on abdomen with a white spot (fig. 26.428) .... *Spodoptera sunia* (p. 574)

42′. Dark, dorsal patches on abdomen without a white spot (figs. 26.425a,b,c) 43

43(42′). Dark, dorsal patches on abdominal segments with thin, white longitudinal line extending through each patch (figs. 26.425a,b,c) 44

43′. Dark, dorsal patches on abdominal segment lacking thin, white lines 45

44(43). Head with distinct reticulation (fig. 73) .... *Spodoptera praefica* (p. 573)

44′. Head pattern infuscated, reticulation obscured, coronal stripes indistinct (fig. 74) .... (figs. 26.425a,b,c) *Spodoptera ornithogalli* (p. 573)

45(43′). Dark, dorsal patch on T2 (fig. 26.424) subequal in size to patch on dorsum of A8 .... *Spodoptera dolichos* (p. 573)

45′. Dark, dorsal patch on T2 (fig. 26.426) distinctly smaller than patch on dorsum of A8 .... *Spodoptera latifascia* (p. 574)

46(38′). Dorsum of A7 and A8 bearing paired, dark triangular patches (fig. 75) .... *Spaelotis clandestina* (p. 568)

46′. Dorsum of A7 and A8 lacking triangular patches, either pale, mottled or uniformly dark 47

47(46′). Segmental series of whitish yellow middorsal spots on abdomen (fig. 26.407) .... *Peridroma saucia* (p. 568)

47′. No pale middorsal spots on abdomen 48

48(47′). Body velvety dark brown to black; subdorsal lines and lateral areas on abdomen white, contrasting .... *Actebia fennica* (p. 567)

48′. Body pale brown to gray; dorsum faintly pigmented (fig. 76), subdorsal lines not contrasting on abdomen .... *Loxagrotis albicosta* (p. 566)

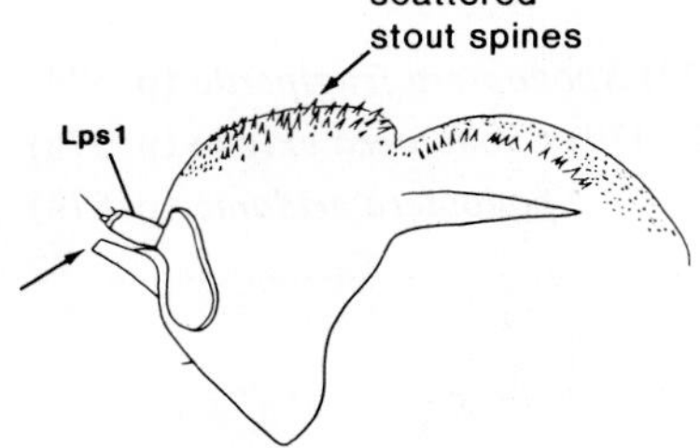

Figure 42

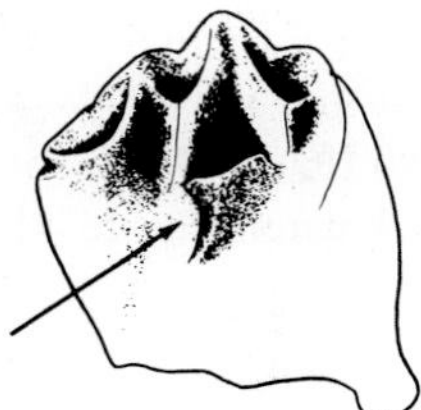

Figure 47

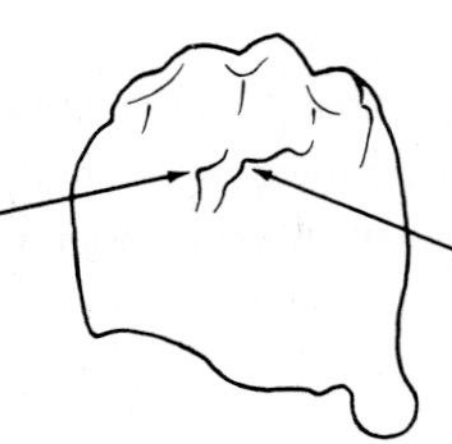

Figure 48

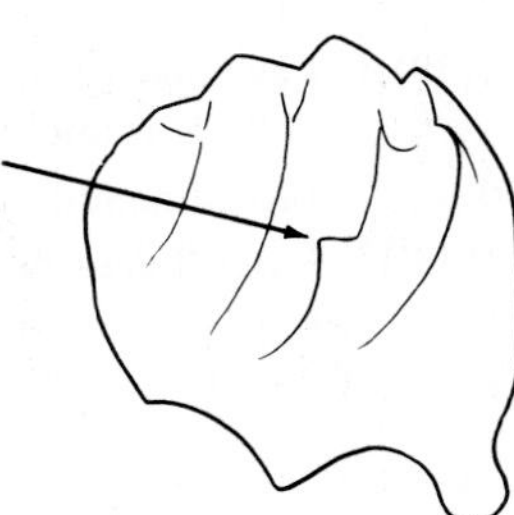

Figure 63

Figure 64

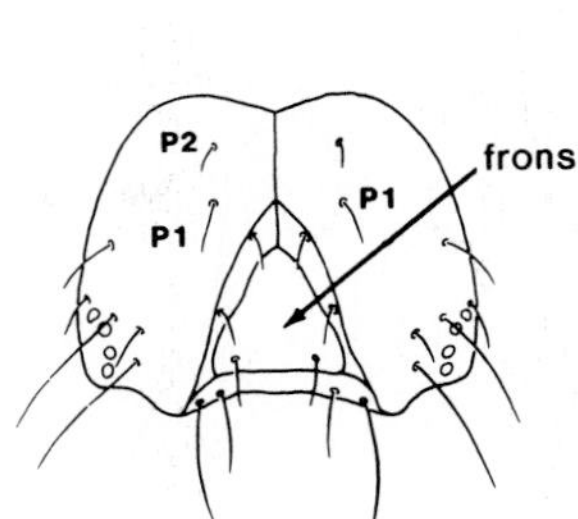

Figure 66

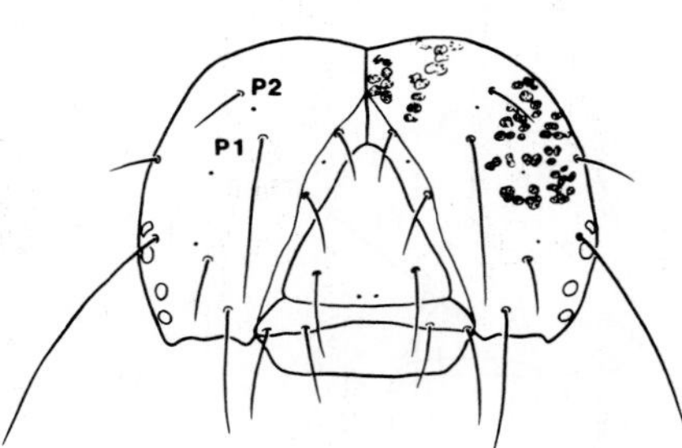

Figure 67

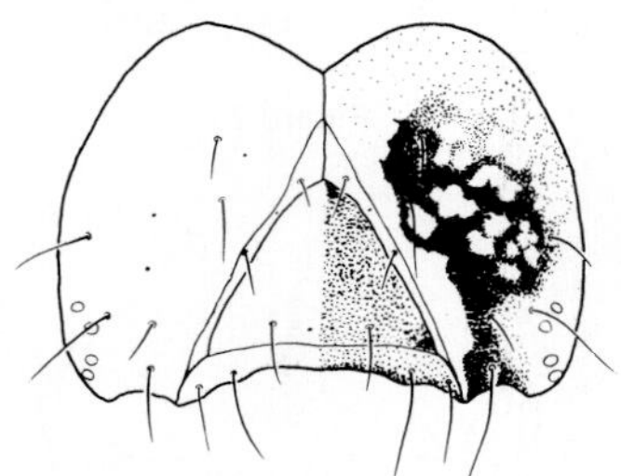

Figure 73

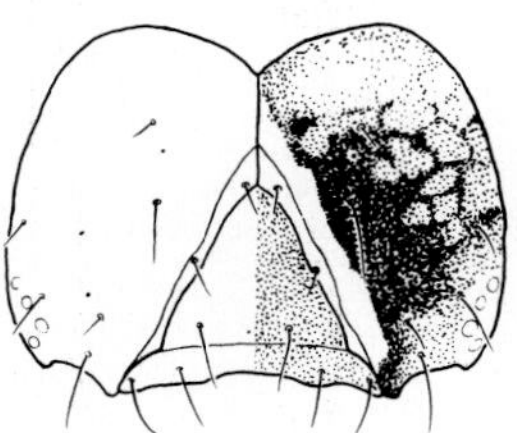

Figure 74

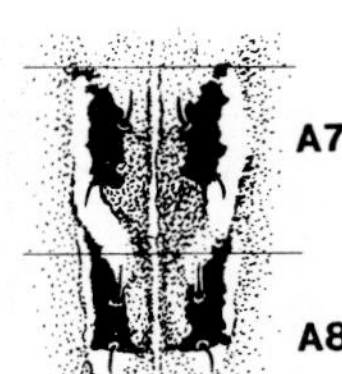

Figure 75

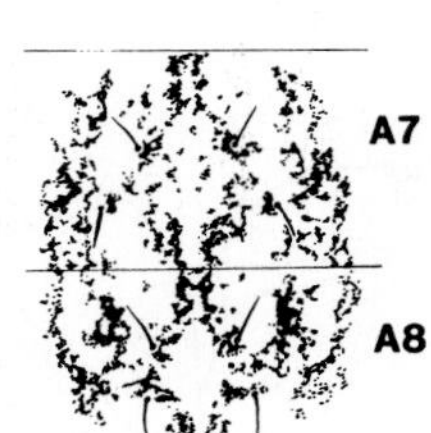

Figure 76

49(32′). Spinneret attenuated apically, drawn to fine point (fig. 77); mandible with two inner teeth (fig. 48) .... *Lithophane laticinerea* (p. 571)

49′. Spinneret, if tapering, blunt or rounded apically (fig. 78); mandible with one inner tooth (figs. 47, 63) or first inner ridge thickened basally and transversely connected to second inner ridge (figs. 64, 79) 50

50(49′). First inner ridge of mandible thickened basally and transversely connected to second inner ridge (figs. 64, 79) 51

50′. First inner ridge of mandible with triangular or quadrate tooth (figs. 47, 63) 52

51(50). Seta P1 of head shorter than width of frons (fig. 66); body not glossy .... *Faronta diffusa* (p. 569)

51′. Seta P1 of head longer than width of frons (fig. 67); body glossy ...... (fig. 26.412) *Trichordestra legitima* (p. 569)

52(50′). Subdorsal area velvety black, interrupted by thin, distinct, vertical, white lines (fig. 26.411) .... *Melanchra picta* (p. 569)

52′. Subdorsal area unicolorous, mottled or bearing white flecks, no vertical lines present (figs. 26.408,26.413a) 53

53(52′). Spinneret subequal to length of first segment of labial palpus (Lps1) (fig. 42) 54

53′. Spinneret surpassing length of first segment of labial palpus (Lps1) (fig. 80) .... *Mamestra configurata* (p. 569)

54(53). Dorsum of abdomen with distinct, white, middorsal line, no subdorsal paired black markings (figs. 26.413a,b) .... (fig. 26.413a,b) *Orthosia hibisci* (p. 570)

54′. Dorsum of abdomen with middorsal line absent or indistinct; subdorsal paired black markings present (fig. 26.408) .... *Xestia spp.* (p. 568)

55(18′). Setae on first segment of maxillary palpus spatulate (fig. 81) .... (fig. 26.417) *Crymodes devastator* (p. 571)

55′. Setae on first segment of maxillary palpus not spatulate (fig. 82) 56

56(55′). Abdominal integument without granules .... (fig. 26.407) *Peridroma saucia* (p. 568)

56′. Abdominal integument with granules visible at 50× 57

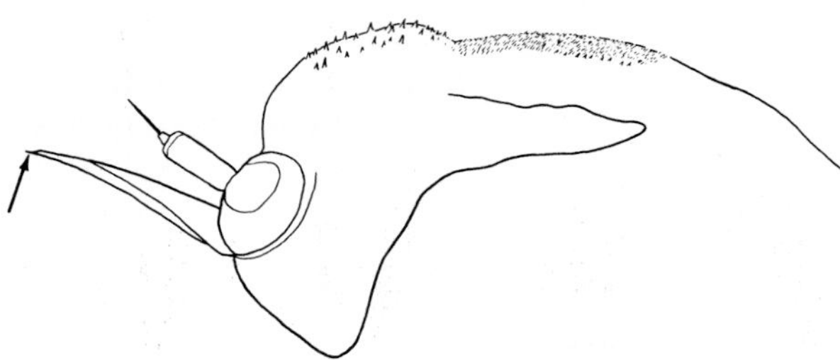

Figure 77

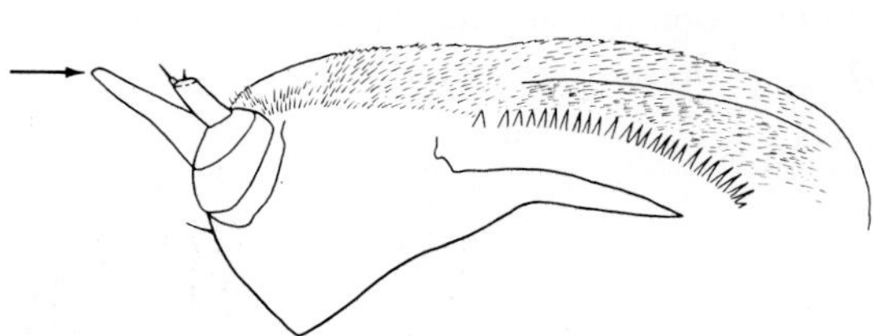

Figure 78

Figure 79

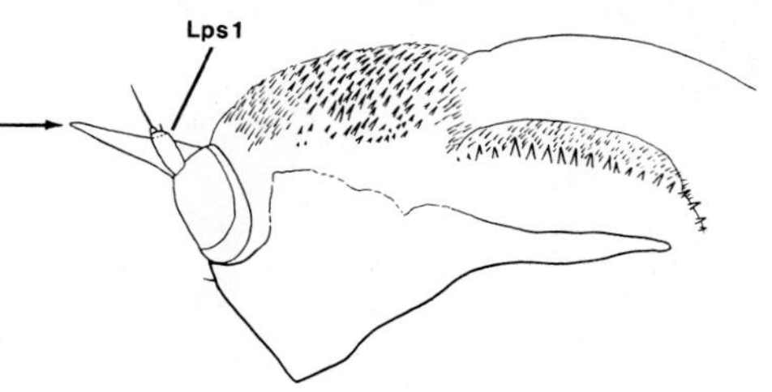

Figure 80

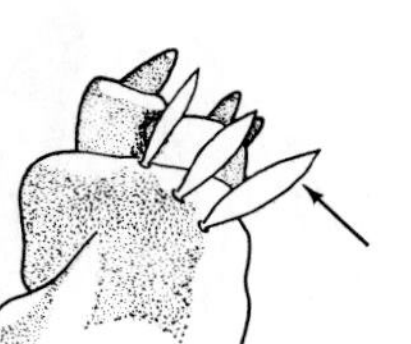

Figure 81

Figure 82

57(56′). Abdominal granules small, slightly convex, set pavementlike without secondary granules (fig. 20) 58

57′. Abdominal granules coarse, isolated, strongly convex or bluntly conical and interspersed irregularly with small secondary granules (figs. 18, 19) 61

58(57). Head marked with pale to dark freckles (fig. 83) ..... (fig. 26.400) *Euxoa auxiliaris* (p. 566)

58′. Head marked with pale brown coronal stripes (fig. 50) and/or distinct reticulation (fig. 73) or greatly infuscated 59

59(58′). Head with brown coronal stripes ..... *Agrotis orthogonia* (p. 567)

59′. Head with distinct reticulate pattern 60

60(59′). Prolegs on A1 each with about 12 crochets; tubercles D1 and D2 on A1–8 subequal ..... *Agrotis malefida* (p. 567)

60′. Prolegs on A1 each with about 6–8 crochets; tubercle D1 about ½ width of D2 on A1–8 ..... (fig. 26.402) *Agrotis gladiaria* (p. 566)

61(57′). Largest abdominal granules conical to retrorse (fig. 19) ..... *Agrotis subterranea* (p. 567)

61′. Largest abdominal granules merely convex, not retrorse (fig. 18) 62

62(61′). Tubercles D1 and D2 subequal on A1–8; subdorsal line darkest on anterior half of each abdominal segment (fig. 26.405) ..... (fig. 26.405) *Feltia jaculifera-subgothica* complex (p. 567)

62′. Tubercle D1 about ½ width of tubercle D2 on A1–8; subdorsal line indistinct or of equal intensity for its length on A1–8 (fig. 26.404) ..... *Agrotis ipsilon* (p. 567)

63(6′). Crochets of prolegs biordinal (fig. 17) ..... *Anagrapha falcifera* (p. 576)

63′. Crochets of prolegs uniordinal (fig. 21) 64

64(63′). T2 with bases of D1 and D2 raised, contiguous; setae D1 and D2 setaceous (fig. 84) 65

64′. T2 with bases of D1 and D2 raised or flat, but usually separated (fig. 85), if contiguous then all D1 and D2 setae stout with rounded tips (fig. 86) 66

65(64). Setal bases of head and body green ..... (fig. 26.398) *Plathypena scabra* (p. 566)

65′. Setal bases of head and body black ..... *Hypena humuli* (p. 566)

66(64′). SD2 on T1 on the shield (fig. 87) (Herminiinae) 67

66′. SD2 on T1 not on the shield (fig. 88) (Catocalinae) 68

67(66). Dorsal setae long, setaceous (fig. 26.440) ..... *Scolecocampa liburna* (p. 578)

67′. Dorsal setae stout (fig. 86) ..... *Idia americalis* (p. 578)

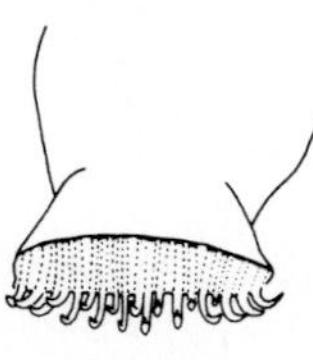

Figure 17

Figure 18

Figure 19

Figure 20

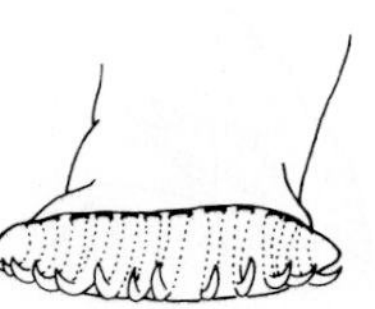

Figure 21

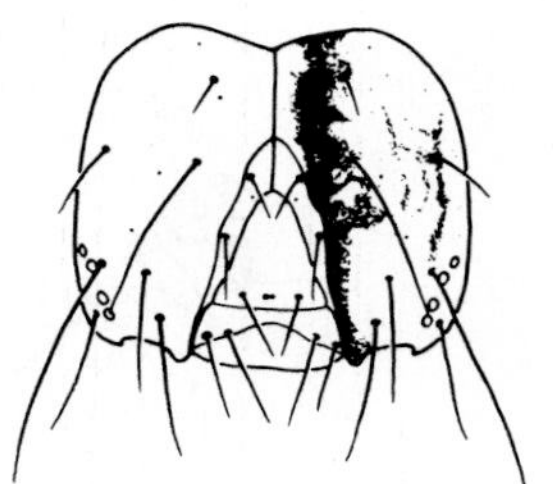

Figure 50

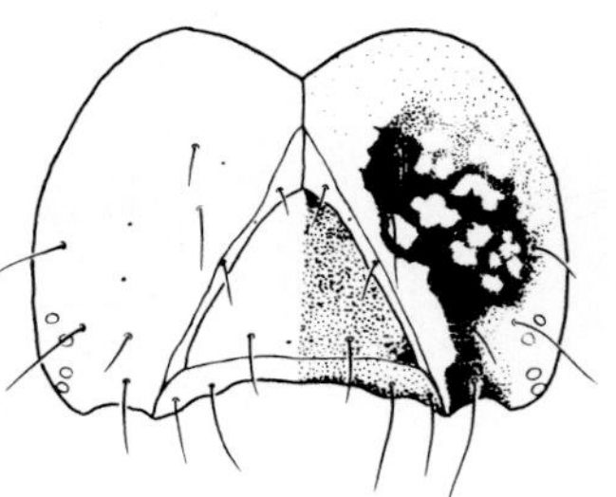

Figure 73

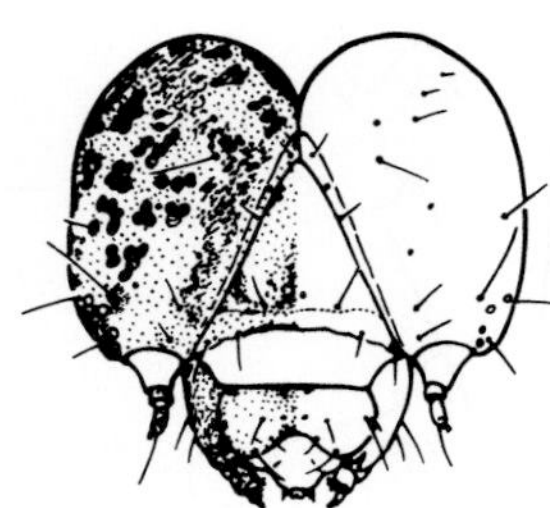

Figure 83

68(66′). Prolegs present only on A5, A6 and A10 (fig. 26.437a) .... *Caenurgina erechtea* (p. 577)

68′. Prolegs present on A3–6 and A10; pair on A3 may be reduced 69

69(68′). Four setae between middorsal line and top of spiracle on A1–6 (fig. 89) .... (fig. 26.439) *Alabama argillacea* (p. 578)

69′. Three setae between middorsal line and top of spiracle on A1–6 (fig. 90) 70

70(69′). Body without subventral abdominal fringes (figs. 90, 26.438a,b) .... *Anticarsia gemmatalis* (p. 578)

70′. Body with subventral abdominal fringes (figs. 91, 26.436a,b) .... *Catocala relicta* (p. 577)

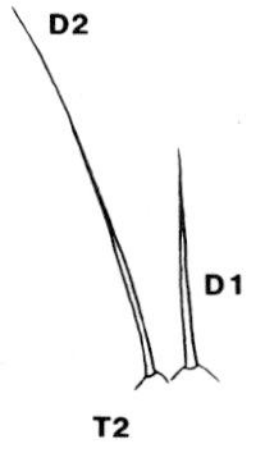

Figure 84

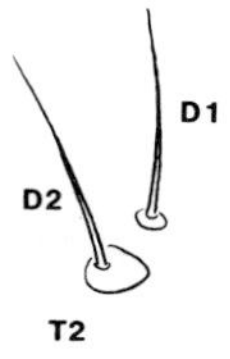

Figure 85

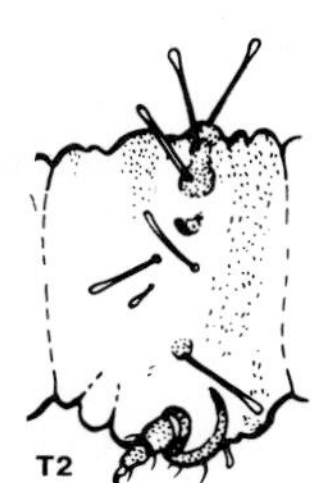

Figure 86

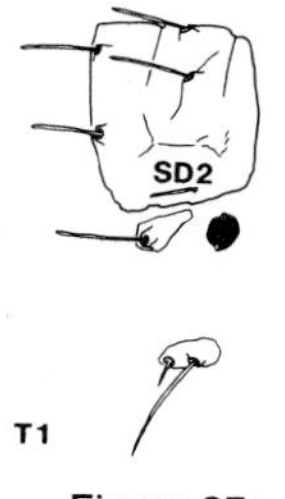

Figure 87

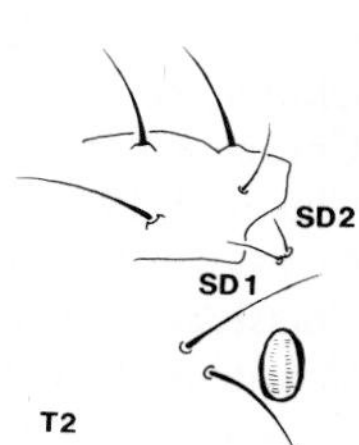

Figure 88

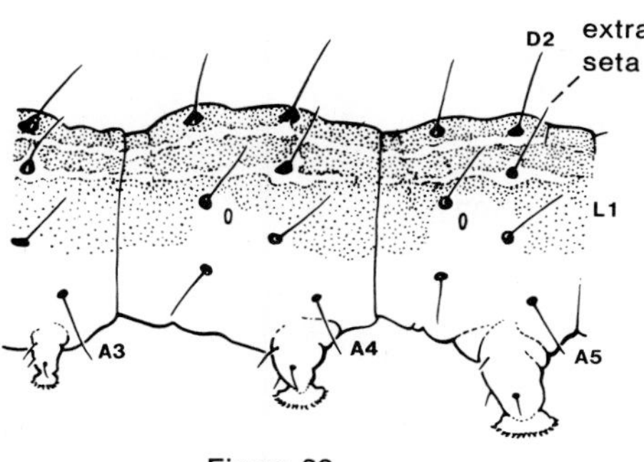

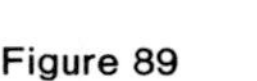
Figure 89

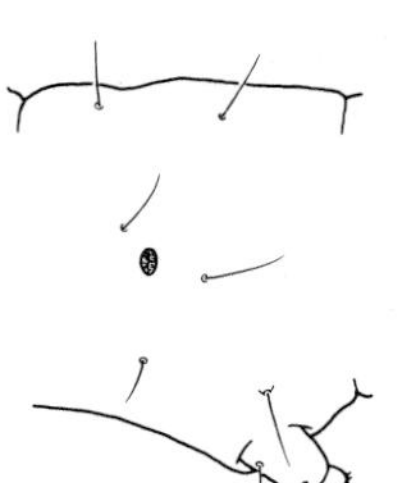

Figure 90

Figure 91

Selected Bibliography (General Coverage)

Claassen 1921.
Common 1970.
Comstock and Dammers 1934.
Darlington 1952.
Edwards and Elliot 1883.
Essig 1958.
Forbes 1960.
Harris 1841.
Kimball 1965.
Mosher 1916a.
Okumura 1961.
Peterson 1948, 1961.
Phipps 1930.
Prentice [coordinator] 1962.
Riley 1870a.
Tietz 1972.
Wilde and Bell 1980.
Wolcott 1936.

Selected Bibliography (Restricted Coverage)

Andrews 1879.
Angulo and Weigert 1975a, 1975b.
Barnes and McDunnough 1918.
Beck 1960.
Bethune 1873.
Beutenmüller 1901, 1902.
Bird 1898.
Brittain 1915, 1918.
Brown and Dewhurst 1975.
Burleigh 1972.
Chapman and Lienk 1974.
Chittenden 1901, 1907, 1913.
Chittenden and Russell 1909.
Chu et al. 1963.
Comstock, J. A. 1939, 1944, 1947.
Comstock, J. A. and Dammers 1942.
Comstock, J. H. 1879, 1880.
Cook 1930.
Cooley 1916.
Crumb 1926a, 1926b, 1927, 1929, 1932, 1934, 1956.
Davis 1912.
Davis and Satterthwait 1916.
Decker 1930, 1931.
Dyar 1894b, 1899b, 1899c, 1904.
Eichlin 1975a, 1975b.
Eichlin and Cunningham 1969, 1978.
Felt 1912.
Forbes, S. A. 1904.
Forbes and Hart, 1900.
Forbes, W. T. M. 1954.
Forbes and Franclemont 1954.
Ford et al. 1975.
Franclemont 1980.
Franklin 1928.
French 1884.
Gardner 1946, 1947, 1948.
Garman 1895.
Gibson 1909, 1915.
Godfrey 1971, 1972, 1980, 1981.
Godfrey and Stehr 1985.
Griswold 1934.
Hardwick 1965.
Hawley 1918.
Heitzman and Habeck 1979(81).
Herzog 1980.
Hill 1925.
Hinds 1912, 1930.
Hinks and Byers 1976.
Hoerner 1948.
Jones 1918.
Khalsa et al. 1979.
King 1928.
Klots 1967b.
Knight 1916.
Knutson 1944.
Kogan et al. 1978.
Lange 1945.
Levine and Chandler 1976.
Levy and Habeck 1976.
Luginbill 1928.
MacNay [compiler] 1959.
Marsh 1913b.
Mason 1922.
Matthewman 1937.
McCabe 1980.
McElvare 1941.
McGuffin 1958a.
Merzheyevskaya 1967.
Muka 1976.
Neunzig 1969.
Ogunwolu and Habeck 1979.
Okumura 1960.
Oliver and Chapin 1981.
Parker et al. 1921.
Pedigo et al. 1973.
Peterson 1964.
Phipps 1927.
Pruess 1961.
Riley 1870b, 1874, 1883a, 1883b, 1883c, 1885, 1892a.
Rings 1969, 1970, 1977.
Rings and Arnold 1977.
Rings et al. 1978.
Rings and Johnson 1976, 1977.
Rings and Musick 1976.
Ripley 1923.
Robertson-Miller 1923.
Salkeld 1975, 1984.
Sargent 1976.
Silver 1933.
Slingerland 1893.
Smith F. 1941.
Smith and Dyar 1898.
Smith R. C. 1924.
Sorenson and Thornley 1941.
Strickland 1916.
Sukhareva 1978 [1979].
Thompson 1943.
Todd 1978.
Todd and Poole 1980.
Van den Bosch and Smith 1955.
Vickery 1915.
Walkden 1950.
Washburn 1903, 1911.
Watson 1916.
Webster 1898.
Weigert and Angulo 1977.
Whelan 1935.
Wood et al. 1954.

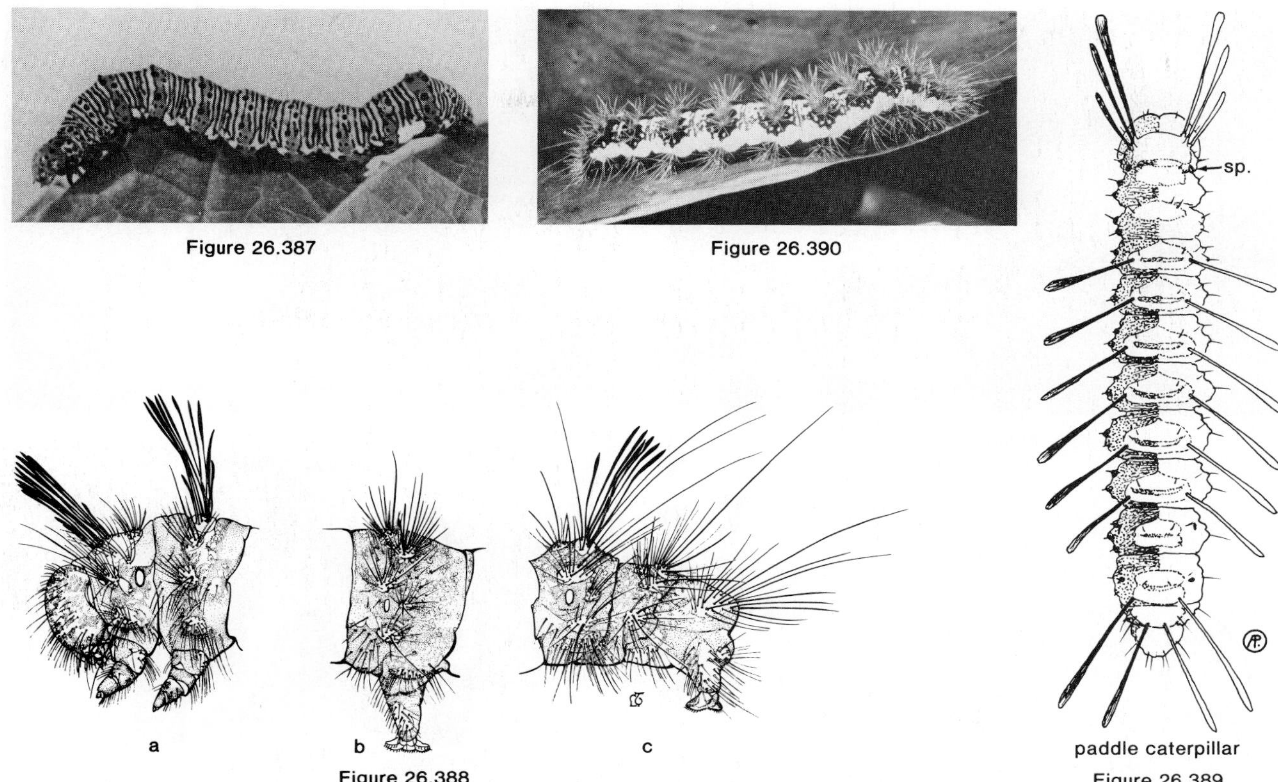

Figure 26.387

Figure 26.390

Figure 26.388

Figure 26.389

Figure 26.387. Noctuidae, Agaristinae. *Alypia octomaculata* (Fabricius), eight-spotted forester. Also see noctuid key fig. 15, head. f.g.l. 30–35 mm. A1–8 with orange band between D1 and D2 tubercles bordered front and rear by alternating black lines and bluish-white stripes; setae white. Head (noctuid key fig. 15) orange, setal bases black. *Diagnosis:* distinct transverse, abdominal stripes and lines plus larval association with Vitaceae separate it (and related agaristines) from other noctuids. In vineyards and on buildings covered with vitaceous vines. *Hosts:* Boston ivy, Virginia creeper, and grape. One and sometimes 2 broods. Throughout the East, from southern Canada to Georgia and Texas. Riley (1874), Washburn 1903), Crumb (1956), and Forbes (1960). Photo by G. L. Godfrey.

Figure 26.388a-c. Noctuidae, Pantheinae. *Panthea furcilla* (Packard). a. head, T1, T2; b. A4; c. A8–10. f.g.l. 40± mm. Dull brown to orange brown with long, black hair-pencils, or blackish with white hair-pencils. Head pale orange, with brown freckles. *Diagnosis:* pale coronal stripes on head join to form "U" above frons; majority of setae in largest dorsal hair-pencils on T1, T2, A1, and A8 long and spatulate. Arboreal. *Hosts:* tamarack, spruce, and pine. Univoltine, mature larvae in late summer. Eastern half of continent, extending to Rocky Mountains in Alberta. Smith and Dyar (1898), Forbes (1954), Crumb (1956), Prentice (1962), and Klots, (1967). Drawings by L. M. Tackett, courtesy of F. W. Stehr.

Figure 26.389. Noctuidae, Acronictinae. *Acronicta funeralis* Grote & Robinson, paddle caterpillar. f.g.l. 25–30 mm. Head and body dark brown to black; dorsum with pale yellow, transverse callouses. *Diagnosis:* setae XD1, XD2, and D2 on T2 and setae D2 on A1–6, A8, and A9 long, black, and spatulate. Arboreal. *Hosts:* deciduous trees and shrubs, including maple, alder, birch, hickory, dogwood, hawthorne, *Holodiscus* (ocean-spray), apple, poplar, cherry, *Rubus* (raspberry, etc.), willow, and elm. Univoltine, larva in August. Throughout the East, south to Mississippi, west to British Columbia. Smith and Dyar (1898), Forbes (1954), Crumb (1956), Prentice (1962) (fig. from Peterson 1948).

Figure 26.390. Noctuidae, Acronictinae. *Acronicta oblinita* (J. E. Smith), smeared dagger moth or smartweed caterpillar. f.g.l. 40–50 mm. Marked with contrasting black and bright yellow, and sometimes red; D1 and D2 tubercles red or black, bearing numerous, bristly setae (yellow or brown); spinules on integument. Head red and black, may also be nearly solid red or black. *Diagnosis:* the wide, bright yellow lateral area, and absence of pale contrasting coronal head stripes separate it from *Simyra henrici* (Grote). Associated with wet habitats, including banks of streams, edges of ponds and lakes, and low-lying areas in open fields. *Hosts:* a general feeder recorded from about 26 genera. The most frequently stated are: alder, *Cephalanthus* (buttonbush), *Epilobium* (fireweed), strawberry, smartweed, raspberry, willow, and cattail. Larva in mid-summer to mid-fall; bivoltine according to some. Transcontinental from southern Canada to Florida and Texas. Smith and Dyar (1898), Beutenmüller (1901), Forbes (1954), Crumb (1956), Prentice (1962). Photo courtesy of D. J. Voegtlin, I.N.H.S.

Figure 26.391

a

P2

b

Figure 26.392

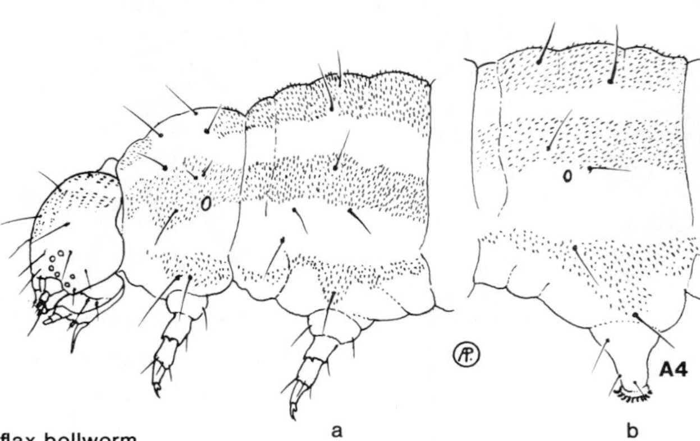

flax bollworm

Figure 26.393

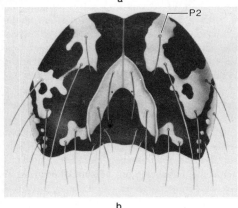

Figure 26.391. Noctuidae, Acronictinae. *Acronicta americana* (Harris), American dagger moth. f.g.l. 50± mm. Densely covered with white to pale yellow setae; dorsal black hair-pencils, 2 each on A1 and A3, 1 on A8. Head black with pale yellow adfrontal ecdysial lines. *Diagnosis:* black hair-pencils about 2X as long as dense, white setae; integument bearing pavement granules. Arboreal. *Hosts:* general feeder on deciduous trees (and a few shrubs). Univoltine in the northern part of range, mature larva in midsummer to late September; possible second generation southward. Common east of the Rocky Mountains from southern Canada to Florida and Texas. Smith and Dyar (1898), Forbes (1954), Crumb (1956), Kimball (1965), and Prentice (1962). Photo courtesy of R. W. Rings and Glenn Berkey, O.A.R.D.C.

Figure 26.392a,b. Noctuidae, Acronictinae. *Simyra henrici* (Grote), cattail caterpillar. a. larva; b. head, showing relationship of coronal stripes and P2 setae. f.g.l. 35-45 mm. Colorfully marked with yellow, orange, and black; subdorsal line and lateral area below spiracles yellow; D1 and D2 tubercles orange, bearing numerous bristly, black and white setae; dorsum black; subdorsal and ventral areas with varying degrees of black mottling; spinules on integument. Head black with white adfrontal areas and coronal stripes (fig. 26.392b). *Diagnosis:* contrasting white adfrontal areas form a distinct inverted "V"; base of P2 on head contained within white coronal stripe. Most frequently associated with semiaquatic plants and large monocots. *Hosts: Cephalanthus* (buttonbush), soybean (infrequently), grasses (unspecified), smartweed, poplar, willow, sedges (unspecified), cattail, and corn (occasionally damaging corn silk). Bivoltine. Southern Canada, west to Alberta and south throughout the U.S. to Florida and southern California. Beutenmüler (1901), Classen (1921), Comstock and Dammers (1934), Comstock (1944), and Crumb (1956). Photo by W. D. Zehr, I.N.H.S.

Figure 26.393a,b. Noctuidae, Heliothinae. *Heliothis ononis* (Fabricius) flax bollworm. Also see noctuid key fig. 22, larva (dorsum) and noctuid key fig. 23, A-8, showing D1 and D2 setal insertions. f.g.l. 30-32 mm. Green, dusky from dark spinules, subdorsal line and lateral area white to pale yellow; crochets biordinal. Head pale green with blackish freckles, especially distinct above P2 setae. *Diagnosis:* insertions of body setae flat, unlike other common *Heliothis* species, which have large tubercles. Prairies and open fields. *Host:* principally on developing seeds of flax, also reported from fruiting structures of vetch, sweet clover, and *Vaccaria* (cow cockle). Univoltine. Sporadically abundant in upper Great Plains. Holarctic. MacNay (1959) and Todd (1978). (figs. from Peterson 1948).

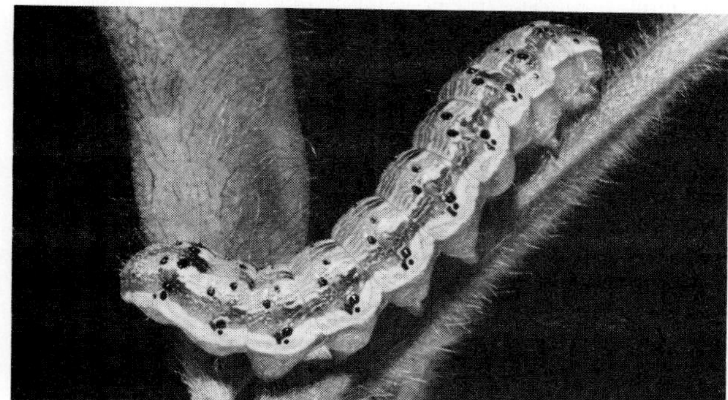

Figure 26.394

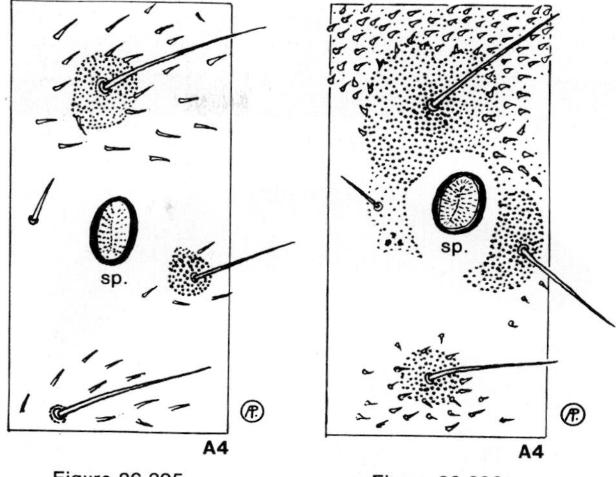

Figure 26.395

Figure 26.396

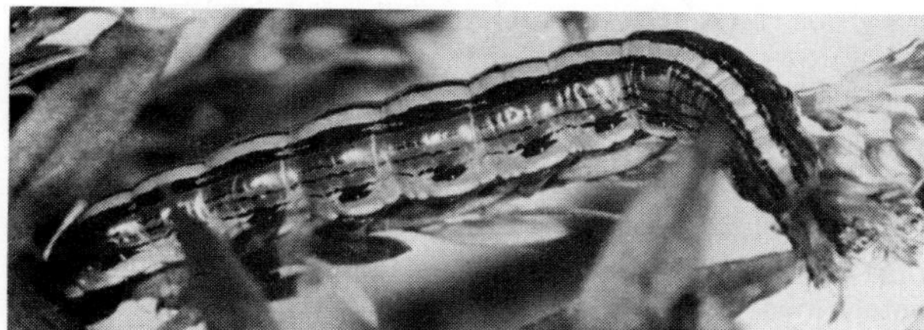

Figure 26.397

Figure 26.394. Noctuidae, Heliothinae. *Heliothis zea* (Boddie), bollworm, corn earworm, tomato fruitworm. Also see noctuid key figs. 25, A8 showing absence of spinules on D1 tubercle, and fig. 26, left mandible (oral view). f.g.l. 32–40 mm. Color quite variable; dorsal half yellow, orange, red, maroon or green, sometimes extensively black; covered with spinules; dorsal tubercles large; crochets biordinal. Head green, pale brown freckles and whitish reticulation may be present. *Diagnosis:* no spinules on dorsal tubercles of A1, A2, and A8; mandible lacks distinct tooth on inner surface. Common in many cultivated crops. *Hosts:* polyphagous, most frequently associated with flowering and fruiting structures of soybean, cotton, tomato, tobacco, corn. Multivoltine in warmer climates. Throughout North America as far north as southern Canada, extending into South America. Crumb (1926b, 1956), Hardwick (1965), Neunzig (1969), and Kogan *et al.* (1978). Photo by W. D. Zehr, I.N.H.S.

Figure 26.395. Noctuidae, Heliothinae. *Heliothis virescens* (Fabricius), tobacco budworm. Also see noctuid key figs. 27, A8 showing spinules on D1 tubercle; fig. 28, left mandible (oral view) with inner tooth, and fig. 29, left mandible with broken inner tooth. f.g.l. 30–37 mm. Commonly green, sometimes red or maroon; covered with spinules; dorsal tubercles large; crochets biordinal. Head green, immaculate or with faint brown freckles and whitish reticulation. *Diagnosis:* spinules on dorsal tubercles of A1, A2, and A8; mandible with distinct tooth (or scar of tooth) on inner surface; differing from *H. subflexus* by the relatively smaller tubercles of SD1, SD2, L1, and L2. Occurs frequently in certain cultivated crops. *Hosts:* polyphagous, feeds on buds, flowers, fruiting structures, and foliage; commonly associated with tobacco; other common hosts include soybean, cotton, and *Geranium.* Range of hosts uncertain because the larvae of *Heliothis virescens, H. subflexus* (Guenée), and *H. zea* frequently are misidentified. Multivoltine. Transcontinental, similar to that of *H. zea.* McElvare (1941), Crumb (1926b, 1956), Neunzig (1969), Todd (1978), and Kogan et al. (1978). (fig. from Peterson 1948).

Figure 26.396. Noctuidae, Heliothinae. *Heliothis subflexus* Guenée. Similar to preceding species. Larval specimens not seen. McElvare (1941), and Peterson (1948) placed emphasis on the large prominent SD1, SD2, L1, and L2 tubercles (fig. 26.396) of the abdominal segments to separate *H. subflexus* from *H. virescens* which has relatively smaller, corresponding tubercles (fig. 26.395). Some authors consider *H. subflexus* to be a synonym or subspecies of *H. virescens.* However, Todd (1978) treated it as a distinct species. Past differences in opinions have led to confusion in the host plant associations of *H. subflexus* and *H. virescens.* McElvare (1941) listed *Solanum nigrum* L. (deadly nightshade) and *Physalis* spp. (ground cherry) as the hosts for *Heliothis subflexus.* (fig. from Peterson 1948).

***Description* 26.396.1.** Noctuidae, Acontiinae. *Acontia dacia* Druce, brown cotton leafworm. f.g.l. 32± mm. Orange brown mottled with white; middorsal line wide, darker brown than rest of dorsal area from which it is separated by thin, white lines; lateral area extending downward on A5 and A6 onto upper portions of prolegs; setae D1 and D2 on A1, A2, A3, and A8 on prominent tubercles, sites of SD1 and L also may be tuberculate, especially on A1, A2, and A3; diffuse, dark transverse bands may be present on dorsa of A1, A2, and A3. Prolegs only on A5, A6, and A10. Head orange brown with brown reticulation and coronal stripes. *Diagnosis:* the brownish body color, an inner mandibular tooth, and the presence of an anal fork (noctuid key fig. 9) distinguish this semilooping noctuid from the common plusiines that may occur on cotton. *Host:* cotton. Reported from Arkansas, Louisiana, and Texas. Okumura (1961); Thomas (pers. comm., 1981) stated that he has not seen this species in Texas since 1957.

Noctuid key fig. 9, *Acontia aprica* (Hübner), showing A10 anal fork that is very similar to that of *A. dacia.*

Figure 26.397. Noctuidae, Cuculliinae. *Cucullia asteroides* Guenée. Also see noctuid key fig. 31, A8–10 showing extension of spiracular line to anal shield, and noctuid key fig. 30, subequal D1, D2, and SD1 setae on A9. f.g.l. 40± mm. Body marked with 2 shades of glossy yellow, green, or reddish brown and 6–7 pairs of thin but distinct black longitudinal lines that separate the shades; middorsal area and lateral area yellow or striped yellow and white in all forms. Head green with small brown freckles. Body and head with pavement granules. *Diagnosis:* the glossy (from the pavement granules) and distinctly striped body make it quite recognizable in the field. In open meadows. *Hosts:* flowers of *Aster* and goldenrod. Univoltine. New Brunswick to Saskatchewan, south to Georgia and Colorado. Forbes (1954) and Crumb (1956). Photo by D. J. Voegtlin, I.N.H.S.

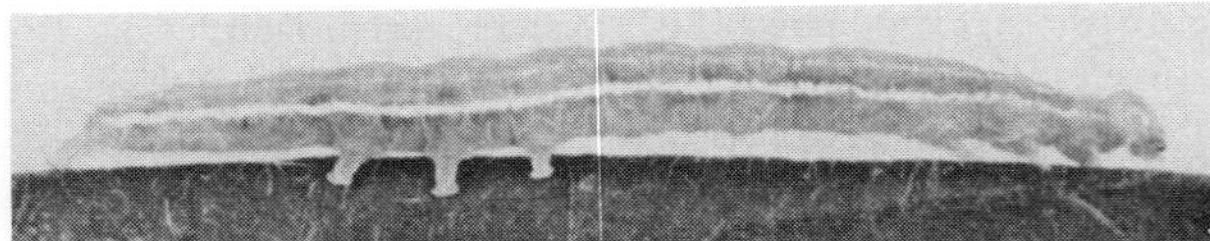

Figure 26.398

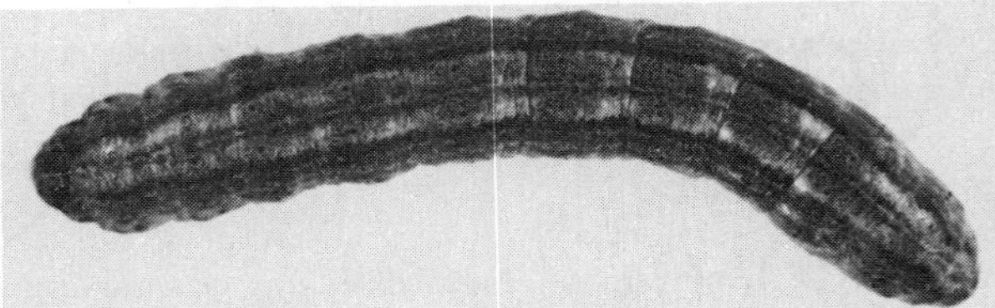

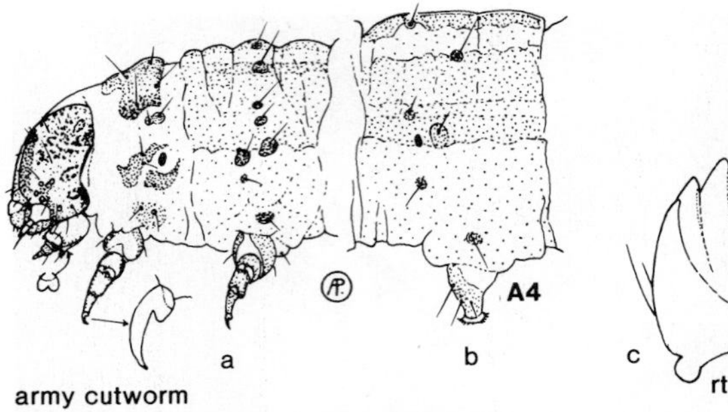

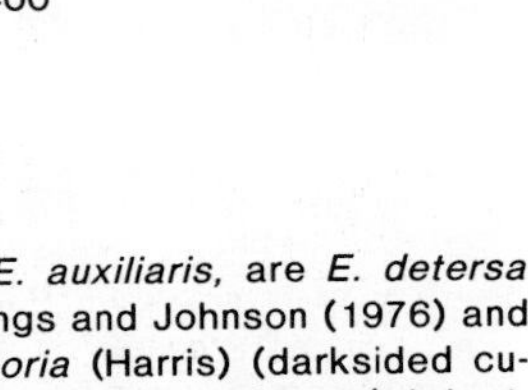

Figure 26.400

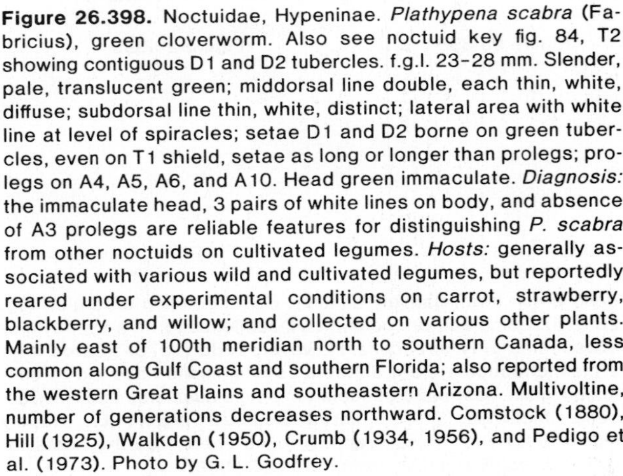

Figure 26.402

Figure 26.398. Noctuidae, Hypeninae. *Plathypena scabra* (Fabricius), green cloverworm. Also see noctuid key fig. 84, T2 showing contiguous D1 and D2 tubercles. f.g.l. 23–28 mm. Slender, pale, translucent green; middorsal line double, each thin, white, diffuse; subdorsal line thin, white, distinct; lateral area with white line at level of spiracles; setae D1 and D2 borne on green tubercles, even on T1 shield, setae as long or longer than prolegs; prolegs on A4, A5, A6, and A10. Head green immaculate. *Diagnosis:* the immaculate head, 3 pairs of white lines on body, and absence of A3 prolegs are reliable features for distinguishing *P. scabra* from other noctuids on cultivated legumes. *Hosts:* generally associated with various wild and cultivated legumes, but reportedly reared under experimental conditions on carrot, strawberry, blackberry, and willow; and collected on various other plants. Mainly east of 100th meridian north to southern Canada, less common along Gulf Coast and southern Florida; also reported from the western Great Plains and southeastern Arizona. Multivoltine, number of generations decreases northward. Comstock (1880), Hill (1925), Walkden (1950), Crumb (1934, 1956), and Pedigo et al. (1973). Photo by G. L. Godfrey.

***Description* 26.399.1.** Noctuidae, Hypeninae. *Hypena humuli* Harris, hop looper (no figure). f.g.l. 18–25 mm. Slender, dark olive green dorsally, paler ventrally; double, thin, white, diffuse middorsal lines; subdorsal line white, distinct; setae long, on black tubercles; prolegs on A4, A5, A6, and A10; minute spinules covering body. Head green with dusky freckles on vertex, bases of setae black. *Diagnosis:* black setal bases on green head and body, and absence of A3 prolegs readily separate *H. humuli* from other noctuids likely to be found on hop or related plants. *Hosts: Humulus* (hop) and *Urtica* (stinging nettle). Quebec to British Columbia, south to California, Colorado, Kansas, Alabama, and Virginia. Bivoltine, larvae in June-July and August-September. Bethune (1873), Dyar (1904), Hawley (1918), Crumb (1934, 1956), and Forbes (1954).

Figure 26.400a-c. Noctuidae, Noctuinae. *Euxoa auxiliaris* (Grote), army cutworm. a. head, T1, T2; b. A4; c. right mandible (oral view). Also see noctuid key fig. 83, head, showing maculation on right side. f.g.l. 38–50 mm. Greenish brown to greenish gray, dorsal half darker than ventral; mid- and subdorsal lines pale, the latter wider than middorsal line; band of white splotches below spiracles; points of muscle attachments on T1 shield brown; small pavement granules. Head pale brown with brown to dark brown freckles; coronal stripes brown or absent; stemmatal area dark brown. Abundant at times in field crops and gardens; feeds above surface. *Hosts:* general, but Strickland (1916) reported a preference for *Sisymbrium incisum* Englem. (tansy mustard) and *Thlaspi arvense* L. (stinkweed) over other plants. Univoltine; larva overwinters, final instar attained the following April-May. Widely distributed from southern Canada to northern Mexico west of the Mississippi River, moths occasionally collected in Illinois. Strickland (1916), Cooley (1916), and Whelan (1935). **Note:** The genus *Euxoa* is a large complex containing several economically important cutworms that, in addition to *E. auxiliaris,* are *E. detersa* (Walker) (sandhill cutworm) (see Rings and Johnson (1976) and Rings and Arnold (1977)), *E. messoria* (Harris) (darksided cutworm) (see Rings et al. 1978)), *E. ochrogaster* (Guenée) (redbacked cutworm), *E. scandens* (Riley) (white cutworm) (see Rings (1977)), and *E. tessellata* (Harris) (striped cutworm). Unfortunately these are very difficult to separate if working with preserved specimens as Crumb (1956) pointed out.

The larvae of *Euxoa* can be recognized quite reliably by the adrontal ecdysial lines that reach or nearly reach the epicranial notch (noctuid key fig. 83), the usually present freckled pattern on the head, and the darkly pigmented points of muscle attachments on the T1 shield. Persons needing specific identifications of *Euxoa* larvae are advised to rear them and obtain determinations of the associated moths. Rearing procedures were developed by Hinks and Byers (1976). By establishing local collections of reared and associated larvae and adults, it may be possible to develop usable identification keys for the *Euxoa* larvae in geographically limited regions. (figs. from Peterson 1948).

***Description* 26.401.1.** Noctuidae, Noctuinae. *Loxagrotis albicosta* (Smith), western bean cutworm. See noctuid key fig. 76, showing dorsal maculation on A7 and A8. f.g.l. 38–40± mm. Pale tan, grayish brown, or pinkish gray with slightly roughened integument; obscure, diamond-shaped pattern on dorsum of abdominal segments; pale brown mid- and subdorsal lines evident on T1 shield, obscure elsewhere; abdominal tubercles small. Head brown, reticulation slightly darker. *Diagnosis:* the dark T1 shield and contrasting paler mid- and subdorsal line will help separate *L. albicosta* from other western cutworms. *Hosts:* green, lima, and pinto beans (leaves and pods) and corn. Univoltine. Alberta to Mexico, especially in the high plains proximate to Rocky Mountains; reported as far east as Iowa. Hoerner (1948), Crumb (1956), and Wilde and Bell (1980).

Figure 26.402. Noctuidae, Noctuinae. *Agrotis gladiaria* Morrison, claybacked cutworm. f.g.l. 33–37 mm. Grayish, dorsal area on T2-A9 distinctly paler than rest of body, varying from pale pinkish gray to pale tan; middorsal line thin, white, distinct on T1 shield; subdorsal line uniformly darker gray; subdorsal area with diffuse white mottling; dorsal tubercles dark brown, D2s on A1–8 ca. 2× as wide as D1s; small, pavement granules. Head brown with black coronal stripes and reticulation; frons infuscated brown. *Diagnosis:* the pale, contrasting dorsal area, relatively larger D2 tubercles on A1–8, and 6–8 crochets on each A3 proleg distinguish it from other common, early-season cutworms. Inhabits tunnels in pastures and poorly cultivated croplands. *Hosts:* a very general feeder, occasionally abundant and destructive to seedling corn, unspecified clover, and various garden plants; also reported on new settings of strawberry, and canes of blackberry and raspberry. Univoltine, larva from early spring to May. Southeastern Canada to Florida, as far west as S. Dakota, Utah, and Arizona. Garman (1895), Forbes (1904), Crumb (1929, 1956), Whelan (1935), Forbes (1954), Rings *et al.* (1978), Rings (1977), and Rings and Arnold (1977). Photo by G. L. Godfrey.

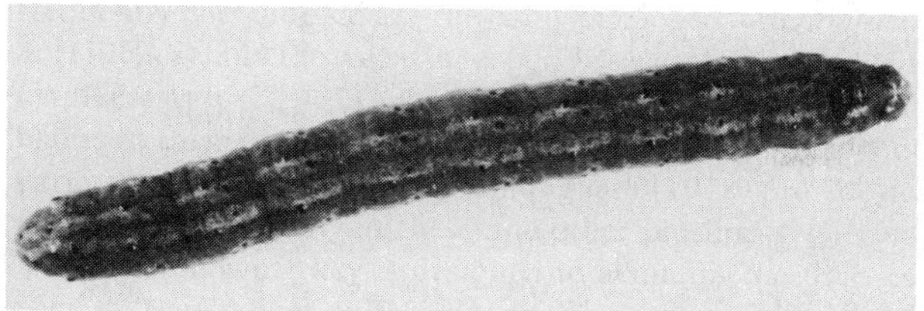

Figure 26.404

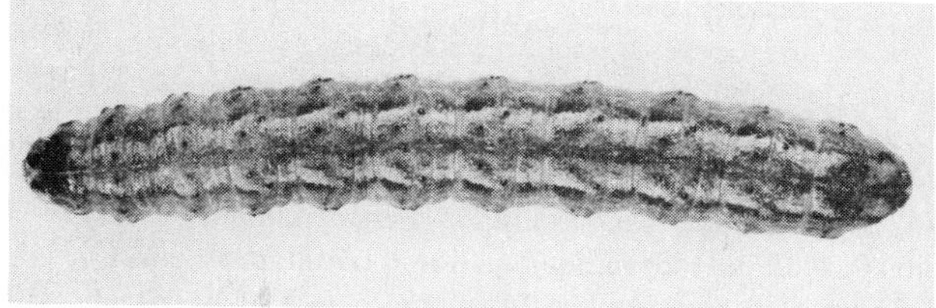

Figure 26.405

Description **26.403.1.** Noctuidae, Noctuinae. *Agrotis orthogonia* Morrison, pale western cutworm (no figure). f.g.l. 31-38 mm. Pale yellowish gray; middorsal line white on pale brown T1 shield; dark dorsal vessel visible on T2-A8; subdorsal line brownish, lateral area whitish, both markings fade during last instar; tubercles small, greenish brown, ringed by pale circle; small pavement granules present. Head pale brown; coronal stripes and stemmatal area dark brown. *Diagnosis:* the distinct, white middorsal line, prominent dark brown coronal stripes and stemmatal areas, and absence of reticulation on the head distinguish *A. orthogonia. Hosts:* a rather general feeder on stems (below soil surface) and roots of herbs; destructive to cultivated cereal crops, e.g., oat, barley, rye, spring and winter wheat, and corn. Univoltine, peak larval feeding from May to mid-June. Primarily east of the Rocky Mountains from southern Alberta to Arizona and New Mexico, extending east to western N. Dakota, S. Dakota, Nebraska, Kansas, and the panhandles of Oklahoma and Texas. Parker, et al. (1921), Cook (1930), Whelan (1935), and Sorenson and Thornley (1941).

Description **26.403.2.** Noctuidae, Noctuinae. *Agrotis malefida* Guenée, palesided cutworm. (no figure). f.g.l. 35-40± mm. Mottled, pale to dark gray dorsally and subdorsally, paler laterally and ventrally with white splotches; T1 shield browish with pale middorsal line; seta SD1 on A1-8 arising from fairly distinct white patch; tubercles small, D1 subequal to D2 on A1-8; small pavement granules. Head pale grayish brown, coronal stripes and reticulation dark brown, reticulation weak. *Diagnosis:* the combination of white SD1 abdominal patch, subequal D1 and D2 abdominal tubercles, small pavement granules, and 12-15 crochets on each A3 proleg help separate *A. malefida* from other common *Agrotis* species. Larva reportedly inhabits shallow, underground tunnels into which it drags plant material. *Hosts:* a general feeder on cultivated row crops and herbs. Multivoltine. Reported from Gulf Coast states as far north as Tennessee. Riley (1885), Crumb (1929, 1956), and Forbes (1954).

Description **26.403.3.** Noctuidae, Noctuinae. *Agrotis subterranea* (Fabricius), granulate cutworm. See noctuid key fig. 19, showing conical and retorse abdominal granules, adapted from Crumb (1929). f.g.l. 30-40± mm. Grayish, mottled; dorsal area medially with segmental series of poorly defined, brownish rhomboid figures on A1-8; lateral margins of dorsal area yellowish gray; T1 shield brown with pale middorsal line; lateral and ventral areas with abundant white flecks and splotches; D1 and D2 tubercles on A1-8 black, large; D2 2× width of D1; body granules heterogenous, conical to retorse (noctuid key fig. 19). Head brownish, coronal stripes black, reticulation faint. *Diagnosis:* similar to *Agrotis ipsilon* except the largest body granules of *A. subterranea* are more pronounced and elongate. A surface feeding cutworm that occasionally will climb to obtain fresh food and reportedly will hide under cut leaves. *Hosts:* general feeder on cultivated crops. Multivoltine. Apparently transcontinental as far north as Massachusetts and S. Dakota. Forbes (1904), Jones (1918), Crumb (1929, 1956), Whelan (1935), and Forbes (1954).

Figure 26.404. Noctuidae, Noctuinae. *Agrotis ipsilon* (Hufnagel), black cutworm. Also see noctuid key fig. 18, convex abdominal granules, adapted from Crumb (1929), and noctuid key fig. 35, head showing adfrontal ecdysial lines reaching epicranial notch. f.g.l. 40-46 mm. Varying from pale gray to black, commonly uniform on dorsal half, some specimens have dorsal area defined by tannish lateral edges (fig. 26.404); abdominal tubercles large, black, D2s on A1-8 at least 2× width of D1s; mid- and subdorsal lines pale, faintly visible on black T1 shield; heterogenous convex granules. Head brown, coronal stripes and reticulation black, the latter frequently obscured by extensive infuscation; early instars bear dark brown freckles in place of coronal stripes and dark reticulation. *Diagnosis:* the heterogenous, convex granules, and relatively large D2 tubercles are distinctive. Commonly in weedy, poorly cultivated crops; appears to be associated with primary weedy invaders of tilled lands, especially if drainage is a problem. Full-grown larva inhabits subterranean tunnel but known to leave to feed. *Hosts:* extremely general on cultivated crops and numerous herbs. Multivoltine. Cosmopolitan, expands northward during spring and summer months. Forbes (1904), Crumb (1929, 1956), Whelan (1935), Walkden (1950), Rings and Musick (1976), Rings (1977), and Rings et al. (1978). Photo by G. L. Godfrey.

Figure 26.405. Noctuidae, Noctuinae. *Feltia "jaculifera-subgothica"* complex, dingy cutworm. *Feltia jaculifera* (Guenée) and *F. subgothica* (Haworth) have very similar larvae, and there are not sufficient data to enable one to distinguish between the larvae. Therefore, the following description is generalized to include both: f.g.l. 32-40 mm. Mottled grayish brown; dorsal area yellowish brown laterally; on A1-8, faint dusky, oblique lines extending mediocephalad from D2s on A1-8 form faint chevrons; subdorsal lines darkly intensified on anterior halves of A1-8; tubercles D1 and D2 dark brown, distinct, subequal; abdominal granules heterogenous, convex. Head brownish with blackish coronal stripes and reticulation, the latter sometimes faint or absent. *Diagnosis: F. jaculifera* and *F. subgothica* may be distinguished from other common spring-feeding cutworms by their heterogenous, convex granules, subequal D1 and D2 tubercles, and the faint dorsal chevrons that point cephalad. Occasionally common in poorly cultivated croplands. *Hosts:* general on cultivated crops, also known to climb and feed on buds of fruit trees. Univoltine, larva overwinters, resumes feeding in early spring and continues until about mid-June. Rather general across southern Canada and northern half of U.S. Forbes (1904), Crumb (1929, 1956), Whelan (1935), Forbes (1954), Rings et al. (1978), Rings and Musick (1976), and Rings (1977). Photo courtesy of R. W. Rings and Glenn Berkey, O.A.R.D.C.

Description **26.406.1.** Noctuidae, Noctuinae. *Actebia fennica* (Tauscher), black army cutworm (no figure). f.g.l. 32-42± mm. Velvety brownish black to black; dorsal area blackish medially, laterally paler with narrow band of scattered white dots and dashes bordered by thin, broken but contrasting white subdorsal line; lateral area white with reddish brown median; T1 shield glossy black, marked by thin, whitish middorsal line. Head blackish, becoming yellow brown laterad. *Diagnosis:* the 4 white lines composed of subdorsal lines and lateral areas contrasting with the dark brown to black dorsal half of body and glossy T1 shield are very distinctive of *A. fennica.* An occasional pest of blueberry. *Hosts:* buds and newer growth of numerous shrubs and trees, also reported from herbs; Phipps (1927, 1930) noted it from over 20 plant families. Univoltine, present from April-early June. Transcontinental, southern Canada; in U.S. reported from Alaska and the northern contiguous states; some records include California (presumably northern) and New Jersey; also northern Eurasia. Phipps (1927, 1930), Crumb (1932, 1956), Darlington (1952), and Forbes (1954).

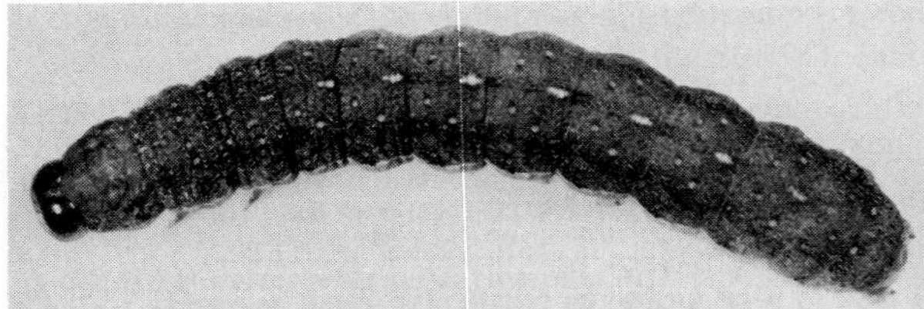

Figure 26.407

Figure 26.408

Figure 26.409

***Description* 26.406.2.** Noctuidae, Noctuinae. *Spaelotis clandestina* (Harris), w-marked cutworm. See noctuid key fig. 75, dorsal maculation on A7 and A8. f.g.l. 28± mm. Gray; segmental series of paired black patches on A1-8, approximating triangles on A8, patches bordered laterally by whitish yellow subdorsal lines; ventral margin of subdorsal area dark above each spiracle; lateral area with white flecks but scarcely paler than subdorsal or ventral areas. Head yellowish brown, coronal stripes and reticulation black. *Diagnosis:* superficially similar to some *Spodoptera* and *Xestia* because of segmental series of paired, dark, dorsal patches. However, the hypopharynx of *Spaelotis clandestina* has distinct spines on the proximolateral lobe whereas the *Spodoptera* hypopharynx lacks such spines. The absence of a tooth on the inner surface of the mandible separates *Spaelotis clandestina* from *Xestia* spp. that possess this structure. A climbing cutworm frequently associated with buds of shrubs and trees. *Hosts:* a general feeder; various cultivated fruit crops, e.g., apple and blueberry, vegetables and grasses (unspecified). Univoltine in New Brunswick, possibly double-brooded farther south. Transcontinental across southern Canada; recorded as far south as Kentucky in U.S. Crumb (1929, 1956), Phipps (1930), and Wood et al. (1954).

Figure 26.407. Noctuidae, Noctuinae. *Peridroma saucia* (Hübner), variegated cutworm. Also see noctuid key fig. 45, spinneret (dorsal), and noctuid key fig. 82, left maxillary palpus. f.g.l. 35-46 mm. Mottled brown or gray, pale or dark, often showing shades of yellow, orange, or pink; segmental series of whitish yellow middorsal spots on A1-8; subdorsal line a segmental series of frosted black dashes; subdorsal area sometimes with oblique frosted black marks extending dorsocephalad from spiracles. Head yellow brown with dark brown reticulation and coronal stripes. *Diagnosis:* the segmental series of whitish yellow, middorsal spots on the abdomen is sufficient to recognize *P. saucia;* the spots frequently are more evident on anterior abdominal segments. A rather common species in vegetable and flower gardens and many field crops. *Hosts:* a general feeder, mostly recorded from herbaceous plants. Multivoltine. General throughout the Americas; nearly cosmopolitan. Crumb (1929, 1956), Walkden (1950), Rings et al. (1978), Rings and Musick (1976), and Rings (1977). Photo courtesy of R. W. Rings and Glenn Berkey, O.A.R.D.C.

Figure 26.408. Noctuidae, Noctuinae. *Xestia "c-nigrum"* complex, spotted cutworm. The Old World species, *Xestia c-nigrum* (Linnaeus), also called *Amathes* or *Noctua c-nigrum,* was thought until recently to occur in North America and commonly was referred to as the "spotted cutworm," but Franclemont (1980) pointed out that *Xestia c-nigrum* is represented in North America by 2 species which he described as *X. dolosa* and *X. adela.* I examined the larvae of both species and could find no substantial characters to separate them. Therefore, the following description is generalized to encompass both: f.g.l. 30-42 mm. Dull gray or brown; dorsal area of abdomen with segmental series of paired black chevrons bordering subdorsal lines, markings small on anterior segments, increasing in size caudad, largest on A7 and A8; lateral area sordid white with shades of pink and orange; spiracles whitish; dorsum of A8 somewhat angulate in lateral view. Head yellow brown, reticulation and coronal stripes dark brown. *Diagnosis:* may be confused with some *Spodoptera,* but unlike them *Xestia* species possess a tooth (or teeth) on the oral surface of the mandible. *Hosts:* general feeder of herbs, field crops, and occasionally on deciduous shrubs and trees. Bivoltine. The combined distributions of *Xestia dolosa* and *X. adela* cover most of North America. See Franclemont (1980) for specific information. Washburn (1903), Franklin (1928), Crumb (1929, 1956), Rings and Musick (1976), Rings (1977), and Rings and Johnson (1977). Photo courtesy of R. W. Rings & Glenn Berkey, O.A.R.D.C.

Figure 26.409. Noctuidae, Hadeninae. *Nephelodes minians* Guenée, bronzed cutworm. f.g.l. 38-45 mm. Thick bodied. Dark brown with bronzy sheen; mid- and subdorsal lines wide, continuous (fig. 26.409), solid white or cream-colored on black cervical shield, cream-colored with reddish brown center elsewhere; lateral area cream-colored with reddish brown mottling; small pavement granules (noctuid key fig. 20). Head tannish with fuscous reticulation. *Diagnosis:* very distinctive with its large head bearing short P1 setae (noctuid key fig. 34); T1 has but 1 SD seta (noctuid key fig. 36). In grassy habitats. *Hosts:* principally on grasses but other plants occasionally reported. Univoltine. East of Rocky Mountains from southern Canada to about Oklahoma and Carolina. Gibson (1915), Crumb (1926a, 1956), Godfrey (1972), Rings and Musick (1976), Rings (1977), and Rings et al. (1978). Photo by G. L. Godfrey.

***Description* 26.410.1.** Noctuidae, Hadeninae. *Dicestra trifolii* (Hufnagel), clover cutworm. f.g.l. 26-30 mm. Body commonly dark or blackish green with white to pale yellow, thin middorsal and distinct subdorsal lines; lateral area bright yellow or yellow tinged with orange; some larvae yellowish green, brown, or gray; segmental series of black bars form lateral margins of dorsal area in some specimens. Head color variable. *Diagnosis:* the bright lateral area contrasting with darker subdorsal and dorsal areas plus the elongate, tapering spinneret (noctuid key fig. 68) is a rather unique combination of characters amongst cutworms associated with herbaceous plants. Capacity for multiple broods. A sporadic or occasional pest on foliage of beets, various legumes, and cabbage. *Hosts:* polyphagous on low herbs, but commonly reported from beets and *Chenopodium* (lamb's-quarters). Transcontinental and Europe. Riley (1883a), Forbes and Hart (1900), Marsh (1913b), Crumb (1956), and Godfrey (1972).

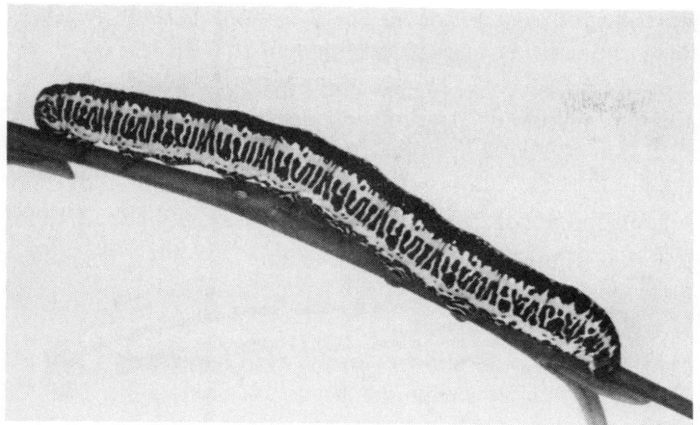

Figure 26.411

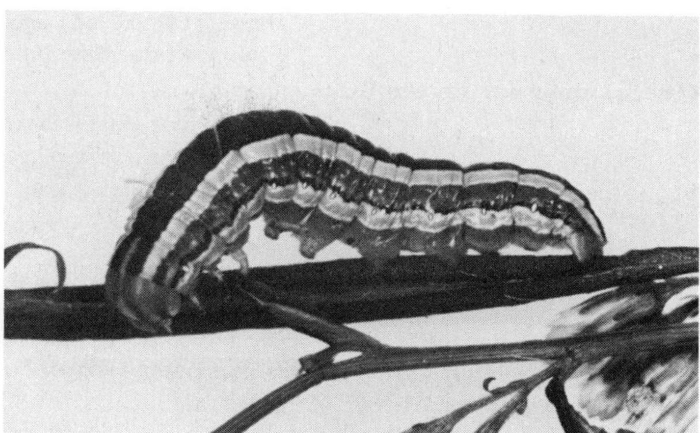

Figure 26.412

Description **26.410.2.** Noctuidae, Hadeninae. *Dargida procincta* (Grote) (no figure). f.g.l. 30–37 mm. Variable, pale green to blackish olive green; mid- and subdorsal lines, pale, narrow, continuous, generally distinct; lateral area with 3 narrow pale lines. Head color and reticulation variable; seta P1 shorter than transverse width of frons. *Diagnosis:* 3 distinct pale lines on a dark lateral area separate *D. procincta* from the *Faronta* species. *Hosts:* various grasses (wild and cultivated), wheat, unspecified sedges and rushes, and *Lupinus* (lupine). Overwinters as larva; more common in mountain meadows than lower. In Pacific Northwest, east to Alberta and south to Utah and California. Thompson (1943), Crumb (1956), Essig (1958), and Godfrey (1972).

Description **26.410.3.** Noctuidae, Hadeninae. *Faronta diffusa* (Walker), wheat head armyworm. f.g.l. 26–43 mm. Moderately slender, head about as wide as anterior abdominal segments. Body varying from dark green, green, yellowish green, to brown; longitudinal stripes and lines dominate pattern; mid- and subdorsal lines, thin, continuous, yellowish; ventral stripe of subdorsal area black, contrasting with very distinct, white lateral area; ventral area greenish becoming white mediad. Head concolorous with body, coronal stripes and reticulation present or absent; seta P1 shorter than transverse width of frons (noctuid key fig. 66). *Diagnosis: F. diffusa* shares affinities with *Dargida procincta,* but the mandible of *Faronta diffusa* has interconnected first and second inner ridges (noctuid key fig. 64). *Hosts:* primarily on heads of grasses and cultivated grains. Bivoltine. Nova Scotia to Alberta and Oregon, south to Arizona, the central Great Plains, and eastern states; not reported from Florida. Washburn (1911), Walkden (1950), Crumb (1956), Tietz (1972), Godfrey (1972), and Wilde and Bell (1980).

Figure 26.411. Noctuidae, Hadeninae. *Melanchra picta* (Harris), zebra caterpillar. f.g.l. 31–46 mm. Body with bold black, yellow, and white markings; dorsal area solid velvety black; middorsal line thin, white on T1 shield, absent or thinly present elsewhere; subdorsal lines wide, yellow; subdorsal area velvety black with thin, distinct, vertical white lines; lateral area yellow plus black spotting; prolegs pinkish orange. Head orange to reddish. *Diagnosis:* alternating black and white vertical lines in subdorsal area are very distinctive. *Hosts:* willow is the most frequently listed host and cabbage second; recorded from over 38 species of dicotyledonous herbs and shrubs, with asparagus being the only listed monocot. Probably bivoltine in most of its range. Southern Canada, west to Alberta and across the northern contiguous U.S., south to California, Kentucky, and Virginia. Harris (1841), Riley (1870a, 1883c), Washburn (1903), Comstock and Dammers (1942), Crumb (1956), Godfrey (1972), and McCabe (1980). Photo courtesy of J. G. Franclemont, Cornell University.

Figure 26.412. Noctuidae, Hadeninae. *Trichordestra legitima* (Grote), striped garden caterpillar. f.g.l. 30–35 mm. Boldly striped and slightly glossy; dorsal area maroonish brown with thin blackish lateral margins, middorsal line present only on T1 shield; subdorsal lines yellow, wide; subdorsal area paler than dorsal area; lateral area pale yellow. Head yellow brown, coronal stripes diffuse or absent, reticulation indistinct. *Diagnosis:* the 2 interconnected inner teeth of the mandible (noctuid key fig. 79) and the yellow and maroonish brown striped body provide ready recognition of *T. legitima. Hosts:* reported plants indicate a general feeder on herbs, common at times on asparagus and flowers of goldenrod; also on some woody species, e.g., blackberry and willow; Crumb (1956) indicated that the normal host may be slender grasses. Univoltine, larva observed September-November. Across southern Canada, south to Florida, Texas, Utah, and California. Chittenden (1907), Crumb (1929, 1956), Godfrey (1972), and McCabe (1980). Photo courtesy of J. G. Franclemont, Cornell University.

Description **26.412.1.** Noctuidae, Hadeninae. *Mamestra configurata* Walker, Bertha armyworm. f.g.l. 32± mm. Highly variable, gray, brown, green, or yellowish with different amounts of black; dorsum with or without segmental series of velvety black shield-shaped marks; mid- and subdorsal lines thin, bright yellow, usually distinct but obscured in specimens nearly solid black dorsally and subdorsally; lateral area pale yellow with reddish tinge, distinct in dark forms, less so in paler forms. Head yellow brown, coronal stripes and reticulation dark, present or absent. *Diagnosis: M. configurata* differs from other common armyworms by the presence of a prominent tooth on the oral mandibular surface and a minute, first segment of the labial palpus (Lps 1) (noctuid key fig. 80). An economic pest in the northern prairies. *Hosts:* very general; common on Chenopodiaceae and Brassicaceae, but reported from a wide range of dicotyledonous plants plus wheat and corn. One larval brood annually in Canada, undetermined elsewhere. Manitoba to British Columbia, south into Mexico. King (1928), Crumb (1956), and Godfrey (1972).

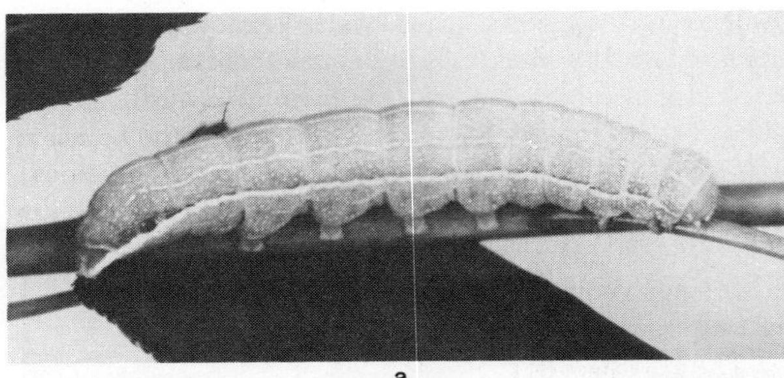

a

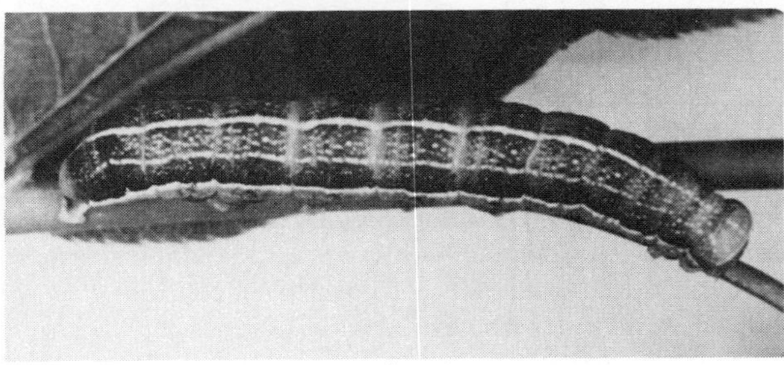

b

Figure 26.413

Figure 26.414

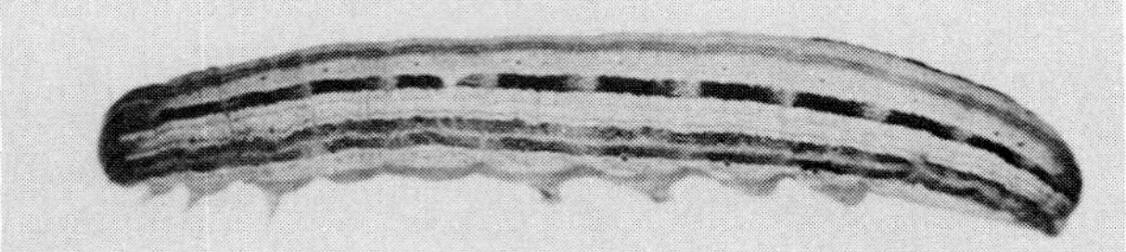

Figure 26.415

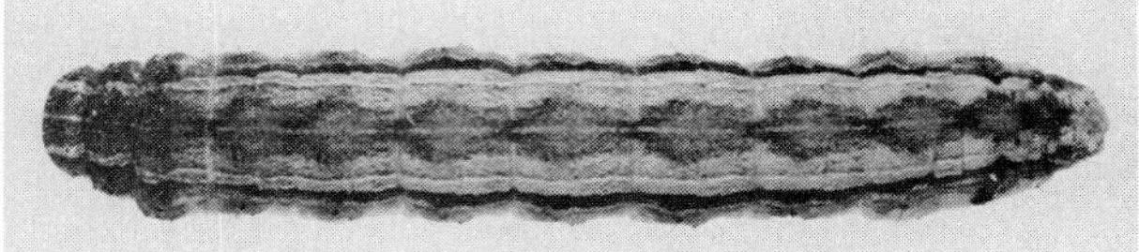

Figure 26.416

Figure 26.413a,b. Noctuidae, Hadeninae. *Orthosia hibisci* (Guenée), a green fruitworm. a. larva (light form); b. larva (dark form). f.g.l. 30–35 mm. Whitish green to blackish green, covered extensively with numerous white flecks; mid- and subdorsal lines white, the former wide and distinct, the latter thin, indistinct; bases of D1 on A1–8 white; lateral area translucent green except for distinct whitish dorsal margin. Head pale green with or without pale brown reticulation. *Diagnosis:* the prominent white, middorsal line (fig. 26.413b), white dorsal margin of lateral area (fig. 26.413a), and strongly developed inner mandibular tooth (noctuid key fig. 47) readily distinguish *O. hibisci* from other common green fruitworms or climbing cutworms. Common in deciduous woodlands. *Hosts:* a rather general feeder on deciduous shrubs and trees (foliage, buds, and young fruit), occasionally reported from conifers, more rarely on herbs. Univoltine, larva during April-June. Newfoundland to Alaska, south to California and Florida. Crumb (1956), McGuffin (1958a), Rings (1970), Godfrey (1972), and Chapman and Lienk (1974). Photos courtesy J. G. Franclemont, Cornell University.

Figure 26.414. Noctuidae, Hadeninae. *Pseudaletia unipuncta* (Haworth), armyworm. f.g.l. 37–41 mm. Brownish with varying degrees of blackish mottling and white flecks; mid- and subdorsal lines thin, white; subdorsal area divided into dark brown dorsal stripe and violet gray ventral stripe; lateral area reddish brown. Head yellow brown with dark brown coronal stripes and brown reticulation. *Diagnosis:* frequently mistaken for fall armyworm, *Spodoptera frugiperda. Pseudaletia unipuncta* is characterized by a deeply cleft and brushy hypopharynx (noctuid key fig. 38) and fringed spinneret (noctuid key fig. 39). The fall armyworm has none of these characters. Also the subdorsal area of *P. unipuncta* has a dark dorsal half and pale ventral half; the opposite is true of fall armyworm. Frequently found in cereal crops and lawns. *Hosts:* mainly grasses. Multivoltine in warmer latitudes. General throughout the continent into southern Canada. Vickery (1915), Davis and Satterthwait (1916), Knight (1916), Crumb (1927, 1956), Franklin (1928), Godfrey (1972), and Rings and Musick (1976). Photo courtesy of R. W. Rings and Glenn Berkey, O.A.R.D.C.

Figure 26.415. Noctuidae, Hadeninae. *Leucania phragmatidicola* Guenée. f.g.l. 28–35 mm. Gray to grayish yellow; dorsal area with thin, white middorsal line and wide, black lateral margins; dorsal half of subdorsal area brownish yellow with 2 thin pinkish lines; ventral half of subdorsal area grayish; lateral area whitish yellow. Head yellow brown with dark brown coronal stripes and brown reticulation. *Diagnosis:* placement of the Fa punctures below the transverse line formed by the F1 setal insertions (noctuid key fig. 50) and the black lateral margins of dorsal area separate *Leucania phragmatidicola* from the allied genera *Pseudaletia* and *Aletia.* Godfrey (1972) may be consulted for distinguishing *Leucania phragmatidicola* from other *Leucania* species. Larva known to occur near lakes and ponds. *Hosts:* grasses and sedges. Bivoltine. East of Rocky Mountains from Manitoba south to Texas and Florida. Crumb (1927, 1956), Knutson (1944), and Godfrey (1972). Photo courtesy of J. G. Franclemont, Cornell University.

Figure 26.416. Noctuidae, Hadeninae. *Lacinipolia renigera* (Stephens), bristly cutworm. f.g.l. 21–31 mm. and Grayish brown; mid- and subdorsal lines pinkish brown; dorsal area with segmental series of diffuse blackish diamonds; subdorsal area with dark brown dorsal stripe; lateral area pinkish brown; small convex granules; setae simple but slightly stout. Head brown, reticulation and coronal stripes black. *Diagnosis:* the combination of the segmental series of dorsal diamonds, prominent black stripe paralleling subdorsal line, and the stiff-appearing dorsal setae are distinctive of *L. renigera.* Found near the soil surface and on low herbs and shrubs in meadows and pastures; larva occasionally seen in the early spring in alfalfa fields. *Hosts:* polyphagous, reported from a wide variety of cultivated garden crops and weed species. Bivoltine in some areas with the second generation overwintering as partially grown larvae. Transcontinental. Crumb (1929, 1932, 1956), Godfrey (1972), Rings and Musick (1976), Rings (1977), and Rings et al. (1978). Photo courtesy of R. W. Rings and Glenn Berkey, O.A.R.D.C.

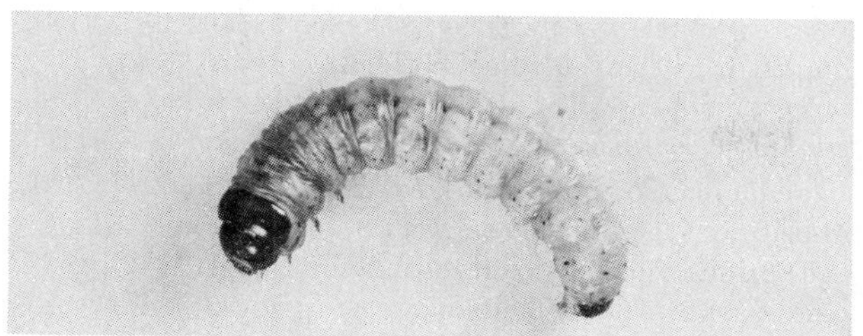

Figure 26.417

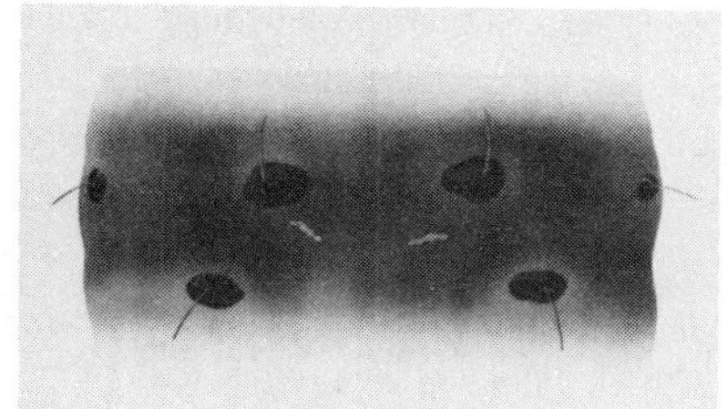

Figure 26.420

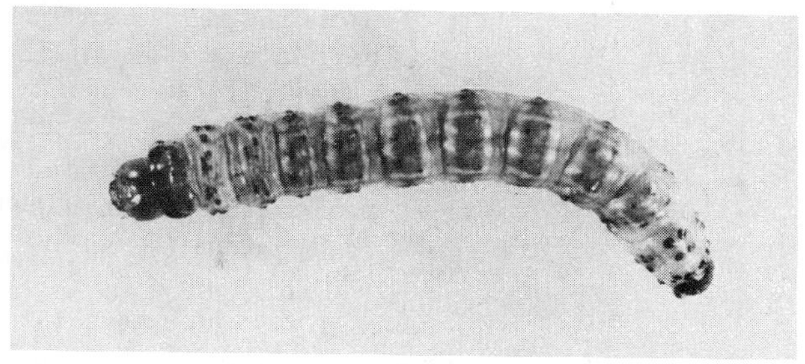

a

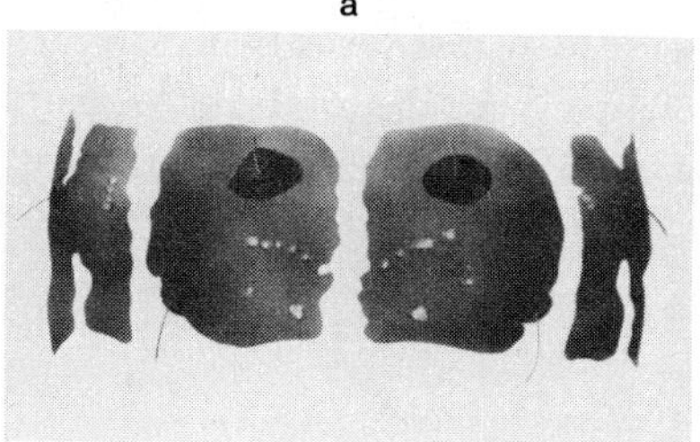

b

Figure 26.419

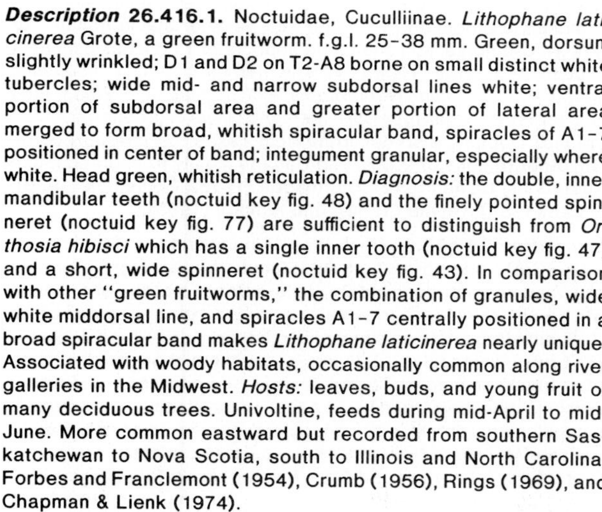

***Description* 26.416.1.** Noctuidae, Cuculliinae. *Lithophane laticinerea* Grote, a green fruitworm. f.g.l. 25–38 mm. Green, dorsum slightly wrinkled; D1 and D2 on T2-A8 borne on small distinct white tubercles; wide mid- and narrow subdorsal lines white; ventral portion of subdorsal area and greater portion of lateral area merged to form broad, whitish spiracular band, spiracles of A1-7 positioned in center of band; integument granular, especially where white. Head green, whitish reticulation. *Diagnosis:* the double, inner mandibular teeth (noctuid key fig. 48) and the finely pointed spinneret (noctuid key fig. 77) are sufficient to distinguish from *Orthosia hibisci* which has a single inner tooth (noctuid key fig. 47) and a short, wide spinneret (noctuid key fig. 43). In comparison with other "green fruitworms," the combination of granules, wide white middorsal line, and spiracles A1-7 centrally positioned in a broad spiracular band makes *Lithophane laticinerea* nearly unique. Associated with woody habitats, occasionally common along river galleries in the Midwest. *Hosts:* leaves, buds, and young fruit of many deciduous trees. Univoltine, feeds during mid-April to mid-June. More common eastward but recorded from southern Saskatchewan to Nova Scotia, south to Illinois and North Carolina. Forbes and Franclemont (1954), Crumb (1956), Rings (1969), and Chapman & Lienk (1974).

Figure 26.417. Noctuidae, Amphipyrinae. *Crymodes devastator* (Brace), glassy cutworm. f.g.l. 35–40 mm. Translucent, may be pale green; shield dark brown, glossy; anal shield brownish to brownish black; tubercles small, may be slightly pigmented. Head reddish brown, darker coronal stripes and reticulation faintly visible. *Diagnosis:* spatulate setae on first segment of maxillary palpus (noctuid key fig. 81) separate *C. devastator* from other cutworms; living specimens may be recognized by the translucent body and contrasting reddish brown head. Found in subterranean burrows in grassy habitats. *Hosts:* roots of grasses, including corn, wheat, etc.; also will feed on other herbs planted in recent grassy areas; once reported damaging seedling peach. Univoltine, larva overwinters, completes development in spring. Transcontinental across southern Canada and contiguous U.S. Riley (1885), Forbes (1904), Crumb (1956), and Rings et al. (1978). Photo courtesy of R. W. Rings and Glenn Berkey, O.A.R.D.C.

***Description* 26.418.1.** Noctuidae, Amphipyrinae. *Luperina stipata* (Morrison) (no figure). f.g.l. 23–31 mm. Body pale yellow alternating with pale reddish brown lines from middorsal line to lateral area; lateral and ventral areas pale yellow only; T1 and anal shields yellow. Head yellow. *Diagnosis:* posterior margin of anal shield crenulate in dorsal view. Late instar larva forms shallow subterranean tunnel. *Hosts:* stem and root borer, prefers *Spartina* (slough grass) but also reported from corn, various other grasses, and sedge. Univoltine, present from April-July. Northeastern U.S., adjacent provinces of Canada, west to Iowa. Decker (1930) and Crumb (1956).

Figure 26.419a,b. Noctuidae, Amphipyrinae. *Hydraecia immanis* Guenée, hop vine borer. a. larva; b. A3 dorsum. f.g.l. 29–35 mm. Sordid white with violet transverse bands (fading close to prepupal stage) on T2-A8 (fig. 26.419a), bands interrupted by thin, white mid- and subdorsal lines (fig. 26.419b), tubercles dark brown to black; T1 shield glossy, pale brownish orange, anterior margin black. Head orange to reddish brown. *Diagnosis:* interrupted, transverse dorsal bands on T2-A8 (fig. 26.419b), and acutely pointed spinneret (noctuid key fig. 41), distinguishes *H. immanis* from *H. micacea*. In areas of coarse grasses. *Hosts:* stem borer of corn, various unspecified grasses, *Humulus* (hop), and *Silphium*. Univoltine, found from late spring to mid-summer. Reportedly from New Brunswick to Manitoba, south to Virginia and Utah; larval records from eastern and midwestern states show it no farther south than New Jersey and central Illinois. Hawley (1918) and Godfrey (1981). Photo by G. L. Godfrey.

Figure 26.420. Noctuidae, Amphipyrinae. *Hydraecia micacea* (Esper), potato stem borer. A3 dorsum. f.g.l. 25–31 mm. Sordid white with reddish, continuous, transverse dorsal bands (fig. 26.420), tubercles dark brown to black; T1 shield, glossy, pale brownish with black anterior margin. Head reddish brown. *Diagnosis:* continuous, transverse dorsal bands on T2-A8 (fig. 26.420) and obtusely rounded tip of spinneret (noctuid key fig. 62) sets *H. micacea* apart from *H. immanis*. *Hosts:* polyphagous, stem borer. Univoltine, biology similar to that of *H. immanis*. Palearctic, introduced into the maritime provinces of Canada around turn of the century; now in Ontario, western New York and Wisconsin. Gibson (1909), Brittain (1915, 1918), Beck (1960), Muka (1976), and Godfrey (1981).

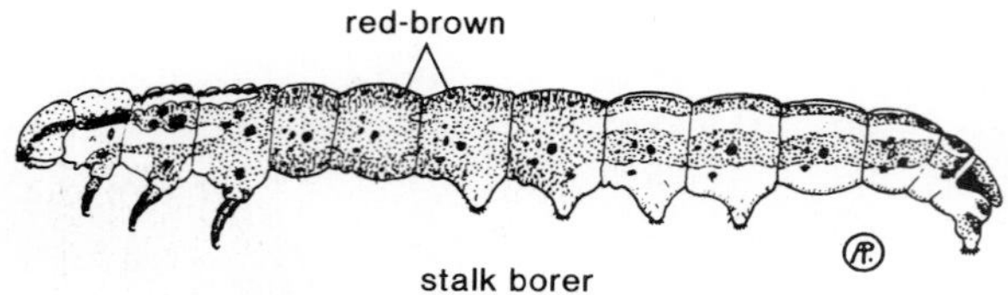

Figure 26.422

Description 26.421.1. Noctuidae, Amphipyrinae. *Papaipema leucostigma* (Harris), columbine borer (no figure). f.g.l. 33-40 mm. Pale purple to salmon (apparently fading prior to pupation) with pale mid- and subdorsal lines; subdorsal lines not present on A1-3; middorsal line, thin, continuous; T1 shield yellow with dark brown lateral margins; abdominal tubercles dark brown. Head yellow with faint, dark brown lateral stripe from stemmata to T1 shield. *Diagnosis:* seta L1 on A7 about at same height as spiracle; middorsal line quite thin. *Hosts:* stem and root borer of wild and cultivated species of *Aquilegia* (columbine). Univoltine. Southern Canada, south to New York, west to Colorado. Bird (1898), Dyar (1899c), Griswold (1934), Matthewman (1937), Knutson (1944), and Forbes (1954).

Description 26.421.2. Noctuidae, Amphipyrinae. *Papaipema cataphracta* (Grote), burdock borer (no figure). f.g.l. 32-43 mm. Pale violet with wide, white mid- and subdorsal lines, the latter continuous on T2-A9; T1 shield yellow with wide, dark brown to black lateral margins; abdominal tubercles large, dark brown to black; D1 tubercles touching middorsal line on A1 and A2. Head yellow with black lateral stripe from stemmata to T1 shield. *Diagnosis:* the combined characters of subdorsal lines on A1-3, D1 tubercles on A1 touching middorsal line, and unmarked yellow head may be used to distinguish *P. cataphracta* from other *Papaipema* species. In pastures and meadows with coarse herbs and in gardens. *Hosts:* a stem borer, general, many species of large-stemmed herbs, e.g., burdock and thistle, and some woody plants, including gooseberry and currant and rose, new growth. Ontario and Quebec, south to New Jersey, west to Minnesota and northern Illinois. Bird (1898), Forbes (1954), and Crumb (1956).

Figure 26.422. Noctuidae, Amphipyrinae. *Papaipema nebris* (Guenée), stalk borer. f.g.l. 30± mm. Purple (fading prior to pupation) with wide, white mid- and subdorsal lines; subdorsal lines absent on A1-3; middorsal line continuous from T2-A9; T1 shield yellow with wide, dark brown or black lateral margins; abdominal tubercles dark brown to black, variable in size. Head orange brown with black lateral stripe from stemmata to T1 shield. *Diagnosis:* no subdorsal lines on A1-3; seta L2 on A7 set caudoventrad of spiracle; middorsal line on A5-8 ca. one-third as wide as dorsal area width. Commonly found in cultivated row crops and gardens. *Hosts:* a stem or stalk borer, polyphagous, recorded from about 44 plant families and nearly 200 species, quite common at times in corn. Univoltine. East of Rocky Mountains, from southern Canada to Texas and S. Carolina. Decker (1931), Knutson (1944). (fig. from Peterson 1948).

Description 26.423.1. Noctuidae, Amphipyrinae. *Callopistria floridensis* (Guenée), Florida fern caterpillar. f.g.l. 24-28 mm. Occurs in pale, intermediate, and dark forms. Pale form: pale green; thin black line traversing anterior margin of T1 shield, same line extending laterocaudad and joining with black, ventral half of lateral area at L1 and L2 on T1; dark portion of lateral area extends to T2, absent or present as separate dashes on T3-A8. Intermediate form: pale green, dorsal areas of T2-A9 each with black transverse bar including D1 and D2 tubercles; black, ventral half of lateral area more prominent. Dark form: velvety black; 2, pale, sinuous middorsal lines; pale, sinuous, subdorsal line; D1 and D2 tubercles black, encompassed by thin, pale circles; lateral area with dorsal half forming continuous pale stripe from T1-A8, ventral half black, continous. Head brownish, black reticulation (dark forms only) and coronal stripes present. *Diagnosis:* serrated cutting edge on mandible (noctuid key fig. 59) and peculiar black, handle-bar mark across dorsal and subdorsal aspects of T1. Occasional pest of cultivated ferns. *Hosts:* various ferns, including *Adiantum, Blechnum, Cyrtomium, Nephrolepis, Polypodium,* and *Pteris;* also reported from *Asparagus sprengeri* and *Smilax* (greenbrier). Multivoltine. Tropical, also occurring from Texas to Florida, records from other regions of U.S.A. and Canada mainly from greenhouse infestations. Davis (1912), Chittenden (1913), Comstock (1939), Forbes (1954), Crumb (1956), and Kimball (1965).

Description 26.423.2. Noctuidae, Amphipyrinae. *Macronoctua onusta* Grote, iris borer. f.g.l. 38-50 mm. Whitish pink; T1 shield brown, sometimes dark brown along anterior margin; anal shield pale brown; spiracles narrow, elongate; tubercles of L1 and L2 on A1-8 quite large, brown; small pavement granules. Head dark reddish brown, immaculate. *Diagnosis:* combination of narrow, elongate, black rimmed spiracles (noctuid key fig. 58), pavement granules, absence of transverse bands or longitudinal body stripes, unmodified anal shield, and immaculate, dark, reddish brown head separate *M. onusta* from other borers. *Hosts:* rhizome and crown borer of *Iris;* also reported from *Belamcanada* (blackberry lily), various other lilies, *Gladiolus,* and corn. Univoltine, mature larva from late July to September. Northeastern U.S.A. and adjacent portions of Canada, south to Washington, D.C., west to Iowa and Minnesota. Felt (1912), Griswold (1934), Knutson (1944), Forbes (1954), and Crumb (1956).

Description 26.423.3. Noctuidae, Amphipyrinae. *Proxenus mindara* (Barnes & McDunnough), roughskinned cutworm (no figure). f.g.l. 25± mm. Tan to brown, dorsal area with thin, white middorsal line, broken, intensified on posterior halves of segments; lateral area white to pale pink with black dorsal margin; small, heterogenous, convex granules. Head pale brown; coronal stripes and reticulation black. *Diagnosis:* uniordinal crochets distinguish *P. mindara* from transcontinental *P. miranda* (Grote) (see Dyar (1899b) and Crumb (1956)) which has biordinal crochets; lacks dorsal diamonds and stiff setae of *Lacinipolia renigera*. *Hosts:* cantaloupe, strawberry, sugar beet, and sweetpotato. California Okumura (1960).

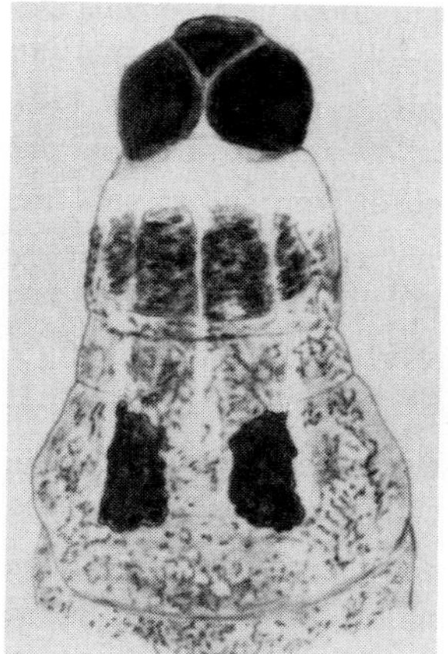

Figure 26.424

a

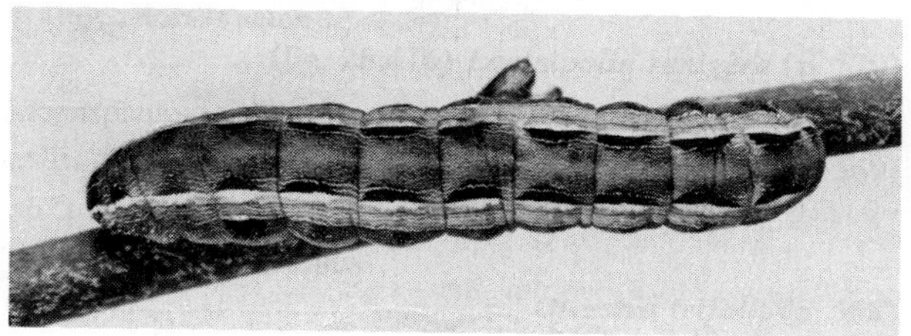

b

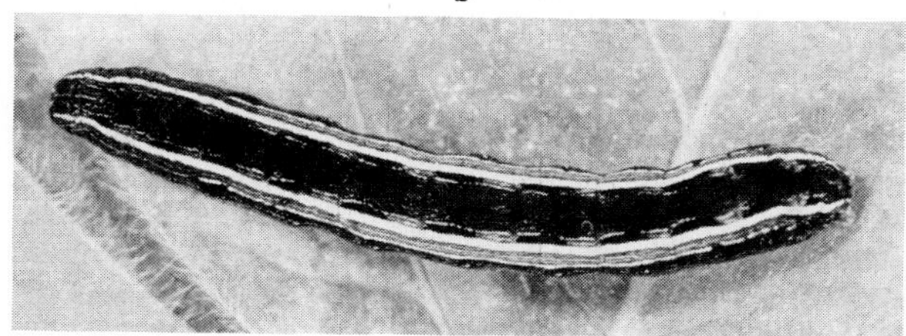

c

Figure 26.425

Figure 26.424. Noctuidae, Amphipyrinae. *Spodoptera dolichos* (Fabricius), large cotton cutworm. Head and dorsa of T1-3, adapted from Levy and Habeck (1976). Also see noctuid key fig. 44, spinneret (dorsal view), adapted from Crumb (1929). f.g.l. 35-43 mm. Pale to dark gray; dorsal area with segmental series of paired velvety black patches bordered by whitish yellow subdorsal lines; black patches on T2 semicircular, about as prominent as patches on A8; middorsal line orange, weakly defined on T1-3, more indistinct elsewhere; subdorsal area grayish with thin indistinct yellow median line, with dark spot on A1; lateral area orangish; ventral area gray brown with white flecks. Head brown, dark brown reticulation masked by infuscation; adfrontal areas white. *Diagnosis: S. dolichos* is recognizable by the subequally developed black patches on T2 and A8. *Hosts:* a general feeder on cultivated and weedy herbs. Multivoltine. Gulf Coast states and sporadically in adjoining states. Chittenden (1901), Crumb (1929, 1956), Levy and Habeck (1976), Todd and Poole (1980), and Oliver and Chapin (1981).

***Description* 26.424.1.** Noctuidae, Amphipyrinae. *Spodoptera praefica* (Grote), western yellowstriped armyworm. f.g.l. 35-46 mm. Whitish, grayish brown, or nearly black; darker colors distributed as fine lines underlaid by white creating mottled effect; dorsal area with or without segmental series of paired black triangles adjacent to subdorsal lines on T2-A9; middorsal line whitish, thin, distinct on T1-3; subdorsal lines yellow, distinct on all segments; dorsal half of subdorsal area paler than ventral half; ventral half of subdorsal area on A1 with black splotch including spiracle; lateral area white, tinged with orange; ventral area translucent grayish brown with white flecks. Head brownish or brownish orange, reticulation dark brown, adfrontal areas pale yellow, forming inverted "V" (noctuid key fig. 73). *Diagnosis: S. praefica* may be recognized by the yellow, subdorsal lines of the body, and the contrasting pale yellow, inverted "V" of the head. Van den Bosch and Smith (1955) observed that the head pattern of *S. praefica* is more heavily reticulated than that of *S. ornithogalli.* An occasional pest of cultivated row crops. *Hosts:* a general feeder, frequently associated with foliage of alfalfa. Multivoltine. Throughout California (more common north of Techachapi Mountains); also recorded (probably adults) from Oregon, Montana, and Great Basin. Van den Bosch and Smith (1955), Crumb (1956), and Todd and Poole (1980).

Figure 26.425a-c. Noctuidae, Amphipyrinae. *Spodoptera ornithogalli* (Guenée), yellowstriped armyworm. a, b. pale form; c. dark form. f.g.l. 30-37 mm. Coloration and maculation very similar to *S. praefica.* Head brownish with dark brown reticulation obscured by infuscation; adfrontal area pale yellow, forming inverted "V" (noctuid key fig. 74). *Diagnosis: S. ornithogalli* has distinct yellow, subdorsal lines, thin white, longitudinal lines passing through the dark dorsal triangles, and a pale yellow or whitish inverted "V" bordering the frons. May be confused with *S. praefica* where the distributions overlap in southern California. *S. ornithogalli* cautiously may be recognized in this area by the less extensive and more obscured reticulation on the head. Also, *S. ornithogalli* in California does not occur north of the Techachapi Mountains. A common pest of field and garden crops in the southern states. *Hosts:* polyphagous, primarily herbs. Multivoltine. A tropical and subtropical species that moves northward as conditions permit. Throughout eastern half of North America north to southern Canada; also Arizona, New Mexico, and extreme southern California. Crumb (1929, 1956), Rings and Musick (1976), Rings (1977), and Todd and Poole (1980). Figs. 26.425a and b courtesy of R. W. Rings and Glenn Berkey, O.A.R.D.C.; fig. 26.425c by G. L. Godfrey.

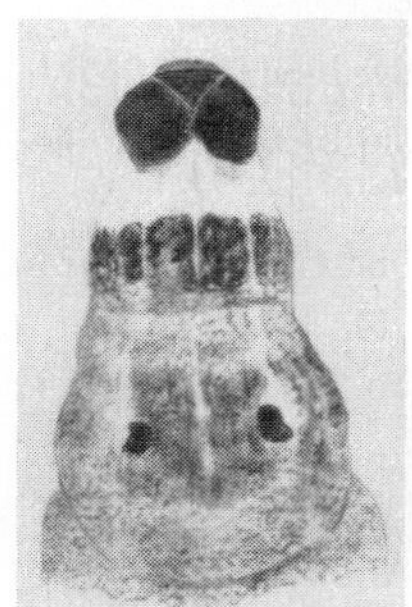

Figure 26.426

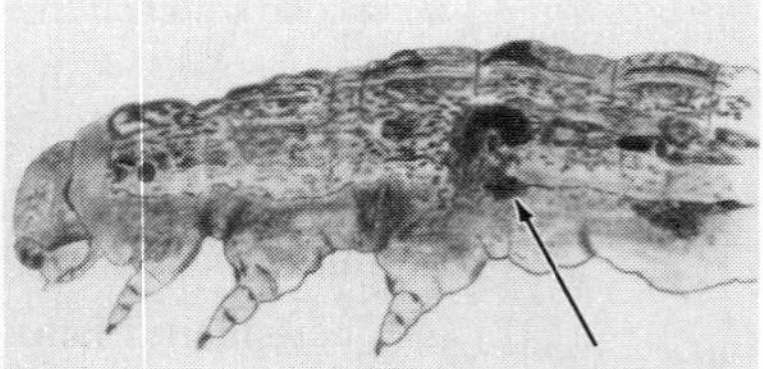

Figure 26.427

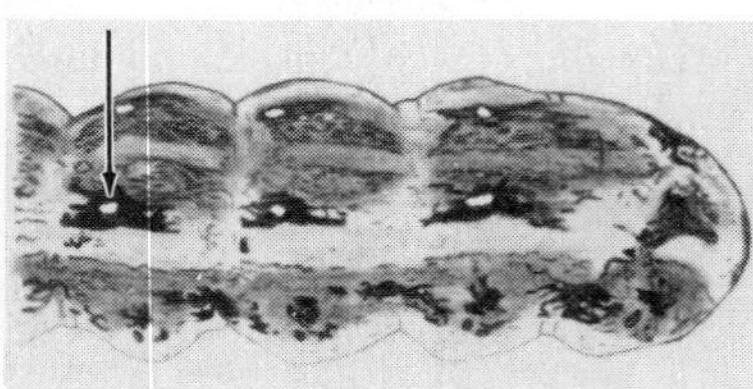

Figure 26.428

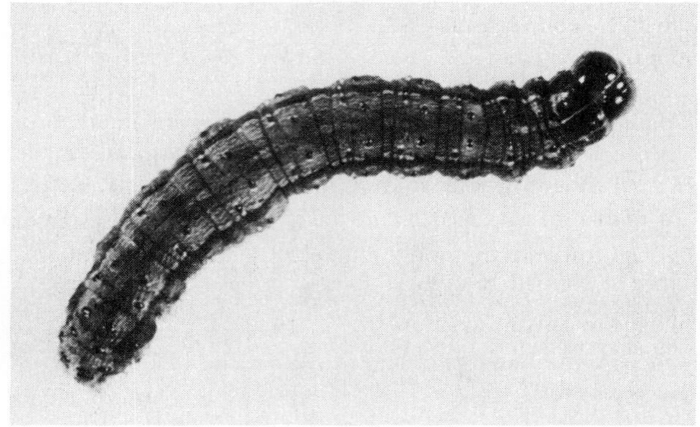

Figure 26.429

Figure 26.426. Noctuidae, Amphipyrinae. *Spodoptera latifascia* (Walker). Head and dorsa of T1-3, adapted from Levy and Habeck (1976). f.g.l. 48± mm. Grayish, blue green to dark reddish brown; dorsal area with segmental series of paired black patches bordering subdorsal lines, patches increase in size caudad from T2-A8, may be absent on T3 and A1, patch on T2 much smaller than on A8; middorsal line whitish pink, indistinct; subdorsal area divided longitudinally by yellow line, dorsal half streaked with thin white lines, ventral half of subdorsal area flecked with white; blackish spots in subdorsal area above spiracles on A1-8, largest on A1; lateral area margined dorsally by yellow line. Head yellowish brown, reticulation and coronal stripes brown, reticulation slightly infuscated; adfrontal area whitish yellow. *Diagnosis: S. latifascia* is distinguished by having the black, dorsal patches on T2 only half the size of the patches on A8 (on *S. dolichos* the 2 pairs of patches are subequal) and by lacking the distinct, contrasting yellow, subdorsal lines that are noticeable on *S. ornithogalli. Hosts:* a general feeder, recorded from onion, grapefruit, soybean, tomato, alfalfa, and tobacco. Gulf Coast states and south. Mason (1922), Wolcott (1936), Levy and Habeck (1976), and Todd and Poole (1980).

Figure 26.427. Noctuidae, Amphipyrinae. *Spodoptera eridania* (Cramer), southern armyworm. Head T1-3, A1, A2, adapted from Levy and Habeck (1976). f.g.l. 25-38 mm. Grayish, may have olive or pink hue; dorsal area with or without segmental series of paired black triangles adjacent to subdorsal lines on T2-A9; mid- and subdorsal lines white, tinged with pink or orange, interrupted on A1 by dark patch (fig. 26.427); ventral area slightly pink or orange, flecked with white. Head yellow with brown coronal stripes and diffuse reticulation. *Diagnosis:* interruption of the distinct lateral area by a large dark patch on A1 separates *S. eridania* from other *Spodoptera* species. Economically important in the South and the tropics. *Hosts:* the early literature associates the larva with *Phytolacca americana* L. (pokeweed) and *Amaranthus spinosus* L. (spiny amaranth); polyphagous on numerous herbs and cultivated crops. Multivoltine. A tropical species reported as far north as Virginia to Texas and New Mexico. Chittenden and Russell (1909), Crumb (1929, 1956), and Todd and Poole (1980).

Figure 26.428. Noctuidae, Amphipyrinae. *Spodoptera sunia* (Guenée). Posterior abdominal segments (dorsolateral view), adapted from Levy and Habeck (1976). f.g.l. 38± mm. Color variable, gray, yellow brown, or blackish; dorsal and subdorsal area appearing mottled or speckled; dorsal area with segmental series of paired blackish triangles on T2, T3 and A1-8; dorsal triangles on A1-7 each with single white spot; mid- and subdorsal lines wide on A1-8; lateral area brownish with yellow dorsal margin, white specks. Head yellowish brown, reticulation and coronal stripes blackish. *Diagnosis:* the white spot in the black dorsal triangles and the wide middorsal line may be used to identify it. *Hosts:* destructive to Swiss chard and potato in Puerto Rico; also reported from celery, asparagus, cotton, alfalfa, tobacco, and pea. Southern Florida and south into the tropics. Wolcott (1936) and Levy and Habeck (1976).

Figure 26.429. Noctuidae, Amphipyrinae. *Spodoptera frugiperda* (J. E. Smith), fall armyworm. f.g.l. 30-38 mm. Color variable, yellowish brown to dark brown; dorsal area brown with faint white mottling, sharply margined by whitish subdorsal lines; middorsal line continuous, yellowish brown, subdorsal and middorsal lines distinct on dark brown T1 shield; D1 and D2 tubercles usually dark brown and prominent; dorsal stripe of subdorsal area uniformly dark brown, ventral stripe paler, mottled; lateral area whitish, continuous; ventral area greenish yellow; small pavement granules. Head pale brown with dark brown coronal stripes and reticulation. *Diagnosis: S. frugiperda* frequently is misidentified as *Heliothis zea* or *Pseudaletia unipuncta. Heliothis zea* has spinules on the integument of the body whereas *Spodoptera frugiperda* has small pavement granules. The striping along the subdorsal area may be used to distinguish between *S. frugiperda* and *Pseudaletia unipuncta.* On *Spodoptera frugiperda* the dorsal stripe of the subdorsal area is darker than the ventral stripe. The opposite is true for *Pseudaletia unipuncta.* If doubt remains compare the hypopharyngeal complex of *Spodoptera frugiperda* (noctuid key fig. 70) with that of *Pseudaletia unipuncta* (noctuid key fig. 38). A common pest of cultivated crops. *Hosts:* polyphagous, but shows preference for Poaceae, including corn, *Sorghum,* etc.; feeds in the whorls of corn. Multivoltine in warmer climates. A tropical species that migrates northward as conditions permit. Throughout much of U.S., common in Great Plains. Chittenden (1901), Franklin (1928), Luginbill (1928), Whelan (1935), Walkden (1950), Crumb (1956), and Rings and Musick (1976). Photo by G. L. Godfrey.

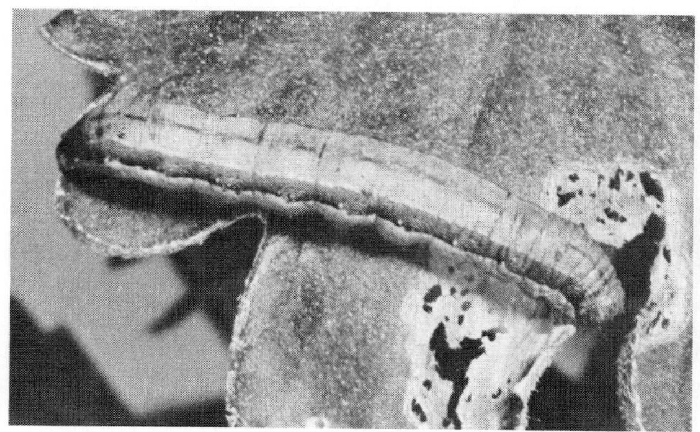

Figure 26.430

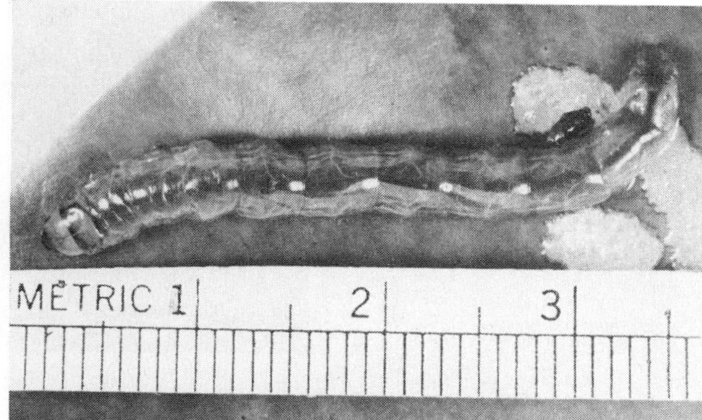

Figure 26.431

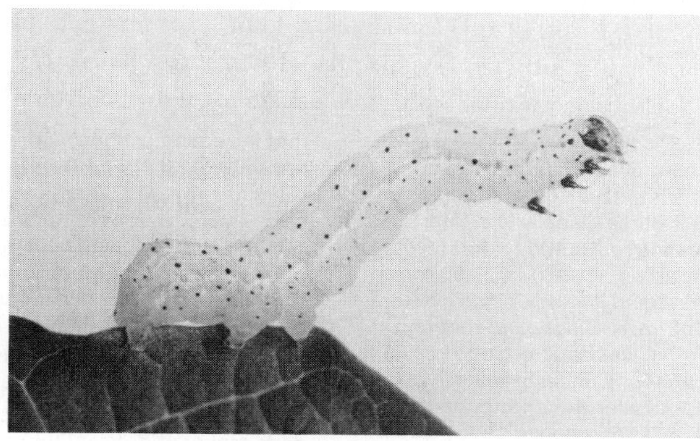

Figure 26.432

Figure 26.430. Noctuidae, Amphipyrinae. *Spodoptera exigua* (Hübner), beet armyworm. f.g.l. 25–30 mm. Highly variable; greenish, brownish, and blackish green; darker ground color on dorsal and subdorsal areas distributed as fine crenulated lines and dashes, underlaid by white; dorsal area paler than subdorsal area; well-marked specimens with blackish patch obscuring SD1 on subdorsal area of T2; mid- and subdorsal lines white, middorsal line thin, obscured by dorsal vessel; lateral area white; ventral area translucent greenish with white flecks. Head with brown to blackish brown reticulation and coronal stripes, sometimes nearly solid blackish brown. *Diagnosis:* the absence of granules and the overall uniformly colored, dorsal area contrasting with the darker subdorsal area (fig. 26.430) is distinctive of *S. exigua*. Common at times in cultivated crops. *Hosts:* polyphagous, principally herbs; infrequently reported from woody plants. Capacity for multiple generations. Introduced from Old World, occurring in southern and southwestern U.S.A., in northern states and Canada under favorable conditions. Dyar (1894b), Comstock and Dammers (1942), Walkden (1950), Crumb (1956), Brown and Dewhurst (1975), and Todd and Poole (1980). Photo by W. D. Zehr, I.N.H.S.

Figure 26.431. Noctuidae, Amphipyrinae. *Bellura gortynoides* Walker. f.g.l. 40–42± mm. Translucent olive brown dorsally and yellowish gray ventrally, trachae visible (fig. 26.431); T1 shield brown. Head orange. *Diagnosis:* spiracles on A8 positioned dorsocaudad (noctuid key fig. 51). Aquatic, shallow lakes and ponds. *Host:* mature larvae in stems of *Nuphar* (yellow pond lily). Bivoltine. Southeastern Canada, south to Florida, west to Illinois. Robertson-Miller (1923), Forbes (1954), Levine and Chandler (1976), and Heitzman and Habeck (1981). Photo by Glenn Berkey, O.A.R.D.C., courtesy of Eli Levine.

***Description* 26.431.1.** Noctuidae, Amphipyrinae. *Achatodes zeae* (Harris), elder shoot borer. f.g.l. 38–45 mm. White with black tubercles; T1 and anal shields black. Head black. *Diagnosis:* anal shield rugose, bearing 6 lobes on posterior margin (noctuid key fig. 56). *Hosts:* principally a borer in stems of *Sambucus* (elder), but also reported from corn, *Dahlia,* and grasses (unspecified, probably secondary hosts). Univoltine, mid-March to early July. General, Minnesota to Louisiana and east. Silver (1933), Knutson (1944), and Crumb (1956).

Figure 26.432. Noctuidae, Plusininae. *Pseudoplusia includens* (Walker), soybean looper. f.g.l. 30± mm. Green; abdominal tubercles, especially D1 and D2, black in well-marked specimens; paired middorsal and paired subdorsal lines white and thin; upper edge of lateral area white; prolegs fully developed on A5, A6, and A10 with biordinal crochets; vestigial prolegs on A3 and A4 lacking crochets. Head green with black setal bases, especially P1, P2, A3, and L; darker specimens with black lateral stripe extending from posterior margin of head to stemma 6. *Diagnosis:* separable from other common plusiines by absence of SV tubercle on A1, the presence of vestigial prolegs on A3 and A4, and the presence of subterminal processes on the second and third inner mandibular ridges (noctuid key fig. 12). Associated with herbs in open fields, sometimes becoming a problem in greenhouses and field crops. *Hosts:* a polyphagous defoliator of many herbs from at least 28 families, but apparently preferring legumes. As many as 4 generations in Gulf Coast states, less farther north. Adults principally along the Gulf and East Coast, westward to Quebec, Illinois, and Arkansas; recorded from southern California, and southward into S. America. Larvae abundant in warmer latitudes. Crumb (1956), Burleigh (1972), Eichlin and Cunningham (1969, 1978), Eichlin (1975a, 1975b), and Herzog (1980). Photo by G. L. Godfrey.

Figure 26.434

Description **26.433.1.** Noctuidae Plusiinae. *Trichoplusia ni* (Hübner), cabbage looper. f.g.l. 30 mm. Green; abdominal tubercles D1, D2, and SD1 pale green or white; SD1 tubercle may be black; paired middorsal and paired subdorsal lines white, thin; lateral area white although white sometimes restricted to dorsal and ventral borders; prolegs fully developed on A5, A6, and A10; crochets biordinal; vestigial prolegs on A3 and A4 lacking crochets. Head immaculate, green to greenish brown. *Diagnosis:* recognizable by absence of SV2 tubercle on A1, vestigial prolegs on A3 and A4 (noctuid key fig. 10), and absence of subterminal processes on second and third inner mandibular ridges (noctuid key fig. 13). A common garden, field crop, and greenhouse pest. *Hosts:* a very general feeder on herbs (a few woody species) from 26 plant families but showing a preference for cabbage, etc. and related crucifers. Multivoltine in subtropics and tropics. Cosmopolitan; north to southern Canada. Crumb (1956), Eichlin and Cunningham (1969, 1978), and Eichlin (1975a, 1975b).

Description **26.433.2.** Noctuidae, Plusiinae. *Autoplusia egena* (Guenée), bean leafskeletonizer (no figure). f.g.l. 30± mm. Green; abdominal tubercles D1 and D2 green or black, SD1 black; pale, paired middorsal lines; pale, sinuate subdorsal line; lateral area white along dorsal border; prolegs fully developed on A5, A6, and A10 with biordinal crochets; no vestigial prolegs on A3 or A4. Head green with black setal bases. *Diagnosis:* the absence of vestigial prolegs on A3 and A4 distinguishes this species from *Pseudoplusia includens* which also may have black setal bases on the head; differs from *Autoplusia californica* in that SV1 and SV2 tubercles on A2 are not confluent. An occasional pest of field crops. *Hosts:* foliage and green pods of beans and soybean; also reported from *Agapanthus,* hollyhock, celery, cole crops, *Chrysanthemum,* carrot, larkspur, mallow, mint, spearmint, *Senecio* (groundsel), *Symphytum* (comfrey), marigold, and *Verbena.* Tropical, north to California, New Mexico, and Florida. Lange (1945), Crumb (1956), Eichlin (1975a, 1975b), and Eichlin and Cunningham (1978).

Figure 26.434. Noctuidae, Plusiinae. *Autographa precationis* (Guenée).f.g.l. 30± mm. Green; tubercles D1 and D2 white, SD1 black on well-marked specimens; paired middorsal and paired subdorsal lines white, thin; ventral border or subdorsal area sometimes blackish green; lateral area white, especially along dorsal border; prolegs fully developed with biordinal crochets on A5, A6, and A10: no vestigial prolegs on A3 or A4. Head green with broad, black lateral stripe encompassing all stemmata. *Diagnosis:* the broad, black lateral stripe on the head in combination with the subterminal processes on the second and third inner mandibular ridges reliably distinguishes *A. precationis* from *A. californica.* The black lateral head stripe, alone, separates it from *Autoplusia egena.* Little is known about the field biology except that it frequently shows up in various field crops but rarely causes appreciable damage. *Hosts:* a general defoliator of herbs. From a limited range of plants in laboratory tests, it prefered Asteraceae over Fabaceae, Solanaceae, and Malvaceae (Khalsa et al., 1979). Probably bivoltine throughout its range. Common in the East, especially southeastern Canada and northeastern U.S.A., west to Nebraska, south to North Carolina. Forbes (1954), Crumb (1956), Eichlin and Cunningham (1978), and Khalsa et al. (1979). Photo by G. L. Godfrey.

Description **26.434.1.** Noctuidae, Plusiinae. *Autographa californica* (Speyer), alfalfa looper (no figure). f.g.l. 30–35 mm. Pale to dark green; abdominal tubercles white, each accented by darker basal circle, especially SD1s; paired middorsal and paired subdorsal lines white, thin; ventral border of subdorsal area sometimes dark green, contrasting with white dorsal border of lateral area; prolegs fully developed on A5, A6, and A10 with biordinal crochets; no vestigial prolegs on A3 and A4. Head greenish, variably marked; lateral stripe present or absent, dominating head in some specimens; black freckles present or absent. *Diagnosis:* a western species separable from the economically important plusiines with overlapping distributions as follows: *A. californica,* unlike *Pseudoplusia includens* and *Trichoplusia ni,* has no vestigial prolegs on A3 and A4; the lateral head stripe and confluent SV1 and SV2 tubercles on A2 distinguish it from *Autoplusia egena; Anagrapha falcifera* possesses 3 SV tubercles on A1 while *Autographa californica* has only 2 SV tubercles on A1. Occasional in cultivated fields and greenhouses. *Hosts:* polyphagous, mainly a defoliator of herbs but reported from a few woody plants. Multivoltine. Western N. America to southern British Columbia and Alberta; also from Kansas and Nebraska. Crumb (1956), Eichlin (1975a, 1975b), and Eichlin and Cunningham (1978).

Description **26.435.1.** Noctuidae, Plusiinae. *Anagrapha falcifera* (Kirby), celery looper. f.g.l. ca. 30 mm. Green; dorsal abdominal tubercles white or green, ventral tubercles encircled by dusky spinules; paired middorsal and subdorsal lines white and thin; ventral border of subdorsal area dusky green; lateral area white, especially along dorsal border; prolegs fully developed on A5, A6, and A10 with biordinal crochets (noctuid key fig. 17); no vestigial prolegs on A3 and A4 (noctuid key fig. 11). Head green, immaculate. *Diagnosis:* unlike other plusiines treated here, *A. falcifera* possesses 3 SV tubercles on A1. Common and may cause economic damage to field crops. *Hosts:* polyphagous on herbs, including: hollyhock, celery, burdock, beet, cole crops, thistle, carrot, lettuce, beans, plantain, dandelion, stinging nettle, and corn; reported woody plants are cranberry and *Viburnum.* Transcontinental, to southern Canada. Forbes (1954), Crumb (1956), Eichlin (1975a, 1975b), and Eichlin and Cunningham (1978).

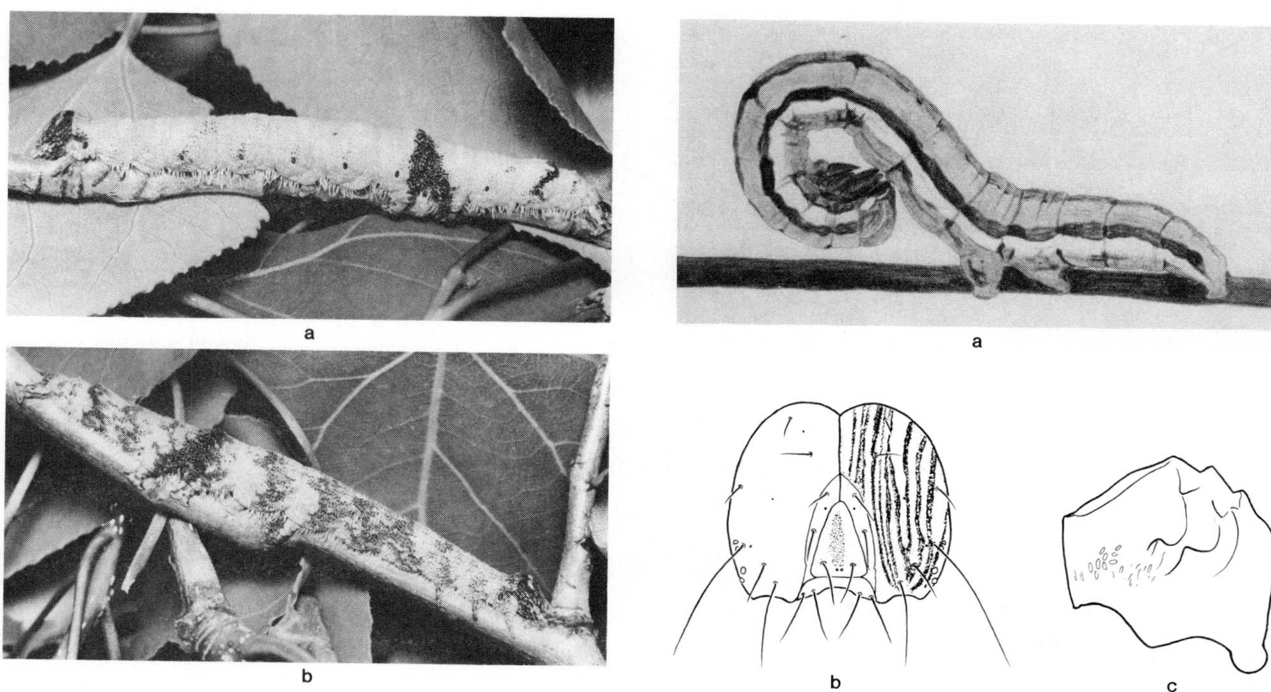

Figure 26.436

Figure 26.437

***Description* 26.435.2.** Noctuidae, Acontiinae. *Lithacodia carneola* (Guenée). f.g.l. 12–15 mm. Green; mid- and subdorsal lines whitish yellow; subdorsal lines 2X width of middorsal line; D1 & D2 setae on all abdominal segments inserted in distinct whitish yellow spots; lateral area whitish yellow; T1 shield marked medially and laterally with thin, dark brown lines. Head pale yellow to yellowish brown with brown freckles on epicranial and lateral areas (noctuid key fig. 67). *Diagnosis:* the apparently restrictive association with *Polygonum* and *Rumex,* the whitish yellow basal spots of setae D1 and D2, and the relatively minute labial palpus (noctuid key fig. 69) distinguish *Lithacodia carneola* from other small noctuids. *Hosts: Rumex* (dock) and *Polygonum* (smartweed). Multivoltine. Southern Canada, west to Saskatchewan, south to Colorado and Georgia. Forbes (1954) and Godfrey (1971).

Figure 26.436a,b. Noctuidae, Catocalinae. *Catocala relicta* Walker. a. pale form; b. dark form. Also see noctuid key fig. 91, A5 showing subventral fringe. f.g.l. 52–62 mm. Stout, dorsum of A5 slightly humped; white subventral fringes present. Whitish gray with slight orange cast; covered with profusion of diffuse, pale brown freckles (dark form with predominance of black freckles on all segments); all forms with black saddle on A5 extending from hump to proleg and expanding ventrocaudad to include spiracle and anterior portion of proleg on A6; A8 with black saddle extending obliquely from middorsum caudad of D2's to spiracle; hints of saddles on other abdominal segments (saddles less conspicuous on dark forms); large, black midventral spots on T2, T3 and A3-A7. Head with distinct dorsal, orange crescent on each half; reticulation black, forming pattern of vertical stripes. *Diagnosis:* the association with salicaceous hosts, orange crescents on dorsum of head, and prominent black saddle between A5 and A6 help identify it. *Hosts: Populus* (especially quaking aspen) and willow. Univoltine, feeds from May-early August. Across southern Canada and northern portions of U.S., as far south as Kentucky and extending into Arizona. Barnes and McDunnough (1918), Forbes (1954), Crumb (1956), Prentice (1962), and Sargent (1976). Photos courtesy of J. E. Rawlins, Carnegie Museum of Natural History.

Figure 26.437a-c. Noctuidae, Catocalinae. *Caenurgina erechtea* (Cramer), forage looper (and *C. crassiuscula* (Haworth), clover looper). a. larva; b. head showing maculation on left side; c. left mandible (oral view). Crumb (1956) and this author do not know of any larval characters that can be used to separate these 2 species. Therefore, the following description of *C. erechtea* also generally applies to *C. crassiuscula:* f.g.l. 35–43 mm. Body and head distinctly longitudinally striped and lined, basically pale yellow brown to gray, alternating with brown and blackish brown; dorsal area with 3 subequally wide stripes, the middorsal pale, the laterals dark; paired intersegmental black spots dorsally between A1 and A2, A2 and A3 visible when larva arches; subdorsal area pale with 3 fine longitudinal lines; dorsal half of lateral area dark brown to black; venter pale with blackish midventral stripe; prolegs only on A5, A6, and A10. Head elongated, striped (fig. 26.437b). *Diagnosis: Caenurgina* species are similar to plusiines by lacking functional prolegs on A3 and A4, but the distinctly striped body and paired inner mandibular teeth (fig. 26.437c) readily distinguish them from the plusiines; they are very similar to species of *Mocis* and *Ptichodis* (not included here), see Crumb (1956) and Ogunwolu and Habeck (1979). *Hosts:* grasses (unspecified), clover (unspecified), soybean, lupine, and alfalfa. Multivoltine. Rather widely distributed. French (1884), Slingerland (1893), Webster (1898), Smith (1924), and Crumb (1956). Fig. 26.437a drawn from photo.

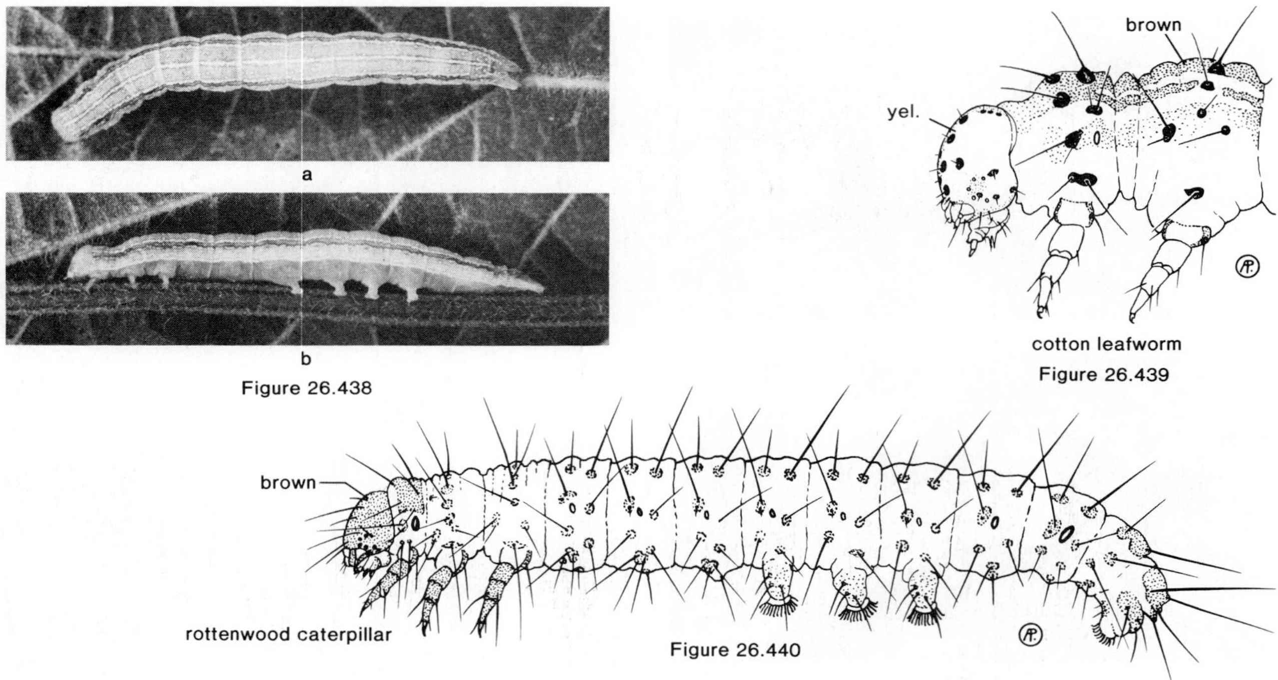

Figure 26.438

Figure 26.439

Figure 26.440

Figure 26.438a,b. Noctuidae, Catocalinae. *Anticarsia gemmatalis* Hübner, velvetbean caterpillar. a. dorsal; b. lateral. f.g.l. 38–48 mm. Slender, varying from yellow green, olive green, to dark green; mid- and subdorsal lines, white, distinct, continuous; D1 and D2 tubercles white, small but size increasing through A9, basal circles of tubercles may be blackish, dorsal stripe of subdorsal area with solid dark margins, ventral stripe with broken white margins; lateral area yellowish with white ventral margin; ventral area concolorous with dorsal area or paler. Head yellow green, setal bases may be small black spots. *Diagnosis:* a slender, smooth skinned, active, legume-associated caterpillar; differing from superficially similar species by the distinct, white, mid- and subdorsal lines and prolegs on A3-A6 and A10 (*Plathypena scabra* lacks prolegs on A3; plusiines have none on A3 and A4 although vestigial pairs may be present on A3 and A4). *Hosts:* principally on herbaceous legumes, especially soybean and *Stizolobium* (velvetbean) in the U.S., but also reported from new growth of black locust; records of nonleguminous hosts are few and probably incidental. Multivoltine. A subtropical and tropical species, larva occasionally as far north as Maryland, Delaware, and Oklahoma, but more common southward; not reported from southern California; abundant in eastern and western Mexico, the Antilles, and most of Central and South America. Watson (1916), Hinds (1930), Crumb (1956), and Ford et al. (1975). Photos by G. L. Godfrey.

Figure 26.439. Noctuidae, Catocalinae. *Alabama argillacea* (Hübner), cotton leafworm. Head, T1, T2. f.g.l. 35–40 mm. Yellow green with varying degrees of black, all forms with prominent white mid- and subdorsal lines; dorsal area of paler forms with distinct black D1 and D2 tubercles encircled by white, black borders along middorsal line; dorsal area of darker form with increasing amount of black, some solid black except for white mid- and subdorsal lines; subdorsal area of darker forms with whitish ventral margin; venter yellow green in pale forms, pale yellow brown in dark forms; an extra seta situated between D2 and L1 on A1-7 (noctuid key fig. 89); crochet-bearing prolegs on A3 and A4 reduced. Head pale orange brown with distinct black setal bases (fig. 26.439). *Diagnosis:* the extra abdominal seta (noctuid key fig. 89), conspicuously spotted head, reduced prolegs on A3 and A4, (noctuid key fig. 89), and smooth skin distinguishes *A. argillacea* from *Heliothis* species; the crochet-bearing prolegs on A3 and A4 separates it from any plusiine that may be found on cotton. *Host:* early instars on foliage of cotton, later instars may also feed on flowering structures and boles. Multivoltine. In cotton growing areas of southern and southwestern U.S. Riley (1870b), Comstock (1879), Hinds (1912), and Crumb (1956). (figs. 26.439, noctuid key fig. 89 from Peterson 1948).

Figure 26.440. Noctuidae, Catocalinae. *Scolecocampa liburna* (Geyer), rottenwood caterpillar. f.g.l. 35–50 mm. Sordid white, partially translucent; T1 and anal shields rugose, brown; T1 shield extending laterad, including SD2; tubercles brown; D2 setae on A1-8 longer than prolegs, conspicuous; prolegs on A4, A5, A6, and A10. Head dark brown to black, somewhat granulated on vertex. *Diagnosis:* the long setae, large spiracles on T1 and A8, and contrastingly dark dorsal tubercles distinguish *S. liburna* from other scavenging noctuid larvae. Inhabits tunnels in moist stumps or fallen logs of hardwood trees in advance stages of decay, feeds either as scavenger or on fungi. New York to Florida and west to Missouri and Texas. Univoltine. Andrews (1879), Edwards and Elliot (1883), Beutenmüller (1902), Crumb (1934, 1956), and Forbes (1954), (fig. from Peterson 1948).

***Description* 26.441.1.** Noctuidae, Herminiinae. *Idia americalis* (Guenée). See noctuid key fig. 86, T2; and fig. 4, A3 and A4 (lateral view). f.g.l. 18± mm. Pale gray with slight pinkish cast, no stripes; T1 shield brownish, including SD2 setal base (as in noctuid key fig. 87); setae on small tubercles, tubercles on T1-T3 brownish, tips of setae blunt; prolegs on A4-A6 and A10; small, heterogenous, slightly stellate granules. Head brown, uniformly covered with small, rounded, convex granules. *Diagnosis:* large sets of characters are needed to recognize species of *Idia*. Therefore, see Crumb (1934 or 1956). Inhabits nests of "*Formica rufa*" according to some authors. Once reared from moldy "bird seed". *Hosts:* probably a scavenger. Transcontinental, northern U.S.A. and contiguous southern Canada; also reported in New Mexico, Texas, and Florida. Riley (1883c, 1892a), Crumb (1934, 1956), Smith (1941), and Forbes (1954). Key figs. 4 and 86 redrawn from Peterson (1948).

BIBLIOGRAPHY*

*Abbot, J., and J. E. Smith. 1797. The natural history of the rarer lepidopterous insects of Georgia, vol. 2. London. pp. 105–214.

Abbott, C. E. 1927. The reaction of *Datana* larvae to sounds. Psyche 34:129–33.

Abrahamson, L., and C. Klass. 1982. Gypsy moth. Cornell Coop. Ext. Inform. Bull. 188. 13 pp.

Aikenhead, P., and C. R. B. Baker. 1964. The larvae of British Hepialidae. Entomologist 97:25–38.

Ainslie, C. N. 1910. The New Mexico range caterpillar. USDA, Bur. Ent. Bull. 85(5):59–96.

Aitken, A. D. 1963. A key to larvae of some species of Phycitinae associated with stored products and some related species. Bull. Ent. Res. 54:175–88.

Allen, D. C. 1973. Fecundity of the saddled prominent, *Heterocampa guttivitta*. Ann. Ent. Soc. Amer. 66:1181–83.

Allen, D. C. 1979. Observations on biology and natural control of the orangehumped mapleworm, *Symmerista leucitys* (Lepidoptera: Notodontidae), in New York. Can. Ent. 111:703–708.

Allen, D. C., and D. G. Grimble. 1970. Identification of the larval instars of *Heterocampa guttivitta* with notes on their feeding behavior. J. Econ. Ent. 63:1201–03.

Allyson, S. 1976. North American larvae of the genus *Loxostege* Hübner. Can. Ent. 108:89–104. (Keys to, and descriptions of, 13 species)

Allyson, S. 1977a. A study of some North American larvae of the genus *Tetralopha* (Lepidoptera: Pyralidae: Epipaschiinae). Can. Ent. 109:329–36. (Keys to, and descriptions of, nine species)

Allyson, S. 1977b. The larvae of two species of Macrothecinae (Lepidoptera: Pyralidae). Can. Ent. 109:839–42.

Allyson, S. 1980. Last instar larva of the gooseberry fruitworm, *Zophodia convolutella* (Lepidoptera: Pyralidae: Phycitinae). Can. Ent. 112:43–45.

Allyson, S. 1981a. Description of the last instar larva of the cabbage webworm, *Hellula rogatalis* (Lepidoptera: Pyralidae), with a key to larvae of North American species of *Hellula*. Can. Ent. 113:361–64.

Allyson, S. 1981b. Last instar larvae of Pyraustini of America North of Mexico (Lepidoptera: Pyralidae). Can. Ent. 113:463–518.

Allyson, S. 1984. Description of last instar larvae of 22 species of North American Spilomelini (Lepidoptera: Pyralidae: Pyraustinae), with a key to species. Can. Ent. 116:1301–34.

Andrews, W. V. 1879. *Scolecocampa liburna*. *In* B. Pickman Mann. Descriptions of some larvae of Lepidoptera, respecting Sphingidae especially, p. 272. Psyche 2:265–72.

Angulo, A. O., and C. Th. Weigert. 1975a. Estados inmaduros de lepidópteros nóctuidos de importancia económica en Chile y claves para su determinación (Lepidoptera: Noctuidae). Soc. Biol. Concepcion, Publ. Espec. 2. 153 pp.

Angulo, A. O., and C. Th. Weigert. 1975b. Noctuidae (Lepidoptera) de interes economico del valle de Ica, Peru: clave para estados inmaduros. Rev. Per. Ent. 18:98–103.

Arbogast, R. T., G. L. LeCato, and R. V. Byrd. 1980. External morphology of some eggs of stored-product moths (Lepidoptera: Pyralidae, Gelechiidae, Tineidae). Int. J. Insect Morphol. and Embryol. 9:165–77.

Ayre, G. L. 1958. Notes on insects found in or near nests of *Formica subnitens* Creighton (Hymenoptera: Formicidae) in British Colombia. Ins. Soc. 5(1):1–7.

Baerg, W. J. 1924. On the life history and the poison apparatus of the white flannel moth, *Lagoa crispata* Packard. Ann. Ent. Soc. Amer. 17:403–12.

Bailey, T. E., and L. T. Kok. 1982. Biology of *Frumenta nundinella* (Lepidoptera: Gelechiidae) on horsenettle in Virginia. Can. Ent. 114:139–44.

*Publications so marked contain significant numbers of color illustrations.

Baker, W. L. 1972. Eastern forest insects. U.S. Dept. Agr. Misc. Publ. 1175. 642 pp.

Balch, R. E. 1932. The fir tussock moth (*Hemerocampa pseudotsugata* McD.). J. Econ. Ent. 25:1143–48.

Balduf, W. V. 1931. Carnivorous moths and butterflies. Trans. Ill. Acad. Sci. 24:156–64.

Barnes, W. J., and A. W. Lindsey. 1921. Pterophoridae of America, north of Mexico. Contrib. Nat. Hist. Lep. N. Amer. 4:281–478. (Appearance of larvae and pupae of some species)

Barnes, W. J., and J. H. McDunnough. 1911. Revision of the Cossidae of North America. Contrib. Nat. Hist. Lep. N. Amer. 1(1):1–35.

Barnes, W. J., and J. H. McDunnough. 1912. On the larval stages of certain Arctian species. Can. Ent. 44:132–63.

Barnes, W. J., and J. H. McDunnough. 1913. The North American species of the liparid genus *Olene*. Contrib. Nat. Hist. Lep. N. Amer. 2(2):45–91.

Barnes, W. J., and J. H. McDunnough. 1918. Illustrations of the North American species of the genus *Catocala*. Mem. Amer. Mus. Nat. Hist. (N.S.) 3:1–47.

Beck, H. 1960. Die Larvalsystematik der Eulen (Noctuidae). Abhandl. Larvalsyst. Ins. (Akademie-Verlag, Berlin) 4:1–406.

Beckwith, R. C. 1978a. Biology of the insect, pp. 25–30. *In* M. H. Brookes, R. W. Stark, and R. W. Campbell (eds.). The Douglas-fir tussock moth: a synthesis. USDA For. Serv., Sci. & Educat. Agency Tech. Bull. 1585.

Beckwith, R. C. 1978b. Larval instars of the Douglas-fir tussock moth. USDA, Combined For. Pest Res. Dev. Prog. Agr. Handbook 536. 15 pp.

Berg, K. 1941. Contributions to the biology of the aquatic moth, *Acentropus niveus* (Oliv.). Vidensk. Medd. fra Dansk Naturh. Foren. 105:60–139.

Bethune, C. J. S. 1873. Insects affecting the hop. (Third) Rept. Ent. Soc. Ontario, 1872:27–34.

Beutenmüller, W. 1888. Descriptions of some lepidopterous larvae. Can. Ent. 20:134–36.

Beutenmüller, W. 1898. Descriptive catalogue of the bombycine moths found within fifty miles of New York City. Bull. Amer. Mus. Nat. Hist. 10:353–448. (9 pls.)

Beutenmüller W. 1901. Descriptive catalogue of the Noctuidae found within fifty miles of New York City. Bull. Amer. Mus. Nat. Hist. 14:229–312. (4 pls.)

Beutenmüller, W. 1902. Descriptive catalogue of the Noctuidae found within fifty miles of New York City. Bull. Amer. Mus. Nat. Hist. 16:413–58. (4 pls.)

Bird, H. 1898. Notes on the noctuid genus *Hydroecia*. Can. Ent. 30:126–33.

Birket-Smith, S. J. R. 1984. Prolegs, legs and wings of Insects. Entomonograph. 5:1–128. Copenhagen, Denmark.

Bishop, F. C. 1923. The puss caterpillar and the effects of its sting on man. USDA, Circ. 288. 14 pp.

Bjerke, J. M., T. P. Freeman, and A. W. Anderson. 1979. A new method of preparing insects for scanning electron microscopy. Stain Technology 54:29–31.

Bottimer, L. J. 1926. Notes on some Lepidoptera from eastern Texas. J. Agr. Res. 33:797–819.

Bourquin, F. 1933. Notas biologicas de la *Castnia archon* Burm. Rev. Soc. Ent. Arg. 24:295–98.

Bourquin F. 1951. Notas sobre la metamorfosis de *Antispastis clarkei* Pastrana 1951 (Lepidoptera: Glyphipterygidae). Acta Zool. Lilloana 12:523–26. (1 pl.)

Bradley, J. D. 1966. A comparative study of the coconut flat moth (*Agonoxena argaula* Meyrick) and its allies, including a new species (Lepidoptera: Agonoxenidae). Bull. Ent. Res. 56:453–72.

Bradley, J. D., W. G. Tremewan, and A. Smith. 1973. British tortricoid moths. Cochylidae and Tortricidae: Tortricinae. London: Ray Society. 260 pp. (21 pls.)

Bradley, J. D., W. G. Tremewan, and A. Smith. 1979. British tortricoid moths. Tortricidae: Olethreutinae. London: Ray Society. 344 pp. (21 pls.)

Bratley, H. E. 1932. The oleander caterpillar, *Syntomeida epilais* Walker. Fla. Ent. 15(4):57–64.

Brandhorst, C. T. 1962. The microcommunity associated with the gall of *Walshia amorphella* (Lep.: Cosmopterygidae) on *Amorpha fruticosa*. Ann. Ent. Soc. Amer. 55:476–79.

Braun, A. F. 1908. Revision of the North American species of the genus *Lithocolletis* Hübner. Trans. Amer. Ent. Soc. 34:268–357. (5 pls.)

Braun, A. F. 1917. Nepticulidae of North America. Trans. Amer. Ent. Soc. 48:155–209. (4 pls.)

Braun, A. F. 1948. Elachistidae of North America (Microlepidoptera). Mem. Amer. Ent. Soc. No. 13. 110 pp.

Braun, A. F. 1963. The genus *Bucculatrix* in America North of Mexico (Microlepidoptera). Mem. Amer. Ent. Soc. No. 18. 208 pp. (45pl.)

Braun, A. F. 1972. Tischeriidae of America north of Mexico (Microlepidoptera). Mem. Amer. Ent. Soc. No. 28. 148 pp. (146 figs.)

Brigham, A. R. and D. D. Herlong. 1982. Aquatic and semiaquatic Lepidoptera, 36 pp. *In* A. R. Brigham, W. U. Brigham, and A. Gnilka. Midwest Enterprises, Mahomet, Ill. Aquatic insects and oligochaetes of North and South Carolina. 837 pp. (Includes tentative keys to genera and some spp. of aquatic pyralid larvae)

Brittain, W. H. 1915. *Hydroecia micacea* as a garden pest. Proc. Nova Scotia Ent. Soc. 1:96–97.

Brittain, W. H. 1918. Notes on two unusual garden pests in Nova Scotia. 48th Ann. Rept. Ent. Soc. Ontario, 1917:94–99.

Brittain, W. H., and H. G. Payne. 1919. Some notes on *Olene vagans* B. & McD. in Nova Scotia. Proc. Ent. Soc. Nova Scotia 4:62–68.

Britton, W. E., and G. A. Crombie. 1911. The leopard moth. Conn. Agr. Expt. Bull. 169. 24 pp.

Brown, E. S., and C. F. Dewhurst. 1975. The genus *Spodoptera* (Lepidoptera, Noctuidae) in Africa and the Near East. Bull. Ent. Res. 65:221–62.

Brown, L. R., and C. O. Eads. 1965. A technical study of insects affecting the oak tree in southern California. Calif. Agr. Expt. Sta. Bull. 818. Berkeley. 38 pp.

Bryant, D., and A. Raske. 1975. Defoliation of white birch by the birch casebearer, *Coleophora fuscedinella* (Lepidoptera: Coleophoridae). Can. Ent. 107(2):217–23.

*Buckler, William. 1887–1899. The larvae of British butterflies and moths. 9 vols. London: The Ray Society.

Burdick, D. J., and J. A. Powell. 1960. Studies on the early stages of two California moths which feed in the staminate cones of digger pine. Can. Ent. 92:310–19.

Burgess, A. F., and W. L. Baker. 1938. The gypsy moth and brown-tail moths and their control. USDA, Circ. 464. 38 pp.

Burgess, A. F., and S. S. Crossman. 1927. The satin moth, a recently introduced pest. USDA, Bull. 1469. 23 pp.

Burleigh, J. G. 1972. Population dynamics and biotic controls of the soybean looper in Louisiana. Environ. Ent. 1:290–94.

Burmeister, H. 1879. Atlas de la description physique de la République Argentine. Contenant des vues pittoresques et des figures d'histoire naturelle. Cinquième section, seconde partie. Lépidoptères. Buenos Aires, Paul-Emile Coni. 66 pp. (Larva of *Dalcera flava* Walker, p. 53)

Burns, J. M. 1964. Evolution in skipper butterflies of the genus *Erynnis*. Univ. Calif. Publ. Ent. 37. 216 pp.

Busck, A. 1907. Revision of the American moths of the genus *Argyresthia*. Proc. U.S. Nat. Mus. 32:5–24. (2 pls.) (Host records, taxonomy)

Busck, A. 1909. Notes on Microlepidoptera, with descriptions of new North American species. Proc. Ent. Soc. Wash. 11:87–103. (Taxonomy)

Busck, A. 1917. The pink bollworm, *Pectinophora gossypiella*. J. Agr. Res. 9:343–70.

Busck, A. 1934. Microlepidoptera of Cuba. Entomologica Amer. 13:151–203.

Byers, J. R. and C. F. Hinks. 1973. The surface sculpturing of the integument of lepidopterous larvae and its adaptive significance. Can. J. Zool. 51:1171–79.

Callaghan, C. 1977. Studies on Restinga butterflies. I. Life cycle and immature stages of *Menander felsina* (Riodinidae), a myrmecophilous metalmark. J. Lepid. Soc. 31:173–82. (References to life histories of tropical taxa)

Campbell, R. W. 1975. The gypsy moth and its natural enemies. USDA, For. Serv., Agr. Inform. Bull. 381. 27 pp.

Candrashoff, S. F. 1962. A description of the immature stages of *Phyllocnistis populiella* Chambers (Lepidoptera: Gracillariidae). Can. Ent. 94:902–909.

Capps, H. W. 1939. Keys for the identification of some lepidopterous larvae frequently intercepted at quarantines. USDA, Bur. Ent. and P. Q. E–475 (revised 1963 as ARS–33–20–1). 37 pp.

Capps, H. W. 1946. Description of the larva of *Keiferia peniculo* Heinrich, with a key to the larvae of related species attacking eggplant, pepper, potato and tomato in the United States (Lepidoptera: Gelechiidae). Ann. Ent. Soc. Amer. 39:561–63.

Capps, H. W. 1958. An illustrated key for identification of larvae of the cotton-pest species of *Pectinophora* Busck and *Platyedra* Meyrick (Lepidoptera: Gelechiidae). Bull. Ent. Res. 49(4):631–32.

Capps, H. W. 1967. Review of some species of *Loxostege* Hübner and descriptions of new species (Lepidoptera: Pyraustidae: Pyraustinae). Proc. U.S. Nat. Mus. 120:1–83. (Mostly adults, but includes descriptions of some larvae of *Hahncappsia* (*Loxostege*))

Carolin, V. M., Jr. 1971. Extended diapause in *Coloradia pandora* Blake (Lepidoptera: Saturniidae). Pan-Pac. Ent. 47:19–23.

Carroll, M. R., and W.H. Kearby. 1978. Microlepidopterous oak leaftiers (Lepidoptera: Gelechioidea) in central Missouri. J. Kans. Ent. Soc. 51:453–71.

Carroll, M. R., M. T. Wooster, W. H. Kearby, and D. C. Allen. 1979. Biological observations on three oak leaftiers: *Psilocorsis quercicella, P. reflexella,* and *P. cryptolechiella* in Massachusetts and Missouri. Ann. Ent. Soc. Amer. 72(3):441–47.

Carter, D. J., and J. S. Dugdale. 1982. Notes on collecting and rearing *Micropterix* (Lepidoptera: Micropterigidae) larvae in England. Ent. Gazette 33:43–47. (2 pls.)

Carey, P., S. Cameron, L. Cerda, and R. Gava. 1978. Ciclo estacional de un minador subcortical de coigue (*Nothofagus dombeyi*). Turrialba, 28(2):151–53.

Caulfield, F. B. 1878. Notes on the larva of *Samia columbia* Smith. Can. Ent. 10:41–42.

Chapman, T. A. 1917. *Micropteryx* entitled to ordinal rank: Order Zeugloptera. Trans. Roy. Ent. Soc. London. pp. 310–14.

*Chapman, P. J., and S. E. Lienk. 1971. Tortricid fauna of apple in New York. N.Y. Agr. Exp. Sta., Geneva, Spec. Publ. 122 pp. (Color paintings)

*Chapman, P. J., and S. E. Lienk. 1974. Green fruitworms. New York Food & Life Sci. Bull., No. 49 (50), Oct. (Nov.) 1974 (Plant Sci. Ent., Geneva), No. 6. 15 pp. (Color illustrations of larvae)

Chauvin, J. T., and G. Chauvin. 1980. Formation des reliefs externes de l'oeuf de *Micropteryx* [sic] *calthella* L. (Lepidoptera: Micropterigidae). Can. J. Zool. 58(5):761–66.

Chittenden, F. H. 1901. The fall armyworm and variegated cutworm. USDA, Div. Ent., Bull. 29 (N.S.). 64 pp.

Chittenden, F. H. 1907. The striped garden caterpillar, (*Mamestra legitima* Grote.). *In* Some insects injurious to truck crops. Pt. 3, USDA, Bur. Ent., Bull. 66. pp. 28–32.

Chittenden, F. H. 1913. The Florida fern caterpillar. USDA, Bur. Ent., Bull. 125. 11 pp.

Chittenden, F. H., and H. M. Russell. 1909. The semitropical armyworm, (*Prodenia eridania* Cram.). *In* Some insects injurious to truck crops. Pt. 4. USDA, Bur. Ent., Bull. 66. pp. 53–70.

Chopra, R. L. 1925. On the structure, life-history, economic importance and distribution of the cocksfoot moth, *Glyphipteryx [sic] fischeriella*, Zell. [=*G. simpliciella*]. Ann. Appl. Biol. 12:359–97. (Biology)

Chu, H. F., C. L. Fang, and L. Y. Wang. 1963. Fauna of Chinese economic insects. Fasc. 7. Noctuidae (immature stages). Peking: Science Press. 120 pp. (31 pls.) (Includes 304 line drawings of eggs, larvae, and pupae. In Chinese)

Clark, G. C., and C. G. C. Dickson. 1956a. The honey glands and tubercles of larvae of the Lycaenidae. Lepid. News 10(1–2):37–40.

Clark, G. C., and C. G. C. Dickson. 1956b. Proposed classification of South African Lycaenidae from the early stages. J. Ent. Soc. So. Afr. 19(2):195–215.

Clark, G. C., and C. G. C. Dickson. 1971. Life histories of the South African lycaenid butterflies. Purnell, Cape Town. 284 pp.

Clarke, A. H. 1926. Carnivorous butterflies. Smithsonian Rpt. for 1925:439–508.

Clarke, J. F. G. 1941. Revision of the North American moths of the family Oecophoridae, with descriptions of new genera and species. Proc. U.S. Nat. Mus. 90:33–286.

Clarke, J. F. G. 1943. A new pest of *Albizzia* in the District of Columbia (Lepidoptera: Glyphipterygidae). Proc. U.S. Nat. Mus. 95: 205–208. (5 pls.) (Biology, chaetotaxy)

Clarke, J. F. G. 1952. Host relationships of moths of the genera *Depressaria* and *Agonopterix*, with descriptions of new species. Smithsonian Misc. Coll. 117:1–20.

Clarke, J. F. G. 1960. A new species of moth injurious to pine (Lepidoptera: Blastobasidae). Fla. Ent. 43:115–17.

Claassen, P. W. 1921. *Typha* insects: their ecological relationships. Cornell Agr. Expt. Sta. Mem. 47:459–531.

Clausen, C. P. 1940. Entomophagous insects. New York: McGraw Hill. 608 pp. (Reprinted by Hafner Publ. Co., N.Y., 1972)

Clench, H. K. 1955. Revised classification of the butterfly family Lycaenidae and its allies. Ann. Carnegie Mus. 33:261–74.

Clench, H. K. 1965. Superfamily Lycaenoidea, pp. 267–403. *In* R. M. Fox et al. (eds.). The butterflies of Liberia. Memoirs Amer. Ent. Soc., No. 19, pp. 1–438.

Cockerell, T. D. A. 1889. The larva of *Gnophaela vermiculata* G. & R. Ent. Amer. 5(3):57–58.

Cockerell, T. D. A. 1911. An *Aleyrodes* on *Euphorbia*, and its parasite (Rhynch., Hym.). Ent. News 22:462–64. (Reports rearing *Doa ampla* from *Euphorbia robusta* in Boulder, Colorado; larvae from this rearing described by Dyar, 1912)

Collins, C. W. 1926. Observations on recurring outbreaks of *Heterocampa guttivitta* Walker and natural enemies controlling it. J. Agr. Res. 32:689–99.

Collins, M. M., and R. D. Weast. 1961. Wild silk moths of the United States. Cedar Rapids, Ia.: Collins Radio Company. 142 pp.

Common, I. F. B. 1963. A revision of the Australian Cnephasiini (Lepidoptera: Tortricidae: Tortricinae). Austral. J. Zool. 11:81–151.

Common, I. F. B. 1970. Lepidoptera (moths and butterflies), pp. 765–866. *In* I. M. Mackerras (ed.). The insects of Australia. Melbourne: Melbourne Univ. Press. 1029 pp. (Brief descriptions of immatures, biology, and classification)

Common, I. F. B. 1975. Evolution and classification of the Lepidoptera. Ann. Rev. Ent. 20:183–203.

Common, I. F.B., and E. D. Edwards. 1981. The life history and early stages of *Synemon magnifica* Strand (Lepidoptera: Castniidae). J. Austral. Ent. Soc. 20:295–302.

Comstock, J. A. 1927. Butterflies of California. Los Angeles: Priv. Publ. 334 pp. (63 pls.)

Comstock, J. A. 1939. Studies in Pacific Coast Lepidoptera. Bull. So. Calif. Acad. Sci. 38:34–35.

Comstock, J. A. 1940. Four new California moths with notes on early stages. Bull. So. Calif. Acad. Sci. 38:172–82. (Biology)

Comstock, J. A. 1944. Four California moths associated with cat-tails. Bull. So. Calif. Acad. Sci. 43:81–83.

Comstock, J. A. 1947. Notes on the life history of *Orthodes accurata* Hy. Edwards. Bull. So. Calif. Acad. Sci. 46:124–26.

Comstock, J. A. 1956a. Notes on the life histories of two southern Arizona butterflies. Bull. So. Calif. Acad. Sci. 55:171.

Comstock, J. A. 1956b. Notes of the metamorphoses of the giant skippers (Megathymidae) and the life history of an Arizona species. Bull. So. Calif. Acad. Sci. 55:19–27. (Descriptions, biology, bibliography)

Comstock, J. A. 1959a. Rare or common! With notes on the life history of the southwestern moths. Bull. So. Calif. Acad. Sci. 58:155–61.

Comstock, J. A. 1959b. Scientific notes. Bull. So. Calif. Acad. Sci. 58:53. (Egg of *Dalcerides ingenita* (Hy. Edw.))

Comstock, J. A., and C. M. Dammers. 1934. Notes on the early stages of three butterflies and five moths from California. Bull. So. Calif. Acad. Sci. 33:137–51.

Comstock, J. A., and C. M. Dammers. 1938. Studies in the metamorphoses of six California moths. Bull. So. Calif. Acad. Sci. 37: 105–28.

Comstock, J. A., and C. M. Dammers. 1942. Notes on the life histories of two common moths. Bull. So. Calif. Acad. Sci. 41:172–78.

Comstock, J. A., and C. Henne. 1967. Early stages of *Lycomorpha regulus* Grinnell, with notes on the imago (Lepidoptera: Arctiidae). J. Res. Lepid. 6(4):275–80.

Comstock, J. H. 1879. The cotton-cutworm, pp. 11–284. *In* J. H. Comstock. Report upon cotton insects. Washington, D.C.: Government Printing Office.

Comstock, J. H. 1880. The green clover-worm (*Plathypena scabra*, Fabr.), p. 252. *In* J. H. Comstock, Report of the entomologist of the U.S.D.A. for the year 1879. USDA, Ann. Rept. Year 1879. Washington, D.C.: Government Printing Office.

Cook, W. C. 1930. Field studies of the pale western cutworm (*Porosagrotis orthogonia* Morr.). Mont. Agr. Expt. Sta. Bull. 225. 79 pp.

Cooley, R. A. 1916. Observations on the life history of the army cutworm, *Chorizagrotis auxiliaris*. J. Agr. Res. 6:871–81.

Corrette, K. B., and H. H. Neunzig. 1979. Descriptions of and notes on larval habits of four immature phycitines in the southeastern United States (Lepidoptera: Pyralidae). Ann. Ent. Soc. Amer. 72:690–99.

Costa Lima, A. da. 1945. Insectos do Brasil. Vol. 5., p. 180. (Larva of *Zadalcera fumata* (Schaus))

Cottrell, C. 1984. Aphytophagy in butterflies: its relationship to myrmecophily. Zool. J. Linnean Soc. 79:1–57.

Crumb, S. E. 1926a. The bronzed cutworm (*Nephelodes emmedonia* Cramer) (Lepidoptera). Proc. Ent. Soc. Wash. 28:201–207.

Crumb, S. E. 1926b. The Nearctic budworms of the lepidopterous genus *Heliothis*. Proc. U.S. Nat. Mus. 68(16):1–8.

Crumb, S. E. 1927. The army worms. Bull. Brook. Ent. Soc. 22:41–55.

Crumb, S. E. 1929. Tobacco cutworms. USDA Tech. Bull. 88. 179 pp.

Crumb, S. E. 1932. The more important climbing cutworms. Bull. Brook. Ent. Soc. 27:73–100.

Crumb, S. E. 1934. A classification of some noctuid larvae of the subfamily Hypeninae. Ent. Amer. 14:133–97.

Crumb, S. E. 1956. The larvae of the Phalaenidae. USDA Tech. Bull. 1135. 356 pp.

Cuscianna, N. 1927. Note morfologiche e biologiche sulla *Simaethis nemorana* Hb. Boll. Lab. Zool. Agr. Portici (Naples) 20:17–34. (Biology, chaetotaxy)

Dampf, A. 1910. Zur Kenntnis gehausetragender Lepidopterenlarven. Zool. Jb. (Suppl.) 12:513–608.

Darlington, E. P. 1952. Notes on blueberry Lepidoptera in New Jersey. Trans. Amer. Ent. Soc. 78:33–37.

Davenport, D., and V. G. Dethier. 1937. Bibliography of the described life-histories of the Rhopalocera of America north of Mexico, 1889–1937. Ent. Amer. 17:155–94.

Daviault, L. 1942. Description et biologie de deus lepidopteres nuisibles a l'orme. Le Naturaliste Canadien. 69:145–56.

Davis, D. R. 1964. Bagworm moths of the Western Hemisphere (Lepidoptera: Psychidae). U.S. Nat. Mus. Bull. 244. 233 pp.

Davis, D. R. 1967. A revision of the moths of the subfamily Prodoxinae (Lepidoptera: Incurvariidae). Bull. U.S. Nat. Mus. 255:1–170.

Davis, D. R. 1969. A revision of the American moths of the family Carposinidae (Lepidoptera: Carposinoidea). U.S. Nat. Mus. Bull. 289:1–105. (Biology, taxonomy)

Davis, D. R. 1972. *Careospina quercivora,* a new genus and species of moth infesting live oaks in California. Proc. Ent. Soc. Wash. 23:137–47. (Incurvariidae)

Davis, D. R. 1975a. A review of Ochsenheimeriidae and the introduction of the cereal stem moth *Ochsenheimeria vacculella* into the United States (Lepidoptera: Tineoidea). Smithsonian Contr. Zool. No. 192. 20 pp.

Davis, D.R. 1975b. A review of the West Indian moths of the family Psychidae with descriptions of new taxa and immature stages. Smithsonian Contr. Zool. No. 188. 66 pp.

Davis, D. R. 1978a. A revision of the North American moths of the superfamily Eriocranioidea with the proposal of a new family Acanthopteroctetidae (Lepidoptera). Smithsonian Contr. Zool. No. 251. pp. 1–131.

Davis, D. R. 1978b. The North American moths of the genera *Phaeoses, Opogona,* and *Oinophila,* with a discussion of their supergeneric affinities (Lepidoptera: Tineidae). Smithsonian Cont. Zool., No. 282. 39 pp. (128 figs.)

Davis, D. R. 1978c. Two new genera of North American Incurvariinae moths. Pan-Pac. Ent. 54:147–53.

Davis, D. R. 1983. Family Tineidae, pp. 5–7. Gracillariidae, pp. 9–11. *In* R. Hodges et al., A check list of the Lepidoptera of America north of Mexico. E. W. Classey Ltd. and The Wedge Ent. Res. Found., London.

Davis, D. R. and S. Faeth. 1986. A new oak mining eriocraniid moth from southeastern United States (Lepidoptera: Eriocraniidae). Proc. Ent. Soc. Wash. 88:145–53.

Davis, J. J. 1912. The southern fern cutworm *Callopistria floridensis* Guen. (*Eriopus floridensis*), pp. 88–91. *In* Twenty-seventh report of the state entomologist on the noxious and beneficial insects of the state of Illinois. Springfield, Ill.

Davis, J. J., and A. F. Satterthwait. 1916. Life-history studies of *Cirphis unipuncta,* the true army worm. J. Agr. Res. 6:799–812. (1 pl.)

Day, G. O. 1912. Notes on *Schizura unicornis,* Smith & Abbot. Proc. Brit. Columbia Ent. Soc. 2(N.S.):40–41.

DeBenedictis, J. A. 1984. On the taxonomic position of *Ellabella* Busck, with descriptions of the larva and pupa of *E. bayensis* (Lepidoptera: Copromorphidae). J. Res. Lepid. 23:74–82. (Biology, chaetotaxy)

Decker, G. C. 1930. The biology of the four-lined borer *Luperina stipata* (Morr.). Iowa Agr. Expt. Sta. Res. Bull. 125, pp. 125–64.

Decker, G. C. 1931. The biology of the stalk borer, *Papaipema nebris* (Gn.). Iowa Agr. Expt. Sta. Res. Bull. 143, pp. 289–351.

DeGryse, J. J. 1916. The hypermetamorphism of the lepidopterous sapfeeders. Proc. Ent. Soc. Wash. 18:164–68.

Dethier, V. G. 1937. Cannibalism among lepidopterous larvae. Psyche 44:110–15.

Dethier, V. G. 1939. Further notes on cannibalism among larvae. Psyche 46:29–35.

Dethier, V. G. 1941. The antennae of lepidopterous larvae. Bull. Mus. Com. Zool. 87:455–507.

Dethier, V. G. 1942a. Abdominal glands of Hesperiinae. J. N.Y. Ent. Soc. 50:203–206.

Dethier, V. G. 1942b. Notes on the life histories of five common Geometridae. Can. Ent. 74:225–34.

Dethier, V. G. 1946. Supplement to the bibliography of described life histories of the Rhopalocera of America north of Mexico. Psyche 53:15–20.

Dethier, V. G. 1980. The world of tentmakers; a natural history of the eastern test caterpillar. Amherst: Univ. of Mass. Press. 148 pp.

Detwiler, J. D. 1922. The ventral prothoracic gland of the red-humped apple caterpillar (*Schizura concinna* Smith & Abbot). Can. Ent. 54:175–91.

Diakonoff, A. 1954. Microlepidoptera of New Guinea. Results of the third Archbold Expedition (American-Netherlands Indian Expedition 1938–1939). Part III. Verh. Koninkl. Nederl. Akad. Wetenschap. Afd. Natuurk. (Amsterdam) (Tweede Reeks) 49(4):1–164. (Adult keys)

Diakonoff, A., and Y. Arita. 1981. The early stages of *Thaumatographa eremnotorna* Diakonoff & Arita with remarks on the status of the Hilarographini (Lepidoptera: Tortricoidea). Ent. Berich. 41:56–60.

Dietz, W. G. 1910. Revision of the Blastobasidae of North America. Trans. Amer. Ent. Soc. 36:1–72.

Doerksen, G. P., and H. H. Neunzig. 1975. Descriptions of some immature *Nephopterix* in the eastern United States (Lepidoptera: Pyralidae: Phycitinae). Ann. Ent. Soc. Amer. 68:623–44. (Keys to, and descriptions of, seven species)

Dominick, R. B. 1972. Practical freeze-drying and vacuum dehydration of caterpillars. J. Lepid. Soc. 26(2):69–79.

Doring, E. 1955. Zur Morphologie der Schmetterlingseier. Berlin: Academie-Verlag. 154 pp. (61 pls.)

Downey, J. C., and A. C. Allyn, Jr. 1979. Morphology and biology of the immature stages of *Leptotes cassius theonus* (Lucas) (Lepid.: Lycaenidae). Bull. Allyn Mus., No. 55, pp. 1–27.

Downey, J. C., and A. C. Allyn, Jr. 1980. Eggs of Riodinidae. J. Lepid. Soc. 34:133–45.

Drew, W. A. 1962. Oklahoma Arctiidae (Lepidoptera). Proc. Okla. Acad. Sci. 42:93–100.

Duckworth, W. D. 1964. North American Stenomidae (Lepidoptera: Gelechioidea). Proc. U.S. Nat. Mus. 116:23–72.

Duckworth, W. D. 1965. North American moths of the genus *Swammerdamia* (Lepidoptera: Yponomeutidae). Proc. U.S. Nat. Mus. 116:549–55. (Biology, taxonomy)

Duckworth, W. D. 1969. A new species of Aegeriidae from Venezuela predaceous on scale insects (Lepidoptera: Yponomeutoidea). Proc. Ent. Soc. Wash. 71:487–90. (Biology)

Duckworth, W. D. 1973. The old world Stenomidae: a preliminary survey of the fauna; notes and relationships and revision of the genus *Eriogenes*. Smithsonian Contr. Zool., No. 147. 21 pp.

Duckworth, W. D., and T. D. Eichlin. 1974. Clearwing moths of Australia and New Zealand (Lepidoptera: Sesiidae). Smithsonian Contr. Zool. No. 180. 45 pp. (Biology, taxonomy)

Duckworth, W. D., and T. D. Eichlin. 1977. A classification of the Sesiidae of America north of Mexico (Lepidoptera: Sesioidea). Occas. Pap. Ent. (Sacramento) 26:1–54. (Host records, taxonomy)

Duckworth, W. D., and T. D. Eichlin. 1978. The clearwing moths of California (Lepidoptera: Sesiidae). Occas. Pap. Ent. (Sacramento) No. 27. 80 pp. (8 pls.) (Biology, taxonomy)

Duckworth, W. D., and T. D. Eichlin. 1981. Sesioidea. Sesiidae. *In* R. B. Dominick, et al. (eds.). The moths of America north of Mexico. Fasc. 8.4. Classey, London. (in prep.). (Biology, taxonomy)

Dugdale, J. S. 1961. Larval characters of taxonomic significance of New Zealand ennomines (Lepidoptera: Geometridae). Trans. Roy. Soc. New Zealand, Zoology 1:215–33.

Dugdale, J. S. 1974. Female genital configuration in the classification of Lepidoptera. New Zealand J. Zool. 1(2):127–46.

Dyar, H. G. 1891a. Preparatory stages of *Datana perspicua* G. & R., and *Cerura borealis* Boisd. Can. Ent. 23:82–87.

Dyar, H. G. 1891b. Preparatory stages of *Pheosia dimidiata* H.S. Psyche 6:194–96.

Dyar, H. G. 1891c. A correction. Psyche 6:197.

Dyar, H. G. 1892a. Life history of *Orgyia cana* Hy. Edw. Psyche 6:203–205.

Dyar, H. G. 1892b. Preparatory stages of *Pheosia portlandia* Hy. Edw. Psyche 6:351–53.

Dyar, H. G. 1893a. Life histories of some bombycid moths. Proc. Boston Soc. Nat. Hist. 26:153–66.

Dyar, H. G. 1893b. Life history of *Orgyia gulosa* Hy. Edw. Psyche 6:438–40.

Dyar, H. G. 1893c. On the differences between the larvae of *Cecrita bilineata* and *Heterocampa manteo.* Ent. News 4:262–63.

Dyar, H. G. 1894a. A classification of lepidopterous larvae. Ann. N.Y. Acad. Sci. 8:194–232.

Dyar, H. G. 1894b. Preparatory stages of *Laphygma flavimaculata* Harv. and other notes. Can. Ent. 26:65–69.

Dyar, H. G. 1895. Notes on drepanid larvae. J. N.Y. Ent. Soc. 3:66–69.

Dyar, H. G. 1895–1899. Life histories of the New York slug caterpillars. J. N.Y. Ent. Soc., Vols. 3–7. (Eighteen papers, some with E. L. Morton)

Dyar, H. G. 1896a. Final notes on *Orgyia.* Psyche 7:340–42.

Dyar, H. G. 1896b. Note on the head setae of lepidopterous larvae. Ann. N.Y. Acad. Sci. 8:194–232.

Dyar, H. G. 1896c. On the probable origin of the Pericopidae: *Composia fidelissima* H.-S. J. N.Y. Ent. Soc. 4:68–72.

Dyar, H. G. 1897a. A comparative study of seven young arctians. J. N.Y. Ent. Soc. 5:130–33.

Dyar, H. G. 1897b. Note on larvae of *Gynaephora groenlandica* and *G. rossii.* Psyche 8:153.

Dyar, H. G. 1899a. A parallel evolution in a certain larval character between the Syntomidae and the Pericopidae. Proc. Ent. Soc. Wash. 4(4):407–409.

Dyar, H. G. 1899b. Descriptions of the larvae of fifty North American Noctuidae. Proc. Ent. Soc. Wash. 4:315–32.

Dyar, H. G. 1899c. Note on two *Hydroecia* larvae. J. N.Y. Ent. Soc. 7: 70. (1 pl., 3 figs.)

Dyar, H. G. 1899–1905. Life histories of North American Geometridae. Psyche, Vols. 8–12 (14 papers).

Dyar, H. G. 1900a. Notes on the larval-cases of Lacosomidae (Perophoridae) and life history of *Lacosoma chiridota* Grt. J. N.Y. Ent. Soc. 8:177–80.

Dyar, H. G. 1900b. Preliminary notes on the larvae of the genus *Arctia.* J. N.Y. Ent. Soc. 8:34–47.

Dyar, H. G. 1901. Descriptions of some pyralid larvae from southern Florida. J. N.Y. Ent. Soc. 9:19–24.

Dyar, H. G. 1902. Descriptions of the larvae of some moths from Colorado. Proc. U.S. Nat. Mus. 25:369–412.

Dyar, H. G. 1904. The Lepidoptera of the Kootenai District of British Columbia. Proc. U.S. Nat. Mus. 27:779–938.

Dyar, H. G. 1910. Notes on the family Dalceridae. Proc. Ent. Soc. Wash. 12:113–21. (Quotes from Burmeister and Jones & Moore regarding larvae)

Dyar, H. G. 1911. Descriptions of the larvae of some Mexican Lepidoptera. Proc. Ent. Soc. Wash. 13:227–32. (Describes larva, cocoon, and pupa of *Doa raspa* (new combination), which he places in the "Hypsidae (Pericopidae)").

Dyar, H. G. 1912. [Untitled note describing larva, cocoon, and pupa of *Doa ampla,* based on material supplied by Cockerell (1911).] Proc. Ent. Soc. Wash. 14:14–15.

Dyar, H. G. 1913. The larva of *Trichostibus parvula.* Ins. Insec. Menstr. 1:49–50. (Biology)

Dyar, H. G. 1914. The pericopid larvae in the national museum. Ins. Insec. Menstr. 2(4):62–64.

Dyar, H. G. 1925. A note on the larvae of the Dalceridae (Lepidoptera). Ins. Insec. Menstr. 13:44–47. (Larvae of *Acraga coa* Schaus and *Paracraga argentea* Schaus, cocoon of *Dalcerides ingenita*)

Eaton, C. B., and G. R. Struble. 1957. The Douglas-fir tussock moth in California (Lepidoptera: Liparidae). Pan-Pac. Ent. 33:105–108.

Edwards, H. 1873. Pacific Coast Lepidoptera, No. 8—On the transformations of some species of *Heterocera,* not previously described. Proc. Calif. Acad. Sci. 5:367–72.

Edwards, H., and S. L. Elliot. 1883. On the transformations of some species of Lepidoptera. Papilio 3:125–36.

Edwards, W. H. 1868–1897. The Butterflies of North America. Three vols. Boston: Houghton-Mifflin. (Descriptions, biology)

Edwards, W. H. 1881a. Description of the preparatory stages of *Libythea bachmanni* Kirtland. Can. Ent. 13:226–29.

Edwards, W. H. 1881b. Notes on the Pacific Coast species of *Orgyia,* with descriptions of larvae and new forms. Papilio 1:60–62.

Edwards, W. H. 1889. Bibliographic catalogue of the described transformations of North American Lepidoptera. Bull. U.S. Nat. Mus. 35:1–147.

Ehrhardt, R. 1914. Über die Biologie und Histologie der myrmekophilen Organe von *Lycaena orion.* Ber. Naturf. Ges. Freiburg, i. Br. 20:40–68.

Ehrlich, P. R. 1958. The comparative morphology, phylogeny and higher classification of the butterflies (Lepidoptera: Papilionoidea). Univ. Kans. Sci. Bull. 39:305–70.

Eichlin, T. D. 1975a. Some looper moth pests of garden and greenhouse. Nat. Pest. Cont. Operator's News 35(6):12–15.

Eichlin, T. D. 1975b. Guide to the adult and larval Plusiinae of California (Lepidoptera: Noctuidae). Occasional Papers Ent. (State of Calif., Dept. Food & Agr., Div. Plant Ind.), No. 21. 73 pp.

Eichlin, T. D., and H. B. Cunningham. 1969. Characters for identification of some common plusiine caterpillars of the southeastern United States. Ann. Ent. Soc. Amer. 62:507–10.

Eichlin, T. D., and H. B. Cunningham. 1978. The Plusiinae (Lepidoptera: Noctuidae) of America north of Mexico, emphasizing genitalic and larval morphology. USDA, Tech. Bull. 1567. 122 pp.

Eidt, D. C., and D. G. Embree. 1968. Distinguishing larvae and pupae of the winter moth, *Operophtera brummata,* and the Bruce spanworm, *O. bruceata* (Lepidoptera: Geometridae). Can. Ent. 100:536–39.

Eisner, T., A. F. Kluge, J. C. Carrel, and J. Meinwald. 1972. Defense mechanisms of arthropods. XXXIV. Formic acid and acyclic ketones in the spray of a caterpillar. Ann. Ent. Soc. Amer. 65:765–66.

Eisner, T., T. E. Pliske, M. Ikeda, D. F. Owen, L. Vasquez, H. Perez, J. G. Franclemont, and J. Meinwald. 1970. Defense mechanisms of arthropods. XXVII. Osmeterial secretions of papilionid caterpillars (*Baronia, Papilio, Eurytides*). Ann. Ent. Soc. Amer. 63:914–15.

Eliot, I. M., and C. G. Soule. 1902. Caterpillars and their moths. New York: The Century Co. 302 pp.

Eliot, J. 1973. The higher classification of the Lycaenidae (Lepidoptera): a tentative arrangement. Bull. British Mus. Nat. Hist. (Ent.) 28:373–505.

Ellis, W. O. 1925. Some lepidopterous larvae resembling the European corn borer. J. Agr. Res. 30:777.

Eltringhan, H. 1921. On the larvae and pupae of Lepidoptera, chiefly Lycaenidae, collected by C. O. Farquharson, W. A. Lamborn and the Rev. Canon K. St. A. Rogers. Trans. Ent. Soc. Lond. 69:473–89, pls. 7, 8. (Plates are the ones often reproduced elsewhere)

Embree, D. G. 1958. The external morphology of the immature stages of the beech leaf tier, *Psilocorsis faginella* (Chamb.) (Lep.: Oecophoridae), with notes on its biology in Nova Scotia. Can. Ent. 90:166–74.

Embree, D. G. 1965. The population dynamics of the winter moth in Nova Scotia, 1954–1962. Mem. Ent. Soc. Can. 46:1–57. (Geometridae)

Emmel, T. C., and J. F. Emmel. 1973. The butterflies of southern California. Nat. Hist. Mus. Los Angeles Co. Sci. Ser. 26. 160 pp.

Emmet, A. M. 1976. Nepticulidae, pp. 171–267. Heliozelidae, pp. 300–306. Tischeriidae, pp. 272–276. *In* J. Heath (ed.). The moths and butterflies of Great Britain and Ireland. Vol. 1, Micropterigidae-Heliozelidae. London. 343 pp.

Empson, D. W. 1956. Cocksfoot moth investigations. Plant Pathol. 5: 12–18. (1 pl.) (Biology)

Engelhardt, G. P. 1946. The North American clear-wing moths of the family Aegeriidae. U.S. Nat. Mus. Bull. 190:1–222. (32 pls.) (Biology, taxonomy)

Esquivel, E. A. 1980. The giant cane borer, *Castnia licus* Drury, and its integrated control. Proc. Sec. Inter-American Sugar Cane Sem. 70–84.

Essig, E. O. 1958. Insects and mites of western North America. New York: The Macmillan Co. 1064 pp.

Evans, W. H. 1951–1955. A catalogue of the American Hesperiidae in the British Museum (Natural History). Part I (1951): Introduction and Pyrrhopyginae, 102 pp. (9 pls.) Part II (1952): Pyrginae, Sect. 1., 178 pp.; Part III (1953): Pyrginae Sect. 2, 252 pp.; Part IV: Hesperiinae. British Museum (Nat. Hist.), Norwich, London: Jarrold & Sons, Ltd.

Eyer, J. R. 1963. A pictorial key to the North American moths of the family Opostegidae. J. Lepid. Soc. 17(4):237–42.

Eyer, J. R., and C. H. Menke. 1914. *Adelocephala bisecta* (Lepid., Family Ceratocampidae). Ent. News 25:156–57. (1 pl.)

Farquharson, C. O. 1922. Five years' observations (1914–1918) on the bionomics of southern Nigerian insects, chiefly directed to the investigation of lycaenid life-histories and to the relation of Lycaenidae, Diptera, and other insects to ants. Trans. Ent. Soc. Lond. 1921:319–448.

Farrier, M. H., and O. E. Tauber. 1953. *Dioryctria disclusa* n. sp. and its parasites in Iowa. J. Sci. Iowa State Coll. 27:495–507. (Pyralidae) (Includes description and figures of last stage larva)

Farris, M. E., and J. E. Appleby. 1980(1981). Field observations of larval behavior of *Datana integerrima* (Notodontidae) in Illinois. J. Lepid. Soc. 34:368–71.

Felt, E. P. 1905. Insects affecting park and woodland trees. Mem. 8, N.Y. St. Mus. 1:1-332; 2:333–755.

Felt, E. P. 1912. Iris borer *Macronoctua onusta* Grote. *In* 27th Report of the State Entomologist, 1911. Albany, N.Y.: N.Y. State Mus. Bull. 155, pp. 52–54.

Felt, E. P. 1917. Apple and thorn skeletonizer (*Hemerophila pariana* Clerck). J. Econ. Ent. 10:502. (U.S. introduction report) (Choreutidae)

Felt, E. P. 1918. Key to American insect galls. N.Y. State Mus. Bull. 200. 310 pp.

Ferguson, D. C. 1963. Immature stages of four Nearctic Notodontidae (Lepidoptera). Can. Ent. 95:946–53.

Ferguson, D. C. 1969. A revision of the moths of the subfamily Geometrinae of America north of Mexico (Insects, Lepidoptera). Peabody Mus. Natur. Hist. Yale Univ. Bull. 29. 250 pp. (49 pls.) (All stages, biology, only adults illustrated)

Ferguson, D. C. 1971–1972. *In* R. B. Dominick et al. The moths of America North of Mexico. Fasc. 20A and 20B, Bombycoidea (part): Saturniidae. London: E. W. Classey Ltd. and R. B. D. Publications, Inc. 275 pp.

Ferguson, D. C. 1975. Host records for Lepidoptera reared in eastern North America. USDA, Tech. Bull. 1521. 49 pp.

Ferguson, D. C. 1978. *In* R. B. Dominick et al. The moths of America north of Mexico, Fasc. 22.2, Noctuoidea (in part): Lymantriidae. London: E. W. Classey Ltd. and The Wedge Ent. Res. Found. 120 pp.

Ferguson, D. C. 1983. Lymantriidae. *In* R. W. Hodges et al. (eds.). A check list of the Lepidoptera of America north of Mexico. London: E. W. Classey Ltd. and The Wedge Ent. Res. Found. 284 pp.

Ferguson, D. C. 1985a. *In* R. B. Dominick et al. The moths of America North of Mexico. Fasc. 18.1 Geometroidea (Geometridae (in part)). Washington: Wedge Fnt. Res. Found. 131 pp.

Ferguson, D. C. 1985b. Contributions toward reclassification of the world genera of the tribe Arctiini. Part 1–Introduction and a revision of the *Neoarctia–Grammia* group. (Lepidoptera: Arctiidae; Arctiinae). Entomographa 3:181–275.

Fernald, C. H. 1878. On the early stages of *Samia columbia* Smith. Can. Ent. 10:43–48.

Ferris, C. D. 1981a. Subfamily Megathyminae Holland, 1900, pp. 141–44. *In* C. D. Ferris and F. M. Brown (eds.). Butterflies of the Rocky Mountain states. Norman, Okla.: Univ. of Okla. Press. 461 pp.

Ferris, C. D. 1981b. Superfamily Nymphaloidea Swainson, 1927 (Monarchs, Long Wings, Brushfooted Butterflies) [Danaidae, Nymphalidae, Heliconiidae, Libytheidae], pp. 290–302, 315–60. *In* C. D. Ferris and F. M. Brown (eds.). Butterflies of the Rocky Mountain states. Norman, Okla.: Univ. of Okla. Press. 461 pp.

Ferris, C. D. 1981c. Superfamily Papilionoidea Latreille, 1809 (Parnassians) [Parnassiinae], pp. 175–78. *In* C. D. Ferris and F. M. Brown (eds.). Butterflies of the Rocky Mountain states. Norman, Okla.: Univ. of Okla. Press, 461 pp.

Ferris, C. D. 1981d. Superfamily Papilionoidea Latreille, 1809 (Whites, Orange Tips, Marbles, Sulphurs) [Pieridae], pp. 145–74. *In* C. D. Ferris and F. M. Brown (eds.). Butterflies of the Rocky Mountain states. Norman, Okla.: Univ. of Okla. Press. 461 pp.

Ferris, C. D., and F. M. Brown (eds.). 1981. Butterflies of the Rocky Mountain states. Norman, Okla.: Univ. of Okla. Press. 461 pp.

Ferris, C. D., and J. D. Eff. 1981. Superfamily Nymphaloidea Swainson, 1827 [Genus *Speyeria,* Family Nymphalidae] pp. 302–15. *In* C. D. Ferris and F. M. Brown (eds.). Butterflies of the Rocky Mountain states. Norman, Okla.: Univ. of Okla. Press. 461 pp.

Fiance, S. B., and R. E. Moeller. 1977. Immature stages and ecological observations of *Eoparargyractis plevie* (Pyralidae: Nymphulinae). J. Lepid. Soc. 31:81–88.

Fibiger, N., and N. P. Kristensen. 1974. The Sesiidae (Lepidoptera) of Fennoscandia and Denmark. Fauna Ent. Scandinavica 2:1–91. (Biology, taxonomy)

Finnegan, R. 1965. The pine needle miner, *Exoteleia pinifoliella* (Chamb.) (Lepidoptera: Gelechiidae), in Quebec. Can. Ent. 97(7):744–50.

Fisher, M. S. 1981. Superfamily Papilionoidea Latreille, 1809 (Swallowtails) [Papilioninae], pp. 178–193. *In* C. D. Ferris and F. M. Brown (eds.). Butterflies of the Rocky Mountain states. Norman, Okla.: Univ. of Okla. Press. 461 pp.

Fitzgerald, T. D., and J. B. Simeone. 1971. Description of the immature stages of the sap feeder *Marmara fraxinicola* (Lepidoptera: Gracillariidae). Ann. Ent. Soc. Amer. 64(4):765–70.

Fletcher, T. B. 1933. Life histories of Indian microlepidoptera (2nd series). Imp. Counc. Agr. Res. Sci. Monogr. 4:1–87.

Forbes, S. A. 1904. The more important insect injuries to corn. Univ. Ill. Agr. Expt. Sta. Bull. 95:331–99.

Forbes, S. A., and C. A. Hart. 1900. The economic entomology of the sugar beet. Univ. Ill. Agr. Expt. Sta. Bull. 60:397–532. (Contains life history information about *Scotogramma trifolii*).

Forbes, W. T. M. 1906. Field tables of Lepidoptera. Worcester, Mass. 141 pp.

Forbes, W. T. M. 1910a. A structural study of some caterpillars. Ann. Ent. Soc. Amer. 3:94–132.

Forbes, W. T. M. 1910b. The aquatic caterpillars of Lake Quinsigamond. Psyche 17:219–27.

Forbes, W. T. M. 1911a. A structural study of some caterpillars. II. The Sphingidae. Ann. Ent. Soc. Amer. 4:261–79.

Forbes, W. T. M. 1911b. Another aquatic caterpillar (*Elophila*). Psyche 18:120–21.

Forbes, W. T. M. 1916. On certain caterpillar homologies. J. N.Y. Ent. Soc. 34:137–42.

Forbes, W. T. M. 1923. The Lepidoptera of New York and neighboring states. Part I. Primitive forms, microlepidoptera, pyraloids, bombyces. Cornell Univ. Agr. Expt. Sta. Mem. 68. 729 pp.

Forbes, W. T. M. 1948. The Lepidoptera of New York and neighboring states. Part II. Geometridae, Sphingidae, Notodontidae, Lymantriidae. Cornell Univ. Agr. Expt. Sta. Mem. 274. 263 pp.

Forbes, W. T. M. 1954. The Lepidoptera of New York and neighboring states. Part III. Noctuidae. Cornell Univ. Agr. Expt. Sta. Mem. 329. 433 pp.

Forbes, W. T. M. 1960. The Lepidoptera of New York and neighboring states. Part IV. Agaristidae through Nymphalidae, including butterflies. Cornell Univ. Agr. Expt. Sta. Mem. 371. 188 pp.

Forbes, W. T. M., and J. G. Franclemont. 1954. Subfamily 3. Cuculliinae, pp. 116–64. *In* Forbes, W. T. M., Lepidoptera of New York and neighboring states. Part III. Cornell Univ. Agr. Exp. Sta. Mem. 329.

Ford, B. J., J. R. Strayer, J. Reid, and G. L. Godfrey. 1975. The literature of arthropods associated with soybeans. IV. A bibliography of the velvetbean caterpillar *Anticarsia gemmatalis* Hübner (Lepidoptera: Noctuidae). Ill. Nat. Hist. Surv. Biol. Notes 92. 15 pp.

Ford, L. T. 1954. The Glyphipterygidae and allied families. Proc. South London Ent. Nat. Hist. Soc. 1952–53:90–99. (8 pls.) (Biology)

Fracker, S. B. 1915. The classification of lepidopterous larvae. Ill. Biol. Mono. 2(1):1–169.

Franclemont, J. G. 1946. A revision of the species of *Symmerista* Hübner known to occur north of the Mexican border (Lepidoptera, Notodontidae). Can. Ent. 78:96–103.

Franclemont, J. G. 1948. *Symmerista* Hübner (*Edema* Walker), pp. 222–24. *In* W. T. M. Forbes. Lepidoptera of New York and neighboring states. Part 2. Cornell Univ. Agr. Expt. Sta. Mem. 274.

*Franclemont, J. G. 1973. The moths of America north of Mexico. Fasc. 20.1, Mimallonoidea (Mimallonidae) and Bombycoidea (Apatelodidae, Bombycidae, Lasiocampidae). London: E. W. Classey and R. B. D. Publ., 86 pp. (11 color pls.)

Franclemont, J. G. 1980. *"Noctua c-nigrum"* in eastern North America, the description of two new species of *Xestia* Hübner (Lepidoptera: Noctuidae: Noctuinae). Proc. Ent. Soc. Wash. 82:576–86.

Franclemont, J. G. 1983. Notodontidae, pp. 112–14. Arctiidae, pp. 114–19. *In* R. W. Hodges et al., (eds.). Check list of the Lepidoptera of America north of Mexico. London: E. W. Classey Ltd. and the Wedge Ent. Res. Found.

Franklin, H. J. 1928. Cape Cod cranberry insects. Mass. Agr. Expt. Sta. Bull. 239:1–67. (3pls.) (Color figs. of *Pseudaletia unipuncta, Spodoptera frugiperda,* and *Xestia* sp.)

Franklin, H. J. 1948. Cranberry insects in Massachusetts. Bull. Mass. Agr. Expt. Sta. 445. 64 pp.

Frediani, D. 1954. Ricerche morfo-biologiche sull *'Acrolepia assectella* Zell. (Lep. Plutellidae) nell'Italia centrale. Redia 39(2):187–249. (Biology, morphology)

Freeman, H. A. 1969. Systematic review of the Megathymidae. J. Lepid. Soc. 23, Suppl. 1:1–59.

Freeman, T. N. 1960. Needle-mining Lepidoptera of pine in North America. Can. Ent. 92, Suppl. 19:1–51. (1 pl.) (Biology)

Freeman, T. N. 1972. The coniferous feeding species of *Argyresthia* in Canada (Lepidoptera: Yponomeutidae). Can. Ent. 104:687–97. (Biology, taxonomy)

French, G. H. 1884. Preparatory stages of *Drasteria erichtea* (sic) Cramer. Papilio 4:148–49.

French, G. H. 1885. Preparatory stages of *Icthyura palla* French, with notes on the species. Can. Ent. 17:41–44.

French, G. H. 1886. Larva of *Seirodonta bilineata* Pack. Can. Ent. 18:49–50.

French, G. H. 1889. Preparatory stages of *Leptarctia californiae* Walker, with notes on the genus. Can. Ent. 21:210–13, 221–26.

Frick, K. F. 1964. *Leucoptera spartifoliella,* an introduced enemy of Scotch broom in the western United States. J. Econ. Ent. 57:589–91.

Friese, G. 1960. Revision der palaarktischen Yponomeutidae unter besonderer Berücksichtigung der Genitalien (Lepidoptera). Beitr. Ent. (Berlin) 10:1–131. (Host records, taxonomy)

Friese, G. 1969. Beiträge zur Insekten-Fauna der DDR: Lepidoptera-Argyresthiidae. Beitr. Ent. (Berlin) 19:693–752. (2 pls.) (Host records, taxonomy European species)

Frost, S. W. 1972. Notes on *Urodus parvula* (Henry Edwards) (Lepidoptera: Yponomeutidae). J. Lepid. Soc. 26:173–77. (Biology)

Furniss, R. L., and V. M. Carolin. 1977. Western forest insects. U.S. Dept. Agr., For. Ser. Misc. Publ. 1339. 654 pp.

Gaedike, R. 1966. Die genitalien der europäischen Epermeniidae (Lepidoptera: Epermeniidae). Beitr. Ent. (Berlin) 16:633–92. (Host plant records)

Gaedike, R. 1970. Revision der paläarktischen Acrolepiidae (Lepidoptera). Ent. Abh. (Dresden) 38:1–54. (Biology, taxonomy palearctic species)

Gaedike, R. 1974. Revision der paläarktischen Douglasiidae (Lepidoptera). Acta Faun. Ent. Mus. Nat. Prag. 15:79–101. (Host records; taxonomy of Palearctic species)

Gaedike, R. 1977. Revision der nearktischen und neotropischen Epermeniidae (Lepidoptera). Beitr. Ent. (Berlin) 27:301–12. (Biology, taxonomy)

Gaedike, R. 1978. Versuch der phylogenitischen Gleiderung der Epermeniidae der Welt (Lepidoptera). Beitr. Ent. (Berlin) 28:201–209. (Phylogeny)

Gardiner, B. O. C. 1974. *Pieris brassicae* L. established in Chile; another Palearctic pest crosses the Atlantic (Pieridae). J. Lepid. Soc. 28:269–77. (Description, biology)

Gardner, J. C. M. 1943. Immature stages of Indian Lepidoptera (5). Indian J. Ent. 5:89–102.

Gardner, J. C. M. 1946. On larvae of the Noctuidae (Lepidoptera). II. Trans. Roy. Ent. Soc. Lond. 97:237–52.

Gardner, J. C. M. 1947. On the larvae of the Noctuidae. III. Trans. Roy. Ent. Soc. Lond. 98:59–89.

Gardner, J. C. M. 1948. On larvae of the Noctuidae (Lepidoptera). IV. Trans. Roy. Ent. Soc. Lond. 99:291–318.

Garman, H. 1895. Cutworms in Kentucky. Kentucky Agr. Expt. Sta. Bull. 58:89–109.

Geiger, H. J. 1980 (1981). Enzyme electrophoretic studies on the genetic relationships of pierid butterflies (Lepidoptera: Pieridae). I. European taxa. J. Res.Lepid. 19:181–95.

Genung, W. G. 1959. Notes on the syntomid moth *Lymire edwardsii* (Grote) and its control as a pest of *Ficus* in south Florida. Fla. Ent. 42(1):39–42.

Gerasimov, A. M. 1935. Zur Frage der Homodyamie der Borsten von Schmetterlingsraupen. Zool. Anz. 112:117–94.

Gerasimov, A. M. 1937. Beitrag zur Systematik der Psychiden auf Grund der Erforschung der Raupen. Zool. Anz. 120(1–2):7–17.

Gerasimov, A. M. 1952. Lepidoptera, Part 1. Caterpillars. Zool. Inst. Acad. Nauk USSR. New Ser. No. 56, Fauna USSR. Insects-Lepidoptera. 2(1):1–338. (In Russian, written before 1942 when Gerasimov died). (Introduction of 120 pages followed by biology of caterpillars and coverage of the more primitive groups through part of Psychidae.) English translation by R. Ericson, 1954, for U.S. Dept. Agr. Syst. Ent. Lab.

Gibbs, G. W. 1983. Evolution of Micropterygidae (Lepidoptera) in the SW Pacific. GeoJournal 7.6:505–10.

Gibson, A. 1902. An interesting caterpillar. Ottawa Nat. 16:161.

Gibson, A. 1903. Notes on Canadian species of the genus *Apantesis* (Arctia), with special reference to the larvae. Can. Ent. 35(5):111–23, 143–54.

Gibson, A. 1909. *Hydroecia micacea* Esper in Canada. 39th Ann. Rept. Ent. Soc. Ontario, 1908:49–51.

Gibson, A. 1915. Cutworms and their control. Can. Dept. Agr. Ent. Bull. 10. 31 pp.

Godfrey, G. L. 1971. The larvae of *Lithacodia muscosula, L. carneola,* and *Neoerastria apicosa*. J. Kan. Ent. Soc. 44:390–97.

Godfrey, G. L. 1972. A review and reclassification of larvae of the subfamily Hadeninae (Lepidoptera: Noctuidae) of America north of Mexico. USDA, Tech. Bull. 1450. 265 pp.

Godfrey, G. L. 1980. Larval descriptions of *Renia hutsoni, R. rigida,* and *R. mortualis* with a key to larvae of *Renia* (Lepidoptera: Noctuidae). Proc. Ent. Soc. Wash. 82:457–68.

Godfrey, G. L. 1981. Identification and descriptions of the ultimate instar larvae of *Hydraecia immanis* (hop vine borer) and *H. micacea* (potato stem borer) (Lepidoptera: Noctuidae). Ill. Nat. Hist. Surv. Biol. Notes 114. 8 pp.

Godfrey, G. L. 1984. Notes on the larva of *Cargida pyrrha* (Notodontidae). J. Lepid. Soc. 38:88–91.

Godfrey, G. L. and F. W. Stehr. 1985. Note on Crumb's "liberae et confluentae" couplet (Noctuidae). J. Lepid. Soc. 39:57–59.

Gomes, J. G., and C. H. Reiniger. 1939. Nota previa sobre uma nova praga da lararijeira. Bol. Soc. Bras. Agron. 11:26. (Larva of *Zadalcera fumata*)

Gould, E. 1931. The pistol-case bearer. W. Va. Univ. Agr. Expt. Sta. Bull. 246. 12 pp.

Gozmany, L. A., and L. Vari. 1973. The Tineidae of the Ethiopian region. Transvaal Museum Memoir No. 18. 244 pp. (570 figures)

Graf, J. E. 1917. The potato tuber moth *Phthorimaea operculella* Zeller. USDA, Bur. Ent. Bull. 427. 56 pp.

Grandi, G. 1932–33. Morfologia ed ecologia comparate di insetti a regime specializzato. IV. La morfologia comparata di vari stati larvali di 30 Microlepidopteri minatori appartenenti a 15 generi ed a 11 famiglie. Boll. Lab. Ent. Bologna 5:143–307. (129 figs.)

Grimble, D. G., and J. D. Kasile. 1974. A sequential sampling plan for saddled prominent eggs. AFRI Res. Rept. No. 15. State Univ. Coll. Envir. Sci. For., Syracuse Univ., Syracuse, N.Y. 15 pp.

Grimble, D. G., and R. G. Newell. 1972a. Saddled prominent pupal sampling and related studies in New York State. AFRI Res. Rept. No. 11. State Univ. Coll. Envir. Sci. For., Syracuse University, Syracuse, N.Y. 31 pp.

Grimble, D. G., and R. G. Newell. 1972b. Saddled prominent oviposition studies in New York and adjacent states. AFRI Res. Rept. No. 12. State Univ. Coll. Envir. Sci. For., Syracuse University, Syracuse, N.Y. 42 pp.

Grimble, D. G., and R. G. Newell. 1973. Saddled prominent moth emergence and oviposition period. AFRI Res. Rept. No. 13. State Univ. Coll. Envir. Sci. For., Syracuse Univ., Syracuse, N.Y. 9 pp.

Grimes, L. R. and H. H. Neunzig. 1984. The larvae and pupae of three phycitine species (Lepidoptera: Pyralidae) that occur in Florida. Proc. Ent. Soc. Wash. 86: 411–21.

Griswold, G. H. 1934. Oviposition in the columbine borer, *Papaipema purpurifascia* (G. & R.) and the iris borer, *Macronoctua onusta* Grt. Ann. Ent. Soc. Amer. 27:545–49.

Grote, R. A., and C. T. Robinson. 1866. Lepidopterological notes and descriptions, No. 2. Proc. Ent. Soc. Phil. 6:1–30.

Grünberg, K. 1909. Zur Metamorphose von *Castnia acraeoides* Gray (Lep.). Deut. Ent. Zeit. 1909:127–30.

Guppy, J. 1904. Notes on the habits and early stages of some Trinidad butterflies. Trans. Ent. Soc. Lond. 52:225–28. (2 pls.) (Describes myrmecophilous glands in some riodinids)

Gustafsson, B. 1981. Characters of systematic importance in European Nepticulidae larvae (Lepidoptera). Ent. Scand. 12:109–16.

Habeck, D. H., and F. W. Mead. 1982. Edward's wasp moth, *Lymire edwardsii* (Grote), (Ctenuchidae: Lepidoptera). Fla. Dept. Agr. Cons. Serv. Ent. Circ. 234. 2 pp.

Hain, F., and W. Wallner. 1973. The life history, biology, and parasites of the pine candle moth, *Exoteleia nepheos* (Lepidoptera: Gelechiidae), on Scotch pine in Michigan. Can. Ent. 105(1):157–65.

Harcourt, D. G. 1956. Biology of the diamondback moth, *Plutella maculipennis* (Curt.) (Lepidoptera: Plutellidae), in eastern Ontario. I. Distribution, economic history, synonymy, and general descriptions. 37th Rep. Quebec Soc. Prot. Plants 1955:155–60. (Biology)

Harcourt, D. G. 1957. Biology of the diamondback moth, *Plutella maculipennis* (Curt.) (Lepidoptera: Plutellidae), in eastern Ontario. II. Life-history, behaviour, and host relationships. Can. Ent. 89:554–64. (Biology)

Hardwick, D. F. 1965. The corn earworm complex. Mem. Ent. Soc. Can. 40:1–247.

Harris, T. W. 1841. A report on the insects of Massachusetts injurious to vegetation. Cambridge. 459 pp.

Harrison, J. 1908. The glands of pierid larvae. Ent. Rec. 20:253–54. (Protrusible larval glands)

Harrison, J. O. 1963. On the biology of three banana pests in Costa Rica (Lepidoptera: Limacodidae, Nymphalidae). Ann. Ent. Soc. Amer. 56:87–94.

Hasenfuss, I. 1960. Die Larvalsystematik der Zünsler (Pyralidae). Akad.-Verlag. Berlin. 263 pp. (Includes keys to subfamilies (tribes) based on European species)

Hawley, I. M. 1918. Insects injurious to the hop in New York with special reference to the hop grub and the hop redbug. Cornell Univ. Agr. Expt. Sta. Mem. 15:141–224.

Heath, J. 1976a. Eriocraniidae. *In* J. Heath (ed.). The moths and butterflies of Great Britain and Ireland, vol. 1: Micropterigidae-Heliozelidae. London. 343 pp.

Heath, J. 1976b. Micropterigidae. *In* J. Heath (ed.). The moths and butterflies of Great Britain and Ireland, vol. 1: Micropterigidae-Heliozelidae. London. 343 pp.

Heath, J., and E. C. Pelham-Clinton. 1976. Incurvariidae, pp. 277–300. *In* J. Heath (ed.). The moths and butterflies of Great Britain and Ireland, vol. 1: Micropterigidae-Heliozelidae. London. 343 pp.

Heinemann, H. 1870. Die Schmetterlinge Deutschlands und der Schweiz. 2d Ab., vol. 2 (pts. 1–2) (Klein-Schmetterlinge), Braunschweig. 927 pp.

Heinrich, C. 1916. On the taxonomic value of some larval characters in the Lepidoptera. Proc. Ent. Soc. Wash. 18:154–64.

Heinrich, C. 1918. On the lepidopterous genus *Opostega,* and its larval affinities. Proc. Ent. Soc. Wash. 20(2):27–34.

Heinrich, C. 1919. Note on the European corn borer (*Pyrausta nubilalis*) and its nearest American allies with descriptions of larvae, pupae, and one new species. J. Agr. Res. 18:171–78.

Heinrich, C. 1921a. On some forest Lepidoptera with descriptions of some new species, larvae and pupae. Proc. U.S. Nat. Mus. 57:53–96.

Heinrich, C. 1921b. Some Lepidoptera likely to be confused with the pink bollworm. J. Agr. Res. 20:807–36.

Heinrich, C. 1923. Revision of the North American moths of the subfamily Eucosminae of the family Olethreutidae. U.S. Nat. Mus. Bull. 123. 302 pp. (59 pls.)

Heinrich, C. 1926. Revision of the North American moths of the subfamilies Laspeyresiinae and Olethreutinae. U.S. Nat. Mus. Bull. 132. 221 pp. (76 pls.)

Heinrich, C. 1956. American moths of the subfamily Phycitinae. U.S. Nat. Mus. Bull. 207:1–581. (Some information on immatures, particularly cactus-feeding species)

Heinrich, C., and J. J. DeGryse. 1915. On *Acrocercops strigifinitelia* Clemens. Proc. Ent. Soc. Wash. 17:6–23.

Heitzman, R. 1961. The life history of *Adelocephala quadrilineata* (Saturniidae). J. Lepid. Soc. 15:233–34.

Heitzman, R. L., and D. H. Habeck. 1979(1981). Taxonomic and biological notes on *Bellura gortynoides* Walker (Noctuidae). J. Res. Lep. 18:228–35.

Heppner, J. B. 1977. The status of the Glyphipterigidae and a reassessment of relationships in yponomeutoid families and ditrysian superfamilies. J. Lepid. Soc. 31:124–34. (Phylogeny)

Heppner, J. B. 1978. Transfers of some Nearctic genera and species of Glyphipterigidae (*auctorum*) to Oecophoridae, Copromorphidae, Plutellidae, and Tortricidae (Lepidoptera). Pan-Pac. Ent. 54:48–55. (Phylogeny)

Heppner, J. B. 1979a. Brachodidae, a new family name for Atychiidae (Lepidoptera: Sesioidea). Ent. Ber. (Amsterdam) 39:127–28. (Taxonomy)

Heppner, J. B. 1979b. *Eutromula pariana* (Clerck) (Lepidoptera: Choreutidae), the correct name of the apple-and-thorn skeletonizer. J. Ent. Soc. Br. Columbia 75:40–41. (Taxonomy)

Heppner, J. B. 1980. Revision of the new genus *Diploschizia* (Lepidoptera: Glyphipterigidae) for North America. Fla. Ent. 64:309–336. (Taxonomy, biology, chaetotaxy)

Heppner, J. B. 1982a. Synopsis of the Hilarographini (Lepidoptera: Tortricidae) of the world. Proc. Ent. Soc. Wash. 84:704–15.

Heppner, J. B. 1982b. Millierinae, a new subfamily of Choreutidae, with new taxa from Chile and the United States (Lepidoptera: Sesioidea). Smithsonian Contr. Zool. No. 370. 27 pp. (Biology, chaetotaxy)

Heppner, J. B. 1984. Revision of the Oriental and Nearctic genus *Ellabella* (Lepidoptera: Copromorphidae). J. Res. Lepid. 23:50–73.

Heppner, J. B. 1985. The sedge moths of North America (Lepidoptera: Glyphipterigidae). Gainesville: Flora and Fauna Publ. (handbook No. 1). 254 pp.

Heppner, J. B. 1986. Revision of the New World genus *Lotisma* (Lepidoptera: Copromorphidae). Pan-Pac. Ent. (In press) (Biology, chaetotaxy)

Heppner, J. B., and G. W. Dekle. 1975. Mimosa webworm, *Homadaula anisocentra* Meyrick (Lepidoptera: Plutellidae). Fla. Dept. Agr. Ent. Circ. 157:1–2. (Biology, taxonomy)

Heppner, J. B., and W. D. Duckworth. 1981. Classification of the superfamily Sesioidea (Lepidoptera: Ditrysia). Smithsonian Cont. Zool. No. 314. 185 pp. (Biology, chaetotaxy, taxonomy)

Hering, E. M. 1951. Biology of the leafminers. Gravenhage: W. Junk. 420 pp.

Herrebout, W. M., P. J. Kuyten, and J. T. Wiebes. 1976. Small ermine moths and their host relationships. Symp. Biol. Hung. 16:91–94.

Herrick, G. W., and J. D. Detwiler. 1919. Notes on the repugnatorial glands of certain notodontid caterpillars. Ann. Ent. Soc. Amer. 12:44–48.

Herzog, D. C. 1980. Sampling soybean looper on soybean, pp. 141–68. *In* M. Kogan and D. C. Herzog (eds.). Sampling methods in soybean entomology. New York: Springer-Verlag.

Hill, A. R. 1952. The bionomics of *Lampronia rubiella* (Bjerkander), the raspberry moth, in Scotland. J. Hort. Sci. 27:1–13. (1 pl.) (Incurvariidae)

Hill, C. C. 1925. Biological studies of the green cloverworm. USDA Bull. 1336. 19 pp.

Hinds, W. E. 1912. Cotton worm or "caterpillar". Alabama Agr. Expt. Sta. Bull. 164:134–60.

Hinds, W. E. 1930. The occurrence of *Anticarsia gemmatilis* (sic) as a soybean pest in Louisiana in 1929. J. Econ. Ent. 23:711–14. (3 pls.)

Hinks, C. F., and J. R. Byers. 1976. Biosystematics of the genus *Euxoa* (Lepidoptera: Noctuidae). V. Rearing procedures, and life cycles of 36 species. Can. Ent. 108:1345–57.

Hinton, H. E. 1942. Notes on the larvae of three common injurious species of *Ephestia*. Bull. Ent. Res. 33:21–25. (Pyralidae)

Hinton, H. E. 1943. The larvae of Lepidoptera associated with stored products. Bull. Ent. Res. 34:163–212. (Includes key to, and descriptions of, Pyralidae (and other families) feeding in stored products; also includes a few species not associated with stored products)

Hinton, H. E. 1946. On the homology and nomenclature of the setae of lepidopterous larvae, with some notes on the phylogeny of the Lepidoptera. Trans. Roy. Ent. Soc. Lond. 97:1–37.

Hinton, H. E. 1947. The dorsal cranial areas of caterpillars. Ann. Mag. Nat. Hist. 14:843–52.

Hinton, H. E. 1951. Myrmecophilous Lycaenidae and other Lepidoptera—a summary. Proc. South Lond. Ent. Nat. Hist. Soc. 1949–50:111–75.

Hinton, H. E. 1952. The structure of the larval prolegs of the Lepidoptera and their value in the classification of the major groups. Lepid. News. 6:1–6.

Hinton, H. E. 1955a. On the structure, function, and distribution of the prolegs of the Panorpoidea, with a criticism of the Berlese-Imms theory. Trans. Roy. Ent. Soc. 106(13):455–541.

Hinton, H. E. 1955b. On the taxonomic position of the Acrolophinae, with a description of the larva of *Acrolophus rupestris* Walsingham (Lepidoptera: Tineidae). Trans. Roy. Ent. Soc. Lond. 105:227–31.

Hinton, H. E. 1956. The larvae of the species of Tineidae of economic importance. Bull. Ent. Res. 47(2):251–346.

Hinton, H. E. 1958. The phylogeny of the panorpoid orders. Ann. Rev. Ent. 3:181–206.

Hinton, H. E. 1979. Biology of insect eggs, vol. 1, structure and biology, pp. 1–474; vol. 2, eggs of insect families, pp. 475–778; vol. 3, bibliography and species, author and subject indices, pp. 779–1113. New York: Pergamon Press.

Hintze, C. 1969. Histologische Untersuchungen am Wehrsekretbeutel von *Cerura vinula* L. und *Notodonta anceps* Goeze (Notodontidae, Lepidoptera). Zeit. Morphol. Tiere 64:1–8.

Hitchcock, S. W. 1958. The orange-striped oakworm. Conn. Agr. Expt. Sta. Circ. 204. 8 pp.

Hitchcock, S. W. 1961. Caterpillars on oaks. Conn. Agr. Expt. Sta. Bull. 641. 15 pp.

Hixson, E. 1941. The walnut *Datana*. Oklahoma Agric. Expt. Sta. Bull. B–246. 29 pp.

Hodges, R. W. 1962. A revision of the Cosmopterygidae of America north of Mexico, with a definition of the Momphidae and Walshiidae (Lepidoptera: Gelechioidea). Ent. Amer. 42(N.S.):1–166.

Hodges, R. W. 1964. A review of the North American moths of the family Walshiidae (Lepidoptera: Gelechioidea). Proc. U.S. Nat. Mus. 115:289–330.

Hodges, R. W. 1966. Review of the new world species of *Batrachedra* with description of three new genera (Lepidoptera: Gelechioidea). Trans. Amer. Ent. Soc. 92:585–651.

Hodges, R. W. 1971. The moths of America north of Mexico. Fasc. 21. Sphingoidea. London: E. W. Classey and R. B. D. Publ., Inc. 157 pp. (14 color plates of adults)

Hodges, R. W. 1974. The moths of America north of Mexico. Fasc. 6.2, Gelechioidea (Oecophoridae). London: E. W. Classey and R. B. D. Publications. 142 pp. (7 color pls.)

Hodges, R. W. 1978. The moths of America north of Mexico. Fasc. 6.1, Gelechioidea (Cosmopterigidae). London: E. W. Classey and The Wedge Ent. Res. Found., 166 pp. (6 color pls. of adults)

Hodges, R. W., et al. 1983. Check list of the Lepidoptera of America north of Mexico. The Wedge Ent. Res. Found., Nat. Mus. Nat. Hist., Wash., D.C. 284 pp.

Hoerner, J. L. 1948. The cutworm *Loxagrotis albicosta* on beans. J. Econ. Ent. 41:631–35.

Hogue, C. L., F. P. Sala, N. McFarland, and C. Henne. 1965(1966). Systematics and life history of *Saturnia* (*Calosaturnia*) *albofasciata* in California (Saturniidae). J. Res. Lepid. 4:173–84.

*Holland, W. J. 1898 (1931). The butterfly book (rev. ed.). New York: Doubleday and Co. 436 pp. (77 pls.) (Descriptions, biology)

*Holland, W. J. 1903. (1916) (later reprints). The moth book. New York: Doubleday, Page & Co. 479 pp. (48 pls.)

Holloway, J. D. 1983. The moths of Borneo: family Notodontidae. Malay. Nat. J. 37:1–107.

Holloway, T. E., W. E. Haley, U. C. Loftin, and C. Heinrich. 1928. The sugar cane moth borer in the United States. USDA, Tech. Bull. 41. 77 pp. (Includes descriptions of immatures)

Hopp, W. 1928. Beitrag zur Kenntnis der Dalceriden. Dt. Ent. Ztschr. Iris 42:284–87. (Molting of dalcerid larvae, especially *Acraga flava*)

Hooker, W. A. 1908. Injury to oak forests in Texas by *Heterocampa manteo* Doubleday. Proc. Ent. Soc. Wash. 10:8–9.

Horn, D. J. 1969. A larva of *Citheronia sepulchralis* (sic) from New Jersey. J. Lepid. Soc. 23:25.

Houlbert, C. 1918. Revision monographique de las Sous Famille des Castniinae. Etudes Lep. Comp., XV. 746 pp.

Howard, L. O., and F. H. Chittenden. 1909. The green-striped maple worm (*Anisota rubicunda* Fab.). USDA, Bur. Ent. Circ. 110. 7 pp.

Howe, W. H. (ed.) 1975. The butterflies of North America. Garden City, N.Y.: Doubleday and Co., Inc. 646 pp. (97 pls.) (Descriptions, biology)

Hudon, M., and J. P. Perron. 1970. First record of *Cisseps fulvicollis* (Lep.: Amatidae) as an economic destructive insect on grain corn in Canada. Can. Ent. 102:1052–54.

Hukusima, S. 1953. Ecological studies on the peach fruit moth, *Carposina niponensis* Walsingham. 1: On the diurnal rhythm of adult. Oyo-Dobutsugaku-Zasshi [Jap. Soc. Appl. Zool.] 18(1–2):55–60. (Biology) (Carposinidae)

*Igarishi, S. 1979. Papilionidae and their early stages. 2 Vols. 218 pp. (102 pls.)

Imms, A. D. 1970. Superfamily Noctuoidea, pp. 564–568. *In* A. D. Imms. A general textbook of entomology. London: Methuen & Co., Ltd.

*Issiki, S., A. Mutuura, Y. Yamamoto, I. Hattori, H. Kuroko, T. Kodama, T. Yasuda, S. Moriuti, and T. Saito. 1965 (vol. 1), 1969 (vol. 2). Early stages of Japanese moths in colour. 1:237 pp., (lower taxa); 2:237 pp., (higher taxa); Hoikuska, Osaka, Japan. (Larvae and adults) (In Japanese)

Ives, W. G. H. 1973. Heat units and outbreaks of the forest tent caterpillar, *Malacosoma disstria* (Lepidoptera: Lasiocampidae). Can. Ent. 105:529–43.

Jäckh, E. 1955. Schutzvorrichtung zum Bau des Vertuppungskekens bei Arten der Gattungen *Bucculatrix* Z. und Lyonetia Hb. (Lep., Lyonetiidae). Zeits. Ent. Gesell. Wiener. 40:118–21.

Janzen, D. H. 1982. Guia para la identificacion de mariposas nocturnas de la familia Saturniidae del Parque Nacional Santa Rosa, Guanacaste, Costa Rica. Brenesia 19/20: 255–99.

Jayewickreme, S. H. 1940. A comparative study of the larval morphology of leafmining Lepidoptera in Britain. Trans. Roy. Ent. Soc. Lond., pp. 63–105.

*Johnson, W. T., and H. H. Lyon. 1976. Insects that feed on trees and shrubs. Ithaca, N.Y.: Cornell Univ. Press. 463 pp. (212 color pls.) (Life histories, damage)

Jones, E. D., and F. Moore. 1882. Metamorphoses of Lepidoptera from Santo Paulo, Brazil, in the Free Public Museum, Liverpool. With nomenclature and descriptions of new forms. And an introductory note. Proc. Lit. Philo. Soc. Liverpool 36:325–77. (Larva of *Pinconia ochracea* Moore, now *Acraga moorei* Dyar, p. 364)

Jones, F. M. 1927. The mating of Psychidae. Trans. Amer. Ent. Soc. 53:293–312.

Jones, F. M., and H. Parks. 1928. The bagworms of Texas. Texas Exp. Stat. Bull. 382. pp. 1–26.

Jones, P. R. 1909. The grape-leaf skeletonizer. USDA, Bur. Ent. Bull. 68:77–90. (Zygaenidae)

Jones, T. H. 1918. II. The granulated cutworm, an important enemy of vegetable crops in Louisiana, pp. 7–14. *In* T. H. Jones. Miscellaneous truck-crop insects in Louisiana. USDA, Bull. 703. Washington, D.C.: Government Printing Office.

Jörgensen, P. 1930. Las especies de Castniidae de la Argentina y Paraguay (Lepidoptera). Rev. de la S. E. A. 14:175–80.

Karsholt, D. and Nielsen, E. S. 1984. A taxonomic review of the stem moths, *Ochsenheimeria* Hübner, of northern Europe (Lepidoptera: Ochsenheimeriidae) Ent. Scand. 15:233–47.

Kato, M. 1940. A monograph of Epipyropidae (Lepidoptera). Ent. World 8(72):67–94. (4 pl.)

Kearby, W. H. 1975. Variable oakleaf caterpillar larvae secrete formic acid that causes skin lesions (Lepidoptera: Notodontidae). J. Kan. Ent. Soc. 48:280–82.

Kearfott, W. D. 1903. Descriptions of new Tineoidea. J. N.Y. Ent. Soc. 11:145–65. (Biology, taxonomy)

Keen, F. P. 1952. Insect enemies of western forests (rev. ed.). U.S. Dept. Agr. Misc. Publ. 273. 280 pp. (Replaced by Furniss & Carolin, 1977).

Kellogg, V. L., and F. J. Jack. 1896. The California phryganidian (*Phryganidia californica* Pack.). Proc. Calif. Acad. Sci. (2d ser.) 5:562–70.

Kendall, R. 1976. Larval foodplants and life history notes for some metalmarks (Lepidoptera: Riodinidae) from Texas and Mexico. Bull. Allyn Mus. No. 32. 12 pp.

Kendall, R. O., and C. A. Kendall. 1971. Lepidoptera in the unpublished field notes of Howard George Lacey, naturalist (1856–1929). J. Lepid. Soc. 25:29–44.

Khalsa, M. S., M. Kogan, and W. H. Luckmann. 1979. *Autographa precationis* in relation to soybean: life history, and food intake and utilization under controlled conditions. Environ. Ent. 8:117–22.

Khare, B. P., and R. B. Mills. 1968. Development of angoumois grain moths in kernels of wheat, sorghum and corn as affected by site of feeding. J. Econ. Ent. 61:450–52.

Keifer, H. H. 1937. California Microlepidoptera, XII. Bull. Calif. Dept. Agr. 26:334–38. (Biology, chaetotaxy)

Keifer, H. H. 1943. Systematic entomology. *In* W. J. Cecil, et al. (eds.), Bull. Dept. Agr. Calif. (Sacramento) 32:256–60.

Kimball, C. P. 1965. Arthropods of Florida and neighboring land areas. 1. Lepidoptera of Florida, an annotated checklist. Div. Plant Industry, State of Florida. Gainesville: Department of Agriculture. 363 pp.

King, K. M. 1928. *Barathra configurata* Wlk., an armyworm with important potentialities on the northern prairies. J. Econ. Ent. 21:279–93.

Kinser, P. D., and H. H. Neunzig. 1981. Description of the immature stages and biology of *Synclita tinealis* Munroe (Lepidoptera: Pyralidae: Nymphulinae). J. Lepid. Soc. 35:137–46.

Kirkpatrick, T. W. 1947. Notes on a species of Epipyropidae (Lepidoptera) parasitic on *Metaphaena* species (Hemiptera: Fulgoridae) at Amani, Tanganyika. Proc. Roy. Ent. Soc. Lond. (Ser. A) 22(4–6):35–64.

Kirkpatrick, T. W. 1953. Notes on minor insect pests of cacao in Trinidad. A report on cacao research. Imp. Coll. Trop. Agr., Trinidad. pp. 67–72. (Immatures of *Acraga ochracea* Walker)

Kitching, R. L. 1983. Myrmecophilous organs of the larva and pupa of the lycaenid butterfly *Jalmenus evagoras* (Donovan). J. Nat. Hist. 17:474–81.

Klots, A. B. 1933. A generic revision of the Pieridae (Lepidoptera). Ent. Amer. (n.s.) 12:139–242.

*Klots, A. B. 1951. A field guide to the butterflies of North America, east of the Great Plains. Boston: Houghton Mifflin Co. 349 pp. (Descriptions, biology)

Klots, A. B. 1966. Life history notes on *Lagoa laceyi* (Barnes & McDunnough). J. N.Y. Ent. Soc. 74:140–42.

Klots, A. B. 1967a. Larval dimorphism and other characters of *Heterocampa pulverea* (Grote & Robinson) (Lepidoptera: Notodontidae). J. N.Y. Ent. Soc. 75:62–67.

Klots, A. B. 1967b. The adaptive feeding habit of a pine caterpillar. J. N.Y. Ent. Soc. 75:43–44.

Klots, A. B. 1969. Audition by *Cerura* larvae (Lepidoptera: Notodontidae). J. N.Y. Ent. Soc. 77:10–11.

Knight, H. H. 1916. The army-worm in New York in 1914. Cornell Univ. Agr. Expt. Sta. Bull. 376:751–65.

Knutson, H. 1944. Minnesota Phalaenidae (Noctuidae). Minn. Agr. Expt. Sta. Tech. Bull. 165:1–128.

Kodama, T. 1961. The larvae of Glyphipterygidae (Lepidoptera) in Japan (I). Osaka Fac. Agr. Ent. Lab. 6:35–45. (Chaetotaxy)

Koehler, C. S., and M. Tauber. 1964. *Periploca nigra,* a major cause of dieback on ornamental juniper in California. J. Econ. Ent. 57:563–66. (Cosmopterigidae)

Kogan, J., D. K. Sell, R. E. Stinner, J. R. Bradley, Jr., and M. Kogan. 1978. The literature of arthropods associated with soybean. V. A bibliography of *Heliothis zea* (Boddie) and *H. virescens* (F.) (Lepidoptera: Noctuidae). INTSOY (University of Illinois) Series, No. 17. 242 pp.

Korytkowski, C. A., and E. A. Ruiz. 1980a. Estado actual de las plagas de Palma Aceitera (*Elaesis quineensis* Jacquin) en Tananta (Hullaga Central, San Martin, Peru). Rev. Peru Ent. 22(1):17–20. (Castniidae)

Korytkowski, C. A., and E. A. Ruiz. 1980b. El Barreno de los racimos de la Palma Aceitera, *Castnia daedalus* (Cramer) Lepidoptera: Castniidae, en la plantacion de Tacocache. Rev. Peru Ent. 22(1):49–62.

Kozhantshikov, I. V. 1956. Fauna SSSR. Nasekomye Cheshuerkrylye. Chekhlonosy-Meshechnitsy (sem. Psychidae) 3(2):1–516. (In Russian, English translation available from U.S. Dept. of Commerce, Springfield, Va.)

Krishnamurti, B. 1933. On the biology and morphology of *Epipyrops eurybrachydis* Fletcher (Lepidoptera). J. Bombay Nat. Hist. Soc. 36(4):944–49. (2 pls.)

Kristensen, N. P. 1971. The systematic position of the Zeugloptera in the light of recent anatomical investigations. Proc. XIII Int. Cong. Ent. 1:261.

Kristensen, N. P. 1976. Remarks on the family level phylogeny of the butterflies (Insecta: Lepidoptera, Rhopalocera). Z. Zool. Syst. Evol.-forsch. 14:25–33.

Kristensen, N. P. 1978. A new family of Hepialoidea from South America, with remarks on the phylogeny of the suborder Exoporia (Lepidoptera). Ent. Germ. 4:272–94.

Kristensen, N. P. 1984. Studies on the morphology and systematics of primitive Lepidoptera (Insecta) Steenstrupia. 10:141–91.

Kristensen, N. P., and E. S. Nielsen. 1979. A new subfamily of micropterigid moths from South America. A contribution to the morphology and phylogeny of the Micropterigidae, with a generic catalogue of the family (Lepidoptera: Zeugloptera). Steenstrupia 5(7):69–147.

Kristensen, N. P., and E. S. Nielsen. 1983. The *Heterobathmia* life history elucidated: immature stages contradict assignment to suborder Zeugloptera (Insecta, Lepidoptera). Sonderdruck aus Zeitschrift für Zool. Systematik u Evolutionsforschung, 21:101–24.

Kumata, T. 1978. A new stem-miner of alder in Japan, with a review of the larval transformation in the Gracillariidae (Lepidoptera). Ins. Matsum. New Series 13:1–27. (8 figs.)

Kuroko, H. 1961. The life history of *Nemophora vaddei* Rebel (Lepidoptera: Adelidae). Sci. Bull. Fac. Agr. Kyushu Univ. 18(4):323–34. (2 pls.)

Kuroko, H. 1964. Revisional studies on the family Lyonetiidae of Japan (Lepidoptera). Esakia, No. 4. 61 pp. (17 pls.)

Kuznetsov, V. I., and A. A. Stekolnikov. 1973. Phylogenetic relationships in the family Tortricidae (Lepidoptera) treated on the base of study of functional morphology of genital apparatus. Tr. Vses. Entomol. Obshch. 56:18–43.

Kuznetsov, V. I., and A. A. Stekolnikov. 1977. Functional morphology of the male genitalia and phylogenetic relationships of some tribes in the family Tortricidae (Lepidoptera) of the fauna of the Far East. Tr. Zool. Inst. Akad. Nauk SSSR 70:65–97.

LaFontaine, J. D., J. G. Franclemont, and D. C. Ferguson. 1982. Classification and life history of *Acsala anomala* (Arctiidae: Lithosiinae). J. Lepid. Soc. 36(3):218–26.

Lange, W. H. 1941. The artichoke plume moth and other pests injurious to the globe artichoke. Calif. Agr. Expt. Sta. Bull. 653:1–71.

Lange, W. H. 1945. *Autographa egena* (Guen.) a periodic pest of beans in California. Pan-Pac. Ent. 21:13.

Lange, W. H. 1950. Biology and systematics of plume moths of the genus *Platyptilia* in California. Hilgardia 19:561–668. (Eggs, larvae and pupae)

Lange, W. H. 1956a. A generic revision of the aquatic moths of North America (Lepidoptera: Pyralidae, Nymphulinae). Wasmann J. Biol. 14:59–144. (Includes information mostly on adults, but has a brief key to genera of known larvae)

Lange, W. H. 1956b. Aquatic Lepidoptera. *In* Usinger, R. L. Aquatic insects of California. Berkeley: Univ. Calif. Press. 508 pp. (Deals primarily with the adults, but includes notes on eggs, pupae and larvae)

Lange, W. H. 1984. Aquatic and semiaquatic Lepidoptera, pp. 348–360. *In* R. W. Merritt and K. W. Cummins (eds.). An introduction to the aquatic insects of North America, 2d. ed. Dubuque, Ia.: Kendall/Hunt Publ. Co. 734 pp.

Langston, R. L., and O. J. Smith. 1953. Notes on the zygaenid genus *Harrisina* Packard, with special reference to *Harrisina metallica* Stretch. Ent. News 64:253–55.

Lawrence, D. A., and J. C. Downey. 1966. Morphology of the immature stages of *Everes comyntas* Godt. (Lycaenidae). J. Res. Lepid. 5(2):61–96.

Lawrence, J. F., and J. A. Powell. 1969. Host relationships in North American fungus-feeding moths (Oecophoridae, Oinophilidae, Tineidae) Bull. Mus. Comp. Zool. 138:29–51.

Leach, W. E. 1815. *In* Brewster (ed.). The Edinburgh Encyclopedia, vol. 9, Edinburgh. 172 pp.

Learned, E. T. 1926. Notes on the early stages of *Estigmene prima* (Lepidoptera: Arctiidae). Can. Ent. 58:1–2.

Levine, E., and L. Chandler. 1976. Biology of *Bellura gortynoides* (Lepidoptera: Noctuidae), a yellow water lily borer, in Indiana. Ann. Ent. Soc. Amer. 68:405–14.

Levy, R., and D. H. Habeck. 1976. Descriptions of the larvae of *Spodoptera sunia* and *S. latifascia* with a key to the mature *Spodoptera* larvae of the eastern United States (Lepidoptera: Noctuidae). Ann. Ent. Soc. Amer. 68:585–88.

Liebherr, J. 1977. Life history and descriptions of the immature stages of *Macrotheca unipuncta* (Lepidoptera: Pyralidae). Great Lakes Ent. 10:201–207.

Lindsey, A. W., E. L. Bell, and R. C. Williams, Jr. 1931. The Hesperioidea of North America. Denison Univ. Bull., J. Sci. Lab. 26:1–142.

Lindquist, O. H. 1982. Keys to lepidopterous larvae associated with the spruce budworm in northeastern North America. Gt. Lakes For. Res. Centre, Sault Ste. Marie, Ont. 18 pp.

Lindquist, O., and R. Bowser. 1966. A biological study of the leaf miner, *Chrysopeleia ostryaella* Chambers (Lepidoptera: Cosmopterygidae), on ironwood in Ontario. Can. Ent. 98(3):252–58.

Lloyd, J. T. 1914. Lepidopterous larvae from rapid streams. J. N.Y. Ent. Soc. 22:145–52. (Larvae and pupae of *Petrophila* (as *Elophila*)) (Pyralidae)

Loesch, A., and J. Foran. 1979. An outbreak of the saddled prominent caterpillar, *Heterocampa guttivitta* (Lepidoptera: Notodontidae) on Beaver Island. Great Lakes Ent. 12:45–47.

Lorenz, R. E. 1961. Biologie und Morphologie von *Micropterix calthella* (L.). Deut. Ent. Zeits. 8:1–23.

Luff, M. L. 1964. Note on larvae of *Micropteryx* [sic]. Proc. Roy. Ent. Soc. Lond. (C) 29:6.

Lugger, O. 1898. Butterflies and moths injurious to our fruit-producing plants. Univ. Minn. Agr. Expt. Sta. Bull. 61, pp. 55–334. (24 pls.)

Luginbill, P. 1928. The fall army worm. USDA, Tech. Bull. 34. 92 pp.

MacKay, M. R. 1943. The spruce foliage worm and the spruce cone worm. Can. Ent. 75:91–98. (Detailed description of immatures and adult of *Dioryctria reniculelloides*)

MacKay, M. R. 1959. Larvae of the North American Olethreutidae (Lepidoptera). Can. Ent., Suppl. 10:1–338.

MacKay, M. R. 1962. Larvae of the North American Tortricinae (Lepidoptera: Tortricidae). Can. Ent., Suppl. 28:1–182.

MacKay, M. R. 1963. Problems in naming the setae of lepidopterous larvae. Can. Ent. 95:996–99.
MacKay, M. R. 1964. The relationship of form and function of minute characters of lepidopterous larvae, and its importance in life-history studies. Can. Ent. 96:991–1004.
MacKay, M. R. 1968. The North American Aegeriidae (Lepidoptera); a revision based on late-instar larvae. Mem. Ent. Soc. Can. 58:1–112. (Chaetotaxy, taxonomy)
MacKay, M. R. 1972. Larval sketches of some Microlepidoptera, chiefly North American. Mem. Ent. Soc. Can. 88:1–83.
MacNay, C. G. (compiler). 1959. Insects affecting cereal, foliage, and special field crops. Can. Ins. Pest Rev. 37:215–35. (Contains life history information on *Heliothis ononis* (Denis and Schiffermüller))
MacNeill, C. D. 1964. The skippers of the genus *Hesperia* in western North America, with special reference to California (Lepidoptera: Hesperiidae). Univ. Calif. Publ. Ent. 35. 234 pp.
MacNeill, C. D. 1975. Family Hesperiidae. The skippers, pp. 423–578. *In* W. H. Howe (ed.). The butterflies of North America. Garden City, N.Y.: Doubleday and Co., Inc.
Madden, A. H., and F. S. Chamberlin. 1945. Biology of the tobacco hornworm in the southern cigar-tobacco district. U.S. Dept. Agr. Tech. Bull. 896. 51 pp.
Malicky, H. 1970. New aspects on the association between lycaenid larvae (Lycaenidae) and ants (Formicidae, Hymenoptera). J. Lepid. Soc. 24(3):190–202.
Manglitz, G. R., H. J. Gorz, and H. J. Stevens. 1971. Biology of the sweetclover root borer. J. Econ. Ent. 64:1154–58. (Cosmopterigidae)
Marlatt, C. L. 1905. The giant sugar-cane borer (*Castnia licus* Fabr.). USDA, Bur. Ent. Bull. 54:71–75.
Marsh, H. O. 1913a. The horse-radish webworm. USDA, Bull. 109: 71–76. (Biology) (Plutellidae)
Marsh, H. O. 1913b. The striped beet caterpillar. USDA, Bur. Ent. Bull. 127, pt. 2. 18 pp.
Marshall, A. T. 1970. External parasitisation and blood-feeding by the lepidopterous larva *Epipyrops anomala* Westwood. Proc. Roy. Ent. Soc. Lond. (A) 45(10–12):137–40. (3 pls.)
Marshall, A. T., C. T. Lewis, and G. Parry. 1974. Paraffin tubules secreted by the cuticle of a insect *Epipyrops anomala* (Epipyropidae: Lepidoptera). J. Ultrastructure Res. 47:41–60.
Martin, J. E. H. (ed.). 1977. The insects and arachnids of Canada. Pt. 1. Collecting, preparing, and preserving insects, mites, and spiders. Can. Dept. of Agr., Publ. 1643. 182 pp.
Martinova, E. F. 1950. On the structure of the larvae of *Micropteryx* [sic] (Lepidoptera, Micropterygidae [sic]). Ent. Obozrenie 31:142–50. (in Russian).
Martouret, D., P. Real, Y. Arambourg, and A. S. Balachowsky. 1966. Hyponomeutidae. *In* A. S. Balachowsky (ed.), Traité d'Entomologie applique à l'agriculture. Tome II:99–249. Paris: Masson and Editeurs. (Biology)
Mason, A. C. 1922. A new citrus insect. Fla. Ent. 6:43–44. (Noctuidae)
Mason, R. R., and J. W. Baxter. 1970. Food preference in a natural population of the Douglas-fir tussock moth. J. Econ. Ent. 63:1257–59.
Matheny, E. L., and E. A. Heinrichs. 1972. Chorion characteristics of sod webworm eggs. Ann. Ent. Soc. Amer. 65:238–46. (Pyralidae)
Mathur, R. N. 1954. Immature stages of Indian Lepidoptera—Pyralidae, subfamily Pyraustinae. Indian For. Rec. 8:241–65.
Mathur, R. N., and P. Singh. 1956. Immature stages of Indian Lepidoptera 13—Pyralidae, subfamily Pyraustinae. Indian For. Rec. 10:117–48.
Matthewman, W. G. 1937. Observations on the life history and habits of the columbine borer. 67th Ann. Rept. Ent. Soc. Ontario, 1936:69–72.
McCabe, T. L. 1980. A reclassification of the *Polia* complex for North America (Lepidoptera: Noctuidae). N.Y. State Mus. Bull. 432. 141 pp.
McCabe, T. L. 1981. *Clemensia albata,* an algal feeding arctiid. J. Lepid. Soc. 35(1):34–40.
McCorkle, D. V. and P. C. Hammond 1985. Observations on the biology of *Parnassius clodius* (Papilionidae) in the Pacific Northwest. J. Lepid. Soc. 39(3):156–62.
McDaniel, E. I. 1933. Important leaf-feeding and gall-making insects infesting Michigan's deciduous trees and shrubs. Mich. Agr. Expt. Sta. Spec. Bull. 243. 70 pp.
McDunnough, J. 1921. A new British Columbia tussock moth, *Hemerocampa pseudotsugata.* Can. Ent. 53:53–56.
McDunnough, J. 1927. Contribution toward a knowledge of our Canadian plume moths (Lepidoptera). Trans. Roy. Soc. Can. 5:175–89. (Brief information on appearance of a few larvae and figures of several pterophorid pupae)
McDunnough, J. 1936. Further notes on Canadian plume moths—Pterophoridae. Can. Ent. 68:63–69. (Larval description of *Oidaematophorus guttatus*)
McDunnough, J. 1938. Check list of the Lepidoptera of Canada and the United States of America. Part 1. Macrolepidoptera. Mem. So. Calif. Acad. Sci. 1:1–272.
McDunnough, J. 1939. Check list of the Lepidoptera of Canada and the United States of America. Part 2. Microlepidoptera. Mem. So. Calif. Acad. Sci. 2:1–171.
McElvare, R. R. 1941. Validity of the species *Heliothis subflexa* (Gn.) (Lepidoptera). Bull. Brooklyn Ent. Soc. 36:29–30.
McFarland, N. 1961. Notes on the early stages of *Daritis (howardi?)* (Pericopidae) from Cabezon Peak, New Mexico. J. Lepid. Soc. 15(3):172–74.
McFarland, N. 1964. Notes on collecting, rearing, and preserving larvae of Macrolepidoptera. J. Lepid. Soc. 18:201–10.
McFarland, N. 1967(1968). Spring moths (Macroheterocera) of a natural area in northeastern Kansas. J. Res. Lepid. 6:1–18.
McFarland, N. 1971(1973). Egg photographs depicting 40 species of southern Australian moths. J. Res. Lepid. 10:215–47.
McFarland, N. 1979. Annotated list of larval foodplant records for 280 species of Australian moths. J. Lepid. Soc. 33, Suppl. to No. 3. 72 pp.
McGiffin, K. C. and H. H. Neunzig. 1985. A guide to the identification and biology of insects feeding on muscadine and bunch grapes in North Carolina. N. Carolina Agr. Res. Ser. Bull. 470. 93 pp.
McGuffin, W. C. 1958a. Biological and descriptive notes on noctuid larvae. Can. Ent. 90:114–24.
McGuffin, W. C. 1958b. Larvae of the Nearctic Larentiinae (Lepidoptera: Geometridae). Can. Ent. 90(8):1–104. (Larvae only)
McGuffin, W. C. 1964. Setal patterns of the anterior abdominal segments of larvae of the Geometridae (Lepidoptera). Can. Ent. 96:841–49.
McGuffin, W. C. 1967. Guide to the Geometridae of Canada (Lepidoptera). I. Subfamily Sterrhinae. Mem. Ent. Soc. Can. No. 50. 67 pp. (All stages, biology)
McGuffin, W. C. 1972. Guide to the Geometridae of Canada (Lepidoptera). II. Subfamily Ennominae, 1. Mem. Ent. Soc. Can. No. 86. 159 pp.
McGuffin, W. C. 1977. Guide to the Geometridae of Canada (Lepidoptera). II. Subfamily Ennominae, 2. Mem. Ent. Soc. Can. No. 101. 191 pp. (All stages, biology)
McGuffin, W. C. 1981. Guide to the Geometridae of Canada. III. Subfamily Ennominae. 3. Mem. Ent. Soc. Can. No. 117. 153 pp. (All stages, biology)
McIndoo, N. E. 1919. The olfactory sense of lepidopterous larvae. Ann. Ent. Soc. Amer. 12:65–84.
McLeod, J. 1969. On the habits of a jack pine needleminer, *Eucordylea canusella* (Lepidoptera: Gelechiidae), with special reference to its association with a fungus, *Aureobasidium pullans* (Maniliales (Deuteromycetes) Dematiaceae). Can. Ent. 101(2):166–80.
Menken, S. B. J. 1981. Host races and sympatric speciation in small ermine moths, Yponomeutidae. Ent. Exp. and Appl. 30:280–92.
Merzheyevskaya, O. I. 1967. The larvae of Noctuidae, their biology and morphology (with keys). Academy of Science, Byelorussia, USSR (Dept. Zool. and Parasit.), Science and Technology, Minsk. 452 pp. (In Russian).

Michener, C. D. 1952. The Saturniidae (Lepidoptera) of the Western Hemisphere, morphology, phylogeny, and classification. Bull. Amer. Mus. Nat. Hist. 98(5):335–502.

Miller, J. Y. 1980. Studies in the Castniidae. II. Descriptions of three new species of *Castnia,* s.l. Bull. Allyn Mus., No. 34. 13 pp.

Miller, J. Y. 1986. The taxonomy, phylogeny, and zoogeography of the Neotropical moth subfamily Castniinae (Lepidoptera: Castnioidea: Castniidae). Ph.D. dissertation, Univ. of Florida, Gainesville, 571 pp.

Miller, L. D. 1981. Superfamily Nymphaloidea Swainson, 1827 (Satyrs) (Satyridae), pp. 267–90. *In* C. D. Ferris and F. M. Brown (eds.). Butterflies of the Rocky Mountain states. Norman, Okla.: Univ. of Okla. Press.

Miller, L. D., and F. M. Brown. 1981. A catalogue/checklist of the butterflies of America north of Mexico. Mem. Lepid. Soc. 2:1–280.

Millers, I., and W. E. Wallner. 1975. The redhumped oakworm. USDA, Forest Pest Leaflet 153. 6 pp.

Milne, L., and M. Milne. 1980. The Audubon Society field guide to North American insects and spiders. New York: Alfred A. Knopf. 989 pp.

Miyashita, K., E. Kawamura, and S. Ikeuchi. 1955. Studies on the seasonal behavior of the peach fruit moth (*Carposina niponensis* Walsingham), 1: On the periods of appearance of the peach fruit moth. Res. Bull. Hokkaido Nat. Agr. Expt. Sta. 68:71–78. (Biology, Carposinidae)

Montgomery, S. L. 1982. Biogeography of the moth genus *Eupithecia* in Oceania and the evolution of ambush predation in Hawaiian caterpillars (Lepidoptera: Geometridae). Entomologia Generalis 8(1):27–34.

Moriuti, S. 1961. Three important species of *Acrolepia* (Lepidoptera: Acrolepiidae) in Japan. Univ. Osaka Pref. Ent. Lab., Publ. 6:23–33. (Biology, chaetotaxy)

Moriuti, S. 1963. Remarks on the *Paraprays anisocentra* (Meyrick, 1922) (Plutellidae), with descriptions of its larva and pupa. Trans. Ent. Soc. Japan 14:52–59. (Biology, chaetotaxy)

Moriuti, S. 1965. Studies on the Yponomeutoidea (XII). *Argyresthia* species (Lepidoptera: Argyresthiidae) attacking coniferous plants in Japan. Bull. Univ. Osaka Pref. (Ser. B) 16:65–80. (Biology, chaetotaxy)

Moriuti, S. 1969. Argyresthiidae (Lepidoptera) of Japan. Bull. Univ. Osaka Pref. (Ser. B) 21:1–50. (Biology, chaetotaxy)

Moriuti, S. 1977. Yponomeutidae *s. lat.* (Insecta: Lepidoptera). *In* Fauna Japonica. Tokyo: Keigaku Publ. Co. 327 pp. (96 pls.; Biology, chaetotaxy, taxonomy)

**Mosher, E. 1914. The classification of the pupae of the Ceratocampidae and Hemileucidae. Ann. Ent. Soc. Amer. 7:277–300.

**Mosher, E. 1916a. A classification of the Lepidoptera based on characters of the pupa. Bull. Ill. State Lab. Nat. Hist. 12, pp. 17–159. (27 pls.) (Reprinted, 1969, Ent. Reprint Specialists, E. Lansing, Mich. (Los Angeles)

**Mosher, E. 1916b. The classification of the pupae of the Saturniidae. Ann. Ent. Soc. Amer. 9:136–58.

**Mosher, E. 1917. Pupae of some Maine species of Notodontoidea. Maine Agr. Expt. Sta. Bull. 259, pp. 29–84.

**Mosher, E. 1918. Pupae of common Sphingidae of eastern North America. Ann. Ent. Soc. Amer. 11:403–22.

Moss, A. M. 1945. The *Castnia* of Para with notes on others (Lep., Castniidae). Proc. Roy. Ent. Soc. Lond., Ser. B, 14:48–52.

Muka, A. A. 1976. A new corn pest is south of the border. Hoard's Dairyman 121:688.

Muller, W. 1886. Sudamerikanischer Nymphalidenraupen. Versuch eines naturlichen Systems der Nymphaliden. Zool. Jahrb. fur Syst. 1:417–678. (4 pls.)

Munroe, E. G. 1949. An unnoticed character in the Saturnioidea (Lepidoptera). Ent. News 60:60–65.

**Reprinted as part of "Lepidoptera pupae: collected works on the pupae of North American Lepidoptera" by Edna Mosher. 1969. Ent. Reprint Specialists, E. Lansing, Mich. 323 pp.

Munroe, E. G. 1961. The classification of the Papilionidae (Lepidoptera). Can. Ent. Suppl. 17:1–51.

Munroe, E. G. 1972–1976. *In* R. B. Dominick et al. (eds.). The moths of America north of Mexico. Pyraloidea. Fas. 13.1A; 13.1B, 13.1C; 13.2A; 13.2B. 482 pp. London: E. W. Classey and R. B. D. Publ., Inc. (Deals mostly with adults, but also includes information on immatures)

Murtfeldt, M. E. 1900. New Tineidae, with life-histories. Can. Ent. 32:161–66. (Biology)

Mutuura, A. 1956. On the homology of the body areas in the thorax and abdomen and a new system of the setae of lepidopterous larvae. Bull. Univ. Osaka Pref. (Ser. B) 6:93–122.

Mutuura, A. 1980. Morphological relations of sclerotized and pigmented areas of lepidopterous larvae to muscle attachments, with applications to larval taxonomy. Can. Ent.12:697–724.

Namba, R. 1956. Description of the immature states and notes on the biology of *Ithome concolorella* (Chambers) (Lepidoptera: Cosmopterygidae), a pest of kiawe in the Hawaiian Islands. Proc. Hawaiian Ent. Soc. 16:95–100.

Naumann, C. M. 1971. Untersuchungen zur Systematik und Phylogenese der holarktischen Sesiiden (Insecta, Lepidoptera). Bonner Zool. Monog. (Bonn) 1:1–190. (English trans. 1977: Studies on the systematics and phylogeny of Holarctic Sesiidae (Insecta, Lepidoptera). Smithsonian Inst., Wash, D.C. 208 pp.) (Taxonomy)

Needham, J. G., S. W. Frost, and B. H. Tothill. 1928. Leaf-mining insects. Baltimore: The Williams and Wilkins Co. 351 pp.

Neumoegen, B., and H. G. Dyar. 1893–94. A preliminary revision of the Bombyces of America north of Mexico. J. N.Y. Ent. Soc. 1 (1893):97–118; 153–80. 2 (1894):1–30; 57–76; 109–32; 147–74. (*Doa* treated on 1:155, as *Coscinia ampla; Doa dora,* new genus and new species, and *Doa ampla* (new combination) discussed on 2:171).

Neunzig, H. H. 1969. The biology of the tobacco budworm and the corn earworm in North Carolina. N. Carolina Agr. Expt. Sta. Tech. Bull. 196. 76 pp.

Neunzig, H. H. 1972. Taxonomy of eastern North American larvae and pupae of *Acrobasis.* USDA, Tech. Bull. 1457. 158 pp. (Keys to, and descriptions of 26 species) (Pyralidae)

Neunzig, H. H. 1979. Systematics of immature phycitines (Lepidoptera: Pyralidae) associated with leguminous plants in the southern United States. USDA, Tech. Bull. 1589. 119 pp. (Keys to, and descriptions of, 21 species of larvae and 20 species of pupae of Phycitinae)

Neunzig, H. H., and E. P. Merkel. 1967. A taxonomic study of the pupae of the genus *Dioryctria* in the southeastern United States. Ann. Ent. Soc. Amer. 60:801–08. (Keys to, and descriptions of, six species) (Pyralidae)

Neunzig, H. H., R. L. Rabb, B. H. Ebel, and E. P. Merkel. 1964. Larvae of the genus *Dioryctria* in the southeastern United States. Ann. Ent. Soc. Amer. 57:693–700. (Keys to, and descriptions of, six species) (Pyralidae)

Newcomer, E. J. 1912. Some observations on the relations of ants and lycaenid caterpillars, and a description of the relational organs of the latter. J. N.Y. Ent. Soc. 20:31–36.

Newton, P. J. and C. Wilkinson. 1982. A taxonomic revision of the North American species of *Stigmella* (Lepidoptera: Nepticulidae). Syst. Ent. 7:367–463.

Nielsen, E. S. 1980. A cladistic analysis of the Holarctic genera of adelid moths (Lepidoptera: Incurvarioidea [sic]). Ent. Scand. 11:161–78.

Nielsen, E. S. 1982a. Incurvariidae and Prodoxidae from the Himalayan area (Lepidoptera: Incurvarioidea). Insecta Matsumurana 26:187–200.

Nielsen, E. S. 1982b. The maple leaf-cutter moth and its allies: a revision of *Paraclemensia* (Incurvariidae, s. str.). Syst. Ent. 7:217–38.

Nielsen, E. S. 1985. A taxonomic review of the adelid genus *Nematopogon* Zeller (Lepidoptera: Incurvarioidea). Ent. Scand. Suppl. 25:1–66.

Nielsen, E. S., and D. R. Davis. 1981. A revision of the Neotropical Incurvariidae s. str., with description of two new genera and two new species (Lepidoptera: Incurvarioidea) Steenstrupia 7(3):25–57.

Nielsen, E. S., and D. R. Davis. 1985. The first southern hemisphere prodoxid, *Prodoxoids asymmetra,* n. gen., n. sp., and the phylogeny of the Incurvarioidea (Lepidoptera). Syst. Ent. 10:307–322.

*Novak, V., F. Hrozinka and B. Stary. 1976. Atlas of insects harmful to forest trees, vol. 1. New York: Elsevier.

Ogunwolu, E. O., and D. H. Habeck. 1979. Descriptions and keys to larvae and pupae of the grass loopers, *Mocis* spp., in Florida (Lepidoptera: Noctuidae). Fla. Ent. 62:402–07.

Oku, T. 1964. On *Prays* sp. boring into the petiole and the bud of the ash-tree. Tohoku Konchu Kenkyu 1:67–69. (Biology) (Plutellidae)

Okumura, G. T. 1960. *Proxenus mindara* B. & McD. (Lepidoptera: Noctuidae) a new pest of strawberries in California. Calif. Dept. Agr. Bull. 49:246–51.

Okumura, G. T. 1961. Identification of lepidopterous larvae attacking cotton, with illustrated key. Calif. Dept. Agr., Bur. Ent. Spec. Publ. 282. 80 pp.

Oliver, A. D. 1963. A behavioral study of two races of the fall webworm, *Hyphantria cunea* (Lepidoptera: Arctiidae), in Louisiana. Ann. Ent. Soc. Amer. 57:192–94.

*Oliver, A. D. and J. B. Chapin. 1981. Biology and illustrated key for the identification of twenty species of economically important noctuid perts. Louisiana Agr. Expt. Sta. Bull. 733. 26 pp.

Oliver, D. R., P. S. Corbet, and J. A. Downes. 1964. Studies in Arctic insects: the Lake Hazen project. Can. Ent. 96:128–39.

Opler, P. A. 1974. Biology, ecology, and host specificity of Microlepidoptera associated with *Quercus agrifolia* (Fagaceae). Univ. Calif. Publ. Ent. 75:1–83.

Opler, P. A., and D. R. Davis. 1981. The leafmining moths of the genus *Cameraria* associated with Fagaceae in California (Lepidoptera: Gracillariidae). Smithsonian Contr. Zool. No. 333. 58 pp. (131 figs., 9 maps)

Opler, P. A., and G. O. Krizek. 1984. Butterflies east of the Great Plains. Baltimore: Johns Hopkins Univ. Press. 360 pp.

Packard, A. S. 1890a. Insects injurious to forest and shade trees. Fifth Rept. U.S. Entomol. Comm. Washington, D.C.: Government Printing Office. 965 pp. (40 pls.)

Packard, A. S. 1890b. Hints on the evolution of the bristles, spines, and tubercles of certain caterpillars, apparently resulting from a change from low-feeding to arboreal habits; illustrated by life-histories of some notodontians. Proc. Boston Soc. Nat. Hist. 24:494–561. (pls. 3–4)

Packard, A. S. 1893. Notes on the life-histories of some Notodontidae. J. N.Y. Ent. Soc. 1:22–28 (part 1); 57–76 (part 2).

Packard, A. S. 1894. A study of the transformations and anatomy of *Lagoa crispata,* a bombycine moth. Proc. Amer. Phil. Soc. 33:275–92.

*Packard, A. S. 1895. Monograph of the bombycine moths of America north of Mexico including their transformations and origin of the larval markings and armature. Part 1: Family 1, Notodontidae. Mem. Nat. Acad. Sci. 7:1–390. (49 pls., maps)

*Packard, A. S. 1905. Monograph of the bombycine moths of North America, including their transformations and origin of the larval markings and armature. Part 2. Family Ceratocampidae, subfamily Ceratocampinae. Mem. Nat. Acad. Sci. 9:1–272. (60 pls.)

*Packard, A. S. 1914. Monograph of the bombycine moths of North America, including their transformations and origin of the larval markings and armature. Part 3. Families Ceratocampidae (excluding Ceratocampinae), Saturniidae, Hemileucidae, and Brahmaeidae (T. D. A. Cockerell, ed.). Mem. Nat. Acad. Sci. 12:1–277, 503–16. (113 pls.)

Parker, H. L. 1964. Life history of *Leucoptera spartifoliella* with results of host transfer tests conducted in France. J. Econ. Ent. 57(4):566–69.

Parker, J. R., A. L. Strand, and H. L. Seamans. 1921. Pale western cutworm (*Porosagrotis orthogonia* Morr.). J. Agr. Res. 22:289–322.

Patch, E. M. 1908. The saddled prominent, *Heterocampa guttivitta* (Wlkr.). Maine Agr. Expt. Sta. Bull. 161:311–50.

Patch, E. M. 1921. A meadow caterpillar, "The adventurer" (*Ctenucha virginica*). Maine Agr. Exp. Sta. Bull. 302:309–20.

Patterson, J. E. 1929. The pandora moth, a periodic pest of western pine forests. USDA, Tech. Bull. 137. 19 pp.

Pavlov, I. F. 1961. Ecology of the stem moth *Ochsenheimeria vaculella* [sic] F.R. (Lepidoptera: Tineoidea). Ent. Rev. 40(4):461–66. (English translation of Entomologischeskoya Obozrenige 40(4):818–27)

Pease, R. W. 1961. A study of first instar larvae of the Saturniidae with special reference to Nearctic genera. J. Lepid. Soc. 14:89–111.

Pedigo, L. P., J. D. Stone, and G. L. Lentz. 1973. Biological synopsis of the green cloverworm in central Iowa. J. Econ. Ent. 66:665–73.

Peiu, M., and E. Patrascanu. 1970. Contribution to the study of the morphology and biology of the insect *Anthophila pariana* Cl. (Lep. Glyphipterygidae), a new apple tree ravager in Romania. Communcari Zool. 1970:179–86. (in Romanian) (Biology)

Penman, D. R. 1978. Biology of *Euciodes suturalis* (Coleoptera: Anthribidae) infesting cocksfoot in Canterbury [a competitor of the cocksfoot moth]. New Zealand Ent. 6(4):414–25. (Biology)

Percy, J., and J. A. MacDonald. 1979. Cells of the thoracic gland of the red-humped caterpillar, *Schizura concinna* . . . ultrastructural observations. Can. J. Zool. 57:80–94.

Perkins, R. C. L. 1905. Leaf-hoppers and their natural enemies (Pt. II Epipyropidae) Lepidoptera. Rpt. Exp. Sta. Haw. Sugar Plant. Assn. Div. Ent., Bull. 1, pt. 2, pp. 79–81.

Peterson, A. 1948 (and 1951, 1956 and subsequent reprintings through 1984). Larvae of insects. An introduction to Nearctic species. Part 1. Lepidoptera and plant infesting Hymenoptera. Printed for the author by Edwards Bros., Ann Arbor, Mich., 315 pp.

Peterson, A. 1953. A manual of entomological techniques. Printed for the author by Edwards Bros., Ann Arbor, Mich. 367 pp.

Peterson, A. 1961. Some types of eggs deposited by moths, Heterocera—Lepidoptera. Fla. Ent. 44:107–14.

Peterson, A. 1963a. Egg types among moths of the Pyralidae and Phycitidae—Lepidoptera. Fla. Ent. 46, Suppl. 1:1–9.

Peterson, A. 1963b. Some eggs of moths among the Amatidae, Arctiidae, and Notodontidae—Lepidoptera. Fla. Ent. 46:169–82.

Peterson, A. 1964. Egg types among moths of the Noctuidae (Lepidoptera). Fla. Ent. 47:71–91.

Peterson, A. 1967. Some eggs of moths from several families of Microlepidoptera. Fla. Ent. 50:125–32.

Petterson, M. A., and R. S. Wielgus. 1973 (1974). Acceptance of artificial diet by *Megathymus streckeri* (Skinner) (Megathymidae). J. Res. Lepid. 12:197–98.

Philippi, R. A. 1863. Metamorphose von *Castnia.* Stett. Ent. Zeit. 24:337–41.

Phipps, C. R. 1927. The black army cutworm. Maine Agr. Expt. Sta. Bull. 340:201–16.

Phipps, C. R. 1930. Blueberry and huckleberry insects. Maine Agr. Expt. Sta. Bull. 356:107–232.

Pierce, N. E., and P. S. Mead. 1981. Parasitoids as selective agents in the symbiosis between lycaenid butterfly larvae and ants. Science 211:1185–87.

Pinhey, E. C. G. 1975. Moths of Southern Africa. Tafelburg Pub. Ltd. 273 pp. (Classification)

Popenoe, A. M., C. L. Marlatt, and S. C. Mason. 1889. The sycamore forktail, *Heterocampa unicolor* Pack., pp. 199–202. *In* A. M. Popenoe et al. Report of the Department of Horticulture and Entomology, First Ann. Rept. Kan. [Agr.] Expt. Sta., for the year 1888.

Popescu-Gorj, A., E. Niculescu, and A. Alexinschi. 1958. Lepidoptera. Familia Aegeriidae. *In* Fauna Republicii Populare Romine. Insecta. Vol. XI. Fasc. 1 (Bucharest). 195 pp. (5 pls.) (Biology, taxonomy)

Porter, B. A., and P. Garman. 1923. The apple and thorn skeletonizer. Conn. Agr. Expt. Sta. Bull. 246:247–64. (4 pls.) (Biology)

Pottinger, R. P., and E. J. LeRoux. 1971. The biology and dynamics of *Lithocolletis blancardella* (Lepidoptera: Gracillariidae) on apple in Quebec. Mem. Ent. Soc. Can. 77. 437 pp.

Povel, G. D. E., and M. M. J. Beckers. 1982. The prothoracic "defense" gland of *Yponomeuta* larvae. Proc. Kon. Ned. Acad. Wet. (c) 85(3):393–98.

Powell, J. A. 1963. Observations on larval and pupal habits of the juniper cone moth, *Periploca atrata* Hodges (Lep.: Gelechioidea). Pan-Pac. Ent. 39:177–81.

Powell, J. A. 1964. Biological and taxonomic studies on tortricine moths, with reference to the species in California. Univ. Calif. Publ. Ent. 32:1–317.

Powell, J. A. 1971. Biological studies on moths of the genus *Ethmia* in California (Gelechioidea). J. Lepid. Soc. (Suppl.) 25(3):1–67.

Powell, J. A. 1973. A systematic monograph of new world ethmiid moths (Lepidoptera: Gelechioidea). Smith. Contrib. Zool., No. 120. 302 pp.

Powell, J. A. 1976a. A remarkable new genus of brachypterous moth from coastal sand dunes in California (Lepidoptera: Gelechioidea, Scythrididae). Ann. Ent. Soc. Amer. 69(2):325–39.

Powell, J. A. 1976b. The giant blastobasid moths of *Yucca* (Gelechioidea). J. Lepid. Soc. 30:219–29.

Powell, J. A. 1980. Evolution of larval food preferences in Microlepidoptera. Ann. Rev. Ent. 25:133–59.

Powell, J. A. 1984. Biological relationships of moths and *Yucca schottii.* Univ. Calif. Publ. Ent. 93 pp. (49 figs.) (Incurvariidae, Cochylidae, Blastobasidae).

Powell, J. A., and R. A. Mackie. 1966. Biological interrelationships of moths and *Yucca whipplei* (Lepidoptera: Gelechiidae, Blastobasidae, Prodoxidae). Univ. Calif. Publ. Ent. 42:1–46.

Powell, J. A., J. A. Comstock, and C. F. Harbison. 1973. Biology, geographical distribution, and status of *Atteva exquisita* (Lepidoptera: Yponomeutidae). Trans. San Diego Soc. Nat. Hist. 17:175–86. (Biology)

*Powell, J. A., and Charles L. Hogue. 1979. California insects. Berkeley: University of California Press. 388 p.

Prentice, R. M. [coordinator]. 1958–1965. Forest Lepidoptera of Canada recorded by the Forest Survey, 4 vols. Can. Dept. For. Vol. 1, 1958, Publ. 1034, Papilionidae to Arctiidae, pp. 1–76; Vol. 2, 1962, Bull. 128, Notodontidae, Noctuidae, Liparidae, pp. 77–281; Vol. 3, 1963, Publ. 1013, Lasiocampidae, Thyatiridae, Drepanidae, Geometridae, pp. 283–541; Vol. 4, 1965, Publ. 1142, Microlepidoptera, pp. 546–840.

Pritchard, A. E., and J. A. Powell. 1959. *Pyramidobela angelarum* Keifer on ornamental *Buddleia* in the San Francisco Bay area. Pan-Pac. Ent. 35:82. (Oecophoridae: Ethmiinae)

Pruess, K. P. 1961. Oviposition response of the army cutworm, *Chorizagrotis auxiliaris,* to different media. J. Econ. Ent. 54:273–77.

*Pyle, R. M. 1981. The Audubon Society field guide to North American butterflies. New York: Alfred A. Knopf. 916 pp. (Descriptions, biology)

Rae, R. M., and C. W. Clifford. 1976. Freeze-drying of spiders and immature insects using commercial equipment. Ann. Ent. Soc. Amer. 69:497–99.

Raine, J. 1966. Life history of *Dasystoma salicellum* Hübner (Lepidoptera: Oecophoridae), a new pest of blueberries in British Columbia. Can. Ent. 98:331–34.

Raske, A. 1976. Complexities in the number of larval instars of the birch casebearer in Newfoundland (Lepidoptera: Coleophoridae). Can. Ent. 108(4):401–07.

Razowski, J. 1965. The Palaearctic Cnephasiini (Lepidoptera, Tortricidae). Acta Zool. Cracov. 10:199–342. (pls. XII-XXVI)

Razowski, J. 1966. World fauna of the Tortricini (Lepidoptera, Tortricidae). Krakow: Panstwowe Wydawnictwo Naukowe. 576 pp. (41 pls.)

Razowski, J. 1976. Phylogeny and system of Tortricidae (Lepidoptera). Acta Zool. Cracov. 21:73–118.

Razowski, J. 1977. Monograph of the genus *Archips* Hübner (Lepidoptera, Tortricidae). Acta Zool. Cracov. 22:56–205.

Razowski, J., and J. Wojtusiak. 1978a. Family group taxa of the Adeloidea (Lepidoptera). Polskie Pismo Ent. 48:3–18.

Razowski, J., and J. Wojtusiak. 1978b. Polish genera of the Adelidae (Lepidoptera) Polskie Pismo Ent. 48:19–33.

Real, P., and A. S. Balachowsky. 1966. Famille des Glyphipterygidae. *In* A. S. Balachowsky (ed.). Traité d'entomologie appliqué a l'agriculture. Tome II. Lepidoptera, vol. 1., pp. 278–87. Paris: Masson and Editeurs. (Biology)

Reinhard, H. J. 1929. The cotton-square borer. Tex. Agr. Expt. Sta. Bull. 401. 36 pp. (Lycaenidae)

Reinhard, H. J. 1938. The sorghum webworm (*Celama sorghiella* Riley). Tex. Agr. Expt. Sta. Bull. 559. 35 pp. (Nolidae)

Remington, J. E. 1963. Laboratory mass rearing of *Cisseps fulvicollis* (Ctenuchidae), with notes on fertility, fecundity, and biology. J. Lepid. Soc. 17:65–80.

Riley, C. V. 1870a. Second annual report on the noxious, beneficial and other insects of the state of Missouri. Jefferson City, Mo. 135 pp.

Riley, C. V. 1870b. The army worm *Leucania unipuncta,* Haw., pp. 37–56. *In* C. V. Riley. Second annual report on the noxious, beneficial and other insects of the state of Missouri. Jefferson City, Mo.

Riley, C. V. 1874. The blue caterpillars of the vine, pp. 87–96. *In* C. V. Riley. Sixth annual report on the noxious, beneficial, and other insects of the state of Missouri. Jefferson City, Mo.

Riley, C. V. 1876. Notes on the yucca borer, *Megathymus yuccae* (Walk.). Trans. St. Louis Acad. Sci. 3:323–44. (Descriptions, biology)

Riley, C. V. 1883a. A myrmecophilous lepidopteron. Amer. Nat. 17:1070.

Riley, C. V. 1883b. The cabbage Mamestra (*Mamestra chenopodii* Albin.) Order Lepidoptera: family Noctuidae. pp. 123–24. *In* C. V. Riley, Report of the Commissioner of Agriculture for the year 1883. Washington, D.C.

Riley, C. V. 1883c. The cabbage worm (*Ceramica picta* Harris) Order Lepidoptera: family Noctuidae, pp. 124–25. *In* C. V. Riley, Report of the Commissioner of Agriculture for the year 1883. Washington, D.C.

Riley, C. V. 1885. Cabbage cut-worms. Order Lepidoptera: family Noctuidae, pp. 289–98. *In* C. V. Riley. Report of the Entomologist of the U.S.D.A. for the year 1884. U.S.D.A. Ann. Rept. Year 1884. Washington, D.C.

Riley, C. V. 1889. The Japanese peach fruit-worm. Insect Life 2:64–66. (Biology)

Riley, C. V. 1892a. A new herbarium pest. Ins. Life 4:108–13.

Riley, C. V. 1892b. The yucca moth and *Yucca* pollination. Third Ann. Rep. Missouri Bot. Garden 3:99–158. (Prodoxidae)

Rings, R. W. 1969. Contributions to the bionomics of the green fruitworms: the life history of *Lithophane laticinerea.* J. Econ. Ent. 62:1388–93.

Rings, R. W. 1970. Contributions to the bionomics of the green fruitworms: the life history of *Orthosia hibisci.* J. Econ. Ent. 63:1562–68. (Other papers in vols. 64 and 65)

Rings, R. W. 1977. A pictorial field key to the armyworms and cutworms attacking vegetables in the north central states. Ohio Agr. Res. Dev. Ctr. (Wooster). Res. Circ. 231. 36 pp.

Rings, R. W., and F. J. Arnold. 1977. Geographical distribution and economic importance of the clay-backed, dingy, dusky, and sandhill cutworms. USDA, Coop. Plant Pest Rept. 2(48–52):881–86.

Rings, R. W., and B. A. Johnson. 1976. An annotated bibliography of the sandhill cutworm *Euxoa detersa* (Walker). Ohio Agr. Res. Dev. Ctr. (Wooster). Res. Circ. 215. 16 pp.

Rings, R. W., and B. A. Johnson. 1977. An annotated bibliography of the spotted cutworm. Ohio Agr. Res. Dev. Ctr. (Wooster), Res. Circ. 235. 50 pp.

Rings, R. W., and G. J. Musick. 1976. A pictorial field key to the armyworms and cutworms attacking corn in the north central states. Ohio Agr. Res. Dev. Ctr. (Wooster), Res. Circ. 221. 36 pp.

Rings, R. W., F. J. Arnold, and B. A. Johnson. 1978. Supplemental annotated bibliographies of the black cutworm, glassy cutworm, bronzed cutworm, bristly cutworm, dingy cutworm, dark-sided cutworm, clay-backed cutworm, dusky cutworm, and variegated cutworm. Ohio Agr. Res. Dev. Ctr. (Wooster), Res. Circ. 238. 59 pp.

Riotte, J. C. E. 1969. Rearing and descriptions of the early stages of the Nearctic species of *Peridea,* with special reference to *P. basitriens* (Lepidoptera: Notodontidae). Mich. Ent. 1:351–56.

Riotte, J. C. E., and R. S. Peigler. 1980 (1981). A revision of the American genus *Anisota* (Saturniidae). J. Res. Lepid. 19:101–80.

Ripley, L. B. 1923. The external morphology and postembryology of noctuid larvae. Ill. Biol. Mon. 8(4):1–102, 243–344. (8 pls.)

Robertson-Miller, E. 1923. Observations on the *Bellura.* Ann Ent. Soc. 16:374–86. (Noctuidae)

Robinson, G. S. 1979. Clothes-moths of the *Tinea pellionella* complex: a revision of the world's species (Lepidoptera: Tineidae). Bull. Brit. Mus. (Natural History). Ent. Series 38(3):57–128.

Roeder, K. D., and A. E. Treat. 1962. The detection and evasion of bats by moths. Smithsonian Rpt. for 1961, pp. 455–64.

Roever, K. 1964a. Bionomics of *Agathymus* (Megathymidae). J. Res. Lepid. 3:103–20.

Roever, K. 1964b (1965). Instar determination of *Agathymus* larvae. J. Res. Lepid. 3:148–50. (Megathymidae)

Roever, K. 1975. Family Megathymidae (the giant skippers), pp. 411–22. *In* W. H. Howe (ed.). The butterflies of North America. Garden City, N.Y.: Doubleday and Co., Inc.

Ross, D. A. 1958. The maple leaf cutter, *Paraclemensia acerifoliella* (Fitch) (Lepidoptera: Incurvariidae), description of stages. Can. Ent. 90(9):544–55.

Ross, D. A. 1959. Abdominal characters of *Dioryctria* pupae from British Columbia. Can. Ent. 91:731–34. (Pyralidae)

Ross, G. N. 1966. Life history studies on Mexican butterflies. IV. The ecology and ethology of *Anatole rossi,* a myrmecophilous metalmark (Lep.: Riodinidae). Ann. Ent. Soc. Amer. 59:985–1004.

Salkeld, E. H. 1975. Biosystematics of the genus *Euxoa* Lepidoptera: Noctuidae). IV. Eggs of the subgenus *Euxoa* Hbn. Can. Ent. 107:1137–52.

Salkeld, E. H. 1983. A catalogue of the eggs of some Canadian Geometridae (Lepidoptera), with comments. Mem. Ent. Soc. Can., No. 126. 271 pp.

Salkeld, E. H. 1984. A catalogue of the eggs of some Canadian Noctuidae (Lepidoptera). Mem. Ent. Soc. Can., No. 127. 167 pp.

Salt, G. 1929. *Castniomera humboldti* (Boisduval), a pest of bananas. Bull. Ent. Res. 20:187–93.

Sargent, T. D. 1976. Legion of night. Amherst: Univ. of Mass. Press. 235 pp.

Sarlet, L. 1964. Iconographie des oeufs de Lépidoptères Belges (Rhopalocera—Heterocera) Pt. 2, Heterocera: Bombycides—Sphingides. 172 pp. (15 figs.)

*Schaus, W. 1927. Lymantriidae, pp. 535–64. *In* A. Seitz, The Macrolepidoptera of the world, vol. 6. The American Bombyces and Sphinges. (English edition incomplete) (198 pls.) Stuttgart: Alfred Kernen Verlag. 1328 pp. (*Leuculodes* and *Doa* treated on pp. 559–60 and 563–64, respectively, both genera transferred to Pericopidae on p. 563).

Schoonhoven, L. M. 1952. Plant recognition by lepidopterous larvae, pp. 87–99. *In* Insect/plant relationships. Sym. Royal Ent. Soc. Lond. 6. London: Blackwell Sci. Publ.

Scoble, M. J. 1980. A new incurvariine leaf-miner from South Africa, with comments on structure, life-history, phylogeny, and the binomial system of nomenclature (Lepidoptera: Incurvariidae). J. Ent. Soc. So. Africa 43(1):7–88.

Scoble, M. J. 1983. A revised cladistic classification of the Nepticulidae (Lepidoptera) with descriptions of new taxa mainly from South Africa. Transvaal Museum Monograph No. 2. 105 pp.

Scoble, M. J. and C. H. Scholtz. 1984. A new gall-feeding moth (Lyonetiidae: Bucculatricinae) from South Africa with comments on larval habits and phylogenetic relationships. Syst. Ent. 9:83–94.

Scudder, S. H. 1889. The butterflies of the eastern United States and Canada, with special reference to New England, three vols. Cambridge, Mass. (Descriptions, biology)

*Seitz, A. 1913. The Macrolepidoptera of the American region, vol. 6. The American Bombyces and Sphinges, pp. 5–7. Stuttgart: Alfred Kernen Verlag.

Shepard, J. H., and S. S. Shepard. 1975. Subfamily Parnassinae, pp. 403–409. *In* W. H. Howe (ed.). The butterflies of North America. Garden City, N.Y.: Doubleday and Co., Inc.

Shields, O. A., J. F. Emmel, and D. E. Breedlove. 1969 (1970). Butterfly larval foodplant records and a procedure for reporting foodplants. J. Res. Lepid. 8:21–36.

*Shirozu, T., and A. Hara. 1960 (vol. 1), 1962 (vol. 2) (and later printings). Early stages of Japanese butterflies in colour. Osaka: Hoikusha Publ. Co. (1: 141 pp., 2: 139 pp.) (Eggs, larvae, pupae, adults; species of all families in both volumes) (In Japanese)

Silver, J. C. 1933. Biology and morphology of the spindle worm, or elder borer. USDA, Tech. Bull. 345. 19 pp.

Singh, B. 1953. Immature stages of Indian Lepidoptera. No. 8—Geometridae. Indian For. Rec. (Ent.) (N.S.) 8(1951):67–158. (Chaetotaxy)

Singh, B. 1956. Some more Indian geometrid larvae (Lepidoptera), with a note on the identity of components of various groups of setae. Indian For. Rec. 9:131–34. (Chaetotaxy)

Singh, P. 1977. Artificial diets for insects, mites, and spiders. New York: Plenum Publ. Corp. 606 pp.

Slingerland, M. V. 1893. *Drasteria erechtea.* Ins. Life 5:87–88.

Smith, F. 1941. A note on noctuid larvae found in ants' nests (Lepidoptera: Hymenoptera: Formicidae). Ent. News 52:109.

Smith, J. B., and H. C. Dyar, 1898. Contributions toward a monograph of the lepidopterous family Noctuidae of boreal North America. A revision of the species of *Acronycta* (Ochsenheimer) and of certain allied genera. Proc. U.S. Nat. Mus. 21(1140):1–194.

Smith, O. J., and R. L. Langston. 1953. Continuous laboratory propagation of western grape leaf skeletonizer and parasites by prevention of diapause. J. Econ. Ent. 46:477–84.

Smith, R. C. 1924. *Caenurgina erechtea* (Cram.) (Noctuidae) as an alfalfa pest in Kansas. J. Econ. Ent. 17:312–19.

Snider, R., J. Shaddy, and J. W. Butcher. 1969. Culture techniques for rearing soil arthropods. Mich. Ent. 1:357–62.

Snodgrass, R. E. 1922. The resplendent shield-bearer and the ribbed-cocoon-maker. Smithsonian Inst. Ann. Rept. 1920, pp. 485–510. (3 pls.)

Snodgrass, R. E. 1935. Principles of insect morphology. New York: McGraw-Hill. 667 pp.

Snodgrass, R. E. 1954. Insect metamorphosis. Smithsonian Misc. Coll. 122:1–124. Washington, D.C.

Snow, F. H. 1885. Preparatory stages of *Hyperchiria zephyria* Grote. Trans. Ann. Meet. Kansas Acad. Sci. (1883–84) 9:61–62. (Saturniidae)

Solomon, J. D. 1966. Artificial rearing of the carpenterworm, *Prionoxystus robiniae* (Lepidoptera: Cossidae), and observations of its development. Ann. Ent. Soc. Amer. 59(6):1197–1200.

Solomon, J. D. 1967. Rearing the carpenterworm, *Prionoxystus robiniae* in the forest (Lepidoptera: Cossidae). Ann. Ent. Soc. Amer. 60(1):283–84.

Solomon, J. D. 1968. Gallery construction by the carpenterworm, *Prionoxystus robiniae*, in overcup oak (Lepidoptera: Cossidae). Ann. Ent. Soc. Amer. 61:72–74.

Solomon, J. D. 1977. Frass characteristics for identifying insect borers (Lepidoptera: Cossidae and Sesiidae; Coleoptera: Cerambycidae) in living hardwoods. Can. Ent. 104:295–303.

Sorenson, C. J., and H. F. Thornley. 1941. The pale western cutworm. (*Agrotis orthogonia* Morrison) in Utah. Utah Agr. Expt. Sta. Bull. 297. 23 pp.

*Spuler, A. 1910. Die Raupen der Schmettlerlinge Europas, vol. 4. Stuttgart: E. Schweizerbart.

Staines, C. L., Jr. 1977. Observations on the variable oak leaf caterpillar, *Heterocampa manteo* (Doubleday) (Lepidoptera: Notodontidae) in Maryland. Proc. Ent. Soc. Wash. 79:343–45.

Stainton, H. T. 1872. The Tineina of North America by Dr. Brackenridge Clemens. London, 297 pp.

Stanford, R. E. 1981. Superfamily Hesperioidea Latreille, 1802 (Skippers), pp. 67–141. *In* C. D. Ferris and F. M. Brown (eds.). Butterflies of the Rocky Mountain states. Norman, Okla.: Univ. of Okla. Press.

Stark, R. W. 1954. Distribution and life history of the lodgepole needle miner (*Recurvaria* sp.) (Lepidoptera: Gelechiidae) in Canadian Rocky Mountain parks. Can. Ent. 86:1–12.

Steel, W. O. 1970. The larvae of the genera of the Omaliinae (Coleoptera: Staphylinidae) with particular reference to the British fauna. Trans. Roy. Ent. Soc. Lond. 122(1):12–47 (3 pl.).

*Stehr, F. W., and E. F. Cook. 1968. A revision of the genus *Malacosoma* Hübner in North America (Lepidoptera: Lasiocampidae): Systematics, biology, immatures, and parasites. U.S. Nat. Mus., Bull. 276. 321 pp.

Stehr, F. W. and N. McFarland. 1985. Crochets on abdominal segments 2 and 7 of dalcerid caterpillars: "missing link" or anomaly? Bull. Ent. Soc. Amer. Spring, 1985. pp. 35–36.

Stehr, F. W., and H. H. Neunzig. 1981. A simplified terminology for the tonofibrillary structures associated with the muscles of Lepidoptera larvae. Can. Ent. 113:1107–12.

Sternburg, J. G., G. P. Waldbauer, and A. G. Scarborough. 1981. Distribution of *Cecropia* moth (Saturniidae) in central Illinois: a study in urban ecology. J. Lepid. Soc. 35:304–20.

Stichel, H. 1930–31. Lepidopterorum Catalogus. Pars 40, 41 and 44. Berlin: W. Junk.

Still, G. N., and H. R. Wong. 1973. Life history and habits of a leaf miner, *Cameraria macrocarpae*, on bur oak in Manitoba (Lepidoptera: Gracillariidae). Can. Ent. 105:239–44.

*Stoeke, W. J. 1944. The caterpillars of British butterflies. London: F. Warne and Co.

*Stoeke, W. J. 1948. The caterpillars of British moths. 2 vols. London: F. Warne and Co.

Stone, M. W. 1965. Biology and control of the limabean pod borer in southern California. USDA, Tech. Bull. 1321. 46 pp. (Pyralidae)

Strickland, E. H. 1916. The army cutworm *Euxoa (Chroizagrotis) auxiliaris* Grote. Dom. Can. Dept. Agr. Ent. Br. Bull. 13. 31 pp.

Sugden, B. A. 1957. A brief history of outbreaks of the Douglas-fir tussock moth, *Hemerocampa pseudotsugata* McD., in British Columbia. Proc. Ent. Soc. Brit. Columbia 54:37–39.

Sukhareva, I. L. 1978 (1979). Biological features of moths of the subfamily Hadeninae (Lepidoptera: Noctuidae). Ent. Rev. 57:342–47. (Translated from Russian)

Surgeoner, G. A., and W. E. Wallner. 1975. Determination of larval instars of *Heterocampa manteo* and reduction of larval head capsule size by the parasitoid *Diradops bethunei*. Ann. Ent. Soc. Amer. 68:1061–62.

Surgeoner, G. A., and W. E. Wallner. 1978. Evidence of prolonged diapause in prepupae of the variable oakleaf caterpillar, *Heterocampa manteo*. Environ. Ent. 7:186–88.

Swaine, J. M. 1909. Notes on the larva and pupa of *Sthenopsis thule* Strecker. Can. Ent. 41:337–43. (Hepialidae)

Swatschek, B. 1958. Die Larvalsystematik der Wickler (Tortricidae und Carposinidae). Abh. Larvalsyst. Ins. 3:1–269. (Chaetotaxy, etc.)

Thompson, B. G. 1943. Cutworm control in Oregon. Oregon Agr. Expt. Sta. Circ. 147. 4 pp.

Tietz, H. M. 1972. An index to the described life histories, early stages and hosts of the Macrolepidoptera of the continental United States and Canada, 2 vols. Sarasota, Fla.: Allyn Museum of Entomology. 1041 pp.

Tillyard, R. J. 1923. On the larva and pupa of the genus *Sabatinca* (Order Lepidoptera, Family Micropterygidae [sic]). Trans. Ent. Soc. Lond., pp. 437–53.

Tillyard, R. J. 1926. The insects of Australia and New Zealand. Angus and Robertson Pub. 560 pp. (Classification)

Tindale, N. B. 1928. Preliminary note on the life history of *Synemon* (Lepidoptera, Fam. Castniidae). Rec. S. Aust. Mus. IV:143–44.

Todd, E. L. 1976. A revision of the genus *Dunama* Schaus (Notodontidae). J. Lepid. Soc. 30:188–96.

Todd, E. L. 1978. A checklist of species of *Heliothis* Ochsenheimer (Lepidoptera: Noctuidae). Proc. Ent. Soc. Wash. 80:1–14.

Todd, E. L., and R. W. Poole. 1980. Keys and illustrations of the armyworm moths of the genus *Spodoptera* Guenée from the Western Hemisphere. Ann. Ent. Soc. Amer. 73:722–38.

Townsend, C. H. T. 1893. *Schizura ipomeae* [sic] Doubl. Ent. News 4:158.

Trägardh, I. 1913. Contributions towards the comparative morphology of the trophi of the lepidopterous leaf miners. Arkiv for Zoologi 8(9):1–49. (67 figs.)

Traugott-Olsen, E., and E. S. Nielsen. 1977. The Elachistidae (Lepidoptera) of Fennoscandia and Denmark. Fauna Entomologica Scandinavica 6:1–299.

Tuskes, P. M. 1976. A key to the last instar larvae of West Coast Saturniidae. J. Lepid. Soc. 30:272–76.

Tuskes, P. M. 1984 (1985). The biology and distribution of California Hemileucinae (Saturniidae). J. Lepid. Soc. 38:281–309.

Tuskes, P. M., and N. J. Smith. 1984. The life history and behavior of *Epimartyria pardella* (Micropterigidae). J. Lepid. Soc. 38:40–46.

Tyler, H. A. 1975. The swallowtail butterflies of North America. Healdsburg, Calif.: Naturegraph Pub., Inc. 200 pp.

Urquhart, F. A. 1970. Mechanism of cremaster withdrawal and attachment in pendant ropalocerous pupae (Lepidoptera). Can. Ent. 102:1579–82.

U.S.D.A. 1985. Insects of eastern forests. U.S. Dept. Agr., For. Ser. Misc. Publ. 1426. Washington, D.C., 608 pp.

Valley, K., and A. G. Wheeler. 1976. Biology and immature stages of *Stomopteryx palpilineella* (Lepidoptera: Gelechiidae), a leafminer and leaftier of crownvetch. Ann. Ent. Soc. Amer. 69(2):317–24.

Valley, K., J. F. Stimmel, and A. G. Wheeler. 1981. Biology of a plume moth, *Platyptilia pica* (Lepidoptera: Pterophoridae), a pest of geraniums in Pennsylvania greenhouses. Ann. Ent. Soc. Amer. 74:209–12. (Includes description of immature stage)

Van den Bosch, R., and R. F. Smith. 1955. A taxonomic and distributional study of the species of *Prodenia* occurring in California (Lepidoptera: Phalaenidae). Pan-Pac. Ent. 31:21–28.

Vanderzant, E. S. 1974. Development, significance, and application of artificial diets for insects. Ann. Rev. Ent. 19:139–60.

Vari, L. 1961. South African Lepidoptera. Vol. 1: Lithocolletidae. Transvaal Museum Mem. No. 12. 256 pp. (112 pls.)

Vickery, R. A. 1915. Notes on three species of *Heliophila* which injure cereal and forage crops at Brownsville, Texas. J. Econ. Ent. 8:389–92.

Viette, P. 1958. Un nouveau microlepidoptere parasite du cafeier (Lep. Orneodidae). Bull. Fran. A. N. (Ser. A) 20:457–59. (Biology)

Walkden, H. H. 1950. Cutworms, armyworms and related species attacking cereal and forage crops in the central Great Plains. USDA, Circ. 849. 52 pp.

Washburn, F. L. 1903. Injurious insects of 1903. Minn. Agr. Expt. Sta. Bull. 84. 192 pp. (1 color pl.) (Color figs. of the larvae of *Alypia octomaculata, Melanchra picta,* and *Xestia* sp.)

Washburn, F. L. 1911. Cutworms, army worms and grasshoppers. Minn. Agr. Expt. Sta. Bull. 123:63–84.

Watson, J. R. 1916. Life-history of the velvet-bean caterpillar (*Anticarsia gemmatilis* (sic) Hübner). J. Econ. Ent. 9:521–28.

Watts, J. G. and T. D. Everett. 1976. Biology and behavior of the range caterpillar. N. Mex. Agr. Expt. Sta. Bull. 646. 32 pp. (Saturniidae)

Weatherston, J., J. E. Percy, L. M. MacDonald, and J. A. MacDonald. 1979. Morphology of the prothoracic defensive gland of *Schizura concinna* (J. E. Smith) (Lepidoptera: Notodontidae) and the nature of its secretion. J. Chem. Ecology 5:165–77.

Webster, F. M. 1898. Notes on the development of *Drasteria erechtea* (Cramer). J. N.Y. Ent. Soc. 6:27–33. (pls. 4, 5)

Weigert, G. Th., and A. O. Angulo. 1977. Nuevos tipos de huevos en noctuidos chilenos (Lepidoptera: Noctuidae). Bol. Soc. Biol. Concepcion 51:289–98.

Wellington, W. G. 1965. Some maternal influences on progeny quality in the western tent caterpillar, *Malacosoma pluviale* (Dyar). Can Ent. 97(1):1–14. (Lasiocampidae)

Werner, K. 1958. Die Larvalsystematik einiger Kleinschmetterlings-familien (Hyponomeutidae, Orthotelidae [sic], Acrolepiidae, Tineidae, Incurvariidae, und Adelidae). Abh. Larvalsystematik Insekten 2:1–145. (Chaetotaxy)

Whalley, P. 1978. New taxa of fossil and recent Micropterigidae with a discussion of their evolution and a comment on the evolution of Lepidoptera (Insecta). Ann. Transvaal Mus. 31(8):71–86. (pls. 11–14).

Weisman, D. M. 1986. Keys for the identification of some frequently intercepted lepidopterous larvae. U.S.D.A., APHIS 81–47, 64 pp.

Whelan, D. B. 1935. A key to the Nebraska cutworms and armyworms that attack corn. Nebr. Agr. Expt. Sta. Res. Bull. 81. 27 pp.

Wickman, B. E. 1971. California oakworm. U.S.D.A. Forest Pest Leaflet 72. 4 pp. (Dioptidae)

Wickman, B. E., and R. C. Beckwith. 1978. Life history and habits, pp. 30–37. *In* M. H. Brookes, R. W. Stark and R. W. Campbell (eds.). The Douglas-fir tussock moth: a synthesis. USDA, For. Serv., Sci. & Ed. Agency, Tech. Bull. 1585.

Wielgus, R. S. 1974 (1975). Artificial diet: the key to mass rearing of *Megathymus* larvae. J. Res. Lepid. 13:271–77. (Megathymidae)

Wielgus, R. S., and D. B. Stallings. 1974. The laboratory biology of *Megathymus streckeri* and *Megathymus texanus texanus* (Megathymidae) with associated field observations. Bull. Allyn Mus. 23:1–15.

Wielgus, R. S., and D. Wielgus. 1974. A new sandy-desert subspecies of *Megathymus coloradensis* (Megathymidae) from extreme northern Arizona. Bull. Allyn Mus. 17:1–16.

Wilde, G., and K. O. Bell. 1980. Identifying caterpillars in field crops. Kansas Agr. Expt. Sta. Bull. 632. 19 pp.

Wilkinson, C. and P. J. Newton. 1981. The microlepidopteran genus *Ectoedemia* Busck (Nepticulidae) in North America. Tijds. voor Ent. 124:27–92.

Wilkinson, C., and M. Scoble. 1979. The Nepticulidae (Lepidoptera) of Canada. Mem. Ent. Soc. Can., No. 107, pp. 1–129. (61 text figs., 10 pls.)

Wille, T. J. E. 1952. Entomologia agricola del Peru. Manual para entomologos, ingenieros agronomos, agricultores & estudiantes de agricultura. Peru Minist. Agr., Lima. (Biology)

Williams, J. R. 1951. The bionomics and morphology of *Brenthia leptocosma* Meyrick (Lep.: Glyphipterygidae). Bull. Ent. Res. 41:629–35. (Biology, chaetotaxy)

Wilson, L. 1974. Life history and habits of a leaf tier, *Aroga argutiola* (Lepidoptera: Gelechiidae), on sweet fern in Michigan. Can. Ent. 106(9):991–95.

Wilson, L. F., and G. A. Surgeoner. 1979. Variable oakleaf caterpillar. USDA, For. Ins. & Disease Leaflet 76. 4 pp.

Wolcott, G. N. 1936. "Insectae Borinquensis". J. Agr. Univ. Puerto Rico 20:1–627.

Wood, G. W., W. T. A. Nelson, C. W. Maxwell, and J. A. McKeil. 1954. Life-history studies of *Spaelotis clandestina* (Harr.) and *Polia purpurissata* (Grt.) (Lepidoptera: Phalaenidae) in low-bush blueberry areas in New Brunswick. Can. Ent. 86:169–73.

Woodroffe, G. E. 1951a. A life-history study of the brown house moth, *Hofmannophila pseudospretella* (Stainton) (Lep.: Oecophoridae). Bull. Ent. Res. 41:529–53. (3 pl.)

Woodroffe, G. E. 1951b. A life-history study of *Endrosis lactella* (Schiff.) (Lep.: Oecophoridae). Bull. Ent. Res. 41:749–60. (3 pl.)

Wygant, N. D. 1941. An infestation of the pandora moth, *Coloradia pandora* Blake, in lodgepole pine in Colorado. J. Econ. Ent. 34:697–702.

Yano, K. 1959. On the larva and pupa of *Commatarcha palaeosema* Meyrick, with its biological notes (Lepidoptera: Carposinidae). Kontyu 27:214–17. (2 pl.) (Biology, immature stages)

Yano, K. 1963. Taxonomic and biological studies of Pterophoridae of Japan. Pac. Ins. 5:65–209. (Detailed treatment of larvae and pupae including descriptions of *P. taprobanes* (Felder), which apparently occurs in North America)

Yasuda, T. 1962. On the larva and pupa of *Neomicropteryx* [sic] *nipponensis* Issiki, with its biological notes (Lepidoptera: Micropterygidae [sic]). Kontyu 30(2):130–36. pls. 6–8.

Zagulajev, A. K. 1960. Fauna of USSR. Lepidoptera, vol. 4, no. 3, True moths (Tineidae), pt. 3, subfamily Tineinae. Zool. Inst. Acad. Sci. USSR. 266 pp. (231 figs., 3 col. pls.) (In Russian; English Translation, 1975 (TT70–57763))

Zagulajev, A. K. 1964. Fauna of USSR. Lepidoptera, vol. 4, no. 2, Tineidae, pt. 2, Nemapogoninae. Zool. Inst. Acad. Sci. USSR. 424 pp. (385 figs., 2 col. pls.) (In Russian; English translation, 1968 (TT67–51342))

Zagulajev, A. K. 1971. The stem moths—pests of grasses. Plant Protection [Zachchita Rastenii], pp. 40–42. (1 col. pl.) (In Russian)

Zagulajev, A. K. 1973. Fauna of USSR. Lepidoptera, vol. 4, no. 4, Tineidae, pt. 3, Scardiinae. Zool. Inst. Acad. Sci. USSR. 126 pp. (99 figs., 2 col. pls.) (In Russian)

Zagulajev, A. K. 1975. Fauna of USSR. Lepidoptera, vol. 4, no. 5, Tineidae, pt. 4, Myrmecozelinae. Zool. Inst. Acad. Sci. USSR. 426 pp. (319 figs. 6 col. pls.) (In Russian)

Zimmerman, E. C. 1958a. Insects of Hawaii, vol. 7: Macrolepidoptera. Geometridae, Noctuidae, Sphingidae, Pieridae, Nymphalidae. Danaidae, Lycaenidae. Honolulu: Univ. Hawaii. 542 pp.

Zimmerman, E. C. 1958b. Insects of Hawaii, vol. 8, Lepidoptera: Pyraloidea (Pyralidae, Pterophoridae, Alucitidae). Honolulu: Univ. Hawaii. 456 pp.

Zimmerman, E. C. 1978. Insects of Hawaii, vol. 9. Microlepidoptera, Part 1, Monotrysia, Tineoidea, Tortricoidea, Gracillarioidea, Yponomeutoidea and Alucitoidea, pp. 1–882. Part II, Gelechioidea, pp. 883–1903. Honolulu: Univ. Hawaii.

Order Hymenoptera

27

Howard E. Evans, Coordinator,
Colorado State University

SAWFLIES, PARASITOIDS, ANTS, WASPS, AND BEES

Larvae of the majority of Hymenoptera feed in concealed habitats, either within the body of a host, in a nest, in wood or stems, or in galls or leaf mines. Many members of the suborder Symphyta (sawflies) do, however, feed externally on leaves, and many parasitoids feed externally on their host. Larval structure varies greatly and is correlated with larval feeding behavior and the ability to spin a cocoon. In general, there is progressive reduction from free-living, eruciform sawfly larvae through wood and stem-boring Symphyta to Apocrita, which lack thoracic legs, eyes, and antennae and palpi of more than one segment. For the most part, larvae of Aculeata (wasps, ants, and bees) show less reduction of the head and mouthparts than do those of parasitoids (Parasitica), living as the former do in a nest in which mastication of food and often some limited body movements occur. Larvae of parasitoids often show greatly reduced structure; in some cases they have few functional spiracles, the maxillae and labium are reduced to simple lobes, and mandibles are largely unsclerotized and unpigmented. However, a more or less complete head capsule, though sometimes sunken into the thorax, is always present. In general, it is necessary to study larvae of Apocrita under a compound microscope, using cleared material mounted on slides. Techniques are discussed further below.

Hypermetamorphosis is common among parasitoids and occurs in a few Aculeata. Only last-instar larvae are considered here. Clausen (1940) and DeBach (1964) provide useful references for the earlier instars of parasitoids; some illustrations of first instars are provided on pp. 45–46.

DIAGNOSIS

Collectively, the last-instar larvae of Hymenoptera may be separated from those of other orders by the following set of characters: (1) head capsule complete or nearly so, sometimes unsclerotized and unpigmented, sometimes sunken into thorax; (2) mandibles present, working in a horizontal plane; (3) lateral eyes consisting of a single pair of simple lenses, or absent; (4) spinneret present or absent on labium, and when present of variable structure; if protruding, generally in the form of transverse lips or paired projections; (5) ten pairs of spiracles often present, but with fewer pairs in several groups; (6) thoracic legs present or absent; (7) prolegs present or absent, when present, without terminal hooklets and in six to eight pairs.

DESCRIPTION

Body elongate, in Symphyta usually cylindrical, in Apocrita usually tapered posteriorly and in many cases anteriorly. Spiracles typically ten pairs: one pair on the prothorax, one between the meso- and metathorax, and one pair on each of the first eight abdominal segments. Reduction in the number of functional spiracles is common, and the second thoracic spiracles are especially likely to be reduced or absent, as are occasionally some of the abdominal spiracles. Spiracular structure is especially useful in the classification of Aculeata (fig. 27.4). There are typically ten distinct abdominal segments, although not all are clearly visible in some parasitoids. Some or all body segments are frequently divided into annulets dorsally. Body setae may be present or absent. A color pattern is present in many sawflies, which may also have six to eight pairs of prolegs (without crochets) on the abdomen.

Head

Epicranium complete or nearly so, in Symphyta sclerotized and often pigmented; in Apocrita usually weakly sclerotized and lightly if at all pigmented. Y-shaped cleavage line usually present, though often incomplete, and often not evident in smaller larvae. In Symphyta a single simple eye is usually present on each side of the head. In Apocrita there is sometimes a pair of pigmented streaks on the upper part of the head capsule; sometimes termed "optic plates", they are of unknown function and are better termed parietal bands. The antennae arise from circular or ovoid orbits on the lower sides of the head, or sometimes fairly high on the head. In the Symphyta, the number of antennal segments varies from one to seven. In many Apocrita the orbits simply contain membranes, which may be flat or elevated, usually containing several small sensilla. In some Apocrita a simple, one-segmented antennal papilla is present (probably a secondary development, not strictly homologous with the antenna of Symphyta). The clypeus may or may not be separated from the frons by a distinct suture, the epistomal suture; the clypeolabral suture, marking off the base of the labrum, may also be present or absent.

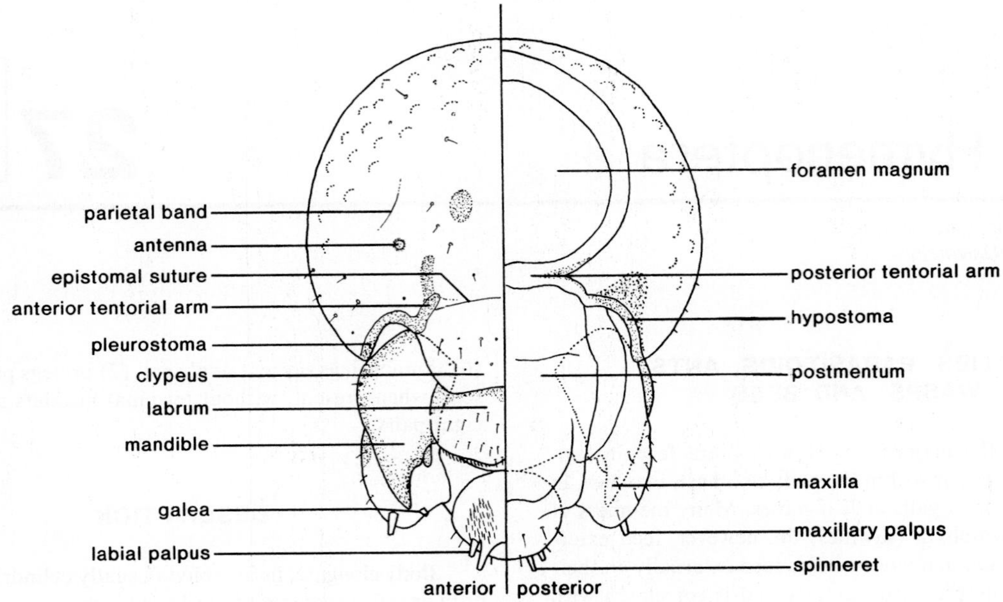

Figure 27.1

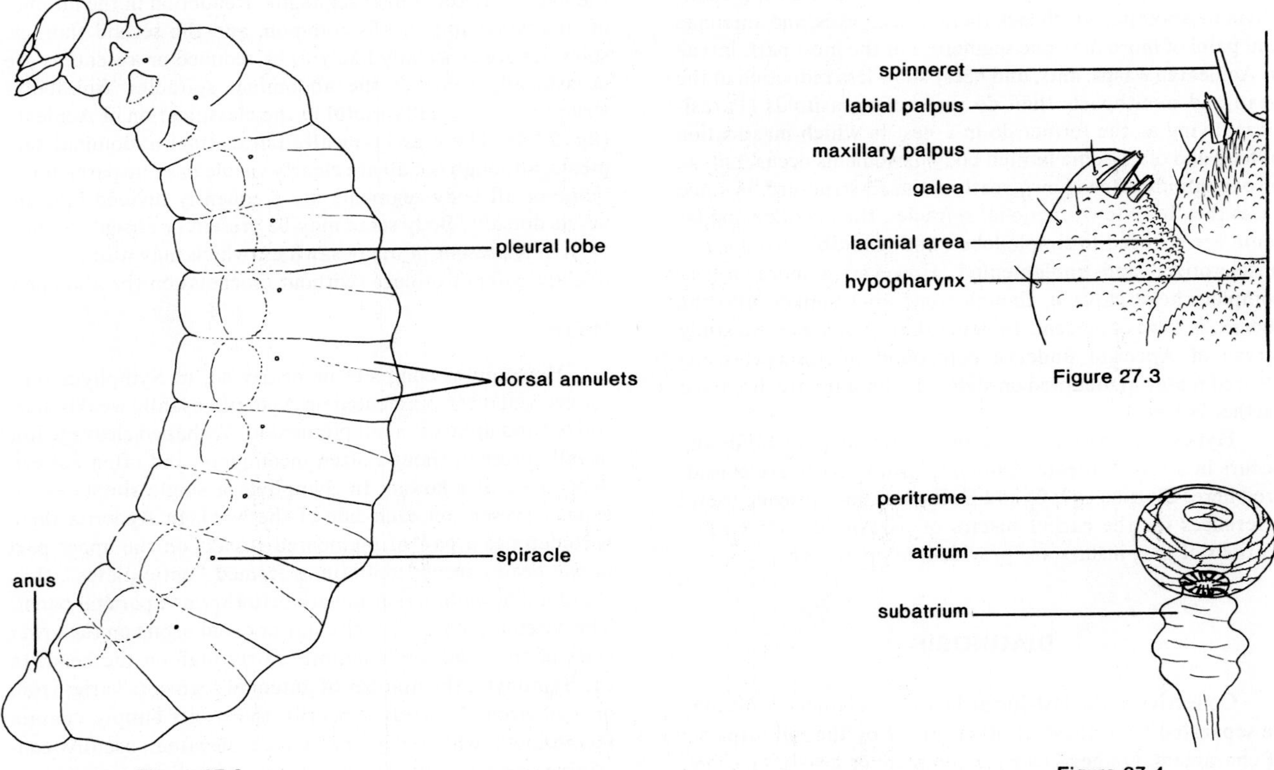

Figure 27.2

Figure 27.3

Figure 27.4

Figures 27.1–27.4. Structures of final-instar larvae of Aculeata (Sphecidae).

Figure 27.1. Head of *Tachytes crassus* Patton, anterior (left) and posterior (right) aspects.

Figure 27.2. Same, lateral aspect of larva.

Figure 27.3. Maxilla, labium, and hypopharynx of *Miscophus evansi* (Krombein), upper (oral) aspect of left side.

Figure 27.4. Anterior thoracic spiracle of *Podium luctuosum* Smith. (From Evans 1964b).

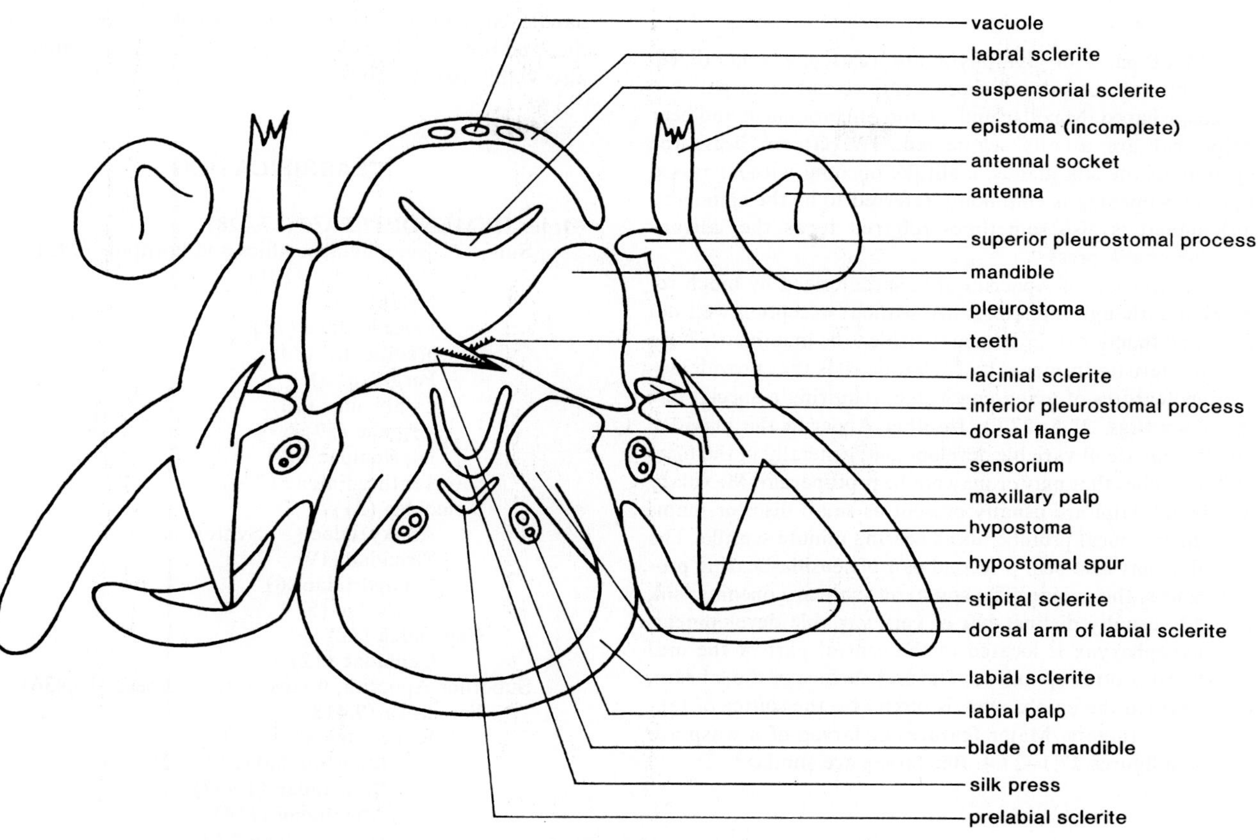

Figure 27.5. Composite diagram of cephalic sclerites of final-instar ichneumonid larva. (From Finlayson and Hagen 1977)

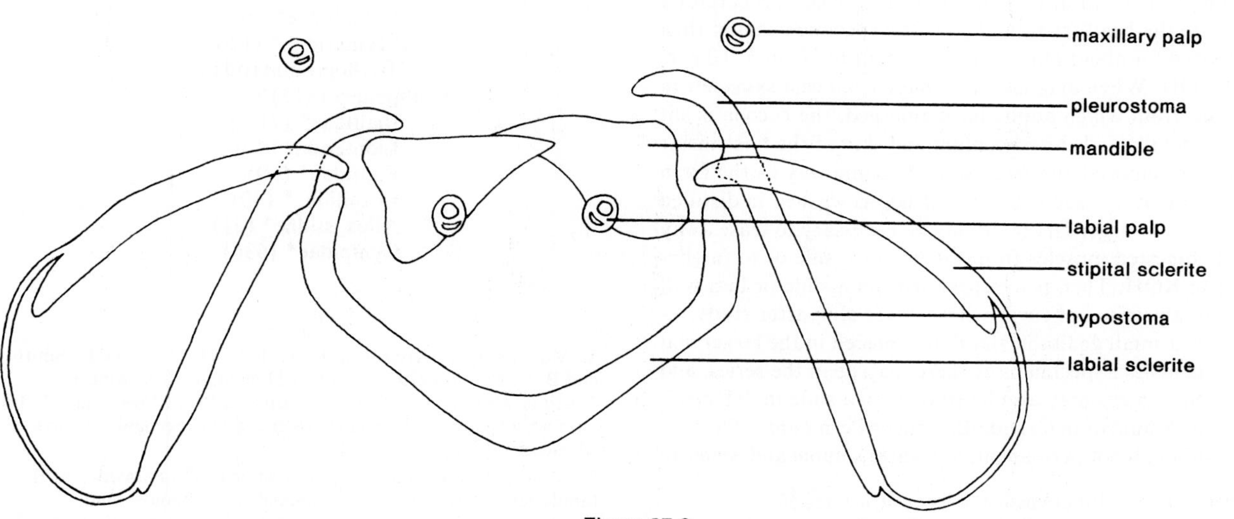

Figure 27.6. Diagram of cephalic sclerites of final-instar aphidiid larva. (From Finlayson and Hagen 1977)

Mouthparts

Mouthparts of Symphyta are relatively generalized. Mandibles are broad and short, dentate, and often palmate or scoop-shaped. Maxillae and labium are prominent and bear palpi that are usually segmented. The labium bears the opening of the silk glands, a slitlike opening without raised lips. This opening is commonly referred to as the spinneret, although it is also sometimes referred to as the salivary opening or silk press.

Mouthparts of Apocrita are characterized by much reduction, although mandibles are without exception well defined and functional. In Ichneumonoidea the mouthparts have a characteristic framework of sclerotic rods that provide numerous features of taxonomic value, requiring a special terminology (figs. 27.5, 27.6). In other Apocrita the maxillae and labium are of variable development, generally in the form of fleshy lobes that may or may not be protuberant. Maxillary and labial palpi are usually present as small discs or mammiform or conical protuberances bearing minute sensilla. The maxillae may also bear a second pair of sensilla-bearing protuberances, the galeae. The spinneret (salivary opening, silk press) is usually distinct but of very variable development. The hypopharynx is located in the central part of the oral cavity and is often spinulose. The oral surface of the labrum, often termed the epipharynx, is often also the source of taxonomic characters. Major features of larvae of a wasp are shown in figures 27.1–27.4. Bee larvae are similar.[1]

TECHNIQUES

It is incorrect to assume that because many of the higher Hymenoptera larvae are small and largely unpigmented, there are no taxonomic characters. There are many; but careful study of the head, spiracles, and body integument is often necessary. Using fresh material or specimens preserved in alcohol, one should first either make a slit in the midventral region of the larva or (in the case of larger larvae) carefully separate the head from the body. The specimen should then be heated for about ten minutes at about 65°C in 5–10 percent KOH. When diagnoses are made from cast skins left in cocoons from which adults have emerged, the cocoon is slit longitudinally and the meconium and skin of the final-instar larva are removed and treated in the same way as the fresh larvae. In either case the softened skin is washed in distilled water. With larger larvae, it may be necessary to tease away partly digested muscles from the head capsule or to further clear in KOH. Then place specimens on a slide or in a well slide in glycerin and cover with a cover slip; after study remove to a small genitalia vial that is placed in the larger vial with such other specimens as there may be in the series, and label. Specimens may also be studied on a slide in deFaure's or Hoyer's fluid or in Canada Balsam or Permount.® The first two, although not permanent, are water soluble and easier to handle. Specimens with well-developed epicrania, e.g., sawflies and bees, should not be permanently slide mounted. See also Wahl (1984) and Torchio and Torchio (1975).

1. For more detailed discussion of larval structures of Hymenoptera, see especially Yuasa (1923), Peterson (1948) (Symphyta), Michener (1953), Torchio and Torchio (1975) McGinley (1981) (bees), and Wheeler and Wheeler (1979) (social Hymenoptera).

CLASSIFICATION

Order **HYMENOPTERA**[2,3] (17,428)
- Suborder Symphyta, sawflies and horntails (992)
  - Megalodontoidea (101)
    - Xyelidae (29)
    - Pamphiliidae (72)
  - Tenthredinoidea (848)
    - Pergidae (5)
    - Cimbicidae (12)
    - Argidae (59)
    - Diprionidae (41)
    - Tenthredinidae (731)
  - Siricoidea (31)
    - Anaxyelidae (= Syntexidae) (1)
    - Siricidae (19)
    - Xiphydriidae (6)
    - Orussidae (5)
  - Cephoidea (12)
    - Cephidae (12)
- Suborder Apocrita, wasps, ants, and bees (16,436)
  - Parasitica (9,418)
    - Ichneumonoidea (5,375)
      - Ichneumonidae (3,322)
      - Braconidae (1,937)
      - Aphidiidae (114)
      - Hybrizontidae* (2)
    - Chalcidoidea (2,223)
      - Torymidae* (175)
      - Pteromalidae* (395)
      - Eurytomidae* (244)
      - Chalcididae* (102)
      - Leucospididae* (5)
      - Eucharitidae* (28)
      - Eupelmidae* (95)
      - Encyrtidae* (509)
      - Eulophidae* (507)
      - Mymaridae* (120)
      - Trichogrammatidae* (43)
    - Cynipoidea (817)
      - Ibaliidae* (7)
      - Liopteridae* (3)
      - Figitidae* (60)
      - Eucoilidae* (80)
      - Alloxystidae* (31)
      - Cynipidae* (636)

2. Modified from: Krombein, K. V., P. D. Hurd, Jr., D. R. Smith, and B. D. Burks. 1979. Catalog of Hymenoptera in America North of Mexico. Smithsonian Institution Press. Vols. 1 and 2. The number following the taxon shows the species recognized north of Mexico.

3. It has not been possible to cover some of the Parasitica at the family level, especially the Chalcidoidea, Cynipoidea and Proctotrupoidea. In addition, the Sclerogibbidae and Embolemidae are not covered, and the Sierolomorphidae (larvae unknown) and the Pelecinidae (larvae unknown?) are not. Families not covered individually are marked with an asterisk (*).

- Proctotrupoidea (938)
  - Vanhorniidae* (1)
  - Roproniidae* (3)
  - Heloridae* (1)
  - Proctotrupidae* (54)
  - Ceraphronidae* (48)
  - Megaspilidae* (61)
  - Diapriidae* (303)
  - Scelionidae* (275)
  - Platygastridae* (192)
- Evanioidea (60)
  - Gasteruptiidae (15)
  - Aulacidae (28)
  - Evaniidae (11)
  - Stephanidae (6)
- Pelecinoidea (1)
  - Pelecinidae* (1)
- Trigonaloidea (4)
  - Trigonalidae (4)

Aculeata (7,018)

- Chrysidoidea (474)
  - Bethylidae (169)
  - Sclerogibbidae* (1)
  - Dryinidae (139)
  - Chrysididae (163)
  - Embolemidae* (2)
- Scolioidea (747)
  - Scoliidae (22)
  - Tiphiidae (236)
  - Sierolomorphidae* (6)
  - Mutillidae (467)
  - Sapygidae (16)
- Formicoidea (580)
  - Formicidae (580)
- Pompiloidea (290)
  - Rhopalosomatidae (2)
  - Pompilidae (288)
- Vespoidea (323)
  - Vespidae (323)
- Sphecoidea (1,139)
  - Sphecidae (1,139)
- Apoidea (3,465)
  - Colletidae (153)
  - Oxaeidae (4)
  - Halictidae (502)
  - Andrenidae (1,199)
  - Melittidae (30)
  - Megachilidae (610)
  - Anthophoridae (920)
  - Apidae (47)

Selected Bibliography (General)

Clausen 1940.
DeBach 1964.
Finalyson and Hagen 1977.
Krombein et al. 1979.
McGinley In press.
Michener 1953.
Peterson 1948.
Torchio and Torchio 1975.
Wahl 1984.
Wheeler and Wheeler 1976, 1979.
Yuasa 1923 (1922).

KEY TO FAMILIES OF HYMENOPTERA (MATURE LARVAE)

Howard E. Evans, *Colorado State University*
Thelma Finlayson, *Simon Fraser University*
Ronald J. McGinley, *Smithsonian Institution*
Woodrow W. Middlekauff, *University of California, Berkeley*
David R. Smith, *Systematic Entomology Laboratory, ARS, USDA*

1. Maxillary and labial palpi distinctly segmented; if 1-segmented, then apex of abdomen with a median, sclerotized postcornus (fig. 1), **or** thoracic legs indicated by sclerotized discs, and each body segment with several stout spines dorsally (fig. 2); stemma (lateral eye) usually present, but absent in most larvae having a postcornus; thoracic legs and prolegs present, sometimes reduced; prolegs sometimes absent in a few tunneling or leaf-mining species .... (suborder SYMPHYTA) 2

1′. Maxillary and labial palpi, if present, disc-like or papilliform, never distinctly segmented; postcornus absent; no sclerotized discs indicating obsolete legs on thorax, and no spines dorsally on each body segment; stemma (lateral eye) never present; thoracic legs and prolegs absent .... (suborder APOCRITA) 14

2(1). Thoracic legs 4- to 5-segmented, and usually with tarsal claw on at least meso- and metathoracic legs; postcornus absent; mostly free-feeding or living in webs, galls, or leafrolls 3

2′. Thoracic legs reduced, represented by indistinctly segmented sclerotized discs, and without tarsal claws, or reduced to clawless, nonsegmented, mammaform protuberances (fig. 3); postcornus (fig. 1) present in stem and wood borers, absent in leaf miners; internal borers in stems, twigs, or wood, and some leaf miners 9

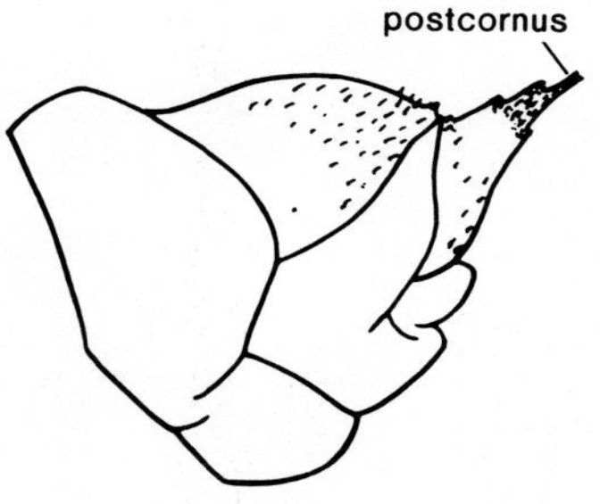

postcornus

Figure 1

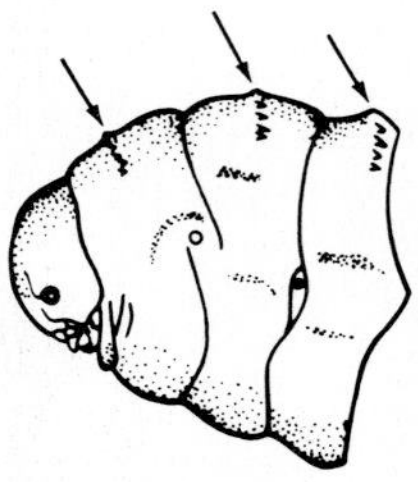

Figure 2

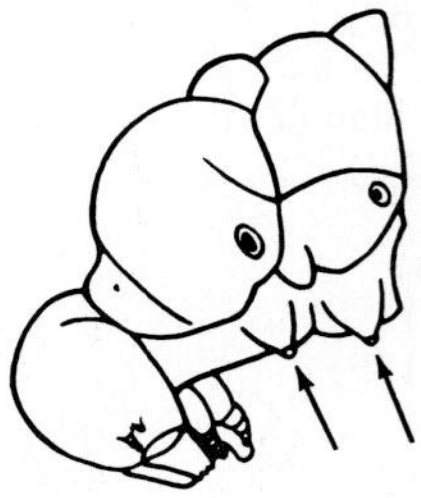

Figure 3

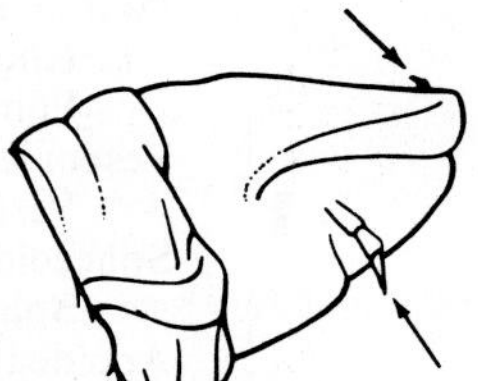

Figure 4

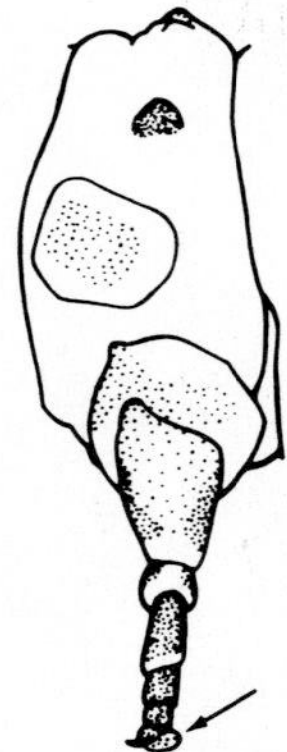

Figure 5

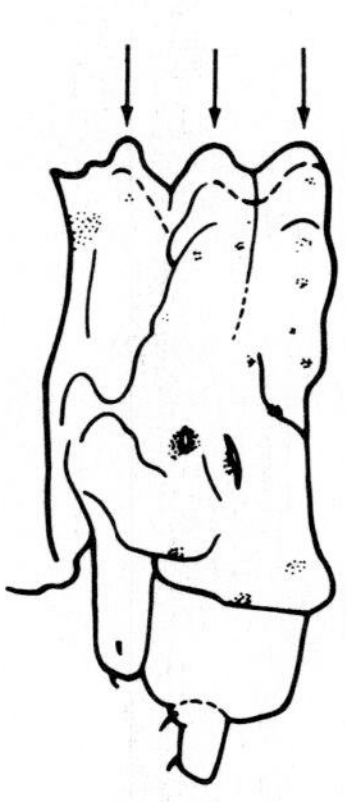

Figure 6

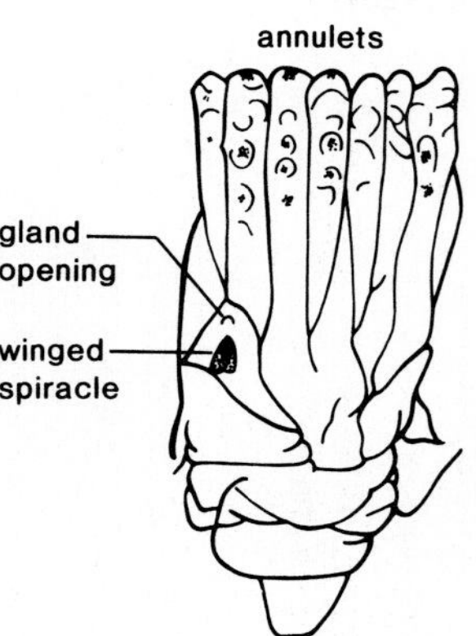

Figure 7

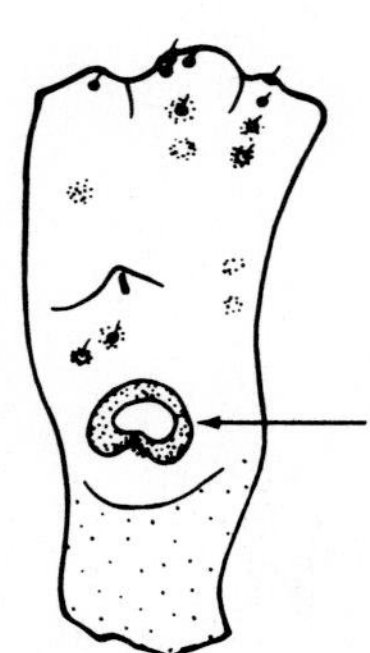

Figure 8

3(2). Antenna 6- or 7-segmented; abdominal prolegs absent, reduced to blunt swellings, or present on *each* abdominal segment in some Xyelidae 4

3′. Antenna with 5 or fewer segments; abdominal prolegs usually distinct but *not on each* abdominal segment 5

4(3). Tenth abdominal segment with pair of segmented, subanal appendages (fig. 4); small sclerotized hook on 10th tergum (fig. 4); thoracic legs slender, elongate, and straight; prolegs absent .... ***Pamphiliidae*** (p. 621)

4′. Tenth segment without subanal appendages and suranal hook; thoracic legs stout; prolegs, if present, appear as blunt swellings or are distinct .... ***Xyelidae*** (p. 619)

5(3′). Thoracic legs with a distinct pad or divergent lobe (empodium) adjacent to claw (fig. 5); prothoracic tarsal claw sometimes absent; abdominal segments with no more than 3 dorsal annulets (fig. 6) 6

5′. Thoracic legs without an enlarged empodium; tarsal claw always present on each thoracic leg; abdominal segments with variable number of annulations, usually more than 3 (fig. 7) 7

6(5). Suckerlike lateral protuberance on abdominal segments 2–4 or 2–5 and 8 (fig. 8); thoracic legs always with tarsal claws; antenna flat, with 4 circular sclerites (fig. 9) .... ***Pergidae*** (p. 622)

6′. Suckerlike lateral protuberances absent on abdomen; thoracic legs with tarsal claws, or with prothoracic tarsal claw absent; antenna 1- or 2-segmented, low and moundlike, or peglike (figs. 10, 11) .... ***Argidae*** (p. 625)

7(5′). Supraspiracular gland openings present (fig. 7); labrum with secondary longitudinal sutures (fig. 12); antenna 1-segmented, low and moundlike (fig. 13); abdominal segments 1–8 each with seven dorsal annulets (fig. 7) .... ***Cimbicidae*** (p. 623)

7′. Supraspiracular gland openings absent; labrum without secondary sutures; antenna 2- to 5-segmented, commonly conical or with apical segment peglike (fig. 14); annulation of abdominal segments variable, if with 7 dorsal annulets, then antenna conical and 5-segmented 8

8(7′). Antenna 3-segmented, 1st and 2nd segments crescentric and flattened, 3rd segment peglike (figs. 14, 15); prolegs present on abdominal segments 2–8 and 10; abdominal segments 1–8 each with 6 dorsal annulets (fig. 16) .... ***Diprionidae*** (p. 628)

8′. Antenna 2- to 5-segmented, usually conical and 5-segmented, sometimes flat, but never with 3rd (apical) segment peglike; number of prolegs and annulations of abdominal segments variable .... (in part) ***Tenthredinidae*** (p. 631)

9(2′). Tenth tergum without postcornus 10

9′. Tenth tergum with postcornus (fig. 17) 11

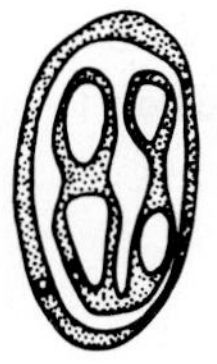

Figure 9

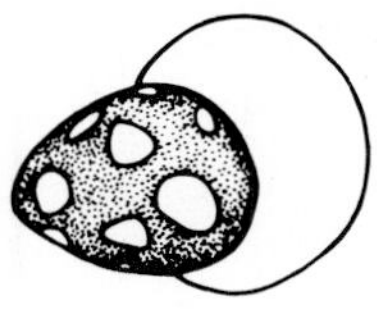

Figure 10

Figure 11

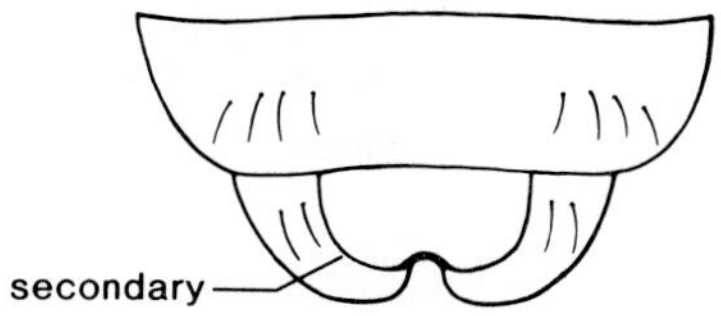

Figure 12

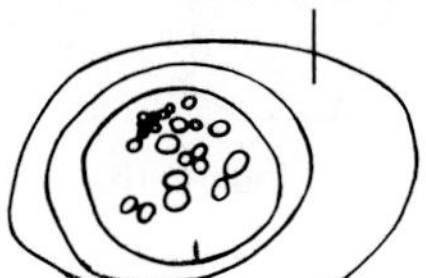

Figure 13

Figure 14

Figure 15

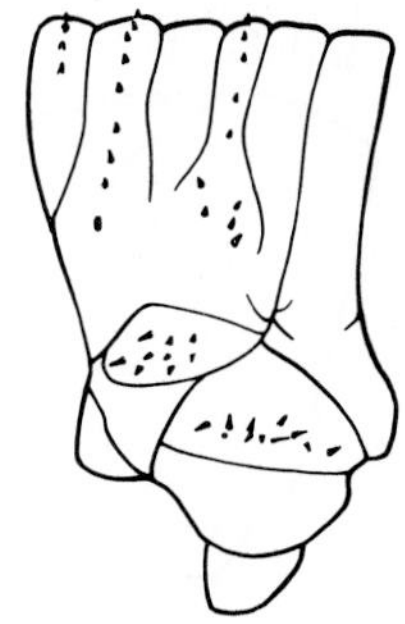

Figure 16

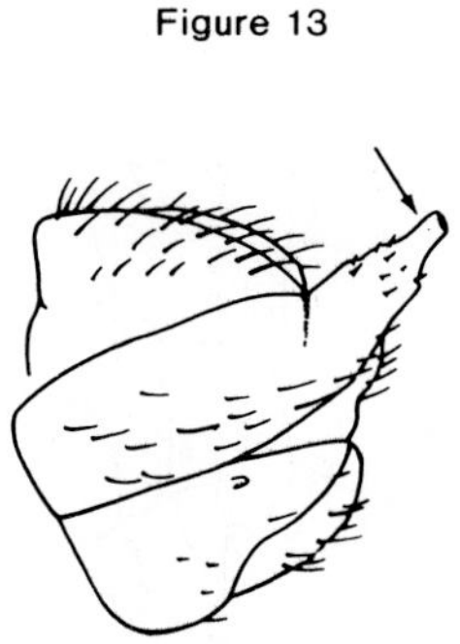

Figure 17

10(9).	Body segments lacking dorsal transverse row of stout spines; thoracic legs indistinctly 4-segmented and usually without claws; maxillary and labial palpi segmented; leaf miners	(in part) ***Tenthredinidae*** (p. 621)
10′.	Body segments each with dorsal transverse row of stout spines (fig. 2); thoracic legs reduced (fig. 2) represented at most by sclerotized discs; maxillary and labial palpi unsegmented, represented by lobes; found in wood	***Orussidae*** (p. 647)
11(9′).	Small, 1- or 2-segmented subanal appendages on 10th sternum (fig. 18); stemma present (fig. 19); stem or twig borers	***Cephidae*** (p. 649)
11′.	Subanal appendages absent on 10th sternum; stemma absent; wood borers	12
12(11′).	Metathoracic spiracle vestigial or absent; postcornus with dorsal tuft of hairs at base (fig. 20); antenna 4- or 5-segmented (fig. 21)	***Xiphydriidae*** (p. 645)
12′.	Metathoracic spiracle conspicuous (fig. 22); postcornus without dorsal tuft of hairs at base (fig. 23); antenna 2- or 3-segmented (fig. 24)	13
13(12′).	First antennal segment setose (fig. 24); vertical furrows on head usually visible; postcornus directed somewhat posteriorly; no small, fleshy protuberance on each side at base of postcornus; mature larva usually over 20 mm long	***Siricidae*** (p. 643)
13′.	First antennal segment bare (fig. 25); vertical furrows on head not visible; postcornus directed more dorsally; a small, fleshy protuberance present on each side at base of postcornus; mature larva not exceeding 15 mm	***Anaxyelidae*** (p. 645)
14(1′).	Maxillae each with 1 papilla (the palpus), with none at all, or the maxillae not clearly distinguishable	15
14′.	Maxillae each with 2 papillae (palpus and galea) (fig. 26) (galea may be quite small, especially in some bees, requiring careful study under high magnification)	72
15(14).	Maxillae and labium not projecting as distinct and separate lobes, being either fused, separated merely by a sclerotic bar, or either or both indistinguishable	16
15′.	Maxillae and labium projecting as separate lobes (fig. 26)	48
16(15).	Mandibles bidentate or tridentate	17
16′.	Mandibles simple (unidentate) (figs. 27, 28) (may be denticulate), or vestigial	23
17(16).	Mandibles bidentate	18
17′.	Mandibles tridentate (figs. 29, 30)	19

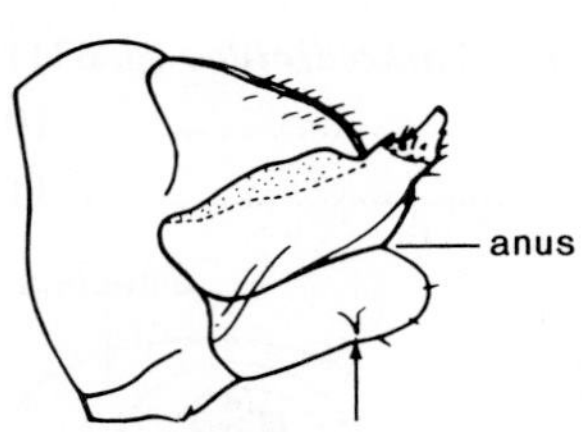

Figure 18

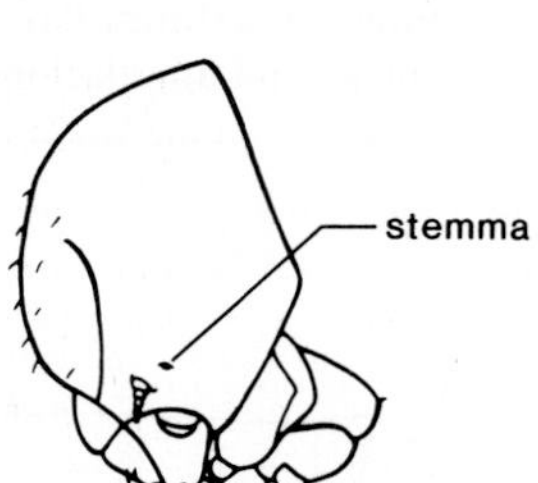

Figure 19

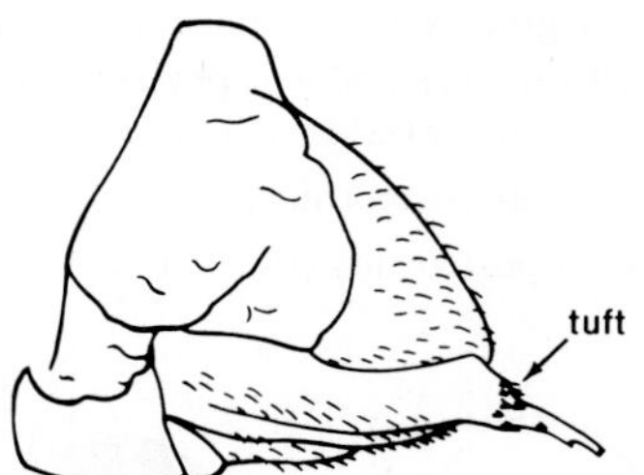

Figure 20

Figure 21

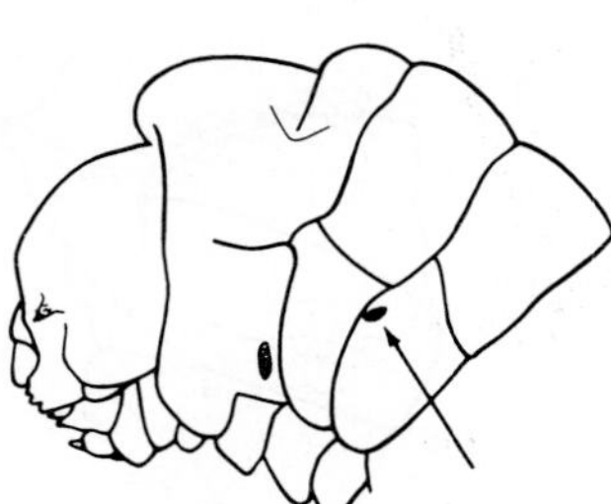
Figure 22

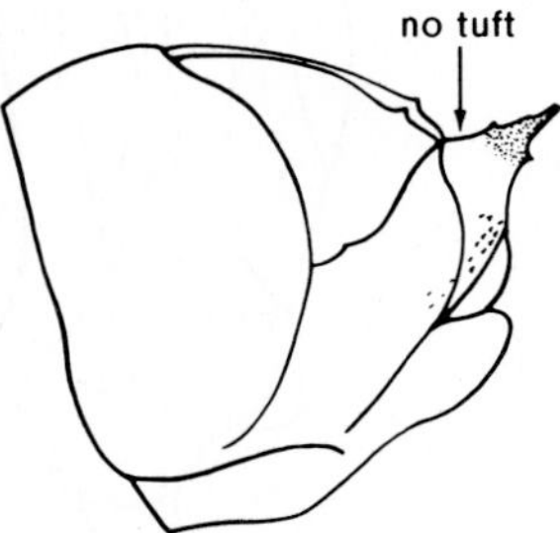

Figure 23

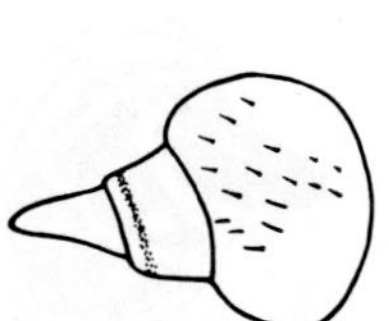
Figure 24

Figure 25

18(17). Head with large setae; second mandibular tooth very much smaller than major tooth .... some Chalcidoidea (***Eurytomidae***) (p. 664)
18′. Head with inconspicuous setae or none; both mandibular teeth well developed .... many parasitic Cynipoidea (p. 665)
19(17′). Body with paired, dorsal, conical tubercles on some segments .... (***Figitidae;*** **Anacharitinae**) some Cynipoidea (p. 665)
19′. Body without dorsal, conical tubercles 20
20(19′). Body with lateral, circular, or oval areas that are thickly studded with spines .... (***Ibaliidae***) some Cynipoidea (p. 665)
20′. Body without such lateral spinose areas 21
21(20′). Mandibles broad, not much tapered apically (fig. 29); antennae not projecting, sometimes difficult to observe .... (gall-forming ***Cynipidae***) some Cynipoidea (p. 665)
21′. Mandibles narrow, tapered (figs. 30, 31); antennae conical or tuberculate 22
22(21′). Maxillary and labial palpi present, though not projecting; mouthparts without a suctorial disc .... ***Stephanidae*** (p. 669)
22′. Maxillary and labial palpi not distinct, mouthparts behind mandibles forming a suctorial disc (figs. 32, 33) .... ***Trigonalidae*** (p. 670)
23(16′). Head with a melanized band encircling lateral, posterior margins (fig. 34) .... ***Heloridae*** (p. 666)
23′. Head without such a melanized band 24

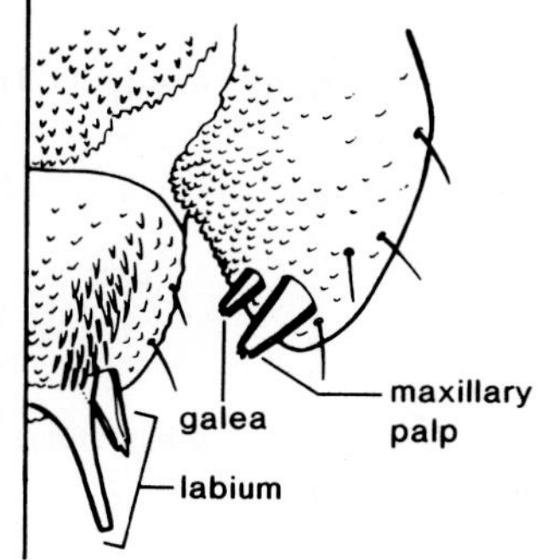

Figure 26

Figure 27

Figure 28

Figure 29

Figure 30

Figure 31

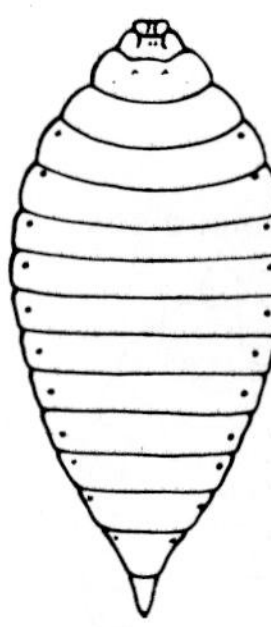
Figure 32

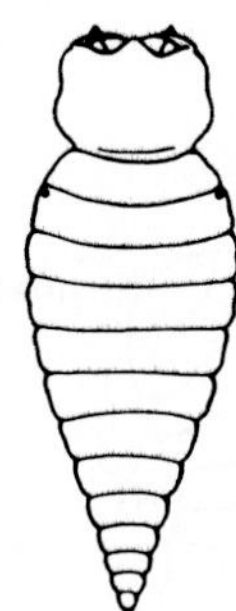
Figure 33

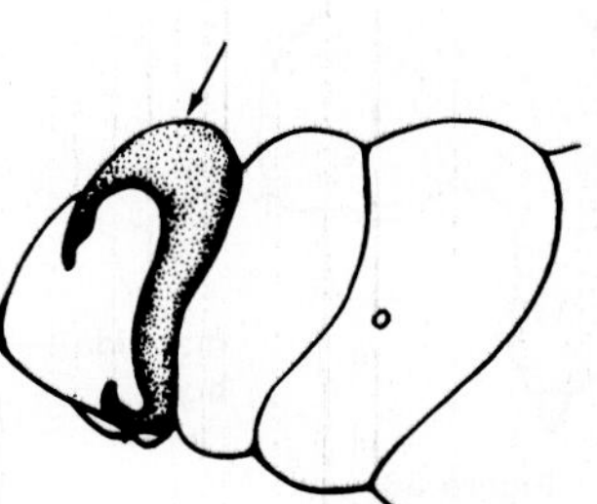
Figure 34

24(23′). Mouthparts with mandibles often sclerotized and pigmented but mouthparts otherwise largely fleshy, parts often difficult to distinguish; pleurostoma present in some Chalcidoidea (fig. 35) 25

24′. Mouthparts with a sclerotic framework that includes the pleurostoma and hypostoma (latter often with a spur), and often the epistoma and labial sclerite (figs. 36, 37) (except that mandibles and head structures unsclerotized and only visible with staining in a few Ichneumonidae) 29

25(24). Body with conical tubercles or sclerotized rings on some segments (fig. 38) .... ***Ceraphronidae*** (p. 666)

25′. Body without tubercles or sclerotized rings 26

26(25′). With discoidal body on side of first abdominal segment (fig. 39) .... ***Platygastridae*** (p. 666)

26′. Without such discoidal body 27

27(26′). Mandibles, maxillae and labium usually distinguishable; 6–9 pairs of spiracles present .... some parasitic Cynipoidea (p. 665)

27′. Only the mandibles well defined in most species; spiracles quite variable in number 28

28(27′). With large antennal discs or protruding antennae in most species (fig. 40); maxillae sometimes distinguishable; spinneret present or absent .... many Proctotrupoidea (p. 666)

28′. Antennae and maxillae usually not distinguishable; spinneret absent .... most Chalcidoidea (p. 664)

29(24′). Epistoma with 2 sclerotized dorsolateral projections (fig. 41) .... ***Aulacidae*** (p. 668)

29′. Epistoma without such sclerotized projections 30

30(29′). Mandibles and head structures unsclerotized and only visible with staining .. **(Collyriinae)** ***Ichneumonidae*** (p. 650)

30′. Mandibles and head structures sclerotized 31

31(30′). Each thoracic and abdominal spiracle at least 5 times longer than width at widest point and usually without obvious atrium and closing apparatus (fig. 42) .... (in part) ***Aphidiidae*** (p. 663)

31′. Each thoracic and abdominal spiracle less than 5 times longer than width at widest point, or if 5 times longer, then atrium is present 32

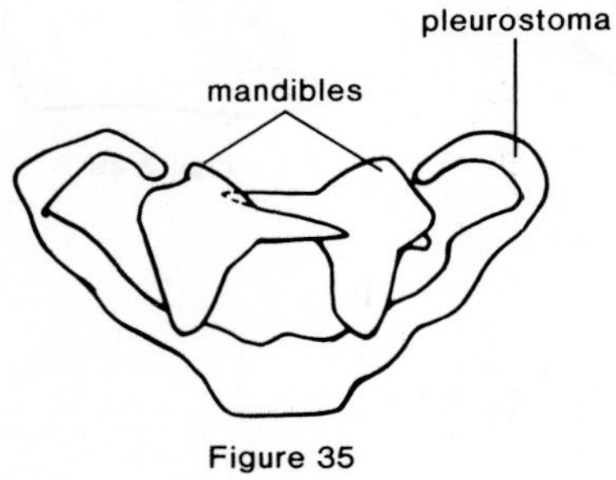

Figure 35

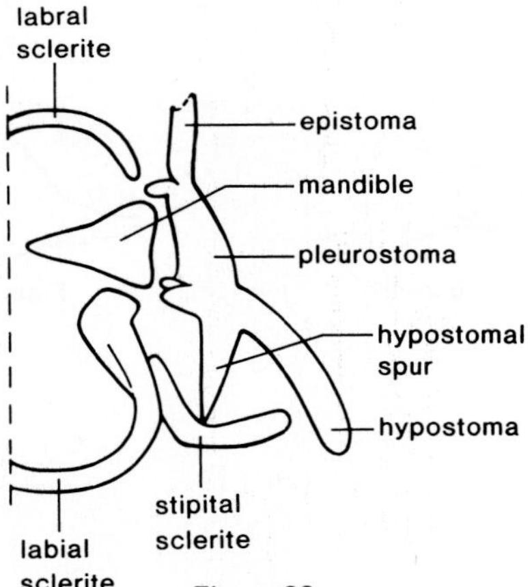

Figure 36

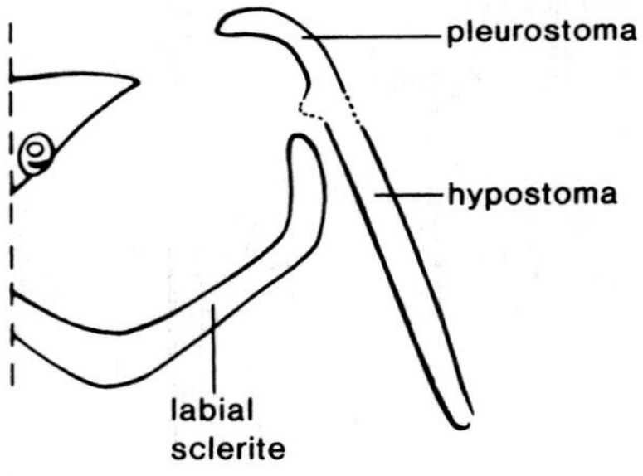

Figure 37

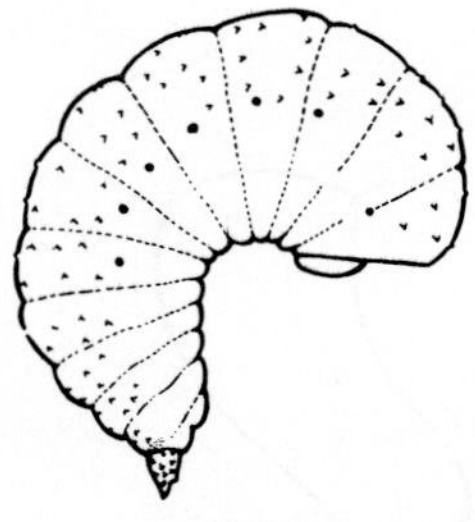

Figure 38

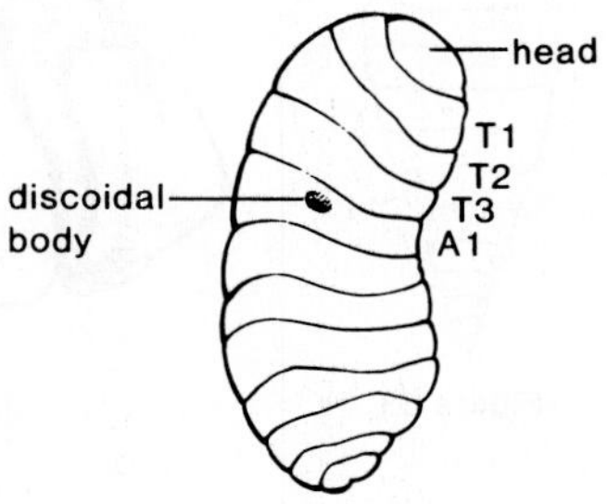

Figure 39

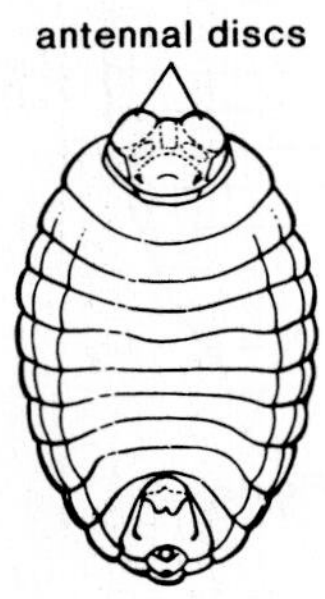

Figure 40

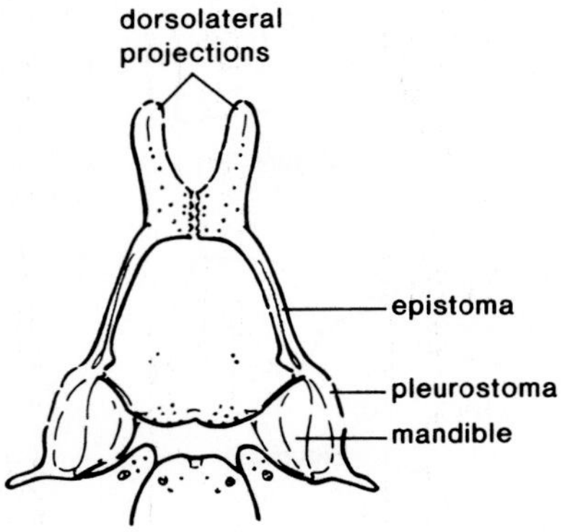

Figure 41

32(31′). Blade of mandible with 1 or 2 rows of teeth (fig. 43) 33

32′. Blade of mandible without teeth 37

33(32). Blade of mandible with 2 rows of teeth .... (in part) ***Ichneumonidae*** (p. 650)

33′. Blade of mandible with 1 row of teeth 34

34(33′). Mandible with a greatly elongated blade with row of teeth on upper edge only, and often with a long, sharp terminal tooth (fig. 44) .... (in part) ***Braconidae*** (p. 657)

34′. Mandible with a relatively short blade with row of teeth on upper or lower edge 35

35(34′). Hypostomal spur absent .... (**Anomaloninae:** *Anomalon*) ***Ichneumonidae*** (p. 650)

35′. Hypostomal spur present (fig. 36) 36

36(35′). Hypostoma extending lateroventrally, usually beyond stipital sclerite, either touching (fig. 45) or passing behind it (fig. 46) .... (in part) ***Braconidae*** (p. 657)

36′. Hypostoma not extending lateroventrally beyond stipital sclerite and not touching or passing behind it .... (in part) ***Ichneumonidae*** (p. 650)

37(32′). Hypostomal spur present (fig. 36) 38

37′. Hypostomal spur absent 41

38(37). Hypostomal spur free or joined to stipital sclerite only (figs. 47, 48, 49) .... (in part) ***Braconidae*** (p. 657)

38′. Hypostomal spur not free or not joined to stipital sclerite only 39

39(38′). Hypostomal spur joined to hypostoma only (fig. 36) .... (in part) ***Ichneumonidae*** (p. 650)

39′. Hypostomal spur joined to both stipital sclerite and hypostoma 40

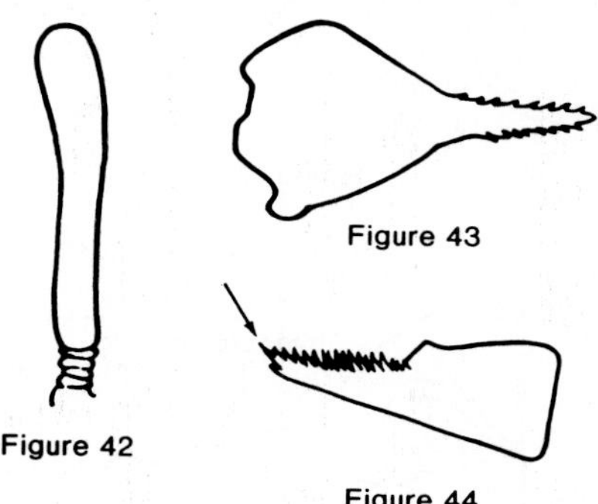

Figure 42

Figure 43

Figure 44

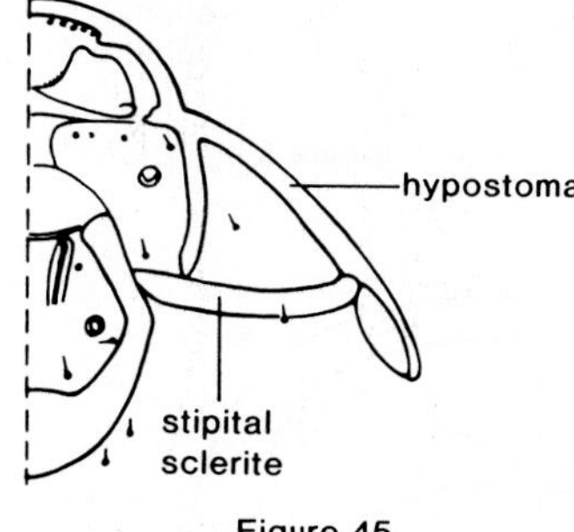

Figure 45

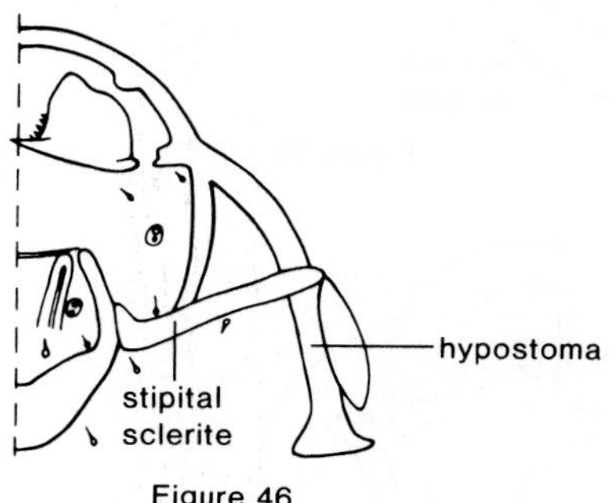

Figure 46

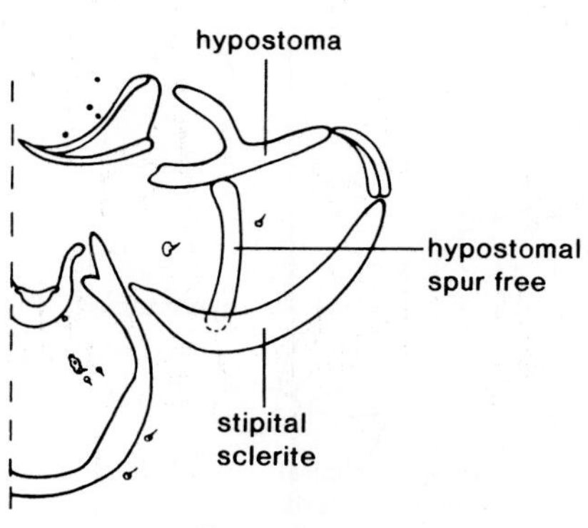

Figure 47

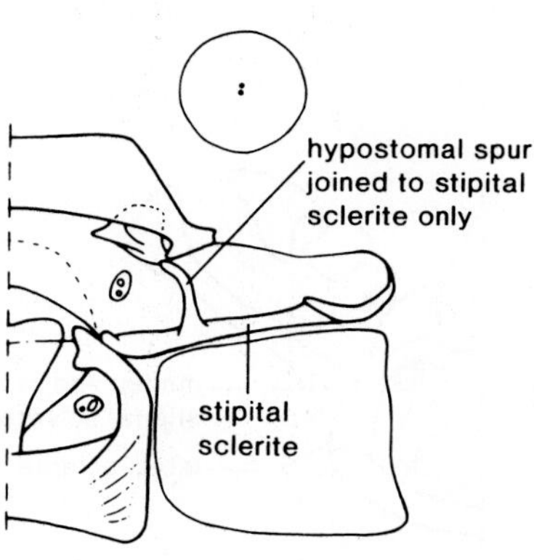

Figure 48

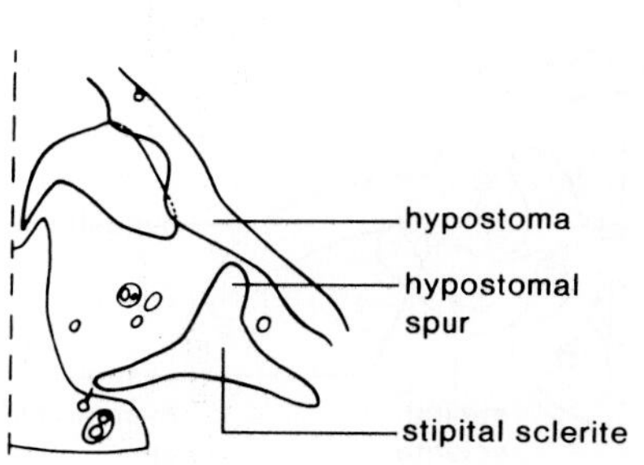

Figure 49

40(39′). Hypostomal spur joined to lateral end of stipital sclerite (figs. 50, 51) .... (in part) ***Ichneumonidae*** (p. 650)

40′. Hypostomal spur joined to stipital sclerite at some point along its length, but not at its lateral end (fig. 52) .... (in part) ***Braconidae*** (p. 657)

41(37′). Stipital sclerite absent (fig. 53) 42

41′. Stipital sclerite present (fig. 55) 43

42(41). Hypostoma present (fig. 53) .... **(Ichneumoninae)** ***Ichneumonidae*** (p. 650)

42′. Hypostoma absent (fig. 54) .... (in part) ***Braconidae*** (p. 657)

43(41′). Stipital sclerite curving dorsally to rest its lateral end on hypostoma (figs. 55, 56) .... **(Anomaloninae** except *Anomalon,* **Metopiinae)** ***Ichneumonidae*** (p. 650)

43′. Stipital sclerite not curving dorsally to rest its lateral end on hypostoma 44

44(43′). Medial end of stipital sclerite not situated at dorsal end of lateral arm of labial sclerite (fig. 57) .... (in part) ***Braconidae*** (p. 657)

44′. Medial end of stipital sclerite situated at dorsal end of lateral arm of labial sclerite (figs. 58, 59, 60, 61) 45

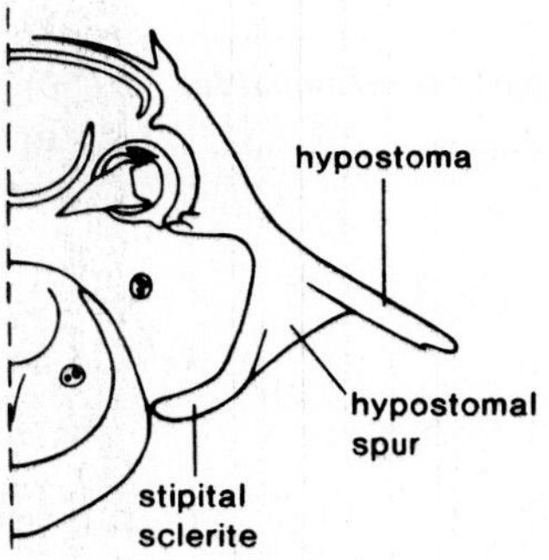

Figure 50

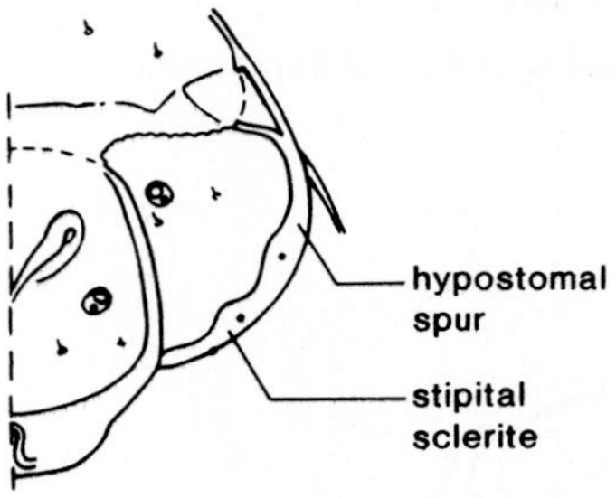

Figure 51

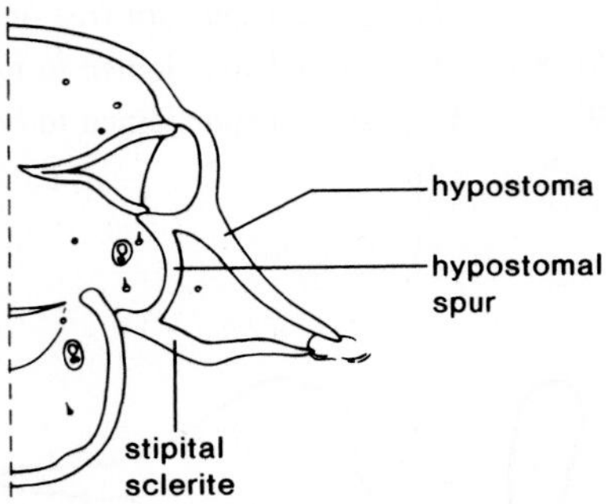

Figure 52

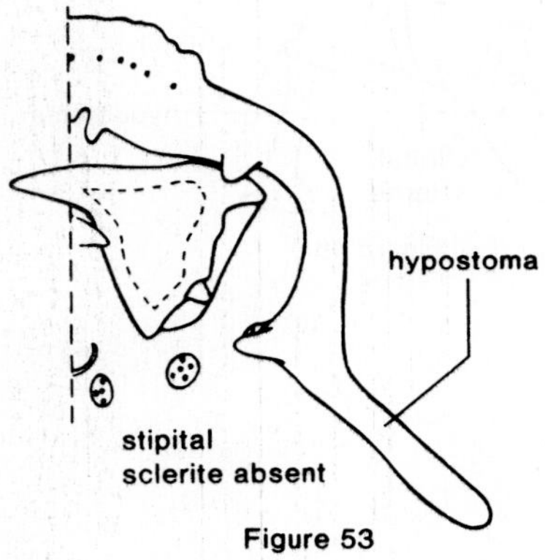

Figure 53

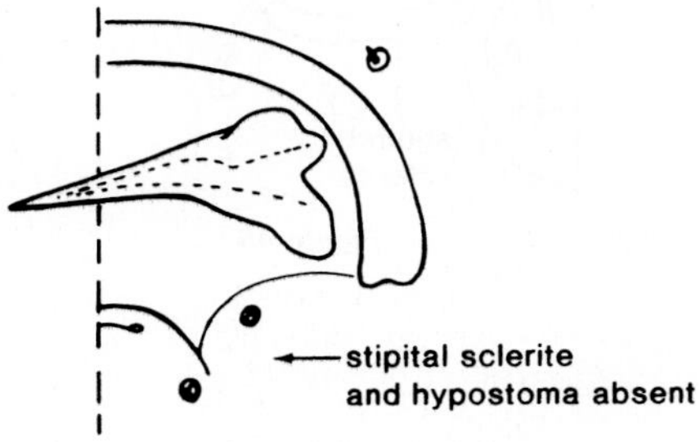

Figure 54

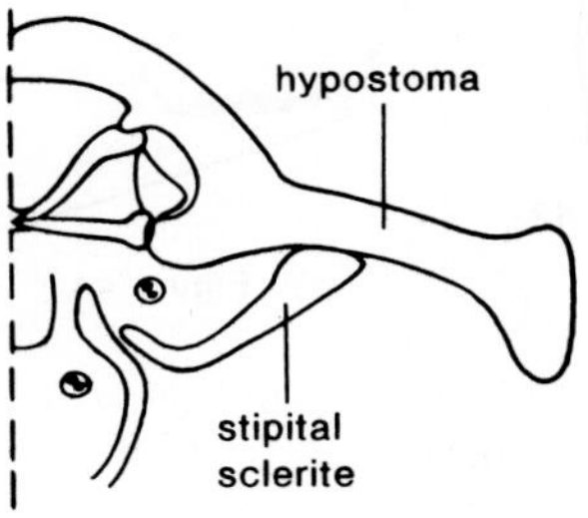

Figure 55

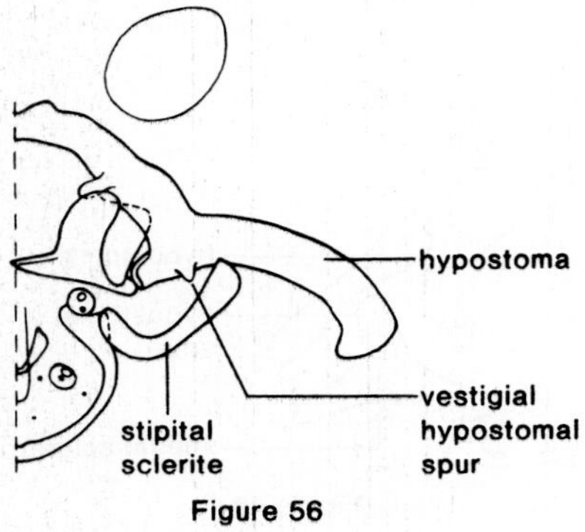

Figure 56

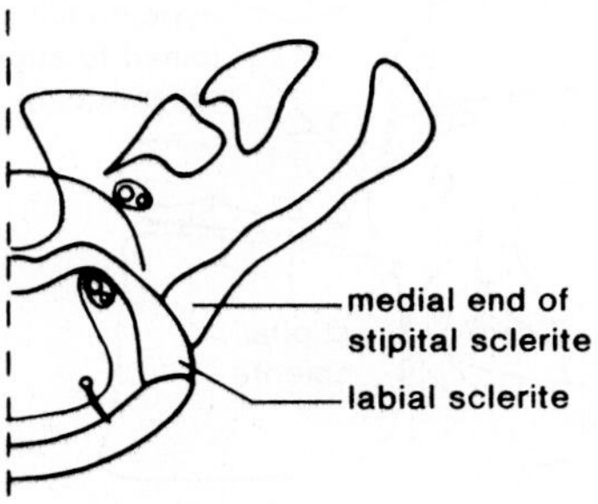

Figure 57

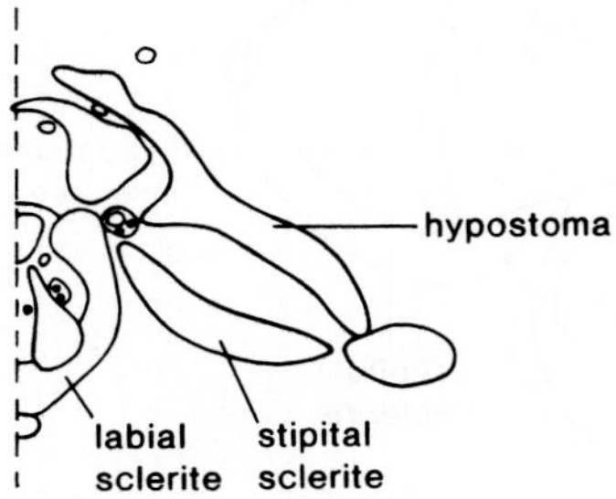

Figure 58

← Fig. 36

45(44′). Hypostoma present (fig. 58) 46
45′. Hypostoma absent or vestigial (fig. 59) 47
46(45). Stipital sclerite either crossing over hypostoma anteriorly (fig. 61) or joined to it by a ring of sclerite (fig. 60) .... (in part) ***Aphidiidae*** (p. 663)
46′. Stipital sclerite not crossing over hypostoma anteriorly nor joined to it by a ring of sclerite (fig. 58) .... (in part) ***Braconidae*** (p. 657)
47(45′). Stipital sclerite enlarging to a paddlelike structure at distal end (fig. 59) .... (in part) ***Aphidiidae*** (p. 663)
47′. Stipital sclerite without a paddlelike enlargement at distal end .... (in part) ***Braconidae*** (p. 657)
48(15′). Mandibles distinctly tridentate (figs. 62, 63) 49
48′. Mandibles simple (fig. 64), denticulate (fig. 65), cuspidate (fig. 66), or bidentate 52
49(48). With 10 pairs of spiracles 50
49′. With 9 pairs of spiracles 51
50(49). Head prognathous (fig. 67); labrum convex apically .... (in part) ***Bethylidae*** (p. 670)
50′. Head hypognathous (fig. 68); labrum bilobed .... (in part) ***Chrysididae*** (p. 673)
51(49′). Antennal papilla present (fig. 69); mandibles longer than broad (fig. 62) .... ***Gasteruptiidae*** (p. 667)
51′. Antennal papilla absent; mandibles about as broad basally as long (fig. 63) .... ***Evaniidae*** (p. 668)

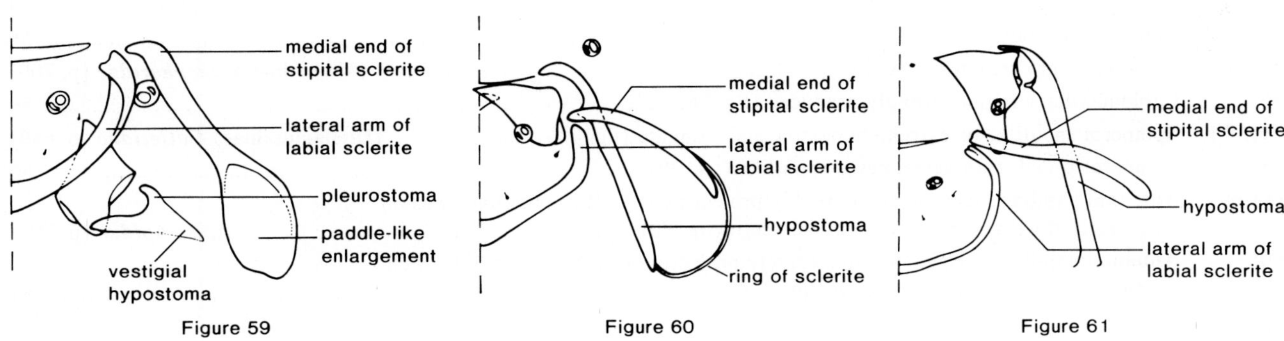

Figure 59 Figure 60 Figure 61

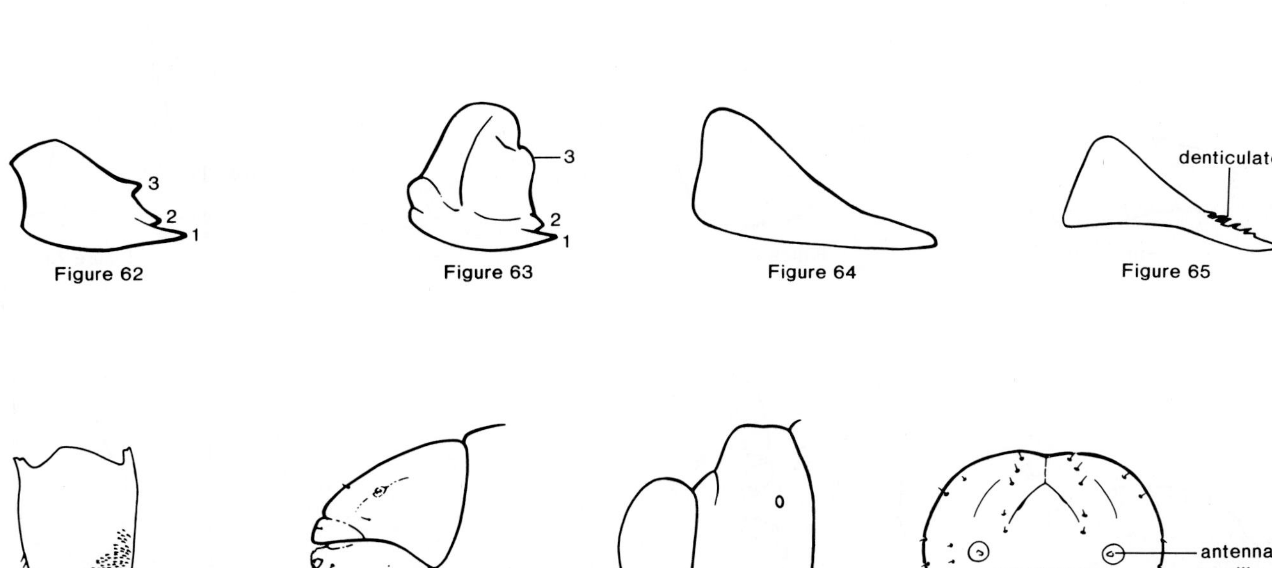

Figure 62 Figure 63 Figure 64 Figure 65

Figure 66 Figure 67 Figure 68 Figure 69

52(48′). Head extremely prognathous (figs. 70, 71); maxillae and labium large, making up more than half the total area of the head in lateral view; mandibles acuminate, either simple or denticulate (fig. 65); parasitoids 53

52′. Head moderately prognathous (figs. 72, 73); maxillae and labium relatively smaller than above; mandibles highly variable in shape and dentition; pollen feeders (look for pollen grains around the mouthparts or in the alimentary tract) (Apoidea, in part) 54

53(52). Body U-shaped prior to leaving host; maxillary palpi unusually large and projecting (fig. 71) .... ***Dryinidae*** (p. 672)

53′. Body not U-shaped (fig. 70); maxillary palpi short (fig. 67) .... (in part) ***Bethylidae*** (p. 670)

54(52′). Spinneret strongly projecting, conspicuous (may be unpigmented) (fig. 72) 55

54′. Spinneret absent or very weakly projecting, inconspicuous (fig. 73) 62

55(54). Thoracic dorsal tubercles minute, conical, darkly pigmented (fig. 74) .... (Bombini) ***Apidae*** (p. 703)

55′. Thoracic dorsal tubercles rounded, transverse or absent, not minute, conical or pigmented 56

56(55′). Body conspicuously slender, often with median dorsal tubercles present (fig. 75) .... (Melitomini) ***Anthophoridae*** (p. 697)

56′. Body more robust, not conspicuously slender (fig. 76); median dorsal tubercles usually absent 57

57(56′). Conspicuous body setae present (fig. 76) .... ***Megachilidae*** (p. 696)

57′. Conspicuous body setae absent 58

58(57′). Labiomaxillary region recessed (fig. 77) .... (Melectini) ***Anthophoridae*** (p. 700)

58′. Labiomaxillary region strongly produced (fig. 78) 59

59(58′). Spinneret spoutlike or extremely narrow and elongate (fig. 78) .... **(Diphaglossinae)** ***Colletidae*** (p. 689)

59′. Spinneret broad, transverse, never spoutlike (fig. 79) 60

60(59′). Antennal papillae inconspicuous, weakly projecting (fig. 79); mandibles weakly sclerotized with a few inconspicuous teeth (fig. 80) .... (*Apis*) ***Apidae*** (p. 703)

60′. Antennal papillae conspicuous, moderately projecting (fig. 81), mandibles well sclerotized with numerous teeth (fig. 82) 61

Figure 70

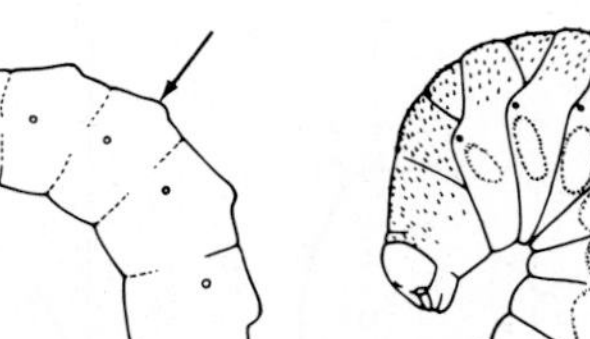

Figure 71

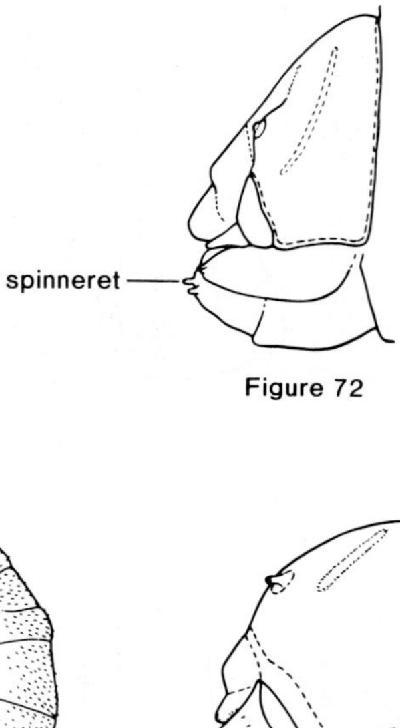

Figure 72

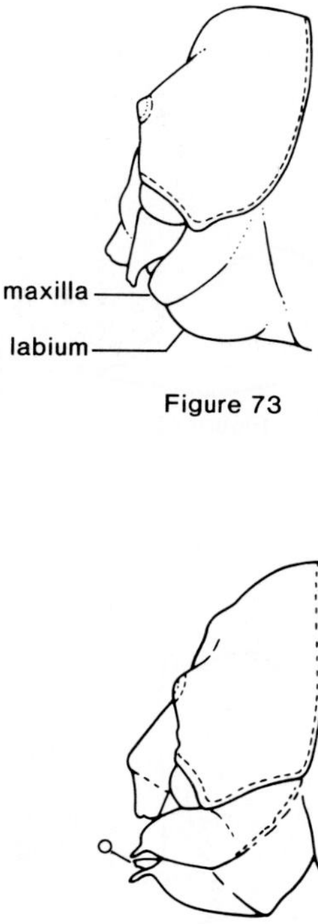

Figure 73

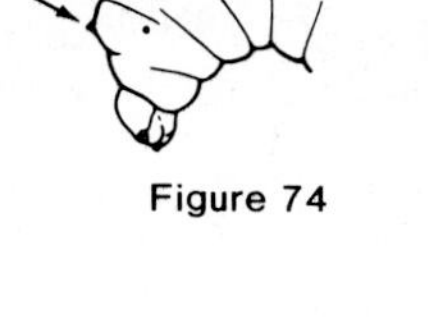

Figure 74

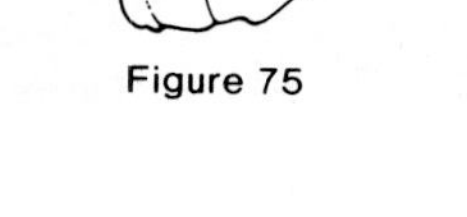

Figure 75

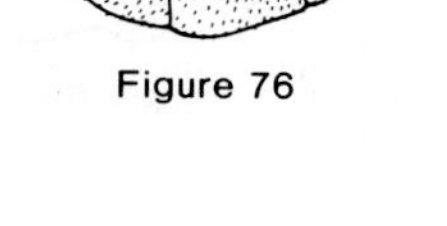

Figure 76

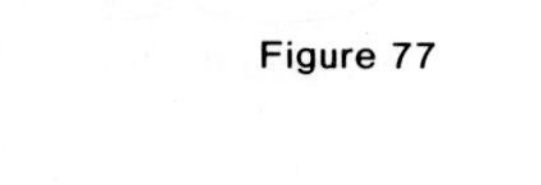

Figure 77

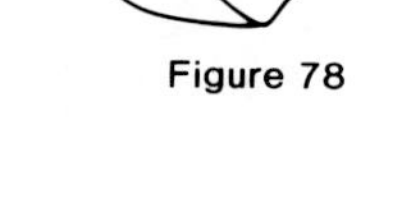

Figure 78

61(60′). Mandibular apex simple, subapical tooth absent (fig. 83) .... **(Dufoureinae)** ***Halictidae*** (p. 691)

61′. Mandibular apex bidentate (fig. 82) .... (*Exomalopsis,* in part) ***Anthophoridae*** (p. 697)

62(54′). Salivary opening positioned dorsally on labium, not visible in frontal view ... **(Xylocopinae)** ***Anthophoridae*** (p. 702)

62′. Salivary opening positioned apically on labium (normal position), visible in frontal view (fig. 84) 63

63(62′). Labiomaxillary region extremely recessed (figs. 85, 86), maxilla usually not projecting beyond hypostomal ridge, or if longer, maxilla greatly exceeded by labium (fig. 86); posterior tentorial pit located anterior to posterior thickening of head capsule and often below hypostomal ridge (figs. 85, 86) 64

63′. Labiomaxillary region produced or moderately recessed (figs. 87, 88), maxilla always exceeding hypostomal ridge and never greatly exceeded by labium; posterior tentorial pit normal in position, located at junction of hypostomal ridge and posterior thickening of head capsule (fig. 88) 65

64(63). Labrum cleft apically (fig. 89); epistomal ridge well developed mesiad of anterior tentorial pits (fig. 89) .... ***Oxaeidae*** (p. 689)

64′. Labrum rounded apically; epistomal ridge incomplete or absent mesiad of anterior tentorial pits .... **(Nomadinae)** ***Anthophoridae*** (p. 700)

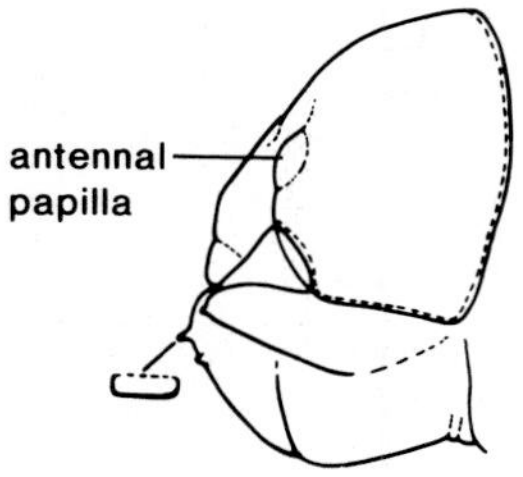

Figure 79

Figure 80

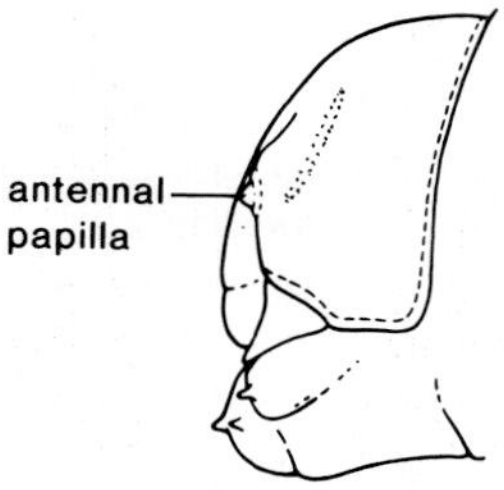

Figure 81

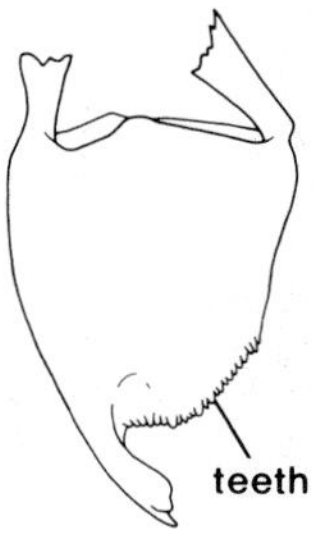

Figure 82

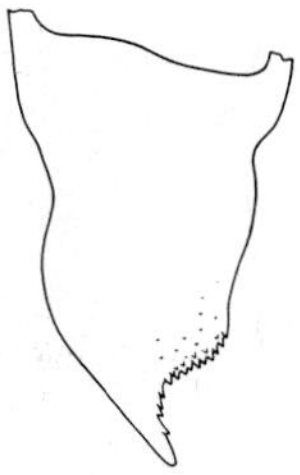

Figure 83

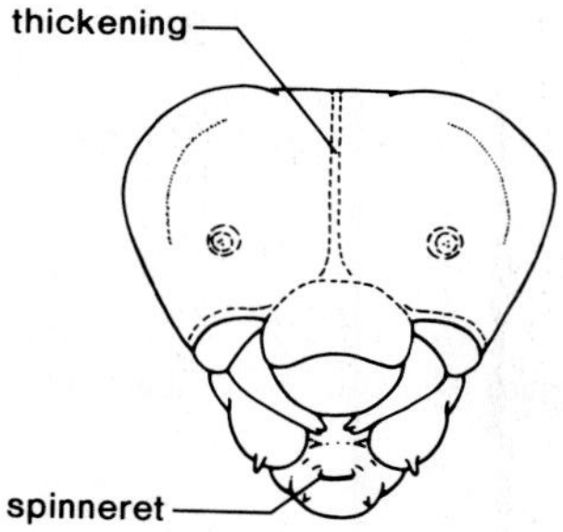

Figure 84

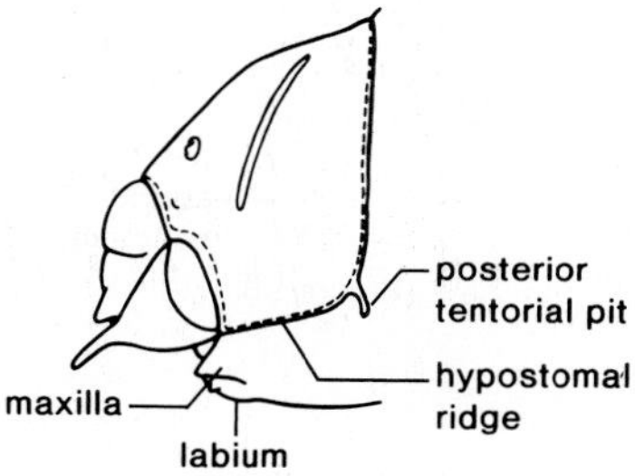

Figure 85

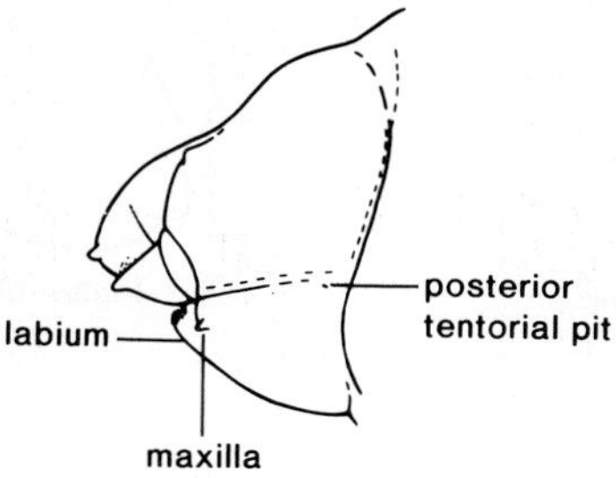

Figure 86

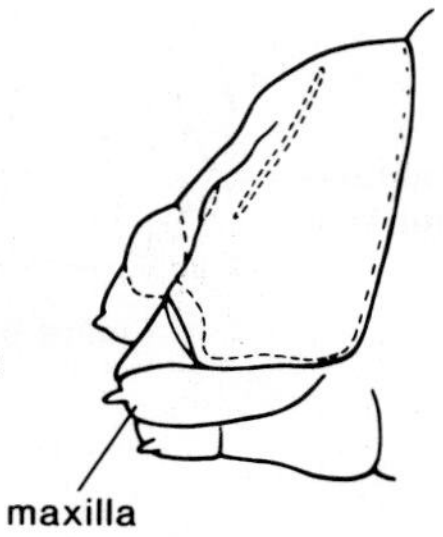

Figure 87

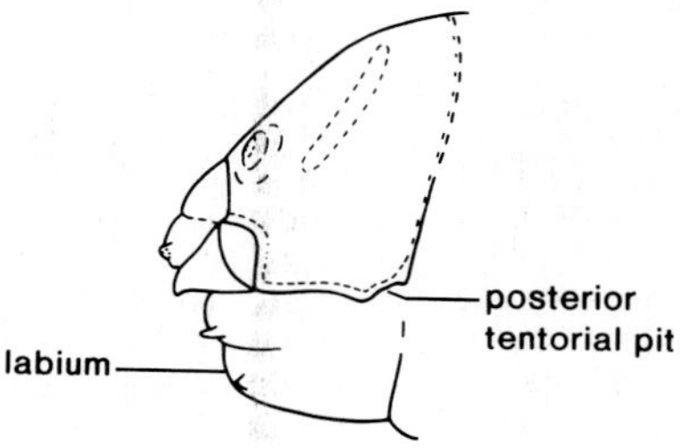

Figure 88

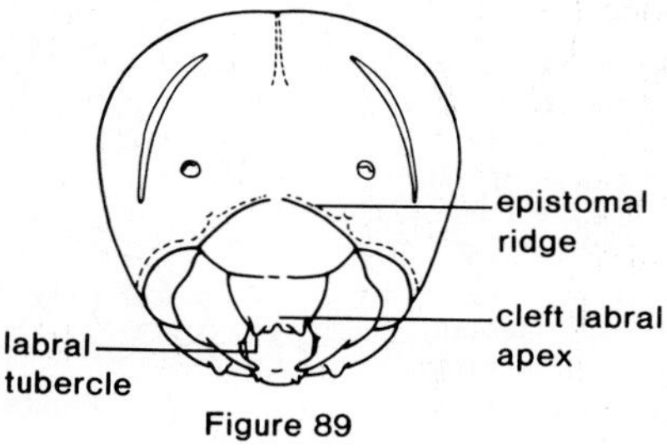

Figure 89

65(63′). Epistomal ridge or epistomal depression extending dorsally beyond ventral margin of antennal discs (see horizontal line in fig. 90) .... (*Colletes, Hylaeus*) ***Colletidae*** (p. 689)

65′. Epistomal ridge or depression extending dorsally at most to ventral margin of antennal discs (fig. 91) 66

66(65′). Labral tubercles distinct (figs. 89, 90, 94, 95) 67

66′. Labral tubercles absent (fig. 91) or possibly evident as broad, poorly defined apicolateral swellings (fig. 92) 69

67(66). Dorsal tubercles of body absent or evident as low, transverse bands, not rounded or conical in shape; antennal elevations weakly developed (fig. 94). Mandibular apex, in adoral view, broadly rounded or truncate (fig. 96); mandibular apical concavity scooplike (fig. 96) .... (Anthophorini) ***Anthophoridae*** (p. 697)

67′. Dorsal tubercles of body well defined, conical, or rounded laterally (fig. 93); antennal elevations strongly developed (fig. 95). Mandibular apex, in adoral view, acute, narrow (fig. 97); mandibular apical concavity not scooplike (fig. 97) 68

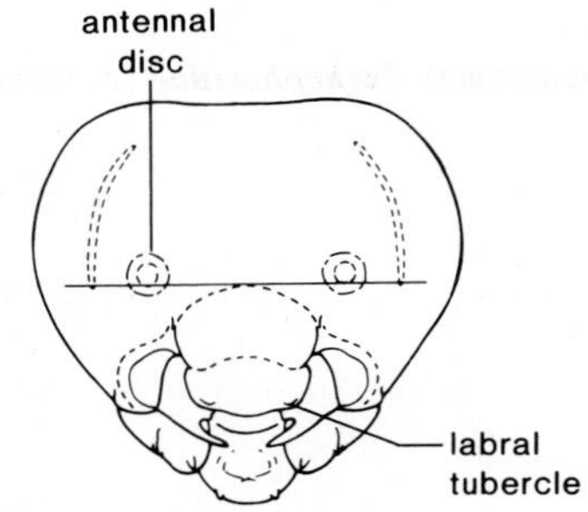

antennal disc
labral tubercle

Figure 90

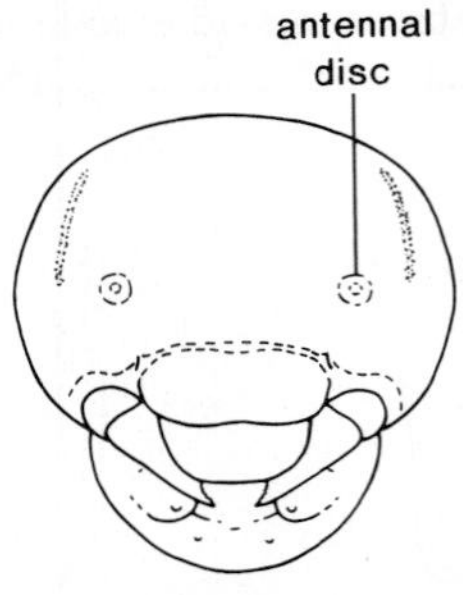

antennal disc

Figure 91

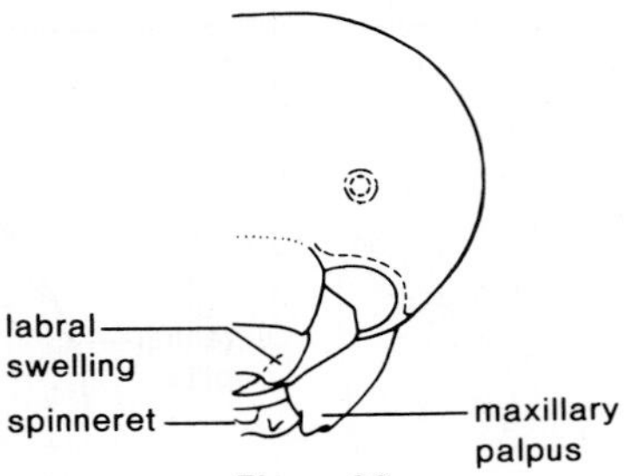

labral swelling
spinneret
maxillary palpus

Figure 92

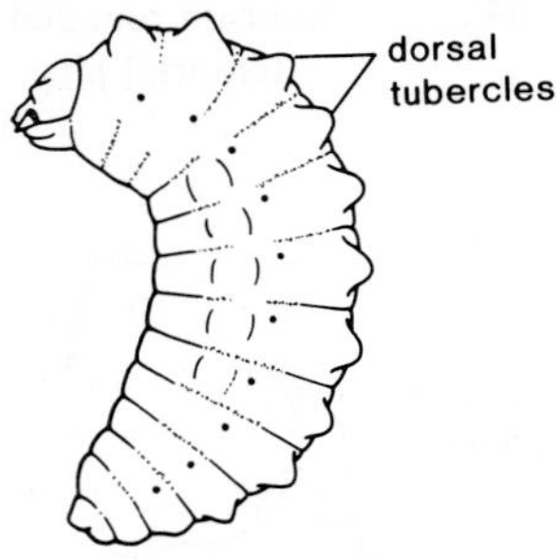

dorsal tubercles

Figure 93

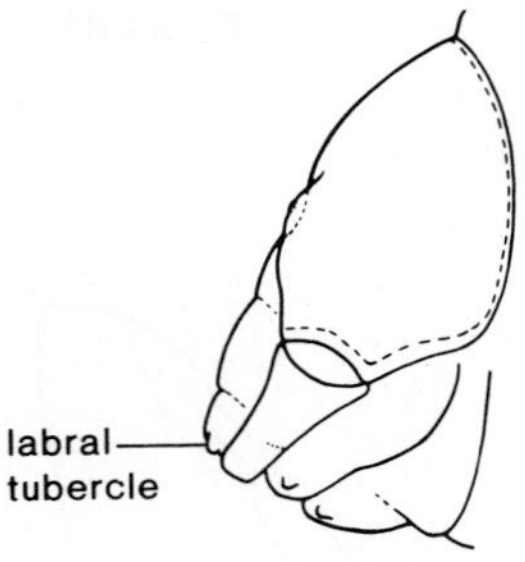

labral tubercle

Figure 94

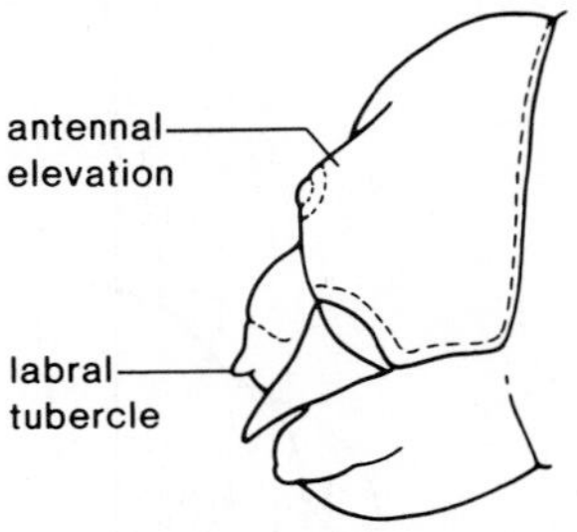

antennal elevation
labral tubercle

Figure 95

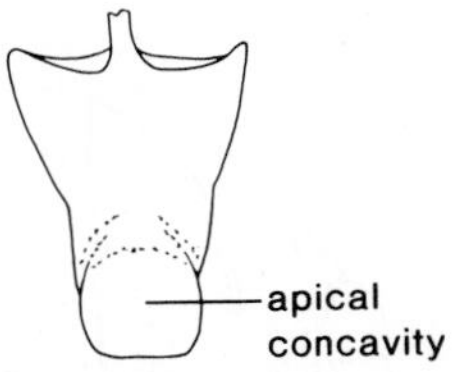

apical concavity

Figure 96

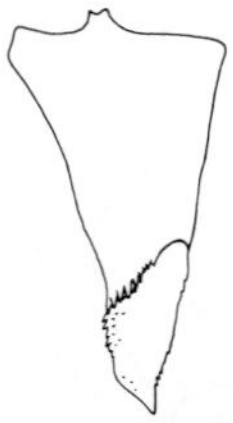

Figure 97

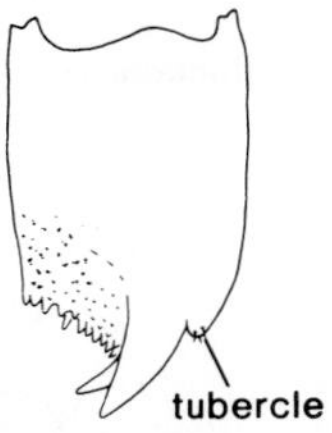

tubercle

Figure 98

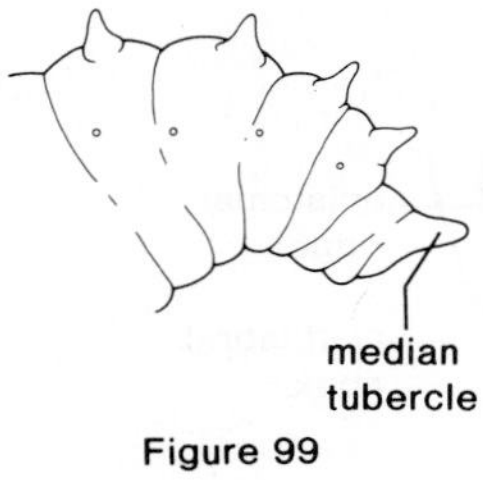

median tubercle

Figure 99

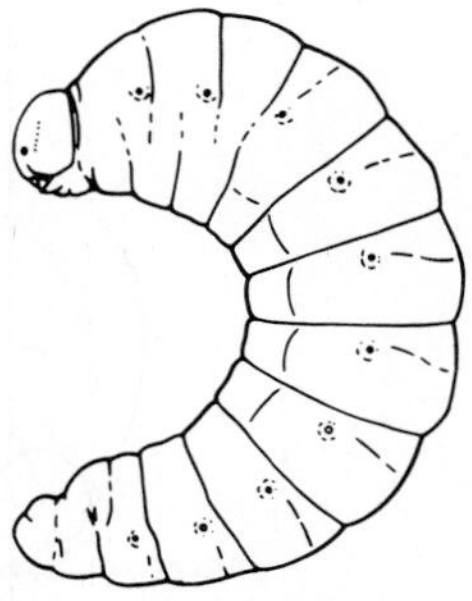

Figure 100

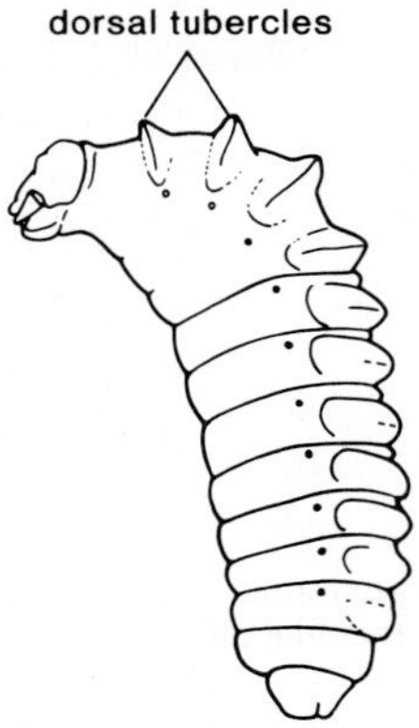

dorsal tubercles

Figure 101

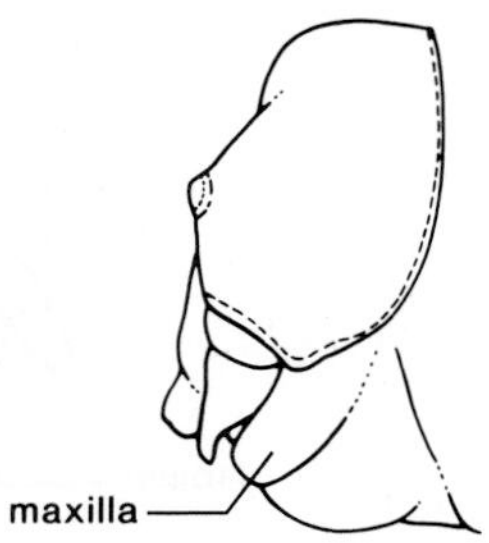

maxilla

Figure 102

68(67′). Outer surfaces of mandible with large, seta-bearing tubercle (fig. 98); salivary slit moderately curved upward .... **(Nomiinae)** ***Halictidae*** (p. 691)

68′. Outer surface of mandible with tubercle absent or very weakly developed; apparent salivary slit strongly curved upward, lateral edges often reaching hypopharyngeal groove (fig. 92); median tubercle on abdominal segment 10 present in most *Perdita* spp. (fig. 99) .... **(Panurginae)** ***Andrenidae*** (p. 693)

69(66′). Body with dorsal tubercles completely absent (fig. 100); mandibular apices simple, never bifid .... (*Hesperapis*) ***Melittidae*** (p. 695)

69′. Body with dorsal tubercles present, varying from low transverse bands to conspicuously developed and rounded lobes (fig. 101); forms with extremely low dorsal tubercles (*Exomalopsis*) have bifid mandibular apices 70

70(69′). Maxillary palpus apparently absent (fig. 102) .... **(Halictinae)** ***Halictidae*** (p. 691)

70′. Maxillary palpus present (fig. 92) 71

71(70′). Venter of abdominal segment 9 nonprotuberant (fig. 103); mandibular apex simple, not bifid .... (*Andrena*) ***Andrenidae*** (p. 693)

71′. Venter of abdominal segment 9 protuberant (fig. 104); mandibular apex bifid .... (*Exomalopsis,* in part) ***Anthophoridae*** (p. 697)

72(14′). Spinneret in the form of paired, projecting processes (fig. 105) (look closely, it may be faint) 73

72′. Spinneret transverse and slitlike, sometimes more prominent laterally than medially but not forming paired projections (figs. 106, 107) 74

73(72). Antennal orbits minute, located somewhat above middle of head capsule (fig. 105); mandibles with a thin, denticulate biting margin set off by a groove (fig. 108) .... ***Rhopalosomatidae*** (p. 679)

73′. Antennal orbit larger, located below middle of head capsule (figs. 109, 110); mandibles without a denticulate margin set off by a groove (figs. 111–115) .... (in part) ***Sphecidae*** (p. 686)

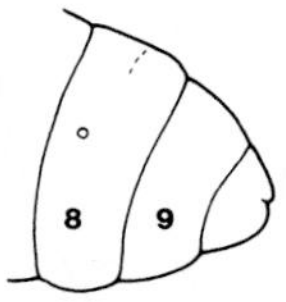

Figure 103

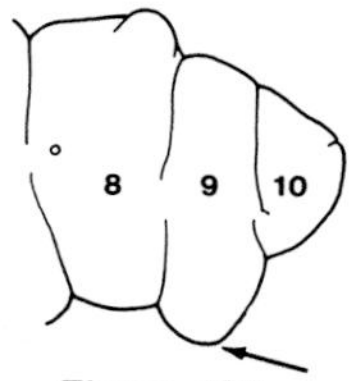

Figure 104

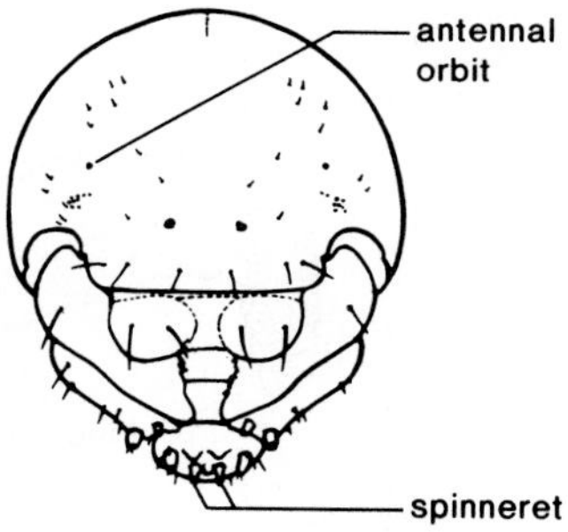

Figure 105

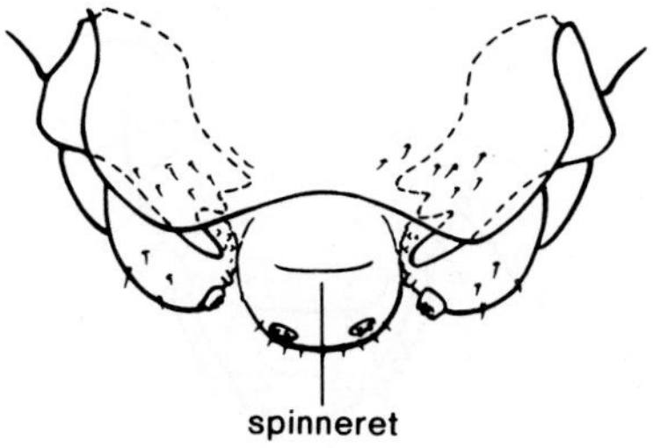

Figure 106

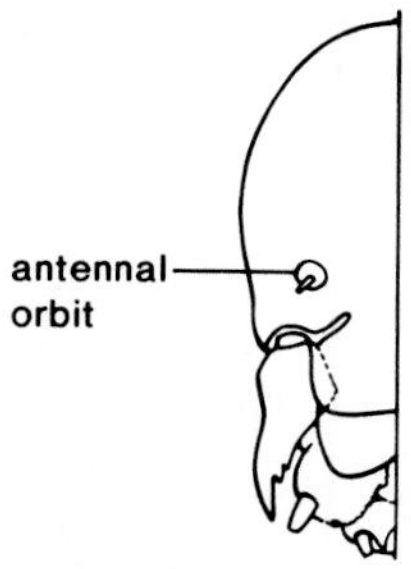

Figure 109

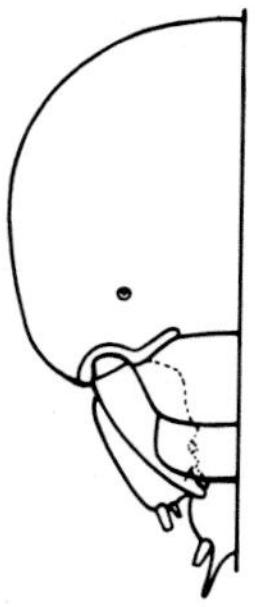

Figure 110

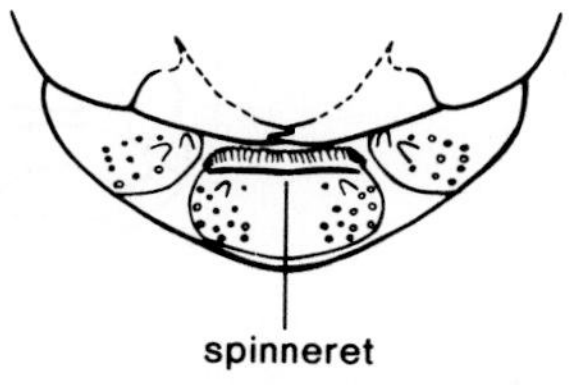

Figure 107

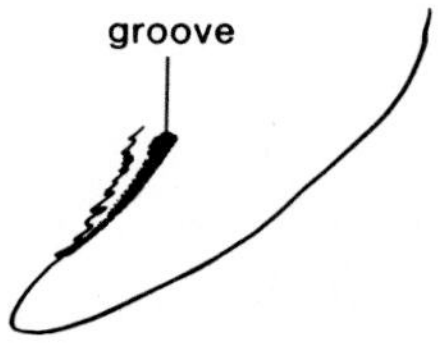

Figure 108

Figure 111

Figure 112

Figure 113

Figure 114

Figure 115

74(72′). Antennal orbits situated at or above middle of head capsule (fig. 116); body often densely hairy, the hairs sometimes branched or otherwise modified (fig. 116a); larvae living in nest chambers, not in individual cells .... ***Formicidae*** (p. 678)

74′. Antennal orbits usually situated below middle of head; body generally bare or with small, sparse setae; larvae in individual cells 75

75(74′). Second pair of spiracles (metathoracic) absent, vestigial, or at least very much smaller than remaining 9 pairs (fig. 117) 76

75′. Second pair of spiracles (metathoracic) slightly if at all smaller than remaining pairs, i.e., there are 10 pairs of clearly defined and apparently functional spiracles 78

76(75). Antennal papilla present; mandibles stout, the teeth tending to surround a concavity; labrum often trilobed (fig. 118) .... ***Pompilidae*** (p. 680)

76′. Antennal papilla absent (orbit may be roundly elevated); mandibular teeth in somewhat the same plane; labrum usually somewhat bilobed 77

77(76′). Maxillary palpi and galeae papillalike, longer than their basal width (fig. 119) .... ***Tiphiidae*** (p. 675)

77′. Maxillary palpi and galeae disclike or mammiform, shorter than their basal width (fig. 120) .... ***Mutillidae*** (p. 676)

78(75′). Maxillary palpi and galeae disclike or mammiform, shorter than their basal width (fig. 121) .... (in part) ***Chrysididae*** (p. 673)

78′. Maxillary palpi and galeae papilliform, longer than their basal width 79

79(78′). Antenna without distinct, protruding papilla, but with 3 or more small sensilla set in membrane of orbit (fig. 121) 80

79′. Antenna with distinct, protruding papilla (figs. 109, 122) 87

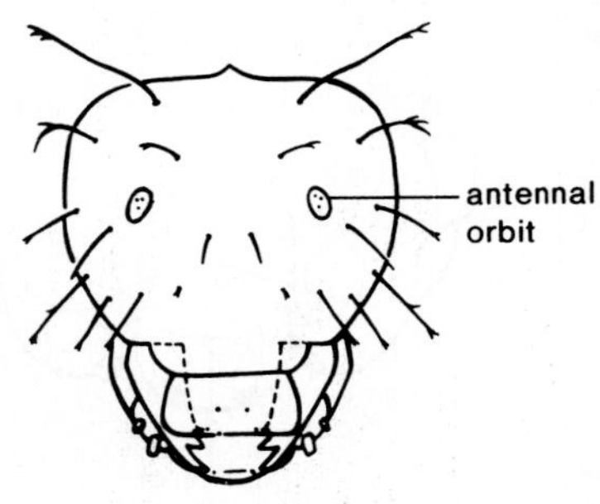

Figure 116

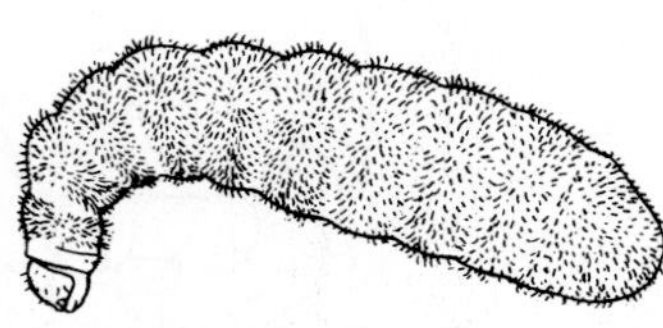
Figure 116a

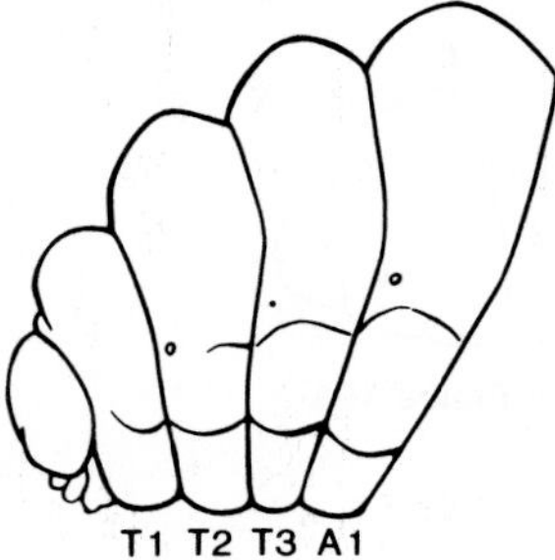

Figure 117

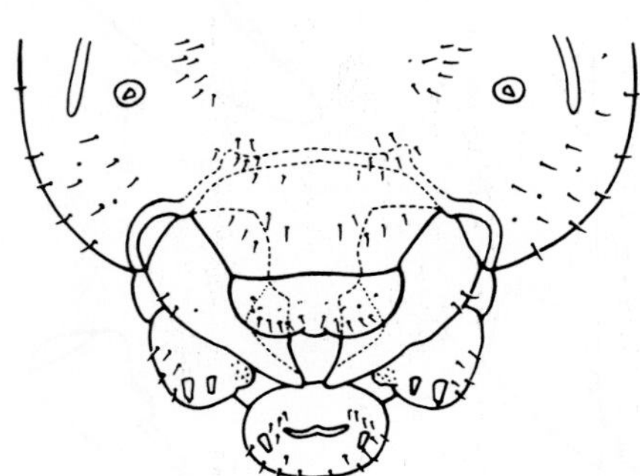
Figure 118

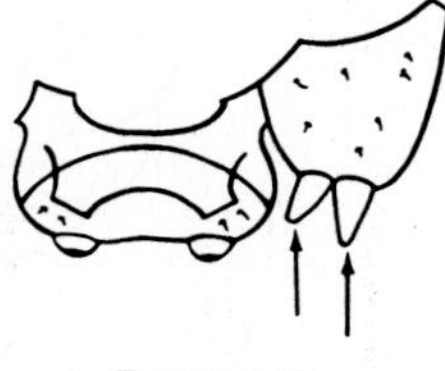
Figure 119

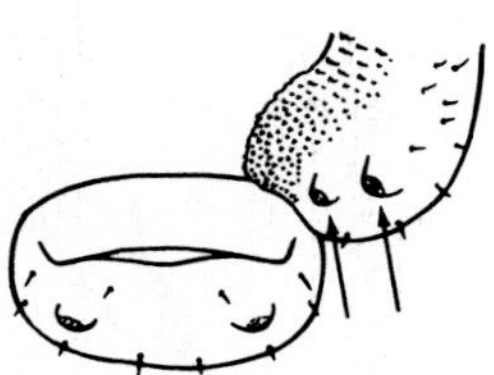
Figure 120

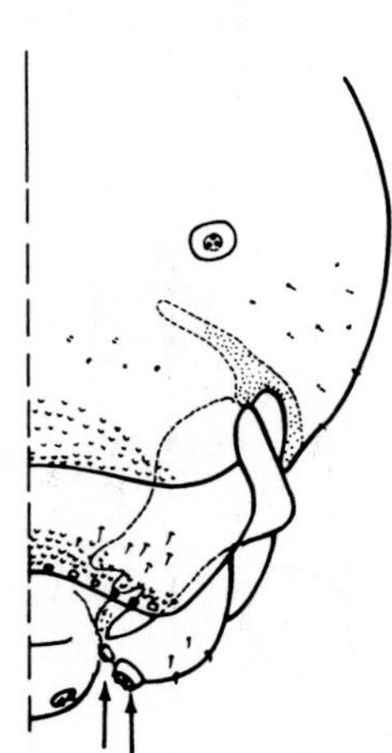
Figure 121

80(79). Median point of clypeolabral suture on or above a line drawn between dorsal points of insertion of mandibles (figs. 123, 124) 81

80′. All of clypeolabral suture well below a line drawn between dorsal points of insertion of mandibles (figs. 125, 126) 82

81(80). Head capsule strongly indented mediodorsally; parietal bands very slender; antennal orbits unusually large (fig. 127); labrum narrowed where it joins clypeus .... (**Masarinae**) ***Vespidae*** (p. 682)

81′. Head capsule at most weakly indented mediodorsally; parietal bands very prominent; antennal orbits not unusually large (fig. 128); labrum not or little narrowed where it joins clypeus .... (**Polistinae**) ***Vespidae*** (p. 683)

82(80′). Mandibles simple, rather bluntly pointed, cusp weakly developed (fig. 129); maxilla with galea minute .... (*Apis*) ***Apidae*** (p. 703)

82′. Mandibles not as above, usually dentate (fig. 130), if weakly so, cusp distinctly cuspidate (fig. 131); maxilla with galea distinct (fig. 132) 83

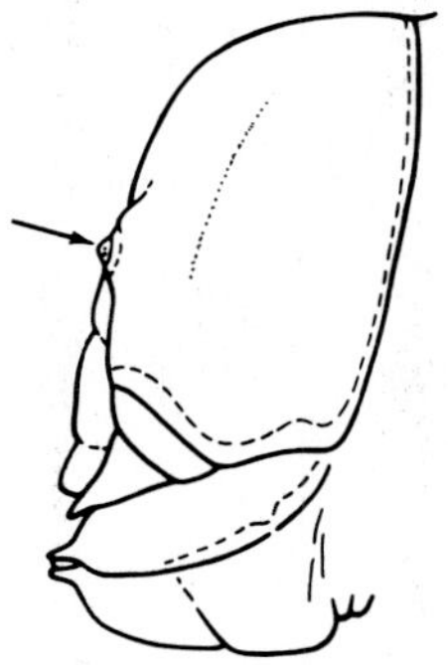

Figure 122

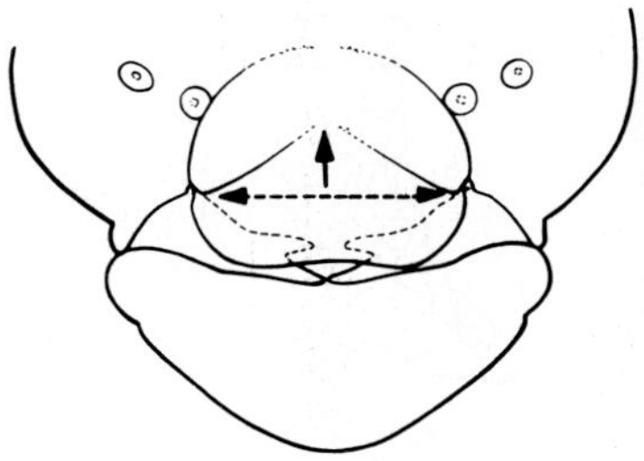

Figure 123

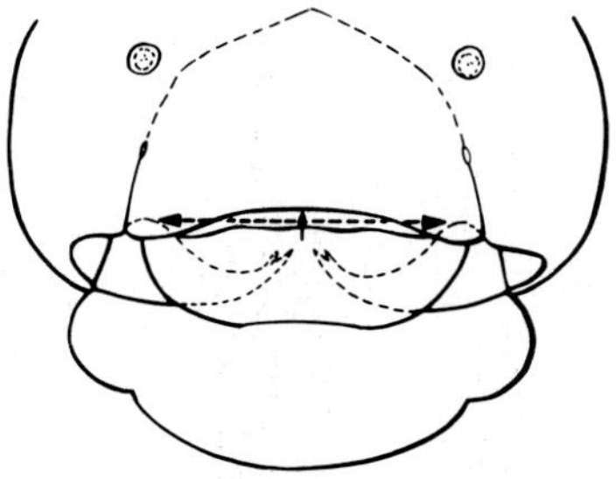

Figure 124

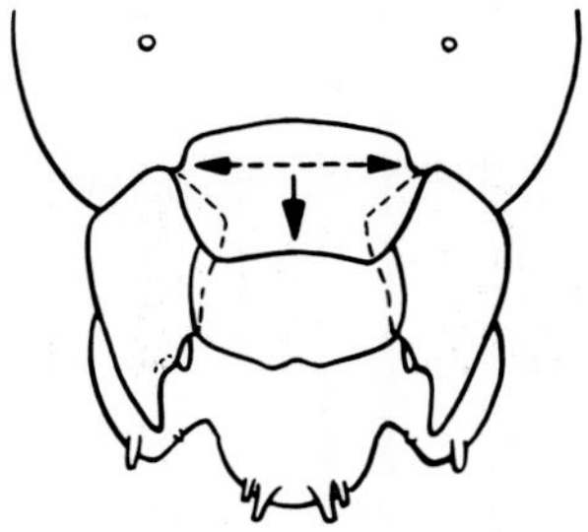

Figure 125

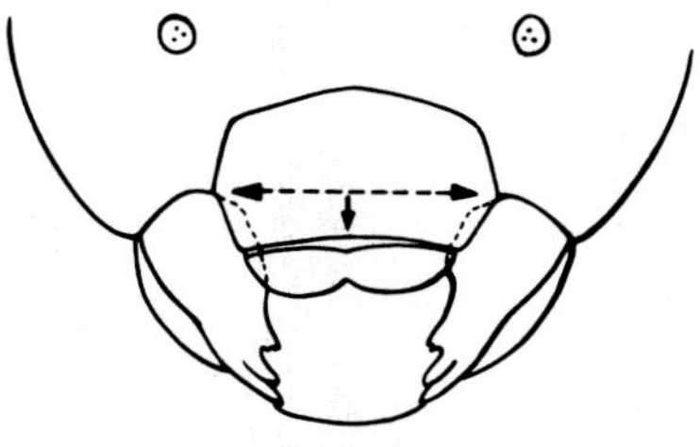

Figure 126

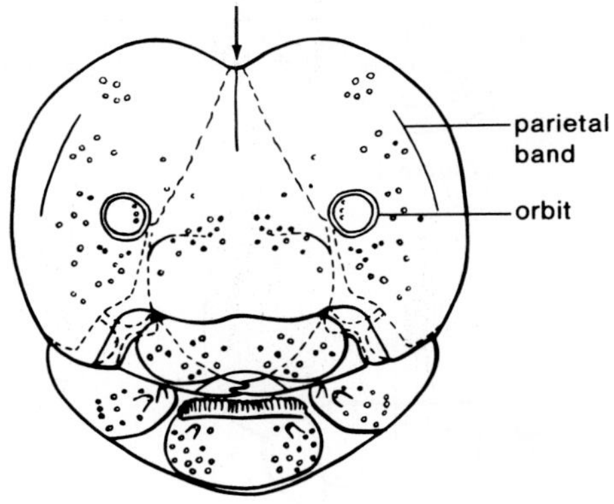

parietal band

orbit

Figure 127

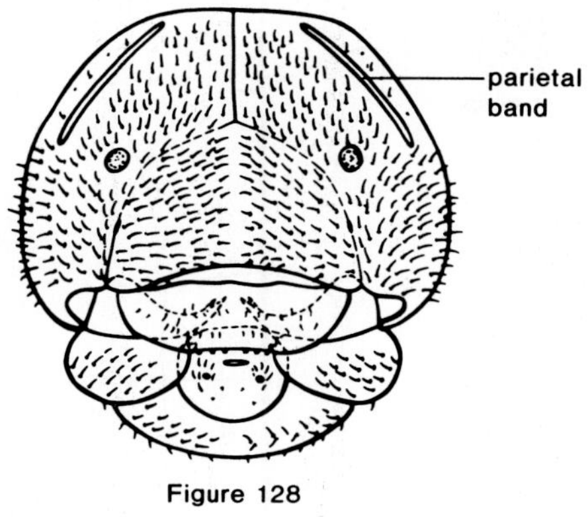

parietal band

Figure 128

Figure 129

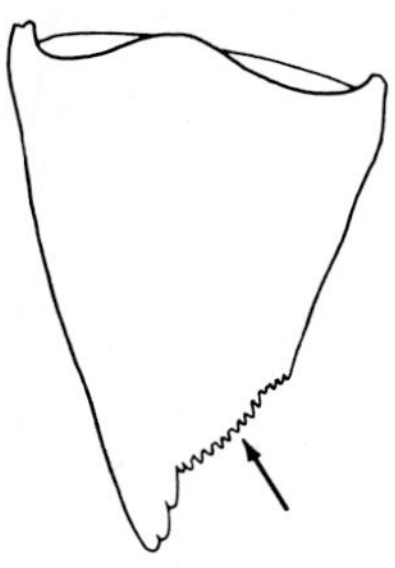

hairlike spicules

cusp

Figure 130

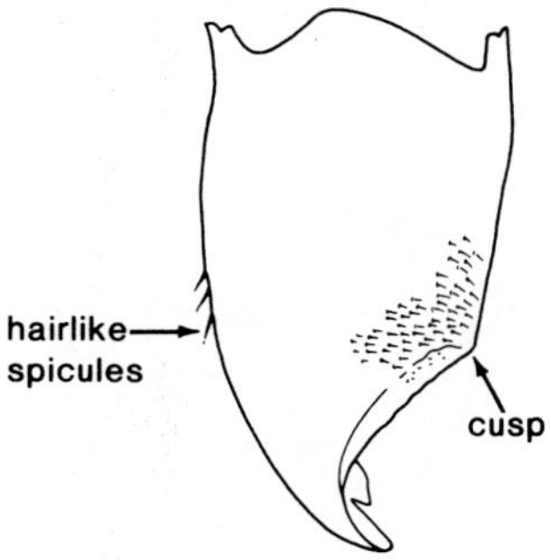

Figure 131

83(82′). Anterior part of body tapered to a small head, which is less than 0.2X as wide as maximum body width (figs. 133, 134); clypeus usually slightly emarginate apically 84

83′. Anterior part of body less tapered, head larger, its width at least 0.3X maximum body width; clypeus usually slightly produced medioapically 85

84(83). Antennal orbits each bearing several (more than 3) sensilla (fig. 135); mandibles tridentate .... ***Scoliidae*** (p. 674)

84′. Antennal orbits each with 3 sensilla; mandibles 4-toothed or at least with a roughened margin basad of third tooth (fig. 136) .... (**Sphecinae** and **Ampulicinae**) ***Sphecidae*** (p. 686)

85(83′). Maximum width of labrum always less than width of labrum where it joins the clypeus (social wasps) (fig. 137) .... (**Vespinae**) ***Vespidae*** (p. 684)

85′. Maximum width of labrum as great as, or greater than, width of labrum where it joins the clypeus (solitary wasps) (fig. 138) 86

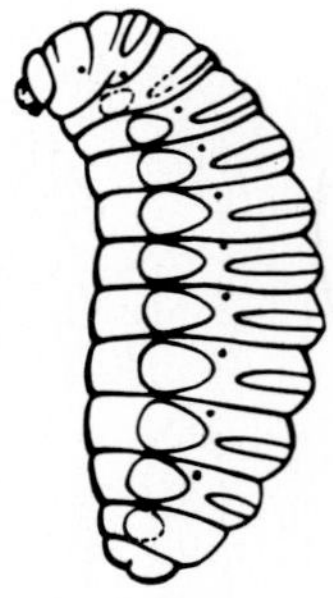

Figure 133

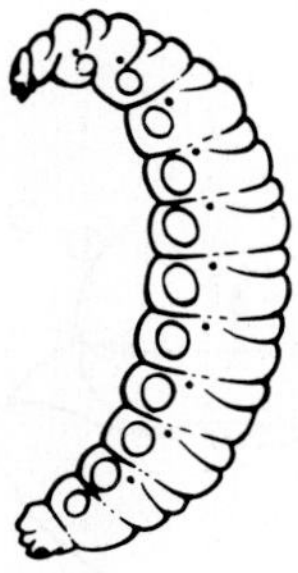

Figure 134

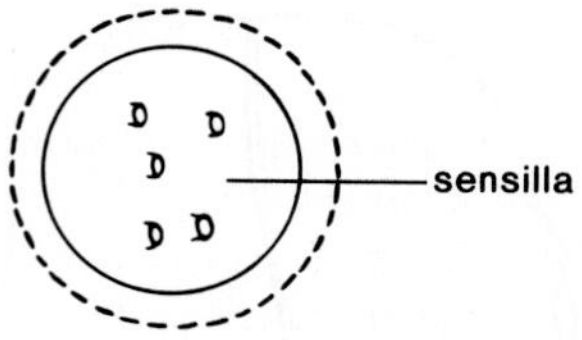

Figure 135

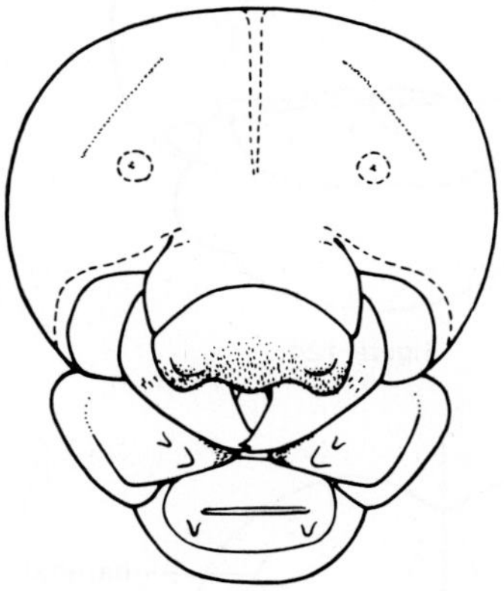

Figure 132

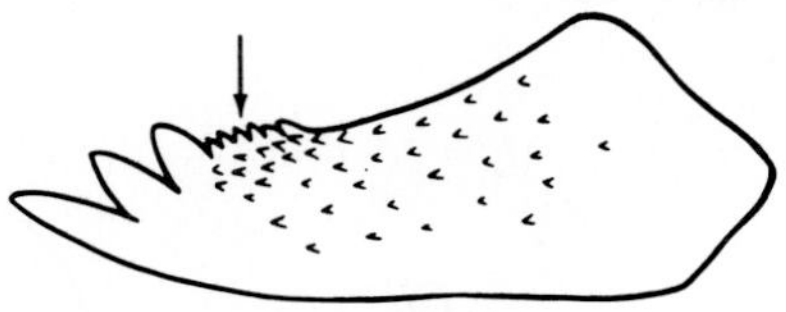

Figure 136

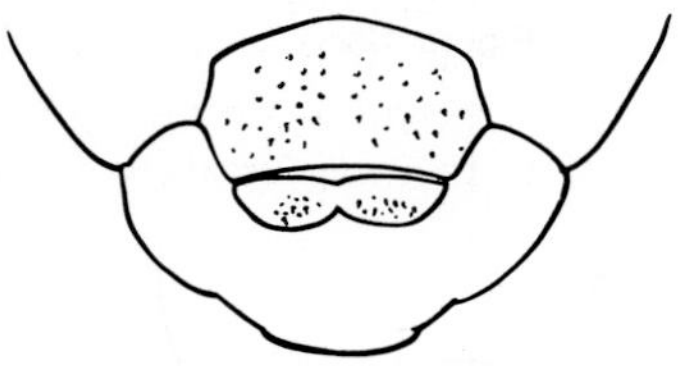

Figure 137

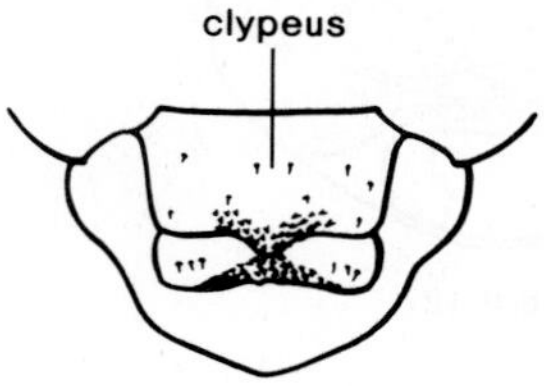

Figure 138

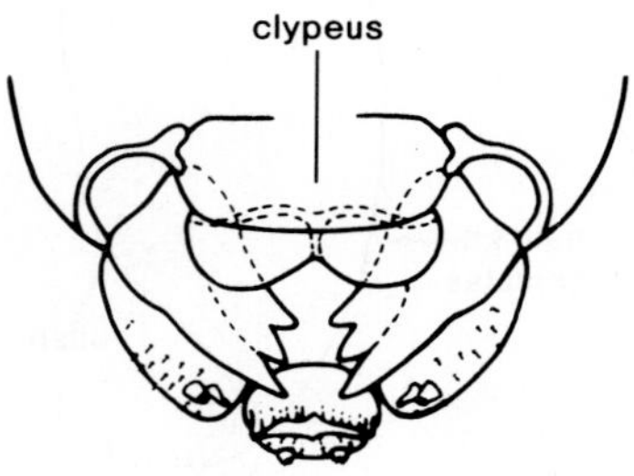

Figure 139

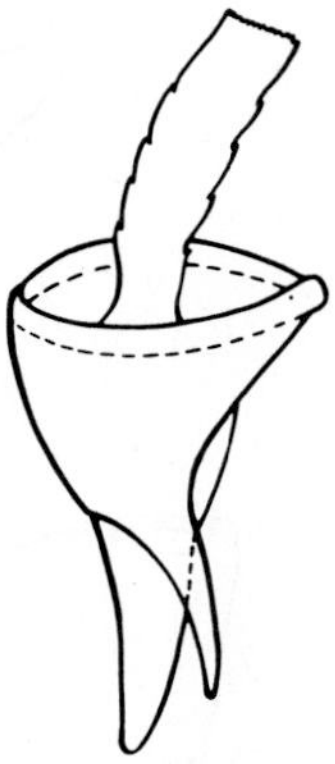

Figure 140

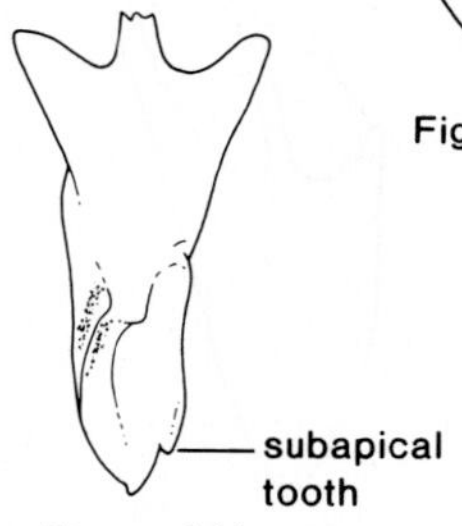

Figure 141

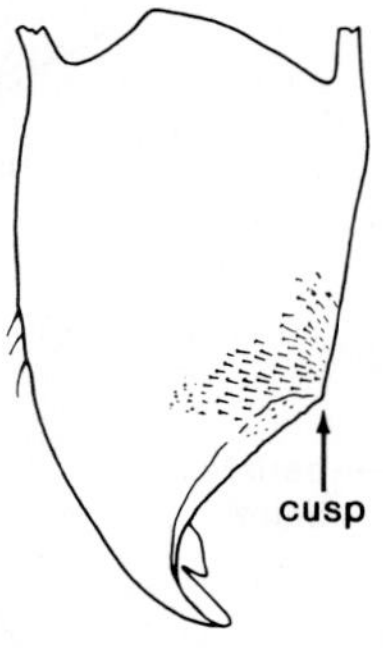

Figure 142

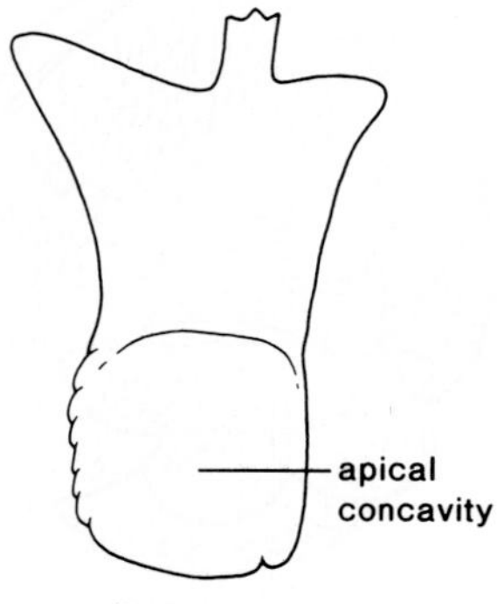

Figure 143

86(85′). Clypeus subquadrate, with nearly parallel sides, anterior margin abruptly produced and somewhat tuberculate medially (fig. 138); labrum with ventral margin notched .... **(Euparagiinae)** ***Vespidae*** (p. 681)

86′. Clypeus with sides somewhat convergent toward apex, apical margin not tuberculate medially (fig. 139); labrum with ventral margin bilobed .... **(Eumeninae)** ***Vespidae*** (p. 682)

87(79′). Mandible strongly bifid, cusp and apical concavity weakly developed or absent (fig. 140) .... ***Sapygidae*** (p. 677)

87′. Mandible at most weakly bifid, subapical tooth, if present, small (fig. 141); mandibular cusp well developed (fig. 142), or if not, apical concavity very well developed and scooplike (fig. 143) (Apoidea, in part) 88

88(87′). Dorsal tubercles of body forming well defined transverse bands, produced and rounded laterally (fig. 144); maxillary apex rounded, not conspicuously projecting mesiad (fig. 145); salivary plate present (fig. 145); abdominal segment 10 moderate in length, rounded with anus apically positioned (fig. 144) .... (*Macropis, Melitta*) ***Melittidae*** (p. 695)

88′. Dorsal tubercles of body absent or evident as very low transverse bands, not produced or rounded laterally (figs. 146, 147); maxillary apex truncate, conspicuously projecting mesiad (fig. 148); salivary plate absent; abdominal segment 10 very short with anus distinctly dorsal in position (fig. 146) or abdominal segment 10 moderately elongate, pointed apically with anus nearly apical in position (fig. 147) .... (in part) ***Anthophoridae*** ... 89 (p. 697)

89(88′). Labral tubercles present (fig. 148); mandible moderately narrow in adoral view, apical concavity not scooplike (fig. 141); dorsal surface of mandible with numerous hairlike spicules (fig. 142); tenth abdominal segment moderately elongate, apically pointed, with anus only slightly dorsal in position (fig. 147) .... (Eucerini) ***Anthophoridae*** (p. 697)

89′. Labral tubercles absent; mandible broad in adoral view, apical concavity scooplike (fig. 143); surface of mandible with fine spicules present, but hairlike spicules absent; tenth abdominal segment very short, not pointed apically, with anus in extreme dorsal position (fig. 149) .... (*Centris*) ***Anthophoridae*** (p. 697)

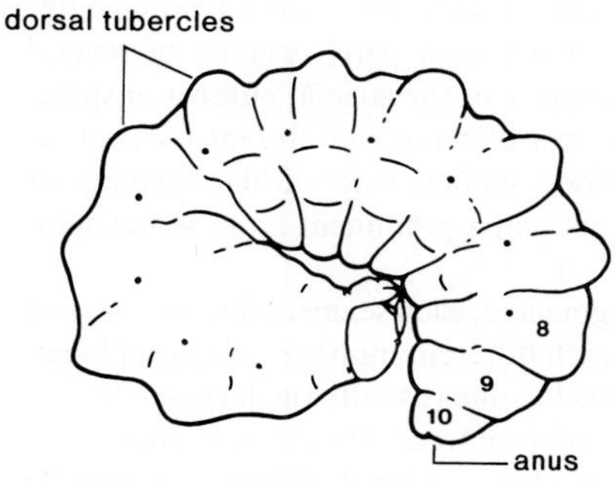

Figure 144

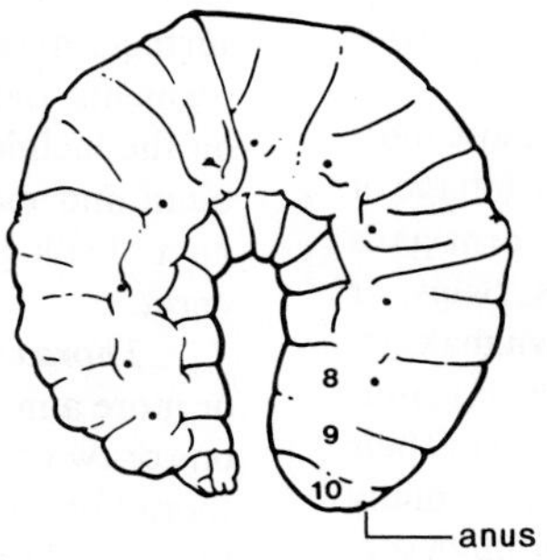

Figure 146

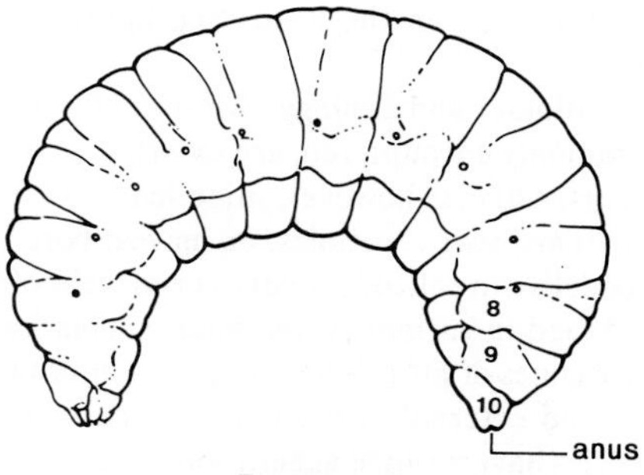

Figure 147

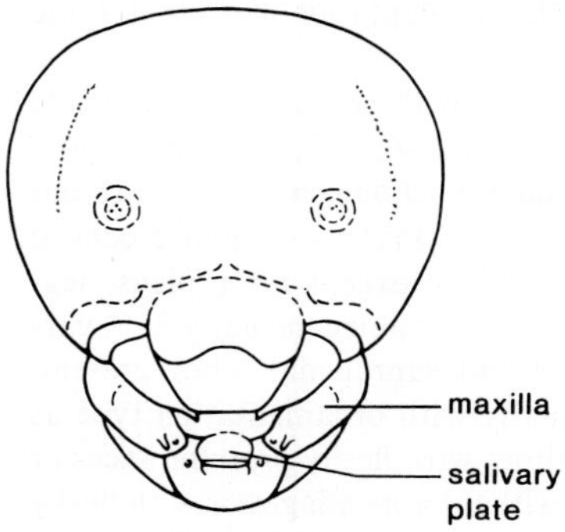

Figure 145

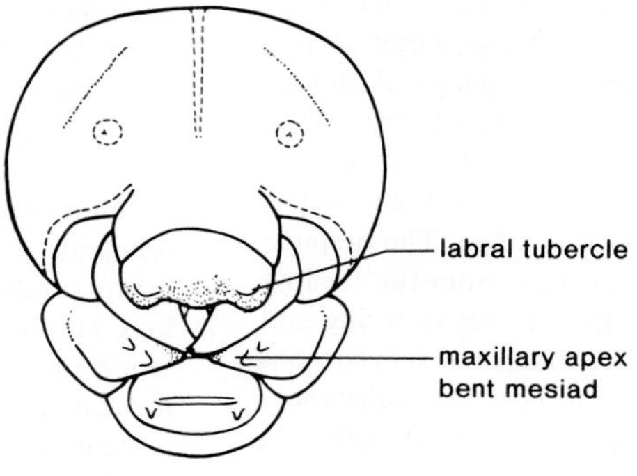

Figure 148

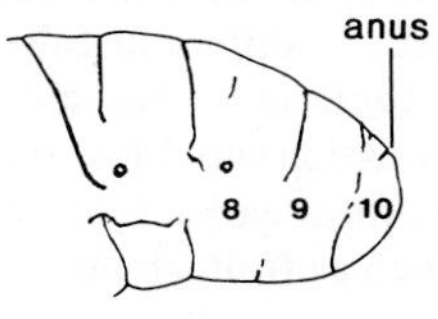

Figure 149

Suborder SYMPHYTA

David R. Smith, *Systematic Entomology Laboratory, ARS, USDA*
Woodrow W. Middlekauff, *University of California, Berkeley*

Sawflies

Relationships and Diagnosis: Symphyta are separated from larvae of Apocrita by the characters in the first couplet of the preceding key to families. The long couplet is necessary for a few exceptional and rarely encountered larvae. The larva of Orussidae (fig. 27.70a) is the main exception: the palpi are not segmented, the thoracic legs are at most represented only by sclerotized discs, and each body segment has a transverse row of stout spines dorsally. Stemmata almost always present, the only exceptions being several families of wood borers in which they are apparently absent, but these families have a sclerotized horn (postcornus) at the apex of the tenth tergum, a structure not found in Apocrita. Thoracic legs are normally segmented, but they may be reduced and lack tarsal claws in some of the leaf miners and wood borers. Most sawfly larvae are caterpillarlike, cylindrical, with three body regions, the head, three-segmented thorax, and ten-segmented abdomen, with the body segments commonly subdivided into divisions called annulets. The mandibles are well developed, opposable, usually with dorsal and ventral cutting edges, and with distinct teeth. Although most sawfly larvae resemble Lepidoptera larvae rather than other Hymenoptera larvae, they are separated from Lepidoptera larvae by the single stemma on each side of the head, lack of a protruding spinneret on the labium, lack of crochets on the prolegs, and lack of adfrontal areas on the head. See figures 27.11a–e for terminology.

Biology and Ecology: Most sawfly larvae, and those most commonly encountered, are external feeders on the foliage of plants. Others, however, mine leaves or petioles; form galls on stems, twigs, petioles, or leaves; bore in stems, twigs, or wood; live in rolled leaf edges or in webs of their own making; and feed in various fruits, buds, shoots, catkins, or in staminate cones of pine. Some may feed internally for part of their life and externally for another part. Life cycles vary: many species have a single generation a year; others go through several generations; and some, such as some Siricidae, may take several years or more to complete their life cycle. Oviposition is usually in the leaf tissue after the female makes an incision in the leaf using her "saw" or ovipositor; twigs, stems, or petioles may be the oviposition site for some species. Although some species feed gregariously during the early larval stages, spreading out later in search of new food, many species feed singly. Larvae go through a number of instars, then molt to the final one, the prepupa or nonfeeding stage. The prepupa usually differs in coloration and structure from the feeding stages, and it searches for and forms an overwintering and pupation site. Most species enter the ground and spin cocoons or form earthen cells; others prefer to bore into substances such as fruit, stems, bark, or rotting wood to form a cell, and some leaf miners remain in fallen leaves. In this cocoon or cell, the larva remains until time for pupation and emergence of an adult.

Hosts are named and other biological notes are given for the different families below. Many species are of economic importance in agricultural crops, in nurseries and ornamental plantings, and in forests. Notes on those particularly important in forests are given in Baker (1972) and Furniss and Carolin (1977).

Description: *Head:* Globose and hypognathous (fig. 27.7a) or dorsoventrally flattened and prognathous (fig. 27.50); deeply or lightly pigmented, with surface smooth to granulate and with various amounts of hairs or setae; ⅄-shaped epicranial suture present (except some wood-boring families) and vertical furrows usually present on each side of head (dorsum) (fig. 27.11c). Frontal area (frons) distinct, bounded by arms of epicranial suture (frontal sutures) dorsally and laterally and by clypeal (or epistomal) suture ventrally. Clypeus transverse, broader than long, usually with definite setal pattern; labrum shallowly or deeply emarginate on anterior margin, usually with several setae, and symmetrical or asymmetrical. Epipharynx with various patterns of spines on each half, usually arranged in semicircular fashion. One stemma (simple eye) on each side of the head, located dorsal to or lateral to antenna (indistinct in some wood borers), clear and convex and usually surrounded by a pigmented spot (ocularium). Antenna located dorsal to mandible, and of various shapes from flat to conical, the base a subcircular or subquadrate ring or antennal suture (antennarium) and within a whitish membrane (antacorium) containing the segments, which may be partial or complete sclerotized rings, flat, or peglike. Mandibles opposable, and asymmetrical, with various dentition, and sometimes with one or more setae on outer surface; maxilla and labium with palpi, usually of several segments, their size, structure of the lacinia, number of spines on the lacinia, and number of setae on different parts of labium and maxilla various; labium with slitlike opening for duct of silk gland but without a prominent protuberant spinneret.

Thorax: Three-segmented, each segment divided into two or more annulets and with different numbers of lateral lobes; one or two pairs of spiracles, one (usually the largest) on prothorax between meso- and prothorax; the other, if present, on the metathorax between meso- and metathorax; legs usually segmented and with claws, but reduced, mammaform, and/or without claws in some internally living species; empodium of claws reduced or enlarged; ornamentation, if present, same as for abdomen.

Abdomen: Ten-segmented, with spiracles on segments 1–8; prolegs present on segments 2–6, 2–7, 2–8, or 4–8 and 10, or on all segments in some Xyelidae, reduced or absent in some internally feeding species. Prolegs without crochets; segments 1–8 similar in structure except for prolegs, segments subdivided into 2–7 dorsal annulets, usually 3–7; various numbers of spiracular and suprapedal lobes present. Tenth tergum rounded, usually with ornamentation type as for rest of abdomen, sometimes with fleshy protuberances or postcornus; tenth sternum without appendages or with fleshy lobes or simple or segmented appendages. Ornamentation usually in a constant pattern for species and may consist of setae, hairs, tubercles, warts, dark plates, or stout or slender simple or branched spines; ornamentation reduced or absent in most internally feeding species; other gland openings such as supraspiracular glands or midventral eversible glands sometimes present; lateral suckerlike protuberances present in Pergidae.

Comments: In North America the Symphyta are represented by four superfamilies, 12 families, 139 genera, and over 1000 species. They are found in all regions from north of the Arctic Circle southward, though most common in temperate areas where vegetation is most diverse and luxuriant. The classification used here follows that of Smith (1978, 1979b).

The only thorough treatment of North American sawfly larvae is that of Yuasa (1923) and, though the nomenclature is out of date, it is still useful if used cautiously. The larvae of European Symphyta were treated by Lorenz and Kraus (1957), and the keys are useful because many of the genera are also North American. More recently, larvae for only certain groups have been described or keyed, mostly in conjunction with revisions; these are referred to in each section.

Those groups with species of economic importance are best known in the larval stage, but still three-fourths or more remain to be found and described. Rearing and adult-larval associations are necessary before many genera and species can be characterized. The feeding stages are the most commonly encountered, and all descriptions and illustrations here are based on the last or late feeding stage. Many characters used at the generic or specific level are not dealt with; these include amount of hairs or setae on the head, clypeus, labrum, and mouthparts; mandible dentition; and type of ornamentation and arrangement of ornamentation on the body. Maxwell (1955) described the internal larval anatomy of many representatives of each family. Pertinent literature is given under each section; additional literature can be found in recent treatments.

XYELIDAE (MEGALODONTOIDEA)

David R. Smith, *Systematic Entomology Laboratory, ARS, USDA*
Woodrow W. Middlekauff, *University of California, Berkeley*

Xyelid Sawflies

Figures 27.7–27.10

Relationships and Diagnosis: Larvae are distinguished from those of other families of Symphyta by the long, conical, usually six-segmented antenna (fig. 27.8d) (if seven-segmented, the apical segment is extremely small and difficult to see); the position of the ocularium, which is nearly adjacent to the antacorium on either the dorsolateral or ventrolateral margin (figs. 27.8a, 27.10a); lack of segmented subanal appendages on the tenth sternum (fig. 27.7c); presence of three or four dorsal annulets on abdominal segments 1–8 (figs. 27.8b, 27.9b); prolegs, if present, on each abdominal segment; and shape of the tenth tergum, which is completely and lightly or darkly sclerotized and rounded or knoblike in appearance (figs. 27.7c, 27.8c, 27.9c). The only other family with more than a five-segmented antenna is the Pamphiliidae, but its larvae (fig. 27.11a) always lack prolegs, have a small hook on the tenth tergum, have three-segmented subanal appendages, and always lack tarsal claws. In all other families, the antenna is five-segmented or less, and the ocularium is removed some distance from the antacorium. The diverse types of xyelid larvae reflect their diverse habits. Larvae of *Macroxyela* and *Megaxyela* are external feeders; they are typically sawflylike with well-developed thoracic legs and prolegs and they commonly have some sort of body ornamentation. Larvae of *Xyela, Pleroneura,* and *Xyelecia* are modified for an internal existence; prolegs are reduced or lacking, tarsal claws are absent in *Xyelecia* and *Pleroneura,* and all larvae are mostly whitish; the larvae of *Pleroneura* are grublike, whereas those of *Xyela* and *Xyelecia* are more cylindrical.

Biology and Ecology: Larvae of *Macroxyela* and *Megaxyela* feed externally on the foliage of deciduous trees such as *Carya, Juglans,* and *Ulmus. Macroxyela ferruginea* (Say) (figs. 27.9a-d) is sometimes abundant on elms, and *Megaxyela major* (Cresson) (figs. 27.10a-d) is an occasional pest of pecan. Larvae of *Xyela* live and feed in staminate cones of pines, the only known exception being *X. gallicaulis* Smith, which forms galls on new shoots of *Pinus taeda, P. echinata,* and *P. elliottii* in the Southeast. Sometimes larvae of *Xyela* are so abundant on pines that they become a nuisance around homes when they drop from trees to go into the soil for overwintering and pupation. Larvae of *Pleroneura* and *Xyelecia* feed in the developing buds and new shoots of fir, *Abies* spp., and can damage new growth. Biological data for *Xyela* are found in Burdick (1961) and for species of *Pleroneura* in Webb and Forbes (1951) and Ohmart and Dahlsten (1977, 1978). Adults are among the earliest sawflies to appear in the spring and may be found on their host plant; adult *Xyela* are commonly attracted to willow or alder catkins in the vicinity of pines. After feeding, larvae enter the soil and form earthen pupal cells, remaining there through the summer and winter. All Xyelidae apparently have a single generation a year.

Description: *Head:* Antenna six- or seven-segmented, conical, if seven-segmented then apical segment difficult to distinguish; ocularium nearly adjacent to antacorium at either dorsolateral or ventrolateral margin; mandible usually linear, without distinct dorsal and ventral cutting margins. *Thorax:* Legs five-segmented, each with tarsal claw in *Macroxyela, Megaxyela* and *Xyela;* without tarsal claw in *Xyelecia* and *Pleroneura.* Ornamentation, if present, same as for abdomen. *Abdomen:* Segments 1–8 with three or four dorsal annulets; prolegs present or absent, or reduced to swellings; if present, then on each abdominal segment; tenth tergum lightly or darkly sclerotized, rounded or knoblike in appearance; tenth segment without suranal hook or segmented subanal appendages; ornamentation absent, only consisting of stout minute setae on venter, or with sclerotized tubercles and/or setae on the annulets and spiracular and surpedal lobes.

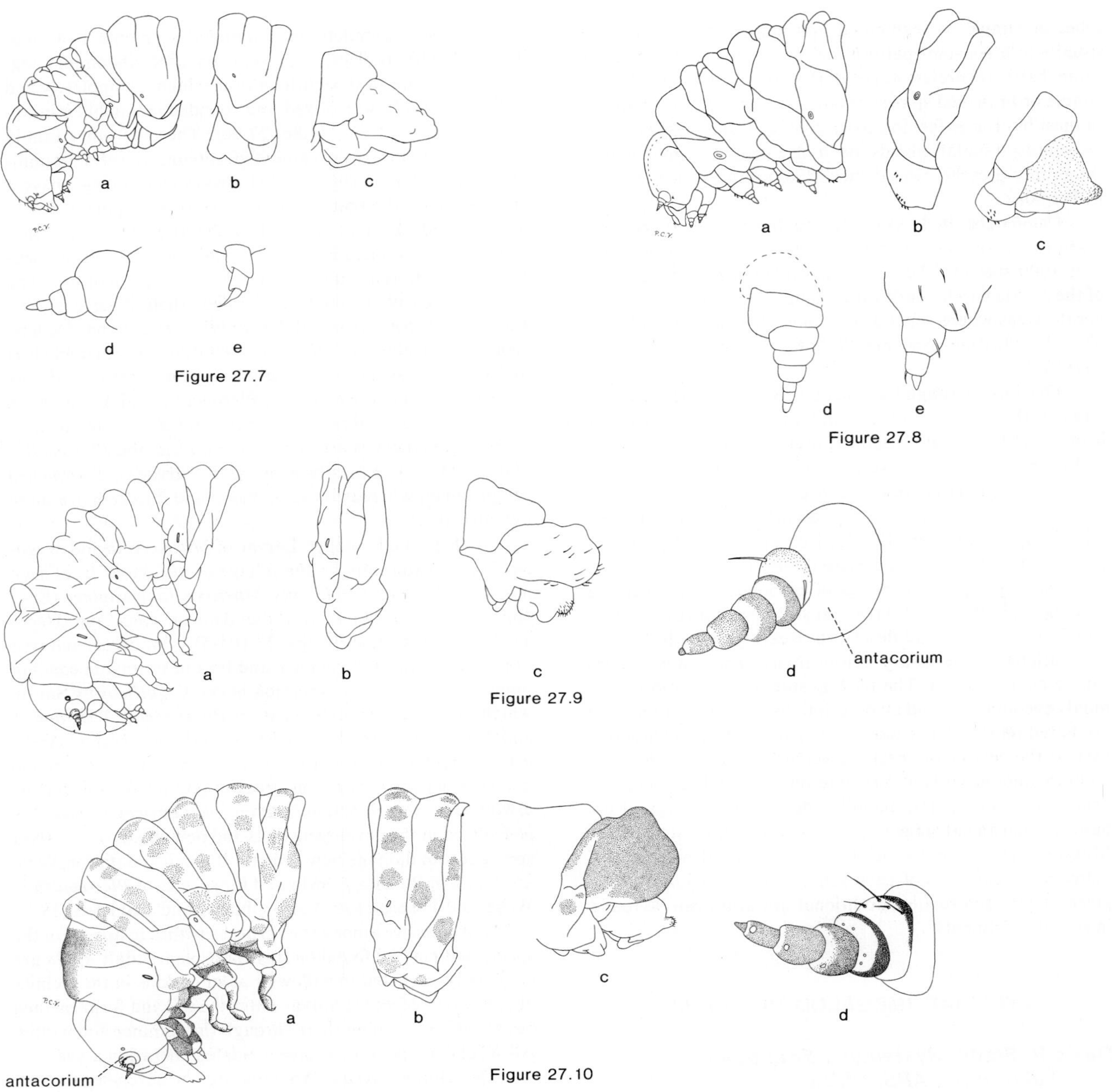

Figure 27.7a-e. Xyelidae. *Xyela* sp. a. head, thorax, A1 segment; b. A3 segment; c. caudal segments; d. antenna; e. T2 leg. f.g.l. 4-6 mm. White with very light brown head and 10th tergum; prolegs reduced; thoracic legs each with claws. Larvae live and feed in staminate cones of pines. Found throughout North America wherever pines occur (Smith 1967).

Figure 27.8a-e. Xyelidae. *Pleroneura* sp. a. head, thorax, A1 segment; b. A3 segment; c. caudal segments; d. antenna; e. A1 leg. f.g.l. 6-10 mm. Entirely whitish; thoracic legs each without tarsal claw; prolegs absent; grublike in appearance. Bud-mining sawflies that feed on the buds and expanding shoots of true firs, *Abies* spp. Occurs in Canada and northern and western U.S. (Smith 1967, Smith et al. 1977, Ohmart and Dahlsten 1977, 1978).

Figure 27.9a-d. Xyelidae. *Macroxyela ferruginea* (Say). a. head, thorax, A1 segment; b. A3 segment; c. caudal segments; d. antenna. f.g.l. 14-18 mm. Creamy white with light brown head and 10th tergum; small setae on annulets of body segments; thoracic legs each with tarsal claw; prolegs present on each abdominal segment, smaller on A1 and A9. Feeds on foliage of elms in eastern North America (Smith 1967).

Figure 27.10a-d. Xyelidae. *Megaxyela major* (Cresson). a. head, thorax, A1 segment; b. A3 segment; c. caudal segments; d. antenna. f.g.l. 22-28 mm. Creamy white with head, 10th tergum, and plates on annulets of body segments dark brown; thoracic legs each with tarsal claw; prolegs present on each abdominal segment, smaller on A1 and A9. Feeds on foliage of pecan in eastern and southeastern U.S. (Smith 1967).

Comments: Five genera and about 28 species of xyelids occur in North America. *Xyela* is the largest genus with 16 species and is coextensive with the distribution of pines from Canada into Mexico. *Pleroneura* with five species is transcontinental in Canada and northern U.S., with southern extensions in the Rockies and Pacific states. *Xyelecia nearctica* Ross, the only species of the genus in North America, is found from the Rockies westward. Both *Megaxyela* and *Macroxyela* are most common in the eastern deciduous forests. Burdick (1961) revised the North American *Xyela,* and Smith et al. (1977) revised *Pleroneura.* References to literature and host lists are found in Smith (1978, 1979b).

Larvae are known for each genus, and keys to genera are found in Smith (1967, 1970b). Specific characters have not been found for *Pleroneura* and *Xyela* larvae; most appear very uniform and larval-adult associations are known for very few.

Selected Bibliography

Burdick 1961.
Ohmart and Dahlsten 1977, 1978.
Smith 1967, 1970b, 1978, 1979b.
Smith et al. 1977.
Webb and Forbes 1951.

PAMPHILIIDAE (MEGALODONTOIDEA)

David R. Smith, *Systematic Entomology Laboratory, ARS, USDA*
Woodrow W. Middlekauff, *University of California, Berkeley*

Webspinning Sawflies

Figures 27.11a-e

Relationships and Diagnosis: Larvae are separated from other families of Symphyta by the long, setiform, seven-segmented antenna (fig. 27.11c); setiform, five-segmented thoracic legs without tarsal claws (fig. 27.11d); absence of prolegs; presence of a small hook on the tenth tergum (fig. 27.11a); and presence of a pair of three-segmented appendages on the tenth sternum (fig. 27.11e). The seven-segmented antenna is similar to that of Xyelidae, but Xyelidae lack the suranal hook and subanal appendages of the tenth segment, some have tarsal claws, and some have prolegs. Larvae of no other North American family of Symphyta have an antenna with more than five segments and a suranal hook and a pair of subanal segmented appendages on the 10th segment. Mature pamphiliid larvae are small to medium in size, 15–24 mm long, slender to robust, subcylindrical, and somewhat flattened ventrally. Segmentation of the abdominal segments is distinct, and there are obvious pleural lobes on the ventrolateral margins of the abdomen. Coloration varies; the head may be pale or dark or with black or brown patterns, the body is usually greenish, sometimes with dark longitudinal stripes, and the tenth tergum may be pale or dark brown.

Biology and Ecology: Larvae live either solitarily or gregariously in a web or rolled leaf. Little is known about the habits of many species, but some are of occasional economic importance, especially those on conifers. *Acantholyda erythrocephala* (Linnaeus), the pine false webworm, an introduced species, damages pines in the East; *A. burkei* Middlekauff and *A. verticalis* (Cresson) are important pests of Monterey pine in California; species of *Cephalcia* may damage spruce, pines, and other conifers; *Pamphilius persicum* MacGillivray is a pest in peach orchards; *P. phyllisae* Middlekauff has severely defoliated large areas of oaks; and species of *Neurotoma,* especially the plum webspinning sawfly, *N. inconspicua* (Norton), are sometimes pests of cherry and plum.

The Cephalciinae (*Acantholyda, Cephalcia*) are associated with conifers, and the Pamphiliinae (*Pamphilius, Onycholyda, Neurotoma*) feed on angiospermous trees and shrubs. The following plant genera have been recorded as suitable hosts: ***Acantholyda*** (*Abies; Picea; Pinus; Pseudotsuga; Tsuga*); ***Cephalcia*** (*Abies; Picea; Pinus; Tsuga*); ***Pamphilius*** (*Amelanchier; Cornus; Corylus; Populus; Prunus; Quercus; Rosa; Viburnum*); ***Onycholyda*** (*Cornus; Rubus*); and ***Neurotoma*** (*Crataegus; Prunus*).

Eggs of all pamphiliids are partially inserted in slits in needles or leaves; the slit pinches the chorion to form a knob-like process that holds the egg in place. Larvae build a characteristic shelter that may be diagnostic for the species. All species on conifers sever needles near the base and consume them from the base, often not finishing one before cutting another. The cut needle is usually taken back to the larval shelter to be eaten. Gregarious species make large unsightly silk nests, which are filled with discarded food, cast skins, and frass. Of those pamphiliids that have been reared, all species with both males and females have an extra instar in the female, and fully grown larvae can be sexed by the larger head of females. Mature larvae form cells in the soil where they may spend more than one winter before pupating. There is usually one generation a year.

Locomotion in pamphiliid larvae is very specialized. With its back to the substrate, the larva prepares a zig-zag trellis of silk under which it wriggles forward. They are dependent on the silk to keep from falling and can only return to a feeding site by preparing a new silk trellis. Larvae of all genera have no prehensile structures other than the mandibles, which are not used in locomotion. The antennae, thoracic legs, and subanal appendages are setiform, and the latter two at least are used for forward locomotion in silk. Fleshy protuberances (pleural lobes) on the abdominal segments may assist in locomotion. The larva wriggles with a peristaltic action and these appendages catch on the silk, forcing the larva ahead.

Pamphiliid larvae differ from other sawfly larvae except for the Palearctic Megalondontidae, by this habit, and the combination of locomotory structures is unique. The suranal hook is also characteristic of Megalodontidae and Pamphiliidae (Lorenz and Kraus 1957). When alarmed, the larva convulses rapidly, the suranal hook catches on the silk, and the larva is propelled backward into its shelter.

Description: *Head:* Semiglobose, sparsely setiferous, as wide as thorax; antenna long, setiform, seven-segmented. *Thorax:* Legs setiform, five-segmented, with elongated distal segments, each without tarsal claw. *Abdomen:* Prolegs absent; segments 1–8 each with four dorsal annulets, with or

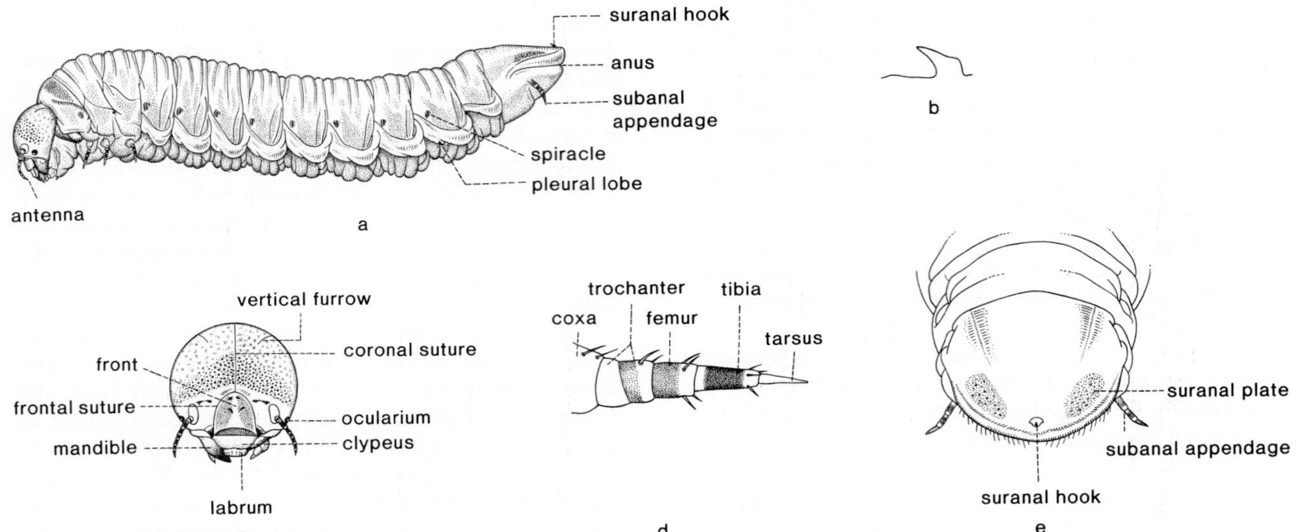

Figure 27.11a-e. Pamphiliidae. *Acantholyda apicalis* (Westwood). a. entire larva; b. suranal hook; c. head; d. T1 leg; e. caudal segments, dorsal view. f.g.l. 23 mm. Characteristics as those given for family. Lives in webs on loblolly pine in southeastern U.S. (Middlekauff 1958).

without setae or other ornamentation; pleural lobes protuberant; venter flattened; tenth segment rounded on caudal margin, somewhat depressed on dorsum and with small suranal hook near caudal margin; tenth sternum with a pair of setiform, three-segmented, subanal appendages.

Comments: The Pamphiliidae are found throughout North America south into Mexico and include five genera and 71 species. *Acantholyda* (33 species) and *Cephalcia* (10 species) are coextensive with the coniferous forests; some species can kill the host. *Pamphilius* (16 species), *Onycholyda* (8 species), and *Neurotoma* (4 species) are most numerous in the eastern deciduous forests; only a few species of *Pamphilius* and *Onycholyda* occur in the West. Several species of *Acantholyda* are known from Mexico, but the family does not occur farther south. The Nearctic fauna was revised by Middlekauff (1958, 1964a) with a more recent treatment of adults, larvae, and life history data of *Cephalcia* by Eidt (1969). Middlekauff (1940) keyed several species of *Neurotoma* larvae. References to more recent biological papers are found in the North American catalog (Smith 1979b).

Although larvae are easily placed to family, there is a lack of morphological characters for generic and specific diagnoses. Keys to genera based on larvae are not available. Most species are unknown in the larval stage, and there is a lack of adult-larval associations.

Selected Bibliography

Eidt 1969.
Lorenz and Kraus 1957.
Middlekauff 1940, 1958, 1964a.
Smith 1979b.

PERGIDAE (TENTHREDINOIDEA)

David R. Smith, *Systematic Entomology Laboratory, ARS, USDA*
Woodrow W. Middlekauff, *University of California, Berkeley*

Pergid Sawflies

Figures 27.12a-e

Relationships and Diagnosis: Pergids are related to the free-feeding sawflylike larvae of Xyelidae, Argidae, Cimbicidae, Diprionidae, and Tenthredinidae. They are small, usually less than 12 mm long, normally greenish when alive with various darker spots, subcylindrical, tapering caudally with the venter flattened, thorax slightly swollen, and the thoracic legs directed laterally. The suckerlike lateral protuberances of abdominal segments 2–4 or 2–5 and 8 (figs. 27.12b, d) and the flat antenna, composed of four circular sclerotized rings (fig. 27.12e) are unique. In addition, the thoracic legs are five-segmented, each with tarsal claw and with empodium enlarged and lobelike adjacent to the claw (fig. 27.12a). Abdominal segments 1–8 each have three dorsal annulets, sometimes indistinct, and each annulet bears ornamentation (fig. 27.12b); and prolegs are present but reduced on abdominal segments 2–7 and 10. The enlarged empodium of the tarsal claws and three-annulate abdominal segments place the Pergidae close to the Argidae.

Biology and Ecology: All pergid sawflies feed externally on foliage, usually on the underside of the leaf, skeletonizing it. They are not considered economically important but may be abundant enough on oaks to attract attention. Hosts for species of the only North American genus, *Acordulecera*, are *Quercus*, *Carya*, *Juglans*, and *Castanea*. Little is known of their life history, but there is apparently one generation a year, and overwintering and pupation are in a cell or very feeble cocoon in the soil. Adults are found in early spring in the eastern deciduous forests.

Description: *Head:* Globose, with numerous hairs; antenna flat, consisting of four sclerotized circular rings. *Thorax:* Legs five-segmented, directed laterally, each with tarsal claw and with empodium as a large fleshy lobe adjacent to claw; prothoracic leg smaller than others; ornamentation as for that of abdomen. *Abdomen:* Segments 1–8 each with three dorsal annulets, sometimes indistinct; each annulet and spiracular and surpedal lobes with setae or darkened tubercles; prolegs reduced to swellings on segments 2–7 and 10; segments 2–4 or 2–5 and 8 each with circular, lateral, suckerlike protuberance; tenth tergum rounded; tenth segment without caudal protuberances or appendages.

Comments: The Pergidae is a large family in the Australian and Neotropical regions. The family characterization given here relates to the only North American genus, *Acordulecera*, and its separation from other North American Symphyta. About six species of *Acordulecera* are known in America north of Mexico, and they are most common in the eastern deciduous forests but have been collected as far west as Arizona. It is a much larger and more diverse genus in the Neotropics. The world catalog of Pergidae (Smith 1978) gives literature references and hosts of species. Though larvae can be identified to *Acordulecera*, specific identification is not yet possible because of a lack of rearings and larval-adult associations. Middleton (1922) described several larvae.

Selected Bibliography

Middleton 1922.
Smith 1978.

CIMBICIDAE (TENTHREDINOIDEA)

David R. Smith, *Systematic Entomology Laboratory, ARS, USDA*
Woodrow W. Middlekauff, *University of California, Berkeley*

Cimbicid Sawflies

Figures 27.13–27.15

Relationships and Diagnosis: Cimbicidae are related to other typically sawflylike and free-feeding larvae of the Xyelidae, Pergidae, Argidae, Diprionidae, and Tenthredinidae, but are separated by the following combination of characters: antenna one-segmented (fig. 27.13e), moundlike or peglike; labrum with secondary longitudinal sutures (fig. 27.13g); thoracic legs five-segmented, each with tarsal claw (fig. 27.13a); prolegs present on abdominal segments 2–8 and 10 (fig. 27.13b); abdominal segments 1–8 each with seven dorsal annulets (fig. 27.13b); a lunate supraspiracular gland opening dorsal to each spiracle on abdominal segments 2–8 (fig. 27.14b). The only other sawfly larvae with seven dorsal annulets on abdominal segments 1–8 are those of the Selandriinae and Tenthredininae of the Tenthredinidae; however, larvae of those subfamilies have conical, five-segmented antennae, lack supraspiracular gland openings, and lack secondary longitudinal sutures on the labrum. Larvae of Cimbicidae are subcylindrical and among the largest of sawflies. Mature larvae of *Cimbex* may be 45–50 mm long. They are mostly greenish when alive, but there may be dark spots or longitudinal stripes on the body, and the head may be unicolorous or with a dark median stripe.

Biology and Ecology: All cimbicid larvae feed externally on foliage. They are usually curled, commonly covered with a whitish, waxy bloom; when disturbed they may eject a yellowish fluid from the supraspiracular glands. A few species may be of economic importance, particularly the elm sawfly, *Cimbex americana* Leach (figs. 27.14a-g), which may severely defoliate elms and other trees, and the honeysuckle sawfly, *Zaraea inflata* Norton.

Host plant genera recorded for North American cimbicid genera are as follows: ***Cimbex*** (*Ulmus; Salix; Alnus; Tilia; Betula; Acer; Populus; Ostrya; Prunus*); ***Trichiosoma*** (*Alnus; Betula; Prunus; Ulmus; Populus; Salix; Fraxinus*); ***Zaraea*** (*Lonicera; Symphoricarpos*).

The elm sawfly, *C. americana*, is a widespread species commonly reported defoliating elms and willows (Stein 1974). Eggs are deposited in the leaf tissue and appear as tiny, pale blisters on the underside. The developing larva feeds and rests in a characteristic coil or helix configuration on the underside of the leaf and readily drops when disturbed. A dorsomedian narrow black line contrasting with the pale greenish-white body is distinctive. When mature, the larva spins a tough parchmentlike cocoon at the base of a tree or shrub in which it passes the winter. There is one generation a year. Adults sometimes chew the tips of twigs and cause partial or complete girdling.

Zaraea americana Cresson, a western species, and its eastern counterpart, *Z. inflata*, are common and occasionally defoliate various species of honeysuckle. Britton (1925) and Middlekauff (1956) gave some biological data. Larvae spin cocoons similar to those of *Cimbex*, and there is a single generation a year.

Description: *Head:* Globose, with numerous or scattered setae or hairs; antenna one-segmented, moundlike or peglike; labrum with secondary longitudinal sutures. *Thorax:* Legs five-segmented, each with tarsal claw, without enlarged empodium adjacent to claw; ornamentation same type as for abdomen. *Abdomen:* Segments 1–8 each with seven dorsal annulets; annulets 2, 4, and 7, and sometimes 3, with setae, tubercles, or wartlike protuberances as are also present on spiracular and surpedal lobes; lunate supraspiracular gland openings on segments 2–8; spiracles winged; prolegs on segments 2–8 and 10; tenth tergum rounded, tenth segment without protuberances or appendages.

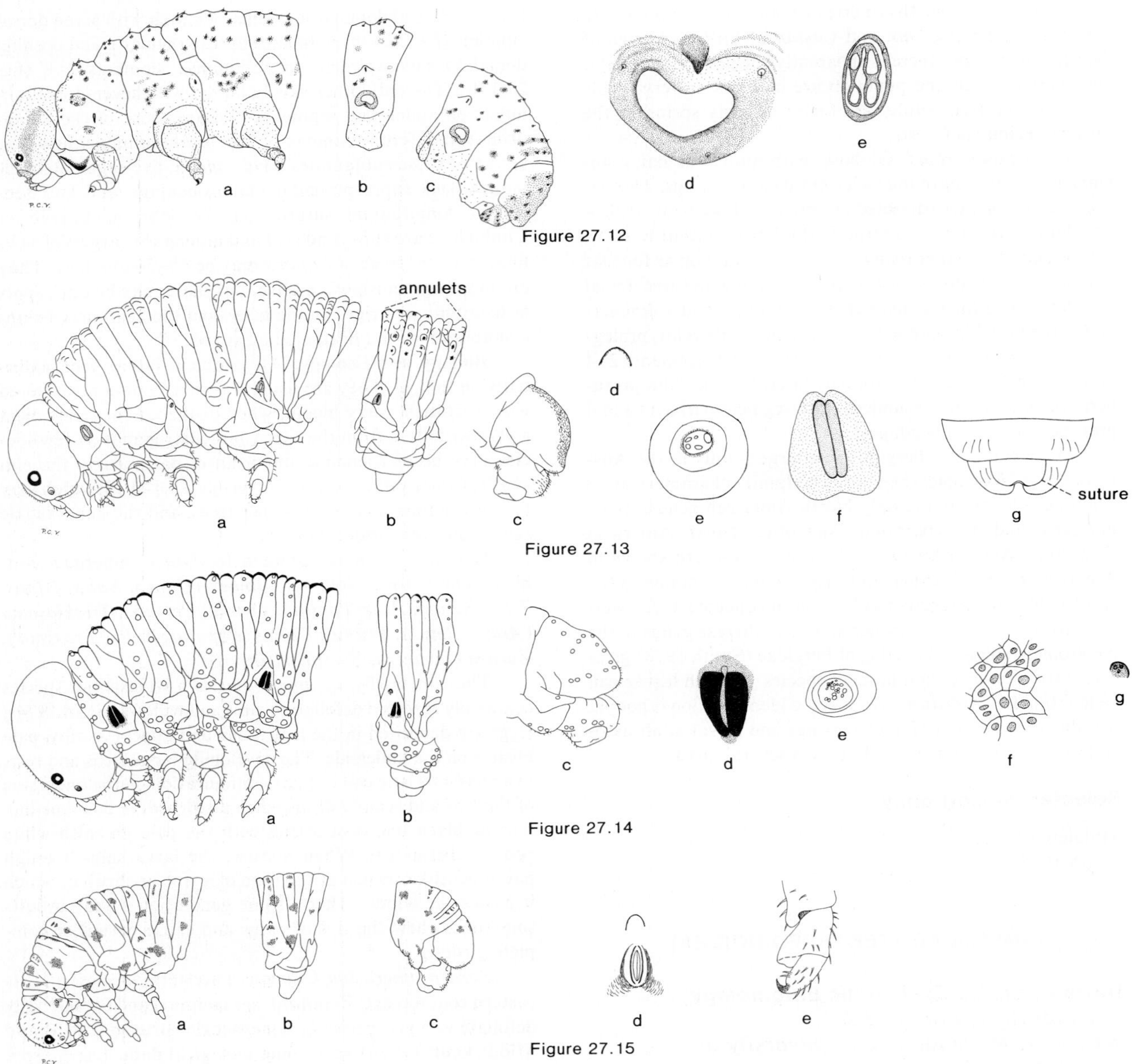

Figure 27.12a-e. Pergidae. *Acordulecera* sp. a. head, thorax, A1 segment; b. A3 segment; c. caudal segments; d. suckerlike protuberance; e. antenna. f.g.l. 8-10 mm. Greenish when alive with slightly darker head; lateral suckerlike protuberances on A2-A4 and A8; each thoracic leg with tarsal claw and enlarged empodium; antenna composed of 4 flat circular sclerites; body somewhat flattened. Gregarious, feeding on underside of leaves of deciduous plants, including oaks, in eastern North America.

Figure 27.13a-g. Cimbicidae. *Trichiosoma* sp. a. head, thorax, A1 segment; b. A3 segment; c. caudal segments; d. semicircular gland orifice (A3 segment); e. antenna; f. A3 spiracle; g. clypeus, labrum. f.g.l. 32-38 mm. Mostly greenish when alive, without longitudinal stripes or dark spots; annulets of body segments each with small lobes. Body may be covered with waxy bloom when alive. Feeds on foliage of alder, birch, willow, and other deciduous trees throughout North America.

Figure 27.14a-g. Cimbicidae. *Cimbex americana* Leach, elm sawfly. a. head, thorax, A1 segment; b. A3 segment; c. caudal segments; d. A3 spiracle; e. antenna; f. reticulated head pattern; g. gland orifice, A3 segment. f.g.l. 45-50 mm. Grayish to yellowish when alive with conspicuous black dorsal stripe, eyespots, and wings of spiracles. Skin of body wartlike or pebbly in texture, and may be covered with waxy bloom. Larvae are leaf-edge feeders, commonly resting in a coiled position on the leaf. Feeds mainly on elm and willow, but many other hosts recorded; occurs throughout North America (Stein 1974).

Figure 27.15a-e. Cimbicidae. *Zaraea americana* Cresson. a. head, thorax, A1 segment; b. A3 segment; c. caudal segments; d. A3 spiracle and semicircular gland orifice; e. T1 leg. f.g.l. 22-28 mm. Greenish with black, yellow, and whitish spots on body and covered with waxy bloom, and with light sublateral band extending length of body. Feeds on foliage of *Lonicera* and *Symphoricarpos*. When disturbed, larva coils into a tight helix. Occurs transcontinentally in North America (Middlekauff 1956).

Comments: The North American fauna is small, with only three genera, *Cimbex, Trichiosoma,* and *Zaraea,* and four species listed for each (Smith 1979b). Each genus is found throughout North America north of Mexico. The only key to genera based on larvae is that by Yuasa (1923) who also described representative larvae for each genus. Identification of larvae must await clarification of species limits, the answers for which may necessitate a combined study of the adults and immatures.

Selected Bibliography

Britton 1925.
Middlekauff 1956.
Smith 1979b.
Stein 1974.
Yuasa 1923.

ARGIDAE (TENTHREDINOIDEA)

David R. Smith, ***Systematic Entomology Laboratory, ARS, USDA,***
Woodrow W. Middlekauff, ***University of California, Berkeley***

Argid Sawflies

Figures 27.16–27.25

Relationships and Diagnosis: Argidae are related to the typically sawflylike, free-feeding larvae of Diprionidae, Cimbicidae, Pergidae, Xyelidae, and Tenthredinidae, but separated by the following combination of characters: antenna one-segmented, rarely two-segmented, flat, rounded, or peglike (figs. 27.16d, 27.18d); thoracic legs five-segmented, tarsal claw present on each leg or only on meso- and metathoracic legs, and with large divergent empodium adjacent to claw (fig. 27.16e); abdominal segments 1–9 each with two or three dorsal annulets (figs. 27.16b, 27.18b); prolegs present on abdominal segments 2–7 or 8 and 10 (fig. 27.23a), on 4–8 and 10 only in *Atomacera* (fig. 27.20). Coloration pattern variable; most species with some sort of body ornamentation in the form of setae, tubular glands, darkened tubercles, or various combinations.

Biology and Ecology: Most larvae are external feeders on foliage, the only known exception in North America being *Schizocerella pilicornis* (Holmgren), which spends part of its feeding stage as a leaf miner of *Portulaca* (fig. 27.19a-i). Several species are economically important. *Arge pectoralis* (Leach), the birch sawfly, may defoliate extensive areas of birch and willow; *Atomacera decepta* Rohwer is common on ornamental *Hibiscus* (figs. 27.21a-c); *Neoptilia malvacearum* (Cockerell) occurs on hollyhock (figs. 27.24a-c); *Schizocerella pilicornis* is common on purslane; and *Sphacophilus cellularis* (Say) (figs. 27.23a-c) has been reported as a pest on sweet potato. Two species have been considered as biological control agents, *Schizocerella pilicornis* to help control purslane, and *Arge humeralis* (Beauvois) (figs. 27.16a-e) to help control poison ivy in Bermuda (Regas-Williams and Habeck 1979).

A broad range of hosts has been recorded for North American Argidae, summarized by genera as follows: ***Arge*** (*Amelanchier, Alnus, Rhododendron, Betula, Carpinus, Corylus, Crataegus, Prunus, Salix, Rhus, Rosa, Quercus*); ***Atomacera*** (*Desmodium, Hibiscus*); ***Aprosthema*** (*Hosackia*); ***Neoptilia*** (*Althaea, Malvastrum, Malva, Sphaeralcea*); ***Ptenus*** (*Acacia, Desmodium ?*); ***Schizocerella*** (*Portulaca*); ***Sphacophilus*** (*Ipomoea, Convolvulus, Psoralea, Desmodium, Petalostemum*).

Life histories vary between genera and species. Eggs are usually laid singly within the leaf tissue, usually on a larger vein or next to the midrib. Some species are leaf skeletonizers; others may consume more or less the entire leaf, whereas early instars of *S. pilicornis* hollow out the leaf, leaving only the cuticle. Larvae may feed gregariously or singly. Pupation usually occurs in a brown, parchmentlike cocoon, usually in ground litter, but occasionally on the host or adjacent plant. Generations per year is variable, from one in some species of *Arge* to as many as six in *Atomacera.*

Description: *Head:* Globose, with some or numerous hairs or setae; antenna usually one-segmented, rarely two-segmented, flat, rounded or peglike. *Thorax:* Legs five-segmented, prothoracic leg commonly shorter than others; empodium an enlarged and conspicuous lobe adjacent to tarsal claw (reduced in *Atomacera*); tarsal claw present on each thoracic leg, or absent only on prothoracic leg in *Schizocerella, Ptenus, Sphacophilus,* and *Neoptilia;* ornamentation as for that of abdomen. *Abdomen:* Segments 1–9 each with two or three dorsal annulets; prolegs present on segments 4–8 and 10 in *Atomacera,* on segments 2–7 or 8 and 10 in other genera; ornamentation various, consisting of tubular spines, hairs, setae, dark tubercles or plates, or various combinations, usually on each annulet, tenth tergum, and spiracular and surpedal lobes; apex of tenth tergum sometimes with fleshy, short protuberances, setae, or stout, stiff setae longer than others.

Comments: Argidae are found throughout North America where 8 genera and 59 species are recorded north of Mexico. These have been treated by Smith (1969d, 1970a, 1971b, 1979b). Most genera are found throughout North America and south to Central and South America, but *Atomacera* is primarily eastern, *Aprosthema* is western; and *Neoptilia, Ptenus,* and *Sphacophilus* are primarily southwestern. Most species occur in the Southwest, and a much larger and more diverse fauna of Argidae occurs from Mexico to South America. *Sericoceros krugii* (Cresson) (fig. 27.25) is the only sawfly in Puerto Rico, U.S. Virgin Islands, and the Dominican Republic.

Keys to genera and some species of argid larvae are found in Smith (1972a); representatives of each North America genus are keyed and described. However, larvae are not known for most species, especially those of *Sterictiphora, Ptenus,* and *Sphacophilus,* and knowledge of other immatures may alter the generic key. The larvae of *Arge* have not been treated except to genus due to taxonomic difficulties related to adults. A combined study of the immatures and adults of *Arge* may help resolve some of the taxonomic problems in the genus.

Selected Bibliography

Martorell 1941.
Regas-Williams and Habeck 1979.
Smith 1969d, 1970a, 1971b, 1972a, 1979b.

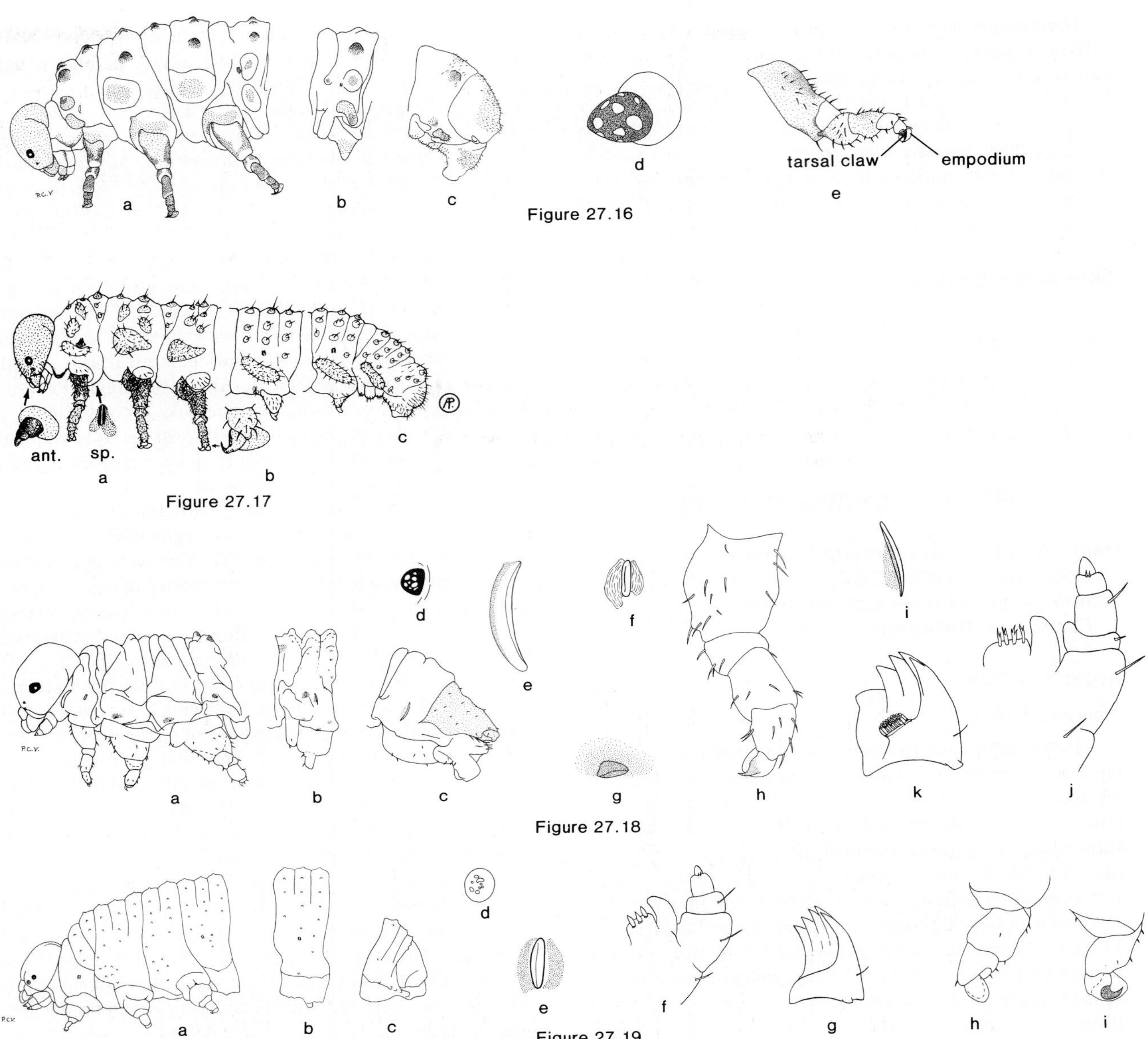

Figure 27.16a-e. Argidae. *Arge humeralis* (Beauvois). a. head, thorax, A1 segment; b. A3 segment; c. caudal segments; d. antenna; e. T1 leg. f.g.l. 15–18 mm. Initially lavender, then turning to bright pink when alive, with lateral lobes mostly bright orange, tipped with black and dorsal lobes black; thoracic legs each with tarsal claw and large, fleshy empodium; abdominal segments 1-8 each 2-annulate. Feeds on foliage of poison ivy in eastern North America (Regas-Williams and Habeck 1979).

Figure 27.17a-c. Argidae. *Arge clavicornis* complex. a. head, thorax; b. A3 segment; c. caudal segments. f.g.l. 15–18 mm. Head reddish-brown to near black; body pale yellowish-green with 4 distinct rows of brown spots on dorsum and an irregular row above and adjacent to spiracles; prominent diagonal projections ventrad of spiracles on abdomen are pigmented and bear numerous setae; base of thoracic legs deeply pigmented, each terminates in claw and distinct fleshy empodium; spiracles usually winged. Food plants *Amelanchier, Azalea, Betula, Carpinus, Corylus, Crataegus, Prunus, Salix* (from Peterson 1948).

Figures 27.18a-k. Argidae. *Aprosthema brunniventre* (Cresson). a. head, thorax, A1 segment; b. A3 segment; c. caudal segments; d. antenna; e. T1 gland; f. A3 spiracle; g. T2 gland; h. T1 leg; i. T3 gland; j. maxilla; k. left mandible, ventral view. f.g.l. 12–14 mm. Preserved specimens whitish with wings on spiracles, spot on postspiracular lobe of each abdominal segment and 10th tergum light brown; thoracic legs each with tarsal claw, empodium not distinctly enlarged; body with wartlike protuberances; 10th tergum with 4 short, fleshy caudal protuberances. Feeds on lotus and occurs in the Pacific Coast states (Smith 1972a).

Figure 27.19a-i. Argidae. *Schizocerella pilicornis* (Holmgren). a. head, thorax, A1 segment; b. A3 segment; c. caudal segments; d. antenna; e. A1 spiracle; f. maxilla; g. left mandible, ventral view; h. T1 leg; i. T2 leg. f.g.l. 8–12 mm. Preserved specimens whitish with body tubercles, head, wings of spiracles, thoracic legs, and 10th tergum amber; T1 without tarsal claw T2 and T3 each with tarsal claw; body with small, rounded tubercles. Larvae mine leaves of purslane. Occurs throughout the western hemisphere (Smith 1972a).

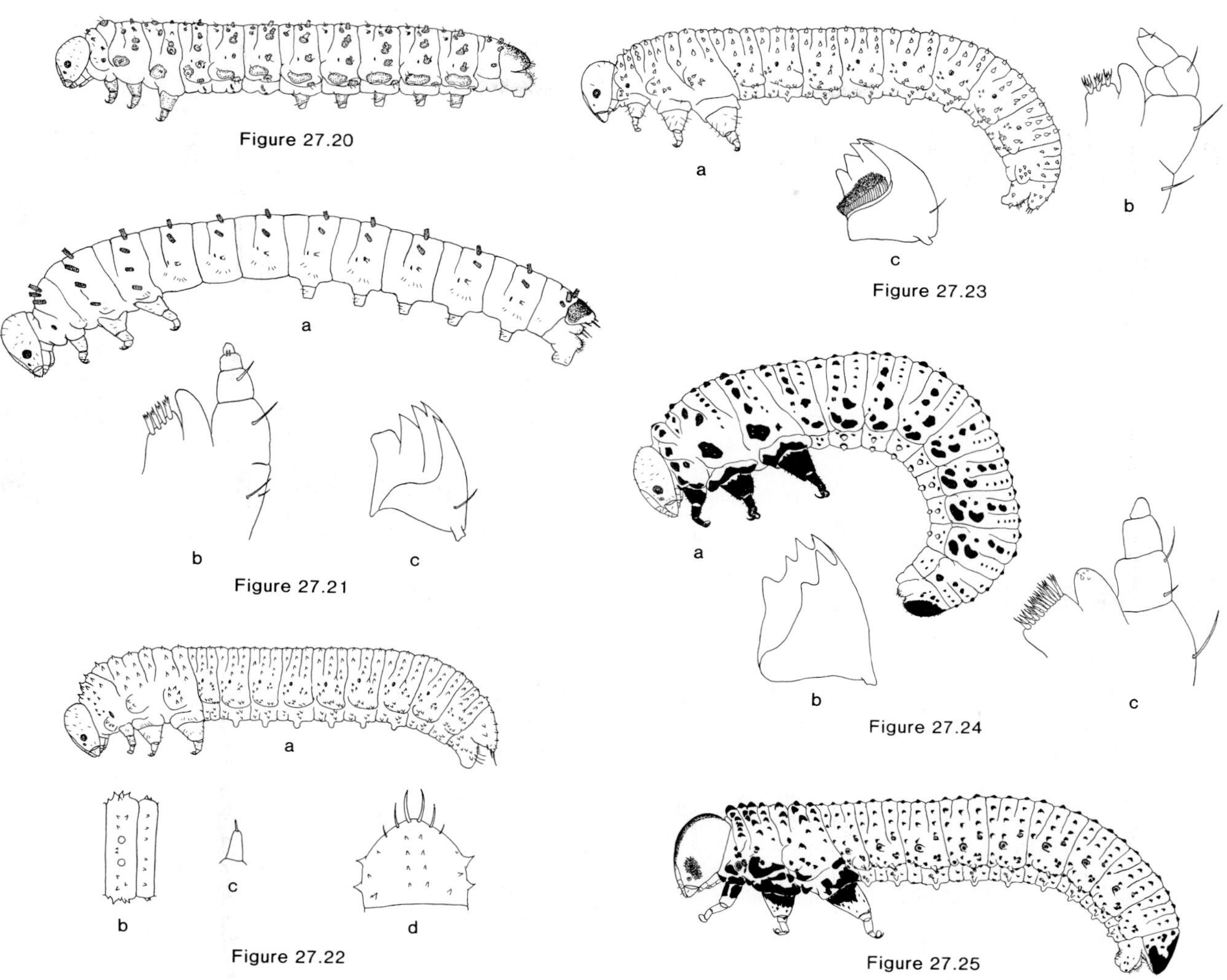

Figure 27.20. Argidae. *Atomacera debilis* Say, entire larva. f.g.l. 8-11 mm. Preserved specimens whitish with head amber and body armature light to dark brown; each thoracic leg with tarsal claw; prolegs on A4-A8 and A10; body armature of tubular glands and setiferous plates. Feeds on foliage of *Desmodium* in eastern North America (Smith 1972a).

Figure 27.21a-c. Argidae. *Atomacera decepta* Rohwer. a. entire larva; b. maxilla; c. left mandible, ventral view. f.g.l. 9-12 mm. Preserved specimens whitish with head, tubular glands, and 10th tergum dark brown; each thoracic leg with tarsal claw; prolegs on A4-A8 and A10; conspicuous tubular glands on body. Commonly found feeding on cultivated *Hibiscus* in eastern North America (Smith 1972a).

Figure 27.22a-d. Argidae. *Sphacophilus nigriceps* (Konow). a. entire larva; b. A3 segment, ventral view; c. body tubercle; d. 10th tergum. f.g.l. 15-20 mm. Preserved specimens whitish with head amber, and apex of tubercles and wings of spiracles brownish; T1 without tarsal claw, T2 and T3 each with tarsal claw; body tubercles conical, each with minute apical segment; apex of 10th tergum with 4 stout, long setae. Feeds on *Petalostemum* and occurs mostly in the midwestern states from Manitoba south to Texas (Smith 1972a).

Figure 27.23a-c. Argidae. *Sphacophilus cellularis* (Say). a. entire larva; b. maxilla; c. left mandible, ventral view. f.g.l. 13-18 mm. Preserved specimens whitish with head, thoracic legs, and body tubercles brown; T1 without tarsal claw, T2 and T3 each with tarsal claw; body tubercles conical, each with minute apical segment. Larvae feed on foliage of sweet potato in eastern North America (Smith 1972a).

Figure 27.24a-c. Argidae. *Neoptilia malvacearum* (Cockerell). a. entire larva; b. left mandible, ventral view; c. maxilla. f.g.l. 15-19 mm. Preserved specimens whitish with head amber and thoracic legs, body plates, and most of 10th tergum dark brown; T1 without tarsal claw, T2 and T3 each with tarsal claw; body armature of low rounded tubercles and plates. Feeds on hollyhock and other plants and occurs from Texas to Arizona (Smith 1972a).

Figure 27.25. Argidae. *Sericoceros krugii* (Cresson). f.g.l. 25± mm. Preserved specimens light with numerous small brown spots on all segments; living late instars pinkish with light green line along the dorsomeson; numerous small black spots scattered over all segments; head reddish with conspicuous black stripe on meson and dark ocellar area; thoracic legs pinkish with dark spots, T1 leg without claw, T2 and T3 legs with single claw and fleshy empodium. Food, seagrape, *Coccoloba uvifera,* common in lowlands. This is the only sawfly in Puerto Rico, the U.S. Virgin Islands, and the Dominican Republic (Martorell 1941, Smith 1972a).

DIPRIONIDAE (TENTHREDINOIDEA)

David R. Smith, *Systematic Entomology Laboratory, ARS, USDA*
Woodrow W. Middlekauff, *University of California, Berkeley*

Conifer Sawflies

Figures 27.26–27.34

Relationships and Diagnosis: Larvae are typically sawflylike and similar to other free-feeding larvae of Cimbicidae, Pergidae, Xyelidae, Argidae, and Tenthredinidae, but are separated by the following combination of characters: Antenna three-segmented, third segment peglike, first and second segments flat and circular (figs. 27.26a, 27.34e); thoracic legs five-segmented, each with claw, without an enlarged empodium adjacent to claw; prolegs on abdominal segments 2–8 and 10; abdominal segments 1–8 each with six dorsal annulets (fig. 27.26b); usually with setae or protuberances on annulets 2 and 4, sometimes 1. The peglike third antennal segment and association with coniferous trees are especially significant in separating Diprionidae larvae. Most larvae are green when alive with different dark patterns such as longitudinal stripes or spots on the body; the head may be black or orange. Mature larvae are 18–40 mm long.

Biology and Ecology: All larvae feed externally on foliage except for *Augomonoctenus libocedrii* Rohwer, which lives and feeds inside developing cones of incense cedar. Larvae are usually gregarious, feeding together at least during the early instars, but later spreading out. Diprionidae are among the most economically important sawflies. Extensive outbreaks may occur in forests, and damage may result in growth loss and sometimes in tree mortality. All Diprionidae are considered destructive to conifers, but the greatest damage is attributed to four species that are adventive: the European pine sawfly, *Neodiprion sertifer* (Geoffroy) (figs. 27.28a-c), the introduced pine sawfly, *Diprion similis* (Hartig) (figs. 27.32a-d), the European spruce sawfly, *Gilpinia hercyniae* (figs. 27.31a-c), and *G. frutetorum* (Fabricius). Important endemic species are the redheaded pine sawfly, *Neodiprion lecontei* (Fitch) (figs. 27.27a-d), the slash pine sawfly, *N. merkeli* Ross, the white pine sawfly, *N. pinetum* (Norton) (figs. 27.26a-d), the jack pine sawfly, *N. pratti banksianae* Rohwer, the Swaine jack pine sawfly, *N. swainei* Middleton, the balsam fir sawfly, *N. abietis* (Harris) (figs. 27.29a-d), the red pine sawfly, *N. nanulus nanulus* Schedl, and the hemlock sawfly, *N. tsugae* Middleton.

Members of the Monocteninae (*Monoctenus,* (figs. 27.33a-g), *Augomonoctenus* (figs. 27.34a-e)) are associated with Cupressaceae, and members of the Diprioninae (*Zadiprion, Neodiprion, Diprion, Gilpinia*) are associated with Pinaceae. Host genera for North America genera are: ***Monoctenus*** (*Juniperus, Thuja*); ***Augomonoctenus*** (*Calocedrus*); ***Zadiprion*** (*Pinus*); ***Neodiprion*** (*Pinus, Abies, Picea, Tsuga, Pseudotsuga*); ***Diprion*** (*Pinus*); ***Gilpinia*** (*Pinus, Picea*).

The life cycle of most diprionid sawflies is fairly uniform, especially in the three most important genera, *Gilpinia, Diprion,* and *Neodiprion.* Eggs are laid in pockets excavated by the ovipositor in living needles. Older needles are usually preferred, but some species such as *N. swainei, N. edulicolus* Ross, and *N. tsugae* choose current needles. Many species overwinter in the egg stage, but some such as *N. lecontei* overwinter in cocoons in the soil. Species in which the eggs undergo diapause have only a single generation per year, whereas all other species with or without a prepupal diapause are capable, under the right conditions, of producing several generations per year.

After hatching, gregarious species form feeding colonies, young larvae at first eating only the parenchymatous tissue, the larger larvae generally consuming the whole needle. Larvae go through four or more instars in the male and an additional one in the female, which is much larger when mature. Tough silken cocoons are spun in the soil where some species overwinter. Prepupal diapause can last several years as in *G. hercyniae* or can be completely absent in one strain of *N. abietis.* Generally it is shorter in northern populations than in southern ones of the same species.

Description: *Head:* Globose, with numerous setae or hairs; antenna three-segmented, segments 1 and 2 flat and circular, segment 3 erect and peglike. *Thorax:* Legs five-segmented, with claw, without enlarged empodium adjacent to claw; prothoracic leg usually smaller than others; legs usually with stiff setae; ornamentation as for that of abdomen. *Abdomen:* Segments 1–8 each with six dorsal annulets, stout setae, hairs, and/or tubercles present on annulets 1, 2, and 4 or on 2 and 4, as well as on spiracular and surpedal lobes; prolegs present on segments 2–8 and 10; tenth tergum rounded, tenth segment without protuberances or appendages.

Comments: Diprionidae are found throughout North America south to Honduras and Cuba wherever coniferous trees occur. Six genera and about 48 species are recorded. *Monoctenus* is eastern with one species in Mexico; *Augomonoctenus* is known only from Oregon and California; *Zadiprion* occurs from Nebraska west to California, south to Guatemala; *Diprion* and *Gilpinia,* both represented by introduced species, are eastern; and *Neodiprion* is widespread throughout the continent south to Central America. Notes on certain species in eastern and western forests are found in Baker (1972) and Furniss and Carolin (1977). Literature and host data are found in Smith (1979b), and a key to genera and world list of species was given by Smith (1974b). Species revisions have been published for *Neodiprion* (Ross 1955) and *Zadiprion* (Smith 1971a, 1974b). There are no keys to genera or species of North American larvae. Most are identified by using host data and comparison with illustrations such as those by Rose and Lindquist (1973), though Smith (1971a) described the larvae of two *Zadiprion* species. Because of their economic importance, many individual species have been covered in papers on biology or life history, but most of this information has not been brought together. Larvae of many species are variable in coloration, especially between the early and late instars, and the adult taxonomy of some groups needs to be settled. However, combinations of adult and larval characters may provide answers to some of the enigmatic taxonomic problems in genera such as *Neodiprion* where apparently identical larval colonies give rise to adults that are

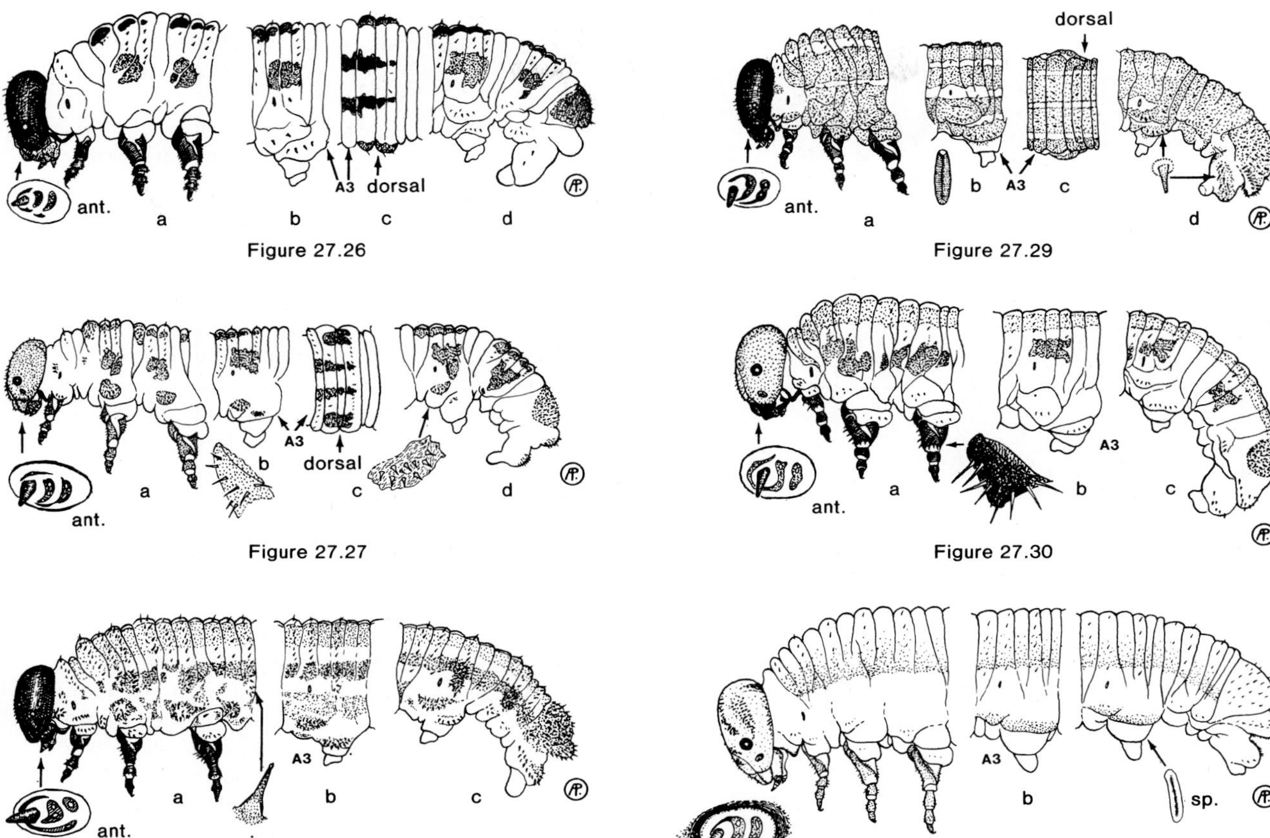

Figure 27.26a-d. Diprionidae. *Neodiprion pinetum* (Norton), white pine sawfly. a. head, thorax; b. A3 lateral; c. A3 dorsal; d. caudal segments. f.g.l. 23-25 mm. Jet black head and yellowish body with 4 longitudinal rows of dark spots dorsad of spiracles; subdorsal spots elongate and supraspiracular spots nearly square; no spots on abdomen below spiracles; suranal area with black patch divided on meson. Feeds on pine needles; destructive on young trees; distinctly gregarious. White pine preferred, but also feeds on pitch, shortleaf, red, and Mugho (Atwood and Peck 1943, Rose and Lindquist 1973; from Peterson 1948).

Figure 27.27a-d. Diprionidae. *Neodiprion lecontei* (Fitch), red-headed pine sawfly. a. head, thorax; b. A3 lateral; c. A3 dorsal; d. caudal segments. f.g.l. 25± mm. Reddish orange to brown head and pale to deep yellow body with 4 rows of conspicuous black spots dorsad of spiracles; subdorsal spots elongate and angular, spuraspiracular spots more square but irregular; on each abdominal segment 2 smaller spots between spiracles and base of prolegs. Feeds gregariously on old needles of pines, especially red, white, and jack (Atwood and Peck 1943, Rose and Lindquist 1973; from Peterson 1948).

Figure 27.28a-c. Diprionidae. *Neodiprion sertifer* (Geoffroy), European pine sawfly. a. head, thorax; b. A3 segment; c. caudal segments. f.g.l. 22± mm. Body dirty gray to green with narrow, middorsal near white line bordered by broad moderately light gray areas that merge into a distinct dark green supraspiracular stripe that is darker in spots, especially on caudal segments; spiracular stripe light and conspicuous bordered ventrally by broad darker area; head, thoracic legs, and dorsum of anal segment black. Food plants are red, Scotch, Japanese red, jack, Swiss mountain, Mugho, and other pines (Rose and Lindquist 1973; from Peterson 1948).

Figure 27.29a-d. Diprionidae. *Neodiprion abietis* (Harris), balsam fir sawfly. a. head, thorax; b. A3 lateral; c. A3 dorsal; d. caudal segments. f.g.l. 15± mm. Yellow to black head and dull greenish body with 6 dark stripes separated by pale dorsal stripe and paler subdorsal and spiracular stripes; thoracic legs black, prolegs pale yellow with proximal portions dark green. Food plants are fir, spruce, hemlock, Douglas fir (from Peterson 1948).

Figure 27.30a-c. Diprionidae. *Neodiprion* sp. from pine. a. head, thorax; b. A3 segment; c. caudal segments. f.g.l. 20-25 mm. Body pale green with 2 dark grayish-green longitudinal lines on dorsum bordering lighter middorsal stripe; between the dark lines and spiracles a conspicuous irregular dark spot occurs on each segment; head reddish-brown; legs black, dorsum of anal segment near black. Food plants are loblolly and shortleaf pines (from Peterson 1948).

Figure 27.31a-c. Diprionidae. *Gilpinia hercyniae* (Hartig), European spruce sawfly. a. head, thorax; b. A3 segment; c. caudal segments. f.g.l. 15-20 mm. Body dark green with 5 longitudinal white lines on dorsum and lateral aspects in fourth and fifth instars and none in last instar; light spiracular stripe broad, extending ventrad to darker areas on prolegs; head yellowish-brown, flecked with brown spots; thoracic legs yellowish; spines on segments very inconspicuous. Larvae feed singly on old needles of spruce (from Peterson 1948).

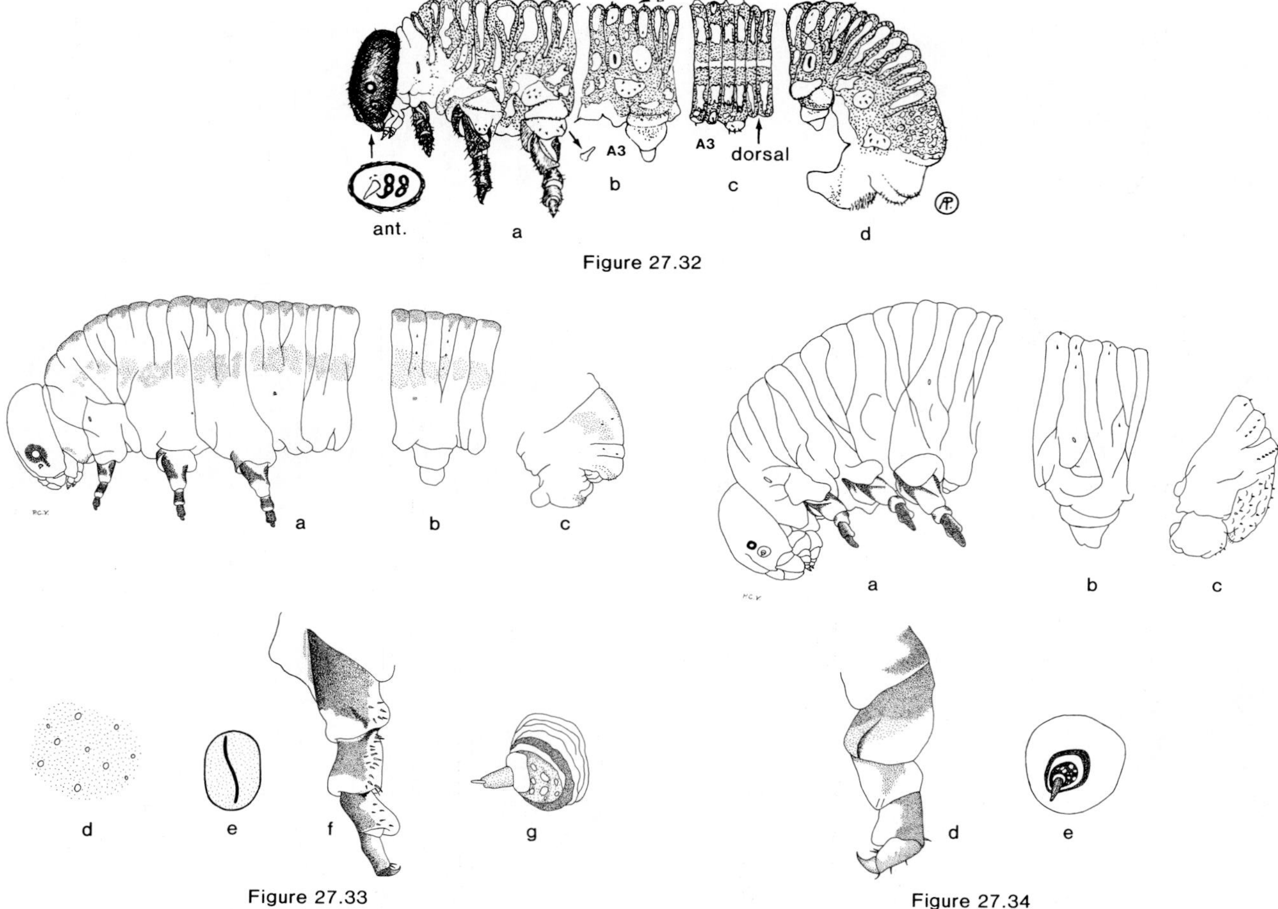

Figure 27.32

Figure 27.33

Figure 27.34

Figure 27.32a-d. Diprionidae. *Diprion similis* (Hartig), introduced pine sawfly. a. head, thorax; b. A3 lateral; c. A3 dorsal; d. caudal segments. f.g.l. 30± mm. Yellowish green on dorsum with 2 black or deep brown stripes adjacent to meson; on lateral aspects depressed portions of segments near black and elevated parts yellow, producing a spotted larva; venter pale yellow to near white; head and thoracic legs near black. Food plants, pines (Rose and Lindquist 1973; from Peterson 1948).

Figure 27.33a-g. Diprionidae. *Monoctenus* sp. a. head, thorax, A1 segment; b. A3 segment; c. caudal segments; d. cuticular pattern of head; e. abdominal spiracle; f. T2 leg; g. antenna. f.g.l. 18-25 mm. Characteristics as for family. Species feed on *Juniperus* and *Thuja* in eastern North America.

Figure 27.34a-e. Diprionidae. *Augomonoctenus libocedrii* Rohwer a. head, thorax, A1 segment; b. A3 segment; c. caudal segments; d. T2 leg; e. antenna. f.g.l. 18-25 mm. Greenish when alive, without pattern on body; thoracic legs light brown. Larvae live and feed in cones of incense cedar; known from Oregon and California.

demonstrably different, and different-appearing larvae produce virtually identical adults (Ross 1955). Repeated rearings of *Neodiprion* colonies have resolved some of the confusion, but many difficulties remain. Judging from these rearings, several species of *Neodiprion* are distinct biological and genetic units, yet they can be diagnosed only on the basis of differences in larval color pattern or details of life history. In the case of certain others, it is difficult to ascertain whether the units involved are species or merely races or strains of the same species. Hence, those studying coniferous sawflies should keep series of each instar, record complete host and biological data, and rear adults.

Selected Bibliography

Atwood and Peck 1943.
Baker 1972.
Coppel and Benjamin 1965.
Furniss and Carolin 1977.
Rose and Lindquist 1973.
Ross 1955.
Smith 1971a, 1974b, 1979b.

TENTHREDINIDAE (TENTHREDINOIDEA)

David R. Smith, *Systematic Entomology Laboratory, ARS, USDA*
Woodrow W. Middlekauff, *University of California, Berkeley*

Tenthredinid Sawflies

Figures 27.35–27.65

Relationships and Diagnosis: Larvae of Tenthredinidae are diverse in habits and morphological forms and difficult to briefly characterize. Most are typically sawflylike and related to Xyelidae, Pergidae, Cimbicidae, Diprionidae, and Argidae. They are separated from the wood and stem borers of the Siricidae, Xiphydriidae, Anaxyelidae, and Cephidae by the lack of a postcornus, and from the Pamphiliidae by the lack of segmented appendages on the tenth sternum, the lack of a hook on the tenth tergum, and having five or fewer antennal segments. The Xyelidae have six- to seven-segmented antennae; the Pergidae have an antenna composed of four flat sclerotized rings and have lateral suckerlike protuberances on the abdomen. The Cimbicidae have one- or two-segmented antennae, secondary longitudinal sutures on the labrum, and supraspiracular gland openings. The Diprionidae have three-segmented antennae with the third segment peglike. The Argidae usually have an enlarged empodium adjacent to each tarsal claw, sometimes lack a tarsal claw on the prothoracic leg, and have three dorsal annulations on the abdominal segments. In the Tenthredinidae, the antenna is flat or conical, with five or fewer segments, the third segment never peglike (figs. 27.35d, 27.37e), there are no secondary longitudinal sutures on the labrum, thoracic legs are usually five-segmented with tarsal claws (except for a few leaf mining forms) and the empodium is not enlarged, annulation of the abdominal segments is distinct, usually with four to seven annulets, and there are no lateral suckerlike protuberances or supraspiracular gland openings. The leaf mining larvae are usually dorsoventrally flattened (fig. 27.50), have reduced thoracic legs and prolegs, and are commonly prognathous; such larvae do not occur in other families of Symphyta.

Biology and Ecology: Most larvae are external feeders on foliage, either leaf edge feeders or feeding on the underside of the leaf and skeletonizing it. However, a number are leaf miners, petiole miners, gall formers on stems, twigs, petioles, leaves, or buds, and some are shoot borers. Additional data are found in the subfamily treatments.

Description: *Head:* Globose and hypognathous or dorsoventrally flattened and prognathous; antenna one- to five-segmented, usually four- or five-segmented, conical, rounded, or flat; labrum without secondary longitudinal sutures. *Thorax:* Legs present, usually five-segmented, each with tarsal claw, and without an enlarged empodium adjacent to claw; a few leaf miners have reduced legs without a claw. *Abdomen:* Segments 1–8 each with two to seven dorsal annulets, usually four to seven in free-feeding larvae, annulation usually distinct; hairs, setae, tubercles, dark plates, or simple or branched spines usually present in a constant pattern on annulets, spiracular and surpedal lobes, and tenth tergum; prolegs present on segments 2–7 or 2–8 and 10, those on 10 sometimes fused, prolegs reduced or absent in some leaf miners; supraspiracular glands and lateral suckerlike protuberances absent; tenth segment without postcornus or segmented appendages, sometimes with two or more fleshy protuberances at apex of tenth tergum.

Comments: The prepupa, or last nonfeeding stage, of the larva may differ in color and structure from the feeding stages. The mandibles of the prepupa are linear, without dorsal and ventral cutting edges, and are similar in appearance, usually without different dentition. Those larvae with spines in the feeding stages lack them in the prepupal stage. The prepupal stage cannot be used in the following key to subfamilies and therefore a couplet separating this stage is given.

The Tenthredinidae is the largest family of sawflies in North America with 95 genera and about 735 species. Representatives are found throughout the continent. Eight subfamilies are recognized, and, because the larvae are distinctive, they are keyed to subfamily and treated separately.

Selected Bibliography

Baker 1972.
Furniss and Carolin 1977.
MacGillivray 1914.
Maxwell 1955.

KEY TO SUBFAMILIES OF TENTHREDINIDAE LARVAE

1. Dorsoventrally flattened; head usually prognathous (fig. 27.50); prolegs reduced or absent; thoracic legs reduced, with or without tarsal claw, and commonly directed laterally; antenna low, one- or two-segmented; sometimes with sclerotized plates on pronotum, thoracic sterna, basal abdominal sterna, or surrounding each proleg; leaf miners .... (in part) **Heterarthrinae** (p. 638)

1′. Cylindrical, typically sawflylike or sluglike; head globose and hypognathous (fig. 27.35a); prolegs present; thoracic legs usually five-segmented, with tarsal claw, usually directed ventrally (fig. 27.35a); antenna four- or five-segmented, low, flat, or conical (figs. 27.35d, 27.39d); ornamentation of thorax and abdomen various; mostly free feeding larvae 2

2(1). Mandible with 3 or 4 teeth in a linear row, both mandibles similar, though sometimes one with more teeth than the other .... prepupae

2′. Mandible with ventral and dorsal cutting edges and sometimes with mesal ridge(s); dentition of each mandible different .... (feeding instars) 3

3(2′). Prolegs present on abdominal segments 2–7 (rarely 2–6) and 10; ventral eversible glands present on abdominal segments 1–8 (fig. 27.43a); antenna flat or conical; 3–6 dorsal annulets on abdominal segments 1–8 .... **Nematinae** (p. 634)

3′. Prolegs present on abdominal segments 2–8 and 10; ventral eversible glands absent; antenna various, usually conical; dorsal annulets of abdominal segments 1–8 various, usually 4–7 4

4(3′). Abdominal segments 1–8 each with 7 dorsal annulets (fig. 27.35b) 5

4′. Abdominal segments 1–8 each with 6 or fewer dorsal annulets 6

5(4). Left mandible with 5 lateral teeth and 2 mesal ridges (fig. 27.35e); tibiae with 10 or fewer setae .... **Selandriinae** (p. 633)

5′. Left mandible with 3–4 lateral teeth and 1 mesal ridge (fig. 27.63e); tibiae with 13–18 setae .... **Tenthredininae** (p. 643)

6(4′). Sclerotized molar process present at floor of mouth between mandibles (visible when mandibles open); labrum distinctly asymmetrical, one lobe much longer and broader than the other (fig. 27.36d) .... **Dolerinae** (p. 633)

6′. Sclerotized molar process absent; labrum symmetrical or nearly so 7

7(6′). Right mandible with large, circular depression between the 2 dorsal teeth (fig. 27.37d); mandible without mesal ridge; each mandible with 1 seta on outer surface (abdominal segments 1–8 each 6-annulate); on Cupressaceae .... **Susaninae** (p. 634)

7′. Right mandible with different type of depression separating teeth or with more than 2 dorsal teeth; mandible with or without mesal ridge(s) and with or without setae on outer surface; annulation of abdominal segments various; host plants other than Cupressaceae 8

8(7′). Sluglike, with thorax enlarged and body tapering toward apex; dark or transparent slime covering body in life; fingerlike projection extending forward from prothoracic leg *or* each mandible with a ventral row of 4 long, rectangular teeth (fig. 27.52d) .... (in part) **Heterarthrinae** (p. 638)

8′. Caterpillarlike, cylindrical, body not distinctly tapering toward apex; slime absent, though body may be covered with whitish bloom; prothoracic leg without fingerlike projection; mandible without row of 4 rectangular teeth 9

9(8′). Left mandible with mesal ridge and elevated mesal area, ridge connected to ventral cutting ridge of mandible (fig. 27.58d); abdominal segments 1–8 each with 6 dorsal annulets, with small setae or tubercles on annulets 2 and 4, sometimes on 1 (figs. 27.59b, 27.61b); body without large dark tubercles or simple or branched spines, only *Dimorphopteryx* larvae with long fleshy projections on pronotum and on posterior margin of tenth tergum (figs. 27.58a,b) .... **Allantinae** (p. 639)

9′. Left mandible with or without mesal ridge, if present, without elevated mesal area and not connected to ventral cutting edge; abdominal segments 1–8 each with 4–6 dorsal annulets, tubercles or other ornamentation only on annulets 2 and 4 (fig. 27.55b); body sometimes with sclerotized tubercles or long simple or branched spines (fig. 27.55a) .... **Blennocampinae** (p. 639)

Subfamily Selandriinae (Tenthredinidae)

Figure 27.35

Relationships and Diagnosis: Larvae are typically sawflylike with conical, five-segmented antenna (fig. 27.35d); five-segmented thoracic legs with only eight to ten setae on the tibiae, prolegs on abdominal segments 2–8 and 10, and abdominal segments 1–8 each with seven dorsal annulets (fig. 27.35b). The seven-annulate abdominal segments are also found in the Tenthredininae and Cimbicidae, but the Selandriinae have a conical, five-segmented antenna and only eight to ten setae on the tibiae. The left mandible of Selandriinae is also characteristic, with five lateral teeth and two mesal ridges (fig. 27.35e), whereas that of Tenthredininae has four or fewer lateral teeth and one mesal ridge.

Biology and Ecology: All larvae feed externally on foliage except for species of *Heptamelus,* which bore in the stems of ferns. The most common species feed on ferns, species of *Eriocampidea, Strongylogaster,* and *Aneugmenus* are associated with *Pteridium,* and *Hemitaxonus* feeds on *Onoclea* and *Athyrium;* species of *Brachythops* feed on sedges, and species of *Birka* have been associated with grasses. Oviposition of fern feeders is in the fronds and pupation in the soil or sometimes in other substances such as stems of plants, bark, or rotting wood; a cocoon is not spun. Most species have one generation a year, but some *Eriocampidea* and *Hemitaxonus* have at least two. Notes on the biology of several species are found in Beer (1955), Hogh (1966), and Smith and Lawton (1980).

Description: *Head:* Globose, with scattered or numerous setae or hairs; antenna conical, five-segmented (three-segmented in *Heptamelus*); labrum symmetrical; molar process absent at floor of mouth; left mandible with five lateral teeth and two mesal ridges. *Thorax:* Legs five-segmented, with tarsal claw; ornamentation same as for abdomen; each tibia with eight to ten setae. *Abdomen:* Prolegs on segments 2–8 and 10; segments 1–8 each with seven dorsal annulets; setae or short tubercles on annulets 2 and 4, sometimes others, and on spiracular and surpedal lobes and tenth tergum; tenth tergum rounded, without caudal protuberances or appendages.

Comments: The Selandriinae are poorly represented in North America where only 11 genera and 27 species have been recorded (Smith 1969b). *Strongylogaster, Aneugmenus,* and *Hemitaxonus* are the most widespread; *Brachythops* and *Birka* are primarily northern; and *Eriocampidea, Stromboceridea,* and *Eustromboceros* are southwestern, being northern extensions of Neotropical genera.

Very few larvae have been associated with adults and few can be reliably identified to species. Generic characters are not yet determined. Difficulties in rearing and their noneconomic importance reflect the poor knowledge of the larvae.

Selected Bibliography

Beer 1955.
Hogh 1966.
Smith 1969b.
Smith and Lawton 1980.

Subfamily Dolerinae (Tenthredinidae)

Figure 27.36

Relationships and Diagnosis: Larvae are typically sawflylike with conical five-segmented antenna, five-segmented thoracic legs each with a tarsal claw (fig. 27.36a), prolegs on abdominal segments 2–8 and 10, and six dorsal annulets on abdominal segments 1–8 (fig. 27.36b). Two characters that immediately separate Dolerinae larvae from those of other Tenthredinidae are the asymmetrical labrum (fig. 27.36d) and the presence of a sclerotized molar process located at the floor of the mouth between the mandibles (visible with mandibles open). Some other tenthredinid larvae have an asymmetrical labrum, though not so decidedly asymmetrical, but the molar process is not known in other subfamilies. Most larvae are green with various dark patterns on the head and body. Ornamentation of the body consists of small tubercles or setae.

Biology and Ecology: All larvae are external feeders and most are not considered of great importance. Several species, such as *Dolerus nitens* Zaddach, which has been reported on fescue, feed on cultivated grass crops. Species of *Dolerus* have been recorded from various Cyperaceae, Poaceae, and *Equisetum,* and species of *Loderus* only on *Equisetum.* Biology and ecology were reported by Ross (1931). Little information is available except that there is apparently a single generation a year and larvae prefer to bore into some soft substance to pupate.

Description: *Head:* Globose, with numerous or scattered setae or hairs; antenna conical, five-segmented; labrum asymmetrical, one lobe decidedly longer, sometimes broader, than the other; sclerotized molar process present at floor of mouth between mandibles. *Thorax:* Legs five-segmented, each with tarsal claw; ornamentation as for the abdomen. *Abdomen:* Prolegs on segments 2–8 and 10; segments 1–8 each with six dorsal annulets; small tubercles or setae usually on annulets 2 and 4, spiracular and surpedal lobes, and tenth tergum; tenth segment without caudal protuberances.

Comments: The two genera, *Dolerus* and *Loderus,* are found throughout North America. About 41 species of *Dolerus* and 5 species of *Loderus* are recorded (Smith 1979b, 1980). They are most common in the temperate regions, and adults are commonly swept from fields of sedges, grasses, or *Equisetum* in very early spring. Difficulties in rearing and their noneconomic importance reflect the poor knowledge of the larvae; none can, with certainty, be identified to species, and larval characters to separate the two genera are not known.

Selected Bibliography

Ross 1931.
Smith 1979b, 1980.

Subfamily Susaninae (Tenthredinidae)

Figure 27.37

Relationships and Diagnosis: Larvae are typically sawflylike, with conical five-segmented antenna (fig. 27.37e), five-segmented thoracic legs each with a tarsal claw, prolegs on abdominal segments 2–8 and 10, and six dorsal annulets on abdominal segments 1–8 with minute tubercles on annulets 2 and 4 (fig. 27.37b). They are similar to larvae of Allantinae and those of the Blennocampinae that lack spines, but are distinguished from those groups by the mandibles, especially the right mandible, which has a large circular depression between the two dorsal teeth. Larvae are most easily separated by their host, species of Cupressaceae, and by comparison with the mandibular dentition in fig. 27.37d.

Biology and Ecology: Larvae are free leaf feeders on *Cupressus, Juniperus,* and *Thuja; Susana cupressi* Rohwer and Middleton is sometimes a pest of cypress in California. Life history information is not available.

Description: *Head:* Globose, with few scattered setae and hairs; antenna conical, five-segmented; clypeus with two setae on each side; each mandible with one seta on outer angle, without mesal ridges; right mandible with large circular depression between two dorsal teeth. *Thorax:* Legs five-segmented, each with tarsal claw; metathoracic legs largest; pair of setiferous lobes on each thoracic sternum. *Abdomen:* Prolegs on segments 2–8 and 10; segments 1–8 each with six dorsal annulets; small inconspicuous tubercles on annulets 2 and 4, spiracular lobes and surpedal lobes; tenth segment without protuberances; ventral eversible glands absent.

Comments: The single genus, *Susana,* with six species occurs in western North America from British Columbia south to California and New Mexico. Larvae of four species have been described, *S. cupressi, S. oregonensis* Smith, and *S. juniperi* (Rohwer) by Smith (1969c), and *S. fuscala* Wong and Milliron from juniper in British Columbia by Wong and Milliron (1972).

Selected Bibliography

Smith 1969c.
Wong and Milliron 1972.

Subfamily Nematinae (Tenthredinidae)

Figures 27.38–27.47

Relationships and Diagnosis: The following features separate nematine larvae from those of other Tenthredinidae: prolegs present on abdominal segments 2–7 and 10; midventral eversible glands present on abdominal segments 1–7 (not always everted) (fig. 27.43a); and antenna four-segmented, rarely five-segmented, and either flat or conical (fig. 27.43c). In addition, there are usually four or five annulets on abdominal segments 1–8, rarely six, and never seven. Ventral eversible abdominal glands are not known in other Tenthredinidae, and most other tenthredinid larvae where prolegs are evident have them on abdominal segments 2–8 and 10.

Biology and Ecology: The habits of Nematinae are very diverse, many are external leaf feeders, but some feed in fruit, form galls on stems, twigs, petioles, buds, or leaves, mine in leaf petioles, or mine in leaves. Following are the more important genera with their host plants and important species.

Cladius: The bristly rose slug, *C. difformis* (Panzer) (figs. 27.39a-d), is a common pest of cultivated roses throughout North America.

Priophorus: *Alnus, Betula, Crataegus, Populus, Prunus, Rubus,* and *Salix* are recorded hosts. One species, *P. morio* (Lepeletier) (figs. 27.40a-c), was introduced into Hawaii, without success, to aid in biological control of *Rubus.*

Trichiocampus: External leaf feeders on *Populus* and *Salix.*

Hoplocampa: Larvae (fig. 27.46a-c) live and feed in developing fruit of *Amelanchier, Pyrus, Prunus, Crataegus,* and *Malus,* and cause premature drop of the fruit. Important species are the cherry fruit sawfly, *H. cookei* (Clarke) in the West, and two introduced species, the European apple sawfly, *H. testudinea* (Klug), and *H. brevis* (Klug) on pear.

Caulocampus: Larvae of *C. acericaulis* (MacGillivray) mine in the petioles of *Acer,* causing early leaf drop.

Hemichroa: External leaf feeders of *Alnus, Amelanchier, Prunus.* The striped alder sawfly, *H. crocea* (Geoffroy) may severely defoliate alder.

Craterocercus: External leaf feeders on *Quercus.*

Fallocampus: External leaf feeders on *Populus.*

Anoplonyx: External leaf feeders and occasional pests of *Larix.*

Neopareophora: On *Vaccinium; N. litura* (Klug) is a pest of low-bush blueberry.

Nematinus: External leaf feeders on *Betula.*

Pachynematus: External feeders of *Carex, Festuca, Salix,* and wheat. Several species may be of importance in cultivated crops such as wheat and fescue.

Pikonema: The yellowheaded spruce sawfly, *P. alaskensis* (Rohwer) (figs. 27.47a-d) and the greenheaded spruce sawfly, *P. dimmockii* (Cresson) are serious defoliators of *Picea.*

Pristiphora: External leaf feeders on *Pyrus, Salix, Aquilegia, Vaccinium, Spiraea, Betula, Quercus, Larix, Sorbus, Picea, Ribes, Alnus, Rubus,* and *Crataegus.* The more important species are the California pear sawfly, *P. abbreviata* (Hartig), the larch sawfly, *P. erichsonii* (Hartig) (figs. 27.44a-c), the mountainash sawfly, *P. geniculata* (Hartig) (figs. 27.45a-c), and a sawfly on *Ribes, P. rufipes* Lepeletier.

Croesus: External leaf feeders on *Castanea, Corylus, Betula,* and *Alnus,* including the dusky birch sawfly *C. latitarsus* Norton.

Nematus: External leaf feeders on *Robinia, Betula, Populus, Salix, Alnus, Carpinus, Corylus, Rhododendron, Vaccinium, Gleditsia, Ribes,* and *Ostrya.* The willow sawfly, *N. ventralis* Say (figs. 27.42a-c), the imported currantworm, *N. ribesii* (Scopoli) (figs. 27.41a-c), and *N. tibialis* Newman on locust are among the most common species.

Pontania: Forms various types of leaf galls on *Salix.*

Euura: Forms stem, twig, petiole, or bud galls on *Salix* (figs. 27.38a-c).

Phyllocolpa: Forms and lives in rolled leaf edges or rolled leaves of *Salix* and *Populus.*

Amauronematus: External leaf feeders on *Salix, Alnus,* and *Rhododendron;* a few species are catkin feeders.

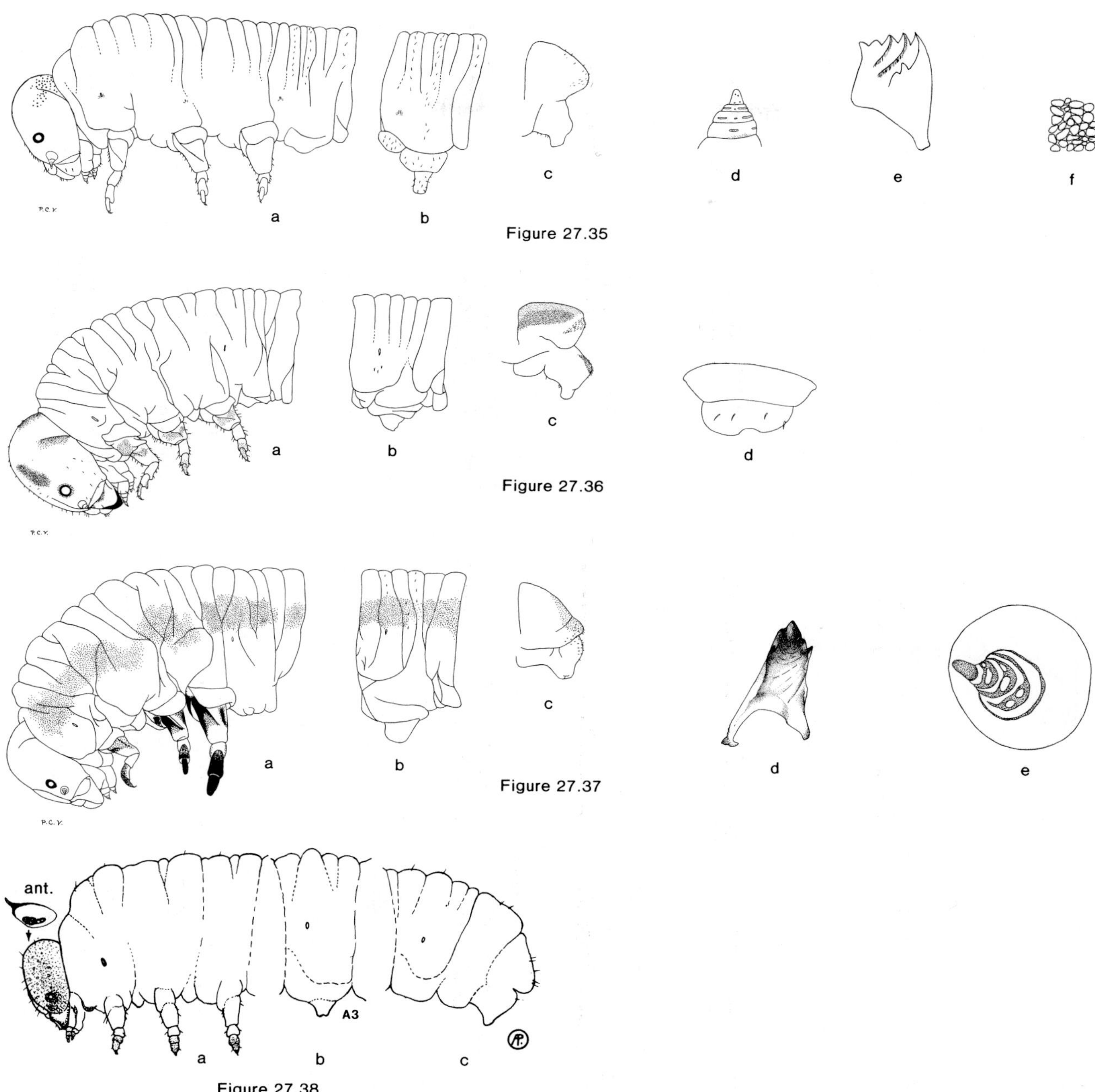

Figure 27.35

Figure 27.36

Figure 27.37

Figure 27.38

Figure 27.35a-f. Tenthredinidae: Selandrinae. *Strongylogaster* sp. a. head, thorax, A1 segment; b. A3 segment; c. caudal segments; d. antenna; e. left mandible, ventral view; f. head reticulation. f.g.l. 18–28 mm. Mostly greenish when alive; abdominal segments 1–8 each 7-annulate; antenna conical. Feeds on fronds of ferns, *Pteridium.*

Figure 27.36a-d. Tenthredinidae: Dolerinae. *Dolerus* sp. a. head, thorax, A1 segment; b. A3 segment; c. caudal segments; d. clypeus and labrum. f.g.l. 18–28 mm. Greenish when alive, usually with various dark patterns on head and sometimes with longitudinal stripes or spots on body; abdominal segments 1–8 each 6-annulate; labrum decidedly asymmetrical. Feeds on grasses, sedges, *Equisetum;* some species damage cultivated crops such as wheat.

Figure 27.37a-e. Tenthredinidae: Susaninae. *Susana* sp. a. head, thorax, A1 segment; b. A3 segment; c. caudal segments; d. right mandible, innerface; e. antenna. f.g.l. 20–28 mm. Greenish when alive, usually with dark thoracic legs and dark longitudinal stripes on body; mandible dentition as described and figured is characteristic. Feeds on cypress and juniper in western North America (Smith 1969c).

Figure 27.38a-c. Tenthredinidae: Nematinae. *Euura* sp. A willow stem gall sawfly. a. head, thorax; b. A3 segment; c. caudal segments. f.g.l. 9± mm. Body cylindrical and near white to cream, without setae on most parts; head light yellow, possessing a few setae, antenna not segmented; suranal area faintly pigmented. Produces stem galls on willows (from Peterson 1948).

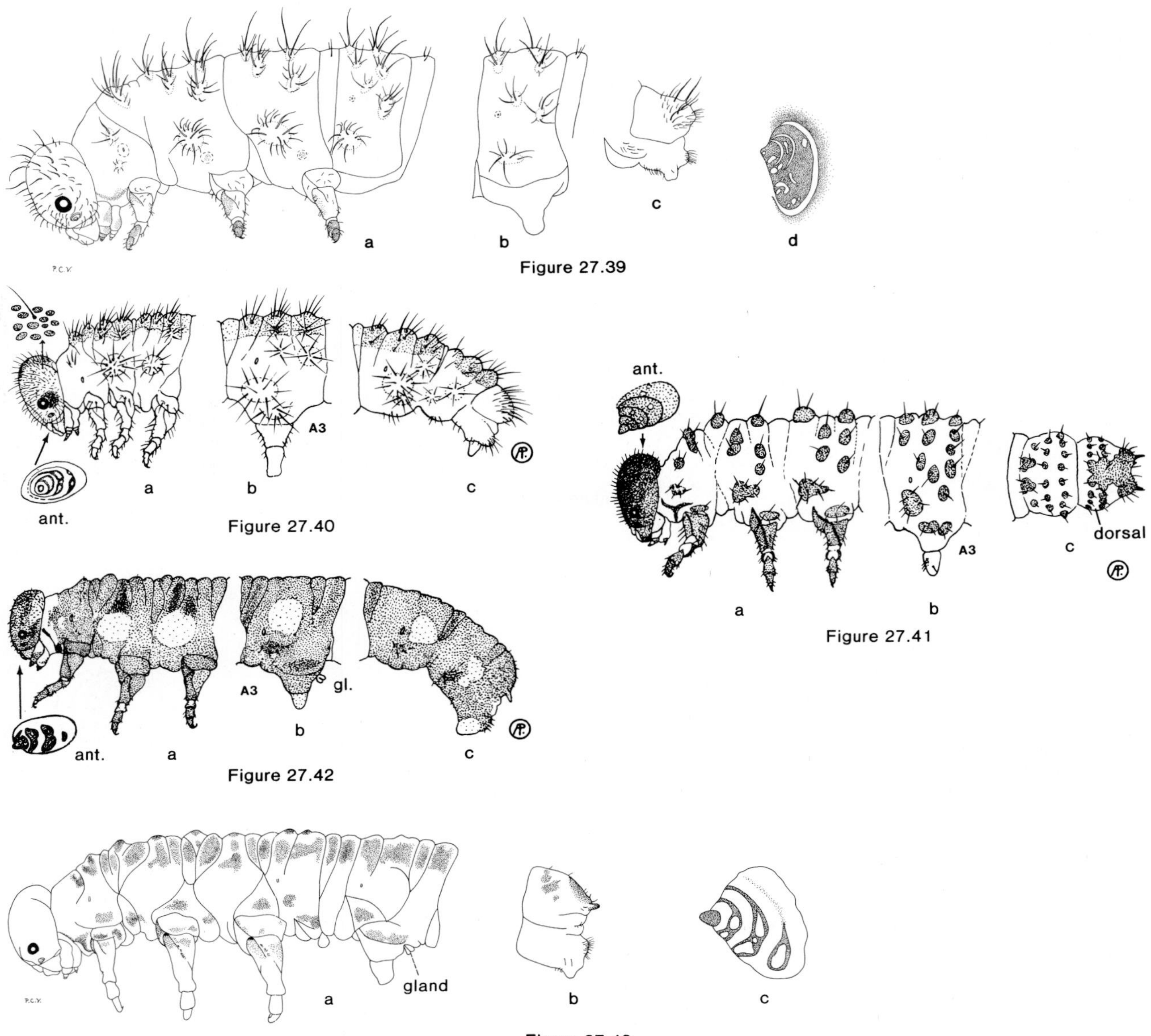

Figure 27.39a-d. Tenthredinidae: Nematinae. *Cladius difformis* (Panzer), bristly roseslug. a. head, thorax, A1 segment; b. A3 segment; c. caudal segments; d. antenna. f.g.l. 10-15 mm. Greenish when alive; prolegs on abdominal segments 2-7 and 10; abdominal segments 1-8 each with 3 dorsal annulets; armature of low plates bearing long setae; head with long hairs. Commonly found feeding on foliage of cultivated roses, throughout North America (Smith 1974a).

Figure 27.40a-c. Tenthredinidae: Nematinae. *Priophorus morio* (Lepeletier). a. head, thorax; b. A3 segment; c. caudal segments. f.g.l. 14-19 mm. Greenish when alive with dark dorsal stripe on body and head; prolegs on A2-A7 and A10; long flexuous setae on head and body. Feeds on *Rubus* throughout North America and may have several generations a year (Smith 1974a; from Peterson 1948).

Figure 27.41a-c. Tenthredininae: Nematinae. *Nematus ribesii* (Scopoli), imported currantworm. a. head and thorax; b. A3 segment; c. caudal segments. f.g.l. 18-23 mm. Greenish with dark spots on body; head, thoracic legs, and 10th tergum blackish. Feeds on currant and gooseberry and is common throughout North America (from Peterson 1948).

Figure 27.42a-c. Tenthredinidae: Nematinae. *Nematus ventralis* Say, willow sawfly. a. head and thorax; b. A3 segment; c. caudal segments. f.g.l. 20-24 mm. Body blackish with conspicuous lateral yellow spots; prolegs yellowish; head and thoracic legs dark. Note caudal protuberances on A10 tergum, present in most species of *Nematus*. Feeds on foliage of willows in the eastern half of the continent; a similarly colored larva but a different species is found in the West (from Peterson 1948).

Figure 27.43a-c. Tenthredinidae: Nematinae. *Nematus* sp. a. head, thorax, A1 and A2 segments; b. caudal segments; c. antenna; f.g.l. 15-30 mm. Coloration varies among species; prolegs on abdominal segments 2-7 and 10; 10th tergum with pair of fleshy protuberances at apex. Species of *Nematus* feed on foliage of many trees and shrubs, particularly birch, poplar, and willow.

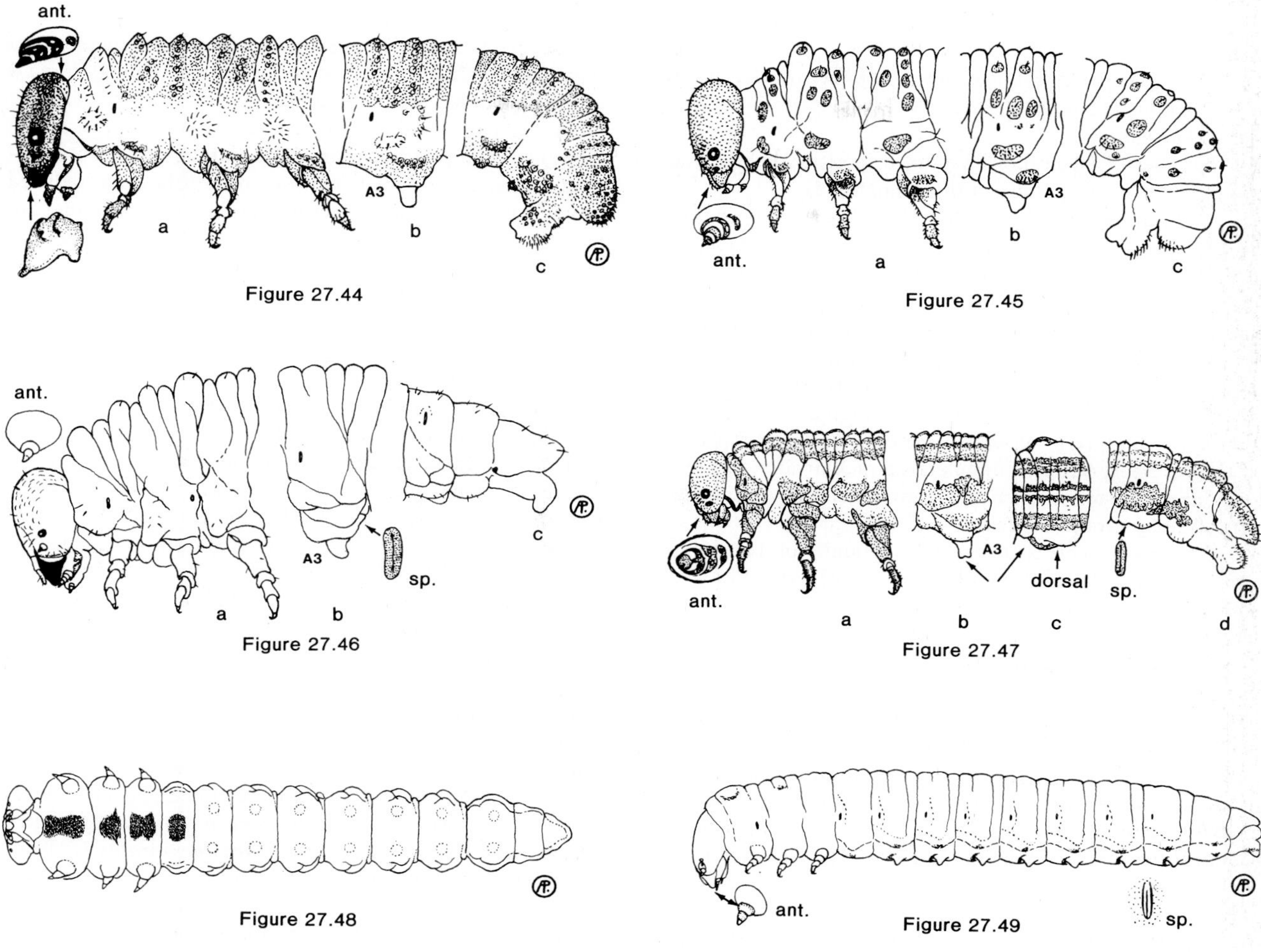

Figure 27.44a-c. Tenthredinidae: Nematinae. *Pristiphora erichsonii* (Hartig), larch sawfly. a. head, thorax; b. A3 segment; c. caudal segments. f.g.l. 19-24 mm. Body greenish with head and thoracic legs dark brown. A serious pest of larches in the United States and Canada; transcontinental (Wong 1963; from Peterson 1948).

Figure 27.45a-c. Tenthredinidae: Nematinae. *Pristiphora geniculata* (Hartig), mountain ash sawfly. a. head, thorax; b. A3 segment; c. caudal segments. f.g.l. 16-20 mm. Greenish with dark brown spots on body; head and thoracic legs yellow to orange. Feeds on mountain ash in eastern North America (from Peterson 1948).

Figure 27.46a-c. Tenthredinidae: Nematinae. *Hoplocampa* sp. a. head and thorax; b. A3 segment; c. caudal segments. f.g.l. 10-14 mm. Late instar white with amber head; early instars may have dark colored head and dark plates on terga A8-A10. Larvae live and feed within fruits of rosaceus plants; the species in pear and apple are especially important. The genus is found throughout North America (Bird 1927; Smith, 1966b; from Peterson 1948).

Figure 27.47a-d. Tenthredinidae: Nematinae. *Pikonema alaskensis* (Rohwer), yellowheaded spruce sawfly. a. head and thorax; b. A3 lateral; c. A3 dorsal; d. caudal segments. f.g.l. 18-23 mm. Body greenish with longitudinal dark stripes; head yellow. Feeds on species of spruce and is transcontinental in northern United States and Canada (Ross 1938; from Peterson 1948).

Figure 27.48. Tenthredinidae: Heterarthrinae. *Fenusa pusilla* (Lepeletier), birch leafminer, entire larva, ventral view. f.g.l. 5-8 mm. White with head amber and dark plates on thoracic sterna and sternum of A1. Transcontinental in North America and may be a serious pest of birch (Smith 1971c; from Peterson 1948).

Figure 27.49. Tenthredinidae: Heterarthrinae. *Profenusa canadensis* (Marlatt). f.g.l. 7± mm. Entirely near white or cream, late-feeding instars possess crescent-shaped pigment areas on prolegs on segments A2-8 and pigment spots immediately above each proleg; also pigment spots on venter of A1 and A9 and a nearly complete pigment circle around the single anal proleg; head somewhat depressed; antennae short and indistinctly 3-segmented; short, rounded, paired prolegs on segments A2-8 and vestigial, lightly sclerotized remnants of prolegs on A1 and A9; 1 median proleg on 10th segment. Food plant hawthorn (*Crataegus*) and cherry. Produces blotch mines along margins of foliage early in growing season (Smith 1971c; from Peterson 1948).

Pseudodineura: One species, *P. parva* (Norton) is a leaf miner of *Hepatica*.

Kerita: *Kerita fidala* Ross is a leaf miner of *Mertensia*.

Pristola: The larva of *P. macnabi* Ross feeds in the berries of *Vaccinium*.

Number of generations a year, oviposition habits, and feeding habits vary greatly in the Nematinae.

Description: *Head:* Globose, with scattered or numerous setae or long hairs; antenna four- to five-segmented, usually four-segmented, conical or flat with segments as sclerotized rings. *Thorax:* Legs five-segmented, each with tarsal claw; ornamentation as for the abdomen. *Abdomen:* Prolegs on segments 2–7 and 10; midventral eversible glands on segments 1–7; segments 1–8 each with three to six dorsal annulets, sometimes not clearly differentiated; various combinations and patterns of setae, tubercles, warts, dark plates, or long or short hairs on annulets, spiracular and surpedal lobes, and tenth tergum; tenth tergum sometimes with caudal pair of fleshy protuberances.

Comments: This is the largest subfamily of Tenthredinidae in North America and includes 27 genera and about 360 species. It is the dominant group in the arctic and subarctic regions with numbers of individuals and species decreasing southward. Very few are known from the southern tier of states.

No generic key to larvae is available, although Yuasa (1923) may be used with caution, and Lorenz and Kraus (1957) may be of some help. Individual larvae have been treated in some papers, such as Bird (1927) and Smith (1966b) for *Hoplocampa;* Lindquist and Miller (1970) for some species feeding on birch and alder in Ontario; E. L. Smith (1968, 1970) for some gall forming species; Wong (1955) for *Anoplonyx;* Ross (1938) for *Pikonema;* and Smith (1972b, 1974a, 1975a) for some other genera. Wong (1963) described the external morphology of the larch sawfly larva. For other references to taxonomic and biological literature, see the North American catalog (Smith 1979b).

More larval-adult associations are necessary before adequate generic characters can be defined.

Selected Bibliography

Bird 1927.
Lindquist and Miller 1970.
Lorenz and Kraus 1957.
Ross 1938.
Smith, D. R. 1966b, 1972b, 1974a, 1975a, 1979b.
Smith, E. L. 1968, 1970.
Wong 1955, 1963.
Yuasa 1923.

Subfamily Heterarthrinae (Tenthredinidae)

Figures 27.48–27.52

Relationships and Diagnosis: Larvae are free feeding or leaf mining; three basic types exist: (1) The typically sawflylike larva of *Endelomyia*, separated from other tenthredinid larvae by the four subtruncate teeth on the ventral margin of each mandible (fig. 27.52d); other characteristics are the conical, five-segmented antenna, prolegs on abdominal segments 2–8 and 10, six dorsal annulets on abdominal segments 1–8, and tubercles on the thorax, annulets 2 and 4 of the abdominal segments, and the spiracular and surpedal lobes; (2) The sluglike or tadpolelike larvae of *Caliroa* (fig. 27.51) have an enlarged thorax, the anterior margin overlapping the head, the body narrowing toward the apex, reduced thoracic legs, and a large, fleshy protuberance stemming anteriorly from each prothoracic leg; (3) The modified leaf mining larvae (figs. 27.48–27.50), which are dorsoventrally flattened, usually with a prognathous head, reduced thoracic legs that sometimes lack tarsal claws, various number and arrangement of sclerotized plates on the pronotum, thoracic sterna, and basal abdominal sterna, and sometimes surrounding prolegs, and reduced or absent prolegs.

Biology and Ecology: The three types of larvae differ in their habits, those of *Endelomyia* and *Caliroa* are external leaf feeders and those of the third type are leaf miners. Larvae of *Caliroa* are closely appressed to the underside of the leaf and are usually covered with a dark or transparent slime that they secrete; they are leaf skeletonizers.

Some of the important Heterarthrinae are the roseslug, *Endelomyia aethiops* (Fabricius) (figs. 27.52a-d), a widespread defoliator of cultivated roses; the pear sawfly, *Caliroa cerasi* (Linnaeus) (fig 27.51), common throughout the continent in pear orchards; *C. liturata* MacGillivray, which defoliates peach and plum; the scarlet oak sawfly, *C. quercuscoccineae* (Dyar), which has caused extensive defoliation of oaks in the East; the birch leafminer, *Fenusa pusilla* (Lepeletier) (fig. 27.48), and *Profenusa thomsoni* (Konow), *Messa nana* (Klug), and *Heterarthrus nemoratus* (Fallén), which are other leaf miners of birch; species of *Profenusa* which are leaf miners of oaks; the European alder leafminer, *Fenusa dohrnii* (Tischbein); and the elm leafminer, *F. ulmi* Sundevall. Host genera for some of the Heterarthrinae genera are as follows: ***Endelomyia*** (*Rosa*); ***Caliroa*** (*Betula, Pyrus, Prunus, Crataegus, Sorbus, Quercus, Salix, Castanea, Nyssa*); ***Heterarthrus*** (*Betula*); ***Messa*** (*Populus, Betula, Salix*); ***Setabara*** (*Prunus*); ***Bidigitus*** (*Platanus*); ***Profenusa*** (*Quercus, Betula, Crataegus, Prunus*); ***Nefusa*** (*Viola*); ***Fenusa*** (*Betula, Alnus, Ulmus*), ***Fenella*** (*Potentilla*).

Some *Caliroa* species may have several generations a year, overwintering and pupating in the soil. *Endelomyia* and most leaf miners have a single generation a year, and overwintering and pupation may be either in the soil or the prepupa may remain in the fallen leaf.

Description: *Head:* Globose and hypognathous or dorsoventrally flattened and prognathous, with scattered or numerous setae or hairs; antenna conical and five-segmented or one- to two-segmented and low and rounded. *Thorax:* Legs five-segmented, each with tarsal claw, or reduced and two-segmented without tarsal claw in some leaf miners; in *Caliroa* legs directed laterally and with a fleshy protuberance stemming anteriorly from prothoracic leg; in leaf miners sclerotized plates sometimes present on pronotum and thoracic sterna, and stiff setae may be present on inner surface of each coxa. *Abdomen:* Prolegs present on segments 2–8 and 10, 2–7 and 10 in some leaf miners, or reduced and absent in some leaf miners; abdominal segments 1–8 each with two to six dorsal annulets, fewer in leaf miners, with small protuberances on annulets 2 and 4 and on spiracular and surpedal lobes in *Endelomyia,* but without ornamentation in most other species; some leaf miners have darkened sclerotized plates surrounding prolegs and anal proleg and on basal sterna; tenth segment without caudal projections or appendages.

Comments: Twelve genera and 38 species of Heterarthrinae occur in North America. *Endelomyia* and *Caliroa* occur throughout the continent, as well as some leaf-mining genera, but most species are eastern. Because of their economic importance, larvae are known for a higher percentage of species than most sawflies. Keys to larvae, including separate keys to the leaf miners of oak and birch, life history data, and references to literature are found in Smith (1971c).

Selected Bibliography

Smith 1971c.

Subfamily Blennocampinae (Tenthredinidae)

Figures 27.53–27.57

Relationships and Diagnosis: All larvae are typically sawflylike except for those of *Ardis,* which are internal shoot borers and have reduced prolegs. Larvae are separated from other tenthredinid larvae by the conical, five-segmented antenna, absence of a mesal ridge on the left mandible (if present then not attached to other ridges), five-segmented thoracic legs each with a tarsal claw, prolegs on abdominal segments 2–8 and 10, reduced in *Ardis,* abdominal segments 1–8 each with four to six dorsal annulets, with setae, tubercles, or spines only on annulets 2 and 4, and tenth segment without caudal protuberances or projections except for ornamentation similar to rest of abdomen. Larvae of some groups have long, simple or furcate spines (figs. 27.55a, 27.56a, 27.57a), not found in other larvae of Tenthredinidae. Those without spines are most similar to larvae of Allantinae, but Allantinae larvae have a mesal ridge on the left mandible and usually a raised mesal area, with the ridge connected to the ventral cutting margin, six-annulate abdominal segments, and some have setae or tubercles on the first annulet of the abdominal segments in addition to annulets 2 and 4.

Biology and Ecology: Blennocampinae are external leaf feeders except for species of *Ardis,* which are shoot borers, and *Blennogeneris* and *Lycaota,* which live and feed in the buds of their hosts. Some important species are the brownheaded ash sawfly, *Tomostethus multicinctus* (Rohwer) (figs. 27.54a-c), the blackheaded ash sawfly, *Tethida barda* (Say), the grape sawfly, *Erythraspides vitis* (Harris) (figs. 27.56a-c), and the raspberry sawfly, *Monophadnoides geniculatus* (Hartig) (figs. 27.57a-c). Host plant genera for some genera of Blennocampinae are as follows: ***Blennogeneris*** (*Symphoricarpos*); ***Tomostethus*** (*Fraxinus*); ***Tethida*** (*Fraxinus*); ***Phymatocera*** (*Smilacina*); ***Paracharactus*** (*Carex*); ***Rhadinoceraea*** (*Veratrum, Calochortus*); ***Lagonis*** (*Sambucus*); ***Monophadnus*** (*Ranunculus*); ***Stethomostus*** (*Ranunculus*); ***Eutomostethus*** (*Poa, Juncus*); ***Ardis*** (*Rosa*); ***Apareophora*** (*Spiraea*); ***Eupareophora*** (*Fraxinus*); ***Periclista*** (*Quercus, Carya*); ***Monophadnoides*** (*Rubus*); ***Erythraspides*** (*Oenothera, Vitis*); ***Halidamia*** (*Galium*).

Most species have a single generation a year. Overwintering and pupation is in a cell in the ground, sometimes a papery cocoon is spun.

Description: *Head:* Globose, with scattered or numerous hairs or setae; antenna five-segmented, conical; left mandible with or without mesal ridge, if present, then not connected at base to dorsal or ventral cutting edges. *Thorax:* Legs five-segmented, each with tarsal claw; prothoracic surpedal lobe of several species protuberant; ornamentation as for abdomen. *Abdomen:* Prolegs on segments 2–8 and 10, reduced in *Ardis;* segments 1–8 each with four to six dorsal annulets, usually six, but indistinctly four to five in some species that have long spines on the body; setae, tubercles, simple or branched spines only on annulets 2 and 4, spiracular and surpedal lobes, and tenth segment; tenth segment without appendages. Those species that have spines in the feeding stages lack spines in the prepupal stage.

Comments: Twenty-one genera and 72 species occur in North America; representatives are found throughout the continent. Larvae are known for most genera, but only a small percentage have adult associations. Smith (1966a) identified three shoot boring sawfly larvae of roses, and Smith (1969a) revised the subfamily for North America, giving keys to larvae, notes on life histories, and literature.

Selected Bibliography

Smith 1966a, 1969a.

Subfamily Allantinae (Tenthredinidae)

Figures 27.58–27.62

Relationships and Diagnosis: Larvae are typically sawflylike and separated from other tenthredinid larvae by the conical, five-segmented antenna, mesal ridge and usually raised mesal area of the left mandible, the ridge connected with the ventral cutting edge (fig. 27.58d), five-segmented thoracic legs each with a tarsal claw (fig. 27.61a), prolegs on abdominal segments 2–8 and 10, abdominal segments 1–8 each with six dorsal annulets, with setae or small tubercles on annulets 2 and 4 and sometimes 1 (fig. 27.59b). Allantinae larvae are closest to the Blennocampinae that lack long spinelike armature on the body, but the Allantinae have the mesal ridge on the left mandible as described above, and sometimes have setae or tubercles on annulet 1 of the abdominal segments. Larvae of Blennocampinae never have setae or tubercles on the first annulet.

Biology and Ecology: All larvae are external leaf feeders. They commonly secrete a white bloom that covers the body and blends in with the foliage. The bloom of *Eriocampa* larvae is extensive and may be in the form of long, white, flaky material sometimes exceeding the width of the larva itself. Few species spin cocoons, most boring into a substance to form a pupal cell. Because of this habit, they are sometimes more destructive to this secondary host than the plant on which they actually feed. Such is the case with the dock sawfly, *Ametastegia glabrata* (Fallén) (figs. 27.62a-c) and some other species of *Ametastegia;* if their host, *Rumex,* is near apple orchards, they commonly bore in apples to form a pupal cell, a habit of more concern than their feeding on the host plant.

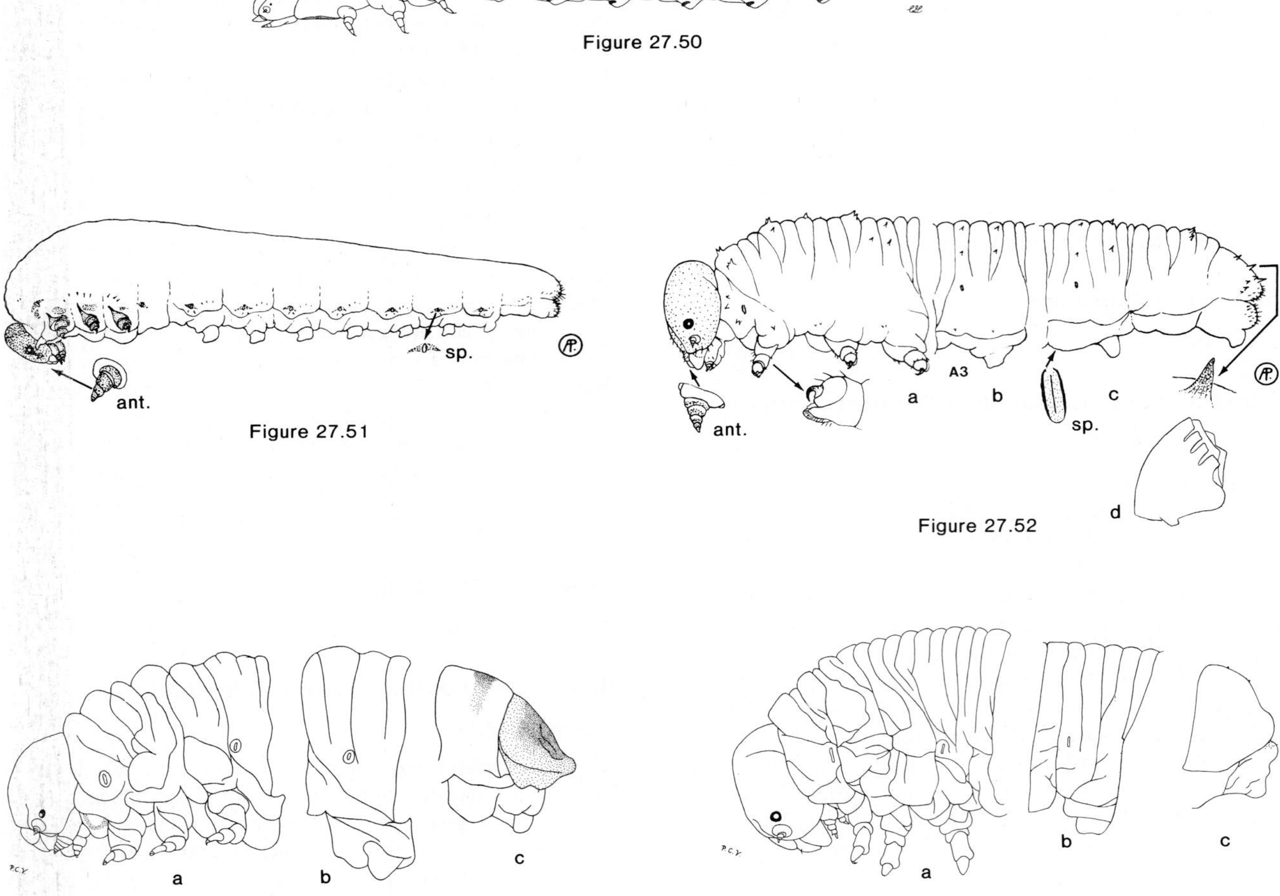

Figure 27.50

Figure 27.51

Figure 27.52

Figure 27.53

Figure 27.54

Figure 27.50. Tenthredinidae: Heterarthrinae. *Messa leucostoma* (Rohwer). f.g.l. 7-9 mm. A leaf miner probably of *Populus;* transcontinental in North America. Typical habitus of leaf-mining larvae; dorsoventrally flattened, head prognathous; white with dark, sclerotized plates on thoracic terga and sterna and sometimes surrounding prolegs; prolegs reduced or absent; abdominal segments sometimes without annulations (Smith 1971c).

Figure 27.51. Tenthredinidae: Heterarthrinae. *Caliroa cerasi* (Linnaeus), pear sawfly. f.g.l. 10-12 mm. Sluglike in habitus, thorax enlarged, tapering caudally; anterior margin of thorax overlaps posterior margin of head; fingerlike protuberance extending anteriorly from prothoracic leg; body covered with dark slimelike substance. Feeds on underside of leaves, skeletonizing pear, cherry, and other rosaceous plants, sometimes destructive in orchards. Occurs throughout North America (Smith 1971c; from Peterson 1948).

Figure 27.52a-d. Tenthredinidae: Heterarthrinae. *Endelomyia aethiops* (Fabr.), roseslug. a. head, thorax; b. A3 segment; c. caudal segments; d. right mandible, ventral view. f.g.l. 8-10 mm. Yellowish-green; mandibles distinctive, right mandible with 4 truncate teeth on ventral cutting edge; a few, short, distinct, conical projections on some annulets of all segments, including A10. Skeletonizes leaflet, usually feeding on under side. Feeds on most varieties of roses (Smith 1971c) (figs. 27.52a,b,c, from Peterson 1948).

Figure 27.53a-c. Tenthredinidae: Blennocampinae. *Ardis brunniventris* (Hartig). a. head, thorax, A1 segment; b. A3 segment; c. caudal segments. f.g.l. 10-15 mm. Whitish with head and 10th tergum amber. A shoot-boring larva of wild and cultivated roses, atypical of the Blennocampinae in that prolegs are absent. Transcontinental in North America (Smith 1966a, 1969a).

Figure 27.54a-c. Tenthredinidae: Blennocampinae. *Tomostethus multicinctus* (Rohwer), brownheaded ash sawfly. a. head, thorax, A1 segment; b. A3 segment; c. caudal segments. f.g.l. 14-18 mm. Greenish when alive, with brown head. Feeds on foliage of ashes and occurs in eastern North America and Oregon and California (Smith 1969a).

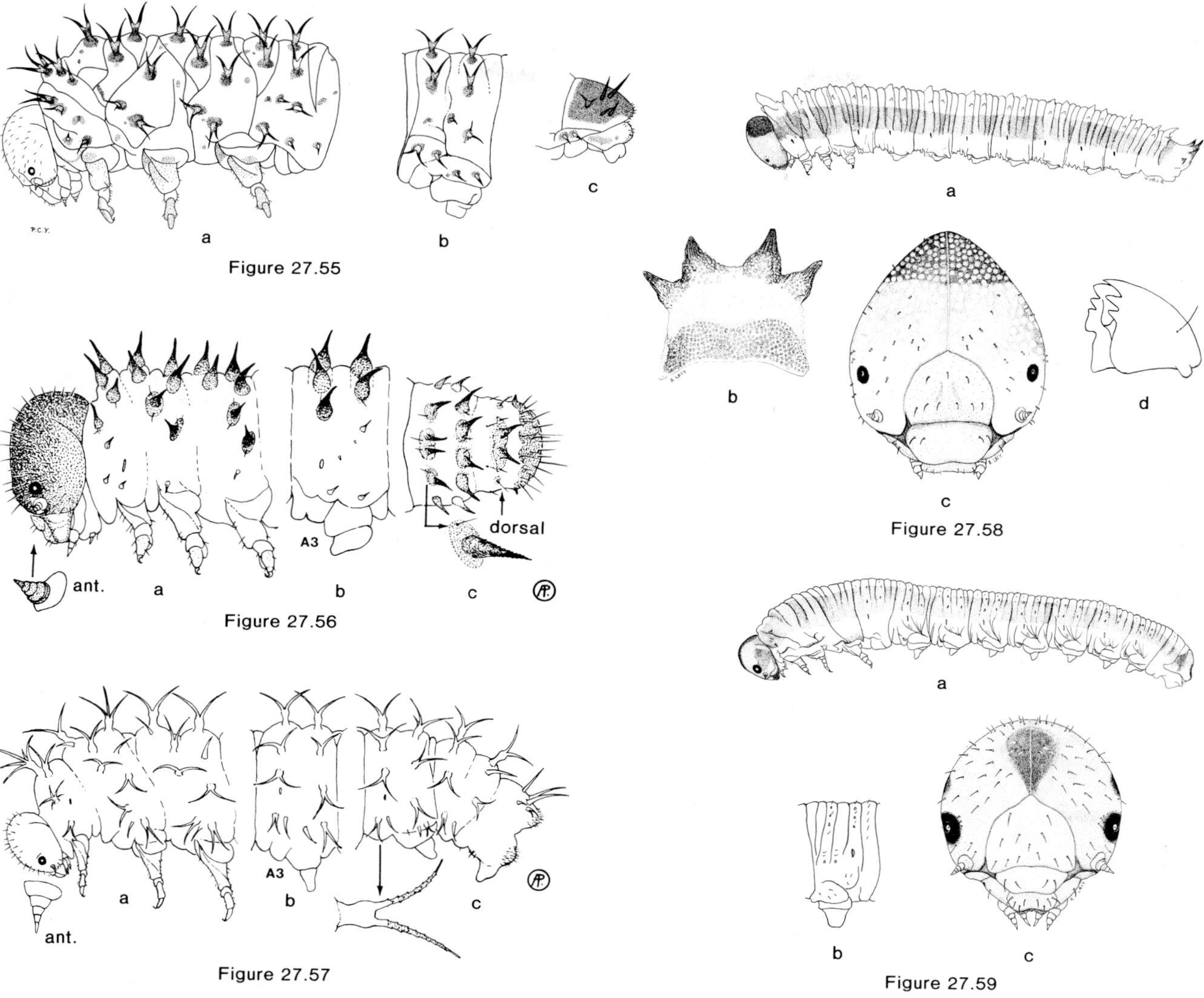

Figure 27.55a-c. Tenthredinidae: Blennocampinae. *Periclista linea* Stannard. a. head, thorax, A1 segment; b. A3 segment; c. caudal segments. f.g.l. 10-15 mm. Body armature of branched spines, arranged in definite pattern on body, note the 1 bifurcate and 1 simple spine on the subspiracular lobe and the 2 simple spines on the surpedal lobe, which are typical for *Periclista* larvae and help to separate them from other Blennocampinae with spine-like armature. Feeds on foliage of oaks in Oregon and California; other *Periclista* species are on oaks throughout North America (Smith 1969a).

Figure 27.56a-c. Tenthredinidae: Blennocampinae. *Erythraspides vitis* (Harris), grape sawfly. a. head and thorax; b. A3 segment; c. caudal segments, dorsal view. f.g.l. 12-16 mm. Greenish with blackish head and body tubercles. Body tubercles are simple, conical spines, mostly located on the dorsum of the body. Larvae are gregarious, feeding on foliage of grape, and occur in eastern North America (Smith 1969a, from Peterson 1948).

Figure 27.57a-c. Tenthredinidae: Blennocampinae. *Monophadnoides geniculatus* (Hartig), raspberry sawfly. a. head and thorax; b. A3 segment; c. caudal segments. f.g.l. 18-22 mm. All greenish. Long-branched spines on body; note subspiracular lobe and surpedal lobe, each with 1 branched and 1 simple spine. Feeds on various species of *Rubus* and occurs throughout North America (Smith 1969a, from Peterson 1948).

Figure 27.58a-d. Tenthredinidae: Allantinae. *Dimorphopteryx virginicus* Rohwer. a. entire larva; b. 10th tergum, dorsal; c. head; d. left mandible, ventral view. f.g.l. 13-17 mm. Greenish when alive with dark lateral band and 10th tergum, apical protuberances of 10th tergum darker; head amber with black spot above; T1 and T2 with protuberances; 4 conical protuberances on 10th tergum; head conical in frontal view. Feeds on foliage of chestnut in eastern North America. A typical larva of the genus *Dimorphopteryx (Smith 1979a).*

Figure 27.59a-c. Tenthredinidae: Allantinae. *Empria multicolor* (Norton). a. entire larva; b. A3 segment; c. head. f.g.l. 12-16 mm. Body greenish when alive with light brown lateral band, 10th tergum light brown, and head amber with dark spot above frons and one posterior to each eye; A1-A8 each 6-annulate with short setae on annulets 1, 2, and 4. Feeds on foliage of birch and alder and is transcontinental in North America. The presence of setae on annulet 1 helps to separate larvae of *Empria* (Smith 1979a).

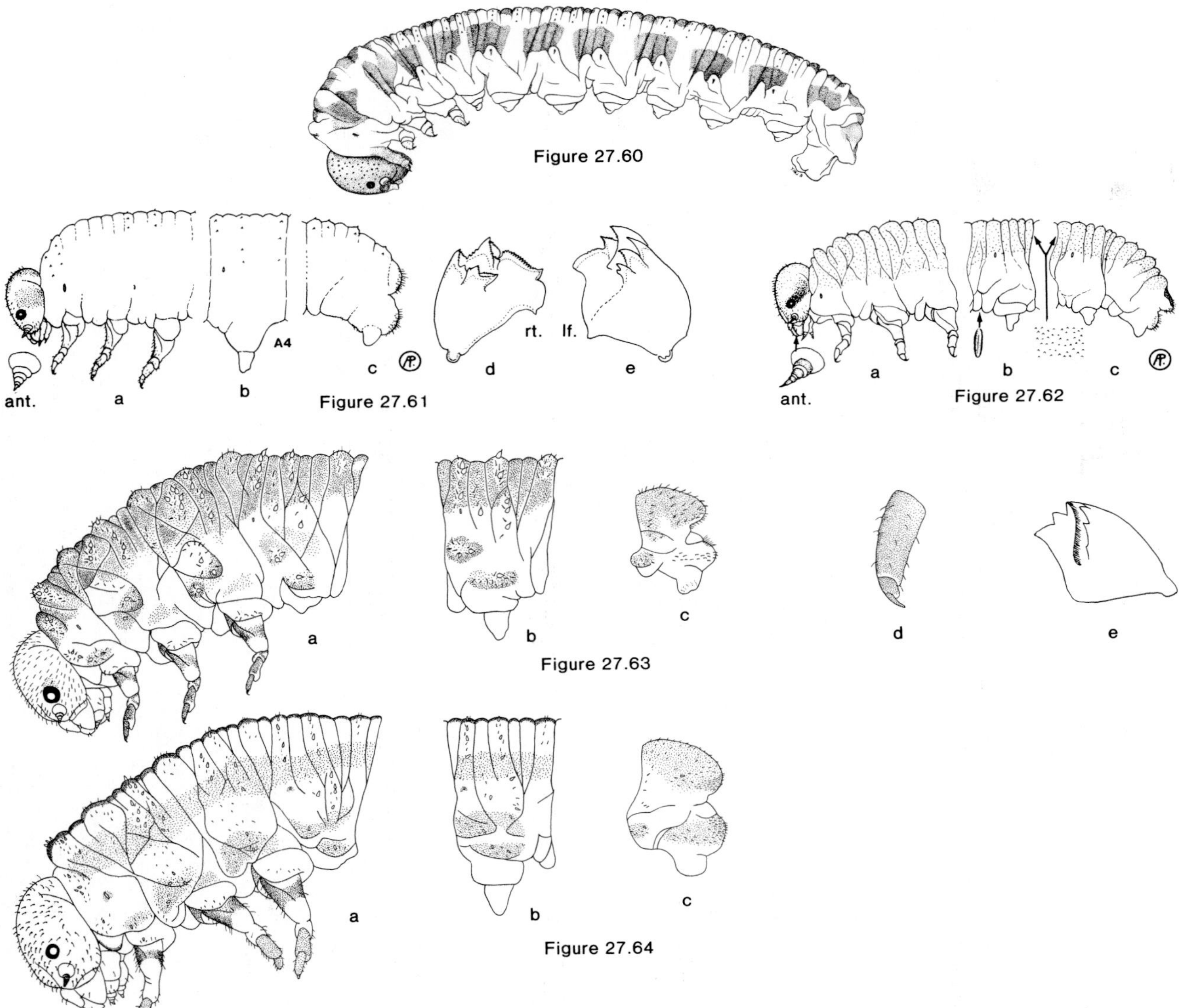

Figure 27.60. Tenthredinidae: Allantinae. *Macremphytus testaceus* (Norton), entire prepupa. f.g.l. 22–30 mm. Prepupa greenish when alive with brown to black dorsal and lateral spots, 10th tergum brownish, and head dark brown; setae on annulets 2 and 4 of A1-A8; head with scattered pits. Late-feeding stage larvae are unicolorous greenish with head dark brown to black, and head lacking pits. Feeds on foliage of dogwood in eastern North America; commonly bores in wood to form a pupal cell (Smith 1979a).

Figure 27.61a-e. Tenthredinidae: Allantinae. *Allantus cinctus* (L.) curled rose sawfly. a. head, thorax; b. A4 segment; c. caudal segments; d. right mandible; e. left mandible, ventral views. f.g.l. 15± mm. Yellow-green, in a coiled position when feeding; structure of the mandibles separates this species from *Endelomyia aethiops* (fig. 27.52d). Without prominent setae. Tiny conelike projections occur on annulets 1, 2 and 4 of A1-9. Early instars skeletonize patches on under side of leaf. Later instars eat holes in leaves and may consume entire leaflet except for largest veins. Feeds on wide variety of roses (Smith 1979a; from Peterson 1948).

Figure 27.62a-c. Tenthredinidae: Allantinae. *Ametastegia glabrata* (Fallén), dock sawfly. a. head, thorax; b. A4 segment; c. caudal segments. f.g.l. 14–17 mm. Slender, setae inconspicuous, olive green to blue-green dorsad of spiracles with rows of setigerous, usually whitish tubercles on some annulets; venter light yellow; pigment spot on A10; head light brown, minutely punctured and with transverse, darker bands extending between stemmata and caudal margin of head. Food plants primarily docks and sorrels (*Rumex*). Skeletonizes and also may consume large portions of leaves. It may bore into apples for a pupation site. (Smith 1979a; from Peterson 1948).

Figure 27.63a-e. Tenthredinidae: Tenthredininae. *Tenthredo* sp. a. head, thorax, A1 segment; b. A3 segment; c. caudal segment; d. T1 tibia and tarsal claw; e. left mandible, ventral view. f.g.l. usually greater than 20 mm. Typical larva of the genus, note especially the 7-annulate abdominal segments, commonly with setae or small tubercles on annulets 2, 3, 4, and 7. The genus is found throughout North America, hosts being known for very few.

Figure 27.64a-c. Tenthredinidae: Tenthredininae. *Filacus pluricinctus* (Norton). a. head, thorax, A1 segment; b. A3 segment; c. caudal segment. f.g.l. 18–24 mm. Greenish when alive with darker dorsal and lateral bands, complete or broken; A1-A8 each 7-annulate with setae and/or tubercles on annulets 2, 4, and 7. Feeds on *Phacelia* and occurs in California, Arizona, and northern Mexico.

Some species of *Macremphytus* are known to bore into sound wood to form a pupal cell. Other species of importance are *Empria maculata* (Norton) and *E. obscurata* (Cresson), which feed on strawberry foliage; the violet sawfly, *Ametastegia pallipes* (Spinola); and the curled rose sawfly, *Allantus cinctus* (Linnaeus) (figs. 27.61a-e). Host plant genera for some genera of Allantinae are as follows: ***Eriocampa*** (*Juglans, Alnus*); ***Pseudosiobla*** (*Cephalanthus*); ***Dimorphopteryx*** (*Prunus, Pyrus, Amelanchier, Crataegus, Betula, Quercus, Alnus, Castanea*); ***Empria*** (*Betula, Corylus, Salix, Fragaria, Potentilla, Rubus, Alnus*); ***Phrontosoma*** (*Cornus*); ***Ametastegia*** (*Rumex, Polygonum, Viola, Salix*); ***Monosoma*** (*Alnus*); ***Monostegia*** (*Lysimachia*); ***Aphilodyctium*** (*Rosa*); ***Allantus*** (*Rosa, Fragaria, Rubus, Betula*); ***Macremphytus*** (*Cornus*); ***Taxonus*** (*Fragaria, Rubus*).

Species have one or more generations a year. Oviposition may be in the stems or foliage of the host. After feeding, larvae may make earthen cells in the soil, or make cells in fruit, stems, bark, wood, or most any other nearby suitable substance.

Description: *Head:* Globose, with numerous or scattered setae or hairs, rounded in front view, but conical in *Dimorphopteryx* (fig. 27.58c); antenna conical, five-segmented; left mandible with mesal ridge and elevated mesal area, ridge connected to ventral cutting edge. *Thorax:* Legs five-segmented, each with tarsal claw; ornamentation as for abdomen; in *Dimorphopteryx*, long fleshy protuberances on pro- and mesothorax. *Abdomen:* Prolegs on segments 2–8 and 10; segments 1–8 each with six dorsal annulets; small tubercles or setae on annulets 2 and 4, sometimes 1, and on spiracular and surpedal lobes and tenth tergum, never with long simple or branched spines; tenth tergum rounded, without protuberances or appendages, only *Dimorphopteryx* with four fleshy caudal protuberances.

The prepupal stage often changes in color and structure. In addition to the different mandibular dentition, the prepupa of *Macremphytus* has very different coloration, and the head has scattered shallow pits, which are absent in the feeding stages.

Comments: There are 15 genera and 64 species in North America, the largest number occurring in the East. Representative larvae for most genera are known, but immatures of most species are not known. For descriptions and keys to larvae, life history data, hosts, and references to literature, see Smith (1979a).

Selected Bibliography

Smith 1979a.

Subfamily Tenthredininae (Tenthredinidae)

Figures 27.63–27.65

Relationships and Diagnosis: Larvae have seven dorsal annulets on abdominal segments 1–8 (figs. 27.63b, 27.64b, 27.65b), and, by this, are separated from all other Symphyta except for Selandriinae and Cimbicidae. From the Cimbicidae, tenthredinine larvae are separated by the conical, usually five-segmented antenna and lack of secondary longitudinal sutures on the labrum; from the Selandriinae, they are separated by the 13–18 setae on the tibiae (fig. 27.64a) (ten or fewer in Selandriinae) and by the left mandible, which has three or four lateral teeth and one mesal ridge (fig. 27.63e) (five lateral teeth and two mesal ridges in Selandriinae). All larvae are typically sawflylike and usually greenish with various dark stripes or spots on the head and body.

Biology and Ecology: All are external leaf feeders, but few are considered important. Hosts and biologies are known for very few species, a very unusual circumstance for the second largest subfamily of Tenthredinidae. Hosts for some Holarctic species are recorded, but these records are from the European literature. Hosts recorded in North America are *Podophyllum, Sambucus,* and *Impatiens* for species of ***Aglaostigma;*** *Sambucus* for a species of ***Lagium;*** *Populus* for species of ***Rhogogaster.*** For ***Macrophya,*** Gibson (1980) recorded plants of the genera *Rudbeckia, Carya, Prunus, Fraxinus, Ligustrum, Syringa, Castanea, Viburnum,* and *Sambucus.* Smith and Kido (1949) gave some biological data for *Tenthredo xantha* Norton on *Rubus* in California. When known, there is a single generation a year, and overwintering and pupation is in an earthen cell or cocoon in the soil.

Description: *Head:* Globose, usually with numerous setae; antenna conical, usually five-segmented; left mandible with three or four lateral teeth and one mesal ridge. *Thorax:* Legs five-segmented, each with a tarsal claw; tibiae with 13–18 setae; ornamentation as for abdomen. *Abdomen:* Prolegs present on segments 2–8 and 10; segments 1–8 each with seven dorsal annulets; ornamentation may consist of hairs, setae, or small tubercles; tenth segment without caudal protuberances.

Comments: In North America, ten genera and about 150 species have been listed (Smith 1979b). *Aglaostigma, Rhogogaster, Tenthredo,* and *Macrophya* are the largest and most widespread genera. Representatives are found throughout the continent; one or two species extend into Mexico, but none is known from south of Mexico.

Larvae are difficult to identify to genus because of the lack of reared material and paucity of adult associations. Larval keys to genera or known species have yet to be worked out.

Selected Bibliography

Gibson 1980.
Smith 1979b.
Smith and Kido 1949.

SIRICIDAE (SIRICOIDEA)

David R. Smith, ***Systematic Entomology Laboratory, ARS, USDA***
Woodrow W. Middlekauff, ***University of California, Berkeley***

Horntails

Figures 27.66, 27.67

Relationships and Diagnosis: Larvae of Siricidae, Xiphydriidae, Anaxyelidae, and Cephidae are similar in that each has reduced, mammaform thoracic legs each without a claw (fig. 27.67b), lacks prolegs (fig. 27.66b), and has a postcornus, a sclerotized, suranal, caudal projection on the tenth tergum (fig. 27.66b). Siricidae larvae are distinguished from the other three families by the following: Antenna one- to three-segmented (figs. 27.66a, 27.67a), with numerous short setae on basal segment; metathoracic spiracle present (figs. 27.66b, 27.67b); vertical furrow indistinct; epicranial sutures absent; abdominal segments without annulations (figs. 27.66b, 27.67b); without subanal appendages; and lack of a ring of hairs at the base of the postcornus. Mature larvae are large, 30–40 mm or more long, usually much larger than those of other internally boring larvae, cylindrical, sometimes weakly S-shaped, and white with postcornus dark brown.

Biology and Ecology: All larvae are woodborers in coniferous or angiospermous trees; the Siricinae (*Sirex, Urocerus, Xeris*) in coniferous trees, and the Tremecinae (*Tremex, Eriotremex*) in angiospermous trees. Siricinae have been taken from *Larix, Juniperus, Cupressus, Sequoia, Pinus, Calocedrus, Pseudotsuga, Chamaecyparis, Abies, Tsuga,* and *Picea;* Tremecinae have been recorded from *Quercus, Malus, Fagus, Betula, Acer, Ulmus, Carya, Pyrus, Platanus, Celtis,* and *Cephalanthus.*

Although the literature on the biology of siricids is widely scattered, much of it has been devoted to a few species that have been imported, especially into New Zealand and Australia. Little study has been devoted to North American species (Morgan 1968).

Most adults emerge in late summer and early autumn when there is usually an abundant supply of suitable hosts resulting from fires, disease, forest insect outbreaks, or other stress factors that kill or weaken forest trees. Females occasionally select a tree that to all appearances is still healthy. Most species lay their eggs after drilling 2–15 mm into the xylem, but others may oviposit in wood or bark. The eggs of *Tremex* are black, those of other genera are white.

Larvae bore in the wood until they are nearly mature, at which time they construct a pupal chamber near the surface. The mines of *Sirex* and *Urocerus* are 5–20 cm long, those of *Tremex* much longer, up to 3 m in length. Larvae have a variable number of instars (5–11) depending upon conditions. The life cycle of some species may be completed in one year under favorable conditions; other species require two or more years. A number of generations may be present in a single log at any one time, which greatly complicates understanding the composition of the population.

Horntails are of additional interest because of the symbiotic relationship of *Sirex, Urocerus,* and *Tremex* with certain wood-destroying fungi. The fungi are commonly found apart from the siricids, but with the exception of *Xeris* (Francke-Grosmann 1939), all adult females examined have had associated fungi. No fungal sacs have been found in adult or larval males. It is believed the eggs are deposited with oidea of a wood-destroying symbiotic fungus upon which the larvae subsist. Larvae do not ingest wood, apparently extracting available nutrients from the fungal mycelium, which is destroyed by secreted saliva. The salivary secretions and dissolved nutrients are ingested by larvae that pass the fragments of wood along the outside of their smooth bodies to join the accumulation of frass behind them. The fungi are carried in paired epidermal glands (hypopleural organs) between abdominal segments 1 and 2 of female larvae (Parkin 1942, Stillwell 1965).

In the United States, siricids are not as a rule considered to be primary enemies of sound timber because they usually do not attack healthy trees. They will, however, readily attack fire-damaged or injured or dying trees and ones recently felled. Female *Sirex areolatus* (Cresson) have even been reported ovipositing in recently sawed redwood lumber. Larval and adult borings reduce the quality of the lumber, but possibly their greatest economic impact in North America occurs when adults of these large and impressive insects emerge in newly built houses. Having completed their larval life in lumber such as wall studs and subflooring, they bore through plaster walls, hardwood floors or other impediments to freedom. Use of infested salvage lumber, frequently available as a result of forest fires, exacerbates the problem.

Description: *Head:* Globular, with scattered short setae; antenna one- to three-segmented, with numerous setae on 1st segment; vertical furrows present but indistinct; epicranial sutures absent; stemmata absent; maxillary palp one-segmented, labial palp one- to two-segmented; each mandible similar. *Thorax:* Legs mammaform, unsegmented, clawless; metathoracic spiracle present, as large as those on abdomen. *Abdomen:* Prolegs absent; segments not annulated; sclerotized postcornus at apex of tenth tergum; subanal appendages absent; without ring of hairs at base of postcornus; with or without small fleshy protuberances on each side of longitudinal groove on tenth tergum at base of postcornus; tenth segment with various amounts of short, stout setae.

Comments: The Siricidae are widespread and found as far south as Guatemala and Chile. Five genera and about 15 species occur in North America. The Siricinae are coextensive with the coniferous forests and the Tremecinae are mainly associated with the eastern deciduous forests. *Eriotremex formosanus* (Matsumura), adventive from southeastern Asia, is known only from the Southeast.

There is no complete recent revision of North American species. Bradley (1913), though out of date, is the only one available. More recent treatments are on the California fauna by Middlekauff (1960), a world catalog including host lists and literature by Smith (1978), the North American catalog by Smith (1979b), and a key to North American genera by Smith (1975b). Smith (1975b) also recorded *Eriotremex* from North America.

Because of their habits of living in wood and sometimes taking several years to complete their life cycle, Siricidae are subject to distribution by commerce throughout the world. Larvae are commonly intercepted by U.S. quarantine inspectors in crates, dunnage, and other wood products from abroad. Though larvae are commonly found, characters have not been found to separate genera or species, and much of the difficulty is due to the fact that very few adult associations have been made.

Selected Bibliography

Bradley 1913.
Francke-Grosmann 1939.
Middlekauff 1960.
Morgan 1968.
Parkin 1942.
Smith 1975b, 1978, 1979b.
Stilwell 1965.

XIPHYDRIIDAE (SIRICOIDEA)

David R. Smith, ***Systematic Entomology Laboratory, ARS, USDA***
Woodrow W. Middlekauff, ***University of California, Berkeley***

Xiphydriid Wood Wasps

Figure 27.68

Relationships and Diagnosis: Larvae are similar to those of other internally boring families, Siricidae, Anaxyelidae, and Cephidae, but separated by the following: antenna three- or four-segmented, first segment without setae (fig. 27.68a); epicranial sutures indistinct; metathoracic spiracle vestigial (fig. 27.68b); indistinct annulations on basal three abdominal segments, remaining segments without annulets (fig. 27.68b); ring of hairs at base of postcornus (fig. 27.68c); without small fleshy protuberances on tenth tergum at each side of furrow at base of postcornus; and without subanal appendages. Larvae are whitish with the head sometimes amber and the postcornus dark brown, cylindrical and slightly S-shaped, and usually more slender than the Siricidae. Like other families of wood and stem borers, the thoracic legs are mammaform, each without a claw, and prolegs are absent (fig. 27.68b).

Biology and Ecology: All are wood borers in angiospermous trees and shrubs, though there is a questionable record from conifers. Hosts for the single North American genus, *Xiphydria* are *Tilia, Carpinus, Carya, Ulmus, Acer, Malus, Betula, Fraxinus, Quercus, Rhus, Fagus, Prunus,* and *Populus.*

Xiphydriids are considered to be of little economic importance because the larvae bore in dead and dying branches or small limbs of shrubs and trees, and are probably secondary to other factors causing mortality. Larger limbs and trunks are apparently rarely attacked. Adults occasionally arouse concern, however, when found in buildings, mostly the result of emerging from firewood.

Little is known concerning the life history of North American species. A symbiotic fungus has been associated with larvae of *Xiphydria* (Buchner 1965) and raises the interesting possibility that *Xiphydria,* like siricids, do not consume and digest wood, but rather subsist upon fungi.

Description: *Head:* Semiglobose, sometimes slightly oblong in frontal view; antenna conical, three- to four-segmented, without setae on first segment; labial palp three-segmented, maxillary palp two-segmented; epicranial sutures indistinct; vertical furrows absent; stemmata absent, without pigmented spot. *Thorax:* Legs mammaform, unsegmented, clawless; metathoracic spiracle vestigial. *Abdomen:* Prolegs absent; segments without annulations though sometimes two or three indistinct annulets on basal 3 segments; without appendages on tenth sternum; with postcornus and with ring of hairs at base of postcornus; tenth tergum with longitudinal groove; without fleshy protuberances at base of postcornus on each side of groove.

Comments: The family was revised by Smith (1976) and includes a single genus, *Xiphydria,* with nine species in North America. All are associated with the eastern deciduous forests; only one species, *X. mellipes* Harris, extends west to British Columbia and Oregon. Species identification of larvae is not yet possible due to inadequate adult associations.

Selected Bibliography

Buchner 1965.
Smith 1976.

ANAXYELIDAE (SIRICOIDEA)

David R. Smith, ***Systematic Entomology Laboratory, ARS, USDA***
Woodrow W. Middlekauff, ***University of California, Berkeley***

Syntexid Wood Wasps

Figure 27.69

Relationships and Diagnosis: Related to larvae of other wood boring families that possess a postcornus, Siricidae, Xiphydriidae, and Cephidae, but separated by the following: antenna three-segmented without setae on basal segment; frontal sutures indistinct (fig. 27.69d); stemmata absent; metathoracic spiracle present (fig. 27.69a); tenth tergum with pair of small fleshy protuberances one on each side of a furrow at base of postcornus; subanal appendages absent. The mature larva is small, less than 15 mm long, white except for the dark brown postcornus, cylindrical, weakly S-shaped, lacks prolegs, and has mammaform thoracic legs.

Biology and Ecology: The single species, *Syntexis libocedrii* Rohwer, is a wood borer of *Calocedrus, Juniperus,* and *Thuja.* Adults were rare in collections until 1963 when Wickman (Middlekauff 1964b) discovered numerous adults hovering near and ovipositing in the trunks of *Calocedrus* following a forest fire in northern California. The tree trunks were still warm and smoking at the time the collections were made. This appears to be a characteristic behavior pattern and helps explain the paucity of specimens in collections. With this information on their biology, additional specimens were readily taken by Wickman and others in California and subsequently by Westcott in Oregon. Biological notes were published by Wickman (1967) and Westcott (1971).

For over 50 years it has been assumed that the incense cedar, *Calocedrus decurrens,* is the sole host of *Syntexis;* however, Westcott (1971) reared specimens from naturally infested western juniper, *Juniperus occidentalis,* and additionally found females ovipositing in western red cedar, *Thuja plicata.*

Description: *Head:* Subglobose with sparse minute setae; epicranial suture apparently obsolete, vertical furrow not visible; stemmata absent, in dried specimens a faint, brown ocularium may be visible; antenna three-segmented, without setae on basal segment; labial palp two-segmented; maxillary palp three-segmented; right mandible with three teeth (fig. 27.69c), left mandible with four teeth. *Thorax:* Swollen, prothorax covers back portion of head; legs mammaform, clawless, with few minute setae; metathoracic spiracle present. *Abdomen:* Prolegs absent; integument smooth, microscopically densely spinulate on venter; tenth tergum depressed by median furrow;

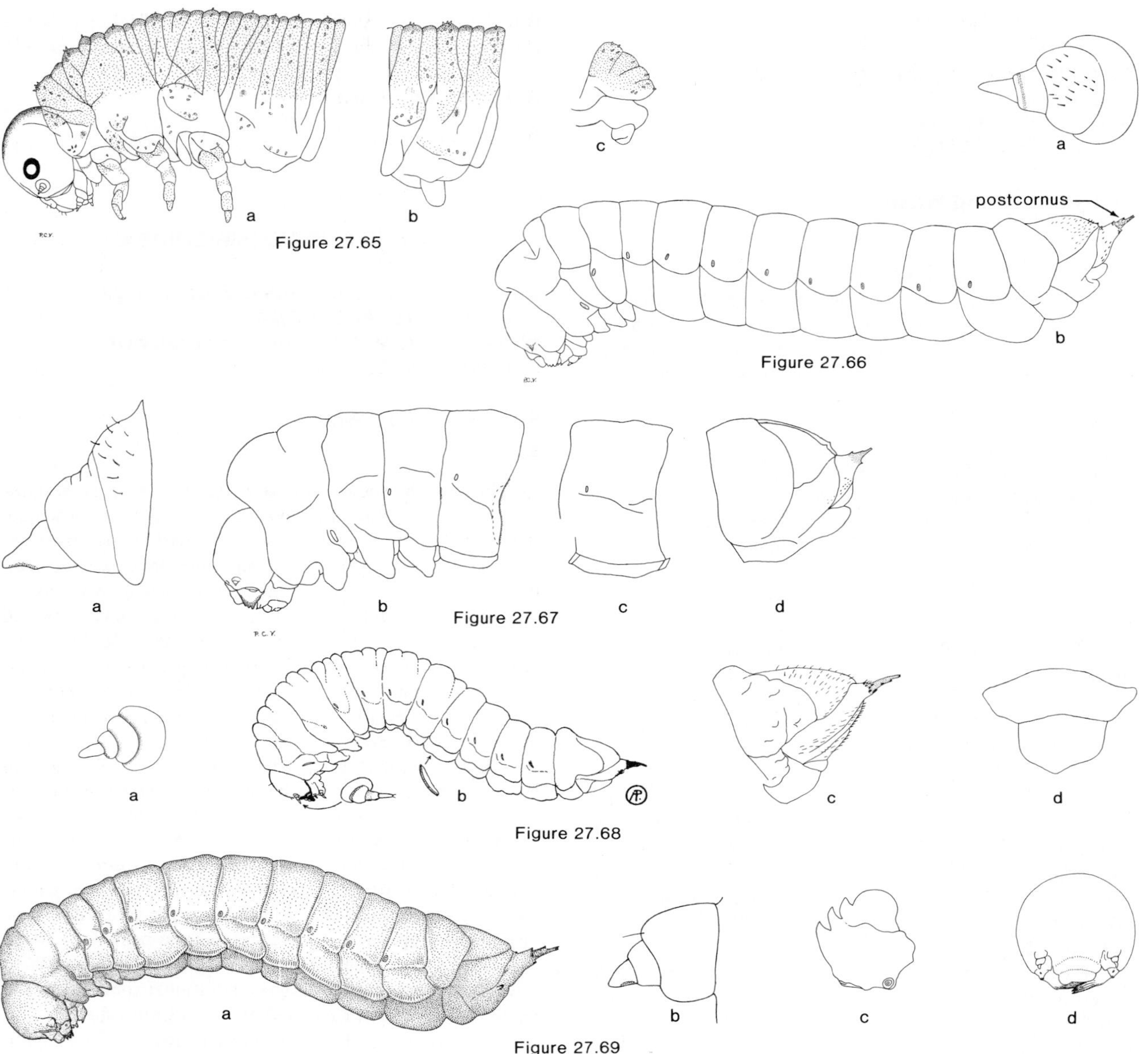

Figure 27.65a-c. Tenthredinidae: Tenthredininae. *Aglaostigma* sp. a. head, thorax, A1 segment; b. A3 segment; c. caudal segment. Typical of larvae of the Tenthredininae, with A1-A8 each 7-annulate; *Aglaostigma* larvae can be separated from most other Tenthredinine larvae by having only 1 seta on the outer surface of each mandible instead of 2. The genus is found throughout North America, known hosts are *Sambucus, Impatiens* and *Podophyllum.*

Figure 27.66a,b. Siricidae. *Sirex cyaneus* Fabricius, blue horntail. a. antenna; b. entire larva. f.g.l. 30-40 mm. Larva typical of Siricidae. *Sirex cyaneus,* a wood borer, has been recorded from fir, larch, spruce, pine and Douglas fir, and is widespread in North America.

Figure 27.67a-d. Siricidae. *Tremex columba* (Linnaeus), pigeon tremex. a. antenna; b. head, thorax, A1 segment; c. A3 segment; d. caudal segments. f.g.l. 30-40 mm. Larva typical of Siricidae. A wood borer of deciduous trees and occurs in most of North America except for the far western states.

Figure 27.68a-d. Xiphydriidae. *Xiphydria maculata* Say. a. antenna; b. entire larva; c. caudal segments; d. clypeus and labrum. f.g.l. 20-30 mm. Larva typical for Xiphydriidae. This species is most commonly a wood borer in small branches and limbs of maples in eastern North America (Smith 1976; fig. 27.68b from Peterson 1948).

Figure 27.69a-d. Anaxyelidae. *Syntexis libocedrii* Rohwer, incense-cedar wasp. a. entire larva; b. antenna; c. right mandible, inner face; d. head. f.g.l. 10-15 mm. Wood borer in incense-cedar and juniper in Oregon and California. See discussion of family (Middlekauff 1974).

no hypopleural organs (as in Siricidae); no subanal appendages; postcornus directed dorsally, without ring of hairs at base; tenth tergum with small protuberances at base of postcornus, one on each side of longitudinal furrow; segments without annulations.

Comments: *Syntexis libocedrii* occurs in California, Oregon, and Idaho. It is the only extant member of an otherwise fossil family (Smith 1978). The larva was described by Middlekauff (1974).

Selected Bibliography

Middlekauff 1964b, 1974.
Smith 1978.
Westcott 1971.
Wickman 1967.

ORUSSIDAE (SIRICOIDEA)

David R. Smith, ***Systematic Entomology Laboratory, ARS, USDA***
Woodrow W. Middlekauff, ***University of California, Berkeley***

Orussid Wood Wasps

Figure 27.70

Relationships and Diagnosis: Morphologically similar to the internally boring larvae of Siricidae, Xiphydriidae, Anaxyelidae, and Cephidae, but the few available larvae of Orussidae (fig. 27.70a) are stouter, about one-third as thick as long, slightly depressed, swollen in the middle, with slightly upcurved caudal end, lack a sclerotized postcornus, lack thoracic legs (represented only by sclerotized discs), have reduced lobelike labium and maxilla without palpi (fig. 27.70b), lack annulations on the abdominal segments, and have transverse rows of short bristles on the thoracic and abdominal terga (fig. 27.70a). The larvae are most likely confused with those of Apocrita, but the thoracic legs, represented by sclerotized discs, and the rows of bristles on the terga distinguish the Orussidae.

Biology and Ecology: Most larvae have been found in galleries of wood-boring insects, and much of the interest in the Orussidae stems from the still unresolved question concerning the alleged entomophagous habits of the larvae. Burke (1918) found two larvae of *Orussus* attacking buprestid larvae and others in pupal cells or near larvae and concluded that *Orussus* is parasitic. Gourlay (1951) stated that the Australian orussid, *Guiglia schauinslandi* (Ashmead), was a parasite of the introduced *Sirex noctilio* Fabricius. He found pupae of *Guiglia* inside the larval remains of *Sirex*. He acknowledged that at the time *Guiglia* oviposits, no young *Sirex* larvae would be present, and thought that the orussid female might hyperparasitize the ichneumonid, *Rhyssa*. Rawlings (1957) dissected logs in which *Guiglia* females had inserted their ovipositors and found that *Sirex* larvae had been punctured and each larva had an egg upon it. He then placed a *Sirex* larva and a *Guiglia* egg in a glass tube. The *Guiglia* larva lived externally for two instars and then penetrated the putrid appearing host where it pupated and eventually emerged as an adult.

Cooper (1953), studying the eastern North American *Orussus terminalis* (Newman), misidentified as *O. sayii* (Westwood), followed the ovipositor's path and found a frass-filled buprestid mine as the target. Subsequent examination disclosed two twisted, water resistant, delicate threads of nonwoody texture, which he believed to be the dried chorions of two eggs. He then stated that, based upon his studies and those of others, it seems possible that *Orussus* does not normally oviposit into larval beetles, or even close to larvae that are working open, fresh mines in the wood. Rather, it seems possible that the normal oviposition site is the core of frass that chokes the old abandoned larval mine. Cooper further stated that the hypothesis that *Orussus* larvae feed upon wood frass, perhaps infested with fungi, yeasts, or other microorganisms, is in accord with the facts of oviposition and the habits of *Orussus'* close living relatives (the Siricidae). Although the evidence appears to be conclusive that the Australian *Guiglia* is parasitic, this may well not be the case with *Orussus*. Evidence (Cooper 1953, Powell and Turner 1975) seems to indicate that early instar feeding of *Orussus* occurs in frass-packed galleries and feeding by at least later instars takes place in the form of an internal parasitoid on larvae and pupae of wood boring insects.

Description: *Head:* Short anteroposteriorly, less than half as broad as greatest diameter of body; antenna one-segmented; labium and maxilla reduced to fleshy lobes, without palpi; stemmata absent; each mandible tridentate. *Thorax:* Legs of no more than sclerotized discs; spiracles on meso- and metathorax (fig. 27.70a); each notum with transverse row of stout spines. *Abdomen:* Prolegs absent; postcornus absent; tenth segment without projections or appendages; segments not annulated; transverse row of short, stout, backward projecting spines on each tergum; spiracles on segments 1–8 visible from dorsal aspect.

Comments: Only four genera and nine species occur in North America (Middlekauff 1983); *Orussus* is most widespread; *Ophrynopus Ophrynon,* and *Ophrynella* are found only in several southern localities extending into Mexico, Central and South America, with one species, *tomentosus,* found in Arizona and California. *Ophrynopus* is the largest genus in South America. Larvae are not commonly found. The above description and illustration are based on the larva described by Rohwer and Cushman (1918) and a single specimen from an unknown Californian locality. Because of lack of material, the family diagnosis is likely to change when more larvae are known.

Selected Bibliography

Burke 1918.
Cooper 1953.
Gourlay 1951.
Middlekauff 1983.
Powell and Turner 1975.
Rawlings 1957.
Rohwer and Cushman 1918.

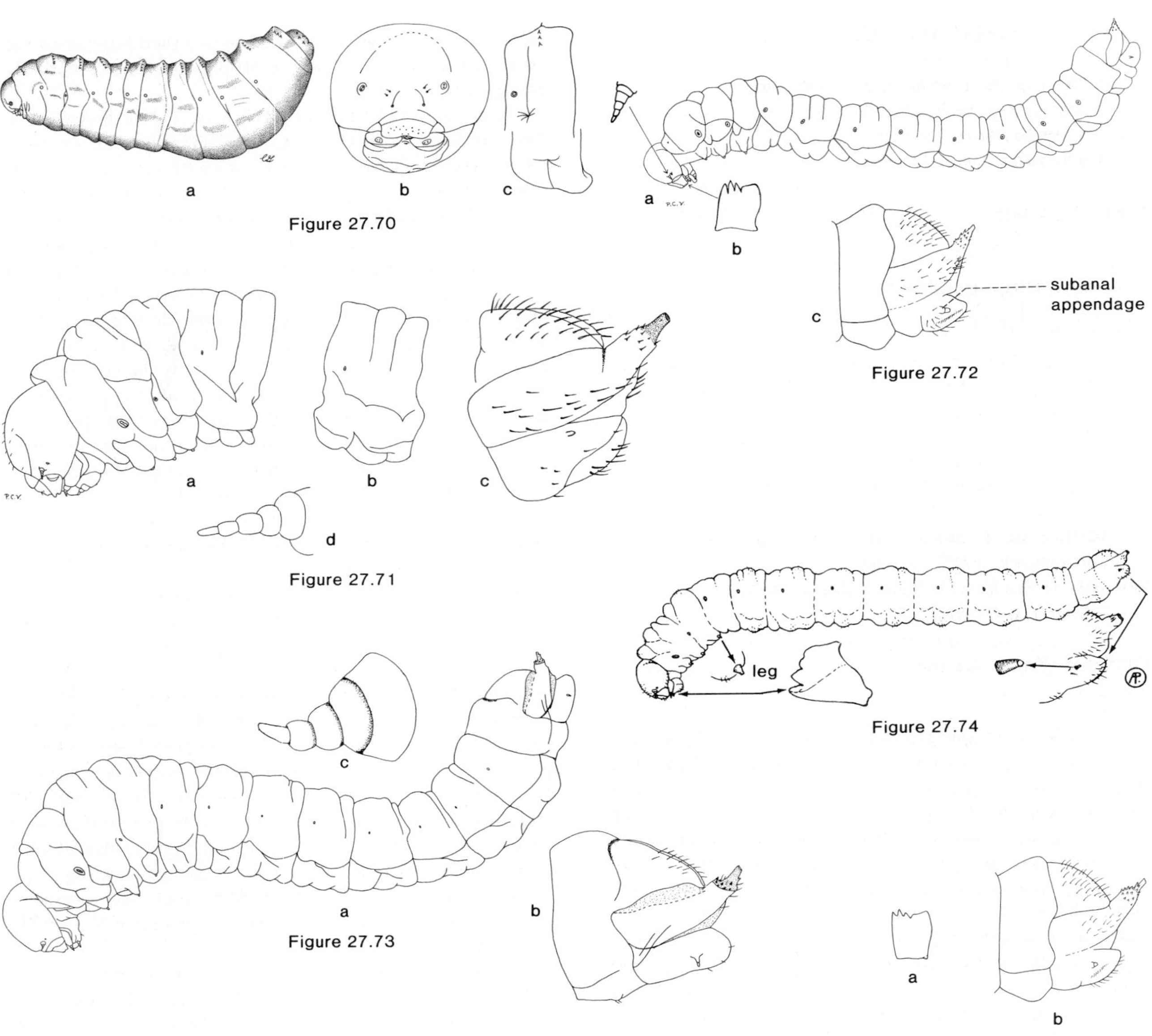

Figure 27.70a-c. Orussidae. *Orussus occidentalis* (Cresson). a. entire larva; b. head; c. A3 segment. Atypical of sawfly larvae; note absence of legs and presence of transverse row of spines on each segment. Usually found in wood, in burrows of other wood-boring insects. *Orussus* is found throughout North America, *O. occidentalis* is western. See text for discussion (Rohwer and Cushman 1918).

Figure 27.71a-d. Cephidae. *Cephus cinctus* Norton, wheat stem sawfly. a. head, thorax, A1 segment; b. A3 segment; c. caudal segment; d. antenna. f.g.l. 9-14 mm. Typical for Cephidae; note the 1-segmented subanal appendage. A serious pest of wheat in the central Canadian provinces and central states of the U.S.; the species occurs from the Midwest westwards (Middleton 1918, Wallace and McNeal 1966).

Figure 27.72a-c. Cephidae. *Hartigia cressonii* (Kirby). a. entire larva; b. left mandible, outer face; c. caudal segments. f.g.l. 19-24 mm. Typical larva of Cephidae; note the 2-segmented subanal appendage. Larva bores in canes of *Rosa* and *Rubus* in Western North America (Middleton 1918).

Figure 27.73a-c. Cephidae. *Janus abbreviatus* (Say), willow shoot sawfly. a. entire larva; b. caudal segments; c. antenna. f.g.l. 11-14 mm. Typical for Cephidae, with 2-segmented subanal appendage. Larva bores in shoots and twigs of willow and poplar in eastern North America (Middleton 1918).

Figure 27.74. Cephidae. *Cephus pygmaeus* (L.) European wheat stem sawfly. f.g.l. 10-14 mm. Lateral view; similar in appearance to *C. cinctus*, fig. 27.71, but caudal segment terminating in caudodorsal nonsclerotized projection except for sclerotized ring at tip (not tubelike as in *C. cinctus*); food plants wheat, barley, rye. Bores in stems and hibernates in stubble near ground level (Middleton 1918; Middlekauff 1979; from Peterson 1948).

Figure 27.75a,b. Cephidae. *Hartigia trimaculata* (Say). a. left mandible, outer face; b. caudal segments. f.g.l. 19-24 mm. Larva typical for Cephidae with 2-segmented subanal appendage. Larvae bores in canes of rose and *Rubus;* the species is transcontinental in North America (Middleton 1918).

CEPHIDAE (CEPHOIDEA)

David R. Smith, *Systematic Entomology Laboratory, ARS, USDA*
Woodrow W. Middlekauff, *University of California, Berkeley*

Stem Sawflies

Figures 27.71–27.75

Relationships and Diagnosis: Distinguished from other internally boring Symphyta larvae with a postcornus, Siricidae, Xiphydriidae, and Anaxyelidae, by the following: antenna conical, four- to five-segmented (figs. 27.71d, 27.72a); stemmata pigmented; tenth sternum with a pair of short one- or two-segmented appendages (fig. 27.73b); abdominal segments 1–8 each with two or three dorsal annulets (figs. 27.72a, 27.73a); and their habit of boring in grass stems or small twigs of shrubs and trees. Larvae are cylindrical, white, and more strongly S-shaped than boring larvae of other families.

Biology and Ecology: Larvae bore and feed internally in grass stems or small twigs of woody plants, never in larger branches or trunks of trees. Species of *Hartigia* and *Janus* are associated with *Rosa, Rubus, Populus, Salix, Viburnum,* and *Ribes. Janus* contains the willow shoot sawfly, *J. abbreviatus* (Say), and the currant stem girdler, *J. integer* (Norton). The single species of *Caenocephus* may also be found in twigs, but the host is unknown. Larvae of *Cephus, Calameuta,* and *Trachelus* are stem borers in grasses and include the most economically important species, the wheat stem sawfly, *Cephus cinctus* Norton (figs. 27.71a-d), the European wheat stem sawfly, *C. pygmaeus* (Linnaeus) (fig. 27.74) and the black grain stem sawfly, *Trachelus tabidus* (Fabricius). Hosts include important cultivated grass crops such as wheat, rye, barley, timothy, and oats. One or more species in the following genera have been recorded as suitable hosts: *Agropyron, Beckmannia, Bromus, Calamagrostis, Calamovilfa, Deschampsia, Elymus, Festuca, Hordeum, Phleum,* and *Stipa.*

Losses of wheat caused by *Cephus cinctus* may range from ten to twenty percent of the crop. Resistant varieties have been developed in Canada and the United States, but since they do not yield as well as certain nonresistant strains they have not been widely accepted. For an excellent account and bibliography see Wallace and McNeal (1966).

Description: *Head:* Globular, with few hairs or setae; antenna conical, four- or five-segmented; labial palp three- or four-segmented; stemmata pigmented; frontal sutures present. *Thorax:* Legs mammaform, each without claw (fig. 27.71a); metathoracic spiracle present (fig. 27.71a). *Abdomen:* Segments 1–8 each with two or three dorsal annulets; prolegs absent; short, tubular postcornus at apex of tenth tergum; tenth sternum with pair of short one- or two-segmented appendages; tenth tergum with longitudinal groove.

Comments: Six genera and 12 species occur in North America; *Caenocephus* and *Calameuta* are western, the other genera are transcontinental with only one species extending into Mexico. Ries (1937) and Middlekauff (1969) treated the adults and gave biological notes and literature. Because of their economic importance, larvae of most species are known; however, the only treatment is that by Middleton (1918). A key to separate larvae of the three common pests of wheat and other grain crops, *Cephus cinctus, C. pygmaeus,* and *Trachelus tabidus,* was given by Gahan (1920) and Wallace and McNeal (1966). Larvae of *Caenocephus* and *Calameuta* are not known, and specific identification of larvae of *Hartigia* must await clarification of the adult taxonomy.

Selected References

Gahan 1920.
Middlekauff 1969.
Middleton 1918.
Ries 1937.
Wallace and McNeal 1966.

ICHNEUMONOIDEA

Thelma Finlayson, *Simon Fraser University*

Parasitic Wasps

Figures 27.5, 27.6, 27.76–27.129a

Relationships and Diagnosis: Larvae of the Ichneumonidae and Braconidae have been studied extensively, those of the Aphidiidae less so. Available information indicates they can be distinguished from other groups of parasitic Hymenoptera by having (1) fully formed mandibles, each with a blade that is not tridentate, (2) maxillary or labial appendages, or sensilla arranged in definitely circumscribed maxillary and labial areas, (3) a silk press, and (4) cocoons are usually formed (Beirne 1941).

Biology and Ecology: Ichneumonoids are mainly "parasites" of other insects and therefore can be a major factor in the regulation of insect populations. The term "parasitoid," sometimes used to distinguish entomophagous insects from parasites as understood in the medical or veterinary context, will be used here. Parasitoid larval instars develop upon or within a single host, which eventually dies as a result. Adults are free living.

Parasitoids can be internal or external feeders; in some cases they are internal in early instars but external in later ones. Eggs may be deposited in insect eggs, in any of the larval instars, in pupae, or, in a few instances in adults. Hosts of parasitoids include most insects and some spiders and pseudoscorpions. Though a few parasitoids attack only one host species, most parasitize a number of different species. Parasitoids may be attracted to their hosts by location, shape, motion, or odor. The host may influence the rate of development, longevity, size or fecundity of the parasitoid. Some species paralyze their hosts by injecting venom; others require a blood meal from the host before eggs mature. Diapause in some stage of the life cycle often is necessary to synchronize development between the host and its parasitoid. Polyembryony occurs in some Ichneumonoidea, notably, in the Braconidae, where one egg divides many times to form numerous individuals; a trophamnion formed from the polar bodies protects and nourishes the developing eggs (Doutt 1959).

Larvae of ectoparasitic species are hatched with a peripneustic tracheal system (spiracles in a row along each side of the body; abdominal spiracles open but metathoracic spiracles closed) with four to nine pairs of spiracles (Hagen 1964). Larvae of endoparasitic species usually hatch with an apneustic tracheal system (none of the spiracles are functional and air enters the closed tracheal system by diffusion through the body surface or gills), which in later instars almost always changes to another type (Hagen 1964). Thorpe (1932) stated that first-instar endoparasitic larvae are haemophagous but later become carnivorous, at which time the spiracles open on the surface of the body. Hagen (1964) suggested that oxygen is obtained either from the surrounding host fluids through the integument or through some mechanism for obtaining air from outside the host.

The larval stomach is a blind sac until the final instar, when it opens to the hind gut. The accumulated fecal contents are evacuated at the conclusion of the larval stage as the meconium, and the skin of the final-instar larva is usually caught in it.

Pupation is usually in a silken cocoon, spun either within the host or nearby after the fully fed larva has emerged from the host. Species that do not make a cocoon pupate within the hollowed-out larval or pupal skin of the host; in these cases a small disc of silk frequently separates the remainder of the host from the pupating parasitoid.

Description: The typical mature ichneumonoid larva (fig. 27.76) is white with translucent skin and has an ovoid, maggotlike form. It has a reduced head that may be sunk into the prothorax, three thoracic and ten (occasionally nine) abdominal segments. There are no thoracic or abdominal appendages. Eyes are absent and antennae and maxillary and labial palpi are papillalike, disclike, or absent. Dark hairs or setae are often present on the skin. First-instar larvae of species that deposit their eggs and hatch apart from the host (e.g., Ichneumonidae: *Euceros,* Finlayson 1960a) may be covered with broad, tough, flat spines that protect them from desiccation or injury while awaiting hosts. Some endoparasitic larvae have a tail-like process in the first instar (*see* p. 46) that is reduced in size after each molt and is absent or rudimentary in the final instar (Hagen 1964); however, species of the genera *Phobocampe* and *Meloboris* (Ichneumonidae) have a caudal appendage in the final instar (Finlayson 1964).

Terminology of cephalic structures of final-instar larvae in this section follows Finlayson (1960a, 1975) and Gillespie and Finlayson (1983), which is based on and is slightly modified from that of Beirne (1941) and Short (1952) (fig. 27.5), and of Mackauer and Finlayson (1967) (fig. 27.6).

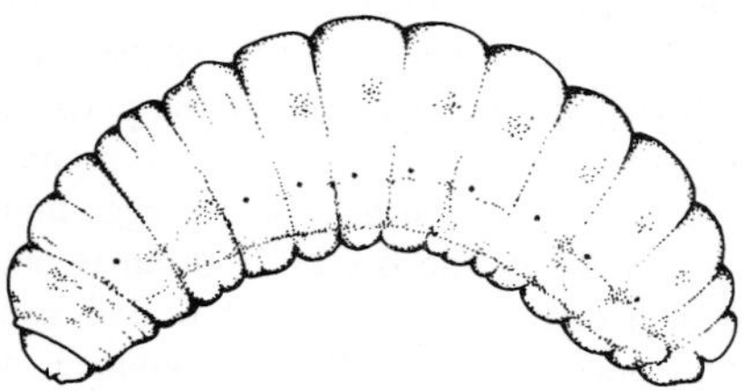

Figure 27.76. Final-instar larva of Braconidae (Orgilinae) (from Finlayson and Hagen 1977).

Comments: The aphid parasitoids of the family Aphidiidae are considered by some workers to be a subfamily, Aphidiinae, of the Braconidae. They are regarded here as having family status because Mackauer (1961) believed the adults formed a separate group and because the immatures are distinctly different from those of the Braconidae.

No single character or group of characters distinguishes final-instar larvae of the families Braconidae, Ichneumonidae, or Aphidiidae. However, they can be separated by using the preceding key to families of Apocrita.

Selected Bibliography (General)

Beirne 1941.
Doutt 1959.
Finlayson 1960a, 1964, 1975.
Finlayson and Hagen 1977 (revised 1979).
Gillespie and Finlayson 1983.
Hagen 1964.
Mackauer 1961.
Mackauer and Finlayson 1967.
Marsh 1979b.
Short 1952, 1959.
Thorpe 1932.

ICHNEUMONIDAE (ICHNEUMONOIDEA)

Thelma Finlayson, *Simon Fraser University*

The Ichneumons

Figures 27.5, 27.77–27.103; First instars pp. 44–46

Most ichneumonid larvae have three thoracic and ten abdominal segments with nine pairs of functional spiracles. Ectoparasitoids have teeth on the lower edge of the blade of the mandible and sometimes on the upper edge as well, and except in *Adelognathus,* possess a labral sclerite (Beirne 1941). The hypostomal spur, if present, is fused to the hypostoma, except in a very few species where it is also fused to the lateral end of the stipital sclerite. A prelabial sclerite may or may not be present. There are accessory longitudinal tracheal commissures in the thorax (Beirne 1941). The parasitoids are solitary, that is, only one emerges from each host. The following key to subfamilies is partly based on those of Beirne (1941) and Short (1959, 1970). Nomenclature conforms to that suggested by Fitton and Gauld (1976, 1978) and Carlson (1979). The division between the Tryphoninae and Cryptinae is difficult to define with certainty and there may be some overlap between these two subfamilies. The key goes to genus where these entities will not fit the usual diagnoses for the subfamily. Descriptions and illustrations in published papers, as listed by Finlayson and Hagen (1977), were used to construct the key. Short's (1978) book on classification of final instar larvae of the Ichneumonidae brings together the published information on this family.

KEY TO SUBFAMILIES OF FINAL-INSTAR LARVAE OF ICHNEUMONIDAE

1. Mandibles and other cephalic structures not sclerotized and only visible with staining; parasitoids of Hymenoptera: Cephidae (stem sawflies) (fig. 27.77) .... **Collyriinae**

1′. Mandibles and other cephalic structures sclerotized, or if unsclerotized they are not parasitoids of wheat-stem sawflies 2

2(1′). Last abdominal segment with 2 hooks .... **Agriotypinae**

2′. Last abdominal segment without 2 hooks 3

3(2′). Labial sclerite sclerotized and attenuated ventrally with a sharp, spinelike projection; remaining sclerites and mandibles very lightly sclerotized and barely visible; parasitoids of Syrphidae (fig. 27.78) .... **Diplazontinae**

3′. Labial sclerite, if present, not attenuated ventrally with a sharp, spinelike projection, or absent; some or all of the remaining sclerites sclerotized 4

4(3′). Hypostomal spur rests on medial end of stipital sclerite where it touches the labial sclerite (fig. 27.79) .... **Pimplinae**

4′. Hypostomal spur, when present, does not rest on medial end of stipital sclerite where it touches the labial sclerite, or absent 5

5(4′). Labral sclerite present 6

5′. Labral sclerite absent 11

6(5). Hypostomal spur fused with lateral end of stipital sclerite (fig. 27.80) .... **Acaenitinae**[1]

6′. Hypostomal spur, if present, not fused with lateral end of stipital sclerite, or absent 7

7(6′). Medial end of stipital sclerite meets labial sclerite at dorsal end of its lateral arm (fig. 27.82) .... (part) (*Opheltes, Perilissus, Protarchus*) **Ctenopelmatinae**

7′. Medial end of stipital sclerite does not meet labial sclerite at dorsal end of its lateral arm 8

8(7′). Epistoma unsclerotized; antenna usually papilliform or absent 9

8′. Epistoma lightly sclerotized; antenna usually disc-shaped or with small papilla (fig. 27.83) .... **Tryphoninae**[2]

9(8). Labral sclerite broken dorsally (fig. 27.85) .... **Labeninae**[3]

9′. Labral sclerite not broken dorsally 10

10(9′). Mandible blade with 2 rows of large teeth on elevations on the dorsal surface (fig. 27.86) .... **Xoridinae**[4]

10′. Mandible blade without teeth or with teeth on dorsal or ventral surfaces, or both (fig. 27.87) .... **Cryptinae**[5]

11(5′). Stipital sclerite absent 12

11′. Stipital sclerite present 13

12(11). Mandible blades strongly curved, and with either a prominent tooth on the medial surface or ridges on the upper edge (fig. 27.88) .... **Orthopelmatinae**

12′. Mandible blades not strongly curved and without a prominent tooth or ridges (fig. 27.89) .... **Ichneumoninae**

13(11′). Hypostomal spur fused with lateral end of stipital sclerite (fig. 27.90) .... **Oxytorinae**

13′. Hypostomal spur not fused with lateral end of stipital sclerite 14

14(13′). Mandible base sclerotized for all or part of the periphery; or if not, then the smaller sensillum of the labial palp seta-like (fig. 27.91) .... **Mesochorinae**

14′. Mandible with fully sclerotized base 15

1. Short's (1959) illustration of *Arotes formosus* Cress. does not show a labral sclerite.
2. Except *Idiogramma* (Short 1959), *Exenterus* (Finlayson 1960 a,b)
3. Townes (1969) and Carlson (1979) placed the genus *Brachycyrtus* in the Labeninae but Short (1959) placed it in the Cryptinae, where it appears to belong on the basis of final-instar larval characters.
4. *Odontocolon geniculatus* (Kreich.) appears to have a complete epistoma (Spradbery 1970).
5. Except *Gambrus* (Short 1959).

15(14′). Stipital sclerite curves in arc from dorsal end of lateral arm of labial sclerite to rest its lateral end on hypostoma (fig. 27.93); hypostomal spur vestigial or absent (fig. 27.93) 16

15′. Stipital sclerite does not curve in arc from dorsal end of lateral arm of labial sclerite to rest its lateral end on hypostoma; hypostomal spur present 17

16(15). Setae present on prelabium (fig. 27.93) .... **Metopiinae**

16′. Setae absent on prelabium (fig. 27.92) .... **Anomaloninae**[6]

17(15′). Hypostoma extends not, or very little laterally beyond its point of junction with hypostomal spur (fig. 27.94) .... (part) (Lissonotanini) **Banchinae**

17′. Hypostoma extends laterally beyond its point of junction with hypostomal spur 18

18(17′). Struts of inferior pleurostomal processes elongated to meet, or almost meet, behind mandibles (fig. 27.95) .... (part) (Glyptini) **Banchinae**

18′. Struts of inferior pleurostomal processes not elongated to meet behind mandibles 19

19(18′). Medial end of stipital sclerite not meeting labial sclerite at dorsal end of its lateral arm 20

19′. Medial end of stipital sclerite meeting labial sclerite at dorsal end of its lateral arm (fig. 27.100) 24

20(19). Prelabium with 6 or more, usually many more, setae (fig. 27.96) .... (part) (Banchini) **Banchinae**

20′. Prelabium with 4 or fewer setae 21

21(20′). Blade of mandible without teeth; epistoma complete 22

21′. Blade of mandible with teeth; epistoma incomplete 23

22(21). Prelabial sclerite present (fig. 27.102) .... **Eucerotinae**

22′. Prelabial sclerite absent (fig. 27.84) .... (*Exenterus*) **Tryphoninae**

23(21′). Prelabial sclerite present (fig. 27.98) .... **Adelognathinae**

23′. Prelabial sclerite absent (fig. 27.99) .... **Lycorininae**

24(19′). Prelabium with many more than 6 setae (fig. 27.100) .... **Ophioninae**

24′. Prelabium with 6 or fewer setae 25

25(24′). Spiracle at least 3 times longer than its width at widest point, and usually much longer (fig. 27.81) .... (part) **Ctenopelmatinae**[7]

25′. Spiracle less than 3 times longer than its width at widest point 26

26(25′). Dorsal flange present on lateral arm of labial sclerite (fig. 27.101) .... **Campopleginae**

26′. Dorsal flange not present on lateral arm of labial sclerite 27

27(26′). Small tooth present on tip of blade of mandible, or tip very sharp (fig. 27.97) .... **Tersilochinae**

27′. Small tooth not present on tip of blade of mandible, or tip not unduly sharp (fig. 27.103) .... **Cremastinae**

Selected Bibliography

For a comprehensive bibliography see Finlayson and Hagen (1977, revised 1979).

Beirne 1941.
Carlson 1979.
Finlayson 1960a, 1960b, 1964, 1967, 1970, 1975, 1976.
Fitton and Gauld 1976, 1978.
Sechser 1970.
Short 1959, 1970, 1978.
Spradbery 1970.
Townes 1969.

6. Except *Anomalon* (Short 1959) and *Agrypon* as illustrated by Sechser (1970), although *Agrypon* as illustrated by Short (1959) falls into this couplet.
7. Except those Ctenopelmatinae that key out in couplet 7.

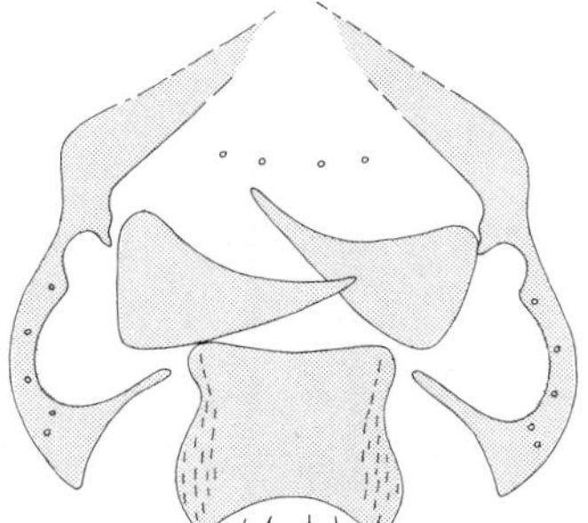
Figure 27.77

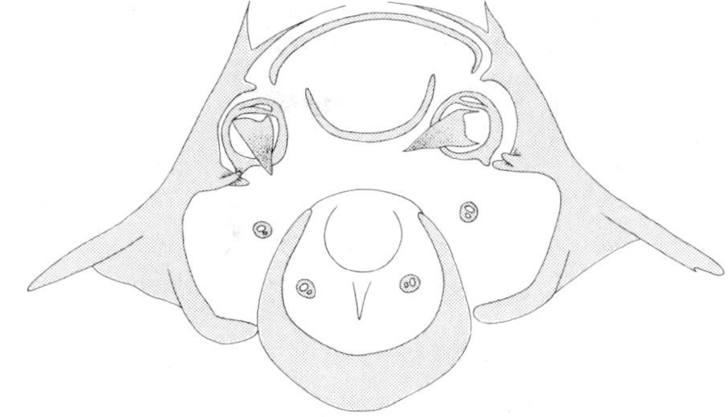
Figure 27.80

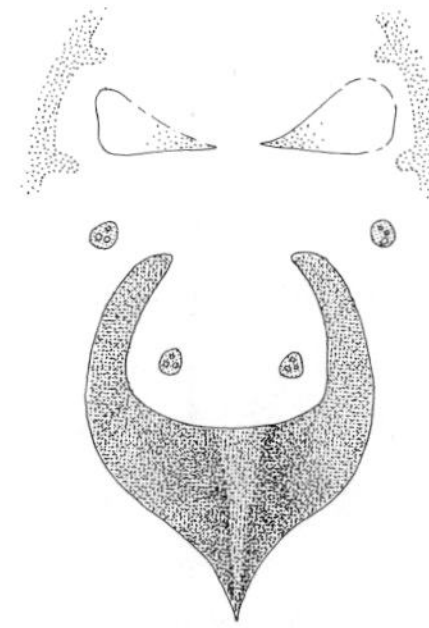
Figure 27.78

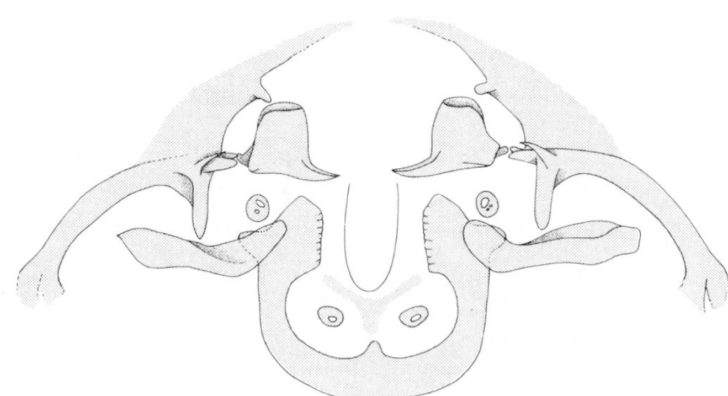
Figure 27.81

Figure 27.79

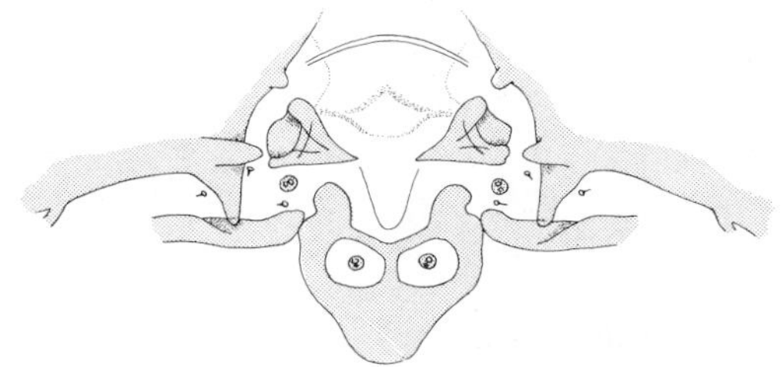
Figure 27.82

Figures 27.77–27.82. Cephalic structures of final-instar ichneumonid larvae.

Figure 27.77. Collyriinae (*Collyria coxator* (Villers)) (redrawn from Beirne 1941).

Figure 27.78. Diplazontinae (*Syrphoctonus* (= *Homocidus*) *tarsatorius* (Panz.)) (redrawn from Beirne 1941).

Figure 27.79. Pimplinae (*Coccygomimus pedalis* (Cresson)) (from Finlayson 1960b).

Figure 27.80. Acaenitinae (*Coleocentrus rufus* Prov.) (from Finlayson 1970).

Figure 27.81. Ctenopelmatinae (*Mesoleius tenthredinis* Morl.) (from Finlayson 1960b).

Figure 27.82. Ctenopelmatinae (*Protarchus testatorius* (Thunb.)) (redrawn from Beirne 1941). (From Finlayson and Hagen 1977).

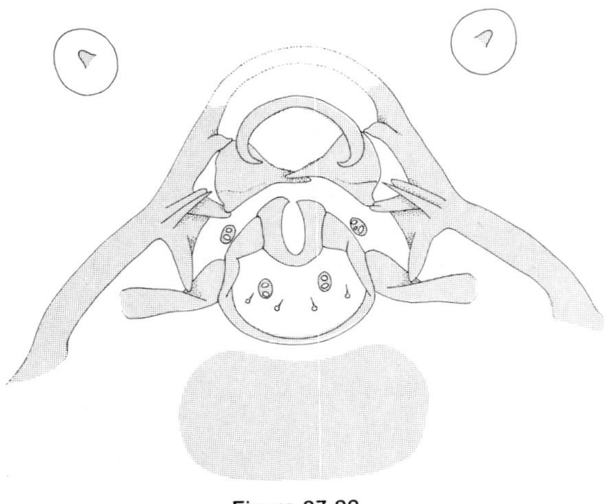

Figure 27.83

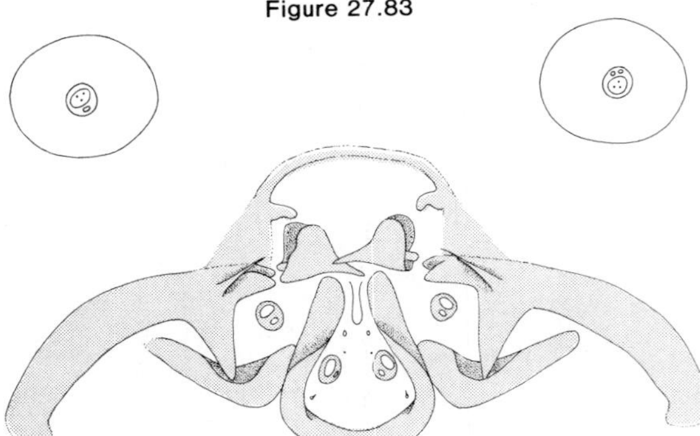

Figure 27.84

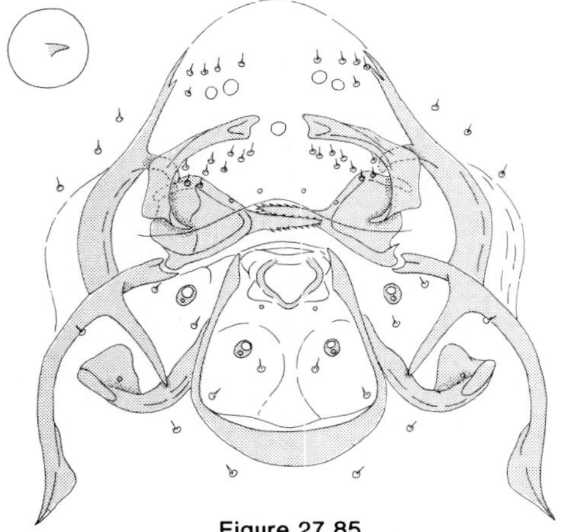

Figure 27.85

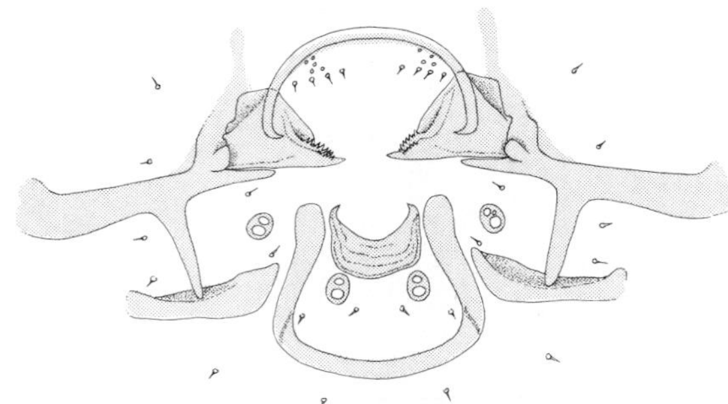

Figure 27.86

Figure 27.87

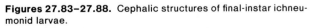

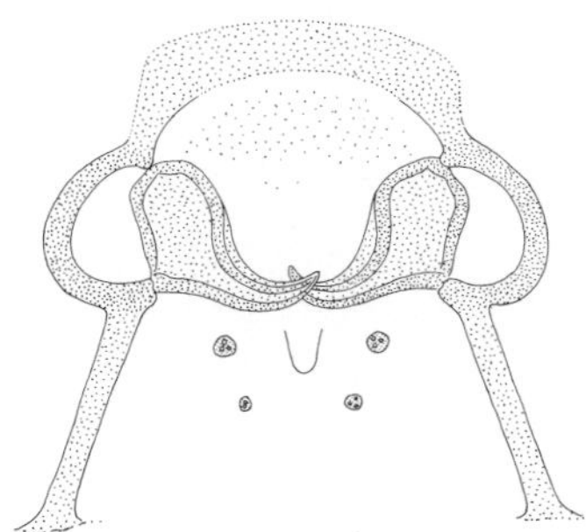

Figure 27.88

Figures 27.83–27.88. Cephalic structures of final-instar ichneumonid larvae.

Figure 27.83. Tryphoninae (*Phytodietus burgessi* (Cress.)) (from Finlayson 1967).

Figure 27.84. Tryphoninae (*Exenterus vellicatus* Cush.) (from Finlayson 1960b).

Figure 27.85. Labeninae (*Labena grallator* (Say)) (redrawn from Short 1959).

Figure 27.86. Xoridinae (*Ischnoceros rusticus* (Geof.)) (redrawn from Beirne 1941).

Figure 27.87. Cryptinae (*Agrothereutes abbreviator iridescens* (Cress.)) (from Finlayson 1960b).

Figure 27.88. Orthopelmatinae (*Orthopelma luteolator* (Grav.)) (redrawn from Beirne 1941).
(From Finlayson and Hagen 1977).

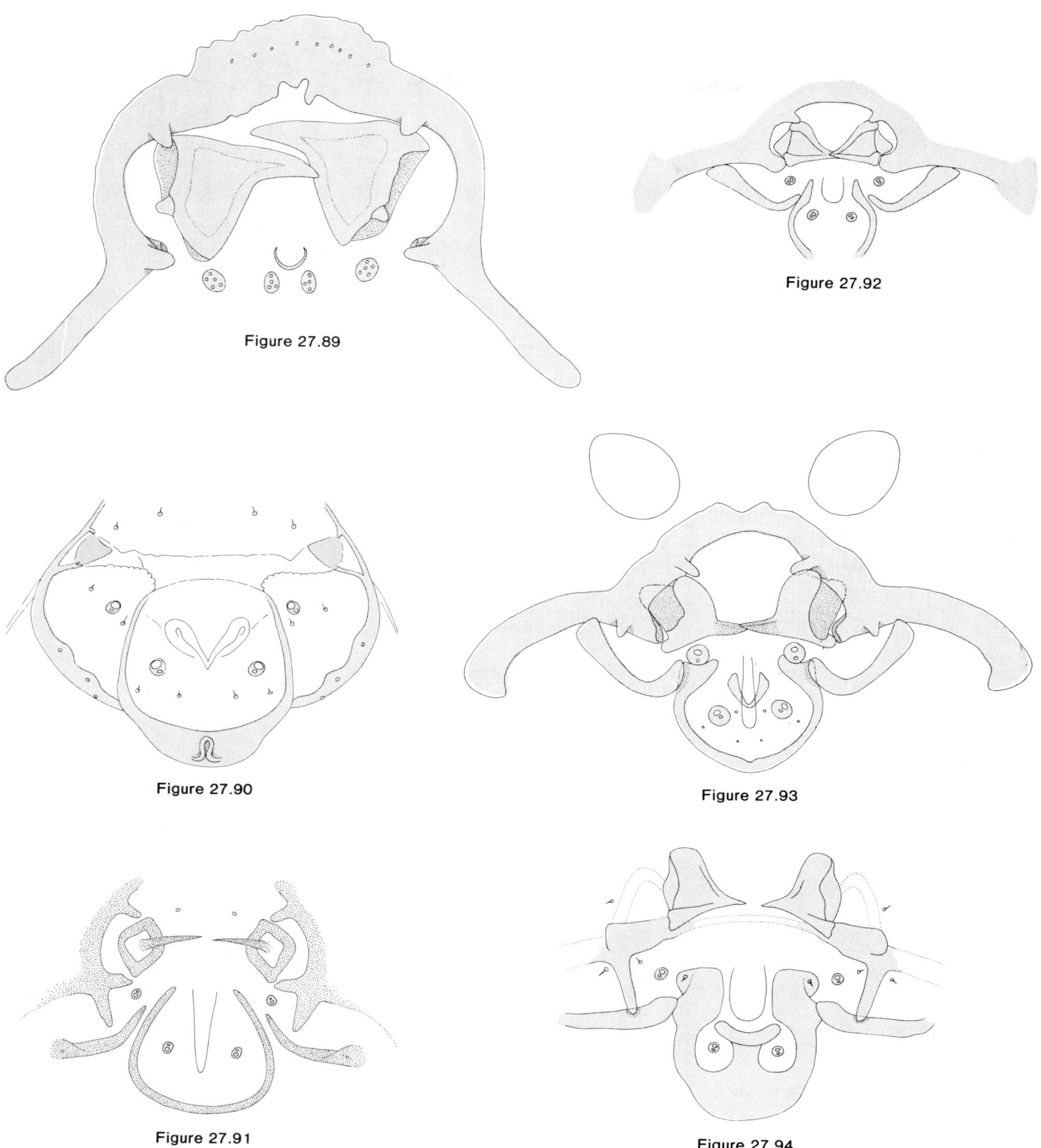

Figures 27.89–27.94. Cephalic structures of final-instar ichneumonid larvae.

Figure 27.89. Ichneumoninae (*Phaeogenes phycidis* Ashm.) (from Finlayson 1967).

Figure 27.90. Oxytorinae (*Megastylus* sp.) (redrawn from Short 1959).

Figure 27.91. Mesochorinae (*Mesochorus fulgurans* Curt.) (redrawn from Beirne 1941).

Figure 27.92. Anomaloninae (*Habronyx nigricornis* (Wesm.)) (redrawn from Beirne 1941).

Figure 27.93. Metopiinae (*Seticornuta apicalis* (Cress.)) (from Finlayson 1967).

Figure 27.94. Banchinae:Lissonotanini (*Lissonota* (= *Meniscus*) *catenator* (Panz.)) (redrawn from Beirne 1941) (from Finlayson and Hagen 1977).

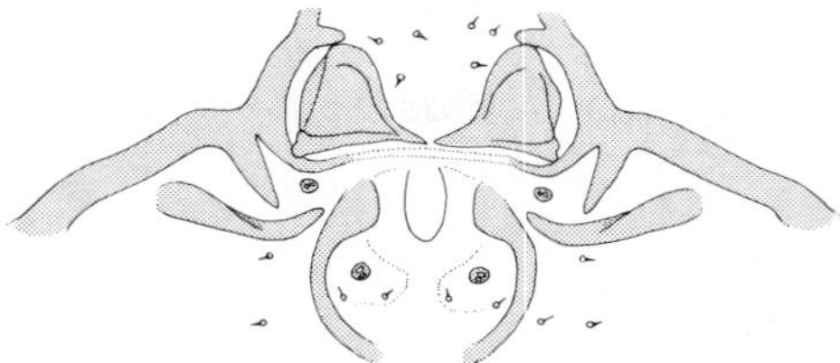

Figure 27.95

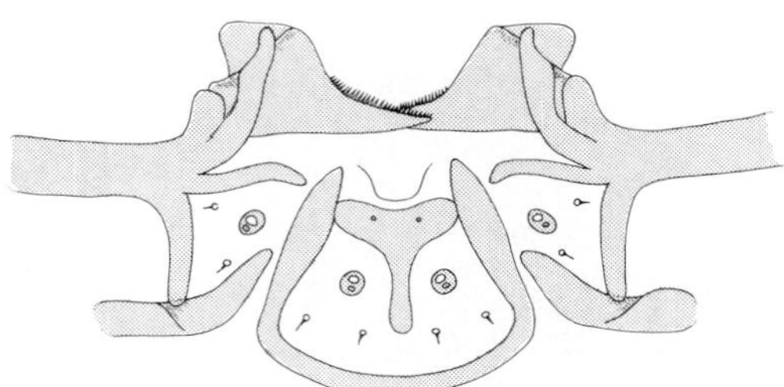

Figure 27.98

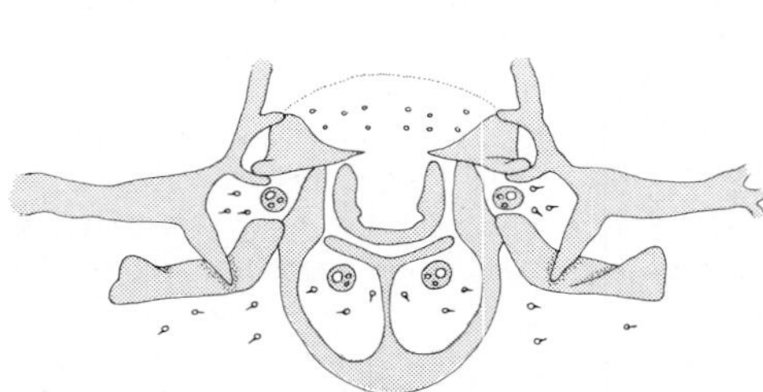

Figure 27.96

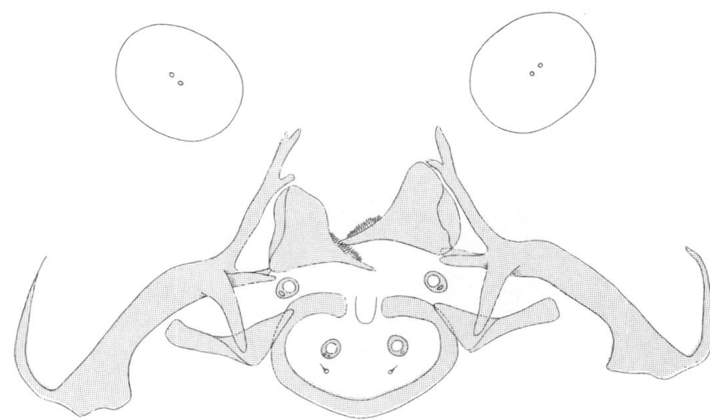

Figure 27.99

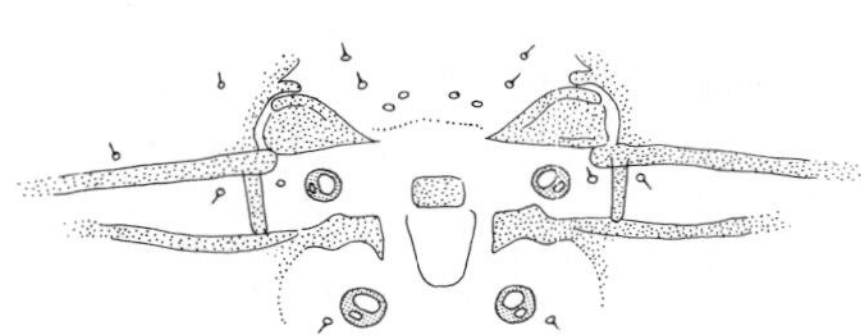

Figure 27.97

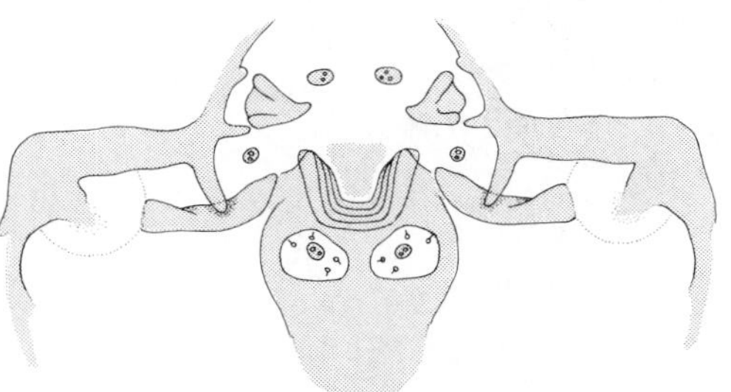

Figure 27.100

Figures 27.95–27.100. Cephalic structures of final-instar ichneumonid larvae.

Figure 27.95. Banchinae: Glyptini (*Glypta parvicaudata* Bridg.) (redrawn from Beirne 1941).

Figure 27.96. Banchinae: Banchini (*Banchus femoralis* Thoms.) (redrawn from Beirne 1941).

Figure 27.97. Tersilochinae (*Tersilochus* sp.) (redrawn from Beirne 1941).

Figure 27.98. Adelognathinae (*Adelognathus* sp.) (redrawn from Beirne 1941).

Figure 27.99. Lycorininae (*Toxophoroides apicalis* (Cress.)) (from Finlayson 1976).

Figure 27.100. Ophioninae (*Enicospilus* (= *Allocamptus*) *macrurus* (Drury)) (redrawn from Beirne 1941). (From Finlayson and Hagen 1977).

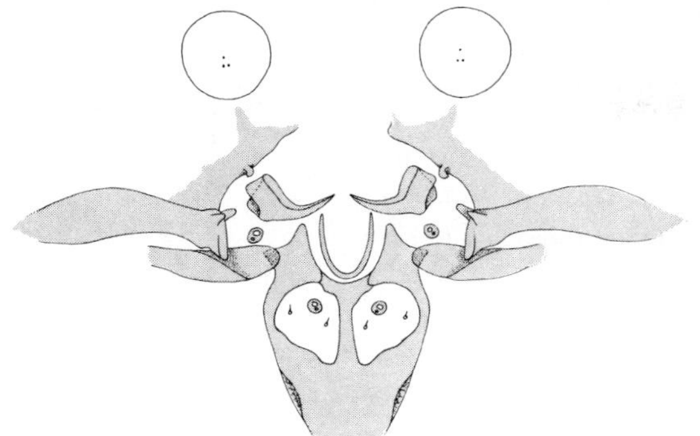

Figure 27.101

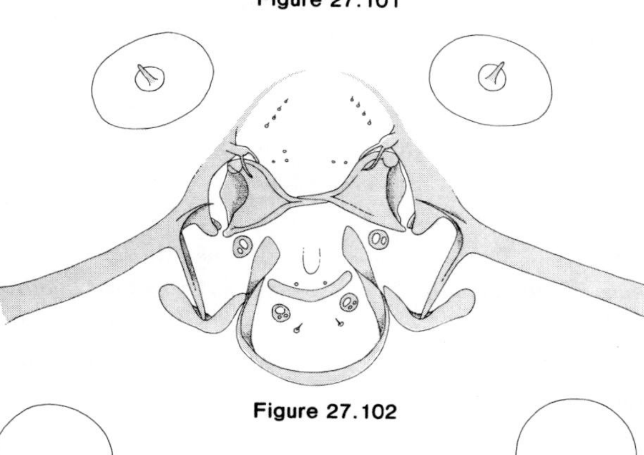

Figure 27.102

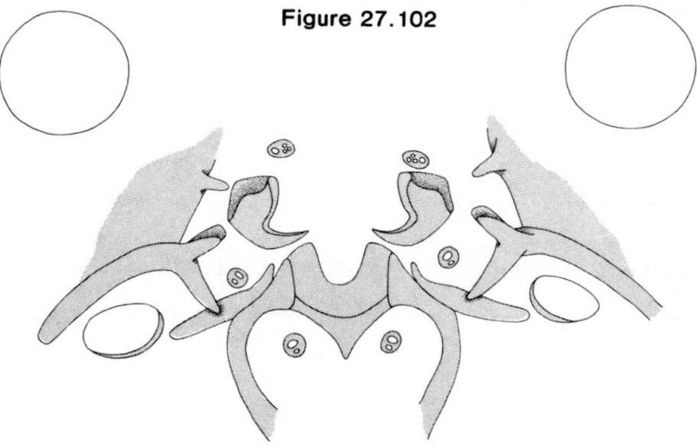

Figure 27.103

Figures 27.101–27.103. Cephalic structures of final-instar ichneumonid larvae.

Figure 27.101. Campopleginae (*Dusona laticincta* (Cress.)) (from Finlayson 1975).

Figure 27.102. Eucerotinae (*Euceros frigidus* Cress) (from Finlayson 1960a).

Figure 27.103. Cremastinae (*Trathala plesia* (Cush.)) (from Finlayson 1967).

BRACONIDAE (ICHNEUMONOIDEA)

Thelma Finlayson, *Simon Fraser University*

The Braconids

Figures 27.76, 27.104–27.127;
First instars p. 46

The braconids exhibit great diversity in the forms of the cephalic structures of the final-instar larvae so it is difficult to find characters to distinguish all the subfamilies. Certain features are characteristic of some groups, e.g., a mandible with an elongated blade bearing a single row of teeth on the upper surface in the Agathidinae (fig. 27.106), and the attachment of the hypostomal spur to the stipital sclerite only in the Euphorinae (fig. 27.114). Nomenclature follows Marsh (1979a) except where that of other authors more closely fits the groupings based on immature characters. These exceptions necessitated moving the following genera from the subfamilies noted in parentheses: *Colastes, Oncophanes, Phanomeris, Cedria, Hormius,* and *Gnaptodon* (Exothecinae) to the Braconinae (Čapek 1970); *Pygostolus, Centistes,* and *Syrrhizus* (Blacinae) to the Euphorinae (Čapek 1973; van Achterberg 1976); and *Zele* (Zeleinae) and *Charmon* [*Eubadizon*] (Blacinae) to the Macrocentrinae (Čapek 1970). To construct the key, descriptions and illustrations in published papers were used (see Finlayson and Hagen 1977 or 1979). Some illustrations that were poor and not accompanied by descriptions in the text could not be used. An illustration of a species that was obviously different from the majority of species within a group was disregarded because of the possibility that the species had been misidentified or that the cast skin examined came from a cocoon mistakenly pinned with the wrong specimen.

KEY TO SUBFAMILIES OF FINAL-INSTAR LARVAE OF BRACONIDAE

1. Mandible with large, heavy-set, often triangular body, with short blade bearing a few teeth at its base; antenna papilliform (figs. 27.104, 27.105) .... **Braconinae, Doryctinae**

1′. Mandible without large, heavy-set body with short blade bearing a few teeth at its base; antenna disclike or absent 2

2(1′). Hypostomal spur joined to hypostoma and also either joined to, or appearing to be joined to, the stipital sclerite 3

2′. Hypostomal spur, if present, not joined to both hypostoma and stipital sclerite, or absent 5

3(2). Mandibular blade elongate, with 1 row of teeth on upper surface (fig. 27.106) .... **Agathidinae**[8]

3′. Mandibular blade without teeth 4

4(3′). Epistoma unsclerotized (fig. 27.107) .... **Opiinae**[9]; **Alysiinae:** *Alysia*

4′. Epistoma sclerotized (fig. 27.112) .... **Rogadinae:** *Batotheca, Pelecystoma harissinae* (Ashm.)

5(2′). Hypostomal spur present, but not attached to either hypostoma or stipital sclerite 6

5′. Hypostomal spur, if present, attached to either hypostoma or stipital sclerite, or absent 9

6(5). Blade of mandible with teeth 7

6′. Blade of mandible without teeth 8

7(6). Hypostoma present (fig. 27.109) .... **Helconinae:** *Aspidocolpus*

7′. Hypostoma absent (fig. 27.123) .... **Blacinae:** *Microtypus, Stantonia;* **Adeliinae:** *Adelius*

8(6′). Pleurostomae present (fig.27.111) .... **Rogadinae**[10]

8′. Pleurostomae absent (fig. 27.113) .... **Neoneurinae**

9(5′). Hypostomal spur attached to stipital sclerite only, often appearing as a vestigial thickening on stipital sclerite 10

9′. Hypostomal spur, if present, not attached to stipital sclerite only, or absent 12

10(9). Labial sclerite absent (fig. 27.127) .... **Dacnusinae:** *Ectilis*

10′. Labial sclerite present 11

11(10′). Mandibular blade without teeth (figs. 27.114, 27.116) .... **Euphorinae**[11]

11′. Mandibular blade with teeth (fig. 27.110) .... **Meteorideinae:** *Meteoridea;* **Helconinae:** *Cenocoelius;* **Agathidinae:** *Cremnops*

12(9′). Hypostomal spur attached to hypostoma only 13

12′. Hypostomal spur absent 18

13(12). Maxillary and labial palpi papilliform (fig. 27.117) .... **Microgastrinae**

13′. Maxillary and labial palpi disclike 14

14(13′). Mandible heavy-set and triangular, with short blade bearing teeth; length of mandible only slightly greater than its width at base (fig. 27.118) .... **Ichneutinae**

14′. Mandible not heavy-set, with long blade bearing teeth; length of mandible more than twice its width at base 15

15(14′). Labial sclerite with 2 ventral extensions (fig. 27.119) .... **Macrocentrinae; Helconinae:** *Diospilus*

15′. Labial sclerite without 2 ventral extensions 16

16(15′). Hypostoma equal to or shorter than hypostomal spur 17

16′. Hypostoma longer than hypostomal spur (fig. 27.120) .... Microgastrinae: Cardiochilini

17(16). Maxillary and labial palpi elongated, with 2 large, widely separated sensoria (fig. 27.125) .... **Blacinae:** *Orgilus*

8. Except *Cremnops* (Simmonds 1947).

9. Except *Opius variegatus* (Szépl.), *Ademon niger* (Ashm.) (Čapek 1970).

10. Except *Batotheca* (Short 1952), *Pelecystoma* (Čapek 1970; Smith et al. 1955), *Rogas aciculatus* (Cress.) (Guppy and Miller 1970), *R. terminalis* (Cress.) (Guppy and Miller 1970). Dowden (1938) indicated that *Aleiodes pallidata* (Thunb.) [=*Rogas unicolor* (Wesm.)] has teeth on the blades of mandibles, but Čapek (1970) stated that teeth were lacking in Rogadini.

11. Except *Syntretus* (Cole 1959), *Wesmaelia* (Čapek 1970, Hendrick and Stern 1970).

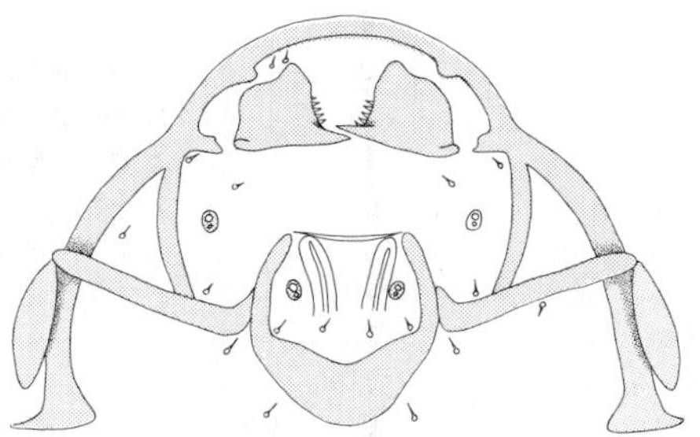

Figure 27.104

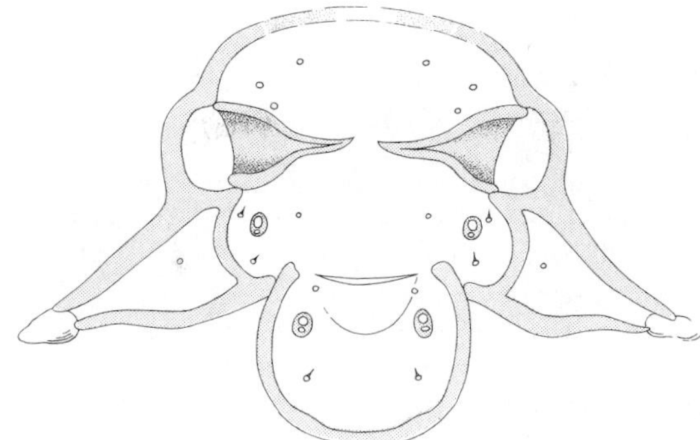

Figure 27.107

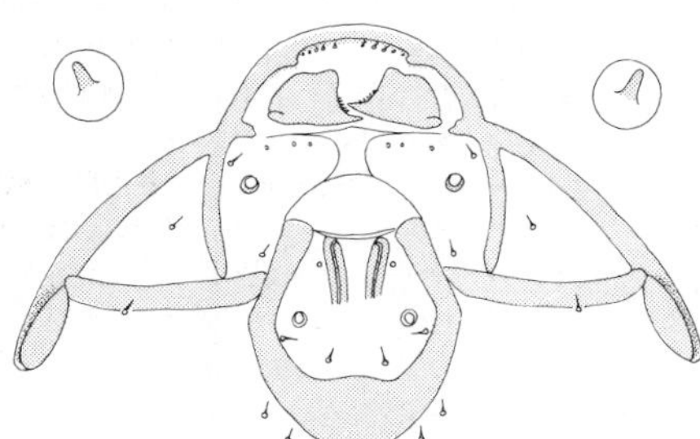

Figure 27.105

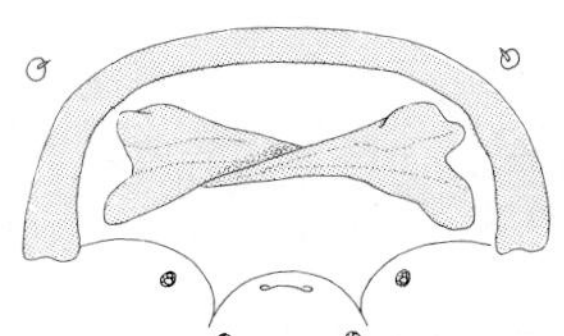

Figure 27.108

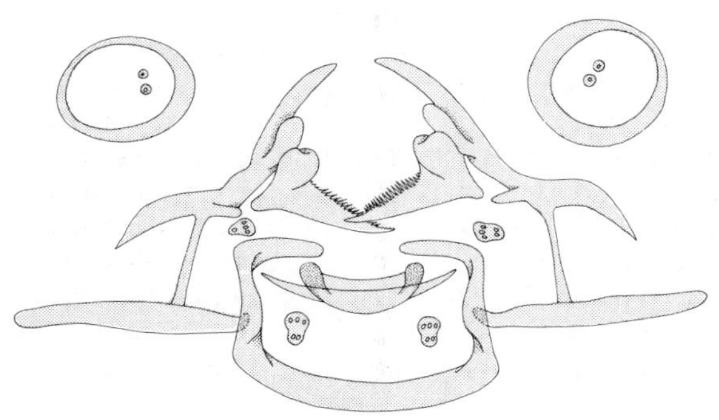

Figure 27.106

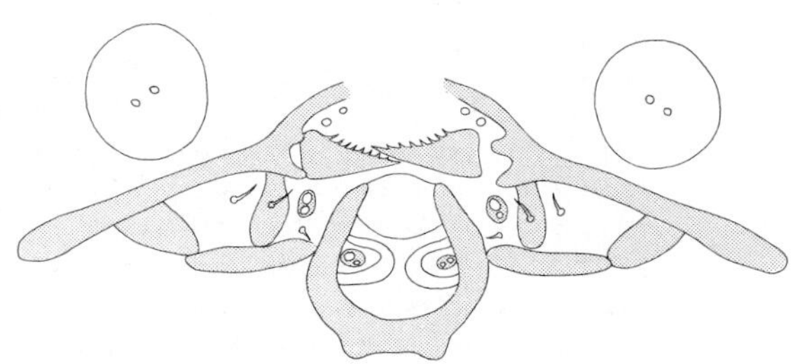

Figure 27.109

Figures 27.104–27.109. Cephalic structures of final-instar braconid larvae.

Figure 27.104. Braconinae (*Bracon terebella* Wesm.) (redrawn from Short 1952).

Figure 27.105. Doryctinae (*Dendrosoter middendorffii* (Ratz.)) (redrawn from Short 1952).

Figure 27.106. Agathidinae (*Agathis calcarata* (Cress.)) (from Finlayson 1967).

Figure 27.107. Opiinae (*Biosteres carbonarius* (Nees)) (redrawn from Short 1952).

Figure 27.108. Alysiinae (*Aspilota vesparum* Stel.) (redrawn from Short 1952).

Figure 27.109. Helconinae (*Aspidocolpus carinator* Ns.) (redrawn from Čapek 1970).
(From Finlayson and Hagen 1977.)

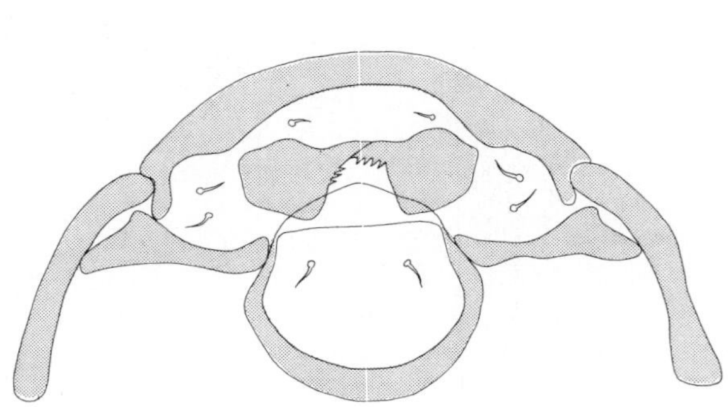

Figure 27.110

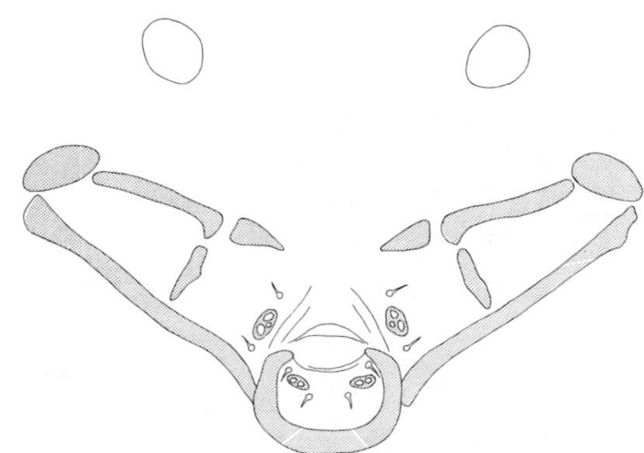

Figure 27.113

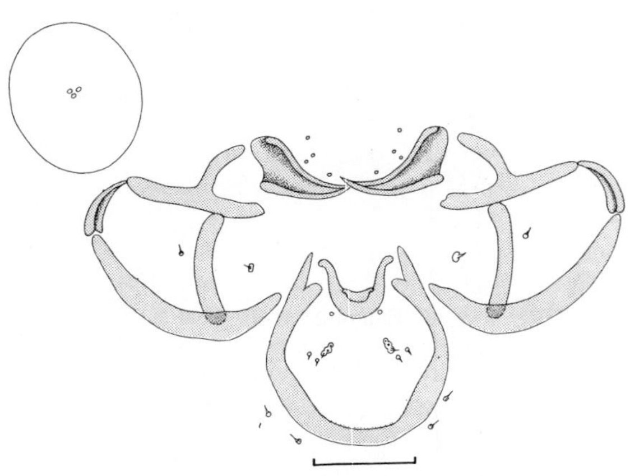

Figure 27.111

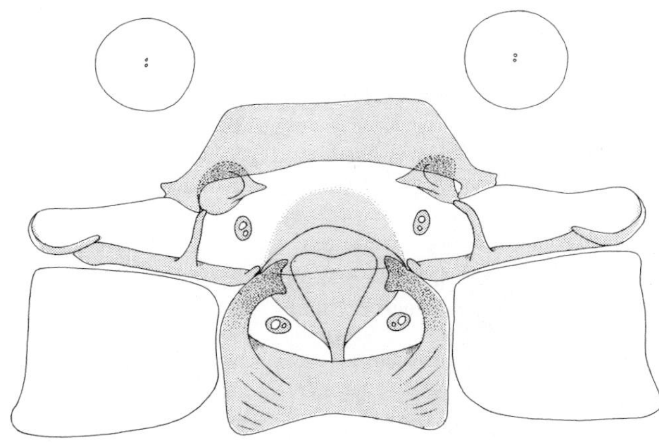

Figure 27.114

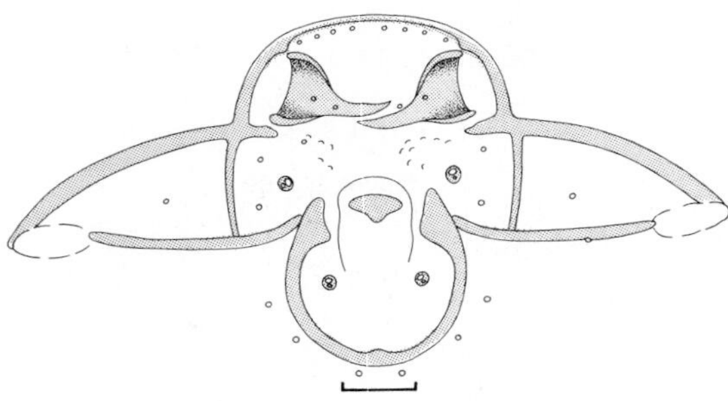

Figure 27.112

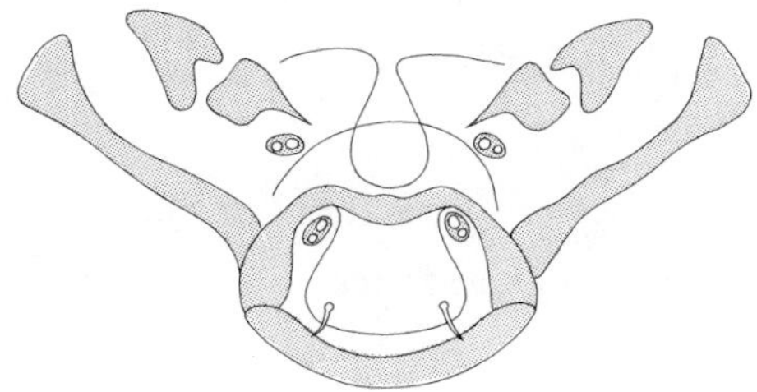

Figure 27.115

Figures 27.110–27.115. Cephalic structures of final-instar braconid larvae.

Figure 27.110. Meteorideinae (*Meteoridea japonensis* (Shen. and Mues.)) (redrawn from Čapek 1970).

Figure 27.111. Rogadinae (*Rogas circumscriptus* Nees) (redrawn from Short 1952).

Figure 27.112. Rogadinae (*Batotheca* sp.) (redrawn from Short 1952).

Figure 27.113. Neoneurinae (*Neoneurus auctus* (Thoms.)) (redrawn from Čapek 1970).

Figure 27.114. Euphorinae (*Meteorus indagator* (Riley)) (from Finlayson 1967).

Figure 27.115. Euphorinae (*Wesmaelia pendula* Foerster) (redrawn from Čapek 1970). (From Finlayson and Hagen 1977).

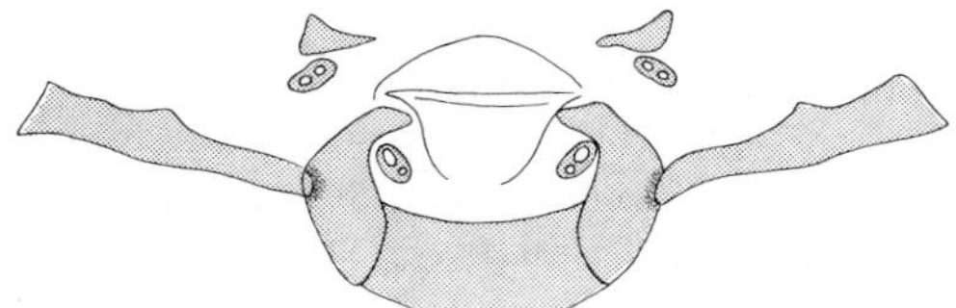

Figure 27.116

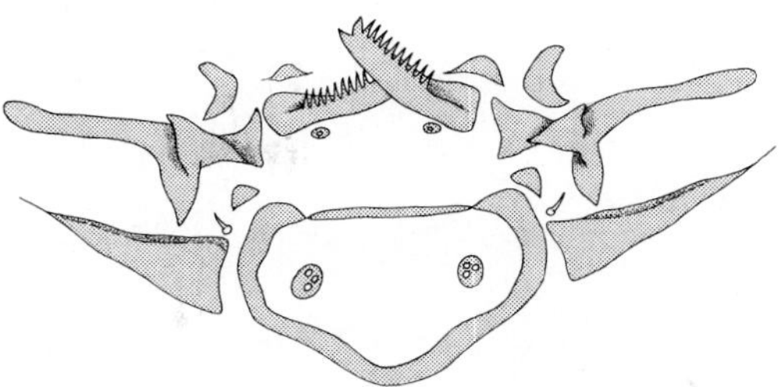

Figure 27.120

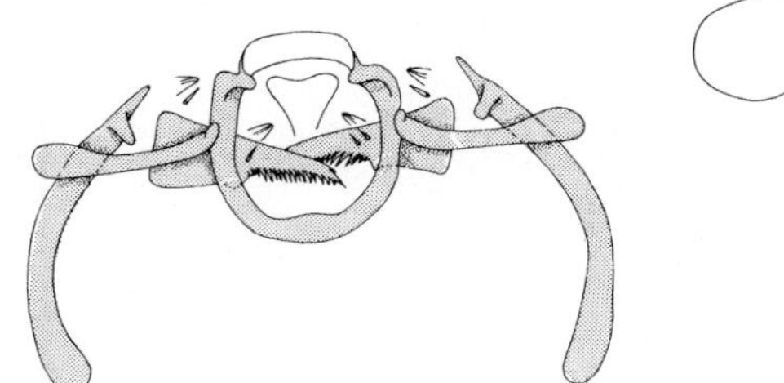

Figure 27.117

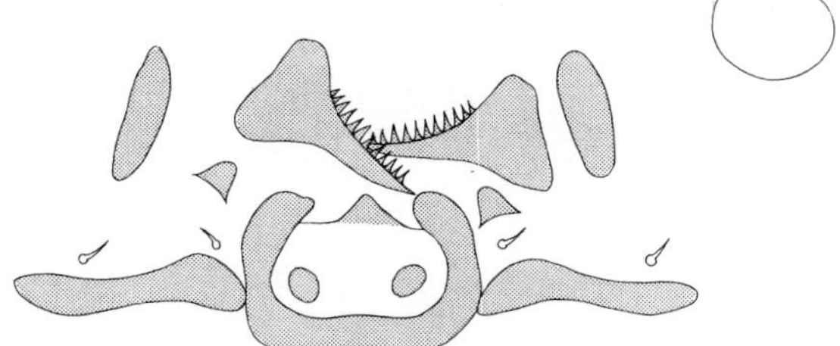

Figure 27.121

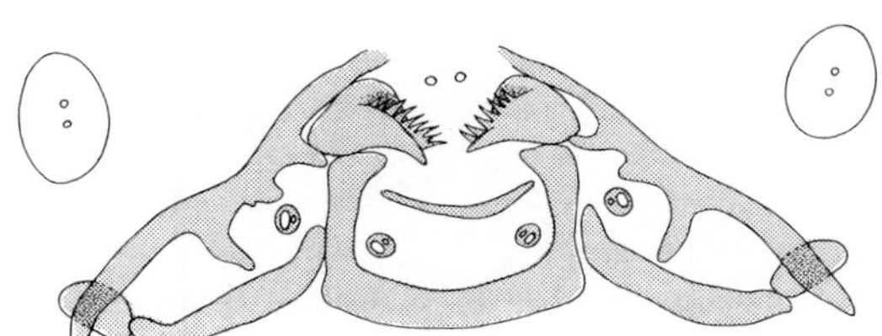

Figure 27.118

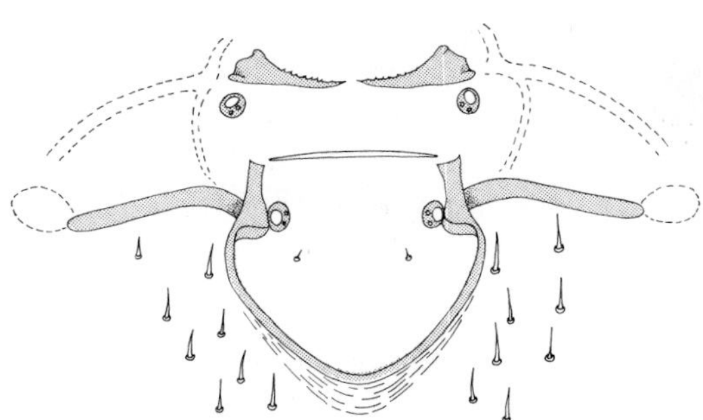

Figure 27.122

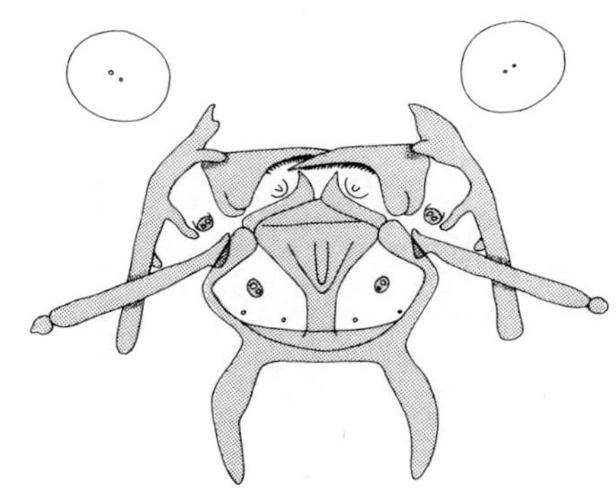

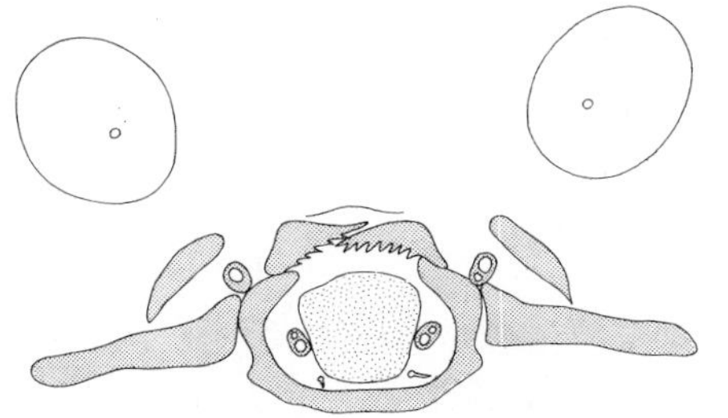

Figure 27.123

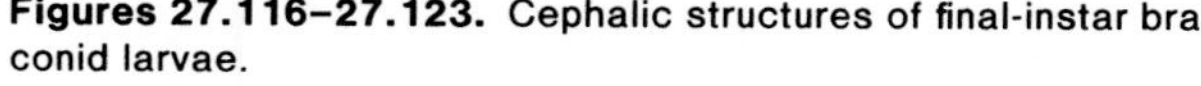

Figure 27.119

Figures 27.116–27.123. Cephalic structures of final-instar braconid larvae.

Figure 27.116. Euphorinae (*Centistes* sp.) (redrawn from Čapek 1970).

Figure 27.117. Microgastrinae (*Apanteles* sp.) (from Finlayson 1967).

Figure 27.118. Ichneutinae (*Ichneutes pikonematis* Mason) (redrawn from Čapek 1970).

Figure 27.119. Macrocentrinae (*Macrocentrus instabilis* Mues.) (from Finlayson 1967).

Figure 27.120. Microgastrinae: Cardiochilini (*Cardiochiles magnus* Mao) (redrawn from Čapek 1970).

Figure 27.121. Sigalphinae (*Sigalphus bicolor* (Cress.)) (redrawn from Čapek 1970).

Figure 27.122. Blacinae (*Eubazus pallipes* Nees) (redrawn from Short 1952).

Figure 27.123. Blacinae (*Microtypus* sp.) (redrawn from Čapek 1970).
(From Finlayson and Hagen 1977.)

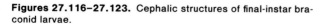

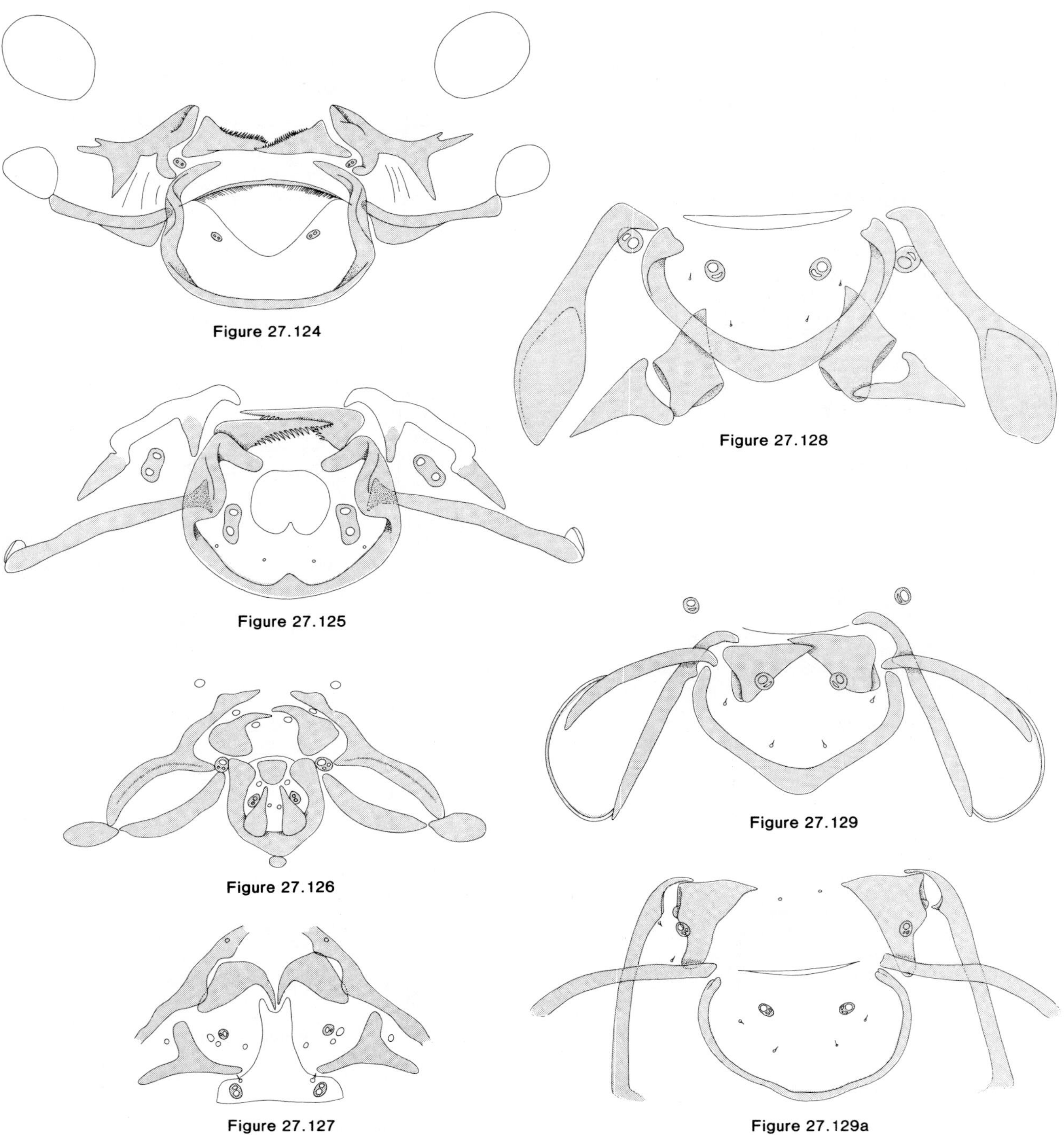

Figures 27.124–27.127. Cephalic structures of final-instar braconid larvae.

Figure 27.124. Cheloninae (*Phanerotoma tibialis* (Hald.)) (from Finlayson 1967).

Figure 27.125. Blacinae (*Orgilus lateralis* (Cress.)) (from Finlayson 1967).

Figure 27.126. Rogadinae (*Pelecystoma luteum* (Nees.)) (redrawn from Čapek 1970).

Figure 27.127. Dacnusinae (*Ectilis semirugosus* Hal.) (redrawn from Čapek 1970) (from Finlayson and Hagen 1977).

Figures 27.128–27.129a. Cephalic structures of final instar aphidiid larvae.

Figures 27.128. Aphidiinae (*Aphidius ervi ervi* Hal.).

Figure 27.129. Prainae (*Praon pequodorum* Vier.).

Figure 27.129a. Ephedrinae (*Ephedrus californicus* Baker). (figs. 27.128–27.129a from MacKauer and Finlayson 1967).

17′. Maxillary and labial palpi small, round, with 2 or more closely spaced sensoria (fig. 27.124) .... **Cheloninae**
18(12′). Mandibular blade with teeth 19
18′. Mandibular blade without teeth 20
19(18). Maxillary palpi papilliform (fig. 27.121) .... **Sigalphinae**
19′. Maxillary palpi not papilliform (fig. 27.122) .... **Blacinae; Helconinae:** *Wroughtonia, Taphaeus*
20(18′). Labial sclerite absent (fig. 27.108) .... **Alysiinae**[12]**; Opiinae:** *Opius variegatus* Szépl., *Ademon*
20′. Labial sclerite present 21
21(20′). Hypostoma absent (fig. 27.115) .... **Euphorinae:** *Syntretus, Wesmaelia*
21′. Hypostoma present (fig. 27.126) .... **Dacnusinae**[13]**; Rogadinae:** *Pelecystoma luteum* Ns., *Rogas aciculatus* (Cress.), *Rogas terminalis* (Cress.)

12. Except *Alysia* (Altson 1920; Evans 1933; Short 1952).
13. Except *Ectilis* (Čapek 1970).

Selected Bibliography

For a comprehensive bibliography see Finlayson and Hagen (1977, revised 1979).

Altson 1920.
Čapek 1970, 1973.
Cole 1959.
Dowden 1938.
Evans 1933.
Finlayson 1967.
Guppy and Miller 1970.
Hendrick and Stern 1970.
Marsh 1979a.
Short 1952.
Simmonds 1947.
Smith et al. 1955.
van Achterberg 1976.

APHIDIIDAE (ICHNEUMONOIDEA)

Thelma Finlayson, *Simon Fraser University*

The Aphidiids

Figures 27.6, 27.128–27.129a

As noted above (Ichneumonoidea, Comments), the Aphidiinae (Braconidae) of some authors is treated here as a family of the Ichneumonoidea (Marsh 1979b). All are parasitoids of aphids. Relationships of the cephalic sclerites of final-instar larvae are shown in fig. 27.6. All have the following characters: blades of mandibles without teeth; epistoma, hypostomal spurs and antennae absent; spiracles about five times or more longer than width at widest point and without obvious atrium and closing apparatus; and cocoon spun within or beneath the dead aphid host or mummy. Information on final-instar larvae is limited, and the key below is based mainly on the work of Mackauer and Finlayson (1967) and on unpublished work by the author. For additional references see Finlayson and Hagen (1977 or 1979).

Selected Bibliography

Finlayson and Hagen 1977 (revised 1979)
Mackauer and Finlayson 1967
Marsh 1979b

KEY TO SUBFAMILIES OF FINAL-INSTAR LARVAE OF APHIDIIDAE

1. Stipital sclerite expanded to form paddle-like enlargement at distal end (fig. 27.128) .... **Aphidiinae**
Stipital sclerite not expanded to form paddle-like enlargement at distal end 2
2. Stipital sclerite crossing over pleurostomal-hypostomal sclerite anteriorly in the dorsal third or not at all; stipital and pleurostomal-hypostomal sclerites joined with a band of sclerite (fig. 27.129) .... **Prainae**
Stipital sclerite crossing over pleurostomal-hypostomal sclerite about mid-point of its length; stipital and pleurostomal-hypostomal sclerites not joined with a band of sclerite (fig. 27.129a) .... **Ephedrinae**

CHALCIDOIDEA

Thelma Finlayson, *Simon Fraser University*

The Chalcidoids

Figures 27.130-27.132; first instars pp. 45-46

Relationships and Diagnosis: The number of families in the superfamily Chalcidoidea ranges from about 9 to more than 20, depending upon the authority; Riek (1970) listed 9, Peck (1963) recognized 21, and Krombein et al. (1979) included 11. Characteristics common to larvae of most are: mandibles fully formed, sclerotized, triangular in form, and with a short blade; each maxilla with 1 papilla or none; maxillae and labium not projecting as separate lobes but either completely fused and indistinguishable or partly fused; labial palpi not visible; silk press absent; and cocoons not formed (except in *Euplectrus* spp. (Eulophidae), Thomsen 1927).

Biology and Ecology: Most chalcidoids are parasitoids in or on immature Homoptera, and the eggs, larvae and, infrequently, adults of Coleoptera, Diptera and Lepidoptera. Many species are hyperparasitoids of parasitic Hymenoptera. A few chalcidoids are phytophagous, feeding internally in seeds and stems, and some are gall formers.

Chalcidoid eggs may be laid singly (solitary), or in groups (gregarious) in or on a single host, or on foliage quite apart from the host, in which case the larvae, usually planidia, must depend upon a chance encounter with the host. In some Eulophidae and Torymidae an egg may be laid in the host egg, and the adult emerges from a late larval instar or from the pupa. Some Encyrtidae are polyembryonic (Marchal 1904), and Doutt (1947) suggested that the absorption of food by the "feeding embryos" of these species is through a trophamnion; the encyrtid egg is deposited in the developing egg of a lepidopteran, and numerous adults emerge from the fully fed host larva (Hagen 1964).

The number of larval instars varies from one to five, with most having three or four. Many larvae are endoparasitic in early instars and ectoparasitic in later ones. Most endoparasitoids do not have open spiracles (*Anastatus* (Eupelmidae) is an exception, Parker and Thompson 1925), obtaining oxygen from the surrounding host fluids through their integument as in the early instars of the Encyrtidae, and in all instars of the Mymaridae and Trichogrammatidae (Hagen 1964). Some Encyrtidae may respire through the egg stalk in the first instar. Nourishment may be absorbed through the skin, as in some Mymaridae (Hagen 1964), but feeding usually occurs through the mouth. Mid- and hindgut are separated until the end of larval feeding when they become joined, at which time the accumulated fecal material (meconium) is voided.

Pupation does not occur within a cocoon but, depending upon the species, within the host larval skin, pupa or cocoon. In some polyembryonic Encyrtidae the pupae are separated by sheaths that are thought to be cast larval cuticle (Marchal 1904); or a product of the host formed mainly from phagocytic blood cells surrounding the parasitoid (Thorpe 1936); or a cocoon-forming substance from ileac glands (Flanders 1938). A true cocoon is formed by species of *Euplectrus*, as noted above, but the silk is produced by the Malpighian tubules and spun through the anus (Noble 1938). For additional information on the biology of chalcidoids see Gordh (1979).

Description: The typical final-instar chalcidoid larva (fig. 27.130) is hymenopteriform, whitish in colour, with a reduced head, three thoracic and ten abdominal segments. Segmentation is indistinct in the Mymaridae, and in most Eucharitidae there is no external segmentation except for the three principal parts of the body. Some ectoparasitic Eulophidae and Pteromalidae have dorsal and/or ventral tubercles in the intersegmental zones that may be used for locomotion.

Terminology of cephalic structures of final-instar larvae follows Finlayson and Hagen (1977) (fig. 27.5). The complexity of these structures is greatly reduced. The only consistent feature is the presence of sclerotized mandibles, each of which almost always has a short blade without teeth (fig. 27.132); a very few species have a row of fine teeth on each edge of the blade, and *Eurytoma* (Eurytomidae) has a bidentate blade (Parker 1924). The pleurostoma and superior

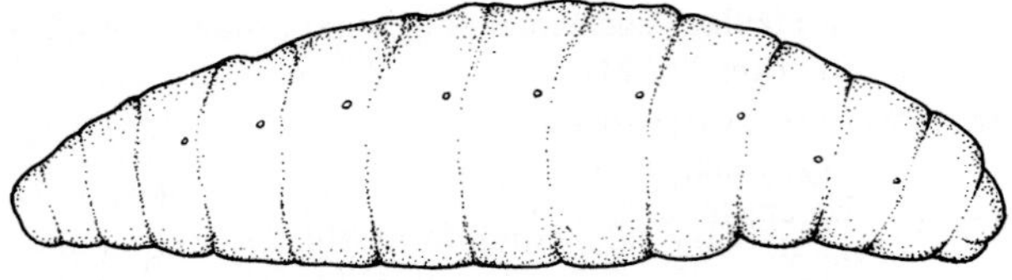

Figure 27.130

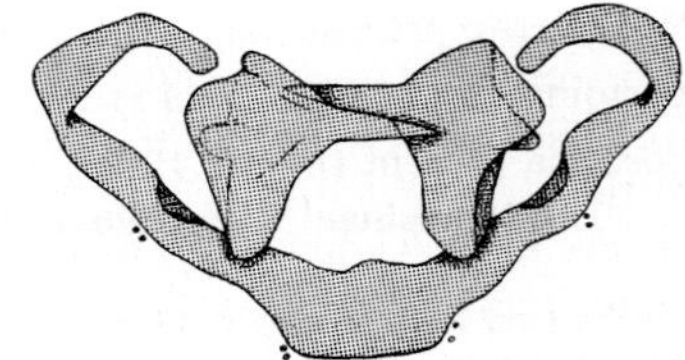

Figure 27.131

Figure 27.132

Figures 27.130–27.132. Final-instar larvae of Chalcidoidea.

Figure 27.130. Chalcididae. Larva of *Brachymeria*.

Figure 27.131. Pteromalidae. Cephalic structures of *Perilampus fulvicornis* Ashm.

Figure 27.132. Pteromalidae. Cephalic structures of *Habrocytus phycidis* Ashm. (Figs. 27.131 and 27.132 from Finlayson 1967.)

and inferior pleurostomal processes are present in some, e.g., *Perilampus* (Pteromalidae) (fig. 27.131) (Finlayson 1967), *Tritneptis, Psychophagus* (Pteromalidae); *Monodontomerus* (Torymidae); and *Tetrastichus* (Eulophidae). Infrequently the epistoma is visible. Antennae are usually not distinguishable. There are usually nine pairs of spiracles except in the Encyrtidae where the number and position is variable (Hagen 1964).

There are several types of eggs, e.g., stalked (Mymaridae, Trichogrammatidae, Pteromalidae, Encyrtidae, Eupelmidae, Torymidae, Eurytomidae, Leucospididae, Chalcididae, Eucharitidae), pedicellate (Eulophidae, Encyrtidae: Aphelininae), encyrtiform (Encyrtidae), and hymenopteriform (Chalcididae, Torymidae) (Hagen 1964).

First-instar larvae are of at least six recognized types: caudate (Encyrtidae), planidium (Pteromalidae: Perilampinae, Eucharitidae), mymariform (Mymaridae, Trichogrammatidae), vesiculate (Encyrtidae), encyrtiform (Encyrtidae), and sacciform (Mymaridae) (Hagen 1964; Clausen 1940). There are four or five pairs of spiracles in the first instar and nine pairs in later instars, whereas ichneumonoids have nine pairs in the first and later instars (Hagen 1964).

Comments: Clausen (1940, pp. 98–238) has summarized descriptions, biology and habits of most of the families of the Chalcidoidea. The greatest differences are found in first instar larvae; these often indicate adaptations to their specific environments rather than phylogenetic relationships, making it difficult to use them taxonomically. No consistent, significant characters have yet been found in final-instar larvae to permit their separation into families, except for the Eupelmidae, which have a heavily sclerotized labrum bearing teeth. It has not been possible to construct a key to separate the families.

Selected Bibliography

For a more complete bibliography see Clausen (1940), Hagen (1964), and Finlayson and Hagen (1977, revised 1979).

Clausen 1940.
Doutt 1947.
Finlayson 1967.
Finlayson and Hagen 1977 (revised 1979).
Flanders 1938.
Gordh 1979.
Hagen 1964.
Krombein et. al. 1979.
Marchal 1904.
Noble 1938.
Parker 1924.
Parker and Thompson 1925.
Peck 1963.
Riek 1970.
Thomsen 1927.
Thorpe 1936.

CYNIPOIDEA

Howard E. Evans, *Colorado State University*

Cynipids, Gall Wasps, and Others

Figures 27.133–27.136; first instar, p. 46

Relationships and Diagnosis: Cynipoidea constitute a large superfamily presumably somewhat related to Chalcidoidea. Relatively few larvae have been described. Gall-formers (Cynipinae) so far as known have tridentate mandibles (fig. 27.135), and some of the parasitoid groups have tridentate or bidentate mandibles; thus these should be separable from the otherwise similar larvae of most Chalcidoidea, which have simple untoothed mandibles. Those having simple mandibles appear to be separable from Chalcidoidea on the basis of the unusually large antennal orbits and by having a line of demarcation between the maxillae and labium.

Biology and Ecology: Cynipinae are gall-makers or inquilines in galls. Ibaliidae, Figitidae, and some Cynipidae (Eucoilinae and Charipinae) are internal parasitoids. Ibaliidae attack siricid larvae; Eucoilinae and most Figitidae attack larvae of Diptera; Charipinae are hyperparasitoids in larvae of Aphidiidae. Many of the parasitoids have a pronounced hypermetamorphosis, and first and second instar larvae are often of unusual form. Most species do not appear to spin cocoons.

Description: Mature larva of a typical gall-former somewhat fusiform, rather tapered posteriorly, anus small, terminal; pleural lobes well developed. Nine pairs of spiracles usually present, two pairs on thorax and one pair on each of first seven abdominal segments; atrium simple, subatrium somewhat tapered from atrium to trachea. Integument smooth, without spines or setae.

Head: Relatively large, wholly unpigmented except for apical half of mandibles, which are very dark; parietal bands absent; antennal orbits small, papilla absent; head devoid of punctures and setae.

Mouthparts: Labrum rounded apically; mandibles broad basally, much less than twice as long as broad, with three large apical teeth; maxillae with a few lateral setae, palpus very short, galea very small and close to palpus; labium well separated from maxilla though extending only slightly beyond it, with very small palpi and an indistinct spinneret.

Comments: The description and figures are based on a series of *Antistrophus pisum* Ashmead, reared from galls of *Lygodesmia juncea* at Fort Collins, Colorado. Short (1952) figured the head of another gall-maker, *Rhodites rosae* L. According to Short, this species has parietal bands and slightly larger antennal orbits than *A. pisum.* There are four large setae on the labrum and evidently no galeae; the spinneret is shown as being very small, circular. In other respects these two larvae resemble one another closely.

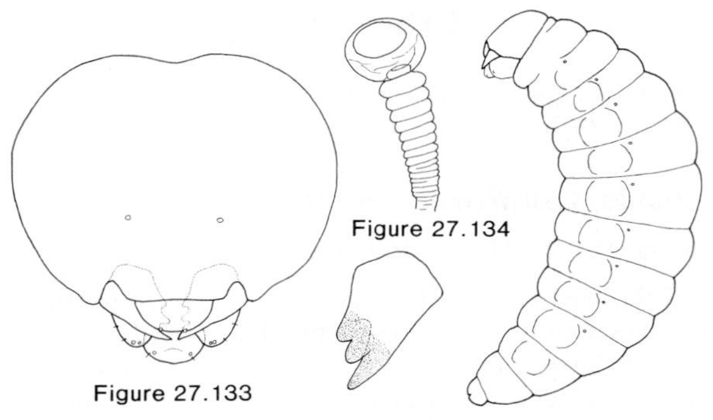

Figures 27.133–27.136. Cynipidae. Larva of *Antistrophus lygodesmiaepisum* Walsh (Cynipinae).

Figure 27.133. Head, anterior view.

Figure 27.134. Anterior thoracic spiracle.

Figure 27.135. Mandible.

Figure 27.136. Body, lateral view.

Among the parasitoid Cynipoidea, tridentate mandibles are reported for *Ibalia* (Ibaliidae) and for Anacharitinae (Figitidae); bidentate mandibles are reported for *Figites* (Figitidae) and for *Trybliographa* and *Hexacola* (Eucoilinae) and *Charips* (Charipinae). Keilin and Baume Pluvinel (1913) state that the mandibles of *Eucoila keilini* Kieffer are simple, unidentate. Whereas most cynipoid larvae have nine pairs of spiracles, those of *Ibalia* are reported to have ten pairs (but last pair vestigial), those of *Hexacola* eight pairs, and those of *Charips* six pairs. Exceedingly large antennal orbits are reported for most of the parasitoid genera that have been studied; however, the antennal orbits of *Ibalia* are very small.

Selected Bibliography

Chrystal 1930.
James 1928.
Keilin and Baume Pluvinel 1913.
Short 1952.
Simmonds 1953.
Wishart and Monteith 1954.

PROCTOTRUPOIDEA

Howard E. Evans, *Colorado State University*

Proctotrupoid Wasps

Figures 27.137–27.141; first instars p. 46

Relationships and Diagnosis: Proctotrupoidea are diverse in adult structure and also quite diverse in larval structure, though chiefly with respect to the degree of reduction of certain features (number of visible body segments, number of functional spiracles, presence or absence of spinneret, etc.). A few groups with unusual features are keyed separately in the family key (Heloridae, Platygastridae, Ceraphronidae).

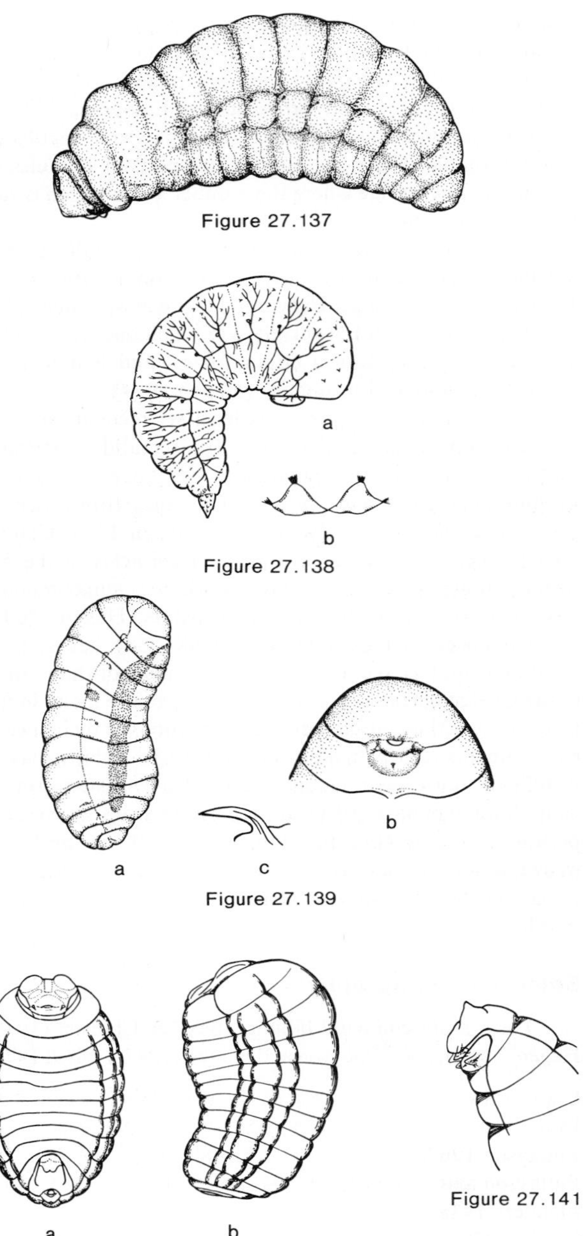

Figures 27.137–27.141. Larvae of Proctotrupoidea.

Figure 27.137. Heloridae. *Helorus paradoxus* (Provancher), lateral view (from Clancy, 1946).

Figure 27.138. Ceraphronidae. a. *Lygocerus* sp., lateral view (redrawn from Haviland 1921); b. mandibles of same (Haviland 1921).

Figure 27.139. Platygastridae. a. *Platygaster herrickii* Fitch (redrawn from Hill and Emery 1937), lateral view; b. ventral aspect of head; c. mandible.

Figure 27.140. Scelionidae. a. *Hadronotus ajax* Girault (from Schell 1943), ventral view; b. lateral view.

Figure 27.141. Diapriidae. *Loxotropa tritoma* (Thomson), lateral view of anterior end (redrawn from Wright, Geering, and Ashby 1946).

The remaining families of this complex cannot easily be separated from Chalcidoidea, and the characters used in the key cannot be fully relied upon. This does not necessarily mean that the two superfamilies are closely related; it simply means that larvae of both groups, being small internal feeders, have undergone similar reductions in morphology. I am unable to present a detailed comprehensive description of last-instar larvae of the superfamily, as relatively few have been described and these show much diversity. Some of this diversity is summarized under "Comments."

Biology and Ecology: With rare exceptions, Proctotrupoidea are primary internal parasitoids of the immature stages of insects, occasionally of the eggs of spiders. Eggs or larvae of diverse orders of Exopterygota and Endopterygota are attacked; a few attack puparia of Diptera; and a few Ceraphronidae are secondary parasitoids. Polyembryony has been well studied in *Platygaster* (Platygastridae). Many Proctotrupoidea exhibit hypermetamorphosis; much information on early instars is summarized in Clausen (1940). Cocoons are spun by many but not all Proctotrupoidea.

Description: Small to minute larvae, up to about 5 mm long (though those of *Pelecinus* may prove to be much larger); with rare exceptions entire body and head unpigmented, whitish or colored like blood of host (tips of mandibles commonly darkened). From 10 to 13 body segments visible behind the head; number of open and apparently functional spiracles varies from one to ten pairs. Integument of body smooth and without spines or setae except in Ceraphronidae. Spiracular atria simple, without ridges or spines.

Head: Unsclerotized (though with a pigmented band in Heloridae); antennae usually visible as small to large, discs on front of head, projecting from surface of head in some Diapriidae.

Mouthparts: Mandibles simple, acute, widely spaced, often minute; maxillae usually discernible as small lobes laterad of the mandibles, often bearing minute sensilla; labium reduced, with or without an evident opening of the salivary glands or spinneret.

Comments: Body form varies considerably; the head is sometimes retracted into the thorax; the posterior end is often blunt, occasionally with a caudal projection. The recorded number of pairs of open spiracles is as follows: one (some Scelionidae and Platygastridae), two (some Proctotrupidae, Scelionidae), three (Diapriidae, some Platygastridae) six to eight (Ceraphronidae, Heloridae), nine (*Telenomus,* in the Scelionidae), ten (*Phaenoserphus,* in the Proctotrupidae). In a number of cases tracheal branches leading to apparently nonfunctional spiracles have been reported. Antennal discs may be absent (*Lygocerus*) or in the form of discs, sometimes large and elevated (*Hadronotus*) or actually projecting (*Loxotropa*). Mandibles vary from vestigial (*Helorus*) to fairly long and sickle-shaped (*Platygaster*); maxillary palpi are said to be present in *Allotropa* (Platygasteridae), but usually the maxillae are mere small, fleshy lobes. The spinneret, when present, is small and circular, or in the form of a small slit.

It seems unwise to attempt to summarize other reported variations in larvae of the superfamily, as much further study is needed. On the basis of available information, it appears that at least most Platygastridae may be recognized by the discoidal bodies on the sides of the first abdominal segment (fig. 27.139a), Heloridae by the melanized band encircling the head posteriorly (fig. 27.137), Ceraphronidae by the conical tubercles on the body segments (*Lygocerus,* fig. 27.138a) or the sclerotized rings on the abdominal segments (*Conostigmus*). It is possible that future study will make it possible to present a key to the families of this diverse complex.

Selected Bibliography

Clancy 1946.
Clausen 1940.
Eastham 1929.
Haviland 1921.
Hill and Emery 1937.
Schell 1943.
Wright et al. 1946.

GASTERUPTIIDAE (EVANIOIDEA)

Howard E. Evans, *Colorado State University*

The Gasteruptiids

Figure 27.142

Relationships and Diagnosis: Relationships of this family are not well understood. The larvae are much more similar to those of wasps than to those of any parasitoid groups. The prominent body setae, numerous head setae (fig. 27.142b), tridentate mandibles (fig. 27.142c), and protruding maxillary palpi but disclike labial palpi serve to separate the known gasteruptiid larvae from those of other groups.

Biology and Ecology: So far as known, these insects are parasitoids of various solitary bees and wasps. A cocoon is spun within the cell of the host.

Description: Body with well-developed setae that are approximately two-thirds as long as length of mandibles. Nine pairs of spiracles present, on second thoracic and first eight abdominal segments; atrium simple, subatrium elongate.

Head: Slightly wider than high, with epicranial suture and parietal bands present though not strongly developed; antennal orbits circular, below middle of head, antennal papilla present; pleurostoma and hypostoma well sclerotized; hypostomal spurs, stipital sclerites, and labial sclerite absent.

Mouthparts: Mandibles well sclerotized, tridentate; maxillae well formed, bearing several setae and a conical palpus; labium broad, bearing several setae and disc-shaped palpi, spinneret somewhat circular.

Comments: This description is largely paraphrased from Short (1978); it is based on *Gasteruption assectator* L.

Selected Bibliography

Short 1978.

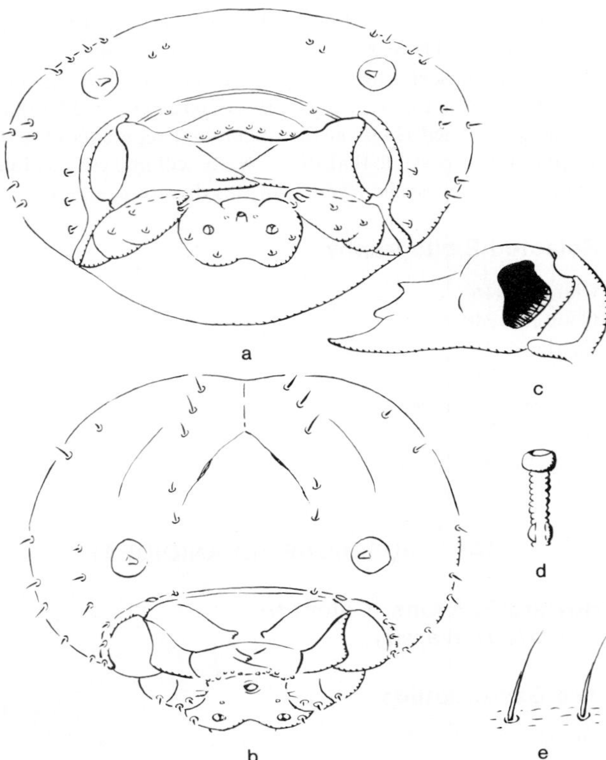

Figure 27.142a-e. Gasteruptiidae. Mature larva of *Gasteruption assectator* L. a. head, ventral aspect; b. head, anterior aspect; c. posterior view of mandibles; d. spiracle; e. body setae (from Short 1978).

AULACIDAE (EVANIOIDEA)

Howard E. Evans, *Colorado State University*

The Aulacids

Figure 27.143

Relationships and Diagnosis: Aulacids have often been grouped with the Gasteruptiidae, sometimes as a subfamily. However, the larvae are not closely similar and the biology is not similar. It appears that the Evanioidea as usually construed is an unnatural group. Aulacid larvae are most readily identified by the paired, sclerotized projections from the epistoma onto the frontal area (fig. 27.143a).

Biology and Ecology: Adult Aulacidae are commonly encountered on tree trunks. The larvae are parasitoids of larvae of wood-boring Coleoptera and Hymenoptera. A silken cocoon is spun within the burrow of the host.

Description: Body smooth, with only minute setae; spiracular atrium simple, with a thin-walled closing apparatus adjoining atrium.

Head: Antennae disc-shaped, without sensilla in the orbit; two sclerotized dorsal projections extending from dorsolateral parts of epistoma onto frontal area; anterior tentorial pits prominent; four sensilla present on clypeus.

Mouthparts: Some 14 sensilla on labrum; pleurostoma and hypostoma unsclerotized; hypostomal spurs, stipital sclerite, and labial sclerite absent; mandible broad basally, blade short, simple, sclerotized; maxillary and labial palpi disc-shaped, each with two sensilla.

Comments: This description is based on *Aulacus striatus* Jurine and is paraphrased from Short (1978).

Selected Bibliography

Short 1978.

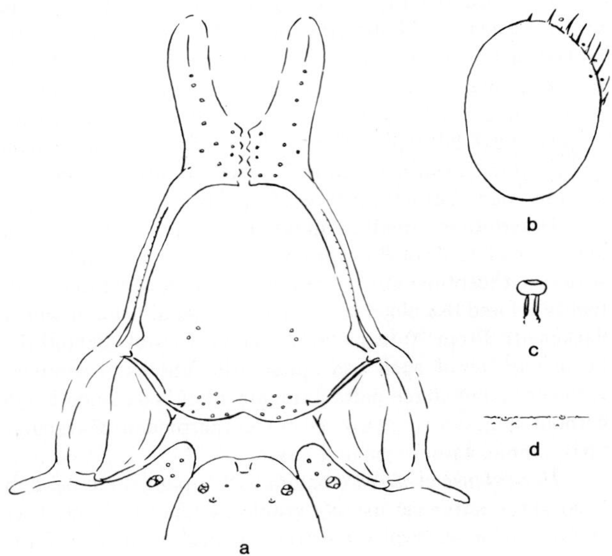

Figure 27.143a-d. Aulacidae. Mature larva of *Aulacus striatus* Jurine: a. mouthparts and frontal region of head; b. antenna; c. spiracle; d. body surface (from Short 1978).

EVANIIDAE (EVANIOIDEA)

Howard E. Evans, *Colorado State University*

Ensign Wasps

Figure 27.144

Relationships and Diagnosis: The evaniids are an isolated group, and neither adults or larvae bear a close resemblance to those of any other group. The larvae have more features in common with wasps than parasitoids; however, the antennal orbits bear no sensilla or papillae, and the maxillae and labium are not strongly differentiated and bear palpi that are disc-shaped and not protuberant. The nine pairs of spiracles and tridentate mandibles suggest a possible relationship with the Gasteruptiidae.

Biology and Ecology: Ensign wasps are parasitoids of cockroach egg capsules. The egg is deposited in the oötheca before the covering is hardened; the larva commonly consumes several eggs. Pupation occurs within the oötheca, which serves as a cocoon.

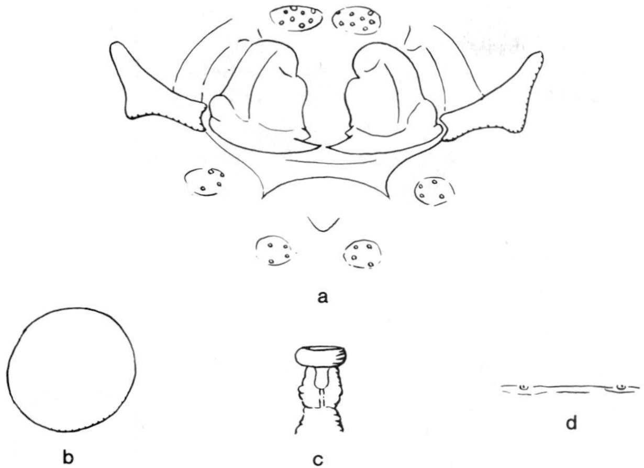

Figure 27.144a-d. Evaniidae. Mature larva of *Brachygaster minuta* Olivier. a. mouthparts; b. antenna; c. spiracle; d. body surface (from Short 1978).

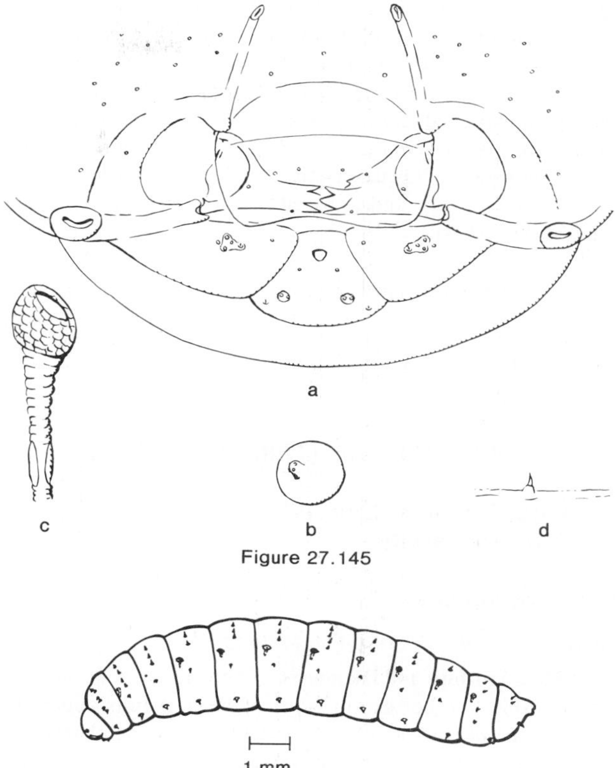

Figure 27.145

Figure 27.146

Figure 27.145a-d. Stephanidae. Mature larva of *Schlettererius cinctipes* Cresson. a. mouthparts; b. antenna; c. spiracle; d. spine on body surface (from Short 1978).

Figure 27.146. Stephanidae. Mature larva of *Schlettererius cinctipes* Cresson, lateral view (from Taylor 1967).

Description: Body robust, grublike, integument with minute setae; nine pairs of spiracles present, atrium simple, with closing apparatus adjoining atrium.

Head: Wider than high; parietal bands absent; antennal orbits large, circular, without sensilla or papillae; head without conspicuous setae. Epistoma unsclerotized, pleurostoma lightly sclerotized, hypostoma short and well sclerotized.

Mouthparts: Labrum with two disc-shaped areas, each with several small sensilla; mandibles broad, tridentate; maxillary and labial palpi disclike, each bearing four sensilla, maxillae and labium not protuberant, partially separated by a line at least in some species.

Comments: This description is based in part on that of Short (1978) based on *Brachygaster minuta* Olivier, and in part on that of Cameron (1957) based on *Evania appendigaster* L. There appear to be no major differences between these two species, and Genieys' (1924) description of the larva of *Zeuxevania splendidula* Costa is also similar.

Selected Bibliography

Cameron 1957.
Genieys 1924.
Short 1978.

STEPHANIDAE (EVANIOIDEA)

Howard E. Evans, *Colorado State University*

The Stephanids

Figures 27.145, 27.146

Relationships and Diagnosis: The family Stephanidae is often placed in the Ichneumonoidea, but the larvae are much more like those of Aculeata. The form of the maxillae and labium is not unlike that of Evanioidea, and the mandibles and antennae resemble those of Gasteruptiidae, so there seems to be no reason not to place the family in the Evanioidea, especially since the biology resembles that of Aulacidae. Each body segment bears several small, sclerotized spines in a transverse series (fig. 27.146). This feature, plus the characteristically shaped mandibles, will serve to separate this family from other Evanioidea and Aculeata.

Biology and Ecology: Adults occur on dead or dying trees and are said to attack larvae of wood-boring Coleoptera. However, the only detailed life history study, that of Taylor (1967), indicates that at least one species is an ectoparasitoid of the larva of *Sirex* (Hymenoptera: Siricidae). Pupation is within the burrow of the host and no cocoon is spun.

Description: Body cylindrical, abruptly tapered at both ends; each body segment with a transverse row of small, sclerotized spines, one ventrally, one pair laterally, below each spiracle, and from two to four dorsally. Nine pairs of functional spiracles present, the second thoracic pair evidently vestigial or absent; atrium with a network of ridges, subatrium elongate. Head wider than high; parietal bands absent; antennal orbits circular, situated rather low on head, each with a short papilla. Labrum transverse, slightly emarginate; pleurostoma and hypostoma very lightly sclerotized; mandibles broad basally, tapered to a tridentate apex; maxillae and labium separated, but neither protuberant; maxillary palpi flattened and triangular, each bearing three sensilla and a seta; labial palpi disc-shaped, each with two sensilla.

Comments: This description is based on *Schletterius cinctipes* (Cresson) and is based on those of Taylor (1967) and Short (1978).

Selected Bibliography

Short 1978.
Taylor 1967.

TRIGONALIDAE (TRIGONALOIDEA)

Howard E. Evans, *Colorado State University*

The Trigonalids

Figures 27.147, 27.148; first instar p. 46

Relationships and Diagnosis: Trigonalidae bear no close relationship to other extant Hymenoptera. Adults appear to share some characters of the Aculeata and some characters of generalized parasitoids. The larvae are more characteristic of parasitoids and are best recognized by the very large head and sickle-shaped mandibles of the third instar (fig. 27.148e) and by the characteristically shaped 3-toothed mandibles of the final instar (fig. 27.148d). Detailed information on larval features is not available.

Biology and Ecology: A few trigonalids are primary parasitoids of sawfly or vespoid larvae, whereas others appear to be secondary parasitoids of sawflies or of Lepidoptera larvae by way of their ichneumonoid or tachinid primary parasitoids. Large numbers of minute eggs are laid, often on the underside of leaves, and in at least some cases these must evidently be eaten by a caterpillar or sawfly larva in order to hatch. Those species attacking vespoid wasps evidently gain access to the nests when the caterpillars are carried to the nest by the wasps. The first three instars develop internally, the last two externally. There is much cannibalism, and only one trigonalid larva survives to pupation, which in some cases occurs inside the host's cocoon.

Description: Mature larvae 9–17 mm long, integument pale but body sometimes appearing brown because contents of the digestive tract show through; head sometimes pigmented; body smooth, without setae or spines, spindle-shaped and tapered at both ends (especially posteriorly) (*Poecilogonalos*) or more cylindrical and less tapered (*Lycogaster*). Eight pairs of spiracles present, two pairs on the thorax and six on the first six abdominal segments.

Head: Large, truncate; antennae conical, each with two small sensilla.

Mouthparts: Represented by a pair of sclerotized mandibles and adjacent pleurostoma and stipital sclerite; mandibles broad basally but strongly tapered, slightly expanded to three acute teeth apically (confirmed for several genera). Maxillae and labium not described in detail, but apparently without prominent palpi or spinneret. Clausen (1929) reports that the mouthparts contain a "suctorial disc," but it is not clear what this represents.

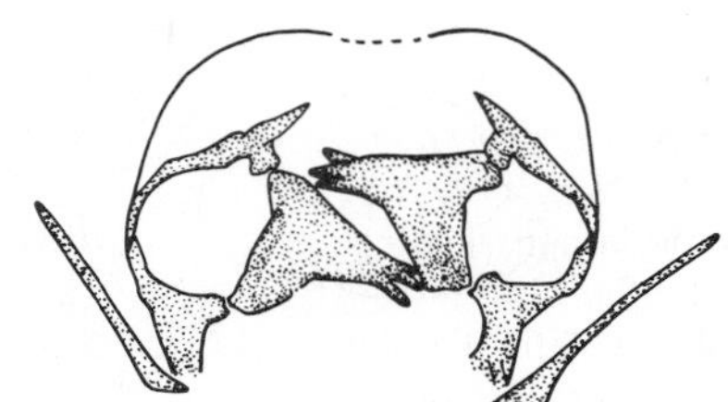

Figure 27.147

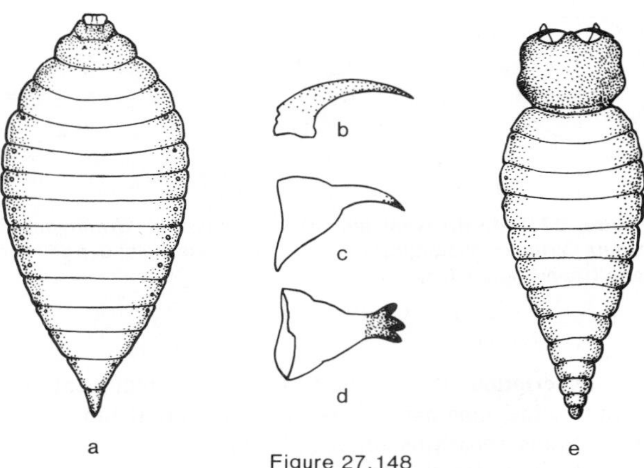

Figure 27.148

Figure 27.147. Trigonalidae. Sclerotized portion of head of *Taeniogonalos* sp (from Riek 1962).

Figure 27.148a-e. Trigonalidae. *Poecilogonalos henicospili* Rohwer. a. body; b. same, mandibles of third instar; c. same, mandibles of fourth instar; d. same, mandibles of fifth (final) instar; e. same, third instar larva (from Clausen 1929).

Selected Bibliography

Clausen 1929.
Cooper 1954.
Riek 1962.

BETHYLIDAE (CHRYSIDOIDEA)

Howard E. Evans, *Colorado State University*

The Bethylids

Figure 27.149

Relationships and Diagnosis: Larvae of Bethylidae feed externally, often gregariously, on larvae of Coleoptera or Lepidoptera, usually on species which themselves feed internally. Published descriptions differ in some details, and this diagnosis is based on an original study of two species of *Goniozus*, with discussion of published descriptions following. There are ten pairs of spiracles, the most anterior pair considerably larger than the others; the head is prognathous (fig. 27.149d), the labium tending to project well beyond the other mouthparts (fig. 27.149b); the mandibles are broad basally, apically denticulate (fig. 27.149e); galeae are absent; the spinneret is an arcuate slit, without raised lips (fig. 27.149e).

Biology and Ecology: Bethylidae are parasitoids of larvae of Coleoptera and Lepidoptera; a few species, however, are reported to drag their paralyzed hosts to a crevice before oviposition. Female Bethylidae often subdue hosts larger than themselves by repeated stinging; they may remain on the host for some time, feeding on body fluids, before laying several eggs. Maternal care is well developed in species of *Scleroderma*. Egg position is variable but tends to be characteristic of genera. Larvae feed with their mandibles through the integument of the host, but in the last instar the anterior third of the body may be inserted through a circular hole in the host's body. Last instar larvae may stand more or less erect on the host. Cocoons are silken, usually whitish, and are often spun next to the shrunken body of the host. Most bethylids do not appear to be strongly host-specific. For reviews, see Clausen (1940), Evans (1964a), and Gordh (1976).

Description: Very small larvae of pale coloration, unsclerotized and unpigmented except for mandibles and sclerotic braces of maxillae and labium. Body spindle-shaped, widest at abdominal segments 6–7; anus small, terminal; pleural lobes not well developed; middle body segments weakly produced dorsally, more anterior segments with indistinct division into two annulets dorsally; body segmentation somewhat indistinct. Ten pairs of apparently functional spiracles, but the most anterior pair the largest, second pair with atrium only about one-third the diameter of first pair, remaining pairs about two-thirds the diameter of first pair; atrium simple except for some weak lines on circumference; opening into subatrium unarmed; subatrium tapering, with a series of swellings. Integument smooth except thoracic segments with bands of spinules that are directed caudad; these spinules located dorsally, laterally, and ventrally.

Head: Prognathous, posterior part sunken into prothorax and not completely formed; coronal suture and parietal bands absent. Antennae represented by two sensory cones each, these on a prominence but not in a distinct orbit. Head with two pairs of setae only. In lateral view, the head appears completely divided into an upper cranium and a low portion consisting of the large maxillary and labial bases; the tentorial arms are located at the extreme base of this excision. Clypeus not distinct, clypeolabral suture barely discernible.

Mouthparts: Labrum broadly rounded, with a thickened margin that is slightly spinulose laterally; surface with two setae but without sensilla; epipharynx apparently smooth. Mandibles strongly tapered from a broad base, denticulate on the oral margin, not capable of protruding much if any beyond margin of labrum. Maxilla with a strong but very short palpus that bears five sensilla, three of them much larger than the remaining two; a single seta located basad of the palpus; galea absent; tip of maxilla with several rows of minute spinules. Labium, at least in spinning larvae, protruding far beyond other mouthparts; labial palpi similar to maxillary palpi; labium with two pairs of setae; spinneret a slightly arched slit without raised lips.

Comments: This description and the figures are based entirely on two species of *Goniozus* that appear to have identical larvae: *G. gordhi* Evans (Gordh 1976, identified as *gallicola* Fouts) and *G. angulata* (Muesebeck) (not labeled as to locality, but believed to be from near Sydney, Australia; *see* Muesebeck 1933, *Proc. Ent. Soc. Wash.* 35:54). Silvestri (1923) described the larva of *G. gallicola* (Kieffer), which is evidently quite similar. Silvestri reported the mandibles of the last-instar larva to be simple, but it should be noted that the mandibles of larvae of *G. gordhi* and *G. angulata* also appear simple when viewed at certain angles.

Laelius anthrenivorus Trani (Vance and Parker 1932) is reported to have only one pair of spiracles, on the thorax. This species also has large antennal orbits without sensilla, an emarginate labrum, and maxillae that lack spinules. In other respects it is similar to *Goniozus:* the mandibles are denticulate, the spinneret is similar, the galeae lacking, and the palpi very short.

Williams (1919) found that the mandibles of the last instar of *Epyris extraneus* Bridwell had three large apical teeth,

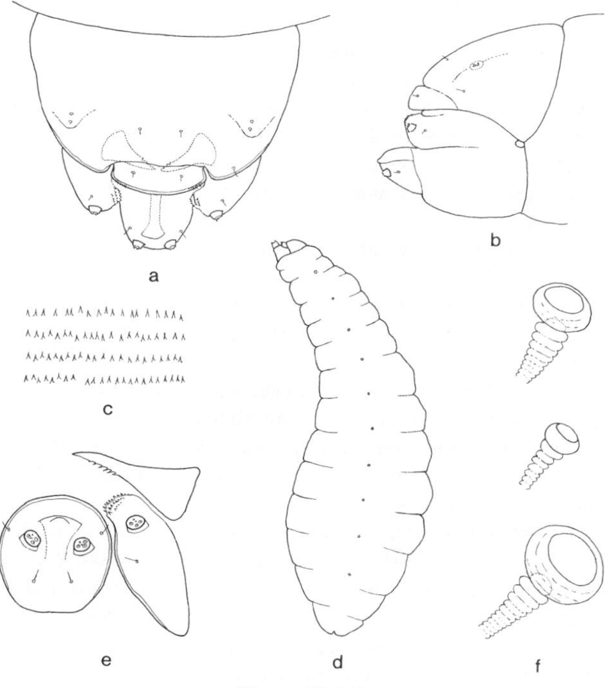

Figure 27.149

Figure 27.149a-f. Bethylidae. *Goniozus gordhi* Evans: a. head, dorsal view; b. head, lateral view; c. spinules on thoracic venter; d. body, lateral view; e. labium, maxilla, and mandible; f. spiracles of first three pairs, drawn to same scale, most anterior spiracle below, second pair above it, third pair at top.

very different from those of *Goniozus* and *Laelius*. Feeding larvae of this and several other species have the apical abdominal segment tapering and nipplelike. Clearly there is still much to be learned regarding larval characters in this family.

Selected Bibliography

Clausen 1940.
Evans 1964a.
Gordh 1976.
Silvestri 1923.
Vance and Parker 1932.
Williams 1919.

DRYINIDAE (CHRYSIDOIDEA)

Howard E. Evans, *Colorado State University*

The Dryinids

Figure 27.150

Relationships and Diagnosis: Last-instar larvae of Dryinidae are U-shaped (fig. 27.150c), and contained in a sac consisting of darkened, cast larval skins; this sac projects from between the abdominal segments of their hosts, Homoptera Auchenorrhyncha. The head is remarkably large, especially the mouthparts, which however are weakly sclerotized except for the mandibles. Antennal orbits are minute and have two sensilla; the mandibles are simple or denticulate; the maxillary palpi are very large, but the labial palpi barely protrude, and there are no galeae. Position of this family in the Chrysidoidea is confirmed by larval characters as there are many features in common with Bethylidae (antennae, labrum, mandibles, absence of galeae, form of labial palpi and spinneret).

Biology and Ecology: Female Dryinidae grasp the host with raptorial fore legs and lay the egg internally. The first-instar larvae also develop internally, but the following instars are external, being enclosed in a sac and feeding through the intersegmental membranes of the abdomen. Mature larvae leave the host and spin a silken cocoon nearby or in the soil. Reviews of the biology have been presented by Fenton (1918) and by Clausen (1940).

Description: Small larvae, not exceeding 5 mm in length, body U-shaped, whitish, but enclosed in a brownish sac formed of partially cast larva skins. Body segments distinct or indistinct; anus not at all conspicuous; pleural lobes absent. Spiracles minute, but first pair slightly larger than remaining eight pairs, atrium simple, only slightly wider than tracheae. Integument smooth, either without spinules and setae or (in *Aphelopus*) with segmentally arranged bands of spines and large setae.

Head: Very large, at least half as high as maximum height of body (lateral view), distinctly prognathous. Coronal suture and parietal bands absent; antennal orbits small and indistinct, each with two minute sensilla; anterior part of head capsule with several to many large, curved setae. Clypeus not distinct.

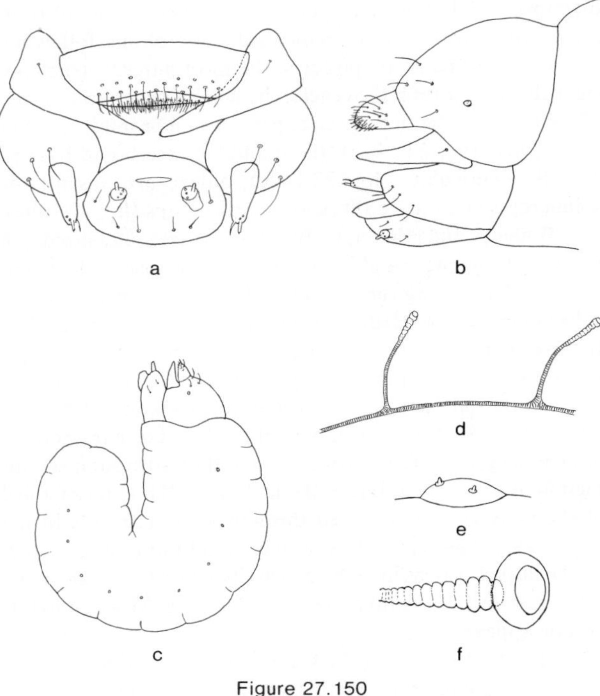

Figure 27.150

Figure 27.150a-f. Dryinidae: Gonatopodini. a. mouthparts; b. head, lateral view; c. body, lateral view; d. portion of tracheal system, showing two spiracles; e. antenna, lateral view; f. spiracle.

Mouthparts: Large, occupying a considerable part of the total area of the head. Labrum transverse, rounded apically, bearing numerous long, curved setae; epipharynx with several setae close to the apical margin and with many small, slender spinules. Mandibles tapering from a wide base, usually somewhat pigmented apically, simple or denticulate. Maxillary palpi unusually large and projecting; galeae absent; inner margin of maxilla not spinulose. Labial palpi very short, like the maxillary palpi with a somewhat fingerlike apical process (at least in Gonatopodini); spinneret a transverse slit without raised lips.

Comments: The specimens from which the drawings were made were unidentified to genus and were taken from *Delphacodes lutulenta* adults from Jasper City, Illinois. At a later date, Dr. Paul Freytag sent me specimens of a dryinid from the same host and they appeared to be identical to the specimens I had drawn. These were from a culture of *Pseudogonatopus similis* Freytag.

The larva of *Gonatopus pilosus* Thoms., figured by Mik (1882) has many similarities to that described above, including the large, simple mandibles and elongate maxillary palpi. Ponomarenko (1975) indicates ten pairs of spiracles in Gonatopodini, the most anterior pair being much below the level of the others; Clausen (1940) also states that some Dryinidae have ten pairs of spiracles. Ponomarenko also figures larvae of the genera *Aphelopus* and *Anteon,* which belong to tribes other than Gonatopodini, but there appear to be no differences in major features. Fenton (1918) reported a fleshy "ventral process" in *Gonatopus,* and this is also figured by Ponomarenko in an unknown Gonatopodini.

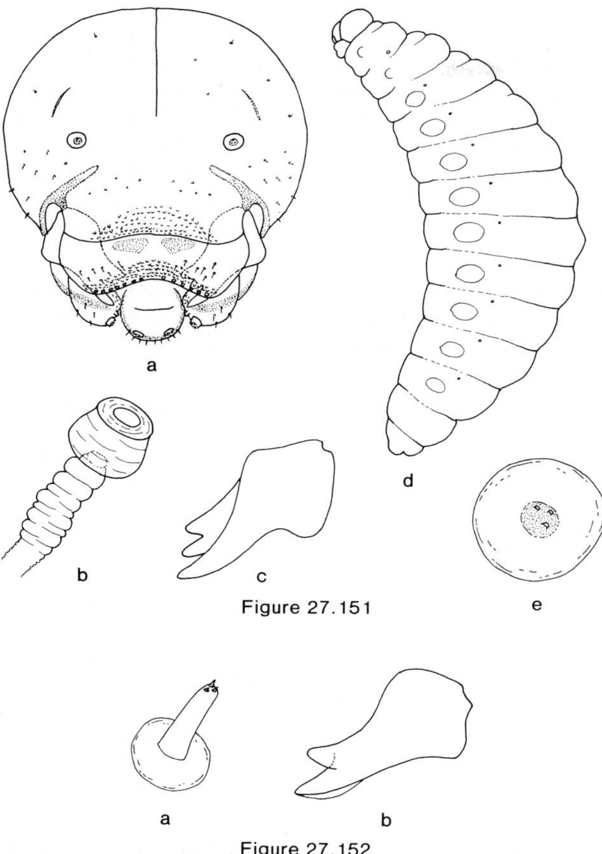

Figure 27.151a-e. Chrysididae. *Chrysis smaragdula* Fabr. a. head, anterior view; b. spiracle of most anterior pair; c. mandible, ventral aspect; d. body, lateral view; e. antenna.

Figure 27.152a,b. Chrysididae. *Omalus aeneus* Fabr. a. antenna b. mandible, dorsal aspect.

Selected Bibliography

Clausen 1940.
Fenton 1918.
Mik 1882.
Perkins 1905.
Ponomarenko 1975.

CHRYSIDIDAE (CHRYSIDOIDEA)

Howard E. Evans, *Colorado State University*

Cuckoo Wasps

Figures 27.151, 27.152

Relationships and Diagnosis: Despite the specializations of adult cuckoo wasps, the mature larvae are relatively generalized Aculeata. The body form is unspecialized and there are ten pairs of spiracles; the mandibles are broad, with two to four teeth, the spinneret a transverse slit. Galeae are apparently lacking in some Chrysididae, a specialized feature, but small galeae occur in most species. Small to rather long antennal papillae occur in most species, but in at least two species of *Chrysis,* the orbits each have three small sensilla set in the membrane. The palpi are short, broader than long. First instar larvae have sickle-shaped mandibles and spines and elevations on the body as well, apparently helping them to find and destroy the egg or small larva of the host; these larvae also have long antennal papillae, which may be absent in the mature larvae.

Biology and Ecology: Most Chrysididae are cleptoparasites of bees and wasps. Females lay their eggs in nest-cells of the host, and the first instar larva destroys the egg or small larva of the host before consuming the provisions in the cell. However, some species are true parasitoids, attacking prepupae of the host. At least one species is an external parasitoid of Lepidoptera larvae (Parker 1936). Reviews of the biology have been presented by Clausen (1940) and by Krombein (1967).

Description: Body of mature larva robust, broadest near middle of body, anus small, terminal. Pleural lobes moderately developed; some body segments, in *Chrysis,* divided into two annulets by a transverse crease; in *Omalus* middle body segments somewhat humped dorsolaterally. Integument apparently smooth, with only sparse, minute setae. Spiracles relatively large and conspicuous, first pair slightly larger than remaining nine pairs; peritreme well developed; atrium simple, lined with weak ridges or none at all, opening into subatrium unarmed.

Head: Small, only about 0.25–0.30 × maximum body width in most species; wider than high, vertex rounded; coronal suture distinct in most species, but parietal bands weak or more commonly absent. Antennal orbits circular, located near or below middle of head, with three small sensilla set in membrane of orbit or (more commonly) at the terminus of a short to rather long papilla. Head with sparse setigerous punctures, all setae very short. Clypeolabral suture located well below mandibular articulations, straight or slightly emarginate.

Mouthparts: Labrum broad, strongly emarginate, bearing some short setae and a series of marginal sensilla that may be numerous but quite small (*Chrysis*) or few in number and strongly protuberant (*Omalus*). Mandibles broad basally, though more slender apically and nearly or about twice as long as their basal width, without lateral setae or sensilla; apex tridentate in *Chrysis,* in *Omalus* bidentate or multidentate. Maxillae with a few lateral setae, mesal margin not spinulose but sometimes papillose; palpi large but wider than long; galea very small, sometimes absent. Labial palpi rather flat, not protuberant; spinneret a transverse slit. Maxillae and labium supported by strong, pigmented and apparently sclerotic bands.

Comments: I have studied the larvae of several species of Chrysididae, and Grandi (1959) has provided detailed descriptions and figures of two species. Other species have been described by Parker (1936) and by Janvier (1933). In *Chrysis,* three-toothed mandibles appear to be consistently present, and small galeae are present in the eight species I have seen; however, galeae are said to be absent in *C. dichroa* Dahlb. (Grandi 1959) and are not figured for *C. shanghaiensis* Smith (Parker

1936). In this genus, antennal papillae are absent in these two species, but short papillae are present in the following species: *carinata* Say, *cembricola* Krombein, *chalcopyga* Mocsary, *coerulans* Fabr., *inflata* Aaron, *texana* Gribodo, and *verticalis* Patton. In *Omalus,* the antennal orbits are situated quite low, and a well-formed papilla is present; however, the mandibles have four or five teeth (*auratus* L., Grandi 1959) or are bidentate, the lower tooth with a rather thin outer margin toward the apex (*aeneus* Fabr., fig. 27.152b). Obviously, there is much variation in major larval features which does not appear to concord closely with generic characters based on adults.

Selected Bibliography

Grandi 1959.
Janvier 1933.
Krombein 1967.
Parker 1936.

SCOLIIDAE (SCOLIOIDEA)

Howard E. Evans, *Colorado State University*

The Scoliids

Figure 27.154

Relationships and Diagnosis: Scoliid larvae are fairly typical of the superfamily, having subcutaneous, sclerotic rods supporting the maxillae, rather slender mandibles with three teeth in about the same plane, the spinneret transverse but much more prominent laterally than medially. In contrast to Tiphiidae and Mutillidae, the parietal bands are strong (fig. 27.154a) and there are ten pairs of well-developed spiracles. A unique feature is the presence of five or six sensilla in each antennal orbit (fig. 27.154d) in contrast to the usual three.

Biology and Ecology: These wasps attack the larvae of Scarabaeidae and Curculionidae, chiefly soil-inhabiting species. The females burrow in the soil in search of their hosts, which are stung into permanent paralysis. The grub is then usually dragged more deeply into the soil and a crude cell prepared; the egg is laid perpendicularly on the median ventral line of the third or fourth abdominal segment. Larval feeding requires about a week in most species, and a brownish cocoon is then spun. Emergence from the cocoon is through a circular cap at the anterior end.

Description: Described larvae are between 22 and 34 mm long when fully extended, though smaller larvae are to be expected. Body whitish, head capsule somewhat brownish in color. Pleural lobes well developed, each body segment except at extreme anterior and posterior ends with a transverse crease (in the nondiapausing larva of *Scolia hirta*) or with a transverse welt that is interrupted on the middorsal line (*Scolia ruficornis*). Spiracles darkly pigmented, all ten pairs about the same size; atrium shallow, with a large, circular opening to the exterior, outer rim elevated slightly above surface of integument; atrium lined throughout with strong spines, those close to the opening into the subatrium essentially filling the entire opening. Integument with fine wrinkles but virtually devoid of spines and setae.

Head: Width less than 0.2 that of maximum body width of diapausing larva. Head wider than high, with strong parietal bands and a short coronal suture. Antennal orbits circular, located well below middle of head, each orbit with four to six sensilla. Frontoclypeal suture weakly defined or absent; clypeolabral suture located well below level of mandibular articulations. Head with an abundance of minute punctures, some bearing minute setae, the punctures grouped in several clusters, especially mesad of the antennae and on the clypeus.

Mouthparts: Labrum broad, weakly bilobed, apical half with many small punctures, some bearing minute setae, apical margin weakly spinulose laterally, also with a series of 16–24 sensory cones; epipharynx minutely spinulose except medially, where there are scattered sensory pores. Mandibles approximately twice as long as their maximum width, without setae or roughened areas, apically with three large teeth in the same plane. Maxillae weakly setose laterally, with a sclerotic, pigmented brace ventrally; palpi and galeae conical, subequal in length; mesal margin slightly produced, spinulose. Labium broad, with a few weak setae, palpi conical, nearly as long as maxillary palpi; spinneret a transverse slit,

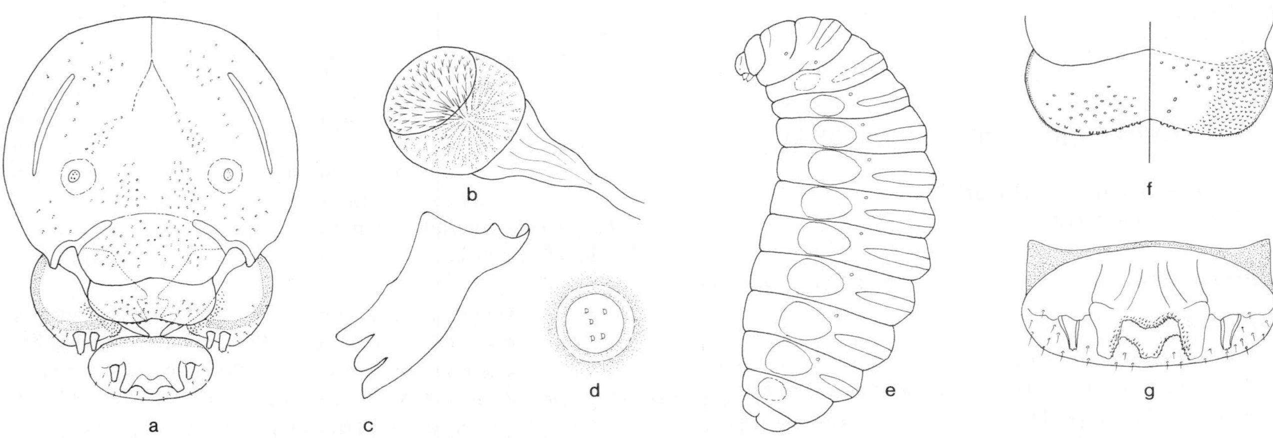

Figure 27.154a-g. Scoliidae. *Scolia ruficornis* Fabr. a. head, anterior view; b. spiracle of most anterior pair; c. mandible; d. antenna; e. body, lateral view; f. labrum (left side) and epipharynx (right side); g. labium.

much more prominent laterally than medially, or produced both laterally and medially and thus biemarginate.

Comments: This description is a composite of that of Grandi (1940) on *Scolia hirta* Schrank and my own study of *S. ruficornis* Fabr., which is figured here. I am not aware that the larvae of any other Scoliidae have been studied in detail. Members of other genera may well differ in some details from those of the genus *Scolia*. The specimens of *S. ruficornis* I studied were collected on the Palau Islands, February 1959, by R. Owen (U.S. National Museum).

Selected Bibliography

Grandi 1940.

TIPHIIDAE (SCOLIOIDEA)

Howard E. Evans, *Colorado State University*

The Tiphiids

Figures 27.156-27.158

Relationships and Diagnosis: Tiphiidae are regarded as among the more generalized Aculeata, and the larvae are generalized with respect to the simple body form, four-toothed mandibles, well-developed galeae, and transverse spinneret. However, the second pair of thoracic spiracles is reduced or even apparently absent, as in Mutillidae and Pompilidae. The antennal orbits are not much if any elevated, the three sensilla resting in the membrane of the orbit. The head and body have only minute, scattered setae.

Biology and Ecology: Tiphiids attack the larvae of Coleoptera that occur in the soil or in rotting wood. The prey is stung into a usually temporary paralysis and the egg is laid on the body. There is no nest, the grub being left in place or less commonly dragged into a cavity. After feeding, the larva spins a brownish, silken cocoon. A good review of biology is provided by Clausen (1940) under the headings Tiphiidae, Thynnidae, and Methocidae, the latter two groups now usually being regarded as subfamilies of Tiphiidae.

Description: Known larvae 10–15 mm long, whitish in color, with the head capsule weakly or not at all pigmented except for parts of the mouthparts. Body elongate-fusiform; pleural lobes well formed, some species with transverse dorsolateral welts on middle body segments; anus terminal, transverse. Spiracles small, circular, nine pairs distinct, second pair of thoracic spiracles very much reduced and probably nonfunctional (may be absent in *Methocha*). Spiracular atrium lined with ridges that are often spinose; opening into subatrium armed with spines or unarmed. Integument apparently smooth but under high power seen to be minutely

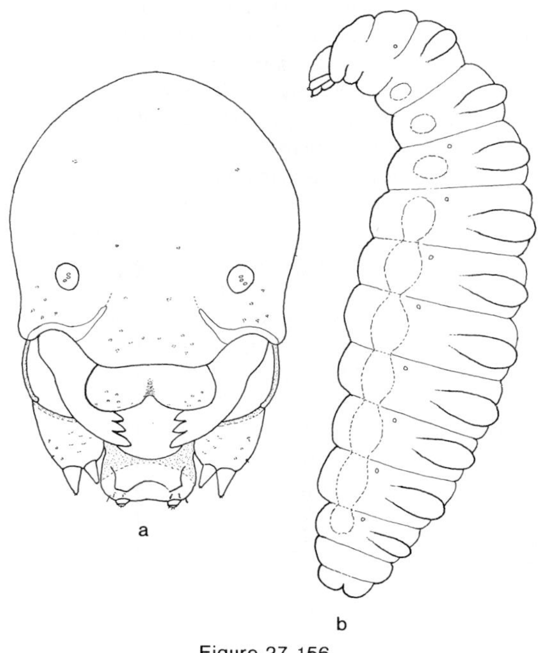

Figure 27.156

Figure 27.156a,b. Tiphiidae. *Pterombrus piceus* Krombein. a. head; b. same, lateral view of body.

Figure 27.157a-c. Tiphiidae. a. *Methocha stygia* (Say). a. spiracle of most anterior pair; b. same, mandible; c. same, labrum (left side) and epipharynx (right side).

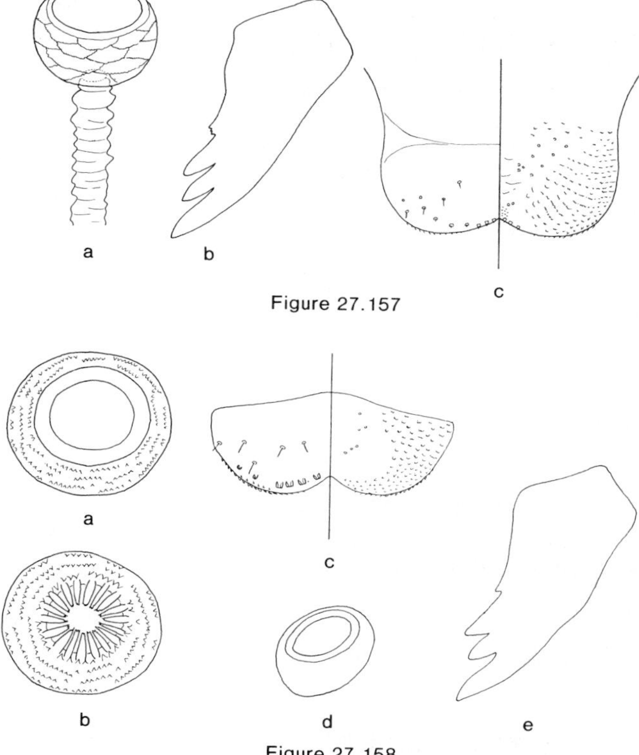

Figure 27.158

Figure 27.158a-e. Tiphiidae. *Tiphia micropunctata* Allen. a. spiracles of anterior pair at plane of peritreme; b. same, spiracle of anterior pair at entrance to subatrium; c. same, labrum (left side) and epipharynx (right side); d. same, vestigial second thoracic spiracle (drawn to same scale as a and b); e. same, mandible.

spinulose or granulose; body setae short, sparse or sometimes fairly numerous on certain parts of body.

Head: Width not more than about one-fourth that of maximum body width of nondiapausing larvae; head subcircular in anterior view, in some species vertex depressed medially. Antennal orbits circular or nearly so, located below middle of head, neither orbits nor membrane within them notably projecting, center of orbit with three small sensilla. Parietal bands absent so far as known. Frontoclypeal suture not well defined; clypeolabral suture well below mandibular articulations. Head setae sparse and short.

Mouthparts: Labrum weakly bilobed, with a few short setae and several barrel-shaped sensoria near the margin, which is at most weakly bristly; epipharynx with numerous minute spinules except laterally. Mandibles slightly more than twice as long as basal width, without setae or roughened areas, with four apical teeth in about the same plane, most basal tooth quite small and sometimes subdivided (mandibles tridentate in *Elaphroptera,* according to Janvier 1933). Maxillae projecting, with large, cone-shaped palpi and galeae, setose laterally but with at most weak spinules on mesal margin. Labium rather broad, palpi short, spinneret transverse, much more prominent laterally than medially. The maxillae are supported by pigmented sclerotic rods, as in other Scolioidea.

Comments: Larvae of Tiphiidae have not been well studied, but as of this writing there seems to be little diversity within the family. Evans (1965) speculated that the larvae of the primitive subfamily Anthoboscinae might have a full complement of well-developed spiracles, but that has proved not to be the case (Brothers 1975). Janvier (1933) figures the spiracles of two species of *Elaphroptera* (Thynninae) as being on the first nine abdominal segments, but surely that is an error.

Selected Bibliography

Brothers 1975.
Clausen 1940.
Evans 1965.
Janvier 1933.
Krombein and Evans 1976.
Ridsdill Smith 1970.
Rivers et al. 1979.

MUTILLIDAE (SCOLIOIDEA)

Howard E. Evans, *Colorado State University*

The Mutillids, Velvet Ants

Figure 27.155,

Relationships and Diagnosis: Mutillidae are closely related to Tiphiidae, and certain groups (e.g., Myrmosinae) have occasionally been shifted from one group to the other. As in Tiphiidae, the second pair of thoracic spiracles is greatly reduced and the mandibles are quadridentate, with a tendency toward modification or reduction of the most basal tooth. On the basis of the few species that have been described, the best separation from Tiphiidae appears to be on the basis of the somewhat mammiform antennae, palpi, and galeae (fig. 27.155a).

Biology and Ecology: Mutillidae are external parasitoids of immature bees and wasps, less commonly of muscoid Diptera and chrysomelid beetles. General reviews have been presented by Mickel (1928) and by Clausen (1940); the most detailed study of a single species is that of Brothers (1972) on *Pseudomethoca frigida* (Smith), which attacks halictid bees. Adult female mutillids are wingless and run over the ground in search of their hosts, most of which occur in nests in the soil; females commonly feed on pupae or prepupae of the host, and certain Old World species are reported to feed on adult honey bees and digger wasps. The egg is usually laid in the nest cell on the mature larva or pupa, which may or may not be stung and paralyzed; after the host forms a cocoon or puparium the female chews a hole in the wall for oviposition and this is later closed up. The fully grown larva spins a silken cocoon, sometimes within the host cocoon.

Description: Mature larva 5–20 mm long, maximum width from about one-third to more than one-half the length, diapausing larvae especially broad because of the prominent pleural lobes. Middle body segments convex dorsally, with faint indication of a transverse crease; anal segment rounded above and below the transverse anal slit. Ten pairs of spiracles present, but the second pair is much smaller than the others and apparently nonfunctional; other spiracles with the atrium lined with only very weak ridges, opening into subatrium unarmed, subatrium long, irregularly expanded before narrowing to the trachea. Integument covered with sharp, minute spinules, especially dorsally, also with a few very small setae, more especially on the pleural lobes.

Head: Width between one-fourth and one-third that of maximum body width, wider than high, with a more or less straight vertex; parietal bands absent or faintly indicated; antennal orbits located below middle of head, located on mammiform elevations, without a distinct papilla but the three sensoria arising from a rounded elevation. Frontoclypeal suture and clypeolabral sutures distinct, the latter well below mandibular articulations. Head setae very sparse. Head unpigmented except for antennal elevations, tentorial arms, hypostomal thickenings, and parts of mouthparts.

Mouthparts: Labrum weakly bilobed, shallowly emarginate; surface with a transverse row of punctures, partly setigerous, and an apical row of conical sensoria; margin very weakly spinulose; epipharynx weakly spinulose apically and laterally, with scattered sensory pores toward the midline. Mandibles robust, 1.5–2 $\times$ as long as wide, without setae or roughened areas, with four apical teeth, the most basal tooth either shorter or more slender than the others, sometimes divided into denticles apically. Maxillae setose laterally, mesal margin slightly produced and minutely spinulose; palpi wider than long, with four apical sensilla, galeae somewhat smaller, also wider than long, with two apical sensilla. Maxillae supported by a pair of curved, sclerotic rods close to the margin but concealed by the membranous portion. Labium with the spinneret a prominent transverse slit with raised lips; palpi very short, with four apical sensilla.

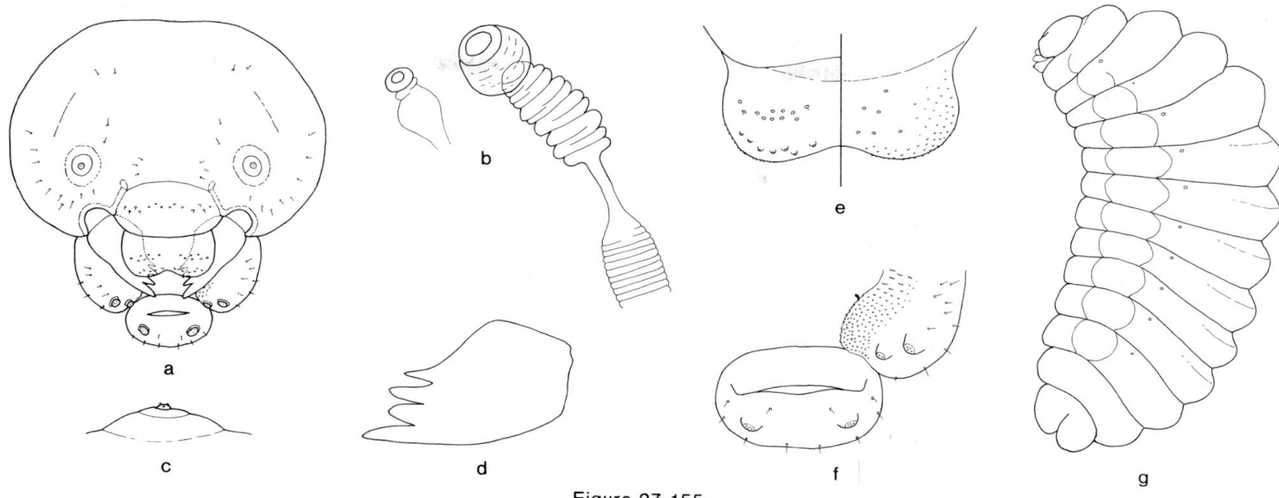

Figure 27.155

Figure 27.155a-g. Mutillidae. *Dasymutilla ursula* Cresson. a. head, anterior view; b. second thoracic spiracle (left) and first thoracic spiracle (right), drawn to same scale; c. antenna, lateral view; d. mandible; e. labrum (left side) and epipharynx (right side); f. labium and maxilla; g. body, lateral view.

Comments: This description is a composite of that provided by Grandi (1954) for *Smicromyrme rufipes* Fabr., that provided by Brothers (1972) for *Pseudomethoca frigida* Smith, and my own study of *Dasymutilla ursula* Cresson. The figures provided are those of a specimen of *D. ursula* found in a fragile cocoon inside a cocoon of *Bembix pruinosa* Fox dug from a sand dune near LaJoya, Socorro County, New Mexico, by W. L. Rubink, who reared several adults from these cocoons.

These three descriptions are detailed and in reasonable agreement except for minor, probably generic differences. Grandi (1954) actually does not mention the spiracles, but as Brothers (1972) has pointed out, Maréchal had earlier stated that the larva of *Smicromyrme* has reduced second thoracic spiracles. Briefer and earlier descriptions are not always in such close accord. Janvier (1933) reports that two Chilean species have tridentate mandibles; both species have the vertex rather strongly depressed medially. Clausen (1940) states that the larva of an African species has 11 pairs of spiracles, but I am inclined to disregard this statement, since no hymenopterous larvae are otherwise known to have more than ten, and no mutillid larvae more than nine that are well developed.

The larva of *Myrmosula parvula* Fox (Myrmosinae) has recently been described by Brothers (1978), whose description largely confirms the placement of this group in the Mutillidae. Differences from the above description are as follows: anus subterminal, apical segment projecting conically above the anus; antennal orbits located above middle of head, each with a very short but distinct papilla; most basal of the four mandibular teeth about the same size as the others.

Selected Bibliography

Brothers 1972, 1978.
Clausen 1940.
Grandi 1954.
Janvier 1933.
Mickel 1928.

SAPYGIDAE (SCOLIOIDEA)

Howard E. Evans, *Colorado State University*

The Sapygids

Figure 27.159

Relationships and Diagnosis: Larvae of Sapygidae bear no close resemblance to those of other Scolioidea. There are ten pairs of well-developed spiracles; the clypeolabral suture is located at the level of the mandibular insertions and passes upward and is weakened medially; the mandibles are strongly bidentate and have some short lateral setae. The antennae are mammiform, having a short papilla bearing two sensilla; the palpi and galeae are quite short. Features of the antennae, palpi, and galeae suggest Mutillidae, but this may be the result of convergence, as both groups are parasitic on other Aculeata.

Biology and Ecology: Sapygidae are cleptoparasites of bees. The female oviposits in freshly completed cells, and the newly emerged larva feeds on any other eggs in the cell before consuming the provisions and spinning a brownish silken cocoon inside the cell. Hosts include various bees that nest in twigs and wood, chiefly Megachilidae and Xylocopinae (Anthophoridae).

Description: Diapausing larva robust, C-shaped, broadest in midbody region; dorsolateral tubercles present on most body segments; pleural lobes well developed. Ten pairs of spiracles present, subequal in size; inner walls of atrium irregularly ridged, opening into subatrium unarmed. Integument finely striate, spiculate, with fairly numerous short setal sensory pegs. Anus a simple, transverse, terminal slit.

Head: Broader than high, vertex depressed medially, coronal suture absent; parietal bands short, curved. Antennal orbits located below middle of head, membrane of orbit elevated, mammiform, crowned with a pigmented papilla that is wider than high and has two terminal sensilla. Front protuberant, clypeus not strongly marked off medially; clypeolabral suture starting at mandibular insertions and passing upward and weakened medially. Head with scattered short setae.

Mouthparts: Labrum protuberant, with lateroapical angles produced into low, broad tubercles; apical margin broadly, shallowly emarginate; surface with numerous setae and peglike sensilla; epipharynx also with numerous hairlike and peglike sensilla. Mandibles heavily sclerotized, broad and cylindrical basally, narrowed apically into two heavily pigmented teeth, the two teeth tending to surround a concavity, shorter tooth with several short setae laterally. Maxillae directed mesad, with several lateral setae; palpi large, conical, projecting, galeae much smaller but longer than wide. Labium with the palpi broader than long; spinneret a transverse slit with raised, fluted lips.

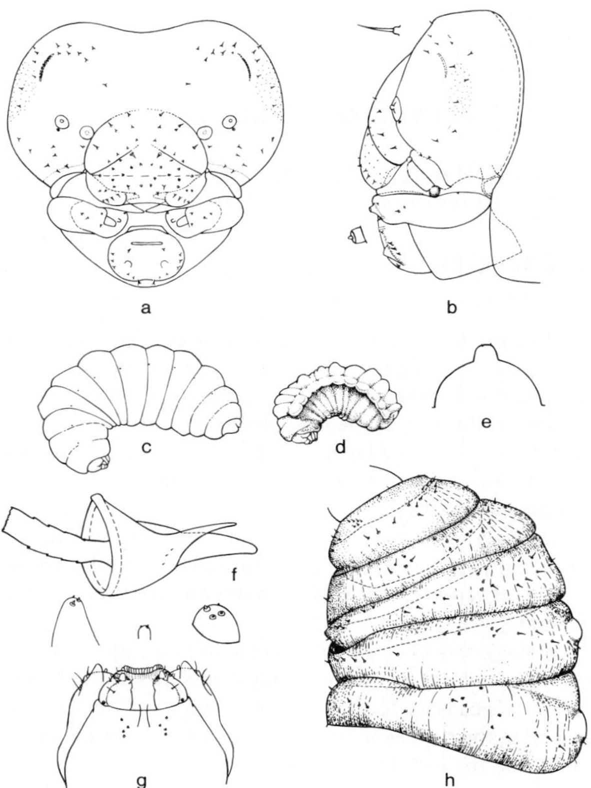

Figure 27.159a-h. Sapygidae. *Sapyga pumila* Cresson. a. head, anterior view; b. head, lateral view; c. body, lateral view before spinning cocoon; d. body from cocoon; e. antenna, lateral view; f. mandible, dorsal view; g. labium and maxilla, ventral view; h. anterior body segments, lateral view (from Torchio 1972).

Comments: This description is based on *Sapyga pumila* Cresson and is paraphrased and abbreviated from Torchio (1972), who has also described the first-instar larva and the egg and pupa. His paper also includes references to other papers on biology and immature stages.

Selected Bibliography

Torchio 1972.

FORMICIDAE (FORMICOIDEA)

Howard E. Evans, *Colorado State University*

Ants

Figures 27.160-27.163

Relationships and Diagnosis: Ant larvae are basically similar to those of generalized wasps such as Scoliidae. Wheeler and Wheeler (1976) have discussed the probable evolution of ant larvae from those of wasps by way of the primitive ant genus *Myrmecia* (Australia). Because of the great variation in larval characters among the ants, it is difficult to present a diagnosis to invariably separate ants from wasps. The Wheelers suggest the following: (1) antennae high on cranium (mostly at or above the middle); (2) thorax often attenuated rather abruptly to form an obvious neck; (3) spiracles usually small and simple; (4) hairs usually abundant and moderately long, often branched or hooked; (5) larvae living in nest chambers of the colony, not in individual cells. Wasplike features worthy of special note include the presence of galeae and of a transverse, slitlike spinneret.

Biology and Ecology: All ants are social and form nest chambers in soil or wood in which larvae are reared in groups, larvae of different sizes often in different chambers. Eggs, the product of the queen, are also usually kept in a separate chamber, as are pupae. Cocoons are brownish and parchmentlike, but naked pupae occur in some ants, in which case the larval spinneret is reduced. Much has been published on the ecology and behavior of ants. Wheeler's book *Ants* (1910) contains much information on biology and development that is still useful, and Wilson's book *The Insect Societies* (1971) provides a more up-to-date reference to ant behavior, with an extensive bibliography.

Description: Length usually 5–15 mm. Body grublike, usually whitish, usually somewhat curved, and often with the thorax considerably narrower than the abdomen; anus a transverse slit, terminal or ventral. Body segmentation distinct; pleural lobes usually well developed. Ten pairs of spiracles present, either of uniform size or one or more of the anterior-most pairs larger than the others; atrium simple or with lines of minute spinules on inner surface; opening into subatrium simple, unarmed. Body hairs abundant and conspicuous, usually rather evenly distributed; hairs may be simple, branched, coiled, plumose, or spatulate. Integument also often with spinules or granules underlying the hairs.

Head: Relatively small, commonly 0.25–0.35 × maximum body width; hypognathous, weakly sclerotized, usually largely or wholly unpigmented, usually wider than high. Coronal suture and parietal bands absent. Antennal orbits located at or more usually above middle of head, membrane of orbit convex, mammiform, bearing three sensilla; in some Ponerinae there is a distinct papilla bearing the sensilla at its apex. Head setae conspicuous, sparse to abundant. Clypeus not set off from front; clypeolabral suture distinct.

Mouthparts: Labrum usually bilobed, but in some genera anterior margin straight, convex, or trilobed; surface with or without setae, usually with a series of subapical sensilla; epipharynx with dense small spinules in rows in most genera. Mandibles from 0.8 to 3.4 × as long as wide, usually about twice as long as wide basally, apex simple, denticulate, or with two to four distinct teeth. Maxillae with palpi and galeae projecting, cone-shaped or digitiform; inner margin of maxilla spinulose in many genera. Labial palpi also projecting but often much shorter than maxillary palpi; spinneret a transverse slit, best developed and projecting in genera that spin cocoons.

Comments: This description was abstracted from Wheeler and Wheeler (1976), who provide much more detail than is possible here, in addition to aids for identification of subfamilies and genera.

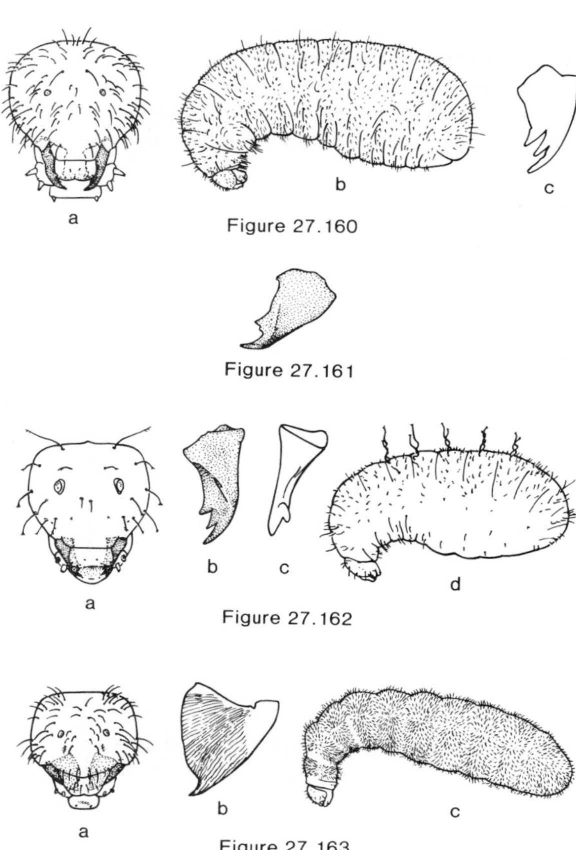

Figures 27.160–27.163. Formicidae. Representative ant larvae.

Figure 27.160a-c. *Pogonomyrmex barbatus* (Smith) a. head; b. same, lateral view of body; c. same, mandible.

Figure 27.161. *Paramyrmica colax* Cole, mandible.

Figure 27.162a-d. *Tetramorium caespitum L.* a. head; b. same, mandible, anterior view; c. same, mandible, mesal view; d. same, lateral view of body.

Figure 27.163a-c. *Myrmecocystus melliger* Forel. a. head; b. same, mandible; c. same, lateral view of body (from Wheeler and Wheeler 1976).

Selected Bibliography

Wheeler and Wheeler 1976 (Includes a list of the many papers by these authors and a bibliography of other important papers.), 1979.
Wheeler 1910 (reprinted 1926, 1960).
Wilson 1971.

RHOPALOSOMATIDAE (POMPILOIDEA)

Howard E. Evans, *Colorado State University*

The Rhopalosomatids

Figure 27.164

Relationships and Diagnosis: This family is of somewhat doubtful taxonomic position, but recent workers have tended to group it with the Pompilidae on the basis of adult structure. The larvae, however, have little in common with those of Pompilidae other than a general resemblance of the labrum and maxillae. Judging from the single available description, that of Gurney (1953), there are ten pairs of well-developed spiracles, no parietal bands, paired, projecting

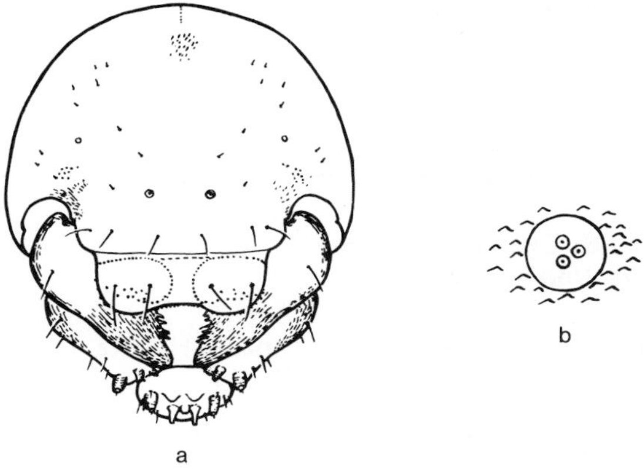

Figure 27.164a,b. Rhopalosomatidae. *Rhopalosoma nearcticum* Brues. a. head; b. antenna greatly enlarged (from Gurney 1953).

spinnerets, and mandibles more like those of bees than of other wasps. In general, larval characters do not seem helpful in placing this family in the scheme of Hymenoptera.

Biology and Ecology: *Rhopalosoma nearcticum* Brues is known to be an ectoparasite of bush crickets of the genera *Orocharis* and *Hapithus*. The egg is apparently laid externally on a cricket, and the larva attaches along the side of the abdomen and feeds through the body wall near the posterior end of the abdomen. The mature larva falls to the ground and spins a brownish, parchmentlike cocoon in the soil. There is only one other recognized species of Rhopalosomatidae in North America, *Olixon banksii* (Brues), which is a brachypterous form usually found running over the ground in sandy places. It may well prove to have quite different biology and immature stages.

Description: Mature larva 8–9 mm long, grublike, widest at midabdominal segments, whitish or yellowish in color, head weakly sclerotized and unpigmented except for mouthparts. Body segmentation somewhat distinct posteriorly; ten pairs of spiracles clearly visible; anus transverse, terminal.

Head: Subcircular in anterior view, without parietal bands or a well-developed coronal suture; antennal orbits small, without projecting papillae, with three small sensory cones set in the orbit; head setae short, sparse; a presumed sensory area consisting of several pigmented spots and minute setae located above the mandibular bases.

Mouthparts: Labrum bilobed, broadly emarginate medially, each lobe with two large setae and with an irregular transverse row of sensoria, margin with a weak fringe of short spinules. Mandibles robust, each with a single large, lateral seta; biting margin thin, sharply and delicately toothed, set off from remainder of mandible by a groove. Maxillae with several lateral setae, mesal margin spinulose and somewhat produced; palpi and galeae elongate, of about equal length. Labial palpi about the same length; spinneret paired, sharply projecting.

Comments: This description is extracted from Gurney (1953), who provides further details on the final as well as earlier instars.

Selected Bibliography

Gurney 1953.

POMPILIDAE (POMPILOIDEA)

Howard E. Evans, *Colorado State University*

Pompilids, Spider Wasps, Tarantula Hawks

Figure 27.153

Relationships and Diagnosis: Larvae of Pompilidae have been little studied, but available information indicates that the vestigial second thoracic spiracles, distinct antennal papillae (fig. 27.153b), and somewhat trilobed clypeus (fig. 27.153a) together will separate most larvae from those of other families. Reduced second thoracic spiracles are also characteristic of some Scolioidea, but it is probable that reduction occurred independently in the two groups. Pompilid larvae bear much resemblance to generalized Sphecidae, and the family might be placed in the Sphecoidea on the basis of larval features alone.

Biology and Ecology: Most larvae occur in nests prepared by the female in the ground or in cavities in wood, or occasionally in mud cells. A few Pompilidae are ectoparasitic and the larvae develop on free-living spiders. In all cases a single, somewhat sausage-shaped egg is deposited on the abdomen of the spider prey (in the book-lungs in *Ceropales*, a cleptoparasite of other Pompilidae). Pupation is within a silken cocoon, usually brownish with a roughly fibrous exterior.

Description: Mature larvae 5–25 mm long, whitish in color, with a weakly sclerotized and mostly unpigmented head. Body grublike, cylindrical but of greatest diameter in the midabdominal region; each body segment except the last weakly divided into two annulets by a transverse dorsal crease; anus terminal, slitlike. Integument apparently smooth, but fine wrinkles, spinules, and short setae can be detected in most species under high magnification. Second pair of thoracic spiracles vestigial, much smaller than the other nine pairs and

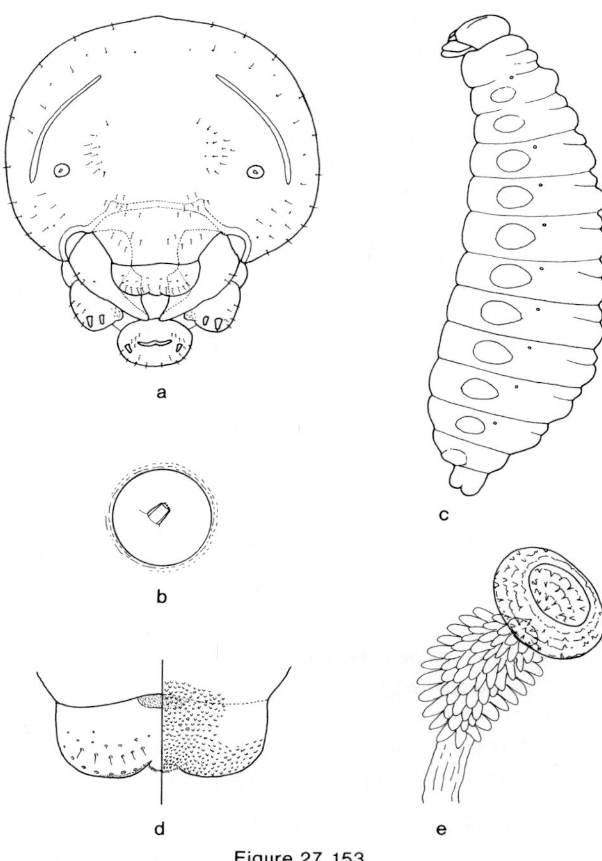

Figure 27.153

Figure 27.153a-e. Pompilidae. *Anoplius marginalis* Banks. a. head, anterior view; b. antenna; c. body, lateral view; d. labrum (left side) and epipharynx (right side); e. spiracle.

sometimes difficult to detect at all; other spiracles of complex structure, differing greatly in different genera, but typically with a very narrow peritreme and a spinose atrium.

Head: Wider than high; parietal bands long and prominent, often brown in color; frontoclypeal suture not distinct; coronal suture weak or absent. Antennal orbits of moderate size, subcircular; each orbit with a projecting papilla that terminates in three small sensory cones. Head setae abundant, short to moderately long.

Mouthparts: Labrum with two large lateral lobes and a smaller median lobe, its margin weakly if at all bristly, but with a series of fairly prominent sensory cones; labrum with an irregular transverse row of small setae; epipharynx with dense, small spinules on the median lobe, elsewhere with rather evenly distributed larger spinules or papillae. Mandibles broad, not more than twice as long as their maximum width, usually more or less pigmented; base usually with one or two lateral setae; apex typically with three teeth that tend to surround a ventral concavity, apical two teeth sharp, basal tooth broad and blunt, in some genera the basal two teeth combined to form a broad cutting edge. Hypopharynx spinulose, bilobed. Maxillae setose laterally, more or less spinulose and often somewhat produced on inner apical margin; palpi and galeae prominent, elongate, of about equal length. Labium with some ventral and apical setae, palpi slightly shorter than maxillary palpi, spinneret a transverse opening with raised lips, more prominent on the sides than medially.

Comments: This description is based on ground-nesting species of generalized behavior. Ectoparasites and cleptoparasites may well be found to have specializations that deviate from this description. There are many papers dealing with pompilid biology, only a few of which are cited here.

Selected Bibliography

Evans 1959b.
Evans and Yoshimoto 1962.
Grandi 1961.
Iwata 1976.
Kaston 1959.
Krombein 1967.
Williams 1956.

VESPIDAE (VESPOIDEA)

Howard E. Evans, *Colorado State University*

The Vespids

Figures 27.165–27.169

Relationships and Diagnosis: Larvae of Vespoidea are extremely diverse, some being relatively generalized and others quite unlike other wasp larvae. The diversity of adult structure and behavior is also great, such that many persons have recognized several famlies within the superfamily, in recent years usually three (Richards 1962). Carpenter (1982) has convincingly argued for recognizing a single family with six subfamilies. One of these, the Stenogastrinae, does not occur in the Americas; Spradbery (1975) has described the larva of a member of this subfamily. The remaining five subfamilies are treated separately here, since they differ considerably in larval structure, biology, and ecology. There seem to be no characters by means of which larvae of Vespidae can always be separated from those of other wasps, and the subfamilies are therefore entered separately in the key.

Biology and Ecology: So far as known, nearly all Vespidae use soil or paper in some aspect of nest construction. Prey consists of diverse insects that are either stung and placed whole into the nest cells or, in eusocial forms, macerated and fed directly to larvae. However, the Masarinae are exceptional in provisioning their nests with pollen and nectar. Euparagiinae, Masarinae, and Eumeninae are solitary wasps, though sometimes nesting gregariously; the remaining three subfamilies show various forms of eusociality. Further details are presented under the subfamilies.

Description: Mature larvae 7–35 mm long, whitish, head weakly to strongly pigmented; pleural lobes well developed; anus terminal; ten pairs of spiracles well developed.

Head: Rather large; parietal bands present; antennal papillae absent in the American species (present in Stenogastrinae).

Mouthparts: Mandibles without lateral setae or sensoria; palpi and galeae projecting, somewhat cone-shaped; spinneret a transverse slit with or without raised lips.

Selected Bibliography

Carpenter 1982.
Reid 1942.
Richards 1962, 1978.
Spradbery 1975.

Subfamily EUPARAGIINAE (VESPIDAE)

The Euparagiines

Figures 27.166a-d

Relationships and Diagnosis: The larva of only one species of this subfamily has been described. These appear to be among the most generalized Vespidae, the larvae bearing much resemblance to those of Eumeninae and differing in no major ways from those of Pompilidae and Scoliidae. Characteristic features include the broad, well-defined clypeus, the anterior margin of which is produced medially; the broad, truncate labrum; and the tridentate mandibles.

Biology and Ecology: *Euparagia scutellaris* Cresson nests gregariously in hard-packed soil, making a vertical burrow that may terminate in two or more branches containing cells. The burrow is topped with a curved chimney built of material excavated from the burrow. Prey consists of weevil larvae. The egg is laid after the cell is provisioned. The cocoon is brownish, very thin and parchmentlike.

Description: Diapausing larva with intersegmental lines complete; pleural lobes strong; cuticle plicate.

Head: Subcircular in anterior view; antennal orbits situated well below middle of head, each containing three minute sensilla; head setae sparse and minute. Clypeus large, its anterior margin far below articulation of mandibles, its surface spinulose apically and somewhat produced medially.

Mouthparts: Labrum transverse, truncate except with a median notch, surface spinulose medially; mandibles robust, with three blunt apical teeth that tend to surround a concavity; galeae somewhat longer and more slender than maxillary palpi; labial palpi rather short; spinneret a simple transverse slit.

Comments: A fuller description of this larva, that of *Euparagia scutellaris* Cresson, is provided by Torchio (1970), from whose paper these notes are taken.

Selected Bibliography

Clement and Grissell 1968.
Torchio 1970.

Subfamily MASARINAE (VESPIDAE)

The Masarids

Figure 27.167a-d

Relationships and Diagnosis: As noted by Torchio (1970), the known Masarinae larvae do not resemble those of Euparagiinae closely but suggest a relationship with Polistinae. All North American Masarinae belong to the genus *Pseudomasaris,* the larva of one species of which has been described in detail by Torchio. This larva is unique in having a deeply emarginate vertex (fig. 27.167a), bidentate mandibles, and a clypeus with its anterior margin even with the articulations of the mandibles.

Biology and Ecology: These wasps are unique in that the cells are provisioned with pollen and water. Cylindrical cells are built of soil and are attached to rocks or other objects in the open; they are usually clustered, up to 13 in a group, and the walls of the nest become very hard. The egg is attached to the cell wall near its base before provisions are brought in. The thin-walled, translucent cocoon consists of salivary secretion and silken strands. Members of the genus *Pseudomasaris* appear to be specialists on plants of the genera *Phacelia* and *Pentstemon.*

Description: Diapausing larva with intersegmental lines complete, pleural lobes absent.

Head: With a strong dorsal concavity; antennal orbits large, near middle of head; head with numerous sensoria; clypeus extending upward nearly as far as antennae, its anterior margin even with upper mandibular articulations.

Mouthparts: Labrum transverse, broadly emarginate medially, without spinules; mandibles elongate, bidentate, inner tooth shorter and sharper than apex; galeae and maxillary palpi subequal in length, labial palpi shorter; lips of spinneret serrate.

Comments: These descriptions are drawn from Torchio (1970), who provides fuller details. Zucchi, Yamane, and Sakagami (1976) have described the larva of the Neotropical species *Trimeria howardi* Bertoni (Masarinae). This larva agrees in many respects with that of *Pseudomasaris,* but the mandibles are tridentate (as in *Euparagia,* but more slender), the vertex only slightly concave, the labrum narrower than the clypeus and deeply emarginate. Curiously, there are seven antennal sensilla in a somewhat elevated orbital membrane.

Selected Bibliography

Cooper 1952.
Torchio 1970.
Zucchi et al. 1976.

Subfamily EUMENINAE (VESPIDAE)

The Eumenids, Mason Wasps, Potter Wasps

Figure 27.165a-f

Relationships and Diagnosis: Eumeninae larvae are rather generalized Aculeata. The anterior margin of the clypeus is well below the articulations of the mandibles and the labrum is about as wide as the clypeus, as in Euparagiinae; the mandibles are tridentate, as in that subfamily, but the teeth are more prominent. The spinneret is in the form of a transverse slit with prominently raised lips.

Biology and Ecology: Larvae occur in nests in the ground, in hollow stems, or in free mud nests, which may be jug-shaped or of rather irregular form. Ground-nesters may also employ mud for barriers between cells or for turrets at the entrance; twig-nesters separate cells with mud barriers. Mud is made by mixing earth with water that is carried in the crop. An egg, which is laid in each cell before provisioning begins, is often suspended by a fine filament from the roof or sides. Most species spend the pupal stage in a delicate silken cocoon, although a few species do not spin cocoons. Prey consists of caterpillars or other types of eruciform larvae.

Description: Mature larvae 7–20 mm long, whitish, with a weakly sclerotized and weakly or scarcely pigmented head except mouthparts, tentorial pits, and adjacent thickenings usually dark in color. Body grublike, usually widest somewhat behind the middle; body segments except those toward the anterior and posterior ends weakly divided into two dorsal annulets by a transverse crease. Pleural lobes weakly developed; anus terminal, transverse. Integument smooth, but in most species with sparse to moderately abundant, minute setae. All ten pairs of spiracles well developed; spiracular atria reinforced with fine lines or ridges that may or may not be spinose; opening into subatrium with or without a circlet of spines.

Head: Subcircular in anterior view, in most species wider than high (exclusive of labrum); parietal bands well developed, rather long and often pigmented; coronal suture distinct. Antennal orbits subcircular, fairly large, each bearing three small sensory cones in the membrane of the orbit. Head setae usually present, sparse, usually quite short. Anterior margin of clypeus convex, or often somewhat produced medially, located well below level of upper insertions of mandibles.

Mouthparts: Labrum about as wide as clypeus, weakly to rather strongly bilobed, emarginate medially; surface of labrum with a variable number of small setae and sensilla; epipharynx covered with small spinules, at least apically and laterally, also with a pair of submedian sensory areas, bare of spinules but sometimes pigmented, each with several sensoria. Mandibles robust, less than twice as long as their maximum width, without setae; apices tridentate (rarely weakly quadridentate), the teeth tending to surround a concavity. Maxillae setose laterally, mesal margin produced and spinulose apically; palpi and galeae both prominent, commonly subequal in length but the galeae more slender and terminating in fewer sensilla. Labium broad, the spinneret a transverse slit with raised lips nearly as wide as the labium itself; palpi prominent, but typically shorter than maxillary palpi.

Comments: There has been no comparative study of eumenid larvae, and available descriptions suggest that there is little diversity within the family. Publications on the biology of Eumeninae are numerous but mostly short; only a few are cited here.

Selected Bibliography

Clement 1972.
Cooper 1953, 1955, 1957.
Evans 1977.
Grandi 1961.
Grissell 1975.
Iwata 1976.
Janvier 1930.
Krombein 1967.

Subfamily POLISTINAE (VESPIDAE)

Paper Wasps

Figure 27.168a–f

Relationships and Diagnosis: The larvae of Polistinae are quite dissimilar to those of other Vespidae, though the form of the clypeus and mandibles suggests the Masarinae, a group with which they otherwise seem to have little in common. The head is dark colored and the mandibles slender and tapering, either simple, bifid, or with a small third tooth; the maxillae are swollen laterally and the postlabium large, swollen.

Biology and Ecology: In genera occurring north of Mexico, larvae are reared in naked paper combs suspended by a pedicel from beams, rocks, branches, or cavities in the soil or in wood. Nests are founded in the spring by fertilized females that have overwintered; in some cases there are associations of foundresses, one of whom becomes the effective queen. Larvae are fed with macerated insects, chiefly caterpillars. Full-grown larvae spin a thick silken cap over the cell and pupate within. Nest structure of tropical Polistinae is exceedingly diverse.

Description: Mature larvae 15–25 mm long, body stout, rather straight, widest at or about the second abdominal segment, head directed forward (actually downward, in the nest), large, approximately half as wide as maximum body width. Pleural lobes well developed; middle body segments with two dorsal annulets, the anterior part of the posterior annulet often elevated to form a transverse welt. Ten distinct pairs of spiracles, walls of atrium with weak, anastomosing ridges; opening into subatrium simple, unarmed. Integument covered with minute spinules and with areas of fairly long setae on the venter of the more anterior segments.

Head: Wider than high, with a distinct coronal suture and strong parietal bands; head capsule usually brown in color, often considerably darker than mouthparts. Antennal orbits near or slightly above level of middle of head, each with three minute sensoria. Clypeus evidently quite large, although the frontoclypeal suture is indistinct except laterally; clypeolabral suture strong, located at or somewhat above level of upper mandibular articulations. Head with an abundance of moderately long setae in many *Polistes,* much less strongly setose in many other Polistinae.

Mouthparts: Labrum transverse, at most weakly emarginate, sparsely setose and with several apical or subapical sensilla; epipharynx largely smooth, with scattered porelike sensilla and often some minute lateral spinules. Mandibles slender, not strongly sclerotized, tapering to a very slender, pointed apex, which may be simple (*Mischocyttarus* and several tropical genera), bifid, or with a very small third tooth (*Polistes* and some tropical genera). Maxillae with palpi and galeae both well developed, but directed inward and upward so that they (like the mandibles) are largely concealed by other mouthparts; lateral portions of maxillae inflated, projecting. Apical part of labium subcircular, with small palpi; postlabium (submentum) large, swollen.

Comments: Social wasps are commonly encountered in nature and are the subjects of much current research. Richards (1978, pp. 14–19) has presented a key to larvae of all major groups of Vespoidea and to the genera of Polistinae, most of which are tropical in distribution. Richards' book contains a wealth of information on larval characters, nest types, and general biology of American Polistinae; Nelson (1982) has presented a key to the species of *Polistes* occurring in the United States. The two genera occurring north of Mexico may be separated by the following key.

Selected Bibliography

Kojima and Yamane 1984.
Parker 1943.
Nelson 1982.
Richards 1978.

KEY TO NEARCTIC GENERA OF POLISTINAE

1. Ventral side of first abdominal segment simple, not greatly produced; terminal abdominal segment with a fingerlike process; mandibles bidentate .... *Polistes* Latreille

1′. Ventral surface of first abdominal segment developed into a large process that is bifid and is directed forward; terminal segment not produced as above; mandibles simple, acuminate .... *Mischocyttarus* Saussure

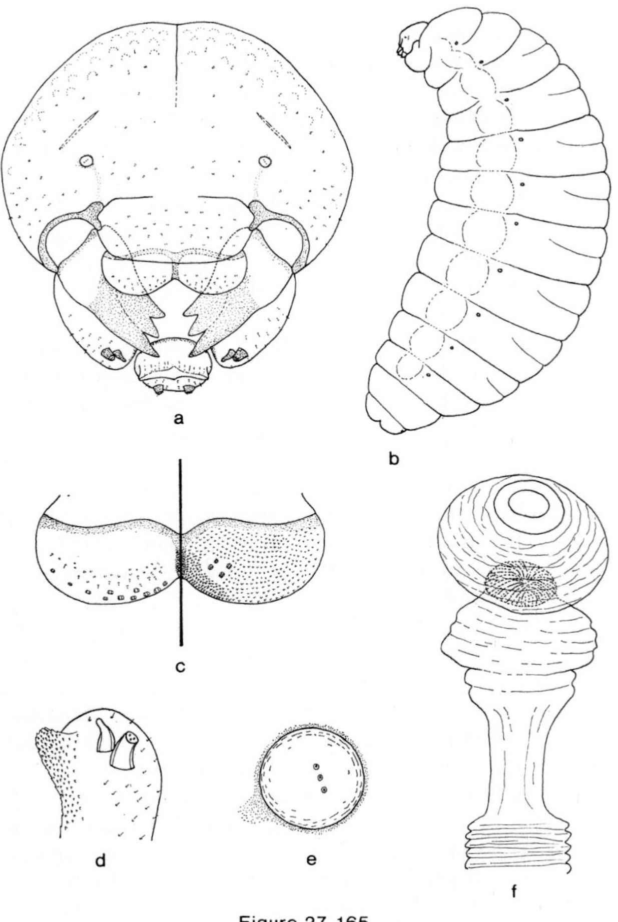

Figure 27.165

Figure 27.165a-f. Vespidae: Eumeninae. Larva of *Pterocheilus quinquefasciatus* Say. a. head; b. body, lateral view; c. labrum (left side) and epipharynx (right side); d. maxilla; e. antenna, greatly enlarged; f. anterior thoracic spiracle, greatly enlarged (from Evans 1977).

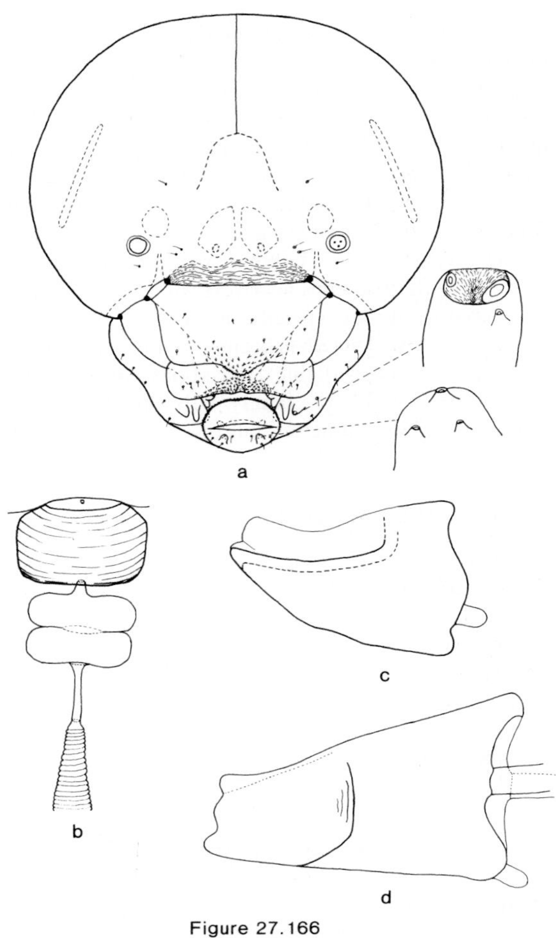

Figure 27.166

Figure 27.166a-d. Vespidae: Euparagiinae. Larvae (prepupae) of *Euparagia scutellaris* Cresson. a. head; b. spiracle; c. mandible, ventral view; d. mandible, mesal view (from Torchio 1970).

Subfamily VESPINAE (VESPIDAE)

Hornets and Yellowjackets

Figure 27.169a-f

Relationships and Diagnosis: Although the social behavior of Vespinae appears more advanced than that of Polistinae, the larvae are more generalized, differing from those of Eumeninae chiefly in the more robust body form and in having the greatest width of the labrum less than the width of the clypeus where it joins the labrum.

Biology and Ecology: Hornets and yellowjackets build paper combs surrounded by a paper envelope; within the nests the combs are in tiers facing downward. Nests are built in trees, bushes, on buildings, or in cavities in the soil. Nests are annual and contain a single queen and numerous workers; late in the mating season males and virgin queens are produced. Following mating, queens overwinter and start new nests in the spring. The larvae are fed with macerated insects, at first by the queen, later by the workers. Mature larvae spin a cocoon that is very thin at the base of the cell but with a thick cap at the opening.

Description: Mature larvae 15–35 mm long, head of moderate size, roughly one third the maximum body width. Head weakly pigmented except for parts of mouthparts and supporting structures. Body fusiform, rather straight, widest at or about the second abdominal segment; pleural lobes well developed; middle body segments divided into two annulets by a transverse crease; anus a terminal, transverse slit, terminal segment sometimes produced above the anus. Ten pairs of spiracles distinct; walls of atrium smooth or with small ridges or spines, opening into the subatrium armed with spines, which may be short or quite long, sometimes branched. Integument apparently smooth, under high power seen to be covered with minute spinules in considerable part; body hairs sparse, minute.

Head: Subcircular in anterior view, often wider than high, sometimes with the vertex depressed; parietal bands strong;

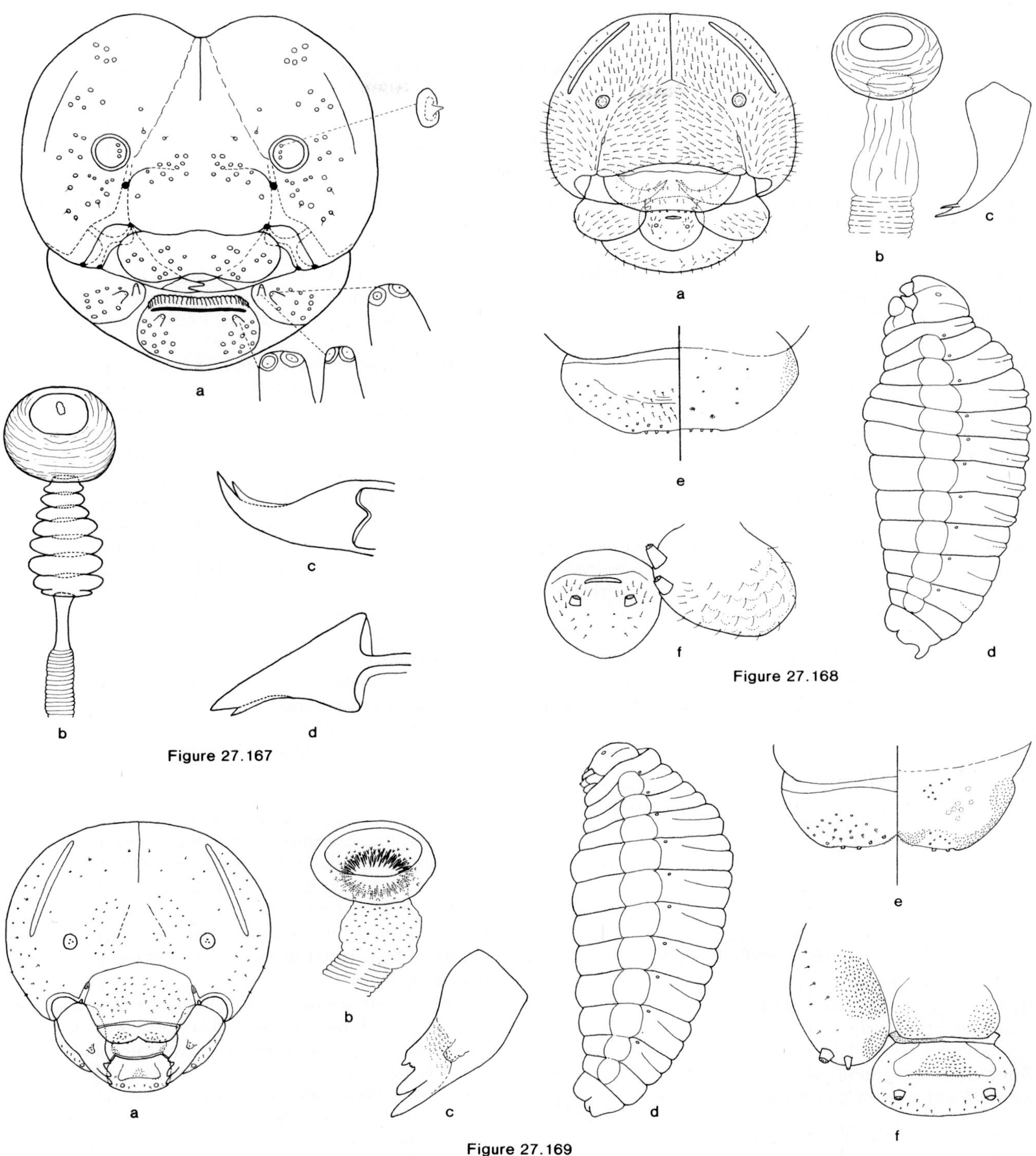

Figure 27.167a-d. Vespidae: Masarinae. Larvae (prepupae) of *Pseudomasaris edwardsii* (Cresson). a. head; b. spiracle; c. mandible, ventral view; d. mandible, mesal view (from Torchio 1970).

Figure 27.168a-f. Vespidae: Polistinae. *Polistes fuscatus* Fabr. a. head, anterior view; b. spiracle of most anterior pair; c. mandible; d. body, lateral view; e. labrum (left side) and epipharynx (right side); f. labium and maxilla.

Figure 27.169a-f. Vespidae: Vespinae. *Vespula atropilosa* Sladen. a. head, anterior view; b. spiracle of most anterior pair; c. mandible, dorsal aspect; d. body, lateral view; e. labrum (left side) and epipharynx (right side); f. maxilla and labium.

Key to Nearctic Genera of Vespinae*

1.	Mandibles with a single apical tooth above which the margin is serrated or cusplike	*Dolichovespula* Rohwer
1′.	Mandibles tridentate, upper tooth sometimes subdivided	2
2(1′).	Spines at opening into subatrium long, thick, and branched	*Vespa* Linnaeus
2′.	Spines at opening into subatrium short, or if longer, then unbranched	*Vespula* Thompson

*Modified from Yamane, 1976. I have studied larvae of only a few Nearctic species and cannot be certain that this key will always work for our species.

antennal orbits situated below middle of head, each bearing three minute sensilla; frontoclypeal suture usually more or less distinct; clypeolabral suture located well below mandibular articulations. Head with sparse punctures, some of which bear minute setae.

Mouthparts: Labrum not as wide as clypeus, somewhat bilobed, bearing short setae and an apical row of small sensilla; epipharynx weakly spinulose apically and laterally, with patches of small sensilla. Mandibles typically tridentate, but the most basal tooth often subdivided into two or more small denticles; but in *Dolichovespula* only the apical tooth is distinct, the margin otherwise being serrate or irregular; frequently the mandibles bear a roughened elevation well back from the apex. Maxillae with the palpi and galeae protruding apically, inner margin with small spinules. Labial palpi short, spinneret a long slit with a somewhat irregular margin.

Comments: Good general reviews of the biology of the Vespinae are those of Spradbery (1973) and Akre (1981). Akre et al. (1974) published a world list of literature on yellowjacket biology. The three genera occurring in North America may be separated by the above key.

Selected Bibliography

Akre et al. 1974, 1981.
Duncan 1939.
Spradbery 1973.
Yamane 1976.

SPHECIDAE (SPHECOIDEA)

Howard E. Evans, ***Colorado State University***

Sphecids, Digger Wasps, Sand Wasps, Mud Daubers

Figures 27.1–27.4, 27.170–27.179

Relationships and Diagnosis: Larvae of Sphecidae are very diverse, some of the more generalized groups (Sphecinae) bearing considerable resemblance to Eumeninae (Vespidae) and Pompilidae, the more specialized groups being nearly unique in having the spinnerets paired and projecting. Sphecid larvae have ten pairs of well-developed spiracles, mandibles with one or more lateral setae or sensilla, labrum usually somewhat bilobed, maxillae with both palpus and galea well developed (galea small in a few genera, said to be absent in one). Larval characters suggest that the family is divisible into eight well-defined subfamilies, and since some authors grant these family status, I have presented below a subfamily key.

Biology and Ecology: Sphecid larvae occur in cells prepared by the female, usually in the ground but sometimes in hollow twigs, burrows in soft, decaying wood, or in mud cells. The egg is deposited in the cell on one of the prey, or occasionally in the empty cell before prey is introduced. Pupation occurs in a cocoon that varies greatly in different subfamilies and genera. Bohart and Menke (1976) have presented a brief survey of the biology of various genera in their generic revision, and Evans (1966) has discussed one subfamily (Nyssoninae) in much detail. Iwata (1942, 1976) has described and figured the shape and position of the egg in several species and summarized many aspects of the behavior. Literature on the biology of these wasps is extensive.

Description: Mature larvae 5–40 mm long, whitish in color, head of similar color except mandibles and other mouth structures darker, portions of head capsule sometimes pigmented. Body usually tapered at both ends and somewhat curved, but more straight and cylindrical in twig-nesters; pleural lobes well developed in most species; most body segments divided into two annulets by a transverse dorsal fold in most species; anus transverse, either terminal or located ventrally on the last segment. Integument smooth, granulose, or with sparse to quite dense, short setae. All ten pairs of spiracles present, second pair at most slightly smaller than remainder; spiracular atria lined with fine ridges that may be spinulose, openings into subatria armed or unarmed.

Head: Subcircular in anterior view or somewhat wider than high or higher than wide, vertex rounded. Parietal bands present or absent; antennal orbits located below middle of head, either with three small sensilla set in orbit or with a short or fairly long, projecting papilla; head capsule with small punctures, some of which are setigerous, setae short to fairly long. Frontoclypeal suture located somewhat above level of mandibular articulations, but ill defined in most genera; clypeolabral suture nearly straight or slightly concave.

Mouthparts: Anterior margin of labrum straight or medially emarginate, with or without short bristles but with a series of somewhat barrel-shaped sensilla set close to anterior margin; epipharynx spinulose or papillose, or in part smooth, always with a group of small sensoria on each side of the midline. Mandibles robust, usually pigmented at least apically, with one or more setae or punctures laterally; apically with from two to five teeth, either in the same plane or tending to surround a concavity (mandibles simple and acute in a few

Key to Subfamilies of Sphecidae

1. Spinneret a long transverse slit with raised lips (fig. 27.170a); antennal papillae absent 2

1′. Spinnerets paired, each opening at the end of a short to fairly long process (figs. 27.171, 27.172a); antennal papillae present or absent 3

2(1). Anal segment slightly produced, anus located on ventral side of this process; epipharynx weakly spinulose laterally but not medially ..... **Ampulicinae**

2′. Anal segment simple, anus terminal; epipharynx with an abundance of spinules laterally and medially ..... **Sphecinae**

3(1′). Anus directed ventrad and situated well before apex of abdomen, the anal segment forming a rounded or conical lobe beyond the anus (fig. 27.172b); antennal papillae absent 4

3′. Anus directed caudad and at the apex of the abdomen or nearly so (as in fig. 27.170b); antennal papillae present except in most Pemphredonini 6

4(3). Body with pleural lobes and a series of dorsolateral prominences conically produced, dark-tipped, and armed with short, stout spinules; oral surface of labium not spinulose ..... **Mellininae**

4′. Body without conical processes that are dark-tipped and armed with stout spinules; oral surface of labium at least partially spinulose 5

5(4′). Inner margin of maxilla with a projecting lobe that is densely spinulose ..... **Larrinae**

5′. Inner margin of maxilla spinulose apically, but rounded and not projecting ..... **Crabroninae**

6(3′). Maxillae with their apices directed mesad, closely associated with the labium and hypopharynx; epipharynx often spinulose laterally, but the spinules not or barely reaching the midline 7

6′. Maxillae projecting as lobes that are free from close association with the labium or hypopharynx (fig. 27.173); epipharynx spinulose (rarely papillose) over much of its surface 8

7(6). Fourth abdominal segment strongly humped; parietal bands present; antennal papillae present ..... **Astatinae**

7′. Fourth abdominal segment no more humped than any other; parietal bands absent or very weak; antennal papillae present or absent ..... **Pemphredoninae**

8(6′). Integument smooth or with setae or spinules in restricted areas; apical abdominal segment simple ..... **Nyssoninae**

8′. Integument clothed nearly everywhere with dense, small spinules; posterior end of body in form of a pseudopod, either the supra-anal or subanal lobes protuberant or the two lobes together forming a tubular process ..... **Philanthinae**

genera). Maxillae setose laterally, mesal margin apically either spinulose or papillose; palpi conical, terminating in about five sensilla; galeae also conical, usually terminating in two sensilla (galeae very short in a few genera, present only as minute knobs and possibly absent in smaller Pemphredonini). Labial palpi conical; spinneret in the form of either a transverse slit with raised lips, or a pair of processes that may be connected basally.

Comments: Larvae of this family are unusually diverse, and a key to subfamilies is presented above. Evans (1959a) has presented an artificial key to Nearctic genera that may in some cases be easier to use than the subfamily key presented here. Evans (1964c) has discussed the classification and evolution of the group as suggested by larval characters.

Added Note: Bohart and Menke (1976) consider the Mellininae only a tribe of Nyssoninae, although the larvae are very different. Evans (1964c) suggested that the Larrinae and Crabroninae be combined as a result of larval similarities, but this was not accepted by Bohart and Menke. These authors place *Xenosphex* in a subfamily of its own, but the larvae of this genus have not been described.

Selected Bibliography

Bohart and Menke 1976.
Evans and Lin 1956.
Evans 1957–1959a, 1964b, 1964c, 1966.
Grandi 1961.
Iida 1967.
Iwata 1942, 1976.

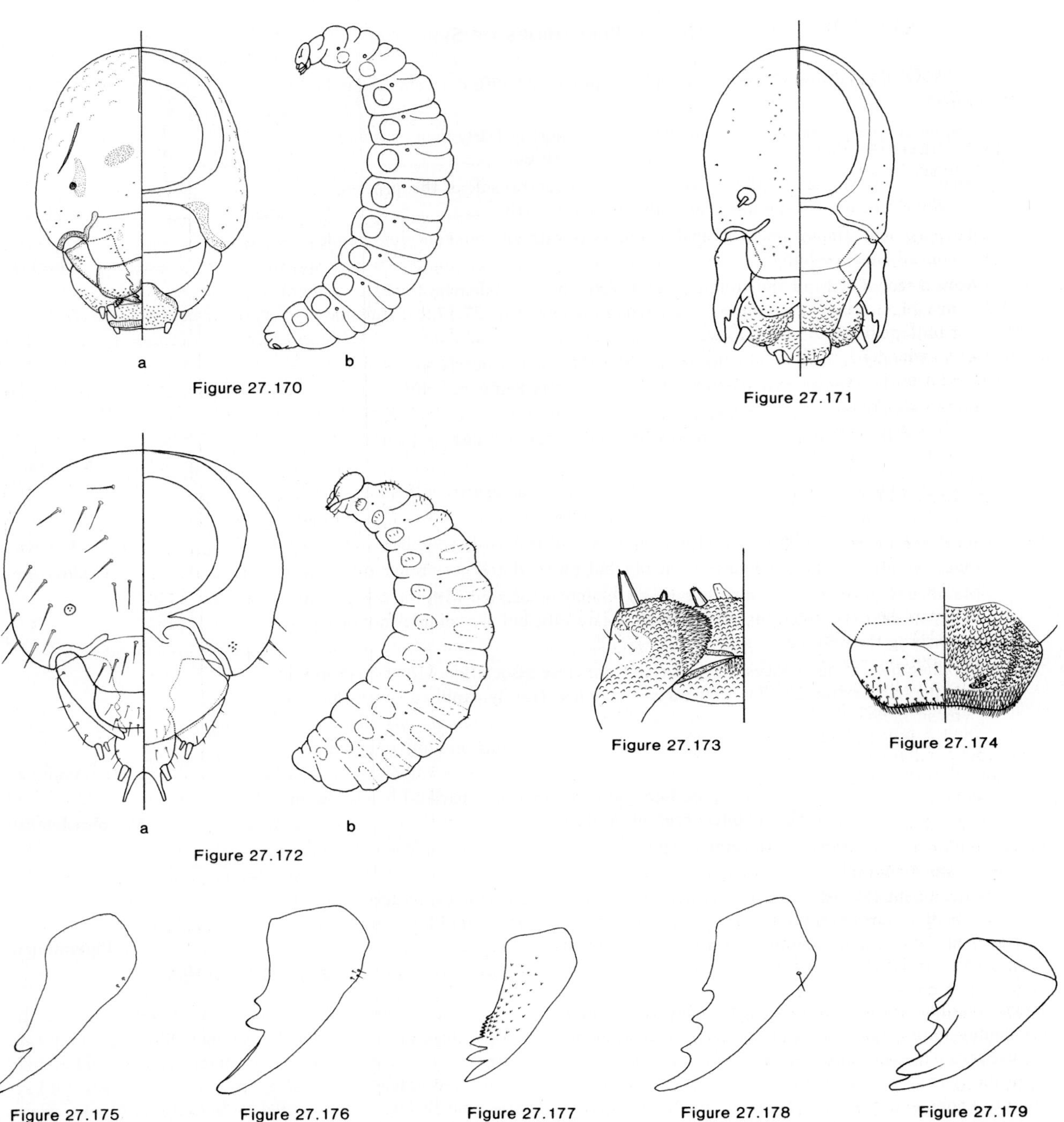

Figure 27.170

Figure 27.171

Figure 27.172

Figure 27.173

Figure 27.174

Figure 27.175

Figure 27.176

Figure 27.177

Figure 27.178

Figure 27.179

Figures 27.170–27.179. Larvae of selected Sphecidae.

Figure 27.170a,b. *Sphex tepanicus* (Saussure) (Sphecinae). a. head (anterior view left side, posterior view right side); b. body.

Figure 27.171. *Clypeadon evansi* Bohart (Philanthinae), head.

Figure 27.172a,b. *Miscophus evansi* (Krombein) (Larrinae). a. head; b. body.

Figure 27.173. *Philanthus bicinctus* Mickel (Philanthinae), maxilla and labium, oral surface.

Figure 27.174. *Steniolia longirostra* (Say) (Nyssoninae), labrum (left side) and epipharynx (right side).

Figure 27.175. *Eucerceris pimarum* Rohwer, mandible (Philanthinae).

Figure 27.176. *Bembecinus mexicanus* (Handlirsch), mandible (Nyssoninae).

Figure 27.177. *Podalonia robusta* (Cresson), mandible (Sphecinae).

Figure 27.178. *Moniaecera asperata* (Fox), mandible (Crabroninae).

Figure 27.179. *Miscophus evansi* (Krombein), mandible (Larrinae).

COLLETIDAE (APOIDEA)

Ronald J. McGinley, ***Smithsonian Institution***

Colletids, Plasterer Bees, Yellow-Faced Bees

Figures 27.180–27.182

Relationships and Diagnosis: Colletids are currently thought by most workers to represent the most primitive group of extant bees, largely due to the sphecoidlike, bifid glossa of the adults (see McGinley (1980) for an alternative viewpoint). Their phylogenetic relationships to other bee families remain poorly understood, as do the interrelationships of most short-tongued bees, i.e., Colletidae, Halictidae, Andrenidae, Oxaeidae, and Melittidae.

Due to their great morphological diversity, it is not possible to give a useful diagnosis for colletid larvae as a whole. Therefore, the three North American subfamilies are treated separately. Larvae of Diphaglossinae (*Caupolicana, Ptiloglossa*) are extremely distinctive, being very large (up to 35 mm in length) and characterized by their unique spinneret, which is long, cylindrical, and spoutlike (figs. 27.180a, b). Larvae of Colletinae (*Colletes*) can be distinguished from all other bee larvae by their very large, unproduced spiracles (figs. 27.180f, 27.181f) in combination with their moderately recessed labiomaxillary region and well-developed maxillary palpi and labral tubercles (fig. 27.181b). Larvae of Hylaeinae (*Hylaeus*) (figs. 27.182a-c) are usually small and slender, about 8 mm in length. The combination of a produced labiomaxillary region and absence of a spinneret is found only in *Hylaeus* larvae and those of the Xylocopinae. Xylocopine larvae can be distinguished by the presence of a subapical mandibular tooth and distinct, projecting antennal papillae (both features lacking in *Hylaeus*).

Biology and Ecology: Colletids are solitary bees that nest in the ground (diphaglossines, *Colletes*) or in pithy stems (*Hylaeus*). Adult females apply a cellophanelike lining to larval cells and portions of the nest burrows (Hefetz et al. 1979). This material can be helpful in identifying a nest as belonging to a colletid.

Description: Mature larvae variable in size and shape (*Hylaeus* small, slender, usually not exceeding 10 mm; *Colletes* moderate in size, up to 20 mm; diphaglossines large, robust, up to 35 mm in length).

Head: Vertex, in lateral view, evenly rounded. Conspicuous setae absent. Epistomal ridge complete or absent (*Hylaeus*), arching dorsally to level of antennal discs (fig. 27.181a). Antennal prominence absent or weakly developed (diphaglossines, fig. 27.180b); antennal papilla usually weakly developed (moderately projecting and robust basally only in diphaglossines).

Mouthparts: Epipharyngeal spiculation present or absent (*Colletes*); hypopharyngeal spiculation present; maxillary spiculation present or absent (*Colletes*). Labral apex variable in shape, rounded to deeply emarginate (*Ptiloglossa*, fig. 27.180a); labral tubercles present (fig. 27.181b). Mandible usually moderately robust, conspicuously attenuate in *Colletes* (fig. 27.181d); mandibular apex acute, subapical tooth absent; cusp well defined (diphaglossines, fig. 27.180c) to absent (*Colletes*, fig. 27.181c); multidentate or nondentate (*Colletes*); teeth on dorsal apical edge present (fig. 27.180c); apical concavity weakly to moderately developed (fig. 27.180d), never scooplike. Labiomaxillary region produced or moderately recessed (*Colletes* (fig. 27.181b)). Inner maxillary surface rounded, not produced mesiad; maxillary palpus present, variable in size and shape; galea absent. Labial palpus present, variable in size and shape. Spinneret absent or present as a spoutlike projection (Diphaglossinae).

Body: Integument spiculate or nonspiculate (*Hylaeus*); conspicuous setae absent. Dorsal tubercles present, transverse, prominently rounded laterally; venter of segment 9 not protuberant; anus apically positioned or slightly dorsal (*Colletes*). Unlike most other bee larvae, spiracular atria of diphaglossines and *Colletes* conspicuously large in diameter (figs. 27.180f, 27.181f).

Comments: Although colletids are cosmopolitan in distribution, they are best represented in the Southern Hemisphere, being especially abundant in Australia. In North America this is a relatively small family that includes about 153 species in five genera. Larvae are known for all North American genera except the southwestern *Eulonchopria* (Colletinae).

Selected Bibliography

Batra 1980.
Hefetz et al. 1979.
McGinley 1980, 1981.
Rajotte 1979.
Rozen 1984.
Rozen and Favreau 1968.
Torchio 1965.

OXAEIDAE (APOIDEA)

Ronald J. McGinley, ***Smithsonian Institution***

The Oxaeids

Figure 27.183a-e

Relationships and Diagnosis: The phylogenetic relationships of this enigmatic family have been the subject of much discussion. Hurd and Linsley (1976) provide a review of this subject.

Oxaeid larvae look superficially like those of the Nomadinae (Anthophoridae). These two groups can be distinguished from all known bee larvae by their extremely recessed labiomaxillary region and the peculiar position of the posterior tentorial pits below the hypostomal ridge (figs. 27.183b, 27.201b). The cleft labral apex (fig. 27.183a) and slitlike primary tracheal opening (figs. 27.183d, e) of oxaeids will separate them from nomadines.

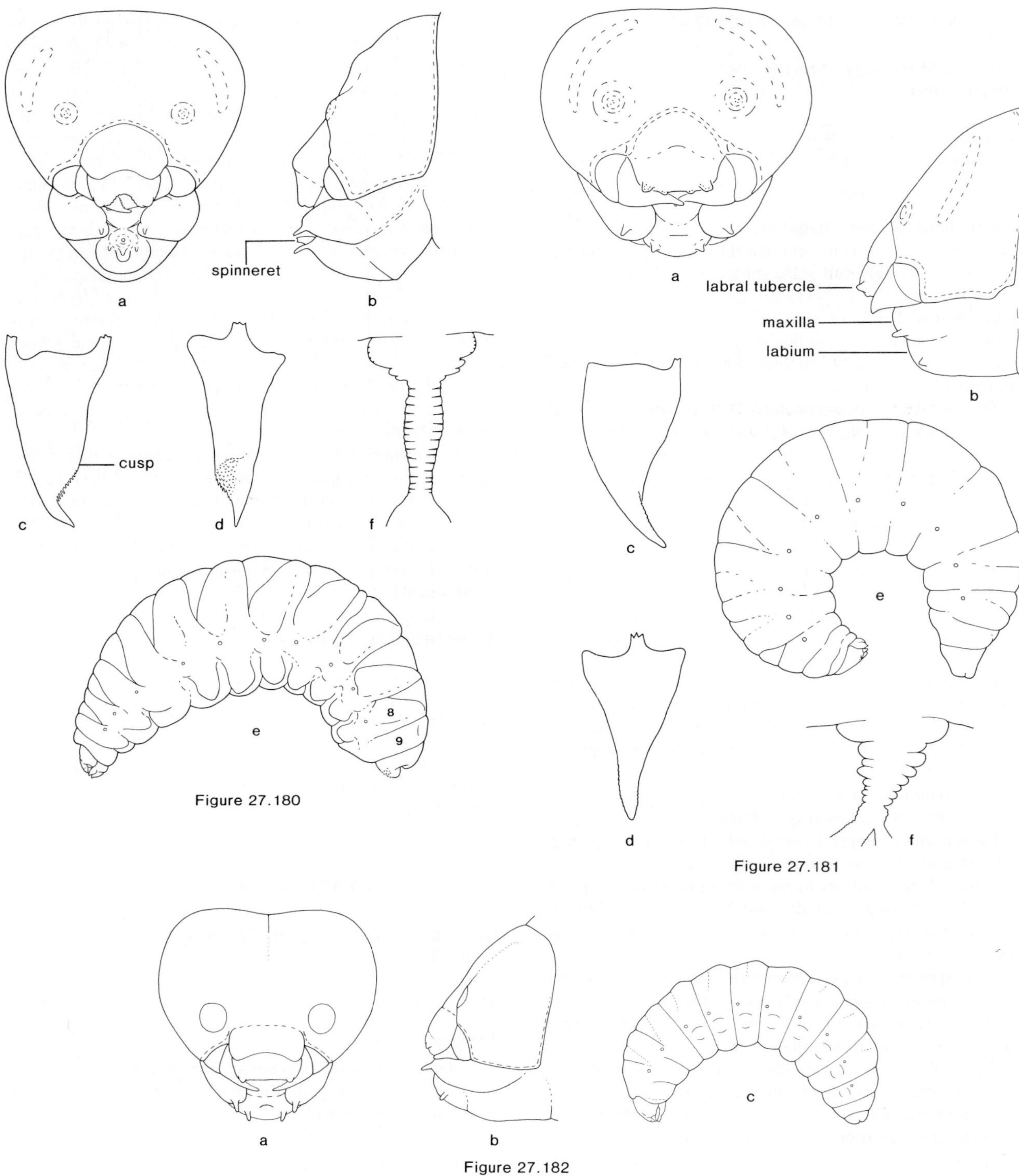

Figures 27.180–27.182.

Figure 27.180a-f. Colletidae: Diphaglossinae. *Ptiloglossa fulvopilosa* (Cameron). a. head, frontal view; b. lateral view; c. mandible, dorsal view; d. adoral view; e. body; f. spiracle, sagittal view. (redrawn from McGinley 1981).

Figure 27.181a-f. Colletidae: Colletinae. *Colletes thoracicus* (Smith). a. head, frontal view; b. lateral view; c. mandible, dorsal view; d. adoral view; e. body; f. spiracle, sagittal view (redrawn from McGinley 1981).

Figure 27.182a-c. Colletidae: Hylaeinae. *Hylaeus* sp. a. head capsule, frontal view; b. lateral view; c. body (redrawn from Michener 1953).

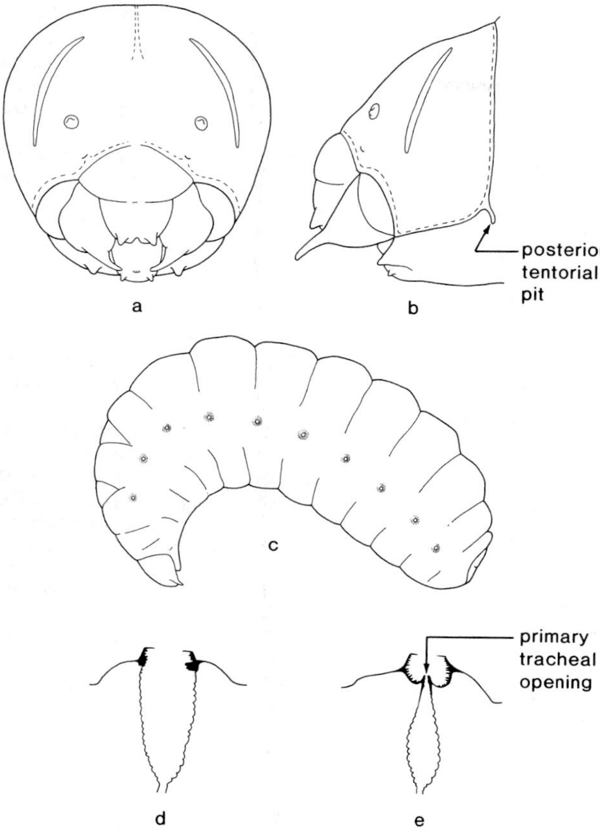

Figure 27.183a-e. Oxaeidae. *Protoxaea gloriosa* Fox. a. head capsule, frontal view; b. lateral view; c. body; d. spiracle, anterior view; e. ventral view (redrawn from Rozen 1964).

Biology and Ecology: Oxaeids are ground-nesting bees that make very deep burrows in the soil. All forms studied are solitary nesters. An excellent review of oxaeid biology can be found in Hurd and Linsley (1976).

Description: Mature larvae moderately large in size (fig. 27.183c); unlike most bee larvae, body integument yellowish with head capsule darkly pigmented (pale or white in most other bee larvae).

Head: Vertex rounded in lateral view, not projecting. Conspicuous setae absent. Epistomal ridge complete (fig. 27.183a), well below level of antennal discs. Antennal prominence absent; antennal papilla weakly developed, small and rounded. Unlike all other known bee larvae except nomadines (Anthophoridae), posterior tentorial pit below juncture of posterior thickening of head capsule and hypostomal ridge (connected to posterior thickening by sclerotic spur (fig. 27.183b)).

Mouthparts: Epipharyngeal and hypopharyngeal spiculation present; maxillary spiculation absent. Labral apex deeply cleft (fig. 27.183a); labral tubercles present, well developed, narrowly rounded apically (fig. 27.183b). Mandible elongate, broad basally in adoral view, tapering to an attenuate, bladelike apical portion (fig. 27.183b); apex acute; subapical tooth absent; cusp weakly defined, multidentate; teeth on dorsal apical edge present; apical concavity weakly developed, not scooplike. Labiomaxillary region extremely recessed (fig. 27.183b). Inner maxillary surface rounded, not produced mesiad; maxillary palpus well developed; galea absent. Labial palpus small. Spinneret absent.

Body: Integument nonspiculate; conspicuous setae absent. Dorsal tubercles very weakly developed, inconspicuous; venter of abdominal segment 9 nonprotuberant; anus apically positioned. Unlike all other bee larvae except those of some Nomadinae, spiracles rest on elevated, pigmented, sclerotic rings (fig. 27.183c), and unlike all other known bee larvae, primary tracheal opening slitlike (figs. 27.183d, e) (circular in other forms).

Comments: This is a small family of bees with 20 known species confined to the Western Hemisphere. Four species in two genera are found in southwestern United States. Larvae of *Mesoxaea* are unknown.

Selected Bibliography

Hurd and Linsley 1976.
Roberts 1973.
Rozen 1964, 1965a.

HALICTIDAE (APOIDEA)

Ronald J. McGinley, Smithsonian Institution

Halictids, Sweat Bees

Figures 27.184–27.187

Relationships and Diagnosis: The phylogenetic relationships of this family are not known, as is the case for other short-tongued bee families. Whereas the monophyletic basis of the Halictidae is well supported (Michener and Greenberg 1985), the three subfamilies included are so distinct that diagnosis of the family based on mature larvae is difficult. Due to the presence of antennal prominences (figs. 27.184b, 27.185b), well-developed dorsal abdominal tubercles (figs. 27.184e, 27.186b), absence of spinnerets and moderately recessed labiomaxillary regions, halictine and nomiine larvae look very much like those of andrenids (figs. 27.188–27.190). In addition to the above characteristics, halictine larvae share the lack of distinct labral tubercles with *Andrena* larvae (tubercles possibly represented by apicolateral swellings, figs. 27.184a, b). Halictines can be distinguished from *Andrena* and virtually all other bee larvae by their apparent lack of maxillary palpi (fig. 27.184b) (Rozen, personal comm., thinks that the palpi are possibly extremely large and nonprotuberant, rather than absent). This condition is found elsewhere only in certain nomadine larvae (*Neopasites, Neolarra*), which can be identified by the unusual position of the posterior tentorial pits that are anterior to the posterior margin of the head (fig. 27.201b).

In addition to these general halictid-andrenid similarities, nomiine and panurgine larvae possess distinct labral tubercles. The very well-developed mandibular tubercles (fig. 27.185c) of nomiines will distinguish these two groups.

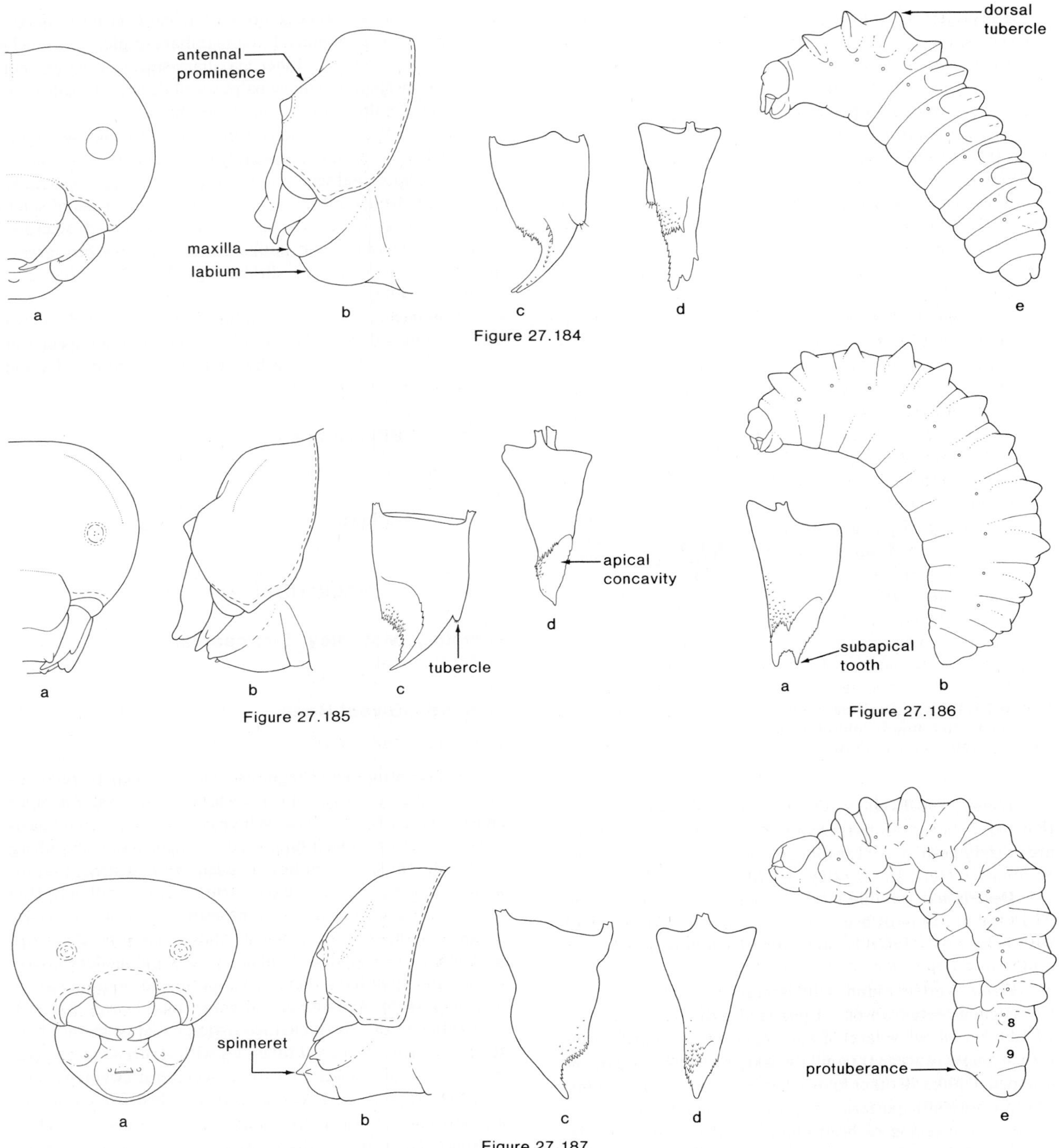

Figure 27.184a-e. Halictidae: Halictinae. *Evylaeus kincaidii* (Cockerell). a. head capsule, frontal view; b. lateral view; c. mandible, ventral view; d. adoral view; e. body. (redrawn from Michener 1953).

Figure 27.185a-d. Halictidae: Nomiinae. *Nomia melanderi* Cockerell. a. head capsule, frontal view; b. lateral view; c. mandible, ventral view; d. adoral view. (redrawn from Michener 1953).

Figure 27.186a,b. Halictidae: Nomiinae. *Nomia nevadensis* Cresson. a. mandible, adoral view; b. body (redrawn from Michener 1953).

Figure 27.187a-e. Halictidae: Dufoureinae. *Dufourea mulleri* (Cockerell). a. head capsule, frontal view; b. lateral view; c. mandible, dorsal view; d. adoral view; e. body.

The projecting labiomaxillary region, transverse, strongly produced spinneret (fig. 27.187b), well-developed dorsal abdominal tubercles, and projecting venter of abdominal segment 9 (fig. 27.187c) will distinguish the larvae of the Dufoureinae from all other known bee larvae except those of *Exomalopsis* (Anthophoridae). These taxa can be separated by the presence of a large subapical mandibular tooth in *Exomalopsis*.

Biology and Ecology: Halictids make their nests in the soil or in rotting wood. Although most species are probably solitary, communal behavior has been reported for all three subfamilies. The Halictinae display a range of behavior from solitary to primitively eusocial forms (Michener 1974).

Description: Mature larvae moderate in size and form (usually under 15 mm in length).

Head: Vertex rounded to moderately produced in lateral view (fig. 27.184b). Setae usually present. Epistomal ridge incomplete or absent, well below level of antennal discs. Antennal prominence usually strongly developed (figs. 27.184b, 27.185b); weakly developed in *Halictus, Dialictus* and some *Sphecodes*); antennal papilla moderately developed, robust basally, and moderately projecting.

Mouthparts: Epipharyngeal, hypopharyngeal and maxillary spiculation present. Labral apex rounded or shallowly emarginate; labral tubercles absent (Dufoureinae), present (Nomiinae, figs. 27.185a, b) or poorly defined, possibly represented by apicolateral swellings (Halictinae, figs. 27.184a, b). Mandible moderate in length to somewhat elongate, slender in form or moderately robust (*Nomia,* figs. 27.185c, d; 27.186a); unlike most bee larvae, mandibular tubercle (figs. 27.184c, 27.185c) usually very well developed (absent in dufoureines); mandibular apex acute; subapical tooth present or absent (figs. 27.185d, 27.186a); cusp usually defined (not developed in some *Sphecodes*), multidentate; teeth on dorsal apical edge present; development of apical concavity variable, never scooplike. Labiomaxillary region strongly produced (Dufoureinae, fig. 27.187b) or moderately recessed (Halictinae, Nomiinae). Inner maxillary surface rounded, not produced mesiad; maxillary palpus well developed or apparently absent (Halictinae, fig. 27.184b), galea absent. Labial palpus elongate (Dufoureinae), small (Nomiinae) or absent (Halictinae). Spinneret transverse, strongly projecting (Dufoureinae, fig. 27.187b), or absent.

Body: Integument usually spiculate, without conspicuous setae. Dorsal abdominal tubercles well developed and transverse, prominently rounded laterally (somewhat conical in Dufoureinae); venter of abdominal segment 9 protuberant only in Dufoureinae (fig. 27.187e) and some *Sphecodes;* anus apical or slightly dorsal in position.

Comments: This is a very large family, distributed worldwide, with about 500 species found in the continental United States. Halictid larvae are poorly known; of the 22 halictid genera occurring in this area, representatives of only 8 genera have been described. In the Dufoureinae, *Conanthalictus* and *Xeralictus* larvae have been collected but await description. Larvae of the other dufoureine genera (*Micralictoides, Protodufourea, Sphecodosoma, Michenerula*) are apparently unknown. Among halictines, larvae of *Augochloropsis, Augochlorella, Pseudaugochloropsis, Temnosoma, Lasioglossum, Sphecodogastra* and *Paralictus* have not been described from the continental United States. Much more material is needed for other genera as well, e.g., from our area, the larvae of only one species of *Sphecodes* has been described (Torchio 1975).

Michener's (1953) treatment of halictid larvae remains the most general account for the family. Eickwort and Eickwort (1973, and papers cited in their bibliography) have contributed significantly to our understanding of many halictid larvae.

Selected Bibliography

Abrams and Eickwort 1980.
Daly 1961.
Eickwort and Eickwort 1973.
Eickwort et al. 1986.
Michener 1953, 1974.
Rozen and McGinley 1976.
Torchio 1975.
Torchio et al. 1967.

ANDRENIDAE (APOIDEA)

Ronald J. McGinley, *Smithsonian Institution*

The Andrenids

Figures 27.188–27.190

Relationships and Diagnosis: Like other short-tongued bees, the relationships of this family to other bees remains poorly understood. Adult and larval characters do, however, support the monophyly of the Andrenidae. Larvae of both subfamilies, Andreninae and Panurginae, appear to have salivary openings that are U-shaped (figs. 27.188a, 27.189a), the ends of which reach the hypopharyngeal groove (Torchio (1975) reports that this U-shaped structure is actually a cross-sectional view of a greatly expanded salivary tube and that no opening to the cuticle exists). This condition is found in all known andrenid larvae except those of some *Perdita* and the Old World *Panurgus;* it is not known to occur in other aculeate larvae. *Perdita* and *Panurgus,* like other panurgines, can be recognized by their conical dorsal abdominal tubercles (figs. 27.189c, 27.190) and well-developed labral tubercles (fig. 27.189b). *Perdita* larvae also have a characteristic median tubercle on the last abdominal segment (fig. 27.190).

Biology and Ecology: Andrenids are ground-nesting bees. Although most are solitary, communal nesting behavior is not uncommon. Floral preferences range from broad polylecty to many cases of narrow oligolecty. Schrader and LaBerge (1978) provide an excellent account of the biologies of *Andrena regularis* and *A. carlini* and include references to earlier works on *Andrena* biology. Rozen (1967a) does the same for the biology of panurgine bees.

Description: Form moderately slender (many Panurginae) to moderately robust.

Head: Vertex usually evenly rounded as seen in lateral view (vertex produced in some *Andrena* and *Perdita*). Setae present. Epistomal ridge usually absent mesiad of anterior

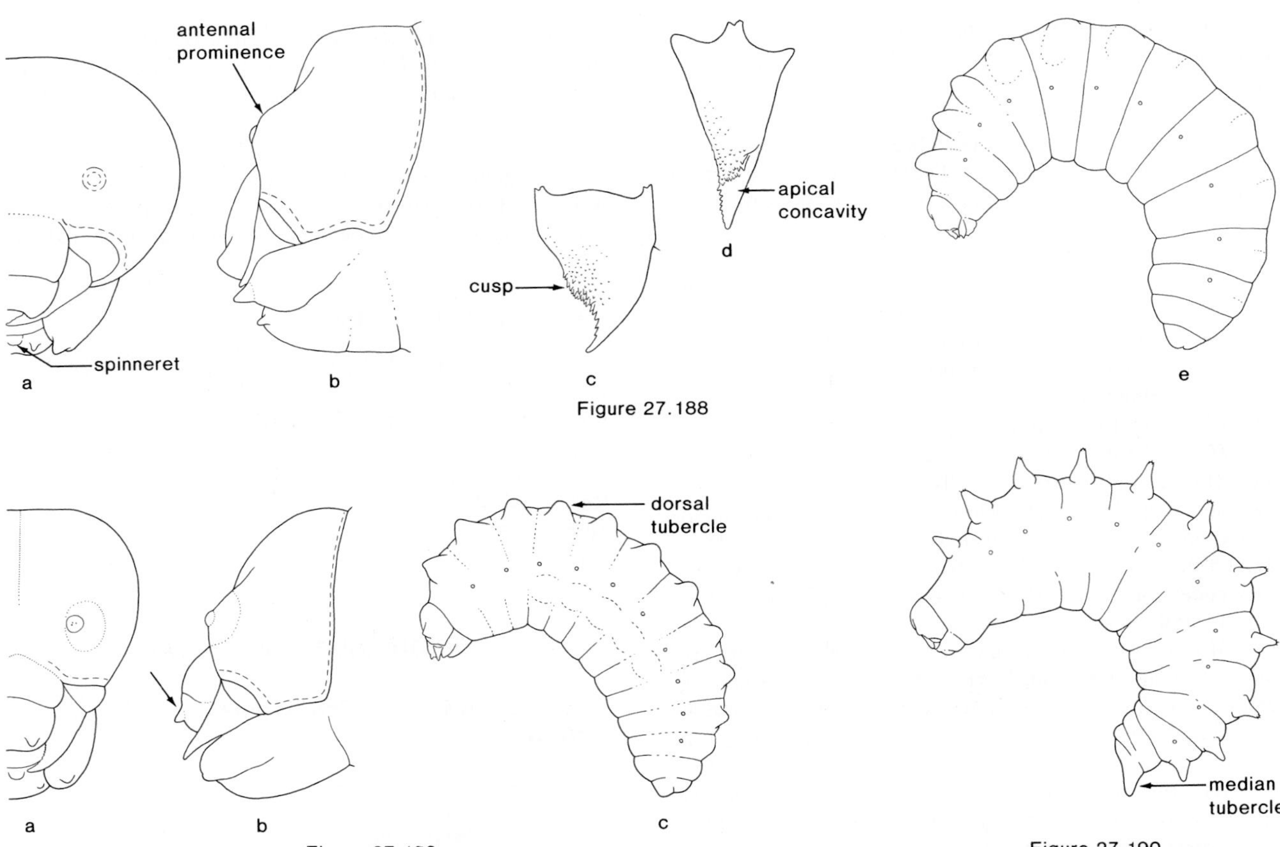

Figure 27.188a-e. Andrenidae: Andreninae. *Andrena* sp. a. head capsule, frontal view; b. lateral view; c. mandible, ventral view; d. adoral view; e. body. (redrawn from Michener 1953).

Figure 27.189a-c. Andrenidae: Panurginae. *Calliopsis andreniformis* Smith. a. head capsule, frontal view; b. lateral view; c. body. (redrawn from Michener 1953).

Figure 27.190. Andrenidae: Panurginae. *Perdita halictoides* Smith. (redrawn from Eickwort 1977).

tentorial pits (complete in *Panurginus*); epistomal ridge or depression well below level of antennal discs. Antennal prominence usually well developed (fig. 27.189b); absent or weakly developed in most *Perdita*; antennal papilla weakly to moderately developed.

Mouthparts: Epipharynx, hypopharynx, and inner maxillary surface spiculate. Labral apex rounded or shallowly emarginate; labral tubercles well developed (Panurginae) or poorly defined, possibly indicated by apicolateral swellings (*Andrena*, figs. 27.188a, b). Mandible moderate to slender as seen in adoral view; mandibular apex acute, with or without subapical tooth (present in some panurgines); mandibular cusp moderately defined to virtually absent; apical concavity weakly developed, never scooplike. Labiomaxillary region moderately recessed; inner maxillary surface rounded, not projecting mesiad; maxillary palpus present, small to moderately well developed and robust; galea absent. Labial palpus small to moderately well developed (absent in some panurgines). Spinneret absent.

Body: Integument spiculate, without conspicuous setae. Dorsal tubercles transverse, prominently rounded laterally (*Andrena*, fig. 27.188e) or distinctly rounded to conical in shape (panurgines, figs. 27.189c, 27.190); venter of abdominal segment 9 nonprotuberant; anus variable in position.

Comments: Andrenids are found on all continents except Australia. It is the largest family of bees in the United States with 1,196 species classified in 15 genera. In this area, larvae of the following andrenid genera remain unknown: *Megandrena, Ancylandrena, Anthemurgus, Metapsaenythia* and *Xenopanurgus.*

Selected Bibliography

Eickwort 1977.
Rozen 1966a, 1967a, 1970, 1973a.
Schrader and LaBerge 1978.
Torchio 1975.
Youssef and Bohart 1968.

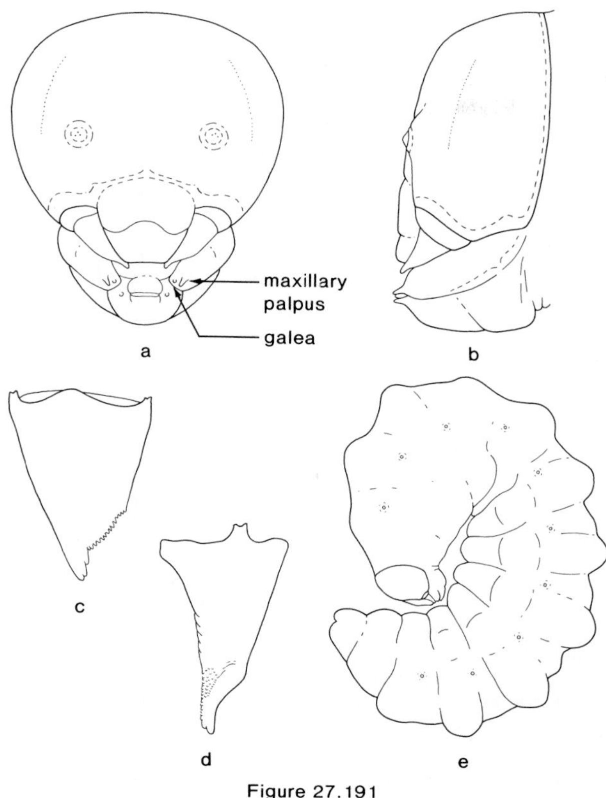

Figure 27.191a-e. Melittidae. *Macropis europaea* Weke. a. head capsule, frontal view; b. lateral view; c. mandible, dorsal view; d. adoral view; e. body. (redrawn from Rozen and McGinley 1974a).

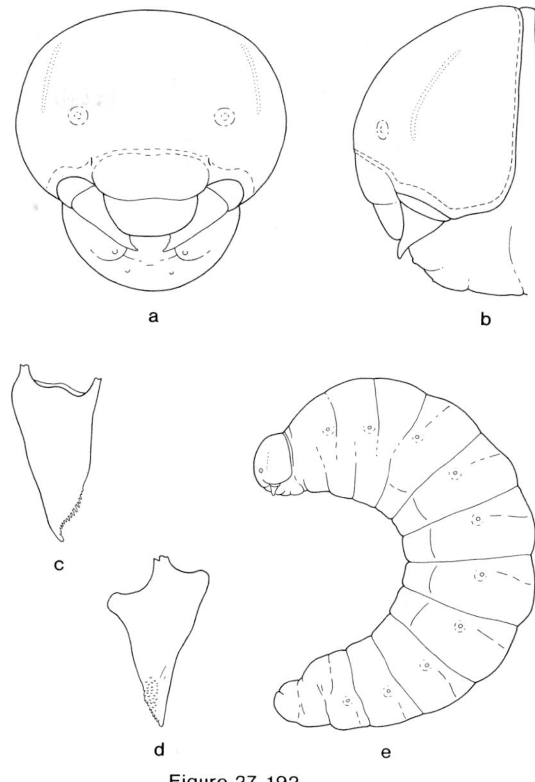

Figure 27.192a-e. Mellitidae. *Hesperapis pellucida* Cockerell. a. head capsule, frontal view; b. lateral view; c. mandible, dorsal view; d. adoral view; e. body (redrawn from Rozen and McGinley 1974a).

MELITTIDAE (APOIDEA)

Ronald J. McGinley, *Smithsonian Institution*

The Melittids

Figures 27.191, 27.192

Relationships and Diagnosis: Recent work by Michener and Greenberg (1980) indicates that melittids are the sister group of the long-tongued bees, i.e., Megachilidae, Anthophoridae, Apidae, and the recently recognized Old World Ctenoplectridae.

The cocoon-spinning melittids, *Macropis* and *Melitta,* are quite generalized in appearance, superficially very similar to cocoon-spinning forms in other families such as *Dufourea* (Halictidae) and *Exomalopsis* (Anthophoridae). *Macropis* and *Melitta,* along with the Eucerini and *Centris* (both Anthophoridae), are the only North American bee larvae to have conspicuous maxillary galeae (fig. 27.191a) (galeae are present in *Apis* but are small and inconspicuous). The two melittid genera are easily distinguished from the above mentioned anthophorids by their possession of dorsal abdominal tubercles (fig. 27.191e) and maxillary apices that do not project mesiad.

Hesperapis larvae (figs. 27.192a, b) do not spin cocoons and therefore look very unlike the larvae of *Melitta* and *Macropis.* The recession of the labiomaxillary region in combination with the absence of labral and abdominal tubercles will distinguish *Hesperapis* from all other known bee larvae.

Biology and Ecology: North American melittids are solitary ground-nesting bees. Nests of *Macropis* have been only recently studied (Rozen and Jacobson 1980). Stage (1966) has worked extensively on the biology of *Hesperapis.* The nesting biology of American *Melitta* species remains unknown.

Description: The following description includes an account of the European *Melitta leporina* (Panzer). Mature larvae moderate in size and form, up to 15 mm in length.

Head: Vertex, in lateral view, rounded. Setae present (*Macropis, Melitta*) or absent. Epistomal ridge complete or fading medially (some *Hesperapis*), well below level of antennal discs. Antennal prominence weakly developed (fig. 27.191b) to absent (most *Hesperapis*); antennal papilla small (*Hesperapis*) to moderately large, robust basally.

Mouthparts: Epipharyngeal and hypopharyngeal spiculation present; maxillary spiculation present or absent. Labral apex variable in shape, truncate, rounded or shallowly emarginate; distinct labral tubercles absent. Mandible moderate in length and form to somewhat short and broad (fig. 27.191c); mandibular apex acute; subapical tooth present only in *Melitta;* cusp moderately defined, multidentate; teeth present on dorsal apical edge; apical concavity moderately to weakly developed (*Hesperapis*), never scooplike. Labiomaxillary region produced (fig. 27.191b) or recessed (*Hesperapis,* fig. 27.192b). Inner maxillary surface rounded, not produced mesiad; maxillary palpus present, very small (*Hesperapis*) or elongate and slender; galea present or absent. Labial palpus minute (*Hesperapis*) or elongate and slender. Spinneret absent or strongly projecting and transverse (fig. 27.191b).

Body: Integument spiculate or nonspiculate, without conspicuous setae. Dorsal tubercles absent (*Hesperapis,* fig. 27.192e) or present as transverse, rounded lobes (fig. 27.191e); venter of segment 9 protuberant (some *Hesperapis*) or nonprotuberant; anus apically positioned.

Comments: Melittids comprise a small bee family that is primarily Holarctic in distribution. In the continental United States it is represented by three genera with approximately 30 species. *Melitta* larvae remain unknown from our area.

Rozen and McGinley (1974a) analysed the interrelationships of melittid genera based on all available mature larvae. This work has been extended by Rozen (1978).

Selected Bibliography

Burdick and Torchio 1959.
Michener and Greenberg 1980.
Rozen 1978.
Rozen and Jacobson 1980.
Rozen and McGinley 1974a.
Stage 1966.

MEGACHILIDAE (APOIDEA)

Ronald J. McGinley, *Smithsonian Institution*

Megachilids, Leafcutting Bees

Figure 27.193

Relationships and Diagnosis: With the anthophorids and apids, megachilids form a monophyletic group usually referred to as the long-tongued bees (Michener and Greenberg 1980). Whereas the interrelationship of these families are not well understood, the megachilids are traditionally thought to be the sister group of the other two families.

Megachilid larvae can be distinguished from other bee larvae by their elongate, conspicuous body setae (fig. 27.193e). Similar setae occur in some *Bombus* (Apidae) (which have highly distinctive conical thoracic tubercles, fig. 27.206d) and some Melitomini (anthophorids which are extremely slender in body form). Other helpful identifying characteristics of megachilid larvae are their absence of labral tubercles, emargination of the labral apex, virtual absence of paired dorsal abdominal tubercles, and in some species, the presence of median dorsal body tubercles.

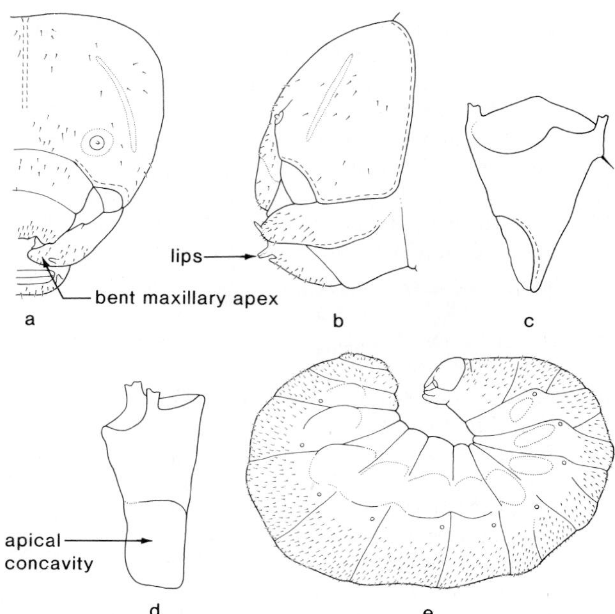

Figure 27.193a-e. Megachilidae. *Trachusa perdita* Cockerell. a. head capsule, frontal view; b. lateral view; c. mandible, ventral view; d. adoral view; e. body (redrawn from Michener 1953).

Biology and Ecology: Megachilids nest in the soil, in stems, in cavities, or in nests made of mud or resin, often with pebbles, attached to stems or rocks. The nests are characteristic in having the cells and burrows lined with leaves, flower petals, resin, mud, or plant hairs. See Eickwort et al. (1981) for a review of megachilid nesting biology.

Although the majority of megachilids are solitary, a few species in a variety of genera are communal. *Coelioxys, Dioxys, Heterostelis,* and *Stelis* are cleptoparasites of other megachilids (*Coelioxys* has been associated with other families in the Old World). See Baker (1971), Rozen (1966b, 1967b), Rust and Thorp (1973), and Thorp (1966) for information on megachilid cleptoparasitic biologies.

Description: The body form of most megachilids is robust, often with a noticeable posterior broadening of the body (fig. 27.193e).

Head: Vertex, in lateral view, evenly rounded. Conspicuous setae present. Epistomal ridge incomplete or absent mesiad of anterior tentorial pits; epistomal depression often arching to level of antennal discs. Antennal prominence absent or weakly developed; antennal papilla usually slender and elongate (extremely large in *Dioxys;* weakly projecting in *Lithurge, Hoplitis*).

Mouthparts: Epipharynx, hypopharynx, and maxilla usually nonspiculate (hypopharynx spiculate in *Trachusa;* maxilla spiculate in *Lithurge*). Labral apex shallowly to deeply emarginate; labral tubercles absent. Mandible usually robust (slender in *Dioxys* and some *Stelis*); mandibular apex usually acute (truncate in *Trachusa;* narrowly rounded in *Coelioxys*) with large subapical tooth (absent in *Trachusa* and some *Stelis*); ventral tooth larger than dorsal tooth; cusp usually weakly defined or absent (well defined in *Lithurge,*

Megachile); cusp nondentate; teeth on dorsal apical edge present or absent; apical concavity moderately to strongly developed (weakly developed in *Dioxys*); apical concavity broad and somewhat scooplike (fig. 27.193d). Labiomaxillary region usually strongly produced (slightly recessed in some *Hoplitis*); inner apical margin of maxilla strongly bent mesiad (fig. 27.193a); maxillary palpus elongate; galea absent. Labial palpus elongate (moderately elongate in *Ashmeadiella*). Spinneret well developed, strongly projecting (fig. 27.193b).

Body: Integument spiculate or nonspiculate, usually with conspicuous setae (fig. 27.193e) (reduced in some *Ashmeadiella*). Paired dorsal tubercles absent or inconspicuous; unlike most other bee larvae, median dorsal tubercles often present; venter of abdominal segment 9 not protuberant; anus often dorsal in position.

Comments: This is a very large family, distributed worldwide. In the United States it is the third largest family, comprising 613 species in 24 genera. Larvae of the following seven genera have not been described from our area: *Heteranthidium, Paranthidium, Adanthidium, Anthidium, Callanthidium, Anthocopa.*

Michener (1953) provides the most comprehensive account of megachilid larvae, and Rozen (1973b) gives a family description based on mature larvae.

Selected Bibliography

Baker 1971.
Baker et al. 1985.
Clement 1976.
Eickwort 1973.
Eickwort et al. 1981.
Matthews 1965.
Michener 1953.
Michener and Greenberg 1980.
Rozen 1966b, 1967b, 1973b.
Rust and Thorp 1973.
Thorp 1966, 1968.

ANTHOPHORIDAE (APOIDEA)

Ronald J. McGinley, *Smithsonian Institution*

Anthophorids, Carpenter Bees

Figures 27.194–27.205

The size (45 genera with approximately 914 species in North America) and biological diversity of the Anthophoridae are reflected in the great morphological variability found among its larvae. Because of this, treatment of the family as a whole proved impractical and the following groups will be discussed separately: pollen-collecting Anthophorinae, parasitic Anthophorinae, Nomadinae and the Xylocopinae.

POLLEN-COLLECTING ANTHOPHORINAE

Anthophorini, Centridini, Eucerini, Exomalopsini, Melitomini

Figures 27.194–27.199

Diagnosis: The presence of galeae (figs. 27.198a, 27.199a) and lack of well-developed dorsal abdominal tubercles will distinguish the larvae of the Eucerini and Centridini from all other known bee larvae except those of *Apis. Apis,* unlike anthophorids, has weakly sclerotized mandibles and extremely broad, weakly projecting antennal papillae. The Eucerini can be separated from the Centridini by possessing labral tubercles (fig. 27.198a) (absent in *Centris*) and an elongate, somewhat pointed tenth abdominal segment (fig. 27.198e) (short and rounded in *Centris,* fig. 27.199d).

Larvae of the Anthophorini are rather unusual among bee larvae in having relatively massive mandibles with scooplike apical concavities (figs. 27.197c, d) combined with the absence of a spinneret. Only *Xylocopa* larvae have this combination of characters but are themselves highly unusual in having a hidden salivary opening (on dorsal surface of labium) and a characteristic ventral labial lobe (fig. 27.204b).

Exomalopsini and Melitomini larvae are similar in their slender, elongate body form (figs. 27.194e, 27.196) and protuberant ninth abdominal venter. The Melitomini differ from the Exomalopsini in having a much more slender body form and maxillae that are strongly bent mesiad. Not all *Exomalopsis* larvae spin cocoons (Rozen, pers. comm.), hence some species have strongly produced labiomaxillary regions and spinnerets, whereas other species lack these features.

Biology and Ecology: Although the majority of pollen-collecting anthophorines nest in the soil, a few species nest in wood. Communal nesting behavior is found in some species of Eucerini and is common but not universal in the Exomalopsini.

Description: Body form extremely slender to robust.

Head: Vertex evenly rounded in lateral view. Conspicuous setae usually present (apparently absent in Anthophorini, *Diadasia*). Epistomal ridge complete or incomplete mesiad of anterior tentorial pits, not arching to level of antennal discs. Antennal prominence absent or weakly developed (Anthophorini, Eucerini); antennal papillae strongly projecting (Melitomini, fig. 27.195b), moderately projecting (*Exomalopsis*) or weakly projecting.

Mouthparts: Epipharynx usually spiculate (nonspiculate in Anthophorini, *Melitoma*); hypopharynx spiculate or nonspiculate; maxillary surface spiculate. Labral apex variable in form, rounded to deeply emarginate; labral tubercles usually present and well defined (absent in *Centris* and *Exomalopsis*). Mandibles massive, moderately narrow in adoral view (Eucerini, *Exomalopsis,* figs. 27.194d, 27.198d) or broad (*Centris,* Melitomini, Anthophorini, figs. 27.195d, 27.197d, 27.199c); apex acute (Eucerini, *Exomalopsis,* some Melitomini) or broadly rounded (*Centris,* Anthophorini, some

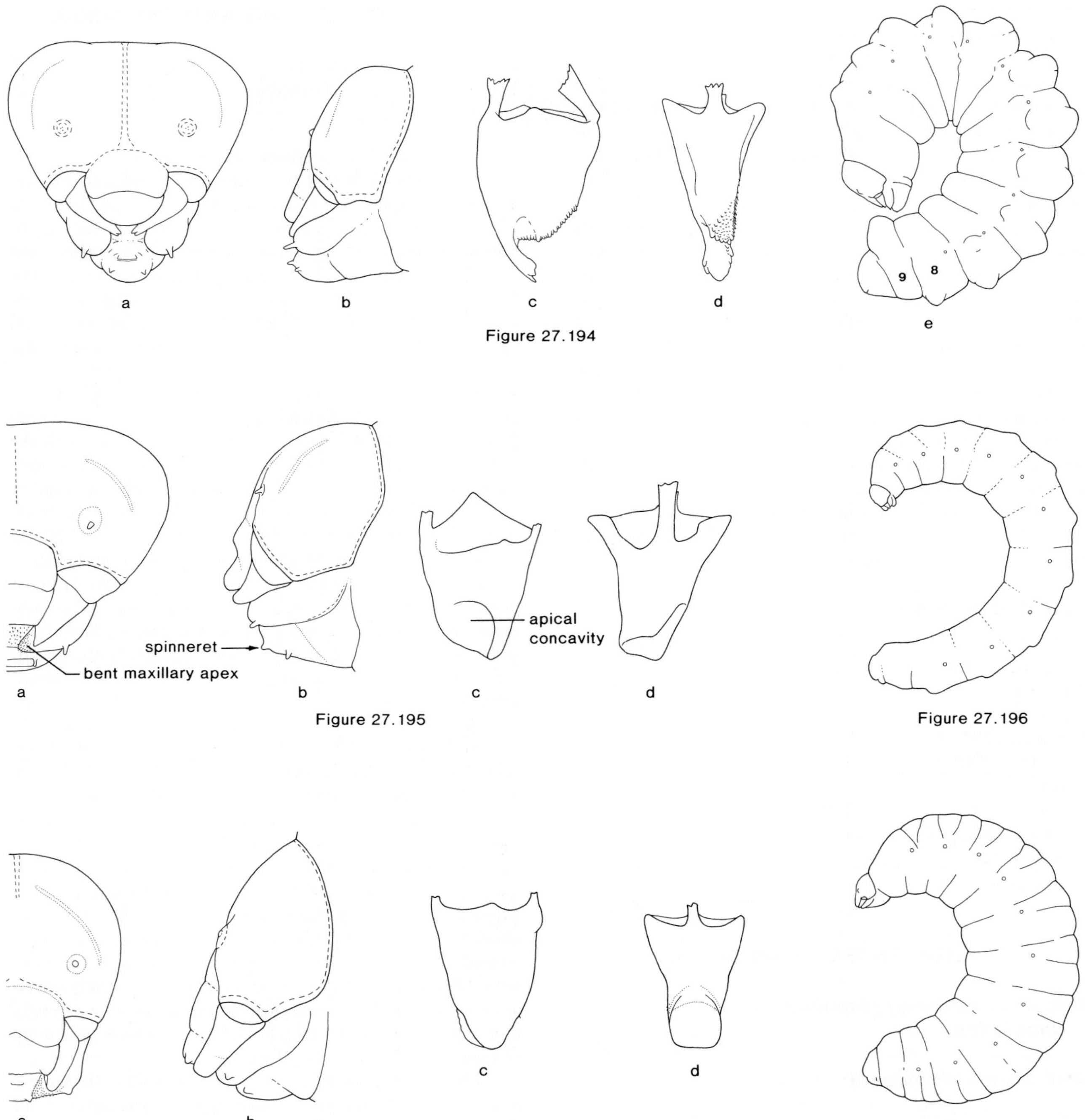

Figure 27.194a-e. Anthophoridae: Anthophorinae. Exomalopsini: *Exomalopsis chionura* (Cockerell). a. head capsule, frontal view; b. lateral view; c. mandible, dorsal view; d. adoral view; e. body (redrawn from Rozen 1957).

Figure 27.195a-d. Anthophoridae: Anthophorinae: Melitomini. *Diadasia enavata* (Cresson). a. head capsule, frontal view; b. lateral view; c. mandible, ventral view; d. adoral view. (redrawn from Michener 1953).

Figure 27.196. Anthophoridae: Anthophorinae: Melitomini. *Ptilothrix bombiformis* (Cresson).

Figure 27.197a-e. Anthophoridae: Anthophorinae: Anthophorini: *Anthophora bomboides stanfordiana* Cockerell. a. head capsule, frontal view; b. lateral view; c. mandible, ventral view; d. adoral view; e. body (redrawn from Michener 1953).

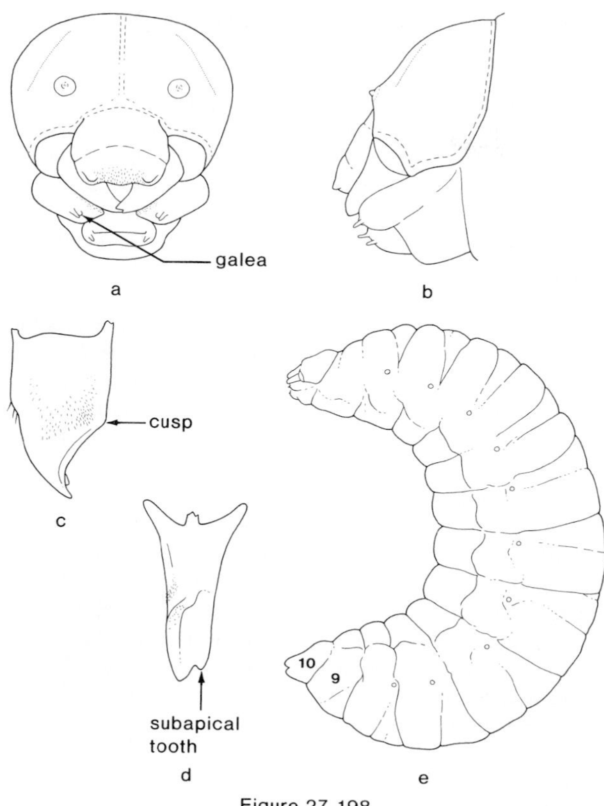

Figure 27.198

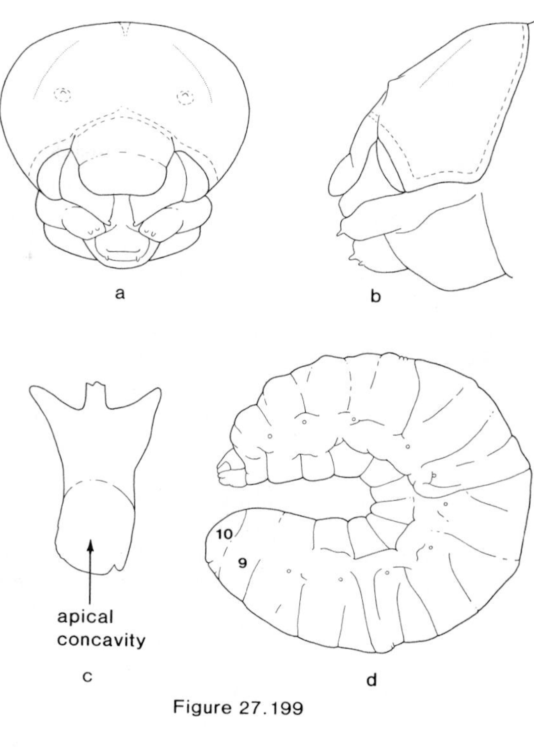

Figure 27.199

Figure 27.198a-e. Anthophoridae: Anthophorinae: Eucerini: *Xenoglossa strenua* (Cresson). a. head capsule, frontal view; b. lateral view; c. mandible, dorsal view; d. adoral view; e. body.

Figure 27.199a-d. Anthophoridae: Anthophorinae: Centridini: *Centris derasa* Lepeletier. a. head capsule, frontal view; b. lateral view; c. mandible, adoral view; d. body (all redrawn from Rozen 1965b).

Melitomini); subapical tooth present (Eucerini, *Exomalopsis, Centris, Emphoropsis,* some Melitomini) or absent (unlike all other known bee larvae except those of the Xylocopinae, dorsal tooth larger than ventral tooth); well-defined cusp in Eucerini, some Melitomini, weakly defined or absent in other taxa, cusp distinctly dentate only in *Exomalopsis* (fig. 27.194d); dorsal apical edge dentate (*Centris,* some Anthophorini, *Exomalopsis*) or nondentate; apical concavity well developed in all forms except *Exomalopsis;* apical concavity scooplike in *Centris,* Anthophorini, and to a lesser extent in Melitomini (apical concavity oblique in Eucerini, *Exomalopsis*). Labiomaxillary region usually strongly produced (slightly recessed in Anthophorini and some Melitomini); inner maxillary margin (fig. 27.195a) usually produced mesiad (rounded in *Centris*); maxillary palpus usually well developed and elongate (moderately short in Anthophorini); galea present (Eucerini, *Centris,* figs. 27.198a, 27.199a) or absent. Labial palpus usually elongate (short in Anthophorini, some Melitomini). Spinneret usually strongly projecting (greatly reduced or absent in some *Exomalopsis* and Anthophorini).

Body: Integument usually nonspiculate (spiculate in Melitomini, *Exomalopsis*); scattered, inconspicuous setae present in many forms (conspicuous in some Melitomini); dorsal tubercles usually absent, moderately well developed in *Exomalopsis;* unlike most other bee larvae, faint, paired dorsal sclerites on thoracic segments present in some Eucerini and *Centris* (fig. 27.199d; known elsewhere only in *Rathymus* and *Acanthopus*); venter of abdominal segment 9 protuberant in *Exomalopsis* and some Melitomini (figs. 27.194e, 27.196); anus usually dorsal in position.

Comments: Among the pollen-collecting Anthophorinae, larvae of the following genera have not been described from our area: *Ancyloscelis, Synhalonia, Syntrichalonia, Cemolobus, Anthedonia, Xenoglossodes, Gaesischia, Simanthedon, Martinapis,* and *Melissoptila.*

The papers by Rozen (1957, 1965b), Torchio (1971, 1974), and Torchio and Stephen (1961) provide the best introduction to the immatures of this group of bees.

Selected Bibliography

Eickwort et al. 1977.
Linsley et al. 1956.
Linsley et al. 1980.
Rozen 1957, 1965b.
Thorp 1969.
Torchio 1971, 1974.
Torchio and Stephen 1961.

PARASITIC ANTHOPHORINAE

Melectini

Figure 27.200

Diagnosis: The recessed (*Melecta*) to strongly recessed (*Zacosmia*) labiomaxillary region, together with a strongly produced spinneret, distinguishes Melectini larvae from those of other known apoids. The only group that might be confused with the Melectini are the Melitomini (figs. 27.195, 27.196) but they can be recognized by their conspicuously slender body form and maxillae that are strongly bent mesiad.

Biology and Ecology: The Melectini are cleptoparasites in the nests of other anthophorids. Host records for North America include *Anthophora* and *Emphoropsis*.

Description: Body form moderately elongate but not conspicuously slender.

Head: Vertex in lateral view, evenly rounded. Conspicuous setae apparently absent. Epistomal ridge nearly complete (*Melecta*) or fading toward median line (*Zacosmia*), not arching to level of antennal discs. Antennal prominences absent; well-developed antennal papillae, elongate (*Melecta*), or only moderately developed (*Zacosmia*).

Mouthparts: Epipharynx and maxillae spiculate (*Melecta*) or nonspiculate; hypopharynx spiculate. Labral apex emarginate (*Melecta*) or narrowly rounded (*Zacosmia*); labral tubercles absent. Mandible (figs. 27.200c, d), in adoral view, broad, massive basally; mandibular apex broadly rounded (*Melecta*) or acute (*Zacosmia*); subapical tooth absent; cusp defined, moderately produced, dentate, or nondentate; apical concavity broad and scooplike (somewhat narrow in *Zacosmia*). Labiomaxillary region moderately (*Melecta*) to strongly (*Zacosmia*) recessed; inner maxillary margin rounded, not produced mesiad; well-developed maxillary palpus, moderately short (*Zacosmia*) to elongate (*Melecta*); maxillary palpus preapical in position as seen in lateral view (*Melecta*, fig. 27.200b); galea absent. Labial palpus moderately elongate (*Melecta*) to somewhat short (*Zacosmia*). Spinneret well developed and strongly projecting (fig. 27.200b).

Body: Integument nonspiculate, conspicuous setae absent. Dorsal tubercles absent or only weakly produced (fig. 27.200e); venter of abdominal segment 9 nonprotuberant; anus dorsally positioned on terminal body segment.

Comments: The other tribe of parasitic Anthophorinae, Eriocrocidini, is represented in the United States by two species of *Ericrocis* and by *Mesoplia dugesi;* all three are presumably cleptoparasites of *Centris*. Rozen (1969) described the larvae of the related *Acanthopus* and *M. rufipes* from Trinidad but the North American larvae remain unknown.

Selected Bibliography

Rozen 1969.
Torchio and Youssef 1968.

Subfamily NOMADINAE (ANTHOPHORIDAE)

Figures 27.201–27.203

Diagnosis: The posterior tentorial pit in nearly all bee larvae is located at the juncture of the hypostomal ridge and the posterior thickening of the head capsule (fig. 27.200b). Nomadines (including *Protepeolus*) are almost unique in having this pit displaced from its typical position, either below the hypostomal ridge, anterior to the posterior thickening of the head capsule or both (fig. 27.201d); oxaeid larvae have a similarly displaced pit but can be recognized by their deeply cleft labral apices (fig. 27.183a). Other useful characters for identifying nomadine larvae (except those of *Protepeolus*) are the extremely recessed labiomaxillary region, rigid integument of postdefecating forms, virtual absence of labial palpi, well-developed labral tubercles, and the protuberant hypopharynx (fig. 27.201b) found in most forms.

Biology and Ecology: Nomadines are cleptoparasites in the nests of all North American bee families except the Megachilidae and Apidae. The majority of nomadine genera restrict their association to one host genus or group of closely related genera. The two major exceptions are *Triepeolus,* which primarily attacks the nests of the Eucerini, and *Nomada,* primarily associated with *Andrena*. Species of both genera are, however, known to parasitize the nests of bees from three additional families.

Tengö and Bergström (1977) have shown that some *Nomada* cleptoparasites secrete compounds that may mimic the nest odor of their *Andrena* hosts.

Description: Mature larvae slender to moderately robust; integument of postdefecating form often rigid; color often faintly yellowish to distinctly yellow.

Head: Vertex, in lateral view, usually evenly rounded (produced only in *Melanomada, Triopasites,* and *Paranomada*). Conspicuous setae present (*Protepeolus*) or absent. Epistomal ridge incomplete or absent mesiad of anterior tentorial pits; epistomal depression below level of antennal discs. Antennal prominence absent, or at most, weakly developed; antennal papilla usually weakly developed (strongly projecting only in *Protepeolus*).

Mouthparts: Epipharyngeal, hypopharyngeal, and maxillary spiculation present or absent. Unlike most other bee larvae, labrum often strongly protuberant, projecting well beyond clypeus in lateral view (fig. 27.201b); labral apex rounded, never emarginate; distinct, narrowly rounded labral tubercles present (absent in *Protepeolus*). Mandible short; in adoral view, mandibular base broad to very broad, tapering to a simple, acute apex (mandible somewhat slender and elongate in *Melanomada* and *Paranomada*); mandibular cusp usually not produced (weakly produced in *Protepeolus, Nomada, Triepeolus*); cusp usually nondentate (dentate in some *Nomada*); dorsal apical edge dentate or serrate, figs. 27.201c, d (smooth in *Triepeolus, Epeolus,* figs. 27.203a, b); apical concavity weakly defined or absent. Labiomaxillary region strongly recessed (moderately produced in *Protepeolus*). Inner maxillary surface rounded apically, produced mesiad only in *Protepeolus;* maxillary palpus present or absent (*Neopasites,*

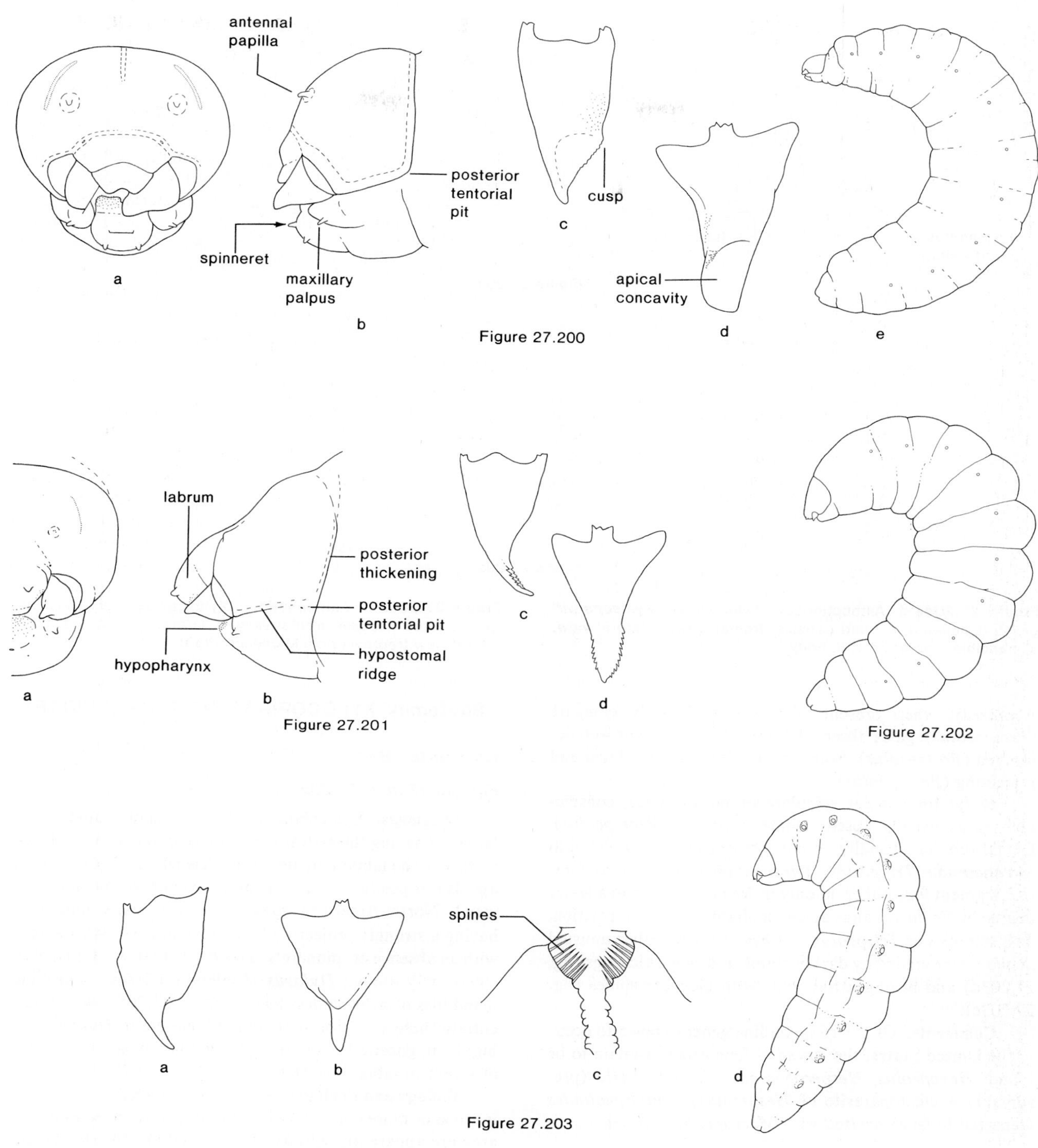

Figure 27.200a-e. Anthophoridae: Anthophorinae: Melectini. *Melecta separata callura* (Cockerell). a. head capsule, frontal view; b. lateral view; c. mandible, dorsal view; d. adoral view; e. body (redrawn from Rozen 1969).

Figure 27.201a-d. Anthophoridae: Nomadinae. *Nomada fowleri* Cockerell. a. head capsule, frontal view; b. lateral view; c. mandible, dorsal view; d. adoral view.

Figure 27.202. Anthophoridae: Nomadinae. *Nomada* species A.

Figure 27.203a-d. Anthophoridae: Nomadinae. *Triepeolus* species A. a. mandible, dorsal view; b. adoral view; c. spiracle, sagittal view; d. body; (redrawn from Rozen 1966c).

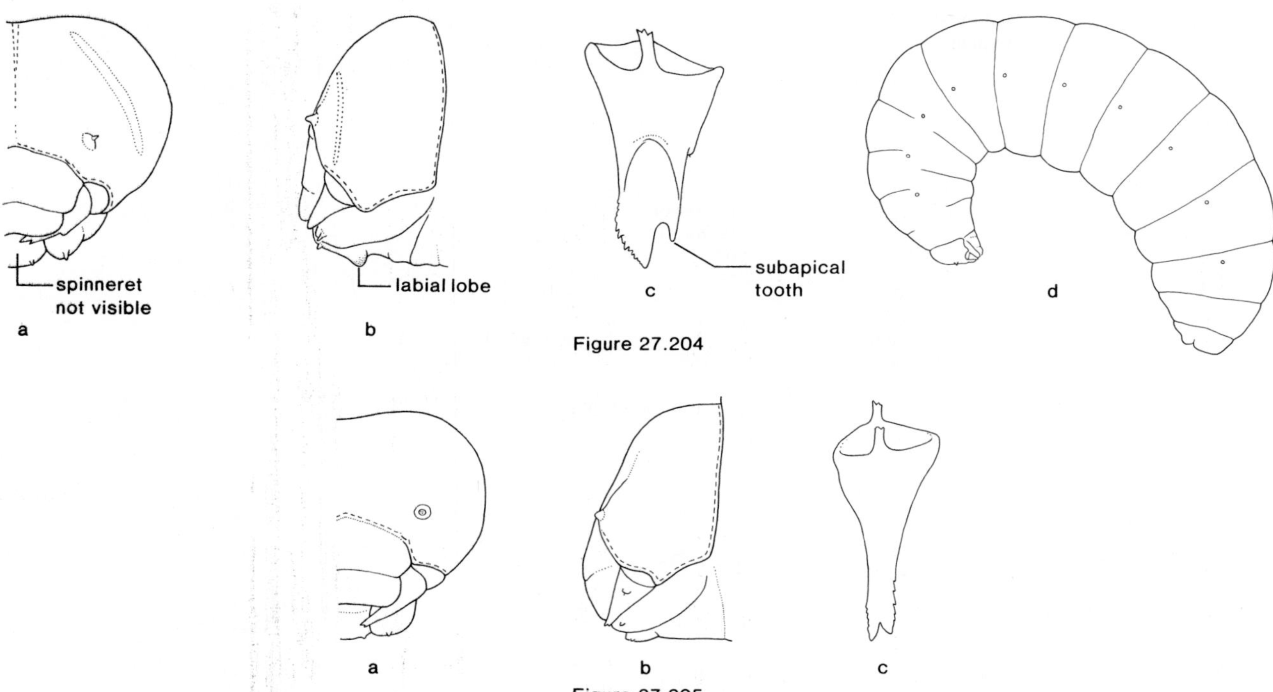

Figure 27.204a-d. Anthophoridae: Xylocopinae. *Xylocopa virginica* (Linnaeus). a. head capsule, frontal view; b. lateral view; c. mandible, adoral view; d. body.

Figure 27.205a-c. Anthophoridae: Xylocopinae. *Ceratina dupla* Say. a. head capsule, frontal view; b. lateral view; c. mandible, adoral view (redrawn from Michener 1953).

Neolarra); when present, short or moderately elongate (*Protepeolus*); galea absent. Labial palpus absent or well developed (*Protepeolus*). Spinneret absent or well defined and projecting (*Protepeolus*).

Body: Integument spiculate or nonspiculate; conspicuous setae usually absent (some present in *Protepeolus*). Dorsal tubercles usually absent (present and well defined in *Melanomada, Triopasites, Paranomada*); venter of abdominal segment 9 protuberant only in *Neopasites* and to a lesser degree in *Neolarra;* anus apical or slightly dorsal in position. The spiracles of *Triepeolus* and *Epeolus* are rather unusual in being surrounded by darkly pigmented, sclerotic rings (fig. 27.203d) and having atrial walls with elongate spines (fig. 27.203c).

Comments: Of the 15 nomadine genera known to occur in the United States, the larvae of four genera remain to be found: *Hexepeolus, Hesperonomada, Townsendiella* (presumably a cleptoparasite of *Hesperapis*), and *Epeoloides* (reported to be associated with *Macropis* in Europe).

Selected Bibliography

Rozen 1966c, 1977.
Rozen et al. 1978.
Rozen and McGinley 1974b.
Tengö and Bergström 1977.

Subfamily XYLOCOPINAE (ANTHOPHORIDAE)

Carpenter Bees

Figures 27.204, 27.205

Diagnosis: Xylocopine larvae are unique among bee larvae in having the salivary opening positioned on the dorsal surface of the labium, hidden from view (fig. 27.204a; in other apoids it is positioned apically and is clearly visible in frontal view). North American xylocopines are nearly unique in having a strongly projecting labiomaxillary region combined with an absence of spinnerets. This combination is found elsewhere only among *Hylaeus* (Colletidae) larvae. The bifid mandibles of xylocopines (figs. 27.204c, 27.205c) will differentiate these two groups (mandibles simple in *Hylaeus*). A highly diagnostic feature of *Xylocopa* larvae is the presence of a ventral labial lobe (fig. 27.204b).

Biology and Ecology: North American xylocopines nest in wood or more commonly in pithy stems. All species in this area are apparently solitary. The related Old World genera, referred to as allodapine bees, contain mostly subsocial and primitively eusocial species. Their larvae, among the most interesting in the Apoidea, have been extensively studied by Michener (1975, 1976) and Houston (1976).

Description: Body form slender (*Ceratina*) to robust (*Xylocopa*).

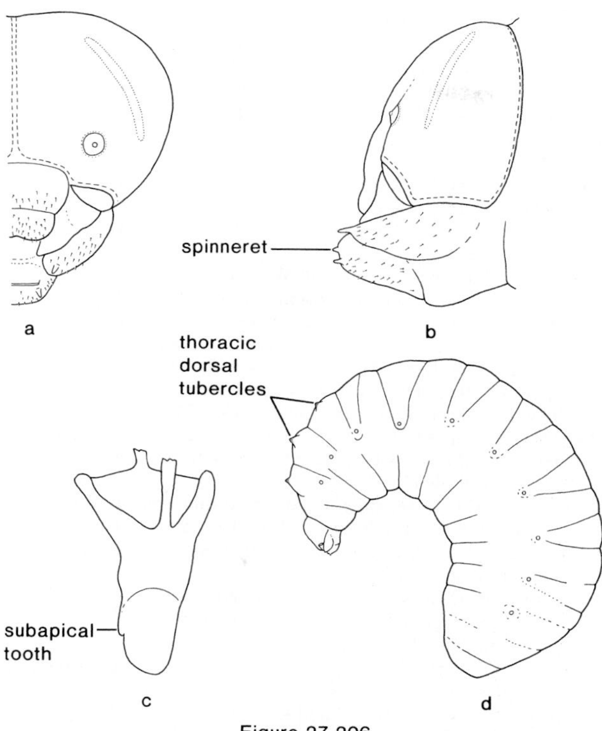

Figure 27.206

Figure 27.206a-d. Apidae. *Bombus pennsylvanicus* (Degeer). a. head capsule, frontal view; b. lateral view; c. mandible, adoral view; d. body.

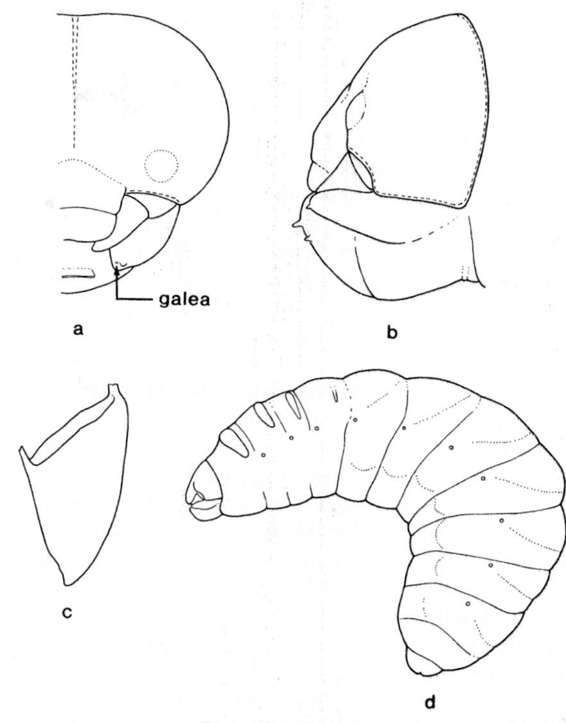

Figure 27.207

Figure 27.207a-d. Apidae. *Apis mellifera* Linnaeus, the honey bee. a. head capsule, frontal view; b. lateral view; c. mandible, ventral view; d. body (redrawn from Michener 1953).

Head: Vertex evenly rounded in lateral view. Distinct setae present. Epistomal ridge incomplete or absent mesiad of anterior tentorial pits, arching to level of antennal discs. Antennal prominence weakly developed; well developed antennal papillae either strongly projecting (*Xylocopa*) or only moderately projecting (*Ceratina*).

Mouthparts: Epipharynx, hypopharynx, and maxillae nonspiculate. Labral apex truncate to shallowly emarginate; labral tubercles absent. Mandible moderately broad in adoral view (*Xylocopa*, fig. 27.204c) or slender and attenuate (*Ceratina*, fig. 27.205c); mandibular apex acute; subapical tooth present (subapical tooth large in *Ceratina*); cusp weakly defined to virtually absent (*Ceratina*), nondentate; dorsal apical edge of mandible dentate; apical concavity well developed, somewhat scooplike (*Xylocopa*, fig. 27.204c) or virtually absent (*Ceratina*). Labiomaxillary region produced. Inner maxillary surface rounded apically, not produced mesiad; maxillary palpus moderately well developed (*Xylocopa*) to somewhat small (*Ceratina*); galea absent. Labial palpus small; labium with a unique, ventral swelling in *Xylocopa* (fig. 27.204b). Spinneret absent; unlike all other known bee larvae, salivary opening greatly recessed on upper labial surface, not visible in frontal view (fig. 27.204a).

Body: Integument spiculate or nonspiculate; conspicuous setae absent. Dorsal tubercles absent; venter of abdominal segment 9 nonprotuberant; anus apically positioned or slightly ventral (*Xylocopa*).

Selected Bibliography

Houston 1976.
Lucas de Oliveira 1974.
Michener 1975, 1976.

APIDAE (APOIDEA)

Ronald J. McGinley, *Smithsonian Institution*

Honey Bees, Bumble Bees

Figures 27.206, 27.207

Relationships and Diagnosis: Apids are usually thought to be most closely related to the Anthophoridae. However, the cladistic interrelationship between these two families remains poorly understood.

Bumble bees (*Bombus, Psithyrus*) can be distinguished from all other bee larvae by their possession of small, darkly pigmented, conical dorsal tubercles on the thoracic segments (fig. 27.206d). Honey bee larvae (*Apis*) can be recognized by the strongly produced labiomaxillary region, well-developed spinneret (both features typical of cocoon-spinning larvae) combined with weakly sclerotized mandibles and extremely broad, weakly projecting antennal papillae (fig. 27.207b).

Apis, along with Eucerini, *Centris,* and some melittids, is unusual among bee larvae in possessing galeae (fig. 27.207a; these are very small in *Apis* and may be overlooked without careful examination under high magnification).

Biology and Ecology: Apids display the greatest spectrum of social behavior found among bees, ranging from the solitary and communal euglossines to the primitively eusocial bumble bees and the highly eusocial honey bees. The family also includes a number of social parasites including the parasitic bumble bee genus *Psithyrus,* represented by six species in the United States.

Bumble bees and honey bees construct nest cells or combs out of wax or a wax-pollen mixture. Nests are built in concealed cavities or under vegetation (see Michener (1974) for an excellent review of this subject).

Description: Body form relatively robust.

Head: Vertex evenly rounded in lateral view. Setae present. Epistomal ridge complete (*Bombus*) or absent mesiad of anterior tentorial pits; epistomal ridge or depression arching dorsally to level of antennal discs. Antennal prominence absent; antennal papilla well developed and slender to broadly rounded and weakly projecting (*Apis*). Epipharyngeal spiculation absent; hypopharyngeal spiculation present; maxillary spiculation present. Labial apex shallowly emarginate; labral tubercles absent. Mandibles weakly sclerotic (*Apis,* fig. 27.207c) or moderately massive, highly sclerotic (*Bombus,* fig. 27.206c); apex narrowly rounded (*Apis*) or broadly rounded (most *Bombus*) (Ritcher (1933) indicates that the apices of *Psithyrus* mandibles are sharply pointed); subapical tooth present (*Bombus*) or absent (*Apis*); mandibular cusp weakly developed to absent; cusp nondentate; dorsal apical edge nondentate or with only few inconspicuous teeth (*Apis*); apical concavity strongly developed, somewhat scooplike (*Bombus*) or virtually absent (*Apis*). Labiomaxillary region strongly produced; inner maxillary margin rounded, not bent mesiad; maxillary palpus elongate (*Bombus*) or short (*Apis*); galea weakly developed (*Apis,* fig. 27.207a) or absent (*Bombus*). Labial palpus elongate (*Bombus*) or short (*Apis*). Spinneret well developed, strongly projecting.

Body: Integument spiculate; conspicuous setae present (some *Bombus*) or absent (*Apis*). Dorsal tubercles absent on most abdominal segments (*Bombus*) to present but weakly defined (*Apis,* fig. 27.207d); thoracic segments with small, darkly pigmented, conical tubercles (*Bombus,* fig. 27.206d) or small, rounded tubercles (*Apis,* fig. 27.207d); venter of abdominal segment 9 nonprotuberant; anus slightly dorsal in position.

Comments: Apids are a cosmopolitan group, represented in the United States primarily by bumble bees (*Bombus*) and the introduced honey bee (*Apis*). Stephen and Koontz (1973) provide an account of certain North American *Bombus* larvae. Torchio and Torchio (1975) give a detailed description of *Apis* species.

Selected Bibliography

Michener 1974.
Ritcher 1933.
Stephen and Koontz 1973.
Torchio and Torchio 1975.

BIBLIOGRAPHY

For a more complete bibliography of parasitoid literature, see Clausen (1940), Hagen (1964), and Finlayson and Hagen (1977, revised 1979).

Abrams, J., and G. C. Eickwort. 1980. Biology of the communal sweat bee *Agapostemon virescens* (Hymenoptera: Halictidae) in New York state. Search (Cornell Univ. Agr. Exp. Sta.) No. 1:1–20.

Akre, R. D., J. F. MacDonald, and W. B. Hill. 1974. Yellowjacket literature (Hymenoptera: Vespinae). Melanderia 18:67–93.

Akre, R. D., A. Greene, J. F. MacDonald, P. S. Landolt, and H. G. Davis. 1981. The yellowjackets of America north of Mexico. U.S. Dept. Agr. Handbook 552. 102 pp.

Altson, A. M. 1920. The life history and habits of two parasites of blow-flies. Proc. Zool. Soc. Lond. 2:195–243.

Atwood, C. E., and O. Peck. 1943. Some native sawflies of the genus *Neodiprion* attacking pines in eastern Canada. Can. J. Res. 21:109–44.

Baker, J. R. 1971. Development and sexual dimorphism of larvae of the bee genus *Coelioxys.* J. Kan. Ent. Soc. 44:225–35.

Baker, W. L. 1972. Eastern forest insects. USDA, Misc. Publ. 1175. 642 pp.

Baker, J. R., E. D. Kuhn, and S. B. Bambara. 1985. Nests and immature stages of leafcutter bees (Hymenoptera: Megachilidae). J. Kan. Ent. Soc. 58:290–313.

Batra, S. W. T. 1980. Ecology, behavior, pheromones, parasites and management of the sympatric vernal bees *Colletes inaequalis, C. thoracicus* and *C. validus.* J. Kan. Ent. Soc. 53(3):509–38.

Beer, R. E. 1955. Biological studies and taxonomic notes on the genus *Strongylogaster* Dahlbom (Hymenoptera: Tenthredinidae). Univ. Kans. Sci. Bull. 37:223–49.

Beirne, B. P. 1941. A consideration of the cephalic structures and spiracles of the final instar larvae of the Ichneumonidae (Hym.). Trans. Soc. Br. Ent. 7:123–90.

Bird, R. D. 1927. The external anatomy of the larva of *Hoplocampa halcyon* Nort. with a key to the instars and to those of related species (Hymenoptera: Tenthredinidae). Ann. Ent. Soc. Am. 20:481–86.

Bohart, R. M., and A. S. Menke. 1976. Sphecid wasps of the world. A generic revision. Berkeley and Los Angeles: Univ. Calif. Press. 695 pp.

Bradley, J. C. 1913. The Siricidae of North America. J. Ent. Zool. 5:1–35.

Britton, W. E. 1925. Sawfly larvae defoliating honeysuckle. Conn. St. Agric. Exp. Stn. Bull. 265:341.

Brothers, D. J. 1972. Biology and immature stages of *Pseudomethoca f. frigida,* with notes on other species (Hymenoptera: Mutillidae). Univ. Kan. Sci. Bull. 50:1–38.

Brothers, D. J. 1975. Phylogeny and classification of the aculeate Hymenoptera, with special reference to Mutillidae. Univ. Kan. Sci. Bull. 50:483–648.

Brothers, D. J. 1978. Biology and immature stages of *Myrmosula parvula* (Hymenoptera: Mutillidae). J. Kan. Ent. Soc. 51:698–710.

Buchner, P. 1965. Siricids, pp. 83–92. *In* Endosymbiosis of animals with plant micro-organisms. New York: J. Wiley. 909 pp. (Revised English version.)

Burdick, D. J. 1961. A taxonomic and biological study of the genus *Xyela* Dalman in North America (Hymenoptera: Xyelidae). Univ. Calif. Publs. Ent. 17:285–356.

Burdick, D. J., and P. F. Torchio. 1959. Notes on the biology of *Hesperapis regularis* (Cresson) (Hymenoptera: Melittidae). J. Kan. Ent. Soc. 32:83–87.

Burke, H. E. 1918 (1917). *Oryssus* is parasitic. Proc. Ent. Soc. Wash. 19:87–89.

Cameron, E. 1957. On the parasites and predators of the cockroach. II. *Evania appendigaster* (L.). Bull. Ent. Res. 48:199–209.

Čapek, M. 1970. A new classification of the Braconidae (Hymenoptera) based on the cephalic structures of the final instar larva and biological evidence. Can. Ent. 102:846–75.

Čapek, M. 1973. Key to the final instar larvae of the Braconidae (Hymenoptera). Acta Inst. For. Zvolen., pp. 258–68.

Carlson, R. W. 1979. Family Ichneumonidae, pp. 315–740. *In* K. V. Krombein et al. Catalog of Hymenoptera in America north of Mexico. Vol. 1. Smithsonian Institution Press.

Carpenter, J. M. 1982. The phylogenetic relationships and natural classification of the Vespoidea (Hymenoptera). Syst. Ent. 7:11–38.

Chrystal, R. N. 1930. Studies of the *Sirex* parasites. The biology and post-embryonic development of *Ibalia leucospoides* Hochenw. (Hymenoptera-Cynipoidea). Oxford For. Mem., No. 11. 63 pp.

Clancy, D. W. 1946. The insect parasites of the Chrysopidae (Neuroptera). Univ. Calif. Publ. Ent. 7:403–96.

Clausen, C. P. 1929. Biological studies on *Poecilogonalos thwaitsii* (Westw.), parasitic in the cocoons of *Henicospilus* (Hymen.: Trigonalidae). Proc. Ent. Soc. Wash. 31:67–79.

Clausen, C. P. 1940. Entomophagous insects. New York and London: McGraw-Hill. 688 pp.

Clement, S. L. 1972. Notes on the biology and larval morphology of *Stenodynerus canus canus* (Hymenoptera: Eumenidae). Pan. Pac. Ent. 48:271–76.

Clement, S. L. 1976. The biology of *Dianthidium heterulkei* Schwarz, with a description of the larva (Hymenoptera, Megachilidae). Wasmann J. Biol. 34:9–22.

Clement, S. L., and E. E. Grissell. 1968. Observations on the nesting habits of *Euparagia scutellaris* Cresson (Hymenoptera: Masaridae). Pan-Pac. Ent. 44:34–37.

Cole, L. R. 1959. On a new species of *Syntretus* Förster (Hym., Braconidae) parasitic on an adult ichneumonid, with a description of the larva and notes on its life history and that of its host, *Phaeogenes invisor* (Thunberg). Ent. Mon. Mag. 95:18–21.

Cooper, K. W. 1952. Records and flower preferences of masarid wasps. II. Polytrophy or oligotrophy in *Pseudomasaris?* (Hymenoptera: Vespidae). Amer. Mid. Nat. 48:103–10.

Cooper, K. W. 1953. Egg gigantism, oviposition, and genital anatomy: Their bearing on the biology and phylogenetic position of *Orussus* (Hymenoptera: Orussidae). Proc. Rochester Acad. Sci. 10:38–68.

Cooper, K. W. 1953, 1955. Biology of eumenine wasps. Parts I and II. Trans. Amer. Ent. Soc. 79:13–35, 80:117–74.

Cooper, K. W. 1954. Biology of eumenine wasps. IV. A trigonalid wasp parasitic on *Rygchium rugosum* (Saussure) (Hymenoptera, Trigonalidae). Proc. Ent. Soc. Wash. 56:280–88.

Cooper, K. W. 1957. Biology of eumenine wasps. Part V. Digital communication in wasps. J. Exp. Zool. 134:469–514.

Coppel, H. C., and D. M. Benjamin. 1965. Bionomics of the Nearctic pine-feeding diprionids. Ann. Rev. Ent. 10:69–96.

Daly, H. V. 1961. Biological observations on *Hemihalictus lustrans,* with a description of the larva (Hymenoptera: Halictidae). J. Kan. Ent. Soc. 34:134–41.

DeBach, P. 1964. Biological control of insect pests and weeds. New York: Reinhold. 844 pp.

Doutt, R. L. 1947. Polyembryony in *Copidosoma koehleri* Blanchard. Amer. Natur. 81:435–53.

Doutt, R. L. 1959. The biology of parasitic Hymenoptera. Ann. Rev. Ent. 4:161–82.

Dowden, P. B. 1938. *Rogas unicolor* (Wesm.), a braconid parasite of the satin moth. J. Agric. Res. 56:523–36.

Duncan, C. D. 1939. A contribution to the biology of North American vespine wasps. Stanford Univ. Publ. Biol. Sci. 8(1):1–272. (Larva of *Vespula pensylvanica* Sauss.)

Eastham, L. E. S. 1929. The postembryonic development of *Phaenoserphus viator* Hal. (Proctotrypoidea), a parasite of the larva of *Pterostichus niger* (Carabidae), with notes on the anatomy of the larva. Parasitology 21:5–21.

Eickwort, G. C. 1973. Biology of the European mason bee, *Hoplitis anthocopoides* (Hymenoptera: Megachilidae), in New York state. Search (Cornell Univ. Agr. Exp. Sta.) 3:1–31.

Eickwort, G. C. 1977. Aspects of the nesting biology and descriptions of immature stages of *Perdita octomaculata* and *P. halictoides* (Hymenoptera: Andrenidae). J. Kan. Ent. Soc. 50:577–99.

Eickwort, G. C., and K. R. Eickwort. 1973. Notes on the nests of three wood-dwelling species of *Augochlora* from Costa Rica. J. Kan. Ent. Soc. 46:17–22.

Eickwort, G. C., K. R. Eickwort, and E. G. Linsley. 1977. Observations on nest aggregations of the bees *Diadasia olivacea* and *D. diminuta*. J. Kan. Ent. Soc. 50:1–17.

Eickwort, G. C., R. W. Matthews, and J. Carpenter. 1981. Observations on the nesting biology of *Megachile rubi* and *M. texana* with a discussion of the significance of soil nesting in the evolution of megachilid bees (Hymenoptera: Megachilidae). J. Kan. Ent. Soc. 54:557–70.

Eickwort, G. C., P. F. Kukuk, and F. R. Wesley. 1986. The nesting biology of *Dufourea novaeangliae* (Hymenoptera:Halictidae) and the systematic position of the Dufoureinae based on behavior and development. J. Kan. Ent. Soc. 59:103–120.

Eidt, D. C. 1969. The life histories, distribution, and immature forms of the North American sawflies of the genus *Cephalcia* (Hymenoptera: Pamphiliidae). Mem. Ent. Soc. Can. 59. 56 pp.

Evans, A. C. 1933. Comparative observations on the morphology and biology of some hymenopterous parasites of carrion-infesting Diptera. Bull. Ent. Res. 24:385–405.

Evans, H. E. 1957–1959a. Studies on the larvae of digger wasps (Hymenoptera: Sphecidae). Parts III, IV, V. Trans. Amer. Ent. Soc. 83:79–117; 84:109–39; 85:137–91.

Evans, H. E. 1959b. The larvae of Pompilidae (Hymenoptera). Ann. Ent. Soc. Amer. 53:430–44.

Evans, H. E. 1964a. A synopsis of the American Bethylidae (Hymenoptera, Aculeata). Bull. Mus. Comp. Zool. Harvard 132:1–222.

Evans, H. E. 1964b. Further studies on the larvae of digger wasps Hymenoptera: Sphecidae). Trans. Amer. Ent. Soc. 90:235–99.

Evans, H. E. 1964c. The classification and evolution of digger wasps as suggested by larval characters (Hymenoptera: Sphecoidea). Ent. News 75:225–37.

Evans, H. E. 1965. A description of the larva of *Methocha stygia* (Say), with notes on other Tiphiidae. Proc. Ent. Soc. Wash. 67:88–95.

Evans, H. E. 1966. The comparative ethology and evolution of the sand wasps. Cambridge, Mass.: Harvard Univ. Press. 526 pp.

Evans, H. E. 1977. Notes on the nesting behavior and immature stages of two species of *Pterocheilus* (Hymenoptera: Eumenidae). J. Kan. Ent. Soc. 50:329–34.

Evans, H. E., and C. S. Lin. 1956. Studies on the larvae of digger wasps (Hymenoptera: Sphecidae). Parts I and II. Trans. Amer. Ent. Soc. 81:131–53; 82: 35–66.

Evans, H. E., and C. M. Yoshimoto. 1962. The ecology and nesting behavior of the Pompilidae (Hymenoptera) of the northeastern United States. Misc. Publ. Ent. Soc. Amer. 3:67–119.

Fenton, F. A. 1918. The parasites of leaf-hoppers. With special reference to the biology of the Anteoninae, Part I. Ohio J. Sci. 18:177–212.

Finlayson, T. 1960a. Taxonomy of cocoons and puparia, and their contents, of Canadian parasites of *Neodiprion sertifer* (Geoff.) (Hymenoptera: Diprionidae). Can. Ent. 92:20–47.

Finlayson, T. 1960b. Taxonomy of cocoons and puparia, and their contents, of Canadian parasites of *Diprion hercyniae* (Htg.) (Hymenoptera: Diprionidae). Can. Ent. 92:922–41.

Finlayson, T. 1964. The caudal appendage of final-instar larvae of some Porizontinae (Hymenoptera: Ichneumonidae). Can. Ent. 96:1155–58.

Finlayson, T. 1967. Taxonomy of final-instar larvae of the hymenopterous and dipterous parasites of *Acrobasis* spp. (Lepidoptera: Phycitidae) in the Ottawa region. Can. Ent. 99:1233–71.

Finlayson, T. 1970. Final-instar larval characteristics of *Coleocentrus rufus* (Hymenoptera:Ichneumonidae). Can. Ent. 102:905–07.

Finlayson, T. 1975. The cephalic structures and spiracles of final-instar larvae of the subfamily Campopleginae, tribe Campoplegini (Hymenoptera: Ichneumonidae). Mem. Ent. Soc. Can. No. 94. 137 pp.

Finlayson, T. 1976. Cephalic structures and spiracles of final-instar larvae of the genus *Toxophoroides* (Hymenoptera: Ichneumonidae:Lycorininae). Can. Ent. 108:981–84.

Finlayson, T., and K. S. Hagen. 1977. (Revised 1979). Final-instar larvae of parasitic Hymenoptera. Pest Management Paper No. 10, Simon Fraser University, Burnaby, B.C. 111 pp.

Fitton, M. G., and I. D. Gauld. 1976. The family-group names of the Ichneumonidae (excluding Ichneumoninae) (Hymenoptera). Syst. Ent. 1:247–58.

Fitton, M. G., and I. D. Gauld. 1978. Further notes on family-group names of Ichneumonidae (Hymenoptera). Syst. Ent. 3:245–47.

Flanders, S. E. 1938. Cocoon formation in endoparasitic chalcidoids. Ann. Ent. Soc. Amer. 31:167–80.

Francke-Grosmann, H. 1939. Ueber das Zusammenleben von Holzwespen (Siricinae) mit Pilzen. Z. Angew. Ent. 25:647–80.

Furniss, R. L., and V. M. Carolin. 1977. Western forest insects. USDA, Misc. Publ. 1339. 654 pp.

Gahan, A. B. 1920. Black grain stem sawfly of Europe in the United States. USDA, Bull. 834. 18 pp.

Genieys, P. 1924. Contributions à l'étude des Evaniidae: *Zeuxevania splendidula* Costa. Bull. Biol. France et Belg. 58:482–94.

Gibson, G. A. P. 1980. A revision of the genus *Macrophya* Dahlbom (Hymenoptera: Symphyta, Tenthredinidae) of North America. Mem. Ent. Soc. Can. 114. 167 pp.

Gillespie, D. R., and T. Finlayson. 1983. Classification of final-instar larvae of the Ichneumoninae (Hymenoptera: Ichneumonidae). Mem. Ent. Soc. Can. No. 124. 81 pp.

Gordh, G. 1976. *Goniozus gallicola* Fouts, a parasite of moth larvae, with notes on other bethylids (Hymenoptera: Bethylidae, Lepidoptera: Gelechiidae). Tech. Bull. U.S. Dept. Agri., No. 1524, 27 pp.

Gordh, G. 1979. Superfamily Chalcidoidea, pp. 743–748. *In* K. V. Krombein et al. Catalog of Hymenoptera in America north of Mexico. Vol. 1. Smithsonian Institution Press.

Gourlay, E. S. 1951. Notes on insects associated with *Pinus radiata* in New Zealand. Bull. Ent. Res. 42:21–22.

Grandi, G. 1934. La constituzione morfologica delle larve di alcuni Vespidi ed Apidi sociali; suoi rapporti con le modalità di assunzione del cibo e con altri comportamenti etologica. Mem. Accad. Sci. Inst. Bologna 1(9):73–79, 2 pls.

Grandi, G. 1940. *Scolia (Scolioides) hirta* Schrk. XVIII. Contributa allo conoscenza degli Imenotteri Aculeati. Mem. Accad. Sci. Inst. Bologna 7(9):165–67, 3pls.

Grandi, G. 1954. Contributi alla conoscenza degli Imenotteri Aculeati. XXVI. Boll. Inst. Ent. Univ. Bologna 20:81–255.

Grandi, G. 1959. Contributi alla conoscenza degli Imenotteri Aculeati. XXVIII. Bol. Inst. Ent. Univ. Bologna 23:239–92.

Grandi, G. 1961. Studi di un Entomologo sugli Imenotteri Superiori. Boll. Inst. Ent. Univ. Bologna 25:1–659. (Biology and immature stages of several European species.)

Grissell, E. E. 1975. Ethology and larva of *Pterocheilus texanus* (Hymenoptera: Eumenidae). J. Kan. Ent. Soc. 48:244–53.

Gurney, A. B. 1953. Notes on the biology and immature stages of a cricket parasite of the genus *Rhopalosoma*. Proc. U.S. Nat. Mus. 103:19–34.

Guppy, J. C., and C. D. F. Miller. 1970. Identifications of cocoons and last-instar larval remains of some hymenopterous parasitoids of the armyworm, *Pseudaletia unipuncta*, in eastern Ontario. Can. Ent. 102:1320–37.

Hagen, K. S. 1964. Developmental stages of parasites, pp. 168–246. *In* P. DeBach (ed.). Biological control of insect pests and weeds. Chapman and Hall, London; or Reinhold, New York.

Haviland, M. D. 1921. On the bionomics and development of *Lygocerus testaceimanus* Kieffer and *Lygocerus cameroni* Kieffer (Proctotrypoidea—Ceraphronidae), parasites of *Aphidius* (Braconidae). Quart. J. Microsc. Sci. 65:101–27.

Hefetz, A., H. M. Fales, and S. W. T. Batra. 1979. Natural polyesters: Dufour's gland macrocyclic lactones form brood cell laminesters in *Colletes* bees. Science 204:415–17.

Hendrick, R. H., and V. M. Stern. 1970. Biological studies of three parasites of *Nabis americoferus* (Hemiptera: Nabidae) in southern California. Ann. Ent. Soc. Amer. 63:382–91.

Hill, C. C., and W. T. Emery. 1937. The biology of *Platygaster herrickii*, a parasite of the Hessian fly. J. Agri. Res. 55:199–213.

Hogh, G. 1966. Local distribution of a fern sawfly, *Strongylogaster multicinctus* (Hymenoptera: Tenthredinidae). J. Kans. Ent. Soc. 39:347–54.

Houston, T. F. 1976. New Australian allodapine bees (subgenus *Exoneurella* Michener) and their immatures (Hymenoptera: Anthophoridae). Trans. Roy. Soc. Australia 100:15–28.

Hurd, P. D., Jr., and E. G. Linsley. 1976. The bee family Oxaeidae with a revision of the North American species (Hymenoptera: Apoidea). Smithsonian. Contrib. Zool. No. 220. 75 pp.

Iida, T. 1967. A study on the larvae of the genus *Sphex* in Japan (Hymenoptera: Sphecidae). Etizenia, Biol. Lab. Fukui Univ., No. 19. 8 pp.

Iwata, K. 1942. Comparative studies on the habits of solitary wasps. Tenthredo 4:1–146.

Iwata, K. 1976. Evolution of instinct: comparative ethology of Hymenoptera. Amerind Publ. Co., New Delhi (translated by A. Gopal and publ. for the Smithsonian Inst.). 535 pp. (General review of biology.)

James, H. C. 1928. On the life-histories and economic status of certain cynipid parasites of dipterous larvae, with descriptions of some new larval forms. Ann. Appl. Biol. 15:287–316.

Janvier, H. 1930. Recherches biologiques sur les prédateurs du Chili. Ann. Sçi. Nat., Zool. 13(10):235–354.

Janvier, H. 1933. Étude biologique de quelques Hyménoptères du Chili. Ann. Sci. Nat. Zool. 10(16):209–356.

Kaston, B. J. 1959. Notes on pompilid wasps that do not dig burrows to bury their spider prey. Bull. Brooklyn Ent. Soc. 54:103–13.

Keilin, D., and G. Baume Pluvinel. 1913. Formes larvaires et biologie d'un cynipide entomophage, *Eucoila keilini* Kieffer. Bull. Sci. France et Belgique (7) 47:88–104, 2 pls.

Kojima, J., and S. Yamane. 1984. Systematic study of the mature larvae of Oriental polistine wasps (Hymenoptera: Vespidae). (I) species of *Ropalidia* and *Polistes* from Sumatra and Java islands. Reports Fac. Sci. Kagoshima Univ. 17:103–127.

Krombein, K. V. 1967. Trap-nesting wasps and bees: life histories, nests, and associates. Smithsonian Press, Washington, D.C. Publ. 4670. 570 pp. (Biology of twig-nesting species.)

Krombein, K. V., and H. E. Evans. 1976. Three new Neotropical *Pterombrus* with description of the diapausing larva (Hymenoptera: Tiphiidae). Proc. Ent. Soc. Wash. 78:361–68.

Krombein, K. V., P. D. Hurd, Jr., D. R. Smith, and B. D. Burks editors. 1979. Catalog of Hymenoptera in America north of Mexico. Smithsonian Institution Press. Vol. 1. Symphyta and Apocrita (Parasitica), pp. 1–1198; Vol. 2. Apocrita (Aculeata), pp. 1199–2210; Vol. 3. Indexes, pp. 2211–2735.

Lindquist, O. H., and W. J. Miller. 1970. A key to free-feeding sawfly larvae on birch and alder in Ontario. Proc. Ent. Soc. Ont. 100:117–23.

Linsley, E. G., J. W. MacSwain, and C. D. Michener. 1980. Nesting biology and associates of *Melitoma* (Hymenoptera, Anthophoridae). Univ. Calif. Publ. Ent. 90:1–45.

Linsley, E. G., J. W. MacSwain, and R. F. Smith. 1956. Biological observations on *Ptilothrix sumichrasti* (Cresson) and some related groups of emphorine bees. Bull. So. Calif. Acad. Sci. 55:83–101.

Lorenz, H., and M. Kraus. 1957. Die Larvalsystematik der Blattwespen (Tenthredinoidea und Megalodontoidea). Akad.-Verlag, Berlin. 339 pp.

Lucas de Oliveira, B. 1974. Estadios imaturos de algumas *Xylocopa neotropicais* (Hymenoptera: Apoidea). Acta Biol. Parana. 3:93–112.

MacGillivray, A. D. 1914(1913). The immature stages of the Tenthredinoidea. Ann. Rept. Ent. Soc. Ont. 44:54–75.

Mackauer, M. 1961. Die Gattungen der Familie Aphidiidae und ihre verwandtschaftliche Zuordnung (Hymenoptera: Ichneumonoidea). Beitr. Ent. 11:792–803.

Mackauer, M., and T. Finlayson. 1967. The hymenopterous parasites (Hymenoptera: Aphidiidae et Aphelinidae) of the pea aphid in eastern North America. Can. Ent. 99:1051–82.

Marchal, P. 1904. Recherches sur la biologie et le développement des hyménoptères parasites. I. La polyembryonie spécifique ou germinogonie. Arch. Zool. Exp. et Gen. 2:257–335.

Marsh, P. M. 1979a. Family Braconidae, pp. 144–295. *In* Catalog of Hymenoptera in America north of Mexico. Vol. 1. Smithsonian Institution Press.

Marsh, P. M. 1979b. Family Aphidiidae, pp. 295–313. *In* Catalog of Hymenoptera in America north of Mexico. Vol. 1. Smithsonian Institution Press.

Martorell, L. F. 1941. Biological notes on the sea-grape sawfly, *Schizocera krugii* Cresson, in Puerto Rico. Caribbean Forester. 2:141–44.

Matthews, R. W. 1965. The biology of *Heriades carinata* Cresson (Hymenoptera, Megachilidae). Amer. Ent. Inst. Contrib. 1:1–33.

Maxwell, D. E. 1955. The comparative internal larval anatomy of sawflies (Hymenoptera: Tenthredinidae). Can. Ent. suppl. 1, Vol. 87, 132 pp.

McGinley, R. J. 1980. Glossal morphology of the Colletidae and recognition of the Stenotritidae at the family level (Hymenoptera: Apoidea). J. Kan. Ent. Soc. 53:539–52.

McGinley, R. J. 1981. Systematics of the Colletidae based on mature larvae with phenetic analysis of apoid larvae (Hymenoptera: Apoidea). Univ. Calif. Pub. Ent. 91:1–307.

McGinley, R. J. In press. A catalog and review of immature Apoidea (Hymenoptera). Smithsonian Contr. Zool.

Mertens, J. W. 1980. Life history and behavior of *Laelius pedatus*, a gregarious bethylid ectoparasitoid of *Anthrenus verbasci*. Ann. Ent. Soc. Amer. 73:686–93.

Michener, C. D. 1953. Comparative morphological and systematic studies of bee larvae with a key to the families of hymenopterous larvae. Univ. Kan. Sci. Bull. 35:987–1102.

Michener, C. D. 1974. The social behavior of the bees. Cambridge, Mass.: Harvard Univ. Press. 404 pp.

Michener, C. D. 1975. Larvae of African allodapine bees. 1. The genus *Allodape*, 2. *Braunsapis* and *Nasutapis*, 3. The genera *Allodapula* and *Eucondylops*. J. Ent. Soc. Southern Africa 38:1–12, 223–42, 243–50.

Michener, C. D. 1976. Larvae of African allodapine bees. 4. *Halterapis*, *Compsomelissa*, *Macrogalea*, and a key to African genera. J. Ent. Soc. Southern Africa 39:33–37.

Michener, C. D., and L. Greenberg. 1980. Ctenoplectridae and the origin of long-tongued bees. Zool. J. Linn. Soc. 69:183–203.

Michener, C. D. and L. Greenberg. 1985. The fate of the lacinia in the Halictidae and Oxaeidae (Hymenoptera: Apoidea). J. Kan. Ent. Soc. 58:137–41.

Mickel, C. E. 1928. Biological and taxonomic investigations on the mutillid wasps. Bull. U.S. Nat. Mus. 143:1–351.

Middlekauff, W. W. 1940. The Nearctic sawflies of the genus *Neurotoma* (Hymenoptera: Pamphiliidae). Can. Ent. 70:201–06.

Middlekauff, W. W. 1956. A cimbicid sawfly feeding on an ornamental honeysuckle. J. Econ. Ent. 49:701.

Middlekauff, W. W. 1958. The North American sawflies of the genera *Acantholyda*, *Cephalcia*, and *Neurotoma* (Hymenoptera: Pamphiliidae). Univ. Calif. Publs. Ent. 14:51–168.

Middlekauff, W. W. 1960. The siricid wood wasps of California (Hymenoptera: Siricidae). Bull. Calif. Insect Surv. 6:59–77.

Middlekauff, W. W. 1964a. The North American sawflies of the genus *Pamphilius* (Hymenoptera: Pamphiliidae). Univ. Calif. Publs. Ent. 38:1–80.

Middlekauff, W. W. 1964b. Notes and description of the previously unknown male of *Syntexis libocedrii* (Hymenoptera: Syntexidae). Pan-Pac. Ent. 40:255–58.

Middlekauff, W. W. 1969. The cephid stem borers of California (Hymenoptera: Cephidae). Bull. Calif. Insect Surv. 11:1–17.

Middlekauff, W. W. 1974. Larva of the wood-boring *Syntexis libocedrii* Rohwer (Hymenoptera: Syntexidae). Pan-Pac. Ent. 50:288–90.

Middlekauff, W. W. 1983. A revision of the sawfly family Orussidae for North and Central America (Hymenoptera: Symphyta, Orussidae). Univ. Calif. Publ. Ent. 101:1–46.

Middleton, W. M. 1918(1917). Notes on the larvae of some Cephidae. Proc. Ent. Soc. Wash. 19:174–79.

Middleton, W. M. 1922. Descriptions of some North American sawfly larvae. Proc. U.S. Nat. Mus. 61:1–31.

Mik, J. 1882. Zür Biologie von *Gonatopus pilosus* Thoms. Wien. Ent. Zeitschr. 1:215–21.

Morgan, F. D. 1968. Bionomics of Siricidae. Ann. Rev. Ent. 13:239–56.

Muesebeck, C. F. W. 1967. Family Braconidae. pp. 27–60. *In* Hymenoptera of America north of Mexico. Synoptic catalog by K. V. Krombein, B. D. Burks and others. USDA Agric. Monogr. 2, Suppl. 2.

Nelson, J. M. 1982. External morphology of *Polistes* (paper wasp) larvae in the United States. Melanderia 38:1–29.

Noble, N. S. 1938. *Euplectrus agaristae* Craw., a parasite of the grape vine moth (*Phalaenoides glycine* Lew.). Bull. N.S. Wales Dept. Agric., No. 63. 27 pp.

Ohmart, C. P., and D. L. Dahlsten. 1977. Biological studies of bud mining sawflies, *Pleroneura* spp. (Hymenoptera: Xyelidae), on white fir in the central Sierra Nevada of California. I. Life cycles, niche utilization, and interaction between larval feeding and tree growth. Can. Ent. 109:1001–07.

Ohmart, C. P., and D. L. Dahlsten. 1978. Biological studies of bud mining sawflies, *Pleroneura* spp. (Hymenoptera: Xyelidae), on white fir in the central Sierra Nevada of California. II. Larval distribution within tree crowns. Can. Ent. 110:583–90.

Parker, D. E. 1936. *Chrysis shanghaiensis* Smith, a parasite of the Oriental moth. J. Agri. Res. 52:449–58.

Parker, H. L. 1924. Récherches sur les formes post-embryonnaires des Chalcidiens. Ann. Soc. Ent. Fr. 93:261–379.

Parker, H. L. 1943. Gross anatomy of the larva of the wasp *Polistes gallicus* (L.) Ann. Ent. Soc. Amer. 36:619–24.

Parker, H. L., and W. R. Thompson. 1925. Notes on the larvae of the Chalcidoidea. Ann. Ent. Soc. Amer. 18: 384–98.

Parkin, L. A. 1942. Symbiosis and siricid woodwasps. Ann. Appl. Biol. 29:268–74.

Peck, O. 1963. A catalogue of the Nearctic Chalcidoidea (Insecta: Hymenoptera). Can. Ent. Suppl., No. 30. 1092 pp.

Perkins, R. C. L. 1905. Leaf-hoppers and their natural enemies. Bull. Hawaiian Sugar Planters' Assoc. 1:1–69.

Peterson, A. 1948. Larvae of insects. Part 1. Lepidoptera and plant infesting Hymenoptera. Printed for the author by Edwards Bros., Ann Arbor, Mich. 315 pp.

Ponomarenko, N. G. 1975. Characteristics of larval development in the Dryinidae (Hymenoptera). Ent. Obr. 54:534–40. (English translation, 1976. Ent. Rev. pp. 36–39.)

Powell, J. A., and W. J. Turner. 1975. Observations on oviposition behavior and host selection in *Orussus occidentalis* (Hymenoptera: Siricoidea). J. Kans. Ent. Soc. 48:299–307.

Rajotte, E. G. 1979. Nesting, foraging and pheromone response of the bee *Colletes validus* Cresson and its association with lowbush blueberries (Hymenoptera: Colletidae) (Ericaceae: *Vaccinium*). J. Kan. Ent. Soc. 52:349–61.

Rawlings, G. B. 1957. *Guiglia schauinslandi* (Ashmead) (Hymenoptera: Orussidae) a parasite of *Sirex noctilio* (Fabricius) in New Zealand. Entomologist 90:35–36.

Regas-Williams, K. A., and D. H. Habeck. 1979. Life history of a poison-ivy sawfly, *Arge humeralis* (Beauvois) (Hymenoptera: Argidae). Fla. Ent. 62:356–63.

Reid, J. A. 1942. On the classification of the larvae of the Vespidae (Hymenoptera). Trans. R. Ent. Soc. London 92:285–331. (Eumenidae considered a subfamily of Vespidae.)

Richards, O. W. 1962. A revisional study of the masarid wasps (Hymenoptera, Vespoidea). British Museum (Natural History), London. 294 pp.

Richards, O. W. 1978. The social wasps of the Americas, excluding the Vespinae. British Museum (Natural History), London. 580 pp.

Ridsdill Smith, T. J. 1970. The biology of *Hemithynnus hyalinatus* (Hymenoptera: Tiphiidae), a parasite on scarabaeid larvae. J. Australian Ent. Soc. 9:183–95.

Riek, E. F. 1962. A trigonalid wasp (Hymenoptera, Trigonalidae) from an anthelid cocoon (Lepidoptera, Anthelidae). Proc. Linn. Soc. N.S. Wales 87:148–50.

Riek, E. F. 1970. Hymenoptera. pp. 867–959. *In* Insects of Australia. CSIRO, Canberra. Melbourne Univ. Press. 1029 pp.

Ries, D. T. 1937. A revision of the Nearctic Cephidae. Trans. Am. Ent. Soc. 63:259–324.

Ritcher, P. O. 1933. The external morphology of larval Bremidae and key to certain species. Ann. Ent. Soc. Amer. 26:53–63.

Rivers, R. L., Z. B. Mayo, and T. J. Helms. 1979. Biology, behavior and description of *Tiphia berbereti* (Hymenoptera: Tiphiidae), a parasite of *Phyllophaga anxia* (Coleoptera: Scarabaeidae). J. Kan. Ent. Soc. 52:362–72.

Roberts, R. B. 1973. Nest architecture and immature stages of the bee *Oxaea flavescens* and the status of Oxaeidae (Hymenoptera). J. Kan. Ent. Soc. 46:437–46.

Rohwer, S. A., and R. A. Cushman. 1918(1917). Idiogastra, a new suborder of Hymenoptera, with notes on the immature stages of *Oryssus*. Proc. Ent. Soc. Wash. 19:89–98.

Rose, A. H., and O. H. Lindquist, 1973. Insects of eastern pines. Dept. Environ., Can. Forestry Serv., publ. 1313. 127 pp.

Ross, H. H. 1931. Sawflies of the sub-family Dolerinae of America north of Mexico. Ill. Biol. Monogr. 12. 116 pp.

Ross, H. H. 1938. The Nearctic species of *Pikonema*, a genus of spruce sawflies (Hymenoptera, Tenthredinidae). Proc. Ent. Soc. Wash. 40:17–20.

Ross, H. H. 1955. The taxonomy and evolution of the sawfly genus *Neodiprion*. Forest. Sci. 1:196–209.

Rozen, J. G., Jr. 1957. External morphological description of the larva of *Exomalopsis chionura* (Ckll.), including a comparison with other anthophorids (Hymenoptera: Apoidea). Ann. Ent. Soc. Amer. 50:469–75.

Rozen, J. G., Jr. 1964. Phylogenetic-taxonomic significance of last instar of *Protoxaea gloriosa* Fox, with descriptions of first and last instars (Hymenoptera: Apoidea). J. N.Y. Ent. Soc. 72:223–30.

Rozen, J. G., Jr. 1965a. The biology and immature stages of *Melitturga clavicornis* (Latreille) and of *Sphecodes albilabris* (Kirby) and the recognition of the Oxaeidae at the family level (Hymenoptera, Apoidea). Amer. Mus. Novitates, No. 2224, pp. 1–18.

Rozen, J. G., Jr. 1965b. The larvae of the Anthophoridae (Hymenoptera, Apoidea). Part 1. Introduction, Eucerini, and Centridini (Anthophorinae). Amer. Mus. Novitates, No. 2233, pp. 1–27.

Rozen, J. G., Jr. 1966a. Systematics of the larvae of North American panurgine bees (Hymenoptera, Apoidea). Amer. Mus. Novitates, No. 2259, pp. 1–22.

Rozen, J. G., Jr. 1966b. Taxonomic descriptions of the immature stages of the parasitic bee, *Stelis (Odontostelis) bilineolata* (Spinola) (Hymenoptera; Apoidea; Megachilidae). J. N.Y. Ent. Soc. 74: 84–91.

Rozen, J.G., Jr. 1966c. The larvae of the Anthophoridae (Hymenoptera, Apoidea). Part 2. The Nomadinae. Amer. Mus. Novitates, No. 2244, pp. 1–38.

Rozen, J. G., Jr. 1967a. Review of the biology of panurgine bees, with observations on North American forms (Hymenoptera, Andrenidae). Amer. Mus. Novitates, No. 2297, pp. 1–44.

Rozen, J. G., Jr. 1967b. The immature instars of the cleptoparasitic genus *Dioxys* (Hymenoptera; Megachilidae). J. N.Y. Ent. Soc. 75:236–48.

Rozen, J. G., Jr. 1969. The larvae of the Anthophoridae (Hymenoptera, Apoidea). Part 3. The Melectini, Ericrocini, and Rhathymini. Amer. Mus. Novitates, No. 2382, pp. 1–24.

Rozen, J. G., Jr. 1970. Biology and immature stages of the panurgine bee genera *Hypomacrotera* and *Psaenythia* (Hymenoptera, Apoidea). Amer. Mus. Novitates, No. 2416, pp. 1–16.

Rozen, J. G., Jr. 1973a. Biology notes on the bee *Andrena accepta* Viereck (Hymenoptera, Andrenidae). J. N.Y. Ent. Soc. 81:54–61.

Rozen, J. G., Jr. 1973b. Immature stages of lithurgine bees with descriptions of the Megachilidae and Fideliidae based on mature larvae (Hymenoptera, Apoidea). Amer. Mus. Novitates, No. 2527, pp. 1–14.

Rozen, J. G., Jr. 1977. Immature stages of and ethological observations on the cleptoparasitic bee tribe Nomadini (Apoidea, Anthophoridae). Amer. Mus. Novitates, No. 2638, pp. 1–16.

Rozen, J. G., Jr. 1978. The relationships of the bee subfamily Ctenoplectrinae as revealed by its biology and mature larva (Apoidea: Melittidae). J. Kan. Ent. Soc. 51:637–52.

Rozen, J. G., Jr. 1984. Nesting biology of diphaglossine bees (Hymenoptera: Colletidae). Amer. Mus. Novitates, No. 2786, pp. 1–33.

Rozen, J. G., Jr., K. R. Eickwort, and G. C. Eickwort. 1978. The bionomics and immature stages of the cleptoparasitic bee genus *Protepeolus* (Anthophoridae, Nomadinae). Amer. Mus. Novitates, No. 2640, pp. 1–24.

Rozen, J. G., Jr., and M. S. Favreau. 1968. Biological notes on *Colletes compactus compactus* and its cuckoo bee, *Epeolus pusillus* (Hymenoptera: Colletidae and Anthophoridae). J. N.Y. Ent. Soc. 76:106–11.

Rozen, J. G., Jr., and N. R. Jacobson. 1980. Biology and immature stages of *Macropis nuda*, including comparisons to related bees (Apoidea, Melittidae). Amer. Mus. Novitates, No. 2702, pp. 1–11.

Rozen, J. G., Jr., and R. J. McGinley. 1974a. Phylogeny and systematics of Melittidae based on the mature larvae (Insecta, Hymenoptera, Apoidea). Amer. Mus. Novitates, No. 2545, pp. 1–31.

Rozen, J. G., Jr., and R. J. McGinley. 1974b. Systematics of ammobatine bees based on their mature larvae and pupae (Hymenoptera, Anthophoridae, Nomadinae). Amer. Mus. Novitates, No. 2551, pp. 1–16.

Rozen, J. G., Jr., and R. J. McGinley. 1976. Biology of the bee genus *Conanthalictus* (Halictidae, Dufoureinae). Amer. Mus. Novitates, No. 2602, pp. 1–6.

Rust, R. W., and R. W. Thorp. 1973. The biology of *Stelis chlorocyanea*, a parasite of *Osmia nigrifrons* (Hymenoptera: Megachilidae). J. Kan. Ent. Soc. 46:548–62.

Schell, S. C. 1943. The biology of *Hadronotus ajax* Girault (Hymenoptera—Scelionidae), a parasite in the eggs of squash-bug (*Anasa tristis* DeGeer). Ann. Ent. Soc. Amer. 36:625–35.

Schrader, M. N., and W. E. LaBerge. 1978. The nest biology of the bees *Andrena (Melandrena) regularis* Malloch and *Andrena (Melandrena) carlini* Cockerell (Hymenoptera: Andrenidae). Ill. Nat. Hist. Surv. Biol. Notes, No. 108:1–23.

Sechser, B. 1970. Der Parasitenkomplex des kleinen Frostspanners (*Operophtera brumata* L.) (Lep., Geometridae) unter besonderer Berücksichtigung der Kokonparasiten. I. Z. angew. Ent. 66:1–35.

Short, J. R. T. 1952. The morphology of the head of larval Hymenoptera with special reference to the head of Ichneumonoidea, including a classification of the final instar larvae of the Braconidae. Trans. R. Ent. Soc. Lond. 103:27–84.

Short, J. R. T. 1959. A description and classification of the final instar larvae of the Ichneumonidae (Insecta, Hymenoptera). Proc. U.S. Nat. Mus. 110 (3419):391–511.

Short, J. R. T. 1970. On the classification of the final instar larvae of the Ichneumonidae (Hymenoptera). Suppl. Trans. R. Ent. Soc. Lond. 112:185–210.

Short, J. R. T. 1978. The final larval instars of the Ichneumonidae. Mem. Amer. Ent. Inst., No. 25. 508 pp.

Silvestri, F. 1923. Contribuzioni alla conoscenza dei tortricidi delle querce. Bol. Portici Scuola Super. Agr. Lab. Zool. Gen. e Agr. 17:41–107.

Simmonds. F. J. 1947. The biology of the parasites of *Loxostege sticticalis* L. in North America—*Bracon vulgaris* (Cress.) (Braconidae, Agathidinae). Bull. Ent. Res. 38:145–55.

Simmonds, F. J. 1953. Parasites of the frit-fly, *Oscinella frit* (L.) in eastern North America. Bull. Ent. Res. 43:503–42.

Smith, D. R. 1966a. The identity of three shoot-boring sawfly larvae of roses (Hymenoptera: Tenthredinidae). Ann. Ent. Soc. Am. 59:1292.

Smith, D. R. 1966b. Recognition of the European apple sawfly and the pear sawfly (Hymenoptera: Tenthredinidae). Coop. Econ. Insect Rep. 16:228–30.

Smith, D. R. 1967. A review of the larvae of Xyelidae, with notes on the family classification (Hymenoptera). Ann. Ent. Soc. Am. 60:376–84.

Smith, D. R. 1969a. Nearctic sawflies I. Blennocampinae: adults and larvae (Hymenoptera: Tenthredinidae). USDA, Tech. Bull. 1397. 198 pp.

Smith, D. R. 1969b. Nearctic sawflies II. Selandriinae: adults (Hymenoptera: Tenthredinidae). USDA, Tech. Bull. 1398. 48 pp.

Smith, D. R. 1969c. The genus *Susana* Rohwer and Middleton (Hymenoptera: Tenthredinidae). Proc. Ent. Soc. Wash. 71:13–23.

Smith, D. R. 1969d. Key to genera of Nearctic Argidae (Hymenoptera), with revisions of the genera *Atomacera* Say and *Sterictiphora* Billberg. Trans. Am. Ent. Soc. 95:439–57.

Smith, D. R. 1970a. Nearctic species of the genus *Ptenus* Kirby (Hymenoptera: Argidae). Trans. Am. Ent. Soc. 96:79–100.

Smith, D. R. 1970b. A new Nearctic *Xyela* causing galls on *Pinus* spp. (Hymenoptera: Xyelidae). J. Ga. Ent. Soc. 5:69–72.

Smith, D. R. 1971a. The genus *Zadiprion* Rohwer (Hymenoptera: Diprionidae). Proc. Ent. Soc. Wash. 73:187–97.

Smith, D. R. 1971b. Nearctic sawflies of the genera *Neoptilia* Ashmead, *Aprosthema* Konow, *Schizocerella* Forsius, and *Sphacophilus* Provancher (Hymenoptera: Argidae). Trans. Am. Ent. Soc. 97:537–94.

Smith, D. R. 1971c. Nearctic sawflies III. Heterarthrinae: adults and larvae (Hymenoptera: Tenthredinidae). USDA, Tech. Bull. 1420. 84 pp.

Smith, D. R. 1972a. North American sawfly larvae of the family Argidae (Hymenoptera). Trans. Am. Ent. Soc. 98:163–84.

Smith, D. R. 1972b. Sawflies of the genus *Croesus* Leach in North America (Hymenoptera: Tenthredinidae). Proc. Ent. Soc. Wash. 74:169–80.

Smith, D. R. 1974a. Sawflies of the tribe Cladiini in North America (Hymenoptera: Tenthredinidae: Nematinae). Trans. Am. Ent. Soc. 100:1–28.

Smith, D. R. 1974b. Conifer sawflies, Diprionidae: key to North American genera, checklist of world species, and new species from Mexico (Hymenoptera). Proc. Ent. Soc. Wash. 76:409–18.

Smith, D. R. 1975a. The sawfly genus *Hemichroa* Stephens: a review of species (Hymenoptera: Tenthredinidae). Ent. Scand. 6:297–302.

Smith, D. R. 1975b. *Eriotremex formosanus* (Matsumura), an Asian horntail in North America (Hymenoptera: Siricidae). Coop. Econ. Insect Rep. 25:851–54.

Smith, D. R. 1976. The xiphydriid woodwasps of North America (Hymenoptera: Xiphydriidae). Trans. Am. Ent. Soc. 102:101–31.

Smith, D. R. 1978. Suborder Symphyta (Xyelidae, Parachexyelidae, Parapamphiliidae, Xyelydidae, Karatavitidae, Gigasiricidae, Sepulcidae, Pseudosiricidae, Anaxyelidae, Siricidae, Xiphydriidae, Paroryssidae, Xyelotomidae, Blasticotomidae, Pergidae). *In* van der Vecht, J. and R. D. Shenefelt, eds., Hymenopterorum Catalogus, Pars 14. 193 pp. The Hague: W. Junk.

Smith, D. R. 1979a. Nearctic sawflies IV: Allantinae: Adults and larvae (Hymenoptera: Tenthredinidae). USDA, Tech. Bull. 1595. 172 pp.

Smith, D. R. 1979b. Symphyta, pp. 3–137. *In* Krombein, K. V., et al. (eds.). Catalog of Hymenoptera in America North of Mexico, vol. 1. Washington, D.C.: Smithsonian Institution Press.

Smith, D. R. 1980. Notes on sawflies (Hymenoptera: Symphyta) with two new species and a key to North American *Loderus*. Proc. Ent. Soc. Wash. 82:482–87.

Smith, D. R., and J. H. Lawton. 1980. Review of the sawfly genus *Eriocampidea* (Hymenoptera: Tenthredinidae). Proc. Ent. Soc. Wash. 82:447–53.

Smith, D. R., C. P. Ohmart, and D. L. Dahlsten. 1977. The fir shoot-boring sawflies of the genus *Pleroneura* in North America (Hymenoptera: Xyelidae). Ann. Ent. Soc. Am. 70:761–67.

Smith, E. L. 1968. Biosystematics and morphology of Symphyta. I. Stem-galling *Euura* of the California region, and a new female genitalic nomenclature. Ann. Ent. Soc. Am. 61:1389–1407.

Smith, E. L. 1970. Biosystematics and morphology of Symphyta. II. Biology of gall-making Nematinae sawflies in the California region. Ann. Ent. Soc. Am. 63:36–51.

Smith, L. M., and G. S. Kido. 1949. The raspberry sawfly. Hilgardia 19:45–54.

Smith, O. J., A. G. Diboll, and J. H. Rosenberger. 1955. Laboratory studies of *Pelecystoma harrisinae* (Ashmead), an adventive braconid parasite of the western grape leaf skeletonizer. Ann. Ent. Soc. Amer. 48:232–37.

Spradbery, J. P. 1970. The immature stages of European ichneumonid parasites of siricine woodwasps. Proc. R. Ent. Soc. Lond. (A) 45:14–28.

Spradbery, J. P. 1973. Wasps: an account of the biology and natural history of social and solitary wasps. Seattle: Univ. Wash. Press. 408 pp.

Spradbery, J. P. 1975. The biology of *Stenogaster concinna* van de Vecht with comments on the phylogeny of Stenogastrinae (Hymenoptera: Vespidae). J. Australian Ent. Soc. 14:309–18. (Includes description of larva).

Stage, G. I. 1966. Biology and systematics of the American species of the genus *Hesperapis* Cockerell. Ph.D. diss., University of California, Berkeley. 464 pp.

Stein, J. D. 1974. Elm sawfly. USDA, Forest Serv., Forest Pest Leaflet 142. 6 pp.

Stephen, W. P., and T. Koontz. 1973. The larvae of the Bombini. Part I. Interspecific variation in larval head characteristics of *Bombus* (Hymenoptera: Apoidea). Part II. Developmental changes in the preadult stages of *Bombus griseocollis* (DeGeer). Melanderia 13:1–29.

Stilwell, M. A. 1965. Hypopleural organs of the woodwasp larva *Tremex columba* (L.) containing the fungus *Daedalea unicolor* Bull. ex Fries. Can. Ent. 97:783–84.

Taylor, K. L. 1967. Parasitism of *Sirex noctilio* F. by *Schlettererius cinctipes* (Cresson) (Hymenoptera: Stephanidae). J. Australian Ent. Soc. 6:13–19.

Tengö, J., and G. Bergström. 1977. Cleptoparasitism and odor mimetism in bees: do *Nomada* males imitate the odor of *Andrena* females? Science 196:1117–19.

Thomsen, M. 1927. Some observations on the biology and anatomy of a cocoon-making chalcid larva, *Euplectrus bicolor* Swed. Vid. Meddel. Dansk. Natur. For. 84:73–89.

Thorp. R. W. 1966. Synopsis of the genus *Heterostelis* Timberlake (Hymenoptera; Megachilidae). J. Kan. Ent. Soc. 39:131–46.

Thorp, R. W. 1968. Ecology of a *Proteriades* and its *Chrysura* parasite, with larval descriptions (Hymenoptera: Megachilidae: Chrysididae). J. Kan. Ent. Soc. 41:324–31.

Thorp, R. W. 1969. Ecology and behavior of *Anthophora edwardsii*. Amer. Midland Nat. 82:321–37.

Thorpe, W. H. 1932. Experiments upon respiration in the larvae of certain parasitic Hymenoptera. Proc. R. Soc. Lond. (B) 109:450–71.

Thorpe, W. H. 1936. On a new type of respiratory interrelation between an insect (chalcid) parasite and its host (Coccidae). Parasitology 28:517–40.

Torchio, P. F. 1965. Observations on the biology of *Colletes ciliatoides* (Hymenoptera: Apoidea, Colletidae). J. Kan. Ent. Soc. 38:182–87.

Torchio, P. F. 1970. The ethology of the wasp, *Pseudomasaris edwardsii* (Cresson), and a description of its immature forms (Hymenoptera: Vespoidea, Masaridae). Los Angeles Co. Mus., Contr. Sci., No. 202. 32 pp. (Includes description of *Euparagia* larva.)

Torchio, P. F. 1971. The biology of *Anthophora (Micranthophora) peritomae* Cockerell. Los Angeles Co. Mus., Contr. Sci. 206:1–14.

Torchio, P. F. 1972. *Sapyga pumila* Cresson, a parasite of *Megachile rotundata* (F.) (Hymenoptera: Sapygidae; Megachilidae). I: Biology and description of immature stages. Melanderia 10:1–22.

Torchio, P. F. 1974. Notes on the biology of *Ancyloscelis armata* Smith and comparisons with other anthophorine bees. J. Kan. Ent. Soc. 47:54–63.

Torchio, P. F. 1975. The biology of *Perdita nuda* and descriptions of its immature forms and those of its *Sphecodes* parasite (Hymenoptera: Apoidea). J. Kan. Ent. Soc. 48:257–79.

Torchio, P. F., and W. P. Stephen. 1961. Description of the larva and pupa of *Emphoropsis miserabilis* (Cresson) and comparisons with other anthophorids (Hymenoptera: Apoidea). Ann. Ent. Soc. Amer. 54:687–92.

Torchio, P. F., and D. M. Torchio. 1975. Larvae of the Apidae (Hymenoptera, Apoidea). Part I. Apini, *Apis*. Agri. Exp. Station, Utah State Univ. Res. Rep. 20:1–35.

Torchio, P. F., and N. N. Youssef. 1968. The biology of *Anthophora (Micranthophora) flexipes* and its cleptoparasite, *Zacosmia maculata,* including a description of the immature stages of the parasite (Hymenoptera: Apoidea, Anthophoridae). J. Kan. Ent. Soc. 41:289–302.

Torchio, P. F., J. G. Rozen, Jr., G. E. Bohart and M. S. Favreau. 1967. Biology of *Dufourea* and of its cleptoparasite, *Neopasites* (Hymenoptera: Apoidea). J. N.Y. Ent. Soc. 75:132–46.

Townes, H. 1969. The genera of Ichneumonidae, Part 1. Mem. Amer. Ent. Inst., No. 11. 300 pp.

van Achterberg, C. 1976. A preliminary key to the subfamilies of the Braconidae (Hymenoptera). Tijds. voor Ent., Deel 119:33–78.

Vance, A. M., and H. L. Parker. 1932. *Laelius anthrenivorus* Trani, an interesting bethylid parasite of *Anthrenus verbasci* L. in France. Proc. Ent. Soc. Wash. 34:1–7.

Wahl, D. 1984. An improved method for preparing exuviae of parasitic Hymenoptera. Ent. News. 95:227–28.

Wallace, L. L., and F. H. McNeal. 1966. Stem sawflies of economic importance in grain crops in the United States. USDA, Tech. Bull. 1350. 50 pp.

Webb, F. B., and R. S. Forbes. 1951. Notes on the biology of *Pleroneura borealis* Felt (Hymenoptera: Xyelidae). Can. Ent. 83:181–83.

Westcott, R. L. 1971. New host and distribution records for three western wood-boring Hymenoptera. Pan-Pacif. Ent. 47:310.

Wheeler, G. C., and J. Wheeler. 1976. Ant larvae: review and synthesis. Mem. Ent. Soc. Wash., No. 7. 108 pp. (Includes a list of the many papers by these authors and a bibliography of other important papers.)

Wheeler, G. C. and J. Wheeler. 1979. Chap. 7. Larvae of social Hymenoptera. *In* Social Insects. Vol. 1. New York: Academic Press.

Wheeler, W. M. 1910. (reprinted 1926, 1960). Ants: their structure, development and behavior. New York: Columbia Univ. Press. 663 pp.

Wickman, B. E. 1967. Life history of the incense-cedar wood wasp, *Syntexis libocedrii*. Ann. Ent. Soc. Am. 60:1291–95.

Williams, F. X. 1919. *Epyris extraneus* Bridwell (Bethylidae), a fossorial wasp that preys on the larva of the tenebrionid beetle, *Gonocephalum seriatum* (Boisduval). Proc. Haw. Ent. Soc. 4:55–63.

Williams, F. X. 1956. Life history studies of *Pepsis* and *Hemipepsis* wasps in California (Hymenoptera, Pompilidae). Ann. Ent. Soc. Amer. 49:447–66.

Wilson, E. O. 1971. The insect societies. Cambridge, Mass.: Harvard Univ. Press. 548 pp.

Wishart, G., and E. Monteith. 1954. *Trybliographa rapae* (Westw.) (Hymenoptera: Cynipidae), a parasite of *Hylemya* spp. (Diptera: Anthomyiidae). Can. Ent. 86:145–54.

Wong, H. R. 1955. Larvae of the Nearctic species of *Anoplonyx* (Hymenoptera: Tenthredinidae). Can. Ent. 87:224–27.

Wong, H. R. 1963. The external morphology of the adults and ultimate larval instars of the larch sawfly *Pristiphora erichsonii* (Htg.) (Hymenoptera: Tenthredinidae). Can. Ent. 95:897–921.

Wong, H. R., and H. E. Milliron. 1972. A Canadian species of *Susana* on western juniper (Hymenoptera: Tenthredinidae). Can. Ent. 104:1025–28.

Wright, D. W., Q. A. Geering, and D. G. Ashby. 1946. The insect parasites of the carrot fly, *Psila rosae* Fab. Bull. Ent. Res. 37:507–29.

Yamane, S. 1976. Morphological and taxonomic studies on vespine larvae, with reference to the phylogeny of the subfamily Vespinae (Hymenoptera: Vespidae). Insecta Matsumurana (n.s.) 8:1–45.

Youssef, N. N., and G. E. Bohart. 1968. The nesting habits and immature stages of *Andrena (Thysandrena) candida* Smith (Hymenoptera, Apoidea). J. Kan. Ent. Soc. 41:442–55.

Yuasa, H. 1923 (1922). A classification of the larvae of the Tenthredinoidea. Ill. Biol. Monogr. 7. 172 pp.

Zucchi, R., S. Yamane, and S. F. Sakagami. 1976. Preliminary notes on the habits of *Trimeria howardi,* a Neotropical communal masarid wasp, with description of the mature larva (Hymenoptera: Vespoidea). Insecta Matsumurana (n.s.) 8:47–57.

Glossary

Accessory Process (of antenna). A short or long secondary process usually arising from the tip of the second or, more rarely, the first segment.

Aculea (-ae). Minute spines.

Aculeate. With aculeae (Lepidoptera); with a sting (Hymenoptera).

Acuminate. Tapering to a long point.

Addorsal Line. A longitudinal line a little to one side of the mid-dorsal line.

Adecticous Pupa. A pupa in which the mandibles are immobile; not used by the pharate adult.

Adfrontal Area(s). An oblique sclerite on each side of the frons, usually extending from the base of the antennae to the epicranial suture where they meet, or to the epicranial notch if they do not meet (Lepidoptera).

Adfrontal Setae. A pair of setae on the adfrontal areas, the more dorsal one being AF1. (Lepidoptera).

Adfrontal (= ecdysial) Suture. The suture separating the adfrontal area from the cranium (Lepidoptera).

Aedeagus. The male intromittent organ; penis plus associated structures.

Aerenchymatous. Cells or tissues of aquatic plants that contain air.

Alveolus (-i). A cuplike depression in the body wall from which a seta arises.

Ambulatorial Warts. Bumplike protuberances used for locomotion; usually ventrally or dorsally on the abdomen (Lepidoptera and others).

Ametabolous. Without metamorphosis.

Amphipneustic. A type of respiratory system in which only the first thoracic and the last or last two abdominal pairs of spiracles are functional.

Amphitoky. A type of parthenogenesis in which both males and females are produced.

Ampulla (-ae). *See* Ambulatorial Warts.

Anal Comb. The mesal sclerotized prong ventrad of the anal plate and adjacent to the anus; used to eject frass (Lepidoptera). A single or double row of fine setae above the ventrolateral setae on the tenth abdominal segment of flea larvae.

Anal Fork. *See* Anal Comb.

Anal Horn. A spinelike mesal horn on the eighth abdominal segment (of most sphingid larvae and the domestic silkworm).

Anal Lobes. Any and various protrusions near the anus.

Anal Plate. In caterpillars and other larvae, the dorsal shieldlike covering of the last abdominal segment (= suranal plate, anal shield, epiproct).

Anal Prolegs. Prolegs on the last abdominal segment.

Anal Shield. *See* Anal Plate.

Anal Slit. Narrow anal opening that may be transverse or vertical.

Anal Struts. Downward- and backward-directed appendages on the tenth abdominal or terminal anal segment of flea larvae.

Anamorphosis. Increase in the number of segments after emergence from the egg (Protura).

Annular Spiracle. A simple, circular or oval spiracle with one opening.

Annulate. Ringed, but not truly segmented.

Annulets. Small secondary rings into which a segment or appendage is divided; a partial dorsal subdivision of a body segment formed by transverse creases in the integument.

Annuliform. In the form of rings, or ringlike.

Annulus (-i). Ring(s) encircling a segment or other structure.

Antacorium (-a). Ring of membrane connecting antenna with head or segmental membrane of segments.

Anteapical. Just proximad of the apex.

Anteclypeus. An anterior and usually more lightly sclerotized part of the clypeus to which the labrum is attached.

Antennal Fossa. A cavity or depression in which the antenna is located.

Antennal Groove. A groove in the head capsule into which part of the antenna fits.

Antennal Papilla. A short, unsegmented appendage arising from the antennal orbit, similar to a small, one-segmented antenna but acquired secondarily.

Antennarium (-a). Antennal suture or ring at base of antenna; annular sclerite forming the edge of the antacoria.

Antepenultimate. Second before the last.

Anterad. Toward the front.

Anterior. Front; in front of.

Anterodorsal. In front of and toward the back or upper part.

Anteroventral. In front of and toward the venter or lower part.

Anteroventral Seta. A smaller seta anterior to the ventrolateral setae of flea larvae on the tenth abdominal segment.

Apical. At the tip or end.

Apneustic. Without functional spiracles; respiration through the cuticle or tracheal gills.

Apodous. Legless.

Apoid. Resembling honey bee larvae.

Apolysis. The process of retraction of the epidermis as the new cuticle is laid down.

Apomorphy, -phic, -phous. The relatively derived state (more recent) of a sequence of homologous characters. *Compare* Plesiomorphy.

Apotome. A sclerotized part, usually of the head capsule, separated at ecdysis; a functional, rather than homologous subdivision.

Approximate. Close together.

Apterous. Wingless.

Apterygota. Primitively wingless subclass of the Insecta.

Armature. Setae, spines, or sclerotized processes.

Arolium (-a). The terminal pad between the tarsal claws.

Arrhenotoky. Facultative parthenogenesis in which fertilized eggs give rise to females and unfertilized eggs to males.

Article. A segment.

Articulation. A connection or joint between two structures.

Asperities. Small spinelike or peglike structures, frequently in rows or patches (microspines, microtrichiae, spinules, spinulae).

Atrial Fine Structure. Sclerotized ornamentation and projections associated with the inner margins of the peritreme and extending into the atrium of the spiracle.

Atrium (of spiracle). A cup-shaped, heavy-walled space immediately beneath the outer spiracular opening; the chamber between the spiracular opening and the trachea.

Atrophied. Rudimentary, reduced.

Attenuated. Slender and gradually tapering.

Attingent. Touching, in contact with.

Basal Articulating Membrane (of antenna). *See* Antacoria.

Beak. The protruding mouthparts of a piercing-sucking or sucking insect.

Bifid. Divided into two parts; forked.

Biforous. Spiracular opening separated into two adjacent openings by a septum.

Bifurcate. Forked, two-branched.

Biordinal. Crochets arranged in a single row of two alternating lengths (Lepidoptera).

Bipartite. Divided into two parts.

Biramous. Having two branches or appendages.

Biserial. Crochets with the bases arranged in two rows (Lepidoptera).

Bisetose. Two setae.

Bivoltine. Having two generations per year.

Blood Channel. An internal duct or external groove in the mandible or combined mandible/maxilla of certain predacious larvae (especially Neuroptera and some Coleoptera).

Brachyptery (brachypterous). In adults, having short wings, not reaching tip of abdomen and usually much shorter.

Breastbone. A sclerotized process (spatula) on the venter of the thorax caudad of the mouth of cecidomyiid larvae (Diptera).

Buccal Cavity. Mouth or oral cavity.

Caducous. Deciduous, easily detached or shed.

Calcar (-ia). A movable spur or spinelike process; specifically the spines at the apex of a tibia. (Orthoptera)

Callus. A rounded swelling.

Campodeiform Larva. An elongate larva, with well-developed thoracic legs and prognathous head, usually active and predacious.

Capitate. With terminal knoblike enlargement; capitate antenna.

Cardo (-ines). The basal segment of a maxilla between the head and stipes.

Carina (-ae). A ridge or keel, not necessarily high or acute.

Carinate. With carinae.

Caste. A form of adult in a social insect (termites, ants).

Caterpillar. Typically a larva with a conspicuous head, three pairs of thoracic legs, and prolegs; the larva of a butterfly, moth, sawfly, or scorpionfly (= eruciform).

Cauda (= tail). Structure located above the anal opening and representing the ninth abdominal segment; length and shape quite variable (Homoptera: Aphididae).

Caudad. Toward the tail or posterior end.

Caudal. Pertaining to the tail or posterior end.

Caudal Filament. A long, often median, tapered process at the end of the body.

Cephalad. Toward the head or anterior end.

Cephalic. Pertaining to the head; anterior.

Cephalo-pharyngeal skeleton. Among muscoid larvae, the heavily sclerotized mouthpart structures withdrawn into the cephalic segments.

Cephalothorax. A body region consisting of head and thorax (Crustacea and Arachnida).

Cercus (-i). Paired appendages of tenth abdominal segment. Appendage(s) at the end of the abdomen—usually slender, filamentous, segmented and paired; sometimes unsegmented (as in Orthoptera, Phasmatodea, Dermaptera). Sometimes misapplied to appendages of segment 9 in Coleoptera larvae. *See* Urogomphus.

Cervical. Pertaining to the neck or cervix.

Cervical Gland. Ventral gland on the prothorax of some caterpillars.

Cervical Triangle. *See* Epicranial Notch.

Cervix. Membranous region between the head and prothorax (neck).

Chaetotaxy. The arrangement and nomenclature of setae. *See* Setal Map.

Chalaza (-ae). A simple, sclerotized, elevated projection, usually bearing a single seta (Lepidoptera primarily).

Chelate. Pincerlike.

Chelicera (-ae). One of the anterior pair of appendages in arachnids.

Cheloniform. Shaped like a turtle, head concealed; sometimes flattened.

Chewing Mouthparts. With opposable, nonsucking mouthparts.

Chitin. A nitrogenous polysaccharide occurring in the cuticle of arthropods.

Chitinized. "Sclerotized," toughened.

Chorion. Outer shell of an insect egg.

Chrysalis (Chrysalids). The pupa of a butterfly.

Cibarium. The area of the mouth cavity just behind the epipharynx and above the hypopharynx.

Ciliate. Fringed or lined with fine hairs.

Clavate. Clublike, or enlarged toward the tip; clavate antennae.

Cleavage Line. The Y-shaped line of weakness on the head capsule along which the integument splits at time of molting (= ecdysial line, epicranial suture).

Cleft. Split or forked.

Cleptoparasite. One that feeds on the food stored for the host larvae.

Clubbed. With the distal part enlarged; clubbed antennae.

Clunium. Fused abdominal terga 9 and 10, or 8, 9, and 10 (Psocoptera).

Clypeus. The sclerite between the frons and labrum.

Coarctate. The third phase of hypermetamorphic larval development in Meloidae, generally equivalent to instar 6, in which the larva is strongly sclerotized, has rudimentary appendages, and is immobile.

Coarctate Pupa. A pupa enclosed in a hardened case formed by the last larval skin (Diptera).

Cocoon. A silken structure spun by the larva in which the pupa is formed.

Colleterial Gland. A glandular structure accessory to the oviduct, secreting the viscous material used to cement eggs together or produce an oötheca.

Collophore. A tubelike structure on the venter of the first abdominal segment of Collembola.

Collum. Necklike constriction at the base of the head capsule, tergite of the first segment (Diplopoda).

Condyle. A knoblike process forming part of an articulation by fitting into a depression (i.e., mandibular condyle).

Contiguous. Touching.

Convergent. Becoming closer together at one end.

Corneous. Sclerotized or horny.

Corniculum. A small process similar to a chalaza, but without a seta.

Coronal Suture. Dorsal, median arm of the cleavage line; a longitudinal suture along the midline of the head, extending from the epicranial notch to the apex of the clypeus (the stem of the frequently Y-shaped epicranial suture). *See* Epicranial Suture.

Coxa (-ae). The proximal segment of the leg.

Coxites. Paired lateral plates of abdominal sterna of Microcoryphia and Thysanura.

Cranium. The sclerotized part of the head capsule.

Crawler. Active first instar of a scale insect.

Cremaster. The terminal spined or hooked process of a pupa, used for attachment or movement (Lepidoptera).

Crenulate. Wavy or scalloped.

Cribriform. Sievelike openings in spiracles, especially among Scarabaeoidea.

Cristate. Crested.

Crochet(s). Sclerotized, hooklike structures, usually arranged in rows or circles on the prolegs of Lepidoptera larvae (and a few others).

Ctenidium. Comb-like row of bristles.

Cultriform. Shaped like a pruning knife.

Cursorial. With legs for running.

Cyphosomatic. With the dorsum curved or humped and the venter flat (common in Chrysomelidae).

Deciduous. Having part(s) that may naturally fall off, be shed, or break away.

Decticous Pupa. A pupa with articulated mandibles; they can be used by the pharate adult.

Decumbent. Bent downward.

Decurved. Curved downward.

Dehiscence, line of. A suture on the dorsum of an insect's body along which the cuticle splits during ecdysis to permit exit of the insect.

Dentate. Toothed.

Denticulate. With minute toothlike projections or edges; with little teeth or notches.

Denticle. Small tooth on inner margin of claw.

Dermal Papillae. Numerous small projections on surface of body.

Deutonymph. Third instar of a mite.

Dextral. In Trichoptera, applied to the direction of coiling of the case in certain species, the coil ascending from left to right of the observer when the anterior opening is held uppermost. Antonym of sinistral.

Diapause. A period of arrested development and reduced metabolic rate, during which growth, differentiation, and metamorphosis cease; a period of dormancy not immediately referable to adverse environmental conditions.

Dicondylous. With two condyles, or two processes articulating the mandible to head capsule.

Digits. Fingerlike lobes, especially of the prothoracic spiracles of maggots (Diptera).

Dilated. Widened.

Dimorphic. Occurring in two distinctive forms.

Disc. The central portion of a structure.

Distal. Toward the tip or end; farthest from the body.

Diurnal. Active during daylight.

Dorsad. Toward the top or back.

Dorsal. At the top or back or above.

Dorsal Acetabulum. A cavity on the dorsal side of the mandibular base that articulates with a condyle (dorsal mandibular articulation) on the cranium.

Dorsal Flange. Enlarged portion of dorsolateral angle of each lateral arm of the labial sclerite (Hymenoptera: Apocrita mouthparts).

Dorsal Line. A longitudinal line along the dorsomeson.

Dorsolateral. At the top and to the side.

Dorsomesal. At the top and near the midline.

Dorsomeson. The middle of the back or top.

Dorsum. The top (dorsal) side or back.

Ecdysial Suture. Where the head capsule usually splits during molting; laterad of the adfrontal areas in Lepidoptera larvae.

Ecdysis (-ses). Molting, shedding the cuticle.

Eclosion. Hatching from the egg.

Ectognathous. With mouthparts exserted or exposed.

Ectoparasite. A parasite that lives on the outside of its host.

Ectoparasitoid. A parasitoid that lives on the outside of its host.

Egg-Burster. A point or ridge on the head used in rupturing the egg shell when hatching.

Elateriform Larva. Resembling a wireworm—slender, heavily sclerotized, nearly naked, and with three pairs of thoracic legs.

Emarginate. With a dent or notch in the margin.

Empodium (-ia). A pad or bristlelike structure between the claws at the tip of the last tarsal segment.

Endognathous. Mandibles and maxillae almost entirely enclosed by fusion of lateral margins of labium with cranial folds, e.g., Protura.

Endoparasitoid. A parasitoid that lives inside its host.

Endopterygote. With the wings developing internally; holometabolous insects.

Ensiform. Resembling the shape of a sword.

Entognathous. *See* Endognathous.

Entomophagous. Feeding on insects.

Epaulet. Dorsal, flattened extension on each side of prothorax in Odonata.

Epicranial Notch. The V-shaped dorsomedial space delimited laterally by the cranial halves (= cervical triangle, verticle triangle).

Epicranial Suture. Inverted, Y-shaped, V-shaped, U-shaped or lyriform suture on face of head with arms diverging ventrad (= ecdysial line, cleavage line); the suture that separates the epicranial halves on the dorsum of the head. Anteriorly (ventrally) it is usually forked into the epicranial arms, on either side of the frons.

Epicranium. The cranium above the frons.

Epimeron. That area of the thoracic pleuron immediately adjacent to the coxa and posterior to the pleural suture.

Epipharynx. Inner surface of labrum.

Epiproct. Dorsal lobe of terminal abdominal segment. *See* Suranal Plate.

Episternum. That region of the thoracic pleuron immediately adjacent to the coxa and anterior to the pleural suture.

Epistoma. Upper part of the frame surrounding the opening of the buccal cavity, continuous with pleurostoma on each side and hypostoma below; area just behind or above the labrum; used by various authors as a synonym of anteclypeus, clypeus, postclypeus, or anterior margin of frons.

Epistomal (Frontoclypeal) Suture. The suture between the frons and the clypeus.

Eruciform Larva. Caterpillarlike; a larva with a well-developed head, thoracic legs, and abdominal prolegs.

Eucephalous. With a well-developed head.

Eversible. Capable of being everted or projected outward.

Ex. From, or out of. (Ex pupa = reared from the pupa).

Exarate Pupa. A type in which the legs and wings are free from the body.

Exopterygote. The wings developing externally—winged insects with hemimetabolous metamorphosis.

Exserted. Projecting from the body; protruding.

Exuvia (-ae). The cast skin. In Latin "exuviae" means clothes, booty, spoils of war. There was (is) no singular, but "exuvia" is the correct derived singular. "Exuvium" is not a correct singular form.

Falciform. Sickle-shaped.

False Legs. Without true segments. *See* Prolegs.

Fastigium (-ia). The extreme point or front of the vertex of the head.

Femur (-ora). The third and usually largest segment of an insect leg.

Filament. A long, slender structure.

Filiform. Hairlike or threadlike; slender and of equal diameter.

First Grub. The second phase of hypermetamorphic larval development in Meloidae, generally incorporating instars 2–5, in which the larva is grublike (scarabaeiform) in appearance.

Flabellate. With fanlike lobes or processes (flabellate antenna).

Flagellate. Whiplike.

Flagellomere. A segment or division of the flagellum or antenna beyond its two basal segments.

Flanking Seta. Flanking hair (of paraproct). A seta immediately above or below the hyaline cone (q.v.) of the distal paraproctal margin (Psocoptera).

Foliaceous. Leaf-like, resembling a leaf.

Fontanelle. A circular, pale gland opening between the eyes (Isoptera).

Fossorial. Structured for digging.

Frass. The pelletlike excrement of insects, especially from caterpillars and wood borers.

Frons. Median sclerite on face of head delimited above by epicranial arms and below by frontoclypeal suture; sometimes termed frontal area.

Front. *See* Frons.

Frontal Seta. A seta on the frons. One of two setae on the frons of caterpillars.

Frontal Suture. The V- or U-shaped suture formed by arms of epicranial suture that delimit frons; the two branches of the epicranial suture on either side of the frons.

Frontoclypeal Region. Area occupied by the frons and clypeus.

Frontoclypeal Suture. The suture separating the frons from the clypeus.

Fungivorous. Feeding on or in fungi.

Furca. A forked structure, frequently internal in the thorax of winged insects.

Furcate. Forked.

Furcula. The springing structure of Collembola.

Fusiform. Tapered at both ends and widened in the middle (spindlelike).

Galea. The outer lobe of the maxilla.

Gena (-ae). The part of the head dorsad (or caudad) of the mandibles; the cheek.

Geniculate. Elbowed.

Gibbose. Bearing one or more swellings or protuberances.

Gill Tuft. Group of filamentous gills joined to a common stem.

Glabrous. Smooth, without setae.

Globose. Approximately spherical.

Glossa (-ae). The inner pair of lobes at the tip of the labium between the paraglossae.

Gressorial. Having legs fitted for walking.

Grub. Relatively inactive U- or C-shaped larvae, especially Coleoptera and Hymenoptera.

Gula. A sclerite on the venter of the head between the labium and the neck.

Gular Suture(s). Longitudinal suture(s), usually one on each side of the gula.

Hastisetae. Spear-headed setae arising from the tergites of some dermestid larvae.

Haustellate. Mouthparts for sucking or piercing-sucking.

Hemimetabolous. A type of metamorphosis in which the form of the immature gradually approaches that of the adult through successive instars; also, metamorphosis in which the immatures (naiads) are aquatic (Odonata, Plecoptera, and Ephemeroptera) and do not resemble the adults as closely as terrestrial groups do (Comstock 1918).

Hemipneustic (Respiratory System). One with spiracles on the thorax and abdominal segments 1–7.

Heteroideous. A mesoseries of crochets with smaller or rudimentary crochets at both ends (many arctiids).

Hibernaculum. A protective retreat made out of silk or other material, in which a larva hides or hibernates; an overwintering retreat or shelter, usually applied to early instars.

Holometabolous. With complete metamorphosis.

Holopneustic. All the spiracles open and functional.

Homoideous. With all crochets the same length.

Horn. A stiff, pointed cuticular process.

Hornworm. A caterpillar with a dorsal spine or horn near the end of the abdomen (Sphingidae).

Host. The organism in or on which a parasitoid or parasite lives; the plant on which an insect feeds.

Hyaline. Translucent but not necessarily flexible or membranous.

Hyaline Cone (of paraproct). A projection on distal margin of paraproct, often bifid at its tip, sometimes double (Psocoptera).

Hygropetric. Rocks with a thin layer of water trickling over them, e.g., a roadside seepage.

Hypermetamorphosis. Larval development involving passage through two or more distinct phases, as in Meloidae; a type of complete metamorphosis in which a larva passes through two or more very different appearing instars.

Hyperparasitoid. A parasitoid whose host is another parasitoid.

Hypognathous. With the head vertical and mouthparts directed ventrad. *Compare* Prognathous.

Hypopharyngeal Filament. A sclerotized filament running from sitophore sclerite to lingual sclerite of hypopharynx (Psocoptera).

Hypopharynx. The median inner mouthpart structure anterior (in a hypognathous head) to the labium (dorsal in a prognathous head) that may bear taxonomically useful structures.

Hyporheic. Subterranean zone beneath the stream bottom.

Hypostoma. Each hypostoma extends laterally on either side of the cephalic structure from the ventral portion of the pleurostoma (Hymenoptera: Parasitica).

Hypostomal Spur. A sclerite extending from each hypostoma across the stipes; this is an abbreviation of Short's (1952) term, "sclerotic spur of the hypostoma" (Hymenoptera: Parasitica).

Imago. An adult.

Imbricated. Overlapping like shingles.

Immaculate. Not marked or spotted.

Inquiline. An animal that lives in the home or nest of a different species.

Instar. The stage between molts.

Integument. The outer body wall.

Interstitial. Referring to the spaces between objects, like leaves and sticks (leaf litter), sand grains, or soil particles. An interstitial organism is of such dimensions that it can move freely within the spaces; thus in substrates composed of fine particles, only minute organisms can be called interstitial, located between two structures.

Kappa Group. The prespiracular wart or group of setae on the prothorax of Lepidoptera larvae (Fracker). (The L group of Hinton)

Keeled. With a distinct ridge (usually ventrally).

Labial Palpus (-pi); Palp (-s). A pair of small, segmented sensory structures arising on the distolateral portions of the labium.

Labial Sclerite. A sclerotized ring, usually broken dorsally, and occasionally ventrally, which represents the marginal sclerotization of the prelabium (Hymenoptera: Parasitica).

Labial Stipites. *See* Prementum.

Labium. The lower lip, basically consisting of submentum, mentum, prementum, ligula, and labial palpi.

Labrum. The upper lip; sclerite attached to the anterior edge of the clypeus or frontoclypeal region.

Labrum-Epipharynx. The combined labrum and epipharynx.

Lacinia (-ae). The inner lobe of the maxilla adjacent to the galea.

Lacinia Mobilus. A small platelike or blade-shaped appendage near the base of the mandible (Coleoptera).

Lamella (-ae). A thin plate or leaflike process.

Lamellate. Made up of a number of lamellae.

Lanceolate. Spear-shaped.

Larva (-ae). The stages between the egg and pupa of those insects having complete metamorphosis; the stages between the egg and adult of those insects not having complete metamorphosis (also known as nymphs); also, the six-legged first instar of Acarina.

Larva I. The second instar of Protura.

Larva II. The third instar of Protura.

Larviform. Shaped like a larva.

Laterad. Toward the side, away from the midline.

Lateral. To the side, or at the side of.

Lateral Penellipse. An incomplete circle of crochets closed laterally and open mesially (Lepidoptera).

Leaf Miner. An insect that feeds and usually lives between the upper and lower surfaces of a leaf, often forming distinctive patterns.

Lentic. Inhabiting still waters, such as lakes, ponds, or swamps.

Ligula. A terminal lobe at the apex of the labium that represents the fused or unfused glossae and/or paraglossae.

Lingual Sclerite. One of a pair of long-oval sclerites situated to each side of midline at distal end of hypopharynx (Psocoptera).

Looper. A caterpillar with some prolegs reduced or missing, usually on abdominal segments 3, 4, and 5, and that moves by looping its body forward (Geometridae; some Noctuidae).

Lotic. Inhabiting running waters, such as streams or rivers.

Lyriform, Lyrate, Lyre- Shaped. Shaped like a lyre; applied to an epicranial suture with the frontal arms sinuate, that is, bowed outwardly and then inwardly again, before curving outwardly toward the antennal insertions.

Maculate. Marked (spotted) with pigmented areas of varying shape.

Maggot. Most larvae of higher Diptera; legless, "headless", and peglike. A wormlike, legless larva with little or no head capsule.

Mala (-ae). A single lobe at the apex of the maxilla, usually representing the fused galea and lacinia; may also be the galea alone.

Mandibular Articulations. The paired dorsal condyles and ventral acetabula at the ends of the pleurostomata, with which the mandibles articulate.

Margined. With a sharp edge.

Mask. The protrusible, hinged labium of Odonata larvae (nymphs).

Maturus Junior. The fourth instar of Protura.

Maxilla (-ae). The paired mouthparts posterior to the mandibles.

Maxillary Palp. A small, segmented sensory structure arising from the maxilla.

Meconium. The liquid waste products excreted by the adult after emergence from the pupa.

Median. In the middle.

Mentum. That part of the labium lying between the submentum and prementum; a term sometimes used for the proximal sclerite of the labium (postmentum), when there are only two labial sclerites.

Mesad. Toward the meson.

Mesal. At, or pertaining to the meson.

Mesal Penellipse. An incomplete circle of crochets closed mesally and open laterally (Lepidoptera).

Meson. An imaginary vertical middle plane of the body.

Mesonotum. The dorsal part of the mesothorax.

Mesopleuron. The lateral part of the mesothorax.

Mesoseries. A longitudinal row of crochets on the mesal side of a proleg; if curved, less than two-thirds of a circle.

Mesosternum. The ventral part of the mesothorax.

Metamorphosis. Change in form during development.

Metanotum. The dorsal part of the metathorax.

Metapleuron. The lateral part of the metathorax.

Metapneustic (Respiratory System). With only the last pair of abdominal spiracles open.

Metasternum. The ventral part of the metathorax.

Metatarsus. The first (basal) tarsal segment.

Micropyle. The minute opening(s) in the insect egg chorion through which sperm enter.

Microspines. Minute spines on the body, usually visible only under magnification (aculeae, spinules).

Microtrichia. Minute, ciliate processes of the cuticle.

Mola. The usually elevated basal grinding surface of a mandible (Coleoptera).

Molar Process. Sclerotized, rounded process at floor of mouth, visible when mandibles open; found in Dolerinae (Tenthredinidae).

Moniliform. With round segments like a string of beads.

Monocondylous. With one condyle articulating mandible to head capsule.

Monophagous. Feeding on only one species.

Mouthhooks. The parallel clawlike mouthparts of the larvae of higher Diptera.

Multiforous. Spiracle with three or more openings within or adjacent to the peritreme.

Multiordinal. Crochets arising in a single row and in many alternating lengths (Lepidoptera).

Multiserial. Crochets arising in several rows (Lepidoptera).

Multiserial Circle. Crochets arising in three or more concentric circles (Lepidoptera).

Muricate. With sharp, rigid points.

Muscidiform. Shaped like the peglike larvae of higher Diptera.

Myiasis. A disease of animals caused by maggots.

Naiad. The aquatic larva (nymph) of Odonata, Plecoptera, and Ephemeroptera.

Nasale. An anterior, median projection from the frons (especially some Coleoptera larvae).

Nasute (-i) (Isoptera). A type of soldier with an elongated forward projection of the head capsule through which are propelled the secretions of the frontal gland. Typical of the Nasutitermitinae (Termitidae).

Navicular. Shaped like a boat, i.e., tapered at both ends.

Neotenic (Isoptera). An individual that becomes a functional reproductive without ever becoming an alate. May be apterous (lacking wing pads) or brachypterous (possessing wing pads). Sometimes termed "supplementary reproductives" because they are functional in the colony in which they originate.

Neoteny. Sexual maturity in a nonadult stage.

Nodate. With node(s) or enlargement(s).

Node. Knotlike swelling.

Nodose. Knotty.

Notum (-a). The tergum or major plate of a thoracic dorsum.

Nuchal. Of the neck.

Nuchal Constriction. A neck-like constriction near the base of the head capsule.

Nymph. The larva of a hemimetabolous insect; the eight-legged second stage of Acarina.

Nymph, Apterous (Isoptera). An immature individual lacking wing pads or other specialized morphological features. Termed "larva" by some.

Nymph, Brachypterous (Isoptera). An immature individual possessing wing pads. Termed "nymphs" by some, in which case apterous nymphs are termed "larvae."

Obtect Pupa. One in which the appendages are tightly appressed to the body (most Lepidoptera, for example).

Obtuse. Blunt, not pointed or acute.

Occipital Foramen. The main posterior opening of the head capsule, leading to the cervical region or prothorax.

Ocellus (-i). A simple eye of an adult insect or of a larva (nymph) not having complete metamorphosis. *Compare* Stemmata.

Ocularium. Elevated or pigmented area bearing or surrounding the stemma or simple eye.

Oligopod. An active larva with well-developed thoracic legs.

Ommatidium (-a). A single section of a compound eye; the eye elements of Mecoptera larvae.

Onisciform Larva. A flattened, ovate larva, resembling some terrestrial isopods.

Oötheca (-ae). An egg case (Mantodea and Blattodea).

Opercular Hooks. Paired hooks attached to the operculum in some aquatic beetle larvae.

Operculum. A circular or ovoid lid formed from the ninth tergum or tenth sternum in some beetle larvae.

Opisthognathous. With the mouthparts directed downward and backward (as in Blattodea).

Opisthosoma. The posterior region of the body, the part behind the legs (Arachnida).

Opposable. Capable of being opposed or meeting one another, as in the mandibles of chewing insects.

Ordinal. Referring to the lengths of crochets (*See* Uni-, Bi- and Triordinal).

Orthosomatic. With surfaces of body subparallel; a straight body.

Osmeterium (-a). A fleshy, eversible, odoriferous, usually forked glandular process (eversible through a dorsal, median slit in the prothorax of papilionid caterpillars (including *Parnassius*)).

Oviparous. Laying eggs.

Oviposit. To lay eggs.

Ovipositor. The egg-laying structure.

Ovoviviparous. Producing living young by the hatching of the ovum while still within the female.

Paedogenesis. Reproduction by an immature stage.

Palmate Mandible. One with a concave mesal surface.

Palp (Palps). Also Palpus (Palpi). A paired segmented, sensory structure arising from the maxilla or labium.

Palpifer (Palpiger). The lobe of the maxilla or labium that bears the palp.

Papilla (-ae). A soft minute bump or projection.

Palpus (-i). *See* Palp.

Paraglossa (-ae). One of a pair of lobes at the tip of the labium, lateral to the glossae.

Paraproct. One of a pair of lobes bordering the anus lateroventrally. Lateral lobe of terminal (eleventh) abdominal segment (Psocoptera).

Paraproctal Prong. A stout spine on distal margin of paraproct in adult males of certain psocids.

Parasite. An animal that lives in or on its host, at least during a part of its life cycle, feeding on it, but usually not killing it (*Compare* Parasitoid).

Parasitic. Living as a parasite or parasitoid.

Parasitoid. An animal that feeds in or on a *single* host, eventually killing it (*Compare* Parasite, Predator).

Parietal Bands. Dorsolateral paired streaks on front of head of some Hymenoptera, slightly differentiated and often pigmented.

Parietals. The lateral areas of the cranium, between the frontal and occipital areas.

Parthenogenesis. Reproduction without fertilization.

Pectinate. With branches like a comb.

Pedal Lobe. A fleshy, bumplike, nonsegmented leg rudiment.

Pedicel. The second segment of an antenna.

Pedipalps. The second pair of appendages of an arachnid.

Pedunculate. On a slender stalk.

Penellipse. Crochets arranged in an incomplete oval or circle. *See* Lateral and Mesal Penellipse.

Penicillum, Penicillus (penicilli). A small pencil or brush of hairs near the base of the mandible (Coleoptera).

Penultimate. Next to last.

Perforate Mandible. One with an internal channel extending from the base to the apex.

Peripneustic (Respiratory System). With one pair of thoracic and all eight abdominal spiracles functional.

Peristoma. The mouth frame, comprised of the epistomal ridge, pleurostomata, and hypostomata.

Peritreme. A sclerotic ring surrounding the outer spiracular opening.

Petiolate. Attached with a narrow stalk or petiole.

Pharate. The condition in which the new (next) stage is enclosed by the cuticle of the present one; i.e., a "pharate" adult is often visible beneath the pupal skin in endopterygotes and may be capable of locomotion in some.

Pharyngeal Skeleton. In higher Diptera, the conspicuous sclerotized structure articulating with the mouthhooks.

Pharynx. The area behind the cibarium; an enlargement of the anterior part of the esophagus.

Phoresy. A relationship (usually temporary) in which an individual of one species is transported by another species, but does not feed on it or otherwise injure it.

Phoretic. Exhibiting phoresy (q.v.).

Physogastric. With an enlarged or swollen abdomen.

Phytophagous. Feeding on plants.

Pi Group. The group of setae just above the thoracic legs or prolegs of Lepidoptera larvae (Fracker). (SV group of Hinton)

Pilose. Covered with soft hair.

Pinaculum (-a). A small, flat, or slightly elevated chitinized area bearing seta(e) (especially Lepidoptera).

Pinaculum Ring. A modified pinaculum forming a relatively broad pigmented band encircling or partially encircling a pale area surrounding a seta (Lepidoptera, especially Pyralidae).

Planidiform (larva). First active stage of certain parasitoids (Diptera and Hymenoptera).

Planidium (-a). A type of first-instar larva that undergoes hypermetamorphosis, a larva that is legless and somewhat flattened (some Diptera and Hymenoptera).

Planta. Distal part of a proleg bearing the crochets but never bearing setae (Lepidoptera).

Plantula. A lobe of the tarsus, a climbing cushion.

Plastron. A permanent physical gill consisting of a gas layer of constant volume, held in place by a hydrofuge mesh that prevents the entry of water under pressure, and a large air-water interface.

Plate. A larger sclerotized area of the body. *See* Anal or Suranal Plate; *compare* Pinaculum).

Plesiomorphy (-phic, -phous). The relatively primitive state (older) of a sequence of homologous characters. *Compare* Apomorphy.

Pleural Lobes. Lateral, rounded lobes on each segment of many Hymenoptera larvae and others.

Pleuron (-a). The lateral area of a segment.

Pleurostoma. Lateral extension of the epistoma on either side of the cephalic structure separating the clypeus from the front of the face and usually continuous with the hypostoma (Hymenoptera: Parasitica).

Plica (-ae). An integumental fold, especially on the dorsum in soft-bodied larvae.

Plumose. With many small branches; featherlike.

Polyembryony. An egg developing into more than one embryo.

Polymorphic. Having many forms.

Polyphagous. Feeding on several species.

Polypod Larva. One with legs on most segments.

Postclypeus. The major and more posterior region of a divided clypeus; the caudal (dorsal) portion of a transversely divided clypeus.

Postcornus. Sclerotized, horny process at apex of tenth abdominal tergum of many boring sawfly larvae and a few others.

Postembryonic. The period after hatching.

Posterad. Toward the rear. *See* Caudad.

Posterior. Caudal or rear.

Posterior Spiracular Plate. The caudal plate bearing the spiracles.

Postgena (-ae). The region immediately behind the gena.

Postmentum. The proximal sclerite of the labium; sometimes divided into a submentum and mentum.

Postspiracular. Caudad of the spiracles.

Preapical. Just before the apex.

Predator. An animal that kills its hosts (prey), requiring more than one to complete its life cycle. *Compare* Parasite, Parasitoid.

Prehensile. Adapted for wrapping around.

Preimago. In males of Acerentomidae (Protura), an instar between the maturus junior and adult.

Prelabial Sclerite. A sclerite below the orifice of the silk press within the labial sclerite; it may be attached to the dorsal arms and/or ventral portion of the labial sclerite or may be free (Hymenoptera: Parasitica).

Prementum. The distal part of the labium, bearing the palps, glossae (and sometimes paraglossae), or the ligula.

Prelarva. The first instar of Protura.

Prepupa. The nonfeeding portion of the last instar preceding the pupa in which the larva is shorter, thicker, and relatively inactive. The waxy-white third-instar flea larva that has expelled all contents from alimentary canal; the prepupal period is postdefecation to pupal molt.

Prespiracular Wart. The L group of setae anterior to the spiracle on the prothorax of Lepidoptera larvae (Kappa group of Fracker).

Pretarsus. The apical segment of a six-segmented larval leg, consisting of a single claw or ungulus, or paired claws (Coleoptera).

Primary Setae. Those setae with definite locations and numbers and found on all instars (Lepidoptera).

Proboscis. Tubelike or beaklike mouthparts.

Produced. Extended or projecting.

Prognathous. With the head horizontal and mouthparts directed forward.

Prolegs. Fleshy, unjointed abdominal legs with or without crochets; also termed larvapods or false legs.

Propneustic (Respiratory System). With only the first (anterior) pair of spiracles functional.

Prosoma. The anterior region of the body—the fused head and thorax, bearing the legs, usually the cephalothorax (Arachnida).

Prostheca. A hyaline or partly sclerotized, rigid or flexibile, simple or complex lobe or process, or a group of hairs or specialized setae, arising from the mesal surface of the mandible just distad of the mola (Coleoptera).

Protonymph. A second instar mite.

Protracted. Extended forward; with the head not withdrawn into the prothorax. Mouthparts—arising near the anterior margin of a prognathous head (Coleoptera).

Protuberance. Any projection with or without setae.

Proximal. Nearest to the point of attachment.

Pseudocercus. *See* Urogomphus.

Pseudoculus. In Protura, one of a pair of disc-shaped sensory organs located on the head subdorsally.

Pseudopod. A soft, footlike appendage (common on many Diptera larvae).

Pterygote. Winged.

Pubescent. With soft, short, fine hairs.

Pulvilliform. Lobelike or padlike; shaped like a pulvillus; pulvilliform empodium.

Pulvillus (-i). A lobe or pad beneath the base of the tarsal claw.

Puncture. Depression, pit or perforation.

Pupa (-ae). The relatively inactive, transformation stage between the larva and the adult of holometabolous insects.

Pupa, Complete. This form is present in the higher Ditrysia (Lepidoptera) and has two or fewer movable abdominal segments in both males and females, no cocoon cutter on the head, and usually no caudally directed spines on the abdomen. These pupae almost never protrude from the cocoon when the adult emerges.

Pupa, Incomplete. This form is typical of the more primitive Lepidoptera with obtect pupae and is characterized by possessing three or more freely movable abdominal segments in the male and two or more in the female, usually a ridge or some other kind of cocoon cutter on the head, and nearly always some caudally directed spines on the dorsum of the abdomen.

Pupal Horns. Hornlike respiratory processes of some Diptera pupae (Culicidae, Syrphidae).

Puparium (-ia). The hardened, barrellike last larval skin in which the pupa is formed (higher Diptera).

Pygidium (-a). The ninth tergum, especially when this is sclerotized; also a sclerotized area, sometimes a concave disc, which is set off from the rest of the ninth tergum (Coleoptera).

Radula. *See* Raster.

Ramose (Setae). Setae with branches, usually arising from the base.

Raptorial. Adapted for seizing prey.

Raster. Complex of bare areas, hairs, and spines on ventral surface of last abdominal segment, in front of anus, in Scarabaeoidea; comprised of septula, palidium, teges, tegillum (in some groups), and campus (Coleoptera).

Recurved. Curved backward.

Reniform. Kidney-shaped.

Respiratory Siphon. A breathing tube of aquatic larvae.

Respiratory Trumpet. Prothoracic respiratory protuberance, usually paired, of the pupa of some aquatic Diptera.

Retinaculum. A projection or toothlike structure on the oral surface of a mandible; usually formed on mandibles without a basal mola (especially Coleoptera).

Retracted. With the head withdrawn into the prothorax. Mouthparts—arising near the rear of a prognathous head.

Retractile. Capable of being drawn in.

Retrorse. Bent backward or downward.

Rostrum. The snout.

Rudimentary. Reduced in size, barely developed.

Rugose. Roughly wrinkled.

Rugosites. Small or minute elevations that are close together.

Saccoid. Swollen; saclike.

Salivary Opening (slit, or lips). Opening of the silk or salivary glands on the extremity of the labium.

Saltatorial. Adapted for leaping or jumping.

Saprophagous. Feeding on dead or decaying materials.

Scale. A highly modified seta, which is somewhat expanded and usually flattened above.

Scansorial Wart. *See* Ambulatorial Warts.

Scarabaeiform, Scarabaeoid. U- or C-shaped larva, with a distinct head and thoracic legs, and resembling a white grub of the Scarabaeidae.

Sclerite. A hardened body plate.

Sclerome. Any heavily sclerotized structure, not platelike.

Sclerotized. Hardened and tanned, so that it is yellow to black in color.

Scolus (-i). A spinose, usually branched projection of the body wall, each branch bearing a stout seta at its tip (Lepidoptera, Mecoptera, Coleoptera).

Scutellum. The third dorsal sclerite of the meso- and metathorax.

Scutum. The middle dorsal sclerite of the meso- and metathorax.

Second Grub. The fourth phase of hypermetamorphic larval development in Meloidae, in which, following the coarctate phase, the larva again becomes grublike.

Secondary Setae. Those setae with indefinite locations and numbers, and usually not present on the first instar.

Seminivorous. Feeding upon seeds of grasses.

Sensilla Styloconicum. A conelike or peglike sense organ.

Sensillum. A sense organ.

Sensoria. Fleshy, peglike sensory organs.

Serial. Referring to the number of rows of the bases of the crochets (Lepidoptera).

Serrate. Sawlike.

Sessile. Attached, incapable of moving from place to place.

Seta (-ae). A hairlike projection of the body wall that is articulated in a socket.

Setal Map. A flat, diagrammatic drawing of the arrangement of the setae on one side of a larva (usually the left side with the head to the left).

Setiferous, Setaceous, Setose. Bearing setae.

Shagreened. Roughened with numerous minute bumps.

Shield. A sclerotized plate covering part of the dorsum of a segment.

Sieve Plate. The perforated plate covering the opening of a cribriform spiracle.

Sigmoid. S-shaped.

Silk Press. Salivary opening associated with silk glands.

Simple Eyes. *See* Stemmata and Ocelli.

Sinistral. Antonym of dextral (q.v.), the coil ascending from right to left of the observer when the anterior opening is held uppermost.

Sinuous. Curved in and out; winding.

Sinus. A cavity or depression.

Siphon. A tubular external structure, usually a breathing tube.

Sitophore Sclerite. Basal sclerite of hypopharynx, bearing a median depression (Psocoptera).

Soldier (Isoptera). A type of individual which is morphologically specialized for defense.

Soldier-Nymph (Isoptera). The presoldier stage that undergoes one molt to become the mature soldier. These individuals have the general form of the mature soldier but are smaller, unsclerotized, and unpigmented.

Spatulate. Spatula-shaped, spoon-shaped; flat, rounded, and broad at tip, narrowed at the base.

Spicule. A minute pointed spine or process.

Spindlelike. Pointed at both ends and swollen in the middle (fusiform).

Spine. An unarticulated thornlike process arising from the cuticle. (*Compare* Spur and Seta).

Spinneret. A structure from which silk is spun, usually located on the labium of larvae, and often protruding as lips or elongate process(es).

Spinule (spinules). Minute spines (microspines).

Spinulose. Bearing little spines or spinules.

Spiracular Line. The line coinciding with or near the spiracles.

Spiracular Lobes. Lobes immediately below or immediately posterior to the lobe bearing the spiracle.

Spur. A movable spine.

Spurious. False.

Squama Genitalis. The partly sclerotized, protrusible genitalia of Protura.

Stadium (-a). The time period between molts.

Stellate. Star-shaped.

Stemapoda. Elongated anal prolegs of some notodontid larvae.

Stemma (-ata). A simple eye of holometabolous larvae. *See* Ocellus.

Sternite. A sclerotized plate on the venter of a thoracic or abdominal segment.

Sternum (-a). The main sclerite or area of the venter of a segment.

Stilettiform. Shaped like a stiletto or style.

Stipes (Stipites). The part of a maxilla distad to the cardo, bearing the palp, galea, and lacinia.

Stipital Sclerite. A sclerotized bar extending from the cardo along the posterior edge of the maxilla on either side of the cephalic structure; homologous with the sclerotized part of the stipes in primitive insects (Hymenoptera: Parasitica).

Striate. With grooves (and ridges).

Stridulate. To make a sound by rubbing two structures together.

Stridulatory Organ. A structure producing sound by moving one sclerotized part, the plectrum, over another, the stridulitrum.

Struma (-ae). A moundlike projection of the body wall bearing a few chalazae.

Strut Setae. Small setae at the base of the anal struts on the tenth abdominal segment of flea larvae.

Stylate. Stylelike.

Stylet (-s). One of the piercing structures in piercing-sucking mouthparts. *Compare* Stylus.

Stylus (-i) Style (-s). A short, slender process; articulated process on posterior border of abdominal sternum or on sternal plates in primitively wingless insects; also on coxae of some Microcoryphia.

Subanal. Below the anus.

Subcuticular Pigment. Pigment granules of the epidermis of thysanopterans of all instars and responsible, in part, for body color. Because it varies intraspecifically, its use in identification is limited.

Subdorsal Line. A line below the dorsal or addorsal line (if present).

Subequal. Nearly equal.

Subimago. The first winged stage of a mayfly after it emerges from the water (the "dun" of fly fishing).

Submentum. The basal part attaching the labium to the head.

Subparallel. Nearly parallel.

Subprimary Setae. Those setae having a definite position, but appearing in the second instar (not present in the first instar); otherwise similar to primary setae.

Subspiracular Line. A line below the spiracles.

Subventral Line. A line just above the bases of the prolegs.

Sulcus (-i). A groove or furrow.

Supplemental Process. An accessory sensory process of the preapical antennal segment.

Supraspiracular Line. Above the spiracles.

Suranal. Above the anus, = supraanal.

Suranal Plate. The area on the dorsum of the last abdominal segment, frequently rather heavily sclerotized (= anal plate, anal shield).

Suranal Process. A sclerotized projection on the meson of the suranal plate (postcornu).

Surpedal Lobes. Lobes dorsal to prolegs and ventral to spiracular lobes.

Symplesiomorphy (-phic, -phous). The sharing by two or more taxa of the more primitive of a homologous pair or sequence of character states. *Compare* synapomorphy, and *see* plesiomorphy and apomorphy.

Synapomorphy, -phic, -phous. The sharing by two or more taxa of the more recent (apomorphic) of a sequence of homologous characters.

Tarsungulus (-i). Fused tarsus and tarsal claw in many larval Coleoptera.

Tarsus (-i). The fifth segment of the thoracic leg which usually is one- to five- segmented and usually bears one or two claws. If holometabolous larvae have tarsal segments, most have only one segment.

Tegmen (Tegmina). Leathery fore wing (especially orthopterans).

Tenaculum. A minute structure on the venter of the third abdominal segment that clasps the furcula (Collembola).

Teneral. Refers to pale, soft-bodied, recently molted immatures or adults.

Tentorium. The internal skeleton of the head.

Tergite. A sclerotized plate on the dorsum of a thoracic or abdominal tergum.

Tergum. The dorsal portion of a body segment (= notum on thorax).

Terminal Filament. Long and slender medial prolongation of last abdominal segment.

Thamnophilous. Living in thickets or dense shrubbery.

Thelytoky (Thelytokous). Parthenogenetic reproduction in which males are absent or extremely rare (Hymenoptera).

Thigmotactic. Contact-loving.

Tibiotarsus (-i). Term applied to the tibia of Polyphaga (Coleoptera), when it is assumed that it represents a fusion of the tibia and tarsus.

Tonofibrillary Platelet. A small, external, flattened, sclerotized area of the integument associated with the tonofibrillae (cuticular fibrils) connecting the muscle fibers with the cuticula (Lepidoptera).

Torma (-ae). One of a pair of dark scleromes originating at each end of the clypeolabral suture and extending mesally and/or posteriorly, sometimes asymmetrical or complex with accessory structures (Coleoptera).

Tormogen Cell. An epidermal cell forming the setal membrane or socket (ring base).

Tracheated. Supplied with tracheae.

Transverse Band. Crochets arranged transversely to the long axis of the body (Lepidoptera).

Trichobothria. Sensory hairs, especially very long ones.

Tricuspidate. Having three cusps or teeth.

Trident. Having three teeth, processes or points.

Trifid. Divided into three parts.

Triordinal. Crochets arranged so the tips are of three alternating lengths.

Triquetral. Triangular in section, with three flat sides.

Triungulin. The active, hypermetamorphic, campodeiform first-instar or primary larva of Meloidae, so named because of paired setae on the tarsungulus, which suggest the presence of "three claws". The term has also been applied to primary larvae of Rhipiphoridae, Strepsiptera, and others, although the term triungulinid is preferable for these other groups without "three claws." *See* Planidium.

Triungulinid. *See* Triungulin.

Trochantin. A sclerite derived from the anterior part of the thoracic pleuron, usually articulated ventrally with the anterior margin of the coxa; of taxonomic value in larval Trichoptera, where it is usually recognizable only on the prothorax.

Truncate. Square at the end.

Tubercle. An abrupt elevation of varying form.

Tuft. A group of setae.

Tympanum (-a). A vibrating membrane; an eardrum.

Uncus (-i). A toothlike or hooklike process, or occasionally a setiferous process at the inner edge of malar apex or at the apex of the lacinia in various beetle larvae.

Uniforous. Spiracle with a single entrance leading into the atrium.

Uniordinal. Crochets arranged so they are of a single length or slightly shorter toward the ends of the row (Lepidoptera). *See* Bi- and Triordinal.

Uniserial. Crochets arranged in a single row with their bases in line (Lepidoptera).

Univoltine. Having one generation per year.

Urogomphus (-i). A dorsal process, usually paired and sclerotized, projecting from the rear of the ninth abdominal segment (may be jointed and movable or unjointed and immovable) (Coleoptera and others).

Urticating Hairs. Hairs or setae connected to poison glands that cause irritation; barbed hairs also may cause irritation without poison.

Venter. The entire under surface of the body.

Ventral Abdominal Vesicle. A transverse, blisterlike swelling bordering both sides of an intersegmental membrane ventrally on abdomen, occurring in several psocid families.

Ventral Condyle. The rounded process on the ventral edge of the mandible that articulates with the head.

Ventral Prolegs. All prolegs on any abdominal segment except the last, which are the anal prolegs.

Ventrolateral Setae. A few large setae anterior to the anal struts on the tenth abdominal segment of flea larvae.

Vermiform. Wormlike.

Verruca (-ae). A somewhat elevated area of the cuticle, bearing setae pointing in many directions like a pincushion.

Verricule. A dense tuft of upright, parallel setae.

Vertex. The "top" of the head.

Vertical Furrow. Furrow on each side of head near dorsal aspect.

Vertical Triangle. *See* Epicranial Notch.

Vestibulum (-a). An external genital cavity formed above the seventh abdominal sternum, when the latter extends beyond the eighth (Snodgrass).

Vestigial. Weakly developed or degenerate; only a remnant left.

Vitta (-ae). A broad stripe.

Vittate. Striped.

Viviparous. Bearing living young.

Wart. A general term for any relatively small, sclerotized area that is usually somewhat raised and bears setae (Lepidoptera).

Wireworm. A larva that is slender, heavily sclerotized, with thoracic legs, without prolegs, and nearly naked; a phytophagous click beetle larva.

Worker (Isoptera). A relatively morphologically unspecialized individual that performs the "work" of the colony; caring for the reproductives, eggs, young, soldiers and other workers; extending and repairing the colony workings, and foraging.

Xylophagous. Feeding on wood.

Host Plant and Substrate Index

NOTE: Because the primary coverage of this book is at the family level, with some coverage below, there are many hosts given in general terms, e. g. "fungi", "grasses", "lichens","sedges", etc. This kind of information may not be of value to many users, but it is included for those who find it helpful.

Many species, especially noctuids, pyralids, and others that feed on deciduous trees, shrubs and herbs are polyphagous; it is not feasible to provide hosts for these species except in terms such as "general feeder on herbs", "polyphagous on woody shrubs", etc. Such terms are not indexed.

No entries are made for the substrates or food of the aquatic orders Ephemeroptera, Odonata, Plecoptera and Trichoptera.

All kinds of names are entered, including common names, family names, and scientific names. Where feasible, combined entries for common name and scientific name are given, even though only one of them may have been used at a given point in the text. For example, text entries for "pecan" are also entered under "*Carya*".

Animal names are entered in the general index.

Index